THIRD EDITION

HANDBOOK OF CAPILLARY and MICROCHIP ELECTROPHORESIS and ASSOCIATED MICROTECHNIQUES

THIRD EDITION

HANDBOOK OF CAPILLARY and MICROCHIP ELECTROPHORESIS and ASSOCIATED MICROTECHNIQUES

EDITED BY JAMES P. LANDERS

CRC Press
Taylor & Francis Group
Boca Raton London New York

CRC Press is an imprint of the
Taylor & Francis Group, an **informa** business

CRC Press
Taylor & Francis Group
6000 Broken Sound Parkway NW, Suite 300
Boca Raton, FL 33487-2742

Printed in the United States of America on acid-free paper
10 9 8 7 6 5 4 3 2 1

International Standard Book Number-13: 978-0-8493-3329-3 (Hardcover)

Library of Congress Cataloging-in-Publication Data

Handbook of capillary and microchip electrophoresis and associated microtechniques / editor, James P. Landers. -- 3rd ed.
p. ; cm.
Rev ed. of: Handbook of capillary electrophoresis. c1997. "CRC title."
Includes bibliographical references and index.
ISBN 978-0-8493-3329-3 (hardcover : alk. paper)
1. Capillary electrophoresis. I. Landers, James P. II. Handbook of capillary electrophoresis. III. Title.
[DNLM: 1. Electrophoresis, Capillary--methods. 2. Electrophoresis, Microchip--methods. QU 25 H2355 2008]

QP519.9.C36H35 2008
543'.4--dc22 2007021969

Visit the Taylor & Francis Web site at
http://www.taylorandfrancis.com

and the CRC Press Web site at
http://www.crcpress.com

Dedication

This undertaking would not have been possible without the support of my family and colleagues. Dedication is, first and foremost, to family—first, my wife, Lianne, and my children, Miranda, Kate, and Ben—who not only recognize my passion for these academic endeavors but also fully encourage me to pursue them. I appreciate the sacrifices they made en route to completion of this undertaking. Second, this laboratory compendium would not exist without the immeasurable effort from the scientists who authored the chapters—they are leaders in their particular fields, both nationally and internationally—I am grateful that they saw value in contributing here. In recognition of their willingness to contribute to this handbook, even in the midst of their already overextended commitments, this book is also dedicated to them.

Contents

Foreword

The esteemed Chinese philosopher Confucius placed works such as this in succinct context when he wrote "*You cannot open a book without learning something.*" Each chapter of this book describes remarkable advancements in one of the most controversial and yet, powerful, analytical tools in separation sciences. Even the most accomplished practitioners in separation science will extract new information on methods and applications that can serve unmet needs in science and medicine.

Even after more than two-and-a-half decades of research and development, many scientists debate whether capillary electrophoresis (CE) technology has provided the speed, resolving power, peak capacity, sensitivity, robustness and cost reduction promised by the pioneers of CE during its genesis in the 1980s and 1990s. Thousands of researchers, focusing on some of the most widespread and practical analytical goals ever considered (that utilize microscale volumes of sample), have demonstrated the utility of CE on a wide range of applications across many disciplines, such as forensic science, medical diagnostics, pharmaceutical science, biotechnology, and environmental science. The last 7 years have seen CE technology evolve quickly from the research laboratory into practical applications in many fields, and the total body of international CE literature continues to grow at an ever-increasing rate.

The first flush of excitement came with the key role that the CE technology played in completion of the Human Genome Project. The project was successfully completed ahead of schedule and at a fraction of the predicted cost. This monumental breakthrough allowed the placement of CE-based DNA sequencers in the most prestigious institutions providing medical-legal testing and in numerous research centers worldwide working on different DNA projects. In addition, an official compendium for the analysis of erythropoietin is now part of a monograph of the European Pharmacopeia. Consequently, CE is now part of the quality control that precedes the release of commercial batches of this critically important drug. Other major biopharmaceutical companies also have official CE-based quality control methods filed with the FDA. In fact, the growth of widely used biotechnology compounds now rivals the growth of conventional pharmaceutical products and, thus, stricter regulations have been applied to biomolecule development due to the complexity of their physicochemical properties.

Each contributor to this book has been actively engaged in CE or microchip research and is a leader in the field. Accordingly, each author contributes various aspects of their experience, specifically the detailed descriptions and awareness of the practical problems inherent in their particular area of research and application. Furthermore, contributors provide sound advice on how to overcome practical problems, thus enabling readers to pursue novel applications equipped with much of the practical knowledge and wisdom possessed by the leading researchers in the field.

The chapters in this book indicate the extraordinary breadth and scope of the work carried out in capillary electrophoresis and microchip technology over the last few years. The core section on "Fundamentals and Methodologies" contains 10 chapters focused on CE technology. These chapters provide an introduction to CE in general, followed by in-depth descriptions of the most widely used modes of CE and their associated detection methods. Several important applications related to protein, nucleic acid, and carbohydrate research are presented, as well as an overview of the emerging field of fabrication and use of microchips. The next 20 or so specialized chapters continue to describe specific applications and methodologies that can be addressed with capillary systems, including studies on the analysis of cellular and subcellular entities, food analysis, environmental science, clinical and forensic applications, solid-phase microextraction/online sample preconcentration, and extreme resolution CE.

This bandwidth of applications is illustrative of the exploding world of microchip systems that perform chemical and biochemical analysis. It is difficult, if not impossible, to determine whether microfabricated analytical devices were predominantly inspired by capillary systems, or by visionaries who recognized the potential of microfabrication techniques (originally designed for microelectronics) to revolutionize chemical, biomolecular, and cellular analysis. Clearly, inspiration has flowed in both directions, and the revolution is upon us. Techniques to exploit the precision, scalability, and unique microscale phenomena enabled by microchip technology now span much of the routine practice of biochemistry and biology, including cell manipulation, protein separations, forensic science, molecular and nanomaterial synthesis, etc. The second part of this book explores core microsystem technologies, such as immunoassays, flow control, device fabrication and reliability. These core technologies are widely used in many applications, such as those outlined in the specialized chapters that conclude this book, and they span everything from on-chip sample preparation to genetic analysis to devices coupled with mass spectrometry.

It would appear that the days of categorizing capillary and microchip electrophoresis as narrow endeavors with limited impact are long gone. The confluence of chemistry, biology, and microengineering undoubtedly will continue to produce new tools for scientific discovery, as well as for forensic applications and clinical diagnostics. These latter applications are poised for a technology revolution that will paradigm shift their respective fields. Microdevices that accept and analyze the components of whole blood—ions, proteins, nucleic acids—using integrated sample preparation domains fluidically interfaced with the analytic process are coming to fruition and will soon become available. Moreover, microfluidics offers the unique capability to control and manipulate (sub)nanoliter volumes of liquids in unique, precise, and reproducible ways that were previously unimaginable in capillary systems. This fact, together with innumerable new developments and insight on cell manipulation and analysis, and recent progress in (inexpensive) polymer microfabrication combine to create exciting opportunities for future analytical microsystems that can be applied as disposable diagnostic tools for site-of-analysis testing. It is our hope that readers will find this book as helpful as the last two editions have been. This work is an outstanding contribution to the development of increasingly powerful analytical tools that will continue to have a positive impact on science and society in the years ahead.

Norberto A. Guzman, Ph.D.
Senior Research Fellow
Bioanalysis, Drug Metabolism, and Drug Toxicity
Johnson & Johnson Pharmaceutical R&D
Raritan, New Jersey

Prof. Dr. Albert van den Berg
BIOS Lab-on-a-Chip group
MESA+ Institute for Nanotechnology
University of Twente
The Netherlands

Matthew R. Begley, Ph.D.
Associate Professor
Department of Mechanical Engineering
Department of Materials Science & Engineering
University of Virginia, Charlottesville, Virginia

Editor

James P. Landers is professor of chemistry and professor of mechanical engineering at the University of Virginia, as well as an associate professor of pathology at the University of Virginia Health System.

Professor Landers received a Bachelor of Science degree in biochemistry with a minor in biomedicine from the University of Guelph in Ontario, Canada, in 1984. He earned his Ph.D. in biochemistry from the same department in 1988. After a one-year postdoctoral fellowship at the Banting Institute at the University of Toronto School of Medicine, he was awarded a Canadian Medical Research Council Fellowship to study cancer biology and diagnostics under Dr. Thomas Spelsberg, a breast cancer biochemist at the Mayo Clinic. He launched and directed Mayo Clinic's Clinical Capillary Electrophoresis Facility in the Department of Laboratory Medicine and Pathology developing clinical assays based on capillary electrophoretic technology—some are still on-board at Mayo today. Beginning as an assistant professor of analytical chemistry at the University of Pittsburgh in 1997, he forayed into analytical microfluidic systems with the goal of developing the next-generation molecular diagnostics platform. These efforts were bolstered by a move to the University of Virginia, where access to a dedicated class-100 cleanroom for microchip fabrication allowed his group to rapidly prototype microdevices for separations, DNA purification, and DNA amplification. In addition to editing the first two editions of this book, he has authored more than 175 papers and 25 book chapters on receptor biochemistry, capillary electrophoretic method development, microchip fabrication, and integrated microfluidic systems for application in the clinical and forensic arenas.

Contributors

Kevin Altria
Research and Development
GlaxoSmithKline
Harlow, Essex, United Kingdom

Imee G. Arcibal
Department of Chemistry
Pennsylvania State University
University Park, Pennsylvania

Edgar A. Arriaga
Department of Chemistry
University of Minnesota
Minneapolis, Minnesota

Yoshinobu Baba
Department of Applied Chemistry and
MEXT Innovative Research Center for
Preventive Medical Engineering
Nagoya University
Nagoya, Japan
and
Health Technology Research Center
National Institute of Advanced
Industrial Science and Technology
Takamatsu, Japan

Annelise E. Barron
Department of Bioengineering
Stanford University
Stanford, California

David J. Beebe
Department of Biomedical Engineering
University of Wisconsin
Madison, Wisconsin

Matthew R. Begley
Department of Mechanical and Aerospace
Engineering
and
Department of Materials Science and Engineering
University of Virginia
Charlottesville, Virginia

Maxim V. Berezovski
Department of Chemistry
York University
Toronto, Ontario, Canada
and
Campbell Family Institute for
Breast Cancer Research
University of Toronto
Toronto, Ontario, Canada

Rajiv Bharadwaj
Microfluidics Group
Caliper Life Sciences
Mountain View, California

Joan M. Bienvenue
Department of Defense DNA Registry
Armed Forces DNA Identification Laboratory
Armed Forces Institute of Pathology
Rockville, Maryland

Ryan Bonn
Department of Chemistry
University of Washington
Seattle, Washington

Michael T. Bowser
Department of Chemistry
University of Minnesota
Minneapolis, Minnesota

Michael C. Breadmore
Australian Centre for Research on Separation Science
School of Chemistry
University of Tasmania
Hobart, Australia

Dean S. Burgi
Molecular Diagnostics
Affymetrix, Inc.
Santa Clara, California

Emanuel Carrilho
Department of Chemistry and Molecular Physics
Institute of Chemistry at São Carlos
University of São Paulo
São Carlos, Brazil

Wendell Karlos Tomazelli Coltro
Department of Chemistry and Molecular Physics
Institute of Chemistry at São Carlos
University of São Paulo
São Carlos, Brazil

Jennifer A. Coyne
Department of Chemical Engineering
Stanford University
Stanford, California

Christopher J. Cullen
Department of Chemistry
Imperial College London
London, United Kingdom

Mercedes de Frutos
Institute of Organic Chemistry
Madrid, Spain

Andrew J. deMello
Department of Chemistry
Imperial College London
London, United Kingdom

Don L. DeVoe
Department of Mechanical Engineering
University of Maryland
College Park, Maryland

Jose Carlos Diez-Masa
Institute of Organic Chemistry
Madrid, Spain

Sheila Donegan
Department of Chemical and Life Sciences
Waterford Institute of Technology
Waterford, Ireland

Kevin D. Dorfman
Department of Chemical Engineering and Material Science
University of Minnesota
Minneapolis, Minnesota

Norman J. Dovichi
Department of Chemistry
University of Washington
Seattle, Washington

Brian M. Dressen
Department of Chemistry
Colorado State University
Fort Collins, Colorado

Christopher J. Easley
Department of Molecular Physiology and Biophysics
Vanderbilt University Medical Center
Nashville, Tennesse

Simon Ekström
Department of Electrical Measurements
Division of Nanobiotechnology
Lund University
Lund, Sweden

Charles A. Emrich
Department of Chemistry and Biophysics Graduate Group
University of California
Berkeley, California

Christopher J. Evenhuis
Australian Centre for Research on Separation Science
School of Chemistry
University of Tasmania
Hobart, Australia

Andrew G. Ewing
Department of Chemistry
Pennsylvania State University
University Park, Pennsylvania

Jerome P. Ferrance
Department of Chemistry
University of Virginia
Charlottesville, Virginia

Barbara A. Fogarty
Tyndall National Institute
Lee Maltings
Cork, Ireland

Robin Fortt
Department of Chemistry
Imperial College London
London, United Kingdom

Hernan V. Fuentes
Department of Chemistry and Biochemistry
Brigham Young University
Provo, Utah

Braden C. Giordano
Nova Research Inc.
Alexandria, Virginia

Michel Girard
Centre for Biologics Research
Health Canada
Ottawa, Ontario, Canada

Carlos F. Gonzalez
Department of Chemistry
Oregon State University
Corvallis, Oregon

Rosanne M. Guijt
Australian Centre for Research on Separation Science
School of Chemistry
University of Tasmania
Hobart, Australia

Christelle Guillo
Department of Chemistry and Biochemistry
Florida State University
Tallahassee, Florida

András Guttman
Horváth Laboratory of Bioseparation Sciences
University of Innsbruck
Innsbruck, Austria

Paul R. Haddad
Australian Centre for Research on Separation Science
School of Chemistry
University of Tasmania
Hobart, Australia

James M. Hempe
Children's Hospital Research Institute for Children and Department of Pediatrics
Louisiana State University Health Sciences Center
New Orleans, Louisiana

Wm. Hampton Henley
Department of Chemistry
The University of North Carolina at Chapel Hill
Chapel Hill, North Carolina

Charles S. Henry
Department of Chemistry
Colorado State University
Fort Collins, Colorado

Ole Hindsgaul
Carlsberg Laboratory
Valby Copenhagen, Denmark

Katie M. Horsman
Department of Chemistry
University of Virginia
Charlottesville, Virginia

Tiemin Huang
Convergent Bioscience Ltd.
Toronto, Ontario, Canada

David E. Huber
Microfluidics Department
Sandia National Laboratories
Livermore, California

Haleem J. Issaq
Laboratory of Proteomics and Analytical Technologies
SAIC Frederick Inc.
Frederick, Maryland

James W. Jorgenson
Department of Chemistry
The University of North Carolina at Chapel Hill
Chapel Hill, North Carolina

Noritada Kaji
Department of Applied Chemistry and MEXT Innovative Research Center for Preventive Medical Engineering
Nagoya University
Nagoya, Japan

James M. Karlinsey
Department of Chemistry
Penn State Berks
The Pennsylvania State University
Reading, Pennsylvania

Thomas M. Keenan
Department of Biomedical Engineering
University of Wisconsin
Madison, Wisconsin

Julia Khandurina
Anadys Pharmaceuticals
San Diego, California

Tarun Khurana
Mechanical Engineering
Stanford University
Palo Alto, California

Takehiko Kitamori
Department of Applied Chemistry
Nagoya University
Aichi, Japan
and
Micro Chemistry Group
Kanagawa Academy of Science and Technology
Kanagawa, Japan

Myra T. Koesdjojo
Department of Chemistry
Oregon State University
Corvallis, Oregon

James Kraly
Department of Chemistry
University of Washington
Seattle, Washington

Sergey N. Krylov
Department of Chemistry
York University
Toronto, Ontario, Canada

Izaskun Lacunza
Institute of Organic Chemistry
Madrid, Spain

James P. Landers
Department of Chemistry
University of Virginia
Charlottesville, Virginia

Theodore Lapainis
Department of Chemistry
University of Illinois at Urbana-Champaign
Urbana, Illinois

Thomas Laurell
Department of Electrical Measurements
Division of Nanobiotechnology
Lund University
Lund, Sweden

Iulia M. Lazar
Virginia Bioinformatics Institute and Department of Biological Sciences
Virginia Polytechnic Institute and State University
Blacksburg, Virginia

Cheng S. Lee
Department of Chemistry and Biochemistry
University of Maryland
College Park, Maryland

Lindsay A. Legendre
Department of Chemistry
University of Virginia
Charlottesville, Virginia

Jennifer S. Lin
Department of Chemical and Biological Engineering
Northwestern University
Evanston, Illinois

Herbert H. Lindner
Division of Clinical Biochemistry
Innsbruck Medical University
Innsbruck, Austria

Susan M. Lunte
Departments of Chemistry and Pharmaceutical Chemistry
Ralph N. Adams Institute of Bioanalytical Chemistry
University of Kansas
Lawrence, Kansas

Miroslav Macka
School of Chemical Sciences
Dublin City University
Dublin, Ireland

Laili Mahmoudian
Department of Applied Chemistry
Nagoya University
Nagoya, Japan

Danqian Mao
Department of Chemistry
University of Washington
Seattle, Washington

György Marko-Varga
Department of Analytical Chemistry
Lund University
Lund, Sweden

Philip J. Marriott
Australian Centre for Research on Separation Science
School of Applied Sciences
RMIT University
Melbourne, Australia

Alex Marsh
Research and Development
GlaxoSmithKline
Harlow, Essex, United Kingdom

María Gabriela Vargas Martínez
Universidad Nacional Autónama de México
Depto. de Química
Cuautitlán, Edo. México, México

Richard A. Mathies
Chemistry Department
University of California
Berkeley, California

Bruce McCord
International Forensic Research Institute
Florida International University
Miami, Florida

Eamon McEvoy
Department of Chemical and Life Sciences
Waterford Institute of Technology
Waterford, Ireland

David Michels
Amgen
Seattle, Washington

Mohamad Reza Mohamadi
Department of Applied Chemistry
Nagoya University
Nagoya, Japan

Jennifer Monahan
Birck Nanotechnology Center
Purdue University
Lafayette, Indiana

Lilliana I. Moreno
International Forensic Research Institute
Florida International University
Miami, Florida

Renee K. Mosing
Department of Chemistry
University of Minnesota
Minneapolis, Minnesota

Pradyot Nandi
Department of Pharmaceutical Chemistry
Ralph N. Adams Institute of Bioanalytical Chemistry
University of Kansas
Lawrence, Kansas

Johan Nilsson
Department of Electrical Measurements
Division of Nanobiotechnology
Lund University
Lund, Sweden

Monica Palcic
Carlsberg Laboratory
Valby Copenhagen, Denmark

Janusz Pawliszyn
Department of Chemistry
University of Waterloo
Waterloo, Ontario, Canada

Stephen L. Pentoney, Jr.
Advanced Technology Center
Beckman Coulter, Inc.
Fullerton, California

Bobby G. Poe
Chemistry Department
University of Minnesota
Minneapolis, Minnesota

Joe Power
Department of Chemical and Life Sciences
Waterford Institute of Technology
Waterford, Ireland

Vincent T. Remcho
Department of Chemistry
Oregon State University
Corvallis, Oregon

Michael G. Roper
Department of Chemistry and Biochemistry
Florida State University
Tallahassee, Florida

Juan G. Santiago
Department of Mechanical Engineering
Stanford University
Stanford, California

Michael F. Santillo
Department of Chemistry
Pennsylvania State University
University Park, Pennsylvania

Bettina Sarg
Division of Clinical Biochemistry
Innsbruck Medical University
Innsbruck, Austria

Kiichi Sato
Department of Applied Biological Chemistry
The University of Tokyo
Tokyo, Japan

Rachel Sauvageot
Advanced Technology Center
Beckman Coulter, Inc.
Fullerton, California

Cory Scanlan
Department of Chemistry
University of Illinois at Urbana-Champaign
Urbana, Illinois

Zak K. Shihabi
Department of Pathology
Wake Forest University School of Medicine
Winston-Salem, North Carolina

Kimia Sobhani
Department of Chemistry
University of Washington
Seattle, Washington

Dana M. Spence
Department of Chemistry
Michigan State University
East Lansing, Michigan

Timothy B. Stachowiak
Department of Chemical Engineering
University of California
Berkeley, California

Frantisek Svec
The Molecular Foundry
Lawrence Berkeley National Laboratory
Berkeley, California

Jonathan V. Sweedler
Department of Chemistry
University of Illinois at Urbana-Champaign
Urbana, Illinois

Eszter Szántai
Institute of Medical Chemistry, Molecular Biology, and Pathobiology
Semmelweis University
Budapest, Hungary

Marina F.M. Tavares
Institute of Chemistry
University of São Paulo
São Paulo, Brazil

Shigeru Terabe
University of Hyogo
Hyogo, Japan

Wolfgang Thormann
Department of Clinical Pharmacology
University of Bern
Bern, Switzerland

Manabu Tokeshi
Department of Applied Chemistry
Nagoya University
Aichi, Japan
and
Micro Chemistry Group
Kanagawa Academy of Science and Technology
Kanagawa, Japan

Fernando G. Tonin
Institute of Chemistry
University of São Paulo
São Paulo, Brazil

Emily Turner
Department of Chemistry
University of Washington
Seattle, Washington

Belinda Vallejo-Cordoba
Centro de Investigation en Alimentacion y Desarrolla, A.C.
Hermosillo, Sonora, Mexico

Paul M. van Midwoud
Pharmaceutical Analysis Group
Groningen Research Institute of Pharmacy
University of Groningen
Groningen, the Netherlands

Timothy D. Veenstra
Laboratory of Proteomics and Analytical Technologies
SAIC Frederick Inc.
Frederick, Maryland

B. Jill Venton
Department of Chemistry
University of Virginia
Charlottesville, Virginia

Elisabeth Verpoorte
Pharmaceutical Analysis Group
Groningen Research Institute of Pharmacy
University of Groningen
Groningen, the Netherlands

Jean-Louis Viovy
Laboratoire Physicochimie-Curie
Section de Recherche, Institut Curie
Paris, France

Stephen G. Weber
Department of Chemistry
University of Pittsburgh
Pittsburgh, Pennsylvania

Colin Whitmore
Department of Chemistry
University of Washington
Seattle, Washington

Adam T. Woolley
Department of Chemistry and Biochemistry
Brigham Young University
Provo, Utah

Robert C.R. Wootton
Department of Pharmacy and Chemistry
Liverpool John Moores University
Liverpool, United Kingdom

Jiaqi Wu
Convergent Bioscience Ltd.
Toronto, Ontario, Canada

David L. Yang
Advanced Technology Center
Beckman Coulter, Inc.
Fullerton, California

Part I

Fundamentals and Methodologies

1 Introduction to Capillary Electrophoresis

James P. Landers

CONTENTS

1.1 INTRODUCTION

While the term "electrophoresis" was coined in 1909 by Michaelis,[1] it was the pioneering experiments of Tiselius in 1937[2] that first showed the separation of serum proteins—albumin, and α-, β-, and γ-globulins—by "moving boundary electrophoresis"; this provided the first intimation of the potential use of electrophoretic analysis for biologically active molecules. However, this approach to electrophoresis was limited by the incomplete separation of proteins, the relatively large sample volume needed, and the necessity of relatively low electrical fields due to the convection currents generated by Joule heating, even in the presence of dense sucrose solutions. The historical events that followed in electrophoretic method development were paradigm shifting—for historical detail, the reader is referred to two reviews in the journal *Electrophoresis*, one by Hjertén[3] in the late 1980s and the other by Rilbe[4] in the mid-1990s. The 1940s saw major efforts directed at improving anticonvective media for zone electrophoresis, while paper electrophoresis was shown to be applicable to the analysis of a wide variety of molecules.[5,6] Gels formed from starch[7] and, later, agarose[8,9] were developed for the analysis of peptides, proteins, and oligonucleotides. In the 1950s, Kolin[10] attempted isoelectric focusing (IEF), but the absence of "ampholytes" led to short pH gradients that had poor stability over time. It was in this period that the pioneering work of Hjertén[3] laid the groundwork for the capillary electrophoresis (CE)analysis of diverse analytes, ranging from small molecules (inorganic ions, nucleotides) to proteins and viruses. Much of this work was conducted with a functional CE-like instrument, albeit with 3 mm tubes that Hjertén constructed as early as 1959 (see Figure 1.1).

It was not until the late 1960s that Vesterberg[11] synthesized ampholines to create stable pH gradients that allowed for effective IEF to be accomplished and take its place among the routine molecular tools for biochemical and biomedical analysis. The functionality and acceptance of starch gels was followed by the introduction of acrylamide as a sieving matrix in 1959 where, for the first time, control over pore size and stability was possible.[12] Ornstein[13] and Davis[14] independently introduced disc gel electrophoresis in 1964, which was followed by the introduction of sodium dodecyl sulfate (SDS) as a denaturing agent for protein separation in 1969. Building on 30 years of development, the 1970s bought the concept of stacking gel together with the use of SDS for the separation of the components in T4 phage,[15] which laid the groundwork for the seminal work by two-dimensional electrophoresis [combination of IEF separation followed by polyacrylamide gel electrophoresis (PAGE)] that was described by Dale and Latner[16] and Macko and Stegemann.[17] Stegemann introduced IEF in polyacrylamide gels followed by SDS–PAGE,[18] which was developed

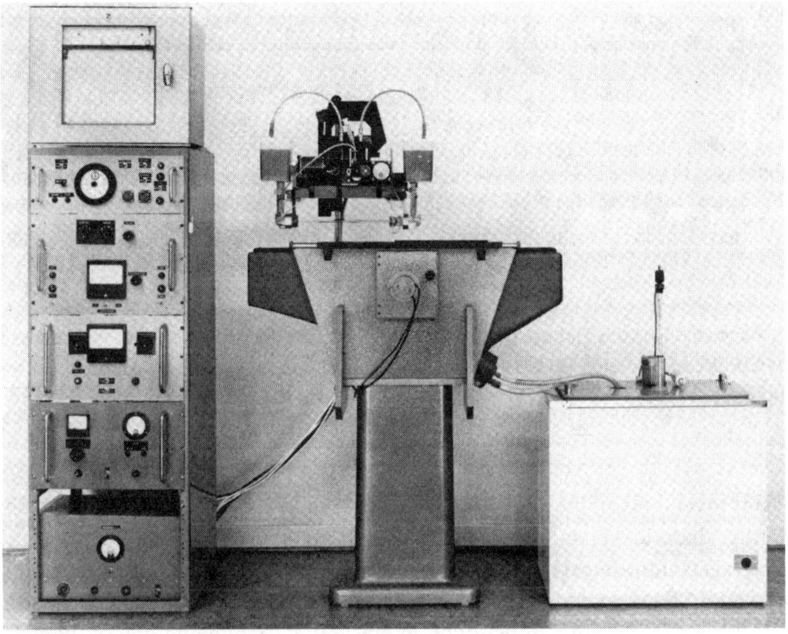

FIGURE 1.1 Photograph of the CE system designed by Hjertén. On the left are stacked the high voltage supply and electronic components for the detector, topped off by a strip-chart recorder. In the center is the carriage with the capillary and the electrode vessels above the immersion bath, while the cooling reservoir flanks it on the right.

into a fine art by O'Farrell[19] in 1975. Interestingly, around this time, Virtanen[20] followed the work of Hjertén a decade earlier with the use of smaller internal diameter (0.2 mm) tubes, which eliminated convection problems and simplified instrumental design. In numerous ways, all of this work seeded the later advances that included development of sequencing gels in 1977, agarose gels a short time later, and then pulsed-field gel electrophoresis in 1983. It was in the very same year (1983) that Jorgenson and Lukacs[21] defined electrophoresis in micron-scale capillaries.

The concurrent development of advanced separation technology in the form of high-performance liquid chromatography (HPLC) made speed, high resolution, quantitative results, and automation a reality. This platform saw widespread adoption in industrial and clinical laboratory settings, filling the need where electrophoresis could not to meet the demands of quantitative analysis, preparative isolation, or automation. These advantages proved HPLC effective for the analysis of small molecules, oligonucleotides, peptides, and small proteins, but with the disadvantage that larger analytes (e.g., structural proteins) were problematic. Moreover, more stringent regulations on waste resulted in increased cost for waste disposal, leading to organic solvent waste generated by HPLC considered as a major contributor to operating costs. Micro LC, in both open-tubular and packed formats,[22–24] were shown to have promise,[25] but technical difficulties in the design of high-pressure, low-volume solvent delivery systems and operational challenges in handling the severe pressure drop in packed column chromatography deterred commercial development.

It was with this historical backdrop of increasing demands for high resolution, quantitative precision of biopharmaceuticals, and control of waste management costs that CE arrived on the analytical scene. The pioneering work of Hjertén[3] and Virtanen[20] preceded the demonstration that CE was shown to be a viable analytical technique by Mikkers et al.[26] and Jorgenson and Lukacs.[21] CE demonstrated the potential for producing high-resolution separations of biopolymers, as well as smaller pharmaceutical agents, and used miniscule amounts of both sample and reagents. In 1989, a decade after Mikkers et al.[26] carried out their defining CE experiments, CE was ready for primetime

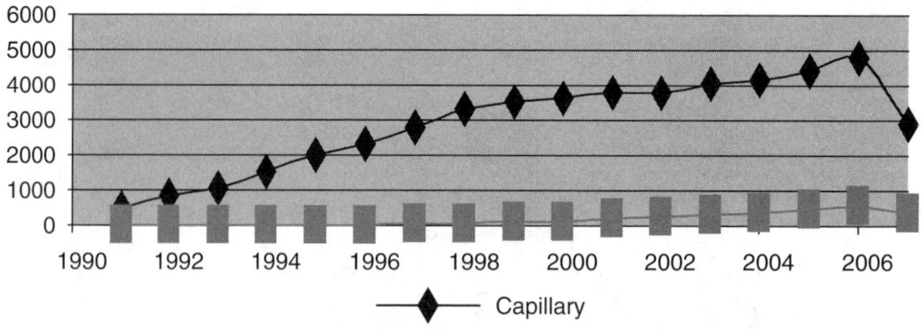

FIGURE 1.2 Growth of the CE and microchip literature. The ISI Web of Science database was searched with the subject keywords of "CE" on a year-by-year basis beginning in 1983 through March 2007 and plotted as a function of publication year.

as a result of improvements in the sensitivity of detectors, advances in automation technology, and, most importantly, the widespread availability of high-quality narrow-bore capillary silica tubing. Beckman Instruments introduced the scientific community to CE with the commercial launch of the first fully automated CE instrument in the form of the *P/ACE*™ 2000. The result of sound engineering is that some of these models can still be found in labs today (ours being one of them).

Within a few years, research groups throughout the world had expanded the horizons of CE. Hjertén,[3] Mazzeo and Krull,[27] and Wehr[28] were early in the pursuit of improvements in capillary coatings to prevent analyte adsorption to the capillary surface while techniques, for casting polyacrylamide and agarose gels in capillaries were developed and applied to protein[29,30] and DNA[31–33] separations. These have now evolved into more sophisticated coatings involving phospholipids,[34] lipoproteins,[35] cholesterol,[36] and polyelectrolytes.[37] Terabe et al.[38,39] pioneered micellar electrokinetic chromatography (MEKC), with the use of micellar solutions to separate neutral and charged species, while IEF was adapted to the capillary format,[40–42] using chemical mobilization methods to move the focused protein zones past the detector. In addition Zhu et al.,[43] Bruin et al.,[44,45] and Ganzler et al.,[46] developed non-gel sieving matrices for size-based separation of biopolymers, which began a flurry of activity defining linear polymers for DNA and proteins analysis by CE.[47] Theoretical studies have led to understanding the problems associated with the technique, and the methods required to overcome them. Productization of more effective and sophisticated instrumentation has made room for the development of new detectors[48–50] and mass spectrometer interfaces.[51] Many of these topics are covered in detail in the chapters that follow in this book.

Beginning with four publications in 1983, growth of the CE literature was exponential for roughly the first decade (not surprising with such small numbers), linearized from 1993 to 1998, and then slowed (but did not plateau) over the next 5 years (1999–2003) (Figure 1.2). While these types of plots often seem superfluous, the point is this—despite the slowing of growth in the CE field around the turn of the century, interest in CE has not stalled. In fact, there has been modest resurgence since then. Method development still continues, but the application focus has intensified in the last half decade. Diversity is evident by application of the technique to a wide spectrum of analyses in a variety of disciplines ranging from the detection and quantitation of priority pollutants in environmental samples, to the analysis of the components of a single cell, the screening for abnormal proteins or DNA fragments indicative of disease or specific typing an individual.

Consistent with the theme of the handbook, this chapter has a practice-oriented focus, with the presentation of select theoretical aspects of CE limited to the basic principles needed for understanding how molecules separate in an applied voltage, and the factors that affect the separation. Appendix 2 provides the reader with a guide to troubleshooting typical problems that may be encountered with CE.

1.2 CAPILLARY ELECTROPHORESIS

1.2.1 Why Electrophoresis in a Capillary?

As introduced in Section 1.1, electrophoresis has been one of the most widely utilized techniques for the separation and analysis of ionic substances. Almost all modes of electrophoresis utilize some form of solid support to prevent convectional distortion of the analyte bands. This has been in the form of paper (high voltage separation of amino acids and other small organic molecules) or, more commonly, polyacrylamide or agarose gels, which have been used extensively for both protein and deoxyribonucleic acid/ribonucleic acid (DNA/RNA) analysis. In light of this, there is little doubt that gel electrophoresis has been an invaluable analytical tool for modern biochemical research. However, despite the ability to resolve the components of complex systems, traditional forms of electrophoresis suffer from several disadvantages, most of which scientists had endured for lack of an effective alternative. Perhaps the most obvious disadvantage is the speed of separation, which is ultimately limited by Joule heating (the heating of a conducting medium as current flows through it). The relatively poor dissipation of Joule heat in slab systems limits their use to low potential electric fields. Moreover, from a methodological perspective, the entire process is a series of cumbersome, time-consuming tasks, from the casting of the gel, preparation and loading of samples, electrophoretic resolution of the ionic/molecular species, to the final stage where the gel is stained, and the results obtained. Other problems include poor reproducibility, particularly with two-dimensional analysis, analyte-dependent differences in staining, which makes quantitative accuracy difficult to achieve (e.g., glycosylated proteins have different dye-binding properties than their unglycosylated counterparts), and the cumbersome methodology involved in modern gel electrophoresis that makes it virtually impossible for the entire electrophoretic procedure to be automated.

The use of capillaries as an electromigration channel for separation of a diverse array of analytes, not only presents a unique approach to separation, but is also associated with several advantages over the standard solid supports used during the evolution of electrophoresis. In particular, the physical characteristics of narrow-bore capillaries make them ideal for electrophoresis. Fused-silica capillaries employed in CE typically have an internal diameter (i.d.) of 20–100 μm (375 μm outside diameter, o.d.), lengths of 20–100 cm, and are externally coated with a polymeric substance, polyimide, which imparts tremendous flexibility to a capillary that would otherwise be very fragile (Figure 1.3). The high surface-to-volume ratio of capillaries with these dimensions allows for very efficient dissipation of Joule heat generated from large applied fields. This is illustrated in Table 1.1, which compares the surface-to-volume ratio of a standard analytical slab gel system with standard capillaries used in CE. From this table, it is clear why the slab gel is limited to fields in the range of $\approx$15–40 V/cm whereas up to 800 V/cm can be applied to a capillary containing the same type of gel matrix shown early on by Dovichi and coworkers.[52] As a result of this ability to dissipate heat, electrophoretic separations can easily be performed at up to 30,000 V with the external capillary environmental thermostatted at ambient temperature.

In addition to the ability to dissipate Joule heat efficiently, the use of capillaries for electrophoresis is associated with many advantages. With typical electrophoresis capillaries, the small dimensions yield total column volumes in the microliter range, thus requiring the use of only milliliter quantities of buffer. Moreover, adhering to the chromatographic rule of thumb restricting sample volume to 1–5% of the total capillary volume, sample volumes introduced into the capillary are in the nanoliter range (as low as 0.2 nL). As a result, as little as a 5 μL of sample will suffice for repetitive analysis on some commercial instruments. Considering the small reagent (buffer) and sample requirements, as well as the rapid analysis times associated with the application of high fields (30,000 V), it is clear why CE has been applied to, and has provided the solution to, a diverse number of analytical problems.

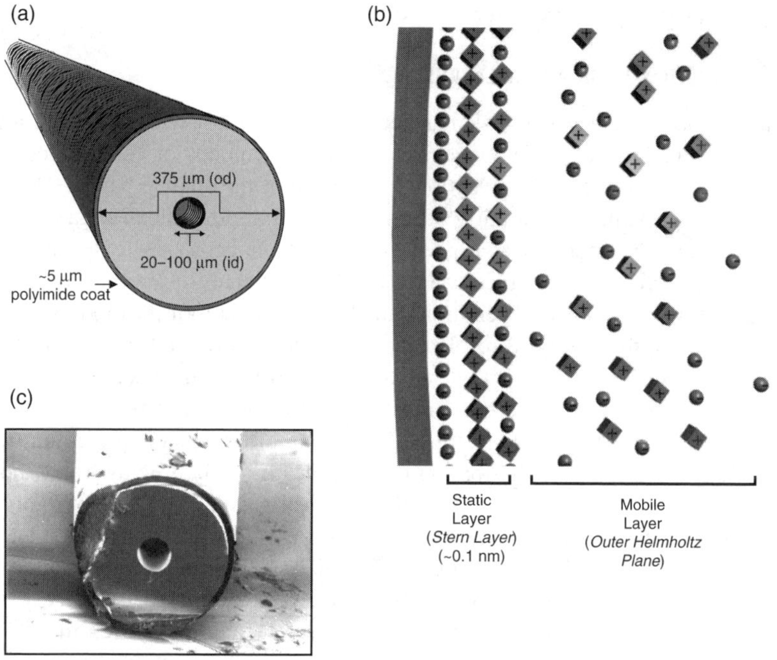

FIGURE 1.3 Diagram of the capillary, its inner surface and the ionic layer. (a) Shows an end view of a capillary with a 375 μm o.d. and a 50 μm i.d. (b) An illustration of the double-ionic layer formed in bare silica capillaries critical in generating endoosmotic flow. (c) A scanning electron photomicrograph of a 375 μm (o.d.) × 75 μm (i.d.) capillary enlarged 170 times. The polyimide coating which grants the capillary flexibility is visible and the "etch" used to break the capillary is clearly seen on the left.

TABLE 1.1
A Comparison of Surface-to-Volume Ratios for an Analytical Slab Gel and a 57 cm Capillary Having a Varied Internal Diameter

	Surface Area (mm^2)	Volume (μL)	Surface-to-Volume Ratio
Slab gel (14 × 11.5 × 0.15 cm)	32,200	24,150	1.3
20 mm i.d. capillary	35.81	0.179	200
50 mm i.d. capillary	89.53	1.119	80
75 mm i.d. capillary	134.3	2.518	53
100 mm i.d. capillary	179.1	4.477	40
200 mm i.d. capillary	258.1	17.907	20

1.2.2 The Family of CE Modes

In much the same way that standard gel electrophoretic techniques diversified, so did CE. This has resulted in a family of specialized modes that collectively constitute "CE." The main modes of CE that have been developed and are presently being exploited include capillary zone electrophoresis (CZE), MEKC, capillary electrokinetic chromatography capillary (CEC), capillary isoelectric focusing (CIEF), capillary gel electrophoresis (CGE), and capillary isotachophoresis (CITP). Over the last decade or so, this latter mode has been primarily used as an on-capillary preconcentration technique.

TABLE 1.2
Modes Used for Analysis of Various Classes of Analytes

CZE	MEKC	CEC	CIEF	CGE
Ions	Small molecules	Small molecules	Peptides	Nucleic acids
Small molecules	Peptides	Peptides	Proteins	
Peptides		Proteins		
Proteins		Carbohydrates		
Carbohydrates				

CZE is the most universal of the techniques, having been shown to be useful for the separation of a diverse array of analytes varying in size and character. Although a brief description of CZE is given in this chapter, various aspects of the technique will be discussed in a number of chapters to follow in this handbook. The other CE modes are covered exclusively, and in detail, in separate chapters dedicated specifically to the theoretical and practical aspects of those modes, along with examples illustrating applications amenable to that particular mode. Table 1.2 provides a reference table categorizing the modes used for the analysis of various classes of analytes.

1.2.3 Capillary Zone Electrophoresis

Capillary zone electrophoresis is not only the simplest form of CE, but also the most commonly utilized. Discussion of this mode permits the presentation of a generic design for the instrumentation for CE. The addition of specialized reagents to the separation buffer readily allows the same instrumentation to be used with the other modes mentioned in the previous section: addition of surfactants with MEKC, ampholines for CIEF and a sieving matrix (linear polymers, entangled matrices) for CGE. The discussion on CZE in the following subsections allows for analysis of some of the basic principles governing analyte separation by this technique.

1.2.3.1 Instrumentation and CE Analysis

A diagrammatic representation of a CE instrument is presented in Figure 1.4. The basic components include a high voltage power supply (0–30 kV; possibly 60 kV if you are in Jorgenson's lab), a polyimide-coated capillary with an internal diameter $\leq$200 μm, two buffer reservoirs that can accommodate both the capillary and the electrodes connected to the power supply, and a detector. As will be discussed later in this chapter, thermostatting of the capillary is critical to efficient and reproducible separations and, hence, some type of capillary thermostatting system should be used. To perform a capillary zone electrophoretic separation, the capillary is filled with an appropriate separation buffer at the desired pH and sample is introduced at the inlet. Both ends of the capillary and the electrodes from the high voltage power supply are placed into buffer reservoirs and up to 30,000 V applied to the system. The ionic species in the sample plug migrate with an electrophoretic mobility (direction and velocity) determined by their charge and mass, and eventually pass a detector where information is collected and stored by a data acquisition/analysis system.

Electrophoretic mobility (μ) of a charged molecular species can be approximated from the Debye–Huckel–Henry theory

$$\mu = q/6\pi\eta r \quad (1.1)$$

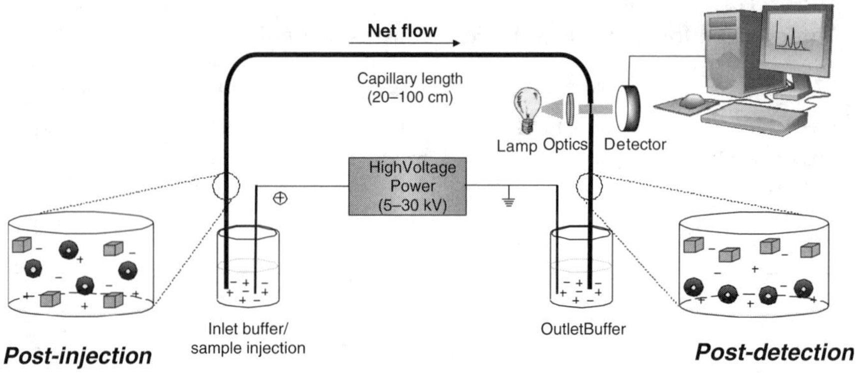

FIGURE 1.4 General schematic of a CE instrument.

where q is the charge on the particle, η is the viscosity of the buffer, and r is the Stokes' radius of the particle. The mass of the particle may be related to the Stokes' radius by $M = (4/3)\pi r^3 V$, where V is the particle specific volume of the solute. Although one might infer the direct proportionality of mass and radius of the particle, empirical data suggest modifications of Equation 1.1 to allow for the nonspherical shape, the counterion effects, and nonideal behavior of proteins and biological molecules.[53]

With CZE, the "normal" polarity is considered to be [inlet—(+), detector—(−) outlet] as shown in Figures 1.4 and 1.5. As electrophoresis ensues, the analytes separate according to their individual electrophoretic mobilities and pass the detector as "analyte zones" (hence, the term capillary zone electrophoresis or CZE). The fact that, under appropriate conditions, all species (net positive, net negative, or neutral) pass the detector indicates that a force other than electrophoretic mobility is involved. If the applied field were the only force acting on the ions, net positively charged (cationic) substances would pass the detector while neutral components would remain static (i.e., at the inlet) and anionic components would be driven away from the detector. It is clear that, if this were the case, CE would be of limited use. Fortuitously, there is another force, "electroosmotic flow" (EOF), driving the movement of all components in the capillary towards the detector when under an applied field (and a normal polarity). EOF plays a principle role in many of the modes of CE and most certainly in CZE. This is discussed briefly in the next section.

1.2.3.2 Role of EOF in CE Analysis

Electroosmotic flow was first identified in the late 1800s when Helmoltz[54] conducted experiments involving the application of an electrical field to a horizontal glass tube containing an aqueous salt solution. Curious about the ionic character of the inner wall and the movement of ions, he found that the silica imparted a layer of negative charge to the inner surface of the tube, which under an applied electric field, led to the net movement of fluid toward the cathode. More than a century later, this phenomenon still plays the fundamental role in CE analysis. Moreover, the importance of the control of EOF has been realized and has become the focus of several research groups.

As a continuation of the pioneering work of Helmholtz, the basic principles governing EOF have been evaluated extensively. As shown by the expanded region of the inner wall of a capillary in Figure 1.3, the ionized silanol groups (SIO) of the capillary wall attract cationic species from the buffer. Obviously, the buffer pH will determine the fraction of the silanol groups that will be ionized; understanding the amorphous nature of silica and the pK_a range (4–6) associated with the various types of silanol groups[55] is key. The ionic layer that is formed has a positive charge density that

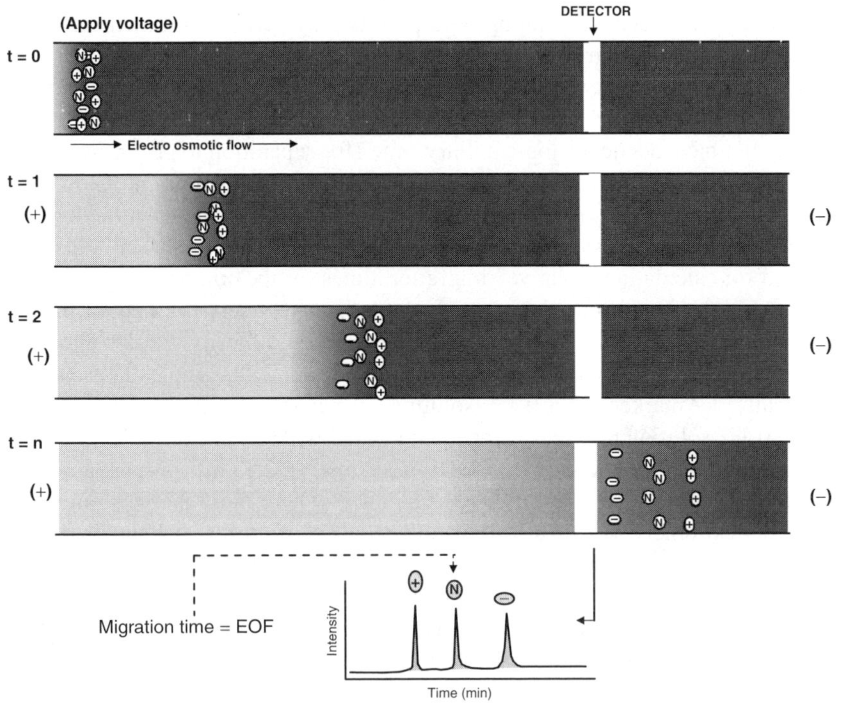

FIGURE 1.5 Mobility of charged and uncharged molecules in an applied field.

decreases exponentially as the distance from the wall increases. The double layer formed closest to the surface is termed the "Inner Helmholtz" or Stern" layer and is essentially static. A more diffuse layer formed distal to the Stern Layer is termed the "Outer Helmholtz Plane" (OHP). Under an applied field, cations in the OHP migrate in the direction of the cathode carrying waters of hydration with them. Because of the cohesive nature of the hydrogen bonding of the waters of hydration to the water molecules of the bulk solution, the entire buffer solution is pulled toward the cathode. This EOF or "bulk flow" acts as a pumping mechanism to propel all molecules (cationic, neutral and anionic) toward the detector with separation ultimately being determined by differences in the electrophoretic migration of the individual analytes. The importance of EOF in capillary electrophoretic analysis is highlighted in Figure 1.5. The buffer entering the capillary inlet behind the sample plug is represented by a "graded shading" for illustrative purposes, and is identical to the buffer preceding the sample. As electrophoretic migration occurs, all analytes are swept towards the detector by bulk flow. Provided that the EOF is adequate but not too strong, the respective electrophoretic mobilities of each of the analytes leads to the formation of discrete zones by the time they pass the detector. If the EOF is low, diffusion of the analyte zones could result in substantial band broadening and, under conditions of very low EOF, some of the analytes may not reach the detector within a reasonable analysis time.

As previously mentioned, the EOF is pH dependent and can be quite strong. For example, in 20 mM borate buffer at pH 9.0, the EOF is ≈2 mm/s, which translates to a flow of ≈4 nL/s in a 50 μm i.d. capillary. The inclusion of EOF into the calculation of velocity is essential and results in

$$v_i = \mu_{\text{app}} E = (\mu_{\text{ep}} \mu_{\text{eo}}) E \tag{1.2}$$

where μ_{ep} is the mobility due to the applied electric potential and μ_{eo} is the mobility due to EOF. From a practical perspective, EOF acts as an electric field-driven pump that may be considered analogous to the mechanical pump used in HPLC. While the simplicity of this pumping system has obvious advantages over mechanical pumps (e.g., no moving parts, no mechanical wear) one does not have the same degree of control associated with mechanical systems.

It is important to have some idea of the magnitude of EOF under the conditions used for a given separation for two reasons. First, under conditions where EOF is very fast, components of the mixture may not have adequate "on-capillary time" for separation to occur. Second, it is useful to know where neutral compounds migrate in the obtained electropherogram. Knowing this, information about the charge character of the sample components is obtained on the basis of whether they migrate faster (cationic) or slower (anionic) than EOF. The EOF marker is also useful as an internal standard for calculating "relative" migration times for the components of the sample from the apparent migration times. Some compounds that adequately serve as neutral markers include dimethyl formamide (DMF), dimethyl sulfoxide (DMSO), and mesityl oxide, although, because of rapid volatilization, the latter compound is of limited use with samples kept at ambient temperature. Typically, a "marker" is a 0.1% solution in water; a 1-s hydrostatic (0.5 psi) sample injection of the 0.1% solution provides an adequate signal with ultraviolet (UV) detection at either 214 or 200 nm. Fluorescent markers (e.g., BODIPY) can serve the same purpose for LIF detection

These markers are not of use with low pH separations where most analytes migrate faster than the EOF, which is extremely low. If a peak is still required as an internal standard for correction of migration time, any compound with a fast cathodic electrophoretic mobility will suffice as a "frontal marker." We have found that a synthetic peptide containing seven lysine residues and a single tryptophan (K_3WK_4) functions adequately for this purpose at pH 2.5. A frontal marker may also be useful at higher pH (e.g., when the neutral marker comigrates with a species of interest), where cationic species such as normetanephrine can suffice.

It is interesting that, since its inception, the holy grail in CZE has been a mechanism to control EOF. The application of radial field in a manner that controls the magnitude of the "zeta potential" (or the thickness of the double layer) on the capillary wall would, in theory, allow one to simply dial up the desired magnitude of EOF. This mechanism has been illusive for the better part of 15 years, still has not been determined, but clearly still is of interest.[56]

1.2.3.3 A Description of the Electrophoretic Process

In the early 1980s, Jorgenson and Lukacs[21] discussed the theory and basis for electrophoretic separations in capillaries. The following brief discussion develops, albeit at a basic level, the theoretical concepts describing the movement of charged molecules in a buffer-filled capillary under an applied electrical field and the shape of the zone during the electrophoretic process. The separation process is discussed in terms of resolution (how adequately two components are separated) and efficiency (how long the separation takes). Finally, we discuss factors that affect the resolution. For more detailed evaluation of these processes, the reader is referred to several excellent discussions on this subject.[57–59]

1.2.3.3.1 Mobility of an Analyte in a Capillary

A charged particle in solution will become mobile when placed in an electric field. The velocity, v_i, acquired by the solute under the influence of the applied voltage H, is the product of μ_{app}, the apparent solute mobility, and the applied field E ($E = H/L$, where L is the length of the field)[21]

$$v_i = \mu_{app}E \tag{1.3}$$

μ_{app} is a property of the particle, and proportional to its charge, and inversely proportional to the frictional forces acting upon it in solution. The electrical force can be given by

$$F_{el} = qE \tag{1.4}$$

and the frictional force on a spherical ion is

$$F_{\text{fr}} = -6\pi\eta r v_i. \tag{1.5}$$

During the electrophoresis, a steady state is attained, where the two forces are equal, but in opposite directions:

$$qE_{\text{r}} = -6\pi\eta r v_i \tag{1.6}$$

Solving for velocity (or E) and substituting Equation 1.6 into Equation 1.3 and rearranging for μ, we obtain

$$\mu_{\text{app}} = q/6\pi\eta r \tag{1.7}$$

From this equation, it is evident that the mobility of the analyte is a property of both the charge and size; a small, highly charged particle will have a high mobility, while a large, minimally charged species will have a low mobility. Refer to Appendix 1 for example on calculating mobility.

1.2.3.3.2 Shape of the Analyte Zone

Under the initial conditions of electrophoresis, the boundary between the buffer solution and the solute mixture forms a zone of infinitesimal thinness at right angles to the direction of applied current and migration. As migration proceeds, this initially sharp boundary will undergo a progressive deterioration in shape. The most important influence on this process is diffusion, as the initial conditions impart a severe concentration gradient across the zonal boundary. By applying Fick's second law of diffusion, Weber[57] has shown that the variation of solute concentration in the direction of migration is given by the equation

$$C_{x,t} = \frac{k}{2(D_i\pi t)^{1/2}} e^{\frac{-(x-v_i t)^2}{4D_i t}} \tag{1.8}$$

where C is the solute concentration at a distance x from the initial position after time t, v_i is the electrophoretic velocity of the solute, D_i is the diffusion coefficient of the solute, and k is a constant. By integration, initially with the boundary conditions at $t = 0$, at distant $x = \infty$, $C = 0$, and a second time with $t = 0$, $x = 0$ and $\mathrm{d}C/\mathrm{d}x = 0$, we obtain a mathematical description of the concentration profile. As the solute migrates a distance x from the origin after time t, the zone may be described by a Gaussian curve.

The characteristics of a Gaussian profile describe the peak maximum and the peak width. The maximum is dependent on initial concentration of solute. The width depends on the length of time from initial conditions (i.e., application of sample in to the system) and the diffusion constant, D_i. The width may be given by the distance between the inflection points, in our case, $2/(2D_i t)^{1/2}$. From analogy with probability calculations, the width of a Gaussian curve is termed the standard deviation (σ)

$$\sigma = (2D_i t)^{1/2} \tag{1.9}$$

and the square of standard deviation is called the variance

$$\sigma^2 = 2D_i t. \tag{1.10}$$

Remembering that

$$t = L_d/v_i = L_d L_t/\mu_{app} V = L_d/\mu_{app} E, \tag{1.11}$$

where L_d is the distance from the origin to the detector, and L_t is the total length of the capillary (i.e., the length of the applied electric field), we may substitute into Equation 1.10, to obtain

$$\sigma^2 = 2D_i L_d/(\mu_{app} E). \tag{1.12}$$

One should remember that the discussion to this point has only dealt with diffusion as a dispersive phenomenon affecting peak shape. Other factors that contribute to what is observed on the electropherogram as band broadening, the practical result of variance, will be discussed later.

1.2.3.3.3 Resolution and Efficiency

The simplest way to characterize the separation of two components is to divide the difference in migration distance by the average peak width to obtain resolution (Res)

$$\text{Res} = 2(x_{i2} - x_{i1})/(w_1 + w_2) \tag{1.13}$$

where x_i is the migration distance of the analyte i, and the subscript 2 denotes the slower moving component, and w is the width of the peak at the baseline.[57] We can readily see that the position of a peak, x_t, is determined by the electrophoretic mobility. The peak width, w, is determined by diffusion and other dispersive phenomena. For two neighboring peaks, $w_1 = w_2$, and

$$\text{Res} = (x_{i2} - x_{i1})/w_2. \tag{1.14}$$

From the equation describing a Gaussian curve, the two peaks touch at baseline when $\Delta x_i = x_2 = 4\sigma$, and Res = 1, or

$$\text{Res} = \Delta x_i/4\sigma. \tag{1.15}$$

Remembering that distance is equal to velocity multiplied by time ($x_i = v_t t$) and substituting for σ from Equation 1.9 into Equation 1.15, we obtain

$$\text{Res} = (\Delta\mu_{app} E)t/4(2D_{i,avg} t)^{1/2} \tag{1.16}$$

where $\Delta\mu_{app}$ is the difference in apparent electrophoretic mobility of the two solutes and $D_{i,avg}$ is the average diffusion of the two solutes.

To obtain a measure of efficiency for the process, we use probability theory.[58] For a random walk process of length L, made of n steps, the variance is given by

$$\sigma = l(n)^{1/2} \tag{1.17}$$

where l is the length of each step. If each step is independent of any other step, each contributes to the total variance of the process[59]

$$\sigma_{tot}^2 = \sum \sigma_i^2. \tag{1.18}$$

Substituting from Equation 1.17 and rearranging

$$1/n = L^2/\sigma_{tot}^2 = N. \tag{1.19}$$

The number of steps in the random process, n, is inversely related to the number of theoretical plates, N, a measure of efficiency for the process.

We may substitute for L and σ in Equation 1.19 to express

$$N = (\mu_{avg}E)^2t^2/(2D_it) \tag{1.20}$$

where μ_{avg} is the average mobility of the two solutes. By comparing the expression for Res (Equation 1.16) with the definition of N (Equation 1.20), we obtain an expression relating resolution to the number of theoretical plates

$$\text{Res} = (1/4)(\Delta\mu_{app}/\mu_{avg})N^{1/2}. \tag{1.21}$$

The utility of Equation 1.21 is that it permits one to independently assess the two factors that affect resolution, selectivity, and efficiency. The selectivity is reflected in the mobility of the analyte(s), while the efficiency of the separation process is indicated by N.

If Res = 1, then

$$N = 16/(\Delta\mu_{app}/\mu_{avg})^2. \tag{1.22}$$

Another expression for N is derived from Equation 1.19, using the width at half-height of a Gaussian peak

$$N = 5.54(L/w_{1/2})^2, \tag{1.23}$$

where $5.54 = 8 \ln 2$, and $w_{1/2}$ is the peak width at half-height.[60]

At this point, it is important to note that it is, in fact, misleading to discuss theoretical plates in electrophoresis. The concept is a carry-over from chromatographic theory, where a true partition equilibrium between two phases is the physical basis of separation. In electrophoresis, separation of the components of a mixture is determined by their relative mobilities in the applied electric field, which is a function of their charge, mass and shape. The theoretical plate is merely a convenient concept to describe the analyte peak shape, and to assess the factors that affect separation. Refer to Appendix 1 for examples on calculating resolution and efficiency.

1.2.3.3.4 Source of Variance

While N is a useful concept to compare the efficiency of separation among columns, or between laboratories, it is difficult to use to assess the factors that affect that efficiency. This is due to the fact that it refers to the behavior of a single component during the separation process, and is unsuited to describing the separation of two components or the resolving power of a capillary. A more useful parameter is the height equivalent of a theoretical plate (HETP).[58]

$$\text{HETP} = L/N = \sigma_{tot}^2/L. \tag{1.24}$$

HETP might be thought of as the fraction of the capillary occupied by the analyte. It is more practical to measure HETP as an index of separation efficiency, rather than N, as the individual components that contribute to HETP may be individually evaluated and combined to determine an overall value. The variance of multiple dispersive phenomena on the analyte may be summed:

$$\sigma_{tot}^2 = \sigma_{diff}^2 + \sigma_T^2 + \sigma_{int}^2 + \sigma_{wall}^2. \tag{1.25}$$

A consideration of all the factors influencing σ_{tot}^2 should include not only diffusion, but also differences in mobility or diffusion generated by Joule heating, the reality that the sample is not introduced as a thin disk but as a plug of finite dimensions, and interaction of analytes with the capillary wall. Each of these factors will be addressed separately in a simplistic manner. Theoretical derivation will be given only to emphasize the importance of various parameters, and how they contribute to the overall result. References shall direct those desiring more information to the appropriate literature.

Two studies have addressed the variance due to a variety of sources. Huang et al.[61] investigated small molecules; Jones et al.,[62] studied proteins as well as amino acids and a neutral marker. The two studies demonstrate that the band broadening observed in CE is in excess over calculated values of diffusion and analyte interaction. Huang et al.[61] and Jones et al.[62] attribute the excess variance to sample introductory practices.

1.2.3.3.4.1 Variance Caused by Temperature Temperature control is an issue, because it may be effected by the efficiency with which the capillary is thermostatted, or by choice of buffer ionic species or ionic strength. Thus, from a practical point of view, we discuss the variance caused by temperature to impress upon the practitioner the importance of controlling temperature in obtaining reproducible electropherograms.

Current passing through a conducting solution generates heat. A sample way of looking at the problem is to compare the equivalent expressions for the applied potential, H, and heat generation, W

$$H = i/\kappa\pi r^2 \tag{1.26}$$

$$W = i^2/\kappa(\pi r^2)^2 \tag{1.27}$$

where i is the current through the electrolyte solution, πr^2 the cross-sectional area of the capillary, and κ is the specific conductance of the buffer. By combining Equations 1.26 and 1.27, to take the ratio we obtain

$$H/W = (i/\kappa\pi r^2)(\kappa(\pi r^2)^2/i^2) = \pi r^2/I \tag{1.28}$$

From Equation 1.28, it can be readily seen that the reduction of heat production can be achieved by reducing the current density in the capillary. This may be accomplished by increasing the cross-sectional area of the capillary or by reducing the current. The latter is preferred, since increasing the diameter of the capillary results in a reduction of the surface-to-volume ratio, and leads to less efficient heat dissipation (see Table 1.1). The choices to reduce the sysetm current lie in carrying out the separation at a lower voltage, or reducing the ionic strength of the separation buffer.

To develop the quantitative expression for thermal heating, we need to describe the electrophoretic front, and its behavior under a thermal gradient. According to the Poiseuille equation, which describes the parabolic flow due to pressure,

$$v_z = \Delta P r^2/8L\eta. \tag{1.29}$$

Joule heating of the electrolyte solution creates a similar flow profile, the equivalent expression being

$$v_z = v(1 + E^2\Lambda C_b B r^2/4k_b T^2)[1 - (r_x/r)^2] \tag{1.30}$$

where $E^2\Lambda$ is the rate of heat generation per unit volume, C_b is the buffer concentration, k_b is the thermal conductivity of the electrolyte solution. Λ is the equivalent conductanceof the electrolyte solution, T is the absolute temperature, r_x is the radial position (which varies from 0 at the center of the capillary to r at the capillary wall), and B is a buffer-related viscosity constant.[63] The variance caused by dispersion in a parabolic velocity profile is given by

$$\sigma_{\mathrm{T}}^2 = 2D_it + \left(r^6E^4C_{\mathrm{b}}^2B^2\Lambda^2t\right)\Big/\left[24D_i\left(8k_{\mathrm{b}}T^2 - E^2\Lambda C_{\mathrm{b}}Br^2\right)^2\right]. \tag{1.31}$$

For most capillary applications, where radius is small (25–50 μm), and E is kV, $E^2\Lambda C_{\mathrm{b}}Br^2$ is $\ll 8k_{\mathrm{b}}T^2$ and the second term reduces to

$$\sigma_T^2 = r^6E^4C_{\mathrm{b}}^2B^2\Lambda^2t/1536D_ik_{\mathrm{b}}^2T^4. \tag{1.32}$$

The strong dependence of σ_{T}^2 on the radial dimension of the capillary (r^6) and the field strength (E^4) demonstrate the importance of performing high voltage electrophoretic separations in narrow-bore capillaries. It also highlights the necessity of obtaining efficient capillary cooling, to prevent thermal effects from not only affecting sample liability, but also to avoid affecting solute mobility. For small molecules with relatively larger diffusion constants (on the order of 10^{-5} cm^2/s) Grushka et al.[63] have calculated a maximum radius of 65 μm with a 0.1 M buffer solution at 30 kV, and 130 mm with 10 μM buffer for less than 5% increase in dispersion. With larger molecules, having diffusion constants on the order of 10^{-6} cm^2/s, these dimensions were halved. Jones et al.[62] have introduced a novel method to study Joule heating effects and nonideal plug flow contributions to analyte dispersion, with polarity reversal at constant voltage. By recording the variance (band broadening) of the peaks over time and plotting the slope of the obtained line versus applied potential, these authors have developed a measure of nonideal behavior. A comparison of charged analyte behavior with neutral solute variance (which should be unaffected by electrophoretic mobility effects of thermal heating, but affected equally by the temperature-dependent effects on a diffusion and viscosity) allows an estimate of the Joule heating effect. With adequate thermostatting, one need not worry about thermal dispersion under normal operating conditions (see Reference 64), but one needs to be aware of how temperature generated within the capillary can affect the separation and resolution of the analytes.

1.2.3.3.4.2 Variance Caused by Finite Sample Introduction Volume In the mobility section above, we assume the solute was initially present as a think disk that dispersed into a zone or band. In reality, the sample is introduced into the capillary as a cylindrical plug. Sternberg[60] derived the commonly used equation describing variance due to sample introduction.

$$\sigma_{\mathrm{int}}^2 = l_{\mathrm{int}}^2/12 \tag{1.33}$$

where l_{int} is the length of the sample introduction plug. This formula assumes that the sample is introduced as a rectangular plug, which is a close approximation for CE. As the plug has a finite volume, it also has measurable length. Huang et al.[61] and Jones et al.[62] have investigated the relative contribution of introduction plug length (by hydrodynamic or electrokinetic methods) to the observed peak width, and concluded that $l_{\mathrm{int}}/L_{\mathrm{d}}$ of less than 3% does not lead to excessive band broadening. Only at very small plug lengths (<200 μm) does one observe variance due to diffusion of the sample plug; typical sample introduction plug lengths of 300–6000 μm (0.3–6 mm) obscure this effect. Huang et al.[61] and Jones et al.[62] conclude that sample introduction volume, that is, plug length, is the most significant factor in excessive band broadening observed in CE.

1.2.3.3.4.3 Variance Caused by Analyte–Wall Interactions The interactions of analyte and capillary wall, or components within the sample solution are numerous, complex and sample specific. The best approach to understanding the band broadening due to adsorption is the approach of McManigill and Swedberg.[64] The general equation of Giddings[59] may be adapted to include an adsorption term

$$\mathrm{HETP} = l^2/12L + 2D_{i,T}/u + \left[k'\mu \Big/ (1+k')^2\right]\left[r_0^2/4D_{i,\mathrm{T}} + 2\big/k_\mathrm{d}\right], \qquad (1.34)$$

where l is the length of the injected plug, L is the total length of the capillary, u is the flow rate, k' the capacity factor, r_0 is the radius of the capillary, $D_{i,T}$ is the diffusion coefficient in the designated solvent at the specific temperature, and k_d in the first-order dissociation constant off the surface.

Through carefully designated and executed experiments, a value of for k' may be determined at differing flow rates, and the effect on plate height calculated. McManigill and Swedberg[64] have shown that the value for k' is not zero under any conditions, and that a value for k' as low as 0.001 can affect the shape of the peak, and, consequently, the efficiency of separation.

To obtain estimates of the relative contribution to overall variance, Jones et al.[62] used a voltage interruption/polarity reversal method. Huang et al.[61] attribute part of the excessive peak width observed in their experiments to analyte–wall interaction. One may minimize such interactions by the appropriate choice of buffer pH, ionic strength, or buffer additives (see Section 1.3).

The variance of multiple dispersive phenomena on the analyte may be summed as

$$\sigma^2_\mathrm{tot} = \sigma^2_\mathrm{diff} + \sigma^2_T + \sigma^2_\mathrm{int} + \sigma^2_\mathrm{wall}. \qquad (1.35)$$

To enable the reader to visualize the magnitude of the contributions by each of these variances, we include some typical calculations.

For a small molecule, $D_i = 10^{-5}$ cm^2/s, time of analysis might be 10 min (600 s), introduction plug of 1.2 nL into a 50 μm by 47 cm capillary, 40 cm to the detector, results in a plug length of 0.6 cm

$$\sigma^2_\mathrm{diff} = 12 \times 10^{-3},$$

$$\sigma^2_\mathrm{T} = \sigma^2_\mathrm{diff} = 12 \times 10^{-3},$$

$$\sigma^2_\mathrm{int} = 3.0 \times 10^{-2},$$

$$\sigma^2_\mathrm{tot}(\mathrm{exp.\ observed}) = 5 \times 10^{-2}.$$

$$\sigma^2_\mathrm{tot}(\mathrm{calculated}) = 1.2 \times 10^{-2} + 1.2 \times 10^{-2} + 3.0 \times 10^{-2} + \sigma^2_\mathrm{wall}$$

$$= 5.4 \times 10^{-2}$$

These figures indicate σ^2_int is, by far, the largest contributor to the band broadening with a small, highly diffusible analyte.

For a protein, $D_i = 10^{-6}$ cm^2/s, time of analysis might be 10 min (600 s), introduction plug of 1.2 nL into a 50 μm by 47 cm capillary, 40 cm to the detector, results in a plug length of 0.6 cm

$$\sigma^2_{\text{diff}} = 12 \times 10^{-4},$$

$$\sigma^2_{\text{T}} = \sigma^2_{\text{diff}} = 12 \times 10^{-4},$$

$$\sigma^2_{\text{int}} = 3.0 \times 10^{-2},$$

$$\sigma^2_{\text{tot}}(\text{exp. observed}) = 5 \times 10^{-2}.$$

$$\begin{aligned}\sigma^2_{\text{tot}}(\text{calculated}) &= 1.2 \times 10^{-3} + 1.2 \times 10^{-3} + 3.0 \times 10^{-2} + \sigma^2_{\text{wall}} \\ &= 5.0 \times 10^{-2}\end{aligned}$$

The injection plug is the largest contributor to the band broadening observed, but the wall interaction has nearly equal contribution.

1.2.3.4 The Capillary

Having the discussed EOF and its dependence on the capillary wall surface, one can understand how the capillary and its inner sufrace condition can have an impact on the efficiency and reproducibility of CE analyses. Hence, the treatment of the capillary before, during, and following electrophoretic separations is crucial. Microbore capillaries employed in CE can be purchased from a number of companies in either bare silica or coated format. As shown in Figure 1.3, the standard fused-silica capillary typically has an internal diameter of 50–100 μm, although a range of internal and external diameters are commercially available (<1μm–>1 mm i.d.; 60 μm–>1 mm o.d.) in shapes that are commonly cylindrical, but also rectangular. The following sections describe some pertinent points regarding capillary preparation, conditioning before use, maintenance, and storage respectively.

1.2.3.4.1 Preparation for CE

For use in CE, the capillary is cut to the appropriate length with a ceramic knife, or an ampoule file, so that both ends are square and flat. This is of particular importance with the inlet end of the capillary where sample is to be introduced. Closer to the outlet, a "window" is created through removal of the polyimide coating so that online detection is possible. The window should ideally be approximately 0.3 cm in length (no longer than 1.0 cm) and can easily be made by burning off the polyimide with a flame and wiping the surface with an ethanol-soaked lens tissue. Another, more labor-intensive method is to scrape off the polyimide surface, being careful not to scratch the silica, or break the capillary. Several mechanical devices have been devised to accomplish this task.[65,66] Commercial instruments are available for carrying this out (e.g., Capital Analytical, UK) Caution must be taken when handling the capillary once the window is created since the window area is extremely fragile. It is important to note that with internally coated capillaries, detector windows cannot be created by burning off the polyimide since the heat required will damage the internal coating. Alternatively, dropwise addition of 103°C sulfuric acid[33] or hot concentrated KOH[67] from a burette or glass pipette have been reported to remove the polyimide coating to create a window without damaging the internal coating. We use an adaptation of the hot sulfuric acid method. A drop of concentrated sulfuric acid is placed on the capillary where the window is desired, and the tip of a hot soldering iron is briefly touched to the drop. This readily removes the polyimide coating without overheating the interior surface, and prevent the necessity of heating a larger volume of acid than would otherwise be needed. The excess acid must be washed off the exterior surface, especially if the capillary is to be placed into a cartridge where the acid would damage the cartridge housing.

When the capillary is installed, it is useful to determine the "capillary fill time." This information will be useful for determining the length of time for each rinse step (NaOH, separation buffer or rinse solution), as well as for diagnosing potential capillary problem, such as partial or complete obstruction. One approach for accomplishing this (with ultraviolet, UV detection) is to pressure rinse the capillary with water, zero the detector, and follow with a rise with 100 mM, NaOH. The time required for a maximum change in absorbance to occur is the "fill time" to the detector. Factoring in the length between the detector and the capillary outlet yields the "capillary fill time."

Before using new bare fused-silica capillary for analysis, the capillary should be "preconditioned" by rinsing with 5–10 column volumes of 100 mM NaOH followed by 5–10 column volumes of water, before rinsing with 3–5 column volumes of separation buffer. If the capillary is coated, preconditioning should be performed as per the protocols recommended by the manufacturer or as deemed necessary by the surface coating.

1.2.3.4.2 Conditioning of the Capillary

When using a newly installed capillary or changing to a new separation buffer, the capillary must be adequately equilibrated with the separation buffer, a process termed "conditioning." Equilibration is particularly important when a phosphate-containing buffer is involved. For acceptable reproducibility, a phosphate-containing buffer should be equilibrated in the capillary for a minimum of 4 h before use.[68] This process can be slightly accelerated by applying a separation-scale voltage to the system during the equilibration period. When preparing to use a new separation buffer in a capillary that has been equilibrated with a phosphate-containing buffer, extensive washing and equilibration of the capillary is typically required. The capillary should be equilibrated with the desired separation buffer for at least several hours before use and, if possible, conditioned overnight. It is for this reason that "dedication" of capillaries to individual buffer systems may be highly recommended. In this manner, no pre-equilibration time may be required, since the capillary can be stored in the buffer to which it is dedicated. However, users often report reproducibility problem that are suspect of surface chemistry deviation, particularly in the buffer ph range of 4–7. Watzig et al.[69] describe preliminary results using x-ray photoelectron spectroscopy to probe the Si-C landscape of fused-silica surfaces. This led the authors to suggest that, in order to yield capillaries that will provide stable migration times, especially in the pH range 4–7, preconditioned for longer than 1 h is required.

1.2.3.4.3 Regeneration of the Capillary Surface

Capillary maintenance plays a critical role in attaining reproducible results with CE. As with any untreated silica surface, ionized silanol groups are ideal for interaction with charged analytes, particularly peptides and proteins in neutral/basic pH buffers. Hence, following each separation, the capillary surface must be "regenerated" or "reconditioned", that is, cleansed of any wall-adsorpted material. This is accomplished by following each run with a 3–5 column-volume rinse with 100 mM NaOH, followed by flushing with 5–8 column volumes of fresh separation buffer. This, of course, should be optimized for the particular buffer system used. However, the users should heed the results of Watzig and coworkers.[69] Alternative solutions for cleaning the capillary are 100 mM sodium tetraborate, pH 11, or mM trisodium phosphate, particularly for use with phosphate containing buffers.

When using acidic buffers, rinsing with NaOH may be a disadvantage since the drastic changes in pH may induce the requirement for extensive rinsing with the separation buffer for adequate pH re-equilibration of the capillary. Under these conditions, it may be advisable to follow the NaOH trines with a brief rinse with a concentrated separation buffer (10X or more), followed by the normal rinse with the IX separation buffer. Alternatively, depending on the sample, the 10X separation buffer rinse alone may be adequate for regenerating the capillary and, hence, the NaOH rinse may be avoided completely. For peptide separations in phosphate buffer, pH 2.5, 1.0 M phosphoric acid solutions may be used for capillary regeneration. Another good protein cleaning solutions is 1 M HNO_3.

Occasionally, after extensive use with protein solutions, the capillary will become fouled and will need to be cleaned to remove strongly adsorpted material. This is usually noticed with migration times that are excessive, with peaks distinctly asymmetrical with extensive tailing, or filling times become noticeably longer. A rinse with sodium ethoxide solution (blended from equal volumes of 1 M NaOH and 95% ethanol) for 5 min, followed by extensive (10–20 column volumes) water rinses usually suffices. A partially plugged capillary may be unplugged by filling with sodium ethoxide and allowing the solution to stand for 5 min before flushing with fresh solution, and following with an extensive water rinse before buffer equilibration. It should be noted that the sodium ethoxide will etch a new surface by removing the surface layer of silica, eventually making the capillary brittle. While we have used this procedure in our laboratory, keeping some capillaries in use for up to one year, this solution should NOT be used with coated capillaries, as it will remove the coating.

1.2.3.4.4 Storage

Because of the small dimensions of the capillary, plugging is always a potential problem. This may occur as a result of solvent evaporation at the capillary ends, which leads to salt crystal formation. This may be avoided in several ways. Dedicated capillaries to be stored for a short time (less than a week) should be rinsed with 100 mM NaOH, re-equilibrated with separation buffer and stored with the ends immersed in buffer/distilled water, or capped with silicone rubber stoppers. Optionally, the capillary may also be stored dry using the procedure below. If the capillary is to be stored for longer periods of time (i.e., >1 week), it should be washed with 100 mM NaOH, rinsed thoroughly with distilled water, purged with nitrogen or air (by pressure) and stored dry. If a capillary has been used with a detergent-/surfactant-containing solution, all traces of the additive must be removed before storage. If this is not done, the capillary will require longer than normal re-equilibration time upon reuse owing to the presence of the residual additive.

To place a stored capillary back into service, one should follow the same procedure outlined for a new capillary. If the capillary has been stored dry, we have found it useful to rehydrate the lumenal surface with a 5 min rinse of distilled water before rinsing with 100 mM NaOH and re-equilibration with buffer.

Before storing coated capillaries, they should be thoroughly rinsed with the appropriate buffers or solvents. For example, we have found that it is best to rinse hydrophobically coated capillaries with one-column volume 100 mM NaOH, then extensively with water (5–10 column volumes), and finally with methanol (5–10 column volumes), followed by drying and storage. Capillaries used with physical gels should be rinsed, filled with fresh buffer, and stored with both ends tightly capped. Capillary coating for particular CE applications can be of paramount importance because of the affinity of some analytes, particularly proteins, for interaction with the wall. A number of methodologies have been described in the literature for coating the internal surface of the capillary.[70] The pH stability, as well as the effectiveness for preventing adsorption of analyte to the wall (i.e., separation efficiency), varies with the chemical nature of the coating. Some precoated capillaries are commercially-available from several suppliers.

1.2.3.4.5 Basic Aspects of a Typical CE Method

As a means of collating the above information into practical form, the following provides what might be considered to be a rudimentary CE method.

1. **Mount capillary in the instrument playing close attention to alignment with the aperature/detection optics**. Once the new uncoated fused-silica capillary has been conditioned as described above, the capillary is mounted in the instrument and properly equilibrated.
2. **Determine the capillary fill time**. First fill the capillary with some low-absorbing solution (buffer or dH_2O) and then zero the detector. Using the rinse mode, fill the capillary with a strongly absorbing solution (e.g., 0.1 M NaOH) noting the amount of time to reach

the detector. Knowing the ratio of the effective length (inlet → detector) and the total length (inlet → outlet), calculate the time required to fill the entire capillary.

3. **Conditioning the capillary as described in Section 1.2.3.4.2.**
4. **Make sure capillary is equilibrated.** The capillary should be equilibrated preanalysis with a minimum of a two capillary volume rinse with *unelectrophoresed separation* buffer, that is, a separate vial of separation buffer that has *not been electrophoresed*. This assumes that the capillary is well conditioned and equilibrated if new, or well re-equilibrated if it is not the first run.
5. **Inject sample.** Sample is injected either by hydrostatic pressure or electrokinetically depending on the nature of the analyte and the sample matrix. Ideally, the injection should be followed by a second injection of separation buffer (equivalent of 1–2 s at 0.5 psi) to avoid any loss of sample into the inlet vial during the first few seconds of voltage application. This is particularly important with sample matrices that end to generate significant Joule heat.
6. **Apply voltage to the system.** With most applications, it is recommended that a low voltage is applied for a short period of time (e.g., 1 kV for 1 min) before application of the higher separation voltage. If the desired resolution is not obtained, one may consider examining voltage ramping.
7. **Regenerate the capillary surface.** The capillary should be regenerated with a minimum two capillary volume rinse with regenerating solution before re-equilibration with separation buffer. For neutral to high pH separations, 0.1–1.0 M NaOH is effective while 0.1–0.5 M phosphoric acid (or the acid equivalent of your separation buffer) is adequate for low pH separations. This should be followed with a short (one capillary volume) rinse with distilled water.
8. **Re-equilibrate the capillary surface.** The capillary should be re-equilibrated with at least 8–10 capillary volumes of *unelectrophoresed* separation buffer. If this is not the first run, this can be shortened to 5–6 capillary volumes since the first step in the method (step 4 above) precedes injection with a 2–3 capillary volume rinse with separation buffer. This is also an appropriate end point following the last run since the capillary remains in separation buffer (and not 0.1 *M* NaOH) overnight.

It should be clear that these are basic guidelines for setting up a CE method. As emphasized throughout this book, no method is universally applicable to all separations. Each method is likely to require the incorporation of idiosyncratic steps specific to the mode, sample, type of capillary, buffer, pH, etc. used.

1.3 METHOD DEVELOPMENT*

As with HPLC, there will be no universal CE conditions that will be appropriate for the analysis of all types of samples/analytes. In fact, it has been shown that one set of conditions is not sufficient for the analysis of one class of proteins, for example, glycoproteins,[71,72] or even variants of the same proteins from different species.[73] One of the first steps in designing a method is to determine, on the basis of the type or "class" of analyte involved, which of the CE modes is best suited to the sample (Table 1.2). Once the appropriate CE mode has been identified, analysis is carried out and separation optimization initiated. Optimal separation of the components of any sample requires a logical approach to sample solubilization and/or dealing with diverse sample matrices as well as identification and utilization of the correct combination of CE operating parameters. Each of these will have distinct effects on the resultant separation and resolution. There is a massive literature on

*Based on information given in the p/ACE 2000 Customer Training Manual. With permission.

CZE method development, not so much in the form of papers that address this per se, but more in the papers that describe a novel separation of a previously unstudied analyte. In these it is likely that fairly extensive details are provided on the developed method.[74–80] In addition, several reviews have been written on aspects of this (e.g., References 81 and 82).

1.3.1 Steps in Designing a Method

A general strategy for CE analysis of a sample is diagrammed in Figure 1.6. The first step is an obvious but important one: choosing the optimum wavelength for detection. This may not necessarily be the most sensitive wavelength (i.e., 200 nm for proteins), but more importantly, one that specifically enhances the detectability of the substance of interest and minimizes the absorbance of background components. Choice of a buffer system is of the utmost importance. The buffer needs to be selected with strict attention to not only the pH requirements of the sample from a stability perspective, but also to the wavelength restrictions of the buffer (i.e., λ_{min} for detection; see Table 1.3 in this chapter). If the use of a specific high ionic strength (high conductivity) buffer system is required (e.g., for preservation of bioactivity), the maximum working voltage should be determined by an Ohm's Law plot (discussed in Section 1.3.3.5 on buffers) to avoid buffer overheating problems. A review of the literature will provide a general idea of the basic parameters for an initial test run. A quick trial should establish the appropriateness of the sample introduction method, initial field strength, and capillary length. Optimization of the separation and resolution should include modification of the field strength, modification of the buffer by either altering pH or using additives to alter analyte mobility or EOF, and changing the length of the capillary as necessary. If peak resolution is poor, one might consider increasing the length of the capillary and/or the frequency of data collection. The remainder of the steps rely largely on the experience and intuition of the operation based on the characteristics of the sample/analyte. Once adequate separation of the sample components is attained, sample introduction should be optimized for a maximum signal-to-noise ratio. This step might include, where possible, altering the sample concentration or the buffer composition (ionic strength, type, pH, etc.). Typical starting conditions for analysis of unknowns (e.g., proteins or some drugs) in our laboratory are 50 μm i.d. $\times$ 30 or 40 cm in effective separation length capillary, 100 mM borate, pH 8.3, 25 kV, with a 3 s pressure injection at 0.5 psi. We find 200 nm to be a good wavelength for UV detection of most organic compounds with most buffers; due to the absorbance of conjugated bonds, there is a 5–10-fold enhancement of the signal over that observed at 280 nm with most proteins and peptides. A reasonable starting concentration for protein samples is 0.1–1.0 mg/mL, usually in 10 mM borate, pH 8.3 or 10–20 mM Tris, pH 7.5.

1.3.2 Sample Parameters to Consider

Clearly, analysis of a sample requires that it be solubilized. Although it is not always possible to know the physicochemical properties of the sample to be analyzed, it is important to be familiar with at least some of the physical properties of the specific analyte of interest in the sample. Some physicochemical properties may ultimately have to be determined empirically. Different questions should be posed depending on whether the sample involved is lyophilized or already solubilized, as in a biological fluid. In either case, it is important to have some knowledge of the detectability, purity, and stability of the sample/component of interest. A number of questions need to be addressed. How complex is the sample? Is the substance of interest a major or minor component? What other substances in the sample might interfere with the detection or separation of the analyte of interest? What is its λ_{max}? Is it thermally stable? If structural information is available, what are the pK values of the ionizable groups? If the sample contains proteins or peptides, what pIs are involved? If the sample is not in solution, there are additional questions that must be considered. Is it soluble in water or low ionic strength buffer at mg/mL concentrations? Does it require extremes of pH for solubility and, if so, is it stable at this pH? Is solubility enhanced by buffer additives such as urea, methanol,

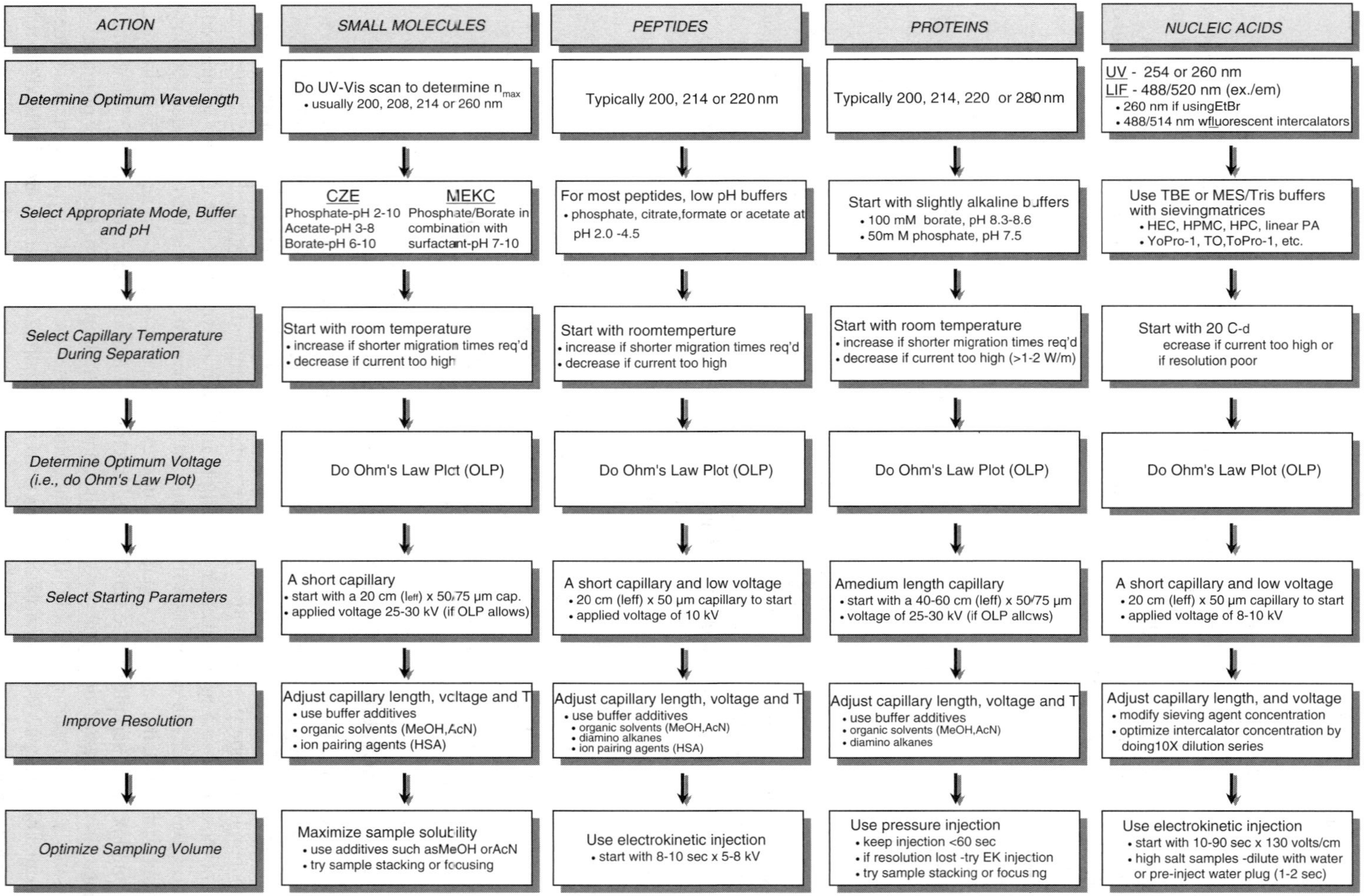

FIGURE 1.6 Flow diagram for CE method development.

TABLE 1.3
Commonly Used CE Buffers and Their Associated Properties

Buffer	Useful pH Range	Minimum Useful λ (nm)
Phosphate	1.14–3.14	195
Formate	2.75–4.75	200
Acetate	3.76–5.76	200
Citrate	3.77–4.77	200
MES*	5.15–7.15	230
Citrate	5.40–7.40	200
PIPES*	5.80–7.80	215
Phosphate	6.20–8.20	195
HEPES*	6.55–8.55	230
Tricine*	7.15–9.15	230
Tris	7.30–9.30	220
Borate	8014–10.14	180
CAPS*	9.70–11.10	220
Phytic	1.9–9.5	200

* Zwitterionic buffers.

acetonitrile, hexane sulfonic acid (HSA), SDS, and so forth? It is these types of questions that will allow for a sound methodological approach to methods development.

A final point worthy of note in this section is the importance of the sample matrix. As per the discussion in Chapter 13 by Burgi and Giordano, the constituents of the sample matrix will play an important role in attaining optimal separation of sample components. The conditions required for solubilization may not provide and adequate sample matrix for introduction to analysis. Compromises may have to be made between complete solubility and an ideal sample matrix. Moreover, it is crucial to view the sample matrix in relation to the separation buffer. Ideally the sample matrix should be approximately 10- to 200-fold less in total ionic strength than the separation buffer, in order that the sample does not contribute to the EOF. This disparity will also enhance the possibility of adequate detection, since the lower ionic strength sample matrix will lead to sample stacking and, hence, on-capillary concentration.

Once solubility has been attained in a reasonable sample matrix, the sample should be filtered through a 0.45 μm filter to remove unsolubilized particulates. A trial analysis is then conducted and the efficiency of the separation optimized. This is accomplished by altering/adjusting a multitude of variables (discussed below) in search of the combination that provide the best separation of the nextsample components within a reasonable analysis time.

1.3.3 Separation Parameters to Consider

With the relatively large number of parameters to contend with in CE methods development, it is beneficial to understand how a change in one parameter will affect others. Figure 1.7 provides a quick reference chart for evaluation of the relationships between the variables discussed below. In this table, the direction of the effect of altering effects on the parameters described on the vertical axis on those in the horizontal is given (↕, signifies the effect is bidirectional). Parameters not covered in Figure 1.7 include the composition of the sample matrix and the presence of buffer additives that either interact with the sample components (e.g., surfactants) or affect the chemical nature of the wall (through dynamic/covalent modification). It is clear that, with any particular sample, each of these variables should be examined systematically for effective methods development.

Increasing these ↓ / Affect these →	pH	Temperature	Injection volume	Viscosity	Electroosmotic flow	Sensitivity/resolution	Diffusion	System current	Ionic strength	Electrophoretic mobility	Capillary surface	Analyte charge	Analyte-wall interaction
pH					↓	↕				↕	↕	↕	↕
Temperature	↕		↑	↑	↓	↕	↑	↑	↑	↑			
Field strength		↑			↓	↕							
Viscosity			↑		↑		↕			↑			
Molecular size/shape						↕	↕			↕		↕	↕
Electroosmotic flow						↕	↑						
Diffusion						↑							
Analyte-wall interaction					↕	↑					↕		
Current		↑		↑	↓		↑						
Ionic Strength		↑	↕		↑	↕		↓			↕		↑
Capillary diameter		↓	↓		↑	↕		↑					
Capillary length		↑	↑			↕		↑					
Surface negative charge					↓						↕		↕
Analyte charge			↕							↕			↕
Electrophoretic mobility			↕			↕							

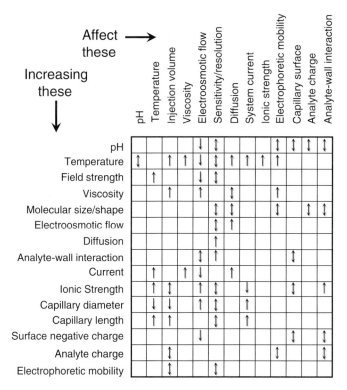

FIGURE 1.7 A reference chart detailing the relationships between variables influencing performance in CE.

The following sections describe the qualitative effects on separation that result from alteration of the parameters described above. For example illustrating the effects of many of these parameters on CE analysis of mixtures, the reader is referred to an excellent review by McLaughlin et al.[83]

1.3.3.1 Electrode Polarity

Establishing the electrode polarity is of paramount importance in CZE and, obviously, one of the first consideration before beginning analysis. As mentioned earlier, the normal polarity for CE is to have the anode (+) at the inlet and cathode (−) at the outlet. In this format, EOF is toward the cathode (detector/outlet). This is the standard polarity for most modes of CE utilizing a bare fused silica capillary. If set in reverse polarity (cathode—inlet: anode—outlet) by mistake, the direction of EOF is away from the detector and only negatively charged analytes with electrophoretic mobility greater than EOF will pass the detector. This format is typically used with capillaries that are coated with substances that reverse the net charge of the inner wall (and subsequently reverse EOF), or when all analytes are net negatively charged (e.g., DNA or SDS–complexed proteins).

1.3.3.2 Applied Voltage

It is advisable to begin analysis with a mid-range voltage (10–20 kV). Increasing the voltage will have a number of effects. While it will increase sample migration and EOF rate, as well as shorten analysis time, it may increase the sharpness of the peaks and improve resolution. However, the advantages associated with increasing the voltage may be lost if the sample matrix ionic strength is much greater than the running buffer ionic strength such that the increased production of Joule heat cannot be efficiently dissipated. Joule heating of the capillary results in a decreased solution viscosity. This leads to a further increase in EOF, ion mobility, and analyte diffusion, which may ultimately result in band broadening. An excellent example of the effects of increased voltage on the separation of five vitamins has been provided by McLaughlin et al.[83] and shown in Figure 1.8. Decreasing the applied voltage from 30 to 10 kV has the expected effect of increasing the migration time of all analytes in the mixture. At 10 kV the separation is inconveniently long and peak 5 had not reached the detector after 33 min. However, adequate separation at higher applied voltages is obtained at the expense of both resolution and efficiency (theoretical plates) (Figure 1.8 inset).

1.3.3.3 Capillary Temperature

Separations should initially be attempted with the capillary thermostatted at close to ambient temperature. The capillary temperature can be increased on most commercial CE units to as high as 60°C without substantially increasing current with most buffers. When this is done using the same applied voltage, decreased buffer viscosity leads to an increase in analyte electrophoretic mobility, thus, decreasing separation times. Also, it is important to note that, when sample introduction is hydrostatic (same pressure/vacuum and time), increased capillary temperature will lead to an increase in the injected sample volume as a result of decreased buffer viscosity.[2,84–89] Sensitivity may not necessarily be increased. However, Undesirable effects include concurrent changes in buffer pH, band broadening due to increased diffusion, and possible thermal denaturation of the sample.

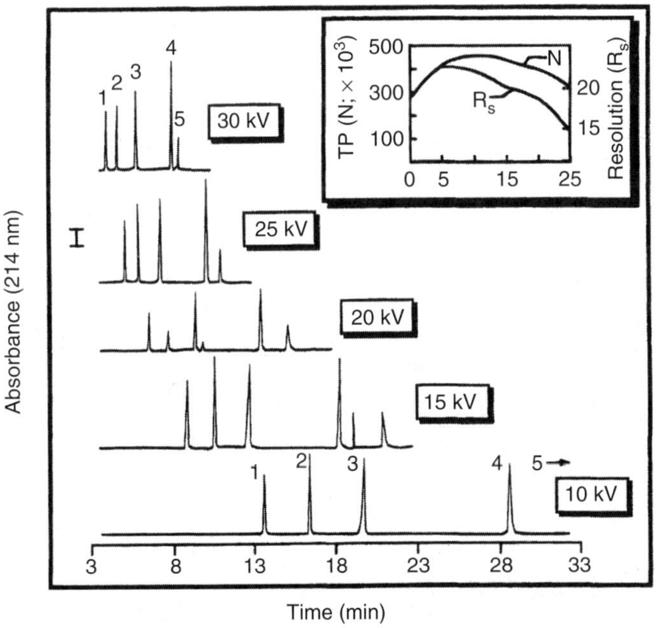

FIGURE 1.8 Effects of voltage on separation efficiency and resolution. A mixture of five vitamins (1) niacinamide, (2) cyanocobalamine (vitamin B_{12}), (3) pyridoxine (vitamin B_6) niacin, and (5) thiamin (vitamin B_1). Separation conditions were as follows: capillary: 75 μm × 50 cm (effective length) fused-silica; *T*: 25°C; buffer: 60 mM borate, 60 mM SDS, 15% methanol, pH 8.92; voltage varied as indicated to the right of each electropherogram. (Modified from McLaughlin, G. M. et al., *J. Liq. Chromatogr.*, 15, 961, 1992. With permission.)

Increasing the capillary temperature can have both positive effects and negative effects on separation. In some cases, the effects of elevated capillary temperatures are not only advantageous but necessary. A study by Guttman et al.[89] showed that separation of a mixture of five proteins in a physical gel was poorest at 20°C and optimized at 50°C. Apparently, higher temperatures were needed to obtain the appropriate structural conformation of the physical gel required for sieving of the proteins. Another positive effect of increasing capillary temperature is the substantial decrease in analysis time (which may or may not be associated with an increase in resolution). Figure 1.9 shows the effect of increasing temperature on the separation of 4-hydroxy-7-nitro-2,1,3-benzoxadiazole (NBD-OH) from the derivatized valine (NBD-valine) at temperatures ranging over 20°C. The temperature shift from 25°C to 45°C was not only associated with the expected decrease in analysis time (reduced by roughly half), but also with a change in the fluorescence signal intensity for the peaks, lower fluorescence intensity observed at lower temperatures. Other effects can include the negative effects of elevated capillary temperature on the stability of the sample. This has been shown with the temperature-dependent CE analysis of α-lactalbumin[87] where separations were optimal at 20°C and 50°C; however, the authors demonstrate that the faster migration time is not solely due to increased EOF, but also to a temperature-induced conformational change in the protein. Hence, while elevated capillary temperatures shorten analysis times, one should be cognizant of the potential for adverse effects on analyte stability. More recently, Giordano et al.[90] defined a new approach for calculating intracapillary temperatures. In an interesting twist, they define a potential positive outcome of Joule heating that overcomes the thermostatting capability of the instrument—the ability to exploit the elevated temperatures for fluorescent labeling proteins.

In the first edition of this CE handbook, Nelson and Cooke[91] addressed the developmental issues associated with capillary temperature control in CE systems. This topic is revisited anew by Haddad in Chapter 18.

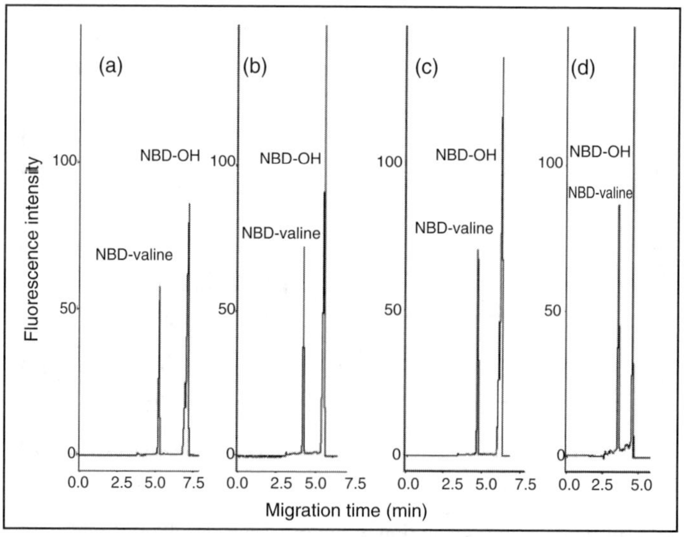

FIGURE 1.9 Effect of temperature on migration time and fluorescence intensity (RFU) with the separation of 4-hydroxy-7-nitro-2,1,3-benzoxadiazole (NBD-OH) and the derivatized valine (NBD-valine). Temperatures were (a) 25°C, (b) 30°C, (c) 35°C, and (d) 45°C. The conditions of the separation were15 kV applied voltage, 40/47 cm effective/separation length, 20 mM borate buffer pH 10.0, injection of 5.56×10^{-5} M valine for 4 s, followed by absolute ethanol for 1 s, and then 5.24×10^{-2} M NBD-F for 2 s. (Modified from Zhang et al. *Anal. Bioanalyt Chem.*, 2006, 386, 1387–1394. With permission.)

1.3.3.4 Capillary Dimensions

1.3.3.4.1 Internal Diameter

The main advantage resulting from increasing the capillary i.d. is the enhancement in detection sensitivity due to increasing the path length. However, a decrease in the surface–volume ratio accompanies an increase in diameter (see Table 1.1). This may lead to less efficient dissipation of Joule heat, which then results in a temperature gradient across the capillary and band broadening. While a narrower capillary will accommodate a smaller sample volume (keeping $L_{int}/L_t < 3\%$), owing to the concentrating phenomenon during electrophoresis, the narrower diameter allows detection a lower initial sample concentration.

1.3.3.4.2 Length

It is recommended that initial separations be performed on shorter capillaries (20–30 cm to detector), which provide short analysis times for determining sample introduction mode and buffer choice, and so forth. Once successful analysis has been achieved, the capillary length can then be increased to improve resolution of closely migrating species since an increased on-capillary time allows for subtle differences in analyte electrophoretic mobility to separate the components. Also, as the length increases, there will be a concomitant decrease in the electrical field strength at constant voltage and, hence, higher voltage may be used. There is a practical limit on capillary length, where the trade-off with increased resolution is overcome by the decrease in sensitivity due to band broadening effects, or the limits of the applied field. This practical limit appears to be about 100 cm.

1.3.3.5 Buffers

The choice of a buffer is critical to obtaining successful CE separation of the analytes. Once the optimal wavelength for detection has been established, a buffer must be selected that does not interfere with the ability to detect the analytes of interest, maintains solubility of the analytes, maintain buffering capacity through the analysis, and produces the desired separation. These topics are covered in more detail in the following sections, with examples highlighting the importance of each parameter. Refer to Appendix 3 for some exemplary buffer systems and associated conditions for separation of various analytes.

1.3.3.5.1 Selection and Preparation of a Separation Buffer

A wide variety of electrolytes can be used to prepare buffers for CE separations. When using absorbance detection, a major requirement of any component used in the buffer system is a low UV absorbance at the wavelength used for detection. This restriction substantially limits the choices to a moderate number of non-UV absorbing electrolytes. However, a number of UV-absorbing organic buffers have been used with success at low concentrations to minimize background absorbance. For low pH buffers phosphate, acetate, formate, and citrate have commonly been used effectively. For buffers in the basic pH range, Tris, Tricine, borate, and 3-(cyclohexylamino)propane-1-sulphonic acid (CAPS) are acceptable electrolytes. For detection modes other than UV absorbance, a number of other electrolytes can be utilized. In fact, for indirect detection, one might desire a buffer with high background absorbance to enable detection limits of analytes. However, if one is considering methods utilizing electrochemical detection, the buffer must not only be compatible with the analytes, but present a stable background conductivity against which the analytes can be detected. This is covered by Lunte in Chapter 48 on electrochemical detection. Table 1.3 presents a list of useful electrolytes for preparing separation buffers for CE. In addition, an indication of their useful pH range and minimum functional wavelength (for absorbance detection) is given.

Perhaps one of the most important aspects of buffers used for CE is the requirement of pure reagents. It is recommended that the water used for buffer preparation be of Milli-Q (or equivalent) purified quality and that the reagents be highly purified or ultrapure (Gold Label) quality; the small

TABLE 1.4
Common Buffer Additives in CE and Their Effects

Additives	Example	Function
Inorganic salts	NaCl, $CaCl_2$, K_2SO_4	Modification of EOF; protein conformational changes; protein hydration
Organic solvents	Methanol, acetonitrile, ethylene glycol	Modification of EOF; analyte solubilization; analyte salvation
Organic additives	Urea	Modification of EOF; protein solubilization
	Pyrenebutanoate	Denaturation of oligonucleotides dynamic modification of protein
Inorganic additives	Borate	Complex *cis*-diols; carbohydrate or glycoprotein separations
Zwitterionic additives	Z1-Methyl	Reduce wall interaction; augment EOF
Sulfonic acids	hexane, heptane, octane, or nonane analogs	Analyte ion-pairing; hydrophobic interaction
Divalent amines	diaminoalkanes; hexamethonium bromide; decamethonium bromide	Modification of EOF; charge neutralization; analyte interaction
Cationic surfactants	Dodecyltrimethylammonium bromide (DTAB); cetyltrimethylammonium bromide (CTAB); tetradecyltrimethylammonium chloride (TTAC)	Charge reversal on-capillary wall; hydrophobic interaction
Cellulose derivatives	Hydroxyethyl cellulose; methyl cellulose; hydroxypropyl methylcellulose	Reduce EOF; provide sieving medium
Miscellaneous polymers	Polyethylene glycol	Protein stability; reduce wall interaction
	Dextran	Manipulate the electrophoretic mobilities
	Polyvinylpyrrolidone	Enhance resolution of peptides and amino acids

amounts required for CE do not entail great expense. Passage of buffers through an 0.22 or 0.45 μm filter is recommended before use to remove any particulate matter. For buffers stored in 4°C, it is imperative that they be brought to room temperature and thoroughly degassed before use.

Buffer additives may also be used to obtain enhanced or differential selectivity in a separation. Several classes of additives have been identified as applicable to CE. Table 1.4 outlines some common buffer additives and their mode of action. As can be seen from the table, the additives are multifunctional, not only suppressing analyte–wall interactions, but also affecting analyte solubility, and in some cases affecting selectivity. Some examples of the use of additives for enhancing/optimizing CE separations are given later in this chapter.

1.3.3.5.1.1 pH Assuming that information about the sample is available, it is advisable to choose a buffer pH that approximates the pK of the solute mixture. This choice is simple with pure or partially purified preparations. With crude biological mixtures, the pK_{avg} will typically be close to neutrality.

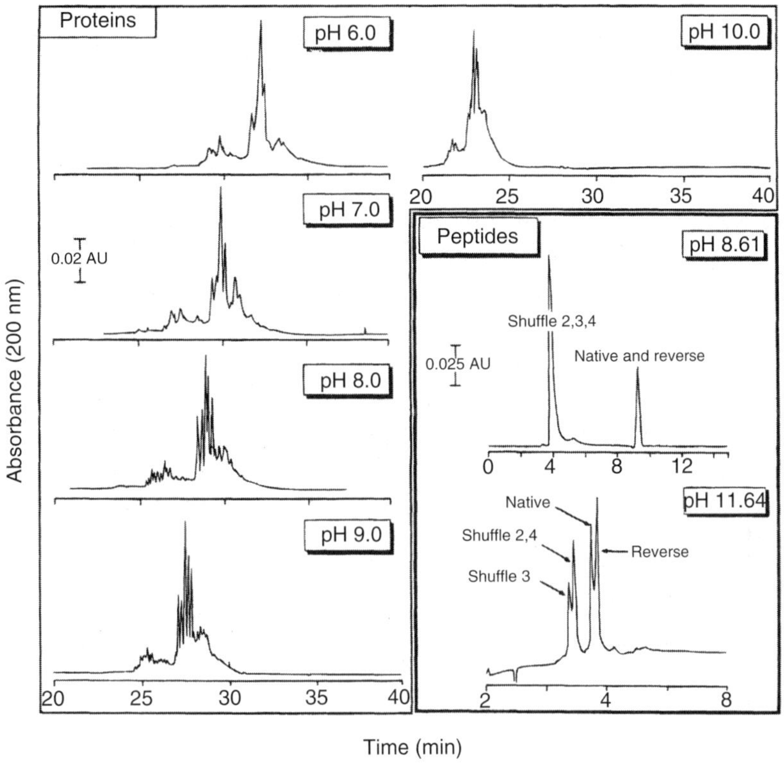

FIGURE 1.10 Dependence of electrophoretic separation and resolution on buffer pH. Proteins: Ovalbumin [1.0 mg/mL, in 100 mM borate, pH 8.3; 3 s pressure (0.5 psi) injection] was introduced into a pre-equilibrated capillary containing 100 mM borate buffer at the appropriate pH, indicated above each trace. Analysis performed on a Beckman P/ACE System 2050. Separation conditions: capillary: 50 μm × 80 cm (effective length), 87 cm total length bare fused silica; *T*: 28°C; voltage: 25 kV; detection: 20 nm. (Modified from Landers, J. P. et al., *Anal. Biochem.*, 205, 115, 1992. With permission.) Peptides: Five peptides containing the same 12 amino acids but in varied sequence were analyzed under different pH conditions as labeled. Sequence of native, reverse, and shuffled peptides are provided in the inset table. Separation conditions: capillary: 50 μm × 20 cm (effective length), 27 cm total length bare fused silica; *T*: 20°C; voltage: 25 kV; detection: 20 nm. Separation buffer: 100 mM borate, plus 10 mM diaminopentane, adjusted to the appropriate pH with NaOH.

Remember that increasing the pH between 4 and 9 results in an increase in the EOF. Also remember that the buffer pH may be altered in a secondary manner by other parameters such as temperature, ion depletion of the buffer (caused by repetitive use of the same separation buffer) and organic additives. Examples of the effect of buffer pH on CE separation are given in Figure 1.8 for both peptides and proteins. With the separation of ovalbumin isoforms, the expected decrease in migration time with increasing pH is observed and optimal separation is observed at pH 9.0. Separation of a mixture of five peptide "isomers" over a wide range of pH is illustrated in the lower right of Figure 1.10. At pH 2.5, all five peptides, which are identical in amino acid composition but vary in sequence, comigrate as a broad peak (not shown). At pH 8.61, the peptides migrate as two groups while at 11.64, resolution of 4 of the 5 peptides is observed.

1.3.3.5.1.2 Ionic Species As discussed above, the choice of buffer for a particular CE separation is important from the perspective of Joule heat generation. Obviously, the pH required for the separation will limit the candidate buffer systems amenable for use. However, it is important to note that, at a given pH, the type of buffer can have dramatic effects on resolution. Figure 1.11 (left)

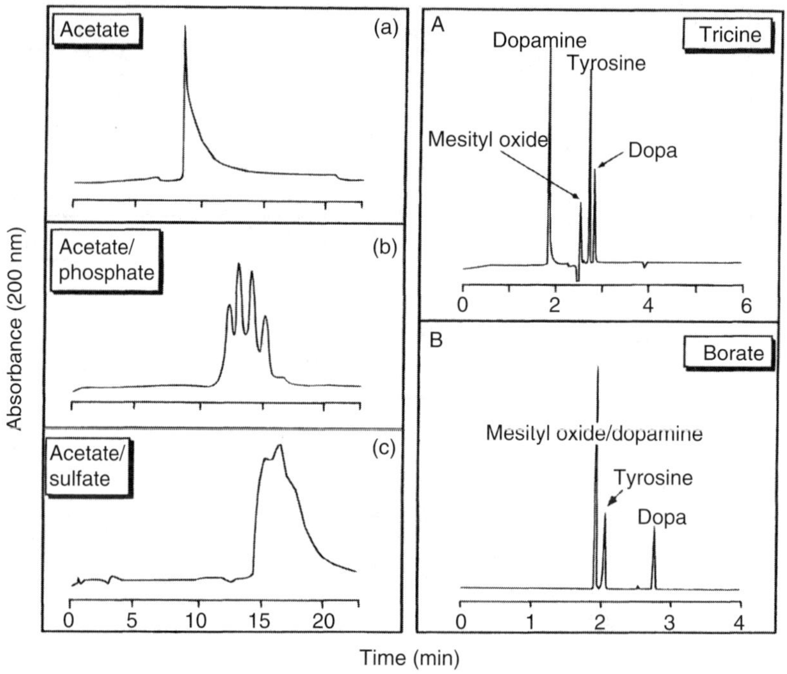

FIGURE 1.11 Effect of buffer type (ionic species) on CE separation. (A) CE analysis of r-hEPO. Separation conditions: capillary: 75 μm × 20 cm (effective length) fused silica; voltage: 10 kV; detection: 214 nm. Buffers were of identical pH (4.0) and 100 mM concentration; (a) acetate buffer (30 μA); (b) acetate-phosphate (120 μA); and (c) acetate–sulfate (200 μA). (Modified from Tran, A. D. et al., *J. Chromatogr.*, 542, 469, 1991. With permission.) (B) Separation of dopamine, tyrosine, dopa [0.5 mg/mL each, in water; 3 s pressure (0.5 psi) injection] and mesityl oxide [neutral marker, 3 s pressure (0.5 psi) injection]. Analysis performed on a Beckman P/ACE System 2050. Separation conditions: capillary: 50 μm × 50 cm (effective length), 57 cm (total length) bare fused silica; *T*: 28°C; voltage: 25 kV; detection: 200 nm. (A) Separation in 100 mM tricine, pH 8.3. (B) Separation in 100 mM borate, pH 8.3. (Modified from Landers, J. P. et al., *Anal. Chem.*, 64, 2846, 1992. With permission.)

illustrates the importance of buffer choice on the CE analysis of recombinant human erythropoietin (r-hEPO).[68] Although all buffers were of identical pH (4.0) and at a 100 mM concentration, resolution of the glycoforms was only observed with acetate–phosphate buffer and not with the acetate alone or acetate–sulfate solutions.

Other buffer systems that may have a dramatic effect on the separation are those capable of interaction with specific analytes. The ability of borate to form stable complexes with *cis*-diols has long as been known and this property exploited in chromatography.[92] Several studies have shown this property to be particularly advantageous in CE since, at a slightly basic pH, borate complexation imparts an additional negative charge to the molecule (due to ionization of one of the free hydroxyl groups; $pK_a = 9.14$).[93–95] Figure 1.11 (right) highlights the importance of borate for the separation of *cis*-diol-containing compounds from those similar in structure, but lacking the functional group in the separation of dopamine, tyrosine, dopa, and mesityl oxide (neutral marker).[96] In tricine at pH 8.3, dopamine is positively-charged and migrates faster than EOF (mesityl oxide) while both tyrosine and dopa migrate slower than EOF (dopa more negatively-charged than tyrosine). In borate buffer at the same pH, borate complexed with dopamine negates the positive charge of the amine, while dopa becomes more negatively-charged. This example emphasizes the importance of buffer selection for enhancing selectivity in CE separations.

1.3.3.5.2 Buffer Concentration: Ohm's Law Plot

Useful buffer concentration is restricted by several parameters including the capillary length and internal diameter, the applied electrical field strength and the efficiency of the capillary thermostatting/cooling system. Generally, use of moderately high ionic strength buffers is desirable for suppression of ion-exchange effects between the charged analyte ions and the ionized silanol groups on the capillary wall. However, the current (Joule heat) associated with buffer concentrations greater than 100 mM may overcome the capillary thermostatting capability of the system at higher applied voltages. Excessive Joule heating can have desirable effects on both resolution and analyte stability. Buffers that may be problematic in this respect include those containing high mobility electrolytes such as chloride, citrate and sulfate. The excessive Joule heating associated with high concentration buffers can be circumvented in two ways. The easiest solution is to lengthen the capillary so that a tolerable current is maintained. However, this will increase the effective capillary length and may compromise resolution. A reduction in the cross-sectional area of the capillary will also reduce heating by decreasing the current density, and allows for more effective dissipation of heat due to a greater surface-to-volume ratio. Alternatively, buffers that run at relatively low current (and Joule heat) can be used. One such buffer in the pH 7–9 range is borate buffer that has been shown to be an excellent CE buffer at concentrations as high as 500 mM.[97] An added advantage of using higher ionic strength separation buffer is the increased sample loading capacity that results owing to the on-capillary stacking effect that will be described in a later section, and by Burgi and Giordano in Chapter 13.

A simple method has been described by Nelson et al.,[98] termed the "Ohm's Law Plot," that allows for easy determination of the "functional" buffer concentration and the maximum voltage that can be utilized with the particular buffer system (i.e., the functional limit for capillary thermostatting). After filling the capillary with the desired buffer and allowing for adequate equilibration, voltage is applied to the system for short intervals (e.g., 1 min) and the current recorded at each voltage. Linearity in a plot of observed current vs. voltage is an indication that the capillary temperature is being adequately maintained (i.e., the generated Joule heat is being effectively dissipated). At the point where linearity is lost, the thermostatting capacity of the system has been exceeded. One should strive for heat generation of <1 W/m (watts per meter) for optimum separation[99] and should not exceed 5 W/m. An example of the Ohm's Law plot is given in Figure 1.12a for 100 mM concentrations of each of three buffers: phosphate, pH 2.5; borate, pH 8.3; and CAPS, pH 11.0.[100] At 20 kV, the current is lowest with borate ($\approx$10 μA) while CAPS ($\approx$100 μA) and phosphate (150 μA) are dramatically higher. The power associated with each buffer at 25 kV is 0.58, 10.07 and 5.88 W/m for borate, phosphate and CAPS, respectively. Refer to Appendix 1 for example on calculating power (W/m). The recommended limits have clearly been exceeded with the latter two buffers. More importantly, linearity in the relationship is lost with the phosphate and CAPS buffer at relatively low applied voltages in comparison with borate; the 100 mM phosphate and CAPS buffers should not be used at voltages greater than 10 and 15 kV, respectively. In contrast, borate buffer (inset in Figure 1.12a) exhibits a linear relationship between current and the applied voltage, even at 30 kV. Figure 1.12b shows that this determination can be simplified if the software used for data acquisition and analysis has the ability to directly monitor/plot current as a function of time. The figure presents a direct (on screen) plot from System Gold Version 7.1 for 100 mM CAPS buffer where voltage is incremented by 2.5 kV/min. This provides the same information without having to manually plot current versus voltage.

1.3.3.5.2.1 Ionic Strength Increasing the ionic strength of the separation buffer increases the thickness of the ionic double layer, and has the effect of decreasing EOF, hence, increasing the analysis time. As mentioned earlier, increasing the ionic strength will also increase the current at a constant voltage and, hence, adequate thermostatting of the capillary becomes a concern. An advantage of increasing ionic strength, in addition to the obvious improvement in buffer capacity,

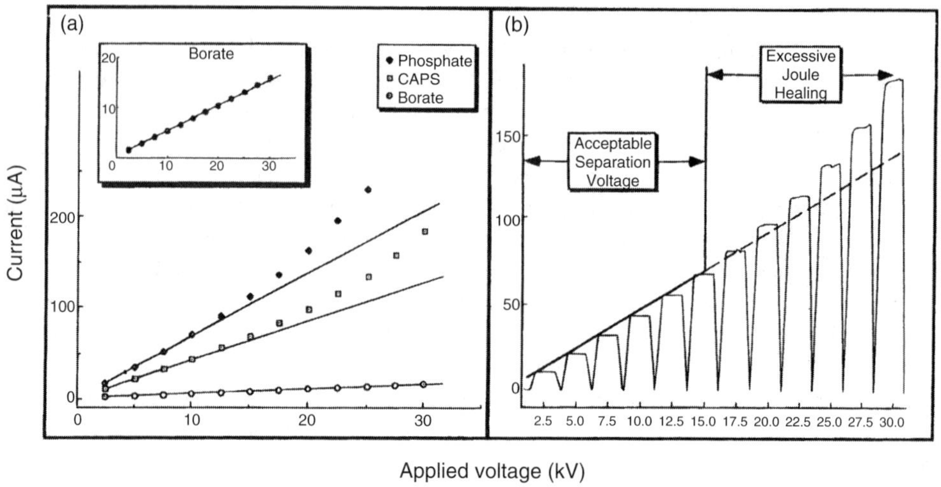

FIGURE 1.12 The Ohm's Law Plot. (a) Plot of observed current vs. applied voltage for each of three buffers. The determinations were carried out on a Beckman P/ACE 2050. The voltage was incremented 2.5 kV/min, and the current recorded through System Gold software, Version 7.1. Buffers were 100 mM phosphate, pH 2.5, made by dilution of phosphoric acid and titration with NaOH; 100 mM borate, pH 8.3, made by titrating 25 mM sodium tetraborate with 100 mM boric acid; and 100 mM CAPS, pH 11.0, made by titration of the appropriate concentration dissolved in water with NaOH. The inset shows the borate data plotted on an expanded scale. (b) Direct plot of current versus applied voltage for 100 mM CAPS, pH 11.0. A direct plot of observed current versus applied voltage obtained through System Gold Version 7.1. Voltage is incremented by 2.5 kV/min. A straight line drawn through the front edge of the plateau illustrates the ability of the cooling system to dissipate the heat generated by the passage of current. The departure from linearity indicates the excessive increase in current at the applied voltage, and is a reflection of the increase in the capillary temperature.

will be to decrease analyte–wall interactions.[100] The net effect on the separation, therefore, will be to increase resolution, provided that capillary thermosetting capability is not overcome and that unwanted analyte dissociative processes (e.g., ligand-peptide/protein multimer dissociation) do not occur. On the other hand, an increase in ionic strength might improve resolution in mixtures by decreasing nonspecific analyte–analyte interaction. One example is with protein–DNA interactions. Figure 1.13 illustrates the effect of increased ionic strength on oligonucleotides electrophoresed in the presence of varying concentrations of urea. With urea, which maintains the oligonucleotides in a denatured form at the thermostatted temperature, there is an obvious improvement in resolution when increasing the urea concentration from 0 to 2 M. However, an optimum is reached and further increase (from 2 to 4 M) results in decreased resolution and a concomitant decrease in the S/N ratio with increasing urea. Other effects have been reported exploiting ionic strength in combination with bovine serum albumin (BSA) to minimize nonspecific interaction with DNA restriction fragments.[100] Addition of BSA to the sample obliterated resolution of the double-stranded DNA-stranded (dsDNA fragments ranging from 72 to 310 base pairs. However, the simple addition of 50 mM NaCl to the 89 mM Tris-borate-ethylenediaminetetraacetic acid (Tris-borate-EDTA) separation buffer negated these effects. While the voltage had to be decreased to accommodate the increased Joule heating effects of the added ionic strength, there was a corresponding improvement in the detection of the low molecular weight DNA fragments, most likely due to a stacking effect produced between the lower ionic strength sample buffer and the higher ionic strength buffer.

Increasing the ionic strength is usually associated with increased Joule heating due to higher current flux. In the above example provided by Figure 1.13, the voltage had to be decreased in the separation that contained salt. One way to increase ionic strength without increasing Joule heating is to use zwitterionic additives.[101] Even at molar levels, zwitterionic agents, such as betaine or

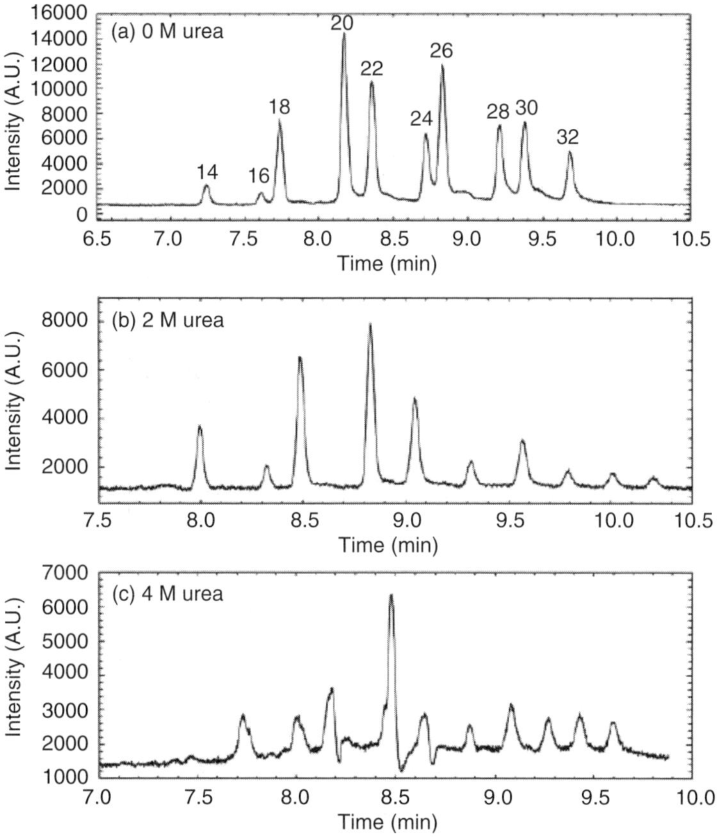

FIGURE 1.13 The effect of increasing ionic strength on the CE separation of oligonucleotides. Dye was mixed with Pluronic F127 solution and the capillary filled by syringe. 25% w/v F127 in 1XTBE buffer as a separation medium at room temperature. Conditions: l (length to detector) = 7 cm, L (column length) = 13 cm, E (field strength) = 200 V/cm, capillary i.d. = 50 mm, o.d. = 360 mm, Electrokinetic injection at 300 V/cm for 1 s. Peak assignments are as indicated in the figures. (Modified from Landers, J. P. et al., *Biotechniques*, 14, 98, 1993. With permission.)

sarcosine, do not contribute to the conductivity of the buffer, and therefore, do not raise the current or induce Joule heating.

1.3.3.5.2.2 Organic Salts The addition of organic modifiers to the separation buffer will have differing effects dependent on the nature of the additive. The positive effects on the separation usually result from the alteration of the solvation properties of the buffer or the diminished EOF due to changes in the thickness of the double-ionic layer (and hence the zeta potential) or the viscosity of the buffer. One affect often results is a change in EOF. For example, the addition 1,4-diaminobutane to the buffer has been proposed to enhance resolution by slowing EOF through a dynamic modification of the capillary wall.[71,72] Ion-pairing agents, such as the alkyl sulfonic acid salts, have been used with success to resolve peptides.[102] Figure 1.14 demonstrates the influence of HSA on the separation of a series of peptides of identical amino acid composition but differing sequence. Here the resolution appears to be due to the differential solvation and/or the microenvironment of nearest neighbor effects on pK values of the charged amino acids in the peptides.

1.3.3.5.2.3 Organic Solvents Organic solvents such as methanol or acetonitrile have the effect of decreasing both the conductivity of the buffer and EOF through their ability to disrupt the ordered

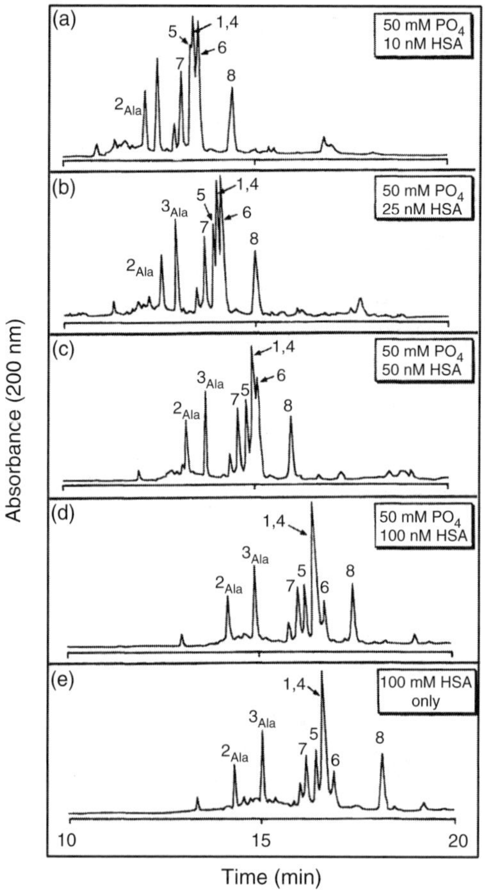

FIGURE 1.14 Effect of organic salts on separation. The separation of eight peptides of which six peptides contain the same 12 amino acids but a varied sequence, and two contain alanine substations for lysine residues. Sequence of native peptide: KTNYCTKPQKSY. Separation conditions: capillary: 50 μm × 50 cm (effective length), 57 cm total length, bare fused silica: T: 28°C; voltage: 15 kV; detection: 200 nm. Separation buffer: 50 mM phosphate, pH 2.0. (a) 50 mM phosphoric acid, pH 2.0, containing 10 mM HSA. (b) 50 mM phosphoric acid, pH 2.0, containing 25 mM HSA. (c) 50 mM phosphoric acid, pH 2.0, containing 50 mM HSA. (d) 50 mM phosphoric acid, pH 2.0, containing 100 mM HSA. (e) 100 mM HSA in distilled water, titrated to pH 2.0 with sulfuric acid. (From Oda, R. P. et al., *J. Chromatogr.*, 680, 341, 1994. With permission.)

structure of the water molecules. In such cases, the subsequent enhancement in resolution may result from a combination of the decreased EOF (i.e., increased on-capillary time), decreased thermal diffusion, and improved analyte solubility. Figure 1.15 shows the effect of acetonitrile concentration on the solvation and resolution of the same series of peptides depicted in Figure 1.12.[102] More recently, Lookhart et al.[103] showed that the addition of acetonitrile to a phosphate–glycine buffer aided in the separation of storage proteins in maize. In another example illustrating the importance of analyte solvation, Gordon et al.[104] demonstrated that adding ethylene glycol to a borate buffer enabled resolution of serum proteins. They found, however, that the presence of the ethylene glycol in the sample, not in the buffer, led to greater sensitivity and reproducibility.

1.3.3.5.2.4 Modifiers of EOF Two approaches can be taken to modify EOF in an attempt to enhance resolution: covalent coverage or dynamic modification of the capillary wall. Many examples of both covalent[105–107] and dynamic[106–113] coatings are found in the literature, and

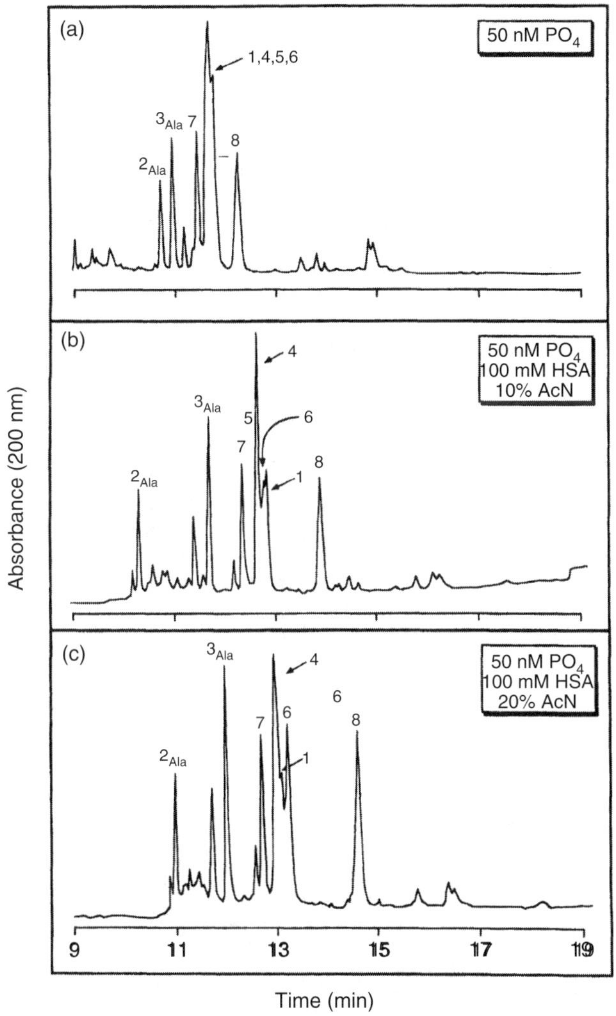

FIGURE 1.15 Effect of organic solvents on CE separation. The separation of the same peptides from Figure 1.12. Separation conditions: capillary: 50 μm × 50 cm (effective length), 57 cm total length, bare fused silica; T: 28°C; voltage: 15 kV; detection: 200 nm. Separation buffer: 50 mM phosphate, pH 2.0. (a) Buffer without any additive, (b) buffer containing 10% acetonitrile, and (c) buffer containing 20% acetonitrile. (From Oda, R. P. et al., *J. Chromatogr.*, 680, 341, 1994. With permission.)

while specific to capillary, surface coating is covered in more detail in Chapter 52 by Henry. A study by Swedberg[105] is an excellent example of the importance of capillary coating for preventing protein–wall interactions during CE separations. A mixture containing seven proteins was clearly resolved (with efficiency >500,000 theoretical plates per meter) in a capillary coated with a pentafluoroaryl compound, whereas separation in bare fused-silica resulted in only a few broad peaks barely greater than baseline. The EOF was not abolished by this coating, suggesting that not all of the silanol groups were reacted, but that the bulky aryl groups prevented protein–wall interaction. As would be expected when protein–wall interactions are diminished, the migration times are markedly shortened. Figure 1.16 highlights the effects of capillary coating on the separation of DNA fragments observed by Chu & coworkers. They describe two new methods for coating capillaries with polyvinyl alcohol (PVA) on characterize the adsorption of PVA onto the wall using atomic

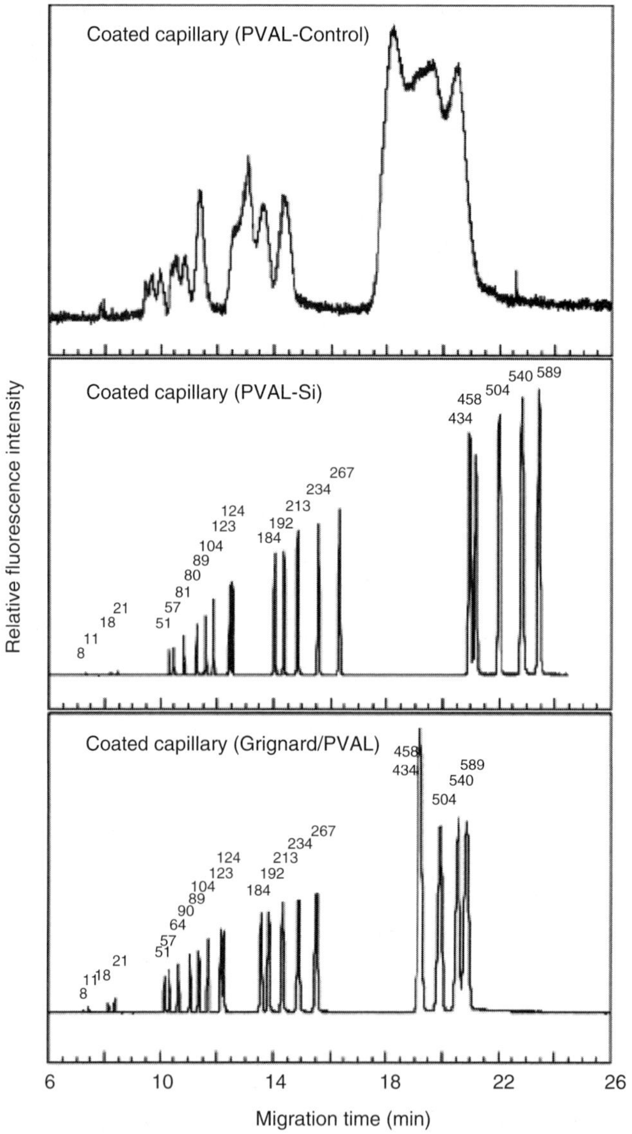

FIGURE 1.16 The effect of capillary surface coating on the resolution of DNA fragments. Capillaries were coated with polyvinyl alcohol and a pBR322 HaeIII digest used to evaluate resolution (fragment lengths from 8 to 589 bp[). Capillary: 13/10 cm total/effective length, 75 m i.d.; medium, 1 TTE solution of polyvinyl alcohol (Mr 125 000; DH = 88 mol%; concentration = 5 wt% for a and c, 6 wt% for b); DNA: pBR322 HaeIII digest, 10 g/mL; buffer, 1 TTE, pH 8.3, including ethidium bromide at 3 g/mL; injection, 0.65 kV, 3 s; electrophoresis: 3.9 kV, 2.6 A, nominal 25C; detection, LIF using an Ar-ion laser beam at = 488 nm. (Modified from Moritani et al. *Electrophoresis*, 2003, 24, 2764–2771. With permission.)

force nueroscopy. The authors describe performance for tens of runs but don't comment on separation reproducibility beyond that.

The other approach to minimizing analyte–wall interactions through modification of the capillary wall is the use o additives that provide a dynamic (noncovalent) coating. Diaminoalkanes are one class of buffer additives that appear to enhance the resolution of protein mixtures by coating the capillary wall,[110,111] and modifying the zeta potential, although the mechanism is poorly understood.[71,72] Figure 1.17 shows the electropherograms resulting from the CE analysis of several

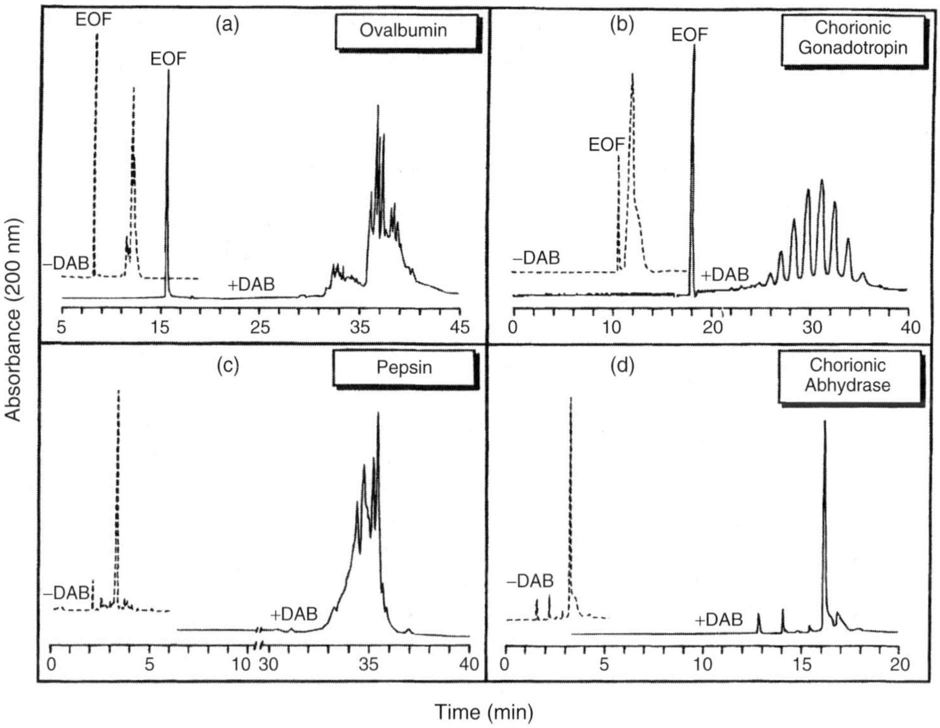

FIGURE 1.17 Effect of modifiers of EOF on separation. (a) The effect of 1 mM 1,4-diaminobutane (DAB) on the resolution of ovalbumin glycoforms. Ovalbumin [1 mg/mL, in water; 3 s pressure (0.5 psi) injection]. Analysis was carried out on a Beckman P/ACE System 2050. Separation conditions: capillary: 50 μm × 80 cm (effective length), 87 cm total length bare fused silica; *T*: 28°C; voltage: 25 kV; buffer: mM borate, pH 8.3; detection: 200 nm. (b) The effect of 5 mM 1,3-diaminopropane on the resolution of hCG glycoforms. hCG [4 mg/mL, in water; 2 s vacuum (–127 mmHg) injection]. Analysis was carried out on a ABI 270A. Separation conditions: capillary: 50 μm × 80 cm (effective length), 100 cm total length bare fused silica: *T*: 28°C; voltage: 25 kV; buffer: 25 mM borate, pH 8.8; detection: 200 nm. (Modified from Morbeck, D. E. et al., *J. Chromatogr.*, 680, 217, 1994. With permission.) (c) Partial resolution of pepsin glycoforms in the presence of 1 mM DAB. Pepsin [0.5 mg/mL, in water; 3 s pressure (0.5 psi) injection]. Analysis was carried out on a Beckman P/ACE System 2050. Separation conditions: capillary: 50 μm × 80 cm (effective length), 87 cm total length bare fused silica; *T*: 28°C; voltage: 25 kV; buffer: 100 mM borate, pH 9.0; detection: 200 nm. (Modified from Landers, J. P. et al., *Anal. Biochem.*, 205, 115, 1992. With permission.) (d) Lack of effect of DAB on carbonic anhydrase separation. Bovine carbonic anhydrase [1.0 mg/mL, in water; 3 s pressure (0.5 psi) injection]. Analysis was carried out on a Beckman P/ACE System 2050. Separation conditions: capillary: 50 μm × 80 cm (effective length), 87 cm total length bare fused silica; *T*: 28°C; voltage: 25 kV; buffer: 100 mM borate, pH 8.3; detection: 200 nm.

glycoproteins [ovalbumin, human chorionic gonadotropin (hCG), and pepsin] known to be microheterogeneous with respect to glycan content, in the absence of any modifier, and in the presence of a diaminoalkane.[72,112] Ovalbumin is resolved into two groups of nine peaks on the basis of protein phosphorylation states; the multitude of peaks observed in the presence of 1 mM diaminobutane are presumed to represent the glycoforms of the protein. A similar effect has been observed by Morbeck, et al.[112] with the CE separation of urinary hCG, another glycoprotein known to be of microheterogeneous character. The presence of 5 mM diaminopropane in the buffer results in a reduction in EOF (as determined by reduced mobility of the neutral marker, DMF) and leads to resolution of eight hCG glycoforms that cannot be completely resolved by any other method. Pepsin glycoforms were only partially resolved in 1 mM diaminobutane. The nonglycosylated protein carbonic anhydrase indicates no microheterogeneity in the presence of 1 mM diaminobutane.[72]

A second method of dynamic coating involves the use of materials to alter the viscosity of the solution at the capillary wall. Hjertén[3] suggest that EOF is inversely dependent on the viscosity of the solution in the double layer. If the viscosity approached infinity, the EOF will fall to zero. Thus polymers that adsorb to the wall should reduce or eliminate EOF, and prevent protein adsorption by blocking the access to the reactive silanes. Various cellulose derivatives[33,42–45] and polyhydroxy compounds such as polyvinyl alcohols[113,114] have been used with varying degrees of success for the separation of peptides and proteins.

1.3.3.5.3 Buffer Depletion

Since the electrophoretic process occurring within a buffer-filled capillary under an applied electrical field involves the movement of ions, the separation buffer can eventually be "depleted" if electrophoresis continues for extensive periods of time. This is not surprising in light of the fact that slab gels are used for electrophoresis only once and buffers up to two or three times. Use of "depleted" buffer for CE analysis can result in deterioration of resolution and poor run-to-run reproducibility.[116] This phenomenon is not isolated to free solution buffer systems, but also problematic in gel-based systems. For this reason buffers should be replenished after a limited number of runs (inlet and outlet every 5–8 runs; 10–15 min analysis; 4–5 mL buffer reservoirs). One method that may be useful for extending the use of a buffer without adverse effects on reproducibility is a "limited replenishment" system in which two inlet buffer reservoirs are used—one for electrophoresis and the other for rinsing the capillary with buffer that has not undergone any changes due to electrophoresis. This "limited replenishment" approach appears to substantially increase the number of runs (with the same electrophoresis buffers) before the onset of buffer-associated reproducibility problems.[117] Trushina et al.[118] have provided one example of pushing CE to the limit with respect to reproducibility under extreme buffer depletion conditions. They used a phosphate buffer containing 10 mM HCl and 250 mM KCl to separate glutathione, nitrosylated glutathione, and glutathione disulfide solubilized in a 2 M HCl sample matrix. Using 11 kV for separation, the system was pushed to the maximum current limit tolerated by the CE (249 μA) exceeding a total power of 7 W/m. Despite the dramatic change in pH observed in the inlet and outlet reservoirs over several runs, reproducible separations of the thiols (injection-to-injection) could be obtained provided that buffer depletion issues were addressed with buffer replenishment after every run.

1.4 INTRODUCTION OF SAMPLE INTO THE CAPILLARY

As a result of the small dimensions of capillaries used in CE, the total capillary volume is typically in the microliter range (see Table 1.1). Adhering to the chromatography "rule of thumb" restricting sample volume to 1–5% of the total capillary volume, sample volumes must be in the low nanoliter range if overloading is to be avoided.[119] The technology for small sample volume introduction into a capillary has converged at three accepted methods. Introduction, or as is sometimes referred, "injection" or "loading" of sample into the capillary can be accomplished hydrostatically, by siphoning, or electrokinetically. With hydrostatic injection, sample is introduced by immersing the capillary inlet into a vial containing the sample and either pressurizing the inlet vial containing the sample or applying a vacuum to the outlet vial (opposite end). Sample introduction by gravity, which is more commonly used with noncommercial systems, relies on the siphoning of sample into the capillary by elevating the injection (inlet) end of the capillary relative to the outlet end. With electrokinetic loading, the injection (inlet) end of the capillary is immersed in the sample, the outlet in the separation buffer and a low voltage (1–10 kV) applied for durations of 1–99 s, depending on the capillary length and i.d.

With electrokinetic introduction, the quantity of sample introduced into the capillary is dependent on a number of parameters, the most important of which are the electrophoretic mobilities of the sample components and the EOF. Other parameters that can effect this mode of loading, therefore, may not be a quantitative representation of the sample components. In contrast, electrokinetic loading may be advantageous if the analyte of interest is a small percentage of the sample, but has a much higher electrophoretic mobility than other constituents. Under these conditions, electrokinetic loading provides a positive sample loading bias and may enhance the detectability of the component of interest. This is containly the case with smaller molecules weight DNA fragments. Another situation where electrokinetic introduction is advantageous is where the capillary contains a polymerized or cross-linked matrix, where pressure injection cannot be used conveniently. Examples of samples advantageously electrokinetically loaded are DNA and SDS–protein complexes. It is important to bear in mid that sample matrices containing significant concentrations of electrolyte are not efficiently electrokinetcally introduced.

This last point highlights a recurring theme in many of the chapters to follow, that is, the sample matrix plays a critical role in obtaining efficient separations. The problems associated with high-salt sample matrices and approaches to manipulating the matrix for adequate CE analysis, are addressed by Shihabi in Chapter 26 and by Burgi and Gordano in chapter 13. In contrast, under the appropriate conditions, the sample matrix can have a positive effect on a sample introduction, enhancing the detectability of the components of a dilute sample by on-capillary preconcentration. This is particularly important in light of the miniscule sample volume capacity inherent with CE and is described in the next section.

1.5 ON-CAPILLARY SAMPLE CONCENTRATION TECHNIQUES

One of the main drawbacks of UV detection CE is the low sensitivity resulting from the inherently small dimensions of the flow cell (i.e., the inner diameter of the capillary) and the sample volume capacity. As mentioned above, introduction of sample volumes larger than 1%–5% of the total capillary volume can be detrimental to resolution. Therefore, while CE is a good mass detector, it is a relatively poor concentration detector. This limits the use of CE as an analytical technique to samples having nominally high concentrations (10 μg/mL or greater). For this reason, the development of several approaches for "on-capillary sample concentration" has been pivotal to the acceptance of CE as a microanalytical technique. Since an entire chapter is dedicated to this subject (Chapter 13 by Burgi and Giordano), these techniques are discussed here in only the briefest of detail.

1.5.1 Sample Stacking

One of the simplest methods for sample preconcentration is to induce "stacking" of the sample components of this is easily accomplished by exploiting the ionic strength differences between the sample matrix and separation buffer.[120,121] Stacking results from the fact that sample ions have an enhanced electrophoretic mobility in a lower conductivity environment. When voltage is applied to the system, ions in the sample plug instantaneously accelerate toward the adjacent separation buffer zone. On crossing the boundary, the higher conductivity environment induces a decrease in electrophoretic velocity and subsequent "stacking" of the sample components into a smaller buffer zone than the original sample plug. Within a short time, the ionic strength gradient dissipates and the charged analyte molecules begin to move from the "stacked" sample zone toward the cathode. Stacking can be utilized with either hydrostatic or electrokinetic injection and can typically yield a 10-fold enhancement in a sample concentration and, thence, sensitivity. An example is given in Figure 1.18, which shows separation of the standard peptide mixture in 50 mM phosphate separation buffer, pH 2.5. There is an obvious enhancement in detectability when the sample matrix is 5 mM phosphate in comparison with 50 mM separation buffer. The undesirable effects of high ionic strength

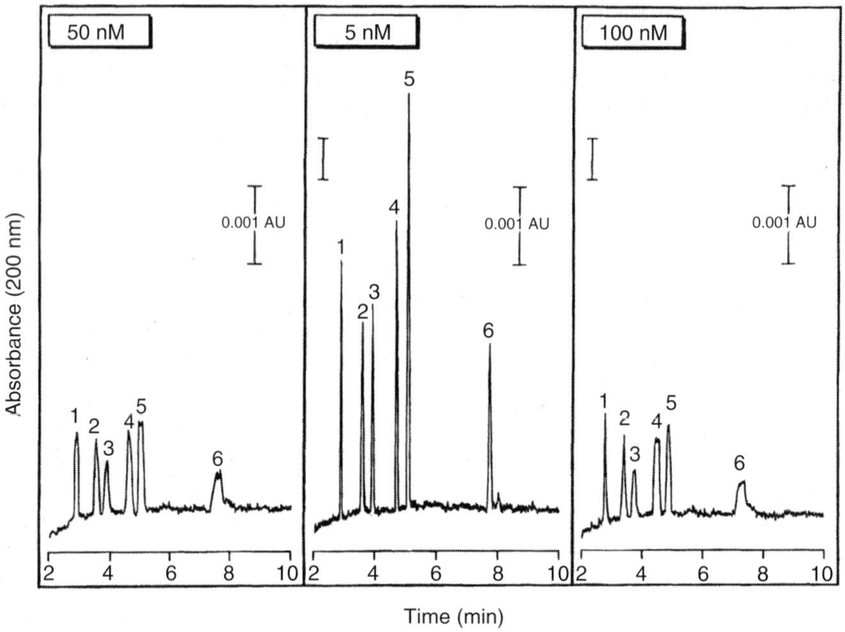

FIGURE 1.18 Sample "stacking." Stacking of peptides in phosphate buffer. Separation of a standard peptide mixture (peptide calibration kit, Bio-Rad, Richmond, CA). Peptides were (1) bradykinin, (2) angiotensin, (3) a-melanin stimulating hormone, (4) thyrotropin-releasing hormone, (5) leutinizing hormone releasing hormone, (6) leucine enkephalin, (7) bombesin, (8) methionine enkephalin, and (9) oxytocin, 50 μg/mL each. (a) Sample was dissolved in 50 mM phosphate buffer, pH 2.5. (b) Sample diluted in 5 mM phosphate buffer, pH 2.5. (c) Sample dissolved in 100 mM phosphate buffer, pH 2.5. Analyses were performed on a Beckman P/ACE System Model 2100. Separation conditions were as follows: capillary: 50 μm × 20 cm (effective length), 27 cm total length, bare fused silica; T: 20°C; voltage: 10 kV; separation buffer: 50 mM phosphate, pH 2.5; 5 s hydrostatic injection; detection: 200 nm.

(100 mM phosphate buffer) sample matrix are clearly illustrated in the right panel. Detailed discussion of sample matrix effects on separation can be found in Chapter 13 and Chapter 26.

When possible, dissolving the sample in dilute separation buffer may be more advisable than in water since the dramatic differences in EOF between the sample plug and separation buffer may cause laminar flow within the capillary and, hence, lead to peak broadening.[63,84,85,122] Moreover, excessive heat produced in the sample plug may denature sample components.[84,85,88,89] Therefore, it may be advisable to "stack" at lower applied voltages (e.g., 1-2 KV) or "ramp" to the separation voltage over several minutes.

1.5.2 Sample Focusing

An alternative approach to sample preconcentration by stacking in on-capillary "focusing" and is based on pH differences between the sample plug and separation buffer. This has been shown to be very useful for the analysis of peptides, mainly because of their relative stability over a wide pH range.[119] Focusing is easily accomplished by increasing the pH of the sample above that of the net pI (isoelectric point) of the sample. The high pH sample plug is flanked between low pH separation buffer zones (i.e., an equivalent volume of low pH separation buffer following introduction of the sample plug) and, on applying a voltage, the negatively charged peptides in the initial sample zone migrate toward the anode. Upon entering the lower pH separation buffer, a pH-induced changed in their charge state causes a reversal in the direction of mobility resulting in a "focusing" on the peptides at the interface of the sample (high pH) and low pH buffer plugs. After the pH gradient dissipates, the peptides, again positively charged, migrate toward the cathode as a sharp zone. This approach, limited to samples that can withstand the inherent changes in pH without substantial denaturation, may yield as much as a fivefold enhancement in sample concentration and, hence sensitivity. The practical and some theoretical aspects of these techniques are addressed in Chapter 26 by Shihabi and Chapter 13 by Burgi and Gierdano.

1.5.3 Isotachoporetic Sample Enrichment

Isotachoporetic (ITP) is a specialized technique for sample application first described for sample concentration by Foret et al.[123] This technique has been covered in detail by Burgi and Chein[124] in a previous edition of this handbook. However, more recently Williams and colleagues[125] utilized this in a very clever way to exploit the original on-capillary sample concentration[123] to enhance injection of DNA into microchannels for higher sensitivity fluorescence detection.

1.5.4 Online Concentration

The development of capillaries with a small "column" or plug of adsorptive resin at the inlet side of the capillary has attracted much attention owing to its ability to allow the many-fold concentration of dilute samples from large volumes. It is the simplicity of the solid-phase extraction (SPE) concept that makes it attractive—load sample by flow through the phase where the analyte of interest is adsorpted onto the resin, wash the resin to remove loosely bound or unwanted components, and then desorb the analytes from the resin and separate by electrophoresis. The notion of online sample concentrate before separation was pioneered by Guzman[126,127] in the early 1990s coincident with the effort of Fuchs et al. [128] A number of groups honed in on the potential utility of this approach for solving sensitivity problems rooted in low concentration analytes in large volume biological samples, and shown the applicability of this method to the analysis of drugs,[129,130] peptides,[131,132] and proteins.[133,134] More recently, this has been applied to DNA extraction[135–139] and immunoaffinity capture.[140–142] This technique is discussed in detail in a various chapters in this edition of the CE handbook including Chapter 43 by Bienvenue.

1.6 CONCLUDING REMARKS

Capillary zone electrophoresis (CZE) is a powerful technique, magnetic in its analytical personality as a result of the simple way diverse analytes can be resolved rapidly and with high efficiency. The attraction is easy to understand—an electrophoretic technique with as much bandwidth as (and complementary to) HPLC and multiple modes of separation available by simply changing the buffer system. Yet within the simple instrumental framework that is, at its root, a power supply, a capillary and a detector, lies the capability to analyze drugs, peptides, carbohydrates, and proteins in sample matrices as simple as buffer or as complex as serum. That power is the magnet that draws people in.

In this chapter, I hope to have imparted to the reader a taste of CE, and some practical suggestions that will be useful in developing a method for the analysis of a specific compound(s). By using the basic principles associated with electrophoretic separation as discussed here, and applying a systematic approach to assess the factors that influence the resolution of the analyte(s) under study, it is likely a successful analytical method can be developed for just about any analyte of interest. In many of the core chapters that follow this chapter, specific methods are covered in detail with an analyte-specific focus—however, method development of some kind will be found in the vast majority one of the 55 chapters that constitute this handbook.

REFERENCES

1. Michaelis, L., Electric transport of enzymes, malt distaste and pepsin, *Biochemische Zeitschrift*, v 17, 1909, pp. 231–234.
2. Tiselius, A., A new apparatus for electrophoretic analysis of colloidal mixtures, *Trans. Faraday Soc.*, 33, 524, 1937.
3. Hjertén, S., High performance electrophoresis, *Chromatogr. Rev.*, 9, 122, 1967.
4. Rilbe, H., Some reminiscences of the history of electrophoresis, *Electrophoresis*, 1995, 16, 1354–1359.
5. Consden, R., Gordon, A. H., and Martin, A. J. P., Quantitative analysis of proteins: partition chromatographic method using paper, *Biochem. J.*, 38, 224, 1944.
6. Wieland, T. and Fischer, E., Electrophoresis onto filter paper, *Naturwissenschaften*, 35, 29, 1948.
7. Smithies, O., Zone electrophoresis in starch gels: group variations in the serum proteins of normal human adults, *Biochem. J.*, 61, 629, 1955.
8. Graber, P. and Williams, C. A., Methode permittant l'etude conjugee des proprietes electrophoretiques et immunichimiques d'un mélanges de proteins: application au serum sanguine, *Biochem. Biophys. Acta*, 10, 193, 1953.
9. Hjerten, S., Agarose as an anticonvection agent in zone electrophoresis, *Biochem. Biophys. Acta*, 53, 514, 1961.
10. Kolin, A., Separation and concentration of proteins in a pH field combined with an electric field, *J. Chem. Phys.*, 22, 1628, 1954.
11. Vesterberg, O., Synthesis and isoelectric fractionation of carrier ampholytes, *Acta Chem. Scand.*, 23, 2653, 1969.
12. Raymond, S. and Weintraub, L., Acrylamide gel as a supporting medium for zone electrophoresis, *Science*, 130, 711, 1959.
13. Ornstein, L., Disc electrophoresis. I. Background and theory, *Ann. NY Acad. Sci.*, 121, 321–349, 1964.
14. Davis, B. J., Disc electrophoresis. II. Method and application to human serum proteins, *Ann NY Acad Sci.*, 121, 404–427, 1964.
15. Laemmli, U. K., Cleavage of structural proteins during the assembly of the head of bacteriophage T4, *Nature*, 227, 680–685, 1970.
16. Dale, G. and Latner, A. L., Isoelectric focusing of serum proteins in a acrylamide gels followed by electrophoresis, *Clin. Chem. Acta*, 24, 61, 1969.
17. Macko, V. and Stegemann, H., Mapping of potato proteins by combined electro-focusing and electrophoresis. Identification of varieties, *Hoppe–Seyler's Z. Physiol. Chem.*, 350, 917, 1969.
18. Stegemann, H., Protein mapping in polyacrylamide and its application to genetic analysis in plants, *Angew. Chem. (Internat. Ed.)*, 9, 643, 1970.
19. O'Farrell, P. H., High resolution two-dimensional electrophoresis of proteins, *J. Biol. Chem.*, 250, 4007, 1975.
20. Virtanen, R., Zone electrophoresis in a narrow-bore tube employing potentionmetric detection, *Acta Polytechnica Scand.*, 123, 1, 1974.
21. Jorgenson, J. W. and Lukacs, K. D., Zone electrophoresis in open tubular glass capillaries, *Anal. Chem.*, 53, 1298, 1981.
22. Ishii, J., A study of high performance liquid chromatography. I. Development of technique for miniaturization of high performance liquid chromatography, *J. Chromatogr.*, 144, 157, 1978.
23. Hirata, Y. and Novotny, M., Techniques of capillary liquid chromatography, *J. Chromatogr.*, 186, 521, 1979.
24. Kucera, P. and Guiochon, G., Use of open-tubular columns in liquid chromatography, *J. Chromatogr.*, 283, 1, 1984.
25. Schomburg, G., Problems and achievements in the instrumentation and column technology for chromatography and capillary electrophoresis, *Chromatographia*, 30, 500, 1990.
26. Mikkers, F. E. P., Everaerts, F. M., and Verheggen, T. P. E. M., High performance zone electrophoresis, *J. Chromatogr.*, 169, 11, 1979.
27. Mazzeo, J. R. and Krull, I. S., Coated capillaries and additives for the separation of proteins by capillary zone electrophoresis and capillary isoelectric focusing, *Biotechniques*, 10, 638, 1991.
28. Wehr, T., Recent advances in capillary electrophoresis columns, *LC-GC Mag.*, 11, 14, 1993.

29. Cohen, A. S. and Karger, B. L., High performance sodium dodecyl sulfate polyacrylamide gel capillary electrophoresis of peptides and proteins, *J. Chromatogr.*, 397, 409, 1987.
30. Cohen, A. S., Paulus, A., and Karger, B. L., High performance capillary electrophoresis using open tubes and gels, *Chromatographia*, 24, 15, 1987.
31. Cohen, A. S., Najarian, D. R., Paulus, A., Guttman, A., Smith, J. A., and Karger, B. L., Rapid separation and purification of oligonucleotides by high-performance capillary gel electrophoresis, *Proc. Natl. Acad. Sci. USA*, 85, 9660, 1988.
32. Drossman, H., Luckey, J. A., Kostichka, A., D'Cunha, J., and Smith, L. M., High-speed separation of DNA sequencing reactions by capillary electrophoresis, *Anal. Chem.*, 62, 900, 1990.
33. Bocek, P. and Chrambach, A., Capillary electrophoresis of DNA in agarose solutions at 40°C, *Electrophoresis*, 12, 1059, 1991.
34. Mansfield, E., Ross, E. E., and Aspinwall, C. A., Preparation and characterization of cross-linked phospholipid bilayer capillary coatings for protein separations, *Anal. Chem.*, 79, 3135–3141, 2007.
35. Kuldvee, R., Wiedmer, S. K., Oorni, K., and Riekkola, M. L., Human low-density lipoprotein-coated capillaries in electrochromatography, *Anal. Chem.*, 15, 77, 3401–3405, 2005.
36. Linden, M. V., Holopainen, J. M., Laukkanen, A., Riekkola, M. L., and Wiedmer, S. K., Cholesterol-rich membrane coatings for interaction studies in capillary electrophoresis: application to red blood cell lipid extracts, *Electrophoresis*, 27, 3988–3998, 2006.
37. Danger, G., Ramonda, M., and Cottet, H., Control of the EOF in CE using polyelectrolytes of different charge densities, *Electrophoresis*, 16, 28, 925–931, 2007.
38. Terabe, S., Otsuka, K., Ichikawa, K., Tsuchiya, A., and Ando, T., Electrokinetic separations with micellar solution and open-tubular capillaries, *Anal. Chem.*, 56, 111, 1984.
39. Terabe, S., Otsuka, K., and Ando, T., Electrokinetic chromatography with miceller solution and open-tubular capillary, *Anal. Chem.*, 57, 834, 1985.
40. Hjerten, S. and Zhu, M., Adaptation of the equipment for high-performance electrophoresis to isoelectric focusing, *J. Chromatogr.*, 346, 265, 1985.
41. Hjerten, S., Liao, J. L., and Yao, K., Theoretical and experimental study of high-performance electrophoretic mobilization of isoelectrically focused protein zones, *J. Chromatogr.*, 387, 127, 1987.
42. Hjerten, S., Elenbring, K., Kilar, F., Liao, J. L., Chen, A. J. C., Siebert, C. J., and Zhu, M., Carrier-free zone electrophoresis, displacement electrophoresis and isoelectric focusing in a high-performance electrophoresis apparatus, *J. Chromatogr.*, 403, 47, 1987.
43. Zhu, M., Hansen, D. L., Burd, S., and Gannon, F., Factors affecting free zone electrophoresis and isoelectric focusing in capillary electrophoresis, *J. Chromatogr.*, 480, 311, 1989.
44. Bruin, G. J. M., Chang, J. P., Kuhlman, R. H., Zegers, K., Kraak, J. C., and Poppe, H., Capillary zone electrophoretic separation of proteins in polyethylene glycol-modified capillaries, *J. Chromatogr.*, 471, 429, 1989.
45. Bruin, G. J. M., Huisden, R., Kraak, J. C., and Poppe, H., Performance of carbohydrate-modified fused-silica capillaries for the separation of proteins by zone electrophoresis, *J. Chromatogr.*, 480, 339, 1989.
46. Ganzler, K., Greve, K. S., Cohen, A. S., Karger, B. L., Guttman, A., and Cooke, N. C., High performance capillary electrophoresis of SDS–protein complexes using UV-transparent polymer networks, *Anal. Chem.*, 64, 2665, 1992.
47. Miksik, I., Sedlakova, P., Mikulikova, K., Eckhardt, A., Cserhati, T., and Horvath, T., Matrices for capillary gel electrophoresis—a brief overview of uncommon gels, *Biomed. Chromatogr.*, 20, 458–465, 2006.
48. Rech, I., Cova, S., Restelli, A., Ghioni, M., Chiari, M., and Cretich, M., Microchips and single-photon avalanche diodes for DNA separation with high sensitivity, *Electrophoresis*, 27, 3797–3804, 2006.
49. Gras, R., Luong, J., Monagle, M., and Winniford, B., Gas chromatographic applications with the dielectric barrier discharge detector, *J. Chromatogr. Sci.*, 44, 101–107, 2006.
50. Castano-Alvarez, M., Fernandez-Abedul, M. T., and Costa-Garcia, A., Amperometric detector designs for capillary electrophoresis microchips, *J. Chromatogr. A*, 24, 1109, 291–299, 2006.

51. Cappiello, A., Famiglini, G., Palma, P., and Siviero, A., Liquid chromatography-electron ionization mass spectrometry: fields of application and evaluation of the performance of a Direct–EI interface, *Mass Spectrom. Rev.*, 24, 978–989, 2005.
52. Rocheleau, M. J. and Dovichi, N. J., Separation of DNA sequencing fragments at 53 bases per minute by capillary gel electrophoresis, *J. Microcol. Sep.*, 4, 449, 1992.
53. Rickard, E. C., Strohl, M. M., and Nielsen, R. G., Correlation of electrophoretic mobilities from capillary electrophoresis with physicochemical properties of proteins and peptides, *Anal. Biochem.*, 197, 197, 1991.
54. Helmholtz, H. Z., About electrical interfaces (translated title), *Ann. Phys. Chem.*, 7, 337, 1879.
55. Tian, H., Hühmer, A. F. R., and Landers, J. P., Evaluation of silica resins for the direct and efficient extraction of DNA from complex biological matrices in a miniaturized format, *Anal. Biochem.*, 283, 175–191, 2000.
56. Sazelova, P., Kasicka, V., Koval, D., Prusik, Z., Fanali, S., Aturki, Z., Control of EOF in CE by different ways of application of radial electric field. *Electrophoresis*, 28, 756–766, 2007.
57. Weber, R., Concerning theories of electrophoretic separations in porous tubes, *Helv. Chim. Acta.*, 36, 424, 1953.
58. Snyder, L. R. and Kirkland, J. J., *Introduction to Modern Chromatography*, 2nd ed., Wiley Interscience, New York, 1979.
59. Giddings, C. J., Generation of variance, "theoretical plates", resolution, and peak capacity in electrophoresis and sedimentation, *Sep. Sci.*, 4, 181, 1969.
60. Sternberg, J. C., Extracolumn contribution to chromatographic band broadening, *Adv. Chromatogr.*, 2, 206, 1996.
61. Huang, X., Coleman, W. F., and Zare, R. N., Analysis of factors causing peak broadening in capillary zone electrophoresis, *J. Chromatogr.*, 480, 95, 1989.
62. Jones, H. K., Nguyen, N. T., and Smith, R. D., Variance contributions to band spreading in capillary zone electrophoresis, *J. Chromatogr.*, 504, 1, 1990.
63. Grushka, E., McCormick, R. M., and Kirkland, J. J., Effect of temperature gradients on the efficiency of capillary zone electrophoresis separations, *Anal. Chem.*, 61, 241, 1989.
64. McManigill, D. and Swedberg, S. A., Factors affecting plate height in high performance zonal capillary electrophoresis (HPZCE), in *Techniques in Protein Chemistry*, Hugli, T., Ed., Academic Press, San Diego, CA, p. 468, 1989.
65. Lux, J. A., Hausig, U., and Schomburg, G., Production of windows in fused silica capillaries for in-column detection of UV-absorption or fluorescence in capillary electrophoresis or HPLC, *J. High Resol. Chromatogr.*, 13, 373, 1990.
66. McCormick, R. M. and Zagursky, R. J., Polymide stripping device for producing detection windows on fused-silica tubing used in capillary electrophoresis, *Anal. Chem.*, 63, 750, 1991.
67. Bruno, A. E., Gassmann, E., Pericles, N., and Anton, K., On-column capillary flow cell utilizing optical waveguides for chromatographic application, *Anal. Chem.*, 61, 876, 1989.
68. Tran, A. D., Park, S., Lisi, P. J., Huynh, O. T., Ryall, R. R., and Lane, P. A., Separation of carbohydrate-mediated microheterogeneity of recombinant human erythropoetin by free solution capillary electrophoresis, *J. Chromatogr.*, 542, 459, 1991.
69. Kaupp, S., Bubert, H., Baur, L., Nelson, G., Watzig, H., Unexpected surface chemistry in capillaries for electrophoresis, *J. Chromatogr. A*, 894, 73–77, 2000.
70. Dolnik, V. Wall coating for capillary electrophoresis on microchips, *Electrophoresis*, 25, 3589–3601, 2004.
71. Taverna, M., Baillet, A., Biou, D., Schluter, M., Werner, R., and Ferrier, D., Analysis of carbohydrate-mediated heterogeneity and characterization of N-linked oligosaccharides of glycoproteins by high performance capillary electrophoresis, *Electrophoresis*, 13, 359, 1992.
72. Landers, J. P., Oda, R. P., Madden, B. J., and Spelsberg, T. C., High-performance capillary electrophoresis of glycoproteins: the use of modifiers of electroosmotic flow for analysis of microheterogeneity, *Anal. Biochem.*, 205, 115, 1992.
73. Oda, R. P. and Landers, J. P., Effect of cationic buffer additives on capillary electrophoretic separation of serum transferrin from different species, *Electrophoresis*, 17, 331, 1996.
74. Dabek-Zlotorzynska, E., Aranda-Rodriguez, R., and Buykx, S. E., Development and validation of capillary electrophoresis for the determination of selected metal ions in airborne particulate matter after sequential extraction. *Anal. Bioanal. Chem.*, 372, 467–472, 2002.

75. Wang, M., Qu, F., Shan, X. Q., and Lin, J. M., Development and optimization of a method for the analysis of low-molecular-mass organic acids in plants by capillary electrophoresis with indirect UV detection, *J. Chromatogr. A*, 989, 285–292, 2003.
76. Jamali, B. and Lehmann, S., Development and validation of a high-resolution capillary electrophoresis method for multi-analysis of ragaglitazar and arginine in active pharmaceutical ingredients and low-dose tablets, *J. Pharm. Biomed. Anal.*, 34, 463–472, 2004.
77. Sanger-van de Griend, C. E., Ek, A. G., Widahl-Nasman, M. E., and Andersson, E. K., Method development for the enantiomeric purity determination of low concentrations of adrenaline in local anaesthetic solutions by capillary electrophoresis, *J. Pharm. Biomed. Anal.*, 11, 41, 77–83, 2006.
78. Kamoda, S., Ishikawa, R., and Kakehi, K., Capillary electrophoresis with laser-induced fluorescence detection for detailed studies on N-linked oligosaccharide profile of therapeutic recombinant monoclonal antibodies, *J. Chromatogr. A*, 1133, 332–339, 2006.
79. Iqbal, J., Scapozza, L., Folkers, G., and Muller CE., Development and validation of a capillary electrophoresis method for the characterization of herpes simplex virus type 1 (HSV-1) thymidine kinase substrates and inhibitors. *J. Chromatogr. B–Anal. Technol. Biomed. Life Sci.*, 846, 281–290, 2007.
80. Wang, W., Tang, J., Wang, S., Zhou, L., and Hu, Z., Method development for the determination of coumarin compounds by capillary electrophoresis with indirect laser-induced fluorescence detection, *J. Chromatogr. A*, 2007. Vol 1148(1): 108–114.
81. Li, P., Li, S. P., and Wang, Y. T., Optimization of CZE for analysis of phytochemical bioactive compounds, *Electrophoresis*, 27, 4808–4819, 2006.
82. Kamoda, S. and Kakehi, K., Capillary electrophoresis for the analysis of glycoprotein pharmaceuticals, *Electrophoresis*, 27, 24, 2006.
83. McLaughlin, G. M., Nolan, J. A., Lindahl, J. L., Palmieri, R. H., Anderson, K. W., Morris, S. C., Morrison, J. A., and Bronzert, T. J., Pharmaceutical drug separations by HPCE: practical guidelines, *J. Liq. Chromatogr.*, 15, 961, 1992.
84. Gobie, W. A. and Ivory, C. F., Thermal model of capillary electrophoresis and a method of counteracting thermal band broadening, *J. Chromatogr.*, 516, 191, 1990.
85. Davis, J. M., Influence of thermal variations of diffusion coefficient on non-equilibrium plate height in capillary zone electrophoresis, *J. Chromatogr.*, 517, 521, 1990.
86. Terabe, S., Otsuka, K., and Ando, T., Electrokinetic chromatography with micellar solution ad open-tubular capillary, *Anal. Chem.*, 57, 834, 1985.
87. Rush, R. S., Cohen, A. S., and Karger, B. L., Influence of column temperature on the electrophoretic behavior of myoglobin and α-lactalbumin in high-performance capillary electrophoresis, *Anal. Chem.*, 63, 1346, 1991.
88. Kurosu, Y., Hibi, K., Sasaki, T., and Saito, M., Influence of temperature control in capillary electrophoresis, *J. High Resolut. Chromatogr.*, 14, 200, 1991.
89. Guttman, A., Horvath, J., and Cooke, N., Influence of temperature on the sieving effect of different polymer matrices in capillary SDS gel electrophoresis of proteins, *Anal. Chem.*, 65, 199, 193.
90. Giordano, B. C., Horsman, K. M., Burgi, D. S., Ferrance, J. P., and Landers, J. P., Method for determining intracapillary solution temperatures: application to sample zone heating for enhanced fluorescent labeling of proteins, *Electrophoresis*, 28, 714, 2007.
91. Nelson and Cooke, *Handbook of Capillary Electrophorosis*, 1st ed., CRC Press.
92. Moyer, T. P., Jiang, N.-S., Tyce, G. M., and Sheps, S. G., Analysis for urinary catecholamines by liquid chromatography with amperometric detection: methodological and clinical interpretation of results, *Clin. Chem.*, 25, 256, 1979.
93. Wallingford, R. A. and Ewing, A. G., Separation of serotonin from catechols by capillary zone electrophoresis with electrochemical detection, *Anal. Chem.*, 61, 98, 1989.
94. Lui, J., Shirota, O., and Novotny, M., Capillary electrophoresis of amino sugars with laser-induced fluorescence detection, *Anal. Chem.*, 63, 413, 1991.
95. Honda, S., Makino, A., Suzuki, S., and Kakehi, K., Analysis of the oligosaccharides of ovalbumin by high performance capillary electrophoresis, *Anal. Biochem.*, 191, 228, 1990.
96. Landers, J. P., Oda, R. P., and Schuchard, M. D., Separation of boron-complexed diol compounds using high-performance capillary electrophoresis, *Anal. Chem.*, 64, 2846, 1992.
97. Chen, F. A., Kelly, L., Palmieri, R., Biehler, R., and Schwartz, H., Use of high ionic strength buffers for the separation of proteins and peptides with capillary electrophoresis, *J. Liq. Chromatogr.*, 15, 1143, 1992.

98. Nelson, R. J., Paulus, A., Cohen, A. S., Guttman, A., and Karger, B. L., Use of Peltier thermoelectric devices to control column temperature in high-performance capillary electrophoresis, *J. Chromatogr.*, 480, 111, 1989.
99. Sepaniak, M. J. and Cole, R. O., Column efficiencies in micellar electrokinetic capillary chromatography, *Anal. Chem.*, 59, 472, 1987.
100. Landers, J. P., Oda, R. P., Spelsberg, T. C., Nolan, J. A., and Ulfelder, K. J., Capillary electrophoresis: A powerful microanalytical technique for biologically active molecules, *Biotechnique*, 14, 98, 1993.
101. Bushey, M. and Jorgenson, J., Capillary electrophoresis of proteins in buffers containing high concentrations of zwitterionic salts, *J. Chromatogr.*, 480, 301, 1989.
102. Oda, R. P., Madden, B. J., Morris, J. C., Spelsberg, T. C., and Landers, J. P., Multiple-buffer-additive strategies for enhanced electrophoretic separation of peptides, *J. Chromatogr.*, 680, 341, 1994.
103. Bean, S. R., Lookhart, G. L., and Bietz, J. A., Acetonitrile as a buffer additive for free zone capillary electrophoresis separation and characterization of maize (*Zeamays* L.) and sorghum (*Sorghum bicolor* L. Moench) storage proteins. *J. Agric. Food Chem.*, 48, 318–327, 2000.
104. Gordon, M. J., Lee, K.-J., Arias, A. A., and Zarc, R. N., Protocol for resolving protein mixtures in capillary zone electrophoresis, *Anal. Chem.*, 63, 69, 1991.
105. Swedberg, S. A., Characterization of protein behavior in high-performance capillary electrophoresis using a novel capillary system, *Anal. Biochem.*, 185, 51, 1990.
106. Bolger, C. A., Zhu, M., Rodriguez, R., and Wehr, T., Performance of uncoated and coated capillaries in free zone electrophoresis and isoelectric focusing of proteins, *J. Liq. Chromatogr.*, 14, 895, 1991.
107. Wiktorowicz, J. E. and Colburn, J. C., Separation of cationic proteins via charge reversal in capillary electrophoresis, *Electrophoresis*, 11, 769, 1990.
108. Towns, J. K. and Regnier, F. E., Capillary electrophoretic separation of proteins using nonionic surfactant coatings, *Anal. Chem.*, 63, 1126, 1991.
109. Bullock, J. A. and Yuen, L.-C., Free solution capillary electrophoresis of basic proteins in uncoated fused silica capillary tubing, *J. Microcol. Sep.*, 3, 241, 1991.
110. Song, L., Ou, Q., and Yu, W., Study on phosphate of ethylenediamine, 1,3-diaminopropane and 1,4-diaminobutane as carrying electrolyte in open-tubular capillary electrophoresis, *J. Liq. Chromatogr.*, 17, 1953, 1994.
111. Oda, R. P., Madden, B. J., Spelsberg, T. C., and Landers, J. P., α, β-Bis-quarternary ammonium alkanes as effective buffer additives for enchanced capillary electrophoretic separation of glycoproteins, *J. Chromatogr.*, 680, 85, 1994.
112. Morbeck, D. E., Madden, B. J., Charlesworth, C. C., and McCormick, D. J., High resolution separation of human choriogonadotropin (hCG) isoforms using capillary electrophoresis, *J. Chromatogr.*, 680, 217, 1994.
113. Gilges, M., Husmann, H., Kleemiß, M. H., Motsch, S. R., and Schomberg, G., CZE separation of basic proteins at low pH in fused silica capillaries with surfaces modified by silane derivatization and/or adsorption of polar polymers, *J. High Res. Chromatogr.*, 15, 452, 1992.
114. Schomberg, G., Belder, D., Gilges, M., Kleemiß, M. H., and Motsch, S. R., Wall chemistry in capillary electrophoresis: the importance of maximal efficiency in analytical CE, Abstract 101, a lecture presented at HPCE '94, San Diego, CA, 1994.
115. Figeys, D., Renborg, A., and Dovichi, N. J., Spatial and temporal depletion of ions from noncrosslinked denaturing polyacrylamide in capillary electrophoresis, *Electrophoresis*, 15, 1512–1517, 1994.
116. Strege, M. A. and Lagu, A. L., Studies of migration time reproducibility of CE proteins separations, *J. Liq. Chromatogr.*, 16, 51, 1993.
117. Landers, J. P., Oda, R. P., Madden, B., Sismelich, T. P., and Spelsberg, T. C., Reproducibility of sample separation using liquid or forced air convection themostated high performance capillary electrophoresis, *J. High Res. Chromatogr.*, 15, 517, 1992.
118. Trushina, E. V., Oda, R. P., McMurray, C. T., and Landers, J. P., Effective and reproducible capillary electrophoretic separation of thiols under conditions where exceptionally high current is generated. *Anal Chem.*, 71, 5569–5573, 1999.
119. Abersold, R. and Morrison, H. D., Analysis of dilute peptide samples by capillary zone electrophoresis, *J. Chromatogr.*, 516, 79, 1990.
120. Moring, S. E., Colburn, J. C., Grossman, P.D., and Lauer, H. H., Analytical aspects of an automated capillary electrophoresis system, *LC-GC*, 8, 34, 1989.

121. Chien, R.-L. and Burgi, D. S., On-column sample concentration using filed amplification in CZE, *Anal. Chem.*, 64, 489A, 1992.
122. Chein, R.-L. and Helmer, J. C., Electroosmotic properties and peak broadening in field-amplified capillary electrophoresis, *Anal. Chem.*, 63, 1354, 1990.
123. Foret, F., Sustacek, V., and Bocek, B. J., On-line isotachophoretic sample preconcentration for enhancement of zone detectability in capillary zone electrophoresis, *J. Microcol. Sep.*, 2, 229, 1990.
124. Burgi, D. and Chien, R.-L., *Handbook of Capillary Electrophoresis* (2nd edition), Ed. J.P. Landers, CRC Press, Boca Raton, FL, Chapter 16, 1996.
125. Kurnik, R. T., Boone, T. D., Nguyen, U., Ricco, A. J., and Williams, S. J., Use of floating electrodes in transient isotachophoresis to increase the sensitivity of detection, *Lab Chip*, 3, 86–92, 2003.
126. Guzman, N. A., Trebilcok, M. A., and Advis, J. P., The use of a concentration step to collect urinary components separated by capillary electrophoresis and further characterization of collected analytes by mass spectrometry, *J. Chromatogr.*, 14, 997, 1991.
127. Guzman, N. A., Automated capillary electrophoresis apparatus, U.S. Patent #5,202,010, 1993.
128. Fuchs, M. and Merion, M., Apparatus for effecting capillary electrophoresis, U.S. Patent #5,246,577, 1993.
129. Strausbauch, M. A., Xu, S. Z., Ferguson, J. E., Nunez, M., Machacek, D., Lawson, G. M., Wettstein, P. J., and Landers, J. P., Concentration and analysis of hypoglycemic drugs using solid phase extraction-capillary electrophoresis, *J. Chromatogr.*, 717, 279–291, 1995.
130. Tomlinson, A. J., Benson, L. M., Braddock, W. D., Oda, R. P., and Naylor, S., On-line preconcentration-capillary electrophoresis-mass spectrometry (PC-EC-MS), *J. High Resolut. Chromatogr.*, 17, 729, 1994.
131. Strausbauch, M., Landers, J. P., and Wettstein, P. J., Mechanism of peptide separations by solid phase extraction-capillary electrophoresis (SPE-CE) at low pH, *Anal. Chem.*, 68, 306–314, 1996.
132. Strausbauch, M. A., Madden, B. J., Wettstein, P. J., and Landers, J. P., Sensitivity enhancement and second dimensional information from the SPE-CE analysis of entire HPLC fractions, *Electrophoresis*, 16, 541, 1995.
133. Beattie, J. H., Self, R., and Richards, M. P., The use of solid phase concentrators for on-line preconcentration of metallothionen prior to isoform separation by capillary zone electrophoresis, *Electrophoresis*, 16, 322, 1995.
134. Roche, M. E., Anderson, M. A., Oda, R. P., Riggs, L. B., Strausbauch, M. A., Okazaki, R., Wettstein, P. J., and Landers, J. P., Capillary electrophoresis of insulin-like growth factors: enhanced UV detection using dynamically-coated capillaries and on-line solid phase extraction (SPE-CE), *Anal. Biochem.*, 258, 87–95, 1998.
135. Cao, W., Easley, C. J., Ferrance, J. P., and Landers, J. P., Chitosan as a novel polymer for pH-induced DNA purification in a totally aqueous system: I. High density open-channel pattern for DNA purification, *Anal. Chem.*, 78, 7222–7228, 2006.
136. Wolfe, K. A., Breadmore, M. C., Ferrance, J. P., Power, M. E., Conroy, J. F., Norris, P. M., and Landers, J. P., Toward a microchip-based solid-phase extraction method for isolation of nucleic acids, *Electrophoresis*, 23, 727–733, 2002.
137. Breadmore, M. C., Wolfe, K. A., Arcibal, I. G., Leung, W. K., Dicks, D., Giordano, B. C., Power, M. E. et al., Microchip-based purification of DNA from biological samples, *Anal. Chem.*, 75, 1880–1886, 2003.
138. Wu, Q., Bienvenue, J. M., Giordano, B. C., Hassan, B. J., Kwok, Y. C., Norris, P. M., Landers, J. P., and Ferrance, J. P., A microchip-based macroporous silica sol-gel monolith for efficient isolation of DNA from clinical samples, *Anal. Chem.*, 78, 5704–5710, 2006.
139. Wen, J., Guillo, C., Ferrance, J. P., and Landers, J. P., DNA extraction using a tetramethyl orthosilicate-grafted photopolymerized monolithic solid phase, *Anal. Chem.*, 78, 1673–1681, 2006.
140. Guzman, N. A. and Phillips, T. M., Immunoaffinity CE for proteomics studies, *Anal. Chem.*, 77, 61A–67A, 2005.
141. Guzman, N. A., Improved solid-phase microextraction device for use in on-line immunoaffinity capillary electrophoresis, *Electrophoresis*, 24, 3718–3727, 2003.
142. Vizioli, N. M., Carducci, C. N., Pena, C., and Guzman, N. A., Monitoring the purity of a synthetic peptide by capillary electrophoresis: utilization of an on-line preconcentration method for improved separation and detection sensitivity, *J. Capill. Electrophor. Microchip Technol.*, 6, 109–118, 1999.

APPENDIX 1

CALCULATIONS OF PRACTICAL USE

In this section, we utilize an exemplary electropherogram (Figure A1.1) and the corresponding data obtained using the data management software (Figure A1.1, inset) to take the reader through a variety of calculations of practical use.

A.1 MOBILITY

The velocity of an analyte, ν_i, is the distance traveled during the time of electric field application. Velocity is related to mobility, μ_i, by field strength, $E = H/L$. Thus,

$$\nu_i = \mu_{\text{app}} H/L = \mu_{\text{app}} E$$

Peak Number	Migration Time	Component	Peak Area	Peak Height	Apparent Mobility
1	1.723	Mesityl oxide	0.10762	0.00688	7.2735
2	2.642	p-OH ϕ acet.	0.10624	0.00463	4.7446
3	2.818	p-OH benz.	0.18887	0.00819	4.4477
4	2.924	benzoate	0.14680	0.00616	4.2868
			0.54952	0.02565	

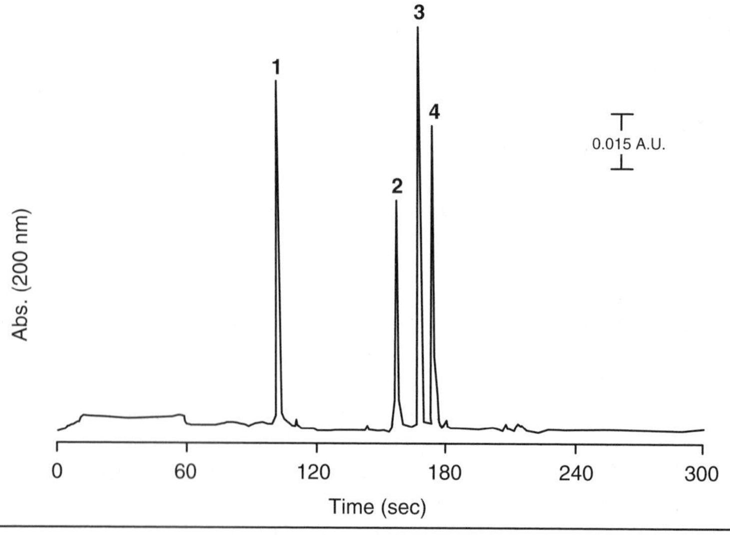

FIGURE A1.1 Simple separation of four small molecules for calculations of practical use. Separation of mesityl oxide [an EOF (neutral) marker], *p*-hydroxyphenylacetic acid, *p*-hydroxybenzoic acid, and benzoic acid [in borate buffer, pH 8.3; 3-s pressure (@0.5 psi) injection]. Analysis was carried out on a Beckman P/ACE System 2050. Separation conditions: capillary, 50 μm $\times$ 40 cm (effective length), 47 cm total length bare fused silica; T, 28°C; voltage, 25 kV; buffer, 100 mM borate, pH 8.3; detection, 200 nm. Inset: data derived from System Gold (vs7.1) data management software.

The apparent mobility μ_{app} measured from an electropherogram is the sum of the mobility of the analyte and that due to electroosmotic flow (EOF):

$$\begin{aligned}\mu_{app,i} &= \mu_{\mathrm{ep}} + \mu_{\mathrm{eo}} \\ &= L_{\mathrm{d}}L_{\mathrm{t}}/t_{\mathrm{app},i}H \\ &= L_{\mathrm{d}}L_{\mathrm{t}}/H(1/t_{\mathrm{app},i} - 1/t_{\mathrm{ref}}) + \mu_{\mathrm{ref}}\end{aligned}$$

where L_d is length to the detector, L_t is total length of the capillary, and, if the reference peak is a neutral marker,

$$\mu_{\mathrm{ref}} = 0.$$

To calculate μ_{ref} for a charged reference peak, the above equation may be used, substituting $t_{\mathrm{app},i}$ for the charged reference and t_{ref} for the neutral marker, with $\mu_{\mathrm{ref}} = 0$.

A.1.1 Example

From Figure A1.1, $L_d = 40$ cm, $L_t = 47$ cm, and $t_1 = 1.723$ min:

$$\begin{aligned}\nu &= L_{\mathrm{d}}/t_1 = 0.4\mathrm{m}/(1.723\ \mathrm{min})(60\mathrm{s/min}) \\ &= 0.003869\ \mathrm{m/s} \\ &= \mu_{\mathrm{app}}E = \mu_{\mathrm{app}}H/L_t\end{aligned}$$

or apparent mobility $= \mu_{\mathrm{app},1} = \nu L_{\mathrm{t}}/H$

$$\begin{aligned}&= (0.003869\ \mathrm{m/s})(0.47\ \mathrm{m})/25\ \mathrm{kV} \\ &= (0.001818/25) = 7.273 \times 10^{-4}\ \mathrm{cm^2/(V \cdot s)}\end{aligned}$$

Similarly for the other peaks,

$$\begin{aligned}\mu_{\mathrm{app},i} &= (0.4)(0.47)/(60)(25 \times 10^3)t_i = 12.533 \times 10^{-4}/t_i\mathrm{cm}^2/(V \cdot s) \\ \mu_{\mathrm{app},2} &= 12.533 \times 10^{-4}/2.642 = 4.744 \times 10^{-4}\mathrm{cm}^2/(V \cdot s) \\ \mu_{\mathrm{app},3} &= 12.533 \times 10^{-4}/2.818 = 4.448 \times 10^{-4}\mathrm{cm}^2/(V \cdot s) \\ \mu_{\mathrm{app},4} &= 12.533 \times 10^{-4}/2.924 = 4.286 \times 10^{-4}\mathrm{cm}^2/(V \cdot s)\end{aligned}$$

Using

$$\mu_{\mathrm{app},i} = L_{\mathrm{d}}L_{\mathrm{t}}/H(1/t_{\mathrm{app},i} - 1/t_{\mathrm{ref}}) + \mu_{\mathrm{ref}}$$

we many make the same calculation using $\mu_{ref} = \mu_{eo} = 7.273 \times 10^{-4}$ cm^2/(V · s):

$$\begin{aligned}
\mu_{app,i} &= [(0.4)(0.47)/25 \times 10^3](1/t_{app,i} - 1/1.723) + 7.273 \\
\mu_{app,2} &= 12.533(1/2.642 - 1/1.723) + 7.273 \times 10^{-4} \\
&= 12.533(0.3785 - 0.5804) + 7.273 \times 10^{-4} \\
&= -2.530 + 7.273 \times 10^{-4}\ \text{cm}^2/(\text{V} \cdot \text{s}) = 4.743 \times 10^{-4}\ \text{cm}^2/(\text{V} \cdot \text{s}) \\
\mu_{app,3} &= 12.533 \times 10^{-4}(0.3549 - 0.5804) + 7.273 \times 10^{-4} \\
&= -2.826 \times 10^{-4} + 7.273 \times 10^{-4} = 4.448 \times 10^{-4}\ \text{cm}^2/(\text{V} \cdot \text{s}) \\
\mu_{app,4} &= 12.533 \times 10^{-4}(0.3420 - 0.5804) + 7.273 \times 10^{-4} \\
&= -2.988 + 7.273 \times 10^{-4} = 4.285 \times 10^{-4}\ \text{cm}^2/(\text{V} \cdot \text{s})
\end{aligned}$$

The electrophoretic mobility, μ_{ep}, may be calculated from

$$\mu_{app,i} = \mu_{ep} + \mu_{eo}$$

as peak 1 in Figure A1.1 is a neutral marker, $\mu_{app,1} = \mu_{eo} = 7.273 \times 10^{-4}$ cm^2/(V · s)

$$\begin{aligned}
\mu_{ep,2} &= 4.744 - 7.273 = -2.529 \times 10^{-4}\ \text{cm}^2/(\text{V} \cdot \text{s}) \\
\mu_{ep,3} &= 4.448 - 7.273 = -2.825 \times 10^{-4}\ \text{cm}^2/(\text{V} \cdot \text{s}) \\
\mu_{ep,4} &= 4.286 - 7.273 = -2.987 \times 10^{-4}\ \text{cm}^2/(\text{V} \cdot \text{s})
\end{aligned}$$

A.2 CORRECTED PEAK AREA

Integrators typically present peak area in dimensions of (response) (time). This calculation is a simple transformation to obtain area in (response) (width).

$$\begin{aligned}
A_{corr} &= A_i(\text{mAU} \times \text{min})v_i \\
&= A_i(\text{mAU} \times \text{min})L_d(\text{cm})/t_m(\text{min}) \\
&= A_i(\text{mAU} \times \text{cm})
\end{aligned}$$

This can usually be achieved through the data collection and software system.

A.2.1 EXAMPLE

$$\begin{aligned}
A_{corr,1} &= (0.10762)(40)/1.723 \\
&= 2.498\ \text{mAU cm}
\end{aligned}$$

Similarly,

$$\begin{aligned}
A_{corr,2} &= (0.10624)(40)/2.642 = 1.608 \\
A_{corr,3} &= (0.18887)(40)/2.818 = 2.681 \\
A_{corr,4} &= (0.14680)(40)/2.924 = 2.008
\end{aligned}$$

Dividing the corrected area by peak height gives peak width, which approximates the peak width at half-height, which may be used to calculate N (see calculations, Section A.5).

$$w_{1/2,1} = 2.498 \text{ mAU}/0.00668\ \mu\text{AU} = 374\ \mu\text{m}$$
$$w_{1/2,2} = 1.608 \text{ mAU}/0.00463\ \mu\text{AU} = 347\ \mu\text{m}$$
$$w_{1/2,3} = 2.681 \text{ mAU}/0.00819\ \mu\text{AU} = 327\ \mu\text{m}$$
$$w_{1/2,4} = 2.008 \text{ mAU}/0.00616\ \mu\text{AU} = 326\ \mu\text{m}$$

A.3 QUANTITY OF SAMPLE INTRODUCED INTO THE CAPILLARY

We assume

$$\begin{aligned} Q &= (\text{volume})_{\text{int}}(\text{concentration}) \\ &= \pi r^2 l \quad [C_i] \end{aligned}$$

where r is the internal radius of the capillary, l is the length of the sample plug, and $[C_i]$ is the concentration of the sample.

A.3.1 Hydrodynamic Injection

$$Q = \pi r^2[\Delta P r^2 t_{\text{int}}/(8\eta L)][C_i]$$

where ΔP is the pressure difference, r is the capillary inner radius, t_{int} is the introduction time, η is the viscosity of the sample solution, and L is the length of the column.

A.3.1.1 Example

Assuming typical values for the constants,

$$\text{Column dimensions: } 50\ \mu\text{m i.d.} \times 47 \text{ cm}, 40 \text{ cm to the detector}$$
$$\eta = 0.9548 \text{ cP} = 9.548 \times 10^{-4} \text{ N/s/m}^2 \text{ at22°C}$$
$$\Delta P = 0.5 \text{ psi} = 3.435 \times 10^3 \text{ N/m}^2$$
$$t_{\text{int}} = 3 \text{ s}$$

Then

$$\begin{aligned} \text{vol}_{\text{int}} &= (3.1416)(25 \times 10^{-6})^2 \left\{\left[\left(3.435 \times 10^3\right)\left(25 \times 10^{-6}\right)^2 3\right]/\left[8\left(9.548 \times 10^{-4}\right)0.47\right]\right\} \\ &= \left(19.635 \times 10^{-10}\right)\left(17.94 \times 10^{-4}\right) \\ &= 3.523 \times 10^{-12} \text{ m}^3 = 3.523 \text{ nL} \end{aligned}$$

which correlates with $\simeq$ 1.2 nL/s injection.

A.3.2 Electrokinetic Introduction

$$Q = [\pi r^2(\mu_{\text{app}})V_{\text{int}}t_{\text{int}}/L][C_i]$$

where μ_{app} is the mobility of the analyte, μ_{eo} is the electroosmotic mobility, and V_{int} is the introduction voltage.

A.3.2.1 Example

Again assuming typical values,

$$\text{Column dimensions: } 50\mu\text{m i.d.} \times 47\text{ cm}, 40\text{ cm to the detector}$$

$$t_{\text{int}} = 20\text{ s}$$

$$\mu_{\text{app}} = 4.744 \times 10^{-8}\text{ m}^2/\text{V/s}$$

$$\mu_{\text{eo}} = 7.273 \times 10^{-8}\text{ m}^2/\text{V/s}$$

$$V_{\text{int}} = 1.5\text{ kV}$$

Then

$$\begin{aligned}\text{vol}_{\text{int}} &= (3.1416)\left(25 \times 10^{-6}\right)^2\left[\left(4.744 \times 10^{-8}\right)\left(1.5 \times 10^3\right)20/0.47\right]\text{m}^3\\ &= \left(19.635 \times 10^{-10}\right)\left(302.81 \times 10^{-5}\right)\text{m}^3\\ &= 5.945 \times 10^{-15}\text{ m}^3 = 5.945\text{ nL}\end{aligned}$$

However, if we calculate the volume of fluid injected, which is due solely to μ_{eo},

$$\begin{aligned}\text{vol}_{\text{int}} &= (3.1416)\,(3.1416)\,(x)^2\left[\left(7.273 \times 10^{-8}\right)\left(1.5 \times 10^3\right)20/(0.47)\right]\text{m}^3\\ &= \left(19.635 \times 10^{-10}\right)\left(464.2 \times 10^{-5}\right)\text{m}^3\\ &= 9115 \times 10^{-15}\text{ m}^3 = 9.115\text{ nL}\end{aligned}$$

With electrokinetic introduction, the amount of sampled injected is proportional to the analyte mobility. It is typically used when the sample has greater mobility than EOF. If one has a highly mobile analyte in relatively low ionic strength sample buffer ($1/x$ times separation buffer) one may load an x-fold greater portion onto the column by electrokinetic introduction without adversely affecting separation resolution. In the above example, if the sample were in 0.1 times separation buffer, we could have loaded the equivalent of 59.5 nL of sample.

A.4 RESOLUTION

Using the equation for resolution, we obtain

$$\text{Res} = 2(x_{i2} - x_{i1})/(w_1 + w_2)$$

A.4.1 Example

$$\begin{aligned}\text{Res}_{2-1} &= 2(2.818 - 2.642)/(0.02295 + 0.02306)\\ &= 0.352/0.04601\\ &= 7.650\\ \text{Res}_{3-2} &= 2(2.924 - 2.818)/(0.02306 + 0.02383)\\ &= 0.212/0.04689\\ &= 4.521\end{aligned}$$

Using the criterion that peaks are resolved when Res = 1, we could state that all three peaks are resolved.

A.5 EFFICIENCY

From the above example, we may calculate the efficiency (from Equation 1.17) of the separation of the peaks in Figure A1.1.

$$N = 5.54(L_d/w_{1/2})^2$$

where L_d, the length to the detector, is 40 cm, and the peak width is (peak area/peak height), both of which must be in the same units, either in time (min) or distance (cm). We shall calculate the efficiency using both time and distance.

A.5.1 Example 1

For time calculations,

$$N_i = 5.54(L_d/w_{1/2})^2 = 5.54[t_i/(\text{area}_i/\text{peak height}_i)]^2$$

From the data in the inset of Figure A1.1, peak area is given in mAU × min, and peak height is in AU. Thus,

$$\begin{aligned}N_1 &= 5.54\,\{(1.732)/[0.10762(0.00668)\,(1000)]\}^2\\ &= 5.54(1.732/0.01611)^2 = 5.54(106.95)^2 = 63,368\\ N_2 &= 5.54(2.642/0.02295)^2 = 5.54(115.14)^2 = 73,445\\ N_3 &= 5.54(2.818/0.02306)^2 = 5.54(122.20)^2 = 82,728\\ N_4 &= 5.54(2.942/0.02383)^2 = 5.54(122.70)^2 = 83,406\end{aligned}$$

A.5.2 Example 2

For distance calculations,

$$N_1 = 5.54(L_d/w_{1/2})^2 = 5.54[L_d/(\text{area}_{\text{con}}/\text{peak height})]^2$$

From calculations, Section A.2, corrected peak area above,

$$w_{1/2,1} = 2.498 \text{ mAU}/0.00668\ \mu\text{AU} = 374\ \mu\text{m}$$

$$w_{1/2,2} = 1.608 \text{ mAU}/0.00463\ \mu\text{AU} = 347\ \mu\text{m}$$

$$w_{1/2,3} = 2.681 \text{ mAU}/0.00819\ \mu\text{AU} = 327\ \mu\text{m}$$

$$w_{1/2,4} = 2.008 \text{ mAU}/0.00616\ \mu\text{AU} = 326\ \mu\text{m}$$

Therefore,

$$N_1 = 5.54(40/0.374)^2$$

$$= 5.54(106.95)^2 = 63,370$$

Likewise,

$$N_2 = 5.54(40/0.347)^2 = 73,616$$

$$N_3 = 5.54(40/0.327)^2 = 82,896$$

$$N_4 = 5.54(40/0.326)^2 = 83,405$$

A.6 JOULE HEATING

To calculate the Joule heating of a buffer, one should run an Ohm's law plot as outlined in Section 1.3.3.5.2, under Buffers. To calculate watts, we use the following:

$$\text{Watts/m} = (\text{voltage})(\text{amperage})/(\text{column length, cm})\ 1000$$

A.6.1 EXAMPLE

For 100 mM borate, pH 8.3: (25 kV)(13.17 μA)/(57 cm) 1000 = 0.58 W/m
For 100 mM CAPS, pH 11.0: (25 kV)(134 μA)/(57 cm) 1000 = 5.88 W/m
For 20 mM CAPS, pH 11.0: (25)(30.03)/57,000 = 1.32 W/m
For 100 mM PO_4, pH 2.5: (25)(229.5)/57,000 = 10.06 W/m
For 50 mM PO_4, pH 2.5: (25)(116.9)/57,000 = 5.13 W/m
For 25 mM PO_4, pH 2.5: (25)(70.03)/57,000 = 3.07 W/m

APPENDIX 2

TROUBLESHOOTING

This table summarizes problems commonly encountered in capillary electrophoresis (CE) experiments, and plausible solutions to these problems.

Problem	Cause	Remedy
I Peak-Associated Problems		
A. No peak observed		
1. Baseline on scale	Inappropriate data collection scale	• Reset scale for appropriate absorbance parameter
	Inappropriate detector range	• Reset detector range
	Inappropriate detector wavelength	• Reset wavelength
	Separation time too short	• Lengthen analysis time • Lengthen capillary
	Reasonable current	• See Section 1.2.1
	Flow through capillary?	• See Section 1.5
	Integrity of sample	• Check sample level • Check for air bubble in bottom of sample vial • Increase sample introductory time • Check sample caps for leakage • See Section 1.5
	Using voltage introduction	• See Section 1.2.1
2. Baseline off scale	Offset baseline	• Rezero detector
	Inappropriate detector wavelength	• Reset wavelength
B. Peaks present		
1. Too many peaks	Random peaks may be caused by microbubbles	• Warm buffer to room temperature • Reduce voltage
	Solid contaminants in sample	• Filter sample (0.45 μm pore size filter)
	Solid contaminants in buffer	• Filter buffer
	Residue from previous analysis	• Wash capillary
	Sample degradation	• Replace sample; check temperature of capillary chamber
2. Too few peaks	Proper wavelength used	• Reset detector wavelength
	Sufficient time	• Lengthen analysis time • Lengthen capillary
	Unreasonable current	• See Section 1.2.1
	Unreasonable voltage	• See Section 1.2.1
	Inappropriate temperature	• Reset temperature control
	Analyte–wall interactions	• Wash capillary • Check running conditions
	Fluid flow inadequate	• See Section 1.5
	Inadequate separation time	• Lengthen analysis time • Lengthen capillary • Reduce electroosmotic flow (EOF)
	Similar charge/mass ratio (not resolving components)	• Alter buffer pH • Reduce EOF • Try alternative separation mode: micellar electrokinetic chromatography (MEKC), capillary isotachophoresis (CITP), capillary isoelectric focusing (CIEF)
3. Distorted peak shape		
a. Small peaks	Improper wavelength	• Reset wavelength
	Inappropriate range on detector	• Reset detector
	Improper integration	• Reset integrator parameters
	Inappropriate sampling time	• Longer sample introduction time

Continued

Problem	Cause	Remedy
b. Flat-topped peaks	Sample too concentrated	• Reduce introduction time • Dilute sample
c. Peak tailing	Current too high	• Ionic strength of separation buffer too high. Check composition; dilute • Reduce voltage
	Sample buffer ionic strength too high	• Dilute sample buffer
	Analyte–wall interactions	*Modify separation buffer*: • Increase ionic strength • Add organic solvents • Add cationic compounds • Add zwitterions (phosphorylethanolamine) • Use coated capillary • Consider different CE mode (e.g., MEKC)
II Instrument-Based Problems		
A. Current		
1. Fluctuating current during separation	Vial levels low	• Replenish buffer
	Loose electrode connections	• Tighten connections
	Plugged capillary	• Rinse/replace
	Sample matrix effects	• Check ionic strength of matrix
2. No current	Safety interlock off	• Reset interlock
	Plugged capillary	• Wash capillary
	Broken capillary	• Replace capillary
	Buffer depletion	• Replace buffers
	Empty capillary	• Fill reservoirs
3. Abrupt loss of current during separation	Short in system (buffer on reservoir cap)	• Dry cap
4. High current	High ionic strength separation buffer	• Decrease ionic strength
B. Baseline drift	Contaminated capillary	• Wash capillary
	Contaminated aperture on detector	• Clean aperture
	Bad capillary alignment	• Check alignment; realign or replace
	Detector instability	• Give adequate warm-up time • Replace lamp
	Unstable capillary temperature	• Check oven/bath temperature • Replace thermostat
	Unstable room temperature	• Stabilize room temperature • Deflect drafts away from instrument • Move instrument
C. Data analysis		
Peaks observed: not analyzed	Inappropriate integration parameters	• Reset integrator attenuation
III Sample Introduction		
A. Electrokinetic		
No peaks	Anodic sample	• Revere polarity • Change separation buffer pH
	Sample ionic strength too high	• Dilute sample buffer
	Sample pH too high	• Adjust pH
	Sample ionic strength too low	• Raise ionic strength
B. Hydrodynamic		
1. No peaks	Poor seal with sample vial	• Replace seal
	Pinched pressure/vacuum line	• Replace line

Problem	Cause	Remedy
	Depleted pressure source	• Replace
	Anodic sample	• Reverse polarity • Change separation buffer pH
2. Irreproducible peak height/area	Poor seal	• Replace cap
	Pinched pressure/vacuum line	• Replace line • Decrease percentage volatile
	Sample matrix volatility	• Decrease solvent concentration
IV Poor Quantitative Reproducibility		
A. Migration time		
1. Unstable temperature	Ionic strength too high	• Dilute buffer
	Voltage too high	• Decrease voltage
	Temperature too high	• Decrease thermostatted temperature (see Section 1.2.2)
	Sample matrix ionic strength too high	• Dilute sample
2. Others	Buffer depletion	• Replenish with fresh buffer
	Analyte–wall interactions	• See Section 1.3.3.5
	Buffer siphoning	• Adjust level in inlet/outlet reservoirs[1]
	Contaminated capillary	• Rinse capillary extensively
	Inadequate rinse steps	• Increase rinse time with rinse solution/buffer
B. Peak height/area	Analyte–wall interactions	• Increase ionic strength • Add dynamic coating agents (e.g., diaminobutane)
	Current instability	• See Section 1.2.1.1
	Fouled capillary surface	• See Section 1.2.3.4.3
V Capillary-Associated Problems[2]		
A. No peaks—proper flow through capillary	Anodic sample	• Reverse polarity
	Sample buffer too viscous	• Dilute sample
B. Reduced flow through capillary	Partially occluded capillary	• Wash capillary. Check sample, buffers for particulates; filter if necessary
	Pinched pressure/vacuum line	• Replace line
	Depleted pressure source	• Replace
C. No flow	Broken capillary	• Replace capillary
D. Peaks irreproducible	Partially occluded capillary	• Unblock chemically (e.g., sodium ethoxide) or with pressure
	Fouled capillary surface	• Recondition capillary (e.g., NaOH or sodium ethoxide)
	Poorly cut capillary inlet	• Retrim inlet or replace capillary

[1]This is unlikely to be significant with narrow diameter capillaries (i.e., <75 μm) since the hydrostatic head differential must be in range of 5 cm of water in order for this phenomenon to contribute appreciably.

[2]Be certain that there is adequate capillary flow. This can be accomplished as described in Chapter 1, Section 1.2.3.4. Alternatively, one may check the flow by pressure rinsing and observing droplets forming at the outlet end. If this must be done manually, a 6-cc syringe with a yellow Eppendorf pipette tip will produce enough pressure to create one drop every 20–30 s. with a 50 μm × 57 cm capillary.

APPENDIX 3

EFFECTIVE SEPARATION CONDITIONS FOR EACH CLASS OF ANALYTES

The following tables provide examples of buffer systems, some containing additives or using surface-modified capillaries, that have provided successful separation conditions for the select analytes. Culled from the capillary electrophoresis (CE) literature, this list is not comprehensive, but is intended to provide the practitioner with a guide for starting conditions for method development for a particular type of analytes—no set of conditions is universally effective for any given class of analytes. For convenience, individual tables are provided for the analyte classes—ions, small molecules, peptides, proteins, and nucleic acids—in both bare silica and with coated surfaces, and with the chromatographic hybrid, micellar electrokinetic chromatography. Full details on the separation conditions can, of course, be found in the original reference.

A.3.1 Ions

Buffer	Additives	Analytes/Sample	References
5 mM chromate or phthalate, pH 10.00	0.5 mM Nice-Pak OFM Anion-BT	Organic/inorganic ions	Romano et al. (1991)
5 mM chromate, pH 8.0	0.4 mM OFM Anion-BT	Organic/inorganic ions	Jones and Jandik (1992)
5 mM chromate, pH 11.0 (LiOH)	None		
30 mM creatinine, pH 4.8 (acetic acid)	8 mM hydroxyisobutyric acid	Metal ions	Lu and Cassidy (1993)
100 mM borate	50 mM tetrabutyl ammonium hydroxide	Inorganic/organic ions	Audalovic et al. (1993)

A.3.2 Small Molecules: Charged and Neutral

Buffer	Additives	Analytes/Sample	Reference
CZE			
50 mM tetraethylammonium perchlorate	10 mM HCl in acetonitrile	Organic bases	Walbroehl (1984)
20–125 mM sodium phosphate, pH 2.5, 7.0–9.2	None	Cinnamic acid and analogs	Fujiwara and Honda (1986)
50 mM sodium phosphate, pH 7.0	50% acetonitrile (v/v)	Substituted benzoic acid isomers	Fujiwara (1987)
50–100 mM sodium acetate, pH 3.9–4.5	0.1% hydroxypropyl cellulose (v/v)	Isotopic benzoic acids	Terabe et al. (1988)
16 mM sodium sulfate, 5 mM MES, pH 6.7	30% methanol (v/v)	Methotrexate analysis	Roach et al. (1988)
20–125 mM sodium phosphate, pH 2.5, 7.0–9.2	None	Pharmaceuticals	Altria and Simpson (1988)
25 mM tetrahexylammonium perchlorate	50% acetonitrile (v/v)	Neutral organic molecules (e.g., polycyclic aromatic hydrocarbons)	Walbroehl and Jorger (1988)
25 mM MES, pH 5.5–5.65	10–20% 2-propanol (v/v)	Catecholamine analysis	Wallingford and Ewing (1989)
10 mM MES/His, pH 6.0	0.5 mM tetradecyltrimethyl-ammonium bromide	Carboxylic acids	Huang et al. (1989)
5 mM sodium borate, pH 9.0	2% SDS (w/v), 0–1% ethylene (v/v) diamine, and 5% ethylene glycol (v/v)	Polyamines	Tsuda et al. (1990)
100 mM CAPS, pH 10.5	None	Nucleotides	Nguyen et al. (1990)
20 mM sodium citrate or 100 mM acetic acid, pH 2.0 or 2.9	None	Shellfish poisons	Thibault et al. (1991)

Buffer	Additives	Analytes/Sample	Reference
150 mM sodium dihydrogen phosphate, pH 2.98 or 5.98	Replacing water with D_2O, pD = 2.98 or 5.98	Aniline derivatives	Okafo et al. (1991)
20 mM phosphoric acid/20% KOH, pH 7.00	None	Cimetidine in pharmaceutical preparations	Arrowood and Hoyt (1991)
50 mM CAPSO and 12.5 mM NAOH, pH 9.55	None	Tricyclic antidepressants	Salomon and Romano (1992)
10 mM sodium tetraborate and 50 mM boric acid or 40 mM NaOH, pH 9.55	Acetonitrile:water 50:50	*cis/trans* Isomers of butenedioic and retinoid acids	Chadwick and Hsieh (1991)
20 mM sodium phosphate or 3–5 mM imidazole or 50 mM acetic acid, pH 4.5–7.0	0.05% ethylene glycol (v/v)	Organic/inorganic cations and anions	Beck and Engelhardt (1992)
100 mM borate, pH 8.3	None	Chloramphenicol acetyl transferase (CAT) enzyme substrates and products	Landers et al. (1992a)
20 mM imidazolium acetate, pH 7.0	None	Sulfonamides	Ackermans et al. (1992)
20–50 mM phosphate, pH 6.8	0.02% hydroxypropylcellulose (w/v) or 0.15% CPDAPS (w/v) and 20% methanol (v/v) or 0.5% polybrene (w/v)	Cefixime and its metabolites	Honda et al. (1992)
6 mM sorbate, pH 12.1	None	Carbohydrates	Vorndran et al. (1992)
100 mM borate, pH 8.4	None	cis-diol containing compounds	Landers et al. (1992b)
100 mM tricine, pH 8.4	None	Norepinephrine, dopamine and metabolites	
8 mM sodium carbonate, 10 mM NaOH, pH 12.0	None	Simple carbohydrates	O'Shea et al. (1993)
5 mM phthalate, and 50 mM MES (1: pH 5.2; 2: pH 5.0)	0.5 mM TTAB	Organic acid counter ions (1: succinic and maleic; 2: trifluoroacetic)	(1) Altria et al. (1997) (2) Little et al. (2007)
25 mM phosphate (pH 2.3) or 0.2 M formic acid (pH 2.15)		Antifungal compounds	Crego et al. (2001)
15 mM borax-sodium phosphate monobasic (pH 8.3)	0.1 mM TAR	Uranium(VI) and other transition metal ions	Evans and Collins (2001)
50 mM borate (pH 8.15)	10 mM phytic acid	Phycobiliproteins	Viskari and Colyer (2002)
30 mM phosphate (pH 5.2)		HVA and VMA (metabolites of catecholamines)	Li et al. (2002)
10 mM sulfated γ-cyclodextrin w/ 50 mM phosphate (pH 2.6)		Amphetamine type stimulants	Iwata et al. (2002)
30 mM phosphate (pH 9.8)		Riboflavin, flavin	Cataldi et al. (2003)
100 mM acetate (pH 4.5)		Glycosaminoglycan monosaccharides	Ruis-Calero et al. (2003)
40 mM phosphate and 10 mM borate (pH 9.0)	40 mM SDS	*E. coli* K4 polysaccharide	Volpi (2004)
10 mM glycine (pH 2.4)	5 mM QA-β-CD (quaternary ammonium β-cyclodextrin)	Highly negative enatiomers	Liu et al. (2004)
10 mM borate (pH 9.5)	20 mM SDS, 30% acetonitrile, 5% ethanol	Antiretroviral agents	Pereira et al. (2005)

Continued

Buffer	Additives	Analytes/Sample	Reference
0.1 M phosphate (pH 3.23)	15 mM β-cyclodextrin	Metabolites of heroin	Qi et al. (2005)
10 mM MES and MOPSO (pH 5.5)	0.25 mM TTAB	EDTA, EDDS, IDS in cosmetics	Katata et al. (2006)
20 mM phosphate (pH 7.2)	60% acetone	Amitriptyline, doxepin, and chlorpromazine in urine	Li et al. (2006)
10 mM DBA, 2 mM 18-crown-6, and 8 mM lactic acid (pH 4.65)		Cations in beverages	Fung and Lau (2006)
15 mM chromate (pH 8.5)	1 mM CTAB	Cyanoacrylate adhesives	Whitaker et al. (2007)
32 mM borate (pH 9.2)	4.5 mM SDS	Monohydroxyl stereoisomers of flavanones	Wang et al. (2007)
Borate	Borate	Corticosteroids	Palmer et al. (1998)
MEKC	MEKC separations		
100 mM phosphate, pH 7.0	25 mM CM-β-cyclodextrin	Cresol isomers	Terabe et al. (1983)
25 mM tetraborate/50 mM phosphate, pH 7.0	1 mM SDS	Phenols	Terabe et al. (1984)
50 mM phosphate/125 mM tetraborate, pH 7.0	50 mM SDS	PTH-amino acids	Otsuka et al. (1985)
100 mM tris-HCL, pH 7.0	50 mM DTAB		
20–50 mM phosphate, pH 7.0–8.0	10–50 mM SDS	Pharmaceuticals	Fujiwara and Honda (1987)
10 mM disodium phosphate and 6 mM borate, pH 7.0–9.0	50 mM SDS or 50 mM dodecyltrimethylammonium chloride or 50 mM STS	Phenols and polycyclic aromatic hydrocarbons	Burton et al. (1987)
10 mM phosphate/6 m*M* borate, pH 7.0	10 mM SDS	Catechols and catecholamines	Wallingford and Ewing (1989)
20 mM phosphate and 20 mM borate, pH 9.0	50 mM SDS and 20–60 mM tetralkylammonium salts	Vitamins and pharmaceuticals	Nishi et al. (1989)
29 mM phosphate-borate, pH 9.0	50 mM sodium cholate	Corticosteroids	Nishi and Terabe (1990)
20 mM sodium dihydrogen phosphate and 20 mM borate, pH 9.2	100 mM SDS and 10% acetonitrile (v/v)	Nucleosides and nucleotide-3-monophosphates	Lecoq et al. (1991)
50 mM phosphate, pH 6.0	50 mM SDS and 5% 2-propanol (v/v)	Creatinine and uric acid and polycyclic aromatic hydrocarbons	Mikaye et al. (1991)
50 mM phosphate and 100 mM borate, pH 8.09	8 mM α-cyclodextrin, 1 mM β-cyclodextrin, and 1 mM γ- cyclodextrin	Plant growth regulators	Yeo et al. (1991)
2.5–5.0 mM borate, pH 7.8–8.9	10–50 mM SDS	Organic gunshot and explosives	Northrop et al. (1991)
10 mM borate-phosphate, pH 8.7	50 mM SDS, 6 M urea, 20% methanol (v/v)	Benzothiazole sulfenamides	Nielsen and Mensink (1991)
8.5 mM phosphate and 8.5 mM borate, pH 8.5	85 mM SDS and 15% (v/v) acetonitrile	Acidic and neutral heroin impurities	Weinberger and Lurie (1991)
100 mM borate and 50 mM phosphate, pH 7.6	30 mM SDS, 3% 2-propanol (v/v) or 3 mM γ-cyclodextrin	Vitamins	Ong et al. (1991)
10 mM borate, pH 9.5	75 mM SDS and 10% methanol (v/v)	Hydroquinone and related molecules in skin toning cream	Sakudinskaya et al. (1992)
50 mM phosphate and 100 mM borate, pH 6.0 or 7.0	3 mM β-cyclodextrin or 2 mM γ-cyclodextrin and 10 mM SDS	Sulfonamides and polycyclic aromatic hydrocarbons	Ng et al. (1992)

Buffer	Additives	Analytes/Sample	Reference
10 mM phosphate and 10 mM borate, pH 9.0	100 mM SDS and 20% methanol (v/v)	Enantiomers of amphetamine, methamphetamine, and their hydroxyphenethylamine precursors	Lurie (1992)
18 mM sodium tetraborate and 30 mM phosphate, pH 7.0	50 mM CTAB	Glucosinolates and their desulfo-derivatives	Morin et al. (1992)
60 mM tris/phosphate, pH 2.5	20 mM cyclodextrin (α,β,γ)	Basic drugs	Nielsen (1993)
10 mM sodium phosphate	10% acetonitrile	Morphine-3-glucuronide	Wernly and Thormann (1993)
6 mM borate, pH 9.2	75 mM SDS		
10 mM sodium phosphate	7 mM cyclodextrin (α,β,γ), acetonitrile (0–15%)	Mycotoxins	Holland and Sepaniak (1993)
6 mM sodium borate	50 mM SDS, 50 mM deoxycholate		
50 mM borate pH 8.5, 100 mM sodium cholate	MeOH 15%	Phthalates/soil	Guo et al. (2005)
20 mM phosphate pH 2.5, 50 mM SDS	ACN 15%, 5% THF	Flavonoids	Tonin et al. (2005)
25 mM borate pH 9.5, 30 mM SDS, 10 mM Triton X-100		Chloropropham aniline metabolites / potatoes	Orejuela et al. (2005)
20 mM borate, 20 mM phosphate pH 9, 80 mM sodium deoxycholate		Nitrofuran antibiotics	Wickramanayake et al. (2006)
50 mM borate pH 9.3, 50 mM SDS	γ–CD 20 mM	Doxorubicin, doxorubicinol	Eder et al. (2006)
20 mM borate, 150 mM CTAB, 10% IPA in MeOH	31.5% H_2O	Biphenyl nitrile derivatives	Gong et al. (2006)
15 mM borate pH 10.2, 40 mM SDS		Ibuprofen, tetrazepam/urine	Nevado et al. (2006)
20 mM Tris pH 9, 300 mM SDS	MeOH 18%	Cefepime, vancomycin/plasma, CSF	Yang et.al. (2007)
100 mM borate pH 9, 60 mM SDS	IPA 2%	Pesticides/wine	Molina-Mayo et al. (2007)
20 mM borate pH 9.3, 50 mM sodium cholate, 20 mM Brij-35	β–CD 5 mM	Amine metabolites/human biofluids	Tseng et al. (2007)
5 mM borate pH 9.3, 20 mM SDS	ACN 20%	Sudan dyes/chili powder	Mejia et al. (2007)

A.3.3 Peptides

Buffer	Additives	Analytes/Sample	Reference
150 mM phosphate, pH 3.0	None	Angiotensin II octapeptides	McCormick (1988)
10 mM tricine, pH 8.0–8.1	5.8–45 mM morpholine and 20 mM NaCl or KCl	LGH tryptic digest	Neilsen et al. (1989a)
100 mM phosphate, pH 2.5	30 mM $ZnSO_4$	DL-His-DL-His	Mosher (1990)
25 mM tris/25 mM phosphate, pH 7.05	50 mM HTAB	Angiotensin analogs	Liu et al. (1990)
0.5 mol/L acetic acid, pH 2.6	None	Di- and triglycine, synthetic growth hormone-releasing peptide	Prusik et al. (1990)

Continued

Buffer	Additives	Analytes/Sample	Reference
250 mM borate, pH 7.0	1% ethylene glycol (v/v) and 7% acetonitrile (v/v)	Proteinase-digested horse myoglobin	Tanaka et al. (1991)
20 mM citric acid pH 2.5	None	Motilin and synthetic peptides	Florance et al. (1991)
20 or 150 mM sodium dihydrogen phosphate, pH 2.93, 7.93, or 7.95	Replacing water with D_2O, pD = 2.93, 7.93, or 7.95	Simple peptides, tryptic digest of calcitonin, glucagons, and cytochrome *c*	Camilleri et al. (1991)
40–80 mM tris and tricine, pH 8.1–8.2	None	β–Casein tryptic and ACTH-endoproteinase Arg C digests	Krueger et al. (1991)
25 mM phosphate, pH 2.2 (KOH)	None	Adrenocorticotropic hormone (ACTH) peptide fragments	Van de Goor et al. (1991)
20 mM formate, pH 3.8	Alanine		
20 mM ε -aminocaproate, pH 4.4 (Acetic acid)	None		
20 mM histidine, pH 6.2	MES		
40 mM imidazole, pH 7.5	MOPS		
100 mM borate, pH 8.3 (KOH)	None		
50 mM phosphate, pH 2.5	40% acetonitrile (v/v)	Multiple antigen peptides	Tanaka et al. (1991)
50 mM sodium dihydrogen phosphate, pH 3.93	0.1% TFA (v/v) and 0.05% hydroxymethylpropyl cellulose (v/v)	α and β species of CGRP	Saria (1992)
20 mM citrate buffer, pH 2.50	None	Peptide monomers and dimmers	Landers et al. (1993)
50 mM formic acid, pH 2.5	10 mM sodium chloride	Basic peptides	Gaus et al. (1993)
10 mM tris and 10 mM disodium, pH 7.05	50 mM dodecyltrimethyl ammonium bromide	Angiotensin analogs	Novotny et al. (1990)
50 mM borate and 20–25 mM tris and 10–25 mM disodium hydrogenphosphate, pH 9.50 or 7.05	50 mM SDS or 20 mM β-cyclodextrin and 1% THF (v/v) or 15% methanol (v/v) or additives such as 0.05 M HTAB or 2–50 mM DTAB	Derivatized peptides and angiotensin analogs	Liu and Novotny (1990)
10 mM sodium phosphate	7 mM cyclodextrin (α,β,γ), acetonitrile (0–15%)	Mycotoxins	Holland and Sepaniak (1993)
6 mM sodium borate	50 mM SDS, 50 mM deoxycholate		
200 mM borate buffer at pH 7.4	None	Human serum proteins: albumin, 1-globulin, 2-globulin, β-globulin, and gamma-globulin	Bossuyta et al. (1998)
5 mM ammonium acetate, pH 4	(Coupled to mass spectrometry)	Phytochelatins in cell extracts	Mounicou et al. (2001)
15 mM sodium citrate (pH 2.1)	0.05% Tylose (methylhydroxyethylcellulose, 30,000 P, that is, viscosity of a 2% w/v solution in H_2O at 20°C)	β–lactoglobulin and α-lactalbumin in milk product isolates	De Block et al. (2003)
50 mM Tricine pH 8.0	20 mM NaCl and 2.5 mM 1,4-diaminobutane (DAB)	Recombinant human granulocyte colony-stimulating factor, glycosylated and nonglycosylated isoforms	Zhoua et al. (2004)
10 mM Tricine, pH 5.5	0.01 M NaCl, 0.01 M sodium acetate, 7 M urea, and 3.9 mM 1,4-diaminobutane	Human alpha-1-acid glycoprotein	Lacunza et al. (2006)

Buffer	Additives	Analytes/Sample	Reference
40 mM sodium phosphate (pH 2.65)	None	Cleavage products of peptide substrates for botulinum neurotoxin	Purcell and Hoard-Fruchey (2007)
Peptides (Coated Capillaries)			
150 mM phosphoric acid, pH 1.5 in polyvinylpyrrolidone-coated capillary	None	Dipeptides	McCormick (1988)
1 M Acetate and 20% MeOH, Polybrene coated capillary	MeOH	Recombinant human erythropoietin (rhEPO)	Balaguer and Neusü (2006)

A.3.4 Proteins

Buffer	Additives	Analytes/Sample	Reference
20 mM borate, pH 8.25	None	Model proteins	Lauer and McManigill (1986)
10 mM tricine, pH 8.22	20 mM KCl		
10 mM tricine, pH 8.0	5.8 mM morpholine and 20 mM NaCl	Biosynthetic human insulin derivatives and growth hormone	Neilsen et al. (1989b)
50 mM glutamine-triethylamine, pH 9.5	Vinyl-bound polyacrylamide-coated capillaries	Model proteins	Cobb et al. (1990)
50 mM sodium phosphate, pH 7.0	Applied Biosystems coating reagent	Multiacetylated histone H4 proteins	Wiktorowicz and Colburn (1990)
50 mM sodium borate decahydrate, pH 10.0	None	Human serum proteins	Gordon et al. (1991)
50 mM tricine, pH 8.0	30% methanol (v/v)	Glycorproteins: recombinant human	Tran et al. (1991)
50 mM MES, pH 6.0	30% ethylene glycol (v/v)	Erythropoietin	
10 mM phosphate, pH 7.0 in capillaries coated with either Tween 20/alkylsilane or Brij	None	Basic proteins	Towns and Regnier (1991)
10 mM citrate, pH 3.0 in BSA-coated capillary	4 mM CHAPSO	Standard proteins	Swedberg (1993)
30 mM citrate, pH 3.0 in a polyacrylamide-coated capillary (uncrosslinked)	None	Standard proteins	Kohr and Engelhard (1991)
50 mM sodium tetraborate, pH 8.3	25–50 mM LiCl	Fluorescamine-labeled interferon and other proteins	Guzman et al. (1992)
100 mM borate, pH 9.0	1 mM diaminobutane	Ovalbumin	Landers et al. (1992c)
20 mM phosphate, pH 3.0	30 mM NaCl, 0.05% polyvinyl alcohol (w/w)	Basic proteins	Gilges et al. (1992)
TES (NaOH), pH 7.0	0.2% SDS linear acrylamide (CGE)	Proteins (3–205 kDa)	Werner et al. (1993)
SDS polyacrylamide gel		Proteins	Tsuji (1994a,b) Shieh et al. (1994)
Iminodiacetic acid (IDA), 50 mM	10% Trifluoroethanol (TFE), 0.5% hydroxyethylcellulose (HEC); or 0.5% HEC and 6-8 M urea	Proteins and peptides	Bossi and Righetti (1997)

Continued

Buffer	Additives	Analytes/Sample	Reference
50 mM aspartic acid	0.5% hydroxyethyl cellulose, 5% trifluoroethanol, 1 zwitterionic detergent (CHAPS)	Alpha and beta-globin chains from human adult Hb	Capelli et al. (1997)
Acetic acid	Acetonitrile and hexane sulfonic acid	Heptathelical membrane proteins	Dong et al. (1997)
Phytic acid		Proteins	Veraart et al. (1997)
Ammonium formate	Trimethylamine (TEA)	Proteins	Lee and Desiderio (1997)
Phosphate buffer	Guaran (neutral polysaccharide)	Basic proteins and drugs	Liu et al. (1991)
50 mM phosphate with 11 mM sodium pentasulfate	11 mM sodium pentasulfate (SPS)	Protein content of a single cell	Zhang et al. (2000)
Acetic acid	10 mM hydroquinone	Protein digests	Moini et al. (1999)
Methanolic buffer (methanol/formic acid/water = 60:20:20)		Enzymic digest peptides	Katayama et al. (2000)
Glutamic acid	1 mM oligoamine (tetraethylene pentamine)	Proteins/muscle acylphosphatase (AcP)	Verzola et al. (2000)
Iminodiacetic acid or aspartic acid	TEPA (tetraethylenepentamine)	Human alpha and beta globin chains	Olivieri et al. (2000)
Borate buffer	SDS and PEO	Proteins: microheterogeneites and isoforms	Tseng et al. (2002)
Acetic acid OR ammonium acetate OR ammonium bicarbonate OR sodium bicarbonate/carbonate OR phosphate OR tris OR ammonium bicarbonate	Acetonitrile	Human high density apolipoproteins	Deterding et al. (2002)
Linear polyacrylamide (sieving)		Proteins	Gomis et al. (2003)
Run buffer: Tris, 0.035 M aspartic acid, 0.1% SDS, 4% acrylamide			
1-Alkyl-3-methylimidazolium-based ionic liquids		Basic proteins	Jiang et al. (2003)
90-mM 1-ethyl-3-methylimidazolium tetrafluoroborate (1E-3MI-TFB)			
25 mM Borax pH 9.4	200 nM SSB (single-strand binding protein) for NECEEM	Basic proteins, separation via Nonequilibrium Capillary Electrophoresis of Equilibrium Mixtures (NECEEM)	Berezovski et al. (2003)
Acidic buffers	Tetraalkylammonium and tetrabutylammonium cations	Peptides and proteins	Quang et al. (2003)
6 mM phosphate buffer	Mexiletine (chiral CE)	Human Serum Ablumin: conformational change	Xu et al. (2004)

Buffer	Additives	Analytes/Sample	Reference
SDS-pullulan (capillary sieving electrophoresis) followed by MECC		Protein fingerprinting of single mammal cells	Hu et al. (2004)
Formic acid (1 M, pH 1.78) for CZE followed by 125 mM ammonium formate (pH 4) for tCITP		Human metal binding proteins	Stutz et al. (2004)
Phosphate buffer (50 mM)	Heptanesulfonic acid (ion pairing agent)	Proteins and peptides	Miksik et al. (2004)
Formic acid (pH 2.4, 50 mM), sodium phosphate buffer, ammonium formate and ammonium borate		Intact proteins	Eriksson et al. (2004)
20 mM phosphate buffer pH 2.5	Pluronic F-127	Enzymic digests of insol. Matrix proteins	Miksik et al. (2004)
Diethylenetriamine (DIEN) phosphate buffer		Basic proteins	Corradini et al. (2005)
Ternary nonaqueous buffer: 60/30/10 v/v methanol/acetonitrile/acetic acid	12.5 mM ammonium acetate	Tryptic digests	Assuncao et al. (2005)
Triethanolamine and phosphoric acid, pH 2.5		Microheterogeneity of intact recombinant glycoprotein	Berkowitz and Zhong (2005)
Replaceable cross-linked polyacrylamide (rCPA)			Lu et al. (2005)
MCRB (Moving Chem. Reaction Boundary): weak acidic run buffer with alk. Sample buffer		Basic proteins	Cao et al. (2005)
Carrier ampholytes		Tryptic digests	Busnel et al. (2006)
Tris-phosphate	0.05% polyethyleneimine	Proteins	Sedlakova et al. (2006a)
0.1 M Phosphate pH 2.5	Pluronic F 127	Proteins, peptides, digests	Sedlakova et al. (2006b)
Acid buffers: ammonium acetate, ammonium hydroxyacetate, phosphate	DDAB capillary coating	Basic proteins	Mohabbati and Westerlund (2006)
Borate pH 8.3	D-PEG dedecylpoly(ethylene glycol ether)	Capsid proteins from rhinovirus	Kremser et al. (2006)
Tris-HCl pH 8.2	"Smart aptamer" affinity probes	MutS protein	Drabovich and Krylov (2006)
Tris-borate pH 10	SDS or PEO	Microheterogeneous proteins	Huang et al. (2006)
50 mM Tris-HCl pH 8.5	Artificial gel antibodies	Transferrin	Takatsy et al. (2006)
30 mM ammonium acetate pH 5.5		Mixtures of drugs, peptides, tryptic digests of proteins, biological fluids	Fanali et al. (2006)
MEKC run buffer: 20 mM CAPS, 60 mM sodium cholate, 20% v/v ACN		Urinary CPIII and CPI	Li and Huie (2006)

Continued

Buffer	Additives	Analytes/Sample	Reference
0.01 M Tricine, 0.01 M NaCl, 0.01 M sodium acetate, 7 M urea		Intact forms of Alpha-1-acid glycoprotein (AGP)	Lacunza et al. (2006)
20 mM Borax in heavy water	Note: Online CE FTIR detection	Model putrescine proteins	Kulka et al. (2006)
Barium tetraborate		Tryptic digests	Mendieta et al. (2006)
10 mM phosphate	Gold nanoparticles	Acidic and basic proteins	Yu et al. (2006)
Poly (*N*,*N*-dimethylacrylamide)-grafted polyacrylamide (self-coating sieving polymer)		Native proteins	Zhang et al. (2006)
Tris-phosphate	0.05% (w/v) polyethlyeneimine	Proteins	Sedlakova and Miksik (2006)
0.1 M Phosphate pH 2.5	Pluronic F 127	Proteins, peptides, digests	Sedlakova et al. (2006)
1. TRIS-CHES-SDS-Dextran (*sieving*) 2. TRIS-CHES-SDS (*MECC*)		Proteins and biogenic amines from mouse AtT-20 cell line	Chen et al. (2007)
10% acetic acid	0.5% chitosan (high molecular weight) with 0.1% acetic acid	Complex peptide mixtures	Busnel et al. (2007)
100 mM borate pH 8.3	6 mM diaminobutate (DAB)	Transferrin	Bortolotti et al. (2007)
100 mM phosphate	High concentration PDDAC: Poly(diallyldimethyl-ammonium chloride)	Cationic proteins Anionic proteins	Lin et al. (2007)

A.3.5 Nucleic Acids

Buffer	Additives	Analytes/Sample	Reference
50 mM phosphate, pH 7.0	5% ethylene glycol (v/v)	Oligo dT ladder	Kasper et al. (1988)
1 mM borate, pH 9.1 (CGE)	30% hydrolink (polymerized)	Restriction fragments	
100 mM trizma	Cesium hydroxide	Forensic DNA	McCord et al. (1993)
100 mM boric acid, pH 8.7 (CGE)	0.1 mM EDTA		
	0.5% hydroxyethylcellulose		
29 mM Tris, 68 mM HEPES (pH 7.2); 15 mM Tris, 27 mM MES, 0.5 M EDTA (pH 6.4); 167 mM Tris, 33 mM boric acid (pH 9.0)	0.5% HPMC (pretreatment); 6% short-chain and linear polyacrylamide (SLPA)	Single-stranded DNA	Ren and Ueland (1999)
20 mM TAPS, pH 7.5	7 M urea; 4% HEC	RNA analysis	Saevels et al. (1999)
89 mM Tris-base [pH 8.3], 89 mM boric acid, 2 mM EDTA	4 M urea; 1% PVP	RNA analysis	Khandurina et al. (2002)
20 mM Tris, 10 mM phosphoric acid, pH 7.3	4.5% HPC	DNA analysis/determination of genetically modified organisms	Giovannoli et al. (2004)
50 mM Tris-borate buffer	5% glycerol; 2% methylcellulose	Single-strand conformation polymorphism analysis	Endo et al. (2005)
100 mM TB buffer (pH 8.0)	0.36 μg/mL CTAB; 0.5% PEO	Double-stranded DNA	Lin and Chang (2006)

Buffer	Additives	Analytes/Sample	Reference
50 mM Tris-acetate (pH 8.2)	2 mM $MgCl_2$ (gel free separation)	SNP analysis	Drabovich and Krylov (2006)
40 mM Tris, 60 mM Mes, and 2 mM EDTA, pH 6.11	1.5% poly(*N*-isopropylacrylamide) (PNIPAM)	DNA analysis	Yu et al. (2006)
30 mM Trizma® Base, 100 mM TAPS, 1 M EDTA, pH 8.0	7 M urea; 8% PVP or 7% PVP, 1% PDMA	DNA sequencing	Ekstrøm, and Bjørheim (2006)
20 mM Tris-HCl, 2 mM KCl	0.03% Triton X-100	Nucleic acid aptamers	Li et al. (2007)
50 mM sodium bicarbonate, pH 9.0	70 mM SDS	DNA analysis	Hua and Naganuma (2007)
20 mM Tris, 9.5 mM orthophosphoric acid, 2 mM EDTA, pH 7.3 (CGE)	4% HEC	DNA analysis/determination of genetically modified organisms	Sánchez et al. (2007)

REFERENCES

Ackermans, M. T., Beckers, J. L., Everaerts, F. M., Koogland, H., and Tomassen, T. J. H. Determination of sulphonamides in pork meat extracts by capillary zone electrophoresis, *J. Chromatogr.*, 596, 101, 1992.

Altria, K. and Simpson, C. F. Analysis of some pharmaceuticals by high voltage capillary zone electrophoresis. *J. Pharm. Biochem. Med. Anal.*, 6, 801, 1988.

Altria, K. D., Assi, K. H., Bryant, S. M., and Clark, B. J. Determination of organic acid drug counter-ions by capillary electrophoresis, *Chromatographia*, 44, 367, 1997.

Arrowood, S. and Hoyt, A. M. Determination of cimetidine in pharmaceutical preparanons by capillary zone electrophoresis, *J. Chromatogr.*, 586, 177, 1991.

Audalovic, N., Pohl, C. A., Rocklin, R. D., and Stillian, J. R. Determination of cations and anions by capillary electrophoresis combined with suppressed conductivity detection, *Anal. Chem.*, 65, 1470–1475, 1993.

Balaguer, E. and Neusü, C. Intact glycoform characterization of erythropoietin-alpha and erythropoietin-beta by CZE-ESI-TOF-MS, *Chromatographia*, 64, 351, 2006.

Beck, W. and Engelhardt, H. Capillary electrophoresis of organic and inorganic cations with indirect UV detection, *Chromatographia*, 33, 313, 1992.

Bossi, A. and Righetti, P. G. Generation of peptide maps by capillary zone electrophoresis in isoelectric iminodiacetic acid. *Electrophoresis*, 18, 2012, 1997.

Bossuyta, X., Schiettekatte, G., Bogaerts, A., and Blanckaer, N. Serum protein electrophoresis by CZE 2000 clinical capillary electrophoresis system, *Clin. Chem.*, 44: 749–759, 1998.

Burton, D. E., Sepaniak, M. J., and Maskarinec, M. P. Evaluation of the use of various surfactants in micellar electrokinetic capillary chromatography, *J. Chromatogr. Sci.*, 25, 514, 1987.

Camilleri, P., Okafo, G. N., Southan, C., and Brown, R. Analytical and micropreparative capillary electrophoresis of peptides from calcitonin, *Anal. Biochem.*, 198, 36, 1991.

Capelli, L., Stoyanov, A. V., Wajcman, H., and Righetti, P. G. Generation of tryptic maps of α-and β-globin chains by capillary electrophoresis in isoelectric buffers. *J. Chromatogr. A*, 791, 313, 1997.

Cataldi, T. R. I., Nardiello, D., Carrara, V., Ciriello, R., and DeBenedetto, G. E. Assessment of riboflavin and flavin content in common food samples by capillary electrophoresis with laser-induced fluorescence detection, *Food Chem.*, 82, 309, 2003.

Chadwick, R. R. and Hsieh, J. C. Separation of cis and trans double bond isomers using capillary zone electrophoresis, *Anal. Chem.*, 63, 2377, 1991.

Cobb, K. A., Dolnik, V., and Novotny, M. Electrophoretic separations of protein in capillaries with hydrolytically stable surface structures, *Anal. Chem.*, 62, 2478, 1990.

Cohen, A. S. High performance capillary electrophoresis of bases, nucleosides and oligonucleotides: retention manipulation via micellar solutions and metal additives, *Anal. Chem.*, 59, 1021, 1987.

Cohen, A. S., Najarian, D., Smith, J. A., and Karger, B. L. Rapid separation of DNA restriction fragments usiny capillary electrophoresis, *J. Chromatogr.*, 458, 323, 1988.

Crego, A. L., Marina, M. L., and Lavandera, J. L. Optimization of the separation of a group of antifungals by capillary zone electrophoresis, *J. Chromatogr. A*, 917, 337, 2001.

De Block J. et al. Monitoring nutritional quality of milk powders: capillary electrophoresis of the whey protein fraction compared with other methods. *Intl. Dairy J.*, 13, 87, 2003.

Drabovich, A. P. and Krylov, S. N. Identification of base pairs in single-nucleotide polymorphisms by muts protein-mediated capillary electrophoresis. *Anal. Chem.*, 78, 2035, 2006.

Eder, A. R., Chen, J. S., and Arriaga, E. A. Separation of doxorubicin and doxorubicinol by cyclodextrin-modified micellar electrokinetic capillary chromatography, *Electrophoresis*, 27, 3263, 2006.

Ekstrøm, P. O. and Bjørheim, J. Evaluation of sieving matrices used to separate alleles by cycling temperature capillary electrophoresis, *Electrophoresis*, 27, 1878, 2006.

Endo, Y., Zhang, L., Katashima, R., Itakura, M., Doherty, E. A. S., Barron, A. E., and Baba, Y. Effect of polymer matrix and glycerol on rapid single-strand conformation polymorphism analysis by capillary and microchip electrophoresis for detection of mutations in *K-ras* gene. *Electrophoresis*, 26, 3380, 2005.

Evans III, L. and Collins, G. E. Separation of uranium (VI) and transition metal ions with 4-(2-thiazolylazo) resorcinol by capillary electrophoresis, *J. Chromatogr. A*, 911, 127, 2001.

Florance, J. R., Konteatis, Z. D., Macielag, M. J., Lessor, R. A., and Galdes, A. Capillary zone electrophoresis studies of motilin peptides, effect of charge, hydrophobicity, secondary structure and length, *J. Chromatogr.*, 559, 391, 1991.

Fujiwara, S. and Honda, S. Determination of cinnamic acid and its analogues silica capillary tube. *Anal. Chem.*, 58, 1811, 1986.

Fujiwara, S. Effect of addition of organic solvent and the separation of positional isomers in high voltage capillary zone electrophoresis, *Anal. Chem.*, 59, 487, 1987.

Fung, Y. S. and Lau, K. M. Separation and determination of cations in beverage products by capillary zone electrophoresis, *J. Chromatogr. A*, 1118, 144, 2006.

Gaus, H.-J., Beck-Sickinger, A. G., and Bayer, E. Optimization of capillary electrophoresis of mixtures of basic peptides and comparison with HPL, *Anal. Chem.*, 65, 1399–1405, 1993.

Gilges, M., Hasmann, H., Kleemib, M. H., Motsch, S. R., and Schomburg, G. CZE separations of basic proteins at low pH in fused silica capillaries with surfaces modified by silane derivitization and/or adsorption of polar polymers, *J. High Resolut. Chromatogr.*, 15, 452, 1992.

Giovannoli, C., Anfossi, L., Tozzi C., Giraudi, G., and Vanni, A. DNA separation by capillary electrophoresis with hydrophilic substituted celluloses as coating and sieving polymers. Application to the analysis of genetically modified meals, *J. Sep. Sci.*, 27, 1551, 2004.

Gong, S. X., Liu, F. J., Li, W., Gao, F., Gao, C. J., Liao, Y. P., and Liu, H. W. Separation of hydrophobic solutes by organic-solvent-based micellar electrokinetic chromatography using cation surfactants, *J. Chromatogr. A*, 1121, 274, 2006.

Gordon, M. J., Zare, R. N., Lee, K. J., and Arias, A. A. Protocol for resolving protein mixtures in capillary zone electrophoresis, *Anal. Chem.*, 63, 69, 1991.

Guo, B. Y., Wen, B., Shan, X. Q., Zhang, S. Z., and Lin, J. M. Separation and determination of phthalates by micellar electrokinetic chromatography, *J. Chromatogr. A*, 1095, 189, 2005.

Guzman, N. A., Moschera, J., Bailey, C. A., Iqbal, K., and Malick, A. W. Assay of protein drug substances present in solution mixtures by fluorescence denvitisation and capillary electrophoresis, *J. Chromatogr*, 598, 123, 1992.

Holland, R. D. and Sepaniak, M. J. Qualitative analysis of mycotoxins using micellar electrokinetic capillary chromatography, *Anal. Chem.*, 65, 1140–1146, 1993.

Honda, S., Taga, A., Kakehi, K., Koda, S., and Okamoto, Y. Determination of cefixime and its metabolites by high performance capillary electrophoresis, *J. Chromatogr.*, 590, 364, 1992.

Hua, N. and Naganuma T. Application of CE for determination of DNA base composition. *Electrophoresis*, 28, 366, 2007.

Huang, X., Luckey, J. A., Gordon, M. J., and Zare, R. N. Quantitative analysis of low molecular weight carboxylic acids by capillary zone electrophoresis/conductivity detection, *Anal. Chem.*, 61, 766, 1989.

Iwata, Y. T., Garica, A., Kanamori, T., Inoue, H., Kishi, T., and Lurie, I. S. The use of a highly sulfated cyclodextrin for the simultaneous chiral separation of amphetamine-type stimulants by capillary electrophoresis. *Electrophoresis*, 23, 1328, 2002.

Jones, W. R. and Jandik, P. Various approaches to analysis of difficult sample matrices of anions using capillary ion electrophoresis, *J. Chromatogr.*, 608, 385–393, 1992.

Kasper, T. J., Melera, M., Gozel, P., and Browniee, R. G. Separation and detection of DNA by capillary electrophoresis, *J. Chromatogr.*, 458, 303, 1988.

Katata, L., Nagaraju, V., and Crouch, A. M. Determination of ethylenediaminetetraacetic acid, ethylenediaminedisuccinic acid and iminodisuccinic acid in cosmetic products by capillary electrophoresis and high performance liquid chromatography, *Analytica Chimica Acta*, 579, 177, 2006.

Khandurina, J., Chang, H., Wanders, B., and Guttman, A. Automated High-throughput RNA analysis by capillary electrophoresis, *Biotechniques*, 32, 1226, 2002.

Kohr, H. and Englehardt, H. Capillary electrophoresis with surface coated capillaries, *J. Microcolumn Sep.*, 3, 491, 1991.

Krueger, R. J., Hobbs, T. R., Mihal, K. A., Therani, J., and Zeece, M. G. Analysis of endoproteinase ArgC action on adrenoconicotrophic hormone by capillary electrophoresis and reverse phase HPLC, *J. Chromatogr.*, 543, 451, 1991.

Lacunza I. et al. CZE of human alpha-1-acid glycoprotein for qualitative and quantitative comparison of samples from different pathological conditions, *Electrophoresis*, 27, 4205, 2006.

Landers, J. P., Schuchard, M D., Subramaniam, S., Sismelich, T. and Spelsberg, T. C. High-performance capillary electrophoresis analysis of chloramphenicol acetyl transferase activity, *J. Chromatogr.*, 603, 247–257, 1992a.

Landers, J. P., Oda, R. P., and Schuchard, M. D. Separation of boron-complexed diol compounds using high-performance capillary electrophoresis, *Anal. Chem.*, 64, 2846–2851, 1992b.

Landers, J. P., Oda, R. P., Madden, B. J., and Spelsberg, T. C. High-performance capillary electrophoresis of glycoproteins: the use of modifiers of electroosmotic flow for analysis of microheterogeneity, *Anal. Biochem.*, 205, 115–124, 1992c.

Landers, J. P., Oda, R. P., Liebenow, J. A., and Spelsberg, T. C. Utility of high performance capillary electrophoresis for monitoring peptide homo- and hetero-dimer formation, *J. Chromatogr.*, 652, 109–117, 1993.

Lauer, H. H. and McMangill, D. Capillary zone electrophoresis of proteins in untreated fused silica tubing, *Anal. Chem.*, 58, 166, 1986.

Lecoq, F., Leuratti, C., Marafante, E., and DiBase, S. Analysis of nucleic acid derivatives by micellar electrokinetic capillary chromatography, *J. High Resolut. Chromatogr.*, 14, 667, 1991.

Li, J., Zhao, F., and Ju, H. Simultaneous determination of psychotropic drugs in human urine by capillary electrophoresis with electrochemiluminescence detection, *Analytica Chimica Acta*, 575, 57, 2006.

Li, T., Li, B., and Dong, S. Aptamer-based label-free method for hemin recognition and DNA assay by capillary electrophoresis with chemiluminescence detection. *Anal. Bioanal. Chem.*, [Epub ahead of print], 2007.

Li, X., Jin, W., and Weng, Q. Separation and determination of homovanillic acid and vanillylmandelic acid by capillary electrophoresis with electrochemical detection, *Analytica Chimica Acta*, 461, 123, 2002.

Lin, Y. W. and Chang, H. T. Analysis of double-stranded DNA by capillary electrophoresis using poly(ethylene oxide) in the presence of hexadecyltrimethylammonium bromide. *J. Chromatogr. A*, 1130, 206, 2006.

Little, M. J., Aubry, N., Beaudoin, M. E., Goudreau, N., and LaPlante, S. R. Quantifying trifluoroacetic acid as a counterion in drug discovery by 19F NMR and capillary electrophoresis, *J. Pharm. Biomed. Anal.*, 43, 1324, 2007.

Liu, J. and Novotny, M. Capillary electrophoretic separation of peptides using micelleforming compounds and cyclodextrins as additives, *J. Chromatogr.*, 519, 189, 1990.

Liu, Q., Inoue, T., Kirchhoff, J. R., Huang, C., Tillekeratne, L. M. V., Olmstead, K., and Hudson, R. A. Chiral separation of highly negatively charged enantiomers by capillary electrophoresis, *J. Chromatogr. A*, 1033, 349, 2004.

Liu, Y. and Chan, K. F. J. High performance capillary electrophoresis of gangliosides, *Electrophoresis*, 12, 402, 1991.

Lu, W. and Cassidy, R. M. Evaluation of ultramicroelectrodes for the detection of metal ions separated by capillary electrophoresis, *Anal. Chem.*, 65, 1649–1653, 1993.

Lurie, I. S. Micellar electrokinetic capillary chromatography of the enantiomers of amphetamine, methamphetamine and their hydroxyphenethylamine precursors, *J. Chromatogr.*, 605, 269, 1992.

McCord, B. R., Jung, J. M., and Holleran, E. S. High resolution CE of forensic DNA using a non-gel sieving buffer, *J. Liquid Chromatogr.*, 16, 1963–1981, 1993.

McCormick, R. M. Capillary zone electrophoresis of peptides and proteins using low pH buffers in modified silica capillaries, *Anal. Chem.*, 60, 2322, 1988.

Mejia, E., Ding, Y. S., Mora, M. F., and Garcia, C. D. Determination of banned sudan dyes in chili powder by capillary electrophoresis, *Food Chem.*, 102, 1027, 2007.

Michaelis, L., Electric transport of enzymes, malt distaste and pepsin, *Biochemische Zeitschrift*, 17, 1909, pp. 231–234.

Mikaye, M., Shibukawa, A., and Nakasgawa, T. Simultaneous determination of creatinine and uric acid in human plasma and urine by micellar electrokinetic chromatography, *J. High Resolut. Chromatogr.*, 14, 181, 1991.

Molina-Mayo, C., Hernandez-Borges, J., Borges-Miguel, T. M., and Rodriguez-Delgado, M. A. Determination of pesticides in wine using micellar electrokinetic chromatography with UV detection and sample stacking, *J. Chromatogr. A*, 1150, 348, 2007.

Morin, Ph., Villard, F., Quinsac, A., and Dreux, M. Micellar electrokinetic capillary chromatography of glucosinolates and desulfoglucosinolates with a cationic surfactant, *J. High Resolut. Chromatogr.*, 371, 1992.

Mosher, R. A. The use of metal ion-supplemented buffers to eliminate the resolution of peptides in capillary zone electrophoresis, *Electrophoresis*, 11, 765, 1990.

Mounicou, S. et al. Determination of phytochelatins by capillary zone electrophoresis with electrospray tandem mass spectrometry detection (CZE-ES MS/MS), *Analyst*, 126, 624, 2001.

Neilsen, R. G., Riggin, R. M., and Richards, E. C. Capillary zone electrophoresis of peptide fragments from trypsin digestion of biosynthetic human growth hormone, *J. Chromatogr.*, 480, 343, 1989a.

Neilsen, R. G., Sittampalam, G. S., and Richard, E. C. Capillary zone electrophoresis of insulin and growth hormone, *Anal. Biochem.*, 177, 20, 1989b.

Nevado, J. J. B., Flores, J. R., Penalvo, G. C., and Dorado, R. M. R. Determination of ibuprofen and tetrazepam in human urine by micellar electrokinetic capillary chromatography, *Anal. Bioanal. Chem.*, 384, 208, 2006.

Ng, C. L., Lee, N. H., and Li, S. F. Y. Systematic optimization of capillary electrophoretic separation of sulphonamides, *J. Chromatogr.*, 598, 133, 1992.

Nguyen, A. L., Luong, J. H. T., and Masson, G. Determination of nucleotides in fish tissue using capillary electrophoresis, *Anal. Chem.*, 62, 2490, 1990.

Nielsen, M. W. F. Chiral separation of basic drugs using cyclodextrin-modified capillary zone electrophoresis, *Anal. Chem.*, 65, 885–893, 1993.

Nielsen, M. W. and Mensink, M. J. A. Separation of benzothiazole sultenamides using micellar electrokinetic capillary chromatography, *J. High Resolut. Chromatogr.*, 14, 417, 1991.

Nishi, H. and Terabe, S. Applications of micellar electrokinetic chromatography to pharmaceutical analysis, *Electrophoresis*, 11, 691, 1990.

Nishi, H., Tsumagari, N., and Terabe, S. Effects of tetraalkylammonium salts on micellar electrokinetic chromatography of ionic substances, *Anal. Chem.*, 61, 2434, 1989.

Northrop, D. M., Martine, D. E., and MacClehan, W. A. Separation and identification of organic gunshot and explosive constituents by micellar electrokinetic capillary chromatography, *Anal. Chem.*, 63, 10338, 1991.

Novotny, M. V., Cobb, K. A., and Liu, J. Recent advances in capillary electrophoresis of proteins, peptides and amino acids, *Electrophoresis*, 11, 735, 1990.

O'Shea, T. J., Lunte, S. M., and LaCourse, W. R. Detection of carbohydrates by CE with pulsed amperometric detection, *Anal. Chem.*, 65, 948–951, 1993.

Okafo, G. N., Brown, R., and Camilleri, P. Some physico-chemical properties that make D.O-based buffer solutions useful media for capillary electrophoresis, *J. Chem. Soc. Chem. Commun.*, 864, 1991.

Ong, L. P., Ng, C. L., Lee, N. H., and Li, S. F. Y. Determination of antihistamines in pharmaceutical by capillary electrophoresis, *J. Chromatogr.*, 588, 335, 1991.

Orejuela, E. and Silva, M. Rapid determination of aniline metabolites of chlorpropham in potatoes by micellar electrokinetic chromatography using negative-charged mixed micelles and laser-induced fluorescence detection, *Electrophoresis*, 26, 2991, 2005.

Otsuka, K., Terabe, S., and Ando, T. Electrokinetic chromatography with micellar solutions: separation of phenylthiohydantoin amino acids, *J. Chromatogr.*, 332, 219, 1985.

Palmer, J., Atkinson, S., Yoshida, W. Y., Stalcup, A. M. and Landers, J. P. Charged chelate-capillary electrophoresis of endogenous corticosteroids, *Electrophoresis*, 19, 3045–3051, 1998.

Pereira, E. A., Micke, G. A., and Tavares, M. F. M. Determination of antiretroviral agents in human serum by capillary electrophoresis, *J. Chromatogr. A*, 1091, 169, 2005.

Prusik, Z., Kasicka, V., Mudra, P., Stepanek, J., Smekal, O., and Hlavacek, J. Correlation of capillary zone electrophoresis with continuous flow zone electrophoresis: application to the analysis and purification of synthetic growth hormone releasing peptide, *Electrophoresis*, 932, 1990.

Purcell, A. L. and Hoard-Fruchey, H. M. A capillary electrophoresis method to assay catalytic activity of botulinum neurotoxin serotypes: Implications for substrate specificity, *Anal. Biochem.*, 366, 207, 2007.

Qi, X. H., Mi, J. Q., Zhang, X. X., and Chang, W. B. Design and preparation of novel antibody system and application for the determination of heroin metabolites in urine by capillary electrophoresis, *Analytica Chimica Acta*, 551, 115, 2005.

Ren, J. and Ueland, P. M. Temperature and pH effects on single-strand conformation polymorphism analysis by capillary electrophoresis, *Hum. Mutat.*, 13, 458, 1999.

Roach, M. C., Gozel, P., and Zare, R. N. Determination of methotrexate and its major metabolite, 7-hydroxymethotreaxate using capillary zone electrophoresis and laser induced fluorescence detection, *J. Chromatograph.*, 426, 129, 1988.

Romano, J., Jandik, P., Jones, W. R., and Jackson, P. E. Optimisation of inorganic capillary electrophoresis for the analysis of anionic solutes in real sample, *J. Chromatogr.*, 545, 411, 1991.

Ruis-Calero, V., Puignou, L., and Galceran, M. T. Determination of glycosaminoglycan monosaccharides by capillary electrophoresis using laser-induced fluorescence detection, *J. Chromatogr. B*, 791, 193, 2003.

Saevels, J., Van Schepdael, A., and Hoogmartens, J. Capillary Electrophoresis of RNA Oligonucleotides: Catalytic Activity of a Hammerhead Ribozyme, *Anal. BioChem.*, 266, 93, 1999.

Sakudinskaya, I. K., Desirodero, C., Nardi, A., and Fanali, S. Micellar electrokinetic chromatographic study of hydroquinone and some of its ethers. Determination of hydroquinone in skin toning cream, *J. Chromatogr.*, 596, 95, 1992.

Salomon, D. R. and Romano, J. Applications of capillary ion electrophoresis in the pulp and paper industry, *J. Chromatogr.*, 602, 219, 1992.

Sánchez, L., González, R., Crego, A. L., and Cifuentes, A. A simple capillary gel electrophoresis approach for efficient and reproducible DNA separations. Analysis of genetically modified soy and maize, *J. Sep. Sci.*, 30, 579, 2007.

Saria, A. Identification of alpha and beta series of calcitonin gene-related peptide in the rat amygdala after separation with capillary zone electrophoresis. 573, 219, 1992.

Sedlakova, P., Svobodova, J., Miksik, I., and Tomas, H. Separation of poly(amidoamine) (PAMAM) dendrimer generations by dynamic coating capillary electrophoresis. *J. Chromatogr. B*, 841, 135–139, 2006a.

Sedlakova, P., Svobodova, J., and Miksik, I. Capillary electrophoresis of peptides and proteins with plug of Pluronic gel. *J. Chromatogr. B*, 839, 112–117, 2006b.

Shieh, P. C. H., Hoang, D., Guttman, A., and Cooke, N. Capillary sodium dodecyl sulfate gel electrophoresis of proteins. I. Reproducibility and stability. *J. Chromatogr. A*, 676, 219, 1994.

Swedberg, S. The impact of column technology on protein separations in OT capillary electrophoresis: past lessons, future promises, *CRC Handbook of Capillary Electrophoresis: A Practical Approach*, Chapter 19, 1993.

Tanaka, H. W., Kanabe, T., Yameda, Y., and Semba, T. Capillary electrophoretic monitoring for C-terminal fragment identification, *J. High Resolut. Chromatogr.*, 14, 491, 1991.

Terabe, S., Otsuka, K., Ichikawa, K., Tsuchiva, A., and Ando, T. Electrokinetic separations with micellar solutions and tubular capillaries, *Anal. Chem.*, 56, 113, 1984.

Terabe, S., Ozuki, H., Otsuka, K., and Ando, T. Electrokinetic chromatography with 2-0-carboxymethyl-cyclodextrin as a moving stationary phase, *J. Chromatogr.*, 332, 211, 1983.

Terabe, S., Yashima, T., Tanaka, M., and Araki, M. Separation of oxygen isotopic benzoic acids by capillary zone electrophoresis based on isotope effects on the dissociation of the carboxyl group, *Anal. Chem.*, 60, 1673, 1988.

Thibault, P., Pleasance, S., and Laycock, M. V. Analysis of paralytic shellfish poisons by capillary electrophoresis, *J. Chromatogr.*, 542, 483, 1991.

Tonin, F. G., Jager, A. V., Micke, G. A., Farah, J. R. S., and Tavares, M. F. M. Optimization of the separation of flavonoids using solvent-modified micellar electrokinetic chromatography, *Electrophoresis*, 26, 3387, 2005.

Towns, J. K. and Regnier, P. E. Capillary electrophoretic separations of proteins using non-ionic surfactant coatings, *Anal. Chem.*, 63, 1126, 1991.

Tran, A. D., Pak, S., List, P. J., Huynh, O. T., Ryall, R. R., and Lane, P. A. Separation of carbohydrate-mediated microheterogeneity of recombinant human erythropoietin by free solution capillary electrophoresis, *J. Chromatogr.*, 542, 459, 1991.

Tseng, H. M., Li, Y., and Barrett, D. A. Profiling of amine metabolites in human biofluids by micellar electrokinetic chromatography with laser-induced fluorescence detection, *Anal. Bioanal. Chem.*, 388, 433, 2007.

Tsuda, T., Kobayashi, Y., Hori, A., Matsumoto, T., and Suzuki, O. Separation of polyamines in rat tissue by capital electrophoresis, *J. Microcol. Sep.*, 2, 21, 1990.

Tsuji, K. Sodium dodecyl sulfate polyacrylamide gel- and replaceable polymer-filled capillary electrophoresis for molecular mass determination of proteins of interest. *J. Chromatogr. A*, 666, 294, 1994a.

Tsuji, K. Factors affecting the performance of SDS gel-filled capillary electrophoresis. *J. Chromatogr. A*, 661, 257, 1994b.

Van de Goor, T. A. A. M., Janssen, P. S. L., Van Nispen, J. W., Van Zeeland, M. J. M., and Everaerts, F. M. Capillary electrophoresis of peptides: analysis of adrenocorticotropic hormone-related fragments, 545, 379, 1991.

Viskari, P. J. and Colyer, C. L. Separation and quantitation of phycobiliproteins using phytic acid in capillary electrophoresis with laser-induced fluorescence detection, *J. Chromatogr. A*, 972, 269, 2002.

Volpi, N. Application of high-performance capillary electrophoresis to the purification process of *Escherichia coli* K4 polysaccharide, *J. Chromatogr. B*, 811, 253, 2004.

Vorndran, A. E., Oefner, P. J., Scherz, H., and Bonn, G. K. Indirect UV detection of carbohydrates in capillary zone electrophoresis, *Chromatographia*, 33, 163, 1992.

Walbroehl, Y. and Jorgenson, J. W. Capillary zone electrophoresis of neutral molecules by solvophobic association with tetraalkylammonium ions, *Anal. Chem.*, 58, 479, 1988.

Walbroehl, Y. On-column UV absorption detector for open tubular capillary zone electrophoresis, *J. Chromatogr.*, 315, 135, 1984.

Wallingford, R. A. and Ewing, A. G. Separation of serotonin from catechols by capillary zone electrophoresis using electrochemical detection. *Anal. Chem.*, 61, 98, 1989.

Wang, S. P., Fu, M. D., and Wang, M. H. Separation mechanism and determination of flavanones with capillary electrophoresis and high-performance liquid chromatography, *J. Chromatogr. A*, 1164, 306, 2007.

Weinberger, R. and Lurie, I. S. Micellar electrokinetic capillary chromatography of illicit drug substances, *Anal. Chem.*, 63, 827, 1991.

Werner, W. E., Demorgst, D. M., Stevens, J., and Wiktorewicz, J. E. Site dependent separation of proteins denatured in SDS by CE using a replaceable sieving matrix, *Anal. Biochem.*, 212, 253–258, 1993.

Wernly, P. and Thormann, W. Determination of morphin-3-glucuronide in human urine by capillary zone electrophoresis and micellar electrokinetic capillary chromatography, *J. Chromatogr.*, 616, 305–310, 1993.

Whitaker, G., Kincaid, B. J., VanHoof, N., Regan, F., Smyth, M. R., and Leonard, R. G. An investigation into the sample preparation procedure and analysis of cyanoacrylate adhesives using capillary electrophoresis, *Intl. J. Adh. Adhes.*, 27, 604, 2007.

Wickramanayake, P. U., Tran, T. C., Hughes, J. G., Macka, M., Simpson, N., and Marriott, P. J. Simultaneous separation of nitrofuran antibiotics and their metabolites by using micellar electrokinetic capillary chromatography, *Electrophoresis*, 27, 4069, 2006.

Wiktorowicz, J. E. and Colburn, J. C. Separation of cationic proteins via charge reversal in capillary electrophoresis, *Electrophoresis*, 11, 769, 1990.

Xu, H., Yu, X., and Chen, H. Enantiomeric separation of basic drugs with partially filled serum albumin as chiral selector in capillary electrophoresis. *Anal. Sci.*, 20, 1409–1413, 2004.

Yang, Y. H., Wu, W. Y., Yeh, H. H., and Chen, S. H. Simultaneous determination of cefepime and vancomycin in plasma and cerebrospinal fluid by MEKC with direct sample injection and application for bacterial meningitis, *Electrophoresis*, 28, 1788, 2007.

Yeo, S. K., Ong, C. P., and Li, S. F. Y. Optimisation of high performance capillary electrophoresis of plant growth regulators usiny the overlapping resolution mapping scheme, *Anal. Chem.*, 63, 2222, 1991.

Yu, S. B., Zhou, P., Feng, A. R., Shen, X. D., Xhang, Z., and Hu, J. M. pH effect on dynamic coating for capillary electrophoresis of DNA. *Anal. Bioanal. Chem.*, 385,730, 2006.

Zhang, Z., Krylov, S., Arriaga, E. A., Polakowski, R., and Dovichi, N. J. One-dimensional protein analysis of an HT29 human colon adenocarcinoma cell. *Anal. Chem.*, 72, 318–322, 2000.

Zhoua, G.-H. et al. Characterization of recombinant human granulocyte colony stimulating factor (rHuG-CSF) by capillary zone electrophoresis, capillary isoelectric focusing electrophoresis and electrospray ionization mass spectrometry. *J. Pharm. Biomed. Anal.*, 35, 425, 2004002E.

2 Protein Analysis by Capillary Electrophoresis

James M. Hempe

CONTENTS

2.1 INTRODUCTION

Almost any analytical protein application can be accomplished using capillary electrophoresis (CE). CE can be used to characterize protein properties, identify specific protein isoforms, quantify expression levels, study protein interactions with other biomolecules or map the distribution of proteins in one or more analytical dimension. Yet, protein analysis arguably remains the least widely used major CE application. Compared to analysis of nucleotides or small organic compounds, protein analysis by CE has faced greater technical challenges and more persistent competition from alternative separation and assay techniques. Although many protein applications are exquisitely or uniquely accomplished with CE, others are still better achieved using conventional analytical methods (e.g., gel electrophoresis, enzyme-linked immunosorbent assay (ELISA), or column chromatography) or other advanced technologies (e.g., high-performance liquid chromatography (HPLC), mass spectrometry (MS), or protein microarray). If an investigator has a small number of specific protein applications to perform over and over again, CE's automation and low reagent costs make it a very attractive alternative. If an investigator wants a versatile system that can be used to analyze more than just proteins, opportunities abound on multimode open analytical CE platforms simply by changing the capillary, separation medium, and/or detector. High resolution, versatility, automation, small sample volume requirements, and minimal waste production are the hallmarks of CE for all applications. Poor reproducibility and lack of sensitivity and specificity are traditional weaknesses of CE for protein analysis. The purpose of this chapter is to provide an overview of how CE is used for protein analysis and how it might be used in the future. CE instrument and method development will be strongly influenced by systems biology and the search for biomarkers of complex diseases. Recent technological advances that improve capillary performance and enhance detection of proteins suggest that CE will play a major role in the evolving field of proteomics.

2.2 BACKGROUND

Capillary electromigration techniques can be effectively applied for protein analysis in all areas of biology. Basic theoretical principles and applications of CE are discussed here in the context of human biology and disease but are similar whether the analytical focus is plants or other animals. Terminology used to describe the separation of proteins by CE will conform to the recommendations of the Analytical Chemistry Division of the International Union of Pure and Applied Chemistry (IUPAC) [1] except that CE will be used according to tradition as a collective term for all capillary electromigration techniques. Recent progress in the use of CE for a wide variety of protein applications has been documented in a periodic series of reviews by Dolnik and colleagues [2–4]. Other recent reviews specifically related to proteins [5], application of specific CE techniques to proteins and other molecules [6–10], and the use of CE in systems biology and proteomics [11,12] are also recommended.

2.2.1 Current Status of CE for Protein Analysis

Capillary electrophoresis has found widespread use for protein analysis in the pharmaceutical [13,14] and food science [15–17] industries, primarily for process quality control in determining the purity of natural and synthetic proteins, peptides, and monoclonal antibodies. Dedicated CE instruments are commercially available for clinical diagnostics based on hemoglobin variants or serum and urine proteins [18–22]. Future development of CE in the pharmaceutical, food science, and diagnostic industries appears strongly linked to advances in microchip CE technology and its promise of high-throughput, low-cost analyses [23–25]. CE is not a standard tool in most biology or biochemistry research laboratories. One reason is that competing technologies were already entrenched when CE was first developed. Another is that CE has low UV/Vis detection sensitivity and lacks specificity for analysis of proteins in complex biological matrices. CE is also consistently outperformed by conventional techniques in applications that require fractionation and recovery of proteins for further study [26].

Significant advances have been made in commercial CE instrumentation since the last edition of this book was published. Open analytical systems are more reliable and dedicated microchip systems are commercially available. High-throughput, multicapillary platforms with dedicated proteomic applications are either on the market or nearing final development. Engineering advances include better microfluidics, better optics and electronics, wider range of detection options, and improved software for instrument control and for data processing, analysis and reporting. Newer coated capillaries allow for better control of electroosmotic flow (EOF) and reduce protein–wall interactions in ways that markedly improve the reproducibility and efficiency of CE analyses. The development of multidimensional assay systems and advances in protein detection by MS and fluorescence has improved the sensitivity and specificity of CE for analysis of proteins in the complex biological matrices that are the targets of proteomic studies. More widespread use of CE for protein analysis might be predicted as biomedical research turns more from the study of individual proteins to large-scale assessment of the human proteome. What role CE will play in the evolving field of proteomics depends on how effectively CE competes with or complements other advanced protein separation technologies, and on how proteomics is applied to the study of complex biological processes.

Some CE techniques are better suited than others for analysis of proteins. A wide variety of applications have been developed for protein analysis by capillary zone electrophoresis (CZE), capillary isoelectric focusing (CIEF), capillary sieving electrophoresis (CSE), and micellar electrokinetic chromatography (MEKC). All of these techniques are excellent for high-resolution analysis of specific polypeptides in uncomplicated matrices, which makes CE ideal for use in process quality control and characterization of purified proteins. CZE and CIEF are also excellent for the analysis of specific proteins in complex biological matrices if the protein or proteins of interest are naturally abundant or readily extracted from biological samples before analysis. Hemoglobin variants, for example, can be directly analyzed in crude hemolysates using commercially available on-capillary detection systems because globins are abundant and do not precipitate at the higher concentrations required for UV/Vis detection. Analysis of less abundant or less soluble proteins often require sample prefractionation, concentration and/or derivatization to enhance detection sensitivity. Like open analytical HPLC platforms, a single multimode open analytical CE platform can be used to accomplish different protein applications depending on the choice of capillary, separation medium and detection system. Capillaries, reagents, and detectors can be readily interchanged, making it simple to switch from one protein application to another. In general, CE offers superior separation efficiency compared to more conventional protein separation techniques when using CZE, CIEF or MEKC. CSE performed on an open analytical CE platform offers relatively fewer advantages over conventional size separation methods beyond those generally characteristic of all CE techniques (i.e., automation, low sample volume requirement, etc.). In contrast, automated microchip CSE represents a significant time- and labor-saving improvement over conventional gel electrophoresis systems but the advantages come at a price. High instrument start-up costs and the need for more advanced operator expertise are obstacles to more widespread CE use as is the fact that CE methods are essentially nonpreparative and fraction collection remains technically challenging.

A significant barrier to the use of CE for protein analysis in many biology and biochemistry research laboratories is the lack of specificity and flexibility provided by conventional polyacrylamide gel electrophoresis, especially when combined with Western blotting. Commercially available reagents and prepoured native, denaturing, isoelectric focusing and two-dimensional (2D) gels offer a wide and usually adequate range of separation efficiencies at relatively low cost per sample, and it requires little expense or expertise to introduce conventional gel electrophoresis into any laboratory. Western blotting is generally as specific as the primary antibody employed, but that functionality is conserved in immunoassays performed by capillary affinity electrophoresis (CAE). A main difference between CAE and Western blotting is that Western blotting is usually performed noncompetitively with enzyme-linked secondary antibody detection, and the target protein can be easily changed by switching from one primary antibody to another. Other than the relatively straightforward need to optimally titrate primary and secondary antibody concentrations, most other analytical aspects of Western blotting are consistent regardless of the protein of interest. Moreover, antibodies and reagent kits are widely and inexpensively available commercially. So are gel documentation systems that feature sensitive chemiluminescence and/or phosphorescence detection and software that makes data analysis simple. In contrast, most CAE protein immunoassays are performed noncompetitively using fluorescent-labeled primary antibodies or competitively using labeled peptides as competing antigens [27], and it is not as easy to switch from one protein immunoassay to another. Automated immunoassay of specific proteins in a microchip or multicapillary format offers many advantages over traditional Western blotting in terms of separation efficiency and throughput potential. However, ELISA and protein microarrays [28] competitively offer similar advantages for many protein applications.

A PubMed search of research literature published over the last 10 years using CE as a search term consistently identified approximately 1000 articles during each year of that period. The proportion of articles identified by adding "protein" as a search term was consistently about one-fourth of that total, suggesting sustained interest in CE for protein analysis. Continued evolution of CE immunoassays, microchip platforms and multidimensional systems that incorporate CE with MS and other advanced detection technologies can be expected to increase the use of CE for protein analysis as biomedical research turns its attention from genomics to proteomics and demand for high-throughput protein analysis increases.

2.2.2 The Challenges of Systems Biology and Proteomics

To study proteins is to study biology, the focus of which is increasingly shifting from individual molecules to systems. Systems biology uses top-down and bottom-up approaches to integrate life science information and better understand the control and operation of biological processes in health and disease [29]. Top-down systems biology is systemic-data driven and identifies molecular interaction networks on the basis of correlated molecular behavior by integrating data from the genome, transcriptome, proteome, and metabolome. Bottom-up systems biology examines the mechanisms through which functional molecular properties arise and constructs detailed mechanistic models of biological processes that can be used to predict system properties. The systems biology approach is more than just a neologistic exercise that combines "–omics" with existing words. It is a conceptual framework that permits consideration of complex systems in a way that begins to encompass the enormous amount of biological variation inherent to living organisms.

Proteomics is the systematic inventory of protein expression levels, posttranslational modifications and interactions with other biomolecules that along with genomics, transcriptomics and metabolomics is an integral part of the systems biology approach. Unlike the relatively static genome, the proteome is a dynamic entity that changes in different physiological states (e.g., health versus disease) and biological compartments (e.g., fluids, cells, tissues, or organisms). The human genome is estimated to contain 30,000–40,000 genes that by alternative splicing or other means code for the production of around 100,000 different proteins [30]. These include enzymes, structural proteins, peptide hormones, molecular transporters, signal transduction elements, transcription activators, and a host of other functionally different proteins. More than two million polymorphic sites are predicted in the human genome [31]. Consequently, many or most of the proteins produced by the human genome exist in multiple primary sequence isoforms in human populations due to allelic polymorphism. Most, if not all, also exist in multiple chemically modified isoforms that are alternatively glycated, phosphorylated, methylated, oxidized and/or otherwise enzymatically or nonenzymatically posttranslationally modified at one or more sites on the protein. Since many of the 100,000 proteins produced by the human genome exist as primary sequence isoforms that also differ by how and where they are posttranslationally modified, the actual number of *chemically distinct proteins* in the human proteome is undoubtedly in the order of millions. The sheer magnitude and diversity of the human proteome assures that the probability of any two organisms of the same species having exactly identical proteomes is vanishingly small, as is the probability that any single individual's proteome will be identical in normal and diseased states. Identifying, characterizing, and quantifying chemically distinct proteins in this vast human proteome is the monumental challenge that proteomics presents to analytical science.

Why is it important to accept this challenge? A current and near future priority of biomedical research is to understand and treat complex diseases like diabetes, heart disease, and cancer that have only recently become prevalent sources of morbidity and mortality in human populations. A major goal of systems biology and proteomics is to identify and quantify specific proteins for use as predictive diagnostic and/or prognostic biomarkers of disease [32]. Biomarker discovery is complicated, however, by the fact that individual response to complex diseases is heritable and highly variable. Sickle cell disease, for example, is caused by a Glu→Val substitution at the sixth position of beta globin in all patients with the disease [33]. But the clinical presentation of sickle cell disease varies from mild to severe, depending on the genetic background of each individual patient [34]. Heritable variation is also observed in human responses to other complex diseases [35,36] and in resistance to drugs used to treat disease [37]. Indeed, remarkably few human traits are truly Mendelian due, in part, to the activities of modifier genes that alter the expression of other genes and obscure regular patterns of inheritance [38,39]. Instead, inheritance of most quantitative traits, including disease risk, is polygenic and attributable to interactions between two or more genes and the environment. As a result, genetic variation and genetic background strongly influence both disease risk and the composition of each individual organism's proteome. Protein expression levels and interactions with other biomolecules are also influenced by environmental factors, like pathogen exposure, nutritional status, or drug intake. Consequently, the types and concentrations of chemically distinct proteins in the human proteome vary widely both between individuals and within individuals over time. The specific challenge that systems biology presents to proteomics is to identify biomarkers of disease within this vast framework of biological variation in the human proteome.

The role CE will play in proteomics will depend in part on how systems biology is applied to biomarker discovery and how biomarkers are applied in the diagnosis and treatment of disease. Given the likelihood that millions of chemically distinct proteins exist in the human proteome, it is clear that comprehensive characterization of an individual organism's entire proteome is beyond the reach of present and near future technology. Given the wide dynamic range of abundance of different proteins in the proteome, and the practical demand for high-throughput, low-cost diagnostic testing, comprehensive characterization of even a limited subset of the human proteome, like the plasma proteome, represents a significant analytical challenge. The estimated dynamic concentration range of proteins in the plasma proteome is 10^8, ranging from 0.5 ng/mL to 50 mg/mL [40]. Even if it were possible to cost-effectively assess all proteins in just the plasma proteome, organizing and understanding the plethora of data produced represents a colossal biomedical and computational challenge because it requires mathematical comparison of the clustering of patterns in proteomic phenotypes (characterized on the basis of the types and concentrations of chemically distinct proteins) with clustering of patterns in clinical phenotypes (characterized on the basis of observable traits or symptoms associated with health and disease). Associating proteomic phenotypes with disease phenotypes is further complicated by the fact that clinical phenotypes of disease can range from mild to severe within and between individuals with the same disease. At present, biomarker discovery appears to be in a sort of Catch-22 situation: biomedical science is technically unable to assess complex disease risk because analytical science does not cost-effectively provide the information needed to discriminate between high and low risk disease phenotypes; but analytical science does not provide this information primarily because biomedical science does not yet know which of the millions of genes, transcripts, proteins and metabolites in an organism are best analyzed in order to characterize normal and disease phenotypes and subphenotypes.

The solution to this situation will require cooperation between clinical and basic scientists and combined expertise in the areas of medicine, biology, chemistry, engineering, and bioinformatics. National and international sample repositories are already being created and categorized to make samples from case controls and subjects with known diseases available for analysis [11]. A plausible top-down approach to biomarker discovery is to develop a general analytical foundation for proteome characterization and later extract functional or diagnostic utility for various proteins on the basis of the results of large-scale systematic data collection [40]. This work is well underway and CE appears poised to be a major contributor, much as it was in the effort to characterize the human genome. A corollary bottom-up approach to biomarker discovery is to comprehensively evaluate narrow proteomic subsets. Characterizing how the levels and functions of a limited number of proteins change in specific cells or specific biochemical pathways associated with disease processes could help differentiate "normal" phenotypes from mild or severe disease phenotypes.

As systems biology progresses toward these goals it is important to emphasize that a range of "normal" human phenotypes exist in part because genotypic and phenotypic variation give human populations the ability to adapt and survive under different environmental and physiological conditions. Furthermore, hereditary risk for mild or severe susceptibility to complex disease is ultimately a function of the interplay between "normal" genetic variation and environmental factors that cause disease. Biochemical and physiological traits often tend to differ in a continuum between individuals in normal populations and within individuals over time or as they progress from normal to diseased states. The fact that phenotypic traits segue almost imperceptibly across a continuum is the ultimate challenge that nature poses to systems biology as it attempts to parse out complex trait patterns associated with normal and disease phenotypes. The advancement of systems biology and proteomics promises to complicate clinical diagnostics due to the increased influx of complex information, but it also holds promise for early detection of complex disease risk and more effective treatment of disease on a case by case basis. The systems biology approach appears indispensable for understanding the clinical heterogeneity of complex diseases, the role of heredity and genetic background in susceptibility to disease, what proteins or metabolites might be effectively used as risk biomarkers for specific diseases, and how biological processes can be manipulated to treat disease. What role CE will play in proteome analysis will depend on how proteomics is applied as part of the systems biology approach. It also depends on how CE instrumentation and method development evolve in competition with other advanced separation technologies whose evolution will also be driven by the demand for more and better information about very complex biological systems and their roles in very complex human diseases.

2.3 THEORETICAL ASPECTS

Capillary electromigration techniques can be used to analyze proteins in a wide variety of ways, all of which depend on principles of chemistry and physics as they apply to the movement of proteins in narrow-bore capillaries. Different CE techniques separate proteins on the basis of different protein properties like charge, mass, isoelectric point and/or hydrophobicity. Different CE techniques also use different detector systems and variously employ hydrodynamic, electrophoretic and/or electroosmotic forces for separation and mobilization. The efficiency and specificity of any CE separation is a function of fluid dynamics, electrical field strength, analyte diffusion and other analytical system characteristics that affect peak migration, resolution, and detection. The term *analytical system* is used here to describe the specific CE technique employed and all properties pertaining to its use, including the CE platform, detector type, capillary type and cooling system temperature, separation media composition and properties (e.g., pH, ionic strength, and viscosity), voltage and other run conditions like duration and capillary conditioning. All of these analytical parameters interactively determine how CE can be effectively applied for different protein applications. The sheer number and complexity of interactions make the theoretical aspects of CE for protein analysis difficult to fathom but it is also the reason that proteins can be analyzed by CE in so many different ways, often with distinct advantages over other separation techniques. Many theoretical aspects regarding the electrophoretic separation of molecules by CE are the same for all analytes and are discussed in greater detail elsewhere in this book. Theoretical aspects of CE most important for protein analysis are discussed next.

2.3.1 PROTEIN PROPERTIES

Proteins are amino acid polymers (polypeptides) with distinctive charge and shape characteristics that can change depending on their microenvironment. Selection of a specific CE analytical system for a specific protein application depends on the objectives of the separation and how the protein or proteins of interest behave in any particular analytical system. The specificity and selectivity of any CE analytical system depends in part on the physical and chemical properties of the protein itself, including concentration, amino acid composition, charge, mass, shape, hydrophobicity, absorption and emission characteristics and interactions with other biomolecules. These properties determine how a protein interacts with separation medium constituents and the capillary or other physical components of the analytical system. The Debye–Hückel–Henry theory can be used to mathematically predict the electrophoretic behavior of proteins [41] but often does not apply to predicting protein mobility in CE due to protein interactions with charged constituents of the capillary wall [42]. Predicting protein mobility is also complicated by inter- and intramolecular chemical interactions that influence a protein's physical properties and electrophoretic migration.

How proteins behave in a particular CE analytical system is strongly influenced by the pH (negative log of the hydrogen ion concentration) and ionic strength (concentration of all ions) of the sample and electrolyte solutions. Both markedly influence the charge properties of ionizable functional groups, especially the amino and carboxyl termini and the basic and acidic side chains of amino acids like lysine or glutamate. In general, polypeptides become more negatively charged as pH increases due to progressive deprotonation of acidic carboxyl and basic ammonium groups. As pH increases and hydrogen ion concentrations decrease, carboxyl groups are converted into carboxylate anions (R-COOH to $R\text{-}COO^-$) and ammonium groups into amino groups ($R\text{-}NH_3^+$ to $R\text{-}NH_2$). Decreasing pH (i.e., increasing hydrogen ion concentration) has the opposite effect. The dissociation constants (pK_a) of the α-carboxyl groups of different amino acids range from about 1.8 for histidine to 2.5 for tryptophan. The pK_a of α-amino groups range from about 8.8 for asparagine to 10.6 for proline. Consequently the carboxyl termini of proteins tend to be negatively charged at pH greater than 2.5 and the amino termini of proteins are positively charged at $pH < 8.8$. The pK_a of carboxyl and amino side chains range from about 3.9 for aspartic acid to 12.5 for arginine. The isoelectric point (pI) of a protein is the pH at which the molecule has zero net surface charge, that is, the effective number of positive charges equals the effective number of negative charges. The absolute number of ionizable groups in a specific polypeptide molecule is a function of its primary amino acid sequence. The actual number is a function of (a) pH, (b) the pK_a of the ionizable functional groups present, (c) posttranslational chemical modification of ionizable functional groups, and (d) inter- and intra-molecular protein interactions that determine which amino acids are present on the surface of the molecule.

2.3.2 CE Separation Techniques

Capillary electromigration techniques collectively represent a family of related microanalytical methods that effect molecular separations using narrow-bore capillaries and high electric field strengths. Operational differences between different modes of separation are primarily a function of the composition of the electrolyte or buffer solutions and whether coated or uncoated capillaries are used. Because different CE techniques variously employ hydrodynamic, electrophoretic and/or electroosmotic forces to mobilize and separate proteins, peak profiles and separation efficiencies differ between techniques based largely on the relative contributions of these forces. For most separations samples are introduced into the capillary either hydrodynamically using pressure or vacuum, or electrokinetically using voltage. The IUPAC subdivides capillary electromigration techniques into capillary electrophoretic techniques and electrically driven capillary chromatographic techniques while recognizing significant overlap between these two classifications [1]. The CE techniques most commonly used for protein analysis include CZE, CIEF, CSE, and MEKC. CAE is an electrophoretic technique where separation is effected by interaction between an analyte of interest and a specific affinity reagent, for example, a protein-specific antibody that changes the electrophoretic migration of a target protein. Other CE techniques are less frequently used for protein analysis. Capillary isotachophoresis (CITP) is used in conjunction with CZE to automate online sample concentration before analysis. Capillary electrochromatography (CEC) is a special case of capillary liquid chromatography where movement of a mobile phase through a capillary packed or coated with a stationary phase is achieved by EOF. CEC is also most often used in conjunction with CZE for online analyte concentration or prefractionation.

Smaller proteins move faster in an electric field than larger proteins with similar charge because acceleration is inversely related to mass to charge ratio (m/z) at any applied force. Protein separation based on differences in m/z in is the fundamental principle of CZE that uses both electrophoretic and electroosmotic forces to mobilize and separate proteins of interest. Since these two forces can act in opposite directions, directional movement of proteins through the capillary depends on the relative direction and magnitude of each force. CZE is the simplest and easiest CE separation mode to use and can be employed in a wide variety of applications. Anode and cathode electrolyte solutions are usually identical and the efficiency and selectivity of protein separations can be modified simply by adjusting pH and ionic strength. Since the pI of most proteins are between pH 3 and 10, most proteins are positively charged in solutions at pH 3 or below and negatively charged at pH 10 or above. The relationship between the pI of a protein and the pH of the separation medium can be exploited to adjust assay selectivity. CZE performed using electrolyte solutions at low or high pH are nonselective because most proteins will be positively or negatively charged, respectively, and migrate to the oppositely charged electrode when voltage is applied. One reason low pH (e.g., 2.5) phosphate buffer is a popular electrolyte solution for exploratory studies is that essentially all proteins introduced into the anode end of the capillary will be positively charged and migrate toward the cathode in CZE. Another reason is that the internal surfaces of fused-silica capillaries bind less protein at low pH. The down side of low pH buffers includes reduced EOF and longer run times. Higher pH buffers increase EOF but also increase binding of proteins with higher pI to capillary walls such that capillaries with fixed or dynamic coatings are often used for analysis of more basic proteins. More selective CZE separations can be achieved, however, by using electrolyte solutions with pH slightly above or below the pI of the protein or proteins of interest. Under appropriate separation conditions proteins with charge opposite that of the protein of interest will be excluded from the separation. The same principle applies to the pH of sample solutions and selective introduction of proteins into the capillary by electrokinetic injection.

Smaller proteins also move faster than larger proteins when electrophoretically mobilized through media containing a selective physical barrier like polyethylene oxide [43] or dextran [44]. Protein separation based on frictional and other forces influenced by protein mass and hydrodynamic radius is the fundamental principle of CSE. CSE uses fixed or replaceable sieving media to differentially retard migration of proteins or protein complexes with different molecular weights. CSE can be effectively used to determine molecular weight in single or multidimensional analyses, or in mobility shift assays that separate a protein from its complex with an antibody or nucleotide. Capillary gel electrophoresis (CGE) is a special case of CSE performed in capillaries containing cross-linked gels. The use of CGE for protein analysis has declined with the advent of noncross-linked soluble polymers that are replaced between runs and afford more reproducible and reliable CSE separations [45].

CIEF [8,46–52] separates proteins in a pH gradient on the basis of surface charge and is ideal for identification and quantification of protein isoforms since different sequence or posttranslationally modified isoforms of the same protein often have different pI. The pH gradients established by including ampholytes in CIEF separation media progressively range from lower pH regions at the anode to higher pH regions at the cathode when voltage is applied. When sample is introduced into the anode end of the capillary, proteins with higher pI migrate further toward the cathode through higher and higher pH regions. When balance is eventually achieved between positive and negative surface charges proteins become neutral and those with similar pI focus into stationary zones. Proteins with lower pI migrate shorter distances and focus in zones closer to the anode. The protein bands remain focused because diffusional movement in either direction puts the protein into a higher or lower pH region that adds positive or negative charge to the molecule. Addition of charge induces electrophoretic repulsions and attractions that cause the protein to return to a pH region where it either loses or gains protons and becomes neutral again. The focused zones may or may not be fully immobilized, depending on the absence or presence of residual EOF. EOF is often suppressed in CIEF separations by using neutral coated capillaries and/or including EOF-suppressing compounds in the separation media. Proteins with higher pI have shorter migration times because they have shorter distances to travel when mobilized past a fixed detector placed near the cathode end of the capillary. Migration times are progressively longer for proteins with progressively lower pI. Wide (e.g., pH 3–10) or narrow (e.g., pH 6.7–7.7) range ampholytes are commercially available and can be used alone or in combination to modify selectivity and separation efficiency. Variations in CIEF operating parameters primarily include the use of pressure or electromigration to mobilize focused protein zones past the detector when using fixed window on-capillary detection. Additives like methylcellulose are often included in CIEF separation media to regulate viscosity and flow rates in separations that hydrodynamically mobilize focused protein zones.

Analyte separation based on hydrophobicity is the fundamental principle of MEKC [53–55]. MEKC is a special case of electrokinetic chromatography (EKC) that separates analytes on the basis of a combination of electroosmosis, electrophoresis and interactions between analytes and surfactants or other additives to the electrolyte solution [1]. At sufficiently high concentrations, detergent molecules like sodium dodecyl sulfate (SDS) self-associate in pseudostationary arrangements called micelles, which have hydrophobic inner cores and hydrophilic outer surfaces. Micelles constantly associate and dissociate and have electrophoretic mobility opposite that of normal EOF. During separation, proteins with differing hydrophobicities distribute differently between the hydrophobic inner core and the hydrophilic solution phase. Hydrophilic (polar) proteins that are completely insoluble in the inner core of the micelle migrate at the velocity of EOF. Hydrophobic (nonpolar) proteins that completely associate with the micelle migrate at the velocity of the micelle. Proteins with intermediate levels of hydrophobicity will spend varying amounts of time between the hydrophobic and hydrophilic phases of the separation media and have migrations times intermediate to those of the micelle and EOF. Variation in protein hydrophobicity is a function of the relative content of nonpolar (e.g., alanine, glycine, leucine, valine, etc.) and polar (e.g., lysine, arginine, asparagines, glutamine, etc.) amino acids. Because separations are based on hydrophobicity, both charged and neutral proteins can be analyzed by MEKC, which is increasingly used to map one dimension of protein distribution in multidimensional analyses [12,56–59].

2.3.3 Instruments

No single CE instrument is ideal for all protein applications. Multicapillary instruments have more throughput capacity than single capillary instruments. Some CE instruments have interchangeable detection systems while others do not and are limited to more specific applications. How different CE instruments are used for protein analysis is primarily a function of what detector systems are available for a specific platform, how many and what types of capillaries can be used, sample throughput capacity and the functionality of the system's software for automation of sampling and data collection. Research laboratories at the forefront of CE instrument development use a wide variety of in-house manufactured platforms and software that are not yet commercially available. Most of these incorporate unusual on- or off-capillary detectors into the analytical system or are designed to permit online pre- or postseparation sample treatment or fraction collection. Perhaps the fastest growing segment of advanced instrument development with relevance to proteomics is that of multidimensional systems that analyze proteins on the basis of more than one protein property, that is, charge, mass and/or hydrophobicity [59]. Systems have been developed for protein separation using CE in two dimensions, HPLC in one dimension and CE in the other, or CE followed by MS, which is both a separation and detection technique.

The advent of microfluidic devices like the Agilent 2100 Bioanalyzer (Agilent Technologies, Sanata Clara, CA) has diverse implications for CE-based protein analysis. This commercially available microchip CE platform is a dedicated instrument that uses manufacturer-supplied reagent kits for the analysis of DNA, RNA and proteins. Analyses are quick (~30 min) and simple with digital control and output that makes it easy to manage and store data. Up to 10 samples can be analyzed simultaneously and the results can be viewed as either electropherograms or gel-like images. The system uses noncross-linked linear polymers, fluorescent protein dye and laser-induced fluorescence (LIF) to separate and detect proteins on the basis of size, and is useful for protein quantification or molecular weight determination. This dedicated analyzer is a feasible alternative to labor intensive conventional electrophoresis methods with advantages superior to that of SDS–polyacrylamide gel electrophoresis (SDS–PAGE) in large part due to automation. One benefit for many biology and biochemistry laboratories is that the system can be used for both proteins and nucleic acids, essentially replacing multiple conventional electrophoresis applications in a single platform. Disadvantages include the inability to recover fractionated proteins and high capital equipment and reagent costs. How microchip systems will be used in proteomic applications will depend on how these systems evolve and their ability to characterize diagnostic subsets of the human proteome.

A relatively limited number of other commercial CE systems is available. Multimode open analytical CE platforms, like those available from Beckman-Coulter (Fullerton, CA) and Agilent Technologies, can perform a wide variety of protein applications because many different separation modes and detection systems can be used. Sebia Electrophoresis (Norcross, GA) and Beckman-Coulter manufacture dedicated platforms and reagent kits for automated clinical analysis of hemoglobin and/or plasma and urine proteins. Convergent Biosciences (Toronto, Canada) markets a CIEF system for whole-column image detection (WCID) of proteins. The systems biology demand for faster, more specific and more sensitive protein analysis has encouraged new companies to enter the CE instrument market with systems specifically designed for proteomic applications. Cell Biosciences (Palo Alto, CA) is developing a dedicated CIEF system specifically for immunodetection of proteins. The instrument is reportedly capable of analyzing 96 samples in 8–12 h and could replace Western blotting for use in some protein applications, especially for analysis of low abundance proteins in an automated format. CombiSep (Ames, IA) manufactures a high-throughput, multicapillary instrument capable of determining the molecular mass of proteins in 96 samples in 30 min [60].

2.3.4 Capillaries

Poor between-run analytical reproducibility has been a major impediment to the wider use of CE in protein applications. Variable capillary performance continues to be the single most important source of analytical variation. Uncoated fused-silica capillaries from different manufacturers or different production lots from the same manufacturer can have significantly different chemical properties. The internal surfaces of uncoated fused-silica capillaries chemically interact with proteins directly and with components of electrolyte solutions in ways that markedly influence protein recovery, mobility and separation efficiency. The need for consistent starting conditions for each analytical run is one reason uncoated capillaries are typically rinsed with sodium hydroxide and/or other regenerating solutions between analytical runs. Capillaries are also usually rinsed between runs with separation media in an effort to establish a reproducible internal chemical equilibrium.

The single most important capillary property influencing protein analysis is the charge characteristics of the internal surface which can be positively charged, neutral or negatively charged. The charge properties of the capillary wall depend on the pH and chemical composition of the electrolyte solution and whether the capillary is uncoated or chemically modified. The surface silanol groups of uncoated fused-silica capillaries are negatively charged at $pH > 2$ and are progressively deprotonated at higher pH. The negatively charged wall attracts a layer of hydrated cations from the electrolyte solution that move toward the cathode when an electric field is applied. This generates the EOF characteristic of many CE separations, which greatly influences the resolution and efficiency of protein separations. When it exists, EOF is always in the direction of the electrode with the same charge as the capillary surface. EOF in uncoated fused-silica capillaries is progressively higher at higher pH and decreases with increasing ionic strength. Protein–wall interactions in uncoated fused-silica capillaries also increase at higher pH, which can decrease protein recovery and degrade capillary performance.

Because the surface of a fused-silica capillary is chemically reactive it can be chemically modified to better control, minimize, or even reverse surface charge. Many of the problems associated with protein analysis in uncoated fused-silica capillaries have been largely overcome by the development of methods to chemically modify the internal surface. Chemical modification produces more consistent starting conditions between analytical runs and decreases protein–wall interactions. Coated capillaries are particularly useful for the analysis of basic proteins, that is, proteins with greater positive charge, because basic proteins bind more readily to the negatively charged surface of uncoated fused-silica capillaries. Typical coating materials include polyacrylamide, polyvinyl alcohol, polyethylene glycol, polyvinylpyrolidone, polyamines, and cationic detergents. Capillaries with covalently-attached (fixed) coatings that are acidic, neutral, basic, hydrophobic, or hydrophilic are commercially available. The performance of capillaries with fixed coatings can deteriorate over time and the capillaries must be replaced periodically due to degradation of the coating material or accumulation of proteins or other foreign matter that corrupts the quality of the separation. Problems encountered with fixed coatings can be overcome somewhat by the use of "dynamic" coatings where the coating agent is applied internally before a separation and then is stripped and replaced between runs. Coating reagents like polyamines or cationic detergents bind to negatively charged fused-silica capillaries and create a positively charged internal surface with reverse EOF (directed toward the anode). MicroSolve Technology Corporation (Eatontown, NJ) markets a family of quality-controlled CE reagents (CElixir) that includes multiple electrolyte solutions with a variety of pH that have been optimized for use with different dynamic coating solutions.

Because low abundance proteins can be difficult to detect by absorption spectroscopy, light transmittance and path length are important properties that also must be considered in capillary selection. Although most capillaries used for protein analyses are constructed of fused-silica, borosilicate glass, quartz, Teflon and other materials are also infrequently used (quartz transmits ultraviolet light better than glass at wavelengths used for protein detection). The internal cavities of most CE capillaries are round but rectangular capillaries have also been used [61]. Typical internal diameters range from 2 to 100 μm. Optical path length and detection sensitivity both increase with diameter but separation efficiency is usually better in narrower capillaries. Most capillaries have an opaque polyimide coating on the outside that provides tensile strength and limits breakage. For analytical systems using on-capillary spectrophotometric detection, the polyimide coating is usually removed by heating or with acid to create a window through which light will pass. Transparent external coatings that provide tensile strength and also optically transmit light have also been developed but are not widely used. Modified capillaries with extended light paths (e.g., bubble- or z-shaped) have been developed to increase detection sensitivity, which can be especially important for low abundance proteins. The utility of capillaries with extended light paths for protein analysis has diminished somewhat with the expanded use of protein-specific derivatizing agents and fluorescence detection.

2.3.5 Separation Media

Individual electrons normally cannot pass through solutions from one electrode to another. Instead, the electrons are carried between electrodes in solution by electrolytes that make protein electrophoresis possible. When a capillary is placed between two separated vials of electrolyte solution and voltage is applied, chemical reactions occur that consume electrons and generate negative ions at the cathode and also produce electrons and generate positive ions at the anode. As negative and positive charge builds up at the cathode and anode, respectively, ions in the electrolyte solutions and sample move through the capillary toward the oppositely charged electrode. This neutralizes charge differences between the electrolyte vials and allows continued electron flow through the electrodes and charged molecule flow through the capillary. Conducting multiple analytical runs without replacing electrolyte solutions can lead to ion depletion, especially when using low ionic strength electrolyte solutions and small volume reservoirs.

As previously noted, the pH and ionic strength of electrolyte solutions markedly influence the charge and migration of proteins due to their effects on both the protein and the internal surface of the capillary. EOF is positively related to pH and inversely related to ionic strength. Although higher ionic strength electrolyte solutions can suppress ionic interactions between charged proteins and ionized silanol groups, they can also generate excessive heat at high voltages that may surpass the cooling capacity of the analytical system. Overheating can alter the stability of proteins, corrupt capillary coatings and degrade the quality and reproducibility of CE separations. CZE often uses the same simple electrolyte solution (e.g., phosphate or borate) at both the anode and cathode ends of the capillary. Other separation techniques use more complex electrolyte solutions that are buffered or contain additives that (a) selectively or nonselectively modify proteins to alter their charge or conformation (e.g., affinity reagents or detergents), (b) selectively or nonselectively modify proteins to increase signal strength (e.g., fluorescent antibodies or nonspecific derivatizing agents), or (c) selectively impede the mobility of proteins through the capillary on the basis of their size, shape, or hydrophobicity (e.g., linear polymers or micelles).

2.3.6 Detection

Various types of detectors are commonly or uncommonly used for protein analysis in CE analytical systems. In all cases, detection is based on characteristic properties of the protein or proteins of interest in their modified or unmodified state. Most commercial platforms use on-capillary absorption or fluorescence spectrophotometers that measure light as it passes transversely through the lumen of the capillary at a fixed narrow window through which analytes sequentially pass during electromigration. In contrast, WCID systems monitor absorbance or fluorescence of all proteins simultaneously over the entire length of the capillary [62,63]. WCID is ideal for use with CIEF because it eliminates band distortions caused by mobilization of narrow focused protein zones. WCID systems are also advantageous for real time on-capillary study of dynamic processes like biomolecular interactions or how proteins migrate under different analytical conditions. In-house constructed CE platforms use a variety of less common on- and off-capillary detection systems (Chapter 9). These include UV/Vis or fluorescence detection using unusual optical systems, polarization, total internal reflection, correlation spectroscopy, and multiphoton excitation. Electrochemical or amperometric CE detectors measure analyte concentrations on the basis of change in current caused by oxidation or reduction of chemical reactants or products at an electrode [64]. Kuijt et al. [65] demonstrated the use of quenched phosphorescence for CE peptide detection using separation media containing 1-bromo-4-naphthalenesulfonic acid whose phosphorescence and background signal is quenched by electron transfer from peptide amino groups at higher pH.

Absorption spectroscopy using ultraviolet or visible light is the detection system most commonly used in CE protein applications. Most commercially available UV/Vis CE detectors use narrow band filters to limit transmittance to specific wavelengths but multiwavelength diode array detectors are also available. Choice of wavelength for UV/Vis protein detection depends on the abundance of the protein of interest and the absorption characteristics of the protein, capillary and electrolyte solution. Nearly all proteins absorb light at 214 and 280 nm due to molecular absorption of electromagnetic energy at these wavelengths by peptide bonds and aromatic amino acids, respectively. Absorbance at any given protein concentration is usually greater at lower UV wavelengths but many separation media constituents, like some ampholytes used in CIEF, also absorb low-frequency UV light and can interfere with detection. The absorption characteristics of naturally-occurring protein prosthetic groups can be used to increase detection specificity for some proteins. Hemoglobin variants, for example, are usually detected by absorbance at 415 nm because the attached heme groups absorb light energy at this wavelength and thus impart some degree of detection specificity. Proteins containing aromatic amino acids like tryptophan, tyrosine, and phenylalanine absorb electromagnetic energy when excited at wavelengths between 250 and 300 nm and naturally fluoresce at wavelengths between 300 and 400 nm. Native fluorescence detection can be used to enhance analytical specificity for proteins with high aromatic amino acid content [66–68].

The obligatory use of narrow-bore capillaries in CE limits permissible optical path lengths. This makes on-capillary UV/Vis absorption spectrophotometry relatively insensitive for analysis of low abundance proteins or when conventional detection systems are used for native fluorescence applications. Optical path length is of less concern when using high-intensity fluorescence detection. Fluorescence spectroscopy measures luminescence generated when molecular absorption of a photon induces the emission of another photon with a longer wavelength. The use of derivatizing agents to enhance the sensitivity of fluorescence detection has greatly expanded the utility of CE for some protein applications [27,69]. LIF detectors using argon ion (488 nm) laser sources have been available for some time on commercial multimode open analytical CE platforms. Fluorescein isothiocyanate (FITC) and other fluorescein derivatives are often used with argon ion lasers because fluorescein has absorption and emission maxima at 490 and 514 nm, respectively. Other high-intensity, monochromatic excitation sources are also used for CE fluorescence detection, including other laser sources and less expensive UV light-emitting diodes [66,67].

The need for greater detection sensitivity and the availability of a wider range of high-intensity excitation wavelength options have combined to promote development and use of novel fluorophores for derivatization of a variety of analytes [70]. Derivatizing agents are used to nonspecifically label carboxyl, amine or sulfhydryl groups of proteins or peptides. Examples include FITC and Cy3.5 [71], 3-(2-furoyl)quinoline-2-carboxaldehyde (FQ) [56,57,72], *o*-phthalic dicarboxaldehyde/beta-mercaptoethanol [66], monobromobimane [73], dipyrrometheneboron difluoride (BODIPY) [74], and MitoTracker Green [75]. Fluorophores are classified as either fluorescent or fluorogenic depending on their emission properties [69]. Fluorescent derivatizing agents emit photons when excited regardless of their binding state while fluorogenic agents only fluoresce when bound to their target. Background fluorescence produced by fluorescent derivatizing agents can limit fluorescence detection sensitivity.

The number of derivatization sites on a protein and the chemical nature of the selected derivatizing agent can influence protein speciation and quantification [69,76]. For example, lysine is a primary target for many derivatizing agents and most proteins have one or more lysine residues. Modification of lysine amino groups or other ionizable functional groups by uncharged derivatizing agents can alter protein surface charge and electrophoretic mobility. Incomplete labeling of ionizable functional groups can produce multiple fluorescent species of the same protein each with different charge properties. Incomplete labeling can also corrupt the relationship between fluorescence signal strength and protein concentration thus jeopardizing accurate quantification. Cationic amine derivatizing agents can be used to apply a fluorescence tag to a protein while maintaining its charge characteristics, but all cationic fluorophores are fluorescent rather than fluorogenic and produce background signal that can obscure signals of interest [69]. Proteins of interest that are not readily separated from contaminating proteins are difficult to specifically detect using nonspecific fluorescence labeling techniques in one-dimensional CE analyses. Selectivity and sensitivity can both be enhanced by the use of analyte-specific fluorescence labeling like that afforded using fluorescent antibodies [27,77–79] or when using fluorescent proteins or peptides as competing antigens [80] in CAE applications.

MS is increasingly the detection method of choice for many proteomic applications because it provides high sensitivity detection and information about protein structure (Chapter 8). Conventional two-dimensional SDS–PAGE has traditionally been the workhorse off-line separation system used in conjunction with MS for the study of proteins [12]. An investigator can run a 2D gel, use an automated spot-cutting system to collect a sample separated in two property dimensions by pI and molecular weight, then process and analyze the sample using a variety of single or tandem MS techniques, or HPLC-MS. Samples can also be analyzed directly using reversed phase, ion exchange or other HPLC-MS technique. Coupling CE with MS permits high efficiency protein separation based on one or more protein properties in an automated format that eliminates post-separation sample handling. CZE is most commonly combined with MS using soft ionization methods due to the simplicity and compatibility of many CZE electrolyte solutions. However, the process of transferring separated proteins from the capillary outlet to the mass spectrometer still presents obstacles to more widespread CE-MS use due to issues related to solvent compatibility and the placement of the CE electrode at the CE–MS interface. The amount of sample that CE can deliver to the MS also remains a major limitation for CE-MS detection of low abundance proteins. Greater detail regarding the advantages and disadvantages of CE-MS for protein analysis is provided in Chapter 8 and recent reviews [6,59,81,82].

2.3.7 Sample Preparation and Fraction Collection

Proteins that are abundant in complex biological matrices or have unique spectral properties can be readily detected and quantified with little or no sample preparation. Little sample preparation may also be needed to analyze proteins in less complicated matrices or where the protein of interest is effectively separated from all potential contaminants. However, many proteins are present in biological samples at concentrations that are lower than the limit of detection of many CE analytical systems. If fluorescence derivatization is not compatible with the desired application, analysis of low abundance proteins may require sample pretreatment to concentrate the protein or proteins of interest or remove interfering compounds. Pre-analytical sample concentration or fractionation can be performed on- or off-column. If sample availability and volume is not severely limited, the same off-line methods commonly used to concentrate or fractionate samples for HPLC or gel electrophoresis can be used to treat samples for CE analysis. Dilute samples can be concentrated by ultrafiltration or reconstituted after lyophilization. Buffer exchange by ultrafiltration or dialysis can be used to remove salts or other compounds that may interfere with a separation. A variety of common off-column fractionation methods from affinity chromatography to acid precipitation can be used to select specific proteins or remove interfering proteins with similar charge, size or hydrophobicity. Development of advanced robotic instruments for automated sample treatment using a variety of different modalities makes off-column sample pretreatment an increasingly feasible option for high-throughput CE protein analysis. CE techniques like CITP or CAC are frequently used for on-column sample pretreatment to concentrate or fractionate proteins before separation, which can be particularly useful in CE-MS for automated detection of low abundance proteins.

Fraction collection by CE is possible but difficult to accomplish in yields sufficient for further study of separated proteins even when using CE methods that are considered preparative. CE is consistently outperformed by other protein separation techniques when it comes to preparative operation or fraction collection for post-separation analysis [26]. Because CE sample injection volumes are very low, the concentrations of separated proteins eluting from the capillary are also very low. Typical CE injection volumes are usually < 50 nL and only nanogram quantities of specific proteins are isolated with each analytical run. Peaks from multiple CE runs can be pooled to increase sample recovery but consistent fraction collection requires highly reproducible run to run separations and accurate prediction of post-detector elution from the capillary. Fraction collection can be performed by adsorption of analytes to moving blotting membranes as they exit the capillary [83]. Inclusion of magnetic beads coated with immobilized antibodies represents an innovative new approach to on-column fraction collection [84]. CE fraction collection is desirable because CE is often able to separate analytes that cannot be resolved by other techniques and because CE separation in free solution often allows preservation of analytes in their native form. A key advantage to preparative CE compared to preparative HPLC is that analyses can be performed using very small sample volumes like those available from single cells.

2.4 PRACTICAL APPLICATIONS

This section details characteristic examples of protein analysis by CZE, CSE, CIEF, and MEKC. The selected examples were chosen to demonstrate some of the more important theoretical aspects of protein separation by each of these techniques while also highlighting the application of CE for characterization of protein properties, isoform identification, use in mobility shift assays, and multidimensional protein analysis. Table 2.1 lists recent references for specific proteins analyzed by different CE methods and show how CE analysis of proteins can be used in a variety of biological applications.

2.5 ANALYSIS OF PROTEIN PROPERTIES BY CZE AND CSE

Information about a protein's physical and chemical properties can help determine its biological function and can be useful in a bottom-up approach to understanding a biological system. The biological activities and binding properties of proteins are strongly influenced by the type and number of ionizable functional groups that are available to participate in chemical reactions with other molecules. Information about a protein's size, shape, and hydrophobicity can also be used to help deduce transport properties, compartmental distribution or other biological characteristics.

CZE separates proteins in narrow-bore capillaries under the combined influence of electrophoretic and electroosmotic forces. Electrophoretic mobility is a function of charge and frictional forces and thus distinguishes proteins with different mass to charge ratios (m/z). Analysis of protein ladders by Sharma and Carbeck [85] elegantly demonstrates the fundamental theoretical principles that permit the separation of proteins by CZE. In these experiments, net charge and hydrodynamic radius ladders were prepared from purified lysozyme. Positively charged functional groups were progressively chemically modified to produce a range of molecules with precisely defined charge and shape characteristics. Figure 2.1 is an idealized schematic showing how charge and radius ladders are synthesized and the net effects of progressive derivatization of one or more functional groups on a protein's charge, hydrodynamic radius and electrophoretic mobility.

To study purified lysozyme from chicken egg white (EC 3.2.1.17, 14.3 kDa, 6 lysine residues, pI 10.9) the authors produced a charge ladder by chemically derivatizing primary amines (RNH_2) at pH 12 with acetic anhydride. Two radius ladders were produced by derivatizing the same amines with two different sizes of polyethylene glycol N-hydroxysuccinimide ester (PEG-NHS, with 2 or 5 kDa PEG chains). The products of multiple reactions were combined to produce complete ladders in a single sample since single reactions did not always produce all derivatized forms. The high lysine ($pK_a = 10.5$) content of unmodified lysozyme makes it a very basic protein that is positively charged at the hydrogen ion concentration of the separation media (pH 8.4). Because the protein would be expected to bind to the negatively charged surface of an uncoated fused-silica capillary the internal surfaces of the capillaries used in these experiments were coated with a cationic polymer to reverse internal surface charge and reduce protein–wall interactions. Because the internal surface charge of the capillary was the reverse of normal, EOF was also the reverse of normal, that is, toward the anode instead of the cathode, and the separations were conducted in reverse polarity where samples were introduced at the cathode end of the capillary with the detector oriented toward the anode.

Under the conditions of these experiments, electroosmosis is a more or less uniform directional force toward the anode on all proteins in the sample. In contrast, positively charged protein molecules are electrophoretically attracted to the cathode and move in a direction opposite that of EOF at rates inversely related to m/z. Electrophoretic mobility toward the cathode is inversely related to m/z because increasing positive charge increases electrostatic attraction while increasing mass increases frictional forces that retard velocity. When electroosmotic force is greater than the opposing electrophoretic force, a protein will migrate in the direction of the detector. Migration times will therefore be greater for proteins with higher electrophoretic mobility (lower m/z) because the difference between electroosmotic and electrophoretic forces increases as electrophoretic mobility increases.

TABLE 2.1
Recent Applications of CE for Protein Analysis

Protein	Purpose	CE Technique	Detection	References
Akt (protein kinase B)	Enzyme activity	CZE	UV	[99]
Albumin	Heparin affinity	Microchip CAE	LIF	[100]
Albumin; alpha(1)-acid glycoprotein	Protein–drug interactions	CZE	UV	[101]
Albumin; lysozyme; ribonuclease a	Protein characterization	CIEF	UV-WCID	[102]
Amyloid precursor protein	Glycosaminoglycan affinity	CZE	UV	[103]
Beta lactoglobulin	Allergen characterization	CZE	LIF	[104]
Botulinum neurotoxin	Enzyme activity	CZE	LIF	[105]
Erythropoietin	Glycoform characterization	CZE	ESI-TOF-MS	[106,107]
Erythropoietin	Glycoform characterization	CZE	UV	[108]
Hemoglobin variants, mouse	Isoform identification and quantification	CIEF	UV	[109]
Hemoglobin variants, human	Isoform identification and quantification	CIEF	UV	[87]
Hemoglobin, globin chains	Isoform identification and quantification	CZE	UV/Vis	[110]
HIV-1 reverse transcriptase	DNA-binding affinity	CZE	LIF	[111]
Insulin; trypsin	Protein–protein interactions	Multicapillary CZE	CCD-LIF	[112]
Metallothionein; superoxide dismutase	Isoform characterization	CZE	ICP-MS	[113]
Methionine-enkephalin	Immunoassay	CZE-CAE	LIF	[114]
Prion protein	Immunoassay	CZE	LIF	[115]
Prion protein	Immunoassay	CZE-CAE/CIEF-CAE	LIF	[116]
Serum proteins	Immunotyping	CZE	UV	[117]
Soybean proteins	Protein profiling	CZE	UV	[118]
Transferrin	Isoform characterization	CZE	UV/ICP-MS	[119]
Transferrin	Isoform quantification	CZE	UV	[120,121]
Whey proteins	Protein profiling	CZE	LIF	[72]

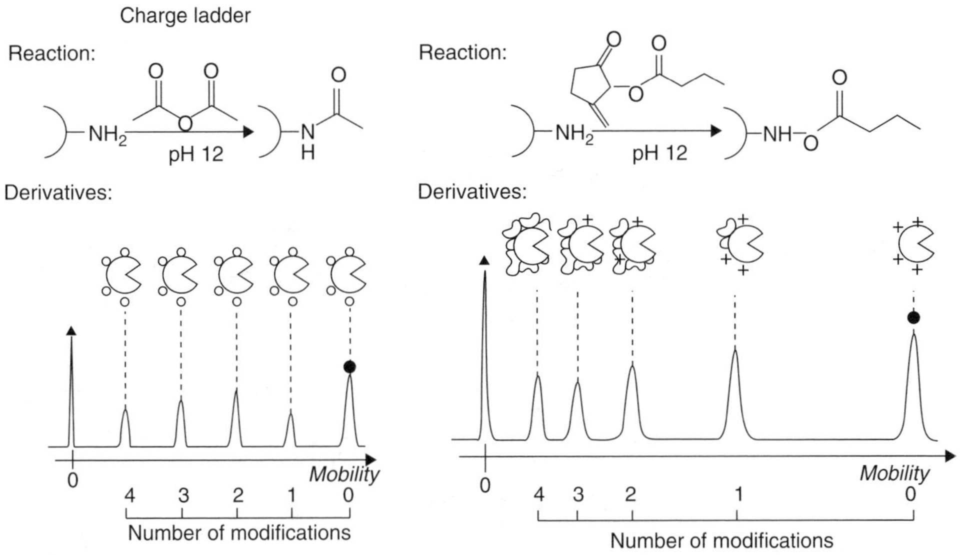

FIGURE 2.1 Schematic illustration of the synthesis of protein charge and radius ladders and their characterization by CZE. Ladders are synthesized by the partial modification of primary amino groups on a protein. Charge ladders are produced using acetic anhydride and radius ladders are produced using PEG-NHS esters. The rungs of a charge ladder contain proteins that differ incrementally only in net charge from those in neighboring peaks and are thus equally spaced. The rungs of a radius ladder contain proteins that differ in both net charge and hydrodynamic radius from those in neighboring peaks. (From Sharma U, Carbeck JD. *Electrophoresis* 2005;26:2086–91. With permission.)

Figure 2.2 shows the effect of progressive derivatization of lysozyme amino groups with acetic anhydride or PEG-NHS on electrophoretic mobility. Mobility was empirically calculated for each observed peak as $\mu = (L_T L_D / V)((1/t_{nm}) - (1/t_x))$, where L_T is the total length of the capillary, L_D is the length of the capillary to the detector window, V is the applied voltage, t_{nm} is the migration time of a neutral marker, and t_x is the peak migration time. The constant influence of electroosmotic force on peak migration is accounted for by including the migration time of the neutral marker in the equation. Since all variables besides t_x are constants in each analytical run, $1/t_x$ decreases as peak migration time increases and the difference between $1/t_{nm}$ and $1/t_x$ increases. Consequently, peak mobility is greater for molecules with lower *m*/*z* and longer migration times.

The precisely defined charge and shape characteristics of the lysozyme molecules in the charge and radius ladders clearly demonstrate these principles. In both types of ladders, unmodified lysozyme had the highest positive charge and the lowest *m*/*z* so it also had the longest migration times and highest calculated peak mobility. In the charge ladder, progressive neutralization of charged lysozyme amino groups by incremental addition of acetic anhydride progressively decreased charge but had little influence on the mass of the molecule. Consequently, the magnitude of the increase in *m*/*z* was approximately equal with each additional amino group modification as evidenced by the even distribution of the differently charged peaks. In contrast, incremental addition of either size PEG-NHS to lysozyme amino groups progressively increased hydrodynamic radius but also decreased charge. Since both mass and charge were modified with each additional amino group modification, the increase in *m*/*z* was disproportional as evidenced by the unequal peak spacing of lysozyme molecules with different numbers of PEG-NHS-modified amino groups.

Separation of purified lysozyme molecules with varying but well-defined charge and mass properties clearly demonstrates the theoretical aspects of how proteins are separated on the basis of these characteristics by CZE. It shows that electroosmotic force must be considered in all separations and

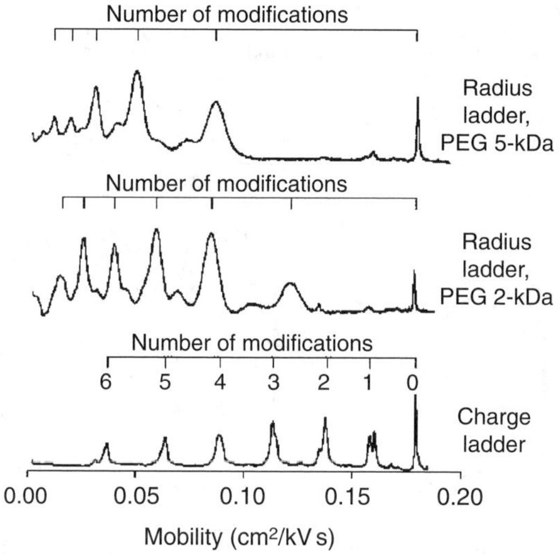

FIGURE 2.2 Analysis of lysozyme charge and radius ladders by CZE. The analytical system included a Beckman-Coulter P/ACE MDQ instrument operated in reverse polarity at 20 kV with UV detection at 214 nm and capillary cooling at 25°C. Derivatized lysozyme molecules with different hydrodynamic radii and/or charge or were separated by CZE on the basis of *m/z* using a positively charged, 60 cm long (50 cm to detector) by 50 μm ID fused-silica capillary (cationic polymer poly-(diallyldimethylammonium chloride) was applied to suppress adsorption of lysozyme). Separation media contained 25 mM Tris and 192 mM glycine (pH 8.4, ionic strength 7.9 mM). The bottom panel shows peak mobility for the charge ladder constituents while the upper two panels show peak mobility for the two radius ladders. (From Sharma U, Carbeck JD. *Electrophoresis* 2005;26:2086–91. With permission.)

can be modified to tailor separations based on the pI of the protein of interest. It also shows how electrophoretic mobility increases with protein charge and decreases with increased mass/hydrodynamic radius. Besides aptly demonstrating the principles of CZE, protein charge and radius ladders can be applied in a variety of practical ways. Sharma and Carbeck [85] used lysozyme radius ladders to measure partitioning of different sized molecules in polymer hydrogels and also suggested that radius ladders can be used to measure the effects of hydrodynamic size on the transport properties of a protein. Ebersold and Zydney [86] used CZE and myoglobin charge ladders to quantify electrostatic interactions between proteins and membranes during ultrafiltration.

Like CZE, CSE also separates proteins in narrow-bore capillaries under the combined influence of electrophoresis and electroosmosis (if present) but markedly increases the influence of frictional forces by including sieving polymers in the separation media. Since frictional force is largely a function of protein mass and shape, CSE primarily separates proteins on the basis of molecular weight if charge differences between proteins are normalized by addition of detergent. SDS–PAGE is the analytical technique most commonly used to determine protein molecular weight in most biology and biochemistry research laboratories. Size exclusion HPLC and MS can also accomplish this task with varying degrees of sensitivity, specificity, and automation.

Betgovargez et al. [58] produced a multidimensional profile of the human serum proteome and used CSE to map one dimension of protein distribution by molecular weight. Their approach is characteristic of how CE techniques can be incorporated in multidimensional analyses for proteome characterization—in this case, by first using an off-line 2D chromatographic system before analysis of the resulting fractions by both CSE and CIEF. Figure 2.3 was selected for inclusion in this section because it very clearly demonstrates the use of CSE for molecular weight determination. In this figure, upper panel shows the electropherogram obtained from a sample containing purified transferrin supplemented with a 10 kDa internal standard and lower panel shows the electropherogram obtained

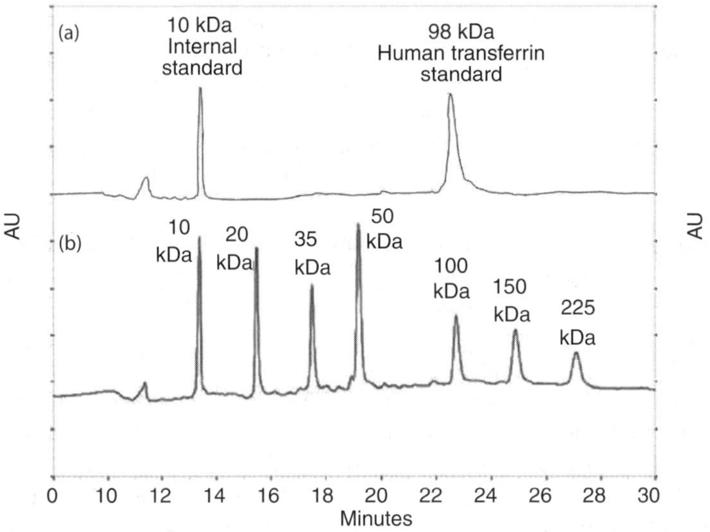

FIGURE 2.3 Protein molecular weight determination by CSE. The analytical system included a Beckman-Coulter ProteomeLab PA 800 operated in reverse polarity at 15 kV with photo diode array UV detection at 220 nm and capillary cooling at 25°C. SDS-treated proteins were separated by CSE on the basis of molecular weight using a negatively charged, 30.2 cm long (20.2 cm to detector) by 50 μm ID uncoated fused-silica-capillary. The separation media included a formulation of polymers optimized for separation of proteins over a wide molecular weight range that is marketed in a kit as part of the system. Analysis of purified transferrin is shown in the upper panel while molecular weight standards are shown the lower panel. (From Betgovargez E, et al. *J Biomol Tech* 2005;16:306–10. With permission.)

from a molecular weight standard containing eight different proteins with known molecular weights (10–225 kDa). Because all samples were treated with SDS, which was also included in the separation media, all proteins were negatively charged and CSE was performed in reverse polarity with the detector toward the anode. SDS also normalizes charge effects on proteins such that separation was a function of molecular weight rather than native m/z and smaller proteins moved more rapidly through the sieving media than larger proteins.

The molecular weights of unknown proteins in test samples can be determined using a calibration curve calculated on the basis of the relationship between the log of the molecular weight standards and 1/mobility. CSE can be performed under denaturing or nondenaturing conditions using replaceable soluble polymers or fixed cross-linked polymers. CSE can be used to determine molecular weight, for direct quantification of abundant proteins in uncomplicated matrices, or for quantitative analysis by immunoassay where bound and free antibodies are identified and quantified on the basis of differences in molecular weight.

2.5.1 Identification of Beta Globin Isoforms by CIEF

Allelic polymorphism in a specific gene can produce nearly identical proteins with primary sequences that differ by one or more amino acids. Primary sequence isoforms can be readily separated by CIEF on the basis of their pI if the amino acid substitution changes the surface charge of the protein. Similarly, enzymatic or nonenzymatic posttranslational chemical modification of ionizable protein functional groups can produce chemically distinct isoforms of the same protein with detectably different surface charge and pI. Analysis of hemoglobin by CIEF is an excellent example of how this technique can be applied for the identification and quantification of a family of related proteins that represent a concise subset of the human proteome. Members of this family include many primary sequence isoforms associated with monogenic diseases but also includes posttranslationally modified isoforms whose chemical composition reflects the concentrations of metabolomic constituents and disease status. The principles and approach used to evaluate primary sequence and posttranslationally modified beta globin isoforms described below can be applied to the study of other proteins, only not as easily since hemoglobin is unusually amenable to one-dimensional analysis by CIEF due to its high abundance, high solubility, and the UV/Vis detection specificity afforded by the attached heme groups.

Hemoglobin is a tetrameric oxygen transport protein composed of two mixed pairs of alpha and beta globin subunits. Hemoglobin tetramers readily dissociate into $\alpha\beta$ dimers that migrate as a unit when analyzed by most charge-based separation techniques [33]. Polymorphism and mutation in globin genes are responsible for the synthesis of hundreds of different primary sequence isoforms of both alpha and beta globin, including the beta globin isoforms responsible for the anemia associated with sickle cell disease and hemoglobin C disease. The most prevalent (normal) beta globin isoform has a glutamic acid residue at position 6 (βGlu6), which imparts negative surface charge to the $\alpha\beta$ dimer at pH > 4.1 due to deprotonation of its carboxyl side chain. Sickle beta globin has a valine residue at the same beta globin position (βVal6) which does not have an ionizable functional group and thus imparts no additional negative or positive charge regardless of pH. The C beta globin isoform has a lysine residue at the same beta globin position (βLys6), which imparts positive surface charge to the $\alpha\beta$ dimer at pH less than ~10.5 due to protonation of its amino side chain.

Homozygous sickle cell disease and heterozygous S/C disease are hereditary hemoglobinopathies prevalent among persons of African ancestry. Figure 2.4a demonstrates the application of CIEF in my laboratory for diagnosis of congenital human hemoglobinopathies [87,88] using pH 6.7–7.7 ampholytes. The sample shown in the figure was prepared for use as a hemoglobin pI standard by pooling erythrocyte hemolysates from multiple subjects that were homozygous or heterozygous for globin alleles that produce normal, sickle and/or C beta globin isoforms. The figure shows that $\alpha\beta$ dimers containing the more positively charged C beta globin isoform ($\alpha\beta^{\text{Lys6+}}$, pI = 7.44) eluted first, followed by $\alpha\beta$ dimers containing the more neutral sickle beta globin isoform ($\alpha\beta^{\text{Val6}}$, pI = 7.21), followed in turn by $\alpha\beta$ dimers containing the more negatively charged normal beta globin isoform ($\alpha\beta^{\text{Glu6}-}$, pI = 6.97). The peak between $\alpha\beta^{\text{Val6}}$ and $\alpha\beta^{\text{Glu6}-}$ contains fetal hemoglobin where gamma globin replaces beta globin in the hemoglobin dimer ($\alpha\gamma$, pI = 7.06). Multiple sequence variations between gamma and beta globin are responsible for the pI difference between $\alpha\gamma$ and $\alpha\beta$ dimers. The proportions of different hemoglobin variants present in a sample can be determined on the basis of peak area at 415 nm. Excellent peak resolution and capacity for automated peak identification based on calculated pI make CIEF a highly effective analytical method for quantitative analysis of primary sequence globin isoforms.

CIEF can also be used for quantification and identification of posttranslationally modified hemoglobin isoforms. Virtually all human beta globin isoforms have a cysteine residue at position 93 (βCys93) whose sulfhydryl group ($pK_a \sim 8.4$) participates in nitric oxide transport. βCys93 also nonenzymatically forms a mixed disulfide with glutathione *in vivo* under conditions of oxidative stress; one physiological role of glutathione is to protect protein sulfhydryl groups from irreversible oxidation. The mixed disulfide formed between βCys93 and glutathione is called *S*-glutathionyl hemoglobin and has been recommended for use as a biomarker of oxidative stress [89–92]. The sample shown in Figure 2.4b was prepared *in vitro* from erythrocytes obtained from a patient with homozygous sickle cell disease, and consequently, only the sickle beta globin isoform and $\alpha\beta^{\text{Val6}}$ dimers were present. A proportion of the dimers were converted into *S*-glutathionyl hemoglobin by disulfide exchange *in vitro* with glutathione disulfide (oxidized glutathione). Glutathione is a tripeptide (Glu–Cys–Gly) that is negatively charged at pH > 4.1 due to deprotonation of the carboxyl side chain of its *N*-terminal glutamic acid. When the mixed disulfide is formed between βCys93 and the glutathione cysteine residue, the $\alpha\beta^{\text{Val6}}$ dimer acquires negative charge that lowers its pI, increases its migration time and produces a new peak ($\alpha\beta^{\text{Val6,Cys93}-SG}$, ~40% of total hemoglobin based on peak area at 415 nm) that was not detected in the untreated hemolysate. Naturally occurring

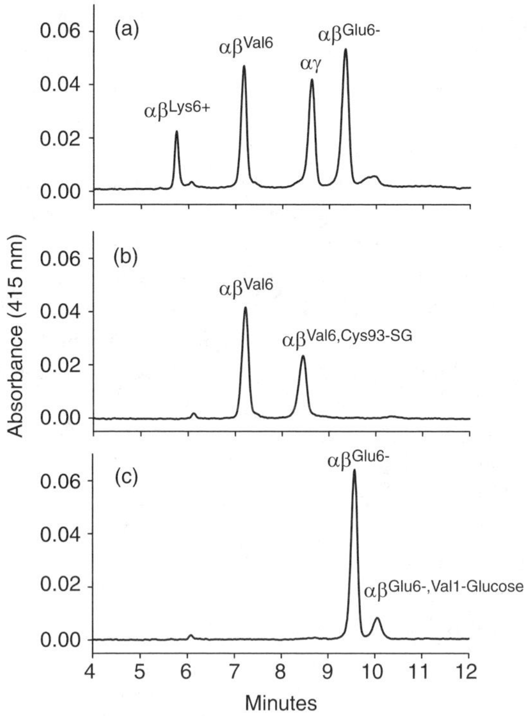

FIGURE 2.4 Analysis of primary sequence and posttranslationally modified beta globin isoforms by CIEF. The analytical system included a Beckman-Coulter P/ACE MDQ instrument operated in normal polarity at 30 kV with UV/Vis detection at 415 nm and capillary cooling at 20°C. Hemoglobin $\alpha\beta$ dimers with different beta globin primary sequences or chemical modifications were separated by CIEF on the basis of pI using a neutral, 31 cm long (21 cm to detector) by 50 μm ID fused-silica capillary (DB-1, coated with dimethylpolysiloxane to suppress EOF). Hemolysates were prepared by adding 10 μl of erythrocytes to 200 μl of hemolyzing reagent (10 mM KCN, 5 mM EDTA) and pressure injected at the anode for 10 s at 1.0 psi. Separation media contained 0.375% methylcellulose and 2% pH 6.7–7.7 ampholytes in deionized water. Cathode solution was 80 mM borate, pH 10.25. Anode solution was 100 mM phosphoric acid in 0.375% methylcellulose. Voltage was applied without pressure for 3 min before focused zones were mobilized past the detector using pressure at 0.5 psi with constant voltage. Samples were (a) mixed hemolysate containing Hbs C, S, F and A; (b) hemolysate containing HbS and S-glutathionyl HbS; (c) hemolysate from a patient with diabetes and elevated glycated hemoglobin.

S-glutathionyl hemoglobin levels *in vivo* measured by MS or ion exchange HPLC are estimated to be $<3\%$ of total hemoglobin in normal individuals and up to $\sim 9\%$ in patients undergoing peritoneal dialysis [89,91,92].

Enzymatic and nonenzymatic protein glycation are posttranslational biochemical processes that also occur naturally in many cells. Attachment of sugars to specific serine, asparagine, and hydroxylysine residues is under strict enzymatic control and serves a multitude of biological functions, such as facilitating protein secretion or prolonging protein survival in the circulatory system [33]. In contrast, nonenzymatic glycation is a concentration- and oxidation-dependent condensation reaction between protein amino groups and the carbonyl groups of aldehydes, ketones, and reducing sugars, also known as the "browning" or Maillard reaction. Nonenzymatically glycated proteins initially form Schiff bases that can chemically rearrange to form Amadori products that can rearrange further to form advanced glycation end (AGE) products that disrupt normal protein function and play an important role in the etiology of vascular complications associated with diabetes and other chronic diseases [93]. Glycated hemoglobin levels increase in proportion to mean blood glucose that remains elevated in diabetes patients with poor metabolic control. Glycated hemoglobin measurements (including HbA1c) are widely used to clinically monitor metabolic status and therapeutic success in diabetes patients.

The *N*-terminal valine of beta globin (βVal1) is particularly susceptible to nonenzymatic posttranslational chemical modification by glucose, glycolytic intermediates, and glucose oxidation/fragmentation products. Chemical modification of the βVal1 α-amino group inhibits its protonation and lowers both the pK_a of the α-amino group and the pI of hemoglobin $\alpha\beta$ dimers. Glucose and other molecules react with other alpha and beta globin amino groups but glycation at these sites apparently does not alter the surface charge of $\alpha\beta$ dimers [33], probably because the charge on these groups is not significantly altered by glycation at pH optimal for separation by charge-based separation techniques. The sample shown in Figure 2.4c was prepared from erythrocytes obtained from a patient with type 1 diabetes and elevated glycated hemoglobin levels. The larger of the two major CIEF peaks ($\alpha\beta^{\text{Glu6}-}$) contains $\alpha\beta$ dimers that are not posttranslationally modified in ways that alter pI, for example, on βVal1. The smaller peak ($\alpha\beta^{\text{Glu6}-,\text{Val1-Glucose}}$, ~15% of total hemoglobin) with longer migration time contains $\alpha\beta$ dimers with lower pI that are posttranslationally modified on β Val1 by glucose and other biomolecules.

These examples show how CIEF can be used to identify and quantify a family of sequence and posttranslationally modified beta globin isoforms based on pI. They also show how oxidative stress and hyperglycemia both influence the speciation of hemoglobin into a variety of chemically distinct isoforms that differ by how and where they are posttranslationally modified. Because hemoglobin is nonenzymatically modified by compounds like glutathione and glucose, isoform levels measured by CIEF reflects the concentrations of these and other metabolites and can be used to quantitatively assess a narrow subset of the human proteome that provides diagnostic information in a more or less bottom-up approach to biomarker discovery. Comprehensive assessment of the hemoglobinome, if you will, holds promise for identifying normal and disease risk phenotypes because beta globins are posttranslationally modified in characteristic patterns associated with characteristic patterns in disease phenotypes, that is, oxidative stress and diabetes. Many other proteins also undergo similar nonenzymatic and/or enzymatic posttranslational modifications with glutathione, glucose and other metabolites. Assessment of posttranslationally modified protein isoforms, especially long-lived proteins like collagen where modifications accumulate irreversibly, could provide information about both the proteome and the metabolome since the chemically distinct protein isoforms present are a function metabolite exposure over time.

2.5.2 SP 1 Mobility Shift Assay by CAE

Capillary affinity electrophoresis is often used to enhance selectivity or study binding interactions between proteins and other biomolecules. CAE applications rely on the principle that the electrophoretic mobilities of bound and unbound forms of a protein of interest differ and can be identified and quantified by CE, most often using CZE or CSE. One example is electrophoretic mobility shift assays, which are used to study protein–protein, protein–antibody, or protein–DNA interactions. A study by Ronai et al. [94] demonstrated the advantages of a CE mobility shift assay (CEMSA) for analysis of SP 1 binding capacity based on affinity between SP 1 and a fluorescent-labeled synthetic oligonucleotide. SP 1 is a sequence-specific DNA-binding protein that plays an important role in the transcriptional regulation of many genes. Figure 2.5 demonstrates the principle of the assay using a fluorescent DNA probe and HeLa nuclear extracts that contained SP 1. In this example, DNA probes were prepared from the 5′ upstream regulatory region of the human dopamine D4 receptor (DRD4) gene by polymerase chain reaction (PCR) amplification using fluorescein labeled primers. The probes (at 40-, 100-, or 400-fold dilutions) were then incubated with or without HeLa nuclear extract and analyzed by CE with fluorescence detection.

The m/z of the free DNA probe was lower than that of the DNA–protein complex and consequently had greater electrophoretic mobility and shorter migration time when analyzed by CZE. The analysis was performed in normal polarity but with the detector window closer to the inlet

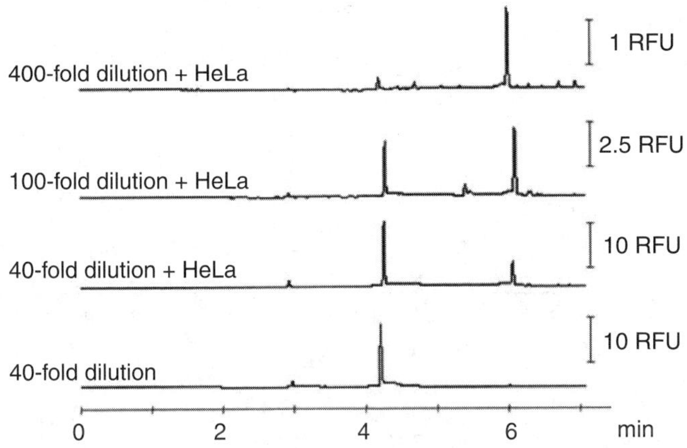

FIGURE 2.5 SP 1 capillary electrophoresis mobility shift assay (CEMSA). The analytical system included a Beckman-Coulter P/ACE MDQ instrument operated in normal polarity at 200 V/cm with laser-induced fluorescence detection (488 nm excitation, 520 nm emission) and capillary cooling at 20°C. Free DNA (~4.2 min) and DNA–protein complex (~6 min) were separated by CZE based on m/z using a neutral coated, 50 cm long (10 cm to detector) by 50 μm ID fused-silica capillary. Separation media (TBE buffer) contained 89 mM Tris HCl, 89 mM boric acid, and 2 mM disodium EDTA, pH 8.3. Samples containing various dilutions of fluorescein labeled DNA probe with or without a fixed amount of HeLa nuclear extract (containing SP 1) were pressure injected for 5 s at 3 psi. (From Ronai Z, et al. *Curr Med Chem* 2004;11:1023–9. With permission.)

(10 cm) in order to shorten the migration time while still using a longer capillary (50 cm total length) and high voltage. The proportion of DNA–protein complex present was a function of the relative concentrations of the DNA probe and nuclear extract. In the absence of HeLa nuclear extract (bottom electropherogram), a single large peak was observed at about 4.2 min representing the unbound (free) DNA probe. Addition of HeLa nuclear extract with the same 40-fold dilution of the same DNA probe (second electropherogram from the bottom) resulted in the appearance of a second peak at approximately 6 min representing the SP 1-DNA probe complex. Decreasing the concentration of the DNA probe while holding the concentration of HeLa extract constant (top two electropherograms) resulted in a proportional increase of bound vs. free DNA probe. Competition assays using nonfluorescent DNA probes with the consensus SP 1 binding sequence or a mutated version of the consensus sequence showed that the effect of the HeLa extract was specific for SP 1. The authors used this CEMSA to show that Sp 1 binding to the DRD4gene was influenced by the number of SP 1 binding 120 base pair repeat sequences present in the highly polymorphic 5′ upstream regulatory region.

2.5.3 Analysis of Free Prostate-Specific Antigen Isoforms by Capillary Zone Electrophoresis

Prostate-specific antigen (PSA) is a single-chain glycoprotein that belongs to the kallikrein family of serine proteases [95]. Overproduction of PSA, for example, in prostate hyperplasia, cancer or prostitis, causes more of the protein to enter the circulatory system such that serum PSA can be used as a biomarker for prostate-related diseases. PSA has one known glycosylation site that is chemically modified by a diantennary N-linked oligosaccharide that itself can vary on the basis of the presence or absence of fucose, one or two sialic acids, or the glycan content of fucose and *N*-acetyl-d-galactosamine. Free PSA thus exists as chemically distinct posttranslational isoforms. Genetic variants of PSA have also been reported. In addition, PSA in serum binds to several serine protease inhibitors, including α1-antichymotrypsin, α2-macroglobulin and α1-antitrypsin. Consequently, PSA in serum exists as both free PSA and complexed PSA, primary sequence

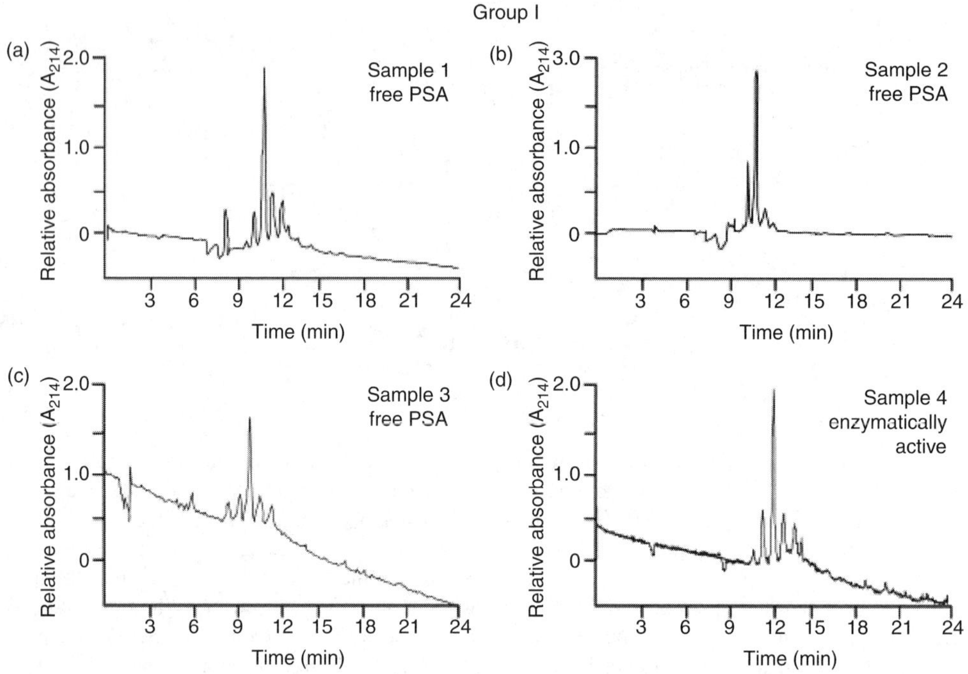

FIGURE 2.6 Analysis of PSA heterogeneity by CZE. The analytical system included a Model 310 Thermo Capillary Electrophoresis Crystal CE instrument operated in normal polarity at 25 kV with UV detection at 214 nm. Free PSA isoforms were separated on the basis of *m/z* using a 60 cm long by 50 μm ID fused-silica capillary. Separation media contained 20 mM sodium borate, 5 mM 1,3-diaminopropane, pH 8.0. Samples were pressure injected for 3 s at 200 mbar. Panels (a)–(d) show the separation of PSA isoforms in purified PSA obtained from four different commercial vendors.

isoforms, and in posttranslationally modified isoforms. Free PSA isoforms were previously reported to have pI between 5.7 and 7.2. Characterization of all structural variants/isoforms is important since modification may differentially modulate PSA biological activity and serve as biomarkers of disease.

Donohue et al. [95] developed a CZE method for the analysis of PSA and compared the profiles of seven commercially available purified PSA products. Since different manufacturers use different proprietary methods to produce purified PSA for research use, their goal was to evaluate the heterogeneity of commercially available PSA. The capillary used in these experiments was dynamically coated by adding 1,3-diaminopropane to the buffer as an EOF modifier to enhance glycoprotein separation and reproducibility. Figure 2.6 shows the heterogeneity observed in four of the seven samples analyzed. Four to seven peaks were observed with one dominant peak in each case representing 51%–70% of the total free PSA (complexed forms and genetic variants are not expected in purified samples). The dominant peak in the sample shown in panel c was not the same as the dominant peaks in panels a and b on the basis of peak retention time, which had intra-assay and interassay relative standard deviations of 0.6% and 5.0%, respectively. The author suggested that the dominant peak is most likely fucosylated disialylated diantennary PSA as demonstrated by other investigators using LC-MS. The majority of the lower-abundance isoforms had longer migration times indicating that these species are more electronegative and likely attributable to alteration of the oligosaccharide structure. Since the profiles of all seven commercially available samples differed, the authors concluded that variation in purification methods strongly influence the isoform composition and heterogeneity of purified PSA products.

2.5.4 Analysis of Histone H1.5 Isoforms by Offline Multidimensional RPHPLC-CZE

The fundamental subunit of chromatin is the nucleosome core, which consists of DNA wrapped around an octamer of core histones [96]. H1 histones are linker histones associated with the core histone–DNA complex and with the linker DNA between adjacent nucleosomes. Histone H1 phosphorylation is a function of the cell cycle, lowest in G_1 phase, rising during S and G_2, and reaching a maximum in M phase. H1 phosphorylation appears to be involved in chromatin decondensation, destabilizing chromatin structure, and weakening binding to DNA. Individual H1 subtypes differ in their degree of phosphorylation during the cell cycle.

To better understand H1 phosphorylation, Sarg et al. [96] incorporated CZE in a multidimensional analysis to show that human lymphoblastic T-cells have unambiguous site specificity for histone H1 phosphorylation. Human lymphoblastic T-cells were synchronized in culture and labeled with ^{32}P. Lyophilized perchloric acid cell extracts containing ^{32}P-labeled H1 histones ($\sim$500 μg) were first separated by reversed phase HPLC (RP-HPLC) on a Nucleosil 300-5 C4 column. Figure 2.7a shows that H1.5, the main component histone and the most highly phosphorylated during interphase and mitosis, was clearly separated from residual histones H1.2, H1.3, and H1.4. The extent of H1.5 phosphorylation was then determined by analyzing the H1.5 fraction by CZE (Figure 2.7b), which identified four peaks representing nonphosphorylated (p0), monophosphorylated (p1), diphsophorylated (p2), and triphosphorylated (p3) H1.5. A similar H1.5 sample pretreated with alkaline phosphatase showed only two peaks, a large p0 peak and a smaller p1 peak but no p2 or p3, confirming the identity of the phosphorylated isoforms. Phosphate groups are negatively charged such that their addition to H1.5 decreases surface charge and alters m/z.

2.5.5 Online Multidimensional Protein Analysis by CSE-MEKC

Multidimensional separation of proteins based on more than one physical property can greatly expand the number of proteins that can be qualitatively or quantitatively identified among the complement of proteins present in any proteomic compartment. The process can be automated by coupling multiple HPLC, CE and/or MS systems in a variety of hybrid/hyphenated combinations. Interfacing any two analytical systems online, even two different CE techniques, requires the use of compatible buffers/electrolytes and transfer of protein-containing fractions from the outlet of the first dimension to the inlet of the second dimension [59]. For complex biological samples, multidimensional analyses can be quantitative if MS is used but is at best semi-quantitative if based on absorbance or derivatized fluorescence intensity across a landscape of intersecting dimensional coordinates (e.g., pI vs. molecular weight). Compared to HPLC, CE provides exceptionally high separation efficiency but is also uniquely suited for analysis of proteins in single cells due to the very low sample volume requirement.

Harwood et al. [56] used CSE and MEKC to map protein distribution in single mouse embryos in two dimensions based on size and hydrophobicity (Figure 2.8). Although CE can be used to inject, lyse, and analyze whole cells completely online [68,69,77,97], mouse embryos are relatively large ($\sim$75 μm) and would require the use of greater diameter capillaries where Joule heating might compromise separation efficiency. Embryo extracts were therefore prepared off-line and protein and nonprotein primary amines were nonspecifically labeled with 3-(2-furoyl)quinoline-2-carboxaldehyde (FQ) before analysis by CSE-MEKC. Over 100 mouse embryo components were identified with fluorescence intensities over 10 times greater than the standard deviation of the background fluorescence.

On average, mouse embryos contained about 27 ng of protein each and yielded approximately 1 μL of lysate. In this study, approximately 3 nL (90 pg of protein or 0.3% of total protein) was injected such that analytical reproducibility could be tested on the basis of the results of multiple analyses. Data collection and processing were performed using in-house developed software that

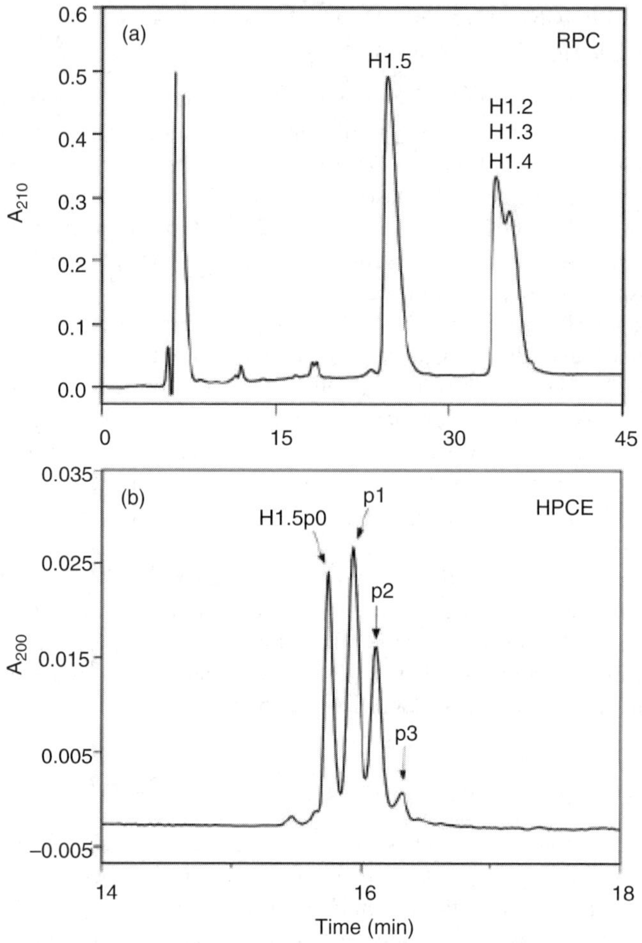

FIGURE 2.7 Isoform analysis of phosphorylated H1 histone by multidimensional HPLC-CZE. Panel (a) shows the separation of H1 histones by RP-HPLC on a Nucleosil 300-5 C4 column. Panel (b) shows the separation of the H1.5 RP-HPLC fraction by CZE. The CE analytical system included a Beckman-Coulter P/ACE 2100 instrument operated in normal polarity at 12 kV with UV detection at 200 nm and capillary cooling at 25°C. Nonphosphorylated (p0) and mono-, di-, and tri-phosphorylated (p1, p2, p3) H1.5 isoforms were separated on the basis of *m/z* using a 57 cm long (50 cm to detector) by 75 μm ID untreated fused-silica capillary. Separation media contained 0.1 M sodium phosphate, 0.02% hydroxypropylmethylcellulose, pH 2.0. Samples were injected for 2 s. (From Sarg B, et al. *J Biol Chem* 2006;281:6573–80. With permission.)

produced gel-like images or landscape images like those shown in Figure 2.8. Each analytical run consisted of approximately 200 cycles where fluorescence intensity was measured between 0 and 17 s as 200 different CSE fractions were analyzed by MEKC. The reproducibility of the separations in replicate analyses were determined in the first and second dimensions by comparing cycle number and seconds, respectively, for 50 sample constituents with the highest fluorescence intensities. The average standard deviations were 4.7 cycles and 0.30 s in the CSE and MEKC dimensions, respectively.

This example shows that CSE-MEKC can separate and detect over 100 different cellular components in a single mouse embryo and could potentially be used to semiquantitatively compare protein expression levels in individual cells from a heterogeneous cell population. Though nonspecific, the use of fluorescence derivatization and detection greatly increased the sensitivity and dynamic range

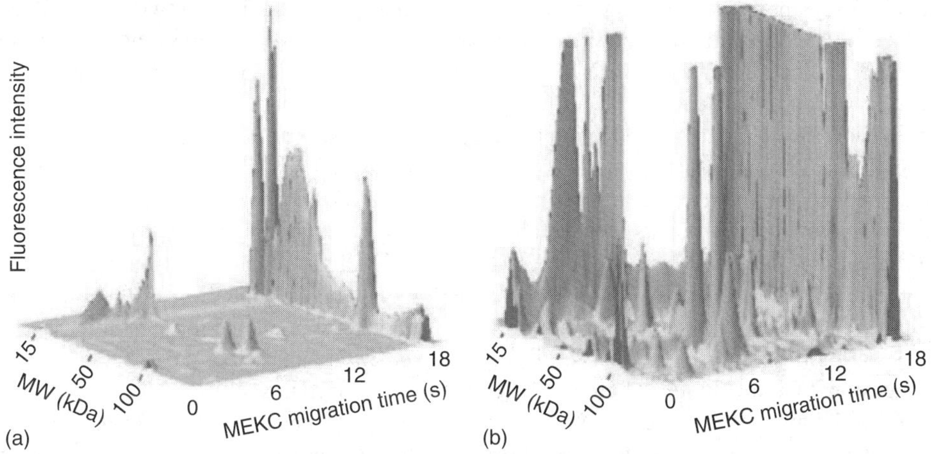

FIGURE 2.8 2D CSE-MEKC separation of proteins in a single mouse embryo. The analytical system included an in-house constructed multidimensional CE system operated in reverse polarity with postcolumn laser-induced fluorescence detection in a sheath-flow cuvette (473 nm excitation, 580 nm emission). Both CSE and MEKC separations used 20 cm long by 30 μm ID fused-silica capillaries dynamically coated with EoTrol LN (Target Discoveries, Palo Alto, CA) to stabilize normal EOF. Samples were injected at the inlet of the CSE capillary for 5 s at 11 kPa using negative pressure. CSE was performed at 1000 V/cm in separation media containing 100 mM Tris, 3.5 mM SDS and 5% dextran polymer (513 kDa) at pH 8.7. Analytes were transferred sequentially into the second capillary where MEKC was performed at 900 V/cm in separation media containing 100 mM CHES, 100 mM Tris and 15 mM SDS at pH 8.7. The net electric field across the first capillary was held at 0 V/cm during analyte separation in the second capillary, a process that was repeated for approximately 200 cycles where each cycle was composed of a 1 s transfer and a 17 s separation in the second dimension. Data collection and processing were performed using in-house developed software that can produce the landscape images shown here at full scale (a) or ten-times expanded scale (b). (From Harwood MM, et al. *J Chromatogr A* 2006;1130:190–4. With permission.)

of protein detection. Multidimensional resolution was better than that attainable in a single CE dimension, and was perhaps similar to that attainable using conventional 2D SDS–PAGE. However, analyzing the protein content of embryos by conventional methods like 2D SDS–PAGE and tandem MS requires pooling extracts from many cells in order to acquire enough sample for analysis. Such results only reflect average protein expression levels rather than expression levels in individual cells as described in the example presented here.

Proteomic analysis of individual cells or specific tissues has important implications for diagnosis of disease since comparing normal and diseased cells can help identify proteomic phenotypes characteristic of disease. For example, Kraly et al. [98] used CSE-MEKC to analyze tissue homogenates from esophageal and stomach biopsy samples collected from patients with Barrett's Syndrome. They identified 18 features from the homogenate profiles as biogenic amines and amino acids that differed in diseased and normal tissues. The results suggest that two-dimensional CE might be useful for rapid characterization of endoscopic and surgical biopsies for this and other diseases.

2.6 CONCLUDING REMARKS

The probability that all chemically distinct proteins in the vast human proteome will be identified and characterized any time soon is vanishingly small. However, research and development toward this goal can only improve the quality and quantity of proteomic data that is available and the ability of biomedical science to use that information to improve the human condition. CE appears poised to play a significant role in the collection of proteomic data for use in both top-down and bottom-up systems biology approaches to biomarker discovery. Electrophoresis in narrow-bore capillaries is capable of extremely high efficiency protein separations based on one or more protein properties. Some of the application examples included in this chapter show how one-dimensional CE techniques can be used to identify and quantify specific protein isoforms or study their physical properties. Other examples show how CE can be used in multidimensional analyses to separate proteins on the basis of multiple physical properties. The efficiency of multidimensional protein separations is approaching the scale needed for relatively comprehensive analysis of the more abundant proteins present in discrete proteomic compartments like the plasma proteome or individual cell proteomes. Many of the reproducibility, sensitivity and specificity issues traditionally associated with analysis of proteins in a capillary format have been overcome by technological advances in CE instrument and method development. MS is and will likely remain the detection method of choice for most microanalytical protein separation techniques due to its high degree of sensitivity and selectivity; characteristics that one-dimensional CE techniques often lack when it comes to protein analysis.

Commercial availability of CE-MS analytical systems has opened the door for high efficiency, multidimensional protein analysis to those who can afford it. Commercialization of multidimensional systems that combine multiple CE techniques, or CE with HPLC, could lead to even greater ability to separate, identify and quantify proteins in complex biological samples. However, multidimensional map coordinates of the distribution of proteins based on charge, size, and/or hydrophobicity provide relatively low quality data because complex biological samples simply contain too many proteins with similar multidimensional properties. Proteomic phenotyping based on limited protein information can only be used to divide individuals in a population into relatively indistinct groups and provides little or no mechanistic information. Higher quality proteome data quantifies specific proteins, as many as possible, and can be used to quantitatively group individuals on the basis of discrete patterns in the proteomic distribution of chemically distinct proteins.

Understanding and classifying biological variation in the vast human proteome is a major undertaking and a challenge that nature and systems biology have presented to biomedical and analytical science. Capacity to cost-effectively and simultaneously profile the concentrations of many chemically distinct proteins in selected proteomes is what biomedical science most urgently needs for biomarker discovery and management of complex diseases. By design or by good fortune, analytical systems for protein analysis by CE are well-positioned to advance in both bottom-up and top-down approaches to the collection of proteomic data. High-throughput microchip and multicapillary CE instruments, and high-resolution multidimensional systems than incorporate CE, can greatly expand both the quantity and quality of protein data that can be acquired. What remains to be determined is how well CE will compete with other advanced protein separation technologies or assay techniques whose development will also be influenced by the emergence of systems biology and biomarker discovery as driving forces in the evolution of analytical instrumentation.

ACKNOWLEDGMENT

The author would like to thank Dr. Mark Richards for his kind and considered review of the manuscript for this chapter.

REFERENCES

1. Riekkola ML, Jonsson L, Smith RM. Terminology for analytical capillary electromigration techniques. *Pure Appl Chem* 2004;76:443–51.
2. Dolnik V, Hutterer KM. Capillary electrophoresis of proteins 1999–2001. *Electrophoresis* 2001; 22:4163–78.
3. Hutterer K, Dolnik V. Capillary electrophoresis of proteins 2001–2003. *Electrophoresis* 2003;24: 3998–4012.

4. Dolnik V. Capillary electrophoresis of proteins 2003–2005. *Electrophoresis* 2006;27:126–41.
5. Huang YF, Huang CC, Hu CC, Chang HT. Capillary electrophoresis-based separation techniques for the analysis of proteins. *Electrophoresis* 2006;27:3503–22.
6. Klampfl CW. Recent advances in the application of capillary electrophoresis with mass spectrometric detection. *Electrophoresis* 2006;27:3–34.
7. Kasicka V. Recent advances in capillary electrophoresis and capillary electrochromatography of peptides. *Electrophoresis* 2003;24:4013–46.
8. Kilar F. Recent applications of capillary isoelectric focusing. *Electrophoresis* 2003;24:3908–16.
9. Eeltink S, Kok WT. Recent applications in capillary electrochromatography. *Electrophoresis* 2006;27:84–96.
10. Schou C, Heegaard NH. Recent applications of affinity interactions in capillary electrophoresis. *Electrophoresis* 2006;27:44–59.
11. Babu CV, Song EJ, Babar SM, Wi MH, Yoo YS. Capillary electrophoresis at the omics level: towards systems biology. *Electrophoresis* 2006;27:97–110.
12. Cooper JW, Wang Y, Lee CS. Recent advances in capillary separations for proteomics. *Electrophoresis* 2004;25:3913–26.
13. Kamoda S, Kakehi K. Capillary electrophoresis for the analysis of glycoprotein pharmaceuticals. *Electrophoresis* 2006;27:2495–504.
14. Little MJ, Paquette DM, Roos PK. Electrophoresis of pharmaceutical proteins: status quo. *Electrophoresis* 2006;27:2477–85.
15. Simo C, Barbas C, Cifuentes A. Capillary electrophoresis-mass spectrometry in food analysis. *Electrophoresis* 2005;26:1306–18.
16. Lookhart GL, Bean SR. Capillary electrophoresis of cereal proteins: an overview. *J Capillary Electrophor* 2004;9:23–30.
17. Frazier RA, Papadopoulou A. Recent advances in the application of capillary electrophoresis for food analysis. *Electrophoresis* 2003;24:4095–105.
18. Jabeen R, Payne D, Wiktorowcz J, Mohammad A, Petersen J. Capillary electrophoresis and the clinical laboratory. *Electrophoresis* 2006;27:2413–38.
19. Fliser D, Wittke S, Mischak H. Capillary electrophoresis coupled to mass spectrometry for clinical diagnostic purposes. *Electrophoresis* 2005;26:2708–16.
20. Kolch W, Neususs C, Pelzing M, Mischak H. Capillary electrophoresis-mass spectrometry as a powerful tool in clinical diagnosis and biomarker discovery. *Mass Spectrom Rev* 2005;24: 959–77.
21. Bossuyt X, Lissoir B, Marien G, Maisin D, Vunckx J, Blanckaert N, Wallemacq P. Automated serum protein electrophoresis by Capillarys. *Clin Chem Lab Med* 2003;41:704–10.
22. Petersen JR, Okorodudu AO, Mohammad A, Payne DA. Capillary electrophoresis and its application in the clinical laboratory. *Clin Chim Acta* 2003;330:1–30.
23. Landers JP. Molecular diagnostics on electrophoretic microchips. *Anal Chem* 2003;75:2919–27.
24. Dolnik V, Liu S. Applications of capillary electrophoresis on microchip. *J Sep Sci* 2005; 28:1994–2009.
25. Fortina P, Surrey S, Kricka LJ. Molecular diagnostics: hurdles for clinical implementation. *Trends Mol Med* 2002;8:264–6.
26. McLaren DG, Chen DD. A quantitative study of continuous flow-counterbalanced capillary electrophoresis for sample purification. *Electrophoresis* 2003;24:2887–95.
27. Lacroix M, Poinsot V, Fournier C, Couderc F. Laser-induced fluorescence detection schemes for the analysis of proteins and peptides using capillary electrophoresis. *Electrophoresis* 2005;26:2608–21.
28. Sanchez-Carbayo M. Antibody arrays: technical considerations and clinical applications in cancer. *Clin Chem* 2006;52:1651–9.
29. Bruggeman FJ, Westerhoff HV. The nature of systems biology. *Trends Microbiol* 2007;15:45–50.
30. Prentice H, Webster KA. Genomic and proteomic profiles of heart disease. *Trends Cardiovasc Med* 2004;14:282–8.
31. Bloom FE. What does it all mean to you? *J Neurosci* 2001;21:8304–5.
32. Muddiman DC. "Lewis and Clark" proteomics. *J Proteome Res* 2006;5:221–2.
33. Bunn HF, Forget BG. *Hemoglobin: Molecular, Genetic and Clinical Aspects*. Philadelphia, PA: W.B. Saunders Company, 1986.

34. Patrinos GR, Kollia P, Papadakis MN. Molecular diagnosis of inherited disorders: Lessons from hemoglobinopathies. *Hum Mutat* 2005;26:399–412.
35. Fu Y, Lee AS. Glucose regulated proteins in cancer progression, drug resistance and immunotherapy. *Cancer Biol Ther* 2006;5:741–4.
36. Becker JC, Kirkwood JM, Agarwala SS, Dummer R, Schrama D, Hauschild A. Molecularly targeted therapy for melanoma: current reality and future options. *Cancer* 2006;107:2317–27.
37. Zimmermann GR, Lehar J, Keith CT. Multi-target therapeutics: when the whole is greater than the sum of the parts. *Drug Discov Today* 2007;12:34–42.
38. Nadeau JH. Genetics. Modifying the message. *Science* 2003;301:927–8.
39. Nadeau JH. Listening to genetic background noise. *N Engl J Med* 2005;352:1598–9.
40. Anderson NL, Anderson NG. The human plasma proteome: history, character, and diagnostic prospects. *Mol Cell Proteomics* 2002;1:845–67.
41. Mosher RA, Dewey D, Thormann W, Saville DA, Bier M. Computer simulation and experimental validation of the electrophoretic behavior of proteins. *Anal Chem* 1989;61:362–6.
42. Pritchett T, Robey FA. Capillary electrophoresis of proteins. In: Landers JP, ed. *Handbook of Capillary Electrophoresis*. Boca Raton, FL: CRC Press, 1997:260–90.
43. He Y, Yeung ES. Rapid determination of protein molecular weight by the Ferguson method and multiplexed capillary electrophoresis. *J Proteome Res* 2002;1:273–7.
44. Karim MR, Janson JC, Takagi T. Size-dependent separation of proteins in the presence of sodium dodecyl sulfate and dextran in capillary electrophoresis: effect of molecular weight of dextran. *Electrophoresis* 1994;15:1531–4.
45. Guttman A. Gel and polymer-solution mediated separation of biopolymers by capillary electrophoresis. *J Chromatogr Sci* 2003;41:449–59.
46. Liu Z, Lemma T, Pawliszyn J. Capillary isoelectric focusing coupled with dynamic imaging detection: A one-dimensional separation for two-dimensional protein characterization. *J Proteome Res* 2006;5:1246–51.
47. Kuroda Y, Yukinaga H, Kitano M, Noguchi T, Nemati M, Shibukawa A, Nakagawa T, Matsuzaki K. On-line capillary isoelectric focusing-mass spectrometry for quantitative analysis of peptides and proteins. *J Pharm Biomed Anal* 2005;37:423–8.
48. Graf M, Watzig H. Capillary isoelectric focusing—reproducibility and protein adsorption. *Electrophoresis* 2004;25:2959–64.
49. Martinovic S, Pasa-Tolic L, Smith RD. Capillary isoelectric focusing—mass spectrometry of proteins and protein complexes. *Methods Mol Biol* 2004;276:291–304.
50. Storms HF, van der HR, Tjaden UR, van der GJ. Capillary isoelectric focusing-mass spectrometry for shotgun approach in proteomics. *Electrophoresis* 2004;25:3461–7.
51. Shimura K. Recent advances in capillary isoelectric focusing: 1997–2001. *Electrophoresis* 2002;23:3847–57.
52. Righetti PG, Gelfi C, Bossi A, Olivieri E, Castelletti L, Verzola B, Stoyanov AV. Capillary electrophoresis of peptides and proteins in isoelectric buffers: an update. *Electrophoresis* 2000;21:4046–53.
53. Holland LA, Leigh AM. Bilayered phospholipid micelles and capillary electrophoresis: a new additive for electrokinetic chromatography. *Electrophoresis* 2003;24:2935–9.
54. Maichel B, Kenndler E. Recent innovation in capillary electrokinetic chromatography with replaceable charged pseudostationary phases or additives. *Electrophoresis* 2000;21:3160–73.
55. Roman GT, McDaniel K, Culbertson CT. High efficiency micellar electrokinetic chromatography of hydrophobic analytes on poly(dimethylsiloxane) microchips. *Analyst* 2006;131:194–201.
56. Harwood MM, Christians ES, Fazal MA, Dovichi NJ. Single-cell protein analysis of a single mouse embryo by two-dimensional capillary electrophoresis. *J Chromatogr A* 2006;1130:190–4.
57. Michels DA, Hu S, Dambrowitz KA, Eggertson MJ, Lauterbach K, Dovichi NJ. Capillary sieving electrophoresis-micellar electrokinetic chromatography fully automated two-dimensional capillary electrophoresis analysis of Deinococcus radiodurans protein homogenate. *Electrophoresis* 2004;25:3098–105.
58. Betgovargez E, Knudson V, Simonian MH. Characterization of proteins in the human serum proteome. *J Biomol Tech* 2005;16:306–10.

59. Stroink T, Ortiz MC, Bult A, Lingeman H, de Jong GJ, Underberg WJ. On-line multidimensional liquid chromatography and capillary electrophoresis systems for peptides and proteins. *J Chromatogr B Analyt Technol Biomed Life Sci* 2005;817:49–66.
60. Luo S, Feng J, Pang HM. High-throughput protein analysis by multiplexed sodium dodecyl sulfate capillary gel electrophoresis with UV absorption detection. *J Chromatogr A* 2004;1051:131–4.
61. Cifuentes A, Poppe H. Rectangular capillary electrophoresis: some theoretical considerations. *Chromatographia* 1994;39:391–404.
62. Wu J, Wu XZ, Huang T, Pawliszyn J. Analysis of proteins by CE, CIEF, and microfluidic devices with whole-column-imaging detection. *Methods Mol Biol* 2004;276:229–52.
63. Wu XZ, Pawliszyn J. Whole-column fluorescence-imaged capillary electrophoresis. *Electrophoresis* 2004;25:3820–4.
64. Pasas SA, Lacher NA, Davies MI, Lunte SM. Detection of homocysteine by conventional and microchip capillary electrophoresis/electrochemistry. *Electrophoresis* 2002;23:759–66.
65. Kuijt J, van Teylingen R, Nijbacker T, Ariese F, Brinkman UA, Gooijer C. Detection of nonderivatized peptides in capillary electrophoresis using quenched phosphorescence. *Anal Chem* 2001;73:5026–9.
66. Hapuarachchi S, Janaway GA, Aspinwall CA. Capillary electrophoresis with a UV light-emitting diode source for chemical monitoring of native and derivatized fluorescent compounds. *Electrophoresis* 2006;27:4052–9.
67. Sluszny C, He Y, Yeung ES. Light-emitting diode-induced fluorescence detection of native proteins in capillary electrophoresis. *Electrophoresis* 2005;26:4197–203.
68. Yeung ES. Study of single cells by using capillary electrophoresis and native fluorescence detection. *J Chromatogr A* 1999;830:243–62.
69. Dovichi NJ, Hu S. Chemical cytometry. *Curr Opin Chem Biol* 2003;7:603–8.
70. Underberg WJ, Waterval JC. Derivatization trends in capillary electrophoresis: an update. *Electrophoresis* 2002;23:3922–33.
71. Kremser L, Petsch M, Blaas D, Kenndler E. Capillary electrophoresis of affinity complexes between subviral 80S particles of human rhinovirus and monoclonal antibody 2G2. *Electrophoresis* 2006;27:2630–7.
72. Veledo MT, de Frutos M, Diez-Masa JC. Development of a method for quantitative analysis of the major whey proteins by capillary electrophoresis with on-capillary derivatization and laser-induced fluorescence detection. *J Sep Sci* 2005;28:935–40.
73. Zinellu A, Sotgia S, Usai MF, Chessa R, Deiana L, Carru C. Thiol redox status evaluation in red blood cells by capillary electrophoresis-laser induced fluorescence detection. *Electrophoresis* 2005;26:1963–8.
74. Welder F, McCorquodale EM, Colyer CL. Proteinase assay by capillary electrophoresis employing fluorescence-quenched protein-dye conjugates. *Electrophoresis* 2002;23:1585–90.
75. Presley AD, Fuller KM, Arriaga EA. MitoTracker Green labeling of mitochondrial proteins and their subsequent analysis by capillary electrophoresis with laser-induced fluorescence detection. *J Chromatogr B Analyt Technol Biomed Life Sci* 2003;793:141–50.
76. Stoyanov AV, Ahmadzadeh H, Krylov SN. Heterogeneity of protein labeling with a fluorogenic reagent, 3-(2-furoyl)quinoline-2-carboxaldehyde. *J Chromatogr B Analyt Technol Biomed Life Sci* 2002;780:283–7.
77. Zhang H, Jin W. Single-cell analysis by intracellular immuno-reaction and capillary electrophoresis with laser-induced fluorescence detection. *J Chromatogr A* 2006;1104:346–51.
78. Sowell J, Salon J, Strekowski L, Patonay G. Covalent and noncovalent labeling schemes for near-infrared dyes in capillary electrophoresis protein applications. *Methods Mol Biol* 2004;276:39–75.
79. Heegaard NH, Nilsson S, Guzman NA. Affinity capillary electrophoresis: important application areas and some recent developments. *J Chromatogr B Biomed Sci Appl* 1998;715:29–54.
80. Schultz NM, Huang L, Kennedy RT. Capillary electrophoresis-based immunoassay to determine insulin content and insulin secretion from single islets of Langerhans. *Anal Chem* 1995;67:924–9.
81. Simpson DC, Smith RD. Combining capillary electrophoresis with mass spectrometry for applications in proteomics. *Electrophoresis* 2005;26:1291–305.
82. Stutz H. Advances in the analysis of proteins and peptides by capillary electrophoresis with matrix-assisted laser desorption/ionization and electrospray-mass spectrometry detection. *Electrophoresis* 2005;26:1254–90.

83. Eriksson KO, Palm A, Hjerten S. Preparative capillary electrophoresis based on adsorption of the solutes (proteins) onto a moving blotting membrane as they migrate out of the capillary. *Anal Biochem* 1992;201:211–5.
84. Kaneta T, Inoue J, Koizumi M, Imasaka T. On-column capture of a specific protein in capillary electrophoresis using magnetic beads. *Electrophoresis* 2006;27:3218–23.
85. Sharma U, Carbeck JD. Hydrodynamic radius ladders of proteins. *Electrophoresis* 2005;26:2086–91.
86. Ebersold MF, Zydney AL. Use of protein charge ladders to study electrostatic interactions during protein ultrafiltration. *Biotechnol Bioeng* 2004;85:166–76.
87. Hempe JM, Vargas A, Craver RD. Clinical analysis of structural hemoglobin variants and Hb A1c by capillary isoelectric focusing. In: Petersen JR, Mohammad AA, eds. *Clinical and Forensic Applications of Capillary Electrophoresis*. Totowa, NJ: Humana Press, 2001:145–63.
88. Hempe JM, Craver RD. Laboratory diagnosis of structural hemoglobinopathies and thalassemias by capillary isoelectric focusing. In: Palfrey SM, ed. *Clinical Applications of Capillary Electrophoresis*. Totowa, NJ: Humana Press, 1999:81–98.
89. Biroccio A, Urbani A, Massoud R, di Ilio C, Sacchetta P, Bernardini S, Cortese C, Federici G. A quantitative method for the analysis of glycated and glutathionylated hemoglobin by matrix-assisted laser desorption ionization-time of flight mass spectrometry. *Anal Biochem* 2005;336:279–88.
90. Pastore A, Francesca MA, Tozzi G, Maria GL, Federici G, Bertini E, Lo Russo A, Mannucci L, Piemonte F. Determination of glutathionyl-hemoglobin in human erythrocytes by cation-exchange high-performance liquid chromatography. *Anal Biochem* 2003;312:85–90.
91. Takayama F, Tsutsui S, Horie M, Shimokata K, Niwa T. Glutathionyl hemoglobin in uremic patients undergoing hemodialysis and continuous ambulatory peritoneal dialysis. *Kidney Int Suppl* 2001;78:S155–8.
92. Niwa T, Naito C, Mawjood AH, Imai K. Increased glutathionyl hemoglobin in diabetes mellitus and hyperlipidemia demonstrated by liquid chromatography/electrospray ionization-mass spectrometry. *Clin Chem* 2000;46:82–8.
93. Jakus V, Rietbrock N. Advanced glycation end-products and the progress of diabetic vascular complications. *Physiol Res* 2004;53:131–42.
94. Ronai Z, Guttman A, Keszler G, Sasvari-Szekely M. Capillary electrophoresis study on DNA–protein complex formation in the polymorphic 5′ upstream region of the dopamine D4 receptor (DRD4) gene. *Curr Med Chem* 2004;11:1023–9.
95. Donohue MJ, Satterfield MB, Dalluge JJ, Welch MJ, Girard JE, Bunk DM. Capillary electrophoresis for the investigation of prostate-specific antigen heterogeneity. *Anal Biochem* 2005;339:318–27.
96. Sarg B, Helliger W, Talasz H, Forg B, Lindner HH. Histone H1 phosphorylation occurs site-specifically during interphase and mitosis: identification of a novel phosphorylation site on histone H1. *J Biol Chem* 2006;281:6573–80.
97. Woods LA, Roddy TP, Ewing AG. Capillary electrophoresis of single mammalian cells. *Electrophoresis* 2004;25:1181–7.
98. Kraly JR, Jones MR, Gomez DG, Dickerson JA, Harwood MM, Eggertson M, Paulson TG, et al. Reproducible two-dimensional capillary electrophoresis analysis of Barrett's esophagus tissues. *Anal Chem* 2006;78:5977–86.
99. Suresh Babu CV, Cho SG, Yoo YS. Method development and measurements of endogenous serine/threonine Akt phosphorylation using capillary electrophoresis for systems biology. *Electrophoresis* 2005;26:3765–72.
100. Liu X, Liu X, Liang A, Shen Z, Zhang Y, Dai Z, Xiong B, Lin B. Studying protein–drug interaction by microfluidic chip affinity capillary electrophoresis with indirect laser-induced fluorescence detection. *Electrophoresis* 2006;27:3125–8.
101. Martinez-Gomez MA, Sagrado S, Villanueva-Camanas RM, Medina-Hernandez MJ. Characterization of basic drug-human serum protein interactions by capillary electrophoresis. *Electrophoresis* 2006;27:3410–9.
102. Wu XZ, Zhang LH, Onoda K. Isoelectric focusing sample injection for capillary electrophoresis of proteins. *Electrophoresis* 2005;26:563–70.
103. McKeon J, Holland LA. Determination of dissociation constants for a heparin-binding domain of amyloid precursor protein and heparins or heparan sulfate by affinity capillary electrophoresis. *Electrophoresis* 2004;25:1243–8.

104. Veledo MT, de Frutos M, Diez-Masa JC. Analysis of trace amounts of bovine beta-lactoglobulin in infant formulas by capillary electrophoresis with on-capillary derivatization and laser-induced fluorescence detection. *J Sep Sci* 2005;28:941–7.
105. Laing TD, Marenco AJ, Moore DM, Moore GJ, Mah DC, Lee WE. Capillary electrophoresis laser-induced fluorescence for screening combinatorial peptide libraries in assays of botulinum neurotoxin A. *J Chromatogr B Analyt Technol Biomed Life Sci* 2006;843:240–6.
106. Balaguer E, Demelbauer U, Pelzing M, Sanz-Nebot V, Barbosa J, Neususs C. Glycoform characterization of erythropoietin combining glycan and intact protein analysis by capillary electrophoresis—electrospray—time-of-flight mass spectrometry. *Electrophoresis* 2006;27:2638–50.
107. Neususs C, Demelbauer U, Pelzing M. Glycoform characterization of intact erythropoietin by capillary electrophoresis-electrospray-time of flight-mass spectrometry. *Electrophoresis* 2005;26:1442–50.
108. Kinoshita M, Murakami E, Oda Y, Funakubo T, Kawakami D, Kakehi K, Kawasaki N, Morimoto K, Hayakawa T. Comparative studies on the analysis of glycosylation heterogeneity of sialic acid-containing glycoproteins using capillary electrophoresis. *J Chromatogr A* 2000;866:261–71.
109. Hempe JM, Ory-Ascani J, Hsia D. Genetic variation in mouse beta globin cysteine content modifies glutathione metabolism: implications for the use of mouse models. *Exp Biol Med* 2007;232:437–44.
110. Shihabi ZK, Hinsdale ME. Simplified hemoglobin chain detection by capillary electrophoresis. *Electrophoresis* 2005;26:581–5.
111. Fu H, Guthrie JW, Le XC. Study of binding stoichiometries of the human immunodeficiency virus type 1 reverse transcriptase by capillary electrophoresis and laser-induced fluorescence polarization using aptamers as probes. *Electrophoresis* 2006;27:433–41.
112. Shimura K, Waki T, Okada M, Toda T, Kimoto I, Kasai K. Analysis of protein–protein interactions with a multi-capillary electrophoresis instrument. *Electrophoresis* 2006;27:1886–94.
113. Montes-Bayon M, Profrock D, Sanz-Medel A, Prange A. Direct comparison of capillary electrophoresis and capillary liquid chromatography hyphenated to collision-cell inductively coupled plasma mass spectrometry for the investigation of Cd-, Cu- and Zn-containing metalloproteins. *J Chromatogr A* 2006;1114:138–44.
114. Babu CV, Chung BC, Lho DS, Yoo YS. Capillary electrophoretic competitive immunoassay with laser-induced fluorescence detection for methionine-enkephalin. *J Chromatogr A* 2006;1111:133–8.
115. Jackman R, Everest DJ, Schmerr MJ, Khawaja M, Keep P, Docherty J. Evaluation of a preclinical blood test for scrapie in sheep using immunocapillary electrophoresis. *J AOAC Int* 2006;89:720–7.
116. Li G, Zhou X, Wang Y, El Shafey A, Chiu NH, Krull IS. Capillary isoelectric focusing and affinity capillary electrophoresis approaches for the determination of binding constants for antibodies to the prion protein. *J Chromatogr A* 2004;1053:253–62.
117. Roudiere L, Boularan AM, Bonardet A, Vallat C, Cristol JP, Dupuy AM. Evaluation of a capillary zone electrophoresis system versus a conventional agarose gel system for routine serum protein separation and monoclonal component typing. *Clin Lab* 2006;52:19–27.
118. Garcia-Ruiz C, Garcia MA, Garcia MC, Marina ML. Development of a capillary electrophoresis method for the determination of soybean proteins in soybean-rice gluten-free dietary products. *Electrophoresis* 2006;27:452–60.
119. Arizaga RS, Blanco GE, Alvarez LG, Montes-Bayon M, Sanz-Medel A. Detection of transferrin isoforms in human serum: comparison of UV and ICP-MS detection after CZE and HPLC separations. *Anal Bioanal Chem* 2005;383:390–7.
120. Joneli J, Lanz C, Thormann W. Capillary zone electrophoresis determination of carbohydrate-deficient transferrin using the new CEofix reagents under high-resolution conditions. *J Chromatogr A* 2006;1130:272–80.
121. Lanz C, Kuhn M, Deiss V, Thormann W. Improved capillary electrophoresis method for the determination of carbohydrate-deficient transferrin in patient sera. *Electrophoresis* 2004;25:2309–18.

3 Micellar Electrokinetic Chromatography

Shigeru Terabe

CONTENTS

3.1 INTRODUCTION

Electrophoresis is a separation technique based on the migration of the analyte in a solution under the influence of electric field. In electrophoretic separation it is essential for the analyte to be charged or ionic. The migration velocity of the analyte is primarily depends on the charge and the size of the analyte under a homogeneous electric field but it can also be modified by several chemical or physical interactions between the analyte and the electrophoretic media including additives or

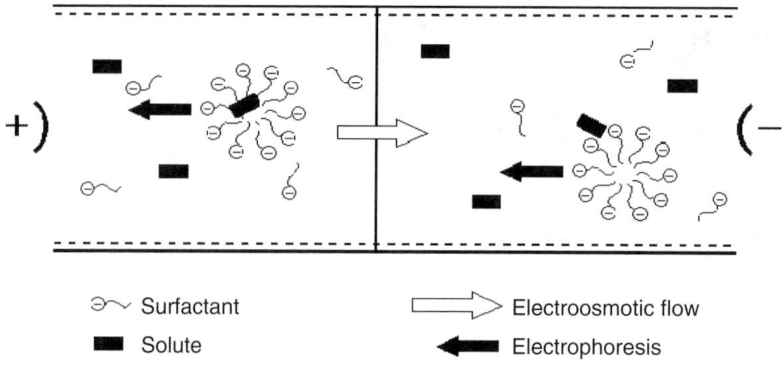

FIGURE 3.1 Schematic illustration of the principle of MEKC.

polymer networks. Neutral molecules were not considered to be separated by electrophoresis until micellar electrokinetic chromatography (MEKC) was developed by Terabe et al. in 1984 [1]. The idea using ionic micelles in electrophoresis was suggested by Nakagawa in 1981 as described by Terabe et al. [1]. The neutral analyte will gain an apparent electrophoretic mobility when it is incorporated into the ionic micelle and will migrate at the same velocity as the micelle under electrophoretic conditions. Since the distribution equilibrium of the analyte between the micelle and the surrounding aqueous phase can be quickly established, the apparent electrophoretic mobility or migration velocity is directly related to the distribution coefficient. The more the analyte is incorporated, the higher the mobility. The analyte also migrates by electroosmotic flow (EOF) under capillary electrophoresis (CE) conditions, although EOF does not contribute to the separation at all. Thus, the migration time in MEKC is a function of electrophoretic velocity of the micelle, distribution ratio, and EOF velocity. A schematic principle of MEKC is shown in Figure 3.1, where EOF is stronger than the electrophoretic migration of the anionic micelle and hence the anionic micelle migrates toward cathode at a retarded velocity. The neutral analyte migrates at the velocity between EOF velocity and that of the micelle and the migration velocity depends on how much fraction of the analyte is incorporated into the micelle. It should be noted that there is optimum distribution of the analyte to the micelle for good separation, that is, totally incorporated analyte by the micelle or free from the micelle cannot be separated by MEKC. MEKC introduced the chromatographic principle into CE by adding ionic micelles into the electrophoretic solution. MEKC is a branch of CE and extended significantly the applicability of CE to a wide range of analytes, particularly small molecules. Since MEKC also belongs to a chromatographic technique, it is familiar for most chromatographers to design separation conditions.

MEKC can be easily performed just by adding ionic micelles to the electrophoretic solution or background electrolyte solution in capillary zone electrophoresis (CZE), which is the simplest separation mode of CE, without any instrumental modification. The ionic micelle added to the background solution (BGS) is called pseudostationary phase because it plays a role as stationary phase in chromatography, although it migrates inside the capillary. It should be noted that micelles are classified as a pseudophase because the micelles exist only in equilibrium with surfactant molecules in solution and cannot be isolated. The idea of separation of neutral analytes by electrophoresis was extended to the use of other pseudostationary phases such as microemulsions, liposomes, charged cyclodextrins (CDs), dendrimers, charged polymers, and so forth, and the technique can be generally named as EKC [2].

Although MEKC has been widely accepted as a CE technique for the separation of neutral analytes, it is also a powerful technique in improving selectivity of charged analytes in CE. MEKC is almost 25 years old and popular among separation scientists as pointed out by a large number of papers published, more than 2600 by 2005 [3], and coverage of the technique in the first edition of this handbook. Therefore, this chapter does not describe much about the fundamental characteristics of MEKC but the main target is to give the guiding principles for full availability of the technique.

There are many books, book chapters, or review articles on fundamental MEKC [4–11]. In particular, online sample preconcentration techniques are becoming popular in CZE to improve concentration detection sensitivity and so are in MEKC, which will be introduced rather in detail. Although it is generally stated that CE requires minimal amounts of sample (certainly true), it should be mentioned that the volume of sample solution required is much greater than that actually introduced into the capillary because the injection end of the capillary must be dipped into the sample solution—this is typically at least a few microliters.

3.2 BACKGROUND

3.2.1 Instrumentation

No special modification of conventional CE instrumentation is required to perform MEKC. However, the micellar solutions, prepared by dissolving a surfactant into a buffer solution, have higher electric conductivity than the conventional CZE buffers and the current is usually high in MEKC. Therefore, care must be taken to apply voltages that avoid excessive current (e.g., preferably less than 50 μA). Otherwise, the separation efficiency will be deteriorated due to high Joule heating. Temperature of capillary should be controlled higher than the Krafft point of the surfactant, which is the temperature where the surfactant solubility increases rapidly with an increase in temperature. Krafft points change depending on the solution, for example, the Krafft point of sodium dodecyl sulfate (SDS) is 15°C in pure water, but it will be lower in buffer solutions.

Native fused-silica capillary with an inside diameter (ID) of 50 or 75 μm is generally used in MEKC. To suppress EOF acidic conditions are employed. The use of surface-treated capillaries, such as polyacrylamide or neutral coating, is effective to suppress EOF. Charged surface coating such as successive multiple ionic layer (SMIL) coating [12] increases EOF even under acidic conditions or reverses the direction of EOF. For example, the capillary is first coated with polybrene after rinsing the surface followed by the coating with dextran sulfate, producing stable negatively charged surface even at low pH. An additional third coating of polybrene produces stable positively charged surface.

Photometric detectors are the most popular in CE instruments including diode array detectors. Laser-induced fluorescence (LIF) detection and electric conductivity detectors are also popular. LIF is particularly sensitive and powerful for detecting low concentration analytes. However, most analytes are not natively fluorescent and some derivatizations are necessary. Conductivity detector is useful for the detection of non-ultraviolet (non-UV) absorbing analytes such as inorganic ions or fatty acids. Both LIF detection and conductivity detectors are commercially available and easy to interface with conventional CE instruments. Electrochemical detectors are also useful for selective high-sensitivity detection. Several techniques have been developed to circumvent the problem of strong effects of electrophoretic field on electrochemical detection, but despite this, commercial electrochemical detectors are not used extensively.

3.2.2 Mass Spectrometric Detection

Mass spectrometry (MS) is becoming an indispensable detection method in every separation analysis. High-performance liquid chromatography (HPLC)-MS is now a practical and robust technique with electrospray ionization (ESI) or atmospheric pressure chemical ionization (APCI) interface to generate analyte ions directly from the liquid phase. The same instrument can be utilized for CE-MS with minor modifications. Since the liquid flow rate is significantly different between HPLC and CE, some modifications are necessary; to keep the supply of the liquid to ESI interface a sheath liquid is added through the coaxial ESI spray nozzle, which is also helpful to keep electrical connection for CE run. A mixture of methanol and water-containing acetic acid or formic acid is usually used as a sheath liquid at a flow rate around few μL/min in the positive ESI mode. The ESI voltage and CE voltage are simultaneously applied at the spray nozzle but the CE current and ESI current are not the same and the mismatch may cause some troubles interrupting current due to bubble formation or unstable CE current. It is preferable to maintain the BGS at a low conductivity to minimize the current mismatch. In some instruments, the spray nozzle can be grounded, thus solving the problem. Since the presence of a high concentration of salts adversely affect the ionization efficiency in ESI, volatile electrolytes such as ammonium formate or acetate are widely employed in CE buffer, which minimizes contamination of the ion source. However, addition of the surfactant to BGS is essential in MEKC, and most surfactants are nonvolatile contaminating the interface, and strong electrolytes reduce the ionization efficiency in ESI. Several techniques have been developed to circumvent this problem: partial-filling technique, reverse migration of the micelle, or the use of volatile surfactant. The partial-filling technique is very effective for avoiding entrance of nonvolatile surfactant into the ion source [13–17], but it is not a mature technique for MEKC-MS as of yet. Other techniques are less popular and will not be introduced here. The more effective procedure for MEKC-MS is to perform conventional MEKC with nonvolatile surfactant and electrolytes, noting that concentrations of SDS and inorganic electrolytes are to be kept low (20 and 10 mM, respectively) [18,19]. Atmospheric pressure photoionization (APPI) interface is a promising interface for MEKC, where up to 50 mM SDS can be used [20].

3.2.3 Chemicals

3.2.3.1 Surfactants

Although a large number of surfactants are commercially available, limited numbers of surfactants are widely utilized in MEKC. SDS is the most popular because it has many advantages over other surfactants—specifically, it has high stability, relatively low Krafft point, low UV absorbance, high solubilizing capability, and high quality reagent is readily available. Among several different surfactants frequently used in MEKC, bile salts such as sodium cholate and sodium deoxycholate, and cetyltrimethylammonium bromide or cetyltrimethylammonium chloride (CTAB or CTAC) are also useful to obtain different selectivity from that obtained with SDS. The effect of surfactant structures on selectivity will be discussed later. The concentration of the surfactant must be higher than critical micelle concentration (CMC), above which the surfactant forms micelles. The CMC of SDS is 8 mM in water, but it will be lower (e.g., down to 3 mM) in buffer solutions. In most applications, SDS concentrations are between 10 and 50 mM, but even 100 mM can be employed if the applied voltage does not cause excessive current. Different surfactants will provide different selectivity for separation and, hence, surfactant choice is a means of selectivity manipulation. The mixed micelle, which consists of two or more surfactants, can provide different selectivity in separation; in particular, the mixed micelle between an ionic and a nonionic surfactant is considered of value because selectivity is closely related to the surface structure of the micelle. Some surfactants such as bile salts and amino acid derived surfactants are chiral and can be used for enantiomers separations.

3.2.3.2 Additives

As will be discussed later, selectivity can be manipulated by changing several parameters in addition to using different surfactants. The choice of additives to modify distribution equilibrium can follow the analogy to reversed-phase HPLC. Water miscible organic solvents such as methanol, acetonitrile, 2-propanol, and so forth, used as mobile phase modifiers in reversed-phase HPLC, are also effective in MEKC. For hydrophobic or less water-soluble analytes, the organic solvent addition is useful to reduce the distribution of the analyte to the micelle. The content of the organic solvents are usually less than 30% to avoid the dissociation of the micelle. However, much higher concentration can be used for separation, although it is not certain that the micelle is still stable. It should be noted that the addition of the organic solvent to BGS affects viscosity; in particular, methanol or 2-propanol addition increases viscosity significantly, and EOF is accordingly decreased because EOF mobility is inversely proportional to viscosity. The addition of hydrophobic solvent may change the micellar structure even if the concentration is low; for example, 1-butanol or 1-propanol may partly be incorporated into the micelle as a cosurfactant, which causes a change in selectivity. Another advantage of the use of an organic solvent is that the adsorption of the analyte on the capillary wall is minimized.

Cyclodextrins (CDs) are useful additives in MEKC to reduce the distribution of the analyte to the micelle, improving the separation of highly hydrophobic analytes. CD addition can also change selectivity, particularly for aromatic isomers. Since CDs are chiral, CD addition makes enantiomer separation possible, and this is particularly useful for the separation of neutral enantiomers. In most applications, β-CD or γ-CD is used at relatively high concentrations (e.g., 10–40 mM). The surfactant molecule may be included into the cavity of the CD and, therefore, the inclusion complex formation constant between the analyte and CD may be different from that observed in the absence of the surfactant. Ion-pair reagents are useful in MEKC in a manner analogous to that in reversed-phase HPLC for the separation of ionic analytes. However, the mechanism is slightly different between MEKC and HPLC, because the micelle is charged [21]. A high concentration of urea [22] or glucose [23] can be used to improve the separation of highly hydrophobic analytes or to change selectivity.

3.2.3.3 The Electrophoresis (Separation or Running) Solution

The micellar solution prepared by dissolving an ionic surfactant into a buffer solution is used as a "running" solution in MEKC. Popular buffer solutions often used to prepare the micellar solution are phosphate, borate, or tris(hydroxymethyl)aminomethane (Tris). It is well known the pH of the buffer significantly affects the EOF; that is, EOF is almost completely suppressed below pH 2, low at ~pH 5 and strong (and relatively constant) above pH 7 when a native capillary is employed. The concentration of buffer electrolytes is usually 20–50 mM, and while higher concentrations give higher buffer capacity, this generally generates high current in addition to the contribution of the ionic surfactant to conductivity. It should be noted that the counter ion of the ionic surfactant is exchanged by the counter ion of the buffer electrolyte and, consequently, the character of the micelle may be changed. The use of potassium-containing electrolytes must be avoided when using SDS because potassium dodecyl sulfate has a higher Krafft point at room temperature, and the surfactant will be precipitated out.

3.2.4 Operating Conditions

No special care is needed to perform MEKC compared with the CZE operation described in Chapter 1. The difference between the two techniques is only the electrophoretic solution or running solution. The current tends to be higher in MEKC, requiring a more frequent replenishment of the running solution to avoid the pH change in running solution associated with buffer depletion. When a cationic surfactant such as CTAB is employed, the direction of EOF is reversed and regeneration of the surface needs to be accomplished through sufficient rinsing of the capillary. It should be noted that CTAB in the anodic vial will generate bromine by electrolysis which ultimately contaminates the running solution.

3.3 THEORETICAL ASPECTS

3.3.1 Retention Factor and Resolution Equation

The migration behaviors of the analyte, the micelle marker, and the EOF marker are shown in Figure 3.2a, where the corresponding electropherogram is also shown (Figure 3.2b). The EOF marker migrates at the velocity of EOF and the micelle at the velocity of difference between EOF and the electrophoretic velocity of the micelle, which is in the opposite direction of EOF. The neutral analyte migrates at the velocity between the two extremes: the analyte is assumed to be equally distributed between the micelle and the aqueous phase in Figure 3.2. EOF mobility is assumed to be

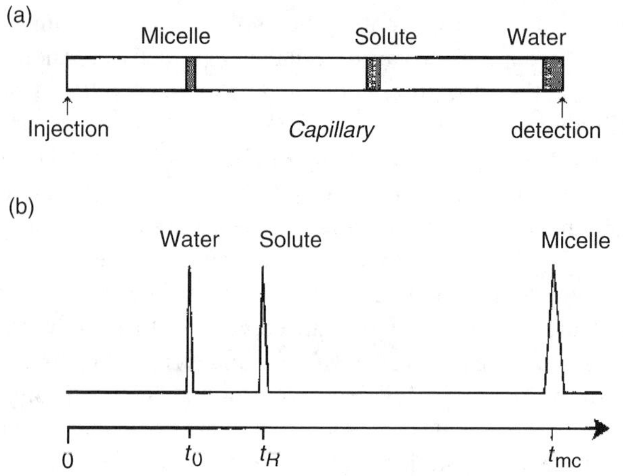

FIGURE 3.2 Schematic illustration of the migration behavior in MEKC. (From Terabe, S. et al., *Anal. Chem.*, 57, 834, 1985. With permission.)

4/3 times larger than electrophoretic mobility of the micelle in absolute values or the migration time of the micelle, t_{mc}, is four times longer than that of the EOF marker, t_0. The EOF marker is free from the micelle throughout the progress and methanol is often used as a marker. Although methanol does not absorb UV, it can be detected due to the fluctuation of the reflective index inside the capillary. Sudan III or IV is often used as a micelle marker, because it is assumed to be totally incorporated into the micelle. Some cationic hydrophobic compounds such as quinine sulfate can be employed as an SDS marker, because quinine interacts strongly with the SDS micelle due to electrostatic and hydrophobic interaction.

The retention factor, k, can be defined as

$$k = \frac{n_{mc}}{n_{aq}}, \tag{3.1}$$

where $n_{aq}$and n_{mc} are the amount of the analyte in the aqueous phase and that incorporated into the micelle. It can be calculated from the migration times t_R of the analyte, t_0, and t_{mc} as follows [1]:

$$k = \frac{t_R - t_0}{t_0\left(1 - t_{R/t_{mc}}\right)}. \tag{3.2}$$

The resolution equation in MEKC can be given by [24]

$$R_s = \frac{\sqrt{N}}{4}\left(\frac{\alpha - 1}{\alpha}\right)\left(\frac{k_2}{1 + k_2}\right)\left(\frac{1 - (t_0/t_{mc})}{1 + (t_0/t_{mc})\,k_1}\right), \tag{3.3}$$

where N is the number of theoretical plates and α is the selectivity factor equal to k_2/k_1, which is assumed to be larger than 1, where k_2 and k_1 are retention factors of analyte 1 and analyte 2, respectively. Equations 3.2 and 3.3 are similar to those of the conventional chromatography but different due to the limited migration time window for the analyte between t_0 and t_{mc}. The last factor (parameter) in right-hand side of Equation 3.3 is not part of the conventional resolution equation, and is ascribed to the change in effective capillary length utilized for separation. In other words, the micelle is not stationary in the capillary but migrates during the separation process, which makes the length of the micelle interacting with the analyte in the capillary shorter than the physical length by the distance the micelle migrates during t_R. We may define this parameter as the migration time window contribution, although the last factor becomes null when EOF is absent or the micelle is stationary. The effect of each factor in Equation 3.3 is briefly discussed below for reasonable optimization of MEKC separation.

3.3.1.1 Number of Theoretical Plates

Column efficiency, expressed as the number of theoretical plates (N: plate numbers), has been shown to be as high as 100,000–200,000 in MEKC. The most important factor contributing to the plate height is the longitudinal diffusion as in CZE—the shorter the migration time, the higher plate number. Since the diffusion coefficient of the micelle is usually one order lower than that of the small molecule, higher plate numbers can be expected for analytes having larger retention factors or those more incorporated into the micelle. The next important factor to be considered is extracolumn effects; in particular, the injection sample volume or sample plug length. In general, the sample plug length should be restricted to less than 1% of the capillary length unless sample preconcentration occurs. The use of a longer capillary alleviates the large volume injection problem, although the plate number is not proportional to the capillary length, as in other CE modes. It should be noted that if the initial sample zone length is minimal, extremely high efficiency can be expected. Temperature effects are not serious if current is lower than 50 μA, and it should be noted that the adverse effects on efficiency often observed are due to the sample adsorption on the capillary wall. Thorough rinsing of the capillary or an addition of a small amount of organic solvent may solve the problem. If the plate number is lower than 100,000, care must be taken to identify the reason of low efficiency; overloading of the sample or adsorption of the analyte on the capillary wall should be considered.

3.3.1.2 Selectivity Factor

As in chromatography, selectivity factor is the most important parameter in optimizing the separation. Since the micelle corresponds to the stationary phase in chromatography, the selection of the surfactant is of primary importance. SDS is a good initial choice and, if not successful, bile salts or CTAB/CTAC should be the second choice. If other surfactants are available, they may be tested. However, the number of different surfactants available for MEKC is rather limited and, hence, the selection of additives to modify selectivity should be considered when expected separation is not obtained by changing the surfactant. Some of the additives suggested above should be considered, as well as the use of mixed micelles. In MEKC, the efficiency is much higher than that of HPLC, the minimum α that provides successful separation may be as low as 1.02.

3.3.1.3 Retention Factor

The retention factor is contained in the third (retention factor) and fourth (migration time window) parameter in Equation 3.3. Therefore, the optimum k value, k_{opt}, that gives the maximum value for the product of third and fourth parameters is a function of k, t_0, and t_{mc}, and given by [25]

$$k_{opt} = \sqrt{t_{mc}/t_0}. \tag{3.4}$$

When SDS is used under neutral or alkaline conditions in native or uncoated capillary, the value t_{mc}/t_0 is usually 3–4 and the retention factor should be adjusted to about 1.7–2.0. Since the retention factor is related to the surfactant concentration by [24]

$$k = \frac{KV_{mc}}{V_{aq}} \cong K\bar{v}\,(C_{sf} - c_{mc}), \tag{3.5}$$

where K is the distribution coefficient of the analyte between the micelle and the aqueous phase, V_{mc} and V_{aq} are volume of the micelle and the aqueous phase, respectively, $\bar{v}$ partial specific volume of the micelle, and C_{sf} is the surfactant concentration. Equation 3.5 suggests that k is nearly proportional to the surfactant concentration; this means that the phase ratio V_{mc}/V_{aq} is adjusted by changing the surfactant concentration. Therefore, the optimization of k is straightforward provided that the specific volume and CMC are known; the $\bar{v}$ value for an SDS micelle is about 0.85 mL/g. Even if these values are not known, it is easy to adjust k by increasing or decreasing the surfactant concentration considering the linear dependence of k on C_{sf}.

3.3.1.4 Migration Time Window

The EOF velocity is difficult to control precisely and depends on the surface charge on the inside wall of the capillary. If the wall is negatively charged, EOF is cathodic and vice versa. The factors affecting EOF are described above. In general, if the EOF velocity is decreased, the migration time window becomes wide and the resolution will be improved at the expense of migration time. The detailed discussion on the effect of EOF on resolution is published by Zhang et al. [26].

3.3.2 Interaction Mechanism between the Micelle and the Analyte

Three types of interaction mechanisms are known between the micelle and the analyte as shown in Figure 3.3: (1) incorporation of the analyte into the hydrophobic core, (2) adsorption of the analyte on the surface or on the palisade layer, and (3) incorporation of the analyte as a cosurfactant. Highly hydrophobic and nonpolar analytes such as aromatic hydrocarbons will be incorporated into the core of the micelle. The selectivity may not be very different among long alkyl-chain surfactants for this class of analyte but the distribution coefficient will be increased with longer alkyl-chain surfactants. Thus, selectivity will not be altered significantly for nonpolar hydrophobic analytes, even when different surfactants are used. However, bile salts may provide substantially different selectivity in comparison with long-alkyl chain surfactants, even for nonpolar hydrophobic analytes.

Most analytes are considered to be incorporated by the micelle on the surface or on the palisade layer [Figure 3.3a (1)] or partly incorporated into the core at the hydrophobic group and on the surface at the polar group [Figure 3.3a (2)]. Therefore, the polar group of the surfactant significantly affects selectivity. The palisade layer will be much different in mixed micelles consisting of an ionic and a nonionic surfactant and, hence, selectivity will be significantly different from the single surfactant micelle (Figure 3.3b).

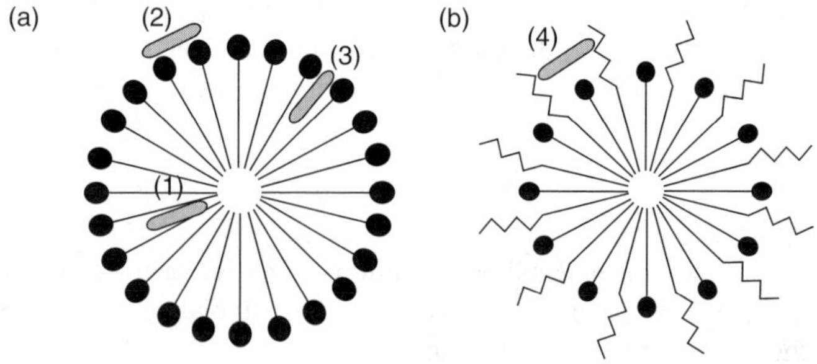

FIGURE 3.3 Schematic illustration of micellar solubilization. (a) Ionic micelle and (b) mixed micelle of ionic and nonionic surfactants interacting (1) with the hydrophobic core, (2) on the surface, (3) as a cosurfactant, and (4) with nonionic surface. (From Terabe, S., *Anal. Chem.*, 76, 240A, 2004. With permission.)

The interaction mechanism between the micelle and the analyte or chemical selectivity in MEKC has been investigated by several groups. Linear solvation energy selectivity relationship (LSER) can be utilized, where structural descriptors such as size, dipolarity, and hydrogen-bonding capabilities are employed to evaluate their contribution to interaction (retention factor) using many different types of solutes and different surfactants [27]. As expected, the molecular size or hydrophobicity causes the strongest positive interaction due to unfavorable energy term for the formation of properly sized cavity in the solvent system for solute accommodation. The second most important term in most MEKC systems is the solute hydrogen bond acceptor capability, but this interaction negatively affects the interaction. Since hydrophobicity affects the interaction almost equally among different MEKC systems, fine tuning in selectivity is performed by the hydrogen bond acceptor capability or basicity of the solute or acidity of the surfactant. Similar conclusions have been obtained by other groups [28–30].

3.3.3 Online Sample Preconcentration

Poor concentration detection sensitivity in CE is a serious problem in almost all CE modes. The main reasons are due to the small amount of sample introduced into the capillary and a short optical path length for photometric detectors. To solve the problem, three major solutions are available: sample pretreatment such as solid-phase extraction (SPE), the use of high sensitivity detectors such as LIF, and the implementation of online sample preconcentration techniques. In addition, extended path length cells such as bubble cell or Z-type cell are commercially available for photometric detection, enabling a 3- to 10-fold enhancement in sensitivity. Most online sample preconcentration techniques are designed not to require any modification of the instrument and to be compatible with any detection methods because the concentration of the sample is increased prior to separation. Several techniques have been developed: transient isotachophoresis (t-ITP), field-enhanced (amplified) sample stacking, sweeping, and dynamic pH junction. The sample zone injected as a much longer plug than normal can be focused to the length less than that of the normal injection. These techniques are based on the migration velocity change of the analytes, between the sample solution zone and BGS, and the analyte must have an electrophoretic *mobility even if it is apparent mobility*. Among these methods, t-ITP and dynamic pH junction are applicable only to ionic or ionizable analytes. Therefore, field-enhanced sample stacking and sweeping are the main techniques described.

3.3.3.1 Field-Enhanced Sample Stacking

The principle of the technique is simple and is based on the fact that the electrophoretic migration velocity is proportional to the field strength. Thus, the sample solution is prepared in a lower conductivity solution than that of the BGS; then the analyte ion migrates rapidly in the sample zone when an electric field is applied. The analyte ion quickly migrates until it reaches the boundary between the sample zone and the BGS zone, where the migration velocity decreases upon entering the BGS zone due to low electric field as shown in Figure 3.4. The larger the difference in electric fields between the sample zone and BGS, the higher the focusing efficiency. However, it is noteworthy that EOF velocity is also proportional to the field strength, but the liquid flow by EOF must be homogeneous throughout the capillary. That is, if the EOF velocity in the two zones is mismatched, mixing of the two zones at the boundary must occur, which can seriously deteriorate the concentration efficiency. Therefore, it is desirable to suppress EOF when online sample stacking is performed. There are several techniques based on the same principle introduced by Chien and Burgi [31].

In principle, field-enhanced stacking is a technique for the concentration of charged analytes using discontinuous solutions in the capillary. Neutral analytes are not concentrated by this method because they do not have electrophoretic mobility. However, by utilizing charged micelles, an apparent electrophoretic mobility can be imparted to a neutral analyte according to the principles of MEKC. As in conventional field-enhanced sample stacking, sample solutions are prepared in a

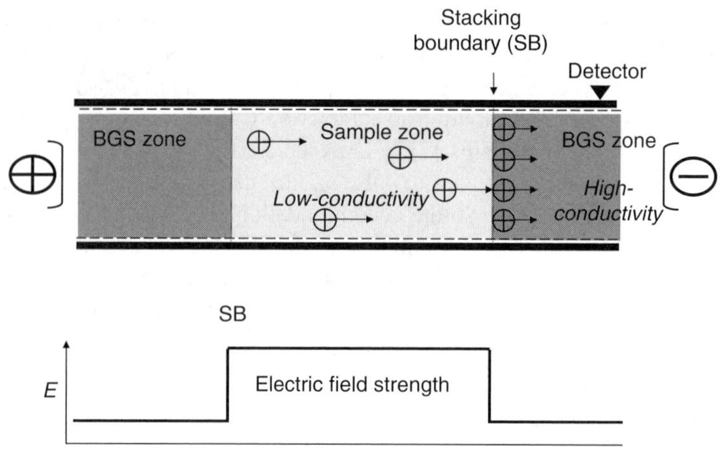

FIGURE 3.4 Schematic illustration of the principle of field-enhanced sample stacking.

low electrical conductivity solution, with or without adding micelles, and a conventional micellar solution is used as the BGS. When the sample solution is prepared without the micelle, the analyte is incorporated by the micelle that enters the sample zone from the BGS zone when voltage is applied. The analyte having a high k that will induce extensive incorporation into the micelle can be efficiently concentrated by field-enhanced stacking because it is apparently highly charged. It should be noted that the micellar concentration is low in the sample zone even if the sample solution contains the micelle, because the migration velocity of the micelle in the sample zone is faster than that in the BGS. As described in Chapter 1, there are two sample injection techniques amenable to field-enhanced stacking: pressurized/hydrostatic injection and electrokinetic injection. With either technique, an EOF suppressed condition will yield a higher concentration efficiency. Larger amounts of sample can, therefore, be injected by electrokinetic injection, while with pressurized injection the sample, volume is limited to that substantially less than the capillary volume. In summary, either injection method will yield a maximum concentration efficiency of less than 100-fold under favorable conditions [32,33].

3.3.3.2 Sweeping

Sweeping is an online concentration technique developed for MEKC where the neutral analyte is efficiently concentrated [34,35]. The principle of sweeping is shown in Figure 3.5 under suppressed EOF conditions. Sweeping seems similar to field-enhanced sample stacking but the difference is that no field enhancement in the sample zone is required in sweeping with a field strength equal to or lower [36] than that of BGS. The sample solution is prepared devoid of the micelle but at a conductivity similar to that of BGS (acidic to suppress EOF) and injected hydrodynamically as a long plug (Figure 3.5a). When the voltage (negative for anionic micelle such as SDS) is applied, the micelle enters the sample zone and collects analytes at the front end of the entering micellar zone (Figure 3.5b) until the front end reaches the original boundary between the sample zone and the BGS (Figure 3.5c). The original sample matrix migrates at the velocity of the micelle and the micelle vacancy zone migrates ahead of the swept sample zone (Figure 3.5c). The length of the swept analyte zone is given by [34]

$$l_{\text{sweep}} = l_{\text{inj}} \left(\frac{1}{1+k} \right), \tag{3.6}$$

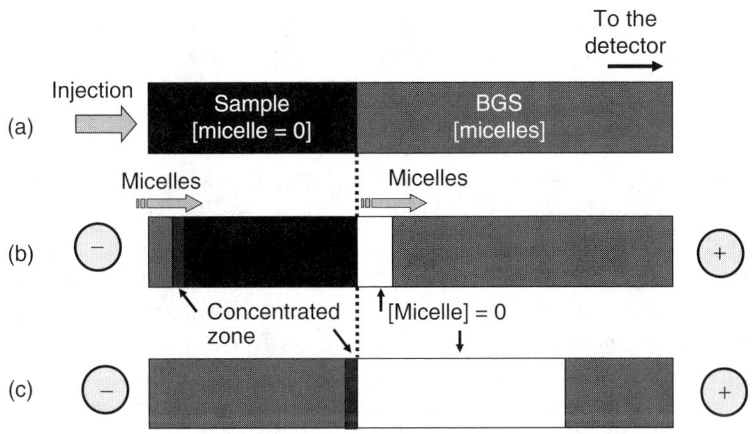

FIGURE 3.5 Schematic illustration of the principle of sweeping under suppressed EOF conditions. The sample solution is prepared devoid of the micelle and BGS is the micellar solution. (a) Sample injection as a long plug. (b) Start of sweeping by application of a voltage. (c) End of sweeping and start of MEKC separation.

where l_{sweep} and l_{inj} are lengths of the sample zone after and before the end of sweeping, respectively. According to Equation 3.6, to obtain high concentration efficiency, k should be maximized. It should be noted that k in Equation 3.6 is the value in the sample matrix; that is, it is different from that in BGS. Therefore, the additive to BGS such as an organic solvent or CD used to improve resolution should not be added to the sample matrix. The zone length of the swept zone of a hydrophobic analyte was extremely narrow as observed when the sweeping process was traced with the peak shape of the sample zone at different positions in the microchannel [37]. Under favorable conditions, 64 cm of 70 cm of effective capillary length was filled with a dilute sample solution and 6 cm of 70 cm was used for the MEKC separation, yet good separation and high efficiency was obtained because the sample zone length after sweeping was very narrow and the time spent for the separation was short. EOF is not related to the sweeping mechanism, but higher concentration efficiency was obtained under suppressed EOF conditions [35]. Consequently, sweeping gives high concentration efficiency with improvements in detection sensitivity on the order of 1000-fold greater than the conventional injection/separation.

Although sweeping is an online sample preconcentration technique for neutral analytes, charged analytes, or ions can be more efficiently concentrated if the micelle employed has a polarity opposite to that of the analyte, because the retention factor will be large due to a strong electrostatic interaction. Therefore, the SDS micelle is convenient for concentrating cationic analytes, while cationic micelles (e.g., CTAB) are suitable for anionic analytes such as aromatic sulfonates or carboxylates [38]. Sweeping is effective in other EKC modes that use pseudostationary phases such as charged CDs, microemulsions, and so forth. In addition, the sweeping principle is applicable to online concentration of metal ions and sugars by utilizing an in-capillary complexation reaction, sweeping of metal ions by ethylenediaminetetraacetic acid (EDTA) [39], or sweeping of sugars (*cis*-diols) using borate ions [40].

As mentioned before, when pressure-based sample injection is employed, the maximum injection volume must be less than the effective length of the capillary. At least 10% of the effective capillary length must remain available for separation. To inject a larger sample volume, electrokinetic injection must be employed. In most online sample preconcentration techniques, the maximum amount that can be injected without loss of separation efficiency is certainly less than the capillary volume. However, with a large volume sample injection under cathodic EOF conditions in SDS MEKC, a sample devoid of the micelle can be continuously electrokinetically injected for a volume equivalent to seven times the capillary volume without significant loss of separation efficiency under favorable conditions. Here, the analyte in the sample is continuously concentrated when it reaches the swept zone [41].

3.3.3.3 Combination of Different Preconcentration Methods

A combination of the different online sample preconcentration techniques enables much higher concentration efficiency or simultaneous concentration of different types of analytes. First, cationic (or anionic) analytes can be electrokinetically injected for a extensive periods of time to cause overloading. Second, the long preconcentrated sample zone resulting from the injection can be further concentrated by sweeping using anionic (or cationic) micelle. This combination of field-enhanced sample injection (FESI) and sweeping is termed cation-selective (or anion-selective) exhaustive injection-sweeping (CSEI/ASEI-sweeping) and can generate extraordinarily high concentration efficiency with close to a million-fold increase in sensitivity, enabling the analysis of samples with analytes present in as low as parts per trillion (ppt) levels [42,43].

Dynamic pH junction is a technique suitable for online sample preconcentration of weakly acidic or basic compounds, with the principle based on the change in electrophoretic velocity between that of the ionized and neutral state of the analyte [44]. The sample solution is prepared in an acidic medium for the acidic analyte with the BGS a basic buffer having a pH higher than the pK_a of the analyte. When electrophoretic process begins, the acidic sample zone is titrated by the alkaline BGS and the pH of the sample zone increases, causing deprotonation of the analyte. The ionized analyte migrates toward the anode against the EOF and focuses at the boundary of pH change. The technique is very effective for the selective concentration of acidic or basic analytes having a range of pK_a values. When the sample contains both neutral and weak acidic compounds, the combination of sweeping and dynamic pH junction is a useful online sample preconcentration method. For example, with urine or plasma containing a combination of riboflavin, flavin mononucleotide (FMN), and flavin adenine dinucleotide (FAD), an enhancement in detection sensitivity of more than 1000-fold was observed for these analytes by dynamic pH junction-sweeping with the sample diluted in a 75 mM phosphate buffer pH 6.0 (and BGS of 140 mM borate) because almost neutral riboflavin was concentrated by sweeping and other FMN and FAD were concentrated by dynamic pH junction simultaneously [45].

3.4 PRACTICAL APPLICATIONS

More than 2000 articles involving “MEKC” have been published since 1990 according to Zare [3]. Therefore, it is impossible to survey all papers to determine which provide the best practical applications of MEKC—consequently, only few practical applications have been arbitrarily selected by the author for inclusion here. For more detailed examples, it is recommended that the reader consult more comprehensive reviews [7,46–48] or relevant chapters of a recently published book [6].

3.4.1 Pharmaceutical Analysis

MEKC is probably most widely accepted in pharmaceutical analysis and several review articles have detailed these applications [49–53]; therefore, only a few recent publications are introduced here. In addition, the reader may refer to Chapter 4 in this book contributed by Altria and colleagues.

As mentioned above, MS is becoming an indispensable detection technique, in particular, for the detection and identification of impurities in drugs. An ESI has been successfully interfaced between MEKC and a single quadrupole or ion-trap MS instrument using a running solution containing 20 mM SDS and 10 mM sodium phosphate buffer (pH 7.5) without any devices to avoid SDS and inorganic electrolytes from entering into the MS ion source [18]. In this report, it was demonstrated that sub-μg/mL levels of mebeverine and related compounds could be detected in full-scan mode, while

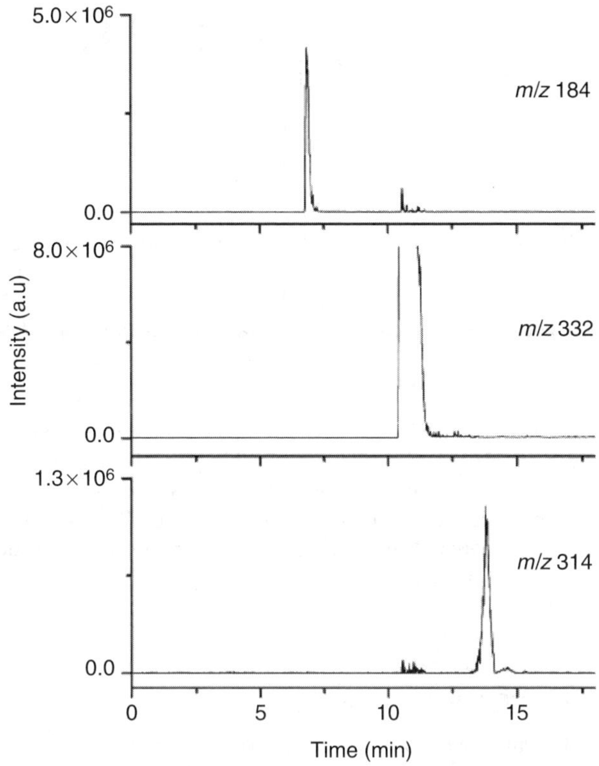

FIGURE 3.6 Extracted-ion chromatograms obtained during MEKC–ESI-MS of a heat-treated solution of ipratropium (1 mg/mL) using a BGE containing 10 mM sodium phosphate (pH 7.5), 20 mM SDS, and 12.5% acetonitrile. Capillary length, 57 cm. Peaks: *m*/*z* 332, ipratropium; *m*/*z* 184 and *m*/*z* 314, degradation products. (From Mol, R., et al., *J. Chromatogr. B*, 843, 283, 2006. With permission.)

detection limits are in the 10–50 ng/mL range when selected ion monitoring was applied. It was also shown that 0.1% (w/w) levels of potential impurities in mebeverine could be detected in full-scan MS. The technique was also successfully applied to the analysis of a galantamine sample containing a number of related impurities [19]. Two degradation products were detected and identified by MS/MS scan in a heat-stressed ipratropium sample as shown in Figure 3.6.

In the pharmaceutical industry, rapid developments and validations of suitable methods for assessment of purity, potency, and stability of new drug substances and drug products are critical to providing appropriate data for early project development decisions. For example, a simple, fast, and selective MEKC method for the simultaneous assay of ketorolac tromethamine and its identified related impurities in both the drug substance and coated tablets was developed by Orlandini et al. [54]. To optimize separation conditions, a response surface study was performed. Flufenamic acid (FL) and tolmetin (TL) were chosen as internal standards to quantify ketorolac tromethamine and the impurities, respectively. The optimized BGS consisted of a mixture of 13 mM boric acid and phosphoric acid (equal mixture), adjusted to pH 9.1 with 1 M sodium hydroxide, containing 73 mM SDS. Optimal temperature and voltage were 30°C and 27 kV. Under these conditions, all compounds were resolved in about 6 min as shown in Figure 3.7. The related substances could be quantified up to the 0.1% (w/w) level. Method validation was performed by evaluating selectivity, robustness, linearity and range, precision, accuracy, detection sensitivity, and quantification limits, as well as system suitability for drug substances/drug product and a satisfactory method performance was obtained.

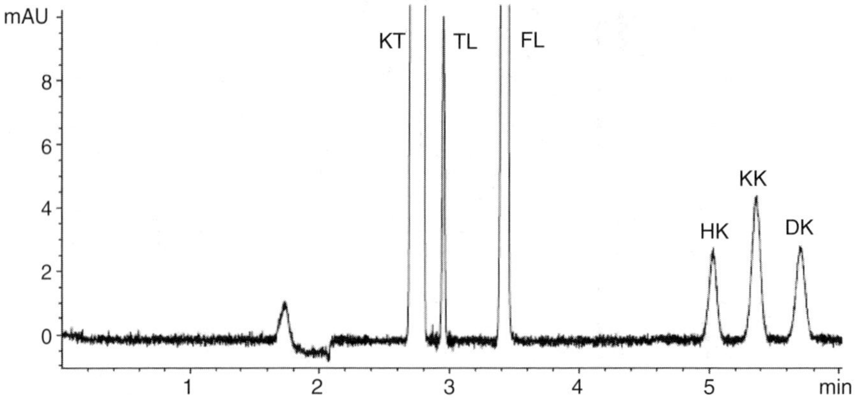

FIGURE 3.7 MEKC electropherogram of a real sample of Lixidol tablets using 73 mM SDS in 13 mM borate-phosphate (equal molar) buffer (pH 9.1). FL, flufenamic acid; KT, 1:1 mixture of ketorolac tromethamine and 2-amino-2-(hydroxymethyl)-1,3-propanediol; DK, decarboxylated ketorolac; HK, 1-hydroxy analog of ketorolac; KK, 1-keto analog of ketorolac; TL, tolmetin. Capillary, 50 μm ID × 48 cm (39.5 cm to the detector); applied voltage, 27 kV; temperature, 30°C; hydrodynamic injection: 50 mbar, 5 s; detection wavelength, 323 nm. (From Orlandini, S., et al., *J. Chromatogr. A*, 1032, 253, 2004. With permission.)

3.4.2 Clinical and Body Fluid Analysis

Development of the analytical separation methods for biomarker discovery and metabolite research is an important domain for the early detection of certain diseases, as well as in pharmaceutical investigations. Recent applications of MEKC and CZE in the field of urinary biomarkers have been summarized by Iadarola et al. [55]. Most target analytes are present in body fluids at low concentrations and, consequently, high sensitivity analysis is essential. The use of high sensitivity detectors or implementation of online sample concentration techniques is essential. Sample pretreatment is also critical in the analysis of body fluids for cleaning up and concentrating the target analytes, and for this several techniques are available.

Direct injection of plasma samples in MEKC was first reported by Nakagawa et al. [56]. Determination of cefotaxime (C) and its deacetylated metabolite (DA) in human plasma were investigated by direct injection of plasma in both CZE and MEKC [57]. MEKC was shown to be superior with regard to simplicity, rapidity, precision, and sensitivity relative to CZE. In MEKC, plasma samples spiked with C, DA, and theobromine (as internal standard) were directly injected after dilution with water, and analyzed using a phosphate buffer (pH 8.0) containing 165 mM SDS as a running solution. The interday precision ($n = 4$ days) was 1.49% in RSD when theobromine was used as internal standard. A satisfactory interday precision between slopes was also obtained with MEKC, even without the use of an internal standard (RSD = 4.38%). Detection limits ($S/N = 3$) were about 1 mg/L in plasma for C and DA. The advantage of MEKC is due to the fact that SDS solubilizes the proteins, and this liberates the drugs from protein-binding sites. The SDS–protein complexes migrate with the micelles and, thus, do not interfere with the detection of the analytes.

An MEKC-sweeping technique for the simultaneous determination of flunitrazepam and its major metabolites, 7-aminoflunitrazepam and *N*-desmethylflunitrazepam, was reported by Huang et al. [58]. The optimized conditions involved a running solution of 25 mM borate (pH 9.5) containing 50 mM CTAB and 30% methanol (v/v) with a 60 cm (50 cm to the detector) × 50 μm ID capillary and sample injected as a 151-mm plug. The limits of detection were 13.4, 5.6, and 12.0 ng/mL for flunitrazepam, 7-aminoflunitrazepam, and *N*-desmethylflunitrazepam, respectively, and the sensitivity enhancement for each compound was within the range of 110- to 200-fold. The method was applied

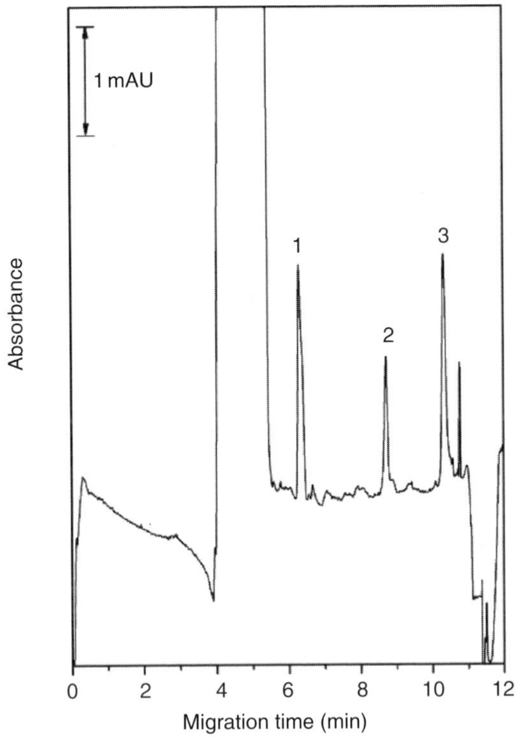

FIGURE 3.8 Sweeping MEKC electropherogram of a spiked urine sample. Peak identification: 1, 7-aminoflunitrazepam; 2, flunitrazepam; 3, *N*-desmethylflunitrazepam. Analyte concentration: 0.3 μg spiked in a 3-mL urine sample before SPE extraction. Conditions: capillary, 60 cm long (50 cm to detector), 50 μm ID; running solution: 25 mM borate buffer (pH 9.5) containing 50 mM CTAB and 30% methanol (v/v), conductivity 7.28 mS/cm; sample matrix: 25 mM borate buffer (pH 9.5); applied voltage: -25 kV; UV detection at 240 nm. (From Huang, C.-W., et al., *J. Chromatogr. A*, 1110, 240, 2006. With permission.)

to a spiked urine sample in combination with SPE — an example of electropherogram is shown in Figure 3.8.

Ewing's group has developed ultramicro separation technique using CE with an electrochemical detector and a narrow bore capillary of 13 μm ID. They applied the technique to the separation of amine metabolite in the fruit fly, *Drosophila melanogaster*, and 14 biogenic amines and metabolites were successfully resolved by MEKC using 25 mM borate buffer (pH 9.5) containing 50 mM SDS and 2% 1-propanol [59]. Flies were snap-frozen in a liquid nitrogen bath, the separated fly heads treated to yield a 3 μL sample solution that could be analyzed for biogenic amines as shown in Figure 3.9. Quantitative results were, for example, 37.2, 311.0, 142.0, 79.5, 732.8 fmol (1×10^{-15} mol) per fly head of a wild-type *Drosophila*, for tyramine, octopamine, *N*-acetyloctopamine, dopamine, and *N*-acetyldopamine, respectively. MEKC with electrochemical detection gave limits of detection as low as 3.4 amol for octopamine, and made the technique useful for volume-limited sample analysis.

Kennedy's group used microdialysis sampling to monitor extracellular dopamine concentration in the brains of rats. The dialysate, mixed online with 6 mM naphthalene-2,3-dicarboxaldehye and 10 mM potassium cyanide in a reaction capillary, was periodically analyzed by MEKC at 90-s intervals [60]. The MEKC system consisted of a 10 μm ID, 369 μm OD, and 16 cm (14.5 cm to the detector) capillary, 30 mM phosphate buffer (pH 7.4) containing 6.5 mM SDS and 2 mM 2-hydoxypropyl-β-CD, with LIF detection using the 413-nm line of a 14-mW diode-pumped laser, and an electric field of 850 V/cm. The detection limit for dopamine was 2 nM when sampling by microdialysis. The separation capability of the developed method is illustrated in Figure 3.10 where

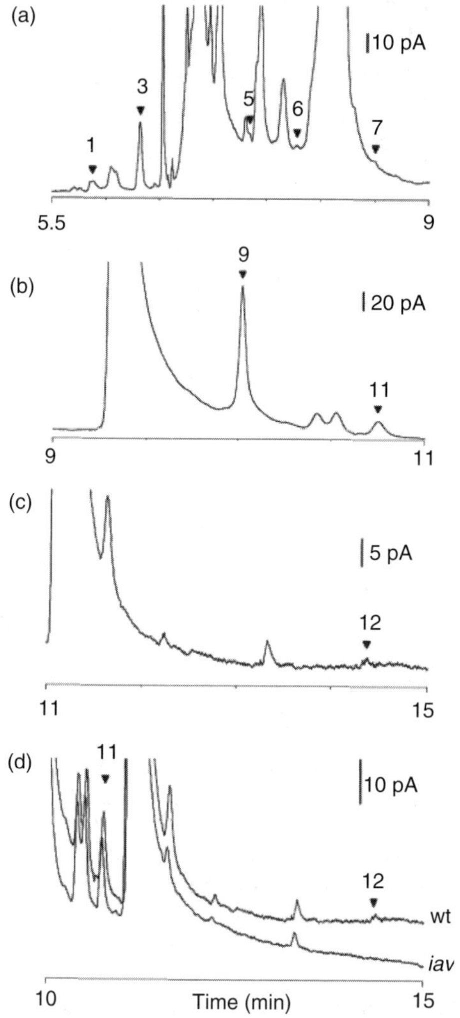

FIGURE 3.9 MEKC-EC separation of a *Drosophila* head homogenate. (a) Enlargement of 5.5 9 min high lighting dopamine (1), *N*-acetyloctopamine (3), octopamine (5), *N*-acetylserotonin (6), and *N*-acetyldopamine (7). (b) Enlargement of 9–11 min emphasizing peaks for 4-dihydroxyphenylalanine (9) and catechol (11). (c) Enlargement of 11–15 min showing tyramine (12). (d) Comparison of Canton-S (wt) and iav^1 *Drosophila* head homogenates highlighting the internal standard catechol (11) and tyramine (12). Conditions: running solution, 25 mM borate buffer (pH 9.5) containing 50 mM SDS and 2% 1-propanol; capillary, a fused silica capillary of 148 μm OD and 13 μm ID, 50 cm long; Field strength, 333 V/cm. The working electrode was held at +750 mV vs. a Ag/AgCl reference electrode. (From Paxon, T.L., et al., *Anal. Chem.*, 77, 5349, 2005. With permission.)

the basal dopamine concentration in dialysates collected from the striatum of anesthetized rats was 18 ± 3 nM ($n = 12$).

3.4.3 Food Analysis

Applications of CE techniques to food analysis is rapidly increasing and recent progresses have been reviewed [61,62]—a detailed discussion of this topic can be found in Chapter 30 by Vargas and Cordoba. The analysis of food components or nutrients does not require high sensitivity but does

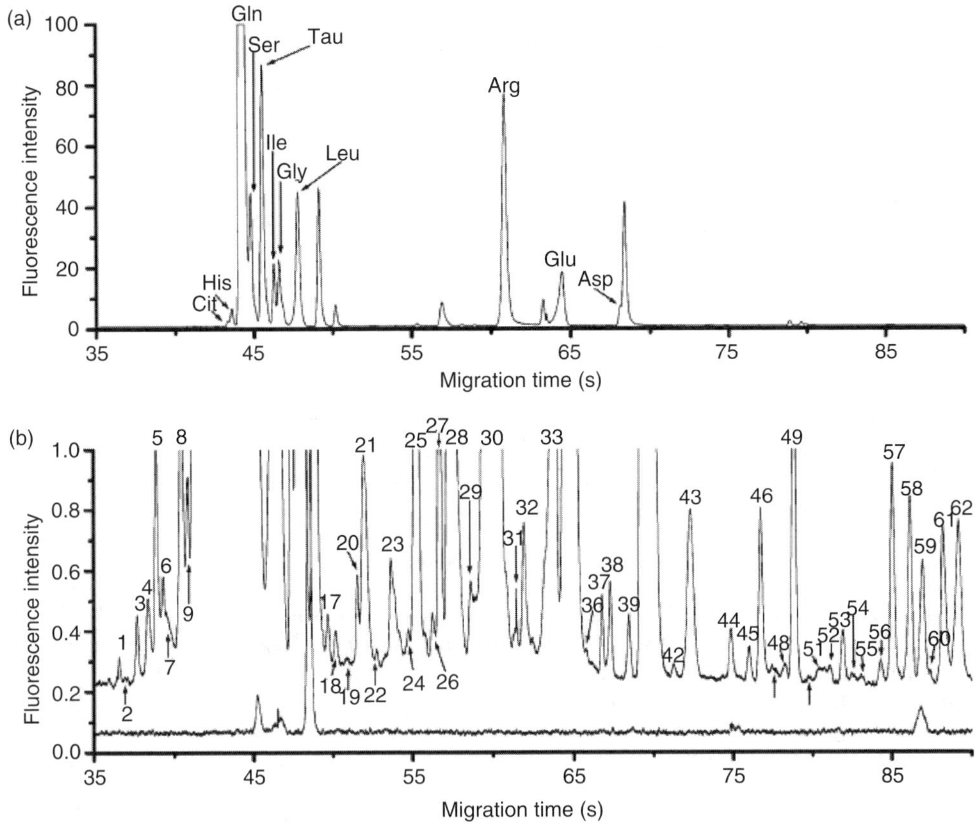

FIGURE 3.10 Sample electropherogram collected from the striatum of an anesthetized rat illustrating peak capacity of the method. (a) Full scale plot with amino acids identified by migration time matching marked. (b) Expanded scale for electropherogram collected from control samples (derivatization reagents sampling artificial cerebral spinal fluid *in vitro*, lower trace) and *in vivo* samples (upper trace) at 10-fold higher gain illustrating smaller peaks detected. Fluorescence units are equivalent in the plots. Peaks that were consistently observed *in vivo* and not observed in control samples were counted. Numbers for off-scale peaks are not shown. Peak 55 is dopamine. (From Shou, M., et al., *Anal. Chem.*, 78, 6717, 2006. With permission.)

require good resolution, in particular for the analysis of plant nutrients; however, for the analysis of contaminants or agrochemical residues, high sensitivity is essential in combination with sample pretreatments.

The catechin content of tea is often analyzed by CE techniques. Matcha is a special powdered green tea used in the Japanese tea ceremony and catechins contained in it have been quantified by MEKC [63]. MEKC conditions are conventional; a 50 μm ID bubble capillary of 77 cm effective length, 25 mM phosphate buffer (pH 7.0) containing 20 mM SDS, applied voltage of 27 kV, and detection at 200 nm. Water and methanol extracts of matcha were compared with water extracts of a popular green tea. An electropherogram of matcha extract is shown in Figure 3.11. The concentration of epigallocatechin gallate (EGCG) available from drinking matcha is 137-times greater than the amount of EGCG available from China Green Tips green tea, and at least three times higher than the largest value reported in the literature for other green teas.

The residue analysis of pesticides or herbicides in food is an important social concern. The contents of these agrochemicals are very low and the sample preparations are time consuming and labor intensive. Therefore, development of simple, inexpensive, and efficient total analysis system is urgently needed. The analysis of pesticides in fruits and vegetables was described by Juan-García

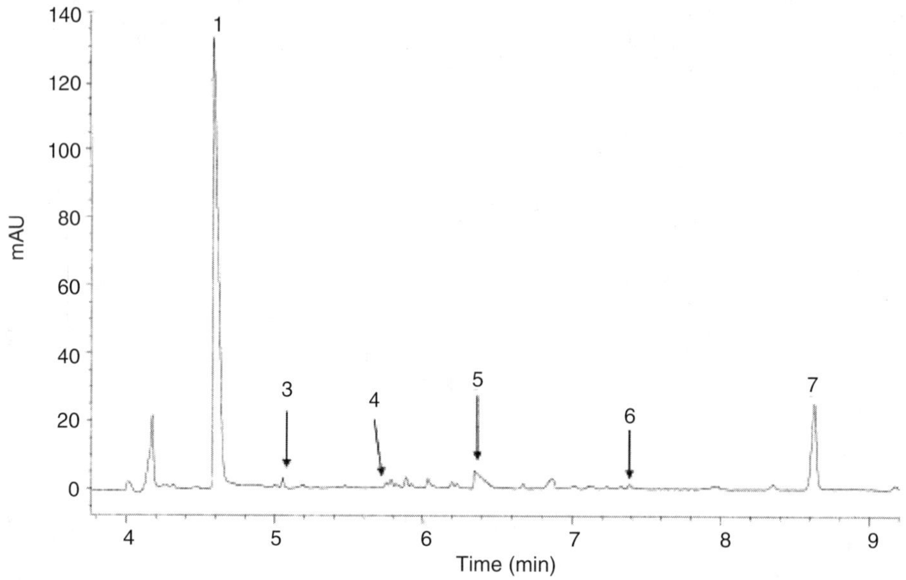

FIGURE 3.11 Electropherogram of a matcha tea sample prepared in deionized water using the traditional Japanese method. Conditions: capillary, 80.5 cm (77.0 cm effective length), 50 μm ID with bubble (three times extended path length); running solution, 25 mM phosphate buffer (pH 7.0) containing 20 mM SDS; applied voltage, 27 kV; detection, at 200 nm. Peak identification: (1) caffeine, (2) catechin, (3) epigallocatechin, (4) epigallocatechin gallate (EGCG), (5) epicatechin, (6) epicatechin gallate, and (7) internal standard (4-amino-2-hydroxybenzoic acid). (From Weiss, D.J. and Anderton, C.R., *J. Chromatogr. A*, 1011, 173, 2003. With permission.)

et al. [64] using SPE and stir-bar sorptive extraction (SBSE), in combination with MEKC. The recoveries obtained by SPE ranged from 40% to 106% with RSD from 10% to 19%, whereas the SBSE method was associated with recoveries in the 12–47% range with RSDs of 3–17%. It is noteworthy that the limits of quantification were much better by SPE (0.2–0.5 mg/kg depending on the processed sample amount) than those obtained by SBSE (1 mg/kg for each compound). MEKC conditions using 75 mM sodium cholate in 6 mM tetraborate buffer (pH 9.2) was employed and exemplary electropherograms are shown in Figure 3.12, where pesticide residues in lettuce were analyzed.

3.4.4 Environmental Analysis

Most target analytes in environmental analysis are low in concentration, although the amount of sample is not very limited. As a result, CE is not an ideal technique for environmental analysis. A combination of a sample pretreatment such as SPE or liquid–liquid extraction, and online sample preconcentration is clearly necessary here. There are several review articles published on the topic [65–67] and a detailed discussion of this topic can be found in Chapter 31 by Tonin and Tavares.

Determination of pesticides in water has been performed by automatic online SPE in combination with MEKC [68]. A C_{18} solid-phase minicolumn was used for the preconcentration, allowing a 12-fold enrichment (as an average value) of the pesticides from fortified water samples. MEKC was performed with 10 mM phosphate buffer (pH 9.5) containing 60 mM SDS and 8% acetonitrile. Pesticides mixtures spiked in water were detected down to a concentration of 50 μg/L in less than 13 min as shown in Figure 3.13.

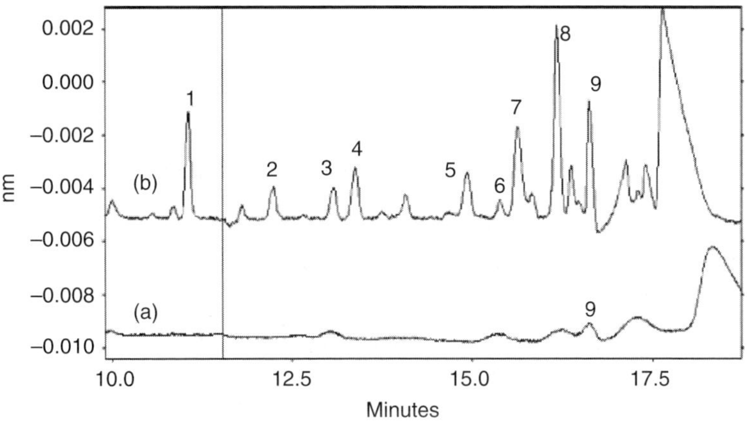

FIGURE 3.12 Electropherograms of SPE extracts from 15 g sample of (a) lettuce that contains pyriproxyfen at 0.2 mg/kg sample and (b) lettuce sample spiked with the pesticides at 0.5 mg/kg levels. Peak identification: (1) flutriafol, (2) cyproconazole I, (3) cyproconazole II, (4) myclobutanil, (5) tebuconazole, (6) acrinathrin, (7) bitertanol, (8) fludioxonil, and (9) pyriproxyfen. MEKC conditions: capillary, 75 μm ID × 57 cm (50 cm to the detector); running solution, 6 mM sodium tetraborate (pH 9.2) containing 75 mM sodium cholate; detection, 214 nm. (From Juan-García, A., et al., *J. Chromatogr. A*, 1073, 229, 2005. With permission.)

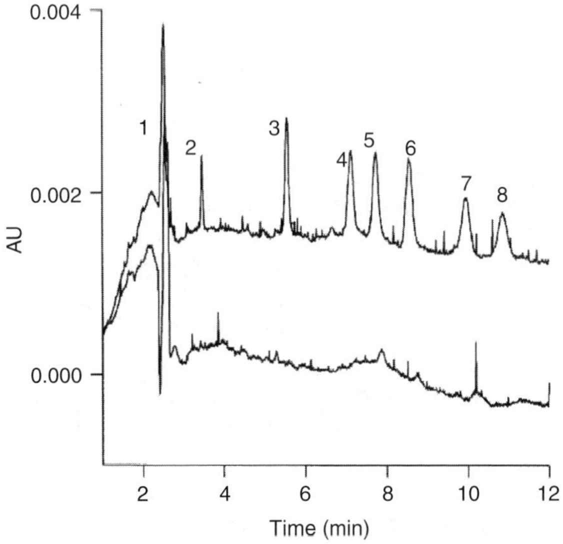

FIGURE 3.13 Electropherograms of an unspiked and spiked river water sample (0.25 μg/mL). (1) EOF, (2) fenuron, (3) simazine, (4) atrazine, (5) carbaryl, (6) ametryn, (7) prometryn, and (8) terbutryn. MEKC conditions: capillary, 75 μm ID × 47 cm; running solution, 10 mM phosphate buffer (pH 9.5) containing 60 mM SDS and 8% acetonitrile; detection, 226 nm. (From Hinsmann, P., et al., *J. Chromatogr. A*, 866, 137, 2000. With permission.)

3.5 METHODS DEVELOPMENT GUIDELINES

Introductory conditions for initial MEKC separations of neutral or weakly ionizable analytes are provided in Table 3.1 [4]. The choice of the running solution is rather arbitrary, but a weakly alkaline borate buffer is recommended because the electrophoretic mobility of the borate ion is rather low and, thus, the current can be kept low. The sample solution can be prepared in any solvents provided

TABLE 3.1
Suggested Standard Operating Conditions

Capillary	50–75 μm ID × 20–50 cm (to the detector) uncoated fused silica capillary
Running solution	50 mM SDS in 50 mM borate buffer (pH 8.5–9.0) or 50 mM phosphate buffer (pH 7.0) (use sodium salts)
Applied voltage	10–25 kV (keep current below 50 μA)
Temperature	25°C or ambient
Sample solvent	Water or methanol (other water miscible organic solvents are usable)
Sample concentration	0.1–1 mg/mL (lower concentrations are acceptable if detectable)
Injection method	Hydrodynamic or siphoning at anodic end
Injection volume	Less than 1% of the capillary length
Detection	Absorbance at 200–210 nm (higher wavelengths are possible if detectable)

Source: Modified from Table 3 of Terabe, S., *Micellar Electrokinetic Chromatography*, Beckman, Fullerton, 1992.

they are miscible with water. However, water-rich solution is advisable if the analyte is water soluble. When the sample contains high concentrations of an organic solvent, each peak may split due to incomplete mixing of the sample solution with the running solution. When online sample preconcentration is to be utilized, the sample solution must be optimized for each concentration technique. For a preliminary run, a relatively high concentration of the analyte will be favorable for easy detection. The separation time depends on the EOF velocity and effective capillary length (length to the detector) but under the conditions listed in Table 3.1 the migration time of the micelle may be less than 20 min.

When a preliminary run shows insufficient separation, several parameters may need to be adjusted in order to obtain optimized separation [9]. First, estimate the retention factors of analytes by Equation 3.1 using measured t_0 and t_{mc}. Should measuring the t_{mc} prove to be nontrivial due to the difficulty observing the micelle marker peak, simply assume t_{mc} is four times longer than t_0. If k values are between 0.5 and 10, k values should be optimized—alternatively, adjust to about 2 by changing the micelle (or surfactant) concentration. If k values are too high ($k > 10$), several options are available; decrease the micelle concentration by adding an organic solvent (methanol or acetonitrile), add a CD, or use another surfactant (e.g., a bile salt). If k values are too small ($k < 0.5$), limited options are available; increase surfactant concentration if the analyte is neutral, add an ion-pairing reagent (a tetraalkylammonium salt for anionic analytes and an alkanesulfonate for cationic analytes) or use cationic surfactant such as CTAB instead of SDS if the analytes are anionic.

If optimum k values do not provide acceptable resolution, further tuning may be required; the use of additives such as organic solvents or CDs may be needed, or the use of alternate surfactants may prove effective. The selection of additives is extensive, but it is recommended that addition of a low concentration of an organic solvent be trialed initially. If the unresolved analytes have closely related structures, addition of a CD derivative may prove effective. There are many different CD derivatives available, but simply β-CD or γ-CD can be tested initially. Further tuning can be performed by selecting other CD derivatives (e.g., sulfated) if necessary. Another choice is the modification of the micelle by using mixed micelles, in particular, addition of a nonionic surfactant such as Brij-35 or Tween 60, adding cosurfactants such as 1-propanol or 1-butanol, or by adding an organic counter ion such as tetraalkylammonium salts for SDS.

The above-mentioned guideline describes practical procedures based on the migration behavior [9]. Computer-assisted modeling, predictions, and multifactor optimization strategy are proposed based on physicochemical models describing the migration behavior for ionizable analytes [69–71]. Other than the factors mentioned above, several experimental parameters that include temperature, applied voltage, buffer concentration, pH, and so forth affect resolution and many parameters can be altered in a manner that provides a fine tuning via statistical approaches, using fitting procedures of polynomial equations (chemometrics). There are several techniques reported to optimize MEKC separation conditions. It is beyond the scope of this chapter to describe chemometric techniques in detail, and only some references are mentioned [72–76]. Although the approaches based on statistical designs are helpful to find the optimum values for many parameters, they should not be used as a black box system [77]. Analysts need to play a crucial role in optimizing separation conditions by carefully considering separation models.

3.6 CONCLUDING REMARKS

Most fundamental characteristics on MEKC were investigated by mid-1990s and, hence, this chapter is not much different from that in second edition of the book [78] in fundamental theory and major procedures. New parts of the chapter are mainly MS detection, online sample preconcentration, and new applications. Unlike a decade ago, MEKC is now a mature technique in CE arsenal, and while a few new, innovative MEKC-related techniques may not arise, applications of MEKC is likely to be widely expanded—this is certainly true in the bioanalytical fields and in the ever-expanding microscale separations including those in microfabricated devices. Although MEKC is considered to be suitable for the separation of small molecules, it is also useful for the separation of large molecules such as peptides and proteins, and applications to large molecules will certainly be explored more extensively in the future. The latest progress in MEKC has been recently reviewed [79].

One of the issues that still remains unsolved in MEKC is the improvement in reproducibility in quantitative analysis, including migration time and peak height or peak area—a problem that is not characteristic of MEKC only, but general to every CE technique. To verify the validity of the MEKC method, intercompany cross-validation exercises have been performed and the results have shown acceptable method performance in terms of relative migration time precision [80]. While reproducible EOF is an issue, it is so in every CE separation modes (except for gel electrophoresis); however, MEKC has more parameters to consider to maintain acceptable reproducibility. First, some characteristic issues in MEKC are discussed to improve quantitative results. CMC and distribution coefficient are both temperature sensitive: an increase in temperature causes an increase in CMC and, hence, a decrease in the micellar concentration, which reduces the retention factor. On the other hand, an increase in temperature causes a decrease in distribution coefficient and, hence, the retention factor. Therefore, the temperature increase will significantly decrease the retention factor from the viewpoint of CMC and distribution coefficient. It is well known that an increase in temperature reduces viscosity of the running solution in any CE mode, which increases both electrophoretic and EOF velocities, reducing the migration time. Thus, temperature and migration time have an inverse relationship where a decrease in migration time accompanies an increase in temperature, and the effect will be larger in MEKC than in CZE. Consequently, temperature must be controlled with diligence in MEKC, relative to other CE modes, in order to obtain high reproducibility. To maintain a low current is also a good strategy, particularly since MEKC usually produces a higher current than CZE due to the addition of the surfactant. A compromise is, therefore, necessary between high Joule heating and fast separation time. It should be noted that an added bonus with the use of micelles is that they may prevent analyte adsorption on the capillary wall, which may contribute to good reproducibility.

Minimizing the temperature effects discussed above could be obtained with the use of polymer micelles or polymer surfactants [81–83], whose CMC is zero, and even in nonaqueous solvent, the micelle is stable. Although several polymer surfactants are commercially available, no such surfactant is widely accepted, probably because SDS, CTAB, or CTAC, and bile salts are superior to polymer surfactants as the pseudostationary phase in MEKC. Although microemulsion electrokinetic chromatography (MEEKC) is not discussed in this chapter but covered in Chapter 4 by Altria and colleagues, a similar optimization strategy to that in MEKC applies to MEEKC [84–86]. Since microemulsion (oil-in-water or o/w) consists of four components, surfactant, oil (organic solvent), cosurfactant, and buffer, available parameters for optimizing selectivity are more extensive than that in MEKC. For most analytes, the choice of the surfactant is most important because the surface of the microemulsion affects significantly selectivity as in the case of micelles. For nonpolar hydrophobic analytes, the choice of the organic solvent is effective but conditions that confer stability to the microemulsion limit the range of organic solvent composition. An advantage of MEEKC over MEKC is the ability to vary the migration time window by changing the concentration of SDS in microemulsion [84]. It should be noted that the composition of microemulsion tends to change with evaporation of the organic solvent, particularly noteworthy if a volatile organic solvent such as hexane is used as the core oil.

Finally, the author would like to emphasize that, although MEKC gives similar separation selectivity to that observed with reversed-phase HPLC, the choice of selectivity manipulations are more versatile and separation efficiency is much higher than that of HPLC. Detection sensitivity, in terms of concentration, has been considered inferior to that of HPLC, but sensitivity is not an issue if high sensitivity detectors are implemented or if online sample preconcentration is employed. One more important feature of MEKC, as a mode of CE, is the ability to use small amounts of sample and almost no need of organic solvents. Therefore, the basis for the apparent lower popularity of MEKC in industrial routine analysis is, at least partially, because CE techniques are not widely accepted among those who are engaged in industrial analysis based largely on the popular HPLC platform. However, it is strongly encouraged that the instrument companies explore the development new CE instruments that are more user-friendly and easy to operate.

REFERENCES

1. Terabe, S. et al., Electrokinetic separations with micellar solutions and open-tubular capillaries, *Anal. Chem.*, 56, 111, 1984.
2. Terabe, S., Electrokinetic chromatography: an interface between electrophoresis and chromatography, *Trends Anal. Chem.*, 8, 129, 1989.
3. Zare, R.N., Tribute to Professor S. Terabe, *J. Chromatogr. A*, 1106, 4, 2006.
4. Terabe, S., *Micellar Electrokinetic Chromatography*, Beckman, Fullerton, 1992.
5. Vindevogel, J. and Sandra, P., *Introduction to Micellar Electrokinetic Chromatography*, Hüthig, Heidelberg, 1992.
6. Pyell, U., Ed., *Electrokinetic Chromatography*, John Wiley, Chichester, 2006.
7. Terabe, S., Chen, N., and Otsuka, K., Micellar electrokinetic chromatography, *Adv. Electrophoresis*, 7, 87, 1994.
8. Khaledi, M.G., Micellar electrokinetic chromatography, in *High Performance Capillary Electrophoresis*, Khaledi, M.G., Ed., Wiley-Interscience, New York, 1998, Chapter 3.
9. Terabe, S., Selectivity manipulation on micellar electrokinetic chromatography, *J. Pharm. Biomed. Anal.*, 10, 705, 1992.
10. Otsuka, K. and Terabe, S., Micellar electrokinetic chromatography, *Bull. Chem. Soc. Jpn.*, 71, 2465, 1989.
11. Terabe, S., Micellar electrokinetic chromatography, *Anal. Chem.*, 76, 240A, 2004.
12. Katayama, H., Ishihama, Y., and Asakawa, N., Stable capillary coating with successive multiple ionic polymer layers, *Anal. Chem.*, 70, 2254, 1998.
13. Nelson, W.M. et al., On-line partial filling micellar electrokinetic chromatography-electrospray ionization mass spectrometry, *J. Chromatogr. A*, 749, 219, 1996.
14. Koezuka, K. et al., Separation and detection of closely related peptides by micellar electrokinetic chromatography coupled with electrospray ionization mass spectrometry using the partial filling technique, *J. Chromatogr. B*, 689, 3, 1997.
15. Muijselaar, P.G., Otsuka, K., and Terabe, S., On-line coupling of partial-filling micellar electrokinetic chromatography with mass spectrometry, *J. Chromatogr. A*, 802, 3, 1998.

16. Suomi, J. et al., Determination of iridoid glycosides by micellar electrokinetic capillary chromatography-mass spectrometry with use of the partial filling technique, *Electrophoresis*, 22, 2580, 2001.
17. Stubberud, K. et al., Partial filling micellar electrokinetic chromatography optimization studies of ibuprofen, codeine and degradation products, and coupling to mass spectrometry, *Electrophoresis*, 23, 572, 2002.
18. Somsen, G.W., Mol, R., and de Jong, G.J., On-line micellar electrokinetic chromatography–mass spectrometry: feasibility of direct introduction of non-volatile buffer and surfactant into the electrospray interface, *J. Chromatogr. A*, 1000, 953, 2003.
19. Mol, R. et al., Micellar electrokinetic chromatography–electrospray ionization mass spectrometry for the identification of drug impurities, *J. Chromatogr. B*, 843, 283, 2006.
20. Mol, R., de Jong, G.J., and Somsen, G.W., Atmospheric pressure photoionization for enhanced compatibility in on-line micellar electrokinetic chromatography-mass spectrometry, *Anal. Chem.*, 77, 5277, 2005.
21. Nishi, N., Tsumagari, N., and Terabe, S., Effect of tetraalkylammonium salts on micellar electrokinetic chromatography of ionic substances, *Anal. Chem.*, 61, 2434, 1989.
22. Terabe, S. et al., Effect of urea addition in micellar electrokinetic chromatography, *J. Chromatogr.*, 545, 359, 1991.
23. Kaneta, T. et al., Effect of addition of glucose on micellar electrokinetic capillary chromatography with sodium dodecyl sulphate, *J. Chromatogr.*, 609, 369, 1992.
24. Terabe, S., Otsuka, K., and Ando, T., Electrokinetic chromatography with micellar solution and open-tubular capillary, *Anal. Chem.*, 57, 834, 1985.
25. Foley, J. P., Optimization of micellar electrokinetic chromatography, *Anal. Chem.*, 62, 1302, 1990.
26. Zhang, C.-X., Sun, Z.-P., and Ling, D.-K., Micellar electrokinetic capillary chromatography theory based on conventional chromatography, *J. Chromatogr. A*, 655, 309, 1993.
27. Yang, S. and Khaledi, M.G., Chemical selectivity in micellar electrokinetic chromatography: characterization of solute-micelle interactions for classification of surfactants, *Anal. Chem.*, 67, 499, 1995.
28. Chen, N. et al., Effect of physico-chemical properties and molecular structure on the micelle-water partition coefficient in micellar electrokinetic chromatography, *J. Chromatogr. A*, 678, 327, 1994.
29. Poole, C.F., Poole, S.K., and Abraham, M.H., Recommendations for the determination of selectivity in micellar electrokinetic chromatography, *J. Chromatogr. A*, 798, 207, 1998.
30. Muijselaar, P.G., Claessens, H.A., and Cramers, C.A., Characterization of pseudostationary phases in micellar electrokinetic chromatography by applying linear solvation energy relationships and retention indexes, *Anal. Chem.*, 69, 1184, 1997.
31. Chien, R.L. and Burgi, D.S., Optimization in sample stacking for high-performance capillary electrophoresis, *Anal. Chem.*, 63, 2042, 1991.
32. Quirino, J.P. and Terabe, S., On-line concentration of neutral analytes for micellar electrokinetic chromatography. 3. Stacking with reverse migrating micelles, *Anal. Chem.*, 70,149, 1998.
33. Quirino, J.P. and Terabe, S., On-line concentration of neutral analytes for micellar electrokinetic chromatography. 5. Field-enhanced sample injection with reverse migrating micelles, *Anal. Chem.*, 70, 1893, 1998.
34. Quirino, J.P. and Terabe, S., Exceeding 5000-fold concentration of dilute analytes in micellar electrokinetic chromatography, *Science*, 282, 465, 1998.
35. Quirino, J.P. and Terabe, S., Sweeping of analyte zone in electrokinetic chromatography, *Anal. Chem.*, 71, 1638, 1999.
36. Palmer, J., Munro, N.J., and Landers, J.P., A universal concept for stacking neutral analytes in micellar capillary electrophoresis, *Anal. Chem.*, 71, 1679, 1999.
37. Sera, Y. et al., Sweeping on a microchip: concentration profiles of the focused zone in micellar electrokinetic chromatography, *Electrophoresis*, 22, 3509, 2001.
38. Kim, J.-B. et al., On-line sample concentration in MEKC using cationic surfactants, *J. Chromatogr. A*, 916, 123, 2001.
39. Isoo, K. and Terabe, S., Analysis of metal ions by sweeping via dynamic complexation and cation-selective exhaustive injection in capillary electrophoresis, *Anal. Chem.*, 75, 6789, 2003.

40. Quirino, J.P. and Terabe, S., Sweeping of neutral analytes via complexation with borate in capillary zone electrophoresis, *Chromatographia*, 53, 285, 2001.
41. Palmer, J., Burgi, D.S., and Landers, J.P., Electrokinetic stacking injection of neutral analytes under continuous conductivity conditions, *Anal. Chem.*, 74, 632, 2002.
42. Quirino, J.P. and Terabe, S., Approaching a million-fold sensitivity increase in capillary electrophoresis with direct ultraviolet detection: cation-selective exhaustive injection and sweeping, *Anal. Chem.*, 72, 1023, 2000.
43. Kim, J.-B., Otsuka, K., and Terabe, S., Anion selective exhaustive injection-sweep–micellar electrokinetic chromatography, *J. Chromatogr. A*, 932, 129, 2001.
44. Britz-McKibbin, P. and Chen, D.D.Y., Selective focusing of catecholamines and weakly acidic compounds by capillary electrophoresis using a dynamic pH junction, *Anal. Chem.*, 72, 1242, 2000.
45. Britz-McKibbin, P., Otsuka, K., and Terabe, S., On-line focusing of flavin derivatives using dynamic pH junction-sweeping capillary electrophoresis with laser-induced fluorescence detection, *Anal. Chem.*, 74, 3736, 2000.
46. Molina, M. and Silva, M., Micellar electrokinetic chromatography: current developments and future, *Electrophoresis*, 23, 3907, 2002.
47. Pappas, T.J., Gayton-Ely, M., and Holland, L.A., Recent advances in micellar electrokinetic chromatography, *Electrophoresis*, 26, 719, 2005.
48. Huie, C.W., Recent applications of microemulsion electrokinetic chromatography, *Electrophoresis*, 27,60, 2006.
49. Nishi, H. and Terabe, S., Micellar electrokinetic chromatography: perspectives in drug analysis, *J. Chromatogr. A*, 735, 3, 1996.
50. Wätzig, H., Degenhardt, M., and Kunkel, A., Strategies for capillary electrophoresis: method development and validation for pharmaceutical and biological applications, *Electrophoresis*, 19, 2695, 1998.
51. Garcia-Ruiz, C. and Marina, M.L., Recent advances in the analysis of antibiotics by capillary electrophoresis, *Electrophoresis*, 27, 266, 2006.
52. Sung, W.-C. and Chen, S.-H., Pharmacokinetic applications of capillary electrophoresis: a review on recent progress, *Electrophoresis*, 27, 257, 2006.
53. Ha, P.T.T., Hoogmartens, J., and Van Schepdael, A., Recent advances in pharmaceutical applications of chiral capillary electrophoresis, *J. Pharm. Biomed. Anal.*, 41, 1, 2006.
54. Orlandini, S. et al., Micellar electrokinetic chromatography for the simultaneous determination of ketorolac tromethamine and its impurities multivariate optimization and validation, *J. Chromatogr. A*, 1032, 253, 2004.
55. Iadarola, P. et al., Micellar electrokinetic chromatographic and capillary zone electrophoretic methods for screening urinary biomarkers of human disorders: a critical review of the state-of-the-art, *Electrophoresis*, 26, 752, 2005.
56. Nakagawa, T. et al., Electrokinetic chromatography for drug analysis. Separation and determination of cefpiramide in human plasma, *Chem. Pharm. Bull.*, 37, 707, 1989.
57. Penalvo, G.C. et al., Evaluation of capillary zone electrophoresis and micellar electrokinetic capillary chromatography with direct injection of plasma for the determination of cefotaxime and its metabolite, *Anal. Chem.*, 69, 1364, 1997.
58. Huang, C.-W. et al., Sweeping technique combined with micellar electrokinetic chromatography for the simultaneous determination of flunitrazepam and its major metabolites, *J. Chromatogr. A*, 1110, 240, 2006.
59. Paxon, T.L. et al., Microcolumn separation of amine metabolites in the fruit fly, *Anal. Chem.*, 77, 5349, 2005.
60. Shou, M. et al., Monitoring dopamine *in vivo* by microdialysis sampling and on-line CE-laser-induced fluorescence, *Anal. Chem.*, 78, 6717, 2006.
61. Frazier, R.A. and Papadopoulou, A., Recent advances in the application of capillary electrophoresis for food analysis, *Electrophoresis*, 24, 4095, 2003.
62. Cifuentes, A., Recent advances in the application of capillary electromigration methods for food analysis, *Electrophoresis*, 27, 283, 2006.
63. Weiss, D.J. and Anderton, C.R., Determination of catechins in matcha green tea by micellar electrokinetic chromatography, *J. Chromatogr. A*, 1011, 173, 2003.

64. Juan-García, A., Picó, Y., and Font, G., Capillary electrophoresis for analyzing pesticides in fruits and vegetables using solid-phase extraction and stir-bar sorptive extraction, *J. Chromatogr. A*, 1073, 229, 2005.
65. Dabek-Zlotorzynska, E., Capillary electrophoresis in the determination of pollutants, *Electrophoresis*, 18, 2453, 1997.
66. Menzinger, F. et al., Analysis of agrochemicals by capillary electrophoresis, *J. Chromatogr. A*, 891, 45, 2000.
67. Dabek-Zlotorzynska, E. and Celo, V., Recent advances in capillary electrophoresis and capillary electrochromatography of pollutants, *Electrophoresis*, 27, 304, 2006.
68. Hinsmann, P. et al., Determination of pesticides in waters by automatic on-line solid-phase extraction–capillary electrophoresis, *J. Chromatogr. A*, 866, 137, 2000.
69. Khaledi, M.G., Smith, S.C., and Strasters, J.K., Micellar electrokinetic capillary chromatography of acidic solutes: migration behavior and optimization strategies, *Anal. Chem.*, 63, 1820, 1991.
70. Strasters, J.K. and Khaledi, M.G., Migration behavior of cationic solutes in micellar electrokinetic capillary chromatography, *Anal. Chem.*, 63, 2503, 1991.
71. Quang, C., Wasters, J.K., and Khaledi, M.G., Computer-assisted modeling, prediction, and multifactor optimization in micellar electrokinetic chromatography of ionizable compounds, *Anal. Chem.*, 66, 1646, 1994.
72. Vindevogel, J. and Sandra, P., Resolution optimization in micellar electrokinetic chromatography: use of Plackett-Burman statistical design for the analysis of testosterone esters, *Anal. Chem.*, 63, 1530, 1991.
73. He, Y. and Lee, H.K., Orthogonal array design experiments for optimizing the separation of various pesticides by cyclodextrin-modified micellar electrokinetic chromatography, *J. Chromatogr. A*, 793, 331, 1998.
74. Persson-Stubberud, K. and Åström, O., Separation of ibuprofen, codeine phosphate, their degradation products and impurities by capillary electrophoresis I. Method development and optimization with fractional factorial design, *J. Chromatogr. A*, 798, 307, 1998.
75. Mikaeli, S., Thorsén, G., and Karlberg, B., Optimisation of resolution in micellar electrokinetic chromatography by multivariate evaluation of electrolytes, *J. Chromatogr. A*, 907, 267, 2001.
76. Wan, H., Öhman, M., and Blomberg, R.G., Chemometric modeling of neurotransmitter amino acid separation in normal and reversed migration micellar electrokinetic chromatography, *J. Chromatogr. A*, 916, 255, 2001.
77. Corstjens, H. et al., Optimisation of selectivity in capillary electrophoresis with emphasis on micellar electrokinetic capillary chromatography, *J. Chromatogr. A*, 715, 1, 1995.
78. Mazzeo, J.R., Micellar electrokinetic chromatography, in *Handbook of Capillary Electrophoresis*, 2nd ed., Landers, J.P., Ed., CRC Press, Boca Raton, FL, 1996, Chapter 2.
79. Silva, M., MEKC: An update focusing on practical aspects, *Electrophoresis*, 28, 174, 2007.
80. Altria, K.D. et al., An inter-company cross-validation exercise on capillary electrophoresis testing of dose uniformity of paracetamol content in formulations, *Chromatographia*, 39, 180, 1994.
81. Palmer, C.P. and Terabe, S., Micelle polymers as pseudostationary phases in MEKC: chromatographic performance and chemical selectivity, *Anal. Chem.*, 69, 1852, 1997.
82. Palmer, C.P., Recent progress in the development, characterization and application of polymeric pseudophases for electrokinetic chromatography, *Electrophoresis*, 23, 3993, 2002.
83. Palmer, C.P. and McCarney, J.P., Recent progress in the use of soluble ionic polymers as pseudostationary phases for electrokinetic chromatography, *Electrophoresis*, 25, 4086, 2004.
84. Terabe, S. et al., Microemulsion electrokinetic chromatography: comparison with micellar electrokinetic chromatography, *J. Chromatogr.*, 608, 23, 1992.
85. Altria, K.D., Highly efficient and selective separations of a wide range of analytes obtained by an optimised microemulsion electrokinetic chromatography method, *Chromatographia*, 49, 457, 1999.
86. Altria, K.D., Clark, B.J., and Mahuzier, P.-E., The effect of operating variables in microemulsion electrokinetic capillary chromatography, *Chromatographia*, 52, 758, 2000.

4 Capillary Electrophoresis for Pharmaceutical Analysis

Eamon McEvoy, Alex Marsh, Kevin Altria, Sheila Donegan, and Joe Power

CONTENTS

4.1 INTRODUCTION

The use of capillary electrophoresis (CE) as an analytical technique has become more widespread and popular in recent years and has established itself as a method of choice for many applications. The ability to analyze small ions and organic molecules has made it the mainstay in many industries such as the food and beverage, biotechnology, and pharmaceutical industries, many of which have incorporated CE into their research as well as their quality assurance (QA) and quality control (QC) departments.

Pharmaceutical analysis is dominated by high-performance liquid chromatography (HPLC) with thin layer chromatography (TLC) and gas chromatography (GC) also used to a lesser extent, as their quantitative abilities are not widespread as HPLC. CE, however, is becoming a more widely used technique and is recognized by several regulatory authorities as a reliable routine analytical technique. Although CE was initially heralded for its high speed and low sample volume, resolving power, and versatility, the technique is also valuable because it is quantitative, can be automated, and can separate compounds that have been traditionally difficult to handle by HPLC. Qualitative analysis is also possible by capillary electrophoresis-mass spectrometry (CE-MS).

The small sample volumes required by CE can be an advantage over HPLC if sample supply is limited, for example, in drug discovery, however, with short path lengths using traditional ultraviolet (UV) detection methods, the technique can suffer from low sensitivity. A number of solutions to this problem have been introduced including "Bubble Cell" by Agilent, which increases the path length at the point of detection. A "High Sensitivity Detection Cell" also by Agilent improves detection sensitivity 10-fold over normal capillaries. Other detection methods such as "Laser-Induced Fluorescence Detection" (LIFD) and CE-MS also give better sensitivity and have led to CE being on a par with HPLC for many pharmaceutical applications and indeed CE is superior to HPLC for many separations.

The advantages of CE for pharmaceutical analysis include its speed and cost of analysis, reductions in solvent consumption, and disposal and the possibility of rapid method development. CE also offers the possibility of using a single set of operating conditions for a wide range of separations that can improve laboratory efficiency. CE instruments can be coupled to a variety of detector types including mass spectrometers, indirect UV detectors, LIFD's, and low UV wavelength detectors.

Separation efficiencies achieved by CE can be an order of magnitude better than competing HPLC methods, which highlights the potential resolving power of CE for complex and difficult separations, which is a valuable advantage in pharmaceutical analysis. The use of an open tubular capillary improves resolution relative to that of packed HPLC columns by eliminating the multiple path term (A) in the following Van Deemter equation. CE reduces plate height further by also eliminating the mass transfer term (Cu_x) that comes from the finite time needed for solutes to equilibrate between the mobile and stationary phases in liquid chromatography (LC). This stationary phase is absent in CE [1] [with the exception of micellar electrokinetic chromatography (MEKC) and microemulsion electrokinetic chromatography (MEEKC) where a "pseudo-stationary phase" forms part of the separation mechanism, these CE modes will be discussed later in the chapter].

$$\text{van Deemter equation for plate height} \quad H = A + \frac{B}{u_X} + \frac{C}{u_X},$$

where A is the multiple path term which is zero for CE capillaries, B/u_X is the longitudinal diffusion term and is the only contributor to band broadening in CE, and C/u_X is the equilibration time (mass transfer) term also zero for CE.

High-speed separations are possible with CE, sometimes as quick as a matter of seconds. Because of this it is an ideal technique for "high-throughput" analyses such as forensic testing, DNA sequencing, and, in particular, screening of candidate drug compounds. Multiplexed CE instruments can be used to perform multiple parallel separations, dramatically increasing sample throughput. For example, the "CombiSep cePRO 9600™" utilizes multiple capillaries and can perform 96 separations in parallel.

An important aspect of CE for pharmaceutical analysis is its ability to separate all molecules of pharmaceutical interest such as small and large synthetic and natural drugs, proteins and peptides, oligonucleotides, carbohydrates and polysaccharides, inorganic ions, and so forth.

This chapter will look at the use of CE for pharmaceutical analysis and will include descriptions of the various modes of CE and their suitability for quantitative and qualitative analysis of pharmaceutical compounds. Practical applications of CE for the analysis of pharmaceuticals will be covered, these applications include drug assay, impurity determination, physicochemical measurements, chiral separations, and the analysis of small molecules. A section covering the approach to CE method development for pharmaceutical analysis will include guidelines to selecting the best mode of CE for an intended separation. Extensive data will be provided on successful pharmaceutical separations with references to extra source material for the interested reader. This chapter will provide a comprehensive and up to date view of the role and importance of CE for the analysis of pharmaceuticals and will provide the reader with practical information and real data that will help them to decide if CE is suitable for an intended separation.

4.2 BACKGROUND

In the 1930s, Tiselius developed the first electrophoretic apparatus to perform separations based on differential migration rates in an electrical field. Up until the early 1980s, most of the research carried out using this technique was by biochemists and molecular biologists on the separation, isolation, and analysis of proteins and other biological macromolecules. Since its introduction in the 1980s [2–5] and with the appearance of commercial instruments for performing analytical electrophoresis on a microscale in capillary columns, CE has become established as a fast, reliable, and highly efficient alternative and complementary method to LC. While CE can perform separations that may be difficult by traditional LC methods and provides different migration patterns, combined use of these two techniques can be a powerful analytical tool for the pharmaceutical analyst. Over the past number of years since the early 1990s, there have been many review papers [6–12] and books [13,14] published concerning CE for pharmaceutical analysis. Since the first applications of CE to pharmaceutical analysis by Altria and Simpson [15] in 1987, the number of reports of CE for a range of pharmaceutical applications now is in hundreds and the number of pharmaceutical compounds analyzed by CE is far greater [13]. There have also been a number of specialized reviews of CE for pharmaceutical analysis including chiral CE [16–20], CE with electrochemical detection [21] and conductivity detection [22], LIFD [23], CE coupled with electrospray-ionization MS (CE-ESI-MS) [24], and CE determination of acid dissociation constants [25].

When deciding on the optimum CE conditions for an intended separation, the pharmaceutical analyst must take a different approach to that of the analyst using HPLC techniques. The vast range of HPLC methods routinely used for pharmaceutical analysis would lead the liquid chromatographer to use literature databases and pharmacopoeia methods along with his/her own experience to decide on initial HPLC conditions. Properties such as sample solubility, log P values, sample polarity, stationary phase characteristics, and mobile phase compositions, among others, all are important to the LC analyst when deciding on the initial chromatographic conditions and can be easily obtained for most applications. The CE analyst, however, must consider the problem separation in a different manner. The depth of knowledge and literature associated with CE method development is in its infancy when compared to LC. As CE is an electro-driven separation technique, the approach to method development must look at factors such as compound pK_a data, functional groups, electrolyte pH, and applied voltage. Compound solubility will also determine which mode of CE is applicable to the required separation. CE method development guidelines will be given more consideration in a later section of this chapter.

4.3 THEORETICAL ASPECTS (ELECTROPHORETIC MODES)

There are several electrophoretic modes that can and have been be used to analyze pharmaceuticals. Each mode offers different possibilities to the CE analyst and the choice of CE mode is the first step when developing a separation method. The mode of CE that best suits the sample to be analyzed must be decided. Some questions that the analyst must ask include are all of the components to be separated soluble in an aqueous electrolyte; are the components charged, neutral, or both; are physicochemical measurements available for the analytes, that is, pK_a and/or log P values; what are the properties of the sample matrix; and so forth. This section will provide the reader with a brief introduction to the various modes of CE, which have been utilized for pharmaceutical analysis and give tips on deciding the mode and detection method for an intended separation/analysis. The detailed theoretical aspects of each CE mode are too broad to be covered in this chapter so the reader will be directed to other sources for more in depth theory.

4.3.1 FREE SOLUTION CAPILLARY ELECTROPHORESIS

Free solution CE (FSCE) or capillary zone electrophoresis (CZE), as it is sometimes called, is the simplest and one of the most widely used forms of CE and involves the use of buffered aqueous solutions as the carrier electrolytes. FSCE is principally used to separate charged, water-soluble analytes and separations rely principally on the pH-controlled dissociation of acidic groups or the protonation of basic functions on the solute. These ionic species are separated based on differences in their charge-to-mass ratios. For example, basic drugs are separated at low pH as cations while acidic drugs are separated as anions at high pH. In FSCE all neutral compounds are swept, unresolved, through the detector together. Under the influence of an applied electric field, sample ions will move toward their appropriate electrode. Cations migrate toward the cathode and anions toward the anode. The speed of their movement toward the electrode is governed by their size and number of appropriate charges. Smaller molecules with a large number of charges will move more quickly than larger or less charged compounds. The speed of movement, known as the electrophoretic mobility, is characteristic of each solute. The mobility of a species can also be changed by complexing the ion as it migrates along the capillary. For example, additives such as cyclodextrins (CDs) can be used to complex with drug enantiomers to achieve chiral separations. The presence of an electroosmotic flow (EOF) allows the separation and detection of both cations and anions within a single analysis, as above pH 7 the EOF is sufficiently strong to sweep anions to the cathode regardless of their charge.

There are a number of variables in FSCE that can be used in the optimization of FSCE methods. These include the operating pH, electrolyte type and concentration, capillary dimensions, temperature, and injection volume. Electrolyte additives such as ion-pair reagents and chiral substances can also be employed in order to manipulate selectivity.

As the majority of pharmaceutical compounds are basic, the highly polar nature of these compounds can make separating them complex by LC methods. Ion-pairing reagents and column regeneration is often necessary to reduce nonspecific ionic interactions in the LC column. With CE, these highly functional groups can be exploited to enable separation. These basic drugs and their impurities can be separated using FSCE in two ways. First, at low pH, the capillary surface is essentially neutral resulting in a suppression of the EOF in the capillary and the protonated analytes migrate toward the detector with little or no influence from the EOF. The second and less desirable method is to use an amine inner capillary surface to repel cationic interactions within the capillary wall, which allows separation over a wide pH range. Using a low pH phosphate buffer, Hudson and coworkers [26] reported the use of FSCE to analyze over 500 basic pharmaceutical compounds.

To separate mixtures of acidic and basic drugs (or any cationic and anionic species), FSCE at high pH can be used [27]. At pH 7 or greater, the EOF generated by the applied current is sufficiently strong to sweep anions to the detector. Analyte migration time is dependent on solute charge type and density, strongly cationic species migrate first while small highly charged anions attempt to migrate against the EOF and are detected last. Neutrals will not be resolved and migrate with the EOF. Readers who require a more detailed description of the theoretical aspects of FSCE will find a number of books dealing with the subject [28–31]. A review paper by Smith and Evans [32] outlines the use of FSCE in pharmaceutical and biomedical analysis.

4.3.2 Nonaqueous Capillary Electrophoresis

Nonaqueous CE (NACE) is used for the separation of water-insoluble or sparingly soluble pharmaceuticals. NACE employs electrolytes composed of organic solvents and has been successfully utilized [33–38]. NACE is also useful for the resolution of water-soluble charged solutes as the selectivity obtained can be different to aqueous-based separations. The viscosity and dielectric constants of organic solvents can have an effect on both sample ion mobility and the level of EOF. Resolution, efficiency, and migration times are critically affected by the nature of the organic solvent, the electrolyte composition, its concentration, and temperature. Because of their different physical and chemical properties (viscosity, dielectric constant, polarity, autoprotolysis constant, etc.), methanol and acetonitrile are the most frequently used solvents for NACE. In particular, methanol–acetonitrile mixture containing 25 mM ammonium acetate and 1 M acetic acid is considered to be the appropriate electrolyte solution for the separation of a large variety of basic drugs [39]. The low currents present in NACE not only allow the use of higher electrolyte salt concentrations and higher electric field strengths but also the sample load can be scaled-up by employing capillaries with wider inside diameter. In addition, an effective sample introduction to a mass spectrometer, in terms of volatility, surface tension, flow rate, and ionization can be expected to further extend the use of NACE [40]. NACE exploits the vastly different physicochemical properties of organic solvents to control EOF and analyte migration. The ability of organic solvents to accept protons from the silanol groups of the capillary wall appears to play an important role in the development of an EOF. Although an EOF may not be required or may be completely undesired in a few electromigration capillary techniques, it plays a significant role in separations in free solution CE. However, the purity of the solvents may affect the EOF as they may contain foreign ionic species. The velocity of EOF is higher in pure solvents than in electrolyte solutions, which may be beneficial when fast liquid transport is desired. Salt-free solvents may also be advantageous in mass spectrometric detection [40].

One of the most attractive features of organic solvents is that their physical and chemical properties differ widely, both from each other and from water. Accordingly, simply changing the organic solvent or varying the proportions of two solvents can achieve selectivity manipulation in NACE. pK_a values in organic solvents can be significantly different from those in water allowing separations that are difficult to achieve in aqueous electrolytes. It has been shown that all solvents in which a measurable EOF is developed in the capillary are either amphiprotic or aprotic solvents. In addition to self-dissociation, amphiprotic solvents act as proton donors or acceptors if there are other proton donors or acceptors in the separation system. However, there are differences in their proton donor and acceptor capabilities. Methanol, similar to water, has an equal tendency to donate and accept protons, but basic amide-type solvents are worse proton donors than proton acceptors. Aprotic solvents, such as acetonitrile, can only accept protons. Inert solvents are capable of neither autoprotolysis nor donation–acceptance of protons to a considerable extent, which makes them less suitable for NACE [40].

A number of review papers have been published which look at detection methods [41], selectivity manipulation [42], and pH of the background electrolyte [43] in NACE. This mode of CE has been reported to be used for a range of pharmaceutical separations; separation of a number of opium alkaloids [44], a mixture of cationic drugs [45], a range of tropane alkaloids [46], a range of beta-blockers [47], tricyclic antidepressants [37], and different basic drugs [48]. NACE has also been used to separate polar acidic and basic drugs [49] and to perform chiral separation of pharmaceutical amines [50].

4.3.3 Micellar Electrokinetic Chromatography

Micellar electrokinetic chromatography (MEKC) was developed by Terabe et al. [4], and combines the separation mechanism of chromatography with the electrophoretic and electroosmotic movement of solutes and solutions for the separation of constituents in a sample. This mode of CE is covered in detail in Chapter 3 by Terabe, and utilizes surfactant micelles as a "pseudo-stationary phase" in the carrier electrolyte. Surfactants are molecules with detergent properties, which are composed of a hydrophilic water-soluble head group and a hydrophobic water-insoluble hydrocarbon chain group [e.g., sodium dodecyl sulfate (SDS), $CH_3–(CH_2)_{11}–O–SO_3^-Na^+$]. When anionic surfactant molecules are present in a solution at a concentration above their critical micelle concentration (CMC) they aggregate to form negatively charged micelles. These are generally spherical in shape with the hydrophilic head groups oriented in the aqueous buffer and the hydrocarbon chains in the centre. Figure 4.1 shows the arrangement of the molecules to form a micelle.

MEKC uses the same instrument set-up as in conventional CE and uses uncoated capillaries. When a water-insoluble hydrophobic compound is added to a micellar solution, it partitions into the core of the micelle and is solubilized. Conversely, if a water-soluble hydrophilic compound is added to the solution it will solubilize in the aqueous phase and will not partition into the micelle. Neutral compounds of intermediate water solubility will partition between the aqueous phase and the micelle core and the extent of the partitioning depends on the hydrophobicity of the compound. When high pH micellar solutions are used as the carrier electrolyte in CE, the negatively charged micelles migrate in the opposite direction to the EOF but the more powerful EOF carries them to the cathode,

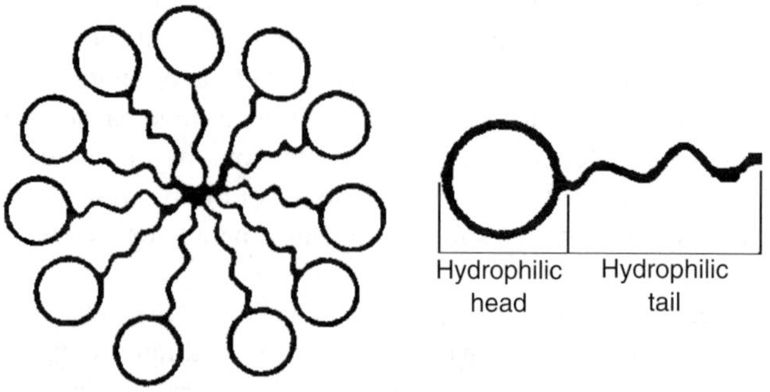

FIGURE 4.1 The arrangement of surfactant molecules when a micelle is formed. Depending on the surfactant, surfactants will form micelles with different numbers of molecules (Aggregation Number, AN). In the case of SDS its AN is 62 and its CMC is 8.27 mM.

through the detector. Very water-soluble solutes will remain in the aqueous phase and will separate due to differences in their electrophoretic mobilities. Solutes, which are completely solubilized by the micelle, will migrate and reach the detector at the same time as the micelle. The main advantage in employing MEKC is that uncharged neutral compounds will partition between the aqueous phase of the electrolyte and the micelle and will separate based on the difference in micelle/water partitioning and migration times are proportional to their micelle/water partition coefficients, log P_{mw}.

The composition of the micellar buffer solution can be changed in many ways in order to optimize the separations. The nature of the surfactant, that is, charge and concentration, the use of additives such as organic solvents, urea, and CDs can be altered in order to manipulate the separation of an analyte mixture. Books by Pyell [51] and Baker [31] dedicate chapters to more detailed discussions of MEKC theory and practice. Review papers by Pappas et al. [52], Molina and Silva [53], and Pyell [54] also cover many MEKC applications and development options.

4.3.4 Microemulsion Electrokinetic Chromatography

Microemulsion electrokinetic chromatography (MEEKC), first presented in 1991 by Watari [55], is an extension of MEKC and is probably the most versatile mode of CE that offers the possibility of highly efficient separations of both charged and neutral solutes with a wide range of water solubilities. The range of pharmaceutical applications of MEEKC has grown rapidly in recent years with chiral analysis, pharmaceutical assay, impurity determination, and log P measurements among the most common. MEKC and MEEKC differ in the type of pseudostationary phase used, as discussed in the previous section, MEKC electrolytes consist of micellar solutions while MEEKC electrolytes are composed of microemulsions, with solutes partitioning between the microemulsion droplets and electrolyte buffer.

Microemulsions are optically clear, thermodynamically stable suspensions of <10 nm diameter droplets of an immiscible liquid dispersed in another liquid, usually oil droplets in an aqueous continuous phase. These are referred to as oil-in-water (O/W) microemulsions and are the most commonly used in MEEKC applications; however, water-in-oil (W/O) microemulsions have also been used for MEEKC analysis of pharmaceuticals [56,57]. The microemulsion droplets are stabilized by the presence of a surfactant, for example, SDS, which was described in Section 4.3.3. The hydrocarbon chain of the surfactant resides in the oil droplet while the hydrophilic head remains in the aqueous phase, which reduces the interfacial tension at the oil–water interface. By incorporating a cosurfactant (usually a short chain alcohol), which bridges the oil–water interface in a manner similar to the surfactant, the interfacial tension is reduced further to almost zero resulting in a thermodynamically stable system. This arrangement is shown in Figure 4.2. SDS, an anionic surfactant, is the most commonly used surfactant in MEEKC and results in a negatively charged droplet. The aqueous continuous phase of the microemulsion usually contains additives, for example, pH buffers and/or ion-pair reagents, CDs, or organic modifiers to provide optimum separation conditions.

The separation mechanism in MEEKC is similar to that in MEKC where solutes are separated by both electrophoretic mechanisms in the capillary and by chromatographic interaction between the microemulsion droplets and the aqueous phase. Similar to MEKC, hydrophobic solutes favor inclusion in the oil droplet while hydrophilic solutes will favor the aqueous phase, neutral solutes will be separated according to differences in their oil/water partitioning coefficients (log P_{ow}) and migration times can be directly related to a neutral solutes log P value. Generally, high pH buffers such as phosphate or borate are used in MEEKC and when a voltage is applied across the capillary, these buffers generate a high EOF toward the cathode end of the capillary. The negatively charged oil droplets attempt to migrate through the capillary to the anode but the EOF sweeps them through the capillary to the cathode. If charged solutes interact with the charged droplet this will also influence their migration time. Separation, therefore, is dependent on solute size, charge, and hydrophobicity, which allows the separation of neutral components.

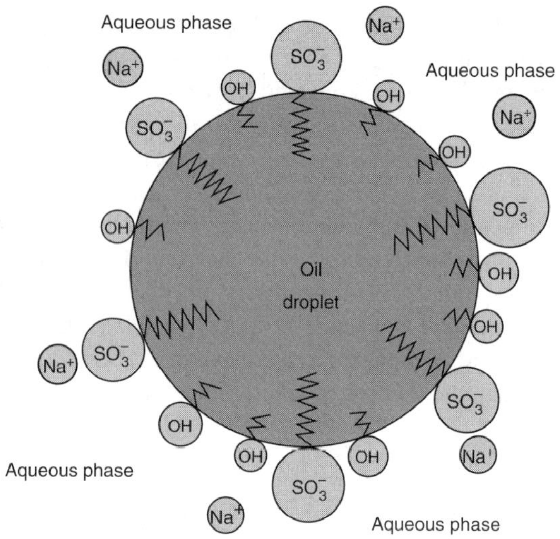

FIGURE 4.2 Schematic representation of an o/w microemulsion droplet.

MEEKC is the most versatile mode of CE for pharmaceutical analysis and offers many advantages over MEKC. It can be applied to a wider range of solutes than MEKC as solutes can penetrate the surface of a microemulsion droplet more easily than a more rigid micelle and offers greater separation capability for water-insoluble compounds [58]. MEEKC has been found to provide superior separation efficiency than MEKC, probably due to improved mass transfer between the microemulsion droplet and the aqueous phase [59]. MEEKC also provides a larger separation window, the size of which can be controlled and therefore offers greater separation capability for hydrophobic compounds [58,60]. Owing to the unique ability of microemulsions to solubilize water-soluble and water-insoluble, charged and neutral compounds, MEEKC is particularly suited to performing complex pharmaceutical separations. The many variables and operating parameters involved in MEEKC provide a large number of method development options for difficult pharmaceutical separations. Other advantages of utilizing MEEKC include the potential elimination or reduction in sample preparation steps resulting in rapid analysis times for difficult water-insoluble samples. This is due to the solubilizing power of microemulsions for water-soluble and water-insoluble compounds. Samples can be simply dissolved in the microemulsion and injected directly onto the capillary with little or no pretreatment. This was demonstrated by Broderick et al. [56] and Altria et al. [57] when utilizing W/O MEEKC for the analysis of a number of pharmaceutical preparations and sunscreen lotions. As the microemulsion droplets are less than 10 nm in diameter and optically transparent, they are very well suited to the use of low wavelength detection. This can be advantageous when detecting analytes with weak chromophores or when the sample matrix interferes with detection at higher wavelengths. This was demonstrated by McEvoy et al. [61] when detecting ibuprofen at 190 nm using O/W microemulsion liquid chromatography (MELC). Many reports have been published detailing the use of MEEKC for pharmaceutical applications and there have been a number of review papers [62–66] written, which describe further examples of MEEKC applications and advancements in method development with descriptions of operating parameter effects. An application-based research paper by Altria [58] highlights the suitability of MEEKC for a broad range of pharmaceutical applications. Readers can also find detailed theoretical aspects of MEEKC in a paper by Watari [66] and *Electrokinetic Chromatography* by Pyell [51].

4.3.5 Capillary Electrophoresis-Mass Spectrometry

Capillary electrophoresis-mass spectrometry combines the short analysis time and high separation efficiency of CE with the molecular weight and structural information from the MS to provide a powerful tool that can be utilized to quantify unknown compounds and impurities. Currently, ESI serves as the most common interface between CE and MS, as it can produce ions directly from liquids at atmospheric pressure, and with high sensitivity and selectivity for a wide range of analytes of pharmaceutical significance. It can also be applied to the detection of a wide range of analytes without derivatization and gives the information necessary to determine the structural formula of the analytes of interest.

The most common detection technique in CE is on-column UV absorbance detection, which can be applied to most pharmaceutical applications as most organic compounds display some UV absorbance. A drawback of UV detection is its relatively low sensitivity due to the short optical path length of the capillary detection window. The marriage of CE and MS instrumental analytical techniques results in an extremely powerful and highly sensitive tool for the separation, identification, and characterization of a wide range of molecules, especially pharmaceuticals. Smith et al. [67] first introduced CE-MS in 1988 and since then there has been an increased utility of CE-MS in pharmaceutical analysis. A chapter covering the coupling of CE to MS [51] describes in more detail the interfacing of the two methods, in particular electrokinetic chromatography-MS (EKC-MS). A number of CE-MS reviews have also been published [67–74], which cover the many applications of this technique and deal with operational, development issues, and quantification [69].

4.3.6 Multiplexed Capillary Electrophoresis

The demand for high-throughput analytical methods to support the evolution of parallel synthetic technologies in drug discovery applications has led to the development of multiplexed CE to characterize libraries of compounds and alleviate backlogs in the discovery process. Multiplexed or parallel analysis is achieved in CE through the use of instruments containing bundles (arrays) of capillaries. The capability to use the same instrument to analyze multiple samples simultaneously, and the diverse separation conditions that are possible, allows rapid turnaround times. In the past 10 years, multiplexed CE instruments have become commercially available with a number of companies such as CombiSep, Beckman Coulter, and Genteon offering multiplexed systems with up to 384 capillaries coupled to laser-induced fluoroscence (LIF) and UV detectors. Applications of multiplexed CE in pharmaceutical analysis include physicochemical profiling [76–78], chiral analysis [79–81], and organic reaction monitoring [82]. In a recent paper, Marsh and Altria [83] reported the use of multiplexed CE for the measurement of log P and pK_a values, tablet assay, and impurity determination. While still in its infancy, the use of multiplexed CE in pharmaceutical applications shows clear advantages over traditional screening techniques. The ability to analyze multiple samples or vary experimental conditions in parallel offers a unique and powerful tool for drug discovery. Multiplexed CE has a broad range of high-throughput applications spanning the pharmaceutical, fine chemical, agrochemical, and biotechnology industries. The multiplexed CE format provides the flexibility to simultaneously vary separation conditions to speed method development processes. As this CE approach is relatively new, there are few sources of literature dedicated to the topic. The papers previously mentioned [76–84] cover the use of multiplexed CE for various pharmaceutical applications and give brief discussions of method development and instrumentation. A review paper by Pang et al. [85] gives more detailed information about the instrumentation and a wider range of applications.

4.4 METHODS OF DETECTION

Detection methods for CE analysis are as diverse as those used for HPLC. The most widely used detection methods used in CE include ultraviolet/visible (UV/Vis) absorbance, LIF, mass spectrometry, conductivity, amperometry, radiometric, and refractive index. When deciding which detection method is best suited to an intended application, the analyst must first know if the compounds to be separated can be detected using a certain type of detector, that is, does the analyte have a chromophore, does it fluoresce, and so forth. If selective detection is desirable then choosing a detector that will only detect the compound of interest or classes of compound may be required, especially if components reach the detector simultaneously. Should quantitative and detailed qualitative data be required, then mass spectrometric detection would be the method of choice. The same criteria for detectors in HPLC apply in CE including sensitivity, selectivity, linear range, and the ability to provide qualitative data.

This section will only discuss the most common detection methods used for CE analysis of pharmaceuticals, that is, UV/Vis absorbance, LIFD, mass spectrometry, and electrochemical detection. In the tables of applications, the method of detection is indicated for each application. This should aid the interested reader and provide information on detector suitability for intended applications.

4.4.1 UV/Vis Absorbance

UV/Vis absorbance is by far the most common detection mode utilized in CE analysis of pharmaceuticals. This is mainly due to the fact that most pharmaceutical compounds are organic in nature and contain one or more chromophores, which leads the analyst to choose this detection method first. This reason along with the fact that UV/Vis detectors are the most common detectors in the majority of laboratories and are easily coupled to the CE instrumentation has led to UV/Vis detection being the usual method of choice in CE analysis of pharmaceuticals. The most advantageous type of UV/Vis detectors are diode array detectors (DADs), which utilize multiple wavelength detection to acquire spectral data that allows purity determination and some peak identification. A drawback of UV/Vis detection for CE analysis is the reduction on concentration sensitivity compared to its use in HPLC. With absorbance detection, minimum detectable concentrations in CE are 10–100 times higher than in HPLC, this is due to the lower path length of the cell. With CE the path length is the inner diameter of the capillary, which is typically 50–100 μm, compared to path lengths of 10 mm for HPLC detectors.

Sensitivity can be enhanced through the use of larger inner diameter capillaries; however, larger capillary diameters produce a higher current, which leads to joule heating and separation efficiency decreases dramatically when the capillary diameter is increased above 100 μm. There have been a number of approaches to enhance sensitivity by increasing the path length without increasing the inner diameter of the capillary. Moring et al. [85] used a “z-cell” with a 3 mm capillary section for path length, which gave a 10-fold increase in sensitivity. Tsuda et al. [86] used a 1000 μm path length “rectangular capillary” to increase sensitivity, however, their application is limited by commercial availability and cost. Wang et al. [87] used a “multireflection cell” to effectively increase the path length more than 40-fold and reported a 40-fold increase in peak height compared to a single pass cell. A “bubble cell” can also be used to increase the path length by only increasing the diameter of the capillary where the light beam goes through it. This was demonstrated by Heiger [88] who reported a threefold increase in peak height when using a 50 μm ID capillary with a 150 μm “bubble.” Some manufacturers supply specialized detection cells such as Agilents high sensitivity detection cell, which provides a 10-fold increase in sensitivity and an extended linear range.

4.4.2 Laser-Induced Fluorescence Detection

Laser-induced fluorescence is the optical emission from molecules that have been excited to higher energy levels by absorption of electromagnetic radiation. Laser-induced fluorescence detection (LIFD) can be regarded as the most sensitive optical detection technique to be coupled to CE with sensitivity proportional to the intensity of emitted light. The main advantage of fluorescence detection compared to absorption measurements is the greater sensitivity achievable because the fluorescence signal has a very low background. Intense light is available from the laser and this light can be efficiently focused into the narrow channel of the capillary in the detection window. LIFD is not a universal detection method and solutes must possess native fluorescence or must be able to be derivatized to generate a fluorophore. The pharmaceutical analyst, when deciding on the suitability of LIFD to their separation must consider this. There are several commercially available LIF detectors designed for use with CE. Beckman Coulter offers a dual wavelength LIF detector that works with a 488 nm (Ar^+ laser) and a 633 nm (HeNe laser). The Picometrics ZETALIF detector is a single wavelength excitation LIF detector which is a modular design and can be used externally to the CE instrument. Picometrics also offer another commercial detector, which can be used inside a cassette of any CE instrument.

A number of publications are available that describe the theoretical aspects of LIFD [51,89], which will provide the reader with a more detailed coverage of this topic. A number of reviews have also been published which cover the use of CE-LIF detection for the analysis of a range of compounds including pharmaceuticals [90–94]. Zhang et al. [94] also show the use of wavelength-resolved LIFD for quantitative and semi-qualitative analysis of target solutes.

4.4.3 Mass Spectrometric Detection

Section 3.5 covered briefly the use of CE-MS for pharmaceutical applications. The suitability of CE-MS to pharmaceutical applications is due to its appeal as a universal and selective detector, which provides very good sensitivity. Significantly, CE-MS can provide quantitative data and where required, conclusive qualitative information.

4.4.4 Electrochemical Detection

The two types of electrochemical detection used most often in CE are amperometry and conductivity. Although not as widely used in pharmaceutical CE as UV/Vis or LIFD, these methods are used when alternative detection methods are not suitable, that is, solutes lack chromophores, fluorophores or if sensitivity is poor using optical detection methods. Amperometric detection measures the current that results from oxidation or reduction of electroactive solutes at a working electrode surface. Oxidation occurs when an electron is transferred from a solute molecule to the working electrode in the amperometric cell, during oxidation the charge on the solute increases. Reduction occurs when electron transfer is in the opposite direction, from the working electrode to the solute molecule, and the charge on the solute is reduced. Oxidation and reduction of the solutes is caused by a potential being applied across a supporting electrolyte between a working and reference electrode. The current that flows through the working electrode is proportional to the number of electron transfers taking place, and therefore to the concentration of the solute. A review by Lui et al. [95] gives a more detailed account of amperometric detection in CE. A number of recent applications of amperometric detection for CE analysis of pharmaceuticals include determination of the hydrolysis rate constants and activation energy of aesculin [96] and analysis of acyclovir [97]. The basis of conductivity detection is the change in electrical conductivity of a solution when an ionic solute is introduced into it. A conductivity detector has two electrodes in the cell and a high frequency alternating current is applied to the electrodes. The buffer from the capillary flows between the electrodes and when an ionic solute comes into the cell it decreases the electrical resistance of the solution, thereby increasing the electrical conductivity. The measured conductivity is proportional to the concentration of the solute. As only ionic solutes cause a change in conductivity, this method cannot be used with MEKC or MEEKC for the detection of separated neutral solutes. As with amperometry, there have been relatively few applications of conductivity detection for pharmaceutical analysis by CE, compared to optical detection methods. Bowman et al. [98] used this technique for the analysis of small amines in pharmaceuticals. Readers are encouraged to consult reviews by de Silva [99], Baldwin [100], Holland and Leigh [101], and Wang and Fang [102] who discuss in more detail electrochemical detection in CE analysis.

TABLE 4.1
Features and Limits of Detection of CE Detection Methods

Detection Method	Detection Requirements	Selective	Universal	Qualitative	On-column Detection	Approximate LOD (Molarity)
UV/Vis absorbance	• Chromophore groups • Absorb UV above 190 nm	Yes	No	Some, using DADs	Yes	10^{-5}–10^{-7}
LIF	• Fluorophore groups • Derivatized to contain fluorophores	Yes	No	No	Yes	10^{-13}–10^{-16}
MS	• No special requirements	Yes	Yes	Yes	No	10^{-8}–10^{-10}
Amperometry	• No special requirements	Yes	No	No	No	10^{-7}–10^{-10}
Conductivity	• Solutes need to be charged	No	Yes	No	No	10^{-7}–10^{-9}

4.4.5 Selecting the Most Suitable Detector

When deciding which detection method is most suitable to an application, the pharmaceutical analyst needs some fundamental information about their sample and also the analytical requirements. The main points to be considered when choosing a detection method are sensitivity and selectivity requirements and the ability to provide qualitative information. If these requirements are known then the list of available detectors is reduced. Knowledge of the chemical structure of target analytes is indispensable when choosing an appropriate detector as information about chromophore and fluorophore groups, solute charge, molecular weight, and so forth narrows the options available to the analyst. Table 4.1 lists the most common detectors as discussed in this section and shows the limits of detection, detection requirements, and so forth for each type that will aid in detector selection. Readers are encouraged to consult reviews by Xu et al. [92] and Swinney and Bornhop [103], which cover CE detectors in more detail.

4.5 SELECTING THE MOST SUITABLE MODE OF CE

As previously discussed in Sections 3.1 through 3.5, the various modes of CE offer various possibilities to the pharmaceutical analyst. Depending on the complexity of the sample, the nature of its components, and the intended application, the nature of pharmaceutical analytes varies enormously from highly water-soluble compounds to species only soluble in pure organic solvents. Each of these modes of CE will provide various advantages and disadvantages for the separation and detection of different classes of compounds. FSCE is well reported for many pharmaceutical assays and impurity determinations, mainly for water-soluble and some water-insoluble acidic and basic compounds. NACE is suited to the separation of highly water-insoluble charged compounds and the volatile nature of the solvents used makes it ideal for coupling to a MS. NACE also offers different selectivity to FSCE for the separation of pharmaceuticals [45,46]. For complex separations, MEEKC is the method of choice as it gives the analyst more method development options, it has a wider application range than MEKC and unlike FSCE and NACE, this mode of CE will separate both charged and neutral compounds. The use of W/O MEEKC has shown applicability of this method for the analysis of highly water-insoluble compounds in hydrophobic matrices without the need for lengthy sample pretreatment or expensive organic solvents [56,57]. For rapid screening, enantiomeric separations and physicochemical measurements of pharmaceuticals, multiplexed CE has been used [77–85]. The main pharmaceutical application areas of CE are chiral separations, pharmaceutical assay, impurity determinations, physicochemical measurements, and the analysis of small molecules and ions. Each of these areas will be discussed in the next section and a selected number of applications will be detailed in tabular form showing the mode of CE used. As the numbers of applications are too numerous to be included in this chapter, sources of further applications will be cited for the interested reader.

4.6 PHARMACEUTICAL APPLICATIONS OF CAPILLARY ELECTROPHORESIS

Since the first application of CE to pharmaceutical analysis in 1988 [15], the number of reports describing the analysis of numerous different pharmaceutical compounds and formulations by CE has expanded to be in the hundreds. Lunn [13] has covered in detail the CE separation methods for over 700 pharmaceutical compounds taken from hundreds of publications. The objective of this applications section is to provide the reader with an introduction to the main pharmaceutical application areas with examples of successful separations including sources of comprehensive reviews for each area. Within each application area there may be more specialized applications (e.g., chiral MEEKC) for enantiomeric separations and some aspects may be cross-linked with other applications or detection methods to comprehensively cover the topic of this chapter. Sources of extra reading material in the form of review papers will be cited where relevant.

4.6.1 Chiral Pharmaceutical Analysis

Chiral analysis has become one of the most studied areas of CE, as it is a powerful analytical tool for separating chiral compounds, which is of major importance in pharmaceutical applications. The two most important analytical techniques used in chiral separations still are LC and CE, followed by GC and capillary electrochromatography (CEC). Compared to other techniques, CE has several advantages, the high resolving power, and low consumption of sample, solvent, and chiral selector. In method development, the chiral selector is added to the background electrolyte instead of using a range of expensive chiral LC columns. Chiral CE provides high flexibility in choosing and changing types of selectors. The possibility of low wavelength UV detection for CE also allows the separation and detection of analytes with poor chromophores, which are difficult to detect by UV/high-performance liquid chromatography (UV/HPLC).

This section will look at recently developed chiral selectors and chiral CE separating techniques. Readers are referred to papers by Rizzi [104] and Gubitz and Schmid [105], which deal with the fundamental aspects and principles of chiral CE. A book dedicated to CE for chiral analysis by Chankvetadze [106] covers a number of aspects of chiral CE including the use of chiral metal complexes, macrocyclic antibiotics, crown ethers, and chiral MEKC. Review papers by Gubitz and Schmid [107], Hoogmartens and coworkers [16], and Amini [108] cover recent applications of chiral CE. The following paragraph gives a very brief introduction to the basic theory of chiral CE separations.

At low pH, basic drugs are positively charged and their migration toward the cathode can be retarded by a chirally selective complexing agent, resulting in the separation of enantiomers of differing affinity for the agent. This principle has been demonstrated for the resolution of chiral basic drugs using CDs as a chiral selector [109]. At high pH, chiral acidic drugs are negatively charged and migrate against the EOF toward the anode. Neutral chiral selector agents are swept along the capillary with the EOF toward the detector; thus, complexation reduces the migration time of the drug and results in enantioseparation [110].

There are many types of chiral selectors that have been applied to the separation of enantiomers by CE, but the most common are native and derivatized CDs. Other chiral selectors, which have been applied to CE separations, include natural and synthetic chiral micelles, crown ethers, chiral ligands, proteins, peptides, carbohydrates, and macrocyclic antibiotics [105,111–114]. A review by Blanco and Valverde [114] describes the separation capabilities of various chiral selectors and provides criteria for their choice in terms of molecular size, charge, and the presence of specific functional groups or substructures in the analytes.

Chiral CE using mass spectrometric detection has also been utilized for detailed qualitative analysis of pharmaceuticals, a review by Shamsi [115] covers the various modes of CE-MS including the use of neutral CDs for FSCE-MS and MEKC-MS.

4.6.1.1 Cyclodextrins

Native and derivatized CDs are employed routinely for enantiomeric separations. These naturally occurring carbohydrates have a bucket-like shape, which allows analytes to become included into the CD cavity by complexation. The migration time of the analytes is dependent on their mobilities and degree of interaction with the CD. Enantiomeric separations are brought about by the difference in the stabilities of the complexes formed between each enantiomer and the CD molecule. The chiral hydroxyl groups around the rim of the CD can interact enantioselectively with chiral analytes, which can fit into the CD cavity, leading to the separation of enantiomers with differing binding constants. There are three types of CDs, α, β, and γ, each differing in the number of glucose units they are composed of. β-CDs are the least soluble in water but their solubility can be improved by the addition of urea to the background electrolyte. Because enantioselection is based on the formation of inclusion complexes between the CD host and the chiral solute, the type of CD chosen is a major factor for achieving efficient resolution of enantiomers. Recent reviews [20,117] have covered the background to the use of CDs in chiral CE. Systematic approaches to the development of chiral CE methods using sulfated CDs for the separation of acidic, basic, neutral, and zwitterionic species have also been reported [118,119]. Systematic method development approaches for several selected compounds were performed by modifying method parameters, such as the concentration of the chiral selectors, buffer pH, type of organic modifiers, buffer type, temperature, and applied voltage. Many practical aspects were also discussed through several specific examples in order to demonstrate how to develop and validate a precise, sensitive, accurate, and rugged separation. Perrin et al. [75] studied the robustness of enantiomeric separations of acidic, basic, and neutral compounds using highly sulfated CDs in a low pH phosphate buffer with short-end injection. An eight-factor experimental design was used to study the robustness of chiral separations of propanolol (basic), praziquantel (neutral), and warfarin (acidic). These factors included electrolyte pH, CD and electrolyte concentration, capillary temperature, and applied voltage. Results showed that control of pH is critical, especially for the separation of acidic enantiomers while CD concentration changes did not adversely affect the robustness of the separations.

4.6.1.2 Crown Ethers

Crown ethers have been developed and synthesized for use in Chiral CE. Kuhn et al. [119] first utilized chiral crown ethers for the enantiomeric separation of drugs and amino acids in 1992. Since then there have been a number of applications of these chiral selectors in CE analysis, 18-crown-6 tetracarboxylic acid [18C6H(4)] is to date, the only chiral crown ether that has been used for chiral

CE separations and forms complexes with chiral analytes in a way similar to CDs, based on differences in complex formation energies. Two new 18-crown-6 diaza derivatives were investigated as chiral selectors [120]. These derivatives did not show any chiral selectivity toward the investigated analytes, they could serve well as an additive to improve the chiral resolution in combination with CDs.

4.6.1.3 Macrocyclic Antibiotics

Although CDs and their derivatives are present in the majority of chiral CE applications, the use of macrocyclic antibiotics can be observed with increasing frequency [121,122]. The number of macrocyclic antibiotics utilized for chiral CE has now exceeded 10 and includes four main groups: glycopeptides, polypeptides, ansamycins, and aminoglycosides (although this group is not always considered as macrocyclic). Thanks to their macrocyclic structure and the diversity in chemical groups, they exhibit a variety of interactions (inclusion, electrostatic, hydrogen bond, hydrophilic-lipophilic, or other Van der Waals bond type), which enables them to achieve high chiral resolution with a wider range of analytes (acidic or basic, with large or small molecular sizes, etc.) [124]. Macrocyclic antibiotics are enantioselective for positively charged solutes using ansamycins and enantioselective for anionic compounds using the glycopeptides. Within a given class of antibiotics such as the glycopeptides, enantioselectivity may also be altered by use of micelles, uncoated vs. coated capillaries, or manipulation of operating parameters such as pH or organic modifiers. In a review paper, Aboul-Enein and Ali [123] describe the chemistry of these antibiotics, the effect of chromatographic conditions on enantioselectivity, the mechanism of resolution, the applications and limitations of the compounds in LC and CE.

4.6.1.4 Oligo- and Polysaccharides

Apart from CDs, many other linear and cyclic oligo- and polysaccharides have been used as chiral selectors [127] (e.g., monosaccharides as D-glucose, D-mannose, or polysaccharides as dextrins, dextrans, and many others). Park et al. [125] used highly sulfated cyclosophoraoses, the sulfated derivatives of chiral unbranched cyclic β-(1→2)-D-glucans, and successfully used them to separate five basic chiral drugs. These derivatives exhibited higher resolution than the original cyclosophoraoses. Another new charged polysaccharide, *N*-(3-sulfo, 3-carboxy)-propionylchitosan, was studied by Budanova et al. [126]. This appeared to have a different chiral recognition mechanism from the charged polysaccharides. The use of polysaccharides combined with CDs has also been reported [128].

4.6.1.5 Proteins

A protein or glycoprotein consists of amino acids or amino acids and sugars, both of which are chiral. Therefore, proteins have the possibility to discriminate between the enantiomers of a chiral molecule. CE methods using proteins as the immobilized ligands or running buffer additives are attractive for the separation of enantiomeric mixtures. Although separation efficiencies by CE are generally somewhat higher than those obtained with HPLC, chiral CE methods based on proteins have a disadvantage of low efficiencies in addition to low loadability. Proteins such as bovine serum albumin (BSA) and human serum albumin (HSA) and several additional proteins have been used as chiral selectors in CE. Reviews by Haginaka [111] and Millot [129] provide a more in-depth coverage of recent chiral CE separations using proteins as chiral selectors. Although proteins display very good chiral discrimination for pharmaceutical enantiomers, their high background UV absorbance limits their usefulness in chiral CE. More can be gleaned on protein analysis by CE from Chapter 2 by Hempe.

4.6.1.6 Chiral MEKC

Enantiomeric separation by MEKC involves the addition of a chiral surfactant or a chiral selective agent to the background electrolyte. A number of selectors have been used in chiral MEKC including crown ethers and CDs. These selectors are usually used in combination with chiral or achiral micelles, for example, CD modified MEKC (CD-MEKC) [130]. MEKC usually utilizes negatively charged micelles formed from anionic surfactants such as SDS, which constitutes the pseudostationary phase. The separation is achieved by differential partitioning of analytes between the pseudostationary phase and the bulk aqueous phase. Chiral surfactants used for chiral MEKC separations include natural surfactants (bile salts, amino acids, and glucose), monomeric synthetic surfactants, and polymeric surfactants [131–133]. Chiral separation in MEKC is affected by the affinity of the enantiomers toward the micelles and the concentration of the micellar phase, which depends on the aggregation properties if the chiral surfactants or chirally modified surfactants. Dobashi et al. [131] first reported the use of chiral mixed micelles to obtain optical separation of enantiomers in 1989 and since then, there have been many reports on enantiomer separations by MEKC. Otsuka and Terabe [134] has reviewed the use of natural and synthetic chiral surfactants in MEKC, in a review by Ha et al. [16], recent applications, and aspects of chiral MEKC are covered in more detail.

4.6.1.7 Chiral MEEKC

Chiral separation by MEEKC was first demonstrated in 1993 [135] using (2*R*,3*R*)-di-*n*-butyl tartrate as a water immiscible chiral selector in the microemulsion electrolyte that successfully separated ephedrine enantiomers. Chiral MEEKC offers increased method development flexibility, the ability to custom tune chiral resolution through the increased method development options, and unique solubility capabilities for both analytes and additives when compared to typical capillary electrophoretic techniques. Chiral MEEKC offers the possibility to simultaneously determine drug enantiomers of more than one compound along with chemical impurities/additives. Importantly, the greatest advantage of this technique is the ability to separate more hydrophobic racemic components for which CE is not currently a preferred methodology. Only in recent years, more research into this area of chiral separations has been carried out by Pascoe and Foley [136], Mertzman and Foley [137,138], and Zheng et al. [139] among others. Chiral separations in MEEKC can be achieved by utilizing a number of chiral agents such as a chiral surfactant dodecoxycarbonylvaline (DDCV) [136], chiral alcohols as cosurfactants [139], a combination of both of these chiral components [140], and the use of CDs in CD modified MEEKC as the only chiral agent and in combination with both SDS and DDCV as surfactants [137]. Although no literature in the form of reviews and few books are dedicated to chiral electrokinetic chromatography (MEKC and MEEKC), Pyell [51] includes a chapter dealing with enantiomer separations by electrokinetic chromatography. There are a number of reviews by Ha et al. [16], Marsh et al. [63], and Huie [64], which cover these subjects in some detail and give examples of recent applications.

4.6.1.8 Pharmaceutical Applications of Chiral CE

Chiral selectors are routinely used in industry for a wide variety of applications and there have been many reports published describing their use in CE for the enantiomeric separations of pharmaceutical compounds. A quick search of scientific databases will yield a far greater number of publications dealing with chiral CE than is possible to cover in this chapter; however, Table 4.2 contains a selection of such applications and details of carrier electrolytes used. Readers are referred to review papers [16,63,64,116,107,141–144], which comprehensively cover recent chiral applications of CE and provide more detailed theory about chiral separations and detection methods.

TABLE 4.2
Selected Chiral Separations of Pharmaceuticals

Application	Chiral Selector	Mode	References
Enantiomeric purity methods for three pharmaceutical compounds	A range of α-, β-, and γ-cyclodextrins, highly and singly sulfated	FSCE	[153]
Evaluation of chiral purity of frovatriptan	Sulfobutyl ether β-cyclodextrin	FSCE	[154]
Chiral separation of four fluoroquinolone compounds	Hydroxypropyl-β-cyclodextrin	FSCE	[155]
Chiral separation of bupivacaine enantiomers	Human serum albumin	FSCE	[156]
Chiral separations of a range of pharmaceuticals	Sulfated β-cyclodextrin (β-CD-(SO_4^-)$_4$	NACE	[157]
Dopa enantiomers	(+)-(18-crown-6)-2,3,11,12-tetracarboxylic acid	FSCE	[147]
Chiral separation of *N′*-nitrosonornicotine in tobacco	Hydroxypropyl-β-cyclodextrin	FSCE	[145]
Enantioseparation of *erythro*-mefloquine and its analogues	Heptakis-(2,3, 6-tri-*O*-methyl)-β-CD Heptakis-(2,3-di-*O*-methyl-6-sulfo)-β-CD Randomly sulfated β-CD	FSCE	[158]
Enantioseparation of nine racemic arylglycine amides	Highly sulfated β-cyclodextrin	FSCE	[159]
Enantioseparation of basic pharmaceutical compounds	Sulfated cyclodextrins	FSCE	[160]
Enantiomeric separation of compounds containing primary amine groups	(+)-(18-crown-6)-2,3,11,12-tetracarboxylic acid	CE-MS	[161]
Separation of 4 diastereomers of an antiviral agent with 2 chiral centers	Highly sulfated α, β, and γ-CDs	FSCE	[162]
Separation of (*R*-(−) and *S*-(+) citalopram) in tablets	Carboxymethyl-γ-cyclodrextrin (CM-γ-CD)	FSCE	[205]
Lisuride bulk substance	Acidic electrolyte with the addition of gamma-cyclodextrin (γ-CD)	FSCE	[206]
Identification and quantification of *cis*-ketoconazole impurity in tablets and gel	Heptakis-(2,3, 6-tri-*O*-methyl)-beta-cyclodextrin	CE-MS	[207]
Separation of omeprazole enantiomers	40 mM phosphate, pH 2.2, 30 mM β-CD, and 5 mM sodium disulfide	FSCE	[208]

4.6.1.9 Examples of Chiral CE and Validated Methods

N′-nitrosonornicotine (NNN) is a nitrosamine found in cured tobacco and is believed to be one of the main carcinogenic constituents of tobacco. Racemic NNN is present as (*R*) and (*S*) enantiomers and it is known that (*S*)-NNN undergoes more 2′ hydroxylation than (*R*)-NNN. For rapid and efficient enantiomeric separation of this nitrosamine, McCorquodale et al. [145] developed a simple CE method, employing a citric acid buffer with hydroxypropyl-β-CD (HP-β-CD). Separation was achieved in 4 min.

A nonaqueous chiral CE method was developed and validated by Olsson et al. [146] for the enantiomeric separation of omeprazole an antiulcer drug and its metabolite 5-hydroxyomeprazole. Heptakis-(2,3-di-*O*-methyl-6-*O*-sulfo)-β-CD (HDMS-β-CD) was chosen as the chiral selector in an

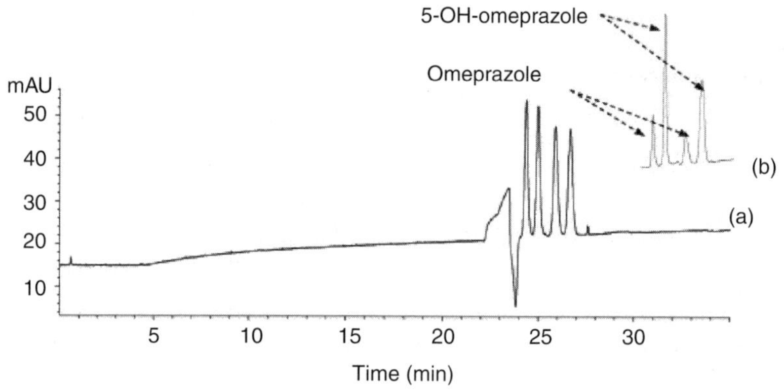

FIGURE 4.3 Enantiomeric separation of (a) omeprazole and 5-hydroxyomeprazole at a concentration of 1 mM and (b) omeprazole and 5-hydroxyomeprazole at a concentration of 1 and 3 mM, respectively, using 30 mM HDMS-β-CD, 1 M formic acid, 30 mM ammonium acetate, 25 kV, 5 s injection at 50 mbar, capillary and tray temperature 16°C. (From Olsson J., et al., *J. Chromatogr. A*, 1129, 291–295, 2006.)

ammonium acetate buffer acidified with formic acid in methanol. The separation is illustrated in Figure 4.3.

The addition of a chiral crown ether [(+)-(18-crown-6)-2,3,11,12-tetracarboxylic acid] to the background electrolyte allowed Blanco and Valverde [147] to separate the enantiomers of benserazide and determine the enantiomeric purity of Dopa (3,4-dihydroxyphenyl-alanine). Levodopa is the main active ingredient in the pharmaceutical formulation Madopar®, which is used to treat Parkinson's disease, while dextradopa causes unwanted side effects. The dextradopa impurity was clearly resolved from the main peak and determined to be 0.5%.

The validation criteria for chiral CE methods are similar to those employed for the validation of chiral HPLC methods and include limits of detection, detector linearity, recovery, precision, and robustness. A number of validated chiral CE methods have been reported, as described previously Olssen et al. [146] developed and validated a chiral NACE method for the enantiomeric determination of an antiulcer drug and its metabolite. Enantiomeric determination of a local anesthetic, ropivacaine in a pharmaceutical formulation showed the required limit of quantification of 0.1% of the impurity [148]. Jimidar et al. [149] reported a validated CE method for the chiral separation of an Alzheimer's treatment, galantamine hydrobromide, using an α-CD chiral selector and the method was successfully included in a New Drug Application. Song et al. [150] achieved a LOD of 0.05% for the undesired enantiomer of an M3 agonist using a highly sulfated γ-CD in a low pH buffer. Other validated chiral CE methods are also cited [151,152].

4.6.2 Pharmaceutical Assay

Assay of pharmaceutical substances and formulated products is one of the most important and regulated activities in the pharmaceutical analysts laboratory. Regulatory authorities require strict validation standards to show that analytical assay methods are robust, accurate, repeatable, and suitable for their intended purpose. In the pharmaceutical industry, HPLC has dominated most analytical assay determinations and it is well established as the method of choice in most laboratories. In addition, pharmacopoeia monographs specify HPLC and titrimetric methods for the majority of pharmaceutical assays. To date, CE is not used extensively in QC work despite displaying excellent efficiencies, resolution, asymmetry factors, and signal-to-noise ratio. This is mainly due to the fact that CE can suffer from insufficient sensitivity and repeatability to control impurities in pharmaceutical substances at the levels required. These issues have been addressed somewhat with sensitivity improvements being reported through the use of alternative detection techniques such as mass spectrometry and LIFD and the use of high sensitivity detection cells for UV detection. When conducting CE analyses, the electrophoretic conditions inside the capillary can sometimes vary slightly between injections, which can lead to greater variability in peak migration times. Poor injection precision due to the very small volumes injected can also cause variations in peak areas. A number of approaches can be made to overcome these problems. Migration times and peak areas can be calculated relative to an internal standard, which leads to improvements in repeatability [164]. Greater migration time reproducibility can also be achieved by applying the separation voltage across the capillary for a very short time prior to injection and separation [165]. The factors affecting CE reproducibility and efforts to address the problem have been covered in more detail by Shihabi and Hinsdale [166], Schaeper and Sepaniak [167], and Mayer [168]. Many factors are involved in reproducibility. Some of these, such as temperature control, voltage control, and sample injection precision, are inherent in the design of the instrument. Other factors, such as the quality of the reagents used and the manner in which the instrument is programmed and operated, are completely in the hands of the user. These two factors, the quality of the instrument and the quality of the operation, are both required in order to achieve reproducible results. A commercially available capillary treatment system of proprietary buffers and rinse solutions has been shown to improve CE repeatability as the capillary is coated with a bilayer of surfactants ensuring that the surface coverage and the EOF is consistent between injections and between capillaries [169]. Figure 4.4 highlights the consistency of the EOF when using the buffer coating system compared to a standard phosphate buffer when using a capillary composed of 19 different channels. In Figure 4.4a, the peaks in the channels have different speeds and the separation is poor while Figure 4.4b shows a single peak due to a consistent EOF in each channel, which makes the peaks move at the same speed.

There is a time lag before new analytical methods such as CE are accepted by regulatory authorities in the submission dossiers for new medicines and pharmaceutical substances. Regulatory authorities are, however, beginning to recognize CE with general monographs appearing in the BP, EP, and Japanese Pharmacopoeia using CE for identification tests. For example, a test for related substances of levocabastine hydrochloride by MEKC has been included in the British Pharmacopoeia 2005 [163].

One of the major advantages of using CE as an alternative to HPLC methods for pharmaceutical assay is the relatively small solvent consumption, that is, milliliters of CE electrolyte compared to liters of HPLC mobile phase, increased efficiency, reduced analysis time, and the possibility of fewer sample pretreatment steps. The ability to quantify a range of sample types using a single set of CE conditions is another strong feature and can contribute to considerable savings in analysis and system setup times. This is particularly true when using multiplexed CE systems where high sample throughput is possible. The merits of using CE compared to HPLC for pharmaceutical analysis will be discussed in more detail in a later section.

Some recent examples of pharmaceutical assay by CE include a MEEKC method for the determination of folic acid in tablets giving a precision of $<1.2\%$ relative standard deviation (RSD) and recovery of $99.8 \pm 1.8\%$ at three concentration levels [170]. Lehmann and Bergholdt [171] developed and validated a high-precision CE method for the main component assay of ragaglitazar, which met the acceptance criteria that are set for HPLC main component assays. In a separate study, Jamali and Lehmann [172] used a FSCE method to analyze ragaglitazar and its counterion arginine in active pharmaceutical ingredients (APIs) and low dose tablets. The method was suitable for the assay and identification of ragaglitazar and arginine, chiral purity of ragaglitazar, and the purity of ragaglitazar, with percentage recovery found to be 101–106% for ragaglitazar and 101–125%. Pajchel et al. [173] used a phosphate buffer supplemented with SDS to develop a selective and precise assay method for quantitative determination of benzylpenicillin, procaine, benzathine, and clemizole. The separation is shown in Figure 4.5.

A number of validated CE methods for pharmaceutical assay have been reported; a selection of recent methods along with the mode of CE used and the run buffer are shown in Table 4.3.

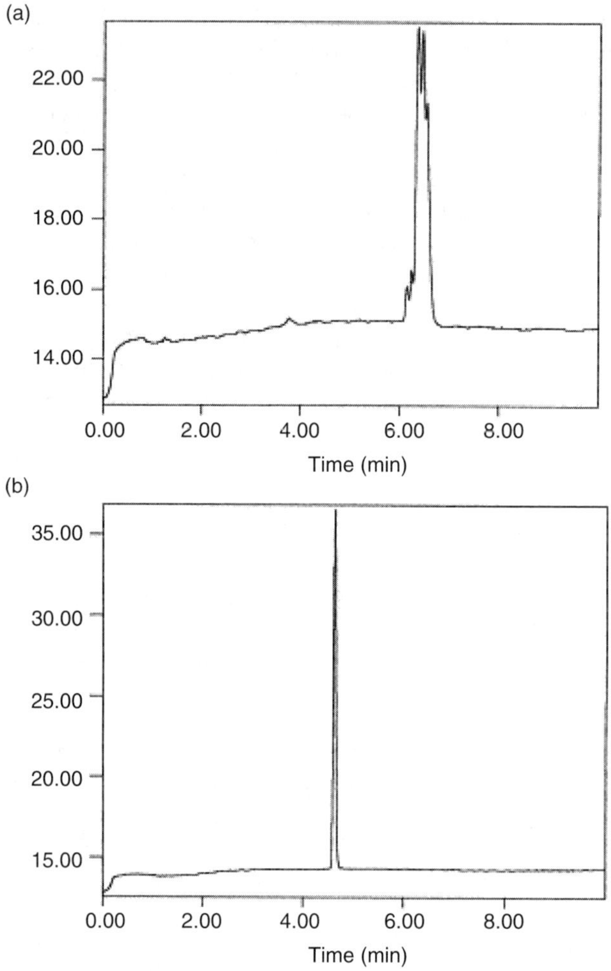

FIGURE 4.4 Separations on a multibore capillary using phosphate buffer or Celixir buffer. (a) Separation using phosphate buffer: 50 mM phosphate 2.5, multibore capillary 19 × 25 μm channels, 27 cm long, 130 μA, +5 kV, 30°C, detection at 200 nm, sample salbutamol 1 mg/mL, 1 s injection. (b) Separation using Celixir buffer: 50 mM phosphate 2.5, multibore capillary 19 × 25 μm channels, 27 cm long, 90 μA, +5 kV, 30°C, detection at 200 nm, sample salbutamol 1 mg/mL, 1 s injection. Multibore capillary 19 × 25 μm channels, 27 cm long, +5 kV, 30°C, Elixir buffer pH 2.5, 90 μA, 200 nm. (From Altria K.D., *J. Pharm. Biomed. Anal.*, 31, 447–453, 2003.)

4.6.3 Impurity Profiling of Pharmaceuticals

Impurities in pharmaceuticals are the unwanted chemicals that remain with the APIs, or develop during formulation, or upon aging of both the API and formulated pharmaceutical product. The presence of these unwanted chemicals even in small amounts may influence the efficacy and safety of the pharmaceutical products. Impurity profiling (i.e., the identity as well as the quantity of impurity in the pharmaceuticals) is another important aspect of the pharmaceutical analysts work and must meet strict regulatory requirements. The different pharmacopoeias, such as the British Pharmacopoeia (BP) and the United States Pharmacopoeia (USP) have tests for related substances incorporated into most monographs for pharmaceutical compounds and formulations. In addition, The International Conference on Harmonization (ICH) has published guidelines on impurities in new drug substances [190]. Impurity profiling is generally carried out by HPLC and cross-correlated with other

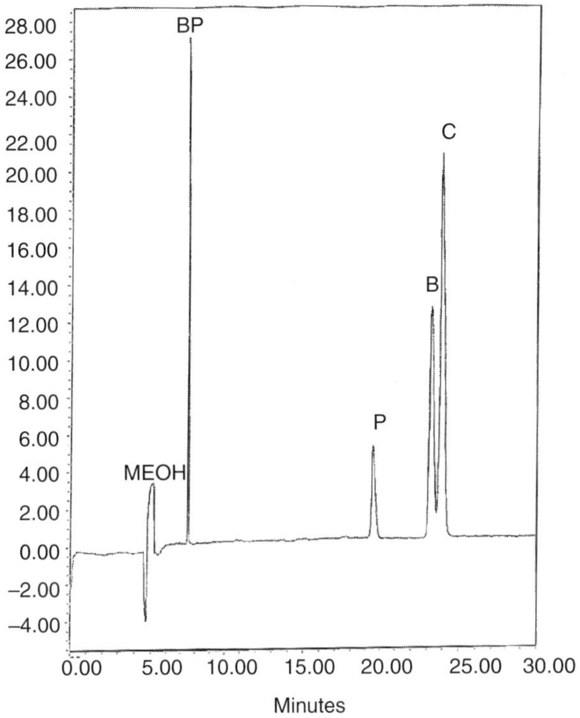

FIGURE 4.5 Electropherogram of benzylpenicillin and procaine, benzathine, and clemizole. Buffer phosphate–borate (pH 8.7), supplemented with 14.4 g/L SDS. Detection wavelength was 214 nm. Separations were performed in 60 cm (52 cm effective length) × 75 μm ID fused-silica capillary coated with polyimide (AccuSep capillaries, Waters) thermo regulated at 25°C, with voltage of 18 kV applied (current about 140 μA). Hydrodynamic injection by gravity-driven siphoning 10 s. P, procaine; B, benzathine; C, clemizole; BP, benzylpenicillin; MEOH, methanol (EOF marker). (From Pajchel G., et al., *J. Chromatogr. A*, 1032, 265–272, 2004.)

chromatographic methods such as TLC or an alternative HPLC method. The overriding requirements in impurity determination methods are that all likely synthetic and degradative impurities are resolved from the main compound and these can be quantified at 0.1% and lower levels. This quantification is possible using commercial CE instrumentation with standard capillaries. The structural impurities of a drug will often possess similar structural properties of the main component, which makes achieving resolution of the compounds challenging. The high separation efficiencies and different separation mechanism of CE often allows easier resolution of the main component and related substances than when using HPLC or TLC. The ability to easily alter separation parameters and the applicability of one CE method to a range of compounds can make CE a cheaper and more rapid complementary method to LC methods. Because LC and CE utilize different separation mechanisms, the analyst can have a high degree of confidence in the cross-correlation of agreeable results.

Various modes of CE have been used for impurity determination with CE. Hansen described the comparison of CZE (FSCE), MEKC, MEEKC, and NACE for the determination of impurities in bromazepam [193] and found that NACE provided the best technique for the poorly water-soluble compounds with impurities determined to be 0.05%. Chiral CE methods can be used to determine enantiomeric impurities [191]. Readers are referred to a paper by Sokoliess and Koller [192] describing method development for chiral purity testing in CE. FSCE is the most widely used as most drugs and impurities are acidic or basic. Low pH buffers are used for basic drug impurities and high pH buffers for acidic compounds.

TABLE 4.3
Selected Validated CE Methods for Pharmaceutical Assay

Application	Buffer	Mode	References
Determination of sertaconazole in pharmaceutical preparations	20 mM phosphate + 40% acetonitrile	FSCE	[174]
Development and validation of a quantitative assay for raloxifene	20 mM acetate (pH 4.5)	FSCE	[175]
Analysis of atorvastatin calcium using capillary electrophoresis and microchip electrophoresis	25 mM sodium acetate (pH 6)	FSCE	[176]
Determination of ibuprofen and flurbiprofen	20 mM *N*-(2-acetamido)-2-aminoethanesulfonic acid (ACES) with 20 mM imidazole and 10 mM alpha-cyclodextrin of pH 7.3	FSCE	[177]
Separation and quantification of fibrate-type antihyperlipidemic drugs	1/15 M phosphate buffer (pH 10)	FSCE	[178]
Simultaneous determination of six angiotensin-II-receptor antagonists.	60 mM phosphate (pH 2.5) 55 mM phosphate 15 mM SDS (pH 6.5)	FSCE MEKC	[179,180]
Simultaneous analysis of four atypical antipsychotics	80 mM sodium phosphate buffer (pH 3.5)	FSCE	[181]
Determination of cyclizine hydrochloride in tablets and suppositories	50 mM phosphate (pH 2.3)	FSCE	[182]
Simultaneous determination of six corticosteroids in commercial pharmaceuticals	Borate buffer containing sodium cholate and sodium deoxycholate	MEKC	[183]
Simultaneous determination of penicillin G, procaine, and dihydrostreptomycin in veterinary drugs	0.08 M borate (pH 8)	FSCE	[184]
Quantitative analysis of quinolizidine alkaloids in Chinese herbs	1% acetic acid, 50 mM ammonium acetate, 20% acetonitrile in methanol	NACE	[185]
Simultaneous determination of four tricyclic antidepressants	50 mM ammonium acetate, 1 M acetic acid in acetonitrile	NACE	[37]
Analysis of benzodiazepines	Ceofix™ Buffer system, using dynamically coated capillaries	CE-MS CE-MS^2 CE-DAD	[187]
Determination of fluoxetine and its main metabolite norfluoxetine in human urine	7:3 methanol–acetonitrile containing 15 mM ammonium acetate	NACE	[186]
Determination of heroin, basic impurities, and adulterants with capillary electrophoresis	Celixir™ Buffer capillary coating system plus added cyclodextrins	FSCE	[188]
Determination of various benzylpenicillin salts	Phosphate–borate buffer supplemented with SDS 14.4 g/L	MEKC	[173]
Simultaneous determination of ingredients in a cold medicine by CD-MEEKC	0.81% pentane, 6.61% 1-butanol, 2% 2-propanol, 4.47% SDS, 86.11% 10 mM sodium tetraborate, 3 mM 2,6-di-*O*-methyl-beta-CD	MEEKC	[189]
Folic acid in tablets	0.5% ethyl acetate, 1.2% butanol, 0.6% SDS, 15% propanol, 82.7% 10 mM tetraborate (pH 9.2)	MEEKC	[170]

A method was developed and validated using a low pH buffer for the determination of ranitidine and potential related impurities in bulk drug and formulations [194]. This method gave detection limits of 0.03% diamine, 0.04% oxime, 0.1% Bis, and 0.24% nitroacetamide but importantly, it detected a number of peaks, which were not resolved by either HPLC or TLC.

High pH buffers such as phosphate or borate are employed in the analysis of acidic components. At high pH, acidic components migrate against the EOF, thus maximizing mobility differences. A high pH borate buffer of 9.2 was developed and validated for the determination of homotaurine as an impurity in calcium acamprosate by FSCE, detection limits of between 0.01% and 0.15% homotaurine were reported [195]. A high pH CE method was used to quantitatively profile the chloromethylated, monomethylated, and hydroxylated impurities of a new compound (LAS 35917). The CE method allowed the quantification of the impurities at levels of 0.04–0.08% of the parent drug, while HPLC failed to resolve the impurities [196].

MEKC and MEEKC have the unique capability to separate both charged and neutral substances and have been employed for the quantification of a number of pharmaceutical compounds and related impurities. A MEKC method was developed and validated for the determination of didanosine, an anti-HIV treatment, which was separated from 13 of its potential impurities [197]. A simple, fast, and selective MEKC method was used for the determination of ketorolac tromethamine and its known impurities in 6 min and gave a quantification limit of 0.1% for impurities [198]. MEEKC is similar to MEKC in its ability to separate and quantify charged and neutral compounds (see Sections 3.3 and 3.4) and has been used to a lesser extent for impurity studies than MEKC, mainly due to the relatively recent introduction of this mode of CE. Future impurity profiling should see an increased utilization of MEEKC methodology. Further to the determination of ketorolac and its impurities using MEKC, Furlanetto et al. [199] used a mixture design in the optimization of a MEEKC method for the same analysis, complete resolution of the analytes was obtained in 3 min. Wen et al. [200] utilized MEEKC to simultaneously separate 17 species of heroin, amphetamine, and their basic impurities and adulterants in under 10 min. Readers are referred to a review by Hilhorst et al. [204] covering impurity profiling of drugs by CE. Some application-based papers will also provide the reader with more information on profiling by CE-MS [201,202] and CE-LIFD [203] (see also Table 4.4).

4.6.4 Physicochemical Profiling

During the early phase and later phase of drug development, knowledge of physiochemical properties of pharmaceutical compounds is important in order to predict their bioavailability and blood-brain barrier distribution to help in formulation design and drug delivery. Physicochemical properties such as acid dissociation constant (pK_a), octanol–water partition-coefficient ($\log P_{ow}$), solubility, permeability, and protein binding are closely related to drug absorption, distribution, metabolism, and excretion. About 30% of drug candidate molecules are rejected due to pharmacokinetic-related failures [220]. When poor pharmaceutical properties are discovered in development, the costs of bringing a potent but poorly absorbable molecule to a product stage can become very high. Fast and reliable *in vitro* prediction strategies are needed to filter out problematic molecules at the earliest stages of discovery. In drug discovery, there can be a vast number of compounds requiring physicochemical screening mainly because of the high volumes of syntheses that can be carried out by combinatorial chemistry. Consequently, there is a need for rapid and reliable methods of physicochemical profiling to maintain high throughput and efficiency. CE is a simple, versatile, automated, and powerful separation technique and widely applied in physicochemical profiling for pharmaceuticals. It has advantages over traditional potentiometric, spectrophotometric, chromatographic, and other methods, as CE requires very small amounts of sample and can measure compounds with impurities and low aqueous solubility. The advent of multiplexed CE instruments for physicochemical measurements can allow for the simultaneous analysis of up to 96 separate compounds using just

TABLE 4.4
Pharmaceutical Impurity Applications

Application	Buffer	Mode	References
Heroin and its basic impurities	100 mM DM-β-CD in Celixir reagent B (pH 2.5); 100 mM HP-β-CD in Celixir reagent B (pH 2.5); 103.2 mM SDS, 50 mM phosphate–borate (pH 6.5)	FSCE	[188]
Separation of penicillin V and its impurities	SDS 20.0 g/L (69 mM) and pentanesulfonic acid sodium salt 2.2 g/L adjusted to pH 6.3	MEKC	[209]
Ranitidine hydrochloride and related substances	190 mM trisodium citrate (pH 2.6)	FSCE	[194]
Aminopyridines and related substances	50 mM phosphate buffer (pH 2.5)	FSCE	[210]
Homotaurine as an impurity in calcium acamprosate	40 mM borate (pH 9.2)	FSCE	[195]
New substance (LAS 35917)	60 mM tetraborate (pH 9.2)	FSCE	[196]
5-Aminosalicylic acid and its major impurities	120 mM CAPS buffer (pH 10.2), 65 mM SDS, 55 mM TBAB, 5% MeOH	MEKC	[211]
Ketorolac tromethamine and its known related impurities	13 mM boric acid, phosphoric acid (pH 9.1) with 1 M sodium hydroxide, 73 mM (SDS), 90.0% 10 mM borate, 2.0% *n*-heptane, 8.0% SDS/*n*-butanol in 1:2 ratio	MEKC MEEKC	[198,199]
Vancomycin and related impurities	120 mM Tris-phosphate buffer (pH 5.2) containing 50 mM CTAC	MEKC	[212]
Ximelagatran thrombin inhibitor and related substances in drug substance and tablet formulation	Phosphate buffer (pH 1.9), 22% (v/v) MeCN, 11 mM hydroxypropyl β-CD	FSCE	[213]
Penicillin and related impurities	20 mM ammonium acetate (pH 6.5) In aqueous and nonaqueous electrolyte	FSCE and NACE	[214]
N-acetylcysteine and its impurities	100 mM borate (pH 8.40)	FSCE	[215]
Ciprofloxacin and its impurities	Phosphate buffer (pH 6.0), 0.075 M pentane–1-sulfonic acid Na salt	FSCE	[216]
Metacycline and its related substances	160 mM sodium carbonate + 1 mM EDTA (pH 10.35), 13% v/v MeOH	FSCE	[217]
Loratadine and related impurities	100 mM H_3PO_4 (pH 2.5), 10% acetonitrile	FSCE	[218]
Rofecoxib and photodegradation impurities	25 mM borate, 15 mM SDS, 10% acetonitrile	MEKC	[219]
Galantamine impurities by CE-MS	50:25:25 (v/v/v) 100 mM ammonium acetate/acetonitrile/methanol	CE-MS	[202]

Abbreviations: CAPS, 3-(Cyclohexylamino)-1-propanesulfonic acid; CTAC, cetyltrimethylammonium chloride; EDTA, ethylenediaminetetraacetic acid; TBAB, tetrabutylammonium bromide.

one instrument [76–79], which can offer huge savings in time costs compared to the more traditional methods mentioned previously. In a review paper by Jia [221], the principles and applications of CE in profiling various physicochemical properties are covered.

Of the physicochemical properties mentioned previously in this section, only capillary electrophoretic measurements of pK_a and $\log P_{ow}$ values will be covered here as they have received the most attention in physicochemical profiling by CE.

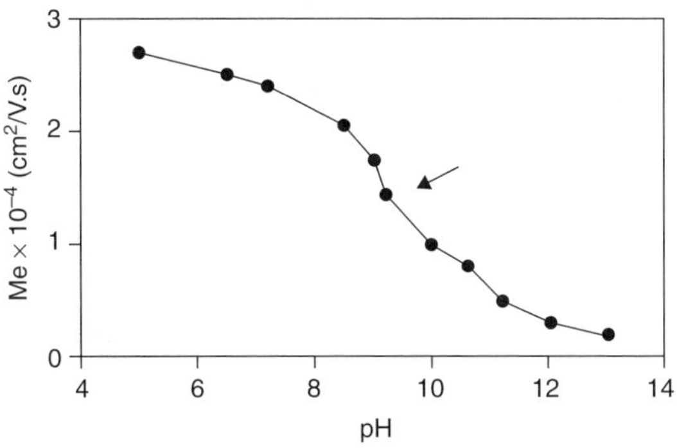

FIGURE 4.6 Dependence of the effective mobilities of a monovalent compound (LY334370) on pH. Arrow indicates the pH equal to the pK_a. (From Caliaro G.A., Herbots C.A., *J. Pharm. Biomed. Anal.*, 26, 427–434, 2001.)

4.6.4.1 pK_a Measurements

The majority of pharmaceutical compounds are either acidic or basic and are therefore ionizable. Other physicochemical properties, such as lipophilicity and solubility, are pK_a dependent; therefore, pK_a is one of the fundamental parameters of a drug molecule. The pK_a determination of acids and bases by CE is based on measuring the electrophoretic mobility of charged species associated with the acid–base equilibria as a function of pH. The detailed theory of pK_a and log P_{ow} measurements is beyond the scope of this chapter and readers are referred to a review by Jia [221], which describes in more detail the theoretical aspects of physicochemical profiling by CE methods including FSCE, NACE, MEKC, and MEEKC.

The pK_a values of pharmaceutical compounds can be determined from migration time data obtained by running the compound with free solution CE electrolytes at a range of pH values. The mobility of the solute at each pH can be calculated from its migration time and the EOF (measured against a neutral marker such as methanol), and a plot of mobility versus pH can be constructed (see Figure 4.6). The pK_a value can be calculated mathematically or obtained from the plot.

The pK_a values of water-insoluble and sparingly soluble compounds can be determined by NACE using methanol as the background electrolyte [48], or a 50% methanol/water electrolyte as used by de Nogales et al. [226] for the measurements of acid dissociation constants of several hydrophobic drugs. The most common method for pK_a measurements, however, is FSCE (see Table 4.5).

CE instruments are highly automated and can be used for high-throughput applications, particularly with the use of capillary array instruments. CE compares favorably to other methods of pK_a measurements [48] and unlike titration methods, precise information of sample concentration is not required as only analyte mobilities are used to calculate dissociation constants by CE. Sparingly soluble compounds are easily analyzed and only very small amounts of material are required, which is useful for screening of newly synthesized compounds where only small quantities may exist or where the molecules are present in only very small amounts. By using a "sample stacking" technique along with pressure assistance and short-end injection for rapid analysis, Wan et al. [224] were able to successfully measure the pK_a values of pharmaceutical compounds with concentrations as low as 2 μM. Techniques commonly used for pK_a measurement such as potentiometric or UV-Vis spectroscopy do not differentiate between the analyte of interest and any other degradant or impurity present, which can cause problems if the analyte is not highly pure or unstable in solution. Because CE is a separating technique, it can measure the pK_a values of impure or unstable compounds where electrolyte purity is not essential. This was demonstrated by Ornskov et al. [225] where a CE method of pK_a measurement was used successfully for a set of labile drug compounds that were unstable in solution. Determining the EOF intensity during pK_a measurements may be time consuming,

TABLE 4.5
Selected CE Applications for pK_a Measurements

Application	Buffer	References
pK_a* values of 21 basic drugs with aliphatic or aromatic amino groups	pH* range 4.9–9.7 Methanol and sodium acetate buffer + acetic acid	[48]
Medium-throughput pK_a of 48 pharmaceutical compounds—acidic, basic, and multivalent	pH range 2.5–11.0 Electrolytes of 0.1 M ionic strength composed of phosphate, acetate, and borate buffers adjusted to required pH with phosphoric, acetic, or boric acid or NaOH Pressure-assisted CE at 2 psi	[227]
Rapid pK_a screening of 26 acidic, basic, and multivalent pharmaceuticals	pH range 2.5–11.0 Electrolytes of 0.05 M ionic strength composed of 0.5 M phosphate and 1 M acetate buffers mixed to obtain required pH PACE at 25 mbar, "short-end" injection	[224]
pK_a determination of labile drug compounds	PH range 2.0–12.0 Electrolytes of 0.05 M ionic strength composed of 1 M phosphate, 0.1 M borate, and 1 M acetate buffers mixed and adjusted with phosphoric acid, acetic acid, and NaOH to obtain required pH "Short-end" injection	[225]
2-Amino-2-oxazolines (antihypertensive agents)	pH 4.77–9.69	[228]
Cephalosporins (antibiotics)	pH 2.0–9.0	[229]
pK_a determination of ^{99m}Technetium radiopharmaceuticals	pH range 1.3–6.6 50 mbar PACE Citric acid adjusted with NaOH	[230]
Dissociation constants of anthrocyclines (antibiotics)	Phosphoric acid, disodium phosphate, and monosodium phosphate (pH 4.20–8.20)	[231]
Dissociation constants of cytokinins (phytohormones)	Phosphate and acetate buffers of ionic strength 0.015 M (pH 1.5–6.0)	[232]
Determination of pK_a values Dihydrofolate reductase inhibitors	pH 2.1–4.5 50 mM phosphate buffer adjusted with phosphoric acid/NaOH	[233]
Quinolones (antibacterial)	pH 2.0–11.0	[234]
Dissociation constants of amino- and guanidinopurine nucleotide analogs and related compounds by CZE	pH 3.50–11.25, ionic strength (25 mM)	[235]
Determination of pK_a values with dynamically coated capillaries	Dynamic capillary coating	[236]
pK_a of organic bases in aq. acetonitrile (0–70%, v/v)	pH range 4.76–9.5 Tris, ethanolamine and acetate buffers	[237]
pK_a of *N*-imidazole derivative aromatase inhibitors	pH range 3.88–9.16 25 mM phosphate buffer adjusted with triethylamine	[238]

especially at a low pH. Geiser et al. [236] overcame this drawback by using a dynamic capillary coating procedure to increase the EOF and thus to reduce the analysis time. In addition, this coating procedure enhanced migration time stability. Table 4.5 lists selected recent applications of CE to the measurement of dissociation constants.

4.6.4.2 Log P_{ow} Measurements

Hydrophobic interaction (or liquid–liquid partitioning) of pharmaceutical compounds in the body plays a significant role in partitioning of drugs into lipid bilayers of biomembranes, bioavailability, and pharmacokinetics. As with pK_a values, liquid–liquid partition coefficient measurements are extremely important during drug discovery, screening, and formulation processes. Solute hydrophobicity is usually expressed by the octanol–water partition coefficient (log P_{ow}) that is defined as the ratio of the concentrations of a species in the two phases at equilibrium. A number of methods to measure log P_{ow} are available including the shake flask method, potentiometric titration, and liquid chromatographic separation methods [241] and extensive data collections of log P_{ow} values can be found in literature [239,240]. CE techniques using pseudostationary phases in the background electrolyte, that is, MEKC and MEEKC, allow the measurement of log P_{ow} values because of the partitioning of solutes between the MEKC micelle and microemulsion droplet (MEEKC). Early work using MEKC showed that the extent of solutes partitioning with micelles was related to the solutes solubility [242,243], the rise in popularity of MEEKC, however, demonstrated its applicability to a wider range of solutes and this technique has been used extensively in log P_{ow} measurements. Examples of using MEEKC for log P_{ow} measurements are cited [78,244,245]. MEEKC review papers by Hansen [62], Marsh et al. [63], and Huie [64] comprehensively cover the many applications in this area of physicochemical profiling.

Using MEEKC, the compounds solubility is assessed by bracketing it with neutral marker compounds of known log P_{ow} values, which are used to create a calibration graph of log P_{ow} against time or log k (log of the retention factor) (see Figure 4.7). The log P_{ow} of the analyte of interest can be calculated by its migration time or retention factor using the graph. The higher the compounds log P_{ow} value, the more it partitions into the microemulsion droplet and the longer it takes to migrate. Table 4.6 contains a number of selected CE applications for log P_{ow} measurements, the interested reader should consult recent reviews [62–64], which contain a comprehensive collection of all recent physicochemical measurements by CE.

4.6.5 Analysis of Small Molecules and Ions (Pharmaceutical)

The separation and detection of small organic and inorganic ions is an important activity in the pharmaceutical industry. CE is routinely used for ion analysis in the pharmaceutical industry for a number of applications, these include counterion determination, stoichiometry, salts, and excipients in drug formulations. Most pharmaceutical molecules are charged and are commonly manufactured

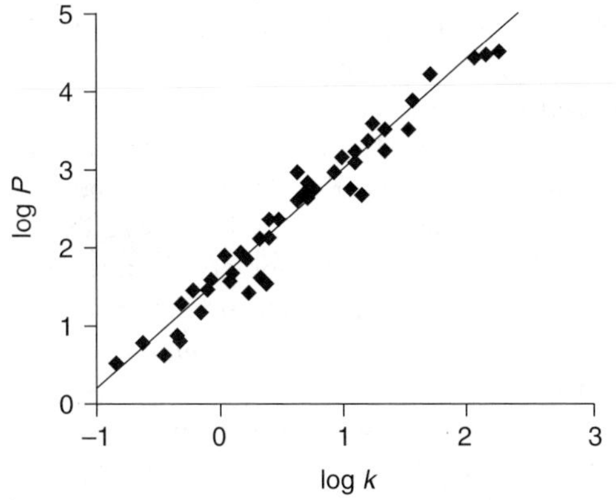

FIGURE 4.7 Plot of log P against the retention factor (log k) from MEEKC on a dynamically coated capillary column for compounds of varied structure. (From Poole S.K., et al., *J. Chromatogr. B*, 793, 265–274, 2003.)

TABLE 4.6
Selected Log *P* Measurements Using MEKC and MEEKC

Application	Buffer	Mode	References
Neutral and weakly acidic compounds using dynamically coated capillary columns	SDS (1.4% w/v), *n*-butanol (8% v/v) and *n*-heptane (1.2% v/v) in 90 mL acidic buffer	MEEKC	[246]
Determination of octanol–water partition coefficients of 80 pesticides	2.16% SDS (w/w), 0.82% heptane (w/w), 6.49% 1-butanol (w/w) in 0.05 M sodium phosphate–0.1 M borate at pH 7	MEEKC	[245]
Estimation of octanol–water partition coefficients of anticancer platinum(II)-complexes	Heptane/SDS/butanol in phosphate buffer at pH 7.4	MEEKC	[247]
Log P_{ow} measurement using multiplexed MEEKC	6.61% (w/v) 1-butanol, 0.81% (w/v) *n*-heptane, 3.31% (w/v) SDS, 800 mL of 68 mM CAPS buffer (pH 10.3)	MEEKC	[248]
Neutral and basic compounds by microchip MEEKC	2% w/v SDS, 1.2% v/v *n*-heptane, 8% v/v 1-butanol, 50 mM CAPS (pH 10.4)	MEEKC	[249]
Neutral pharmaceuticals	25 mM borate buffer (pH 8.5), 20–150 mM SDS	MEKC	[242]
Neutral pharmaceuticals and steroids	0.05 M phosphate, pH 7 (pH 9 for steroids), 80 mM sodium cholate, 100 mM SDS or 100 mM CTAB	MEKC	[250]
Determination of drug in phospholipids	50 mM phosphate buffer (pH 7.4) containing different concentrations of sodium cholate C (50, 60, 80, 125 mM)	MEKC	[251]

with a counterion, commonly a metal cation for acidic drugs and an ionic salt or small organic acid for basic drugs. During development of a new drug, a range of different counterions may be synthesized to compare pharmaceutical properties such as solubility, stability, and crystallinity of the different salts. The ratio of the drug to counterion is known as the drug stoichiometry and this needs to be characterized analytically. The typical stoichiometry is a 1:1 drug–counterion mixture; however, frequently 2:1 and 1:2 compositions are manufactured depending on the ionic nature of the drug and/or counterion. There is a clear analytical need to quantify drug counterion levels to demonstrate that the correct salt version has been manufactured and that the required stoichiometry can be reliably achieved batch-to-batch when the final drug salt has been selected. The counterion of basic drugs includes inorganics such as sulfate and chloride or organic acids such as maleate, fumarate, acetate, or succinate. Cations analyzed involve a range of metal ions including Na^+, K^+, Mg^{2+}, Ca^{2+}, and simple low molecular weight amines. These analytes possess little or no chromophore, which generally necessitates use of indirect UV detection. However, some larger anionic counterions such as benzoates and simple organic acids can possess sufficient UV activity to allow direct UV detection. Alternatively, metal ions may be complexed "on capillary" to form metal chelates that can then be detected by direct UV measurement. Alternative detection methods such as conductivity detection have also been used [268] to detect potassium counterion and other inorganic cationic impurities in pharmaceutical drug substances.

Popular techniques for the analysis of small ions include ion-exchange chromatography and flame atomic absorption spectrometry but CE is becoming more popular for such applications due to its simplicity and speed of method development and analysis, elimination of the need for specialized columns, high resolving power, and simple sample treatment steps (typically, the sample just needs to be diluted in the background electrolyte and injected onto the capillary). Commercial ion analysis

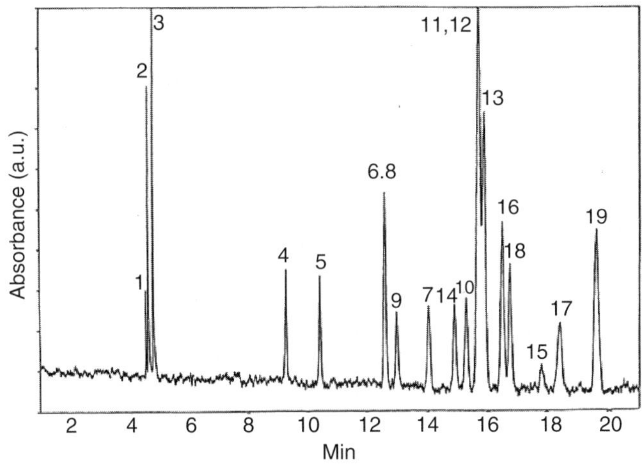

FIGURE 4.8 Electropherogram of 20 common amino acids. 50 mM Ethanesulfonic acid, pH 2.8; applied voltage, 30 kV; injection time, 10 s. Peaks: 1 = Lys; 2 = Arg; 3 = His; 4 = Gly; 5 = Ala; 6 = Val; 7 = Ser; 8 = Ile; 9 = Leu; 10 = Thr; 11 = Asn; 12 = Met; 13 = Gln; 14 = Trp; 15 = Glu; 16 = Phe; 17 = Pro; 18 = Tyr; 19 = Cys. (From Fritz J.S., *J. Chromatogr. A*, 884, 261–275, 2000.)

kits are available that contain predefined method conditions, reagents and buffers that simplify method development and allow analysis times of 2–10 min, comparing favorably to ion-exchange chromatography methods.

Pharmaceutical excipients such as SDS or alginic acid can be analyzed as raw materials or when present in formulations [263]. Kelly et al. [252] developed a reliable quantitative CE method for the determination of SDS in a cefuroxime axetil pharmaceutical preparation. Separation of amino acids can be troublesome and complicated using HPLC methods as they first need to be derivatized to provide a chromophore for detection. Performing these separations using CE and low pH electrolytes, however, can be relatively simplistic as lower detection wavelengths of 185 nm can be used to detect zwitterionic amino acids, which become cations at low pH. A background electrolyte of 50 mM ethanesulfonic acid (pH 2.8) was used to resolve a number of amino acids, which were detected at 185 nm without any sample pretreatment [253], as shown in Figure 4.8.

Simple organic acids often possess chromophores, which makes direct UV detection possible, even at low wavelengths. CE can measure the presence of small ion contaminant impurities in drug substances. For example, an NACE method with indirect UV detection was used to monitor ammonium ion contaminant in pharmaceutical preparations with a limit of detection of 50 ppb [254]. A number of reviews have been published that will provide the reader with a comprehensive coverage of the status of CE for the analysis of ions and small molecules, including detection methods, quantification, and stoichiometric determinations [8,9,11,253,255,256]. Table 4.7 shows some recent analysis applications of CE for ion analysis.

4.7 METHOD DEVELOPMENT AND VALIDATION

As with any other analytical technique developing a CE method involves some basic steps. As a general guide to method development, the following items should be considered: the objectives of the intended separation, instrumental requirements, sample characteristics, sample pretreatment, data handling, and reporting. The objectives may be straightforward and require the detection and quantification of a single charged compound of which there is much information available and separation can be achieved using simple aqueous buffers. They may be more complex and require the separation, identification, and quantification of a number of charged and neutral components,

TABLE 4.7
Selected CE Applications for the Analysis of Small Molecules and Ions (Pharmaceutical)

Application	Electrolyte	References
Anions—indirect detection		
SDS levels in tablets	3–10 mM barbital buffer, pH ~9	[252]
Determination of Br, Cl, and SO_4 as impurities in calcium acamprosate	10 mM potassium chromate + 1 mM borate, pH 9.15	[257]
Acetate counterion in an antifungal lipopeptide	4.0 mM 4-hydroxybenzoic acid, myristyltrimethylammonium bromide, as an EOF modifier	[258]
Sodium acetate in antisense oligonucleotides	Phthalate, myristyltrimethylammonium bromide type EOF modifier	[259]
Drug inorganic counter ion determination	Chromate, tetradecyltrimethylammoniumbromide (TTAB)	[260]
Drug organic acid counter ion determination	Phthalate, 4-morpholineethanesulfonic acid (MES), tetradecyltrimethylammoniumbromide (TTAB)	[261]
Anion screening for drugs and intermediates	Chromate, myristyltrimethylammonium bromide type EOF modifier	[262]
Anions—direct detection		
Quantitative determination of alginic acid in pharmaceutical formulations	12.5 mM borate/boric acid buffer (pH 8.3)	[263]
Determination of residual Br in excess of chloride for local anaesthetic analysis	60:40 MeCN:methanesulfonic acid buffer (pH 1.3)	[264]
Arginine counterion determination for ragaglitazone	10:90 ACN:(2% SB-β-CD and 0.7% DM-β-CD in 25 mM phosphate buffer, pH 8.0)	[172]
Cations—indirect detection		
Ammonium ion impurities	NACE method, imidazole	[254]
Ca, Li, K, and Na counterions of glycosaminoglycans	4-Aminopyridine buffer (pH 9)	[265]
Ca in calcium acamprosate drug substance	10 mM imidazole containing 1 mM tetrabutylammonium sulfate (pH 4.5)	[266]
Sodium in drug substance	Imidazole, formic acid	[267]
Quarternary amine residues in drug substance	Quinine, THF	[267]
Piperazine in pharmaceutical drug substances	50 mM benzylamine (pH 8.7)	[269]
Cations—direct detection		
K counterion and inorganic cationic impurities of pharmaceutical drug substances by conductivity detection	30 mM creatinine, 30 mM acetic acid, 4.5 mM 18-Crown-6	[268]

which may be only possible using MEKC or MEEKC. The compound of interest may be present in a number of isomeric forms, which requires chiral selectors added to the background electrolyte or may be present in trace quantities as degradation products or impurities, which require separation and quantification with high sensitivity. Other items to be considered when deciding the objectives are as follows: do all of the sample components need to be separated/quantified/identified and what is the sample matrix. Identifying these goals helps to determine the instrumental requirements. For example, if qualitative analysis is required then a DAD or MS detector will be required. To achieve lower limits of detection, LIFD or UV sensitivity enhancement techniques may be used. The characteristics of the sample and sample matrix are helpful in selecting the mode of CE. If the sample and/or sample matrix is water insoluble, then NACE, MEEKC may be the CE mode of choice. Data handling depends on the type of the intended analysis, quantitative or qualitative and whether internal or external standardization is used for quantification. If the analyst possesses enough knowledge and experience of CE separations, he/she may be able to decide on initial CE conditions and optimize the separation from there. This, however, is not always the case particularly as the in-depth knowledge and experience of pharmaceutical analysis by CE is not as widely established as it is for other separation methods such as HPLC, TLC, and GC. Once the analyst identifies the separation objectives, a literature search of published reports to find sources of developed and/or validated methods for the same or similar separations should be carried out. A variety of literature sources are available. Sources of reviews and publications for each application area, CE mode and detection method described in this chapter are cited toward the end of each individual section to direct the reader to relevant literature sources. A number of tables of selected applications are also presented. Lunn [13] provides detailed method information for several hundred applications. In addition to searching published application literature, a number of commercial CE instrument and consumable manufacturers and suppliers offer generic and sample methods for a range of applications. Beckman Coulter, for example, has published a generic method development strategy for chiral CE, previously published in *LCGC Europe* [283].

Developing a method involves selecting and optimizing a number of variables

- *Mode of CE*: as discussed previously the mode of CE must suit the separation.
- *Capillary*: fused-silica capillaries are the standard type used. Capillary length will affect the speed of separations. Small internal diameter capillaries allow higher electric field strengths to be applied resulting in more efficient separations. Typically, capillaries of 50 μm inner diameter and 0.5–1 m in length are used.
- *Capillary conditions*: these are affected by the capillary pretreatment reagents (e.g., NaOH) or dynamic capillary coating systems to achieve better reproducibility.
- Applied voltage and operating current.
- Injection mode and volume.
- Capillary temperature.
- *Detector*: as discussed previously, the detector type is governed by quantitative/qualitative and suitability requirements.
- *Buffer composition*: this is one of the most important variables to consider, as the type of compounds to be analyzed will dictate the pH, polarity, concentration, additives, and so on of the buffer.

A number of books and reviews provide a more detailed insight and coverage of the operating parameters of CE and their effects in method development, such as those mentioned above. This chapter does not have the scope to cover these in detail and readers are encouraged to consult the publications [13,14,34–37].

If there are no previous published methods for a particular pharmaceutical compound or class of compound then evaluation of some of the properties and functionalities must be considered to determine the initial method conditions. If pK_a data are available for the solute of interest and other species that need to be resolved then an appropriate electrolyte pH can be selected. If the compound is present as a racemic mixture then selection of an appropriate chiral selector can be made based on the compounds functional groups and other chiral separation applications. For basic compounds, a 50 mM phosphate (Na_2HPO_4) buffer adjusted to pH 2.5 with phosphoric acid, in a standard capillary with an applied voltage of +20 kV and UV detection at 200 nm is a good starting point.

Samples should be dissolved in an appropriate solvent at a concentration of ~1 mg ml^{-1}. This set of operating conditions is appropriate for a wide range of basic compounds. Additional optimization can be achieved using additives such as CDs, ion-pairing reagents, and organic solvents.

For the separation of acidic compounds, the buffer can be replaced with a 15 mM borate solution with a natural pH ~9.3. The same operating parameters as those used for the separation of basic compounds can be used to separate a wide range of acidic drugs and buffer additives similar to those used for basic separations can be used.

For the separation of neutral compounds using MEKC, 15 mM Na_2HPO_4 containing 50 mM SDS as the micellar electrolyte can be employed. Using different surfactants such as bile salts can help if the solutes are particularly water insoluble. Optimization can be carried out by adjusting the surfactant concentration and type, adding organic solvents, ion-pairing reagents, CDs, and so forth. Reviews by Pappas et al. [52], Molina and Silva [53], and Pyell [54] discuss MEKC operating parameters and applications in detail.

MEEKC offers more selectivity options than MEKC and can be used to separate a wider range of solutes. Owing to the composition of the microemulsion electrolytes, MEEKC can be used for more complex separations of water-soluble, water-insoluble, charged and neutral compounds, particularly where water-insoluble matrices are involved. A standard O/W microemulsion composed of 3.3% SDS, 0.8% octane, 6.6% butanol in 10 mM borate with fused-silica capillaries are used for most applications and this is a good starting point for any intended MEEKC separation. With a range of buffer additives available such as organic modifiers, CDs, ion-pairing reagents along with the various microemulsion parameters, MEEKC offers a wider selection of optimization choices than any other form of CE for pharmaceutical analysis. The background theory and operating parameters of MEEKC have been reviewed by Altria [58,270,271] and Klampfl [272], which detail the effects of each microemulsion and instrument variable on separation selectivity.

4.7.1 Method Optimization Using Experimental Design

Although the approach to CE method development is similar to HPLC methods, the number of variables involved can sometimes make the optimization of separation conditions a difficult task. This is particularly true when dealing with MEKC and MEEKC where pH, nature, and concentration of the surfactant, buffer, and organic modifiers combined with instrument variables such as applied voltage, temperature, and injection volume all have a great influence on solute separation. Ideally, a method should be optimized to find the best separation conditions in as short a time as possible with the minimum number of experiments. Statistical experimental design is a powerful tool to quantify the effect of one or more variables on a set of measured responses. It provides a methodological framework for changing operating parameters simultaneously by the help of experimental designs. These approaches involve the smallest possible number of useful experiments and provide maximum information. The use of experimental design strategies for CE methods involving univariate (studying one factor at a time) and multivariate (studying multiple factors simultaneously) techniques has evolved as a rapid method development and optimization option. Multivariate techniques are probably the least time consuming and are recommended. Multivariate (chemometric) experimental designs have been reviewed by Altria et al. [273] and Sentellas and Saurina [274,275]. Veuthey and Rudaz [276] have also covered the use of statistical and chemometric tools for pharmaceutical analysis. The use of artificial neural networks (ANNs) in separation science has also been reviewed [277]. Pyell [51] includes a detailed chapter containing many examples of experimental work (MEKC and MEEKC) using experimental design.

4.7.2 Method Validation Guidelines for Pharmaceutical Applications

Validation of pharmaceutical CE methods requires similar considerations to those examined during HPLC validation procedures. A useful guide for method validation in analytical chemistry is the

Eurachem guide [285], which discusses when, why, and how methods should be validated. However, for the pharmaceutical industry, the main reference source is the *ICH Guidelines* [286], which provides recommendations on the various characteristics to be tested for the most common types of analytical procedures developed in a pharmaceutical laboratory. The main characteristics of any analytical method to be tested are specificity, linearity, accuracy, precision, solution stability, limits of detection and quantification, and robustness. Specific aspects should be considered for a CE method including method transfer between instrument manufacturers, reagent purity and source, electrolyte stability, capillary treatment and variations in new capillaries, and buffer depletion. Fabre and Altria [284] discuss CE method validation in more detail and include a number of examples of validated CE methods for pharmaceutical analysis. Included in Table 4.3 are a number of validated pharmaceutical assay methods.

4.8 COMPARISON OF HPLC, GC, AND CE FOR PHARMACEUTICAL ANALYSIS

As HPLC is the most widely used pharmaceutical analysis technique among current chromatographic methods, a short comparison of HPLC and CE should be very informative to the reader. Some comparisons will also be made to GC. This section will illustrate the advantages, disadvantages, and complementary natures of HPLC and CE.

The range of CE applications in pharmaceutical analysis is at least as extensive as that of HPLC. Adopting CE testing provides several distinct advantages, including faster analysis and method development, lower consumable expenses, and easier operation. All of these factors are important in pharmaceutical separations where high throughput is becoming ever important. The disadvantages of CE are not to be forgotten; they include poorer injection precision (hence the need to incorporate internal standards) and the limited number of staff members who are trained and experienced in using CE compared with the number who are competent with HPLC. The highly complementary nature of the two techniques is a valuable tool for the pharmaceutical analyst where "cross-validation" using the two methods can give a high degree of confidence in results.

4.8.1 Efficiency

As solute peaks in HPLC and GC chromatograms are integrated and measured in the same way as peaks in CE electropherograms, direct comparisons of peak efficiency for each method is possible. Typical CE capillary dimensions are 30–100 cm long (although when using "short-end injection" the effective length can be reduced further) with internal diameters of between 50 and 100 μm and can generate ~400,000 theoretical plates, compared to ~150,000 plates for a typical wall coated open tubular GC column 0.18 mm internal diameter and 20 m long. HPLC, however, can generate at best 30,000 plates for a typical 150 mm C18 column. The main reason for the higher efficiency possible in CE compared to HPLC is that the solutes move through the column with a "plug" or flat flow profile, while they move through HPLC columns with a "laminar flow" profile. The other reasons for higher plate numbers in CE are the absence of multipath and longitudinal diffusion and slow equilibration of the solute in the capillary. Turbulent flow, often associated with GC columns, is also absent from CE capillaries due to the electrodriven nature of the EOF.

4.8.2 Sample Types

Similar to HPLC, CE is well suited for the analysis of both polar and nonpolar compounds but unlike HPLC columns, the same capillary can be used for most CE applications, that is, chiral separations, small ions, and so forth. Sample preparation for CE analyses is straightforward and usually requires sample dissolution in the buffer or water, sometimes with the addition of a small volume of organic solvent to help dissolve the solute. CE buffers can be used over a wider pH range than HPLC mobile phases, as most HPLC columns can only be used within a specified pH limits without damaging the column.

4.8.3 Sample Volume

Relatively small sample volumes are required for CE analyses compared to HPLC and GC, only a few nanoliters are injected compared to microliters used in GC and HPLC. This is due to the small volumes of the capillaries, typically 0.98 μl for a 50 cm × 50 μm ID capillary. This is advantageous when only limited amounts of the sample is available.

4.8.4 Sensitivity

As mentioned in Section 4, detection limits in CE are generally poorer than HPLC or GC due to the narrow detection path length for UV/Vis detection and the nanoliter injection volumes. Various approaches to improve CE sensitivity such as the use of LIFD detectors, low wavelength detection, and so forth have been covered in Section 4.4. Some comparisons of CE and HPLC sensitivity for pharmaceutical analysis are illustrated in Table 4.8.

4.8.5 Precision

Because of the small injection volumes used in CE, peak area reproducibility is poorer than HPLC. The use of high sample concentrations and internal standards makes CE precision values more comparable to HPLC. The precision of solute migration times can also suffer because of variations within the capillary between injections, affecting the EOF. This can be alleviated by careful preparation of buffers and samples and the use of dynamic capillary coating systems.

TABLE 4.8
Comparisons of Recent HPLC and CE Applications

Application	Method	Analysis Time	LOD	Precision RSD%	References
Determination of four parabens in cosmetics	HPLC CE	21 min 16 min	0.02–0.05 $\mu g\ ml^{-1}$ 0.16–0.21 $\mu g\ ml^{-1}$	0.41–1.14 0.77–2.17	[279]
Determination of glycyrrhizinic acid in pharmaceutical preparations	HPLC CE	7 min 15.5 min	n/a n/a	0.9–1.7 3.2–4.0	[287]
Stereoselective analysis of carvedilol in serum	HPLC CE	27 min 15 min	1 $ng\ ml^{-1}$ 50 $ng\ ml^{-1}$	n/a n/a	[280]
Determination of ketoconazole in drug formulations	HPLC CE	3 min 5 min	HPLC = 2.5 times less than CE	1.91 2.95	[281]
Determination of diazepam in pharmaceutical tablets	HPLC CE	4.9 min 4.1 min	1.44 $\mu g\ ml^{-1}$ 4.24 $\mu g\ ml^{-1}$	0.98 1.62	[282]

4.8.6 Analysis Times

Typical analysis times for both HPLC and CE are between 5 and 30 min. With CE, however, a number of options are available to shorten analysis times; pressure-assisted CE where an external pressure is applied to the capillary in conjunction with the EOF helps speed up separations. Short-end injection, where the sample is introduced into the detector end of the capillary can also reduce analysis times, these methods were demonstrated by Mahuzier et al. [278]. A shorter capillary can also be used to achieve shorter analysis times. As discussed in Section 4.3.6, the use of multiplexed or capillary array CE instruments can provide the pharmaceutical laboratory with the capability to conduct multiple analyses simultaneously, offering a significant advantage in terms of time and cost savings. This is particularly advantageous where the screening of large numbers of new pharmaceutical compounds is required due to combinatorial organic synthesis in drug discovery.

4.8.7 Reagents and Consumables

This area provides CE with many advantages over HPLC methods for a number of reasons. Owing to the applicability of the same CE instrumentation and capillaries to a wide range of applications only the buffer composition needs to be changed, whereas a variety of normal and reversed-phase, chiral and ion-exchange, and so forth HPLC columns are available with a wide range of packing material for different classes of compounds and separations. CE capillaries are also only a fraction of the cost of HPLC columns, particularly chiral columns, which can cost in excess of 1200, a standard fused-silica capillary that costs a few euro can be used with milligrams of a chiral selector to perform the same separation. Relatively small amounts of buffers and reagents are used in CE, milliliters compared to liters for HPLC mobile phases. This offers significant cost savings for both the purchase and disposal of expensive chemicals.

As discussed in Section 4.3.6, the use of multiplexed CE instruments can provide the pharmaceutical laboratory with the capability to conduct multiple analyses simultaneously, offering a significant advantage in terms of time and cost savings.

4.9 CONCLUSIONS

Capillary electrophoresis has been shown as a suitable analytical method for a wide range of pharmaceutical applications, indeed for principal applications such as pharmaceutical assay, physicochemical measurements, and chiral analysis CE can be superior to HPLC in terms of speed and range of method development options, cost efficiency, speed of analysis, ease of use, selectivity, peak efficiency, and the possibility of implementation of a single set of method conditions for the analysis of several different samples. The various advantages of CE have been highlighted which include low cost of consumables, speed of analysis, the use of standard instruments and capillaries for most applications, high-throughput analysis using capillary array instruments, superior efficiency, and low wavelength detection. The main disadvantages of CE are its poor sensitivity due to the narrow detection path length using UV/Vis detection methods and poor injection precision. Measures to improve detection limits include more sensitive detection methods such as LIFD, while CE-MS offers the possibility of highly sensitive qualitative analysis. Quantitative methods such as the use of internal standards serve to improve injection precision to levels approaching that of HPLC.

The various modes of CE available for pharmaceutical analysis means that effectively all types of drug compounds can be separated using standard instrumentation, that is, charged, neutral, polar and nonpolar, large and small molecules, and counterions. MEKC and MEEKC have been shown to be particularly suitable for the separation of complex mixtures of both charged and neutral compounds, in particular MEEKC, which can be applied to a wider range of solutes of a broadly varying water solubility using both W/O and O/W microemulsions as the carrier electrolyte.

The range of pharmaceutical applications of CE is very extensive with hundreds of reports published and this chapter only serves to give a brief introduction to each application area with tabulated examples of proven methods. Many validated methods are cited which indicate the usefulness of CE in the analytical laboratory, indeed the use of CE in pharmaceutical laboratories is now very widespread and regulatory authorities have begun to accept CE for a number of new drug product submissions. CE identification tests in general monographs for drug substances have been included in the British and European Pharmacopoeia.

CE is regularly used to provide complementary information to HPLC or other methods as well as being a technique of choice for certain applications. As further research is carried out and the use of CE becomes more widespread in pharmaceutical laboratories, the use of CE is certain to become more popular and possibly take over from traditional analytical methods such as HPLC and titrimetric techniques for routine testing of pharmaceuticals.

4.10 DEDICATION

To the memory of Johnny McEvoy, a great man who left us with many great memories.

REFERENCES

1. Harris D.C., *Quantitative Chemical Analysis*, 4th ed. (W. H. Freeman and Company, New York, 1995), pp. 643, 713.
2. Jorgenson J.W., Lukacs K.D., Zone electrophoresis in open tubular glass capillaries. *Anal. Chem.*, 53, 1289–1302 (1981).
3. Mikkers F.E.P., Everaerts F.M., Verheggen T.P.E.M., High-performance zone electrophoresis. *J. Chromatogr.*, 169, 11–20 (1979).
4. Terabe S., Otsuka K., Ando T., Electrokinetic chromatography with micellar solution and open tubular capillary. *Anal. Chem.*, 57, 834–841 (1985).
5. Terabe S., Electrokinetic chromatography: an interface between electrophoresis and chromatography. *Trends Anal. Chem.*, 8, 129–134 (1989).
6. Altria K.D., Quantitative aspects of the application of capillary electrophoresis to the analysis of pharmaceuticals and drug related impurities. *J. Chromatogr.*, 646, 245–257 (1993).
7. Holland L.A., Chetwyn N.P., Perkins M.D., Lunte S.M., Capillary electrophoresis in pharmaceutical analysis. *Pharm. Res.*, 14, 372–387 (1997).
8. Altria K.D., Kelly M.A., Clark B.J., Current applications in the analysis of pharmaceuticals by capillary electrophoresis. I. *Trends Analyt. Chem.*, 17, 204–214 (1998).
9. Altria K.D., Kelly M.A., Clark B.J., Current applications in the analysis of pharmaceuticals by capillary electrophoresis. II. *Trends Analyt. Chem.*, 17, 214–226 (1998).
10. Altria K.D., Overview of capillary electrophoresis and capillary electrochromatography. *J. Chromatogr. A*, 856, 443–463 (1999).
11. Altria K.D., Elder D., Overview of the status and applications of capillary electrophoresis to the analysis of small molecules. *J. Chromatogr. A*, 1023, 1–14 (2004).
12. Natishan T.K., Recent progress in the analysis of pharmaceuticals by capillary electrophoresis. *J. Liq. Chromatogr. Relat. Technol.*, 28, 1115–1160 (2005).
13. Lunn G., *Capillary Electrophoresis Methods for Pharmaceutical Analysis* (John Wiley & Sons, New York, 2000).
14. Landers J.P. (Ed.), *Handbook of Capillary Electrophoresis*, 2nd ed. (CRC Press, FL, 1997).
15. Altria K.D., Simpson C.F., Analysis of some pharmaceuticals by high voltage capillary zone electrophoresis. *J. Pharm. Biomed. Anal.*, 6, 801–807 (1988).
16. Ha P.T.T., Hoogmartens J., Van Schepdael A., Recent advances in pharmaceutical applications of chiral capillary electrophoresis. *J. Pharm. Biomed. Anal.*, 41, 1–11 (2006).
17. Scriba G.K.E., Selected fundamental aspects of chiral electromigration techniques and their application to pharmaceutical and biomedical analysis. *J. Pharm. Biomed. Anal.*, 27, 373–399 (2002).

18. Scriba G.K.E., Pharmaceutical and biomedical applications of chiral capillary electrophoresis and capillary electrochromatography: an update. *Electrophoresis*, 24, 2409–2421 (2003).
19. Amini A., Recent developments in chiral capillary electrophoresis and applications of this technique to pharmaceutical and biomedical analysis. *Electrophoresis*, 22, 3107–3130 (2001).
20. de Boer T., de Zeeuw R.A., de Jong G.J., Ensing K., Recent innovations in the use of charged cyclodextrins in capillary electrophoresis for chiral separations in pharmaceutical analysis. *Electrophoresis*, 21, 3220–3239 (2000).
21. Wang A.B., Fang Y.Z., Applications of capillary electrophoresis with electrochemical detection in pharmaceutical and biomedical analyses. *Electrophoresis*, 21, 1281–1290 (2000).
22. Zemann A.J., Conductivity detection in capillary electrophoresis. *Trends Analyt. Chem.*, 20, 346–354 (2001).
23. Lacroix M., Poinsot V., Fournier C., Couderc F., Laser-induced fluorescence detection schemes for the analysis of proteins and peptides using capillary electrophoresis. *Electrophoresis*, 26, 2608–2601 (2005).
24. Smyth W.F., Recent applications of capillary electrophoresis-electrospray ionisation-mass spectrometry in drug analysis. *Electrophoresis*, 27, 2051–2062 (2006).
25. Poole S.K., Patel S., Dehring K., Workman H., Poole C.F., Determination of acid dissociation constants by capillary electrophoresis. *J. Chromatogr. A*, 1037, 445–454 (2004).
26. Hudson J., Golin M., Malcolm M., Whiting C., Capillary zone electrophoresis in a comprehensive screen for drugs of forensic interest in whole blood. *.Can. Soc. Forens. Sci. J.*, 31, 1–19 (1998).
27. Altria K.D., Bryant S.M., Hadgett T., Validated CE method for the assay of a range of acidic drugs and excipients. *J. Pharm. Biomed. Anal.*, 15, 1091–1101 (1997).
28. Khaledi Morteza G. (Ed.), *High-Performance Capillary Electrophoresis: Theory, Techniques, and Applications*. (John Wiley & Sons, New York 1998).
29. Foret F., Bocek P., Krivankova L. *Capillary Zone Electrophoresis*. (John Wiley and Sons, New York, 1993).
30. Altria K.D. (Ed.), *Capillary Electrophoresis Guidebook: Principles, Operations and Applications*. (Humana Pr. 1996).
31. Baker Dale R., *Capillary Electrophoresis*. (John Wiley & Sons, New York, 1995).
32. Smith N.W., Evans M.B., Capillary zone electrophoresis in pharmaceutical and biomedical analysis. *J. Pharm. Biomed. Anal.*, 12, 579–611 (1994).
33. Esaka Y., Okumura N., Uno B., Goto M., Non-aqueous capillary electrophoresis of p-quinone anion radicals. *Anal. Sci..*, 17, 99–102 (2001).
34. Yuqin L., Shuya C., Yuqiao C., Xingguo C., Zhide H., Application of nonaqueous capillary electrophoresis for quantitative analysis of quinolizidine alkaloids in Chinese herbs. *Anal. Chim. Acta*. 2004, 508, 17–22.
35. Leung G. N. W., Tang, H. P. O., Tso, T. S. C., Wan, T.S. M., Separation of basic drugs with non-aqueous capillary electrophoresis. *J. Chromatogr. A*, 738, 141–154 (1996).
36. Cherkaoui S., Varesio E., Christen P., Veuthey J.L., Selectivity manipulation using nonaqueous capillary electrophoresis. Application to tropane alkaloids and amphetamine derivatives. *Electrophoresis*, 19, 2900–2906 (1998).
37. Cantú M. D., Hillebrand S., Costa Queiroz M. E., Lanças F. M., Carrilho E., Validation of non-aqueous capillary electrophoresis for simultaneous determination of four tricyclic antidepressants in pharmaceutical formulations and plasma samples. *J. Chromatogr. B*, 799, 127–132 (2004).
38. Anurukvorakun O., Suntornsuk W., Suntornsuk L., Factorial design applied to a non-aqueous capillary electrophoresis method for the separation of β-agonists. *J. Chromatogr. A*, 1134, 326–332 (2006).
39. Cherkaoui S., Veuthey J.L., Separation of selected anesthetic drugs by non-aqueous capillary electrophoresis. *Analysis*, 27, 765–771 (1999).
40. Riekkola M.L., Jisilli M., Porras S.P., Valko I.E., Non Aqueous capillary electrophoresis. *J. Chrom. A*, 892, 155–170 (2000).
41. Matysik F.M., Special aspects of detection methodology in nonaqueous capillary electrophoresis. *Electrophoresis*, 23, 400–407 (2002).
42. Fillet M., Servais A.C., Crommen J., Effects of background electrolyte composition and addition of selectors on separation selectivity in nonaqueous capillary electrophoresis. *Electrophoresis*, 24, 1499–1507 (2003).

43. Porras S.P., Kenndler E., Capillary zone electrophoresis in non-aqueous solutions: pH of the background electrolyte. *J. Chromatogr. A*, 1037, 455–465 (2004).
44. Bjornsdottir I., Hansen S.H., Determination of opium alkaloids in pure opium using non-aqueous capillary electrophoresis. *J. Pharm. Biomed. Anal.*, 13, 1473–1481 (1995).
45. Bjornsdottir I., Hansen S.H., Comparison of the separation selectivity in aqueous and non-aqueous CE. *J. Chromatogr. A*, 711, 313–319 (1995).
46. Cherkaoui S., Varesio E., Christen P., Veuthey J.L., Selectivity manipulation using non-aqueous CE. Application to tropane alkaloids and amphetamine derivatives. *Electrophoresis*, 19, 2900–2906 (1998).
47. Servais A.C., Fillet M., Chiap P., Abushoffa A.M., Hubert P., Crommen J., Optimisation of the separation of beta-blockers by ion-pair CE in non-aqueous media using univariate and multivariate approaches. *J. Sep. Sci.*, 25, 1087–1095 (2002).
48. Porras S.P., Reikkola M.-L., Capillary zone electrophoresis of basic analytes in methanol as non-aqueous solvent: Mobility and ionisation constant. *J. Chromatogr. A*, 905, 259–268 (2001).
49. Siren H., Hiissa T., Min Y., Aqueous and non-aqueous capillary electrophoresis of polar drugs. *Analyst*, 9, 1561–1568 (2000).
50. Hedeland Y., Hedeland M., Bondesson U., Petterson C., Chiral separations of amines with *N*-benzoxycarbonylglycil-L-proline as selector in non-aqueous CE using methanol and 1,2-dichloroethane in the background electrolyte. *J. Chromatogr. A*, 984, 261–271 (2003).
51. Pyell U. (Ed.), *Electrokinetic Chromatography: Theory, Instrumentation and Applications* (John Wiley & Sons, England, 2006).
52. Pappas T.J., Gayton-Ely M., Holland L.A., Recent advances in micellar electrokinetic chromatography. *Electrophoresis*, 26, 719–734 (2005).
53. Molina M., Silva M., Micellar electrokinetic chromatography: Current developments and future. *Electrophoresis*, 23, 3907–3921 (2002).
54. Pyell U., Micellar electrokinetic chromatography—From theoretical concepts to real samples (Review). *Fresenius J. Anal. Chem.*, 371, 691–703 (2001).
55. Watari H., Microemulsion capillary electrophoresis. *Chem. Lett.*, 3, 391–394 (1991).
56. Broderick M., Donegan S., Power J., Altria K.D., Optimisation of water-in-oil MEEKC in pharmaceutical analysis. *J. Pharm. Biomed. Anal.*, 37, 877–884 (2005).
57. Altria K.D., Broderick M.F., Donegan S., Power J., The use of novel water-in-oil microemulsions in microemulsion electrokinetic chromatography. *Electrophoresis*, 25, 645–652 (2004).
58. Altria K.D., Application of microemulsion electrokinetic chromatography to the analysis of a wide range of pharmaceuticals and excipients. *J. Chromatogr. A*, 844, 371–386 (1999).
59. Hansen S.H., Gabel-Jenson C., Mohammed El-Sherbiny D., Pedersen-Bjergaard S., Microemulsion electrokinetic chromatography—or solvent modified micellar electrokinetic chromatography. *Trends Anal. Chem.*, 20, 614–619 (2001).
60. Cherkaoui S., Veuthey J.-L., Micellar and microemulsion electrokinetic chromatography of selected anaesthetic drugs. *J. Sep. Sci.*, 25, 1073–1078 (2002).
61. Wilson I.D., Poole C.F., (Eds). Liquid chromatography: Microemulsion based HPLC methods. Encyclopedia of Separation Science—Online Update ISBN: 978-0-12-226770-3 (Elsevier, 2007).
62. Hansen S.H., Recent applications of microemulsion electrokinetic chromatography. *Electrophoresis*, 24, 3900–3907 (2003).
63. Marsh A., Broderick M., Altria K.D., Clark B., Power J., Donegan S., Recent advances in microemulsion electrokinetic chromatography. *Electrophoresis*, 25, 3970–3980 (2004).
64. Huie C.W., Recent applications of microemulsion electrokinetic chromatography. *Electrophoresis*, 27, 60–75 (2006).
65. McEvoy E., Marsh A., Altria K.D., Donegan S., Power J., Recent advances in the development and applications of microemulsion electrokinetic chromatography. *Electrophoresis*, 28, 193–207 (2007).
66. Watari H., Microemulsions in separation sciences. *J. Chromatogr. A*, 780, 93–102 (1997).
67. Smith R.D., Barinaga C.J., Udseth H.R., Improved electrospray ionisation interface for capillary zone electrophoresis-mass spectrometry. *Anal. Chem.*, 60, 1948–1952 (1988).
68. Smyth W.F., Recent applications of capillary electrophoresis-electrospray ionisation-mass spectrometry in drug analysis. *Electrophoresis*, 27, 2051–2062 (2006).
69. Ohnesorge J., Neususs C., Watsig H., Quantitation in capillary electrophoresis-mass spectrometry. *Electrophoresis*, 26, 3973–3987 (2005).

70. Schmitt-Kopplin P., Englmann M., Capillary electrophoresis-mass spectrometry: Survey on developments and applications 2003–2004. *Electrophoresis*, 26, 1209–1220 (2005).
71. Smyth W.F., Recent applications of capillary electrophoresis-electrospray ionisation-mass spectrometry in drug analysis. *Electrophoresis*, 26, 1334–1357 (2005).
72. Shamsi S.A., Miller B.E., Capillary electrophoresis-mass spectrometry: Recent advances to the analysis of small achiral and chiral solutes. *Electrophoresis*, 25, 3927–3961 (2005).
73. Issaq H.J., Janini G.M., Chan K.C., Veenstra T.D., Sheathless electrospray ionization interfaces for capillary electrophoresis-mass spectrometric detection—Advantages and limitations. *J. Chromatogr. A*, 1053, 37–42 (2004).
74. Schmitt-Kopplin P., Frommberger M., Capillary electrophoresis-mass spectrometry: 15 years of developments and applications. *Electrophoresis*, 24, 3737–3867 (2003).
75. Tu J., Halsall H.B., Seliskar C.J., Limbach P.A., Wehmeyer K.R., Heineman W.R., Estimation of log P_{ow} values for neutral and basic compounds by microchip microemulsion electrokinetic chromatography with indirect fluorimetric detection (mu MEEKC-IFD). *J. Pharm. Biomed. Anal.*, 38, 1–7 (2005).
76. Zhou C.X., Yin Y.K., Kenseth J.R., Stella M., Wehmeyer K.R., Heineman W.R., Rapid pK_a estimation using vacuum-assisted multiplexed capillary electrophoresis (VAMCE) with ultraviolet detection. *J. Pharm. Biomed. Anal.*, 94, 576–589 (2005).
77. Wong K.S., Kenseth J., Strasburg R., Validation and long-term assessment of an approach for the high throughput determination of lipophilicity (log *P*-OW) values using multiplexed, absorbance-based capillary electrophoresis. *J. Pharm. Sci.*, 93, 916–931 (2004).
78. Wehmeyer K.R., Tu J., Yin Y.K., King S., Stella M., Stanton D.T., Kenseth J., Wong K.S., The application of multiplexed microemulsion electrokinetic chromatography for the rapid determination of log P_{ow} values for neutral and basic compounds. *LC GC North America* 90–95 Suppl. S, (2004).
79. Kenseth J., Bastin A., High-throughput chiral separations using the cePRO 9600 (TM) multiplexed capillary electrophoresis system. *LC-GC North America*, 28–29 Suppl. S (2004).
80. Reetz M.T., Kuhling K.M., Deege A., Hinrichs H., Belder D., Super-high-throughput screening of enantioselective catalysts by using capillary array electrophoresis. *Angew. Chem. Int. Ed.*, 39, 3891–3893 (2000).
81. Zhong W., Yeung E.S., Combinatorial enantiomeric separation of diverse compounds using capillary array electrophoresis. *Electrophoresis*, 23, 2996–3005 (2002).
82. Zhang Y.H., Gong X.Y., Zhang H.M., Larock R.C., Yeung E.S., Combinatorial screening of homogeneous catalysis and reaction optimization based on multiplexed capillary electrophoresis. *J. Combinat. Chem.* 2(5), 450–452 (2000).
83. Marsh A., Altria K.D., Use of multiplexed capillary electrophoresis for pharmaceutical analysis. *Chromatographia*, 64, 327–333 (2006).
84. Pang H.M., Kenseth J., Coldiron S., High-throughput multiplexed capillary electrophoresis in drug discovery. *Drug Discov. Today*, 9, 1072–1080 (2004).
85. Moring S.E., Reel R.T., Van Soest R.E.J., Optical improvements of a Z shaped cell for high sensitivity UV absorbency detection in capillary electrophoresis. *Anal. Chem.*, 62, 3454–3459 (1993).
86. Tsuda T., Sweedler J.V., Zare R.N., Rectangular capillaries for capillary zone electrophoresis. *Anal. Chem.*, 62, 2149–2152 (1990).
87. Wang T.S., Aiken J.H., Huie C.W., Hartwick R.A., Nanoliter-scale multireflection cell for absorption detection in capillary electrophoresis. *Anal. Chem.*, 63, 1372–1376 (1991).
88. Heiger D.N., *High Performance Capillary Electrophoresis—An Introduction*. (Hewlett-Packard Co., France), 1992.
89. Mathies R., Peck K., Optimisation of high sensitivity fluorescence detection. *Anal. Chem.*, 62, 1786–1791 (1990).
90. Dittrich P.S., Manz A., Single-molecule fluorescence detection in microfluidic channels—The Holy Grail in mu TAS? *Anal. Bioanal. Chem.*, 382, 1771–1782 (2005).
91. Johnson M.E., Landers J.P., Fundamentals and practice for ultrasensitive laser-induced fluorescence detection in microanalytical systems. *Electrophoresis*, 25, 3513–3527 (2004).
92. Xu C.L., Li B.X., Zhang Z.J., Recent developments in detectors for micro total analysis systems. *Chin. J. Anal Chem.*, 31, 1520–1526 (2003).

93. McWhorter S., Soper S.A., Near-infrared laser-induced fluorescence detection in capillary electrophoresis. *Electrophoresis*, 21, 1267–1280 (2000).
94. Zhang X., Stuart J.N., Sweedler J.V., Capillary electrophoresis with wavelength-resolved laser-induced fluorescence detection. *Anal. Bioanal. Chem.*, 373, 332–343 (2002).
95. Lui J.F., Yang X.R., Wang E.K., Progress in amperometric detection for capillary electrophoresis. *Chin. J. Anal Chem.*, 30, 748–753 (2002).
96. Zang L., Tong P., Chen G.N., Determination of the hydrolysis rate constants and activation energy of aesculin with capillary electrophoresis end-column amperometric detection. *J. Chromatogr. A*, 1098, 194–198 (2005).
97. Zhang S.S., Yuan Z.B., Liu H.X., Zou H., Xiong H., Wu Y.J., Analysis of acyclovir by high performance capillary electrophoresis with on-column amperometric detection. *Electrophoresis*, 21, 2995–2998 (2000).
98. Bowman J., Tang L.Y., Silverman C.E., Analysis of small amines in pharmaceuticals by capillary ion electrophoresis with conductivity detection. *J. Pharm. Biomed. Anal.*, 23, 663–669 (2000).
99. de Silva J.A.F., Electrochemical detection in capillary electrophoresis. *Quimica Nova.*, 26, 56–64 (2003).
100. Baldwin R.P., Recent advances in electrochemical detection in capillary electrophoresis. *Electrophoresis*, 21, 4017–4028 (2000).
101. Holland L.A., Leigh A.M., Amperometric and voltametric detection for capillary electrophoresis. *Electrophoresis*, 23, 3649–3658 (2002).
102. Wang A.B., Fang Y.Z., Applications of capillary electrophoresis with electrochemical detection in pharmaceutical and biomedical analyses. *Electrophoresis*, 21, 1281–1290 (2000).
103. Swinney K., Bornhop D.J., Detection in capillary electrophoresis. *Electrophoresis*, 21, 1239–1250 (2000).
104. Rizzi A., Fundamental aspects of chiral capillary electrophoresis. *Electrophoresis*, 22, 3079–3106 (2001).
105. Gubitz G., Schmid M.G., Chiral separation principles in chromatographic and electromigration techniques. *Mol. Biotechnol.*, 32, 159–179 (2006).
106. Chankvetadze, B. *Capillary Electrophoresis in Chiral Analysis.* (John Wiley & Sons, New York, 1997).
107. Gubitz G., Schmid M.G., Recent advances in chiral separation principles in capillary electrophoresis and capillary electrochromatography. *Electrophoresis*, 25, 3981–3996 (2004).
108. Amini A., Recent developments in chiral capillary electrophoresis and applications of this technique to pharmaceutical and biomedical analysis. *Electrophoresis*, 22, 3107–3130 (2001).
109. Denola N.L., Quiming N.S., Catabay A.P., Saito Y., Jinno K., Optimization of capillary electrophoretic enantioseparation for basic drugs with native beta-CD as a chiral selector. *Electrophoresis*, 27, 2367–2375 (2006).
110. Koppenhoefer B., Zhu X., Jakob A., Wuerthner S., Lin B., Separation of drug enantiomers by capillary electrophoresis in the presence of neutral cyclodextrins. *J. Chrom. A*, 875, 135–161 (2000).
111. Haginaka J., Enantiomer separation of drugs by capillary electrophoresis using proteins as chiral selectors. *J. Chromatogr. A*, 875, 235–254 (2000).
112. Chiou C.-S., Shih J.-S., Application of crown ethers as modifiers for the separation of amines by capillary electrophoresis. *Anal. Chim. Acta*, 360, 69–76 (1998).
113. Fanali S., Aturki Z., Desiderio C., Righetti P.G., Use of MDL 63 246 (Hepta-Tyr) antibiotic in capillary zone electrophoresis: II. Chiral resolution of α-hydroxy acids. *J. Chromatogr. A*, 838, 223–235 (1999).
114. Blanco M., Valverde I., Choice of chiral selector for enantioseparation by capillary electrophoresis. *Trends Anal. Chem.*, 22, 428–439 (2003).
115. Shamsi S.A., Chiral capillary electrophoresis-mass spectrometry: Modes and applications. *Electrophoresis*, 23, 4036–4051 (2003).
116. Fillet M., Hubert P., Crommen J., Method development strategies for the enantioseparation of drugs by capillary electrophoresis using cyclodextrins as chiral additives. *Electrophoresis*, 19, 2834–2840 (1998).
117. Evans C.E., Stalcup A.M., Comprehensive strategy for chiral separations using sulfated cyclodextrins in capillary electrophoresis. *Chirality*, 15, 709–723 (2003).

118. Zhou L., Thompson R., Song S., Ellison D., Wyvratt J.M., A strategic approach to the development of capillary electrophoresis chiral methods for pharmaceutical basic compounds using sulfated cyclodextrins. *J. Pharm. Biomed. Anal.*, 27, 541–553 (2002).
119. Kuhn R., Stoecklin F., Erni F., Chiral separations by host-guest complexation with cyclodextrin and crown ether in capillary zone electrophoresis. *Chromatographia*, 33, 32–36 (1992).
120. Iványi T., Pal K., Lázár I., Massart D.L., Vander Heyden Y., Application of tetraoxadiaza-crown ether derivatives as chiral selector modifiers in capillary electrophoresis. *J. Chromatogr. A*, 1028, 325–332 (2004).
121. Fanali S., Cartoni C., Desiderio C., Chiral separation of newly synthesized arylpropionic acids by capillary electrophoresis using cyclodextrins or a glycopeptide antibiotic as chiral selectors. *Chromatographia*, 54, 87–92 (2001).
122. Desederio C., Fanali S., Chiral analysis by capillary electrophoresis using antibiotics as chiral selector. *J. Chromatogr. A*, 807, 37–56 (1998).
123. Aboul-Enein H.Y., Ali I., Macrocyclic antibiotics as effective chiral selectors for enantiomeric resolution by liquid chromatography and capillary electrophoresis. *Chromatographia*, 52, 679–691 (2000).
124. Ha P.T.H., Van Schepdael A., Roets E., Hoogmartens J., Investigating the potential of erythromycin and derivatives as chiral selector in capillary electrophoresis. *J. Pharm. Biomed. Anal.*, 34, 861–870 (2004).
125. Park H., Lee S., Kang S., Jung Y., Jung S., Enantioseparation using sulfated cyclosophoraoses as a novel chiral additive in capillary electrophoresis. *Electrophoresis*, 25, 2671–2674 (2004).
126. Budanova N., Shapovalova E., Lopatin S., Varlamov V., Shpigun O., *N*-(3-sulfo, 3-carboxy)-propionylchitosan as new chiral selector for enantioresolution of basic drugs by capillary electrophoresis. *Chromatographia*, 59, 709–713 (2004).
127. Bortocan R., Bonato P.S., Enantioselective analysis of primaquine and its metabolite carboxyprimaquine by capillary electrophoresis. *Electrophoresis*, 25, 2848–2853 (2004).
128. Quan Z., Song Y., Feng Y., LeBlanc M.H., Liu Y.M., Detection of D-serine in neural samples by saccharide enhanced chiral capillary electrophoresis. *Anal. Chim. Acta*, 528, 101–106 (2005).
129. Millot M.C., Separation of drug enantiomers by liquid chromatography and capillary electrophoresis, using immobilized proteins as chiral selectors. *J. Chromatogr. B*, 797, 131–159 (2003).
130. Tesarova E., Sevcik J., Gag B., Armstrong D.W., Effects of partial/asymmetrical filling of micelles and chiral selectors on capillary electrophoresis enantiomeric separation: Generation of a gradient. *Electrophoresis*, 25, 2693–2700 (2004).
131. Dobashi A., Ono T., Hara S., Yamaguchi J., Optical resolution of enantiomers with chiral mixed micelles by electrokinetic chromatography. *Anal. Chem.*, 61, 1984–1986 (1989).
132. Terabe S., Shibata M., Miyashita Y.J., Chiral separation by electrokinetic chromatography with bile salt micelles. *J. Chromatogr. A*, 480, 403–411 (1989).
133. Tickle D.C., Okafo G.N., Camilleri P., Jones R.F.D., Kirby J., Glucopyranoside-based surfactants as pseudostationary phases for chiral separations in capillary electrophoresis. *Anal. Chem.*, 66, 4121–4126 (1994).
134. Otsuka K., Terabe S., Enantiomer separation of drugs by micellar electrokinetic chromatography using chiral surfactants. *J. Chromatogr. A*, 875, 163–178 (2000).
135. Aiken J.H., Huie C.W., Use of a microemulsion system to incorporate a lipophilic chiral selector in electrokinetic capillary chromatography. *Chromatographia*, 35, 448–450 (1993).
136. Pascoe R., Foley J., Rapid separation of pharmaceutical enantiomers using electrokinetic chromatography with a novel chiral microemulsion. *Analyst*, 127, 710–714 (2002).
137. Mertzman M.D., Foley J.P., Chiral cyclodextrin modified microemulsion electrokinetic chromatography. *Electrophoresis*, 25, 1188–1200 (2004).
138. Mertzman M.D., Foley J.P., Effect of surfactant concentration and buffer selection on chromatographic figures of merit in chiral microemulsion electrokinetic chromatography. *Electrophoresis*, 25, 3247–3256 (2004).
139. Zheng Z.X., Lin J.M., Chan W.H., Lee A.W.M., Huie C.W., Separation of enantiomers in microemulsion electrokinetic chromatography using chiral alcohols as cosurfactants. *Electrophoresis*, 25, 3263–3269 (2004).

140. Kahle K.A., Foley J.P., Chiral microemulsion electrokinetic chromatography with two chiral components: Improved separations via synergies between a chiral surfactant and a chiral cosurfactant. *Electrophoresis*, 27, 896–904 (2006).
141. Garcia-Ruiz C., Marina M.L., Sensitive chiral analysis by capillary electrophoresis. *Electrophoresis*, 27, 195–212 (2006).
142. Shamsi S.A., Miller B.E., Capillary electrophoresis-mass spectrometry: Recent advances to the analysis of small achiral and chiral solutes. *Electrophoresis*, 25, 3927–3961 (2004).
143. Gubitz G., Schmid M.G., Recent advances in chiral separation principles in capillary electrophoresis and capillary electrochromatography. *Electrophoresis*, 25, 3981–3996 (2004).
144. Gubitz G., Schmid M.G., Chiral separation by chromatographic and electromigration techniques. A review. *Biopharm. Drug Dispos.* 22, 291–336 (2001).
145. McCorquodale M.E., Boutrid H., Coyler C.L., Enantiomeric separation of N'-nitrosonornictine by capillary electrophoresis. *Anal. Chim. Acta*, 496, 177–184 (2003).
146. Olsson J., Stegander F., Marlin N., Wan H., Blomberg L.G., Enantiomeric separation of omeprazole and its metabolite 5-hydroxyomeprazole using non-aqueous capillary electrophoresis. *J. Chromatogr. A*, 1129, 291–295 (2006).
147. Blanco M., Valverde I., Chiral and non chiral determination of Dopa by capillary electrophoresis. *J. Pharm. Biomed. Anal.*, 31, 431–438 (2003).
148. Sänger-van de Griend C.E., Wahlström H., Gröningsson K., Widahl-Näsman M., A chiral capillary electrophoresis method for ropivacaine hydrochloride in pharmaceutical formulations: Validation and comparison with chiral liquid chromatography. *J. Pharm. Biomed. Anal.*, 15, 1051–1061 (1997).
149. Jimidar M., Van Ael W., De Smet M., Cockaerts P., Method validation and robustness testing of an enantioselective CE method for chemical quality control. *LCGC Europe*, 15, 230–243 (2002).
150. Song S., Zhou L., Thompson R., Yang M., Ellison D., Wyvratt J.M., Comparison of capillary electrophoresis and reversed-phase liquid chromatography for determination of the enantiomeric purity of an M3 antagonist. *J. Chromatogr. A*, 959, 299–308 (2002).
151. Kuo C.Y., Wu H.L., Wu S.M., Enantiomeric analysis of methotrexate in pharmaceuticals by cyclodextrin-modified capillary electrophoresis. *Anal. Chim. Acta*, 471, 211–217 (2002).
152. Zhang B., Krull I.S., Cohen A., Smisek D.L., Kloss A., Wang B., Bourque A.J., Separation of quaternary ammonium diastereomeric oligomers by capillary electrophoresis. *J. Chromatogr. A*, 1034, 213–220 (2004).
153. Ramstad T., Enantiomeric purity methods for three pharmaceutical compounds by electrokinetic capillary chromatography utilizing highly sulfated-γ-cyclodextrin as the chiral selector. *J. Chromatogr. A*, 1127, 286–294 (2006).
154. Khan M., Viswanathan B., Sreenivas R.D., Sreenivas R.G., A validated chiral CE method for Frovatriptan, using cyclodextrin as chiral selector. *J. Pharm. Biomed. Anal.*, 41, 1447–1452 (2006).
155. Zhou S., Ouyang J., Baeyens W.R.G., Zhao H., Yang Y., Chiral separation of four fluoroquinolone compounds using capillary electrophoresis with hydroxypropyl-β-cyclodextrin as chiral selector. *J. Chromatogr. A*, 1130, 296–301 (2006).
156. Martínez-Pla J.J., Martín-Biosca Y., Sagrado S., Villanueva-Camañas R.M., Medina-Hernández M.J., Chiral separation of bupivacaine enantiomers by capillary electrophoresis partial-filling technique with human serum albumin as chiral selector. *J. Chromatogr. A*, 1048, 111–118 (2004).
157. Wang F., Khaledi M.G., Non-aqueous capillary electrophoresis chiral separations with sulfated β-cyclodextrin. *J. Chromatogr. B: Biomed. Sci. Appl.*, 731, 187–197 (1999).
158. Chankvetadze B., Burjanadze N., Blaschke G., Enantioseparation of *erythro*-mefloquine and its analogues in capillary electrophoresis. *J. Pharm. Biomed. Anal.*, 32, 41–49 (2003).
159. Guo L., Lin S.J., Yang Y.F., Qi L., Wang M.X., Chen Y., Fast enantioseparation of arylglycine amides by capillary electrophoresis with highly sulfated-β-cyclodextrin as a chiral selector. *J. Chromatogr. A*, 998, 221–228 (2003).
160. Gómez-Gomar A., Ortega E., Calvet C., Andaluz B., Mercé R., Frigola J., Enantioseparation of basic pharmaceutical compounds by capillary electrophoresis using sulfated cyclodextrins: Application to E-6006, a novel antidepressant. *J. Chromatogr. A*, 990, 91–98 (2003).
161. Tanaka Y., Otsuka K., Terabe S., Separation of enantiomers by capillary electrophoresis-mass spectrometry employing a partial filling technique with a chiral crown ether. *J. Chromatogr. A*, 875, 323–330 (2000).

162. Lipka E., Selouane A., Postel D., Len C., Vaccher M.P., Bonte J.P., Vaccher C., Enantioseparation of four *cis* and *trans* diastereomers of 2′,3′-didehydro-2′,3′-dideoxythymidine analogs, by high-performance liquid chromatography and capillary electrophoresis. *J. Chromatogr. A*, 1034, 161–167 (2004).
163. *British Pharmacopoeia* 2005, Volume 2, p. 1177.
164. Dose E.V., Guiochon G.A., Internal standardisation technique for capillary zone electrophoresis. *Anal. Chem.*, 63, 1154–1158 (1991).
165. Ross G.A., Voltage pre-conditioning technique for optimisation of migration time reproducibility in capillary electrophoresis. *J. Chromatogr. A*, 718, 444–447 (1995).
166. Shihabi Z.K., Hinsdale M.E., Some variables affecting reproducibility in capillary electrophoresis. *Electrophoresis*, 16, 2159–2163 (1995).
167. Schaeper J.P., Sepaniak M.J., Parameters affecting reproducibility in capillary electrophoresis. *Electrophoresis*, 21, 1421–1429 (2000).
168. Mayer B.X., How to increase precision in capillary electrophoresis. *J. Chromatogr. A*, 907, 21–37 (2001).
169. Altria K.D., Enhanced pharmaceutical analysis by CE using dynamic surface coating system. *J. Pharm. Biomed. Anal.*, 31, 447–453 (2003).
170. Aurora-Prado M.S., Silva C.A., Tavares M.F.M., Altria K.D., Determination of folic acid in tablets by microemulsion electrokinetic chromatography. *J. Chromatogr. A*, 1051, 291–296 (2004).
171. Lehmann S.V., Bergholdt A.B., Development and validation of a high-precision capillary electrophoresis method for main component assay of ragaglitazar. *J. Chromatogr. A*, 1011, 203–211 (2003).
172. Jamali B., Lehmann S.V., Development and validation of a high-resolution CE method for multi-analysis of ragaglitazar and arginine active pharmaceutical ingredients and low-dose tablets. *J. Pharm. Biomed. Anal.*, 34, 463–472 (2004).
173. Pajchel G., Michalska K., Tyski S., Application of capillary electrophoresis to the determination of various benzylpenicillin salts. *J. Chromatogr. A*, 1032, 265–272 (2004).
174. Nemutlu E., Yardimci C., Ozaltin N., Determination of sertaconazole in pharmaceutical preparations by capillary electrophoresis. *Anal. Chim. Acta*, 547, 83–88 (2005).
175. Perez-Ruiz T., Martinez-Lozano C., Sanz A., Bravo E., Development and validation of a quantitative assay for raloxifene by capillary electrophoresis. *J. Pharm. Biomed. Anal.*, 34, 891–897 (2004).
176. Guihen E., Sisk G.D., Scully N.M., Glennon J.D., Rapid analysis of atorvastatin calcium using capillary electrophoresis and microchip electrophoresis. *Electrophoresis*, 27, 2338–2347 (2006).
177. Hamoudova R., Pospisilova M., Determination of ibuprofen and flurbiprofen in pharmaceuticals by capillary zone electrophoresis. *J. Pharm. Biomed. Anal.*, 41, 1463–1467 (2006).
178. Komsta L., Misztal G., Majchrzak E., Hauzer A., Separation of fibrate-type antihyperlipidemic drugs by capillary electrophoresis and their quantitation in pharmaceuticals. *J. Pharm. Biomed. Anal.*, 41, 408–414 (2006).
179. Hillaert S., Van den Bossche W., Simultaneous determination of hydrochlorothiazide and several angiotensin-II-receptor antagonists by capillary electrophoresis. *J. Pharm. Biomed. Anal.*, 31, 329–339 (2006).
180. Hillaert S., De Beer T.R., De Beer J.O., Van den Bossche W., Optimisation and validation of a micellar electrokinetic chromatographic method for the analysis of several angiotensin-II-receptor antagonists. *J. Chromatogr. A*, 984, 135–146 (2003).
181. Hillaert S., Snoeck L., Van den Bossche W., Optimization and validation of a capillary zone electrophoretic method for the simultaneous analysis of four atypical antipsychotics. *J. Chromatogr. A*, 1033, 357–362 (2004).
182. Mohammadi A., Kanfer I., Walker R.B., A capillary zone electrophoresis (CZE) method for the determination of cyclizine hydrochloride in tablets and suppositories. *J. Pharm. Biomed. Anal.*, 35, 233–239 (2004).
183. Kou C.Y., Wu S.M., Micellar electrokinetic chromatography for simultaneous determination of six corticosteroids in commercial pharmaceuticals. *J. Sep. Sci.*, 28, 144–148.
184. Michalska K., Pajchel G., Tyski S., Capillary electrophoresis method for simultaneous determination of penicillin G, procaine and dihydrostreptomycin in veterinary drugs. *J. Chromatogr. B*, 800, 203–209 (2004).

185. Li Y., Cui S., Cheng Y., Chen X., Hu Z., Application of nonaqueous capillary electrophoresis for quantitative analysis of quinolizidine alkaloids in Chinese herbs. *Anal. Chim. Acta*, 508, 17–22 (2004).
186. Flores J.R., Nevado J.J.B., Penalvo G.C., Diez N.M., Development and validation method for determination of fluoxetine and its main metabolite norfluoxetine by nonaqueous capillary electrophoresis in human urine. *Talanta*, 65, 163–171 (2005).
187. Vanhoenacker G., de l'Escaille F., De Keukeleire D., Sandra P., Analysis of benzodiazepines in dynamically coated capillaries by CE-DAD, CE-MS and CE-MS^2. *J. Pharm. Biomed. Anal.*, 34, 595–606 (2004).
188. Lurie I.S., Hays P.A., Garcia A.E., Panicker S., Use of dynamically coated capillaries for the determination of heroin, basic impurities and adulterants with capillary electrophoresis. *J. Chromatogr. A*, 1034, 227–235 (2004).
189. Okamoto H., Nakajima T., Ito Y., Aketo T., Shimada K., Yamato S., Simultaneous determination of ingredients in a cold medicine by cyclodextrin-modified microemulsion electrokinetic chromatography. *J. Pharm. Biomed. Anal.*, 37, 517–528 (2005).
190. International Conferences on Harmonization, Draft revised guidance on impurities in new drug substances. Q3A(R). *Fed. Regist.*, 65, 45085–45090 (2000).
191. Cruz L.A., Hall R., Enantiomeric purity assay of moxifloxacin hydrochloride by capillary electrophoresis. *J. Pharm. Biomed. Anal.*, 38, 8–13 (2005).
192. Sokoliess T., Koller G., Approach to method development and validation in capillary electrophoresis for enantiomeric purity testing of active basic pharmaceutical ingredients. *Electrophoresis*, 26, 2330–2341 (2005).
193. Hansen S.H., Sheribah Z.A., Comparison of CZE, MEKC, MEEKC and non-aqueous capillary electrophoresis for the determination of impurities in bromazepam. *J. Pharm. Biomed. Anal.*, 39, 322–327 (2005).
194. Kelly M.A., Altria K.D., Grace C., Clark B.J., Optimisation, validation and application of a capillary electrophoresis method for the determination of ranitidine hydrochloride and related substances. *J. Chromatogr. A*, 789, 297–306 (1998).
195. Fabre H., Perrin C., Bosc N., Determination of homotaurine as impurity in calcium acamprosate by capillary zone electrophoresis. *J. Chromatogr. A*, 853, 421–430 (1999).
196. Toro I., Dulsat J.F., Fábregas J.L., Claramunt J., Development and validation of a capillary electrophoresis method with ultraviolet detection for the determination of the related substances in a pharmaceutical compound. *J. Chromatogr. A*, 1043, 303–315 (2004).
197. Mallampati S., Leonard S., De Vulder S., Hoogmartens J., Van Schepdael A., Method development and validation for the analysis of didanosine using micellar electrokinetic capillary chromatography. *Electrophoresis*, 26, 4079–4088 (2005).
198. Orlandini S., Fanali S., Furlanetto S., Marras A.M., Pinzauti S., Micellar electrokinetic chromatography for the simultaneous determination of ketorolac tromethamine and its impurities—Multivariate optimization and validation. *J. Chromatogr. A*, 1032, 253–263 (2004).
199. Furlanetto S., Orlandini S., Marras A.M., Mura P., Pinzauti S., Mixture design in the optimization of a microemulsion system for the electrokinetic chromatographic determination of ketorolac and its impurities: Method development and validation. *Electrophoresis*, 27, 805–818 (2006).
200. Wen T., Zhao X., Luo G.A., Wang J., Wang Y.M., Li P., Zhu J., Yu Z.S., Simultaneous determination of heroin, amphetamine and their basic impurities and adulterants using microemulsion electrokinetic, chromatography. *Chin. Chem. Lett.*, 16, 1499–1502 (2005).
201. Vassort A., Barrett D.A., Shaw P.N., Ferguson P.D., Szucs R., A generic approach to the impurity profiling of drugs using standardised and independent capillary zone electrophoresis methods coupled to electrospray ionisation mass spectrometry. *Electrophoresis*, 26, 1712–1723 (2005).
202. Visky D., Jimidar I., Van Ael W., Vennekens T., Redlich D., De Smet M., Capillary electrophoresis-mass spectrometry in impurity profiling of pharmaceutical products. *Electrophoresis*, 26, 1541–1549 (2005).
203. Wan H., Schmidt S., Carlsson L., Blomberg L.G., Some factors affecting enantiomeric impurity determination by capillary electrophoresis using ultraviolet and laser-induced fluorescence detection. *Electrophoresis*, 20, 2705–2714 (1999).
204. Hilhorst M.J., Somsen G.W., de Jong G.J., Capillary electrokinetic separation techniques for profiling of drugs and related products. *Electrophoresis*, 22, 2542–2564 (2001).

205. Berzas Nevado J.J., Cabanillas C.G., Villaseñor Llerena M.J., Rodríguez R.V., Enantiomeric determination, validation and robustness studies of racemic citalopram in pharmaceutical formulations by capillary electrophoresis. *J. Chromatogr. A*, 1072, 249–257 (2005).
206. Kvasnièka F., Bíba B., Cvak L., Capillary zone electrophoresis separation of enantiomers of lisuride. *J. Chromatogr. A*, 1066, 255–258 (2005).
207. Castro-Puyana M., Garcia-Ruiz C., Cifuentes A., Crego A.L., Marina M.L., Identification and quantitation of *cis*-ketoconazole impurity by capillary zone electrophoresis-mass spectrometry. *Electrophoresis*, 1114, 170–177 (2006).
208. Berzas Nevado J.J., Castañeda Peñalvo G., Rodríguez Dorado R.M., Method development and validation for the separation and determination of omeprazole enantiomers in pharmaceutical preparations by capillary electrophoresis. *Anal. Chim. Acta*, 533, 127–133 (2005).
209. Pajchel G., Michalska K., Tyski S., Analysis of phenoxymethylpenicillin potassium by capillary electrophoresis. *J. Chromatogr. A*, 1087, 197–202 (2005).
210. Sabbah S., Scriba G.K.E., Development and validation of a capillary electrophoresis assay for the determination of 3,4-diaminopyridine and 4-aminopyridine including related substances. *J. Chromatogr. A*, 907, 321–328 (2001).
211. Gotti R., Pomponio R., Bertucci C., Cavrini V., Determination of 5-aminosalicylic acid related impurities by micellar electrokinetic chromatography with an ion-pair reagent. *J. Chromatogr. A*, 916, 175–183 (2001).
212. Kang J.W., Van Schepdael A., Roets E., Hoogmartens J., Analysis of vancomycin and related impurities by micellar electrokinetic capillary chromatography. Method development and validation. *Electrophoresis*, 22, 2588–2592 (2001).
213. Owens P.K., Wikström H., Någård S., Karlsson L., Development and validation of a capillary electrophoresis method for ximelagatran assay and related substance determination in drug substance and tablets. *J. Pharm. Biomed. Anal.*, 27, 587–598 (2002).
214. Hilder E.F., Klampfl C.W., Buchberger W., Haddad P.R., Comparison of aqueous and non-aqueous carrier electrolytes for the separation of penicillin V and related substances by capillary electrophoresis with UV and mass spectrometric detection. *Electrophoresis*, 23, 413–420 (2002).
215. Jaworska M., Szulinska G., Wilk M., Tautt J., Capillary electrophoretic separation of *N*-acetylcysteine and its impurities as a method for quality control of pharmaceuticals. *J. Chromatogr. A*, 853, 479–485 (1999).
216. Michalska K., Pajchel G., Tyski S., Determination of ciprofloxacin and its impurities by capillary zone electrophoresis. *J. Chromatogr. A*, 1051, 267–272 (2004).
217. Gil E.C., Dehouck P., Van Schepdael A., Roets E., Hoogmartens J., Analysis of metacycline by capillary electrophoresis. *Electrophoresis*, 22, 497–502 (2001).
218. Fernández H., Rupérez F.J., Barbas C., Capillary electrophoresis determination of loratadine and related impurities. *J. Pharm. Biomed. Anal.*, 31, 499–506 (2003).
219. Nemutlu E., Ozaltin N., Altinoz S., Determination of rofecoxib, in the presence of its photodegradation product, in pharmaceutical preparations by micellar electrokinetic capillary chromatography. *Anal. Bioanal. Chem.*, 378, 504–509 (2004).
220. Avdeef A., Physicochemical profiling (solubility, permeability and charge state). *Curr. Top. Med. Chem.*, 1, 277–351 (2001).
221. Jia Z.J., Physicochemical profiling by capillary electrophoresis. *Curr. Pharm. Anal.*, 1, 41–56 (2005).
222. Caliaro G.A., Herbots C.A., Determination of pK_a values of basic new drug substances by CE. *J. Pharm. Biomed. Anal.*, 26, 427–434 (2001).
223. Poole S.K., Patel S., Dehring K., Workman H., Poole C.F., Determination of acid dissociation constants by capillary electrophoresis. *J. Chromatogr. A*, 1037, 445–454 (2004).
224. Wan H., Holmén A., Någård M., Lindberg W., Rapid screening of pK_a values of pharmaceuticals by pressure-assisted capillary electrophoresis combined with short-end injection. *J. Chromatogr. A*, 979, 369–377 (2002).
225. Ornskov E., Linusson A., Folestad S., Determination of dissociation constants of labile drug compounds by capillary electrophoresis. *J. Pharm. Biomed. Anal.*, 33, 379–391 (2003).
226. de Nogales V., Ruiz R., Rosés M., Ràfols C., Bosch E., Background electrolytes in 50% methanol/water for the determination of acidity constants of basic drugs by capillary zone electrophoresis. *J. Chromatogr. A*, 1123, 113–120 (2006).

227. Jia Z.J., Ramstad T., Zhong M., Medium-throughput pK_a screening of pharmaceuticals by pressure-assisted capillary electrophoresis. *Electrophoresis*, 22, 1112–1118 (2001).
228. Matoga M., Laborde-Kummer E., Langlois M.H., Dallet P., Bosc J.J., Jarry C., Dubost J.P., Determination of pK_a values of 2-amino-2-oxazolines by capillary electrophoresis. *J. Chromatogr. A*, 984, 253–260 (2003).
229. Mrestani Y., Neubert R., Munk A., Wiese M., Determination of dissociation constants of cephalosporins by capillary zone electrophoresis. *J. Chromatogr. A*, 803, 273–278 (1998).
230. Jankowsky R., Friebe M., Noll B., Johannsen B., Determination of dissociation constants of ^{99m}Technetium radiopharmaceuticals by capillary electrophoresis. *J. Chromatogr. A*, 833, 83–96 (1999).
231. Hu Q., Hu G., Zhou T., Fang Y., Determination of dissociation constants of anthrocycline by capillary zone electrophoresis with amperometric detection. *J. Pharm. Biomed. Anal.*, 31, 679–684 (2003).
232. Barták P., Bednár P., Stránský Z., Bocek P., Vespalec R., Determination of dissociation constants of cytokinins by capillary zone electrophoresis. *J. Chromatogr. A*, 878, 249–259 (2000).
233. Cao J., Cross R.F., The separation of dihydrofolate reductase inhibitors and the determination of pK_a values by capillary zone electrophoresis. *J. Chromatogr. A*, 695, 297–308 (1995).
234. Jiménez-Lozano E., Marqués I., Barrón D., Beltrán J.L., Barbosa J., Determination of pK_a values of quinolones from mobility and spectroscopic data obtained by capillary electrophoresis and a diode array detector. *Anal. Chim. Acta*, 464, 37–45 (2002).
235. Solinova V., Kasicka V., Koval D., Cesnek M., Holy A., Determination of acid–base dissociation constants of amino- and guanidinopurine nucleotide analogs and related compounds by capillary zone electrophoresis. *Electrophoresis*, 27, 1006–1019 (2006).
236. Geiser L., Henchoz Y., Galland A., Carrupt P.A., Veuthey J.L., Determination of pK_a values by capillary zone electrophoresis with a dynamic coating procedure. *J. Sep. Sci.*, 28, 2374–2380 (2005).
237. Buckenmaier S.M.C., McCalley D.V., Euerby M.R., Determination of pK_a values of organic bases in aqueous acetonitrile solutions using capillary electrophoresis. *J. Chromatogr. A*, 1004, 71–79 (2003).
238. Foulon C., Danel C., Vaccher C., Yous S., Bonte J.-P., Goossens J.-F., Determination of ionization constants of *N*-imidazole derivatives, aromatase inhibitors, using capillary electrophoresis and influence of substituents on pK_a shifts. *J. Chromatogr. A*, 1035, 131–136 (2004).
239. Hansch C., Leo A., *Substituent Constants for Correlation Analysis in Chemistry and Biology*. (John Wiley and Sons, New York, 1979).
240. Sangster J., *Octanol-water partition coefficients for simple organic compounds. J. Phys. Chem. Ref. Data*, 18, 1111–1229 (1989).
241. Berthod A., Carda-Broch S., Determination of liquid–liquid partition coefficients by separation methods. *J. Chromatogr. A*, 1037, 3–14 (2004).
242. Chen N., Zhang Y., Terabe S., Nakagawa T., Effect of physico-chemical properties and molecular structure on the micelle–water partition coefficient in micellar electrokinetic chromatography. *J. Chromatogr. A*, 678, 327–332 (1994).
243. Herbert B.J., Dorsey J.G., *n*-Octanol water partition-coefficient estimation by micellar electrokinetic capillary chromatography. *Anal. Chem.*, 67, 744–749 (1995).
244. Lucangioli S.E., Carducci C.N., Scioscia S.L., Carlucci A., Bregni C., Kenndler E., Comparison of the retention characteristics of different pseudostationary phases for microemulsion and micellar electrokinetic chromatography of betamethasone and derivatives. *Electrophoresis*, 24, 984–991 (2003).
245. Klotz W.L., Schure M.R., Foley J.P., Determination of octanol–water partition coefficients of pesticides by microemulsion electrokinetic chromatography. *J. Chromatogr. A*, 930, 145–154 (2001).
246. Poole S.K., Patel S., Dehring K., Workman H., Dong J., Estimation of octanol–water partition coefficients for neutral and weakly acidic compounds by microemulsion electrokinetic chromatography using dynamically coated capillary columns. *J. Chromatogr. B*, 793, 265–274 (2003).
247. Rappel C., Galanski M., Yasemi A., Habala L., Keppler B.K., Analysis of anticancer platinum(II)-complexes by microemulsion electrokinetic chromatography: Separation of diastereomers and estimation of octanol–water partition coefficients. *Electrophoresis*, 26, 878–884 (2005).
248. Wong K.S., Kenseth J., Strasburg R., Validation and long-term assessment of an approach for the high throughput determination of lipophilicity (log *P*-OW) values using multiplexed, absorbance-based capillary electrophoresis. *J. Pharm. Sci.*, 93, 916–931 (2004).

249. Tu J., Halsall H.B., Seliskar C.J., Limbach P.A., Arias F., Wehmeyer K.R., Heineman W.R., Estimation of log P_{OW} values for neutral and basic compounds by microchip microemulsion electrokinetic chromatography with indirect fluorimetric detection (μMEEKC-IFD). *J. Pharm. Biomed. Anal.*, 38, 1–7 (2005).
250. Yang S., Bumgarner J.G., Kruk L.F.R., Khaledi M.G., Quantitative structure–activity relationships studies with micellar electrokinetic chromatography influence of surfactant type and mixed micelles on estimation of hydrophobicity and bioavailability. *J. Chromatogr. A*, 721, 323–335 (1996).
251. Quaglia M.G., Fanali S., Barbato F., Donati E., Micellar electrokinetic chromatography for determination of drug partition in phospholipids. *Il Farmaco*, 60, 77–83 (2005).
252. Kelly M.A., Altria K.D., Clark B.J., Quantitative analysis of sodium dodecyl sulphate by capillary electrophoresis. *J.Chromatogr. A*, 781, 67–71 (1997).
253. Fritz J.S., Recent developments in the separation of inorganic and small organic ions by capillary electrophoresis. *J. Chromatogr. A*, 884, 261–275 (2000).
254. Gong X.Y., Shen Y., Mao B., Quantitation of ammonium cations in pharmaceutical samples by non-aqueous capillary electrophoresis with indirect UV detection. *J. Liq. Chrom. Rel. Tech.*, 27, 661–675 (2004).
255. Padarauskas A., CE determination of small ions: Methods and techniques. *Anal. Bioanal. Chem.*, 384, 132–144 (2006).
256. Timerbaev A.R., Capillary electrophoresis of inorganic ions: An update. *Electrophoresis*, 25, 4008–4031 (2004).
257. Fabre H., Blanchin M.D., Bosc N., Capillary electrophoresis for the determination of bromide, chloride and sulfate as impurities in calcium acamprosate. *Anal. Chim. Acta*, 381, 29–37 (1999).
258. Zhou L., Dovletoglou A., Practical capillary electrophoresis method for the quantitation of the acetate counter-ion in a novel antifungal lipopeptide. *J. Chromatogr. A*, 763, 279–284 (1997).
259. Chen D.H., Klopchin P., Parsons J., Srivatsa G.S., Determination of sodium acetate in antisense oligonucleotides by capillary zone electrophoresis. *J. Liq. Chrom. Rel. Tech.*, 20, 1185–1195 (1997).
260. Altria K.D., Goodall D.M., Rogan M.M., Quantitative determination of drug counter-ion stoichiometry by capillary electrophoresis. *Chromatographia*, 38, 637–642 (1994).
261. Altria K.D., Assi K.H., Bryant S.M., Clark B.J., Determination of organic acid drug counter-ions by capillary electrophoresis, *Chromatographia*, 44, 367–371 (1997).
262. Nair J.B., Izzo C.J., Anion screening for drugs and intermediates by capillary ion electrophoresis. *J. Chromatogr. A*, 640, 445–461 (1993).
263. Moore D.E., Miao W.G., Benikos C., Quantitative determination of alginic acid in pharmaceutical formulations using capillary electrophoresis. *J. Pharm. Biomed. Anal.*, 34, 233–238 (2004).
264. Stålberg O., Sander K., Sänger-van de Griend C., The determination of bromide in a local anaesthetic hydrochloride by capillary electrophoresis using direct UV detection. *J. Chromatogr. A*, 977, 265–275 (2002).
265. Malsch R., Harenberg J., Purity of glycosaminoglycan-related compounds using capillary electrophoresis. *Electrophoresis*, 17, 401–405 (1996).
266. Fabre H., Blanchin M.D., Julien E., Segonds C., Mandrou B., Bosc N., Validation of a capillary electrophoresis procedure for the determination of calcium in calcium acamprosate. *J. Chromatogr. A*, 772, 265–269 (1997).
267. Johnson B.D., Grinberg N., Bicker G., Ellison D., The quantitation of a residual quaternary amine in bulk drug and process streams using capillary electrophoresis. *J. Liq. Chrom. Rel. Tech.*, 20, 257–272 (1997).
268. Williams R.C., Boucher R.J., Analysis of potassium counter ion and inorganic cation impurities in pharmaceutical drug substance by capillary electrophoresis with conductivity detection. *J. Pharm. Biomed. Anal.*, 22, 115–122 (2000).
269. Denis C.M., Baryla N.E., Determination of piperazine in pharmaceutical drug substances using capillary electrophoresis with indirect UV detection. *J. Chromatogr. A*, 1110, 268–271 (2006).
270. Altria K.D., Background theory and applications of microemulsion electrokinetic chromatography. *J. Chromatogr. A*, 892, 171–186 (2000).
271. Altria K.D., Mahuzier P.E., Clark B.J., Background and operating parameters in microemulsion electrokinetic chromatography. *Electrophoresis*, 24, 315–324 (2003).

272. Klampfl C.W., Solvent effects in microemulsion electrokinetic chromatography. *Electrophoresis*, 24, 1537–1543 (2003).
273. Altria K.D., Clark B.J., Filbey S.D., Kelly M.A., Rudd D.R., Application of chemometric experimental designs in capillary electrophoresis—A review. *Electrophoresis*, 16, 2413–2148 (1995).
274. Sentellas S., Saurina J., Chemometrics in capillary electrophoresis. Part A: Methods for optimization. *J. Sep. Sci.*, 26, 875–885 (2003).
275. Sentellas S., Saurina J., Chemometrics in capillary electrophoresis. Part B: Methods for data analysis. *J. Sep. Sci.*, 26, 1395–1402 (2003).
276. Veuthey J.L., Rudaz S., Statistical and chemometric tools applied to pharmaceutical analysis. *Chimia*, 59, 326–330 (2005).
277. Marengo E., Robotti E., Bobba M., Liparota M.C., Artificial neural networks applications in the field of separation science optimisation. *Curr. Anal. Chem.*, 2, 181–194 (2006).
278. Mahuzier P.E., Clark B.J., Bryant S.M., Altria K.D., High-speed microemulsion electrokinetic chromatography. *Electrophoresis*, 22, 3819–3823 (2001).
279. Labat L., Kummer E., Dallet P., Dubost J.P., Comparison of high-performance liquid chromatography and capillary zone electrophoresis for the determination of parabens in a cosmetic product. *J. Pharm. Biomed. Anal.*, 23, 763–769 (2000).
280. Clohs L., McErlane K.M., Comparison between capillary electrophoresis and high-performance liquid chromatography for the stereoselective analysis of carvedilol in serum. *J. Pharm. Biomed. Anal.*, 31, 407–412 (2003).
281. Velikinac I., Cudina I., Jankovic I., Agbaba D., Vladimirov S., Comparison of capillary zone electrophoresis and high performance liquid chromatography methods for quantitative determination of ketoconazole in drug formulations. *Il Farmaco*, 59, 419–424 (2004).
282. Aurora-Prado M.S., Steppe M., Tavares M.F.M, Kedor-Hackmann E.R.M, Santoro M.I.R.M., Comparison of capillary electrophoresis and reversed-phase liquid chromatography methodologies for determination of diazepam in pharmaceutical tablets. *J. Pharm. Biomed. Anal.*, 37, 273–279 (2005).
283. Chapman J., Chen F.T.A., Implementing a Generic Methods Development Strategy for Enantiomer Analysis. *LCGC* 19, 427–431 (2001).
284. Fabre H., Altria K.D., Validating CE methods for pharmaceutical analysis. *LCGC* 14, 302–310 (2001).
285. Eurachem Guide, 'The Fitness for Purpose of Analytical Methods' (1998).
286. Text on validation of analytical procedures: methodology. *International Conference on Harmonization of Technical Requirements for the Registration of Pharmaceuticals for Human Use (ICH)*, IFPMA Ed., (Geneva, Switzerland, 1996).
287. Hansen H.K., Hansen S.H., Kraunsøe M., Petersen G.M., Comparison of high-performance liquid chromatography and capillary electrophoresis methods for quantitative determination of glycyrrhizinic acid in pharmaceutical preparations. *Eur. J. Pharm. Sci.*, 9, 41–46 (1999).

5 Principles and Practice of Capillary Electrochromatography

Myra T. Koesdjojo, Carlos F. Gonzalez, and Vincent T. Remcho

CONTENTS

5.1 INTRODUCTION

Capillary electrochromatography (CEC) is a hybrid of capillary electrophoresis (CE) and capillary high-performance liquid chromatography (HPLC). As such, it has been characterized by some as exhibiting the best of both technologies and by others as a composite of the worst attributes of both—the reality lies somewhere between the extremes. Shifting fortunes and interest in CEC are well represented in Figure 5.1, which illustrates a best estimate of the number of publications on the topic over the past decade. What is evident is (1) a boom attributable to the promise of highly efficient liquid phase chromatographic separations and (2) a decline in interest due to practical issues such as the irreproducibility of electroosmotic flow (EOF) (also a factor in the limited market impact of CE)

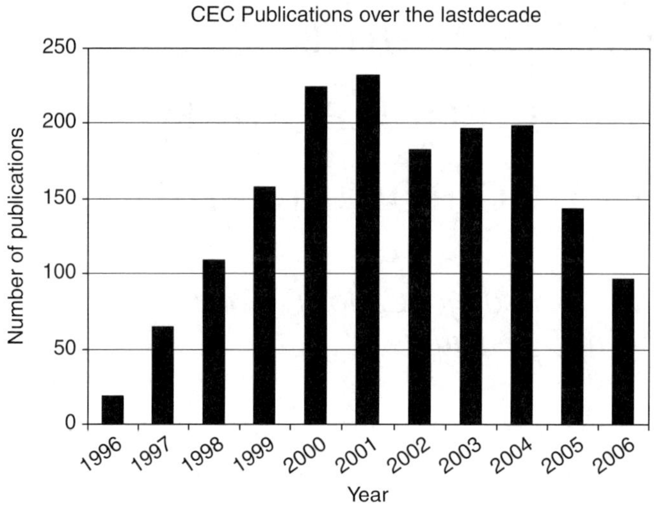

FIGURE 5.1 A bar chart showing the number of publications on the topic of CEC per year for the last decade.

and a rediscovery of ultra high pressure liquid chromatography (UPLC). It is possible that there will be a resurgence of interest arising at least in part from the growth of microchip-based separations technology. CEC is a good fit on the microfluidic platform owing to its compatibility with small-pore-size monolithic media and fine particulate packings (due to the lack of backpressure limitations in the absence of pressure-driven bulk flow), its relative independence from mechanical valves and pumps, the inherent ease of miniaturizing electrical and electronic components, and its facility to be interfaced to other chip-based unit operations such as solid phase extraction, filtration, optical spectroscopic detection, and electrochemistry. This chapter describes the basic theoretical concepts and operational principles of CEC. The discussion is intended to provide sufficient background on CEC instrumentation, the principles of CEC operation, and on the range of applications of CEC to equip the reader to employ CEC in the laboratory.

5.2 THEORY

Capillary electrochromatography is a hybrid separation method that couples CE with HPLC. It combines the high separation efficiency that capillary zone electrophoresis (CZE) offers with the wide range of parameters that can be manipulated in HPLC, particularly the wide range of stationary phases from which to choose. Subsequently, CEC has become a powerful technique that has gained interest in the last few years.

5.2.1 Electroosmotic Flow

Capillary electrochromatography utilizes an electric field rather than a pressure gradient to propel the mobile phase through a packed bed. High efficiencies can be achieved in CEC mainly because of the use of EOF. The flow profile of EOF is pluglike, unlike the parabolic flow of pressure-driven HPLC, making for less dispersion and therefore higher efficiency than pressure-driven separations under otherwise identical conditions. Another benefit of EOF is that it generates no pressure drop, thus, small diameter packings (0.5–2 μm) can be used, thereby further enhancing efficiency. In CEC, separations are generally carried out using aqueous buffers with organic modifiers in fused-silica capillaries of 25–100 μm internal diameter (i.d.) packed with small particles (<5 μm) in an applied electric field.

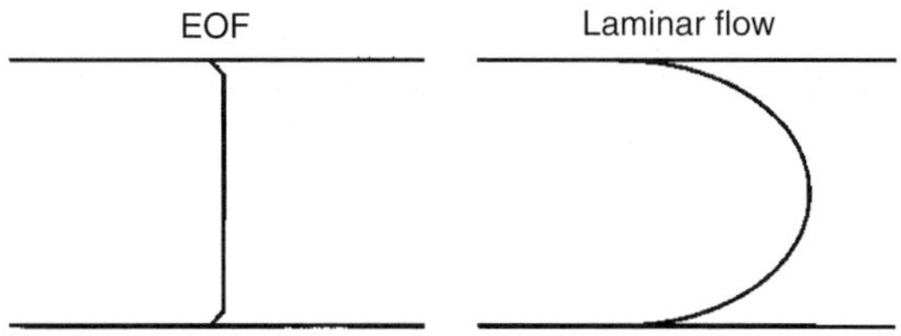

FIGURE 5.2 Flow profiles of EOF and laminar flow.

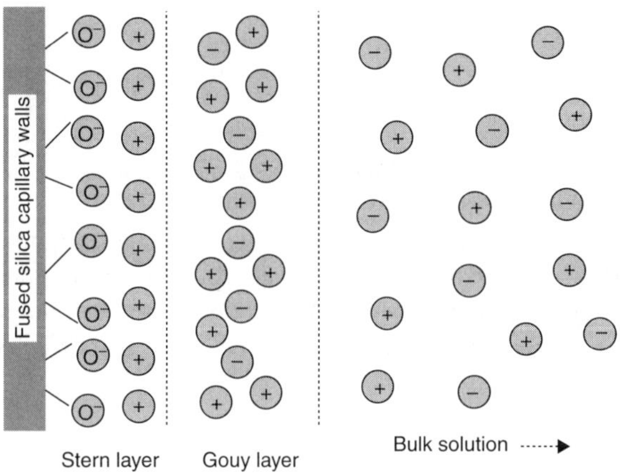

FIGURE 5.3 Schematic representation of the electrical double layer at a negatively charged capillary wall.

CEC was first introduced more than 30 years ago (1974) when Pretorius et al. [1] first demonstrated the use of EOF to drive a mobile phase through a liquid chromatography (LC) column. Pretorius used 75–125 μm particles in a 1-mm i.d. glass tube and was able to show that band broadening with EOF was considerably smaller than with pressure-driven flow. In 1981, Jorgenson and Lukacs [2] further proved the principle of electrically driven chromatography using small-diameter particles in fused-silica capillaries. Although the efficiencies were relatively low (~60,000 plates m^{-1}), they were able to separate 9-methylanthracene from perylene on a 170-μm i.d. capillary packed with 10-μm reverse-phase packing material. The work that finally resurrected the interest in CEC, however, was pioneered by Knox and Grant [3] in 1987 and 1991 with their theoretical and practical approach that demonstrated the practical feasibility of CEC.

The stark contrast of EOF to the parabolic laminar flow profile generated by an external pump used in HPLC is shown in Figure 5.2. The flat profile of EOF originates in the evenly distributed charge on the capillary wall, or packing material surface, at which an electrochemical double layer is formed. On application of a longitudinal potential, a shear plane develops and ions, their solvation layer, and adjacent species slip toward the outlet with a uniform velocity cross section. This contrasts with pressure-driven flow, as in HPLC, in which frictional forces at the column walls lead to laminar flow profile. That yields broad peaks due to the dispersive nature of the broad range in flow velocities. With a flat profile, zone broadening is minimized, leading to high separation efficiencies in CEC. The uniform EOF velocity (in channels ~0.1 to ~150 μm in diameter) is independent of the channel width.

To understand the generation of EOF inside the capillary, Stern's model, as presented in Figure 5.3, is discussed. The inner wall of a fused-silica capillary possesses ionizable silanol groups on the surface. In a buffer filled capillary, these silanol groups dissociate to give the capillary wall a negative charge (SiO^-). This is centered by cations from the buffer solution, giving rise to an electrical double layer. The layer of ions closest to the surface is called the "Stern layer" and is essentially rigid or immobile. A more diffuse layer is formed distal to the Stern layer and is known as the "Gouy layer." This mobile layer is rich in cations and decays into the bulk liquid. The double layer is formed at the capillary wall and the surface of the particulate packing material.

The formation of this electrical double layer gives rise to a potential that falls off exponentially as a function of distance from the capillary surface. This zeta potential (ζ) has values ranging from 0 to 100 mV. The distance between the Stern layer and a point in the bulk liquid at which the potential is 0.37 times the potential at the interface between Stern and diffuse layer is defined as the thickness of the double layer (δ).

The equation describing δ (the thickness of double layer) g [3] is

$$\delta = [\varepsilon_r \varepsilon_0 RT/2cF^2]^{1/2}, \tag{5.1}$$

where δ is the thickness of double layer, ε_r the dielectric constant or relative permittivity of the eluent, ε_0 the permittivity of a vacuum, R the universal gas constant, T the absolute temperature, c the molar concentration, and F the Faraday constant.

Solving Equation 5.1, for a monovalent electrolyte at a concentration of 0.001 M in water ($\varepsilon_r = 80$), the thickness of the electrical double layer would be 10 nm, and at a concentration of 0.1 M it would be 1 nm for the eluent/sorbent and eluent/capillary wall interfaces.

The zeta potential (at the shear plane) is dependent on δ and the charge density σ [4]

$$\zeta = \frac{\delta\sigma}{\varepsilon_0 \varepsilon_r}. \tag{5.2}$$

The relationship between the EOF linear velocity (u_{eo}) and ζ is given by the Smoluchowski equation

$$u_{eo} = \frac{\varepsilon_0 \varepsilon_r \zeta E}{\eta}, \tag{5.3}$$

where E is the electric field strength and η is the viscosity of the solvent.

It is now apparent that the EOF depends upon the surface charge density, the field strength, the thickness of the electrical double layer, and the viscosity of the separation medium, which in turn is dependent upon the temperature. In a packed capillary, the vast majority of the total surface area is contributed by the packing material, thus it is reasonable to approximate the surface charge density of the system (both the capillary wall and the packing) using the value for the sorbent.

When a voltage is applied longitudinally along the capillary, cations in the diffuse (Gouy) layer are free to migrate toward the cathode, carrying the bulk solution with them. The result is a net flow in the direction of the cathode.

A main factor affecting electroosmotic mobility is buffer pH. EOF will be significantly greater at high pH compared to low pH. At high pH (pH > 9), silanols on the capillary surface are completely ionized and thus, EOF mobility is at its greatest. At low pH (pH < 4), however, the ionization of silanols is low and the EOF mobility is nearly insignificant. The ionic strength of the buffer will also affect mobility.

A solute's apparent or effective mobility (μ_x) takes into account both its individual electrophoretic mobility (μ_{ep}) and EOF mobility (μ_{eo}):

$$\mu_x = \mu_{ep} + \mu_{eo}. \tag{5.4}$$

Under "normal" polarity with a fused-silica capillary, samples are introduced at the anode and EOF migrates toward the cathode. In this case, cations have positive μ_{ep}, neutrals have zero μ_{ep}, and

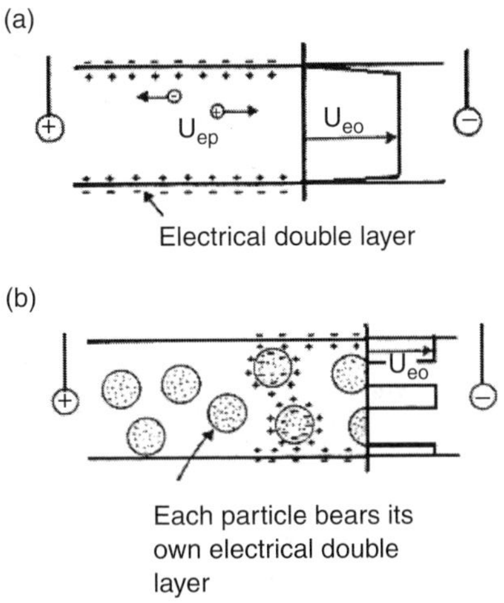

FIGURE 5.4 Schematic representation of (a) CZE and (b) CEC inside a capillary column. (From Knox, J.H. and Grant, I.H., *Chromatographia*, 26, 329, 1988. With permission.)

anions have negative μ_{ep}. In other words, cations migrate faster than the EOF and anions migrate more slowly than the EOF, whereas neutrals migrate with the same velocity as the EOF. Thus, to determine EOF velocity experimentally, a neutral and chromatographically unretained marker can be injected into the capillary under a given set of conditions. Often, acetone is used as a marker in reversed-phase CEC. By measuring the time that it takes for the neutral unretained marker to reach the detector, EOF can be determined on the basis of the distance traveled.

As discussed above, the origin of EOF is the electrical double layer that exists at the interface of a charged surface in contact with an electrolyte solution, as in the case of CZE (Figure 5.4a). In a capillary packed with silica particles, the surface of both the capillary wall and of the particles are negatively charged owing to the dissociation of silanol groups (Figure 5.4b). Again, the surface areas of silica-based packing materials are much greater than that of the capillary wall, thus most of the EOF is generated by the surface silanol groups of the packing material. When an axial electrical field is applied to the column, ions in the diffuse section of the double layer migrate towards the cathode moving the bulk solution by viscous drag.

Looking back at Equation 5.3, in an open tubular (OT) capillary with thin double layer and in the absence of significant polarization, ζ is the zeta potential of the wall ζ_w, and is defined as potential on a hypothetical surface of shear close to the tube wall. For porous/nonporous packing particles that are nonconducting, ζ is zeta potential at the particle surface ζ_p.

For OT fused-silica capillaries, the ζ potential has been determined to be between 20 and 120 mV depending on the pH. The potential ζ in capillaries made of polymeric materials (i.e., polytetrafluorethylene, polyethylene, or polyvinylchloride) ranges from 0 to 60 mV at pH 6.0 and 10, respectively. The potential ζ of C18 derivatized particles can be assumed to be in the same range as that of polymeric materials [5]. Figure 5.5 shows the EOF velocities versus the electric field strength measured for a C18 packing material (CEC Hypersil, C18, 3 μm). Depending on the surface charge of the packing materials and the pH of the buffer, linear flow rates of as much as 2.5 mm s^{-1} can be achieved in a 35-cm long column.

5.2.2 Zone Broadening: CEC versus HPLC

Electroosmotic flow-driven chromatography yields higher separation efficiencies than HPLC because of the use of small particles and reduction of plate heights as a result of the plug-flow profile.

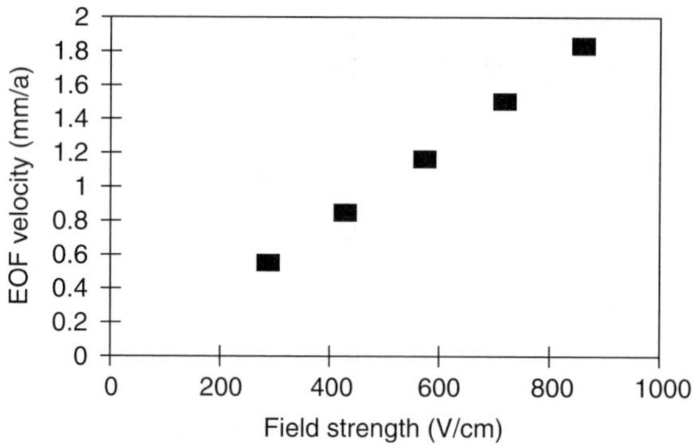

FIGURE 5.5 Plot of EOF velocity vs. the applied electric strength. Conditions: 250 (335) mM × 0.1 mM CEC Hypersil C18, 2.5 μm, 80% ACN/20% MES 25 mM pH 6, 10 bar pressure applied to both ends of capillary, 20°C. (Reprinted from Landers, J.P, *Handbook of Capillary Electrophoresis*, 2nd ed., CRC Press, FL, 1997, Chapter 5. With permission.)

The linear velocity in a pressure-driven system is given by the following equation described by Knox and Grant [3]:

$$u = \frac{d_p^2 \Delta p}{\Phi \eta L}, \tag{5.5}$$

where u is the linear velocity, d_p the particle diameter, Φ the flow resistance parameter for packed columns, Δp the pressure drop across the column, L the column length, and η the viscosity of the solvent.

On the other hand, the linear velocity through a packed capillary under an applied electric field is given by the Smoluchowski equation, as discussed above in Equation 5.3.

By comparing the two Equations 5.3 and 5.5, it can be seen that the linear velocity u is proportional to d_p^2 in a pressure-driven system while it is independent of the particle diameter in an electrically driven system. Since plate height values are generally lowered as a result of using small diameter particles, it is possible in electrically driven systems to use very small diameter packing materials and still maintain high linear velocities to yield rapid and very efficient separations.

Since electromatography uses a stationary phase just like in conventional LC, the principles of band broadening in CEC and LC are similar. The plate height can be expressed using a modified Van Deemter equation:

$$H = \mathrm{A}d_p \left[\frac{ud_p}{D_m}\right]^{1/3} + \frac{\mathrm{B}}{u} + \frac{\mathrm{C}ud_p^2}{D_m}, \tag{5.6}$$

where A, B, and C are constants, and D_m is the diffusion coefficient.

The A term refers to eddy diffusion that arises from the different flow paths the solute molecules may traverse through the packed bed. Band broadening occurs because molecules in different flow streams are moving at different velocities. Solute molecules move faster (on average) in wider paths than they do in narrow paths. This term is much more significant in a pressure-driven system, where the flow rate varies from one channel to another, while in an electrically driven system like CEC, the contribution of the A term to zone broadening is significantly lower, because of the closely matched velocities between the channels (in the absence of electrochemical double layer overlap). The C term represents the contribution to the plate height resulting from resistance to mass transfer between the mobile phase and the stationary phase. The effect is greater as the mobile phase velocity increases

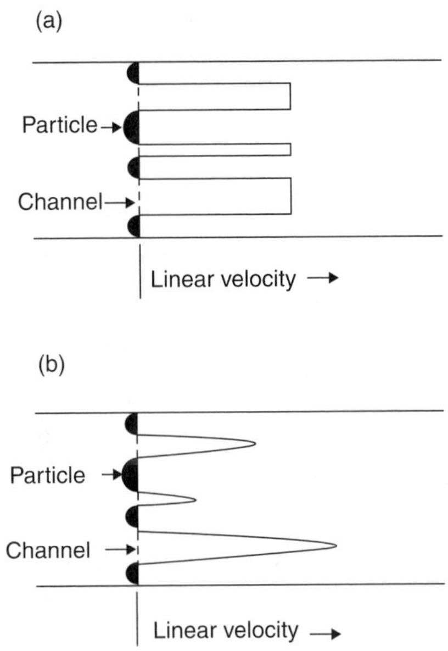

FIGURE 5.6 Schematic representation of (a) electroosmotic and (b) pressure-driven flow in a CEC packed column. (From Knox, J.H. and Grant, I.H., *Chromatographia*, 26, 329, 1988. With permission.)

owing to the diminished equilibration time. Just as the use of small diameter particles reduces the A term contribution to zone broadening, the C term contribution can be minimized by using smaller packing materials.

In a recent study by Horváth and coworkers [6], band broadening between electrically driven and pressure-driven flows in several packed capillaries were examined. Their results suggested that detection and injection contributions to band spreading are negligible. The experimental data were examined and fitted to the simplified Van Deemter equation to evaluate the eddy diffusion (A) and the mass transfer (C) parameters of each mode of flow:

$$H = \mathrm{A} + \frac{\mathrm{B}}{u} + \mathrm{C}u. \tag{5.7}$$

As expected, their column studies showed that the A term was much smaller in CEC than in HPLC. This was attributed to the plug-like EOF flow profile in CEC as shown in Figure 5.6. This results in reduced multipath dispersion effects by a factor 2–4. The C term in HPLC was shown to be greater than for CEC using packing materials with pore sizes $\geq$300 Å, but was not observed in packing materials with average pore size of 80 Å. This was attributed to the absence of EOF transport through the pores of the particles because of double layer overlap. The intraparticle resistance to mass transfer contribution to plate height, H_i, was evaluated using

$$H_i = \left[\frac{\theta(k_0 + k + k_0 k)^2 d}{30\, k_0(1 + k_0)^2(1 + k)}\right] \frac{D_{\mathrm{eff}}}{D_{\mathrm{app}}} \frac{\mu d}{D_{\mathrm{m}}}, \tag{5.8}$$

where θ is the tortuosity of the support, d is the diameter of particle, D_{m} is the diffusivity of the solute in the mobile phase, μ is the interstitial mobile phase velocity, k is the retention factor, D_{eff} is the effective molecular diffusivity in the pores, and D_{app} is the apparent diffusivity for transport in the porous particle by diffusion and by intraparticle convective transport. The retention factor, k,

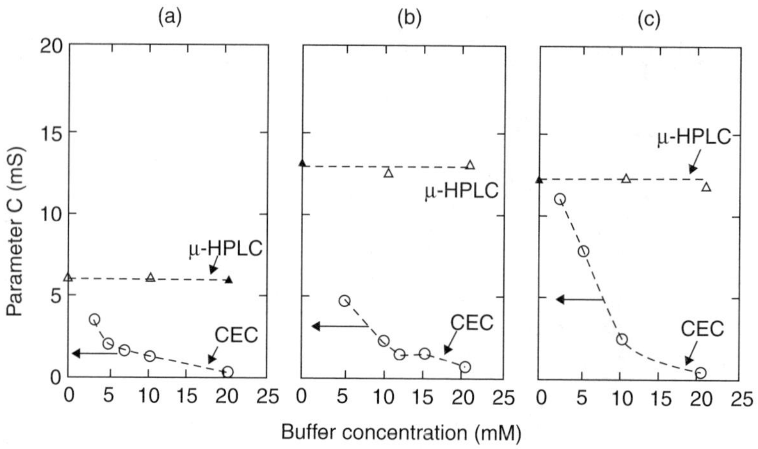

FIGURE 5.7 C term vs. the concentration of electrolyte in solution evaluated for HPLC and CEC. (Reprinted from Wen, E., et al., *J. Chromatogr. A*, 855, 349–366, 1999. With permission from Elsevier.)

is given by

$$k = \frac{\varepsilon_i(1 - \varepsilon_i)}{\varepsilon_\theta}, \tag{5.9}$$

where ε_i is the intraparticle porosity and ε_θ is interstitial porosity. Figure 5.7 shows how the C term can be reduced by intraparticle EOF for large particles. It was also shown that the contribution of the C term is reduced as the concentration of electrolytes in solution is increased.

In an additional study on the differences of zone broadening in CEC versus HPLC, Dittmann and coworkers modified the Horvath model for HPLC to account for the effects of EOF [3,7].

$$H = H_{\text{disp}} + H_{\text{e,diff}} + H_{\text{i,diff}} + H_{\text{t,diff}} + H_{\text{kin}}, \tag{5.10}$$

where H_{disp} is the plate height increment due to the axial dispersion of the solute in interstitial space, $H_{\text{e,diff}}$ is the plate height increment resulting from film resistance at the particle boundary, $H_{\text{i,diff}}$ is the plate height contribution from interparticle diffusion, $H_{\text{t,diff}}$ is the plate height contribution of transchannel mass transfer, and H_{kin} is the plate height contribution from the interaction between the solute and the stationary phase. The sources of band broadening from $H_{\text{e,diff}}$, $H_{\text{i,diff}}$, and H_{kin}, are shown in Figure 5.8 [8]. In Equation 5.10, height equivalent to a theoretical plate (HETP) is expressed as the sum of independent terms that contribute to zone broadening. In a packed column, dispersion in the axial direction is expressed as

$$H_{\text{disp}} = H_{\text{a,diff}} + H_{\text{eddy,diff}}, \tag{5.11}$$

where $H_{\text{a,diff}}$ is the plate height contribution arising from static diffusion in the axial direction, and $H_{\text{eddy,diff}}$ is the plate height contribution due to the differences in axial transport velocity (eddy diffusion).

In Equation 5.10, it is assumed that the contributions of all individual terms are independent of flow profile except for the $H_{\text{eddy,diff}}$ and $H_{\text{t,diff}}$ terms. Because of the plug-like flow velocity profile, the contribution of transchannel diffusion is much smaller in an EOF-driven system (Figure 5.6). The contribution of transchannel diffusion to the total HETP in a packed column is very small owing to the small channel diameter; approximately one-sixth of the particle diameter. Flow velocity in the channels between particles determines the magnitude of the eddy diffusion term contribution. In

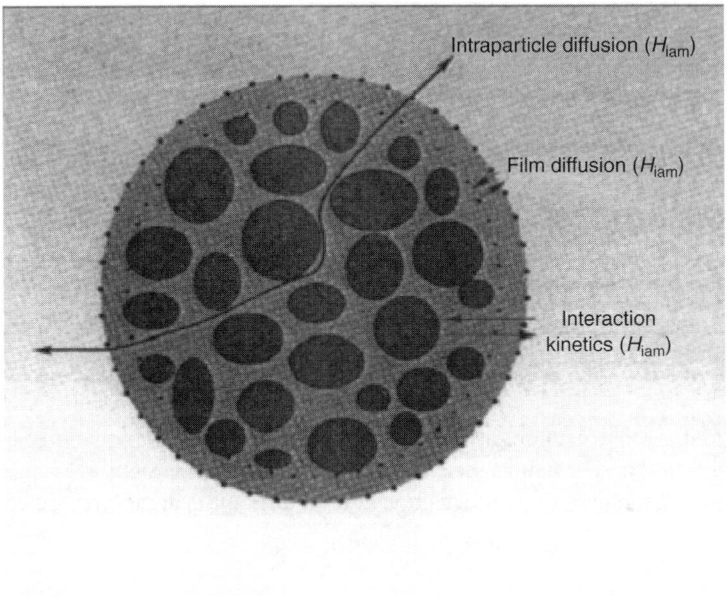

FIGURE 5.8 Sources of dispersion in liquid chromatography. (Reprinted from Bartle, K.D. and Myers, P., *J. Chromatogr. A*, 916, 3–23, 2001. With permission from Elsevier.)

a pressure-driven system velocity varies with the diameter of the channels, while in an electrically driven system, flow velocity is largely independent of the width of the channel. Therefore, solutes exchanging between one channel and another do not usually experience a difference in flow velocity as they would in a pressure-driven system, resulting in much lower dispersion.

5.3 COLUMNS

The packing in a CEC column plays the key role of providing sites for the sorptive interactions (as in HPLC) as well as supporting EOF. The development of both particulate packings having properties tuned for CEC, and alternative column technologies such as monoliths, is emerging rapidly.

CEC columns are generally made of fused-silica tubing, usually packed with the appropriate stationary phase. Today, the most commonly used CEC columns have i.d. of 100 μm or less, with 50 and 75 μm i.d. being the most popular. The stationary phase is retained in the column by two frits. Column designs can be categorized into two major types: OT columns and packed structures, which include packed columns, monolithic columns, and microfabricated structures (open or continuous beds). Packed capillary columns are most commonly used, as has been demonstrated in numerous papers [9–11]. They can be subdivided into three different categories: columns packed with particles, columns with continuous beds fabricated *in situ* creating a rod-like monolithic structure, and columns with immobilized or entrapped particulate materials.

5.3.1 Open Tubes

Open tubular CEC (OTCEC) is similar to CE, with the exception of having a stationary phase attached on the wall of the fused-silica capillary. OT columns offer some advantages over packed capillaries [9]. OT columns with inner diameters of 10 μm or less can generate plate height lower than those for packed columns with the same i.d. The limited band broadening in OT columns results from the absence of packing material and end frits. The small diameter OT column also allows a faster separation with higher field strength without causing significant Joule heating. The drawbacks of

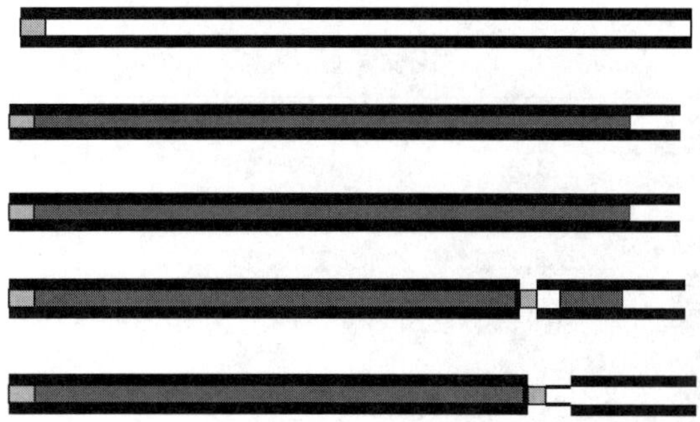

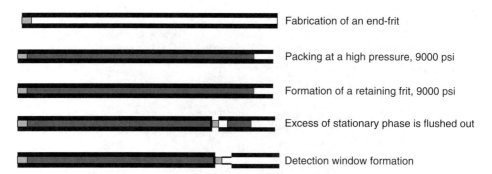

FIGURE 5.9 Schematic of capillary packing procedure.

OTCEC columns include difficulties associated with sample injection volume (due to the extremely small loading capacity) and detection. Because of the low nL to pL sample size range, and the small i.d. and therefore path length, optical detection can be difficult. This has restricted the growth of application of OTCEC.

5.3.2 Packed Columns

Capillary electrochromatography capillaries are usually packed with 3–5 μm C_{18} or C_8 packing materials. A packed column typically consists of two parts, a packed and an OT section as illustrated in Figure 5.9. The preparation of these columns includes the fabrication of retaining frits within a capillary and the packing of small diameter particles into narrow-bore capillaries. The retaining frits are typically made by filling a small section of the column with silica, then, sintering to produce two porous plugs. After this temporary frit is formed, stationary phase is slurry packed into the column at high pressure. Once packed, a second retaining frit is sintered in place with a heating element, and the excess stationary phase is flushed out by applying pressure in the reverse direction. Packed columns provide higher column capacity than the OT columns. Most columns for CEC have been packed with reversed-phase materials.

Both these steps, fritting and packing, present technical difficulties and therefore reproducibility remains problematic. The use of retaining frits in fabricating a packed column presents several issues. The fabrication of frits causes the removal of the protective polyimide coating and causes the column to be fragile at the frit. Nonreproducibility in manufacturing the frits is also another drawback of this approach. Moreover, the heat applied in the process of making the frits changes the characteristics of the packing materials at the frits. The difference in surface chemistry leads to bubble formation at the interface between the frits and the unpacked segments of the capillary. These limitations have led to the development of various alternative approaches.

5.3.3 Monolithic Columns

Fritless monolithic columns, which emerged during the last decade [4,12–19], have proven to be a viable alternative to packed capillaries for CEC. They are prepared by *in situ* polymerization to form a continuous rod-like porous bed, thereby, eliminating the difficulties encountered with packed columns. The porous monoliths can be either in rigid structures [12,16,19–31] or soft gels [4,13,24,32–38]. Monolithic columns have special characteristics and present several advantages, mainly in the ease of their preparation [39–41]. These advantages include the following: (1) the polymerization process is simple and can be performed directly within the confines of a capillary or a microfluidic chip, thus avoiding the problems related to both frit formation and packing; (2) columns of virtually any length and shape can be fabricated; (3) the polymerization mixture can be prepared using a wide variety of monomers, allowing a nearly unlimited choice of

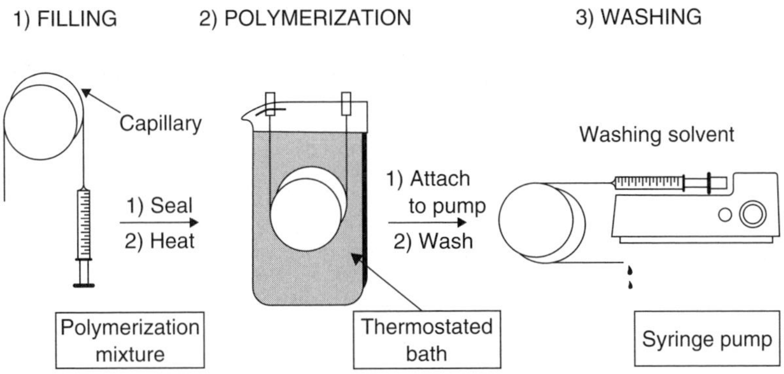

FIGURE 5.10 Schematic representation of the preparation of a monolithic column. (Reprinted from Svec, F., et al., *J. Chromatogr. A*, 887, 3–29, 2000. With permission from Elsevier.)

both matrix and surface chemistries; and (4) the polymerization process can be easily controlled, which enables optimization of the porous properties of the monolith, and consequently the flow rate and chromatographic efficiency of the system. Studies conducted by Luo and Andrade [42] suggest that continuous polymeric beds with submicron channels may be the ideal packing structure for CEC. With greatly reduced geometric tortuosity and more favorable EOF, this approach should simplify column technology and alleviate other problems associated with CEC (i.e., bubble formation) [43].

Polymeric monolithic columns have been widely developed and successfully applied for CEC. As can be seen in Figure 5.10 [44], the preparation of a monolithic porous polymer sorbent is a simple and straightforward process. The steps consist of (1) modification of the capillary wall in order to provide functional groups on the surface that will participate in the subsequent polymerization process; (2) filling the capillary with the homogeneous polymerization mixture consisting of the monomers, initiator, and porogenic solvent; (3) initiation of polymerization thermally or by exposure to ultraviolet (UV) radiation to obtain a rigid monolithic porous polymer; and (4) the removal of unreacted components, such as the porogenic solvents and any other soluble compounds that remain in the polymer.

As mentioned before, a large number of available monomers with a wide variety of functionalities are available giving unlimited choices of surface chemistry. They may be used directly for the preparation of polymer monoliths and some are shown in Figure 5.11.

5.3.3.1 Molecularly Imprinted Polymer Sorbents for CEC

Molecularly imprinted polymers (MIPs) are tailor-made materials able to selectively recognize an analyte. These stationary phases have been successfully employed in CEC, particularly when a high degree of selectivity is required. The first applications of MIPs were as stationary phases in affinity chromatography [45,46]. Extensive peak broadening and tailing, however, were observed, especially for the more retained compound (usually the template). It was reported that this was due to the heterogeneity of binding sites, in terms of both affinity and accessibility, and different association and dissociation kinetics.

The use of MIPs for CEC gained considerable interest, most likely owing to the need for a highly effective liquid-based separation system with unique selectivity for predetermined molecular species in a miniaturized format. This is especially attractive since CEC is known to have more efficiency than conventional HPLC, and thus CEC-mode separations should lead to improved performance of imprinted polymers compared to that achieved in conventional LC.

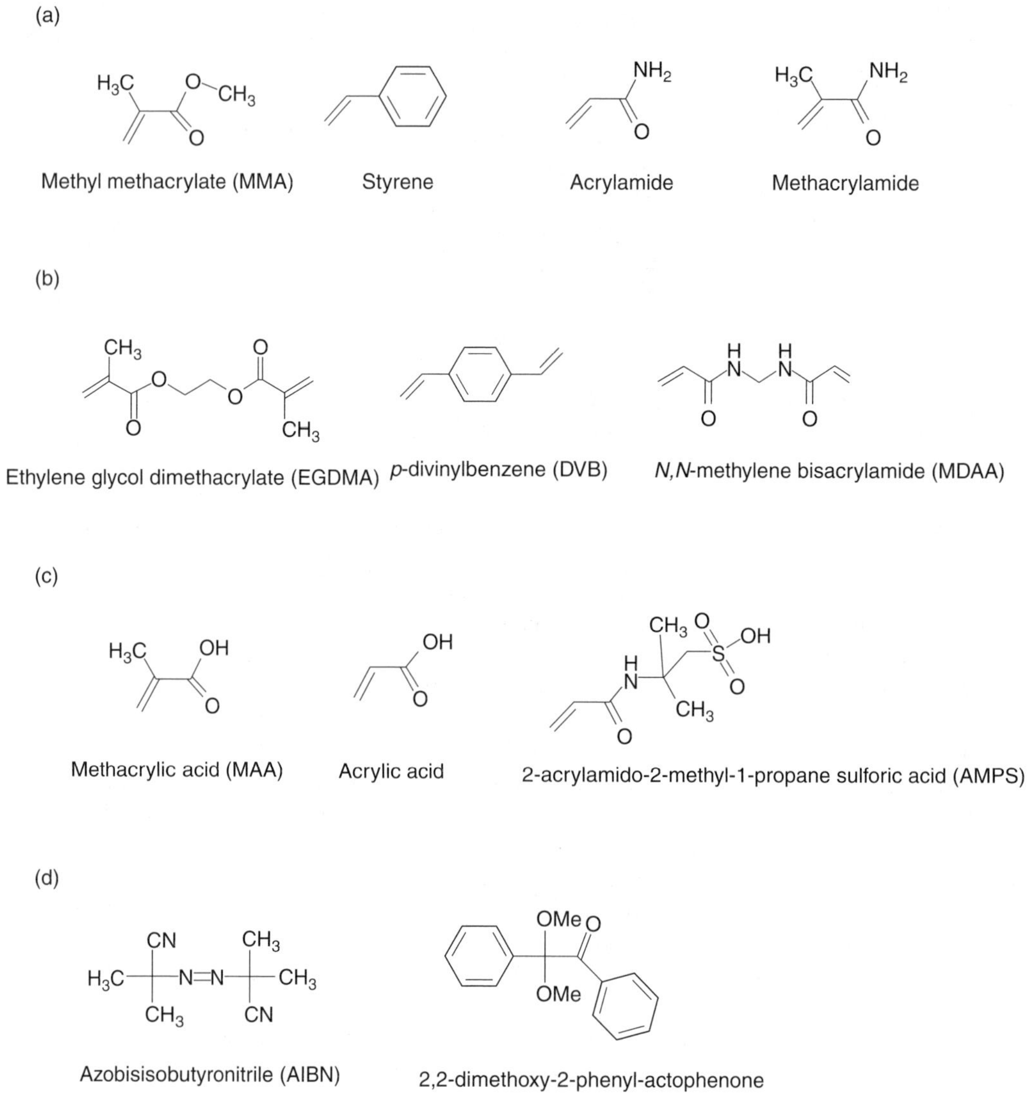

FIGURE 5.11 Selection of (a) monomers with functional groups, (b) cross-linkers, (c) charged monomers for EOF generation, and (d) chemical initiators used for the preparation of polymer monolithic columns.

The first requirement for MIP–CEC is therefore the adaptation of the molecular imprinting technology to the capillary format. This is not a simple task and the straightforward approach taken from HPLC experiments by packing irregular MIP particles to prepare MIP sorbents presented some technical difficulties in the capillary format. Thus, most approaches have reported the synthesis of MIP stationary phases within the confines of the capillary column. This monolithic format is attractive but deceptively challenging. While laborious packing procedures, such as frit making as well as bulk polymerization of MIP with subsequent crushing and sieving steps, are avoided, other challenges abound.

The first attempt to adapt MIPs to capillary columns was reported using an *in situ* dispersion polymerization, which involved the formation of about 10 μm sized agglomerates of 0.5–4 μm particles [47]. These MIP capillaries were imprinted against L-phenylanaline anilide, pentamidine, and benzamidine. The results showed low selectivity of these columns and only pentamidine showed

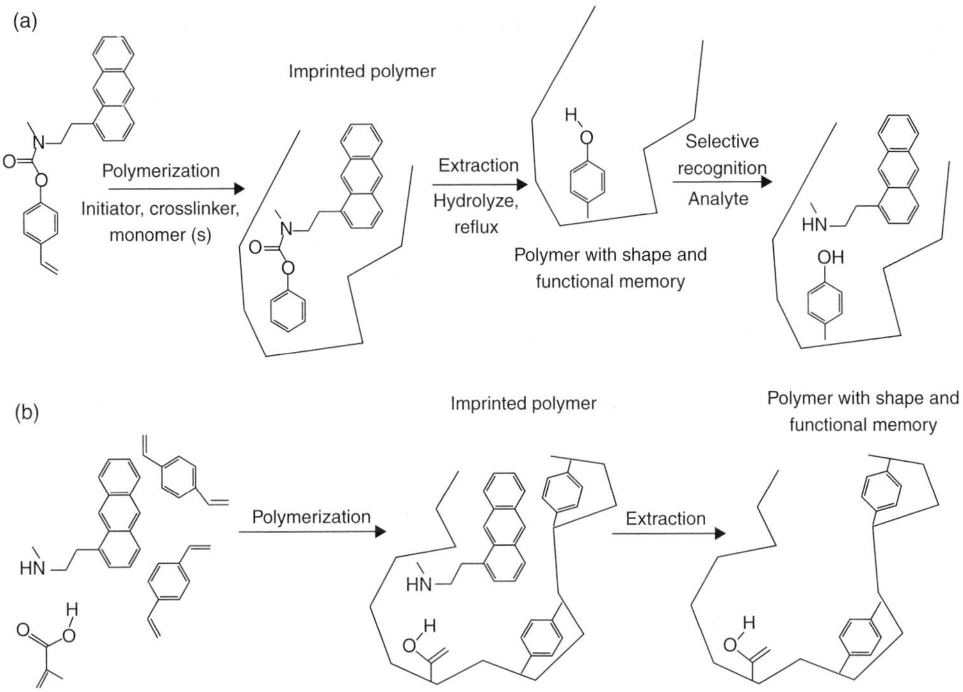

FIGURE 5.12 Schematic illustration of (a) covalent and (b) noncovalent imprinting procedures.

a pH-dependent retention that might originate from imprinting. Despite the poor performance of these columns, this work is of high importance in indicating the feasibility of synthesizing MIPs *in situ*.

As shown in Figure 5.12, the preparation of MIPs includes the polymerization of suitable monomers around a template in the presence of an appropriate cross-linker and solvent (porogen). Following polymerization, the template molecule is removed to leave cavities complementary in size, shape, and chemical functionality to the analyte. Accordingly, the polymer should be able to selectively rebind the target analyte.

Two main approaches are used to produce MIPs: the noncovalent [48] and the covalent [49] approach. In the covalent approach (Figure 5.12a), the functional monomer is covalently bonded to the template molecule before polymerization. When polymerization is complete, the covalent bonds between the template molecule and the polymer are cleaved and the template molecule is extracted. The resulting imprint is then able to recognize and rebind the imprinted analyte *via* reversible covalent bonds. However, this technique suffers from lack of generality owing to the difficulties of finding suitable monomers.

A more general technique is the noncovalent approach. So far, only the noncovalent imprinting technique has been used for the preparation of MIP phases for CEC. In the noncovalent approach, monomers interact with the template molecule *via* hydrogen bonding, ionic bonding, Van der Waals forces, hydrophobic effects, and so forth. The noncovalent imprinting technique has been dominant because of its generality and simplicity. A broad range of species have been imprinted during the past several years [50]. Figure 5.12b shows the procedure for preparation of an MIP using the noncovalent approach.

With the noncovalent approach, the selectivity of the MIP can be altered by optimizing the types and quantities of functional monomers, cross-linking monomers and solvents used. Careful selection of a porogenic solvent is necessary in order to produce a polymer with enough porosity to assure good permeability. The degree of cross-linking is also important in achievement of stability of the template–monomer complex during polymerization, and will also affect polymer porosity.

Figure 5.11 shows some of the most common monomers, cross-linkers, and initiators used for the preparation of monolithic MIPs.

5.3.3.2 Particle Entrapment

In the entrapped monolithic column approach, particles are immobilized inside the capillary column in a sol–gel matrix [14,16,17,19,51] or by sintering the particles with heat [25] instead of using retaining frits. The entrapment approach eliminates the problems associated with retaining frits in conventional packed columns. Moreover, column longevity improves significantly because the problems caused by the failure of frits can be avoided.

Columns fabricated via the sintering method have reproducible separations with ~125,000 plates m^{-1}. The column preparation involves packing the capillary with silica particles (premodified with the desired stationary phase), a washing step, drying with nitrogen gas, and a two-step heat treatment at different temperatures (120°C and 360°C). To reintroduce stationary phase that might have been damaged during the process, *in situ* resilanization is required. The multistep fabrication and stationary phase reattachment process can be time consuming.

There are two means of immobilizing particles in sol–gel matrix. One is to prepare a sol–gel solution (typically a mixture containing alkoxysilanes, ethanol, and hydrochloric acid), followed by addition of particulate materials containing the stationary phase to form a suspension. The suspension is introduced into the column by vacuum or pressure, and then dried to result in an immobilized packed bed. Studies have shown that columns prepared using this method generate ~80,000 plates m^{-1}. Fabrication of these columns is problematic owing to irreproducibilities in the finished products.

The second approach to particle entrapment is to introduce the solution sol gelled after the column has been packed. Chirica and Remcho [16] used high pressure to slurry pack the column. Tang et al. [20,27,51] used supercritical CO_2 to pack the column. After packing, the sol–gel solution was forced into the column and dried. If temporary retaining frits are used, they are eliminated after drying. These methods generated columns with efficiencies of ~125,000 plates m^{-1}. In comparison to the "one-step" method, less entrapment matrix is required and higher reproducibility and homogeneity were noted.

5.4 DEVELOPING A CEC METHOD

Developing a CEC method can seem like a daunting task, but by focusing on some key items and understanding how they are interrelated, the process can become reasonably straightforward. The easiest path is of course to find and use a preexisting method. These may be found by either conducting a literature search or contacting a column manufacture. If a method cannot be located, then the analyst will need to take into consideration: the sorbent, separation buffer, organic modifiers, temperature, and voltage. Table 5.1 is a compilation of CEC applications by compound class, suitable stationary phase, particle size, and reference where the method can be found, and may serve as a good starting point.

5.4.1 SORBENTS

One item to consider when developing a new method is that there are many capillary columns available to the consumer, and even though the stationary phases may be similar there will be variations in EOF and selectivity between them. This effect was studied by Dittmann and Rozing [201] for five reverse-phase sorbents anchored onto silica particles. Figure 5.13 shows the EOF variation for five C18 stationary phases from the work conducted by Dittmann and Rozing. Since variations in EOF exist between different columns with similar stationary phases, it is important to be aware that methods may need to be modified when using a column that is different in even a seemingly minor way than the one used in creating the original method.

TABLE 5.1
Applications of CEC Classified on the Type of Compounds Separated

Compound Class	Suitable Stationary Phase	Particle Size	References
Pollutants			
PAHs	C18 and silica	3 and 1 µm	[52]
PAHs	Nonporous C18	1.5 µm	[53]
PAHs	Methacrylate monolith	—	[54,55]
PAHs	Entrapped C18	5 µm	[16,56,57]
PAHs	4000 Å Nucleosil C18	7 µm	[58]
Triazine herbicides	Hypersil C18 Hypersil C8 Spherisorb C6/SCX	3 µm for all three	[59]
Cinosulfuron and byproducts	C18	3 µm	[60]
Carbonyl 2,4 dinitrophenylhydrazones	Spherisorb ODS 1	3 µm	[61]
Carbamate insecticides	ODS	3 µm	[62]
Insecticidal pyrethrin esters	Hypersil C 18	3 µm	[63]
Pyrethroid insecticides	Nucleosil C18 Zorbax C8 J.T. Baker C18	5 µm for all three	[64]
Carbamate and pyrethroid insecticides	Nucleosil C18 Zorbax C8 J.T. Baker C18	5 µm for all three	[65]
Pyrethroid pesticides	Hypersil ODS	5 µm	[66]
Pirimicarb and azoxystrobin pesticides	Hypersil C18	3 µm	[67]
Phenols	C18	3 µm	[68]
Mono- and dichlorophenols	Silica with methylated β-CDs	—	[69]
Pentachlorophenol	Spherisorb ODS 1	3 µm	[70]
Phenols in tobacco smoke	Hypersil C 18	3 µm	[71]
Nitroaromatic and nitramine explosives	Nonporous C18	1.5 µm	[72]
Nitroaromatic and nitramine explosives	Hypersil C 18 Nonporous C18	3 µm 1.5 µm	[73]
Amino acids, peptide, proteins			
PTH amino acids	Hypersil ODS	3 µm	[74]
PTH amino acids	Zorbax C18	3.5 µm	[75]
PTH amino acids	Zorbax C18 sintered	6 µm	[25]
Dansylated amino acids	ECTFE	10 µm	[76]
Tryptophan, tyrosine	OT CEC DNA aptamers	—	[77]
NDA amino acids	Methacrylate monolith	—	[78]
PTH amino acids			
Peptides (2–7 amino acids)			
Trp-Arg, Arg-Trp	OT CEC DNA aptamers	—	[79]
Dipeptides	Methacylate monolith	—	[80]
Tripeptides	OT CEC porphyrin	—	[81]
Peptides (2–3 amino acids)	Methacrylate monolith	—	[82]
Peptides (2–4 amino acids)	Sperisorb ODS SAX/C18	3 µm 5 µm	[83]
Peptides (2–5 amino acids)	Methacrylate monolith	—	[30,84]
Peptides (5 amino acids)	Hypersil C18	3 µm	[85,86]
Peptides (4–5 amino acids)	Gigaporous polymeric SCX	8 µm	[87]

Continued

TABLE 5.1
(Continued)

Compound Class	Suitable Stationary Phase	Particle Size	References
Peptides (4 amino acids)	Polyacrylamide poly(ethyleneglycol) monolith	—	[13]
Synthetic peptides	Hypersil C18	3 μm	[88]
Synthetic peptides dipeptides	Spherisorb SCX	5 μm	[89]
Basic peptides	Spherisorb Si	5 μm	[90]
Angiotensins	Acrylate monolith	—	[91]
Angiotensins	PS/DVB monolith	—	[15]
Lysozymes, angiotensins	OT CEC C18	—	[92]
Cytochrome c digest	Vydac C18	3 μm	[93]
Chicken albumin digest, angiotensins			
L-threonine, L-tyrosine	(IBMA-EDMA-AMPS) monolith	—	[94]
Myoglobin (horse heart), transferrin (human), α-lactalbumin (bovine milk)	C4 monolith	—	[95]
β-Lactoglobulin	Vydac C18	3 μm	[96]
Tryptic digest of transferrin	OT CEC Si	—	[97]
Tryptic digest of ovalbumin	Monolith	—	[98]
Map of cytochrome c	Gromsil ODS2	1.5 μm	[99]
Tryptic digest of cytochrome c	Spherisorb SAX/C6	3 μm	[88]
Carbonic anhydrase, α-lactalbumin, trypsin inhibitor, ovalbumn, conalbumin, haemoglobin variants	Spherisorb S5- W SAX	5 μm	[100]
Ribonuclease, insulin, α-lactalbumin	Acrylate monolith	—	[91]
Lysozyme, α-chymotrypsinogen, ribonuclease A, cytochrome c	C18 monolith	—	[37]
Cytochrome c mixture (horse, tuna, chicken and bovine)	OT CEC cholesteryl	—	[101]
Cytochrome c mixture (horse, tuna, chicken and bovine)	OT CEC C18	—	[102]
Trypsinogen, α-chymotrypsinogen, ribonuclease A, cytochrome c	OT CEC polyaspartic acid	—	[103]
Lysozyme, α-chymotrypsinogen, ribonuclease A, cytochrome c	OT CEC DVB	—	[104]
Carbohydrates			
Sucrose, saccharin	C18	3 μm	[105]
Sucralose and related carbohydrates	C18	3 μm	[106]
Glucose-maltohexose	Polyacrylamide poly(ethyleneglycol) monolith	—	[13]
p-Nitrophenyl labeled glucopyranosides, maltooligosaccharides	Zorbax C18	5 μm	[107]
α- and β-anomers of glucopyranoside			
Phenyl-methyl-pyrazolone labeled monosaccharides	Hypersil ODS I	5 μm	[108]
Phenyl-methyl-pyrazolone labeled mono and disaccharides	Hypersil ODS I	5 μm	[109]
Phenyl-methyl-pyrazolone labeled aldopentose and monosaccharides	Aminopropylated Si octadecylammonium Si	—	[110]
Nucleotides			
AMP, ADP and ATP	Nucleosil C18	5 μm	[111]

TABLE 5.1
(Continued)

Compound Class	Suitable Stationary Phase	Particle Size	References
AMP, CMP, GMP, and UMP	OT: 4, 8, 12, 18, 22, 26-hexaaza-1, 15-dioxacyclooctaeicosane ([28]ane-N_6O_2) derivatized	—	[112]
Adenine, cytosine, guanine, thymine	9-ethyladenine MIP	—	[113]
Adenosine, cytidine, uridine, guanosine, thymidine	Hypersil phenyl	3 µm	[114]
Adenosine, cytidine, inosine, uridine, guanosine, thymidine	CEC Hypersil C18	3 µm	[115]
Thymine, cytosine, adenine, guanine, adenosine	OT-cholesteryl-undecanoate and cyanopentoxybiphenyl derivatized	—	[116]
Nucleic acids, mono-, di- and triphosphonucleotides	ODSS ODSS	10 µm 5 µm	[117]
Dinucleotides	Nonporous ODSS	2 µm	
t RNAs			
Purine and pyrimidine bases and their nucleotides	ODSS	10 µm	[118,119]
Primicarb and related pyrimidines	Hypersil C18	3 µm	[120]
Synthetic nucleoside	Spherisorb ODS1	3 µm	[121]
PAH-DNA adduct products of in vitro reactions	ODS	3 µm	[122,123]
Miscellaneous			
Vitamin E in vegetable oils	C18	3 µm	[124]
Flavonoids (hespederin, hesperetin)	Hypersil C8	3 µm	[125]
Flavonoids (hop acids)	Hypersil C18	3 µm	[126]
Antraquionones in Rhubarb	Hypersil C18	3 µm	[127]
Antraquionones in Rhubarb	C18	5 µm	[128]
Triglycerides of vegetable and fish oil	CEC Hypersil C18	3 µm	[129–131]
Fatty acids of vegetable and fish oil	CEC Hypersil C18	3 µm	[130]
Unsaturated fatty acids methyl esters	GROM-SIL ODS	3 µm	[132]
Glycosphingolipids	Porous ODSS	5 µm	[133]
Cannabinnoids	Hypersil C18 Hypersil C8	3 µm 3 µm	[134]
Retinyl esters	Nucleosil C18	5 µm	[135]
Retinyl esters	7 µm Nucleosil C18	7 µm	[136]
Retinyl esters	C30	5 µm	[137]
Carotenoid isomers	C30	3 µm	[138]
N-Nitrosodiethanolamine in cosmetics	OT-CEC C18	—	[139]
Aloins, and related constituents of aloe	C18	3 µm	[140]
Food colorants and aromatic glucoronides	Nucleosil C18	5 µm	[141]
Azo and antraquinone textile dyes	Hypersil C18	3 µm	[142]
Alkaloids	Nucleosil C18	7 µm	[20]
Fullurenes C_{60} and C_{70}	Vydac C18	3 µm	[143]
Polystyrene standards	Methacrylate monolith	—	[144]

Continued

TABLE 5.1
(Continued)

Compound Class	Suitable Stationary Phase	Particle Size	References
Celluloses	Nucleosil silica	5 μm	[145]
Pharmaceuticals			
Steroids			
Fluticasone propionate and synthesis impurities	Hypersil C18 Spherisorb ODS-I	3 μm 3 μm	[146] [147]
Triamcinolone, hydrocortisone, prednisolone, cortisone, methylprednisolone, betamethasone, dexamethasone, adrenosterone, fluocortolone, triamcinolone acetonide	Hypersil C18	3 μm	[71,148]
Tipredane and related substances	Spherisorb ODS-I	3 μm	[121]
Corticosteroids			
Corticosterone, testosterone, androsten-3, 17-dione, androstan-3, 17-dione, pregnan-3, 20-dione	Zorbax ODS	6 μm	[75]
Aldosterone, hydrocortisone, testosterone	Hypersil C18	3 μm	[149]
Digoxigenin, gitoxigenin, cinobufatalin, digitoxigenin, cinobufagin, bufalin	Spherisorb ODS-I	3 μm	[150]
Hydrocortisone, testosterone, 17-α-methyltestosterone, progesterone	Nonporous Chromspher-ODS	1.5 μm	[151]
Estriol, hydrocortisone, estradiol, estrone, testosterone, 17-α-methyltestosterone, 4-pregnen-20α-ol-3-one, progesterone	Zorbax ODS	1.8 μm	[152]
Deesterified steroid, budesonide, steroid A	Spherisorb ODS-I	3 μm	[153]
Hydrocortisone, prednisolone, hydrocortisone 21-acetate, testosterone	Poly (AMPS-co-IPPAm) hydrogel	—	[38]
Hydrocortisone, prednisolone, betamethasone, betamethasone dipropionate, clobetasol butyrate, fluticasone propionate, clobetase butyrate, betamethasone-17-valerate	Hypersil C18	3 μm	[154]
Aldosterone, dexamethasone, β-estradiol, testosterone	Hypersil C18 NPS ODS 2	3 μm 1.5 μm	[155]
Desogestrel and analogs, tibolon and analogs	Hypersil C18	3 μm	[156]
Dexamethasone, betamethasone valerate, fluticasone propionate	Hypersil C18/SCX	3 μm	[157]
Neutral and conjugated steroids	Macroporous monolith, C12	—	[158]

TABLE 5.1
(Continued)

Compound Class	Suitable Stationary Phase	Particle Size	References
Corticosteroids and esters (hydrocortisone, hydrocortisone 17-butyrate, hydrocortisone 21-acetate, hydrocortisone 17-valerate, hydrocortisone 21-caprylate, hydrocortisone 21-cypionate, hydrocortisone 21-hemisuccinate)	Spherisorb ODS-I	3 μm	[159]
Cholesterol and ester derivatives	Hypersil C18	3 μm	[160]
Corticosteroids (ouabain, strophantidin, 4-pregnene-6b,11b,21-triol-3,20-dione)	Spherisorb small pore ODS/SCX, sol-gel bonded	3 μm	[20]
Bile acids and conjugates	Macroporous monolith, amino (normal phase) Macroporous monolith, C12 (reverse phase)	—	[161]
Levonorgestrel and racemic norgestrel	Cellulose tris(3,5-dichlorophenylcarbamate coated onto spherical Daisogel	5 μm	[162]
Estrogens (diethylstilbestrol, hexestrol, dienestrol)	ODS	3 μm	[163,164]
Estrogens (estriol, estradiol, equiline, estrone)	GROM-SIL ODS-0 AB	3 μm	[132]
Benzodiazepines			
Nitrazepam and diazepam	Hypersil C18	3 μm	[148]
Nitrazepam, diazepam, triazolam	3-(1,8-Naphthalimido)propyl-modified silyl silica gel	—	[165]
Oxazepam, lorzepam, temazepam, diazepam, tofizepam	phenyl	3 μm	[166]
Nitrazepam, nimetazepam, estazolam, brotizolam, clonazepam, axazolam, haloxazolam, cloxazolam, medazepam	Cholesteryl bonded silica OT, cholesteryl modified capillary wall	6.5 μm	[166]
Temazepam, oxazepam, clonazepam, diazepam, nitrazepam	OT fused silica etched with NH_4HF_2, chemically bonded cholesteryl-10-undecanoate	—	[116]
Flunitrazepam, temazepam, diazepam, oxazepam, lorazepam, clonazepam, nitrazepam	PEM-coated capillary	—	[168]
Flunitrazepam, temazepam, diazepam, oxazepam, lorazepam, clonazepam, nitrazepam	Reliasil C18	3 μm	[169]
Nonsteroidal anti-inflammatory drugs			
Fenoprofen, ibuprofen, indoprofen, ketoprofen, naproxen, suprofen	LiChrospher 100 RP-18	5 μm	[170]

Continued

TABLE 5.1
(Continued)

Compound Class	Suitable Stationary Phase	Particle Size	References
Indoprofen, ketoprofen, naproxen, ibuprofen, fenoprofen, flurbiprofen, suprofen	OT histidine coated capillary	—	[171]
Indoprofen, suprofen, tiaprofen, ketoprofen, naproxen, fenoprofen, carprofen, flurbiprofen, cicloprofen, ibuprofen	LiChrospher 100 RP-18	5 μm	[172]
Etodolac and five metabolites	LiChrospher 100 RP-18	5 μm	[173]
Acetaminophen, caffeine	Nucleosil C18, sol-gel bonded	5 mm	[57]
Ibuprofen, indoprofen, fenoprofen, ketoprofen, suprofen, diclofenac, metenamic acid	Acrylamide-based monoliths	—	[174]
Ibuprofen, naproxen, ketoprofen, suprofen	Methacrylate-based macroporous SAX monolith	—	[175]
Tricyclic antidepressants			
Nortriptyline and amitriptyline	Nucleosil 50 Nucleosil 100 Nucleosil 300 Nucleosil 1000 Nucleosil 4000	5 μm for all	[176]
Bendroflumethiazide, nortriptyline, chlomipramine, methdilazine, imipramine, desipramine	Spherisorb ODS-I Spherisorb SCX	3 μm 3 μm	[147]
Nortriptyline, *N*-methyl-amitriptyline, amitriptyline, imipramine, clomipramine, *N*,*N*-dipropyl-protriptyline	Spherisorb ODS-I	3 μm	[177]
Nortriptyline, amitriptyline, *N*-methyl-amitriptyline, desipramine, imipramine, chlomipramine, *N*, *N*-dimethyl-protriptyline, *N*,*N*-dipropyl-protriptyline	Nucleosil SCX Zorbax SCX	5 μm 5 μm	[141]
Nortriptyline, imipramine, amitriptyline, clomipramine	Spherisorb SCX	3 μm	[142]
Nortriptyline, *N*-methyl-amitriptyline, amitriptyline	Continuous-bedpolyacrylamide with various contents of isopropyl and sulfonate ligands	—	[178]
Nortriptyline, doxepin, imipramine, amitriptyline, trimipramine, clomipramine	Nortriptyline MIP	—	[179]
Various pharmaceuticals			
Prostaglandins and relates impurities	Spherisorb ODS-I Zorbax SBC8	3 μm 1.8 μm	[146]

TABLE 5.1
(Continued)

Compound Class	Suitable Stationary Phase	Particle Size	References
Neutral related *S*-oxidation compounds	Hypersil C18	3 μm	[180]
2-Phenylmethyl-1-naphthol	Hypersil C18	3 μm	[181]
p-Hydroxybenzoic acid, bumetanide, flurbiprofen	Hypersil C18	3 μm	[121]
Thomapyrin: containing acetaminophen, caffeine and acetylsalicylic acid	GROM-SIL 100 ODS-0 AB	3 μm	[182]
Antiviral drug suramin	Nucleosil 100 C18	5 μm	[111]
Amino group containing drugs: codeine phosphate, ephedrine hydrochroride, thebaine, berberine, hydrochloride, jatrorrizine hydrochloride, cocaine hydrochloride	Micra bare silica	3 μm	[183]
Isradepin and by-products	Hypersil C18	3 μm	[184]
Morphine alkaloids	Nucleosil 100 C18	5 μm	[111]
Barbiturates (barbital, phenobarbital, secobarbital, thiopental)	3-(1,8-Naphthalimido)propyl-modified silyl silica gel	—	[165]
Antiepileptic drugs: ethosuccinimide, primidon, CBZ-10, 11 diol, CBZ-10, 11-epoxid, phenytoin, carbamazepine (CBZ)	Spherisorb ODS-I	3 μm	[125]
Macrocyclic lactone, S541 Factor B from Streptamyces S541	CEC Hypersil	3 μm	[128]
Sulfanilamide, sulfaflurazol, sulfadicramide	Nucleosil 100-5C8	5 μm	[185]
Cardiac glycosides: digoxigenin, digoxin, digitoxigenin	NPS ODS II	1.5 μm	[155]
Tetracyclines	Etched and modified C18 OT column	—	[186]
2-Phenylethylamine derivatives: epinephrine, Dopa, 2-amino-3-hydroxy-3-phenyl-propanol, ephedrine	Nucleosil 5C8	5 μm	[125]
Vitamin D2 and D3	GROM-SIL 100 ODS-0 AB	3 μm	[132]
Doxorubicin	Luna C18	3 μm	[187]
Thalidomide and hydroxylated metabolites	Aminopropyl coated with amylose and cellulose derivatives	5 μm	[188]
Thalidomide and metabolites	LiChrospher 100 RP-18	5 μm	[189]
Methylamphetamine and impurities	NPS ODS II	1.5 μm	[190]
Related opiate compounds (morphine, hydromorphone, nalorphine, codeine, oxycodone, diacetylmorphine)	NPS ODS II	1.5 μm	[191]

Continued

TABLE 5.1
(Continued)

Compound Class	Suitable Stationary Phase	Particle Size	References
Theophylline, caffeine, sulfanilamide	C18	3 μm	[105]
Theophylline, caffeine, aminophylline, theobromine, β-hydrocyethyltheophylline, phenylbutazone, hydrochlorothiazide, acetominophen	Silica	3 μm	[192]
Fluvoxamine and possible isomers	Spherisorb ODS-I	3 μm	[193]
Hydroquinone and ethers	LiChrospher 100 RP-18	5 μm	[194]
Serotonin and metabolites	OT, fused silica etched with NH_4HF_2, chemically bonded 4-cyano-4'-pentoxybiphenyl	—	[116]
Neurotransmitters (5-hydroxytryptamine, dopamine, norepinephrine, epinephrine, DOPA)	OT, chemically-modified wall with macrocyclic dioxopolyamine (dioxo[13]aneN_4)	—	[195]
Clenbuterol, salbutamol, methadone	Hypersil MOS	3 μm	[196]
Salbutamol, salmeterol	Spherisorb C18	3 μm	[142]
	Spherisorb C6/SCX	3 μm	
	Spherisorb SCX	3 μm	
Carbovir, ranitidine, ondansetron, imipramine, amitriptyline, clomipramine	Symmetry Shield RP-8	3.5 μm	[197]
Benzylamine, nortriptyline, diphen-hydramine, terbutalin, procainamide	Hypersil C18	3 μm	[198]
Benzylamine, nortriptyline, diphen-hydramine, procainamide	Hypersil silica	3 μm	[199]
	Hypersil BDS silica	3 μm	
	Hypersil silica	3 μm	
Basic and acidic drugs (metoclopramide, timolol, procain, ambroxol, antipyrine, naproxen)	Spherisorb ODS-I, Hypersil C18, Hypersil C8	3 μm	[200]
Amphetamine, methamphetamine, procaine, cocaine, quinine, heroin, noscapine, Phenobarbital, methaqualone, diazepam, testosterone, testosterone propionate, CBN, d9-THC, d9-THC acid-A	Hypersil C8	3 μm	[134]

5.4.2 Separation Buffer

Proper selection of a separation buffer is crucial because it will help resist changes in pH, is used to dissolve the sample, will mitigate the interactions between the sample and bonded phase, affect the EOF, and is the electrical connection between the cathode and anode when voltage is applied during a separation. There are some important features of the separation buffer that must be considered: pH, ionic strength, and organic modifiers. Of these, pH and ionic strength will be covered in this section, and organic modifiers will be discussed separately.

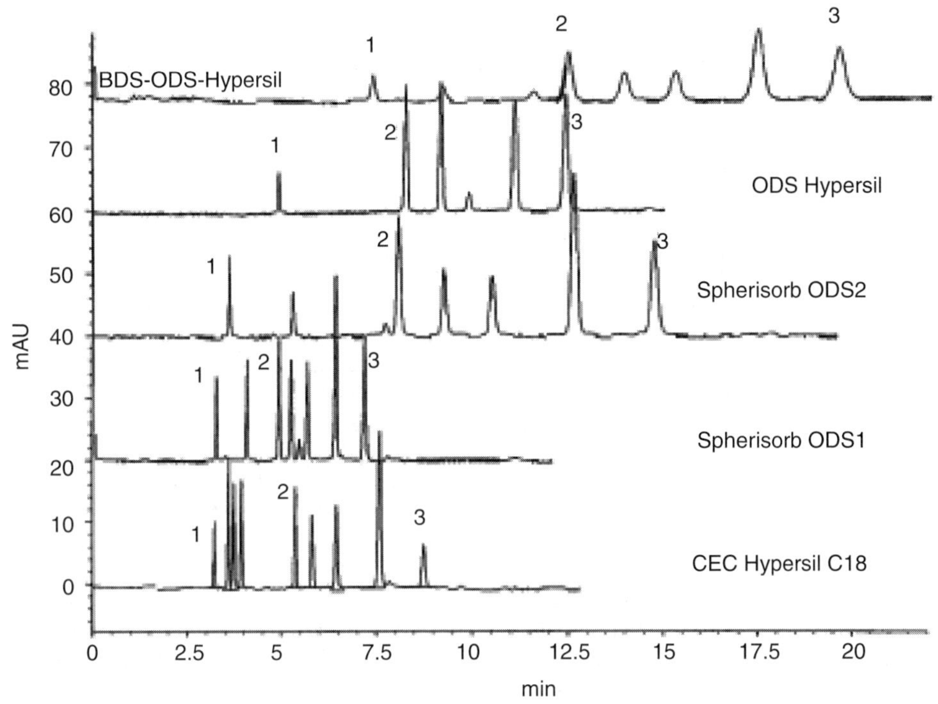

FIGURE 5.13 Separation of PAHs on five reversed-phase C18 stationary phases. Column 250 (335) mm × 0.1 mm, 3 μm, mobile phase 80% ACN 20% 50 mM Tris-HCl, pH 8, 20 kV, temperature 20°C, 10 bar pressure applied to both ends of capillary, 20°C. Samples were not identical but all contained (1) thiourea, (2) napthalene, and (3) fluoranthene. (Reprinted from Dittmann, M.M. and Rozing, G.P., *J. Microcol. Sep.*, 9, 399, 1997. With permission.)

The separation buffer in the column and inlet/outlet vials should also be replaced—ideally after each run—to prevent buffer depletion. Buffer depletion occurs as the cation and anion components of the buffer migrate toward cathode and anode respectively during a run. Over time, the concentration of anions at the cathode and cations at the anode will be reduced. This reduction of ions will also occur within the buffer in the column itself. When this occurs to an extreme, major performance degradation is the result.

5.4.2.1 pH

In selecting a separation buffer, the sample to be dissolved must be considered first. It is important to consider the pK_as of the components of the sample as the pH will determine whether the sample components are charged or neutral. Euerby et al. [121] showed that by changing buffer pH, and using a 25-cm effective length 3-μm particle size Hypersil phenyl stationary phase, the migration order of some of the barbiturates in their mixture changed when the pH was increased from 6.1 to 7.8, owing to a change in analyte ionization state. For biological samples the pH is not only important in dictating the charge of the analyte, but also critical in preventing sample denaturation.

EOF is also pH dependent. For example, in columns having silica support particles, ionizable silanols will cause these columns to exhibit a more pronounced dependency of EOF on pH than do other columns. The silanols begin to ionize at pH 2 and will continue to ionize progressively (based on the regiospecificity of a given silanols' pK_a) until approximately pH 8–9, at which point the preponderance of silanols are fully ionized. The zeta potential will therefore increase, and it is for this reason that the electroosmotic velocity increases with increasingly alkaline solutions.

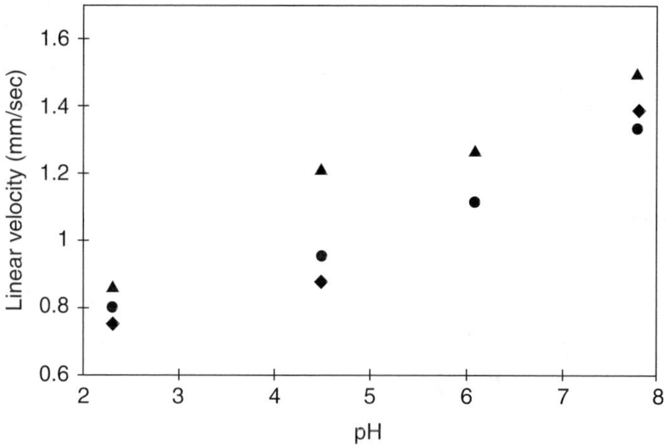

FIGURE 5.14 Effect of mobile phase pH on the linear velocity of the EOF. Comparison of CEC Hypersil C18 (♦), C8 (●), and phenyl (▲) packing materials. (Reprinted from Euerby, M.R., et al., *J. Microcol. Sep.*, 11, 305, 1999. With permission.)

Euerby et al. [121] demonstrated that the linear velocities for three different commercially available columns increased with rising pH, as shown in Figure 5.14. These Agilent columns were packed with Hypersil C18, Hypersil C8, and Hypersil phenyl.

A possible solution to the low electroosmotic velocity encountered with standard reverse-phase columns is to use a mixed-bed column. A mixed-bed column contains both normal-phase (NP) and reverse-phase (RP) components, and when working at lower pH regions the NP component will yield EOF from the surface charge associated with the anionic or cationic NP sorbent. Tang and Lee [20] compared electroosmotic velocity over the pH range of 2–9 for reverse-phase and mixed-bed columns; both columns were sol–gel bonded continuous-bed columns. The reverse-phase column contained an ODS1 stationary phase whereas the mixed-bed column was an ODS/SCX mixture. They noted that the mixed-bed column exhibited a nearly constant EOF for the entire pH range studied, unlike the reverse-phase column that continually increased as the pH climbed. These findings are shown in Figure 5.15.

5.4.2.2 Ionic Strength

Another variable to be optimized is the ionic strength of the separation buffer. The ionic strength not only determines the buffering capacity of the buffer being used, but also affects EOF and Joule heating.

From Equation 5.3, it seems that there is no relation between EOF velocity and double layer thickness, but there is a direct relation to the zeta potential. To find the relation between EOF and ionic strength, Equation 5.1 must be substituted into Equation 5.2 to obtain

$$\zeta = \frac{[(\varepsilon_0\varepsilon_r RT/2F^2cz^2)^{1/2}]\sigma}{\varepsilon_0\varepsilon_\mathrm{r}}. \tag{5.12}$$

Thus the zeta potential will have an inverse relation to the square root of the electrolyte concentration. Likewise, it can be seen that EOF velocity will be dependent on the inverse square root of the ionic strength of the electrolyte.

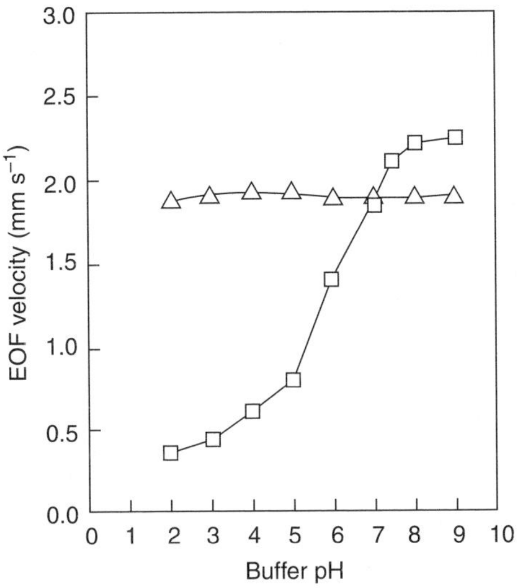

FIGURE 5.15 Plots of EOF velocity vs. mobile phase pH for sol–gel bonded continuous-bed columns. Conditions: 25/34 cm × 75 μm i.d. continuous-bed columns containing sol-gel bonded (□) 3 μm, 80 Å ODS1, and (Δ) 3 μm, 80 Å ODS/SCX; 70: 25: 5 (v/v/v) ACN/H_2O/50 mM phosphate buffer; 5 kV × 2 s injection; 30 kV applied voltage; 0.3 mM thiourea used as EOF marker. (Reprinted from Tang, Q. and Lee, M.L., *J. High Resol. Chromatogr.*, 23, 73, 2000. With permission.)

Heating is also a concern when working with high ionic strength buffer solutions. As the ionic strength of the solution is increased, the current will likewise increase. Utilizing the following equation for estimating Joule heating (Q)

$$Q = E^2 \Lambda c, \tag{5.13}$$

where E^2 is the voltage gradient, Λ is molar conductivity of the electrolyte, and c is the electrolyte molar concentration, it can be determined that heating will increase by raising the molar conductivity of the electrolyte. The changes brought on by heating will be discussed in more detail in Section 5.4.4.

5.4.3 Organic Modifier Effects

The effect of organic modifiers such as acetonitrile (ACN), ethanol, isopropanol, methanol, and tetrahydrofuran on CEC separations is quite complicated. Organic modifiers added to an aqueous buffer can increase the solubility of neutral and nonpolar compounds, but at the same time yield some EOF effects in concert with changes in retention factors.

Addition of a modifier will alter the viscosity of the run buffer depending on its interactions with the aqueous buffer. According to Equation 5.3, there is an inverse relation between electroosmotic velocity and solution viscosity, and therefore, the electroosmotic velocity will change accordingly when a modifier is used. In addition, the dielectric constant of the buffer solution will change with increasing organic solvent concentration, which will affect the zeta potential inversely thus altering EOF.

An example of addition of organic modifiers decreasing EOF was presented by Kanitsar et al. [202] for the effect of various organic additives on the separation of three carboxylic acids. Separation buffers consisting of 15% v/v solvent to pH 6.0, 25 mM KH_2PO_4 and 0.75 mM TTAB buffer were used in conjunction with benzyl alcohol, as the neutral marker, for EOF determination. For the

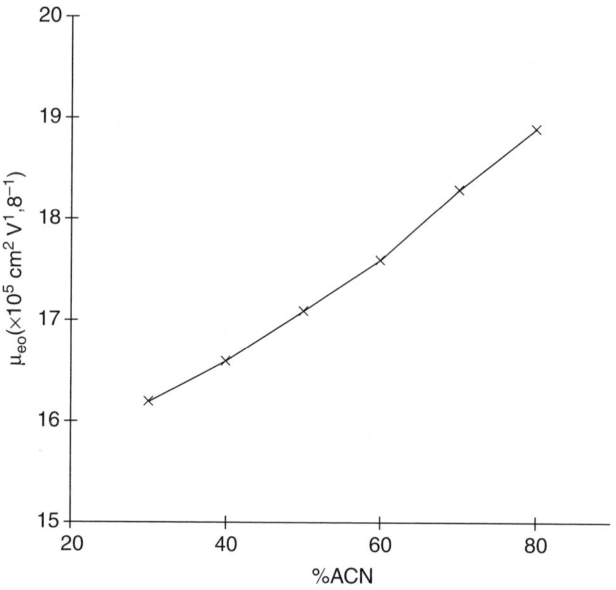

FIGURE 5.16 Effect of the ACN content on EOF mobility in CEC. Electropak phenyl column, EP-75-40-3-PH [40 cm packed (47 cm total) 75 μm i.d.]; electrolyte: Tris HCl (pH 8)–ACN mixture (ionic strength 5 mM); temperature: 25°C; applied voltage: +20 kV; UV detection: 220 nm; electrokinetic injection: 4 s (+10 kV); EOF marker: thiourea. (Reprinted from Cahours, X., et al., *J. Chromatogr. A*, 845, 203–216, 1999. With permission from Elsevier.)

solvent systems studied, the respective EOF values were determined to be 4.03×10^{-4} cm^2 V^{-1} s^{-1} for water, 2.71×10^{-4} cm^2 V^{-1} s^{-1} for ethanol, 2.56×10^{-4} cm^2 V^{-1} s^{-1} for 1-propanol, 2.29×10^{-4} cm^2 V^{-1} s^{-1} for ACN, 2.24×10^{-4} cm^2 V^{-1} s^{-1} for methanol, and 0.69×10^{-4} cm^2 V^{-1} s^{-1} for ethylene glycol. An example of increasing EOF was given by Cahours et al. [166] for addition of ACN to a 5 mM, pH 8 Tris–HCl buffer. They noticed a 17% increase in EOF velocity when the ACN percentage was increased from 30% to 80%, and a plot of their findings is shown in Figure 5.16.

As with HPLC, organic modifiers will of course affect interactions between analytes and the stationary phase. With reverse-phase materials, addition of an organic modifier to a buffer will lower the partition coefficient of an analyte. At an appropriate concentration, it is possible to optimize the partition function for a given analyte and tune the retention factor (k'). Zhang et al. [91] showed that by increasing the percentage of ACN in their mobile phase that the retention times of four proteins separated by RP CEC were not altered significantly. What they did notice, however, is that as the ACN percentage was increased a reduction in retention attributed to chromatographic interactions occurred, and at a sufficiently high ACN percentage the separation was largely electrophoretic in nature. Electrochromatograms for this separation showing ACN percentages in the range of 20%–50% are given in Figure 5.17.

The organic modifier/buffer species and organic modifier/surface interactions are not fully understood, and as such there is no easy way to summarize the effect of organic modifiers on EOF. These effects are specific to a given stationary phase/buffer species/buffer concentration combination, and therefore experimentation and optimization are required.

5.4.4 Influence of Temperature

Joule heating leads to zone spreading and both qualitative and quantitative irreproducibility; to attenuate Joule heating the column is housed in a temperature-controlled environment. Unregulated

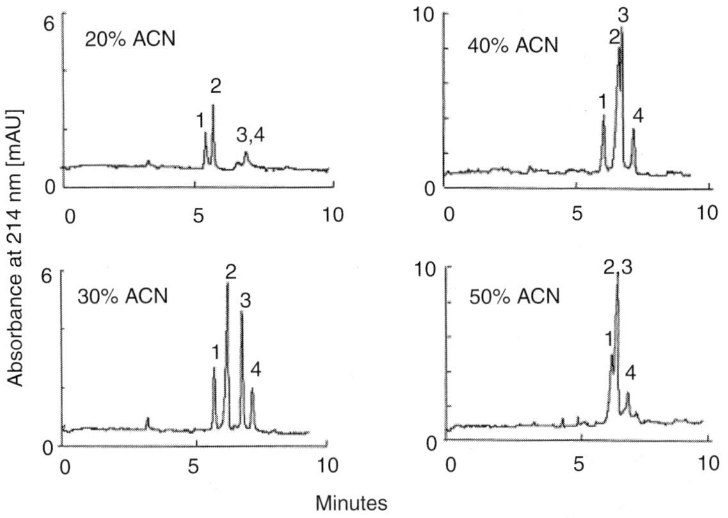

FIGURE 5.17 Effect of ACN concentration in the eluent on the separation of four proteins. Column, 39 cm (effective length 29 cm) × 50 μm i.d., fused-silica capillary with porous methacrylic monolith having tertiary amino functions; mobile phase, ACN (%, v/v) in 60 mM aqueous sodium phosphate, pH 2.5; applied voltage, −25 kV; detection, 214 nm; sample: (1) ribonuclease A, (2) insulin, (3) α-lactalbumin, and (4) myoglobin. (Reprinted from Zhang, S., et al., *J. Chromatogr. A*, 887, 465–477, 2000. With permission from Elsevier.)

Joule heating leads to a cycle of increased resistivity and increased heating. In the worst case scenario, the buffer will heat until ultimately out-gassing occurs. Most of the CE instrumentation available today is equipped with heating or cooling capability, but when dealing with a homebuilt system temperature control should be considered.

EOF velocity will also be changed by fluctuations in temperature. With an increase in temperature the EOF velocity will increase, owing to a decrease in the viscosity of the buffer system in accordance with Equation 5.3. Zhang et al. [203] showed that an increase in temperature from 25°C to 55°C increased the velocity of the EOF as shown in Figure 5.18. Also, the kinetics of partitioning of analytes between the stationary and mobile phases will be more rapid which can aid in minimizing band broadening effects.

5.4.5 Voltage

As with any other separation method, the ultimate goal of developing a method is to have the shortest analysis time that affords the desired resolution between sample components. In CEC, reduced analysis times can be gained by working at higher voltages because of the linear relation between applied voltage and both electrophoretic and electroosmotic velocity. One concern about working with elevated voltages is that there is increased possibility of excessive Joule heating. Li and Remcho [204] demonstrated the linear relationship between field strength and the EOF velocity using acetone as an unretained neutral marker in a packed Nucleosil C18 column (shown in Figure 5.19) for 3, 5, and 7 μm particles.

5.4.6 Summary

There are numerous variables that can be altered in developing a CEC method, and the vast majority of them are interrelated. This contributed greatly to the decline in interest in CEC in the late 1990s. For example a change in buffer pH may change EOF, the charge state of an analyte, and

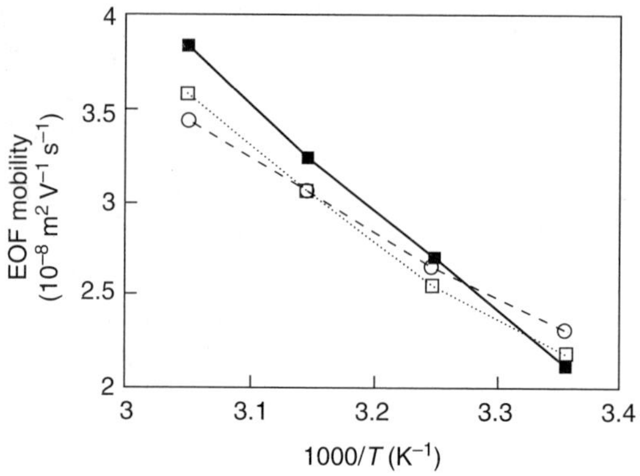

FIGURE 5.18 Plots of the EOF mobility, measured with DMSO as the unretained neutral marker, against the reciprocal absolute temperature. Mobile phase (■) 10% ACN, (□) 20% ACN, (○) 30% ACN in 60 mM sodium phosphate buffer, pH 2.5; temperature, 25°C; applied voltage, −30 kV; detection, 214 nm. Column, 40 cm (effective length 30 cm) × 75 μm, fused silica with styrenic monolith having quaternary ammonium functions. (Reprinted from Zhang, S., et al., *J. Chromatogr. A*, 914, 189–200, 2001. With permission from Elsevier.)

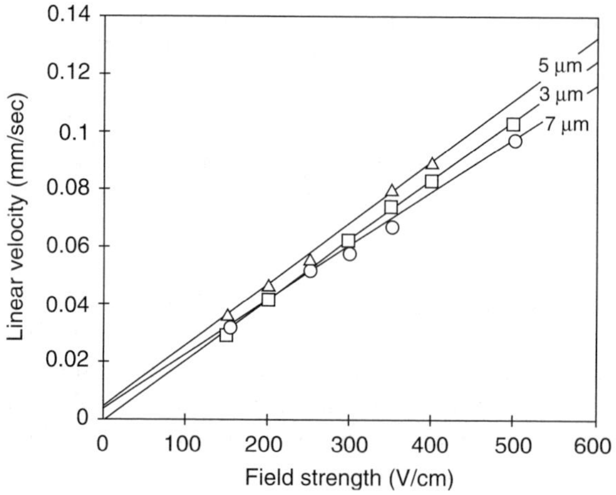

FIGURE 5.19 Linear velocity as a function of field strength for 3-, 5-, and 7-μm-diameter Nucleosil C18 particles. (Reprinted from Li, D.M. and Remcho, V.T., *J. Microcol. Sep.*, 9, 389, 1997. With permission.)

its partitioning behavior. The ionic strength of a mobile phase can also change the EOF, and current will increase as salts are added to the system. If the current is sufficiently high (owing to the mobile phase ionic strength) heating effects may be encountered, which can lower the partition coefficients of analytes and the mobile phase viscosity, and may ultimately lead to bubble formation. Therefore, it is important to consider how a change in one component of the method can alter other components.

5.5 APPLICATIONS

Capillary electrochromatography can be an effective separation tool for a wide range of sample types, as evidenced in Table 5.1. In this section, examples of separations utilizing CEC will be highlighted.

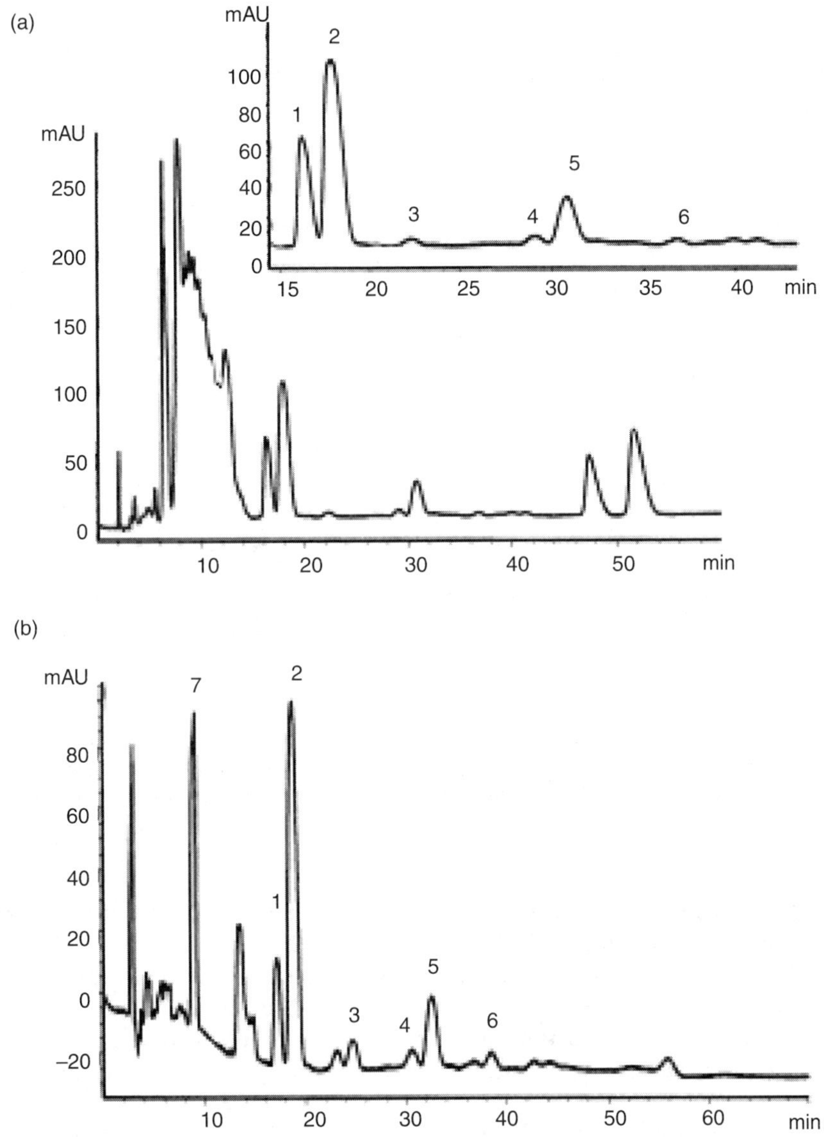

FIGURE 5.20 CEC of (a) pyrethrin dip and (b) flea and tick mist. Conditions: MeCN-25 mM Tris–THF (55:35:10, pH 9). (1) Cinerin II, (2) pyrethrin II, (3) jasmolin II, (4) cinerin I, (5) pyrethrin I, (6) jasmolin I, and (7) piperonyl butoxide. (Reprinted from Henry, C.W., et al., *J. Chromatogr. A*, 905, 319–327, 2001. With permission from Elsevier.)

A CEC separation of insecticidal pyrethrin esters was accomplished by Henry et al. [63]. Here it was shown that it is possible to identify pyrethrin esters (cinerin I, pyrethrin I, jasmolin I, cinerin II, pyrethrin II, and jasmolin II) in a commercially available pyrethrin dip and flea and tick mist (Figure 5.20). The column inner and outer diameters were 100 and 350 μm, respectively, and the total and effective lengths were 33 and 25 cm, respectively. The column was packed with 3 μm particle size Hypersil C18. The mobile phase was composed of 55:35:10, pH 9, ACN-25 mM Tris-tetrahydrofuran. Other parameters were an applied voltage of 30 kV, column temperature of 25°C, and UV detection at 254 nm.

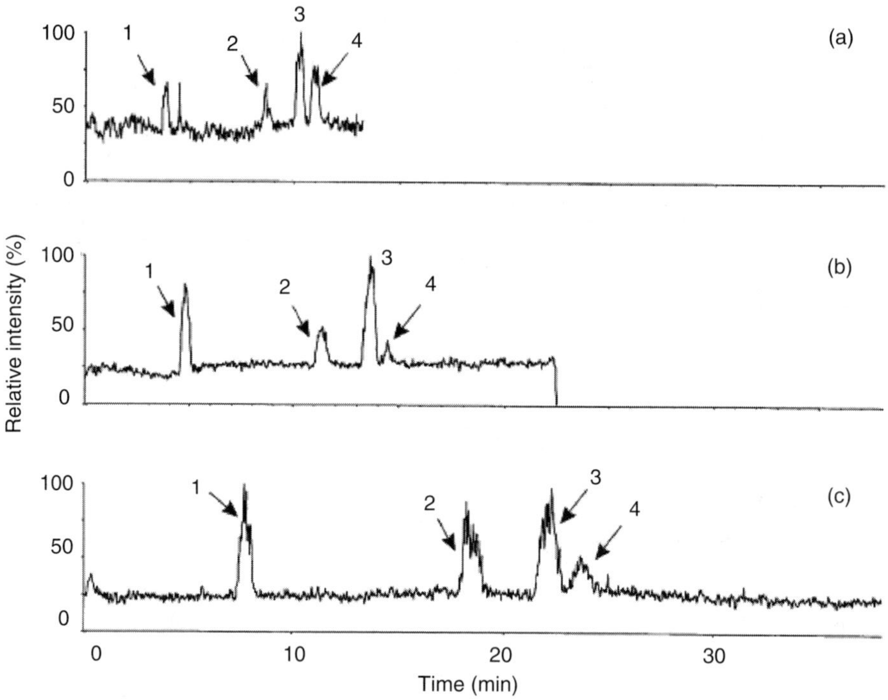

FIGURE 5.21 Separation of estrogenic compounds: (1) estriol, (2) estradiol, (3) equiline, and (4) estrone. (a) With pCEC-CIS-MS: 80 bar and +15 kV applied, (b) pCEC-CIS-MS: 80 bar and +5 kV applied, and (c) CHPLC-CIS-MS: 80 bar applied. (Reprinted from Rentel, C., et al., *Electrophoresis*, 20, 2329, 1999. With permission.)

An example of a separation of estrogens (estriol, estradiol, equiline, and estrone) with pressurized CEC (pCEC) was shown by Rentel et al. [132]. Capillaries with inner diameters of 100 μm and outer diameters of 360 μm were used. The GROM-SIL ODS-0 AB packing material used had a particle size of 3 μm. Detection of the estrogenic analytes was accomplished with coordination ion spray mass spectrometry (CIS-MS). Other experimental parameters were: pH 9, 4 mM ammonium acetate in 5:95 water–ACN mobile phase, applied pressure of 80 bar, and applied voltage of 15 (Figure 5.21a) and 5 kV (Figure 5.21b).

Tricyclic antidepressants were analyzed by Vallano et al. [179] via an MIP-CEC separation. In this separation, doxepin, imipramine, amitriptyline, trimipramine, and clomipramine were separated from the template molecule nortriptyline (Figure 5.22). The capillary inner diameter and total length were 100 μm and 33 cm, respectively, and the total length of the monolithic bed was 22.5 cm. The eluent was 92:2 ACN: 10 mM sodium acetate pH 3.0 to which 0.02% trifluoracetic acid and 0.015% triethylamine (TEA) (v/v) was added. A constant applied voltage of 30 kV was utilized.

CEC has also been used to successfully separate nucleotides. For example, Helboe and Hansen [115] separated cytidine, uridine, inosine, guanosine, thymidine, adenosine, and thiourea, with a CEC Hypersil C18 column. Column specifics are as follows: particle size of 3 μm, inner diameter of 100 μm, and bed length of 25 cm. The mobile phase was 98% 5 mM acetic acid, 3 mM TEA, pH 5: 2% ACN. Figure 5.23, shows the optimization of temperature and voltage, 25°C and 20 kV for 5.4A, 25°C and 25 kV for 5.4B, and 20°C and 25 kV for 5.4C. Nucleotides were also examined by Cahours et al. [114]. Figure 5.24 illustrates the effect ionic strength has on the migration order of adenosine (A), cytidine (C), guanosine (G), uridine (U), and thymidine (T), and thiourea, an EOF marker. For this separation, the column dimensions were 27 cm total length × 75 μm inner diameter.

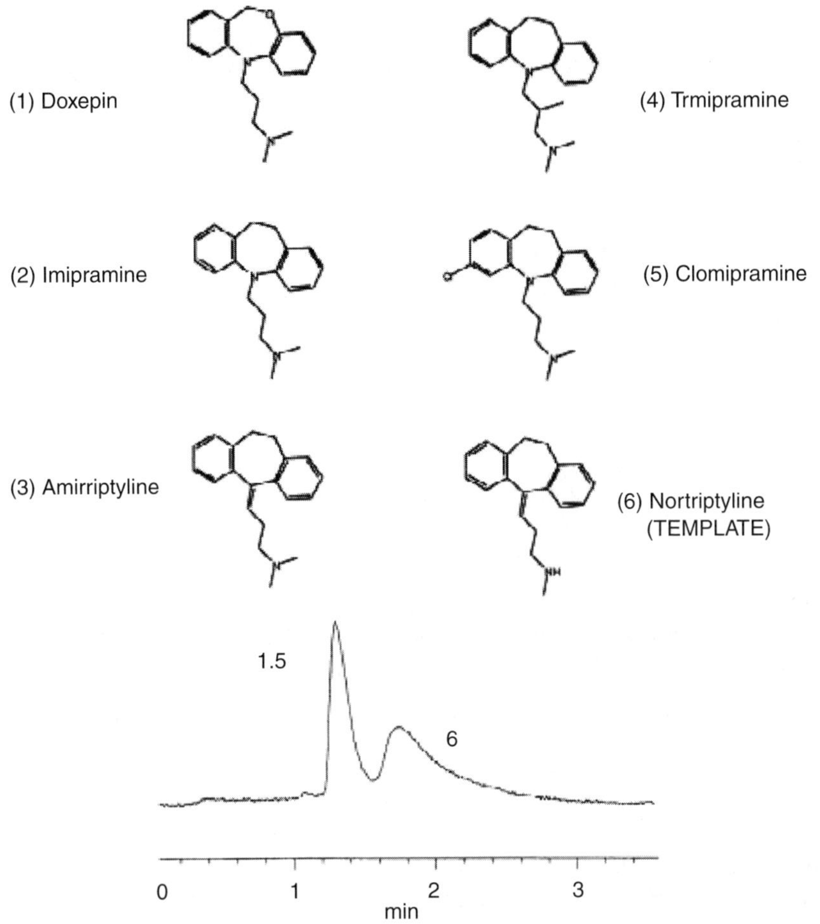

FIGURE 5.22 MIP-CEC separation of a simulated combinatorial library consisting of several tricyclic antidepressants. Conditions: capillary i.d. 100 μm; Ltot: 33 cm; Lbed: 22.5 cm; eluent:ACN: 10 mM Na acetate pH 3.0 (98:2) with 0.02% trifluoracetic acid and 0.015% TEA (v/v); voltage +30 kV constant; injection: +2 kV, 2 s; column temperature: 50°C. (Reprinted from Vallano, P.T. and Remcho, V.T., *J. Chromatogr. A*, 887, 125-135, 2000, with permission from Elsevier.)

The effective length of the column was 20 cm and it was packed with 3 μm Phenyl Hypersil. The separation temperature and voltage were 20°C and 20 kV, respectively, and the detection wavelength was 254 nm. The electrolyte was composed of pH 5 acetic acid/ammonia-ACN (95/5), and the ionic strengths used were 10 and 5 mM, respectively.

Pharmaceuticals have also been separated by CEC. Taylor and Teale [148] conducted gradient CEC separations of drug mixtures. In their study, a mixture of corticosteroids consisting of triamcinolone, hydrocortisone, prednisolone, cortisone, methylprednisolone, betamethasone, dexamethasone, adrenosterone, fluocortolone, and tramcinolone acetonide was examined. This mixture was separated using a 3-μm Hypersil ODS. The packed length, length to detection window, and total length of this column were 30, 30.1, and 42 cm, respectively. The applied voltage was 30 kV, and the detection wavelength used was 240 nm. Gradient elution was used as follows: initially, 5 mM ammonium acetate in ACN–water (17:83) for the first 3 min, then for 15 min; the ACN concentration was ramped to 38% and kept constant for the remainder of the analysis. The flow rate for this analysis was 10 μL min^{-1} for the first 3 min, and was then elevated to 100 μL min^{-1}. This separation is shown in Figure 5.25.

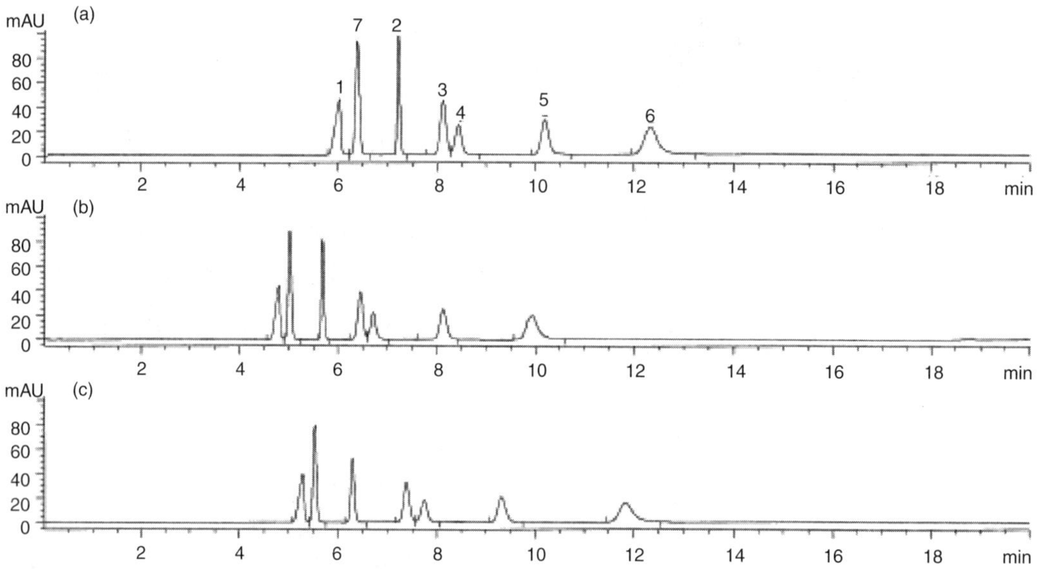

FIGURE 5.23 Final optimization of temperature and voltage. Order of peaks: (1) cytidine, (7) thiourea, (2) uridine, (3) inosine, (4) guanosine, (5) thymidine, and (6) adenosine. Conditions: injection, 10 kV for 3 s; 25 cm × 100 μm column (CEC Hypersil C_{18}, 3 μm); mobile phase (5 mM acetic acid, 3 mM TEA, pH 5)–ACN (92:8, v/v). (a) 25°C and 20 kV; (b) 25°C and 25 kV; and (c) 20°C and 25 kV. (Reprinted from Helboe, T. and Hansen, S.H., *J. Chromatogr. A*, 836, *J. Chromatogr. A*, 315–324, 1999. With permission from Elsevier.)

An example of a CEC analysis of an illicit drug was shown by Lurie and collaborators [134]. In this study, a standard mixture of seven cannabinoids was separated, and this test mixture was then compared with concentrated hashish and marijuana extracts (Figure 5.26). The cannabinoid standards used were cannabigerol (CBG), cannabidiol (CBD), cannabinol (CBN), Δ-9-tetrahydrocannabinol (d9-THC), Δ-8-tetrahydrocannabinol (d8-THC), cannabichromene (CDB), and Δ-9-tetrahydrocannabinolic acid (d9-THCA-A), with dimethyl sulfoxide (DMSO) as the neutral marker. A Hypersil C18 column (3 μm d_p) was used. The column inner diameter was 100 μm and had a total length of 49 cm, and the effective length was 40 cm. UV detection was conducted at 210 nm. The run buffer was 75% ACN and 25% 25 mM phosphate buffer pH 2.57. The applied voltage was 30 kV, and the column temperature was kept at 20°C.

Carbohydrate separations too have been accomplished using CEC. Such an example was presented by Zhao and Johnson [106], where they studied sucralose and related carbohydrates, as shown in Figure 5.27. The 100 μm inner diameter column was packed with 3 μm ODS particles, had a total bed length of 25 cm, and a total length of 33 cm. During each run, an external pressure of six bars was applied to the inlet and outlet vials to eliminate bubble formation. The compounds were detected at 195 nm for this experiment. Different mobile phase compositions were studied, and for Figure 5.28 the eluent was: 35:65 ACN/4 mM borate (Figure 5.28a), 30:70 ACN/4 mM borate, and 25:75 ACN/4 mM borate. For the three runs shown in Figure 5.28, the applied voltage was 15 kV. Note that sucralose was successfully separated from related compounds A and B, a system peak, and ethyl acetate (EA) traces from compound A.

Clearly, CEC is a flexible and efficient separation technique that, with careful attention in method development, is capable of addressing analytical needs across a broad spectrum of sample types.

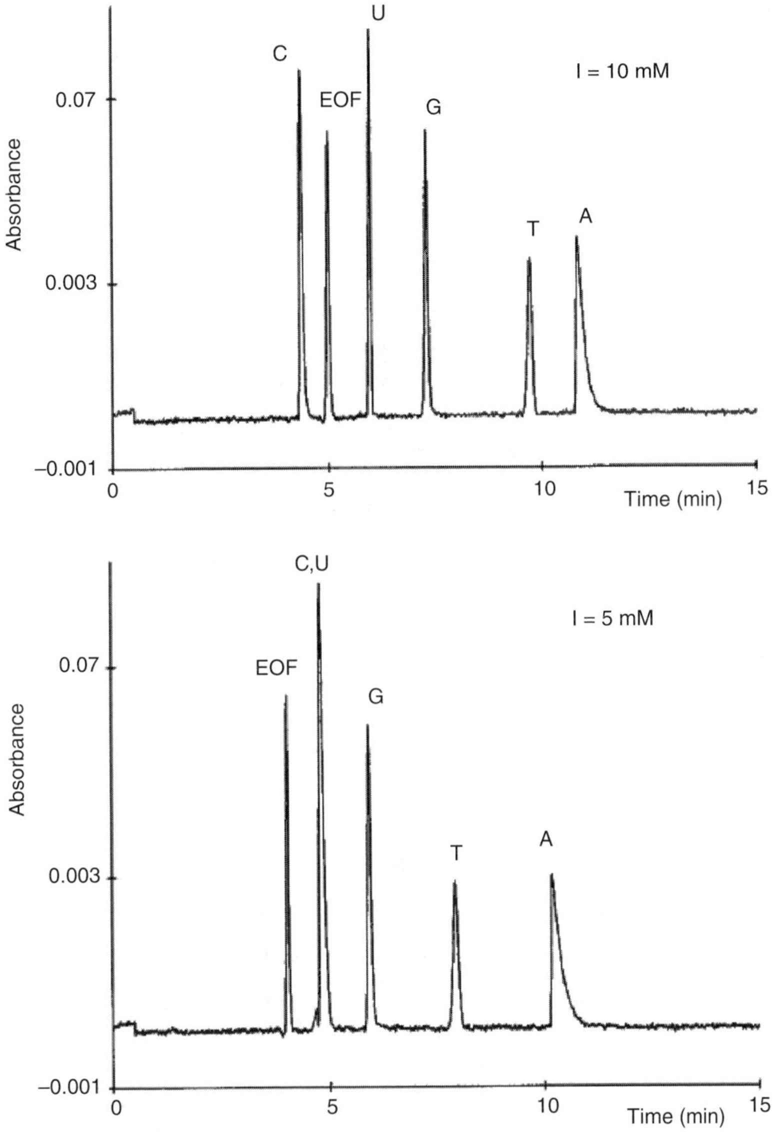

FIGURE 5.24 Effect of the ionic strength on the migration order of nucleosides in CEC. Column: 27 cm × 75 μm i.d., bed length 20 cm, packing: 3 μm Phenyl Hypersil; electrolyte: acetic acid/ammonia (pH 5)/ACN (95/5); temperature: 20°C; applied voltage: +20 kV; UV detection: 254 nm; electrokinetic injection: 4 s (+10 kV); nucleoside concentration: 50 ppm. (Reprinted from Cahours, X., et al., *J. High Resol. Chromatogr.*, 23, 138, 2000. With permission.)

5.6 CONCLUSIONS

What does the future hold for CEC? It is impossible to predict the path the technique will follow. It is of course unlikely that it will ever enjoy the great breadth of application of HPLC and GC. That said, it is entirely possible that the technique will find one or more niches in which it will thrive. Likely areas of application include: field-portable, low-cost, chip-based analytical instruments; disposable components in medical diagnostic/assay devices; and consumables for clinical diagnostic tools.

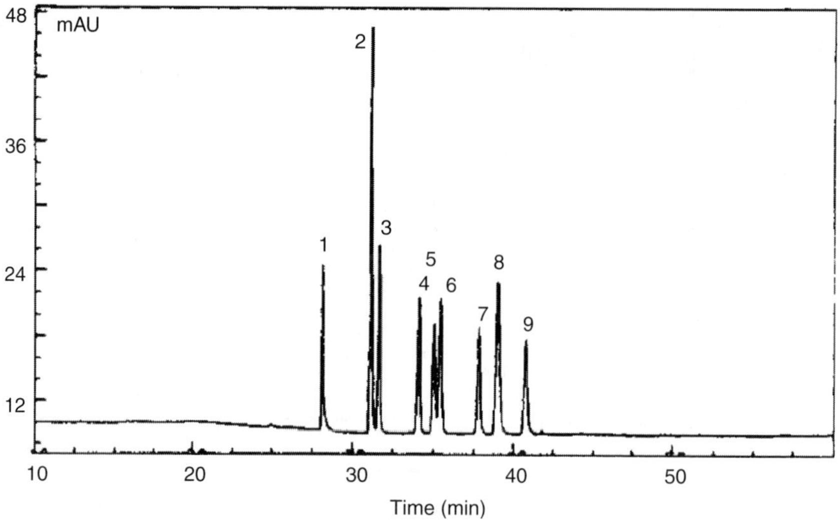

FIGURE 5.25 CEC-UV chromatogram (240 nm) of a mixture of 10 corticosteroids (100 μg mL^{-1}) using a linear gradient elution program. Voltage = 30 kV, HPLC injection volume = 10 μL, flowrate = 10 μL min^{-1} for 3 min then increased to 100 μL min^{-1}. Gradient program = initial: ammonium acetate, 5 mM, in ACN–water (17:83), held for 3 min then ramped to 38% acetronitrile at 15 min and maintained to end of run. Column = Hypersil ODS, 3 μm, 42 cm total length, 30 cm packed length, 30.1 cm to window, (1) triamcinolone, (2) hydrocortisone and prednisolone coeluting, (3) cortisone, (4) methylprednisolone, (5) betamethasone, (6) dexamethasone, (7) adrenosterone, (8) fluocortolone, and (9) triamcinotone acetonide. (Reprinted from Taylor, M.R. and Teale, P., *J. Chromatogr. A*, 768, 89–95, 1997. With permission from Elsevier.)

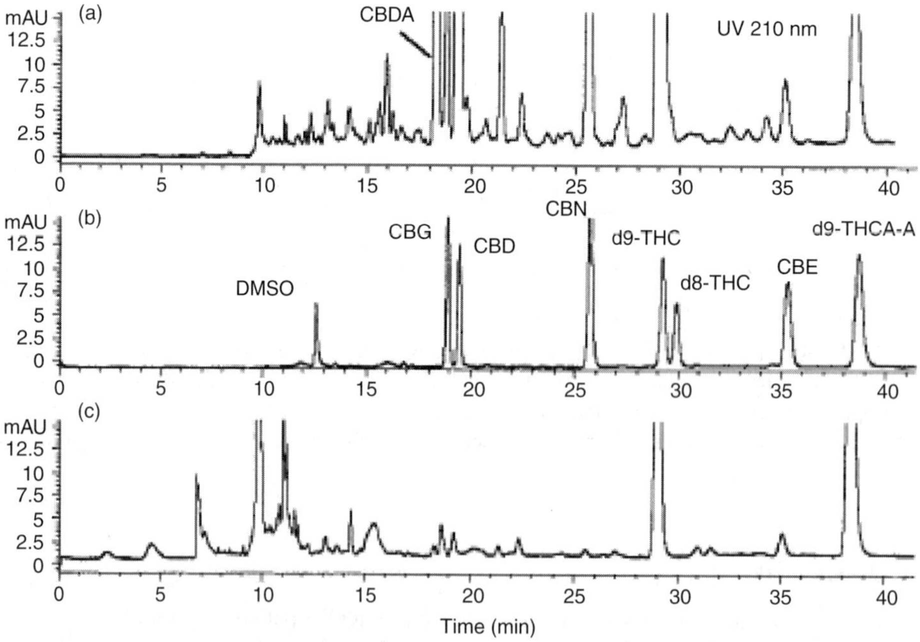

FIGURE 5.26 CEC of (a) concentrated hashish extract, (b) standard mixture of cannabinoids, and (c) concentrated marijuana extract. Conditions: 75% ACN/25 mM phosphate buffer pH 2.57 with voltage 30 kV and temperature 20°C. A Hypersil C18, 3-μm [100 μm × 49 cm (40-cm length to detector)] column is used. Electrokinetic injections of 8.0 s at 5.0 kV are used. (Reprinted from Lurie, I.S., et al., *Anal. Chem.*, 70, 3255, 1998. With permission.)

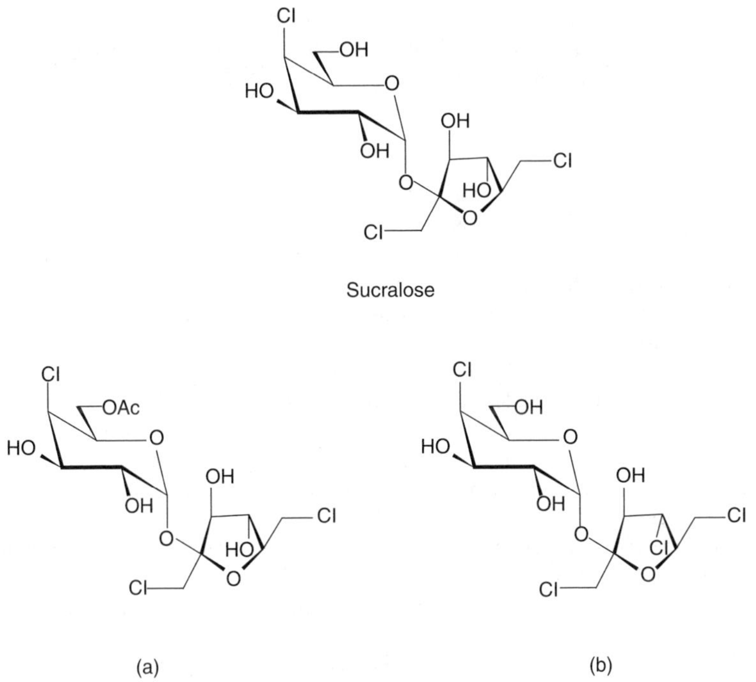

FIGURE 5.27 Structures of sucralose and related carbohydrate compounds (a) and (b). (Copyright 2000 from CEC: analysis of sucralose and related carbohydrate compounds by Zhao, R.R. and Johnson, B.P. Reproduced by permission of Taylor & Francis Group, LLC., http://www.taylorandfrancis.com)

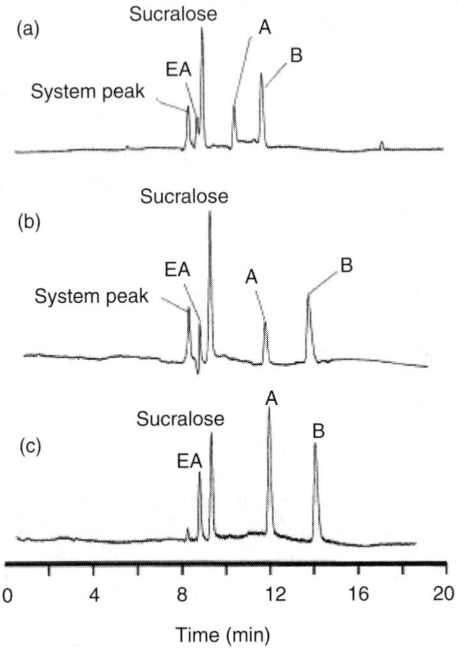

FIGURE 5.28 Electrochromatograms of sucralose and related compounds in different mobile phases at 15 kV run: (a) 35/65 ACN/4 mM borate, (b) 30/70 ACN/4 mM borate, and (c) 25/75 ACN/4 mM borate. (Copyright 2000 from CEC: analysis of sucralose and related carbohydrate compounds by Zhao, R.R. and Johnson, B.P. Reproduced by permission of Taylor & Francis Group, LLC., http://www.taylorandfrancis.com)

CEC is a technique that must compete in an increasingly crowded field of technologies, yet it has a sufficient quantity of positive attributes—detailed in this chapter—to be worthy of consideration. It will be intriguing to see where this road leads.

REFERENCES

1. Pretorius, V., Hopkins, B.J., and Schieke, J.D., Electro-osmosis: A new concept for high-speed liquid chromatography, *J. Chromatogr. A*, 99, 23, 1974.
2. Jorgenson, J.W. and Lukacs, K.D., Zone electrophoresis in open-tubular glass capillaries: Preliminary data on performance, *J. High Res. Chrom.*, 4, 230, 1981.
3. Knox, J.H. and Grant, I.H., Thermal effects and band spreading in capillary electro-separation, *Chromatographia*, 26, 329, 1988.
4. Schweitz, L., Andersson, L.I., and Nilsson, S., Capillary electrochromatography with predetermined selectivity obtained through molecular imprinting, *Anal. Chem.*, 69, 1179, 1997.
5. Landers, J.P, *Handbook of Capillary Electrophoresis*, 2nd ed., CRC Press, FL, 1997, Chapter 5.
6. Wen, E., Asiaie, R., and Horváth, C., Dynamics of capillary electrochromatography: II. Comparison of column efficiency parameters in microscale high-performance liquid chromatography and capillary electrochromatography, *J. Chromatogr. A*, 855, 349, 1999.
7. Yan, C., Schaufelberger, D., and Erni, F., Electrochromatography and micro high-performance liquid chromatography with 320 μm I.D. packed columns, *J. Chromatogr. A*, 670, 15, 1994.
8. Bartle, K.D. and Myers, P., Theory of capillary electrochromatography, *J. Chromatogr. A*, 916, 3, 2001.
9. Colon, L.A., et al., Recent progress in capillary electrochromatography, *Electrophoresis*, 21, 3965, 2000.
10. Vanhoenacker, G., et al., Recent applications of capillary electrochromatography, *Electrophoresis*, 22, 4064, 2001.
11. Cikalo, M.G., et al., Capillary electrochromatography. Tutorial Review, *Analyst*, 123, 87R, 1998.
12. Peters, E.C., et al., Molded rigid polymer monoliths as separation media for capillary electrochromatography, *Anal. Chem.*, 69, 3646, 1997.
13. Palm, A. and Novotny, M.V., Macroporous polyacrylamide/poly(ethylene glycol) matrixes as stationary phases in capillary electrochromatography, *Anal. Chem.*, 69, 4499, 1997.
14. Dulay, M.T., Kulkarni, R.P., and Zare, R.N., Preparation and characterization of monolithic porous capillary columns loaded with chromatographic particles, *Anal. Chem.*, 70, 5103, 1998.
15. Gusev, I., Huang, X., and Horváth, C., Capillary columns with *in situ* formed porous monolithic packing for micro high-performance liquid chromatography and capillary electrochromatography, *J. Chromatogr. A*, 855, 273, 1999.
16. Chirica, G. and Remcho, V.T., Silicate entrapped columns—New columns designed for capillary electrochromatography, *Electrophoresis*, 20, 50, 1999.
17. Tang, Q.L., Xin, B.M., and Lee, M.L., Monolithic columns containing sol–gel bonded octadecylsilica for capillary electrochromatography, *J. Chromatogr. A*, 837, 35, 1999.
18. Tanaka, N. et al., Monolithic silica columns for HPLC, micro-HPLC, and CEC, *J. High Resol. Chromatogr.*, 23, 111, 2000.
19. Ratnayake, C.K., Oh, C.S., and Henry, M.P., Particle loaded monolithic sol–gel columns for capillary electrochromatography: A new dimension for high performance liquid chromatography, *J. High Resol. Chromatogr.*, 23, 81, 2000.
20. Tang, Q. and Lee, M.L., Continuous-bed columns containing sol-gel bonded packing materials for capillary electrochromatography, *J. High Resol. Chromatogr.*, 23, 73, 2000.
21. Peters, E.C., et al., Molded rigid polymer monoliths as separation media for capillary electrochromatography. 1. Fine control of porous properties and surface chemistry, *Anal. Chem.*, 70, 2288, 1998.
22. Peters, E.C., et al., Molded rigid polymer monoliths as separation media for capillary electrochromatography. 2. Effect of chromatographic conditions on the separation, *Anal. Chem.*, 70, 2296, 1998.
23. Ishizuka, N., et al., Performance of a monolithic silica column in a capillary under pressure-driven and electrodriven conditions, *Anal. Chem.*, 72, 1275, 2000.

24. Fujimoto, C., Preparation of fritless packed silica columns for capillary electrochromatography, *J. High Resol. Chromatogr.*, 23, 89, 2000.
25. Asiaie, R., et al., Sintered octadecylsilica as monolithic column packing in capillary electrochromatography and micro high-performance liquid chromatography, *J. Chromatogr. A*, 806, 251, 1998.
26. Dulay, M.T., et al., Preparation and characterization of monolithic porous capillary columns loaded with chromatographic particles, *Anal. Chem.*, 70, 5103, 1998.
27. Tang, Q., Wu, N., and Lee, M.L., Continuous bed columns containing sol–gel bonded large-pore octadecylsilica for capillary electrochromatography, *J. Microcol. Sep.*, 11, 550, 1999.
28. Schweitz, L., Andersson, L.I., and Nilsson, S., Molecular imprint-based stationary phases for capillary electrochromatography, *J. Chromatogr. A*, 817, 5, 1998.
29. Xiong, B. et al, Capillary electrochromatography with monolithic poly(styrene-co-divinylbenzene-co-methacrylic acid) as the stationary phase, *J. High Resol. Chromatogr.*, 23, 67, 2000.
30. Yu, C., Svec, F., and Frechet, J.M., Towards stationary phases for chromatography on a microchip: Molded porous polymer monoliths prepared in capillaries by photoinitiated *in situ* polymerization as separation media for electrochromatography, *Electrophoresis*, 21, 120, 2000.
31. Remcho, V.T. and Tan, Z.J., MIPs as chromatographic stationary phases for molecular recognition. *Anal. Chem.*, 71, 248A, 1999.
32. Liao, J.L., et al. Preparation of continuous beds derivatized with one-step alkyl and sulfonate groups for capillary electrochromatography, *Anal. Chem.*, 68, 3468, 1996.
33. Schure, M.R., et al., High-performance capillary gel electrochromatography with replaceable media, *Anal. Chem.*, 70, 4985, 1998.
34. Fujimoto, C., Charged polyacrylamide gels for capillary electrochromatographic separations of uncharged, low molecular weight compounds, *Anal. Chem.*, 67, 2050, 1995.
35. Ericson, C., et al., Preparation of continuous beds for electrochromatography and reversed-phase liquid chromatography of low-molecular-mass compounds, *J. Chromatogr. A*, 767, 33, 1997.
36. Fujimoto, C., Sakurai, M., and Muranaka, Y., PEEK columns for open-tubular liquid chromatography with electroosmotic flow, *J. Microcol. Sep.*, 11, 693, 1999.
37. Ericson, C. and Hjerten, S., Reversed-phase electrochromatography of proteins on modified continuous beds using normal-flow and counterflow gradients. Theoretical and practical considerations, *Anal. Chem.*, 71, 1621, 1999.
38. Fujimoto, C., Fujise, Y., and Matsuzawa, E., Fritless packed columns for capillary electrochromatography: Separation of uncharged compounds on hydrophobic hydrogels, *Anal. Chem.*, 68, 2753, 1996.
39. Svec, F., Tennikova, T.B., and Deyl, Z., Monolithic materials: Preparation, properties, and applications, *Anal. Bio. Chem.*, 379, 8, 2004.
40. Svec, F., Preparation and HPLC applications of rigid macroporous organic polymer monoliths, *J. Sep. Sci.*, 27, 747, 2004.
41. Allen, D. and El Rassi, Z., Silica-based monoliths for capillary electrochromatography: Methods of fabrication and their applications in analytical separations, *Electrophoresis*, 24, 3962, 2003.
42. Luo, Q.L. and Andrade, J.D., Electrokinetic flow through packed capillary columns, *J. Microcol.*, 11, 682, 1999.
43. Svec, F., Monolithic stationary phases for capillary electrochromatography based on synthetic polymers: Designs and applications, *J. High Resol. Chromatogr.*, 23, 3, 2000.
44. Svec, F., et al., Design of the monolithic polymers used in capillary electrochromatography columns, *J. Chromatogr. A*, 887, 3, 2000.
45. Nishi, H. and Terabe, S., Optical resolution of drugs by capillary electrophoretic techniques, *J. Chromatogr. A*, 694, 276, 1994.
46. Sellergen, B., Imprinted chiral stationary phases in high-performance liquid chromatography, *J. Chromatogr. A*, 906, 227, 2001.
47. Nilsson, K., et al., Imprinted polymers as antibody mimetics and new affinity gels for selective separations in capillary electrophoresis, *J. Chromatogr. A*, 680, 57, 1994.
48. Arshady, R. and Mosbach, K., Synthesis of substrate-selective polymers by host–guest polymerization, *Macromol. Chem.*, 182, 687, 1981.

49. Wulff, G. and Sarhan, A., Über die Anwendung von enzymanalog gebauten Polymeren zur Racemattrennung, *Angew. Chem.*, 84, 364, 1972.
50. Sellergren, B., *Molecularly Imprinted Polymers—Man Made Mimics of Antibodies and Their Applications in Analytical Chemistry*, Elsevier, Amsterdam, 2001, pp. 113–184.
51. Tang, Q., Xin, B., and Lee, M.L., Monolithic columns containing sol–gel bonded octadecylsilica for capillary electrochromatography, *J. Chromatogr. A*, 837, 35, 1999.
52. Yan, C., et al., Capillary electrochromatography: Analysis of polycyclic aromatic hydrocarbons, *Anal. Chem.*, 67, 2026, 1995.
53. Dadoo, R. et al., Advances in capillary electrochromatography: Rapid and high-efficiency separations of PAHs, *Anal. Chem.*, 70, 4787, 1998.
54. Ngola, S.M., et al., Conduct-as-cast polymer monoliths as separation media for capillary electrochromatography, *Anal. Chem.*, 73, 849, 2001.
55. Fintschenko, Y., et al., Chip electrochromatography of polycyclic aromatic hydrocarbons on an acrylate-based UV-initiated porous polymer monolith, *Fresenius J. Anal. Chem.*, 371, 174, 2001.
56. Chirica, G.S. and Remcho, V.T., A simple procedure for the preparation of fritless columns by entrapping conventional high performance liquid chromatography sorbents, *Electrophoresis*, 21, 3093, 2000.
57. Chirica, G.S. and Remcho, V.T., Fritless capillary columns for HPLC and CEC prepared by immobilizing the stationary phase in an organic polymer matrix, *Anal. Chem.*, 72, 3605, 2000.
58. Vallano, P.T. and Remcho, V. T., Modeling interparticle and intraparticle (perfusive) electroosmotic flow in capillary electrochromatography, *Anal. Chem.*, 72, 4255, 2000.
59. Dittman, M.M. and Rozing, R., Capillary electrochromatography: Investigation of the influence of mobile phase and stationary phase properties on electroosmotic velocity, retention, and selectivity, *J. Microcol. Sep.*, 9, 399, 1997.
60. Mayer, M., et al., Fritless capillary electrochromatography, *Electrophoresis*, 20, 43, 1999.
61. Engelhardt, H. and Hafner, F.T., Porous and non-porous stationary phases for capillary electrochromatography under conditions of reversed phase, *Chromatography*, 52, 769–776, 2000.
62. Wu, X., et al., Rapid separation and determination of carbamate insecticides using isocratic elution pressurized capillary electrochromatography, *Electrophoresis*, 27, 768, 2006.
63. Henry, C.W., McCarroll, M.E., and Warner, I.M., Separation of the insecticidal pyrethrin esters by capillary electrochromatography, *J. Chromatogr. A*, 905, 319, 2001.
64. Tegeler, T. and El Rassi, Z., Surfactant-mediated capillary electrochromatography with octadecylsilica- packed capillary columns for the separation of nonpolar compounds. Case of pyrethroid insecticides, *Electrophoresis*, 23, 1217, 2002.
65. Tegeler, T. and El Rassi, Z., On-column trace enrichment by sequential frontal and elution electrochromatography: II. Enhancement of sensitivity by segmented capillaries with z-cell configuration—Application to the detection of dilute samples of moderately polar and nonpolar pesticides, *J. Chromatogr. A*, 945, 267, 2002.
66. O'Mahony, T., et al., Monitoring the supercritical fluid extraction of pyrethroid pesticides using capillary electrochromatography. *Int. J. Environ. An. Ch.*, 83, 681, 2003.
67. Cooper, P.A., Jessop, K.M., and Moffatt, F., Capillary electrochromatography for pesticide analysis: Effects of environmental matrices, *Electrophoresis*, 21, 1574, 2000.
68. Fung, Y.S. and Long, Y.H., Determination of phenols in soil by supercritical fluid extraction–capillary electrochromatography, *J. Chromatogr. A*, 907, 301, 2001.
69. Araki, T., Chiba, M., Tsunoi, S., and Tanaka, M., Separation of chlorophenols by HPLC and capillary electrochromatography using β-cyclodextrin-bonded stationary phases, *Anal. Sci.*, 16, 412, 2000.
70. Cassells, N.P., et al., Microtox® testing of pentachlorophenol in soil extracts and quantification by capillary electrochromatography (CEC)—A rapid screening approach for contaminated land, *Chemosphere*, 40, 609, 2000.
71. Saeed, M., et al., Application of capillary electrochromatography (CEC) for the analysis of phenols in mainstream and sidestream tobacco smoke, *Chromatographia*, 49, 391, 1999.
72. Bailey, C.G. and Yan, C., Separation of explosives using capillary electrochromatography, *Anal. Chem.*, 70, 3275, 1998.
73. Hilmi, A. and Luong, J.H.T., In-line coupling capillary electrochromatography with amperometric detection for analysis of explosive compounds, *Electrophoresis*, 21, 1395, 2000.

74. Mueller, T., Posch, U.C., and Lindner, H.H., Separation of phenylthiohydantoin amino acids by capillary electrochromatography, *Electrophoresis*, 25, 578, 2004.
75. Huber, C., Choudhary, G., and Horvath, C., Capillary electrochromatography with gradient elution, *Anal. Chem.*, 69, 4429, 1997.
76. Alicea-Maldonado, R. and Colon, L., Capillary electrochromatography using a fluoropolymer as the chromatographic support material, *Electrophoresis*, 20, 37, 1999.
77. Kotia, R.B., Li, L., and McGown, L.B., Separation of nontarget compounds by DNA aptamers, *Anal. Chem.*, 72, 827, 2000.
78. Shediac, R., et al., Reversed-phase electrochromatography of amino acids and peptides using porous polymer monoliths, *J. Chromatogr. A*, 925, 251, 2001.
79. Charles, J.A.M. and McGown, L.B., Separation of Trp–Arg and Arg–Trp using G-quartet-forming DNA oligonucleotides in open-tubular capillary electrochromatography, *Electrophoresis*, 23, 1599, 2002.
80. Wu, R., et al., Separation of peptides on mixed mode of reversed-phase and ion-exchange capillary electrochromatography with a monolithic column, *Electrophoresis*, 23, 1239, 2002.
81. Charvatova, J., Kral, V. and Deyl, Z., Capillary electrochromatographic separation of aromatic amino acids possessing peptides using porphyrin derivatives as the inner wall modifiers, *J. Chromatogr. B*, 770, 155, 2002.
82. Wu, R., et al., Capillary electrochromatography for separation of peptides driven with electrophoretic mobility on monolithic column, *Anal. Chem.*, 73, 4918, 2001.
83. Steiner, F. and Scherer, B., Separation of small peptides by electrochromatography on silica-based reversed phases and hydrophobic anion exchange phases, *Electrophoresis*, 26, 1996, 2005.
84. Svec, F., et al., Monolithic stationary phases for capillary electrochromatography based on synthetic polymers: Designs and applications, *J. High Resol. Chromatogr.*, 23, 3, 2000.
85. Walhagen, K., Unger K.K., and Hearn, M.T.W., Capillary electrochromatography analysis of hormonal cyclic and linear peptides, *Anal. Chem.*, 73, 4924, 2001.
86. Gucek, M., et al., Capillary electrochromatography/nanoelectrospray mass spectrometry for attomole characterization of peptides, *Rapid Commun. Mass Spectrom.*, 14, 1448, 2000.
87. Ludtke, S., Adam, T., and Unger, K.K., Application of 0.5-μm porous silanized silica beads in electrochromatography, *J. Chromatogr. A*, 786, 229, 1997.
88. Adam, T. and Unger, K.K., Comparative study of capillary electroendosmotic chromatography and electrically assisted gradient nano-liquid chromatography for the separation of peptides, *J. Chromatogr. A*, 894, 241, 2000.
89. Ye, M.L., et al., Separation of peptides by strong cation-exchange capillary electrochromatography, *J. Chromatogr. A*, 869, 385, 2000.
90. Ye, M.L., et al., Capillary electrochromatography with a silica column with a dynamically modified cationic surfactant, *J. Chromatogr. A*, 855, 137, 1999.
91. Zhang, S., et al., Capillary electrochromatography of proteins and peptides with a cationic acrylic monolith, *J. Chromatogr. A*, 887, 465, 2000.
92. Pesek, J.J., et al., Synthesis and characterization of alkyl bonded phases from a silica hydride via hydrosilation with free radical initiation, *J. Chromatogr. A*, 786, 219, 1997.
93. Wu, J.T., et al., Protein digest analysis by pressurized capillary electrochromatography using an ion trap storage/reflectron time-of-flight mass detector, *Anal. Chem.*, 69, 2908, 1997.
94. Lin, Z., et al., On-column coaxial flow chemiluminescence detection for underivatized amino acids by pressurized capillary electrochromatography using a monolithic column, *Anal. Chem.*, 78, 5322, 2006.
95. Bandilla, D. and Skinner, C.D., Protein separation by monolithic capillary electrochromatography, *J. Chromatogr. A*, 1004, 167, 2003.
96. Huang, P., Wu, J.T., and Lubman, D.M., Separation of tryptic digests using a modified buffer in pressurized capillary electrochromatography with an ion trap storage/reflectron time-of-flight mass spectrometer, *Anal. Chem.*, 70, 3003, 1998.
97. Pesek, J., et al., Protein and peptide separations on high surface area capillaries, *Electrophoresis*, 20, 2343, 1999.
98. He, B., Ji, J., and Regnier, F., Capillary electrochromatography of peptides in a microfabricated system, *J. Chromatogr. A*, 853, 257, 1999.

99. Behnke, B. and Metzger, J. W., Tryptic digest mapping by gradient capillary electrochromatography, *Electrophoresis*, 20, 80, 1999.
100. Zhang, J., et al., Capillary electrochromatography of proteins on an anion-exchanger column, *Anal. Chem.*, 72, 3022, 2000.
101. Matyska, M.T., et al., Electrochromatographic characterization of etched chemically-modified capillaries with small synthetic peptides, *J. Chromatogr. A*, 924, 211, 2001.
102. Pesek. J.J., Matyska, M.T., and Cho, J., Open tubular capillary electrochromatography in etched, chemically modified 20 μm I.D. capillaries, *J. Chromatogr. A*, 845, 237, 1999.
103. Xu, W. and Regnier, F., Electrokinetically-driven cation-exchange chromatography of proteins and its comparison with pressure-driven high-performance liquid chromatography, *J. Chromatogr. A*, 853, 243, 1999.
104. Huang, X., Zhang, J., and Horvath, C., Capillary electrochromatography of proteins and peptides with porous-layer open-tubular columns, *J. Chromatogr. A*, 858, 91, 1999.
105. Guo, W., Koropchak, J.A., and Yan, C., Sensitive, universal detection for capillary electrochromatography using condensation nucleation light scattering detection, *J. Chromatogr. A*, 849, 587, 1999.
106. Zhao, R.R. and Johnson, B.P., Capillary electrochromatography: analysis of sucralose and related carbohydrate compounds, *J. Liq. Chromatogr. Rel. Technol.*, 23, 1851, 2000.
107. Yang, C. and El Rassi, Z., Capillary electrochromatography of derivatized mono- and oligosaccharides, *Electrophoresis*, 19, 2061, 1998.
108. Suzuki, S., et al., Separation of 1-phenyl-3-methyl-5-pyrazolone derivatives of monosaccharides by capillary electrochromatography, *Electrophoresis*, 19, 2682, 1998.
109. Gucek, M. and Pilar, B., Capillary electrochromatography of 1-phenyl-3-methyl-5-pyrazolone derivatives of some mono- and disaccharides, *Chromatographia*, 51, S139, 2000.
110. Suzuki, S., et al., Preparation of various silica-based columns for capillary electrochromatography by in-column derivatization, *J. Chromatogr. A*, 873, 247, 2000.
111. Verheij, E.R., et al., Pseudo-electrochromatography–mass spectrometry: A new alternative, *J. Chromatogr. A*, 554, 339, 1991.
112. Lin, S.Y. and Liu, C.Y., An insight into the phenomena involved in a multiple-function stationary phase for the capillary electrochromatographic separation of 2′-, 3′-, and 5′-monophosphorylated nucleoside isomers, *Electrophoresis*, 24, 2973, 2003.
113. Huang, Y.C., Lin, C.C., and Liu, C.Y., Preparation and evaluation of molecularly imprinted polymers based on 9-ethyladenine for the recognition of nucleotide bases in capillary electrochromatography, *Electrophoresis*, 25, 554, 2004.
114. Cahours, X., et al., Fast separation of nucleosides by capillary electrochromatography on non-endcapped phenyl-bonded silica phase using short-end injection method, *J. High Resol. Chromatogr.*, 23, 138, 2000.
115. Helboe, T. and Hansen, S.H., Separation of nucleosides using capillary electrochromatography, *J. Chromatogr. A*, 836, 315, 1999.
116. Matyska, M.T., Pesek, J.J., and Katrekar, A., Open tubular capillary electrochromatography using etched fused-silica tubing modified with chemically bonded liquid crystals, *Anal. Chem.*, 71, 5508, 1999.
117. Zhang, M., Yang, C., and El Rassi, Z., Capillary electrochromatography with novel stationary phases. 3. Retention behavior of small and large nucleic acids on octadecyl-sulfonated-silica, *Anal. Chem.*, 71, 3277, 1999.
118. Zhang, M. and El Rassi, Z., Capillary electrochromatography with novel stationary phases. I. Preparation and characterization of octadecyl-sulfonated silica, *Electrophoresis*, 19, 2068, 1998.
119. Zhang, M. and El Rassi, Z., Capillary electrochromatography with novel stationary phases: II. Studies of the retention behavior of nucleosides and bases on capillaries packed with octadecyl-sulfonated-silica microparticles, *Electrophoresis*, 20, 31, 1999.
120. Moffatt, F., Cooper, P.A., and Jessop, K.M., Comparison of capillary electrochromatography with high-performance liquid chromatography for the analysis of pirimicarb and related compounds, *J. Chromatogr. A*, 855, 215, 1999.

121. Euerby, M.R., et al., Solvent and stationary phase selectivity in capillary electrochromatography method development: Comparison of C18, C8, and phenyl-bonded phases for the separation of a series of substituted barbiturates, *J. Microcol. Sep.*, 11, 305, 1999.
122. Ding, J. and Vorous, P., Capillary electrochromatography–mass spectrometry for the separation and identification of isomeric polyaromatic hydrocarbon DNA adducts derived from in vitro reactions, *J. Chromatogr. A*, 887, 103, 2000.
123. Ding, J. and Vourous, P., Capillary electrochromatography and capillary electrochromatography–mass spectrometry for the analysis of dna adduct mixtures, *Anal. Chem.*, 69, 379, 1997.
124. Aturki, Z., D'Orazio, G., and Fanali, S., Rapid assay of vitamin E in vegetable oils by reversed-phase capillary electrochromatography, *Electrophoresis*, 26, 798, 2005.
125. Eimer, T., Unger, K.K., and Van der Greef, J., Selectivity tuning in pressurized-flow electrochromatography, *Trends Anal. Chem.*, 15, 463, 1996.
126. Vanhoenacker, G., et al., Single-run capillary electrochromatographic analysis of hop acids and prenylated hop flavonoids, *J. Sep. Sci.*, 24, 55, 2001.
127. Li, Y., et al., Optimized separation of pharmacologically active anthraquinones in Rhubarb by capillary electrochromatography, *Electrophoresis*, 21, 3109, 2000.
128. Lane, S. J., et al., Evaluation of a new capillary electrochromatography/mass spectrometry interface using short columns and high field strengths for rapid and rfficient analyses, *Rapid Commun. Mass Spectrom.*, 10, 733, 1996.
129. Sandra, P., et al., Analysis of triglycerides by capillary electrochromatography, *J. Microcol. Sep.*, 9, 409, 1997.
130. Dermaux, A., et al., Analysis of the triglycerides and the free and derivatized fatty acids in fish oil by capillary electrochromatography, *J. High Resol. Chromatogr.*, 1998, 21, 545.
131. Dermaux, A., et al., Elucidation of the triglycerides in fish oil by packed-column supercritical fluid chromatography fractionation followed by capillary electrochromatography and electrospray mass spectrometry, *J. Microcol. Sep.*, 11, 451, 1999.
132. Rentel, C., Gfroerer, P., and Bayer, E., Coupling of capillary electrochromatography to coordination ion spray mass spectrometry, a novel detection method, *Electrophoresis*, 20, 2329, 1999.
133. Zhang, M.Q., Ostrander, G.K., and El Rassi, Z., Capillary electrochromatography with novel stationary phases: IV. Retention behavior of glycosphingolipids on porous and non-porous octadecyl sulfonated silica, *J. Chromatogr. A*, 887, 287, 2000.
134. Lurie, I.S., Meyers, R.P., and Conver, T., Capillary electrochromatography of cannabinoids, *Anal. Chem.*, 70, 3255, 1998.
135. Roed, L., Lundanes, E., and Greigrokk, T., Nonaqueous electrochromatography on continuous bed columns of sol-gel bonded large-pore C30 material: Separation of retinyl esters, *J. Microcol. Sep.*, 12, 561, 2000.
136. Roed, L., Lundanes, E., and Greibrokk, T., Nonaqueous electrochromatography on continuous bed columns of sol-gel bonded large-pore C_{18} material: separation of retinyl esters, *J. Chromatogr. A*, 890, 347, 2000.
137. Roed, L., Lundanes, E., and Greigrokk, T., Nonaqueous electrochromatography on C30 columns: Separation of retinyl esters, *Electrophoresis*, 20, 2373, 1999.
138. Sander, L.C., et al., Separation of carotenoid isomers by capillary electrochromatography with C_{30} stationary phases, *Anal. Chem.*, 71, 3477, 1999.
139. Matyska, M., Pesek, J., and Yang, L., Screening method for determining the presence of *N*-nitrosodiethanolamine in cosmetics by open-tubular capillary electrochromatography, *J. Chromatogr. A*, 887, 497, 2000.
140. Girelli, A.M., et al., Reversed-phase capillary electrochromatography of aloins and related constituents of aloe, *Chromatographia*, 53, S284, 2001.
141. Enlund, A.M., Isaksson, R., and Westerlund, D., Capillary electrochromatography of tricyclic antidepressants on strong cation exchangers with different pore sizes, *J. Chromatogr. A*, 918, 211, 2001.
142. Spikmans, V., et al., Automated capillary electrochromatography tandem mass spectrometry using mixed mode reversed-phase ion-exchange chromatography columns, *Rapid Commun. Mass Spectrom.*, 13, 141, 1999.

143. Whitaker, K.W. and Sepaniak, M.J., Nonaqueous packed capillary electrokinetic chromatographic separations of large polycyclic aromatic hydrocarbons and fullerenes, *Electrophoresis*, 15, 1341, 1994.
144. Peters, E.C., et al., Molded rigid polymer monoliths as separation media for capillary electrochromatography. 2. Effect of chromatographic conditions on the separation, *Anal. Chem.*, 70, 2296, 1998.
145. Stol, R., et al., Application of size exclusion electrochromatography to the microanalytical determination of the molecular mass distribution of celluloses from objects of cultural and historical value, *Anal Chem.*, 74, 2314, 2002.
146. Smith, N.W. and Evans, M.B., The analysis of pharmaceutical compounds using electrochromatography, *Chromatographia*, 38, 649, 1994.
147. Smith, N.W. and Evans, M.B., The efficient analysis of neutral and highly polar pharmaceutical compounds using reversed-phase and ion-exchange electrochromatography, *Chromatographia*, 41, 197, 1995.
148. Taylor, M.R. and Teale, P., Gradient capillary electrochromatography of drug mixtures with UV and electrospray ionisation mass spectrometric detection, *J. Chromatogr. A*, 768, 89, 1997.
149. Gordon, D.B., Lord, G.A., and Jones, D.S., Development of packed capillary column electrochromatography/mass spectrometry, *Rapid Commun. Mass Spectrom.*, 8, 544, 1994.
150. Lord, G.A., et al., Tapers and restrictors for capillary electrochromatography and capillary electrochromatography–mass spectrometry, *J. Chromatogr. A*, 768, 9, 1997.
151. Seilar, R.M., et al., Capillary electrochromatography with 1.5 μm ODS-modified non-porous silica spheres, *Chromatographia*, 46, 131, 1997.
152. Seilar, R.M., et al., Capillary electrochromatography with 1.8-μm ODS-modified porous silica particles, *J. Chromatogr. A*, 808, 71, 1998.
153. Euerby, M.R., et al., "Short-end injection" rapid analysis capillary electrochromatography, *Chromatographia*, 47, 135, 1998.
154. Frame, L.A., Robinson, M.L., and Lough, W.J., Simplification of capillary electrochromatography procedures, *J. Chromatogr. A*, 798, 243, 1998.
155. Mayer, M., et al., Fritless capillary electrochromatography, *Electrophoresis*, 20, 43, 1999.
156. Hilhorst, M.J., Somsen, G.W., and de Jong, G.H., Sensitivity enhancement in capillary electrochromatography by on-column preconcentration, *Chromatographia*, 53, 190, 2001.
157. Spikmans, V., Lane, S.J., and Smith, N.W., Capillary electrochromatography of complex plasma matrix on a C_{18}/SCX column using UV-vis and mass spectrometric detection, *Chromatographia*, 51, 18, 2001.
158. Que, A.H., et al., Steroid profiles determined by capillary electrochromatography, laser-induced fluorescence detection and electrospray–mass spectrometry, *J. Chromatogr. A*, 887, 379, 2000.
159. Djordjevic, N.M., et al., High temperature and temperature programming in capillary electrochromatography, *J. Chromatogr. A*, 887, 245, 2000.
160. Thiam, S., et al., Capillary electrochromatography of cholesterol and its ester derivatives, *Anal. Chem.*, 72, 2541, 2000.
161. Que, A.H., et al., Analysis of bile acids and their conjugates by capillary electrochromatography/electrospray ion trap mass spectrometry, *Anal. Chem.*, 72, 2703, 2000.
162. Chankvetadze, B., et al., Comparative study on the application of capillary liquid chromatography and capillary electrochromatography for investigation of enantiomeric purity of the contraceptive drug levonorgestrel, *J. Pharm. Biomed. Anal.*, 30, 1897, 2003.
163. Liu, S., et al., Separation of structurally related estrogens using isocratic elution pressurized capillary electrochromatography, *J. Chromatogr. A*, 1092, 258, 2005.
164. Liu, S., et al., On-line coupling of pressurized capillary electrochromatography with end-column amperometric detection for analysis of estrogens, *Electrophoresis*, 26, 2342, 2005.
165. Ohyama, K., et al., Rapid separation of barbiturates and benzodiazepines by capillary electrochromatography with 3-(1,8-naphthalimido)propyl-modified silyl silica gel, *Biomed. Chromatogr.*, 18, 396, 2004.
166. Cahours, X., Morin, P., and Dreux, M., Influence of ionic strength and organic modifier on performance in capillary electrochromatography on phenyl silica stationary phase, *J. Chromatogr. A*, 845, 203, 1999.

167. Jinno, K., et al., Comparison of separation behavior of benzodiazepines in packed capillary electrochromatography and open-tubular capillary electrochromatography, *J. Chromatogr. A*, 887, 479, 2000.
168. Kapnissi, C., et al., Analytical separations using molecular micelles in open-tubular capillary electrochromatography, *Anal. Chem.*, 74, 2328, 2002.
169. Kapnissi, C.P. and Warner, I.M., Separation of benzodiazepines using capillary electrochromatography, *J. Chromatogr. Sci.*, 42, 238, 2004.
170. De Rossi, A. and Desiderio, C., Separation of negatively charged nonsteroidal anti-inflammatory drugs by reversed-phase capillary electrochromatography, *J. Chromatogr. A*, 984, 283, 2003.
171. Pai, Y.F., Lin, C.C., and Liu, C.Y., Optimization of sample stacking for the simultaneous determination of nonsteroidal anti-inflammatory drugs with a wall-coated histidine capillary column, *Electrophoresis*, 25, 569, 2004.
172. Desiderio, C. and Fanali, S., Capillary electrochromatography and capillary electrochromatography–electrospray mass spectrometry for the separation of non-steroidal anti-inflammatory drugs, *J. Chromatogr. A*, 895, 123, 2000.
173. Strickmann, D.B. and Blaschke, G., Capillary electrochromatography–electrospray ionization mass spectrometry for the qualitative investigation of the drug etodolac and its metabolites in biological samples, *J. Chromatogr. B*, 748, 213, 2000.
174. Hoegger, D. and Freitag, R., Acrylamide-based monoliths as robust stationary phases for capillary electrochromatography, *J. Chromatogr. A*, 914, 211, 2001.
175. Lammerhofer, M., et al., Capillary electrochromatography in anion-exchange and normal-phase mode using monolithic stationary phases, *J. Chromatogr. A*, 925, 265, 2001.
176. Steiner, F. and Lobert, T., Capillary electrochromatography with bare silicas of different pore sizes as stationary phases, *J. Sep. Sci.*, 26, 1589, 2003.
177. Endund, A.M. and Westerlund, D., Effects of aliphatic amines on capillary electrochromatographic performance of tricyclic antidepressants on octadecylsilica, *J. Chromatogr. A*, 895, 17, 2000.
178. Enlund, A.M., et al., Capillary electrochromatography of hydrophobic amines on continuous beds, *Electrophoresis*, 22, 511, 2001.
179. Vallano, P.T. and Remcho, V.T., Highly selective separations by capillary electrochromatography: Molecular imprint polymer sorbents, *J. Chromatogr. A*, 887, 125, 2000.
180. Miyawa, J.H., Alasandro, M.S., and Riley, C.M., Application of a modified central composite design to optimize the capillary electrochromatographic separation of related S-oxidation compounds, *J. Chromatogr. A*, 769, 145, 1997.
181. Miyawa, J.H., Lloyd, K.D., and Alasandro, M.S., Capillary electrochromatography as a method development tool for the liquid chromatographic separation of DUP 654 and related substances, *J. High Resol. Chromatogr.*, 21, 161, 1998.
182. Gfrorer, P., et al., Gradient elution capillary electrochromatography and hyphenation with nuclear magnetic resonance, *Electrophoresis*, 20, 3, 1999.
183. Wei, W., et al., Capillary electrochromatographic separation of basic compounds with bare silica as stationary phase, *J. Chromatogr. A*, 817, 65, 1998.
184. Yamamoto, H., Baumann, J., and Erni, F., Electrokinetic reversed-phase chromatography with packed capillaries, *J. Chromatogr. A*, 593, 313, 1992.
185. Dekkers, S.E.G., Tjaden, U.R., and Van der Greef, J., Development of an instrumental configuration for pseudo-electrochromatography-electrospray mass spectrometry, *J. Chromatogr. A*, 712, 201, 1995.
186. Pesek, J.J. and Matyska, M.T., Separation of tetracyclines by high-performance capillary electrophoresis and capillary electrochromatography, *J. Chromatogr. A*, 736, 313, 1996.
187. Nagaraj, S. and Karnes, H.T., Visible diode laser induced fluorescence detection of doxorubicin in plasma using pressurized capillary electrochromatography, *Biomed. Chromatogr.*, 14, 234, 2000.
188. Meyring, M., Chankvetadze, B., and Blaschke, G., Simultaneous separation and enantioseparation of thalidomide and its hydroxylated metabolites using high-performance liquid chromatography in common-size columns, capillary liquid chromatography and nonaqueous capillary electrochromatography, *J. Chromatogr. A*, 876, 157, 2000.
189. Meyring, M., et al., Investigation of the in vitro biotransformation of *R*-(+)-thalidomide by HPLC, nano-HPLC, CEC and HPLC–APCI-MS, *J. Chromatogr. B*, 723, 255, 1999.

190. Lurie, I.S., et al., Profiling of impurities in illicit methamphetamine by high-performance liquid chromatography and capillary electrochromatography, *J. Chromatogr. A*, 870, 53, 2000.
191. Lim, J.T., et al., Separation of related opiate compounds using capillary electrochromatography, *Electrophoresis*, 21, 737, 2000.
192. Lai, E.P.C. and Dabek-Zlotorzynska, E., Separation of theophylline, caffeine and related drugs by normal-phase capillary electrochromatography, *Electrophoresis*, 20, 2366, 1999.
193. Hilhorst, M.J., Somsen, G.W., and de Jong, G.H., Capillary electrochromatography of basic compounds using octadecyl-silica stationary phases with an amine-containing mobile phase, *J. Chromatogr. A*, 872, 315, 2000.
194. Desiderio, C., Ossicini, L., and Fanali, S., Analysis of hydroquinone and some of its ethers by using capillary electrochromatography, *J. Chromatogr. A*, 887, 489, 2000.
195. Guan, N., et al., Open tubular capillary electrochromatography in fused-silica capillaries chemically bonded with macrocyclic dioxopolyamine, *Anal. Chim. Acta*, 418, 145, 2000.
196. Stol, R., et al., Pseudo-electrokinetic packing of high efficiency columns for capillary electrochromatography, *J. Chromatogr. A*, 873, 293, 2000.
197. Smith, N.W., Comparison of aqueous and non-aqueous capillary electrochromatography for the separation of basic solutes, *J. Chromatogr. A*, 887, 233, 2000.
198. Gillot, N.C., et al., The analysis of pharmaceutical bases on a silica stationary phase by capillary electrochromatography using aqueous mobile phases, *Chromatographia*, 51, 167, 2000.
199. McKeown, A.P., et al., An evaluation of unbonded silica stationary phases for the separation of basic analytes using capillary electrochromatography, *Chromatographia*, 52, 777, 2000.
200. Dittmann, M.M., Masuch, K., and Rozing, G.P., Separation of basic solutes by reversed-phase capillary electrochromatography, *J. Chromatogr. A*, 887, 209, 2000.
201. Dittmann, M.M. and Rozing, G.P., Capillary electrochromatography: Investigation of the influence of mobile phase and stationary phase properties on electroosmotic velocity, retention, and selectivity, *J. Microcol. Sep.*, 9, 399, 1997.
202. Kanitsar, K., et al., Influence of organic modifiers on the separation of carboxylic acids using co-EOF capillary electrophoresis, *J. Liquid Chromatogr.*, 26, 455, 2003.
203. Zhang, S., Zhang, J., and Horvath, C., Rapid separation of peptides and proteins by isocratic capillary electrochromatography at elevated temperature, *J. Chromatogr. A*, 914, 189, 2001.
204. Li, D.M. and Remcho, V.T., Perfusive electroosmotic transport in packed capillary electrochromatography: Mechanism and utility, *J. Microcol. Sep.*, 9, 389, 1997.

6 Capillary Electrophoresis of Nucleic Acids

Eszter Szántai and András Guttman

CONTENTS

6.1 INTRODUCTION

Since Francis Crick and James Watson discovered the double helical structure of deoxyribonucleic acid in 1953 [1], which finding—with the understanding of the basic rules of inheritance suggested by Gregor Mendel in the nineteenth century—became the theoretical basis of contemporary genetic studies, there has been an urgent need for rapid, precise, sensitive, and high-throughput analysis of nucleic acids. In the mid-1960s, electrophoresis-based methods employing polyacrylamide and agarose gels were rapidly developed; however, these were fairly time-consuming and labor-intensive techniques. The use of gels as sieving matrices for DNA electrophoresis was necessary to provide and appropriate anticonvective media and to make electric field mediated separation of nucleic acid molecules size based. Please note that the constant linear charge density of DNA chains results in practically equal charge-to-mass ratio for the different lengths oligo- and polynucleotide molecules. Slab gels are still widely used in DNA analysis in spite of the fact that they are not quantitative, they lack online detection option and the analysis usually takes a long time. The next generation of DNA separation techniques transpired in the late 1980s by the application of capillaries to electrophoresis. This resulted in higher separation speed, as the large surface area-to-volume ratio allowed effective Joule heat dissipation (i.e., higher voltages could be applied) [2]. In the past two decades, capillary electrophoresis (CE) became a powerful separation technique featuring automation, ease of use, and high efficiency. Novel detection methods, such as laser-induced fluorescence (LIF), increased the sensitivity of the technique, thus advancing low-level DNA analysis from small amount of samples. The next big step toward large-scale and automated DNA analysis was the introduction of low viscosity so-called replaceable sieving polymer networks, which provided much higher flexibility compared with high viscosity cross-linked gels [3].

Capillary gel electrophoresis (CGE) also shed light to the possibility of miniaturization (lab-on-a-chip), as microfluidic devices can be manufactured similar to that of semiconductor chips, and existing CE methods can be readily transferred from the capillary to microchip electrophoresis [4]. Thanks to the rapid technical developments of CE in the past 15 years [5], the identification of the three billion base pair of the human genome was completed well ahead of schedule [6,7] enabling new high-throughput genomics based approaches to be manifested in biomedicine, especially in clinical diagnostics.

6.2 BACKGROUND

Slab gel electrophoresis is a well established and still very popular separation method for nucleic acid analysis. Polyacrylamide and agarose are the most frequently used matrices for horizontal and vertical slab gel electrophoresis [8]. DNA fragments can be detected after the separation either by the application of a fluorescent dye, such as ethidium bromide, or by autoradiographic techniques. Besides the traditionally used slab gel electrophoresis, analysis of DNA molecules in the beginning of the twenty-first century is also based on high-performance liquid chromatography (HPLC) and CE. Liquid chromatography partitions analytes between a very hydrophobic stationary phase and a relevant mobile phase. As the diffusion characteristic of the analyte is an important issue, electrophoresis is advantageous over HPLC when large molecules, such as longer DNA fragments, are separated. The mass transport of DNA molecules is slow in the HPLC mobile phase, resulting in increased band broadening. In CGE this does not represent a problem. CE instrumentation is also less complex and capillaries are cheaper than HPLC columns. A good critical comparison of the two methods is described in Reference 9.

Slab gels feature the benefit of separating multiple samples simultaneously in multilane format; however, the emergence of multicapillary array instruments prevailed over this advantage [10]. Today, mostly 96 capillary units (ABI, Molecular Dynamics) but even 384-capillary array electrophoresis instruments (Genteon) are commercially available, allowing high-throughput analysis primarily for industrial sequencing and genotyping laboratories. As these instruments are quite expensive and the change of a single damaged capillary represents a major issue (i.e., in most instances replacement of the entire array), smaller molecular diagnostic laboratories cannot afford to use them. Microchip-based units (Agilent, Biorad) or small cartridge-based multicapillary systems (eGene) with fluorescence-based nucleic acid detection represent a good alternative to slab gel electrophoresis for smaller molecular biology or clinical laboratories. In the past few years several reviews have been published on DNA separation by CE discussing theoretical issues, separation matrices and detection modes [11–16], and genetic diagnostic [17–23] and forensic applications [24,25].

6.3 THEORETICAL ASPECTS

6.3.1 ELECTROPHORETIC MIGRATION OF NUCLEIC ACIDS

Applying a uniform electric field (E) to an oligo- or polynucleotide molecule with a net charge of Q, the electrical force (F_e) is defined as

$$F_e = QE. \tag{6.1}$$

In a gel or polymer network solution, a frictional force (F_f) acts in the opposite direction

$$F_f = f\left(\frac{dx}{dt}\right), \tag{6.2}$$

where f is the translational friction coefficient and dx and dt are the distance and time increments, respectively. Differences in shape, size, and overall charge of the solute molecules result in variances in electrophoretic mobilities providing the basis of the electrophoretic separation. Under steady-state conditions, F_e and F_f are counterbalanced, thus the solute migrates with a steady-state velocity of v

$$v = \frac{dx}{dt} = \frac{EQ}{f}. \tag{6.3}$$

The electrophoretic mobility (μ) is defined as the velocity per unit field strength

$$\mu = \frac{v}{E}. \tag{6.4}$$

Retardation of the solute molecules in gel-filled or polymer-filled capillaries is a function of the separation matrix concentration (P) and its physical interactions with the molecules subject to electromigration is defined by the retardation coefficient (K_R)

$$\mu_{app} = \mu_0 \exp(-K_R P), \tag{6.5}$$

where μ_{app} is the apparent electrophoretic mobility and μ_0 is the free solution mobility of the analyte (i.e., with no sieving matrix) [8].

When the average pore size of the matrix is in the same size range as the hydrodynamic radius of the migrating analyte molecule, the classical sieving theory applies (Ogston regime) [26,27], that is, at constant polymer concentration the retardation coefficient (K_R) is an apparent logarifunction of the molecular weight (MW) of the migrating analyte [28]:

$$\mu \sim \exp(-\mathrm{MW}). \tag{6.6}$$

In this instance, the so called Ferguson plots [29] cross each other at zero gel concentration.

The Ogston theory assumes that the migrating solute behaves as an unperturbed spherical object with comparable size to the pores of the gel. However, DNA molecules can migrate through polymer networks with pores significantly smaller than their size [30] by the phenomenon referred to as reptation, suggesting "snakelike" motion for large biopolymers through the much smaller gel pores [31–33]. The reptation model implies an inverse relationship between the size (MW) and the mobility of the analyte molecules as shown in the following equation:

$$\mu \sim \frac{1}{\mathrm{MW}}. \tag{6.7}$$

At extremely high electric field strengths, reptation turns into biased reptation and the resulting mobility of the analyte is described by

$$\mu \sim \left(\frac{1}{\mathrm{MW}} + bE^{\mathrm{a}}\right), \tag{6.8}$$

where b is a function of the mesh size of the sieving matrix, the charge, and segment length of the migrating DNA molecules, and $1 < (a) < 2$.

6.3.2 Efficiency and Resolution

In CGE one of the major contributors to band broadening, besides the injection and detection extra column effects, is the longitudinal diffusion of the solute molecules in the capillary tube [34]. The theoretical plate number (N) is characteristic of column efficiency

$$N = \mu \frac{E \cdot \ell}{2D}, \tag{6.9}$$

where μ is the electrophoretic mobility, D is the diffusion coefficient of the solute in the separation gel-buffer system, and ℓ is the effective column length.

Resolution (R_s) between two peaks can be calculated from the differences of their electrophoretic mobilities ($\Delta\mu$) [35]

$$R_{\mathrm{s}} = 0.18 \cdot \Delta\mu \sqrt{\frac{E \cdot \ell}{D \cdot \bar{\mu}}}, \tag{6.10}$$

where $\bar{\mu}$ is the mean mobility of the sample components of interest. As one can see, Equations 6.9 and 6.10 suggest that higher applied electric field and lower solute diffusion coefficient would result in higher separation efficiency (N) and concomitantly higher resolution (R_s). One of the limiting factors is the so-called Joule heat (Q_j), generated by the applied power ($P = V \times I$) [2]

$$Q_{\mathrm{j}} = \frac{P}{r^2 \cdot I \cdot L}, \tag{6.11}$$

where I is the current, L is the total column (electrode to electrode) length, and r is the inside radius of the capillary. Owing to the temperature dependence of electrophoretic mobility, efficient temperature control during CE separation is important in order to attain good reproducibility. Modern, automated CE instruments are equipped with effective liquid- or air-cooling systems to address temperature change-related problems.

6.4 PRACTICAL APPLICATIONS OF CAPILLARY ELECTROPHORESIS ANALYSIS OF DNA MOLECULES

6.4.1 Polymerase Chain Reaction Product Analysis

6.4.1.1 Mutation Detection and Polymorphism Studies

Polymerase chain reaction (PCR) is an *in vitro* DNA replication and amplification technique that revolutionized nucleic acid analysis [36]. It enables small amount of nucleic acid molecules to be exponentially amplified (i.e., to generate enough material for their analysis and sequencing). PCR is a commonly used technique in biomedical, molecular biology, and clinical diagnostics laboratories accommodating a variety of tasks, such as detection of hereditary diseases, identification of genetic fingerprints, diagnosis of infectious diseases, cloning of genes, and paternity testing. Electrophoresis is a crucial part of this method (except in case of real-time PCR) and CE—especially combined with LIF—more and more replaces classical electrophoretic techniques in many fields.

The first research groups to investigate the utility of CE for PCR product and in general for DNA analysis were Brownlee's and Karger's in the late 1980s [37–39]. In the years, novel high-resolution polymer networks and capillary coatings were developed and CE conditions were optimized for better DNA separation performance [12,40–42]. At the present time, many DNA-related applications are still being reported by using similar conditions and capillaries. In this section, we list a few examples from several important reports on PCR-based methods that greatly helped the fields of molecular genetics. Chromosome 18q allelic loss has been reported to have prognostic significance in stage II colorectal carcinoma. Erill et al. [43] have developed a robust and reliable fluorescent and multiplex PCR assay to analyze five microsatellite markers for the allelic loss at the long arm of chromosome 18. Amplicon detection and evaluation was accomplished by an ABI 310 Genetic Analyzer (Applied Biosystems).

The most frequently used methods for DNA length polymorphism analysis usually consists of two major steps. First is the amplification of the genome region of interest followed by electrophoresis-based separation of the resulting fragments. Hyytia-Trees et al. [44] developed a subtyping method for Shiga toxin-producing *Escherichia coli* O157 strains, which was based on the analysis of several variable number of tandem repeats in the bacterial genome. PCR products were sized using a multicapillary electrophoresis-based sequencing system (CEQ™8000; Beckman Coulter) with the following conditions: injection at 2.0 kV after 15 s and separation at 6.0 kV after 60 min.

One of the largest groups of genetic polymorphisms is referred to as single nucleotide polymorphisms (SNPs), which were conventionally detected by restriction fragment analysis. This method is based on the cleavage of molecules by sequence-specific endonucleases. Interrogation of the size of the resulting DNA fragments is usually electrophoresis separation based. The C677T mutation of the methylenetetrahydrofolate (*MTHFR*) gene is a nutrient-oriented, "eco" genetic mutation, which is associated with elevated levels of homocysteine and an increased risk for coronary heart disease. Sell and Lugemwa [45] reported on an automated assay by means of the Hinf I RFLP to detect this mutation. The resulting restriction fragments were analyzed on an ABI PRISM 310 single capillary-based Genetic Analyzer.

Allele-specific PCR is another very popular genotyping method and is also referred to as amplification refractory mutation system. The technique is based on the utilization of an allele-specific primer as its 3′ end hybridizes to the SNP site followed by amplification using a DNA polymerase that lacks 3′ exonuclease activity. Amplification in this case occurs only if the primer sequence perfectly matches with the template sequence. Carrera et al. [46] describe refractory mutation system analysis of point mutations by CE. The first application of a multiplex multicolor assay for the simultaneous detection of three of the most frequent mutations related to hereditary haemochromatosis was presented by Gomez-Llorente et al. [47]. One of the described methods was allele-specific PCR and CE analysis of the amplified products that enabled easy, rapid, unambiguous, and high-resolution discrimination between wild-type and mutant alleles. Fluorescently labeled products were analyzed on an ABI PRISM 310 Genetic Analyzer using POP-4 polymer and fused-silica capillaries of 47-cm length and 50-μm diameter [47].

Ligase detection reaction (LDR) developed by Barany [48] provides an elegant technique for multiplexed typing of SNPs, deletions, and insertions. It utilizes the ability of the DNA ligase to preferentially seal adjacent oligonucleotides hybridized to a target DNA with perfect complementation at a nick junction. In their paper, Thomas et al. [49] reported on the use of CE and microchip electrophoresis format for detecting single base mutations in selected gene fragments with high diagnostic value in colorectal cancer using LDR. The electrophoretic separations were carried out for the single-stranded DNA products generated by allele-specific ligation assay to screen for a single base mutation in the *K-ras* oncogene. Various separation matrices were investigated in CGE.

LDR products (44 and 51 bp) were analyzed by a cross-linked polyacrylamide gel (5%T/5%C) in approximately 45 min having a 1000-fold molar excess of the LDR primers (25 bp). Interestingly, when linear polyacrylamide gels were used, these same fragments could not be detected because of significant electrokinetic bias during the electrokinetic injection process.

Another frequently used genotyping technique is based on primer extension. In this method, the 3′ end of the applied primer anneals just before the SNP site as the first step of a quasi-minisequencing reaction, which is carried out using dideoxy-ribonucleotide triphosphate terminators. For detection purposes, either the primers or the dideoxy nucleotide terminators are labeled by an appropriate fluorophor. Another version of this technique is allele-specific primer extension, which also represents a good method of choice for efficient genotyping. This technique is based on the sequence specific extension of two allele-specific primers that differ at their 3′ end defining the allele. Single base extension (SBE) assay—next to allele-specific PCR—was used for the detection of three mutations related to hereditary haemochromatosis [47]. ABI PRISM 310 Genetic Analyzer using POP-4 polymer and fused-silica capillaries of 47-cm length and 50-μm diameter was applied for separation.

Brazill and Kuhr [50] presented a model system to show possible advantages in combining SBE technique with CGE and electrochemical detection (Figure 6.1). An electrochemically labeled primer, with ferrocene acetate covalently attached to its 5′ end, was used. The complementary dideoxynucleotide (ddNTP) extended the primer by a single nucleotide and the reaction mixture

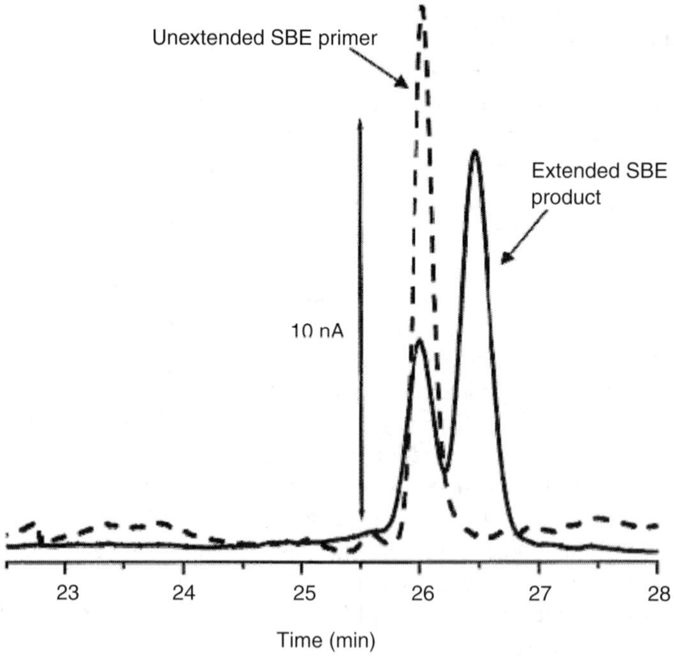

FIGURE 6.1 CGE coupled to sinusoidal voltammetric detection of SBE product. Separation capillary: 25-μm ID/25-cm length; Coating and separation medium: POP-4. Separation conditions: injection: 6 s at –4 kV; Separation voltage: –4 kV. The SV detection employed a 21 Hz sine wave scanning from –200 mV to 800 mV versus Ag/AgCl. Both of the time courses shown in this figure are from the second harmonic or 42 Hz. The solid trace is from the injection of the successful SBE utilizing ddATP terminator. The first peak corresponds to migration of the unextended primer, whereas the late-eluting peak corresponds to the extended product. The dotted trace represents the control reaction utilizing ddCTP. (From Brazill, S.A. and Kuhr, W.G., *Anal Chem*, 74, 3421, 2002. With permission.)

was subsequently separated by CGE in a 25-cm-long fused-silica capillary filled with POP-4 sieving medium. The ferrocene-tagged fragments were detected with sinusoidal voltammetry.

Heteroduplex analysis is another very useful approach to analyze known or unknown genetic variations. The method is based on the differential migration of heteroduplex DNA molecules (formed between a wild-type gene segment and the corresponding homologous segment containing an induced mutation or a naturally occurring SNP), compared to the corresponding homoduplex in nondenaturing polyacrylamide gel. Heteroduplexes migrate slower than their corresponding homoduplexes due to their quasi open configuration surrounding the mismatched bases. Heteroduplex analysis is usually carried out by slab gel electrophoresis, CE, or denaturing HPLC (dHPLC); however, it can also be performed by microchip electrophoresis. Kozlowski et al. [51] demonstrated the influence of a number of parameters on the electrophoretic properties of DNA duplexes such as temperature, presence of glycerol, capillary length, as well as polymer (GeneScan) concentration, and evaluated their contribution to the overall analysis time. Their study was carried out on an ABI 310 apparatus equipped with argon-ion laser. The time required for the detection of two typical *BRCA1* gene deleterious mutations by heteroduplex analysis could be significantly decreased through careful optimization of analysis conditions. The effect of different silanizing reagents, polymeric coatings, and polymer networks were investigated for detecting DNA mutations in the *BRCA1* gene via heteroduplex analysis by Landers' group [52]. Figure 6.2 shows the results of analyzing one heterozygous *BRCA1* mutant using capillaries coated with different silanizing reagents and polymers and how they compare with commercial coated capillaries. In general, the electroosmotic flow (EOF) was found to be 19- to 76-fold lower than that of with bare fused-silica capillary. Optimal performance was observed using the chlorodimethyloctylsilane (OCT)- and poly(vinylpyrrolidone) (PVP)-coated capillary and hydroxyethylcellulose (HEC) as the polymer network. More recently, Weber et al. [53] proposed a novel method for the detection of unknown mutations called enhanced mismatch mutation analysis, which is also based on electrophoretic heteroduplex analysis. Their experimental results showed that the combination of high-resolution block-copolymer sieving matrix poly(acrylamide-γ-polydimethylacrylamide), and nucleosides as additives in the electrophoretic medium increased the resolution significantly between the homoduplex and heteroduplex peaks. The enhanced mismatch mutation analysis method was compared to denaturing HPLC in a large-scale mutation study of the breast cancer-associated gene *BRCA2* and the success rate of detection of both methods was comparable (94%).

Single-strand conformation polymorphism (SSCP) is a simple and versatile method combining PCR amplification, denaturation of DNA molecules, and the analysis of denatured fragments by electrophoresis. This technique, originally developed by Orita et al. [54], is based on subtle sequence differences (often just a single base pair) that can result in a different three-dimensional conformation and concomitantly measurable differences in their electrophoresis mobility. Hofman-Bang et al. [55] developed a multiplex CE-based SSCP screening protocol on an automated genotyping platform for mutation detection in the *SCN5A* gene coding for the alpha-subunit of the cardiac Na^+ ion channel. The separation was carried out by using a commercial polymer at 18°C and 30°C. Disease-causing mutations were scattered over the DNA sequence, making it difficult to screen for specific mutations that could cause several diseases, such as long QT syndrome, Brugada syndrome, idiopathic ventricular fibrillation, sick sinus node syndrome, progressive conduction disease, dilated cardiomyopathy, and atrial standstill. These diseases exhibited variable expressivity and identification of gene carriers was clinically important, particularly in sudden infant and adult death syndromes. The method was highly efficient with a false positive rate of 0.5% of the analyzed amplicons. Holmila and Husgafvel-Pursiainen [56] compared CE-SSCP, denaturant gradient gel electrophoresis (DGGE), and direct sequencing to investigate the benefits and sensitivity of each of the methods for the detection of unknown TP53 mutations in human lung cancer. Their study revealed that direct sequencing performed less well in finding mutations than the other two methods, and CE-SSCP was found to be a fast and highly reproducible method, also considerably less laborious compared to DGGE, for screening of unknown TP53 mutations. The CE-SSCP analysis was performed using ABI PRISM

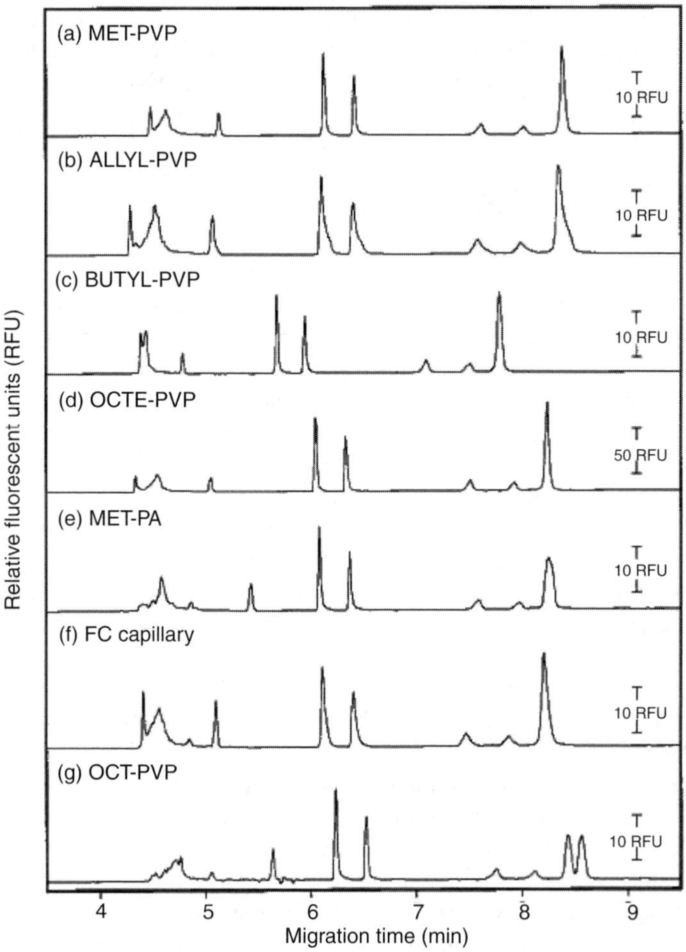

FIGURE 6.2 Effect of different coatings on CE-based heteroduplex analysis. The PCR products were amplified from a heterozygous individual containing 1294del40 mutation in the *BRCA1* gene and analyzed under the following conditions: Injection: 20 s at 370 V/cm. Separation: 370 V/cm (current 27.0–27.6 μA), using reversed polarity. EOF of the bare fused-silica capillary was 3.7×10^{-4} $cm^2/V{\cdot}s$ under the conditions used. The different silanizing reagents and polymers used for modifying the surface of the capillary were as follows: (a) $CH_2 = C(CH_3)COO(CH_2)_3Si(OCH_3)_3$ (MET), PVP, EOF is 8.5×10^{-6} $cm^2/V \cdot s$; 24.8 μA. (b) $CH_2 = C(CH_2)Si(CH_3)_2(Cl)$ (ALLYL), PVP, EOF is $8.9 \times 10^{-6} cm^2/V \cdot s$; 25.0 μA. (c) $Cl(CH_2)_4Si(CH_3)_2(Cl)$ (BUTYL), PVP, EOF is $19.7 \times 10^{-6} cm^2/V \cdot s$; 25.5 μA. (d) $CH_2 = CH - (CH_2)_6Si(OCH_3)_3$ (OCTE), PVP, EOF is $7.3 \times 10^{-6} cm^2/V \cdot s$; 25.4 μA. (e) $CH_2 = C(CH_3)COO(CH_2)_3Si(OCH_3)_3$ (MET), PA, EOF is $4.9 \times 10^{-6} cm^2/V \cdot s$; 25.3 μA. (f) FC capillary (J&W), EOF is $6.7 \times 10^{-6} cm^2/V \cdot s$; 24.9 μA. (g) $CH_3(CH_2)_7Si(CH_3)_2(Cl)$ (OCT), PVP, EOF is $8.1 \times 10^{-6} cm^2/V \cdot s$; 24.7 μA. (From Tian, H., et al., *Anal Chem*, 72, 5483, 2000. With permission.)

310 capillary sequencer at 30°C, and with ABI PRISM 3100 Avant capillary sequencer at five different analysis temperatures (18°C, 25°C, 30°C, 35°C, and 40°C). Electrophoresis time was 30 min with the applied voltage of 13 kV. The separation media were 5% GeneScan Polymer in 1× Tris-borate- EDTA (TBE) buffer containing 10% glycerol. Endo et al. [57] studied SSCP parameters that might affect the mutation analysis in the *K-ras* gene—such as electric field strength, separation temperature, polymer, and additive in the sieving matrix (high concentration methyl cellulose)—by CE and microchip electrophoresis. Figure 6.3 shows the effect of glycerol concentration in the buffer

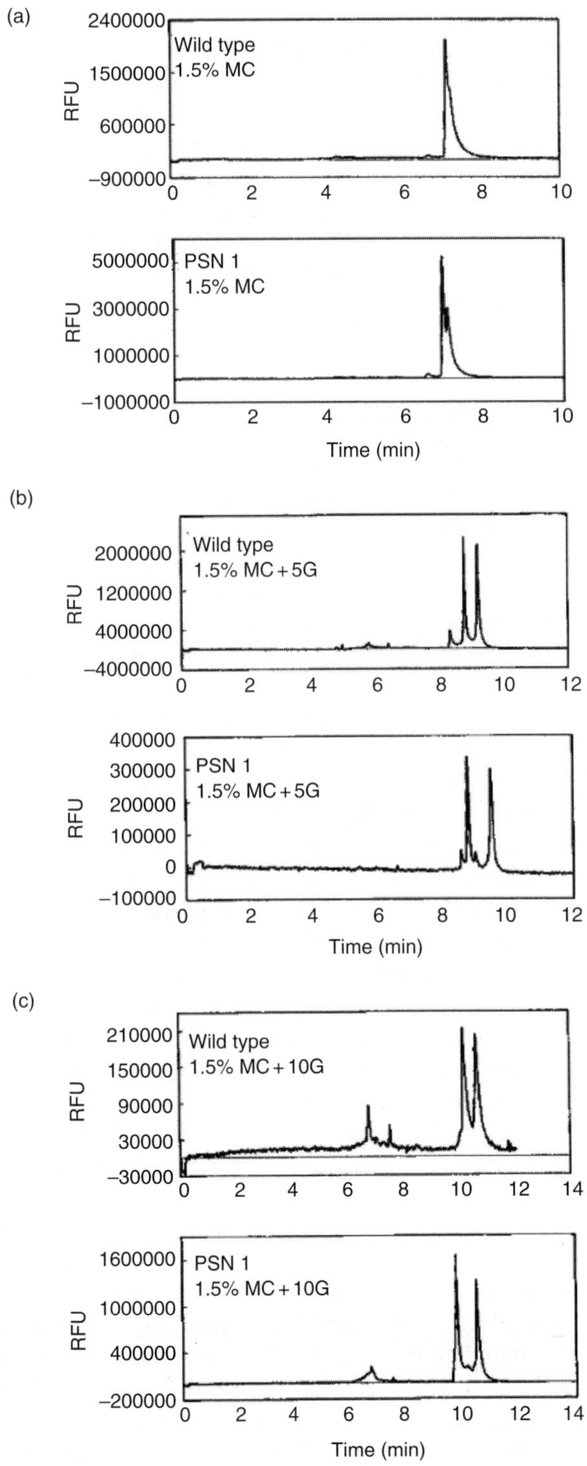

FIGURE 6.3 Effects of glycerol concentration in the buffer on CE-SSCP analysis of PSN1 and the wild-type of *K-ras* gene. Experimental conditions: 1.5% MC in 50 mM Trisborate buffer; (a) without glycerol; (b) with 5% glycerol; (c) with 10% glycerol; electric field: 300 V/cm; samples were labeled with FAM. RFU represents relative fluorescence unit. (From Endo, Y., et al., *Electrophoresis*, 26, 3380, 2005. With permission.)

on CE-SSCP analysis of PSN1 and the wild-type *K-ras* gene. Better resolution of single-stranded DNAs was obtained with the addition of 5% glycerol to the buffer system. However, as it can be seen in Figure 6.3c, the separation time was increased with increasing glycerol concentration. Analysis of seven mutants of the *K-ras* gene was accomplished in 10 min by CE and in 1 min by microchip electrophoresis under optimized separation conditions. Culiat et al. [58] identified point mutations in the mouse genome induced by *N*-ethyl-*N*-nitrosourea using a new high-throughput mutation-scanning technique: temperature-gradient CE (TGCE). TGCE detects the presence of partially denatured heteroduplex molecules that are resolved from homoduplexes through their differential mobilities during CE conducted with a closely controlled temperature gradient. All TGCE analyses were conducted using the SCE9610 Genetic Analysis System (Spectrumedix). The default running condition was a broad 60–68°C temperature ramp with 5 kV injection voltage and 60 s injection time.

6.4.1.2 Forensic Application/Identity Testing

Individual DNA fingerprints can be provided by microsatellite- or SNP analysis and can be used in forensic testing, paternity testing, and identification of suitable recipients for organ transplantation, just to mention a few. A recent review summarizes the most relevant examples of analytical applications of CE techniques in the forensic field from 2001 till 2004 [25]. In an earlier review, Butler et al. [24] detailed the characteristics of DNA typing with short tandem repeat (STR) markers widely used for a variety of applications including human identification. Y-chromosome STRs (Y-STRs) provided valuable information in cases of rape and questioned paternity. This test also allowed genetic identification of male lineages. In a study of Johnson et al. [59], a Y-STR 10-plex on a commercially available automated genetic analyzer was validated for use in forensic and paternity laboratories. In a comparison study, Dixon et al. [60] used artificially degraded samples and demonstrably obtained a profile using SNPs instead of STRs for certain sample types with greater likelihood. A method for simultaneous analysis of SNPs has been developed and validated to analyze highly degraded and low copy number DNA templates (i.e., $<$100 pg) for scenarios including mass disaster identification. The multiplex assay interrogated 20 autosomal noncoding loci—not in linkage disequilibrium—and Amelogenin for sex determination. The samples were amplified in a single-tube PCR and analyzed by an automated CE-based sequencer. Inagaki et al. [61] developed a novel 39-plex typing system for SNP study. Thirty-seven genomic DNA fragments containing a total of 38 SNPs and one sex-discriminating site were amplified in a multiplex PCR. Following the reaction, single nucleotide primer extension reaction was performed by dividing these SNP loci into five groups. The SNP type of each of the 39 loci was determined in the same run by an ABI PRISM™ 310 Genetic Analyzer using POP-4 polymer in 47 cm $\times$ 0.050 mm ID capillaries applying a newly designed multi-injection method. In this novel technique, the collection time was set to 1 or 2 min for the first four groups and 20 min for the last group. The method was applied in forensic cases of paternity testing and personal identification.

An additional interesting forensic application of CE is the identification of differences to determine the species of botanical evidence found at a crime site or to associate a sample with a source. Amplified fragment length polymorphism (AFLP) analysis of botanical forensic evidence provided a means of obtaining a reproducible DNA profile in a relatively short period of time for species with no available sequence information. Bless et al. [62] obtained AFLP profiles of 40 Acer rubrum trees using an automated DNA fragment analyzer. This information could be used to link a piece of evidence with a particular location or a suspect. Another important application for investigating an accident or a crime is dog DNA profiling, as dogs are intensely integrated in human social life. Several STR markers were analyzed for the individualization of dogs using an ABI PRISM 3100 Genetic Analyzer with POP-4 gel in 36-cm capillary arrays using default instrument settings [63].

Graft rejection is one of the most severe complications after allogenic transplants in leukemia. Detection of increasing quantities of lymphoid and myeloid host cells might be predictive of graft rejection and leukemia relapse, respectively. A sensitive and automatic method of quantifying the degree of mixed chimerism after allogenic stem cell transplantation was based on PCR assays of polymorphic STRs. CE allowed quantification of the recipient's cells relative to the donor's cells by calculating ratios between STR alleles [64].

6.4.1.3 Diagnosis of Infectious Diseases

Identification of bacteria, viruses, and fungi is crucial for diagnostic and therapeutic purposes; however, microbiologists often have problems with microorganisms that are difficult to culture. Molecular methods on the other hand enable detection of these infectious agents without requiring the presence of viable organisms to allow appropriate identification. PCR can be used to target conserved stretches of DNA, similar to the region encoding 16S ribosomal RNA. These conserved sequences enable robust applications for the identification of unknown or known infectious agents. Several groups reported case histories in instances where infectious agents were identified based on the sequence analysis of *16S rRNA* gene (*rDNA*). Sequencing was carried out using the ABI Big Dye cycle sequencing reaction kit with AmpliTaq FS DNA polymerase and the electrophoresis was performed on an ABI 310 capillary analyzer. Analysis of the sequences and clustering was performed by GeneCompar, version 2.0 (Applied Maths, Kortrijk, Belgium) [65,66]. In some cases, 16S rDNA sequence determination does not seem to be sufficient for species identification just like in the case of some *Helicobacter* species commonly found in dogs and cats. Phylogenetically, these species are highly related to each other, thus their *16S rRNA* gene sequences show >99% similarity. A multiplex PCR method was described based on the tRNA intergenic spacers and in the urease gene, combined with CE in the study. The PCR products were separated using an automated genetic analyzer and the tRNA intergenic spacers amplified in tDNA-PCR were sequenced by using the BigDye Terminator cycle sequencing kit (Applied Biosystems). The sequencing products were electrophoretically separated by a commercially available multicapillary genetic analyzer. Their procedure was shown to be very useful in determining the species identity of "*Helicobacter heilmannii*"-like organisms observed in human stomachs and would facilitate research concerning their possible zoonotic importance [67]. CE also offers great versatility in studying viral systems. Krylova et al. [68] demonstrated the utility of a P/ACE MDQ instrument for monitoring DNA release from virus particles in a running buffer of 25 mM sodium tetraborate containing 10 mM SDS. Drug development targeting viral propagation requires fast and sensitive methods for *in vitro* monitoring of viral DNA release. A T5 bacteriophage/*Escherichia coli* K-12 model was reported to study molecular mechanisms of viral infections and to evaluate antiviral drug candidates. The identification of clinically significant fungi species was demonstrated by the amplification of noncoding internal transcribed spacer regions of the *rRNA* gene and sequencing. All PCR/sequencing identifications from positive broths were in agreement with the final species identification of the isolates grown from subculture. Early identification of fungi by PCR/sequencing method may facilitate prompt and more appropriate antifungal therapy [69].

An earlier review by Righetti and Gelfi [70] describes a number of applications using capillary zone electrophoresis with sieving liquid polymers (mostly linear polyacrylamides and celluloses) for the analysis of PCR products of clinically relevant, diagnostic DNA samples. The fields of microbiology and virology were also covered, just like human genetics, quantitative gene dosage, forensic medicine, and therapeutic DNA.

It has to be taken into account that microorganisms can also be separated, identified, and characterized by CE without the necessity of using PCR. A comprehensive review of Desai and Armstrong [71] gives details about these possibilities, such as the recent paper of Gao et al. [72] who detected *Staphylococcus aureus* by a combination of monoclonal antibody-coated latex and CE. CZE separations were performed on a Beckman P/ACE MDQ System equipped with a photodiode array detector. A 27-cm-long capillary column was used for the separation (20 cm to the detection window) applying 215 kV at a constant temperature of 25°C.

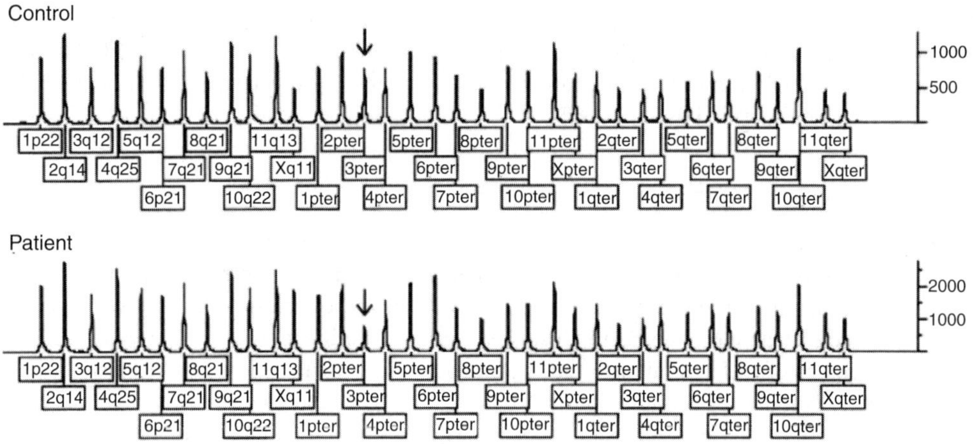

FIGURE 6.4 CE trace showing the deletion of the 3p telomere. The peak area of the 3p telomere is reduced in the patient sample when compared to the control sample. (From Rooms, L., et al., *Hum Mutat*, 23, 17, 2004. With permission.)

6.4.1.4 Molecular Karyotyping

Conventional karyotyping (chromosome banding) is one of the most widely used techniques in routine cytogenetics and has been invaluable in the search for chromosomal aberrations causally related to, for example, congenital mental retardation and malformation syndromes. It requires culturing of cells from viable tissues but many clinical specimens may fail to provide adequate cellularity for karyotyping and it also has a rather limited resolution (5–10 million bp). To solve these problems, molecular karyotyping assays have been developed that allow sensitive and specific detection of single copy number changes of submicroscopic chromosomal regions throughout the entire human genome. Among such methods are array-based comparative genomic hybridization (array CGH), fluorescent *in situ* hybridization (FISH), and multiplex ligation-dependent probe amplification (MLPA). MLPA is a recently developed technique, which is based on PCR amplification of ligated probes hybridized to chromosome regions in question using unique primers attached to the probes, making multiplexing easier. Only probes hybridized to a target sequence will be ligated and subsequently amplified in the PCR. After separation by CGE, the peak area of each amplification product reflects the relative copy number of that target sequence. Sequences that are deleted or duplicated can also be easily identified, such as the deletion of the 3p telomere in Figure 6.4 [73,74]. However, it should be noted that balanced translocations could not be detected by CE methods at this time.

6.4.1.5 Quantification of Cellular mRNA/Expression Analysis

Quantitative transcript analysis is usually carried out by microarray techniques, real-time PCR, or even with serial analysis of gene expression (SAGE) [75–78]. An alternative of these techniques is competitive template reverse transcriptase method described by Gilliland et al. [79] in which the PCR mix contains a known amount of competitor DNA that differs from the cDNA of interest only by having, for example, a mutated internal restriction enzyme site. Therefore, the competitor and the unknown amount of target DNA compete for the same primers. The ratio of products remains constant through the coamplification and can be readily quantified by CE and LIF [80,81]. In the paper of Warner et al. [82] bronchogenic carcinoma was diagnosed with the help of standardized reverse transcriptase PCR (StaRT-PCR)—which is a modification of competitive template reverse transcriptase method—since morphological analysis of cytologic samples obtained by fine-needle aspirate or bronchoscopy was not sensitive enough (65–80%). They measured the c-myc × E2F-1/p21 index in cDNA samples and found that the index may augment cytomorphologic diagnosis of bronchogenic carcinoma biopsy samples, particularly those considered nondiagnostic by cytomorphologic criteria. After amplification, each PCR product was analyzed by microchip electrophoresis on an Agilent 2100 Bioanalyzer instrument [82]. Relative quantification of mutated and normal mRNAs was carried out by amplification refractory mutation sequencing PCR assay using 5-carboxyfluorescein dye-labeled mutation-specific primers and a commercial DNA analyzer with standard protocols [83]. A method was designed and validated to detect a point mutation in the *Janus tyrosine kinase 2* gene in patients with chronic myeloproliferative disorders. The proposed method might complement current technologies based on genomic DNA analysis. Spyres et al. [84] described a semiautomated PCR-based technology, called quantitative rapid analysis of gene expression (Q-RAGE), which provided fast measurements of mRNA abundance with extremely high sensitivity using fluorescent detection of the specific products. An array of sixteen 36-cm-long fused-silica capillaries were used in an ABI 3100 instrument employing POP-4 separation polymer. They claimed that their method was more sensitive compared to SAGE and its throughput was higher than that of real-time PCR or StaRT-PCR. The flexibility of Q-RAGE makes it eminently well suited for large-scale validation of microarray results and for more directed quantitative studies of medium-sized sets of related genes, especially when low-abundance transcripts are of interest.

6.4.2 DNA Sequencing

DNA sequencing enables to yield the greatest amount of information by identifying the order of each deoxynucleotide base in a particular DNA molecule. With this knowledge, for example, one can locate regulatory elements and gene sequences, make comparisons between homologous genes across species, and identify mutations. In the 1970s, two DNA-sequencing methods were independently developed. Maxam and Gilbert [85] used a "chemical cleavage protocol," while Sanger [86] designed a procedure similar to the natural process of DNA replication. Even though both teams shared the 1980 Nobel Prize, the Sanger method became the industry standard because of its practicality. Sanger's method is based on dideoxy chain termination using ddNTPs in addition to the normal nucleotides (dNTPs) in appropriate concentrations. When these ddNTPs are integrated into the newly built-up nucleotide chain, they prohibit the addition of further nucleotides due to the lack of their phosphodiester bond forming OH group, thus the DNA chain is terminated. When different fluorescent labels are attached to each of the four reaction products (primer or terminator), DNA sequencing can be carried out in one test tube as multispectral imaging can readily determine which fragment ends with what base by simple fluorescent signal based differentiation. In the beginning of the various Genome Projects (Human, Mice, Rice, etc.), most laboratories did DNA sequencing by means of very labor-intensive and time-consuming slab gel electrophoresis, also facing difficulties with automation. The first paper on high-resolution single base separation of DNA fragments using CGE was published in 1988 with the promise that the method was applicable for automated DNA sequencing [87]. In this early work, cross-linked polyacrylamide gels were utilized within the capillary, in analogy to slab gel electrophoresis [38]. Later, it was found that the cross-linked matrix could not withstand osmotic shock occurring when salt plugs migrated through the gel-filled capillary. Therefore, it became clear that non-cross-linked (so-called physical gels) had much better characteristics in this respect. Initially, covalent coatings were utilized to minimize EOF [88]; more recently, dynamic (i.e., adsorbed) coatings prevailed [89]. It was clear that if CE were to be selected as the technology to sequence the Human Genome, researchers had to find the means for automatic replenishment of the polymer matrix, as no one was interested in reusing a polymer matrix due to fear of cross-contamination. Changing columns after each run

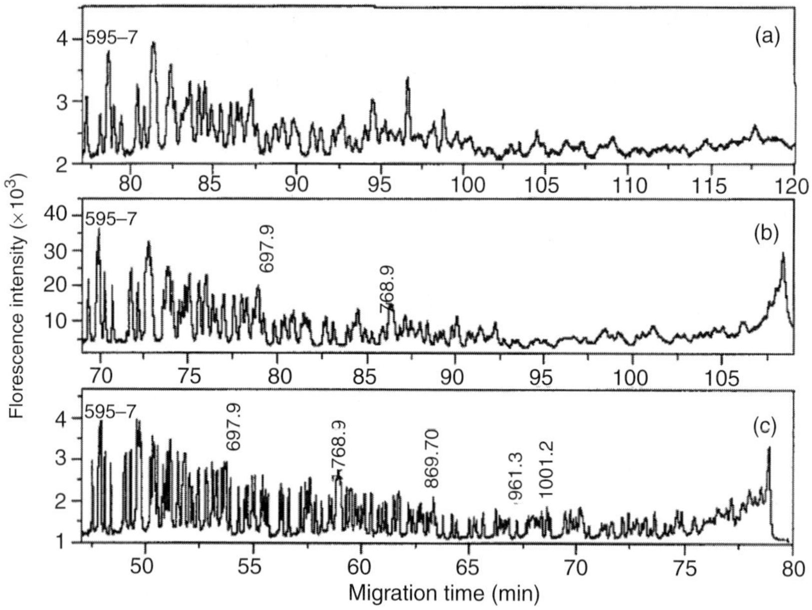

FIGURE 6.5 Separation of single-color DNA sequencing of larger than 590 bases by using (a) POP-6, (b) MegaBACE matrix, and (c) quasi-IPN. Temperature: 60°C. Capillary, 40 cm effective length, 75 mm ID. DNA injection, 41 V/cm and 30 s for POP-6, 50 V/cm and 40 s for MegaBACE matrix, 75 V/cm and 8 s for quasi-IPN; running electric field strength, 200 V/cm for POP-6, 150 V/cm for MegaBACE matrix and quasi-IPN. (From Wang, Y., et al., *Electrophoresis*, 26, 126, 2005. With permission.)

was also not a workable strategy. In 1990, Guttman [3] demonstrated that, through the use of solutions of linear polyacrylamide and high pressure, it was possible to automatically replace the polymer after each run and still maintain the high separation efficiency. This was a key step forward, because it proved the potential of CE to become the basis of an automated DNA-sequencing instrument. The technology was further developed for long read-length sequencing by Karger and coworkers [90] in the late 1990s. It was also recognized that with longer DNA chains, reptation-like migration became more significant and this stretching phenomenon reduced chain length-based electrophoretic mobility differences to the point where separation was unsuccessful. After numerous optimization and improvement steps by increasing the molecular weight of the sieving polymer, 1000 base read lengths were achieved in 1996 at high column temperatures (50°C) [91]. Today, high-molecular-weight linear polymers are used routinely for DNA sequencing and further developments on DNA separation media are carried out by the use of copolymers, possessing high sieving ability, low viscosity, and dynamic coating ability [92]. As it was difficult for homopolymers to possess all the properties that can meet the different challenges, copolymers were used to combine the desirable properties of the different monomers. On the basis of this principle, a range of DNA separation matrices were developed featuring dynamic coating ability, low viscosity, and/or adjustable viscosity covering a wide range of read length. The latest results are summarized in the paper of Wang et al. [93] and Figure 6.5 compares the separation efficiency of two polymers and a copolymer.

Another significant improvement in capillary-based DNA sequencing was multiplexing (i.e., using arrays of capillaries) [10]. This topic has been extensively reviewed by a number of groups, just to list a few [5,94–97]; however, it is important to highlight the work of Kheterpal and Mathies [98], Li and Yeung [99], Dovichi and Zhang [100], and Kamahori and Kambara [101] for their contributions on automated multicapillary instrumentation.

6.4.3 Purity Control of Synthetic Nucleotides

6.4.4 Oligonucleotides

Oligonucleotides are commonly used for DNA amplification in PCR mixtures as primers, for *in situ* hybridization techniques as probes for detecting PCR products, and for antiviral and anticancer treatment as antisense oligonucleotide therapeutic agents, just to list the most important applications. In all of these instances, the oligonucleotides have to be of high purity, so their quality control after synthesis is very important. All synthetic oligonucleotides should be tested for a defect in length or sequence. The study of Willems et al. [102] showed that the combination of capillary zone electrophoresis and electrospray ionization quadrupole-time of flight mass spectrometry allowed the identification of oligonucleotides differing in length by only one nucleotide, and also any misincorporated nucleotide by measuring tiny mass differences. Another approach for such tests was published by Olson et al. [103] who modified an automated multicapillary DNA sequencer and applied noncovalent fluorescent labeling.

6.4.4.1 Antisense DNA

Antisense therapeutics are synthetic oligonucleotides that have a complementary base sequence to a target messenger RNA which encodes for disease-causing proteins or to the double-stranded DNA which the mRNA is transcribed from. These molecules inactivate the genetic message, thus inhibiting gene expression. Since oligonucleotides with a phosphodiester backbone are very susceptible to nuclease degradation, DNA analogs with phosphorus-modified backbone (e.g., phosphorothioates and methylphosphonates) are of high interest as they exhibit increased resistance against degradation. Another type of antisense DNA is called as peptide nucleic acid. In these molecules, the deoxyribose-phosphate backbone is substituted by a pseudopeptide backbone. Synthetic peptide nucleic acid oligomers have been used in recent years in molecular biology procedures, diagnostic assays and therapies. Some of these applications typically require stringent purity criteria for the antisense DNA agents.

A paper by Chen and Gallo [104] on the instrumental and technological aspects of analyzing antisense DNA in biological samples recommends the use of CGE to reveal the pharmacokinetics of these molecules. An earlier review by DeDionisio and Lloyd [105] described a CGE-based method for the purity analysis of antisense oligonucleotides after their chemical synthesis and for a range of applications in the field of antisense-based technology. In addition, a CGE-based analysis method using eCAP™ ssDNA 100-R gel in a Beckman PACE 5510 instrument was presented by the same group for the analysis of a N3′ → P5′ phosphoramidate/phosphorothioate chimera, a second generation antisense oligonucleotide, prior and after purification [106]. More recently, Malek and Khaledi [107] used CE with LIF (CE-LIF) detection to evaluate the effectiveness of delivery and fate of a model 25mer DNA-based phosphorothioate antisense drug in HeLa cells. Belenky et al. [108] described a CE method for sequence determination of antisense DNA analogs of unknown base composition, while Froim et al. [109] developed a method of phosphorothioate antisense DNA-sequencing analysis using UV detection (i.e., with no fluorescent or other labeling requirement). Vilenchik et al. [110] applied high-performance CE for low level and rapid monitoring of phosphorothioate DNA that modulated *in vitro* gene expression of human immunodeficiency virus (HIV). The method was based on Watson-Crick hybridization between phosphodiester and the target DNA. Various techniques of staining or labeling were investigated to improve detection limits (0.1 ng/mL).

6.4.5 Separation of Large DNA Molecules (>2 kb)

Initial efforts in large DNA fragment analysis tended to apply constant electric fields. The size limit of high resolution was demonstrably about 20,000 bp [111,112], while in case of low-resolution requirement it was as high as 48.5 kb using linear polyacrylamide in CE [113]. DNA fragments (72 bp to 23.1 kb) were separated in ultradilute HEC polymer solutions (<0.002% w/w) using high voltage. On the basis of their results, Barron et al. [114] developed a transient entanglement coupling mechanism theory for DNA separation, which suggested that there was no a priori upper size limit of DNA separation by CE at a constant electric field. More recently, Chiu and Chang [115] demonstrated that HEC solution at concentrations higher than its entanglement threshold provided good separation for large DNA fragments in the presence of EOF at high pH. Separation of DNA fragments ranging in size from 5 to 40 kb was completed in 6 min using 1.5% HEC, prepared in 20 mM methylamine-borate (pH 12.0). A novel separation mechanism was demonstrated by Zheng and Yeung [116], which was based on radial migration in CE with additionally applied hydrodynamic flow. Without the need for gel/polymer or complex instrumentation, this separation technique—while also applicable for proteins, cells, and so on—was complementary to CGE and field-flow fractionation techniques.

Plasmid analysis is often used in recombinant DNA technology. A new separation matrix, consisting of poly(*N*-isopropylacrylamide) (PNIPAM) polymer and mannitol as small molecule additive, was used for CE-based plasmid DNA separation. Supercoiled, linear, and nicked conformers of lambda plasmid were separated in 1% PNIPAM + 6% mannitol. The effect of the applied mannitol concentration on the separation quality is shown in Figure 6.6 [117].

Application of pulsed electric fields in conjunction with dilute polymer solutions are one way to extend the size limit of DNA fragment analysis in CE. In the early 1990s, Heiger et al. [40] demonstrated the use of pulsed-field electrophoresis to enhance the separation of large DNA molecules using linear polyacrylamide matrix. Later, Kim and Morris [118] separated long-chain double-stranded DNA (<10 kb and >1.5 Mb) in less than 4 min in buffers containing ultradilute polymer solutions. They concluded that field inversion with higher peak amplitude in the forward direction than in the reverse, but with equal pulse durations, provided the best resolution if running time was an issue. Kabatek et al. [119] examined the behavior of large DNA fragments in pulsed-field CE under various temperatures in short capillaries. Tseng et al. [120] reported the analysis of long DNA molecules by nanoparticle-filled CE under the influences of hydrodynamic and electrokinetic forces. The polymer composite prepared from gold nanoparticles and poly(ethyleneoxide) was filled into the capillary column as separation matrix. Separations of lambda-DNA (0.12–23.1 kb) and high-molecular-weight DNA markers (8.27–48.5 kb) by nanoparticle-filled CE, under the electric field strength of 140 V/cm with the additional hydrodynamic flow velocity of 554 μm/s, were accomplished within 5 min. Electrophoresis of very long DNA molecules (T4 DNA: 166 kb; *S. pombe* chromosomal DNA: 3–6 Mb) in linear polyacrylamide solutions was investigated by fluorescence microscopy and CE [121]. It was found that at higher polymer concentrations, the shape of the migrating DNA changed from U shape to linear shape, and it was possible to separate the DNA molecules under this linear shape motion. Actually, Mb-sized DNAs were well separated within 5 min in the region of the linear shape motion by means of CE with a d.c. field. Considering that it usually takes 20 h to separate Mb-sized DNAs by standard pulsed-field gel electrophoresis, this result proved extremely useful for the separation of giant DNA molecules.

6.4.6 Analysis of Mononucleotides and Nucleosides

Deoxyribonucleic acid and ribonucleic acid, as well as their nucleotides, nucleosides, and base constituents play an important role in many vital biochemical processes of medical interest. To better understand these processes, fundamental investigations into the structure, occurrence, search for modifications, and biochemical impact of structural variation are required. Thus, reliable high-resolution analytical methods for the separation and identification of the nucleic acid constituents (often at extremely low concentration levels) had to be developed. Chromatography (including reversed-phase liquid chromatography), ion exchange chromatography, dHPLC, and electrophoresis

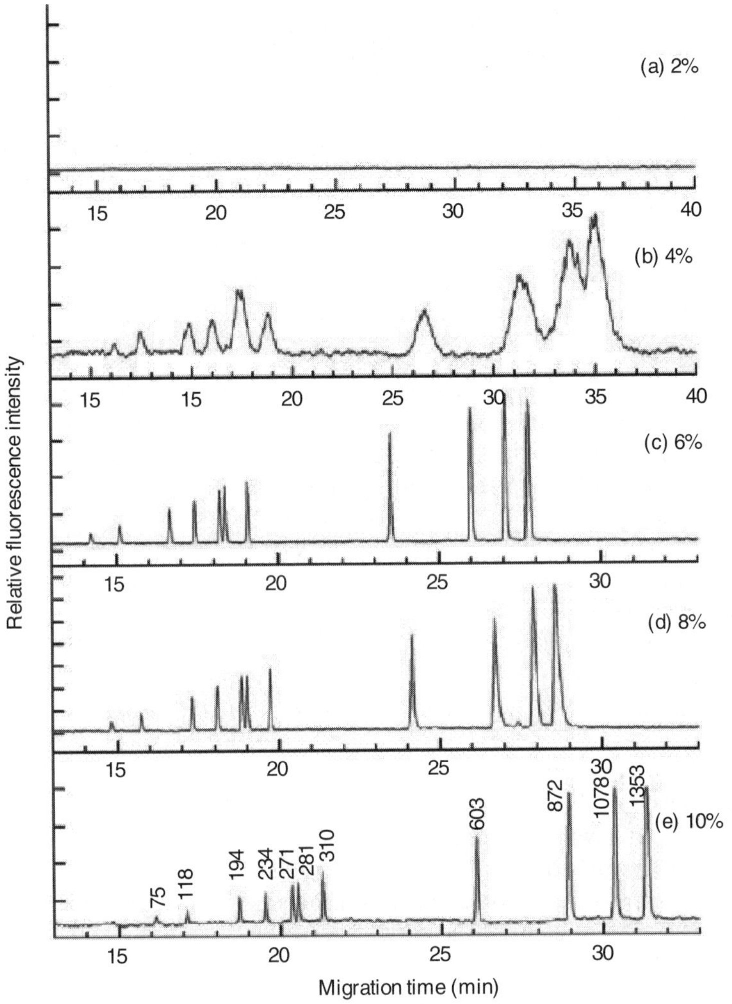

FIGURE 6.6 Separation of _X174/*Hae*III DNA by CE with: (a) 1.5% + PNIPAM + 2% mannitol; (b) 1.5% PNIPAM4% mannitol; (c) 1.5% PNIPAM + 6% mannitol; (d) 1.5% PNIPAM + 8% mannitol; (e) 1.5% PNIPAM + 10% mannitol. Capillary, 75/365 μm i.d./o.d., and 32/40 cm efficient/total length; inject, 5 s at −8 kV; separation electric field strength 220 V/cm. (From Zhou, P., et al., *J Chromatogr A*, 1083, 173, 2005. With permission.)

(slab gel and other modes) were the major techniques that have been attempted for the analysis of nucleic acids, nucleotides, nucleosides, and bases.

6.4.6.1 Nucleotides

Nucleotides are among the most important metabolites in a cell. Besides being the monomeric units of the major nucleic acids of RNA and DNA, nucleotides are also required for numerous other important functions within cells. These functions include (1) acting as energy carriers in phosphate transfer reactions [adenosine 5′-triphosphate (ATP), guanosine 5′-triphosphate (GTP)]; (2) supplying the basic structures for important compounds, such as cofactors (NAD^+, $NADP^+$, FAD, and coenzyme A); (3) serving as mediators of several important cellular processes such as secondary messengers in signal transduction events [cyclic adenosine 5′-monophosphate (cAMP), cyclic guanosine

5′-monophosphate (cGMP)]; (4) controlling many enzymatic reactions through allosteric effects on enzyme activity; and (5) acting as activated intermediates in numerous biosynthetic reactions (*S*-adenosylmethionine, also referred to as *S*-AdoMet). Nucleotide analyses by CE were reportedly carried out in biochemical, biomedical, and pharmacological studies, as well as in the food industry [122–128]. A new method based on pressure-assisted CE coupled to electrospray ionization mass spectrometry was recently developed by Harada et al. [124] for anionic metabolome analysis, including nucleotides. The key step of the method was that CE polarity was inverted from conventional CE analysis for anions with the use of uncoated fused-silica capillary.

6.4.6.2 Nucleoside Analogs

Several nucleoside analogs are chemically synthesized and used as therapeutics, for example, to inhibit specific enzymatic activities. Such analogs can be cytostatics (e.g., 6-mercaptopurine, 5-fluorouracil, 5-iodo-2'-deoxyuridine, and 6-thioguaninc), bccausc they interfere with DNA synthesis, thus preferentially kill rapidly dividing cells like tumor cells. Another family of nucleoside analogs has been used as therapeutics for viral infection treatment, such as for HIV. Azidothymidine (AZT) was the first of such analogs, but followed by many others: ddI (dideoxyinosine), ddC (dideoxycytidine), d4T (didehydro-deoxythymidine), and so forth. Nucleoside analogs are also used to suppress the immune system after organ transplantation and to reduce the likelihood of transplant rejection by the host. CZE has proved to be a suitable and useful method for the determination of acid–base dissociation constants (pK_a) of new synthetic compounds of amino- and (amino)guanidinopurine nucleoside analogs, such as acyclic nucleoside phosphonate, acyclic nucleoside phosphonate diesters, and other related compounds [129]. Determination of this characteristic seemed to be very important in the optimization stage of new drug development projects to understand their passage and metabolism.

In an earlier study, Balayiannis et al. [130] developed a capillary zone electrophoresis method for the analysis of a series of novel synthetic dideoxynucleoside analogs with potential anti-HIV activity. The method was readily applied for purity testing as well as to resolve *cis* and *trans* diastereomers. The purity and separation of diastereomers of such analogs are of great importance for further testing of their biological and pharmacological activity.

6.4.6.3 DNA Adducts

The initial step in chemical carcinogenesis is believed to be the covalent attachment of a chemical to DNA to form DNA adducts, like the extensively studied 8-hydroxyguanine. DNA adducts are most often used as molecular dosimeters of exposure. If not repaired correctly, these modifications may lead to mutations and eventually to cancer, in particular if the adduct is located in an oncogene or in a tumor suppressor gene. Therefore, significant efforts are being made to understand how the diversity in DNA adduct conformations can affect cellular responses to DNA damage. Analysis of DNA adducts can be carried out on the nucleotide or nucleoside level after enzymatic digestion. Analytical methods used for the detection of DNA adducts must have excellent sensitivity, since only 1 in 10^6 to 1 in 10^{12} bases are usually modified that represents a great challenge. Next to immunoassays and fluorescence-based assays, the most common method is ^{32}P-postlabeling, but CE-MS holds the potential to replace it in the near future. Markushin et al. [131] recently reported that catechol estrogen quinines-derived DNA adducts are present in urine samples from subjects with prostate cancer. Their finding suggests that if it can be repeated in larger sample sizes, the presence of depurinating adducts in human urine samples may be used as prostate cancer biomarker.

6.5 FUTURE PROSPECTIVE

Recent technological developments in nucleic acid analysis by electric field mediated separation techniques are tending toward miniaturization with the goal to provide high-speed and low-cost clinical analysis tools. CE, and more recently microchip electrophoresis, features the use of small sample volumes, low reagent consumption, and equally important in a high-throughput clinical environment, almost negligible waste. Another essential impact of miniaturization in capillary and microchip-based techniques is the potential for system integration of several processing steps, such as sample preparation, desalting, PCR amplification, restriction digestion, separation, fraction collection, and so forth. The speed of CE and microchip electrophoresis is significantly higher than that of any other conventional electrophoresis methods, and its cost is also favorable, especially in a disposable plastic chip format. Multicapillary and microchip-based bioanalytical methods will probably play a significant role in the future of biomedical and clinical applications [4].

ACKNOWLEDGMENTS

This work was supported by the European Community Marie Curie Chair 006733. Eszter Szantai gratefully acknowledges the grant of the Austrian Ministry of Arts, Science and Education's GenAu program.

REFERENCES

1. Watson, J.D. and Crick, F.H., Molecular structure of nucleic acids: a structure for deoxyribose nucleic acid, *Nature*, 171, 737, 1953.
2. Nelson, R.J., et al., Use of peltier thermoelectric devices to control column temperature in high-performance capillary electrophoresis, *J Chromatogr A*, 480, 111, 1989.
3. Guttman, A., US patent #5,332,481, Capillary Electrophoresis Using Replaceable Gels, 1994.
4. Heller, M. and Guttman, A. (eds), *Integrated Microfabricated BioDevices: Advanced Technologies for Genomics, Drug Discovery, Bioanalysis, and Clinical Diagnostics*, Marcel Dekker, New York, 2001.
5. Karger, B.L. and Guttman, A., Capillary electrophoresis and the human genome project, *Genomic/Proteomic Technol.*, 3, 14, 2003.
6. Lander, E.S., et al., Initial sequencing and analysis of the human genome, *Nature*, 409, 860, 2001.
7. Venter, J.C., et al., The sequence of the human genome, *Science*, 291, 1304, 2001.
8. Andrews, A.T., *Electrophoresis*, Claredon Press, Oxford, 1986.
9. Gelfi, C., et al., Analysis of antisense oligonucleotides by capillary electrophoresis, gel-slab electrophoresis, and HPLC: a comparison, *Antisense Nucleic Acid Drug Dev*, 6, 47, 1996.
10. Mathies, R.A. and Huang, X.C., Capillary array electrophoresis: an approach to high-speed, high-throughput DNA sequencing, *Nature*, 359, 167, 1992.
11. Barbier, V. and Viovy, J.L., Advanced polymers for DNA separation, *Curr Opin Biotechnol*, 14, 51, 2003.
12. Guttman, A., Gel and polymer-solution mediated separation of biopolymers by capillary electrophoresis, *J Chromatogr Sci*, 41, 449, 2003.
13. Lin, Y.W., Chiu, T.C., and Chang, H.T., Laser-induced fluorescence technique for DNA and proteins separated by capillary electrophoresis, *J Chromatogr B Analyt Technol Biomed Life Sci*, 793, 37, 2003.
14. Quigley, W.W. and Dovichi, N.J., Capillary electrophoresis for the analysis of biopolymers, *Anal Chem*, 76, 4645, 2004.
15. Slater, G.W., et al., The theory of DNA separation by capillary electrophoresis, *Curr Opin Biotechnol*, 14, 58, 2003.
16. Xu, F. and Baba, Y., Polymer solutions and entropic-based systems for double-stranded DNA capillary electrophoresis and microchip electrophoresis, *Electrophoresis*, 25, 2332, 2004.
17. Andersen, P.S., et al., Capillary electrophoresis-based single strand DNA conformation analysis in high-throughput mutation screening, *Hum Mutat*, 21, 455, 2003.
18. Babu, C.V.S., et al., Capillary electrophoresis at the omics level: towards systems biology, *Electrophoresis*, 27, 97, 2006.
19. Godde, R., et al., Electrophoresis of DNA in human genetic diagnostics—state-of-the-art, alternatives and future prospects, *Electrophoresis*, 27, 939, 2006.
20. Hempel, G., Biomedical applications of capillary electrophoresis, *Clin Chem Lab Med*, 41, 720, 2003.

21. Jabeen, R., et al., Capillary electrophoresis and the clinical laboratory, *Electrophoresis*, 27, 2413, 2006.
22. Landers, J.P., Clinical capillary electrophoresis, *Clin Chem*, 41, 495, 1995.
23. Mitchelson, K.R., The use of capillary electrophoresis for DNA polymorphism analysis, *Mol Biotechnol*, 24, 41, 2003.
24. Butler, J.M., et al., Forensic DNA typing by capillary electrophoresis using the ABI Prism 310 and 3100 genetic analyzers for STR analysis, *Electrophoresis*, 25, 1397, 2004.
25. Tagliaro, F. and Bortolotti, F., Recent advances in the applications of CE to forensic sciences (2001–2004), *Electrophoresis*, 27, 231, 2006.
26. Guttman, A., On the separation mechanism of capillary SDS gel electrophoresis of proteins, *Electrophoresis*, 16, 611, 1995.
27. Ogston, A.G., The spaces in a uniform random suspension of fibres, *Trans Faraday Soc*, 54, 1754, 1958.
28. Grossman, P.D., Menchen, S., and Hershey, D., Quantitative analysis of DNA-sequencing electrophoresis, *GATA*, 9, 9, 1992.
29. Ferguson, K.A., Starch-gel electrophoresis: application to the classification of pituitary proteins and polypeptides, *Metab Clin Exp*, 13, 985, 1964.
30. Lumpkin, O.J., Dejardin, P., and Zimm, B.H., Theory of gel electrophoresis of DNA, *Biopolymers*, 24, 1573, 1985.
31. De Gennes, P.G., *Scaling Concept in Polymer Physics*, Cornell University Press, Ithaca, NY, 1979.
32. Slater, G.W. and Noolandi, J., The biased reptation model of DNA gel electrophoresis: mobility vs. molecular size and gel concentration, *Biopolymers*, 28, 1781, 1989.
33. Viovy, J.L. and Duke, T., DNA electrophoresis in polymer solutions: Ogston sieving, reptation and constraint release, *Electrophoresis*, 14, 322, 1993.
34. Terabe, S., Otsuka, K., and Ando, T., Band broadening in electrokinetic chromatography with micellar solutions and open-tubular capillaries, *Anal Chem*, 61, 2434, 1989.
35. Karger, B.L., Cohen, A.S., and Guttman, A., High performance capillary electrophoresis in the biological sciences, *J Chromatogr*, 492, 585, 1989.
36. Mullis, K.B., Target amplification for DNA analysis by the polymerase chain reaction, *Ann Biol Clin (Paris)*, 48, 579, 1990.
37. Cohen, A.S., et al., Rapid separation of DNA restriction fragments using capillary electrophoresis, *J Chromatogr*, 458, 323, 1988.
38. Cohen, A.S., et al., Rapid separation and purification of oligonucleotides by high-performance capillary gel electrophoresis, *Proc Natl Acad Sci USA*, 85, 9660, 1988.
39. Kasper, T.J., et al., Separation and detection of DNA by capillary electrophoresis, *J Chromatogr*, 458, 303, 1988.
40. Heiger, D.N., Cohen, A.S., and Karger, B.L., Separation of DNA restriction fragments by high performance capillary electrophoresis with low and zero crosslinked polyacrylamide using continuous and pulsed electric fields, *J Chromatogr*, 516, 33, 1990.
41. Schwartz, H.E., et al., Analysis of DNA restriction fragments and polymerase chain reaction products towards detection of the AIDS (HIV-1) virus in blood, *J Chromatogr*, 559, 267, 1991.
42. Zhu, M., et al., Factors affecting free zone electrophoresis and isoelectric focusing in capillary electrophoresis, *J Chromatogr A*, 480, 311, 1989.
43. Erill, N., et al., A novel multiplexing, polymerase chain reaction-based assay for the analysis of chromosome 18q status in colorectal cancer, *J Mol Diagn*, 7, 478, 2005.
44. Hyytia-Trees, E., et al., Second generation subtyping: a proposed PulseNet protocol for multiple-locus variable-number tandem repeat analysis of Shiga toxin-producing *Escherichia coli* O157 (STEC O157), *Foodborne Pathog Dis*, 3, 118, 2006.
45. Sell, S.M. and Lugemwa, P.R., Development of a highly accurate, rapid PCR-RFLP genotyping assay for the methylenetetrahydrofolate reductase gene, *Genet Test*, 3, 287, 1999.
46. Carrera, P., et al., Amplification refractory mutation system analysis of point mutations by capillary electrophoresis, *Methods Mol Biol*, 163, 95, 2001.
47. Gomez-Llorente, C., et al., Multiplex analysis of the most common mutations related to hereditary haemochromatosis: two methods combining specific amplification with capillary electrophoresis, *Eur J Haematol*, 72, 121, 2004.

48. Barany, F., The ligase chain reaction in a PCR world, *PCR Methods Appl*, 1, 5, 1991.
49. Thomas, G., et al., Capillary and microelectrophoretic separations of ligase detection reaction products produced from low-abundant point mutations in genomic DNA, *Electrophoresis*, 25, 1668, 2004.
50. Brazill, S.A. and Kuhr, W.G., A single base extension technique for the analysis of known mutations utilizing capillary gel electrophoresis with electrochemical detection, *Anal Chem*, 74, 3421, 2002.
51. Kozlowski, P., Olejniczak, M., and Krzyzosiak, W.J., Rapid heteroduplex analysis by capillary electrophoresis, *Clin Chim Acta*, 353, 209, 2005.
52. Tian, H., et al., Effective capillary electrophoresis-based heteroduplex analysis through optimization of surface coating and polymer networks, *Anal Chem*, 72, 5483, 2000.
53. Weber, J., et al., Improving sensitivity of electrophoretic heteroduplex analysis using nucleosides as additives: application to the breast cancer predisposition gene BRCA2, *Electrophoresis*, 27, 1444, 2006.
54. Orita, M., et al., Detection of polymorphisms of human DNA by gel electrophoresis as single-strand conformation polymorphisms, *Proc Natl Acad Sci USA*, 86, 2766, 1989.
55. Hofman-Bang, J., et al., High-efficiency multiplex capillary electrophoresis single strand conformation polymorphism (multi-CE-SSCP) mutation screening of SCN5A: a rapid genetic approach to cardiac arrhythmia, *Clin Genet*, 69, 504, 2006.
56. Holmila, R. and Husgafvel-Pursiainen, K., Analysis of TP53 gene mutations in human lung cancer: comparison of capillary electrophoresis single strand conformation polymorphism assay with denaturing gradient gel electrophoresis and direct sequencing, *Cancer Detect Prev*, 30, 1, 2006.
57. Endo, Y., et al., Effect of polymer matrix and glycerol on rapid single-strand conformation polymorphism analysis by capillary and microchip electrophoresis for detection of mutations in *K-ras* gene, *Electrophoresis*, 26, 3380, 2005.
58. Culiat, C.T., et al., Identification of mutations from phenotype-driven ENU mutagenesis in mouse chromosome 7, *Mamm Genome*, 16, 555, 2005.
59. Johnson, C.L., et al., Validation and uses of a Y-chromosome STR 10-plex for forensic and paternity laboratories, *J Forensic Sci*, 48, 1260, 2003.
60. Dixon, L.A., et al., Validation of a 21-locus autosomal SNP multiplex for forensic identification purposes, *Forensic Sci Int*, 154, 62, 2005.
61. Inagaki, S., et al., A new 39-plex analysis method for SNPs including 15 blood group loci, *Forensic Sci Int*, 144, 45, 2004.
62. Bless, C., Palmeter, H., and Wallace, M.M., Identification of *Acer rubrum* using amplified fragment length polymorphism, *J Forensic Sci*, 51, 31, 2006.
63. Eichmann, C., et al., Estimating the probability of identity in a random dog population using 15 highly polymorphic canine STR markers, *Forensic Sci Int*, 151, 37, 2005.
64. Fernandez-Aviles, F., et al., Serial quantification of lymphoid and myeloid mixed chimerism using multiplex PCR amplification of short tandem repeat-markers predicts graft rejection and relapse, respectively, after allogeneic transplantation of CD34+ selected cells from peripheral blood, *Leukemia*, 17, 613, 2003.
65. De Baere, T., et al., Bacteremic infection with *Pantoea ananatis*, *J Clin Microbiol*, 42, 4393, 2004.
66. Vaneechoutte, M., et al., Isolation of *Moraxella canis* from an ulcerated metastatic lymph node, *J Clin Microbiol*, 38, 3870, 2000.
67. Baele, M., et al., Multiplex PCR assay for differentiation of *Helicobacter felis*, *H. bizzozeronii*, and *H. salomonis*, *J Clin Microbiol*, 42, 1115, 2004.
68. Krylova, S.M., et al., Monitoring viral DNA release with capillary electrophoresis, *Analyst*, 129, 1234, 2004.
69. Pryce, T.M., et al., Rapid identification of fungal pathogens in BacT/ALERT, BACTEC, and BBL MGIT media using polymerase chain reaction and DNA sequencing of the internal transcribed spacer regions, *Diagn Microbiol Infect Dis*, 54, 289, 2006.
70. Righetti, P.G. and Gelfi, C., Capillary electrophoresis of DNA for molecular diagnostics, *Electrophoresis*, 18, 1709, 1997.
71. Desai, M.J. and Armstrong, D.W., Separation, identification, and characterization of microorganisms by capillary electrophoresis, *Microbiol Mol Biol Rev*, 67, 38, 2003.
72. Gao, P., et al., Rapid detection of *Staphylococcus aureus* by a combination of monoclonal antibody-coated latex and capillary electrophoresis, *Electrophoresis*, 27, 1784, 2006.

73. Rooms, L., et al., Subtelomeric deletions detected in patients with idiopathic mental retardation using multiplex ligation-dependent probe amplification (MLPA), *Hum Mutat*, 23, 17, 2004.
74. Slater, H.R., et al., Rapid, high throughput prenatal detection of aneuploidy using a novel quantitative method (MLPA), *J Med Genet*, 40, 907, 2003.
75. Cai, Y., et al., Abnormal expression of Smurf2 during the process of rat liver fibrosis, *Chin J Dig Dis*, 7, 237, 2006.
76. Lee, S., et al., Gene expression profiles in acute myeloid leukemia with common translocations using SAGE, *Proc Natl Acad Sci USA*, 103, 1030, 2006.
77. Nakamura, N., et al., Isolation and expression profiling of genes upregulated in bone marrow-derived mononuclear cells of rheumatoid arthritis patients, *DNA Res*, 13, 169, 2006.
78. Sanada, Y., et al., Down-regulation of the claudin-18 gene, identified through serial analysis of gene expression data analysis, in gastric cancer with an intestinal phenotype, *J Pathol*, 208, 633, 2006.
79. Gilliland, G., et al., Analysis of cytokine mRNA and DNA: detection and quantitation by competitive polymerase chain reaction, *Proc Natl Acad Sci USA*, 87, 2725, 1990.
80. Borson, N.D., et al., Direct quantitation of RNA transcripts by competitive single-tube RT–PCR and capillary electrophoresis, *Biotechniques*, 25, 130, 1998.
81. Williams, S.J. and Williams, P.M., Quantitation of mRNA by competitive PCR using capillary electrophoresis, *Methods Mol Biol*, 163, 243, 2001.
82. Warner, K.A., et al., The c-myc × E2F-1/p21 interactive gene expression index augments cytomorphologic diagnosis of lung cancer in fine-needle aspirate specimens, *J Mol Diagn*, 5, 176, 2003.
83. Vannucchi, A.M., et al., A quantitative assay for JAK2(V617F) mutation in myeloproliferative disorders by ARMS-PCR and capillary electrophoresis, *Leukemia*, 20, 1055, 2006.
84. Spyres, L., et al., Quantitative high-throughput measurement of gene expression with sub-zeptomole sensitivity by capillary electrophoresis, *Anal Biochem*, 345, 284, 2005.
85. Maxam, A.M. and Gilbert, W., A new method for sequencing DNA, *Proc Natl Acad Sci USA*, 74, 560, 1977.
86. Sanger, F. and Coulson, A.R., A rapid method for determining sequences in DNA by primed synthesis with DNA polymerase, *J Mol Biol*, 94, 441, 1975.
87. Guttman, A., et al., Capillary Gel Electrophoresis: High Resolution and Micropreparative Applications, in *Electrophoresis'88*, Schafer-Nielsen, C. (Ed.), VCH, Weinheim, Germany, 1988, p. 151.
88. Hjerten, S., High-performance electrophoresis: elimination of electroendosmosis and solute adsorption, *J Chromatogr*, 347, 191, 1985.
89. Madabhushi, R.S., Separation of 4-color DNA sequencing extension products in noncovalently coated capillaries using low viscosity polymer solutions, *Electrophoresis*, 19, 224, 1998.
90. Goetzinger, W., et al., Characterization of high molecular mass linear polyacrylamide powder prepared by emulsion polymerization as a replaceable polymer matrix for DNA sequencing by capillary electrophoresis, *Electrophoresis*, 19, 242, 1998.
91. Carrilho, E., et al., Rapid DNA sequencing of more than 1000 bases per run by capillary electrophoresis using replaceable linear polyacrylamide solutions, *Anal Chem*, 68, 3305, 1996.
92. Barbier, V., et al., Comb-like copolymers as self-coating, low-viscosity and high-resolution matrices for DNA sequencing, *Electrophoresis*, 23, 1441, 2002.
93. Wang, Y., et al., Quasi-interpenetrating network formed by polyacrylamide and poly(*N*,*N*-dimethylacrylamide) used in high-performance DNA sequencing analysis by capillary electrophoresis, *Electrophoresis*, 26, 126, 2005.
94. Carrilho, E., DNA sequencing by capillary array electrophoresis and microfabricated array systems, *Electrophoresis*, 21, 55, 2000.
95. Dovichi, N.J., DNA sequencing by capillary electrophoresis, *Electrophoresis*, 18, 2393, 1997.
96. Dovichi, N.J. and Zhang, J.-Z., *Methods in Molecular Biology*, Totowa, N.J. (Ed.), United States, 2001, p. 225.
97. Yeung, E.S. and Li, Q., *Chemical Analysis*, New York, 1998, 767.
98. Kheterpal, I. and Mathies, R.A., Capillary array electrophoresis DNA sequencing, *Anal Chem*, 71, 31A, 1999.
99. Li, Q. and Yeung, E.W., Simple two-color base-calling schemes for DNA sequencing based on standard four-label Sanger chemistry, *Appl Spectrosc*, 49, 1528, 1995.

100. Dovichi, N.J. and Zhang, J., How capillary electrophoresis sequenced the human genome this essay is based on a lecture given at the Analytica 2000 conference in Munich (Germany) on the occasion of the Heinrich-Emanuel-Merck prize presentation, *Angew Chem Int Ed Engl*, 39, 4463, 2000.
101. Kamahori, M. and Kambara, H., Capillary array electrophoresis analyzer, *Methods Mol Biol*, 163, 271, 2001.
102. Willems, A.V., et al., Development of a quality control method for the characterization of oligonucleotides by capillary zone electrophoresis-electrospray ionization-quadrupole time of flight-mass spectrometry, *Electrophoresis*, 26, 1412, 2005.
103. Olson, N.A., Khandurina, J., and Guttman, A., DNA profiling by capillary array electrophoresis with non-covalent fluorescent labeling, *J Chromatogr A*, 1051, 155, 2004.
104. Chen, S.H. and Gallo, J.M., Use of capillary electrophoresis methods to characterize the pharmacokinetics of antisense drugs, *Electrophoresis*, 19, 2861, 1998.
105. DeDionisio, L.A. and Lloyd, D.H., *Progress in HPLC-HPCE*, VSP, Hayward, 1997, p. 293.
106. DeDionisio, L.A., Raible, A.M., and Nelson, J.S., Analysis of an oligonucleotide N3′ →P5′ phosphoramidate/phosphorothioate chimera with capillary gel electrophoresis, *Electrophoresis*, 19, 2935, 1998.
107. Malek, A.H. and Khaledi, M.G., Monitoring liposome-mediated delivery and fate of an antisense drug in cell extracts and in single cells by capillary electrophoresis with laser-induced fluorescence, *Electrophoresis*, 24, 1054, 2003.
108. Belenky, A., Smisek, D.L., and Cohen, A.S., Sequencing of antisense DNA analogues by capillary gel electrophoresis with laser-induced fluorescence detection, *J Chromatogr A*, 700, 137, 1995.
109. Froim, D., et al., Method for phosphorothioate antisense DNA sequencing by capillary electrophoresis with UV detection, *Nucleic Acids Res*, 25, 4219, 1997.
110. Vilenchik, M., Belenky, A., and Cohen, A.S., Monitoring and analysis of antisense DNA by high-performance capillary gel electrophoresis, *J Chromatogr A*, 663, 105, 1994.
111. Pariat, Y.F., et al., Separation of DNA fragments by capillary electrophoresis using replaceable linear polyacrylamide matrices, *J Chromatogr A*, 652, 57, 1993.
112. Strege, M. and Lagu, A., Separation of DNA restriction fragments by capillary electrophoresis using coated fused silica capillaries, *Anal Chem*, 63, 1233, 1991.
113. Guszczynski, T., et al., Capillary zone electrophoresis of large DNA, *Electrophoresis*, 14, 523, 1993.
114. Barron, A.E., Blanch, H.W., and Soane, D.S., A transient entanglement coupling mechanism for DNA separation by capillary electrophoresis in ultradilute polymer solutions, *Electrophoresis*, 15, 597, 1994.
115. Chiu, T.C. and Chang, H.T., Comparison of the separation of large DNA fragments in the presence and absence of electroosmotic flow at high pH, *J Chromatogr A*, 979, 299, 2002.
116. Zheng, J. and Yeung, E.S., Mechanism for the separation of large molecules based on radial migration in capillary electrophoresis, *Anal Chem*, 75, 3675, 2003.
117. Zhou, P., et al., Electrophoretic separation of DNA using a new matrix in uncoated capillaries, *J Chromatogr A*, 1083, 173, 2005.
118. Kim, Y. and Morris, M.D., Ultrafast high resolution separation of large DNA fragments by pulsed-field capillary electrophoresis, *Electrophoresis*, 17, 152, 1996.
119. Kabatek, Z., Kleparnik, K., and Gas, B., Effect of temperature on the separation of long DNA fragments in polymer solution, *J Chromatogr A*, 916, 305, 2001.
120. Tseng, W.L., et al., Nanoparticle-filled capillary electrophoresis for the separation of long DNA molecules in the presence of hydrodynamic and electrokinetic forces, *Electrophoresis*, 26, 3069, 2005.
121. Ueda, M., et al., Electrophoresis of long DNA molecules in linear polyacrylamide solutions, *Biophys Chem*, 71, 113, 1998.
122. Geldart, S.E. and Brown, P.R., Analysis of nucleotides by capillary electrophoresis, *J Chromatogr A*, 828, 317, 1998.
123. Grob, M.K., et al., Optimization of cellular nucleotide extraction and sample preparation for nucleotide pool analyses using capillary electrophoresis, *J Chromatogr B Analyt Technol Biomed Life Sci*, 788, 103, 2003.
124. Harada, K., Fukusaki, E., and Kobayashi, A., Pressure-assisted capillary electrophoresis mass spectrometry using combination of polarity reversion and electroosmotic flow for metabolomics anion analysis, *J Biosci Bioeng*, 101, 403, 2006.

125. Huang, M., et al., High resolution separation and quantitation of ribonucleotides using capillary electrophoresis, *Anal Biochem*, 207, 231, 1992.
126. Nguyen, A.L., Luong, J.H., and Masson, C., Determination of nucleotides in fish tissues using capillary electrophoresis, *Anal Chem*, 62, 2490, 1990.
127. Soga, T., et al., Quantitative metabolome analysis using capillary electrophoresis mass spectrometry, *J Proteome Res*, 2, 488, 2003.
128. Soga, T., et al., Simultaneous determination of anionic intermediates for *Bacillus subtilis* metabolic pathways by capillary electrophoresis electrospray ionization mass spectrometry, *Anal Chem*, 74, 2233, 2002.
129. Solinova, V., et al., Determination of acid-base dissociation constants of amino- and guanidinopurine nucleotide analogs and related compounds by capillary zone electrophoresis, *Electrophoresis*, 27, 1006, 2006.
130. Balayiannis, G., Papaioannou, D., and Karamanos, N.K., Application of capillary electrophoresis in the analysis of novel synthetic dideoxynucleoside analogues with potential anti-HIV activity, *Biomed Chromatogr*, 15, 271, 2001.
131. Markushin, Y., et al., Potential biomarker for early risk assessment of prostate cancer, *Prostate*, 2006.

7 Analysis of Carbohydrates by Capillary Electrophoresis

Julia Khandurina

CONTENTS

7.1 INTRODUCTION

Carbohydrates are widely distributed in nature. Generally speaking, they represent a class of polyhydroxylated compounds, which contain a carbonyl functional group, aldehyde or ketone, often as semiacetal or hemiketal. The abundance of hydroxyl groups and a carbonyl functionality result in the formation of diverse oligomers and macromolecules. Carbohydrate biopolymers, oligo- and polysaccharides, are often very structurally complex comprising monosaccharides linked by glycosidic bonds. The most common examples are α- and β-1,4-linked polyglucans, for example, cellulose, the main constituent of plant cell walls, or starch, which is present in large quantities in plant seeds and roots. Many fruits contain high concentration of oligosaccharides based on glucose, fructose, and raffinose building blocks. On the other hand, bacterial cell walls are built of complex glycans containing glucosamine, muramic acid, and oligosaccharide derivatives of various monosaccharides. Various kinds of carbohydrates play important roles in animal and human tissues supporting their living activities. For example, glucose is a major constituent of blood that generates energy via TCA metabolic cycle (tricarboxylic acid cycle, also known as citric acid cycle, or the Krebs cycle). Glycogen, a highly branched glucose polymer, is the principal energy storage in animal and human cells. Abnormal increase of glucose in blood is a symptom of diabetes; therefore, rapid estimation of its level is an important task in clinical applications. Polymerized glucosamine constructs hard shell animal and insect substances. Besides, carbohydrates exist in living organisms in various forms of glycoconjugates. Glycan moieties in glycoconjugates, where oligosaccharide chains are covalently linked to an aglycon (e.g., protein or lipid), are very structurally and functionally diverse. They are responsible for key functions in protein folding and stability, biological recognition, and cell–cell interaction processes [1–3]. Glycoconjugates act as receptors on the cell surface. Structures of glycomoieties range from relatively simple, as in glycosphingolipids (GSLs) and some glycoproteins, to extraordinary complex mucins and proteoglycans (PGs). Saccharides have an enormous potential to carry information, far exceeding that of proteins and nucleic acids [4], due to diversity of monomer building blocks and linkages between them, as well as substitute groups. Recent advances in the research of biological roles of glycans in glycoconjugates call for the development of more sensitive and reproducible analytical methods to determine their compositions and structures.

In spite of the apparent significance, carbohydrate separation and analysis technology has not been as universal in life sciences and biotechnology as nucleic acid and protein analysis. Heterogeneity of oligo- and polysaccharides in molecular weight, primary sequence and branching, linkage, variety of sugar structural isoforms, and lack of chromophore/fluorophore or active functional groups have presented a challenge. A number of analytical methods have been applied to carbohydrate analysis, including paper and thin-layer chromatography [5,6], gas chromatography [8], high-performance liquid chromatography (HPLC) [7], high-pH anion-exchange chromatography (HPAEC) [9–12], supercritical fluid chromatography [13], affinity chromatography [14], polyacrylamide slab gel electrophoresis and capillary electrophoresis (CE) as separation, and mass spectrometry and nuclear magnetic resonance spectroscopy as structural characterization techniques [15]. CE features the highest separation efficiency in carbohydrate analysis compared with the other reported methods. A number of excellent reviews on carbohydrate analysis by CE have been published in the past [16–21], which described the method in detail.

This chapter is intended to summarize most significant advances in carbohydrate analysis by CE-based techniques during the past decade. Clearly, it is no longer possible to write a comprehensive review of the massive literature published recently, which includes a variety of different aspects and applications involving carbohydrate analysis. Therefore, the focus has been made on the most significant and interesting, in the author's opinion, developments in high separation efficiency of CE, hyphenated techniques, in particular, CE-mass spectrometry (CE-MS) and analysis of complex polysaccharides and glycoconjugates. In addition, special attention has been paid to specific applications, recently emerging in pharmaceutical and bioindustrial arenas. These include analysis of polysaccharides and glycoconjugates in drugs, analysis of traditional Chinese medicines, and approaches to high-throughput carbohydrate profiling at the large industrial scale, particularly, problems associated with the processes of biomass conversion to fermentable sugars. Recent trends toward miniaturization of CE carbohydrate analysis have also been reviewed.

7.2 BACKGROUND: HIGH-PERFORMANCE CE SEPARATIONS OF CARBOHYDRATES

Numerous research papers and reviews on carbohydrate separations by CE have been written for the past several years. Researches have successfully addressed problems, such as tremendous diversity and complexity of this class of compounds, polar and neutral nature of most carbohydrates, their low ultraviolet (UV) extinction coefficients, and lack of functional groups. In the previous edition of this book, Olechno and Nolan [16] published a comprehensive overview of the CE separation techniques, attempted and developed for intact and derivatized carbohydrates, charged and neutral, as well as detection approaches by UV, indirect fluorescence, electrochemical (e.g., amperometric) detection, refractive index, and laser-induced fluorescence (LIF). A variety of buffer systems were introduced, depending on the nature of mono- and oligosaccharides of interest: acidic and basic, using normal or reversed polarity, either with electroosmotic flow (EOF) present or suppressed. For more details and specific examples, the reader is directed to Reference 16. In brief, the most popular buffers utilized have been phosphates (pH 3 and 9), borates (pH 8–9), mixed phosphate-borate systems, acetates, and so forth [16]. Addition of surfactants, such as sodium dodecyl sulfate (SDS) or cetyltrimethylammonium bromide (CTAB), has often improved the peak efficiency due to micellar electrokinetic chromatography (MEKC) effect. Other additives have been exploited as well, for example, thriethylamine to improve run-to-run reproducibility, or polyethyleneoxide (PEO) to enhance resolution, most likely, by imposing additional sugar–polymer interactions based on hydrogen bonding. Owing to weak acidic properties, neutral sugars become ionized at very high-pH values (pH 12–13) and can be separated according to their pK_a differences under strongly alkaline conditions (analogous to HPAEC). Borate-assisted CE has been widely used for the analysis of both native and derivatized sugars. Borate ions are known to form anionic complexes with polyols (e.g., sugars) and more stable adducts are typically formed with *cis*-oriented hydroxyl groups. Therefore, in the presence of borate anions the resolving power of CE can be significantly enhanced, and very structurally similar carbohydrate molecules can be separated.

In most cases, carbohydrate molecules must be derivatized to render charge, prior to or during electrophoretic separation, as majority of carbohydrates are neutral. This can be achieved by interaction with oxoacid [22] or metal ions [23], or introduction of ionic tags [24,25]. On the other hand, electrokinetic chromatography approach using hydrophilic monolithic columns, developed by Novotny and coworkers [26], extended CE applicability to neutral sugar molecules. Certain surfactants, such as SDS, for example, form negatively charged micelles in the running buffer and serve as pseudostationary phase that incorporates carbohydrate molecules with a hydrophobic tag [e.g., 1-phenyl-3-methyl-5-pyrazolone (PMP)] [27] to a various extent, resulting in differential migration and separation.

Absence of chromophore or fluorophore groups and very low extinction coefficients of most carbohydrate molecules called for application of alternative (to UV and fluorescence) detection methodologies. Electrochemical detection and refractometry allow direct measurements although the sensitivity is relatively low. Carbohydrates are very weak acids, and their hydroxyl groups get ionized at very high pH values (>12–13), enabling CE analysis under such strong alkaline conditions using gold or transition metal electrodes [28,29] in electrochemical detection settings.

With respect to UV or fluorescent detection, a number of methods have been developed to introduce UV absorbing and/or fluorescent tags into carbohydrate molecules. Derivatization of carbohydrates, to render chromophore/fluorophore group and/or charge, was first demonstrated in CE by Honda [22] who used 2-aminoacridone (first introduced by Hase for HPLC analysis [30]). Since then, different derivatization strategies have been explored. Most typical of these are reductive amination [e.g., 2-aminoacridone, 2-aminopyridine, 4-aminobenzoate, *p*-aminobenzonitrile, APTS, 8-amino-naphtalene-1,3,6-trisulfonic acid (ANTS), etc.]; reaction of amino sugars with 1,2-dioxo aromatic compounds to form isoindoles [e.g., *o*-phthaldehyde, 3-(4-carboxybenzoil)-2-quinoline-carboxaldehyde]; reaction of aldoses with pyrazalones (e.g., PMP); reaction of dioxo sugars with 1,2-diamonoaromatic compounds; amidation of sugar carboxylic acids with aromatic amines through carbodiimide effect [16]. PMP tag allows quantitative labeling under mild conditions, in both pre-column mode and *in migratio* [31]. The most effective fluorescent derivatization agents have been APTS and methylglycamine-4-nitro-2,1,3-benzoxadiazole (MG-NBD) [17,32]. These methods are based on reduction amination, which results in quantitative and uniform labeling of different carbohydrates, and utilize most common Ar-ion laser for excitation (488 nm) [33–36]. The details of these and other derivatization approaches are well summarized in several articles [18–19,32].

High-performance and resolving power of CE enable distinguishing between very similar structural isomers of short oligosaccharides, which have different anomeric and positional configurations of sugar monomers. Typical separations of various APTS-labeled homooligosaccharide series are exemplified in Figure 7.1 [37]. Traces A and B correspond to oligomers, composed of glucose

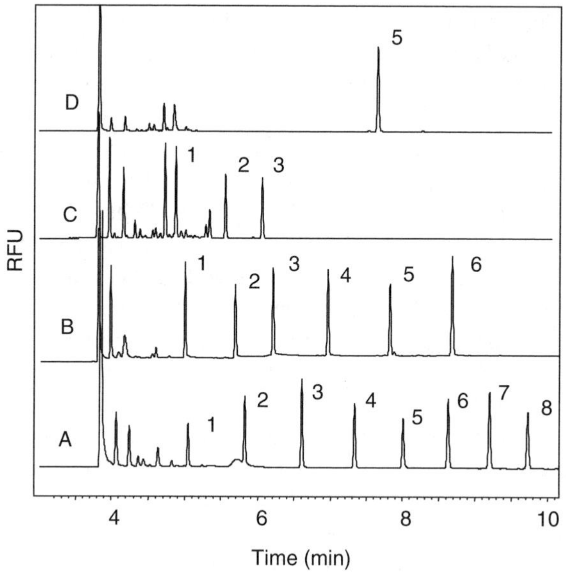

FIGURE 7.1 High-resolution CE separation of oligosaccharide structural isomers. Traces from the bottom: (A) maltooligosaccharide oligomers (DP 1–8); (B) cellooligosaccharide oligomers (DP 1–6); (C) xylooligosaccharides (DP 1–3); and (D) xylopentaose. Numbers above the peaks correspond to the DP. Conditions: fused-silica capillary, 60 cm × 50 μm ID, effective separation length 50 cm; running buffer 25 mM lithium acetate (pH 5); dynamic coating, 1% PEO; injection 5 s at 0.5 psi (3447 Pa); applied electric field 500 V/cm; separation temperature 20°C. (Adapted from Khandurina, J. and Guttman, A. *Chromatographia*, 62, S37, 2005. With permission.)

molecules linked via α-1,4 (maltooligosaccharides) and β-1,4(cellooligosaccharides) bonds, respectively. Numbers above the peaks denote the degree of polymerization (DP). As one can see, species with the same DP but different anomeric configuration between the monomer units (α or β) have different electrophoretic mobilities, and can be readily separated in CE, using 25 mM lithium acetate (pH 5) buffer in combination with a dynamically coated column and reversed polarity. The shorter cello-oligomers migrated faster than the corresponding malto-analogs (DP 2–5 peaks in traces A and B, Figure 7.1). However, the migration order was reversed above DP 5, as the cellohexaose migrated slower than maltohaxaose. Most likely, the differences in the spatial configuration, rigidity and hydrodynamic radius of the short α- and β-glucopyranose oligosaccharides in the solution caused this change in the migration order. Traces C and D in Figure 7.1 depict the separation of xylose series (xylose, xylobiose, xylotriose, and xylopentaose), which possess higher electrophoretic mobility compared to glucose series (traces A and B, Figure 7.1) with the same DP, due to more compact structure of pentose-based sugars versus hexose ones. Very close structural isomers of di- and trisaccharides can be resolved by CE, under suppressed EOF conditions, as demonstrated in Figure 7.2. Here, maltose and isomaltose (α-1,4 and α-1,6 linkages) are baseline resolved (peaks 1 and 2). On the other hand, trisaccharide analogs (maltotriose and panose, peaks 3 and 4) could be separated only with the addition of the polymer additive, PEO. The PEO additive is capable of selectively interacting with sugar hydroxyls via the formation of hydrogen bonds, thus, enhancing (or decreasing) separation selectivity. As it is illustrated in Figure 7.2, the described CE approach can provide an excellent tool for fine tuning of the separation selectivity of the carbohydrate molecules of interest. CE profile of a complex mixture of 10 mono- and oligosaccharides, constituents of plant cell walls, is presented in Figure 7.3. An excellent resolution of the three different galactobiose isoforms (β-1,4; α-1,4; and α-1,3 linkages) was achieved in capillary zone electrophoresis (CZE) format [37].

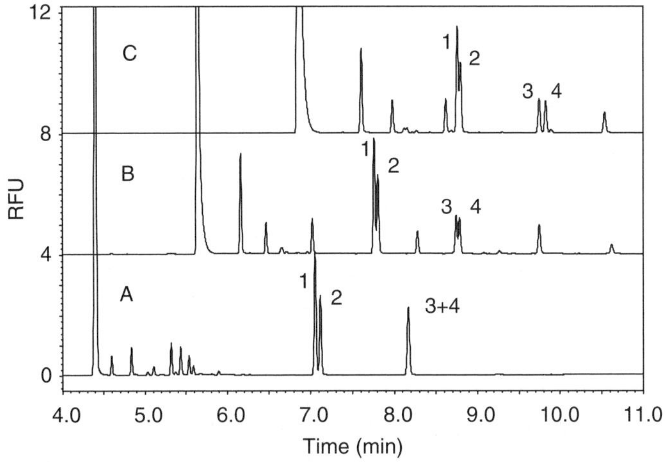

FIGURE 7.2 CE analysis of close structural di- and trisaccharide isomers. Peaks: (1) maltose, (2) isomaltose, (3) panose, and (4) maltotriose. Conditions: 0.5% PEO (B) and 1% PEO (C) was added to the running buffer (no PEO additive in (A). Separation temperature 15°C. Other conditions were the same as in Figure 7.1. (Adapted from Khandurina, J. and Guttman, A. *Chromatographia*, 62, S37, 2005. With permission.)

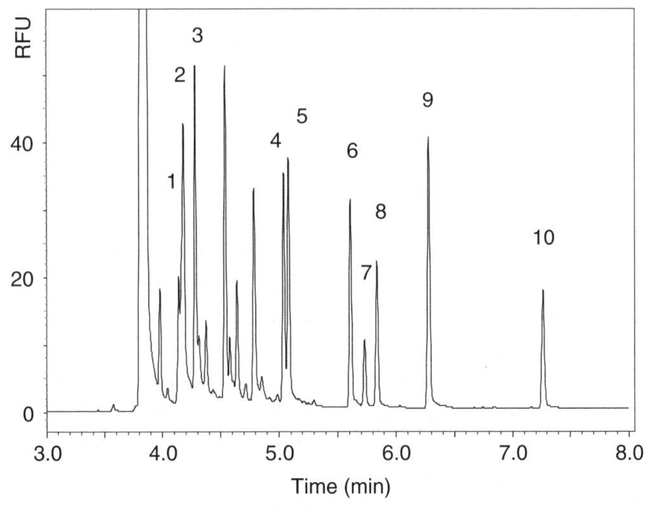

FIGURE 7.3 High-resolution analysis of a combined mono- and oligosaccharide mixture representing fragments of major cell wall components of bioindustrial interest. (1) Galacturonic acid, (2) digalacturonic acid, (3) trigalacturonic acid, (4) rhamnose, (5) galactose, (6) galactobiose β-1,4 linkage, (7) galactobiose α-1,4 linkage, (8) galactobiose α-1,3 linkage, (9) galactotriose α-1,3; β-1,4, and (10) galactotetraose α-1,3; β-1,4; α-1,3. Conditions were the same as in Figure 7.1. (Adapted from Khandurina, J. and Guttman, A. *Chromatographia*, 62, S37, 2005. With permission.)

7.3 THEORETICAL AND PRACTICAL ASPECTS OF MODERN CARBOHYDRATE ANALYSIS: CE-MS

7.3.1 Basic Underlying Principles

The popularity of CE-MS hyphenation has been a fast growing analytical approach, since its introduction in the late 1980s. Over the years the technique has advanced to the commercial ready-to-use level. Still, there is a lot of development to be done toward optimization of CE-MS instrumentations, in particular, improving its robustness and efficiency. A variety of analytical applications in different fields have been investigated [38]. With respect to carbohydrate analysis, MS detection offers a way to overcome detection sensitivity problems associated with the lack of chromophore groups, as well as structural elucidation of complex carbohydrate molecules. Structural characterization of carbohydrates includes the determination of molecular weight, type, and number of saccharide components, sequence and branching, glycosidic attachment sites, anomeric configurations, conformation of glycosyl ring, and secondary structure. Electrospray ionization (ESI) and matrix-assisted laser desorption ionization (MALDI) mass spectrometry, especially in the tandem MS^n mode, are the most popular approaches that offer high ionization efficiency of saccharides and possibility of their sequencing. In particular, nano-ESI-MS/MS, in both positive and negative mode, enable elaborate structural determination of carbohydrates from biological samples using only sub-pmole amounts [39–46]. Basically, three major types of ESI interfaces have been used for CE-MS coupling: coaxial sheath-flow, sheathless, and liquid junction, as schematically presented in Figure 7.4 [47]. Different modern mass analyzers e.g., ion trap (IT), quadrupole time-of-flight (QTOF), TOF–TOF, and even Fourier transform ion cyclotron resonance (FT-ICR) can provide highly informative fragmentation spectra of oligosaccharides [26,48–53]. One of the challenges in carbohydrate characterization in complex mixtures is associated with the presence of overlapping isobaric peaks that complicates spectra interpretation. In addition, modified and substituted sugar molecules are prone to in-source fragmentation due to labile nature of acetyl, phospho-, sulfo-, and other functional groups, causing loss of structural information of intact molecules. Moreover, most carbohydrates, unlike peptides and proteins, do not form multiply charged ions causing difficulties analyzing oligo- and polysaccharides of high molecular weight, which often exceeds MS detection capabilities [4].

MS analysis of complex biological mixtures of carbohydrates often follows a liquid phase separation method, such as chromatography (HPLC), CE, or capillary electrochromatography (CEC). The advantages of mass spectrometry, combined with superior resolving power of CE, have boosted glycoscreening development in biomedical research [4]. Very efficient and fast separations, automation and capabilities of miniaturization, already proven for small molecules, peptides, proteins, and nucleic acids [54–57], represent a promising analytical alternative for all types of carbohydrates. CE-MS combination can be carried out either off-line or online. The former approach was often realized by CE fraction collection followed by ESI-MS characterization of individual fractions, which contain a single carbohydrate species or mixtures of significantly reduced complexity. However, fraction dilution associated with this approach (collection of nanoliter fractions into microliter total volume) imposes additional detection sensitivity challenges, especially with limited amounts of glycoconjugate samples extracted from biological sources, and therefore, is of limited use. On the other hand, online CE-MS interface has become truly widespread during the past decade. This methodology allows minimum sample handling and consequently maximum efficiency in MS detection. A limitation though is the necessity to select ESI-MS compatible volatile buffers, which often do not possess the highest resolving power characteristics for CE. Besides, high sensitivity and high speed of data acquisition, especially for online CE-MS/MS, and certain restrictive conditions of ion formation also exist [58]. In spite of these limitations, CE-MS technique is invaluable, providing carbohydrate composition identification, quantification, and structural insights. To decipher molecular composition, MS/MS measurements are typically conducted using low-energy collision-induced dissociation (CID) in ESI applications or MALDI postsource decay (PSD). These types of fragmentation preferentially cause glycosidic bond cleavage. Sugar ring cut, on the other hand, resulting in the formation of mostly C and Y type alcoholate ions [39], provide information on the branching pattern. The latest developments in CE-MS analysis of carbohydrates are highlighted and summarized in a number of comprehensive reviews (e.g., by Zamfir and Peter-Katalinic [4], Campa et al. [59], and Kamoda and Kakehi [60]). Here, we briefly outline examples of CE-MS applications to the analysis of carbohydrates, including complex glycoconjugate mixtures, neutral and negatively

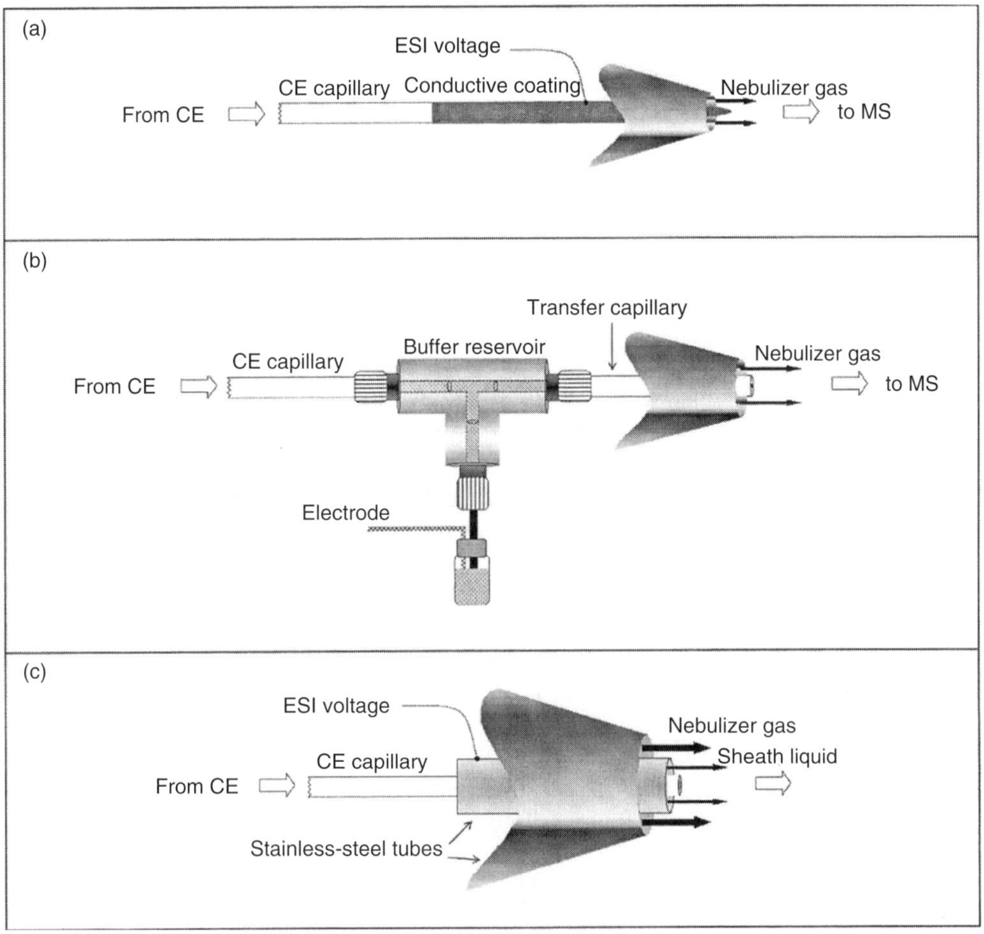

FIGURE 7.4 Interfaces for the direct coupling of CE to ESI-MS using (A) sheathless; (B) liquid-junction; and (C) coaxial sheath-flow designs. (Adapted from Simo, C. et al. *Electrophoresis*, 26, 1306, 2005. With permission.)

charged glycopeptides and glycoproteins, glycosaminoglycans (GAGs), lipopolysaccharides (LPS), and gangliosides.

7.3.2 CE-MS Characterization of Simple Sugars and Glycoconjugates

CE-MS analysis of simple sugar mixtures is typically conducted using either derivatized neutral sugars or underivatized carbohydrates, containing ionizable groups (e.g., carrageenan oligosaccharides) [61]. Derivatization is typically accomplished through reductive amination coupling to negatively charged tags, such as aminobenzoic acid (ABA) [62], 8-aminonaphtalene-1,3,6-trisulfonic acid (ANTS) [59], APTS [63], and 7-amino-1,3-naphtalene-disulfonic acid (ANDSA) [64]. Positively charged tags, for example, 3-(acetylamino)6-aminoacridone [65] or rhodamine 123 [59], are sometimes used as well. Labeling results in enhanced ionization efficiency and MS response, as well as CE separation selectivity [48], for most carbohydrates regardless of their native charge. Simple sugars are often used as standard mixtures to assess performance and optimize CE-MS conditions prior to characterization of more complex samples.

Glycoproteins, a very important type of glycoconjugates, are formed during co- and posttranslational modification in eukaryotic cells. Typically, N-glycans are linked to asparagines, while O-glycan moieties to serine or threonine. Protein glycosylation is characterized by pronounced microheterogeneity of glycosylation sites, which poses challenges in controlled manufacturing of recombinant glycoprotein therapeutics [66]. CE-MS analysis has been successfully applied to characterization of sugar moieties, native or derivatized, using direct or reversed polarity CE in combination with ESI-MS or MALDI. Since typically only low concentrations of glycans are available from glycoconjugates, various derivatization approaches have been implemented to increase MS response and facilitate interpretation of MS/MS spectra of glycans. Some research groups developed new derivatizing agents with improved physicochemical properties and better suited for CE-MS applications [65,67]. Min et al. [68] reported a new interesting labeling approach based on enzymatic reaction rather than chemical derivatization. In this work a fluorescent acceptor, naphthalene-2,3-dicarboxaldehide-asparagine-*N*-acetylglucosamine, was used. Asn-linked glycan moieties were enzymatically transferred from ovalbumin to the tag reagent. This strategy can help alleviate degradation problems sometimes observed in reductive amination, especially with unstable glycan residues (e.g., sialylated carbohydrates). A number of research groups used APTS and ANTS labeling in conjunction with CE-MS/MS analysis of N-glycans following their enzyme-assisted release from glycoproteins [69–71]. CE-UV and CE-MS electropherograms, obtained for the same oligosaccharide mixture, were compared under optimized analysis conditions in Reference 69. As expected, separation efficiency and selectivity were lower in CE-MS, primarily due to sheath liquid dilution and certain dead volume in the capillary outlet-MS interface. In addition, application of an MS-friendly buffer (6-aminocapronic acid, pH 4.1), instead of 10 mM citric acid (pH 3) used in CE-UV settings, could have contributed to the decrease of selectivity. Nevertheless, the use of powerful mass spectrometers can counterbalance the loss of separation efficiency, as it was demonstrated by successful structural characterization of a number of *N*-linked carbohydrates and other glycoconjugates [4,72,73]. The MS^n capacity of ion-trap spectrometer has been particularly useful in elucidation of fine structural details.

Analysis of high-mannose-type oligosaccharides by a combination of micro-liquid chromatography (μLC)-MS and CE was reported by Koller et al. [70]. The authors separated glycopeptides and high-mannose-type oligosaccharides, derived and digested from a recombinant enzyme phospholipase C expressed in *Pichia pastoris* yeast. The glycopeptides were subjected to μLC-ESI-MS and μLC-ESI-MS/MS that revealed variation in high-mannose structures in the range between $Man_7GlcNAc_2$ and $Man_{14}GlcNAc_2$. Then, high-performance CE was applied to identify possible positional isomers of the high-mannose structures. One $Man_9GlcNAc_2$, two $Man_{10}GlcNAc_2$, three $Man_{11}GlcNAc_2$, $Man_{12}GlcNAc_2$, and $Man_{13}GlcNAc_2$, and two $Man_{14}GlcNAc_2$ were identified, while no $Man_7GlcNAc_2$ and $Man_8GlcNAc_2$ were observed (Figure 7.5). It was found that CE results provided complementary information to the μLC-ESI-MS/MS data, revealing the exact number of positional isomers in the glycan pool.

Cell surface and extracellular proteins are typically O-glycosylated, and the most abundant type of O-glycosylation is the attachment of *N*-acetylgalactosamine (GalNAc) to serine or threonine in the protein sequence by *a*-glycosidic linkage [74]. Many eukaryotic cytoplasmic and nuclear proteins, containing such carbohydrate moiety, exhibit reciprocal glycosylation and phosphorylation during cell cycle, stimulation, and growth. In the group of O-GalNAc glycosylated proteins, musins are the most ubiquitous. They are especially interesting with respect to cancer research, understanding of cell adhesion, and metastasis processes. The distribution of O-glycans in human tissues and sera is important for diagnostic and prognostic studies [74,75]. Mucin-like O-glycans are characterized by high degree of structural variation, even at the core level, as well as in the chain elongation arising from the core extension (Figure 7.6). The carbohydrate chains often contain repeating units of *N*-acetyllactosamine, terminal sialic groups, and also noncarbohydrate substituents. Less investigated types of O-glycosylation are O-fucosylation, O-mannosylation, and O-glucosylation. These types of modifications are vital for physiological functions of proteins. O-glycans are traditionally detached

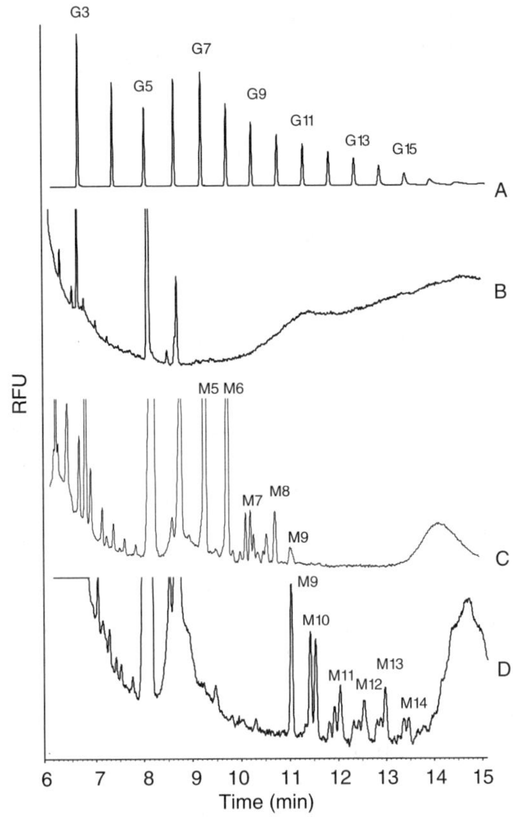

FIGURE 7.5 High-performance CE separation of the APTS-labeled high-mannose type oligosaccharides released by PNGase F from bovine ribonuclease B (C) and PLC (D). (A) Electropherogram of the maltooligosaccharide ladder; (B) derivatization background peaks. Conditions: 48 cm neutrally coated column (58 cm total length), 50-μm ID. LIF detection: Ar-ion laser, 488 nm excitation, 520 nm emission. Separation buffer: 25 mM acetate (pH 5); 25°C capillary cartridge temperature; pressure injection 10 s at 0.5 (A) and 2 psi (B-D). (Adapted from Koller, A. et al. *Electrophoresis*, 25, 2003, 2004. With permission.)

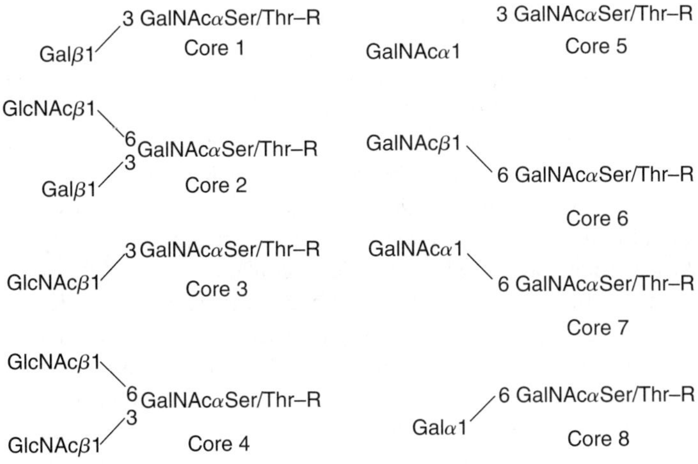

FIGURE 7.6 Eight types of O-glycan core structures. (Adapted from Peter-Katalinic, J. *Methods Enzymol*, 405, 139, 2005. With permission.)

from the parent protein by β-elimination using strong bases (e.g., NaOH or hydrazine). Hydrozinolysis does not always result in the selective cleavage of O-glycans, and often is accompanied by N-glycan release, as well as modification of the reducing terminus [74] and references therein. β-Elimination by NaOH under reductive conditions allows chemical derivatization of the reducing end (e.g., with a fluorescent label). Nonreductive release of O-glycans by β-elimination is also possible [74], as well as an enzymatic release, but the latter options are limited by high substrate specificity of O-glycosidases available. Comprehensive profiling of complex glycan and glycopeptide mixtures requires sophisticated analysis methods to characterize glycosylation patterns and structures of individual components, and the coupling of CE separation devices to ESI-MS or MALDI-MS is a powerful approach. Off-line CE-MALDI analysis has been used for the characterization of musin carbohydrate chains, comprising up to six GalNAc moieties [75]. CE-MS/MS interface enabled deducting molecular structure of individual glycopeptide components based on specific fragmentation, especially when micro- and nano-ESI sources are used. O-glycopeptides in urine of patients, suffering from a hereditary *N*-acetylhexosaminidase deficiency, were profiled by this approach, using high-pH and MS-compatible buffer for CE separation, sheathless CE-ESI-MS interface, and QTOF-MS equipped with automated high-speed MS-MS/MS switching [75]. The MS/MS fragmentation of glycans results mainly in the formation of Y and C types of ions, which are indicative of the specific structures of the parent ions. The sheathless CE-ESI/QTOF design proved to be one of the most suited designs for glycomics, providing high coverage of detection and identification of CE-MS data, MS-MS/MS switching, electrospray stability, and acquisition rate [74,75].

The glycan MS fragmentation nomenclature introduced by Costello, Domon, and Vath [76–79] is a widely accepted system used by most of the MS and glycoconjugate community. The simplest fragmentation of the carbohydrate moiety of glycoconjugates and glycosides results from the cleavage of the glycosidic bond and therefore provides information on the sugar sequence. More complex processes involving the fragmentation of the sugar ring can occur, particularly, in CID-MS/MS spectra. These ions are more difficult to assign, although they contain important structural information. Figure 7.7 schematically illustrates the nomenclature. Fragments are designated A_i, B_i, and C_i, where i represents the number of the glycosidic bond cleaved, counted from nonreducing end. Ions containing aglycone, or the reducing sugar unit in case of oligosaccharides, are labeled as X_j, Y_j, and Z_j, where j is the number of interglycosidic bond counted from aglycone, or reducing end. The glycosidic bond linking to aglycone is 0. More complex fragmentations involve cleavage of C–C bonds of the sugar ring. The product ions are designated as A_i and X_j. Since several ring cleavage pathways are possible, two additional superscripts are used (e.g., $^{k,1}A_j$) to indicate the sugar ring bonds that have been broken [78]. This system, described in more detail elsewhere [78,80], comprises types of cleavages, observed in collision-induced decomposition MS/MS spectra of carbohydrate moiety of glycoconjugates, and is applicable to both positive and negative ionization modes.

Glycolipids are carbohydrate-attached lipids (e.g., phospholipids) found on the outer exoplasmic surface of all eukaryotic cell membranes. Their biological role is to provide energy and serve as markers for cellular recognition. One subtype is LPSs, which are major components of the cell membrane of Gram-negative bacteria, contributing greatly to the structural integrity, and protecting the membrane from certain kinds of chemical attacks. LPS are endotoxins and induce a strong response from normal animal immune systems. It comprises three parts: polysaccharide (O) side chains, core polysaccharides, and lipid A. A lipid A contains certain fatty acids (e.g., hydroxymyristic acid), and is inserted into the outer membrane while the rest of an LPS projects from the surface. A core polysaccharide contains unusual sugars, such as keto-deoxyoctulonate and heptulose. It also contains two glucosamine sugar derivatives, each having three fatty acids with phosphate or pyrophosphate attached. The core polysaccharide is attached to lipid A, which is also in part responsible for the toxicity of Gram-negative bacteria. The polysaccharide side chain is referred as the O-antigen of the bacteria. O side chain (O-antigen) is also a polysaccharide chain that extends from the core polysaccharide. The composition of the O side chain varies between different Gram-negative bacterial strains. O side chains are easily recognized by the host antibodies, however, the nature of

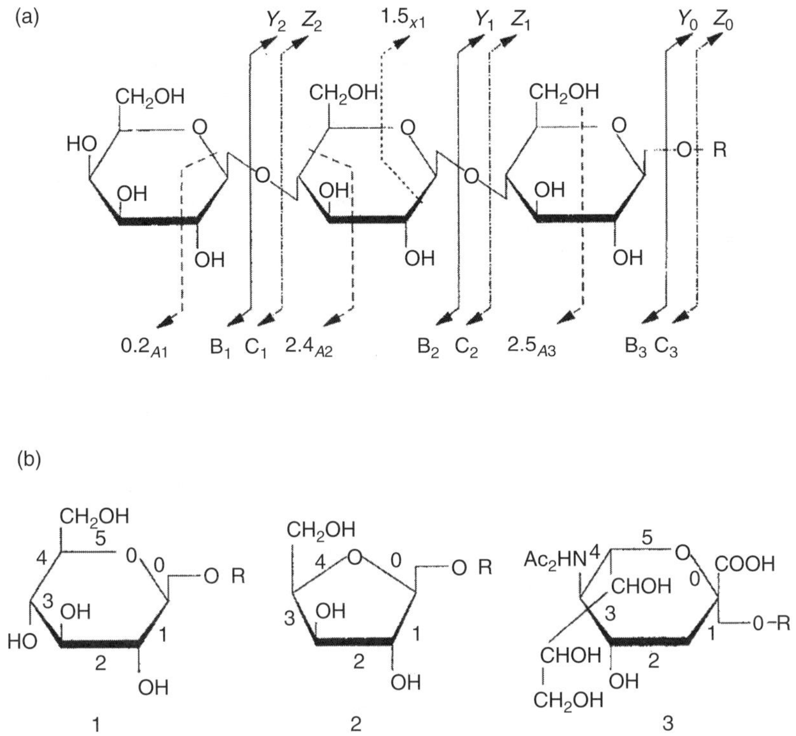

FIGURE 7.7 Types of carbohydrate fragmentation. (Adapted from Domon, B. and Costello, C. E. *Glycoconjugate J*, 5, 397, 1988. With permission.)

the chain can easily be modified by the bacteria to avoid detection. LPS also increases the negative charge of the cell wall and helps stabilize the overall membrane structure [81,82].

A comprehensive discussion on the application of CE-MS to the analysis of complex LPS can be found in Reference 83 and references therein. Various research groups developed electrophoretic conditions to conduct trace-level enrichment and separation of closely related glycoforms and isoforms. Sensitive detection of glycolipids from as little as five bacterial colonies has been demonstrated. Mixed MS scanning functions can assist in the identification of specific LPS functionalities [e.g., pyrophosphoethanolamine, phosphocholine, and *N*-acetylneuraminic acid (Neu5Ac)]. High resolving power of CE combined with sensitivity of tandem MS provides a unique analytical tool to probe subtle structural differences and location of oligosaccharide isoforms of LPS. Correlation of structural changes in bacterial strains and isogenic mutants allow establishing functional gene relationships, using CE-MS screening capabilities and wide dynamic range.

Figure 7.8 shows structure of the oligosaccharide region of LPS from *Neisseria meningitidis* and *Haemophilus influenzae*. LPS of *H. influenzae* contains a common L-glycero-D-manno-heptose inner-core trisaccharide unit attached to the lipid A via a phosphorylated 2-keto-3-deoxyoctulosonic acid (KDO) residue (Figure 7.8A). Similarly, *N. meningitidis* (Figure 7.8B) has a conserved inner core structure with two KDOs, two heptoses (Hep), and one *N*-acetylglucosamine (GlcNAc). Further carbohydrate extension can occur at each heptose. In addition, various noncarbohydrate substituents can be incorporated, for example, acetyl (Ac), phosphate (P), phosphoethanolamine (PE), pyrophosphoethanolamine (PPE), phosphocholine (PC), as well as different aminoacids (Ala, Ser, Thr, Lys). Details on typical LPS extraction and preparation methods are well described in Reference 83. Hydrozinolysis can be utilized to release *O*-linked fatty acids, and, if necessary, enzyme sialidase to remove Neu5Ac residues. Li et al. [83] described interfacing of a CE instrument with a triple quad

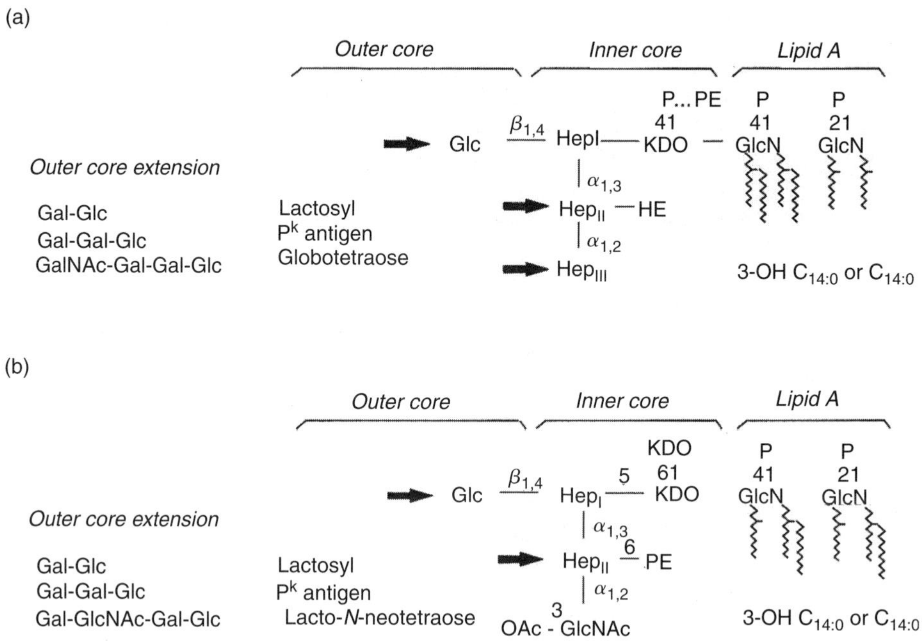

FIGURE 7.8 Structure of the conserved inner core oligosaccharide regions of LPS from *N. meningitidis* and *H. influenzae*. (A) *Haemophilus influenzae* capsular and noncapsular strains. (B) *Neisseria meningitidis* strain. (Adapted from Li, J. et al. *Methods Enzymol*, 405, 369, 2005. With permission.)

or QTOF mass spectrometers via a microionspray assembly, using isopropanol–methanol sheath liquid, a bare fused-silica capillary, and morpholine–formic acid (or ammonium acetate) buffer (pH 9) with 5% methanol. To maintain high separation efficiency of CE ($N > 150{,}000$ plates/m), the injection volume should be kept below 2% of the total column volume (e.g., 40 nL for 1 m × 50 μm ID capillary) [84]. MS identification of bacterial glycolipids can be achieved using 100–300 pg on column loadings, which typically translates into low μg/mL detection limit when using ESI. Different techniques to improve LPS sample loading have been explored, including preconcentration via transient isotachophoresis, hydrophobic adsorption membrane at the sample inlet, and field amplification stacking using a microdialysis interface junction and polarity switching ([83] and references therein).

CE-MS has been employed for the structural elucidation of an O-chain polysaccharide (LPS component exposed on bacteria cell surface and mediating host–cell invasion and other cell–cell interaction processes) from *Aeromonas salmonicida* strains [85,86]. The authors used in-source fragmentation approach for the analysis of these complex macromolecules. As mentioned above, saccharides typically form singly charged ions during the ESI process. Therefore, the masses of these oligosaccharides often exceed the mass limit of common mass spectrometers (e.g., quadrupole or ion-trap types). The in-source fragmentation technique produces fragments that can be readily detected by these instruments.

LPS are often presented by a complex distribution of closely related glycoforms varying by length and site of attachment. Structural characterization of these diverse populations is important for the development of protein therapeutics—antibodies specifically targeted toward these immunodeterminate structures. Glycoforms in the different families of glycolipids are identified based on the progressive extension of Hex residues on the core structure, and concurrent increase in molecular weight and decrease in electrophoretic mobility. This approach is described in great detail by Li et al.

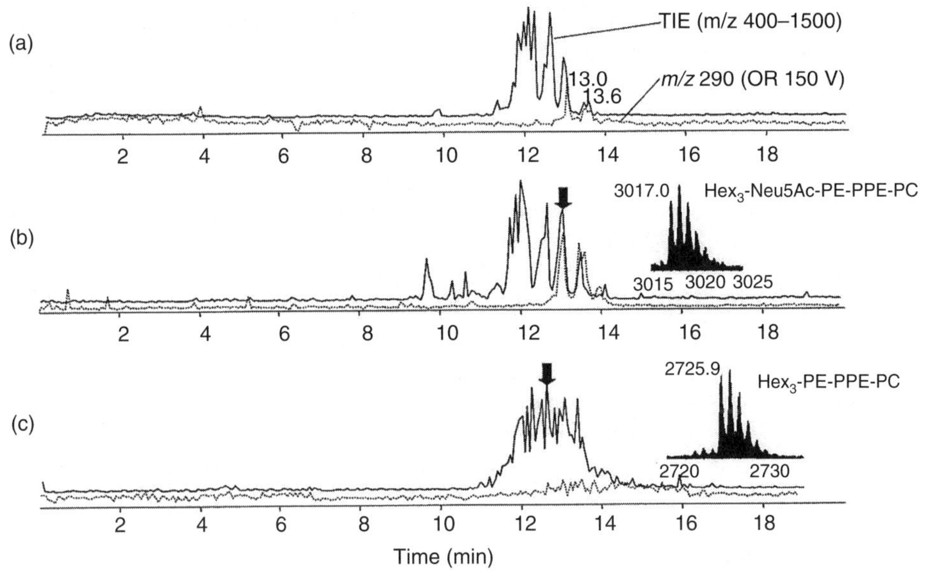

FIGURE 7.9 Analysis of sialylated O-deacylated LPS from *H. influenzae* 375. (A) Results for *H. influenzae* 375 wild type strain. (B) Strain grown in presence of CMP Neu5Ac. (C) Strain following incubation with α-2,3 sialidase. The total ion electropherogram (solid line) is shown together with the fragment anion m/z 290 characteristic of Neu5Ac. The arrows indicate the migration of the glycolipid shown as an inset on the right in B and C. (Adapted from Li, J. et al. *Methods Enzymol*, 405, 369, 2005. With permission.)

[83]. In addition to isoform analysis, monitoring of trace-level glycolipids present as a small subset in a bacterial extract, is often required. To probe characteristic functional groups and residues, such as P, PE, PPE, PC, and Neu5Ac, CE-ESI-MS has been successfully applied [83] using in-source fragmentation and selected ion monitoring (SIM) mode. The ability to probe specific functionalities can also be an important analytical tool in monitoring incorporation of monosaccharides by glycosyltranseferases or their removal by glycosidases [83], as illustrated in Figure 7.9 for Neu5Ac residue in *H. influezae* 375 strain. The precise location of a modification or the assignment of branching site on the glycolipid structure typically requires using tandem MS/MS. Higher order MS (MS^n with FT-ICR, for example) helps rationalize the observation of characteristic fragment ions or assess fragmentation pathways.

Glycosphingolipids (GSLs) are a subtype of glycolipids, containing aminoalcohol sphingosine, which include cerebrosides, gangliosides, and globosides. GSLs are ubiquitously distributed among all eukaryotic species and some bacteria. Since GSLs are secondary metabolites, their direct and comprehensive analyses must be considered an essential complement to genomic and proteomic approaches and establishing the structural repertoire and functional roles of these compounds within an organism. Gangliosides, for example, are composed of a GSL (ceramide and oligosaccharide) with one or more sialic acids linked on the sugar chain. It is a component in the cell plasma membrane, which modulates cell signal transduction events. They have recently been found to be of great importance in immunology. Natural and semisynthetic gangliosides are considered possible therapeutics for neurodegenerative disorders. A detailed review on GSLs and their structural analysis, mostly by various MS techniques, can be found in Reference 80. Several research groups [87–89] applied CE-MS technique to ganglioside analysis. Online glycoscreening of gangliosides by CE-nano-ESI-QTOF-MS [87] is feasible, provided suitable buffer systems, compatible with MS and ensuring acceptable CE resolution, are developed. Ju et al. [90] demonstrated good sensitivity and electrophoretic resolution of a number of gangliosides. MEKC has been coupled with ESI-MS by Tseng et al. [91] who used nonvolatile buffer systems.

Bindila et al. [72] reported a method for the analysis of underivatized glycoconjugates by CE-TOF-MS. The method was first optimized using simple monosaccharides followed by the application to the studies of urine samples from patients with *N*-acetylglucosaminidase deficiency syndrome. Identification of nearly 50 major and minor species, O-glycosylated sialylated amino acids and peptides in patient's urine, using a CE buffer of 6.4% formic acid in 60% methanol at pH 2.8 (adjusted with NH_3), was demonstrated. Another research group applied CE-MS approach to the analysis of APTS-labeled structural isomers of glucooligosaccharides [63]. The known properties of borate ions to form complexes with sugars, changing their hydrodynamic characteristics and therefore improving the resolving power of CE, were also exploited in the CE-MS format. Isomeric LPS and their isomeric glycoforms from *N. meningitidis* were distinguished and characterized by Li et al. [91], employing CE-MS interface in either positive or negative mode.

More examples on CE-MS analysis of various glycoconjugates, including PGs, LPSs, gangliosides, highly glycosylated peptides, and so forth, can be found in References 4, 59,60, and 83.

7.4 APPLICATIONS

7.4.1 Complex Polysaccharides

Complex polysaccharides, glucosaminoglycans (GAGs), are a class of ubiquitous macromolecules exhibiting a wide range of biological functions. They exist as side chains of PGs in the cells and extracellular matrix. The recent development of analytical tools for their study has spurred a virtual explosion in the field of glycomics. A number of electrophoretic separation techniques, including agarose gel, CE, and fluorophore-assisted carbohydrate electrophoresis (FACE), have been employed for the structural analysis and quantification of hyaluronic acid (HA), chondroitin sulfate (CS), dermatan sulfate (DS), keratan sulfate (KS), heparan sulfate (HS), heparin (Hep), and acidic bacterial polysaccharides [93]. Moreover, certain diagnostic analytical applications have been developed to detect fine structural and compositional changes of GAGs at various pathological conditions. These have been accomplished by analyzing oligosaccharides derived from GAGs by enzymatic or chemical degradation. The major analytical advances in this field in the past decade have been reviewed in Reference 93.

GAGs are linear, complex, and polydisperse polysaccharides [94–99]. Most of them, except for KS, consist of alternating copolymers of uronic acids and amino sugars, and their structures are commonly represented by disaccharide sequences. KS, CS, DS, HS, and Hep are sulfated heteropolysaccharides with different degrees of density and positional distribution of sulfate groups. These compounds are very heterogeneous not only by structure but also in biological and pharmacological activities. With the exception of HA, GAG chains are covalently attached at their reducing end through an *O*-glycosidic linkage to a serine residue or *N*-linked to asparagine (e.g., KS) in a core protein [100–103]. Some glycobiology studies suggest that PGs are not only structural components but also regulate many cellular and physiological processes (e.g., cell proliferation and differentiation) [101,104–106]. Most GAGs are obtained from animal sources by extraction and purification processes [99]. Recently, these natural substances have also been chemically modified to get synthetic analogues and GAGs-based drugs [107,108].

Figure 7.10 shows structures of some typical GAGs. For example, HA is a linear polysaccharide composed of alternating residues of the monosaccharides D-glucuronic acid and *N*-acetyl-D-glucosamine linked by β-1,3 bonds in repeating units; and these disaccharides are connected via β-1,4 linkages [109,110]. On the other hand, the repeatable blocks in KS molecules are *N*-acetyl-lactosamine fragments of -β-1,3-[D-galactose-β-1,4-*N*-acetyl-D-glucosamine]-β-1,3-, which are typically sulfated at C6 position of acetylglucosamine and sometimes at C6 of galactose [111]. Structures of some other typical GAGs are schematically shown in Figure 7.10 and described in more detail in Reference 93 and references therein.

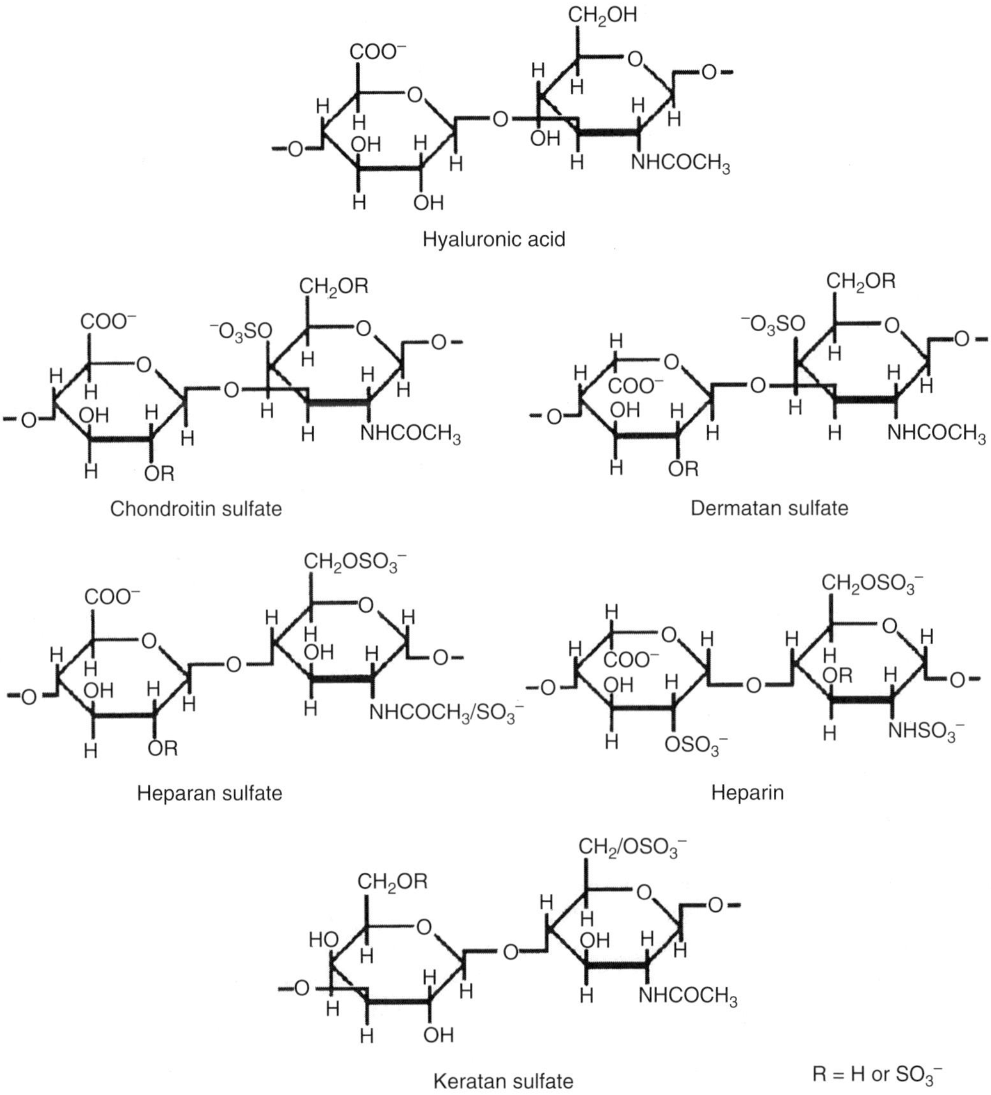

FIGURE 7.10 Structures of disaccharides forming GAGs. Major modifications for each structure are illustrated (R = H or SO_3^-) but minor variations are possible. (Adapted from Volpi, N. and Maccari, F. *J Chromatogr B Analyt Technol Biomed Life Sci*, 834, 1, 2006. With permission.)

Separation and quantification of GAGs compounds in mixed samples have been conducted by one- and two-dimensional electrophoresis on cellulose acetate strips, nitrocellulose membranes, agarose gels, and polyacrylamide gels (PAGE), using staining dyes for detection (alcian blue, toluidine blue, azure A, methylene blue, combined azure blue, and silver staining) [93]. CZE in uncoated fused-silica capillaries, as well as MEKC, have been successfully exploited for GAGs analysis. In the latter case, SDS or CTAB was used in the basic borate buffers [112–114]. In addition, microemulsion electrokinetic capillary chromatography mode was demonstrated, where GAGs species can be separated based on partitioning into oil droplets moving in the running buffer [93]. Different modes of CE were realized with either normal or reversed polarity, under basic or acidic (suppressed EOF) conditions, respectively. Proven benefits of CE, such as ease of use, fast analysis, very small sample amount requirements, high resolution and sensitivity, and reproducible quantification, worked very

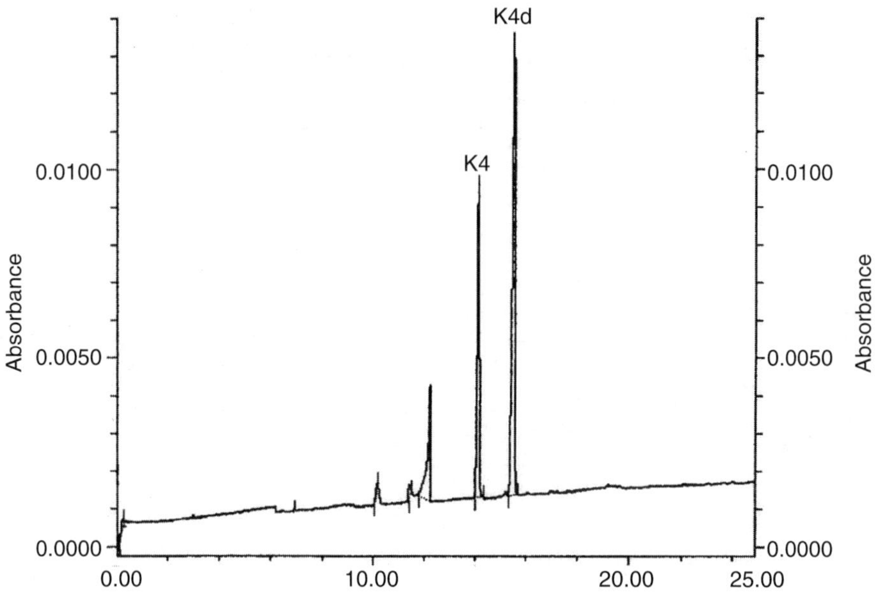

FIGURE 7.11 HPCE electropherograms of unsaturated disaccharide ΔHexAFrc-GalNAc, forming polysaccharide K4 (K4), and ΔHexA-GalNAc of defructosylated K4 product (K4d). (Adapted from Volpi, N. *Electrophoresis*, 24, 1063, 2003. With permission.)

well especially for nonsulfated GAGs, that is, HA and bacterial polysaccharides, oligosaccharides, and disaccharides generated by enzymatic digestion reactions [112–114]. The possibility to label GAGs polysaccharides or their depolymerized short fragments with chromophores or fluorophores, to significantly boost sensitivity, is another attractive feature of CE analysis. FACE, based on 2-aminoacridone (2-AMAC) derivatization, has been applied to the analysis of plant cell wall polysaccharides [115], and HA and GAGs disaccharides [116]. As an example, Figure 7.11 presents CE separation of two disaccharides, products of chondroitinase digestion of polysaccharides from uropathogenic *Escherichia coli* K4 bacteria and its defructosylated product [117]. SDS MEKC mode in uncoated fused-silica capillary under normal polarity and UV detection (230 nm) was used in this work. However, this approach required enzymatic treatment of bacterial polysaccharides with lyases. In the other study of the same group [118], direct CE analysis of native K4 and defructosylated K4 has been developed. The two polyanions were separated and detected within 30 min, as depicted in Figure 7.12. Linear region of quantification was observed in the 30–210 ng range. Other bacteria polysaccharides and membrane LPS have been analyzed by high-performance CE [119,120] and the results were invaluable for characterization and quantification of these complex polymers in biological and medical research.

A method for the determination of HA oligomers by CE-MS was reported by Kuhn et al. [121]. Oligosaccharides of 4–16 DP were generated by enzymatic digestion of HA with bacterial hyaluronidase, followed by CE-MS analysis. Structural information was obtained in MS^n experiments using an ion-trap instrument (IT-MS). Another research group [122] described a homemade sheathless ESI interface for the highly sensitive analysis of GAGs by CE-MS: CS and dermatan oligosaccharides of extended chain length and increased degree of sulfonation from decorin transfected human embryonic kidney cells were detected. More examples of CE-MS characterization of complex polysaccharides are presented in Sections 7.3.2 and 7.4.2, as well reviewed in References 4, 59, and 93.

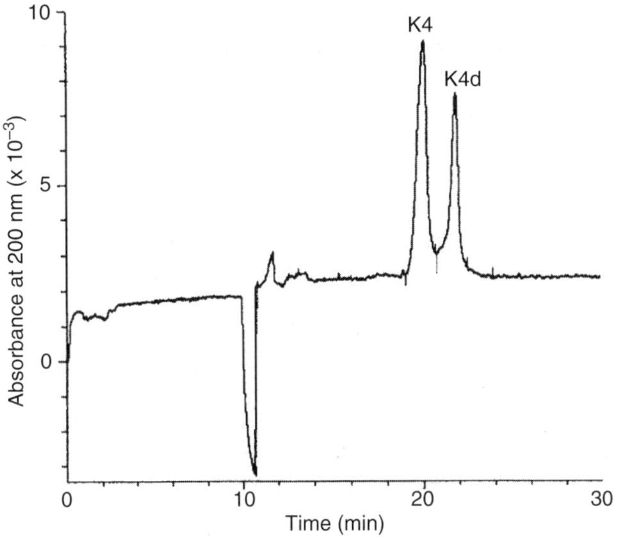

FIGURE 7.12 HPCE electropherograms of the polysaccharide K4 (K4) and defructosylated K4 polysaccharide (K4d). (Adapted from Volpi, N. *Electrophoresis*, 25, 692, 2004. With permission.)

7.4.2 Carbohydrates in Pharmaceuticals

In drug development and clinical and forensic applications, the popularity of CE and CE-MS technologies have grown due to high discrimination power of CE and possibility to directly analyze complex biological matrices, such as serum, urine, or other fluids [123,124]. This section overviews most typical examples related to carbohydrate analysis in biopharmaceutical research. More details and references on the topic can be found in recent comprehensive reviews [38,59,60,123,125]. Overall, implementation of CE in pharmaceutical bioanalysis has tremendous potential by providing additional detailed information on the structure of oligosaccharide fragments in glycoconjugates of pharmaceutical interest and ultimately contributing to the development of more effective and safe medicines.

7.4.2.1 Glycoproteins

Glycosylation is one of the most common co- and post-translational modifications of the proteins. Carbohydrate moieties in glycoproteins have direct effect on their biological activity and cellular functions, such as protein folding, recognition, signaling, immune response, differentiation, and so forth [126–128]. Modern recombinant technologies enable production of protein pharmaceuticals in living cells. However, inherent structural heterogeneity in glycosylation patterns affect their efficacy, pharmacokinetic and pharmacodynamic properties [129,130], and therefore should be carefully controlled. The structure and composition of carbohydrate chains depend on the expression conditions in the host cells and purification steps involved in the production processes. Characterization of complex heterogeneous carbohydrate chains is a challenging, yet important, task in production optimization, regulatory submissions, and quality control. Glycosylation patterns and heterogeneity are not directly controlled by genes, therefore cannot be fully predicted by protein expression studies. Stand-alone CE and CE-MS^n have become methods of choice in analysis of glycoforms of protein and antibody therapeutics in biopharmaceutical research and development. High sensitivity and resolution of CE permit detection of fmol and amol amounts of oligosaccharides, when using an appropriate fluorescent-labeling method and LIF detection. Remarkably, oligosaccharide analysis at a single cell level has been reported [131].

Recombinant human erythropoietin (rhuEPO), one of the most successful recombinant protein therapeutics on the market, has three *N*-linked and one *O*-linked glycosylation sites, and each glycan can have variable number of sialic acid residues [132]. Isoelectrofocusing (IEF) in capillary format has been successfully demonstrated to resolve rhuEPO glycoforms, replacing time- and labor-consuming slab gels IEF [58,133–135]. Noticeably, this methodology is now included in *European Pharmacopoeia* [136]. The separation was performed in a bare fused-silica capillary using a buffer containing putresine and urea. To overcome irreproducibility issues [58,135,136], capillary surface modification through covalent and/or dynamic coating, was utilized [137]. Monitoring changes in glycoprotein heterogeneity by CE separation of glycoforms is now in use in many quality control labs for release and stability testing.

Structural analysis of *N*-linked oligosaccharides is typically conducted by their enzymatic release from the protein, derivatization with a suitable fluorescent tag, and CE separation. A variety of oligosaccharide standards have been developed to facilitate identification of unknown species based on migration time comparison [60,138]. For further structural characterization, glycans can be fractionated and subjected to LC-ESI-MS or MALDI-TOF-MS analysis [139–141]. Alternatively, online CE-MS analysis is feasible as described in Section 7.3.1.

Sequencing of oligosaccharides can be accomplished by their digestion with various exoglycosidase enzymes, followed by CE analysis and peak shift monitoring [60,142]. Specific glycosidase cleavage is useful to distinguish positional isomers and glycosidic linkages. For example, Galβ-1,4GlcNAc and Galβ-1,3GlcNAc at nonreducing end can be discriminated using β-1,4 galactosidase or β-1,3 galactosidase. In glycan profiling of therapeutic antibody rituximab, two positional isomers of biantennary *N*-linked oligosaccharides (with one galactose at nonreducing terminal of different arms) were separated in CE, as shown in Figure 7.13 (peaks 2 and 3) [143]. The oligosaccharides can be identified using α-1,2 and α-1,3 mannosidase digestion following *N*-acetylhexosaminidase treatment [144]. Such subtle structural differences could not be resolved by MS or MS/MS methods.

Capillary affinity electrophoresis (CAE) based on specific carbohydrate–protein interaction, is a valuable method in carbohydrate characterization. A variety of carbohydrate-binding proteins (e.g., lectins) specific to certain oligosaccharide structures are available and have been exploited by different research groups to characterize glycoprotein pharmaceuticals [145,146].

In recombinant antibodies [147], oligosaccharide moieties are attached to the Fc region of antibodies and often control their biological activity and stability. CE-LIF analysis of oligosaccharides derived from recombinant monoclonal IgGs has been used by different researchers and exhibited high degree of reproducibility and accuracy in the determination of composition of the released glycan isoforms. A number of examples of application of CE-based technology to glycan profiling of therapeutic antibodies can be found in References 59, 60, 144, and 142.

7.4.2.2 Analysis of Complex Polysaccharides in Drugs

Certain GAGs and other sulfated polysaccharides are known to possess anti-inflammatory properties. Some of them are also used as anticlotting and anti-arthritis agents. Recent studies by Liang et al. [148–150], using CZE, demonstrated that heparin, carrageenan, and dextran sulfate interact with a hematopoietic growth factor, granulocyte colony-stimulating factor (G-CSF), and have potential therapeutic effect on cancers through inhibition of the growth and induction of the differentiation of the leukemia cells. Such phenomenon might be of great importance for the treatment of patients under radio- and chemotherapy, since heparin, for instance, appears to protect G-CSF from degradation, therefore, increase its circulation half-life. It was found that these interactions are dependent on the polysaccharide chain length and sulfate content. For example, separation, identification, and interaction of heparin with G-CSF were accomplished by CE, following enzymatic digestion of heparin with haparinase to render smaller oligosaccharides [150]. The oligosaccharides were well separated in CE using 50 mM phosphate buffer (pH 9). The smaller di- and trisaccharides were also identified by CE-ESI-MS.

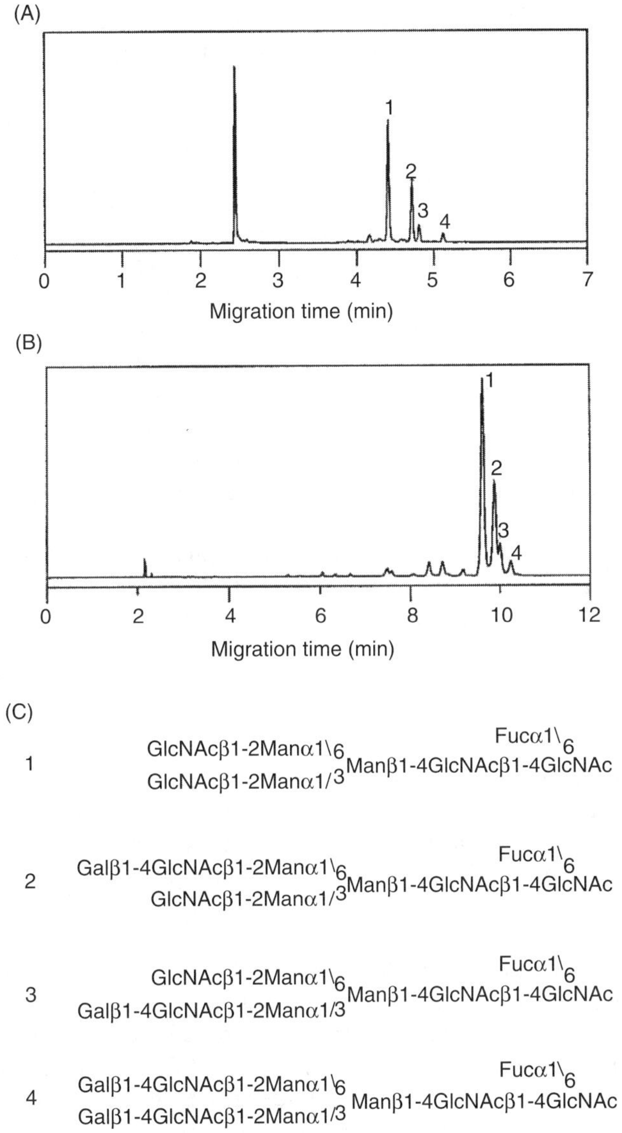

FIGURE 7.13 Oligosaccharide maps of trastuzumab by CE. (A) Oligosaccharides derivatized with APTS, (B) oligosaccharides derivatized with 3-AA. Analytical conditions: (a) capillary DB-1 (50 μm ID, 20 cm effective length, 30 cm total length); running buffer 50 mM Tris–acetate buffer (pH 7.0) containing 0.5% PEG70000; injection 0.5 psi for 5 s; applied voltage, 18 kV at 25°C. (b) Capillary DB-1 (100 μm ID, 20 cm effective length, 30 cm total length); running buffer 100 mM Tris–borate buffer (pH 8.3) containing 10% PEG70000; injection 1.0 psi for 10 s; applied voltage 25 kV at 25°C. LIF detection 405 nm emission/325 nm ex (He-Cd laser). (C) Structures of major oligosaccharides in rituximab, peaks 1–4. (Adapted from Kamoda, S. et al. *J Chromatogr A*, 1050, 211, 2004. With permission.)

The presence of native ionizable groups in GAGs and some other polysaccharides make them particularly suited for both CE and MS, without need for derivatization. A number of examples of the CE separation and CE-MS characterization of GAGs oligosaccharides can be found in literature [59,87–89,121,151,152]. Both positive and negative ionization modes, as well as normal and reversed CE polarities, were tested for analysis of heparin oligomers [151], although negative ionization was particularly useful for complex mixtures. In many cases, use of low pressure in conjunction with CE voltage, especially in the reversed separation mode, helped boost electromigration, and electrospray stability [121,151]. MS/MS fragmentation enabled qualitative and quantitative identification of coeluting species, for example, disaccharides IIS-IIIS and ΔUAS(1→4)α-d-GlcNAc6s (IIA)-ΔUA2S(1→4)α-d-GlcNAc(IIIA) [152]. MS/MS fragmentation results in loss of sulfate groups in di- and trisulfated disaccharides, and loss of water and ring cleavage in mono- and nonsulfated ones. It was shown that limits of detection (LODs) were comparable (low-μM range) for CE-UV and CE-MS [152]. A coated capillary was used to enhance and stabilize EOF and characterize oligosaccharides of HA, up to 16mer, under positive CE polarity and negative ESI [152].

Degree of sulfation of GAGs, in particular oversulfation, affect their biological activity [87–89]; therefore, structural studies are important in the analysis of these compounds. Zamfir et al. [87–89] have extensively studied DS and CS using CE-ESI-MS, both off-line and online. Mild MS/MS fragmentation conditions allowed preventing sulfate cleavage and more effectively detecting multiply charge ions, while higher collision energy produced smaller fragments and was more useful for structure elucidation. Introduction of sheathless CE-MS interface with nano-ESI probe resulted in improved overall performance of the analysis and LOD in pmol range [89]. Structural characterization of CS/DS derived from skin fibroblast secretion lit more light toward understanding of binding activity to fibroblast growth factor-2 (FGF2) [88], which is related to epithelial cell proliferation and growth of new blood vessels (i.e., angiogenesis). Overall, CE-MS/MS approach seems to represent a powerful tool for structural characterization of sulfated GAGs.

7.4.2.3 Glucuronidation of Drugs

Glucuronidation and *N*-glucuronidation are major inactivating pathways for a vast variety of endogenous and exogenous molecules, such as drugs. Therefore, analysis of glucuronidated drug metabolites is of importance for drug development and clinical and forensic applications. Owing to ionic nature of glucoronidated compounds, CE and MS are often the methods of choice for their identification and characterization.

For example, detection of opoids in body fluids is needed in clinical and forensic toxicology because of their widespread use for both therapeutic and illicit purposes. CE-MS analysis of urine proved to be a very efficient aproach in the detection of opoids' metabolites [153–155]. Urinary samples of codeine, dihydrocodein, morphine and their glucuronides, and oxycodone and its metabolites were successfully analyzed by Wey et al. [153–155], utilizing CZE-based immunoassay, as well as CE-UV and CE-ion trap MS. Theses researchers investigated effects of different pretreatments and found that solid-phase extraction (SPE) and liquid–liquid extraction improved detection limits [154]. CE separation before MS detection versus direct MS has been beneficial, since it enabled resolution of the compounds with identical fragmentation patterns (e.g., morphine-2-glucuronide and morphine-6-glucuronide). Besides, direct MS analysis required significantly larger amount of samples.

CE-ESI-MS^n has been used for the analysis of entacapone and tolcapone and their metabolites [156]. These drugs are administered in combination with anti-Parkinson's levodopa. The main metabolites found were 3-*O*-glucuronide conjugates, and the analytical approach proved to be reproducible and very sensitive. The latter should facilitate direct metabolite identification in complex biological matrices.

CE and CE-ESI-MS were exploited for the analysis of lorazepam (3-hydroxy-1,4-benzodiazepine) and its metabolites in urine [157]. This drug is commonly used for the treatment of anxiety and as a sedative and hypnotic agent. The authors found that 75% of administered lorazepam is excreted in urine as its 3-*O*-glucuronide. Interestingly, since glucuronidation occurs at the chiral center of the molecule, two diastereoisomers can be formed, which was confirmed by MEKC analysis of urine extracts and also *in vitro*, via incubation of the drug with human liver microsomes and 5′-diphospho-glucuronic acid as coenzyme. The evidence of stereoselectivity of lorazepam glucuronidation, with one diastereoisomer being formed preferentially than the other, was found as well. Achiral analysis was performed with a running buffer composed of 6 mM $Na_2B_4O_7$, 10 mM Na_2PO_4 (pH 9.1), and 75 mM SDS. For enantioselective analysis, the same buffer was diluted with isopropanol (2.5%), and 2-hydroxypropyl-cyclodextrin stereoselector was added. CE-MS experiments were performed in negative ESI mode with ammonium acetate as a separation buffer (pH 9 adjusted with concentrated ammonia) and a sheath liquid containing 50% methanol with 0.1% concentrated ammonia to ensure the formation of negatively charged ions. CE-MS^2 and CE-MS^3 experiments were conducted to confirm the presence of lorazepam and its 3-*O*-glucronide, respectively.

7.4.2.4 Characterization of Oligosaccharides in Pathogenic Bacteria

Pathogenic bacteria contain LPS and capsular polysaccharides (CPS) in their outer membranes. These compounds play important role in virulent properties of bacterial species, that is, invasion into human immune system through molecular mimicry mechanisms [158]. LPS functions have been under experimental research for several years due to their role in activating many transcriptional factors, which become active after stimulation with LPS. LPS also produces many types of mediators involved in septic shock. Structural characterization of the core oligosaccharide domain and O-chain polysaccharide components help identify the infecting pathogens and classify new strains, and therefore, contributes to the development of antibacterial drugs. For example, *Pseudomonas aeruginosa* is a Gram-negative bacterium affecting individuals suffering from cystic fibrosis, certain cancers, and immunodeficiency. The wild type of this pathogen expresses a heterogeneous population of O-PSs, and LPSs obtained from cystic fibrosis patients have been found to be deficient of O-chain extension [159].

To facilitate analysis of LPS/CPS, a partial hydrolysis is typically employed, either chemical or enzymatic, which is specific for each type of polysaccharide. Hydrolysis is usually followed by purification by a chromatographic means. Taking into consideration these time-consuming sample preparation steps, online coupling of separation and structural characterization methods represent advantage of potential minimization of sample pretreatment efforts. CE-MS have been exploited in this type of applications by several research groups. Li et al. [86] used ESI-MS in-source fragmentation technique, following CE separation, to break down CPSs and O-PS obtained from *Aeromonas pleuropneumoniae*, instead of conventional chemical degradation pretreatment. This approach helped to avoid extra sample preparation steps while keeping O-PS structure intact [85,86]. The latter also allowed acquiring information on O-acetylation of the molecules. In addition, online preconcentration methods coupled with CE-MS have been attempted [159].

CE-MS was used for the characterization of LPSs at different structural levels, gaining more understanding of mechanisms of action of the pathogens and, consequently, effecting drug design. *Moraxella catarrhalis* is a pathogen affecting the respiratory tract and middle ear causing ottitis. CE-MS/MS analysis of oligosaccharides derived from this bacterium revealed both basic (glucosamine) and acidic groups (deoxy-manno-octulosonic acid) [158]. Mycobacteria contain lipoarabinomannans (LAMs) as their capsular components, which comprise arabinimannan core and a hydrophobic anchor, (e.g., phosphatidyl-*myo*-inositol). The immunological activity of LAMs depends on both cap and anchor structure: arabinofuranosyl side chains can be substituted with either mannose (ManLAMs) or phosphatidyl-*myo*-inositol (PI-LAMs) [59,160]. Nonvirulent mycobacteria have been shown to have preferentially PI-LAMs, while ManLAMs were found in pathogenic *Mycobacterium tuberculosis*, *Mycobacterium bovis*, and *Mycobacterium leprae*. Mannooligosaccharides can be hydrolytically released, derivatized by reductive amination with APTS, and undergo CE, coupled online or off-line with MS, using either nano-ESI source or MALDI-TOF [160]. More examples on characterization of oligosaccharides derived from bacterial sources can be found in Reference 59.

7.4.2.5 Sugar Analysis in Traditional Chinese Drugs

There has been growing attention to traditional Chinese drugs (TCD) in the world, due to their effectiveness in a number of disease therapies. Various sugars, including mono-, oligo-, and polysaccharides, are significant constituents of traditional Chinese herbs, reaching in some cases up to 80–90% of the whole plant dry weight [125]. There are over 300 kinds of polysaccharides extracted from natural plants. Among those, water-soluble polysaccharides typically possess most valuable pharmaceutical properties. The mechanisms of pharmaceutical effects of bioactive sugars have been extensively studied, and the number of their therapeutic applications have been increasing [161,162].

Most promising areas are anticancer activity [163–165], immune system stimulation [166–168] through the production of interleukins and T and B cell regeneration, anti-HIV effects (polysaccharides inhibiting binding of HIV to T cells) [169], blood sugar reduction (enhancing secretion of insulin and carbohydrate metabolism) [125,170], and anti-inflammatory [171,172].

The analysis of pharmaceutical polysaccharides and other sugars in TCD typically involves extraction, purification, and characterization steps. Water and weak basic solutions, as well as some organic solvents, at different temperatures, are typically used to extract sugars [125]. Purification can be accomplished by serial precipitation and chromatography steps (e.g., size exclusion, gel permeation, ion exchange, etc.) [125]. Electromigration separation methods for the analysis of sugar constituents of TCD have been gaining popularity in the past decade, due to their high separation efficiency and compatibility with UV, LIF, refraction index (RI), and electrochemical detection techniques, as well as hyphenation with MS and NMR. CE occupies a special niche in carbohydrate analysis, and TCD are not exception. In addition to high performance of CE, emphasized throughout this chapter, simplicity of instrumentation and low sample and all the reagents consumption are the main reasons for that. TCD sugars are usually separated and identified by CE as mono- and oligosaccharides, from which the composition of polysaccharides can be derived. Thus, complex polysaccharides are typically hydrolyzed to render simpler oligomers and monomers. Electrochemical detection and LIF of underivatized and fluorescently labeled saccharides, respectively, are highly sensitive. For example, when using amperometric detection approach μM sensitivity level for monosaccharides can be achieved with copper or gold electrodes [173–175]. Different aqueous separation buffers have been utilized, including highly basic ones ($pH > 11–12$), when most neutral sugars become negatively charged due to their weak acidic properties. Sodium hydroxide based running buffers, borate (pH 10), phosphate–borate buffers, used in combination with uncoated fused-silica capillaries, have most often been mentioned. Certain buffer additives, for example, surfactants (SDS, CTAB), THF, tryptophan, can often improve separation efficiency. Examples and references to analysis of carbohydrates in traditional Chinese medicines can be found in the recent review article by Wang and Fang [125].

One promising approach to be exploited more in the future with respect to TCD analysis is microchip-based CE. It is expected to enable ultimate integrated and highly sensitive systems to be built [176–178]. CE-MS coupling represents another area of fast development in carbohydrate analysis of TCD, as it has advanced in other applications in research, biopharmaceutical, and bioindustrial fields.

7.4.3 Carbohydrates in Foods

Carbohydrate composition of foods and feeds allow characterization of the quality, ripeness, possible adulterations, and so forth. CE-MS approach has been widely used in analysis of foods. With a few exceptions [179], food carbohydrates are analyzed without any derivatization. Arabinose, ribose, xylose, inositol, galactose, fructose, and mannitol were monitored by CE-ESI-MS, using highly alkaline CE running buffer and negative ionization mode [180], reaching LOD 0.5–30 μg/mL. Figure 7.14 exemplifies different carbohydrate patterns in red and white wines. The higher contents for most sugars were found in white wines. CEC-ESI-MS approach was exploited for the analysis of

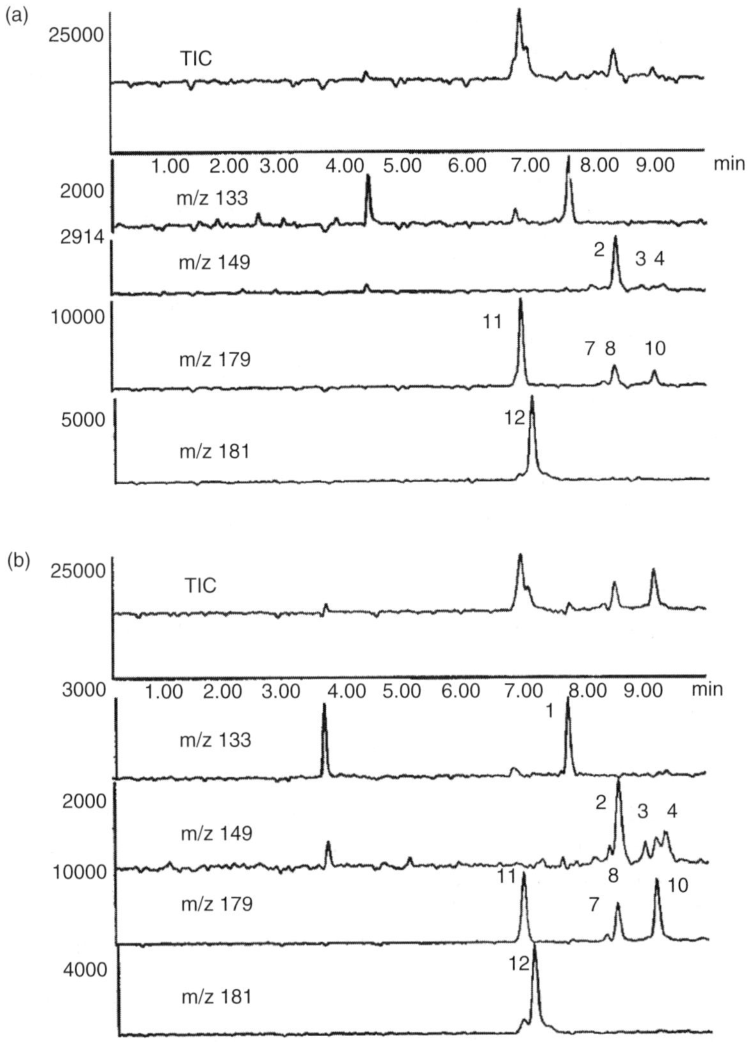

FIGURE 7.14 CE-MS total ion current (TIC) and SIM electropherograms for (A) a red wine and (B) white wine sample. CE conditions: fused-silica capillary (70 cm length to detection, 50 mm ID); running buffer 300 mM diethylamine (DEA); running voltage 20 kV; injection at 50 mbar for 9 s. ESI-quadrupole MS conditions: negative ion mode; capillary voltage 5 kV; sheath liquid 2-propanol:water (80:20 v/v) with 0.25% DEA at a flow rate of 4 mL/min; drying gas N_2 at a flow rate of 1.4 L/min and a temperature of 1507°C; nebulizer pressure 0 psi. Peaks: 1: deoxyribose; 2: arabinose; 3: ribose; 4: xylose; 5: galactose; 6: glucose; 7: fructose; 8: inositol; 9: mannitol. (Adapted from Klampfl, C. W. and Buchberger, W. *Electrophoresis*, 22, 2737, 2001. With permission.)

enzymatically digested GAGs [151] and other derivatized and underivatized glycans [69], complex oligosaccharide mixtures using hydrophilic monolithic columns [26], and FT-MS to achieve high mass resolution [53]. MEKC-ESI-MS was attempted to the analysis of iridoid glycosides in plant samples from different species [181]; in this case, the partial filling technique was used to prevent surfactants from entering the MS interface. Gellan gum, a complex sugar polymer food additive, was detected based on the occurrence of a characteristic tetrasaccharide released in the process of enzymatic digestion by a specific gellan degrading enzyme [182]. Low molecular weight sugar-based acids (gluconic and isosaccharinic) present in ale were unambiguously identified by CE-MS, which helped overcome the problem associated with comigration of the two acids by SIM application during acquisition of MS spectra [183]. Glycoalkoloids of vegetable origin are also compounds of interest due to their toxic properties and occasional presence in food products. CE-MS was used to separate some typical glycoalkoloids, as well as related aglycones, in potatoes (α-chaconine, α-solanine, solanidine) and tomatoes (α-tomatine, tomatidine) [184]. In this case, nonaqueous CE coupled with SIM or SRM (selected reaction monitoring) MS mode resulted in excellent separation and identification of all the investigated compounds. Moreover, nonaqueous buffer was found to significantly improve ionization efficiency. Another interesting application of CE-MS was reported by Bringmann et al. [185] who characterized glucosinolates from seeds, the compunds of anticarcinogenic effects. The authors employed IT analyzer for structural elucidation, while TOF MS was used for information on elemental and isotopic composition. Higher analysis efficiency was achieved compared to chromatographic methods, resulting in not only identification of known compounds but also discovery of some new ones.

7.4.4 Bioindustrial Applications

Enzymatic saccharification of lignocellulosic biomass to fermentable sugars is of great importance for a number of agroindustrial processes. These include conversion of cellulose and hemicelluloses to fuels and other important chemicals, delignification of paper pulp, digestibility enhancement of animal feed stock, clarification of juices, to name just a few. Chemical pretreatment methods in conjunction with proper cost-effective enzyme combinations (e.g., hemicellulases or cellulases) tailored to each specific biomass will enable production of fuel additives, such as ethanol, and other valuable chemicals by fermentation from this vast renewable resource [186–190]. Various agricultural residues, such as corn fiber and stover, wheat and rice straw, and so forth, contain up to 40% hemicellulose, the second most abundant polysaccharide in nature after cellulose. Cost-effective conversion of these materials to valuable chemicals is a current issue for the industry [188]. Lignocellulosic biomass structure is very complex [186,187]. Industrial corn fiber contains cellulose chains, assembled into highly crystalline and aggregated structures, microfibrils and fibers. The cellulose fibers are imbedded in a matrix of hemicellulose (highly branched heteropolysaccharides) and lignin. Cellulose and hemicellulose degrading enzymes are found in nature either as enzyme complexes or free in solution [191]. Different enzyme classes are involved in the biological depolymerization of cellulose and hemicellulose components: cellulases (endoglucanases, cellobiohydrolases, and β-glucosidases); xylanases to degrade the hemicellulose backbone, as well as accessory enzymes, for example, arabinofuranosidases, mannanases, galactosidases, glucuronidases, ferulic acid esterases and acetyl-xylan esterases. It is desirable to find high-specific activity enzymes that effectively hydrolyze biomass fiber in synergistic fashion [189]. Evaluation of different enzymes and their combinations, optimization of the reaction conditions (temperature, pH, reagent concentration, reaction time, etc.) are critical. Therefore, hundreds and thousands of samples are to be analyzed to screen the various enzyme combinations, tune reaction parameters, and characterize novel expressed recombinant enzymes and microorganisms.

Implementation of efficient and robust industrial scale bioconversion processes requires high-throughput analysis of mono- and oligosaccharide products, released by enzymatic digestion of various biomasses, to evaluate activity of enzymes and microorganisms. CE with LIF detection offers excellent resolving power and high sensitivity, superior to many other analytical approaches [192]. High-resolution separations of complex carbohydrate mixtures, simple sugars, glycoforms of glycoproteins, and glycolipid oligosaccharides by CE have been demonstrated for the past decade [19,48,193]. Using suitable fluorescent-labeling procedures, charged derivatives of various carbohydrates can be formed enabling attomole detection level in CE [193]. The majority of carbohydrate derivatization methods utilize tagging of the reducing termini with a fluorophore via reductive amination [34]. APTS is one of the most popular of such derivatization agents, providing very bright fluorescence and good spectral characteristics suitable for the commonly used Ar-ion laser [195].

There is a great demand for large, industrial-scale carbohydrate profiling methods to meet the current growing needs to screen enzymatic activity of recombinant enzyme and microorganism libraries. A few advances have been made to date to establish reliable high-throughput methodology for automated oligosaccharide analysis. Morell et al. [196] adapted a polyacrylamide slab gel based DNA-sequencing device to parallel multilane analysis of oligosaccharide products of enzymatic digests of starches. Although, good resolution up to DP 75 and sensitivity (1 fmol per oligosaccharide) were achieved, the technique had limitations with respect to reproducibility, precise tracking of gel lanes, resolution of subtle differences in profiles, and labor consumption. Another research group [197] utilized a similar DNA-sequencing instrument using 32 parallel lanes, in combination with sample preparation, derivatization and cleanup steps, to accomplish ultrasensitive profiling and sequencing of *N*-linked glycans. On the other hand, CE demonstrated superior performance enabling fast and reproducible separation in the range of DP 1–100, providing at the same time flexibility of separation formats for the development of various assays [48]. A multicapillary CE-based DNA analyzer combined with 96-well-plate-based sample preparation method and a thermocycler were used to perform high-throughput clinical analysis of *N*-linked oligosaccharides from serum protein of liver disease patients [141].

Khandurina et al. [198] demonstrated the capabilities of a single CE instrument to provide automated high resolution and quantitative analysis of mono- and oligosaccharide products of enzymatic digestion of cellohexaose, as a model substrate, and corn fiber biomass in bioindustrial settings. The authors developed various protocols to accommodate different assays needs. Unattended batch sample processing from 96-well plates after the one-step derivatization reaction enabled reliable industrial-scale carbohydrate profiling. A single CE system (P/ACE MDQ; Beckman Coulter) was employed, with the cathode on the inlet side and anode on the outlet side (reversed polarity), since the negatively charged APTS-labeled oligosaccharides migrate toward the anode under applied electric field. For the CE analysis, either bare fused-silica (Polymicro Technologies, Phoenix, AZ) or coated capillary columns (eCAP™ N-CHO Capillary; Beckman Coulter) of 50 μm ID were used, with the effective separation length either 48 cm to achieve high resolution or 10 cm in fast screening experiments. 25 mM lithium acetate (pH 5) buffer was used as running buffer. Samples were injected by vacuum for 5–10 s at 0.5 psi (3447 Pa) and separated at 500 V/cm electric field strength. In case of uncoated capillaries, a dynamic coating by 1% polyethylene oxide (MW 600,000 Da) was utilized to suppress EOF. Figure 7.15 shows high-resolution oligosaccharide profiling by CE using 48 cm effective separation length. Traces A and B depict the electropherograms of a maltooligosaccharide ladder and cello-oligomer standards, respectively. There are well-observable differences in migration times of the oligosaccharides with the same DP but different linkages between glucose monomer units (e.g., maltose vs. cellobiose, maltotriose vs. cellotriose, etc.). These differences result from the changes in structure and, therefore, hydrodynamic radius of the molecules, having either α-1,4 or β-1,4 glycosidic bonds. Profiles C–J in Figure 7.15 show CE analyses of the reaction products of cellohexaose digestion (trace C) using different cellulases. Precise monitoring of the differences in cellulase enzymatic activity, revealing a variety of combinations of glucose, cellobiose, cellotriose, cellotetraose, and cellopentaose, was readily achieved by CE. The excellent migration time reproducibility allowed easy identification of the cello-oligomer products of the cellohexaose enzymatic hydrolysis.

Separation of monosaccharides is a challenging task, particularly due to very minor differences between most common sugar molecules, representing the building blocks of lignocellulosic biomasses. Monosaccharide analysis by CE has been previously demonstrated using borate-based buffer systems at high pH (>9), taking advantage of the resolution enhancement due to anionic complex formation between the polyhydroxy compounds (sugars) and borate [199,200]. Good separation performance was demonstrated using CZE [200] or MEKC [27] in bare fused-silica capillaries under EOF conditions. However, in most cases an internal standard was required to account for possible migration time shifts. The authors [198] have found that comparable separation performance of some monosaccharides (major building blocks of hemicellulose) can be achieved using the same CE

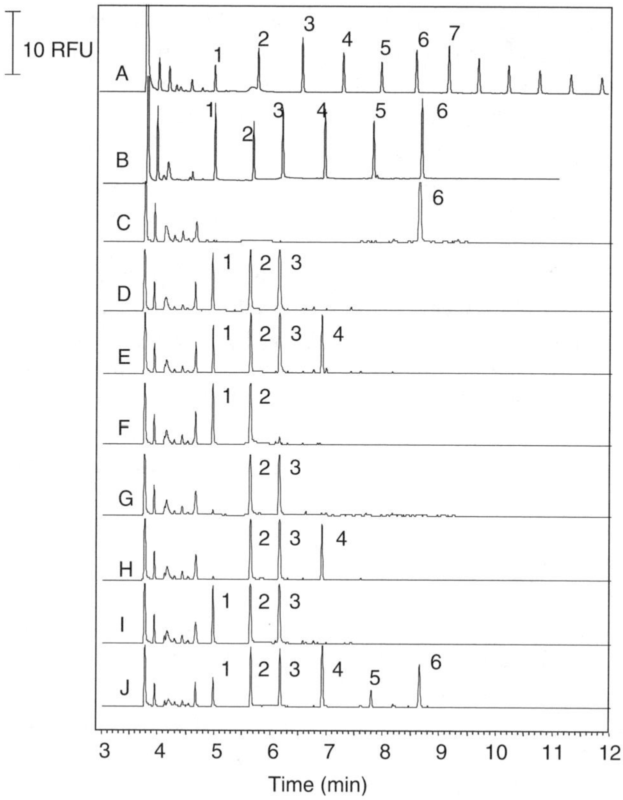

FIGURE 7.15 High-resolution oligosaccharide profiling by CE. Traces: (A) Maltooligosaccharide ladder; (B) cellooligosaccharide ladder; (C) model substrate cellohexaose; (D–J) products of enzymatic digestion of cellohexaose by seven different enzymes with cellulase activity. Numbers above peaks correspond to DP of oligosaccharides. Conditions: eCAP neutral coated capillary (50 μm ID); effective separation length 48 cm; electric field strength 500 V/cm; running buffer 25 mM lithium acetate (pH 5); injection 5 s at 0.5 psi (3447 Pa). (Adapted from Khandurina, J. et al. *Electrophoresis*, 25, 2326, 2004. With permission.)

conditions, as described above for oligosaccharide profiling. Moreover, the absence of EOF typically renders more reproducible performance and elimination of additional rinsing steps, otherwise required.

Trace A in Figure 7.16 shows a well-resolved separation of a mixture of the five monosaccharides of interest in just 5 min, using 25 mM lithium acetate (pH 5) running buffer under apparently zero EOF conditions. Glucuronic acid (1) is the fastest migrating sugar, due to its additional charge, followed by the pentoses of xylose (2) and arabinose (3), and the hexoses of glucose (4) and galactose (5), respectively. By this means, monosaccharide reaction products of corn fiber hydrolysis, either chemical or enzymatic, can be easily identified, assuming very similar labeling efficiency for all the released monosaccharides [34,196]. Traces B and C in Figure 7.3 depict separations of sugars released from equal amounts of corn fiber by acidic hydrolysis, for 2 h at 100°C, with 1 M sulfuric acid and 2 M tetrafluoroacetic acid (TFA), respectively. The glucose (4), most likely, comes from the cellulose domains of corn fiber. To identify the peaks precisely, in ambiguous cases, the samples were spiked with proper amounts of standards. Trace D in Figure 7.16 delineates decomposition products of corn fiber under harsh hydrolysis conditions, started with 72% H_2SO_4 for 1 h followed by dilution to 4%, autoclaving at 120°C for 1 h, and storing for 4 months after the treatment. This procedure apparently resulted in the release of significantly more monosaccharides, as one can observe from the increased peak areas of this particular trace. In addition, the relative glucose concentration was

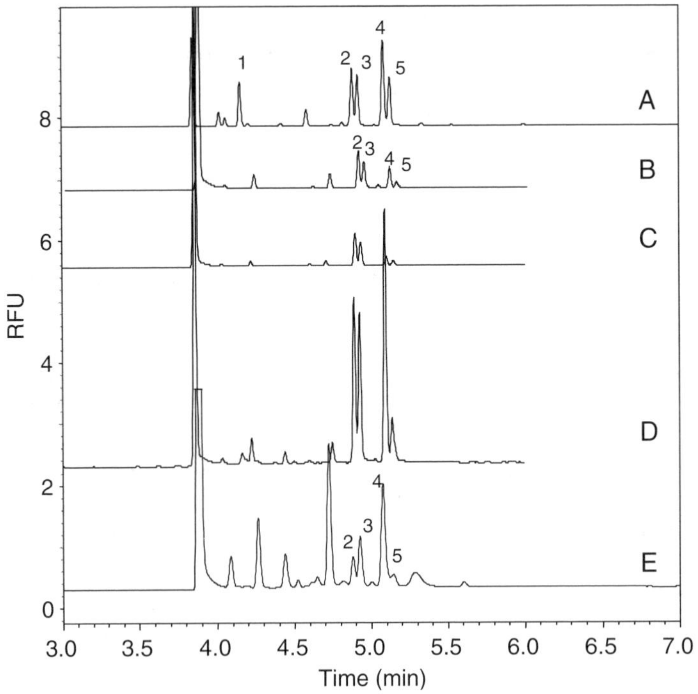

FIGURE 7.16 Monosaccharide analysis by CE under zero EOF condition. Traces: (A) five standard sugar mix (1: glucuronic acid, 2: xylose, 3: arabinose, 4: glucose, and 5: galactose); (B) corn fiber acid hydrolysates—1 M sulfuric acid; (C) 2 M TFA; (D) 72% sulfuric acid; (E) corn fiber enzymatic digestion. Conditions: same as in previous figure, except a bare fused-silica capillary (50μm ID, 48 cm effective separation length) was used with dynamic PEO coating. (Adapted from Khandurina, J. et al. *Electrophoresis*, 25, 2326, 2004. With permission.)

substantially elevated (peak 4) compared with milder hydrolysis conditions. An illustrative example of the enzymatic digestion of corn fiber by a fungal culture medium is shown in Trace E (Figure 7.16). The lower relative content of the released xylose, compared with the acidic hydrolysates, was caused by lack of xylanase activity in the enzyme system used in the experiment. Most of the other earlier migrating peaks in the electropherogram are probably associated with background and matrix effects originating from the labeling reaction and constituents of the enzyme medium.

The same research group introduced a 96-capillary array electrophoresis (CAE) approach for large-scale mono- and oligosaccharide analysis and characterization [201,202]. In this work, a DNA sequencer, MegaBACE1000 (Amersham Biosciences) was adapted for carbohydrate screening by developing and optimizing operational protocols and data processing tools. This approach brings high-performance carbohydrate analysis within reach of life sciences laboratories without the need for additional costly equipment, and helps expedite the pace of discovery and development of biomass conversion processes, as well as glycobiology studies. Carbohydrate samples (1 μL) were labeled in 96-well plate format, through reductive amination by the addition of 2 μL of 0.2 M APTS in 15% acetic acid and 2 μL of 1 M $NaBH_3CN$ in tetrahydrofuran [19,34,48,193,194]. The plates were incubated for 1 h at 75°C, followed by the addition of 100 μL water to stop the reaction. Before CE analysis, the APTS-labeled samples were further diluted in water (up to three orders of magnitude depending on the initial concentration of carbohydrates), resulting in low nM concentration range of each derivatized carbohydrate species. The authors found that uncoated fused-silica capillary arrays (The Gel Company, San Francisco, CA) in combination with dynamic PEO coating render more consistent performance and significantly longer lifetime compared to original covalently coated capillary arrays, designed for DNA-sequencing applications. The dynamic coating is easily regenerated before each run by sequential washing with hydrochloric acid, water, PEO solution in the running buffer, and finally lithium acetate running buffer. It is important to emphasize that carbohydrate analysis was performed in CZE (open tube) mode and, therefore, more sensitive to capillary surface quality and any residual EOF, unlike in DNA sequencing and genotyping, where viscous sieving media can compensate for some coating nonuniformity and deterioration. To further speed up the process, the dynamic coating polymer can be directly added to the running buffer. One important problem to avoid in carbohydrate CAE is a possible siphoning effect caused by uneven liquid levels in inlet and outlet buffer reservoirs [202]. Capillary-to-capillary and run-to-run variation in migration time and signal intensity necessitated the development of data normalization tools. Internal bracketing fluorescent standards were incorporated into the analysis enabling CAE trace alignment. A four-color detection system allowed a choice of different fluorophores to spectrally separate the sample peaks and internal standards in the corresponding detection channels. An excess of unreacted APTS, remaining in the samples after derivatization and migrating faster than the labeled sugars, was used as leading bracketing standard. A fluorescent dye sulforhodamine B (SRB), emitting in the red spectral region (λ_{max} = 586 nm) and migrating with similar mobility as the maltooligosaccharide of DP 16, served as a terminating bracketing standard. A set of programming tools were developed to process the data and properly normalize migration time for each electropherograms based on internal bracketing standards. The procedure is described in detail in Reference 201. Figure 7.17 presents a set of raw (left) and aligned (right) electropherograms, utilizing the developed algorithm. APTS-derivatized maltoologosaccharide ladder mixed with the terminating bracketing standard (SRB) was used in this experiment. As one can see, all corresponding oligosaccharide peaks match very well in the aligned data set in the right panel. The lower panel in Figure 7.17 delineates the migration time of APTS peak across the entire 96-CAE before and after the normalization procedure. A mixture of four monosaccharides, xylose, arabinose, glucose, and galactose were also successfully separated by CAE under low-pH conditions (Figure 7.18). The data were normalized, similar to the maltooligosaccharide ladder, enabling an accurate peak identification of closely migrating monosaccharides. Figure 7.18 shows five representative traces of such analysis before (upper panel) and after (lower panel) the data normalization. Albeit, the efficiency of APTS-labeling reaction is very similar for most monosaccharide product and short oligosaccharide product, precise quantification requires correction for possible small differences [201]. This can be accomplished by introduction of another internal standard into the sample derivatization mix, for example, oligosaccharide migrating with no overlapping with the sugars of interest. Relative labeling efficiency coefficients can be derived this way and used for accurate quantification of the sugars of interest in the unknown samples [201]. The described large-scale qualitative and quantitative carbohydrate profiling method based on 96-CAE can accommodate high-throughput demands of the biotechnology industry. The areas of application for the developed methodology include biomass conversion processes, which require massive screening of enzymatic reaction products to evaluate large libraries of recombinant enzymes and microorganisms, as well as protein glycosylation studies.

7.4.5 Miniaturization in Carbohydrate Analysis

The recent fast developing trend for miniaturization of analytical systems has influenced carbohydrate analysis as well. Advances in microfabrication technologies, microfluidics-based devices, and microchip electrophoresis in particular, in combination with available high-sensitivity detection methods, stimulated a series of attempts to implement carbohydrate analysis on-chip. However, advancement of on-chip glycomics has been somewhat limited compared to fast growth of microscale genomic and proteomic applications. A number of limitations, such as lack of charge and functional groups and detection issues, present a challenge. Suzuki and Honda [17] have briefly summarized the latest

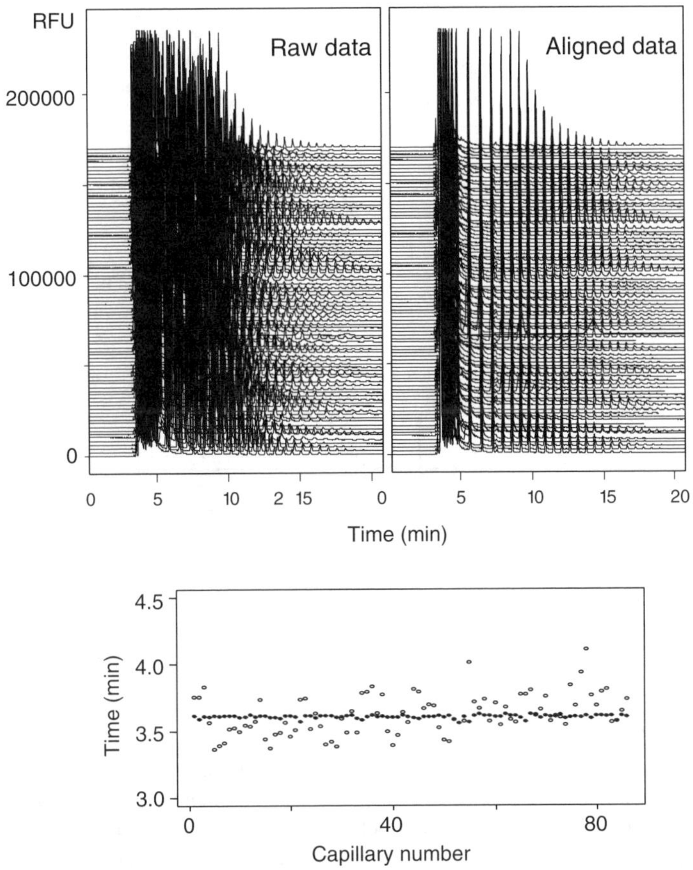

FIGURE 7.17 Migration time normalization and CAE data alignment. Left and right panels represent raw and aligned traces, respectively, for maltooligosaccharide ladder samples containing SRB as internal bracketing standard. The lower panel shows migration time standard deviation for APTS peak before and after normalization procedure. See details of data processing in the text. Conditions: injection 10 s at 5 kV, separation 15 kV (240 V/cm), temperature 30°C, 25 mM lithium acetate (pH 5) separation buffer, uncoated fused-silica capillary arrays dynamically coated with PEO (see details in the text). Arrows and numbers indicate DP of oligosaccharides, for reference. (Adapted from Khandurina, J. et al. *Electrophoresis*, 25, 3122, 2004. With permission.)

developments in this area. Mechref and Novotny [32] described chip approaches in glycomics, both CE and chromatography based, in their latest review.

Miniaturization of sample handling and processing steps should significantly reduce losses associated with adsorption at the surfaces, sample transfers, as well as potential contamination. Integration and concomitant minimization of the analysis steps on-chip, including purification, concentration, separation, and detection, will ultimately enhance sensitivity and efficiency of the carbohydrate analysis, and glycan screening in particular.

Several research groups implemented carbohydrate analysis on-chip with direct detection of underivatized sugar molecules. Electrochemical detection is the most attractive approach, as it offers reasonable sensitivity and selectivity, and it is ideally suited for microchip format. Schwarz et al. [203] developed amperometric detection of sugars using microfabricated copper electrode. They separated fructose, sucrose, and galactose in 70 s on a glass chip with 50-μm wide and 20-μm deep microchannel and double tee injection geometry. The detection was based on Teflon-coated platinum wire plated with copper and inserted in the end of the separation channel etched in a conical shape. The detection limit down to 1 μM was achieved. Hebert and coworkers [204] reported an

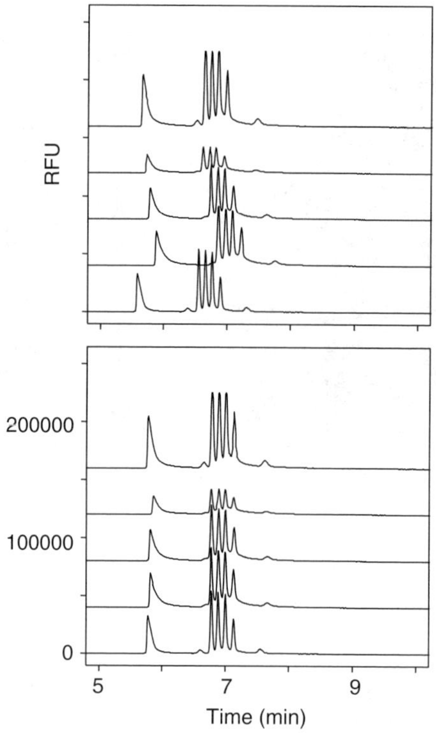

FIGURE 7.18 CAE analysis and data alignment for monosaccharides. Only five traces representing wells # 60–65 of a sample plate are shown to better visualize the separation and data processing performance. Raw and normalized data are in the upper and lower panels, respectively. The first migrating peak in each trace is unreacted APTS followed by xylose, arabinose, glucose, and galactose, in order of migration. Electrophoretic conditions (except for injection ay 1 kV for 5 s) and internal bracketing standards are the same as in previous figure. (Adapted from Khandurina, J. et al. *Electrophoresis*, 25, 3122, 2004. With permission.)

interesting detection system based on sinusoidal voltammetry, where Fourier transformation was used to convert current data, associated with carbohydrate molecules, into the frequency domain, thus, effectively isolating the analyte signal from background current. A polydimethylsiloxane chip with a planar copper electrode was fabricated, enabling detection limit of 200 amol and efficiency of 10^5 theoretical plates per meter. The signal-to-noise ratio was enhanced by utilizing a digital lock-in method, which allowed selective picking and masking of carbohydrate signals. An on-chip-pulsed amperometric detection system, described by Fanguy and Henry [205], accomplished analysis of glucose, xylose, and maltose in 4 cm separation channel with a platinum electrode built in the end part of the channel. The authors demonstrated a linear response in 20–500 μM concentration range, with 20 μM detection limit. A unique system, combining flow injection and short CE, was constructed by Fu and Fang [206]. Some carbohydrates were monitored by amperometric detection on a horizontal channel connected to a vertical plastic tubing. A solution of glucose and sucrose was continuously introduced by droplets from the tubing into the junction with the channel followed by electrophoretic separation. Sugars were separated in 60 s on a 5-μm ID and 5-cm-long capillary, with a dynamic range of 10–1000 μM.

Refractometry is another approach for direct detection of carbohydrates. An interesting holography-based refractive index detector was implemented by Burggraf et al. [207]. A cyclic channel of 80 mm in circumference and 10-μm deep was fabricated. The detection scheme was constructed having a diode laser beam (670 nm) split into two beams, one passing through the separation channel and the second through the adjacent reference channel. A photodiode array detected changes in holographic optical image. Separation time (17 s) of raffinose, sucrose, and *N*-acetylglucosamine was achieved, with the detection optimum at 10 mM level (600–900 fmol of injected material). Although the analysis sensitivity was quite limited, the described coupling of electrophoresis microchip and refractive index detector can be used for certain application when sensitivity is not required (e.g., measurements of simple sugars in foods).

The most successful microchip separations were achieved with derivatized carbohydrates when a UV absorbing or, better, fluorescent tag is introduced. On-chip UV detection suffered from low sensitivity and allowed only mM detection level using PMP label and free zone electrophoresis in borate buffer [208]. To alleviate poor detection limit, microchip Shimadzu instrument is equipped with the special whole-channel detection system, where the signal is enhanced by simultaneous accumulation and averaging of the absorption data from multiple points. Baba and coworkers [209] analyzed APTS-labeled oligosaccharides on a Hitachi SV1100 microchip system using a blue light emitting diode (LED) source and confocal fluorescent detection configuration. In this work, a plastic polymethylmethacrylate (PMMA) chip was employed, and the detrimental adsorption of APTS to PMMA material was alleviated by dynamic coating with methylcellulose to the running buffer. The maltooligosaccharides were resolved up to DP 10 (maltodecaose) in 2 min. In another work from the same group [210] protein glycosylation structures were analyzed: high-mannose-type glycans were enzymatically released from bovine spleen ribonuclease B and separated on-chip, using asymmetric pinched injection with field-amplified stacking resulting in 20-fold improved sensitivity. Derivatization of sugar reducing ends by reductive amination has been undoubtedly one of the most successful and universal approaches; however, some classes of carbohydrates contain functional groups that are suitable for alternative chemical couplings. For example, hexosamines and their derivatives possess an amino group at the C2 position, and therefore can be readily fluorescently labeled via reactions developed for aminoacids. Suzuki et al. [211] implemented a condensation reaction of hexosamines with 4-nitro-2,1,3-benzoxadiazole 7-fluoride (NBD-F), followed by a microchip electrophoresis on a quartz chip using 33 mm separation distance and Ar-ion LIF detection. The authors separated glucosamine, galactosamine, and their corresponding reduced forms, glucosaminitol and galactosaminitol, in 1 min under free buffer conditions (phenylborate sugar complexes), reaching detection limit of 2.5 μM (0.5 fmol injected amount). The same labeling methodology was successfully applied to *O*-linked glycans enzymatically stripped from glycoproteins (e.g., bovine submaxillary mucin).

Another interesting example of miniaturized carbohydrate analysis is on-chip integration of enzymatic reactions, conventionally used for sugar determination, with electrophoretic separation of the reaction products. Wang and coworkers [174] demonstrated a simultaneous assay of glucose, ascorbic acid, and acetaminophen on a microchip with cross-channels and a gold electrode assembly in the separation channel. The sample containing glucose was mixed with glucose oxidase (GOx) in the channel intersection, followed by the oxidation reaction and separation of the generated hydrogen peroxide from ascorbic acid and acetaminophen. The latter two species migrated more slowly due to their anionic nature. The gold-coated thick-film amperometric detector was positioned downstream in the separation channel, and the glucose level was assessed based on comparison of response with and without GOx (Figure 7.19) [174]. The same group reported an oxidase/dehydrogenase assay, implemented on-chip, for the measurements of glucose and ethanol [175]. The developed assay was applied to the analysis of wine samples.

Integration of microchip CE and MS is a new emerging technology, which is expected to enhance high-throughput glycan mapping and sequencing. The combination of microfluidics, automation/robotization, and software for assignment of MS data has already been implemented by introducing Advion Nanomate dispenser and chip to glycoscreening by coupling it to QTOF and FT-ICR mass analyzers [73]. The fully automated chip-based MS approach for complex carbohydrate system analysis was applied to urine analysis of patient suffering from hereditary diseases. Addition of sample CE preseparation to the ESI-QTOF via Nanomate system has been demonstrated

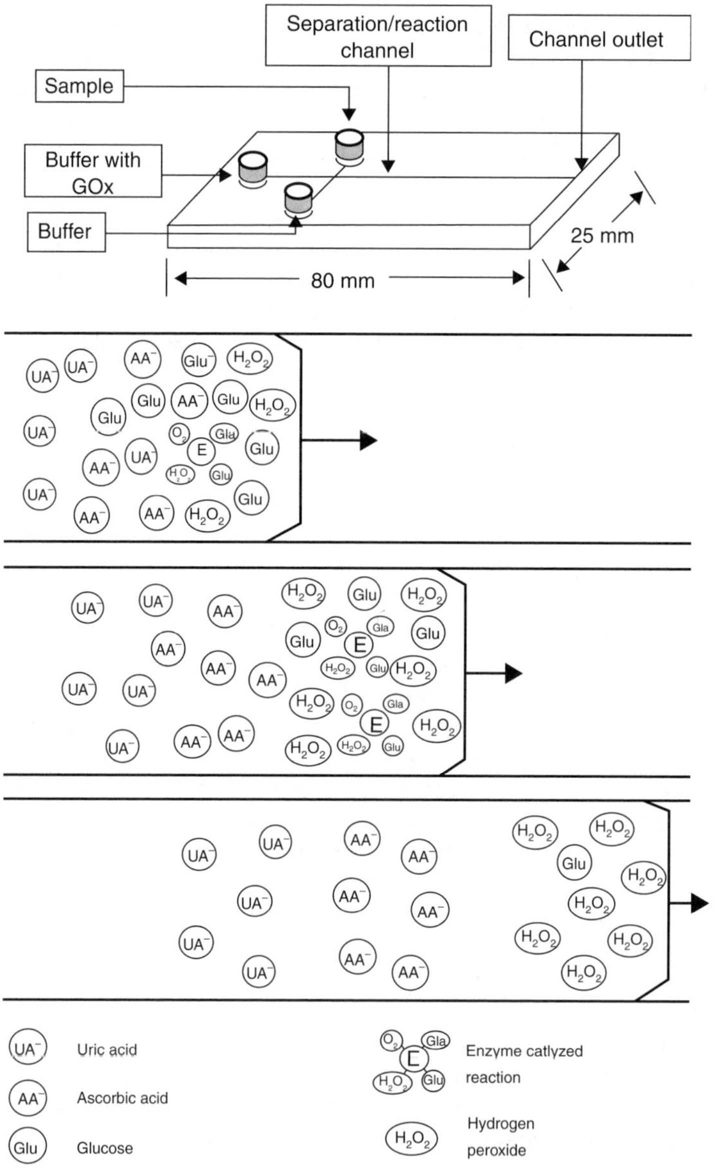

FIGURE 7.19 Layout of the separation/reaction microchip for bioassays of glucose, ascorbic acid, uric acid, and acetaminophen (upper panel). Enzymatic and separation processes along the reaction/separation channel of the CE biochip (lower panel). (Adapted from Wang, J. et al. *Anal Chem*, 72, 2514, 2000. With permission.)

to provide advantages compared to more conventional CE-MS interface, that is, higher ionization yield, decreased in-source fragmentation, and stable spray [212].

Recently, a total serum protein N-glycosylation profiling was attempted on a CE chip by Ehrlich and coworkers [213]. The authors employed a glass chip with a double-tee injector and a 4% linear polyacrylamide sieving medium (analogous to nucleic acid separations). Profiling of serum samples from chronic hepatitis patients identified the differences in N-glycan composition in cirrhotic and noncirrhotic cases, and demonstrated the potential of microchip approach for these types of clinical studies.

Integrated microchips, packed with reversed-chromatographic phase and interfaced with TOF [214] or IT [215] MS, for the analysis of oligosaccharides, *O*- and *N*-glycans of different sources have also been reported.

Although only relatively simple carbohydrate systems have been transferred on-chip so far, primarily due to resolution limitation in short separation distances and sensitivity challenges in microchannels, rapid progress in this area is expected in the nearest future. Column efficiencies observed in microchannels were similar or higher (the latter was due to shorter injection plugs realized in microchip settings) compared to conventional CE format. Shorter separation paths utilized in miniaturized analysis settings limit complexity of analytes, which can be successfully resolved. On the other hand, microfabricated devices offer a number of other attractive analytical advantages. One of them is possibility of integration, that is, combining sample prep and derivatization, separation, and detection steps on a single integrated device [216,217]. Second, disposable microchips can be inexpensively mass produced and utilized in testing analytical kits or point-of-care diagnostic tools in food, environmental, and medical applications.

7.5 METHOD DEVELOPMENT GUIDELINES

As it has been emphasized throughout this chapter, CE in its various modes of separation and detection is suitable for the analysis of a wide variety of carbohydrate molecules, that is, mono-, oligo-, and polysaccharides, glycopeptides, glycoproteins, and glycolipids. This can be credited to the progress made in the capillary column technology, introduction of novel electrolyte systems, as well as the development of various detection approaches, such as indirect UV and LIF detection, electrochemical detection, CE-MS interface, and precolumn labeling with suitable chromophores and fluorophores. Precolumn derivatization proved to be one of the most elegant approaches for the separation and detection of carbohydrates. Among the different reaction schemes, introduced for the labeling of carbohydrates, multiply charged tags (ANDSA, ANTS, and APTS) have become the most popular, not only because of their high detection sensitivity by either UV or LIF, but also because they yield derivatives that are readily separated by CE.

Majority of CE separations are typically accomplished utilizing borate complexation, for both derivatized and underivatized carbohydrates. However, noncomplexing electrolyte systems (phosphate, acetate based buffers, for example) are often successfully used as well. As it has been described in Section 7.2, borate complexation enhances small structural differences of structurally similar molecules, resulting into highly selective separations of multicomponent sugar mixtures. Under a given set of conditions, various sugars, whether charged or neutral, undergo varying degrees of complexation with borate leading to differences in the electrophoretic mobilities of the complexed solutes and hence separation.

Highly alkaline electrolyte solutions, for example, lithium, potassium, or sodium hydroxide at pH > 12 have been proved to be useful in the separation of underivatized saccharides by CZE. The resolution of various saccharides is typically improved with pH increase (up to pH 13) due to increasing ionization of the separated analytes. In addition, the nature of the alkali-metal influences the resolution of the sugar analytes.

The separations at extremely high pH can only be performed in bare fused-silica capillaries, since most coatings undergo hydrolytic degradation under the highly basic conditions. The use of electrolyte systems containing alkaline-earth metals for the separation of neutral carbohydrates by CZE is mainly based on differences in the extent of complexation of the divalent metals with the carbohydrate solutes, as well as the size and shape of the molecule. These systems provided a different selectivity from that of borate buffers, although the resolution is often inferior.

The electrolyte systems in CE can be easily modified and tailored to a specific separation task. For example, most glycolipids are not compatible with aqueous electrolyte solutions but they can be readily separated in their monomeric forms in hydroorganic buffers. Buffer additives, such as surfactants and polymers, often enhance separation efficiency. Examples of various separation buffer systems, suitable for different classes of carbohydrates, and approaches to their development can be found in the above sections, dedicated to specific types of carbohydrate species and their sources, and references therein.

Characterization of glycoconjugates are typically carried out by first cleaving the carbohydrate moieties from glycoproteins, glycolipids, GAGs, or other complex molecules, often followed by further reduction of the structure complexity through enzymatic digestion and/or depolymerization.

Owing to enormous diversity and complexity of carbohydrates and their conjugates with other classes of biomolecules, it is impossible to briefly outline any definitive guidelines for analysis method development. Each type of carbohydrate molecules should be approached separately, based on its specific properties, such as molecular weight, linearity or branching, presence/absence of the reducing end, charge, additional functional groups present, matrix complexity, and so forth. Therefore, the reader is directed to the above sections to find basic modern strategies, as well as specific protocols in the corresponding references, for sample preparation, derivatization, separation, and detection in the analysis and structural characterization of simple and complex saccharides and glycoconjugates in pharmaceuticals, foods, plants, bacteria, animal tissues, and other sources.

7.6 CONCLUSIONS AND OUTLOOK

Progress in carbohydrate research and concomitant advances of analytical methodologies during the past decade have been undoubtedly remarkable. In particularly, the expansion of CE and its various implementations supported the increasing demand for new level of information output, such as integrated systems biology approach.

The future developments in CE-based analysis of carbohydrates are envisioned to bring forward automation and miniaturization to ensure high-throughput and large-scale carbohydrate profiling to keep up with the industrial needs. On the other hand, capabilities of elaborate structural insights into complex carbohydrate molecules should be further developed as well. Hyphenated techniques, such as CE-MS, offer possibility to identify composition of complex polysaccharides and glycoconjugates, as well as their sequence and fine structural details, that is, isomeric configurations, branching, substitutions, linkages between monomer units, and so forth. Glycomics and glycoproteomics deserve special attention, as part of system biology efforts toward complete understanding of the complexity of living organisms. Just like the developments in genomics, transcriptomics, proteomics, and metabolomics, functional glycomics is strongly dependent on the current and future advances in analytical methodologies and instrumentation. Quantitative measurements are essential to better understanding of cellular and molecular interactions associated with different physiological and pathological processes in living systems. CE-based methodologies directly combined with MS are particularly useful in this respect. Electromigration techniques, for example, CZE, MEKC, and CEC, exhibit most superior resolving power enabling separation and identification of close oligosaccharide isomers. LIF is most preferred detection approach for high-sensitivity profiling of glycan mixtures by CE. Coupling these techniques to MS is still a challenging task, although significant progress has been made. Increasing number of manufacturers, offering ready-to-use CE-MS instruments and a variety of available interfaces (especially nanospray sheathless approach), represent a step forward toward a more routine use of carbohydrate analysis by CE-MS. Growing sophistication of MS instrumentation fuels development of new CE-MS applications, involving IT, TOF, TOF-TOF, and FT-ICR. CE-MS^n offers powerful capabilities. However, optimization of this technology, including instrumental solutions, sample preparation, and separation aspects, is still in progress, and robust standardized systems and protocols are to be refined yet. Careful choice of separation buffers is the key, as many salts and other buffer components adversely affect sensitivity of most MS detectors.

It is reasonable to expect that microfluidic and chip-based technology will become widely applicable to glycomics, as it has already become in many aspects of genomics and proteomics. Miniaturized CE-MS technology should develop into reliable task-oriented commercial instrumentation, applicable to investigation of different biological systems under well-designed conditions.

REFERENCES

1. Crocker, P. R. and Feizi, T. Carbohydrate recognition systems: functional triads in cell–cell interactions. *Curr Opin Struct Biol*, 6, 679, 1996.
2. Helenius, A. and Aebi, M. Intracellular functions of *N*-linked glycans. *Science*, 291, 2364, 2001.
3. Karlsson, K. A. Meaning and therapeutic potential of microbial recognition of host glycoconjugates. *Mol Microbiol*, 29, 1, 1998.
4. Zamfir, A. and Peter-Katalinic, J. Capillary electrophoresis-mass spectrometry for glycoscreening in biomedical research. *Electrophoresis*, 25, 1949, 2004.
5. Gordon, H. T., Thornburg, W. W. and Werum, L. N. Rapid paper chromatographic fractionation of complex mixtures of water-soluble substances. *J Chromatogr*, 9, 44, 1962.
6. Ghebregzabher, M., Rufini, S., Monaldi, B. and Lato, M. Thin-layer chromatography of carbohydrates. *J Chromatogr*, 127, 133, 1976.
7. Honda, S. High-performance liquid chromatography of mono- and oligosaccharides. *Anal Biochem*, 140, 1, 1984.
8. Churms, S. C. Recent developments in the chromatographic analysis of carbohydrates. *J Chromatogr*, 500, 555, 1990.
9. Townsend, R. R., Hardy, M. R. and Lee, Y. C. Separation of oligosaccharides using high-performance anion-exchange chromatography with pulsed amperometric detection. *Methods Enzymol*, 179, 65, 1989.
10. Hardy, M. R., Townsend, R. R. and Lee, Y. C. Monosaccharide analysis of glycoconjugates by anion exchange chromatography with pulsed amperometric detection. *Anal Biochem*, 170, 54, 1988.
11. Hermentin, P. et al. The mapping by high-pH anion-exchange chromatography with pulsed amperometric detection and capillary electrophoresis of the carbohydrate moieties of human plasma alpha 1-acid glycoprotein. *Anal Biochem*, 206, 419, 1992.
12. Hermentin, P. et al. A strategy for the mapping of N-glycans by high-pH anion-exchange chromatography with pulsed amperometric detection. *Anal Biochem*, 203, 281, 1992.
13. Hanko, V. P. and Rohrer, J. S. Determination of carbohydrates, sugar alcohols, and glycols in cell cultures and fermentation broths using high-performance anion-exchange chromatography with pulsed amperometric detection. *Anal Biochem*, 283, 192, 2000.
14. Rohrer, J. S. Analyzing sialic acids using high-performance anion-exchange chromatography with pulsed amperometric detection. *Anal Biochem*, 283, 3, 2000.
15. Townsend, R. R. *Techniques in Glycobiology*, Marcel Dekker, New York, 1997.
16. Olechno, J. D. and Nolan, J. A. *Handbook of Capillary Electrophoresis*, ed. Landers, J. p. 297, CRC Press Inc., Boca Raton, FL, 1996.
17. Suzuki, S. and Honda, S. Miniaturization in carbohydrate analysis. *Electrophoresis*, 24, 3577, 2003.
18. Honda, S. Separation of neutral carbohydrates by capillary electrophoresis. *J Chromatogr A*, 720, 337, 1996.
19. Suzuki, S. and Honda, S. A tabulated review of capillary electrophoresis of carbohydrates. *Electrophoresis*, 19, 2539, 1998.
20. Novotny, M. V. and Sudor, J. High-performance capillary electrophoresis of glycoconjugates. *Electrophoresis*, 14, 373, 1993.
21. El Rassi, Z. and Mechref, Y. Recent advances in capillary electrophoresis of carbohydrates. *Electrophoresis*, 17, 275, 1996.
22. Honda, S., Iwase, S., Makino, A. and Fujiwara, S. Simultaneous determination of reducing monosaccharides by capillary zone electrophoresis as the borate complexes of *N*-2-pyridylglycamines. *Anal Biochem*, 176, 72, 1989.

23. Honda, S., Yamamoto, K., Suzuki, S., Ueda, M. and Kakehi, K. High-performance capillary zone electrophoresis of carbohydrates in the presence of alkaline earth metal ions. *J Chromatogr A*, 588, 327, 1991.
24. Grill, E., Huber, C., Oefner, P., Vorndran, A. and Bonn, G. Capillary zone electrophoresis of *p*-aminobenzoic acid derivatives of aldoses, ketoses and uronic acids. *Electrophoresis*, 14, 1004, 1993.
25. Jackson, P. The analysis of fluorophore-labeled glycans by high-resolution polyacrylamide gel electrophoresis. *Anal Biochem*, 216, 243, 1994.
26. Que, A. H. and Novotny, M. V. Separation of neutral saccharide mixtures with capillary electrochromatography using hydrophilic monolithic columns. *Anal Chem*, 74, 5184, 2002.
27. Chiesa, C., Oefner, P. J., Zieske, L. R. and O'Neill, R. A. Micellar electrokinetic chromatography of monosaccharides derivatized with 1-phenyl-3-methyl-2-pyrazolin-5-one. *J Capillary Electrophor*, 2, 175, 1995.
28. O'Shea, T. J., Lunte, S. M. and LaCourse, W. R. Detection of carbohydrates by capillary electrophoresis with pulsed amperometric detection. *Anal Chem*, 65, 948, 1993.
29. Colon, L. A., Dadoo, R. and Zare, R. N. Determination of carbohydrates by capillary zone electrophoresis with amperometric detection at a copper microelectrode. *Anal Chem*, 65, 476, 1993.
30. Hase, S., Hra, S. and Matsushima, Y. Tagging of sugars with a fluorescent compound, 2-aminopyridine. *J Biochem*, 85, 217, 1979.
31. Honda, S., Suzuki, S. and Taga, A. Analysis of carbohydrates as 1-phenyl-3-methyl-5-pyrazolone derivatives by capillary/microchip electrophoresis and capillary electrochromatography. *J Pharm Biomed Anal*, 30, 1689, 2003.
32. Mechref, Y. and Novotny, M. V. Miniaturized separation techniques in glycomic investigations. *J Chromatogr B Analyt Technol Biomed Life Sci*, 841, 65, 2006.
33. Evangelista, R. A., Liu, M.-S. and Chen, F.-T. A. Characterization of 9-aminopyrene-1,4,6-trisulfonate derivatized sugars by capillary electrophoresis with laser-induced fluorescence detection. *Anal Biochem*, 67, 2239, 1995.
34. Guttman, A., Chen, F. T., Evangelista, R. A. and Cooke, N. High-resolution capillary gel electrophoresis of reducing oligosaccharides labeled with 1-aminopyrene-3,6,8-trisulfonate. *Anal Biochem*, 233, 234, 1996.
35. O'Neill, R. A., Hoff, L. B. and Khan, S. H. Method and apparatus for automated carbohydrate mapping and sequencing. *PCT Int Appl WO*, 95, 989, 1995.
36. Guttman, A. and Herrick, S. Effect of the quantity and linkage position of mannose, alpha 1,2 residues in capillary gel electrophoresis of high-mannose-type oligosaccharides. *Anal Biochem*, 235, 236, 1996.
37. Khandurina, J. and Guttman, A. High resolution capillary electrophoresis of oligosaccharide structural isomers. *Chromatographia*, 62, S37, 2005.
38. Klampfl, C. W. Recent advances in the application of capillary electrophoresis with mass spectrometric detection. *Electrophoresis*, 27, 3, 2006.
39. Peter-Katalinic, J. Analysis of glycoconjugates by fast atom bombardment mass spectrometry and related MS techniques. *Mass Spectrom Rev*, 13, 77, 1994.
40. Sagi, D., Peter-Katalinic, J., Conradt, H. S. and Nimtz, M. Sequencing of tri- and tetra-antennary N-glycans containing sialic acid by negative mode ESI QTOF tandem MS. *J Am Soc Mass Spectrom*, 13, 1138, 2002.
41. Beckedorf, A. I., Schaffer, C., Messner, P. and Peter-Katalinic, J. Mapping and sequencing of cardiolipins from Geobacillus stearothermophilus NRS 2004/3a by positive and negative ion nanoESI-QTOF-MS and MS/MS. *J Mass Spectrom*, 37, 1086, 2002.
42. Bahrke, S. et al. Sequence analysis of chitooligosaccharides by matrix-assisted laser desorption ionization postsource decay mass spectrometry. *Biomacromolecules*, 3, 696, 2002.
43. Metelmann, W., Peter-Katalinic, J. and Muthing, J. Gangliosides from human granulocytes: a nano-ESI QTOF mass spectrometry fucosylation study of low abundance species in complex mixtures. *J Am Soc Mass Spectrom*, 12, 964, 2001.
44. Macek, B., Hofsteenge, J. and Peter-Katalinic, J. Direct determination of glycosylation sites in O-fucosylated glycopeptides using nano-electrospray quadrupole time-of-flight mass spectrometry. *Rapid Commun Mass Spectrom*, 15, 771, 2001.

45. Hanisch, F. G., Jovanovic, M. and Peter-Katalinic, J. Glycoprotein identification and localization of O-glycosylation sites by mass spectrometric analysis of deglycosylated/alkylaminylated peptide fragments. *Anal Biochem*, 290, 47, 2001.
46. Viseux, N., Costello, C. E. and Domon, B. Post-source decay mass spectrometry: optimized calibration procedure and structural characterization of permethylated oligosaccharides. *J Mass Spectrom*, 34, 364, 1999.
47. Simo, C., Barbas, C. and Cifuentes, A. Capillary electrophoresis-mass spectrometry in food analysis. *Electrophoresis*, 26, 1306, 2005.
48. El Rassi, Z. *Carbohydrate Analysis by Modern Chromatography and Electrophoresis*, ed. El Rassi, Z., Elsevier, Amsterdam, 2002.
49. Zaia, J. Mass spectrometry of oligosaccharides. *Mass Spectrom Rev*, 23, 161, 2004.
50. Novotny, M. V. and Mechref, Y. New hyphenated methodologies in high-sensitivity glycoprotein analysis. *J Sep Sci*, 28, 1956, 2005.
51. Tang, H., Mechref, Y. and Novotny, M. V. Automated interpretation of MS/MS spectra of oligosaccharides. *Bioinformatics*, 21 (Suppl 1), i431, 2005.
52. Huang, L. and Riggin, R. M. Analysis of nonderivatized neutral and sialylated oligosaccharides by electrospray mass spectrometry. *Anal Chem*, 72, 3539, 2000.
53. Que, A. H. et al. Coupling capillary electrochromatography with electrospray Fourier transform mass spectrometry for characterizing complex oligosaccharide pools. *Anal Chem*, 75, 1684, 2003.
54. Shen, Y. and Smith, R. D. Proteomics based on high-efficiency capillary separations. *Electrophoresis*, 23, 3106, 2002.
55. Bergstrom, S. K., Samskog, J. and Markides, K. E. Development of a poly, dimethylsiloxane interface for on-line capillary column liquid chromatography-capillary electrophoresis coupled to sheathless electrospray ionization time-of-flight mass spectrometry. *Anal Chem*, 75, 5461, 2003.
56. Ivanov, A. R., Horvath, C. and Karger, B. L. High-efficiency peptide analysis on monolithic multimode capillary columns: Pressure-assisted capillary electrochromatography/capillary electrophoresis coupled to UV and electrospray ionization-mass spectrometry. *Electrophoresis*, 24, 3663, 2003.
57. Guetens, G. et al. Nanotechnology in bio/clinical analysis. *J Chromatogr B Biomed Sci Appl*, 739, 139, 2000.
58. Lopez-Soto-Yarritu, P., Diez-Masa, J. C., Cifuentes, A. and de Frutos, M. Improved capillary isoelectric focusing method for recombinant erythropoietin analysis. *J Chromatogr A*, 968, 221, 2002.
59. Campa, C., Coslovi, A., Flamigni, A. and Rossi, M. Overview on advances in capillary electrophoresis-mass spectrometry of carbohydrates: a tabulated review. *Electrophoresis*, 27, 2027, 2006.
60. Kamoda, S. and Kakehi, K. Capillary electrophoresis for the analysis of glycoprotein pharmaceuticals. *Electrophoresis*, 27, 2495, 2006.
61. Hau, J. and Roberts, M. Advantages of pressurization in capillary electrophoresis/electrospray ionization mass spectrometry. *Anal Chem*, 71, 3977, 1999.
62. Li, D. T., Sheen, J. F. and Her, G. R. Structural analysis of chromophore-labeled disaccharides by capillary electrophoresis tandem mass spectrometry using ion trap mass spectrometry. *J Am Soc Mass Spectrom*, 11, 292, 2000.
63. Joucla, G., Brando, T., Remaud-Simeon, M., Monsan, P. and Puzo, G. Capillary electrophoresis analysis of glucooligosaccharide regioisomers. *Electrophoresis*, 25, 861, 2004.
64. Wolff, M. W., Bazin, H. G. and Lindhardt, R. J. Analysis of fluorescently labeled oligosaccharides by capillary electrophoresis and electrospray ionization mass spectrometry. *Biotechnol Tech*, 13, 797, 1999.
65. Charlwood, J. et al. A probe for the versatile analysis and characterization of *N*-linked oligosaccharides. *Anal Chem*, 72, 1453, 2000.
66. Muthing, J. et al. Effects of buffering conditions and culture pH on production rates and glycosylation of clinical phase I anti-melanoma mouse IgG3 monoclonal antibody R24. *Biotechnol Bioeng*, 83, 321, 2003.
67. An, H. J., Franza, A. H. and Lebrilla, C. B. Improved capillary electrophoretic separation and mass spectrometric detection of oligosaccharides. *J Chromatogr*, 1004, 121, 2003.
68. Min, J. Z., Toyo'oka, T., Kato, M. and Fukushima, T. Resolution of *N*-linked oligosaccharides in glycoproteins based upon transglycosylation reaction by CE-TOF-MS. *Chem Commun* (Camb), 27, 3484, 2005.

69. Gennaro, L. A., Delaney, J., Vouros, P., Harvey, D. J. and Domon, B. Capillary electrophoresis/electrospray ion trap mass spectrometry for the analysis of negatively charged derivatized and underivatized glycans. *Rapid Commun Mass Spectrom*, 16, 192, 2002.
70. Koller, A. et al. Analysis of high-mannose-type oligosaccharides by microliquid chromatography-mass spectrometry and capillary electrophoresis. *Electrophoresis*, 25, 2003, 2004.
71. Sandra, K. et al. Characterization of cellobiohydrolase I N-glycans and differentiation of their phosphorylated isomers by capillary electrophoresis-Q-Trap mass spectrometry. *Anal Chem*, 76, 5878, 2004.
72. Bindila, L., Peter-Katalinic, J. and Zamfir, A. Sheathless reverse-polarity capillary electrophoresis-electrospray-mass spectrometry for analysis of underivatized glycoconjugates. *Electrophoresis*, 26, 1488, 2005.
73. Zamfir, A., Konig, S., Althoff, J. and Peter-Katalinc, J. Capillary electrophoresis and off-line capillary electrophoresis-electrospray ionization quadrupole time-of-flight tandem mass spectrometry of carbohydrates. *J Chromatogr A*, 895, 291, 2000.
74. Peter-Katalinic, J. Methods in enzymology: O-glycosylation of proteins. *Methods Enzymol*, 405, 139, 2005.
75. Baldus, S. E. and Hanisch, F. G. Biochemistry and pathological importance of mucin-associated antigens in gastrointestinal neoplasia. *Adv Cancer Res*, 79, 201, 2000.
76. Domon, B., Vath, J. E. and Costello, C. E. Analysis of derivatized ceramides and neutral glycosphingolipids by high-performance tandem mass spectrometry. *Anal Biochem*, 184, 151, 1990.
77. Domon, B., Vath, J. E. and Costello, C. E. Analysis of derivatized ceramides and neutral glycosphingolipids by high-performance tandem mass spectrometry. *Anal Biochem*, 184, 151, 1990.
78. Domon, B. and Costello, C. E. A systematic nomenclature for carbohydrate fragmentations in FABMS/MS of glycoconjugates. *Glycoconjugate J*, 5, 397, 1988.
79. Costello, C. E. and Vath, J. E. Tandem mass spectrometry of glycolipids. *Methods Enzymol*, 193, 738, 1990.
80. Levery, S. B. Glycosphingolipid structural analysis and glycosphingolipidomics. *Methods Enzymol*, 405, 300, 2005.
81. Seydel, U., Oikawa, M., Fukase, K., Kusumoto, S. and Brandenburg, K. Intrinsic conformation of lipid A is responsible for agonistic and antagonistic activity. *Eur J Biochem*, 267, 3032, 2000.
82. Netea, M. G., van Deuren, M., Kullberg, B. J., Cavaillon, J. M. and Van der Meer, J. W. Does the shape of lipid A determine the interaction of LPS with Toll-like receptors? *Trends Immunol*, 23, 135, 2002.
83. Li, J. et al. Mapping bacterial glycolipid complexity using capillary electrophoresis and electrospray mass spectrometry. *Methods Enzymol*, 405, 369, 2005.
84. Krylov, S. N. and Dovichi, N. J. Capillary electrophoresis for the analysis of biopolymers. *Anal Chem*, 72, 111R, 2000.
85. Wang, Z. et al. Structural and serological characterization of the O-chain polysaccharide of *Aeromonas salmonicida* strains A449, 80204 and 80204-1. *Carbohydr Res*, 340, 693, 2005.
86. Li, J., Wang, Z. and Altman, E. In-source fragmentation and analysis of polysaccharides by capillary electrophoresis/mass spectrometry. *Rapid Commun Mass Spectrom*, 19, 1305, 2005.
87. Zamfir, A., Vukelic, Z. and Peter-Katalinic, J. A capillary electrophoresis and off-line capillary electrophoresis/electrospray ionization-quadrupole time of flight-tandem mass spectrometry approach for ganglioside analysis. *Electrophoresis*, 23, 2894, 2002.
88. Zamfir, A., Seidler, D. G., Kresse, H. and Peter-Katalinic, J. Structural characterization of chondroitin/dermatan sulfate oligosaccharides from bovine aorta by capillary electrophoresis and electrospray ionization quadrupole time-of-flight tandem mass spectrometry. *Rapid Commun Mass Spectrom*, 16, 2015, 2002.
89. Zamfir, A., Seidler, D. G., Kresse, H. and Peter-Katalinic, J. Structural investigation of chondroitin/dermatan sulfate oligosaccharides from human skin fibroblast decorin. *Glycobiology*, 13, 733, 2003.
90. Ju, D. D., Lai, C. C. and Her, G. R. Analysis of gangliosides by capillary zone electrophoresis and capillary zone electrophoresis-electrospray mass spectrometry. *J Chromatogr A*, 779, 195, 1997.
91. Tseng, M. C., Chen, Y. R. and Her, G. R. A low-makeup beveled tip capillary electrophoresis /electrospray ionization mass spectrometry interface for micellar electrokinetic chromatography and nonvolatile buffer capillary electrophoresis. *Anal Chem*, 76, 6306, 2004.

92. Li, J., Cox, A. D., Hood, D., Moxon, E. R. and Richards, J. C. Application of capillary electrophoresis-electrospray-mass spectrometry to the separation and characterization of isomeric lipopolysaccharides of Neisseria meningitidis. *Electrophoresis*, 25, 2017, 2004.
93. Volpi, N. and Maccari, F. Electrophoretic approaches to the analysis of complex polysaccharides. *J Chromatogr B Analyt Technol Biomed Life Sci*, 834, 1, 2006.
94. Ofosu, F. A., Danishefsky, I. and Hirsh, J. *Heparin and Related Polysaccharides: Structure and Activities*, eds. Ofosu, F. A., Danishefsky, I. and Hirsh, J., N.Y. Acad. Sci., New York, 1989.
95. Mammen, E. F. *Seminars in Trombosis and Hemostasis*, ed. Mammen, E. F., Theme Medical Pub. Inc., New York, 1991.
96. Lane, D. A. and Lindahl, U. *Heparin: Chemical and Biological Properties, Clinical Applications*, eds. Lane, D. A. and Lindahl, U., Edward Arnold, London, Melbourne, Auckland, 1989.
97. Lane, D. A., Bjork, I. and Lindahl, U. *Advances in Experimental Medicine and Biology*, eds. Lane, D. A., Bjork, I. and Lindahl, U., Plenum Press, New York, 1992.
98. Scott, J. E. *Dermatan Sulfate Proteoglycans—Chemistry, Biology, Chemical Pathology*, ed. Scott, J.E., Portland Press, University of Manchester, UK, 1993.
99. Crescenzi, V., Dea, I. C. M., Paoletti, S., Stivala, S.S. and Sutherland, I. W. *Biomedical and Biotechnological Advances in Industrial Polysaccharides*, eds. Crescenzi, V., Dea, I. C. M., Paoletti, S., Stivala, S.S. and Sutherland, I. W., Gordon and Breach Sc. Pub, New York, 1989.
100. Wight, T. N. and Mecham, R. P. *Biology of Proteoglycans*, eds. Wight, T. N. and Mecham, R. P., Academic Press, New York, 1987.
101. Kjellen, L. and Lindahl, U. Proteoglycans: structures and interactions. *Annu Rev Biochem*, 60, 443, 1991.
102. Ruoslahti, E. Structure and biology of proteoglycans. *Annu Rev Cell Biol*, 4, 229, 1988.
103. Heinegard, D. and Oldberg, A. Structure and biology of cartilage and bone matrix noncollagenous macromolecules. *FASEB J*, 3, 2042, 1989.
104. Salmivirta, M. and Jalkanen, M. Syndecan family of cell surface proteoglycans: developmentally regulated receptors for extracellular effector molecules. *Experientia*, 51, 863, 1995.
105. Iozzo, R. V. The biology of the small leucine-rich proteoglycans. Functional network of interactive proteins. *J Biol Chem*, 274, 18843, 1999.
106. Ruoslahti, E. Proteoglycans in cell regulation. *J Biol Chem*, 264, 13369, 1989.
107. Sasisekharan, R., Shriver, Z., Venkataraman, G. and Narayanasami, U. Roles of heparan-sulphate glycosaminoglycans in cancer. *Nat Rev Cancer*, 2, 521, 2002.
108. Capila, I. and Linhardt, R. J. Heparin–protein interactions. *Angew Chem Int Ed Engl*, 41, 391, 2002.
109. Fraser, J. R., Laurent, T. C. and Laurent, U. B. Hyaluronan: its nature, distribution, functions and turnover. *J Intern Med*, 242, 27, 1997.
110. Laurent, T. C. and Fraser, J. R. Hyaluronan. *FASEB J*, 6, 2397, 1992.
111. Choi, H. U. and Meyer, K. The structure of keratan sulphates from various sources. *Biochem J*, 151, 543, 1975.
112. Koketsu, M. and Linhardt, R. J. Electrophoresis for the analysis of acidic oligosaccharides. *Anal Biochem*, 283, 136, 2000.
113. Vynios, D. H., Karamanos, N. K. and Tsiganos, C. P. Advances in analysis of glycosaminoglycans: its application for the assessment of physiological and pathological states of connective tissues. *J Chromatogr B Analyt Technol Biomed Life Sci*, 781, 21, 2002.
114. Mao, W., Thanawiroon, C. and Linhardt, R. J. Capillary electrophoresis for the analysis of glycosaminoglycans and glycosaminoglycan-derived oligosaccharides. *Biomed Chromatogr*, 16, 77, 2002.
115. Hoffman, P., Hsu, D. and Mashburn, T. A., Jr. Acrylamide gel electrophoresis of acid glycosaminoglycans for the study of molecular size distribution. *Anal Biochem*, 52, 382, 1973.
116. Calabro, A. et al. Fluorophore-assisted carbohydrate electrophoresis, FACE of glycosaminoglycans. *Osteoarthritis Cartilage*, 9 (Suppl A), S16, 2001.
117. Volpi, N. Separation of capsular polysaccharide K4 and defructosylated K4 derived disaccharides by high-performance capillary electrophoresis and high-performance liquid chromatography. *Electrophoresis*, 24, 1063, 2003.

118. Volpi, N. Separation of capsular polysaccharide K4 and defructosylated K4 by high-performance capillary electrophoresis. *Electrophoresis*, 25, 692, 2004.
119. Volpi, N. Separation of *Escherichia coli* 055:B5 lipopolysaccharide and detoxified lipopolysaccharide by high-performance capillary electrophoresis. *Electrophoresis*, 24, 3097, 2003.
120. Volpi, N. Purification of the *Escherichia coli* K5 capsular polysaccharide and use of high-performance capillary electrophoresis to qualitative and quantitative monitor the process. *Electrophoresis*, 25, 3307, 2004.
121. Kuhn, A. V., Ruttinger, H. H., Neubert, R. H. and Raith, K. Identification of hyaluronic acid oligosaccharides by direct coupling of capillary electrophoresis with electrospray ion trap mass spectrometry. *Rapid Commun Mass Spectrom*, 17, 576, 2003.
122. Zamfir, A., Seidler, D. G., Schonherr, E., Kresse, H. and Peter-Katalinic, J. On-line sheathless capillary electrophoresis/nanoelectrospray ionization-tandem mass spectrometry for the analysis of glycosaminoglycan oligosaccharides. *Electrophoresis*, 25, 2010, 2004.
123. Smyth, W. F. Recent applications of capillary electrophoresis-electrospray ionisation-mass spectrometry in drug analysis. *Electrophoresis*, 26, 1334, 2005.
124. Schweda, E. K., Li, J., Moxon, E. R. and Richards, J. C. Structural analysis of lipopolysaccharide oligosaccharide epitopes expressed by non-typeable Haemophilus influenzae strain 176. *Carbohydr Res*, 337, 409, 2002.
125. Wang, Q. and Fang, Y. Analysis of sugars in traditional Chinese drugs. *J Chromatogr B Analyt Technol Biomed Life Sci*, 812, 309, 2004.
126. Muramatsu, T. Carbohydrate signals in metastasis and prognosis of human carcinomas. *Glycobiology*, 3, 291, 1993.
127. Fukuda, M. Possible roles of tumor-associated carbohydrate antigens. *Cancer Res*, 56, 2237, 1996.
128. Zara, J. and Naz, R. K. The role of carbohydrates in mammalian sperm-egg interactions: how important are carbohydrate epitopes? *Front Biosci*, 3, D1028, 1998.
129. Takeuchi, M. et al. Relationship between sugar chain structure and biological activity of recombinant human erythropoietin produced in Chinese hamster ovary cells. *Proc Natl Acad Sci USA*, 86, 7819, 1989.
130. Egrie, J. C., Dwyer, E., Browne, J. K., Hitz, A. and Lykos, M. A. Darbepoetin alfa has a longer circulating half-life and greater in vivo potency than recombinant human erythropoietin. *Exp Hematol*, 31, 290, 2003.
131. Krylov, S. N., Arriaga, E. A., Chan, N. W., Dovichi, N. J. and Palcic, M. M. Metabolic cytometry: monitoring oligosaccharide biosynthesis in single cells by capillary electrophoresis. *Anal Biochem*, 283, 133, 2000.
132. Dordal, M. S., Wang, F. F. and Goldwasser, E. The role of carbohydrate in erythropoietin action. *Endocrinology*, 116, 2293, 1985.
133. Bietlot, H. P. and Girard, M. Analysis of recombinant human erythropoietin in drug formulations by high-performance capillary electrophoresis. *J Chromatogr A*, 759, 177, 1997.
134. Cifuentes, A., Moreno-Arribas, M. V., de Frutos, M. and Diez-Masa, J. C. Capillary isoelectric focusing of erythropoietin glycoforms and its comparison with flat-bed isoelectric focusing and capillary zone electrophoresis. *J Chromatogr A*, 830, 453, 1999.
135. Kinoshita, M. et al. Comparative studies on the analysis of glycosylation heterogeneity of sialic acid-containing glycoproteins using capillary electrophoresis. *J Chromatogr A*, 866, 261, 2000.
136. Sanz-Nebot, V., Benavente, F., Valverdu, A., Guzman, N. and Barbossa, J. Separation of recombinant human erythropoietin glycoforms by capillary electrophoresis using volatile electrolytes. Assessment of mass spectrometry for the characterization of erythropoietin glycoforms. *Eur Pharmacopoeia*, 1316, 1123, 2002.
137. Sanz-Nebot, V., Benavente, F., Vallverdu, A., Guzman, N. A. and Barbosa, J. Separation of recombinant human erythropoietin glycoforms by capillary electrophoresis using volatile electrolytes. Assessment of mass spectrometry for the characterization of erythropoietin glycoforms. *Anal Chem*, 75, 5220, 2003.
138. Kamoda, S., Nakano, M., Ishikawa, R., Suzuki, S. and Kakehi, K. Rapid and sensitive screening of N-glycans as 9-fluorenylmethyl derivatives by high-performance liquid chromatography: a method which can recover free oligosaccharides after analysis. *J Proteome Res*, 4, 146, 2005.

139. Kawasaki, N., Ohta, M., Hyuga, S., Hyuga, M. and Hayakawa, T. Application of liquid chromatography/mass spectrometry and liquid chromatography with tandem mass spectrometry to the analysis of the site-specific carbohydrate heterogeneity in erythropoietin. *Anal Biochem*, 285, 82, 2000.
140. Kawasaki, N. et al. Structural analysis of sulfated *N*-linked oligosaccharides in erythropoietin: application of liquid chromatography/mass spectrometry and liquid chromatography with tandem mass spectrometry to the analysis of the site-specific carbohydrate heterogeneity in erythropoietin. *Glycobiology*, 11, 1043, 2001.
141. Okafo, G. et al. A coordinated high-performance liquid chromatographic, capillary electrophoretic, and mass spectrometric approach for the analysis of oligosaccharide mixtures derivatized with 2-aminoacridone. *Anal Chem*, 68, 4424, 1996.
142. Callewaert, N. et al. Noninvasive diagnosis of liver cirrhosis using DNA sequencer-based total serum protein glycomics. *Nat Med*, 10, 429, 2004.
143. Kamoda, S. et al. Profiling analysis of oligosaccharides in antibody pharmaceuticals by capillary electrophoresis. *J Chromatogr A*, 1050, 211, 2004.
144. Ma, S. and Nashabeh, W. Carbohydrate analysis of a chimeric recombinant monoclonal antibody by capillary electrophoresis with laser-induced fluorescence detection. *Anal Chem*, 71, 5185, 1999.
145. Nakajima, K., Oda, Y., Kinoshita, M. and Kakehi, K. Capillary affinity electrophoresis for the screening of post-translational modification of proteins with carbohydrates. *J Proteome Res*, 2, 81, 2003.
146. Nakajima, K. et al. Screening method of carbohydrate-binding proteins in biological sources by capillary affinity electrophoresis and its application to determination of Tulipa gesneriana agglutinin in tulip bulbs. *Glycobiology*, 14, 793, 2004.
147. Glennie, M. J. and van de Winkel, J. G. Renaissance of cancer therapeutic antibodies. *Drug Discov Today*, 8, 503, 2003.
148. Liang, A. et al. Interactions of dextran sulfates with granulocyte colony-stimulating factor and their effects on leukemia cells. *Electrophoresis*, 27, 3195, 2006.
149. Liang, A. et al. Structural features in carrageenan that interact with a heparin-binding hematopoietic growth factor and modulate its biological activity. *J Chromatogr B Analyt Technol Biomed Life Sci*, 843, 114, 2006.
150. Liang, A. et al. Separation, identification, and interaction of heparin oligosaccharides with granulocyte-colony stimulating factor using capillary electrophoresis and mass spectrometry. *Electrophoresis*, 26, 3460, 2005.
151. Duteil, S. et al. Identification of heparin oligosaccharides by direct coupling of capillary electrophoresis/ionspray-mass spectrometry. *Rapid Commun Mass Spectrom*, 13, 1889, 1999.
152. Ruiz-Calero, V., Moyano, E., Puignou, L. and Galceran, M. T. Pressure-assisted capillary electrophoresis-electrospray ion trap mass spectrometry for the analysis of heparin depolymerised disaccharides. *J Chromatogr A*, 914, 277, 2001.
153. Wey, A. B., Caslavska, J. and Thormann, W. Analysis of codeine, dihydrocodeine and their glucuronides in human urine by electrokinetic capillary immunoassays and capillary electrophoresis-ion trap mass spectrometry. *J Chromatogr A*, 895, 133, 2000.
154. Wey, A. B. and Thormann, W. Capillary electrophoresis-electrospray ionization ion trap mass spectrometry for analysis and confirmation testing of morphine and related compounds in urine. *J Chromatogr A*, 916, 225, 2001.
155. Wey, A. B. and Thormann, W. Capillary electrophoresis and capillary electrophoresis-ion trap multiple-stage mass spectrometry for the differentiation and identification of oxycodone and its major metabolites in human urine. *J Chromatogr B Analyt Technol Biomed Life Sci*, 770, 191, 2002.
156. Keski-Hynnila, H., Raana, K., Taskinen, J. and Kostiainen, R. Direct analysis of nitrocatechol-type glucuronides in urine by capillary electrophoresis-electrospray ionisation mass spectrometry and tandem mass spectrometry. *J Chromatogr B Biomed Sci Appl*, 749, 253, 2000.
157. Baldacci, A. and Thormann, W. Analysis of lorazepam and its 30-glucuronide in human urine by capillary electrophoresis: evidence for the formation of two distinct diastereoisomeric glucuronides. *J Sep Sci*, 29, 153, 2006.
158. Kelly, J., Masoud, H., Perry, M. B., Richards, J. C. and Thibault, P. Separation and characterization of O-deacylated lipooligosaccharides and glycans derived from *Moraxella catarrhalis* using capillary

electrophoresis-electrospray mass spectrometry and tandem mass spectrometry. *Anal Biochem*, 233, 15, 1996.

159. Auriola, S., Thibault, P., Sadovskaya, I. and Altman, E. Enhancement of sample loadings for the analysis of oligosaccharides isolated from *Pseudomonas aeruginosa* using transient isotachophoresis and capillary zone electrophoresis—electrospray—mass spectrometry. *Electrophoresis*, 19, 2665, 1998.
160. Ludwiczak, P., Brando, T., Monsarrat, B. and Puzo, G. Structural characterization of *Mycobacterium tuberculosis* lipoarabinomannans by the combination of capillary electrophoresis and matrix-assisted laser desorption/ionization time-of-flight mass spectrometry. *Anal Chem*, 73, 2323, 2001.
161. Tzianabos, A., Wang, J. Y. and Kasper, D. L. Biological chemistry of immunomodulation by zwitterionic polysaccharides. *Carbohydr Res*, 338, 2531, 2003.
162. Ishii, T. Structure and functions of feruloylated polysaccharides. *Plant Sci*, 127, 111, 1997.
163. Tang, Y.-J. and Zhong, J.-J. Exopolysaccharide biosynthesis and related enzyme activities of the medicinal fungus, Ganoderma lucidum, grown on lactose in a bioreactor. *Biotechnol Lett*, 24, 1023, 2002.
164. Ho, C. Y., Lo, T. W., Leung, K. N., Fung, K. P. and Choy, Y. M. The immunostimulating activities of anti-tumor polysaccharide from K1 capsular, polysaccharide antigen isolated from *Klebsiella pneumoniae*. *Immunopharmacology*, 46, 1, 2000.
165. Wong, C. K., Leung, K. N., Fung, K. P. and Choy, Y. M. The immunostimulating activities of anti-tumor polysaccharides from *Pseudostellaria heterophylla*. *Immunopharmacology*, 28, 47, 1994.
166. Sonoda, Y. et al. Stimulation of interleukin-8 production by acidic polysaccharides from the root of Panax ginseng. *Immunopharmacology*, 38, 287, 1998.
167. Lu, R. et al. Specific biological activities of Chinese lacquer polysaccharides. *Carbohydr Polymers* 43, 47, 2000.
168. Yang, J. and Du, Y. Chemical modification, characterization and bioactivity of Chinese lacquer polysaccharides from lac tree Rhus vernicifera against leukopenia induced by cyclophosphamide. *Carbohydr Polymers*, 52, 405, 2003.
169. Mitsuya, H., Yarchoan, R. and Broder, S. Molecular targets for AIDS therapy. *Science*, 249, 1533, 1990.
170. Yoshida, T. Synthesis of polysaccharides having specific biological activities. *Prog Polymer Sci*, 26, 379, 2001.
171. Xu, H.-X., Lee, S. H. S., Lee, S. F., White, R. L. and Blay, J. Isolation and characterization of an anti-HSV polysaccharide from *Prunella vulgaris*. *Antiviral Res*, 44, 43, 1999.
172. Wang, Y. et al. Protective effect of Fructus Lycii polysaccharides against time and hyperthermia-induced damage in cultured seminiferous epithelium. *J Ethnopharmacol*, 82, 169, 2002.
173. Hu, Q., Zhou, T., Hu, G. and Fang, Y. Determination of sugars in Chinese traditional drugs by CE with amperometric detection. *J Pharm Biomed Anal*, 30, 1047, 2002.
174. Wang, Q., Ding, F., Zhu, N., He, P. and Fang, Y. Determination of the compositions of polysaccharides from Chinese herbs by capillary zone electrophoresis with amperometric detection. *Biomed Chromatogr*, 17, 483, 2003.
175. Wang, Q., Yu, H., Zong, J., He, P. and Fang, Y. Determination of the composition of Chinese ligustrum lucidum polysaccharide by capillary zone electrophoresis with amperometric detection. *J Pharm Biomed Anal*, 31, 473, 2003.
176. Wang, J., Chatrathi, M. P., Tian, B. and Polsky, R. Microfabricated electrophoresis chips for simultaneous bioassays of glucose, uric acid, ascorbic acid, and acetaminophen. *Anal Chem*, 72, 2514, 2000.
177. Wang, J., Chatrathi, M. P. and Tian, B. Microseparation chips for performing multienzymatic dehydrogenase/oxidase assays: simultaneous electrochemical measurement of ethanol and glucose. *Anal Chem*, 73, 1296, 2001.
178. Tseng, K., Liu, J., Lebrilla, C. B., Collins, S. D. and Smith, R. L. Fabrication and design of open microchannels for capillary electrophoresis separations and matrix-assisted laser/desorption mass spectroscopy analysis. *Proc SPIE Int Soc Opt Eng*, 3606, 137, 1999.
179. Larsson, M., Sundberg, R. and Folestad, S. On-line capillary electrophoresis with mass spectrometry detection for the analysis of carbohydrates after derivatization with 8-aminonaphthalene-1,3,6-trisulfonic acid. *J Chromatogr A*, 934, 75, 2001.

180. Klampfl, C. W. and Buchberger, W. Determination of carbohydrates by capillary electrophoresis with electrospray-mass spectrometric detection. *Electrophoresis*, 22, 2737, 2001.
181. Suomi, J., Wiedmer, S. K., Jussila, M. and Riekkola, M. L. Determination of iridoid glycosides by micellar electrokinetic capillary chromatography-mass spectrometry with use of the partial filling technique. *Electrophoresis*, 22, 2580, 2001.
182. Craston, D. H. et al. Determination of gellan gum by capillary electrophoresis and CE-MS. *Food Chem*, 73, 103, 2001.
183. Hagberg, J. Analysis of low-molecular-mass organic acids using capillary zone electrophoresis-electrospray ionization mass spectrometry. *J Chromatogr A*, 988, 127, 2003.
184. Bianco, G., Schmitt-Kopplin, P., De Benedetto, G., Kettrup, A. and Cataldi, T. R. Determination of glycoalkaloids and relative aglycones by nonaqueous capillary electrophoresis coupled with electrospray ionization-ion trap mass spectrometry. *Electrophoresis*, 23, 2904, 2002.
185. Bringmann, G. et al. Analysis of the glucosinolate pattern of Arabidopsis thaliana seeds by capillary zone electrophoresis coupled to electrospray ionization-mass spectrometry. *Electrophoresis*, 26, 1513, 2005.
186. Kirk, T. K. and Cullen, D. *Environmentally Friendly Technologies for the Pulp and Paper Industry*, eds. Young, R. A. and Akhtar, M., John Wiley & Sons, New York, 1998.
187. Viikari, L. et al. *Bioconversion of Forest and Agricultural Plant Residues*, ed. Saddler, J. N., CAB, Oxford, 1993.
188. Saha, B. C. Hemicellulose bioconversion. *J Ind Microbiol Biotechnol*, 30, 279, 2003.
189. Wyman, C. E. Potential synergies and challenges in refining cellulosic biomass to fuels, chemicals and power. *Biotechnol Prog*, 19, 254, 2003.
190. Wyman, C. E. Ethanol from lignocellulosic biomass: technology economics, and opportunities. *Bioresour Technol*, 50, 3, 1994.
191. Felix, C. R. and Ljungdahl, L. G. The cellulosome: the exocellular organelle of Clostridium. *Annu Rev Microbiol*, 47, 791, 1993.
192. Paulus, A. and Klokow-Beck, A. *Analysis of Carbohydrates by Capillary Electrophoresis*, Vieweg Publishers, Wiesbaden, 1999.
193. Mechref, Y. and Novotny, M. V. Structural investigations of glycoconjugates at high sensitivity. *Chem Rev*, 102, 321, 2002.
194. Chiesa, C. and Horvath, C. Capillary zone electrophoresis of malto-oligosaccharides derivatized with 8-aminonaphthalene-1,3,6-trisulfonic acid. *J Chromatogr*, 645, 337, 1993.
195. Guttman, A. High-resolution carbohydrate profiling by capillary gel electrophoresis. *Nature*, 380, 461, 1996.
196. Morell, M. K., Samuel, M. S. and O'Shea, M. G. Analysis of starch structure using fluorophore-assisted carbohydrate electrophoresis. *Electrophoresis*, 19, 2603, 1998.
197. Callewaert, N., Geysens, S., Molemans, F. and Contreras, R. Ultrasensitive profiling and sequencing of *N*-linked oligosaccharides using standard DNA-sequencing equipment. *Glycobiology*, 11, 275, 2001.
198. Khandurina, J., Blum, D. L., Stege, J. T. and Guttman, A. Automated carbohydrate profiling by capillary electrophoresis: a bioindustrial approach. *Electrophoresis*, 25, 2326, 2004.
199. Stefansson, M. and Novotny, M. Electrophoretic resolution of monosaccharide enantiomers in borate-oligosaccharide complexation media. *J Am Chem Soc*, 115, 11573, 1993.
200. Guttman, A. Analysis of monosaccharide composition by capillary electrophoresis. *J Chromatogr A*, 763, 271, 1997.
201. Khandurina, J., Anderson, A. A., Olson, N. A., Stege, J. T. and Guttman, A. Large-scale carbohydrate analysis by capillary array electrophoresis: part 2. Data normalization and quantification. *Electrophoresis*, 25, 3122, 2004.
202. Khandurina, J., Olson, N. A., Anderson, A. A., Gray, K. A. and Guttman, A. Large-scale carbohydrate analysis by capillary array electrophoresis: part 1. Separation and scale-up. *Electrophoresis*, 25, 3117, 2004.
203. Schwarz, M. A., Galliker, B., Fluri, K., Kappes, T. and Hauser, P. C. A two-electrode configuration for simplified amperometric detection in a microfabricated electrophoretic separation device. *Analyst*, 126, 147, 2001.

204. Hebert, N. E., Kuhr, W. G. and Brazill, S. A. Microchip capillary electrophoresis coupled to sinusoidal voltammetry for the detection of native carbohydrates. *Electrophoresis*, 23, 3750, 2002.
205. Fanguy, J. C. and Henry, C. S. Pulsed amperometric detection of carbohydrates on an electrophoretic microchip. *Analyst*, 127, 1021, 2002.
206. Fu, C.-G. and Fang, Z.-L. Combination of flow injection with capillary electrophoresis. *Anal Chim Acta* 422, 71, 2000.
207. Burggraf, N., Krattiger, B., de Mello, A. J., de Rooij, N. F. and Manz, A. Holographic refractive index detector for application in microchip-based separation systems. *Analyst*, 123, 1443, 1998.
208. Suzuki, S., Ishida, Y., Arai, A., Nakanishi, H. and Honda, S. High-speed electrophoretic analysis of 1-phenyl-3-methyl-5-pyrazolone derivatives of monosaccharides on a quartz microchip with whole-channel UV detection. *Electrophoresis*, 24, 3828, 2003.
209. Dang, F., Zhang, L., Hagiwara, H., Mishina, Y. and Baba, Y. Ultrafast analysis of oligosaccharides on microchip with light-emitting diode confocal fluorescence detection. *Electrophoresis*, 24, 714, 2003.
210. Dang, F., Zhang, L., Jabasini, M., Kaji, N. and Baba, Y. Characterization of electrophoretic behavior of sugar isomers by microchip electrophoresis coupled with videomicroscopy. *Anal Chem*, 75, 2433, 2003.
211. Suzuki, S., Shimotsu, N., Honda, S., Arai, A. and Nakanishi, H. Rapid analysis of amino sugars by microchip electrophoresis with laser-induced fluorescence detection. *Electrophoresis*, 22, 4023, 2001.
212. Bindila, L. et al. Off-line capillary electrphoresis/fully automated nanoelectrospray chip quadrupole time-of-flight mass spectrometry and tandem mass spectrometry for glycoconjugate analysis. *J Mass Spectrom*, 39, 1190, 204.
213. Callewaert, N. et al. Total serum protein *N*-glycome profiling on a capillary electrophoresis-microfluidics platform. *Electrophoresis*, 25, 3128, 2004.
214. Ninonuevo, M. et al. Nanoliquid chromatography-mass spectrometry of oligosaccharides employing graphitized carbon chromatography on microchip with a high-accuracy mass analyzer. *Electrophoresis*, 26, 3641, 2005.
215. Kang, P., Mechref, Y., Klouckova, I. and Novotny, M. V. Solid-phase permethylation of glycans for mass spectrometric analysis. *Rapid Commun Mass Spectrom*, 19, 3421, 2005.
216. Khandurina, J. and Guttman, A. Microscale separation and analysis. *Curr Opin Chem Biol*, 7, 595, 2003.
217. Khandurina, J. and Guttman, A. Bioanalysis in microfluidic devices. *J Chromatogr A*, 943, 159, 2002.

8 The Coupling of Capillary Electrophoresis and Mass Spectrometry in Proteomics

Haleem J. Issaq and Timothy D. Veenstra

CONTENTS

8.1 INTRODUCTION

Modern science is firmly entrenched in the "omics" era. Spurred by the human genome project, omics-thinking quickly spread to ribonucleic acid (RNA), proteins, and now metabolites. Although different technologies are required for each specific omic endeavor, they all have similar goals of attempting to garner as much information about a specific class of biomolecules as possible. Another thing that each has in common that often goes overlooked is the need for separation. While very complex mixtures of deoxyribonucleic acid (DNA), RNA, protein, or metabolites are introduced to the appropriate analytical instrumentation, the technology still measures each basic unit (whether it be a base, transcript, or peptide) individually. For example, even though entire transcriptomes are introduced to array platform, the RNA transcripts must be separated via annealing to their complementary strand for a quantitative signal to be deciphered for each. While proteomics, one of the foci of this chapter, is arguably driven by the development of mass spectrometry (MS) technology, separation technologies have been just as instrumental in its development. If proteomics was limited to direct infusion techniques, we would still be floundering in the detection of only the highest abundant proteins present within biological samples.

When integrating a separation technique into a MS-based proteomics study many different aspects need to be considered. The first consideration is if an online or an off-line separation of the proteome sample will be integrated into the research plan. There are a large number of separation techniques available to do off-line.[1] These include many different chromatography [e.g., strong/weak cation exchange (S/WCX), strong/weak anion exchange (S/WAX), reversed-phase (RP), size exclusion, etc.] and electrophoretic gel and capillary electrophoresis (CE) [e.g., capillary zone electrophoresis (CZE), isoelectric focusing (IEF), isotachophoresis (ITC), etc.] methods. Off-line separations are simpler, easier to perform, and require no instrument modification. In addition, a simple solvent exchange step can be administered to each fraction to introduce the correct conditions for downstream MS analysis. RP columns packed with C18 particles are most commonly used to remove salts and other MS incompatible materials from fractions collected via off-line separations. The next consideration is the amount of material available for fractionation. This consideration is where chromatographic methods have a decided advantage over gel-based electrophoretic techniques. Proteins in a proteomic sample are fractionated off-line mainly to simplify the mixture prior to high-performance liquid chromatography/mass spectrometry (HPLC/MS) analysis and are normally done when a significant amount (50–200 μg) of protein is available. Although chromatography columns have the capacity to handle such samples sizes, typical gel electrophoretic methods are unsuitable for these amounts of materials.[2] This aspect is the primary reason that SCX has become the most popular choice of off-line liquid-based separations for MS-based proteomic studies. The final design aspect that needs to be considered is the choice of online separation that will be used with direct MS analysis. The most important consideration for this part of the study is making the solvent conditions compatible with electrospray ionization (ESI).[3] ESI requires the solvent to be volatile, acidic, and contain only low levels of salts and buffers. ESI is almost intolerable to detergents.

The most popular type of separation technique to which MS has been directly coupled is reversed-phase liquid chromatography (RPLC). This chromatography has all the necessary attributes that make it an excellent choice for online MS analysis. What is not optimal, however, is its speed and resolution. RP columns require re-equilibration between analyses when gradient mobile phases are used, leading to an almost 25% downtime when using this type of separation. While many laboratories have implemented column-switching so that one column is being used for online MS analysis while the other is re-equilibrating, this technology is not widely used and has issues related to repositioning of the column within the ESI source. While the resolution of RPLC is quite good, the duty cycles of tandem MS instruments is so short that even higher resolution separations could provide a greater solution for maximizing proteome coverage. It is in these two areas that CE has distinct advantages.

Capillary electrophoresis is an excellent microseparation technique that has been used for the separation of a wide diversity of different molecules.[4] Its separation capabilities extend to ions, small molecules (such as amino acids), and large biomolecules (such as peptides, proteins, and nucleic acids). Indeed the human genome project owes its success, in part, to the use of CE for the separation of DNA bases. In the past, CE has been combined with detection devices such as ultraviolet (UV) and laser-induced fluorescence (LIF) spectrophotometers. The detection of the separated analytes is carried out on column by etching the capillary. Unfortunately, UV detection lacks sensitivity and not every compound of interest will absorb in the UV region of the spectrum. Detection using LIF is sensitive, however, the analytes of interest may require derivatization with a fluorescent tag or have an aromatic amino acid in their structure (e.g., proteins and peptides). An advantage of MS detection that neither UV nor LIF detection provides is the information necessary to directly determine the structure of the detected analyte(s).

Mass spectrometry, however, has high sensitivity for the detection of a wide variety of analytes. When operated in a multistage, MS mode can provide information to assist in the determination of the structural formula of the molecules of interest. A marriage of CE and MS would potentially result in an extremely powerful tool for the separation, identification, and characterization of a wide range of molecules. Unfortunately, the online interfacing of CE with MS is not trivial and presents several technological challenges not present with CE-UV or CE-LIF. Some major issues are the requirements of both the CE continuous electrical circuit for electrophoretic separation and the mass spectrometer electrical contact for efficient ESI. Running buffers normally used in CE, such as sodium phosphate or borate that have low volatility, are not compatible with ESI-MS. These buffers need to be replaced with volatile buffers such as ammonium formate or acetate. This substitution, however, may have a detrimental impact on the quality of the separation.

The resolution afforded with CE is a good match for the duty cycle times of present day tandem MS instruments, particularly in the area of proteomics. In the past, the speed at which mass spectrometers selected, isolated, fragmented, and detected peptide ions did not warrant extremely high resolution separations because some analytes would not be present long enough in the MS source to be surveyed by the instrument. Nowadays, with the speed of mass spectrometers such as the linear-ion traps, peptides that are present for only a few seconds (i.e., <10) still have a high probability of being selected for tandem MS analysis. The biggest drawback in using CE online with MS is the amount of material that can be loaded onto the capillary. Although MS is a sensitive detection technique, it still has its limits. Sample sizes in the range of 1–10 μg are typically loaded onto a capillary RP column for subsequent MS analysis. Unfortunately, CE capillaries have a very limited volume and can only accommodate sample amount in the nanogram (ng) range. Fortunately, the use of on-column preconcentration devices has enabled the loading of greater amounts of material onto CE capillaries.[5]

8.2 ONLINE CE–ESI/MS SYSTEM REQUIREMENTS

An online CE–ESI/MS system is made up of a CE instrument, a mass spectrometer, an interface that will allow the transfer of analytes from the CE column outlet to the MS source, a closed CE electrical circuit for electrophoretic separation of the analyte mixture, and a closed electrical circuit for the generation of continuous, stable and uniform fine spray stream that affords sensitive MS detection. The challenge in coupling CE directly online with ESI-MS is achieving electrical continuity that allows uninterrupted operation of both systems, without affecting the quality of the CE separation or the ESI efficiency (which is directly related to the detection sensitivity). The coupling of CE to MS has been a challenging problem because an ideal interface is one that is constructed in such a way that the CE separation column and the spray tip form a single continuous unit in order to eliminate any dead volume that may lead to diffusion and affect the quality of separation. In addition, the design should preserve the electric circuits of both the CE system and the spray tip. Also, it would be advantageous if no external solvent is added to the system (sheath liquid) that dilutes the analyte concentration and affects the detection sensitivity.

The most popular choice for coupling CE with ESI/MS is the coaxial sheath flow interface. Coaxial sheath flow was introduced as the first CE-MS interface in 1987.[6] This coupling device uses three concentric capillaries in which the CE capillary (used for the actual analyte separation) is innermost and protrudes into the ESI source region. A stainless steel capillary (the ESI needle) is placed around the CE capillary and delivers a sheath liquid. The sheath liquid is required to complete the electrical circuit necessary for both CE separation and ESI. The sheath liquid comes into contact with both the CE buffer and ESI needle. The sheath liquid is usually a mixture of a volatile acid, H_2O, and an organic modifier, such as methanol (CH_3OH) or acetonitrile (ACN). The sheath liquid may also contain an electrolyte since it also acts as the outlet buffer for the CE separation process. The electrolyte concentration must be carefully selected, as both separation and ionization efficiency must be considered. A high electrolyte concentration will provide good separation efficiency but will negatively impact ESI efficiency. The sheath liquid is mixed with the CE buffer at the very tip of the capillary within the ESI source region. It is critical to keep the mixing volume as small as possible to maximize ES stability and sensitivity. Therefore, the distance that the CE capillary extends past the sheath liquid capillary is very important. While the correct distance may need to be determined through experimentation, it is generally in the range of 0.1–0.5 mm. Both of these capillaries are placed within a third concentric stainless steel tube that delivers a gas flow, which is necessary for effective desorption of the electrosprayed ions as well as providing the necessary cooling of the CE capillary.

8.3 SHEATH VERSUS SHEATHLESS INTERFACES

There are many advantages of sheathless interfaces compared to those that require a sheath flow.[7,8] The main difference between sheath and sheathless interface designs is that sheathless interface does not require the external flow of a coaxial sheath liquid to establish electrical contact with the CE effluent to facilitate the ESI process. Introduction of a sheath liquid dilutes the analytes and results in poorer detection limits. Indeed, a comparable sheathless interface can give an order of magnitude greater sensitivity than its counterpart that requires a sheath liquid. One of the major weaknesses of CE is that although the mass detection levels are quite high the amount of sample that may be injected onto the capillary is low. Dilution of the analytes by sheath liquid only serves to compound this deficiency. The construction of an interface with sheath flow is also more complicated, requiring coupling of different capillaries on connectors. In principle, the sheathless interface is the best design for coupling CE online with ESI-MS, since the flow rates required for each technique are compatible and no dilution of the analytes occurs during separation and elution, thereby maximizing the signal recorded by the mass spectrometer.

8.4 SHEATHLESS DESIGN STRATEGIES

Several different types of sheathless designs have been introduced that satisfy the requirement of closing the CE separation capillary circuit while simultaneously providing an electrical potential to the spray tip for ESI. These include the use of a single capillary, two capillaries, and three capillaries.[5]

Many methods for interfacing CE with MS using a single-capillary format whereby uninterrupted electrical contact required for CE and ESI was established by different means have been introduced. An example of such is shown in Figure 8.1a. The first single capillary-sheathless interface was introduced by Wahl et al.[9] who coated the capillary outlet with a conductive metal (i.e., silver deposition) at the cathode end. This coating served to define the CE field strength along the capillary column and provide the necessary ESI potential. One of the best features of this design is that it did not interfere with the CE separation process. Since this work, a number of studies reported using similar strategies for coating the capillary's cathodic end with metal, polymer, or carbon.[10,11] The biggest drawback to these coating methods was their lack of mechanical stability. The coatings would slowly disintegrate and each CE capillary was only useful for a few hours to days. The stability of the metal coating by precoating the capillary with materials such as silanes, silicon oxide, or chromium improved the length of the capillary's lifetime.

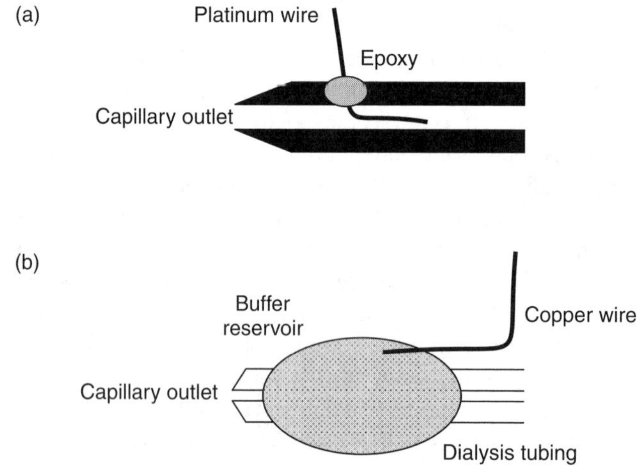

FIGURE 8.1 (a) Schematic diagram of a single capillary design for CE-MS. (From Cao, P. and Moini, M., *J. Am. Soc. Mass Spectrom.*, 8, 561, 1997.) In this design, a small hole is drilled into the capillary through which a platinum wire is inserted. (b) Schematic of a two capillary design for CE-MS in which two pieces of fused-silica capillaries are connected via a section of dialysis tubing. (From Severs, J. C. and Smith, R. D., *Anal. Chem.*, 69, 2154, 1997.) This junction is immersed into the CE buffer and into which a copper electrode is inserted.

Other groups have taken different approaches that require directly inserting a metal wire into the CE capillary. For example, Cao and Moini[10] constructed a sheathless interface by inserting a platinum wire (25 μm diameter by 0.2–0.3 cm in length), through a pinhole that was drilled into a 150-μm o.d. fused-silica capillary. The pinhole was positioned approximately 2 cm from the outlet of the capillary. Epoxy was used to maintain the position of the wire as well as seal the capillary. The platinum wire electrode serves as both the low voltage electrode of the CE electric circuit and as a connection for the ESI voltage. This design proved to be stable over the maximum time studied (i.e., 50 min), however, such capillaries are difficult to construct (requiring a dental drill to create a small opening) and the presence of the wire may generate turbulent flow that affects the resolution of the CE separation. The laboratories of Wahl and Smith[12] developed a similar "micro-hole" method, however, in this case an interface whereby the electrical connection was made was constructed by drilling a small hole 2 cm from the end of the separation capillary. The micro-hole was then sealed with conductive gold epoxy. While this method allows for the necessary electrical conductance, the design is cumbersome and the proper capillaries are not easy to construct.

Our laboratory developed a simple sheathless interface for CE-MS in which the separation column, an electrical porous junction, and the spray tip are integrated within a single piece of a fused-silica capillary, as illustrated in Figure 8.2.[13] The electrical potential to induce ESI is introduced through a porous junction across a 3–4 mm length of fused silica. A 3–4 mm section of the polyimide coating was removed from the capillary approximately 5 cm from end of the spray tip. Electrical conductance between the solute within the capillary and the anode buffer was achieved by etching the exposed section using hydrofluoric acid (HF). This etching reduced the outside diameter of the capillary at this point without affecting the inner diameter. The HF etching requires approximately 6 h and reduces the capillary wall to about 15–20 μm. Since the porous junction created by HF etching is fragile, the capillary must be secured inside the reservoir protecting it from breaking. The reservoir contains the buffer for closing the CE circuit and providing the voltage necessary for ESI. This design proved to be quite rugged as this group reported continuous operation of the same column for over a 2-week period with no evidence of deterioration in separation or ESI performance.

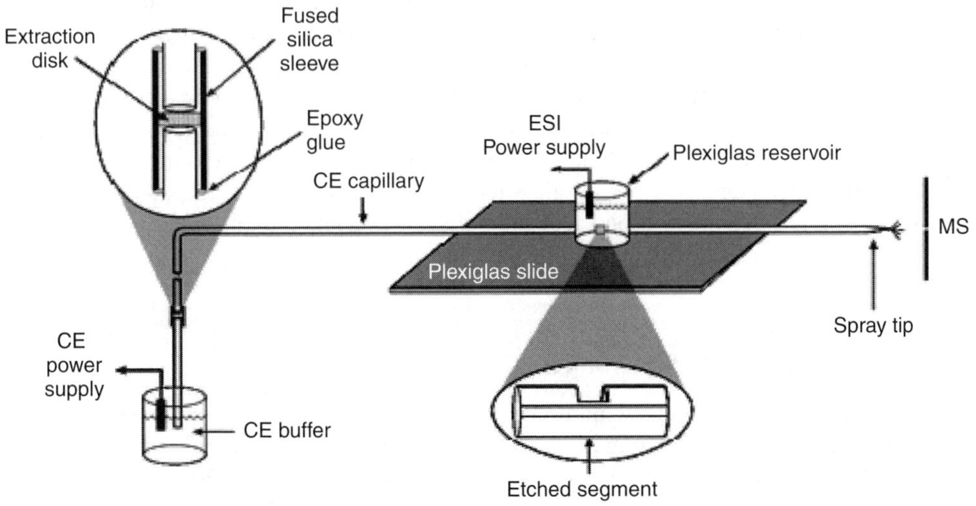

FIGURE 8.2 Schematic diagram of a CE-MS design that incorporates etching of the capillary wall to enable electrical conductance and a solid-phase extraction disk for pre-concentration of the sample prior to CE-MS analysis. (From Janini, G. M., et al., *Anal. Chem.*, 76, 1615, 2003. With permission.)

The same group followed up on this initial design by adding the capability for on-column sample enrichment into the sheathless interface.[4] A small solid-phase extraction cartridge, made of RP material, was attached to the CE capillary near the point of injection. Optimization of the proper capillary diameters showed that they could achieve a mass limit of detection of 500 amol for CE-MS/MS analysis of a standard peptide using a 20-μm i.d. capillary. This design brings to reality a true zero dead-volume sheathless CE—MS interface with the ability to preconcentrate the sample within the same capillary.

Another strategy for coupling CE with MS is a sheathless interface in which two pieces of capillary are used. In this method the CE separation capillary is connected to a short spray tip (which functions as the ESI needle) via a sleeve. The sleeve can be prepared from microdialysis tubing, stainless steel tubing, or a micro-tee junction. Severs and Smith[14] developed a separation capillary that is connected to a 2-cm-long ESI emitter capillary via a 1.5 cm length of dialysis tubing. Epoxy is then applied to maintain the connection between the tubing and capillary pieces, as illustrated in Figure 8.1b. The manufactured capillary is then inserted through a 250 μL Eppendorf pipet tip containing an electrolyte identical to that employed for the CE separation. Connection of the pipet tip to an *x-y-z* stage allows the position of the spray tip relative to the MS orifice for maximum detection sensitivity. A copper wire is inserted in the electrolyte reservoir and connected to a high-voltage power supply to provide the necessary voltage for the CE separation and ESI.

The group of Tong et al.[15] developed a sheathless liquid–metal junction interface for CE-MS by removing approximately 2–3 mm length of polyimide coating from a coated capillary (5–10 cm long), and exposing the fused-silica tip to 48–51% HF. This resulted in etching of the exposed fused silica and the polyimide coating was further burned to expose approximately 3 mm of the tapered tip. The butt ends of the tip and CE capillary were then carefully polished flat. A liquid–metal junction was then created with a polyether ether ketone (PEEK) micro-tee. The side channel of the tee was enlarged to accommodate a 2.5 cm length of 0.06 cm diameter gold wire. The electrospray tip and the separation capillary were carefully butted together along the channel of the micro-tee and connected via microfittings. The gold wire was inserted into the side channel of the micro-tee to supply the electrical connection to supply the required voltage for ESI.

Other researchers have also reported on the design of three-piece capillary/spray tips in which the electrical contact is established through a porous glass joint that is connected to the CE capillary using polytetrafluoroethylene (PTFE) sleeves on one side and the spray tip on the other. A different three-piece designed based on the one previously described, used a porous glass joint. In this design, the polyimide layer was burned off a 2–3 mm piece of a 3-cm-long fused-silica capillary.[16] This exposed portion was etched with HF to reduce the thickness of the capillary wall to <20 μm. This porous glass joint was inserted into a piece of PTFE, in which a small notch was cut through its center. The diameter of this notch matched that of the outer diameter of the capillary. The porous glass joint was coupled to the CE column with the aid of a PTFE sleeve, and the ESI emitter was connected to the other side of the porous glass joint. To conduct online CE separations the porous fused silica was immersed in 1% acetic acid and voltage was applied to the acidic solution to drive both the CE separation and ESI.

8.5 PROTEOMIC ANALYSIS USING CE-MS

Currently there is no greater focus within proteomics than the discovery of biomarkers of various disease states. The number of proteomic techniques and strategies that are being utilized in an attempt to find novel biomarkers is a veritable "who's who" list of available methods. The number of different methods being attempted is related to a number of different factors including the enormity of the experimental question, complexity of human biofluids, and the lack of success achieved using one single design. While the primary separation technologies that have been combined with MS for the discovery of biomarkers are two-dimensional polyacrylamide gel electrophoresis (2D-PAGE) two-dimensional polyacrylamide gel electrophoresis, and RPLC, CE has also played a role in this research area. In a study attempting to find indicators of graft vs. host disease (GvHD) in patients that had received allogeneic hematopoietic stem cell transplants (HSCT), investigators examined the CE-MS profiles of urine obtained from 40 HSCT patients and 5 patients with sepsis.[17] They were able to find 16 differentially urine-excreted peptides that indicated the presence of GvHD. In a cross validation study using the pattern generated by these markers, they correctly diagnose GvHD with a sensitivity and specificity of 82% and 100%, respectively.

While the above study utilized urine, probably no biofluid has attracted the attention of the proteomic biomarker discovery field as serum and plasma. Although the analysis of serum by RPLC-MS has almost become routine, such is not the case for CE-MS. In a proof of principle study to show the efficacy of using CE-MS to find differentially abundant proteins in serum, the group of Sassi et al.[18] analyzed groups of sera that were spiked with different concentrations of known standard peptides. The groups of sera were analyzed by CE-MS and the standard peptides were successfully identified as being differentially abundant with a success rate of 95%.

These first promising data using CE-MS to detect biomarkers in serum/plasma was a prospective, randomized, open-label trial to assess the effect of vitamin C deficiency on oxidative stress and analyze inflammatory markers in hemodialysis patients.[19] The CE-MS analysis of plasma obtained from dialysis patients and those with normal renal functions showed more than 30 polypeptides that were significantly different in abundance between the two groups. This study showed that oral vitamin C supplementation did indeed have an affect on the plasma proteome of dialysis patients.

Another important biofluid that has been utilized in CE-MS biomarker discovery studies is cerebrospinal fluid (CSF). Because of the blood–brain barrier, blood has limited diagnostic potential for nervous system related diseases, such as Alzheimer's disease (AD). CSF is a clear, colorless liquid that provides mechanical protection for the brain. The proteomic content of CSF is similar to that of blood, in that 70% of CSF's protein content consists of isoforms of albumin, transferrin, and immunoglobulins. A study conducted by Wetterhall et al.[20] described a method for analyzing CSF, by first digesting the proteome into tryptic peptides and then analyzing these using CE coupled directly online with FT-ICR MS. In the first step, CSF was analyzed by CE-MS to establish the ability to identify peptides/proteins within this biofluid. Using the mass measurement accuracy capabilities of FTICR-MS, 30 proteins were identified with mass measurement errors of less than 5 parts per million (ppm). In an alternative approach, Wittke et al.[21] obviated the tryptic digestion step, however, they depleted the high abundance proteins from CSF prior to its analysis by CE-MS. They were able to detect 450 different proteins in the molecular mass range of 800–15,000 Da. They proceeded to use this experimental setup to analyze protein patterns within CSF from 4 healthy donors, 8 patients with AD and 7 with schizophrenia. They were able to find peaks that were capable of differentiating AD and schizophrenic patients, but were unable to differentiate these two groups from the controls. While the number of samples analyzed in this study was low, it did show the potential of direct CE-MS as a diagnostic tool for brain-related disorders.

Current literature suggests CE-MS is potentially a powerful tool for the detection of biomarkers in a variety of biofluids. It has the ability to reliably detect peptides within biological samples at a rate far greater than conventional RPLC-MS. There are, however, several challenges that remain. While the ability of CE-MS to provide information on hundreds of peptides in a single sample has been established, and the diagnostic ability of many of patterns of these peptides has been shown, the link to potential physiological function of these markers is still lacking. This deficiency is primarily due to the inability to routinely identify the sequence and therefore the identity of these peptide peaks. At this point, the diagnostic features are simply; just features and not known proteins. To identify these putative biomarkers is extremely difficult when one considers that they must be isolated from an extremely complex mixture (e.g., serum, plasma, etc.) and significant sequencing is required to establish their identity. While CE has been established with MS/MS capable instrumentation, the broad identification of proteins in complex biofluids is still not at the level of RPLC-MS/MS.

ACKNOWLEDGMENTS

This project has been funded in whole or in part with federal funds from the National Cancer Institute, National Institutes of Health, under Contract NO1-CO-12400. The content of this publication does not necessarily reflect the views or policies of the Department of Health and Human Services, nor does the mention of trade names, commercial products, or organizations imply endorsement by the United States Government.

REFERENCES

1. Edwards, E. and Thomas-Oates, J., Hyphenating liquid phase separation techniques with mass spectrometry: on-line or off-line. *Analyst*, 130, 13, 2005.
2. Bonneil, E. and Waldron, K. C., On-line solid-phase preconcentration for sensitivity enhancement in capillary electrophoresis. *J. Capill. Electrophor. Microchip Technol.*, 6, 61, 1999.
3. Canas, B., Lopez-Ferrer, D., Ramos-Fernandez, A., Camafeita, E., and Calvo, E., Mass spectrometry technologies for proteomics. *Brief. Funct. Genomic. Proteomic.*, 4, 295, 2006.
4. Liu, Z. and Pawliszyn, J., Online coupling of solid-phase microextraction and capillary electrophoresis. *J. Chromatogr. Sci.*, 44, 366, 2006.
5. Janini, G. M., Zhou, M., Yu, L. R., Blonder, J., Gignac, M., Conrads, T. P., Issaq, H. J., and Veenstra, T. D., On-column sample enrichment for capillary electrophoresis sheathless electrospray ionization mass spectrometry: evaluation for peptide analysis and protein identification. *Anal. Chem.*, 75, 5984, 2003.
6. Olivares, J. A., Nguyen, N. T., Yonker, C. R., and Smith, R. D., On-line mass spectrometric detection for capillary zone electrophoresis. *Anal. Chem.*, 59, 1230, 1987.
7. Ding, J. and Vouros, P., Advances in CE/MS. *Anal. Chem.*, 71, 378A, 1999.
8. Kelly, J. F., Ramaley, L., and Thibault, P., Development of electrophoretic conditions for the characterization of protein glycoforms by capillary electrophoresis-electrospray mass spectrometry. *Anal. Chem.*, 69, 51, 1997.
9. Wahl, J. H., Gale, D. C., and Smith, R. D., Sheathless capillary electrophoresis-electrospray ionization mass spectrometry using 10 μm I.D. capillaries: analysis of tryptic digests of cytochrome c. *J. Chromatogr. A*, 659, 217, 1994.
10. Cao, P. and Moini, M., A novel sheathless interface for capillary electrophoresis/electrospray ionization mass spectrometry using an in-capillary electrode. *J. Am. Soc. Mass Spectrom.*, 8, 561, 1997.
11. Chang, Y. Z. and Her, G. R., Sheathless capillary electrophoresis/electrospray mass spectrometry using a carbon-coated fused-silica capillary. *Anal. Chem.*, 23, 626, 2000.
12. Wahl, J. H. and Smith, R. D., Comparison of buffer systems and interface designs for capillary electrophoresis-mass spectrometry. *J. Cap. Electrophoresis*, 1, 62, 1994.
13. Janini, G. M., Conrads, T. P., Wilkens, K. L., Issaq, H. J., and Veenstra, T. D., A sheathless nanoflow electrospray interface for on-line capillary electrophoresis mass spectrometry. *Anal. Chem.*, 76, 1615, 2003.
14. Severs, J. C. and Smith, R. D., Characterization of the microdialysis junction interface for capillary electrophoresis/microelectrospray ionization mass spectrometry. *Anal. Chem.*, 69, 2154, 1997.
15. Tong, W., Link, A., Eng, J. K., and Yates, J. R., Identification of proteins in complexes by solid-phase microextraction/multistep elution/capillary electrophoresis/tandem mass spectrometry. *Anal. Chem.*, 71, 2270–2278, 1999.
16. Settlage, R. E., Russo, P. S., Shabanowitz, J., and Hunt, D. F., A novel μ-ESI source for coupling capillary electrophoresis and mass spectrometry: sequence determination of tumor peptides at the attomole level. *J. Microcolumn Sep.*, 10, 281–285, 1998.
17. Kaiser, T., Kamal, H., Rank, A., Kolb, H. J., Holler, E., Ganser, A., Hertenstein, B., Mischak, H., and Weissinger, E. M., Proteomics applied to the clinical follow-up of patients after allogeneic hematopoietic stem cell transplantation. *Blood*, 104, 340, 2004.
18. Sassi, A. P., Andel, F., III, Bitter, H. M., Brown, M. P., Chapman, R. G., Espiritu, J., Greenquist, A. C., et al.. An automated, sheathless capillary electrophoresis-mass spectrometry platform for discovery of biomarkers in human serum. *Electrophoresis*, 26, 1500, 2005.

19. Weissinger, E. M., Nguyen-Khoa, T., Fumeron, C., Saltiel, C., Walden, M., Kaiser, T., Mischak, H., Drueke, T. B., Lacour, B., and Massy, Z. A., Effects of oral vitamin C supplementation in hemodialysis patients: a proteomic assessment. *Proteomics*, 6, 993, 2006.
20. Wetterhall, M., Palmblad, M., Hakansson, P., Markides, K. E., and Bergquist, J., Rapid analysis of tryptically digested cerebrospinal fluid using capillary electrophoresis-electrospray ionization-Fourier transform ion cyclotron resonance-mass spectrometry. *J. Proteome Res.*, 1, 361–366, 2002.
21. Wittke, S., Mischak, H., Walden, M., Kolch, W., Radler, T., and Wiedmann, K., Discovery of biomarkers in human urine and cerebrospinal fluid by capillary electrophoresis coupled to mass spectrometry: towards new diagnostic and therapeutic approaches. *Electrophoresis*, 26, 1476–1487, 2005.

9 Light-Based Detection Methods for Capillary Electrophoresis

Cory Scanlan, Theodore Lapainis, and Jonathan V. Sweedler

CONTENTS

9.1 INTRODUCTION

The small sample volumes used in capillary electrophoresis (CE) make the technique suitable for myriad small-scale biochemical and materials science applications. These small volumes also require detection methods that exhibit high sensitivity and suitable selectivity. Such rigorous demands on detector performance can have important consequences with respect to other aspects of the CE system. For example, it is well known that higher efficiency and faster separations are achieved using smaller inner diameter (ID) capillaries. And yet 50-μm ID capillaries are more common than 5-μm ID capillaries in systems using light-based detection, primarily because the longer optical path length of the larger capillary aids in detection. In this way, the performance of the detection system can dictate the overall performance of a given CE system. Thus, knowledge of the various detection modes available, as well as their respective figures of merit, aids in designing CE-based methods of analysis (see Table 9.1).[1]

In addition to a working knowledge of the various detection schemes available for CE, the sensitivity, selectivity, and detectability requirements of the particular analyte(s) and application should be considered when choosing a suitable detector. In a setting where the CE system will be devoted to a single application, the detector's parameters can be tuned to optimize the detection of the analyte(s) of interest for that application. If, however, the instrument is to be used for a wide variety of applications, then versatility is of primary importance, which may come at the expense of optimal performance for any individual analyte. If none of the available detection methods allows for adequate detection of the analyte(s) of interest, it may be possible to chemically modify the analyte by derivatization in order to render it detectable.[2]

This chapter describes detection methods for CE that involve probing a sample with electromagnetic radiation. While the discussion will emphasize the two most common forms of light-based

TABLE 9.1
Light-Based Detection Modes for CE, and Corresponding Limits of Detection

Detection Principle	Minimum Detectable Concentration (M)	Minimum Detectable Quantity (mol)	Representative Detectable Molecules
Direct absorbance	10^{-5} to 10^{-7}	10^{-13} to 10^{-16}	Most
Indirect absorbance	10^{-4} to 10^{-6}	10^{-12} to 10^{-15}	Small ions, amino acids, fatty acids, polymers
Laser-induced fluorescence	10^{-9} to 10^{-12}	10^{-18} to 10^{-21}	Dyes, amino acids, peptides, proteins, carbohydrates, nucleotides, vitamins, pharmaceuticals
Indirect fluorescence	10^{-6} to 10^{-8}	10^{-14} to 10^{-16}	Small ions, pollutants, amino acids, peptides
Chemiluminescence	10^{-7} to 10^{-9}	10^{-14} to 10^{-16}	Dyes, amino acids, peptides, proteins, nucleotides
Thermooptical absorbance	10^{-5} to 10^{-7}	10^{-15} to 10^{-18}	Dyes, amino acids, proteins, nucleotides, vitamins
Radioactivity	10^{-6} to 10^{-10}	10^{-14} to 10^{-18}	Small ions, amino acids, peptides, proteins, nucleotides, pharmaceuticals
Raman	10^{-3} to 10^{-5}	10^{-12} to 10^{-15}	Dyes, small ions, amino acids, nucleotides

Source: Adapted from Khaledi, M. G., in *Chemical Analysis: A Series of Monographs on Analytical Chemistry and Its Applications*, 1st ed., Winefordner, J. D. (ed.), Wiley-Interscience, Raleigh, 1998, pp. 308–9.

detection—ultraviolet-visible (UV-Vis) absorbance and fluorescence detection—other less common methods, such as thermooptical detection, are discussed as well. Prior to discussion of the individual detection schemes, the basic considerations and requirements of optical detectors for CE are presented.

There have been a number of important developments in light-based detection for CE since the previous edition of this work was published. For example, multichannel detection has become more common for absorbance as well as fluorescence, and Fourier transform infrared (FTIR) and nuclear magnetic resonance spectroscopies are now viable detection modes. Thus, while the purpose of this chapter is to describe the instrumentation and performance of more common optical detectors, newer methods are presented throughout.

Finally, it is worth noting that limits of detection of a system mentioned throughout the chapter depend on the separation performance, preconcentration steps, and mode of operation (i.e., zone electrophoresis, electrochromatography, isotachophoresis, etc.) of the separation system. Thus, one can often optimize a CE method to improve the overall system performance, given the sample stacking and other approaches that may be appropriate for a particular sample type or application.[3–6]

9.2 GENERAL CONSIDERATIONS FOR DETECTORS

9.2.1 The Detector Cell

The capillaries typically used in CE are made up of fused silica, with IDs of 10–100 μm, outer diameters (ODs) of 100–375 μm, and coated with a 10–30 μm protective layer of polyimide (PI). There are several locations relative to this capillary where detection can take place. These detection locations include on-column, end-column, whole-column, and post-column detection.

For on-column detection, a portion of the protective PI coating is removed, either by acid application (hot HNO_3) or by burning. This results in a cylindrical length of bare fused silica, which is used as the detection window. The window is generally placed as close as practicable to the outlet of the capillary in order to maximize the fraction of the capillary used for the separation.

End-column detection is done by coupling the outlet of the capillary to an external detection cell. The properties of this detection cell (path length, material, etc.) can be tailored to allow for high sensitivity and low analyte limits of detection. The careful selection of coupling and transfer fluidics, such as zero dead-volume unions, ensures that the high resolution afforded by CE is maintained within the end-column detector.

In the post-column detection configuration, detection is carried out at a location remote to the outlet of the capillary. For example, in radionuclide detection, the capillary eluent is spotted onto a membrane or collected as fractions, and subsequently exposed to a scintillator for detection.[7,8] Post-column detection allows for the separation and detection to be independently optimized. In the case of radionuclide detection, the detector integration time can be maximized without a corresponding loss in resolution (see Section 9.2.2).

Whole-column imaging can be used to view the entire separation process as it occurs. As an example, this type of detection is often used in capillary isoelectric focusing in order to determine when substances have become sufficiently concentrated for collection and further characterization. To do this, the capillary can be held stationary while mobile scanning systems scan the length of the capillary,[9] or detectors with high spatial resolution such as charge-coupled devices (CCDs) can image the entire capillary at once.[10] These imaging detectors typically measure UV-Vis absorbance or fluorescence.

9.2.2 Resolution and Response Time

In general, the signal-to-noise ratio (SNR) of a detected peak increases with longer detector integration times. With the exception of post-column detectors, the integration time is limited by the velocity of the migrating analyte bands. Typical flow rates are on the order of 10–100 nL/min, resulting in analyte bands that migrate through the detector cell on the order of a second. In peak profile analysis, six or more data points are desired over the width of the peak band in order to describe it adequately. This means that the length of the detection cell should be restricted to one-sixth of the width of the narrowest band; common cell lengths are 100–200 μm. There are several strategies that can be used to increase detector response within these general spatial and temporal restrictions, including reducing the operating voltage of the separation, as well as transient stopped- or reduced-flow detection.[11]

9.2.3 Maximizing Separation Efficiency

The magnitude of detector response for a given amount of analyte is maximized when the analyte is confined within narrow bands. As described in Chapter 1, there are several important considerations for maintaining high efficiency in CE. These include voltage, temperature, and sample introduction. In general, the use of the highest practicable voltages maximizes efficiency. Also, since temperature gradients decrease efficiency, narrow bore capillaries allow for more effective heat dissipation. Peak dispersion can be minimized by ensuring that the concentration of any given analyte is more than two orders of magnitude lower than the concentration of the background electrolyte.[12]

9.2.4 Figures of Merit

Selection of an appropriate detection method for CE is facilitated by a comparison of each method's respective figures of merit. The relevant figures of merit include the limit of detection (LOD), sensitivity, and linear dynamic range.

The LOD is a measure of the smallest amount of analyte that can be detected. Typically, LODs are defined as the concentration or absolute amount (mass or moles) of analyte that yields a signal that is equal to three times the standard deviation of the baseline signal.

Sensitivity describes the ability of a detector to differentiate between different amounts of analyte. When a plot of signal versus amount of analyte generally yields a straight line, the term "sensitivity" refers to the slope of this line.

It is observed that the relationship between signal and concentration is linear over a finite range of concentrations. At higher concentrations, various phenomena (scattering, nonlinear optical processes, "inner filter" effect, etc.) lead to deviations from linearity. The range of concentrations where a linear relationship between signal and concentration holds is known as the linear dynamic range of the detector.

As mentioned in Section 9.1, these figures of merit depend not only on the detection system, but also on other factors as well, including the nature of the analyte, the composition of the separation medium, and the mode of operation of the separation system.

9.2.5 Qualitative versus Quantitative Information

The linear relationship between analyte concentration and detector response can be exploited to gain quantitative information about an "unknown" sample. To do this, solutions of known concentrations are used to generate a standard plot of intensity versus concentration, which is then used to determine the concentration of the unknown.

In addition to this quantitative information, it is also possible to obtain qualitative information about an analyte, such as the presence or absence of specific functional groups and their relative locations or orientations. A subset of the available optical detection methods are capable of providing this information, such as multiwavelength UV-Vis absorbance or fluorescence, nuclear magnetic resonance, and Raman spectroscopies; these techniques will be discussed in more detail in this chapter.

9.2.6 Indirect Detection

Detection methods are typically chosen based on the characteristics of the analyte of interest, for example, absorbance detectors for highly absorbing species, or fluorescence detectors for intrinsic fluorophores with sufficient quantum yields. There are, however, cases where the analyte lacks the properties that enable it to be analyzed by any of the conventional detection methods. If this is the case, and suitable derivatization chemistry is unavailable, it may be possible to detect the analyte indirectly.[13–17] To do this, a signal-generating substance is incorporated into the separation medium, and the presence of the analyte is detected as a decrease in the measured signal. The observed decrease is due to the displacement of the signal-generating species by the analyte; charge-density conservation and solubility changes are mechanisms by which this displacement can occur. Indirect detection has been used in conjunction with absorbance, electrochemical, and fluorescence detection systems.

9.3 ABSORBANCE DETECTION

9.3.1 Theory

The term absorbance refers to a phenomenon whereby a molecule is promoted to an excited state by a photon. Absorbance is normally determined by measuring the transmission of light through a sample. Transmittance, T, is the ratio of intensity of a light beam before and after it passes through a solution, that is, $T \equiv I/I_0$, where I is the intensity after passing through the sample and I_0 is the initial intensity. Absorbance, A, is related to the transmission by $A = -\log(T)$.

A solution's absorbance can be related to its constituent molecular components by the Beer–Lambert law: $A = \varepsilon lc$, where ε is the molar absorptivity (in units of inverse length per M), l is the path length through which the light passes, and c is the concentration of the absorbing species. The molar absorptivity of a substance is dependent on various parameters, including the solution pH, solvent polarity, and the wavelength of light. It can be seen from the Beer–Lambert law that, for a fixed path length and molar absorptivity, the absorbance of a solution is linearly dependent on the concentration of the absorbing species in solution. This relationship allows for the quantification of an absorbing species, if the measured value is within the dynamic range of the instrument (see Section 9.2.4).

There are several considerations with respect to measuring absorbance in CE. First, the decrease in measured light intensity after passing through a solution is due to both stray light/scattering processes and absorbance. Thus, it is important that the stray light/scattering be accounted for when relating absorbance to analyte concentration. This is typically accomplished by making appropriate "blank" measurements of a solution containing all of the pertinent buffers and solvents, but without the analyte of interest. Second, the magnitude of the measured signal scales with the path length. In CE, capillary IDs are kept small in order to minimize the deleterious effects of temperature gradients on resolution. This trade-off between detector performance and separation resolution is characteristic of light-based detection methods for CE and underscores the importance of a working knowledge of both the principles of CE and of the various methods of detection.

9.3.2 Instrumentation

Figure 9.1 shows the layout of a typical absorbance detector for CE. The light generated by the source is collected and focused into a beam by the focusing optics. The spatial profile of this beam can be defined by an aperture, which also serves to reduce background light. The beam passes through the detection cell, where the analyte absorbs a portion of the incident light. The emerging beam is then spectrally filtered by a prism or spectrograph to select the desired wavelength(s) and detected by an appropriate signal transducer, such as a photodiode array or photomultiplier tube. The resulting

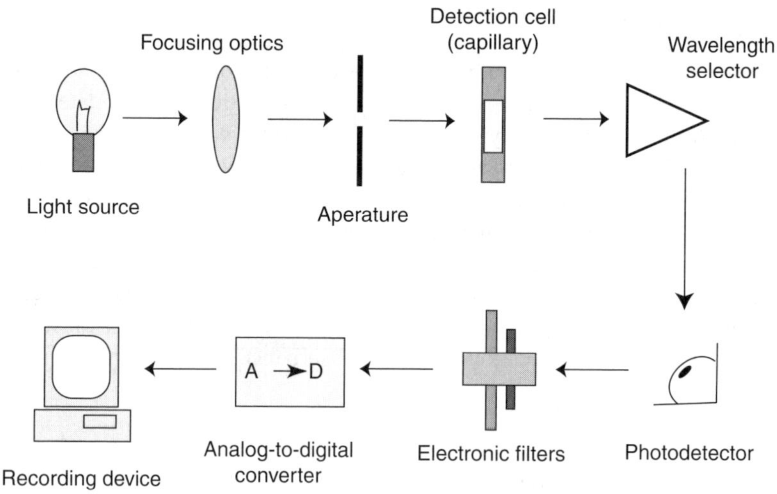

FIGURE 9.1 Flow-diagram of an absorbance detection system for CE.

signal is either displayed using an analogue device or it is converted to a digital signal and sent to a PC for analysis. The following subsections describe some of these essential components in greater detail.

9.3.2.1 Light Source

The primary considerations with respect to light sources for absorbance-based detectors are output stability and spectral characteristics. Because analyte quantification is dependent on the measured intensity of light after it emerges from the detection cell, it is important that the amount of light that enters the cell does not fluctuate in time. Thus, light sources that produce a stable output of light are integral to the satisfactory performance of an absorbance-based detector. In addition to output stability, the spectral characteristics of the output are of importance. Not only must the wavelengths provided match the absorbance bands of the analyte(s) of interest, but also the spectral profile must be suitable as well. Some detectors monitor a single wavelength, so that narrow emission lines are preferred, while wavelength-resolved instruments require a source with a broad spectral output.

Light sources that are suitable for use in absorbance-based detectors include atomic emission lamps, deuterium arc lamps, lasers, and light-emitting diodes (LEDs). In atomic emission lamps, photons are generated when electronically excited atoms relax to their ground states. Atomic emission lamps are available that use transition metals such as mercury, cadmium, and zinc, which generate strong emission lines at 254, 229, and 214 nm, respectively. These line source lamps are a reasonable choice for systems that require only a single (or relatively few) output wavelength(s). In addition, there are a number of blackbody lamps available that emit broadband radiation, such as a tungsten filament, which produces a range of wavelengths from 350 to 2500 nm. Another broadband source is the deuterium arc lamp. In contrast to atomic emission lamps, the UV light provided by the deuterium arc lamp is due to the radiant decay of molecular deuterium. The lamp produces a relatively stable output in the range of 160–400 nm, and is the most commonly used light source for absorbance detectors.

In addition to the various lamps, lasers can be used as light sources for absorbance detectors. The highly directional, spatially coherent emission from lasers can be efficiently collected and focused to a small spot within a capillary. These characteristics make lasers a viable choice for sources in absorbance detection systems (especially for smaller-diameter capillaries), although their use is less common than the various incoherent lamps. More recently, LEDs have been used as light sources for optical detectors for CE, including absorbance detectors.[18–20] The low cost and compact size of LEDs make them well suited for field-deployable sensors and lab-on-a-chip applications.

As mentioned above, output stability is important. All sources exhibit some degree of output fluctuation, and so many instruments include a means by which these fluctuations can be monitored. The most common approach is to use a beam-splitter or displacer to generate a reference beam that can be monitored by an auxiliary detector.[21] The reference channel can serve to minimize the effects of instability on detection limits, although various factors (including shot noise) place restrictions on this ability.

9.3.2.2 Focusing Optics

The performance of an absorbance detector is dependent on the ability to effectively couple the excitation source to the detection cell. This involves collecting and focusing as much source light as practical into the detection zone, as well as minimizing the stray source light that reaches the detector without interacting with the analyte.[22]

Various lenses and lens arrangements are available to collect and focus a large fraction of the incident source light. Commonly used lenses include the plano-convex singlet, as well as the quartz ball lens. The specific lens arrangement that best suits a given detector is dependent on the source used. As an alternative to the use of lenses, fiber optics can be effective in coupling the source light to the detection cell (as well as coupling the detection zone to the detector).[23–25] The use of fiber optics allows for flexibility in the placement of the source and detector and enables the illumination of multiple capillaries or multiple locations on the same capillary, although fiber optics are more difficult to interface for the smallest diameter capillaries.

The amount of stray light that reaches the detector can be minimized by the appropriate use of apertures placed before/after the detection cell.[22] A pre-cell aperture can be used to block extraneous light and allow only a well-defined cross section to pass through the capillary. For applications where high spatial resolution is critical, the lateral dimension of this area (along the capillary) can be decreased. In addition, a second aperture, placed after the detection cell but in front of the detector, can be used to prevent light that is scattered off the capillary walls from reaching the detector.

9.3.2.3 The Detection Cell

The most common illumination geometry used in absorbance detection cells is the transcolumn configuration (Figure 9.2a), in which the light passes through the detection cell along a path that is perpendicular to the length of the capillary. For on-column detection, this sets the maximum optical path length at the ID of the capillary. As mentioned previously, the performance of a light-based detector is directly proportional to this path length. Thus, it is not surprising that numerous approaches to extending the effective path length of optical detectors have been reported in the literature.

Axial-capillary illumination (Figure 9.2b) is an approach to increasing optical path length that involves focusing the source light into the capillary bore and directing it along the capillary axis.[26] In this set-up, the tightly focused lamp light or laser beam is directed into the capillary inlet, and the light exiting the outlet end of the capillary is detected by an appropriate signal transducer. Axial illumination relies on partial internal reflection within the capillary, which, in turn, requires that the refractive index of the separation buffer be greater than that of the capillary walls. The light exiting the capillary contains both the wall-propagated and internally propagated light rays, and so spatial filtering or increasing the refractive index of the separation medium is used to selectively measure the internally propagated rays.

A Z-cell can be constructed by heating the capillary in several places, and bending it into the shape of a Z, as shown in Figure 9.2c. The source light enters the cell at the first of the two bends, propagates axially though the capillary, and exits the cell at the other bend. The use of a Z-shaped cell

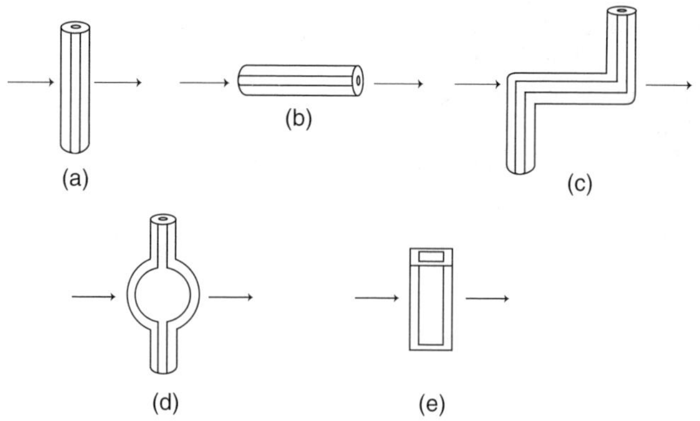

FIGURE 9.2 Different detector cell types used to increase the path length for optical detection. (a) Trans-capillary illumination, (b) axial-capillary illumination, (c) Z-cell capillary, (d) bubble cell capillary, and (e) rectangular capillary trans-illumination.

has been reported to yield up to a 10-fold improvement in detection limits relative to conventional transcolumn illumination.[22]

The bubble cell increases the optical path length by expanding the ID of a small length of capillary that makes up the detection zone relative to the ID of the rest of the capillary, as shown in Figure 9.2d. Path length increases of up to fivefold have been reported using the bubble cell, for example, a 125-μm ID bubble cell formed on a 25-μm ID capillary.[27]

While the most common cross section for fused-silica capillaries is circular, square cross section capillaries have been employed in CE (Figure 9.2e). Square capillaries minimize the scattering caused by curved capillary surfaces, and increase the probed volume relative to a circular capillary of the same width. Originally, these rectangular capillaries were made up of borosilicate glass, which is opaque in the UV.[28] More recently, square fused-silica capillaries have become commercially available and are a viable option for increasing the path length in UV absorbance detectors.[29,30]

9.3.3 Wavelength Selection

Measuring absorbance at 200–325 nm is often satisfactory for general-use absorbance detectors, as most analytes (including many biomolecules) have some absorbance at these wavelengths.[31] There are, however, cases where the analyte(s) of interest do not absorb appreciably in this range.[32] Thus, it may be beneficial to determine a more suitable wavelength for a particular analyte, for example, when dealing with low-abundance analytes or buffers that strongly absorb at these wavelengths. The optimum wavelength is the one that gives the maximum SNR for the analyte. Often, this wavelength is the one where the analyte exhibits maximum absorbance (λ_{max}). A spectrophotometer can be used to generate background-subtracted absorbance spectra, which can in turn be used to identify λ_{max}.

As an alternative to selecting and monitoring absorbance at a single wavelength, it is possible to use a multiwavelength absorbance detection system. These systems involve either rapid scanning gratings[33] used in conjunction with a single channel detector or, more commonly, a fixed spectrograph used with a photodiode array.[34–36] Indeed, a majority of the available commercial CE instruments use wavelength-resolved diode-array absorbance detection. There are a number of reasons for the popularity of these instruments, which have been discussed in detail elsewhere.[37,38] With these instruments, it is not necessary to select an appropriate wavelength for detection and quantification before analysis; the optimal wavelength for each analyte can be determined and used for subsequent data analysis after running the samples of interest. The spectral information provided by these detectors can also be useful for analyte identification or characterization, by comparing the spectral features of a peak of interest to those of standard solutions. Not only is this beneficial for analyzing "unknown" samples, but also it facilitates method development, since differences in separation conditions can alter migration times and even elution order. Moreover, completely unresolved peaks can be identified as such, because the observed spectral characteristics will differ from those of any pure analyte (i.e., peak purity confirmation).[39] For partially resolved components, the multichannel data can be used in conjunction with various computational approaches to enable quantification of the constituent components, without having to change separation conditions and reanalyze samples.[39,40]

9.4 FLUORESCENCE DETECTION

9.4.1 Theory

Fluorescence is a process in which a photon is emitted by an electronically excited molecule as it relaxes to its ground electronic state. Each of the relevant electronic states contains numerous vibrational sublevels, and so fluorescence excitation and emission spectra are relatively broad. The manifold of various vibronic states of a fluorophore results in unique and complex emission spectra, which may enable differentiation between two closely related molecules. In addition, the observed emission spectra are dependent on the local chemical environment of the fluorophore, and can differ depending on factors such as solvent polarity and solution pH. Indeed, this exquisite dependence of emission spectra on surroundings can be used to deduce the relative location of tryptophan residues within a protein. An excellent overview of the qualitative aspects of fluorescence can be found in Lacowicz's text on the subject.[41]

The observed magnitude of fluorescence emission is given by $I_f = cdkI_lQ_f$, where I_f is the intensity of fluorescence (photons cm^{-2} s^{-1}), c is the concentration of the fluorophore (M), d is the path length of excitation (cm), k is the base-ten logarithm of the molar absorptivity (ε, see Section 9.3), I_l is the intensity of the laser, and Q_f is the fluorescence quantum yield (unitless).[42] The dependence of the fluorescence signal on the intensity of the excitation source suggests that high-power light sources can be used. In practice, however, the optimal laser power is also a function of other factors such as analyte photostability, and so should be determined empirically.[43,44]

Fluorescence detection affords the lowest reported detection limits of the various detector types for CE (Table 9.1). The high performance of fluorescence-based detection systems results from a number of characteristics. First, a single fluorophore molecule can cycle repeatedly between excitation and emission, thus generating up to thousands of photons before exiting the detection zone or undergoing photodestruction. Second, the observed emission spectrum is shifted relative to the excitation wavelength(s), so that filters can be used to minimize the amount of scattered excitation radiation that reaches the detector. As a result, single molecule detection is possible using LIF, and LIF-based detection systems for CE are capable of providing LODs in the yoctomole range.[45,46]

9.4.2 Instrumentation

The basic fluorescence detection system in CE consists of several components, including an excitation source, a detection cell, collection optics, and a detector. Within these requirements, there are a wide variety of possible optical configurations, and vastly differing performance specifications between systems. The design constraints and general considerations relevant to the development of a fluorescence detection system have been detailed in the literature.[47–50] The following sections will discuss the major components of a fluorescence detection system and recent innovations.

9.4.2.1 Light Source

There are a number of criteria relevant to the selection of an appropriate light source for a CE fluorescence detection system. As with sources for absorbance detectors, the output wavelength and spectral characteristics are important, as is the stability of the source output. In addition, the intensity of a light source is important for fluorescence detectors, as the observed signal scales with the excitation irradiance. The sources for fluorescence detection can be differentiated into two types, coherent (lasers) and incoherent (lamps). Jorgenson and Lukacs[51–53] built the first fluorescence detector for CE in 1981, and it used a high-pressure mercury lamp for excitation. Such types of incoherent lamps, combined with predetection cell monochromaters to filter out unwanted wavelengths from the emission source, provide flexibility in wavelength selection to match the wavelength of maximum absorbance for an analyte. Stabilized arc lamps are becoming increasingly available that, along with spatial filters and carefully constructed optical setups, allow for sensitive detection for chromophore-labeled analytes, such as detection limits of 10 pM for fluorescein-labeled amino acids.[54,55]

The use of LEDs as sources for fluorescence is becoming more commonplace.[56–59] As mentioned previously (in Section 9.3), the small footprint and low cost of these sources make them a good choice for microchip (e.g., μTAS or lab-on-a-chip) devices. It is likely that these sources will become more commonplace as the output power generated by these sources increases.

Lasers are another source of excitation radiation used in fluorescence detection systems. The high-directional output of a laser maximizes the fraction of total output that can be easily focused down to a spot size compatible with the dimensions of CE detection cells. The output of a laser is also typically monochromatic, or a discrete set of spectrally narrow lines. This type of output makes it relatively easy to filter out low-level incoherent plasma radiation and undesired emission lines without greatly diminishing the overall output power. In addition, many lasers provide flexibility in terms of pulse width and repetition rate, which allows one to optimize excitation with respect to analyte photostability.

The original reported use of lasers in CE with fluorescence detection was described by Gassmann et al.[60] in which laser-induced fluorescence (LIF) was used to distinguish and measure the chiral enantiomers in mixtures of dansylated amino acids. Since that time, lasers have become commonplace in CE systems with fluorescence detection. The most common types of lasers used have been the He–Cd, Ar-ion, and He–Ne lasers, as they are relatively inexpensive and have emission lines that match commonly used fluorescent reagents, such as *o*-phthalaldehyde (OPA) (325 nm) and fluorescamine (354 nm).

The majority of LIF instruments use excitation wavelengths in the visible portion of the spectrum.[61] There is, however, interest in the development of LIF instruments that use near-infrared (NIR) excitation.[62] There are several reasons for this, including the availability of inexpensive NIR diode lasers, as well as the decreased likelihood of photobleaching or significant Raman background at longer wavelengths. The further development of suitable derivatization strategies will increase the number of applications that can be addressed by NIR-LIF.[63,64]

At the opposite end of the spectrum, UV sources for CE-LIF are becoming increasingly popular. UV radiation is capable of inducing fluorescence in many intrinsic fluorophores, including a number of biologically relevant molecules. The frequency-doubled Ar-ion laser (257 nm) was one of the first examples reported of UV-excitation for CE-LIF, and yielded improvements in LODs for a number of substances, such as conalbumin (1.4×10^{-8} M).[65] As another example, a frequency-doubled Kr laser operating at 284 nm has been used for the analysis of neuropeptides and small biomolecules, and exhibited LODs for tryptophan of ~800 zmol.[66] In addition to frequency-doubled ion lasers, a number of relatively inexpensive pulsed lasers such as frequency quadrupled YAG (266 nm),[67] KrF (248 nm),[68] and hollow-cathode metal vapor lasers have appeared, which provide deep-UV excitation (e.g., 224 and 248 nm).[69–73]

Multiphoton excitation is another strategy for exciting the same transitions that UV sources excite, while using NIR or visible lasers and associated optical elements.[74,75] In multiphoton excitation, two or three photons are absorbed by a fluorophore, which is then promoted to a vibronic state with an associated energy that is equal to the sum of the energies of the individual absorbed photons. Because the photons must be absorbed simultaneously (e.g., within ~1 fs), this technique requires high peak power. More specifically, the observed fluorescence emission is proportional to I^n_{peak}, where I_{peak} is the peak intensity of the excitation radiation and n is the number of photons involved in the transition. The small probe volumes dictated by this nonlinear intensity dependence are compatible with the small ID capillaries used in CE; this technique has been used to detect a variety of substances after CE separation, including the aromatic amino acids, nucleotides, and monoamine neurotransmitters such as serotonin.[76,77]

9.4.2.2 The Detection Cell

On-column capillary detection is the most widely employed configuration in CE with fluorescence detection, as shown in Figure 9.3a. A small window is made on the capillary by burning off a section

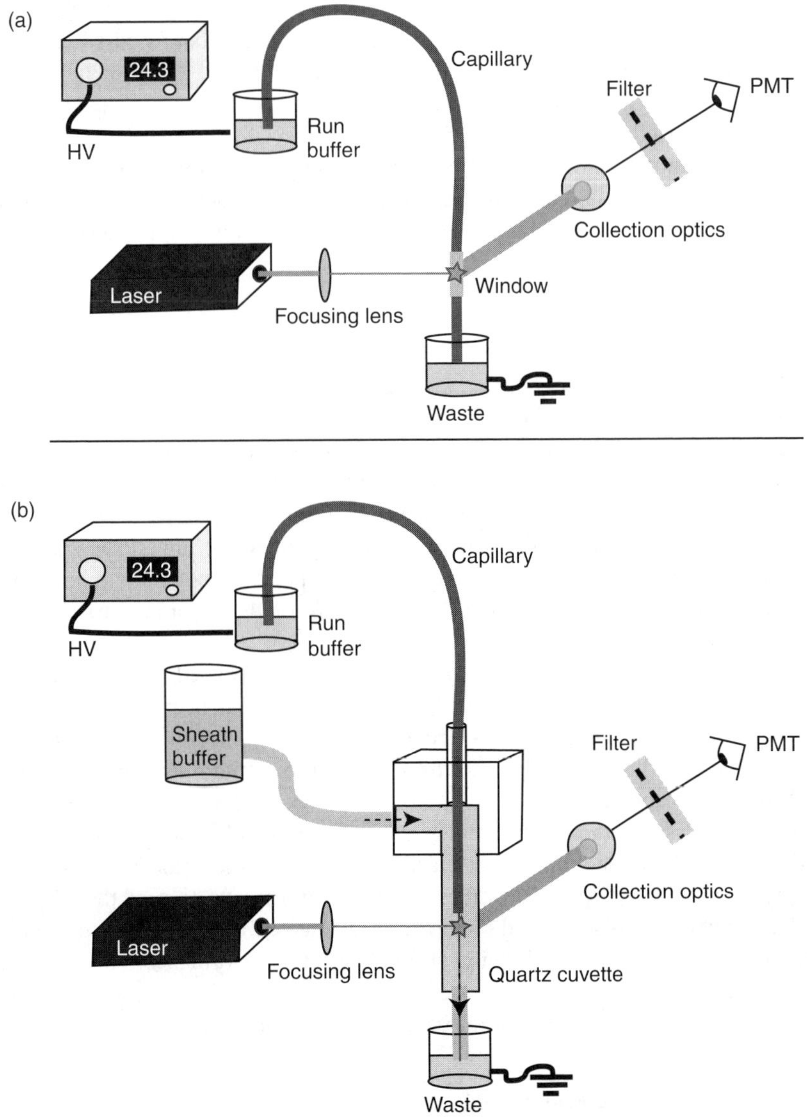

FIGURE 9.3 Schematic diagrams of two CE-LIF detection systems. (a) On-column CE-LIF and (b) end-column sheath-flow CE-LIF.

of the polyimide coating. In most cases, the source light is focused through this window, although axial illumination (directed up the bore of the capillary) has been used as well.[26] In either case, the fluorescence emission is collected through the window. Collection can take place orthogonal to the light source, or coaxial with it, as in the epi-illumination configuration.[49,78] Immersion of the cylindrical detection window in an index-matching solution can be helpful in maximizing excitation and collection efficiency.[79]

End-column detection is another frequently used configuration for fluorescence-based CE detection systems. One example is the sheath-flow cuvette, originally used for flow cytometry adapted to CE by Dovichi and coworkers (Figure 9.3b).[80–82] In sheath-flow post-column detection, the end of the capillary is inserted into a cuvette into which buffer is flowing. The linear flow rate of the sheath buffer is set to match or slightly exceed the electroosmotic flow of the capillary, thus minimizing turbulence and maintaining a narrow sample stream. The laser is focused to a spot immediately below the outlet of the capillary; the use of a cuvette with a square cross section maximizes the amount of excitation light that is transmitted into the probe volume. This type of detection cell is especially useful for CE-LIF systems using UV-excitation, as metal ions present in fused-silica capillary walls give rise to a broadband luminescent background. Impressive limits of detection have been obtained using the sheath-flow cuvette.[83,84] For example, detection limits of 10^{-13} M have been reported for numerous rhodamine-based fluorophores.[85,86]

The post-column configuration has also been used in conjunction with fluorescence. Typically, this involves depositing the eluent on a membrane, or collecting fractions, as is commonly done in HLPC.[87–89] The resulting spots or fractions can then be imaged with an appropriate fluorimeter, fluorescence microscope, or other suitable fluorescence detector.

9.4.2.3 Collection Optics

The ideal collection optic would collect fluorescence from a $4\pi\,sr$ solid angle around the detection zone with 100% efficiency, while being completely inefficient at collecting scattered laser light. Obviously, there exists no such optic, but there are various approaches reported in the literature taken to emulate this ideal. In most cases, either a high numerical aperture microscope objective or a fiber-optic system is used to collect the fluorescence and transmit it to the signal transducer.

Microscope objectives used as collection optics have yielded the lowest reported LODs in CE with fluorescence detection. They can be used alone or as part of a complete microscope setup. A high-numerical-aperture objective collects a large quantity of light from a very small region and focuses it onto the detector.

A number of fiber optic-based systems have been described for CE-LIF systems.[90–92] These systems boast a small footprint, and the terminal of the fiber optic can be placed exceedingly close to the detection window, thus enabling high collection efficiency. Hence, fiber optics are an excellent choice for systems where the overall footprint must be kept to a minimum, such as field-deployable sensors or lab-on-a-chip applications.

Fluorescence-based CE detection systems generally also include a spatial filter and wavelength filter located between the collection optics and the detector, in order to minimize the amount of Rayleigh and Raman scatter that reaches the detector. Careful selection of the filter wavelength range and rejection characteristics, based on the desired detected wavelength range and the spectral output of the source, is important to achieving optimal detector performance. The use of line sources such as lasers is advantageous in this respect, as a narrow pass filter can be used to block excitation radiation while minimizing the amount of fluorescence that is blocked.

9.4.3 Multidimensional Fluorescence

9.4.3.1 Wavelength-Resolved Fluorescence

In wavelength-resolved fluorescence detection, a complete emission spectrum is generated for each time point. The resulting data are plotted as a three-dimensional electropherogram, with axes

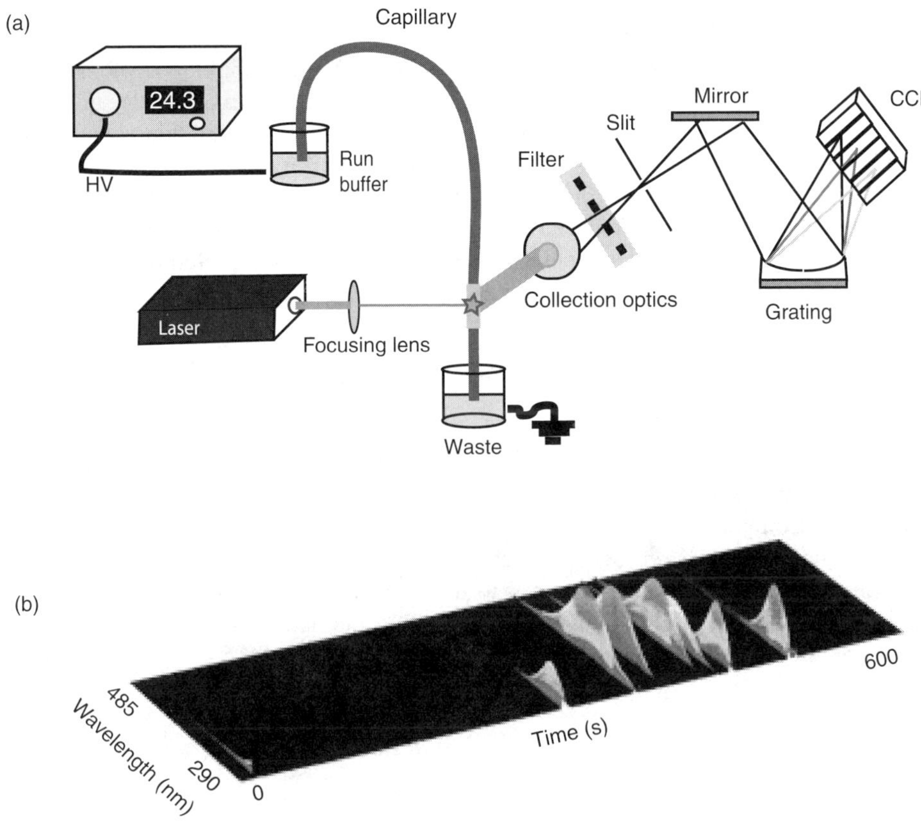

FIGURE 9.4 (a) Schematic diagram of a CCD-based wavelength-resolved CE-LIF instrument. (b) Wavelength-resolved electropherogram of a mixture of seven peptides containing tryptophan and tyrosine residues. (Adapted from Timperman, A. T., et al., *Anal. Chem.*, 67, 3421, 1995. With permission.)

corresponding to time, wavelength, and fluorescence intensity. The first report of wavelength-resolved fluorescence detection for microscale separations was by Swaile and Sepaniak in 1989,[35] in which a spectrograph was used to disperse the fluorescence onto a photodiode array with 4-nm spectral resolution. The advent of higher-sensitivity array detectors, such as CCDs,[93] made this type of detection feasible for low-abundance analytes (see Figure 9.4), as demonstrated by 100 molecule detection limits for sulforhodamine[94] and 3 nM LODs for the neurotransmitter serotonin.[95]

Multichannel fluorescence detection can also be carried out using a series of dichroic beamsplitters and photomultiplier tubes (PMTs) in place of a spectrograph and CCD.[96] Rather than generating a high-resolution emission spectrum, this type of detector divides the collected emission into several channels, each of which corresponds to a relatively broad range of wavelengths. Peaks are identified based on a ratiometric "fingerprint" calculated from the relative intensities of the channels. This "fingerprint" is capable of differentiating two closely related substances, such as various catecholamine neurotransmitters.

Spectrally resolved fluorescence detection for CE has found a variety of applications and is often used for the detection and identification of various biologically relevant intrinsic fluorophores, including amino acids, proteins, neuropeptides, and small molecule neurotransmitters.[66,95] For example, wavelength-resolved fluorescence was used to elucidate various aspects of serotonin catabolism.[97–99] In addition, multichannel fluorescence detection was used to differentiate each of four fluorescently labeled dideoxy bases that terminate replication in a polymerase chain reaction.[48,100] This technique was instrumental in sequencing the human genome, as well as those of other organisms such as *Drosophila melanogaster* and *Caenorhabditis elegans*.

9.4.3.2 Time-Resolved Fluorescence

While wavelength-resolved fluorescence is the most common form of multidimensional fluorescence detection in CE, other forms of multidimensional fluorescence have also been reported. Time-resolved fluorescence involves measuring the temporal evolution of the fluorescence signal, and using this information to differentiate various analytes.[101] For example, high-resolution time-domain data can be used to calculate fluorescence lifetime of an analyte, which in turn can assist in peak identification by discriminating between short-lived and long-lived fluorophores.[102] Moreover, typical fluorescence lifetimes are orders of magnitude longer than scattering lifetimes (both Raleigh and Raman). As a result, the signals resulting from scattering and fluorescence can be effectively deconvolved.

9.4.3.3 Polarization Fluorescence

In addition to wavelength and time-resolved fluorescence techniques, polarization fluorescence can yield important information about an analyte.[41] This is especially true when differentiating between chiral compounds. Combining, for example, a fluorescently tagged antibody immunoassay with polarization detection allows for very sensitive detection limits of chiral enantiomers. Laser-induced fluorescence polarization (LIFP) has been used to detect concentrations as low as 0.9 nM of an antibody-bound cyclosporine A (an immunosuppressive drug) in human blood.[103] A conventional single channel fluorescence detector can be easily modified to perform such measurements, simply by adding the appropriate polarization filters.

9.4.4 Detecting Biomolecules

9.4.4.1 Derivatization

As mentioned previously, fluorescence-based detection exhibits the lowest reported detection limits of any detection mode for CE. And while this is true, the effective use of fluorescence detection is contingent upon the ability to induce fluorescence in the analyte of interest with reasonable quantum efficiency. While native fluorescence is appropriate for a number of analytes, many others (including many of biological relevance) are either not intrinsically fluorescent or exhibit low fluorescence quantum yields. As a result, a significant amount of research has been devoted to appropriate labeling methods, whereby a fluorescent moiety is covalently bound to the analyte of interest in order to enable its detection.[104] While attaching fluorescent probes to molecules is not unique to CE, the small sample volumes involved make the development of suitable derivatization reactions and procedures critically important.

Derivatization can be carried out end-column, on-column, or precolumn, with the latter being the most common. With precolumn derivatization, large starting sample volumes and a large molar excess of derivatizing reagents can be used, and the reaction can be allowed to proceed indefinitely. These steps minimize losses due to dilution or sample handling, and help to ensure that the derivatization reaction proceeds to completion. Precolumn derivatization of lysine-containing neuropeptides with fluorescamine has shown pmol detection limits,[105] and 3-(*p*-carboxybenzoyl) quinoline-2-carboxaldehyde was used to gain a detection limit of 9 zmol (10^{-21}) for arginine.[106] While using large volumes of the sample solution helps to minimize the effects of dilution by the derivatizing reagents, there are times when this is not possible. This is the case in many biological applications, such as single-cell analysis. For example, Oates and coworkers[107,108] demonstrated derivatization of proteins and peptides from a single cell in a 25 nL volume using naphthalene-2,3-dicarboxaldehyde (NDA) as the derivatant. Derivatization of intracellular signaling compounds can be done by using the cell itself as a reaction chamber. Selective derivatization with monobromobimane of thiol-containing compounds within a single red blood cell,[15] as well as selective derivatization of glutathiones by NDA in single neuroblastoma cells were some of the first reports published showing successful derivatization within a single cell.[109]

On-column analyte derivatization is another approach often used. For on-column labeling, it is beneficial to use a reagent that is nonfluorescent in its unreacted form, such as NDA. Gilman and Ewing[110] were able to use on-column labeling with NDA to analyze a whole PC12 cell. To do this, they injected the intact cell followed by a plug of derivatization mixture, and left the voltage off for a brief time interval to facilitate cell lysis and derivatization prior to CE analysis. On-column derivatization can also be carried out using low-volume connectors attached near the outlet of the capillary to introduce derivatization reagent. Carefully constructed transfer fluidics maintain separation efficiency upon introduction of the derivatant. This type of apparatus can be constructed by laser-drilling holes into the separation capillary and introducing the reagent by gravity through a T-connector. The introduction of OPA into the capillary using this arrangement allows for femtomole amounts of amino acids to be detected.[111]

There are cases when it is desirable to carry out derivatization after the separation has taken place. For example, small molecules can be difficult to separate after derivatization, because the size of the label is often large relative to the analyte. In addition, some analytes have multiple reactive sizes (such as amino groups) where derivatization can occur. This can result in a distribution of fluorescence products from a single analyte, each giving rise to a peak on the electropherogram. To address these and other complications, a number of methods have been developed to perform end-column derivatization. Most end-column systems employ a similar set of components, including a micropump to add derivatization reagent to the CE effluent and a secondary column where the labeling reactions occur before the tagged analytes reach the detector. The simplest approach is to use a sheath-flow system to introduce the reactants, such as the coaxial capillary reactor system originally described by Rose and Jorgenson.[112] In this setup, OPA was introduced at the outlet of the capillary tip, allowing for attomole detection limits. Another approach is to use the sheath-flow cuvette (discussed earlier) as a derivatization reactor; the fluorogenic reagent is simply added to the sheath-flow buffer (Figure 9.5).[113,114]

Another end-column derivatization system involves a gap junction reactor, as described by Albin et al.[115] In the gap junction reactor, two capillaries are separated by a small gap in which the fluorophore is introduced. Careful alignment of the two capillaries ensures that high separation efficiency is maintained at the junction. The gap junction interface is an excellent choice when using high-cost derivates, as smaller volumes can be used for reagent introduction.

For all of the various end-column derivatization schemes, it is important that labeling reactions be chosen that exhibit fast kinetics. This ensures that derivatization is complete before the analytes enter the detection zone, which in turn ensures maximum analyte detectability. Common derivates with suitable kinetics for end-column detection include fluorescamine, OPA, and NDA.[2,116,117] In addition to kinetics, the selection of an appropriate derivatization reaction can be influenced by other factors, including the potential side reactions for a given set of reagents and analytes, equilibrium and concentration considerations, and the pH at which the derivatization is carried out.

9.4.4.2 Native Fluorescence

When possible, it can be advantageous to take advantage of an analyte's intrinsic fluorescence. Ideally, native fluorescence would exhibit a high quantum yield and could be excited with readily available excitation sources. Unfortunately, few compounds exhibit fluorescence in the spectral range of the most common lasers, that is, the visible portion of the spectrum. Nonetheless, there are some compounds that fluoresce upon excitation with visible light. Examples include many pharmaceutical compounds,[118–120] porphyrins,[121,122] and vitamin B6 metabolites.[123] In contrast, excitation below 300 nm excites quite a number of biomolecules, including proteins and peptides containing Tyr, Trp, or Phe residues, polycyclic aromatic hydrocarbons, as well as nucleosides and DNA.[95,124–126] For

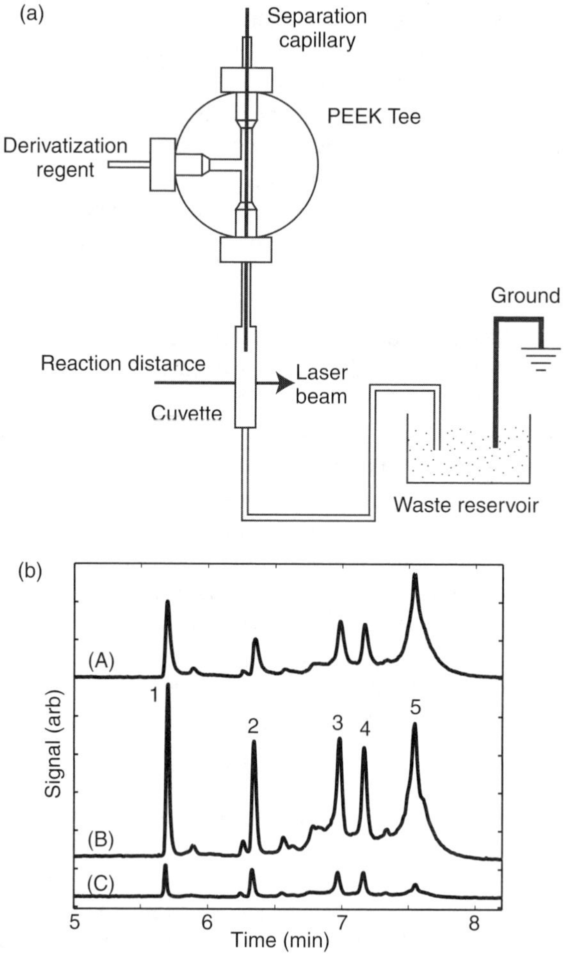

FIGURE 9.5 (a) Schematic diagram of the sheath-flow post-column reactor. (b) Separation of protein standards derivatized with NDA/β-mercaptoethanol at a height difference of (A) 0.5 cm, (B) 3 cm, and (C) 6 cm between the derivatization solution reservoir and waste reservoir. (Adapted from Ye, M. L., *J. Chromatogr. A*, 1022, 201, 2004. With permission.)

example, laser-induced native fluorescence can be used to detect epinephrine and norepinephrine content of single bovine adrenal medullary cells (Figure 9.6).[127]

Most applications of native fluorescence detection involve the direct detection of a fluorescent analyte of interest. However, native fluorescence can also be used indirectly to detect the presence of nonfluorescent species. For example, the presence of a nonfluorescent enzyme can be detected based on its conversion of substrate into natively fluorescent products. This is often used in conjunction with electrophoretically mediated microanalysis, in which differences in buffer mobilities enable on-column mixing and reaction of enzymes with substrates.[128] One such study exploited the intrinsic fluorescence of nicotinamide adenine dinucleotide (NADH) to investigate differences in the reactivity of individual molecules of lactate dehydrogenase.[129]

The quantum efficiencies and photostabilities of most intrinsically fluorescent biomolecules are modest relative to those of the available fluorescent labels. There are, however, several factors that can affect these quantum yields. The ionic strength and pH of the separation conditions, as well as the presence or absence of organic additives, can have a significant effect on the fluorescent properties (i.e., quantum yields and Stokes' shift) of these analytes.[41,130] Thus, if sensitivity is of

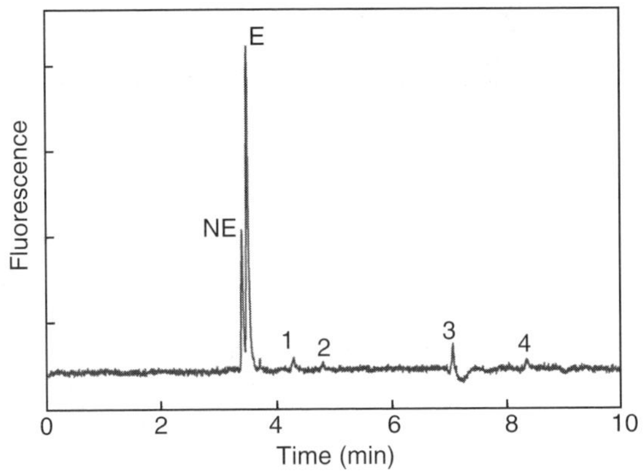

FIGURE 9.6 Electrophoretic separation of intracellular components of a single bovine adrenal medullary cell detected by laser-induced native fluorescence. NE = Norepinephrine; E = Epinephrine; Peaks 1–4 are unknown components. (Reprinted from Chang, H. T. and Yeung, E. S., *Anal. Chem.*, 67, 1079, 1995. With permission.)

utmost importance for a particular application, the separation buffer can be optimized in terms of analyte fluorescence and separation efficiency.

9.5 THERMOOPTICAL DETECTION

Thermooptical detection exploits the deflection of light that occurs as a result of changes in the refractive index of a light-absorbing solution. Although it is a transmittance-based detection method, it is distinct from absorbance detection in that the measured decrease in transmittance is related to changes in the index of refraction of the absorbing solution. This detection mode involves the use of two laser beams that intersect in the detection zone of the capillary. One of the beams, the "probe" beam, passes through the detection cell, and is detected by a photodiode or other appropriate signal transducer. The second beam, the "pump" beam, is selected to match the absorbance profile of the separation medium. This beam is temporally chopped, so that it passes through the detection window only transiently. The index of refraction of the separation medium changes in time as a result of heat-induced density changes in the solution, depending on whether or not the pump beam is passing through the detection cell. As a consequence, the transmittance of the probe beam changes in time, which can be measured by the photodetector. Furthermore, the change in refractive index that occurs is dependent on the absorbing species in the separation medium. Hence, as different substances migrate through the detection zone, the magnitude of the detected signal changes. The electropherogram is generated by plotting the difference in transmittance between "pump on" and "pump off" as a function of time.

Thermooptical detection has been combined with CE and used for protein and peptide Edman degradation sequencing detection, native protein detection, as well as the analysis of derivatized amino acid mixtures.[131–134] Frequency-doubled argon-ion lasers have been employed to supply an UV (257 nm) pump beam for the analysis of dansylated amino acids as well as the analysis of etopside and etopside phosphate in human blood plasma.[135,136] In addition, two-color thermooptical absorbance detectors have been constructed, where each laser serves a dual role of "pump" and "probe"; this system can be used to detect analytes that absorb in differing regions of the electromagnetic spectrum.

The use of a near-field thermal lens system to increase pump beam absorption has recently been reported by Pyell and coworkers.[137] This innovation has resulted in a nearly 30-fold improvement in detection limits for a multitude of nitroaromatic compounds, including a 0.15 mg/L detection limit (compared to a previously reported LOD of 3.4 mg/L) for 2-nitrotoluene.

9.6 CHEMILUMINESCENCE

Chemiluminescence (CL) is a phenomenon whereby the electronically excited state product of a chemical reaction generates optical radiation during relaxation to its ground state. There are two general mechanisms of CL that are employed in the context of detection for CE.[138] In direct CL, the photon(s) are emitted by the excited state reaction product as it relaxes to the ground state. In the second, the relaxation of the excited state takes place via energy transfer to a fluorophore, which subsequently fluoresces; this is referred to as sensitized CL, because a fluorophore with a high quantum yield can be used.

CL is an attractive detection method for CE in part due to its simple instrumentation. An external radiation source is not required, reduces instrument cost, complexity, as well as the overall footprint of the instrument. In addition, since there is no background from an excitation source, there is no need to filter the collected radiation, which increases detection efficiency. In its simplest form, a CE-CL instrument can consist of only a capillary, power supply, and a detector, such as a PMT or photodiode detector. The small footprint of such instrumentation is particularly attractive for microscale total analysis systems (μTAS) or lab-on-a-chip applications.[139,140]

CL detection can be used to probe intrinsic luminescence, such as that found in numerous biological systems (bioluminescence), or analytes of interest can be derivatized by reaction with an isothiocyanate or succinimidyl ester-linked chemiluminescent moiety. The first reported use of CL for CE detection was by Hara et al.[141–143] who used peroxyoxolate as the luminophore to detect a protein–dye complex. Since then, CL has been used successfully to detect a variety of substances after CE separation including metal ions,[144] amino acids,[145] peptides[146] and proteins,[147] and even DNA.[148]

CL involves a chemical reaction between two or more species, and so the magnitude of the detected signal is a function of each of these substances. To facilitate quantification of one of these, the concentrations of the others must be held constant, or otherwise known. Several distinct strategies have emerged to enable this, including coaxial flow on-column detection, off-column detection, and end-column detection.[149,150] One method that has gained considerable attention recently is to generate the luminophore *in situ* by an electron transfer reaction facilitated by an electrode; this specific type of CL is referred to as electrogenerated chemiluminescence (ECL). For example, ruthenium (II) tris(2,2′-bipyridine) can be oxidized to $Ru(bpy)_3^{2+}$, which undergoes a chemiluminescent reaction with various analytes.[151–156] Not only can this luminophore be used with biologically relevant analytes such as amino acids and NADH without their prior derivatization, but the reactive state of the luminophore can be regenerated by oxidation at an electrode in the detection zone.

Many developments in CL detection have taken place since the last edition of this book. There are a number of reviews that highlight current research in the field.[149,150] Given the relative low cost, simplicity, and small footprint of CL detectors, as well as the persistent drive toward miniaturization of analytical instruments, it is likely that research in CL detection for CE will continue to progress at a rapid pace.

9.7 RADIONUCLIDE DETECTION

In radionuclide detection, the presence of a radiolabeled analyte is detected by the response of a secondary substance to the radioactive decay of the radiolabel. The secondary substance is referred to as a scintillator, and it responds to collision with a radioactive particle by the emission of photons. Since it is this emission of photons that is actually measured in radionuclide detection, it is appropriate that this method of detection be included in a chapter covering light-based detection methods. Radionuclide detection is both highly sensitive and selective. The high sensitivity of this technique is due to the ease of detection of the energetic decay events, in conjunction with a very low natural background. The high selectivity of this technique results from the fact that only the radiolabeled sample yields a detectable response. As a consequence, it is possible, for example, to inject a living system with a radiolabeled sample and investigate various metabolic pathways or biochemical changes by following the movement of the radiolabeled nuclei, without interference from unlabeled compounds.

The sensitivity and detection limits for radionuclide detection depend in large part on the particular radioactive nuclei being used. More specifically, the isotope's rate of decay and particle energy determine the lowest attainable detection limit. In addition, the sensitivity is dependent on the observation time of the isotope and the fraction of events detected, which can be expressed as follows:

$$\text{counts} = (\text{DPS})\ (\text{time observed})\ (\text{efficiency of detector})$$

where "counts" represents the number of detector counts recorded over an observed peak, "DPS" (measured in Becquerels, or Bq) represents the number of radioactive decompositions per second from the sample, "time observed" is the amount of time (in seconds) a radioactive molecule spends in the detection volume, and "efficiency of detector" is the average fractional number of events sensed by the detector per total number of events. DPS is directly proportional to the concentration of a particular isotope in a given sample, which is determined by the isotope's rate of decay. The rate of decay for an isotope is described as the amount of time for half the nuclei in a sample to decay, commonly expressed in days or years, also known as the half-life ($t_{1/2}$). These rates can vary widely, from more than 10^{22} years for very stable radionuclides, to 10^{-6} s for highly unstable ones. More reasonable ranges for isotopes used in CE analysis are those of ^{32}P with a half-life of 14.2 days, and ^{14}C with a half-life of 5.73×10^3 years. Because peak times in CE are normally limited to a few seconds, the "time observed" term in the preceding equation limits the counts observed, and hence the overall detection sensitivity when measurements are performed online. In addition, it is evident from the "DPS" term that the shorter the half-life of the isotope, the more likely the detector will see enough events in a given amount of time to obtain a clear signal. Another consideration worth noting is that high-energy decay events are more easily detected than low-energy events. This is due to the decreased probability that an intervening atom will absorb a higher-energy particle before it reaches the scintillator. Common particle energies for isotopes used in online radionuclide detection in CE range from 1.71 MeV for ^{32}P to 0.156 MeV for ^{14}C. The higher-energy particles are generally able to travel a longer distance before being absorbed. Besides the differing particle energies, there are also different forms of decay for radioparticles, the most common in CE detection being the emission of negative (β^-) particles, positive (β^+) particles, and neutral gamma rays (γ).

Both on-column and post-column detection schemes have been developed for radionuclide detection for CE. The most common type used is an on-column configuration, which yields detection limits in the 10^{-9} M range for isotopes such as ^{32}P. Isotachophoretic separations of ^{14}C were among the first examples of online capillary radionuclide detection, performed by Kaniansky et al.[157–159] The associated instrument uses 300-μm ID fluorinated ethylene–propylene copolymer capillary tubing, and the separation eluent flows directly into a plastic scintillator cell between two PMTs. The scintillation events are detected coincidentally between the two PMTs, such that only if both PMTs receive an input within a short time will they register the count as signal. This kind of coincidence detection ensures that nonscintillation photons that come from outside the detection cell and only hit one PMT are not counted. This system exhibits a detection limit of 16 Bq for ^{14}C analytes, with a detector efficiency of 13–15%.

Gamma-emitting radioparticles are typically much higher in energy than beta particles, and so their detection requires a more dense scintillator material. The CE separation and detection of radiopharmaceuticals containing ^{99}Tc was first reported by Altria et al.[160] who constructed an instrument in which the capillary passes through a block of solid scintillator; the solid scintillator emits photons in response to impinging γ-rays.

A similar design was employed by Pentoney et al.[161] in the pursuit of detecting ^{32}P particles, using a γ^- and β^- sensitive Cd-Te detector. The sensitivity of this instrument was augmented by the use of flow-programming—reducing the voltage while an analyte is in the detection window in order to increase residence (and hence integration) time. Similar systems were developed by Klunder et al.[162] to detect fissionable materials, in which the radioactivity detector consists of conical plastic scintillating material, with the capillary passing through the center to provide a near $4\pi\, sr$ detection geometry. This type of system provides an 80% detection efficiency and detects the decay of ^{152}Eu and ^{137}Cs at the nCi level.

In biological studies, two important radionuclides are ^{3}H and ^{35}S, due to their common presence in small molecules and proteins, respectively. Unfortunately, ^{3}H and ^{35}S lack sufficiently energetic decay products to penetrate the capillary wall and so must be sampled outside the capillary tubing if they are to be detected. Tracht et al.[7,8] developed a post-column method to detect these low-energy β^- particle emitters. In this setup, the end of the capillary is painted with a conductive paint to maintain an electrical circuit, and the capillary eluent is deposited directly onto a binding membrane with the aid of a multiaxis translation stage. Scintillation compounds deposited onto the membrane will luminesce from any radioactive decay present in the sample (Figure 9.7). The intensity can be measured with a photodetector, or the plate can be read by a phosphorimager. This post-column CE collection and exposure system has since been used for a number of nanoscale biological analyses, from analysis of a single neuron[163] to observation of synthase activity within a ganglion.[164]

A major advantage unique to the coupling CE with radionuclide experiments is the significant reduction in the volume of hazardous waste generated as compared to what would be generated using other separation methods such as high-pressure liquid chromatography (HPLC). Furthermore, the abundance of potential applications in the biological sciences and environmental monitoring will ensure the continued development of this method of analysis in years to come.

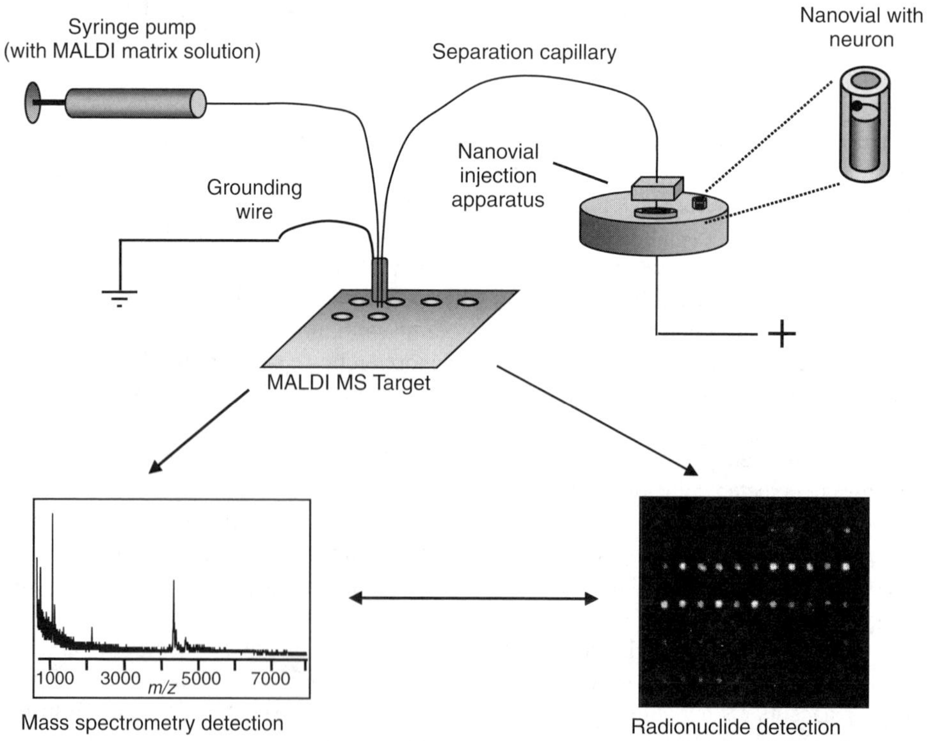

FIGURE 9.7 Post-column radionuclide detection system for CE. The spots can be analyzed by other detection modes, such as mass spectrometry, to gain additional information. (Reprinted from Page, J. S., et al., *Anal. Chem.*, 74, 497, 2002. With permission.)

9.8 OTHER LIGHT-BASED DETECTION METHODS

The set of optical detection techniques presented thus far in this chapter is not comprehensive; emphasis has been placed on those that are most common in CE, the most influential to the field, or both. There are other optical detection modes in various stages of development, each one with its own unique characteristics. Many of these alternative detection modes aim to provide structural information about the analyte being detected. One such method uses FTIR detection. FTIR detection is of interest because it is potentially a nearly universal detection method and provides a wealth of structural information. The nearly universal nature of IR activity also presents challenges, as the capillary itself as well as the separation buffer are often IR-absorbing. Thus, most attempts at FTIR detection for CE have been off-line; the eluting bands are sprayed onto an IR-transparent (e.g., CaF_2) disk using a glass or metal nebulizer for subsequent (off-line) FTIR analysis.[165,166] The instrumentation for online FTIR detection was developed for HPLC before its implementation in CE.[167] The first report of online CE-FTIR appeared some 4 years later, in which adenosine, guanosine, and adenosine-5′-monophosphate were separated, and the online-generated IR spectra were compared with standard IR spectra.[168] The most recent report of CE-FTIR uses mid-IR detection to distinguish between different sugars in fruit juices; this is the first report of using CE-FTIR to differentiate compounds in "real" samples.[169] In these reports, LODs were in the sub- to low-millimolar (>0.2 mM) range; this technique holds great promise for elucidating structural information from samples, and future work in this field will likely enhance the performance of this technique for commercial application.

Raman spectroscopy (RS) is another detection method for CE that is potentially useful for providing structural information about an eluting analyte. The use of special Raman modes, such as resonance Raman spectroscopy (RRS) and surface-enhanced Raman spectroscopy (SERS), offer substantial signal enhancements (10^2–10^{15}) compared to other Raman processes. RRS and SERS have been coupled with CE both online and off-line (so-called at-line); an excellent review has been published recently covering considerations for at-line and online coupling of Raman spectroscopies to liquid separations, including CE.[170] For "at-line" SERS, the CE eluent can be spotted on an appropriate substrate, such as a thin film of gold, before being subjected to Raman spectroscopic analysis.[171,172] Online CE-SERS analysis has also been reported using a silver colloid solution (with appropriate electrolytes) as a separation medium.[173] RRS was coupled online to CE as early as 1991. More recently, online UV CE-RRS was used to obtain spectral information from the separated components of standard mixtures of aromatic sulfonic acids and nucleotides. Through the enhancement factors of RRS and SERS, and recent work toward the adaptation of deep-UV and tunable wavelength lasers to CE-RRS, RS has become an intriguing and useful detection method for obtaining structural information from small-volume separations.

Radio frequency (RF) electromagnetic radiation can be used to induce nuclear spin transitions in the presence of an external magnetic field, thus enabling detection via NMR. NMR can be used to both detect and obtain structural information about analytes that are not electroactive and do not possess a chromophore. Hence, research is ongoing with the aim of improving the online coupling of NMR to microscale separations. The use of NMR on nanoliter volume samples was first enabled by the development of the microcoil probe, which consists of a thin copper wire (rf coil) wrapped around a small diameter (e.g., 50 μm) capillary.[174–176] A variety of approaches have been used in conjunction with the microcoil probe in order to enhance the detected signal. Sample concentration techniques, such as capillary isotachophoresis, have been used to improve analyte detectability and enable trace analysis.[177] The generation of high-resolution NMR spectra can be accomplished using periodic stopped-flow CE to increase the spectral acquisition time.[11,178] Several methods have been developed to enable continuous-flow separations and high resolution, including the dual microcoil split-flow method[179] and a "splitless" method.[180] For example, in the dual microcoil method, multiple capillaries are interconnected such that a separated band can be directed to a secondary capillary and microcoiled, parked, and analyzed, while the separation continues in the primary capillary.[179]

Several relatively recent reviews have been published that discuss the application of NMR detection for microscale separations (including CE) in more detail.[181,182]

Future research will determine whether the potential of these techniques for structure-based analyte identification can be realized in the form of practicable and profitable detection methods for CE.

9.9 CONCLUSIONS

As CE continues to evolve, it is not surprising to see a concomitant development of optical detection methods. Detector innovations have been required to keep up with smaller diameter capillaries, faster separations, parallel capillary systems, and microfluidic-based electrophoretic formats. In parallel to these changes, more flexible and higher performance imaging detectors, UV light sources, and clever applications of these advances continue to push the envelope in terms of light-based detection for CE.

While this chapter summarizes the state-of-the-art in optical detection, it is perhaps worthwhile to speculate as to the potential fruits of this continued development, that is, the future of light-based detection for CE. While photodiode array-based absorbance detection will continue to dominate CE detection, fluorescence detection will be used for an ever-increasing set of CE-applications. For example, the widespread use of fluorescence microscopy in biological and biomedical research ensures that new derivatization chemistries will be developed, allowing a wider range of analytes to be characterized using fluorescence. Further development of low-cost deep-UV sources is likely to lead to the proliferation of native fluorescence detection. This, in turn, will be accompanied by a dramatic increase in the use of multichannel systems to provide information about the identity of eluting analytes. And of course, solid-state light sources will replace lasers in many applications.

Outside of absorbance and fluorescence, microchip CL systems will become more common, likely using ECL with *in situ* regenerated lumophores. Will the next decades finally witness the commercial development of a capillary-scale Raman probe? Of course, the continued hyphenation of UV-absorbance, mass spectrometry, and other detection schemes will push the envelope of information-rich detection systems for small-volume separations.

ACKNOWLEDGMENTS

This material is based upon work supported by the National Science Foundation under Award No. CHE-04-00768, the National Institutes of Health under Award No. R33 DK 070285, and the National Institute on Drug Abuse under Award No. P30 DA 018310 to the UIUC Neuroproteomics Center.

REFERENCES

1. Khaledi, M. G., High-performance capillary electrophoresis: theory, techniques, and applications, in *Chemical Analysis: A Series of Monographs on Analytical Chemistry and its Applications*, 1st ed., Winefordner, J. D. (ed.), Wiley-Interscience, Raleigh, 1998, pp. 308.
2. Szulc, M. E. and Krull, I. S., Improved detection and derivatization in capillary electrophoresis, *J. Chromatogr. A*, 659, 231, 1994.
3. Burgi, D. S. and Chien, R. L., Optimization in sample stacking for high-performance capillary electrophoresis, *Anal. Chem.*, 63, 2042, 1991.
4. Beckers, J. L. and Bocek, P., Sample stacking in capillary zone electrophoresis: Principles, advantages and limitations, *Electrophoresis*, 21, 2747, 2000.
5. Osbourn, D. M., Weiss, D. J., and Lunte, C. E., On-line preconcentration methods for capillary electrophoresis, *Electrophoresis*, 21, 2768, 2000.

6. Urbanek, M., Krivankova, L., and Bocek, P., Stacking phenomena in electromigration: From basic principles to practical procedures, *Electrophoresis*, 24, 466, 2003.
7. Tracht, S. E., Cruz, L., StobbaWiley, C. M., and Sweedler, J. V., Detection of radionuclides in capillary electrophoresis using a phosphor-imaging detector, *Anal. Chem.*, 68, 3922, 1996.
8. Tracht, S., Toma, V., and Sweedler, J. V., Postcolumn radionuclide detection of low-energy beta-emitters in capillary electrophoresis, *Anal. Chem.*, 66, 2382, 1994.
9. Beale, S. C. and Sudmeier, S. J., Spatial-scanning laser fluorescence detection for capillary electrophoresis, *Anal. Chem.*, 67, 3367, 1995.
10. Wu, X. Z. and Pawliszyn, J., Whole-column fluorescence-imaged capillary electrophoresis, *Electrophoresis*, 25, 3820, 2004.
11. Olson, D. L., Lacey, M. E., Webb, A. G., and Sweedler, J. V., Nanoliter-volume 1H NMR detection using periodic stopped-flow capillary electrophoresis, *Anal. Chem.*, 71, 3070, 1999.
12. Sustacek, V., Foret, F., and Bocek, P., Selection of the background electrolyte-composition with respect to electromigration dispersion and detection of weakly absorbing substances in capillary zone electrophoresis, *J. Chromatogr.*, 545, 239, 1991.
13. Foret, F., Fanali, S., Nardi, A., and Bocek, P., Capillary zone electrophoresis of rare earth metals with indirect UV absorbance detection, *Electrophoresis*, 11, 780, 1990.
14. Mala, Z., Vespalec, R., and Bocek, P., Capillary zone electrophoresis with indirect photometric detection in the visible range, *Electrophoresis*, 15, 1526, 1994.
15. Hogan, B. L. and Yeung, E. S., Determination of intracellular species at the level of a single erythrocyte via capillary electrophoresis with direct and indirect fluorescence detection, *Anal. Chem.*, 64, 2841, 1992.
16. Kuhr, W. G. and Yeung, E. S., Indirect fluorescence detection of native amino acids in capillary zone electrophoresis, *Anal. Chem.*, 60, 1832, 1988.
17. Olefirowicz, T. M. and Ewing, A. G., Capillary electrophoresis with indirect amperometric detection, *J. Chromatogr.*, 499, 713, 1990.
18. Dasgupta, P. K., Eom, I. Y., Morris, K. J., and Li, J., Light emitting diode-based detectors: Absorbance, fluorescence and spectroelectrochemical measurements in a planar flow-through cell, *Anal. Chim. Acta*, 500, 337, 2003.
19. Macka, M., Andersson, P., and Haddad, P. R., Linearity evaluation in absorbance detection: The use of light-emitting diodes for on-capillary detection in capillary electrophoresis, *Electrophoresis*, 17, 1898, 1996.
20. Boring, C. B. and Dasgupta, P. K., An affordable high-performance optical absorbance detector for capillary systems, *Anal. Chim. Acta*, 342, 123, 1997.
21. Xue, Y. J. and Yeung, E. S., On-column double-beam laser-absorption detection for capillary electrophoresis, *Anal. Chem.*, 65, 1988, 1993.
22. Bruin, G. J. M., Stegeman, G., Vanasten, A. C., Xu, X., Kraak, J. C., and Poppe, H., Optimization and evaluation of the performance of arrangements for UV detection in high-resolution separations using fused-silica capillaries, *J. Chromatogr.*, 559, 163, 1991.
23. Bruno, A. E., Gassmann, E., Pericles, N., and Anton, K., On-column capillary-flow cell utilizing optical wave-guides for chromatographic applications, *Anal. Chem.*, 61, 876, 1989.
24. Foret, F., Deml, M., Kahle, V., and Bocek, P., Online fiber optic UV detection cell and conductivity cell for capillary zone electrophoresis, *Electrophoresis*, 7, 430, 1986.
25. Ludi, H., Gassmann, E., Grossenbacher, H., and Marki, W., Capillary zone electrophoresis for the analysis of peptides synthesized by recombinant DNA technology, *Anal. Chim. Acta*, 213, 215, 1988.
26. Taylor, J. A. and Yeung, E. S., Axial-beam absorbency detection for capillary electrophoresis, *J. Chromatogr.*, 550, 831, 1991.
27. Heiger, D. N., Kaltenbach, P., and Sievert, H. J. P., Diode-array detection in capillary electrophoresis, *Electrophoresis*, 15, 1234, 1994.
28. Tsuda, T., Sweedler, J. V., and Zare, R. N., Rectangular capillaries for capillary zone electrophoresis, *Anal. Chem.*, 62, 2149, 1990.
29. Dubber, M. J. and Kanfer, I., Application of reverse-flow micellar electrokinetic chromatography for the simultaneous determination of flavonols and terpene trilactones in Ginkgo biloba dosage forms, *J. Chromatogr. A*, 1122, 266, 2006.

30. Wu, J. Q. and Pawliszyn, J., Capillary isoelectric-focusing with imaging detection, *J. Liquid Chromatogr.*, 16, 1891, 1993.
31. Stamler, J. S. and Loscalzo, J., Capillary zone electrophoretic detection of biological thiols and their S-nitrosated derivatives, *Anal. Chem.*, 64, 779, 1992.
32. Oefner, P. J., Vorndran, A. E., Grill, E., Huber, C., and Bonn, G. K., Capillary zone electrophoretic analysis of carbohydrates by direct and indirect UV detection, *Chromatographia*, 34, 308, 1992.
33. Jones, D. H. and Hinman, A. S., Thin-layer ultraviolet visible reflectance spectroelectrochemistry with a spinning-grating monochromator, *Can. J. Chem.*, 68, 2234, 1990.
34. Kobayashi, S., Ueda, T., and Kikumoto, M., Photodiode array detection in high-performance capillary electrophoresis, *J. Chromatogr.*, 480, 179, 1989.
35. Sepaniak, M. J., Swaile, D. F., and Powell, A. C., Instrumental developments in micellar electrokinetic capillary chromatography, *J. Chromatogr.*, 480, 185, 1989.
36. Yeo, S. K., Lee, H. K., and Li, S. F. Y., Separation of antibiotics by high-performance capillary electrophoresis with photodiode-array detection, *J. Chromatogr.*, 585, 133, 1991.
37. Heiger, D. N., Kaltenbach, P., and Sievert, H. J. P., Diode array detection in capillary electrophoresis, *Electrophoresis*, 15, 1234, 1994.
38. Cruz, L., Shippy, S. A., and Sweedler, J. V., Capillary electrophoretic detectors based on light, in *High Performance Capillary Electrophoresis*, Khaledi, M. G. (ed.), Wiley, New York, 1998, p. 303.
39. Lilley, K. A. and Wheat, T. E., Drug identification in biological matrices using capillary electrophoresis and chemometric software, *J. Chromatogr. B: Biomed. Appl.*, 683, 67, 1996.
40. Li, H., Hou, J., Wang, K., and Zhang, F., Resolution of multicomponent overlapped peaks. A comparison of several curve resolution methods, *Talanta*, 70, 336, 2006.
41. Lakowicz, J. R., *Principles of Fluorescence Spectroscopy*, 3rd ed. Springer, New York, 2006.
42. Larson, A. P., Ahlberg, H., and Folestad, S., Semiconductor laser-induced fluorescence detection in picoliter volume flow cells, *Appl. Opt.*, 32, 794, 1993.
43. Peck, K., Stryer, L., Glazer, A. N., and Mathies, R. A., Single-molecule fluorescence detection: Autocorrelation criterion and experimental realization with phycoerythrin, *Proc. Natl Acad. Sci. USA*, 86, 4087, 1989.
44. Mathies, R. A., Peck, K., and Stryer, L., Optimization of high-sensitivity fluorescence detection, *Anal. Chem.*, 62, 1786, 1990.
45. Chen, D. Y. and Dovichi, N. J., Single-molecule detection in capillary electrophoresis: Molecular shot noise as a fundamental limit to chemical analysis, *Anal. Chem.*, 68, 690, 1996.
46. Soper, S. A., Davis, L. M., and Shera, E. B., Detection and identification of single molecules in solution, *J. Opt. Soc. Amer. B-Opt. Phys.*, 9, 1761, 1992.
47. Goodwin, P. M., Ambrose, W. P., and Keller, R. A., Single-molecule detection in liquids by laser-induced fluorescence, *Acc. Chem. Res.*, 29, 607, 1996.
48. Zhang, J. Z., Chen, D. Y., Wu, S., Harke, H. R., and Dovichi, N. J., High-sensitivity laser-induced fluorescence detection for capillary electrophoresis, *Clin. Chem.*, 37, 1492, 1991.
49. Johnson, M. E. and Landers, J. P., Fundamentals and practice for ultrasensitive laser-induced fluorescence detection in microanalytical systems, *Electrophoresis*, 25, 3513, 2004.
50. Fu, J. L., Fang, Q., Zhang, T., Jin, X. H., and Fang, Z. L., Laser-induced fluorescence detection system for microfluidic chips based on an orthogonal optical arrangement, *Anal. Chem.*, 78, 3827, 2006.
51. Jorgenson, J. W. and Lukacs, K. D., High-resolution separations based on electrophoresis and electroosmosis, *J. Chromatogr.*, 218, 209, 1981.
52. Jorgenson, J. W. and Lukacs, K. D., Zone electrophoresis in open-tubular glass-capillaries, *Anal. Chem.*, 53, 1298, 1981.
53. Jorgenson, J. W. and Lukacs, K. D., Capillary zone electrophoresis, *Science*, 222, 266, 1983.
54. Green, J. S. and Jorgenson, J. W., Variable-wavelength on-column fluorescence detector for open-tubular zone electrophoresis, *J. Chromatogr.*, 352, 337, 1986.
55. Arriaga, E., Chen, D. Y., Cheng, X. L., and Dovichi, N. J., High-efficiency filter fluorometer for capillary electrophoresis and its application to fluorescein thiocarbamyl amino-acids, *J. Chromatogr. A*, 652, 347, 1993.
56. Chabinyc, M. L., Chiu, D. T., McDonald, J. C., Stroock, A. D., Christian, J. F., Karger, A. M., and Whitesides, G. M., An integrated fluorescence detection system in poly(dimethylsiloxane) for microfluidic applications, *Anal. Chem.*, 73, 4491, 2001.

57. Hillebrand, S., Schoffen, J. R., Mandaji, M., Termignoni, C., Henrik Grieneisen, H. P., and Ledur Kist, T. B., Performance of an ultraviolet light-emitting diode-induced fluorescence detector in capillary electrophoresis, *Electrophoresis*, 23, 2445, 2002.
58. Kuo, J. S., Kuyper, C. L., Allen, P. B., Fiorini, G. S., and Chiu, D. T., High-power blue/UV light-emitting diodes as excitation sources for sensitive detection, *Electrophoresis*, 25, 3796, 2004.
59. Zhao, S., Yuan, H., and Xiao, D., Optical fiber light-emitting diode-induced fluorescence for capillary electrophoresis, *Electrophoresis*, 27, 461, 2006.
60. Gassmann, E., Kuo, J. E., and Zare, R. N., Electrokinetic separation of chiral compounds, *Science*, 230, 813, 1985.
61. Higashijima, T., Fuchigami, T., Imasaka, T., and Ishibashi, N., Determination of amino-acids by capillary zone electrophoresis based on semiconductor-laser fluorescence detection, *Anal. Chem.*, 64, 711, 1992.
62. Williams, D. C. and Soper, S. A., Ultrasensitive near-IR fluorescence detection for capillary gel-electrophoresis and DNA-sequencing applications, *Anal. Chem.*, 67, 3427, 1995.
63. Shealy, D. B., Lipowska, M., Lipowski, J., Narayanan, N., Sutter, S., Strekowski, L., and Patonay, G., Synthesis, chromatographic-separation, and characterization of near-infrared-labeled DNA oligomers for use in DNA-sequencing, *Anal. Chem.*, 67, 247, 1995.
64. Williams, R. J., Lipowska, M., Patonay, G., and Strekowski, L., Comparison of covalent and noncovalent labeling with near-infrared dyes for the high-performance liquid-chromatographic determination of human serum-albumin, *Anal. Chem.*, 65, 601, 1993.
65. Swaile, D. F. and Sepaniak, M. J., Laser-based fluorometric detection schemes for the analysis of proteins by capillary zone electrophoresis, *J. Liquid Chromatogr.*, 14, 869, 1991.
66. Timperman, A. T., Oldenburg, K. E., and Sweedler, J. V., Native fluorescence detection and spectral differentiation of peptides containing tryptophan and tyrosine in capillary electrophoresis, *Anal. Chem.*, 67, 3421, 1995.
67. Chan, K. C., Muschik, G. M., and Issaq, H. J., Solid-state UV laser-induced fluorescence detection in capillary electrophoresis, *Electrophoresis*, 21, 2062, 2000.
68. Chan, K. C., Janini, G. M., Muschik, G. M., and Issaq, H. J., Laser-induced fluorescence detection of 9-fluorenylmethyl chloroformate derivatized amino acids in capillary electrophoresis, *J. Chromatogr.*, 653, 93, 1993.
69. Miao, H., Rubakhin, S. S., and Sweedler, J. V., Analysis of serotonin release from single neuron soma using capillary electrophoresis and laser-induced fluorescence with a pulsed deep-UV NeCu laser, *Anal. Bioanal. Chem.*, 377, 1007, 2003.
70. McNeil, J. R., Johnson, W. L., and Collins, G. J., Ultraviolet laser action in He–Ag and Ne–Ag mixtures, *Appl. Phys. Lett.*, 29, 172, 1976.
71. Sparrow, M. C., Jackovitz, J. F., Munro, C. H., Hug, W. F., and Asher, S. A., New 224 nm hollow cathode laser-UV Raman spectrometer, *Appl. Spectrosc.*, 55, 66, 2001.
72. Storrie-Lombardi, M. C., Hug, W. F., McDonald, G. D., Tsapin, A. I., and Nealson, K. H., Hollow cathode ion lasers for deep ultraviolet Raman spectroscopy and fluorescence imaging, *Rev. Sci. Instrum.*, 72, 4452, 2001.
73. Warner, B. E., Persson, K. B., and Collins, G. J., Metal-vapor production by sputtering in a hollow-cathode discharge: Theory and experiment, *J. Appl. Phys.*, 50, 5694, 1979.
74. Xu, C., Zipfel, W., Shear, J. B., Williams, R. M., and Webb, W. W., Multiphoton fluorescence excitation: New spectral windows for biological nonlinear microscopy, *Proc. Natl. Acad. Sci. USA*, 93, 10763, 1996.
75. Shear, J. B., Multiphoton-excited fluorescence, *Anal. Chem.*, 71, 598A–605A, 1999.
76. Wise, D. D. and Shear, J. B., Quantitation of nicotinamide and serotonin derivatives and detection of flavins in neuronal extracts using capillary electrophoresis with multiphoton-excited fluorescence, *J. Chromatogr. A*, 1111, 153, 2006.
77. Gostkowski, M. L., McDoniel, J. B., Wei, J., Curey, T. E., and Shear, J. B., Characterizing spectrally diverse biological chromophores using capillary electrophoresis with multiphoton-excited fluorescence, *J. Am. Chem. Soc.*, 120, 18, 1998.
78. Hernandez, L., Marquina, R., Escalona, J., and Guzman, N. A., Detection and quantification of capillary electrophoresis zones by fluorescence microscopy, *J. Chromatogr.*, 502, 247, 1990.

79. Abromson, D. and Bickel, W. S., Fluorescent angular scattering emissions from dye-filled fibers, *Appl. Opt.*, 30, 2980, 1991.
80. Cheng, Y. F. and Dovichi, N. J., Subattomole amino acid analysis by capillary zone electrophoresis and laser-induced fluorescence, *Science*, 242, 562, 1988.
81. Swerdlow, H., Wu, S. L., Harke, H., and Dovichi, N. J., Capillary gel electrophoresis for DNA sequencing. Laser-induced fluorescence detection with the sheath flow cuvette, *J. Chromatogr.*, 516, 61, 1990.
82. Wu, S. and Dovichi, N. J., High-sensitivity fluorescence detection of fluorescein isothiocyanate derivatives of amino acids separated by capillary zone electrophoresis, *J. Chromatogr.*, 480, 141, 1989.
83. Chen, D. Y. and Dovichi, N. J., Yoctomole detection limit by laser-induced fluorescence in capillary electrophoresis, *J. Chromatogr. B*, 657, 265, 1994.
84. Zhao, J. Y., Dovichi, N. J., Hindsgaul, O., Gosselin, S., and Palcic, M. M., Detection of 100 molecules of product formed in a fucosyltransferase reaction, *Glycobiology*, 4, 239, 1994.
85. Chen, D. Y., Swerdlow, H. P., Harke, H. R., Zhang, J. Z., and Dovichi, N. J., Low-cost, high-sensitivity laser-induced fluorescence detection for DNA sequencing by capillary gel electrophoresis, *J. Chromatogr.*, 559, 237, 1991.
86. Swerdlow, H., Zhang, J. Z., Chen, D. Y., Harke, H. R., Grey, R., Wu, S. L., Dovichi, N. J., and Fuller, C., Three DNA sequencing methods using capillary gel electrophoresis and laser-induced fluorescence, *Anal. Chem.*, 63, 2835, 1991.
87. Eriksson, K. O., Palm, A., and Hjerten, S., Preparative capillary electrophoresis based on adsorption of the solutes (proteins) onto a moving blotting membrane as they migrate out of the capillary, *Anal. Biochem.*, 201, 211, 1992.
88. Cheng, Y. F., Fuchs, M., Andrews, D., and Carson, W., Membrane-fraction collection for capillary electrophoresis, *J. Chromatogr.*, 608, 109, 1992.
89. Huang, X. H. and Zare, R. N., Continuous sample collection in capillary zone electrophoresis by coupling the outlet of a capillary to a moving surface, *J. Chromatogr.*, 516, 185, 1990.
90. Mesaros, J. M., Luo, G., Roeraade, J., and Ewing, A. G., Continuous electrophoretic separations in narrow channels coupled to small-bore capillaries, *Anal. Chem.*, 65, 3313, 1993.
91. Quesada, M. A. and Zhang, S., Multiple capillary DNA sequencer that uses fiber-optic illumination and detection, *Electrophoresis*, 17, 1841, 1996.
92. Stokes, D. L., Sepaniak, M. J., and Vo-Dinh, T., Development of a new capillary electrophoresis-based fibre optic sensor, *Biomed. Chromatogr.*, 11, 187, 1997.
93. Sweedler, J. V., Charge-transfer device detectors and their applications to chemical-analysis, *Crit. Rev. Anal. Chem.*, 24, 59, 1993.
94. Timperman, A. T., Khatib, K., and Sweedler, J. V., Wavelength-resolved fluorescence detection in capillary electrophoresis, *Anal. Chem.*, 67, 139, 1995.
95. Fuller, R. R., Moroz, L. L., Gillette, R., and Sweedler, J. V., Single neuron analysis by capillary electrophoresis with fluorescence spectroscopy, *Neuron*, 20, 173, 1998.
96. Lapainis, T., Scanlan, C., Rubakhin, S. S., and Sweedler, J. V., A multichannel native fluorescence detection system for capillary electrophoretic analysis of neurotransmitters in single neurons, *Anal. Bioanal. Chem.*, 387, 97–105, 2007.
97. Stuart, J. N., Zhang, X., Jakubowski, J. A., Romanova, E. V., and Sweedler, J. V., Serotonin catabolism depends upon location of release: characterization of sulfated and gamma-glutamylated serotonin metabolites in *Aplysia californica*, *J. Neurochem.*, 84, 1358, 2003.
98. Stuart, J. N., Ebaugh, J. D., Copes, A. L., Hatcher, N. G., Gillette, R., and Sweedler, J. V., Systemic serotonin sulfate in opisthobranch mollusks, *J. Neurochem.*, 90, 734, 2004.
99. Squires, L. N., Jakubowski, J. A., Stuart, J. N., Rubakhin, S. S., Hatcher, N. G., Kim, W. S., Chen, K., Shih, J. C., Seif, I., and Sweedler, J. V., Serotonin catabolism and the formation and fate of 5-hydroxyindole thiazolidine carboxylic acid, *J. Biol. Chem.*, 281, 13463, 2006.
100. Karger, A. E., Harris, J. M., and Gesteland, R. F., Multiwavelength fluorescence detection for DNA sequencing using capillary electrophoresis, *Nucleic Acids Res.*, 19, 4955, 1991.
101. Bright, F. V. and Munson, C. A., Time-resolved fluorescence spectroscopy for illuminating complex systems, *Anal. Chim. Acta*, 500, 71, 2003.

102. Miller, K. J. and Lytle, F. E., Capillary zone electrophoresis with time-resolved fluorescence detection using a diode-pumped solid-state laser, *J. Chromatogr.*, 648, 245, 1993.
103. Ye, L. W., Le, X. C., Xing, J. Z., Ma, M. S., and Yatscoff, R., Competitive immunoassay for cyclosporine using capillary electrophoresis with laser induced fluorescence polarization detection, *J. Chromatogr. B*, 714, 59, 1998.
104. Haughland, R. P., *The Handbook—A Guide to Fluorescent Probes and Labeling Technologies*, 10th ed. Molecular Probes, Eugene, 2005.
105. Shippy, S. A., Jankowski, J. A., and Sweedler, J. V., Analysis of trace-level peptides using capillary electrophoresis with UV laser-induced fluorescence, *Anal. Chim. Acta*, 307, 163, 1995.
106. Arriaga, E. A., Zhang, Y. N., and Dovichi, N. J., Use of 3-(P-carboxybenzoyl)quinoline-2-carboxaldehyde to label amino-acids for high-sensitivity fluorescence detection in capillary electrophoresis, *Anal. Chim. Acta*, 299, 319, 1995.
107. Oates, M. D., Cooper, B. R., and Jorgenson, J. W., Quantitative amino acid analysis of individual snail neurons by open tubular liquid chromatography, *Anal. Chem.*, 62, 1573, 1990.
108. Oates, M. D. and Jorgenson, J. W., Quantitative amino acid analysis of subnanogram levels of protein by open tubular liquid chromatography, *Anal. Chem.*, 62, 1577, 1990.
109. Orwar, O., Fishman, H. A., Ziv, N. E., Scheller, R. H., and Zare, R. N., Use of 2,3-naphthalenedicarboxaldehyde derivatization for single-cell analysis of glutathione by capillary electrophoresis and histochemical-localization Ion by fluorescence microscopy, *Anal. Chem.*, 67, 4261, 1995.
110. Gilman, S. D. and Ewing, A. G., Analysis of single cells by capillary electrophoresis with on column derivatization and laser-induced fluorescence detection, *Anal. Chem.*, 67, 58, 1995.
111. Pentoney, S. L., Huang, X. H., Burgi, D. S., and Zare, R. N., Online connector for microcolumns—Application to the on-column ortho-phthaldialdehyde derivatization of amino-acids separated by capillary zone electrophoresis, *Anal. Chem.*, 60, 2625, 1988.
112. Rose, D. J. and Jorgenson, J. W., Post-capillary fluorescence detection in capillary zone electrophoresis using ortho-phthaldialdehyde, *J. Chromatogr.*, 447, 117, 1988.
113. Oldenburg, K. E., Xi, X., and Sweedler, J. V., Simple sheath flow reactor for post-column fluorescence derivatization in capillary electrophoresis, *Analyst*, 122, 1581, 1997.
114. Ye, M. L., Hu, S., Quigley, W. W. C., and Dovichi, N. J., Post-column fluorescence derivatization of proteins and peptides in capillary electrophoresis with a sheath flow reactor and 488 nm argon ion laser excitation, *J. Chromatogr. A*, 1022, 201, 2004.
115. Albin, M., Weinberger, R., Sapp, E., and Moring, S., Fluorescence detection in capillary electrophoresis—Evaluation of derivatizing reagents and techniques, *Anal. Chem.*, 63, 417, 1991.
116. Gilman, S. D. and Ewing, A. G., Postcolumn derivatization for capillary electrophoresis using naphthalene-2,3-dicarboxaldehyde and 2-dercaptoethanol, *Anal. Method Instrum.*, 2, 133, 1995.
117. Zhu, R. and Kok, W. T., Postcolumn reaction system for fluorescence detection in capillary electrophoresis, *J. Chromatogr. A*, 716, 123, 1995.
118. Reinhoud, N. J., Tjaden, U. R., Irth, H., and Vandergreef, J., Bioanalysis of some anthracyclines in human plasma by capillary electrophoresis with laser-induced fluorescence detection, *J. Chromatogr. B*, 574, 327, 1992.
119. Roach, M. C., Gozel, P., and Zare, R. N., Determination of methotrexate and its major metabolite, 7-hydroxymethotrexate, using capillary zone electrophoresis and laser-induced fluorescence detection, *J. Chromatogr. B*, 426, 129, 1988.
120. Soini, H., Novotny, M. V., and Riekkola, M. L., Determination of naproxen in serum by capillary electrophoresis with ultraviolet absorbency and laser-induced fluorescence detection, *J. Microcolumn. Sep.*, 4, 313, 1992.
121. Quirino, J. P., Dulay, M. T., Fu, L., Mody, T. D., and Zare, R. N., Capillary electrophoresis separation and native laser-induced fluorescence detection of metallotexaphrins, *J. Sep. Sci.*, 25, 819, 2002.
122. Wu, N., Li, B. H., and Sweedler, J. V., Recent developments in porphyrin separations using capillary electrophoresis with native fluorescence detection, *J. Liquid Chromatogr.*, 17, 1917, 1994.
123. Burton, D. E., Sepaniak, M. J., and Maskarinec, M. P., Analysis of B6 vitamers by micellar electrokinetic capillary chromatography with laser-excited fluorescence detection, *J. Chromatogr. Sci.*, 24, 347, 1986.

124. Lee, T. T. and Yeung, E. S., High-sensitivity laser-induced fluorescence detection of native proteins in capillary electrophoresis, *J. Chromatogr.*, 595, 319, 1992.
125. Mcgregor, D. A. and Yeung, E. S., Detection of DNA fragments separated by capillary electrophoresis based on their native fluorescence inside a sheath flow, *J. Chromatogr. A*, 680, 491, 1994.
126. Milofsky, R. E. and Yeung, E. S., Native fluorescence detection of nucleic-acids and DNA restriction fragments in capillary electrophoresis, *Anal. Chem.*, 65, 153, 1993.
127. Chang, H. T. and Yeung, E. S., Determination of catecholamines in single adrenal-medullary cells by capillary electrophoresis and laser-induced native fluorescence, *Anal. Chem.*, 67, 1079, 1995.
128. Regnier, F. E., Patterson, D. H., and Harmon, B. J., Electrophoretically-mediated microanalysis (Emma), *Trends Anal. Chem.*, 14, 177, 1995.
129. Xue, Q. and Yeung, E. S., Differences in the chemical reactivity of individual molecules of an enzyme, *Nature*, 373, 681, 1995.
130. Park, Y. N., Zhang, X., Rubakhin, S. S., and Sweedler, J. V., Independent optimization of capillary electrophoresis separation and native fluorescence detection conditions for indolamine and catecholamine measurements, *Anal. Chem.*, 71, 4997, 1999.
131. Yu, M. and Dovichi, N. J., Attomole amino acid determination by capillary zone electrophoresis with thermooptical absorbance detection, *Anal. Chem.*, 61, 37, 1989.
132. Saz, J. M., Krattiger, B., Bruno, A. E., Diezmasa, J. C., and Widmer, H. M., Thermooptical absorbency detection of native proteins separated by capillary Electrophoresis in 10μM ID tubes, *J. Chromatogr. A*, 699, 315, 1995.
133. Waldron, K. C. and Dovichi, N. J., Sub-femtomole determination of phenylthiohydantoin-amino acids: Capillary electrophoresis and thermooptic detection, *Anal. Chem.*, 64, 1396, 1992.
134. Yu, M. and Dovichi, N. J., Sub-femtomole determination of dabsyl-amino acids with capillary zone electrophoresis separation and laser-induced thermooptical absorbance detection, *Mikrochim. Acta*, 3, 27, 1988.
135. Bruno, A. E., Paulus, A., and Bornhop, D. J., Thermooptic absorption detection in 25 micrometer ID capillaries—Capillary electrophoresis of Dansyl-amino acids mixtures, *Appl. Spectrosc.*, 45, 462, 1991.
136. Ragozina, N. Y., Putz, M., Heissler, S., Faubel, W., and Pyell, U., Quantification of etoposide and etoposide phosphate in human plasma electrokinetic by micellar chromatography and near-field thermal lens detection, *Anal. Chem.*, 76, 3804, 2004.
137. Ragozina, N. Y., Putz, M., Heissler, S., Faubel, W., and Pyell, U., Quantification of etoposide and etoposide phosphate in human plasma electrokinetic by micellar chromatography and near-field thermal lens detection, *Anal. Chem.*, 76, 3804, 2004.
138. Huang, X. and Ren, J., Chemiluminescence detection for capillary electrophoresis and microchip capillary electrophoresis, *Trends Anal. Chem.*, 25, 155, 2006.
139. Hashimoto, M., Tsukagoshi, K., Nakajima, R., Kondo, K., and Arai, A., Microchip capillary electrophoresis using on-line chemiluminescence detection, *J. Chromatogr. A*, 867, 271, 2000.
140. Mangru, S. D. and Harrison, D. J., Chemiluminescence detection in integrated post-separation reactors for microchip-based capillary electrophoresis and affinity electrophoresis, *Electrophoresis*, 19, 2301, 1998.
141. Hara, T., Nishida, H., Kayama, S., and Nakajima, R., Capillary zone electrophoretic separation of protein labeled with rhodamine B isothiocyanate and its on-line chemiluminescence detection, *Bull. Chem. Soc. Japan*, 67, 1193, 1994.
142. Hara, T., Okamura, S., Kato, S., Yokogi, J., and Nakajima, R., Separation and detection of protein as a supramolecular complex using capillary electrophoresis, *Anal. Sci.*, 7, 261, 1991.
143. Hara, T., Yokogi, J., Okamura, S., Kato, S., and Nakajima, R., On-line chemiluminescence detection of proteins separated by capillary zone electrophoresis, *J. Chromatogr.*, 652, 361, 1993.
144. Liu, Y. M., Liu, E. B., and Cheng, J. K., Ultrasensitive chemiluminescence detection of sub-fM level Co(II) in capillary electrophoresis, *J. Chromatogr. A*, 939, 91, 2001.
145. Zhao, S., Xie, C., Lu, X., Song, Y., and Liu, Y. M., A facile and sensitive chemiluminescence detection of amino acids in biological samples after capillary electrophoretic separation, *Electrophoresis*, 26, 1745, 2005.

146. Hendrickson, H. P., Anderson, P., Wang, X., Pittman, Z., and Bobbitt, D. R., Compositional analysis of small peptides using capillary electrophoresis and Ru(bpy)3/3+-based chemiluminescence detection, *Microchem. J.*, 65, 189, 2000.
147. Tsukagoshi, K., Nakahama, K., and Nakajima, R., Direct detection of biomolecules in a capillary electrophoresis-chemiluminescence detection system, *Anal. Chem.*, 76, 4410, 2004.
148. Tsukagoshi, K., Shikata, Y., Nakajima, R., Murata, M., and Maeda, M., Analysis of a biopolymer by capillary electrophoresis with a chemiluminescence detector using a polymer solution as the separation medium, *Anal. Sci.*, 18, 1195, 2002.
149. Huang, X. and Ren, J., Chemiluminescence detection for capillary electrophoresis and microchip capillary electrophoresis, *Trends Anal. Chem.*, 25, 155, 2006.
150. Huang, X. J. and Fang, Z. L., Chemiluminescence detection in capillary electrophoresis, *Anal. Chim. Acta*, 414, 1, 2000.
151. Forbes, G. A., Nieman, T. A., and Sweedler, J. V., On-line electrogenerated $[Ru(bpy)_3]^{3+}$ chemiluminescent detection of beta-blockers separated with capillary electrophoresis, *Anal. Chim. Acta*, 347, 289, 1997.
152. Li, J., Yan, Q., Gao, Y., and Ju, H., Electrogenerated chemiluminescence detection of amino acids based on precolumn derivatization coupled with capillary electrophoresis separation, *Anal. Chem.*, 78, 2694, 2006.
153. Ding, S. N., Xu, J. J., and Chen, H. Y., Tris(2,2′-bipyridyl)ruthenium(II)-zirconia-Nafion composite films applied as solid-state electrochemiluminescence detector for capillary electrophoresis, *Electrophoresis*, 26, 1737, 2005.
154. Yin, X. B., Dong, S., and Wang, E., Analytical applications of the electrochemiluminescence of tris (2,2/-bipyridyl) ruthenium and its derivatives, *Trends Anal. Chem.*, 23, 432, 2004.
155. Wang, X. and Bobbitt, D. R., *In situ* cell for electrochemically generated Ru(bpy)33+-based chemiluminescence detection in capillary electrophoresis, *Anal. Chim. Acta*, 383, 213, 1999.
156. Chiang, M. T. and Whang, C. W., Tris(2,2/-bipyridyl)ruthenium(III)-based electrochemiluminescence detector with indium/tin oxide working electrode for capillary electrophoresis, *J. Chromatogr. A*, 934, 59, 2001.
157. Kaniansky, D., Marak, J., Rajec, P., Svec, A., Koval, M., Lucka, M., and Sabanos, G., On-column radiometric detector for capillary isotachophoresis and its use in the analysis of C-14-labeled constituents, *J. Chromatogr.*, 470, 139, 1989.
158. Kaniansky, D., Rajec, P., Svec, A., Havasi, P., and Macasek, F., Online radiometric detection in capillary isotachophoresis. 1. Preliminary experiments, *J. Chromatogr.*, 258, 238, 1983.
159. Kaniansky, D., Rajec, P., Svec, A., Marak, J., Koval, M., Lucka, M., Franko, S., and Sabanos, G., On-column radiometric detector for capillary isotachophoresis, *J. Radioanal. Nucl. Chem.*, 129, 305, 1989.
160. Altria, K. D., Simpson, C. F., Bharij, A. K., and Theobald, A. E., A gamma-ray detector for capillary zone electrophoresis and its use in the analysis of some radiopharmaceuticals, *Electrophoresis*, 11, 732, 1990.
161. Pentoney, S. L., Zare, R. N., and Quint, J. F., Online radioisotope detection for capillary electrophoresis, *Anal. Chem.*, 61, 1642, 1989.
162. Klunder, G. L., Andrews, J. E., Grant, P. M., Andresen, B. D., and Russo, R. E., Analysis of fission products using capillary electrophoresis with on-line radioactivity detection, *Anal. Chem.*, 69, 2988, 1997.
163. Page, J. S., Rubakhin, S. S., and Sweedler, J. V., Single-neuron analysis using CE combined with MALDI MS and radionuclide detection, *Anal. Chem.*, 74, 497, 2002.
164. Miao, H., Rubakhin, S. S., Scanlan, C. R., Wang, L. P., and Sweedler, J. V., D-aspartate as a putative cell-cell signaling molecule in the *Aplysia californica* central nervous system, *J. Neurochem.*, 97, 595, 2006.
164. Jarman, J. L., Todebush, R. A., and De Haseth, J. A., Glass nebulizer interface for capillary electrophoresis—Fourier transform infrared spectrometry, *J. Chromatogr. A*, 976, 19, 2002.
165. Todebush, R. A., He, L. T., and De Haseth, J. A., A metal nebulizer capillary electrophoresis/fourier transform infrared spectrometric interface, *Anal. Chem.*, 75, 1393, 2003.

166. Vonach, R., Lendl, B., and Kellner, R., High-performance liquid chromatography with real-time Fourier-transform infrared detection for the determination of carbohydrates, alcohols and organic acids in wines, *J. Chromatogr. A*, 824, 159, 1998.
167. Kolhed, M., Hinsmann, P., Svasek, P., Frank, J., Karlberg, B., and Lendl, B., On-line fourier transform infrared detection in capillary electrophoresis, *Anal. Chem.*, 74, 3843, 2002.
168. Kolhed, M. and Karlberg, B., Capillary electrophoretic separation of sugars in fruit juices using on-line mid infrared Fourier transform detection, *Analyst*, 130, 772, 2005.
169. Dijkstra, R. J., Ariese, F., Gooijer, C., and Brinkman, U. A. T., Raman spectroscopy as a detection method for liquid-separation techniques, *Trends Anal. Chem.*, 24, 304, 2005.
170. Dijkstra, R. J., Gerssen, A., Efremov, E. V., Ariese, F., Brinkman, U. A. T., and Gooijer, C., Substrates for the at-line coupling of capillary electrophoresis and surface-enhanced Raman spectroscopy, *Anal. Chim. Acta*, 508, 127, 2004.
171. He, L., Natan, M. J., and Keating, C. D., Surface-enhanced Raman scattering: A structure-specific detection method for capillary, *Anal. Chem.*, 72, 5348, 2000.
172. Nirode, W. F., Devault, G. L., Sepaniak, M. J., and Cole, R. O., On-column surface-enhanced Raman spectroscopy detection in capillary electrophoresis using running buffers containing silver colloidal solutions, *Anal. Chem.*, 72, 1866, 2000.
173. Olson, D. L., Peck, T. L., Webb, A. G., Magin, R. L., and Sweedler, J. V., High-resolution microcoil 1H-NMR for mass-limited, nanoliter-volume samples, *Science*, 270, 1967, 1995.
174. Peck, T. L., Magin, R. L., and Lauterbur, P. C., Design and analysis of microcoils for NMR microscopy, *J. Magn. Reson. B*, 108, 114, 1995.
175. Wu, N., Peck, T. L., Webb, A. G., Magin, R. L., and Sweedler, J. V., 1H-NMR spectroscopy on the nanoliter scale for static and on-line measurements, *Anal. Chem.*, 66, 3849, 1994.
176. Wolters, A. M., Jayawickrama, D. A., Larive, C. K., and Sweedler, J. V., Capillary isotachophoresis/NMR: Extension to trace impurity analysis and improved instrumental coupling, *Anal. Chem.*, 74, 2306, 2002.
177. Wolters, A. M., Jayawickrama, D. A., Webb, A. G., and Sweedler, J. V., NMR detection with multiple solenoidal microcoils for continuous-flow capillary electrophoresis, *Anal. Chem.*, 74, 5550, 2002.
178. Jayawickrama, D. A. and Sweedler, J. V., Dual microcoil NMR probe coupled to cyclic CE for continuous separation and analyte isolation, *Anal. Chem.*, 76, 4894, 2004.
179. Rapp, E., Jakob, A., Schefer, A. B., Bayer, E., and Albert, K., Splitless on-line coupling of capillary high-performance liquid chromatography, capillary electrochromatography and pressurized capillary electrochromatography with nuclear magnetic resonance spectroscopy, *Anal. Bioanal. Chem.*, 376, 1053, 2003.
180. Jayawickrama, D. A. and Sweedler, J. V., Hyphenation of capillary separations with nuclear magnetic resonance spectroscopy, *J. Chromatogr. A*, 1000, 819, 2003.
181. Webb, A. G., Nuclear magnetic resonance coupled microseparations, *Mag. Reson. Chem.*, 43, 688, 2005.

10 Microfluidic Devices for Electrophoretic Separations: Fabrication and Use

Lindsay A. Legendre, Jerome P. Ferrance, and James P. Landers

CONTENTS

10.1 INTRODUCTION

The first miniaturized analytical device reported was a gas chromatographic analyzer fabricated from silicon in 1979.[1] The work showed the separation of a simple mixture of compounds in only a few seconds. While unperfected, this was an example unlike anything else reported in the literature at that time, and while the scientific community did not immediately grasp the achievements of Terry and coworkers, they did see a glimpse of the future provided in this work with respect to the value of miniaturizing analytical methods. Multiple papers followed this initial work in miniaturization over the next 12 years[2–11]; however, it was not until the concept of a total analysis system (TAS) was first proposed by Manz et al.[12] in 1990 that the concept of miniaturization began to have a significant impact.

The TAS concept was originally motivated by a lack of adequate sensors for the detection of specific species from a complex mixture. It was hypothesized in this seminal paper[12] that, by improving the sample treatment steps, an ultrasensitive sensor would not be required if interfering chemical compounds were removed. A TAS, as proposed, entailed initial sampling, transport of the sample, sample pretreatment steps, and the final detection of the analyte. Miniaturization of all

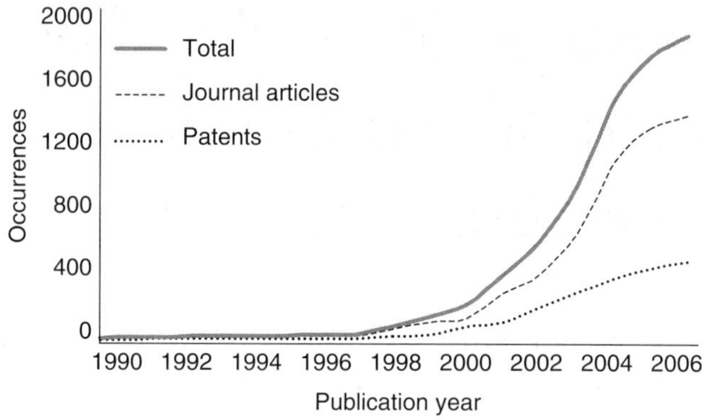

FIGURE 10.1 An increase in microfluidics publications. A search was performed using the term "microfluidics" and the number of publications was graphed as a function of publication year. The data shown represent the total number of journal articles (- - -), patents (· · ·), and the sum of both (—).

of the processes required for total analysis, a micrototal analysis system (μTAS), would allow the development of a TAS that performs these functions at the site of measurement. Additional benefits are also derived in decreasing the size of a TAS, that include decreased volumes of required sample, decreased reagent consumption and the potential for dramatically reduced analysis times.

From its inception, the driving force behind the miniaturization of electrophoretic separation techniques was related to an enhancement in its analytical performance, rather than a reduction in size.[12] Because separation efficiency is a function of the applied voltage, microfluidic devices utilize shorter separation lengths along with the application of high voltages typically used in capillary-based analysis, resulting in fast, efficient separations. In addition, and perhaps most importantly, microchips also offer the opportunity to integrate sample-processing steps onto the same device used for separations, producing the μTAS initially described by Manz and coworkers. Process integration results in a decrease in total analysis time as well as a reduction in potential contamination because sample handling between processes is eliminated.

Microfluidic devices are becoming increasingly more prevalent in the scientific arena, not only revealed by a rise in the number of reports in the literature but also by the release of commercial products. A graph plotting publications versus time, as shown in Figure 10.1, highlights the increased number of reports using these miniaturized devices since the inception of the original TAS concept. The goal of this chapter is to educate the novice microchip user on fabrication methods and "tricks of the trade" regarding glass and polymer [specifically poly(dimethylsiloxane) or PDMS] microdevices with a view to application of these devices.

10.2 BACKGROUND

The initial experiments transitioning the capillary electrophoretic separation method to a microdevice were performed by Manz et al.[13] This paper, which provides details about the process of fabricating the devices, describes the four different processes necessary for standard one-mask microfabrication; these processes included film deposition, photolithography, etching, and bonding. These processes were used to fabricate a fluidic layer in silicon, which was sealed with a glass cover plate, and then the device was used for the electrophoretic separation of amino acids. The techniques outlined in this original paper for microfabrication in glass and silicon are still used today (and detailed below), along with additional advances required to generate the significantly more complex devices now being reported in the literature. The goal of this background section is to cover the basic definitions and concepts, all of which are addressed in more detail in subsequent sections.

Film deposition refers to the deposition of materials, such as photoresist and metals, onto the wafer. These materials can serve as protective layers during etching, or can act as a master for polymeric devices fabricated using molding. Metal layers are usually deposited using sputtering or vacuum deposition, with chrome being used most frequently because of its avid bonding to both glass and silicon. Photoresists are photoactive materials normally coated on wafers using spin coating to deposit a material onto the surface at a desired thickness. An alternative to conventional spin coating of photoresist has been developed using a constant-volume injection method,[14] which provides an increased pattern definition of the photoresist structures under the conditions investigated. All of these methods require cleanroom conditions and extensive instrumentation to provide either the protective surface that can be patterned for etching or for the development of a master used for PDMS molding.

Once the photoresist has been deposited, the channel geometry required for the final device is "imprinted" onto the photoresist. This is normally performed using a photomask containing the microchip design, through which a ultraviolet (UV) light is passed, exposing the photoresist in the appropriate areas, to transfer the pattern from the mask onto the photoresist. Photolithography, a well-established method from the microelectronics industry, is widely utilized but requires initial production of the photomask. Recent reports describe methods for patterning photoresist without a mask, the most promising of which is laser direct-write systems in which the design is transferred to the photoresist using a laser that travels along the desired pattern on the chip surface creating features based on a schematic contained within the software file used to create the design.[15,16]

With the traditional etching method for fabricating glass or silicon devices, the desired pattern is transferred to the photoresist by exposure of the photoresist to UV light in the pattern designated by the design. This UV-induced damage to the photoresist makes it susceptible to removal with a developer, which is essentially a buffered alkaline solution, exposing the underlying metal layers that cover the wafer. Once these are removed, exposure to hydrofluoric acid (HF) etches the glass or silicon at a rate that can be predetermined based on the substrate used and the concentration of HF used in the etch solution. It is noteworthy to mention that HF is a hazardous acid that must be handled with extreme caution, and appropriate countermeasures must be taken upon exposure. In another work, a combination of photoresist and sacrificial metal layers were used that have an increased resistance to HF and, therefore, allowed even deeper glass etching (>300 μm).[17] Wet chemical etching with HF has also been performed without a mask, using patterned PDMS sealed to the glass with HF flowed through the PDMS channels to etch corresponding channels in the glass substrate.[18] The same concept was also applied to a plastic substrate where an etchant, in place of HF, was pumped through the PDMS channels.[19]

Because of the hazards of HF, alternatives to wet chemical etching have also been reported; Belloy et al.[20] used powder blasting to fabricate high-aspect ratio structures in glass. Their technology uses a beam of powder particles that erodes the protective layers patterned on a substrate; this technique allowed for fabrication of free-standing monolithic glass microstructures. A similar method called deep reactive ion etching (DRIE) was used by Ceriotti et al.[21] to fabricate rectangular channels for UV-detection in fused-silica wafers. DRIE, as well as regular reactive ion etching, involves bombarding the material to be etched with highly energetic chemically reactive ions, typically fluoride, to collide with the material's surface and remove surface atoms in the process. All of these etching processes are used to generate channels with depths in the 1–500 μm range. Pushing the limits of structure size, Hibara et al.[22] fabricated channels with only nanometer dimensions in fused silica by fast atom beam etching; this method uses accelerated neutral particles to bombard the surface. An additional method for the generation of nanometer channels was reported by Ionescu et al.,[23] who used enzymes to create nanometer-sized depressions in a protein surface for nanofluidic photolithography. Proteases were delivered with scanning probe microscopy and were able to precisely control the dimensions of the resulting features.[24]

Once channels/chambers have been etched in to the substrate, it is bonded to a flat cover plate to enclose the channels and complete the fluidic architecture of the microchip. With glass microdevices, thermal bonding is used most frequently, with heating to a temperature that approaches their glass transition temperature while under pressure, allowing the glass plates to fuse. For silicon/silicon, and for glass/silicon bonding, anodic bonding is often employed in which a high voltage is applied between the top and bottom layers, causing the two to fuse. Both of these bonding procedures utilize high temperatures, which may not be desirable if metals or other components are an integral part of the device. Other microchip bonding methods have been reported including techniques utilizing silicate solutions,[25–27] UV-curable adhesive,[28,29] and low-temperature methods,[30–33] many of which are carried out at low temperatures (<150°C), and, therefore, are ideal for surfaces that require modification (i.e., metallization) prior to bonding.

10.3 GLASS MICROCHIP FABRICATION

Glass offers some distinct advantages over other substrates that are used to fabricate microdevices. It is optically-transparent through much of the visible and UV spectrum, its surface chemistry is reasonably well-characterized, it is compatible with most solvents, and it can withstand high temperature applications due to its high glass transition temperature. One of the most significant drawbacks associated with using glass as a substrate is the apparent requirement for a cleanroom-based fabrication facility. However, while cleanroom environments are useful if available, they are not a necessity as glass devices can be fabricated outside a cleanroom. In line with this, the need to sputter metals and spin-coat photoresist can be avoided with commercially available silicon and glass wafers that are precoated with the necessary layers in a variety of sizes and material, in many ways eliminating the need for sophisticated facilities. For the photolithography step, a laminar flow hood can be used to protect the surface and mask from dust during the exposure, ensuring exact replication of the mask image in the photoresist. Yellow lighting is also required to prevent unwanted exposure of the photoresist, and all the postexposure steps can be performed in an acid-resistant hood (e.g., polypropylene). More detailed methods for the fabrication steps, beginning with pre-coated wafers, are presented in the following sections.

10.3.1 PHOTOLITHOGRAPHY AND WET ETCHING OF GLASS MICROCHIPS

Glass microchips are traditionally fabricated using standard photolithography techniques exploiting either a film or metal masks (Photosciences, San Jose, CA) for design transfer, and wet chemical etching for structure creation (Figure 10.2). For both types of masks, the chip image is created using software (e.g., AutoCAD or Adobe Illustrator) that contains the design for the channels, chambers, and reservoirs desired in the microfluidic chip. The computer image can be transferred lithographically onto a film by most commercial printing companies using a high-resolution printer (~2500 dpi)—this makes the mask generation inexpensive (a few tens of dollars), allowing the user to quickly and easily modify the design with successive iteration. While the resolution of the printer restricts designs to features greater than about 20 μm (width), in most cases, microfluidic channels range from 40 to 200 μm deep and, thus, film masks prove adequate for most applications. Creating a metal mask is a more costly endeavor (hundreds of dollars per mask) and is typically done when either higher resolution structures are required or the user has defined a final design that is tried and true. This is created through the fabrication of the microfluidic architecture into a glass substrate, followed by a metallization step to increase the robustness of the mask. While these metal-on-glass masks are more expensive, they can be used repeatedly without degradation (unless dropped), and allow for higher resolution structures (<20 μm initial channel width) to be fabricated.

Borosilicate glass has traditionally been used as substrate for glass microchips (e.g., Schott 1.1-mm-thick BOROFLOAT), but quartz[34] and soda lime[35] glass have also been used. Borosilicate

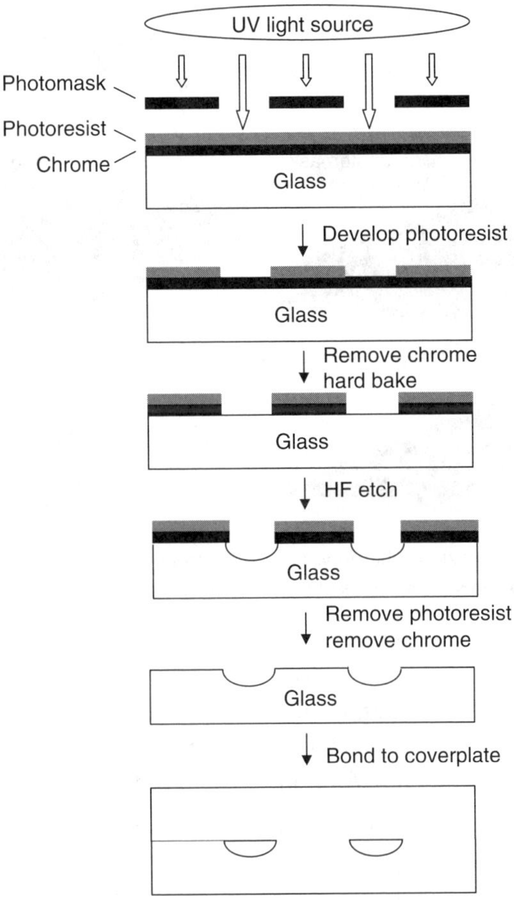

FIGURE 10.2 A diagram illustrating the processes involved in microchip fabrication using standard photolithography and wet chemical etching methods.

glass plates can be obtained commercially, precoated with sacrificial chromium (<1 μm thickness) layered underneath a coating of photoresist. To transfer the image from the photomask to the photoresist, a high power (e.g., 350 W) UV exposure system is normally used, with the exposure time being dependent on the thickness and type of photoresist; a 5 second exposure works well with the commercially coated plates described above. The UV exposure system can be used with a mask aligner, but is only essential if exact alignment is required, or if multiple exposures are to be used; extended low power UV sources can also be sufficient, but could require exposure times. For exposure, the photomask is placed over the coated wafer and the photoresist irradiated with UV light through the mask. If a film mask is used (flimsy in nature), a clean bare glass plate is placed over it to obtain intimate contact with the surface of the photoresist (i.e., to hold it flat) and ensure accurate transfer of the image. The exposed glass is placed in a developer solution and, in the case of a positive photoresist, the photoresist that was exposed to the UV radiation is removed, revealing the underlying chromium layer (see Figure 10.3). In contrast, negative photoresist becomes insoluble where the UV radiation was incident,[36] therefore, the photoresist not exposed to the UV would be removed. In general, either a positive or a negative photoresist can be used as long as the photomask is designed accordingly. For most photoresists, a "hard bake" is required after the photoresist has been developed—this stabilizes it for further processing and eliminates additional light sensitivity. The time and temperature for hard baking varies based on the choice of photoresist, but normally these range around 110°C for 30 min and this can be carried out on a standard laboratory hot plate.

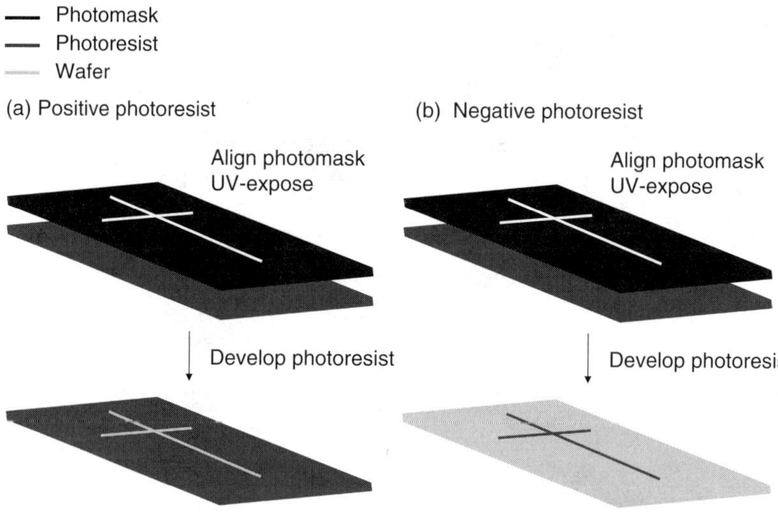

FIGURE 10.3 Positive and negative photoresists. An illustration demonstrating the difference between (a) positive and (b) negative photoresists when used in photolithograpy.

TABLE 10.1
Empirically Determined Glass Etch Rates Based on the Amount of HF Used

%HF	Etch Rate (μm/min)
49	10
45	8
25	3
10	0.6

The chromium layer exposed by removal of the photoresist is dissolved using a chromium etchant solution to reveal the underlying glass; all of the solutions used in the photolithography process (developer, chromium etchant, stripper) are commercially-available from a number of sources. Once the chrome is removed, the glass plate is immersed in an HF solution, which will etch away the exposed glass. The etching rate is dependent on the HF concentration of the solution and the type of glass. Borosilicate glass etches very slowly, and higher concentrations of HF are normally used with this glass; soda lime glass etches rapidly due to the high concentration of ions in the glass, and buffered etch solutions utilizing ammonium fluoride are often used to control the etch rate. The etch rate is also dependent on the final depth required for the features in the device, as long exposures to HF solution begin to consume the photoresist and chrome protective surfaces, causing pitting in the areas where etching was not desired. At the same time, etching at excessively rapid each rates can cause irregular surfaces to be formed on the bottom and walls of the channel, and other components are often included in the etch solution to ameliorate this effect. An exemplary etch solution for borosilicate glass is 10% HF/10% HNO_3 in water and provides an etch rate of approximately 0.6 μm/min, which will decrease as the solution is used. Table 10.1 provides a general guide for achieving different etch rates by varying the concentration of HF. Gentle rocking during the process also provides a cleaner etched surface, as precipitate forms during etching which can build up in the channel structures unless removed. For carrying out the etching, costly acid-resistant containers are often specified (e.g., Nalgene), but better, lower-cost alternatives can be reusable polypropylene food storage containers; the polypropylene is completely HF resistant, and these containers have tight-fitting, spill-proof lids that make it easy to store the HF solutions when not in use. Once the etching is complete, the protective layer of baked photoresist is removed using a photoresist stripper, and the underlying chrome layer is removed using the chromium etchant.

One important aspect to bear in mind with this process is that glass is isotropically-etched by HF meaning that the glass structure is etched laterally at the same time it is etched deep—this results in hemispherical or U-shaped channels. As a result of this phenomenon, side etching of the glass quickly undercuts the protective chrome layer that has been deposited on top of the substrate, resulting in a top channel width that is slightly larger than what would be expected. This widening at the top occurs in concert with depth etching and must be taken into account when determining the depth of the etch and used accordingly to define the width of the channel designed into the mask. While this places limitations on the possible dimensions of structures that can be attained, and does not allow for high aspect ratio structures in finished glass devices, this can be advantageous in two important ways. The first is that isotropic etching provides a simple and inexpensive method for determining the depth of the etched structures. A microscope containing an eyepiece reticule can be used to measure the channel width at the top of the structure (the distance between the chrome flanking the sides of the structure). The etch depth can be calculated by

$$\text{Depth} = (\text{top width of channel} - \text{original width})/2,$$

where the top width of the channel is measured using the microscope and the original width is known from the mask design, or it can be measured prior to etching. This method provides an approximate depth of the channel, and can be used for most applications. Exact measurements of the etch depth can be obtain using a profilometer in cases where knowing if the depth is critical. A traditional profilometer uses a stylus tip which is mobilized across the surface; its vertical deflection, as it "profiles" the surface, provides measurements of surface features in the nano- to micrometer range. More recently, optical profilometers have been developed that provide similar information without the need for the stylus. In our hands, fabrication involves estimating an etch rate based on the chemical ratio of the etch solution or etch rates reported in the literature for the use of etch solutions of a particular composition. After a specific etch time, the glass plate is removed from the etch solution, immersed into a large volume of water to terminate the etching, then rinsed thoroughly with water, and the depth of the structure measured by profilometry. These measurements are used to empirically determine the etch rate for better control of timing to achieve a specific channel depth.

The second advantage of the isotropic etching is the ability to easily form weirs—an abrupt narrowing of the channel—for retaining beads of the appropriate size. Within the channel design itself, a line across the channel is left unexposed. As the exposed part of the channel etches, the chrome line across it is undercut from both sides by the side etching of the HF. As long as the line is less than twice the final depth, it will be completely undercut and removed during the etching process. Careful predetermination of the width of the line allows the final height of the weir to be controlled (i.e., when the etching from the two sides connect). A less reproducible method for generating weirs, but useful for etching different depths on a device with a single exposure, is to cover parts of the exposed glass with an HF resistant tape during the initial part of the etching process. The deeper desired features can be etched part way and the tape removed for the remaining etch time.

10.3.2 Drilling and Dicing Glass Microchips

Once the etched layer containing the desired microchannel structures is completed, access holes at the entrance and exit location of each channel must be fabricated in either the etched plate or a cover plate. The simplest and most popular method involves drilling holes through the glass with a 1.1 mm diamond-tipped drill bit using a simple Dremel tool and miniature drill press, both of which are available at a local hardware store.[37] Slight submersion of the glass plate in water while drilling prevents heating of the drill bit, eliminating wear of the cutting surface and allowing the bits to be reused multiple times (however appropriate precautions must be taken, as this is a power tool connected to a high-voltage power source). Silicon carbide drill bits can also be used for drilling glass and are also readily available in a variety of diameters. More complicated drilling can be achieved using a computer numerical controlled (CNC) mill, which is a more expensive option for drilling holes, but may be a good choice if accurate placement of the hole location is required or if a large number of holes are required in a device.[38] A simple program guides the CNC mill to each of multiple locations on a glass substrate in the X–Y plane, with an accuracy in a simple system down to ~5 μm, as well as controlling the speed and drilled depth (the Z-axis). While less efficient on simple chip designs, or initial prototypes because of the setup time involved (alignment in particular), it is extremely useful for device designs that contain multiple access holes or designs that will be utilized multiple times. We have utilized this system to drill holes in a wafer containing 8 chips each with 20 holes.[39] Alternative methods for generating holes in glass substrates are also available, such as laser ablation,[40] but these are mainly used with substrates that are are not amenable to standard drilling or in situations where very small holes are required.

In addition to drilling, cutting the glass plates is often required. Glass substrates used to fabricate microchips are commonly 4″–6″ in diameter—those etched in our laboratory are normally 5″ × 5″ in size—but the final dimensions of the devices are significantly smaller. Consequently, each mask will contain multiple devices to be fabricated in a single glass plate during the etching process. Each of the individual devices, along with the corresponding cover plates, needs to be diced prior to the bonding step. Scoring and snapping is the traditional method for cutting glass, which utilizes a diamond-tipped scoring tool for scoring the glass on one side, followed by application of pressure to the other side so as to snap it. While this method is simple and easy to employ, it is not foolproof as the glass does not always break along the scored line, and requires both practice and confidence to master. A more accurate and reliable way to dice the microchips utilizes a glass-cutting saw with a thin diamond-edged blade. While wafer saws created for the microelectronics industry are available for this purpose, a simple and inexpensive option is to use a commercial ceramic tile saw (available at hardware stores). The blade on a tile saw is normally designed for grinding away the tile during cutting, but this blade is easily replaced with a thin diamond-edged blade designed for the more expensive wafer saws. This provides a simple method for cutting the glass wafers into individual devices without worrying about cracking or breaking the etched plate at unwanted locations.

10.3.3 Glass Microchip Bonding Techniques

The most common glass bonding technique is thermal bonding, where the etched glass chip and cover plate are properly aligned, brought into intimate contact and placed into a programmable, high temperature oven where heating to 640–690°C for roughly 6–8 hours is required for annealing the glass. While thermal bonding is simple in principle, it is the slowest and, arguably, the most unpredictable step in microchip fabrication procedure. There are two main aspects that play a role in achieving successful thermal bonding, cleanliness and contact. Any organic residue on the glass plates will prevent them from annealing, thus the surfaces of the etched and cover plates are often cleaned using piranha wash (3:1 mixture of concentrated H_2SO_4:H_2O_2) before assembly. An easier method involves simply wiping the plates with an ammonia-based window cleaner to thoroughly clean the glass surface; however, use of this method may result in partially unbonded devices that will never bond completely. Wiping the glass with lint-free wipes prior to assembly is advantageous as any glass particles that might have adhered to the surface during the glass cutting or drilling procedures will be removed. Care must be taken in wiping the glass plates in this manner as the edges can be sharp. Once cleaned, the top and bottom must be aligned and pressed together to achieve intimate contact between the two plates. This is easily detected by the formation and disappearance of "Newton rings" (rainbow circles) that form when the distance between the two plates approaches the wavelengths of visible light. Although the float glass is manufactured to be optically flat, the processing alters the surface enough that significant pressure (under heat) is required to achieve this intimate contact. This can be achieved with small individual weights or a press, all the while, careful not to stress the glass to the point that it breaks. Simply squeezing the glass plates between ones thumbs can often achieve some areas of intimate contact, and a drop of water to wet the surface and keep the plates aligned during oven loading will also increase the chances of success. Dust or glass particles between the plates are easily detected after assembly (Newton rings), at which point, the plates must be disassembled and cleaned. Putting the plates together in a cleanroom or laminar flow hood assists in avoiding some of these problems, but equal success can be achieved directly in the laboratory environment. The assembled device (etched and cover plates) are placed between two polished graphite sheets or ceramic plates in a high temperature furnace, with or without a weight (~5 kg) placed on top to apply additional pressure during the heating process. The bonding cycle is normally performed in four steps, a rapid heating up to about 550°C, a slow heating up to the annealing temperature (640–690°C), a hold at the annealing temperature (3–10 h), and an unforced cool-down step that allows the glass to cool slowly back to room temperature.

Multiple bonding cycles are often used to generate devices that are completely bonded. The annealing temperature in the step program is normally raised 5°C with each subsequent bonding step, as long as the melting temperature of the glass is not exceeded. Two issues to be aware of in the high temperature bonding procedure are that the graphite or ceramic used to sandwich the fabricated devices must be polished, otherwise imperfections or roughness in these may be transferred to the outer surface of the glass devices, giving them a frosted appearance or creating unwanted features in the surface that can interfere with optical detection and visualization of the channels inside. In addition, above certain temperatures (e.g., ~670°C with borofloat glass), the glass will fuse to the ceramic plates during the annealing process. To prevent this, a liquid graphite coating is applied to the ceramic plates, allowed to dry, then polished flat before the glass chips are placed between them.

In addition to high temperature bonding, a number of other bonding procedures have been reported (described in the Section 10.2). These procedures are most often used when additional functionality is to be incorporated into the chip. Most often this is metal sputtered onto the microchip architecture for electrodes or electrochemical detection, but could also include modification to the channel surface or incorporation of materials that cannot withstand the high bonding temperatures. In these cases, alternative cleaning methods may also need to be employed and the protocol for microchip bonding altered.[41]

10.3.4 Postbonding Cleaning of Glass Microdevices

Owing to the high temperatures and long dwell times associated with the thermal bonding steps, the surface of the glass within the channel becomes dehydrated, which will ultimately affect not only the electroosmotic flow (EOF) in the channel, but also reactions on the glass surface important for permanently or dynamically coating the channels. Rehydrating the glass surface involves increasing the ability to wet the glass usually through the formation of hydroxyl groups in an isolated or geminal formation; this results in an increase in reproducible electrophoretic separations and allows for consistent coating of the channels. While many techniques have been suggested as effective for cleaning glass microchannels, adequate rehydration is required after the devices are bonded, and a simple rinse with NaOH is not sufficient for this purpose. Cras et al.[42] evaluated nine different popular techniques for glass cleaning and preparing the glass for silanization. We have had success adapting to borofloat, a technique described by Cras et al. for soda lime glass: a 30-min rinse with a 1:1 mixture of methanol:concentrated HCl, a thorough rinse with water, a 30-min rinse with concentrated H_2SO_4, followed by a second rinse with water.

TABLE 10.2
Comparison of Polymers and Glass Properties

	Glass	PDMS	PMMA	PI
Polymer type	NA	Elastomeric	Thermoplastic	Thermoplastic
Glass temperature (°C)	525	−120	106	285
Useful temperature range (°C)	≤500	−40 to 50	−70 to 100	−73 to 240
Thermal conductivity (W/mK)	1.2	0.17 to 0.3	0.186	0.2
Visible transmittance (%)	>90	91	92	87
Surface charge (native)	Yes	Weak	Yes	No
Chemical resistance				
Acid	Excellent	Fair–good	Good	Fair–good
Solvent	Excellent	Poor	Poor	Fair
Alkalis	Excellent	Poor–fair	Excellent	Fair–good

The information for the polymer properties was adapted from Sun and Kwok[43] and Becker and Gartner.[81] The polymers listed include poly(dimethylsiloxane) (PDMS), polymethyl methacrylate (PMMA) and polyimide (PI). The information for the glass properties was obtained from the Schott website, specifically BOROFLOAT® 33.

10.4 FABRICATION OF POLYMERIC MICRODEVICES

The glass fabrication technique detailed above is used extensively, but for many first time microdevice researchers, the rapid prototyping technique first described by Whitesides and coworkers[49] using PDMS provides an easy starting point. Moreover, glass is not the ideal substrate for many applications, and for this reason, a variety of polymers have been utilized for fabrication of microchips. In general, polymers can be classified into three categories based on their properties: thermoplastics, thermoset, and elastomers.[43] Thermoplastics are a class of polymers that consist of unlinked or weakly cross-linked chains, which soften and flow when heated above their glass transition temperature (Tg). Examples of these polymers include polymethyl methacrylate (PMMA), polycarbonate (PC), and polyimide (PI). Thermoset polymers[44] contrast thermoplastics in that they are heavily cross-linked and do not melt or flow upon heating; examples include phenol formaldehyde and vinyl esters. The last category, elastomeric polymers, consists of weakly cross-linked polymer chains that form structures, which remain flexible and elastic after polymerization; an increasingly popular example for microfluidic device fabrication is PDMS. A described list of polymers from each class that have been used as having utility for microfluidic devices and their associated properties, can be seen in Table 10.2. The choice of substrate is dictated by a range of variables, such as the application of interest, the available materials and fabrication, and the solvents/reagents to be used, to name a few. For example, PI does not have an inherent surface charge and channels in PI would require surface treatment in order to perform separations requiring a substantial EOF. This section will focus on PDMS because of its ease of use; chapters in this volume written by DeVoe and Lee (Chapter 33), Carrilho (Chapter 41), Woolley (Chapter 51) and Henry (Chapter 52) provide details on the fabrication of microdevices from other polymeric substrates.

Microdevices fabricated in PDMS have been used for electrophoresis[45] as well as a variety of other applications ranging from cell sorting[46] to combinatorial screening.[47] PDMS is a heterogeneous polymer exhibiting methyl on the surface, which imparts significant hydrophobicity; its surface structure is illustrated in Figure 10.4. It is flexible, inexpensive, and optically transparent (down to 280 nm) making it compatible with many optical detection methods.[48] Owing to the flexibility of PDMS, it easily forms reversible bonds with another flat surface through van der Waals interactions. The seal can easily be broken by peeling the PDMS from the surface intact, thus yielding a single piece reusable in this manner; irreversible bonds can also be formed between PDMS and a number of other materials including glass and PDMS itself (discussed in detail below).[48,49] Owing to the

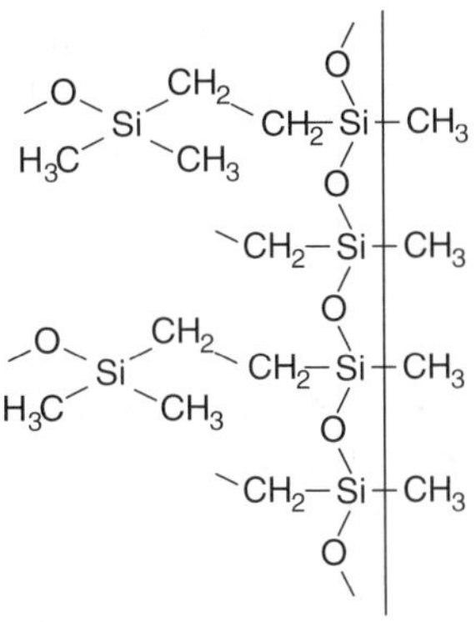

FIGURE 10.4 Chemical structure of PDMS at the surface. (Adapted from http://mrsec.wisc.edu/Edetc/background/PDMS/index.html)

popularity of PDMS for microfluidic devices, a number of reviews have been written[43,48,50] detailing the procedures for making and using these devices. However, one must be cognizant of the drawbacks associated with the hydrophobic nature of the polymer.[51]

Numerous methods for device fabrication, based on the particular polymer selected, have been reported.[43] For fabrication of PDMS microdevices, rapid prototyping[49] and replica molding are two processes that are readily used. These methods allow for fabrication of microdevices in a matter of hours, with the only drawback being the inability to make channels smaller than 20 μm. For rapid prototyping, a photoresist is spin coated onto a wafer in the same way it is for as with the glass fabrication; however, the photoresist for this application must be as thick as the desired depth of the channels and a photoresist created from SU-8 is used most often for this purpose. Photolithography is used to transfer the fluidic design from the mask to the photoresist, but in this case, the SU-8 remains on the wafer where the channels and chambers are desired in the PDMS surface. With this approach the SU-8 forms a master that can be used repeatedly for replica molding to fabricate devices. This involves simply casting the polymer over the master to make the fluidic layer that can be sealed to another piece of PDMS or a glass cover plate.

10.4.1 Fabrication of the Master

A common method for fabricating PDMS microfluidic devices is to use a silicon wafer and SU-8 photoresist. To assist in adhesion of the photoresist to the Si wafer, a cleaning step and/or a surface treatment with hexamethyldisilazane (HMDS) is performed prior to spinning on the photoresist. Placing the wafer into a 10% (v/v) HF bath for roughly 1–2 min will remove surface silanols, rendering the surface hydrophobic; piranha wash has also been used for this purpose.[52] Treatment of Si wafers with HMDS can be performed simply by rinsing the wafer with methanol, spinning a small amount of HMDS onto the surface and baking at 105°C for 5 min before spinning on the photoresist. Either method should provide a surface adequate for adhesion between the Si wafer and SU-8. When spinning the photoresist onto a blank Si wafer, the viscosity of the SU-8 as well as the speed and duration of the spin will dictate the thickness of the SU-8 layer (and therefore the depth of the channels in the PDMS). SU-8 is available in a number of formulations designed specifically for achieving particular photoresist thickness upon spin coating. A soft bake is then performed to evaporate the solvent from the SU-8 layer and improve the adhesion of the resist to the Si wafer—this step can be accomplished on a hot plate and is best performed at 95°C (the heating time is dependent on the thickness of the photoresist). Standard photolithography methods are used to transfer the image from the mask to the photoresist, where the exposed SU-8 is cross-linked by the UV light to make it insoluble in the developer (propyleneglycolmethylether acetate-PGMEA). The master is then placed into the developer to remove the uncrosslinked photoresist, leaving the design that is desired in the PDMS as a raised structure on the surface. The final step involves a hard bake of the master (done at 105°C for at least 15 minutes depending on the thickness of the photoresist) which stabilizes the photoresist. To assist in removal of cured PDMS from the master, the SU-8 master can be silanized through using vapor deposition by simply placing it into a chamber with a tridecafluoro-1,1,2,2-tetrahydrooctyl-1-trichlorosilane solution, essentially making the surface a nonstick-type coating.

For all steps that involve heating, there are several noteworthy details. Attention must be paid to the flatness of the surface used for the soft and hard baking steps, particularly if a hot plate is used. If the surface is not heated evenly, there could be an uneven distribution of the SU-8 layer, causing unwanted variation in structure depth in the PDMS microchips fabricated from the master. The transition rates associated with the heating and cooling are also important and should not occur rapidly. The coefficients of thermal expansion for SU-8 and the Si wafer are very different, and if the heating/cooling step is expedited, the SU-8 layer could delaminate from the wafer. In our experience with using a hot plate for heating, it may be best to use a separate hot plate set to 65°C for an intermediate heating step prior to bringing the master to 95°C on a second hot plate.[52] The master need only be exposed to 65°C for about 1–2 min before transfer to the higher temperature.

This method for fabrication of a master is the traditional method, but does require access to a cleanroom and instrumentation for spin coating the photoresist to the desired thickness as well as exposing the photoresist to UV light. A master for PDMS molding can be fabricated out of any material; however, be aware that there are predefined limitations on the size of the features that are set by the method used to generate the master. Micromachining capabilities can be used to generate a master from metal substrates or other polymers (such as PMMA) that work just as well as the traditional method, but removal of PDMS from the master after the casting process can be an issue. We have found micromachined Teflon to be an excellent substrate for fabrication of masters, particularly when it comes to ease of removal of the PDMS. A distinct advantages of this approach is that the mold can be heated to more rapidly cure the PDMS without destroying or altering the master in the process.

10.4.2 PDMS Device Fabrication

Assuming a master of sufficient quality has been created, fabrication of the PDMS microdevice can be attempted. PDMS is purchased as a two-component system consisting of the elastomeric base and a curing agent, both in liquid form at room temperature. Sylgard 184 (Dow Corning) is the most widely-used PDMS formulation, which uses a base:curing agent ratio of 10:1 (by weight as recommended by the manufacturer). Harder or stickier polymers can be created by varying the ratio of the two as needed for a particular application. Once the two components are mixed thoroughly, the mixture should be degassed to remove any air bubbles; this can be accomplished by placing the PDMS into a vacuum chamber and applying adequate vacuum to gently degas the PDMS. The master should be placed in a container to hold the liquid PDMS, which is then slowly poured over the master (taking care to avoid introducing bubbles) and allowed to cure. At room temperature, at least 18 hours is required for the PDMS to completely crosslink, but this time can be reduced by increasing the temperature. For example, at 40°C the PDMS will cure in less than 4 h, and the process will only consume 2 hours at 70°C. A PDMS cover plate for the device can be fabricated simultaneously, by pouring additional PDMS over a blank Si wafer and allowing it to cure at the same time. Once the PDMS has cured, the device can be easily cut to size with a razor blade, and access holes introduced into the device using a simple hole punch.

10.4.3 PDMS Bonding

Bonding of PDMS devices, in general, is much simpler than the thermal bonding required for glass devices. To obtain a reversible bond between PDMS and any number of complimentary substrates (glass, silicon, or another piece of PDMS), a cleaning step must be performed to ensure that the seal between the fluidic layer and the cover plate is water tight, at least to the extent that it can withstand the pressures associated with flow through the device. Cleaning the surfaces to assure no dust or particles are present on either surface is a must, followed by rinsing the PDMS thoroughly with ethanol and allowing to dry. The layers are then gently pressed together manually, starting at one corner and making sure no air bubbles are trapped between the two surfaces. Because the bonding is reversible, if air bubbles do appear, or if particles remain on the surface, the layers can be separated (by simply peeling them apart), recleaned, and rebonded. A more permanent seal can also be obtained, both by activation of the PDMS surface using either a plasma oxidizer (plasma cleaner)[48,49] or a UV source that generates ozone near the surfaces to be activated. This activation of the surface is immediately followed by manually pressing the layers together, allowing the two surfaces to react and form a permanent bond; curing the devices at 80°C for 2.5 hours aids in improving the avid nature of the bonding.

A clever alternative bonding procedure for sealing PDMS to itself was reported by Quake and coworkers and involves altering the elastomer:base ratio in the two layers.[53] They utilized a ratio of 20:1 for the fluidic layer and 5:1 for the cover layer, so that when the device was oven-bonded, the two layers fused together as a result of the excess curing agent in one layer and excess base in the other. This method of bonding completely eliminates the seam between the two layers, negating the opportunity for fluid leakage between the layers.

10.4.4 PDMS Surface

After the PDMS device has been fabricated, the separation channel (or other structures) may require surface modification to suit the needs of the chemistry desired. Unmodified PDMS is hydrophobic with no charge,[51] therefore yielding a very weak EOF. One approach to decreasing the hydrophobicity in PDMS surface is to invoke the use of the plasma cleaning method prior to bonding. Through the introduction of ionizable silanols created by the plasma, this method temporarily renders the surface hydrophilic, allowing for direct use in this state or providing the option of further surface modification. However, the user should be aware that this effect is only temporary, lasting several tens of minutes to a few hours (depending on the system and the oxidation process)—consequently, expediency in cleaning and assembling is advised.[54] For more information about altering the surface of PDMS, please refer to Chapter 52 by Henry and coworkers.

10.5 PERFORMING MICROCHIP ELECTROPHORESIS

From the earliest inception of the microchip as an analytical platform,[1,55] rapid separations were demonstrated. The major components required for microchip-based electrophoresis (ME) include a detection system, a power supply, and a computer with programmable software for data collection and control of the applied voltages and data collection/manipulation; a schematic diagram illustrating these components can be seen in Figure 10.5. While there are a number of different detection systems that can be utilized with microchips (see Chapter 45 by Karlinsey and coworkers), a fluorescence detection system (a homemade fluorescence microscope at its core), is the most widely utilized. There are several parameters to consider when choosing a power supply, including the number of voltage outputs required and the range of electric fields that are desired. For a standard "cross-t" chip design, a minimum of four voltage outputs are required to control the sample injection and separation. As the complexity of the chip design increases, more outputs may be necessary. One of the benefits of glass devices is the inherent ability to dissipate heat; consequently, reasonably large voltages can be applied and dual-polarity, high-voltage supplies are typically used. In addition, the holes drilled into the microfluidic device may not be large enough to accommodate a sufficient volume of buffer to allow for electrical contact with the platinum electrodes during the separation. The simplest solution

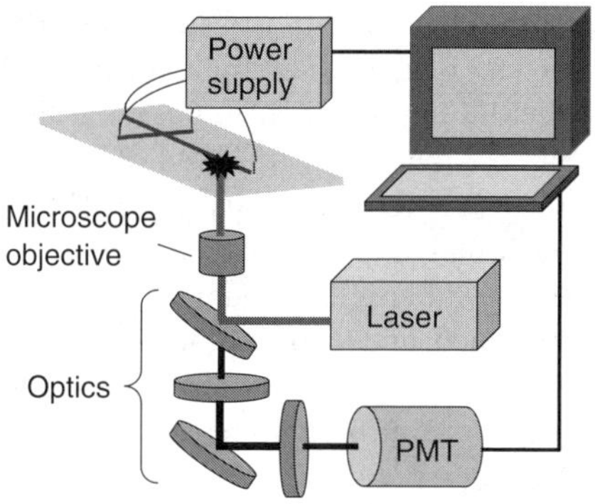

FIGURE 10.5 A schematic diagram of an LIF system. The laser beam is reflected up by a dichroic filter and focused into the separation channel by a microscope objective. The resultant excitation passes back down through the dichroic and bandpass filters, and is directed onto the PMT. The computer is used to collect a voltage output from the PMT proportional to the fluorescence detected. A power supply is used to apply the appropriate voltages to the microchip reservoirs, to perform electrophoresis.

for obtaining expanded volume reservoirs involves cutting the large ends off of pipette tips and gluing them onto the chip to act as reservoirs—more recently, commercially-developed reservoirs for microchips have become available.

A detection system is required to track the progress of the separation, and with the increasing popularity of microchips for separations, a variety of different detection options are now available. By far, the most prevalent detection system described in the literature is laser-induced fluorescence (LIF). Chapter 45 by Karlinsey and coworkers not only provides a guide to selection of the components for LIF, but also easy-to-follow strategies for alignment of the optics in the LIF system (as well as reviewing other detection modes used in microfluidics). The components used in an LIF system are a light detection module, such as a photomultiplier tube or charge-coupled device (CCD) camera, a laser source, optics for alignment (to guide excitation light in and collect emitted light out) for efficient detection, a computer with reasonable processing speed for data collection, and a user-friendly software for interfacing with the hardware. LabVIEW software (National Instruments) is commonly used as the interface, and allows for easy application-specific computer programming to provide software interfaces that provide exquisite control over data collection and control of the detection system.

10.5.1 Electrophoretic Separations

When performing electrophoretic separations, the cross-t design originally proposed by Verheggen et al.[5] for electrokinetic (EK) injection of sample is the most popular; Figure 10.6a shows a typical cross-t design. Using the appropriate buffer for the separation of interest, the channel surface must be conditioned in a manner not too dissimilar to those processes used with capillary systems. If the separation to be performed requires EOF, the rigorous cleaning method suggested in Section 10.3.4 is recommended. All channels of the device are filled with buffer along with the reservoirs to ensure electrical contact between the platinum electrodes and the solution. A larger volume of buffer in the reservoirs minimizes problems with buffer depletion which can have adverse effects on the separation (see Chapter 1 by Landers). Consistent with the scheme in Figure 10.6a, the sample is placed in the reservoir SI and a voltage applied between SI and SO to allow for mobilization of sample from reservoir SI to reservoir SO. After the voltage has been applied for a predetermined amount of time, voltage is applied to the BO and BI reservoirs to electrophorese the sample (down) into the separation channel, thus, completing the injection process. The detection system, in this case, the incoming laser light and the photomultiplier tube (PMT), are positioned at a point in the separation channel

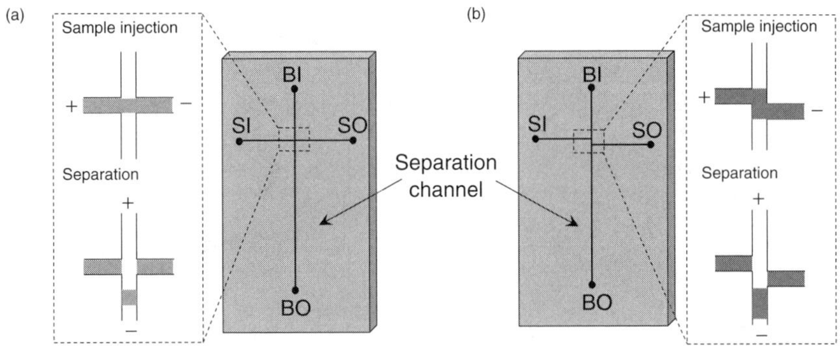

FIGURE 10.6 Schematic representation of microchip electrophoresis device designs (a) shows a traditional cross-t design while (b) shows an offset cross-t. The reservoirs are labeled as follows: SI is the sample inlet, SO is sample outlet, BI is buffer inlet, and BO is buffer outlet. A closer view of the cross-t can be seen in the dotted boxes during sample injection and separation; the sample is in gray and the + and − represent the position and what type of voltage is being applied.

commensurate with providing the necessary effective length for the separation—the sample analytes passing that detection point are excited and the resultant fluorescence collected.

For sample injection on a microdevice, a few different modes of EK injection have been reported. Ramsey and coworkers described a floating injection, in which the BI and BO reservoirs are both floating during the sample injection; this can lead to a variable volume plug due to diffusion.[56] A "pinched" injection allows for a constant-volume injection by also pumping buffer from the separation channel along the sample channel, minimizing the length of the plug.[56] Another injection mode involves slightly modifying the device design to incorporate an offset cross-t.[57] With this design, the incoming and outgoing channels are offset relative to the separation channel, thus elongating the sample plug length to be injected (Figure 10.6b). As with EK injections in capillary systems, there is a bias toward higher mobility analytes when performing EK injections in a microdevice. For example, when performing microdevice-based DNA separations with a sieving matrix, EOF is predominantly suppressed and the applied voltage is used to attract the (anionic) DNA across the sample arm to the sample outlet. However, long injection times (upwards of 60 seconds) are generally needed to obtain a representative sample in the cross-t because the mobility of smaller DNA fragments dominates during the initial phase of the injection process.

10.5.2 Exemplary Separations

In 1992, the first microchip-based electrophoretic separation was performed: two fluorescent dyes, fluorescein and calcein, were separated in a glass/silicon device utilizing LIF as the detection method.[13,58] While the analysis time was not significantly shorter than separations that could be accomplished using capillary electrophoresis (CE), there was a drastic reduction in the consumption of sample and reagents (i.e., electrophoresis buffer). Shortly thereafter, faster chip-based separations began to emerge in the literature. Seiler et al.[59] were able to separate three fluorescein 5-isothiocyanate (FITC)-labeled amino acids—arginine, phenylalanine, and glutamine—in less than 120 seconds using a separation distance of 9.6 cm; the electropherogram and conditions are shown in Figure 10.7. Ramsey and coworkers[60] decreased the separation length to 0.9 mm and were able to baseline resolve fluorescein and rhodamine B in a fraction of a second (150 milliseconds).

The early success of zone electrophoresis in a microchip gave way to DNA separations in microdevices. Mathies and coworkers were among the pioneers in this section of the microfluidics field, successfully resolving the restriction fragments of ΦX174 *Hae*III in 120 seconds, which ranged from 70 to 1000 base-pairs in length.[61] This technology has since been advanced tremendously, with

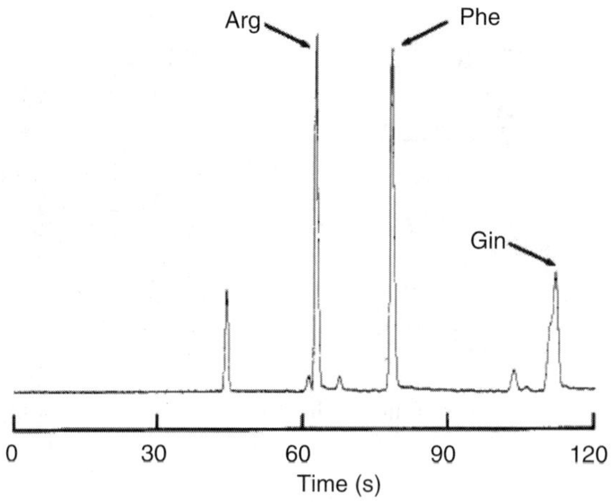

FIGURE 10.7 Microchip electropherogram of three 10 μM FITC-labeled amino acids (Arg, Phe, and Gln in a pH 9.2 carbonate buffer). Electrokinetic injection was performed by applying a voltage of 250 V for 60 s (~1-mm plug length). For the separation, a voltage of 10 kV was applied (6.3 kV between injector and detector). The separation distance from injection to detection was 9.6 cm. (Reprinted from Seiler, K., et al., *Analytical Chemistry*, 65, 1481, 1993. With permission.)

multiple groups having accomplished high-resolution separations (including DNA sequencing) and contributed advancements to this arena.[62,63] High-throughput DNA electrophoresis microchips have also been demonstrated to be functional (shown in Figure 10.8), with Paegel et al.[38] having multiplexed 96 channels on a radial capillary electrophoresis microchannel plate using a four-color rotary confocal fluorescence scanner for detection. Tapered turns have been included in the channel design to gain the effective separation length necessary in order to achieve the high-resolution required for DNA sequencing separations. On the detection front, Karlinsey and Landers[64] have shown the value of multicolor detection using an acoustooptical tunable filter (AOTF), an approach easily capable of detecting tens of different colors without any moving parts—they illustrate this with the detection of biowarfare agents. The electropherogram shown in Figure 10.9 provides an exemplary separation of a PCR-generated fragment amplified from *Bacillus anthracis* labeled with a TET dye and co-injected with a ROX-labeled sizing standard. They also show the utility of the AOTF for rapid channel alignment, sample scanning to collect spectral information from the desired analytes, and displayed the wide range of wavelength detection by performing simultaneous 19 wavelength detection. DNA separations on microdevices have also been successful for decreasing the analysis time associated with many post-amplification separations for molecular diagnostics, such as single-stranded conformation polymorphism (SSCP) for the detection of breast cancer,[65] tandem SSCP/heteroduplex analysis for detecting p53 mutations,[66] and the detection of biowarfare agents using an AOTF to detect multiple fluorescent signals.[64] These topics are covered in more detail in individual chapters in this book.

Electrophoretic separations involving proteins have also been successfully adapted to microfluidic devices utilizing electrophoretic techniques that go beyond capillary zone electrophoresis. Giordano et al.[68] separated partially purified human plasma using a microchip-based capillary electrophoresis-SDS analysis. Figure 10.10 shows a comparison between the glass microchip separation with an 11% acrylamide SDS-PAGE gel separation. Traditionally, this separation had been performed using capillary electrophoresis and UV-detection and suffered from poor detection limits. Through the use of a dynamic labeling dye, they were able to achieve improved sensitivity without a loss of resolution. Liu et al.[69] modified the separation channels in PMMA microchips using an atom-transfer radical polymerization to graft polyethylene glycol to the surface. This passivation

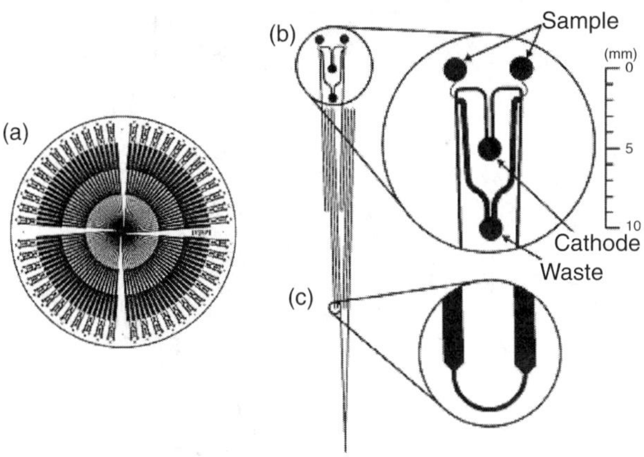

FIGURE 10.8 Overall layout of the (a) 96-lane DNA sequencing microchannel plate. (b) expanded view of the injector. Each doublet features two sample reservoirs and common cathod and waste reservoirs. The arm from the sample to the separation channel is 85-μm wide, and the arm from the waste to the seapration channel is 300-μm wide. The separation channel connecting the central anode and cathode is 200-μm wide (c) expanded view of the hyperturn region. The turns are symmetrically tapered with a tapering length of 100 μm, a turn channel width of 65 μm, and a radius of curvature of 250 μm. Channel widths and lengths are not drawn to scale. (Reprinted from Paegel, B. M. et al., *Proceedings of the National Academy of Science USA* 99, 575, 2002. With permission.)

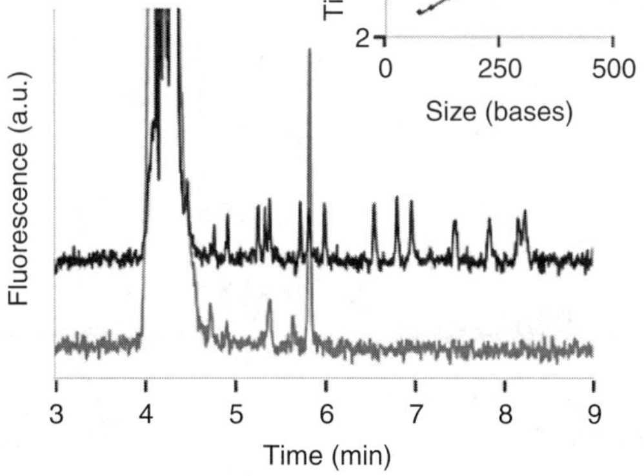

FIGURE 10.9 Amplified TET-anthrax fragment coinjected electrokinetically with a ROX sizing standard and separated under denaturing conditions. Data were obtained by the AOTF between emission wavelengths at 18 Hz, and the electropherograms are offset for clarity. Inset: sizing plot with ladder (diamond) and anthrax (square) peaks. (Reprinted from Karlinsey J. M. and Landers, J. P., *Analytical Chemistry* 78, 5593, 2006. With permission.)

step allowed for a separation with increased resolution of the components of bovine serum albumin (BSA) when compared to an uncoated PMMA micro device; the resultant electropherogram can be seen in Figure 10.11. Roman et al.[70] used micellar-EK chromatography in a PDMS device to provide high-resolution separations of standard proteins and hydrophobic molecules by adding SDS as a dynamic channel coating.[71] Isoelectric focusing has also been utilized successfully in the microformat for the separation of proteins with multiple detection techniques including LIF,[72–75] UV detection[76] and whole-column detection.[76,77] While most microchip IEF methods have used either

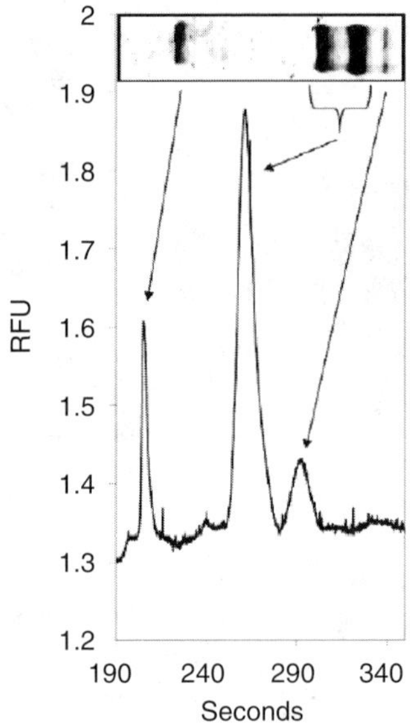

FIGURE 10.10 Comparison of 11% acrylamide SDS–PAGE gel separation of partially purified human plasma with microchip-based CE-SDS analysis. Separation conditions include 25 mM Tris-CHES, 1 mM DTT, 0.1% SDS, 370 V/cm field strength, electrokinetic injection and 0.2% (v/v) NanoOrange (excitation 488 nm and emission 590 nm). (Reprinted from Giordano, B. C. et al., *Analytical Chemistry* 76, 4711, 2004. With permission.)

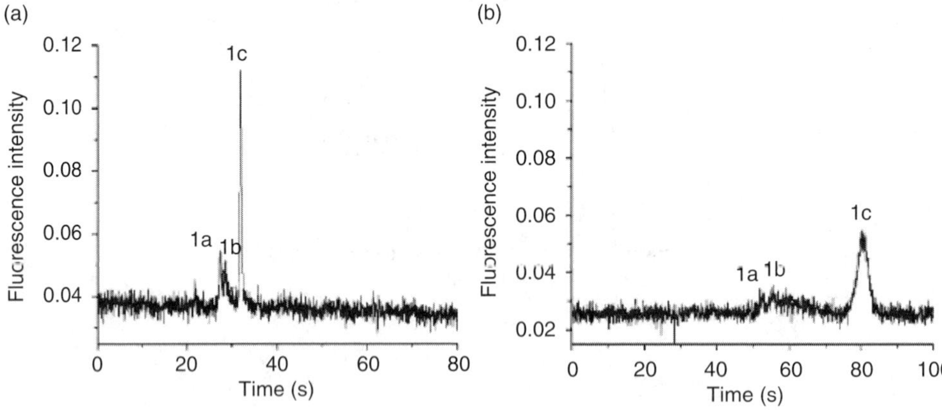

FIGURE 10.11 Microchip CE of FITC-BSA. (a) Polyethylene glycol (PEG)-grafted and (b) untreated chip. Peaks 1a, 1b, and 1c are the three main components of the BSA sample. The PEG was grafted onto the PMMA surface using an atom-transfer radical polymerization. (Reprinted from Liu, J. et al., *Analytical Chemistry* 76, 6953, 2004. With permission.)

EOF- or chemically-induced mobilization of the focused zones past the detector, Guillo et al.[75] illustrated the use of on-chip pumping, generated by elastomeric diaphragm pumps in a three-layer device (one PDMS layer sandwiched by two glass layers), to mobilize the zones. The authors explored the effect of mobilization flow rates on the separation of two amino acids, L-lysine and L-histidine; the electropherograms can be seen in Figure 10.12.

Other biologically-relevant analytes, in addition to DNA and proteins, have been successfully separated on microchip. Figure 10.13 shows the successful electrophoretic separation of 8-aminopyrene-1,3,6-trisulfonate (APTES)-labeled N-linked oligosaccharides from ribonuclease B,

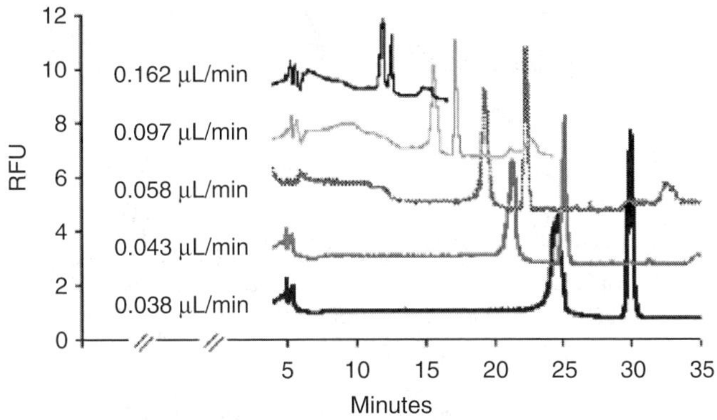

FIGURE 10.12 Effect of mobilization flow rates on the overall separation performance of L-lysine and L-histidine. The glass fluidic layer was coated using a modified Hjerten coating. The samples were labeled with FQ. The focusing step was performed at 7.6 kV for 6 min, and then the mobilization step was initiated. Valve actuation conditions for mobilization: actuation vacuum: 60 kPa; actuation pressure ranging from 3 to 20 kPa. (Reprinted from Guillo, C., et al., *Lab on a Chip* 7, 117, 2007. With permission.)

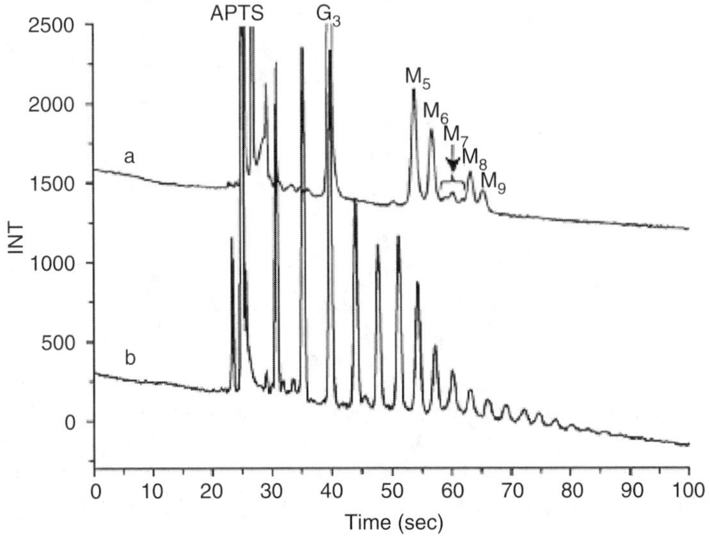

FIGURE 10.13 Microchip electropherograms of 8-aminopyrene-1,3,6-trisulfonate (APTS)-labeled N-linked oligosaccharides from ribonuclease B (A) and APTS-labeled oligosaccharide ladder (B). Experimental conditions: E_{sep} = 300 V/cm, 0.5% methyl cellulose in 20 mM phosphate buffer (pH 6.66). (Reprinted from Dang, F., et al., *Journal of Chromatography A* 1109, 141, 2006. With permission.)

using microchips fabricated in PMMA.[78] The authors achieved separation in a channel that had a 30-mm effective separation length and have also extended this work to include fast profiling of N-linked complex oligosaccharides released from three other glycoproteins. The separation of lipids, an area of growing interest, has also been accomplished on microchips. Lin et al.[79] reported on the separation of anionic phosphoinositides on a single-sipper microdevice (Figure 10.14). Once a method for successful electrophoretic separations was established, the authors used their technology to monitor the activity of lipid-modifying enzymes. They were able to show the applicability of the method to high-throughput screening applications by demonstrating phospholipase A_2 in a 384-well format; this technology offers an attractive alternative to current methods that involve radioactive substrates. In an elegant report, Vrouwe et al.[80] showed the ability to detect lithium in blood samples down

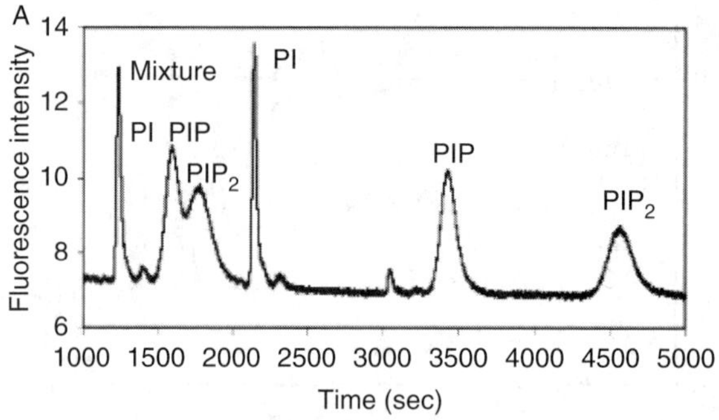

FIGURE 10.14 Separation of anionic phosphoinositides on a single-sipper microchip. Separation conditions: sample injection time, 0.5 s; buffer time, 800 s; field strength: 530 V/cm; pressure, −1.65 psi (vacuum); lipid concentration: 1.3 μM each and labeled with BODIPY FL on the acyl chain. (Reprinted from Lin, S., et al., *Analytical Biochemistry* 314, 101, 2003. With permission.)

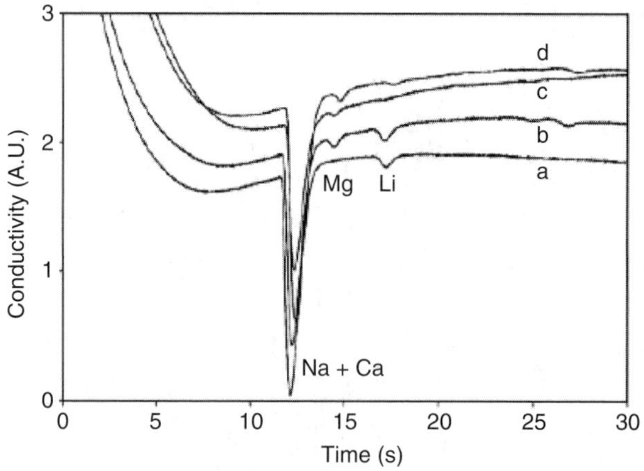

FIGURE 10.15 Electropherogram of an (a) aqueous calibration mixture containing 140 mmol/L sodium and 2 mmol/L lithium, (b) citrated whole blood enriched with 2 mmol/L lithium, (c) whole blood without anticoagulant, and (d) heparinized plasma from a patient on lithium therapy containing 0.62 mmol/L lithium. (Reprinted from Vrouwe, E. X., et al., *Clinical Chemistry* 53, 121, 2007. With permission.)

to levels of 0.15 mmol/L. Figure 10.15 shows an exemplary separation from samples that included whole blood, citrated whole blood, and heparinized whole blood while detecting sodium, calcium, magnesium, and lithium. Through the incorporation of a plexiglass sample collector, the authors have created a functional device capable of point-of-care analysis. To further challenge their device, the plexiglass sample collector was tested with a blinded study where they tested five-patient serum samples on the microchip, and were able to detect lithium concentrations similar to that determined using ion-selective electrodes.

10.6 CONCLUDING REMARKS

The purpose of this chapter has been two-fold. First, to highlight some of the basic methods and substrates that have been successfully utilized to create microfluidic devices. We have attempted to describe these methods, albeit at a rudimentary level, in a manner that allows for someone skilled in CE to fabricate, functionalize and execute microchip-based electrophoresis. In addition to the basic methodologies provided here, the second aim was to provide a few 'tricks of the trade' from our own experience, ones that should enable a novice to create devices of standard cross-t design, or even more complex and innovative designs tuned to their specific applications. It is noteworthy that, this is not a comprehensive methods treatise on chip fabrication, and that every microchip fabricator, regardless of the substrate used, has their own 'tricks of the trade'—methodology preferences that enhance their fabrication success with the substrate and design of choice. It is our goal that the methods given here (and in other chapters in this book) enable the reader to enter into the microchip fabrication arena where they can demonstrate the application of microfluidics in ways that, to-date, have not been revealed. But perhaps more important than enhancing the availability of microfluidics to researchers looking to adapt this analytical platform into their lab, we have enabled the novices of today to become the microfluidics leaders of tomorrow.

REFERENCES

1. Terry, S. C., Herman, J. H., and Angel, J. B., A gas chromatographic air analyzer fabricated on a silicon wafer, *IEEE Trans. Elec. Dev.* 26, 1880, 1979.
2. van der Schoot, B. and Bergveld, P., An ISFET-based microlitre titrator: integration of a chemical sensor-actuator system, *Sensors and Actuators* 8, 11–22, 1985.
3. Shoji, S., Esashi, M., and Matsuo, T., Prototype miniature blood gas analyser fabricated on a silicon wafer, *Sensors and Actuators* 14, 101–107, 1988.
4. van Lintel, H. T. G., van de Pol, F. C. M., and Bouwstra, S., A piezoelectric micropump based on micromachining of silicon, *Sensors and Actuators* 15, 153–167, 1988.
5. Verheggen, T. P. E. M., Beckers, J. L., and Everaerts, F. M., Simple sampling device for capillary isotachophoresis and capillary zone electrophoresis, *Journal of Chromatography* 452, 615–622, 1988.
6. Esashi, M., Shoji, S., and Nakano, A., Normally closed microvalve and micropump fabricated on a silicon wafer, *Sensors and Actuators* 20, 163–169, 1989.
7. van de Pol, F. C. M., Wonnink, D. G. J., Elwenspoek, M., and Fluitman, J. H. J., A thermo-pneumatic actuation principle for a microminiature pump and other micromechanical devices, *Sensors and Actuators* 17, 139–143, 1989.
8. Esashi, M., Integrated micro flow control systems, *Sensors and Actuators, A* 21, 161–167, 1990.
9. Shoji, S., Nakagawa, S., and Esashi, M., Micropump and sample-injector for integrated chemical analyzing systems, *Sensors and Actuators, A* 21, 189–192, 1990.
10. Smits, J. G., Piezoelectric micropump with three valves working peristaltically, *Sensors and Actuators, A* 21, 203–206, 1990.
11. van de Pol, F. C. M., van Lintel, H. T. G., Elwenspoek, M., and Fluitman, J. H. J., A thermopneumatic micropump based on micro-engineering techniques, *Sensors and Actuators, A* 21, 198–202, 1990.
12. Manz, A., Graber, N., and Widmer, H. M., Miniaturized total chemical-analysis systems—a novel concept for chemical sensing, *Sensors and Actuators B-Chemical* 1, 244–248, 1990.
13. Manz, A., Harrison, D. J., Verpoorte, E. M. J., Fettinger, J. C., Paulus, A., Ludi, H., and Widmer, H. M., Planar chips technology for miniaturization and integration of separation techniques into monitoring systems—capillary electrophoresis on a chip, *Journal of Chromatography* 593, 253–258, 1992.
14. Lin, C. H., Lee, G. B., Chang, B. W., and Chang, G. L., A new fabrication process for ultra-thick microfluidic microstructures utilizing SU-8 photoresist, *Journal of Micromechanics and Microengineering* 12, 590–597, 2002.
15. Paufler, J., Kuck, H., Seltmann, R., Doleschal, W., Gehner, A., and Zimmer, G., High-throughput optical direct write lithography, *Solid State Technology* 40, 175–182, 1997.
16. Corbett, S., Strole, J., Johnston, K., Swenson, E. J., and Lu, W. X., Laser direct exposure of photodefinable polymer masks using shaped-beam optics, *IEEE Transactions on Electronics Packaging Manufacturing* 28, 312–321, 2005.
17. Bien, D. C. S., Rainey, P. V., Mitchell, S. J. N., and Gamble, H. S., Characterization of masking materials for deep glass micromachining, *Journal of Micromechanics and Microengineering* 13, S34–S40, 2003.

18. Rodriguez, I., Spicar-Mihalic, P., Kuyper, C. L., Fiorini, G. S., and Chiu, D. T., Rapid prototyping of glass microchannels, *Analytica Chimica Acta* 496, 205–215, 2003.
19. Brister, P. C. and Weston, K. D., Patterned solvent delivery and etching for the fabrication of plastic microfluidic devices, *Analytical Chemistry* 77, 7478–7482, 2005.
20. Belloy, E., Pawlowski, A. G., Sayah, A., and Gijs, M. A. M., Microfabrication of high-aspect ratio and complex monolithic structures in glass, *Journal of Microelectromechanical Systems* 11, 521–527, 2002.
21. Ceriotti, L., Weible, K., de Rooij, N. F., and Verpoorte, E., Rectangular channels for lab-on-a-chip applications, *Microelectronic Engineering* 67–68, 865–871, 2003.
22. Hibara, A., Saito, T., Kim, H. B., Tokeshi, M., Ooi, T., Nakao, M., and Kitamori, T., Nanochannels on a fused-silica microchip and liquid properties investigation by time-resolved fluorescence measurements, *Analytical Chemistry* 74, 6170–6176, 2002.
23. Ionescu, R. E., Marks, R. S., and Gheber, L. A., Nanolithography using protease etching of protein surfaces, *Nano Letters* 3, 1639–1642, 2003.
24. Ionescu, R. E., Marks, R. S., and Gheber, L. A., Manufacturing of nanochannels with controlled dimensions using protease nanolithography, *Nano Letters* 5, 821–827, 2005.
25. Ito, T., Sobue, K., and Ohya, S., Water glass bonding for micro-total analysis system, *Sensors and Actuators B–Chemical* 81, 187–195, 2002.
26. Satoh, A., Water glass bonding, *Sensors and Actuators A–Physical* 72, 160–168, 1999.
27. Khandurina, J., McKnight, T. E., Jacobson, S. C., Waters, L. C., Foote, R. S., and Ramsey, J. M., Integrated system for rapid PCR-based DNA analysis in microfluidic devices, *Analytical Chemistry* 72, 2995–3000, 2000.
28. Satyanarayana, S., Karnik, R. N., and Majumdar, A., Stamp-and-stick room-temperature bonding technique for microdevices, *Journal of Microelectromechanical Systems* 14, 392–399, 2005.
29. Huang, Z. L., Sanders, J. C., Dunsmor, C., Ahmadzadeh, H., and Landers, J. P., A method for UV-bonding in the fabrication of glass electrophoretic microchips, *Electrophoresis* 22, 3924–3929, 2001.
30. Chiem, N., Lockyear-Shultz, L., Andersson, P., Skinner, C., and Harrison, D. J., Room temperature bonding of micromachined glass devices for capillary electrophoresis, *Sensors and Actuators B–Chemical* 63, 147–152, 2000.
31. Pigeon, F., Biasse, B., and Zussy, M., Low-temperature pyrex glass wafer direct bonding, *Electronics Letters* 31, 792–793, 1995.
32. Jia, Z. J., Fang, Q., and Fang, Z. L., Bonding of glass microfluidic chips at room temperatures, *Analytical Chemistry* 76, 5597–5602, 2004.
33. Howlader, M. M. R., Suehara, S., and Suga, T., Room temperature wafer level glass/glass bonding, *Sensors and Actuators A–Physical* 127, 31–36, 2006.
34. Nakanishi, H., Nishimoto, T., Arai, A., Abe, H., Kanai, M., Fujiyama, Y., and Yoshida, T., Fabrication of quartz microchips with optical slit and development of a linear imaging UV detector for microchip electrophoresis systems, *Electrophoresis* 22, 230–234, 2001.
35. Lin, C. H., Lee, G. B., Fu, L. M., and Chen, S. H., Integrated optical-fiber capillary electrophoresis microchips with novel spin-on-glass surface modification, *Biosensors and Bioelectronics* 20, 83–90, 2004.
36. Shaw, J. M., Gelorme, J. D., LaBianca, N. C., Conley, W. E., and Holmes, S. J., Negative photoresists for optical lithography, *IBM Journal of Research and Development* 41, 81–94, 1997.
37. Ferrance, J. P., Wu, Q. R., Giordano, B., Hernandez, C., Kwok, Y., Snow, K., Thibodeau, S., and Landers, J. P., Developments toward a complete micro-total analysis system for Duchenne muscular dystrophy diagnosis, *Analytica Chimica Acta* 500, 223–236, 2003.
38. Paegel, B. M., Emrich, C. A., Weyemayer, G. J., Scherer, J. R., and Mathies, R. A., High throughput DNA sequencing with a microfabricated 96-lane capillary array electrophoresis bioprocessor, *Proceedings of the National Academy of Sciences of the United States of America* 99, 574–579, 2002.
39. Easley, C. J., Karlinsey, J. M., Bienvenue, J. M., Legendre, L. A., Roper, M. G., Feldman, S. H., Hughes, M. A., et al., A fully integrated microfluidic genetic analysis system with sample-in-answer-out capability, *Proceedings of the National Academy of Sciences of the United States of America* 103, 19272–19277, 2006.

40. Rossier, J. S., Schwarz, A., Reymond, F., Ferrigno, R., Bianchi, F., and Girault, H. H., Microchannel networks for electrophoretic separations, *Electrophoresis* 20, 727–731, 1999.
41. Simpson, P. C., Woolley, A. T., and Mathies, R. A., Microfabrication technology for the production of capillary array electrophoresis chips, *Biomedical Microdevices* 1, 7–26, 1998.
42. Cras, J. J., Rowe-Taitt, C. A., Nivens, D. A., and Ligler, F. S., Comparison of chemical cleaning methods of glass in preparation for silanization, *Biosensors and Bioelectronics* 14, 683–688, 1999.
43. Sun, Y. and Kwok, Y. C., Polymeric microfluidic system for DNA analysis, *Analytica Chimica Acta* 556, 80–96, 2006.
44. Holland, A., Cher, C., Rosengarten, G., and Simon, D., Plasma etching of thermoset polymer films for microchannels, *Smart Materials and Structures* 15, S104–S111, 2006.
45. Dou, Y. H., Bao, N., Xu, J. J., and Chen, H. Y., A dynamically modified microfluidic poly(dimethylsiloxane) chip with electrochemical detection for biological analysis, *Electrophoresis* 23, 3558–3566, 2002.
46. Fu, A. Y., Spence, C., Scherer, A., Arnold, F. H., and Quake, S. R., A microfabricated fluorescence-activated cell sorter, *Nature Biotechnology* 17, 1109–1111, 1999.
47. Hansen, C. L., Skordalakes, E., Berger, J. M., and Quake, S. R., A robust and scalable microfluidic metering method that allows protein crystal growth by free interface diffusion, *Proceedings of the National Academy of Sciences of the United States of America* 99, 16531–16536, 2002.
48. McDonald, J. C., Duffy, D. C., Anderson, J. R., Chiu, D. T., Wu, H., Schueller, O. J., and Whitesides, G. M., Fabrication of microfluidic systems in poly(dimethylsiloxane), *Electrophoresis* 21, 27–40, 2000.
49. Duffy, D. C., McDonald, J. C., Schueller, O. J. A., and Whitesides, G. M., Rapid prototyping of microfluidic systems in poly(dimethylsiloxane), *Analytical Chemistry* 70, 4974–4984, 1998.
50. Sia, S. K. and Whitesides, G. M., Microfluidic devices fabricated in poly(dimethylsiloxane) for biological studies, *Electrophoresis* 24, 3563–3576, 2003.
51. Toepke, M. W. and Beebe, D. J., PDMS absorption of small molecules and consequences in microfluidic applications, *Lab on a Chip* 6, 1484–1486, 2006.
52. Vickers, J. A. and Henry, C. S., Simplified current decoupler for microchip capillary electrophoresis with electrochemical and pulsed amperometric detection, *Electrophoresis* 26, 4641–4647, 2005.
53. Liu, J., Enzelberger, M., and Quake, S., A nanoliter rotary device for polymerase chain reaction, *Electrophoresis* 23, 1531–1536, 2002.
54. Vickers, J. A., Caulum, M. M., and Henry, C. S., Generation of hydrophilic poly(dimethylsiloxane) for high-performance microchip electrophoresis, *Analytical Chemistry* 78, 7446–7452, 2006.
55. Harrison, D. J., Fluri, K., Seiler, K., Fan, Z. H., Effenhauser, C. S., and Manz, A., *Science* 261, 895–897, 1993.
56. Jacobson, S. C., Hergenroder, R., Koutny, L. B., Warmack, R. J., and Ramsey, J. M., Effects of injection schemes and column geometry on the performance of microchip electrophoresis devices, *Analytical Chemistry* 66, 1107–1113, 1994.
57. Effenhauser, C. S., Manz, A., and Widmer, H. M., Glass chips for high-speed capillary electrophoresis separations with submicrometer plate heights, *Analytical Chemistry* 65, 2637–2642, 1993.
58. Harrison, D. J., Manz, A., Fan, Z. H., Ludi, H., and Widmer, H. M., Capillary electrophoresis and sample injection systems integrated on a planar glass chip, *Analytical Chemistry* 64, 1926–1932, 1992.
59. Seiler, K., Harrison, D. J., and Manz, A., planar glass chips for capillary electrophoresis—repetitive sample injection, quantitation, and separation efficiency, *Analytical Chemistry* 65, 1481–1488, 1993.
60. Jacobson, S. C., Hergenroder, R., Koutny, L. B., and Ramsey, J. M., High-speed separations on a microchip, *Analytical Chemistry* 66, 1114–1118, 1994.
61. Woolley, A. T. and Mathies, R. A., Ultra-high-speed DNA fragment separations using microfabricated capillary array electrophoresis chips, *Proceedings of the National Academy of Sciences of the United States of America* 91, 11348–11352, 1994.
62. Woolley, A. T. and Mathies, R. A., Ultra-high-speed DNA-sequencing using capillary electrophoresis chips, *Analytical Chemistry* 67, 3676–3680, 1995.
63. Liu, S. R., Ren, H. J., Gao, Q. F., Roach, D. J., Loder, R. T., Armstrong, T. M., Mao, Q. L., Blaga, I., Barker, D. L., and Jovanovich, S. B., Automated parallel DNA sequencing on multiple channel

microchips, *Proceedings of the National Academy of Sciences of the United States of America* 97, 5369–5374, 2000.
64. Karlinsey, J. M. and Landers, J. P., Multicolor fluorescence detection on an electrophoretic microdevice using an acoustooptic tunable filter, *Analytical Chemistry* 78, 5590–5596, 2006.
65. Tian, H. J., Jaquins-Gerstl, A., Munro, N., Trucco, M., Brody, L. C., and Landers, J. P., Single-strand conformation polymorphism analysis by capillary and microchip electrophoresis: a fast, simple method for detection of common mutations in BRCA1 and BRCA2, *Genomics* 63, 25–34, 2000.
66. Hestekin, C. N., Jakupciak, J. P., Chiesl, T. N., Kan, C. W., O'Connell, C. D., and Barron, A. E., An optimized microchip electrophoresis system for mutation detection by tandem SSCP and heteroduplex analysis for p53 gene exons 5–9, *Electrophoresis* 27, 3823–3835, 2006.
67. Giordano, B. C., Jin, L. J., Couch, A. J., Ferrance, J. P., and Landers, J. P., Microchip laser-induced fluorescence detection of proteins at submicrogram per milliliter levels mediated by dynamic Labeling under pseudonative conditions, *Analytical Chemistry* 76, 4705–4714, 2004.
68. Liu, J. K., Pan, T., Woolley, A. T., and Lee, M. L., Surface-modified poly(methyl methacrylate) capillary electrophoresis microchips for protein and peptide analysis, *Analytical Chemistry* 76, 6948–6955, 2004.
69. Roman, G. T., Carroll, S., McDaniel, K., and Culbertson, C. T., Micellar electrokinetic chromatography of fluorescently labeled proteins on poly(dimethylsiloxane)-based microchips, *Electrophoresis* 27, 2933–2939, 2006.
70. Roman, G. T., McDaniel, K., and Culbertson, C. T., High efficiency micellar electrokinetic chromatography of hydrophobic analytes on poly(dimethylsiloxane) microchips, *Analyst* 131, 194–201, 2006.
71. Li, Y., DeVoe, D. L., and Lee, C. S., Dynamic analyte introduction and focusing in plastic microfluidic devices for proteomic analysis, *Electrophoresis* 24, 193–199, 2003.
72. Tan, W., Fan, Z. H., Qiu, C. X., Ricco, A. J., and Gibbons, I., Miniaturized capillary isoelectric focusing in plastic microfluidic devices, *Electrophoresis* 23, 3638–3645, 2002.
73. Han, J. and Singh, A. K., Rapid protein separations in ultra-short microchannels: microchip sodium dodecyl sulfate-polyacrylamide gel electrophoresis and isoelectric focusing, *Journal of Chromatography A* 1049, 205–209, 2004.
74. Guillo, C., Karlinsey, J. M., and Landers, J. P., On-chip pumping for pressure mobilization of the focused zones following isoelectric focusing, *Lab on a Chip* 7, 112–118, 2006.
75. Mao, Q. L. and Pawliszyn, J., Demonstration of isoelectric focusing on an etched quartz chip with UV absorption imaging detection, *Analyst* 124, 637–641, 1999.
76. Huang, T. M., Ertl, P., Wu, X. Z., Mikkelsen, S., and Pawliszyn, J., Microfabrication of microfluidic cartridge for isoelectric focusing by screen printing, *Sensors and Materials* 14, 141–149, 2002.
77. Dang, F. Q., Kakehi, K., Nakajima, K., Shinohara, Y., Ishikawa, M., Kaji, N., Tokeshi, M., and Baba, Y., Rapid analysis of oligosaccharides derived from glycoproteins by microchip electrophoresis, *Journal of Chromatography A* 1109, 138–143, 2006.
78. Lin, S. S., Fischl, A. S., Bi, X. H., and Parce, W., Separation of phospholipids in microfluidic chip device: application to high-throughput screening assays for lipid-modifying enzymes, *Analytical Biochemistry* 314, 97–107, 2003.
79. Vrouwe, E. X., Luttge, R., Vermes, I., and van den Berg, A., Microchip capillary electrophoresis for point-of-care analysis of lithium, *Clinical Chemistry* 53, 117–123, 2007.
80. Becker, H. and Gartner, C., Polymer microfabrication methods for microfluidic analytical applications, *Electrophoresis* 21, 12–26, 2000.

Part IIA

Capillary-Based Systems: Core Methods and Technologies

11 Kinetic Capillary Electrophoresis

Maxim V. Berezovski and Sergey N. Krylov

CONTENTS

11.1 INTRODUCTION AND BACKGROUND OF AFFINITY METHODS

Affinity methods play a crucial role in modern life sciences. In addition to affinity purification, their applications include quantitative analyses of biomolecules, studies of biomolecular interactions, and selection of affinity probes and drug candidates from complex mixtures, such as combinatorial libraries. Conceptually, all affinity methods are based on noncovalent binding of a ligand (L) and a target (T) with the formation of a ligand–target complex (C):

$$L + T \underset{k_{off}}{\overset{k_{on}}{\rightleftarrows}} C, \tag{11.1}$$

where k_{on} and k_{off} are rate constants of complex formation and dissociation, respectively. The stability of the complex is typically described in terms of the equilibrium dissociation constant:

$$K_d = \frac{k_{off}}{k_{on}} \tag{11.2}$$

Methods for the measurement of equilibrium and rate constants can be classified into two categories: mixture-based and separation-based (Figure 11.1). The first category includes light

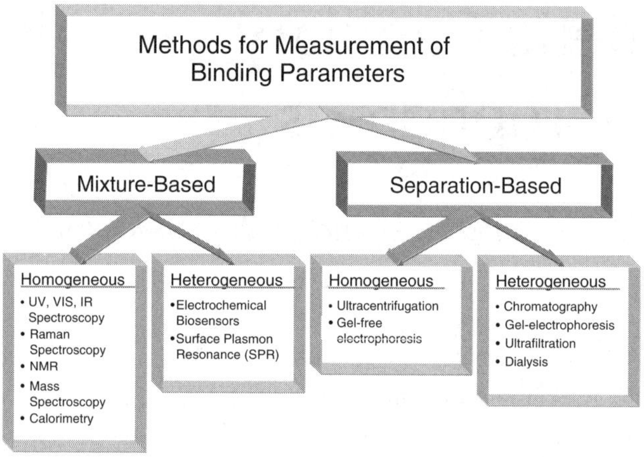

FIGURE 11.1 Classification of affinity methods.

absorption, fluorescence spectroscopy, nuclear magnetic resonance (NMR), Fourier transform infrared spectroscopy (FTIR), mass spectrometry (MS), Raman spectroscopy, potentiometry, and calorimetry. The separation-based methods include dialysis, ultrafiltration, ultracentrifugation, chromatography (liquid chromatography and thin-layer chromatography), and electrophoresis (planar and capillary electrophoresis). Separation-based methods can detect individual interacting components and/or complexes, thus avoiding the interference of other components.

Mixture-based and separation-based methods can be subdivided into two broad categories: heterogeneous and homogeneous binding assays. In heterogeneous assays, T is affixed to a solid substrate, while L is dissolved in a solution and can bind T affixed to the surface. In an advanced heterogeneous mixture-based technique such as SPR, T is affixed to a sensor that can change its optical or electrical signal upon L binding to T.[1,2] In SPR, K_d can be found in a series of equilibrium experiments. The concentration of L in the solution is varied and L and T are allowed to reach equilibrium. The signal from the sensor versus the concentration of L has a characteristic sigmoidal shape and K_d can be found from the curve by identifying the concentration of L at which the signal is equal to half of its maximum amplitude. The k_{off} value can be determined by SPR in a single nonequilibrium experiment in which the equilibrium is disturbed by fast replacing the solution of L with a buffer devoid of L. The complex on the surface dissociates in the absence of L in the solution, and the complex dissociation generates an exponential signal on the sensor.

Heterogeneous binding assays have certain advantages and drawbacks. The most serious drawback is that affixing T to the surface changes the structure of T. The extent of such change depends on the method of immobilization. The change in the structure can potentially affect binding of L to T. This problem is especially severe for L that binds to T through interaction with a large part of T. In addition, the immobilization of T on the surface may be time consuming and expensive. Moreover, nonspecific interactions with the surface are always a concern.

In homogeneous binding assays, T and L are mixed and allowed to form a complex in solution; neither of the molecules affixed to the surface. Complex formation is followed by either monitoring the changing physical–chemical properties of L or T upon binding. Such properties can be optical (absorption, fluorescence, polarization) or separation-related (chromatographic or electrophoretic mobility). Equilibrium experiments with varying concentrations of L can be used similarly to heterogeneous analyses to find K_d. Nonequilibrium stopped-flow experiments, in which L and T are mixed in a fast fashion and the change in spectral properties is monitored, can be used to find k_{on}.

Separation-based affinity methods can also be classified as kinetic or nonkinetic. Kinetic methods are those that do not assume equilibrium in reaction 1 and can thus be used for (1) quantitative affinity analyses with "weak" affinity probes (high k_{off}), (2) measuring k_{on} and k_{off}, and (3) selection of binding ligands with predefined k_{on} and k_{off}. Nonkinetic methods, in contrast, assume equilibrium and, thus, cannot serve for these tasks. The assumption of equilibrium in nonkinetic methods is not conceptually required; moreover, equilibrium cannot be maintained in separation-based affinity methods. Thus, all nonkinetic methods can be converted to kinetic methods by changing conditions and approaches for data analysis.

In general, homogeneous methods are preferable due to their simplicity and kinetic methods are preferable due to their enabling kinetic features. Until recently, the only mixture-based heterogeneous method with comprehensive kinetic capabilities was SPR. In this chapter, kinetic capillary electrophoresis (KCE) is described as a conceptual platform for separation-based homogeneous methods with comprehensive kinetic capabilities, which will find multiple applications in chemistry, biology, medicine, and drug discovery.

11.2 THEORY OF KINETIC CAPILLARY ELECTROPHORESIS

Kinetic capillary electrophoresis is defined as capillary electrophoretic separation of species that interact during electrophoresis.[3–5] Thus, KCE involves two major processes: affinity interaction of L and T, described by Equation 11.1, and separation of L, T, and C based on differences in their electrophoretic velocities, v_L, v_T, and v_C, respectively. These two processes are described by the following general system of partial differential equations:

$$\begin{aligned}
\frac{\partial L(t,x)}{\partial t} + v_L \frac{\partial L(t,x)}{\partial x} &= -k_{on} L(t,x) T(t,x) + k_{off} C(t,x) \\
\frac{\partial T(t,x)}{\partial t} + v_T \frac{\partial T(t,x)}{\partial x} &= -k_{on} L(t,x) T(t,x) + k_{off} C(t,x) \\
\frac{\partial C(t,x)}{\partial t} + v_C \frac{\partial C(t,x)}{\partial x} &= -k_{off} C(t,x) + k_{on} L(t,x) T(t,x)
\end{aligned} \tag{11.3}$$

where L, T, and C are concentrations of L, T, and C, respectively; t is time passed since the beginning of separation; and x is the distance from the injection end of the capillary. System 11.3 describes two basic processes, which are always present in KCE. Depending on species studied and a specific analytical setup, other processes, such as binding with complex stoichiometry, diffusion, adsorption to capillary walls, and so forth, can play significant roles in KCE. In such cases, mathematical terms, describing additional processes, must be added to system 11.3. The solution of system 11.3 depends on the initial and boundary conditions: initial distribution of L, T, and C along the capillary and the way L, T, and C are introduced into the capillary and removed from the capillary during separation. This solution can be found nonnumerically for specific sets of initial and boundary conditions and specific assumptions.[6–9] For KCE to be a generic approach, it is required that system 11.3 be solved for any set of conditions; such solutions can be found only numerically.

11.3 KINETIC CAPILLARY ELECTROPHORESIS METHODS

Every set of qualitatively unique initial and boundary conditions for system 11.3 defines a unique KCE method. Here we will describe six KCE methods: nonequilibrium capillary electrophoresis of the equilibrium mixtures (NECEEM), sweeping CE (SweepCE), continuous NECEEM (cNECEEM), short SweepCE (sSweepCE), plug-plug KCE (ppKCE), and short SweepCE of Equilibrium Mixture (sSweepCEEM). Table 11.1 contains drawings that schematically illustrate initial and boundary conditions and show the mathematical representation of initial and boundary conditions. It also contains representative functions $L(t)$, $T(t)$, and $C(t)$ for a fixed x for each method. The notion of equilibrium mixture (EM) refers to the mixture of L, T, and C at equilibrium, typically prepared

TABLE 11.1
Summary of KCE Methods

KCE method	Schematic representation of initial and boundary conditions	Simulated concentration profiles
NECEEM	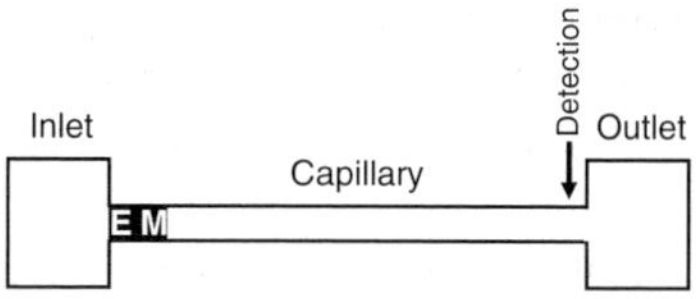	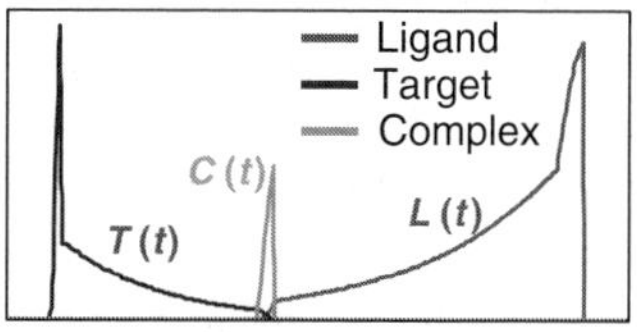
ppKCE	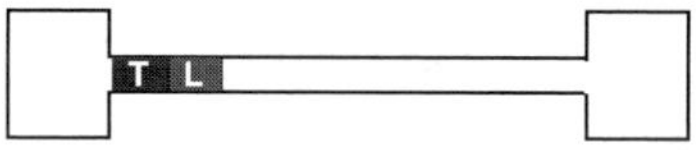	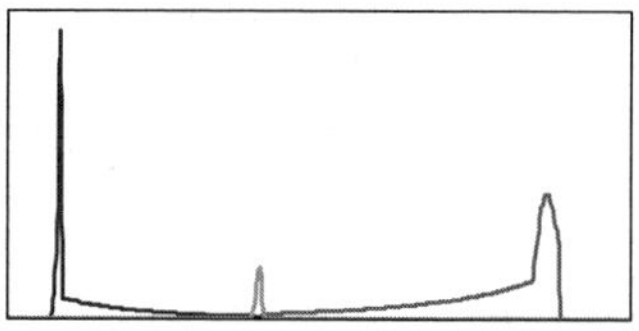
SweepCE		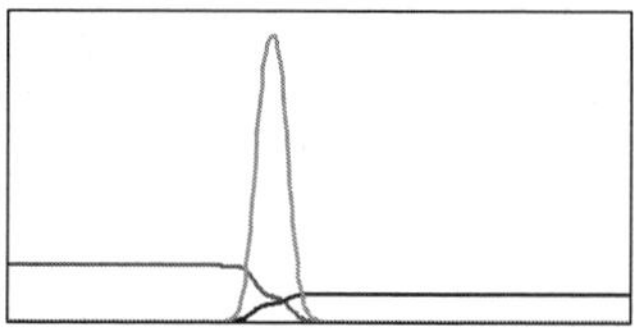
cNECEEM	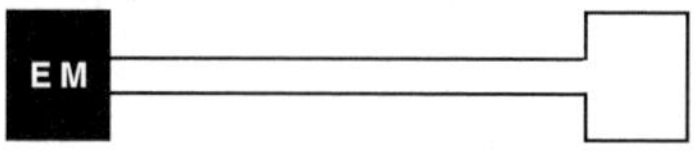	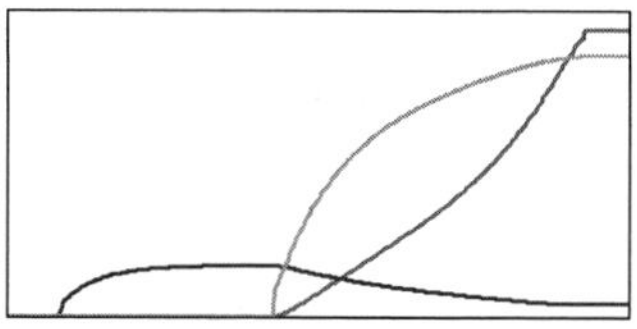
sSweepCE		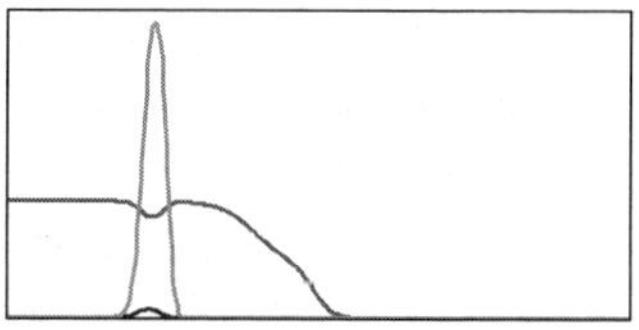
sSweepCEEM		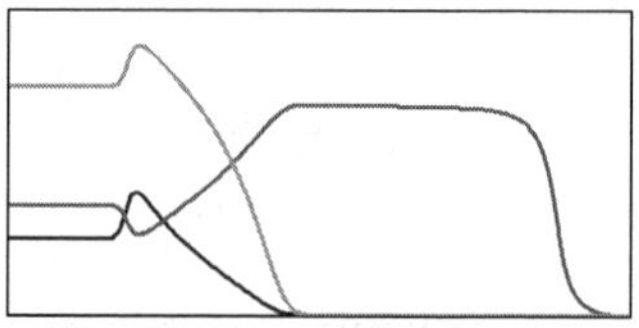
		Migration time to the detector

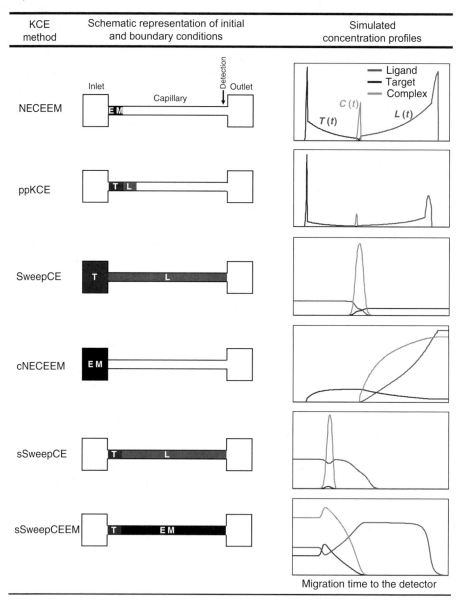

outside the capillary. The concentrations of the three components ($\tilde{T}, \tilde{L}$, and $\tilde{C}$) in the EM are interconnected through the equilibrium dissociation constant, K_d, as $K_d = (\tilde{T}\tilde{L})/\tilde{C}$.

11.3.1 NECEEM

Nonequilibrium capillary electrophoresis of the equilibrium mixtures (NECEEM) is a unique KCE method that facilitates finding both K_d and k_{off} from a single electropherogram.[6] Mechanistically, it is similar to previously introduced affinity probe capillary electrophoresis (APCE)[10–12] and falls into a broad category of capillary zone electrophoresis (CZE).[13–18] APCE was introduced as being applicable to systems having suitably stable complexes where the dissociation rate constants are small in the scale of characteristic separation times. In other words, the characteristic time of dissociation is much longer than the characteristic separation time. That is why, APCE has been applied only to measuring equilibrium constants but not rate constants.

So far, NECEEM was used to study the interaction between several proteins and DNA such as an *Escherichia coli* single-stranded DNA-binding protein (SSB) and a fluorescently labeled oligonucleotide (ssDNA),[6,7,19] *Taq* DNA polymerase and its aptamer,[19] thrombin and its aptamer,[20] Tau protein and single-stranded and double-stranded DNA,[21] protein farnesyltransferase (PFTase) and its aptamer,[22] MutS protein and its aptamer,[23] h-Ras protein and its aptamer,[24,25] and Mef2c protein and double-stranded DNA it naturally binds.[26] It has also been used to study protein–peptide interactions.[12]

In NECEEM, a short plug of the EM is injected into the inlet of the capillary, which is prefilled with the run buffer. Separation is carried out with both inlet and outlet reservoirs containing the run buffer only. C continuously dissociates during electrophoresis. If separation is efficient, re-association of T and L can be neglected. The resulting NECEEM electropherograms contain three peaks of T, C, and L and two exponential "smears" of L and T, which occur from the dissociation of C whose migration times and areas are used to calculate k_{off} and K_{d} (Figure 11.2). The important feature of NECEEM is that a single electropherogram contains data sufficient for finding both K_{d} and k_{off}. NECEEM starts with the EM; therefore, it has a memory of the equilibrium necessary for finding K_{d}. Although a single electropherogram is sufficient to find both K_{d} and k_{off}, for accurate measurements of the constants and their experimental errors several experiments have to be performed. The essential feature of NECEEM electropherograms is that the areas of peaks and smears in them are proportional to the amounts of corresponding species. A single NECEEM electropherogram can be used for finding four measurable parameters required for the determination of K_{d} and k_{off} (Figure 11.3). A_1 is the area of the peak corresponding to L, which was free in the EM. A_3 is the area of the exponential smear left by L dissociated from C during the separation. A_2 is the area of the peak corresponding to C, which remained intact by the time of passing the detector. Finally, t_{C} is the migration time of the complex.

The values of K_{d} and k_{off} can be calculated using the following algebraic formulas[8]:

$$K_{\text{d}} = \frac{[T]_0 \left(1 + A_1/(A_2 + A_3)\right) - [L]_0}{1 + (A_2 + A_3)/A_1}, \tag{11.4}$$

$$k_{\text{off}} = \frac{\ln\left((A_2 + A_3)/A_2\right)}{t_{\text{C}}}, \tag{11.5}$$

where $[T]_0$ and $[L]_0$ are total concentrations of T and L in the EM, which include free components and complexes. Advantageously, areas and migration time associated with a single species only (L in our example) are required. This simplifies the use of fluorescence detection because finding a strategy for labeling a single species is relatively easy. A major step in the method development for NECEEM involves finding conditions for good quality separation of L from C. Figure 11.3 shows an experimental NECEEM electropherogram. In this example, interaction between ssDNA and ssDNA-binding protein was studied. In the experimental electropherogram, the peaks have Gaussian-type shapes rather than rectangular ones presented in a schematic electropherogram in Figure 11.2. While measuring the areas, it is important to accurately define the boundary between them. The boundary between A_1 and A_3 can be found by comparing the peak of free L in the presence and absence of T. It was shown that the uncertainty in defining the boundaries between the areas leads to experimental errors in the range of 10%. This is an acceptable level of experimental errors for most applications. Alternatively, mathematical modeling of a NECEEM electropherogram can be used to find both K_{d} and k_{off} from the nonlinear regression analysis without the need to define the areas. Typically, the area method is used as a simple, fast, and acceptably accurate approach.[8]

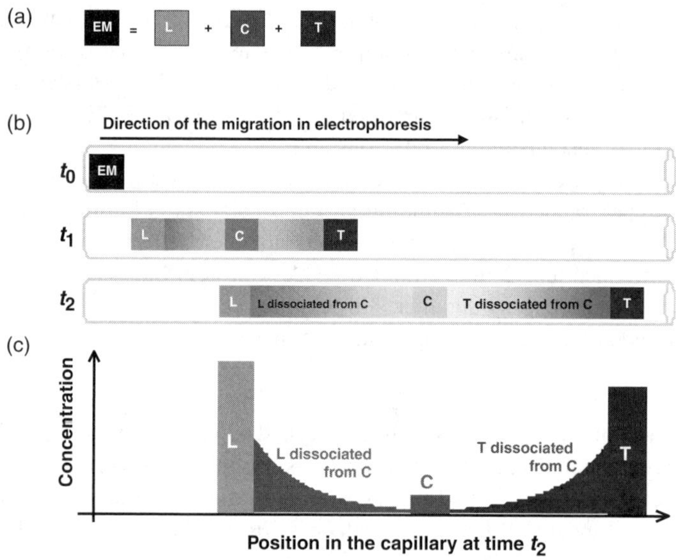

FIGURE 11.2 Conceptual representation of nonequilibrium capillary electrophoresis of equilibrium mixtures (NECEEM). (a) Components of the equilibrium mixture (EM): free ligand (L), free target (T), and the ligand-target complex (C). (b) NECEEM-based separation of L, T, and C. A short plug of the equilibrium mixture is injected into a capillary at time t_0. High voltage is then applied. It is assumed that T migrates faster than L; C typically has an intermediate mobility. Equilibrium fractions of free L and T migrate as individual zones, which do not change in time. The equilibrium fraction of C continuously dissociates during separation (time t_1 and t_2), leaving smears of L and T. By time t_2, only a fraction of C remains intact. (c) The graph shows concentrations of the separated components as functions of the position in the capillary at time t_2.

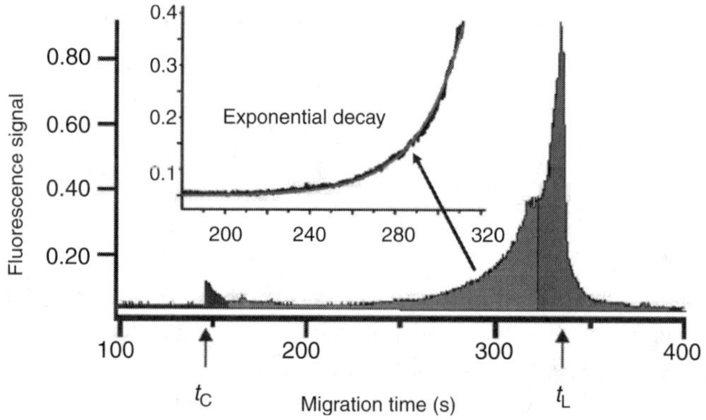

FIGURE 11.3 Example of a NECEEM electropherogram for the interaction between fluorescently labeled ssDNA (ligand) and single-stranded binding (SSB) protein (target). t_C and t_L are the migration time of the complex and the ligand, respectively.

NECEEM-based determination of K_d and k_{off} is fast and accurate, and it has a wide and adjustable dynamic range. The upper limit of K_d values depends on the highest concentration of T available and can be as high as millimolar. This allows for the measurement of K_d values for very low-bulk affinities of naive combinatorial libraries.[24] The lower limit of K_d depends on the concentration limit of detection; for fluorescence detection, it can be as low as picomolar. The dynamic range of k_{off} values is defined by the migration time of the complex, which can be easily regulated by the length of the capillary, electric field, or electro-osmotic velocity. The practically proven dynamic range of k_{off} spans from 10^{-4} to 1 s^{-1}.[6,8,12,19–21,27–29] Although only one electropherogram is required for finding both $K_d$and k_{off}, the concentration of T (if L is used as a detectable species) should be within an order of magnitude from the K_d value. Titration of T with 10-time increments in concentration is recommended as the fastest way of finding suitable T. Furthermore, conducting several experiments may be required to find the experimental deviation of the K_d value. The equilibrium is typically established in the incubation buffer, whereas dissociation occurs in the electrophoresis run buffer. The values of K_d and k_{off} are, thus, measured for the incubation buffer and run buffer, respectively. If the incubation buffer and the electrophoresis run buffer are identical, then K_d and k_{off} are determined under the same conditions, and k_{on} can be calculated as $k_{on} = k_{off}/K_d$. It is typically possible to make the incubation and run buffers identical. An example of when such matching is difficult is when T is the protein, which requires a high salt concentration. CE cannot tolerate high salt concentrations in the run buffer because of the high Joule heating, which can deteriorate the quality of separation.

11.3.1.1 Affinity-Mediated NECEEM

To explain the rationale for affinity-mediated NECEEM, it needs to be emphasized that NECEEM requires good separation of free ligand from the target–ligand complex. If the separation is poor, the accuracy of the method with respect to the determination of rate constants and equilibrium constants decreases. Affinity-mediated NECEEM is based on the insight that adding to the run buffer a background affinity agent, which can bind free ligand but not the target–ligand complex, can improve the separation by changing the mobility of free ligand while not affecting that of the complex. Affinity-mediated NECEEM was demonstrated for interaction between thrombin and its DNA aptamer by using the single-stranded binding (SSB) protein from *E. coli* as a background affinity agent in the run buffer (Figure 11.4).[20,30] Hypothetically, affinity-mediated NECEEM can be also realized with a target-binding affinity agent, such as an antibody, instead of a ligand-binding agent, provided that the agent binds the target but does not bind the target–ligand complex. A serious assumption used in affinity-mediated NECEEM is that the affinity agent does not affect the interaction between the target and a ligand.

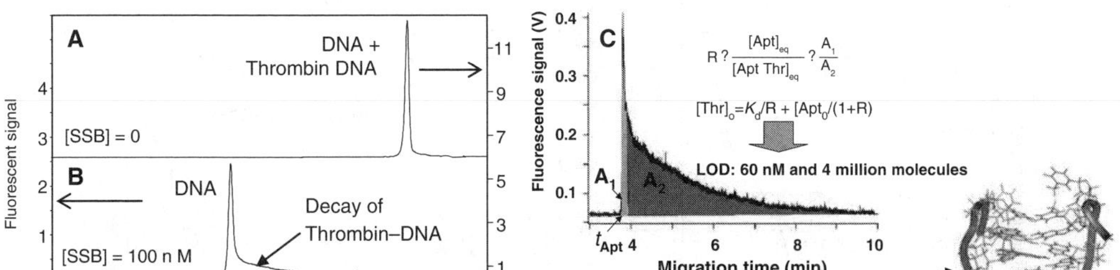

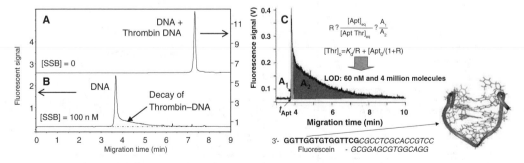

FIGURE 11.4 Affinity-mediated NECEEM. (a) Separation of thrombin and its aptamer (DNA) in the absence of SSB in the run buffer. (b) Separation of thrombin and DNA in the presence of SSB. (c) Quantitative detection of thrombin without observing the peak of the complex and just uses a decay of the complex. The structure of the aptamer (bottom of c) for human thrombin (boldface) with an additional 16-nucleotide sequence and a fluorescein-labeled probe that is complementary to the additional 16mer sequence. Complementary strands are given in italics.

11.3.1.2 Temperature-Controlled NECEEM

Temperature-controlled NECEEM was developed to study thermochemistry of target–ligand interactions (protein–DNA interaction as one of the examples).[19] Knowing how temperature influences kinetic and equilibrium binding parameters of noncovalent protein–DNA interactions is important for understanding fundamental biological processes, such as gene expression and DNA replication.[31–35] It is also essential for developing analytical applications of DNA aptamers and DNA-binding proteins in affinity and hybridization analyses and in optimizing the polymerase chain reaction (PCR).[30,36–38] Conventional methods for thermo-chemical studies of protein–DNA interactions have limitations. Differential scanning calorimetry (DSC) and isothermal titration calorimetry (ITC) are not applicable to finding kinetic parameters.[39–41] SPR can serve to determine equilibrium and kinetic parameters but it is a heterogeneous method, which requires the immobilization of either DNA or protein on the surface of a sensor.[42,43] Being a homogeneous kinetic method, temperature-controlled NECEEM uniquely allows finding temperature dependencies of equilibrium and kinetic parameters of complex formation without the immobilization of the interacting molecules on the surface of a solid substrate. Moreover, it requires much lower quantities of the protein than DSC, ITC, and SPR. Two protein–DNA pairs: (1) *Taq* DNA polymerase with its DNA aptamer and (2) *E. coli* single-stranded DNA-binding protein with a 20-base long ssDNA were analyzed by temperature-controlled NECEEM (Figure 11.5).[19] Temperature dependencies of three parameters were determined: the equilibrium binding constant ($K_b = 1/K_d$), the rate constant of complex dissociation (k_{off}), and the rate constant of complex formation (k_{on}). The $K_b(T)$ functions for both protein–DNA pairs had phase-transition-like points suggesting temperature-dependent conformational changes in structures of the interacting macromolecules. Temperature dependencies of k_{on} and k_{off} provided insights into how the conformational changes affected two opposite processes: binding and dissociation. Finally, thermodynamic parameters, ΔH and ΔS, for complex formation were found for different conformations. With its unique features and potential applicability to other macromolecular interactions, temperature-controlled NECEEM establishes a valuable addition to the arsenal of analytical methods used to study dynamic molecular complexes.

Additionally, there is a bonus application of temperature-controlled NECEEM: a nonspectroscopic approach to determining temperature in CE (Figure 11.6).[44] It is based on measuring a temperature-dependent rate constant of complex dissociation by NECEEM. This work was demonstrated on the dissociation of a protein–DNA complex to show that the new method allows

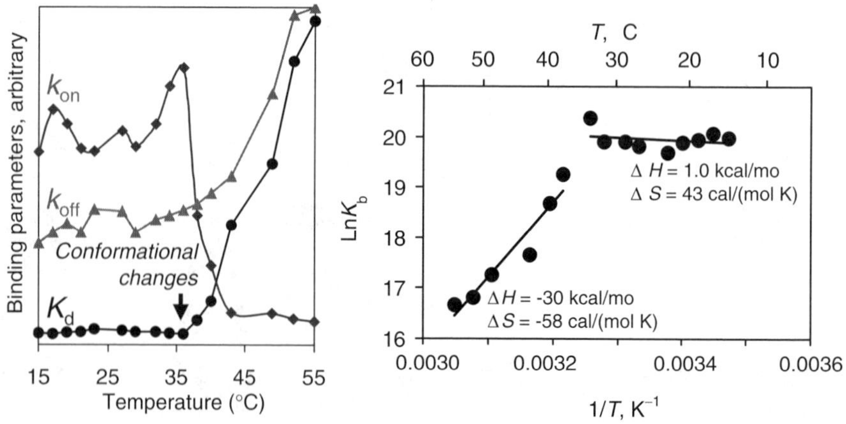

FIGURE 11.5 NECEEM-measured temperature dependencies of equilibrium and kinetic binding parameters and Van't Hoff plots used for the determination of ΔH and ΔS for complex formation for *Taq* DNA polymerase with its DNA aptamer.

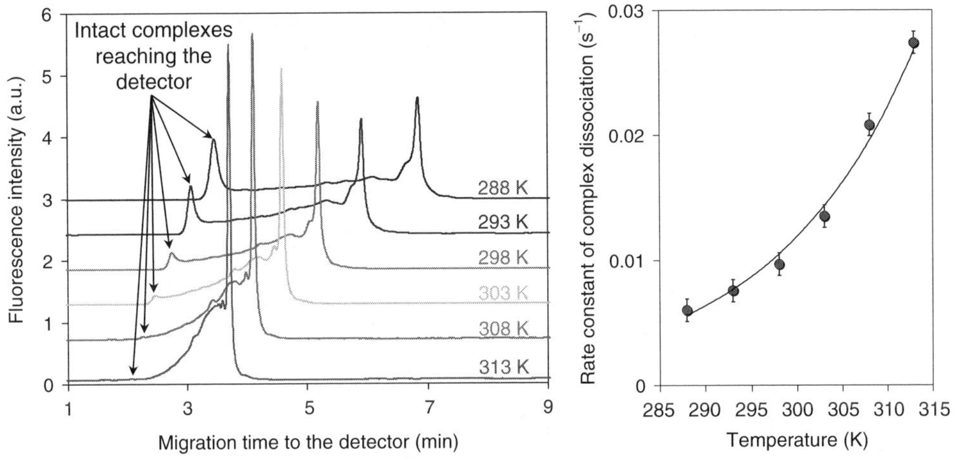

FIGURE 11.6 Temperature dependence of NECEEM electropherograms for SSB-ssDNA interaction (left panel), and calibration curve "k_{off} versus T" (right panel).

for temperature determination in CE with a precision of 2°C. With a number of advantages over conventional spectroscopic approaches, NECEEM-based temperature determination will find practical applications in CE method development for temperature-sensitive analyses, such as hybridization and affinity assays.

11.3.2 SweepCE

The monomolecular rate constant of complex dissociation, k_{off}, can be determined by either SPR,[45,46] or by NECEEM.[6–8] As for the bimolecular rate constant of complex formation, k_{on}, until now, the only technique available for its direct measurements was stopped-flow spectroscopy.[47–49] Stopped-flow spectroscopy relies on the change of spectral properties of either protein or DNA during complex formation. Such changes are often insignificant, which limits the applicability of stopped-flow methods to studies of protein–DNA interactions. SweepCE was introduced for directly measuring k_{on}, and demonstrated for studying protein–DNA interactions.[50] In contrast to stopped-flow spectroscopy, SweepCE does not rely on spectral changes of the protein or DNA upon complex formation. It requires only that electrophoretic mobilities of the protein and DNA be different, which is always achievable.

The concept of SweepCE is based on the sweeping of slowly migrating L by a fast migrating T during electrophoresis. The capillary is filled with L, while the inlet reservoir contains T and the outlet reservoir contains a run buffer. During electrophoresis, T continuously moves through L, causing continuous binding of T to L. Dissociation of C can also contribute to the resulting concentration profiles that contain a single peak of C and plateaus of T and L. The value of k_{on} for complex formation can be determined from the time-profile of T concentration using a simple mathematical model of the sweeping process. Mathematical analysis is an essential part of SweepCE. The method was demonstrated on the phenomenon of DNA sweeping by a DNA-binding protein in CE (Figure 11.7).

11.3.3 ppKCE

In ppKCE, the plugs of L and T are injected into the capillary prefilled with the run buffer. The inlet and outlet reservoirs contain the run buffer as well. During electrophoresis T moves through L causing the formation of C. When the zone of T passes L, C starts to dissociate. ppKCE can be considered as a functional hybrid of NECEEM and SweepCE. The resulting concentration profiles

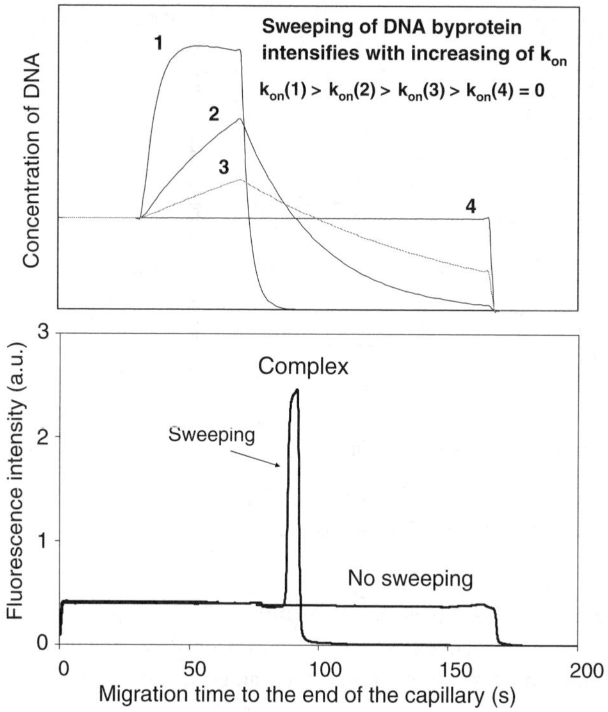

FIGURE 11.7 SweepCE is a first nonspectroscopic method for direct measuring of k_{on}. It is based on the sweeping of a slowly migrating ligand by a fast migrating target during electrophoresis. The value of k_{on} for complex formation can be determined from the time-profile of target concentration using a mathematical model of the sweeping process.

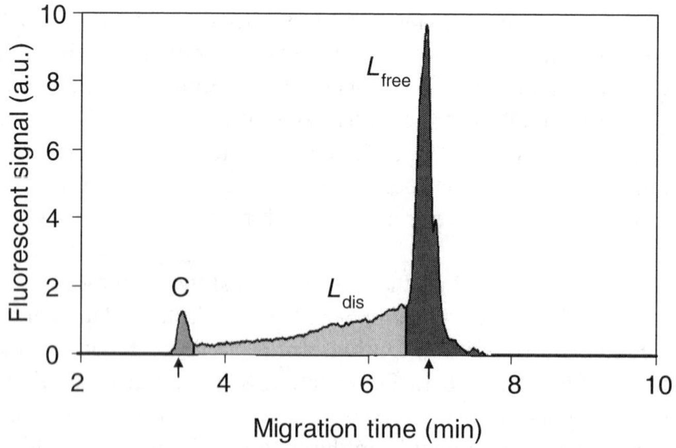

FIGURE 11.8 ppKCE is a method for direct measurement of both rate constants, k_{on} and k_{off}. The resulting concentration profile is very similar to NECEEM with a smaller peak of a complex (area = C), a decay of the complex (area = L_{dis}) and a free ligand (area = L_{free}).

resemble those of NECEEM with a smaller peak of C and "smears" of T and L (Figure 11.8).[3,9] However, since ppKCE does not start with the EM of L and T, the resulting electropherogram does not have a "memory" of K_d but rather has a memory of k_{on} and k_{off}. Both k_{on} and k_{off} can, thus, be calculated from a single ppKCE electropherogram using areas of peaks and smears and migration times of peaks. Calculation of rate constants of complex formation and dissociation from one experimental electropherogram is possible due to the developed simple mathematical approach without performing nonlinear regression analysis so that the method does not require expertise in mathematical modeling. So far, there has not been any method that allowed direct measurement of both rate constants. ppKCE is simple and robust. It requires only nanoliters volumes of reagents and can be readily adjusted for different ranges of both constants. This method will be useful for screening large libraries for drug candidates as well as the development of novel research and diagnostic tools.

11.3.4 Other KCE Methods

In cNECEEM, the inlet reservoir is filled with the EM while the capillary and the outlet reservoir contain the run buffer. During electrophoresis, C is separated from T, which moves faster, and from L, which moves slower. As a result, C continuously dissociates inside the capillary. Although dissociation is a prevalent process in cNECEEM, re-association can also contribute to the resulting concentration profiles, which are represented by smooth functions of $T(t)$, $L(t)$, and $C(t)$ with no pronounced peaks.

In sSweepCE, a short plug of T is injected into the capillary prefilled with L. Both inlet and outlet reservoirs contain the run buffer. T moves through L during electrophoresis causing both association of T and L and dissociation of resulting C to occur. The concentration profiles of T and C are peak-like, while that of L is a smooth function.

In sSweepCEEM, a short plug of T is injected into the capillary prefilled with the EM. Both inlet and outlet reservoirs contain the run buffer. During electrophoresis, an intricate interplay of dissociation of C and association of T and L occur resulting in sophisticated concentration profiles, which contain peaks and plateaus.

11.3.5 Multimethod KCE Toolbox

The degree of formation and dissociation of a complex differ in different KCE methods. KCE methods, therefore, have different accuracies of determination of k_{on} and k_{off}. For example, in NECEEM, complex dissociation prevails over complex formation, thus, making it more "sensitive" to k_{off} than k_{on}. In SweepCE, in contrast, complex formation can prevail over complex dissociation, making it more sensitive to k_{on} than k_{off}. The ppKCE method can be tuned to have comparable accuracy of both k_{on} and k_{off} determination. Therefore, KCE methods, which involve EMs (e.g., NECEEM, cNECEEM, and sSweepCEEM), are expected to be more accurate for the determination of the equilibrium constant, K_d. The most accurate determination of all constants can be achieved if multiple KCE methods are combined in a single kinetic tool. The approach can be used for testing hypotheses about the mechanisms of interaction and finding kinetic parameters of the interaction. Conceptually, experimental electropherograms are obtained by multiple KCE methods first. A hypothetical model of interactions between L and T is suggested and the system of differential equations (system 11.3) is built. The experimental KCE electropherograms are fitted with simulated electropherograms simultaneously to obtain the best fits with one of the criteria used for nonlinear regression analysis (e.g., minimum chi-square). If the quality of fitting is not satisfactory, a new hypothesis is suggested for the interaction. The procedure is repeated until a satisfying hypothesis is found. The best fits for the accepted hypothesis lead to the determination of stoichiometric and kinetic parameters of the interaction. Using the general concept of KCE, other KCE methods can be defined by simply selecting new sets of initial and boundary conditions. Importantly, this approach requires no serendipity but, rather, a rational (or irrational) design of conditions, which can be performed in an intuitive way schematically depicted in Table 11.1. The multimethod KCE toolbox allowed, for example, the determination of kinetic parameters of specific and nonspecific protein–DNA interactions.[3]

KCE establishes a new paradigm: separation methods can be used as comprehensive kinetic tools. The majority of previous attempts to utilize chromatography and electrophoresis for studying biomolecular interactions were limited to assuming equilibrium between interacting molecules.[51] Not only does such an assumption limit applications to measuring equilibrium constants, but also this assumption is conceptually mistaken since separation disturbs equilibrium. Kinetics must be appreciated when separation methods are used for studies of noncovalent interactions. This appreciation can dramatically enrich analytical capabilities of the methods.

11.4 KCE FOR APTAMER SELECTION AND DRUG DISCOVERY

DNA or RNA aptamers are single-stranded oligonucleotides that can bind proteins, small-molecule compounds, and living cells with high affinity and specificity.[52,53] Aptamers are very promising affinity ligands with the potential to change the field of affinity probes and replace antibodies as diagnostic, analytical, and therapeutic reagents.[54–56] Aptamers have indisputable advantages over antibodies due to the simplicity and low cost of production, simplicity of chemical modifications, and simplicity of integration into different analytical schemes. The unique properties of aptamers have led to their application in multiple areas of bioanalytical and biomedical sciences. They have been successfully used in proteomics and development of bioanalytical assays,[57] inhibition of enzymes and receptors,[7,8] development of artificial enzymes (ribozymes and aptazymes),[9] target validation and screening for drug candidates,[58–60] cytometry and imaging of cellular organelles,[61] and development of biosensors.[62] Aptamers are gaining a strengthening reputation as therapeutic reagents for the treatment of different pathologies.[63,64] Their potential medical applications also include gene therapy and drug delivery to therapeutic targets.[65] Despite a great promise and a significant effort in the development of aptamers over a period of 15 years, they have only been obtained for approximately 100 protein targets.[66] Such a slow progress is largely due to limitations of conventional technologies used for aptamer development.

Aptamers are typically selected from large libraries of random DNA (RNA) sequences in a general approach termed Systematic Evolution of Ligands by EXponential enrichment (SELEX).[67,68] In essence, SELEX involves repetitive rounds of two alternating processes: (1) partitioning of aptamers from nonaptamers by separating target-bound DNA from free DNA and (2) amplification of aptamers by the PCR (Figure 11.9a).

Noninstrumental methods of partitioning, such as filtration and gel electrophoresis, were initially used for SELEX; they still dominate the area.[69] Because of high background (the high level of target-nonbound DNA collected along with target-bound DNA), SELEX based on conventional partitioning methods requires a large number of rounds of selection, typically greater than 10. As a result, SELEX based on conventional partitioning methods is a lengthy and resource-consuming process. It often leads to DNA structures that bind to the surfaces of the filters or chromatographic support used in partitioning rather than to the target. Counter selection is successfully employed to eliminate such "surface aptamers"; however, it introduces additional rounds of selection, thus, making the procedure even longer. Another disadvantage of too many rounds of selection is the very limited number of unique aptamer sequences obtained at the output of conventional SELEX. This disadvantage is

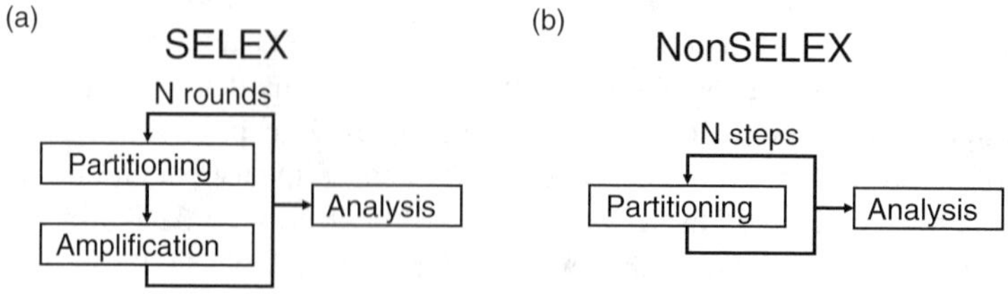

FIGURE 11.9 (a and b) Schematic representation of SELEX vs. nonSELEX.

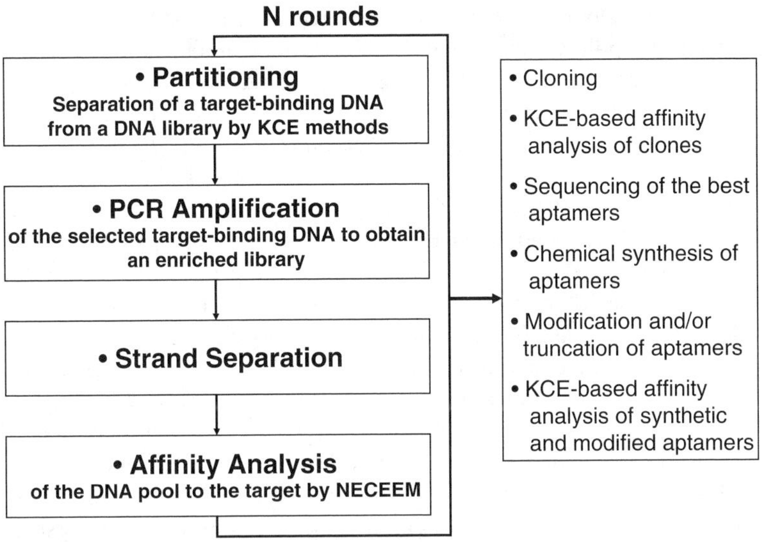

FIGURE 11.10 Schematic representation of KCE-based selection of DNA aptamers. This figure shows a simplified flowchart of *in vitro* selection of DNA aptamers. A random DNA pool containing 10^{11}–10^{15} unique sequences is incubated with a target. The next steps are KCE-based measuring of bulk affinity of DNA library to the target and KCE-based partitioning of a DNA-target complex from free DNA. The collected DNA is then amplified using PCR with a fluorescein-labeled forward primer and a biotin-labeled reverse primer. DNA strands are separated using streptavidin iron particles and a fluorescently labeled strand is collected to yield the affinity-enriched DNA library. This lower diversity pool is incubated with new aliquot of the target to examine its binding affinity using NECEEM. If the affinity is not high enough, a new selection round starts. When desirable affinity is reached, pool of aptamers is cloned into bacteria, clones are screened, and the best aptamers are sequenced. Aptamer sequences can be further modified to improve binding or other specific properties.

especially critical for aptamer-based drug development, which requires as many "lead molecules" as possible. Finally, if the efficiency of partitioning is too low, SELEX can completely fail to select aptamers.

Methods of KCE,[3] which started with pioneering works of Heegaard and Whitesides on affinity capillary electrophoresis (ACE),[15,70] establish a new methodological platform for partitioning of aptamers. Two distinct KCE methods have been used for the selection of aptamers: NECEEM[24,25,29] and ECEEM.[71,72] Bowser and co-authors[73,74] were the first to use NECEEM in SELEX; they called the approach CE-SELEX. The partitioning efficiency of KCE methods exceeds that of conventional partitioning methods, such as filtration and column chromatography, by at least two orders of magnitude.[29] As a result, KCE methods decrease the number of rounds of SELEX from 10 or more (required with conventional partitioning techniques) to 1–3 (Figure 11.10). In addition, KCE methods can be equally used for the selection of aptamers and for measurements of all their binding parameters: K_d, k_{on}, k_{off}, ΔH, and ΔS.[19] KCE methods have been demonstrated to facilitate selection of "smart" aptamers—ligands with predefined binding parameters.

11.4.1 KCE-Based Selection of Smart Aptamers

The designing of advanced aptamer-based diagnostics and therapeutics requires aptamers with predefined kinetic and/or thermodynamic parameters of aptamer–target interaction. Technological limitations of aptamer selection methods have so far precluded selection of such aptamers. Two KCE methods, ECEEM[71,72] and NECEEM,[24,25,29] have been successfully used for the selection of aptamers with predefined K_d and k_{off} of aptamer–target complexes. Conceptually, in ECEEM, a mixture of a target with a DNA (RNA) library is prepared and equilibrated (Figure 11.11). A plug of the EM is injected into a capillary prefilled with a run buffer containing the target at the concentration identical to the target concentration in the EM. The components of the EM are separated by CE while quasi-equilibrium is maintained between the target and aptamers inside the capillary. The unique feature of ECEEM is that aptamers with different K_d values migrate with different and predictable mobilities. Thus, collecting fractions with different mobilities results in aptamers with different and predefined K_d values. As an example, ECEEM was used to select aptamers with predefined K_d for binding MutS protein.[71] Three rounds of ECEEM-based selection were sufficient to obtain aptamers with K_d values approaching theoretically predicted ones (Figure 11.11).

In the NECEEM approach, smart aptamers can be selected with predefined K_d and k_{off} as well. A combinatorial DNA library is equilibrated with a protein target and the components of the EM are separated under nonequilibrium conditions. The nonequilibrium conditions promote the dissociation of the complex during separation. Fractions collected in a time window preceding the DNA library yield pools of DNA sequences capable of binding the target and dissociating from the complex with specific rates (Figure 11.12). Being a homogeneous method with comprehensive kinetic features, NECEEM provides a means for the selection of DNA aptamers with predefined ranges of all binding parameters of complex formation (K_d, k_{off}, k_{on}). First, the selection can be with respect to K_d values by varying the concentration of the protein target $[T]$ in the EM. The ratio between protein bound and unbound ligands changes according to classical equilibrium: the ligands where $K_d < [T]$ are preferentially bound to the protein and selected, while the ligands where $K_d > [T]$ are preferentially unbound and not selected. Second, the selection can be implemented with respect to k_{off} values by varying time windows in which fractions are collected. Finally, selection with respect to k_{on} values can be carried out by varying the time of incubation of the library with the target. To select for a single binding constant, the parameters that control the other two binding constants should be kept unchanged.

NECEEM appears to be a more complicated method than ECEEM, because each collection window may contain ligands with totally different k_{off} values; however, ligands with k_{off} defined in Figure 11.12 are the most abundant and will be predominant after multiple rounds of selection. The concept is different in ECEEM, in which every window theoretically contains only ligands with calculated K_d values, and deviations occur only because of nonspecific interactions during separation, nonzero width of the peaks, and disturbed equilibrium. Thus, NECEEM is a more "evolutionary" method of iterative selection, which requires more than one round to select oligonucleotides with a narrow range of k_{off} values.

11.4.2 NonSELEX Selection of Aptamers

The outstanding partitioning capabilities of KCE methods have motivated the attempt to select aptamers in a procedure that does not include intermediate amplifications steps; this approach is called nonSELEX[24,25] (Figure 11.9b). Excluding repetitive steps of PCR amplification accelerates the procedure of aptamer selection without compromising its efficiency. Omitting repetitive steps of PCR also excludes quantitative errors associated with the exponential nature of PCR amplification, thus making nonSELEX a useful tool for studies of properties of DNA libraries with respect to their interaction with targets. For example, nonSELEX can be used to accurately back-calculate the number of aptamer molecules in the naive DNA library. Further, excluding repetitive steps of PCR allows one to avoid the bias related to differences in PCR efficiency with respect to different oligonucleotide sequences. Finally, nonSELEX can potentially provide a viable alternative to SELEX in the commercial development of aptamers. It should be noted that the implementation of nonSELEX with currently available commercial CE instrumentation has a limitation: only a fraction of the collected ligands can be sampled for the next step of nonSELEX. This limitation requires that the fraction of aptamers in the naive library be no lower than 5×10^{-10} for the parameters used in this protocol. Usually, the abundance of aptamers in the naive library is greater than 5×10^{-10}, thus making

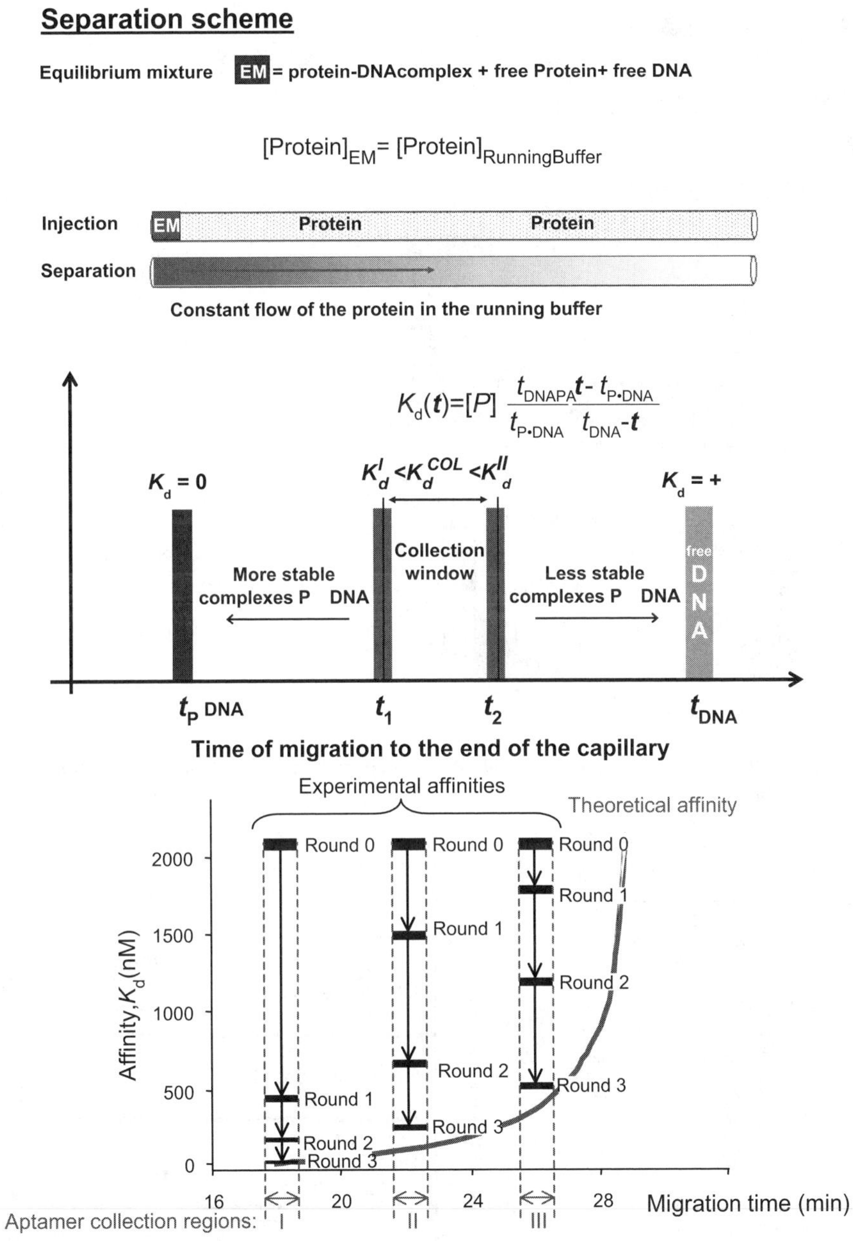

FIGURE 11.11 ECEEM-based selection of smart aptamers with predefined K_d values.

the limitation less important. As a real example, NECEEM-based nonSELEX for the selection of DNA aptamers was shown for h-Ras protein.[24] Three steps of NECEEM-based partitioning in the NonSELEX approach were sufficient to improve the affinity of a DNA library to a target protein by more than four orders of magnitude (Figure 11.13). The resulting affinity was higher than that of the enriched library obtained in three rounds of NECEEM-based SELEX. Remarkably, NECEEM-based NonSELEX selection took only 1 h to complete in contrast to several days or several weeks required for a typical SELEX procedure by conventional partitioning methods. In addition, NECEEM-based NonSELEX allowed to accurately measure the abundance of aptamers in the library. Not only does

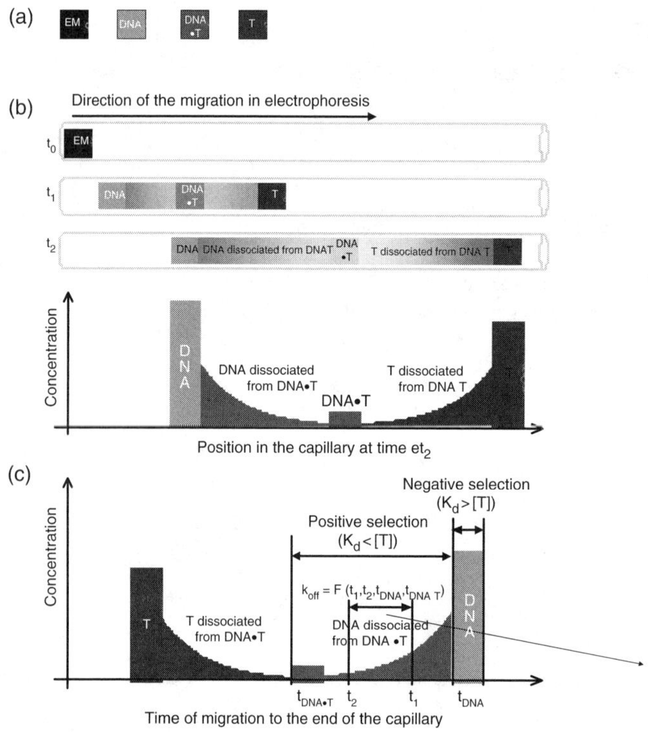

$$k_{off} = \frac{t_{DNA} - t_{DNA\bullet T}}{t_{DNA\bullet T}(t_1 - t_2)} \ln\left(\frac{t_{DNA} - t_2}{t_{DNA} - t_1}\right)$$

FIGURE 11.12 NECEEM-based selection of DNA aptamers with predefined ranges of all binding parameters of complex formation (K_d, k_{off}, k_{on}).

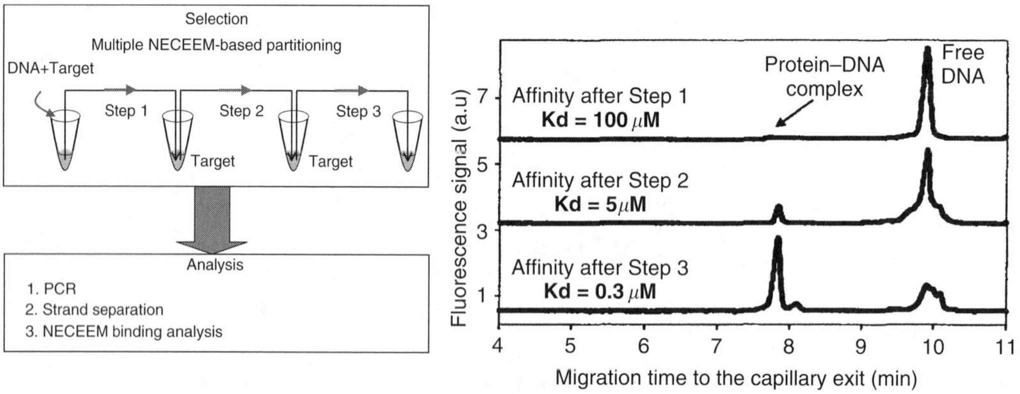

FIGURE 11.13 NonSELEX. Partitioning of aptamers from nonaptamers in nonSELEX by multiple steps of NECEEM with no PCR amplification between the steps.

this work introduce an extremely fast and economical method for aptamer selection, but it also suggests that aptamers may be much more abundant than they are thought to be. The step-by-step protocol for nonSELEX has been published.[25] Even more importantly, this work opens the opportunity for the selection of drug candidates from large nonDNA libraries, which cannot be PCR-amplified and are, thus, not approachable by SELEX.

11.4.3 The Prospective of KCE in Drug Discovery

The nonSELEX concept provides the opportunity for the selection of affinity ligands from DNA-encoded libraries of small molecules and peptides.[75,76] DNA tags in such libraries encode the information, which allows one to identify structures of corresponding small molecules, when the DNA tags are PCR amplified and sequenced. Owing to the small size of the molecules with respect to that of the covalently attached DNA tag, such libraries are expected to have electrophoretic properties identical to those of DNA libraries. SELEX is not applicable to such libraries since small molecules and peptides cannot be amplified by PCR.

11.5 CONCLUSION

KCE is based on the major principle of separation science that a complex can be separated from a target. In this case, the kinetics of molecular interactions must be appreciated. All methods of KCE use a single instrumental platform and a single conceptual platform for solving multiple tasks associated with biomolecular screening. A variety of different KCE methods can be designed by defining different ways of interaction between molecules in CE. Because of their comprehensive analytical capabilities, KCE methods have the potential to become a workhorse of biomolecular screening. This makes KCE methods highly attractive for the pharmaceutical industry as a novel approach to the selection and characterization of drug candidates. To conclude, it is clear that KCE methods will find multiple applications in fundamental studies of biomolecular interactions, designing clinical diagnostics, and the development of affinity probes and drug candidates. New applications will emerge with further development of KCE.

REFERENCES

1. Englebienne, P., Van Hoonacker, A., and Verhas, M. Surface plasmon resonance: Principles, methods and applications in biomedical sciences. *Spectroscopy* 17, 255–273 (2003).
2. Homola, J. Present and future of surface plasmon resonance biosensors. *Anal. Bioanal. Chem.* 377, 528–539 (2003).
3. Petrov, A., Okhonin, V., Berezovski, M., and Krylov, S. N. Kinetic capillary electrophoresis (KCE): A conceptual platform for kinetic homogeneous affinity methods. *J. Am. Chem. Soc.* 127, 17104–17110 (2005).
4. Berezovski, M., Okhonin, V., Petrov, A., and Krylov, S. N. Kinetic methods in capillary electrophoresis and their applications. *Proc. Int. Soc. Opt. Eng.* 5969, 203–215 (2005).
5. Berezovski, M. Kinetic capillary electrophoresis and its applications. *Ph.D. Dissertation*, Department of Chemistry, York University, Toronto (2005).
6. Berezovski, M., and Krylov, S. N. Nonequilibrium capillary electrophoresis of equilibrium mixtures—A single experiment reveals equilibrium and kinetic parameters of protein–DNA interactions. *J. Am. Chem. Soc.* 124, 13764–13765 (2002).
7. Krylov, S. N., and Berezovski, M. Non-equilibrium capillary electrophoresis of equilibrium mixtures—Appreciation of kinetics in capillary electrophoresis. *Analyst* 128, 571–575 (2003).
8. Okhonin, V., Krylova, S. M., and Krylov, S. N. Nonequilibrium capillary electrophoresis of equilibrium mixtures, mathematical model. *Anal. Chem.* 76, 1507–1512 (2004).
9. Okhonin, V., Petrov, A., Berezovski, M., and Krylov, S. N. Plug-plug kinetic capillary electrophoresis: A method for direct determination of rate constants of complex formation and dissociation. *Anal. Chem.* 78, 4803–4810 (2006).
10. Shimura, K., and Karger, B. L. Affinity probe capillary electrophoresis: Analysis of recombinant human growth hormone with a fluorescent labeled antibody fragment. *Anal. Chem.* 66, 9–15 (1994).
11. German, I., Buchanan, D. D., and Kennedy, R. T. Aptamers as ligands in affinity probe capillary electrophoresis. *Anal. Chem.* 70, 4540–4545 (1998).
12. Yang, P., Whelan, R. J., Jameson, E. E., Kurzer, J. H., Argetsinger, L. S., Carter-Su, C., Kabir, A., Malik, A., and Kennedy, R. T. Capillary electrophoresis and fluorescence anisotropy for quantitative

analysis of peptide–protein interactions using JAK2 and SH2-Bb as a model system. *Anal. Chem.* 77, 2482–2489 (2005).

13. Dolnik, V. Recent developments in capillary zone electrophoresis of proteins. *Electrophoresis* 20, 3106–3115 (1999).
14. Busch, M. H. A., Boelens, H. F. M., Kraak, J. C., Poppe, H., Meekel, A. A. P., and Resmini, M. Critical evaluation of the applicability of capillary zone electrophoresis for the study of hapten–antibody complex formation. *J. Chromatogr. A* 744, 195–203 (1996).
15. Heegaard, N. H. H., and Robey, F. A. Use of capillary zone electrophoresis to evaluate the binding of anionic carbohydrates to synthetic peptides derived from human serum amyloid P component. *Anal. Chem.* 64, 2479–2482 (1992).
16. Kraak, J. C., Busch, S., and Poppe, H. Study of protein–drug binding using capillary zone electrophoresis. *J. Chromatogr. A* 608, 257–264 (1992).
17. Wu, S., and Dovichi, N. J. High-sensitivity fluorescence detector for fluorescein isothiocyanate derivatives of amino acids separated by capillary zone electrophoresis. *J. Chromatogr.* 480, 141–155 (1989).
18. Zhao, J. Y., Dovichi, N. J., Hindsgaul, O., Gosselin, S., and Palcic, M. M. Detection of 100 molecules of product formed in a fucosyltransferase reaction. *Glycobiology* 4, 239–242 (1994).
19. Berezovski, M., and Krylov, S. N. Thermochemistry of protein–DNA interaction studied with temperature-controlled nonequilibrium capillary electrophoresis of equilibrium mixtures. *Anal. Chem.* 77, 1526–1529 (2005).
20. Berezovski, M., Nutiu, R., Li, Y., and Krylov, S. N. Affinity analysis of a protein–aptamer complex using nonequilibrium capillary electrophoresis of equilibrium mixtures. *Anal. Chem.* 75, 1382–1386 (2003).
21. Krylova, S. M., Musheev, M., Nutiu, R., Li, Y., Lee, G., and Krylov, S. N. Tau protein binds both double-stranded and single-stranded DNA—The proof obtained with nonequilbrium capillary electrophoresis of equilibrium mixtures. *FEBS Lett.* 579, 1371–1375 (2005).
22. Berezovski, M., Drabovich, A., Krylova, S. M., Musheev, M., Okhonin, V., Petrov, A., and Krylov, S. N. Nonequilibrium Capillary Electrophoresis of Equilibrium Mixtures: A universal tool for development of aptamers. *J. Am. Chem. Soc.* 127, 3165–3171 (2005).
23. Drabovich, A., Berezovski, M., and Krylov, S. N. Selection of smart aptamers by equilibrium capillary electrophoresis of equilibrium mixtures (ECEEM). *J. Am. Chem. Soc.* 127, 11224–11225 (2005).
24. Berezovski, M., Musheev, M., Drabovich, A., and Krylov, S. N. Non-SELEX selection of aptamers. *J. Am. Chem. Soc.* 128, 1410–1411 (2006).
25. Berezovski, M. V., Musheev, M. U., Drabovich, A. P., Jitkova, J., and Krylov, S. N. Non-SELEX: Selection of aptamers without intermediate amplification of candidate oligonucleotides. *Nat. Protoc.* 1, 1359–1369 (2006).
26. Krylov, S., Krylova, S., and Berezovski, M. Non-equilibrium capillary electrophoresis of equilibrium mixtures (NECEEM)-based methods for drug and diagnostic development. *U.S. Pat. Appl. Publ.*, 41 (2005).
27. Berezovski, M., and Krylov, S. N. Using nonequilibrium capillary electrophoresis of equilibrium mixtures for the determination of temperature in capillary electrophoresis. *Anal. Chem.* 76, 7114–7117 (2004).
28. Krylov, S. N. Nonequilibrium capillary electrophoresis of equilibrium mixtures (NECEEM): A novel method for biomolecular screening. *J. Biomol. Screen.* 11, 115–122 (2006).
29. Berezovski, M., Drabovich, A., Krylova, S. M., Musheev, M., Okhonin, V., Petrov, A., and Krylov, S. N. Nonequilibrium capillary electrophoresis of equilibrium mixtures: A universal tool for development of aptamers. *J. Am. Chem. Soc.* 127, 3165–3171 (2005).
30. Berezovski, M., and Krylov, S. N. Using DNA-binding proteins as an analytical tool. *J. Am. Chem. Soc.* 44, 13451–13454 (2003).
31. Ma, H., and Zou, Y. Thermodynamic characterization of the interaction of mutant UvrB protein with damaged DNA. *Biochemistry* 43, 4206–4211 (2004).
32. Schubert, F., Zettl, H., Haefner, W., Krauss, G., and Krausch, G. Comparative thermodynamic analysis of DNA–protein interactions using surface plasmon resonance and fluorescence correlation spectroscopy. *Biochemistry* 42, 10288–10294 (2003).

33. Galletto, R., and Bujalowski, W. Kinetics of the *E. coli* replication factor DnaC protein–nucleotide interactions. II. Fluorescence anisotropy and transient, dynamic quenching stopped-flow studies of the reaction intermediates. *Biochemistry* 41, 8921–8934 (2002).
34. Mosig, G., Gewin, J., Luder, A., Colowick, N., and Vo, D. Two recombination-dependent DNA replication pathways of bacteriophage T4, and their roles in mutagenesis and horizontal gene transfer. *Proc. Natl Acad. Sci. USA* 98, 8306–8311 (2001).
35. Sutani, T., and Yanagida, M. DNA renaturation activity of the SMC complex implicated in chromosome condensation. *Nature* 388, 798–801 (1997).
36. Savran, C. A., Knudsen, S. M., Ellington, A. D., and Manalis, S. R. Micromechanical detection of proteins using aptamer-based receptor molecules. *Anal. Chem.* 76, 3194–3198 (2004).
37. Dang, C., and Jayasena, S. D. Oligonucleotide inhibitors of *Taq* DNA polymerase facilitate detection of low copy number targets by PCR. *J. Mol. Biol.* 264, 268–278 (1996).
38. Yakimovich, O. Y., Alekseev, Ya. I., Maksimenko, A. V., Voronina, O. L., and Lunin, V. G. Influence of DNA aptamer structure on the specificity of binding to *Taq* DNA polymerase. *Biochemistry* (*Moscow*) 68, 228–235 (2003).
39. Carra, J. H., and Privalov, P. L. Energetics of folding and DNA binding of the MAT alpha 2 homeodomain. *Biochemistry* 36, 526–535 (1997).
40. Knapp, S., Karshikoff, A., Berndt, K. D., Christova, P., Atanasov, B., and Ladenstein, R. Thermal unfolding of the DNA-binding protein Sso7d from the hyperthermophile *Sulfolobus solfataricus*. *J. Mol. Biol.* 264, 1132–1144 (1996).
41. Davis, K. G., Plyte, S. E., Robertson, S. R., Cooper, A., and Kneale, G. G. Comparison of Pf1 and Fd gene 5 proteins and their single-stranded DNA complexes by NMR spectroscopy and differential scanning calorimetry. *Biochemistry* 34, 148–154 (1995).
42. Shumaker-Parry, J. S., Aebersold, R., and Campbell, C. T. Parallel, quantitative measurement of protein binding to a 120-element double-stranded DNA array in real time using surface plasmon resonance microscopy. *Anal. Chem.* 76, 2071–2082 (2004).
43. Tsoi, P. Y., and Yang, M. Surface plasmon resonance study of the molecular recognition between polymerase and DNA containing various mismatches and conformational changes of DNA–protein complexes. *Biosens. Bioelectron.* 19, 1209–1218 (2004).
44. Berezovski, M., and Krylov, S. N. A non-spectroscopic method for the determination of temperature in capillary electrophoresis. *Anal. Chem.* 76, 7114–7117 (2004).
45. Imanishi, M., and Sugiura, Y. Artificial DNA-bending six-zinc finger peptides with different charged linkers: Distinct kinetic properties of DNA bindings. *Biochemistry* 41, 1328–1334 (2002).
46. Cheskis, B., and Freedman, L. P. Modulation of nuclear receptor interactions by ligands: Kinetic analysis using surface plasmon resonance. *Biochemistry* 35, 3309–3318 (1996).
47. Kozlov, A. G., and Lohman, T. M. Kinetic mechanism of direct transfer of *Escherichia coli* SSB tetramers between single-stranded DNA molecules. *Biochemistry* 41, 6032–6044 (2002).
48. Noel, A.-J., Wende, W., and Pingoud, A. J. DNA Recognition by the homing endonuclease PI-SceI involves a divalent metal ion cofactor-induced conformational change. *Biol. Chem.* 279, 6794–6804 (2004).
49. Zeeb, M., and Balbach, J. Single-stranded DNA binding of the cold-shock protein CspB from *Bacillus subtilis*: NMR mapping and mutational characterization. *J. Protein Sci.* 12, 112–123 (2003).
50. Okhonin, V., Berezovski, M., and Krylov, S. N. Sweeping capillary electrophoresis: A non-stopped-flow method for measuring bimolecular rate constant of complex formation between protein and DNA. *J. Am. Chem. Soc.* 126, 7166–7167 (2004).
51. Heegaard, N. H. H., and Kennedy, R. Identification, quantitation, and characterization of biomolecules by capillary electrophoretic analysis of binding interactions. *Electrophoresis* 20, 3122–3133 (1999).
52. Wilson, D. S., and Szostak, J. W. *In vitro* selection of functional nucleic acids. *Annu. Rev. Biochem.* 68, 611–647 (1999).
53. Famulok, M., Mayer, G., and Blind, M. Nucleic acid aptamers—from selection *in vitro* to applications *in vivo*. *Acc Chem. Res.* 33, 591–599 (2000).
54. Rimmele, M. Nucleic acid aptamers as tools and drugs: Recent developments. *Chembiochem* 4, 963–971 (2003).

55. Jayasena, S. D. Aptamers: An emerging class of molecules that rival antibodies in diagnostics. *Clin. Chem.* 45, 1628–1650 (1999).
56. Ellington, A. D., and Conrad, R. Aptamers as potential nucleic acid pharmaceuticals. *Biotechnol. Annu. Rev.* 1, 185–214 (1995).
57. Bock, C., Coleman, M., Collins, B., Davis, J., Foulds, G., Gold, L., Greef, C., et al. Photoaptamer arrays applied to multiplexed proteomic analysis. *Proteomics* 4, 609–618 (2004).
58. Mayer, G., and Jenne, A. Aptamers in research and drug development. *BioDrugs* 18, 351–359 (2004).
59. Nutiu, R., Yu, J. M., and Li, Y. Signaling aptamers for monitoring enzymatic activity and for inhibitor screening. *Chembiochem* 5, 1139–1144 (2004).
60. Famulok, M., and Mayer, G. Intramers and aptamers: Applications in protein-function analyses and potential for drug screening. *Chembiochem* 6, 19–26 (2005).
61. Ulrich, H., Martins, A. H., and Pesquero, J. B. RNA and DNA aptamers in cytomics analysis. *Cytometry A* 59, 220–231 (2004).
62. Tombelli, S., Minunni, M., and Mascini, M. Analytical applications of aptamers. *Biosens. Bioelectron.* 20, 2424–2434 (2005).
63. Nimjee, S. M., Rusconi, C. P., and Sullenger, B. A. Aptamers: An emerging class of therapeutics. *Annu. Rev. Med.* 56, 555–583 (2005).
64. Lee, J. F., Stovall, G. M., and Ellington, A. D. Aptamer therapeutics advance. *Curr. Opin. Chem. Biol.* 10, 282–289 (2006).
65. Patil, S. D., Rhodes, D. G., and Burgess, D. J. DNA-based therapeutics and DNA delivery systems: A comprehensive review. *AAPS J.* 7, 61–77 (2005).
66. Lee, J. F., Hesselberth, J. R., Meyers, L. A., and Ellington, A. D. Aptamer database. *Nucleic Acids Res.* 32, D95–D100 (2004).
67. Ellington, A., and Szostak, J. *In vitro* selection of RNA molecules that bind specific ligands. *Nature* 346, 818–822 (1990).
68. Tuerk, C., and Gold, L. Systematic evolution of ligands by exponential enrichment: RNA ligands to bacteriophage T4 DNA polymerase. *Science* 249, 505–510 (1990).
69. Gopinath, S. C. Methods developed for SELEX. *Anal. Bioanal. Chem.* 387, 171–182 (2006).
70. Chu, Y. H., Avila, L. Z., Biebuyck, H. A., and Whitesides, G. M. Use of affinity capillary electrophoresis to measure binding constants of ligands to proteins. *J. Med. Chem.* 35, 2915–2907 (1992).
71. Drabovich, A., Berezovski, M., and Krylov, S. N. Selection of smart aptamers by equilibrium capillary electrophoresis of equilibrium mixtures (ECEEM). *J. Am. Chem. Soc.* 127, 11224–11225 (2005).
72. Drabovich, A. P., Berezovski, M., Okhonin, V., and Krylov, S. N. Selection of smart aptamers by methods of kinetic capillary electrophoresis. *Anal. Chem.* 78, 3171–3178 (2006).
73. Mendonsa, S. D., and Bowser, M. T. *In vitro* evolution of functional DNA using capillary electrophoresis. *J. Am. Chem. Soc.* 126, 20–21 (2004).
74. Mendonsa, S. D., and Bowser, M. T. *In vitro* selection of high-affinity DNA ligands for human IgE using capillary electrophoresis. *Anal. Chem.* 76, 5387–5392 (2004).
75. Kanna, M. W., Rozenman, M. M., Sakurai, K., Snyder, T. M., and Liu, D. R. Reaction discovery enabled by DNA-templated synthesis and *in vitro* selection. *Nature* 431, 545–549 (2004).
76. Levin, A. M., and Weiss, G. A. Optimizing the affinity and specificity of proteins with molecular display. *Mol. Biosyst.* 2, 49–57 (2006).

12 DNA Sequencing and Genotyping by Free-Solution Conjugate Electrophoresis

Jennifer A. Coyne, Jennifer S. Lin, and Annelise E. Barron

CONTENTS

12.1 INTRODUCTION

Most DNA sequencing today is done essentially by the same enzymatic method conceived by Frederick Sanger in the 1970s—while DNA electrophoresis instruments and molecular labels have improved (we now use fluorescent dyes, whereas Sanger used radioactive labeling), the basic approach is the same. When we read sequence from samples produced by the Sanger method, the problem is reduced to biomolecule separation and detection. DNA molecules, each carrying a terminal fluorophore that encodes the identity of the terminal base (A, C, G, or T), must be fractionated according to chain length with single-base resolution, and for the best process efficiency, with the longest "read-length" possible. Ideally at least 600–650 bases of contiguous, high-accuracy sequence is desired per read for the *de novo* sequencing of complex genomes such as the human genome, which has a high content of repetitive DNA sequence that can be difficult to "read through" and assemble correctly if only shorter reads are available.

Accordingly, since 1990 or so, a key aspect of pushing sequencing technology forward has been perfecting the ability of electrophoresis instruments to separate DNA molecules according to size with great efficiency. A major turning point, reached in 1999, was the widespread shift of sequencing from slab gel instruments, which required genome center technicians to carry out manual steps (e.g., pouring very large, ultra-thin polyacrylamide slab gels, loading DNA samples), to automated capillary array electrophoresis (CAE) systems, which separate fluorescently labeled DNA molecules within 50- or 75-μm inner diameter glass capillaries, in fluid, uncrosslinked polyacrylamide solutions rather than crosslinked gels. Although the basic technological approach to sequencing is still essentially "gel electrophoresis," the greater automation provided by CAE along with intensive logistical and computational efforts in academia and private enterprise enabled a draft human genome sequence to be obtained 2 years ahead of schedule, in 2003.[1,2]

The sequencing of that first composite human genome, a good deal of which was done with slab gel systems, cost a total of $3 billion.[1–6] Thanks to improvements to CAE and other technological developments seeded in academic research laboratories with funds granted by the National Institutes of Health (NIH)/National Human Genome Research Institute (NHGRI) as well as in industry, genome centers now analyze genomes 24 h/day by automated CAE of DNA in solutions of entangled, water-soluble polymers.[7] The aforementioned viscous, entangled polymer solutions are at present necessary, consumable, and expensive elements of CAE technology, as they provide the size-based separation of DNA required to read sequence by the Sanger method.

In 2004, the NIH/NHGRI initiated a race to develop new and different technologies that could decrease the cost for sequencing an individual human genome (which comprises 6 billion base pairs, considering both sets of chromosomes) from the now-current (2007) price of $10–20 million to $100K in the near future, and eventually to $1K.[8] Changes to the separation technology could drop the cost a lot, since about 60% of the present cost of sequencing a genome is due to CAE and its required consumables (capillary arrays and polymer solutions). While CAE has transformed both science and society and remains extremely important (as the only technology suitable for the de novo sequencing of complex genomes, such as the human), less expensive, more efficient, and more highly parallel technologies are still greatly needed. In 2006, the National Cancer Institute (NCI) announced "The Cancer Genome Atlas" (TCGA), a new project that aims to map and analyze genetic mutations occurring in different types of human cancers; this will require full sequencing and assembly of ~15,000 human genomes, an effort much greater in scale than the Human Genome Project (HGP).[9,10] A 100-fold reduction in sequencing cost to $100K per genome will be needed if the Cancer Genome Atlas Project is to be accomplished within its presently planned budget.[11] The second 100-fold cost decrease, to a $1K genome, would enable individual human genomes to be sequenced as a new aspect of general medical care. In addition, the recently announced, privately backed Archon X Prize for Genomics promises $10 million to the first group that sequences 100 human genomes in less than 10 days,[12,13] which would represent a giant leap beyond the 2003 completion of the HGP. However, given current sequencing costs and the state of present technologies, old and new, it certainly seems that we have a long way to go before sequencing is cheap enough to make $10 million an attractive prize for sequencing 100 human genomes, so that the Archon Prize can be paid out.[14]

Reducing sequencing costs requires both increased throughput and decreased cost per base. In practical terms, capillary electrophoresis (CE) using polymer matrices for DNA separation is limited to a maximum read-length (the number of contiguous DNA bases that can be sequenced with high accuracy, i.e., with good single-base resolution) of about 800–900 bases, in part because of limitations imposed by the physical mechanism of DNA separation that prevails in entangled polymer solutions.[14,15] A phenomenon known as "biased reptation" occurs as DNA molecules move through a polymer sieving matrix in an electric field; the DNA molecules unstretch and move through the polymer "mesh" head-first in a snake-like fashion, orienting themselves in the direction of the field. The higher the electric field strength, the more DNA molecules—especially long DNA chains greater than 300 bases in length—are likely to reptate with a strong field-bias, and the weaker the dependence of DNA electrophoretic mobility on size for these reptating chains. This causes the size-dependence of DNA mobility to decrease, thereby also decreasing read-length. Therefore, capillary in a polymer matrix electrophoresis requires a balance to be struck between the increased separation speed possible under high electric fields and the decreased read-length.[16] DNA sequencing is also limited by the number of capillaries that can be run in parallel; CAE instruments are currently available with up to 384 capillaries, though instruments with 96 capillaries are much more typical. Polymer sieving matrix costs are a significant percentage of the cost of DNA sequencing consumables, however, and the cost of polymer increases linearly with the number of capillaries. Throughput may be increased by highly parallel separations, but the cost per base is only minimally decreased.

Miniaturized microfluidic electrophoresis devices or "chips," which are still primarily in the research stage for the applications of DNA sequencing and genotyping [which often requires the fractionation of small, single-stranded DNA (ssDNA) molecules] are a natural next step from CE because they decrease the separation time and increase throughput, thus offering to reduce the cost per base of DNA sequencing. DNA has been separated by electrophoresis on both glass[17-21] and plastic[22,23] microchips of varying lengths. However, polymer-sieving matrices that are easily pressure-loaded into capillaries (with pressure limits $>$ 1000 psi) are difficult to load into glass and plastic microchips (which have pressure limits in the range of 50–200 psi). If DNA could be separated by electrophoresis in free solution, that is, without a polymer sieving matrix, DNA sequencing by microchip electrophoresis would be significantly easier. Eliminating the need for polymer matrix in chip electrophoresis could decrease the cost per base even further, relative to microchip electrophoresis using separation matrices.

Woolley and Mathies[24] of U.C. Berkeley published the first four-color DNA sequencing results obtained on microfluidic devices back in 1995. The Mathies group then showed sequencing of 500 bases in 20 min in 1999.[19] Since then, the Jovanovich group at Molecular Dynamics has succeeded in sequencing 450 bases on a chip in 15 min.[21] While these sequencing results are certainly impressive, they are not yet fast enough, or cheap enough, to provide the "$100,000 genome" on their own. Research groups are working to integrate several parts of the sequencing protocol (DNA extraction, purification, amplification, and detection) on a "lab-on-a-chip" microfluidic device[17,25–41] that includes polymer matrix-based CE as the final separation step.

In CE, polymer sieving matrices are necessary to separate DNA molecules according to size, for the reason that DNA chains behave as "free-draining" polymers during electrophoresis and also have a monotonic size-to-charge ratio, and hence elute as one peak in free-solution CE, regardless of molecular size. In free-solution electrophoresis, the electric field pulls negatively charged DNA to the cathode, as the positively charged buffer counterions that surround DNA migrate toward the anode; solvated cations move around and through the DNA coil, giving it its "free-draining" characteristic. The counterions effectively screen hydrodynamic interactions between different parts of the DNA molecule, so the molecule adopts an open conformation, causing the electrophoretic friction coefficient ζ to scale linearly with the number of monomers in the DNA chain M. The total electrophoretic force F applied to the DNA molecule also scales linearly with the number of charges along the DNA. Each of the four deoxynucleotides (dNTPs) carries the same amount of negative charge from identical phosphate groups, causing the number of charges in a DNA molecule to scale with M. The electrophoretic velocity of DNA in free solution is calculated by dividing total force by the friction coefficient: $v = F/\zeta \sim M/M \sim M^0$, thus no size separation is achieved for DNA in the absence of a gel. The sieving matrix used in capillary gel electrophoresis separates DNA molecules by size as they collide with and interact with polymers and their movement is retarded. Small DNA molecules move through the polymer matrix faster because they have fewer collisions, so that DNA molecules to elute in order from smallest to largest. Interaction with polymers to effect this size separation necessarily slows down the migration of DNA.

To accomplish free-solution separations of DNA would obviously require that DNA molecules of different sizes are modified in such a way that they have size-dependent electrophoretic velocities, which could be achieved either by changing the density of charge on each monomer or by changing the dependence of the molecular friction coefficient on M. In free-solution conjugate electrophoresis (FSCE), the linear scaling of DNA's charge-to-friction ratio, described above, is abolished by attaching another molecule (which we call a "drag-tag") to each DNA chain, which perturbs the average charge density per monomer. Noolandi first suggested this general idea in 1992[42] and mentioned the possibility of using a natural protein as a drag-tag in 1993.[43] The idea is potentially very powerful, since free-solution electrophoresis would be much easier to integrate into microfluidic devices than gel electrophoresis. To try to make this concept a concrete reality, one has first to consider what the structure and properties of such a "drag-tag" molecule ought to be. Should it be charged or uncharged? If charged, should it be cationic or anionic? Experiments predict that negatively charged drag-tags might only slightly modify the electrophoretic velocity of DNA while reducing the separation efficiency, read-length, and effective drag, while a positively charged drag-tag could interact electrostatically with the negatively charged DNA and/or the microchannel walls, increasing band broadening. Thus, a neutral drag-tag seems to offer the best size-based separation of DNA; the drag-tag should act as a sort of "parachute" behind the migrating DNA molecule, slowing down the molecule. If we attach the exact same drag-tag to every DNA molecule in a mixture, the largest molecule will migrate fastest because it experiences the smallest retardation from the additional frictional drag. Hence, elution should occur in order from largest to smallest, which is the opposite order of gel electrophoresis. This chapter concerns the present state-of-the-art of this bioconjugate approach to DNA sequencing and genotyping.

Other nonelectrophoresis technologies are being developed in the search for the \$100,000 and \$1000 genomes. Several of these technologies offer promising results and are expected to play a role in the sequencing revolution currently at hand. Specifically, sequencing by synthesis (SBS) and nanopore detection methods are under development as potential nonelectrophoresis technologies.[44–53] SBS is currently limited to average read-lengths of about 120 bases, is not as accurate as Sanger sequencing,[54] and requires difficult and demanding sample preparation, amplification, and genome assembly.[55] CE-MS is also being investigated as a DNA sequencing method, and is likely to have relatively short read-lengths.[56] While these nonelectrophoretic sequencing methods are being rapidly developed, DNA sequencing by FSCE offers one potential avenue to the \$100,000 genome, especially if it is incorporated into an integrated "lab-on-a-chip"[17,25–41] sequencing device. The separation time for FSCE is faster than matrix-based electrophoresis, and the cost per base should be decreased by eliminating the polymer matrix. This chapter will present an overview of the free-solution electrophoresis of bioconjugates, including a brief summary of the theory, a description of the perturbing entities, recently achieved free-solution DNA separations for genotyping and sequencing, and finally, future directions for the development of this potentially powerful new technology.

12.2 FREE-SOLUTION CONJUGATE ELECTROPHORESIS OF DNA

Free-solution electrophoresis of DNA that has been conjugated to a friction-perturbing entity, which we call a "drag-tag," was not feasible until after the development of CE in the 1990s. Free-solution electrophoresis generates a significant amount of heat, which must be dissipated. The narrow channels used in CE have diameters in the order of 25–100 μm, which efficiently remove this "Joule heat." Free-solution conjugate electrophoresis (FSCE) was first examined quantitatively in 1994 by Mayer et al.[57] who developed the theory and named the method end-labeled free-solution electrophoresis (ELFSE). However, we prefer the name FSCE (since the drag-tag is not a "label") and will use that name in this chapter. The original theory presented in 1994 has been further developed and was recently reviewed.[14] A brief summary of the general theory of FSCE will be presented here, along with several "proof-of-concept" experiments.[16,58–61]

12.2.1 Theory of FSCE

If DNA is electrophoresed in an aqueous buffer only (i.e., in "free-solution"), the electrophoretic mobility of an entire DNA molecule is essentially equal to the mobility of just one monomer unit of DNA. Because DNA molecules of all lengths elute at exactly the same time by free-solution electrophoresis, DNA is called a "free-draining" polymer[15,58,62] (as if DNA molecules are in "free fall"). The electrophoretic mobility of DNA μ is defined as the ratio between the electrophoretic velocity v and electric field strength E. For a free-draining molecule, this ratio is equivalent to the ratio of charge Q to friction ζ.[57] As each of the four dNTP monomers has an identical negatively charged phosphate group, the total charge Q of a molecule of DNA scales linearly with the number of monomers in the DNA chain M. The electrophoretic friction coefficient ζ of a free-draining chain also scales linearly with the number of monomers in the chain; hence, the electrophoretic mobility of DNA in free-solution (μ_0) is given by

$$\mu_0 = \frac{v}{E} = \frac{QE/\zeta}{E} = \frac{Q}{\zeta} \propto \frac{M}{M}. \tag{12.1}$$

Because both charge and friction scale linearly with the number of DNA monomers, $\mu(M)$ is not length-dependent. However, when an uncharged drag-tag is attached to the DNA molecule, the modified mobility is size-dependent in free-solution electrophoresis. In 1994, Mayer et al.[57] theorized that the mobility of this bioconjugate could be determined solely by dividing total electrical force by total friction coefficient. Assuming that an uncharged drag-tag is used, they predicted the free-solution mobility of the ssDNA conjugated to the drag-tag as

$$\mu(M) \approx \frac{\rho(M)}{\zeta(M+\alpha)} \approx \mu_0 \frac{M}{M+\alpha}, \tag{12.2}$$

where $\mu_0 = \rho/\zeta$ is the mobility of untagged (free) ssDNA in free-solution electrophoresis and α is the friction of the drag-tag in units of the friction of one base of DNA.

The original assumption made by Mayer et al.[57] concerning how the velocity of the ssDNA + drag-tag conjugate is calculated was found later to be true only in special cases, and to be determined by the hydrodynamic conformation of the ssDNA + drag-tag conjugate.[63] The theory was then further developed to determine the probability that the experimental conditions coincided with the specific hydrodynamic conformation of that special case. Luckily, that special case is actually the most common when separating these charged-uncharged conjugates in free-solution electrophoresis.[14,64] A brief explanation of the theory confirming this is presented below.

The ssDNA + drag-tag conjugate may adopt four basic types of hydrodynamic conformations, depending on the relative sizes of the ssDNA and the drag-tag and the intensity of the electric field E (Figure 12.1).[64] The electrophoretic mobility of the DNA + drag-tag depends on the hydrodynamic conformation of the conjugate when separated by free-solution electrophoresis. At equilibrium, the neutral drag-tag is integrated into the random coil of the negatively charged DNA (Figure 12.1a). When an electric field is applied during electrophoresis, the drag-tag itself does not have any intrinsic mobility because it is not charged. The ssDNA moves through the electric field according to its electrophoretic mobility and the retardation of the drag-tag. If a large enough electric field is applied ($E_1 > E > E_0$), the DNA "engine" will begin to segregate itself away from the drag-tag as it is pulled

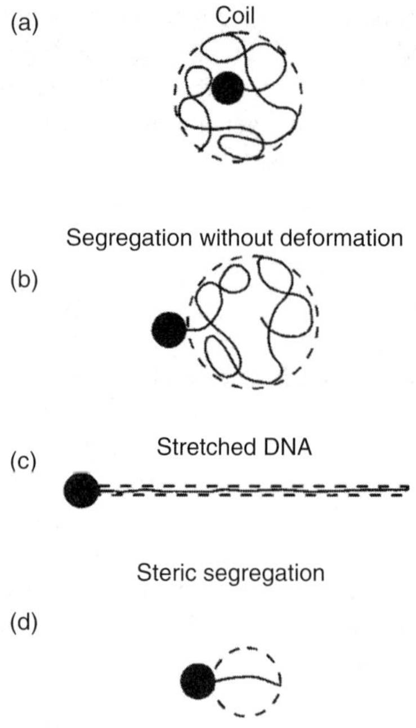

FIGURE 12.1 Schematic representation of four potential conformations of DNA-label bioconjugates: (a) label is part of the random coil of DNA, (b) label and DNA segregate without the DNA being deformed, (c) label and DNA segregate, and DNA is stretched, (d) DNA is too small to form a coil around the label, forcing steric segregation. (Reprinted with permission from Desruisseaux, C., et al. *Macromolecules* 34, 44–52, 2001.)

by the electric field (Figure 12.1b).[64] When the applied electric field has a high intensity ($E > E_1$), the DNA and label are segregated, and the DNA becomes fully stretched in the electric field as it is pulled through the capillary (Figure 12.1c).[64] In extreme cases where the drag-tag is much larger hydrodynamically than the DNA, the DNA is sterically segregated from the drag-tag because it is unable to surround the large label completely (Figure 12.1d). As bioconjugate electrophoresis aims to sequence large pieces of DNA, the theory focuses on determining the critical electric field intensity (E_0) required to move from the random coil hydrodynamic conformation to segregation without deformation.

Before calculating the critical electric field strength E_0 for the transition from the random coil to the hydrodynamic segregation regime, the mobility of labeled DNA $\mu(M)$ in the random coil regime must be determined. The uncharged drag-tag monomers cannot be assumed to have the same hydrodynamic friction per unit as negatively charged DNA. Because each block of the DNA + drag-tag block copolymer has different friction, the so-called blob theory is used to create "supermonomers" within each block that have the same hydrodynamic properties and drag but different charge.[14,65] Taking into account the molecular weights of the "blobs" as determined by Kuhn lengths, the electrophoretic mobility of the DNA-label conjugate simplifies to

$$\mu = \mu_0 \frac{M_c}{M_c + \alpha_1 M_u}, \tag{12.3}$$

where M_c is the number of charged monomers of DNA, M_u is the number of neutral drag-tag monomers, and α_1 is the friction coefficient of each uncharged drag-tag monomer.[14,64–69] Using this equation, the value for E_0 with typical experimental conditions was $E_0 \approx 8$ kV/cm,[14] which is much larger than typical electric fields used for CE (100–300 V/cm). The possibility of segregated and stretched DNA + drag-tag conjugates is highly unlikely under typical electrophoresis conditions; these conjugates are always in the random coil regime during electrophoresis.

The value of $\alpha = \alpha_1 M_u$ is critically important for determining the potential separation efficiency of drag-tags for free-solution electrophoresis. Drag-tags are evaluated by their α-value, and higher α-values correlate to a greater amount of drag imposed on each piece of DNA, indicating that larger DNA sequencing fragments can be resolved by a particular drag-tag. The approximate elution time of a DNA + drag-tag conjugate can be determined using the equation

$$t(M) = \frac{L}{\mu_0 E}\left(1 + \alpha_1 \frac{M_u}{M_c}\right), \tag{12.4}$$

where L is the total length of the capillary channel. This equation clearly shows that longer pieces of DNA (smaller M_u/M_c ratio) elute before shorter ones (larger M_u/M_c ratio and thus larger time), which is the opposite order of traditional CGE. This equation also offers an easy way to calculate $\alpha = \alpha_1 M_u$.

The first iteration of FSCE theory predicted the potential sequencing of up to 2000 bases in less than 1 h under perfect (diffusion-limited) conditions using a drag-tag with an α between 100 and 200.[57] The initial theory, however, assumed that $\mu(M)$ and the diffusion coefficient $D(M)$ could be related using the Nernst–Einstein equation, which it turns out is not a valid assumption under free-solution electrophoresis conditions.[70] This assumption caused Mayer to overestimate the potential performance of DNA sequencing by FSCE. Further development of the theory with a correction for this error led to the following equation that estimates the maximum number of bases that can be sequenced by a drag-tag with a specified α-value:

$$M_c^{1/2}(M_c + \alpha_1 M_u)^{5/4} \cong 215\alpha_1 M_u. \tag{12.5}$$

The protein streptavidin was suggested as a potential drag-tag in the early ELFSE literature.[43,57,71] Proteins are a natural choice for drag-tags because, in principle, they can be completely monodisperse (homogeneous in size and structure) and are large enough to provide appropriate amount of drag. Furthermore, streptavidin was chosen because it is close to electrostatically neutral under standard experimental conditions[14] and can be easily linked to DNA with biotin. The α-value of streptavidin is approximately equivalent to the friction of 30 monomers of ssDNA.[16] With an α of 30, the maximum read-length of DNA sequencing with a streptavidin drag-tag is predicted by theory to be 129 bases.[14]

12.2.2 Proof-of-Concept Experiments

12.2.2.1 Streptavidin

The first confirmation of the potential capabilities of ELFSE was achieved in 1998 by Heller et al.[58] when streptavidin was used as a drag-tag to separate double-stranded DNA (dsDNA).[58] Using biotinylated nucleotides and a polymerase enzyme, either one or both ends of the dsDNA were biotinylated so the dsDNA could be conjugated, end-on, with either one or two streptavidins (one on each end). Double-stranded DNA ladders (100 bp) with one or two terminal biotin moieties were generated and conjugated to streptavidin. FSCE analysis of the conjugates was run in 1× TBE buffer (89 mM Tris, pH 8.0, 89 mM boric acid, 2 mM EDTA), and the resulting electropherograms are presented below (Figure 12.2).

Large α-values are necessary to achieve good resolution of DNA, as shown by the double-labeled dsDNA ladder. When two drag-tags are attached to dsDNA, an α-value of 54 imparts enough drag on the dsDNA so the 100 and 200 bp strands separate over approximately 2 min instead of less than 1 min with only one streptavidin.[58] This first foray into FSCE also confirmed the prediction that increased electric fields lead to higher resolution between peaks.[58] The electrophoretic velocity is

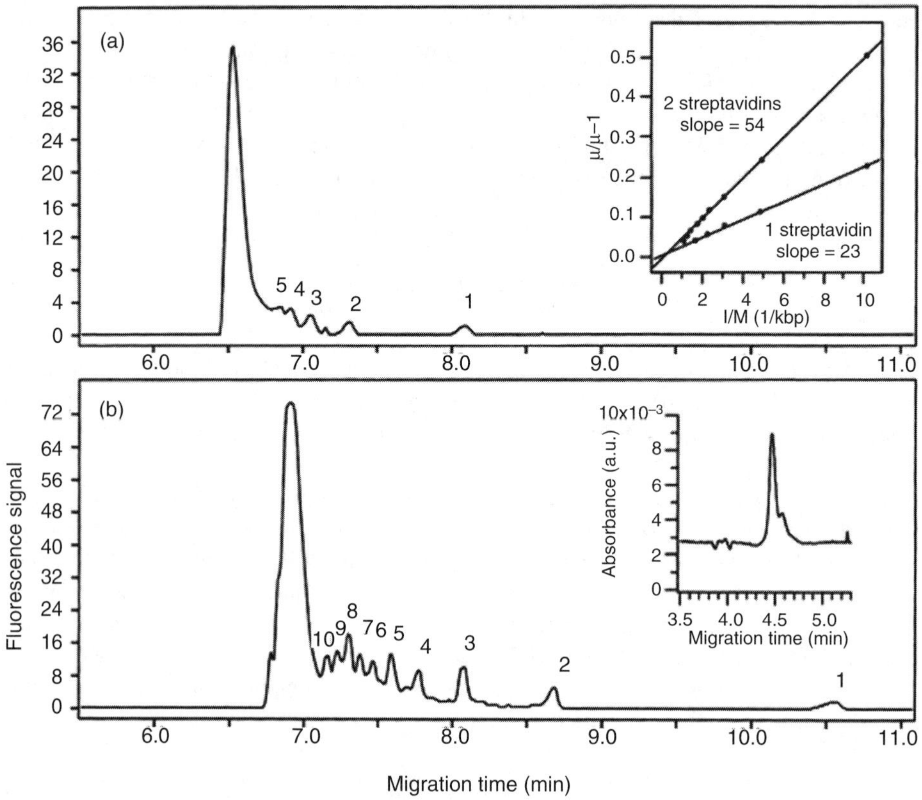

FIGURE 12.2 Free-solution capillary electrophoresis analysis of 100 bp dsDNA ladder labeled on one (a) or both (b) ends with streptavidin. The drag of two streptavidins clearly resolves the dsDNA fragments more efficiently than one streptavidin. Inset of (a) is plot of $\mu_0/\mu 1$ versus $1/M$; the slopes of this plot are equal to the α value of one or two streptavidin drag-tags. Plot inset of (b) shows the polydispersity of the streptavidin used. The large unnumbered peak at the beginning of both electropherograms is the unlabeled dsDNA. (Reprinted with permission from Heller, C., et al. *Journal of Chromatography A* 806, 113–121, 1998.)

also increased with higher electric fields without any loss of resolution, which is an advantage of FSCE over matrix-based CE systems that must balance the increased speed of separations with loss of resolution from increased electric fields.[14,15]

The polydispersity of the streptavidin used to separate the dsDNA (Figure 12.2b, inset) was troublesome because it potentially causes each peak of DNA to split into two or more peaks. The protein is fairly heterogeneous due to differing degrees of proteolysis and glycosylation as well as the fact that it has not one, but four active sites to react with biotin. DNA sequencing requires single-base resolution, which is clearly not possible with a polydisperse drag-tag that generates two or more peaks for each length of ssDNA. When streptavidin was purified using a homebuilt, preparative, nondenaturing polyacrylamide gel electrophoresis instrument, the polydispersity was greatly decreased.[16]

To fully demonstrate the potential of ELFSE, sequencing fragments were generated with a biotinylated primer. The purified streptavidin was attached to the fragments after cycle sequencing. Single-base resolution of the sequencing fragments in four colors was achieved for the first 100–110 bases in 18 min by ELFSE (Figure 12.3).[16] The approximately monodisperse streptavidin used here had an α-value of around 24. As mentioned in Section 12.2.1, single-base resolution of 129 bases of ssDNA is predicted for a drag-tag with an α of 30. Therefore, sequencing 100–110 bases with streptavidin is near the upper limit predicted by theory.

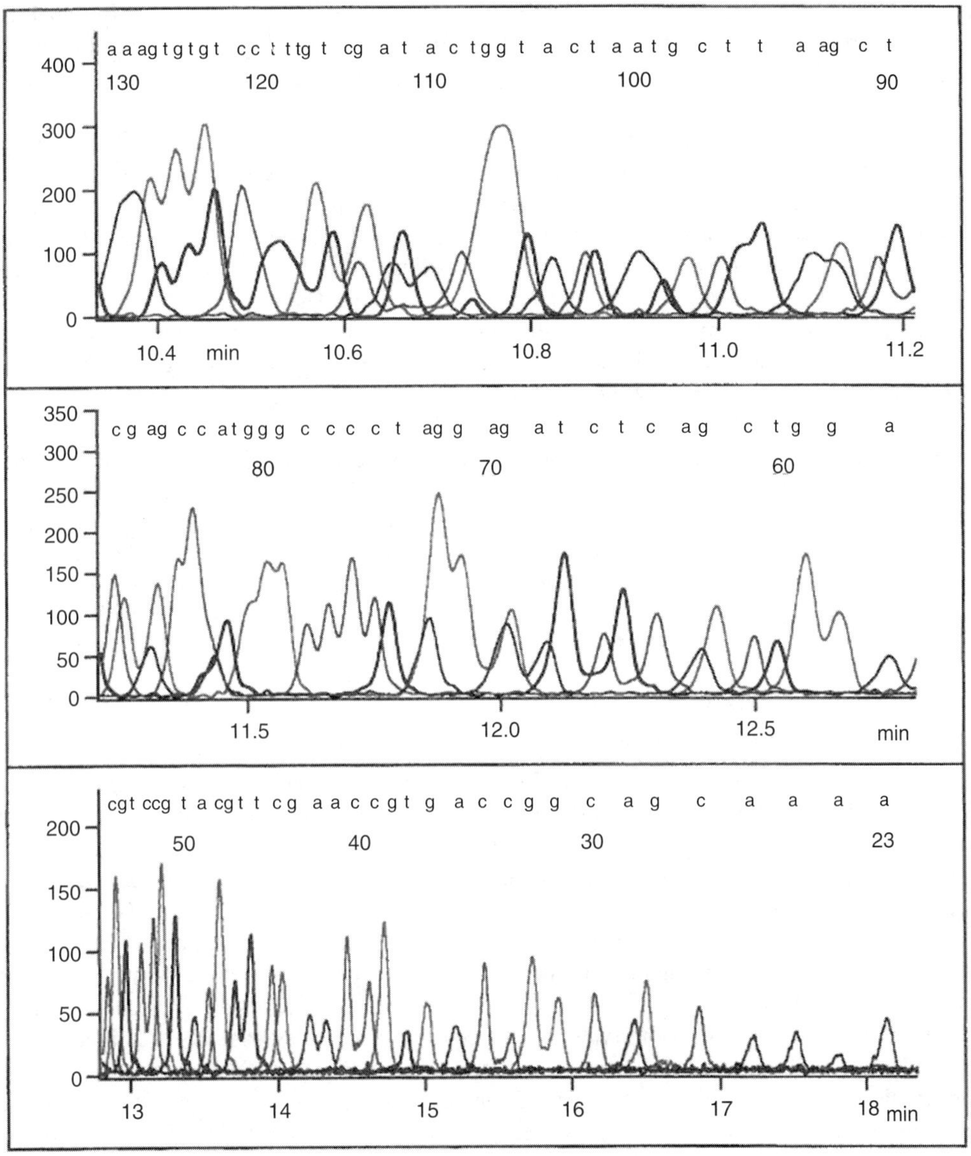

FIGURE 12.3 DNA sequencing of M13mp18 in free-solution electrophoresis using purified streptavidin and a biotinylated sequencing primer. The bases elute in reverse order compared to traditional CGE; the base immediately after the primer elutes at 18 min. Electrophoresis conditions were 1X TAPS buffer (pH 8.5) with 7 M urea and 0.01% POP-6 as a dynamic wall coating, 34 cm effective length (45 cm total length) capillary with 20 μm inner diameter. (Reprinted with permission from Ren, H., et al. *Electrophoresis* 20, 2501–2509, 1999.)

While reading 100–110 bases of DNA by free-solution CE is obviously not yet competitive with current matrix-based and nonelectrophoresis DNA sequencing methods, it clearly shows ELFSE's potential. Based on the streptavidin drag-tag, a drag-tag with an α of 300 is predicted to resolve ssDNA sequencing fragments up to 625 bases long.[16] Separating sequencing fragments by microchip electrophoresis may also be able to increase the read-length by applying a higher electric field than is possible with traditional CE instruments.[16,58] However, several issues must be resolved before that many bases can be separated using ELFSE. First, the mobility of the fluorescent dye labels must be determined to deconvolute the mobility shifts of the sequencing fragments induced by the four dyes. Second, an effective capillary wall coating must be developed to further decrease interactions between the walls and the bioconjugates.

Third, and most critical to long-read sequencing, is the development of a large, uncharged, monodisperse drag-tag. While streptavidin was a natural choice of drag-tag for testing the first DNA separation by FSCE, it will never be able to sequence more than 100–130 bases of DNA. Streptavidin, like almost all natural proteins, folds into a fairly compact sphere under native (nondenaturing) conditions. The frictional coefficient of spheres increases linearly with the radius; however, the radius only increases with the one-third power of the molecular weight. Therefore, streptavidin would need a molecular weight of approximately 30 million Da (a 600-fold increase in mass) to have an α-value of 300.[14] Also, as a natural protein, streptavidin has areas of positive and negative charge on its surface that may adversely affect its ability to be an effective drag-tag. Natural, folded proteins like streptavidin are thus not the best choice of drag-tags for long-read ssDNA sequencing. A monodisperse drag-tag that is uncharged and unstructured with a large α-value should theoretically provide much better sequencing results. Denaturing conditions that unfold proteins cannot be used with streptavidin without compromising the biotin-binding sites. If another method of end-on conjugation is used, proteins of either random coil or α-helical secondary structure may be used. Two different approaches have been taken toward developing appropriate drag-tags, and they will be discussed in Section 12.3.

12.2.2.2 Oligosaccharides

While work on FSCE has mainly focused on separating DNA molecules, the Novotny group of Indiana University has also successfully achieved free-solution separations of end-labeled, charged oligosaccharides.[59–61] Oligosaccharides are similar to DNA in that their charge-to-friction ratio remains constant as length changes.[59] Two different fluorescent labels were used as drag-tags for carbohydrates: one to increase the charge-to-friction ratio of the oligosaccharides and one to decrease it. The fluorescent label 8-aminonaphthalene-1,3,6-trisulfonic acid (ANTS) increased the charge-to-friction ratio and separated the oligosaccharides by free-solution electrophoresis with a migration order from smallest to largest. When the charge-to-friction ratio was decreased by the attachment of 6-aminoquinoline, the migration order was reversed to largest to smallest oligosaccharides, like ELFSE. Both types of labeled oligosaccharides were also run in a 1% linear polyacrylamide (LPA) matrix to determine the impact of sieving matrix on separation. The matrix increased the separation efficiency of the ANTS-labeled oligosaccharides but significantly decreased separation of the oligosaccharides with a decreased charge-to-friction ratio. The LPA slowed the larger oligosaccharides by a much larger percentage than the smaller ones, effectively cutting the separation efficiency into half.[59] The Novotny group has also applied FSCE to the separation of heparin oligosaccharides[61] and the study of how polygalacturonic acid interacts with metal ions.[60]

12.3 DRAG-TAGS FOR FSCE

Selection of an appropriate drag-tag is key to achieving successful DNA sequencing by ELFSE. The ideal drag-tag for many DNA separations of interest would have a large value of α. With a large hydrodynamic drag, higher resolution and performance can be achieved for the separation of large sequencing fragments (greater than 200 bases). In addition to a high α-value, several other properties are desired in the ideal drag-tag[14]

1. Complete monodispersity
2. Water solubility (DNA sequencing is performed under aqueous conditions)

3. Minimal to no electrostatic charge
4. Minimal adsorption to or nonspecific interaction with microchannel walls
5. Ability to be uniquely and stably attach to DNA (to ensure DNA is attached to exactly one drag-tag)

The most important property of an ideal drag-tag is complete monodispersity, where every tag is identical in charge and drag. When analyzed by free-solution CE, a monodisperse DNA + label conjugate should correspond to exactly one peak on the electropherogram. A polydisperse DNA + drag-tag conjugate, though, may show up as two or more ambiguous peaks on an electropherogram, which is not conducive to single-base resolution of ssDNA sequencing fragments. Accurate DNA sequencing would be challenging if not impossible with a polydisperse drag-tag.

Three different types of drag-tags have been examined in the search for appropriate drag-tags for sequencing and genotyping: synthetic polymers such as poly(ethylene glycol) (PEG),[69] poly-*N*-substituted glycines,[68,72,73] and genetically engineered protein polymers.[68,74–76] The dual requirements for complete monodispersity and sufficiently large hydrodynamic drag eliminate many commercially available synthetic polymers from consideration. Although completely monodisperse PEGs are now available, their small sizes result in low hydrodynamic drag that is insufficient for DNA separation. Larger PEG molecules with low polydispersity (PDI = 1.01) are also inadequate. When an "engine" of fluorescently labeled monodisperse ssDNA is conjugated to such PEG samples, they separate into more than 100 peaks in free-solution electrophoresis, with single-monomer resolution of the PEGs with different chain lengths.[69] The development of poly-*N*-substituted glycines and genetically engineered protein polymers for drag-tags will be discussed in the following sections.

To characterize a drag-tag, a completely monodisperse, fluorescently labeled ssDNA molecule is conjugated to a potentially polydisperse drag-tag. The DNA is conjugated to the drag-tag by the reaction between a maleimide-terminated drag-tag and a 5′-thiol-terminated oligomer of DNA (Figure 12.4). In electrophoresis, the monodisperse DNA acts like an "engine" that imparts the exact same amount of force on each drag-tag. Each different drag-tag is represented by a peak on the electropherogram of the free-solution separation; the polydispersity of the drag-tag can be evaluated by the resulting number of peaks. If just one product peak is present, then the drag-tag is actually

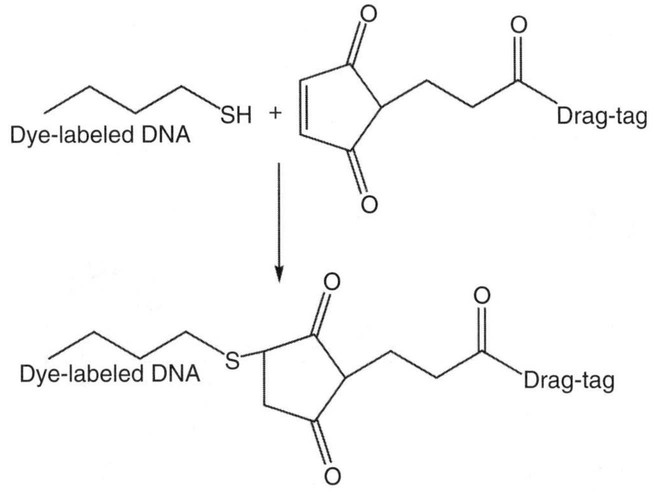

FIGURE 12.4 Conjugation of DNA to drag-tag by reacting dye-labeled DNA terminated with a 5′-thiol group with a maleimide-terminated drag-tag. (Modified from Vreeland, W. N., et al. *Analytical Chemistry* 73, 1795–1803, 2001.)

monodisperse and can be used for DNA sequencing or genotyping. More than one product peak present indicates that the drag-tag has an underlying polydispersity that may not have been identified during a previous purification. Potential drag-tags are evaluated and their effective drag α (measured in units equivalent to the drag from one base of DNA) is determined using Equation 12.3 and the identity $\mu_0/\mu \approx t/t_0$.[14,68,69,73,75–78]

12.3.1 Synthetic Polymers

FSCE was first applied to the evaluation of synthetic polymers as potential drag-tags, specifically the water-soluble polymer (PEG) whose antifouling properties should keep it from interacting with the capillary walls. PEG with a polydispersity index (weight average molecular weight/number average molecular weight) of 1.01 was tested, which is considered "monodisperse" by commercial standards. The PEG-DNA bioconjugate was expected to show up on the electropherogram as essentially one peak. The necessity for absolute monodispersity of a drag-tag was quickly realized upon separating the PEG-DNA bioconjugates by free-solution CE (Figure 12.5). The polydispersity of the "monodisperse" PEG immediately ruled out its potential application as a drag-tag for DNA sequencing. Ever since this first experiment with PEG, FSCE has been the "gold standard" by which the monodispersity of a drag-tag is determined.

12.3.2 Polypeptoids

Polypeptoids are similar to polypeptides except the side chains are attached to the amide nitrogen of the peptide backbone instead of the α-carbon[79–81] (Figure 12.6). Peptoids are synthesized by solid-phase synthesis on a peptide synthesizer using a "submonomer" approach[79,81] that proceeds by adding each "monomer" in a two-step process. With this method, many different side chain groups

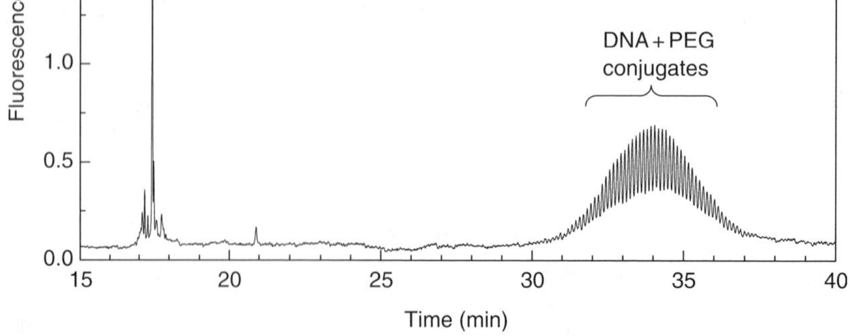

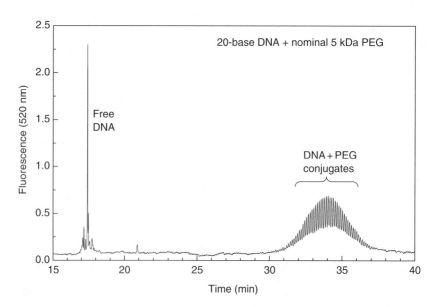

FIGURE 12.5 Electropherogram of FSCE analysis of DNA–PEG bioconjugate. A 20-base ssDNA oligomer was conjugated to a 5 kDa PEG. Separation was achieved in 1× TAPS buffer in a 100 cm (total length) capillary with 25 μm diameter. Pressure injection was used (138 kPa·s), 300 V/cm running field strength with 5.6 μA current. (Modified from Vreeland, W. N., et al. *Analytical Chemistry* 73, 1795–1803, 2001.)

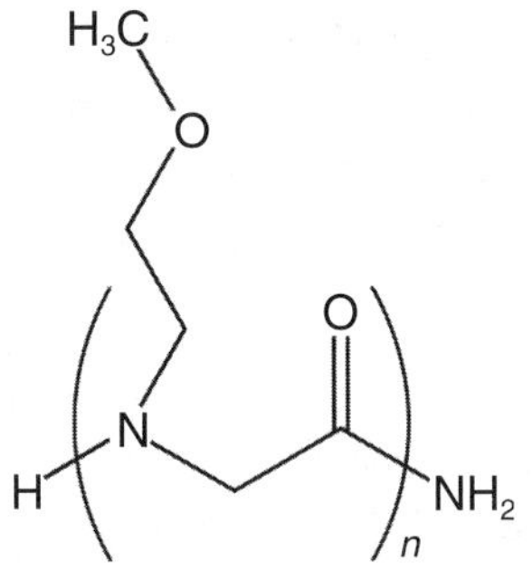

FIGURE 12.6 General structure of polypeptoid, with methoxyethyl as a glycine-like side chain. To attach to ssDNA, a thiol group is part of the R group at the end of the peptoid. (Modified from Haynes, R. D., et al. *Bioconjugate Chemistry* 16, 929–938, 2005; Zuckermann, R. N., et al. *Journal of the American Chemical Society* 114, 10646–10647, 1992.)

can be incorporated into the growing peptoid, including side chain analogues of amino acids. In particular, methoxyethyl has been used as a glycine-like side chain in peptoids for drag-tags because an *N*-methoxyethylglycine (NMEG) drag-tag is water-soluble, uncharged, and shows minimal or no interaction with the capillary walls.

12.3.2.1 Linear Polypeptoids

Peptoid drag-tags are advantageous because they can be purified to monodispersity by reversed-phase high-performance liquid chromatography (RP-HPLC). However, the solid-phase method can only synthesize linear polypeptoids up to 60 NMEG monomers before the yield drastically decreases from $< 100\%$ coupling efficiency of solid-phase synthesis.[69,78] The α-value of peptoid drag-tags is $\alpha \approx 0.2$–0.25 per peptoid monomer.[78] Therefore, a linear peptoid would have to be 100–125 monomers in length to have an α of 25 similar to streptavidin. Because solid-phase synthesis methods are limited to 60 monomers, achieving a large α like 300 for long-read sequencing requires something other than a linear peptoid drag-tag.

12.3.2.2 Branched Polypeptoids

Introducing oligopeptoid branches onto polypeptoid backbones is one approach to increase the effective drag of these synthetic drag-tags. At certain intervals in the synthesis of the peptoid backbone, side chains with a terminal ϵ-amino group are incorporated. Branches with terminal glutamic acid residues can then be attached to the backbone via amide bonds. Using a 30mer peptoid backbone with five evenly spaced "grafting" sites, three differently sized branches were appended to the drag-tag backbone to determine the impact of branch size on the effective drag. The smallest "branch" was an acetyl group, and the other two branches were linear polypeptoid NMEGs with a length of either 4 or 8 monomers. The α-value increased approximately linearly as the molar mass increased, with the effective drag of the octamer-branched drag-tag more than doubling that of the drag-tag with the acetyl groups as "branches."[77] The original FSCE theory did not predict this; branched objects are more compact than if the branches were added serially to the end of a linear drag-tag and theory did not predict that the branch length would make a dramatic impact on the drag. The theory was recently re-examined, though, and now suggests several parameters to take into account when designing branches to help achieve the highest α-value.[14] The search for the most effective drag-tag for long-read sequencing is now focused on finding the largest possible branch for a peptoid backbone.

12.3.2.3 Analysis of Solid-Phase Synthesis Products

FSCE can also show impurities present in solid-phase synthesis products that are not clearly evident by conventional purification methods. RP-HPLC with ultra violet (UV) detection is traditionally used to analyze solid-phase synthesis products such as peptoids and peptides as well as to separate them preparatively away from any deletion products. Since each synthetic monomer addition cycle is less than 100% efficient,[73] products that are one or two monomer units shorter than the desired product must be purified out to ensure monodispersity of the final polypeptide or polypeptoid. The separating power of FSCE was compared with analytical RP-HPLC traces after RP-HPLC purification (Figure 12.7). The one peak in the RP-HPLC trace actually separates into one large peak

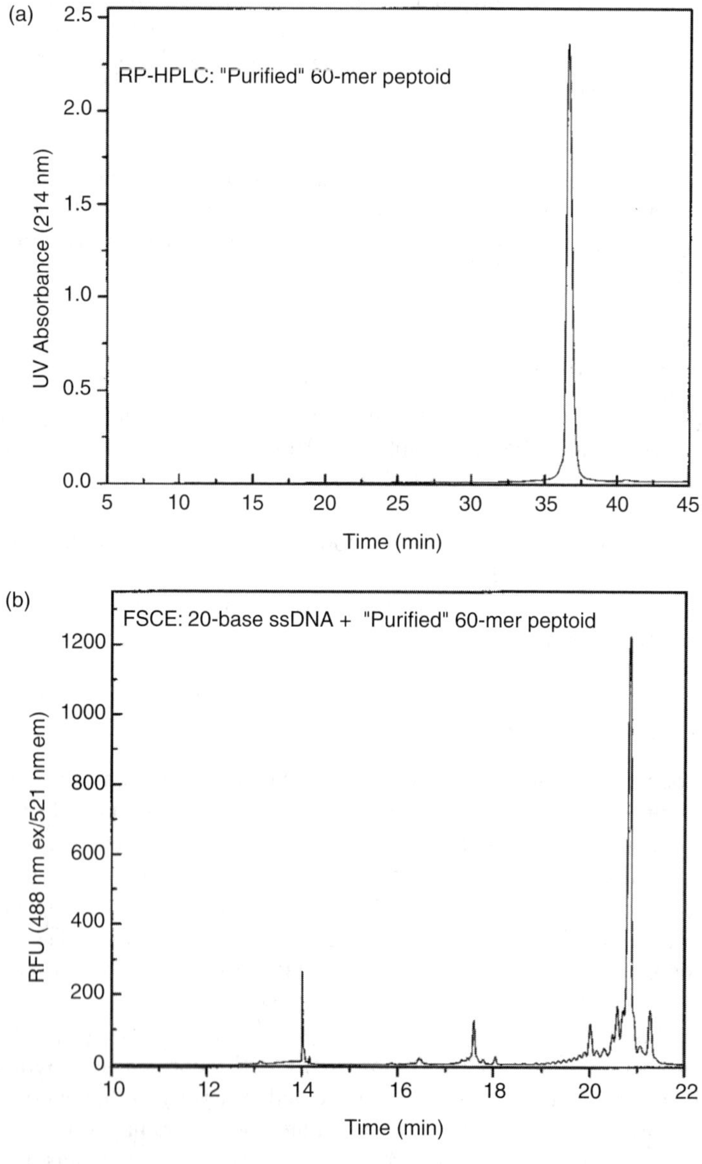

FIGURE 12.7 Analysis of polydispersity of a 60-unit NMEG polypeptoid drag-tag post RP-HPLC purification. RP-HPLC (a) shows one peak, but FSCE (b) shows a series of smaller peaks that may impact the drag-tag's read-length. (Reprinted from Vreeland, W. N., et al. *Bioconjugate Chemistry* 13, 663–670, 2002. With permission.)

with several smaller peaks in FSCE. While this drag-tag might have an α appropriate for sequencing and is practically monodisperse by RP-HPLC, the slight polydispersity might negatively impact the read-length. Overall, FSCE is very useful for evaluating both polydispersity and the α-value of a potential drag-tag and is recommended before any drag-tag is used to separate or sequence DNA.

12.3.3 Genetically Engineered Protein Polymers

In order to meet all of the criteria for an ideal drag-tag and avoid the drawbacks of synthetic polymers or natural proteins, the approach of nonnatural "protein polymers" has been employed. Protein polymers are produced by genetic engineering and consist of a repeating amino acid sequence. In general, this repetitive sequence can be a mimic of a natural sequence motif (e.g., elastin, silk) or highly nonnatural and designed specifically for a particular structure (e.g., α-helix) or function.[82–99] The properties of protein polymers are determined by the selected DNA sequence that codes for the final protein and can be customized by arranging specific amino acids in a desired order. Unlike conventional, synthetic polymerization techniques, protein-based materials produced in biological systems, such as the bacterium *Escherichia coli*, offer much better control over the properties of the final product.[100] Protein engineering is not limited to just the 20 naturally occurring amino acids. Several researchers have successfully incorporated a wide variety of unnatural amino acid analogs into the protein production machinery of living cells.[101–105]

12.3.3.1 Designing the Drag-Tag Sequence

The first and most important step in producing a nonnatural protein polymer is to design the actual sequence. In the genetic code, three DNA bases (a codon) code for either a particular amino acid sequence or a stop codon. The genetic code is degenerate; several different codons can correspond to the same amino acid. When designing the DNA sequence, it is important to use codons preferred by the particular host species because expressing highly repetitive protein sequences can rapidly deplete the available amino acid pool in the cell.

As mentioned previously, the ideal drag-tag characteristics are monodispersity, water solubility, lack of charged residues, unique attachment to DNA, and minimal adsorption to the microchannel walls. Based on these requirements, phenylalanine, isoleucine, tyrosine, and tryptophan were eliminated from consideration due to their strongly hydrophobic aromatic groups. Cysteine also was excluded because its thiol side-chain is highly reactive and may oxidize to form disulfide bonds. Several charged amino acids (lysine, arginine, histidine, aspartate, and glutamate) were similarly eliminated from consideration.

The first-generation of six protein polymer drag-tags that were designed for FSCE sequencing were designated PZ-1 through PZ-6. Glutamic acid was included, sparsely, in four of the sequences (even with the negative charge) to address concerns about water solubility. Valine and leucine were included in PZ-3 and PZ-4 to explore the effects of increasing hydrophobicity in drag-tags. The sequences were designed to generate random-coil structured protein polymers so each residue is exposed to the outside solvent and is not buried inside a sphere (such as in streptavidin). Unfortunately, not all of the proteins expressed well in *E. coli*, providing an important lesson to the designers and indicating the same persistence would be required. Of the proteins that could be expressed, none of these first six sequence designs was suitable as drag-tags for FSCE sequencing for the various reasons summarized in Table 12.1 and discussed in Reference 76. A key issue with the first generation drag-tags was the observed polydispersity that was later attributed to structural degradation of the drag-tag under acidic conditions, and more specifically, to deamidation of the glutamine into negatively charged glutamic acid.[75]

Therefore, the second generation drag-tags (PZ-7 and PZ-8) replaced glutamine with either serine or threonine, but of these only the PZ-8 sequence expressed well. Variants of this sequence containing a few serine-to-arginine mutations at different positions were also produced. The first of

TABLE 12.1
Summary of Initially Studied (Unsuitable) Drag-Tag Sequences

Name	Repeating Sequence	Comments
PZ-1	Gly-Ser-Gly-Gln-Gly-Glu-Ser	Good expressibility and water solubility; too hydrophilic for RP-HPLC purification; contains both glutamine and glutamic acid; similar electrophoretic mobility to DNA makes it poor drag-tag
PZ-2	Gly-Ala-Gly-Gln-Gly-Glu-Ala	
PZ-3	Gly-Val-Gly-Gln-Gly-Glu-Val	Both: not well expressed by *E. coli*; could not be purified in good yield
PZ-4	Gly-Leu-Gly-Gln-Gly-Glu-Leu	
PZ-5	Gly-Ala-Gly-Gln-Gly-Asn-Ala	Only obtained in impure preparation; both asparagine and glutamine included, which turned out to be nonideal; soluble in water only with addition of 1 M urea
PZ-6	Gly-Ala-Gly-Gln-Gly-Ser-Ala	Contains glutamine
PZ-7	Gly-Ala-Gly-Ser-Gly-Ser-Ala	Not well expressed
PZ-8	Gly-Ala-Gly-Thr-Gly-Ser-Ala	Well expressed and well behaved in aqueous solution

these variants has a mutation where one of every nine serines is replaced by a positively charged arginine. This mutant sequence and the original PZ-8 sequence have been expressed in at least three protein chain lengths (127, 253, and 505 amino acids). The average secondary structure of these proteins was profiled by circular dichroism (CD) spectroscopy, and all had spectra characteristic of random coil structures at 25°C in water. These results confirmed that the protein sequence had been successfully designed to form random coil proteins upon expression and purification.

Much effort and discussion has focused on making linear protein polymers for drag-tags but a parallel effort to create branched drag-tags is ongoing. These branched proteins have a sequence similar to the PZ proteins with the one notable exception that they include multiple lysines at varying intervals to be reactive sites for grafting on branches (either protein polymer or peptoid). A branched protein polymer is expected to generate greater hydrodynamic drag than a linear protein polymer of same length, based on the branched peptoid drag-tags.

12.3.3.2 Analysis of Protein Polymers as Drag-Tags

Unfortunately when these proteins are conjugated to DNA primers and analyzed in free-solution CE, all but the two 127-amino-acid proteins show multiple peaks, despite the exclusion of glutamine residues from the sequences. These additional peaks get more numerous as the protein polymer length increases, rendering these proteins useless as drag-tags. The expected effective α-value can still be determined by FSCE assuming the final peak in the electropherogram is the desired protein. Figure 12.8 shows a plot of protein molecular weight versus α for each protein. A clear linear trend exists for both series of proteins tested. The protein containing a few arginine mutations showed higher "effective" α-values than their uncharged counterparts due to the positive charge "pulling" the molecule in the opposite direction of the DNA. No significant detrimental interaction of the positively charged drag-tags with the DNA or microchannel walls was observed. The inclusion of even more positively charged residues may increase the hydrodynamic drag and avoid the need to generate very long, uncharged, drag-tags to achieve high α-values.

Meeting the requirement of a completely monodisperse drag-tag has proven more challenging than originally expected, and this work is still ongoing. Several theories have been explored and eliminated as causes of the heterogeneity, including the possibility that cell lysate proteases degrade the protein during purification. The process of histidine tag removal by formic acid with cyanogen bromide was also eliminated as a possible cause by using an enterokinase cleavage site already

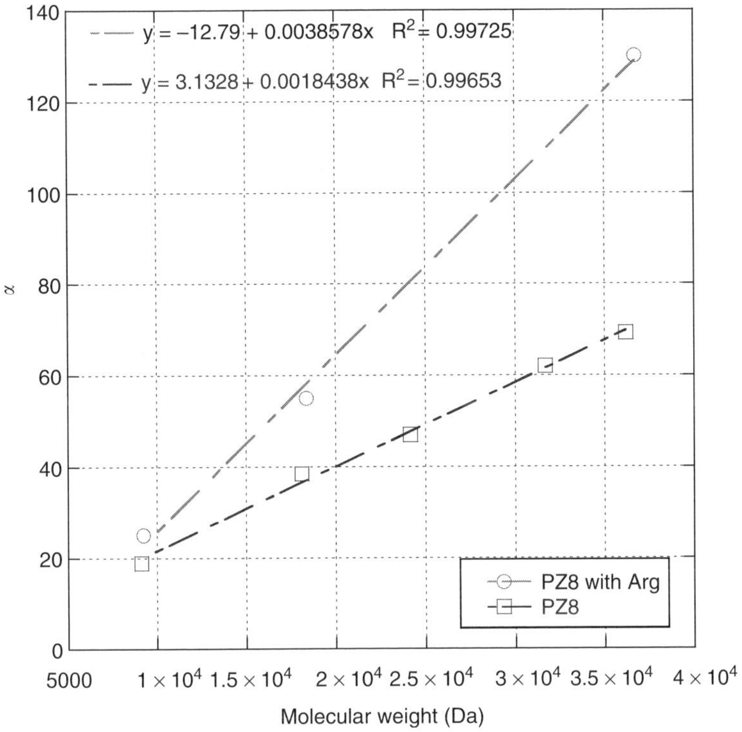

FIGURE 12.8 Chain length and charge of linear protein polymers (PZ-8 variants) versus effective α as determined by FSCE.

coded into the expression vector to allow enzymatic removal of the histidine tag and by shortening the cyanogen bromide reaction time from 24 to 4 h. Also, protein obtained from cells lysed by a freeze/thaw process demonstrated the same polydispersity as those obtained from cells lysed by ultrasonication. Even though MALDI-TOF mass spectrometry appears to only show a single peak, a difference of one amino acid may be what generates additional peaks in the FSCE electropherogram.

12.3.4 Double-Labeled DNA

FSCE can also be used to test different labeling techniques. The dsDNA separations with streptavidin drag-tags back in 1998 revealed an anomaly: the drag from a drag-tag on both ends of the dsDNA was expected to be equivalent to double the drag of a drag-tag on only one end, but as can be seen in Figure 12.2a (inset), the effective drag from dsDNA with drag-tags on both ends is 17% more than double the drag of one drag-tag.[16] Accordingly, the theory of ELFSE was re-examined to determine the cause of this anomaly. The theory is thoroughly developed in Reference 68 to take these "end-effects" into account. The monomer "blobs" at the ends of the bioconjugate molecule were found to be weighted more than those in the middle of the bioconjugate in how they influence the effective drag and electrophoretic mobility of the bioconjugate.[68]

To determine how the theory compared to experimental data, two different lengths of linear polypeptoid NMEG drag-tags were used to test end-effects on ssDNA, and four different drag-tags (one linear polypeptoid NMEG drag-tag, one branched NMEG peptoid drag-tag, and two different protein polymer drag-tags) were used to test the effect of double-labeling dsDNA. One drag-tag on both ends of ssDNA increased the effective drag α 6–9% over one drag-tag that was twice as long. Double-labeling dsDNA increased the α 10–23% over one drag-tag twice as long.[68] Figure 12.9 shows

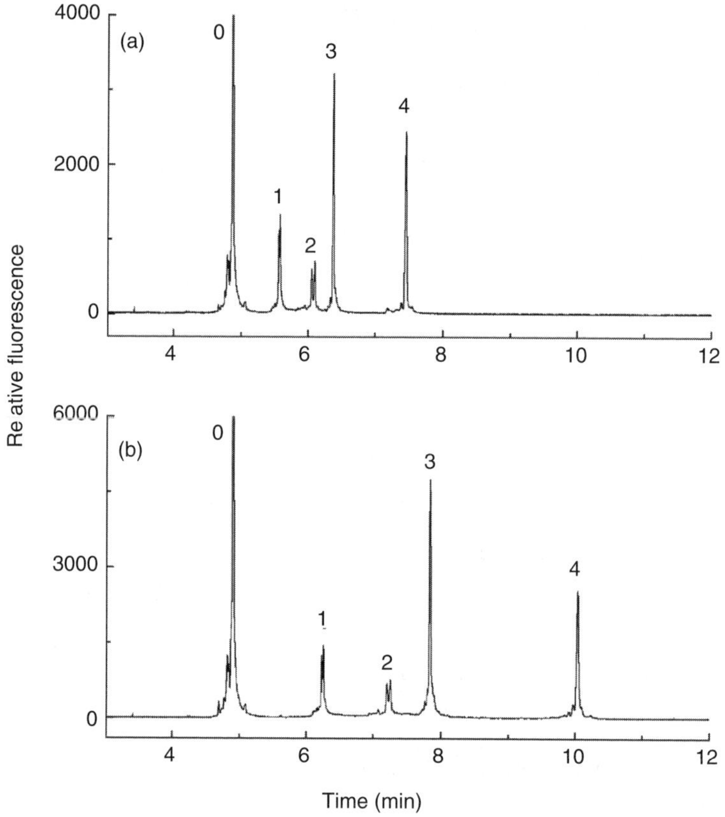

FIGURE 12.9 Free-solution capillary electrophoresis analysis of single- and double-labeled fluorescently tagged ssDNA. (a) Is a separation with 20mer NMEG peptoid drag-tag, and (b) uses a 40mer. Peak 0 is unlabeled, "free" DNA. Peak 1 is 40mer DNA oligomer with one drag-tag; peak 2 is 20mer DNA with one drag-tag; peak 3 is 40mer DNA with two drag-tags; and peak 4 is 20mer DNA with two drag-tags. Separations were achieved in 36 cm capillaries (47 cm total length; 50 μm inner diameter) with 22 V/cm electrokinetic injection for 15 s and current of 15 μA. (Reprinted with permission from Meagher, R. J., et al. *Electrophoresis* 27, 1702–1712, 2006.)

the separation achieved with double-labeled ssDNA. Overall, the use of double-labeled DNA could be a way to increase the effective drag imparted by a particular drag-tag without having to synthesize another, much larger, drag-tag.

12.4 DNA ANALYSIS BY FSCE

12.4.1 DNA GENOTYPING: SINGLE-BASE EXTENSION

While DNA sequencing is often the "gold standard" in detecting disease-causing genetic mutations, it is very expensive and time- and energy intensive. A majority of the genetic mutations responsible for diseases such as sickle cell anemia, Alzheimer's, and most cancers are single-nucleotide polymorphisms (SNPs) that are point mutations in a gene. Specifically, mutations in the *p*53 gene have been implicated as the cause of many cancers; most are missense mutations that cause a change in amino acid.[106–111] Many different methods are currently used for detecting SNPs,[112,113] including single-base extension (SBE) assays. In SBE, the sample is amplified using primers that anneal one base 5′ of the known SNP location in a reaction mixture containing only fluorescently labeled dideoxynucleotide terminators (ddNTPs) and no dNTPs. When the polymerase extends the chain from the 3′ end of the primer, the ddNTP incorporated corresponds to the complement of the SNP.[114]

Bioconjugate electrophoresis was first applied to SBE genotyping in 2002 to separate three SNPs in the *p53* gene in approximately 9 min, testing four different mutants.[72] Free-solution electrophoresis is important to SBE because it eliminates the polymer matrix cost. The degree of multiplexing can also be increased much more easily with FSCE. The elution order and separation efficiency of peaks for unique loci is determined by the length of drag-tag; a longer and more expensive primer is not required. While the genotyping peaks were rather broad in these first separations, the advantages to this method are clear: multiplexed separations are possible in free-solution CE and should be easily achieved in microchip electrophoresis.

SBE-FSCE was further multiplexed in 2006 when 22 samples (21 mutants and 1 wild-type) were tested for mutations at 16 loci.[115] Thermal cycling times were decreased by 90%, and free-solution electrophoresis of the multiplexed SBE-FSCE reaction was performed both in capillaries (Figure 12.10a) and microfluidic devices (Figure 12.10b) with greatly improved separation resolution over the 2002 study. Separations by microchip electrophoresis decreased the separation time from 20 min in capillaries to less than 70 s in an 8-cm-long chip. The SBE-FSCE technique was 96.3% accurate in genotyping by this method; of 16 loci across 22 samples (total of 352 loci), 339 were correctly genotyped, including 5 wild-type/mutant heterozygotes that were later confirmed by direct sequencing.[115]

The "modular" approach of SBE-FSCE shows the promise of genotyping using free-solution bioconjugate electrophoresis. Any drag-tag can be conjugated to any primer, allowing the design of multiple different sets of primers to probe many loci. Because of the abbreviated thermal cycling protocol and rapid separation on the microchip, this assay is a promising candidate to be incorporated into an integrated microfluidic "lab-on-a-chip" device for genotyping.[17,25–41,116]

12.4.2 DNA Sequencing by FSCE

Rapid, long-read DNA sequencing using capillary and/or microchannel electrophoresis is an ultimate goal of FSCE. Free-solution electrophoresis in microchips, in particular, eliminates the difficulties of loading viscous polymer matrix into the microchannels and ensures a higher run-to-run reproducibility. Because polypeptoids do not have sufficient drag capable of long-read sequencing, they have not been tried as drag-tags for sequencing. The branched peptoid drag-tag with five 8mer NMEG branches grafted to it had an α-value of approximately 17,[77] and was used recently to sequence almost 100 bases (unpublished results R. J. Meagher, R. D. Haynes). However, the 8mer NMEG branches did not increase the effective drag of the branched polypeptoid drag-tag enough to make it as effective as the original streptavidin drag-tag.

The recent successful cloning and purification of the PZ-8 protein polymer has allowed sequencing of end-labeled bioconjugates in free-solution to become a reality. The 127-amino-acid PZ-8 protein polymer has been purified almost to monodispersity, as mentioned previously. It has an α-value of approximately 25, which is appropriate for ssDNA sequencing. Both PZ-8 variants have been tested as sequencing drag-tags. The PZ-8 sequence variant with two serine-to-arginine mutations has a higher effective drag than the unmutated PZ-8 (Figure 12.8) and thus was expected to have a longer read-length. To create sequencing fragments, a 5′-thiol terminated M13 primer was first conjugated to the pure, monodisperse 127-amino-acid PZ-8. A typical Sanger sequencing reaction mixture, including the drag-tag-labeled M13 primer, M13 ssDNA template, dNTPs, fluorescent dye-labeled ddNTPs, DNA polymerase, and a polymerase buffer is thermal cycled to create drag-tagged ssDNA sequencing fragments.[54] Notably, the sequencing reaction was carried out with the protein polymer drag-tag already conjugated to the thiolated primer, allowing for easy post-amplification cleanup and sample preparation.[117] The fact that the 127-amino-acid protein polymer conjugated to the 5′ end of the thiolated 17mer M13 primer did not impact the DNA polymerase activity or

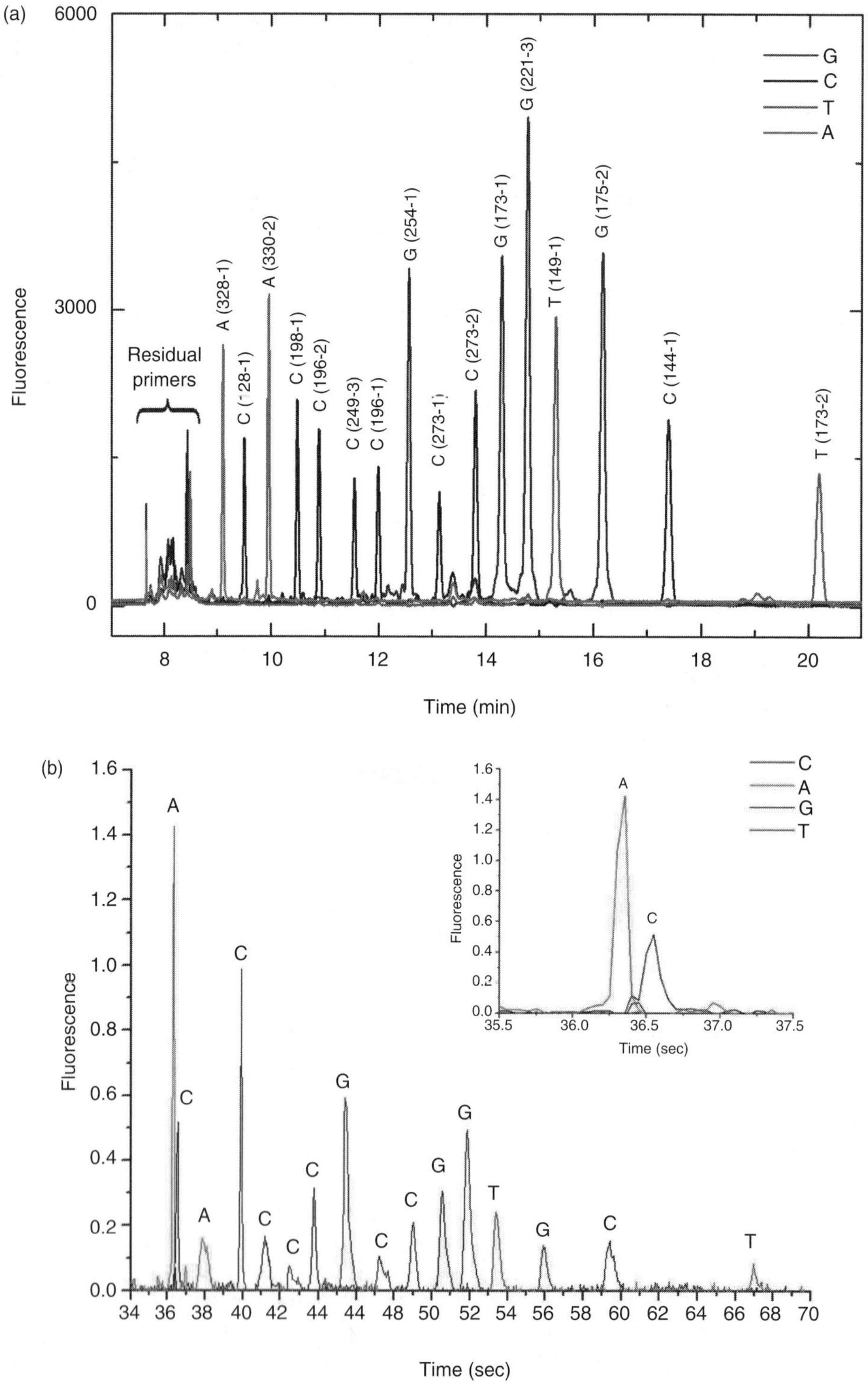

FIGURE 12.10 (**See color insert following page 810.**) Electropherograms of multiplexed SBE-FSCE separation of 16 loci in a wild-type *p*53 gene. (a) CE separation in 36 cm capillary (47 cm total length, 50 μm inner diameter) in 1× TTE (89 mM Tris, 89 mM TAPS, 2 mM EDTA), 7 M urea and 0.5% (v/v) POP-6 for dynamic coating. (b) Separation on 8 cm microchip in 1× TTE (49 mM Tris, 49 mM TAPS, 2 mM EDTA) with 7 M urea on chip coated with poly-*N*-hydroxyethylacrylamide. Both separations were performed at 55°C. (Reprinted with permission from Meagher, R. J., et al. *Analytical Chemistry* 79, 1848–1854, 2007.)

primer–template hybridization is a positive step for the potential application of DNA sequencing by FSCE in genome centers and on integrated microfluidic devices.

The first 120 bases of sequencing of the M13 ssDNA sequencing template were very clearly resolved using the 127-amino-acid PZ-8 drag-tag with two serine-to-arginine mutations, and the next 60 bases could be distinguished with prior knowledge of the sequence (Figure 12.11), giving an effective read-length of nearly 180 bases.[117] Comparing Figure 12.11 with Figure 12.3, the peaks of the ssDNA sequencing analysis using the protein polymer drag-tag are significantly sharper. The 127-amino-acid protein polymer was also used to test the effect of capillary length, electrophoretic velocity, and ionic strength of the buffer on resolution of the separation of sequencing peaks. The longest capillaries allowed slightly narrower peaks with deeper valleys, at the cost of a fourfold increase in separation time, while changing electrophoretic velocity did not drastically impact the separations. Increased buffer concentrations slightly increased α, but higher buffer concentrations caused more heat generation, which may degrade the quality of sequencing.[117]

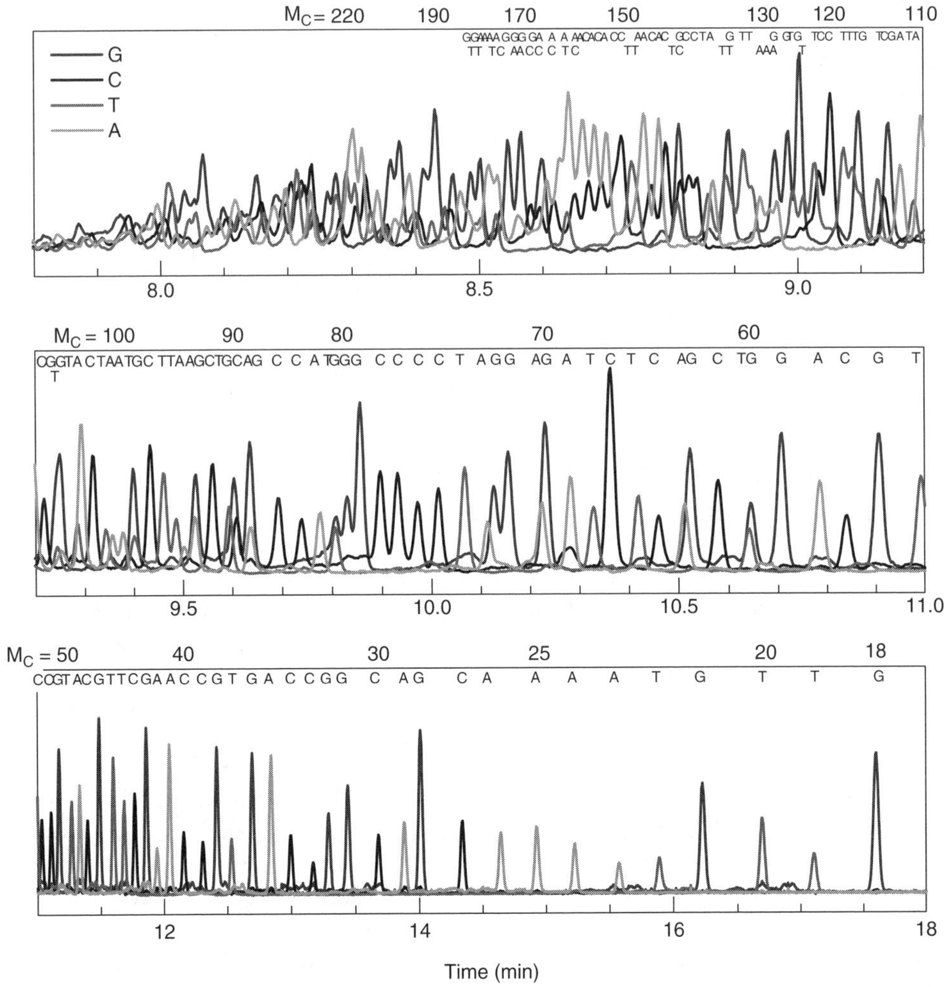

FIGURE 12.11 **(See color insert following page 810.)** DNA sequencing analysis of M13mp18 ssDNA in free-solution with genetically engineering protein polymer drag-tag in 36 cm capillary (47 cm total length, 50 μm inner diameter) with 312 V/cm field strength and electrokinetic injection of 1 kV for 20 s. Capillaries were filled with 1× TTE (89 mM Tris, 89 mM TAPS, 2 mM EDTA) with 7 M urea and 0.5% (v/v) POP-6 to dynamically coat the walls of the capillary. (Reprinted with permission from Meagher, R. J., et al. In preparation 2007.)

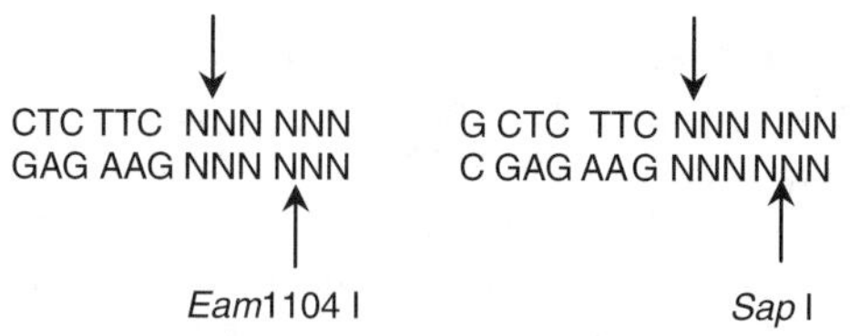

FIGURE 12.12 *Eam*1104 I and *Sap* I restriction enzyme recognition and cleavage sites are shown by the arrows. The *Sap* I enzyme has the additional base requirement of G on the top strand and C on the bottom strand as depicted here.

12.5 METHODS DEVELOPMENT GUIDELINES

12.5.1 DRAG-TAG CLONING AND PRODUCTION

12.5.1.1 Standard Protocol for Cloning and Production of Protein Polymers

Protein polymers are synthesized using biological systems (often *E. coli*) and then purified from the cell lysate by affinity chromatography. Chemical synthesis of peptides is limited to the production of short chains (<50 amino acids), and it is similarly difficult to synthesize long DNA oligonucleotides. The large repetitive gene necessary to express a protein polymer must be generated by joining several DNA "monomers" with compatible end base-pair sequences using a ligase enzyme.

Several cloning methods have been developed over the years to create repetitive genes for protein polymers.[92,98,99,118–121] Nonetheless, obtaining large concatemer genes was still difficult because of their limited yield after DNA ligation. Long concatemers could only be found by chance after laborious screening of hundreds of colonies. In order to consistently create the long DNA concatemers needed for expression of protein polymer drag-tags, a modified version of the seamless cloning method was developed called controlled cloning.[74] This method allows for the construction of large DNA concatemers without any specific sequence requirement using two Type II restriction enzymes, *Eam*1104 I and *Sap* I, which cleave downstream of their nearly identical recognition sites (Figure 12.12). Figure 12.13 illustrates the controlled cloning method for the example of doubling a multimer.

Each "monomer" of the gene is a 100 bp synthetic oligonucleotide that encodes three tandem repeats of a seven amino acid sequence. This oligomer is polymerase chain reaction (PCR)-amplified, enzymatically digested by the *Eam*1104 I enzyme to generate the 63-bp monomer DNA fragments, and then self-ligated to generate a ladder of different sized multimers. The gene is inserted into a cloning vector, pUC18, and colonies obtained after transformation of the plasmid into *E. coli* cells are screened for the desired insert sizes. Controlled cloning can then be used to combine one gene of the desired size with another, which may be the same or different in sequence and size.

After DNA sequencing confirms that the gene is correct, it is then moved into an expression plasmid, pET-19b, encoding an *N*-terminal histidine fusion tag, which is used to purify the desired protein by immobilized metal affinity chromatography (IMAC) from other cellular proteins. The protein is expressed in BLR(DE3) *E. coli* cells using either LB (Luria-Bertani) or TB (Terrific Broth) media. Cells are grown under antibiotic selection at 37°C until they reach mid-log growth phase with an OD_{600} value between 0.6 and 0.8. The cells are then induced to begin protein expression with the addition of the lactose analog, isopropyl-β-D-thiogalactopyranoside (IPTG) at 1 mM, and grown for 3–4 h before harvesting by centrifugation. Cells are resuspended in buffer then lysed using multiple freeze/thaw cycles as well as ultrasonication. The clarified cell lysate is purified using column chromatography under denaturing conditions on either nickel or cobalt-chelated resin, which bind to the histidine tag. The protein with the affinity tag is eluted off the column with the addition of imidazole. The collected fractions are analyzed using sodium dodecyl sulfate–polyacrylamide gel electrophoresis (SDS–PAGE). Elution fractions are dialyzed against water to remove salts and small molecular weight contaminants before being lyophilized into a dry powder. Typical protein

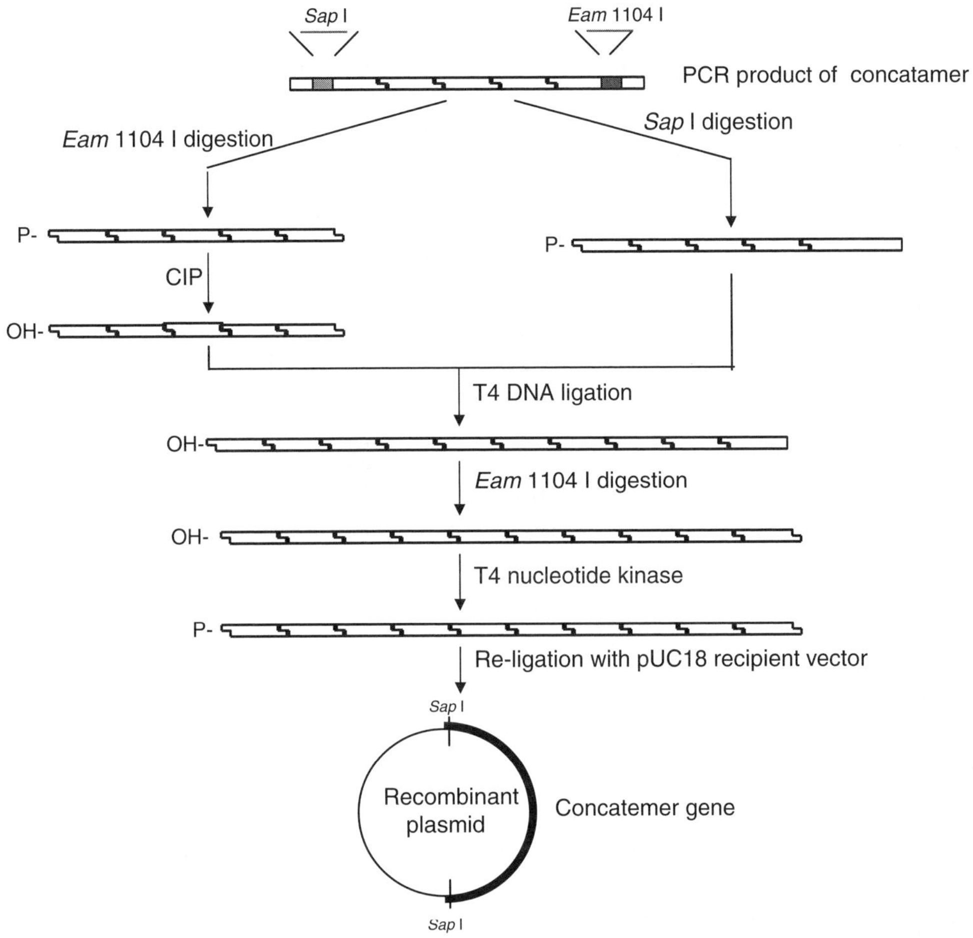

FIGURE 12.13 Controlled cloning method for generation of long concatemer genes. (Reprinted with permission from Won, J. I. and Barron, A. E. *Macromolecules* 35, 8281–8287, 2002.)

polymer yields range from 20 to 30 mg/L of cell culture. The purified protein is then analyzed for purity by RP-HPLC on a C4 column. Peaks are detected by UV absorbance at 220 nm. MALDI-TOF analysis yields the molecular mass of the protein sample which can then be compared to the expected mass. Removal of the affinity tag is achieved by reaction with cyanogen bromide in formic acid for 24 h.[122] Cleavage of the affinity tag is essential as it contains multiple positively and negatively charged amino acids that can adversely affect the drag-tag's properties in free solution.[76]

12.5.1.2 Polypeptoid Synthesis

Polypeptoid synthesis reactions are carried out on automatic peptide synthesizers (ABI 433A; Applied Biosystems, Foster City, CA). Fmoc-rink amide resin (0.25 mmol; Nova Biochem, San Diego, CA) is deprotected by treating with 20% (v/v) piperdine (ABI) in dimethylformamide (DMF; Fisher Scientific, Itasca, IL) in two 15-min treatments. To assemble the polypeptoid chain, alternating cycles of bromoacetylation and amine displacement of the bromine from the *N*-terminal alkyl halide are performed (Figure 12.14). To bromoacetylate, the resin is vortexed for 45 min with 4.3 mL of 1.2 M bromoacetic acid (Aldrich, Milwaukee, WI) in DMF and 1 mL of diisopropylcarbodiimide (DIC; Aldrich). The liquid is drained from the resin, and the resin is washed four times with 7 mL

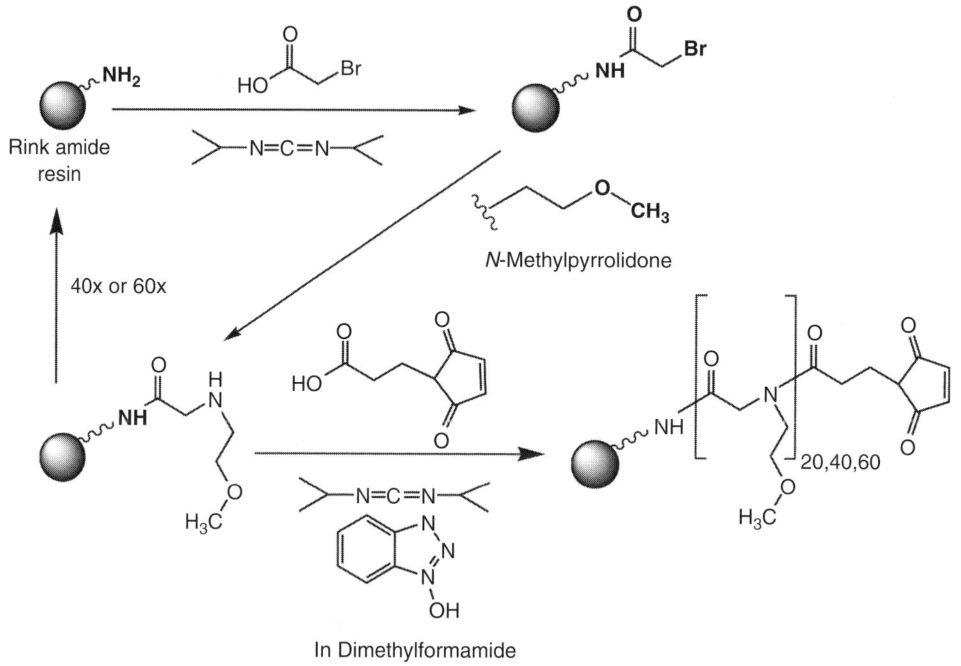

FIGURE 12.14 Reaction scheme for synthesis of linear NMEG polypeptoids of length 20, 40, or 60 monomer units and addition of maleimidopropionic acid. (Reprinted with permission from Vreeland, W. N., et al. *Bioconjugate Chemistry* 13, 663–670, 2002.)

of DMF. After the DMF wash, 4 mL of 1M methoxyethylamine (Aldrich) in *N*-methylpyrrolidone (NMP; Aldrich) is introduced to the reaction vessel and vortexed for 45 min to displace the amine and introduce the methoxyethyl side-chain moiety to the growing chain. The 4× DMF wash is then repeated.[72,73,115]

The peptoid chain is cleaved from the resin by treating it with 95% trifluoroacetic acid (TFA; Aldrich) in water for 30 min. After the peptoid is cleaved from the resin and dissolved in water, RP-HPLC is used to purify the peptoids (C18 packing, Vydac, 5 μm, 300 Å, 2.1 × 250 mm). A linear gradient of 10–60% B in A is typically run at 60°C over 50 min at a flow rate of 0.1 mL/min, where solvent A is 0.1% TFA in water and B is 0.1% TFA in acetonitrile. Polypeptoid concentrations are detected by UV absorbance at 214 nm. To fully purify the peptoids, preparative HPLC (C18 packing, Vydac, 15 μm, 300 Å, 22 × 250 mm) is run using the same solvents, gradient, time, and a flow rate of 12 mL/min.[72,73,115]

12.5.2 DNA + Drag-Tag Conjugation

Conjugation of DNA to a drag-tag can be used for one of two purposes. To test the monodispersity and effective drag of a newly produced drag-tag, a fluorescently labeled DNA oligomer should be used so the bioconjugate can be tested by electrophoresis immediately postconjugation. For a sequencing or genotyping reaction, however, DNA primers should not have a fluorescent label if fluorescently labeled ddNTPs are to be used. DNA and drag-tags are conjugated through the use of a maleimide linker. For DNA to be attached to a maleimide-activated drag-tag, it must first be modified to have a thiol (-SH) functionality on its 5′ terminus. Any DNA–DNA disulfide bonds present must be reduced to ensure the highest degree of conjugation efficiency. To reduce the DNA, 2 nmol of DNA primer are incubated with a 20:1 molar excess of Tris (2-carboxyethyl)phosphine (TCEP; Acros Organics, Morris Plains, NJ) at 40°C for 90 min.[115]

Protein polymer drag-tags are activated with a maleimide by the addition of sulfosuccinimidyl 4(*N*-maleimidomethyl)cyclohexane-1-carboxylate (Sulfo-SMCC; Pierce Biotechnology, Rockford, IL). A mixture of a 10:1 excess of Sulfo-SMCC to 1.2 mg of protein polymer in 80 μL of 100 mM sodium phosphate buffer (pH 7.2) is vortexed for 1 h at room temperature. Excess Sulfo-SMCC is removed from the activated protein polymer drag-tag by gel filtration with a Centri-Sep column (Princeton Separations, Adelphia, NJ). The activated, purified protein polymer is frozen, lyophilized and then resuspended to a concentration of 10 mg/mL.[117] Polypeptoid drag-tags are activated with a maleimide during the solid-phase synthesis on the peptide synthesizer before cleavage. To do this, a mixture of 500 μL of 1.2 M maleimidopropionic acid (Fluka, Buchs, Switzerland) in DMF and 144 μL of DIC is added to the resin and mixed for 45 min, then washed three times with DMF (Figure 12.14).[73] After cleavage and purification by HPLC, the peptoid is frozen, lyophilized to a golden, viscous oil and then dissolved in water to a concentration of 10 mg/mL.

To conjugate the activated drag-tag to the reduced DNA, 90 pmol of DNA is mixed with 2.5 nmol of drag-tag in a total volume of 10 μL of sodium phosphate buffer at pH 7.2 (Figure 12.4). The mixture is then incubated at room temperature for 3–24 h.[115,117] A large excess of drag-tag to DNA is necessary to ensure nearly complete conjugation of drag-tags to each DNA molecule.

12.5.3 Thermal Cycling Protocols

Single-base extension (SBE) reactions are commonly used both for genotyping and also for testing drag-tag + DNA primer conjugations. Because neither the drag-tag nor the DNA primer is fluorescently labeled, one fluorescently labeled ddNTP is added to the DNA so the bioconjugate can be detected by laser-induced fluorescence (LIF). SBE reactions require a sequencing polymerase, polymerase buffer, dye-labeled ddNTPs, DNA primer, and DNA template. To this end, the SNaPshot Multiplex Kit (ABI) has been frequently used for SBE reactions. The SNaPshot kit premix includes a sequencing polymerase, reaction buffer, and dichlororhodamine (dRhodamine) dye-labeled ddNTPs. SBE reactions are prepared by mixing 2.5 μL of the SNaPshot premix, 1 pmol of primer + drag-tag conjugate, 0.025–0.10 pmol of template DNA, 0.5 μL of 125 mM HCl, and sufficient water for a total volume of 5 μL. The mixture is then thermocycled for 5 cycles at 96°C for 2 s (denaturation), 51.5°C for 5 s (annealing), and 60°C for 10 s (extension). Centri-Sep gel filtration columns are then used to remove excess dye terminators and buffer salts from the reaction mixture before electrophoresis.[72,115]

DNA sequencing reactions differ from SBE reactions in that they also include dNTPs in the reaction mixture to extend the chain further than just one base by Sanger sequencing. For ELFSE sequencing, the following reaction mixture is used: 5 μL of SNaPshot premix, 8 nmol of dNTPs, 4.2 pmol of M13 primer (5′-$\mathbf{X}_1$GTTTTCCCA-GTCACGAC, where $\mathbf{X}_1$ is a 5′-C6 thiol linker) conjugated to a drag-tag, 0.16 μg of M13mp18 control DNA template (Amersham Biosciences, Piscataway, NJ), and sufficient water for a total volume of 10 μL. The mixture is then thermocycled for 26 cycles at 96°C for 5 s (denaturation), 50°C for 5 s (annealing), and 60°C for 30 s (extension). For purification, Centri-Sep gel filtration columns are used to remove excess dye terminators, buffer salts, and dNTPs before electrophoresis.[117]

12.5.4 Electrophoresis Conditions

Capillary electrophoresis separations of DNA + drag-tag conjugates are performed in a denaturing buffer: 1× TTE (89 mM Tris, 89 mM TAPS, 2 mM EDTA) with 7 M urea and a 0.5–1% (v/v) POP-5 or POP-6 polymer solution (ABI) for a dynamic wall coating agent to suppress electroosmotic flow (EOF) and prevent adsorption to capillary walls. Filtering the buffer with a 0.45 μm filter before loading it into the capillaries and buffer reservoirs reduces noise. In theory, any capillary array instrument with LIF detection (preferably 4-color) can be used for ELFSE separations. Capillaries with an effective length from inlet to detector of 36, 50, and 80 cm have been used to separate both ELFSE sequencing and genotyping fragments. Before analysis by electrophoresis, samples should be denatured at 95°C for 30–60 s and snap-cooled on ice for 2–5 min. Typical electrophoresis conditions include electrokinetic injection with a potential of 1–2 kV applied for 5–30 s and running voltage of 15 kV, all at 55°C.[68,115,117]

12.6 FUTURE DIRECTIONS

In order for DNA sequencing by FSCE to make an impact on the sequencing field, sequencing of at least 400 bases at a time is necessary. At present, a monodisperse drag-tag with a significantly larger effective drag α is the major obstacle blocking long-read sequencing by free-solution electrophoresis of DNA + drag-tag conjugates. Long protein polymers of the PZ-8 family are currently the most promising large drag-tags because the (GAGTGSA) sequence has proved amenable to this method. Research is also underway to investigate the possibility of achieving drag-tags with appropriate α-values for long-read sequencing by attaching very large branches to peptoid backbones.

As can be seen in Figure 12.11, the final peaks of FSCE sequencing are over-separated while the beginning peaks are under-separated. For long-read sequencing to be efficient and competitive, the smaller molecules (final peaks) need to be less resolved, and the large molecules (first peaks) need better resolution. The over-separation of the last peaks does not limit the read-length of the sequencing but does increase the separation time. The number of bases of sequencing read by FSCE is, however, limited by the under-separation of the first peaks. FSCE separations have always been performed under conditions to minimize EOF; however, strong EOF should cause rapid elution of small DNA fragments with inherently low electroosmotic mobility. Therefore, performing FSCE separations in the presence of strong EOF was predicted to reverse the migration order of DNA fragments so that they elute in order from smallest to largest, like CGE. The possibility of performing FSCE with EOF was examined theoretically to determine if it would also slow down the longer molecules and speed up the smaller ones when the migration order was reversed.[66]

A theoretical study of FSCE with EOF showed that FSCE in the presence of EOF mobility exactly equivalent to the free-solution mobility of DNA (scaled EOF mobility = 1) will more than double the

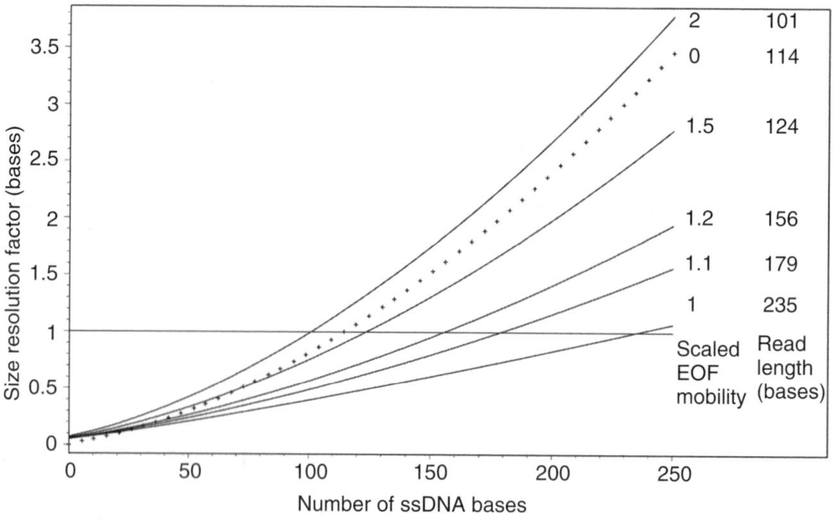

FIGURE 12.15 Graph of size resolution factor as a function of the number of ssDNA bases that are sequenced. When the size resolution factor is ≤1, single-base resolution of ssDNA fragments is possible. The scaled EOF mobility is the EOF mobility divided by the free-solution mobility of DNA. Therefore, the read-length is predicted to be the longest when the EOF mobility exactly equals the free-solution mobility of DNA. (Reprinted with permission from McCormick, L. C. and Slater, G. W. *Electrophoresis* 27, 1693–1701, 2006.)

read-length of sequencing. The counterflow of EOF not only reverses the migration order of DNA fragments but also speeds up small fragments while slowing down long fragments. The impact of the scaled EOF mobility on read-length in bases of ssDNA is shown in Figure 12.15, which used the effective drag of streptavidin as a comparison. According to these results, approximately 100 bases of ssDNA that were sequenced in Figure 12.3 could actually have been 235 bases of ssDNA if the EOF mobility was accurately tuned to be equal to the mobility of DNA in free-solution.[66] According to this theory, FSCE with a scaled EOF mobility of 1 could be a significant contributor in the search for long-read free-solution sequencing. The combination of FSCE with EOF, a drag-tag with a significantly large α-value, and electrophoretic separations on a microfluidic device offers the opportunity for sequencing by FSCE to be competitive with current methods of DNA sequencing. Both glass and plastic microfluidic devices can be easily functionalized with different coatings, allowing EOF to be tuned for FSCE separations to test how experimental data compare to the theory of FSCE with EOF. Once long-read sequencing by FSCE is achieved, FSCE has the potential to be seamlessly integrated into a "lab-on-a-chip" integrated microfluidic device.[25–37,40,41,123,124] If long-read sequencing in free-solution can be incorporated into a fully integrated microfluidic device, it has the chance to be competitive in the race for the $100,000 genome.

ACKNOWLEDGMENT

The authors would like to thank Louisa R. Carr for her help in the preparation of this manuscript.

REFERENCES

1. Lander, E. S. et al. Initial sequencing and analysis of the human genome. *Nature* 409, 860–921 (2001).
2. Venter, J. C. et al. The sequence of the human genome. *Science* 291, 1304–1351 (2001).
3. Kling, J. The search for a sequencing thoroughbred. *Nature Biotechnology* 23, 1333–1335 (2005).
4. Service, R. F. Gene sequencing—The race for the $1000 Genome. *Science* 311, 1544–1546 (2006).
5. Collins, F. S., Lander, E. S., Rogers, J., Waterston, R. H., and Conso, I. H. G. S. Finishing the euchromatic sequence of the human genome. *Nature* 431, 931–945 (2004).
6. Collins, F. S., Green, E. D., Guttmacher, A. E., and Guyer, M. S. A vision for the future of genomics research. *Nature* 422, 835–847 (2003).
7. Zhou, H. H. et al. DNA sequencing up to 1300 bases in two hours by capillary electrophoresis with mixed replaceable linear polyacrylamide solutions. *Analytical Chemistry* 72, 1045–1052 (2000).
8. NIH News Release: NHGRI seeks next generation of sequencing technologies, http://www.genome.gov/12513210 (2004).
9. Bonetta, L. Genome sequencing in the fast lane. *Nature Methods* 3, 141–147 (2006).
10. Project., R. f. a. H. C. G. Report of the Working Group on Biomedical Technology (2005).
11. Recommendation for a Human Cancer Genome Project: Report of the Working Group on Biomedical Technology. http://www.genome.gov/15015123 (2005).
12. Ledford, H. Kudos, not cash, is the real X-factor. *Nature* 443, 733 (2006).
13. Pennisi, E. GENOMICS: On your mark. Get set. Sequence! *Science* 314, 232 (2006).
14. Meagher, R. J. et al. End-labeled free-solution electrophoresis of DNA. *Electrophoresis* 26, 331–350 (2005).
15. Viovy, J. L. Electrophoresis of DNA and other polyelectrolytes: Physical mechanisms. *Reviews of Modern Physics* 72, 813–872 (2000).
16. Ren, H. et al. Separating DNA sequencing fragments without a sieving matrix. *Electrophoresis* 20, 2501–2509 (1999).
17. Paegel, B. M., Blazej, R. G., and Mathies, R. A. Microfluidic devices for DNA sequencing: Sample preparation and electrophoretic analysis. *Current Opinion in Biotechnology* 14, 42–50 (2003).
18. Paegel, B. M., Emrich, C. A., Weyemayer, G. J., Scherer, J. R., and Mathies, R. A. High throughput DNA sequencing with a microfabricated 96-lane capillary array electrophoresis bioprocessor. *Proceedings of the National Academy of Sciences of the United States of America* 99, 574–579 (2002).

19. Liu, S. R., Shi, Y. N., Ja, W. W., and Mathies, R. A. Optimization of high-speed DNA sequencing on microfabricated capillary electrophoresis channels. *Analytical Chemistry* 71, 566–573 (1999).
20. Dolnik, V., Liu, S. R., and Jovanovich, S. Capillary electrophoresis on microchip. *Electrophoresis* 21, 41–54 (2000).
21. Liu, S. et al. Automated parallel DNA sequencing on multiple channel microchips. *Proceedings of the National Academy of Sciences of the United States of America* 97, 5369–5374 (2000).
22. Llopis, S. D., Stryjewski, W., and Soper, S. A. Near-infrared time-resolved fluorescence lifetime determinations in poly(methylmethacrylate) microchip electrophoresis devices. *Electrophoresis* 25, 3810–3819 (2004).
23. Shi, Y. N. and Anderson, R. C. High-resolution single-stranded DNA analysis on 4.5 cm plastic electrophoretic microchannels. *Electrophoresis* 24, 3371–3377 (2003).
24. Woolley, A. T. and Mathies, R. A. Ultra-high-speed DNA-sequencing using capillary electrophoresis chips. *Analytical Chemistry* 67, 3676–3680 (1995).
25. Blazej, R. G., Kumaresan, P., and Mathies, R. A. Microfabricated bioprocessor for integrated nanoliter-scale Sanger DNA sequencing. *Proceedings of the National Academy of Sciences of the United States of America* 103, 7240–7245 (2006).
26. Burns, M. A. et al. An integrated nanoliter DNA analysis device. *Science* 282, 484–487 (1998).
27. Burns, M. A. et al. Microfabricated structures for integrated DNA analysis. *Proceedings of the National Academy of Sciences of the United States of America* 93, 5556–5561 (1996).
28. Dunn, W. C. et al. PCR amplification and analysis of simple sequence length polymorphisms in mouse DNA using a single microchip device. *Analytical Biochemistry* 277, 157–160 (2000).
29. Ferrance, J. P. et al. Developments toward a complete micro-total analysis system for Duchenne muscular dystrophy diagnosis. *Analytica Chimica Acta* 500, 223–236 (2003).
30. Fredlake, C. P., Hert, D. H., Mardis, E. A., and Barron, A. E. What is the future of electrophoresis in large-scale genome sequencing? *Electrophoresis* 27, 3689–3702 (2006).
31. Giordano, B. C., Ferrance, J., Swedberg, S., Huhmer, A. F. R., and Landers, J. P. Polymerase chain reaction in polymeric microchips: DNA amplification in less than 240 seconds. *Analytical Biochemistry* 291, 124–132 (2001).
32. Grover, W. H., Skelley, A. M., Liu, C. N., Lagally, E. T., and Mathies, R. A. Monolithic membrane valves and diaphragm pumps for practical large-scale integration into glass microfluidic devices. *Sensors and Actuators B-Chemical* 89, 315–323 (2003).
33. Kan, C. W., Fredlake, C. P., Doherty, E. A. S., and Barron, A. E. DNA sequencing and genotyping in miniaturized electrophoresis systems. *Electrophoresis* 25, 3564–3588 (2004).
34. Khandurina, J. et al. Integrated system for rapid PCR-based DNA analysis in microfluidic devices. *Analytical Chemistry* 72, 2995–3000 (2000).
35. Koh, C. G., Tan, W., Zhao, M. Q., Ricco, A. J., and Fan, Z. H. Integrating polymerase chain reaction, valving, and electrophoresis in a plastic device for bacterial detection. *Analytical Chemistry* 75, 4591–4598 (2003).
36. Lagally, E. T. and Mathies, R. A. Integrated genetic analysis microsystems. *Journal of Physics D-Applied Physics* 37, R245–R261 (2004).
37. Lagally, E. T. et al. Integrated portable genetic analysis microsystem for pathogen/infectious disease detection. *Analytical Chemistry* 76, 3162–3170 (2004).
38. Legendre, L. A., Bienvenue, J. M., Roper, M. G., Ferrance, J. P., and Landers, J. P. A simple, valveless microfluidic sample preparation device for extraction and amplification of DNA from nanoliter-volume samples. *Analytical Chemistry* 78, 1444–1451 (2006).
39. Paegel, B. M., Yeung, S. H. I., and Mathies, R. A. Microchip bioprocessor for integrated nanovolume sample purification and DNA sequencing. *Analytical Chemistry* 74, 5092–5098 (2002).
40. Ugaz, V. M., Elms, R. D., Lo, R. C., Shaikh, F. A., and Burns, M. A. Microfabricated electrophoresis systems for DNA sequencing and genotyping applications: current technology and future directions. *Philosophical Transactions of the Royal Society of London Series A-Mathematical Physical and Engineering Sciences* 362, 1105–1129 (2004).
41. Woolley, A. T. et al. Functional integration of PCR amplification and capillary electrophoresis in a microfabricated DNA analysis device. *Analytical Chemistry* 68, 4081–4086 (1996).
42. Noolandi, J. A new concept for sequencing DNA by capillary electrophoresis. *Electrophoresis* 13, 394–395 (1992).

43. Noolandi, J. A new concept for separating nucleic-acids by electrophoresis in solution using hybrid synthetic end labeled nucleic-acid molecules. *Electrophoresis* 14, 680–681 (1993).
44. Bai, X. P. et al. Photocleavable nucleotide analogues for DNA analysis. *Biochemistry* 42, 8597–8598 (2003).
45. Ju, J. Y. et al. Design and synthesis of fluorescence energy-transfer dye-labeled primers and their application for DNA-sequencing and analysis. *Analytical Biochemistry* 231, 131–140 (1995).
46. Ju, J. Y., Ruan, C. C., Fuller, C. W., Glazer, A. N., and Mathies, R. A. Fluorescence energy-transfer dye-labeled primers for DNA-sequencing and analysis. *Proceedings of the National Academy of Sciences of the United States of America* 92, 4347–4351 (1995).
47. Li, Z. M. et al. A photocleavable fluorescent nucleotide for DNA sequencing and analysis. *Proceedings of the National Academy of Sciences of the United States of America* 100, 414–419 (2003).
48. Seo, T. S. et al. Four-color DNA sequencing by synthesis on a chip using photocleavable fluorescent nucleotides. *Proceedings of the National Academy of Sciences of the United States of America* 102, 5926–5931 (2005).
49. Seo, T. S. et al. Photocleavable fluorescent nucleotides for DNA sequencing on a chip constructed by site-specific coupling chemistry. *Proceedings of the National Academy of Sciences of the United States of America* 101, 5488–5493 (2004).
50. Margulies, M., Egholm, M., Altman, W. E., Attiya, S., Bader, J. S., Bemben, L. A., Berka, J., et al. Genome sequencing in microfabricated high-density picolitre reactors. *Nature* 437, 376–380 (2005).
51. Astier, Y., Braha, O., and Bayley, H. Toward single molecule DNA sequencing: Direct identification of ribonucleoside and deoxyribonucleoside 5′-monophosphates by using an engineered protein nanopore equipped with a molecular adapter. *Journal of the American Chemical Society* 128, 1705–1710 (2006).
52. Yan, H. and Xu, B. Q. Towards rapid DNA sequencing: Detecting single-stranded DNA with a solid-state nanopore. *Small* 2, 310–312 (2006).
53. Shendure, J. et al. Accurate multiplex polony sequencing of an evolved bacterial genome. *Science* 309, 1728–1732 (2005).
54. Sanger, F., Nicklen, S., and Coulson, A. R. DNA sequencing with chain-terminating inhibitors. *Proceedings of the National Academy of Sciences of the United States of America* 74, 5463–5467 (1977).
55. Rogers, Y. H. and Venter, J. C. Genomics—Massively parallel sequencing. *Nature* 437, 326–327 (2005).
56. Edwards, J. R., Ruparel, H., and Ju, J. Y. Mass-spectrometry DNA sequencing. *Mutation Research-Fundamental and Molecular Mechanisms of Mutagenesis* 573, 3–12 (2005).
57. Mayer, P., Slater, G. W., and Drouin, G. Theory of DNA-sequencing using free-solution electrophoresis of protein–DNA complexes. *Analytical Chemistry* 66, 1777–1780 (1994).
58. Heller, C. et al. Free-solution electrophoresis of DNA. *Journal of Chromatography A* 806, 113–121 (1998).
59. Sudor, J. and Novotny, M. V. End-label, free-solution capillary electrophoresis of highly-charged oligosaccharides. *Analytical Chemistry* 67, 4205–4209 (1995).
60. Wiedmer, S. K., Cassely, A., Hong, M. F., Novotny, M. V., and Riekkola, M. L. Electrophoretic studies of polygalacturonate oligomers and their interactions with metal ions. *Electrophoresis* 21, 3212–3219 (2000).
61. Sudor, J. and Novotny, M. V. End-label free-solution electrophoresis of the low molecular weight heparins. *Analytical Chemistry* 69, 3199–3204 (1997).
62. Olivera, B. M., Baine, P., and Davidson, N. Electrophoresis of the nucleic acids. *Biopolymers* 2, 245–257 (1964).
63. Long, D., Viovy, J. L., and Ajdari, A. Simultaneous action of electric fields and nonelectric forces on a polyelectrolyte: Motion and deformation. *Physical Review Letters* 76, 3858–3861 (1996).
64. Desruisseaux, C., Long, D., Drouin, G., and Slater, G. W. Electrophoresis of composite molecular objects. 1. Relation between friction, charge, and ionic strength in free solution. *Macromolecules* 34, 44–52 (2001).

65. McCormick, L. C. et al. Desruisseaux, C., and Drouin, G. Capillary electrophoretic separation of uncharged polymers using polyelectrolyte engines—Theoretical model. *Journal of Chromatography A* 924, 43–52 (2001).
66. McCormick, L. C. and Slater, G. W. A theoretical study of the possible use of electroosmotic flow to extend the read length of DNA sequencing by end-labeled free solution electrophoresis. *Electrophoresis* 27, 1693–1701 (2006).
67. McCormick, L. C. and Slater, G. W. The molecular end effect and its critical impact on the behavior of charged-uncharged polymer conjugates during free-solution electrophoresis. *Electrophoresis* 26, 1659–1667 (2005).
68. Meagher, R. J. et al. Free-solution electrophoresis of DNA modified with drag-tags at both ends. *Electrophoresis* 27, 1702–1712 (2006).
69. Vreeland, W. N. et al. Molar mass profiling of synthetic polymers by free-solution capillary electrophoresis of DNA-polymer conjugates. *Analytical Chemistry* 73, 1795–1803 (2001).
70. Nkodo, A. E. et al. Diffusion coefficient of DNA molecules during free solution electrophoresis. *Electrophoresis* 22, 2424–2432 (2001).
71. Ulanovsky, L., Drouin, G., and Gilbert, W. DNA trapping electrophoresis. *Nature* 343, 190–192 (1990).
72. Vreeland, W. N., Meagher, R. J., and Barron, A. E. Multiplexed, high-throughput genotyping by single-base extension and end-labeled free-solution electrophoresis. *Analytical Chemistry* 74, 4328–4333 (2002).
73. Vreeland, W. N., Slater, G. W., and Barron, A. E. Profiling solid-phase synthesis products by free-solution conjugate capillary electrophoresis. *Bioconjugate Chemistry* 13, 663–670 (2002).
74. Won, J. I. and Barron, A. E. A new cloning method for the preparation of long repetitive polypeptides without a sequence requirement. *Macromolecules* 35, 8281–8287 (2002).
75. Won, J. I., Meagher, R. J., and Barron, A. E. Characterization of glutamine deamidation in a long, repetitive protein polymer via bioconjugate capillary electrophoresis. *Biomacromolecules* 5, 618–627 (2004).
76. Won, J. I., Meagher, R. J., and Barron, A. E. Protein polymer drag-tags for DNA separations by end-labeled free-solution electrophoresis. *Electrophoresis* 26, 2138–2148 (2005).
77. Haynes, R. D., Meagher, R. J., Won, J. I., Bogdan, F. M., and Barron, A. E. Comblike, monodisperse polypeptoid drag-tags for DNA separations by end-labeled free-solution electrophoresis (ELFSE). *Bioconjugate Chemistry* 16, 929–938 (2005).
78. Vreeland, W. N. and Barron, A. E. Free-solution capillary electrophoresis of polypeptoid-oligonucleotide conjugates. *Polymer Preprints* 41, 1018–1019 (2000).
79. Kirshenbaum, K. et al. Sequence-specific polypeptoids: A diverse family of heteropolymers with stable secondary structure. *Proceedings of the National Academy of Sciences of the United States of America* 95, 4303–4308 (1998).
80. Simon, R. J. et al. Peptoids—A modular approach to drug discovery. *Proceedings of the National Academy of Sciences of the United States of America* 89, 9367–9371 (1992).
81. Zuckermann, R. N., Kerr, J. M., Kent, S. B. H., and Moos, W. H. Efficient method for the preparation of peptoids [oligo(*N*-substituted glycines)] by submonomer solid-phase synthesis. *Journal of the American Chemical Society* 114, 10646–10647 (1992).
82. Welsh, E. R. and Tirrell, D. A. Engineering the extracellular matrix: A novel approach to polymeric biomaterials. I. Control of the physical properties of artificial protein matrices designed to support adhesion of vascular endothelial cells. *Biomacromolecules* 1, 23–30 (2000).
83. Panitch, A., Yamaoka, T., Fournier, M. J., Mason, T. L., and Tirrell, D. A. Design and biosynthesis of elastin-like artificial extracellular matrix proteins containing periodically spaced fibronectin CS5 domains. *Macromolecules* 32, 1701–1703 (1999).
84. Trabbic-Carlson, K., Setton, L. A., and Chilkoti, A. Swelling and mechanical behaviors of chemically cross-linked hydrogels of elastin-like polypeptides. *Biomacromolecules* 4, 572–580 (2003).
85. Zhou, Y. T., Wu, S. X., and Conticello, V. P. Genetically directed synthesis and spectroscopic analysis of a protein polymer derived from a flagelliform silk sequence. *Biomacromolecules* 2, 111–125 (2001).
86. Asakura, T. et al. Synthesis and characterization of chimeric silkworm silk. *Biomacromolecules* 4, 815–820 (2003).

87. Fukushima, Y. Genetically engineered syntheses of tandem repetitive polypeptides consisting of glycine-rich sequence of spider dragline silk. *Biopolymers* 45, 269–279 (1998).
88. Chilkoti, A., Dreher, M. R., and Meyer, D. E. Design of thermally responsive, recombinant polypeptide carriers for targeted drug delivery. *Advanced Drug Delivery Reviews* 54, 1093–1111 (2002).
89. Chilkoti, A., Dreher, M. R., Meyer, D. E., and Raucher, D. Targeted drug delivery by thermally responsive polymers. *Advanced Drug Delivery Reviews* 54, 613–630 (2002).
90. Lee, J., Macosko, C. W., and Urry, D. W. Elastomeric polypentapeptides cross-linked into matrixes and fibers. *Biomacromolecules* 2, 170–179 (2001).
91. Di Zio, K. and Tirrell, D. A. Mechanical properties of artificial protein matrices engineered for control of cell and tissue behavior. *Macromolecules* 36, 1553–1558 (2003).
92. McMillan, R. A. and Conticello, V. P. Synthesis and characterization of elastin-mimetic protein gels derived from a well-defined polypeptide precursor. *Macromolecules* 33, 4809–4821 (2000).
93. Megeed, Z., Cappello, J., and Ghandehari, H. Genetically engineered silk-elastinlike protein polymers for controlled drug delivery. *Advanced Drug Delivery Reviews* 54, 1075–1091 (2002).
94. Nagarsekar, A. et al. Genetic synthesis and characterization of pH- and temperature-sensitive silk-elastinlike protein block copolymers. *Journal of Biomedical Materials Research* 62, 195–203 (2002).
95. Nagarsekar, A. et al. Genetic engineering of stimuli-sensitive silkelastin-like protein block copolymers. *Biomacromolecules* 4, 602–607 (2003).
96. Farmer, R. S., Argust, L. M., Sharp, J. D., and Kiick, K. L. Conformational properties of helical protein polymers with varying densities of chemically reactive groups. *Macromolecules* 39, 162–170 (2006).
97. Farmer, R. S. and Kiick, K. L. Conformational behavior of chemically reactive alanine-rich repetitive protein polymers. *Biomacromolecules* 6, 1531–1539 (2005).
98. McGrath, K. P., Tirrell, D. A., Kawai, M., Mason, T. L., and Fournier, M. J. Chemical and biosynthetic approaches to the production of novel polypeptide materials. *Biotechnology Progress* 6, 188–192 (1990).
99. McGrath, K. P., Fournier, M. J., Mason, T. L., and Tirrell, D. A. Genetically directed syntheses of new polymeric materials—Expression of artificial genes encoding proteins with repeating (Alagly)3proglugly elements. *Journal of the American Chemical Society* 114, 727–733 (1992).
100. Van Hest, J. C. M. and Tirrell, D. A. Protein-based materials, toward a new level of structural control. *Chemical Communications* 1897–1904 (2001).
101. Kiick, K. L., Saxon, E., Tirrell, D. A., and Bertozzi, C. R. Incorporation of azides into recombinant proteins for chemoselective modification by the Staudinger ligation. *Proceedings of the National Academy of Sciences of the United States of America* 99, 19–24 (2002).
102. Link, A. J., Mock, M. L., and Tirrell, D. A. Non-canonical amino acids in protein engineering. *Current Opinion in Biotechnology* 14, 603–609 (2003).
103. Mock, M. L., Michon, T., van Hest, J. C. M., and Tirrell, D. A. Stereoselective incorporation of an unsaturated isoleucine analogue into a protein expressed in *E. coli*. *Chembiochem* 7, 83–87 (2006).
104. Wang, C. C. Y., Seo, T. S., Li, Z. M., Ruparel, H., and Ju, J. Y. Site-specific fluorescent labeling of DNA using Staudinger ligation. *Bioconjugate Chemistry* 14, 697–701 (2003).
105. Wang, J., Xie, J., and Schultz, P. G. A genetically encoded fluorescent amino acid. *Journal of the American Chemical Society* 128, 8738–8739 (2006).
106. Birch, J. M. et al. Relative frequency and morphology of cancers in carriers of germline TP53 mutations. *Oncogene* 20, 4621–4628 (2001).
107. Collins, F. S., Brooks, L. D., and Chakravarti, A. A DNA polymorphism discovery resource for research on human genetic variation. *Genome Research* 8, 1229–1231 (1998).
108. Greenblatt, M. S., Bennett, W. P., Hollstein, M., and Harris, C. C. Mutations in the P53 tumor-suppressor gene—Clues to cancer etiology and molecular pathogenesis. *Cancer Research* 54, 4855–4878 (1994).
109. Kirk, B. W., Feinsod, M., Favis, R., Kliman, R. M., and Barany, F. Single nucleotide polymorphism seeking long term association with complex disease. *Nucleic Acids Research* 30, 3295–3311 (2002).
110. Soussi, T. and Beroud, C. Assessing TP53 status in human tumours to evaluate clinical outcome. *Nature Reviews Cancer* 1, 233–240 (2001).
111. Soussi, T. and Lozano, G. p53 mutation heterogeneity in cancer. *Biochemical and Biophysical Research Communications* 331, 834–842 (2005).

112. Chen, X. and Sullivan, P. F. Single nucleotide polymorphism genotyping: Biochemistry, protocol, cost and throughput. *Pharmacogenomics Journal* 3, 77–96 (2003).
113. Landegren, U., Kaiser, R., Sanders, J., and Hood, L. A ligase-mediated gene detection technique. *Science* 241, 1077–1080 (1988).
114. Pastinen, T., Partanen, J., and Syvanen, A. C. Multiplex, fluorescent, solid-phase minisequencing for efficient screening of DNA sequence variation. *Clinical Chemistry* 42, 1391–1397 (1996).
115. Meagher, R. J. et al. Multiplexed p53 mutation detection by free-solution conjugate microchannel electrophoresis with polyamide drag-tags. *Analytical Chemistry* 79, 1848–1854 (2007).
116. Lagally, E. T., Simpson, P. C., and Mathies, R. A. Monolithic integrated microfluidic DNA amplification and capillary electrophoresis analysis system. *Sensors and Actuators B-Chemical* 63, 138–146 (2000).
117. Meagher, R. J., Won, J. I., and Barron, A. E. Sequencing of DNA by free-solution capillary electrophoresis using a genetically engineered protein polymer drag-tag. Submitted (2006).
118. Mi, L. X. Molecular cloning of protein-based polymers. *Biomacromolecules* 7, 2099–2107 (2006).
119. Cappello, J. et al. Genetic-engineering of structural protein polymers. *Biotechnology Progress* 6, 198–202 (1990).
120. Meyer, D. E. and Chilkoti, A. Genetically encoded synthesis of protein-based polymers with precisely specified molecular weight and sequence by recursive directional ligation: Examples from the elastin-like polypeptide system. *Biomacromolecules* 3, 357–367 (2002).
121. Goeden-Wood, N. L., Conticello, V. P., Muller, S. J., and Keasling, J. D. Improved assembly of multimeric genes for the biosynthetic production of protein polymers. *Biomacromolecules* 3, 874–879 (2002).
122. Gross, E. The cyanogen bromide reaction. *Methods in Enzymology* 11, 238–255 (1967).
123. Tan, H. D. and Yeung, E. S. Integrated on-line system for DNA sequencing by capillary electrophoresis: From template to called bases. *Analytical Chemistry* 69, 664–674 (1997).
124. Tan, H. D. and Yeung, E. S. Automation and integration of multiplexed on-line sample preparation with capillary electrophoresis for high-throughput DNA sequencing. *Analytical Chemistry* 70, 4044–4053 (1998).

13 Online Sample Preconcentration for Capillary Electrophoresis

Dean S. Burgi and Braden C. Giordano

CONTENTS

13.1 INTRODUCTION

The detection of material in a capillary column has proven to be interesting and dynamic over the years.[1] Ultimately, the ability to detect a given analyte comes down to the amount of material in the cross-sectional area of the capillary at the detector location. For example in ultra violet (UV) detection, the absorbance of light, A, is related to three terms in the Beer–Lambert law,[2]

$$A = \varepsilon lc. \tag{13.1}$$

The molar absorptivity, ε, is an intrinsic property of the analyte, the second term, l, is the path length the light will travel through the medium; for capillary electrophoresis (CE), the second term is small (25–100 μm) and cannot be easily change without impacting the resolution of the separation process. Thus, manipulation of the third term, c, the concentration of the analyte of interest, has evolved into a topic of significant interest.

13.1.1 The Kohlrausch Regulation Function and Stacking

While there are a number of methods by which analytes can be stacked or preconcentrated in CE, ultimately, all stacking phenomena require a change in analyte velocity as the analyte transitions from the sample zone to the background electrolyte (BGE).

The capillary can be viewed as a wire, which will carry current when a potential is placed across it. The amount of potential drop is related to the resistivity of the cross-sectional area of the column. In a wire, the free electrons move from one end of the wire to the other and their ability to move freely is a measure of the resistance of the wire. In the liquid domain where CE resides, each ionic species has the ability to carry a part of the current based on its charge, z, and concentration, c. Since each ion has a unique mobility, the sum of all the ionic species in the cross-sectional area is a measure of the resistance. Before the application of potential, the Kohlrausch regulation function (KRF)[3] for a given point in the capillary is defined by the following equation:

$$\omega = \sum_i \frac{c_i z_i}{|\mu_i|}, \tag{13.2}$$

where c_i is the concentration of a given ion, z_i is the relative charge of strong electrolytes, and μ_i is the mobility. For any given point in the capillary, the KRF value, ω, will remain constant. With that in mind, upon application of voltage, species will move in and out of said cross-sectional area and increase (stack) or decrease (destack) in concentration in order to satisfy this function. For example, if the whole column was filled with a single mono electrolytic compound like sodium chloride dissolved in water, Equation 13.2 would allow the calculation of the Kohlrausch value at all points from one end of the capillary to the other like a wire. As different ions are added to the system, the equation becomes a summation of all the ions and their mobilities as they move into and out of the analysis point. It must be noted that each time the electric field is broken; the ionic composition at the time of reconnection will generate a new Kohlrausch value. Thus, an injection potential will have one Kohlrausch value and separation will have a second Kohlrausch value. The delta in values between the two events can help further stack the samples into tighter bands.[4]

13.1.2 Three Solutions to the KRF

13.1.2.1 Continuous Systems

There exist three types of states for the regulation function to operate. State 1 is where the BGE is continuous throughout the capillary. It is assumed that in this state a low analyte concentration is present in the sample plug and the presence of analyte does not affect the Kohlrausch value anywhere throughout the capillary. Put simply, the sample plug has the same Kohlrausch value as the BGE region surrounding it, thus, has no boundary. The flux of ions in and out of the cross-sectional area is uniform. This state is known as zone electrophoresis. Upon application of voltage, the sample compounds move out of the sample region as a function of their intrinsic mobilities and more importantly, no velocity difference exists between analyte in the sample zone and analyte in the BGE. Thus, the back end of the sample maintains equidistance from the front end of the sample. That velocity, v, is a function of the ions mobility and is given in the following equation:

$$v = \mu E, \tag{13.3}$$

where E is the electric field across the capillary. The width of the peak for each compound in the sample would be as narrow as the first original injection plug barring the diffusion process. Figure 13.1 shows several mobility markers from a short electrokinetic injection. The peak widths are becoming wider as the amount of time spent in the column increase due simply to longitudinal diffusion.

The second state is where the sample plug is higher or lower in BGE concentration than that of the surrounding BGE. If the sample plug is lower in concentration, the back of the sample plug will migrate rapidly toward the front of the plug because the Kohlrausch value at the boundary would decrease the sample ion velocity whereas the rest of the sample is still under the original Kohlrausch

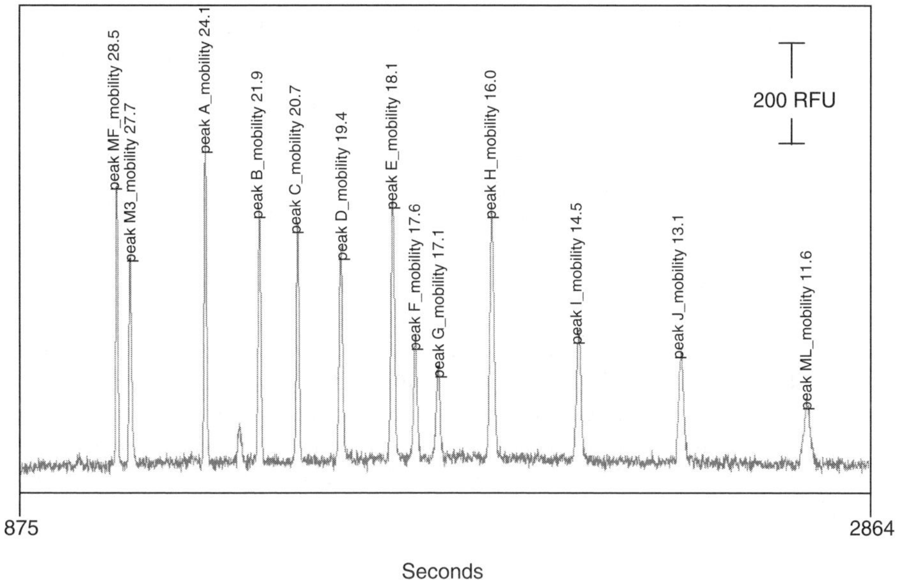

FIGURE 13.1 An electropherogram of 13 mobility markers run on a MegaBACE 4000. Injection voltage 8 kV, 20 s, run voltage 8 kV, 60 min, and oven temperature is 44°C. Mobilities calculated using a 60 cm long capillary, 43 cm to the detector. Concentration of each marker is 5 nM.

value. This difference in velocity is directly related to the difference in the concentrations as seen in the following equation:

$$v = \mu E = \mu IR = \mu I/c. \tag{13.4}$$

The resistance (R) is inversely proportional the concentration (c) of the electric field carrying ions. Thus, sample is stacked into a small region and the concentration increases, which allows a lower concentration sample to be detected. The front boundary (or the detector side of the sample plug) is a stable stationary boundary and does not move; only the sample moves through it. This stacking mechanism (see Figure 13.6) is commonly referred to as field-amplified sample stacking (FASS). If the sample plug is higher in concentration than the surrounding BGE, the plug will spread out, as the front of the plug has a higher velocity than the back end. The front boundary is a function of the Kohlrausch value calculated at the time of electric application and will cause mixing of the sample with the BGE to maintain the Kohlrausch value. This effect is known as electrodispersion and often presents as analyte fronting, and is typically considered undesirable. In continuous systems, the boundaries are stable and stationary and will not move except by electroosmotic flow (EOF).

13.1.2.2 Discontinuous Systems

The third state requires a multi-ion sample matrix (SM). In this system, there are several ions present of varying mobilities at high enough concentrations to affect the electric fields. For example, the column matrix can include an ion with a greater mobility than those in the sample. In effort to satisfy the KRF, this ion with the greatest mobility will have the ion next largest mobility ion to stack up behind it. The third ion and so further will fall in line; the ions are stacked into regions where the concentration/mobility (KRF) equals the leading electrolytes value. Thus, low concentration ions would stack into very narrow bands and higher concentration ions will be broader. This state is commonly known as isotachophoresis (ITP). There are several review articles,[5–7] which explain the various states in more detail. In a discontinuous system, the boundaries between the ion species are moving but stable through the system. They will move from one end of the column to the other and are called system peaks.

Another version of a discontinuous system is called sample self-stacking.[8,9] In this system, a co-ion is placed into the sample matrix, which acts like the leading electrolyte in ITP but only during the injection phase. After injection, the sample will continue to stack until the stacking ion leaves the sample zone. In Reference 9, an equation is presented to calculate a stacking ratio known as a_{crit}.

$$a_{\text{crit}} = [(\mu_b + \mu_r)/(\mu_a + \mu_r)] \times [(\mu_a - \mu_x)/(\mu_x - \mu_b)] \tag{13.5}$$

where μ_x is the mobility of the last ion to be stacked, μ_a is the mobility of the major stacking co-ion, μ_b is the mobility of the sample BGE (which may not be the same as the separation run buffer), and μ_r is the mobility of the counter ion of the system. μ_a is the same as the leading electrolyte in ITP and μ_b is the trailing electrolyte. Thus, a temporary ITP event is setup in the sample during electrokinetic injection and during the first moments of the separation time.[9] a_{crit} is the theoretical value at which the major co-ion in the sample system is proposed to self-stack the slower ions, μ_x, in a transitory ITP fashion.

In addition, in Reference 9, a value related to the major ion concentrations is give.

$$a = C_{\text{s}}/C_{\text{b}} \tag{13.6}$$

where C_{s} is the concentration of the sample-stacking co-ion, C_{b} is the concentration of the sample BGE. The value of "a" will indicate if the sample system is near or at a self-stacking domain. If the "a" value is greater than or equal to a_{crit} then self-stacking can occur. If not, then other stacking or destacking conditions are the driving force for the analysis.

13.2 EXAMPLES OF LONG INJECTION STACKING

In general, sample stacking will occur when a boundary exists between the sample regions and the surrounding electrolyte (BGE). This boundary must be stable for the length of time such that the concentration of the sample of interest can be increased. When the boundary has passed on, stacking will stop and normal zone electrophoresis occurs. Several standard methods for boundary formation and use are described elsewhere[1,4] like FASS, capillary ion electrophoresis,[10] and ITP to name a few.

Several novel stable pseudostationary boundary techniques have appear, one technique dubbed moving chemical reaction boundary[11] forms a neutral zone in the column were the H^+ ions combine with the OH^- ions to make a water zone. This zone is stable and leads to stacking across it. A second method uses EOF balanced against back-pressure to hold the stacking boundary in a fixed location.[12] The stability of the boundary allows for injections up to 3 h to be made.[13] The most robust method for long injects has arguably been electrokinetic injection out of a sample vial in which the column has one Kohlrausch value and the vial a lower value. This method has shown to improve sample injection by 1000 fold.[14]

When considering long electrokinetic injections, it becomes necessary to carefully consider the ionic species included in the sample matrix and the BGE. Figure 13.2 illustrates the importance of matching the composition of the ions in the sample matrix to the BGE. In this case, 4-Morpholinepropanesulfonic acid was added to the sample matrix in order to achieve a desired pH for purposes of a reaction; thus creating a discontinuity between the sample matrix and the BGE (100 mM TAPS). Consequently, a stable moving boundary (or system peak) results between the MOPS in the sample matrix and *N*-[Tris(hydroxymethyl)methyl]-3-aminopropanesulfonic acid in the BGE (noted in Figure 13.2 by arrows). With increasing injection time, this system peak becomes more obvious, and more importantly, significant destacking of analytes that have to migrate through this

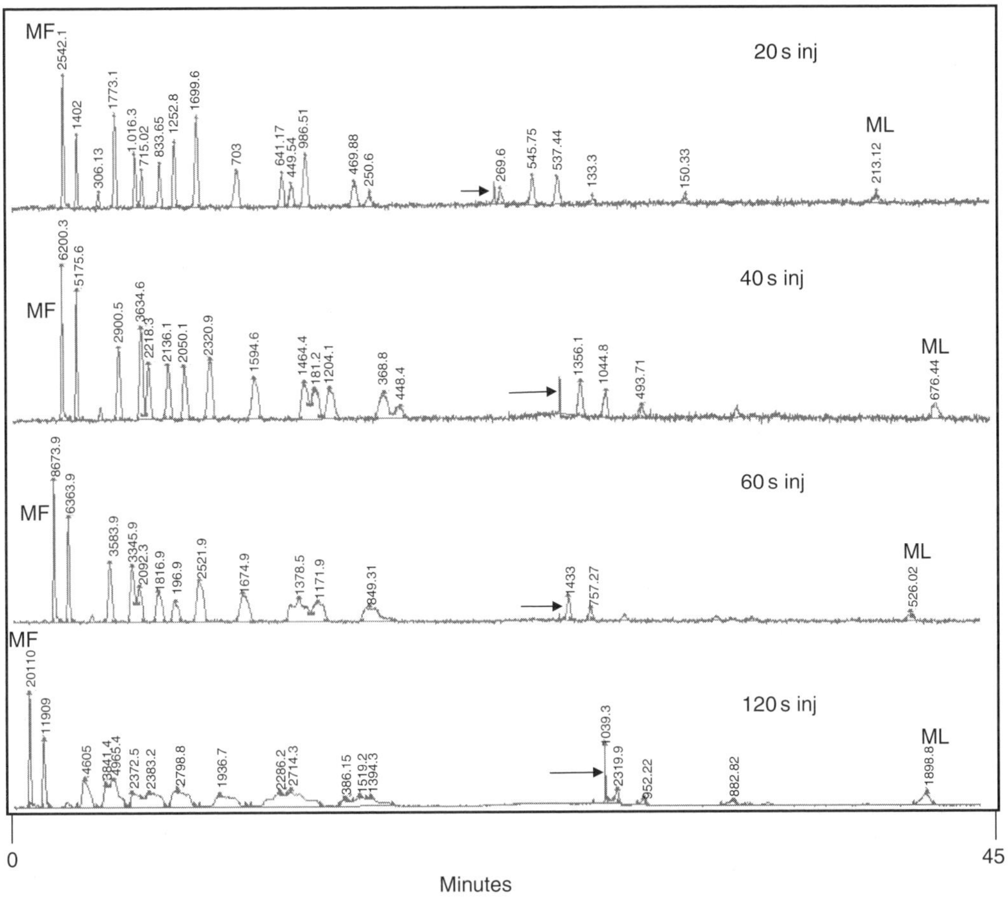

FIGURE 13.2 Injection of mobility markers prepared in a 100 mM MOPS buffer system (pH 7.9). The injection time ranged from 20 s in the top image to 120 s in the bottom image. The separation was done on a MegaBACE 1000, injection voltage was 15 kV, run voltage 15 kV, run time 45 min, oven is set to 44°C. The MOPS system peak is located at about 28 min and indicated by an arrow.

boundary is observed. If the sample matrix composition is changed such that TAPS is in both the sample matrix and BGE (Figure 13.3), the injection time can be increased to 120 s and the system peak is located at the interface of the column and the sample vial.

To further push the limits of long injections a mobility difference between the major current carrying ionic species is required, resulting in the formation of a stable moving boundary; sample will stack in the plug during the time it takes to establish steady-state conditions. The leading or terminating electrolyte serves to define the electric field that each of the analyte zones experience such that band width will adjust itself to have the same concentration per cross section as the leading or terminating ion (isotachophoresis). Using a self sample-stacking ion in the sample plug at the appropriate "a" value can generate a transient ITP event to occur during the injection and the first part of the separation time. The sample compounds isotachophoretically stack as the leading ion is slowed down by the mismatch at the moving boundary between the sample plug and the BGE. After the leading ion moves into the separation part of the column, the self-stacked stops and the sample begins to separate as if under zone electrophoresis and the resolution are preserved.

Close proximity to a_{crit}, as determined by the ratio of the concentration of the sample-stacking co-ion to the sample BGE (Equation 13.6), will result in a steady-state ITP effect before the leading ion is out of the sample zone. For the set of mobility markers shown in this work, a range of a_{crit} can

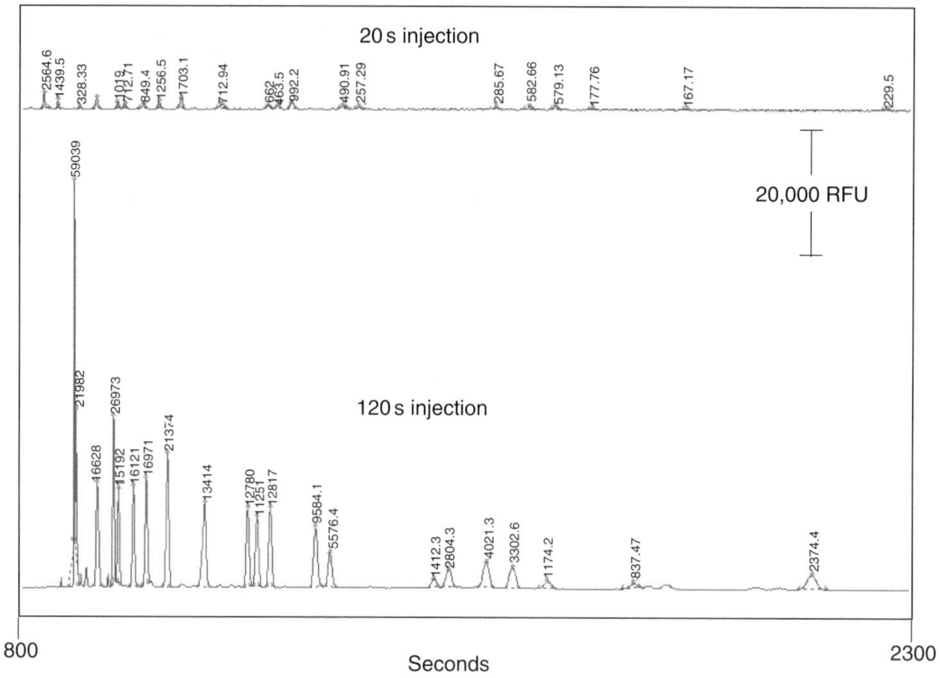

FIGURE 13.3 Comparison of a 20 s injection with a 120 s injection using 100 mM TAPS buffer (pH 8.3) instead of 100 mM MOPS (pH 7.9) buffer in the sample.

be determined from Equation 13.5. On the basis of the use of chloride (from $MgCl_2$) as the leading electrolyte and TAPS as the trailing electrolyte; a range of a_{crit}'s from 1.23 to 34.24 can be calculated. Figure 13.4 illustrates the difference between an "a" value (ratio of chloride concentration to TAPS concentration) close to a_{crit} versus not close to a_{crit}. In the top trace, "a" is 0.5; it is apparent that the mobility markers with a_{crit} values closest to 0.5 are stacked more efficiently (peaks A and B with a_{crit} values of 1.23 and 1.43, respectively). Decreasing the TAPS concentration relative to the chloride concentration to yield an "a" value of 4.0 drastically improves the efficiency of stacking to markers with much higher a_{crit} values, allowing injection times up to 150 s long with no loss in resolution or distortion in peak shape. However, there is a functional limit to the amount of self-stacking that can occur as determined by the time it takes the leading electrolyte to migrate out of the sample zone. If electrokinetic injection continues beyond this point significant, destacking can occur as illustrated by Figure 13.5.

Whether the stacking technique is called field-amplified stacking, transient isotachophoresis, or sample self-stacking, the driving force in on-column sample preconcentration for charges species is the need to satisfy the initial KRF conditions. Under separation and sample matrix conditions where the KRF is difficult to determine, simply calculations such as a_{crit} and "a" offer a rapid evaluation of conditions for optimal transient ITP.

13.3 SAMPLE STACKING IN MICELLAR ELECTROKINETIC CHROMATOGRAPHY

Micellar electrokinetic chromatography (MEKC) is a mode of CE primarily used in the separation of neutral molecules. Since a mixture of neutral molecules will all migrate with EOF, it becomes necessary to include a proxy within the BGE to afford a separation. This proxy is typically a micelle—an

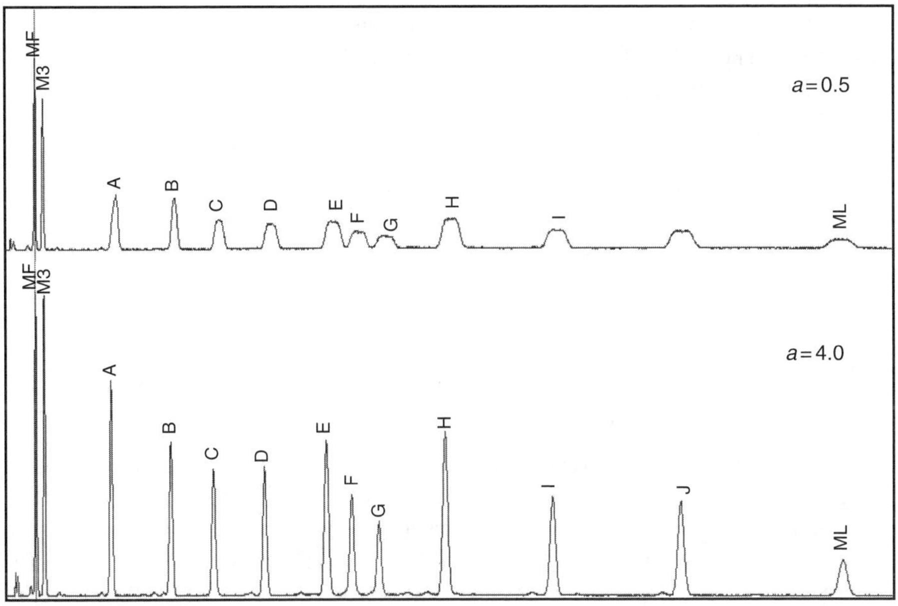

FIGURE 13.4 Electropherogram of the mobility markers run under different sample self-stacking conditions as defined by the concentration of chloride. The background buffer was 100 mM TAPS with 25 mM $MgCl_2$ for an "*a*" value = 0.5 and 12.5 mM TAPS BGE with 25 mM $MgCl_2$ for an "*a*" value = 4.0. The *x*-axis is setup as a relative scale in which the MF is set at zero and the ML peak is set as 1.0. All the other peak mobilities are calculated relative to this unit scale. The data were generated on a MegaBACE 4000 under the same condition as Figure 13.1.

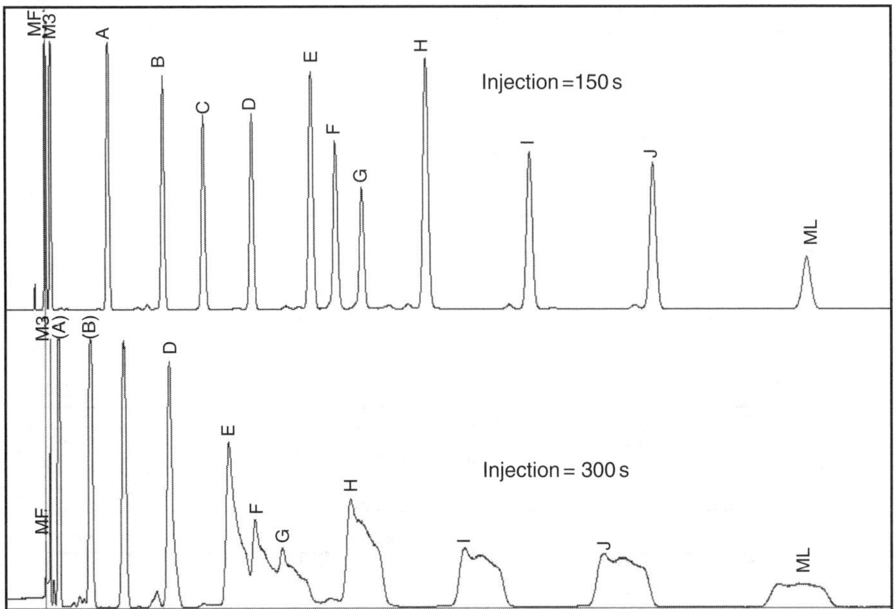

FIGURE 13.5 Electropherogram of the mobility markers at 150 s versus 300 s injection time. Data were generated on a MegaBACE 4000 as per Figure 13.1.

aggregate of surfactant molecules. Neutral analytes migrate into and out of the micelle, and when in the micelle, briefly take on the electrophoretic mobility of the micelle.

In the previous sections, several methods of sample preconcentration were described. Regardless of the mechanism, every example of sample stacking requires a change in the velocity of the analyte as determined by the drive to maintain the KRF value. Unfortunately, neutral analytes do not contribute to the KRF and a different perspective on sample stacking must be taken. There are a number of stacking methods available when using MEKC. This section will describe the fundamental mechanism and conditions by which sample preconcentration of neutral molecules occurs.

13.3.1 Modes of Sample Stacking in MEKC

13.3.1.1 Field-Amplified Stacking

Field-amplified stacking is made possible by preparing sample in diluted surfactant solution such that the conductivity of the sample matrix is lower than that of the BGE.[15] This results in localized higher electric field in the sample zone. When using normal polarity with a high-EOF velocity, this will result in analyte and micelle in the sample zone to rapidly travel toward the negative electrode due to the increased field strength in the sample zone. On entering the BGE, the analyte will experience a lower electric field, resulting in stacking. This mechanism is illustrated in Figure 13.6.

Throughout this section, electropherograms illustrating the effectiveness of the various modes of sample stacking will be presented. Figure 13.7 shows the electropherogram resulting from preparing sample in a dilute micelle solution, thus under FASS-like conditions. The analyte set is a mixture of neutral alkaloids. The bottom trace shows the separation resulting from preparing sample in BGE, thus under normal zone electrophoresis conditions, the middle trace shows the resulting electropherogram when sample is prepared in 0.1× BGE, and the top trace shows the resulting electropherogram when sample is prepared in 0.05× BGE. Experimental details are included in the figure legend. It is clear from the figure that the peak height of several analytes increase, however, some tailing is observed indicating that this method of preconcentration may be inefficient for this specific analyte set.

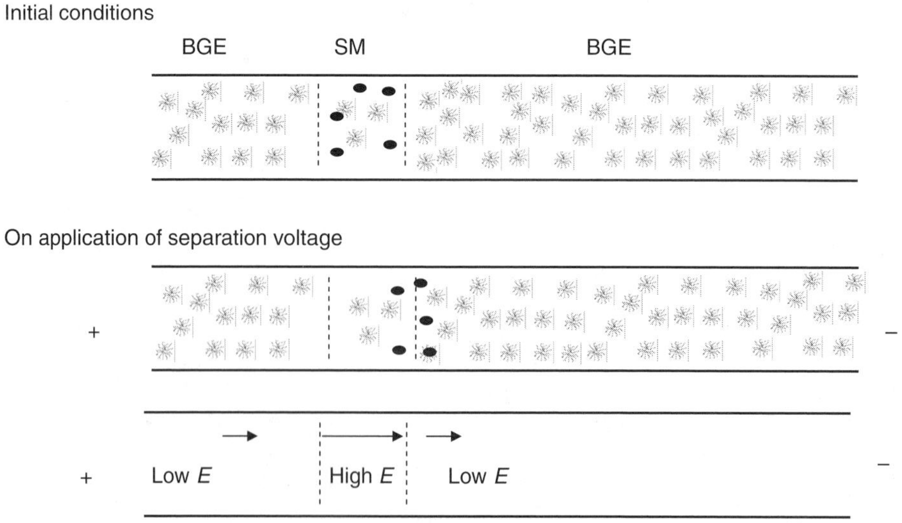

FIGURE 13.6 Illustration of the mechanism of field-amplified stacking. In this case, the SM is dilute BGE typically between 0.05× and 0.1× BGE.

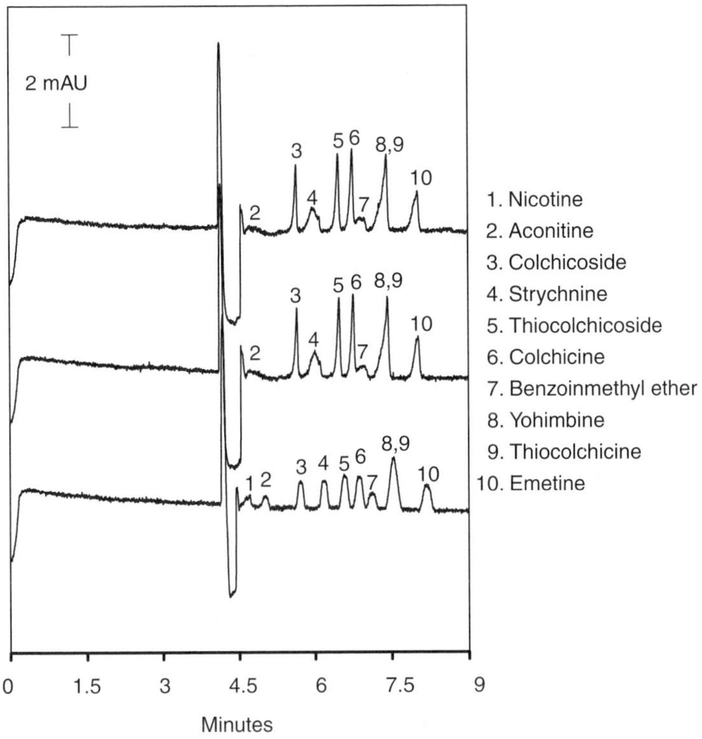

FIGURE 13.7 Separation of 10 alkaloids. Bottom trace: sample matrix is BGE. Middle trace: sample matrix is 0.1× BGE. Top trace: sample matrix is 0.05× BGE. Experimental conditions: 50 μm internal diameter, $L_T = 60$ cm (50 cm effective length), $l_{inj} = 10$ mm (20 s × 0.5 psi), $V = 25$ kV (47 μA), UV absorbance at 214 nm. BGE: 10 mM $Na_2B_4O_7$, 80 mM sodium cholate.

13.3.1.2 Sweeping

Sweeping was originally defined as "the picking and accumulating of analytes by the pseudo stationary phase that fills the sample zone during application of voltage."[16] Sweeping was initially implemented by preparing analytes in a sample matrix of equal conductivity to the BGE, but devoid of surfactant sodium dodecyl sulfate.[16,17] Terabe and coworkers[18] later expanded the definition of "sweeping" to include any situation in which the BGE contains a separation vector and the sample matrix does not. The mechanism of stacking for an equal conductivity sample matrix is presented in Figure 13.8 for a system with negatively charged micelles and normal EOF. Briefly, under a homogenous electric field, neutral analytes are carried toward the negative electrode only to be swept up in the micelles at the boundary of the sample region. The effectiveness of the stacking is dependent on an analytes affinity for the micelle, the higher affinity of the micelle the larger the extent of stacking. Terabe derived the following equation representing the extent of stacking as a function of an analytes affinity for the micelle

$$l_{sweep} = l_{inj}\left(\frac{1}{1+k}\right), \tag{13.7}$$

where l_{sweep} is the length of the swept analyte, l_{inj} is the length of the injected plug, and k is the retention factor for a given analyte. Figure 13.9 shows the effectiveness of sweeping on the alkaloid separation. Note that peaks 1 and 2 are not well stacked. This is expected as they have the lowest affinity (k) for the cholate micelle.

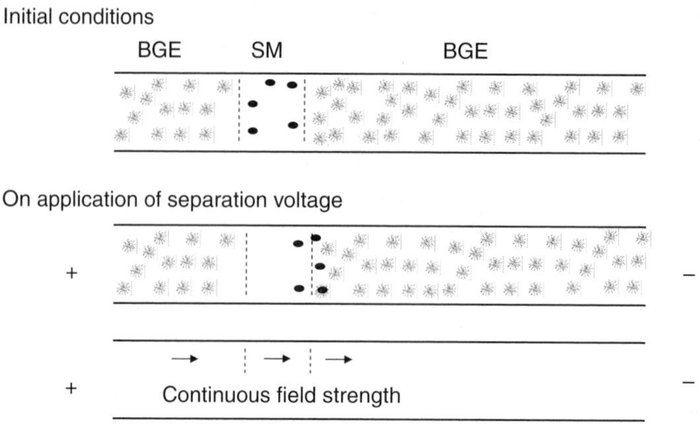

FIGURE 13.8 Illustration of the mechanism of sweeping. In this case, the ionic strength of the sample matrix, devoid of micelle, is adjusted such that a continuous field strength is achieved.

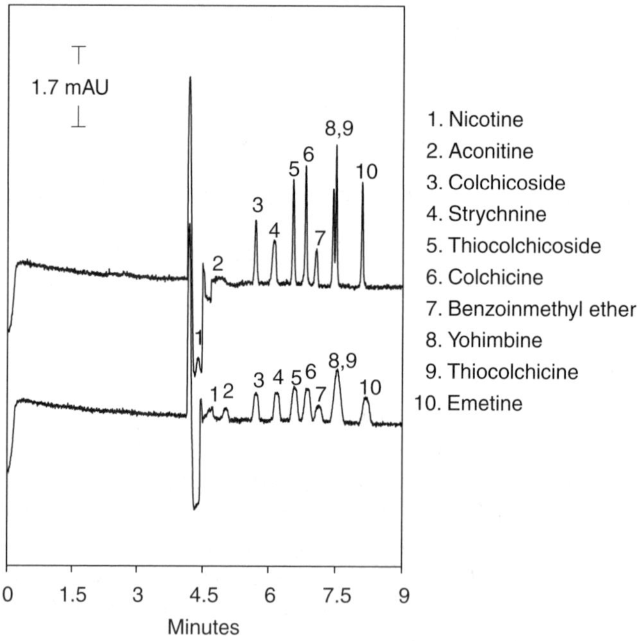

FIGURE 13.9 Separation of 10 alkaloids. Bottom trace: sample matrix is BGE. Top trace: sample matrix is 30 mM $Na_2B_4O_7$. Experimental conditions: 50 μm internal diameter, $L_T = 60$ cm (50 cm effective length), $l_{inj} = 10$ mm (20 s × 0.5 psi), $V = 25$ kV (47 μA), UV absorbance at 214 nm. BGE: 10 mM $Na_2B_4O_7$, 80 mM sodium cholate.

13.3.1.3 High-Salt Stacking

High-salt stacking is a mechanism of analyte preconcentration afforded by the stacking of micelles at the negative electrode side of the sample plug.[19] Analyte migrates into the stacked micelle region experiencing a locally high retention factor due to the high micelle concentration. The nomenclature "high-salt stacking" comes from the fact that in order to stack micelles the conductivity of the sample

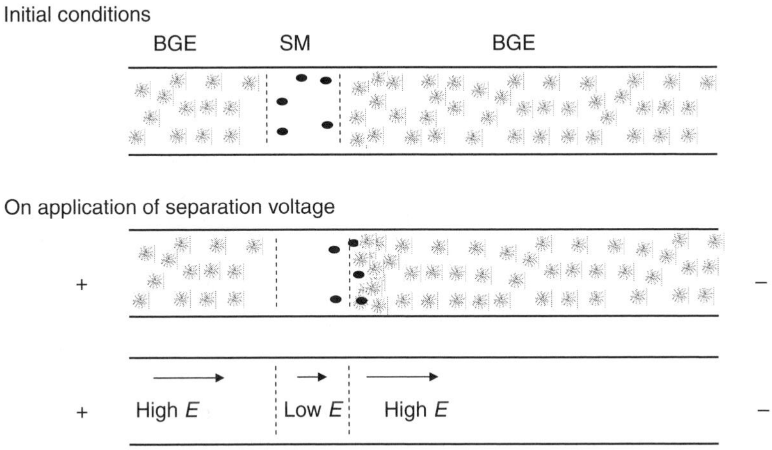

FIGURE 13.10 Illustration of the mechanism of high-salt stacking. In this case the ionic strength of the sample matrix is adjusted such that it is 1.5–2.5 times the ionic strength of the BGE.

zone is typical 1.5–2.5 × that of the BGE. High-salt stacking requires the following conditions be met:

$$\mu_{\text{sample}} E_{\text{sample}} < \mu_{\text{ev}} E_{\text{ev}} \tag{13.8}$$

$$\mu_{\text{sample}} > \mu_{\text{ev}}, \tag{13.9}$$

where μ_{sample} is the mobility of the sample stacking co-ion, E_{sample} is the electric field in the sample zone, μ_{ev} is the mobility of the electrokinetic vector (micelle), and E_{ev} is the field in the BGE. The condition that the mobility of the sample stacking co-ion be greater than the mobility of the micelle requires that the field in the BGE be greater than that in the sample zone, this is the need to significantly increase the conductivity of the sample zone relative to the BGE. The stacking mechanism is summarized in Figure 13.10. Note that the illustration shows micelles concentrating at the detector side of the sample plug upon application of a separation voltage. The effectiveness of high-salt stacking for the analysis of the alkaloid samples is shown in Figure 13.11. In this example, peaks 1 and 2 are well stacked demonstrating the power of high-salt stacking for analytes with low k's.

13.3.1.4 Electrokinetic Injection

Field-amplified stacking, sweeping, and high-salt stacking all require a discrete hydrodynamically injected sample plug. Alternatively, electrokinetic injections can be utilized to introduce sample to the column. Landers and coworkers[20] presented electrokinetic stacking under continuous conductivity conditions. Sample is injected with EOF and interacts with micelles included in the BGE. The analytes concentrate at the sample zone/BGE interface. Provided a large enough deference between the velocity of EOF and the net velocity of the micelle, it is possible to inject the equivalent of multiple column volumes of analyte. This stacking scheme is illustrated in Figure 13.12, and the corresponding example using this method for the alkaloid separation is shown in Figure 13.13.

13.3.1.5 Field-Enhanced Sample Injection

While the above method of electrokinetic injection requires only a single step, it is possible to couple electrokinetic injections with a brief hydrodynamic injection of water.[21] The column is initially full of BGE, followed by a hydrodynamic injection of water. Analyte is then electrokinetically injected until the measured current through the column is approximately 70–90% of the column when filled

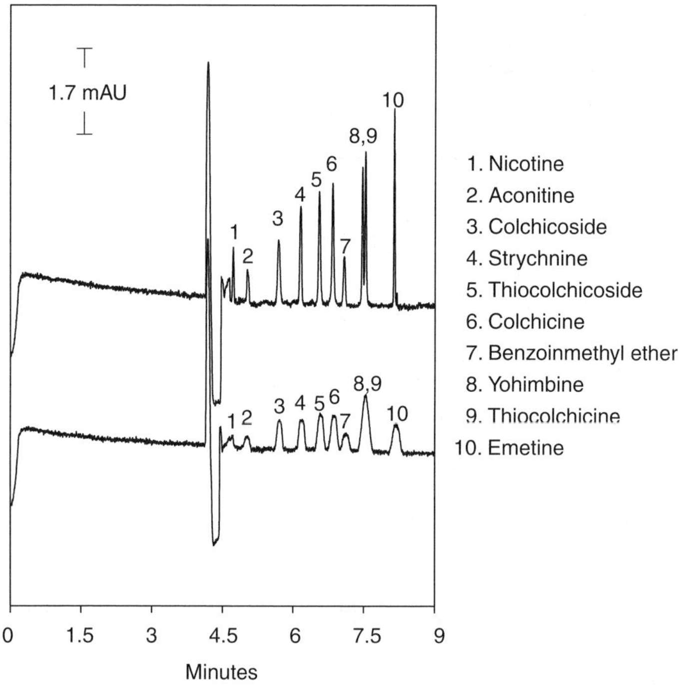

FIGURE 13.11 Separation of 10 alkaloids. Bottom trace: sample matrix is BGE. Top trace: sample matrix is 150 mM NaCl. Experimental conditions: 50 μm internal diameter, $L_T = 60$ cm (50 cm effective length), $l_{inj} = 10$ mm (20 s × 0.5 psi) $V = 25$ kV (47 μA), UV absorbance at 214 nm. BGE: 10 mM $Na_2B_4O_7$, 80 mM sodium cholate.

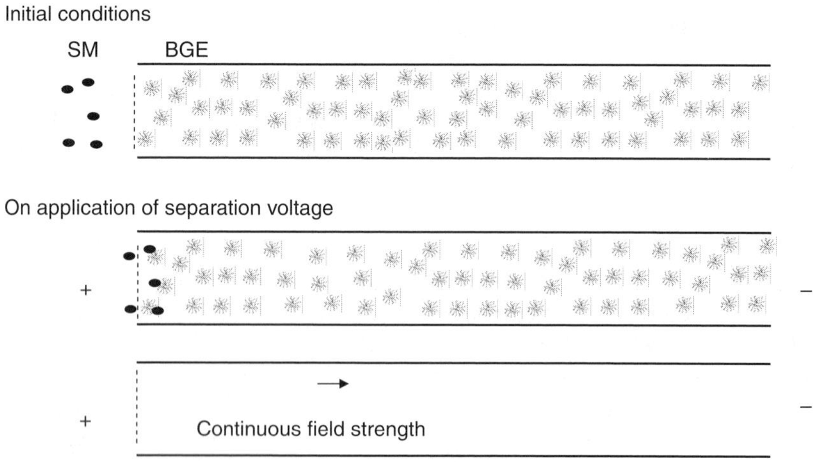

FIGURE 13.12 Illustration of the mechanism of electrokinetic stacking injection. The sample matrix is prepared devoid of micelles to an equal conductivity of the BGE.

exclusively with BGE. Upon application of voltage, analyte migrates into the water plug. As potential continues to be applied across the capillary the water plug compresses—narrowing the analyte plug. This scheme is shown in Figure 13.14. A separation of a mixture of alkylphenyl ketones using this method is shown in Figure 13.15.

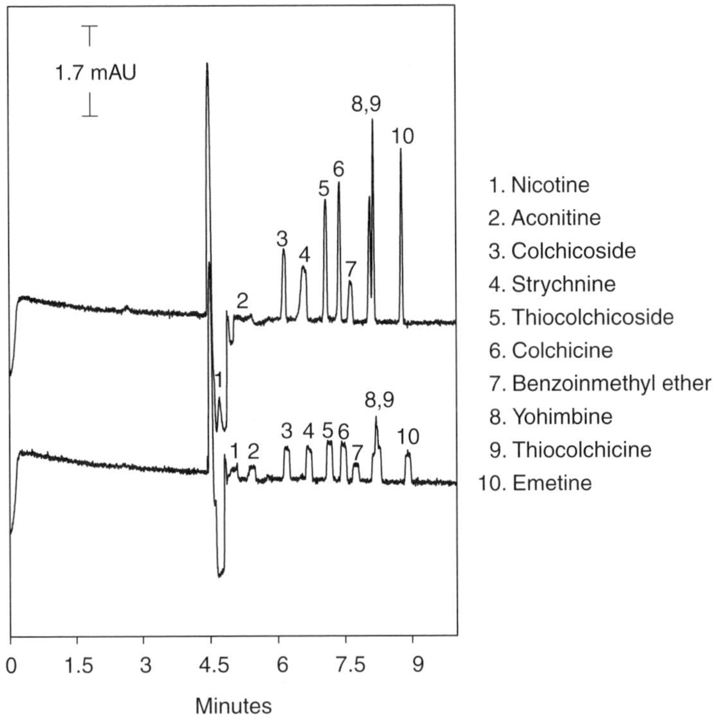

FIGURE 13.13 Separation of 10 alkaloids. Bottom trace: sample matrix is BGE. Top trace: sample matrix is 30 mM $Na_2B_4O_7$. Experimental conditions: 50 µm internal diameter, $L_T = 60$ cm, $l_{inj} = 20$ s at 10 kV, $V = 25$ kV (47µA), UV absorbance at 214 nm. BGE: 10 mM $Na_2B_4O_7$, 80 mM sodium cholate.

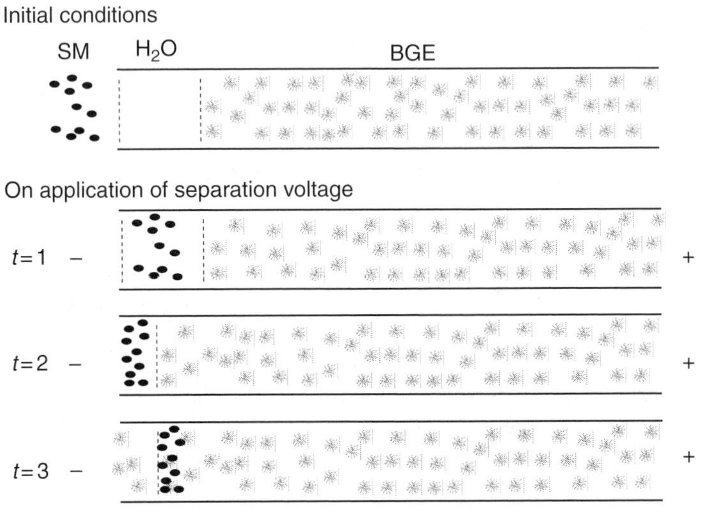

FIGURE 13.14 Illustration of the mechanism or field-enhanced sample injection.

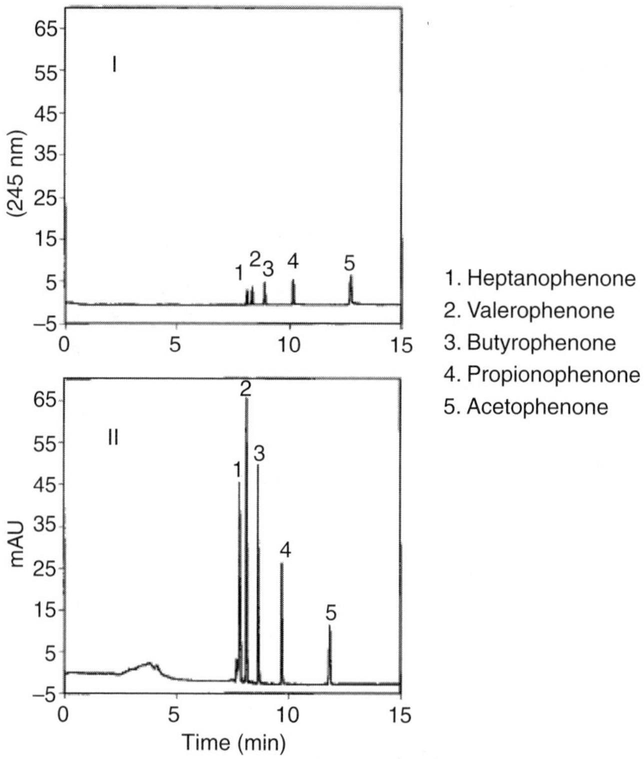

FIGURE 13.15 Separation of five alkyl phenyl ketones. Sample matrix is 8.3 mM SDS, 16.7 mM phosphate buffer (pH 2.5) (I). Water plug length of 8.7 cm for FESI (II). Experimental conditions: 50 μm internal diameter, $L_T = 64.5$ cm (56 cm effective length), $l_{inj} = 2$ s at 50 mbar, $V = -20$ kV, UV absorbance at 245 nm. BGE: 100 mM phosphate buffer (pH 2.25), 50 mM SDS. (Reprinted with permission from Quirino, J.P. and Terabe, S., *Am. Chem. Soc.*, 70, 1893, 1998. Copyright 1998.)

13.4 CONCLUSIONS

The goal of this chapter was to present some of the basic concepts of online sample preconcentration in capillary electrophoresis. The relationship between large volume injections of charged analytes and the Kohlraush regulation function were discussed in Section 13.2. Irrespective of whether one chooses a continuous or discontinuous buffer system, careful consideration of the background electrolyte and sample matrix components and the formation of a long-term stable Kohlrausch boundary are absolutely necessary for efficient large volume injections. Rapid evaluation of the potential for sample stacking is made possible by calculations for "a" and a_{crit}, precluding the sometimes cumbersome KRF calculation. Stacking of neutral analytes, which do not contribute to the KRF, was presented in Section 13.3. The mechanism and effect of several sample stacking modes were presented. The underlying phenomenon common to each mode of sample stacking in MEKC is that the analyte experiences a velocity change on partitioning into the micelle. In this regard, preconcentration efficiency in MEKC is, at the very least, dependent on the analyte affinity for the micelle and the mobility of the micelle. Ultimately, whether analytes are neutral or charged, successful optimization of online preconcentration is made possible with a complete understanding of the system components including the analytes of interest, sample matrix, and background electrolyte.

REFERENCES

1. Landers, J.P., Ed., *Handbook of Capillary Electrophoresis*, 2nd ed., CRC Press, Boca Raton, FL, 1996.
2. Skoog, D.A. and West, D.M., *Principles of Instrumental Analysis*, 2nd ed., Chapter 6, p. 161, Saunders College, Philadelphia, PA, 1980.
3. Kohlrausch, F., Ueber concentrations-verschiebungen durch electrolyse im inneren van Losungen und losungegemischen, *Ann. Phys. Chem.*, 62, 209, 1897.
4. Chien, R.L., Sample stacking revisited: A personal perspective, *Electrophoresis*, 24, 486–497, 2003.
5. Gerbauer, P. and Bocek, P., Theory of zone separation in isotachophoresis: A diffusional approach, *Electrophoresis*, 16, 1999–2007, 1995.
6. Beckers, J.L., Steady-state models in electrophoresis: From isotachophoresis to capillary zone electrophoresis, *Electrophoresis*, 16, 1987–1998, 1995.
7. Urbanek, M., Krivankova, L., and Bocek, P., Stacking phenomena in electromigration: From basic principles to practical procedures, *Electrophoresis*, 24, 466–485, 2003.
8. Gerbauer, P., Thormann, W., and Bocek, P., Sample self-stacking and sample stacking in zone electrophoresis with major sample components of like charge: General model and scheme of possible modes, *Electrophoresis*, 16, 2039–2050, 1995.
9. Gerbauer, P., Krivankova, L., Pantuckova, P., Bocek, P., and Thormann, W., Sample self-stacking in capillary zone electrophoresis: Behavior of samples containing multiple major coionic components, *Electrophoresis*, 21, 2797–2808, 2000.
10. Landers, J.P., Ed., *Handbook of Capillary Electrophoresis*, 1st ed., Chapter 9, p. 209, CRC Press, Boca Raton, FL, 1993.
11. Cao, C.X., Zhang, W., Qin, W.H., Li, S., Zhu, W., and Liu, W., Quantitative predictions to conditions of zwitterionic stacking by transient moving chemical reaction boundary created with weak electrolyte buffers in capillary electrophoresis, *Anal. Chem.*, 77, 955, 2005.
12. Zhu, J. and Feng, Y.L., A powered online enhancement technique for the analysis of nucleotides using capillary zone electrophoresis/mass spectrometry, *Anal. Chem.*, 78, 6608, 2006.
13. Gong, M., Wehmeyer, K.R., Limbach, P.A., and Heimeman, W.R., Unlimited-volume electrokinetic stacking injection in sweeping capillary electrophoresis using a cationic surfactant, *Anal. Chem.*, 78, 6035, 2006.
14. Krivankova, L., Pantuckova, P., Gebauer, P., Bocek, P., Ccaslavska, J., and Thormann, W., Chloride present in biological samples as a tool for enhancement of sensitivity in capillary zone electrophoretic analysis of anionic trace analytes, *Electrophoresis*, 24, 505–517, 2003.
15. Liu, Z., Sam, P., Sirimanne, S.R., McClure, P.C., Grainger, J., and Patterson, D.G., Field-amplified sample stacking in micellar electrokinetic chromatography for on-column sample concentration of neutral molecules, *J. Chromatogr. A*, 673, 125, 1994.
16. Quirino, J.P. and Terabe, S., Exceeding 5000-fold concentration of dilute analytes in micellar electrokinetic chromatography, *Science*, 282, 465, 1998.
17. Quirino, J.P. and Terabe, S., Sweeping of analyte zones in electrokinetic chromatography, *Anal. Chem.*, 71, 1638, 1999.
18. Quirino, J.P., Kim, J.B., and Terabe, S., Sweeping: Concentration mechanism and applications to high-sensitivity analysis in capillary electrophoresis, *J. Chromatogr. A*, 965, 357, 2002.
19. Palmer, J., Munro, N.J., and Landers, J.P., A universal concept for stacking neutral analytes in micellar electrokinetic chromatography, *Anal. Chem.*, 71, 1679, 1999.
20. Palmer, J., Burgi D.S., and Landers, J.P., Electrokinetic stacking injection of neutral analytes under continuous conductivity conditions, *Anal Chem.*, 74, 632, 2002.
21. Quirino, J.P. and Terabe, S., Online concentration of neutral analytes for micellar electrokinetic chromatography. 5. Field-enhanced sample injection with reverse migrating micelles, *Anal. Chem.*, 70, 1893, 1998.

14 Capillary Electrophoresis for the Analysis of Single Cells: Sampling, Detection, and Applications

Imee G. Arcibal, Michael F. Santillo, and Andrew G. Ewing

CONTENTS

14.1 INTRODUCTION

Conventional analysis of biological specimens typically involves sampling from populations of cells (e.g., tissue homogenates). The heterogeneity of these populations, however, often yields information averaged over various types of cells. Single-cell sampling, on the other hand, allows researchers to study fundamental processes, which may not be observed in a heterogeneous cell population. To look at these processes at the single-cell level, analytical techniques must have the capability to utilize small volumes and investigate a variety of compounds concurrently. Obtaining both quantitative and qualitative information from single cells in complex biological environments allows researchers to learn more about basic cellular function as well as the mechanisms of drugs and toxins.

Capillary electrophoresis (CE) is an analytical technique ideally suited for studying the small volume contents of single cells, sampling whole cell and subcellular volumes on the femtoliter scale.

In addition, high separation potentials can be applied when using small inner diameter capillaries with CE since Joule heat can be rapidly dissipated. The use of high potentials results in fast, efficient separations of over 10^6 theoretical plates. Furthermore, CE is compatible with a diverse array of detectors including laser-induced fluorescence (LIF), ultraviolet (UV) absorbance, electrochemical, mass spectrometric, and radiochemical. This chapter builds upon the previous two editions of this book discussing advances in sampling techniques and detectors used for analyzing individual cells with CE. Furthermore, applications of single-cell CE analysis in the areas of neuroscience, immunology, nucleic acids and gene expression, as well as enzymology are also presented. The use of microfabricated devices for cellular analysis has become a popular area of research and is covered elsewhere in this book.

14.2 METHODOLOGY

14.2.1 Sampling Techniques

14.2.1.1 Whole Cell Sampling

Sampling whole cells for CE involves either siphoning the cell by either applying a pressure differential or applying a potential across the capillary to electrokinetically inject the cell. In each case, a drag force is produced by the fluid flow, driving the cell into the capillary for lysis and the separation of its contents.[1] Though these schemes are the most simplistic forms of injection, there have been several recently developed complementary techniques for introduction of cells and their contents into capillaries.

14.2.1.1.1 Continuous Cell Introduction

Individually selecting cells for siphoning or electrokinetic injection is a time-consuming process, thereby limiting the number of samples that can be run in a given period of time. In an effort to increase throughput, Chen and Lillard[2] have created a two capillary scheme capable of continuously introducing cells for injection (Figure 14.1a). The thinner first capillary is immersed in a dilute cell suspension and cells are continuously pumped in single file by electroosmotic flow. Individual cells flow toward the lysis junction connecting the two capillaries where they are lysed by mechanical disruption. The lysate is then siphoned into the second capillary where separation takes place. This system reduces analysis time from 6 to 4 min for each cell[2] and is capable of repeatedly injecting cells for over 40 min before many of the cells descend to the bottom of solution and out of reach of the capillary.

14.2.1.1.2 Laser Cavitation and Electrical Lysis

Because removing a cell from its growth substrate alters many of the biological processes occurring in the cell, its detachment increases the difficulty in analyzing the cell and the variation from its normal function.[3] Thus, in order to obtain the most accurate portrait of the cell's internal environment, it must be sampled quickly after removal from the substrate. Two innovative techniques, laser cavitation and fast electrical lysis, allow for rapid sampling of substrate-bound cells.

Sims et al.[4] have exploited the production of a shockwave created by a laser beam for their laser cavitation scheme (Figure 14.1b). The shockwave generated by laser pulse is focused in close proximity to the cell to be analyzed, forming plasma at the focal point. A cavitation bubble is produced that subsequently collapses and causes a shockwave that ruptures the cell's membrane. Cellular contents can then be loaded rapidly after lysis, reducing the time available for deviation from standard cell function to occur.

To circumvent the expense and complication of using a pulsed laser, a fast electrical lysis procedure for single-cell analysis has also been developed (Figure 14.1c).[3,4] This setup takes advantage of gold conductivity, with both the capillary and the coverslip on which cells are grown coated

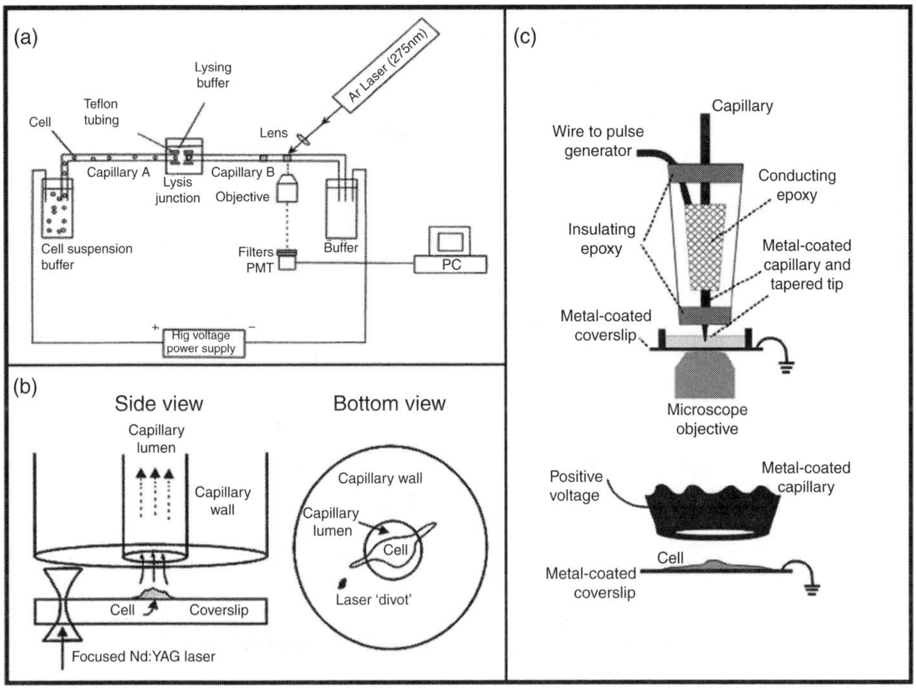

FIGURE 14.1 Recently developed whole cell sampling schemes. (a) Continuous cell introduction. (From Chen, S. and Lillard, S.J., *Anal. Chem.*, 73, 111, 2001. With permission.) (b) Laser cavitation. (From Sims, C.E. et al., *Anal. Chem.*, 70, 4570, 1998. With permission.) (c) Fast electrical lysis. (From Han, F., et al., *Anal. Chem.*, 75, 3688, 2003. With permission.)

with gold prior to analysis. Once the tapered capillary is positioned above a chosen cell, a voltage pulse is applied between the capillary and the coverslip. The applied pulse passes through the cell concurrently lysing and injecting the contents into the capillary for separation.

14.2.1.2 Release from Whole Cells

In addition to quantifying the amount of a compound within a whole cell, exocytosis can be monitored by the sampling, separation, and detection of compounds released following stimulation. Because this release often results in modifications in the environment surrounding the cell by initiating changes in adjacent cells, it is important to have the ability to sample releasate from single cells to further understand the effects of different compounds in a complex cellular environment. Direct monitoring of exocytosis was initially completed by Chen et al.[5] from the giant dopamine cell of the *Planorbis corneus* by manipulating the injection end of the separation capillary near the cell. After stimulation, a plug of secreted compounds was injected into the capillary, which was then returned to the buffer reservoir for separation. Since then, release from single mammalian cells has been achieved by Tong and Yeung.[6] In their setup, single bovine adrenal medullary cells were injected individually and allowed to adhere to the capillary wall. Subsequently, a plug of stimulant was injected and the separation run to quantify the amount of secreted catecholamines.

In order to maintain temporal resolution while obtaining chemical data, Liu et al.[7] have used a dynamic channel electrophoresis scheme to sample release from single cells. As shown in Figure 14.2, the sampling capillary is scanned across the entrance of a separation channel such that the position of the capillary at the channel inlet yields the temporal information of release. Single cells are placed in a "nanoperfusion chamber" for analysis within the sampling capillary (inset of Figure 14.2). Following

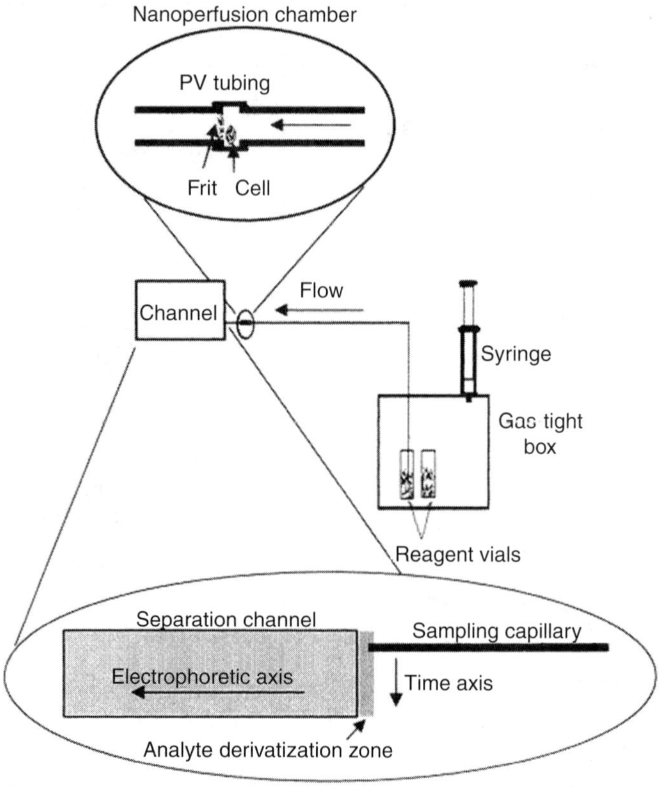

FIGURE 14.2 Dynamic channel electrophoresis instrumentation. (From Liu, Y.-M. et al., *Anal. Chem.*, 71, 28, 1999. With permission.)

release of a plug of stimulant, secreted compounds flow through the frit into the separation channel, minimizing diffusion and increasing time resolution.

14.2.1.3 Subcellular Sampling

Sampling the cellular contents has become more prevalent as capillaries have become smaller and easier to position. Subcellular sampling allows only the cellular contents of interest to be analyzed, and offers the possibility to sample specific subcellular compartments, minimizing interference from the extracellular environment. In this manner, the function or dysfunction of different compartments can be investigated, simplifying the information being obtained and yielding a more basic understanding of the mechanisms occurring during fundamental cellular processes.

14.2.1.3.1 Cytoplasmic Sampling

The synthesis, storage, and metabolism of different compounds are regulated by levels found in cellular cytoplasm. Sampling the cytosol of an intact cell can thus yield about the use of various chemical species. A microinjector having a tip tapering to 8–10 μm o.d. was previously created to sample cytoplasm directly from single neurons in *P. corneus*.[8] The size of this injector, though suitable for large invertebrate cells, is not compatible for use with single mammalian cells, which average 15 μm in diameter.[9] To probe smaller rat adrenal cells, Woods et al.[10] have fabricated microinjectors from 770 nm i.d. capillaries with tips narrowing to 2.5 μm o.d. to use in conjunction with electroporation. As transient pores are created in the cell by electroporation, the microinjector can be inserted with minimal perturbation. The voltage pulse used to electroporate is also used to electrokinetically inject cytoplasm into the capillary before the tip is removed and the separation is carried out.

14.2.1.3.2 Optical Trapping

The utility of optical trapping for manipulating subcellular contents into a capillary was initially demonstrated using single vesicles from the *Aplysia californica*, which were lysed and their contents separated.[11] Optical trapping has since also been employed to maneuver nuclei,[12] mitochondria,[13,14] and various other organelles[15] into capillaries for separation. This noninvasive and sterile manner of sampling subcellular contents utilizes a single beam laser diode focused through a high numerical aperture objective onto the desired specimen.[11,16] The "scattering force" produced from the scattering of the light upon encountering the organelle is able to move the organelle along the direction of light propagation, in this case into the capillary for lysis and separation.[17]

14.2.1.3.3 Organelle Separations

Much of the study of different organelles by single-cell CE has been completed by the Arriaga group, particularly focusing on acidic organelles and nuclei. In both cases, the organelles of interest were labeled with fluorescent tags for ease in detecting the desired species. During the analysis of single acidic organelles, populations of cells were first incubated with fluorescent microspheres, which were endocytosed and transported to these organelles. Cells were then lysed in a buffer solution and the fluorescent "organelle fraction" obtained prior to injection and separation.[18] For the investigation of nuclei, individual nuclei were tagged with a fluorescent nuclear-targeted protein in order to determine their presence with LIF detection. Analyzed cells were subsequently chosen based on observed fluorescence with a microscope and injected individually into the capillary. Following injection, the cell was lysed with digitonin and its cellular components separated.

14.2.2 Detection Methods

CE systems can be coupled to many detectors, allowing a variety of different molecules to be separated and quantified. UV absorbance, LIF, electrochemistry, radiochemistry, mass spectrometry, and refractive index are all detectors that have been successfully coupled to CE. However, electrochemical and LIF detection are the two most commonly used detectors for single-cell CE due to their high sensitivity. There have also been reports of using mass spectrometry and radiochemistry in single-cell CE analyses, though their use is not as widespread.

14.2.2.1 Electrochemistry

Electrochemical detection has found widespread use for single-cell studies with CE owing to both its selectivity and sensitivity. Though once hampered by the presence of a high voltage electric field, innovations such as the porous glass coupler for off-column detection and development of an end-column conical detector have been used to isolate the microelectrode from the potential field used for separation.[19,20] Electrochemical detection is typically carried out in the amperometric mode with the applied voltage held at a constant overpotential versus a reference electrode. Most often, carbon fiber working electrodes are used since they can be easily coupled at the ends of capillaries. As analytes encounter the working electrode surface following separation, they are either oxidized or reduced to produce a measurable current.

Scanning electrochemical detection has also been employed for single-cell CE separations.[21] In contrast to amperometry, scanning electrochemical detection varies the voltage of the working electrode. As separated species pass the working electrode, a voltammogram is obtained for each

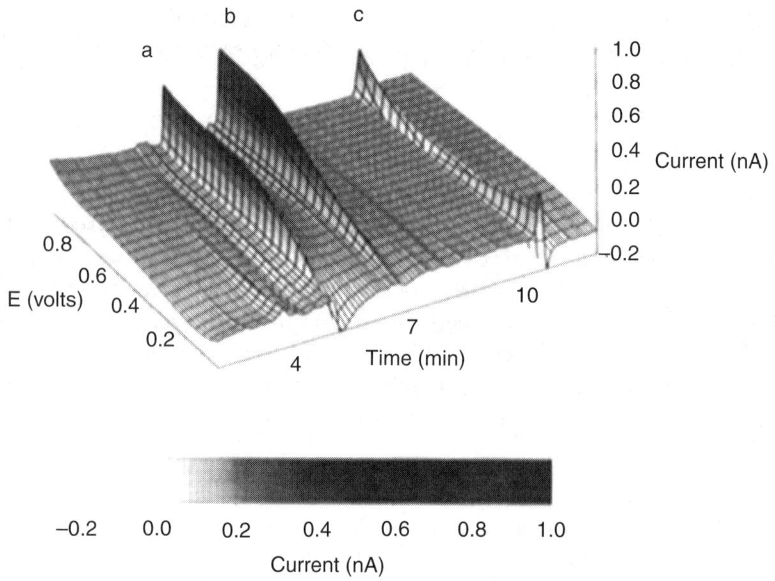

FIGURE 14.3 Electropherogram of the separated contents from a single dopaminergic neuron obtained using CE with scanning electrochemical detection and plotted in three dimensions. Peaks "a" and "b" correspond to dopamine that was immediately released during lysis and dopamine located deeper in the cell, respectively. Peak "c" represents a neutral electroactive species in cellular buffer. (From Swanek, F.D. et al., *Anal. Chem.*, 68, 3912, 1996. With permission.)

analyte and a three-dimensional surface is constructed showing current versus voltage and time (Figure 14.3). This unique detection mode allows eluted substances to be identified according to their characteristic voltammogram as well as retention time. Thus, scanning electrochemical detection offers enhanced selectivity since co-eluting peaks can be further resolved by their unique oxidation potentials. Scanning electrochemical detection has been utilized to detect dopamine at the femtomole level from single *P. corneus* neurons.[22]

14.2.2.2 Laser-Induced Fluorescence

Along with amperometry, LIF is the most popular mode of detection in single-cell CE experiments owing to its high sensitivity. Most CE-LIF systems employ visible argon ion lasers; however, UV lasers have also been used for fluorescence exitation.[23,24] One problem with LIF is that not all species are natively fluorescent, so labeling is often required. Although LIF detection most often involves a single emission wavelength, Sweedler has demonstrated simultaneous multiwavelength detection (Figure 14.4).[25] Over 30 compounds containing aromatic amino acids in an individual neuron were separated and detected using native LIF with a UV laser, and a unique fluorescence emission spectrum was obtained for each eluted species.

14.2.2.3 Mass Spectrometry

Mass spectrometry (MS) has perhaps the most potential as a detector since it can provide detailed chemical information to identify separated species, particularly molecules with high molecular weights like polypeptides. Unfortunately, mass spectrometers have not been widely used in single-cell CE experiments as detectors due to their relatively poor sensitivity compared to LIF. Nonetheless,

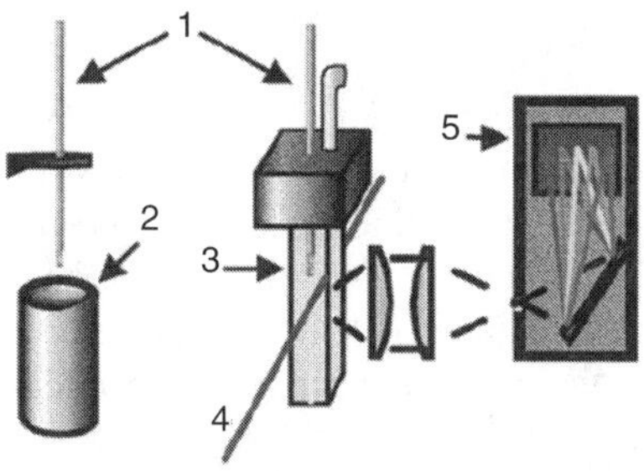

FIGURE 14.4 (**See color insert following page 810.**) CE system with simultaneous multiwavelength fluorescence detection. Cellular samples were electrokinetically drawn into a capillary (1) from a steel microvial (2). The analytes were separated in the capillary, and eluted into a square-cuvette sheath-flow cell (3) where fluorescence was induced with a laser beam (4). The fluorescent light was collected orthogonally and directed into a spectrograph and CCD where a complete spectrum was obtained for each separated analyte (5). (From Fuller, R.R. et al., *Neuron*, 20, 173, 1998. With permission.)

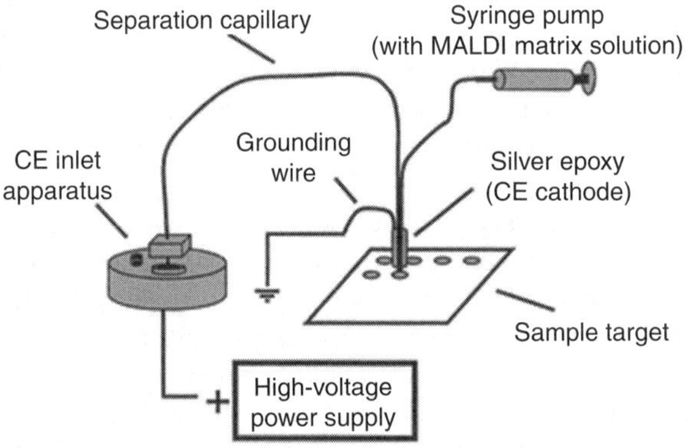

FIGURE 14.5 CE coupled to a MALDI-TOF-MS system for detecting neuropeptides. As the contents of a single cell were separated in the capillary, the separated species were collected as fractions deposited onto a MALDI-MS target. (From Page, J.S. et al., *Analyst*, 125, 555, 2000. With permission.)

there have been a few interesting papers using mass spectrometric detection for single-cell analysis. One early application of CE-MS for single-cell analysis demonstrated the use of electrospray ionization and Fourier transform ion cyclotron resonance mass spectrometry to detect the α and β chains of hemoglobin in a single human erythrocyte.[26]

Sweedler and coworkers[27] have developed a novel system coupling CE to matrix-assisted laser desorption–ionization time-of-flight mass spectrometry (MALDI-TOF-MS) to study peptides (Figure 14.5). In some initial work, the contents of a single cerebral ganglion cell from *A. californica* were separated and mixed with MALDI matrix after exiting the capillary; every 30 s, a new fraction was deposited onto the MALDI target plate for mass spectrometric analysis. In following work, using an instrumentation setup similar to Figure 14.5, MALDI-TOF-MS was coupled with CE to identify neuropeptides and hormones released from individual *A. californica* cells.[28] Although MALDI-TOF-MS is excellent for qualitative chemical identification of high molecular weight species, it lacks the ability for quantification. Sweedler and coworkers,[29] however, complemented CE-MALDI-TOF-MS with radiochemical detection to achieve both chemical identification and successful quantification of hormones and neuropeptides from individual cells.

14.2.2.4 Radiochemistry

Radiochemistry has not been extensively used for single-cell CE systems because of the cumbersome nature of working with radioactive elements and lengthy time involved in collecting an image of radiolabeled eluents. A postcolumn radionucleotide detector has been developed in which radiolabeled eluents are deposited onto a membrane with a scintillator, and then imaged with a charge-coupled device (CCD). This unique system had limits of detection of 88 zmol for ^{32}P-labeled analytes, 17 amol for ^{35}S, and 8 fmol for ^{3}H, which were found in peptides.[30] An improvement of the previous system was made by using a commercial phosphor plate and photomultiplier tube, which yielded even lower limits of detection (4.9 zmol for ^{32}P and 0.13 amol for ^{35}S found in peptides), and was successfully applied to detecting radiolabeled peptides single buccal ganglion cells in *A. californica*.[31]

14.3 APPLICATIONS

14.3.1 NEUROSCIENCE

Single-cell CE experiments in neurochemistry were first performed in the late 1980s with initial cytoplasmic sampling from *P. corneus* completed in 1988[32] and whole cell injection of *Helix aspersia* in 1989 to separate fluorescently labeled amino acids.[33] Since these pioneering studies, many different small molecule neurotransmitters and other related species have been detected in single cells by CE.

Serotonin is a monoamine neurotransmitter believed to regulate physiological functions such as appetite and sleep. Most importantly, decreased serotonin transmission in the brain has been implicated in playing a role in depression and other related mood disorders, making it an important focus of research in the contemporary neuroscience community. In one study, individual *Aplysia* metacerebral cells were electrically stimulated, followed by separation by CE and fluorescence detection with a UV laser.[23] Each cell soma contained approximately 450 fmol of serotonin with a limit of detection for the experiment of 39 nM. A later report[34] investigated serotonin metabolism in *Aplysia*. Single cells from different locations of the *Aplysian* nervous system were incubated with serotonin with two new serotonin catabolites discovered. Interestingly, it was recently shown that electrokinetic injection of serotonin results in the formation of a serotonin dimer artifact, but when hydrodynamic injection is used, the dimer is not present.[35]

Like serotonin, the amino acid D-aspartate is a small molecule neurotransmitter and is present in the central nervous system of both invertebrates and vertebrates. While most proteins and polypeptides contain only amino acids of the L-enantiomer, mounting evidence suggests that the D-enantiomer of aspartate plays a role in intercellular signaling. The addition of cyclodextrin to the separation buffer in CE allows these two enantiomers to be separated and individually quantified.[36–38] When individual *Aplysian* cell processes and soma were analyzed separately by CE-LIF, it was shown that D-aspartate was not only located in the nucleus but also in the processes and the amount of D-aspartate in different processes of the same cell was similar.[37] In a subsequent experiment, CE-LIF was used to show that the amounts of D-aspartate varied among different clusters from single *A. californica* neurons, D-aspartate is synthesized endogenously from its L-enantiomer, and it travels long intracellular distances to its release sites.[38] In addition to aspartate, the addition of cyclodextrin has allowed fluorescently labeled D-glutamate and L-glutamate to be separated from an individual cell of *Aplysia*.[39] Neuropeptides may also contain D-amino acids with the importance of peptides containing D-amino acids illustrated with the discovery of a neuropeptide containing D-tryptophan in a single cell of *A. californica*.[40]

Nitric oxide (NO), a more recently discovered neurotransmitter and regulator of other physiological processes, has been indirectly quantified by CE of its primary metabolites. Because of the ephemeral nature of NO, nitrate and nitrite, two products of NO oxidation, have been assumed to be good indicators of NO synthase activity. These anions in single *Pleurobranchaea californica* and *A. californica* neurons have been separated by CE and quantified with UV absorbance detection.[41]

The limits of detection were less then 200 fmol and the concentrations of nitrite and nitrate were typically 2 and 12 mM, respectively. A more recent report involved CE separations of arginine, citrullene, nitrate, and nitrite with LIF and conductivity detection.[42] Contrary to previous results, it was determined that nitrates are not always reliable indicators of NO synthase activity. Similarly, other metabolites associated with NO—arginine, citrullene, arginosuccinate, ornithine, and arginine phosphate—were quantified in *Pleurobranchaea* and *Aplysia* with limits of detection ranging from 5 nM to 17 μM.[43]

In vivo detection of NO is commonly accomplished using the fluorophore 4,5-diaminofluorescien (DAF-2). When the fluorophore reacts with N_2O_3 (a species related to NO), it is converted to a triazole derivative, DAF-2T. Dehydroascorbic acid present in cells also reacts with DAF-2 and forms a complex with an emission spectrum similar to that of DAF-2T, thereby interfering with NO detection. The separation of the interfering species with CE-LIF has allowed quantification of NO in single cells,[44] along with ascorbic acid and dehydroascorbic acid,[44,45] which both play important roles in cellular metabolism.

14.3.2 Immunology

For years, enzyme-linked immunosorbent assays (ELISA) have been the standard method for investigating antibody–antigen interactions. The time-consuming nature of ELISA, however, has led to the development of single-cell CE for the rapid analysis of the reaction of antibodies and target antigens. One such study has utilized single-cell CE-LIF to analyze human interferon-gamma (IFN-γ), which provides antiviral, anticell proliferation, and immunity adjustment effects as support for other cells.[46] Though initially employing on-capillary immunoreactions, where the contents of a lysed natural killer (NK) cell could interact with fluorescently tagged anti-IFN-γ monoclonal antibody (Ab*) prior to the separation,[46] Zhang and Jin[47] further optimized sample preparation by loading Ab* into cells via electroporation prior to entering the capillary. Single cells were subsequently washed and lysed on-capillary following a short incubation time prior to analysis by CE-LIF. Labeled antibodies for both IFN-γ I and II were detected in the cell electropherograms with increases in the relative fluorescence intensity observed for each form of the protein–antibody complex with the electroporated loaded cells versus those that had undergone the on-capillary reaction due to a decrease in Ab* dilution during the intracellular immunoreaction (Figure 14.6). In addition, the average value determined for total IFN-γ concentration in NK cells (98 pM) corresponded well with the value acquired from an ELISA assay (99 pM),[47] demonstrating the utility of the CE technique in performing immunoreaction assays rapidly while maintaining accuracy.

In another investigation, the role of P-glycoprotein (PGP) in imparting multidrug resistance to tumor cells by functioning as transmembrane drug pump was examined.[48] Two groups of single cells—those that were drug resistant and those drug sensitive—were incubated with primary (JSB-1, a mouse raised antibody that binds to PGP) and secondary [goat antimouse IgG-fluorescein isothiocyanate (GAMIF), an antimouse antibody tagged with fluorescein isothiocyanate (FITC)] antibodies prior to analysis with CE-LIF. Owing to the minimal nonspecific binding of JSB-1, its quantity in a single-cell was taken as the total amount of PGP present on the cell since the antibody reacts with the protein epitope on a 1:1 ratio. When comparing the amount of PGP expressed in each cell group, a larger quantity was present on the drug resistant cell line, with 157,000 molecules of PGP present versus 47,000 molecules present on the drug sensitive cell line.

14.3.3 Nucleic Acids and Gene Expression

A popular technique used to observe gene expression is reverse transcriptase-polymerase chain reaction (RT–PCR). RT–PCR works by reverse transcribing mRNA to a complementary DNA sequence that corresponds to the gene of interest. Several years ago, Zabzdyr and Lillard[49] showed that CE could be used to separate and detect RT-PCR products corresponding to β-actin in individual

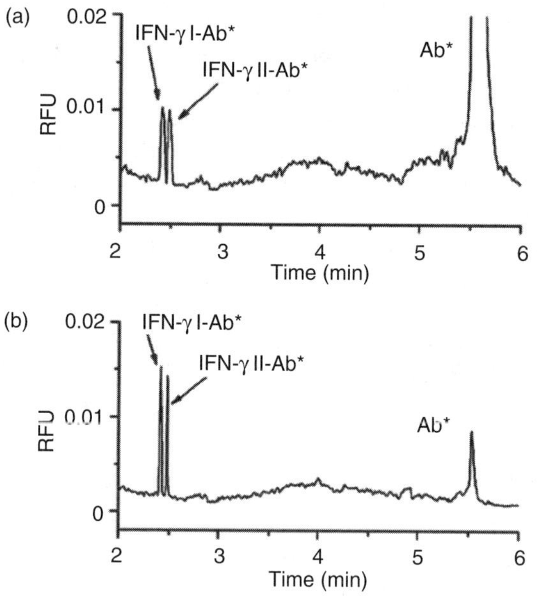

FIGURE 14.6 Electropherograms detailing IFN-γ—antibody complex (IFN-γ-Ab*) content in NK cells using on-capillary immunoreaction (a) and intracellular immunoreaction (b) for both IFN-γ I and II. (From Zhang H. and Jin, W., *J. Chromatogr. A*, 1104, 346, 2006. With permission.)

human prostate carcinoma cells, which was similar to an earlier study done by Li and Yeung[50] who detected PCR products corresponding to β-actin from lymphoblast cells. The cells were lysed with detergent or by a freeze-thaw method and a primer and other PCR reagents were added. The resulting PCR products were complexed with ethidium bromide and were separated by CE-LIF. The freeze-thaw lysing method yielded a better signal-to-noise ratio versus the use of detergent, and limits of detection corresponded to 133 initial molecules of β-actin mRNA per nL. Yeung and coworkers[51] successfully developed a microthermocycler for integrated cell lysis and RT–PCR in which the PCR products of β-actin were measured from a group of 16 cells, demonstrating the possibility of using this novel system for single-cell analyses. Lillard and coworkers[52] more recently used CE-LIF to detect RT–PCR products corresponding to β-actin and the α estrogen receptor simultaneously from single human breast carcinoma cells (Figure 14.7). Surprisingly, only a single round of RT–PCR amplification was necessary, but additives were needed to enhance the low signals for the analytes. It is hypothesized that the signals were low because two primers were used simultaneously, and it is thought that they formed dimers thereby lowering the amplification efficiency.

In addition to using CE for detecting single-cell RT–PCR products, CE has also been used to detect RNA in individual cells without the need for extraction. Lillard and coworkers[53] injected single Chinese hamster ovary cells into a capillary, where the cells were lysed and the RNA complexed with ethidium bromide for LIF detection. Only rRNA and tRNA were detected, which was verified by monitoring changes in the peak patterns after adding RNase. mRNA was not detected because it is more unstable than the other forms of RNA with the buffer conditions used in these experiments. The total amount of RNA in individual cells was calculated to be approximately 10–20 pg.[53] Since aging, carcinogens, and other toxins cause damage to nucleic acids, single-cell CE was also used to observe changes in peak patterns of nucleic acids after single cells were incubated with hydrogen peroxide.[53] In addition to monitoring changes in nucleic acids following chemical damage, single-cell CE-LIF has also been used to look at changes in RNA during cell division.[54] Single Chinese hamster ovary cells were taken from different phases of the cell cycle, injected individually

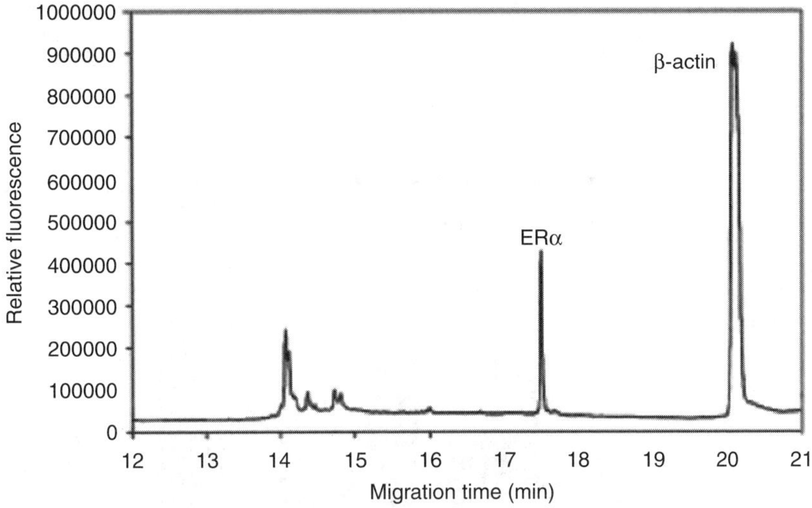

FIGURE 14.7 Electropherogram of multiplex RT–PCR products corresponding to β-actin and the α estrogen receptor from single human breast carcinoma cells. (From Zabzdyr, J.L. and Lillard, S.J., *Electrophoresis*, 26, 137, 2005. With permission.)

via suction into a capillary, followed by lysis and detection with LIF. The total RNA content increased after each phase in the cycle, while the rates of RNA synthesis varied between the steps in the cycle.[54]

14.3.4 Enzymology

Understanding the mechanism behind the interaction between enzymes and substrates in a biological system is inherently difficult because of the variation that exists in enzyme expression and function across a population of cells. The use of single-cell CE, however, is particularly suited to enzymology as the activity of individual cells can be assayed alone. In one study, for example, post-translational enzymes have been investigated at potential targets for cancer therapy. The role of Ras proteins in cell growth and differentiation, as well as the presence of their mutated forms in human cancers, has made them a particularly interesting area of study for understanding how cancer proliferates and spreads. In efforts to target possible mechanisms for thwarting the function of mutated Ras proteins, the activity of three enzymes required to create functional proteins—farnesyltransferase, endoprotease, and methyltransferase—was evaluated.[55] Immortalized mouse cell lines were incubated with a peptide analog of the Ras protein to evaluate the function of each enzyme. Three enzyme products, two found in extracellular media and one intracellularly, were detected using CE-LIF upon lysis of single incubated cells. Disappointingly, comparison of the experimental product migration times did not correspond with those of the known products of the enzymes with the substrate, so no conclusion could be made as to the identity of the experimental compounds. However, the protocol developed was clearly shown to be compatible with single-cell CE.

While simply investigating the action of an enzyme in a cell at a single time point can provide a great deal of information, the ability to monitor how a reaction proceeds over time would also be advantageous. Shoemaker et al.[56] have developed nanoliter reaction vessels to sample the contents of a single lysed cell over time. Intracellular enzymes, such as α-galactosidase, from the lysed cell are capable of interacting with a particular substrate in the vessel. In the case of α-galactosidase, a disaccharide substrate, αGlc(1 → 3)αGlc-TMR, was utilized as the substrate and nanoliter aliquots were sampled for analysis by CE-LIF over several hours to detect the presence of the enzymatic product (Figure 14.8). A heterogeneity in enzyme function was observed with average conversion

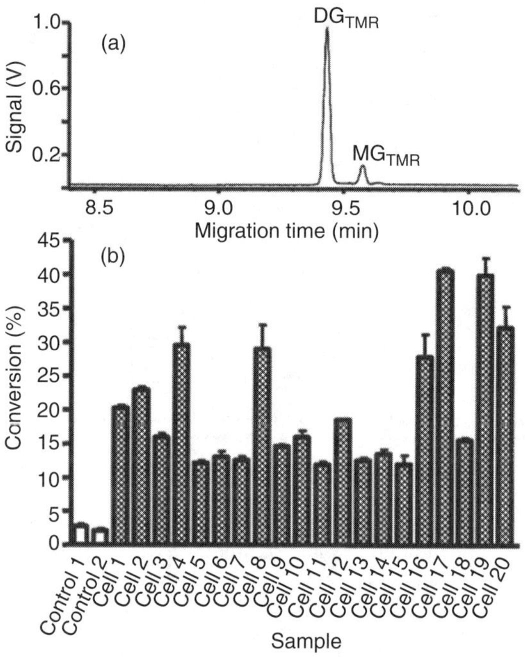

FIGURE 14.8 Electropherogram of an α-glucosidase II single-cell (Cell 7) reaction aliquot taken after incubation (a). Percent conversion of the dissaccharide (DG-TMR) substrate by α-glucosidase II in single cells and control assays (b). Percent conversion was calculated by comparing the DG-TMR peak to that of the monosaccharide product MG-TMR. (From Shoemaker, G.K. et al., *Anal. Chem.*, 77, 3132, 2005. With permission.)

of 20.9% ± 9.4% of substrate to product and is attributable to differences in cell age, cell cycle, and cell size.[55,56]

14.4 FUTURE PROSPECTS

Capillary electrophoresis of single cells has been employed in many scientific areas such as neuroscience, immunology, gene expression, and enzymology in the past 20 years. Though numerous developments have been made, many avenues are still left to pursue. One area that can be developed further is sample acquisition. The Chiu group, for instance, utilizes nanosurgery to choose which subcellular component (organelle) to analyze, using a combination of optical trapping and electroporation[15,57] to remove it from the cell. Other organelle types can be studied by pairing this technique with CE separations to better elucidate their functions. In addition, although continuous sampling from a single cell has already been achieved to study enzyme activity, it would be beneficial to maintain an intact cell while sampling to gain a more accurate picture of what is occurring in terms of cellular function. Further, sampling single cells within a population simultaneously will give clues about intercellular interactions in tissues and organs. In this manner, one can determine how a stimulus to one affects the others.

As always, separation conditions are constantly being optimized to gain the best resolution between analytes. For instance, when analyzing single whole cells, both lipids and proteins are capable of adsorbing to capillary walls, reducing peak resolution, and separation efficiency. Thus, it is imperative to reduce adsorption to the capillary walls to maintain the advantages of using CE over other techniques. Further, to lower detection limits, more sensitive detectors must be developed. Also, since LIF and electrochemistry are most common detection methods, but most analytes are not natively fluorescent or electroactive, derivatization of analytes must be optimized so that all compounds present can be adequately measured.

While new innovations in sampling schemes and detectors will further increase its use for the analysis of single cells, CE has shown great promise as a tool to study single cells. The key advantages of CE are small volume capability combined with rapid and efficient separations. In addition, the diversity of detectors that can be coupled to a CE system further increases its versatility as a variety of analytes can be investigated. With continued use of CE to study individual cells, more information can be learned about the basic function of cells as well as their interactions with each other and different chemical species.

REFERENCES

1. Krylov, S.N. et al., Instrumentation for chemical cytometry, *Anal. Chem.*, 72, 872, 2000.
2. Chen, S. and Lillard, S.J., Continuous cell introduction for the analysis of individual cells by capillary electrophoresis, *Anal. Chem.*, 73, 111, 2001.
3. Han, F. et al., Fast electrical lysis of cells for capillary electrophoresis, *Anal. Chem.*, 75, 3688, 2003.
4. Sims, C.E. et al., Laser-micropipet combination for single-cell analysis, *Anal. Chem.*, 70, 4570, 1998.
5. Chen, G. et al., Observation and quantitation of exocytosis from the cell body of a fully developed neuron in *Planorbis corneus*, *J. Neurosci.*, 15, 7747, 1995.
6. Tong, W. and Yeung, E.S., On-column monitoring of secretion of catecholamines from single bovine adrenal chromaffin cells by capillary electrophoresis, *J. Neurosci. Methods*, 76, 193, 1997.
7. Liu, Y.M., Moroz, T., and Sweedler, J.V., Monitoring cellular release with dynamic channel electrophoresis, *Anal. Chem.*, 71, 28, 1999.
8. Olefirowicz, T.M. and Ewing, A.G., Capillary electrophoresis in 2 and 5 microns diameter capillaries: Application to cytoplasmic analysis, *Anal. Chem.*, 62, 1872, 1990.
9. Lu, X. et al., Recent developments in single-cell analysis, *Anal. Chim. Acta*, 510, 127, 2004.
10. Woods, L.A., Gandhi, P.U., and Ewing, A.G., Electrically assisted sampling across membranes with electrophoresis in nanometer inner diameter capillaries, *Anal. Chem.*, 77, 1819, 2005.
11. Chiu, D.T. et al., Probing single secretory vesicles with capillary electrophoresis, *Science*, 279, 1190, 1998.
12. Liang, H. et al., Giant cell formation in cells exposed to 740 nm and 760 nm optical traps, *Lasers Surg. Med.*, 21, 159, 1997.
13. Fuller, K.M. and Arriaga, E.A., Advances in the analysis of single mitochondria, *Curr. Opin. Biotechnol.*, 14, 35, 2003.
14. He, M. et al., Selective encapsulation of single cells and subcellular organelles into picoliter- and femtoliter-volume droplets, *Anal. Chem.*, 77, 1539, 2005.
15. Shelby, J.P., Edgar, J.S., and Chiu, D.T., Monitoring cell survival after extraction of a single subcellular organelle using optical trapping and pulsed-nitrogen laser ablation, *Photochem. Photobiol.*, 994, 2005.
16. Kuyper, C.L. and Chiu, D.T., Optical trapping: a versatile technique for biomanipulation, *Appl. Spectrosc.*, 56, 300A, 2002.
17. Block, S.M., Making light work with optical tweezers, *Nature*, 360, 493, 1992.
18. Fuller, K.M. and Arriaga, E.A., Analysis of individual acidic organelles by capillary electrophoresis with laser-induced fluorescence detection facilitated by the endocytosis of fluorescently labeled microspheres, *Anal. Chem.*, 75, 2123, 2003.
19. Wallingford, R.A. and Ewing, A.G., Capillary zone electrophoresis with electrochemical detection, *Anal. Chem.*, 59, 1762, 1987.
20. Sloss, S. and Ewing, A.G., Improved method for end-column amperometric detection for capillary electrophoresis, *Anal. Chem.*, 65, 577, 1993.
21. Ferris, S.S., Lou, G., and Ewing, A.G., Scanning electrochemical detection in capillary electrophoresis, *J. Microcol. Sep.*, 6, 263, 1994.
22. Swanek, F.D., Chen, G.Y., and Ewing, A.G., Identification of multiple compartments of dopamine in a single cell by CE with scanning electrochemical detection, *Anal. Chem.*, 68, 3912, 1996.

23. Miao, H., Rubakhin, S.S., and Sweedler, J.V., Analysis of serotonin release from single neuron soma using capillary electrophoresis and laser-induced fluorescence with a pulsed deep-UV NeCu laser, *Anal. Bioanal. Chem.*, 377, 1007, 2003.
24. Zabzdyr, J.L. and Lillard, S.J., UV- and visible-excited fluorescence of nucleic acids separated by capillary electrophoresis, *J. Chromatogr. A*, 911, 269, 2001.
25. Fuller, R.R. et al., Single neuron analysis by capillary electrophoresis with fluorescence spectroscopy, *Neuron*, 20, 173, 1998.
26. Hofstadler, S.A. et al., Analysis of single cells with capillary electrophoresis electrospray ionization Fourier transform ion cyclotron resonance mass spectrometry, *Rapid Commun. Mass Spectrom.*, 10, 919, 1996.
27. Page, J.S., Rubakhin, S.S., and Sweedler, J.V., Direct cellular assays using off-line capillary electrophoresis with matrix-assisted laser desorption/ionization time-of-flight mass spectrometry, *Analyst*, 125, 555, 2000.
28. Rubakhin, S.S. et al., Analysis of cellular release using capillary electrophoresis and matrix assisted laser desorption/ionization-time of flight-mass spectrometry, *Electrophoresis*, 22, 3752, 2001.
29. Page, J.S., Rubakhin, S.S., and Sweedler, J.V., Single-neuron analysis using CE combined with MALDI MS and radionuclide detection, *Anal. Chem.*, 74, 497, 2002.
30. Tracht, S., Toma, V., and Sweedler, J.V., Postcolumn radionuclide detection of low-energy beta emitters in capillary electrophoresis, *Anal. Chem.*, 66, 2382, 1994.
31. Tracht, S.E. et al., Detection of radionuclides in capillary electrophoresis using a phosphor-imaging detector, *Anal. Chem.*, 68, 3922, 1996.
32. Wallingford, R.A. and Ewing, A.G., Capillary zone electrophoresis with electrochemical detection in 12.7 microns diameter columns, *Anal. Chem.*, 60, 1972, 1988.
33. Kennedy, R.T. et al., Microcolumn separations and the analysis of single cells, *Science*, 246, 57, 1989.
34. Stuart, J.N. et al., Serotonin catabolism depends upon location of release: Characterization of sulfated and gamma-glutamylated serotonin metabolites in *Aplysia californica*, *J. Neurochem.*, 84, 1358, 2003.
35. Stuart, J.N. et al., Spurious serotonin dimer formation using electrokinetic injection in capillary electrophoresis from small volume biological samples, *Analyst*, 130, 147, 2005.
36. Miao, H., Rubakhin, S.S., and Sweedler, J.V., Confirmation of peak assignments in capillary electrophoresis using immunoprecipitation. Application to D-aspartate measurements in neurons, *J. Chromatogr. A*, 1106, 56, 2006.
37. Miao, H., Rubakhin, S.S., and Sweedler, J.V., Subcellular analysis of D-aspartate, *Anal. Chem.*, 77, 7190, 2005.
38. Miao, H. et al., D-Aspartate as a putative cell-cell signaling molecule in the *Aplysia californica* central nervous system, *J. Neurochem.*, 97, 595, 2006.
39. Quan, Z. and Liu, Y.M., Capillary electrophoretic separation of glutamate enantiomers in neural samples, *Electrophoresis*, 24, 1092, 2003.
40. Sheeley, S.A. et al., Measuring D-amino acid-containing neuropeptides with capillary electrophoresis, *Analyst*, 130, 1198, 2005.
41. Cruz, L. et al., Nitrite and nitrate levels in individual molluscan neurons: Single-cell capillary electrophoresis analysis, *J. Neurochem.*, 69, 110, 1997.
42. Moroz, L.L. et al., Direct single cell determination of nitric oxide synthase related metabolites in identified nitrergic neurons, *J. Inorg. Biochem.*, 99, 929, 2005.
43. Floyd, P.D. et al., Capillary electrophoresis analysis of nitric oxide synthase related metabolites in single identified neurons, *Anal. Chem.*, 70, 2243, 1998.
44. Kim, W.S. et al., Measuring nitric oxide in single neurons by capillary electrophoresis with laser-induced fluorescence: Use of ascorbate oxidase in diaminofluorescein measurements, *Anal. Chem.*, 78, 1859, 2006.
45. Kim, W.S. et al., Ascorbic acid assays of individual neurons and neuronal tissues using capillary electrophoresis with laser-induced fluorescence detection, *Anal. Chem.*, 74, 5614, 2002.
46. Zhang, H. and Jin, W., Determination of different forms of human interferon-gamma in single natural killer cells by capillary electrophoresis with on-capillary immunoreaction and laser-induced fluorescence detection, *Electrophoresis*, 25, 1090, 2004.
47. Zhang, H. and Jin, W., Single-cell analysis by intracellular immuno-reaction and capillary electrophoresis with laser-induced fluorescence detection, *J. Chromatogr. A*, 1104, 346, 2006.

48. Xiao, H. et al., Immunoassay of P-glycoprotein on single cell by capillary electrophoresis with laser induced fluorescence detection, *Anal. Chim. Acta*, 556, 340, 2006.
49. Zabzdyr, J.L. and Lillard, S.J., Measurement of single-cell gene expression using capillary electrophoresis, *Anal. Chem.*, 73, 5771, 2001.
50. Li, H. and Yeung, E.S., Selective genotyping of individual cells by capillary polymerase chain reaction, *Electrophoresis*, 23, 3372, 2002.
51. Matsunaga, H., Anazawa, T., and Yeung, E.S., Integrated on-capillary instrumentation for gene expression measurement directly from cells, *Electrophoresis*, 24, 458, 2003.
52. Zabzdyr, J.L. and Lillard, S.J., A qualitative look at multiplex gene expression of single cells using capillary electrophoresis, *Electrophoresis*, 26, 137, 2005.
53. Han, F. and Lillard, S.J., *In-situ* sampling and separation of RNA from individual mammalian cells, *Anal. Chem.*, 72, 4073, 2000.
54. Han, F. and Lillard, S.J., Monitoring differential synthesis of RNA in individual cells by capillary electrophoresis, *Anal. Biochem.*, 302, 136, 2002.
55. Arkhipov, S.N. et al., Chemical cytometry for monitoring metabolism of a Ras-mimicking substrate in single cells, *Cytometry A*, 63, 41, 2005.
56. Shoemaker, G.K. et al., Multiple sampling in single-cell enzyme assays using CE-laser-induced fluorescence to monitor reaction progress, *Anal. Chem.*, 77, 3132, 2005.
57. Chiu, D.T., Micro- and nano-scale chemical analysis of individual sub-cellular compartments, *Trends Anal. Chem.*, 22, 528, 2003.

15 Ultrafast Electrophoretic Separations

Michael G. Roper, Christelle Guillo, and B. Jill Venton

CONTENTS

15.1 INTRODUCTION

Many chemical and biological events are dynamic in nature and rapid analytical techniques are required to monitor their occurrence. For example, reaction intermediates may be unstable and only present for milliseconds. Similarly, receptor binding and affinity complex formation or dissociation occurs on a rapid time scale. Neurotransmission and other cellular signaling processes are fast, with rapid release and clearance of signaling compounds. In addition, high-throughput analyses such as DNA analysis or drug screening require repetitive and rapid observations. Fast analytical techniques are therefore needed to better understand the time course of these phenomena.

Separations have often been considered a slow step in analysis, unsuitable for monitoring rapid processes. The first demonstration of an electrophoretic separation was achieved in hours[1] and today, conventional capillary electrophoresis separations are typically completed in around 10 min, depending on the sample. Other analytical methods, such as spectroscopy or electrochemistry, have therefore been used to obtain dynamic information about chemical changes. However, separations provide chemical information that may not be obtained with other techniques. Fast separations that allow simultaneous collection of temporal and chemical information would be advantageous for monitoring many of the real-world examples given above. Electrophoretic separations performed in capillaries or planar substrates have advantages over high-performance liquid chromatography (HPLC) and gel electrophoresis because rapid separations can be achieved without a loss of separation efficiency. Advances in electrophoresis instrumentation have reduced separation times from the minute to the second time scale, and even to microseconds in some applications. The improved temporal resolution is leading to a better understanding of rapid chemical and biological processes. In this chapter, we will outline the theory, technological advances, and practical aspects of performing ultrafast capillary and microfluidic electrophoresis.

15.2 THEORY

When examining the theoretical basis of rapid separations, it is important to balance the considerations of temporal resolution with well-resolved separations. The fundamental equations defining electrophoresis show that it is possible to increase both the speed and efficiency of a separation by increasing the voltage.[2,3] The migration time (t_{mig}) is given by

$$t_{mig} = \frac{L^2}{V \times \mu_{ep}} \tag{15.1}$$

where L is the separation length (and length to the detector), V is the separation voltage, and μ_{ep} is the electrophoretic mobility of the analyte. Because the migration time is inversely proportional to voltage, higher voltages lead to shorter migration times. The equation for theoretical plates (N) is

$$N = \frac{L^2}{\sigma_{tot}^2} = \frac{\mu_{ep} \times V}{2D} \tag{15.2}$$

where D is the diffusion coefficient and σ_{tot}^2 is the total zone variance due to band broadening. If band broadening is limited to diffusion, then the equation simplifies to the term on the right, which shows that N is proportional to voltage. This reveals that more efficient separations will result from high applied voltages. Shorter separation lengths will result in faster separations, given the L^2 proportionality to migration time, but with no loss in separation efficiency since N is not dependent on separation length. In contrast, in chromatography, shorter column lengths lead to faster separations but also result in a lower number of theoretical plates. Therefore, the strategy for fast, efficient electrophoretic separations is clearly high voltages and short separation distances.

The above analysis assumed that the only source of band broadening was diffusion, but other sources of band broadening can occur in an experiment. The total variance is given by

$$\sigma_{tot}^2 = \sigma_{diff}^2 + \sigma_{heat}^2 + \sigma_{inj}^2 + \sigma_{det}^2 + \sigma_{ads}^2 + \sigma_{ED}^2 \tag{15.3}$$

where variances are due to diffusion, joule heating, injection, detection, adsorption, and electromigration dispersion, respectively. Joule heating increases with high voltages, which is problematic for fast separations where electric field strengths are often greater than 1000 V cm^{-1}. Heat is only dissipated through the surface of the separation path and radial temperature gradients between the inside and outside of the separation path result. The temperature gradients can lead to band broadening due to convection or differences in temperature-dependent variables that control the migration rate, such as viscosity. The effect of Joule heating on plate height has been quantified and the parameters can be predicted where heating will have a deleterious effect on efficiency.[4] The temperature difference between the wall and the middle of the separation path is proportional to the radius (r) of the capillary inner diameter (i.d.) to the second power (i.e., r^2), so to reduce the effects of Joule heating most fast electrophoretic studies employ small i.d. capillaries ($<10\,\mu$m i.d. compared to the standard 50–100 μm i.d. capillaries) or shallow microfluidic channels. Joule heating can also be reduced by using low conductivity buffers.[5]

Small i.d. capillaries (or shallow microfluidic channels) present experimental challenges, particularly for injection and detection techniques. Band broadening due to adsorption is also a prominent mode of dispersion in small i.d. capillaries due to high surface area to volume ratios, but covalent or dynamic coatings can be used to reduce this effect.[6,7] When small i.d. capillaries are used, lower amounts of analyte must be injected to avoid band broadening due to sample overloading. Therefore, injection instrumentation must reproducibly inject small plugs of sample; however, if smaller quantities are being injected, then high-sensitivity detectors are needed to detect the analytes. Typically, ultraviolet (UV) detectors are not sensitive enough, and so most fast separations rely on high-sensitivity detection methods such as laser-induced fluorescence (LIF). Advances in injection and detection methodologies for ultrafast separations will be described more in the Section 15.3.

Another parameter to consider in electrophoresis is electroosmotic flow (EOF), which is present in most examples of rapid electrophoretic separations. Therefore, μ_{ep} in Equations 15.1 and 15.2 must be replaced with the sum of electrophoretic mobility and the electroosmotic mobility (μ_{EOF}):

$$t_{mig} = \frac{L^2}{V \times (\mu_{ep} + \mu_{EOF})} \tag{15.4}$$

$$N = \frac{L^2}{\sigma_{tot}^2} = \frac{(\mu_{ep} + \mu_{EOF}) \times V}{2D}. \tag{15.5}$$

At first glance, EOF appears to be favorable for both migration time and efficiency. However, while EOF mobilizes the analytes off the column faster, allowing less time for diffusional band broadening and an improvement in N, it is a nonselective mode of transport. As a result, if EOF is in the same direction as the ion mobility, two analytes have less time to be separated and a loss of resolution results. This fact is shown in the equation for resolution (R_s) of two compounds:

$$R_s = 0.177 \left(\Delta\mu_{ep}\right) \left[\frac{V}{D\left(\bar{\mu} + \mu_{EOF}\right)}\right] \tag{15.6}$$

where $\Delta\mu_{ep}$ is the difference in the electrophoretic mobilities of two analytes and $\bar{\mu}$ is the average electrophoretic mobility of two analytes. Thus, the highest resolution is attained when $\bar{\mu}$ is approximately equal and opposite to μ_{EOF}; however, this situation is at the cost of increased analysis time (see Equation 15.4).[2] Methods to reduce EOF, such as wall coatings and buffer additives, promote better resolution while maintaining fast separations.[8]

15.3 METHODS DEVELOPMENT

Examining the theory for electrophoresis reveals that high voltages and small separation lengths are two strategies primarily involved in ultrafast separations, for both capillary and microfluidic formats. However, most commercial instruments are incompatible with short separation distances and are not equipped with the fast injection devices and sensitive detectors needed for rapid separations. Therefore, methods development for both ultrafast capillary and microfluidic separations has gone hand-in-hand with instrument development. Instrument development has focused on methods to maximize voltages, minimize injection volumes, detect small amounts of analyte, and automate data analysis.

15.3.1 Applied Voltage

Typical separations might employ electric fields from 100 to 300 V cm^{-1} while electric fields in the 1000–5000 V cm^{-1} range are common for fast separations. Ultrafast separations on the millisecond or microsecond timescale can require fields up to MV cm^{-1}. These high voltages are a safety risk and the user should be properly isolated from the voltage. Proper care must also be taken to ensure that there are no additional pathways to ground for the applied voltage besides the capillary.

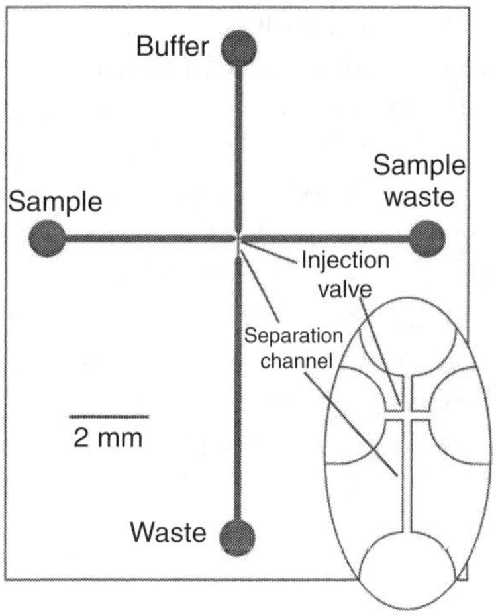

FIGURE 15.1 Schematic representation of a microchip used for high-speed electrophoretic separations. Narrow channels were etched in the injection and separation areas, while wide channels were fabricated for all other sections. These differential channel widths ensured the majority of the potential was applied across the narrow channels. Owing to the high separation field strengths (up to 6.1 V cm^{-1} per volt of applied potential), subsecond separations were possible. (Reproduced from Jacobson, S. C. et al. *Anal. Chem.*, 70, 3476, 1998. With permission from American Chemical Society.)

Extremely high voltages (greater than 30 kV) can lead to breakdown of the material in the separation path and research with ultrahigh voltages has shown that shielding, such as immersion in a weakly conducting solution, will help prevent material breakdown.[9] The Jorgenson lab has applied ultrahigh voltages to increase the temporal response when analyzing tryptic digests of proteins by capillary electrophoresis.[9] High voltages yield faster, more efficient separations but a complicated separation can still take over an hour, even with 120 kV applied. The technology could be applied to making second or even subsecond separations in the future.

Amplified electric fields have been obtained by changing the geometry of a capillary as well. Pulling a capillary to an hourglass shape gives amplified fields in a 50 μm region where the inner diameter is thinnest. Fields of up to 0.15 MV cm^{-1} were obtained by applying 5–20 kV potentials.[10] In the same vein, microfluidic devices have been fabricated that use differential channel widths to apply a large percentage of the separation voltage across the appropriate channel. For example, Ramsey and coworkers[11] applied 8.7 kV across a narrow injection valve design (Figure 15.1) and obtained a 53 kV cm^{-1} electric field in the separation channel, while 71 kV would have been required if all channels had the same cross-sectional area.

15.3.2 Injection Methods

Injection technology is also crucial for achieving efficient separations. A thin plug of analyte must be injected because the band broadening caused by injection is proportional to the width of the sample plug squared. Sample overloading, or electromigration dispersion, can also occur due to differences in the electric field between the sample and the separation buffer if the plug is too wide. Thin plug injections onto small i.d. capillaries or shallow microfluidic channels require that low volumes of sample are loaded. Injection methods for microfluidic devices will be outlined at the end of this section; however, the two most common injection methods for capillary electrophoresis systems are hydrodynamic and electrokinetic injections. Modifications of hydrodynamic injections and electrokinetic methods are necessary to obtain thin plugs.

Injection valves have been used to introduce samples into a capillary electrophoresis instrument.[12] The injection system is similar to a six-port valve used in a conventional HPLC instrument, where the sample can be loaded into a sample loop on the valve, then the valve actuated to inject the plug onto the capillary. The size of the injection is determined by the plug size and flow rate. This system is not the fastest or most reproducible, but it relies on fairly simple technology. Valved injections have been used in conjunction with separation times down to 10 s.[13]

Flow-gated injection is also a common method of injection. Pioneered by Jorgenson, this method involves a cross intersection between the reagent capillary and the separation capillary.[14] During the separation, a cross-flow buffer sweeps the continuously flowing sample out to waste. However, to make an injection, this cross-flow buffer is stopped, the sample builds up in the gap at the middle of the cross, and then sample is injected electrokinetically. The disadvantages are the same as normal electrokinetic injections, as sampling biases occur due to differences in analyte electromigration rates. However, the reproducibility of injections is good, with relative standard deviations of 4% between runs. A range of flow rates can be used and the injection voltage can be varied to control the injection volume. Flow switching on a very rapid scale is difficult to achieve, so flow-gated injection is not useful for the fastest microsecond separations. However, it has been routinely used for multidimensional separations of 2–3 s[15] and coupling to *in vivo* monitoring with 10 s separations.[16,17]

Optical gating is an extremely fast method of injection for CE. The sample is fluorescently tagged and a separation voltage is applied to allow a continuous stream of sample to be electrokinetically injected.[18] An intense gating laser beam near the entrance to the capillary is applied to photobleach the fluorophore and render the analyte undetectable with LIF at the probe beam (Figure 15.2). Only when the gating beam is pulsed "off" is an injection made, allowing tagged molecules to be separated and detected by LIF. The gating laser beam can be pulsed quickly, on the microsecond

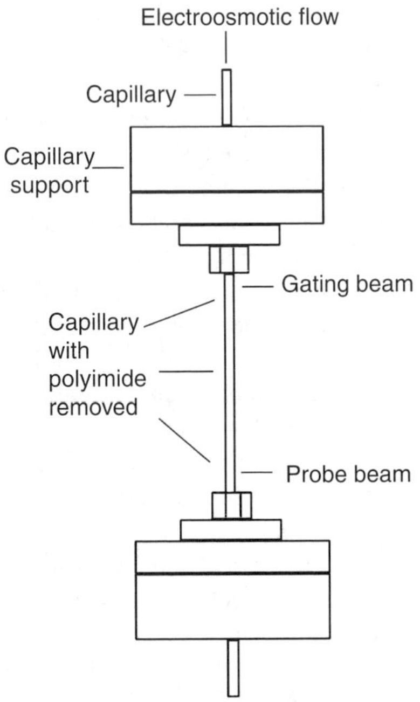

FIGURE 15.2 Instrumental diagram for optical gating. The gating laser beam photobleaches the sample most of the time. To make an injection, the gating beam is pulsed off, and analytes can then be detected by LIF at the probe beam. (Reproduced from Monnig, C. A. and Jorgenson, J. W. *Anal. Chem.* 63, 802, 1991. With permission from American Chemical Society.)

level, making optical gating methods faster than fluid switching methods. However, effectively photobleaching the fluorescent tags can be difficult and incomplete photobleaching can lead to high background light levels.[18] The separation voltage is continuously applied to the capillary, so capacitive charging when applying a voltage does not limit the sample plug sizes. The sample plug is limited by the dimensions of the laser focus, the time for switching and the velocity of the sample migrating through the gate. A concern is that the sample is continuously being injected onto the column. If the sample adsorbs to the capillary or contains high salt levels that alter EOF, poor separations may result. Optical gating has the advantage of being amenable to fast, repeated injections. Therefore, it has been used to apply pulse profiles for multiple injection methods such as Hadamard transform.[19]

A twist on optical gating, which photodestroys most analytes, is optical generation of fluorescent products. For example, multiphoton-excited photoreactions have been used to generate fluorescent products from thermally labile, nonfluorescent reactants.[20] As with optical gating, the laser beams needed for the reaction can be pulsed, so the fluorescent products are only produced for a very short period. Another approach is to generate a fluorescent analyte to photoactivate a caged fluorophore.[21] Caged fluorescein can be photolyzed below 365 nm yielding a fluorescent species that can be detected using the 488 nm line of an Ar ion laser. Therefore, a UV laser beam can be used as a gate to activate a small plug of sample, making it fluorescent and detectable with LIF. This method has lower noise than traditional optical gating because it does not suffer from the background currents caused by incomplete photobleaching.

Injection methods for microfluidic experiments are different, but often based on similar principles. For example, gated injections (either via a perpendicular cross-flow or via an optical gate) that are similar to their CE counterparts are used for high-speed separations on microdevices. Often these injection times are fast, less than 5 s is typical, and since the sample remains close to the injection intersection, serial injections can be easily attained.

A tee intersection,[22] double tee,[23] or a tee intersection operated in a pinched format[24] are the most popular choices for injection schemes in microfluidic formats due to their simplicity and ability to deliver time-independent concentrations of samples. These types of injection formats are unique to microfluidic devices. While short and rapid injections are needed in CE experiments, long loading times are regularly used with these injection systems even with high-speed separations since (ideally) no sample enters the separation channel until analysis begins.

The downside to these methods is that sample often diffuses out of the loading channel during the analysis, which leads to increased band broadening. One method to reduce this effect is to apply "pullback" voltages. In this way, the analyte is driven away from the injection intersection after separation commences to ensure no diffusion of sample into the separation channel. The disadvantage of the pullback method is that the sample is moved away from the injection intersection making it difficult to achieve rapid serial injections. To alleviate this problem, a double-tee format has been used with one tee in the sample channel (allowing the sample to remain close to the injection cross while the pullback voltage was applied), and a second tee in the separation channel (which enabled loading of subsequent injections during the analysis).[25] With this design, repetitive injections up to 10 Hz could be achieved, although the method was not coupled to a separation.

In gated injections the sample is shunted away from the separation channel using a perpendicular flow.[26] To perform an injection, the perpendicular flow is stopped for a specific amount of time and the sample is allowed to fill the injection cross before reinitiating the perpendicular flow. If the sample is being driven by application of an electric field, the injection is biased toward analytes with the highest mobilities. This type of injection is similar to conventional flow-gate injections in capillary systems. Often the injection times are fast, less than 5 s is typical, and since the sample remains close to the injection intersection, highly reproducible serial injections can be easily attained (Figure 15.3).[27]

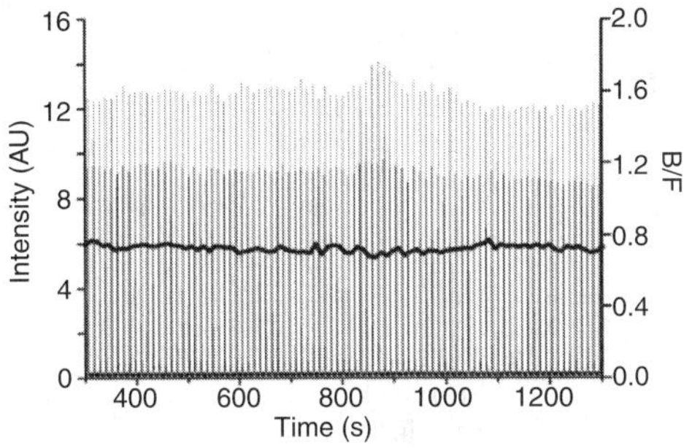

FIGURE 15.3 Series of 110 consecutive separations performed in a microchip to monitor insulin secretion via a competitive immunoassay. Fast monitoring can be achieved by increasing the rate at which a reaction is sampled (e.g., 1 s injections with a 5 s separation time). The gated injection scheme used in these separations enabled rapid sampling, resulting in high temporal resolution of insulin secretion. The black line through the peaks shows the bound/free ratios for each electropherogram; reproducibility ranged from 2% to 6% over the course of 30 min. (Reproduced from Roper, M. G. et al. *Anal. Chem.*, 75, 4711, 2003. With permission from American Chemical Society.)

15.3.3 Detection Methods

Absorbance detection is frequently utilized with capillary electrophoresis but is difficult to use in conjunction with rapid separations. While most molecules absorb in the UV, small capillaries or shallow channels entail small pathlengths, resulting in high limits of detection. Detection cells that lengthen pathlengths, such as Z-shaped cells and bubble cells, are incompatible with fast separations and can lead to band broadening. Signal averaging can boost signal-to-noise (S/N) ratios but is inconsistent with rapid detection because it is time consuming.

The most common detection method for fast separations with CE or microfluidic platforms is LIF. With this method, molecules are usually derivatized with a fluorescent moiety, then a laser is used to induce fluorescence and the resultant light intensity measured. Lasers can be focused into small beams, minimizing band broadening due to the length of the detector. On-column detection is popular because it limits band broadening after the separation. A detection window is created on the capillary by removing the protective polyimide coating; however, scattering of light from the capillary can cause noise and decrease S/N ratios. Dovichi and coworkers[28] designed a sheath-flow cell for off-column detection that concentrates the analyte into a Taylor cone after the capillary. The sheath-flow detection cuvette was square, eliminating the light scattering caused by the round capillary. A schematic diagram of an LIF instrument is shown in Figure 15.4a with a close-up of the sheath-flow cell for LIF detection in Figure 15.4b.[29] Sheath-flow cells are used to increase S/N ratios by a factor of 10 over on-column detection.

Similar to capillaries, the most common detection method for high-speed separations on microdevices is fluorescence, yet few methods have matched the limits of detection found in sheath-flow formats routinely used in capillary systems. This is due in part to the microfluidic substrate being in a planar format and therefore difficult to reproduce the three-dimensional Taylor cone found in sheath-flow capillary systems. The most simple and common method for high-sensitivity detection in microdevices is the use of a confocal detection scheme.[30]

Native-fluorescent LIF has been used to detect proteins with high tryptophan levels[31] and serotonin.[32] However, most molecules are not fluorescent and must be derivatized with a fluorescent moiety in order to be detected. Derivatization has been achieved both with covalent attachments and affinity agents such as aptamers.[33,34] The limitation for LIF as a detection technique for rapid separations is usually the chemistry of the derivatization reaction. First, the kinetics of the tagging reaction should be fast, especially for online analysis systems. Some fluorophores such as fluorescein

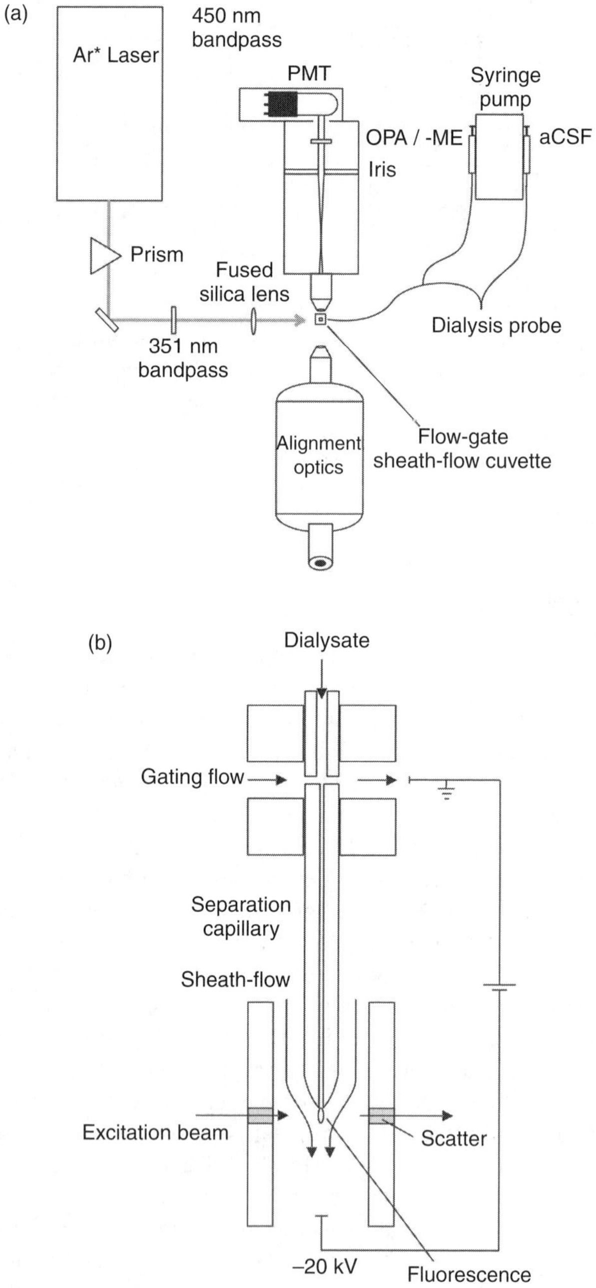

FIGURE 15.4 Instrumental diagram of a flow-gated injection CE instrument for monitoring neurochemicals. (a) Schematic diagram of the optical setup for the detection of analytes after OPA derivatization online. (b) Schematic diagram of the flow-gated interface and sheath-flow cuvette for detection. A cross-flow buffer between the reaction and separation capillary normally sweeps the sample to waste. An injection is made by stopping the cross-flow, allowing sample to build up in the gap, then applying a voltage for an electrokinetic injection. The sheath-flow cuvette has a sheath buffer flowing around the capillary to concentrate the sample into a Taylor cone and facilitate off-column detection. (Reproduced from Bowser, M. T. and Kennedy, R. T. *Electrophoresis*, 22, 3668, 2001. With permission from WILEY-VCH.)

isothiocyanate (FITC) can take 6–12 h to react completely,[35] making them incompatible with methods requiring fast analysis. However, a few tags for primary amines, such as *o*-phthaldialdehyde (OPA)[36] and naphthalene dicarboxylic acid (NDA)[37] can react within a few minutes. The classes of compounds that can be detected are limited by the chemical reactions. Methods for derivatizing primary amines,[33] thiols,[38] and carboxylic acids[39] have been developed, but secondary amines, for example, have slower kinetics and are much more difficult to rapidly detect. Also problematic is that derivatization of real-world samples in complex matrices with nM concentration is difficult.[40] Despite the drawbacks of the chemical reaction, LIF is popular because the spectral properties of the fluorescent tag can be tuned and high mass sensitivity detection can be obtained. Detection of yoctomoles of analyte has been reported using LIF.[41]

Electrochemical detection has also been used for rapid electrophoretic separations.[42,43] The main challenge of electrochemical detection after electrophoresis is separating the electrode from the high voltage applied for the separation. Noise at the electrode from the separation voltage must be minimized for efficient, rapid separations. This is usually accomplished by detecting on-column after a decoupler that grounds the separation voltage before the electrode or by end-column detection after the capillary.[44] Because there is no EOF after the column, care must be taken to avoid band broadening during end-column detection. Electrochemistry can provide high sensitivity and low mass detection limits. Electroactive molecules can be detected without derivatization, an advantage over LIF. Microelectrodes are compatible with small i.d. capillaries and the small size of electrodes makes them particularly attractive for interfacing with microchips where miniaturization is a goal. Electrodes can even be fabricated into chips. Because mass limits of detection are not as good as with LIF, electrochemical detection has been used with separations in the minute time frame[45] but it has not been used for ultrafast, subsecond separations.

Mass spectrometry (MS) is especially useful for monitoring changes in peptides and proteins.[46] MS can be coupled to fast separations;[47] however, there are multiple challenges including detection of low mass samples inherent to rapid electrophoretic separations (see previous section) and coupling the separation eluent to the mass spectrometer for detection. Electrospray detection has primarily been used as the interface method, but the separation voltage and electrospray voltage are of different magnitudes (kV and V, respectively). Coupling of nanospray and CE has been achieved to allow faster separations.[48] Ideal mass analyzers for rapid separations would make measurements several times a second, making ion traps and time-of-flight analyzers preferred. Quadrupoles are more sensitive but have slower scan speeds. MS coupled to CE has not been used for ultrafast microsecond separations, but could be useful for proteomics and other applications that require separations of tens of compounds. These separations could normally take hours, but could be reduced to the minute time scale with rapid capillary electrophoresis procedures outlined here.

15.3.4 Electronics and Data Analysis

Analysis of ultrafast separation data requires a consideration of instrumentation electronics and data analysis. High collection frequencies are needed to sample sharp analyte peaks. In addition, filtering cutoffs and rise times must be ample to allow fast signals to pass.[49] Fast sampling rates, while providing a better description of the peak shape, require larger bandwidths and can lead to greater noise from environmental sources. For subsecond separations, sampling rates greater than 1000 Hz and filter rise times less than 1 ms are needed. Often, high-speed separations are used for monitoring purposes. In these cases, methods to analyze a large number of electropherograms in a reasonable time are indispensable.[50]

A simple approach to increasing S/N ratios for samples, thereby lowering detection limits, is to average multiple electropherograms from the same sample. The difficulty of this approach with rapid separations is that performing multiple separations increases the analysis time and sample consumed. Mathematical multiplexing approaches such as Hadamard transforms allow multiple separations to be performed at once, increasing throughput and improving S/N. Aspinwall has designed a fast Hadamard transform injection profile that when coupled with low volume optical gating injections yields separations with ninefold enhancement in S/N in less than 10 s.[19] This is the first multiplexing method that is amenable to fast monitoring.

15.4 PRACTICAL APPLICATIONS

For ease in presenting literature examples of rapid electrophoretic separations, this section is broken into examples using capillary and microfluidic electrophoresis systems.

15.4.1 Capillary Electrophoresis

The fastest separation to date is the separation of 5-hydroxytryptamine (5-HT) and 5-hydroxytryptophan (5-HTrp) in 20 μs (Figure 15.5).[10] Multiphoton excitation was used to create the fluorescent, highly unstable photoproducts.[20] Short injection pulses were achieved by pulsing the laser on for 200 μs to achieve a femtoliter reaction plug. Separation of these products could be achieved in a 4.7 μm i.d. traditional capillary with a separation distance of 15 μm in under 10 ms. Different reaction products were observed than when the separation products eluted 0.1–1.0 s after pulse, indicating the reaction occurs in less than 0.1 s and the products were very unstable. To achieve microsecond separations, higher electric fields were needed. This was accomplished by pulling capillaries into an hourglass shape to create extremely high electric fields (MV cm^{-1} range) in the pulled region.[10] With a separation distance of 10 μm, separations could be performed in these high fields in 20 μs. This experiment demonstrated that capillary electrophoresis could be used to analyze short-lived reaction products. The intermediates have similar spectroscopic properties, but different electrophoretic mobilities that are probed with CE. The technological advances of both the geometry of the capillary and the width of the injection pulse will allow faster probing of a variety of reactions.

The Kennedy lab has been at the forefront of using rapid capillary electrophoresis to measure neurochemical changes. Optically gated injection of OPA-derivatized amino acids has been achieved in less than 2 s.[33] However, the high salt concentrations of physiological samples, such as cerebral spinal fluid, can reduce EOF, and make separations slower. Still, 10 s temporal resolution for online monitoring of directly sampled and microdialysis samples were obtained. An instrument has been developed in the Kennedy lab for online analysis of microdialysis samples after precolumn

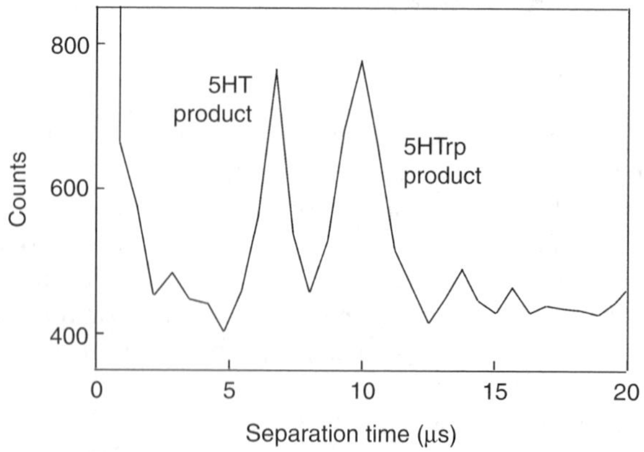

FIGURE 15.5 Microsecond separation of 5-HT and 5-HTrp. The capillary was pulled to an hourglass geometry to achieve an electric field of 0.15 MV cm^{-1} over the 10-μm separation distance. (Reproduced from Plenert, M. L. and Shear, J. B. *Proc. Natl Acad. Sci. USA*, 100, 3853, 2003. With permission.)

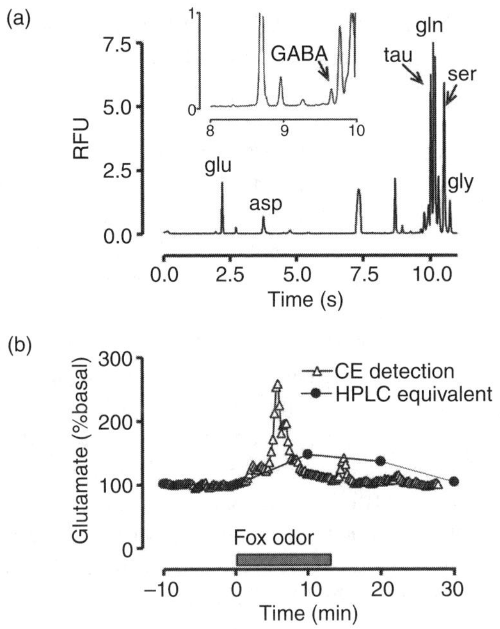

FIGURE 15.6 Use of rapid CE for neurochemical monitoring during a behavioral experiment. (a) Example electropherogram of a rapid separation of OPA-derivatized amino acids in a microdialysis sample. Overlapping injections were used, so glutamate (glu) and aspartate (asp) are from the previous injection. Also separated were taurine (tau), glutamine (gln), serine (ser), and glycine (gly). The enlarged inset shows a clear peak for GABA. (b) Average peak height for glutamate changes during presentation of a predator fox odor to a rat ($n = 14$ animals, fox odor present during gray bar). The CE data (triangles) collected a point every 14 s and revealed that changes in glutamate were fast and large. Averaging the data to show the equivalent 10 min temporal resolution of HPLC experiments show that both the magnitude and the width of the peak are distorted with slower temporal resolution techniques. (Adapted from Venton, B. J. et al. *J. Neurochem.*, 96, 236, 2006. With permission.)

derivatization with OPA or NDA (Figure 15.4).[29,37] With flow-gated injection, short, small i.d. capillaries, and electric field strengths of 2000 V cm^{-1}, separations of amino acids were achieved in less than 15 s. Sheath-flow detection with LIF allowed low nM detection limits of real samples. An example electropherogram from a behavioral experiment using this instrumentation is shown in Figure 15.6a. With overlapping injections, six amino acids were detected every 10 s. Plotting the peak current for glutamate revealed large, fast amino acid changes in response to a predator fox odor (Figure 15.6b).[16] Conventional detection by HPLC would allow only 10 min temporal resolution, and grouping the rapid CE data in 10 min bins to mimic that temporal resolution reveals that fast detection is needed to characterize both the time course and magnitude of the changes. Capillary electrophoresis has also been used to demonstrate that glutamate and γ-aminobutyric acid (GABA) transients correlate with behavioral changes during stereotypic mouse behaviors[51] and fear conditioning.[52] The application of fast CE separations to neurochemical monitoring is allowing a better picture of the dynamics of neurotransmission in the rat brain.

Others have also demonstrated fast separations that could be useful for neurochemical monitoring. Shear's laboratory has detected cyclization reactions of catecholamines to form trihydroxyindoles that allowed them to be detected with two photon fluorescence.[53] Their method allowed separation of norepinephrine and epinephrine within 80 s and cofactors such as FADH and NADH within 4 min. Bowser and coworkers[54] have developed rapid chiral CE separations for D-serine in salamander retinas and rat brain samples. Fast separations of nitrate and nitrite in rat brain perfusates have been achieved using UV detection in under 1.5 min.[55] Multiplexing rapid separation using Hadamard transforms has improved the detection limits of amino acids.[19] Monitoring of amino acid mixtures at low nM concentrations indicates that this method may be beneficial for neurochemical monitoring of compounds such as GABA and dopamine, which are usually present at low concentrations.

Conventional capillary electrophoresis is useful for separating small inorganic ions in a complex mixture. CE with a contactless conductivity detector has been utilized to separate solutions of cations or anions in under 20 s.[13] Together, inorganic ions and cations were detected in under 2 min.[56] The conductivity detector allows detection of ions that are not UV active. Small anions and cations have also been separated from the complex solutions blood serum and milk.[57]

Rapid separations of DNA fragments in conventional capillaries have been demonstrated. Commercial instrumentation commonly used to sequence DNA is not equipped to use small diameter, short length capillaries so extremely fast separations have not been extensively utilized. Boček and coworkers[58] have shown that CE-LIF detection of denatured DNA fragments can be achieved in 2.5 cm capillaries in less than 45 s. Comparable separations on a traditional CE-UV instrument with 50 cm capillaries required 23 min. Short tandem repeat polymorphisms could be detected because baseline separation of DNA fragments differing by two nucleotides was possible in under a minute. In another study, narrow electrophoretic injections were the key to minimizing dispersion and obtaining fast, efficient separations of DNA. Here, short capillaries were used to demonstrate separations of double-stranded fragments of DNA within 30 s.[59] Fast mutation detection was achieved of a single base pair mutation in double-stranded mitochondrial DNA in 72 s.

Separations of peptides and proteins are not routinely performed in under a minute. However, the Jorgenson lab has used rapid CE to investigate isomerization between *cis* and *trans* peptide bonds.[60] Isomerization of proline-containing peptide bonds from *cis* to *trans* can occur in a few seconds. The multiple peaks detected with CE for peptides containing more than one proline are thought to be due to differential migrations of the *cis* and *trans* conformations. Optical gating injections combined with a 6 μm i.d. capillary was used to separate *cis* and *trans* isomers in 2 s, less than the rate of the conformational shift. Longer peptides could be separated in 5 s. These studies show that rapid CE is useful for studying fast isomerization rates for optical isomers.

Affinity probe capillary electrophoresis (APCE) uses a fluorescently labeled ligand to bind a receptor. Separation of the bound and free affinity probe has been achieved with CE with LIF detection. One advantage of the rapid CE method is that if the separation is rapid compared to the dissociation time, affinity constants can be determined.[61] Competition experiments can be performed to observe the conditions of complex formation. For example, a fluorescent GTP analog was used as an affinity probe and G protein receptor subunits as the ligand.[62] The complex and free fluorescent probe were separated within 20 s and Ras-like G protein complexes could be separated in less than 15 s. The rapid separations allowed kinetics of binding to be studied.

APCE has also been used to investigate the fast kinetics of peptide–protein interactions.[63] A fluorescently labeled phosphorylated peptide was mixed with a Src homology 2-domain protein and separations of complex and free probe could be achieved in less than 4 s. These rapid separations were necessary, because no complex peak was observed with 8 s separations, presumably because it had dissociated. Estimated dissociation constants for the complex were obtained from peak areas and results were similar to constants from fluorescence anisotropy data. Aptamers have also been used as affinity ligands for fast separations. Separations as short as 30 s were accomplished with a fluorescently labeled aptamer against IgE.[34] APCE has also been used to detect nM concentrations of digoxin in less than 1 min.[64]

The rapid separations offered by capillary electrophoresis have made it amenable as a detector in hyphenated techniques. For LC-CE, the total analysis time is usually governed by the LC separation, which generally takes minutes. However, capillary electrophoresis detection adds more peak capacity because of a second and orthogonal dimension for separation, and shorter separation conditions for LC can often be tolerated. For example, a 2.5 min reversed-phase liquid chromatography gradient was used in conjunction with 2.5 s CE separations for the detection of a tryptic digest of cytochrome *c*.[65]

The experiment used optical gating for injections, but flow-gated injections were actually developed to interface the two separation techniques. For example, using a capillary LC column, flow-gated injections of LC eluent onto the CE capillary were performed every 30–60 s.[14] Fluorescently labeled phenylalanine and glutamate could be separated in 35 s. A more complicated analysis of fluorescently labeled urine required longer separations, up to 1 min. However, overlapping electrokinetic injections were made before one run finished, allowing 30 s temporal resolution.

15.4.2 Microfluidic Applications

Owing to the numerous examples of microfluidic separations in the literature, we have attempted to limit the applications reviewed to those where the separation was performed in less than 30 s.

As an example of utilizing the gated methodology for rapid, sequential sample introduction, Roper et al.[27] made 1 s injections every 15 s over the course of 30 min to monitor secretion of a peptide hormone from a single islet of Langerhans on a glass microfluidic chip. During the 15 s separation, free FITC-insulin and FITC-insulin bound to an insulin antibody were separated and used to quantify the amount of insulin released from the islet. The Kennedy and coworkers[66] have continued to expand on this initial work by improving the throughput using 0.5 s injections every 5.5 s. These rapid serial injections and separations were used to monitor insulin secretion from an islet for over 2 h.

Gated injections have also been used in several applications where microdialysis sampling was coupled to microfluidic devices for separation. For example, Lunte and coworkers[67] coupled the outlet of a microdialysis sampling probe to a microfluidic chip and used the inherent flow resistances within the device to split a portion of the dialysate to the separation channel. Their device was used to monitor the product of an enzyme reaction using 1 s injections every 30 s and could observe a decrease in the substrate peak with a concomitant increase in the product peak. Sandlin et al.[68] have also coupled a microdialysis probe to a microfluidic chip and used this device to monitor changes in neurotransmitter levels in the brains of anesthetized rats. In this report, the dialysate was mixed and reacted on-chip with OPA, before a 0.1 s gated injection with a 25 s separation. With this protocol, sequential injections were made every 130 s over the time course of 100 min while obtaining high efficiency separations ($N = 156{,}000$ for glutamate).

Lapos and Ewing[69] have performed 80 ms injections every 5 s for the separation of 4-choloro-7-nitrobenzofurazan (NBD)-labeled amino acids. They have continued to improve upon this method by utilizing optical gating to inject analytes in a parallel channel format[70] and have achieved five serial separations of four NBD-labeled amino acids in four parallel channels every 10 s. In addition, they have performed three serial separations of five fluorescein-labeled oligonucleotides (10, 20, 30, 60, and 80 basepairs in length) within 20 s.[71]

Li and coworkers[72] described the separation of FITC and FITC-labeled antihuman IgG by zone electrophoresis in a glass microchip. The two compounds were electrokinetically injected into the separation channel for 2 s and separated in less than 12 s. A comparison between the separation performance in capillary and microchip showed that higher efficiency was achieved in the microchip format due to the shorter separation length and the higher electric field that were applied in microfluidic structures.

Belder and coworkers[73] reported the fast chiral separation of acidic and basic compounds using a commercially available quartz microchip with a simple cross-tee design and linear imaging UV detection. Highly sulfated cyclodextrins were added to the separation medium to improve selectivity and samples were loaded for 60 s using the pinched injection mode. The chiral separation of norephedrine was achieved in 2.5 s, and a mixture of three basic drugs was resolved in 11 s.

The same microchip electrophoresis system was used by the Takeda and coworkers[74] to separate phenolic compounds [bisphenol A, 4-nonylphenol, 4-(1,1,3,3-tetramethylbutyl)phenol and 4-*tert*-butylphenol] by micellar electrokinetic chromatography (MEKC). The samples were loaded for 25 s using a pinched injection scheme. β-Cyclodextrin was added to the MEKC buffer to further improve the separation and the four phenolic compounds were resolved in 13 s in a 25 mm separation channel using an electric field of 450 V cm^{-1}.

The requirement for low sample injection volumes in microchip electrophoresis often leads to problems in detection sensitivity. In cases where the analytes of interest are present in low amounts, sample concentration techniques are typically applied before separation. deMello and coworkers[75] has reported performing in-column field-amplified sample stacking (FASS) for the separation and detection of low concentrations of biogenic amines. A hybrid PDMS/glass device patterned with a narrow sample channel tee-injector design allowed the sample to be directly injected into the separation channel without leakage control. The separation length was 30 mm and the sample was loaded into the separation channel for 10 s. The extent of the sample plug was defined by the chip design, more specifically by the distance between the injection intersection and the inlet buffer reservoir, which enabled reproducible injections. Stacking occurred on application of the separation voltage, and separation of putrescine and tryptamine was performed in less than 20 s. Limits of detection for putrescine and tryptamine using FASS were 20 and 25 pM, respectively.

Several research groups have investigated the use of polymeric materials to fabricate microfluidic devices, which can help reduce production costs compared to glass or quartz materials. Henion and coworkers[76] reported the use of Zeonor (a polymer commonly used to manufacture CDs and DVDs) as the analytical platform for the separation of carnitine derivatives by microchip electrophoresis-MS. Using a double-tee junction design for sample injection (15 s loading time), carnitine, acetylcarnitine, and butyrylcarnitine were separated in a 3.5 cm long separation channel in less than 10 s, with an applied electric field of 2 kV cm^{-1}.

Nagata et al.[77] reported the use of a polyethylene glycol-coated poly(methyl methacrylate) (PMMA) microchip for microchip gel electrophoresis separations. Trypsin inhibitor, bovine serum albumin (BSA) and β-galactosidase (labeled with Alexa Fluor 488) were electrokinetically loaded for 60 s. The sodium dodecyl sulfate (SDS)–protein complexes were then separated and baseline resolved in 8 s using a 5% linear polyacrylamide/0.1% (w/v) SDS-based separation buffer, 3 mm separation channel, and 303 V cm^{-1} electric field.

Fast microchip electrophoresis in polymeric devices was also applied to multiplex enzyme assays, which enabled high-throughput screening of one or several substrates against multiple enzymes targets. One particular advantage of this technique is that the activity, specificity, and cross-reactivity of enzymes for drug candidates may be monitored as well as protein–protein interactions in a single fast separation. Gibbons and coworkers[78] performed a fourplex protein kinase assay in a PMMA chip patterned with a double-tee injection (150 μm offset) design. The separation was performed in a buffer containing 1% polyethylene oxide (PEO) to suppress EOF. Three substrate peaks and four product peaks were separated in less than 25 s.

The fabrication of separation platforms from less common materials has also recently been described. Wirth and coworkers[79] performed electrophoretic separations inside silica colloidal crystals. The advantages of this separation medium include fast mass transport, high electric field resistance, and chemically modifiable surfaces. The silica colloidal crystals were treated and reacted with chloro-dimethyl-octadecyl silane to form a C_{18} stationary phase. The system differed from traditional electrochromatography in that EOF was negligible. In such a system, separation occurred through both adsorption and electrophoresis phenomena. A comparison between the Van Deemter plots obtained from the silica colloidal system (obtained using an unretained dye since there was no flow) and from a commercial monolithic column showed that the colloidal phase produced much smaller plate heights due to smaller A (Eddy diffusion) and C (mass transfer) terms. Fast mass transport was also beneficial to produce fast separations. Three hydrophilic peptides, labeled with rhodamine and electrokinetically injected for 5 s, were separated in less than 10 s using an electric field of 800 V cm^{-1} in a 6-mm-long separation channel.

More complex experimental designs have also been described. Hyphenation of microchip electrophoresis with an orthogonal separation technique was reported by Zhang and coworkers.[80]

A two-dimensional HPLC/electrophoresis device (fabricated in glass) was used to separate FITC-labeled peptides. To interface the two techniques, a microchip design similar to a cross-tee was used, containing a hole in the sample-loading channel that connected the outlet end of the HPLC capillary column to the microchip channels. The HPLC effluent was continuously introduced into the sample-loading channels and was repeatedly injected (pinched injection for 3 s) into the separation channel every 20 s. The peptides were separated in a 40-mm-long separation channel using an electric field of 635 V cm^{-1}. The partial profile of a tryptic digest of BSA was obtained from 10 consecutive injections and separations of the HPLC effluent, with eight FITC-labeled peptides being separated and detected in a total analysis time of 230 s. With these experimental conditions (i.e., sampling and separation times), two peptides were detected per electrophoretic separation (20 s analysis time).

Ramsey and coworkers[11,81] were one of the first groups reporting subsecond separation in a microfluidic device. A simple cross-T design, etched into a planar glass chip, was used to separate rhodamine B and fluorescein in 150 ms, in a 0.9-mm-long separation channel.

Liu et al.[82] also reported the microchip electrophoretic separation of three flavin metabolites (riboflavin, flavin-adenine dinucleotide, flavin mononucleotide) in less than 1 s using a pinched injection scheme. The experimental conditions were optimized to achieve ultrafast separation, by shortening the separation channel length and increasing the electric field. The separation quality was also improved by controlling the sample injection size by optimizing sample pinching dispensing factors.

Subsecond chiral separations of DNS-labeled amino acids were also reported by Belder and coworkers[83] using quartz microchips. Highly sulfated γ-cyclodextrin was added to the separation buffer as a chiral selector and the samples were loaded for 20 s using the pinched injection mode. The separation length and electric field strength (1.5 mm and 2012 V cm^{-1}, respectively) were optimized to achieve baseline resolution of DNS-norleucine enantiomers in 720 ms, and DNS-tryptophan enantiomers in 800 ms (Figure 15.7). The authors also showed the baseline separation of a more complex sample consisting of three amino acids (DNS-phenylalanine, DNS-norvaline, and DNS-glutamic acid) in 3.3 s using a 7 mm separation length and 2012 V cm^{-1} electric field.

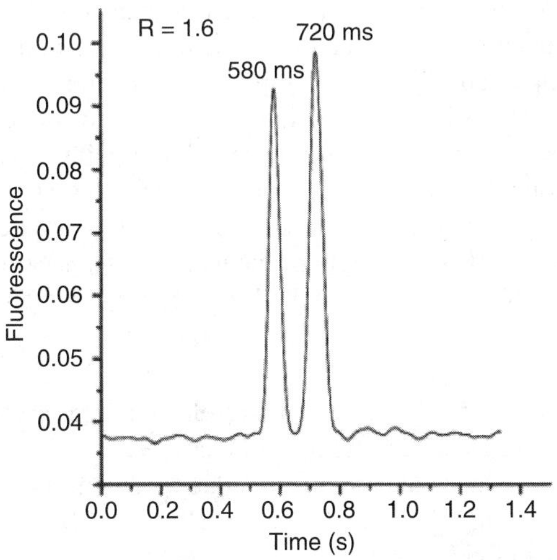

FIGURE 15.7 Subsecond chiral separation of DNS-norleucine achieved using a simple cross-injector quartz microchip. The separation was performed in a 1.5-mm-long channel, using an electric field of 2012 V cm^{-1} and 2% highly sulfated γ-cyclodextrin, 25 mM triethylammonium phosphate buffer, pH 2.5, as the separation buffer. (Reproduced from Piehl, et al. Electrophoresis 25, 3848, 2004.)

15.5 METHODS DEVELOPMENT GUIDELINES

In this section, it is our goal to provide a brief procedure that may be followed to produce successful ultrafast separations.

15.5.1 CAPILLARY ELECTROPHORESIS

Rapid capillary electrophoresis measurements in general require small i.d., short length capillaries with fast injection and sensitive detection techniques. One example of a successful implementation of all these principles is a rapid capillary electrophoresis instrument developed by Bowser and Kennedy[29] to analyze online microdialysis samples for *in vivo* monitoring (Figure 15.4). This instrument used small, 10 μm i.d. capillaries that were 10 cm long. Applied voltages were 20,000 V, or 2000 V cm^{-1}.

Analysis of amino acid neurotransmitters was achieved by online, precolumn derivatization with OPA. The reaction of OPA with primary amines occurs in less than 2 min. An online reactor was created by inserting a smaller capillary with the analyte into a larger capillary through, which the OPA was pumped. After the end of the smaller capillary, the analyte would be mixed with OPA and derivatized. A flow-gated interface was used for injection. The sample was swept to waste by a cross-flow buffer except when an injection was made. Then analyte was allowed to accumulate in the gap (30–50 μm) between the reaction and separation capillaries. A small voltage, less than the separation voltage, was applied for only 200 ms to perform an electrokinetic injection. This allowed very efficient separations with only 30 ms wide peaks.

Separations were performed with 40 mM borate buffer. One key is to filter all solutions so that no dust particles are present that can clog the small capillaries. Hydroxypropyl-β-cyclodextrin was added to the separation buffer to increase the selectivity and was useful for the separation of isomers α- and β-aminobutyric acid from GABA. In addition, side chains of the amino acids differentially interact with the hydroxyl groups on the cyclodextrin, allowing more separations that are efficient. Overall, efficiencies as high as 500,000 theoretical plates were obtained with separation times less than 20 s. Overlapping injections, where the separation is stopped (i.e., the voltage turned off) and more sample injected before the first separation is finished, allowed separations every 10 s (Figure 15.6a). Limits of detection for glutamate were less than 50 nM.

LIF in a sheath-flow cuvette allows detection of small amounts of analyte with high S/N ratios. The sheath-flow cuvette improved S/N ratios 15 times over conventional on-capillary detection. The end of the separation capillary was inserted into a square, quartz cuvette. Buffer was siphoned around the outside of the capillary end, which had been ground to a point. This created a Taylor cone of analyte. For OPA, excitation was with a 351 nm line of an Ar ion laser and emission at 450 nm was collected at a 90 angle using an objective to focus it on a photomultiplier tube. Collection of light at this angle avoids interference from the light of the incident beam. Software programs automated data collection and allowed the analysis of hundreds of electropherograms simultaneously.

To summarize

1. High voltages are needed for rapid separations (electric fields greater than 1000 V cm^{-1}).
2. Short separation distances are crucial for fast analyses.
3. Small i.d. capillaries reduce band broadening due to joule heating.
4. Precolumn derivatization with a fast reaction eliminates band broadening issues with postcolumn derivatization.
5. Short injection times, less than a second, yield narrow peaks and efficient separations.
6. A flow-gated interface allows sampling and injection from a continuous flow stream.
7. LIF is a high-sensitivity detection scheme for low quantities of analyte.
8. A sheath-flow cuvette increases S/N ratios and decreases detection limits.
9. Rapid, batch analysis of hundreds of electropherograms enables fast data analysis.

15.5.2 Microfluidic Chips

As detailed in Section 15.4, there are multiple examples of fast separations performed in microfluidic devices and the channel design and speed of the separation is ultimately dictated by the analytes to be separated. In one example, Roper et al.[27] detected insulin secretion from islets of Langerhans using a combination of several factors designed for rapid separations. These factors included the use of shallow channels, 1 s gated injections, 15 s separations, and high-speed data processing.

Use of shallow channels reduces current (and therefore, Joule heating) by increasing the overall resistance of the device. Unfortunately, these type of channels were also easy to clog with dust particulates, salt, or cellular debris so a thorough cleaning procedure before and after experiments was outlined in a subsequent paper.[67] This cleaning method included flushing the channels with NaOH, followed by multiple buffer rinses. In addition, caps were placed on top of microfluidic reservoirs to reduce dust and particulate entry into the device over a day of experimentation. With these types of precautions in place, separations every 10 s over a period of several hours could be performed.

Gated injections were used in this example since a pullback flow would perturb the insulin concentration upstream that was being monitored. With 1 s injections, the absolute number of analytes being detected was relatively low; however, the concentration being detected, approximately 100 nM FITC-insulin was high for LIF detection. The separation buffer was also optimized for high-speed separations by using a zwitterionic buffer. Zwitterionic buffers or separation buffers with added organic modifiers allow the application of high separation voltages while maintaining a low current, essential for rapid separations.[7] Finally, a software program was written to analyze the large amount of data being produced from these rapid separations.[50] Without this specialized software, data analysis was orders of magnitude longer than the separation time, limiting the utility of the method.

An analogous circuit diagram should be produced while the fluidic architecture is being designed. With this diagram, the distribution of the separation voltage over the channel network can be tested using relative resistance values[84–86] before spending the time and money fabricating the photomask or device. Through careful consideration of the relative resistances, the majority of the separation voltage can be applied to specific areas within the channel network, such as across the separation channel, leading to decreased separation times.

To summarize

1. Fabrication of analogous circuit diagram ensures proper channel lengths and widths (relative resistances) so that if only 3000 V is applied across the entire device, most of the voltage is dropped across the separation channel.
2. Production of a glass microfluidic device with shallow channels. These two features allow for high EOF rates while maintaining a low current.
3. Extensive cleaning of the channel surfaces before and after experiments with NaOH and buffers.
4. Use a high concentration of zwitterionic buffer to allow application of a high voltage while maintaining a low current.
5. One-second gated injections using a computer-controlled high voltage relay followed by a 10–15 s separation allows rapid, serial monitoring.
6. LIF in conjunction with an epifluorescence microscope facilitates sensitive detection of low mass, low volume injections.
7. High-throughput data program allows for efficient postprocessing of collected data.

15.6 CONCLUSIONS

Capillary electrophoresis has continuously demonstrated that short analysis times are compatible with well-resolved separations. With the use of microfluidic devices, more integrated sample processing and handling is possible while maintaining the benefits associated with rapid separations. The major obstacles to performing rapid separations are nondiffusional sources of band broadening (i.e., large injection amounts and finite detection zones). With the proper choice of instrumentation and analytes, separation time scales could be readily achieved below 1 s. The development of commercial instrumentation is needed for widespread implementation of rapid electrophoresis methodologies. It will be interesting to observe if technology development in the areas of injection and detection techniques paves the way for new applications, or if the applications drive the development of new technologies. Certainly now it appears that technology is at the forefront as many examples of high-speed separations are performed on fluorescent dyes. When the reverse is true, the true power of rapid separations may be observed. Thus, the research field of rapid electrophoretic separations is expected to be active for quite some time.

REFERENCES

1. Tiselius, A. A new apparatus for electrophoretic analysis of colloidal mixtures. *Trans. Faraday Soc.* 33, 524, 1934.
2. Jorgenson, J. W. and Lukacs, K. D. Zone electrophoresis in open-tubular glass capillaries. *Anal. Chem.* 53, 1298, 1981.
3. Jorgenson, J. W. and Lukacs, K. D. Capillary zone electrophoresis. *Science* 222, 266, 1983.
4. Liu, J., Dolnik, V., Hsieh, Y.-Z., and Novotny, M. Experimental evaluation of the separation efficiency in capillary electrophoresis using open tubular and gel-filled columns. *Anal. Chem.* 64, 1328, 1992.
5. Hjerten, S., Valtcheva, L., Elenbring, K., and Liao, J. L. Fast, high-resolution (capillary) electrophoresis in buffers designed for high field strengths. *Electrophoresis* 16, 584, 1995.
6. Hjerten, S. and Kubo, K. A new type of pH- and detergent-stable coating for elimination of electroendosmosis and adsorption in (capillary) electrophoresis. *Electrophoresis* 14, 390, 1993.
7. Bushey, M. M. and Jorgenson, J. W. Capillary electrophoresis of proteins in buffers containing high concentrations of zwitterionic salts. *J. Chromatogr.* 480, 301, 1989.
8. Qiao, R. Control of electroosmotic flow by polymer coating: Effects of the electrical double layer. *Langmuir* 22, 7096, 2006.
9. Hutterer, K. M. and Jorgenson, J. W. Ultrahigh-voltage capillary electrophoresis. *Anal. Chem.* 71, 1293, 1999.
10. Plenert, M. L. and Shear, J. B. Microsecond electrophoresis. *Proc. Natl Acad. Sci. USA* 100, 3853, 2003.
11. Jacobson, S. C., Culbertson, C. T., Daler, J. E., and Ramsey, J. M. Microchip structures for submillisecond electrophoresis. *Anal. Chem.* 70, 3476, 1998.
12. Hogam, B. L., Lunte, S. M., Stobaugh, J. F., and Lunte, C. E. On-line coupling of *in vivo* microdialysis sampling with capillary electrophoresis. *Anal. Chem.* 66, 596, 1994.
13. Wuersig, A., Kuban, P., Khaloo, S. S., and Hauser, P. C. Rapid electrophoretic separations in short capillaries using contactless conductivity detection and sequential injection analysis manifold for hydrodynamic sample loading. *Analyst* 131, 944, 2006.
14. Hooker, T. F. and Jorgenson, J. W. A transparent flow gating interface for the coupling of microcolumn LC with CZE in a comprehensive two-dimensional system. *Anal. Chem.* 69, 4134, 1997.
15. Moore, A. W. and Jorgenson, J. W. Rapid comprehensive 2-dimensional separations of peptides via RPLC optically gated capillary zone electrophoresis. *Anal. Chem.* 67, 3448, 1995.
16. Venton, B. J., Robinson, T. E., and Kennedy, R. T. Transient changes in nucleus accumbens amino acid concentrations correlate with individual responsivity to the predator fox odor 2,5-dihydro-2,4,5-trimethylthiazoline. *J. Neurochem.* 96, 236, 2006.
17. Ciriacks, C. M. and Bowser, M. T. Monitoring D-serine dynamics in the rat brain using online microdialysis-capillary electrophoresis. *Anal. Chem.* 76, 6582, 2004.
18. Monnig, C. A. and Jorgenson, J. W. On-column sample gating for high-speed capillary zone electrophoresis. *Anal. Chem.* 63, 802, 1991.
19. Braun, K. L., Hapuarachchi, S., Fernandez, F. M., and Aspinwall, C. A. Fast Hadamard transform capillary electrophoresis for on-line, time-resolved chemical monitoring. *Anal. Chem.* 78, 1628, 2006.
20. Gordon, M. J., Okerberg, E., Gostkowski, M. L., and Shear, J. B. Electrophoretic characterization of transient photochemical reaction products. *J. Am. Chem. Soc.* 123, 10780, 2001.

21. Hapuarachchi, S., Premeau, S. P., and Aspinwall, C. A. High-speed capillary zone electrophoresis with online photolytic optical injection. *Anal. Chem.* 78, 3674, 2006.
22. Harrison, D. J. et al. Capillary electrophoresis and sample injection systems integrated on a planar glass chip. *Anal. Chem.* 64, 1926, 1992.
23. Effenhauser, C. S., Manz, A., and Widmer, H. M. Glass chips for high-speed capillary electrophoresis separations with submicrometer plate heights. *Anal. Chem.* 65, 2637, 1993.
24. Jacobson, S. C. et al. Effects of injection schemes and column geometry on the performance of microchip electrophoresis devices. *Anal. Chem.* 66, 1107, 1994.
25. Thomas, C. D., Jacobson, S. C., and Ramsey, J. M. Strategy for repetitive pinched injections on a microfluidic device. *Anal. Chem.* 76, 6053, 2004.
26. Jacobson, S. C., Hergenroder, R., Moore, A. W., and Ramsey, J. M. Precolumn reactions with electrophoretic analysis integrated on a microchip. *Anal. Chem.* 66, 4127, 1994.
27. Roper, M. G., Shackman, J. G., Dahlgren, G. M., and Kennedy, R. T. Microfluidic chip for continuous monitoring of hormone secretion from living cells using an electrophoresis-based immunoassay. *Anal. Chem.* 75, 4711, 2003.
28. Ye, M., Hu, S., Quigley, W. W., and Dovichi, N. J. Post-column fluorescence derivatization of proteins and peptides in capillary electrophoresis with a sheath flow reactor and 488 nm argon ion laser excitation. *J. Chromatogr. A* 1022, 201, 2004.
29. Bowser, M. T. and Kennedy, R. T. *In vivo* monitoring of amine neurotransmitters using microdialysis with on-line capillary electrophoresis. *Electrophoresis* 22, 3668, 2001.
30. Ocvirk, G., Tang, T., and Harrison, D. J. Optimization of confocal epifluorescence microscopy for microchip-based miniaturized total analysis systems. *Analyst* 123, 1429, 1998.
31. Lee, T. T. and Yeung, E. S. High-sensitivity laser-induced fluorescence detection of native proteins in capillary electrophoresis. *J. Chromatogr.* 595, 319, 1992.
32. Zhang, X., Stuart, J. N., and Sweedler, J. V. Capillary electrophoresis with wavelength-resolved laser-induced fluorescence detection. *Anal. Bioanal. Chem.* 373, 332, 2002.
33. Tao, L., Thompson, J. E., and Kennedy, R. T. Optically gated capillary electrophoresis of *o*-phthaldialdehyde/B-mercaptoethanol derivatized amino acids for chemical monitoring. *Anal. Chem.* 70, 4015, 1998.
34. German, I., Buchanan, D. D., and Kennedy, R. T. Aptamers as ligands in affinity probe capillary electrophoresis. *Anal. Chem.* 70, 4540, 1998.
35. Fukushima, T., Usui, N., Santa, T., and Imai, K. Recent progress in derivatization methods for LC and CE analysis. *J. Pharm. Biomed. Anal.* 30, 1655, 2003.
36. Kutlan, D., Presits, P., and Molnar-Perl, I. Behavior and characteristics of amine derivatives obtained with *o*-phthaldialdehyde/3-mercaptopropionic acid and with *o*-phthaldialdehyde/*N*-acetyl-L-cysteine reagents. *J. Chromatogr. A* 949, 235, 2002.
37. Shou, M. et al. *In vivo* monitoring of amino acids by microdialysis sampling with on-line derivatization by naphthalene-2,3-dicarboxyaldehyde and rapid micellar electrokinetic capillary chromatography. *J. Neurosci. Methods* 138, 189, 2004.
38. Lada, M. W. and Kennedy, R. T. *In vivo* monitoring of glutathione and cysteine in rat caudate nucleus using microdialysis on-line with capillary zone electrophoresis-laser induced fluorescence detection. *J. Neurosci. Methods* 72, 153, 1997.
39. Toyo'oka, T. Use of derivatization to improve the chromatographic properties and detection selectivity of physiologically important carboxylic acids. *J. Chromatogr. B* 671, 91, 1995.
40. de Bee, T., Velthorst, N. H., Brinkman, U. A. T., and Gooijer, C. Laser-based non-fluorescent detection techniques for liquid separation systems. *J. Chromatogr. B* 971, 1, 2002.
41. Chen, D. Y. and Dovichi, N. J. Yoctomole detection limit by laser-induced fluorescence in capillary electrophoresis. *J. Chromatogr. B* 657, 265, 1994.
42. Holland, L. A. and Leigh, A. M. Amperometric and voltammetric detection for capillary electrophoresis. *Electrophoresis* 23, 3649, 2002.
43. Vandaveer, W. R., Pasas, S. A., Martin, R. S., and Lunte, S. M. Recent developments in amperometric detection for microchip capillary electrophoresis. *Electrophoresis* 23, 3667, 2002.
44. Wallenborg, S. R., Nyholm, L., and Lunte, C. E. End-column amperometric detection in capillary electrophoresis: Influence of separation-related parameters on the observed half-wave potential for dopamine and catechol. *Anal. Chem.* 71, 544, 1999.
45. Chen, G., Zhang, L. Y., and Wang, J. Miniaturized capillary electrophoresis system with a carbon nanotube microelectrode for rapid separation and detection of thiols. *Talanta* 64, 1018, 2004.

46. Babu, C. V. et al. Capillary electrophoresis at the omics level: Towards systems biology. *Electrophoresis* 27, 97, 2006.
47. Servais, A. C., Crommen, J., and Fillet, M. Capillary electrophoresis-mass spectrometry, an attractive tool for drug bioanalysis and biomarker discovery. *Electrophoresis* 27, 2616, 2006.
48. Hsieh, F., Baronas, E., Muir, C., and Martin, S. A. A novel nanospray capillary zone electrophoresis/mass spectrometry interface. *Rapid Commun. Mass Spectrom.* 13, 67, 1999.
49. Moore, A. W. and Jorgenson, J. W. Study of zone broadening in optically gated high-speed capillary electrophoresis. *Anal. Chem.* 65, 3550, 1993.
50. Shackman, J. G., Watson, C. J., and Kennedy, R. T. High-throughput automated post-processing of separation data. *J. Chromatogr. A* 1040, 273, 2004.
51. Presti, M. F. et al. Behavior-related alterations of striatal neurochemistry in a mouse model of stereotyped movement disorder. *Pharmacol. Biochem. Behav.* 77, 501, 2004.
52. Venton, B. J., Robinson, T. E., Kennedy, R. T., and Maren, S. Dynamic amino acid increases in the basolateral amygdala during acquisition and expression of conditioned fear. *Eur. J. Neurosci.* 23, 3391, 2006.
53. Gostkowski, M. L., Wei, J., Okerberg, E., and Shear, J. B. Attomole electrophoretic analysis of catecholamines using copper-catalyzed intramolecular cyclization. *Anal. Biochem.* 303, 199, 2002.
54. O'Brien, K. B. et al. A high-throughput on-line microdialysis-capillary electrophoresis assay for D-serine. *Electrophoresis* 24, 1227, 2003.
55. Gao, L. et al. Determination of nitrate and nitrite in rat brain perfusates by capillary electrophoresis. *Electrophoresis* 25, 1264, 2004.
56. Rainelli, A. and Hauser, P. C. Fast electrophoresis in conventional capillaries by employing a rapid injection device and contactless conductivity detection. *Anal. Bioanal. Chem.* 382, 789, 2005.
57. Yang, W. and Zhang, Z. Fast and direct determination of small anions in proteinaceous samples using low voltage-short tube capillary zone electrophoresis system. *Anal. Lett.* 36, 465, 2003.
58. Kleparnik, K. et al. Ultrafast detection of microsatellite repeat polymorphism in endothelin 1 gene by electrophoresis in short capillaries. *Electrophoresis* 21, 238, 2000.
59. Muller, O., Minarik, M., and Foret, F. Ultrafast DNA analysis by capillary electrophoresis/laser-induced fluorescence detection. *Electrophoresis* 19, 1436, 1998.
60. Moore, A. W., Jr. and Jorgenson, J. W. Resolution of cis and trans isomers of peptides containing proline using capillary zone electrophoresis. *Anal. Chem.* 67, 3464, 1995.
61. Heegaard, N. H. and Kennedy, R. T. Antigen-antibody interactions in capillary electrophoresis. *J. Chromatogr. B* 768, 93, 2002.
62. Jameson, E. E. et al. Detection of G proteins by affinity probe capillary electrophoresis using a fluorescently labeled GTP analogue. *Anal. Chem.* 75, 4297, 2003.
63. Yang, P. et al. Capillary electrophoresis and fluorescence anisotropy for quantitative analysis of peptide-protein interactions using JAK2 and SH2-beta as a model system. *Anal. Chem.* 77, 2482, 2005.
64. Hafner, F. T. et al. Noncompetitive immunoassay of small analytes at the femtomolar level by affinity probe capillary electrophoresis: Direct analysis of digoxin using a uniform-labeled scFv immunoreagent. *Anal. Chem.* 72, 5779, 2000.
65. Hutterer, K. M. and Jorgenson, J. W. Separation of hyaluronic acid by ultrahigh-voltage capillary gel electrophoresis. *Electrophoresis* 26, 2027, 2005.
66. Shackman, J. G., Dahlgren, G. M., Peters, J. L., and Kennedy, R. T. Perfusion and chemical monitoring of living cells on a microfluidic chip. *Lab Chip* 5, 56, 2005.
67. Huynh, B. H., Fogarty, B. A., Lunte, S. M., and Martin, R. S. On-line coupling of microdialysis sampling with microchip-based capillary electrophoresis. *Anal. Chem.* 76, 6440, 2004.
68. Sandlin, Z. D., Shou, M., Shackman, J. G., and Kennedy, R. T. Microfluidic electrophoresis chip coupled to microdialysis for *in vivo* monitoring of amino acid neurotransmitters. *Anal. Chem.* 77, 7702, 2005.
69. Lapos, J. A. and Ewing, A. G. Injection of fluorescently labeled analytes into microfabricated chips using optically gated electrophoresis. *Anal. Chem.* 72, 4598, 2000.
70. Xu, H., Roddy, T. P., Lapos, J. A., and Ewing, A. G. Parallel analysis with optically gated sample introduction on a multichannel microchip. *Anal. Chem.* 74, 5517, 2002.
71. Xu, H. et al. Parallel separations of oligonucleotides with optically gated sample introduction on multichannel microchips. *J. Sep. Sci.* 27, 7, 2004.

72. Rodriguez, I., Zhang, Y., Lee, H. K., and Li, S. F. Y. Conventional capillary electrophoresis in comparison with short-capillary capillary electrophoresis and microfabricated glass chip capillary electrophoresis for the analysis of fluorescein isothiocyanate anti-human immunoglobulin G. *J. Chromatogr. A* 781, 287, 1997.
73. Ludwig, M., Kohler, F., and Belder, D. High-speed chiral separations on a microchip with UV detection. *Electrophoresis* 24, 3233, 2003.
74. Wakida, S-I. et al. On-chip micellar electrokinetic chromatographic separation of phenolic chemicals in waters. *J. Chromatogr. A* 1109, 179, 2006.
75. Beard, N. P., Zhang, C-X., and deMello, A. J. In-column field-amplified sample stacking of biogenic amines on microfabricated electrophoresis devices. *Electrophoresis* 24, 732, 2003.
76. Kameoka, J., Craighead, H. G., Zhang, H., and Henion, J. A polymeric microfluidic chip for CE/MS determination of small molecules. *Anal. Chem.* 73, 1935, 2001.
77. Nagata, H., Tabuchi, M., Hirano, K., and Baba, Y. High-speed separation of proteins by microchip electrophoresis using a polyethylene glycol-coated plastic chip with a sodium dodecyl sulfate-linear polyacrylamide solution. *Electrophoresis* 26, 2687, 2005.
78. Xue, Q. F., Wainright, A., Gangakhedkar, S., and Gibbons, I. Multiplexed enzyme assays in capillary electrophoretic single-use microfluidic devices. *Electrophoresis* 22, 4000, 2001.
79. Zheng, S., Ross, E., Legg, M. A., and Wirth, M. J. High-speed electroseparations inside silica colloidal crystals. *J. Am. Chem. Soc.* 128, 9016, 2006.
80. Yang, X. et al. Comprehensive two-dimensional separations based on capillary high-performance liquid chromatography and microchip electrophoresis. *Electrophoresis* 24, 1451, 2003.
81. Jacobson, S. C., Hergenroder, R., Koutny, L. B., and Ramsey, J. M. High-speed separations on a microchip. *Anal. Chem.* 66, 1114, 1994.
82. Liu, B-F., Hisamoto, H., and Terabe, S. Subsecond separation of cellular flavin coenzymes by microchip capillary electrophoresis with laser-induced fluorescence detection. *J. Chromatogr. A* 1021, 201, 2003.
83. Piehl, N., Ludwig, M., and Belder, D. Subsecond chiral separations on a microchip. *Electrophoresis* 25, 3848, 2004.
84. Seller, K., Fan, Z. H., Flurl, K., and Harrison, D. J. Electroosmotic pumping and valveless control of fluid flow within a manifold of capillaries on a glass chip. *Anal. Chem.* 66, 3485, 1994.
85. Attiya, S. et al. Design of an interface to allow microfluidic electrophoresis chips to drink from the fire hose of the external environment. *Electrophoresis* 22, 318, 2001.
86. Kim, D., Chesler, N. C., and Beebe, D. J. A method for dynamic system characterization using hydraulic series resistance. *Lab Chip* 6, 639, 2006.

16 DNA Sequencing by Capillary Electrophoresis

David L. Yang, Rachel Sauvageot, and Stephen L. Pentoney, Jr.

CONTENTS

16.1 INTRODUCTION

The cell is the basic functional unit of all living systems and the instruction set required to manage all cellular activities is contained within the biopolymer known as deoxyribonucleic acid (DNA). The sequential ordering of paired bases within the DNA polymer codes for the exact instructions that define an organism and determine its ability to survive and reproduce. DNA sequencing is the process of determining the unknown locations and ordering of bases within an organism's genome. Identifying and understanding sequence changes or "DNA variations" among individuals can lead

TABLE 16.1
Milestones in DNA Sequencing by Capillary Electrophoresis

Year	References	Description
1981	[204]	Capillary zone electrophoresis revisited
1985	[31]	High-sensitivity CE-LIF detection
1988	[64]	Single base resolution using cross-linked polyacrylamide gel-filled capillary
1990	[15]	Sequencing using a four-color CE-LIF system
1991–1995	[45–48]	Replaceable gels introduced
1994	[172]	Single base resolution demonstrated in a microfluidic system
1995	[90]	Sequencing performed using noncovalent capillary coating
1995	[175]	Four-color separation demonstrated in a microfluidic channel
1996	[13,123]	Multicapillary four-color CE-LIF
1995–1998	*	Three commercial capillary-based DNA sequencers introduced
1998	[76]	Sequencing with low-viscosity polydimethylacrylamide performed using noncovalent capillary coating
2001 and 2003	[10,11]	Sequence and analysis of the Human Genome working draft published

*Applera Corporation, Beckman Coulter, Inc., GE Healthcare.

to significant improvements in the way physicians diagnose, treat, and take measures to prevent the large number of disorders that affect us.

In the late 1980s and early 1990s, when laboratories involved in DNA sequence analysis were still using labor-intensive, manually loaded slab gels, many people in the biotechnology field recognized several potential process improvements offered by capillary electrophoresis (CE). The most highly valued potential improvement areas include increased throughput, process simplification through automation, and sequencing cost reduction achieved through minimizing the consumption of key reagents. The promise of these significant improvements drove substantial R&D investment in CE and as a result, rapid technological advances were made during the 1990s in the development of replaceable gels, capillary surface chemistries, and sensitive optical detection methods (the reader is directed to several informative references [1–9]). Table 16.1 provides an approximate time-line for several of the advances made in the field of CE that would ultimately prove to be key elements to improving the field of automated DNA sequencing.

At a very simple level, current state-of-the-art sequencing methodology may be viewed to involve two fundamental steps. In the first step, an enzyme catalyzed reaction is utilized to create a complicated mixture of molecules that are structurally very similar to one another. Then, in the second step, these molecules must be rapidly size-sorted and accurately identified at low concentration levels. None of the advances made in the field of CE were more important than the development of robust, high-resolution separation matrices required to size-sort these mixtures and the degree of success quickly realized in this area was quite impressive. Figure 16.1 illustrates rapid progress made during the 1990s in our ability to resolve single-stranded DNA fragments of increasing size using CE. Over roughly a 5 year period, reported read-lengths progressed from a "proof-of-principle" few hundred bases to a much more impressive range of 600–900 bases and commercial efforts in CE ultimately played a significant role in the early completion of the human genome sequencing initiative [10,11]. Today, automated DNA sequencing is the largest single commercial application of CE and CE has effectively displaced slab gel based sequencing methodologies. Significant improvements have been demonstrated in all three of the earlier mentioned process improvement areas, though one may argue that significant opportunity remains for the reduction of expensive reagent consumption.

The key technology areas involved in the maturation of CE-based DNA sequencing can be broken down into the following major categories:

- Sequencing reaction chemistry
- Separation matrix and surface coatings

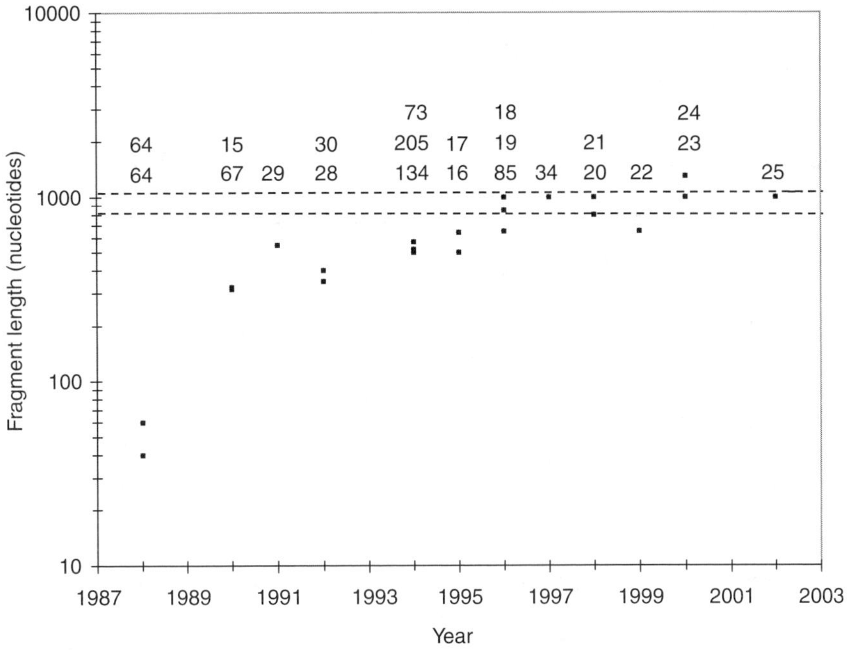

FIGURE 16.1 Graph showing literature reports of single-stranded DNA fragment lengths resolved by capillary gel electrophoresis versus publication year. The area between the two dashed lines identifies the current read-length range typically achieved using commercial capillary-based DNA sequencing systems. Reference numbers are provided above the data points.

- Sample introduction and separation methods
- Excitation/detection systems
- Algorithms

Advances were required in all of these areas in order to make reliable and competitive CE-based DNA sequencing a reality and in the following sections of this chapter we describe fundamental progress made in each of these areas as we survey the related literature.

16.2 CORE CAPILLARY ELECTROPHORESIS SEQUENCING SYSTEM TECHNOLOGIES

16.2.1 DNA Sequencing Reaction Chemistry

16.2.1.1 Sequencing Chemistries

All current CE-based DNA sequencing systems utilize chemistry based on the chain termination method originally described by Sanger et al. [12]. In this approach, a DNA polymerase enzyme is used, in the presence of chain-terminating dideoxynucleoside triphosphates, to catalyze the extension of an oligonucleotide primer hybridized to target DNA, as illustrated in Figure 16.2. As chain extension progresses, a subset of the original target pool is terminated at each base position thus creating a nested set of single-stranded fragments in the denatured reaction product. These fragments range in length from that of the primer (plus one nucleotide) to a few thousand bases, well beyond the upper resolution limit of sequencing matrices developed to date (generally in the range of 1200 bases). Fluorescent reporter molecules are integrated into the sequencing reaction at either the primer or

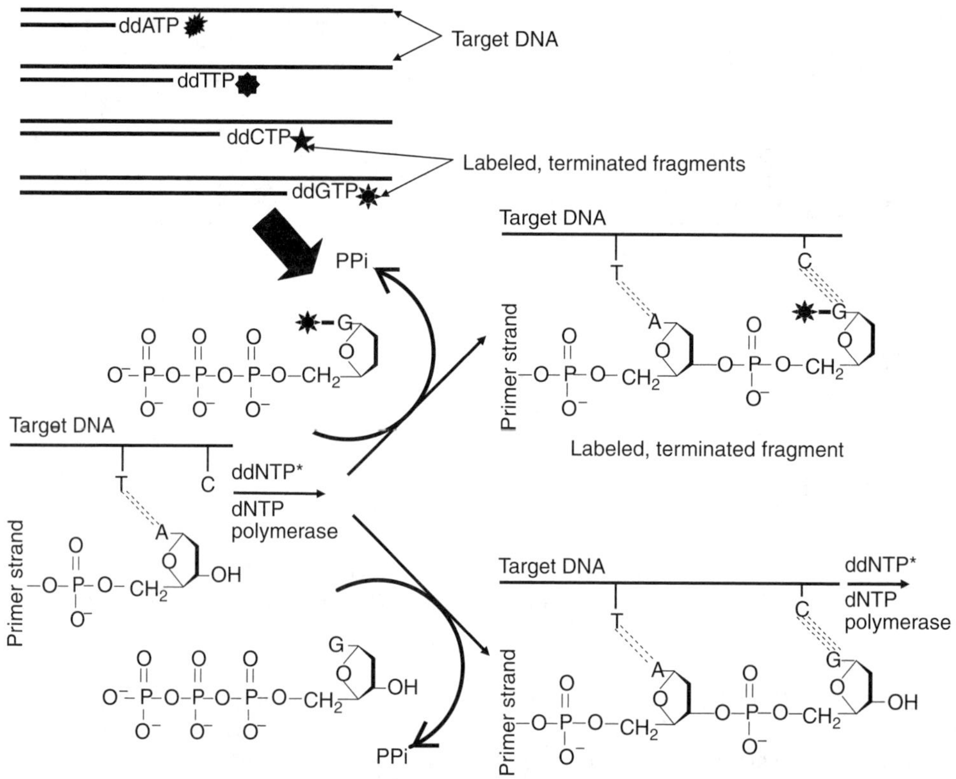

FIGURE 16.2 Graphical depiction of Sanger–Coulson chain terminator sequencing chemistry.

the dideoxy chain terminators. Once the extension reaction is complete, the mixture of fluorescently labeled fragments is purified to remove unincorporated reagents, denatured to free-labeled fragments from the template strand, and loaded into the inlet end of a gel-filled capillary. Under the influence of an applied electric field, the carefully formulated sieving matrix contained within the capillary allows smaller fragments to migrate more freely along the capillary length and so the migrating fragments become size ordered by the time they reach the detection region located near the capillary outlet. The identity of the 3′ terminal base of each migrating fragment is associated with the corresponding fluorescent reporter signal and this signal, combined with the ordering of migration, is used for algorithm-based sequence calling.

During the 1990s, several variations of the chain terminator chemistry were utilized in CE-based sequencing studies. The first of these variations is known as fluor-primer sequencing. Here, a single fluorescently labeled primer is elongated during four separate extension reactions in which only one of the four dideoxy chain terminators is present. The reaction products from each reaction are separated electrophoretically in a capillary, yielding "one-color primer-labeled" sequencing data [13]. The one-color primer sequencing chemistry was widely utilized in separation optimization and gel matrix development studies but was not popular for actual CE-based sequencing of DNA because four separate reactions and separations would have been required to read sequence. In an alternate approach, four separate extension reactions are run, each employing one of the four dideoxy chain terminators and a primer labeled with one of four different fluorophores. The reactions are then combined and separated in a capillary, yielding "four-color primer sequencing" data [14–25]. This approach was more acceptable as it required only one separation be performed, but it still suffered from the requirement of running four separate reactions.

A fundamentally different variation of the dye-primer chain termination chemistry, originally suggested by Tabor and Richardson [26], was explored by several CE groups in the 1990s. Here, relative peak height, rather than the spectral identity of the fluorescent labels, was used to determine sequence. This approach offered the hope of simplified instrumentation and chemistry since only one channel of fluorescence needed to be monitored from a single separation of labeled fragments produced in a single reaction. Tabor and Richardson [26] showed that incorporation of Mn^{2+} into sequencing reaction buffers using T7 DNA polymerase nearly eliminated template sequence-dependent variability in dideoxynucleotide triphosphate (ddNTP) incorporation and thereby resulted in vastly improved peak height uniformity. They exploited this observation in a slab-based sequencing protocol using a single fluorescently labeled primer and different concentrations of the four ddNTPs to determine sequence using relative peak intensity. Ansorge et al. [27] also described using a similar strategy. Using this approach, the advantages of single-lane sequencing were realized without the associated use of four different fluors. Pentoney et al. [28] and Dovichi et al. [29,30] investigated the possibility of adapting the Tabor–Richardson type approach for DNA sequence determination for capillary gel electrophoresis (CGE) with laser-induced fluorescence (LIF) detection [31,32].

Pentoney et al. [28] demonstrated that Tabor–Richardson type sequencing could be effectively utilized in combination with a relatively simple and inexpensive CE/LIF system. Here the authors modified the Tabor–Richardson approach so that only three ddNTPs are included in the sequencing chemistry. Each template was sequenced twice using two different sets of terminator concentrations and both reactions were analyzed simultaneously. The position of the fourth nucleotide in the sequence was determined from the location, size, and shape of gaps appearing in the electropherogram in between peaks produced by the three ddNTP terminators present in the reaction. This simplified the software task of discerning between the different intensity levels and more efficiently utilized the available system dynamic range. The authors found that careful selection of the terminator concentrations and algorithm-based comparison of the two complementary data sets resulted in a robust sequence determination.

The current state-of-the art chemistry approach utilizes four dideoxynucleotides, each labeled with a different fluorophore, to generate fragments from a single reaction [33], which are also analyzed in one capillary (four-color dye-terminator or "single tube" sequencing) [34]. This approach eliminates the need to fluorescently label and purify different primer sets, the sequence of which varies with each target. Note that slab gel based automated sequencers using one or more of these chemistry approaches were originally developed by Pharmacia, Hitachi, EG&G, Applied Biosystems, and Du Pont. CE-based automated DNA sequencers have now been commercialized by Applera Corporation, Beckman Coulter, Inc., and GE Healthcare. All the CE-based systems currently utilize the four-color dye-terminator chemistry in combination with nondiscriminating thermal stable DNA polymerases that allow linear amplification of reaction product via a chemistry protocol known as "cycle sequencing." Cycle sequencing refers to a modified version of the above described Sanger–Coulson sequencing chemistry in which a thermal stable DNA polymerase is used in order to repetitively run the extension reaction in a cyclic fashion. The thermal stable DNA polymerases used today can be repeatedly heated to 95°C and still retain their enzymatic activity. Through the use of repeated temperature cycles (annealing, extension, denaturation), it becomes possible to repeat the extension reaction over and over again in the same reaction vessel, each time generating more reaction product from the same sample target. This cycle sequencing method results in an n-fold increase in reaction product, where n is the number of times the cycle is repeated. Figure 16.3 depicts today's typical DNA sequencing chemistry process flow.

16.2.1.2 Fluorescent Dyes

Fluorescent dye sets used in capillary sequencing attached to either primers or terminators have been developed for gas, solid state, and diode lasers, and they span the spectral range from roughly 488 to 800 nm. The basic requirements of a fluorescently labeled set of four terminators are demanding and include high-extinction coefficients at wavelengths for which reasonably low-cost lasers exist, high-quantum yields for fluorescence, sufficient separation of emission maxima for confident identification, reasonable aqueous buffer solubility, good photostability, minimal impact on enzyme

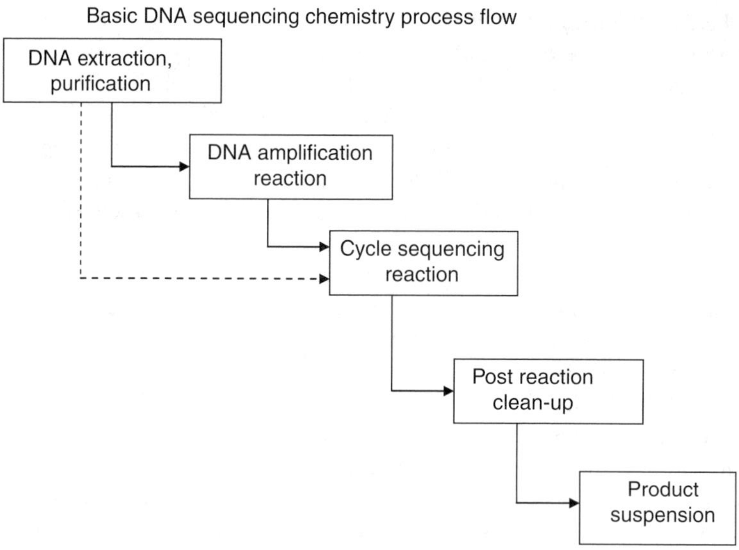

FIGURE 16.3 High-level graphical depiction of the steps involved in running a typical Sanger–Coulson sequencing chemistry.

incorporation efficiency, and uniform mobility effects upon labeled fragments generated in the Sanger–Coulson chemistry. No truly perfect set of fluorophores has been developed for DNA sequencing and algorithm-based data correction is utilized to compensate for some of the above concerns.

Four-color DNA sequence determination involves the accurate recognition of one of four different emission profiles at each base position. In addition, in some instances, such as the sequencing of mixed viral strains, it is necessary to identify the presence of two or more bases at a single-base location in the electropherogram. To minimize spectral cross talk, families of four dyes have generally been selected to have emission maxima that differ by about 30 nm. Two different approaches have been developed to efficiently excite these four-dye families. The first approach is to assign two pairs of dyes to two different laser excitation sources. This allows the excitation wavelength that is used to stimulate each dye pair to be located reasonably near the corresponding dye absorbance maxima. The two laser excitation lines are usually alternately modulated in this scheme. A second approach is to exploit the use of fluorescence energy transfer [16,35–37]. Here, a common absorber molecule is used to efficiently couple each of the four labels with a single laser line and a large fraction of the absorbed laser energy is then rapidly transferred to one of four different acceptor dyes. In this manner, a single laser source can be used to sensitively address four different labels having very different emission maxima.

16.2.2 Surface Coating and Separation Matrix

16.2.2.1 Capillary Characteristics and EOF Suppression

Most CE-based DNA sequencing separations are performed using fused-silica capillaries and some have been reported for channels etched into glass plates. Since DNA sequencing separations are performed at alkaline pH (~8.5) to provide a negatively charged DNA phosphate backbone, the acidic capillary surface silanol groups will also be ionized. This leads to the formation of negatively charged surface silanol groups and a charge imbalance near the capillary inner wall. Mobile, hydrated cationic species attracted to this negatively charged zone define a region of net positive charge density that decreases in magnitude with increasing distance from the capillary wall. Under the influence

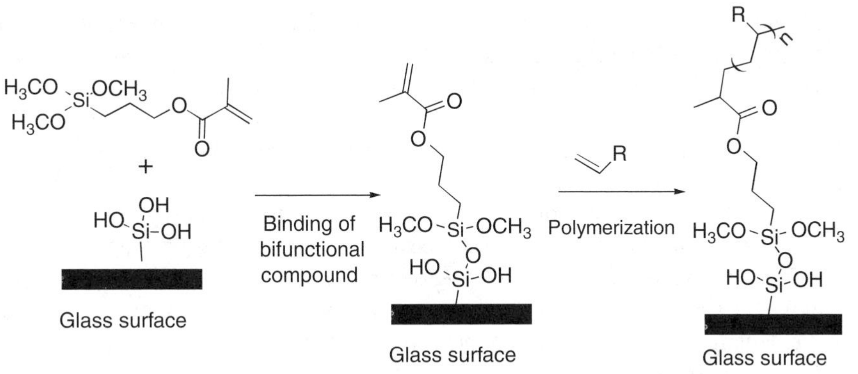

FIGURE 16.4 Inner surface of a capillary derivatized using the chemical method described by Hjertén to eliminate electroendosmosis and surface adsorption of solutes.

of an applied electric field, the hydrated cationic groups near the capillary wall will migrate toward the cathode (negatively charged electrode) and will effectively drag bulk fluid along with them in an interesting pumping phenomenon known as electroendoosmotic flow (EOF) [38]. Unless eliminated, EOF will lead to gel movement and results in very poor separation quality.

Several methods exist to suppress EOF. The earliest method, described by Hjertén in 1985, involved covalent attachment of a hydrophilic masking layer of noncross-linked polyacrylamide to the inner capillary wall in order to prevent both EOF and solute adsorption in free zone and isoelectric focusing separations of proteins. In this original work, the bifunctional compound γ-methacryloxypropyltrimethoxysilane was covalently bonded to the inner surface of a glass capillary and the other end of this bridging molecule was then reacted with acrylamide (AA) monomer undergoing linear polymerization, as illustrated in Figure 16.4 [39]. Several other combinations involving bifunctional bridging groups and noncross-linked polymers may be used in a similar manner. Additional covalent attachment chemistries have since been developed to mask the surface silanol groups [40–43].

A second simpler method that has been used to eliminate EOF is to mask the surface silanol groups using an adsorbed polymer. This "self-coating" or "dynamic-coating" method will be discussed in more depth in a later section of this chapter.

16.2.2.2 Role of the Separation Matrix

The separation matrix truly lies at the core of CE-based DNA sequencing. Since DNA fragments have constant size to charge ratios (each additional base adds another phosphodiester linkage and hence another unit of charge), both the force acting upon each fragment and the frictional drag experienced by each fragment increase linearly with size during electrophoresis performed in free solution. This effect causes the DNA mobility to be size-independent [44], and so a sieving matrix is required in order to separate DNA fragments of differing lengths from one another. We should keep in mind that the basic separation requirement in DNA sequencing is to rapidly resolve single-stranded DNA over a range of approximately 20 bases to more than a thousand bases with single-base resolution. This is a well-defined nucleic acid mixture, but a very demanding separation indeed.

One of the most important technical hurdles to overcome was the development of a practical solution to the capillary separation matrix problem. Early capillary gel work involved the adaptation of cross-linked polyacrylamide gels, used for years in slab-based DNA sequencing separations, to the capillary format. In cross-linked gels, a DNA mixture is sieved through pores formed by a network of polymer branches that are covalently fixed. The size selectivity of this network is tuned through variation of both the monomer (%T) and cross-linker (%C) concentrations. Cross-linked gels were introduced into capillaries by initiating polymerization in an external vessel and then driving the flowable, prepolymerized solution into the capillary with a simple syringe. These "fixed" capillary matrices were capable of resolving DNA fragments but suffered from frequent failure modes including sharp current drops due to intermittent bubble formation. As a result, use-lives were often as short as one or two runs [45,46] and fixed capillary gels proved to be an impractical solution for automated DNA sequencing.

The development of replaceable noncross-linked or linear polyacrylamide gels for DNA sequencing was probably the single most significant advance made in the field of CE during the 1990s [45–48]. In replaceable gels, a dynamic network of entangled linear polymers forms the pores through which the DNA mixture is sieved. The size selectivity of this noncross-linked network is tuned through variation in the length and concentration of the polymers defining the replaceable matrix. Replaceable gel matrices for CE were described in 1991 for the separation of oligodeoxycytidylic acids ("poly-C") 10–15 and 24–36 bases long [47] and analysis of double-stranded restriction digest products [49]. These gel matrices were not only capable of resolving sequencing fragments with single base resolution over a wide size range but also easily replaced, thereby eliminating the need for frequent capillary removal or capillary inlet trimming. A number of reviews provide detailed discussions of the theory of DNA separation in gels as well as a description of the various separation matrices that have been evaluated for DNA sequencing [2,44,50–52], especially in References 53–56.

But what characteristics or physical properties of a replaceable sieving gel are most important in terms of resolving single-stranded DNA over such a wide size range? And what metrics have proven useful in the optimization of gel formulations and separation parameters? Quesada and Menchen [55] have published a helpful theoretical discussion of the parameters affecting the performance of semi-dilute replaceable polymer systems used in capillary-based DNA sequencing. It turns out that for a given separation matrix polymer, the separation speed and resolving power of the entangled gel matrix are generally dependent on polymer concentration. Below a critical or "overlap" concentration c^*, little or no interaction between matrix polymer strands exists. The overlap concentration is generally reached when the space volume $4\pi(R_G^3)/3$ (where R_G is the radius of gyration of the polymer) is occupied by exactly one polymer strand [55,57]. Above c^*, the polymers begin to impinge upon one another and the mesh size of the gel matrix becomes inversely proportional to polymer concentration. Mesh size in an entangled polymer system may be thought of as the average distance between polymer–polymer entanglements. For linear polyacrylamide studied at different concentrations above c^* ($C > c^*$), the mesh size was observed to follow the relationship

$$\xi = 2.09C^{-0.76\pm0.03}, \tag{16.1}$$

where ξ is the mesh size (in angstroms) and C is the polyacrylamide concentration (in gm/mL) [2,58]. As an example, a 3.5% noncross-linked linear polyacrylamide (LPA) gel would correspond to an estimated average pore size of ~27Å using Equation 16.1.

The past 10 years of capillary gel research has indicated that optimum separation efficiency and maximum read-length are achieved through the use of highly entangled systems of hydrophilic, high-average molecular weight polymers. Typical replaceable gel systems utilize polymer molecular weights on the order of 1–8 million Da and concentrations on the order of 2.5–6%. The polymer molecular weight distribution, polymer concentration, column length, field strength, and separation temperature are typically settled upon empirically to yield an optimum compromise between separation quality and analysis time.

Several models exist to describe the migration of DNA fragments through the separation matrix. In the Ogston model [59], often applied to cross-linked gels, the DNA fragments are treated as migrating rigid spheres traversing variously sized pores within the matrix; smaller "spheres" encounter more pores with equal or greater diameters and migrate more quickly than larger "spheres." The DNA

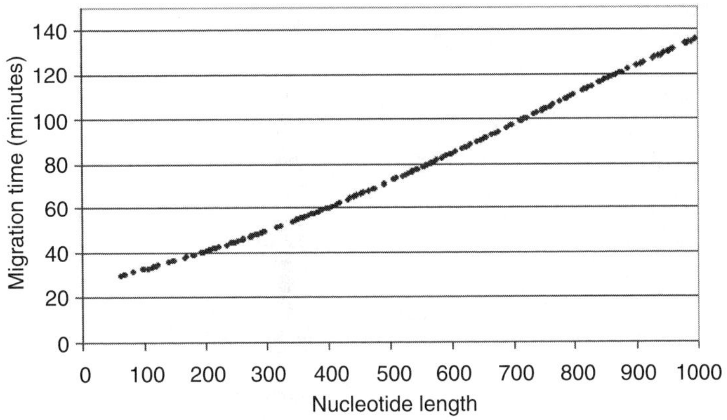

FIGURE 16.5 Graph showing the proportional relationship between migration time and DNA fragment length for a replaceable capillary gel separation of dye-terminator labeled sequencing fragments. (pUC 18 template DNA, 1009 bases called to 98% accuracy, run on a CEQ analysis system from Beckman Coulter).

fragment mobility in the separation matrix is approximated by the following equation:

$$\mu = \mu_0 \exp(-K_r C) \tag{16.2}$$

where μ is the mobility in the separation matrix, μ_0 is the mobility in free solution, K_r is the retardation factor and is proportional to both the radius of the matrix polymers and the radius of gyration of the DNA fragment, and C is the polymer concentration. This model is valid only for DNA fragments in the approximate size range of 50–300 bases in polyacrylamide gels [56,60].

A second model known as biased reptation with fluctuations (BRF) is an extension of the biased reptation model (BRM), and treats the DNA fragment not as a solid sphere but as a long flexible tube that snakes through the pores of the separation matrix [61,62]. The electrophoretic migration times are predicted to be directly proportional to DNA fragment size, as observed experimentally [50] and illustrated in Figure 16.5. A large number of reports have confirmed empirically that DNA fragment mobility in well-formulated and properly run gels is inversely related to molecular weight over a size range of roughly 20–1000 bases, after which mobility asymptotically approaches a minimum and the remaining fragments are observed to comigrate. Interested readers are directed to Viovy [54] for a detailed description of the physical mechanisms of DNA separations.

Another important feature of the separation matrix is its ability to relieve any secondary structure that arises due to intramolecular base pairing in the single-stranded DNA molecules. The presence of such regions of secondary structure leads to increased electrophoretic mobility of DNA fragments and hence congested regions in the electropherogram known as "compressions." These compressed regions make accurate base-calling difficult or impossible. To alleviate this problem gels are typically formulated with the addition of urea, formamide, or other denaturing compounds [25,63]. The use of elevated temperature also helps to disassociate the regions of secondary structure.

16.2.2.3 Evolution of Separation Matrices

16.2.2.3.1 Cross-linked versus Linear Acrylamides

Capillary gel electrophoretic separation of DNA fragments at single base resolution was first demonstrated by Karger and coworkers in 1988 [64] and this work was expanded in 1990 [65]. Extending

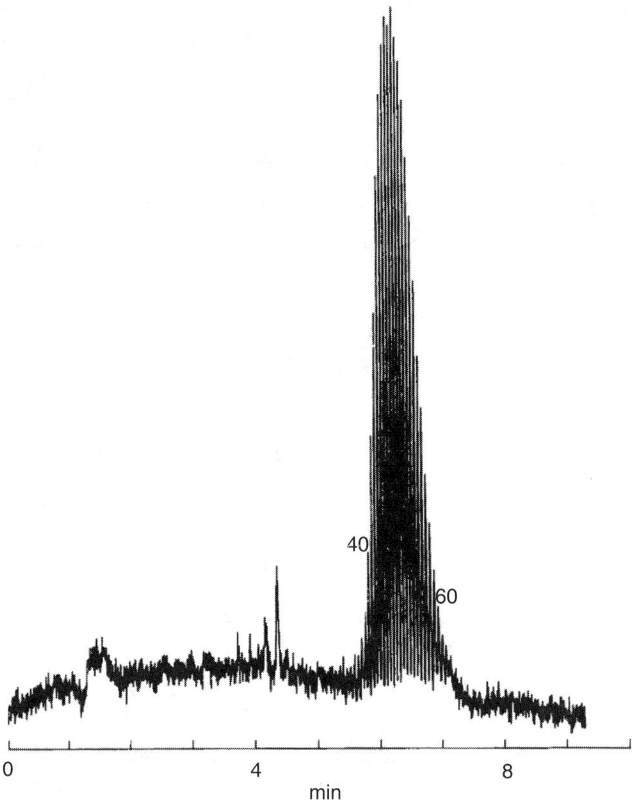

FIGURE 16.6 Capillary electrophoretic separation with single base resolution of poly-A 40–60 bases long. (Reprinted with permission from Cohen, A.S., et al., *Proc. Natl Acad. Sci. USA*, 85, 9660, 1988.)

what had been learned from slab gel work, a cross-linked polyacrylamide gel (T = 7.5%, C = 3.3%) was formed in a 75 μm i.d. column having an effective length of 13 cm. Using an absorbance-based detector, the individual fragments in a sample containing polydeoxyadenylic acid homopolymers (poly-A) 40–60 nt in length were resolved from one another with efficiencies equivalent to 5×10^6 plates per meter in less than 8 min (Figure 16.6). Because of this high-resolving power, the authors accurately predicted that "capillary gel electrophoresis [may] be used as a tool for oligonucleotide sequencing" as well as a tool for "rapid assessment of purity" [65]. Paulus et al. [66] demonstrated single base resolution for samples containing polydeoxythymidilic acid 20–160 nt long using a cross-linked T = 2.5%, C = 3.3% matrix. Subsequent work using fluorescence detection also demonstrated DNA fragments with single base resolution [15,29,67–71]. These separation gels typically consisted of a Tris–borate buffer at pH values >8.0 and with 7–8 M urea used as a denaturant.

However, the cross-linked nature of these separation matrices limited reuse since the polyacrylamide degraded, bubbles or voids tended to form in the gels and impurities injected with the sample accumulated in the capillaries, all leading to instability and unacceptably short average use-life [45,46]. In an attempt to improve capillary use-life, Righetti and coworkers [72] polymerized AA in a capillary without a cross-linker (T = 10%, C = 0%) and then treated the resultant media with cysteine to scavenge any remaining monomer, since this monomer is both toxic and potentially reactive. The walls of the capillary were pretreated such that during matrix polymerization, polyacrylamide strands were also bonded to the inner wall thus eliminating EOF. The authors reported that these matrix-filled capillaries were expected to have an improved use-life of approximately 2 weeks.

The use-life issue was resolved by replacing the medium within the capillaries after each use using a non cross-linked gel. Bocek separated a sample of poly-C 10–15 and 24–36 bases long in a replaceable gel created from a solution containing 10% AA [47] while Karger and coworkers

[45] used replaceable gels to analyze dye-primer labeled DNA sequencing reaction products. In the latter work, Karger and coworkers [45,46] polymerized a degassed solution containing 6% w/v AA monomer in a Tris–borate–EDTA (TBE) buffer containing 7 M urea with ammonium persulfate (APS) and N, N, N', N'-tetramethylethylenediamine (TEMED) in 100 μL syringes. Linear polyacrylamide (LPA) viscosities, ranging from 160 to 3600 centipoise, as measured with a falling ball viscometer, were found to be inversely proportional to the radical initiator concentration while migration times were found to be proportional to viscosity, reflecting the size (or molecular weight) of the LPA strands. LPA in this viscosity range required pressures of the order of 1200 psi to be forced through capillaries 33 cm long with an inner diameter of 75 μm, well within the range of a syringe pump [45], thus allowing the matrix to be replaced after every run. Using LPA with an average molecular weight of approximately 1×10^6 Da (determined via light scatter) and fluorescently labeled primer samples, fragments >370 bases in length could be identified using capillaries that were modified via the covalent masking method of Hjertén [39].

Subsequent work using noncross-linked LPA matrices and run under more optimized conditions led to incremental increases in read-length. Although the separation medium was not replaced after each run, Dovichi and coworkers [73] reported reads of up to 570 bases in less than 140 min when a primer-labeled sample was separated at room temperature using a matrix created from a 6% AA solution polymerized directly in capillaries treated with a solution of [(γ-methacryloyloxy) propryl] trimethoxysilane to bind a surface layer of polyacrylamide to the capillary walls, similar to the method of Hjertén [39]. Read-lengths of up to 640 bases were achieved in 2 h run times at 60°C using a capillary containing a matrix consisting of a 5% AA solution [17].

Karger and coworkers [18] using a replaceable separation matrix introduced into the capillary before each run, performed separations at 50°C at a field strength of 150 V/cm using a 2% AA solution that yielded >1000 bases for a set of dye primer-labeled sequencing reaction fragments. Separations performed in the temperature range 50–60°C were reported to yield higher resolution for large fragments, decreased separation times, and longer read-lengths [19]. A separation matrix created by mixing LPA of two different molecular weight distributions, 0.5% (w/w) of 270 kDa and 2% (w/w) 17 MDa, operated at 70°C with a field of 125 V/cm yielded read-lengths of ~1300 bases in just over 2 h for primer-labeled sequencing reaction fragments [24].

Synthetic procedures incorporating inverse emulsion polymerization were later used to create a fine powder of high-purity LPA with a molecular mass of 9 MDa [20]. Separations performed using a matrix created from a combination of this powder and polymer having a molecular weight of 50 kDa resulted in read-lengths of greater than 1000 bases in runs requiring less than 60 min [21]. More recently, Barron and coworkers [74,75] showed promising results using very sparsely cross-linked "nanogel" matrices. The authors used very low concentrations of methylene bisacrylamide (Bis) cross-linker ($\sim 10^{-4}$ mol%) in an emulsion polymerization of AA, to produce polymers with molecular weights of ~10–11 MDa. The low level of cross-linking was carefully controlled to retain polymer solution fluidity, but was apparently high enough to stabilize pores within the matrix, thereby allowing separations of large DNA fragments to occur with higher resolution than is typically observed using noncross-linked LPA [75]. The authors reported that the improvement in separation quality for the larger single-stranded fragments led to an increase in read-length of approximately 19% [75].

To date, LPA remains a very popular separation matrix for DNA sequencing, and is used on the family of sequencers and genetic analyzers available from both Beckman Coulter and GE Healthcare. Figure 16.7 illustrates a typical separation of dye terminator-labeled DNA sequencing reaction products performed using the CEQ™ system from Beckman Coulter and a linear polyacrylamide based replaceable gel matrix. In this run, 1009 bases were called with 98% accuracy.

16.2.2.3.2 Other Matrix Materials

A gel matrix based upon the use of polydimethylacrylamide (PDMA) has also enjoyed great commercial success [76,77]. Developed by Madabhushi and coworkers at Perkin Elmer Corporation (now Applera Corporation) and used in their family of high-throughput DNA sequencers, PDMA

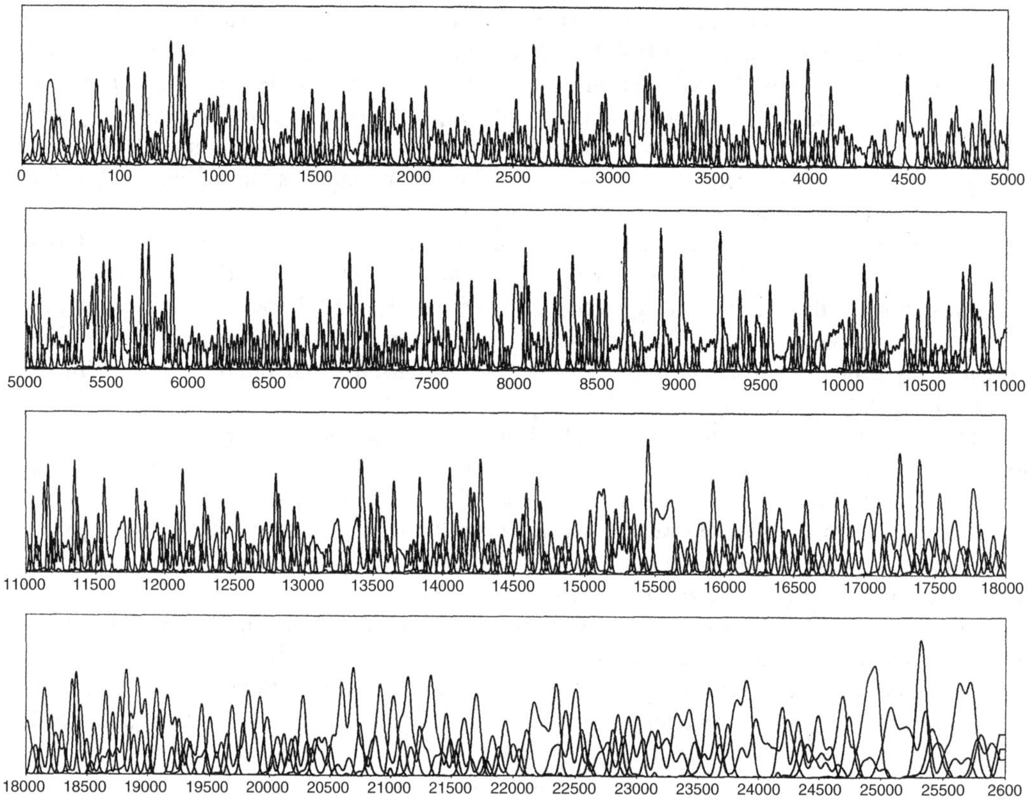

FIGURE 16.7 **(See color insert following page 810.)** Analyzed four-color dye-terminator sequencing from a CEQ analysis system (Beckman Coulter): 1009 bases called to 98% accuracy.

polymers of various molecular weights were created by polymerization of *N*, *N*-dimethylacrylamide (DMA) solutions with APS and TEMED at 50°C and purification by precipitation. The resulting polymer had molecular weights in the range 20–200 kDa. A 6.5% PDMA (98 kDa) solution was reported to have a low viscosity of 75 cP and yielded read-lengths of 600 bases. In addition, since PDMA is reported to be more hydrophobic than LPA, the segmental adsorption energy of PDMA onto fused silica is greater than that of LPA [76]. Consequently, PDMA will adsorb to capillary walls and mask the surface silanol groups, significantly reducing EOF and eliminating the task of having to chemically modify the capillary walls. Madabhushi [76] reported performing over 100 runs, replenishing the gel before each run, without ever having to recondition the capillary. Heller [78,79] performed a detailed study of the utility of PDMA in DNA separations by creating polymers with various molecular weights for use in optimizing polymer concentration, polymer chain length, and electric field strength. Chu and coworkers [80] achieved read-lengths of up to 1000 bases in 96 min at room temperature using 2.5% w/v PDMA with a molecular weight of 5.2 MDa. Although the added analysis time would be impractical, preconditioning of PDMA, by subjecting the gel medium to an electric field strength of 162–320 V/cm for at least 6 h before separation, was reported to improve peak separation and was predicted to extend the read-length [81].

Barron and coworkers [82] reported success using replaceable poly *N*-hydroxyethylacrylamide (PHEA) as both a separation matrix for DNA sequencing and a dynamic coating that eliminated EOF. Using bare fused-silica capillaries, 4–6% solutions of PHEA with a molecular weight of ~5.2 MDa and field strengths ranging from 40 V/cm to 117 V/cm, read-lengths of 445–750 bases were achieved with 98.5% accuracy. Separation times of 1–5 h were reported in this study.

Chu and coworkers [83] combined LPA and PDMA in an attempt to obtain a gel with the best properties of each polymer, namely high-quality separations using uncoated capillaries. They observed that mixtures containing PDMA with high molecular mass or in high concentrations adversely affected the quality of separations. A mixture of 2.5% (w/v) of LPA with mass 2.2 MDa and 0.2% PDMA with mass 8 kDa resulted in a matrix that produced read-lengths of ~700 bases using bare (untreated) capillaries [83]. To more intimately entangle the LPA and PDMA polymers, the group polymerized PDMA monomer in a homogeneous solution of LPA, thus creating a "quasi-interpenetrating network" or quasi-IPN [84]. Using single-color dye-primer separations, they reported sequencing up to 1200 bases in less than 60 min at 65°C using 2% quasi-IPN formed by LPA with a molecular weight of 9.9 MDa and an AA to DMA ratio of ~11 : 1.

Although AA-based matrices dominate the commercial field of DNA sequencing today, other polymers have been tested with varied results. Bashkin et al. [85], Marsh et al. [86], and Dolnik and Gurske [87] reported having some success using hydroxyethylcellulose (HEC), with Bashkin et al. reporting base calls to 500 bases in less than 60 min using a matrix composed of 2% HEC in the molecular weight range 90,000–105,000 using Hjertén-modified capillaries [85]. Kheterpal and Mathies [88] compared the read-lengths achieved using HEC matrix and LPA matrix in sequencing and found LPA read-lengths to be ~1000 bases while that of HEC was limited to approximately 600 bases.

Yeung and coworkers [34,89–93] demonstrated success using poly(ethyleneoxide) (PEO) as a sieving matrix. PEO is an attractive alternative to LPA in the sense that PEO is available commercially over a broad and high-molecular weight range, eliminating the need to synthesize the long chain polymer [89]. Similar to PDMA, PEO polymer adsorbs to the capillary walls and masks the surface silanol groups. However, Yeung's group [90,94] found that periodic capillary regeneration with HCl or polyvinylpyrrolidone (PVP) [93] was required to maintain good performance. The authors reported separations of >900 bases in less than 110 min using a 2.5% solution of PEO with a molecular weight of 8 MDa and separation performed at 40°C using a capillary effective length of 40 cm, an inner capillary diameter of 75 μm and a field strength of 160 V/cm [93]. Yeung and coworkers [95,96] also reported using PVP as the separation matrix in uncoated capillaries and demonstrated read-lengths approaching 350 bases. PVP-based separation media have the dual advantage of masking surface silanol groups and possessing an extremely low viscosity (<30 cP) [76,93].

Dolnik and coworkers [97] used a novel, natural material to create a sieving matrix for capillary sequencing. Guar gum, or guaran, is a polysaccharide originating from the endosperm portion of the legume seed (*Cyamopsis tetragonoloba*) that grows mainly on the Indian subcontinent and in some parts of Texas and Oklahoma and contains 75–85% galactomannan. In this sequencing application, purified 2.1 MDa guaran was dissolved to a final concentration of 15 g/L in a urea/Tris/HEPES buffer, pumped into a capillary array and run using the MegaBACE system available from GE Healthcare. Separation of sequencing fragments using this matrix, as well as matrices using galactomannan from tara gum and locust bean gum exhibited poorer efficiencies than did separations performed using LPA; read-lengths were generally in the range of 600–700 bases (98.5% accuracy) [97].

Menchen et al. [98] created self-assembling polymers for sieving DNA by capping polyethylene glycol (PEG) molecules with micelle forming fluorocarbon tails. In solution just above the critical micelle concentration, these molecules are believed to form intramolecular micellular structures and, at even higher concentrations, intermolecular micellular structures are believed to form. Reasonably good resolution was obtained for fragment lengths up to ~500 bases using a 6% solution of a 1:1 mixture of polymers created from PEG35000 end-capped with $(C_6F_{13})_2$ and PEG35000 end-capped with $(C_8F_{17})_2$ [98].

Several new copolymer formulations have recently been created [99]. In the first type of copolymer, AA monomer was copolymerized with a second monomer type, creating a polymer with units of the two monomers randomly distributed. Chu and coworkers [100] created AA/DMA random copolymers with molecular weights of ~2.2 MDa, starting with 3:1, 2:1, and 1:1 molar ratios of AA and DMA. Using a 2.5% w/v copolymer created with either 3:1 or 2:1 molar ratios, reasonably

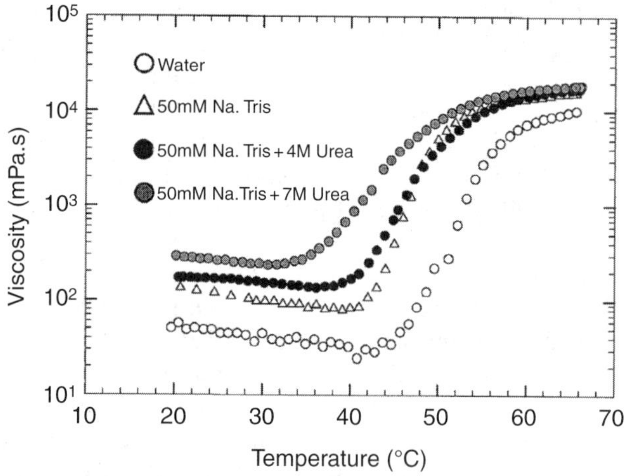

FIGURE 16.8 Viscosity change as a function of temperature for thermal associating gels with different additives. (Reprinted with permission from Sudor, J., et al., *Electrophoresis*, 22, 720, 2001.)

good resolution up to 700 bases, with visual identification of ~900 bases, was achieved in room temperature separations of primer-labeled fragments using uncoated, bare fused-silica capillaries. Barron and coworkers [101] copolymerized DMA with *N*,*N*-diethylacrylamide (DEA) at different molar ratios. These copolymers generally performed poorly in comparison to either LPA or PDMA homopolymers in terms of resolution, with performance suffering as the relative amount of DEA increased, indicating the role that polymer hydrophobicity plays in separation quality [102]. Although performance suffered with increasing amounts of DEA, the viscosities also decreased, possibly indicating a tradeoff in performance and ease-of-use [101].

A second class of copolymer is the "thermo-associating" or "thermo-gelling" copolymer [103–107]. These are made up of two polymers, one of which exhibits a phase change across a critical temperature known as the lower critical solution temperature (LCST). This phase change is characterized by a change in turbidity and/or viscosity [103,105]. These thermo-associating matrix materials have the interesting and useful property of switching from a low-viscosity state to a high-viscosity state upon increasing temperature from ambient (Figure 16.8). This facilitates easy matrix replenishment at ambient temperature while providing a good separation environment at higher temperatures, where DNA sequencing separations are normally performed. Examples include the "comb-like" copolymers consisting of poly (*N*-isopropylacrylamide) (PNIPAM) groups [106] or PDMA groups [107] grafted onto an LPA backbone and copolymers of *N*-ethoxyethylacrylamide and *N*-methoxyethylacrylamide [105]. The thermal-thickening behavior of the comb-like copolymers is primarily attributed to the behavior of the LCST units: at lower temperatures, little interaction between or within the backbones or grafts exists. As the temperature approaches the LCST, the grafted units begin to associate with one another, forming the aggregates that provide sieving during separation (Figure 16.9). Presumably due to interactions of the grafted PNIPAM or PDMA units with the capillary walls [107], the comb-like polymers may be run on bare fused-silica capillaries. Separations made with the PDMA grafts yielded resolutions and migration times comparable to those of LPA [107].

16.2.3 Excitation/Detection Methods: The Road to Four-Color Multicapillary Systems

The detection of DNA fragments in a capillary separation channel is a challenging proposition [5,108]. Injected sample quantities for each (ultimately) resolved band are typically in the sub nanomolar to picomolar range and detection volumes are on the order of a 100 pL. Peak efficiencies

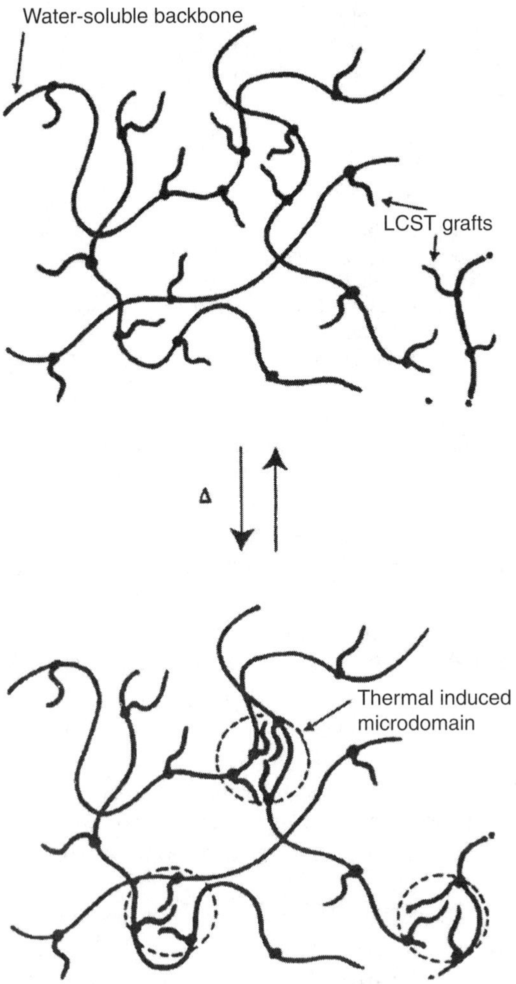

FIGURE 16.9 Simplified depiction of the mechanism for thermo-thickening. Upon heating above the LCST, grafted side chains become less soluble and tend to aggregate with one another thereby forming a transiently cross-linked structure. (Reprinted with permission from Sudor, J., et al., *Electrophoresis*, 22, 720, 2001.)

of $\sim 10^5$–10^6 are required for single base resolution. For a 30 cm capillary, these high-efficiency numbers correspond to peak widths of approximately 500 μm at the detector window. LIF detection [31] is the method of choice for capillary-based DNA sequencing because it extends to extremely low limits of detection, is compatible with the narrow capillary bore confines, is easily extended to multiwavelength monitoring, and is readily compatible with Sanger–Coulson sequencing chemistries (both primer and terminator labeling). For example, the CEQ-family of four-color DNA analysis systems (Beckman Coulter) utilizes diode lasers that are focused to an elliptical spot approximately 60 μm in length along the capillary axis. Within the corresponding 120 pL illuminated volume, approximately 220 molecules of fluor-labeled DNA in 120 pL illuminated volume may be detected at a signal-to-noise ratio (SNR) of 2 using LIF detection (unpublished data).

16.2.3.1 Single-Capillary On-Column Detection Systems

Several groups have designed on-column LIF detection systems for capillary-based sequencing. Excitation and emission light is typically passed through a short portion of the capillary from which

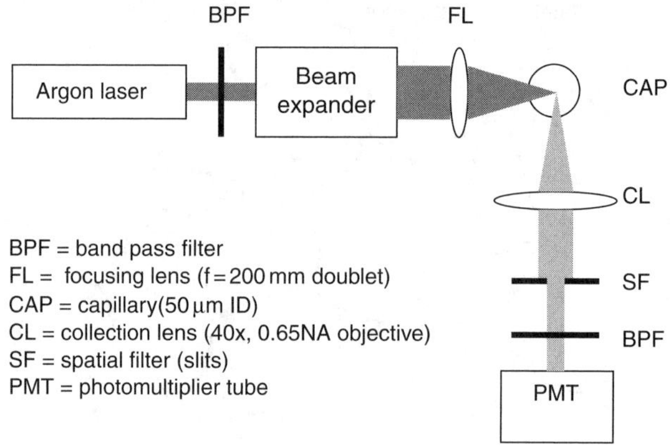

FIGURE 16.10 Single-color on-column CE/LIF detection system. (Reprinted with permission from Drossman, H., et al., *Anal. Chem.*, 62, 900, 1990. Copyright 1990 American Chemical Society.)

the polyimide cladding has been removed, usually by burning or charring. The effective separation length of the column is defined as the distance from inlet tip of the capillary to this detection window.

Zagursky and McCormick [109] provided the first report of DNA sequencing performed in a capillary-like format. The authors modified a commercial automated slab-gel sequencing system (Genesis 2000 DNA Analyzer from E. I. DuPont) to accommodate up to 12 polyacrylamide gel-filled, 530 μm i.d. by 40 cm long (effective length) columns. The separation of dye-terminator labeled sequencing reactions was performed using the Genesis 2000 system equipped with an argon-ion laser and two-color LIF detection optics and sequencing calls to 500 bases with 96% accuracy were performed in 9.5 h. This work was performed using low field strength but conversion to capillary dimension tubing (i.e., <200 μm i.d.) was expected to allow the use of higher voltage with a resulting decrease in separation time.

Smith and coworkers [67] designed one of the first on-column, single-capillary, single-wavelength systems used to detect primer-labeled DNA sequencing reaction fragments (Figure 16.10). The output beam of a 40 mW argon-ion laser was passed through a 488 nm bandpass filter, expanded, and then focused into a 50 μm i.d. capillary to excite fluorescein-labeled DNA sequencing fragments. Emission light was collected normal to the excitation direction using a 40×, 0.65 NA microscope objective. Slits placed at the objective plane reduced background interference from excitation light scattered at the capillary walls. A bandpass filter placed just in front of a photomultiplier tube (PMT) further eliminated scattered light and selected the appropriate detection wavelength range. The detection limit for this system, determined using vacuum injection of known concentrations of fluorescein-labeled primers, was estimated to be 60,000 molecules (at a SNR of 2:1). Swerdlow and Gesteland [68] and Karger and coworkers [69], working in parallel with comparable CE/LIF systems, reported detection limits of 10^{-11} M for a solution of fluorescein, at a SNR of 3:1 and 5:1, respectively. Swerdlow and Gesteland [68] reported focusing the laser to a spot size of approximately 20 μm diameter, corresponding to an illumination volume of 30 pL and a reported detection limit of <200 molecules.

Pentoney et al. [28] used a 543 nm HeNe laser to excite primer-labeled sequencing reaction fragments. The capillary was threaded though a simple aluminum parabolic reflector and the portion of the capillary with the polyimide cladding removed to allow detection was located at the focal point of the reflector (Figure 16.11). The collimated light exiting the reflector passed through several optical filters to eliminate scattered light before detection using a PMT. A scattering block was placed across the mouth of the parabolic reflector to reject the intense plane of scattered light that surrounds the capillary. Detection limits in the low 10^{-11} M range for solutions of fluor-labeled primers were reported [28].

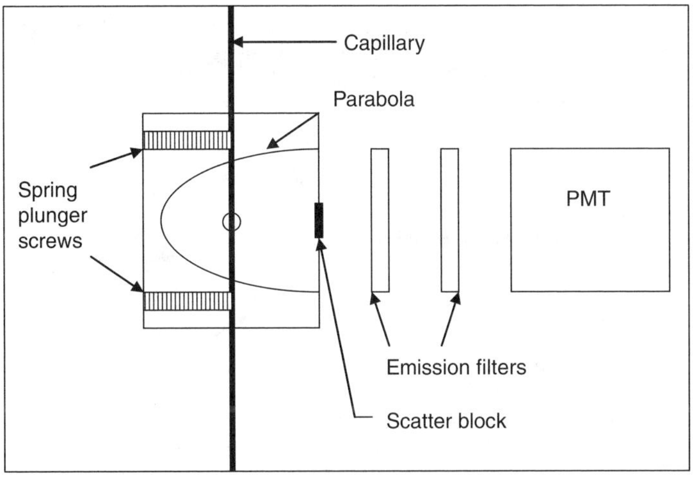

FIGURE 16.11 A single-color CE/LIF detection system utilizing a parabolic reflector for capturing and collimating fluorescence. (Reprinted with permission from Pentoney, S.L., Jr., et al., *Electrophoresis*, 13, 467, 1992.)

Recognizing the need to perform multicolor detection for truly robust DNA sequencing, a number of groups devised methods to perform single-capillary four-color detection. Smith and coworkers [15] expanded their single detector system by incorporating 50/50 mirrors in the emission path in order to split the emission from four fluorophores and direct it toward four optically filtered PMTs (Figure 16.12). The four fluorophores, FAM, JOE, TAMRA, and ROX, from Applied Biosystems (now part of Applera) were coupled to primers and used in four separate chain extension reactions. These four fluorophores can all be stimulated by a common argon-ion laser, allowing detection in all four channels to occur simultaneously. However, since 50/50 mirrors were used in the emission path, only 25% of the original emission intensity for any channel actually reached the detector. Dovichi and coworkers [29] opted for an alternate time sharing approach in a similar system by placing a spinning filterwheel, containing four bandpass filter segments, before the detector. Dovichi used two lasers, an argon-ion laser at 488 nm and a HeNe laser at 543 nm for excitation of the same family of fluorophores used by Luckey [15], because the 543 nm HeNe laser is a better spectral match with the absorbance profile of TAMRA and ROX fluorophores. The two lasers were modulated with a two-sector chopping wheel, with 488 nm light used to stimulate FAM- and JOE-labeled primers while the 543 nm laser was used to stimulate TAMRA- and ROX-labeled primers. Synchrony between the two spinning wheels allowed assignment of the emission collected at the PMT to a particular spectral channel and hence to a specific fluor-labeled primer. Since the single detector is now temporally shared among the four dyes, sensitivity will depend on the integration time for each channel. This, in turn, will depend on the data acquisition rate, typically set to 2 Hz or greater due to peak width and sampling considerations [29].

Gesteland and coworkers [71] eliminated the need to split the fluorescence emission or time-share the detector to obtain multicolor data by using a spectrometer to disperse the emission from the four Applied Biosystems fluors before imaging onto a charge-coupled device (CCD) camera. Quantum yield in the 500–600 nm range was 1.6–2.5× greater for the CCD camera than for the PMT which made this imaging detection system quite attractive. However, the need to place multiple mirrors, with a composite efficiency of $0.94^4 = 0.78$, and the 0.70–0.75 efficiency of the grating employed reduced the light levels directed toward the detector by 50%. Sensitivities using this CCD-based optical system with a 500 ms integration time were reported to be comparable to PMT-based systems [71]. Karger also eliminated the emission splitting/time sharing issue by using a 488 nm laser and a 543 nm laser focused to two different points on the capillary that were separated by approximately

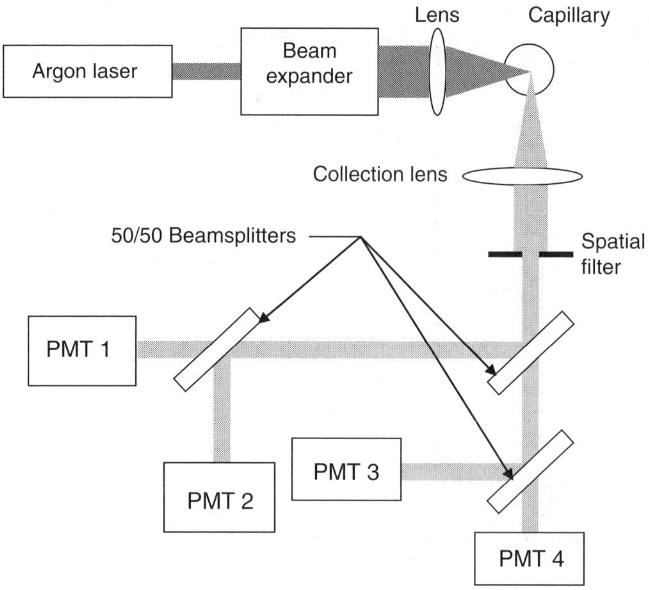

FIGURE 16.12 Four-color on-column detection system utilizing four dedicated PMTs. A 540 nm bandpass filter was placed in front of PMT1; a 560 nm bandpass filter was placed in front of PMT2; a 580 nm bandpass filter was placed in front of PMT3; and a 610 nm bandpass filter was placed in front of PMT4. All pass bands are 10 nm. (Reprinted from Luckey, J.A., et al., *Nucleic Acids Res.*, 18, 4417, 1990. With permission from Oxford University Press.)

2 cm [110]. Two collection objectives were used in this study, one for each detection window, to bring the emission to different regions of a diode array detector. A grating placed before the diode array dispersed the emission from each of the detection windows, with a resolution of 1 nm per pixel in the 555–700 nm range. Monitoring different portions of the array gave excitation wavelength-associated spectral information about the emission and thus about the primer-labeled fragments as they migrated through the capillary.

16.2.3.2 Multicapillary On-Column Detection Systems

The systems described above may all be regarded as important steps or feasibility demonstrations in the development of CE-based optical systems for DNA sequencing. The use of capillaries, with their small cross-sectional areas and high surface-to-volume ratios permitted the use of significantly higher field strengths than had typically been used in slab gel separations but the corresponding increase in separation speed was not so great as to give single-capillary sequencing systems a throughput advantage over slab-based systems where many parallel lanes were run simultaneously. Therefore, in order to significantly improve sequencing throughput over that already realized with slab-based systems, many researchers developed fluorescence optical approaches that allowed simultaneous monitoring of multiple capillaries run in a parallel format. Mathies and coworkers [13,112] and Mathies et al. [111] developed a multicapillary DNA sequencing system in which a linear array of capillaries, placed on a moving stage, is scanned under a stationary confocal microscope head, as displayed in Figure 16.13. In their first report [13], 25 capillaries (100 μm i.d., 200 μm o.d.) were bundled into a linear array. An array holder, mounted to a motor driven translation stage, held the capillary windows in optical alignment. For excitation, the output beam of an argon-ion laser was expanded, collimated, and reflected through a long pass dichroic mirror to a 32×0.4 NA microscope objective, which focused the beam to a 9 μm spot at the plane of the arrayed capillaries. A portion

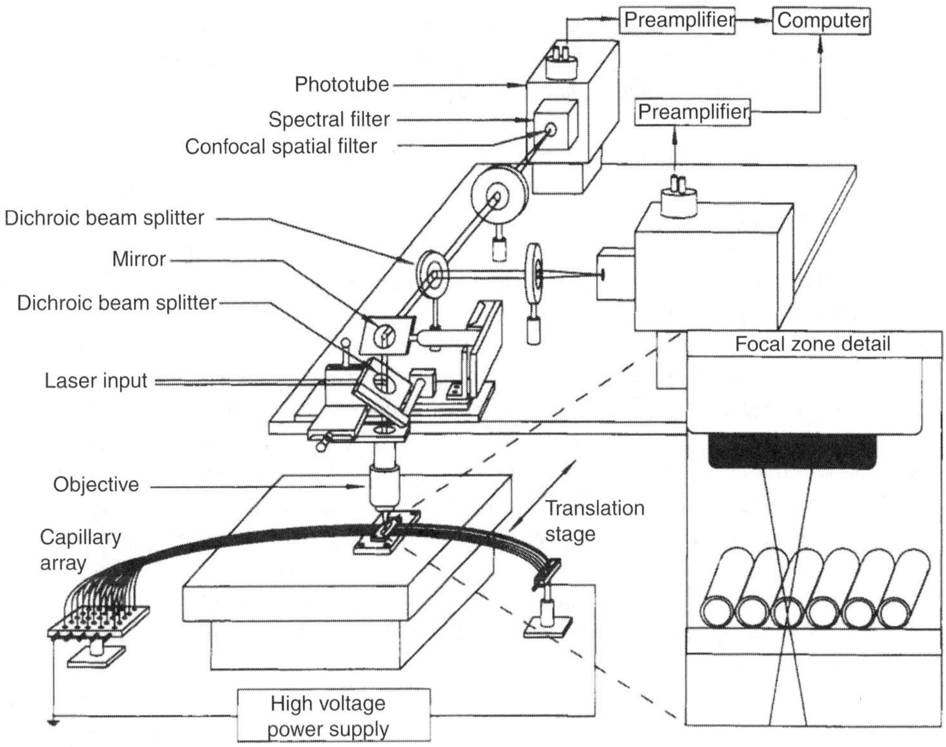

FIGURE 16.13 Multicapillary detection system utilizing a planar capillary array arrangement. Although two detectors are shown, one- and four-color detection systems have also been implemented. (Reprinted with permission from Huang, X.C., et al., *Anal. Chem.*, 64, 2149, 1992. Copyright 1992 American Chemical Society.)

of the fluorescence originating from the capillaries was captured by the same microscope objective used to focus the laser (confocal arrangement) and transmitted back through the long pass dichroic mirror to reduce background laser interference at the detector. The fluorescence emission was focused though a 400 μm pinhole spatial filter, a 488 nm rejection filter, and a long pass filter, all to further reduce laser scatter. The initial system contained a single PMT and monitored one or two colors, but later systems reported by the same group used either two or four detectors for multicolor monitoring [111–113]. In those systems, the emissions from different fluorophores were split and directed to the appropriate PMTs using dichroic beam splitters. The translation stage, with the capillary array holder mounted to it, was scanned beneath the microscope objective at 1 Hz. During a sweep, the velocity of the translational stage was 20 mm/s while data were collected at 2000 Hz; thus, an image resolution of 10 μm per data point was achieved [13]. Since 100 μm i.d. capillaries were used in this work, 10 points defined the inner bore of the capillary. These 10 measurements were co-added to yield a single intensity measurement for that capillary and that scan. A detection limit of $\sim 2 \times 10^{-12}$ M (SNR = 3) was reported using solutions of fluorescein that were flowed through the system (so-called flowing stream characterization measurements) [13].

Bashkin et al. [86,114] modified this system by placing the microscope objective on a translation stage: the position of the 48-capillary array in their system was fixed and the objective was translated above the array. The capillary signals were sampled unidirectionally at a sampling rate of 38 kHz for each of two PMTs, yielding a spatial resolution of 37.5 μm per pixel, and a temporal resolution of 1 Hz. Flowing stream detection limits of 10^{-12} M were reported using solutions of BODIPY505/515 (Molecular probes) prepared in methanol [114]. Automated replacement of the hydroxyethyl cellulose (HEC) gel separation matrix used in their studies was accomplished via a stainless steel pressure manifold/anode assembly [114]. In the upper part, four ports were available for arrays of 16 capillaries to be inserted while one port introduced a common electrode and another port allowed high-pressure nitrogen to be introduced. The lower part of the manifold assembly contained a 25 mL beaker that contained either the separation matrix or water. Pressurization of up to 1000 psi allowed water or matrix to be pushed out the distal end of the capillary. This system was extended to four PMTs and eventually commercialized by Molecular Dynamics (now part of GE Healthcare) as the MegaBACE DNA sequencing system [114,56].

Mathies and coworkers [115] also developed a highly multiplexed system, with the potential to monitor 1000 capillaries. In this "rotary capillary array electrophoresis scanner," the capillaries were supported in grooves on the outside of two coaxial ~4′ diameter cylinders (see Figure 16.14). A microscope objective, designed to deliver excitation light and collect the resulting emission, was placed between the two cylinders and mounted on a central shaft that was driven by a stepper motor to rotate at 2–4 revolutions per second. Light traveled bidirectionally along the axis of the rotating shaft between the dichroic beam splitter located just outside the top of the shaft and a diagonal mirror located within the shaft. Simultaneous replacement of separation matrix in all capillaries was achieved by placing matrix in an O-shaped well, where the capillary tips reside, and pressurizing with helium. During a sequencing run, buffer and the high voltage electrode were placed in this same well.

Yeung and coworkers [116–119] studied several on-column excitation and detection schemes and developed several multicapillary systems. In one scheme [117], the windows of 96 capillaries, with an i.d. of 75 μm, were aligned in a plane oriented at 45° with respect to a CCD camera detector. A sheet of 514 nm excitation light, oriented perpendicular to the viewing axis of the CCD camera was focused into the windows of the capillary array. Emission was collected through a holographic filter to reduce background laser scatter. Two-color fluorescence detection was obtained by directing one image of the array onto a portion of the CCD chip through a 630 nm longpass filter and directing a second image of the array onto another region of the CCD chip through a quartz plate without additional optical filtering. The 1.5 cm range of capillaries imaged covered 300 pixels on the CCD with each individual capillary window encompassing three pixels in one image. An integration time of 300 ms was used to acquire each frame at 1.75 frames per second and good uniformity in migration time was observed. Cross talk in this sheet illumination approach, arising from emission in an adjacent capillary being refracted by the capillary walls into the CCD camera and appearing in the electropherogram of the second capillary, was estimated to be <10%. Automation of matrix replenishment was discussed in this work but not implemented. Heller and coworkers [120] developed a similar system, but used a grating to disperse the emission onto the CCD camera. Intercapillary cross talk was estimated to be 1–2% in their system. This system was coupled to a stacker that held up to forty 96-well mictotiter plates, allowing up to 15,000 samples to be run without reloading. However, no automated matrix replenishment system was described.

Yeung and coworkers [121,122] also developed a side-entry illumination system for exciting up to 24 capillaries. In this work, capillaries with either round or square bores were used. After windowing, the capillaries were packed side-by-side with the windows aligned and fixed into place using epoxy. Fused-silica plates were used to create a cell for the windowed region of the capillaries; this cell was then filled with index-matching fluid to reduce laser light loss at the capillary interfaces. The 514 nm line of an argon-ion laser was focused through the side of the array and transferred through the detection bores of all capillaries. Because of the refractive index-matching fluid, negligible light loss was observed across the array and the authors predicted that greater than 34% of the incident laser power would be available in the 500th capillary of a hypothetical 500-capillary array [122]. Detection was made above the plane of the capillary array with a CCD camera. A 514-nm notch filter was used to reduce scattered light while an image splitter, consisting of a 610- or 630-nm long pass filter and a quartz plate, allowed two-color imaging onto two regions of the CCD chip. Detection limits of 90 pM were reported from long injections of fluorescein dissolved in run buffer, to eliminate stacking. Two-color DNA sequencing was demonstrated in this work.

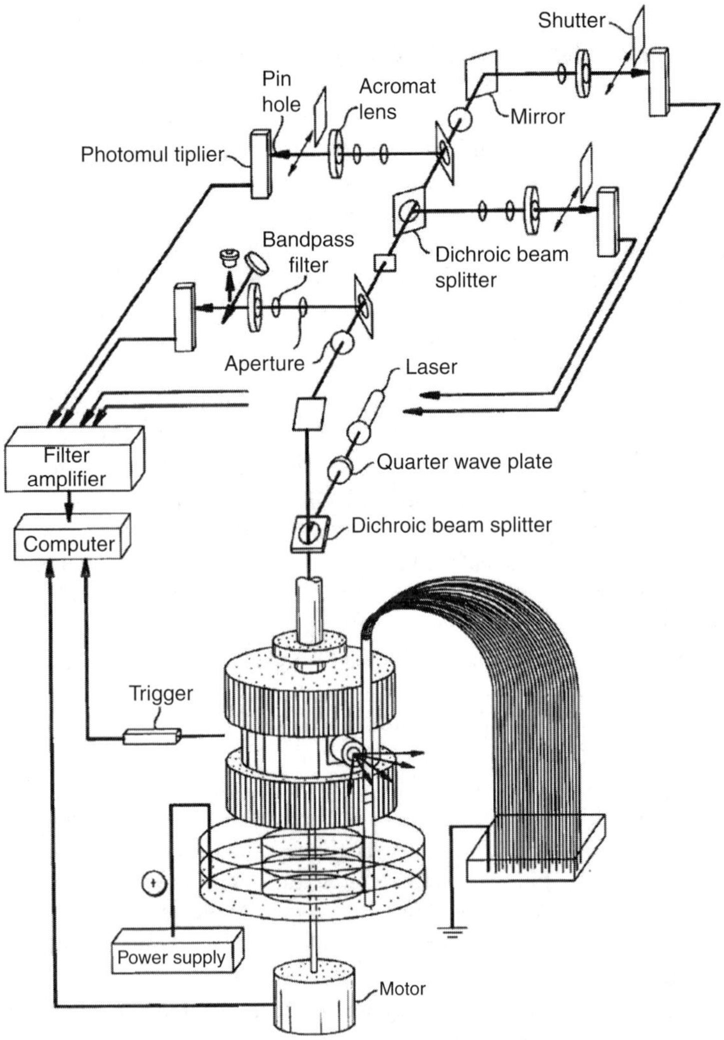

FIGURE 16.14 Illustration showing a 1000 capillary, four-color detection rotary capillary array electrophoresis scanner device. (From Mathies and Scherer, US Patent 7,090,758.)

Quesada and Zhang [123] and Quesada et al. [124] performed detailed ray tracing and also demonstrated success using a side illumination scheme. A planar 12 capillary array was created and the windowed region was placed in index-matching fluid. The output of a 488 nm laser was split into two fiber-optic transmitters and delivered to either end of a planar array, thus reducing variation in laser intensity across the array. Emission was collected with 12 optical fibers, each placed against one capillary detection window, normal to the direction of laser excitation. The emission was dispersed using a spectrograph and imaged onto a CCD camera, thus providing spatial and four-color spectral information. Cross talk between adjacent capillaries was estimated to be ~1–2% while a detection limit of ~3.7 pM at a SNR of 2, derived from flowing stream measurements using fluorescein, was reported. Simultaneous four-color dye-terminator sequencing for all 12 capillaries was demonstrated in this work [124]. Since the number of capillaries that could be used was generally limited by laser attenuation due to beam divergence, glass rods were placed between capillaries by Anazawa et al. [125]. This allowed them to achieve detection limits of <1 pM of flowing fluorescein solution in the 45th capillary, furthest away from the illumination source.

Recently, Roeraade and coworkers [126,127] developed a detection scheme in which each capillary acted as a waveguide to direct the emitted fluorescence to the capillary tip, where it was collected with a CCD camera. Strip illumination of up to 91 capillaries was made in a region where the capillaries are arranged in a planar array; these capillaries were externally coated with a fluoropolymer that did not require removal for use in LIF detection. Approximately 15 cm downstream from the excitation region, the capillaries were cast into a two-dimensional hexagonal array with black epoxy and placed into a container consisting of buffer and the electrode. Fluorescence emitted from the ends of the capillaries transmitted through the glass wall of the container was collimated, passed through two bandpass filters to eliminate 488 nm scattered light, and was then dispersed with a prism onto a cooled-CCD camera. This arrangement yielded spatial and spectral information about the fluorescence emitted from the capillary outlets. Cross talk between capillaries, arising from emission from one capillary being captured and transmitted by a second capillary, was found to be less than 0.4%. A detection limit of 2.7 pM fluorescein measured in flowing stream mode was reported, but this sensitivity measurement was made without use of the prism to disperse the light [126].

The parabolic reflector concept developed by Pentoney et al. [28] was extended to a scanning eight-capillary system, which was commercialized by Beckman Coulter in their family of four-color, eight-capillary DNA analysis systems. In this work, two diode lasers, also used in References 128 and 129, were alternately modulated and the beams directed to a precision rotating mirror positioned on a galvanometer. This mirror sequentially directed the focused laser beams into each capillary window in the planar array. The windows were centered about the focus of a parabolic reflector, which collimated and directed the emission though a four-quadrant spinning filter wheel, a dual-notch filter (designed to reduce scatter from each laser) and then through a focusing lens and onto a PMT (Figure 16.15). The mouth of the parabolic reflector again had a black bar placed across it to block the intense plane of laser scatter. The spinning filter wheel was synchronized with the modulation of the two diode lasers and the galvo scanner such that when the first laser was striking the capillary, the first two filter quadrants passed in front of the detector. Similar interrogations of the capillaries were made for laser 2 and filter wheel quadrants 3 and 4. To ensure high sensitivity, the diode lasers were optically filtered to remove low level emission ("diode wings") overlapping those of the detection wavelengths. Detection limits below 10 pM at a SNR of 2 were determined from flowing stream measurements using proprietary dye-labeled terminator solutions (unpublished data, Beckman Coulter).

An interesting aspect of this step-scanned system is the automated optical alignment feature. To maximize duty cycle, the galvo mirror rapidly advances from one capillary bore to the next and then pauses during data acquisition. In order to predetermine the location of the capillary bores, an automated scan was made before every run. In this prescan, which was completed in less than 2 min, the first laser was scanned across the entire region where the array was believed to reside. An autoalignment photodiode, placed behind the array, facing the laser but off axis to the laser, detected light refracted from the capillaries and an auto-alignment spectrum was obtained, such as the one displayed in Figure 16.16. The peaks seen in this spectrum contain the information required to direct the galvo to position the beams at the capillary bore centers. This process was repeated for the second laser, yielding a second set of eight peaks corresponding to the galvo positions required to send the second laser to each of the capillary bores. The 16 positions (eight-capillary positions for each of the two lasers) were saved in a lookup table and used during the separation to place the first laser in the first capillary, the second laser in the first capillary, the first laser in the second capillary, and so on. Experience has shown that the positions of the capillary bores with respect to the laser positions do not change during the course of the run.

16.2.3.3 Sheath-Flow Detection

Dovichi and coworkers [29,70,130] developed a high-sensitivity fluorescence detection system for DNA sequencing based on a detection scheme borrowed from flow cytometry and previously

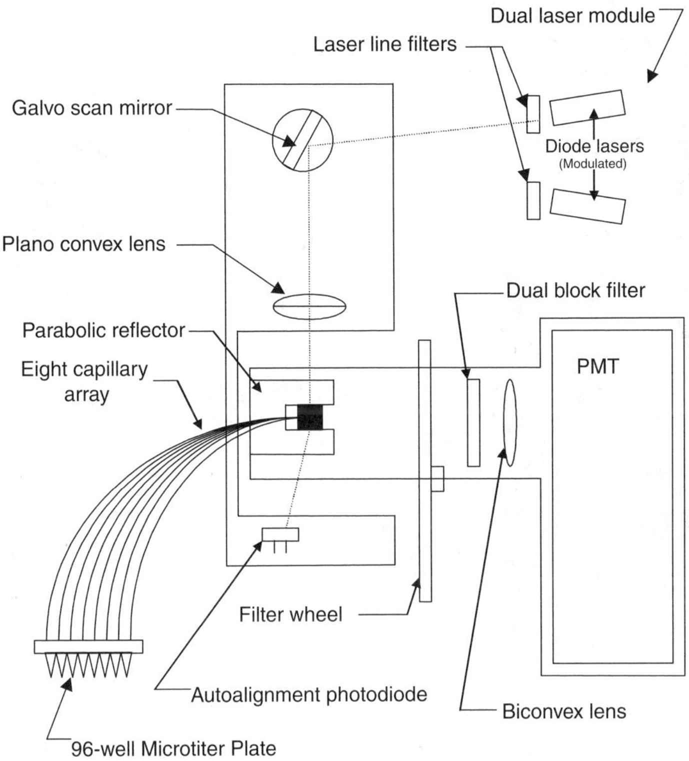

FIGURE 16.15 Illustration of the galvanometer scanner based, eight-capillary DNA sequencer optics employed in the Beckman Coulter DNA analysis system.

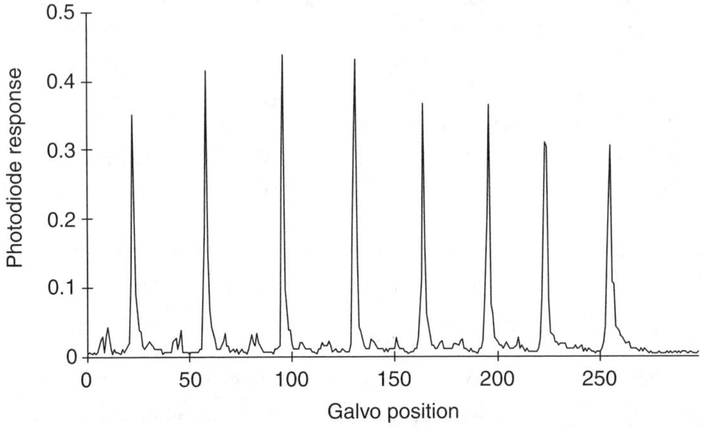

FIGURE 16.16 Capillary auto-alignment spectrum used to determine the capillary bore center positions.

developed by the same group for CE separations of labeled amino acids [131]. To minimize or eliminate background scattered light arising from the excitation source striking the curved capillary surface, the separated DNA fragments were carried away from the capillary outlet before detection. This was accomplished by placing the end of a 1 m $\times$ 50 μm i.d. capillary in a flow cytometer cell (Figure 16.17): the labeled sequencing fragments eluting from the end of the capillary were entrained

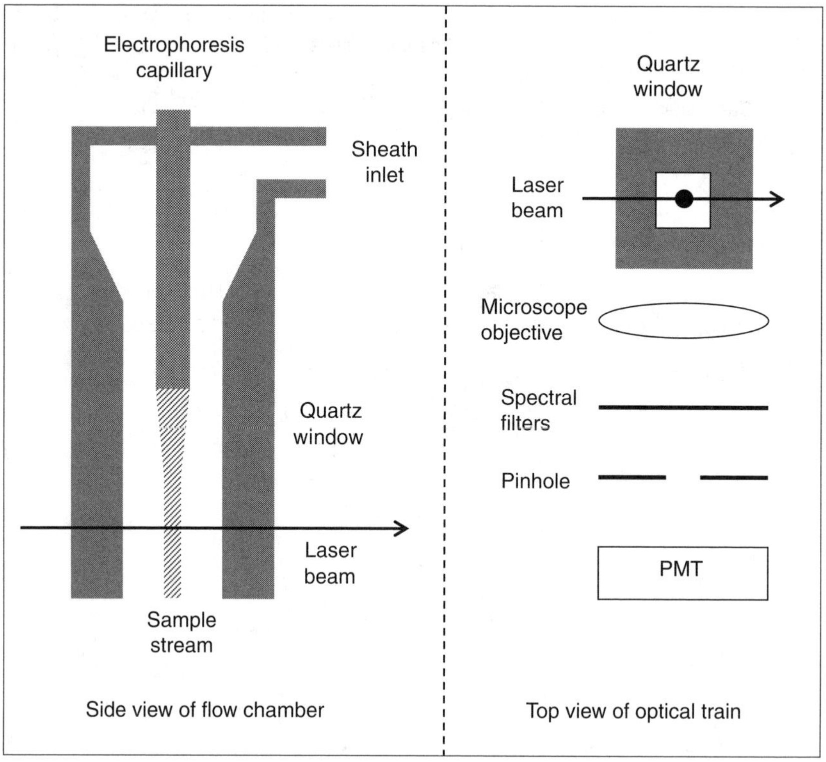

FIGURE 16.17 Postelectrophoresis sheath-flow detection of DNA fragments. (Reprinted with permission from Zhang, J.Z., et al., *Clin. Chem.*, 37/9, 1942, 1991.)

and carried down the flowcell by sheath fluid, flowing at 0.16 mL/h, into a square detection region. The sample stream was illuminated with a focused laser and fluorescence from the labeled sequencing products was collected, filtered to eliminate potentially interfering Rayleigh and Raman scatter and passed on to a PMT [17,29,30,70,73,130,132–134]. Two-color detection was achieved either using a dichroic beam splitter to separate the emission from two fluors [29,30,73,134] or by detection on both sides of the sheath-flow cuvette [30], facilitating two-label peak-height encoded DNA sequencing. A detection limit of less than 60 molecules of Rhodamine 6G injected into the system was reported [133].

Four-color sequencing in a single capillary using this sheath-flow concept was also demonstrated [17]. Two alternately modulated lasers, an argon-ion laser and a green helium-neon (HeNe) laser, were combined using a dichroic filter. The collinear beams were focused onto the sheath flow entraining the separated four-color primer-labeled DNA fragments. To spectrally resolve the emission, a four-sector filter wheel spinning at 2 Hz was placed between the flow cell and a microscope objective, which in turn coupled the emission into an optical fiber mated to an avalanche-photodiode operated in single-photon-counting mode. With the argon-ion laser on, the first two quadrants of the filter wheel, containing bandpass filters centered at 540 nm and 560 nm, respectively, passed in front of the microscope objective. Similarly, excitation with the green HeNe laser was synchronized to detection through the 580 nm and 610 nm quadrants of the spinning filter wheel. Read-lengths of 640 bases in 2 h were reported for fluorescently labeled primers [17]. A multicolumn approach soon followed [22] and with a planar array of five capillaries, the lasers were focused to illuminate across the outlets of the capillary array, exciting all five sheath-entrained samples in the cuvette simultaneously approximately 100 μm below the ends of the capillaries (Figure 16.18). Emission was collected normal to the direction of the sheath flow and excitation laser direction using a microscope objective (20×,

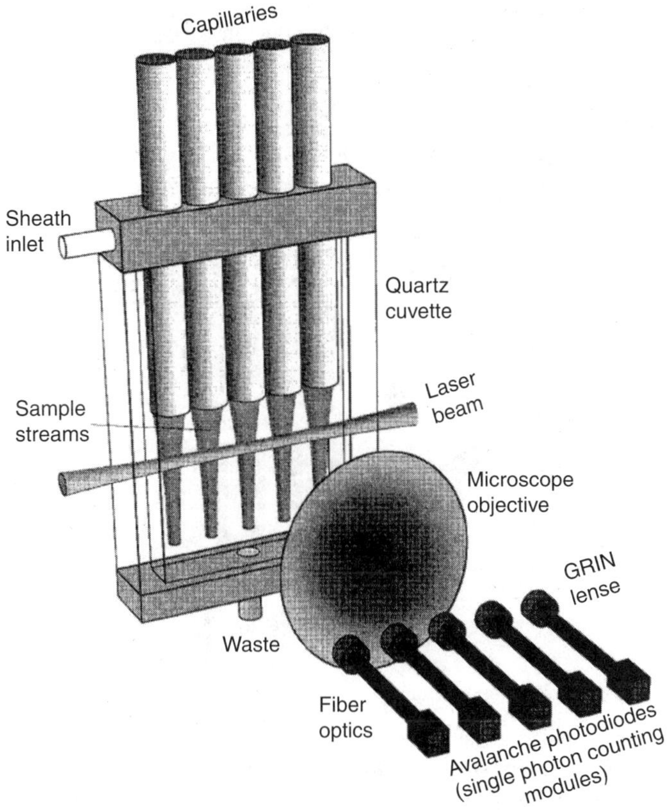

FIGURE 16.18 A five-capillary sheath-flow CE/LIF detection scheme. (Reprinted from Zhang, J., et al., *Nucleic Acid Res.*, 27, e36, 1999. With permission from Oxford University Press.)

0.5 NA) and was passed through a spinning filter wheel. At the image plane, five gradient refractive index lenses, each coupled to a fiber optic, brought the emission from the five spots onto five dedicated avalanche photodiodes [22]. The implementation of a micromachined sheath-flow cuvette for capillary alignment extended multiplexing to 16 capillaries [135] while a two-dimensional capillary array scheme using 32 capillaries, with illumination by a "sheet" of laser light and detection in the sheath stream at the exit tips of the capillaries using a CCD camera, suggested feasible scaling to "several thousand capillaries" [136].

Kambara and Takahashi also employed the sheath-flow technique for the interrogation of 20 capillaries using a CCD camera [137]. In that work, the 20 capillaries were aligned into a linear array, with the outlet end of all capillaries engulfed by a sheathing fluid, in this case, a TBE buffer. No flow cell was used in this work. A laser was directed through the sheath fluid carrying the sample, normal to the direction of the sheath flow, allowing sample from all 20 capillaries to be illuminated simultaneously (Figure 16.19). A cylindrical lens followed by a prism was used to disperse the fluorescence from all 20 capillary bores onto a CCD camera. One axis of the CCD image yielded spatial information about the interrogated capillaries and the other axis of the CCD yielded spectral information. No moving parts was required in this approach. In addition, since excitation and detection for all capillaries and fluors occur simultaneously, no time sharing was required, improving duty cycle and greatly increasing the capability of this system for multiplexing. PE Biosystems Inc. (now part of Applera) has commercialized a similar sheath-flow design on several of their high-throughput DNA sequencing systems [6,56,127].

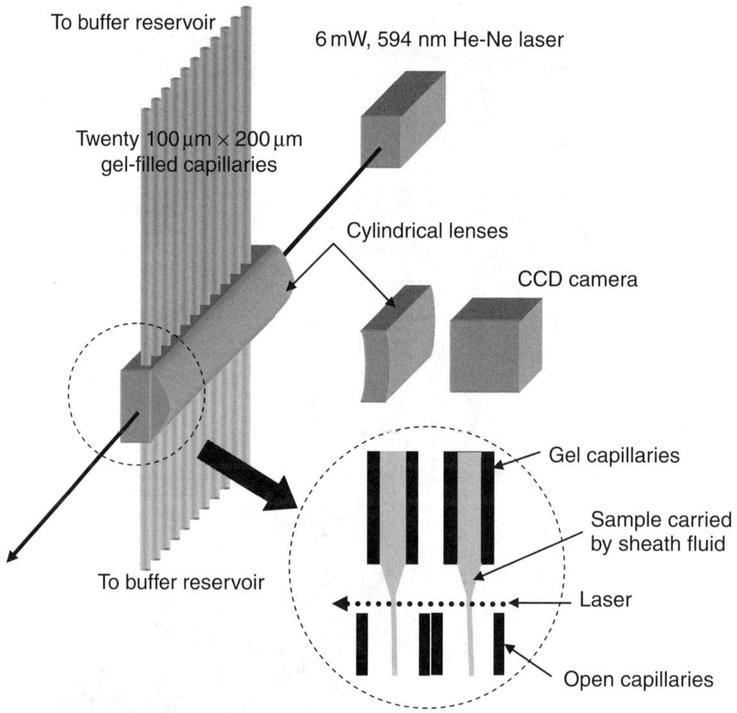

FIGURE 16.19 A twenty-capillary sheath-flow CE/LIF system. Detection is accomplished in the gap region shown in the inset. (Adapted by permission from Macmillan Publishers Ltd: Kambara, H. and Takahashi, S., *Nature*, 361, 565, 1993. Copyright 1993.)

16.2.3.4 Other Detection Schemes Explored

The detection methods thus far described focus on measuring the intensity of fluorophores at one to four wavelengths. For multicolor work, an important requirement is that the fluorophores are spectroscopically unique. To obtain spectroscopic information, the emission is either monitored using one time-shared detector, split among several detectors, or dispersed with a grating or prism, all resulting in reduced light levels and lower SNR.

One alternative method is to use differences in fluorescence lifetime as the metric to discriminate between the fluor-labels [138,139]. Several potential benefits of measuring lifetimes for DNA sequencing have been described [140,141]. Lifetimes may be accurately determined in the absence of spectral discrimination; thus, discrete lifetimes may be determined for mixtures of fluors coeluting in a single peak. Lifetimes are also independent of intensity even at concentrations where relatively poor SNR may lead to an erroneous fluorescence intensity-based call. In addition, since scattered light has an effective lifetime of zero, scatter may be temporally rejected and is thus less of an issue. However, this method requires a family of fluors in which the lifetimes are resolvable [141–143] and the use of relatively complex electronics [141].

There are two methods that have been used to determine fluorescence lifetimes in DNA sequencing. In the first method, known as the frequency-domain or phase-modulation method, the excitation beam is intensity modulated. The a.c. portion of the resulting emission is phase-shifted relative to the laser modulation; this phase-shift contains information about the fluorescence lifetime, or lifetimes if more than one fluor is present [140]. McGown and coworkers [144,145] used this method for four-color sequencing. In that work, 488 nm or 514 nm laser light was electronically modulated with a Pockels cell before being focused onto a capillary column. Detection, made normal to the laser direction, was optically filtered to reduce laser scatter and was focused onto the detector of a fluorescence lifetime instrument (Model 4850 MHF from Spectronics Instruments, Rochester, NY). Fluorescence lifetimes of 1.7, 2.5, 2.9, or 3.5 ns were measured and assigned to primers labeled with either CY3, the energy transfer dye fluorescein-dTMR, rhodamine green, or BODIPY-fluorescein, respectively. Recent advances using acridone dyes with more resolvable lifetimes of 4, 6, 11, and 14 ns have been reported [143].

The second method of determining fluorescence lifetime is to modulate the excitation source with a very short pulse width and measure the decaying fluorescent emission with high-temporal resolution when the laser is off. Soper and Mattingly [146] constructed a near-infrared detection system with picosecond resolution. In this study, an argon-ion pumped Ti:sapphire laser generated 210 femtosecond excitation light pulses at 76 MHz. They found a strong dependence of fluorescence lifetime on sieving matrix material. Detection of separated dye-primer labeled fragments was demonstrated [147]. Wolfrum and coworkers [138] used a single 630 nm diode laser, pulsed at 22 MHz with 500 picosecond on durations, to excite a family of fluor-labeled primers. The lifetimes of the four fluorophores studied were 3.7, 2.9, 2.4, and 1.6 ns. Emission, filtered with a dichroic lens (required in their confocal illumination/detection setup) and a 675 nm bandpass filter, was monitored using an avalanche photodiode. This work was expanded by placing up to 16 capillaries onto a translation stage and scanning them under the confocal excitation/detection system [23]. In all cases, relatively complex curve fitting of the fluorescence decay was required to extract the lifetime values. Detection limits were not reported for any of the fluorescence lifetime-based capillary sequencing studies probably because that was an area of difficulty relative to conventional fluorescence approaches.

Another detection scheme, studied by Gorfinkel and coworkers [148,149] with capillary-based DNA sequencing in mind, was based on single-photon detection. In this approach, the output beams from four lasers were combined in a single optical fiber and used to excite the labeled fragments in the CEQ DNA test sample (Beckman Coulter) as they migrated through the detection window of a capillary. Each laser, selected to preferentially excite one of the four BCI fluors at 635, 675, 750, and 810 nm, was modulated at a different frequency, in the range 1–2 Hz. Collection was made via a fiber receiver connected to a single-photon PMT detector. Four stacked notch filters were used to eliminate scatter from each of the lasers. Since the four dye-matched lasers were modulated at different frequencies, the fluorescence signal arising from each fluor was also modulated at a different frequency. A Fourier spectrum contained a peak or peaks corresponding to the fluorescence frequency of the fluor(s) present. Cross talk, corresponding to the emission from a fluor generated by excitation from a second laser, may be determined by running each fluor independently and determining the relative Fourier amplitude for each of the lasers. Although demonstrated only for sequencing in a single capillary, a multiplexed scheme to monitor separations in 32 capillaries was also described [149].

Metzker and coworkers [150] developed a "color-blind fluorescence detection" system using four fluorophores. In this system, each of the four fluorophores was stimulated primarily by only one of four lasers. The four lasers, at 399, 488, 594, and 685 nm, were modulated using mechanical shutters such that only one laser was striking the capillary at any time. These were used to excite Alexa Fluor 405, BODIPY-FL, 6-ROX, and Cy5.5, respectively. Emission was collected normal to the direction of excitation and was then transmitted through a long pass and several notch filters to block light scattered from any of the lasers before being focused onto a PMT. Since each of the four fluorophores is primarily stimulated by only one laser, fluorescence detected at the PMT was assigned to the DNA fragment labeled with the fluorophore primarily excited by the laser currently striking the capillary. In general, cross talk observed between lasers and fluorophores was in the range of 2–4%. However, Cy5.5 excited by 594 nm yielded an emission that was 20% of the emission generated upon excitation with 685 nm light. Mobility correction was required before good DNA sequence could be called.

An interesting nonoptical detection scheme for CE-based DNA sequencing was proposed by Brennan et al. [151]. In this approach to sequencing, capillary gel electrophoresis was coupled to mass spectrometric detection. The idea was based on the use of four different sulfur isotopes, having masses of 32, 33, 34, and 36 to establish the terminal base identity of Sanger–Coulson sequencing fragments eluting from the capillary. The labels were incorporated in the form of thiophosphate analogues. Described advantages included the possibility of internal labeling, which would produce an increase in signal with increasing fragment length, and the elimination of spectral overlap issues that exist with fluorescent labeling schemes. The novel proposition involved a CE/MS interface in which capillary effluent droplets were created using a piezo-electric dispenser similar to those used in ink-jet printers. The small, picoliter volume droplets were to be combusted and transferred to a mass dedicated spectrophotometer tuned over the small mass range required to differentiate the four sulfur isotopes.

16.2.3.5 Integrated Sample Processing and Detection

The systems described thus far addressed only a portion of the individual tasks required to elucidate DNA sequence. A more efficient process would fully integrate the tasks of DNA extraction, purification, template preparation, and amplification with sequencing reaction chemistry, post reaction cleanup, separation, and detection [152]. Several groups have explored partially integrated systems intended to streamline various steps in this process [152–156].

Swerdlow et al. [153] developed a system for DNA sequencing that integrated the cycle sequencing reaction, post reaction sample cleanup, sample injection, CE separation, and LIF detection. A schematic representation of the system is displayed in Figure 16.20. Cycle sequencing reaction components and sample were loaded by syringe into a temperature programmable rotary valve. Once the cycle sequencing reaction was completed, the reaction product was directed to a gel filtration HPLC column for cleanup. Their system allowed "heart cutting" of the product peak that eluted from the gel filtration column and this purified aliquot of the product peak was then electrokinetically injected into a gel-filled capillary using the "Tee" injection scheme depicted in Figure 16.20. After a 10 s injection, pressure drove a stream of TBE buffer through valve 2 and into the "T" in order to chase away any remaining sample. Once the "T" had been filled with TBE buffer, electrophoresis was initiated. On-column detection was made using a single PMT to monitor one-color fluorescence induced by 532 nm laser excitation. The entire process was completed in 90 min [153].

Yeung and coworkers devised a similar scheme to integrate thermal cycling and purification followed by injection, separation, and detection, first in a single capillary [157] and then in eight capillaries [154]. In the multicapillary work, multiplexed freeze/thaw values (MFTV) were used to simplify the manipulation of samples during the thermal cycling process. In the MFTV, an array of eight capillaries was threaded through two stainless steel tubes; the length of capillary between the tubes was placed in a thermalcycler and was used as the reaction vessels. When liquid nitrogen was flowed through the two stainless steel tubes, the liquid in the capillary portions within the steel tubes froze, thus "closing" the valves and forming an immobile volume of liquid between the two frozen plugs. Passage of warm air through the steel tubes melted the plugs, opening the "valves." Figure 16.21 displays the operation of the MFTV for a single capillary. A pump was used to draw thermalcycling reagents from a well in a microtiter plate into the portion of the capillary within the thermalcycler. With valves 1 and 2 in the closed position, the cycle sequencing reaction was performed, thereby creating the labeled DNA fragments (Figure 16.21a). With 1 and 2 in the open position, and valve 3 in the closed position, the pump drew the sample above the "T" (Figure 16.21b). The sample was then pushed into the size exclusion column for sample purification by closing valves 1 and 2 and opening valve 3 (Figure 16.21c). After this, sample purification and capillary loading/separation occurred in much the same manner as described by Swerdlow et al. [153]. Detection of separated DNA fragments was made via side illumination of all eight capillaries via the method described above [122]. This method was later expanded to include sequence determination directly from single bacterial colonies [155].

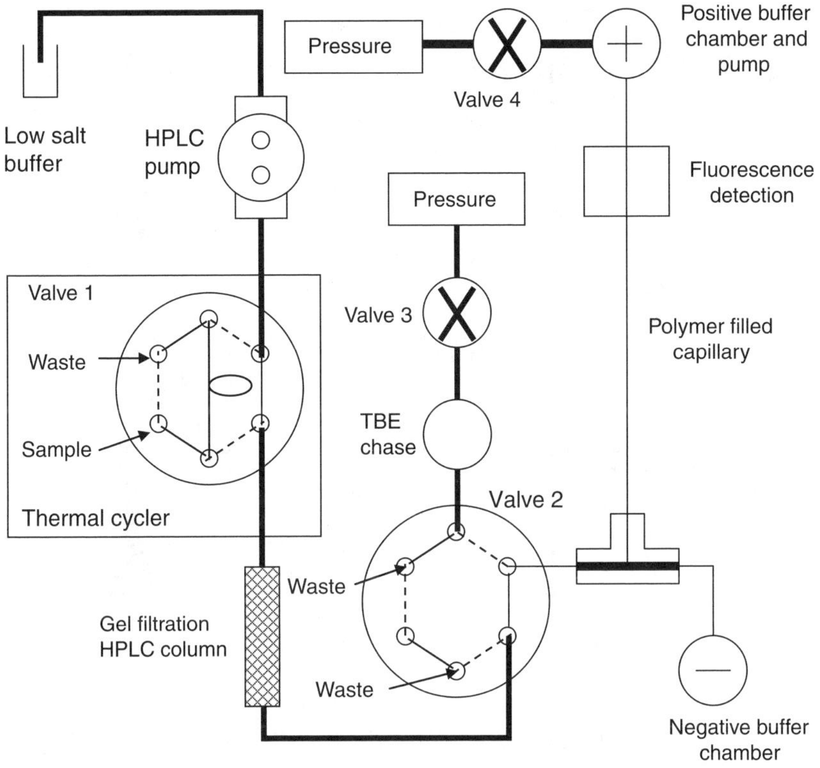

FIGURE 16.20 Integrated cycle sequencing reaction, sample cleanup, separation, and detection system. (Reprinted with permission from Swerdlow, H., et al., *Anal. Chem.*, 69, 848, 1997. Copyright 1997 American Chemical Society.)

16.2.4 Separation Methods

16.2.4.1 Sample Injection

In terms of sample loading, capillary-based DNA sequencing systems have a fairly narrow window of stable operation. This window ranges on the low injection end, from the minimum amount of sample required to produce sufficient signal to reliably call bases throughout the sequencing run (note that signal strength generally declines with increasing base number) to the high injection end of the window, where sample overloading results in resolution losses and/or current instabilities. Several complicating factors can further reduce the width of this stability window including residual template DNA, protein contaminants in the sample, excess salts and unincorporated nucleotides, and the unintentional introduction of air into the gel upon executing electrokinetic injection. The ideal sample loading process, speaking strictly from a separations perspective, would introduce nothing but labeled sequencing fragments onto the gel, and those fragments would be introduced over a column length that contributed nothing to total peak variance at the detector. The fragments would furthermore be introduced onto the gel at concentrations that were electrically transparent relative to the conductive gel medium. Obviously, practical considerations having to do with post sequencing reaction cleanup and low level fluorescence detection prevent us from completely realizing this ideal injection goal.

In 1998, Karger and coworkers [158,159] published a two-part study on the role of sample matrix components and post reaction sequencing product cleanup . In this study, the authors used ultra-filtration to remove template DNA and gel filtration to reduce salt concentration in the sample

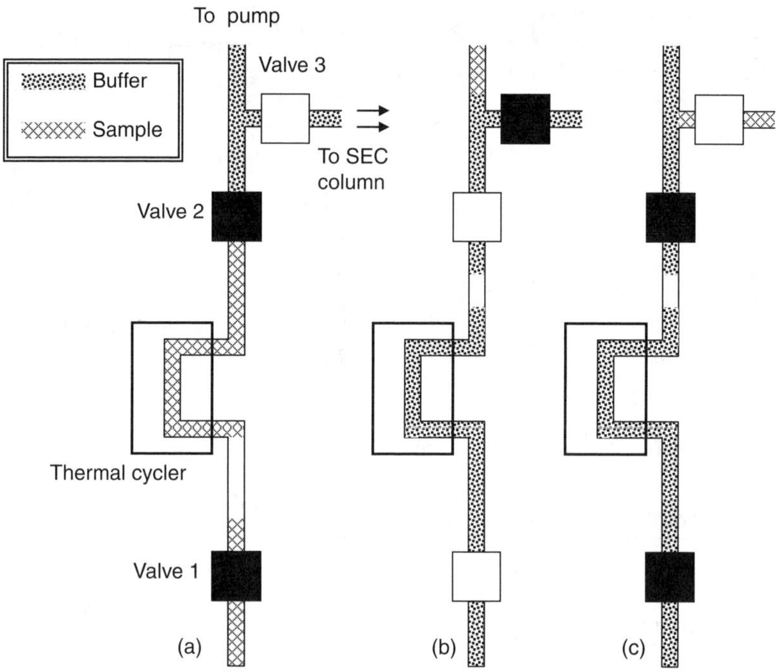

FIGURE 16.21 Multiplexed freeze/thaw valves used to isolate sample during thermalcycling. (Reprinted with permission from Tan, H. and Yeung, E.S., *Anal. Chem.*, 70, 4044, 1998. Copyright 1998 American Chemical Society.)

loading matrix. The cleanup procedure was reported to be very reproducible, resulted in a 10- to 50-fold increase in the amount of sequencing fragments loaded onto the gel, and facilitated the realization of read-lengths in excess of 1000 bases.

Manufacturers of commercial systems generally recommend a spin-column cleanup method or an ethanol precipitation/washing cleanup protocol that must then be followed by suspension of the lyophilized DNA pellet in a sample loading matrix designed to optimize or enhance sample loading. Note that separating DNA sequencing fragments generated from polymerase chain reaction (PCR) amplicons is generally less problematic than when larger template DNA is sequenced, thus supporting the idea that the presence of template DNA in the sample often leads to poor performance.

16.2.4.2 Separation Parameters

Several groups have searched for optimum run parameter settings in an attempt to maximize sequencing rates and read-lengths in both fixed and replaceable gels [15,18,21,24,50,67,73,80,160]. Rather than provide an exhaustive review of those studies here, we describe a few metrics that we have found useful in the optimization of separation conditions. The first of these is the "cross-over plot" in which peak width and adjacent peak spacing are plotted against base number [55]. Figure 16.22 shows cross-over plots for the separation of dye-terminator labeled sequencing fragments run at three different field strengths using a CEQ system (Beckman Coulter) and a linear polyacrylamide gel. A qualitative resolution limit is visible in these plots as the point where the two curves cross. At that point in the separation, the peak-to-peak spacing has dropped to the level where it is equal to the peak widths. Beyond this intersection point, discerning individual bands becomes increasingly difficult. Note that the cross-over point in these figures is observed to occur at decreasing base number as the field strength is increased and so any increase in speed realized by using higher voltage is offset by a loss in read-length.

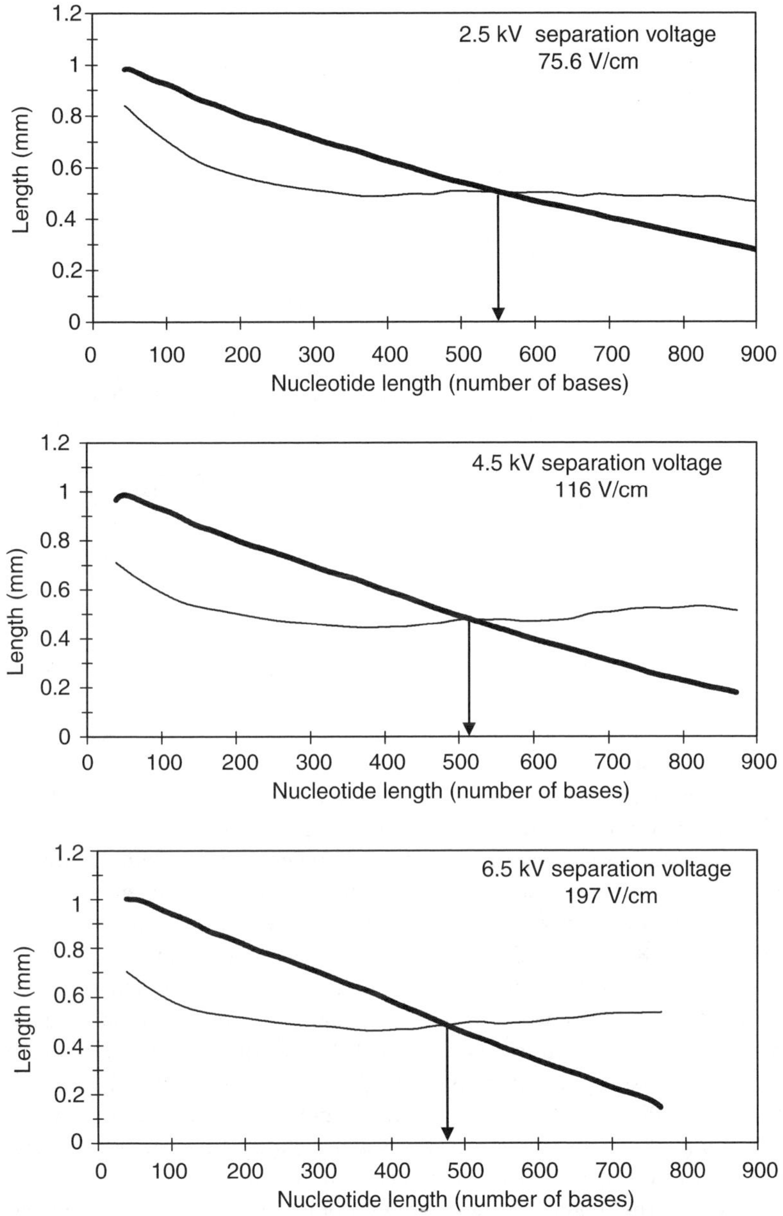

FIGURE 16.22 Peak-to-peak spacing (thick curves) and full-width at half maximum (FWHM) values (thin curves) as a function of nucleotide length for several run voltages.

Another useful, and perhaps more significant, metric to monitor as run conditions are varied is the "98% cutoff" or "length of read" number. This refers to the length of read achieved before the cumulative errors reach a value equal to or greater than a prespecified value of 2%.

A third metric that has proven useful in the optimization of run parameters and gel formulations is the "first base called" (FBC). This is the first position in the sequence that is correctly called and it is located typically 5–15 bases from the 3′ end of the primer used in the extension reaction. Factors affecting the FBC include the gels ability to sieve over a wide range of sizes, including the smaller extension products and congestion that often occurs early in the electropherogram due to coelution of unincorporated terminators, salts, and unextended primer.

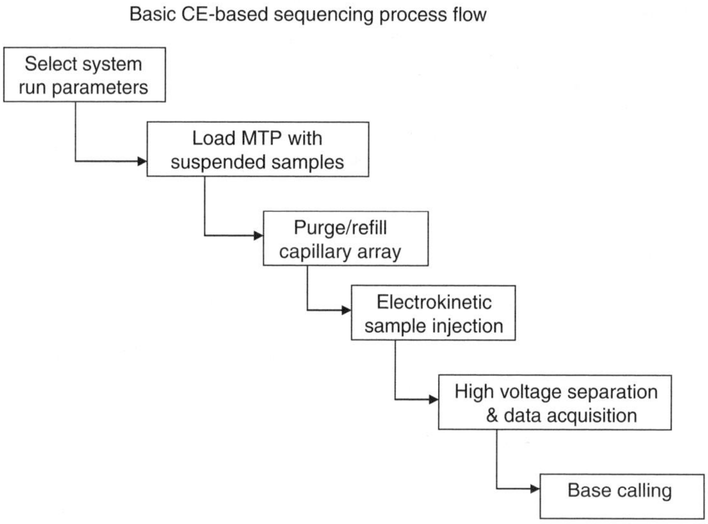

FIGURE 16.23 High-level graphical depiction of the steps involved in CE/LIF-based DNA sequence determination.

A fourth metric that we have found useful is to track resolution and/or base-calling accuracy over short stretches of sequence "known to be difficult" (KD stretches). This term refers, for example, to regions in electropherograms where anomalous migration occurs due to either secondary structure or the presence of salt fronts.

Figure 16.23 illustrates the steps involved in executing a typical CE-based DNA sequencing separation using one of today's automated sequencers. Run parameters such as injection voltage and duration, capillary temperature, denaturation time and temperature, separation voltage and duration, and sample well assignments are pre-programmed using simple graphical user interfaces. After cleanup, sequencing reaction samples are suspended in a suitable loading matrix and are transferred to appropriate wells in a 384- or 96-well microtiter plate. Fresh gel is automatically introduced into the capillaries of the array, optical alignment is verified, samples are denatured and then electrokinetically loaded into the capillaries. Recommended injection parameters vary somewhat but generally involve loading the samples onto the gel at low voltage (~60 V/cm) for approximately 30 s. Separation is initiated, often with the voltage ramped up over a prespecified time interval, to a final separation voltage on the order of 100–200 V/cm. Higher field strengths will result in faster separations but often at the cost of shorter read-lengths and/or reduced system stability. Multicolor detector output signals are acquired at data rates suitable to accurately represent the corresponding peak profiles (generally in the range of 2–4 Hz). Individual capillary current profiles, which are a useful diagnostic, are also often collected throughout the injection and separation process.

16.2.5 Algorithms

16.2.5.1 Basic Sequence Calling Steps

The importance and value of base-calling algorithms that have been developed for automated DNA sequencing cannot be overstated. Figure 16.24 graphically depicts the major steps involved in the analysis of raw detector output signals from today's capillary-based sequencers. Raw LIF signals are collected at a data acquisition rate sufficient to adequately represent the true concentration profiles of the narrow bands migrating through the beam. The raw data is generally smoothed to reduce noise and is baseline corrected to reduce drift. Color separation refers to dealing with the fact that fluorescence emissions typically exhibit some degree of overlap in the spectral channels monitored for

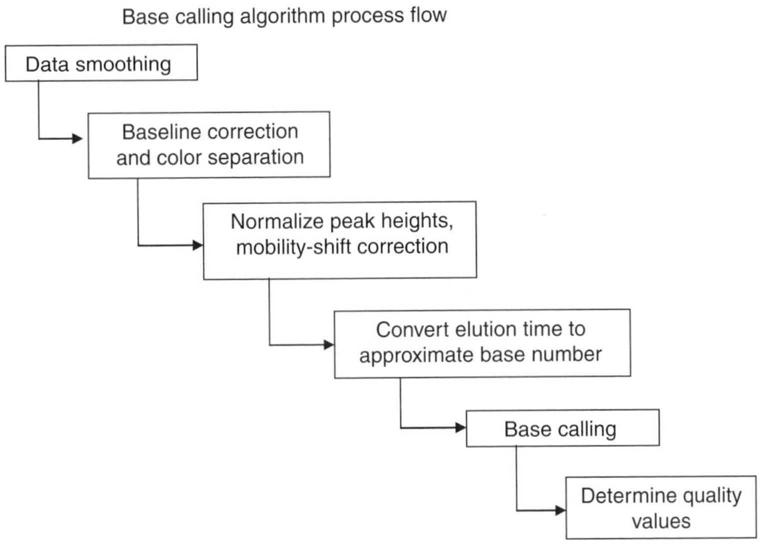

FIGURE 16.24 Graphical depiction of the steps involved in analyzing CE/LIF sequencing data for base calling.

each of the four fluorescent labels. Manufacturers of these systems understand the degree of spectral overlap that exists in these measurements and are able to mathematically eliminate this concern so that an unambiguous base assignment is made at each position. Peak-height uniformity has seen improvements over the years, owing both to improved sequencing reaction chemistry (e.g., the role played by pyrophosphatase in References 26 and 161) and algorithm-based mathematical normalization for differences in response across the dye set and for signal decay with increasing base number. The four different dyes attached to the terminators can also have an impact on fragment mobility and thus can complicate the clear size ordering assumed to be provided by the gel matrix. This issue has also been minimized through both chemistry improvements (similarity of dye structures, common degree of sulfonation, etc.) and mathematical algorithm corrections.

16.2.5.2 Quality Values

Phred is a base-calling algorithm, developed by Ewing and Green [162,163] that reads DNA sequence data files in a variety of formats, makes base call assignments, and also associates a highly accurate, base-specific quality score with each base call. The reported Phred quality scores are logarithmically related to error probabilities as shown in Table 16.2. Manufacturers of commercial sequencing systems also incorporate similar base-specific quality scores into their base-calling algorithms.

16.3 THE NEXT GENERATION: SEPARATIONS IN MICROFLUIDIC SYSTEMS

With replaceable gel CE now having effectively displaced all slab-based DNA sequencing, many have speculated that the next generation of fully automated, high-throughput DNA sequencing systems will be based on the use of microfluidic technologies such as compact plates or discs that are channel etched and micro-plumbed at high density [7,8,164–171]. Using technologies such as photo and chemical etching, originally developed for the microelectronics industry, features much smaller than those required for conventional CE may be created directly in fused silica or other materials. This technology could allow for significant miniaturization of multicolumn systems, more control options

TABLE 16.2
Error Probability for Various Phred Quality Scores

Phred Quality Score	Probability of Error at That Base Call	Basecall Accuracy at That Base (%)
10	1 in 10	90.0000
20	1 in 100	99.0000
30	1 in 1000	99.9000
40	1 in 10,000	99.9900
50	1 in 100,000	99.9990
60	1 in 1,000,000	99.9999

for sample introduction, and simplification of fluid handling in even more highly paralleled systems. The ability to reliably introduce much smaller sample plugs into the separation channels (reduced injection variance) may one day allow even faster separations to be run using significantly shorter separation channels. As demonstrated below, plug lengths as short as 30 μm may be introduced in microfluidic channels. Low signal-to-noise issues anticipated with reduced sample loading may be resolved using innovative solutions such as brighter fluorophores and increasing channel depths in the detection region.

The first DNA separation with single base resolution performed in a microfluidic device was reported by Effenhauser et al. [172]. Using standard photolithographic techniques channels were etched into a polished glass plate over which a second plate containing micromachined holes was thermally bonded (Figure 16.25). Both thin channels (50 μm wide and 12 μm deep) and broad channels (1000 μm wide and 12 μm deep) were created in this study. Before use, all the channels were filled with a 10% AA solution that was allowed to polymerize *in situ*. Pipette tips were glued to the holes in the covering plate and acted as reservoirs. Referring to Figure 16.25, sample was injected from reservoir 1 into the main separation channel by applying a field across 1 and 5. The loaded sample plug was 150 μm long with a volume of 90 pL [164]. Separation of this plug was achieved by switching the field to ports 3 and 7. LIF detection of the fluorescein-labeled fragments was made 3.8 cm downstream of the injection manifold and single base resolution of fragments 10–25 bases long was achieved in less than 45 s [172].

Mathies and coworkers [173] developed a similar system to separate double-stranded DNA restriction fragment digests. Standard photolithography techniques were also employed here [174]. Specifically, the channels were etched into a glass slide by first covering the slide with a photoresist film and then transferring the channel pattern to the film by exposing UV light through a patterned mask. The photoresist film exposed to the UV light dissolved while the remaining film hardened. The exposed glass region was then chemically etched with solutions of NH_4F/HF to form the appropriate channels and the remaining photoresist film was removed with H_2SO_4/H_2O_2. A glass plate with access holes drilled into it was then thermally bonded to the etched plate. Channel widths of 30–120 μm were created. To prepare the channels for electrophoresis and eliminate EOF, the surfaces were treated using a modified version of the Hjertén protocol. A general scheme for sample injection and separation is displayed in Figure 16.26. Low-viscosity HEC separation matrix was placed in the separation channel by applying vacuum to opening 4. The cross channel regions were filled with TBE buffer. Electrodes, buffers, and sample were placed in reservoirs numbered 1 through 4. Two injection types were explored. For "stack" injection, sample was placed in 3 while a field was applied across 3 and 4. For "plug" injection, sample was placed in 3 and a field was applied across 1 and 3 to move sample across the intersecting separation channel. In either case, once sample was injected, separation occurred by applying the field across positions 2 and 4. Initially, single-color detection was made using an optical arrangement similar to that described for multicapillary work, as illustrated in Figures 16.13 and 16.14.

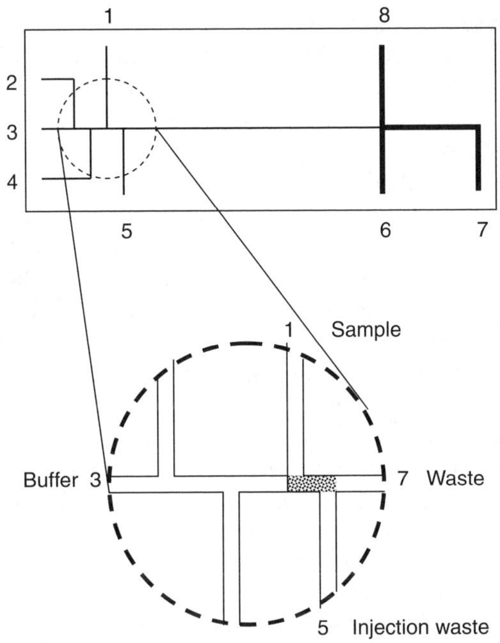

FIGURE 16.25 Microfluidic design for sample injection from a twin-T channel geometry. (Reprinted with permission from Effenhauser, C.S., et al., *Anal. Chem.*, 66, 2949, 1994. Copyright 1994 American Chemical Society.)

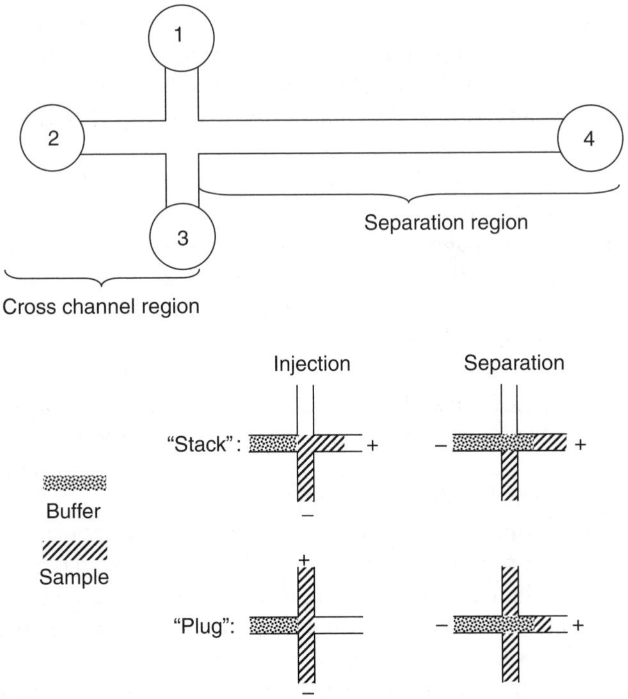

FIGURE 16.26 Schematic diagram describing two different modes of injection in a microfabricated cross channel format. (Reprinted with permission from Woolley, A.T. and Mathies, R.A., *Proc. Natl Acad. Sci. USA*, 91, 11348, 1994. Copyright 1994 National Academy of Sciences, USA.)

The first reported DNA sequencing separation with this system utilized plug injection of primer-labeled fragments onto an LPA matrix in a channel having an effective separation length of 3.5 cm and resulted in a four-color basecall of ~150 bases at 97% accuracy in 540 s [175]. Further work using a replaceable 4% LPA matrix, delivered via a high-pressure loader [176], and the "twin-T" injection scheme used earlier by Effenhauser [172], yielded read-lengths of 500 bases at 99.4% accuracy in less than 20 min [177]. Microfluidic chips containing 12 [178], 16 [179], 96 [180–183], 188 [184], and 384 [185,186] separation channels have since been reported. Linear 96-channel systems were constructed [179,180] as were novel radial microplates for 96- and 384-channel systems [181–183,185,186], although no sequencing was demonstrated with the 384-channel system. Figure 16.27 displays a system with 384 channels etched into a 20 cm diameter wafer; lanes were ~60 μm wide and 30 μm deep, with an effective separation length of 8.0 cm. A multicapillary array loader was created to transfer sample (via pressure) from a microtiter plate to the radial microplate [181]. For detection, a rotary scanning confocal fluorescence detector was created [181,182]. Similar to the linear scan head developed to interrogate a multicapillary array (see Figure 16.13), the objective is placed on a rotating scan head (Figure 16.28) [181]. This system was capable of collecting data at 200 revolutions per second, with 5000 data points collected per revolution for each color. Given that the scanning diameter was 2 cm, a spatial resolution of ~12.6 μm was achieved. A detection limit of ~1 pM at a SNR of 2 was reported using this system [182]. Simultaneous four-color sequencing of primer labeled DNA fragments yielded an average of 400 high-quality bases in each of the 95 channels (one channel failed due to a defect in the photolithography) [183].

Ehrlich and coworkers [187] performed a systematic study of LPA polymer concentration and composition, device temperature, effective channel length, and electric field strength for microfluidic-based DNA sequencing platforms. They determined that separations performed in an 11.5 cm channel (the longest used in this work) at 50°C using a field strength of 125 V/cm and a separation matrix consisting of 3.0% (w/w) 10 MDa plus 1.0% (w/w) 50 kDa LPA were most optimal, yielding a read-length of 640 bases at 98.5% accuracy in ~30 min for primer-labeled DNA fragments. Very long read-lengths required an extension of the column length and separations in 40 cm channels (40 μm deep and 90 μm wide at the top) yielded reads >800 bases at 98% accuracy in 80 min at 50°C [188]. Comparable results required 180 min of run time on the MegaBACE 1000 system (GE Healthcare) [188,189].

In addition to the advantages of performing separations in microfluidic channels presented at the beginning of this section, several others are now evident. For example, in all microfluidic systems in which the sample is loaded through a gel-filled "T" region before separation, "preseparation" is occurring. With judicious injection timing, unwanted materials contained in the sample solution may be prevented from entering the separation zone, thereby increasing separation quality [190,191]. For example, fragments contained in the sample that are larger than the upper resolving limit of the gel normally coelute as a large unresolved "blob" at the end of the electropherogram. This signal can be completely eliminated from the analysis by running sample through the T junction just long enough to allow the largest resolvable fragments to transit the intersection. Once the high-voltage contacts are switched from injection to separation positions, only the resolvable fraction of the sample will be directed on for separation and analysis.

In addition, such injection methods have also been reported to enhance loading of larger fragments relative to direct electrokinetic sample injection onto a gel-filled capillary [179,192]. Finally, because there is no analogy to the polyimide cladding protecting capillaries in microfluidic channels, the entire chip process may be monitored with a CCD camera, yielding information about sample injection and providing flexibility in detection [193,194].

Ehrlich and coworkers [188] found that long channel lengths yielded longer read-lengths, with a channel 40 cm long yielding reads of greater than 800 bases at 98% accuracy in 80 min. However, long straight channels are difficult to create primarily because of the special equipment needed to etch large plates [184,191]. To remedy this situation, some groups have created channels with several turns (i.e., serpentine channels) to increase column length. However, dispersion effects degraded

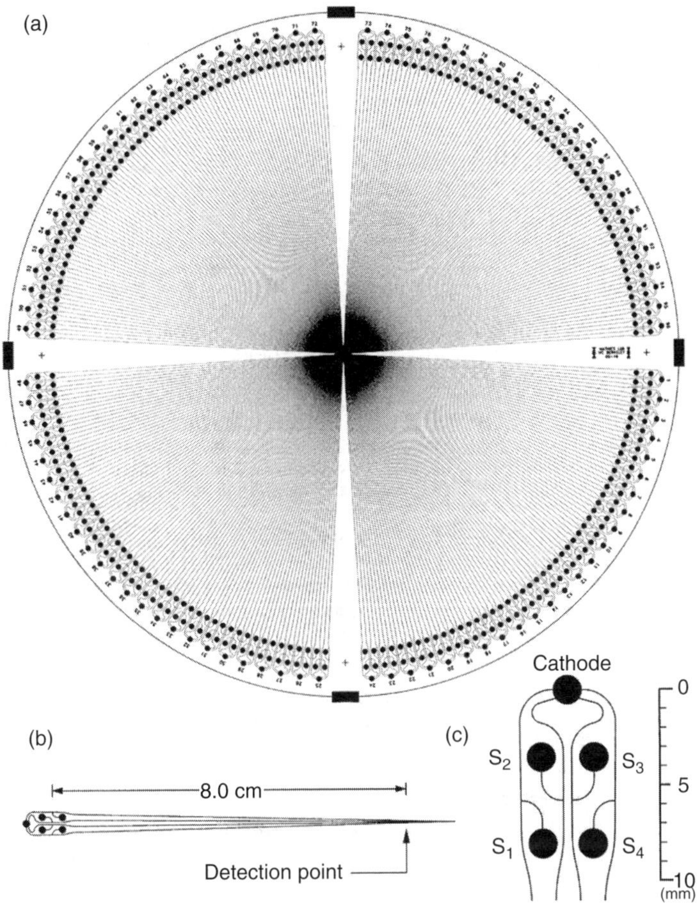

FIGURE 16.27 (a) Layout of 384-channel microfluidic electrophoresis device. (b) Expanded view of a single quartet of channels with their injectors. (c) Expanded view of a quartet of individual sample reservoirs. (Reprinted with permission from Emrich, C.A., et al., *Anal. Chem.*, 74, 5076, 2002. Copyright 2002 American Chemical Society.)

the individual bands because the path-length within the channel varied depending on whether the molecules traveled along the interior or exterior radius [195,196]. New designs implemented with unique channel geometries in the turn regions have minimized this problem [197,198].

To circumvent the drawbacks associated with the etching of long straight channels on large glass plates, Liu [191] created a hybrid capillary/microchip design. In his work, twin-T injectors with round channels as well as round capillary connection channels were created using very narrow line-width isotropic etching (Figure 16.29). The chip was then diced and a 75/200 μm capillary inserted along the chip edge into the connection channel, mating with the twin-T region (Figure 16.29b). Hybrid 16-capillary chips were created and used on a modified MegaBACE 1000 system (GE Healthcare) for multicapillary detection [199]. Sample loaded in the twin-T was then separated in the 40 cm long capillary, resulting in >800 bases read to 98.5% accuracy in 56 min. Comparable capillary-only runs with electrokinetic injection yielded only ~650 bases [191].

The use of microfluidics in a commercial DNA sequencing system has yet to be demonstrated. This is likely due to the lack of a compelling driver. Capillaries are still relatively inexpensive in comparison with etched plates and schemes have been engineered to allow fairly cost effective replacement of subgroups of capillaries, even in the high-throughput systems employing hundreds of

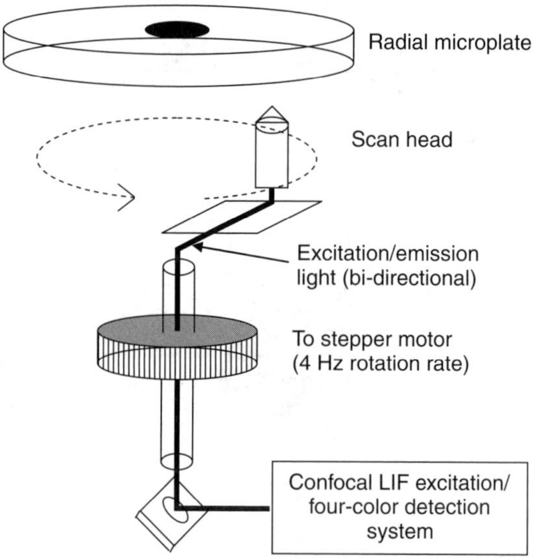

FIGURE 16.28 Rotary confocal fluorescence microplate detector. (Reprinted with permission from Shi, Y., et al., *Anal. Chem.*, 71, 5354, 1999. Copyright 1999 American Chemical Society.)

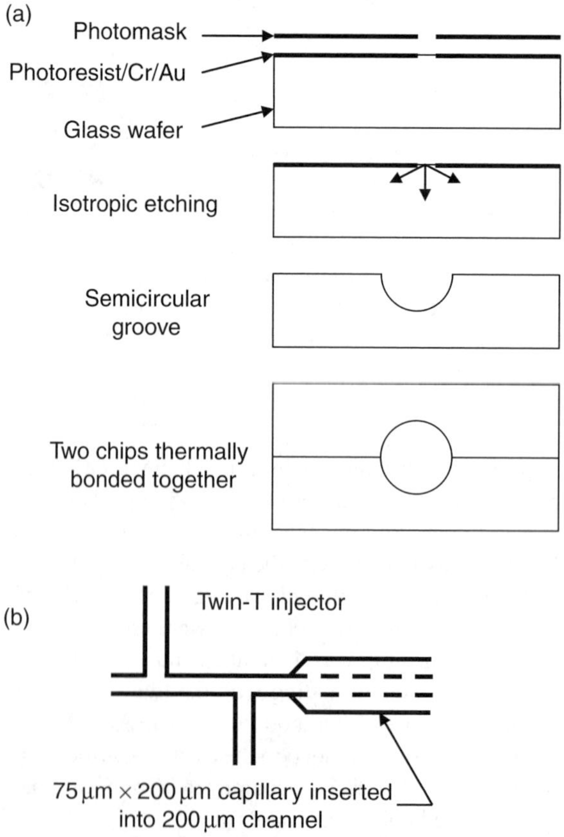

FIGURE 16.29 (a) Isotopic etching to create round channels. (b) Top view of capillary inserted into the connection channel, mating with twin-T region. (Reprinted with permission from Liu, S., *Electrophoresis*, 24, 3755, 2003.)

separation channels. Future improvements in the field of microfluidics may result in a transition from capillaries to etched substrates but competitive sequencing technologies that seem to have finally eliminated the need for electrophoresis are now on the horizon [200–203] and progress in this area will likely redefine the role of CE in DNA sequencing.

16.4 OTHER NUCLEIC ACID APPLICATIONS

Although the focus of this chapter has been on capillary-based DNA sequencing, it is important to point out that the same systems originally developed for DNA sequencing are also capable of supporting a wide variety of other important electrophoresis-based assays. Methods and kits have been developed for applications such as gene expression, heterozygote detection, mutation analysis, allele identification, single nucleotide polymorphism screening, microsatellite instability, and amplified fragment length polymorphism (AFLP) fingerprinting and these applications are now finding widespread acceptance in a variety of fields of science including biomedical research, human and animal diagnostics, and agriculture.

16.5 CONCLUSIONS

Clearly, a remarkable amount of progress has been made over the past two decades in the area of automated DNA sequencing. Early sequencing methodologies were slow, tedious, labor–intensive, and riddled with error-making opportunities. Until the late 1980s, slab gels were manually poured, samples were loaded by hand, and images created using X-ray film were painstakingly read by eye. Automated film readers were introduced to improve the manual reading process but their utility was short lived as better technologies were soon developed. The late 1980s and early 1990s saw important improvements as the manual sequencing process was partially automated. Radio isotopic tags were displaced by the use of fluorescent labels and the process of reading slab gels was automated using multilane fluorescence slab gel readers and powerful base-calling algorithms. The initiative to sequence the human genome drove the development of faster, high-throughput sequencing technologies, and CE emerged in the late 1990s as that decade's winner. Slab gels were displaced by capillary arrays filled with replaceable gels and sensitive fluorescence technologies were successfully adapted to the more challenging multicapillary format. Current large-scale sequencing projects would be slowed unthinkably without automated capillary sequencing machines, which became commercially available in the late 1990s and have now made DNA sequencing much quicker and more reliable. Today, one automated capillary DNA sequencing system can produce data at a rate of approximately 1 million bases of raw sequence per day with as little as 15 min of user intervention required. Without question, the works cited in this chapter and the authors who performed those studies have all contributed to a technology and field that has enjoyed impressive progress.

New technologies involving nonelectrophoretic, massively paralleled sequencing strategies now seem poised to deliver the next significant improvement in the area of lower cost, high-throughput DNA sequencing. These systems are targeting single platform throughput levels of tens of millions of bases per hour and may make whole genome sequencing more affordable.

ACKNOWLEDGMENTS

The authors wish to thank Cynthia Johnson, Clarence Lew, Feng Liu, Lucy Liu, Paul Kraght, Veronica V. Colinayo, and Christopher Pentoney for their assistance in creating this chapter and Dr James C. Osborne for his support.

REFERENCES

1. Dovichi, N.J., DNA sequencing by capillary electrophoresis, *Electrophoresis*, 18, 2393, 1997.
2. Quesada, M.A., Replaceable polymers in DNA sequencing by capillary electrophoresis, *Curr. Opin. Biotechnol.*, 8, 82, 1997.
3. Slater, G.W., Kist, T.B.L., Ren, H., and Drouin, G., Recent developments in DNA electrophoretic separations, *Electrophoresis*, 19, 1525, 1998.
4. Dolnik, V., DNA sequencing by capillary electrophoresis (review), *J. Biochem. Biophys. Methods*, 41, 103, 1999.
5. Timperman, A.T. and Sweedler, J.V., Capillary electrophoresis with wavelength-resolved fluorescence detection, *Analyst*, 121, 45R, 1996.
6. Schmalzing, D., Koutny, L., Salas-Solano, O., Adourian, A., Matsudaira, P., and Ehrlich, D., Recent developments in DNA sequencing by capillary and microdevice electrophoresis, *Electrophoresis*, 20, 3066, 1999.
7. Carrilho, E., DNA sequencing by capillary array electrophoresis and microfabricated array systems, *Electrophoresis*, 21, 55, 2000.
8. Mitnik, L. Novotny, M., Felten, C., Buonocore, S., Koutny, L., and Schmalzing, D., Recent advances in DNA sequencing by capillary and microdevice electrophoresis, *Electrophoresis*, 22, 4104, 2001.
9. Righetti, P.G., Gelfi, C., and D'Acunto, M.R., Recent progress in DNA analysis by capillary electrophoresis, *Electrophoresis*, 23, 1361, 2002.
10. Venter, J.C., et al., The sequence of the human genome, *Science*, 291, 1304, 2001.
11. The International Human Genome Mapping Consortium, A physical map of the human genome, *Nature*, 409, 934, 2001.
12. Sanger, F., Nicklen, S., and Coulson, A.R., DNA sequencing with chain-terminating inhibitors, *Proc. Natl Acad. Sci. USA*, 74, 5463, 1977.
13. Huang, X.C., Quesada, M.A., and Mathies, R.A., DNA sequencing using capillary array electrophoresis, *Anal. Chem.*, 64, 2149, 1992.
14. Smith, L.M., Sanders, J.Z., Kaiser, R.J., Hughes, P., Dodd, C., Connell, C.R., Heiner, C., Kent, S.B.H., and Hood, L.E., Fluorescence detection in automated DNA sequence analysis, *Nature*, 321, 674, 1986.
15. Luckey, J.A., Drossman, H., Kostichka, A.J., Mead, D.A., D'Cunha, J., Norris, T.B., and Smith, L.M., High speed DNA sequencing by capillary electrophoresis, *Nucleic Acids Res.*, 18, 4417, 1990.
16. Ju, J., Kheterpal, I., Scherer, J.R., Ruan, C., Fuller, C.W., Glazer, A.N., and Mathies, R.A., Design and synthesis of fluorescence energy transfer dye-labeled primers and their application for DNA sequencing and analysis, *Anal. Biochem.*, 231, 131, 1995.
17. Zhang, J.Z., Fang, Y., Hou, J.Y., Ren, H.J., Jiang, R., Roos, P., and Dovichi, N.J., Use of non-cross linked polyacrylamide for four-color DNA sequencing by capillary electrophoresis separation of fragments up to 640 bases in length in two hours, *Anal. Chem.*, 67, 4589, 1995.
18. Carrilho, E., Ruiz-Martinez, M.C., Berka, J., Smirnov, I., Goetzinger, W., Miller, A.W., Brady, D., and Karger, B.L., Rapid DNA sequencing of more than 1000 bases per run by capillary electrophoresis using replaceable linear polyacrylamide solutions, *Anal. Chem.*, 68, 3305, 1996.
19. Klepárnik, K., Foret, F., Berka, J., Goetzinger, W., Miller, A.W., and Karger, B.L., The use of elevated column temperature to extend DNA sequencing read lengths in capillary electrophoresis with replaceable polymer matrices, *Electrophoresis*, 17, 1860, 1996.
20. Goetzinger, W., Kotler, L., Carrilho, E., Ruiz-Martinez, M.C., Salas-Solano, O., and Karger, B.L., Characterization of high molecular mass linear polyacrylamide powder prepared by emulsion polymerization as a replaceable polymer matrix for DNA sequencing by capillary electrophoresis, *Electrophoresis*, 19, 242, 1998.
21. Salas-Solano, O., Carrilho, E., Kotler, L., Miller, A.W., Goetzinger, W., Sosic, Z., and Karger, B.L., Routine DNA sequencing of 1000 bases in less than one hour by capillary electrophoresis with replaceable linear polyacrylamide solutions, *Anal. Chem.*, 70, 3996, 1998.
22. Zhang, J., Voss, K.O., Shaw, D.F., Roos, K.P., Lewis, D.F., Yan, J., Jiang, R., Ren, H., Hou, J.Y., Fang, Y., Puyang, X., Ahmadzadeh, H., and Dovichi, N.J., A multiple-capillary electrophoresis system for small-scale DNA sequencing and analysis, *Nucleic Acids Res.*, 27, e36, 1999.
23. Neumann, M., Herten, D.-P., Dietrich, A., Wolfrum, J., and Sauer, M., Capillary array scanner for time-resolved detection and identification of fluorescently labeled DNA fragments, *J. Chromatogr. A*, 871, 299, 2000.

24. Zhou, H., Miller A.W., Sosic, Z., Buchholz, B., Barron, A.E., Kotler, L., and Karger, B.L., DNA sequencing up to 1300 bases in two hours by capillary electrophoresis with mixed replaceable linear polyacrylamide solutions, *Anal. Chem.*, 72, 1045, 2000.
25. Kotler, L., He, H., Miller, A.W., and Karger, B.L., DNA sequencing of close to 1000 bases in 40 minutes by capillary electrophoresis using dimethyl sulfoxide and urea as denaturants in replaceable linear polyacrylamide solutions, *Electrophoresis*, 23, 3062, 2002.
26. Tabor, S. and Richardson, C.C., DNA sequence analysis with a modified bacteriophage T7 DNA polymerase: Effect of pyrophosphorolysis and metal ions, *J. Biol. Chem.*, 265, 8322, 1990.
27. Ansorge, W., Zimmermann, J., Schwager, C., Stegemann, J., Erfle, H., and Voss, H., One label, one tube, Sanger DNA sequencing in one and two lanes on a gel, *Nucleic Acids Res.*, 18, 3419, 1990.
28. Pentoney, S.L., Jr., Konrad, K.D., and Kaye, W., A single-fluor approach to DNA sequence determination using high performance capillary electrophoresis, *Electrophoresis*, 13, 467, 1992.
29. Swerdlow, H., Zhang, J.Z., Chen, D.Y., Harke, H.R., Grey, R., Wu, S., and Dovichi, N.J., Three DNA sequencing methods using capillary gel electrophoresis and laser-induced fluorescence, *Anal. Chem.*, 63, 2835, 1991.
30. Chen, D., Harke, H.R., and Dovichi, N.J., Two-label peak-height encoded DNA sequencing by capillary gel electrophoresis: three examples, *Nucleic Acids Res.*, 20, 4873, 1992.
31. Gassman, E., Kuo, J.E., and Zare, R.N., Electrokinetic separation of chiral compounds, *Nature*, 339, 813, 1985.
32. Gordon, M.J., Huang, X., Pentoney, S.L., Jr., and Zare, R.N., Capillary electrophoresis, *Science*, 242, 224, 1988.
33. Prober, J.M., Trainor, G.L., Dam, R.J., Hobbs, F.W., Robertson, C.W., Zagursky, R.J., Cocuzza, A.J., Jensen, M.A., and Baumeister, K., A system for rapid DNA sequencing with fluorescent chain-terminating dideoxynucleotides, *Science*, 238, 336, 1987.
34. Kim, Y. and Yeung, E.S., Separation of DNA sequencing fragments up to 1000 bases by using poly(ethylene oxide)-filled capillary electrophoresis, *J. Chromatogr. A*, 781, 315, 1997.
35. Ju, J., Ruan, C., Fuller, C.W., Glazer, A.N., and Mathies, R.A., Fluorescence energy transfer dye-labeled primers for DNA sequencing and analysis, *Proc. Ntl. Acad. Sci. USA*, 92, 4347, 1995.
36. Lee, L.G., Spurgeon, S.L., Heiner, C.R., Benson, S.C., Rosenblum, B.B., Menchen, S.M., Graham, R.J., Constantinescu., A., Upadhya, K.G., and Cassel, J.M., New energy transfer dyes for DNA sequencing, *Nucleic Acids Res.*, 25, 2816, 1997.
37. Rosenblum, B.B., Lee, L.G., Spurgeon, S.L., Khan, S.H., Menchen, S.M., Heiner, C.R., and Chen, S.M., New dye-labeled terminators for improved DNA sequencing patterns, *Nucleic Acids Res.*, 25, 4500, 1997.
38. Lee, C.S., Improved capillary electrophoretic separations associated with controlling electroendoosmotic flow, in *Handbook of Capillary Electrophoresis*, 2nd ed., Landers, J.P., ed., CRC Press, Boca Raton, FL, 1996, chap. 25.
39. Hjertén, S., High-performance electrophoresis elimination of electroendosmosis and solute adsorption, *J. Chromatogr.*, 347, 191, 1985.
40. Cobb, K.A., Dolnik, V., and Novotny, M., Electrophoretic separations of proteins in capillaries with hydrolytically stable surface structures, *Anal. Chem.*, 62, 2478, 1990.
41. Schmalzing, D., Piggee, C.A., Foret, F., Carrilho, E., and Karger, B.L., Characterization and performance of a neutral hydrophilic coating for the capillary electrophoretic separation of biopolymers, *J. Chromatogr. A*, 652, 149, 1993.
42. Hjertén, S. and Kubo, K., A new type of pH- and detergent-stable coating for elimination of electroendosmosis and adsorption in (capillary) electrophoresis, *Electrophoresis*, 14, 390, 1993.
43. Moritani, T., Yoon, K., Rafailovich, M., and Chu, B., DNA capillary electrophoresis using poly(vinylalcohol). I. inner capillary coating, *Electrophoresis*, 24, 2764, 2003.
44. Slater, G.W., Kenward, M., McCromick, L.C., and Gauthier, M.G., The theory of DNA separation by capillary electrophoresis, *Curr. Opin. Biotechnol.*, 14, 58, 2003.
45. Ruiz-Martinez, M.C., Berka, J., Belenkii, A., Foret, F., Miller, A.W., and Karger, B.L., DNA sequencing by capillary electrophoresis with replaceable linear polyacrylamide and laser-induced fluorescence detection, *Anal. Chem.*, 65, 2851, 1993.

46. Pariat, Y.F., Berka, J., Heiger, D.N., Schmitt, T., Vilenchik, M., Cohen, A.S., Foret, F., and Karger, B.L., Separation of DNA fragments by capillary electrophoresis using replaceable linear polyacrylamide matrices, *J. Chromatogr. A*, 652, 57, 1993.
47. Sudor, J., Foret, F., and Boček, P., Pressure refilled polyacrylamide columns for the separation of oligonucleotides by capillary electrophoresis, *Electrophoresis*, 12, 1056, 1991.
48. Guttman, A., Capillary electrophoresis using replaceable gels, US Patents 5,332,481, 5,421,980, RE37,606E, and RE37,941E.
49. Guttman, A. and Cooke N., Capillary gel affinity electrophoresis of DNA fragments, *Anal. Chem.*, 63, 2038, 1991.
50. Grossman, P.D. and Soane, D.S., Experimental and theoretical studies of DNA separations by capillary electrophoresis in entangled polymer solutions, *Biopolymers*, 31, 1221, 1991.
51. Karger, B.L., Foret, F., and Berka, J., Capillary electrophoresis with polymer matrices: DNA and protein separation and analysis, *Methods Enzymol.*, 271, 293, 1996.
52. Sunada, W.M. and Blanch, H.W., Polymeric separation media for capillary electrophoresis of nucleic acids, *Electrophoresis*, 18, 2243, 1997.
53. Heller, C., Principles of DNA separation with capillary electrophoresis, *Electrophoresis*, 22, 629, 2001.
54. Viovy, J.-L., Electrophoresis of DNA and other polyelectrolytes: Physical mechanisms, *Rev. Modern Phys.*, 72, 813, 2000.
55. Quesada, M.A. and Menchen, S., Replaceable polymers for DNA sequencing by capillary electrophoresis, *Methods Mol. Biol.*, 162, 139, 2001.
56. Albarghouthi, M.N. and Barron, A.E., Polymeric matrices for DNA sequencing by capillary electrophoresis, *Electrophoresis*, 21, 4096, 2000.
57. Ying, Q. and Chu, B., Overlap concentration of macromolecules in solution, *Macromolecules*, 20, 362, 1987.
58. Wu, C., Quesada, M.A., Schneider, D.K., Farinato, R., Studier, F.W., and Chu, B., Polyacrylamide solutions for DNA sequencing by capillary electrophoresis: Mesh sizes, separation and dispersion, *Electrophoresis*, 17, 1103, 1996.
59. Ogston, A.G., The spaces in a uniform random suspension of fibers, *Trans. Faraday Soc.*, 54, 1754, 1958.
60. Slater, G.W., Mayer, P., and Drouin, G., Migration of DNA through gels, *Methods Enzymol.*, 270, 272, 1996.
61. Slater, G.W. and Noolandi, J., On the reptation theory of gel electrophoresis, *Biopolymers*, 25, 431, 1986.
62. Duke, T., Viovy, J.-L., and Semenov, A.N., Electrophoretic mobility of DNA in gels. I. New biased reptation theory including fluctuations, *Biopolymers*, 34, 239, 1994.
63. Konrad, K.D. and Pentoney, S.L., Jr., Contribution of secondary structure to DNA mobility in capillary gels, *Electrophoresis*, 14, 502, 1993.
64. Cohen, A.S., Najarian, D.R., Paulus, A., Guttman, A., Smith, J.A., and Karger, B.L., Rapid separation and purification of oligonucleotides by high-performance capillary gel electrophoresis, *Proc. Natl Acad. Sci. USA*, 85, 9660, 1988.
65. Guttman, A., Cohen, A.S., Heiger, D.N., and Karger, B.L., Analytical and micropreparative ultrahigh resolution of oligonucleotides by polyacrylamide gel high-performance capillary electrophoresis, *Anal. Chem.*, 62, 137, 1990.
66. Paulus, A., Gassmann, E., and Field, M.J., Calibration of polyacrylamide gel columns for the separation of oligonucleotides by capillary electrophoresis, *Electrophoresis*, 11, 702, 1990.
67. Drossman, H., Luckey, A.J., Kostichka, J., D'Cunha, J., and Smith, L.M., High-speed separations of DNA sequencing reactions by capillary electrophoresis, *Anal. Chem.*, 62, 900, 1990.
68. Swerdlow, H. and Gesteland, R., Capillary gel electrophoresis for rapid, high resolution DNA sequencing, *Nucleic Acids Res.*, 18, 1415, 1990.
69. Cohen, A.S., Najarian, D.R., and Karger, B.L., Separation and analysis of DNA sequence reaction products by capillary gel electrophoresis, *J. Chromatogr.*, 516, 49, 1990.
70. Swerdlow, H., Wu, S., Harke, H., and Dovichi, N.J., Capillary gel electrophoresis for DNA sequencing laser-induced fluorescence detection with the sheath flow cuvette, *J. Chromatogr.*, 516, 61, 1990.

71. Karger, A.E., Harris, J.M., and Gesteland, R.F., Multiwavelength fluorescence detection for DNA sequencing using capillary electrophoresis, *Nucleic Acids Res.*, 19, 4955, 1991.
72. Chiari, M., Nesi, M., Fazio, M., and Righetti, P.G., Capillary electrophoresis of macromolecules in "syrupy" solutions: Facts and misfacts, *Electrophoresis*, 13, 690, 1992.
73. Best, N., Arriga, E., Chen, D.Y., and Dovichi, N.J., Separation of fragments up to 570 bases in length by use of 6% T non-cross-linked polyacrylamide for DNA sequencing in capillary electrophoresis, *Anal. Chem.*, 66, 4063, 1994.
74. Doherty, E.A.S., Kan, C.-W., and Barron, A.E., Sparsely cross-linked "nanogels" for microchannel DNA sequencing, *Electrophoresis*, 24, 4170, 2003.
75. Doherty, E.A.S., Kan, C.-W., Paegel, B.M., Yeung, S.H.I., Cao, S., Mathies, R.A., and Barron, A.E., Sparsely cross-linked "nanogel" matrixes as fluid, mechanically stabilized polymer networks for high-throughput microchannel DNA sequencing, *Anal. Chem.*, 76, 5249, 2004.
76. Madabhushi, R.S., Separation of 4-color DNA sequencing extension products in noncovalently coated capillaries using low-viscosity polymer solutions, *Electrophoresis*, 19, 224, 1998.
77. Madabhushi, R., DNA sequencing in noncovalently coated capillaries using low viscosity polymer solutions, *Methods Mol. Biol.*, 163, 309, 2001.
78. Heller, C., Finding a universal low viscosity polymer for DNA separation (II), *Electrophoresis*, 19, 3114, 1998.
79. Heller, C., Separation of double-stranded and single-stranded DNA in polymer solutions: I. mobility and separation mechanism, *Electrophoresis*, 20, 1962, 1999.
80. Song, L., Liang, D., Fang, D., and Chu, B., Fast DNA sequencing up to 1000 bases by capillary electrophoresis using poly(*N*, *N*-dimethylacrylamide) as a separation medium, *Electrophoresis*, 22, 1987, 2001.
81. Griess, G.A., Hardies, S.C., and Serwer, P., Matrix conditioning for lengthened capillary DNA sequencing, *Electrophoresis*, 26, 102, 2005.
82. Albarghouthi, M.N., Buchholz, B.A., Huiberts, P.J., Stein, T.M., and Barron, A.E., Poly-*N*-hydroxyethylacrylamide (polyduramide): A novel, hydrophilic, self-coating polymer matrix for DNA sequencing by capillary electrophoresis, *Electrophoresis*, 23, 1429, 2002.
83. Song, L., Liang, D., Kielescawa, J., Liang, J., Tjoe, E., Fang, D., and Chu, B., DNA sequencing by capillary electrophoresis using copolymers of acrylamide and *N*, *N*-dimethyl acrylamide, *Electrophoresis*, 22, 729, 2001.
84. Wang, Y., Liang, D., Ying, Q., and Chu, B., Quasi-interpenetrating network formed by polyacrylamide and poly(*N*, *N*-dimethylacrylamide) used in high-performance DNA sequencing analysis by capillary electrophoresis, *Electrophoresis*, 26, 126, 2005.
85. Bashkin, J., Marsh, M., Barker, D., and Johnston, R., DNA sequencing by capillary electrophoresis with a hydroxyethylcellulose sieving buffer, *Appl. Theor. Electrophor.*, 6, 23, 1996.
86. Marsh, M., Tu, O., Dolnik, V., Roach, D., Solomon, N., Bechtol, K., Smietana, P., Wang, L., Li, X., Cartwright, P., Marks, A., Barker, D., Harris, D., and Baskin, J., High-throughput DNA sequencing on a capillary array electrophoresis system, *J. Capillary Electrophor.*, 4, 83, 1997.
87. Dolnik, V. and Gurske, W.A., Capillary electrophoresis in sieving matrices: Selectivity per base, mobility slope, and inflection slope, *Electrophoresis*, 20, 3373, 1999.
88. Kheterpal, I. and Mathies, R.A., Capillary array electrophoresis DNA sequencing, *Anal. Chem.*, 71, 31A, 1999.
89. Chang, H.T. and Yeung, E.S., Poly(ethyleneoxide) for high-resolution and high-speed separation of DNA by capillary electrophoresis, *J. Chromatogr. B*, 669, 113, 1995.
90. Fung, E.N. and Yeung, E.S., High-speed DNA sequencing by using mixed poly(ethylene oxide) solutions in uncoated capillary columns, *Anal. Chem.*, 67, 1913, 1995.
91. Tan, H. and Yeung, E.S., Characterization of dye-induced mobility shifts affecting DNA sequencing in poly(ethylene oxide) sieving matrix, *Electrophoresis*, 18, 2893, 1997.
92. Fung, E.N., Pang, H.-M., and Yeung, E.S., Fast DNA separations using poly(ethylene oxide) in non-denaturing medium with temperature programming, *J. Chromatogr. A*, 806, 157, 1998.
93. Wei, W. and Yeung, E.S., Improvements in DNA sequencing by capillary electrophoresis at elevated temperature using poly(ethylene oxide) as a sieving matrix, *J. Chromatogr. B*, 745, 221, 2000.
94. Preisler, J. and Yeung, E.S., Characterization of nonbonded poly(ethyleneoxide) coating for capillary electrophoresis via continuous monitoring of electroosmotic flow, *Anal. Chem.*, 68, 2885, 1996.

95. Gao, Q. and Yeung, E.S., A matrix for DNA separation: Genotyping and sequencing using poly(vinylpyrrolidone) solution in uncoated capillaries, *Anal. Chem.*, 70, 1382, 1998.
96. Song, J.M. and Yeung, E.S., Optimization of DNA electrophoretic behavior in poly(vinyl pyrrolidone) sieving matrix for DNA sequencing, *Electrophoresis*, 22, 748, 2001.
97. Dolnik, V., Gurske, W.A., and Padua, A., Galactomannans as a sieving matrix in capillary electrophoresis, *Electrophoresis*, 22, 707, 2001.
98. Menchen, S., Johnson, B., Winnik, M.A., and Xu, B., Flowable networks as DNA sequencing media in capillary columns, *Electrophoresis*, 17, 1451, 1996.
99. Chu, B. and Liang, D., Copolymer solutions as separation media for DNA capillary electrophoresis, *J. Chromatogr. A*, 966, 1, 2002.
100. Song, L., Liang, D., Kielescawa, J., Liang, J., Tjoe, E., Fang, D., and Chu, B., DNA sequencing by capillary electrophoresis using copolymers of acrylamide and *N*, *N*-dimethylacrylamide, *Electrophoresis*, 22, 729, 2001.
101. Albarghouthi, M.N., Buchholz, B.A., Doherty, E.A.S., Bogdan, F.M., Zhou, H., and Barron, A.E., Impact of polymer hydrophobicity on the properties and performance of DNA sequencing matrices for capillary electrophoresis, *Electrophoresis*, 22, 737, 2001.
102. He, H., Buchholz, B.A., Kotler, L., Miller, A.W., Barron, A.E., and Karger, B.L., DNA sequencing with hydrophilic and hydrophobic polymers at elevated column temperatures, *Electrophoresis*, 23, 1421, 2002.
103. Buchholz, B.A., Doherty, E.A.S., Albarghouthi, M.N., Bogdan, F.M., Zahn, J.M., and Barron, A.E., Microchannel DNA sequencing matrices with a thermally controlled "viscosity switch," *Anal. Chem.*, 73, 157, 2001.
104. Buchholz, B.A., Shi, W., and Barron, A.E., Microchannel DNA sequencing matrices with switchable viscosities, *Electrophoresis*, 23, 1398, 2002.
105. Kan, C.-W., Doherty, E.A.S, and Barron, A.E., A novel thermogelling matrix for microchannel DNA sequencing based on poly-*N*-alkoxyalkylacrylamide copolymers, *Electrophoresis*, 24, 4161, 2003.
106. Sudor, J., Barbier, V., Thirot, S., Godfrin, D., Hourdet, D., Millequant, M., Blanchard, J., and Viovy, J.-L., New block-copolymer thermoassociating matrices for DNA sequencing: Effect of molecular structure on rheology and resolution, *Electrophoresis*, 22, 720, 2001.
107. Barbier, V., Buchholz, B.A., Barron, A.E., and Viovy, J.-L., Comb-like copolymers as self-coating, low-viscosity and high-resolution matrices for DNA sequencing, *Electrophoresis*, 23, 1441, 2002.
108. Pentoney, Jr., S.L. and Sweedler, J.V., in *Handbook of Capillary Electrophoresis*, 2nd ed., Landers, J.P., ed., CRC Press, Boca Raton, FL, 1996, chap. 12.
109. Zagursky, R.J. and McCormick, R.M., DNA sequencing separations in capillary gels on a modified commercial DNA sequencing instrument, *BioTechniques*, 9, 74, 1990.
110. Carson, S., Cohen, A.S., Belenkii, A., Ruiz-Martinez, M.C., Berka, J., and Karger, B.L., DNA sequencing by capillary electrophoresis: Use of a two-laser-two-window intensified diode array detection system, *Anal. Chem.*, 65, 3219, 1993.
111. Mathies, R.A. and Huang, X.C., Capillary array electrophoresis: An approach to high-speed, high-throughput DNA sequencing, *Nature*, 359, 167, 1992.
112. Huang, X.C., Quesada, M.A., and Mathies, R.A., Capillary array electrophoresis using laser-excited confocal fluorescence detection, *Anal. Chem.*, 64, 967, 1992.
113. Kheterpal, I., Scherer, J.R., Clark, S.M., Radhakrishnan, A., Ju, J., Ginther, C.L., Sensabaugh, G.F., and Mathies, R.A., DNA sequencing using a four-color confocal fluorescence capillary array scanner, *Electrophoresis*, 17, 1852, 1996.
114. Bashkin, J.S., Bartosiewicz, M., Roach, D., Leong, J., Barker, D., Johnston, R., Implementation of a capillary array electrophoresis instrument, *J. Capillary Electrophor.*, 3, 61, 1996.
115. Scherer, J.R., Kheterpal, I., Radhakrishnan, A., Ja, W.W., and Mathies, R.A., Ultra-high throughput rotary capillary array electrophoresis scanner for fluorescent DNA sequencing and analysis, *Electrophoresis*, 20, 1508, 1999.
116. Ueno, K. and Yeung, E.S., Simultaneous monitoring of DNA fragments separated by electrophoresis in a multiplexed array of 100 capillaries, *Anal. Chem.*, 66, 1424, 1994.
117. Pang, H., Pavski, V., and Yeung, E.S., DNA sequencing using 96-capillary array electrophoresis, *J. Biochem. Biophys. Methods*, 41, 121, 1999.

118. Xue, G. and Yeung, E.S., Two-color excitation system for fluorescence detection in DNA sequencing by capillary array electrophoresis, *Electrophoresis*, 23, 1490, 2002.
119. Zhong, W. and Yeung, E.S., Multiplexed capillary electrophoresis for DNA sequencing with ultra violet absorption detection, *J. Chromatogr. A*, 960, 229, 2002.
120. Behr, S., Matzig, M., Levin, A., Eickhoff, H., and Heller, C., A fully automated multicapillary electrophoresis device for DNA analysis, *Electrophoresis*, 20, 1492, 1999.
121. Lu, X. and Yeung, E.S., Optimization of excitation and detection geometry for multiplexed capillary array electrophoresis of DNA fragments, *Appl. Spectrosc.*, 49, 605, 1995.
122. Lu, S.X. and Yeung, E.S., Side-entry excitation and detection of square capillary array electrophoresis for DNA sequencing, *J. Chromatogr. A*, 853, 359, 1999.
123. Quesada, M.A. and Zhang, S., Multiple capillary DNA sequencer that uses fiber-optic illumination and detection, *Electrophoresis*, 17, 1841, 1996.
124. Quesada, M.A., Dhadwal, H.S., Fisk, D., and Studier, F.W., Multi-capillary optical waveguides for DNA sequencing, *Electrophoresis*, 19, 1415, 1998.
125. Anazawa, T., Takahashi, S., and Kambara, H., A capillary-array electrophoresis system using side-entry on-column laser irradiation combined with glass rod lenses, *Electrophoresis*, 20, 539, 1999.
126. Hanning, A., Lindberg, P., Westberg, J., and Roeraade, J., Laser-induced fluorescence detection by liquid core waveguiding applied to DNA sequencing by capillary electrophoresis, *Anal. Chem.*, 72, 3423, 2000.
127. Hanning, A., Westberg, J., and Roeraade, J., A liquid core waveguide fluorescence detector for multicapillary electrophoresis applied to DNA sequencing in a 91-capillary array, *Electrophoresis*, 21, 3290, 2000.
128. Chen, F.-T., Tusak, A., Pentoney, S. Jr., Konrad, K., Lew, C., Koh, E., and Sternberg, J., Semiconductor laser-induced fluorescence detection in capillary electrophoresis using a cyanine dye, *J. Chromatogr. A*, 652, 355, 1993.
129. Mank, A.J.G. and Yeung, E.S., Diode laser-induced fluorescence detection in capillary electrophoresis after pre-column derivatization of amino acids and small peptides, *J. Chromatogr. A*, 708, 309, 1995.
130. Zhang, J.Z., Chen D.Y., Wu, S., Harke, H.R., and Dovichi, N.J., High-sensitivity laser-induced fluorescence detection for capillary electrophoresis, *Clin. Chem.*, 37/9, 1942, 1991.
131. Wu, S. and Dovichi, N.J., High-sensitivity fluorescence detector for fluorescein isothiocyanate derivatives of amino acids separated by capillary zone electrophoresis, *J. Chromatogr.*, 480, 141, 1989.
132. Chen, D.Y., Swerdlow, H.P., Harke, H.R., Zhang, J.Z., and Dovichi, N.J., Low-cost, high-sensitivity laser-induced fluorescence detection for DNA sequencing by capillary gel electrophoresis, *J. Chromatogr.*, 559, 237, 1991.
133. Chen, D.Y. and Dovichi, N.J., Yoctomole detection limit by laser-induced fluorescence in capillary electrophoresis, *J. Chromatogr. B*, 657, 265, 1994.
134. Starke, H.R., Yan, J.Y., Zhang, J.Z., Mühlegger, K., Effgen, K., and Dovichi, N.J., Internal fluorescence labeling with fluorescent deoxynucleotides in two-label peak-height encoded DNA sequencing by capillary electrophoresis, *Nucleic Acids Res.*, 22, 3997, 1994.
135. Crabtree, H.J., Bay, S.J., Lewis, D.F., Zhang, J., Coulson, L.D., Fitzpatrick, G.A., Delinger, S.L., Harrison, D.J., and Dovichi, N.J., Construction and evaluation of capillary array DNA sequencer based on a micromachined sheath-flow cuvette, *Electrophoresis*, 21, 1329, 2000.
136. Zhang, J., Yang, M., Puyang, X., Fang, Y., Cook, L.M., and Dovichi, N.J., Two-dimensional direct-reading fluorescence spectrograph for DNA sequencing by capillary array electrophoresis, *Anal. Chem.*, 73, 1234, 2001.
137. Kambara, H. and Takahashi, S., Multiple-sheathflow capillary array DNA analyzer, *Nature*, 361, 565, 1993.
138. Lieberwirth, U., Arden-Jacob, J., Drexhage, K.H., Herten, D.P., Müller, R., Neumann, M., Schulz, A., Siebert, S., Sagner, G., Klingel, S., Sauer, M., and Wolfrum, J., Multiplex dye DNA sequencing in capillary gel electrophoresis by diode laser-based time-resolved fluorescence detection, *Anal. Chem.*, 70, 4771, 1998.
139. Zhu, L., Stryjewski, W., Lassiter, S., and Soper, S.A., Fluorescence multiplexing with time-resolved and spectral discrimination using a near-IR detector, *Anal. Chem.*, 75, 2280, 2003.

140. Li, L.-C. and McGown, L.B., On-the-fly frequency-domain fluorescence lifetime detection in capillary electrophoresis, *Anal. Chem.*, 68, 2737, 1996.
141. Flanagan, J.H., Jr., Owens, C.V., Romero, S.E., Waddell, E., Kahn, S.H., Hammer, R.P., and Soper, S.A., Near-infrared heavy-atom-modified fluorescent dyes for base-calling in DNA-sequencing applications using temporal discrimination, *Anal. Chem.*, 70, 2676, 1998.
142. Nunnally, B.K., He, H., Li, L.-C., Tucker, S.A., and McGown, L.B., Characterization of visible dyes for four-decay fluorescence detection in DNA sequencing, *Anal. Chem.*, 69, 2392, 1997.
143. Mihindukulasuriya, S.H., Morcone, T.K., and McGown, L.B., Characterization of acridone dyes for use in four-decay detection in DNA sequencing, *Electrophoresis*, 24, 20, 2003.
144. Li, L.-C., He, H., Nunnally, B.K., and McGown, L.B., On-the-fly fluorescence lifetime detection of labeled DNA primers, *J. Chromatogr. B*, 695, 85, 1997.
145. He, H. and McGown, L.B., DNA sequencing by capillary electrophoresis with four-decay fluorescence detection, *Anal. Chem.*, 72, 5865, 2000.
146. Soper, S.A. and Mattingly, Q.L., Steady-state and picosecond laser fluorescence studies of nonradiative pathways in tricarbocyanine dyes: Implications to the design of near-IR fluorochromes with high fluorescence efficiencies, *J. Am. Chem. Soc.*, 116, 3744, 1994.
147. Lassiter, S.J., Stryjewski, W., Owens, C.V., Flanagan, J.H., Jr., Hammer, R.P., Khan, S., and Soper, S.A., Optimization of sequencing conditions using near-infrared lifetime identification methods in capillary gel electrophoresis, *Electrophoresis*, 23, 1480, 2002.
148. Alaverdian, L., Alaverdian, S., Bilenko, O., Bogdanov, I., Filippova, E., Gavrilov, D., Gorbovitski, B., Gouzman, M., Gudkov, G., Domratchev, S., Kosobokova, O., Lifshitz, N., Luryi, S., Ruskovoloshin, V., Stepoukhovitch, A., Tcherevishnick, M., Tyshko, G., and Gorfinkel, V., A family of novel DNA sequencing instruments based on single-photon detection, *Electrophoresis*, 23, 2804, 2002.
149. Gavrilov, D.N., Gorbovitski, B., Gouzman, M., Gudkov, G., Stepoukhovitch, A., Ruskovoloshin, V., Tsuprik, A., Tyshko, G., Bilenko, O., Kosobokova, O., Luryi, S., and Gorfinkel, V., Dynamic range of fluorescence detection and base-calling accuracy in DNA sequencer based on single-photon counting, *Electrophoresis*, 24, 1184, 2003.
150. Lewis, E.K., Haaland, W.C., Nguyen, F., F., Heller, D.A., Allen, M.J., MacGregor, R.R., Berger, C.S., Willingham, B., Burns, L.A., Scott, G.B.I., Kittrell, C., Johnson, B.R., Curl, R.F., and Metzker, M.L., Color-blind fluorescence detection for four-color DNA sequencing, *Proc. Natl Acad. Sci. USA*, 102, 5346, 2005.
151. Brennan, T., Chakel, J., Bente, P., and Field, M., New methods to sequence DNA by mass spectrometry, in *Proceedings of SPIE—Volume 1206 New Technologies in Cytometry and Molecular Biology*, Salzman, G.C., ed., July 1990, pp. 60–77.
152. Soper, S.A., Williams, D.C., Xu, Y., Lassiter, S.J., Zhang, Y., Ford, S.M., Sanger DNA-sequencing reactions performed in a solid phase nanoreactor directly coupled to capillary gel electrophoresis, *Anal. Chem.*, 70, 4036, 1998.
153. Swerdlow, H., Jones, B.J., and Wittwer, C.T., Fully automated DNA reaction and analysis in a fluidic capillary instrument, *Anal. Chem.*, 69, 848, 1997.
154. Tan, H. and Yeung, E.S., Automation and integration of multiplexed on-line sample preparation with capillary electrophoresis for high-throughput DNA sequencing, *Anal. Chem.*, 70, 4044, 1998.
155. Zhang, Y., Tan, H., and Yeung, E.S., Multiplexed automated DNA sequencing directly from single bacterial colonies, *Anal. Chem.*, 71, 5018, 1999.
156. He, Y., Pang, H.-M., and Yeung, E.S., Integrated electroosmotically-driven on-line sample purification system for nanoliter DNA sequencing by capillary electrophoresis, *J. Chromatogr. A*, 894, 179, 2000.
157. Tan, H. and Yeung, E.S., Integrated on-line system for DNA sequencing by capillary electrophoresis: From template to called bases, *Anal. Chem.*, 69, 664, 1997.
158. Ruiz-Martinez, M.C., Salas-Solano, O., Carrilho, E., Kotler, L., and Karger, B.L., A sample purification method for rugged and high-performance DNA sequencing by capillary electrophoresis using replaceable polymer solutions. A. Development of the cleanup protocol, *Anal. Chem.*, 70, 1516, 1998.
159. Salas-Solano, O., Ruiz-Martinez, M.C., Carrilho, E., Kotler, L., and Karger, B.L., A sample purification method for rugged and high-performance DNA sequencing by capillary electrophoresis using replaceable polymer solutions. B. Quantitative determination of the role of sample matrix components on sequencing analysis, *Anal. Chem.*, 70, 1528, 1998.

160. Luckey, J.A. and Smith, L.M., Optimization of electric field strength for DNA sequencing in capillary gel electrophoresis, *Anal. Chem.*, 65, 2841, 1993.
161. Ruan, C.C., Samols, S.B., and Fuller, C.W., Role of pyrophosphorolysis in DNA sequencing, Editorial Comments, Vol. 17, No. 2, p. 1, United States Biochemical Corporation, 1990.
162. Ewing, B., Hillier, L., Wendl, M.C., and Green, P., Base-calling of automated sequencer traces using phred. I. Accuracy assessment, *Genome Res.*, 8, 175, 1998.
163. Ewing, B. and Green, P., Base-calling of automated sequencer traces using phred. II. Error probabilities, *Genome Res.*, 8, 175, 1998.
164. Effenhauser, C.S., Bruin, G.J.M., and Paulus, A., Integrated chip-based capillary electrophoresis, *Electrophoresis*, 18, 2203, 1997.
165. Dolnik, V., Liu, S., and Jovanovich, S., Capillary electrophoresis on microchip, *Electrophoresis*, 21, 41, 2000.
166. Gao, Q., Shi, Y., and Liu, S., Multiple-channel microchips for high-throughput DNA analysis by capillary electrophoresis, *Fresenius J. Anal. Chem.*, 371, 137, 2001.
167. Chen, L. and Ren, J., High-throughput DNA analysis by microchip electrophoresis, *Comb. Chem. High Throughput Screen.*, 7, 29, 2004.
168. Zhu, L., Stryjewski, W.J., and Soper, S.A., Multiplexed fluorescence detection in microfabricated devices with both time-resolved and spectral-discrimination capabilities using near-infrared fluorescence, *Anal. Biochem.*, 330, 206, 2004.
169. Kan, C.-W., Fredlake, C.P., Doherty, E.A.S., and Barton, A.E., DNA sequencing and genotyping in miniaturized electrophoresis systems, *Electrophoresis*, 25, 3564, 2004.
170. Ueberfeld, J., El-Difrawy, S.A., Ramdhanie, K., and Ehrlich, D.J., Solid-support sample loading for DNA sequencing, *Anal. Chem.*, 78, 3632, 2006.
171. Shi, Y., DNA sequencing and multiplex STR analysis on plastic microfluidic devices, *Electrophoresis*, 27, 3703, 2006.
172. Effenhauser, C.S., Paulus, A., Manz, A., and Widmer, H.M., High-speed separation of antisense oligonucleotides on a micromachined capillary electrophoresis device, *Anal. Chem.*, 66, 2949, 1994.
173. Woolley, A.T. and Mathies, R.A., Ultra-high-speed DNA fragment separations using microfabricated capillary array electrophoresis chips, *Proc. Natl Acad. Sci. USA*, 91, 11348, 1994.
174. Simpson, P.C., Woolley, A.T., and Mathies, R.A., Microfabrication technology for the production of capillary array electrophoresis chips, *J. Biomed. Microdevices*, 1, 7, 1998.
175. Woolley, A.T. and Mathies, R.A., Ultra-high-speed DNA sequencing using capillary electrophoresis chips, *Anal. Chem.*, 67, 3676, 1995.
176. Scherer, J.R., Paegel, B.M., Wedemayer, G.J., Emrich, C.A., Lo, J., Medintz, I.L., and Mathies, R.A., High-pressure gel loader for capillary array electrophoresis microchannel plates, *Biotechniques*, 31, 1150, 2001.
177. Liu, S., Shi, Y., Ja, W.W., and Mathies, R.A., Optimization of high-speed DNA sequencing on microfabricated capillary electrophoresis channels, *Anal. Chem.*, 71, 566, 1999.
178. Woolley, A.T., Sensabaugh, G.F., and Mathies, R.A., High-speed DNA genotyping using microfabricated capillary array electrophoresis chips, *Anal. Chem.*, 69, 2181, 1997.
179. Liu, S., Ren, H., Gao, Q., Roach, D.J., Loder, R.T., Jr., Armstrong, T.M., Mao, Q., Blaga, L., Barker, D.L., and Jovanovich, S.B., Automated parallel DNA sequencing on multiple channel microchips, *Proc. Natl. Acad. Sci. USA*, 97, 5369, 2000.
180. Simpson, P.C., Roach, D., Woolley, A.T., Thorsen, T., Johnston, R., Sensabaugh, G.F., and Mathies, R.A., High-throughput genetic analysis using microfabricated 96-sample capillary array electrophoresis microplates, *Proc. Natl Acad. Sci. USA*, 95, 2256, 1998.
181. Shi, Y., Simpson, P.C., Schere, J.R., Wexler, D., Skibola, C., Smith, M.T., and Mathies, R.A., Radial capillary array electrophoresis microplate and scanner for high-performance nucleic acid analysis, *Anal. Chem.*, 71, 5354, 1999.
182. Medintz, I.L., Paegel, B.M., Blazej, R.G., Emrich, C.A., Berti, L., Scherer, J.R., and Mathies, R.A., High-performance genetic analysis using microfabricated capillary array electrophoresis microplates, *Electrophoresis*, 22, 3845, 2001.
183. Paegel, B.M., Emrich, C.A., Wedemayer, G.J., Scherer, J.R., and Mathies, R.A., High throughput DNA sequencing with a microfabricated 96-lane capillary array electrophoresis bioprocessor, *Proc. Natl. Acad. Sci. USA*, 99, 574, 2002.

184. Backhouse, C., Caamano, M., Oaks, F., Nordman, E., Carrillo, A., Johnson, B., and Bay, S., DNA sequencing in a monolithic microchannel device, *Electrophoresis*, 21, 150, 2000.
185. Emrich, C.A., Tian, H., Medintz, I.L., and Mathies, R.A., Microfabricated 384-lane capillary array electrophoresis bioanalyzer for ultrahigh-throughput genetic analysis, *Anal. Chem.*, 74, 5076, 2002.
186. Tian, H., Emrich, C.A., Scherer, J.R., Mathies, R.A., Andersen, P.S., Larsen, L.A., and Christiansen, M., High-throughput single-strand conformation polymorphism analysis on a microfabricated capillary array electrophoresis device, *Electrophoresis*, 26, 1834, 2005.
187. Salas-Solano, O., Schmalzing, D., Koutny, L., Buonocore, S., Adourian, A., Matsudaira, P., and Ehrlich, D., Optimization of high-performance DNA sequencing on short microfabricated electrophoretic devices, *Anal. Chem.*, 72, 3129, 2000.
188. Koutny, L., Schmalzing, D., Salas-Solano, O., El-Difrawy, S., Adourian, A., Buonocore, S., Abbey, K., McEwan, P., Matsudaira, P., and Ehrlich, D., Eight hundred-base sequencing in a microfabricated electrophoretic device, *Anal. Chem.*, 72, 3388, 2000.
189. Swanson, D., The art of the state of nucleic acid sequencing, *Scientist*, 14, 23, 2000.
190. Schmalzing, D., Tsao, N., Koutny, L., Chisholm, D., Srivastava, A., Adourian, A., Linton, L., McEwan, P., Matsudaira, P., and Ehrlich, D., Toward real-world sequencing by microdevice electrophoresis, *Genome Res.*, 9, 853, 1999.
191. Liu, S., A microfabricated hybrid device for DNA sequencing, *Electrophoresis*, 24, 3755, 2003.
192. Schmalzing, D., Adourian, A., Koutny, L., Ziaugra, L., Matsudaira, P., and Ehrlich, D., DNA sequencing on microfabricated electrophoretic devices, *Anal. Chem.*, 70, 2303, 1998.
193. Simpson, J.W., Ruiz-Martinez, M.C., Mulhern, G.T., Berka, J., Latimer, D.R., Ball, J.A., Rothberg, J.M., and Went, G.T., A transmission imaging spectrograph and microfabricated channel system for DNA analysis, *Electrophoresis*, 21, 135, 2000.
194. Haab, B.B. and Mathies, R.A., Single-molecule detection of DNA separations in microfabricated capillary electrophoresis chips employing focused molecular streams, *Anal. Chem.*, 71, 5137, 1999.
195. Jacobson, S.C., Hergenröder, R., Koutny, L.B., Warmack, R.J., and Ramsey, J.M., Effects of injection schemes and column geometry on the performance of microchip electrophoresis devices, *Anal. Chem.*, 66, 1107, 1994.
196. Culbertson, C.T., Jacobson, S.C., and Ramsey, J.M., Dispersion sources for compact geometries on microchips, *Anal. Chem.*, 70, 3781, 1998.
197. Paegel, B.M., Hutt, L.D., Simpson, P.C., and Mathies, R.A., Turn geometry for minimizing band broadening in microfabricated capillary electrophoresis channels, *Anal. Chem.*, 72, 3030, 2000.
198. Molho, J.I., Herr, A.E., Mosier, B.P., Santiago, J.G., and Kenny, T.W., Optimization of turn geometries for microchip electrophoresis, *Anal. Chem.*, 73, 1350, 2001.
199. Liu, S., Elkin, C., and Kapur, H., Sequencing of real-world samples using a microfabricated hybrid device having unconstrained straight separation channels, *Electrophoresis*, 24, 3762, 2003.
200. Shendure, J., Porreca, G.J., Reppas, N.B., Lin, X., McCutcheon, J.P., Rosenbaum, A.M., Wang, M.D., Zhang, K., Mitra, R.D., and Church, G.M., Accurate multiplex colony sequencing of an evolved bacterial genome, *Science*, 309, 1728, 2005.
201. Margulies, M., et al., Genome sequencing in microfabricated high-density picolitre reactors, *Nature*, 437, 376, 2005.
202. Bonetta, L., Genome sequencing in the fast lane, *Nat. Methods*, 3, 141, 2006.
203. Pennisi, E., Cut-rate genomes on the horizon? *Science*, 309, 862, 2005.
204. Jorgenson, J.W. and Lukacs, K.D., Zone electrophoresis in open-tubular glass capillaries, *Anal. Chem.*, 53, 1298, 1981.
205. Manabe, T., Chen, N., Terabe, S., Yohda, M., and Endo, I., Effects of linear polyacrylamide concentrations and applied voltages on the separation of oligonucleotides and DNA sequencing fragments by capillary electrophoresis, *Anal. Chem.*, 66, 4243, 1994.

17 Dynamic Computer Simulation Software for Capillary Electrophoresis

Michael C. Breadmore and Wolfgang Thormann

CONTENTS

17.1 INTRODUCTION

Since the early 1980s, the use of computers within society has increased at a significant rate. They are now involved in almost every aspect of modern day life, and it is therefore not surprising that this prevalence has transferred into the professional field of the modern day scientist. Of particular interest in this chapter is the use in capillary electrophoresis (CE) to model or "simulate" the myriad of chemical and physical processes that occur during an electrophoretic separation. This is achieved through a series of model equations derived from the transport concepts in solution under the influence of a d.c. electric field together with user-inputted experimental conditions, such as concentrations, mobilities, field strengths or currents, capillary lengths, and so on, and a theoretical separation is calculated. One of the potential advantages of a simulation is the possibility of determining appropriate separation conditions well before any laboratory experiments are undertaken, making method development a simpler task, although it must be noted that this can currently only be done with limited capacity. Simulations can also provide detailed insight into the processes involved in the separation allowing the researcher to understand the result of a particular combination of experimental conditions, and with this information design new superior systems, and this has been the major area in which simulation software has been used to date. In addition, simulations can be employed for educational purposes.

The result sought by the researchers will quite often dictate the simulation approach to be used, and in this regard it is necessary to distinguish the different types of computer models for electrophoresis. The most important for CE are (1) the dynamic models providing complete sets of concentration, conductivity, pH, and flow profiles as a function of time and location (e.g., at the point of detection along the capillary) for zone electrophoresis (ZE), moving boundary electrophoresis (MBE), isotachophoresis (ITP), and isoelectric focusing (IEF) [1,2]; and (2) the models for rapid assessment of ZE buffer systems and analyte separability [3–6] with a comprehensive program available for free (Peakmaster 5.1, http://www.natur.cuni.cz/Gaš). An overview of other models, including the steady-state models for ITP and IEF, which can predict the shape but not the evolution of steady-state electrophoretic boundaries, is given in [7], and a survey of training software for electrophoresis is provided in [8]. Computer simulations using dynamic models are more powerful than the other alternative as they are applicable to any buffer configuration and follow the entire separation at every step from start to finish according to the underlying transport laws. This not only yields the end result but also provides snapshots along the way, which can be used to monitor and understand the evolution of a particular process. While this is a huge advantage, it has one main disadvantage, namely that the algorithms are much more complicated and therefore requires more computer power. Furthermore, in contrast to high-performance liquid chromatography (HPLC) [9–12], electrophoretic simulations are much more computationally intense in three distinct ways: (1) electrophoresis usually has more components to deal with, (2) the separation efficiency is much higher requiring that component concentrations be updated more often and at more numerous column positions, and (3) the equations required to describe the processes are more numerous. This is less of an issue today than it was 20 years ago as computer hardware technology has developed significantly since then. Simulations that used to take several days can now be performed within several hours, allowing more complex and realistic simulations to be undertaken. It has only been within the past 5 years, over 20 years since the first reports on computer simulations, that it is possible to simulate simple CE separations under conditions that approximate those routinely employed in the laboratory.

In this chapter, we will discuss the theory underlying the use of dynamic simulation software and the benefits that the use of simulations can provide, illustrated through a number of simulated examples and examples taken from the literature over the past two decades. Finally, the most common computer simulation software will be contrasted and discussed from a practical perspective providing some guidelines as to the best way to approach a simulation problem.

17.2 BACKGROUND

Simulators with dynamic models are based on the description of algebraic acid–base equations and continuity equations, which are partial differential equations in time and space that can only be solved numerically using computers. Such models calculate the transport of each component through the electrophoretic space as a result of electromigration, diffusion, imposed and/or electrically driven bulk flow, solution-based chemical reactions such as protolysis and, if incorporated, also interaction of solutes with the capillary walls. An example is presented in Figure 17.1. A complete analysis of the temporal behavior of an electrophoretic system is thereby obtained and such models are thus often referred to as dynamic or transient state models. The numerical solutions are called dynamic simulations or dynamic modeling. Many dynamic models of various degrees of complexity have been described in the literature [13–46]. Only two of them are mostly used these days, namely the generalized PC-based models developed by Gaš and coworkers [30] and Mosher and coworkers [34–36], models that permit simulation of the transient states in all major electrophoretic modes with predictive values for MBE, ZE, ITP, and IEF and are referred to as SIMUL5 and GENTRANS, respectively, in the remainder of this chapter. Furthermore, the construction of a three-dimensional stochastic simulation model for electrophoretic separations has been reported [47]. In contrast to the dynamic models referred to above, this approach is based on the modeling of the trajectories of each individual molecule, requires extremely powerful computers in order to compute the motion of a statistically significant number of molecules, and is not further considered in this chapter. The

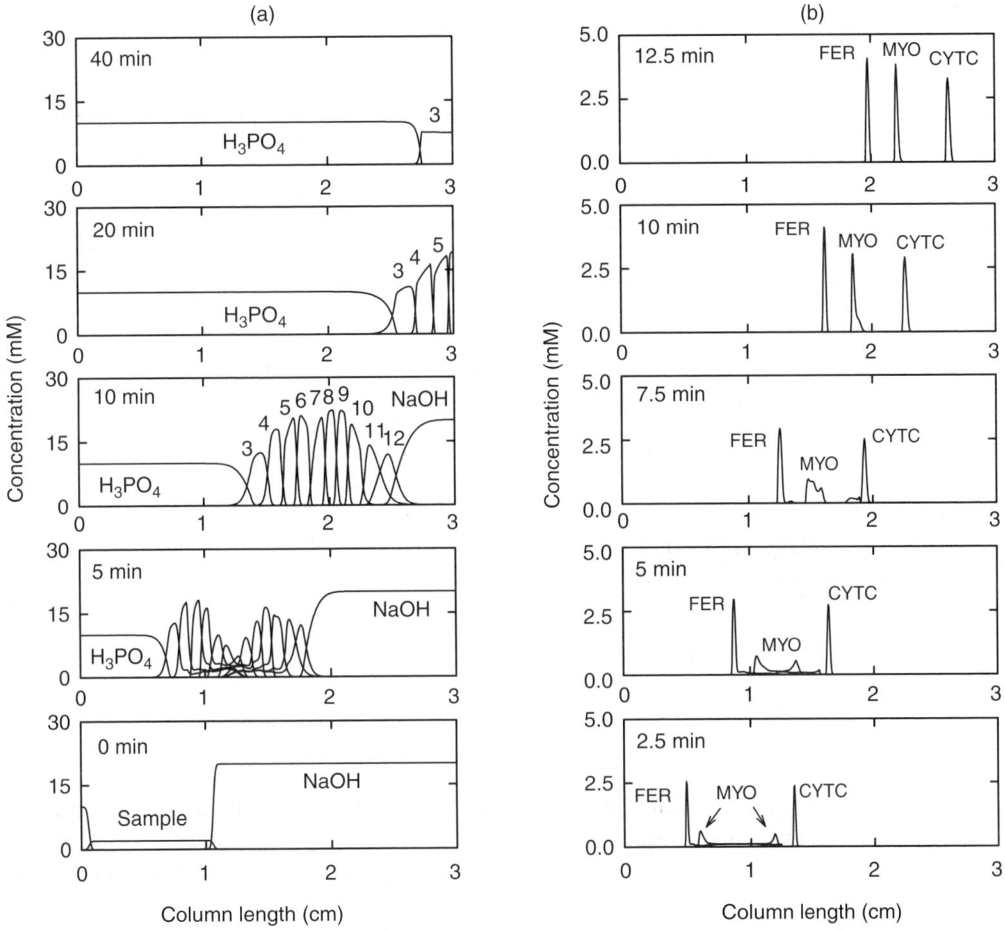

FIGURE 17.1 Computer predicted IEF of 10 carrier components and 3 proteins between 10 mM H_3PO_4 (anolyte) and 20 mM NaOH (catholyte) in a bare fused-silica capillary using GENTRANS with (i) the electroosmosis model that considers electroosmosis as function of pH and ionic strength according to experimental data (electroosmosis model V of Reference 54), (ii) a 3 cm column divided into 600 segments ($\Delta x = 50$ μm), (iii) boundary conditions with open column ends, (iv) a constant voltage of 10 V, and (v) data smoothing. (a) Concentration profiles of carrier ampholytes at 0, 5, 10, 20, and 40 min of current flow. (b) Dynamics of the three proteins at 2.5, 5, 7.5, 10, and 12.5 min (from bottom to top, respectively). (c) Column properties at 10 min (solid lines) and 0 min (dotted lines) for component distributions, ionic strength, pH, conductivity, and electroosmotic flow (from bottom to top, respectively). (d) Temporal behavior of current density, pH at two column locations (70 and 100% of column length), detector profiles for the proteins at the two detector locations, and detector profiles for all carriers and proteins for the two column locations (from top to bottom, respectively). For constructing the detection traces, all analytes had a detector response of 1. The cathode in panels a–c is to the right. The numbers in panel c refer to the pI values of the carrier ampholytes. S in panel c denotes the initial sample and C, F and M in panel (d) refer to CYTC, FER, and MYO, respectively.

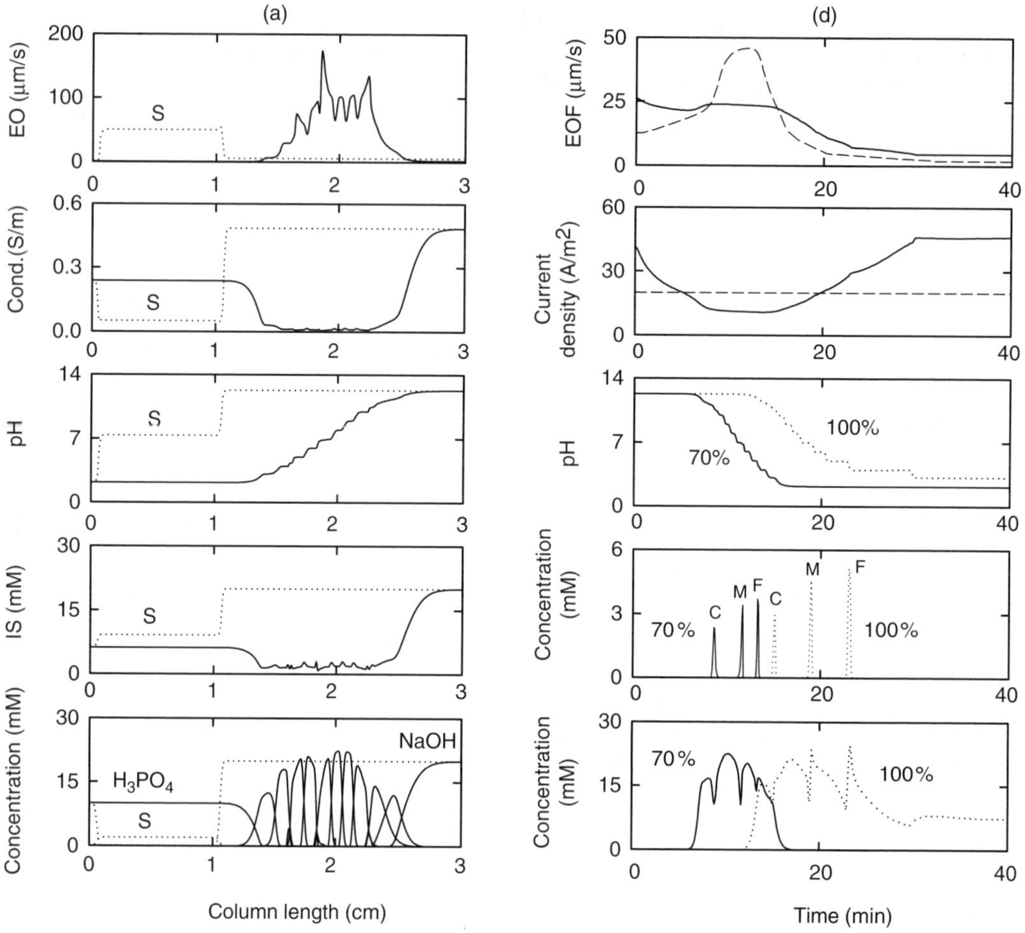

FIGURE 17.1 (Continued)

same is true for all multidimensional models that describe electrokinetically driven mass transport in microfabricated chip devices, such as those of Ermakov et al. [48], Bianchi et al. [49], Chatterjee [50], and Sounart and Baygents [51], as well as predict sample zone distortion in continuous flow electrophoresis [52].

Dynamic simulation provides plentiful data of any given CE system. This is illustrated here with the IEF separation example presented in Figure 17.1, which is performed in a fused-silica capillary and comprises a sample composed of ten hypothetical biprotic carrier ampholytes and three proteins [cytochrome *c* (CYTC, pI of 10.49), myoglobin (MYO, pI of 6.80), and ferritin (FER, pI of 4.40)] bracketed between 10 mM phosphoric acid (anolyte) and 20 mM sodium hydroxide (catholyte). A 3 cm separation space divided into 600 segments of equal length, 33% initial occupation of the sample at the anodic capillary end (panel a, bottom graph), and a constant voltage of 10 V were used. The pI values of the carrier ampholytes uniformly span the range 3–12 ($\Delta pI = 1$). For each ampholyte, ΔpK was 2, the ionic mobility was 3×10^{-8} m^2/Vs and the initial concentration was 2.0 mM. The initial concentrations of the three proteins were 0.1 mM each. Physico-chemical input properties were the same as used before [32,53] and the input data for the calculation of electroosmosis are those for a fused-silica capillary and consideration of the impact of the ionic strength on electroosmosis according to model V [54]. Simulation reveals that the carrier ampholytes are completely separated within about 10 min of current flow (panel a) and the pH gradient is therefore established within that time period (center graph of panel c). The data shown also reveal that carrier ampholytes at the interfaces to anolyte and catholyte (nicely seen with the pI 3 and pI 12 carriers, respectively) become gradually isotachophoretically adjusted and somewhat removed from the gradient. Owing to electroosmosis, the liquid and thereby the whole zone structure is being transported toward the cathode. With time, the catholyte and most carrier compounds are swept out at the cathodic capillary end and the net cathodic transport is decreasing and eventually disappearing completely [53]. A stationary steady-state zone configuration, which is characterized by an equilibrium between anodic isotachophoretic zone transport and cathodic electroosmosis, is produced. Under the employed conditions, the boundary between anolyte (phosphoric acid) and the most acidic carrier ampholyte is predicted to become immobilized at 91% of column length (top graph in panel a). The graphs presented in panel b depict the dynamics of the protein separation and are essentially those that are seen in whole column imaging [36]. The data of panel c represent zone properties predicted for the 10 min time point (solid line graphs) compared to the corresponding initial distributions (dotted line graphs). Data of other time points are not displayed, as this would complicate the figure. The electroosmotic data of each column segment presented in the top graph of panel c illustrate that the IEF gradient functions as the driving fluid pump of the capillary. These data do not reflect a real physical distribution of that property, but provide insight into the pumping activity of each zone. The electroosmotic pumping activity within anolyte and catholyte are predicted to be significantly smaller compared to that within the gradient system. Furthermore, computer-predicted temporal data for the net electroosmotic flow (EOF), current density, and various detector profiles are given in panel d. EOF is thereby shown to vary with time (even when the separation is executed under a constant current density, as is shown with the broken line graph, which was obtained for a constant 20 A/m^2). The same is true for the current density under the simulation conditions of constant voltage. Detector profiles for pH, the proteins (e.g., absorbance at 280 nm), and the carriers and proteins (e.g., absorbance at 200 nm) are given for two column locations, namely 70% of column length (solid line graphs) and the cathodic capillary end (100% of column length, dotted line graphs). The entire IEF zone structure is detected at the first location (and this can be obtained with a detector placed between about 70% and 90% of column length), whereas at the end of the column, the anolyte and the most acidic part of the gradient is not detectable at this location. Information of this kind is useful for designing experiments with detection at the capillary end, most notably with mass spectrometry (MS). It should be appreciated that without the use of computer simulation software, it would be very difficult to discern and delineate the different electrophoretic phenomena involved in the system in Figure 17.1.

17.3 HISTORICAL CONTEXT

Shortly after computers became available to University-based research entities, separation scientists at universities in the Czech Republic (Prague, supervision of J. Vacík), Switzerland (Bern, E. Schumacher), and Arizona, USA (Tucson, M. Bier) began to construct dynamic computer models for electrophoresis with the goal of exploring the basics of electrokinetic separations in solution. The early models of Moore [13], Ryser [14], Gaš [15], and Vacík and coworkers [16–18], although restricted to strong electrolytes, allowed the simulation of ZE, ITP, and MBE configurations and can be regarded as the first dynamic electrophoretic models. These efforts were associated with the CE studies of the 1960s and 1970s, which led to ITP analyses in polymer capillaries. First dynamic models predicting the behavior of strong and weak electrolyte systems were developed in the 1980s, including those of Bier et al. [1,19,20], Radi and Schumacher [21,22], Roberts [23,24], and Schafer-Nielsen [25]. Except for [21,22], they were applied to all basic electrophoretic modes, including IEF. The model of Radi and Schumacher [21,22] is unique in using the kinetic constants of the association and dissociation reactions to describe a chemical equilibrium. The model of Bier et al. [1], which led to a unified view of all basic electrophoretic modes, was extensively used to characterize a large number of electrophoretic configurations (for overviews see [7] and the detailed monograph of Mosher et al. [2]) and also simulates the behavior of proteins using effective and mean square valences that are determined from titration data [26]. Furthermore, the first model with incorporated EOF together with the transport based on electromigration and diffusion is that of Dose and Guiochon [27]. This approach, in which strong electrolytes are considered only and EOF is treated as constant plug flow, has been applied to the modeling of capillary ZE and ITP. The model of Roberts includes two spatial dimensions, whereas all others are one-dimensional approaches [23,24]. It is important to note that all these efforts were undertaken with hybrid or digital mainframe computers.

With the advent of PCs and the increasing popularity of capillary electrophoretic separations and analyses in fused-silica capillaries, many other dynamic simulators were developed in the 1990s, including those of Gaš et al. [28–30], Mosher and Thormann [31–36], Ermakov et al. [37–39], Shimao [40,41], Schafer-Nielsen [42], Martens et al. [43], Beckers and Boček [44], Ikuta and Hirokawa [45], and Sounart and Baygents [46]. Gaš et al. [28,29] extended their early model with incorporation of weak electrolytes, an approach that recently resulted in a comprehensive package that can account for any number of mono- and multivalent electrolytes and ampholytes (SIMUL5, [30]). The PC adapted model of Mosher and Thormann, GENTRANS, which is based on the dynamic simulator of Bier et al. [1,19,20] and was extended for a more realistic treatment of proteins [31], application of plug flow [32] and *in situ* calculation of EOF using wall titration data as input [33,34], can handle strong and weak electrolytes, simple univalent ampholytes, peptides and proteins (currently limited to 150 components total), and be executed at voltage gradients that are typically employed in experimental work [35,36]. It is interesting to note that the models in current use are again promoted by researchers from the same regions from which the early dynamic models originated, namely Prague, Czeck Republic (B. Gaš), Bern, Switzerland (W. Thormann), and Tucson, Arizona, USA (R.A. Mosher).

17.4 THEORETICAL ASPECTS OF SIMUL5 AND GENTRANS

17.4.1 Mathematical Model

The generalized models SIMUL5 and GENTRANS are simulation programs that predict the impact of current flow on a specified distribution of electrolytes. They are composed of a set of balance laws governing the transport of components in electrophoretic separations as was originally developed in the 1980s by Bier et al. [1,19,20], later detailed in the monograph of Mosher et al. [2] and recently reported by Hruška et al. [30]. They comprise a coupled set of nonlinear partial differential equations describing the appropriate balance laws and algebraic equations describing chemical equilibria, which includes an unsteady electromigration–diffusion equation for each component, a charge balance with inclusion of the diffusion current (for importance, see [55]), the electroneutrality approximation, expressions for dissociation–association equilibria of weak electrolytes and amphoteric compounds, and a model for calculating protein mobilities as a function of pH and ionic strength (GENTRANS only). Because dissociation–association reactions are fast compared with the mass transport [22], ion concentrations are constrained by a coupled set of mass action relations, namely the dissociation of water, and the dissociation–association equilibria of the components. Flow (constant or time dependent) is incorporated by adding an additional term to the flux equation. The assumed plug flow with cross-sectional uniformity does not contribute to any boundary and zone dispersion. GENTRANS also includes *in situ* calculation of electroosmosis using wall titration data as input [33,34]. For each column segment, electroosmosis is calculated with the use of a pH and ionic strength dependent electroosmotic mobility and the voltage gradient. Then, in analogy to Darcy's equation, which is valid for pressure-driven flow, the bulk capillary flow is taken to be the average of all of the segment flows.

Both models are one-dimensional and based on the principles of electroneutrality and conservation of mass and charge. Isothermal conditions are assumed and relationships between the concentrations of the various species of a component are described by equilibrium constants. Component fluxes are computed on the basis of electromigration, diffusion, and convection (imposed flow and/or electroosmosis). In GENTRANS, electrophoretic mobilities of small molecular mass components are considered to be independent of the ionic strength and temperature, but vary as a function of pH. SIMUL5 has an optional feature to account for the influence of the ionic strength on ionic mobilities and electrolyte activities. Furthermore, to save computational time, SIMUL5 calculates only the part of the column where considerable changes from the initial values are expected.

17.4.2 Numerical Implementation

The complete partial differential equations are solved numerically. The spatial region is overlaid with a uniform set of grid points or segments of length Δx for which all properties are calculated for each time step. The numerical treatment involves replacing the spatial derivatives in the partial differential equations by finite difference approximations that yields a set of ordinary differential equations with time as the independent variable. These equations are solved using sophisticated algorithms and computer codes that enable one to use variable step sizes for the time discretization and thus control error growth. SIMUL5 solves the equations using the Runge-Kutta and predictor-corrector method by Hamming [30], whereas GENTRANS is based on the fifth order Runge-Kutta-Fehlberg time step and second order central difference spatial discretization using DAREP simulation language [2,20], an approach that was later referred to as CSD algorithm in the work of Sounart and Baygents [46]. In both models, the algebraic equations are treated with the Newton's iteration method. Eventual differences in performance of the two simulators have not been elucidated thus far although there appears to be no major difference in the predictions between the two systems for a relatively simple simulation of the capillary ZE separation of a number of inorganic and organic anions as shown in Figure 17.2. It should be noted, however, that there are some practical differences, namely the actual simulation time using GENTRANS was approximately 40 min, with another 20 min required to process the data and import it to graphing software, while SIMUL5 required approximately 3 h to perform the same simulation, although did not require postsimulation data processing to visualize the results.

17.4.3 Boundary Conditions

For each simulation, the boundary conditions have to be specified. As the models rely on cross-sectional uniformity, there is only one spatial dimension, the separation axis x to be considered. For a constant voltage simulation, the boundary conditions on the potential $\phi\ (x,t)$ are $\phi\ (0,t) = V$ and $\phi\ (L,t) = 0$ where V is the potential and L the column length. Boundary conditions on the component concentrations at the column ends, that is, the permeabilities for the components, vary with electrophoretic technique. Both models feature the conditions for open column ends, which allow mass transport into and out of the separation space. After each step and for each component, the implemented algorithm sets the concentration at the first (last) mesh point equal to that of the second (next-to-last) mesh point. This approach accounts for changing boundary conditions at the column ends that occur when, for example, sample components are leaving the separation space. Simulations with GENTRANS can also be executed with fixed concentrations at column ends. Experimentally, this is equivalent to the presence of large electrolyte reservoirs at each column end. This choice is generally used (1) for ZE, ITP, and MBE simulations in which no sample components reach the column ends; (2) for ITP and MBE in which no buffer constituents from one electrode vessel reach the opposite side of the separation space; and (3) for IEF simulations in the presence of anolyte and catholyte whose concentrations remain unchanged at the column ends during the investigated time interval. Conditions with column ends that are impermeable to any buffer and sample compounds can also be selected. It represents a configuration having the column ends only permeable for OH^- and H^+ at cathode and anode, respectively, and is equivalent to an IEF experimental system in which

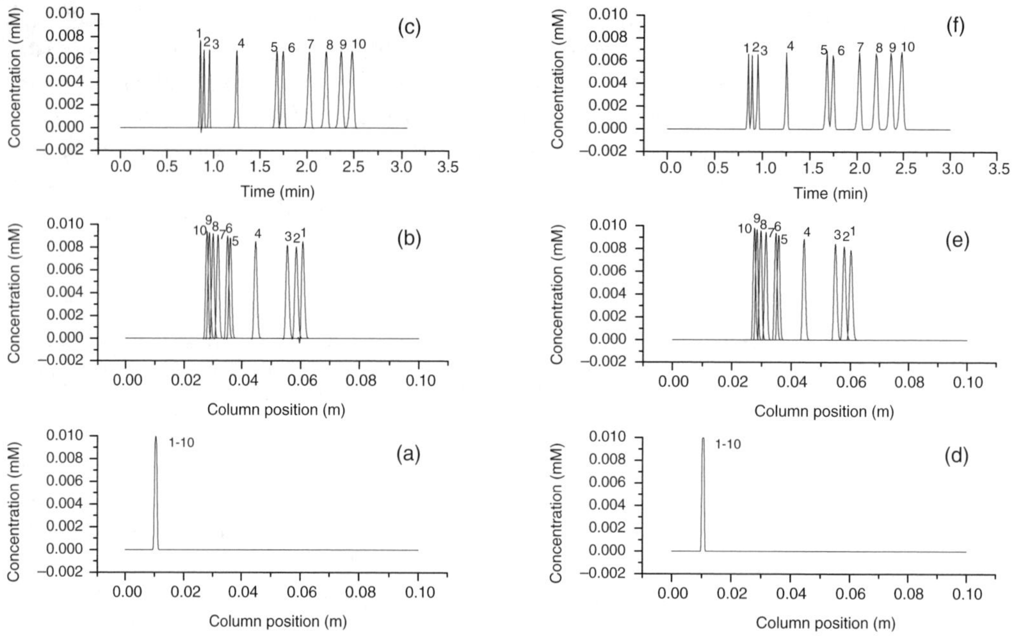

FIGURE 17.2 Simple capillary ZE separation of a mixture of inorganic anions and carboxylic acids in a MES/HIS electrolyte using SIMUL5 (panels a–c) and GENTRANS (panels d–f) under identical conditions (no EOF, same PC). Data were produced using a 10 cm length that was divided into 4000 segments ($\Delta x = 25$ μm) with the simulation in GENTRANS featuring variable column ends and smoothing. A constant voltage of –2000 V (200 V/cm) was applied and the sample initially occupied 1% of the capillary length. Panels a and d show the initial distribution of the inorganic and organic anions. Panels b and e show the column distributions 30 s after application of the voltage. Panels c and f show the detector trace constructed with the detector placed 9.5 cm from the injection end (segment 3800). For constructing the detection trace in GENTRANS, all analytes had a detector response of 1. The electrolyte consisted of 20 mM MES (pK_a = 6.095, $\mu_{ep} = 28 \times 10^{-9}$ m^2/Vs), 20 mM HIS (pK_a = 6.04, $\mu_{ep} = 44.7 \times 10^{-9}$ m^2/Vs), pH 6.06. The sample consisted of 0.01 mM of the following anions prepared as sodium salt ($\mu_{ep} = 51.9 \times 10^{-9}$ m^2/Vs) in BGE. Analytes are 1 = sulphate ($\mu_{ep} = 82.9 \times 10^{-9}$ m^2/Vs), 2 = chloride ($\mu_{ep} = 79.1 \times 10^{-9}$ m^2/Vs), 3 = nitrate ($\mu_{ep} = 75.1 \times 10^{-9}$ m^2/Vs), 4 = formate (pK_a = 3.752, $\mu_{ep} = 56.6 \times 10^{-9}$ m^2/Vs), 5 = iodate ($\mu_{ep} = 42.0 \times 10^{-9}$ m^2/Vs), 6 = acetate (pK_a = 4.756, $\mu_{ep} = 42.4 \times 10^{-9}$ m^2/Vs), 7 = propanoate (pK_a = 4.874, $\mu_{ep} = 37.1 \times 10^{-9}$ m^2/Vs), 8 = butyrate (pK_a = 4.820, $\mu_{ep} = 33.8 \times 10^{-9}$ m^2/Vs), 9 = valerate (pK_a = 4.842, $\mu_{ep} = 31.6 \times 10^{-9}$ m^2/Vs), and 10 = caproate (pK_a = 4.857, $\mu_{ep} = 30.2 \times 10^{-9}$ m^2/Vs).

electrodes are used to define the ends of an electrophoresis chamber. It is also quite similar to an IEF arrangement in which (1) a cation exchange membrane is used to isolate the anode chamber from the separation space, and the anolyte is an acid, and (2) an anion exchange membrane is used to isolate the cathode chamber from the separation space, and the catholyte is a base. Furthermore, GENTRANS also permits the specification of mixed boundary conditions. This selection allows the user to specify a boundary condition, either no transport or free transport of each component at both left and right boundaries.

17.4.4 Execution of a Simulation

Initial conditions that must be specified for a simulation include (1) the distribution of all components, (2) the pK and mobility values of the buffer and sample constituents, (3) the diffusion coefficients and charge tables of the proteins (GENTRANS only), (4) the electroosmotic input data [constant velocity

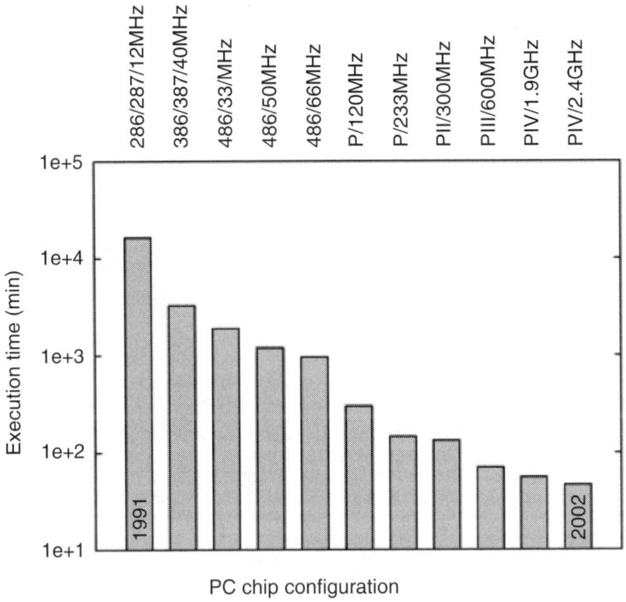

FIGURE 17.3 Execution time of GENTRANS under MS-DOS (MS Fortran version) as function of chip and clock speed assessed with PCs that were equipped with Intel chips for a 40 min simulation of the IEF example of Figure 17.1 under a constant current density of 10 A/m^2, a constant net EOF of 15 μm/s toward the cathode and fixed concentrations at column ends. A 350-fold decrease in execution time was observed with the PCs purchased during the 11-year time period. Using the FTN77 Fortran version on the Pentium PCs, execution time was found to be about half compared to that with the MS Fortran version.

value (SIMUL5 and GENTRANS) or as wall titration data for *in situ* calculation of electroosmosis (GENTRANS only)], (5) the magnitude of constant voltage or constant current density, (6) the duration of power application and the number of time points (frequency) of data storage, (7) the column length as well as its segmentation, and (8) the species permeabilities (boundary conditions) at the ends of the separation space (GENTRANS only). The programs output concentration, pH, conductivity, ionic strength, and flow (GENTRANS only) distributions, and allow the presentation of these data either as profiles along the column at specified time intervals or as temporal data, which would be produced by detectors at specified column locations, that is, segment numbers. Furthermore, they output the current density and the net flow (GENTRANS only) as functions of time. SIMUL5 also has the added option of viewing the results as an animation allowing simple visualization of the progression of the electrophoretic system.

The programs are executed on Pentium computers typically running at 233–3000 MHz. Execution time, which can vary between a few seconds and a full month, mainly depends on the PC's clock speed (Figure 17.3), the complexity of the electrophoretic situation (mainly the number of components), the applied power level, and the number of column segments. SIMUL5, which is free and can be downloaded from the web (http://www.natur.cuni.cz/Gaš), features a comfortable windows environment for data input, data evaluation, and visualization of the ongoing simulation, as illustrated in Figure 17.4. Furthermore, SIMUL5 features an inbuilt database of mobilities and pK_a values of common components as well as visualization of a completed simulation in a movie format. The GENTRANS program is somewhat older, runs in the DOS environment and exists as two versions, a low-resolution version for up to 600 segments (written in MS FORTRAN), and a high-resolution version for up to 100,000 segments (FTN77 FORTRAN). The use of GENTRANS can be learned in our research laboratories at the Universities of Bern and Tasmania.

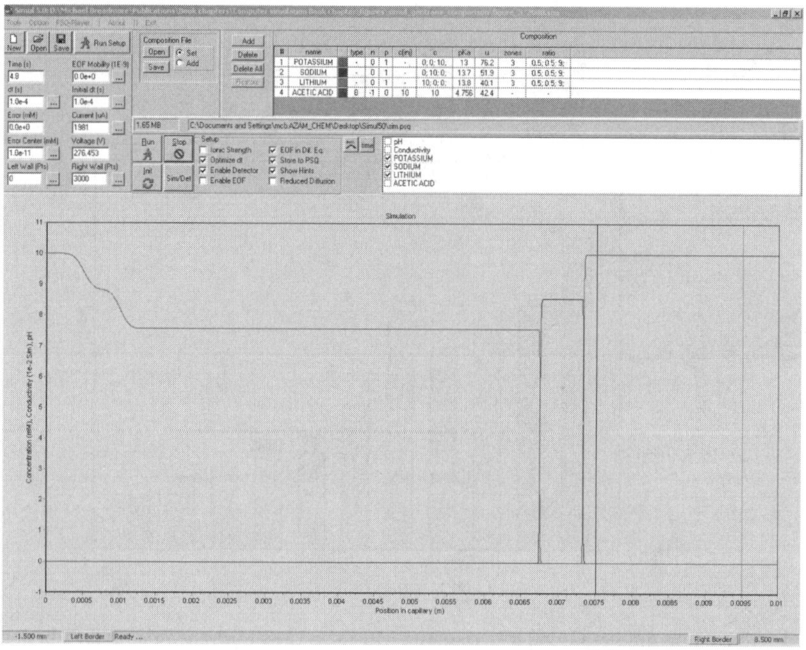

FIGURE 17.4 Screen capture of the windows environment of SIMUL5 performing exactly the same simulation as displayed in Figure 17.11 with a current density of 2000 A/m^2. The main panel displays the current simulation progress, and is currently only displaying the column distributions for potassium, sodium, and lithium. The vertical line at a column position of 0.0075 m is the right-hand calculation boundary, with the left-hand calculation boundary at a position of 0 m. The second vertical line at 0.0095 m is the position of the detector. The top left corner features the simulation parameters, such as time, voltage, current, and so on. The top right displays the physico-chemical properties of the buffer and sample components.

17.5 PRACTICAL APPLICATIONS

A considerable proportion of the published reports on the use of dynamic computer simulation software have been on the development of the simulation software itself and the agreement of the simulation with real experimental data. This is an important and essential part of the development of any theoretical system for without agreement between what is obtained in the laboratory, the knowledge gained from the simulation becomes meaningless. These applications will not be discussed here, instead, we will focus on the way in which computer simulations have been used to increase our understanding of the system and the field of electrophoretic separations. The early work is summarized in [7] and the book of Mosher et al. [2], which appeared in 1992. In examining the literature published thereafter, it has been dominated by application into a number of key areas, particularly in ITP and MBE, IEF and ZE, as well as in understanding various forms of stacking in ZE, and will be discussed under these categories. It is interesting to note that the application of simulation software seemed to peak in the mid 1990s with a number of groups working in this field, with a noticeable decline after this, until a few years ago when there was renewed interest around the world. While one may speculate about the causes, it is worthwhile noting that the past few years have seen an increasing number of applications of computer simulations in electrophoresis for two main reasons. First, with the advent of faster computers, it is possible to simulate more realistic and complex separations thus improving the quantitative prediction ability of the software and the ability to simulate a range of separations that were previously not possible. Second, the development of simple to use windows based software, such as SIMUL5, means that computer simulations are no longer restricted to highly trained researchers, but can be used by almost anyone with a basic knowledge of electrophoresis. It is anticipated that this area will see considerable growth over the next few years as new and more powerful computers are used in conjunction with more sophisticated simulation software.

17.5.1 Isotachophoresis and Moving Boundary Electrophoresis

ITP is one of the oldest electrophoretic separation techniques, and as such there have been a number of simulations on ITP and the behavior of the boundaries, although many of these have been used for the verification of the simulation software and for educational purposes [2,7,8]. In one of the first nontrivial studies on ITP boundaries, Mosher et al. [56] provided the first precise description of enforced migration due to changes in pH between the leading and terminating electrolyte. A boundary with self-sharpening properties was created by a conductivity decrease across the boundary in the direction of migration due to the pH gradient, which could be used to force analytes to migrate between the two electrolytes as a result in changes in ionization (and hence, mobility). They also showed that experimentally observed "bumps" were not necessarily artifacts due to adsorption/desorption phenomena, but were due to changes in conductivity and hence a change in electric field. A similar study was later undertaken by Gaš et al. [28] using an early version of the SIMUL5 software. They also simulated the occurrence of "bumps" due to the presence of H^+ or OH^- and the migration inversion of analytes depending on the pH of the electrolytes used. They concluded that in a well-buffered system, migration inversion could occur only for analytes that had a pK_a difference of approximately 0.3–1.0 pH units.

The impact of hydrodynamic flow upon the formation and stabilization of ITP zones has been examined by Deshmukh and Bier [57]. They modified the GENTRANS software to account for the addition of counter flow and found that it had no significant impact on the shape of the ITP boundary, but did have an impact on the position and could be used to effectively immobilize a particular ITP boundary. This, however, can be a problem when the counter flow is applied at the beginning of the separation as forming zones that migrate with a velocity lower than the counter flow were found to be pushed from the capillary, as illustrated in Figure 17.5. This explained experimentally observed results for the ITP stacking of serum components in which some components were not stacked appropriately. Subsequently, Mosher et al. [33] introduced equations to account for EOF, and in contrast to other available software, to also account for the variation of EOF with pH by using titration data. This was later used by Thormann et al. [58] to examine the effect of EOF on ITP and illustrate the difficulties of performing anionic ITP in unmodified fused-silica capillaries. Particularly interesting was the generation of a stationary anionic ITP boundary in which the boundary velocity was perfectly balanced by the EOF. Subsequent work by Caslavska and Thormann [59] simulated bidirectional ITP in which the EOF was used to mobilize the zones past the detector. The use of high pH electrolytes allowed both cationic and anionic ITP zones to be detected, but using a lower pH caused the formation of stationary anionic ITP zones that could not be detected without the application of hydrodynamic flow. Importantly, simulation data agreed well with the results obtained by experiment. The same was found to be true for the prediction of ITP in polymethylmethacrylate (PMMA) capillaries [54]. Furthermore, very recent work by Caslavska and Thormann [60] revealed that fused-silica capillaries double coated with Polybrene and poly(vinylsulfonate) feature a strong cathodic EOF across the entire pH range such that bidirectional ITP zone patterns at acidic pH reach the detector without the addition of imposed flow.

In a series of work, Mosher et al. [26,31,61] expanded GENTRANS to be applicable for proteins. As proteins are complicated molecules with a large number of ionizable groups, the mobility changes significantly as a function of pH and as such a table of pH titration data and diffusion constants is necessary to calculate the mobility of the protein [26]. After their initial work demonstrating the applicability to proteins, two algorithms to account for the influence of ionic strength on protein mobility were evaluated [31] with the best system subsequently used to examine the feasibility of using titration data generated from the amino acid composition and free amino acid pK_a values

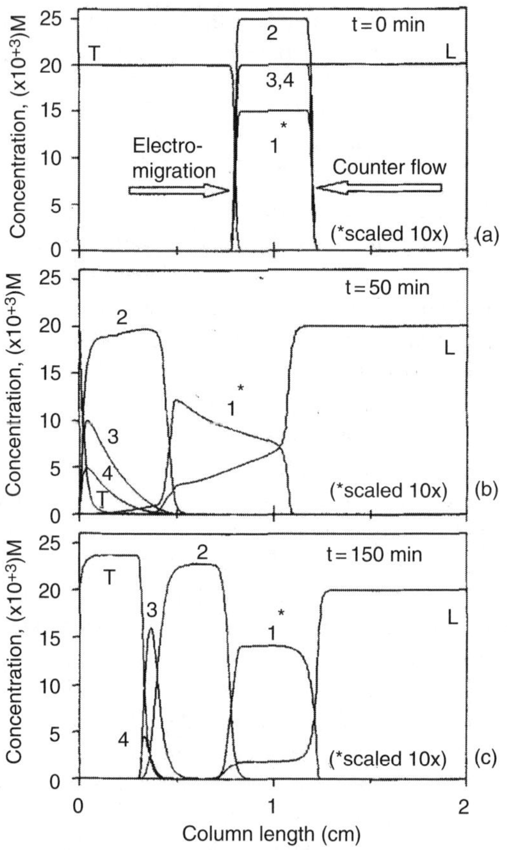

FIGURE 17.5 Selective washout of sample components during ITP due to application of a counter flow. The counter flow applied was sufficient to immobilize the leading front. For the first 50 min, the bands are pushed to the left due to the counter flow, followed by a movement to the right due to ITP adjustment. Leader (L): 20 mM cacodylic acid/Tris, pH 7.1; Terminator (T): 20 mM β-alanine/Tris, pH 9.1. The four sample components were (1) albumin, (2) Gly-Gly, (3) Asn, and (4) Gln. Their initial concentrations were 1.5 mM, 25 mM, 20 mM, and 20 mM, respectively. Simulation was performed under constant current density of 20 A/m^2 with a separation length of 4 cm with 400 segments and an applied counter flow of 0.47×10^{-5} m/s. (Reproduced from Deshmukh, R.R. and Bier, M., *Electrophoresis*, 14, 205, 1993. With permission.)

[61]. When using theoretical titration data, a different ITP order of several proteins was obtained to that observed by experiment, although this could be rectified by shifting the pH data to correct for differences between theoretical and experimental pIs. Thormann and Mosher [62] simulated two different anionic ITP systems and one cationic ITP system for the separation of proteins. In all cases, there was qualitative agreement with experimental data, although the plateau concentrations were typically lower than those observed experimentally. This was attributed to additional protein interactions that are not considered in the model. In subsequent work, Thormann et al. [63] performed simulations of ITP separations of serum albumin and compared these with experimental capillary ITP, continuous flow ITP, and recirculating flow ITP in various instruments. Distortion of protein zones found in the collected fractions of continuous flow ITP, which was not observed for low molecular mass compounds and was not observed in the simulations, was attributed to various instrument properties, such as high density/viscosity gradient formed at the buffer interfaces as well as the design of the outlet ports. This was not observed in recirculating flow ITP where faster fluid flows appeared to alleviate this problem.

In a different application, Dubrovčáková et al. [64] performed simulations of ITP separations in which a neutral complexing agent (a cyclodextrin in this case) was added to the electrolytes. Mechanistic insight into the movement of the ITP boundaries with addition of the cyclodextrin to the sample, leading and terminating electrolyte, and various combinations, was gained. Simulations also provided insight into the optimum concentration of cyclodextrin required to separate two analytes with particular interaction constants.

Assessment and characterization of the evolution of moving and stationary electrophoretic boundaries is another topic of interest in simulation [2,7,44]. As an interesting application, Foret et al. [65] investigated with GENTRANS the formation of a moving boundary induced by the liquid sheath in coupling of CE with electrospray ionization MS. When the buffer in the capillary and the liquid sheath contain different counterions, a boundary migrating from the capillary tip into the capillary is formed. Before reaching the MS interface, analytes will suddenly face another buffer composition (difference in pH, conductivity, ionic strength, etc.) and thus a change in migration that can result in delays, loss of resolution, or inversion of migration order. The authors showed that the migrating boundary can be either diffuse or sharp, depending on whether the effective mobility of the sheath liquid counterion is higher or lower than that of the buffer in the capillary. In the most recent application, Gebauer and Boček [66] used SIMUL to evaluate a new type of boundary that shows both steady-state (sharp) and nonsteady-state (diffuse) properties. After discussing the theory behind the new boundary, simulation results of a simple system comprising picrate and acetate were used to demonstrate the validity of their theory, with the new type of boundary shown in Figure 17.6b. Figure 17.6a depicts conditions in which a normal steady-state boundary is formed.

17.5.2 Isoelectric Focusing

The unique and powerful nature of IEF, which comprises a complex discontinuous configuration, has ensured that there have been a number of simulations of IEF, again with the majority of those being used to validate the simulation software and to ensure that there was reasonable agreement with experimental data. The majority of these simulations were performed with a current density typically much lower than that employed in the laboratory, which provided qualitative information about the system, but could not be used in any quantitative capacity. Nevertheless, these simulations provided new insights into the separation, stabilizing, and destabilizing processes of IEF in closed columns without electrolytes, in configurations with acid and base as anolyte and catholyte, respectively, and in systems with immobilized pH gradients (for a review of that simulation work see Mosher et al. [2]). Simulations also revealed the basics of one-step capillary IEF performed in the presence of electroosmosis in fused-silica capillaries [32,53]. In 2000, Mao et al. [35] presented the first results with a new version of GENSTRANS that could handle up to 150 components and simulate IEF at experimental current densities. This allowed the simulation of a pH gradient spanning 7 pH units from 3 to 10, with the pH gradient made from 140 ampholytes, and low molecular mass amphoteric dyes as sample constituents. These simulation data were used in conjunction with whole column capillary imaging to examine the mechanism and quantitative agreement between the two. Both systems demonstrated the well-known transient "double-peak" approach to focusing with the simulation software predicting complete focusing being achieved 5 min after current application, while experimentally it was observed to be 6 min. This was remarkable agreement given that the input data (pK and mobility values) of carrier ampholytes and the dyes used were not accurately known and that the composition of the commercial ampholyte mixture is largely unknown. A subsequent study by Mosher and Thormann [67] revisited the focusing and stabilization phases of IEF. The simulation results performed at 300 V/cm suggested that focusing is complete within less than 10 min and is followed by a lengthy stabilization phase (up to 7000 min), which is characterized by changes that progress from the column ends toward neutrality and the formation of nonlinear pH gradients, as illustrated in Figure 17.7. The presence of electrolytes at the column ends disrupts the stabilizing phase, with the degree of disruption being dependent on the concentrations of the acid and base

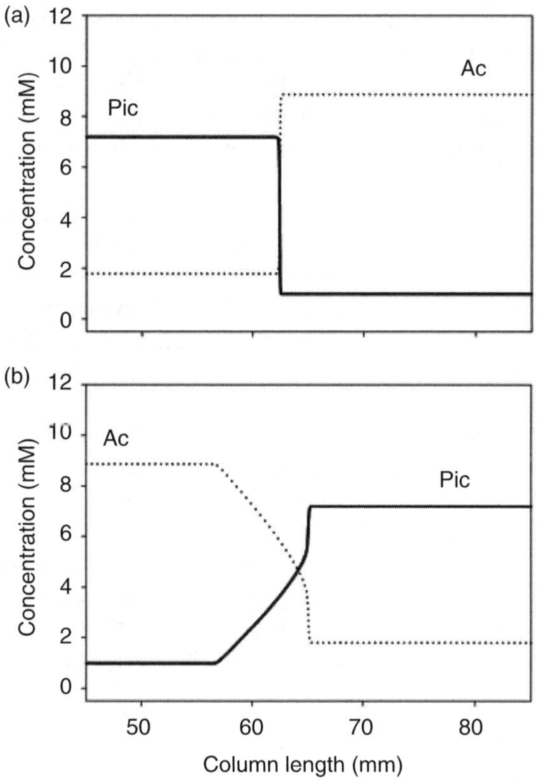

FIGURE 17.6 Simulated picrate (solid line) and acetate (dotted line) concentration profiles evolved after 25 s of electrophoresis from initial sharp boundaries between solutions in which the molar fraction of picrate in the front and rear zone was (a) 0.1 and 0.8 and (b) 0.8 and 0.1, respectively. The simulation was performed with 6000 segments ($\Delta x = 16\ \mu m$) at a current density of 4000 A/m^2. (Reproduced from Gebauer, P. and Boček, P., *Electrophoresis*, 26, 453, 2005. With permission.)

of anolyte and catholyte, respectively. The data suggest that it is impossible to reach a completely stationary steady-state pH gradient when using carrier ampholyte solutions, and that the only way to achieve this is through the use of immobilized pH gradients such as those of Immobiline gels.

As GENTRANS can also handle proteins, Thormann et al. [36] performed simulations of the IEF focusing of hemoglobin variants and compared these with whole column imaging IEF (both approaches at 300 and 600 V/cm). They simulated a broad range pH gradient from 3 to 10 with 20 ampholytes per pH unit (140 ampholytes over the entire gradient) and a short-range pH gradient from 5 to 8 with 40 ampholytes per pH unit. In contrast to IEF simulations of small molecular mass compounds, the number of segments had to be increased from 1000 to 8000 in order to get smooth protein peaks, which was due to the sharper peaks produced by proteins. The authors also noted that in order to establish a smooth stepless pH gradient at experimental field strengths, it was necessary to use 40 ampholytes per pH unit to establish the focusing gradient, while previously most simulations had used at most 20. Again, agreement between experimental results with whole column imaging was obtained supporting the use of this software for studying IEF.

In their latest work, Thormann and Mosher [68] have used GENTRANS to examine IEF when sodium chloride is added to the catholyte at the beginning of the separation. This approach is a single-step approach to post focusing mobilization of the IEF gradient by the addition of salt to the cathode. Results were shown to demonstrate that proteins still focus at a pH near their pI, but that the anion flux from the addition of chloride caused the pH to be slightly lower giving all ampholytes a

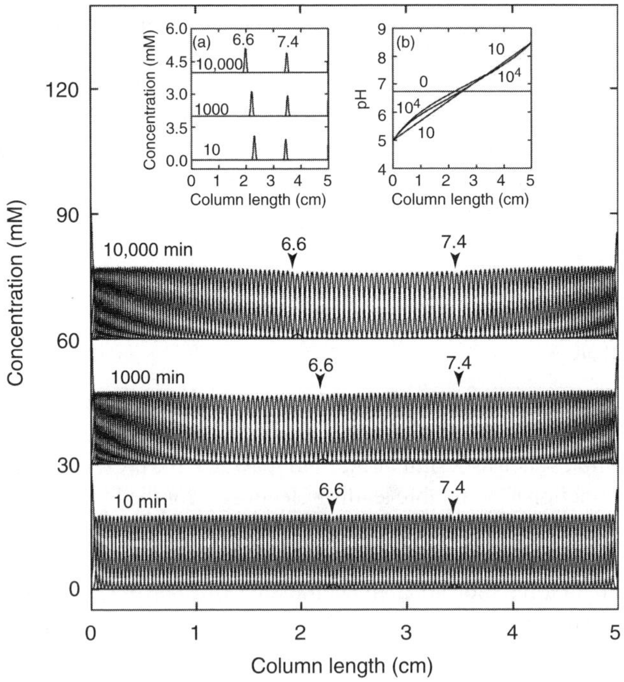

FIGURE 17.7 Computer-simulated distributions of the 140 carrier components and the three dyes for the pH 5–8.5 gradient system after 10, 1000, and 10,000 min of constant voltage application. The numbers refer to the pI values of the dyes and the arrowheads mark their locations. Successive graphs are presented with a y-axis offset of 30 mM. The insets a and b depict the concentration profiles of the pI 6.6 and 7.4 amphoteric dyes and the pH profiles, respectively, at the indicated time points. Simulations were performed with $\Delta x = 50$ μm and having column ends that are impermeable to any sample and carrier compounds. (Modified from Mosher, R.A. and Thormann, W., *Electrophoresis*, 23, 1803, 2002. With permission.)

slight positive charge and hence movement toward the cathode. The rate of movement was found to be dependent on the concentration of chloride added, with high concentrations causing more rapid movement. This also caused a gradual pH loss at the anodic end, while the cathodic end (where the salt was) remained unchanged causing a change in the profile of the pH gradient.

17.5.3 Zone Electrophoresis

Simulations of capillary ZE where the sample is the only discontinuity present have appeared in the literature, but the reports are fewer than those of ITP and IEF, and are dominated by confirming the validity of the simulation model, showing the property changes across ZE zones and demonstrating the effect of electrophoretic sample concentration when the sample is administered in a matrix of reduced ionic strength [2,7,33,34,54]. Simulation of ZE is useful for designing configurations with indirect detection, for the assessment of system zones and the investigation of unusual peak forms or peak shape distortions that cannot be handled with Peakmaster [30,69–72]. Particular worthy of discussion here is the work of Ermakov et al. [73–75] on developing algorithms to describe interaction of analytes with the capillary wall and the effect that this has on peak shapes in capillary ZE. Simulations were undertaken with different levels of adsorption and desorption at different concentrations of analytes and also as a function of the adsorption capacity of the surface. Peaks shapes were then compared with experimental results to gauge the source of interaction of analytes. Small monovalent cations were found to have minimal interaction with the wall (symmetrical peak shapes and no baseline shift) while large polycations were found to have slow desorption and produced a tailed peak with a shift in baseline after the peak, consistent with simulation data in which analytes showed very slow, almost irreversible, desorption. Independent from the work of Ermakov et al., Gaš et al. [76] studied the dynamics of peak dispersion in capillary ZE with wall adsorption using a simulation model that accounts for linear and nonlinear adsorption isotherms together with slow and fast sorption rates. Furthermore, options in electrolyte systems for online combined capillary ITP and ZE were investigated theoretically and verified by computer simulation using GENTRANS [77].

17.5.4 tITP Stacking

The combination of capillary ITP and ZE has long been recognized as an ideal approach to improve the sensitivity of problematic samples, particularly those with significant ionic strength. For biological and environmental samples such as serum, urine, and seawater, the presence of high concentrations of chloride is perfect to function as a suitable leading electrolyte for the ITP step, and all that is required is to ensure that an appropriate terminator is used and that there is sufficient time to complete the ITP stage and then to destack and separate the components by capillary ZE. tITP systems with samples naturally abundant in chloride have been the basis of several significant experimental and computational studies, the most significant belonging to Gebauer et al. [78] for the analysis of biological samples by tITP, or sample self stacking as referred to by these authors. Their first report in 1992 presented a simple theory to provide explicit description of various zone parameters such as detection time, variance, and resolution, which was confirmed by computer simulation using the GENTRANS software. This was later continued to look at the concentrations of components required to induce tITP stacking, with the concentration of the sample macrocomponent or stacking ion required to be above a critical concentration [79]. This concentration is a function of the buffer co-ion concentration in the sample and the mobilities of the buffer co-ion, sample macrocomponent, sample minor component, and the counterion. Simulations were undertaken to show when stacking does and does not occur as a function of this ratio. Results were also presented illustrating that the closer the mobility of the minor component to the stacking macrocomponent, a higher critical concentration of macrocomponent is required to induce efficient stacking. Later work focused on simulating the effect of multiple macrocomponents upon stacking and conditions under which one of them would function as a destacker [80]. There was again found to be a critical concentration of both the stacking and destacking macrocomponents, with this ratio only a function of the mobilities of the ions within the system. Above this ratio, the destacker is actually stacked also facilitating stacking of the minor components. Using the concepts of this system, Křivánková et al. [81] performed simulations to examine the conditions required to induce tITP stacking of minor components in serum (citrate, maleate, and acetoacetate), at various ratios of the destackers (phosphate and lactate), with simulation results for different amounts of salt in the sample shown in Figure 17.8. Simulation data demonstrated that 50 mM Cl or more was sufficient to induce ITP stacking in all possible phosphate and lactate concentrations found in human serum and could therefore be used for the analysis of plasma and serum samples from patients with a number of disorders. Their latest work used SIMUL to examine the properties of borate buffers and ways in which tITP stacking could be induced for samples with a high concentration of NaCl [82]. They examined two approaches, first, the addition of a suitable terminating ion (MOPS) to allow borate to function as the leader, and second, the use of ammonium as the counterion that reduced the mobility of hydroxide, allowing chloride to function as the leader, and borate to function as the terminator. Both approaches were able to provide suitable tITP stacking of minor components and were found to agree well with experimental approaches.

In the area of environmental analysis, Hirokawa et al. [83] used SIMUL to simulate the electrokinetic injection of cations with and without the hydrodynamic injection of potassium acetate. Not

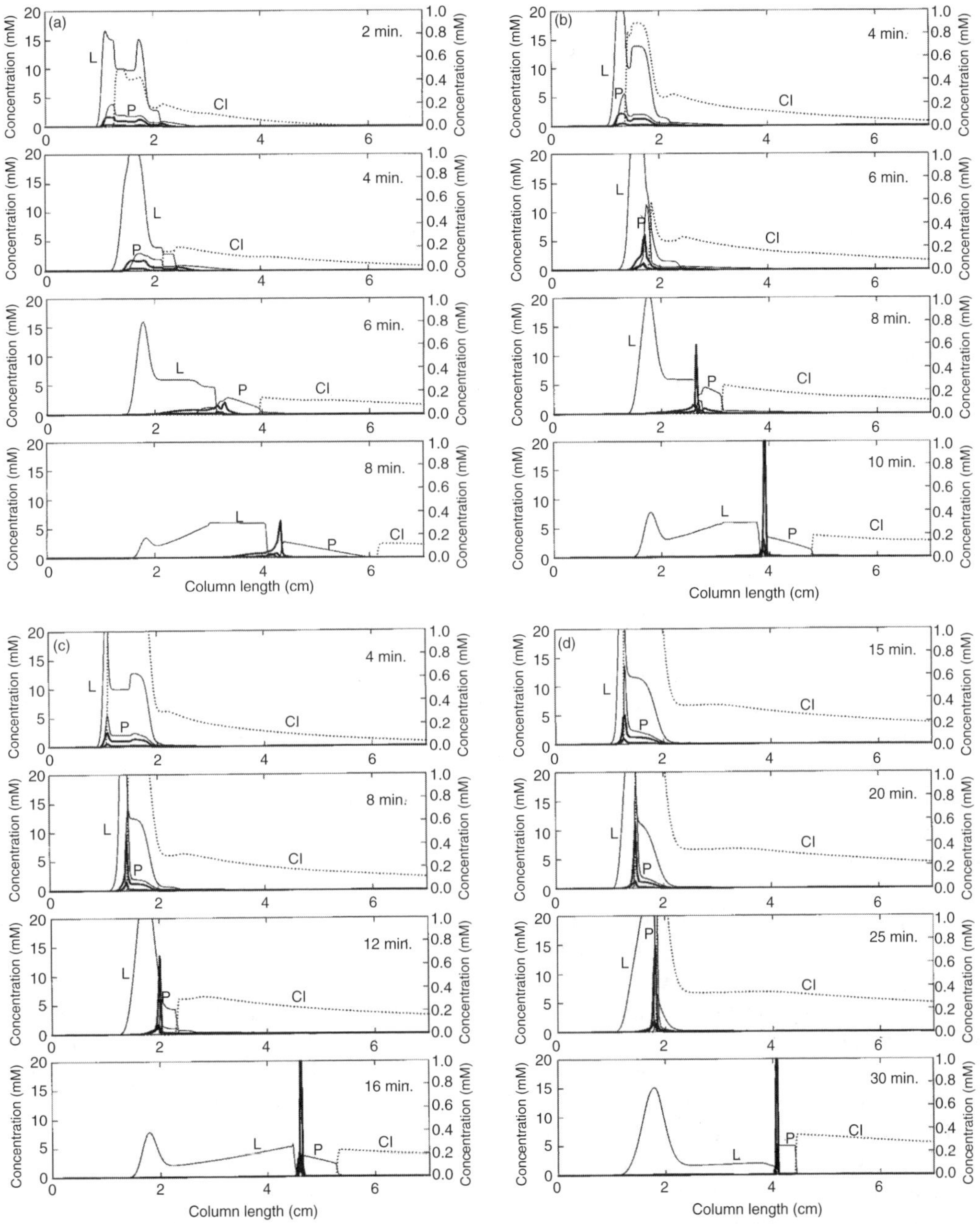

FIGURE 17.8 Simulated concentration profiles at the given four time points for the sample containing (a) 10 mM, (b) 20 mM, (c) 40 mM, and (d) 100 mM chloride. Left concentration axis relates to the macrocomponents chloride (Cl, dotted line), phosphate and lactate (P and L, respectively, both thin lines). Right concentration axis relates to the microcomponents citrate, malate, and acetoacetate (thick lines). Simulations were performed with $\Delta x = 62$ μm at a field strength of 40.54 V/cm. (Reproduced from Křivánková, L., et al., *Electrophoresis*, 24, 505, 2003. With permission.)

surprisingly, much better preconcentration was obtained for cations with a mobility faster than the mobility of the electrolyte co-ion with the injection of potassium as this functioned as a leading ion for tITP. Following this, they presented results for the simulation tITP stacking of iodide in seawater. Simulation data showed that the sample of unbuffered seawater maintains a high pH during ITP and therefore a suitable terminator with a low mobility at this pH was necessary, although it should be noted that practically seawater is actually buffered slightly by the presence of carbonate and borate. On the basis of these results, the authors used 500 mM MES at pH 6.0 as the terminating electrolyte that showed excellent stacking of an analyte with a mobility of 30×10^{-9} m^2/Vs (approximating the mobility of iodide after ion association with CTAC).

The only other report of note in the area of tITP is that by Schwer et al. [29] who used an early version of SIMUL to examine tITP stacking with a discrete pH difference between the leading and terminating electrolytes. Dissipation of the pH difference was used to increase the mobility of the cationic terminator thus shifting the separation from ITP stacking to ZE separation. A simplified system involving the injection of a high pH terminator before the sample using a low pH terminator was developed and showed to be suitable for the stacking of a tryptic digest of α-caesin.

17.5.5 pH Stacking

An alternative approach to stacking components in highly saline samples is though the use of a pH discontinuity between the sample and the electrolyte to stack ionizable analytes. This is similar to the enforced migration method examined by Mosher et al. [56] discussed above, although in this case it is transient in nature and is destroyed before migration by capillary ZE. Kim et al. [84] used SIMUL to simulate the formation and movement of a pH boundary and its influence on stacking weak acids. The pH boundary eventually dissipates allowing stacked analytes to be separated according to their electrophoretic mobility. Simulations also explained the experimental observation of an improvement in stacking of *m*-nitrophenol with increasing injection volume, which was related to the velocity of the stacking pH boundary and the length of time before dissipation and hence, destacking of the analytes.

Shim et al. [85] used their own program for the simulation of stacking weak acids (such as fluorescein) in buffered and unbuffered saline samples. Buffered saline samples showed stacking via tITP, while in unbuffered samples (NaCl in water) the rapid exit of the matrix ions from the sample caused a localized pH change to maintain electroneutrality thus creating an additional pH boundary. This pH boundary was restricted to the initial position of the sample injection and merged with a migrating tITP boundary induced by the high concentration of Cl. Thus, analytes were initially stacked on both sides of the sample zone and merged into a single focused zone in a fashion similar to the transient "double peak" mechanism observed in IEF.

In the above two simulations, stacking was simulated for one analyte and the magnitude of the pH boundary was only several pH units. Breadmore et al. [86] used GENTRANS to simulate the stacking of 24 weak bases of different mobility and pK_a in which the sample had a pH almost 7 units higher than that of the electrolyte. The influence of the addition of a high concentration of sodium chloride to the sample was also examined, with simulation results demonstrating that the movement of the boundary can be influenced by the addition of sodium chloride to the sample. Upon the addition of 100 mM NaCl, a significant reduction in mobility of the boundary was observed allowing bases with a much lower mobility to be focused. However, one of the limitations of lowering the mobility of the boundary in this fashion is that it causes prolonged migration of the boundary itself, which can result in analytes migrating through the entire capillary as focused bands, as shown in Figure 17.9. Care must therefore be taken to allow sufficient time for the boundary to dissipate and the analytes to destack.

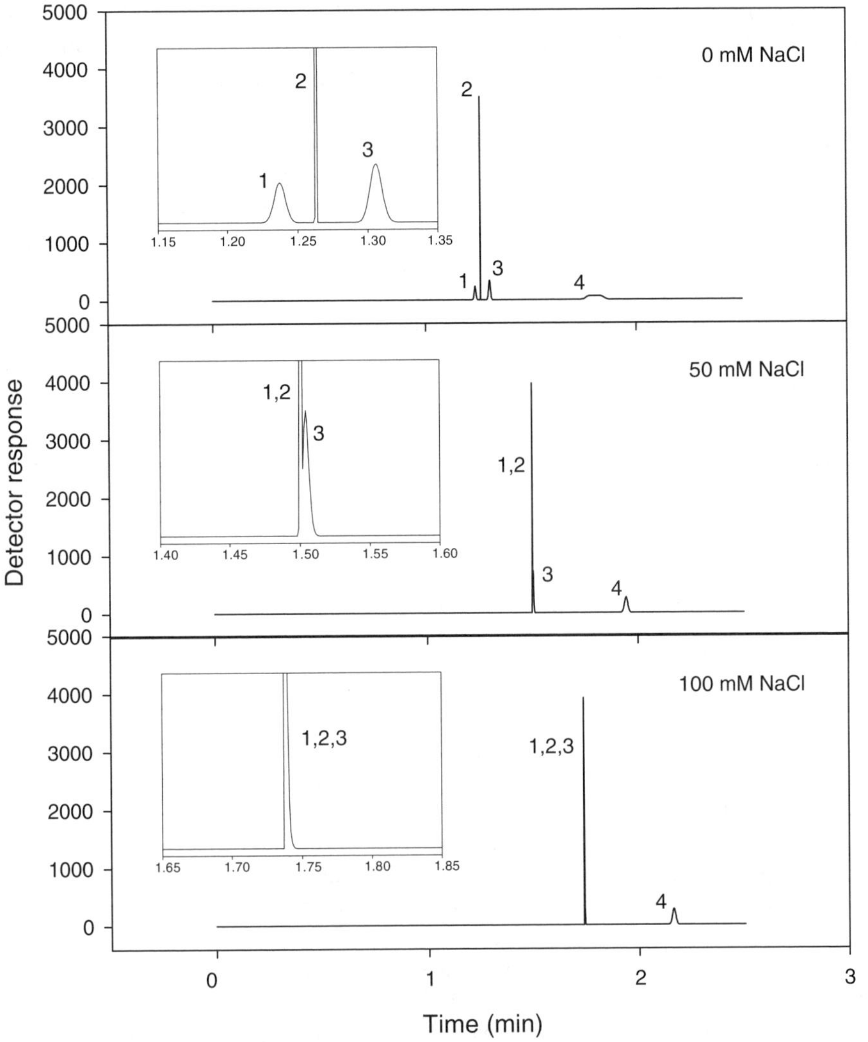

FIGURE 17.9 Simulated detector response for the separation of four weak bases ($pK_a = 4.5$) after stacking with a transient pH boundary without and with the addition of NaCl to the sample zone. Buffer: 65.6 mM formic acid, pH 2.85, with NaOH. Sample: 65.6 mM formic acid, pH 8.60 adjusted with NaOH, 0, 50, or 100 mM NaCl and the four bases (5 μM each). Insets show a close-up view of the peaks 1–3. Voltage 400 V/cm. Detector location 5 cm. Peaks are (1) 25×10^{-9} m^2/Vs, (2) 20×10^{-9} m^2/Vs, (3) 15×10^{-9} m^2/Vs, and (4) 10×10^{-9} m^2/Vs. (Reproduced from Breadmore, M.C., et al., *Anal. Chem.*, 78, 538, 2006. With permission.)

17.5.6 Miscellaneous Applications

Tesařová et al. [87] used a modified version of SIMUL, which they called SIMULMIC to simulate the separation of neutral analytes in a system with a neutral cyclodextrin and anionic micelles. A number of systems were examined in which various combinations of the inlet and outlet vials and the capillary itself were filled with cyclodextrin. Simulation results were used to examine the micellar/cyclodextrin boundary at various times and concentrations although no simulation results for a chiral separation were reported. To the best of our knowledge, this is the only dynamic simulation of an electrokinetic chromatography (EKC) separation to date.

Several reports have been presented to simulate capillary affinity electrophoresis (CAE) under equilibrium (complex is formed with ligand in the electrolyte and migrates in an equilibrium state) and nonequilibrium (ligand is not in the electrolyte allowing the complex to dissociate during separation) conditions [88–90] using binding constants and association/dissociation rate constants. Because of the added complexity of the additional equations, implementation often involves simplification by assuming no changes in the electrolyte macrocomponents and hence no change in pH or conductivity, thus only allowing more limited simulations to be performed. Nevertheless, these simulators can successfully predict peak shapes in CAE, and as was demonstrated by Okhonin et al. [90] can be used to determine the binding and rate constants by using nonlinear regression to fit the experimental separation to that predicted with the CAE software.

SIMUL5 has been used to study oscillation patterns that occur in background electrolytes with complex eigenmobilities when they are subjected to an electric field [91]. For a system composed of 0.21 mM sebacic acid and 0.323 mM imidazole, simulation revealed the production of local 10^{-12} mM perturbations of the imidazole concentration, which exponentially grew as long as the system was in the region of complex eigenmobilities. This unusual behavior could be experimentally validated and is thus not a simulation artifact.

17.6 METHODS DEVELOPMENT GUIDELINES

In performing a simulation, there are some points that the researcher must understand in order to maximize the utility of their simulation. Selection of the simulation parameters, such as the number of segments and the applied power level, is crucial for success. These aspects are visualized with the following simple example, which represents the simulation of the ITP zone formation of sodium between potassium (leader L) and lithium (terminator T) with acetate as the counterion (Figure 17.10). This simulation was performed in a 1 cm column that was divided into 2000 segments of equal length ($\Delta x = 5$ μm) and having a constant current of 200 A/m^2. All initial concentrations are 10 mM and the sample occupies 5% of column length (between 5% and 10%, bottom graph of left panel of Figure 17.10). Upon application of power, the sodium zone becomes adjusted within about 0.1 min (right panel of Figure 17.10) to a concentration of 8.48 mM as it gradually penetrates into the space originally occupied by the leader where it continues to migrate between potassium and lithium. The same applies to lithium, the terminating constituent, which becomes adjusted to a concentration of 7.59 mM. Thereafter, the entire zone structure is migrating at a constant velocity and without change of the zone structure toward the cathode (left panel of Figure 17.10). The corresponding zone distribution after 0.08 min (4.8 s) at 2000 A/m^2 obtained with SIMUL5 is shown in Figure 17.4. The migrating boundaries between the alkali metal ions are sharp, steady-state transitions whose widths are dependent on the applied power level. At a low power, electrophoretic boundaries are rather broad and easy to treat with a relatively low number of segments. Before the availability of fast computers, this was the rule rather than the exception [2,7]. SIMUL5 and GENTRANS are capable of handling high power levels. The example presented in Figure 17.11 illustrates the impact of current density on the ITP zone boundaries in the configuration of Figure 17.10. Having 20 A/m^2 (electric field $E = 1.79$ V/cm) results in broad boundaries and simulation predicts that there is insufficient sodium to produce an ITP zone with a plateau concentration (upper left graph of Figure 17.11). The opposite is true with an increase of the current density to 200 A/m^2 ($E = 17.9$ V/cm, upper right graph), 2000 A/m^2 ($E = 179$ V/cm, lower left graph), and 20,000 A/m^2 ($E = 1787$ V/cm, lower right graph), conditions with much sharper boundaries and 10-, 100- and 1000-fold, respectively, faster solute transport. The latter two examples were simulated at power levels that are typically used in experiments. ITP boundaries between the alkali metal ions of our example are predicted to be in the order of 350, 35, and 3.5 μm, respectively (Figure 17.11, insets depict the sodium–lithium transitions at elongated x-axis scales), that is, boundary width is indirectly proportional to the current density which is in agreement with previous knowledge [92]. The transition between potassium and

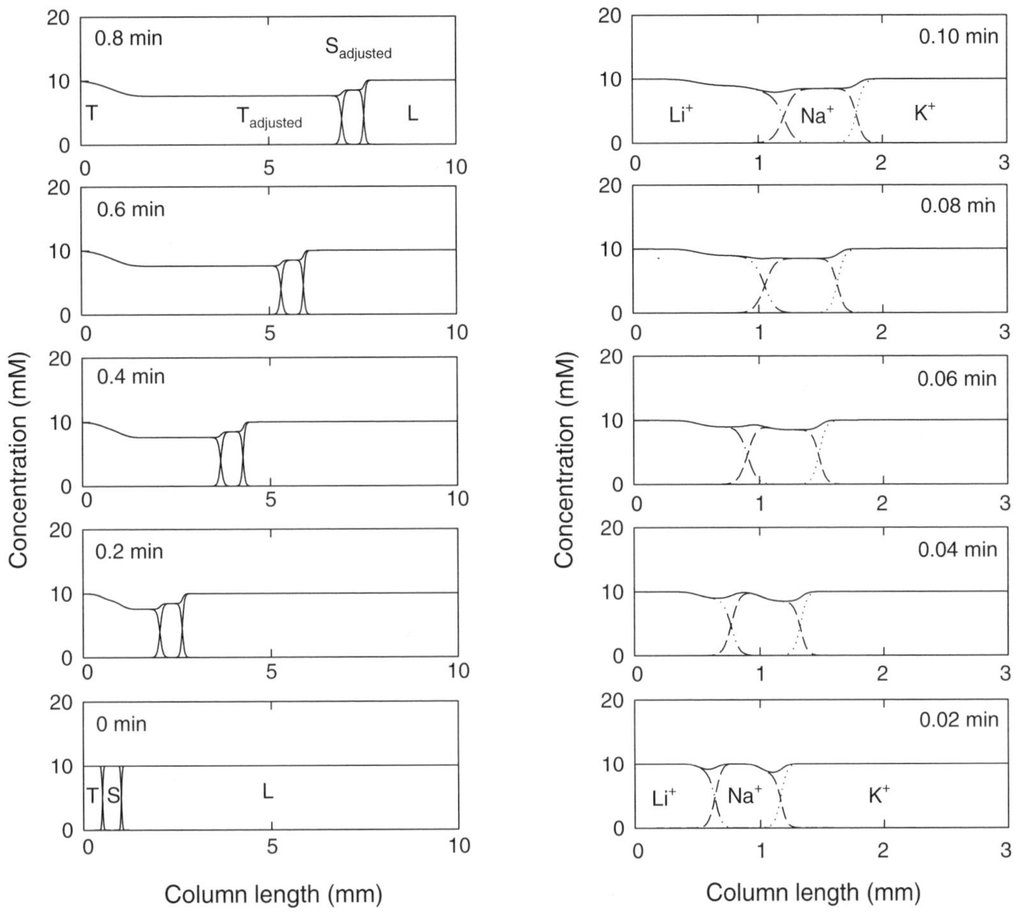

FIGURE 17.10 Computer predicted ITP zone formation with sodium as sample (S) between potassium (leader L) and lithium (terminator T) and having acetate as counter component and no electroosmosis. Data were produced with GENTRANS using a column of 1 cm length that was divided into 2000 segments ($\Delta x = 5\ \mu$m), fixed concentrations at column ends and no data smoothing. A constant current density of 200 A/m^2 was applied and the sample initially occupied 5% of column length. Left-hand graphs depict concentration distributions at 0, 0.2, 0.4, 0.6, and 0.8 min after current application. Right-hand graphs depict the dynamics of the formation of the migrating sodium ITP zone with concentration profiles after 0.02, 0.04, 0.06, 0.08, and 0.1 min. The cathode is to the right. The mobilities of potassium, sodium, and lithium were taken as 7.91×10^{-8} m^2/Vs, 5.19×10^{-8} m^2/Vs, and 4.10×10^{-8} m^2/Vs, respectively. The pK_a and mobility of acetic acid were 4.76 and 4.12×10^{-8} m^2/Vs, respectively.

sodium is somewhat sharper than that between sodium and lithium. The reason for this is the smaller conductivity change across the latter boundary (Figure 17.12, right panel).

Using insufficient number of segments, there is a risk of numerical oscillations that are seen in all profiles, including concentration, conductivity, and pH distributions. This is illustrated with the data presented in Figure 17.12 that were generated with 2000 A/m^2. Strong oscillations are obtained for a simulation with 1000 segments ($\Delta x = 10\ \mu$m, left graph in Figure 17.12). With 2000 segments ($\Delta x = 5\ \mu$m, center graph in Figure 17.12), much smaller oscillations are predicted, whereas 4000 segments led to smooth transitions ($\Delta x = 2.5\ \mu$m, right graphs in Figure 17.12). It is important to realize, however, that this is at the expense of simulation time. For a total of 0.1 min electrophoresis time, the simulations performed on a Pentium IV 2.4 GHz PC lasted 2.45, 7.20, and 20.68 min, respectively, suggesting that doubling the grid requires 2.9-fold longer simulation time intervals.

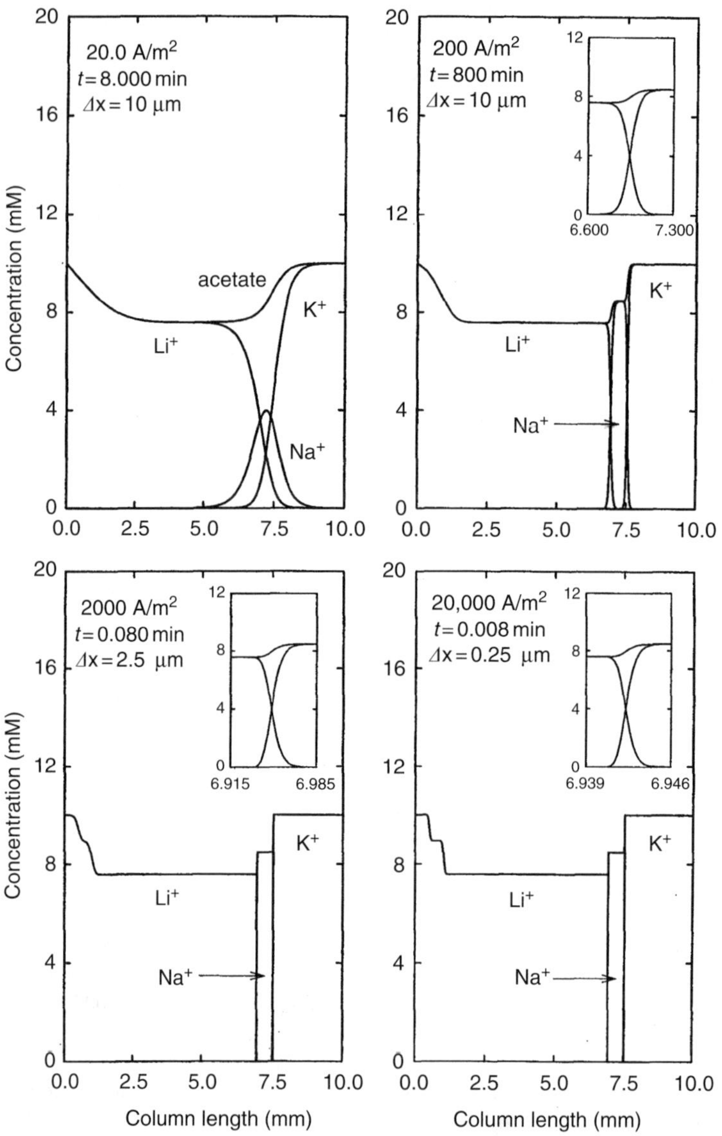

FIGURE 17.11 ITP simulation of the system of Figure 17.10 at different constant current densities of 20 A/m^2 (upper left graph), 200 A/m^2 (upper right graph), 2000 A/m^2 (lower left graph), and 20,000 A/m^2 (lower right graph) with data shown for 8 min, 0.8 min, 0.08 min, and 0.008 min, respectively. The insets depict the sodium-lithium transitions on elongated x-axis scales. Data were produced with GENTRANS without smoothing using 1000, 1000, 4000, and 40,000 segments, respectively.

On the other hand, the simulation at 200 A/m^2 and 1 min of electrophoresis time (Figure 17.10) executed on a Pentium 233 MHz PC with 1000, 2000, and 4000 segments required 0.27, 2.0, and 16.5 h, respectively. For this case, execution time was found to increase about eightfold with doubling of the mesh. This illustrates that simulation time intervals are also hardware dependent. It should also be noted that the alternative to increasing the mesh size is to restrict the magnitude by which the step size dt can change. This can be used successfully to overcome floating-point errors that result from numerical oscillations, however, this also obviously results in an elongation in simulation time.

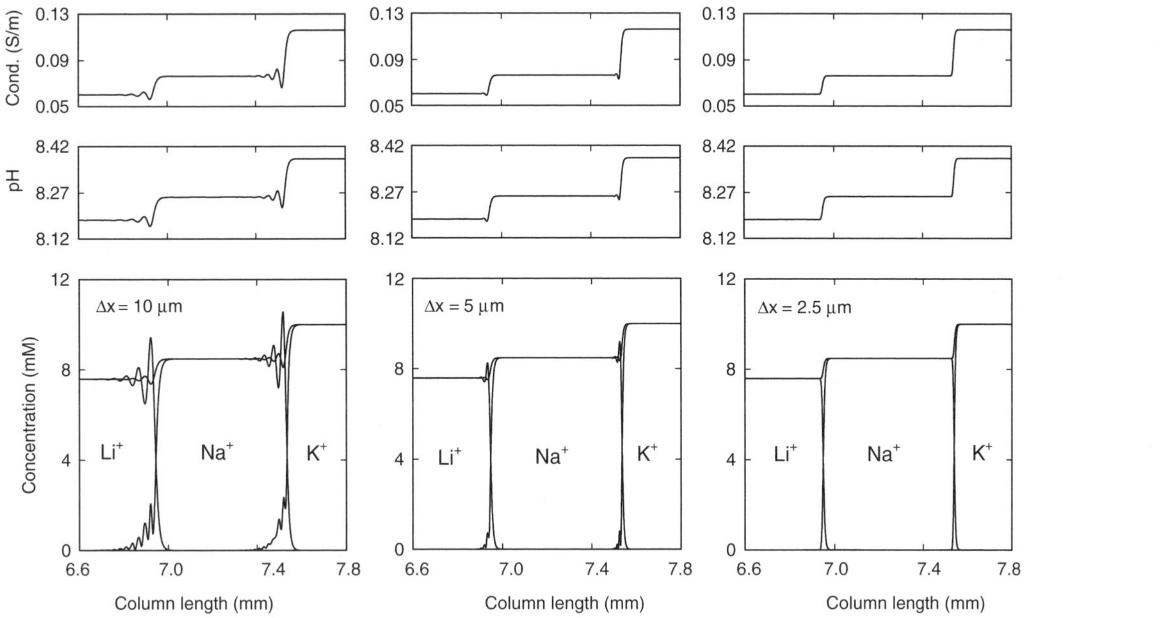

FIGURE 17.12 ITP simulation data at 2000 A/m^2 of the system of Figure 17.10 using 1000 (left graphs), 2000 (center graphs), and 4000 (right graphs) segments. Concentration, pH, and conductivity data shown are for 0.08 min and were obtained with GENTRANS without application of data smoothing.

While no quantitative study has been undertaken, this approach can often be quicker than doubling the number of segments.

Instead of an increased mesh, other algorithms have been tested in various laboratories as is discussed in depth by Sounart and Baygents [46]. None of the proposed procedures, however, was found to provide data without undesired boundary distortions in a much reduced time interval. Instead, GENTRANS is having an optional smoothing algorithm that removes negative concentrations caused by numerical oscillations. Using smoothing will decrease accuracy and creates the potential for nonphysical results to be produced. If smoothing is used and data are obtained that are suspected to be nonphysical, that is, behavior not expected to occur in an experiment, the simulation should be repeated without invoking smoothing, and the results compared to the suspect data. The data presented in Figure 17.13 illustrate the impact of the smoothing option for the simulation of a ZE separation of five dipeptides that are negatively charged in the CHES buffer at pH 9.5. Without smoothing, peptides 2–4 show oscillations with negative concentrations at the rear ends of their peaks (shown for 10 and 20 min of power application, left-hand graphs). These are not observed anymore when smoothing is used (right-hand graphs). Smoothing is thereby shown that such a simulation can be executed with a relatively large mesh size ($\Delta x = 166.7$ μm), that is, a low number of segments that results in a rather fast simulation.

While there has been a significant improvement in computer hardware over the past decade, it is still necessary to note that due to the intimate relationship between the length of the capillary, number of segments, and current density or applied electric field strength, there are still some laboratory experiments that are difficult to simulate. These are typically those that employ very high current densities (approaching 45,000 A/m^2, which translates to a current of 88 μA in a 50 μm ID capillary, or 200 μA in a 75 μm ID capillary) in which there is a self-sharpening discontinuity. These are frequently encountered when trying to simulate various stacking methods, which are difficult to simulate at realistic currents unless short capillaries, typically in the order of several cm, with several thousands (typically 20,000) segments, are used. Longer capillaries or fewer segments

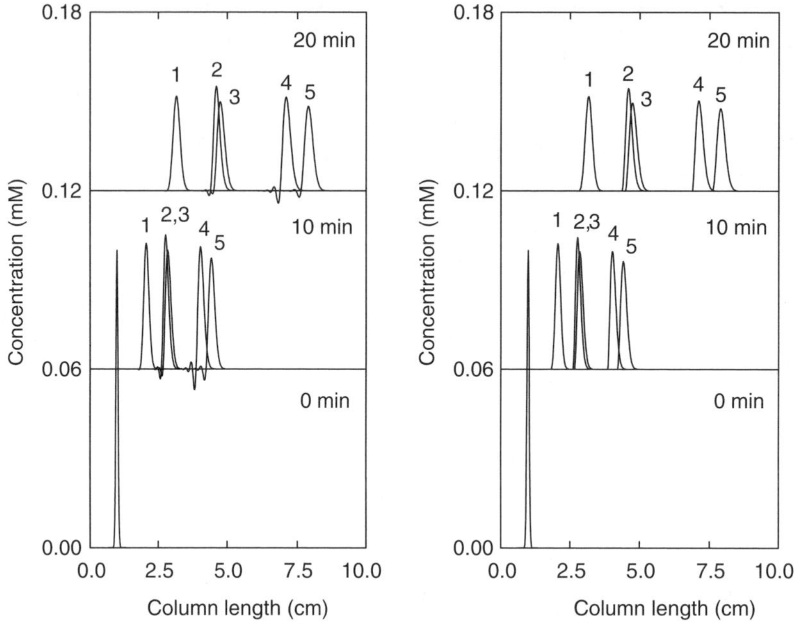

FIGURE 17.13 Simulation of the ZE separation of five dipeptides in a 20.0 mM CHES/9.457 mM NaOH buffer at pH 9.5 and without electroosmosis using GENTRANS without (left-hand graphs) and with (right-hand graphs) data smoothing. Data were produced using a column of 10 cm length that was divided into 600 segments ($\Delta x = 166.7$ μm) and fixed concentrations at column ends. A constant voltage of –145 V (current density of –100 A/m^2) was applied and the sample was applied as Gaussian peak (0.1 mM peak height each, half width of 10 segments) superimposed onto the buffer at 10% of column length. Sample distributions after 0, 10, and 20 min of power application are shown with a *y*-axis offset of 0.06 mM between graphs. The anode is to the right. Key: 1, β-ala-his (pK_a values: 2.73, 6.87, 9.73; mobility: 2.14×10^{-8} m^2/Vs); 2, lys-glu (pK_a values: 2.93, 4.47, 7.75, 10.5; mobility: 1.96×10^{-8} m^2/Vs); 3, his-gly (pK_a values: 2.40, 5.80, 7.82; mobility: 2.26×10^{-8} m^2/Vs); 4, asp-his (pK_a values: 2.45, 3.02, 6.82, 7.98; mobility: 1.81×10^{-8} m^2/Vs); 5, asp-gly (pK_a values: 2.10, 4.53, 9.07; mobility: 2.46×10^{-8} m^2/Vs). The pK_a and mobility values of the weak acid CHES were 9.55 and 2.31×10^{-8} m^2/Vs, respectively. The mobility of sodium was 5.19×10^{-8} m^2/Vs.

cause undesirable oscillations, and while this can be avoided by lowering the current density, it again deviates from the situation encountered in the laboratory. A version of GENTRANS has been created that can handle up to 100,000 segments while SIMUL has no apparent restriction on the number of segments, potentially allowing simulations to be performed with capillary lengths typically used in the laboratory at experimentally used electric field strengths (up to 1000 V/cm), however as noted above, doubling the number of segments increases the computational time by a factor of 4–8, meaning current simulations with 20,000 segments taking 18 days, would stretch out to 72–144 days (about 2.5–5 months!). While it is expected that the advent of newer computers will continue to push the boundaries of simulation software, it will necessitate revision of the software itself, particularly to make use of multiprocessor computers that are now becoming commonly available in desktop computers.

Finally, it should be noted that there are a number of essential requirements that need to be reported to allow others to repeat a particular simulation. Obviously, the software and the version are important as are the input data used to describe the components (pK_a values, mobilities, and concentrations), but it is often the overlooked parameters that can mean the difference between successfully repeating a simulation. The length of the capillary (L) and number of segments (either as the number of segments, n, or the mesh size, Δx), the current density and the voltage, and the initial step size (dt) and whether there are any restraints on these should always be given. Also important is the position and distribution of the components, particularly the presence of boundaries, which should be given with their position as well as their width. In principle, simulations can be performed at constant voltage or constant current density. For simulations comprising boundaries with a steady-state shape, such as ITP, simulations with constant current density rather than constant voltage might be preferred as the width of a steady-state boundary is current dependent.

17.7 CONCLUDING REMARKS

The state of dynamic computer simulation software has progressed significantly over the past two decades. Software is available that will simulate electrophoretic systems including MBE, ZE, ITP, and IEF under almost exactly the same conditions used in the laboratory, and this has been used to show the detailed mechanisms of many of the fundamental phenomena that occur in a simple electrophoretic separation. A detailed insight into a number of electrophoretic processes has been achieved and this is having a significant impact on designing new and improved electrophoretic systems. Noticeably absent is the inability of the current software to simulate EKC and electrochromatography (EC), and while the underlying algorithms of the current software is applicable to these forms of CE, the necessity of adding more equations and constants to describe the additional equilibria that necessitates an increase in computation time, and is likely to be the reason this has not been implemented to date. This will change over the coming years as improvements in computer hardware will result in a decrease in total simulation times and will make it feasible to complicate the system by introducing additional equilibria to account for other modes of electrophoresis making it possible to simulate even more varied and complex CE separations. Furthermore, two- and three-dimensional dynamic simulation models will have to be developed that permit simulations in capillaries and microchannels of different shapes and geometries.

It is important to mention that dynamic electrophoretic simulations are relevant for separations on any scale and instrumental format, including free-fluid preparative, gel, capillary, and chip electrophoresis. Separations in the minute to subsecond time domains (Figure 17.11) can thereby be assessed. Electrophoretic modeling is extremely flexible. In addition to the classical electrophoretic modes (MBE, ITP, ZE, and IEF), it has been used to examine the behavior of systems containing arrays of fixed charges such as immobilized pH gradients or the presence of ion exchange membranes [1,2]. CE performed in capillaries and microchannels (chips) benefits from computer modeling as simulation is well suited (1) to develop/modify buffer systems, including those employed for indirect and conductivity detection (determination of the impact of the sample on background buffer composition); (2) to study the effects of sample composition on sample concentration/migration for different injection procedures; (3) to determine the effect of a pH change on migration/separation; (4) to investigate system peaks, unusual sample peaks, and buffer oscillations; (5) to explore combinations of electrophoretic modes, including ITP/ZE separations and mobilization in IEF; and (6) to elucidate the impact of EOF on separations in discontinuous buffer systems (ITP and IEF), to name a few. Furthermore, it allows to predict the impact of various modifications on the electrophoretic behavior of peptides and proteins, including genetic mutations/modifications, chemical modifications such as deamidation during downstream processing, changes in amino acid composition, stripping of tightly bound, charged small ion cofactors, and binding of metal ions. This and other areas of application, including the prediction of enantioselective and other affinity separations, will be explored on a broad basis by dynamic electrophoretic simulation in the not too distant future.

ACKNOWLEDGMENTS

The authors acknowledge the valuable contributions of Dr R. A. Mosher to the development of GENTRANS and the discussions about SIMUL5 with Dr B. Gaš and V. Hruška. This work was supported by the Swiss National Science Foundation and the Australian Research Council.

REFERENCES

1. Bier, M., Palusinski, O.A., Mosher, R.A., and Saville, D.A., Electrophoresis: Mathematical modeling and computer simulation, *Science*, 219, 1281, 1983.
2. Mosher, R.A., Saville, D.A., and Thormann, W., *The Dynamics of Electrophoresis*, VCH Publishers, Weinheim, 1992.
3. Reijenga, J.C. and Kenndler, E., Computational simulation of migration and dispersion in free capillary zone electrophoresis. I. Description of the theoretical model, *J. Chromatogr.*, 659, 403, 1994.
4. Reijenga, J.C. and Kenndler, E., Computational simulation of migration and dispersion in free capillary zone electrophoresis. II. Results of simulation and comparison with measurements, *J. Chromatogr.*, 659, 417, 1994.
5. Jaroš, M., Hruška, V., Štědrý, M., Zuskova, I., and Gaš, B., Eigenmobilities in background electrolytes for capillary zone electrophoresis: IV. Computer program PeakMaster, *Electrophoresis*, 25, 3080, 2004.
6. Gaš, B., Jaroš, M., Hruška, V., Zuskova, I., and Štědrý, M., PeakMaster—A freeware simulator of capillary zone electrophoresis, *LC-GC Eur*, 18, 282, 2005.
7. Thormann, W. and Mosher, R.A., Theory of electrophoretic transport and separations: The study of electrophoretic boundaries and fundamental separation principles by computer simulation, *Adv. Electrophor.*, 2, 45, 1988.
8. Reijenga, J.C., Martens, J.H.P.A., and Everaerts, F.M., Training software for electrophoresis, *Electrophoresis*, 16, 2008, 1995.
9. Golshanshirazi, S. and Guiochon, G., Comparison between experimental and theoretical band profiles in nonlinear liquid-chromatography with a binary mobile phase, *Anal. Chem.*, 60, 2634, 1988.
10. Golshanshirazi, S., Ghodbane, S., and Guiochon, G., Comparison between experimental and theoretical band profiles in nonlinear liquid-chromatography with a pure mobile phase, *Anal. Chem.*, 60, 2630, 1988.
11. Guiochon, G., Golshanshirazi, S., and Jaulmes, A., Computer-simulation of the propagation of a large-concentration band in liquid-chromatography, *Anal. Chem.*, 60, 1856, 1988.
12. Guiochon, G. and Ghodbane, S., Computer-simulation of the separation of a 2-component mixture in preparative scale liquid-chromatography, *J. Phys. Chem.*, 92, 3682, 1988.
13. Moore, G.T., Theory of isotachophoresis development of concentration boundaries, *J. Chromatogr.*, 106, 1, 1975.
14. Ryser, P., *Omegaphorese: Theorie und Anwendung einer Analysenmethode*, Dissertation, University of Bern, Bern, 1976.
15. Gaš, B., Modeling separation process in capillary isotachophoresis, *Diploma Thesis*, Charles University, Prague, 1975.
16. Vacík, J. and Fidler, V., Dynamics of isotachophoretic separation, in *Analytical Isotachophoresis*, Everaerts, F.M. (Ed.), Elsevier, Amsterdam, 1981, p. 19.
17. Fidler, V., Vacík, J., and Fidler, Z., Dynamics of isotachophoretic separation. 1. Computer-simulation, *J. Chromatogr.*, 320, 167, 1985.
18. Fidler, Z., Fidler, V., and Vacík, J., Dynamics of isotachophoretic separation. 2. Impurities in the electrolytes and sample injection, *J. Chromatogr.*, 320, 175, 1985.
19. Saville, D.A. and Palusinski, O.A., Theory of electrophoretic separations. Part I: formulation of a mathematical model, *AIChE J.*, 32, 207, 1986.
20. Palusinski, O.A., Graham, A., Mosher, R.A., Bier, M., and Saville, D.A., Theory of electrophoretic separations. Part II: Construction of a numerical simulation scheme and its applications, *AIChE J.*, 32, 215, 1986.
21. Radi, P., *Numerische Integration der elektrophoretischen Transportgleichung*, Diploma Thesis, University of Bern, Bern, 1982.
22. Radi, P. and Schumacher, E., Numerical simulation of electrophoresis—The complete solution for 3 isotachophoretic systems, *Electrophoresis*, 6, 195, 1985.
23. Roberts, G.O., *Electrophoresis modeling*, USRA/RAI/ES73 Report, Marshall Space Flight Center, Huntsville, 1984.
24. Roberts, G.O., Rhodes, P.S., and Synder, R.S., Dispersion effects in capillary zone electrophoresis, *J. Chromatogr.*, 480, 35, 1989.

25. Schafer-Nielsen, C., Steady-state gel electrophoresis systems, in *Gel Electrophoresis of Proteins*, Dunn, M.A. (Ed.), Wright, Bristol, 1986, p. 1.
26. Mosher, R.A., Dewey, D., Thormann, W., Saville, D.A., and Bier, M., Computer simulation and experimental validation of the electrophoretic behavior of proteins, *Anal. Chem.*, 61, 362, 1989.
27. Dose, E.V. and Guiochon, G.A., High-resolution modeling of capillary zone electrophoresis and isotachophoresis, *Anal. Chem.*, 63, 1063, 1991.
28. Gaš, B., Vacík, J., and Zelensky, I., Computer-aided simulation of electromigration, *J. Chromatogr.*, 545, 225, 1991.
29. Schwer, C., Gaš, B., Lottspeich, F., and Kenndler, E., Computer simulation and experimental evaluation of on-column sample preconcentration in capillary zone electrophoresis by discontinuous buffer systems, *Anal. Chem.*, 65, 2108, 1993.
30. Hruška, V., Jaroš, M., and Gaš, B., Simul 5—Free dynamic simulator of electrophoresis, *Electrophoresis*, 27, 984, 2006.
31. Mosher, R.A., Gebauer, P., Caslavska, J., and Thormann, W., Computer simulation and experimental validation of the electrophoretic behavior of proteins. 2. Model improvement and application to isotachophoresis, *Anal. Chem.*, 64, 2991, 1992.
32. Thormann, W., Molteni, S., Stoffel, E., Mosher, R.A., and Chmelík, J., Computer modeling and experimental validation of the dynamics of capillary isoelectric focusing with electroosmotic zone displacement, *Anal. Methods Instrum.*, 1, 177, 1993.
33. Mosher, R.A., Zhang, C.-X., Caslavska, J., and Thormann, W., Dynamic simulator for capillary electrophoresis with in situ calculation of electroosmosis, *J. Chromatogr. A*, 716, 17, 1995.
34. Thormann, W., Zhang, C.-X., Caslavska, J., Gebauer, P., and Mosher, R.A., Modeling of the impact of ionic strength on the electroosmotic flow in capillary electrophoresis with uniform and discontinuous buffer systems, *Anal. Chem.*, 70, 549, 1998.
35. Mao, Q., Pawliszyn, J., and Thormann, W., Dynamics of capillary isoelectric focusing in the absence of fluid flow: High-resolution computer simulation and experimental validation with whole column optical imaging, *Anal. Chem.*, 72, 5493, 2000.
36. Thormann, W., Huang, T., Pawliszyn, J., and Mosher, R.A., High-resolution computer simulation of the dynamics of isoelectric focusing of proteins, *Electrophoresis*, 25, 324, 2004.
37. Ermakov, S.V., Finite-difference algorithm for convection diffusion equation applied to electrophoresis problem, *Informatica*, 3, 173, 1992.
38. Ermakov, S.V., Mazhorova, O.S., and Zhukov, M.Y., Computer simulation of transient states in capillary zone electrophoresis and isotachophoresis, *Electrophoresis*, 13, 838, 1992.
39. Ermakov, S.V., Bello, M.S., and Righetti, P.G., Numerical algorithms for capillary electrophoresis, *J. Chromatogr. A*, 661, 265, 1994.
40. Shimao, K., Computer simulation of electrophoretic transport. II. Modified general procedure and application to isoelectric focusing of proteins using immobilised pH gradient, *Jap. J. Electrophor.*, 38, 221, 1994.
41. Shimao, K., Computer simulation of electrophoretic transport. V. Computer simulation of sodium dodecyl sulfate-polyacrylamide gel electrophoresis, *Jap. J. Electrophor.*, 43, 245, 1999.
42. Schafer-Nielsen, C., A computer model for time-based simulation of electrophoresis systems with freely defined initial and boundary conditions, *Electrophoresis*, 16, 1369, 1995.
43. Martens, J.H.P.A., Reijenga, J.C., ten Thije Boonkkamp, J.H.M., Mattheij, R.M.M., and Everaerts, F.M., Transient modeling of capillary electrophoresis. Isotachophoresis, *J. Chromatogr. A*, 772, 49, 1997.
44. Beckers, J.L. and Boček, P., Sample stacking in capillary zone electrophoresis: Principles, advantages and limitations, *Electrophoresis*, 21, 2747, 2000.
45. Ikuta, N. and Hirokawa, T., Numerical simulation for capillary electrophoresis. I. Development of a simulation program with high numerical stability, *J. Chromatogr. A*, 802, 49, 1998.
46. Sounart, T.L. and Baygents, J.C., Simulation of electrophoretic separations by the flux-corrected transport method, *J. Chromatogr. A*, 890, 321, 2000.
47. Hopkins, D.L. and McGuffin, V.L., Three-dimensional molecular simulation of electrophoretic separations, *Anal. Chem.*, 70, 1066, 1998.
48. Ermakov, S.V., Jacobson, S.C., and Ramsey, J.M., Computer simulations of electrokinetic transport in microfabricated channel structures, *Anal. Chem.*, 70, 4494, 1998.

49. Bianchi, F., Ferrigno, A., and Girault, H.H., Finite element simulation of an electroosmotic-driven flow division at a T-junction of microscale dimensions, *Anal. Chem.*, 72, 1987, 2000.
50. Chatterjee, A., Generalized numerical formulations for multi-physics microfluidics-type applications, *J. Micromech. Microeng.*, 13, 758, 2003.
51. Sounart, T.L. and Baygents, J.C., Electrically-driven fluid motion in channels with streamwise gradients of the electrical conductivity, *Colloids Surf. A Physicochem. Eng. Asp.*, 195, 59, 2001.
52. Afonso, J.L. and Clifton, M.J., Optimization of protein separation by continuous-flow electrophoresis: Influence of the operating conditions and the chamber thickness, *Electrophoresis*, 20, 2801, 1999.
53. Steinmann, L., Mosher, R.A., and Thormann, W., Characterization and impact of the temporal behavior of the electroosmotic flow in capillary isoelectric focusing with electroosmotic zone displacement, *J. Chromatogr. A*, 756, 219, 1996.
54. Caslavska, J. and Thormann, W., Electrophoretic separations in PMMA capillaries with uniform and discontinuous buffers, *J. Microcol. Sep.*, 13, 69, 2001.
55. Sounart, T.L. and Baygents, J.C., Simulation of electrophoretic separations: effect of numerical and molecular diffusion on pH calculations in poorly buffered systems, *Electrophoresis*, 21, 2287, 2000.
56. Mosher, R.A., Thormann, W., and Bier, M., Computer aided analysis of electric field gradients within isotachophoretic boundaries between weak electrolytes, *J. Chromatogr.*, 320, 23, 1985.
57. Deshmukh, R.R. and Bier, M., Counterflow in isotachophoresis: Computer simulation and experimental studies, *Electrophoresis*, 14, 205, 1993.
58. Thormann, W., Caslavska, J., and Mosher, R.A., Impact of electroosmosis on isotachophoresis in open-tubular fused-silica capillaries: Analysis of the evolution of a stationary steady-state zone structure by computer simulation and experimental validation, *Electrophoresis*, 16, 2016, 1995.
59. Caslavska, J. and Thormann, W., Bidirectional isotachophoresis in open-tubular, untreated fused-silica capillaries, *J. Chromatogr. A*, 772, 3, 1997.
60. Caslavska, J. and Thormann, W., Isotachophoresis in dynamically double coated fused-silica capillaries, *Electrophoresis*, 27, 4618–4630, 2006.
61. Mosher, R.A., Gebauer, P., and Thormann, W., Computer simulation and experimental validation of the electrophoretic behavior of proteins. III. Use of titration data predicted by the protein's amino acid composition, *J. Chromatogr.*, 638, 155, 1993.
62. Thormann, W. and Mosher, R.A., Experimental and theoretical description of the isotachophoretic behavior of serum albumin, *Electrophoresis*, 11, 292, 1990.
63. Thormann, W., Firestone, M.A., Sloan, J.E., Long, T.D., and Mosher, R.A., Isotachophoretic zone formation of serum albumin in different free-fluid electrophoresis instruments, *Electrophoresis*, 11, 298, 1990.
64. Dubrovčáková, E., Gaš, B., Vacík, J., and Smolková-Keulemansová, E., Electromigration in systems with additives in background electrolytes I. Addition of the neutral complexing agent, *J. Chromatogr.*, 623, 337, 1992.
65. Foret, F., Thompson, T.J., Vouros, P., Karger, B.L., Gebauer, P., and Boček, P., Liquid sheath effects on the separation of proteins in capillary electrophoresis/electrospray mass spectrometry, *Anal. Chem.*, 66, 4450, 1994.
66. Gebauer, P. and Boček, P., A new type of migrating zone boundary in electrophoresis: 1. General description of boundary behavior based on electromigration dispersion velocity profiles, *Electrophoresis*, 26, 453, 2005.
67. Mosher, R.A. and Thormann, W., High-resolution computer simulation of the dynamics of isoelectric focusing using carrier ampholytes: The post-separation stabilizing phase revisited, *Electrophoresis*, 23, 1803, 2002.
68. Thormann, W. and Mosher, R.A., High-resolution computer simulation of the dynamics of isoelectric focusing using carrier ampholytes: Focusing with concurrent electrophoretic mobilization is an isotachophoretic process, *Electrophoresis*, 27, 968, 2006.
69. Ermakov, S.V., Zhukov, M.Y., Capelli, L., and Righetti, P.G., Artifactual peak splitting in capillary electrophoresis. 2. Defocusing phenomena for ampholytes, *Anal. Chem.*, 67, 2957, 1995.
70. Ermakov, S.V., Zhukov, M.Y., Capelli, L., and Righetti, P.G., Experimental and theoretical-study of artifactual peak splitting in capillary electrophoresis, *Anal. Chem.*, 66, 4034, 1994.

71. Williams, R.L., Childs, B., Dose, E.V., Guiochon, G., and Vigh, G., Peak shape distortions during capillary electrophoretic separations of multicomponent samples in two co-ion buffers, *J. Chromatogr. A*, 781, 107, 1997.
72. Williams, R.L., Childs, B., Dose, E.V., Guiochon, G., and Vigh, G., Peak shape distortions in the capillary electrophoretic separations of strong electrolytes when the background electrolyte contains two strong electrolyte co-ions, *Anal. Chem.*, 69, 1347, 1997.
73. Ermakov, S.V. and Righetti, P.G., Computer simulation for capillary zone electrophoresis. A quantitative approach, *J. Chromatogr. A*, 667, 257, 1994.
74. Ermakov, S.V., Zhukov, M.Y., Capelli, L., and Righetti, P.G., Wall adsorption in capillary electrophoresis. Experimental study and computer simulation, *J. Chromatogr. A*, 699, 297, 1995.
75. Zhukov, M.Y., Ermakov, S.V., and Righetti, P.G., Simplified mathematical model of irreversible sample adsorption in capillary zone electrophoresis, *J. Chromatogr. A*, 766, 171, 1997.
76. Gaš, B., Štědrý, M., Rizzi, A., and Kenndler, E., Dynamics of peak dispersion in capillary zone electrophoresis including wall adsorption I. Theoretical model and results of simulation, *Electrophoresis*, 16, 958, 1995.
77. Křivánková, L., Gebauer, P., Thormann, W., Mosher, R.A., and Boček, P., Options in electrolyte systems for online combined capillary isotachophoresis and capillary zone electrophoresis, *J. Chromatogr.*, 638, 119, 1993.
78. Gebauer, P., Thormann, W., and Boček, P., Sample self-stacking in zone electrophoresis. Theoretical description of the zone electrophoretic separation of minor compounds in the presence of bulk amounts of a sample component with high mobility and like charge, *J. Chromatogr.*, 608, 47, 1992.
79. Gebauer, P., Thormann, W., and Boček, P., Sample self-stacking and sample stacking in zone electrophoresis with major sample components of like charge: General model and scheme of possible modes, *Electrophoresis*, 16, 2039, 1995.
80. Gebauer, P., Křivánková, L., Pantůčková, P., Boček, P., and Thormann, W., Sample self-stacking in capillary zone electrophoresis: Behavior of samples containing multiple major coionic components, *Electrophoresis*, 21, 2797, 2000.
81. Křivánková, L., Pantůčková, P., Gebauer, P., Boček, P., Caslavska, J., and Thormann, W., Chloride present in biological samples as a tool for enhancement of sensitivity in capillary zone electrophoretic analysis of anionic trace analytes, *Electrophoresis*, 24, 505, 2003.
82. Křivánková, L., Brezkova, M., Gebauer, P., and Boček, P., Importance of the counterion in optimization of a borate electrolyte system for analyses of anions in samples with complex matrices performed by capillary zone electrophoresis, *Electrophoresis*, 25, 3406, 2004.
83. Hirokawa, T., Okamoto, H., and Gaš, B., High-sensitive capillary zone electrophoresis analysis by electrokinetic injection with transient isotachophoretic preconcentration: Electrokinetic supercharging, *Electrophoresis*, 24, 498, 2003.
84. Kim, J.-B., Britz-McKibbin, P., Hirokawa, T., and Terabe, S., Mechanistic study on analyte focusing by dynamic pH junction in capillary electrophoresis using computer simulation, *Anal. Chem.*, 75, 3986, 2003.
85. Shim, S.-H., Riaz, A., Choi, K., and Chung, D.S., Dual stacking of unbuffered saline samples, transient isotachophoresis plus induced pH junction focusing, *Electrophoresis*, 24, 1603, 2003.
86. Breadmore, M.C., Mosher, R.A., and Thormann, W., High-resolution computer simulations of stacking of weak bases using a transient pH boundary in capillary electrophoresis. 1. Concept and impact of sample ionic strength, *Anal. Chem.*, 78, 538, 2006.
87. Tesařová, E., Ševčík, J., Gaš, B., and Armstrong, D.W., Effects of partial/asymmetrical filling of micelles and chiral selectors on capillary electrophoresis enantiomeric separation: Generation of a gradient, *Electrophoresis*, 25, 2693, 2004.
88. Andreev, V.P., Pliss, N.S., and Righetti, P.G., Computer simulation of affinity capillary electrophoresis, *Electrophoresis*, 23, 889, 2002.
89. Fang, N. and Chen, D.D.Y., General approach to high-efficiency simulation of affinity capillary electrophoresis, *Anal. Chem.*, 77, 840, 2005.

90. Okhonin, V., Krylova, S.M., and Krylov, S.N., Nonequilibrium capillary electrophoresis of equilibrium mixtures, mathematical model, *Anal. Chem.*, 76, 1507, 2004.
91. Hruška, V., Jaroš, M., and Gaš, B., Oscillating electrolytes, *Electrophoresis*, 27, 513, 2006.
92. Thormann, W. and Mosher, R.A., Theoretical and computer aided analysis of steady state moving boundaries in electrophoresis: An analytical solution for the estimation of boundary widths between weak electrolytes, *Trans. Soc. Comput. Simul.*, 1, 83, 1984.

18 Heat Production and Dissipation in Capillary Electrophoresis

Christopher J. Evenhuis, Rosanne M. Guijt, Miroslav Macka, Philip J. Marriott, and Paul R. Haddad

CONTENTS

18.1 INTRODUCTION

Joule heating is an unavoidable phenomenon in capillary electrophoresis (CE) and results from resistive heating that occurs when an electric current (I) flows through the electrolyte when a potential difference is applied. The increase in conductivity with temperature results in a positive feedback effect in which the current increases until a steady state is reached. This may even cause the electrolyte to boil or superheat: this is known as autothermal runaway [1]. Temperature control is usually employed in CE to aid heat dissipation and provide acceptable precision, but measurement of the electrolyte temperature is often overlooked. In some older systems, only ambient temperature operation was available, with or without fan-forced airflow. The temperature of the electrolyte affects its viscosity (η), its dielectric constant (ε_r), and the zeta potential (ζ) [2], which affect the precision of migration times through the effect of η, ε_r, and ζ on the electroosmotic mobility (μ_{eof}) and the electrophoretic mobility (μ_{ep}). Even small changes in temperature can cause significant deviations in migration times [3]. The electrolyte temperature also has a major influence on peak broadening [4–6], such that separation efficiency generally decreases with increasing temperature. Radial temperature differences in the electrolyte result in viscosity differences across the capillary with analytes traveling faster in the warmer, lower viscosity zone near the axis of the capillary than in the cooler zones near the capillary wall [5]. Axial temperature differences that result from the layout of the instrument [7], and differences caused by variations in conductivity as the sample migrates through the electrolyte, also increase dispersion effects [6].

This chapter discusses the phenomena of Joule heating, heat dissipation, estimation of the electrolyte temperature and its effects, and extends the discussion of Nelson and Burgi [3] provided in the first edition of this book. The focus of this chapter is on the use of simple methods to determine the temperature of the electrolyte, free from the influence of Joule heating. An improved understanding of the radial temperature profile that exists during electrophoresis [8] allows a simple technique to be introduced for evaluating the heat transfer coefficient for an instrument.

18.2 HEAT GENERATION

18.2.1 THEORY

A significant number of authors have modeled the theory of heat generation and heat dissipation in CE [4,9–13]. The rate of Joule heating (P), which is the resistive heating power that occurs when an electric current (I) flows through the electrolyte, can be quantified simply using the following equation:

$$P = IV, \tag{18.1}$$

where V is the applied voltage. If the current is measured in microamperes (μA) and the voltage is measured in kilovolts (kV), P will be measured in milliWatts (mW). Instead of P, the power per unit length (P/L) is generally used to take the length of the capillary into account.

Traditionally, it has been suggested that a graph of current versus voltage should be constructed for an electrolyte to characterize the CE system. In Figure 18.1a, an Ohm's Law plot is given for a fused-silica (FS) capillary and for a fluorinated ethylene–propylene (FEP) capillary. The greater the deviation of the curve from the Ohm's Law plot, the less efficient is the cooling system. However, assignment of the voltage at which the deviation of the curve from the Ohm's Law plot becomes excessive is somewhat arbitrary, although this is sometimes used as a guide for maximum operating potential.

A more useful plot to characterize the CE system is a graph of conductance (G) versus the power dissipated per unit length as this can provide information about the increase in the buffer temperature, the internal radius of the capillary, and the cooling efficiency of the instrument. Conductance varies linearly with temperature, making it very useful as a probe for the average electrolyte temperature in the capillary. A detailed description is provided in Section 18.4.2.

Equation 18.2 shows that conductance is the reciprocal of resistance and can be found by dividing the current by the voltage, parameters that can be easily monitored and recorded

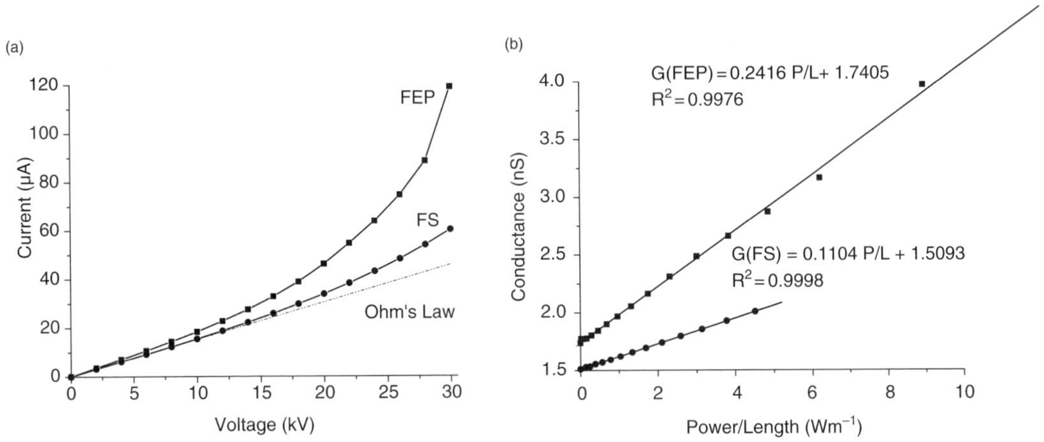

FIGURE 18.1 Graphs of (a) I vs. V and (b) G vs. P/L for different capillaries of equal length. Conditions: $L_{tot} = 41.0$ cm fused-silica (FS) $d_i = 74.0$ μm, $d_o = 362.8$ μm and fluorinated ethylene–propylene copolymer (FEP) d_i 80 μm, $d_o \approx 370$ μm. Electrolyte: 10 mM phosphate buffer at pH 7.21.

during the course of a separation.

$$G = \frac{I}{V}. \tag{18.2}$$

If the current is measured in μA and the voltage is in kV, the conductance is determined in nanoSiemens (nS $= 10^{-9}$ AV^{-1} $= 10^{-9}\Omega^{-1}$). Plots of G versus P/L for a FS capillary and for a fluorinated FEP capillary are given in Figure 18.1b.

The later onset of deviation from Ohm's Law for the FS capillary as voltage increases in the Ohm's Law plot in Figure 18.1a demonstrates the superior heat dissipation of FS over FEP. This is confirmed by the smaller slope of G vs. P/L for the FS capillary in Figure 18.1b. The ratio of the gradient of the G vs. P/L graph to the intercept can be used to estimate the temperature increase of the electrolyte. A more detailed description of temperature calculation will follow in Section 18.4. In addition, the average internal diameter (d_i) of the capillary can be derived from the conductance at zero power (G_0). The intercept, G_0, is proportional to the cross-sectional area of the capillary. Provided that the lengths of the capillaries are equal, this allows one to use Equation 18.3 to calculate the internal diameter of an unknown capillary based on the internal diameter of another capillary [14].

$$\frac{G_{0(\text{FEP})}}{G_{0(\text{FS})}} = \frac{(d_{\text{i FEP}})^2}{(d_{\text{i FS}})^2}. \tag{18.3}$$

Figure 18.1b demonstrates that the FEP capillary has a larger internal diameter, as evident from its larger conductance at zero power (G_0). On the basis of the d_i of the FS capillary, the diameter of the FEP capillary can be calculated as illustrated in the following equation:

$$d_{\text{i FEP}} = \sqrt{\frac{G_{0(\text{FEP})}}{G_{0(\text{Fs})}}} \cdot d_{\text{i Fs}} = \sqrt{\frac{1.7405}{1.5093}} \cdot 74.0\,\mu\text{m} = 79.5\,\mu\text{m}. \tag{18.4}$$

The possibility of determining the electrolyte temperature and the capillary diameter make the G versus P/L graph a superior method to the Ohm's Law plot for characterizing Joule heating.

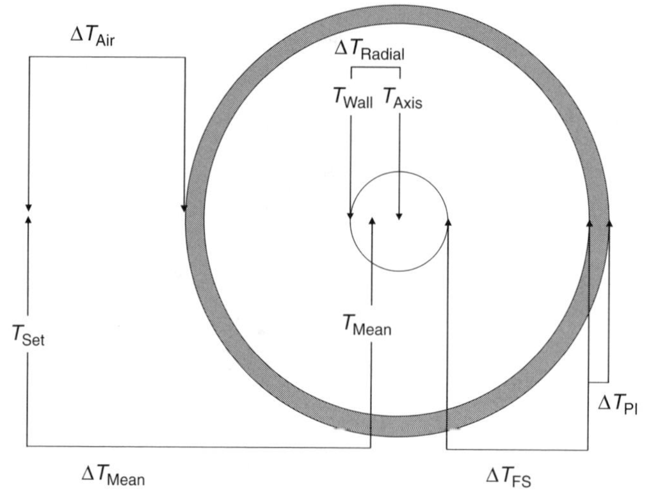

FIGURE 18.2 Schematic diagram showing the locations at which the temperature determinations are made, and definitions of temperature differences. (Reproduced with permission from Evenhuis, C. J., et al., *Anal. Chem.* 2006, *78*, 2684–2693. Copyright 2006 American Chemical Society.)

Grushka et al. [4] have provided a useful rule of thumb that the radial temperature difference in the capillary (ΔT_{Radial}, see Figure 18.2) should not be allowed to exceed 1.5 °C otherwise there is an unacceptable loss of efficiency due to peak broadening.

It has been shown that the radial temperature difference is directly proportional to the power per unit length and is independent of the cooling efficiency of the instrument. The following equation applies for aqueous electrolytes:

$$\Delta T_{\text{Radial}} = \frac{1}{4\pi\lambda} \cdot \frac{P}{L} = 0.1315\ °\text{C W}^{-1}\text{m} \cdot \frac{P}{L}, \tag{18.5}$$

where λ is the thermal conductivity of the electrolyte ($\lambda_{\text{water}} = 0.605\ \text{Wm}^{-1}\ \text{K}^{-1}$). To avoid excessive peak broadening *P*/*L* should not be allowed to exceed 11 Wm^{-1}. For nonaqueous electrolytes, which tend to have lower thermal conductivities, or electrolytes containing organic constituents, it is wise to calculate or estimate the maximum value of *P*/*L*[15]. For example, for methanol ($\lambda_{\text{MeOH}} = 0.202$ $\text{Wm}^{-1}\ \text{K}^{-1}$) the maximum power level would be 3.8 Wm^{-1}.

A number of other problems are associated with working at high values of P/L. McCormick [16,17] observed that if the electrolyte heated up too quickly, part of the sample could actually be expelled from the capillary by the rapid expansion of the buffer. The formation of bubbles or even boiling of the electrolyte can also occur at high power levels, resulting in a sudden decrease of the current. Microbubbles formed by outgassing of dissolved air allow the current to continue but can often be detected as random sharp peaks (spikes). To avoid these spurious peaks, it is good practice to degas the electrolyte in an ultrasonic bath before its use.

18.2.2 Factors Affecting Heat Generation

The flowchart (see Figure 18.3) demonstrates that the capillary, electrolyte, and instrumental parameters all have an influence to a varying degree on the overall temperature rise of the electrolyte (ΔT_{Mean}, see Figure 18.2). An explanation of the terms used is shown in Table 18.1. Key parameters are discussed in detail in the following sections.

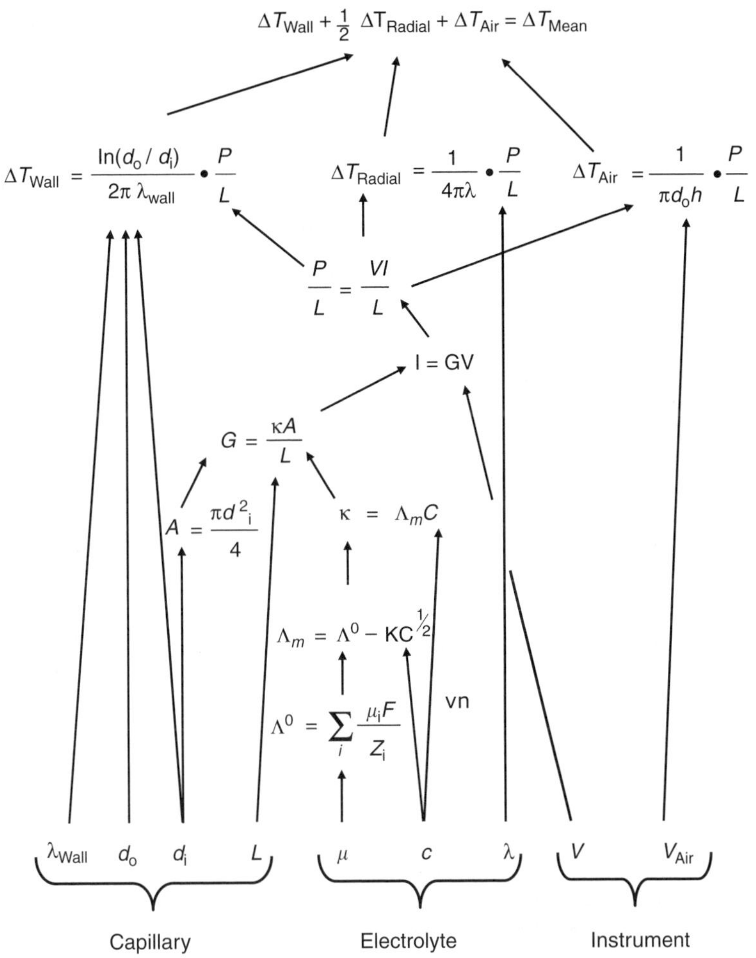

FIGURE 18.3 Flowchart showing the influences of various experimental variables on the electrolyte temperature.

18.2.2.1 Buffer Composition

Figure 18.4 illustrates that ΔT_{Mean} increases linearly with the power per unit length (P/L). The buffer composition and concentration affect its conductivity and therefore the rate of heat generation during a separation. Equation 18.6 illustrates the principle that the use of a lower conductivity buffer results in a smaller increase in temperature for the same electrical field strength (E).

$$\frac{P}{L} = \frac{\kappa A V^2}{L^2} = \kappa A E^2. \tag{18.6}$$

As conductivity depends on the mobility of the ions involved (see Equation 18.7), the use of counterions of lower mobility can reduce the current.

$$\Lambda^0 = \sum_i \frac{\mu_{iF}}{z_i}, \tag{18.7}$$

TABLE 18.1
Explanation of Terms

Abbreviation	Quantity	Unit
ΔT_{Air}	Temperature difference across air layer surrounding capillary (see Figure 18.2)	°C
ΔT_{Wall}	Temperature difference across the capillary wall [sum of ΔT_{FS} and ΔT_{PI} (see Figure 18.2)]	°C
ΔT_{FS}	Temperature difference across fused-silica wall (see Figure 18.2)	°C
ΔT_{Mean}	Change in mean temperature of the electrolyte as a result of Joule heating (see Figure 18.2)	°C
ΔT_{PI}	Temperature difference across poly(imide) coating (see Figure 18.2)	°C
ΔT_{Radial}	Radial temperature difference across electrolyte (see Figure 18.2)	°C
γ	Thermal coefficient of electrical conductivity	$°C^{-1}$
η	Viscosity of electrolyte	$kg\ m^{-1}\ s^{-1}$
Λ^0	Limiting ionic conductivity	$Sm^2\ mol^{-1}$
Λ_m	Molar conductivity	$Sm^2\ mol^{-1}$
ε	Electrical permittivity of the electrolyte	$F\ m^{-1}$
ζ	Zeta potential	V
κ	Electrical conductivity of electrolyte	$S\ m^{-1}$
λ	Thermal conductivity of electrolyte	$Wm^{-1}\ K^{-1}$
λ_{Wall}	Thermal conductivity of wall	$Wm^{-1}\ K^{-1}$
μ	Limiting ionic mobility	$m^2\ s^{-1}\ V^{-1}$
μ_i	Limiting ionic mobility of *i* species	$m^2\ s^{-1}\ V^{-1}$
μ_{ep}	Electrophoretic mobility	$m^2\ s^{-1}\ V^{-1}$
μ_{EOF}	Electroosmotic mobility	$m^2\ s^{-1}\ V^{-1}$
$\mu_{EOF}(0)$	Electroosmotic mobility at zero power	$m^2\ s^{-1}\ V^{-1}$
Φ	Constant relating ΔT_{Mean} to P/L	$°C\ W^{-1}\ m$
A	Cross-sectional area	M^2
c	Molar concentration	$mol\ L^{-1}$
d_i	Internal diameter of capillary	M
d_o	External diameter of capillary	M
F	Faraday's constant	$96{,}487\ C\ mol^{-1}$
G	Conductance	S
$G(0)$	Conductance at zero power	S
h	Heat transfer coefficient	$Wm^{-2}\ K^{-1}$
I	Electric current	A
K	Kohlrausch constant	$Sm^2\ mol^{-3/2}\ L^{-1/2}$
L	Total length of the capillary	M
P	Power	W
P/L	Power per unit length	Wm^{-1}
q	Charge of ion	C
r_h	Hydrodynamic radius	M
V	Applied voltage	V
z_i	Valency of ionic species *i*	Dimensionless

where Λ^0 is the limiting ionic conductivity, μ_i is the limiting ionic mobility of species *i*, and z_i is its valency.

For example, Na^+ should be used in preference to NH_4^+ or K^+ when counterions are being considered. Large organic counterions such as histidine can significantly reduce the current but one needs to be aware that separation selectivity can also be affected. Zwitterionic buffers such as

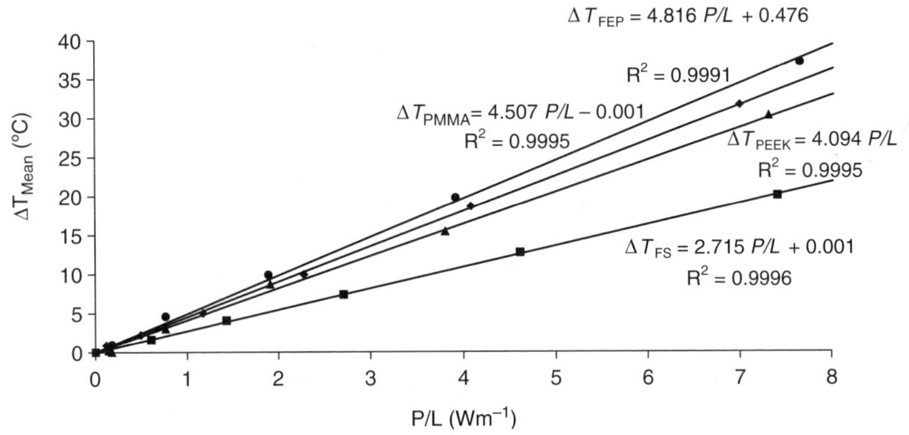

FIGURE 18.4 Variation of mean increase in electrolyte temperature with power per unit length for capillaries made from different materials •, FEP; ♦, PMMA; ▲, PEEK; and ■, FS. Conditions: $h = 376\ \text{Wm}^{-2}\ \text{K}^{-1}$, $L_{tot} = 32.2$ cm, $d_i \approx 75\ \mu\text{m}$, $d_o \approx 365\ \mu\text{m}$, See Reference [14] for more details for each capillary. Electrolyte: 10 mM phosphate, pH = 7.21. (Reproduced with permission from Evenhuis, C. J., et al., *Electrophoresis* 2005, *26*, 4333–4344. Copyright 2005 Wiley VCH.)

3-(*N*-morpholino)propanesulfonic acid (MOPS) and 2-(*N*-morpholino)ethanesulfonic acid (MES) have high buffer capacities with low conductivities; they tend to absorb at low ultraviolet (UV) wavelengths but may be used to advantage for UV detection above 210 nm [3].

18.2.2.2 Applied Voltage

The rate of Joule heating increases with the square of the applied voltage, so doubling the voltage increases *P*/*L* by a factor of 4 (see Equation 18.6). The voltage is usually optimized for maximum efficiency but a compromise between efficiency and heat generation can often lead to shorter analysis times if there is sufficient resolution and the analytes are not influenced by elevated temperatures [3].

18.2.2.3 Capillary Length

The rate of Joule heating (P) is inversely proportional to the length of the capillary, so halving the capillary length increases both *P*/*L* and ΔT_{Radial} by a factor of 4 (see Equation 18.6), but decreases the separation time by a factor greater than 4 due to the increased mobility of the analytes. Resolution can be improved by using a longer capillary if the same electrical field can be maintained [3].

18.2.2.4 Capillary Internal Diameter

As the conductance depends on the cross-sectional area of the capillary (see Equation 18.6), for a constant applied voltage, *P*/*L* increases with the square of the internal diameter (d_i).

$$\frac{P}{L} = \frac{\pi \kappa d_i^2 V^2}{4L^2}. \tag{18.8}$$

Decreasing the internal diameter from 75 μm to 50 μm leads to a 56% reduction in P/L. It follows that the use of smaller bore capillaries enables higher conductivity buffers to be used at similar field strengths, which can significantly improve peak focusing. However, decreasing the internal diameter of the capillary may result in a loss of sensitivity for concentration-sensitive detection techniques, such as UV absorption, due to the decreased detection pathlength.

18.2.2.5 Buffer Viscosity

Increasing the buffer viscosity (η) affects the apparent mobility through the electrophoretic mobility (μ_{ep}, see Equation 18.9) and electroosmotic mobility (μ_{EOF}, see Equation 18.10) by increasing the resistance to movement

$$\mu_{ep} = \frac{q}{6\pi r_h \eta}, \tag{18.9}$$

where q is the charge on the solvated ion, r_h is its hydrodynamic radius, ε is the electrical permittivity of the electrolyte, and ζ is the zeta potential.

$$\mu_{EOF} = -\frac{\varepsilon\zeta}{\eta}. \tag{18.10}$$

To a good approximation, there is an inverse relationship between the viscosity and the resulting current; doubling the buffer viscosity will halve the current and ΔT_{Radial}. In nonaqueous solvents, electrical currents are generally lower than that for the same electrolyte dissolved in water, but predicting ΔT_{Radial} becomes complex as one needs to take into account variations in the mobilities of species, dissociation equilibria, dielectric constant, zeta potential, and thermal conductivity [18].

18.3 HEAT DISSIPATION

18.3.1 THEORY

It is not surprising that since heat energy is generated in the electrolyte and the heat is conducted through the capillary walls, the electrolyte temperature is greater at the axis than at the inside wall. A temperature gradient therefore exists from the central axis to the outside wall of the capillary. The size of the radial temperature difference across the electrolyte (ΔT_{Radial}) and across the capillary wall(s) is dictated by the power generated per unit length (P/L) and the thermal conductivity of each medium, but the overall increase in temperature of the electrolyte is determined by the cooling efficiency, which is characterized by the heat transfer coefficient at the outer surface (h). A relatively straightforward method of finding h for a CE instrument is described in Section 18.5. It has been predicted that an approximately parabolic temperature profile exists within the electrolyte [9,12] and this has been verified using noninvasive methods, such as Raman Spectroscopy [19]. In FS capillaries, an exponential decrease in temperature occurs across the FS and poly(imide) layers. Interestingly, the magnitude of the temperature difference across the thin polymer layer is comparable to that across the FS. However, even when highly efficient cooling systems are in operation, the main temperature drop occurs from the outside surface of the poly(imide) coating to the set temperature [12] (see Figure 18.5).

18.3.2 FACTORS INFLUENCING HEAT DISSIPATION

18.3.2.1 Type of Cooling System

The thermal conductivity of the cooling medium and the speed at which the medium flows over the capillary influence the cooling efficiency. For example, helium has been shown to be about six times more effective than air for capillary cooling due to its greater thermal conductivity [20]. As mentioned earlier, the elevation in electrolyte temperature is determined mainly by how efficiently

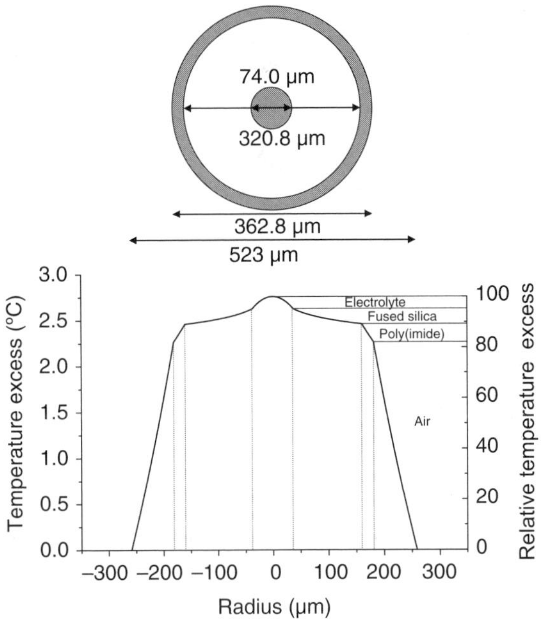

FIGURE 18.5 Calculated average radial temperature profile for a FS capillary. Conditions: $P/L = 1.00\,\mathrm{Wm^{-1}}$, $h = 376\,\mathrm{Wm^{-2}\,K^{-1}}$, $L_{tot} = 32.2$ cm, $d_i = 74.0\,\mu$m, $d_o = 362.8\,\mu$m, thickness of poly(imide) layer $= 21.0\,\mu$m. Electrolyte: 10 mM phosphate, pH = 7.21. (Reproduced with permission from Evenhuis, C. J., et al., *Anal. Chem.* 2006, *78*, 2684–2693. Copyright 2006 American Chemical Society.)

heat is conducted from the outer surface of the capillary, and this is characterized by the heat transfer coefficient (h).

Typical values for h in CE instruments are approximately 50 $\mathrm{Wm^{-2}\,K^{-1}}$ for stagnant air [21], 300–700 $\mathrm{Wm^{-2}\,K^{-1}}$ for fan-forced air depending on the speed of flow and the fraction of the capillary that is temperature controlled [8,21] and between 700 and 1200 $\mathrm{Wm^{-2}\,K^{-1}}$ for liquid cooling [5,22]. The most effective cooling system to be demonstrated for CE used a Peltier thermoelectric device to cool an alumina block. The capillary was housed in a purpose built groove and thermal contact was enhanced using ethylene glycol. A heat transfer coefficient of $h = 2600\ \mathrm{Wm^{-2}\,K^{-1}}$ was reported for this device [10].

Figure 18.6 illustrates that increasing the speed of the airflow becomes less effective at higher airflow rates, but that there is a large change in heat transfer at low flows.

18.3.2.2 Cooling System Design

Axial temperature differences are to be avoided. Ideally, the temperature of the whole capillary should be actively controlled to the same temperature. Unfortunately, the design of some commercial CE instruments is such that there are substantial portions of the capillary without temperature control. These portions of the capillary are surrounded either by stagnant air or by electrolyte in the vials. Without redesigning the instrument, areas with stagnant air cannot be avoided, but many instruments allow regulation of the temperature of the vials containing the sample and electrolyte vials, which allows significant improvements in the reproducibility.

18.3.2.3 Capillary Material

Although poly(imide)-coated FS accounts for the vast majority of applications in CE, polymeric materials have also been studied as materials for CE capillaries since these offer a variety of different

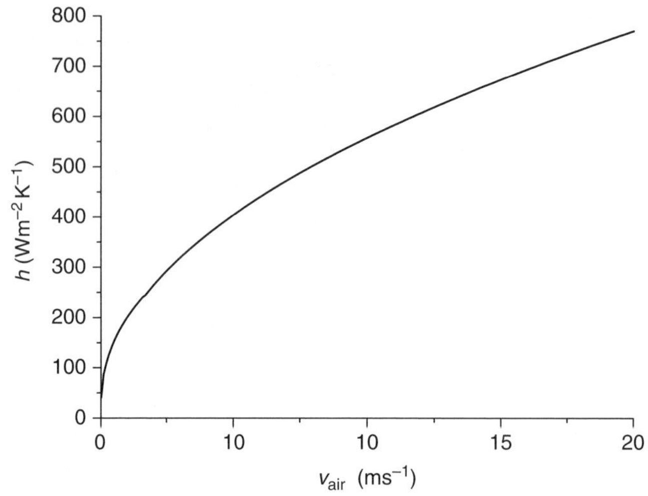

FIGURE 18.6 Theoretical variation of the heat transfer coefficient with the speed of the airflow for a poly(imide) coated FS capillary with $d_i = 75$ μm and $d_o = 375$ μm.

TABLE 18.2
Thermal Conductivity of Various Substances

Material	Thermal Conductivity (λ) at 25 °C (Wm^{-1} K^{-1})[a]	References
Air	0.025	[12]
Fused-silica (FS)	1.40	[12]
Borosilicate glass	1.0	[12]
Water	0.605	[12]
Poly(dimethylsiloxane) (PDMS)	0.18	[46]
Poly(etherether ketone) (PEEK)	0.252	[47]
Poly(tetrafluoroethene) (PTFE)	0.265	[48]
Copolymer of tetrafluoro-ethene and hexafluoropropene (FEP)	0.209	[49]
Poly(fluoroethene)	0.1	[50]
Poly(imide)	0.155	[4]
Poly(iminoadipoyliminohexane-1,6-diyl) (Nylon-6,6)	0.23	[51]
Poly(methyl methacrylate) (PMMA)	0.193	[52]
Poly(phenylethene) (Polystyrene)	0.11	[53]
Teflon AF	0.116	[54]

[a] To convert from Wm^{-1} K^{-1} to cal s^{-1} m^{-1} K^{-1} divide by a factor of 4.184.

surface physical properties. Examples include the following: poly(methylmethacrylate) (PMMA) [23,24], FEP [25], poly(propene) [26], poly(butylene-terephthalate) [26], poly(tetrafluoroethene) (PTFE) [27–30], and poly(etherether ketone) (PEEK) [31].

Generally speaking, the thermal conductivity of polymers is much less than FS (see Table 18.2) so that the temperature difference across the capillary walls and the rise in temperature of the electrolyte are both significantly greater for polymeric capillaries under the same conditions (see Figure 18.5) [14].

18.3.2.4 Capillary Outer Diameter

It has been noted that increasing the external diameter of the capillary by using a thicker layer of FS actually reduces the rise in temperature of the electrolyte [3,10]. Although the temperature gradient across the capillary wall increases, heat is removed more effectively from the larger surface area at the external surface. However, increasing the thickness of the polymer layer causes an increase in the temperature but can lead to a reduction in the optical baseline noise [3]. These authors suggested that the increased temperature of the electrolyte could be counterbalanced by reducing the temperature of the cooling system [3].

18.4 DETERMINATION OF THE AVERAGE ELECTROLYTE TEMPERATURE

A knowledge of the electrolyte temperature is important in CE as temperature changes in the electrolyte influence precision, accuracy, separation efficiency, and method robustness [7,14,32]. During the past two decades, a considerable amount of research has been conducted toward electrolyte temperature measurements in CE [1,14,19,21,32–42]. Early methods have included using the variation of electroosmotic mobility (μ_{EOF}), electrophoretic mobility (μ_{ep}), and electrical conductivity (κ) to measure temperature [38,39]. More recently, techniques such as external thermocouples [21], Raman thermometry [19,40], NMR spectroscopy [32,35], thermochromic probes [41], and the variation in fluorescence response [42] have been used to measure temperatures. Most of these methods require the modification of the existing instrument and/or the purchase of additional equipment.

Two noninvasive methods of temperature measurement based on the electroosmotic mobility and conductance are discussed in further detail below. The linear relationships between both the electroosmotic mobility and the conductance versus *P*/*L* are illustrated in Figure 18.7.

18.4.1 Electroosmotic Flow Method

It should be noted that the electroosmotic mobility (μ_{EOF}) is not a measure of the average electrolyte temperature (T_{Mean}, see Figure 18.2) over the whole cross section; but instead reflects the average temperature of the electrolyte near the inner wall of the capillary (T_{Wall}). This is because the electroosmotic flow is generated at the capillary wall. To determine the average temperature of

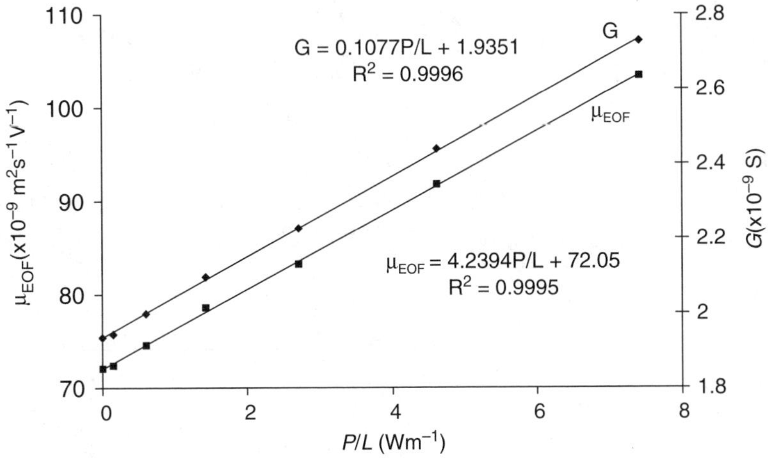

FIGURE 18.7 Variation of electroosmotic mobility and conductance with power per unit length. Conditions as in Figure 18.5.

the electrolyte (T_{Mean}), a correction of $\frac{1}{2}\Delta T_{Radial}$ needs to be added to the wall temperature to take into account the radial temperature profile of the electrolyte [8].

$$T_{Mean} = T_{Wall} + \frac{1}{2}\Delta T_{Radial} = T_{Wall} + 0.066\,\mathrm{KW^{-1}m} \cdot \frac{P}{L} \tag{18.11}$$

The previous methods of using μ_{EOF} as a temperature probe have assumed that changes in μ_{EOF} were due solely to variations in viscosity. More recently, it has been shown that changes to the dielectric constant and zeta potential also influence μ_{EOF} [8]. To obtain the most accurate results, an initial calibration curve for the variation of electroosmotic mobility with temperature is required for a specific electrolyte and capillary material. To make this calibration plot, μ_{EOF} needs to be determined at three or more different values of *P*/*L* for a range of set temperatures using the instrument's temperature control system. At each temperature, the graph of μ_{EOF} versus *P*/*L* is extrapolated to $P/L = 0$ to find μ_{EOF} free of the effects of Joule heating, $\mu_{EOF}(0)$. These values of $\mu_{EOF}(0)$ can be used to prepare the calibration curve for the variation of electroosmotic mobility with temperature, as illustrated in Figure 18.8.

For the example illustrated in Figure 18.8, the electrolyte used was 10 mM phosphate buffer at pH = 7.21. μ_{EOF} was found to increase at 2.22% per °C, that is, the temperature coefficient (or slope of μ_{EOF} versus *T*) was $0.0222°\mathrm{C}^{-1}$. Once the temperature coefficient for electrophoretic mobility is known, the electrolyte temperature can be easily determined for any value of *P*/*L*as long as the same capillary and electrolyte are used.

$$T_{Mean} = T_0 + \frac{\mu_{EOF} - \mu_{EOF}(0)}{0.0222\mu_{EOF}(0)} + 0.066\,\mathrm{KW^{-1}m} \cdot \frac{P}{L} \tag{18.12}$$

where T_0 is the temperature set for the instrument, μ_{EOF} is the electroosmotic mobility measured for the electrophoretic separation, and $\mu_{EOF}(0)$ is the extrapolated value of μ_{EOF} to zero power at ambient temperature.

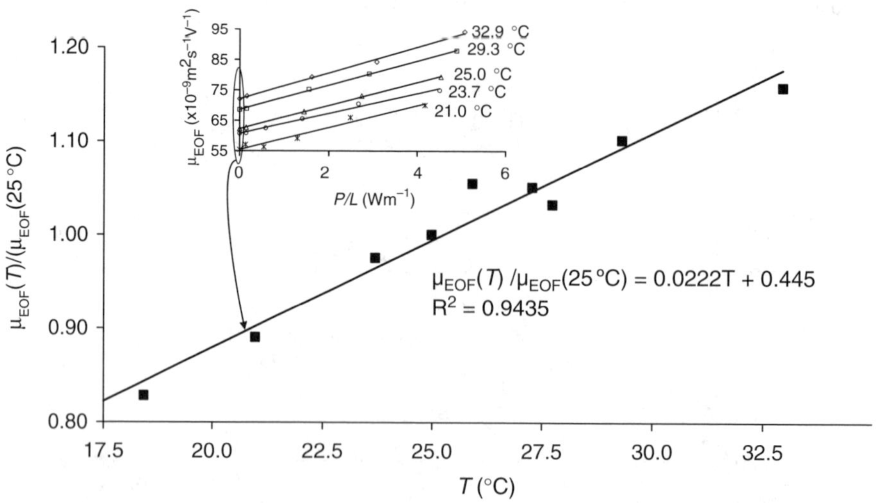

FIGURE 18.8 Calibration curve for μ_{EOF} versus Ambient temperature. Conditions as in Figure 18.5. (Reproduced with permission from Evenhuis, C. J., et al., *Anal. Chem.* 2006, *78*, 2684–2693. Copyright 2006 American Chemical Society.)

Example: If $\mu_{\text{EOF}}(0) = 7.209 \times 10^{-8}\ \text{m}^2\ \text{s}^{-1}\ \text{V}^{-1}$ at 25.00 °C and $\mu_{\text{EOF}}(P/L = 4.625\ \text{Wm}^{-1}) = 9.179 \times 10^{-8}\ \text{m}^2\ \text{s}^{-1}\ \text{V}^{-1}$, then

$$\begin{aligned} T_{\text{Mean}} &= 25.00\ °\text{C} + \frac{9.179 \times 10^{-8} - 7.209 \times 10^{-8}}{0.0222\ °\text{C}^{-1}\left(7.209 \times 10^{-8}\right)} + 0.066\ °\text{CmW}^{-1} \cdot 4.625\ \text{Wm}^{-1} \\ &= 25.00\ °\text{C} + 12.31\ °\text{C} + 0.30\ °\text{C} = 37.61\ °\text{C} \end{aligned}$$

The increase in mean temperature of the electrolyte (ΔT_{Mean}) is directly proportional to P/L.

$$\Delta T_{\text{Mean}} = \Phi \frac{P}{L}, \tag{18.13}$$

where Φ is a constant relating ΔT_{Mean} to the P.

For the example above, $\Delta T_{\text{Mean}} = 12.61$ °C,

$$\Phi = \frac{\Delta T_{\text{Mean}} L}{P} = \frac{12.61\ °\text{C}}{4.625\ \text{Wm}^{-1}} = 2.726\ °\text{C}\,\text{W}^{-1}\text{m}.$$

To obtain the most accurate results, the calibration process described above would need to be applied for each different electrolyte. This makes the method less attractive for routine use.

18.4.2 Conductance Method

Conductance (G) can be calculated by simply dividing the current by the voltage (see Equation 18.2). Conductance can be used as a probe for the mean electrolyte temperature (T_{Mean}). For most electrolytes, the thermal coefficient of electrical conductivity (γ) varies from about 0.019 K^{-1} to 0.021 K^{-1} [43] at the concentrations used for CE. It is possible to calculate γ from first principles using the Debye–Hückel–Onsager equation [8] but this is far from being a trivial process. As a rule of thumb, using $\gamma = 0.020\ \text{K}^{-1}$ gives values of ΔT_{Mean} to within 5%, but for more accurate values a calibration process is necessary.

To measure ΔT_{Mean}, the value of conductance at the ambient temperature is required without the influence of Joule heating [$G(0)$]. This can be obtained by measuring G at 3 or more values of P/L and extrapolating to $P/L = 0$ (see Figure 18.7).

$$\Delta T_{\text{Mean}} = \frac{G - G(0)}{\gamma\, G(0)}. \tag{18.14}$$

For example, 10 mM phosphate buffer at pH = 7.21 has $\gamma = 0.0205\ °\text{C}^{-1}$. Using the data from Figure 18.7, $G(0) = 1.935$ nS and $G(4.625\ \text{Wm}^{-1}) = 2.441$ nS.

This value is within 0.15 °C of the increase in temperature that was calculated above using μ_{EOF}. The overall process is far more straightforward than the method based on μ_{EOF} and is less sensitive to slight changes in wall chemistry that can occur over a period of time.

18.5 DETERMINING THE HEAT TRANSFER COEFFICIENT (*h*)

As illustrated in Figure 18.5, the difference between the outer wall temperature and the ambient temperature (referred to as ΔT_{Air}, see Figure 18.2) can easily account for 80% of the temperature increase of the electrolyte, even in an efficiently cooled instrument. The size of ΔT_{Air} is determined by the size of the heat transfer coefficient, so it useful to be able to determine h experimentally rather than to assume that the instrument is still functioning according to the manufacturer's specifications.

It should be noted that it is not possible to calculate h for an instrument per se as the value calculated depends on the outer diameter of the capillary used. However as most practitioners use capillaries with $d_o \approx 375$ μm, this value will be used in the example below for the purposes of illustration.

h is calculated from the difference in temperature between the outer wall and the ambient temperature of the coolant (ΔT_{Air})

$$h = \frac{1}{\pi d_o \Delta T_{\text{Air}}} \cdot \frac{P}{L}, \tag{18.15}$$

where d_o is the outer diameter of the capillary and P/L is the power dissipated per unit length. ΔT_{Air} can be found from the rise in the mean temperature of the electrolyte and the temperature differences across the FS and poly(imide) layers (see Figure 18.2).

$$\Delta T_{\text{Air}} = \Delta T_{\text{Mean}} - 1/2\, \Delta T_{\text{Radial}} - \Delta T_{\text{FS}} - \Delta T_{\text{PI}}. \tag{18.16}$$

For both FS and poly(imide) capillaries, the temperature difference across the wall is found using the following equation:

$$\Delta T_{\text{AcrossWall}} = \frac{1}{2\pi \lambda_{\text{Wall}}} \ln\left(\frac{d_o}{d_i}\right) \cdot \frac{P}{L}, \tag{18.17}$$

where λ_{Wall} refers to the thermal conductivity of the wall material. The thermal conductivities for FS and poly(imide) are $\lambda_{\text{FS}} = 1.40\ \text{Wm}^{-1}\text{K}^{-1}$ and $\lambda_{\text{PI}} = 0.155\ \text{Wm}^{-1}\text{K}^{-1}$, respectively [12,21].

For convenience, the following calculations use $P/L = 1.00\ \text{Wm}^{-1}$ but the calculations are applicable for any value of P/L. For the purpose of the calculation, the values used are internal diameter, $d_i = 75.0$ μm, diameter of FS $d_{\text{FS}} = 335$ μm, and outer diameter $d_o = 375$ μm (the thickness of the poly(imide) coating is 20 μm).

$$\Delta T_{\text{FS}} = \frac{1}{2\pi \times 1.40\ \text{Wm}^{-1}\text{K}^{-1}} \ln\left(\frac{335\ \mu\text{m}}{75.0\ \mu\text{m}}\right) \cdot 1.00\ \text{Wm}^{-1} = 0.170\ \text{K}$$

$$\Delta T_{\text{PI}} = \frac{1}{2\pi \times 0.155\ \text{Wm}^{-1}\text{K}^{-1}} \ln\left(\frac{375\ \mu\text{m}}{335\ \mu\text{m}}\right) \cdot 1.00\ \text{Wm}^{-1} = 0.116\ \text{K}$$

$$\frac{1}{2}\Delta T_{\text{Radial}} = \frac{1}{2} \times \frac{1}{4\pi \times 0.605\ \text{Wm}^{-1}\text{K}^{-1}} \cdot \frac{P}{L} = 0.066\ \text{KW}^{-1}\text{m} \cdot 1.00\ \text{Wm}^{-1} = 0.066\ \text{K}$$

Using the previously determined value of $\Delta T_{\text{Mean}} = 2.726$ K for $P/L = 1.00\ \text{Wm}^{-1}$ for a capillary with a total length of 32.2 cm,

$$\begin{aligned}\Delta T_{\text{Air}} &= \Delta T_{\text{Mean}} - \frac{1}{2}\Delta T_{\text{Radial}} - \Delta T_{\text{FS}} - \Delta T_{\text{PI}} \\ &= 2.726\ \text{K} - 0.066\ \text{K} - 0.170\ \text{K} - 0.116\ \text{K} = 2.374\ \text{K}\end{aligned}$$

$$\therefore h = \frac{1}{\pi d_o \Delta T_{\text{Air}}} \cdot \frac{P}{L} = \frac{1}{\pi \times 375 \times 10^{-6}\text{m} \times 2.374\ \text{K}} \cdot 1.00\ \text{Wm}^{-1} = 358\ \text{Wm}^{-2}\text{K}^{-1}.$$

It should be noted that the value of h calculated is an average over the whole length of the capillary and that the size of h will depend on the fraction of the capillary that is under thermostat control. Longer capillaries that have a larger fraction of the capillary under thermostat control will tend to give larger values of h.

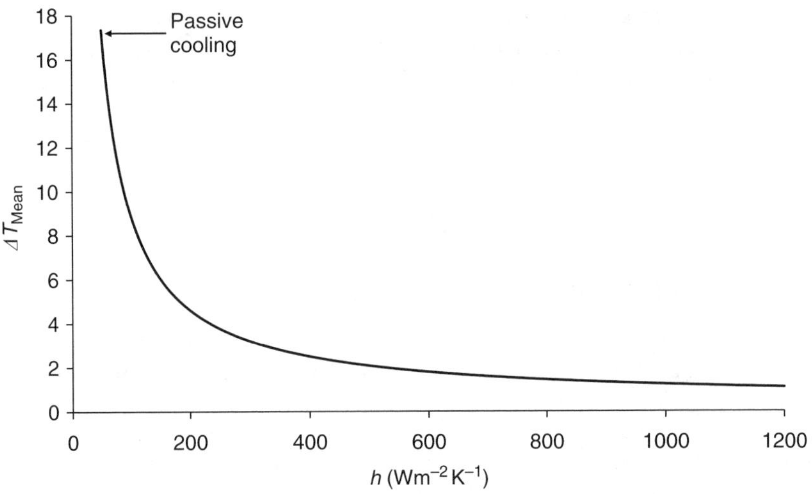

FIGURE 18.9 Variation of the mean rise in electrolyte temperature with heat transfer coefficient for $P/L = 1.00\ \mathrm{Wm}^{-1}$.

18.6 ESTIMATING THE TEMPERATURE INCREASE OF THE ELECTROLYTE

The temperature rise of the electrolyte can be estimated using a combination of the heat transfer coefficient and the power per unit length using Equation 18.18. For the FS capillary described above [$d_i = 75.0\ \mu\mathrm{m}$, $d_o = 375\ \mu\mathrm{m}$, thickness of poly(imide) $= 20\ \mu\mathrm{m}$]:

$$\begin{aligned}\Delta T_{\mathrm{Mean}} &= \Delta T_{\mathrm{Air}} + {}^{1}\!/_{2}\Delta T_{\mathrm{Radial}} + \Delta T_{\mathrm{FS}} + \Delta T_{\mathrm{PI}} = \Delta T_{\mathrm{Air}} + 0.352\ \mathrm{KmW}^{-1}\cdot\frac{P}{L} \\ &= \left(\frac{1}{\pi d_o h} + 0.352\ \mathrm{KmW}^{-1}\right)\frac{P}{L} = \left(\frac{849}{h} + 0.352\ \mathrm{KmW}^{-1}\right)\frac{P}{L}\end{aligned} \qquad (18.18)$$

A graph of Equation 18.18 is shown in Figure 18.9.

For capillaries with $d_o = 375\ \mu\mathrm{m}$, the error introduced by Equation 18.18 is less than 0.4°C for $d_i = 50$–$150\ \mu\mathrm{m}$. For capillaries with narrower bores, the temperature rise is slightly greater than predicted and for wider bore capillaries the predicted temperature rise is smaller.

18.7 CONTROL OF THE CAPILLARY TEMPERATURE

When selecting the desired temperature of the capillary, consideration should be given to reproducibility of sample injection and migration times, optimization of resolution, the possibility of sample decomposition, and the minimization of analysis times [3].

18.7.1 Reproducibility of Migration Times and Injection Volumes

Both the electroosmotic mobility and electrophoretic mobility are temperature-dependent so it is not surprising that controlling the temperature of the capillary leads to improved intraday-, interday-, and interlaboratory-reproducibility of migration times. Nevertheless, it is possible to achieve acceptable run-to-run reproducibility using an instrument without temperature control if experiments are carried out in a laboratory in which the ambient temperature is maintained within a narrow range.

Since the viscosity of the electrolyte decreases with increasing temperature at a rate of approximately 2% per °C, the volume of sample injected using a particular pressure and time combination will increase with the ambient temperature. As a rule of thumb, the product of pressure and time should be reduced by 20% for each 10°C rise in the ambient temperature. Alternatively, peak areas can be normalized by multiplying them by the ratio of the viscosities at the different temperatures [3]. As mentioned earlier, where the facility exists, it is wise to control the temperature of the tray housing the electrolyte vials as this improves reproducibility of both injections and migration times.

18.7.2 Optimizing Resolution

Nelson and Burgi [3] observed that the relative migration times could be affected differently by changes to the ambient electrolyte temperature so that the order in which a range of anions is detected can be changed considerably. Ions that comigrate at one temperature can be well resolved at a different temperature. Although the mechanism for these changes is not well understood, changing the temperature may be an effective strategy to improve resolution. However, it should be noted that generally there is an increase in peak distortion that accompanies increases in temperature [5].

18.7.3 Sample Decomposition

It is well known that biological samples may be sensitive to decomposition and/or conformational changes at elevated temperatures [10,44,45]. Such changes can be avoided by monitoring the electrolyte temperatures using the procedures described earlier or by controlling the electrolyte temperature at a lower value to counteract the effect of Joule heating.

18.7.4 Band Broadening

Joule heating is well established as a source of peak broadening. Petersen et al. [5] showed that a 10°C increase in temperature is associated with a 3.3% decrease in the number of theoretical plates due to increased axial diffusion. Radial temperature differences tend to be more significant; Grushka et al. [4] recommended that the radial temperature difference should not be allowed to exceed 1.5°C. Gobie and Ivory [1] found that thermally induced parabolic distortion of the migration velocity can be countered by applying an opposing laminar flow to the outlet end of the capillary. This method was found to be particularly effective in wider bore capillaries.

18.8 CONCLUSIONS

Although Joule heating is inevitable during electrophoretic separations, its magnitude is simple to quantify. Variation of the electrical conductance (G) with the power per unit length (P/L) is a simple and effective method of measuring the temperature increase of the electrolyte. An experimental method of determining the heat transfer coefficient (h) for a capillary has been outlined, along with a simple equation to calculate the rise in the mean temperature of the electrolyte (ΔT_{Mean}). ΔT_{Mean} increases linearly with the power per unit length and can be minimized by using narrow bore capillaries with low electrical conductivity electrolytes and by increasing the length of the capillary. Finally, the ambient temperature of the capillary during a separation should always be considered as a variable that must be optimized with respect to speed, efficiency, and resolution [3].

REFERENCES

1. Gobie, W. A. and Ivory, C. F., *J. Chromatogr.* 1990, *516*, 191–210.
2. Evenhuis, C. J., Guijt, R. M., Macka, M., Marriott, P. J., and Haddad, P. R., *Electrophoresis* 2005, *27*, 672–676.

3. Nelson, R. J. and Burgi, D. S., in Landers, J. P. (Ed.), *Handbook of Capillary Electrophoresis*, CRC Press, Boca Raton, 1994, pp. 549–562.
4. Grushka, E., McCormick, J. R. M., and Kirkland, J. J., *Anal. Chem* 1989, *61*, 241–246.
5. Petersen, N. J., Nikolajsen, R. P. H., Mogensen, K. B., and Kutter, J. P., *Electrophoresis* 2004, *25*, 253–269.
6. Xuan, X. and Li, D., *Electrophoresis* 2005, *26*, 166–175.
7. Rathore, A. S., *J. Chromatogr. A* 2004, *1037*, 431–443.
8. Evenhuis, C. J., Guijt, R. M., Macka, M., Marriott, P. J., and Haddad, P. R., *Anal. Chem.* 2006, *78*, 2684–2693.
9. Hjerten, S., *Chromatog. Rev.* 1967, *9*, 122–219.
10. Nelson, R. J., Paulus, A., Cohen, A. S., Guttman, A., and Karger, B. L., *J. Chromatogr.* 1989, *480*, 111–127.
11. Bello, M. S. and Righetti, P. G., *J. Chromatogr.* 1992, *606*, 103–111.
12. Knox, J. H., *Chromatographia* 1988, *26*, 329–337.
13. Knox, J. H. and McCormack, K. A., *Chromatographia* 1994, *38*, 215–221.
14. Evenhuis, C. J., Guijt, R. M., Macka, M., Marriott, P. J., and Haddad, P. R., *Electrophoresis* 2005, *26*, 4333–4344.
15. Porras, S. P., Marziali, E., Gas, B., and Kenndler, E., *Electrophoresis* 2003, 24, 1553–1564.
16. McCormick, R. M., *Anal. Chem.* 1988, *60*, 2322–2328.
17. Knox, J. H. and McCormick, K. A., *Chromatographia* 1994, *38*, 279–282.
18. Porras, S. P. and Kendler, E., *Electrophoresis* 2005, *26*, 3203–3220.
19. Liu, K.-L. K., Davis, K. L., and Morris, M. D., *Anal. Chem.* 1994, *66*, 3744–3750.
20. Cifuentes, A., Kok, W., and Poppe, H., *J. Microcolumn Sep.* 1995, *7*, 365–374.
21. Nishikawa, T. and Kambara, H., *Electrophoresis* 1996, *17*, 1115–1120.
22. Bello, M. S. and Righetti, P. G., *J. Chromatogr.* 1992, *606*, 95–102.
23. Caslavska, J. and Thormann, W., *J. Microcol. Sep.* 2001, *13*, 69–83.
24. Chen, S. and Lee, M. L., *J. Microcol. Sep.* 1997, *9* 57–62.
25. Kvasnicka, F., *Electrophoresis* 2003, *24*, 860–864.
26. Liu, P. Z., Malik, A., Kuchar, M. C., Vorkink, W. P., and Lee, M. L., *J. Microcol. Sep.* 1993, *5*, 245–253.
27. Deng, B. and Chan, W.-T., *Electrophoresis* 2001, *22*, 2186–2191.
28. Lukacs, K. D. and Jorgenson, J. W., *HRC & CC* 1985, *8*, 407–411.
29. Macka, M., Yang, W.-C., Zakaria, P., Shitangkoon, A., Hilder, E. F., Andersson, P., Nesterenko, P., and Haddad, P. R., *J. Chromatogr. A* 2004, *1039*, 193–199.
30. Mikkers, F. E. P., Everaerts, F. M., and Verheggen, T. P. E. M., *J. Chromatogr.* 1979, *169*, 11–20.
31. Tanyanyiwa, J., Leuthardt, S., and Hauser, P. C., *J. Chromatogr. A* 2002, *978*, 205–211.
32. Lacey, M. E., Webb, A. G., and Sweedler, J. V., *Anal. Chem.* 2002, *74*, 4583–4587.
33. Bello, M. S. and Righetti, P. G., *J. Chromatogr.* 1989, *606*, 95–102.
34. Bello, M. S. and Righetti, P. G., *J. Chromatogr.* 1989, *606*, 103–111.
35. Lacey, M. E., Webb, A. G., and Sweedler, J. V., *Anal. Chem.* 2000, *72*, 4991–4998.
36. Thormann, W., Zhang, C.-X., Caslavska, J., Gebauer, P., and Mosher, R. A., *Anal. Chem.* 1998, *70*, 549–562.
37. Berezovski, M. and Krylov, S. N., *Anal. Chem.* 2004, *76*, 7114–7117.
38. Burgi, D. S., Salomon, K., and Chien, R.-L., *J.Liq. Chromatogr.* 1991, *14*, 847–867.
39. Knox, J. H. and McCormack, K. A., *Chromatographia* 1994, *38*, 207–214.
40. Davis, K. L., Liu, K.-L. K., Lanan, M., and Morris, M. D., *Anal. Chem* 1993, *65*, 293–298.
41. Wätzig, H., *Chromatographia* 1992, *33*.
42. Ross, D., Gaitan, M., and Locascio, L. E., *Anal. Chem.* 2001, *73*, 4117–4123.
43. Isono, T., *J. Chem.Eng. Data* 1984, *29*, 45–52.
44. Kuldvee, R., Vunder, K., and Kaljurand, M., *Proc. Estonian Acad. Sci. Chem.* 1999, *48*, 119–128.
45. Guttmann, A. and Cooke, N., *J. Chromatogr.* 1991, *559*.
46. Dow-Corning, in 10-898E-01, F. N. (Ed.), http://www.dowcorning.com/DataFiles/090007b5802e201c .pdf, 2005.
47. Boedeker Plastics, I., 2004.
48. Kerbow, D. L. and Sperati, C. A., in Brandrup, J., Immergut, E. H., and Grulke, E. A. (Eds.), *Polymer Handbook 4th Edition*, J. Wiley & Sons, Hoboken, 1999, p. V/36.

49. Kerbow, D. L. and Sperati, C. A., in Brandrup, J., Immergut, E. H., and Grulke, E. A. (Eds.), *Polymer Handbook 4th Edition*, J. Wiley and Sons, Hoboken, 1999, p. V/44.
50. Kerbow, D. L. and Sperati, C. A., in Brandrup, J., Immergut, E. H., and Grulke, E. A. (Eds.), *Polymer Handbook 4th Edition*, J. Wiley and Sons, Hoboken, 1999, p. V/51.
51. Mehta, R. H., in Brandrup, J., Immergut, E. H., and Grulke, E. A. (Eds.), *Polymer Handbook 4th Edition*, J. Wiley & Sons, Hoboken, 1999, p. V/93.
52. Wunderlich, W., in Brandrup, J., Immergut, E. H., and Grulke, E. A. (Eds.), *Polymer Handbook 4th Edition*, J. Wiley & Sons, Hobken, 1999, p. V/88.
53. Schrader, D., in Brandrup, J., Immergut, E. H., and Grulke, E. A. (Eds.), *Polymer Handbook 4th Edition*, J. Wiley & Sons, Hobken, 1999, p. V/93.
54. Kerbow, D. L. and Sperati, C. A., in Brandrup, J., Immergut, E. H., and Grulke, E. A. (Eds.), *Polymer Handbook 4th Edition*, J. Wiley & Sons, Hoboken 1999, p. V/56.

19 Isoelectric Focusing in Capillary Systems

Jiaqi Wu, Tiemin Huang, and Janusz Pawliszyn

CONTENTS

19.1 INTRODUCTION OF CAPILLARY ISOELECTRIC FOCUSING

Isoelectric focusing (IEF) [1] is a powerful electrophoretic method for characterizing proteins and other biopolymers. IEF separates amphoteric substances based on their isoelectric point (pI) differences. In an electric field, a charged particle is subjected to the force, F ($F = QE$, Q is the charge of the particle and E the strength of the electric field). At the pI, where the charge of the particle equals zero, the mobility of the particle should be zero. IEF is a special electrophoresis performed in a pH gradient. The result of an IEF separation is that amphoteric substances in a sample are separated and focused at the positions along the pH gradient where their pI equals the pH.

IEF has been applied mainly to two areas in the biological sciences: the analysis of complex protein mixtures and the characterization of purified proteins [2]. For the first application, IEF is used as a separation dimension in multidimensional electrophoretical methods, such as 2-D polyacrylamide gel electrophoresis (2-D PAGE). The 2-D PAGE is a common tool in proteomics. For the second application, IEF is one of the two most common methods (another one is ion exchange chromatography) to characterize a protein's charge heterogeneity. The charge heterogeneity of a protein is frequently caused by post-translational modifications. Thus, the IEF method is an important tool in the study of these modifications. Pharmaceutical companies utilize IEF methods to monitor product processing, study formulations, and perform quality control for their protein products.

As a major protein characterization tool in biotech laboratories, slab gel IEF is slow, labor intensive, and semiquantitative. In addition, the quality of the results depends on the skill of the analyst. It was recognized that if IEF could be performed in a column format, significant advantages over slab gel IEF could be realized, in terms of automation, speed, and quantification. In 1985, Hjertén and Zhu [3] first reported an IEF method performed in a capillary column format—capillary isoelectric focusing (cIEF). In the early years of cIEF, it was performed on commercial capillary electrophoresis (CE) instruments that were designed for multiple CE modes. cIEF was expected to have both the high resolution of slab gel IEF and the advantages of a column-based separation technology that include automation and quantification.

Despite the inherent appeal of cIEF, since it was first introduced more than 20 years ago, widespread acceptance of cIEF by biotechnology laboratories and the pharmaceutical industry as a substitute for slab gel IEF for protein characterization did not occur. The main factor for this slow acceptance by the industry was related to the difficulty in performing cIEF using general purpose CE instruments (conventional cIEF). Subsequently, the introduction of whole-column detection cIEF [4] and its commercialization [5] revolutionized cIEF technology. It has been quickly adopted by leading pharmaceutical companies and research laboratories. Many analysis methods based on the whole-column detection cIEF have been validated in laboratories regulated by the U.S. Food and Drug Administration (FDA) [6,7]. To our knowledge, the top 10 pharmaceutical companies in the world are using the whole-column detection cIEF technique for protein characterization, for drug discovery, product processing, including formulation development, and quality control (QC).

It has been almost 25 years since CE was first reported [8]. The late 1980s and early 1990s were CE's booming years. During that period, it was widely expected that CE would replace high-performance liquid chromatography (HPLC) for most HPLC applications, although this did not occur. After the 25 years, CE almost disappeared from industrial laboratories. Only two CE technologies survived and were accepted by biotech pharmaceutical companies, including CE-sodium dodecyl sulfate (SDS) and cIEF. CE-SDS could replace SDS–PAGE and cIEF could replace slab gel IEF for many applications, especially in protein characterization and QC. The speed of the acceptance of these two CE technologies in the biotech field has been accelerated in the past 5 years [9] due to the improvements in these two CE technologies.

In this chapter, a review of cIEF applications for protein characterizations performed by biotech and pharmaceutical laboratories will be presented.

19.2 REVIEW OF cIEF TECHNOLOGY

19.2.1 CONVENTIONAL cIEF

When a general purpose CE instrument is used to perform cIEF (conventional cIEF), as shown in Figure 19.1a, the separation column, which is usually a 50–100 μm inner diameter, 25–60 cm long capillary, is filled with a mixture of a protein sample and carrier ampholytes (CAs). After the sample is injected, the two ends of the capillary column are dipped into the catholyte and the anolyte, as

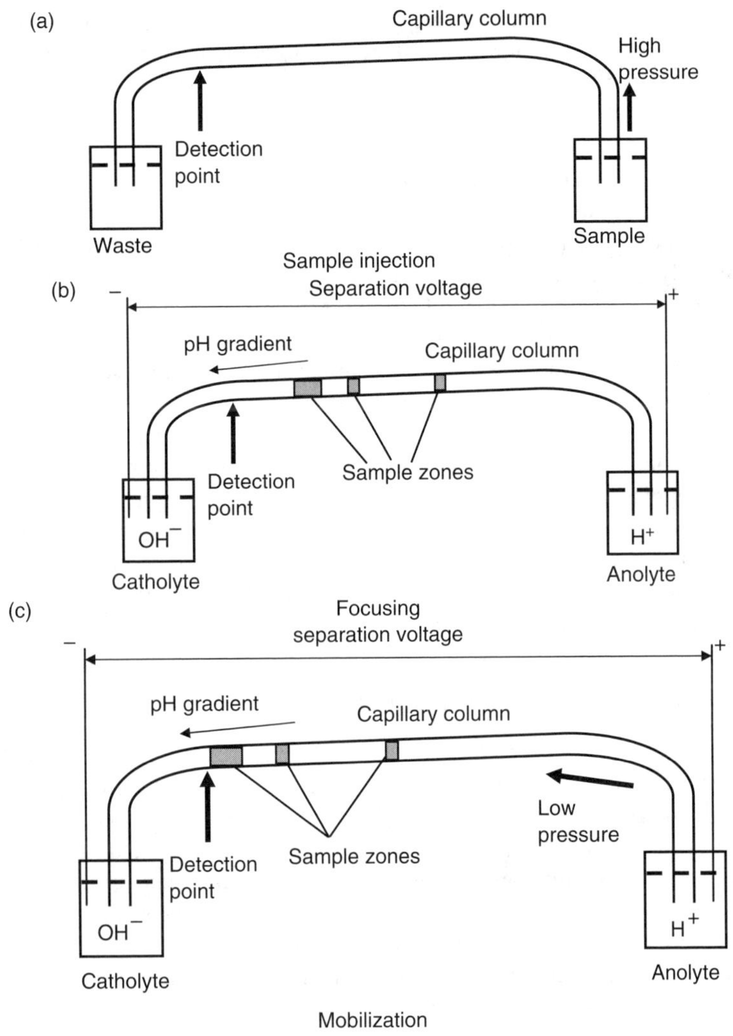

FIGURE 19.1 Steps involved in cIEF using general purpose CE instruments (conventional cIEF). (a) Sample injection, (b) focusing step, and (c) mobilization step.

shown in Figure 19.1b. A separation voltage (usually at 100–500 V/cm) is applied to the two ends of the capillary column. Under the separation voltage, a pH gradient is established along the capillary column from the anodic end of the column to the cathodic end of the column by isoelectric stacking of components in the CAs [10,11]. In the pH gradient, driven by the separation voltage, proteins in the sample migrate to the points within the capillary column where their isoelectric point (pI) values are equivalent to the pH values and the migration then stops. The proteins are focused into very narrow zones at their pI points.

There are three basic requirements of a separation column for cIEF; zero or substantially reduced electroosmotic flow (EOF), hydrodynamic flow, and interaction between protein samples and the column wall. Columns used in cIEF are usually coated on their inner wall with stable, neutral coatings to control for EOF and prevent the protein from interaction with the wall [10]. During the focusing process, the two ends of the separation column should be maintained at the same level to eliminate the hydrodynamic flow. Under these conditions, at the end of the focusing process, all protein zones are stationary or near stationary within the separation column [3].

Since general purpose CE instruments are only equipped with a single point on-column detector, which is usually located close to one end of the capillary column as shown in Figure 19.1c, these stationary protein zones within the column have to be mobilized past the detection point of the single point detector. For cIEF performed on these CE instruments (conventional cIEF), a mobilization process is necessary following the focusing process [10,11]. There are three ways of performing the mobilization: pressure-driven mobilization [11], salt mobilization [3], and EOF-driven mobilization [12]. These three methods can be used in any combination.

Pressure is often used for mobilization. As shown in Figure 19.1c, at the end of the focusing a low pressure (usually about 0.5 psi) is applied to one end of the capillary column to mobilize the entire pH gradient past the detection point. During the mobilization, the separation voltage remains to reduce the sample zone broadening.

With *salt mobilization*, at the end of the focusing process salts are added into one of the electrolytes. The addition of the salts creates a pH shift at this end of the column under the separation voltage, and then the shift gradually progresses deeper into the column. This causes the whole pH gradient within the column to shift toward this end of the column.

In *EOF-driven mobilization*, the EOF of the separation column is not totally eliminated, but is controlled with additives, such as methylcelulose (M.C.). Upon application of the separation voltage, the focusing process is initiated. Concurrently, the whole volume of solution within the column is mobilized toward the cathodic end of the column, driven by the EOF. In this way, the mobilization process is combined with the focusing process.

19.2.2 Applications of the Conventional cIEF

There are a range of applications of conventional cIEF to real-world analysis, including the characterization of protein samples in laboratories of biotech pharmaceutical companies, and clinical analysis. In the pharmaceutical industry, some methods have been developed based on this technology [13,14] and validated [15–17] as the identity assays in regulated pharmaceutical laboratories for the analysis of therapeutic monoclonal antibodies and glycoproteins. cIEF has also been used in clinical analysis for human hemoglobin analysis [18,19] and the analysis of proteins in cerebrospinal fluid [20]. There are several reviews on the applications of the conventional cIEF [10,21]. However, the pace of acceptance of conventional cIEF for the analysis of real-world samples is slow.

19.2.3 Difficulties Associated with Conventional cIEF

The mobilization process in conventional cIEF inhibits its widespread use. The process introduces many problems, such as poor resolution, poor reproducibility, and long sample analysis time (less than 2 samples/h). In conventional cIEF, the dynamic process of IEF within the separation column is not monitored, which makes it difficult to optimize focusing time and mobilization parameters. Focusing time is the most important parameter in cIEF. In addition, the real focusing time in conventional cIEF is equal to the focusing time and the mobilization time, since the separation voltage is always on during the mobilization process. This makes it almost impossible to optimize the focusing time because the mobilization time of a sample component depends on its pI value. It is also difficult to detect problems in the focusing process within the separation column, such as sample aggregation and precipitation.

19.2.4 Whole-Column Detection cIEF

The introduction of whole-column detection cIEF technology [22,23] solved the problems caused by the mobilization process in conventional cIEF. Figure 19.2 illustrates the principle of whole-column detection cIEF. The separation column is a short capillary (typically, 50 mm long, 100 μm inner diameter silica capillary). The inner wall of the capillary is coated to eliminate EOF and prevent

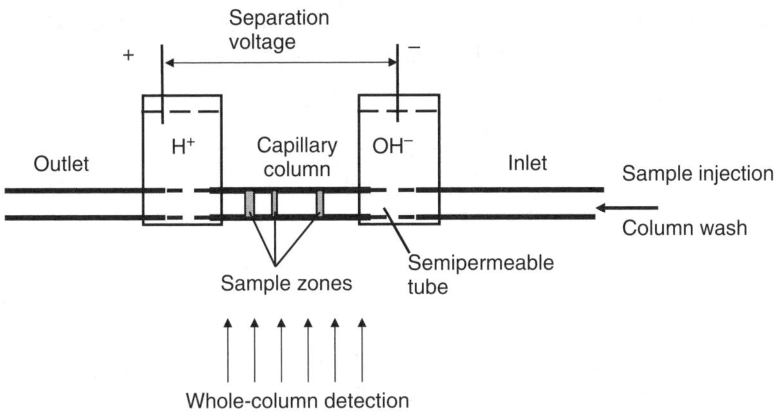

FIGURE 19.2 Principle of whole-column detection cIEF.

proteins from interacting with the wall surface. The whole-column detection system is an ultraviolet (UV) absorption-imaging detector operated at 280 nm. During the operation of cIEF, the two tanks on the cartridge are filled with the catholyte and anolyte, respectively, as shown in Figure 19.2. A protein sample premixed with CAs is injected from the inlet of the column into the separation column between the two semipermeable membrane tubes. After the column is filled with the sample, the sample injection is stopped. A separation voltage, typically, 3 kV (600 V/cm), is applied across the two tanks. IEF only occurs within the separation column between the two semipermeable tubes. The structure of the separation column makes accurate sample injection unnecessary. As long as the separation column between the two semipermeable tubes is filled with the sample solution, the injection is quantitative. The IEF process within the entire length of the separation column is monitored by the whole-column detection system. At the end of the focusing, after recording the last image, the sample is washed out of the column by injecting a wash solution from the inlet, then, the separation column is then ready for the next sample.

Since the whole-column detector monitors the IEF process within the separation column in an online fashion, the focusing time can be determined in a single sample run. At the end of the focusing process, all sample zones within the capillary column are simultaneously recorded by the detector without disturbing the separation resolution. Any sample precipitation and aggregation during focusing can be observed and distinguished while focusing is in process. Sample precipitation and aggregation are the two most common problems in IEF faced while obtaining reproducible results. Different additives may be used to improve peak pattern reproducibility. These features facilitate fast method development. In turn, the commercialization of the whole-column detection cIEF technology accelerated the acceptance of cIEF technology by pharmaceutical companies.

19.3 THEORETICAL ASPECTS OF cIEF

19.3.1 NATURE pH GRADIENT AND STEADY-STATE IEF

The separation of ampholytes in a pH gradient formed by CAs was proposed by Svensson in the early 1960s [1,24–26]. It is known that the pH value of a pure ampholyte solution is approximately its isoelectric point [24]. In a Svensson's IEF system, the electrode reactions must be the electrolysis of water ions, which ensures that protons are produced at the anode and hydroxide ions are produced at the cathode. Before the electric field is applied, components of CA are uniformly distributed throughout the separation channel between the anode and the cathode. The pH along the channel is uniform, representing the average pH of the mixture of all of the components of the CAs. Upon the application of electric field, influenced by the migration of protons from the anode and hydroxide ions from the cathode of the separation channel, the components of the CAs begin to separate. The components with pI values higher than the average pH are positively charged and migrate toward the cathode whereas the components with pI values lower than the average pH are negatively charged and migrate toward the anode. The components cease migrating when their net charge becomes zero, at the region where the pH equals their pIs. Consequently, a smooth and stable pH gradient, which is positive all the way from anode to cathode, can be "naturally" generated.

19.3.2 The Stability of pH Gradient

Svensson [24,25] regarded the natural pH gradient IEF as a true steady-state process and expected its unlimited stability. However, as noted by Svensson, a true steady state was never reached in gel IEF [27]. Instabilitics of pH gradicnt, such as thc platcau phenomenon, cathodic drift, anodic drift, or symmetric drift, were commonly observed [1,27–34]. There have been hypotheses to explain the instability of the pH gradient in IEF by isotachophoresis (ITP) mechanism [35,36] or stationary neutralization reaction boundary equilibriums (SNRBEs) [37]. Hjerten et al. [38] suggested, as he proposed the mechanism for chemical mobilization in cIEF, that pH instability is inherent in natural pH gradient IEF due to the need for electroneutral conditions.

Nevertheless, it is commonly accepted that IEF is a steady-state technique, but this steady state is limited and conditional. To exploit the high-resolution capability of IEF, it is necessary to understand IEF's uniqueness. IEF is a steady state or equilibrium technique in nature. Therefore, factors that affect the steady-state IEF should be avoided. For example, as mentioned in Section 19.2, interferences that disturb the formation and stability of the pH gradient, such as the existence of EOF, hydrodynamic flow within the separation channel, or impurities in the anolyte and catholyte (even adsorption of CO_2), should be controlled.

19.3.3 Resolution

The resolution in IEF is defined by the resolving power on pI (ΔpI) that depends on the diffusion coefficient of the ampholyte, D, the electric field, E, the mobility slope ($-\mathrm{d}u/\mathrm{dpH}$), and the pH gradient ($\mathrm{dpH}/\mathrm{d}x$) [1]

$$\Delta \mathrm{pI} = 3\sqrt{\frac{D\left(\mathrm{dpH}/\mathrm{d}x\right)}{E\left(-\mathrm{d}\mu/\mathrm{dpH}\right)}}.$$

The above equation shows that resolution can be improved by high E, low D, a narrow pH gradient, and a high mobility slope.

Experimentally, electric field, diffusion coefficient, and $\mathrm{dpH}/\mathrm{d}x$ can be manipulated. The rate of change of mobility with a pH near the isoelectric point is intrinsic to the ampholytes and cannot be changed. Performing IEF under higher applied voltages facilitates the development of narrow focused protein zones. However, an increase in the electric field will intensify the Joule heating. As a result, a moderate voltage, which can be determined experimentally, is applied in IEF.

19.4 EXAMPLES OF PRACTICAL APPLICATIONS

In this section, several examples of method development procedures for protein characterization assays used for samples from biotech pharmaceutical companies are presented.

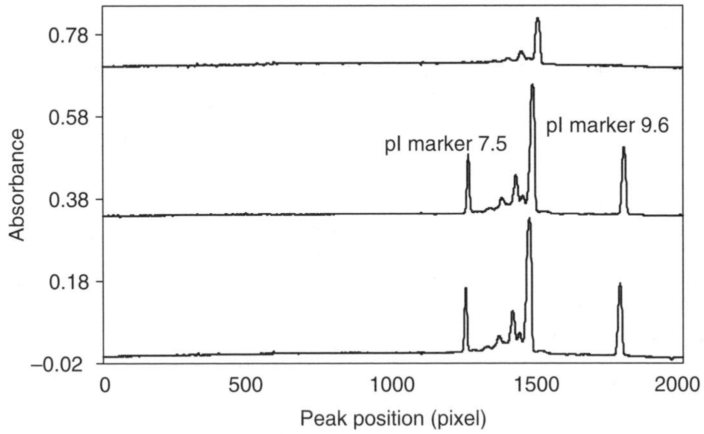

FIGURE 19.3 (Trace 1) Monoclonal antibody 1 at 0.2 mg/mL in 4% Pharmalyte pH 3–10 and 0.35% MC. Focusing time was 6 min at 600 V/cm with a 1 min prefocusing at 300 V/cm. (Traces 2 and 3) Two consecutive runs of the same sample were performed in the same running buffer and under the same focusing conditions as that of trace 1 Sample concentration was 0.6 mg/mL. Two pI markers, pI 7.5 and pI 9.6, were spiked for pI calibration.

19.4.1 Monoclonal Antibody 1

First, following the initial conditions, 4% Pharmalyte (pH 3–10), and 0.2 mg/mL sample concentration were tested for this sample. Under the conditions, as shown in the trace 1 of Figure 19.3, the sample peak height is detected at 0.1 Abs. In the next run, the sample concentration was increased to 0.6 mg/mL. The focusing time was 6 min at 600 V/cm with a 1 min prefocusing at 300 V/cm. Two pI markers were spiked into the sample for pI calibration. The results show good reproducibility in peak pattern (traces 2 and 3 in Figure 19.3). Since the separation resolution under these conditions was satisfactory, no further method development was pursued to enhance the resolution.

19.4.2 Recombinant Protein 1

This protein was expected to be very heterogeneous (multiple peaks). At the beginning, higher than usual sample concentration (0.4 mg/mL) was tested. However, as shown in the trace 1 of Figure 19.4, the peak height of the major peaks was only in 0.03 Abs range. In the next run, the sample concentration was boosted to 2.5 mg/mL in order to increase the peak height to above the 0.1 Abs level. Since the sample was focused into about 15 peaks, higher resolution was required. To enhance the resolution, Pharmalyte mixture (pH 3–10 and pH 5–8 at 1:3 ratio) was tested. As shown in traces 2 and 3 of Figure 19.4, the resolution was improved under these conditions.

19.4.3 Protein Conjugate 1

The purpose of developing an IEF assay for this sample was to monitor the pI values of different production lots of the product. For this sample, it was very difficult to obtain reproducible peak patterns that could match the peak pattern reproducibility of the previous two samples. As shown in Figure 19.5, the sample started to aggregate and was focused into an unstable big spike when the focusing time was longer than 4.5 min. Many additives and denatured conditions were tested, but no solution was found. However, if the focusing time was limited to 4 min (before sample started to aggregate), the peak pattern reproducibility was acceptable. The peak pattern reproducibility is shown in Figure 19.6 for this protein conjugate. This cIEF method is therefore useful in detecting pI changes in different lots.

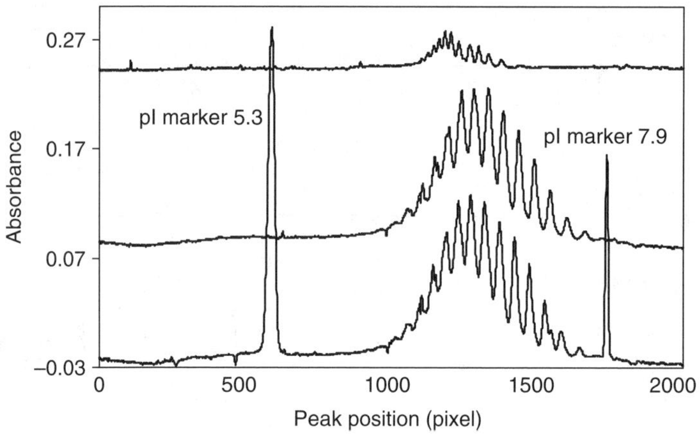

FIGURE 19.4 (Trace 1) Recombinant protein 1 at 0.4 mg/mL in 4% Pharmalyte pH 3–10 and 0.35% MC. Focusing time was 6 min at 600 V/cm with a 1 min prefocusing at 300 V/cm. (Traces 2 and 3) Two consecutive runs of the sample were performed at a concentration of 2.5 mg/mL in 1% Pharmalyte pH 3–10, 3% Pharmalyte pH 5–8, and 0.35% MC. Focusing time was 6 min at 600 V/cm with a 1 min prefocusing at 300 V/cm. Two pI markers, pI 5.3 and pI 7.9, were spiked into the sample for pI calibration.

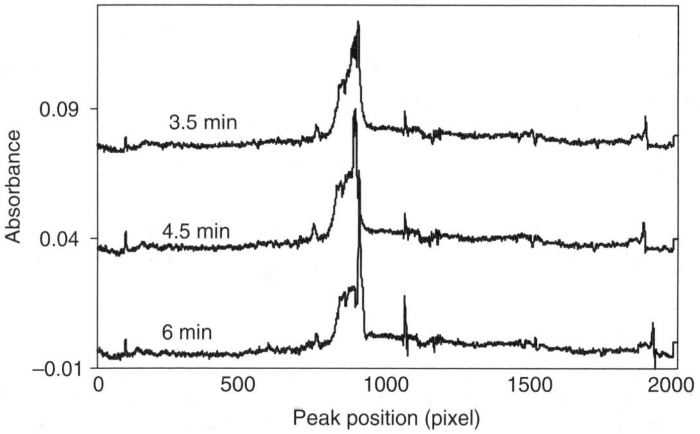

FIGURE 19.5 Effect of focusing time on sample protein conjugate 1's peak pattern.

19.4.4 Deactivated Virus 1

A second sample was also analyzed to determine pI values of different lots in production. As shown in Figure 19.7, when focusing time was limited to 6 min, the peak pattern reproducibility was satisfactory for the application. From our experiences with cIEF method development for virus samples, cIEF may not achieve the peak pattern reproducibility that can match that of monoclonal antibodies or other protein samples. However, in most cases, the reproducibility is satisfactory for the desired applications [39].

19.4.5 Monoclonal Antibody 2

The monocolonal antibody 2 protein started to aggregate before the completion of the focusing process (Figure 19.8). In the early stage of the focusing process, it went smoothly. The protein was focused into well-defined peaks. However, before the completion of the focusing process, a new peak

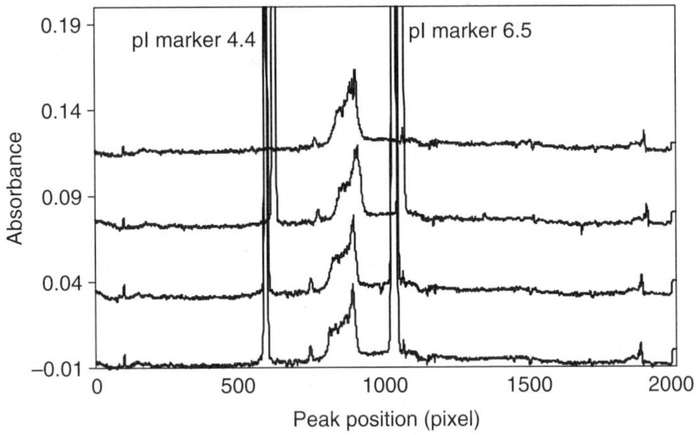

FIGURE 19.6 Four consecutive runs of protein conjugate 1 at 0.25 mg/mL in 4% Pharmalyte pH 3–10 and 0.35% MC. Focusing time was 4 min at 600 V/cm. Two pI markers, pI 4.4 and pI 6.5, were spiked into the sample for pI calibration.

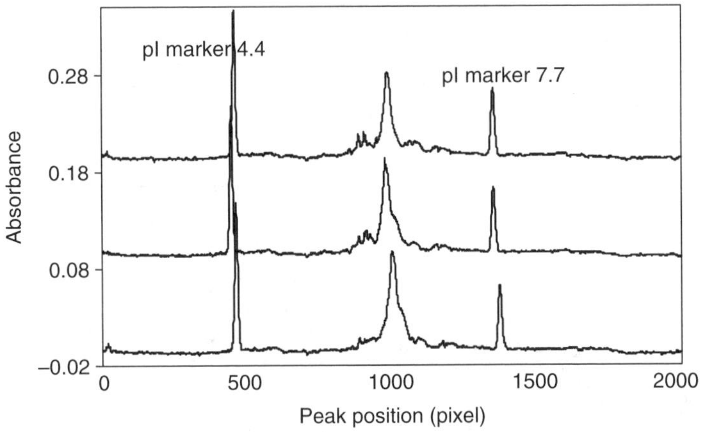

FIGURE 19.7 Three consecutive runs of deactivated virus 1 at 0.15 mg/mL in 4% Pharmalyte pH 3–10 and 0.35% MC. Focusing time was 6 min at 600 V/cm. Two pI markers, pI 4.4 and pI 7.7, were spiked into the sample for pI calibration.

(indicated by an arrow in Figure 19.8) started to appear and developed quickly along the focusing time. This is a typical symptom of sample aggregation created by the focusing process. In order to eliminate the problem, 4 M urea was added to the sample, effectively eliminating the aggregation problem (Figure 19.9), resulting in a reproducible peak pattern. This example illustrates the power of the whole-column detection cIEF for facilitating fast method development.

19.5 METHOD DEVELOPMENT GUIDELINES

In this section, the development of protein characterization methods for samples in typical pharmaceutical laboratories will be examined. All of the method development procedures discussed in this chapter are based on the use of the whole-column detection cIEF instrument, since the IEF process can be observed and the easily associated with optimizing the focusing time. However, the basic conditions described in this chapter are also applicable to conventional cIEF method.

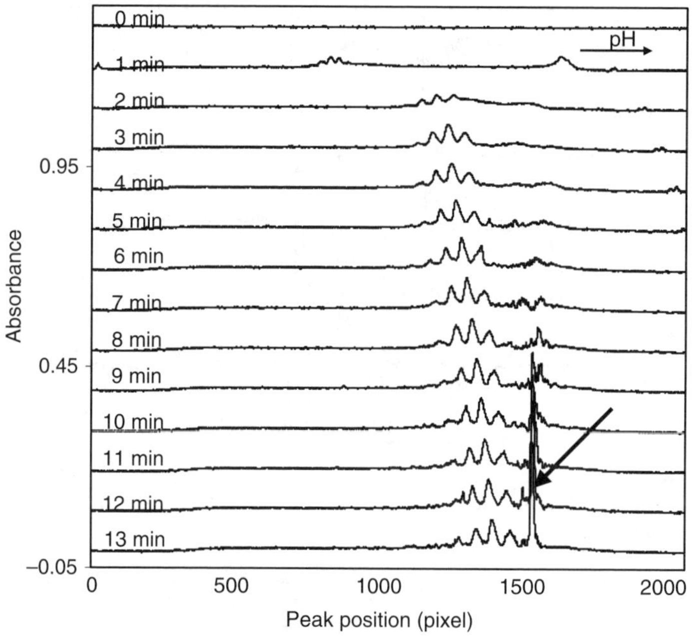

FIGURE 19.8 Focusing process of sample monoclonal antibody 2. The sample concentration was 0.2 mg/mL. Focusing conditions: 4% Pharmalyte pH 5–8 and 600 V/cm.

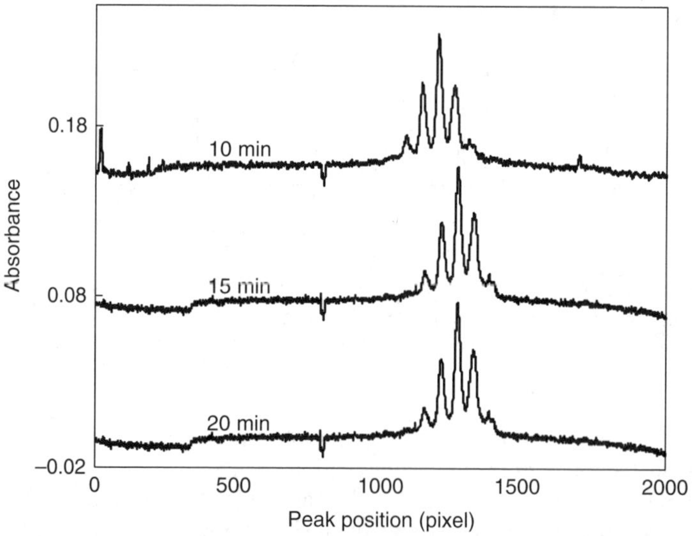

FIGURE 19.9 Focusing process of sample monoclonal antibody 2. The sample concentration was 0.2 mg/mL. Focusing conditions: 4% Pharmalyte pH 5–8 and 600 V/cm; 4 M urea was added to the sample.

19.5.1 Initial Conditions

At the beginning of the development of a cIEF method for a new sample, the initial conditions for the first sample run must be determined. These conditions include brand name, pH range, pH gradient linearity, and concentration of CAs, electrolytes, additives, sample concentration, focusing voltage, and focusing time.

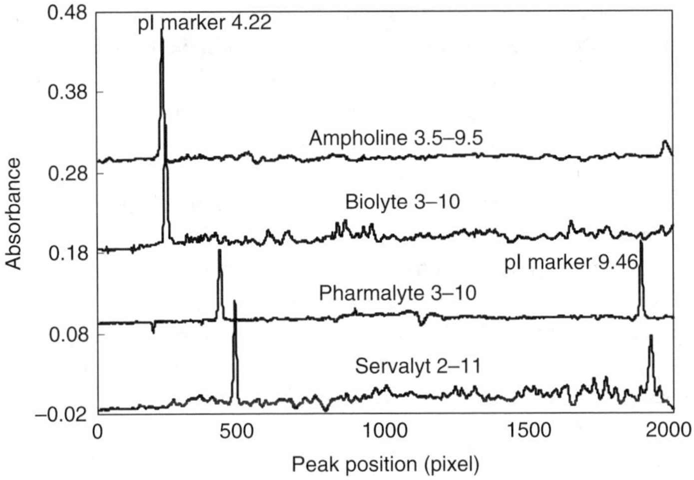

FIGURE 19.10 Electropherograms of four commercially available CAs. Concentrations of the CAs were 8%. Focusing time was 6 min at 600 V/cm. Two internal pI markers, pI 4.22 and pI 9.46, were spiked into the samples.

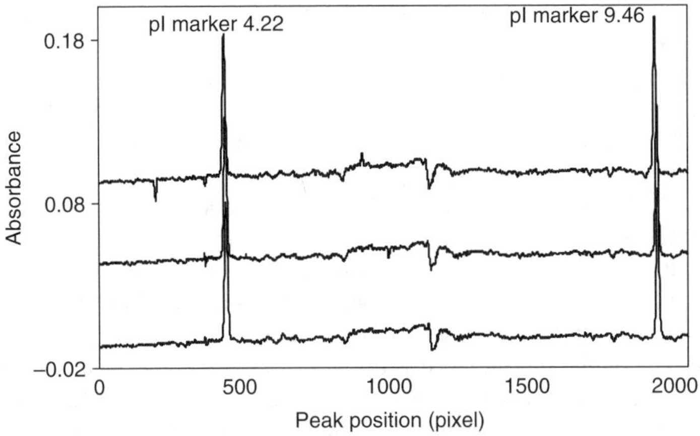

FIGURE 19.11 Three consecutive runs of Pharmalyte pH 3–10. Concentration was 8% and focusing time was 6 min at 600 V/cm. Two internal pI markers, pI 4.22 and pI 9.46, were spiked into the sample.

The most important condition in cIEF is the CA. As mentioned in Section 19.3, CAs are "special buffers" in cIEF. A CA may contain over 900 amphoteric components [2,40]. According to the literature [2] and to our knowledge, there are four different kinds of commercial CAs: Ampholine, Biolyte, Pharmalyte, and Servalyt. Experiments are needed to determine which CA results in the best resolution for a given sample. Thus, the resolution cannot be the first criterion in CA selection because it is unknown for a given protein sample. Usually, a wide pH range CA is the first choice for a sample of unknown pI value. From our experience, the first criterion in selecting the CA for cIEF should be the background UV absorption. All these CAs were designed for slab gel IEF in which the CA's UV absorption was not a concern [41]. Figure 19.10 shows the background absorption of the four CAs at 280 nm along their entire pH gradient. In addition, as shown in Figure 19.11, the background profiles of these CAs are reproducible from run to run and relatively stable for different lots. From the point view of background absorption, Pharmalyte should be the first choice due to its low background along its entire pH gradient, as illustrated in Figures 19.10 and 19.11.

Because of the UV absorption of the CAs, from background point of view, lower the concentration of the CAs is in samples, the better the signal-to-noise ratio will be. However, the buffer capacity of the CAs is determined by their concentration [2]. The sample peaks become broad and the relationship between the peak height and the sample concentration becomes nonlinear when the buffer capacity of the CAs is saturated. There should therefore be a trade-off between the buffer capacity and the background. From the literature and our experiences, a 2–8% Pharmalyte solution (the stock solution is considered to be 100%) results in acceptable background noise, while still maintaining enough buffer capacity for most proteins at working concentrations. Thus, it is recommended that a 4% Pharmalyte solution is a good starting point for method development.

As we will discuss later in this chapter, for many samples, separation resolution can be enhanced with the use of narrow pH range CAs. In addition, if the pI value of a sample is known, a narrow pH range CA can be selected for the initial conditions. It should be noted that there has been a misunderstanding associated with the use of narrow pH range CAs. Traditionally, it is believed that narrow pH range CAs must be mixed with one or more wide pH range CAs, or one or more narrow pH range CAs of different pH ranges. From our experience, this is not necessary. A single narrow pH range CA can be used alone for cIEF. Actually, the best resolution is usually achieved using a single narrow pH range CA. For a given protein sample, as long as the pH range of a single narrow pH range CA coves the pIs of the protein, this CA should be the first choice.

Literature suggests that weak acids and bases should be used as the electrolytes when narrow pH range CAs are used [10,11]. In our experiments, we found that although weak acids and bases gave similar results, if not better, to that when using strong acids and bases; they had to be frequently refreshed in order to keep consistent results for a long batch of sample runs. This is probably due to their weak buffer capacity. In our experiments, the same anolyte (0.08 M H_3PO_4) and catholyte (0.1 M NaOH) are used for all CAs of wide pH ranges and narrow pH ranges between 4% and 8%. The advantage of this approach is the relative simplicity of subsequent method development. Rather, different CAs can be tested without changing the electrolytes.

Focusing time is the second most important parameter for cIEF. Before the focusing time is chosen, the focusing voltage must be determined. Although theoretically, as described in the Section 19.3.3, the separation resolution is higher when a higher voltage is applied, the detection noise also increases along with the voltage [42]. The optimal voltage is around 500 V/cm [42].

Under the focusing voltage, focusing time of different samples can be easily optimized with the use of the whole-column detection cIEF because it monitors the focusing process within the separation column in an online fashion. There is a common misunderstanding related to focusing time in cIEF. The literature suggests that one must determine the end of the focusing process by the focusing current. It is believed that the focusing process is complete when the focusing current stabilizes. However, in our experiments using the whole-column detection cIEF, we found that there is no such relationship between the end of the focusing process and the current. The stabilization of the focusing current only roughly reflects the end of focusing for the CAs. Proteins are usually focused at a slower speed compared to CAs. Some proteins, such as PEGylated proteins, may require an additional 20 min of focusing time after the focusing current stabilizes. Therefore, focusing time has to be determined experimentally for each sample. In addition, the focusing time can be affected by the concentration of salts in the samples. A higher salt concentration will result in a faster focusing process, since the pH gradient created by the CAs is narrowed by the salts [43]. The focusing speed becomes slower when the concentration of CAs is higher. In addition, the focusing speed is much slower when narrow pH range CAs are used. From our experience, for most protein samples, when 4% pH 3–10 Pharmalyte is used, a 6 min focusing time is a good starting point.

For all samples, especially samples in a salt matrix, performing a prefocusing step at a voltage lower than focusing voltage reduces detection noise and increases peak pattern reproducibility. The prefocusing voltage is usually half of the focusing voltage. The step can be 1–2 min long depending on the samples.

At the end of the cIEF process, proteins are focused into narrow zones within the separation column and are concentrated hundreds of times. Confining proteins at their pI points (zero net charge) and at high concentrations for a long time increases the likelihood that they will precipitate or aggregate. Many chemicals can be used as the additives in cIEF [2] to stabilize proteins during IEF, such as sugars, nonionic or zwitterionic surfactants, and urea, thus, effectively reducing the likelihood of protein precipitation and aggregation occur during the focusing process.

For all cIEF methods, polymer solutions are added into the samples, including methyl cellulose (MC) and (hydroxypropyl)methylcellulose (HPMC). These polymers modify the capillary surface [10–12] and enhance separation resolution. The existence of the polymer in the sample solution reduces diffusion coefficients of the proteins, thus, as described in Section 19.3.3, enhancing the separation resolution. In all cIEF applications described in this chapter, 0.35–0.5% of MC are added to the samples.

Because of the UV absorption of the CAs used in cIEF, all UV absorption detections in cIEF are performed at 280 nm. At 280 nm, the absorption of the proteins is determined by the presence of two amino acids in the proteins: tyrosine and tryptophan. Thus, the sensitivity of proteins at 280 nm varies a great deal from protein to protein. It is not easy to determine the initial sample concentration. However, we found that the sensitivity of monoclonal antibodies is relatively even and predictable. For these samples and many other protein samples, 0.2 mg/mL is a good starting point for sample concentration if the samples have only one major peak (the major peak is defined as a peak having >50% of the total protein in a sample). If the samples can be separated into several major peaks (each major peak should be >20% of the total protein in a sample), the starting point can be calculated as follows:

$$\text{Number of major peaks} \times 0.1\,\text{mg/mL}$$

In summary, for an unknown sample, the initial cIEF analysis conditions would include 4% pH 3–10 Pharmalyte, 0.5% MC as the additive, 0.08 M H_3PO_4 as the anolyte, 0.1 M NaOH as the catholyte, 0.2 mg/mL sample concentration, and 6 min focusing time at 500 V/cm voltage with a 1–2 min prefocusing step.

19.5.2 Method Development Flow Chart

Figure 19.12 illustrates a flow chart for the development of the cIEF method. At the beginning, it is recommended that the initial conditions described in Section 19.5.1 are tested. The second step should then involve adjusting the sample concentration. The third step is the critical step in the method development because a high-resolution method is based on good reproducibility. All methods should be attempted to ensure a reproducible peak pattern. Then, the separation resolution should be enhanced with the use of narrow pH range CAs. The final step in the method development is pI determination and sample peak identification.

19.5.3 pI Determination and Peak Identification

Determination of the pI of a protein using cIEF requires two basic conditions: pI markers with accurate pI values and a linear pH gradient created by the chosen CAs. Using pI markers is the only way to characterize a pH gradient within the capillary column created by CAs [44]. As for the second condition, a pI value in a nonlinear pH gradient is difficult to characterize using a limited number of pI markers since the number of pI markers in a given pH region is always limited. In most cases, these pI markers do not distribute evenly within the pH region. Curve fitting in a nonlinear pH gradient using these pI markers involves a large error that is difficult to estimate. The determination of the pI value should be performed in a single CA with good linearity in its pH gradient.

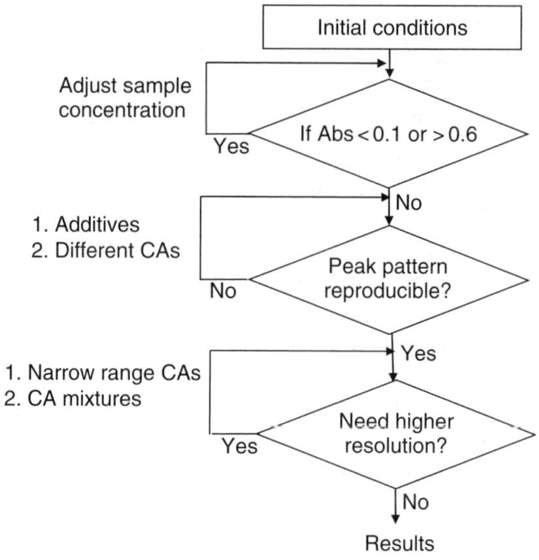

FIGURE 19.12 cIEF method development.

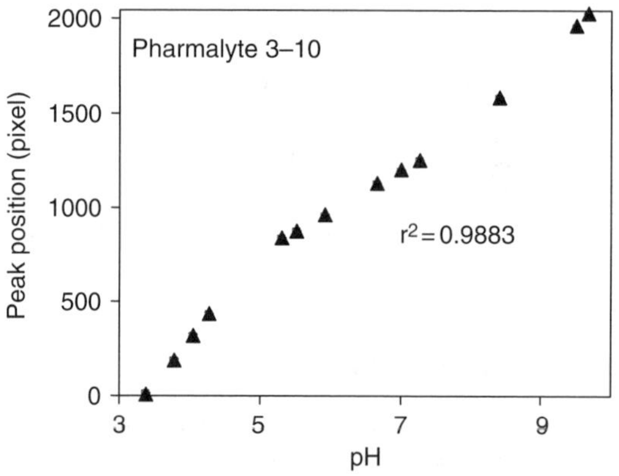

FIGURE 19.13 pH gradient profile of 8% Pharmalyte pH 3–10.

Profiling the pH gradient of different CAs is important for this purpose. Figure 19.13 illustrates an example with a pH gradient for Pharmalyte pH 3–10. The overall linearity of the pH gradient is good (r^2 is about .99). However, the gradient bends at two places, around pH 5 and 8. If pI determination is performed around these two regions, multiple pI markers should be used to ensure good accuracy. However, in other pH regions of this CA, the linearity in the pH gradient is well above $r^2 = .99$. Two pI markers should therefore be adequate to calibrate the pI values.

In most practical applications, the determined pI values of a sample are only used to identify component peaks. The true pI value of a peak is not relevant in the applications, as long as the determined pI value of this peak (component) is reproducible in all samples run under identical experimental conditions. It has been proven that peaks (components) in different samples can be identified using the determined pI values within a standard deviation of 0.1 pH units (<2% RSD) by using two internal pI markers, regardless of the linearity of the pH gradient created by the CAs [43]. This method applies to salt concentrations up to 15 mM in final sample solutions.

19.6 EXPANDING cIEF APPLICATIONS AND FUTURE PROSPECTS OF cIEF

In the above sections, the cIEF applications of protein characterization in biotech laboratories and pharmaceutical companies were discussed. In this section, other applications of the cIEF are presented, as well as new developments in cIEF.

19.6.1 ONLINE cIEF-MASS SPECTROMETER COUPLING

Coupling cIEF to mass spectrometer is a perfect match because cIEF separates proteins only based on their charges while the results of other CE methods are more or less related to the molecular weight of the samples. Early demonstrations of cIEF-MS work were published in 1996 [45,46]. For this application, a mobilization process after the focusing process is necessary to move sample zones into the MS detector. This technology has recently been applied to the field of proteomics [47–50]. One example is the analysis of *Escherichia coli* whole cell lysates using cIEF-MS for the study of intact proteins [50]. A complete discussion of the applications of cIEF in proteomics is discussed in a recent review [48].

The biggest difficulty in coupling cIEF to MS is that high concentration CAs [47] and most of the additives used in cIEF are not compatible with MS. As we have seen in the previous sections, many protein samples require these additives at high concentration in order to achieve reproducible results. This method limitation must be addressed before cIEF-MS technology can be applied as a tool for the analysis of real-world samples.

19.6.2 MULTIPLE DIMENSION SEPARATION

The high speed of separation and detection of the whole-column cIEF makes it attractive as a second separation dimension in a two-dimension (2-D) separation scheme. For example, using a 1-cm-long separation column, the IEF separation, and detection of the separated zones can be performed in 30 s [51]. As the second separation dimension, the whole-column detection cIEF has been successfully coupled to gel filtration chromatography [52]. The whole-column detection cIEF can also be coupled with other CE separation dimensions [53].

19.6.3 FLUORESCENCE WHOLE-COLUMN DETECTION FOR cIEF

Using cIEF as the second separation dimension in a 2-D separation system may require high detection sensitivity for cIEF because after the first dimension separation, the sample concentration can be very low. In this case, UV absorption detection is apparently not enough. Fluorescence is considered to be one of the most sensitive methods in optical detectors. A whole-column fluorescence detection system has been used for cIEF and other CE modes [54]. Two schemes have been proposed for the introduction of an excitation laser light beam into a narrow bore capillary column: side illumination and axial illumination. Axial illumination has the advantages of simple optical alignment and low background noise caused by scattering excitation light beam [55]. These advantages expanded the application of the whole-column cIEF [56].

The difficulty of the fluorescence detector for the cIEF is the lack of a chromophore in most proteins in the visible wavelength region. Labeling proteins with fluorescence dyes may change their pI values. In addition, multiple labeling is unavoidable for high sensitive detection.

19.6.4 New Applications of Whole-Column Detection cIEF

19.6.4.1 Study of Protein Interactions

Since the entire separation column is monitored in an online fashion, the whole-column detection cIEF is an ideal tool for studies of protein–protein interactions and protein interactions with other molecules, such as phospholipids [57,58]. The basic principle of applying whole-column detection cIEF for such studies is based on the pI differences of the proteins and their complexes after interaction with the molecules.

19.6.4.2 Estimation of Protein Molecular Weight Based on Their Diffusion Coefficients Determined by Whole-Column Detection cIEF

At the end of focusing process, if the separation voltage is stopped, the protein zones focused within the cIEF column will start to diffuse. This diffusion is almost one dimensional along the axis of the capillary column because of its narrow inner diameter. The diffusion process can be well monitored by the whole-column detection system. From this process, the protein's diffusion coefficient can be calculated. Once the diffusion coefficient is known, the protein's molecular weight will be estimated [59]. One potential application of this method is the determination of aggregation. If the aggregation peaks of a protein can be separated from its monomer, the aggregation peaks should be identified using the diffusion method since the difference of their molecular weight from that of the monomer will be large.

REFERENCES

1. Righett, P. G. *Isoelectric Focusing: Theory, Methodology and Application*, Elsevier, Amsterdam, 1983.
2. Righetti, P. G., Stoyanov, A. V., Zhukov, M. Y. *The Proteome Revisited*, Elsevier, Amsterdam, 2001.
3. Hjertén, S., Zhu, M. *J. Chromatogr.*, 1985, 346, 265–270.
4. Wu, J., Pawliszyn, J. *Am. Lab.*, 1994, October, 48–52.
5. Wu, J., Watson, A. H., Torres, A. R. *Biotech. Lab.*, 1999, 17, 24–26
6. Janini, G., Saptharishi, N., Waselus, M., Soman, G. *Electrophoresis*, 2002, 23, 1605–1611.
7. Li, N., Kessler, K., Bass, L., Zeng, D. *J. Pharmaceut. Biomed. Anal.*, 2006, doi: 10.1016/jpba.2006.09.024.
8. Jorgenson, J. W., Lukacs, K. D. *Anal. Chem.*, 1981, 53, 1298–1302.
9. Annual conference of CEPharm, see website www.casss.org.
10. Kilár, F. *Electrophoresis*, 2003, 24, 3908–3916.
11. Rodriguez-Diaz, R., Wehr, T., Zhu, M. *Electrophoresis*, 1997, 18, 2134–2144.
12. Mazzeo, J. R., Krull, I. S. *Anal. Chem.*, 1991, 63, 2852–2857.
13. Silverman, C., Komar, M., Shields, K., Diegnan, G., Adamovics, J. *J. Liq. Chromatogr.*, 1992, 15, 207–219.
14. Costello, M. A., Woititz, C., De Feo, J., Stremlo, D., Wen, L.-F. L., Palling, D. J., Iqbal, K., Guzman, N. A. *J. Liq. Chromatogr.*, 1992, 15, 1081–1097.
15. Hunt, G., Hotaling, T., Chen, A. B. *J. Chromatogr. A*, 1998, 800, 355–367.
16. Lasdun, A. M., Kurumbail, R. R., Leimgruber, N. K., Rathore, A. S. *J. Chromatogr. A*, 2001, 917, 147–158.
17. Schmerr, M. J., Cutlip, R. C., Jenny, A. *J. Chromatogr. A*, 1998, 802, 135–141.
18. Hempe, J. M., Craver, R. D. *Clin. Chem.*, 1994, 40, 2288–2295.
19. Mario, N., Baudin, B., Giboudeau, J. *J. Chromatogr. B*, 1998, 706, 123–129.
20. Hiraoka, A., Tominaga, I., Hori, K. *J. Chromatogr. A*, 2002, 961, 147–153.
21. Righetti, P. G., Gelfi, C., Conti, M. *J. Chromatogr. B*, 1997, 699, 91–104.
22. Wu, J., Pawliszyn, J. *Anal. Chem.*, 1992, 15, 224–227.
23. Wu, J., Watson, A. *J. Chromatogr. B*, 1998, 714, 113–118.
24. Svensson, H. *Acta. Chem. Scan.*, 1961, 15, 325–341.

25. Svensson, H. *Acta. Chem. Scan.*, 1962, 16, 456–466.
26. Vesterberg, O. *Acta. Chem. Scan.*, 1969, 23, 2653–2666.
27. Rilbe, H. In: *Electrofocusing and Isotachophoresis*, Radola, B. J., Graesslin, D., eds., Walter de Gruyter, New York, 1977.
28. Chrambach, A., Doerr, P., Finlayson, G. R., Miles, L. E. M., Sherins, R., Rodbard, D. *Ann. NY Acad. Sci.*, 1973, 290, 44–64.
29. Catsimpoolas, N. *Isoelectric Focusing*, Chapter 1, Academic Press, New York, 1976.
30. Murel, A., Kirjanen, I., Kirret, O. *J. Chromatography*, 1979, 174, 1–11.
31. Finlayson, G. R., Chrambach, A. *Anal. Biochem.*, 1971, 40, 292–311.
32. Miles, L. E. M., Simmons, J. E., Chrambach, A. *Anal. Biochem.*, 1972, 49, 109–117.
33. Arosio, P., Gianazza, E., Gighetti, P. G. *J. Chromatography*, 1978, 166, 55–64.
34. Nguyen, N. Y., Chrambach, A. *Anal. Biochem.*, 1976, 74, 145–153.
35. Mosher, R. A., Thormann, W., Bier, M. *J. Chromatogr.*, 1986, 351, 31–38.
36. Mosher, R. A., Thormann, W. *Electrophoresis*, 1990, 11, 717–723.
37. Cao, C. X. *J. Chromatogr.*, 1998, 813, 153–171.
38. Hjerten, S., Liao, J., Yao K. *J. Chromatogr.*, 1987, 387, 127–138.
39. Goodridge, L., Goodridge, C., Wu, J., Griffiths, M., Pawliszyn, J. *Anal. Chem.*, 2004, 76, 48–52.
40. Mao, Q., Pawliszyn, J., Thormann, W. *Anal. Chem.*, 2000, 72, 5493–5502.
41. Chen, A. B., Rickel, A., Flanigan, G., Hunt, G., Moorhouse, K. G. *J. Chromatogr. A*, 1996, 744, 279–284.
42. Wu, J., Pawliszyn, J. *Anal. Chem.*, 1992, 64, 219–224.
43. Wu, J., Huang, T. *Electrophoresis*, 2006, 27, 3584–3590.
44. Righetti, P. G. *J. Chromatogr. A*, 2004, 1037, 491–499.
45. Yang, L., Tang, Q., Harrata, A. K., Lee, C. S. *Anal. Biochem.*, 1996, 243, 140–149.
46. Kirby, D. P., Thorne, J. M., Göetzinger, W. K., Karger, B. L. *Anal. Chem.*, 1996, 68, 4451–4458.
47. Storms, H. F., van der Heijden, R., Tjaden, U. R., van der Greef, J. *Electrophoresis*, 2004, 25, 3461–3467.
48. Simpson, D. C., Smith, R. D. *Electrophoresis*, 2005, 26, 1291–1305.
49. Wang, Y., Rudnick, P. A., Evans, E. L., Li, J., Zhuang, Z., Devoe, D. L., Lee, C. S., Balgley, B. M. *Anal. Chem.*, 2005, 77, 6549–6556.
50. Martinovic, S., Pasa-Tolic, L., Smith, R. D. *Methods Mol. Biol.*, 2004, 276, 291–304.
51. Wu, J., Pawliszyn, J. *Anal. Chem.*, 1995, 34, 2010–2014.
52. Tragas, C., Pawliszyn, J. *Electrophoresis*, 2000, 21, 227–237.
53. Sheng, L., Pawliszyn, J. *Analyst*, 2002, 127, 1159–1163.
54. Wu, X.-Z., Huang, T., Liu, Z., Pawliszyn, J. *Trends Anal. Chem.*, 2005, 24, 369–382.
55. Huang, T., Pawliszyn, J. *Analyst*, 2000, 125, 1231–1233.
56. Liu, Z., Pawliszyn, J. *Anal. Biochem.*, 2005, 336, 94–101.
57. Bo, T., Pawliszyn, J. *Anal. Biochem.*, 2006, 350, 91–98.
58. Bo, T., Pawliszyn, J. *Anal. Chim. Acta.*, 2006, 559, 1–8.
59. Liu, Z., Lemma, T., Pawliszyn, J. *J. Proteome Res.*, 2006, 5, 1246–1251.

Part IIB

Capillary-Based Systems: Specialized Methods and Technologies

20 Subcellular Analysis by Capillary Electrophoresis

Bobby G. Poe and Edgar A. Arriaga

CONTENTS

20.1 INTRODUCTION

Many cellular processes are localized within specific subcellular compartments; therefore, recent bioanalytical research has been aimed at utilizing the natural organization found within cells to reduce the complexity of biological samples. Subcellular compartments include (i) distinct, vesicular organelles, such as nuclei, mitochondria, and acidic organelles; (ii) indistinct, continuous organelles such as the endoplasmic reticulum, Golgi apparatus, plasma membrane, and cytoskeletal network; and (iii) the cytoplasm.

In spite of the reduction in complexity that can be achieved by cellular fractionation, an analytical separation is frequently required to separate one or more components from the cellular milieu. As evidenced throughout this book, capillary electrophoresis (CE) provides high resolution and separation efficiency, both of which are necessary for subcellular analysis. In addition, CE is advantageous because it requires very little sample volume, typically less than a nanoliter, and generally very little sample preparation. Hence, capillaries have been used to directly sample subcellular compartments within neurons,[1] oocytes,[2] and muscle tissue sections.[3] They have also been used to analyze individual organelles from a single cell following on-column lysis.[4,5]

Taking advantage of these attributes, CE has the potential to benefit diverse fields, that study the vital cellular processes that are localized within specific subcellular compartments. In fact, subcellular analysis by CE has already proven useful in different areas such as neurochemistry;[1,6–14] cellular physiology;[2,3,15–23] gene therapy;[24] drug accumulation, metabolism and localization;[25–32] disease diagnosis;[33,34] and proteomics.[35–38]

20.2 BACKGROUND

20.2.1 ORGANELLE PREPARATION

In general, the aim of subcellular analysis is to quantify an analyte within a specific subcellular compartment. Consequently, in most cases, an organelle fraction must be purified before analysis. Cellular fractionation, that is, isolation and purification of organelles, has been indispensable in the biochemical fields and, as evidenced in the literature, has been used pervasively. Since complete reviews can be found in the biochemical literature, we will only briefly describe the principles of cellular fractionation.

20.2.1.1 Cellular Fractionation

To purify an organelle fraction, the first step is cell homogenization, which can be accomplished using diverse methods such as hypotonic shock or mechanical homogenization. Hard tissue samples such as muscle require harsh homogenization techniques (e.g., using a Waring blender), but soft samples such as cells from cultures can be homogenized with gentler techniques like nitrogen cavitation. The choice of a homogenization technique must take into account the sample hardness, volume, and desired organelle intactness.

Following cell homogenization, centrifugation techniques enjoy nearly universal use in the purification of subcellular organelles. Centrifugation is relatively simple and can be performed on several liters to submilliliter volumes of cell homogenate. As shown in Equation 20.1, the sedimentation velocity $\mathrm{d}x/\mathrm{d}t$ per unit centrifugal field ($\omega^2 x$) is dependent on the size of the particle (r^2), the density difference of the particle compared to the medium ($\rho_\mathrm{p} - \rho_\mathrm{m}$), and the viscosity of the medium (η):

$$\frac{\mathrm{d}x/\mathrm{d}t}{\omega^2 x} = \frac{2r^2\left(\rho_\mathrm{p} - \rho_\mathrm{m}\right)}{9\eta} \tag{20.1}$$

With advances in separation media, organelles can now be separated based on their density, size, or both. This makes centrifugation the most versatile and practical cellular fractionation method.

Other techniques can be combined with centrifugation to achieve higher purity. These techniques include immunoisolation and electrophoretic purification. Immunoisolation holds great potential for isolating highly purified organelle fractions, since it relies on the molecular recognition of surface antigens by antibodies. Prerequisites for immunoisolation are (i) an antigen that is highly abundant on the surface of a specific organelle type; (ii) an antibody that recognizes the antigen; and (iii) a monodisperse organelle suspension, which reduces contamination due to aggregation. Frequently, the antibody is attached to magnetic beads to allow more efficient recovery of the organelles.

20.2.1.2 Electrophoretic Preparation of Organelles

Although electrophoretic separations are chiefly preparatory, they provide much of the basis for the separation of intact organelles by CE. Electrophoretic separations were originally developed for proteins, but have been used since the 1970s as a purification technique for intact organelles.[39–43] Biological particles, including organelles, carry a net negative charge at physiological pH, and therefore, are mobilized in the presence of an electric field (see Section 20.3). Common separation modes used in electrophoretic preparations are zone electrophoresis, isotachophoresis, and isoelectric focusing modes. The separation scheme for free-flow electrophoresis (zone electrophoresis mode) is illustrated in Figure 20.1. A laminar fluid flow carries the organelles through the separation chamber, while the electric field mobilizes the organelles perpendicular to the flow. Organelle types are then separated based on their electrophoretic mobility differences. At the end of the chamber the suspension is separated into many fractions and the material can be recovered for further analysis. As such, organelles can be continually introduced into the chamber, allowing for the purification of large quantities of material.

The earliest electrophoretic techniques concentrated mainly on the preparation of mitochondria[39,42,43] and lysosomes,[40,41] but further research has made possible the purification of secretory vesicles;[44] clathrin-coated vesicles;[45] endoplasmic reticulum;[46] early, middle-, and late-stage endosomes;[46] peroxisomes;[47] microsomes;[48] and phagosomes.[49] These preparatory techniques have functioned as the proof-of-principle for analytical separations of intact organelles using CE, by demonstrating that isolated organelles are amenable to electrophoretic separation techniques.

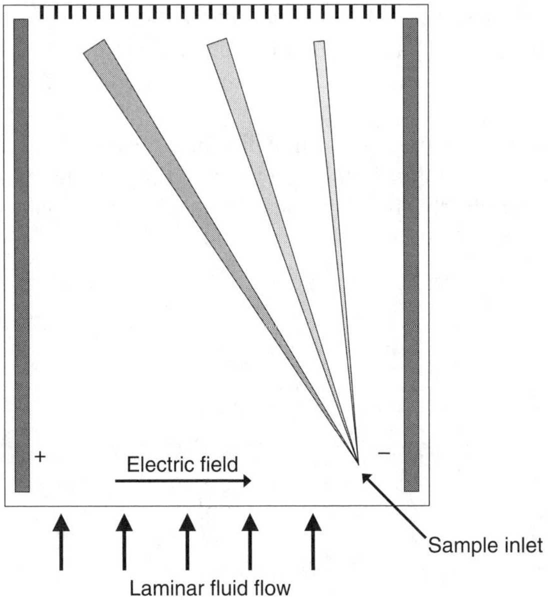

FIGURE 20.1 Free-flow electrophoresis separation. Laminar fluid flow carries the organelles perpendicularly to the applied electric field. Organelles migrate in the opposite direction of the electric field according to their electrophoretic mobility. Sample is introduced continually at the inlet and collected in many fractions following separation.

20.3 THEORETICAL ASPECTS

The theories describing micellar electrokinetic chromatography (MEKC), capillary zone electrophoresis (CZE), and capillary gel electrophoresis (CGE) separations of small molecules and biopolymers are described in other chapters of this book and will not be discussed here. Here, we will briefly touch upon the theoretical aspects of the electrophoretic mobility of organelles, foregoing an in depth discussion of electrokinetic theory that can be found elsewhere in original publications and comprehensive reviews.[50–58] The electrophoretic mobility (μ_E) is defined as

$$\mu_E = \frac{\nu}{E} \tag{20.2}$$

where ν is the velocity of the particle and E is the electric field in V/cm. The electrophoretic mobility of small molecules is as given in the following equation:

$$\mu_E = \frac{q}{6\pi \eta r} \tag{20.3}$$

where q is the net charge, η is the viscosity of the separation medium, and r is the radius of the molecule. As shown, electrophoretic mobility depends on the electrical force and Stokes frictional drag.

On the other hand, the electrophoretic mobility of charged particles (e.g., organelles) is more complex. Figure 20.2 illustrates the forces acting upon a charged particle in an electric field (E). While electrophoretic mobility is still governed in part by the electrical force (F_1) and the Stokes frictional force (F_2), it is also affected by electrophoretic retardation (F_3) and the relaxation effect (F_4). The electrophoretic retardation force and the relaxation effect are both caused by the ionic atmosphere (dashed circle) that surrounds the charged particle and are dependent on the zeta-potential (ζ), which is related to the surface charge. Electrophoretic retardation (F_3) is caused by the force exerted upon the solvent molecules by the oppositely charged ions in the ionic atmosphere, resulting in fluid flow around the particle, which increases frictional drag. In addition, the ionic atmosphere lags slightly behind the charged particle, since its formation requires a finite amount of time, which results in a small electrical force in the opposite direction of the electric field; this is termed the relaxation effect (F_4).

The first solution for the electrophoretic mobility of a charged particle was reported by Von Smoluchowski[59] for the case of a thin double layer (i.e., $\kappa a \gg 1$); where κ is the Debye factor and a is the particle radius. Accordingly, the electrophoretic mobility of a charged particle should follow

$$\mu_E = \frac{\varepsilon\zeta}{\eta} \tag{20.4}$$

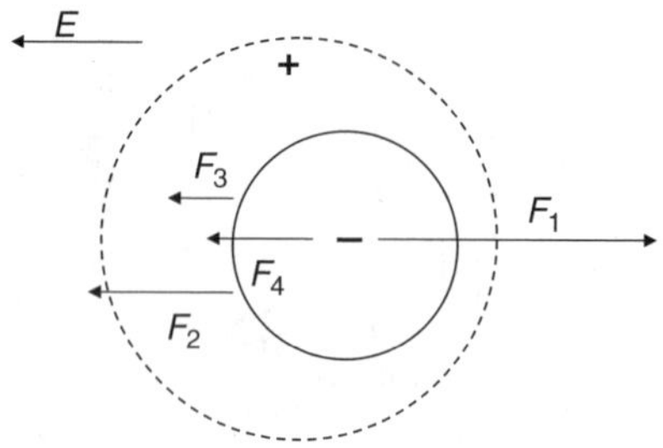

FIGURE 20.2 Forces acting on a charged particle. The particle is negatively charged and surrounded by a positively charged ionic atmosphere, indicated by the dashed circle. F_1 is the electrical force, F_2 is Stokes frictional drag, F_3 is electrophoretic retardation, and F_4 is the relaxation effect.

in which ε and η are the dielectric constant and viscosity of the liquid surrounding the particle, respectively, and ζ is defined as follows:

$$\zeta = \frac{\sigma a}{\varepsilon(1+\kappa a)} \tag{20.5}$$

where σ is the net electrokinetic charge density.

In the opposite case, $\kappa a \ll 1$, the electrophoretic mobility follows the Hückel equation:[60]

$$\mu_E = \frac{2\varepsilon\zeta}{3\eta} \tag{20.6}$$

Henry provided a solution[61] (Equation 20.7) for various values of κa by taking into account the deformation of the electric field lines near the particle:

$$\mu_E = \frac{2\varepsilon\zeta}{3\eta} f_1\,(\kappa a) \tag{20.7}$$

When $\kappa a \rightarrow 0$, $f_1 = 1$ and the Henry equation reduces to Hückel's solution (Equation 20.6). Conversely, when $\kappa a \rightarrow \infty$, $f_1 = 3/2$ and the equation reduces to Smoluchowski's solution (Equation 20.4). The function f_1 was recently given as a single equation,[62] as opposed to two power series as reported by Henry.

Equation 20.7 can be applied when any of the following conditions are met: (i) $\kappa a \gg 1$, (ii) $\kappa a \ll 1$, or (iii) $\zeta \ll 25$ mV. In these cases, the relaxation effect is negligible. More accurate theories that take into account the relaxation effect and electrophoretic retardation have been developed by Booth,[52] Overbeek,[53,54] and O'Brien.[55–58]

Regrettably, the electrokinetic theories described above do not adequately describe the electrophoretic mobility of organelles. This is because the classical colloidal theories assumed the particle was rigid and nonconducting with uniform charge, whereas in reality, organelles are nonspherical, deformable, and the surface charges are mobile, which leads to polarization. This makes it difficult to interpret electrophoretic mobility with the classical theories. However, Hayes' group has aimed to describe the electrophoretic mobility of deformable particles using liposomes as a model.[63,64] It has been demonstrated that models that incorporate the effects of particle deformation, mobile surface charges, and polarizability predict the electrophoretic behavior of liposomes better than do traditional electrokinetic models.[64] Since liposomes more closely resemble biologically relevant particles, this model provides a more complete description of organelle electrophoretic mobility.

20.4 PRACTICAL APPLICATION

Subcellular analysis by CE can be performed in three modes: (i) an organelle fraction can be dissolved or lysed and the analytes found in the fraction analyzed; (ii) intact, isolated organelles can be separated and detected; and (iii) analytes or organelles can be directly sampled from a single cell or tissue section. Each mode of analysis will be illustrated below.

20.4.1 PROTEOMIC ANALYSIS OF DISSOLVED SUBCELLULAR FRACTIONS

Proteomics research has benefited greatly from subcellular fractionation, because reducing the complexity of the entire proteome to smaller organelle proteomes makes it possible to separate and detect low abundance proteins. Furthermore, since the goal of proteomics is not only to learn protein sequences but also the localization and function of proteins, subcellular analysis is advantageous because it provides the subcellular localization. Indeed, the benefits of cellular fractionation before proteomic analysis have been proven with conventional two-dimensional gel electrophoresis and multidimensional liquid chromatography-tandem mass spectrometry (LC-MS/MS).[65–68]

Although two-dimensional gel electrophoresis provides unmatched resolution (~2000 protein spots using conventional gels), it does have some drawbacks: these include the precipitation of highly acidic or basic proteins in the first dimension (isoelectric focusing), limited sensitivity and dynamic range, low throughput, and difficulty in automating time-consuming and tedious tasks. CE, on the other hand, is easily automated, has a very low limit of detection (LOD) when combined with off-column laser-induced fluorescence (LIF) detection, and has a large dynamic range. In addition, isoelectric focusing does not need to be performed in the first dimension, thereby eliminating the precipitation of acidic and basic proteins. Separations based on CE thus show potential for increasing the throughput and sensitivity of proteomic analyses.

Working toward this goal, Dovichi and coworkers[69–71] have developed highly sensitive two-dimensional CE protein separations and automated the analysis. Since two-dimensional CE separations do not resolve as many proteins as traditional two-dimensional gel electrophoresis,[70] in order to increase the resolution and aid in the detection of low abundance proteins, a differential detergent fractionation technique was used to reduce the complexity of cell homogenates before separation.[38,72] Differential detergent fractionation was developed by Ramsby et al.[73] for isolated hepatocyes and is based on the different solubilities of proteins in various buffers. A schematic representation of the fractionation procedure is shown in Figure 20.3 where four fractions are obtained: (i) cytosolic or soluble proteins, (ii) membrane-organelle proteins, (iii) nuclear membrane proteins, and (iv) cytoskeletal-matrix proteins. The identity of each fraction was confirmed and its selectivity measured with immunoblotting using over 20 antibodies.[73]

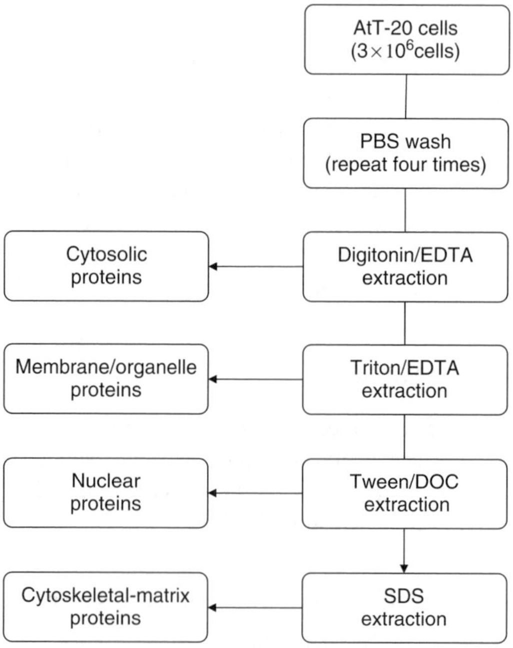

FIGURE 20.3 Schematic representation of the differential detergent fractionation procedure applied to cultured cells. Four fractions are obtained: (1) cytosolic proteins, (2) membrane/organelle proteins, (3) nuclear-membrane proteins, and (4) cytoskeletal proteins. (Modified from Fazal, M. A., et al., *J. Chromatogr. A*, 1130, 182–189, 2006.)

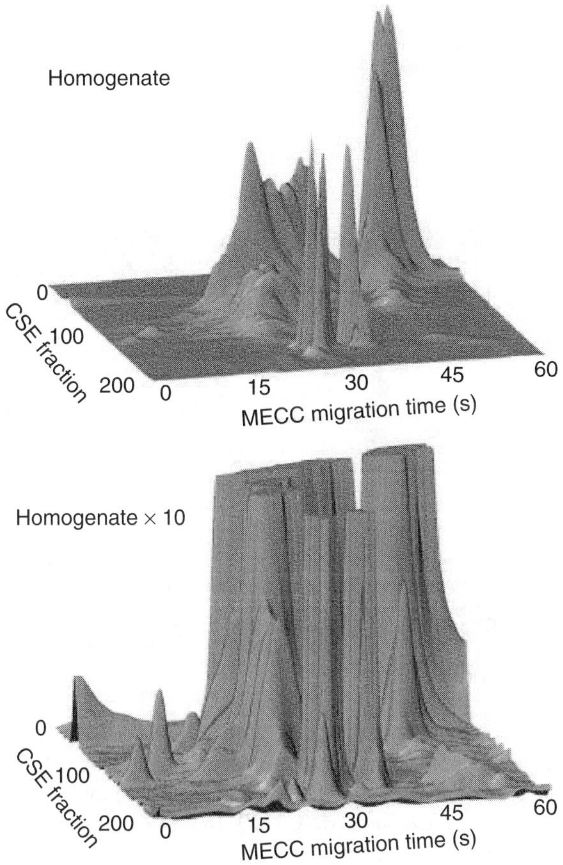

FIGURE 20.4 Two-dimensional electropherogram of whole cell homogenate. AtT-20, mouse adrenal gland cells, were homogenized and the proteins were labeled with FQ. The electropherogram is shown as a landscape image, in which the fluorescence intensity is proportional to the peak height. Proteins were separated first by capillary sieving electrophoresis and then by micellar electrokinetic capillary chromatography in the second dimension. Approximately 150 fractions were transferred from the first capillary to the second capillary. (Reprinted from Fazal, M. A., et al., *J. Chromatogr. A*, 1130, 182–189, 2006. Copyright 2006. With permission from Elsevier.)

Ramsby's procedure was used by Fazal et al.[38] to fractionate the homogenate of AtT-20 cells (mouse adrenal gland cells) before two-dimensional CE separation. Proteins were labeled with 3-(2-furoyl) quinoline-1-carboxaldehyde (FQ) before separation and detected using LIF in a sheath flow cuvette. Capillary sieving electrophoresis was performed in the first dimension to separate the proteins based on size, using dextran in the sieving buffer; this was followed by micellar electrokinetic capillary chromatography in the second dimension to separate proteins based on hydrophobicity. The resulting two-dimensional electropherogram for the whole cell homogenate is displayed in Figure 20.4. The separation detected 51 unique protein peaks, and the origin of the large mesa, shown in the close up view, was determined using the differential detergent fractionation procedure. Figure 20.5 shows all four detergent fractions and indicates that the large mesa in Figure 20.4 was from proteins located in the nuclear protein extract. As anticipated, by performing the fractionation technique before the two-dimensional separation, the number of resolved proteins increased from 51, for the whole cell homogenate, to 231, for the detergent fractions. This increase is due to the detection of low abundance proteins that had been previously masked by highly abundant proteins in the whole

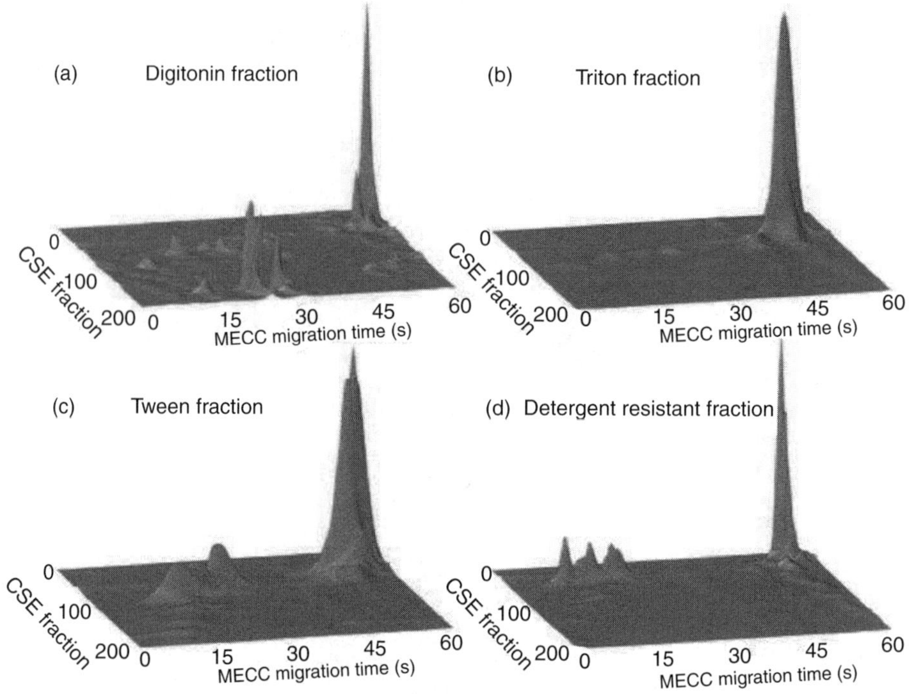

FIGURE 20.5 Two-dimensional electropherograms of each differential detergent fraction. The digitonin fraction (A) is the cytosolic proteins, the Triton fraction (B) is the membrane/organelle proteins, the Tween fraction (C) is the nuclear-membrane proteins, and the detergent resistant fraction (D) is the cytoskeletal proteins. (Reprinted from Fazal, M. A., et al., *J. Chromatogr. A*, 1130, 182–189, 2006. Copyright 2006. With permission from Elsevier.)

cell homogenate. In addition, each detergent fraction had distinct electrophoretic features, indicating that each fraction contained unique proteins.

20.4.2 Quantification of Hydroxychloroquine in the Rat Liver Microsomal Fraction

A more conventional subcellular fractionation approach was taken for the quantification of hydroxychloroquine (HCQ) in the microsomal fraction of mouse liver homogenates.[27] Microsomes are vesicles formed from the disruption of the endoplasmic reticulum and provide a convenient means of studying metabolism by the cytochrome CYP450 enzyme superfamily, which metabolizes the majority of drugs given to humans. HCQ has proven efficacious against rheumatoid arthritis, lupus erythematosus, and has been used as an antimalarial drug. It is a chiral molecule and, as shown in Figure 20.6, forms three chiral metabolites when incubated with rat liver microsomes. Since it is well known that enantiomers show different activities toward drug targets, a method for quantifying HCQ and its metabolites was developed.

Mouse livers were excised and homogenized with a Potter–Elvehjem homogenizer. The microsomal fraction was isolated by centrifuging the homogenate at $9000 \times g$ for 15 min, to sediment nuclei, mitochondria, and membrane debris. The supernatant was centrifuged at $100,000 \times g$ for 60 min to pellet microsomes. The microsomes were further purified by another centrifugation step to remove soluble proteins (e.g., hemoglobin). HCQ was incubated with the microsomes for 2 h. Then a liquid–liquid extraction technique was performed to recover HCQ and its metabolites. The molecules were separated by a commercial CE system with ultraviolet (UV) absorbance detection.

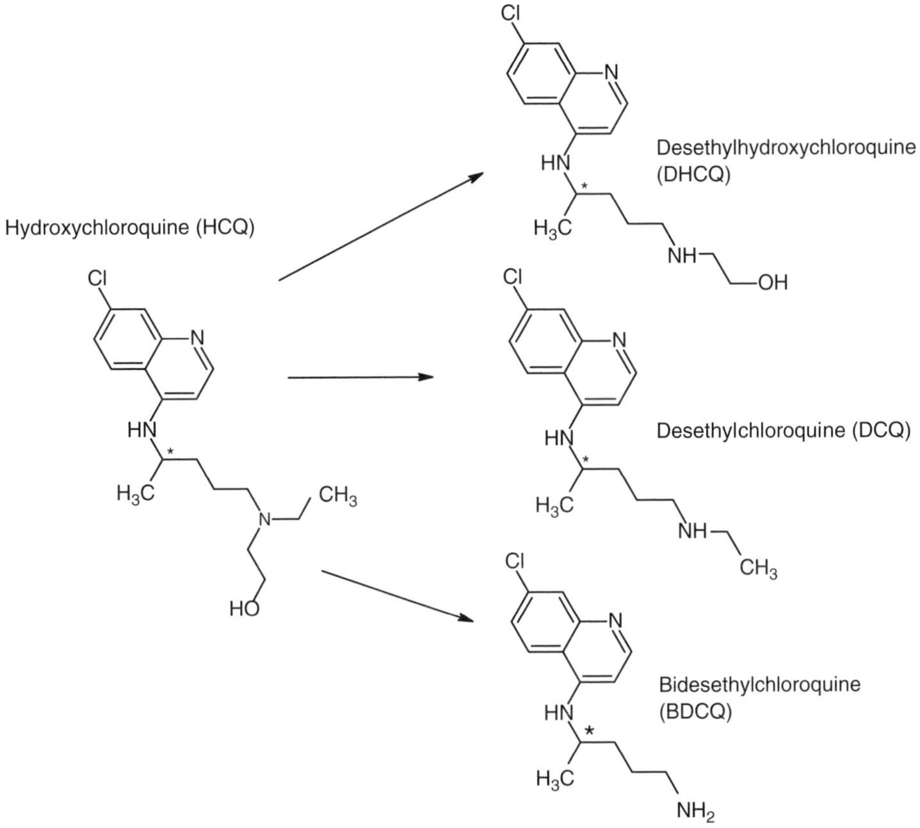

FIGURE 20.6 Illustration of HCQ and its three metabolites. The chiral center in each molecule is indicated by an asterisk.

The resulting electropherogram is displayed in Figure 20.7, which shows the separation of all eight molecules. The separation was performed in Tris buffer that contained sulfated-β-cyclodextrin and hydroxypropyl-β-cyclodextrin, which provided the enantiomeric selectivity. Metabolites were quantified from 0.41 to 0.89 μM with relative standard deviations of ~10%. The subcellular analysis shows the preferential production of (−)-(*R*)-metabolites (Table 20.1). Compared with a previously reported high-performance liquid chromatography (HPLC) method,[74] the CE method allowed these researchers to quantify two more metabolites and required less time.

20.4.3 Measurement of the pH in Individual Acidic Organelles

An alternative approach to separating analytes that have been released from dissolved organelles is to quantify the analytes within intact organelles. The major aims of this type of analysis are to reveal the heterogeneity amongst organelles, determine the subcellular localization of analytes, and quantify the analytes within individual organelles. Arriaga and coworkers successfully analyzed intact nuclei,[4,75,76] mitochondria,[3,15,17,77–80] and acidic organelles[16,28,81] by CE-LIF. The advantages of individual organelle analysis are illustrated below using pH measurements in individual acidic organelles as an example.

Multidrug resistance is a common problem with chemotherapeutic treatments in which cancer cells become insensitive to drug treatment. It has been proposed that multidrug resistance may be related to a larger than normal pH gradient between the cytoplasm and acidic organelles in

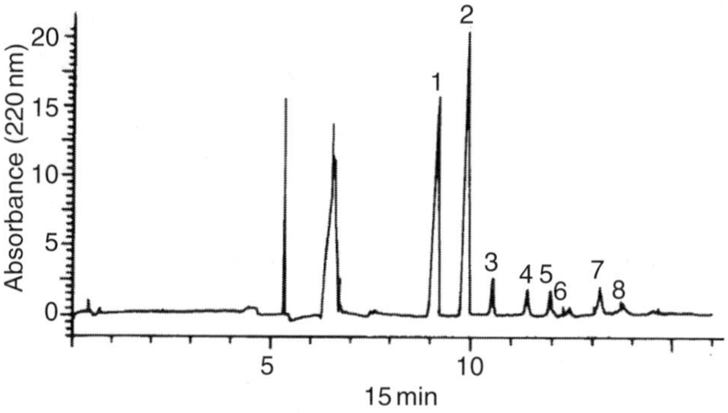

FIGURE 20.7 Chiral resolution of HCQ and its metabolites after incubation with mouse liver microsomes. Peaks are numbered and correspond as follows: (1) (−)-(*R*)-HCQ, (2) (+)-(*S*)-HCQ, (3) (−)-(*R*)-DHCQ, (4) (+)-(*S*)-DHCQ, (5) (+)-(*S*)-DCQ, (6) (−)-(*R*)-BDCQ, (7) (−)-(*R*)-DCQ, and (8) (+)-(*S*)-BDCQ. Electrophoresis was performed in a 50 cm, 50 μm i.d. capillary with a separation voltage of 13 kV in 100 mM Tris/phosphate, 1% sulfated-β-cyclodextrin, 30 mg/mL hydroxypropyl cyclodextrin, pH 9.0. The molecules were detected with UV absorbance at 220 nm. (Reprinted from Cardoso, C. D., et al., *Electrophoresis*, 27, 1248, 2006. With permission.)

TABLE 20.1
HCQ Metabolites Formed by Incubation with Mouse Liver Microsomes

	(−)-(*R*)-DHCQ	(+)-(*S*)-DHCQ	(−)-(*R*)-DCQ	(−)-(*R*)-DCQ	(−)-(*R*)-BDCQ	(+)-(*S*)-BDCQ
Concentration (M)	8.9×10^{-7}	6.5×10^{-7}	7.8×10^{-7}	6.3×10^{-7}	4.9×10^{-7}	4.1×10^{-7}
RSD(%)n = 3	7.6	8.9	10.1	12.9	10.5	11

Source: Reprinted from Cardoso, C. D., et al. *Electrophoresis*, 27, 1248, 2006. With permission.

drug-resistant cells as compared to drug-sensitive cells.[82] The larger pH gradient is hypothesized to result in drug protonation and sequestration in acidic organelles and secretion from the cell (Protonation, Sequestration, and Secretion, PSS model). When the drug is secreted from the cell, it can no longer act to halt cell proliferation, which causes drug resistance. The average pH value of acidic organelles can be determined with fluorescence microscopy and flow cytometry, but neither technique has been used to measure the pH of individual acidic organelles or reveal acidic organelle heterogeneity. To obtain this information, CE-LIF was used to measure, for the first time, the pH of individual acidic organelles in drug-resistant and drug-sensitive cell lines.[81]

Fluorescein tetramethylrhodamine dextran (FRD) was used as a ratiometric probe, which provides a pH-independent signal (tetramethylrhodamine), to compensate for different organelle volumes, and a pH-dependent signal (fluorescein). The ratio of the pH-independent and pH-dependent fluorescence was expected to provide a quantitative measure of the pH. Acidic organelles (lysosomes and endosomes) were specifically labeled by FRD, because the cells endocytose small amounts of the extracellular medium, which is accumulated in endosomes and eventually lysosomes. CE was performed in a poly(*N*-acryloyl aminopropanol, AAP) modified capillary to reduce both adsorption at the capillary surface and the electroosmotic flow. LIF detection was performed off-column using a

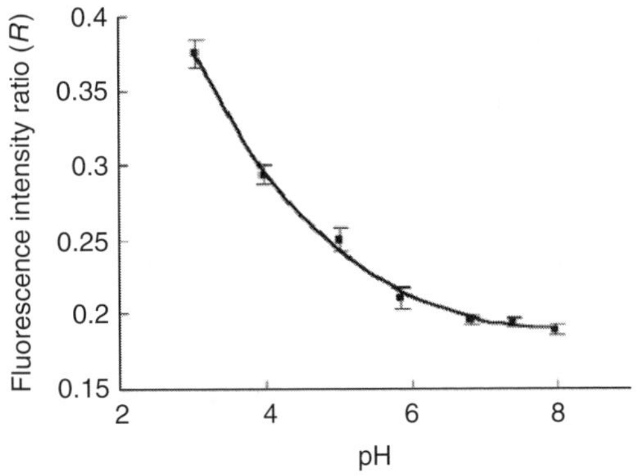

FIGURE 20.8 Calibration of the pH measurement. *R*, the ratio of tetramethylrhodamine and fluorescein fluorescence, versus the pH for liposomes constructed with internal pH values ranging from 3 to 7. The average ratio and standard deviation is shown with a marker and bars at 7 pH values. The calibration curve represents a quadratic model and was determined by multiple regression analysis. (Reprinted from Chen, Y. and Arriaga, E. A., *Anal. Chem.*, 78, 821, 2006. Copyright 2006. With permission from American Chemical Society.)

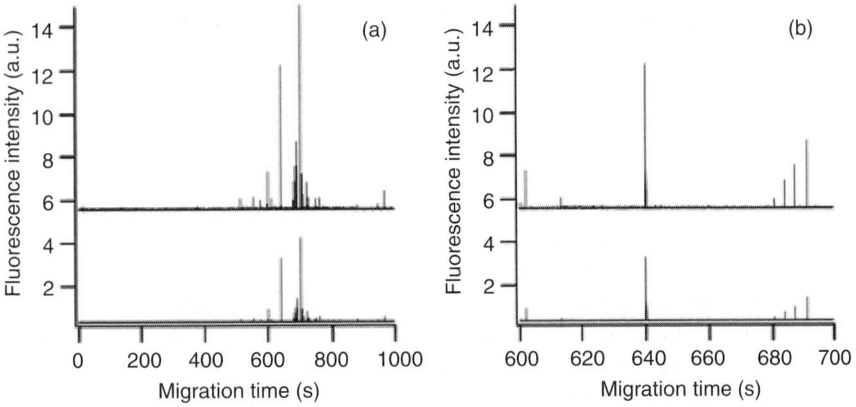

FIGURE 20.9 Electropherogram displaying CE-LIF detection of individual acidic organelles. The entire electropherogram is shown in panel a, where the top trace is the fluorescein channel and the bottom trace is the tetramethylrhodamine channel. Panel b shows an enlarged region of the electropherogram to illustrate detection of acidic organelles simultaneously in each channel. Acidic organelles were injected for 1 s at 11 kPa and separated at –300 V/cm in a poly(AAP) coated capillary. (Reprinted from Chen, Y. and Arriaga, E. A., *Anal. Chem.*, 78, 821, 2006. Copyright 2006. With permission American Chemical Society.)

single excitation source (488 nm argon-ion laser) and two emission wavelengths (λ_1-pH independent and λ_2-pH dependent) that were monitored simultaneously.

To calibrate the CE-LIF detector, liposomes were constructed with internal pH values ranging from 3.1 to 8.0. The liposomes were injected onto the capillary, electrophoretically separated and detected individually following hydrodynamic focusing in a sheath flow cuvette. The calibration curve obtained is shown in Figure 20.8. To test the PSS model, acidic organelles were isolated from drug-resistant and drug-sensitive cell lines using a Dounce homogenizer and differential centrifugation. The acidic organelles were then analyzed in a fashion similar to the liposomes and the resulting electropherogram is shown in Figure 20.9; the sharp spikes correspond to the detection of individual acidic organelles. By tabulating the fluorescence in both channels and comparing the ratios with

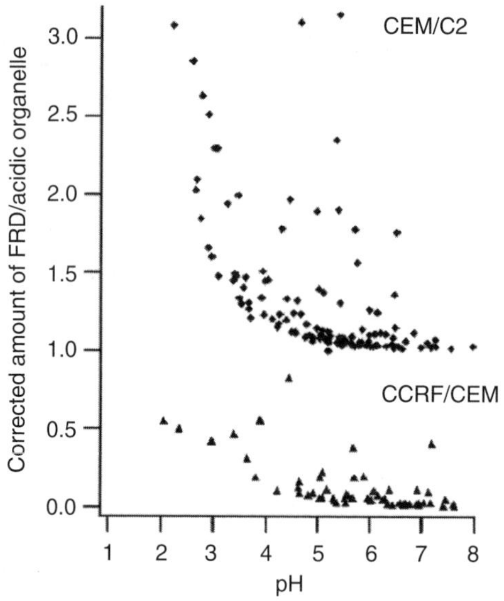

FIGURE 20.10 Amount of FRD per acidic organelle sorted from high to low pH. The *y*-axis is offset 1 unit for the CEM/c2 cells (drug-resistant). (Reprinted from Chen, Y. and Arriaga, E. A., *Anal. Chem.*, 78, 821. Copyright 2006. With permission from American Chemical Society.)

the calibration curve, the pH for each individual acidic organelle was measured. The average pH values were 5.1 ± 0.2 for drug-resistant cells and 6.1 ± 0.9 for drug-sensitive cells, respectively. The average pH values correlated well with fluorescence microscopy values (5.0 ± 0.6 and 6.2 ± 0.7 for drug resistant and drug sensitive, respectively).

Unlike fluorescence microscopy, which measures the average pH value of all the acidic organelles in an entire cell, the individual pH measurements obtained with CE revealed a range of pH values from 3 to 7 in both cell lines, which indicated that several types of acidic organelles were being analyzed; specifically, early endosomes (pH 6–6.5), late endosomes (pH 5.5–6), and lysosomes (pH 4.5–5.5). This was confirmed by plotting the amount of FRD versus pH as shown in Figure 20.10. The more acidic organelles contained more FRD, which is consistent with the maturation of the acidic organelles as they transform from early endosomes to lysosomes, since they fuse with each other to become larger during this process.

20.4.4 Direct Sampling of Dopamine from Mammalian Cell Cytoplasm

The third approach to subcellular analysis involves using the separation capillary to sample subcellular compartments within single cells or from tissue sections. Capillaries are uniquely suited to provide not only high resolution and separation efficiency but also, due to their physical dimensions, sufficient spatial resolution to sample subcellular compartments. Most commonly, the capillary is used to sample the cytoplasm from single cells. However, the capillary can also be used to sample intact organelles from tissue sections[3,15] and analyze organelles released from single cells.[5,76]

In particular, neurochemistry has found CE especially useful for quantifying neurotransmitter concentrations in single cells. Since important cellular processes such as neurotransmitter synthesis, storage and release are affected by the cytoplasmic concentrations of neurotransmitters, methods to sample and quantify cytoplasmic concentrations of neurotransmitters from single invertebrate cells have been developed.[6,7,9–11] Invertebrate neurons are particularly amenable for sampling because they can be quite large, for example, the giant dopamine neuron of pond snails is ~200 μm

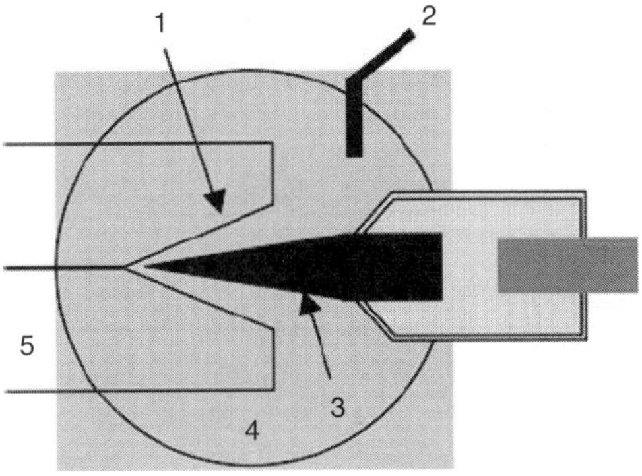

FIGURE 20.11 Schematic representation of the end-column amperometric detection system: (1) the outlet of the separation capillary was etched to 13 μm i.d., (2) Ag/AgCl reference electrode, (3) flame-etched carbon fiber electrode, (4) buffer reservoir, and (5) separation capillary, 770 nm i.d. (Reprinted from Woods, L. A., et al., *Electroanalysis*, 17, 1192, 2005. With permission.)

in diameter. However, sampling from mammalian cells, which are over four times smaller, was impracticable until recently.

To overcome these difficulties, a CE system utilizing a 770 nm i.d. capillary was developed to sample and quantify dopamine from the cytoplasm of single mammalian cells.[7] To sample the cytoplasm, the inlet of the separation capillary was tapered to an outer diameter of 2.5 μm, which is five times smaller than the diameter of the cells. The capillary was inserted through the plasma membrane by applying a small voltage (e.g., 2 kV) to cause the electroporation of the membrane. This voltage also electrokinetically injected the cytoplasm into the capillary, simultaneously. This system allowed as little as 380 fL of cytoplasm to be injected from a single PC12 cell, which amounts to ~40% of the cell cytoplasm. The smallest percentage of cytoplasm that was sampled was 8%.

Such small injection volumes necessitate a very low LOD. This is achieved using an improved end-column amperometric detection scheme,[83] as shown in Figure 20.11; a 2.5 μm flame-etched carbon fiber microelectrode is positioned inside the etched lumen of the separation capillary. The conical tip of the flame-etched microelectrode allows for better positioning of the electrode into the capillary lumen, which reduces the dead volume and hence decreases the LOD. The reported LODs for dopamine and catechol using this detection scheme were 400 ± 100 and 410 ± 80 zmol (10^{-21} moles), respectively. The separation and detection of dopamine sampled from a single cell is displayed in the resulting electropherogram shown in Figure 20.12. The concentration of cytoplasmic dopamine from three cells was 240 ± 60 μM, which indicated the presence of dense core vesicles in the injection plug.

20.4.5 Detection of Individual Mitochondria Sampled from Muscle Tissue Cross sections

In a similar fashion, intact organelles can also be sampled directly from a cell sample. This was first demonstrated by sampling and separating individual mitochondria from skeletal muscle tissue cross sections.[3] Skeletal muscle is a complex and heterogeneous tissue in which thousands of individual fibers extend approximately in parallel throughout the length of the tissue. Each fiber contains multiple nuclei and thousands of mitochondria. Using conventional histochemical stains, the heterogeneity among muscle fibers can be visualized to investigate the effects of disease and aging. However, an analytical method that could reveal heterogeneity among the organelles in a single fiber as well as fiber-to-fiber heterogeneity was desirable.

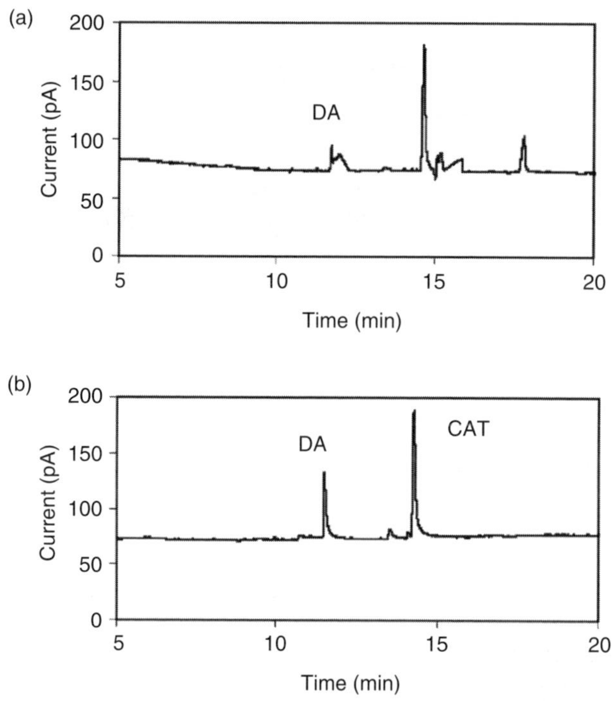

FIGURE 20.12 Separation of dopamine (DA) and catechol (CAT) from a single mammalian cell. (a) Dopamine was separated and detected in a 380 fL injection of PC12 cell cytoplasm and (b) a standard injection of DA and CAT is used to verify the detection of dopamine. Both injections used a 3 kV injection voltage for 5 s and a bare fused-silica capillary. Dopamine was separated at 25 kV in 50 mM TES, pH 7.2 with 2% 1-propanol. (Reprinted from Woods, L. A., et al., *Electroanalysis*, 17, 1192, 2005. With permission.)

Semimembranous muscle tissues from Fisher rats were excised, frozen, and sectioned perpendicular to the fiber length, with a cryostat. The ~10 μm thick cross sections were mounted on microscope slides and frozen before use. Mitochondria were labeled with the mitochondria-specific fluorescent stain 10-*N*-nonyl acridine orange (NAO) by depositing a microdroplet on the tissue cross section and using a capillary, micromanipulator, and a microscope to visualize deposition. Mitochondria were sampled by positioning the separation capillary on top of the labeled fiber and applying suction. Since the cross-sectional area of the 50 μm i.d. capillary was ~2000 μm^2, which is slightly smaller than the ~3000 μm^2 area of a single fiber, the capillary was able to sample mitochondria from a single fiber. The labeling and sampling of mitochondria are shown in Figure 20.13, wherein the capillary lumen can be easily seen above a single muscle fiber. After the mitochondria were introduced into the capillary, the inlet was placed in separation buffer and the separation voltage applied. Mitochondria were electrophoretically separated and detected individually by off-column LIF detection as they migrated out of the capillary. Figure 20.14 displays an electropherogram showing the detection of individual mitochondria sampled from a muscle cross section.

One complicating factor in the analysis of mitochondria from tissue cross sections is that mitochondria are found in two locations in muscle tissue: (i) subsarcolemmal mitochondria are found directly beneath the sarcolemmal membrane and are easily extracted and (ii) interfibrillar mitochondria are located between the myofibrils and require exposure to a protease to be efficiently released. To release these latter mitochondria, a short treatment with trypsin was tested by depositing a microdroplet of trypsin on the fiber before sampling. Table 20.2 compares the analysis of mitochondria from muscle fiber that had undergone trypsin treatment with the analysis of mitochondria from

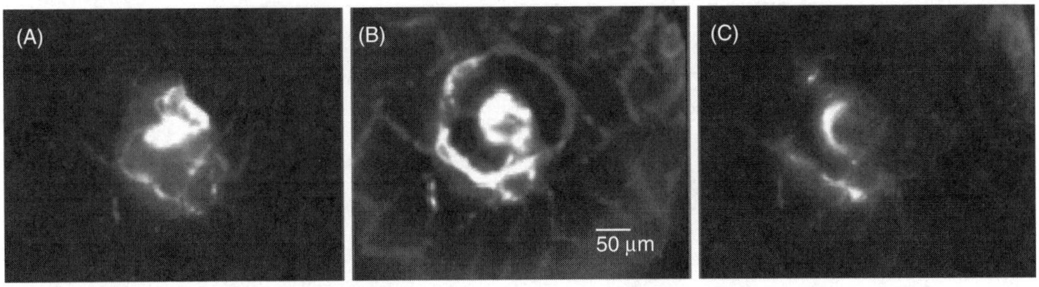

FIGURE 20.13 Sampling mitochondria from a muscle cross section. A series of fluorescence images show (A) the tissue cross section labeled with a microdroplet of 5 μM NAO using a capillary and a micromanipulator, (B) the separation capillary is positioned above the desired fiber and the mitochondria are sampled for 3 s with negative pressure, and (C) the fluorescence image indicates that the sampling was successful since some of the fluorescence in the region above the desired fiber has disappeared. (Reprinted from Ahmadzadeh, H., et al., *Anal. Chem.*, 76, 315, 2004. Copyright 2006. With permission American Chemical Society.)

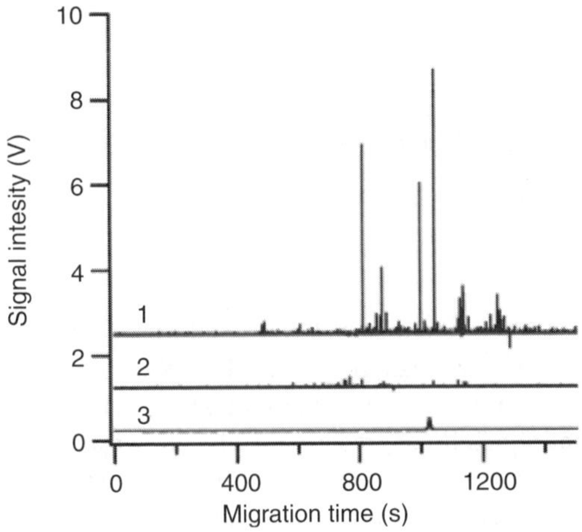

FIGURE 20.14 Detection of individual mitochondria sampled from a muscle tissue cross section. CE-LIF was performed on (1) cross sections that were labeled with 5 μM NAO, (2) an unlabeled cross section, and (3) a solution of the free dye, NAO. The three traces show that mitochondria are detected when labeled with NAO and there is relatively little interference from other cellular components. The small spikes seen in (2) are likely caused by scattering of the excitation source or autofluorescence but only comprise 5% of the peaks detected in (1). Mitochondria were separated at –200 V/cm in a poly(AAP) coated capillary in 250 mM sucrose, 10 mM HEPES, pH 7.4. (Reprinted from Ahmadzadeh, H. et al., *Anal. Chem.*, 76, 315, 2004. Copyright 2006. With permission American Chemical Society.)

untreated muscle fiber. As can be seen, many more mitochondria, presumably interfibrillar, were released from the fiber after exposure to trypsin.

20.5 METHODS DEVELOPMENT GUIDELINES

20.5.1 ISOLATION OF MITOCHONDRIA FROM MAMMALIAN CELL CULTURE

Many complex cellular processes are best studied through the use of an appropriate cell model. Cell cultures provide a relatively stable and virtually unlimited supply of organelles for subcellular analysis. We will discuss organelle preparations that are commonly used in the author's laboratory

TABLE 20.2
Effect of Trypsin on the Number of Mitochondrial Peaks

	Number of Peaks[a]	
Cross Section Number	**Trypsin Treatment**	**No Trypsin Treatment**
1	1782	350
2	2339	256
3	1285	339
Mean	1802	315
RSD (%)	29	16

[a] Conditions for sampling and CE-LIF analysis are the same as for Figure 20.14.

Source: Reprinted from Ahmadzadeh, H., et al., *Anal. Chem.*, 76, 315, 2004. Copyright 2006. With permission American Chemical Society.

for the isolation of mitochondria from human osteosarcoma 143B cells. With minor changes, the procedure detailed below should be applicable to other mammalian cell lines.

20.5.1.1 Cell Lysis

Approximately 5×10^6 143B cells should be harvested and washed three times with an ice-cold, iso-osmotic buffer, for example, 220 mM mannitol, 70 mM sucrose, 0.5 mM EGTA, 2 mM HEPES, pH 7.4 buffer (MSHE buffer). Whenever possible, the sample should be kept on ice and centrifugation performed at 4°C to prevent organelle degradation. Cells are washed by centrifuging at $\sim 700 \times g$ for 10 min, followed by removal of the supernatant and resuspension in MSHE buffer. Approximately 3 mL of cells are then transferred into an ice-cold cell disruption chamber (Parr Instruments, Moline, IL), which is then pressurized to 500 psi. After equilibration for 15 min, the pressure is released as the cells are collected through the discharge port and kept on ice for 5 min. Disruption efficiency is generally >90% and the nitrogen pressure can be adjusted to obtain the desired efficiency. A phase-contrast microscope or trypan blue staining can be used to visually confirm cell disruption.

20.5.1.2 Differential Centrifugation

Following lysis, centrifugation is nearly universally implemented to separate organelle fractions based on their size or density. Centrifugation is popular because it is very versatile; it can be used to fractionate organelles from several liters of tissue homogenates or submilliliter volumes from cell cultures. The simplest centrifugation procedure is differential centrifugation, which produces crudely purified organelle fractions.

Differential centrifugation begins by pelleting nuclei at low centrifugal force (i.e., $600\text{–}1000 \times g$ for 10 min). The resulting nuclear fraction is always contaminated with organelles that have been trapped in the cytoskeletal network, whole cells, and large membrane debris. The postnuclear supernatant is then transferred to a new microcentrifuge tube and the light-organelle fraction can be sedimented at $16{,}000 \times g$ for 10 min, and the supernatant will contain the cytosolic components. The light-organelle fraction is now enriched with mitochondria, but also contains significant amounts of acidic organelles, microsomes, and peroxisomes. For some analyses that use organelle-specific labels, no further purification is necessary. However, if purer organelle fractions are required, the light-organelle fraction can be further fractionated as described below.

20.5.1.3 Density Gradient Centrifugation

The mitochondria in the light-organelle fraction can be further purified using common methods such as continuous gradient and discontinuous gradient centrifugation. While continuous gradients can be used (e.g., sucrose gradient), since the osmolarity changes quite drastically through the gradient, discontinuous gradients are preferred to reduce mitochondrial disruption. Furthermore, the mitochondrial band is harder to distinguish when continuous gradients are used, which reduces the recovery.

Discontinuous density gradient centrifugation separates organelles based on their density and has been described by Storrie and Madden.[84] Unfortunately, the popular medium for density gradient centrifugation, Metrizamide, is no longer produced, and therefore, the existing methods must be redesigned with appropriate new media. We have found success using Histodenz (Sigma, St Louis, MO) following the procedure of Okado-Matsumoto and Fridovich.[85] Briefly, a 50% (w/v) solution of Histodenz is prepared by dissolving in 5 mM Tris–HCl, 1 mM EGTA at pH 7.4. Solutions of 34%, 30%, 25%, 23%, and 20% solutions (w/v) of Histodenz are prepared by diluting the stock solution in 250 mM sucrose, 5 mM Tris−HCl, 1 mM EGTA at pH 7.4. The light-organelle fraction is suspended in ~10 mL of 25% Histodenz and the discontinuous gradient is prepared as shown in Figure 20.15. The samples are then centrifuged at 52,000 × g for 90 min at 4°C. After centrifugation is complete, the bands can be removed sequentially including the band with the purified mitochondria located at the 25/30% interface. Finally, the purity and yield of the mitochondrial preparation can be determined using enzymatic marker assays, such as cytochrome *c* oxidase (mitochondrial marker) and β-galactosidase (lysosomal marker).

20.5.2 Capillary Modifications

Mitochondria interact strongly with the surface of bare fused-silica capillaries, preventing any accurate measurement of electrophoretic mobility. Since mitochondria and the capillary surface both carry a net negative charge, the adsorption is probably not due to any electrostatic interactions, but rather to hydrophobic intermolecular interactions. To prevent the mitochondria from interacting with the capillary surface, coatings can be applied to the bare silica. Among these coatings, AAP has been reported to be more hydrophilic and hydrolytically stable in comparison with acrylamide, dimethyl acrylamide, and *N*-acryloylaminoethoxyethanol.[86] We have found that capillaries

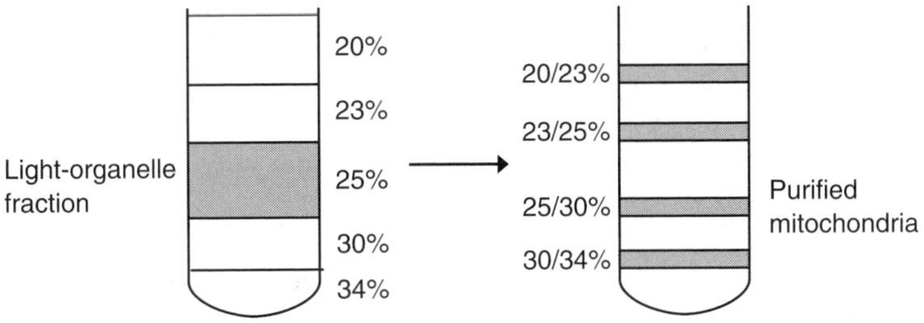

FIGURE 20.15 Density gradient centrifugation. The postnuclear supernatant, suspended in 25% Histodenz (Hz), is layered on top of the 30% and 34% Hz solutions. Then the tube is filled sequentially with the 23% and 20% solutions of Hz. Centrifugation is performed at 52,000×*g* for 90 min at 4°C. The mitochondria will form a band at the 25/30% Hz interface.

coated with poly(AAP), based on Gelfi's procedure,[87] reduce organelle adsorption, and thereby allow reproducible separations; this coating remains stable for several months.

Covalent surface modification with poly(AAP) consist of the following steps:

1. Flush 4–5 m of fused-silica capillary sequentially with 0.1 M KOH for 3 h, 0.1 M HCl for 1 h, MeOH for 1 h with nitrogen pressure (90 psi for all flushes unless stated otherwise)
2. Heat the capillary to 110°C and flush with nitrogen at 10 psi for 8 h
3. Place the capillary in a moisture-free, inert environment (glove box) and do not remove until step (9)
4. Flush the capillary with thionyl chloride for 1 h at 65°C
5. Seal the capillary ends with a gas chromatography (GC) septum and heat for 8 h at 65°C
6. Flush the capillary with freshly distilled tetrahydrofuran (THF) for 30 min at 65°C
7. Prepare a 0.25 M vinylmagnesium bromide solution with freshly distilled THF, flush the capillary with this solution for 1 h and heat for 8 h at 70°C
8. Flush the capillary with freshly distilled THF for 1 h and water for 30 min
9. Cut the capillary into 50 cm pieces
10. Add 5 μL of freshly prepared 10% ammonium persulfate and 2 μL tetramethylethylenediamine (TEMED) to a monomer solution and immediately flush this solution through the capillaries for 5 min
11. Polymerize the capillaries at room temperature for 1 h
12. Flush the capillaries with water for 1 h

Freshly distilled THF must be used in each step to prevent the precipitation of vinylmagnesium bromide in the capillary. To increase the robustness of the capillary modification, steps (4) and (7) may be repeated. The AAP monomer can be synthesized in a single step reaction with >99% yield and >99% purity.[86] The polymerization time and concentrations of ammonium persulfate and TEMED must be optimized for each batch of AAP. The success of the procedure can be checked by measuring the electroosmotic flow, which should be $<0.5 \times 10^{-5}$ cm^2 V^{-1} s^{-1}. The success rate of the procedure described above is ~90%, but decreases dramatically if freshly distilled THF or anhydrous reagents are not used.

As described above, covalently linking AAP to the capillary surface is a tedious and time-consuming task. Procedures that utilize silane chemistry are faster and easier to perform, but the resulting Si–O–Si bonds are prone to nucleophilic cleavage and degrade relatively quickly. Simpler procedures using dynamic coatings or unmodified capillaries would be beneficial and have also been investigated. Under tightly controlled conditions, it was reported that mouse liver mitochondria could be analyzed in unmodified fused-silica capillaries, but this method was not easily adapted to other samples, including rat liver mitochondria, and requires strict attention to buffer composition.[88] Poly(vinyl alcohol) was investigated as a dynamic capillary coating and reduced mitochondrial adsorption to values that were similar to those for poly(AAP) coated capillaries. Although such dynamic coatings show potential, poly(AAP) coated capillaries continue to be the gold standard for organelle separations.

20.5.3 Separation Conditions

To separate analytes from a dissolved organelle fraction, a useful and popular mode of separation is MEKC. As applied to subcellular analysis, the MEKC buffer is often used to dissolve the organelles and acts to separate the components based on their hydrophobicity. A common buffer utilized for this purpose is 10 mM borate, 10 mM sodium dodecyl sulfate (SDS) pH 9.5 (BS buffer), which has been used to separate doxorubicin and its metabolites from nuclear, mitochondrial, and cytosolic-enriched fractions.[25,26] Here, separations were performed from each fraction by injecting a small sample plug that had been dissolved in BS buffer and separating at 400 V/cm in a fused-silica capillary while using BS buffer as the running buffer. To increase the resolution between structurally similar molecules, additive can be used to improve the selectivity. For example, γ-cyclodextrin has been added to BS buffer to resolve doxorubicin and a structurally similar chiral metabolite, doxorubicinol.[89] MEKC buffers that contain detergents (e.g., SDS) also reduce the interactions of molecules with the capillary wall; consequently, capillary surface modifications are rarely needed for an MEKC separation, greatly simplifying the analysis.

To separate intact organelles, iso-osmotic buffers must be used to reduce disruption caused by osmotic shock. For instance, 220 mM mannitol, 70 mM sucrose, 0.5 mM ethylene glycol tetraacetic acid (EGTA), 2 mM 4-(2-hydroxyethyl)-1-piperazineethanesulfonic acid (HEPES), pH 7.4, form a commonly used buffer for isolating mitochondria. The inclusion of mannitol as an osmotic support reduces glycogen binding, whereas the sucrose reduces organelle aggregation. Since organelles will spend several minutes in the running buffer during the separation, it must also be iso-osmotic. Accordingly, a buffer containing 250 mM sucrose, 10 mM HEPES, pH 7.4, has been used in the CE analysis of nuclei,[76] mitochondria,[17] and acidic organelles.[81] For mitochondria, the pH should be adjusted with KOH, as K^+ ions are less permeant to the mitochondrial inner membrane than is Na^+, thereby reducing mitochondrial degradation. Another consideration is that large electric fields (e.g., 600 V/cm) can electrically disrupt organelle membranes.[90] To minimize electric field-induced organelle disruption, smaller electric fields (e.g., –200 V/cm) must then be used to separate organelles.

20.5.4 Organelle Labeling

Instead of performing time-consuming and tedious organelle purification techniques, it is often possible to use organelle-specific fluorescent labels to provide the necessary detection selectivity. Such labels allow crudely purified organelle fractions, prepared from differential centrifugation, to be analyzed. Many organelle-specific fluorescent labels are commercially available and Table 20.3 lists some of the most commonly used labels.

In the biological chemistry field, fusion proteins have recently shown more potential as organelle-specific labels. For subcellular analysis, fusion proteins can be expressed that contain both a fluorescent protein and a subcellular localization sequence. This combination results in organelle-specific labeling that retains the excellent photochemical properties of fluorescent proteins (e.g., intense fluorescence and photostability). We commonly use a commercially available plasmid that codes for a fusion protein of red fluorescent protein (DsRed2) and the mitochondrial targeting sequence from subunit VIII of cytochrome *c* oxidase, which contains the neomycin/kanamycin resistance gene, as described below.

TABLE 20.3
Organelle-Specific Fluorescent Labels

Organelle	Fluorescent Label	Target
Nuclei	Ethidium bromide	Double-stranded DNA
Mitochondria	NAO	Cardiolipin
	Mitotracker Green FM	Reacts with thiol groups in mitochondria
	Rhodamine 123	Membrane potential dependent accumulation
Acidic organelles	Fluorescein dextran conjugate	Endocytosed into acidic compartments
	LysoTracker Green	Weak base is protonated and accumulated
Cytoskeletal network	Fluorescein Phalloidin	Actin
	Oregon Green 488 paclitaxel	Tubulin

Transfection of mammalian cells:

1. Suspend pDsRed2-Mito plasmid (Clontech, Mountain View, CA) in OPTI-MEM medium (Invitrogen, Carlsbad, CA) and incubate with DMRIE-C (Invitrogen) at room temperature for 30 min
2. Wash the 143B human osteosarcoma cells with OPTI-MEM
3. Layer the DMRIE-C-plasmid complex over the cells and incubate for 6 h
4. Add cell culture medium (MEM with 20% calf serum)
5. After 24 h, seed the cells onto a Petri dish at low density and culture in growth medium that contains 500 μg/mL geneticin
6. After 5–6 days, select a single colony of cells that have been successfully transfected
7. Subculture the stably fluorescent transformants in medium [Dulbecco's modified Eagle's medium (DMEM)] containing 250 μg/mL geneticin

In addition to mitochondria, commercial plasmids are also available that code for fusion proteins that incorporate subcellular localization signals for nuclei, endoplasmic reticulum, Golgi apparatus, peroxisomes, and actin filaments.

20.5.5 Detection Methods

Choosing a detection scheme is an important component in designing any CE system, and as shown in the preceding examples (Section 20.4), this can be done in many ways. Owing to the small injection volumes used in CE, often the LOD must be submicromolar or lower; in these instances, achieving the required LOD becomes our chief concern when designing a detection strategy. As illustrated above, LIF, UV absorbance, and electrochemical detection may be employed for subcellular analysis by CE, but other techniques such as mass spectrometry[35] and biological cell sensors[2] have also been used.

20.5.5.1 UV Absorbance

UV absorbance is the most common detection method for any CE system, primarily because nearly all compounds absorb somewhere in the UV region of the spectrum, and also because fused-silica capillaries are transparent to UV radiation. Given these advantages, UV absorbance detectors from HPLC instruments were naturally adopted into CE systems, but a problem arises when these detectors are used for subcellular analysis: subcellular analyses commonly quantify analytes that are below the LOD for UV absorbance. According to Beer's law, the sensitivity of a UV absorbance detector is related to the pathlength of the light. Therefore, UV absorbance has limited applicability to traditional CE separations due to the short path of the typical capillary diameters.

20.5.5.2 Electrochemical Detection

Electrochemical detection is less widely used than UV absorbance or LIF detection, but has been used successfully by Ewing and coworkers to detect neurotransmitters within subcellular structures.[6–11] Electrochemical detection is attractive because it provides a lower LOD than UV absorbance, does not require sample derivatization, and in the case of amperometric detection, can be tuned to specific classes of compounds. Furthermore, unlike UV absorbance in which the sensitivity is dependent on the sample volume (i.e., the pathlength), the LOD of electrochemical detection improves when applied to miniaturized CE systems, since in this case sensitivity is related to contact between the analyte and the electrode surface. In fact, as described above, LODs in the zeptomole range for the subcellular quantification of dopamine have been reported,[7] with the sensitivity of these results making electrochemical detection the rival of many LIF detectors. However, the application of electrochemical detection to subcellular analysis has been limited to the quantification of neurotransmitters.

20.5.5.3 Laser-Induced Fluorescence

Lastly, the most popular detection technique for subcellular analysis is LIF due to the extremely low LODs that can be attained. Mass LODs on the order of attomoles can be achieved with commercial CE-LIF instruments; and custom-built instruments, in conjunction with sheath-flow cuvettes, can reach down to the yoctomole (10^{-24} mole) scale.[91] LIF detection thus allows the quantification of minute amounts of analyte, and is the only detection method that has been used to detect individual organelles. We will therefore describe off-column LIF detection in further detail.

Sheath-flow cuvettes were originally developed for flow cytometry, but were adapted for use with CE by Dovichi.[92–94] Figure 20.16 illustrates LIF detection in conjunction with a sheath-flow cuvette. As displayed, the capillary is surrounded by sheath-flow buffer inside the cuvette. As organelles migrate out of the capillary, they are hydrodynamically focused by the sheath flow into a narrow stream. The excitation source is focused beneath the capillary outlet to excite the organelles once they have been focused. Since the diameter of the sample stream is dependent on the difference between the volumetric flow rates of the sample stream and the sheath flow, the sample stream can be narrowed to the desired width by increasing the sheath volumetric flow.

At this point, the fluorescence from each organelle can be collected by a microscope objective. As with all LIF detectors, care must be taken to choose an objective with high numerical aperture (NA) that can maximize the collection efficiency. The collected fluorescence can then be spatially filtered with a pinhole placed at the image plane to remove out-of-focus scatter from the buffer–cuvette interfaces. The size of the pinhole should be matched with the magnification of the collection objective

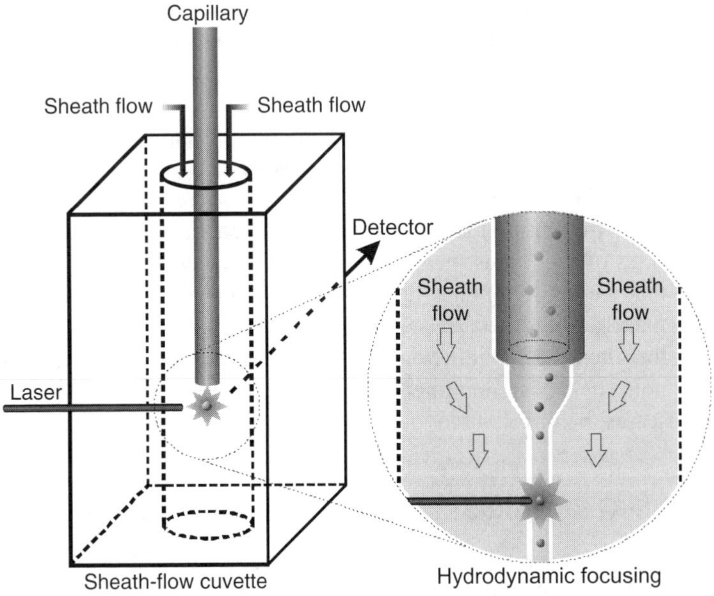

FIGURE 20.16 Diagram of the sheath-flow cuvette used for LIF detection. The capillary is inserted into the sheath-flow cuvette and the laser beam is focused several micrometers away from the outlet. Sheath fluid focuses the sample as it migrates out of the capillary into a narrow stream, in which it is excited by a focused laser beam. The fluorescence is collected at a 90° angle from the excitation source by a high N.A. objective. (Reprinted from Johnson, R. et al., *Anal. Bioanal. Chem.*, In Press. Copyright 2006. With kind permission from Springer Science and Business Media.)

and desired detection area. For instance, if the desired spot diameter is 50 μm and the fluorescence is collected with a 60× objective, a 3-mm diameter pinhole is required. The collected fluorescence must also be spectrally filtered to remove Raman and Raleigh scattering. In general, a long-pass optical filter is used to block Raleigh scattering from the excitation source, and a bandpass filter that encompasses the fluorophore emission to exclude Raman scattering. The Raman scattering bands must be taken into consideration when choosing the emission wavelength range, since a lower LOD may be achieved at wavelengths that are not at the maximum emission of the fluorophore if it coincides with a Raman band. Finally, the fluorescence is detected using a photomultiplier tube or photodiode.

20.5.6 Data Analysis

Electropherograms that contain wide peaks, that is, peaks with baseline widths of several seconds or more, are routinely analyzed. However, in the case of individual organelle detection, the peak widths are much smaller, typically ~80 ms, and the electropherograms nearly always contain wide peaks, baseline drift, and the sharp spikes that correspond to the detection of individual organelles. Special data treatment is required to discern these types of detected events. Toward this end, Jorgenson and coworkers[95] described an important data analysis tool, wherein a moving median filter is used to remove low-frequency background drift from a chromatogram. The median filter is shown mathematically as follows:

$$y_i = \text{median}\,(J_i)$$

$$Ji = \{x_{i-r}, x_{i-r+1}, x_{i-1} x_i, x_{i+1} x_{i+r-1}, x_{i+r}\}$$

$$i = 0, 1, 2, n - 1$$

where J_i is a subset of the input array X (data to be analyzed) centered around the ith element of X, y_i is the ith element of the output array Y, n is the size of the input array, and r is the median filter rank.[95] Consequently, peaks that do not comprise over 50% of the subset are excluded from the data set and only broad peaks remain.

For organelle analysis, median filtering is performed with a large rank (large filter window) so that all the sharp spikes are removed from the electropherogram. Then, the filtered electropherogram, which contains the wide peaks and background drift, is subtracted from the original to yield an electropherogram with a flat baseline and sharp spikes that represent the detection of individual organelles. As reported, the optimum rank value should generally be equal to, or greater than, the baseline peak width of the widest peak of interest.[95] However, in the analysis of individual organelles, we find that the optimum rank value is much greater than the baseline peak width; that is, typically greater than 10 times the average peak width. When the rank is too low, negative peaks will be observed in the filtered data. Following the median filter procedure, organelle events can be discriminated from noise based on their intensities. Frequently, five times the standard deviation in the background is used as the threshold to identify organelles.

20.6 CONCLUDING REMARKS

As shown throughout this chapter, CE is a very versatile and powerful separation technique that has proven useful in subcellular analyses. As in all applications of CE, a trend for subcellular analysis is increasing throughput by using microfluidic devices to reduce separation times. This was recently demonstrated by analyzing fluorescently labeled mitochondria, for the first time, on a microfluidic chip.[96] The analysis allowed a fivefold reduction in analysis time, from 20 to 4 min. To realize the full potential of CE, advances in electrophoretic models must also be made. Models that can accurately predict electrophoretic mobility, as well as explain observed differences, will allow this information to be used more effectively. Finally, as the information that can be gleaned from an electrophoretic separation is limited, the development of complimentary techniques that can be combined with CE will allow more information about subcellular compartments to be discovered.

REFERENCES

1. Miao, H., Rubakhin, S. S., and Sweedler, J. V., Subcellular analysis of D-aspartate, *Anal. Chem.*, 77, 7190–7194, 2005.
2. Sims, C. E., Luzzi, V., and Allbritton, N. L., Localized sampling, electrophoresis, and biosensor analysis of *Xenopus laevis* cytoplasm for subcellular biochemical assays, *Methods Mol. Biol.*, 322, 413–424, 2006.
3. Ahmadzadeh, H., Johnson, R. D., Thompson, L., and Arriaga, E. A., Direct sampling from muscle cross sections for electrophoretic analysis of individual mitochondria, *Anal. Chem.*, 76, 315–321, 2004.
4. Gunasekera, N., Xiong, G., Musier-Forsyth, K., and Arriaga, E., A capillary electrophoretic method for monitoring the presence of alpha-tubulin in nuclear preparations, *Anal. Biochem.*, 330, 1–9, 2004.
5. Johnson, R. D., Navratil, M., Poe, B. G., Xiong, G., Olson, K. J., Ahmadzadeh, H., Andreyev, D., Duffy, C. F., and Arriaga, E. A., Analysis of mitochondria isolated from single cells, *Anal. Bioanal. Chem.*, 387, 107–118, 2006.
6. Woods, L. A., Gandhi, P. U., and Ewing, A. G., Electrically assisted sampling across membranes with electrophoresis in nanometer inner diameter capillaries, *Anal. Chem.*, 77, 1819–1823, 2005.
7. Woods, L. A., Powell, P. R., Paxon, T. L., and Ewing, A. G., Analysis of mammalian cell cytoplasm with electrophoresis in nanometer inner diameter capillaries, *Electroanalysis*, 17, 1192–1197, 2005.
8. Kristensen, H. K., Lau, Y. Y., and Ewing, A. G., Capillary electrophoresis of single cells: Observation of two compartments of neurotransmitter vesicles, *J. Neurosci. Methods*, 51, 183–188, 1994.
9. Chien, J. B., Wallingford, R. A., and Ewing, A. G., Estimation of free dopamine in the cytoplasm of the giant dopamine cell of *Planorbis corneus* by voltammetry and capillary electrophoresis, *J. Neurochem.*, 54, 633–638, 1990.
10. Olefirowicz, T. M. and Ewing, A. G., Dopamine concentration in the cytoplasmic compartment of single neurons determined by capillary electrophoresis, *J. Neurosci. Methods*, 34, 11–15, 1990.
11. Olefirowicz, T. M. and Ewing, A. G., Capillary electrophoresis in 2 and 5 microns diameter capillaries: Application to cytoplasmic analysis, *Anal. Chem.*, 62, 1872–1876, 1990.
12. Ho, A. M. and Yeung, E. S., Capillary electrophoretic study of individual exocytotic events in single mast cells, *J. Chromatogr. A*, 817, 377–382, 1998.
13. Lillard, S. J., Chiu, D. T., Scheller, R. H., Zare, R. N., Rodriguez-Cruz, S. E., Williams, E. R., Orwar, O., Sandberg, M., and Lundqvist, J. A., Separation and characterization of amines from individual atrial gland vesicles of *Aplysia californica*, *Anal. Chem.*, 70, 3517–3524, 1998.
14. Chiu, D. T., Lillard, S. J., Scheller, R. H., Zare, R. N., Rodriguez-Cruz, S. E., Williams, E. R., Orwar, O., Sandberg, M., and Lundqvist, J. A., Probing single secretory vesicles with capillary electrophoresis, *Science*, 279, 1190–1193, 1998.
15. Ahmadzadeh, H., Thompson, L. V., and Arriaga, E. A., On-column labeling for capillary electrophoretic analysis of individual mitochondria directly sampled from tissue cross sections, *Anal. Bioanal. Chem.*, 384, 169–174, 2006.
16. Fuller, K. M. and Arriaga, E. A., Analysis of individual acidic organelles by capillary electrophoresis with laser-induced fluorescence detection facilitated by the endocytosis of fluorescently labeled microspheres, *Anal. Chem.*, 75, 2123–2130, 2003.
17. Fuller, K. M., Duffy, C. F., and Arriaga, E. A., Determination of the cardiolipin content of individual mitochondria by capillary electrophoresis with laser-induced fluorescence detection, *Electrophoresis*, 23, 1571–1576, 2002.
18. Gunasekera, N., Lee, S. W., Kim, S., Musier-Forsyth, K., and Arriaga, E., Nuclear localization of aminoacyl-tRNA synthetases using single-cell capillary electrophoresis laser-induced fluorescence analysis, *Anal. Chem.*, 76, 4741–4746, 2004.
19. Li, H., Sims, C. E., Wu, H. Y., and Allbritton, N. L., Spatial control of cellular measurements with the laser micropipet, *Anal. Chem.*, 73, 4625–4631, 2001.

20. Lochmann, H., Bazzanella, A., and Bachmann, K., Analysis of solutes and metabolites in single plant cell vacuoles by capillary electrophoresis, *J. Chromatogr. A*, 817, 337–343, 1998.
21. Luzzi, V., Lee, C. L., and Allbritton, N. L., Localized sampling of cytoplasm from *Xenopus* oocytes for capillary electrophoresis, *Anal. Chem.*, 69, 4761–4767, 1997.
22. Muscari, C., Pappagallo, M., Ferrari, D., Giordano, E., Capanni, C., Caldarera, C. M., and Guarnieri, C., Simultaneous detection of reduced and oxidized glutathione in tissues and mitochondria by capillary electrophoresis, *J. Chromatogr. B Biomed. Sci. Appl.*, 707, 301–307, 1998.
23. Yoon, S., Ban, E., and Yoo, Y. S., Direct monitoring of the expression of the green fluorescent protein-extracellular signal-regulated kinase 2 fusion protein in transfected cells using capillary electrophoresis with laser-induced fluorescence detection, *J. Chromatogr. A*, 976, 87–93, 2002.
24. McKeon, J. and Khaledi, M. G., Quantitative nuclear and cytoplasmic localization of antisense oligonucleotides by capillary electrophoresis with laser-induced fluorescence detection, *Electrophoresis*, 22, 3765–3770, 2001.
25. Anderson, A. B. and Arriaga, E. A., Subcellular metabolite profiles of the parent ccrf-cem and the derived cem/c2 cell lines after treatment with doxorubicin, *J. Chromatogr. B Analyt. Technol. Biomed. Life Sci.*, 808, 295–302, 2004.
26. Anderson, A. B., Ciriacks, C. M., Fuller, K. M., and Arriaga, E. A., Distribution of zeptomole-abundant doxorubicin metabolites in subcellular fractions by capillary electrophoresis with laser-induced fluorescence detection, *Anal. Chem.*, 75, 8–15, 2003.
27. Cardoso, C. D., Jabor, V. A. P., and Bonato, P. S., Capillary electrophoretic chiral separation of hydroxychloroquine and its metabolites in the microsomal fraction of liver homogenates, *Electrophoresis*, 27, 1248–1254, 2006.
28. Chen, Y., Walsh, R. J., and Arriaga, E. A., Selective determination of the doxorubicin content of individual acidic organelles in impure subcellular fractions, *Anal. Chem.*, 77, 2281–2287, 2005.
29. Clohs, L. and Wong, J., Validation of a capillary electrophoresis assay for assessing the metabolic stability of verapamil in human liver microsomes, *J. Capillary Electrophor.*, 7, 113–117, 2002.
30. Liu, H. C., Wang, N., Yu, Y., and Hou, Y. N., Stereoselectivity in *trans*-tramadol metabolism and *trans-O*-demethyltramadol formation in rat liver microsomes, *Acta Pharmacol. Sin.*, 24, 85–90, 2003.
31. Sakai-Kato, K., Kato, M., and Toyo'oka, T., On-line drug-metabolism system using microsomes encapsulated in a capillary by the sol-gel method and integrated into capillary electrophoresis, *Anal. Biochem.*, 308, 278–284, 2002.
32. Ward, E. M., Smyth, M. R., O'Kennedy, R., and Lunte, C. E., Application of capillary electrophoresis with pH-mediated sample stacking to analysis of coumarin metabolites in microsomal incubations, *J. Pharm. Biomed. Anal.*, 32, 813–822, 2003.
33. Lin, C., Cotton, F., Boutique, C., Dhermy, D., Vertongen, F., and Gulbis, B., Capillary gel electrophoresis: Separation of major erythrocyte membrane proteins, *J. Chromatogr. B Biomed. Sci. Appl.*, 742, 411–419, 2000.
34. Prange, A., Schaumloffel, D., Bratter, P., Richarz, A. N., and Wolf, C., Species analysis of metallothionein isoforms in human brain cytosols by use of capillary electrophoresis hyphenated to inductively coupled plasma-sector field mass spectrometry, *Fresenius J. Anal. Chem.*, 371, 764–774, 2001.
35. Moini, M. and Huang, H., Application of capillary electrophoresis/electrospray ionization-mass spectrometry to subcellular proteomics of *Escherichia coli* ribosomal proteins, *Electrophoresis*, 25, 1981–1987, 2004.
36. Presley, A. D., Fuller, K. M., and Arriaga, E. A., Mitotracker green labeling of mitochondrial proteins and their subsequent analysis by capillary electrophoresis with laser-induced fluorescence detection, *J. Chromatogr. B Analyt. Technol. Biomed. Life Sci.*, 793, 141–150, 2003.
37. Kustos, T., Kustos, I., Gonda, E., Kocsis, B., Szabo, G., and Kilar, F., Capillary electrophoresis study of outer membrane proteins of pseudomonas strains upon antibiotic treatment, *J. Chromatogr. A*, 979, 277–284, 2002.
38. Fazal, M. A., Palmer, V. R., and Dovichi, N. J., Analysis of differential detergent fractions of an AtT-20 cellular homogenate using one- and two-dimensional capillary electrophoresis, *J. Chromatogr. A*, 1130, 182–189, 2006.

39. Heidrich, H. G., Stahn, R., and Hannig, K., The surface charge of rat liver mitochondria and their membranes. Clarification of some controversies concerning mitochondrial structure, *J. Cell Biol.*, 46, 137–150, 1970.
40. Henning, R. and Heidrich, H. G., Membrane lipids of rat liver lysosomes prepared by free-flow electrophoresis, *Biochim. Biophys. Acta*, 345, 326–335, 1974.
41. Stahn, R., Maier, K. P., and Hannig, K., A new method for the preparation of rat liver lysosomes. Separation of cell organelles of rat liver by carrier-free continuous electrophoresis, *J. Cell Biol.*, 46, 576–591, 1970.
42. Valdivia, E., Pease, B., Gabel, C., and Chan, V., Electrophoresis of isolated mitochondria, *Anal. Biochem.*, 51, 146–151, 1973.
43. Kamo, N., Muratsugu, M., Kurihara, K., and Kobatake, Y., Change in surface charge density and membrane potential of intact mitochondria during energization, *FEBS Lett.*, 72, 247–250, 1976.
44. Sengelov, H., Nielsen, M. H., and Borregaard, N., Separation of human neutrophil plasma membrane from intracellular vesicles containing alkaline phosphatase and NADPH oxidase activity by free flow electrophoresis, *J. Biol. Chem.*, 267, 14912–14917, 1992.
45. Gottlieb, M. H., Steer, C. J., Steven, A. C., and Chrambach, A., Applicability of agarose gel electrophoresis to the physical characterization of clathrin-coated vesicles, *Anal. Biochem.*, 147, 353–363, 1985.
46. Tulp, A., Verwoerd, D., Fernandez-Borja, M., Neefjes, J., and Hart, A. A., High resolution density gradient electrophoresis of cellular organelles, *Electrophoresis*, 17, 173–178, 1996.
47. Volkl, A., Mohr, H., Weber, G., and Fahimi, H. D., Isolation of rat hepatic peroxisomes by means of immune free flow electrophoresis, *Electrophoresis*, 18, 774–780, 1997.
48. Weber, G. and Bocek, P., Optimized continuous flow electrophoresis, *Electrophoresis*, 17, 1906–1910, 1996.
49. Hasan, Z. and Pieters, J., Subcellular fractionation by organelle electrophoresis: Separation of phagosomes containing heat-killed yeast particles, *Electrophoresis*, 19, 1179–1184, 1998.
50. Hunter, R. J., *Foundations of Colloid Science*, 2nd ed. Oxford University Press, Oxford, 2001.
51. Overbeek, J. T. G. and Wiersema, P. H., *The Interpretation of Electrophoretic Mobilities*, Academic Press, New York, 1967.
52. Booth, F., Theory of electrokinetic effects, *Nature*, 161, 83–86, 1948.
53. Wiersema, P. H., Loeb, A. L., and Overbeek, J. T. G., Calculation of the electrophoretic mobility of a spherical colloid particle, *J. Coll. Interf. Sci.*, 22, 78–99, 1966.
54. Overbeek, J. T. G., Quantitative interpretation of the electrophoretic velocity of colloids, *Adv. Coll. Sci.*, 3, 97–134, 1950.
55. O'Brien, R. W., The solution of the electrokinetic equations for colloidal particles with thin double layers, *J. Coll. Interf. Sci.*, 92, 204–216, 1983.
56. O'Brien, R. W. and Ward, D. N., The electrophoresis of a spheroid with a thin double layer, *J. Coll. Interf. Sci.*, 121, 402–413, 1988.
57. O'Brien, R. W. and Hunter, R. J., The electrophoretic mobility of large colloidal particles, *Can. J. Chem.*, 59, 1878–1887, 1981.
58. O'Brien, R. W. and White, L. R., Electrophoretic mobility of a spherical colloidal particle, *J. Chem. Soc., Faraday Trans.*, 77, 1607–1626, 1978.
59. Von Smoluchowski, M., *Bull. Acad. Sci. Cracovie*, 182, 182–200, 1903.
60. Hückel, E., Die Kataphorese der Kugel, *Physik Z.*, 25, 204, 1924.
61. Henry, D. C., The cataphoresis of suspended particles. I. The equation of cataphoresis, *Proc. Roy. Soc. Lond.*, A133, 106, 1931.
62. Ohshima, H., A simple expression of Henry's function for the retardation effect in electrophoresis of spherical colloidal particles, *J. Coll. Interf. Sci.*, 168, 269–271, 1994.
63. Pysher, M. D. and Hayes, M. A., Examination of the electrophoretic behavior of liposomes, *Langmuir*, 20, 4369–4375, 2004.
64. Pysher, M. D. and Hayes, M. A., Effects of deformability, uneven surface charge distributions, and multipole moments on biocolloid electrophoretic migration, *Langmuir*, 21, 3572–3577, 2005.
65. Gaucher, S. P., Taylor, S. W., Fahy, E., Zhang, B., Warnock, D. E., Ghosh, S. S., and Gibson, B. W., Expanded coverage of the human heart mitochondrial proteome using multidimensional liquid chromatography coupled with tandem mass spectrometry, *J. Proteome Res.*, 3, 495–505, 2004.

66. Lescuyer, P., Strub, J. M., Luche, S., Diemer, H., Martinez, P., Van Dorsselaer, A., Lunardi, J., and Rabilloud, T., Progress in the definition of a reference human mitochondrial proteome, *Proteomics*, 3, 157–167, 2003.
67. Sickmann, A., Reinders, J., Wagner, Y., Joppich, C., Zahedi, R., Meyer, H. E., Schonfisch, B., et al., The proteome of *Saccharomyces cerevisiae* mitochondria, *Proc. Natl. Acad. Sci. USA*, 100, 13207–13212, 2003.
68. Strong, R., Nakanishi, T., Ross, D., and Fenselau, C., Alterations in the mitochondrial proteome of adriamycin resistant mcf-7 breast cancer cells, *J. Proteome Res.*, 5, 2389–2395, 2006.
69. Hu, S., Michels, D. A., Fazal, M. A., Ratisoontorn, C., Cunningham, M. L., and Dovichi, N. J., Capillary sieving electrophoresis/micellar electrokinetic capillary chromatography for two-dimensional protein fingerprinting of single mammalian cells, *Anal. Chem.*, 76, 4044–4049, 2004.
70. Michels, D. A., Hu, S., Dambrowitz, K. A., Eggertson, M. J., Lauterbach, K., and Dovichi, N. J., Capillary sieving electrophoresis-micellar electrokinetic chromatography fully automated two-dimensional capillary electrophoresis analysis of *Deinococcus radiodurans* protein homogenate, *Electrophoresis*, 25, 3098–3105, 2004.
71. Michels, D. A., Hu, S., Schoenherr, R. M., Eggertson, M. J., and Dovichi, N. J., Fully automated two-dimensional capillary electrophoresis for high sensitivity protein analysis, *Mol. Cell. Proteomics*, 1, 69–74, 2002.
72. Hu, S., Jiang, J., Cook, L. M., Richards, D. P., Horlick, L., Wong, B., and Dovichi, N. J., Capillary sodium dodecyl sulfate-dalt electrophoresis with laser-induced fluorescence detection for size-based analysis of proteins in human colon cancer cells, *Electrophoresis*, 23, 3136–3142, 2002.
73. Ramsby, M. L., Makowski, G. S., and Khairallah, E. A., Differential detergent fractionation of isolated hepatocytes: Biochemical, immunochemical and two-dimensional gel electrophoresis characterization of cytoskeletal and noncytoskeletal compartments, *Electrophoresis*, 15, 265–277, 1994.
74. Cardoso, C. D. and Bonato, P. S., Enantioselective analysis of the metabolites of hydroxychloroquine and application to an *in vitro* metabolic study, *J. Pharm. Biomed. Anal.*, 37, 703–708, 2005.
75. Gunasekera, N., Musier-Forsyth, K., and Arriaga, E., Electrophoretic behavior of individual nuclear species as determined by capillary electrophoresis with laser-induced fluorescence detection, *Electrophoresis*, 23, 2110–2116, 2002.
76. Gunasekera, N., Olson, K. J., Musier-Forsyth, K., and Arriaga, E. A., Capillary electrophoretic separation of nuclei released from single cells, *Anal. Chem.*, 76, 655–662, 2004.
77. Anderson, A. B., Xiong, G., and Arriaga, E. A., Doxorubicin accumulation in individually electrophoresed organelles, *J. Am. Chem. Soc.*, 126, 9168–9169, 2004.
78. Fuller, K. M. and Arriaga, E. A., Capillary electrophoresis monitors changes in the electrophoretic behavior of mitochondrial preparations, *J. Chromatogr. B Analyt. Technol. Biomed. Life Sci.*, 806, 151–159, 2004.
79. Meany, D. L., Poe, B. G., Navratil, M., Moraes, C. T., and Arriaga, E. A., Superoxide released into the mitochondrial matrix, *Free Radic. Biol. Med.*, 41, 950–959, 2006.
80. Strack, A., Duffy, C. F., Malvey, M., and Arriaga, E. A., Individual mitochondrion characterization: A comparison of classical assays to capillary electrophoresis with laser-induced fluorescence detection, *Anal. Biochem.*, 294, 141–147, 2001.
81. Chen, Y. and Arriaga, E. A., Individual acidic organelle pH measurements by capillary electrophoresis, *Anal. Chem.*, 78, 820–826, 2006.
82. Schindler, M., Grabski, S., Hoff, E., and Simon, S. M., Defective pH regulation of acidic compartments in human breast cancer cells (mcf-7) is normalized in adriamycin-resistant cells (mcf-7adr), *Biochemistry*, 35, 2811–2817, 1996.
83. Woods, L. A. and Ewing, A. G., Etched electrochemical detection for electrophoresis in nanometer inner diameter capillaries, *Chemphyschem*, 4, 207–211, 2003.
84. Storrie, B. and Madden, E. A., Isolation of subcellular organelles, *Methods Enzymol.*, 182, 203–225, 1990.
85. Okado-Matsumoto, A. and Fridovich, I., Subcellular distribution of superoxide dismutases (SOD) in rat liver: Cu, Zn-SOD in mitochondria, *J. Biol. Chem.*, 276, 38388–38393, 2001.
86. Simo-Alfonso, E., Gelfi, C., Sebastiano, R., Citterio, A., and Righetti, P. G., Novel acrylamido monomers with higher hydrophilicity and improved hydrolytic stability: I. Synthetic route and product characterization, *Electrophoresis*, 17, 723–731, 1996.

87. Gelfi, C., Curcio, M., Righetti, P. G., Sebastiano, R., Citterio, A., Ahmadzadeh, H., and Dovichi, N. J., Surface modification based on Si-O and Si-C sublayers and a series of *n*-substituted acrylamide top-layers for capillary electrophoresis, *Electrophoresis*, 19, 1677–1682, 1998.
88. Whiting, C. and Arriaga, E. A., CE-LIF analysis of mitochondria using uncoated and dynamically coated capillaries, *Electrophoresis*, 27, 4523–4531, 2006.
89. Eder, A. R., Chen, J. S., and Arriaga, E. A., Separation of doxorubicin and doxorubicinol by cyclodextrin-modified micellar electrokinetic capillary chromatography, *Electrophoresis*, 27, 3263–3270, 2006.
90. Zimmerman, U. and Neil, G., *Electromanipulation of Cells*, CRC Press, New York, 1996.
91. Chen, D. Y. and Dovichi, N. J., Yoctomole detection limit by laser-induced fluorescence in capillary electrophoresis, *J. Chromatogr. B Biomed. Appl.*, 657, 265–269, 1994.
92. Dovichi, N. J., Martin, J. C., Jett, J. H., and Keller, R. A., Attogram detection limit for aqueous dye samples by laser-induced fluorescence, *Science*, 219, 845–847, 1983.
93. Cheng, Y. F. and Dovichi, N. J., Subattomole amino acid analysis by capillary zone electrophoresis and laser-induced fluorescence, *Science*, 242, 562–564, 1988.
94. Cheng, Y. F., Wu, S., Chen, D. Y., and Dovichi, N. J., Interaction of capillary zone electrophoresis with a sheath flow cuvette detector, *Anal. Chem.*, 62, 496–503, 1990.
95. Moore, A. W., Jr. and Jorgenson, J. W., Median filtering for removal of low-frequency background drift, *Anal. Chem.*, 65, 188–191, 1993.
96. Duffy, C. F., MacCraith, B., Diamond, D., O'Kennedy, R., and Arriaga, E. A., Fast electrophoretic analysis of individual mitochondria using microchip capillary electrophoresis with laser induced fluorescence detection, *Lab. Chip*, 6, 1007–1011, 2006.

21 Chemical Cytometry: Capillary Electrophoresis Analysis at the Level of the Single Cell

Colin Whitmore, Kimia Sobhani, Ryan Bonn, Danqian Mao, Emily Turner, James Kraly, David Michels, Monica Palcic, Ole Hindsgaul, and Norman J. Dovichi

CONTENTS

21.1 INTRODUCTION

The cell is the organizing unit of life, and characterization of the composition of cells is of value in both fundamental research and in clinical applications. Classic analytical methods are often used to characterize cellular homogenates produced from thousands to billions of cells. These methods are

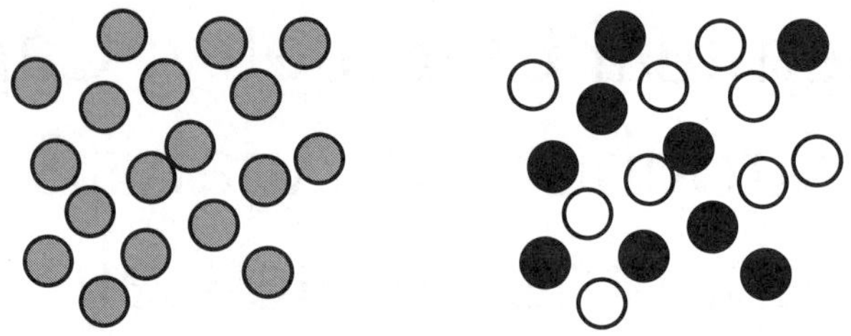

FIGURE 21.1 Comparison of two cellular populations, one homogeneous, with a key component present at intermediate levels, and the other heterogeneous, with some cells lacking the compound while others have that compound in high abundance.

unable to distinguish between a homogeneous population of intermediate composition and a heterogeneous population made up of a mixture of high- and low-expression cells. Figure 21.1 presents two samples that have the same average composition. In one case, the population is homogeneous and in the other is highly heterogeneous; the responses of these populations to a stimulus are likely to be quite different.

In many cases, it is clearly useful to study single cells, rather than an average of the population. Classically, flow and image cytometry are used to measure physical properties and chemical composition of single cells, and are very important in both clinical and fundamental research. In these methods, affinity reagents are usually used to determine the distribution of a handful of components in a cellular population. While extremely powerful, these methods are limited in the number of components resolved. More importantly, classic cytometry methods can only detect those components for which affinity reagents are available; the unexpected is undetectable.

In contrast, chemical cytometry employs powerful ultrasensitive analytical methods to characterize the composition of single cells. In principle, chemical cytometry can resolve hundreds of components from a single cell.

Our work in this field is motivated by three broad applications. First, this technology will be of value whenever characterizing complex tissues, such as neurons and cells of the immune system, where there is a tremendous cell-to-cell heterogeneity in composition. For example, one can imagine providing a detailed cell-by-cell map of selected brain structures or monitoring the evolution in a T-cell population in response to an infection. Second, this technology will be to be of value in characterizing the molecular changes associated with the maturation of stem and precursor cells into their differentiated progeny; in an ideal experiment, an organism such as *Caenorhabditis elegans* would be microdissected so that each cell can be analyzed as the organism develops from a single-cell embryo to an adult. Finally, we believe that chemical cytometry may prove of value in improving the accuracy of cancer prognosis and diagnosis; just as cell-to-cell variation in ploidy can have prognostic value, determination of cell-to-cell heterogeneity in other cellular components may improve prognostic accuracy.

21.2 BACKGROUND

21.2.1 The Challenge

Chemical cytometry employs modern analytical tools to characterize single cells. Analysis of the content of a single cell requires extremely sensitive instrumentation. Consider a typical mammalian cell that has a 10-μm diameter, 500-fL volume, and 500-pg mass. If that cell is 10% protein by weight, then the cell contains 50 pg of protein. Assuming an average molar mass of 30,000 g/mol, a single cell contains about 2 fmol of protein. The fraction of the proteome expressed by a single cell is unknown; perhaps 10,000 different proteins are present in each cell, with an average of 200 zmol

(1 zmol = 10^{-21} mol = 600 copies) per protein. As we show below, there is a large distribution in protein expression within a cell. In addition, the cell contains ~15 fmol of amino acids and other biogenic amines. Chemical cytometry requires exquisite detection sensitivity.

21.2.2 Classic Cytometry

Cytometry is the measurement of physical properties and chemical composition of single cells, and a wonderful review of the field is given by Howard Shapiro.[1] Cytometry was pioneered by Caspersson,[2,3] who developed microspectrophotomers that were used to measure ultraviolet (UV) absorbance of single cells at 260 nm to characterize nucleic acids and at 280 nm to characterize protein content. This work was particularly important because it led to the discovery that DNA content of cells doubled during the cell cycle, strengthening the hypothesis that DNA contained genetic material.

Image cytometry had its genesis in Caspersson's instrumentation, where a scanning stage was used to translate a sample through the focal volume of the microspectrophotometer, recording transmission data from a large number of cells. Since then, many technological improvements have been made to measure transmission, scatter, and fluorescence from cells and tissues.

Flow cytometers record optical and electrical signals from single cells flowing in a stream. The earliest instruments were based on light scatter and electrical impedance and were used for cell counting, replacing the hemocytometer. More sophisticated instruments record fluorescence intensity in one or more spectral bands and can characterize perhaps a half-dozen components in a single cell. The most sophisticated instruments are cell sorters that employ an electrical signal to deflect cells of interest from the flowing stream into a receiving reservoir for subsequent culture or analysis. Modern instruments can measure the signal from 200,000 cells/s and sort 50,000 cells/s.[4] Classic cytometry typically employs fluorescence to characterize a few components per cell; by use of several excitation wavelengths and multiple detection channels, instruments can characterize up to a dozen parameters per cell.[5]

Fluorescence-based assays employ reagents to characterize a wide range of cellular components, including nucleic acids, proteins, biogenic amines, carbohydrates, and lipids. These assays are often based on affinity probes, such as oligonucleotides, antibodies, aptamers, and lectins, which are used to characterize a specific component within the cell. In addition, fluorogenic substrates are used to characterize enzyme activity in single cells.

21.2.3 Chemical Cytometry

Chemical cytometry employs modern analytical tools to characterize the chemical composition of single cells. As an example, mass spectrometry has been used to monitor hemoglobin in single erythrocytes and neuropeptides in giant neurons. The information content of the mass spectrometer's signal is invaluable in component identification. Caprioli and coworkers has reported the use of matrix-assisted laser desorption/ionization time of flight (MALDI-TOF) to analyze tissue samples.[6] Over 800 components are detected from a tissue section.[7] Li and coworkers has reported the detection of hemoglobin from a single erythrocyte by MALDI-TOF mass spectrometry(MS).[8] Each cell contains about 450 amol of hemoglobin, and only extremely highly expressed proteins are detected and identified in a single cell. Sweedler and coworkers[9] has reported the use of MALDI-TOF to analyze neuropeptides in single cells.

Several separation methods have been developed for chemical cytometry. In an early example, silk fibers were used for the electrophoretic determination of 100 pg of RNA contained within a single cell, employing a UV microscope to detect the components.[10] Kennedy et al.[11] inaugurated the modern era of chemical cytometry by using open tubular capillary chromatography for the analysis of biogenic amines in a single giant neuron from a snail.

21.2.4 Chemical Cytometry of Proteins, Biogenic Amines, and Metabolic Cascades by One- and Two-Dimensional Capillary Electrophoresis

21.2.4.1 Introduction

A rich literature has developed for the use of capillary electrophoresis (CE) for chemical cytometry. Wallingford and Ewing[12] reported the use of a capillary to sample the internal contents of a single giant neuron for electrophoresis of biogenic amines. Hogan and Yeung[13] reported the use of a specific label to derivatize thiols in individual erythrocytes; once the derivatization reaction was completed, a single cell was injected into a capillary, lysed, and the contents separated by CE. Gilman and Ewing[14] reported the on-column labeling of amines from a cell that had been injected into and lysed within a capillary; the capillary was used for electrophoretic separation of amines. Allbritton and coworkers[15] reported the use of a pulsed laser to lyse a cell before analysis of kinase activities by CE. Han and Lillard[16] reported the use of cell synchronization based on the shakeoff method for the characterization of RNA synthesis in single cells as a function of cell cycle. Zare and coworkers[17] reported a microfabricated device to automate cell lysis, labeling, and separation.

21.2.4.2 High-Sensitivity Electrophoresis

Capillary electrophoresis employing a postcolumn sheath-flow cuvette has produced outstanding detection limits, approaching single molecule levels in favorable cases.[18,19] We have used chemical cytometry based on CE to study proteins and biogenic amines in a wide range of cell types, including the HT29 colon cancer cell line; the MCF7 breast cancer cell line; the AtT20 adrenal gland cancer cell line; the MC3T3 osteoprecursor cell line; the hTERT telomerase-expressing Barrett's esophagus cell line; the A549 lung cancer cell line; the SupT1 T-cell line; RAW 264.7 macrophage cells; *Deinococcus radiodurans* cells; primary neurons, macrophages, and T-cells; and single-cell *C. elegans* and mouse embryos.

21.2.4.3 Proteins and Biogenic Amines

In chemical cytometry of protein and biogenic amines, cells are lysed and their constituents are labeled with a fluorescent or fluorogenic reagent before separation by CE and detection by laser-induced fluorescence. The use of fluorescent reagents to label proteins will target the most highly abundant proteins in the cell. The best data on the protein expression in a cellular homogenate were generated by Ghaemmaghami and coworkers,[20] who created a tick anticoagulant peptide (TAP) fusion of each open reading frame (ORF) in yeast; these cells were homogenized and assayed by Western blotting. Figure 21.2 presents a cumulative sum of protein abundance, starting from the most abundant protein. A few proteins are present at very high levels, and dominate the total population, while a large fraction (25%) of ORFs generated no detectable signal. The most abundant protein accounts for 3.4% of the total, and over 50% of the total protein content is due to the 100 most abundant proteins in a yeast cell. Chemical cytometry methods that target all proteins in a cell will be dominated by a relatively small number of highly abundant proteins.

We have also downloaded the ORF data for yeast;[21] this database includes the predicted sequence of 6057 ORFs that are 100 residues or longer. Note that these data are for the full-length ORF. As result, all ORFs contain a methionine residue, which corresponds to the start codon. Post-translational proteolysis will remove the leader sequence from the N-terminus of the mature protein. Figure 21.3 presents a histogram of the abundances of each amino acid in those ORFs. The figure also reports the number of ORFs that do not contain that amino acid (n_0) and the percentage of all amino acids. For example, only 9 out of 6057 ORFs contain no lysine residues. The three most abundant amino acids in yeast are leucine, serine, and lysine, which account for 9.6%, 9.0%, and 7.3% of all amino acids, whereas the three least abundant amino acids are tryptophan, cysteine, and methionine, which account for 1.0%, 1.3%, and 2.1% of all amino acids; the methionine abundance in mature proteins

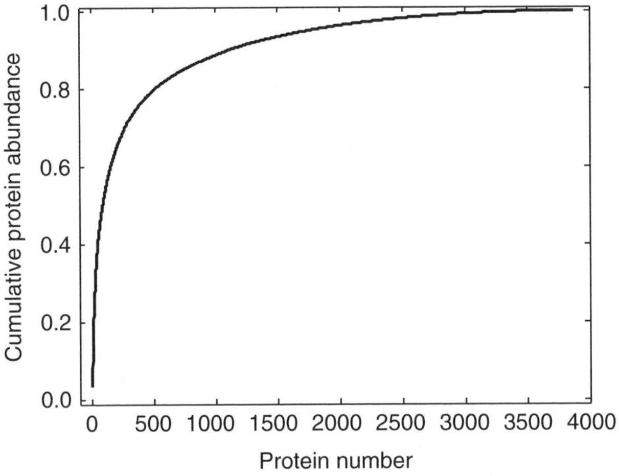

FIGURE 21.2 Cumulative abundance of proteins in yeast, starting from the most abundant protein. The 100 most abundant proteins represent 50% of the total protein content in this organism. Data obtained from Reference 27.

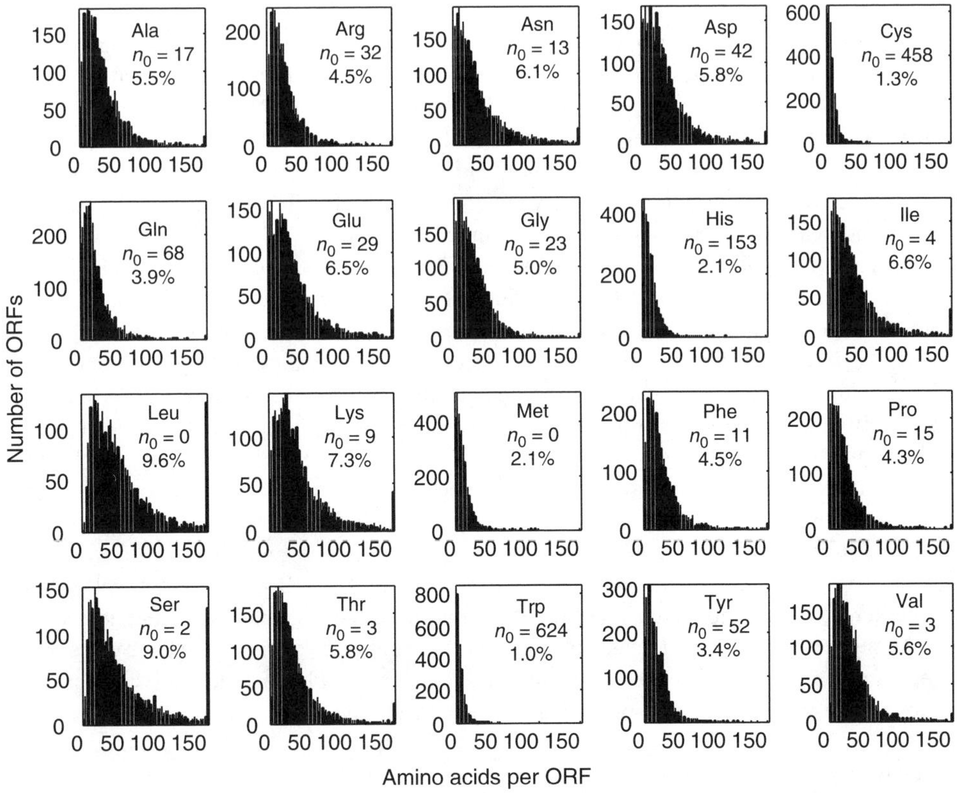

FIGURE 21.3 Amino acid distribution in yeast ORFs. A total of 6057 ORFs coding for 100 or more amino acids were analyzed. The histogram of amino acid abundance in each ORF is plotted. The number of ORFs lacking that amino acid (n_0) is listed. Finally, the percentage of each amino acid in the yeast ORFs is listed as the bottom line.

will be lower than predicted from these data because of the loss of the start codon during cleavage of the signal sequence from the precursor.

Most labeling chemistry targets primary amines, which are the ε-amine of lysine residues and unblocked N-termini. Primary amines are also found on amino acids, aminolipids, aminosugars, neurotransmitters, and other biogenic amines, and understanding their role is of interest in physiology, molecular biology, and cell biology. These compounds make up 5–10% of the content of a cell, and are important mediators of a wide range of cellular processes. Their study is extremely difficult by classic cytometry methods, such as flow cytometry, because of the lack of specific affinity reagents.

21.2.4.4 Fluorescence Labeling

While it is possible to detect proteins based on the native fluorescence of tryptophan, that amino acid is the least abundant in yeast, representing 1.0% of the total amino acid content of yeast. Excitation of tryptophan requires expensive and temperamental UV lasers. Instead, it is usually more convenient to label the protein with a fluorophore that is excited by inexpensive and reliable lasers that operate in the visible portion of the spectrum.

21.2.4.4.1 Chemical Derivatization

As noted above, lysine is one of the most abundant amino acids, and most chemical cytometry experiments rely on reagents that target the primary amine found on lysine residues and the N-terminus. Classic fluorescent reagents, such as fluorescein, rhodamines, Cy-dyes, and bodipy, are not useful in chemical cytometry. While they create highly fluorescent products, unreacted reagent and trace levels of fluorescent impurities create a sea of background signals that can swamp the fluorescence signal from analyte. For example, if 1 nL of a 10^{-4} M labeling reagent is used in an experiment, then there will be 100 amol of impurities present at the 0.1% level, and impurities present at the part-per-million level will contribute 100 zmol, which corresponds to the average protein level in a cell. These trace impurities are inevitably present, many are highly fluorescent, and they create a sea of background peaks that obscures target compounds.

We instead employ fluorogenic reagents, such as 3-(2-furoyl) quinoline-2-carboxaldehyde (FQ), in our experiments. These compounds are nonfluorescent until they react with a primary amine in the presence of a nucleophile. Fluorogenic reagents have very few fluorescent impurities and generate a very low background signal that does not interfere in the chemical cytometry experiment.

There is another subtle issue in fluorescent labeling of proteins. Heroic efforts are required to completely label all lysine residues in a protein, and these efforts are impractical in single-cell analysis.[22] Instead, the reaction targets the most labile residues, and usually creates a complex mixture of products; there are $2^N - 1$ fluorescent products generated from a protein with N primary amines.[23] For example, ovalbumin has 20 lysine residues. Incomplete labeling can result in the production of $2^{20} - 1 = 1048575$ different fluorescent products and each product will have a different mobility during electrophoresis, creating a complex electropherogram.[23–25] Both this group and Whitesides and coworkers[26–28] have noted that addition of an anionic surfactant, such as sodium dodecyl sulfate (SDS), causes the complex set of peaks to collapse into a single peak, which can have extraordinarily high separation efficiency that exceeds 10^5 plates. However, this phenomenon only appears to occur when acylation produces a neutral product and when an anionic surfactant is present.

An alternative strategy is to use very mild reaction conditions, so that few proteins have more than one fluorescent label. However, these conditions lead, by necessity, to inefficient labeling.

21.2.4.4.2 Labeling by Genetic Engineering

Chemical labeling is best used to characterize the most abundant proteins in a cell. Low abundance proteins tend to be buried under the signal generated by high abundance components. Fluorescently labeled antibodies are commonly used to label specific components in flow cytometry, but cellular autofluorescence limits their use to characterize the lowest abundance proteins. In addition, immunoassay seldom provides information on post-translational modifications on the target protein.

Genetic engineering can be used to introduce a label with high specificity. In this case, the label is a fluorescent protein, such as green fluorescent protein (GFP). Labeling is performed by fusing the gene for GFP with the gene of the target protein. Transcription and translation of the target gene leads to the production of a fusion protein that is labeled with GFP.

GFP is a relatively highly fluorescent protein that can be detected with high sensitivity.[29] The study of fusion proteins by microscopy and flow cytometry is common. However, there are three issues associated with study of GFP. First, autofluorescence from endogenous components provides a background fluorescence signal that limits the detection of trace level proteins. Second, GFP fluorescence reflects the structure of GFP itself, and provides no information on post-translational modifications to the target protein, such as phosphorylation or glycosylation. Third, GFP is a relatively proteolysis-resistant protein. As a result, the target protein may have undergone proteolytic destruction while GFP remains intact. Detection of fluorescence from GFP does not necessarily mean that the target protein is present; it simply means that the target protein had been expressed.

The use of chemical cytometry for the study of GFP fusion proteins deals with these issues.[29] Electrophoretic conditions can be manipulated to minimize overlap of GFP and autofluorescent components, so that GFP can be detected on a very low background, which improves detection limits. The migration time of the fusion will reflect any post-translational modifications to the protein, including proteolysis of the target protein.

21.3 PRACTICAL APPLICATIONS

21.3.1 Chemical Cytometry of Biogenic Amines and Proteins Using One-Dimensional Capillary Electrophoresis

Figure 21.4 presents a one-dimensional capillary sieving electrophoresis (CSE) analysis of the proteins from a human primary T-cell. In this separation, the cell was injected into the capillary, lysed, and primary amines were labeled with FQ. Components were separated by CSE. The electropherogram consists of a set of ~25 peaks, some of which are extraordinarily sharp; for example, the component that migrates at 14.2 min generates a peak with 2.8 million theoretical plates (7.7 million plates/m), whereas most peaks generate a few hundred thousand theoretical plates.

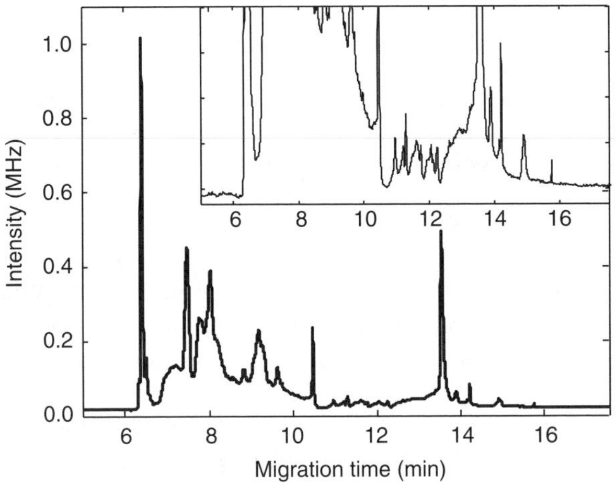

FIGURE 21.4 One-dimensional CSE analysis of a single human T-cell. The inset expands the fluorescent signal to highlight lower amplitude components.

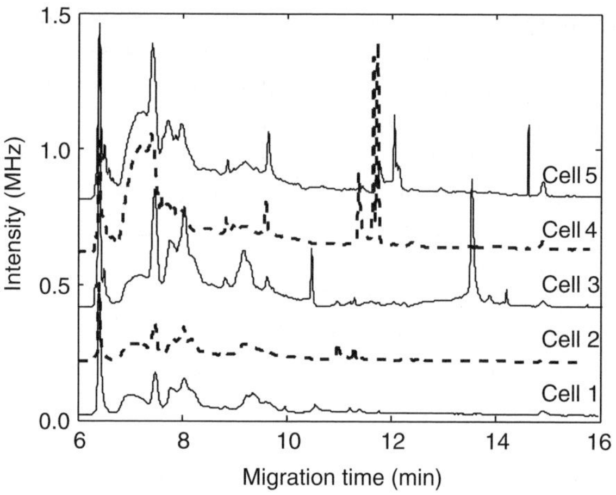

FIGURE 21.5 CSE analysis of five human T-cells. Curves are offset for clarity.

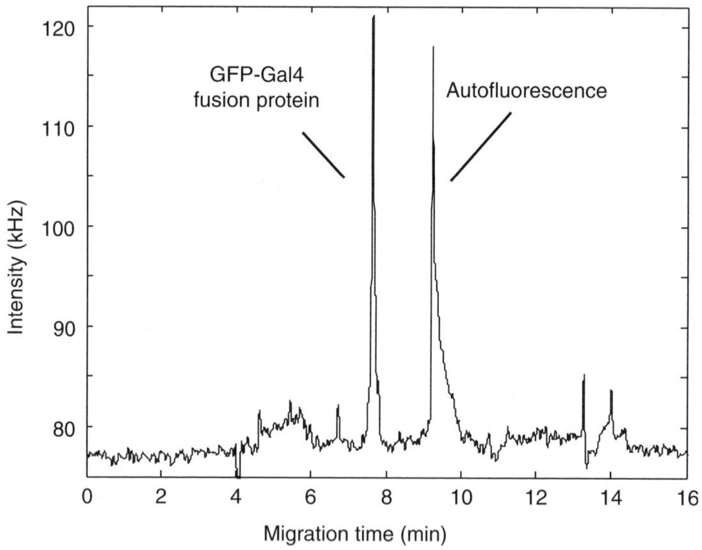

FIGURE 21.6 Chemical cytometry of a single yeast cell expressing the GFP–Gal4 fusion protein. The autofluorescence signal is similar in amplitude to the signal due to the fluorescent protein; without electrophoresis, accurate quantification of fluorescence due to the fusion protein would be difficult.

Figure 21.5 presents the electropherograms generated from five cells. While some cells, such as cells 1 and 2, generate similar electropherograms, there is a fairly wide cell-to-cell heterogeneity in expression patterns. This pattern likely reflects the inherent heterogeneity in this complex primary cell type.

Krylov and coworkers and this group[29,30] have also employed chemical cytometry to characterize GFP in single cells (Figure 21.6). This electropherogram shows the analysis of a single yeast cell that has been engineered to express the Gal4–GFP fusion protein. Figure 21.7 presents the electropherogram generated from a single cell of *D. radiodurans* expressing GFP. This organism is a prokaryote, which is ~1000-fold smaller in mass than typical mammalian cells.

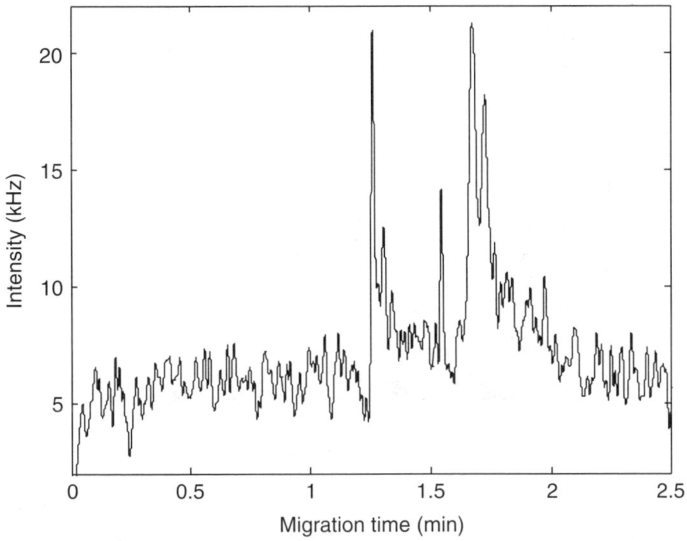

FIGURE 21.7 Chemical cytometry of a single *D. radiodurans* cell expressing GFP. The set of peaks appears to be associated with proteolytic fragments of GFP.

21.3.2 Chemical Cytometry of Biogenic Amines and Proteins Using Two-Dimensional Capillary Electrophoresis

We have developed two-dimensional CE systems for the characterization of proteins and biogenic amines.[27,31,32] The use of this technology for chemical cytometry is similar to the use of one-dimensional electrophoresis: a cell is aspirated into the column, lysed, and its components labeled with FQ. For two-dimensional electrophoresis, components are separated based on CSE in the first-dimension capillary. Fractions are then transferred across an interface to a second capillary, where they undergo additional separation based on micellar electrokinetic chromatography (MECC) before detection by fluorescence. The voltage drop across the first capillary is set to zero during the second dimension separation, holding components stationary. In a typical experiment, ~300 fractions are transferred between capillaries under computer control.

Figure 21.8 presents the raw fluorescence signal generated from the two-dimensional electrophoresis analysis of a single RAW 264.7 macrophage cell. The fluorescence signal consists of a long time series that contains a large number of peaks, and these peaks tend to repeat with 17 s periods, which was the time between transfers of fractions from the first to the second capillary.

The data can be considered a spiral wrapped around a cylinder (Figure 21.9), where the axial distance corresponds to the first dimension separation, the angle corresponds to the second dimension separation, and the density of the image is related to the fluorescence intensity. To generate an image, the cylinder is slit and flattened, creating a gel image of the separation.

Figure 21.10 presents such an image that resembles a silver-stained two-dimensional gel. The image consists of a set of spots, distributed across the surface. The positions of these spots are highly reproducible from cell to cell, with typical precision in spot position being on the order of the size of the spot itself.

The gel image is convenient for comparing spot position. However, the image, particularly when overexposed, is less useful in comparing the amplitude of spots or characterizing the dynamic range of the experiment. Instead, the data can be plotted as a surface in the form of a landscape, where height is proportional to the fluorescence intensity (Figure 21.11).

These data have a very large dynamic range, approaching 10^4 in favorable cases. This dynamic range is suggested by expanding the image by a factor of 10 (Figure 21.12). These separations tend

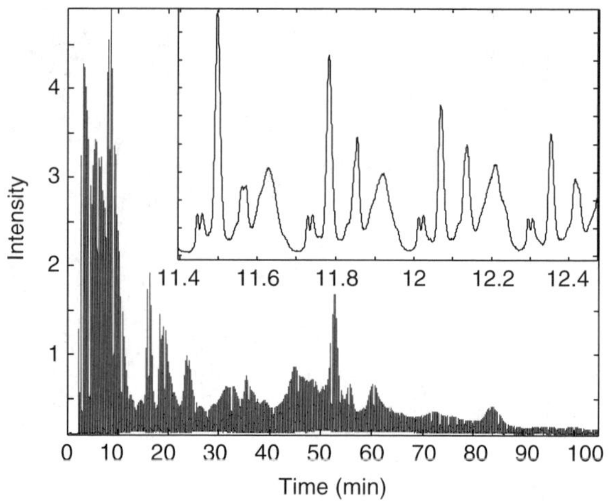

FIGURE 21.8 Raw data generated from a single macrophage cell. Each fraction transfer from the first to the second capillary generates a set of peaks as the first capillary's components are separated in the second capillary. The inset shows a close up of a typical region.

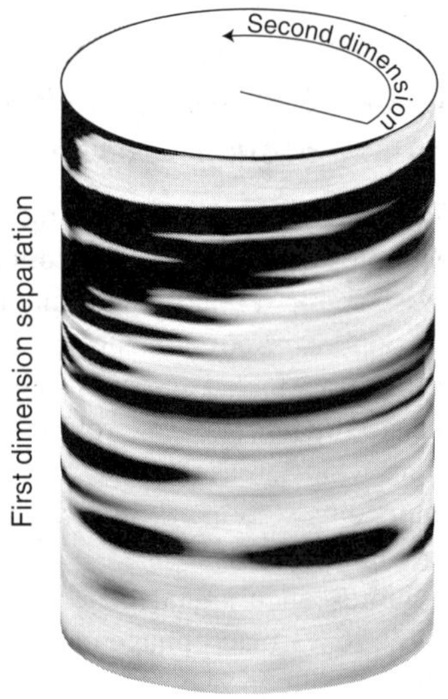

FIGURE 21.9 The raw electrophoresis data of Figure 21.8 are wrapped as a spiral around a cylinder, where the optical density is now related to the fluorescence intensity.

to give excellent peak capacity in the CSE dimension, but poorer capacity in the MECC dimension. The overall spot capacity of this two-dimensional separation is typically between 250 and 500 spots. This form of chemical cytometry has the potential of providing a very rich data set with which to characterize a cellular population.

21.3.3 Metabolic Cytometry

This research group coined the terms chemical cytometry and metabolic cytometry in 1999.[33] Metabolic cytometry is a form of chemical cytometry that monitors biosynthetic and biodegradation enzymatic cascades in a single cell.

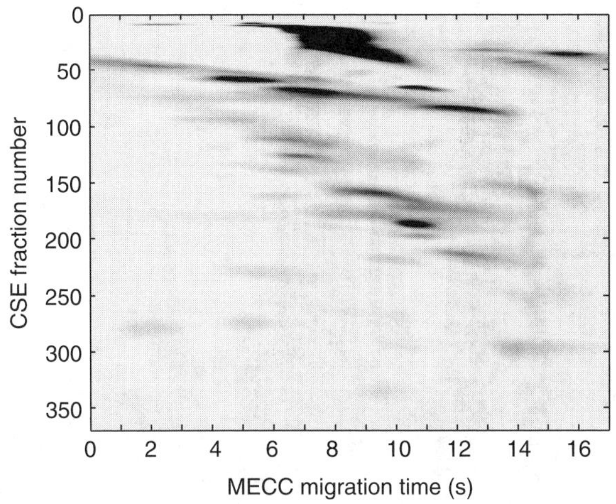

FIGURE 21.10 The cylinder is slit and flattened, creating a gel image, where the optical density is related to fluorescence intensity. The image is overexposed to highlight low-amplitude components.

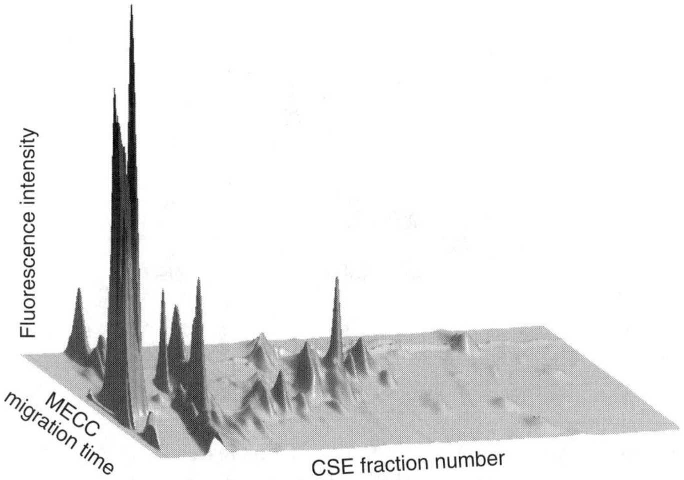

FIGURE 21.11 Landscape view, where height is proportional to fluorescence intensity.

In metabolic cytometry, cells are incubated with a substrate that is tagged with a highly fluorescent dye; we prefer tetramethylrhodamine because of its excellent spectroscopic properties and its compatibility with the frequency-doubled neodymium YAG laser. This substrate is prepared at high concentration and undergoes chromatographic purification to eliminate fluorescent impurities.

As long as the dye is preserved during the transformations, any enzymatic transformation can be monitored with exquisite sensitivity. Figure 21.13 presents an illustrative cartoon, where a substrate, tagged with a fluorescent label (star), undergoes biosynthesis (left path) or biodegradation (right path).

In our first report, we measured DNA ploidy, the uptake of a fluorescent carbohydrate, and that of carbohydrate's biosynthetic and biodegradation products in a single HT29 colon adenocarcinoma cell.[33] We also monitored similar transformations in a single yeast cell.[34]

More recently, we have been studying glycosphingolipid metabolism in single AtT20 cells. Glycosphingolipids are among the most common molecules on neural cell surfaces. These

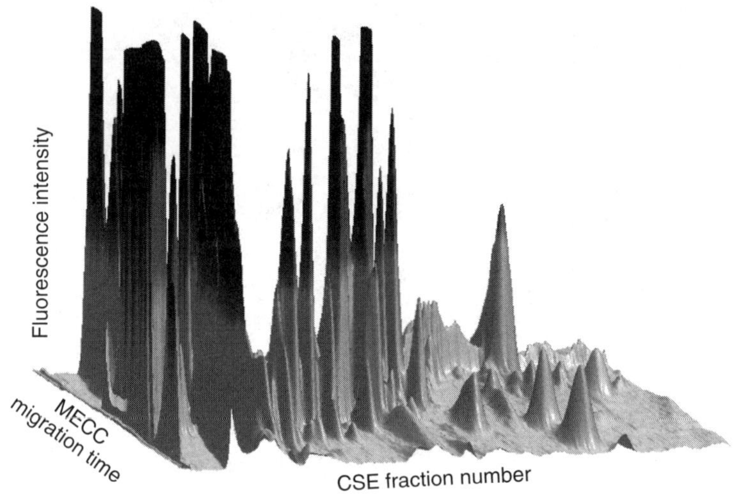

FIGURE 21.12 The image of Figure 21.11 has been expanded by a factor of 10 to highlight low-amplitude components.

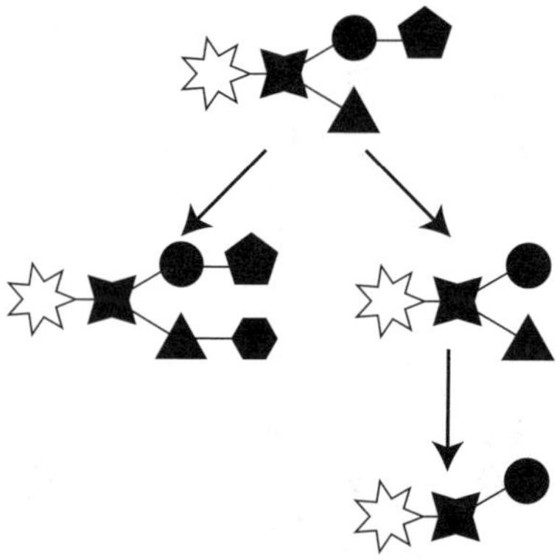

FIGURE 21.13 Metabolic cytometry. A substrate (top) is fluorescently labeled (star). This substrate can undergo biosynthesis (left) or biodegradation (right). As long as the fluorescent tag remains intact, the metabolic products can be monitored by chemical cytometry.

compounds consist of several components. Ceramide is the nonpolar tail that consists of a long-chain amino alcohol linked to a long-chain fatty acid. A chain of saccharide groups attached to the ceramide form the constituent glycosphingolipids, and the presence of one to four sialic acids turns these glycosphingolipids into gangliosides. In the common naming of gangliosides, the subscript letter indicates the quantity of sialic acid groups: M for monosialo, D for disialo, and so forth; A refers to asialo that are glycosphingolipids without sialic acids. The number refers to the order of peak appearance in thin layer chromatography and thus the length of the sugar chain; the sequential removal of zero, one, and two terminal saccharides increases retention to give G_{M1}, G_{M2}, and G_{M3}, respectively.

These glycolipids form a large fraction of neuronal cell membranes, and defects in their metabolism leads to a series of devastating diseases, the best known of which is Tay-Sachs. The compounds are synthesized from the lipid ceramide by successive addition of monosaccharides (Figure 21.14), where the numbers in boxes correspond to the E.C. number for the enzyme responsible

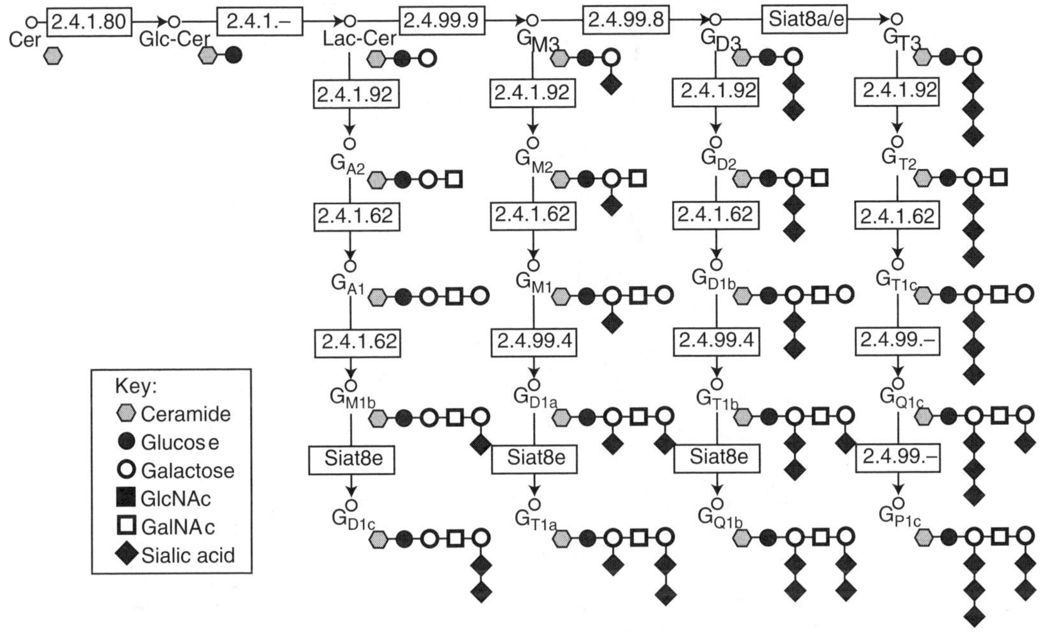

FIGURE 21.14 Partial metabolic pathway associated with sphingolipids. The number in the boxes corresponds to the EC name for the biosynthetic enzyme responsible for the transformation.

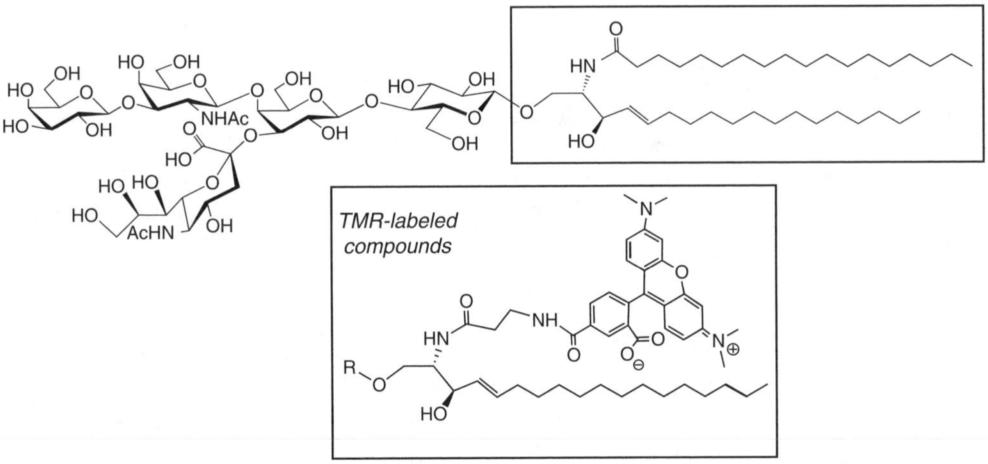

FIGURE 21.15 Structure of G_{M1}. A fluorescently labeled compound has been prepared where the ceramide group (box) is replaced with the TMR-label (bottom box).

for the addition of the corresponding sugar. Biodegradation is not shown; defect in the degradation of G_{M_2} to G_{M_3} leads to accumulation of the former, and is the underlying cause of Tay-Sachs disease.

We have synthesized the fluorescently labeled G_{M1} (Figure 21.15).[35] In the native compound, top, ceramide is shown in the box. In our labeled compound, the fatty acid has been replaced with tetramethylrhodamine.

We have also synthesized the biodegradation products for this compound, including the asialo G_{A1}, and G_{A2}, which are used as standards to tentatively identify metabolic cytometry products based on comigration.

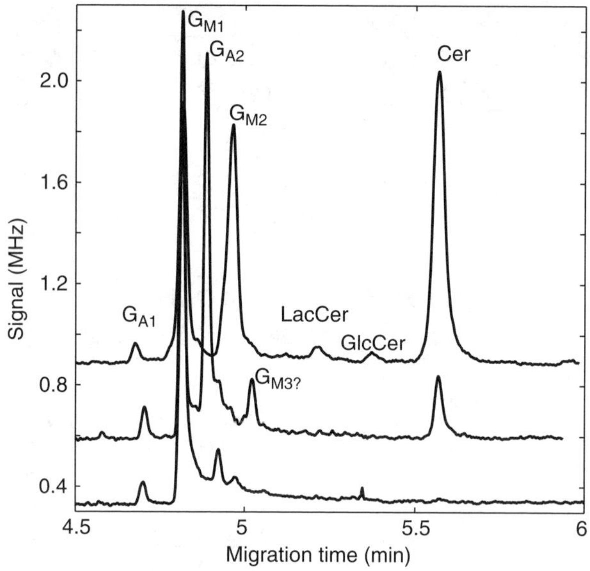

FIGURE 21.16 Metabolic cytometry analysis of three cells. AtT20 cells were incubated with the substrate of Figure 21.15, aspirated into a capillary, lysed, and analyzed by CE.

Figure 21.16 presents metabolic cytometry data generated from single AtT20 cells. In these experiments, a cell is incubated with the substrate. The substrate is taken up, which can be confirmed by fluorescence microscopy. Enzymatic transformations convert the substrate to products. To analyze the products, the cell is aspirated into a capillary, lysed, components are separated by CE, and products detected by laser-induced fluorescence.

These data show remarkable cell-to-cell variations in metabolism. The substrate, G_{M1}, is present in all cells, and all cells show a modest amount of G_{A1}, which requires complex transformation of the substrate. Some cells show very active biodegradation, completely removing the carbohydrates and leaving only the labeled ceramide. Other cells show much less biodegradation and no detectable ceramide.

21.4 METHOD DEVELOPMENT GUIDELINES

Instrumentation for chemical cytometry consists of three parts. The first is an injection module that facilitates aspiration of a cell into the capillary. The second is instrumentation to perform one- or two-dimensional CE. The third is a high sensitivity laser-induced fluorescence detector.

21.4.1 Injection Block

Krylov et al.[36] designed and constructed an injection block that greatly facilitates single-cell analysis. This block consists of three Plexiglas pieces (Figure 21.17). The top piece holds both the capillary and the high-voltage electrode. The second piece contains a hole to connect to pure nitrogen, which is used to pressurize the chamber in order to flush buffer through the capillary. The bottom piece holds a microcentrifuge vial that contains the running buffer.

This injection block is attached to micromanipulators. In operation, the capillary is first flushed with buffer to prepare it for separation. The bottom piece is then removed, and micromanipulators are used to center the capillary over a cell of interest. The cell is aspirated into the capillary, the vial holder is replaced, and high voltage is applied to drive the separation.

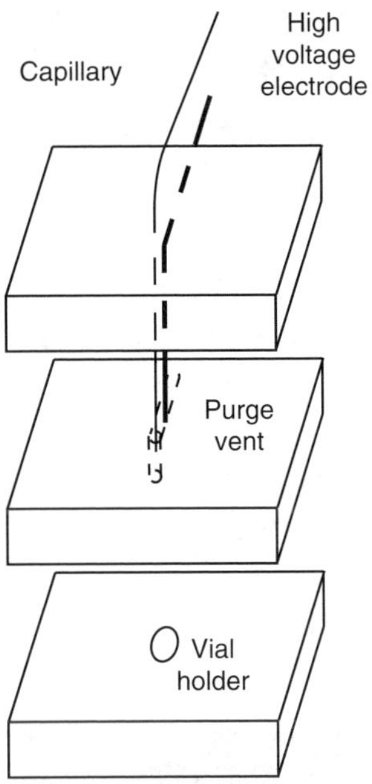

FIGURE 21.17 Injection block. A capillary and electrode are threaded through the top two blocks. The third block forms a gas-tight seal and is used to pressurize a reservoir that pumps reagents through the capillary. The bottom block is removed for injection of a cell.

Aspiration of a single cell typically requires a pulse of ~11 kPa vacuum to the distal end of the capillary for 1 or 2 s. Control of such modest pressure is not trivial. Krylov[36] developed an elegant method for generating an extremely reproducible vacuum. The distal end of the capillary is connected to a sheath-flow cuvette, and the waste stream of the cuvette is connected to a three-way valve. In normal operation, the valve directs the waste to a receiving reservoir. To inject a sample, the valve directs flow to a solenoid valve, which is connected to a water-filled piece of tubing that terminates in a beaker placed 1 m below the level of the injection block. A timer circuit is used to open the solenoid valve for a precisely timed period, during which time the injection end of the capillary is connected to a water column of 1 m high, which applies vacuum necessary to aspirate the cell within the capillary.

21.4.2 Capillary Electrophoresis

Both one- and two-dimensional CE have been used for chemical cytometry. One-dimensional electrophoresis is similar to conventional experiments, albeit with the issues of sample loading, lysis, and on-column labeling.

Two-dimensional CE instrumentation is more complicated, with two power supplies, one for each capillary (Figure 21.18).[37,38] The sample undergoes separation by CSE. Fractions are transferred between capillaries by manipulation of the voltages applied across the two capillaries.

The interface is manufactured using standard macromachining technology. Two capillaries are aligned to be coaxial with a ~50-μm gap in a buffer-filled chamber. Analyte is transferred between capillaries by creating a voltage drop across the interface. Figure 21.19 presents a micrograph of fluorescein being transferred between capillaries in an interface. The fluorescence is centered on the two capillaries. A stagnant zone appears to form an annulus about 10 μm in radius about the inner lumen of the first dimension capillary. However, there is no evidence that analyte leaks from the interface or is lost during a transfer.

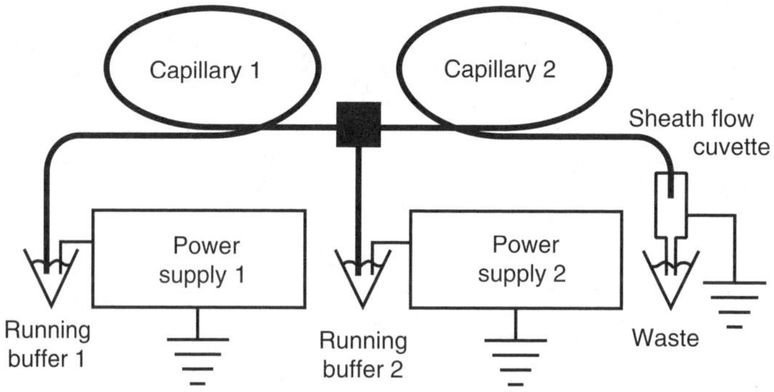

FIGURE 21.18 Instrumentation for two-dimensional CE. Two capillaries are joined through an interface, that allows fractions to be transferred by manipulation of the voltage applied to the capillaries. In our system, the sheath-flow cuvette detector is held at ground potential.

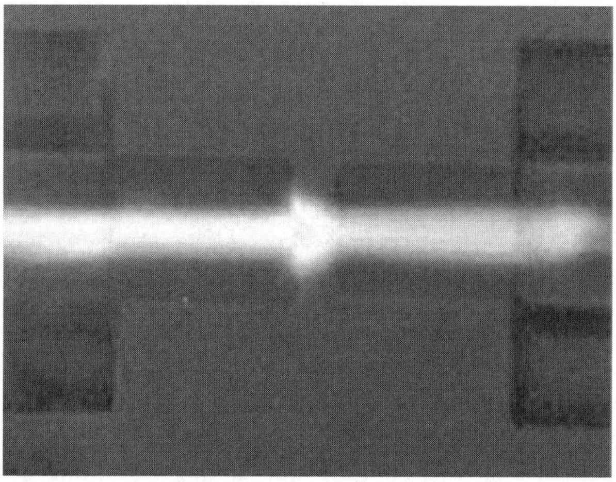

FIGURE 21.19 Transfer of fluorescein between capillaries in the interface used in two-dimensional CE.

This interface differs from Jorgenson and coworkers[39] flow-gated interface between a liquid chromatographic column and a CE column. In that design, analyte continually flows from the chromatographic column; cross-flow is used to flush analyte to waste except during fraction transfer. In contrast, analyte flow is easily stopped in two-dimensional CE by applying equal voltage at both ends of the capillary; as a result, there is no need for crossed flow with its accompanying sample loss.

A typical voltage program is shown in Figure 21.20. After lysis, a preliminary CSE separation is performed so that the fastest moving components approach the exit of the first capillary. Then, a series of fraction transfers and MECC separations is performed. There is a voltage drop across the first capillary during the transfer steps, which continues the CSE separation and which drives analyte to the second dimension capillary. There is also a voltage drop across the second dimension capillary during the transfer, so that separation also continues in that dimension during fraction transfer. There is no voltage drop across the first dimension capillary during the MECC separation step, and analyte does not migrate in that capillary.

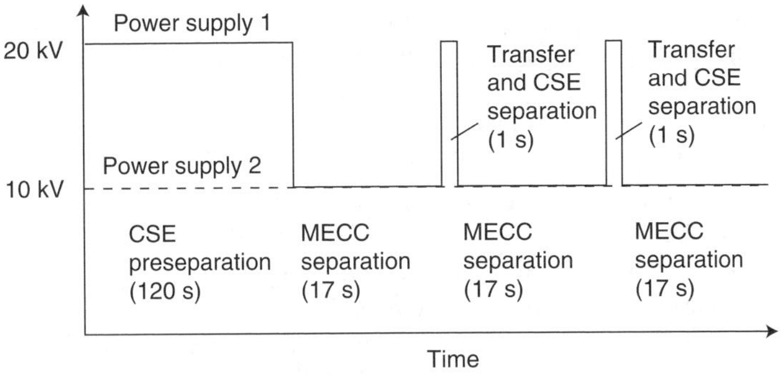

FIGURE 21.20 Typical timing diagram used in two-dimensional CE for chemical cytometry. The distal end of capillary 2 is held in the sheath-flow cuvette, which is at ground potential.

21.4.3 Fluorescence Detection

High-sensitivity fluorescence detection is required for chemical cytometry. We use a postcolumn sheath-flow cuvette for detection. This device is based on a cuvette used in the venerable Ortho cytofluorograph, a flow cytometer developed in the late 1970s. A stream of flowing buffer ensheaths analyte flowing from the separation capillary. Fluorescence is detected in a flow chamber with high optical quality windows, which decreases light scatter and improves the detection limit compared to using the capillary itself as a detection cuvette.

21.4.4 Data Processing

There are a couple useful steps in data processing. First, a morphological opening procedure can be used to remove baseline drift. This process is available in the image processing toolbox in Matlab. Like many other filtering processes, baseline correction is subject to artifact. These artifacts are most obvious when performing the morphological opening procedure on too small an area; in this case, wide peaks and ridges in the data can disappear.

Second, filtering is useful. A number of filters are available for two-dimensional data. The most important is the median filter that effectively removes noise spikes caused by bubbles or particles passing through the laser beam. Next, it is often useful to smooth the data; convolution with a two-dimensional Gaussian filter is particularly useful if the shape of the Gaussian filter is matched to those of typical peaks.

21.5 CONCLUDING REMARKS AND FUTURE WORK

The field of chemical cytometry is in its infancy, and it is clear that there is much work to be done before the technology is widely used. First, instrumentation throughput must be increased. Flow cytometry today can process a 100,000 cells/s. While chemical cytometry will not produce similar throughput, it is reasonable to expect the technology to process 10,000 cells/day in one-dimensional electrophoresis and perhaps 1000 cells/day in two-dimensional electrophoresis. Such instrumentation will likely resemble the multiple-CE systems that have become ubiquitous in DNA sequencing.[37,38,40]

Second, separation efficiency must be improved. Current two-dimensional electrophoresis systems are generating a spot capacity of a few hundred components; an order of magnitude improvement is desirable. Spot capacity is given by the product of the peak capacity in each dimension, so that a threefold improvement in separation efficiency in each dimension will generate the desired improvement. MECC is not an ideal separation mode for proteins, and would be replaced with a technique such as isoelectric focusing, which can provide exquisite resolution. A major hurdle deals with protein labeling. FQ, and presumably other dyes that produce an anionic or neutral reaction product, generate very complex isoelectric focusing electropherograms due to multiple labeling. Anionic surfactants are not compatible with IEF, and it will be necessary to employ a labeling reagent that produces a cationic product. The recently developed chameleon dyes are interesting examples of fluorogenic reagents that produce cationic products.[41,42]

Third, component identification must be improved. Current methods rely on comigration with standards. While practical for metabolic cytometry and biogenic amines, due to the relatively small number of possible compounds, spiking is extremely tedious for protein analysis. Ideally, an online mass spectrometer can be used to monitor compounds as they migrate from the capillary. Although the current generation of mass spectrometers does not have sufficient sensitivity to monitor the minute amounts of compounds present in a single cells, mass spectrometry could be used to study concentrated homogenates prepared from large numbers of cells. Based on excellent run-to-run reproducibility of the two-dimensional electrophoresis system, component identification should be practical in fluorescence-based detection of single cells. As a complication, it would be necessary to use fluorescently labeled proteins, which would need to be considered during interpretation of mass spectra.

ACKNOWLEDGMENTS

This work was funded by grants from National Institutes of Health (NIH) and Department of Energy (DOE).

REFERENCES

1. Shapiro, H.M. *Practical Flow Cytometry*, 4th ed., Hoboken, NJ, Wiley, 2003.
2. Caspersson, T.O. *Cell Growth and Cell Function*, New York, Norton, 1950.
3. Caspersson, T.O. History of the development of cytophotometry from 1935 to the present. *Anal. Quant. Cytol. Histol.* 1987; 9, 2–6.
4. Ibrahim, S.F. and van den Engh, G. High-speed cell sorting: Fundamentals and recent advances. *Curr. Opin. Biotech.* 2003; 14, 5.
5. Baumgarth, N. and Roederer, M. A practical approach to multicolor flow cytometry for immunophenotyping. *J. Immunol. Methods* 2000; 243, 77–97.
6. Xu, B.J., Caprioli, R.M., Sanders, M.E. and Jensen, R.A. Direct analysis of laser capture microdissected cells by MALDI mass spectrometry. *J. Am. Soc. Mass Spectrom.* 2002; 13: 1292–1297.
7. Laurent, C., Levinson, D.F., Schwartz, S.A., Harrington, P.B., Markey, S.P., Caprioli, R.M. and Levitt, P. Direct profiling of the cerebellum by matrix-assisted laser desorption/ionization time-of-flight mass spectrometry: A methodological study in postnatal and adult mouse. *J. Neurosci. Res.* 2005; 81: 613–621.
8. Whittal, R.M., Keller, B.O. and Li, L. Nanoliter chemistry combined with mass spectrometry for peptide mapping of proteins from single mammalian cell lysates. *Anal. Chem.* 1998; 70: 5344–5347.
9. Li, L., Garden, R.W. and Sweedler, J.V. Single-cell MALDI: A new tool for direct peptide profiling. *Trends Biotechnol.* 2000; 18: 151–160.
10. Edstrom, J.E. Nucleotide analysis on the cyto-scale. *Nature* 1953; 172: 809.
11. Kennedy, R.T., Oates, M.D., Cooper, B.R., Nickerson, B. and Jorgenson, J.W. Microcolumn separations and the analysis of single cells. *Science* 1989; 246: 57–63.
12. Wallingford, R. A. and Ewing, A. G. Capillary zone electrophoresis with electrochemical detection in 12.7 microns diameter columns. *Anal. Chem.* 1988; 60: 1972–1975.
13. Hogan, B.L. and Yeung, E.S. Determination of intracellular species at the level of a single erythrocyte via capillary electrophoresis with direct and indirect fluorescence detection. *Anal. Chem.* 1992; 64: 2841–2845.

14. Gilman, S.D. and Ewing, A.G. Analysis of single cells by capillary electrophoresis with on-column derivatization and laser-induced fluorescence detection. *Anal. Chem.* 1995; 67: 58–64.
15. Meredith, G.D., Sims, C.E., Soughayer, J.S. and Allbritton, N.L. Measurement of kinase activation in single mammalian cells. *Nat. Biotechnol.* 2000; 18: 309–312.
16. Han, F. and Lillard, S.J. *In-situ* sampling and separation of RNA from individual mammalian cells. *Anal. Chem.* 2000; 72: 4073–4079.
17. Wu, H., Wheeler, A. and Zare, R. N. Chemical cytometry on a picoliter-scale integrated microfluidic chip. *Proc. Natl Acad. Sci. USA* 2004; 101: 12809–12813.
18. Cheng, Y.F. and Dovichi, N.J. Subattomole amino acid analysis by capillary zone electrophoresis and laser-induced fluorescence. *Science* 1988; 242: 562–564
19. Chen, D.Y. and Dovichi, N.J. Single-molecule detection in capillary electrophoresis: Molecular shot noise as a fundamental limit to chemical analysis. *Anal. Chem.* 1996; 68: 690–696.
20. Ghaemmaghami, S., Huh, W.K., Bower, K., Howson, R.W., Belle, A., Dephoure, N., O'Shea, E.K. and Weissman, J.S. Global analysis of protein expression in yeast. *Nature* 2003; 425: 737–741.
21. Available at: http://yeastgtp.ucsf.edu/allOrfData.txt
22. Liu, H.J., Cho, B.Y., Krull, I.S. and Cohen, S.A. Homogeneous fluorescent derivatization of large proteins. *J. Chromatogr. A* 2001; 927: 77–89.
23. Zhao, J.Y., Waldron, K.C., Miller, J., Zhang, J.Z., Harke, H. and Dovichi, N.J. Attachment of a single fluorescent label to peptides for determination by capillary zone electrophoresis. *J. Chromatogr.* 1992; 608: 239–242.
24. Craig, D.B. and Dovichi, N.J. Multiple labeling of proteins. *Anal. Chem.* 1998; 70, 2493–2494.
25. Richards, D., Stathakis, C., Polakowski, R., Ahmadzadeh, H. and Dovichi, N.J. Labeling effects on the isoelectric point of green fluorescent protein. *J. Chromatogr. A* 1999; 853: 21–25.
26. Pinto, D.M., Arriaga, E.A., Craig, D., Angelova, J., Sharma, N., Ahmadzadeh, H., Dovichi, N.J. and Boulet, C.A. Picomolar assay of native proteins by capillary electrophoresis—Precolumn labeling, sub-micellar separation and laser induced fluorescence detection. *Anal. Chem.* 1997; 69: 3015–3021.
27. Kraly, J.R., Jones, M.R., Gomez, D.G., Dickerson, J.A., Harwood, M.M., Eggertson, M., Paulson, T.G., et al. Rapid and reproducible two-dimensional capillary electrophoresis analysis of Barrett's esophagus tissues. *Anal. Chem.* 2006; 78: 5977–5986.
28. Gudiksen, K.L., Gitlin, I., Moustakas, D.T. and Whitesides, G.M. Increasing the net charge and decreasing the hydrophobicity of bovine carbonic anhydrase decreases the rate of denaturation with sodium dodecyl sulfate. *Biophys. J.* 2006; 91: 298–310.
29. Turner, E.H., Lauterbach, K., Pugsley, H., Palmer, V.R. and Dovichi, N.J. Ultrasensitive green fluorescent protein detection in a single bacterium by capillary electrophoresis with laser-induced fluorescence, *Anal. Chem.* 2007; 79: 778–781.
30. Hu, K., Ahmadzadeh, H. and Krylov, S.N. Asymmetry between sister cells in a cancer cell line revealed by chemical cytometry. *Anal. Chem.* 2004; 76: 3864–3866.
31. Michels, D.A., Hu, S., Schoenherr, R.M., Eggertson, M.J. and Dovichi, N.J. Fully automated two-dimensional capillary electrophoresis for high sensitivity protein analysis. *Mol. Cell. Proteomics* 2002; 1: 69–74.
32. Michels, D.A., Hu, S., Dambrowitz, K.A., Eggertson, M.J., Lauterbach, K. and Dovichi, N.J. Capillary sieving electrophoresis-micellar electrokinetic chromatography fully automated two-dimensional capillary electrophoresis analysis of *Deinococcus radiodurans* protein homogenate. *Electrophoresis* 2004; 25: 3098–3105.
33. Krylov, S.N., Zhang, Z., Chan, N.W.C., Arriaga, E., Palcic, M.M. and Dovichi, N.J. Correlating cell cycle with metabolism in single cells: The combination of image and metabolic cytometry. *Cytometry*, 1999; 37: 15–20.
34. Le, X.C., Tan, W., Scaman, C.H., Szpacenko, A., Arriaga, E., Zhang, Y., Dovichi, N.J., Hindsgaul, O. and Palcic, M.M. Single cell studies of enzymatic hydrolysis of a tetramethylrhodamine labeled trisaccharide in yeast. *Glycobiology* 1999; 9: 219–225.
35. Larsson, E.A., Olsson, U., Whitmore, C.D., Martins, R., Tettamanti, G., Schnaar, R.L., Dovichi, N.J., Palcic, M.M. and Hindsgaul, O. Synthesis of reference standards to enable single cell metabolomic studies of tetramethylrhodamine-labeled ganglioside GM1. *Carbohydr. Res.* 2007; 342: 482–489.
36. Krylov, S.N., Starke, D.A., Arriaga, E.A., Zhang, Z., Chan, N.W., Palcic, M.M. and Dovichi, N.J. Instrumentation for chemical cytometry. *Anal. Chem.* 2000; 72: 872–877.

37. Zhang, J., Voss, K.O., Shaw, D.F., Roos, K.P., Lewis, D.F., Yan, J., Jiang, R., et al. A multiple-capillary electrophoresis system for small-scale DNA sequencing and analysis. *Nucleic Acids Res.* 1999; 27: e36.
38. Crabtree, H.J., Bay, S.J., Lewis, D.F., Zhang, J., Coulson, L.D., Fitzpatrick, G.A., Delinger, S.L., Harrison, D.J. and Dovichi, N.J. Construction and evaluation of a capillary array DNA sequencer based on a micromachined sheath-flow cuvette. *Electrophoresis* 2000; 21: 1329–1335.
39. Opiteck, G.J., Lewis, K.C., Jorgenson, J.W. and Anderegg, R.J. Comprehensive on-line LC/LC/MS of proteins. *Anal. Chem.* 1997; 69: 1518–1524.
40. Zhang, J., Yang, M., Puyang, X., Fang, Y., Cook, L.M. and Dovichi, N.J. Two-dimensional direct-reading fluorescence spectrograph for DNA sequencing by capillary array electrophoresis. *Anal. Chem.* 2001; 73: 1234–1239.
41. Craig, D.B., Wetzl, B.K., Duerkop, A. and Wolfbeis, O.S. Determination of picomolar concentrations of proteins using novel amino reactive chameleon labels and capillary electrophoresis laser-induced fluorescence detection. *Electrophoresis* 2005; 26: 2208–2213.
42. Wetzl, B.K., Yarmoluk, S.M., Craig, D.B. and Wolfbeis, O.S. Chameleon labels for staining and quantifying proteins. *Angew. Chem. Int. Ed. Engl.* 2004; 43: 5400–5402.

22 Glycoprotein Analysis by Capillary Electrophoresis

Michel Girard, Izaskun Lacunza, Jose Carlos Diez-Masa, and Mercedes de Frutos

CONTENTS

22.1 INTRODUCTION

Under the action of enzymes globally called glycosyltransferases, proteins undergo cotranslational and post-translational modifications (PTMs) resulting in the covalent attachment of glycan chains to the polypeptide core. This fundamental process is called glycosylation and results in the formation of glycoproteins. Although many others exist, glycosylation is undoubtedly one of the most prevalent

PTM of eukaryotic proteins. For instance, in humans, glycoproteins account for more than 50% of all proteins. Despite this high frequency of occurrence, no generalized role can be attributed to glycosylation of proteins. In fact, the role of the glycan moiety can vary from one protein to another and have direct involvement in the biological properties of proteins such as transport, cell surface recognition, enzymatic processes, or the immune system. The large size of oligosaccharides may allow them to cover functionally important regions of proteins, to modulate the interactions of glycoconjugates with other molecules, and to affect the rate of the processes that involve conformational changes. Besides these important biological functions, glycosylation also confers physical properties to proteins such as solubility, hydrophilicity, stability, and assists in protein folding.

Just as for the rest of proteins, other PTM events and the existence of genetic variants contribute to the heterogeneity of glycoproteins. All of the heterogeneities are generically termed "forms" of a protein (e.g., deamidated forms), while forms arising from the glycosidic part of the protein are specifically called "glycoforms." The term "sialoform" refers to the glycoforms that have the same number of sialic acid residues. In the field of capillary electrophoresis (CE), "isoform" indicates an electrophoretic peak or band, which can include one or more forms of a protein.

From an analytical point of view, there are several approaches to the study of glycoproteins. The protein can be submitted to partial or total hydrolysis/proteolysis resulting in the formation of glycopeptides, glycans, individual monosaccharides, peptides, or amino acids, which can then be analyzed separately. Alternatively, the different forms of the intact glycoprotein can be separated and studied using one of many suitable techniques providing high resolution of closely related proteins such as sodium dodecylsulfate–polyacrylamide gel electrophoresis (SDS–PAGE), isoelectric focusing (IEF), high-performance liquid chromatography (HPLC), mass spectrometry (MS), or CE. Traditionally, conventional gel electrophoresis methods have been employed to study the intact forms of glycoproteins and their hydrolysis/proteolysis products. However, these methods are laborious to carry out, cannot be automated, and do not consistently provide quantitative results. In contrast, CE analysis has been demonstrated to provide rapid, automated, and quantitative results. In addition, it is now often used as a complementary technique to the more established HPLC in several glycoprotein studies.

In this chapter, CE approaches to the study of intact forms of glycoproteins will be discussed with particular emphasis on factors such as separation conditions and modes of detection. In addition, an overview of practical applications of CE in different sectors of activities such as the pharmaceutical industry and clinical research will be presented. Although examples illustrating various aspects will be mentioned, the primary goal of this chapter is not to provide an extensive review of the literature reported on the subject or details about all of the glycoproteins that have been studied. For in-depth discussions several reviews covering this subject in part or in total, which are mentioned in Section 22.2 of this chapter, can be read.

The general principles related to the different modes of CE and of protein separation by CE have not been considered here as they are the subjects of other chapters in this book.

22.2 BACKGROUND

Glycoprotein formation is a complex process that involves multiple enzymatic reactions occurring in the endoplasmic reticulum and the Golgi apparatus.

A glycoprotein is usually present in an organism as a population of glycoforms, the heterogeneity of which is cell/tissue or organism-dependent, as well as a function of the pathophysiological status [1]. Changes in protein glycosylation occur in both congenital and acquired illnesses [2–4] and the study of these changes is generating considerable interest in the field of biomarkers. The production by recombinant DNA technology of glycoproteins for specific therapeutic uses is another field that has attracted considerable research efforts. In fact, many of the biopharmaceuticals currently produced are glycoproteins [5,6]. The glycosylation of these proteins is crucial in their role

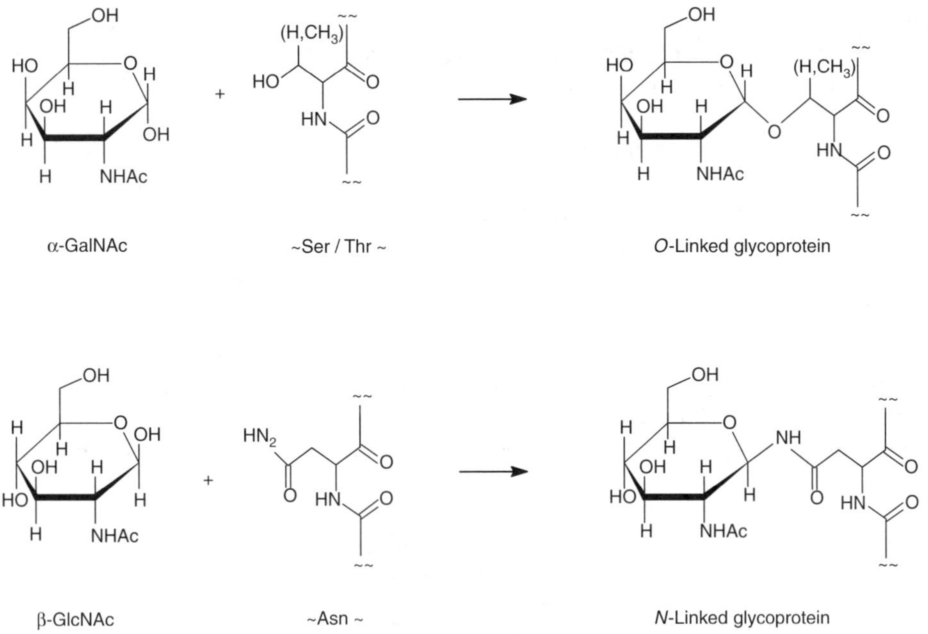

FIGURE 22.1 Representation of O- and N-linked glycosylation from the reaction of α-*N*-acetylgalactosamine with serine/threonine and β-*N*-acetylglucosamine with asparagine, respectively.

as biopharmaceuticals since it will affect, for instance, their biological activities, solubility, and other properties [7,8]. In the case of recombinant glycoproteins, in addition to being cell-dependent, glycosylation is affected by culture media conditions and the purification process [1]. These considerations give an idea of the enormous variety of glycoforms that a glycoprotein can contain and, therefore, the difficult task that the in-depth analysis of all of these structures involves, particularly when the approach selected is the study of intact glycoprotein.

Glycans are attached to proteins through O- and N-linkages at specific amino acids (Figure 22.1). O-linked glycosylation occurs at the hydroxyl functional group of the side chain of serine (Ser) or threonine (Thr) residues and, to a lesser extent, as a secondary modification at hydroxylysine (Hyl) and hydroxyproline (Hyp) residues (e.g., in collagen). N-linked glycosylation takes place at the amide side chain of asparagine (Asn) residues. There are several monosaccharides involved in the glycosylation of proteins: mannose (Man), galactose (Gal), fucose (Fuc), *N*-acetylglucosamine (GlcNAc), *N*-acetylgalactosamine (GalNAc), and sialic acid. The latter is the generic term for several derivatives of neuraminic acid that includes *N*-acetylneuraminic acid (NeuAc), one of the most abundant sialic acids in mammals. All of these sugars are neutral at physiological pH except for sialic acid, which is negatively charged due to the presence of a carboxylic acid functional group.

N-glycans are usually classified into the following three types: complex, high-mannose, and hybrid [9]. Complex-type glycans have several antennae, with biantennary being the most abundant ones, although tri- and tetra-antennary ones are also common (Figure 22.2a) [10], each of which may feature high heterogeneity due to several possible permutations of the sugar residues. High-mannose-type glycans are common in cell-surface proteins and have, in most cases, between 5 and 9 mannose residues (Figure 22.2b). Hybrid glycans feature characteristics common to both high-mannose and complex glycans (Figure 22.2c).

O-glycans are also composed of a variety of common monosaccharides that, in addition to those found in *N*-glycans, include xylose (Xyl), glucose (Glu), and arabinose (Ara) (Figure 22.2d). *O*-glycans are very common in mucines where long polypeptides with highly negatively charged zones are present due to the presence of sialic acids.

The complexity of glycoproteins and the fact that CE compares favorably with conventional electrophoretic techniques are among the reasons for which glycoprotein analysis has been a preeminent feature of CE from its very early days. A monograph dedicated to CE of proteins and peptides by Strege and Lagu [11] is available. Reviews covering the early development of analytical CE methods for proteins, including glycoproteins, for years up to 1997 have been cited by Prichett and Robey [12] in the previous edition of this handbook. Several recent reviews on the analysis of glycoproteins by CE have discussed various aspects [13–26].

22.3 THEORETICAL ASPECTS

Several CE modes that will be considered in detail in Section 22.4 of this chapter have been used in the separation of glycoproteins: capillary zone electrophoresis (CZE), capillary isoelectric focusing (CIEF), capillary electrophoresis in SDS (CE-SDS), micellar electrokinetic chromatography (MEKC), affinity capillary electrophoresis (ACE), and capillary isotachophoresis (CITP). It is out

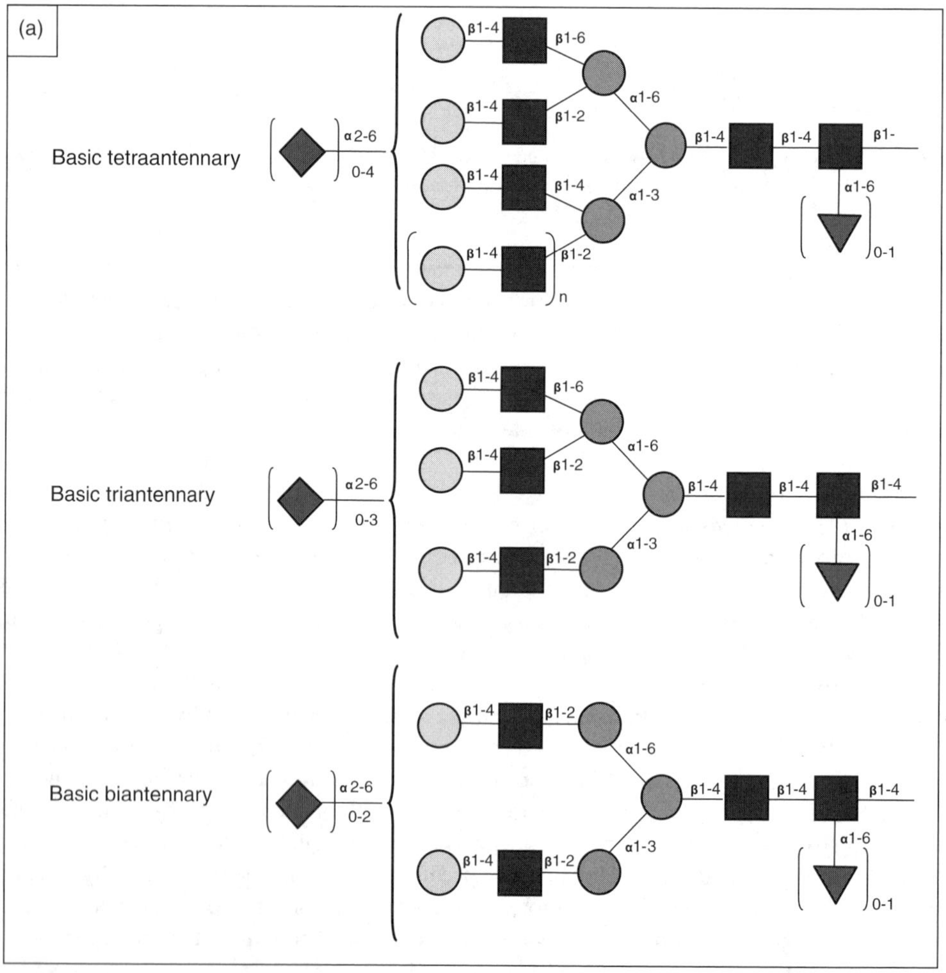

FIGURE 22.2 Typical structures of (a) complex-type N-glycans; (b) high-mannose N-glycans; (c) hybrid-type N-glycans; and (d) core structures of O-glycans. Representation according to Nomenclature Committee of the Consortium for Functional Glycomics (http://www.glycomics.scripps.edu/CFGnomenclature.pdf, accessed on November 2006).

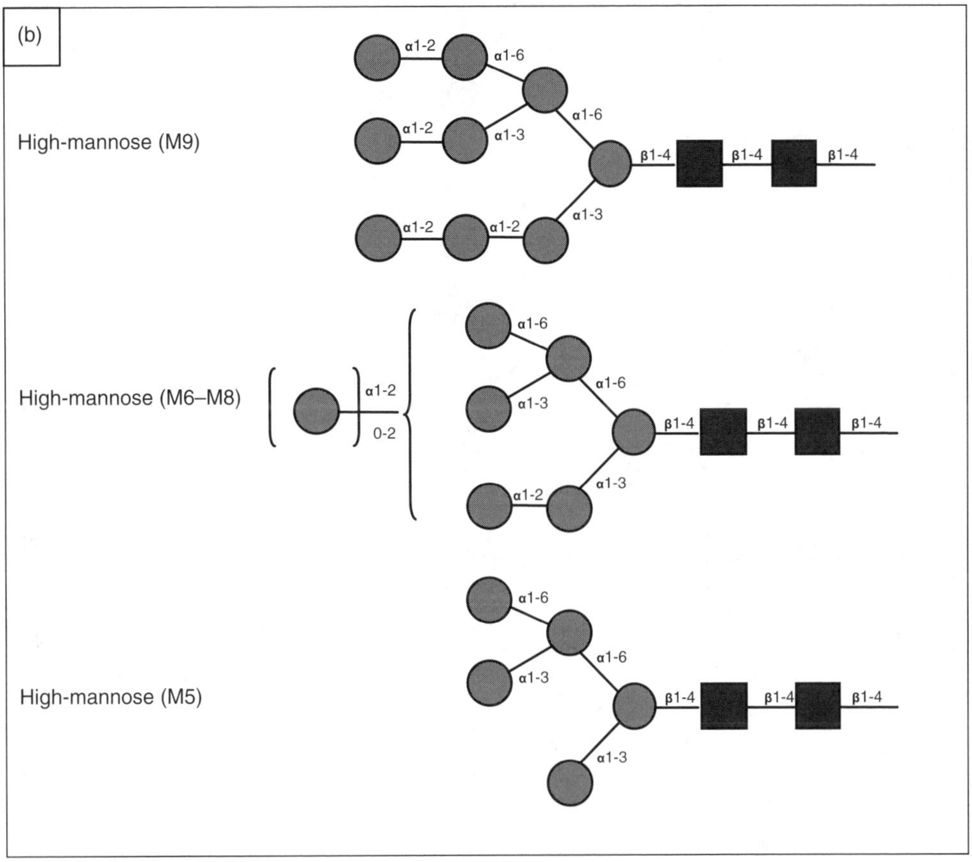

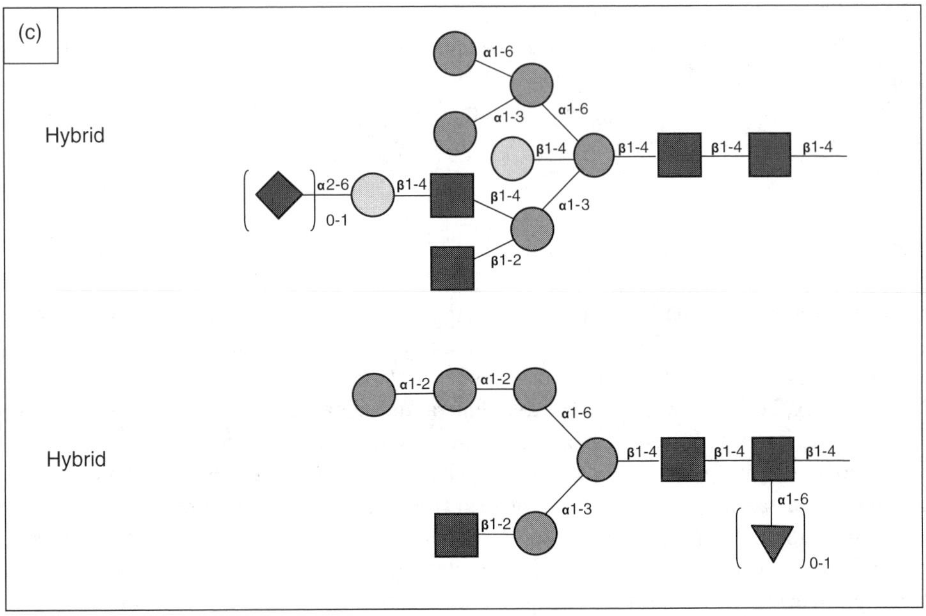

FIGURE 22.2 (Continued)

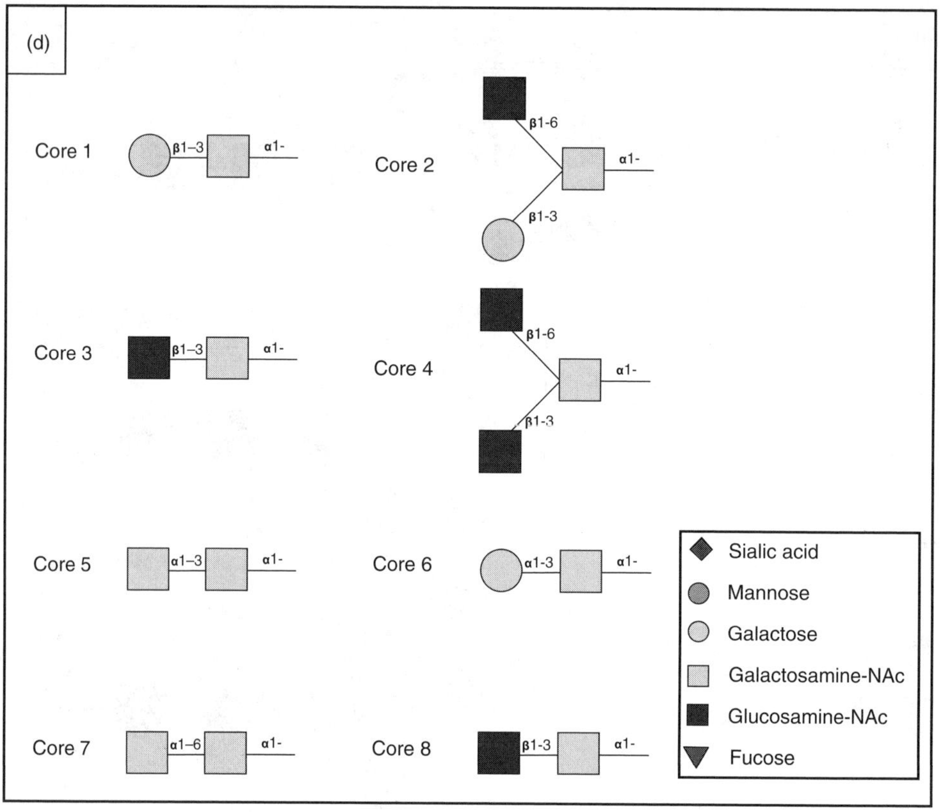

FIGURE 22.2 (Continued)

of the scope of this chapter to review the theoretical aspects of each of these techniques for glycoproteins. In this section, only a few considerations about the separation of proteins with a special emphasis on glycoforms separation by CZE will be presented.

In CZE, proteins can be considered from a theoretical point of view as charged particles. The mobility, μ, can be described in a first approximation by [27]

$$\mu = \frac{q}{f}, \tag{22.1}$$

where q is the charge of the particle and f is its friction coefficient in the separation buffer, which for small molecules is proportional to their respective mass. For proteins, the friction coefficient is also related to the shape of the particle. In fact, studies [12,28] have shown that the separation of proteins in CZE is better explained in terms of hydrodynamic radius, because this parameter describes both the mass and shape of the protein. In a similar way, studies have shown that the displacement of a protein in a given separation buffer is better described by the net charge of the protein and, consequently, q is better expressed as the net charge of the protein (for a review see Reference 29). The net charge includes contributions from all of the electrostatic elements of the protein, such as charged amino acids, metal ions, and cofactors bound to it. The net charge is screened to a certain degree by the microenvironment of the charged group, by the association of the protein with the counterions in solution, and by the characteristics of the protein that influence its attraction to counterions. In conclusion, electrophoretic mobility of proteins can be interpreted in terms of their hydrodynamic radius and net charge [12].

In CZE of glycoproteins the situation is more complicated than for nonglycosylated ones because of additional differences originating from the glycosidic part. For the sake of simplicity, the following will present only a limited discussion of the overall implications of the charge and composition of the glycans on the migration of glycoproteins.

The simplest cases are those in which glycoforms differ only in electric charge. These differences are very often due to the different content in sialic acid of the glycoforms (sialoforms). Sialic acid moieties occur in most cases in the outer part of the protein, in a relatively free position to move around the N- or O-linkage with the rest of the oligosaccharide. Therefore, differences in sialic acid content of these glycoforms produce important variations in the net charge of the proteins resulting in differences in their mobilities. One well-documented example of sialoforms separation by CZE is presented by Balaguer et al. [30]. Using neutral coated capillaries, the authors carried out the separation of recombinant human erythropoietin (rhEPO) isoforms (Figure 22.3).The main peaks were identified using electrospray ionization (ESI) time-of-flight mass spectrometry (ESI-TOF MS) coupled to CZE as several sialoforms (Figure 22.3a). Using the reconstructed extracted ion electropherograms together with the deconvoluted mass of the MS isoform peaks, they observed that, as predicted by the theory, only glycoforms differing in their sialic acid content migrated as separate peaks and that their mobilities were directly related to the number of sialic acids present in each sialoform.

More complex situations arise when different glycoforms have the same charge but the hydrodynamic radius increases with the size of the glycan chain. In the same study mentioned earlier [30], the authors obtained the extracted ion electropherogram of rhEPO glycoforms containing the same number of sialic acids but differing in the number of hexose and *N*-acetylhexosamine residues (Figure 22.3b). An increase in the effective hydrodynamic radius of the protein causing a mobility shift to lower migration times was observed.

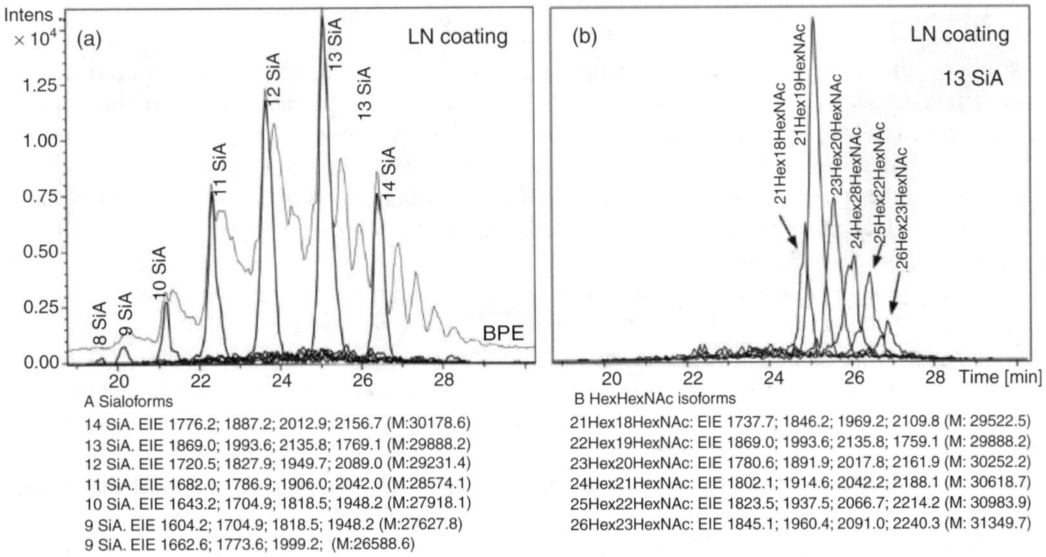

FIGURE 22.3 Electropherogram of rhEPO (BRP European Pharmacopeia, 2.5 μg/μL in water). Separation performed in a capillary 80 cm × 50 μm coated with a neutral polymer (UltraTrol Pre-Coated LN), separation buffer 2 M HAc, and separation voltage at +30 kV. (a) Clear trace represents the base–peak electropherogram (BPE) and darker trace represents the electropherogram reconstructed from the extracted ions together with the deconvoluted mass of the MS peaks of different sialic acid isoforms. (b) Extracted ion electropherogram obtained for glycoforms with different HexHexNAc content and identical sialic acid number. (From Balaguer, E. et al., *Electrophoresis*, 27, 2638, 2006. With permission.)

In some cases, however, a clear effect of either the charge or the size on the glycoform mobility cannot be discerned. This is illustrated in the work of Yim et al. [31] on the study of recombinant human bone morphogenic protein 2 (rhBMP-2). This basic protein (pI > 8.5) is a dimer containing in each of the monomer an N-linked glycan of the high-mannose type. Using neutral hydrophilic capillaries and phosphate buffer at pH 2.5 as background electrolyte (BGE), the 15 possible rhBMP-2 glycoforms were separated into nine peaks. The use of the glycan and peptide maps, together with fast atom bombardment mass spectrum of the generated glycopeptides, enabled the authors to identify these peaks as the nine glycoforms of the structure $(rhBP\text{-}2)_2\text{-}(GlcNAc)_4\text{-}(Man_z)$, where z varies from 10 to 18. As all of the glycoforms have the same charge, the authors concluded that the size of a mannose residue was fundamental for the separation and that the increasing hydrodynamic radius of the protein could control the separation. However, in rhBMP-2, the mannose glycans appear to be large enough to cover part of the polypeptide backbone, a situation that would lead to the shielding of the most charged region of the protein and to a reduction of its effective charge. An estimation of the effect of the mass of a mannose on the mobility of rhBMP-2 in terms of differences in charge and frictional coefficient led to the conclusion that it was not possible to assess whether the effect of the mannose residue was from a reduction of the z potential, from an increase of the frictional coefficient, or from a combination of the two.

From the above discussion, it can be concluded that both the effective charge and the hydrodynamic radius play a major role in the control of the electrophoretic mobility in the separation of proteins by CZE. Since proteins are flexible molecules, the interplay of both factors has to be taken into consideration to explain their mobility in electrophoresis. For glycoproteins, a full understanding of the factors that control their separation in CZE is a complex task because the size and flexibility of the glycans constitute an additional cause of structural variability.

22.4 METHODS DEVELOPMENT GUIDELINES

22.4.1 Sample Preparation

As in any other analytical procedure, sample preparation, in those cases where it is needed, is a key step in the separation of glycoforms by CE. The main goals of the sample preparation step are concentration and purification and, when designing the protocol, the peculiarities of each analyte and of the sample matrix must be taken into account. For protein separations by CE, in addition to aspects common to other types of analysis such as filtration, it is often necessary to eliminate compounds that may interfere either by causing a high electrical current, by adsorbing to the capillary walls, or by comigrating with the analytes.

Without sample preparation, the analysis of a dilute serum sample by CE with ultraviolet (UV) detection leads to the separation and detection of the major serum proteins, γ-globulin, transferrin (Tf), β-lipoproteins, haptoglobin, α_2-macroglobulin, α_1-antitrypsin (AAT), α_1-lipoproteins, human serum albumin (HSA), pre-albumin, and complements [32], as shown in Figure 22.4. These proteins often mask other important analytes that are usually more clinically relevant than the major proteins. In fact, it has been widely mentioned that global approaches to proteome analysis detect less than 0.1% of the protein species present in a sample, which span a range of 10 orders of magnitude between the most abundant and the less abundant proteins [33].

The following discussion will focus on an example of a glycoprotein to which several sample preparation procedures have been reported, that is, erythropoietin (EPO) to which a lot of attention has been devoted for its analysis by CE.

The elimination of low molecular weight (MW) compounds from a sample is widely carried out by using microcentrifugal devices provided with membranes of different MW cutoffs. Clean-up of the rhEPO Biological Reference Preparation (BRP) of the European Pharmacopoeia (EP) was carried out by passing an aqueous solution through a microconcentrator cartridge with a 10 kDa

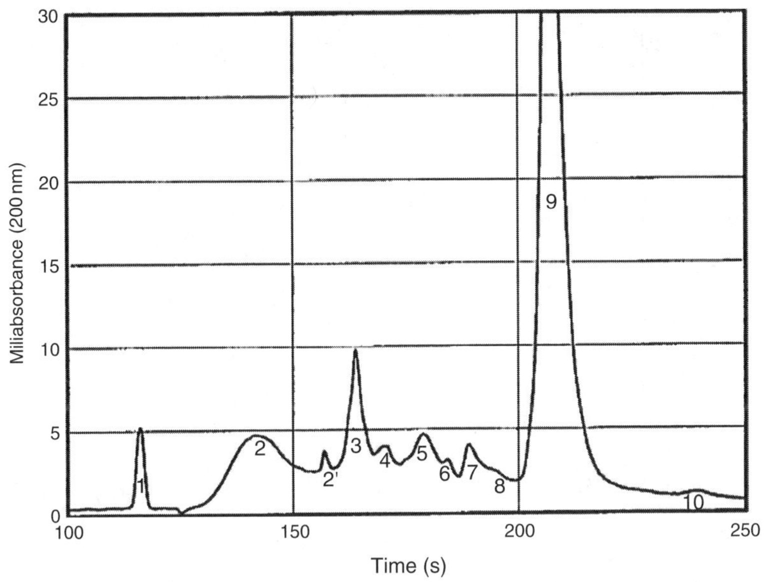

FIGURE 22.4 Electropherogram of a diluted normal control serum. Peaks: 1, neutral marker; 2, γ-globulin; 3, transferrin; 4, β-lipoproteins; 5, haptoglobin; 6, α_2-macroglobulin; 7, α_1-antitrypsin; 8, α_1-lipoproteins; 9, albumin; 10, prealbumin; 2′, complements. Analytical conditions: fused-silica capillary 25 cm × 25 μm i.d.; buffer: Beckman proprietary buffer of pH 10.0; 10 kV; 22°C. (From Chen, F.-T.A., *J. Chromatogr.*, 559, 445, 1991. With permission.)

MW cutoff membrane and followed by two consecutive rinses with 0.2 mL water [34,35]. A more detailed procedure detailing the time and the centrifugal force to be applied as well as an increase in the number of washing steps was subsequently reported [36]. Ways to decrease the preparation time by increasing the centrifugal force [37,38], and to increase the recovery by repeatedly washing the membrane and inverting the filtering device were also described [38]. An intermediate procedure led to an enhanced resolution of rhEPO glycoforms when compared with the analysis performed under identical conditions but without sample pretreatment [30]. As it will be shown later, this type of sample preparation step that eliminates excipients of low MW is especially important when the separation mode is CIEF [36,39]. This same type of procedure for eliminating low MW excipients was also useful to remove mannitol from a standard of human urinary EPO (uEPO) to be analyzed by CZE [40].

When the sample matrix contains high MW components, such as in biological fluids or in pharmaceutical products formulated with protein excipients, strategies other than MW-based sample preparation are often required. Using again rhEPO as an example, glycoform analysis of products containing HSA under the same CZE conditions as those used for samples containing low MW components was unsuccessful. An approach at separation that did not require a sample preparation step was the development of a CZE method involving the selective binding of HSA. This was accomplished on a polyamine-coated capillary by adding nickel ions to the BGE. The metal ions selectively bonded HSA and decreased its mobility leading to separation of the otherwise comigrating rhEPO [41]. Polybrene (PB)-coated capillaries with a buffer containing acetic acid and methanol also allowed the separation of rhEPO from HSA [42]. A different approach relied on performing the immunochromatographic removal of HSA before separation. Figure 22.5 shows the electropherograms of the unresolved peak obtained for an HSA-formulated rhEPO sample before HSA removal and the separated rhEPO bands after the immunochromatographic HSA depletion [43]. Figure 22.6

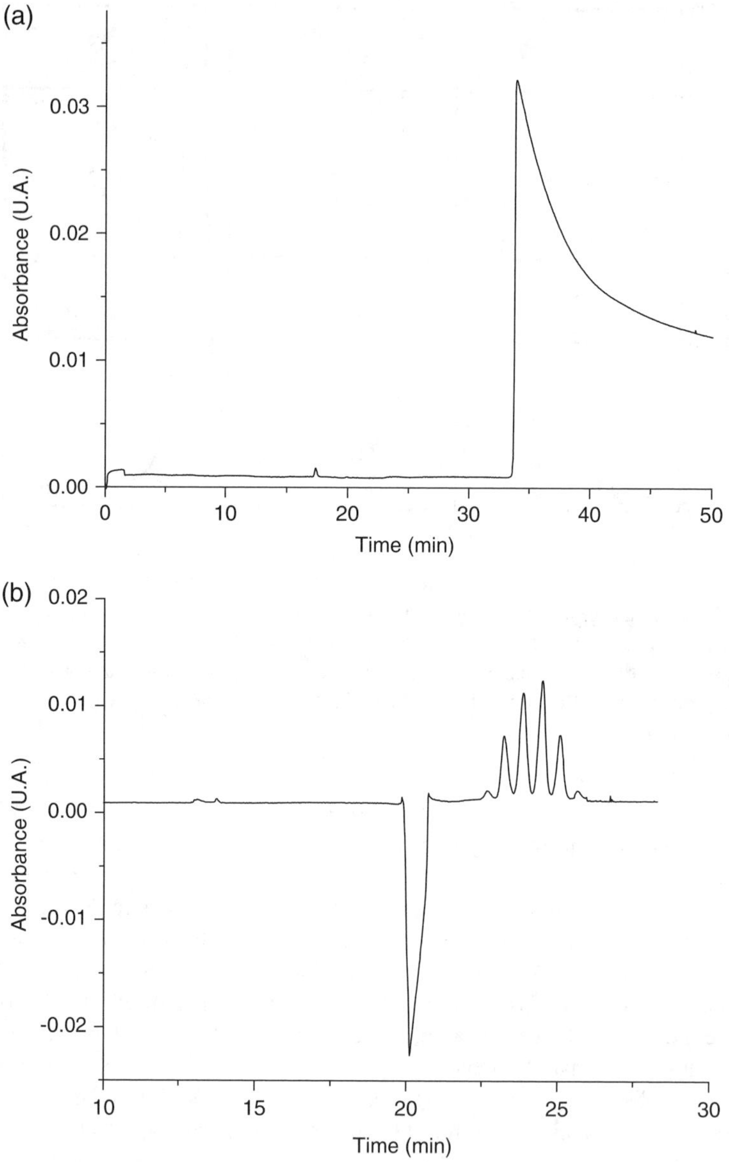

FIGURE 22.5 CZE electropherograms of rhEPO alpha formulated with human serum albumin (HSA), (a) without immunodepletion of HSA, (b) after HSA immunodepletion. Separation conditions: Uncoated fused-silica capillary 87 cm × 50 μm i.d. Separation buffer: 0.01 M Tricine, 0.01 M NaCl, 0.01 M sodium acetate, 7 M urea, and 3.9 mM putrescine, pH 5.5; pressure injection 30 s at 0.5 psi; 25 kV; temperature 35°C; detection at 214 nm. (Adapted from Lara-Quintanar, P. et al., *J. Chromatogr. A*, 1153, 227, 2007. With permission.)

shows the separation of a similar sample on polyamine-coated capillary and nickel ions added to the BGE [41]. Both methods provide separation of glycoforms and while the one performed on coated capillaries does not require a prior sample preparation step, the other method can be performed with bare fused-silica capillaries and provides better glycoform resolution.

Sample preparation steps used for other samples are mainly a function of the matrix, the most complicated ones being for the biological fluids. Specific methods used to prepare some of these samples are detailed in Section 22.5.2.

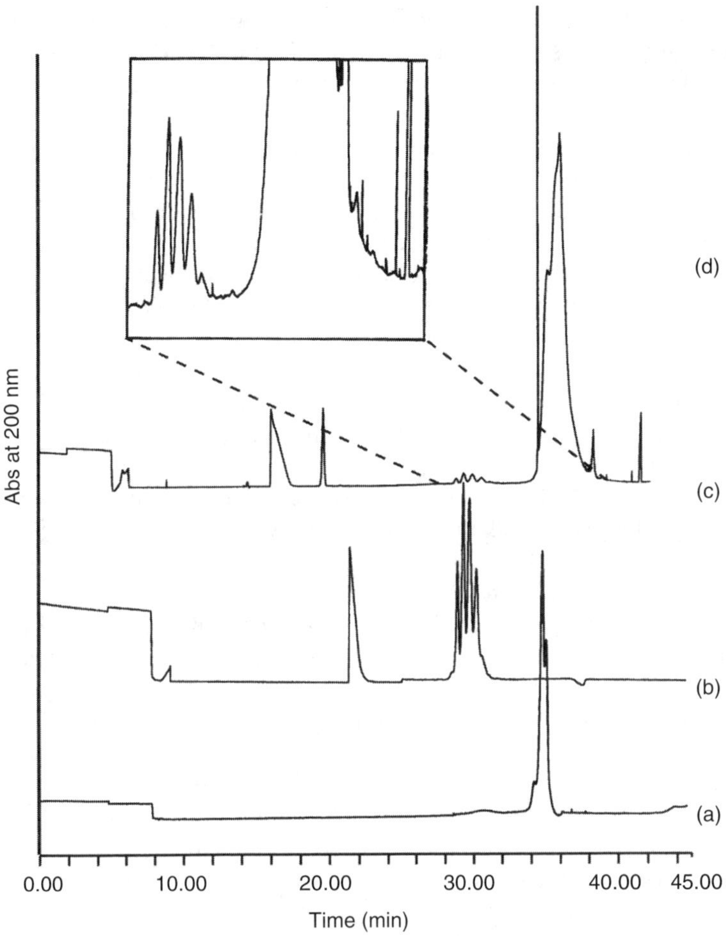

FIGURE 22.6 CZE electropherograms of (a) human serum albumin, (b) rEPO, (c) HSA-formulated rEPO. Inset (d): enlarged view showing rEPO glycoform separation in formulated product. Separation conditions: Beckman eCAP amine capillary, 200 mM sodium phosphate, pH 4.0, 1 mM nickel chloride, –15 kV (75 μA), UV detection at 200 nm. (From Bietlot, H.P. and Girard, M., *J. Chromatogr. A*, 759, 177, 1997. With permission.)

22.4.2 Separation of Glycoforms

22.4.2.1 Solutions to Protein Adsorption Problems

Very high separation efficiency can be achieved when the interaction between a protein and the inner wall of the capillary can be eliminated [44]. This is especially important for glycoproteins where this type of unwanted interaction has to be prevented to achieve the high efficiency required for the separation of glycoforms. Two basic strategies can be utilized to overcome this problem. The first one consists in changing the buffer pH and the second one in modifying the inner capillary wall.

The first, and simplest, approach is to select a buffer pH at which interactions become small. At a pH below 3, the silanol groups on the silica surface are fully protonated and bear no electric charge, thus minimizing interactions since most proteins have pI values above 3. On the other hand, at pH 8 or higher, silanol groups are fully ionized and the silica surface becomes negatively charged. So, at pH 8, a protein having a pI smaller than 8 will have a net negative charge and will be repelled by the surface, while one having a pI higher than 8 (a basic protein) will be charged positively and may be adsorbed on the capillary walls. In such a case, a buffer with a pH higher than the pI of the protein could be used for preventing their adsorption. Although these strategies have been proposed for the separation of proteins since the early days of CE [45,46], they limit the selection of buffer pH as a freely adjustable parameter for optimization. In addition, too large differences between the buffer pH and the protein pI can lead to structural changes resulting in band broadening and low recovery. These limitations have led to other approaches for preventing protein adsorption, which allow a wider selection of buffer pH.

Several methods employing capillary wall modifications have been developed and they can be grouped into several categories: dynamic coatings by the addition of a cationic modifier (usually an amine) to the BGE [47,48], permanent coatings by physical adsorption of a cationic modifier (usually a polymer) [47,48], and permanent coating by covalent bonding of a hydrophilic polymeric layer [49].

22.4.2.1.1 Cationic Buffer Additives

The addition to the buffer of cationic substrates that interact with the silanol groups at the surface of the fused-silica capillary represents the simplest method to improve the separation efficiency in glycoprotein analysis. Several effects have been attributed to these additives, which include decreasing the interaction of the analyte with the inner capillary walls, controlling the electroosmotic flow (EOF), increasing the protein solubility, and enhancing the selectivity of the separation. Although additives of several types, including neutral and ionic substances, have been reported [47,48], amine modifiers have been more frequently used for the separation of protein glycoforms. There are two main groups of amines, one comprising diaminoalkanes and the other polyaminoalkanes.

In the first group, short alkyl chain diamines have been used at millimolar concentrations for the effective separation of glycoproteins such as ovalbumin (OVA) [50,51], rhEPO [52], recombinant factor VIIa [53], granulocyte colony-stimulating factor (rhGCSF) [54], Tf [55], and human chorionic gonadotropin (hCG) [56]. The diaminoalkanes play a role in the separation by affecting both the resolution and the migration time. The separation of hCG glycoforms in uncoated silica capillaries at different concentrations of 1,3 diaminopropane (DAP) in borate buffer (Figure 22.7) shows that an increase in the diamine concentration results in an increase in resolution with a concomitant increase in migration times [56].

Replacing the terminal diamino groups of diaminoalkanes by quaternary ammonium moieties gave rise to more efficient separations [57]. Although a similar resolution was achieved with both types of additives, lower concentrations of quaternary ammonium salts were necessary and shorter migration times were obtained, an indication of effectiveness of these bis-quaternary ammonium compounds. Quaternary ammonium salts such as hexamethonium chloride and bromide and decamethonium bromide (DcBr), having alkyl chains of 6 and 10 carbon atoms, respectively, have been used for the separation of the isoforms of OVA and hCG [57], and Tf [55]. For quaternary ammonium salts, the longer alkyl chains were shown to be more effective and chloride salts led to shorter analysis time than bromide ones.

Polyaminealkanes, such as spermine and spermidine, have been used for the separation of OVA glycoforms [51]. Interestingly, while the efficiency of a polyamine for preventing protein adsorption to the silica walls was shown to increase with the number of amino groups in the chain, this number needed to be higher than three for it to become an effective adsorption inhibitor [58].

22.4.2.1.2 Capillary Coatings by Physical Adsorption

Physical adsorption of polymeric cationic coatings on the inner surface can be easily accomplished by equilibrating the capillary with a solution containing the polymeric additive. In some cases, the polymer remains so tightly bonded to the surface that this does not lose its coating even after performing rinses and, therefore, the polymer does not need to be included in the buffer.

Polybrene (PB) has been used to prepare physically adsorbed polymeric coatings [59,60]. PB is a polycationic polymer, composed of quaternary amines (N,N,N',N'-tetramethyl-1,3-propylenediamine), which strongly adsorbs to the inner surface of the capillary and reverses the

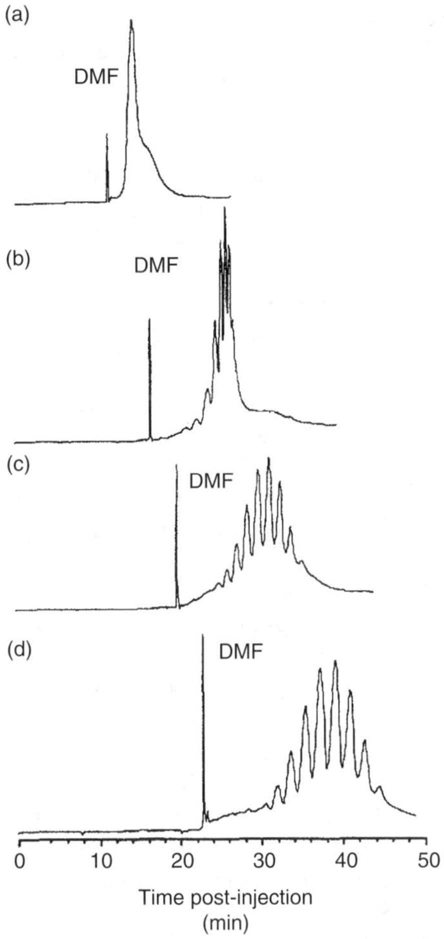

FIGURE 22.7 Effect of various concentrations of diaminopropane on the electrophoretic migration and separation of hCG (4 mg/mL). Separation capillary: 100 cm × 50 μm fused silica. Separation buffer: 25 mM borate (pH 8.8). Other separation conditions: 25 kV, 28°C. Detection: 200 nm. The diaminopropane concentrations used were: 0 (a), 1.0 (b), 2.5 (c), and 5.0 mM (d). (From, Morbeck, D.E. et al., *J. Chromatogr. A*, 680, 217, 1994. With permission.)

EOF in acidic buffers. In these buffers, most proteins are positively charged and are repelled from the surface. This strategy has been used for the separation of rhEPO glycoforms [37], however, with a lower resolution than that obtained with the diaminoalkane putrescine (1,4-diaminobutane, DAB). Its main advantage resides in that this coating allows the use of volatile buffers compatible with MS. Glycoforms of ribonuclease B (RNase B) and horseradish peroxidase (HRP) [61] have also been separated using this strategy. Other polymers prepared from 1,3-propylenediamine and having different charge density have also been reported [62].

PB has a limited stability and is degraded after a few separations. Better coating stability was shown for multiple ionic polymer layers. They are prepared from a first layer made up of a cationic PB coating that is then covered with another physically adsorbed layer from an anionic polymer such as dextran sulfate [63] or poly(vinylsulfonate) [64]. In these coatings the EOF is essentially constant at most pH values. However, because of the negative charge at the capillary surface, positively charged proteins could adsorb to the coating and as such cannot be separated in this type of capillaries. The use of a third layer composed of a cationic polymer, usually PB, has overcome this limitation and has improved the stability of the coating [63,65]. Several groups [66–69] have used commercial reagents that produce a double ionic layer (e.g., CEofix-CDT kit; Analisis, Namur, Belgium) for the analysis of Tf isoforms in human serum as will be shown in detail in Section 22.5.2.

Recently, a new polyacrylamide (PAA) derivative that allows dynamic coating (UltraTrol LN-; Target Discovery, Palo Alto, CA) has been reported for the efficient separation of glycoforms of ATT [70]. A comparison of the ability of PB and UltraTrol LN for the separation of bovine alpha-1-acid glycoprotein (AGP), bovine serum fetuin (BSF), and rhEPO sialoforms and their application coupling with MS has been reported [30,71]. Better resolution of sialoforms was obtained with capillaries coated with UltraTrol-LN in acidic volatile buffers used for coupling CE to ESI-TOF MS.

22.4.2.1.3 Capillary Coatings by Covalent Binding

The covalent binding of polymers to the inner capillary walls leads to permanent coatings with a thin layer of polymeric material. In a way similar to physically adsorbed polymeric coatings, they minimize protein adsorption by shielding the silanol groups at the surface. In addition, these types of coatings eliminate or fix the EOF over a wide range of buffer pH allowing for a better control on the separation. However, good permanent coatings are more difficult to prepare than dynamically prepared coatings. Typically, preparation is carried out in a two-step process. First, a bifunctional compound, usually a silane derivative, is reacted with the silanol groups at the capillary surface through one of its functional groups, giving rise to a first layer and leaving the second functional group available to be covalently bound to the second layer. The second layer is produced by reacting a monomeric unit and polymerizing it. In some cases, an out-of-column synthesized polymer is bonded directly to the first layer. Several hydrophilic polymers such as linear and cross-linked PAA [72,73], methylcellulose and dextran [74], and polyvinyl alcohol (PVA) [75] to cite only a few have been developed for the separation of proteins. Commercial hydrophilic-coated capillaries have also been used for the separation of recombinant human tissue plasminogen activator (rhtPA) glycoforms [76].

The importance of using coated capillaries for glycoform separations has been shown by Thorne et al. [77] who demonstrated that rhtPA does not migrate in uncoated capillaries, even when ε-aminocaproic acid was added to the separation buffer. However, when the same buffer was used with capillaries having hydrophilic coatings, rhtPA could be separated into several isoforms. They also obtained better separations using PVA-coated capillaries than with PAA-coated ones.

The importance of additives to the separation buffer, even for coated capillaries, should be emphasized here. Both zwitterionic buffers and tensioactive additives have been shown to improve the separation and recovery of rhtPA with PVA-coated capillaries [77]. On DB1 capillaries, a dimethylpolysiloxane coating widely used in gas chromatography (GC), optimal separations of fetuin, rhEPO, and AGP isoforms were obtained with acidic acetate buffers containing 0.4–0.5% (w/v) hydroxypropylmethycellulose (HPMC). In this case, the addition of a neutral polymer to the separation buffer makes the EOF almost negligible and protein interactions with the inner surface of the capillary were prevented [78].

The importance of the elimination of protein adsorption to the surface to obtain good glycoform resolution should also be emphasized. This is well illustrated in the separation of rhBMP-2 mentioned earlier, where the use of a commercial hydrophilic-coated capillary with a phosphate buffer without any additive led to efficiencies in excess of 7×10^5 plates/meter and up to nine isoforms differing only in the number of mannose residues being separated (Figure 22.8) [31].

22.4.2.2 Factors Affecting the Resolution of Glycoform Peaks

Owing to the importance of the capillary on resolution, the previous section was devoted to its coating. Besides preventing or diminishing interactions with the silanol groups on the silica surface, the capillary coating modifies the EOF. In this way, the apparent electrophoretic mobilities of the glycoform peaks are modified with respect to their effective mobilities, thus allowing for the modulation of the separation. In addition, some of the compounds used as dynamic coatings have been

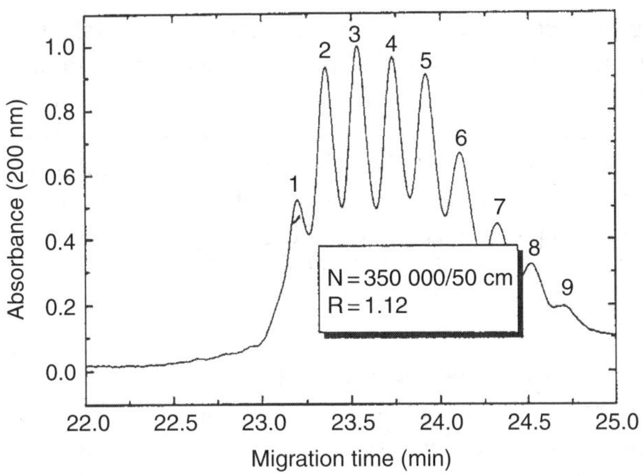

FIGURE 22.8 CZE separation of rhBMP-2 in a 50 cm × 50 μm i.d. Bio-Rad coated capillary. Sample was electroinjected for 4–8 s at 6–12 kV in 0.1 M phosphoric buffer (pH 2.5). Detection: 200 nm. Separation temperature 20°C. (From, Yim, K. et al., *J. Chromatogr. A*, 716, 401, 1995. With permission.)

shown to play a role in the separation of glycoforms through interactions with glycoproteins or their complexes, and in this sense they will be considered in this section.

In principle, each CE mode should be adequate to perform the separation of intact glycoforms as long as they differ in molecular properties that are distinguished in the particular mode used. When taking into account that the change in mass between two glycoforms is often almost negligible compared with the total mass of the protein and that relative changes in charge, when occurring, are more noticeable, it is not surprising that most separations have been achieved by CZE and CIEF. However, the other modes, MEKC, CGE, ACE, and CITP have also proven to be useful in some instances. The respective advantages and drawbacks of different modes have been discussed for some glycoproteins [36,76,77,79,80]. The following discussion will focus on aspects of the effect of factors affecting glycoform resolution that can be helpful for methods development together with providing examples to illustrate them. Although this section is organized according to separation modes, it does not imply that in some instances more than one mechanism may be involved in the separation achieved. Details can also be found in Section 22.5.

22.4.2.2.1 Capillary Zone Electrophoresis

CZE was used for the separations of forms of glycoproteins, namely iron-free Tf and RNase almost two decades ago [79,81]. Different factors are known to influence the CZE separation of glycoforms and, when possible, the influence of individual variables will be considered. However, as it will be mentioned, in several instances an interconnection between the effects of two or more factors exists.

The resolution between two peaks can be described [45] as

$$R_S = \frac{\Delta\mu_e}{4\,(\mu_{EOF} + \mu_e)} N^{1/4}, \tag{22.2}$$

where $\Delta\mu_e$ is the difference in the electrophoretic mobility between two peaks, μ_{EOF} is the electroosmotic mobility, and μ_e is the mean electrophoretic mobility of the two peaks. Consequently, a decrease in the μ_{EOF} should result in an increase in resolution for other factors remaining constant. An option to selectively modify μ_{EOF} is through manipulation of the zeta potential of the capillary and ways of achieving low values of μ_{EOF} by using dynamically or covalently coated capillaries have been described in the previous section of this chapter. Other options to increase resolution would be to increase the capillary plate number, to decrease the mean electrophoretic mobility, to increase the difference in the electrophoretic mobility of the two peaks, or to match the electroosmotic and the mean electrophoretic mobilities of different signs so the vectorial result is close to zero. The following discussion will briefly present the influence of some factors on these parameters. Buffer additives, which can act as capillary coatings, will also be considered in this section with respect to their effects on resolution by a joint action with other additives.

22.4.2.2.1.1 Buffer pH A variable that is typically changed to optimize differences in effective mobilities between glycoforms is the buffer pH. This frequently used approach derives from the potential improvement in resolution when using a buffer at a pH close to the (pI) of the glycoprotein where both its charge and mobility are lower. In other words, the closer the pH and pI values the higher the chances of increasing charge differences between two glycoforms. An example of this effect was shown for rhtPA, a protein with pIs in the range 4.5–6.0, for which better resolution was obtained at pH 3.6 than at pH 2. [82]. The enhancement of resolution is frequently accompanied by an increase in the analysis time as a consequence of the decrease in apparent mobility. For example, enhanced resolution at the cost of doubling the migration time was obtained by decreasing the pH from 5.5 to 4.5 when analyzing the novel erythropoiesis-stimulating protein (NESP), a protein known to be more acidic than rhEPO [83]. The same effect was observed for AGP, a very acidic protein (pI 1.8–3.8), when the buffer pH was changed from 4.5 to 3.5. In this case, the increase in analysis time was compensated by using a shorter capillary, while maintaining baseline resolution (Figure 22.9) [84]. While the previous examples were carried out on bare fused-silica capillaries, a similar phenomenon has been observed on coated capillaries. Resolution of fetuin (pI 3.2–3.8) glycoforms on DB-1 coated capillaries gradually increased for the pH buffer going from 9.0 to 5.0 [78].

In some instances, the choice of the most appropriate buffer pH precludes the need for buffer additives. This is well illustrated in a study of rhGCSF that showed that although additives, mainly DAB, could be effective in enhancing the separation of glycoforms, the pH control in borate buffer

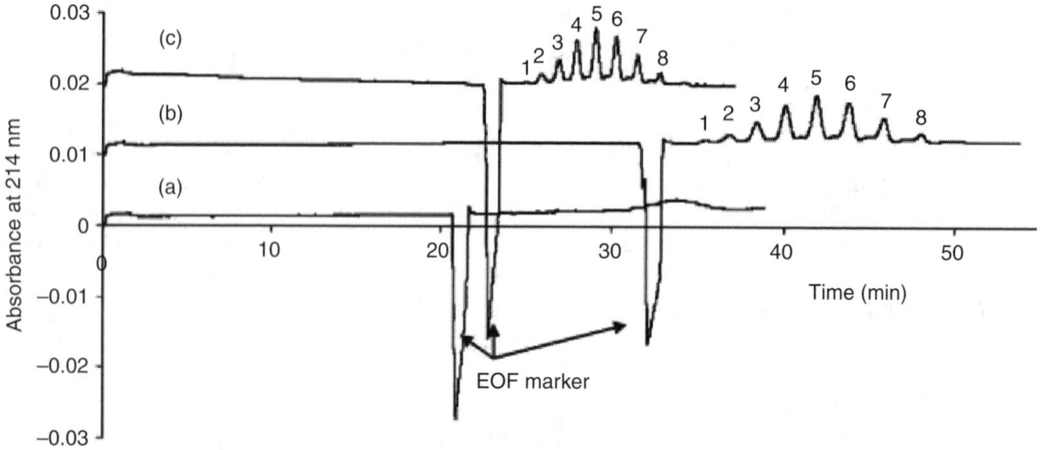

FIGURE 22.9 CZE electropherograms of standard human AGP. Conditions: (a) uncoated capillary, 87 cm × 50 μm; separation buffer: 0.01 M Tricine, 0.01 M NaCl, 0.01 M sodium acetate, 7 M urea and 3.9 mM putrescine, pH 5.5; injection 30 s at 0.5 psi; 25 kV; 35°C; detection at 214 nm; pH of separation buffer 4.5, other conditions as in (a); (c) capillary length 77 cm, other conditions as in (b). (From Lacunza, I. et al., *Electrophoresis*, 27, 4205, 2006. With permission.)

was enough to obtain baseline resolution of the two glycosylated, the nonglycosylated, and the desialylated forms [54].

22.4.2.2.1.2 Buffer Ionic Strength The ionic strength of the separation buffer is another factor to be taken into account when trying to optimize glycoform resolution. Generally, an increase in ionic strength increases resolution by decreasing the EOF and the analyte apparent mobility. However, anomalous results have been observed. This is the case for formic acid used as the separation buffer in the CZE-MS analysis of RNase B on PB-coated capillaries. Increased resolution and decreased EOF were observed when the formic acid concentration was increased from 0.1 M to 2.0 M. Interestingly, at 0.5 M formic acid, resolution was worse than for any other concentration although the EOF was lower than at 0.1 M formic acid. This behavior was interpreted as resulting from conformational changes of the protein [61].

22.4.2.2.1.3 Borate Buffer In view of its ability to resolve peaks not solely based on charge differential borate has become a substance of major importance in the resolution of glycoforms. The formation of complexes between hydroxyl groups and borate ions is at the basis of the separation mechanism, making it possible to resolve peaks of glycoforms with the same charge. Borate complexation is usually favored with *cis* 1,2-diols over *trans* 1,2-diols, but some carbohydrates do not follow this rule [85]. For a constant amount of carbohydrate, the complex concentration increases with increasing borate concentration and pH as a result of higher alkaline borate ion concentration, which is known to be the species that complexes with diols [86]. Asialo fetuin analyzed in DB-1 coated capillaries with Tris-borate buffer at pH 8.5 and containing HPMC is an example. The addition of the cellulose polymer, which shields the capillary surface from interactions with the protein, was necessary to achieve resolution; however, excess HPMC hampered the separation, probably due to higher viscosity [78]. Depending on the polymer concentration a sieving effect could also be in play. A comparison of the separation buffers Tris-glycine, ammonium formate, and borate for the analysis of hCG glycoforms also showed an enhancement in resolution when using the borate buffer [56]. However, borate is not necessarily always the buffer of choice and depending on the glycoproteins and the specific separation conditions, other buffers provide better resolution [87].

22.4.2.2.1.4 Borate Buffer and Diamino Additives The combined action of borate and diamino additives in the separation buffer reported in 1992 by the groups of Taverna and Landers [50,82] has been widely used to achieve separation of glycoforms. In addition to those reported by the two groups, other types of bifunctional alkyl amines or bis-quarternary amines have been reported in separations involving borate-based buffers [56,57]. The necessity for the diamine to exist as a divalent cation was stated in these studies and a mechanism in which the borate-complexed protein interacts with the amino group of the capillary-bound additive was suggested. Such a partition mechanism that would involve CZE and electrochromatography components has been accepted by other authors [88].

22.4.2.2.1.5 Other Components of the Separation Buffer The use of zwitterions as buffer additives has the advantage of not affecting the conductivity of the separation buffer. These additives are thought to enhance glycoform resolution by suppressing protein–wall interactions, by forming ion pairs with the glycoproteins, and by preventing individual protein molecules from interacting with lysine residues on adjacent molecules resulting in precipitation. For example, EACA [76] and other similar ω-amino acid buffers have proven effective in the separation of rhtPA forms [77]. In this instance, the addition of a nonionic detergent such as Tween 80 increased the protein recovery. Modified celluloses added to a borate buffer in DB-17 coated capillaries allowed the separation of Tf sialoforms. Although the separation mechanism is not clear, it is thought that both charge differences and gel sieving are contributing factors [89].

22.4.2.2.1.6 CZE Separation of EPO Glycoforms To illustrate the many factors that can be modified to optimize glycoforms resolution it is interesting to follow the different approaches reported for the CZE separation of rhEPO, starting with the method reported by Watson and Yao [52]. On a bare fused-silica capillary, no separation was possible for a commercial rhEPO sample at pH values between 5 and 10 using buffers of different ionic strengths. However, at pH 6.2, the use of additives that decreased the EOF allowed some resolution, with the best results achieved with 2.5 mM DAB. Further improvement was achieved by adding 7 M urea to the buffer and the optimized conditions consisted of a separation buffer of 10 mM tricine, 10 mM NaCl, 2.5 mM DAB, 7 M urea, pH 6.2, in a 50 cm bare fused-silica capillary and applying 10 kV. Baseline resolution of six peaks was obtained in 35 min. A modified version of this method was developed and used for an international collaborative study to assess the BRP for rhEPO of the EP, which consisted in a mixture of alpha- and beta-epoetin. [34]. The modified method consisted in the addition of 10 mM sodium acetate to the separation buffer, adjusting the pH to 5.5 with 2 M acetic acid, and performing the separation at 30°C in a 107 cm capillary. In the initial study, large discrepancies in migration times were found among participants (between 32 and 120 min). These results were possibly attributable to differences of the capillary walls since capillaries from different manufacturers and even from different batches from the same manufacturer have been shown to have very different EOFs. This method was later implemented as an official method for the identification of rhEPO (see Section 22.5.1.1) [35]. A reduction of migration times from 70 to 36 min while keeping enough resolution between peaks and good repeatability has been achieved by decreasing the DAB concentration to 0.025 mM [36]. In contrast, the increased analysis time at higher DAB concentrations could be reduced by applying a higher voltage, from 15.4 to 30 kV [90]. In this instance, baseline resolution of glycoforms have been performed in less than 30 min, and this method could also be applied, although without baseline resolution, to the NESP [38,43].

The development of several CZE methods for the same glycoprotein, using buffers and capillaries other than those previously mentioned, clearly shows the versatility of CE. As another example, the early work by Tran et al. [91] on the influence of factors such as buffer pH and type, and the addition of organic modifiers showed the possibility of obtaining partial resolution of multiple rhEPO peaks with a 100 mM acetate–phosphate buffer at pH 4 in uncoated fused-silica capillaries. A marked effect on resolution was observed when the pre-equilibration time before using the capillary was increased from 4 to 10 h. The shorter pre-equilibration time and the same separation conditions have been used by others [92]. On the other hand, a different method for rhEPO analysis has been performed on C8-coated capillaries using a phosphate buffer [93]. A 300 μm i.d. tube made up of fluorinated ethylene–propylene (FEP) copolymer was used in a hydrodynamically closed separation system, which allowed enhancing the loadability. A buffer consisting on *N*-(2-hydroxyethyl)piperazin-*N′*-2-(hydroxypropane sulfonic acid) (HEPPSO), 1,3-*bis*[*tris*(hydroxymethyl)-methylamino] propane (BTP), and methylhydroxyethylcellulose (MHEC) 30,000, at pH 7.25 allowed to obtain partial resolution of seven bands for rhEPO [94]. CZE under similar conditions with a borate buffer at pH 8.8 containing MHEC was used to monitor the fractionation of rhEPO by preparative CITP [95].

The coupling of the CZE step to detection systems other than UV has required the development of separation conditions compatible to the detection system used. For instance, the presence of primary amines, such as DAB, in buffers needed to be avoided for compatibility with laser-induced fluorescence (LIF) of compounds derivatized with fluorogenic substrates through their amino groups [90]. Baseline resolution of eight peaks in approximately the same time was achieved by substituting DAB by morpholine and tricine by boric acid (to avoid potential traces of primary amines in the tricine buffer) and by adjusting the concentration of other buffer components to compensate for the increase in electrical current. In the same work, modifications were also required to achieve compatibility with MS detection where nonvolatile salts, urea, and amines should be usually avoided. A physically adsorbed polyethylenimine-coated capillary was used to overcome protein adsorption to the capillary walls in the absence of cationic additives and the use of an acetate buffer at pH 5.05 allowed the partial resolution of at least five bands of rhEPO. Other types of coated capillaries have been used for the analysis of EPO by CE-MS as detailed in Section 22.4.3.3 [30,37,42,62,96].

22.4.2.2.2 *Capillary Isoelectric Focusing*

Almost two decades ago, CIEF was described by Kilar and Hjerten [79,97] for the separation of Tf glycoforms and it has, in principle, some advantages over other CE modes, namely, (1) the large volume of sample introduced in the capillary would result in an increase in sensitivity compared with nanoliters amounts in other modes; (2) the measurement of an important molecular property, the isoelectric point (pI), is determined; and (3) a large number of experimental factors can be modified to improve resolution. With regard to the first two advantages, experimental considerations made it that their importance is not what it should be. With respect to the sensitivity issue, it should be stated that commercial ampholytes are generally not designed to work in a capillary format with UV detection and they show high absorbance at 214 nm. In fact, they are manufactured for classical IEF slab gels with detection at 280 nm, a wavelength at which the extinction coefficient of proteins is lower than at 214 nm. So, in practice, the sensitivity of CIEF has been shown in some instances to be similar to that obtained for the same glycoprotein by CZE [36]. With respect to the pI determination, the value obtained for a given analyte should be considered cautiously since, in many instances, urea or other additives that affect the pI of a protein are used in addition to the fact that the pH gradient is not always linear [98].

CIEF can be performed in two general ways: a one-step or a two-step process. In the one-step process, the capillary retains a residual EOF that allows protein bands to migrate through the detection window, while in the two-step process the absence of EOF is desirable for better focusing of the protein bands being mobilized to pass the detection point in the next step. However, even in the one-step mode a controlled EOF is necessary to achieve enough focusing before reaching the detector. Both modes have advantages and drawbacks as was shown for the effect of different factors in the separation of AGP [99].

In CIEF, the total time elapsed between sample introduction and detection depends on the focusing time and on the time for a band to reach the detection window during mobilization. These two effects can take place either simultaneously or consecutively depending on whether the process is a one-step or two-steps, respectively. So, the term "migration time" is not strictly correct in CIEF but it is used in this chapter for simplicity. It is important to note that the need for mobilization through the detector window can be eliminated in systems where whole-capillary imaging detection is used [100].

Samples are introduced in the capillary either separated or mixed from ampholytes, the latter being the most frequently used. In addition to the sample and ampholytes, additives can also be introduced in the capillary in the so-called sample mixture, as will be shown in later text.

The different types of capillary coatings providing reduced or no EOF have been described in a previous section. A comparison of different capillary coatings for the resolution of rhtPA glycoforms has been performed and, in some cases, capillaries with different degree of EOF reduction have shown similar resolution, a situation that was probably due to the presence of additives controlling the EOF [101]. While similar resolution has been achieved with different coatings, the stability of the coatings in terms of the number of runs performed was evaluated in the analysis of recombinant human antithrombin III (rhATIII) and the recombinant human immunodeficiency virus envelope glycoprotein rgp 160s. Higher stability was observed for dextran-coated [102] and PVA-coated [103] capillaries than for PAA-coated capillaries. Besides the capillary coating, other variables that can be controlled for improving resolution of glycoforms are described in the following text.

22.4.2.2.2.1 Additives for Reducing EOF Besides the use of coated capillaries for controlling the EOF, the addition of substances aimed toward reducing the EOF through an increase in the viscosity is frequently used. This is the case for poly(ethylene oxide) [36,39,77] and HPMC [101,103]. The latter has proven to be effective even when using bare fused-silica capillaries, where it acts as a dynamic coating [104].

22.4.2.2.2.2 Ampholytes This is one of the variables to which a lot of attention has been paid. Both the nature and pH of ampholytes as well as the combination of several pH ranges have been

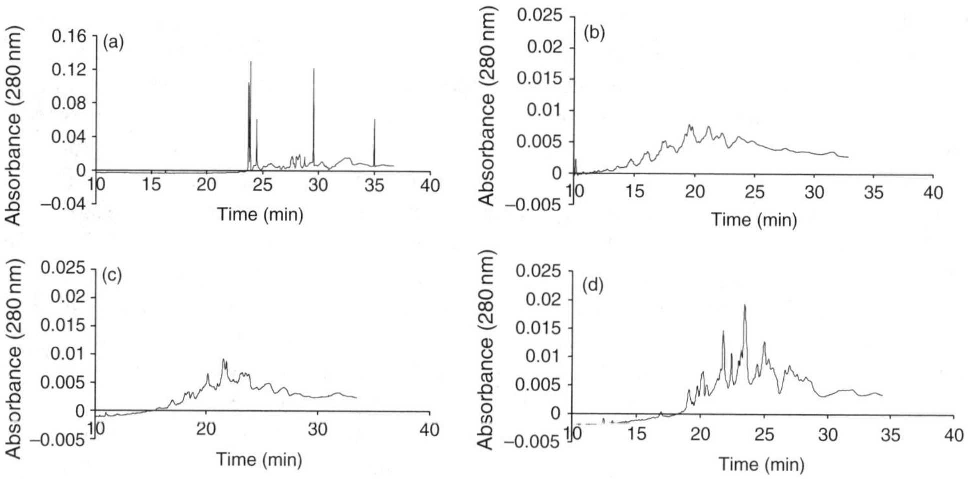

FIGURE 22.10 Effect of types of ampholytes on separation of AGP forms. Total ampholyte concentration in the sample mixture: 6.3% (v/v). Types of ampholytes, specified by their pH range (and ratios (v/v)): (a) range 3–10; (b) mixture of ranges 3–10 and 2.5–5 (1:1); (c) mixture of ranges 3–10, 2.5–5, and 2–4 (1:1:1), (d) mixture of ranges 3–10, 2.5–5, 2–4, and 3–5 (1:1:1:1). Other analytical conditions: PVA-coated capillary, 27 cm (effective length 20 cm). Sample mixture in CIEF gel: 1 mg/mL AGP, 5.6 M urea, 20 mM NaCl, and ampholytes. Catholyte: 20 mM NaOH titrated with H_3PO_4 to pH 11.85. Anolyte: 91 mM H_3PO_4 in CIEF gel. Focusing: 10 min at 20 kV. Mobilization step: 20 kV and 0.5 p.s.i. N_2 pressure. Temperature: 20°C. Detection: 280 nm. (From Lacunza, I. et al., *Electrophoresis*, 28, 1204, 2007. With permission.)

considered. In general, better resolution is attained with mixtures of ampholytes. The influence of the commercial source of ampholytes on resolution has been shown, even when only a wide-range ampholyte solution is used [77]. Similarly, for a given total concentration of ampholytes mixed in constant proportions, the resolution was shown to depend on the nature of the ampholytes [105]. Besides providing different resolutions of rhtPA glycoforms, commercial ampholytes have different optical properties that have an impact on detection [106]. Ampholytes from different sources have been shown to be different in nature and in focusing properties [107,108]. As such, mixtures from different manufacturers are generally preferred, since the different nature of the compounds from the different sources may help to achieve better focused zones by reinforcing the pI range of interest. This effect was clearly seen for the separation of rhEPO [36] and AGP glycoforms [109] (see Figure 22.10).

Apart from the nature of ampholytes, their proportion and total concentration are other factors to be considered. Usually, higher proportions of an ampholyte or narrow-range ampholytes close to the pI of the glycoprotein provide better resolution [103]. For the amount of total ampholyte, a compromise is generally achieved since higher percentages not only provide better resolution but also lead to higher background absorption and, consequently, decreased sensitivity [39,103,109].

22.4.2.2.2.3 Agents to Avoid Precipitation Protein precipitation is favored during the focusing process where a protein is concentrated in a narrow zone close to its pI, and associated salts are separated from it. The addition of nonionic surfactants, organic modifiers, or chaotropes to the sample can be used to decrease chances of precipitation [110]. The presence of urea in the separation buffer has been shown to markedly improve the resolution of AGP an EPO glycoforms [36,100,109], being also used to obtain adequate separation of rhtPA glycoforms [77,104]. Triton X-100 was found to be unnecessary when urea was used to separate rhtPA glycoforms [101]. This detergent was used to avoid precipitation for the analysis of rhATIII [102]. However, it may be necessary to resort to a combination of agents. This was the case for recombinant human immunodeficiency virus envelope glycoprotein where several additives including a chaotrope (urea), a

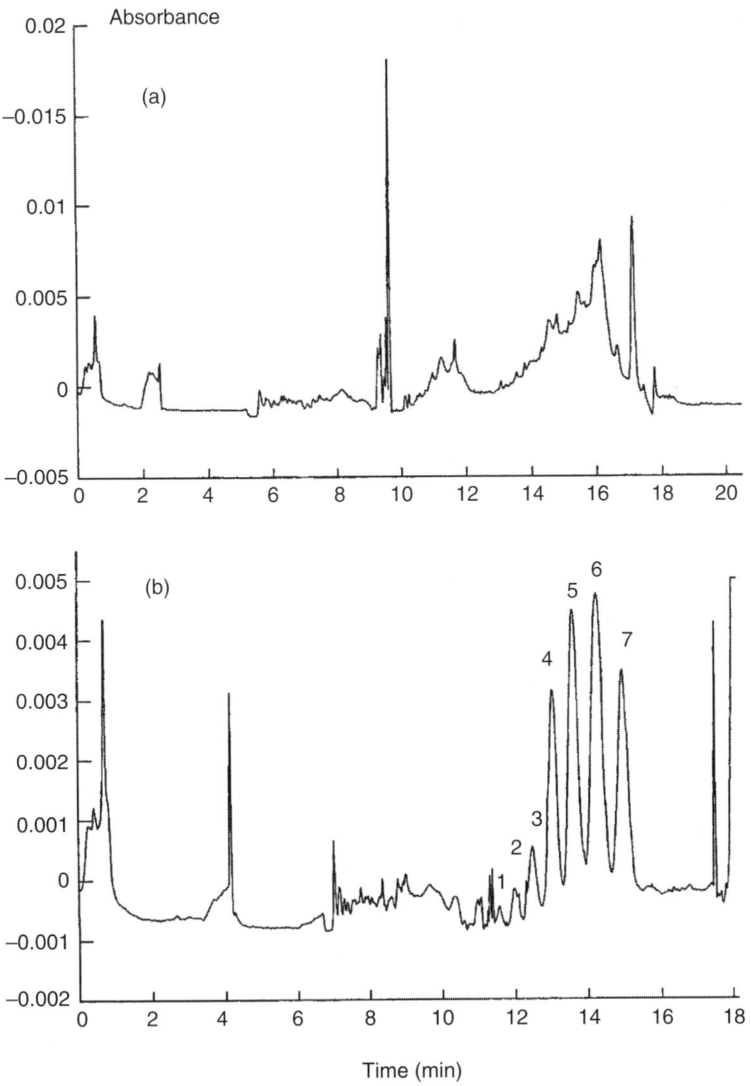

FIGURE 22.11 Effect of urea on CIEF separation of rhEPO. (a) Without urea in the sample mixture, and (b) with 7 M urea. Analytical conditions: polyacrylamide-coated capillary 27 cm (effective length 20 cm) × 50 μm i.d. Focusing: 6 min at 25 kV. Sample mixed with CIEF gel and ampholytes. Ampholyte mixture (1:2, v/v) of pH 3–10 and 2.5–5 ranges. Detection: 280 nm. (From Cifuentes, A. et al., *J. Chromatogr. A*, 830, 453, 1999. With permission.)

zwitterion [3-(cyclohexylamino)-1-propanesulfonic acid, CAPS], and a sugar (saccharose) provided the best results [103].

22.4.2.2.2.4 pH Extenders In order to avoid the focusing of glycoforms past the detection window, a situation that would prevent their detection, it is necessary to add a substance with an appropriate pI to block that zone of the capillary. It is referred to as a pH extender and the most frequently used is *N,N,N′,N′*-tetramethylenediamine (TEMED). It is usually added to the sample mixture at a concentration proportional to the length of capillary to be blocked. The larger the pH extender concentration is the farther away a glycoform will focus from the detection point and, consequently, the greater the migration time will be, as was observed for rhtPA without any marked influence on resolution [105]. The same authors also reported that for TEMED concentrations about 10 times higher, in the range of 3.75–7.5% (v/v), the effect on migration time was not noticeable [101]. Caution should be taken when using TEMED as it has been shown to damage the capillary coating [99,103]. Depending on the pI of the glycoprotein, other less basic extenders, such as alanine, can be used [99].

22.4.2.2.2.5 Focusing Time and Voltage For the one-step CIEF separation of rhtPA glycoforms, an increase in voltage led, as expected, to a decrease in migration times without any influence on resolution [105]. For the two-step process, the two factors involved on the focusing step (i.e., voltage and time) have an impact on resolution. According to the theory [111], increasing voltages up to 25 kV provided slightly better resolution of rhEPO glycoforms; however, loss of resolution was observed at higher voltage, probably due to an excess of Joule heating [36].

It is usually considered that focusing is finished when the current decreases below 10% of its initial value and stabilizes. However, in practice, focusing times as short as 0.5 min and corresponding to 69% of the initial current have been shown to lead to the same resolution of EPO isoforms as for longer focusing times. This can be understood taking into account that during the pressure-mobilization step, the voltage is applied to avoid band broadening resulting in a continuation of focusing [36]. In contrast, focusing times longer than needed may have a detrimental effect not only on the analysis time but also on the quality of the separation. The total depletion of salts due to a long focusing time led to precipitation of monoclonal anti-alpha-1 antitrypsin [100]. In addition, focusing times that are too long may lead to greater anodic drift and can affect the isoform separation of highly acidic proteins such as AGP [109]. Incomplete focusing due to times shorter than needed, however, may lead to double peaks. These examples show that careful optimization of the focusing time is required.

22.4.2.2.2.6 Presence of Salts in the Sample Mixture The presence of salts in the sample mixture remains a controversial point. Some authors have claimed that the ionic strength of the sample should be as low as possible while others have found it to help improve resolution. For the analysis of anti-alpha-1 antitrypsin [100], AGP [109], or rhEPO [36,39], the presence of salt was necessary for the separation of isoforms. This effect is likely related to prevent protein precipitation and denaturation, although other effects such as improved focusing or masking of the silanols of the capillary walls cannot be excluded.

22.4.2.2.2.7 Mobilization Step In the two-step CIEF mode the mobilization step is performed after focusing. Hydrodynamic (more frequently performed by applying pressure rather than by vacuum) and chemical mobilization are the two general ways of driving the focused zones to the detection point.

As indicated earlier, when the mobilization is performed by pressure, a voltage is applied simultaneously, usually at the same strength as the voltage used during the focusing step. However, a higher voltage may be needed in some instances as shown for a monoclonal antibody (mAb) [112]. Chemical mobilization with a strong electrolyte has been shown to lead to better resolution than by using a weak one [97].

Performing hydrodynamic and chemical mobilizations together has been shown to be beneficial for the separation of rhEPO [39] and even necessary for the separation of glycoforms of acidic proteins such as AGP [109] (see Figure 22.12). The addition of phosphoric acid to the catholyte, which consists of sodium hydroxide, at the very beginning of the focusing step promotes chemical mobilization that continues when the pressure is applied at the hydrodynamic mobilization step.

22.4.2.2.3 Micellar Electrokinetic Chromatography

There are many factors involved in the MEKC separation mode that contribute to the complexity of its mechanism. On the other hand, they provide a large number of variables that can be modulated to optimize the separation. MEKC has been used on a few occasions for the separation of glycoproteins,

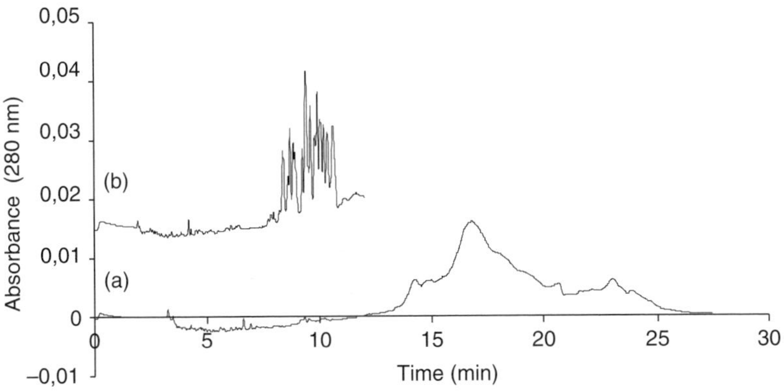

FIGURE 22.12 One-step CIEF of AGP. PAA coated-capillary 27 cm (effective length 7 cm) × 50 μm i.d. Voltage: –20 kV. Temperature: 20°C, Detection 280 nm. Sample mixture: 5.6 M urea, 1.7% (v/v) TEMED, 9.7 % (v/v) ampholytes in the following distribution of pH ranges: 3–10, 2.5–5, 2–4, 3–5 (2:2:3:3) and 1 mg/mL AGP in CIEF gel. (a) anolyte: 91 mM H_3PO_4 in CIEF gel, catholyte: 20 mM NaOH, (b) anolyte: 91 mM H_3PO_4 in CIEF gel, catholyte: 20 mM NaOH titrated to pH 11.85 with H_3PO_4.

usually under conditions involving the use of borate or phosphate salts and SDS as the micellar agent. Factors, such as borate/phosphate and SDS concentrations, and the nature and pH of the buffer can be modified to optimize resolution. RNase B was separated into five peaks, each one of them corresponding to one of the Man5 to Man9 oligomannose structures, in a fused-silica capillary with a buffer consisting of sodium phosphate, SDS, and sodium tetraborate [113]. For recombinant human interferon-γ (rhIFN-γ), over 30 species were resolved giving rise to three groups of peaks, PG1, PG2, and PG3, as shown in Figure 22.13 [114]. The increase in the concentrations of the borate buffer and SDS led to a concomitant increase in separation efficiency and migration times. The comparison between phosphate buffers at pH 6.5 and 7.5 and borate buffers at pH 8.5, at a constant SDS concentration and at buffer concentrations chosen to generate similar currents, showed that the best resolution was obtained with borate buffers. At a higher pH (9.5) and a lower borate concentration than that used at pH 8.5, higher current, slower separation, and lower efficiency were observed. The better resolution obtained with the borate buffer pH 8.5 was likely due to the effect of the complexation between borate anions and diols mentioned earlier. It has been hypothesized that the early migrating glycoforms would be those with the larger glycan structures and that this should be taken into account when designing a separation method or when assigning peaks to specific glycoforms. The reasoning behind this hypothesis is that at high borate concentrations, sugar residues would be extensively complexed and that the resulting negative charge, when added to that from sialic acid residues, would cause repulsion between SDS micelles and the borate-complexed glycoprotein. According to the authors, the repulsion intensity would depend on the glycan size. The same study reported the possibility of separating the glycoforms of RNAse B and HRP, while the separation of BSF led to the separation of only one major peak and three minor, broad ones, and no separation at all of AGP glycoforms. According to the authors, the lack of separation for proteins glycosylated at multiple sites such as BSF and AGP could be due to the fact that the heterogeneity of individual oligosaccharides does not give rise to measurable differences in carbohydrate contents.

22.4.2.2.4 Capillary Electrophoresis in SDS

In conventional polyacrylamide gel electrophoresis (PAGE) as well as in its capillary counterpart CE-SDS, proteins form stable complexes with SDS and are separated on the basis of their size. The separation of glycoproteins is known, however, to deviate from the linear relationship of migration versus MW observed with nonglycosylated proteins. This anomalous behavior arises from the fact that carbohydrate moieties bind with much lower amounts of SDS than expected for a corresponding

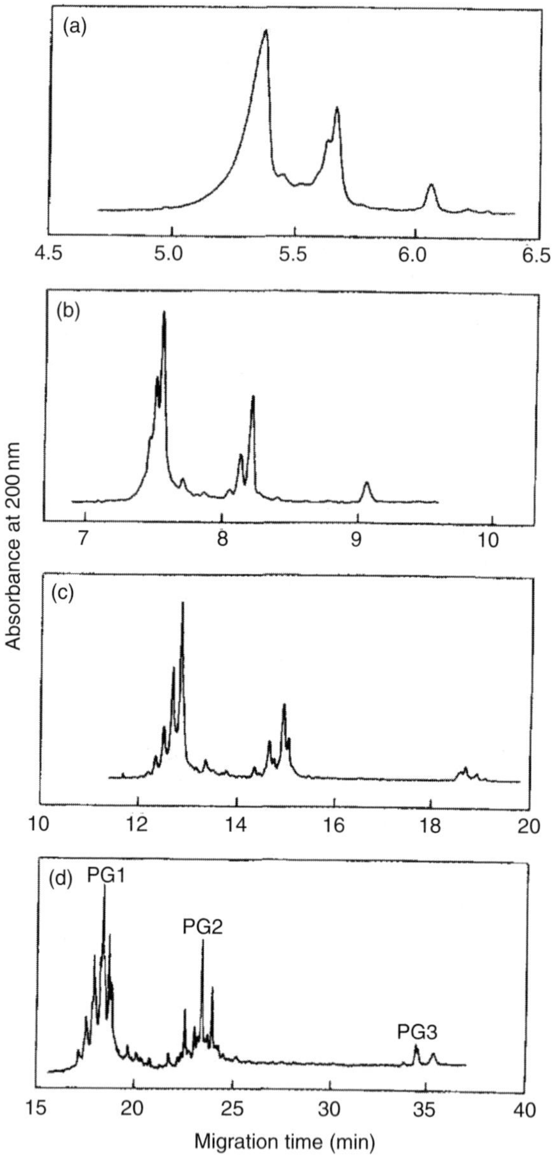

FIGURE 22.13 Optimization of the separation of IFN-γ glycoforms by MEKC. Borate/SDS buffer at the following concentrations: (a) 40 mM borate, 10 mM SDS; (b) 40 mM borate, 100 mM SDS; (c) 400 mM borate, 10 mM SDS; (d) 400 mM borate, 100 mM SDS. Fused-silica column 57 cm × 50 μm i.d. Injection 5 s of 1 mg/mL protein in 50 mM borate, 50 mM SDS, pH 8.5. Voltage: 22 kV. In (d), the main groups are designated PG1, PG2, and PG3 in order of migration. (From James, D.C. et al., *Anal. Biochem.*, 222, 315, 1994. With permission.)

peptide of similar MW. Advantages were taken from this fact in order to differentiate glycosylated fetuin from the nonglycosylated protein produced by enzymatic hydrolysis [115]. The release of the carbohydrate moiety gives rise to a decrease in size that translates into an increase in the charge-to-mass ratio since the charge imparted by SDS to the carbohydrate moiety is low compared with that imparted by the interaction of the surfactant with the peptide core. The resulting differences in mobility make it possible to monitor the deglycosylation reaction. Similarly, rhATIIIα was separated from rhATIIIβ by CE-SDS due to the lack of glycosylation at Asp 135 in the latter [116].

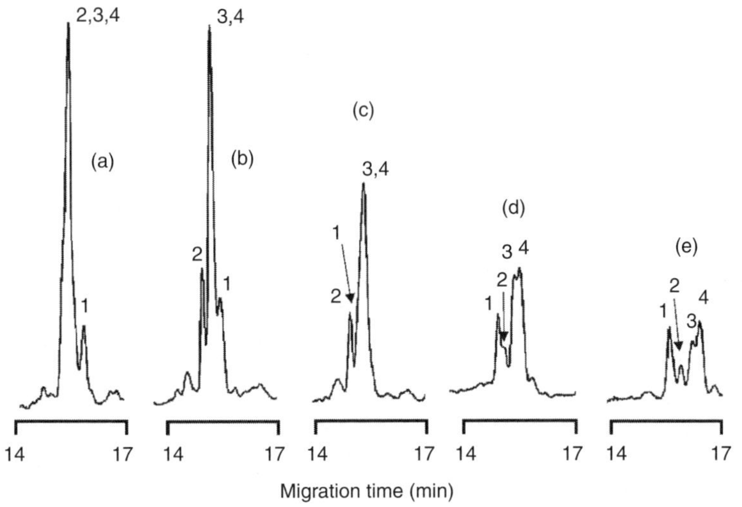

FIGURE 22.14 Changes in the migration times of OVA glycoforms with LCA concentration. Capillary: linear PAA-coated 58 cm × 50 μm i.d; Buffer: 100 mM phosphate pH 6.8, (A) not containing LCA, (b–e) containing 0.4, 0.8, 1.4, and 2.0 mg/mL LCA, respectively. Voltage: 20 kV; Detection: 214 nm. (From Uegaki, K. et al., *Anal. Biochem.*, 309, 269, 2002. With permission.)

22.4.2.2.5 Affinity Capillary Electrophoresis

Glycoform separations by ACE have been carried out by taking advantage of the known interactions between lectins and carbohydrates. The "partial filling technique" in which only a portion of the capillary is filled with an affinity ligand present in the separation buffer before injecting the sample was shown to be useful for separating AGP glycoforms. The separation of two peaks based on the dependence of the strength of the AGP-concanavalin A (Con A) interaction as a function of the content in biantennary glycans was achieved [117]. Differential interactions between lectins and glycans are useful not only to perform glycoform separations but also to estimate the values of the apparent association constant between the lectin and each of the separated glycoform. PB-coated and PAA-coated capillaries completely filled with a buffer containing the lectin *Lens culinaris* agglutinin (LCA) were shown to be suitable for separating glycoforms of RNase B and OVA, respectively, and for calculating the affinity constants without a prior separation of the glycoforms outside the capillary. Figure 22.14 shows the separation obtained for OVA glycoforms and the influence of the LCA concentration [118].

22.4.2.2.6 Capillary Isotachophoresis

Few studies on the use of the CITP electrophoretic mode have been reported for the separation of protein glycoforms. The separation of AGP isoforms by CIEF mentioned previously [109] can be considered to involve an isotachophoretic mechanism taking into account the concurrence of focusing and chemical mobilization [119]. Preparative CITP has been used to fractionate rhEPO with a leading electrolyte containing chloride as anion and BTP as counter ion, and a terminating electrolyte containing glycine as anion and BTP as counter ion. Under these conditions, the fractions contained mixtures of glycoforms enriched in the predominant glycoform of each fraction. The process was monitored by CZE [95].

22.4.2.3 Reproducibility Problems

With respect to the issue of reproducibility, the separation of protein glycoforms by CE is faced with the same kind of problems than for any other protein. However, the consequences of a lack of adequate precision can have a detrimental impact due to the fact that differences in the charge/mass ratio or pI among isoforms are so small that the different peaks usually migrate very close to one another, making it sometimes impossible to compare samples. For example, peaks of rhBMP-2 separated by CZE on the basis of a single mannose residue between adjacent peaks differ in migration times by only 0.15–0.22 min or $0.01 \times 10^{-4}\ cm^2V^{-1}\ s^{-1}$ in their respective mobility [31]. This can give rise to unreliable comparisons of samples since the difference in migration time between two consecutive peaks can be in the same range as the migration time dispersion for a given peak [38]. For instance, this problem made it impossible to correctly compare a reference hCG sample with a sample obtained from the urine of a patient with metastatic choriocarcinoma [56]. Solutions to this type of problems require a two-pronged approach. First, experimental factors affecting reproducibility must be carefully controlled. Once this is achieved, reliable migration parameters must be used for band assignment.

Some experimental factors such as buffer preparation are usually well controlled. However, there are other variables to which not enough attention is paid. One of these, which is of key importance for achieving good intra- and interday repeatability, is capillary conditioning. All of the steps performed for the conditioning of brand-new capillaries as well as the washing steps carried out between runs and before the storage of capillaries are decisive in achieving good precision. This may help in explaining the low reproducibility reported for some methods [17], while the same methods when applied to the same analytes showed highly reproducible results in other studies [36,38,90].

Another factor that has a large influence on migration time reproducibility, especially for coated capillaries, is capillary aging. Buffer components that can alter the coating need to be identified and avoided. However, even when using the appropriate buffers, capillaries gradually degrade and they should be discarded if the resolution achieved is no longer within easily verifiable limits or if peak assignments cannot be reliably performed.

Once experimental factors are controlled as much as possible there are several ways to minimize the problem of migration time reproducibility through the use of migration parameters. One of the more common approaches is the use of internal standards from which it is possible to calculate the migration time of each isoform relative to that of the internal standard. This approach has been used in the analysis of several glycoprotein isoforms [38,39,109,120]. For example, in the CIEF separation of rhEPO glycoforms, the intraday migration time reproducibility was improved from an RSD value of around 1.8–0.3% when using an internal standard [39]. In another approach and using the CZE analysis of rhEPO as a model, a statistical program was developed in order to choose, among others, the best migration parameter to correctly assign peaks and reliably compare different samples. The program estimated the probability of correct assignment for each migration parameter [38] and was used for the comparison of rhEPO from different samples [38,43] and for the comparison of electrophoretic profiles of AGP purified from healthy donors and cancer patients [84]. In the latter publication, it was shown that the electrophoretic mobility of each AGP peak and the migration time of each peak relative to the migration time of the EOF marker were more effective migration parameters than the migration time of each peak.

22.4.3 Detection and Identification of Forms of Glycoproteins

22.4.3.1 UV Detection

The UV absorbance detector is one of the most widely used systems in CE. Detection takes place in the same separation capillary, avoiding band broadening of the separated compounds. Since the internal diameter of the capillary is very small (25–100 μm), the detection pathlength is very short and the concentration sensitivity is typically 10–20 times poorer than that obtained with the same type of detector in other separation techniques (e.g., HPLC).

Glycoproteins, similar to nonglycosylated ones, can be monitored at 280 nm at which wavelength the aromatic residues of the polypeptide chain, tryptophan, tyrosine, and, to a lesser extent, phenylalanine give relatively good absorption. However, the detection limit at this wavelength remains in the low mg/mL range and instead, detection is usually accomplished at 200 nm where proteins present 50- to 100-fold greater absorptivity. At this wavelength, sensitivities in the μg/mL range can be obtained.

While this level of sensitivity may be enough for the quality control of glycoproteins, it is, in most cases, insufficient for the analysis of these proteins in clinical samples. Strategies based on sample concentration or protein derivatization can sometimes be used in these cases and some aspects have already been discussed in other parts of this chapter and will only be considered briefly in this section.

22.4.3.2 Laser-Induced Fluorescence Detection

For glycoproteins present at low sample concentrations (generally lower than 10^{-6} M), such as in clinical samples, fluorescence detection will likely be one of the detection methods to be improved in the future for analysis of these proteins by CE. It should be recalled that, due to the small size of the detection volume in on-column monitoring (usually 5–10 nL), high optical power (>1 mW) from the light source should reach the detection window of the capillary for intense fluorescence excitation [121]. This is more efficiently achieved using a laser as a source of light than a ultraviolet-visible (UV-VIS) lamp and confirmed by practical experience of the fluorescence detection of proteins and glycoproteins by CE.

LIF detection of glycoproteins is not an easy task. Most analytes do not produce fluorescence at the wavelength of the emission of the most frequently used lasers (those with light emission in the 400–600 nm range) and, therefore, native fluorescence cannot be obtained with these light sources. Derivatization with a fluorescent reagent is a useful strategy for LIF detection when native fluorescence is unavailable. Although several fluorogenic compounds with good chemical and spectroscopic characteristics have been described in the literature [122], this approach suffers from several problems when the analyte is a glycoprotein, including the formation of multiple reaction products [123] and the difficulties associated with derivatization of proteins at low concentration (below 10^{-8} M). In the following discussion, a few approaches reported in the literature to make LIF detection feasible for glycoproteins are presented.

Glycoproteins can be directly detected by native fluorescence owing to the fluorescence emission of the tryptophan and tyrosine residues; however, this requires the use of laser with emission in the UV region of the electromagnetic spectrum. This approach has been scarcely used [124] for glycoproteins, probably due to the high cost of such lasers and the expertise required for their maintenance. In the past few years, however, pulsed diode-pumped solid-state (DPSS) lasers with emission in the UV region (266 nm) have become affordable and this has opened the way to a more universal native LIF detection of glycoproteins.

The covalent derivatization of a glycoprotein with a fluorescent agent has been shown to be useful in some cases. As mentioned earlier, the main difficulty with this type of approach is the formation of multiple products due to the fact that each lysine residue in the peptide chain can react to a different extent with the derivatizing agent. As a result, protein molecules having a variable number of covalently bonded fluorescent tags are produced, resulting in a variable number of species with similar MW (the fluorescent moiety has a small MW in relation to that of the protein) but with different electric charges. In these cases, analysis by CZE or CIEF may give rise to broad or even multiple peaks for the same protein. On the other hand, when the separation mode is CE-SDS, narrow peaks with good efficiency are obtained and make LIF a more suitable technique for this separation mode.

Using this approach, a CE-SDS method with LIF detection has been developed and validated as a quality control procedure for the purity determination of a recombinant mAb [125]. 5-Carboxytetramethylrhodamine succinimidyl ester was used as fluorescent reagent and the optimal conditions for derivatization of the nonreduced and reduced (reacted with dithiothreitol) mAb and their respective degradation products were determined. It is worth mentioning that for reduced samples, the proposed method was able to separate the fraction corresponding to the nonglycosylated fragment of the mAb from the glycosylated one. The commercial LIF instrumentation (argon ion laser, λ_{exc} 488 nm, λ_{em} 560 nm) provided excellent limits of detection (around 10 ng/mL) for the intact antibody and its degradation products.

One approach to overcome the formation of multiple derivatives and the difficulties associated with derivatization of proteins at low concentrations was proposed by the group of Dovichi [126,127]. They developed a derivatization procedure for proteins with the fluorogenic reagent, 5′-furoylquinoline-3-carboxaldehyde (FQ), that provided good sensitivity and minimized the band broadening caused by the variable labeling from the reaction of FQ with lysine residues. The derivatization could be carried out before the sample injection or inside the separation column. For the latter, two successive short plugs, one containing the sample and the other containing the reagent, were injected into the column; both plugs were mixed inside the capillary (due to their differences in electrophoretic mobility) and allowed to react. After derivatization, either inside or outside the capillary, the protein separation took place in a buffer containing a*n*-alkylsulfate tensioactive (e.g., sodium dodecylsulfate or sodium pentylsulfate) at submicellar concentrations, which masked the charge differences introduced in the protein by the multiple labeling. The method provided good results in terms of sensitivity, with an assay detection limit of 10^{-10} M for OVA (the authors define the assay detection limit as the minimum amount of protein that can be successively derivatized and detected), and band broadening (around 20,000 theoretical plates) (Figure 22.15). Limitations of the technique included variable protein sensitivity due to the different number of lysine residues and the possibility of masking differences in mobility of the protein glycoforms from the added tensioactive [128]. It should be noted that the reported sensitivity was achieved with postcolumn sheath-flow LIF detection; sensitivity using a commercial on-column LIF detector was about 100 times poorer.

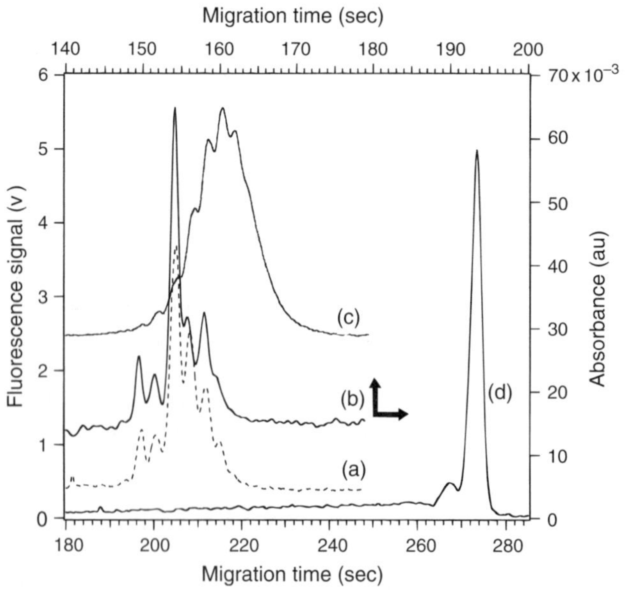

FIGURE 22.15 Comparison of ovalbumin analyzed under various conditions. Running buffer: (a–c) 25 mM tricine, pH 8.0; (d) 25 mM tricine +5 mM SDS. (a) FQ-labelled ovalbumin, 15 s reaction, LIF detection; (b) unlabelled ovalbumin, UV adbsorbance detection; (c) FQ-labelled ovalbumin, 10 min reaction. LIF detection. Note that electropherogram B is plotted versus top *x*-axis and right *y*-axis. (From Pinto, D.M. et al., *Anal. Chem.*, 69, 3015, 1997. With permission.)

Affinity binding is another good alternative for fluorescent detection of glycoproteins. This kind of binding involves several types of interactions between the glycoproteins and the fluorescent reagent (excluding the covalent one) and include electrostatic, van der Waals, hydrophobic, and hydrogen-bonding forces. A large choice of fluorescently labeled biomolecules can be used in this approach. The preparation of a homogeneously labeled biomolecule is in some cases a labor-intensive task involving chromatographic purification of the probe. Bornemann et al. [129] have used affinity derivatization with LIF monitoring to improve the selective detection of rhEPO. They first prepared a monomeric antigen-binding fragment (Fab) from the mAb 5F12 that bonded to a conformationally independent epitope of the N-terminal region of human EPO. This fragment was then labeled with the fluorescent dye, Alexa Fluor 488, yielding a mixture of labeled products (doubly and singly labeled and unlabeled) from which a homogeneous fraction was prepared by HPLC. This fraction was used as affinity reagent in CIEF with LIF detection (argon ion laser λ_{exc} 488 nm, λ_{em} 520 nm) for the detection of less than 100 pmol of rhEPO. Although this detection limit is far from the sensitivity necessary for the determination of this hormone in serum or urine, the authors believe that further refinement of the labeled Fab fragment and decrease of sample volume should permit the detection of smaller amounts of EPO.

Postcolumn derivatization with a fluorescent reagent is another alternative. However, this approach requires the selection of a fast derivatization reaction to achieve good sensitivity and a careful design of the postcolumn reactor that minimizes the loss in resolution achieved in the separation capillary. To this end, affinity interactions of some biomolecules, such as antibodies, with other glycoproteins are especially useful. That has been the case reported by Kelly and Lee [130] who have developed a CE method with online postcapillary affinity LIF detection for the monitoring and quantification of microheterogeneities of a mouse antihuman follicle-stimulating hormone mAb in samples containing culture medium without further purification. The fluorogenic reagent, consisting in fragment B (BF) of protein A conjugated with fluorescein, was mixed with the effluent of the CE column using a homemade postcolumn capillary reactor without causing large band broadening. Affinity binding between the antibody variants and the BF–fluorescein occurred in the reaction capillary and the complexes were monitored by the LIF detector (argon ion laser, λ_{exc} 488 nm, λ_{em} 515 nm) placed in the same capillary. By binding the BF–fluorescein fragment to the antibody, the emission of the fluorescein moiety was enhanced, so that the mAb peaks were detected with a low fluorescence background. The detection of up to five peaks was reported and the authors speculated that this could be due to the presence of different isoforms of the antibodies differing in the sialic acid content of their oligosaccharides.

22.4.3.3 Mass Spectrometry

Recent progress made in the development of MS ionization techniques has enabled its application to the analysis of intact glycoproteins. The most widely used ionization modes are ESI and matrix-assisted laser desorption ionization (MALDI). When considering the gains in sensitivity when compared with conventional UV detection, the structural information that can be derived from its use and the possibility of increasing the selectivity in the selected ion monitoring (SIM) mode, it is not surprising that MS has become a widely used detection system for coupling to CE. In this section, only those systems where the online coupling of CE to mass spectrometry (CE-MS), and for which MS is used as the detection (and information) system for the CE separation of glycoforms will be considered.

Owing to the high heterogeneity and structural complexity of glycoproteins, a prior separation step is generally necessary to obtain qualitatively and quantitatively useful results. Otherwise, only broad signals are obtained. Despite the compatibility of CE with ESI-MS and the relatively easy coupling of the two techniques, appropriate separation buffers, usually consisting of volatile substances, must be employed. This is one of the more difficult issues to the coupling of CE to MS where buffer electrolytes compatibility with MS involves choosing buffers that do not inhibit ionization and that, at the same time, provide reasonable glycoform separation [30]. Formic acid was used as the separation buffer in a pioneering work aimed at the characterization of glycoforms [61]. Analysis required constructing a CE–MS interface that could be adapted to a commercial CE cartridge. Under these separation conditions, the CE-MS system provided analysis of intact RNase B and HRP, a more complex glycoprotein in which the microheterogeneity arises from glycosylation at two residues and variability in the carbohydrate composition at each site. Figure 22.16 shows the CE-UV and the CE-MS electropherograms obtained for RNase B.

Although volatile buffers are usually employed, successful analyses of intact glycoproteins have also been achieved by using low concentrations of nonvolatile compounds in separation buffers with low ionic strength [131]. Using PAA-coated capillaries and a separation buffer consisting of 50 mM β-alanine adjusted to pH 3.5 with acetic acid, two high mannose-containing proteins, RNAse B and rhBMP-2, were analyzed by CE-MS employing a home-built CE instrument. Reporter ions for carbohydrates generated by in-source fragmentation for the glycoforms separated by CE were used to provide information about the degree of glycosylation of the different isoforms.

The presence of urea in the separation buffer was found to be needed to accomplish the separation of antithrombin III (ATIII) glycoforms from a commercial sample purified from plasma [132]. In order to make the CZE separation step compatible with the ESI-MS ion trap, special precautions about the decoupling of the needle were needed to minimize the amount of urea introduced into the spray chamber. Although the resolution was lower than that obtained by CZE-UV, it was enough to allow deconvolution of the multiply charged peaks, even for minor isoforms. It was also possible to distinguish isoforms differing in fucosylation for which small differences in mass for species having similar pIs would not make it possible to be differentiated on the basis of electrophoretic mobilities. The precision of the molecular mass determination was better than 1%. Assignment of particular glycan structures to isoforms was aided by the known canonical structures of the main isoforms of alpha- and beta-ATIII.

Successful characterization of protein glycoforms has been reported using MS-friendly conditions on capillaries with dynamic coatings that generate low-EOF [70]. An acidic BGE containing an organic additive was used (1 M acetic acid/20% methanol at pH 2.4). Under these conditions, partial separation of glycoforms was attained for several glycoproteins including AGP and BSF [71].

By using ionene-coated capillaries and an acetate buffer at pH 4.8, baseline resolution of the three main isoforms of rhEPO and a comparison of rhEPO and uEPO were obtained by CE-ESI-MS [62].

The development of CZE-ESI-MS methods using an orthogonal accelerated time-of-flight (TOF) mass spectrometer has allowed distinguishing glycosylation differences between a reference preparation of rhEPO and a commercially available product sold for research purposes [42]. Membrane concentrators were used to obtain sample concentrations at the required level followed by separation on PB-coated capillaries (the amine coating effectively reverses the EOF). The concentration step provided the additional advantage of removing small MW excipients (e.g., salts, polysorbate), which are known to interfere in the ionization step. Despite the lack of complete resolution in the CE separation step, the MS dimension provided additional resolution and enabled the identification of intact glycoforms varying in the number of HexHexNAc residues, sialic acids, and even the identification of fucosylated, acetylated, and oxidized glycoforms. A total of about 135 isoforms were distinguished in the rhEPO BRP of the EP. A possible composition of the recombinant protein was speculated taking into account the information provided by the CE-MS system and that obtained by other methods. The same method allowed the characterization of the two pharmaceutical products, epoetin-α and epoetin-β [96]. Ultimately, the two products could be distinguished on the basis of the presence of two additional basic (i.e., glycoforms containing less sialic acids) glycoforms in epoetin-β.

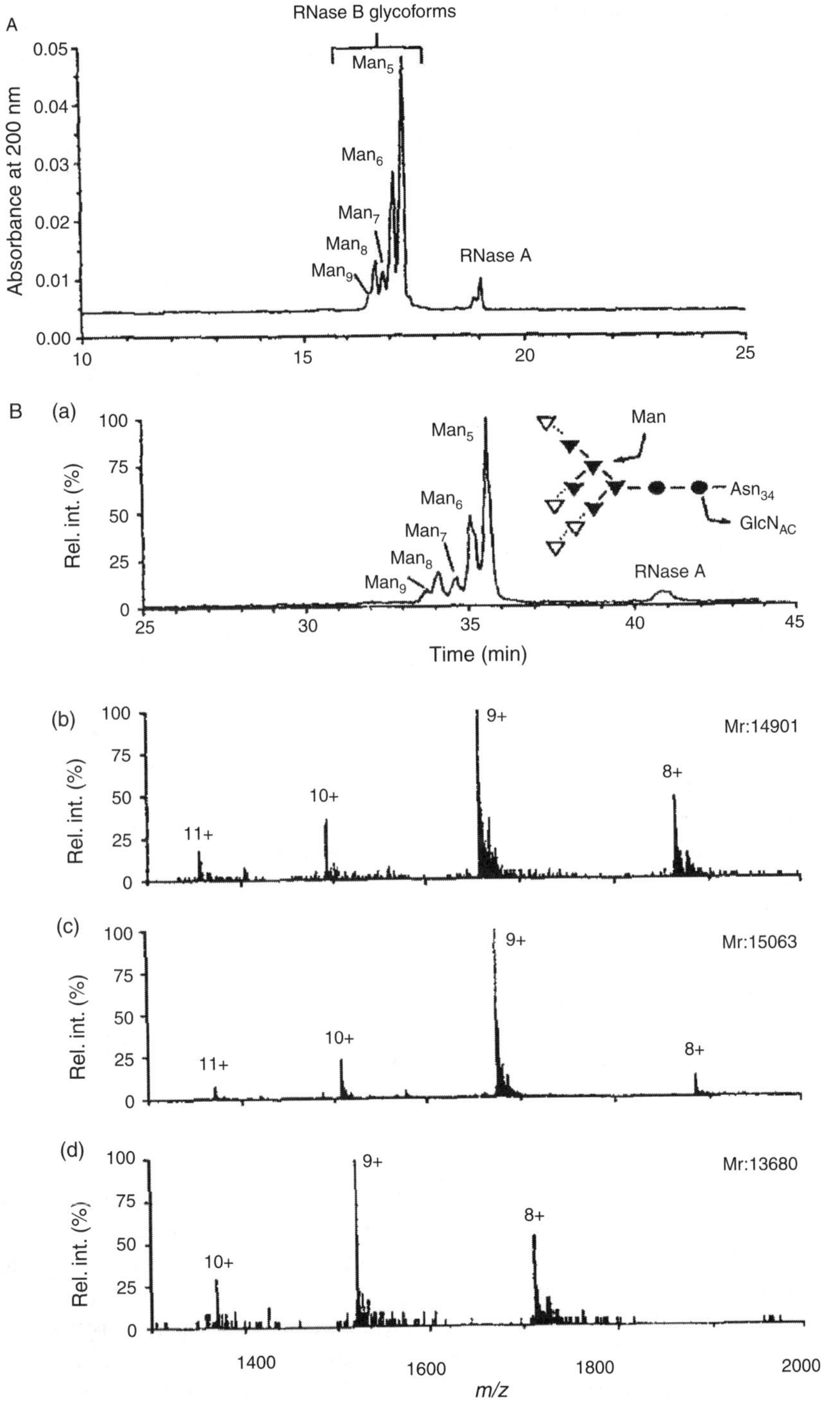

FIGURE 22.16 CZE of RNase B. (A) Fused-silica capillary 80 cm × 50 μm i.d. coated with a solution of polybrene and ethylenglycol; Buffer containing 2M formic acid; injection of 6 pmol of RNase B; UV detection. (B) Fused-silica capillary 110 cm × 50 μm i.d. coated with a solution of polybrene and ethylenglycol; Buffer containing 2 M formic acid; injection of 6 pmol of RNase B; ESMS detection. Sheath flow 5 μL/min of an aqueous solution of 0.2% formic acid and methanol. (a) Total ion electropherogram for the full mass scan acquisition (*m/z* 1300–2000). (b) Extracted mass spectra for peaks migrating at 35.2 min, (c) at 35.6 min, and (d) at 40.8 min. The calculated molecular mass is shown on the right corner of each spectrum. (From Kelly, J.F. et al., *J. Chromatogr. A*, 720, 409, 1996. With permission.)

A comparison of rhEPO by CZE-ESI-MS using capillaries that provided reversal of the EOF (PB coating) or suppressed EOF (UltraTrol Dynamic PreCoat LN) showed, as indicated in a previous section of this chapter, that increased resolution was obtained with the suppressed EOF capillaries (see Figure 22.3) [30]. The improved separation allowed detection of low-level glycoforms. The observed mass differences were attributed to differences in the content of sialic acids, hexoses, and *N*-acetylhexosamines. In addition, the high mass resolution of the system allowed the detection of comigrating substances differing by 42 and 16 Da corresponding to acetylated and oxidized variants, respectively. Similarly to the results obtained for rhEPO, better glycoform resolution was obtained for bovine fetuin and bovine alpha 1-acid glycoprotein standards on capillaries with suppressed EOF [71].

Despite a level of sensitivity about 100 times lower than that obtained by CE-UV, inductively coupled plasma mass spectrometry (ICP-MS) coupled to CZE has been shown to separate and distinguish between isoforms containing 3, 4, and 5 sialic acid residues in standard Tf samples [133].

For more complete details on the use of CE-MS for glycoproteins some recent reviews have also appeared [134–138].

22.4.3.4 Other Detection Modes

Conductivity detection has been used in a CITP instrument for the fractionation of glycoforms of rhEPO mentioned previously in this chapter [95]. The preparative CITP instrument is composed of two columns made out of a fluorinated ethylene copolymer, one acting as a separation column and the other one as a trapping column. Both columns are provided with their respective conductivity detectors. That of the first one served, in addition to monitoring the CITP separation, as a proper timing in the switching of the columns. The other one in the second column provided the isotacopherograms used in the control of the fraction collector placed at the end of this second column, and equipped with a collection valve. Although the sensitivity of conductivity detectors was poor for the monitoring of proteins, in this case sensitivity was not an issue since 100 μg of rhEPO were injected in each run.

22.5 PRACTICAL APPLICATIONS

22.5.1 Characterization and Quality Assessment of Biologicals and Biopharmaceuticals

22.5.1.1 General Considerations

The development of physicochemical methods of analysis based on high-resolution techniques such as CE has had a significant impact on fields of activities dealing with complex biomolecules such as glycoproteins. While undoubtedly variable in magnitude, this impact has been considerable in the development of foods, pharmaceuticals, vaccines, blood and blood products, diagnostics, and for regulatory authorities. In many of these areas, the establishment of more sensitive, precise, and selective methods of analysis has particular relevance since manufacturers are required to provide scientific evidence with regard to major tenets that guide the development of regulatory requirements in most jurisdictions: that products entering the market must be safe and of high quality. More specifically, the assessment of a product through the verification of its identity, the quantitative evaluation of its purity, its stability, or the consistency of its manufacturing is of prime importance. For example, considerations regarding the decision to market a pharmaceutical protein resides not only on the demonstration of its biological activity but also on the demonstration that the manufacturing process can provide a product consistently, at the same level of quality and purity as that used for clinical trials. While the application of these specifications does not affect all areas universally (e.g., purity evaluation may not be a significant factor for a diagnostic product), they nevertheless are now generally well adhered to an industry and applied at various stages of the manufacturing process. Data generated are then incorporated in the preparation of documentation to be submitted to regulatory authorities for evaluation. In certain instances, harmonized guidelines relating to test procedures have been developed. This is the case for biotechnological and biological products where an international initiative, the International Conference on Harmonization (ICH), has enshrined these principles in a series of guidelines of test procedures for the quality evaluation (Q5A–Q5E) and specifications (Q6B) of these products [139].

Traditionally, complex biomolecules such as glycoproteins have been assessed using conventional gel electrophoresis methods. However, these methods are laborious to carry out, cannot be automated, and do not consistently provide quantitative results. In contrast, CE has been demonstrated to provide rapid, automated, and quantitative results. In addition, it is now often used as a complementary technique to the more established HPLC in several fields of activities. This progress has led to the recognition by pharmacopeial authorities in Japan, the United States, and Europe of CE for the generation of data relating to product identification, assay or tests for related protein impurities, and to its use by industry in various other aspects of the development process including drug substance and drug product characterization, in-process monitoring, and stability determination. It is also one of the recommended physicochemical techniques of the ICH Q6B guideline on specifications for biotechnological/biological products mentioned earlier. Underlying this recognition is the necessity to devise validated methods based on widely accepted analytical parameters. Typically, method validation involves the determination of one or more analytical parameters (e.g., accuracy, precision, linearity range, limit of detection, limit of quantitation, robustness, specificity) to demonstrate that the method performs according to its intended purpose. The degree with which a method requires validation varies according to the procedure being implemented as shown in Table 22.1 derived from the ICH guidelines Q2(R1). For example, while an identification test requires only that specificity be demonstrated, the quantitative evaluation of impurities necessitates that all characteristics be determined with the exception of the limit of detection.

The coming of age of CE as a recognized analytical method has been marked by its introduction into pharmacopeial monographs. Given its improved separation and quantification capabilities, a CZE method for rhEPO glycoform separation was found suitable to replace conventional IEF as an identification test and, as previously mentioned, has been incorporated into the EP monograph for rhEPO concentrated solution (i.e., the bulk substance solution before formulation) [35]. Using

TABLE 22.1
ICH Q2(R1) Recommendations for Method Validation of Analytical Test Procedures

Type of Analytical Procedure	Identification	Testing for Impurities	Assay (Content/Potency)
Characteristics		**Quantitation/Limit**	
Accuracy	−	+/−	+
Precision			
Repeatability	−	+/−	+
Interm. precision	−	+/−	+
Specificity	+	+/+	+
Detection limit	−	−/+	−
Quantitation limit	−	+/−	−
Linearity	−	+/−	+
Range	−	+/−	+

Source: http://www.ich.org/LOB/media/MEDIA417.pdf, Accessed November 2006.

conditions adapted from the method of Watson and Yao [52], rhEPO glycoforms are separated on a bare fused silica with a BGE consisting of 0.01 M tricine, 0.01 M sodium chloride, 0.01 M sodium acetate, 7 M urea, and 2.5 mM putrescine at pH 5.55. The method system suitability specifies several required criteria including the separation of the BRP into a number of well-resolved peaks [34] as shown in Figure 22.17. For this particular preparation, the rhEPO BRP batch 1, eight glycoforms are separated with nearly baseline resolution and the order of migration follows the increasing total number of terminal sialic acid residues present on the carbohydrate chains. In other words, glycoforms with the lowest total number of sialic acids migrate first and those with the highest number migrate last. It should be noted here that the BRP was a mixture prepared from known quantities of two commercialized rhEPO forms, rhEPO-α and rhEPO-β. Characterization of these two products had shown that they had different glycoform profiles and distribution, with rhEPO-β containing higher amounts of the more basic glycoforms (i.e., those with lower total amounts of sialic acid residues). The use of a mixture of products for the BRP allowed the preparation of a unified monograph appropriate to analyze both forms of rhEPO that have essentially identical biological activities. Specifications on the distribution of glycoforms as a percentage of the total content in the test solution are provided and constitute an integral part of the identity test (Table 22.2). (At present, a new standard, rhEPO BRP batch 2, has been established with a slightly different electropherogram to that shown in Figure 22.17).

Somatropin, the recombinant version of the 22 kDa form of human growth hormone, is another important biopharmaceutical for which a CZE method has recently been adopted for inclusion into the EP somatropin monographs. Although not a glycoprotein, it is mentioned here to exemplify the increased use of CE for the specifications of proteins. In this case, the test is for the quantitative assessment of charged variants [140]. It is expected that CE will become a regularly used technique in pharmacopeial monographs of biopharmaceuticals in view of its usefulness for the characterization of these structurally complex molecules. In addition, the patent expiration date for several biopharmaceuticals has either passed or is fast approaching and this will undoubtedly lead to several manufacturers wishing to enter this growing market to produce similar products.

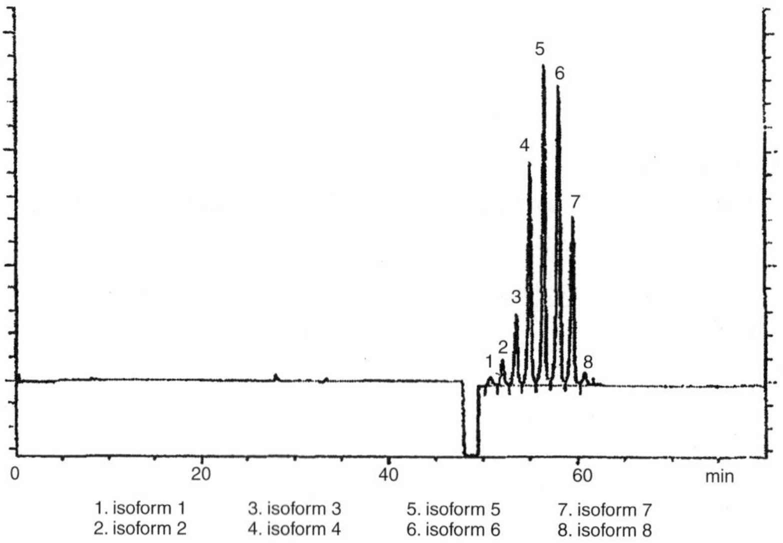

FIGURE 22.17 Electropherogram of the European Pharmacopoeia erythropoietin biological reference preparation batch 1. (From Bristow, A. and Charton, E., *Pharmaeuropa*, 11, 290, 1999. With permission.)

TABLE 22.2
Ranges of Glycoforms Content Specified in the EP Monograph for Erythropoietin

Isoform	Content (Percent)[a]
1	0–15
2	0–15
3	5–20
4	10–35
5	15–40
6	10–35
7	0–20
8	0–15

[a] Ranges for the current EP standard, rhEPO BRP batch 2 have been modified (see Behr-Gross, M.-E., Daas, A. and Bristow, A., *Pharmeuropa Bio*, 2004, 1, 23, 2004).
Source: Erythropoietin concentrated solution, 01/2005:1316, European Pharmacopoeia 5th edition, published by EDQM, June 2004.

From a strategic perspective, product characterization is paramount to the development of identity, purity, or assay procedures since it involves the determination of many of the structural and physicochemical properties of the product under study. Consequently, product characterization is, by necessity, first in the series of test procedures to be developed. For glycoproteins, in addition to the more basic structural properties inherent to proteins such as amino acid sequence and higher-order structural elements (i.e., secondary, tertiary, and quaternary structures), in-depth characterization of carbohydrate-mediated properties are required.

In the following sections, an overview of the applications of CE with respect to the guiding principles and analytical parameters described earlier will be presented. For more in-depth discussions on specific aspects or individual glycoproteins, excellent reviews cited in the Section 22.2 can be read.

22.5.1.2 Product Characterization and Identity Testing

As previously mentioned, glycoproteins exist as mixtures of closely related species that differ in their glycosylation patterns. These differences are often the result of both compositional and sequence variations of the glycan chains. Moreover, the biological activity of glycoproteins is frequently linked to the presence of these carbohydrates and, consequently, the characterization of glycoprotein microheterogeneity represents one of the more challenging tasks for the characterization and identity testing. In both cases, high specificity is required since it is expected that the method will differentiate between the active ingredient, variants, impurities, and contaminants. Identity tests typically involve the determination of one or more intrinsic properties of a molecule, whether physicochemical, biological, and/or immunological and are often carried out by comparison with a reference standard. Apart from performing a simple identity test through coanalysis of the substrate and the reference standard, qualitative and quantitative information derived from CE experiments with respect to specific structural characteristics such as carbohydrate-mediated glycoform profiles, pI, or MW determination can be useful for product characterization and identity testing.

22.5.1.2.1 Recombinant Sialoglycoproteins

Several important biopharmaceuticals that received market authorization around the world in the past decades are recombinant glycoproteins that include such successful products as EPOs (rhEPO-α and -β), darbepoetin-α (also called NESP), rhtPA, interferons (rhIFN-β1, -γ), and colony-stimulating factors (rhGCSF, recombinant human granulocyte macrophage colony-stimulating factor, rhGM-CSF). They are used for the management of diseases not otherwise well treated with small molecule pharmaceuticals. Unequivocally, it is advances in molecular biology and recombinant DNA technology that are largely responsible for the availability of these molecules in large quantities, thus allowing biopharmaceutical companies to introduce them in the markets. Despite these advances, protein glycosylation remains a crucial factor that requires a thorough assessment especially when considering that the glycosylation of recombinant proteins is inherently heterogeneous and rarely identical to that found in humans, a situation that has the potential to result in problems linked to immunogenicity.

The ubiquitous occurrence of sialic acid residues on glycan chains of recombinant glycoproteins has provided an ideal tool for analysis by CE by taking advantage of the presence of the ionizable carboxylic acid group. As indicated in previous sections, both CZE and CIEF have been shown to be particularly sensitive to differences in net charge and, as such, have been widely applicable to characterize sialic acid-mediated glycoprotein profiles. However, as it will become evident in the remainder of this section, the biggest limitation to the application of CE methods is undoubtedly the complex nature of sialoglycoproteins, especially when more than one glycosylation site is present in the molecule. In such cases, the number of possible combinations and permutations of individual carbohydrate residues make it difficult to even anticipate separating all glycoforms.

Studies of relatively simple sialoglycoproteins, that is, containing a single glycosylation site, were among the first examples of the use of CE for characterization of intact glycoproteins. RhGCSF stimulates progenitor cell proliferation, differentiation, and activation of neutrophils and features a single glycosylation site at Thr-133. It was separated into two glycoforms under basic (pH 7–9) [54] or acidic (pH 2.5) conditions [141]. In the latter case, rhGCSF was analyzed in a preparation containing HSA and the addition of HPMC prevented adsorption to the capillary walls and facilitated detection at low pH. In both cases, the order of migration occurred in a predictable fashion according to increasing number of sialic acid residues. Both methods could distinguish between the nonglycosylated and glycosylated products.

A highly efficient CZE method capable of resolving in excess of 30 glycoforms of a simple sialoglycoprotein, a recombinant 24 kDa glycoprotein containing a single glycosylation site, has been reported by Berkowitz et al. [142]. Using a bare fused-silica capillary with triethanolamine/phosphoric acid, pH 2.5 as BGE, the 30 glycoforms were separated on a 27 cm capillary in less than 15 min (Figure 22.18). Lengthening the capillary to 77 cm resolved over 60 peaks in less than 30 min. Partial validation of the method was carried out by examining day-to-day reproducibility and sample matrix effect. For the latter, it became apparent that increased resolution was obtained for samples dissolved in low ionic buffer and low pH. While the effect of low sample ionic strength compared with the BGE is well known and is due to stacking effects, the low sample pH effect was explained in terms of the initial movement of the protein, which if injected from a sample at high pH was toward the anode and, consequently, against the stacking effect since the protein has a relatively high pI ($6.5 < pI < 8.0$). Using a kinetic study of the enzymatic removal of terminal sialic acids in conjunction with fractionation of the glycoprotein by anion exchange, they were able to identify several glycoforms according to their antennarity and sialic acid content as shown by the shaded areas in Figure 22.18. Although not all of the peaks were identified, this approach allowed to demonstrate that parts of the observed microheterogeneity were due to the presence of large amounts of deamidated variants (i.e., >30%). In addition to the qualitative assessment of the product, the results when taken collectively allowed to derive quantitative expressions for determination of the content in deamidated forms, nonglycosylated forms and monoglycosylated forms, biantennary forms, and

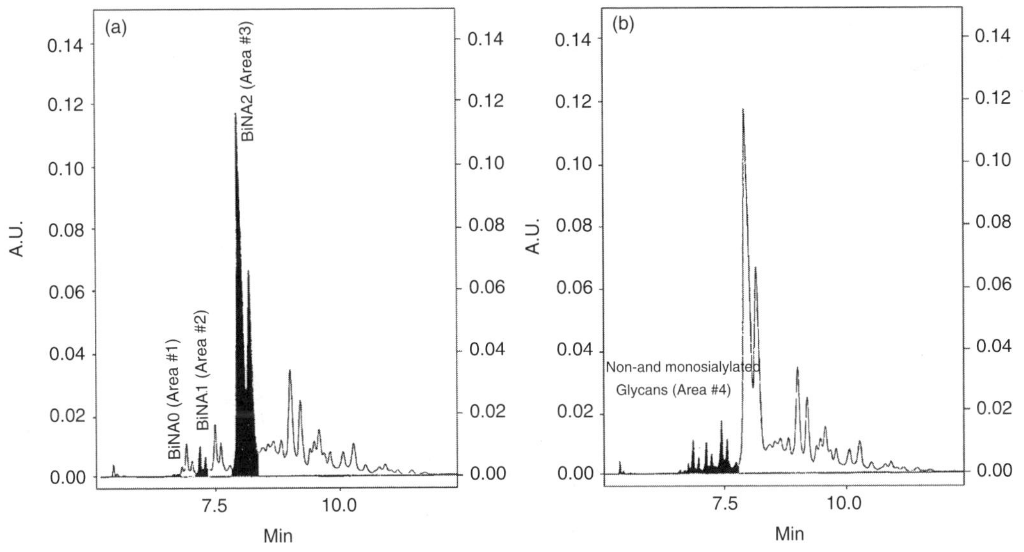

FIGURE 22.18 CZE separation of 24 kDa sialoglycoproteins with peak region assignments (shaded areas). In (a): BiNA0, biantennary oligosaccharides with no terminal sialic acid residues; BiNA1, biantennary oligosaccharides with one sialic acid terminal residue; and BiNA2, biantennary oligosaccharides with two terminal sialic acid residues. In (b): nonglycosylated protein and monosialylated glycoforms (area #4). Separation conditions: fused-silica capillary, 50 μm × 27 cm capillary, 100 mM triethanolamine–phosphoric acid, pH 2.5, 10 kV, 25 °C and UV detection at 214 nm. (From Berkowitz, S.A. et al., *J. Chromatogr. A*, 1079, 254, 2005. With permission.)

finally, sialylation level of biantennary forms. The method was also adequate to monitor the effects of changes in culture conditions such as host strains or growth conditions.

The analysis of complex sialoglycoproteins, that is, containing multiple glycosylation sites has been the subject of many reports. One such example is rhtPA, an important biopharmaceutical used in the treatment of myocardial infarction, which contains three N-glycosylated sites. Among several articles published on CE separation conditions, Thorne et al. [77] reported the separation of rhtPA into no less than eight glycoforms by CIEF. The method showed high reproducibility and precision for migration times and good total recovery of rhtPA from the capillary was demonstrated. CZE conditions using amines as additives to the BGE were also used to separate numerous major and minor glycoforms. In addition, they demonstrated the use of CE-SDS as a potential purity test for rhtPA Type I and Type II, which differ in the level of glycosylation site-occupancy.

RhEPO is another complex sialoglycoprotein, with three N-linked and one O-linked glycosylation sites, that has been extensively characterized by CZE and cIEF as described in previous sections. The comparison of rhEPO and uEPO has been reported on uncoated capillaries [40] and on capillaries dynamically coated with ionene [62]. Although increases in resolution and sensitivity are desirable, and the samples of uEPO corresponded to purified standards, these works showed the potential of discriminating between exogenous and endogenous EPO.

CE was also used to fractionate rhEPO glycoforms as briefly discussed previously. Madajova et al. [95] developed a preparative CITP method, operating in a discontinuous mode, which enabled the separation of 100 μg of rhEPO into six fractions in 30 min. The number of collected fractions corresponded approximately to the number of glycoforms detected in the original product by CZE. Subsequent analysis of the collected fractions by CZE showed that several glycoforms had been substantially enriched by 10- to 100-fold in some cases. Such an approach may provide a desirable way to access specific glycoform levels in relation to their biological activity or for the evaluation of low glycoprotein levels in biological matrices.

Recombinant human deoxyribonuclease (rhDNAse) is a complex sialoglycoprotein whose heterogeneity is due in part to the presence of phosphorylated mannose residues. CE studies of this acidic (3.5 < pI < 4.5) phosphosialoglycoprotein that contains two glycosylation sites have been reported [143,144]. CZE methods were developed under both acidic (pH 4.8) and basic (pH 8.0) conditions. The addition of calcium ions significantly improved resolution at both pH as a consequence of interactions with calcium-binding sites that stabilize the protein and make it more resistant to proteases. The pH-dependent calcium binding provided a means to distinguish between acidic and neutral glycoforms that were identified by a two-dimensional (2D) investigation of neuraminidase and alkaline phosphatase–protein digestions. The acidic pH resolved acidic charge heterogeneity and the basic pH discriminated neutral heterogeneity.

22.5.1.2.2 *Naturally Occurring Sialoglycoproteins*

The need for characterization and identity determination of naturally occurring glycoproteins has been one of the driving forces behind the development of CE methods. Naturally occurring glycoproteins are highly heterogeneous and sometimes present a formidable challenge. Only a few examples of approaches to their characterization will be discussed in this section.

ATIII is a glycoprotein that plays a key role in the inhibition of blood coagulation. Despite the availability of a recombinant version, plasma-derived ATIII has continued to be the subject of recent characterization studies by CE. Two isoform groups isolated from plasma-derived ATIII that differ in the number of glycosylation sites, ATIII-α (glycosylated at four asparagines) and ATIII-β (glycosylated at three asparagines), were studied by a combination of techniques that included gel IEF, 2D-gel electrophoresis, CIEF, and CZE [145]. The two isoform groups were separated by CZE as shown in Figure 22.19. With a lower glycosylation site-occupancy and, consequently less sialic acid residues, the major component of ATIII-β (peak A in Figure 22.19a and d) predictably migrated before the ATIII-α components. The quantitative analysis indicated a content of about 70% ATIII-α main isoform and about 6.6% of ATIII-β in good agreement with published data. The pI values of ATIII determined by CIEF using an internal calibration were in fair agreement with values of the main isoforms measured by 2D-gel electrophoresis. The CZE method provided a definite advantage in terms of quantification and accuracy. In another report, the same group further characterized the separated isoforms by online coupling with MS [132].

The commercially available glycoprotein, OVA, has been extensively used as a model protein for approaches to glycoform separations by CE since the very early years of the technique. Recent studies on OVA have focused on novel separation conditions that included the effect of new additives to the BGE and new dynamic coatings [87,146,147] as well as conditions suitable for online coupling to MS [148]. As indicated in a previous section, an innovative separation approach based on the free-solution (i.e., without immobilization) selective affinity of OVA glycoforms to a common protein, LCA, has been reported [118]. The separation mechanism was explained in terms of the increased affinity of glycoforms having high-mannose type N-glycans for LCA and, consequently, these species were more strongly retarded than other glycoforms. A similar type of approach was recently used to separate AGP glycoforms without biantennary glycans from those having one or more biantennary glycans [117]. Once again, the separation mechanism appeared to involve greater affinity of the ligand for high-mannose species.

22.5.1.2.3 *Glycoproteins without Sialic Acids*

CE separation of glycoforms of proteins where no sialic acid residues are present has been reported. RhBMP-2 is a protein with high-mannose glycan chains that has been unexpectedly separated under CZE conditions [31,131]. Glycoforms were separated on the basis of the number of mannoses a given rhBMP-2 molecule possesses (see Figure 22.8). This type of approach may prove useful for the qualitative and quantitative monitoring of fermentation processes for glycoproteins without substantial charge differences.

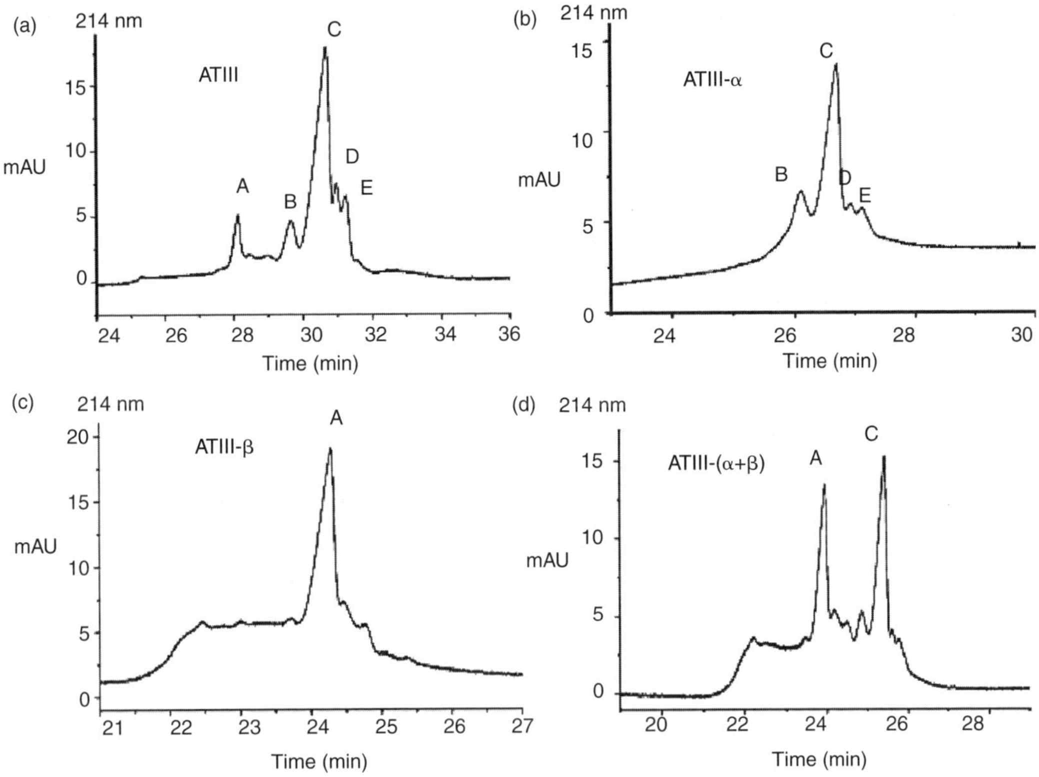

FIGURE 22.19 CZE of (a) ATIII, (b) ATIII-α, (c) ATIII-β, and (d) ATIII-(α + β). Separation conditions: PVA-coated capillary 50 μm i.d. × 64.5 cm (56 cm effective length), 1 M acetic acid containing 4 M urea, 0.1 M sodium chloride added to the aqueous sample solution, 120 kV, UV at 214 nm. (From Kremser, L., et al., *Electrophoresis*, 24, 4282, 2003. With permission.)

Cellobiohydrolase I (CBH-I) is a phosphoglycoprotein that contains a catalytic domain with four N-glycosylation sites and where phosphate groups form phosphodiester links between the anomeric carbon and the C6 of two consecutive Man residues (i.e., Man1-P-6Man). CIEF was used to separate the phospho-isoforms of the catalytic domain of CBH-I into four glycoform species according to the number of phosphate groups, that is, 0, 1, 2, and 3 groups [149]. Since glycosylation of CBH-I is dependent on the strain and growth medium, the method may be useful for the analysis of cellulase glycosylation from different strains, different organisms, mutated organisms, or cellulases expressed in other organisms.

22.5.1.2.4 Monoclonal Antibodies

A number of therapeutic mAbs produced by recombinant DNA technology have been marketed in recent years. Most mAbs exhibit some heterogeneity due to PTM and degradation occurring during the manufacturing process and shelf life of the product. As such, they have been the subject of several studies using CE-based methods and reviewed [150]. Charge heterogeneity is a parameter commonly monitored for mAbs. Common sources of charge heterogeneity include sialylation, deamidation, and C-terminal lysine cleavage. Sialylation can vary widely between mAbs, from being insignificant in some cases to being the major source of charge heterogeneity in other cases. It follows that the sialic acid-mediated heterogeneity is commonly determined from the characterization of the charge heterogeneity before and after enzymatic treatment with sialidase.

Both CIEF and CZE-based methods have been used to study mAbs. A combination of cation exchange chromatography and CIEF was used to study human tumor necrosis factor (hTNF) mAb [151]. Of the four peaks separated by cIEF, three corresponded to C-terminal lysine variants and the remaining peak to sialic acid-mediated charge differences. Similarly, CIEF was used to determine the sialic acidmediated heterogeneity of a IgG_1 mAb that contained N-linked glycosylation in the Fab region at Asn56, with approximately half of the biantennary glycans being sialylated [152]. Using an approach based on the well-known formation of complexes between borate and diols, the carbohydrate heterogeneity of a mAb was demonstrated by CZE [153]. Three partially resolved peaks were observed after optimization of the conditions to 150 mM borate, pH 9.4, on a fused-silica capillary. The method was further used to monitor batch-to-batch consistency and for stability testing.

In addition to providing molecular mass information, CE-SDS has been used for the detection and quantification of low levels of unglycosylated heavy chain (HC) in mAbs. The optimization and validation of generic and quantitative CE-SDS procedures with LIF detection was reported [125,154]. An alkylation step was incorporated to decrease thermally induced fragmentation of nonreduced, labeled mAb samples. The unglycosylated variant was detected under reducing conditions where light chain (LC) and HC are separated (Figure 22.20). A collaborative study report involving multiple laboratories in different organizations has demonstrated the robustness of CE-SDS without labeling and using UV detection for assessing mAbs and the unglycosylated variant [155].

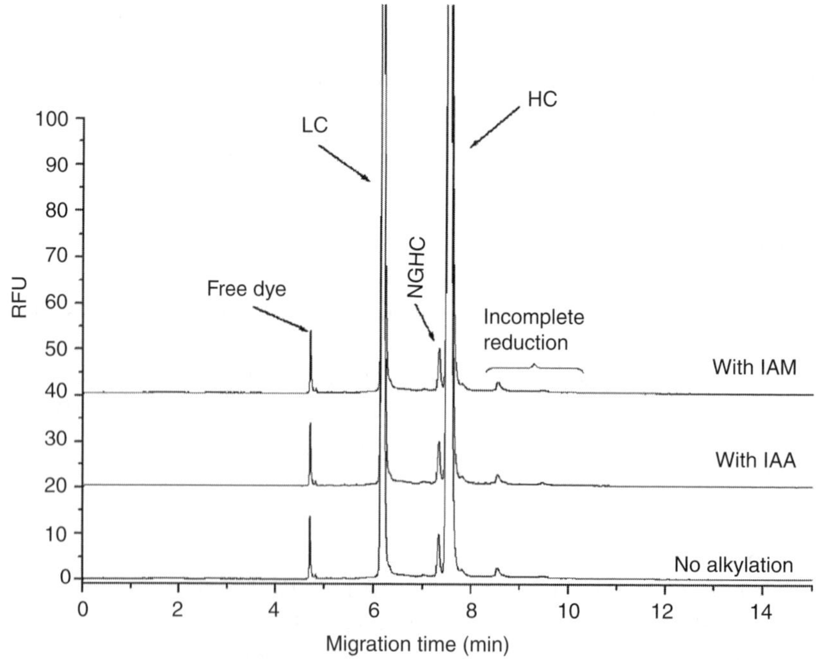

FIGURE 22.20 CE-SDS separations of reduced labeled mAb samples in the presence of different alkylating agents: NGHC, nonglycosylated heavy chain; LC, light chain; HC, heavy chain; IAA, iodoacetic acid; IAM, iodoacetamide. Separation conditions: uncoated fused-silica capillary 50 μm i.d. × 30 cm (effective length 19.4 cm), both anode and cathode buffers were the Bio-Rad SDS running buffer, samples were injected at a constant electric field of 417 V/cm for 15 s and electrophoresed at 625 V/cm (21.2 μA) at 20°C, detection was performed with LIF using a 3.5-mW argon ion laser, 488 nm excitation, 560 ± 20 nm emission. (From Salas-Solano, O. et al., *Anal. Chem.*, 78, 6583, 2006. With permission.)

22.5.1.3 In-Process Monitoring and Product Consistency

Studies of recombinant DNA technology processes have clearly demonstrated that PTM such as glycosylation are not constant throughout culture and that many factors must be examined to obtain and maintain the required glycosylation. They include choice of host cell, genetic engineering of glycan processing, or control of bioprocess parameters such as culture environment, method of cell culture, and culture time [156]. Furthermore, even after a process has been well established, changes may be required from time to time, a situation that may also have an impact on PTM. Examination of pre- and postchange products is then required to ascertain the absence of unwanted changes to the product quality [157]. It follows that monitoring of glycosylation at all stages of a bioprocess has become a necessity to ensure product quality and batch-to-batch product consistency. Practical considerations to the application of an analytical technique for in-process monitoring and product consistency assessment range from the necessity of the method to have high accuracy and sensitivity, to provide high reproducibility and be relatively simple in its application, to be robust enough to sustain minor changes in sample matrix, to be specific and quantitative, and to permit fast development and analysis time. CE has become widely applicable at various stages of the manufacturing process due to its already mentioned advantages regarding the flexibility provided by the various separation modes that can be applied, the low sample volume required, and its capability for automation and fast analysis.

22.5.1.3.1 In-Process Monitoring

RhIFN-γ is, as mentioned in a previous section of this chapter, a sialoglycoprotein with two N-linked glycosylation sites at Asn25 and Asn97, which has been shown to be heterogeneous in the number of species resulting from variable glycosylation site-occupancy (nonglycosylated: 0N; monoglycosylated at Asn25: 1N; and diglycosylated at Asn25 and Asn97: 2N) as well as from variability in the number of terminal sialic acids [114,134,158]. In combination with CIEF, a method based on MEKC conditions discussed previously (see Section 4.2.2.3 and Figure 22.13) was used to monitor rhIFN-γ glycosylation during perfused fluidized-bed production [159]. Glycosylation site-occupancy was monitored by MEKC. A plot of the relative amounts of each of 0N, 1N, and 2N provided evidence of constant levels of glycosylation site-occupancy over the course of the process. In addition, the sialic acid mediated isoform content was monitored by CIEF using a commercial kit over a pI range of 3–10. No less than 11 major variants were detected between pI values of 3.4 and 6.4 (Figure 22.21b). After enzymatic desialylation of this mixture, a decrease in the number of variants and, as expected, a major shift to higher pI values was obtained (Figure 22.21a), indicating that most of the observed heterogeneity resulted from sialic acid variability. The remaining heterogeneity was ascribed to C-terminal truncated variants. The quantitative analysis of sialylated variants indicated a sharp decrease in the mean pI after 200 h of perfusion culture (Figure 22.21c). This suggested that sialylation is increased during the perfusion process to reach a constant level afterward. On the other hand, examination by MEKC of glycosylation site-occupancy for a stirred-tank batch culture showed that the process did not provide constant levels, but gradually declined throughout the fermentation process. They also demonstrated that this decline was at the expense of the 2N variant and that there were generally higher proportions of the doubly glycosylated (2N) rhIFN-γ glycoform and lower proportions of the nonglycosylated (0N) variant in the perfused fluidized-bed process compared with stirred-tank culture, that is, the rhIFN-γ protein was more heavily glycosylated during perfusion culture.

Similar MEKC conditions were used to monitor rhEPO glycosylation during continuous culture of Chinese hamster ovary (CHO) cells in a fluidized-bed bioreactor [160]. While individual glycoforms were not separated under these conditions, clear differences in both peak shapes and migration times were observed when compared with unglycosylated EPO, suggesting that adequate glycosylation was obtained. Analysis of the product by 2D-electrophoresis provided further evidence that the required isoforms had been obtained.

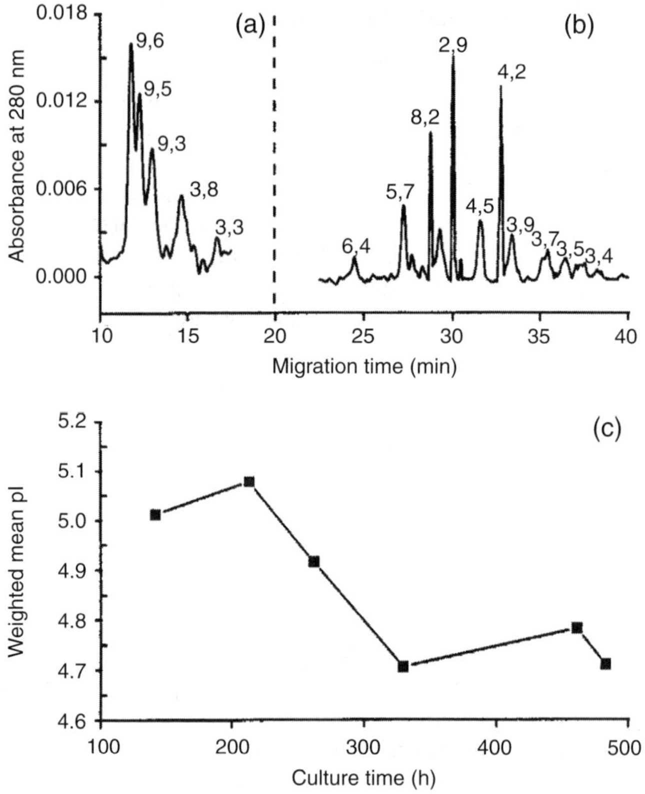

FIGURE 22.21 Monitoring of rh IFN-γ microheterogeneity by capillary isoelectric focusing. (a) rhIFN-γ desialylated with neuraminidase from *A. ureafaciens*. (b) Typical electropherogram of immunoaffinity-purified rhIFN-γ from perfusion culture of CHO cells. (c) The weighted mean pI of rhIFN-γ variants secreted by CHO cells during perfusion culture. (From Goldman, M. H. et al., *Biotech. Bioeng*. 60, 596, 1998. With permission.)

The influence of changing conditions for the fermentation process of rhATIII using a combination of liquid chromatography and CZE has been reported [161]. While the CZE method only partially resolved glycoforms, it provided quantification within 2 min. The use of a 2D–HPLC/CZE design provided additional sensitivity and resolution and quantitative results in 5 min.

Although not developed specifically for glycoproteins, a rapid CE-SDS method for the in-process monitoring of fermentation, hydridoma cell cultivation, and purification has been reported [162]. It was carried out in a total analysis time of less than 5 min.

22.5.1.3.2 Product Consistency

The ability to detect differences of product quality is a key component of the assessment of any manufacturing process. As seen in the preceding section, biological processes are complex and many factors influence the quality of the resulting product. In addition, manufacturing processes are rarely static and more often than not involve continual refinement to improve yields, to obtain higher quality products, or to implement new technologies. In these cases, manufacturing changes are implemented and the significance of these changes on the product consistency needs to be assessed.

In addition to examples already described in previous sections, a CIEF method was used to compare the isoform profiles of IgG_{2a} obtained under different conditions of osmolality and CO_2 partial pressure (pCO_2) in growth media [163]. Significant increases in the pIs of the major peaks as well as in the number of peaks were observed with increases in osmolality and pCO_2 (Figure 22.22). The change in pI was not a consequence of differences in sialic acid content (this particular mAb

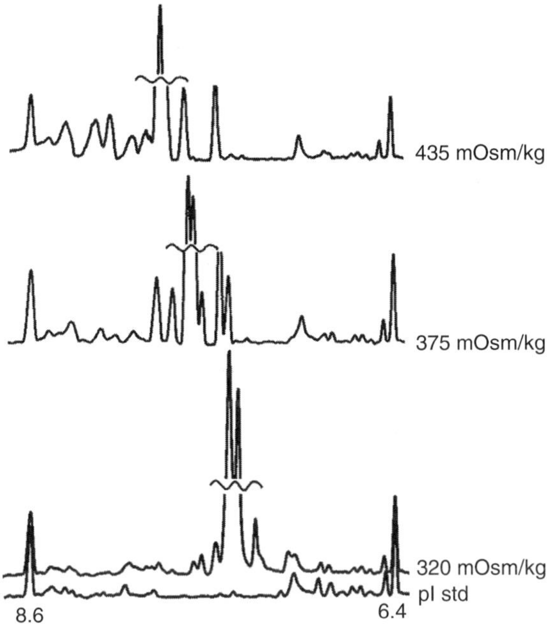

FIGURE 22.22 Representative CIEF separations showing the effect of osmolality on the pI distribution of IgG_{2a} produced in serum-containing medium. The pI standard curve is shown at the bottom of the figure for comparison. A shift in the main pI peaks toward higher pI values with increasing osmolality, as well as an increase in the number of peaks, especially at higher pI values are observed. (From Schmelzer, A. E. and Miller, W.M., *Biotechnol. Prog*., 18, 346, 2002. With permission.)

was not sialylated) and was due to an increase in the level of galactose incorporation, a situation that indicated that hyperglycosylation was occurring, possibly arising from galactosylation at another Asn site.

22.5.1.4 Product Comparability and Analysis of Finished Products

The ability to compare products derived from different manufacturing processes or sources is an important aspect of the regulatory process, for the setting of standards, the evaluation of potential problems related to one specific product or the assessment of differences linked to product characteristics such as biological activity. For example, currently throughout the world, there are biopharmaceuticals for which more than one manufacturer have received marketing authorization and this situation is likely to expand as the expiration dates of patents for several products are fast approaching. In addition, some biopharmaceuticals are being used in situations other than those for which they were marketed, such as their use as performance-enhancing substances by athletes. In many instances, products can be obtained through a number of sources other than conventional pharmacies and may, in some instances, be produced by unauthorized manufacturers. In this context, it has become essential to develop generic methods for the evaluation of multisource products.

22.5.1.4.1 Product Comparability

There are several sources of commercially available OVA, and CZE can be used to distinguish their glycoform profiles [146]. Two batches of turkey egg ovalbumin (tOVA) showed qualitatively similar glycoform profiles with some differences in the levels of a number of glycoforms (Figure 22.23a and b). On the other hand, the glycoform profile of chicken egg ovalbumin (cOVA) (Figure 22.23c)

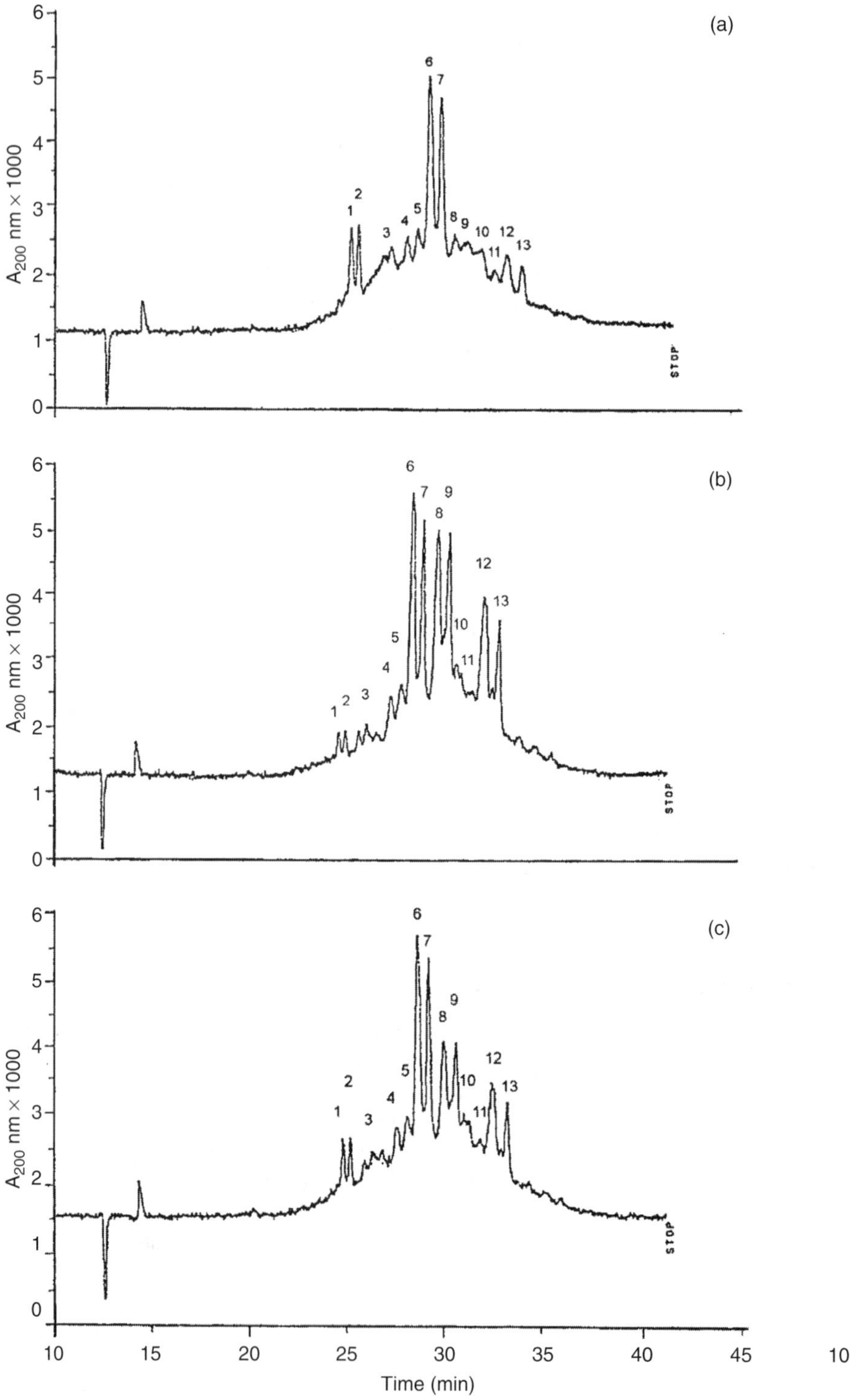

FIGURE 22.23 Electropherograms of (a) and (b): two batches of turkey ovalbumin and (c): chicken ovalbumin. Separation conditions: fused-silica capillary 72 cm (50 cm effective length), 100 mM H_3BO_3/NaOH and 1.8 mM putrescine, pH 8.6, 20 kV, detection at 200 nm, temperature controlled at 30 °C. (From Che, F.Y. et al., *J. Chromatogr. A*, 849, 599, 1999. With permission.)

revealed marked qualitative differences when compared with tOVA with respect to migration times. These differences were a result of both polypeptide and glycan chains structural variations.

Caseinoglycomacropeptide (CGMP) is a small glycoprotein (MW of approx. 7000 Da) derived from bovine κ-casein that has been shown to have a variety of biological activities such as inhibition of gastric secretion, depression of platelet aggregation, growth promoting effect on bifidobacteria, inhibition of oral *Actinomyces* adhesion to red blood cell membranes, inhibition of adhesion of oral *Streptococci* to saliva-coated hydroxyapatite beads, inhibition of adhesion of *Streptococcus sanguis* to human buccal epithelial cells, and inhibition of cholera toxin binding to its receptor. A CZE method that had been previously reported for the separation of CGMP glycoforms [164] and for stability monitoring in cosmetic lotions [165] was also used for the assessment of sialylation levels of commercial batches of CGMP [166]. Results indicated that the quantitative assessment using CZE were in good agreement with those obtained by enzymatic release of sialic acids or colorimetric methods. The CZE method was fully validated for the usual analytical criteria (e.g., linearity, precision, accuracy, LOD, and LOQ). This approach provided a fast, reliable, precise, and cost-effective way to compare products.

Another example of product comparability was reported for prostate-specific antigen (PSA), a single-chain glycoprotein that is used as a biomarker for prostate-related diseases [167]. PSA has one known PTM, a sialylated biantennary N-linked glycan chain attached to Asn45. It is commercially available from different sources. The glycoform profiles of seven free PSA (fPSA) samples from several manufacturers, of which two were specialized, enzymatically active PSA (EA-PSA) and noncomplexing PSA (NC-PSA), were assessed by CZE. Results indicated that PSA samples could be classified into three distinct groups according to glycoform profiles. Figure 22.24 shows

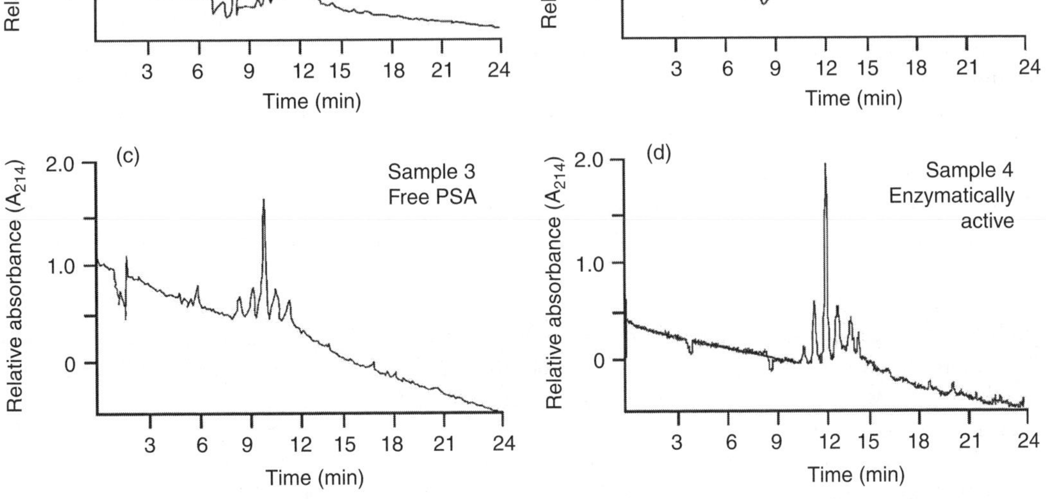

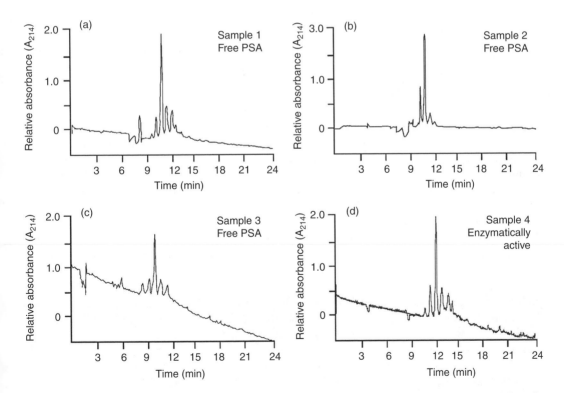

FIGURE 22.24 Electropherograms of protein specific antigen (PSA) group I. (a) Sample 1 (f PSA); (b) Sample 2 (f PSA); (c) Sample 3 (f PSA); (d) Sample 4 (EA-PSA). Separation conditions: 50 μm i.d. × 60 cm fused-silica capillary, 20 mM sodium borate (pH 8.0)/5 mM diaminopropane or 20 mM sodium phosphate (pH 7.0 or 8.0), 25 kV, detection at 214 nm. (From Donohue, M.J. et al., *Anal. Biochem.*, 339, 318, 2005. With permission.)

electropherograms of products from group I that contained 4–7 peaks with one peak dominating and representing 51–70% of the total fPSA in the sample. This study demonstrated that proprietary purification procedures have a major influence on the composition and yield of the number of structural forms observed in a PSA sample.

22.5.1.4.2 Glycoprotein Drug Products

Methods capable of analyzing the active ingredient in a formulated (i.e., finished) drug product are useful to measure product integrity and/or its stability as well as for comparison to other products. However, the approach to their analysis entails careful consideration of a number of specific issues to drug products such as the low amounts of active ingredient, the presence of large amounts of excipients, or the potential of modifying the product through manipulation.

(Glyco)protein drug products generally contain low amounts of the active ingredient since the therapeutic effect can usually be achieved at low concentrations. For example, it is not unusual to find active ingredient at low microgram quantities in a dosage form. In turn, formulations at low protein content usually require the addition of large amounts of excipients to enhance product stability and prevent nonspecific adsorption to the container. Commonly used excipients vary widely in terms of their chemical nature and include inorganic salts, amino acids, sugars, and surfactants such as polysorbate or other proteins such as HSA. Typically, isotonic salt preparations are produced as most of these products are injectables and, consequently, high salt concentrations are present. Furthermore, many of these excipients may be present simultaneously, leading to complex mixtures. They have also been shown in some cases to interfere with traditional assay methodologies such as UV or high-pressure liquid chromatography (HPLC). The choice of CE as a separation technique to be applied to the analysis of protein drug products resides in its ability to provide sufficient selectivity to separate the active ingredient from interfering species.

Studies aiming the direct analysis of finished products with no or minimal prefractionation or manipulations in order to minimize potential loss of product or the generation of product artifacts have been reported. As mentioned previously, a CZE method using an amine-coated capillary to separate rhEPO glycoforms in drug products containing large amounts of HSA and inorganic salts has been reported [41]. The addition of nickel ions to the high ionic strength BGE (200 mM phosphate, pH 4.0) selectively altered the mobility of HSA and allowed the separation of the otherwise comigrating proteins (Figure 22.6). Dosage forms containing HSA/rhEPO at a ratio of 250:1 w/w could be analyzed under these conditions. The method was validated and provided satisfactory results for active ingredient assay, quantification of glycoforms, and comparison of products from different manufacturers. The usefulness of the method was further extended to assess manufacturing changes and to compare rhEPO products formulated with different excipients as shown in Figure 22.25 for different rhEPO-α and rhEPO-β products formulated with HSA or polysorbate [168]. As described in a previous section, a similar comparison of finished rhEPO products was carried out following the immunochromatographic removal of HSA from formulations (see Section 4.1) and enabled the comparison of glycoform profiles of products from different manufacturers and from the BRP of the EP formulated with excipients of low-MW and with HSA [43].

The biophysical properties of two commercially available rhEPO-α products, Epogen and Eprex, both of which are produced under similar conditions by different manufacturers, were studied [169]. Among many techniques used in this study, the analysis by CZE showed that the products had similar glycoform profiles in terms of isoform type and distribution, an indication that the products had similar sialic acid contents.

CZE was also used for the analysis of two preparations of rhGCSF, Granocyte, a glycosylated product formulated with large amounts of HSA, and Gran300, a nonglycosylated product [170]. The two products were compared with a bulk rhGCSF preparation that had been obtained from *Escherichia coli* (nonglycosylated). Analysis with 50 mM Tricine, 20 mM NaCl, 2.5 mM DAB, pH 8.0, as BGE showed the separation of two glycoforms in Granocyte that corresponded to monosialylated and disialylated glycoforms. As expected, the nonglycosylated product, Gran300, showed

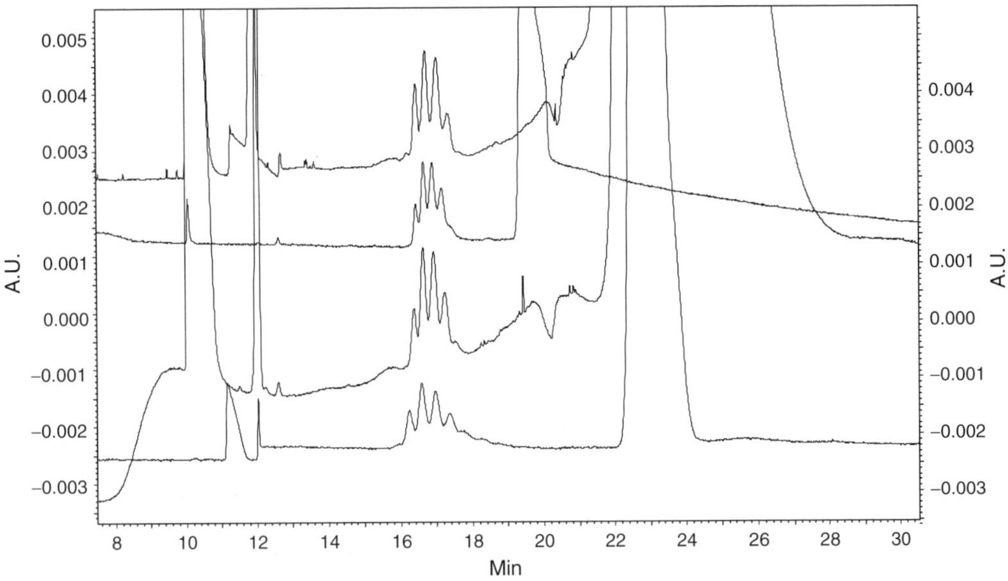

FIGURE 22.25 Typical electropherograms of rhEPO drug products (from top to bottom): rhEPO-α formulated with HSA (product I), rhEPO-α formulated with polysorbate 80 (product II), rhEPO-α formulated with HSA (product III), and rhEPO-β formulated with polysorbate 20 (product IV). Separation conditions: Beckman eCAP amine capillary, 200 mM sodium phosphate/1 mM nickel chloride pH 4.0, −15 kV (75 μA), UV detection at 200 nm. (From Girard, M. et al., Presented at the *7th International symposium on capillary electrophoresis in the biotechnology and pharmaceutical industries*, Montreal, August 2005.)

a single peak with a shorter migration time than that of the two glycosylated counterparts and that corresponded to that of the bulk rhGCSF from *E. coli*.

22.5.1.5 Determination of Biological Activity/Potency

The determination of the biological activity of biopharmaceuticals by physicochemical methods has long been a desirable goal of manufacturers, regulatory authorities, and policy makers alike, driven in part by the ability to replace costly and often imprecise biological assay testing using animals. There are currently a few examples of biopharmaceuticals for which this has occurred. Recombinant insulin and recombinant somatropin (human growth hormone) are two such products, both of which can be described as relatively simple, well-characterized proteins. For both of these two nonglycosylated proteins, physicochemical tests have replaced animal-based bioassays for batch release of bulk products (for somatropin see [140]). In-depth physicochemical characterizations as well as the establishment of a direct correlation with the *in vivo* bioassay were carried out.

For more complex products such as glycoproteins, difficulties in obtaining in-depth characterization of the glycosylation-mediated microheterogeneity have limited the development of this type of approach. However, in a few cases, reports have shown that there exists a correlation between glycoform profiles and biological activity. This is the case for recombinant follicle stimulating hormone (rhFSH) where the CZE-derived glycoform profile has been shown to correlate with the *in vivo* bioassay in a study directed at predicting the biological potency of several preparations [171]. Using CZE conditions consisting of a fused-silica capillary and 100 mM borate, 5 mM DAP at pH 8.9 [172], where rhFSH isoforms migrated as a broad peak between two internal standards, sucrose and the dipeptide Lys-Asp, four highly purified rhFSH preparations differing in their isoform composition and biological potencies were analyzed (Figure 22.26a). The inclusion of internal standards enabled the accurate determination of the median migration time (tm) of the rhFSH peak in each of the four

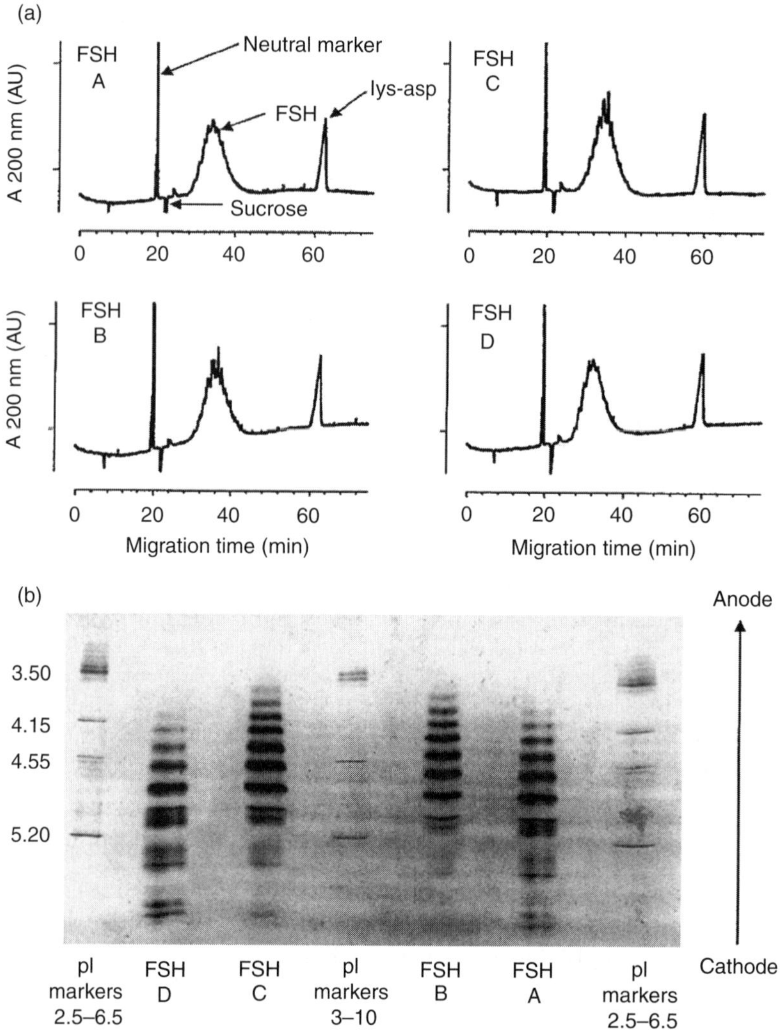

FIGURE 22.26 (a) Electropherograms of rhFSH preparations A–D with different isoform compositions and biological potencies. Separation conditions: fused-silica capillary i.d.: 20 μm × 77 cm (effective length 70.6 cm), 100 mM borate, 5 mM diaminopropane (DAP) at pH 8.9, 20 kV (approx. 3.5 μA), 28 °C, detection at 200 nm. (b) Isoelectric focusing electropherograms of preparations A–D. (From Storring, P.L. et al., *Biologicals*, 30, 217, 2002. With permission.)

preparations; tm correlated directly with the biological potency on the basis that an increase in the content of highly sialylated (the more acidic) isoforms led to a corresponding increase in biological activity. The same four preparations were also analyzed by gel IEF (Figure 22.26b) and the predicted biological potencies were derived and compared with those obtained by CZE. In both cases it was found that the methods were sufficiently accurate, precise, and robust to predict the bioactivity of batches of rhFSH when used in conjunction with a standard preparation.

22.5.2 CE of Isoforms of Intact Glycoproteins in the Clinical Field

As mentioned in the introduction to this chapter, changes in the glycosylation of proteins have been widely related to pathophysiological changes in an individual [1–4]. For instance, a common phenotypic alteration in malignant cells is the transformation of their glycosylation [173]. CE is an attractive technique for the study of glycoproteins with a clinical interest, due to its well-known quantitative results and speed of analysis [21]. However, not many works dealing with this subject have been published. There are many challenges in the application of CE to glycoproteins of clinical interest. One of the problems is that, as for many other important molecules, these glycoproteins often exist at low concentrations in biological matrices where other proteins are usually present in large amounts and interfere in the analysis. Thus, a sample preparation step, sometimes including concentration, is required. Moreover, increasing the detection sensitivity of CE for the analysis of glycoproteins is one of the challenges reviewed in this chapter. Another pitfall of the technique is that it is not easy to develop a single method for the analysis of most of the proteins as it is the case for other analytical techniques such as SDS–PAGE. On the other hand, CE presents the advantage of its high resolving power that enables the study of slight modifications in protein composition such as for glycoproteins. In addition, its small sample volume requirements make it appropriate for the analysis of clinical samples. This section is devoted to a discussion of CE reports on the separation of isoforms of glycoproteins with a clinical interest.

22.5.2.1 Transferrin

The isoforms of Tf, one of the major proteins in serum, have been widely analyzed by CE as alcohol abuse markers. Tf is a glycoprotein synthesized mainly in the liver, which consists, in a single polypeptide chain, of 679 amino acid residues and two N-linked complex-type oligosaccharide chains. It has an important role in iron metabolism and it also acts as a growth factor [174,175]. Human Tf presents several heterogeneities due to its iron content, its genetic polymorphism, and its carbohydrate moieties. With respect to iron content, a Tf molecule can be apotransferrin, monoferric transferrin, or diferric transferrin [174,175]. The influence of this heterogeneity on CE separations of glycoforms can be easily overcome by iron saturation of the sample. On the other hand, more than 38 genetic Tf variants, attributable to substitutions at one or more amino acids, have been described, although only four of them show a prevalence of $\geq 1\%$ [176]. The common type of Tf is called C-type and most Caucasian individuals express this allele [177]. With respect to carbohydrate heterogeneity, each carbohydrate chain can be bi-, tri-, or tetra-antennary and each antenna possesses a terminal sialic acid residue. Therefore, nine sialoforms (from zero to eight sialic acid residues per molecule) may be present in serum, resulting in pI variations of 0.1 pH unit/residue. The most abundant Tf-form in normal sera is tetrasialo-Tf and it has a pI of 5.4. The other forms have higher and lower pI values [178].

Variations in the heterogeneity of Tf are induced by several pathological and physiological conditions, such as rheumatoid arthritis (RA), congenital disorders of glycosylation (CDG), or pregnancy [175,179], even though variations in the Tf pattern due to alcohol abuse has attracted the highest interest [176,178,180]. The first to report abnormal Tf heterogeneity in cerebrospinal fluid (CSF) associated with alcoholic cerebellar degeneration were Stibler and Kjellin [181] and further studies determined the Tf abnormality in serum of alcohol abusers [182,183]. The variation in Tf heterogeneity in sera from alcohol abusers corresponds to increased amounts of asialo-, monosialo-, and disialotransferrrin (generically called carbohydrate-deficient transferrin or CDT). At first, the increased CDT was only attributed to changes in the number of sialic acid residues. However, it was later learnt that the difference was more complex and included changes in some of the neutral carbohydrates [178]. CDT has become the most specific marker for chronic alcohol abuse [66,177,178,184].

The first studies dealing with the quantitative determination of CDT in sera from alcohol abusers were performed by conventional IEF with immunological detection. This method can easily detect genetic variants and has the power to resolve individual isoforms. However, it is laborious and time consuming [178,184]. Since then, a number of CDT methods of analysis, such as anion-exchange chromatography, chromatofocusing, HPLC, and CZE have been developed [176,184].

At the present time, there are several commercial kits for the quantification of CDT based on the fractionation of CDT and non-CDT serum variants by anion-exchange chromatography with immunochemical detection [176]. The main characteristic of almost all of these micro columns is that they do not separate CDT sialoforms and they are quantified as a whole. Therefore, the CDT forms may contain an undetermined amount of trisialo-Tf, which is under debate as to whether it should be considered a CDT form, and that introduces an error in the determination [185,186]. Another drawback of the anion-exchange chromatographic methods is that they fail to detect genetic variants, which may give rise to incorrect determination of CDT or even false-positive results when sera from individuals with genetic Tf variants are analyzed [177,186]. Some authors have claimed that the separation of individual Tf sialoforms is a more accurate procedure than CDT quantification as a whole [68,69,185,186]. One of these commercial tests was compared with HPLC and CZE methods to conclude that the data obtained needed to be systematically confirmed by HPLC or CZE [187]. In spite of these drawbacks, these kits are still convenient for routine analysis due to their speed.

HPLC and CZE methods have been developed to resolve all of the Tf sialoforms present in serum. An HPLC method based on anion-exchange chromatography with direct detection at 420 nm was developed by Jeppsson et al. [188] to individually separate Tf sialoforms in about 16 min. The serum required to be saturated with iron and lipoproteins to be precipitated. Several authors followed the method with minor changes and improvements [186,189,190].

22.5.2.1.1 CZE of Carbohydrate-Deficient Transferrin (CDT)

In general, the CZE identification of Tf sialoforms in the electropherograms has been performed in three ways: first, using anti-Tf to perform immunosubtraction and then comparing the electropherograms of the Tf sample before and after being immunosubtracted; second, the amounts of the different sialoforms can be monitored by CZE analysis after progressive desialylation of Tf with neuraminidase treatment of the sample; and third, by identification from the migration times of each sialoform. With very few exceptions, Tf is iron-saturated before performing CE to eliminate the iron-binding-mediated heterogeneity.

22.5.2.1.1.1 Sample Preparation Commercially available Tf standards were first separated by CIEF and CZE by Kilar and Hjerten [79,97] and Oda and Landers [55]. Iourin et al. [120] purified Tf from sera of one donor and of two CDG type I syndrome by consecutive precipitation of sera with rivanol and ammonium sulfate for their analysis by CZE. On the other hand, immunopurification of Tf from sera was used as sample preparation before CE analysis [89,191]. While these were pioneering attempts at the use of CZE to study CDT as a disease marker, the laborious and time-consuming sample preparation steps required were not conducive to their use in routine settings. Sample preparation of Tf from serum became easier after the development of a procedure that involved only iron saturation and dilution of the serum [192]. This sample preparation method has since then been adopted by most authors with only slight modifications [66–68,177,185,193–201]. In another approach, interfering proteins, such as immunoglobulins, can be eliminated from serum with protein A to enhance CDT detection [202]. Another interesting approach to sample preparation is the direct injection of serum into the capillary, carrying out complexation of Tf with iron during the electrophoretic separation [203].

22.5.2.1.1.2 CZE Separation With respect to CZE performance, many approaches have been published for the separation of Tf sialoforms. Two groups of methods can be distinguished. In the first one, studies were conducted with a commercial buffer system (CEofix-CDT kit; Analisis, Namur, Belgium), as already mentioned in a previous section, widely accepted and used, sometimes with slight modifications to the protocol proposed by the manufacturer [66–69,177,185,195,197–199,203]. This commercial reagent kit offers the advantage of interlaboratory standardization. In the second group of studies, methods developed in-house were used

[55,79,89,97,120,185,191,192,196,200–202]. To the best of our knowledge, all of the reported methods were carried out at basic pH (around 8.5) so that Tf is negatively charged to prevent protein adsorption to the capillary wall.

The commercial buffer system is based on the dynamic double coating of an uncoated, bare fused-silica capillary. A first buffer (the so-called initiator) is used to coat the capillary wall with a polycation and a second buffer containing a polyanion adds the second layer. The separation buffer is usually borate-based. Between injections, the capillary is rinsed with NaOH so that the coating layers are eliminated after each analysis. This procedure ensures a constant EOF and increases the negative charges on the capillary wall for preventing protein adsorption and increasing the speed of analysis (usually between 6 and 13 min, depending on the modifications of the method) [185,203]. Although in one of the first studies with this commercial system, disialo- and trisialo-Tf could not be resolved [203], modifications of the initial method, which included among others offline iron-saturation and dilution of serum, increased capillary length, and injection of an SDS solution before the injection of the sample, led to baseline resolution of tri, tetra, penta, and hexasialo-Tf in nonalcoholic donors and asialo, di, tri, tetra, penta, and hexasialo-Tf in alcoholic patients [177]. It appears that the SDS plug before sample injection is performed to keep β-lipoprotein peaks out of the area of the electropherogram of interest. SDS is now incorporated in the $FeCl_3$ solution used for iron saturation of the commercial reagent set, so it is not necessary to add it separately if the complete reagent kit is used [69].

In the second group of CZE separations, many different approaches have been published. The main goal has been to avoid Tf adsorption to the capillary walls. This problem has been overcome by working with covalently coated capillaries, in some cases with hydroxyethylcellulose in the running buffer [79,89,191,202], or silica capillaries dynamically coated with DAB [185,196,200,202], spermine [185], DcBr [55], or diethylentriamine (DETA) [201]. In general, the methods performed in uncoated capillaries with dynamic coatings are slower than the ones performed with the commercial buffers. On the other hand, the methods using covalently coated capillaries are as fast (usually in less than 10 min of analysis time) as the methods developed with the commercial buffers.

Lanz et al. [185] compared two different methods in fused-silica capillaries with alkaline borate buffers and different dynamic coatings (DAB and spermine, respectively) and with the double-coating method based on commercially available buffers. The latter method provided faster and more reproducible results. In addition, Tf isoforms peaks were more intense when analyzed with the latter method.

The same group [66] studied the precision of a method based on the double-coating buffer kit over a 20-day period. They found out that the method is highly reproducible both in migration times and peak areas of Tf isoforms. Martello et al. [198] ran an interlaboratory comparison of a method based on the dynamic double coating of the capillary wall performed with commercially available reagents. Results from both laboratories showed high correlation.

22.5.2.1.2 Clinical Applications

In general, the study by CZE of CDT as an illness marker is performed by qualitative or quantitative comparison of CDT between two different groups of participants. To the best of our knowledge, CDT reference limits for direct classification of samples are not yet routinely used, although some attempts have been performed. In that sense, not even a common CDT calculation is performed. For example, area of disialo-Tf as a percentage of tetrasialo-Tf [185,192,201,204], area % of disialo-Tf in relation to the sum of all Tf-isoforms [191,202], and CDT isoforms (asialo-, monosialo-, and disialo-Tf) as area % of total Tf isoforms [177,194,203] have been used as CDT measurements. Lanz and Thormann [205] worked on the establishment of normal CDT reference limits using a CZE method based on a dynamic double coating of the capillary with commercially available reagent kits. They found that the reference CDT intervals (in the case of CDT, only the upper reference limit is important) for a group of 54 individuals with no or moderate alcohol consumption was dependent on the complete or incomplete separation of disialo- and trisialo-Tf, on the integration approach and on the applied voltage. The upper reference limit obtained by Lanz et al. [197] was used in a subsequent study where more than 600 samples were analyzed with a CZE method established as a routine method. Later on, the reference intervals found by those authors [205] were revised for a modified version of the commercially available reagent kits [67]. The same samples from the 54 individuals were re-analyzed with the new reagents and the CDT reference limits were found to be comparable, although slightly smaller, to the old ones so that the same upper reference limit was valid for the study. The establishment of reliable reference CDT intervals is critical in the sensitivity (number of false negatives) and specificity (number of false positives) of the method.

22.5.2.1.2.1 Alcohol Abuse Alcohol abuse is an important public health problem, and self-reporting of alcoholism or alcohol abuse is not reliable. Furthermore, diagnosis on the basis of clinical symptoms is not easy [197]. Therefore, reliable alcohol markers are required for the diagnosis of alcoholism and CDT is one of the better markers of alcohol abuse.

Prasad et al. [191] based their CDT calculation of Tf separated by CZE on the ratio of the area of disialo-Tf (they did not detect asialo-Tf) to the area of total Tf in the sample. On the basis of this index and with a control population of social drinkers, they established a CDT cutoff value. This cutoff value was exceeded by the majority of alcohol abusers studied. This CDT measurement was compared with other alcohol abuse markers (e.g., aspartate amino transferase, alanine amino transferase, and others) and it was found to be the most specific one.

In another study, Tagliaro et al. [192] found that by comparing the Tf profile of a control group of 30 participants against a group of 13 alcoholics, the latter showed a significant increase in disialo- and trisialo-Tf (expressed as percentages of the tetrasialo-Tf peak). They did not detect asialo-Tf in alcoholic samples. These results are controversial since several studies have claimed that not only trisialo-Tf is not an alcohol abuse marker [68,195] but also that it interferes with the correct CDT quantification [204]. On the other hand, there are also favorable opinions to include trisialo-Tf in the calculation of CDT or, at least, to establish a debate [177,206,207]. Crivellente et al. [193] improved on their published method [192] and tested it with real samples. Although the presence of asialo-Tf was suspected to be present in the sera of alcoholic individuals, no confirmation was performed. An increase in disialo- and trisialo-Tf (expressed as percentage of the tetrasialo-Tf peak) compared with healthy participants happened in alcoholic individuals.

Another study using CZE for the analysis of CDT as a marker of alcohol abuse was carried out by Giordano et al. [202]. They did not detect asialo-Tf in sera from alcoholic people, but they detected a significant increase in the disialo-Tf (related to total Tf) in those samples in comparison with control (nonalcoholic subjects) samples. In the same direction, and by establishing the Tf index as % area of disialo-Tf in relation to tetrasialo-Tf, Lanz et al. [185] found that a good classification of sera could be performed on the basis of this index increasing in sera of alcoholic individuals.

Interesting CZE studies on CDT isoforms were reported by Legros et al. [68,69]. They found that asialo-Tf was missing in teetotalers and was present in 92% of alcohol abusers, and that disialo-Tf was increased in alcohol abusers. Figure 22.27 shows the comparison of the Tf profiles of an alcoholic and a teetotaler. Under these conditions, C-reactive protein (CRP) comigrated with monosialo-Tf in the alcoholics electropherogram. Samples before and after Tf immunosubtraction are also compared in the figure. They proposed to use the presence of asialo-Tf as alcohol intake marker because it showed the highest sensitivity and specificity when compared with other CDT isoforms or combination of CDT isoforms. The study focused on clearly different groups (teetotalers and alcohol abusers). In their subsequent work [69], they included moderate drinkers in another group. They found that asialo-Tf was able to discriminate between moderate drinkers and alcohol abusers better than other CDT measurements tested.

22.5.2.1.2.2 Congenital Disorders of Glycosylation Congenital disorders of glycosylation (CDG) (previously known as carbohydrate-deficient glycoprotein syndromes) are inherited

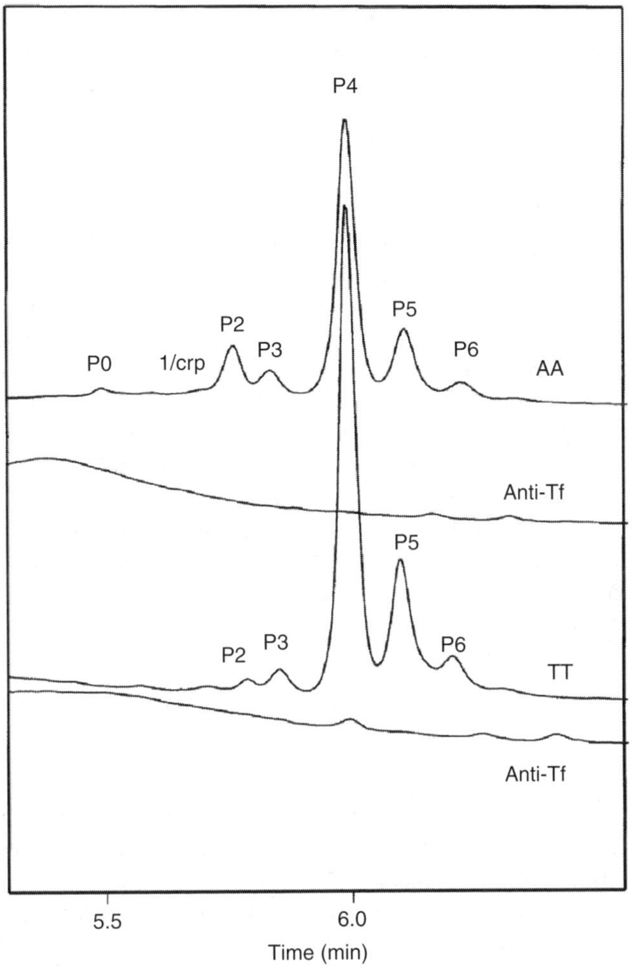

FIGURE 22.27 Comparison between CZE Tf electropherograms of an alcohol abuser (AA) and a teetotaler (TT). In both cases, anti-Tf antibody was added after the first electrophoretic run. P0: asialo-Tf. 1/CRP: comigration of monosialo-Tf and CRP; P2: disialo-Tf; P3: trisialo-Tf; P4: tetrasialo-Tf; P5: pentasialo-Tf; P6: hexasialo-Tf. Analytical conditions: fused-silica capillary 57 cm × 50 μm i.d; buffer: reagents from the CEofix CDT kit. Capillary was first rinsed with a solution of polycation dissolved in malic acid, pH 4.8; then rinsed again with a polyanion dissolved in Tris-borate, pH 8.5. Capillary was rinsed with the same buffer for 0.5 min under low pressure. After 3 s low-pressure injection of SDS, the iron-saturated sample was injected by low pressure for 2 s; Voltage: 28 kV; Temperature: 40°C; Detection: 214 nm. (From Legros, F.J. et al., *Clin. Chem.*, 48, 2177, 2002. With permission.)

conditions that are usually recognized from glycosylation changes of serum proteins [2]. The hypoglycosylation of different proteins and sometimes of other glycoconjugates leads to several symptoms, such as mental retardation. IEF measurement of Tf is the main diagnostic test used for the detection of CDG because these disorders appear to have an influence on its sialylation (and consequently on its pI) [199,208]. There are several studies that have been reported in which the CZE of Tf has been used to study CDG. An increase in disialo- and asialo-Tf in CDG patients has been reported by several authors [89,120,196,203]. Carchon et al. [199] demonstrated that patients with abnormal IEF results and with confirmed CDG could be identified by CZE. However, they warned about the possibility of finding compounds migrating in the Tf-region and when this is suspected in a sample, CZE analysis with immunosubtraction should be performed.

22.5.2.1.2.3 Genetic Variants In addition to the common C-type Tf, other variants such as B and D have been reported. As stated earlier, genetic polymorphism is limited to Caucasians. Most individuals present CC phenotypes (homozygous for Tf with the C gene variant) and, rarely, CB phenotypes. West African, African American, and indigenous American populations, for example, present higher frequency of the D allele [177]. Certain genetic variants are not detected by anion-exchange chromatography-based methods and may result in false positives for alcohol abuse [177,178]. In that sense, CZE is a promising tool for the determination of genetic variants due to its high resolving power. Several workers have published Tf CZE patterns of different genetic variants [66,67,177,197,203]. In general, these patterns are characterized by having two major peaks instead of one (corresponding to the tetrasialo-Tf of both genetic variants). Wuyts et al. [177] found that their CZE method was able to separate (in an alcoholic carrying a CD phenotype) D-asialo-, D-disialo-, D-trisialo-, D-tetrasialo-, and D-pentasialo-Tf in addition to C-tetrasialo-, C-pentasialo-, and C-hexasialo-Tf. In spite of this high resolution power, some of the D-Tf peaks comigrated with C-CDT forms. To solve this problem for the CDT calculation for alcohol consumption diagnostic purposes, the authors proposed a CDT calculation adjusted to CD-phenotypes in order to avoid false positives. Figure 22.28 shows the comparison of Tf patterns from a nonalcohol consumer homozygous (CC), an alcoholic homozygous (CC), a non-alcohol consumer heterozygous (CD), and an alcoholic heterozygous (CD). On the other hand, Lanz et al. [67] analyzed a CD-phenotype alcoholic individual. They detected two peaks for asialo-Tf and two peaks for disialo-Tf. Since C-disialo-Tf was not baseline resolved, the CDT calculation was done by taking twice the area of D-disialo-Tf because both peaks had the same height. The CDT value clearly exceeded the upper CDT reference limit established for normal nonalcoholic-CC phenotypes.

22.5.2.1.2.4 Transferrin and Cancer A study has been recently performed in which the Tf CZE profile of a group of cancer patients who consumed alcohol moderately and a group of cancer patients who were alcoholics has been studied [195]. Their profiles were compared with a group of participants without cancer and that were teetotalers, moderate drinkers, and alcohol abusers. While asialo-Tf was present in 95% of alcohol abusers (with and without cancer), trisialo-Tf was higher in cancer patients. The CZE peak corresponding to this trisialo-Tf was not completely eliminated by Tf immunosubtraction. Some tests on the remaining peak suggested that it might be a polysaccharide.

22.5.2.1.2.5 Interferences in CDT Determination Usually, Tf analysis by CZE is performed after the injection of diluted, iron-saturated serum, which is an extremely complex media. Therefore, possible interferences in the determination of Tf should be taken into account. Generally, interferences are found after Tf immunosubtraction of the sample and re-analysis. Sometimes, immunosubtraction of the interference is also performed. Legros et al. [68] detected comigration of CRP with monosialo-Tf, so this peak was not taken into account in the CDT calculations. CRP was also identified and migrated before disialo-Tf in some samples analyzed by Lanz et al. [66,197]. In some patients with hepatic disorders, a broad interference, assumed to be due to high levels of immunoglobulin A, was found under the Tf peaks [197]. Wuyts et al. [203] added antihemoglobin and anti-C3c (degradation products of complement-C) antibodies to the running buffer to avoid interferences from these compounds. As described before, SDS was injected into the capillary before the sample for methods performed with commercial reagent buffers to avoid β-lipoprotein peaks comigrating with Tf [68,69,177,195]. This modification was later adopted by the manufacturer [69]. Moreover, a compound comigrating with trisialo-Tf (probably, a polysaccharide) in cancer patients was described [195]. Recently, the commercial reagents for the double-coating CZE method have been reformulated to avoid undesired interferences, and CRP comigration is no longer interfering, neither are some paraproteins that interfered in some samples [67].

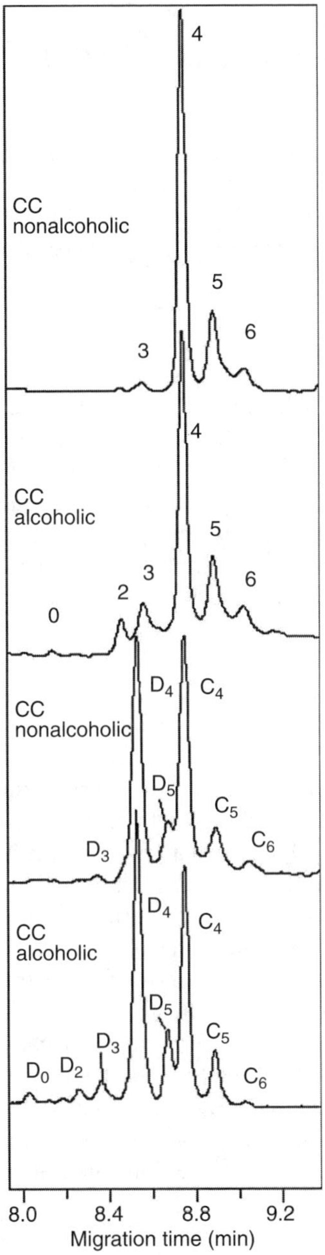

FIGURE 22.28 Comparison between CZE Tf electropherograms of a healthy nonconsuming (CC nonalcoholic) and an alcohol-consuming (CC alcoholic) carrier of the homozygote CC-Tf, and a nonconsuming (CD nonalcoholic) and an alcohol-consuming (CD alcoholic) carrier of the heterozygous CD-Tf. C_0: asialo-C, C_2: disialo-C, C_3: trisialo-C, C_4: tetrasialo-C, C_5: pentasialo-C, C_6: hexasialo-C-Tf. D_2: disialo-D, D_3: trisialo-D, D_4: tetrasialo-D, D_5: pentasialo-D-Tf. Analytical conditions: fused-silica capillary 67 cm × 50 μm i.d; buffer: reagents from the CEofix CDT kit. Capillary was first rinsed with the initiator solution, then rinsed again with the buffer solution containing a polyanion. A plug of SDS was injected during 5 s before the sample. The iron-saturated sample was injected for 1 s. Voltage: 28 kV; detection: 214 nm. (From Wutys, B. et al., *Clin. Chem. Lab. Med.*, 39, 937, 2001. With permission.)

22.5.2.1.3 Concluding Remarks

The CZE analysis of Tf seems to be an alternative to CDT measurements kits, which are prone to inaccurate diagnostics. Normal CDT ranges are method-dependent and, consequently, results must be interpreted by taking into account method-specific cutoff values [176,184]. In any case, several authors support the application of CZE for CDT determination in the clinical laboratory [191] and claim that it represents an alternative to HPLC and that it should be taken into account as a reference method for CDT [67]. In comparison to HPLC, CZE methods have the advantages of easier sample preparation, faster analysis times, higher isoforms resolution, and faster column reconditioning [197].

22.5.2.2 Alpha-1-Acid Glycoprotein

AGP or orosomucoid is synthesized mainly in the liver. It is an acute-phase protein with an unclear function, although it is usually accepted as a natural immunomodulatory and anti-inflammatory agent. It is a 41–43 kDa protein with a pI of 2.8–3.8 with 5 N-complex type glycans. Terminal sialic acid residues are usually present and contribute to the acidity of the protein [209]. Since it is expressed by various alleles at two *loci*, it exhibits genetic polymorphism [210].

Changes in the glycosylation of AGP, as well as in the distribution of genetic AGP forms, have been related to several pathophysiological states, such as cancer [210–213], RA [214,215], and other types of inflammation [216–218].

There are several studies that have described the separation of AGP forms by CE [78,84,87,100,109,117,219,220]. However, only a few authors have carried out CE analysis of AGP samples from patients in order to study if the technique is suitable for distinguishing the changes described earlier.

Kinoshita et al. [78] analyzed AGP purified from rat sera from normal and inflammation states. They separated electrokinetically injected AGP into several nonbaseline resolved peaks in a DB-1 capillary with an acetate buffer (pH 4.1) containing HPMC. They found qualitative differences between the two samples. The sample from rat sera in the inflammation state had more AGP bands moving slower than the AGP sample from sera in normal state. The authors speculated about those bands having decreased sialic acid content. Later on, Kakehi et al. [219] proposed an interesting approach to the analysis of AGP by CZE. They performed electrokinetic injection of desalted serum in a DB-1 coated capillary with a running buffer at pH 4.5. They assumed that almost all serum proteins have pIs above that pH except for AGP, which is negatively charged at that pH. Therefore, a selective electro-injection of AGP was performed. They obtained nonbaseline resolution of 10 AGP forms. They analyzed two serum samples from patients who had acquired methicillin-resistant *Staphylococcus aureus* during hospitalization and one serum sample from a healthy volunteer. However, they did not find a relationship between the abundance of each glycoform and the clinical data of patients.

A completely different approach, based on lectin-affinity CE was used by Bergstrom et al. [117] to separate AGP fractions. As mentioned in a previous section, using Con A as affinity ligand, they separated AGP into two peaks according to the biantennary content of the glycoforms. They applied this methodology to two AGP samples from patients with severe RA and one sample from a healthy donor. They found that the RA samples showed a decrease in the relative peak area of the biantennary peak compared with normal AGP.

In another study, a CZE method capable of baseline separation of up to 11 AGP bands was developed [84]. This method, together with a statistical program that allowed to correctly compare AGP bands from different samples, was used to compare three AGP samples from one healthy donor and two pools of sera from cancer patients. Results showed that samples from cancer patients had one extra AGP band with a higher charge to mass ratio, while the AGP profile from the healthy donor presented one extra band with lower charge to mass ratio. A quantitative study of the samples demonstrated that there were peak area differences for the AGP forms. The cancer patient samples presented higher proportion of forms with higher charge to mass ratio and lower proportion of forms with lower charge to mass ratio when compared with AGP from the healthy donor (Figure 22.29). The same authors also developed a CIEF method to separate AGP isoforms and applied it to the analysis of AGP from sera from ovary cancer patients [109].

The studies discussed earlier represent the beginning of research efforts directed toward the utility of CE analysis of AGP for clinical uses. In all cases, AGP samples from people and animals with different pathophysiological conditions were analyzed as a proof of concept for the methods developed. Studies of large populations are needed to make clinical conclusions.

22.5.2.3 Other Glycoproteins

hCG is a glycohormone of 38,000 Da that consists of two noncovalently bound subunits. Both subunits (α and β, respectively) contain two N-glycans and the β-subunit also contains four O-glycosylated chains. Its role is to avoid the disintegration of the corpus luteum in the ovary. The more acidic hCG glycoforms have increased biological activity. In addition, the glycosylation pattern of this hormone seems to change in patients with trophoblastic disease [56]. As indicated previously, this glycoprotein (in its native heterodimer form) was separated into eight peaks in an uncoated capillary with a borate buffer containing DAP [56]. The method was used to analyze a reference hCG, two samples purified from crude urinary hCG, and one hCG sample from the urine of a patient with metastatic choriocarcinoma. The four hCG electropherograms had the same number of forms, but the relative concentration of each one appeared to vary. However, the hCG sample coming from the cancer patient migrated in a way that did not allow an accurate comparison with the other samples. Those peaks could be aligned with either the previous or subsequent peak, resulting in a different quantitative result.

Hiroaka et al. [221] developed a one-step CIEF method for the analysis of CSF proteins with molecular masses between 10,000 and 50,000 Da. They found differences in the distribution of lipocalin-type prostaglandin D synthase isoforms (a sialic acid-containing glycoprotein) in the CSF of patients with certain neurological disorders. It was speculated that the four different peaks attributable to this protein were due to different numbers of sialic acid residues.

PSA is a glycoprotein that is used as a biomarker for prostate cancer. Its release into blood is increased during the development of prostate cancer. It has a MW of 28,430 Da and contains one N-glycosylation site. A fraction of circulating PSA is bound to plasma proteins. The measurement of unbound PSA increases the diagnosis specificity of total blood PSA [222]. However, these measurements do not take into account the glycosylation pattern of PSA. It was recently stated that when comparing PSA oligosaccharides from healthy donors with that from a prostate tumor cell line, the PSA glycosylation pattern may be different [223]. In that case, the use of CZE for the comparative analysis of PSA glycosylation patterns in samples from healthy persons and prostate cancer patients would contribute to the study of PSA glycosylation as prostate marker. Donohue et al. [167] obtained a separation of PSA forms from several manufacturers in 12 min in an uncoated capillary with a borate buffer containing DAP using UV detection. The commercial PSA samples had been purified from human seminal fluid. As mentioned earlier in this chapter, an interesting finding of this work was that the purification procedure influences the PSA pattern.

22.6 CONCLUDING REMARKS AND FUTURE PROSPECTS

As demonstrated in this chapter, CE has clearly become a widely applicable tool for the analysis of glycoforms of glycoproteins, without which the determination of some significant qualitative and quantitative information would not have been possible. This is aptly exemplified from some of the earliest work in the field on Tf, work which has culminated in the use of CE in clinical settings, to a more recent application of CE-MS from which over 130 glycoforms of rhEPO were

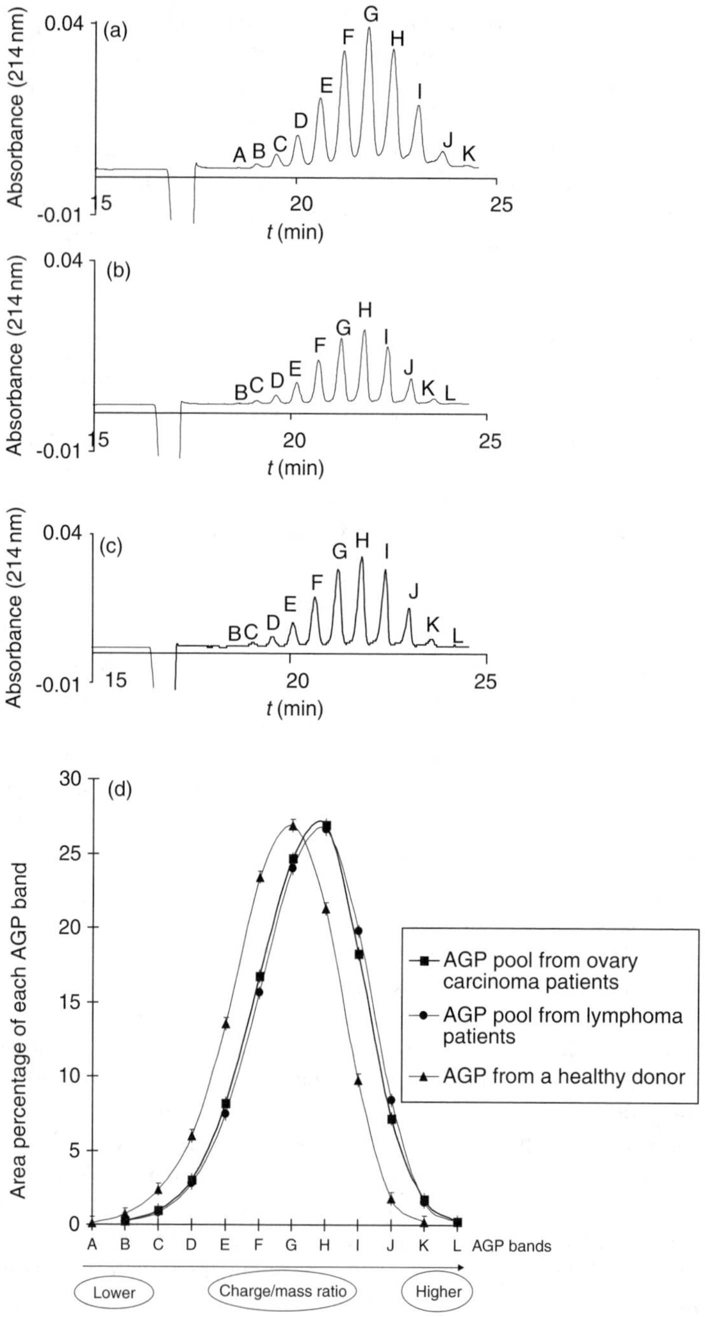

FIGURE 22.29 (a, b, and c): Electropherograms of AGP purified from serum of a healthy donor, a pool of sera from ovary cancer patients, and a pool of sera from lymphoma patients, respectively. A statistical program was used to compare the three electropherograms, showing the different AGP profile between the healthy donor and the cancer patients. Analytical conditions: fused-silica capillary 77 cm × 50 μm i.d; buffer: 0.01 M Tricine, 0.01 M NaCl, 0.01 M sodium acetate, 7 M urea and 3.9 mM putrescine, pH 4.5; voltage: 25 kV, 35°C. (d) Mean percentage areas and standard deviation (indicated as "I" and calculated with standard AGP, $n = 6$) of AGP peaks in real samples (2 injections each). Names of the AGP peaks according to electropherograms (a), (b), and (c). Discontinuous lines are used to link the real values for the sake of clarity of the representation. (Adapted from Lacunza, I. et al., *Electrophoresis*, 27, 4205, 2006. With permission.)

identified. Undoubtedly, efforts at understanding the separation mechanism of glycoforms have made it possible. The applicability of CE at most stages of biopharmaceutical manufacturing processes as well as its usefulness for comparing products from different sources, whether of recombinant or natural origins, has led most manufacturers to endorse this technique. CE methods have also appeared in pharmacopeial monographs to replace the more laborious, conventional gel methods. A list of several proteins for which glycoforms have been separated and the CE method employed in each case can be seen in Table 22.3.

One particular aspect of CE analysis that requires improvement to make its practical application wider is sample preparation. There is little sense in developing simple and fast CE methods if

TABLE 22.3
Separation of Glycoprotein Forms by Capillary Electrophoresis

Protein	Separation Mode	Reference
24 kDa glycoprotein, recombinant	CZE	Berkowitz, S.A. et al., *J. Chromatogr. A*, 1079, 254, 2005
Alpha-1-acid glycoprotein, human (AGP) (orosomucoid)	ACE	Bergstrom, M. et al., *J. Chromatogr. B*, 809, 323, 2004
	CIEF	Wu, J. et al., *J. Chromatogr. A*, 817, 163, 1998
		Lacunza, I. and de Frutos, M., *PACE Setter*, 10, 5, 2006
		Lacunza, I. et al., *Electrophoresis*, 28, 1204, 2007
	MEKC	James, D.C. et al., *Anal. Biochem.*, 222, 315, 1994
	CZE	Kubo, K., *J. Chromatogr. B*, 697, 217, 1997
		Kinoshita, M. et al., *J. Chromatogr. A*, 866, 261, 2000
		Pacakova, V. et al., *Electrophoresis*, 22, 459, 2001
		Kakehi, K. et al., *Anal. Chem.*, 73, 2640, 2001
		Sei, K. et al., *J. Chromatogr. A*, 958, 273, 2002
		Lacunza, I. et al., *Electrophoresis*, 27, 4205, 2006
Alpha-1-acid glycoprotein, bovine (bovine AGP)	CZE	Che, F.-Y. et al., *Electrophoresis*, 20, 2930, 1999
		Kinoshita, M. et al., *J. Chromatogr. A*, 866, 261, 2000
		Balaguer, E. and Neususs, C., *Anal. Chem.*, 78, 5384, 2006
Alpha-1-acid glycoprotein, rat (rat AGP)	CZE	Kinoshita, M. et al., *J. Chromatogr. A*, 866, 261, 2000
Alpha-1-acid glycoprotein, sheep (sheep AGP)	CZE	Kinoshita, M. et al., *J. Chromatogr. A*, 866, 261, 2000
Alpha-1-antitrypsin (AAT)	CZE	Kubo, K., *J. Chromatogr. B*, 697, 217, 1997
		Chang, W.W.P. et al., *Electrophoresis*, 26, 2179, 2005
Antithrombin III, human (ATIII)	CIEF	Kremser, L. et al., *Electrophoresis*, 24, 4282, 2003
	CZE	Demelbauer, U.M. et al., *Electrophoresis*, 25, 2026, 2004
		Kremser, L. et al., *Electrophoresis*, 24, 4282, 2003
Antithrombin III, recombinant human (rhATIII)	CZE	Reif, O.-W. and Freitag, R., *J. Chromatogr. A*, 680, 383, 1994
	CIEF	Reif, O.-W. and Freitag, R., *J. Chromatogr. A*, 680, 383, 1994
		Buchacher, A. et al, *J. Chromatogr. A*, 802, 355, 1998
Antithrombin III β	CE-SDS	Buchacher, A. et al., *J. Chromatogr. A*, 802, 355, 1998
Avidin	CZE	Bateman, K.P. et al., *Methods Mol. Biol.*, 213, 219, 2003
Beta-trace proteins (BTP)	CIEF	Hiraoka, A. et al., *Electrophoresis*, 22, 3433, 2001
Bone morphogenic protein-2, recombinant human (rhBMP-2)	CZE	Yim, K. et al., *J. Chromatogr. A*, 716, 401, 1995
		Yeung, B. et al., *Anal. Chem.*, 69, 2510, 1997
Caseinoglycomacropeptide	CZE	Cherkaoui, S. et al., *J. Chromatogr. A*, 790, 195, 1997
		Cherkaoui, S. et al, *Chromatographia*, 50, 311, 1999
		Daali, Y. et al., *J. Pharm. Biomed. Anal.*, 24, 849, 2001
CD4, recombinant (rCD4)	CZE	Wu, S.-L. et al., *J. Chromatogr.*, 516, 115, 1990

TABLE 22.3
(Continued)

Protein	Separation Mode	Reference
Cellobiohydrolase I (CBH-I)	CIEF	Sandra, K. et al., *J. Chromatogr. A*, 1058, 263, 2004
Cell adhesion molecule (CAM)	CITP	Josic, D. et al., *J. Chromatogr.*, 516, 89, 1990
Chorionic gonadotropin, human (hCG)	CZE	Morbeck, D.E. et al., *J. Chromatogr. A*, 680, 217, 1994
		Oda, R.P. et al, *J. Chromatogr. A*, 680, 85, 1994
Darbepoetin-α (novel erythropoiesis stimulating protein, NESP)	CZE	Lacunza, I. et al., *Electrophoresis,* 25, 1569, 2004
		Sanz-Nebot, V. et al., *Electrophoresis*, 26, 1451, 2005
Desmodus salivary plasminogen activator (DSPAα1)	CZE	Apfel, A. et al., *J. Chromatogr. A*, 717, 41, 1995
		Chakel, J.A. et al., *J. Chromatogr. B*, 689, 215, 1997
DNAse, recombinant human (rhDNAse)	CZE	Felten, C. et al., *J. Chromatogr. A*, 853, 295, 1999
		Quan, C.P. et al., *Chromatographia Suppl.*, 53, S39, 2001
Erythropoietin, recombinant human (rhEPO)	CZE	Tran, A.D. et al., *J. Chromatogr.*, 542, 459, 1991
		Watson, E. and Yao, F., *Anal. Biochem.*, 210, 389, 1993
		Nieto, O. et al., *Anal. Commun.*, 33, 425, 1996
		Bietlot, H.P. and Girard, M., *J. Chromatogr. A*, 759, 177, 1997
		Kaniansky, D. et al., *J. Chromatogr. A*, 772, 103, 1997
		Zhou, G.-H. et al., *Electrophoresis*, 19, 2348, 1998
		Bristow, A. and Charton, E., *Pharmaeuropa*, 11, 290, 1999
		Che, F.-Y. et al., *Electrophoresis*, 20, 2930, 1999
		Kinoshita, M. et al., *J. Chromatogr. A*, 866, 261, 2000
		Lopez-Soto-Yarritu, P. et al., *J. Sep. Sci.*, 25, 1112, 2002
		Sanz-Nebot, V. et al., *Anal. Chem.* 75, 5220, 2003
		European Pharmacopoeia 5th edition, published by EDQM. June 2004
		Lacunza, I. et al., *Electrophoresis,* 25, 1569, 2004
		Neususs, C. et al., *Electrophoresis*, 26, 1442, 2005
		Madajova, V. et al., *Electrophoresis*, 26, 2664, 2005
		Yu, B. et al., *J. Sep. Sci.*, 28, 2390, 2005
		Balaguer, E. and Neususs, C., *Chromatographia*, 64, 351, 2006
		Balaguer, E. and Neususs, C., *Anal. Chem.*, 78, 5384, 2006
		Balaguer, E. et al., *Electrophoresis*, 27, 2638, 2006
		Kamoda, S. and Kakehi, K., *Electrophoresis*, 27, 2495, 2006
		Lara-Quintanar, P. et al., *J. Chromatogr. A*, 1153, 227, 2007
	CIEF	Kubach, J. and Grimm, R., *J. Chromatogr. A*, 737, 281, 1996
		Cifuentes, A. et al., *J. Chromatogr. A*, 830, 453, 1999
		Lopez-Soto-Yarritu, P. et al., *J. Chromatogr. A*, 968, 221, 2002
	CITP	Madajova, V. et al., *Electrophoresis*, 26, 2664, 2005
Erythropoietin, urinary (uEPO)	CZE	de Frutos, M. et al., *Electrophoresis*, 24, 678, 2003
		Yu, B. et al., *J. Sep. Sci.*, 28, 2390, 2005
Factor VIIa, recombinant human	CZE	Klausen, N.K. and Kornfelt, T., *J. Chromatogr. A*, 718, 195, 1995
Factor IX, human	CIEF	Buchacher, A. et al., *J. Chromatogr. A*, 802, 355, 1998
Fetuin, bovine	CZE	Kinoshita, M. et al., *J. Chromatogr. A*, 866, 261, 2000
		Balaguer, E. and Neususs, C., *Anal. Chem.*, 78, 5384, 2006
Fetuin	CE-SDS	Werner, W.E. et al., *Anal. Biochem.*, 212, 253, 1993
	MEKC	James, D.C. et al., *Anal. Biochem.* 222, 315, 1994

Continued

TABLE 22.3
(Continued)

Protein	Separation Mode	Reference
FG basic chimeric glycoprotein, recombinant	CZE	Tsuji, K. and Little, R.J., *J. Chromatogr.*, 594, 317, 1992
Follicle stimulating hormone, recombinant human (rhFSH)	CZE	Mulders, J.W.M. et al., *3rd WCBP meeting*, Washington, DC, 1999 Storring, P.L. et al., *Biologicals*, 30, 217, 2002
Granulocyte colony-stimulating factor, recombinant human (rhG-CSF)	CZE	Watson, E. and Yao, F., *J. Chromatogr.*, 630, 442, 1993 Somerville, L. E. et al., *J. Chromatogr. B*, 732, 81, 1999 Zhou, G.-H. et al., *J. Pharm. Biomed. Anal.*, 35, 425, 2004
Hepatitis C virus highly glycosylated protein, recombinant	MEKC	Kundu, S. et al., *J. Cap. Electrophor.*, 3, 301, 1996
Hirudin, novel O-glycosylated P6 (leech)	CZE	Steiner, V. et al., *Biochemistry*, 31, 2294, 1992
HIV gp120, recombinant	MEKC	Jones, D.H. et al., *Vaccine*, 13, 991, 1995
	CIEF	Tran, N.T. et al., *J. Chromatogr. A*, 866, 121, 2000
Horseradish peroxidase (HRP)	MEKC	James, D.C. et al., *Anal. Biochem.*, 222, 315, 1994
	CZE	Kelly, J.F. et al., *J. Chromatogr. A*, 720, 409, 1996
Interferon-γ, recombinant human (rhIFN-γ)	MEKC	James, D.C. et al., *Anal. Biochem.* 222, 315, 1994 James, D. C. et al., *Prot. Sci.*, 5, 331, 1996 Goldman, M.H. et al., *Biotech. Bioeng.*, 60, 597, 1998 Hooker, A.D. and James, D.C., *Mol. Biotech.*, 14, 241, 2000
Interferon-ω	MEKC	Kopp, K. et al., *Arzneim. Forsch./Drug Res.*, 46, 1191, 1996
Interleukin-2 (IL-2)	CZE	Knuver-Hopf, J. and Mohr, H., *J. Chromatogr. A*, 717, 71, 1995
Lipocalin-type prostaglandin D synthase	CIEF	Hiroaka, A. et al., *Electrophoresis*, 22, 3433, 2001
Monoclonal anti-alpha-1 antitrypsin	CIEF	Wu, J. et al., *J. Chromatogr. A*, 817, 163, 1998
Monoclonal antibody HER2, recombinant humanized (rhuMAbHER2)	CIEF	Hunt, G. et al., *J. Chromatogr. A*, 744, 295, 1996 Harris, R.J., *J. Chromatogr. A*, 705, 129, 1995
	CE-SDS	Hunt, G. et al., *J. Chromatogr. A*, 744, 295, 1996
Monoclonal antibody antihuman follicle stimulating hormone, mouse	CZE	Kelly, J.A. and Lee, C.S., *J. Chromatogr. A*, 790, 207, 1997
Monoclonal antibodies anti-HIV gp-41	CZE	Wenisch, E., et al., *J. Chromatogr.*, 516, 13, 1990
Monoclonal antibody, unspecified (mAb)	CZE	Compton, B.J. *J. Chromatogr.*, 559, 357, 1991
Monoclonal antibody, unspecified (mAb)	CIEF	Kubach, J. and Grimm, R., *J. Chromatogr. A*, 737, 281, 1996
Monoclonal antibody, unspecified (mAb)	CIEF	Schwer, C., *Electrophoresis*, 16, 2121, 1995
Monoclonal antibody, unspecified (mAb)	CE-SDS	Salas-Solano, O. et al., *Anal.Chem.*, 78, 6583, 2006

Continued

TABLE 22.3
(Continued)

Protein	Separation Mode	Reference
Ovalbumin (OVA)	CZE	Bullock, J.A. and Yuan, L.C., *J. Microcol. Sep.*, 3, 241, 1991
		Landers, J.P. et al., *Anal. Biochem.*, 205, 115, 1992
		Taverna, M. et al., *Electrophoresis*, 13, 359, 1992
		Bullock, J., *J. Chromatogr.*, 633, 235, 1993
		Kelly, J.F. et al., *Beckman discovery Series 2,* 1993
		Thibault, P. et al., *HPCE'94*, San Diego, CA 1994, Abstract P-323, p. 126
		Oda, R.P. et al, *J. Chromatogr. A*, 680, 85, 1994
		Legaz, M.E. and Pedrosa, M.M., *J. Chromatogr. A*, 719, 159, 1996
		Oda, R.P. and Landers, J.P., *Mol. Biotechnol.* 5, 165, 1996
		Chakel, J.A. et al., *J. Chromatogr. B*, 689, 215, 1997
		Chen, Y., *J. Chromatogr. A*, 768, 39, 1997
		Pinto, D.M. et al., *Anal. Chem.*, 69, 3015, 1997
		Che, F.Y., et al., *J. Chromatogr. A*, 849, 599, 1999
		Kubo K. and Hattori, A., *Electrophoresis*, 22, 3389, 2001
		Pacakova, V. et al., *Electrophoresis*, 22, 459, 2001
		Catai, J.R. et al., *J. Chromatogr. A*, 1083, 185, 2005
	ACE	Uegaki, K. et al., *Anal. Biochem.*, 309, 269, 2002
Ovalbumin (turkey)	CZE	Che, F.Y., et al., *J. Chromatogr. A*, 849, 599, 1999
Pepsin	CZE	Landers, J.P. et al, *Anal. Biochem.*, 205, 115, 1992
Placental alkaline phosphatase	CZE	Eriksson, H.J.C. et al., *J. Chromatogr. B*, 755, 311, 2001
Pollen allergens	CZE	Pacakova, V. et al., *Electrophoresis*, 22, 459, 2001
Prostate-specific antigen (PSA)	CZE	Donohue, M. J. et al., *Anal. Biochem.*, 339, 318, 2005
Proteinase A (*S. cerevisiae*)	CZE	Pedersen, J. and Biedermann, K., *Biotechnol. Appl. Biochem.* 18, 377, 1993
Ribonuclease B (RNase B)	MEKC	Rudd, P.M. et al., *Glycoconj. J.*, 9, 86, 1992
		James, D.C. et al., *Anal. Biochem.*, 222, 315, 1994
		Rudd, P.M. et al., *Biochemistry*, 33, 17, 1994
	CZE	Grossman, P.D. et al., *Anal. Chem.*, 61, 1186, 1989
		Kelly, J.F. et al., *J. Chromatogr. A*, 720, 409, 1996
		Yeung, B. et al., *Anal. Chem.*, 69, 2510, 1997
		Bateman, K.P. et al., *Meth. Mol. Biol.*, 213, 219, 2003
	ACE	Uegaki, K. et al., *Anal. Biochem.*, 309, 269, 2002
Serum proteins (albumin, globulins, IgG), human	CZE	Kim, J.W. et al., *Clin. Chem.*, 39, 689, 1993
Somatropin, recombinant bovine (rbSt)	CGE	Tsuji, K., *J. Chromatogr. A*, 652, 139, 1993
Superoxide dismutase, recombinant	CZE	Wenisch, E. et al., *J. Chromatogr.*, 516, 13, 1990
Tissue plasminogen activator, recombinant human (rhtPA)	CZE	Wu, S.-L. et al., *J. Chromatogr.*, 516, 115, 1990
		Yim, K., *J. Chromatogr.*, 559, 401, 1991
		Taverna, M. et al., *Electrophoresis*, 13, 359, 1992
		Thorne, J. M. et al., *J. Chromatogr. A*, 744, 155, 1996
	CIEF	Yim, K., *J. Chromatogr.*, 559, 401, 1991
		Taverna, M. et al., *Electrophoresis*, 13, 359, 1992
		Moorhouse, K.G. et al., *J. Chromatogr. A*, 717, 61, 1995
		Moorhouse, K.G. et al., *Electrophoresis*, 17, 423, 1996
		Chen, A.B. et al., *J. Chromatogr. A*, 744, 279, 1996

TABLE 22.3
(Continued)

Protein	Separation Mode	Reference
		Thorne, J.M. et al., *J. Chromatogr. A*, 744, 155, 1996
		Kubach, J. and Grimm, R., *J. Chromatogr. A*, 737, 281, 1996
	CGE-SDS	Thorne, J.M et al., *J. Chromatogr. A*, 744, 155, 1996
Transferrin, human (Tf)	CZE	Kilar, F. and Hjerten, S., *J. Chromatogr.*, 480, 351, 1989
		Bergmann, J. et al., *Pharmazie*, 51, 644, 1996
		Oda, R.P. and Landers, J.P., *Electrophoresis*, 17, 431, 1996
		Iourin, O. et al., *Glycoconj. J.*, 13, 1031, 1996
		Kubo, K., *J. Chromatogr. B*, 697, 217, 1997
		Prasad, R. et al., *Electrophoresis*, 18, 1814, 1997
		Oda, R.P. et al., *Electrophoresis*, 18, 1819, 1997
		Tagliaro, F. et al., *Electrophoresis*, 19, 3033, 1998
		Tagliaro, F. et al., *J. Cap. Electrophor Microchip Tech.*, 5, 137, 1999
		Beisler, A.T. et al., *Anal. Biochem.*, 285, 143, 2000
		Giordano, B.C. et al., *J. Chromatogr. B*, 742, 79, 2000
		Trout, A.L. et al., *Electrophoresis*, 21, 2376, 2000
		Crivellente, F. et al., *J. Chromatogr. B*, 739, 81, 2000
		Wuyts, B. et al., *Clin. Chem. Lab. Med.*, 39, 937, 2001
		Wuyts, B. et al., *Clin. Chem.*, 47, 247, 2001
		Lanz, C. et al., *J. Chromatogr. A*, 979, 43, 2002
		Legros, F.J. et al., *Clin. Chem.*, 48, 2177, 2002
		Ramdani, B. et al., *Clin. Chem.*, 49, 1854, 2003
		Sanz-Nebot, V. et al., *J. Chromatogr. B*, 798, 1, 2003
		Legros, F.J. et al., *Clin. Chem.*, 49, 440, 2003
		Wuyts, B. and Delanghe, J.R., *Clin. Chem. Lab. Med.*, 41, 739, 2003
		Lanz, C. et al., *J. Chromatogr. A*, 1013, 131, 2003
		Lanz, C. and Thormann, W., *Electrophoresis*, 24, 4272, 2003
		Ramdani, B. et al., *Clin. Chem.*, 49, 1854, 2003
		Fermo, I. et al., *Electrophoresis*, 25, 469, 2004
		Lanz, C. et al., *Electrophoresis*, 25, 2309, 2004
		Martello, S. et al., *Forensic Sci. Internat.*, 141, 153, 2004
		Carchon, H.A. et al., *Clin. Chem.*, 50, 101, 2004
		Bortolotti, F. et al., *Clin. Chem.*, 51, 2368, 2005
		Chang, W.W.P. et al., *Electrophoresis*, 26, 2179, 2005
		Joneli, J. et al., *J. Chromatogr. A*, 1130, 272, 2006
	CIEF	Kilar, F. and Hjerten, S., *Electrophoresis*, 10, 23, 1989
		Kilar, F. and Hjerten, S., *J. Chromatogr.*, 480, 351, 1989
		Molteni, S. and Thormann, W., *J. Chromatogr.*, 638, 187, 1993
Trypsin inhibitor, human urinary (UTI)	CZE	Che, F.-Y. et al., *Electrophoresis* 20, 2930, 1999
Tumor necrosis factor receptor fusion protein, recombinant human (rhTNFR:Fc)	CIEF	Jochheim, C. et al., *Chromatographia Suppl.*, 53, S59, 2001

they first require time-consuming and labor-intensive steps for obtaining adequate samples. Additional efforts along the lines of work described in this chapter for developing CE methods that do not require a prior purification step, or developing fast and simple purification methods will be required. In combination to more general purification methods, specific ones could be included to reach this goal. The use of affinity-based systems using lectins, antibodies, or aptamers may provide elegant solutions in this regard. The online combination of an affinity purification step with the CE separation has been demonstrated recently and appears promising even in cases where high throughput is achieved by using multicapillary systems, such as the one described for other analytes [224].

One of the major challenges to be faced in future years will be that of subsequent entry biologics (Canada)/follow-on biologics (USA)/biosimilar medicines (EU), and the ability to establish parameters that will allow the comparison of complex glycoproteins whether it is at the glycoform or isoform levels. Work in this area has been reported in recent years but it is anticipated that with the expiration of patents for several biopharmaceuticals currently on the market, many of them will be produced under a wide variety of manufacturing conditions. This will be even more challenging as it has been shown in this chapter that many parameters can affect the glycosylation profiles. Consequently, the refinement of CE methods will be required in order to improve peak resolution as well as to detect minor glycoforms.

The development of multidimensional, multimodal CE, that is, the online coupling of two or more CE separation modes has been reported several years ago. However, despite its potential for the analysis of complex mixtures such as glycoproteins, its application to practical problems has been limited and it is expected that more applications will be reported.

The field of microchip CE (MCE) has progressed tremendously over the past few years and its application to the analysis of glycoproteins has only been recently reported [225] and will undoubtedly continue to find applications in the coming years.

The majority of the work on CE of glycoforms carried out up to now is related to methods development and the successful demonstration of proof of concept for specific applications. This should lead as a logical next step to its application to a larger number of practical situations, especially in the clinical and diagnostic fields. This will require the methods to be applied to large, well-controlled populations to achieve well-defined conclusions about their validity. However, the application of CE techniques (including MCE) to the clinical field for specific applications such as the analysis of glycoforms as disease markers may be somewhat restrained (if not completely prevented) if more sensitive detection methods are not available. In this sense, newer detection schemes, perhaps involving the development of new fluorescent labeling agents for LIF detection of glycoproteins are required.

ACKNOWLEDGMENTS

Financial support from Spanish Ministry of Education and Science (Projects TIC2003-01906, HH04-33, CTQ2006-05214, DEP2006-56207-CO3-01), Fundacion Ramon Areces (Project Biosensors), Fundacion Domingo Martinez (project Microorganisms), and Comunidad de Madrid (Project S2006/GEN-0247) is acknowledged.

ABBREVIATIONS

ACE Affinity capillary electrophoresis
AGP Alpha-1-acid glycoprotein, orosomucoid
Ara Arabinose
Asn Asparagine
AAT Alpha-1-antitrypsin
ATIII Antithrombin III
BGE Background electrolyte

BRP	Biological Reference Preparation
BSF	Bovine serum fetuin
BTP	1,3-Bis[tris(hydroxymethyl)-methylamino] propane
CAPS	3-(Cyclohexylamino)-1-propanesulfonic acid
CBH-I	Cellobiohydrolase I
CDG	Congenital disorders of glycosylation
CDT	Carbohydrate-deficient transferrin
CE	Capillary electrophoresis
CE-MS	Online coupling of CE to mass spectrometry
CE-SDS	Capillary electrophoresis in sodium dodecylsulfate
CGMP	Caseinoglycomacropeptide
CHO	Chinese hamster ovary
CIEF	Capillary isoelectric focusing
CITP	Capillary isotacophoresis
Con A	Concanavalin A
CRP	C-reactive protein
CSF	Cerebrospinal fluid
CZE	Capillary zone electrophoresis
DAB	1,4-Diaminobutane, putrescine
DAP	1,3-Diaminopropane
DcBr	Decamethonium bromide
EACA	ε-Aminocaproic acid
EOF	Electroosmotic flow
EP	European Pharmacopoeia
EPO	Erythropoietin
ESI	Electrospray ionization
ESI-TOF MS	Electrospray ionization time of flight mass spectrometry
Fuc	Fucose
Gal	Galactose
GalNAc	*N*-Acetylgalactosamine
GlcNAc	*N*-Acetylglucosamine
Glu	Glucose
HC	Heavy chain
hCG	Human chorionic gonadotropin
HEPPSO	*N*-(2-hydroxyethyl)piperazin-*N′*-2-(hydroxypropane sulfonic acid)
HPLC	High performance liquid chromatography
HPMC	Hydroxypropylmethylcellulose
HRP	Horseradish peroxidase
HSA	Human serum albumin
hTNF	Human tumour necrosis factor
Hyl	Hydroxylysine
Hyp	Hydroxyproline
IEF	Isoelectric focusing
LC	Light chain
LCA	*Lens culinaris* agglutinin
LIF	Laser-induced fluorescence
mAb	Monoclonal antibody
MALDI	Matrix-assisted laser desorption ionization
Man	Mannose

MEKC	Micellar electrokinetic chromatography
MHEC	Methylhydroxyethylcellulose
MS	Mass spectrometry
NESP	Novel erythropoiesis-stimulating protein
NeuAc	*N*-Acetylneuraminic acid
OVA	Ovalbumin
PAA	Polyacrylamide
PAGE	Polyacrylamide gel electrophoresis
PB	Polybrene
pI	Isoelectric point
PSA	Prostate specific antigen
PTM	Postranslational modification
PVA	Polyvinyl alcohol
RA	Rheumatoid arthritis
rhATIII	Recombinant human anti-thrombin III
rhBMP-2	Recombinant human bone morphogenic protein 2
rhDNAse	Recombinant human deoxyribonuclease
rhEPO	Recombinant human erythropoietin
rhFSH	Recombinant human follicle-stimulating hormone
rhGCSF	Recombinant human granulocyte colony-stimulating factor
rhGMCSF	Recombinant human granulocyte macrophage colony-stimulating factor
rhIFN-γ	Recombinant human interferon-γ
rhtPA	Recombinant human tissue plasminogen activator
RNase	Ribonuclease
SDS	Sodium dodecylsulfate
SDS–PAGE	Sodium dodecylsulfate polyacrylamide gel electrophoresis
Ser	Serine
SIM	Selected ion monitoring
TEMED	*N*,*N*,*N′*,*N′*-Tetramethylene diamine
Tf	Transferrin
Thr	Threonine
TOF	Time-of-flight
uEPO	Human urinary EPO
UV	Ultraviolet
Xyl	Xylose

REFERENCES

1. Dwek, R.A., Glycobiology: Toward understanding the function of sugars, *Chem. Rev.*, 96, 683, 1996.
2. Taylor, M.E. and Drickamer, K., *Introduction to Glycobiolgy*, Oxford University Press, New York, 2004, chap. 12.
3. Brockhausen, I. and Kuhns, W., *Glycoproteins and Human Disease*, Springer-Verlag, Heidelberg, Germany, 1997, chap. 2.
4. Durand, G. and Seta, N., Protein glycosylation and diseases: Blood and urinary oligosaccharides as markers for diagnosis and therapeutic monitoring, *Clin. Chem.*, 46, 795, 2000.
5. Walsh, G., Pharmaceutical biotechnology products approved within the European Union, *Eur. J. Pharm. Biopharm.*, 55, 3, 2003.
6. Walsh, G., Second-generation biopharmaceuticals, *Eur. J. Pharm. Biopharm.*, 58, 185, 2004.
7. Jenkins, N. and Curling, M.A., Glycosylation of recombinant proteins: Problems and prospects, *Enzyme Microb. Technol.*, 16, 354, 1994.

8. Jung, E. and Williams, K.L., The production of recombinant glycoproteins with special reference to simple eukaryotes including *Dictyostelium discoideum*, *Biotechnol. Appl. Biochem.*, 25, 3, 1997.
9. Varki, A. et al., *Essentials of Glycobiology*, Cold Spring Harbor Laboratory Press, New York, 1999, chap. 2.
10. Taylor, M.E. and Drickamer, K., *Introduction to Glycobiolgy*, Oxford University Press, New York, 2004, chap. 2.
11. Strege, M.A. and Lagu, A.L., *Capillary Electrophoresis of Proteins and Peptides*, *Meth. Mol. Biol.*, Humana Press, Inc., NJ, 2004, Vol. 276.
12. Prichett, T. and Robey, F.A., Capillary electrophoresis of proteins, in *Handbook of Capillary Electrophoresis*, Landers, J.P., Ed., CRC Press, Boca Raton, FL, 1997, chap. 9.
13. Kakehi, K., Kinoshita, M., and Nakano, M., Analysis of glycoproteins and the oligosaccharides thereof by high-performance capillary electrophoresis—Significance in regulatory studies on biopharmaceutical products, *Biomed. Chromatogr.*, 16, 103, 2002.
14. Pantazaki, A., Taverna, M., and Vidal-Madjar, C., Recent advances in the capillary electrophoresis of recombinant glycoproteins, *Anal. Chim. Acta*, 383, 137, 1999.
15. Taverna, M. et al., Electrophoretic methods for process monitoring and the quality assessment of recombinant glycopropteins, *Electrophoresis*, 19, 2572, 1999.
16. Taverna, M., Tran, N.T., and Ferrier, D., Separation of protein glycoforms by capillary electrophoresis, *Meth. Mol. Biol.* 213, 163, 2003.
17. Kamoda, S. and Kakehi, K., Capillary electrophoresis for the analysis of glycoprotein pharmaceuticals, *Electrophoresis*, 27, 2495, 2006.
18. Watzig, H., Degenhardt, M., and Kunkel, A., Strategies for capillary electrophoresis: Method development and validation for pharmaceutical and biological applications, *Electrophoresis*, 19, 2695, 1998.
19. Chen, A.B. and Canova-Davis, E., Capillary electrophoresis in the development of recombinant protein biopharmaceuticals, *Chromatographia Suppl.*, 53, S7, 2001.
20. Patrick, J.S. and Lagu, A.L., Review applications of capillary electrophoresis to the analysis of biotechnology-derived therapeutic proteins, *Electrophoresis*, 22, 4179, 2001.
21. Mechref, Y. and Novotny, M.V., Structural investigations of glycoconjugates at high sensitivity, *Chem. Rev.*, 102, 321, 2002.
22. Little, M.J., Paquette, D.M., and Roos, P.K., Electrophoresis of pharmaceutical proteins: Status quo, *Electrophoresis*, 27, 2477, 2006.
23. Dolnik, V., Capillary electrophoresis of proteins 2003–2005, *Electrophoresis*, 27, 126, 2006.
24. Huck, C.W., Bakry, R., and Bonn, G.K., Progress in capillary electrophoresis of biomarkers and metabolites between 2002 and 2005, *Electrophoresis*, 27, 111, 2006.
25. Kishino, S. and Miyazaki, K., Separation methods for glycoprotein analysis and preparation, *J. Chromatogr. A*, 699, 371, 1997.
26. El Rassi, Z., Recent developments in capillary electrophoresis of carbohydrate species, *Electrophoresis*, 18, 2400, 1997.
27. Grossman, P.D., Free-solution capillary electrophoresis, in *Capillary Electrophoresis. Theory and Practice*, Grossman, P.D. and Colburn J.C., Eds., Academic Press, New York, 1992, chap. 2.
28. Kalman, F. et al., Charge and size effect in capillary zone electrophoresis of nuclease A and variants, *Electrophoresis*, 16, 595, 1995.
29. Gitlin, I., Carbeck, J.D., and Whitesides, G.M., Why are proteins charged? Network of charge-charge interaction in proteins measured by charge ladders and capillary electrophoresis, *Angew. Chem. Int. Ed.*, 45, 3022, 2006.
30. Balaguer, E. et al., Glycoform characterization of erythropoietin combining glycan and intact protein analysis by capillary electrophoresis-electrospray-time-of-flight mass spectrometry, *Electrophoresis*, 27, 2638, 2006.
31. Yim, K., Abrams, J., and Hsu, A., Capillary zone electrophoretic resolution of recombinant human bone morphogenic protein 2 glycoforms. An investigation into the separation mechanisms for an exquisite separation, *J. Chromatogr. A*, 716, 401, 1995.
32. Chen, F.-T.A., Rapid protein analysis by capillary electrophoresis, *J. Chromatogr.*, 559, 445, 1991.

33. van der Greef, J., Stroobant, P., and van der Heijden, R., The role of analytical sciences in medical systems biology, *Curr. Opin. Chem. Biol.*, 8, 559, 2004.
34. Bristow, A. and Charton, E., Assessment of the suitability of a capillary zone electrophoresis method for determining isoform distribution of erythropoietin *Pharmaeuropa*, 11, 290, 1999.
35. Erythropoietin concentrated solution 01/2005:1316, *European Pharmacopoeia* 5th edition, published by EDQM. June 2004.
36. Cifuentes, A. et al., Capillary isoelectric focusing of erythropoietin glycoforms and its comparison with flat-bed isoelectric focusing and capillary zone electrophoresis, *J. Chromatogr. A*, 830, 453, 1999.
37. San-Nebot, V. et al., Separation of recombinant human erythropoietin glycoforms by capillary electrophoresis using volatile electrolytes. Assessment of mass spectrometry for the characterization of erythropoietin glycoforms, *Anal. Chem.*, 75, 5220, 2003.
38. Lacunza, I. et al., Selection of migration parameters for a highly reliable assignment of bands of isoforms of erythropoietin separated by capillary electrophoresis, *Electrophoresis*, 25, 1569, 2004.
39. Lopez-Soto-Yarritu, P. et al., Improved capillary isoelectric focusing method for recombinant erythropoietin analysis, *J. Chromatogr. A*, 968, 221, 2002.
40. de Frutos, M., Cifuentes, A., and Diez-Masa, J.C., Differences in capillary electrophoresis profiles of urinary and recombinant erythropoietin, *Electrophoresis*, 24, 678, 2003.
41. Bietlot, H.P. and Girard, M., Analysis of recombinant human erythropoietin in drug formulations by high performance capillary electrophoresis, *J. Chromatogr. A*, 759, 177, 1997.
42. Neususs, C., Demelbauer, U., and Pelzing, M., Glycoform characterization of intact erythropoietin by capillary electrophoresis-electrospray-time of flight-mass spectrometry, *Electrophoresis*, 26, 1442, 2005.
43. Lara-Quintanar, P. et al., Immunochromatographic removal of albumin in erythropoietin biopharmaceutical formulations for its analysis by capillary electrophoresis, *J. Chromatogr. A*, 1153, 227, 2007.
44. Hjerten, S., Zone broadening in electrophoresis with special reference to the high-performance electrophoresis in capillaries: An interplay between theory and practice, *Electrophoresis*, 11, 665, 1990.
45. Lauer, H.H. and McManigill, D., Capillary zone electrophoresis of proteins in untreated fused silica tubing, *Anal. Chem.*, 58, 166, 1986.
46. McCormick, R.A., Capillary zone electrophoretic separation of peptides and proteins using low pH buffers in modified silica capillaries, *Anal. Chem.*, 6, 2322, 1986.
47. Corradini, D., Buffer additives other than the surfactant sodium dodecyl sulfate for protein separation by capillary electrophoresis, *J. Chromatogr. B*, 699, 21, 1997.
48. Righetti, P.G. et al., The state of the art of dynamic coatings, *Electrophoresis*, 22, 603, 2001.
49. Ahmadzadeh, H., Dovichi, N.J., and Krylov, S. Capillary coating for protein separation on Si—O and Si—C covalent bond formation for capillary electrophoresis with laser-induced fluorescence detection, in *Capillary Electrophoresis of Proteins and Peptides*, Strege, M.A. and Lagu, A.L., Eds., Humana Press, Totowa, NJ, 2004, chap. 2.
50. Landers, J.P. et al., High performance capillary electrophoresis of glycoproteins: The use of modifiers of electroosmotic flow for analysis of microheterogeneity, *Anal. Biochem.*, 205, 115, 1992.
51. Legaz, M.E. and Pedrosa, M.M., Effect of polyamines on the separation of ovalbumin glycoforms by capillary electrophoresis, *J. Chromatogr. A*, 719, 159, 1996.
52. Watson, E. and Yao, F., Capillary electrophoretic separation of human recombinant erythropoietin (r-HuEPO) glycoforms, *Anal. Biochem.*, 210, 389, 1993.
53. Klausen, N.K. and Kornfelt, T., Analysis of the glycoforms of human recombinant factor VIIa by capillary electrophoresis and high-performance liquid chromatography, *J. Chromatogr. A*, 718, 195, 1995.
54. Watson, E. and Yao, F., Capillary electrophoretic separation of recombinant granulocyte-colony-stimulating factor glycoforms, *J. Chromatogr.*, 630, 442, 1993.
55. Oda, R.P. and Landers, J.P., Effect of cationic buffer additives on the capillary electrophoretic separation of serum transferrin from different species, *Electrophoresis*, 17, 431, 1996.
56. Morbeck, D.E., Madden, B.J., and McCormick, D.J., Analysis of the microheterogeneity of the glycoprotein chorionic gonadotropin with high-performance capillary electrophoresis, *J. Chromatogr. A*, 680, 217, 1994.
57. Oda, R.P. et al., α, ω-Bis quaternary ammonium alkanes as effective buffer additives for enhanced capillary electrophoretic separation of glycoproteins, *J. Chromatogr. A*, 680, 85, 1994.

58. Verzola, B., Gelfi, C., and Righetti, P.G., Protein adsorption to the bare silica wall in capillary electrophoresis. Quantitative study on the chemical composition of the background electrolyte for minimizing the phenomenon, *J. Chromatogr. A*, 868, 85, 2000.
59. Wicktorowicz, J.E. and Colburn, J.E., Separation of cationic proteins via charge reversal in capillary electrophoresis, *Electrophoresis*, 11, 769, 1990.
60. Cordova, E., Gao, J., and Whitesides G.M., Noncovalent polycationic coating for capillaries in capillary electrophoresis of proteins, *Anal. Chem.*, 69, 1370, 1997.
61. Kelly, J.F. et al., Development of electrophoretic conditions for the characterization of protein glycoforms by capillary electrophoresis-electrospray mass spectrometry, *J. Chromatogr. A*, 720, 409, 1996.
62. Yu, B. et al., Ionene-dynamically coated capillary for analysis of urinary and recombinant human erythropoietin by capillary electrophoresis and online electrospray ionization mass spectrometry, *J. Sep. Sci.*, 28, 2390, 2005.
63. Katayama, H., Ishihama, Y., and Asakawa, N., Stable capillary coating with successive multiple ionic polymer layers, *Anal. Chem.*, 70, 2254, 1998.
64. Bendahl, L., Hansen, S.H., and Gammelgaard, B., Capillaries modified by noncovalently anionic polymer adsorption for capillary zone electrophoresis, micellar electrokinetic capillary chromatography and capillary electrophoresis, *Electrophoresis*, 22, 2565, 2001.
65. Graul, T.W. and Schlenoff, J.B., Capillaries modified by polyelectrolyte multilayers for electrophoretic separations, *Anal. Chem.*, 71, 4007, 1999.
66. Lanz, C., Marti, U., and Thormann, W., Capillary zone electrophoresis with a dynamic double coating for analysis of carbohydrate-deficient transferrin in human serum. Precision performance and pattern recognition, *J. Chromatogr. A*, 1013, 131, 2003.
67. Joneli, J., Lanz, C., and Thormann, W., Capillary zone electrophoresis determination of carbohydrate-deficient transferrin using the new Ceofix reagents under high-resolution conditions, *J. Chromatogr. A*, 1130, 272, 2006.
68. Legros, F.J. et al., Carbohydrate-deficient transferrin isoforms measured by capillary zone electrophoresis for detection of alcohol abuse, *Clin. Chem.*, 48, 2177, 2002.
69. Legros, F.J. et al., Use of capillary zone electrophoresis for differentiating excessive from moderate alcohol consumption, *Clin. Chem.*, 49, 440, 2003.
70. Chang, W.W.P. et al., Rapid separation of protein isoforms by capillary zone electrophoresis with new dynamic coating, *Electrophoresis*, 26, 2179, 2005.
71. Balaguer, E. and Neususs, C., Glycoprotein characterization combining intact protein and glycan analysis by capillary electrophoresis-electrospray ionization-mass spectrometry, *Anal. Chem.*, 78, 5384, 2006.
72. Hjerten, S., High-performance electrophoresis. Elimination of the electroendosmosis and solute adsorption, *J. Chromatogr.*, 347, 191, 1985.
73. Schmalzing, D. et al., Characterization and performance of a neutral hydrophilic coating for the capillary electrophoretic separation of biopolymers, *J. Chromatogr. A.*, 652, 149, 1993.
74. Hjerten, S. and Kubo, K., A new type of pH- and detergent stable coating for elimination of electroendosmosis and adsorption in (capillary) electrophoresis, *Electrophoresis*, 14, 390, 1993.
75. Gilges, M., Kleemiss, M.H., and Schomburg, G., Capillary zone electrophoresis separation of basic and acidic proteins using poly(vinyl alcohol) coatings in fused silica capillaries, *Anal. Chem.*, 66, 2038, 1994.
76. Yim, K., Fractionation of the human recombinant tissue plasminogen activator (rtPA) glycoforms by high-performance capillary zone electrophoresis and capillary isoelectric focusing, *J. Chromatogr.*, 559, 401, 1991.
77. Thorne, J.M. et al., Examination of capillary zone electrophoresis, capillary isoelectric focusing and sodium dodecyl sulphate capillary electrophoresis for the analysis of recombinant tissue plasminogen activator, *J. Chromatogr. A*, 744, 155, 1996.
78. Kinoshita, M. et al., Comparative studies on the analysis of glycosylation heterogeneity of sialic acid-containing glycoproteins using capillary electrophoresis, *J. Chromatogr. A*, 866, 261, 2000.
79. Kilar, F. and Hjerten, S., Separation of the human transferrin isoforms by carrier-free high-performance zone electrophoresis and isoelectric focusing, *J. Chromatogr. A*, 480, 351, 1989.

80. Hunt, G., Moorhouse, K.G., and Chen, A.B., Capillary isoelectric focusing and sodium dodecyl sulfate-capillary gel electrophoresis of recombinant humanized monoclonal antibody HER2, *J. Chromatogr. A*, 744, 295, 1996.
81. Grossman, P.D. et al., Application of free-solution capillary electrophoresis to the analytical scale separation of proteins and peptides, *Anal. Chem.*, 61, 1186, 1989.
82. Taverna, M. et al., Analysis of carbohydrate-mediated heterogeneity and characterization of N-linked oligosaccharies of glycoproteins by high performance capillary electrophoresis, *Electrophoresis*, 13, 359, 1992.
83. Sanz-Nebot, V. et al., Capillary electrophoresis and matrix-assisted laser desorption/ionization-time of flight-mass spectrometry for analysis of the novel erythropoiesis-stimulating protein (NESP), *Electrophoresis*, 26, 1451, 2005.
84. Lacunza, I. et al., CZE of human alpha-1-acid glycoprotein for qualitative and quantitative comparison of samples from different pathological conditions, *Electrophoresis*, 27, 4205, 2006.
85. Honda, S. et al., Simultaneous determination of reducing monosaccharides by capillary zone electrophoresis as the borate complexes of N-2 pyridylglycamines, *Anal. Biochem.*, 176, 72, 1989.
86. Hoffsteter-Kuhn, S. et al., Influence of borate complexation on the electrophoretic behavior of carbohydrates in capillary electrophoresis, *Anal. Chem.*, 63, 1541, 1991.
87. Pacakova, V. et al., Effects of electrolyte modification and capillary coating on separation of glycoproteins isoforms by capillary electrophoresis, *Electrophoresis*, 22, 459, 2001.
88. Chen, Y., Critical conditions for separating the microheterogeneous components of glycoproteins by capillary electrophoresis, *J. Chromatogr. A*, 768, 39, 1997.
89. Oda, R.P. et al., Capillary electrophoresis-based separation of transferrin sialoforms in patients with carbohydrate-deficient glycoprotein syndrome, *Electrophoresis*, 18, 1819, 1997.
90. Lopez-Soto-Yarritu, P. et al., Comparison of different capillary electrophoresis methods for analysis of recombinant erythropoietin glycoforms, *J. Sep. Sci.*, 25, 1112, 2002.
91. Tran, A.D. et al., Separation of carbohydrate-mediated microheterogeneity of recombinant human erythropoietin by free solution capillary electrophoresis, *J. Chromatogr.*, 542, 459, 1991.
92. Zhou, G.-H. et al., Application of capillary electrophoresis, liquid chromatography, electrospray-mass spectrometry and matrix-assisted laser desorption/ionization-time of flight-mass spectrometry to the characterization of recombinant human erythropoietin, *Electrophoresis*, 19, 2348, 1998.
93. Nieto, O., Hernandez, P., and Hernandez, L., Capillary zone electrophoresis of human recombinant erythropoietin using C_8 coated columns without additives in the running buffer, *Anal. Commun.*, 33, 425, 1996.
94. Kaniansky, D. et al., Capillary zone electrophoresis in a hydrodynamically closed separation system with enhanced sample loadability, *J. Chromatogr. A*, 772, 103, 1997.
95. Madajova, V. et al., Fractionation of glycoforms of recombinant human erythropoietin by preparative capillary isotachophoresis, *Electrophoresis*, 26, 2664, 2005.
96. Balaguer, E. and Neususs, C., Intact glycoform characterization of erythropoietin-α and erythropoietin-β by CZE-ESI-TOF-MS, *Chromatographia*, 64, 351, 2006.
97. Kilar, F. and Hjerten, S., Fast and high resolution analysis of human serum transferrin by high performance isoelectric focusing in capillaries, *Electrophoresis*, 10, 23, 1989.
98. Righetti, P.G., Determination of the isoelectric point of proteins by capillary isoelectric focusing, *J. Chromatogr. A*, 1037, 491, 2004.
99. Lacunza, I. and de Frutos, M., Different capillary isoelectric focusing approaches for analysis of α-1-acid glycoprotein, *PACE Setter*, 10, 5, 2006.
100. Wu, J., Li, S.-C., and Watson, A., Optimizing separation conditions for proteins and peptides using imaged capillary isoelectric focusing, *J. Chromatogr. A*, 817, 163, 1998.
101. Moorhouse, K.G. et al., Rapid one-step capillary isoelectric focusing method to monitor charged glycoforms of recombinant human tissue-type plasminogen activator, *J. Chromatogr. A*, 717, 61, 1995.
102. Reif, O.-W. and Freitag, R., Control of the cultivation process of antithrombin III and its characterization by capillary electrophoresis, *J. Chromatogr. A*, 680, 383, 1994.
103. Tran, N.T. et al., One-step capillary isoelectric focusing for the separation of the recombinant human immunodeficiency virus envelope glycoprotein glycoforms, *J. Chromatogr. A*, 866, 121, 2000.
104. Kubach, J. and Grimm, R., Non-native capillary isoelectric focusing for the analysis of the microheterogeneity of glycoproteins, *J. Chromatogr. A*, 737, 281, 1996.

105. Moorhouse, K.G., Rickel, C.A., and Chen, A.B., Electrophoretic separation of recombinant tissue-type plasminogen activator glycoforms: Validation issues for capillary isoelectric focusing methods, *Electrophoresis*, 17, 423, 1996.
106. Chen, A.B. et al., Comparison of ampholytes used for slab gel and capillary isoelectric focusing of recombinant tissue-type plasminogen activator glycoforms, *J. Chromatogr. A*, 744, 279, 1996.
107. Sebastiano, R. et al., Mass distribution and focusing properties of carrier ampholytes for isoelectric focusing: I. Novel and unexpected results, *Electrophoresis*, 27, 3919, 2006.
108. Simo, C. et al., Mass distribution, polydispersity and focusing properties of carrier ampholytes for IEF II: pH 4-6 intervals, *Electrophoresis*, 27, 4849, 2006.
109. Lacunza, I. et al., CIEF with hydrodynamic and chemical mobilization for the separation of forms of alpha-1-acid glycoprotein, *Electrophoresis*, 28, 1204, 2007.
110. Wehr, T., Rodriguez-Diaz, R., and Zhu, M., Recent advances in capillary isoelectric focusing, *Chromatographia Suppl.*, 53, S45, 2001.
111. Righetti, P.G., *Isoelectric Focusing: Theory, Methodology, and Applications*, Elsevier Biomedical, Amsterdam, 1989, 1.
112. Schwer, C., Capillary isoelectric focusing: A routine method for protein analysis?, *Electrophoresis*, 16, 2121, 1995.
113. Rudd, P.M. et al., Separation and analysis of the glycoform populations of ribonuclease B using capillary electrophoresis, *Glycoconj. J.*, 9, 86, 1992.
114. James, D.C. et al., High-resolution separation of recombinant human interferon-γ glycoforms by micellar electrokinetic capillary chromatography, *Anal. Biochem.*, 222, 315, 1994.
115. Werner, W.E. et al., Size-dependent separation of proteins denatured in SDS by capillary electrophoresis using a replaceable sieving matrix, *Anal. Biochem.*, 212, 253, 1993.
116. Buchacher, A. et al., High-performance capillary electrophoresis for in-process control in the production of antithrombin III and human clotting factor IX, *J. Chromatogr. A*, 802, 355, 1998.
117. Bergstrom, M. et al., Lectin affinity capillary electrophoresis in glycoform analysis applying the partial filling technique, *J. Chromatogr. B*, 809, 323, 2004.
118. Uegaki, K. et al., Simultaneous estimation of the association constants of glycoprotein glycoforms to a common protein by capillary electrophoresis, *Anal. Biochem.*, 309, 269, 2002.
119. Thormann, W. and Mosher, R.A., High-resolution computer simulation of the dynamics of isoelectric focusing using carrier ampholytes: Focusing with concurrent electrophoretic mobilation is an isotachophoretic process, *Electrophoresis*, 27, 968, 2006.
120. Iourin, O. et al., The identification of abnormal glycoforms of serum transferrin in carbohydrate deficient glycoprotein syndrome type I by capillary zone electrophoresis, *Glycoconj. J.*, 13, 1031, 1996.
121. Thibault, P. and Dovichi, N., General instrumentation and detection systems including mass spectrometry, in *Capillary Electrophoresis. Theory and Practice*, Camilleri, P., Ed., CRC Press, Boca Raton, FL, 1998, chap. 2.
122. http://probes.invitrogen.com/handbook/sections/0900.html Accessed January 2007.
123. Craig, D.B. and Dovichi, N.J., Multiple labeling of proteins, *Anal. Chem.*, 70, 2493, 1998.
124. Tseng, W.-L., Lin, Y.-W., and Chang, H.-T., Improved separation of microheterogeinities and isoforms of proteins by capillary electrophoresis using segmental filling with SDS and PEO in the background electrolyte, *Anal. Chem.*, 74, 4828, 2002.
125. Salas-Solano, O. et al., Optimization and validation of a quantitative capillary electrophoresis sodium dodecyl sulfate method for quality control and stability monitoring of monoclonal antibodies, *Anal.Chem.*,78, 6583, 2006.
126. Pinto, D.M. et al., Picomolar assay of native proteins by capillary electrophoresis precolumn. Submicellar separation, and laser-induced fluorescence detection, *Anal. Chem.*, 69, 3015, 1997.
127. Lee, I.H. et al., Picomolar analysis of proteins using electrophoretically mediated microanalysis and capillary electrophoresis with laser-induced fluorescence detection, *Anal. Chem.*, 70, 4546, 1998.
128. Wehr, T., Rodriguez-Diaz, R., and Zhu, M., *Capillary Electrophoresis of Proteins*, Marcell Dekker, New York, 1999, chap. 3.

129. Bornemann, C. et al., Fluorescence-labelled antigen-binding fragments (Fab) from monoclonal antibody 5F12 detect human erythropoietin in immunoaffinity capillary electrophoresis, *Anal. Bioanal. Chem.*, 376, 1074, 2003.
130. Kelly, J.A. and Lee, C.S., Online postcapillary affinity detection of immunoglobulin G subclass and monoclonal antibodies variants for capillary electrophoresis, *J. Chromatogr. A*, 790, 207, 1997.
131. Yeung, B., Porter, T.J., and Vath, J.E., Direct isoform analysis of high-mannose-containing glycoproteins by online capillary electrophoresis electrospray mass spectrometry, *Anal. Chem.*, 69, 2510, 1997.
132. Demelbauer, U.M. et al., Characterization of glyco isoforms in plasma-derived human antithrombin by online capillary zone electrophoresis-electrospray ionization-quadrupole ion trap-mass spectrometry of the intact glycoproteins, *Electrophoresis*, 25, 2026, 2004.
133. Arizaga-Rodriguez, S. et al., Detection of transferrin isoforms in human serum: Comparison of UV and ICP-MS detection after CZE and HPLC separations, *Anal. Bioanal. Chem.*, 383, 390, 2005.
134. Hooker, A.D. and James, D.C., Analysis of glycoprotein heterogeneity by capillary electrophoresis and mass spectrometry, *Mol. Biotech.*, 14, 241, 2000.
135. Morelle, W. and Michalski, J.C., The mass spectrometric analysis of glycoproteins and their glycan structures, *Curr. Anal. Chem.*, 1, 29, 2005.
136. Bateman, K.P. et al., Glycoprotein analysis by capillary zone electrophoresis–electrospray mass spectrometry, *Meth. Mol. Biol.*, 213, 219, 2003.
137. Monton, M.R.N. and Terabe, S., Recent developments in capillary electrophoresis—Mass spectrometry of proteins and peptides, *Anal. Sci.*, 21, 5, 2005.
138. Servais, A.-C., Crommen, J., and Fillet, M., Capillary electrophoresis-mass spectrometry, an attractive tool for drug bioanalysis and biomarker discovery, *Electrophoresis*, 27, 2616, 2006.
139. ICH Guidelines, http://www.ich.org/cache/compo/276-254-1.html, accessed November 2006.
140. Somatropin, 07/2007:0951, European Pharmacopoeia 5th edition, published by EDQM, April 2007.
141. Somerville, L.E., Douglas, A.J., and Irvine, A.E., Discrimination of granulocyte colony-stimulating factor isoforms by high-performance capillary electrophoresis, *J. Chromatogr. B*, 732, 81, 1999.
142. Berkowitz, S.A. et al., Rapid quantitative capillary zone electrophoresis method for monitoring the micro-heterogeneity of an intact recombinant glycoprotein, *J. Chromatogr. A*, 1079, 254, 2005.
143. Felten, C. et al., Use of acidic and basic pH and calcium ion addition in the capillary zone electrophoretic characterization of recombinant human deoxyribonuclease, a complex phosphoglycoprotein, *J. Chromatogr. A*, 853, 295, 1999.
144. Quan, C.P., Canova-Davis, E., and Chen, A.B., Effects of the solution environment on the resolution of recombinant human deoxyribonuclease variants in capillary zone electrophoresis, *Chromatographia Suppl.*, 53, S39, 2001.
145. Kremser, L. et al., Characterization of antithrombin III from human plasma by two-dimensional gel electrophoresis and capillary electrophoretic methods. *Electrophoresis*, 24, 4282, 2003.
146. Che, F.Y. et al., Comparative study on the distribution of ovalbumin glycoforms by capillary electrophoresis, *J. Chromatogr. A*, 849, 599, 1999.
147. Kubo, K. and Hattori, A., Assessment of the capillary zone electrophoretic behaviour of proteins in the presence of electroosmotic modifiers: Protein-polyamine interaction studied using a polyacrylamide-coated capillary, *Electrophoresis*, 22, 3389, 2001.
148. Catai, J.R. et al., Noncovalently bilayer-coated capillaries for efficient and reproducible analysis of proteins by capillary electrophoresis, *J. Chromatogr. A*, 1083, 185, 2005.
149. Sandra, K. et al., Combining gel and capillary electrophoresis, nano-LC and mass spectrometry for the elucidation of posttranslational modifications of *Trichoderma reesei* cellobiohydrolase I., *J. Chromatogr. A*, 1058, 263, 2004.
150. Krull, I.S. et al., HPCE methods for the identification and quantitation of antibodies, their conjugates and complexes, *J. Pharm. Biomed. Anal.*, 16, 377, 1997.
151. Santora, L.C., Krull, I.S., and Grant, K., Characterization of recombinant human monoclonal tissue necrosis factor-α antibody using cation-exchange HPLC and capillary isoelectric focusing, *Anal. Biochem.*, 275, 98, 1999.
152. Harmon, B.J., Imaged capillary isoelectric focusing for the characterization and monitoring of charge heterogeneity of recombinant monoclonal antibodies, in *Proceedings of the 5th Symposium on CE in the Biotechnology and Pharmaceutical Industries*, San Francisco, CA, 2003.

153. Hoffstetter-Kuhn, S., Alt, G., and Kuhn, R. Profiling of oligosaccharide-mediated microheterogeneity of a monoclonal antibody by capillary electrophoresis, *Electrophoresis*, 17, 418, 1996.
154. Ma, S. in *State of the Art Analytical Methods for the Characterization of Biological Products and Assessment of Comparability*, Mire-Sluis, A.R., Ed., Karger, Basel, 2005, pp. 49–68.
155. Nunnally, B. et al., A series of collaborations between various pharmaceutical companies and regulatory authorities concerning the analysis of biomolecules using capillary electrophoresis, *Chromatographia*, 64, 359, 2006.
156. Jenkins, N., Parekh, R.B., and James, D.C., Getting the glycosylation right—Implications for the biotechnology industry, *Nat. Biotechnol.*, 14, 975, 1996.
157. Chirino, A.J. and Mire-Sluis, A., Characterizing biological products and assessing comparability following manufacturing changes, *Nat. Biotechnol.*, 22, 1383, 2004.
158. James, D.C. et al., Posttranslational processing of recombinant human interferon-γ in animal expression systems, *Prot. Sci.*, 5, 331, 1996.
159. Goldman, M.H. et al., Monitoring recombinant human interferon-gamma *N*-glycosylation during perfused fluidized-bed and stirred-tank batch culture of CHO cells, *Biotech. Bioeng.*, 60, 596, 1998.
160. Wang, M.D. et al., Erythropoietin production from CHO cells grown by continuous culture in a fluidized-bed bioreactor, *Biotech. Bioeng.*, 77, 194, 2002.
161. Freitag, R., et al., Production of recombinant h-AT III with mammalian cell cultures using capillary electrophoresis for product monitoring, *Cytotechnology*, 21, 205, 1996.
162. Klyushnichenko, V., Capillary electrophoresis in the analysis and monitoring of biotechnological processes, *Meth. Mol. Biol.*, 276, 77, 2004.
163. Schmelzer, A.E. and Miller, W.M., Hyperosmotic stress and elevated pCO_2 alter monoclonal antibody charge distribution and monosaccharide content, *Biotechnol. Prog.*, 18, 346, 2002.
164. Cherkaoui, S. et al., Development of a capillary zone electrophoresis method for caseinoglycomacropeptide determination, *J. Chromatogr. A*, 790, 195, 1997.
165. Cherkaoui, S. et al., Determination of the stability of caseinoglycomacropeptide in a cosmetic lotion by use of capillary zone electrophoresis with a coated capillary, *Chromatographia*, 50, 311, 1999.
166. Daali, Y., Cherkaoui, S., and Veuthey, J.L., Capillary electrophoresis and high-performance anion exchange chromatography for monitoring caseinoglycomacropeptide sialylation, *J. Pharm. Biomed. Anal.*, 24, 849, 2001.
167. Donohue, M.J. et al., Capillary electrophoresis for the investigation of prostate-specific antigen heterogeneity, *Anal. Biochem.*, 339, 318, 2005.
168. Girard, M., Boucher, S., and Kane, A., Analysis of erythropoietin preparations obtained from various sources worldwide by capillary zone electrophoresis, in *Proceedings of the 7th International symposium on capillary electrophoresis in the biotechnology and pharmaceutical industries*, Montreal, 2005.
169. Deechongkit, S. et al., Biophysical comparability of the same protein from different manufacturers: a case study using epoetin alfa from Epogen® and Eprex®, *J. Pharm. Sci.*, 95, 1931, 2006.
170. Zhou, G.-H. et al., Characterization of recombinant human granulocyte colony stimulating factor (rHuG-CSF) by capillary zone electrophoresis, capillary isoelectric focusing electrophoresis and electrospray ionization mass spectrometry, *J. Pharm. Biomed. Anal.*, 35, 425, 2004.
171. Storring, P.L. et al., Physicochemical Methods for predicting the biological potency of recombinant follicle stimulating hormone: An international collaborative study of isoelectric focusing and capillary zone electrophoresis, *Biologicals*, 30, 217, 2002.
172. Mulders, J.W.M., Wijn, H., and Theunissen, F. Development and validation of a capillary zone electrophoresis method which can be used to predict the *in-vivo* bioactivity of recombinant follicle stimulating hormone, in *Proceedings of the 3rd Symposium on the Analysis of Well Characterized Biotechnology Pharmaceuticals (WCBP)*, Washington, DC, 1999.
173. Kim, Y.J. and Varki, A., Perspectives on the significance of altered glycosylation of glycoproteins in cancer, *Glycoconj. J.*, 14, 569, 1997.
174. de Jong, G. and van Eijk, H.G., Microheterogeneity of human serum transferrin: A biological phenomenon studied by isoelectric focusing in immobilized pH gradients, *Electrophoresis*, 9, 589, 1998.
175. de Jong, G., van Dijk, J.P., and van Eijk, H.G., The biology of transferrin, *Clin. Chim. Acta*, 190, 1, 1990.

176. Arndt, T., Carbohydrate-deficient transferrin as a marker of chronic alcohol abuse: A critical review of preanalysis, analysis, and interpretation, *Clin. Chem.*, 47, 13, 2001.
177. Wuyts, B. et al., Carbohydrate-deficient transferrin and chronic alcohol ingestion in subjects with transferrin CD-variants, *Clin. Chem. Lab. Med.*, 39, 937, 2001.
178. Stibler, H., Carbohydrate-deficient transferrin in serum: A new marker of potentially harmful alcohol consumption reviewed, *Clin. Chem.*, 37, 2029, 1991.
179. Keir, G., Winchester, B.G., and Clayton, P., Carbohydrate-deficient glycoprotein syndromes: inborn errors of protein glycosylation, *Ann. Clin. Biochem.*, 36, 20, 1999.
180. Whitfield, J.B., Transferrin isoform analysis for the diagnosis and management of hazardous or dependent drinking, *Clin. Chem.*, 48, 2095, 2002.
181. Stibler, H. and Kjellin, KG., Isoelectric-focusing and electrophoresis of CSF proteins in tremor of different origins—Study of 38 cases with cerebellar syndrome of chronic-alcoholism, Parkinson's-disease and essential tremor, *J. Neurol. Sci.*, 30, 269, 1976.
182. Stibler, H. et al., Abnormal microheterogeneity of transferrin in serum and cerebrospinal-fluid in alcoholism, *Acta Med. Scand.*, 204, 49, 1978.
183. Stibler, H., Borg, S., and Allgulander, C., Clinical significance of abnormal heterogeneity of transferrin in relation to alcohol-consumption, *Acta Med. Scand.*, 206, 275, 1979.
184. Wuyts, B. and Delanghe, J.R., The analysis of carbohydrate-deficient transferrin, marker of chronic alcoholism, using capillary electrophoresis, *Clin. Chem. Lab. Med.*, 41, 739, 2003.
185. Lanz, C. et al., Evaluation and optimization of capillary zone electrophoresis with different dynamic capillary coatings for the determination of carbohydrate-deficient transferrin in human serum, *J. Chromatogr. A*, 979, 43, 2002.
186. Helander, A. et al., Interference of transferrin isoform types with carbohydrate-deficient transferrin quantification in the identification of alcohol abuse, *Clin. Chem.*, 47, 1225, 2001.
187. Bortolotti, F. et al., Analysis of carbohydrate-deficient transferrin: Comparative evaluation of turbidimetric immunoassay, capillary zone electrophoresis, and HPLC, *Clin. Chem.*, 51, 2368, 2005.
188. Jeppsson, J.-O., Kristensson, H., and Fimiani, C., Carbohydrate-deficient transferrin quantified by HPLC to determine heavy consumption of alcohol, *Clin. Chem.*, 39, 2115, 1993.
189. Turpeinen, U. et al., Comparison of HPLC and small column (CDTect) methods for disialotransferrin, *Clin. Chem.*, 47, 1782, 2001.
190. Helander, A., Husa, A., and Jeppsson, J.-O., Improved HPLC method for carbohydrate-deficient transferrin in serum, *Clin. Chem.*, 49, 1881, 2003.
191. Prasad, R. et al., Analysis of carbohydrate deficient transferrin by capillary zone electrophoresis, *Electrophoresis*, 18, 1814, 1997.
192. Tagliaro, F. et al., Optimized determination of carbohydrate-deficient transferrin isoforms in serum by capillary zone electrophoresis, *Electrophoresis*, 19, 3033, 1998.
193. Crivellente, F. et al., Improved method for carbohydrate-deficient transferrin determination in human serum by capillary zone electrophoresis, *J. Chromatogr. B*, 739, 81, 2000.
194. Trout, A.L. et al., Direct capillary electrophoretic detection of carbohydrate-deficient transferrin in neat serum, *Electrophoresis*, 21, 2376, 2000.
195. Ramdani, B. et al., Analyte comigrating with trisialotransferrin during capillary zone electrophoresis of sera from patients with cancer, *Clin. Chem.*, 49, 1854, 2003.
196. Sanz-Nebot, V. et al., Characterization of human transferrin glycoforms by capillary electrophoresis and electrospray ionization mass spectrometry, *J. Chromatogr. B*, 798, 1, 2003.
197. Lanz, C. et al., Improved capillary electrophoresis method for the determination of carbohydrate-deficient transferrin in patient sera, *Electrophoresis*, 25, 2309, 2004.
198. Martello, S. et al., Determination of carbohydrate deficient transferrin (CDT) with capillary electrophoresis: an inter laboratory comparison, *Forensic Sci. Internat.*, 141, 153, 2004.
199. Carchon, H.A. et al., Diagnosis of congenital disorders of glycosylation by capillary zone electrophoresis of serum transferrin, *Clin. Chem.*, 50, 101, 2004.
200. Tagliaro, F. et al., Carbohydrate-deficient transferrin determination revisited with capillary electrophoresis: A new biochemical marker of chronic alcohol abuse, *J. Cap. Electrophor. Microchip Tech.*, 5, 137, 1999.
201. Fermo, I. et al., Capillary zone electrophoresis for determination of carbohydrate-deficient transferrin in huan serum, *Electrophoresis*, 25, 469, 2004.

202. Giordano, B.C. et al., Dynamically-coated capillaries allow for capillary electrophoretic resolution of transferrin sialoforms via direct analysis of human serum, *J. Chromatogr. B*, 742, 79, 2000.
203. Wuyts, B. et al., Determination of carbohydrate-deficient transferrin using capillary zone electrophoresis, *Clin. Chem.*, 47, 247, 2001.
204. Tagliaro, F. et al., Caveats in carbohydrate-deficient transferrin determination, *Clin. Chem.*, 48, 208, 2001.
205. Lanz, C. and Thormann, W., Capillary zone electrophoresis with a dynamic double coating for analysis of carbohydrate-deficient transferrin in human serum: Impact of resolution between disialo- and trisialotransferrin on reference limits, *Electrophoresis*, 24, 4272, 2003.
206. Vitala, K., Lähdesmäki, K., and Niemela, O., Comparison of the Axis % CDT TIA and the CDTect method as laboratory tests of alcohol abuse, *Clin. Chem.*, 44, 1209, 1998.
207. Delanghe, J.R., Wuyts, B., and De Buyzere, M.L., Caveats in carbohydrate-deficient transferring determination–Response, *Clin. Chem.*, 48, 209, 2002.
208. Marquardt, T. and Dencke, J., Congenital disorders of glycosylation: Review of their molecular bases, clinical presentations and specific therapies, *Eur. J. Pediatr.*, 162, 359, 2003.
209. Fournier, T., Medjoubi-N, N., and Porquet, D., Alpha-1-acid glycoprotein, *Biochim. Biophys. Acta*, 1482, 157, 2000.
210. Duche, J.-C. et al., Expression of the genetic variants of human alpha-1-acid glycoprotein in cancer, *Clin. Chem.*, 33, 197, 2000.
211. Mackiewicz, A. and Mackiewicz, K., Glycoforms of serum α1-acid glycoprotein as markers of inflammation and cancer, *Glycoconj. J.*, 12, 241, 1995.
212. Hashimoto, S. et al., α_1-Glycoprotein fucosylation as a marker of carcinoma progression and prognosis, *Cancer*, 101, 2825, 2004.
213. Kremmer, T. et al., Liquid chromatographic and mass spectrometric analysis of human serum acid alpha-1-glycoprotein, *Biomed. Chromatogr.*, 18, 323, 2004.
214. Hrycaj, P. et al., Micoheterogeneity of alpha 1 acid glycoprotein in rheumatoid arthritis: Dependent on disease duration?, *Ann. Rheum. Dis.*, 52, 138, 1993.
215. Elliott, M.A. et al., Investigation into the Concavalin A reactivity, fucosylation and oligosaccharide microheterogeneity of α_1-acid glycoprotein expressed in the sera of patients with rheumatoid arthritis, *J. Chromatogr. B*, 688, 229, 1997.
216. Iijima, S. et al., Changes of α_1-glycoprotein microheterogeneity in acute inflammation stages analyzed by isoelectric focusing using serum obtained postoperatively, *Electrophoresis*, 21, 753, 2000.
217. Higai, K. et al., Glycosylation of site-specific glycans of α_1-acid glycoprotein and alterations in acute ad chronic inflammation, *Biochim. Biophys. Acta*, 1725, 128, 2005.
218. Higai, K. et al., Altered glycosylation of α_1-acid glycoprotein in patients with inflammation and diabetes mellitus, *Clin. Chim. Acta*, 329, 117, 2003.
219. Kakehi, K. et al., Capillary electrophoresis of sialic acid-containing glycoprotein. Effect of the heterogeneity of carbohydrate chains on glycoform separation using an α_1-acid glycoprotein as a model, *Anal. Chem.*, 73, 2640, 2001.
220. Sei, K. et al., Collection of α_1-acid glycoprotein molecular species by capillary electrophoresis and the analysis of their molecular masses and carbohydrate chains. Basic studies on the analysis of glycoprotein glycoforms, *J. Chromatogr. A*, 958, 273, 2002.
221. Hiroaka, A. et al., Charge microheterogeneity of the β-trace proteins (lipocalin-type prostaglandin D synthase) in the cerebrospinal fluid of patients with neurological disorders analyzed by capillary isoelectric focusing, *Electrophoresis*, 22, 3433, 2001.
222. Lilja, H., Biology of prostate-specific antigen, *Urology*, 62, 27, 2003.
223. Peracaula, R. et al., Altered glycosylation pattern allows the distinction between prostate-specific antigen (PSA) from normal and tumor origins, *Glycobiology*, 13, 457, 2003.
224. Guzman, N.A., Immunoaffinity capillary electrophoresis applications of clinical and pharmaceutical relevance, *Anal. Bioanal. Chem.*, 378, 37, 2004.
225. Mao, X., Chu, I.K., and Lin B., A sheath-flow nanoelectrospray interface of microchip electrophoresis MS for glycoprotein and glycopeptide analysis, *Electrophoresis*, 27, 5059, 2006.

23 Capillary Electrophoresis of Post-Translationally Modified Proteins and Peptides

Bettina Sarg and Herbert H. Lindner

CONTENTS

23.1 INTRODUCTION

A large number of chemical modifications can occur in individual amino acids that can fundamentally affect their physicochemical and functional properties. Over 200 distinct covalent modifications have been reported, with phosphorylation, glycosylation, acetylation, methylation, and ADP-ribosylation being the most common. Some amino acids can be converted into other amino acids, for example, asparagine in aspartic acid or glutamine in glutamic acid by deamidation. Knowledge of these modifications is extremely important because they may alter physical and chemical properties, folding, conformation distribution, stability, activity, and consequently, function of the proteins.

Phosphorylation, principally on serine, threonine, or tyrosine residues, is one of the most important and abundant post-translational modifications (PTMs), with more than 30% of proteins being modified by the covalent attachment of one or more phosphate groups. It plays a critical role in the regulation of various cellular processes including cell cycle, growth, apoptosis, and transmitting extracellular signals to the nucleus. In fact, protein phosphorylation is probably the single most common intracellular signal transduction event. Owing to its biochemical importance, various analytical techniques for the detection and analysis of protein phosphorylation have been described.[1] A widely used approach for the detection of phosphorylation involves metabolic labeling with [^{32}P]orthophosphate followed by one- or two-dimensional polyacrylamide gel electrophoresis. Detection of phosphorylation sites is based on radiography combined with enzymatic digestions of the separated proteins and two-dimensional phosphopeptide mapping. High-performance capillary electrophoresis (HPCE) offers substantial method inherent advantages over such protocols, for example, a more rapid and accurate analysis with less sample consumption, while avoiding hazardous radioactive labeling.

In vivo methylation of the side chains of specific arginines, histidines, and lysines in proteins is a common phenomenon in nature involving numerous classes of proteins in both prokaryotic and eukaryotic cells.[2,3] Methylation has been most well studied in histones, with distinct lysine residues mono-, di-, or tri-methylated playing a major role in the regulation of gene expression, DNA replication, and repair. Methylated amino acids have often been determined in protein and tissue hydrolysates using amino acids analyzers and through cells radiolabeled with [methyl-^{3}H]methionine.

Acetylation of proteins occurs in two different ways. The N-terminal acetylation is one of the most common protein modification reactions in eukaryotes and it is estimated that up to about 85% of all mammalian proteins are affected. It is an irreversible process occurring cotranslationally, unlike the reversible side chain acetylation of internal lysine residues, most famously for histones and transcription factors that affect selective gene transcription and chromatin structure. Despite many hypotheses about the role of N-terminal acetylation, its biological significance is still unclear.

Protein glycosylation is a PTM of eukaryotic proteins and does not occur in prokaryotes. The carbohydrate groups are highly variable with significant effects on protein folding, stability, and activity. The importance of HPCE for the analysis of glycosylated proteins is described in Chapter 22.

Nonenzymatic deamidation of peptides or proteins represents an important degradation reaction occurring *in vitro* in the course of isolation or storage and *in vivo* during development and/or aging of cells.[4,5] Deamidation is a hydrolytic reaction resulting not only in the introduction of negative charges but also in a change in the primary structure of proteins or peptides. Deamidation is a common PTM resulting in the conversion of an asparagine residue to a mixture of isoaspartate and aspartate. Deamidation of glutamine residues can occur but does so at a much lower rate. Detecting and separating deamidated forms from the parent molecules are still problematic aspects of protein analysis.

HPCE continues to become more widely used for the detection, separation, and quantification of peptides and proteins, since the use of capillaries greatly reduces sample volume and analysis time compared to conventional gel electrophoresis. This offers perfect qualification for the analysis of PTMs, but surprisingly only a small number of HPCE applications have been developed during the past years. Since these modifications represent phenomena that occur *in vivo* often in very small amounts, they can be easily missed. This may explain why in certain cases recombinantly expressed proteins have an altered or absent activity compared with the naturally expressed proteins.

Taking this into account, more sensitive methods of sample detection have been developed, for example, special detection cell constructions for ultraviolet (UV) adsorption and laser-induced fluorescence (LIF), and particularly the introduction of mass spectrometry (MS) brought tremendous progress in online and offline characterization not only of modified peptides but also of modified proteins. In this chapter, we try to provide information about capillary electrophoresis (CE) methods developed for the separation of proteins and peptides with various PTMs.

23.2 THEORETICAL ASPECTS

More or less all modification reactions occurring on proteins may alter their overall charge either by introducing or eliminating charges, thereby changing to some extent the m/z ratio and the pI of the parent proteins. For this reason, capillary zone electrophoresis (CZE) and isoelectric focusing (IEF) are CE modes that are especially well suited to separate post-translationally modified forms. However, the ease of use makes CZE the preferentially used method in the laboratory and only very few applications employing other CE modes can be found in the literature.

Depending on the pI of the proteins and the pH of the buffers used for the separation in CZE proteins are either positively or negatively charged (except when pI = pH, where the overall charge of the proteins is zero). In the course of phosphorylation, for example, negatively charged phosphate groups are bound to uncharged amino acids such as Ser, Thr, or Tyr. Also, glycosylation has an impact on protein charge if one or more charged sialic acid molecules are attached (detailed in Chapter 22). In both cases, modification causes either a decrease in the overall positive charge or at appropriate buffer pH conditions an increase in the negative charge of the modified forms.

Acetylation can take place either on the N-terminal amino group of the first amino acid of a protein or at the ε-amino group of lysines. As basic amino groups, forming cationic ammonium ions under usual buffer conditions, become neutral due to amide formation one positive charge for each acetyl group bound will be removed in course of this modification reaction. This charge difference enables the separation of distinctly acetylated proteins from each other and from their unacetylated form.

Another modification of lysine, its mono-, di-, and, tri-methylation, has been the focus of great attention in histones due to its biological importance in gene regulation. As expected, the impact of methylation on charge is not as pronounced as of acetylation; however, electron donor effects of the alkyl group slightly increase the charge density of the nitrogen of the amino group. These differences can be sufficiently high for a CE separation provided the molecular mass of the protein is low.

An essential and in terms of its biological importance as well as of its frequency often underestimated modification reaction of proteins and peptides occurring under both physiological and laboratory conditions involves the nonenzymatic spontaneous deamidation of asparagine and glutamine residues to aspartate and glutamate, respectively. The deamidation of Asn, which takes place much more frequently than that of Gln, follows a rather complex mechanism via a cyclic imide intermediate. End products of its hydrolytic cleavage are the formation of isoaspartic acid as main reaction product and aspartic acid. In certain cases also truncation of the protein backbone can occur.[4] Under buffer conditions where proteins are cations (pH $<$ pI) the presence of both isoaspartic and aspartic acid reduces the overall positive charge of the proteins compared to the Asn containing protein. Moreover, due to minor differences in the pK_a values of iso-Asp and Asp (pK_a isoAsp $<$ pK_a Asp) a separation of these two isomeric protein forms can be achieved if pH of the separation buffer is adjusted thoroughly.

Certain protein classes are known to be multiply modified. Histone proteins are most probably the best investigated representatives in this respect. Under certain biological conditions, they are known to be acetylated, phosphorylated, and methylated, for example, in a definite manner at the same time. Several examples of successful CE separations of even multiply modified proteins are reported in the literature.[6,7]

Problems in CE occurring with the analysis of PTMs are usually not related to the presence of the modifying residues. They are much more associated with problems generally occurring when proteins are analyzed by capillary electrophoresis. Proteins stick to many different surfaces including metals, plastics, and also glass, which are the most commonly used materials for capillaries in CE. Depending on the extent of interaction between proteins and the glass surface of the capillary peak tailing, loss of resolution, reduced sensitivity, or even total adsorption can occur. The primary causes for these detrimental effects are ionic interactions between cationic protein regions and the negatively charged silanol groups of the capillary surface. Moreover, also hydrophobic interactions may contribute to these adsorption phenomena.

A variety of strategies were developed in the past to overcome these troublesome effects. Among these, a few are very simple to accomplish, for example, avoiding very low buffer concentrations. Increasing the buffer concentration decreases protein–wall interactions by reducing the effective

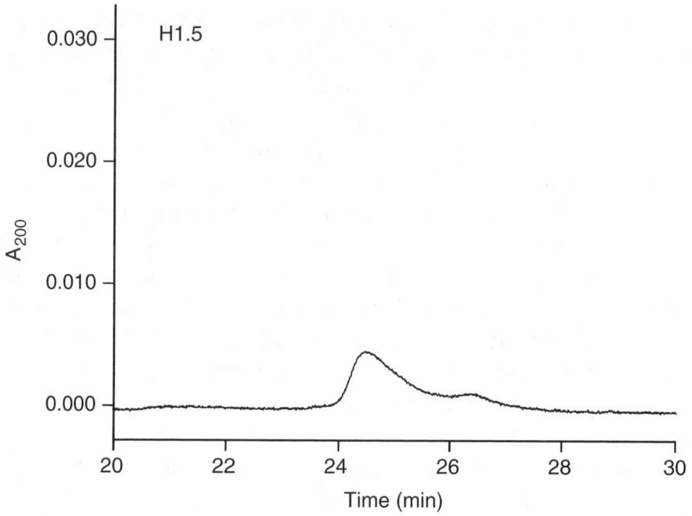

FIGURE 23.1 CZE separation of phosphorylated histone H1.5 from human tumor cells (CCRF-CEM) in a 100 mM phosphate buffer (pH 2.0). Other conditions: voltage 12 kV, temperature 25°C, injection 2 s, detection 200 nm, untreated capillary (50 cm $\times$ 75 μm i.d.).

surface charge and, depending on buffer type also ion-pairing effects come into play. Another approach to limit adsorption is simply by working at pH extremes: At low pH ($\sim$2) silanol groups are protonated, their dissociation is significantly reduced and in this way the interaction of the under these conditions cationic proteins with the now more or less uncharged surface minimized. Conversely, at high pH ($\sim$10) both the silica surface and the proteins will be deprotonated and negatively charged. In this case, adsorption is remarkably reduced as a result of a charge repulsion effect. However, while these approaches may be very successful strategies for the separation of modified forms of peptides and some small proteins (<5 kDa), their application is very limited to the separation of "normal sized" proteins as in this case undesired wall interactions cannot sufficiently enough be suppressed. A striking example is illustrated in Figure 23.1. It shows the separation of a mixture of distinctly phosphorylated and nonphosphorylated human linker histone H1.5 using an uncoated capillary and a sodium phosphate buffer with pH 2.0. Owing to their remarkably basic (pI > 10) and also hydrophobic properties, histone proteins particularly strongly interfere with the inner surface of the silica wall. From Figure 23.1, it is evident that even at this low pH essential electrostatic protein–wall interactions still occur, which are responsible for the poor resolution, broad peaks, and low sensitivity.

In contrast to separations performed at low pH, working at high pH is not always feasible for the following reasons:

1. Artificial protein modification reactions can be induced, for example, deamidation of proteins.
2. At very high pH (>11) dissolution of the silica becomes an issue.

For this reason, distinct coating procedures, often combined with low pH buffers, have been found to be a prerequisite for the successful separation of proteins as well as of their PTMs.

Two basic strategies have been applied in CE to limit protein adsorption.

1. Permanent modification by covalently bonded or physically adhered phases. Basically, hydrophilic polymers, polar functional groups, or even positively charged residues, for

example, amino groups are bound to the silica surface.[8] In general, life time of the permanent coating and batch to batch reproducibility may still be a problem. Therefore, it is not so surprising that most applications for the separation of protein modifications, which can be found in the literature, are based on an alternative approach to suppress undesired protein–surface interactions.

2. It is the concept of dynamically coating the inner wall of the capillary by the addition of suitable additives to the running buffer. These additives themselves show high affinity to the silica surface and act like a shield to prevent positively charged proteins from coming into close contact with the capillary wall. Various compounds have been used for this purpose; for instance, neutral and cationic hydrophilic polymers, cationic hydrophobic polymers, diamines, and so on.[8] Outstanding advantages of this approach are stability of the coating and simplicity of handling. Since the coating agent is in the buffer, the coating layer of the capillary surface is continuously regenerated and no permanent stability is required. Moreover, cleaning of the capillary can be performed easily by rinsing with 0.1 M sodium hydroxide and water. However, potential disadvantages of the dynamic coating approach are effects of the surfactants, for example, influencing protein structure, incompatibility with MS analysis, range of pH applicable can be limited, time for equilibrium needed to obtain reproducible surface coating, surface properties may influence quality and reproducibility of the coating.

As a consequence, both methods have advantages and shortcomings and no method is clearly superior. For this reason, depending on protein primary structure and type of modification method development is still an issue; however, many promising recipes for their successful separation are available in the meantime.

23.3 APPLICATIONS

Many proteins such as histones, ribonucleic acid (RNA) polymerase II, tubulin, myelin basic protein, p53, and tyrosine kinases are post-translationally modified at multiple sites. Among them histones are the most extensively studied group as their hydrophilic N- and C-terminal tail domains are subjected to a great variety of PTMs including phosphorylation, acetylation, methylation, deamidation, ubiquitination, and ADP-ribosylation. Distinct combinations of covalent histone modifications including lysine acetylation, lysine and arginine methylation, and serine phosphorylation form the basis of the histone code hypothesis.[9–11] This hypothesis proposes that a pre-existing modification affects subsequent modifications on histone tails and that these modifications generate unique surfaces for the binding of various proteins or protein complexes responsible for higher-order chromatin organization and gene activation and inactivation. Owing to their biological importance, it is not surprising that a variety of methods were developed particularly for the CE separation of histone modifications, more than for any other protein family.

This chapter summarizes practical applications of HPCE on the analysis of various modified histones, other proteins and, moreover, also some modified peptide separations are described.

23.3.1 PROTEINS

23.3.1.1 Phosphorylation

A set of different microscale techniques of CE exists for analyzing phosphorylated proteins and peptides. The effectiveness of a buffer system containing hydroxypropylmethyl cellulose (HPMC) as dynamic coating agent in combination with low pH preventing undesired interactions of positively charged histone proteins with the silica surface could be first demonstrated by Lindner et al.[12–14] A complex mixture of rat testis H1 histones consisting of eight microsequence variants and, in addition,

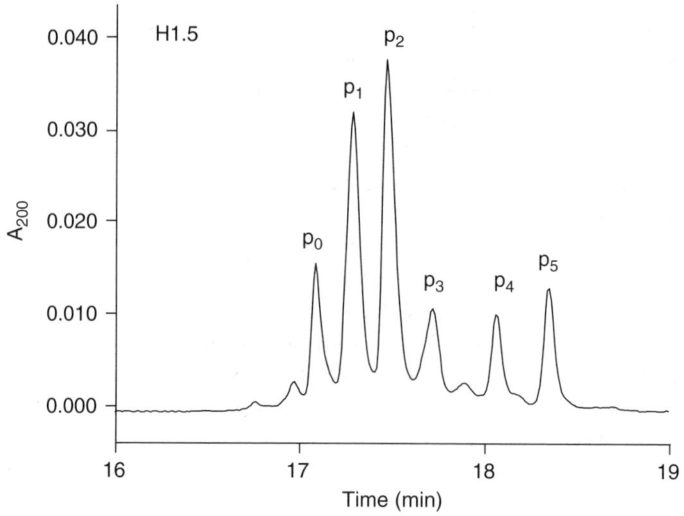

FIGURE 23.2 CE separation of phosphorylated histone H1.5 from human tumor cells (CCRF-CEM) in a 100 mM phosphate buffer (pH 2.0) and in the presence of 0.02% HPMC. Other conditions: voltage 12 kV, temperature 25°C, injection 2 s, detection 200 nm, untreated capillary (50 cm × 75 μm i.d.). Designations p0–p5 = non-, mono-, di-, tri-, tetra-, and pentaphosphorylated histone H1.5.

various phosphorylated forms, was separated using an uncoated capillary and a sodium phosphate buffer (pH 2.0) and 0.03% HPMC. A striking example for the resolving power of this separation buffer system is depicted in Figure 23.2, illustrating the CE separation of the same multiphosphorylated histone H1.5 sample already shown in Figure 23.1. Applying the same CE conditions as described in Figure 23.1, except that 0.02% HPMC was added to the phosphate buffer, all five phosphorylated forms are now clearly separated from each other and from the unphosphorylated parent protein. The nonphosphorylated protein migrates fastest, followed by the distinctly phosphorylated forms, because binding of the negatively charged phosphate groups decrease the overall positive charge of the histone molecule thus diminishing the electrophoretic mobility of the phosphorylated protein species.

Recently, Yoon et al.[15] developed a CE-LIF method to determine phosphorylation levels and to follow the translocation of the green fluorescent protein-extracellular signal regulated protein kinase 2 (GFP-ERK2) from the cytoplasm to the nucleus. CE conditions applied were an untreated capillary and a 100 mM CAPS buffer (pH 11.0) containing 2 M betaine. LIF detection was performed with a 5 mW air-cooled argon ion laser (excitation, 488 nm/emission, 520 nm). Phosphorylated and non-phosphorylated GFP-ERK2 were not separated simultaneously, but in consecutive runs. Significant differences in migration time allowed a clear assignment of the modified and unmodified protein. Owing to separation conditions employed (high pH, high salt buffer), the GFP-ERK2 proteins are negatively charged and, therefore, no dynamic coating agent was added. Phosphorylation causes a further increase in the negative charge; thereby, phosphorylated GFP-ERK2 migrates slower. Compared to conventional Western blotting, the CE method allows the analysis of sample volumes as low as a few nanoliters and does not require a separate sample purification step and radiolabeling.

On the basis of their differences in isoelectric point (pI), phosphorylated proteins can be separated in capillary isoelectric focusing (CIEF). Wei et al.[16] employed a CIEF method using a capillary covalently coated with linear polyacrylamide and a pH gradient from 4 to 6.5 for the resolution of mono- and diphosphorylated ovalbumins. Proteins were detected by their UV absorbance at 280 nm. Additional ovalbumin variants within each of the mono- and diphosphoovalbumins, differing in their amount of glycosylation, were further analyzed by online CIEF-electrospray ionization (ESI)-MS.

In a similar manner, non-, mono-, and diphosphorylated myosin light chain using CIEF, either with UV or LIF detection, was separated by Shiraishi et al.[17] Neutral coated capillaries (eCAP; Beckman Coulter) and HPMC in the ampholyte solution reduced electroosmotic flow (EOF) and protein adsorption. A detection limit of ~1 pg fluorescently labeled myosin light chain/capillary was achieved.

23.3.1.2 Acetylation

Histone H4 with a calculated pI of 11.9 is one of the most basic proteins known. Under certain biological conditions, it can be reversibly acetylated in its N-terminal region at four lysine residues. CE separation of these different acetylated forms was not possible applying chemically coated capillaries. However, when HPMC was used as a dynamic coating agent, the histone H4 sample isolated from whole histones by reversed phase chromatography (RPC) was clearly resolved into the non-, mono-, di-, tri-, and tetra-acetylated forms within about 22 min.[18] Applying a special buffer system consisting of 500 mM formic acid/LiOH/10 mM urea (pH 2.0) with 0.02% HPMC, an ultrafast CE separation (4 min) of the distinctly acetylated H4 proteins could be achieved (shown in Figure 23.3). As acetylation of lysines decreases the positive charge of histone H4, the unacetylated protein migrates fastest.

Using HPMC as buffer additive another core histone, subfraction H2A, which is a single peak in RPC, could be further separated by CE into five peaks consisting of non-and monoacetylated H2A.2a and H2A.2b, respectively, and even a third subtype H2A.3 was resolved.[19]

Wiktorowicz and Colburn[20] separated core histone H4 into its different acetylated forms using a commercially available cationic surfactant (MicroCoat, ABI). Coating of the silica surface was performed by rinsing the capillary with the reagent followed by a wash with running buffer. Like other cationic surfactants such as cetyl trimethylammonium bromide (CTAB), the coating reagent is primarily bound to the silica surface by ionic interaction. In a second step, the surface-bound neutralized surfactant binds additional reagent cations by virtue of hydrophobic interactions between alkyl side groups. In this way, a remarkably stable bilayer is formed and the charge of the capillary wall is reversed from negative to positive. Consequently, the positively charged histone molecules

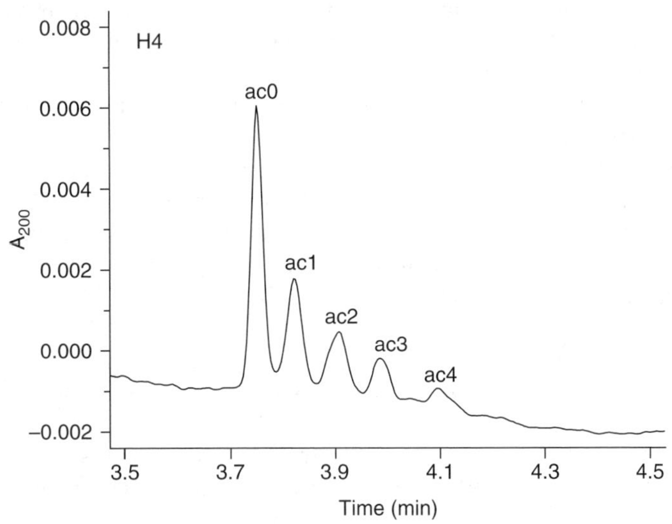

FIGURE 23.3 Electropherogram of multiacetylated histone H4 from mouse tumor cells (NIH) in 500 mM formic acid/LiOH/10 mM urea buffer (pH 2.0) containing 0.02% HPMC. Conditions: voltage 30 kV, injection 1 s. Other conditions are as in Figure 23.2. Designations ac0–ac4 = non-, mono-, di-, tri-, and tetra-acetylated histone H4.

are repelled from the capillary surface and under the influence of an electric field, a reversal of the EOF takes place. It should be mentioned, however, that strictly speaking this particular approach of charge reversal is a special case of the dynamic coating concept, as no coating reagent is included in the running buffer. For this reason, periodic replenishment of the coating by flushing the capillary with the surfactant solution is required.

Besides histones, very little is described about the CE analysis of other acetylated proteins. Some articles report on the separation of N-terminally acetylated and unacetylated forms of metallothioneins (MTs). MTs are structurally unusual proteins due to their small size (6–7 kDa), high cysteine and metal content, and remarkable kinetic lability. Purified rat MT-2 protein was separated by MEKC using an electrolyte of 100 mM sodium borate buffer containing 75 mM SDS at pH 8.4 and showed two peaks.[21] Identification of both peaks was performed with online CE-MS using 100 mM formic acid and 2% methanol. The two peaks differed by a mass equivalent to that of a single acetyl group. Another group established a CE-ESI-time of flight (TOF)-MS method to characterize rabbit liver MT isoforms and was successful in separating several N-acetylated and non-N-acetylated forms using 100 mM acetic acid:100 mM formic acid (pH 2.3).[22]

CIEF has been applied to the separation and quantification of the three main hemoglobin components of umbilical cord blood (fetal, acetylated fetal, and adult hemoglobins).[23] CIEF was performed with a poly (acryloylaminoethoxy-ethonal) [poly(AAEE)] coated capillary and a carrier ampholyte consisting of 5% Ampholine pH 6–8, supplemented with 0.5% TEMED.

An interesting CE application has been the use of so-called charge ladders of proteins for measuring the role of charge in protein stability, protein–ligand binding, and ultrafiltration.[24–26] A protein charge ladder is a collection of derivatives of a protein by converting its charged groups into electrically neutral ones. Charge ladders can be easily generated *in vitro* by the treatment of a variety of model proteins with acetic anhydride. CE has been shown to be an effective tool in the analysis of such charge ladders, as it is used to separate the proteins that constitute the charge ladder into individual "rungs," each rung contains derivatives with the same number of modified groups.

Cordova et al.[27] explored the CE behavior of protein charge ladders obtained by acetylation of lysozyme and carbonic anhydrase II using noncovalent polycationic coated capillaries. Two of them, polyethylenimine and Polybrene, were very effective in preventing the adsorption of positively charged proteins. Conditions used were fused-silica capillary (50 μm i.d. × 38 cm) coated with the polymer by flushing the capillary with a 7.5% (w/v) polymer solution, prepared in 25 mM Tris–192 mM Gly buffer (pH 8.3), for 15 min. The running buffer was 25 mM Tris–192 mM Gly buffer (pH 8.3) in the absence of polymer. Separations were obtained within 5 min.

Carbeck et al.[28] showed CE-ESI MS to be a useful tool for the study of charge ladders of lysozyme, carbonic anhydrase II, and bovine pancreatic trypsin inhibitor and for the examination of the relationship between the properties of proteins in the solution phase and in the gas phase.

An example for a CE separation of such a charge ladder is shown in Figure 23.6 (Section 23.4.3).

Separation of multiply modified proteins place high demands on the method applied. An example for a successful separation of a protein containing differently phosphorylated and acetylated forms in a single run using the HPMC-based CZE method is shown in Figure 23.4.

23.3.1.3 Deamidation

The analysis of recombinant human growth factor (rhGH), one of the first biotechnologically produced proteins in *Escherichia coli*, by CE with UV absorbance and MS detection using bilayer-coated capillaries was demonstrated very recently.[29] The authors present an improved CE method using capillaries noncovalently coated with polybrene and poly(vinyl sulfonic acid) and a background electrolyte of 400 mM Tris phosphate (pH 8.5) to achieve efficient separations of intact rhGH and degradation products like deamidated and oxidated forms.

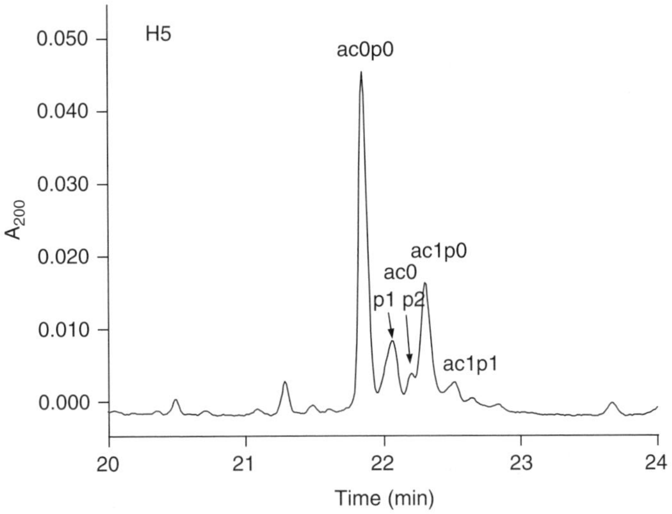

FIGURE 23.4 Electropherogram of histone H5 from chicken erythrocytes. Conditions were the same as in Figure 23.2. Designations ac0p0 = nonacetylated nonphosphorylated H5; ac0p1 and ac0p2 = nonacetylated mono- and diphosphorylated H5; ac1p0 and ac1p1 = monoacetylated non-and monophosphorylated histone H5.

A novel method for the determination of glutamine deamidation in a long repetitive protein polymer via bioconjugate capillary electrophoresis was published recently by Won et al.[30] By conjugating a monodisperse, fluorescently labeled DNA oligomer to long polydisperse nearly electrostatically neutral protein polymers, protein polymers differing in degree of deamidation were successfully separated. For protein polymers with increasing extents of deamidation, the electromotive force of DNA + polypeptide conjugate molecules increases due to the introduced negative charge of deamidated glutamic acid residues. CE analysis reveals increasing differences in the electrophoretic mobilities of conjugate molecules, which qualitatively shows the degree of deamidation. CE of protein–DNA conjugates was performed using a capillary coated with POP-5 polymer and a 50 mM Tris, 50 mM TAPS, 2 mM EDTA buffer containing 7 M urea (pH 8.4) added with 3% POP-5 solution.

Histone $H1^0$ is known to be deamidated in the course of aging.[5] Intact $H1^0$ contains asparagine in position 3, while depending on the age of the organ deamidated forms of $H1^0$ containing either aspartic acid or isoaspartic acid accumulate. The buffer system containing HPMC enables the separation of intact histone $H1^0$ fragment (residues 1–52) as well as its deamidated forms (Figure 23.5).

23.3.1.4 Methylation

Unlike protein modification by acetylation or phosphorylation, methylation does not greatly influence the charge of individual amino acids, thus making the electrophoretic separation of distinct methylated proteins from each other and from the unmethylated parent protein a problematic part. Separation of acetylated histone H4 has already been described by Lindner et al.[18] using CZE with HPMC as dynamic coating reagent. The same method enables the simultaneous separation of mono-, di-, and trimethylated H4, including their distinct acetylated forms.[6] Changes in the modification pattern of H4 from normal tissues, cancer cell lines, and primary tumors were found suggesting that a global loss of monoacetylation and trimethylation of histone H4 is a common hallmark of human cancer.

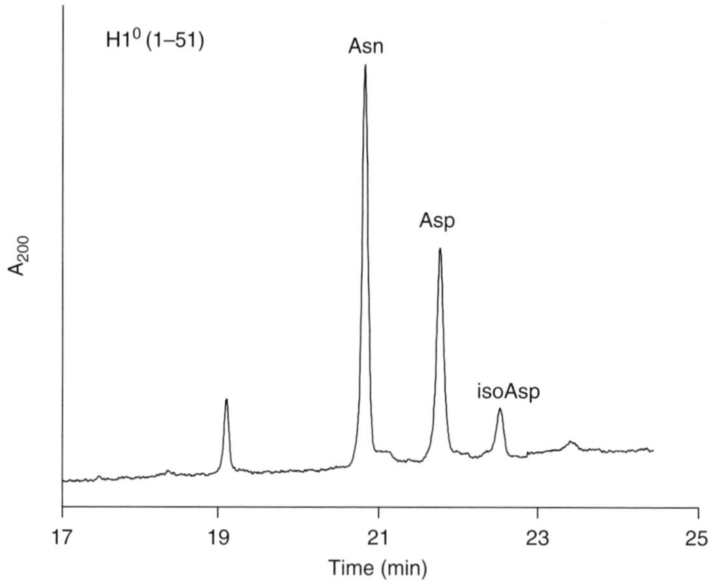

FIGURE 23.5 CE separation of deamidated proteins. Histone $H1^0$ from rat liver was digested with chymotrypsin and the N-terminal fragment obtained by RPC analyzed by CE. Conditions were the same as in Figure 23.2. Designation Asn, Asp, isoAsp = asparagine, aspartic acid, and isoaspartic acid containing fragments.

23.3.2 Peptides

23.3.2.1 Phosphorylation

A CZE procedure for detection and assay of protein kinase and phosphatase activities in complex biological mixtures has been developed by Dawson et al.[31] The phosphorylated and dephosphorylated forms of several peptides were resolved using an uncoated capillary and a 150 mM phosphoric acid buffer at pH 2.0 or pH 5.0. Furthermore, the CZE-based assay was capable of resolving a peptide phosphorylated on different sites and permitted the quantification of each peptide. Since this application, much research has been done on protein kinase assays based on CE.

For the same purpose, Gamble et al.[32] tested both neutral and cationic coated and uncoated capillaries and found that the best conditions for the separation of phosphopeptide isomers is formic acid or phosphate buffer with pH ranging between 5.5 and 6.5 and a PVA-coated capillary. Synthetic peptides were incubated with various protein kinases and phosphatases and the phosphorylation and dephosphorylation was monitored using CZE. These methods enabled the assay of several protein kinases and phosphatases and the determination of the sites of phosphorylation.

Improved CE-based methods to measure kinase activation in single cells have been described recently.[33] Phosphorylated and nonphosphorylated forms of peptide substrates for protein kinase C and calcium-calmodulin activated kinase were separated by CZE using a polydimethylacrylamide (PDMA) coated capillary and buffers containing different concentrations of betaine (0–1 M). The separation system is compatible with a living cell and, therefore, adaptable to the laser micropipet system, a strategy to measure the activation of enzymes in single mammalian cells.

Synthetic peptides containing phosphorylated tyrosine have been shown to be easily separated by a CZE method using a linear polyacrylamide-coated capillary and a 300 mM borate buffer (pH 8.5).[34]

Another separation mode of HPCE, namely MEKC, has been used to separate a mixture of phosphopeptide isomers of the insulin receptor peptide.[35] The resolution in coated and uncoated capillaries was compared using a 50 mM phosphate buffer with 25 mM SDS (pH 6.1). An efficient separation of diphosphorylated isomers could be obtained by using polyacrylamide-coated capillaries.

Mass spectrometry of phosphopeptides has become a powerful tool for phosphorylation site identification. However, proteolytic digests examined by MS are often likely to fail to detect phosphopeptides because the ionization of phosphorylated peptides in positive ion mode MS is generally less efficient compared with the ionization of their nonphosphorylated counterparts resulting in ion suppression effects. A further problem is that phosphopeptides may not be retained by RP chromatography because they are too small and/or hydrophilic to bind to the C18 stationary phase. Therefore, capillary electrophoresis coupled to MS is a powerful method to enhance the detection of phosphoproteins and phosphopeptides due to their efficient separation by CE.

Sandra et al.[36] developed a CE-MS technique to the characterization of the N-glycans from the glycoprotein cellobiohydrolase I (CBH I) with special interest for the phosphorylated species. CBH I shows phosphorylated glycans that are only present when the organism is grown under minimal conditions and could be related to stress response. The glycans were labeled with the negatively charged tag 8-aminopyrene-1,3,6-trisulfonate (APTS) by reductive amination and separated using an untreated capillary and a 25 mM ammonium acetate buffer (pH 4.55), allowing the simultaneous analysis of uncharged and charged glycans and, moreover, the differentiation of phosphorylated isomers.

Detection of phosphorylated proteins and peptides can be enhanced by selectively isolating these species. Online immobilized metal affinity chromatography (IMAC)-CE-ESI-MS is such a powerful analytical tool. The IMAC resin retains and preconcentrates phosphorylated proteins and peptides, CE separates the phosphorylated species and MS/MS identifies the components and their phosphorylation sites. Cao and Stults[37,38] applied this method to the analysis of phosphorylated angiotensin II and tryptic digests of α- and β-casein (CE conditions: buffer, 0.1% acetic acid/10% methanol; uncoated capillary). Beta-casein is a well-characterized protein with five phosphorylation sites and is widely used as a standard for protein phosphorylation studies.

23.3.2.2 Deamidation

The effectiveness of CZE for the separation of deamidated peptides was shown by Ganzler et al.[39] Excellent resolution of peptides containing all three forms, Asn, Asp, and isoAsp, could be achieved by using a 20 mM sodium citrate buffer (pH 2.5).

23.3.2.3 Farnesylation

Protein farnesylation, catalyzed by protein farnesyltransferase, plays important roles in the membrane association and protein–protein interaction of a number of eukaryotic proteins. The enzyme transfers a 15-carbon farnesyl moiety from farnesyl diphosphate (FPP) to the sulfhydryl group of cysteine. The activity of the enzyme was measured by CE with LIF detection, which is a powerful alternative to classical methods involving radiolabeled FPP.[40] LIF detection was performed with an argon ion laser (excitation, 488 nm/emission, 520 nm). A fluorescently labeled pentapeptide that was used as substrate was clearly separated from its farnesylated form under the four CE buffer conditions investigated (e.g., 25 mM borax, 25 mM SDS, pH 9.3, uncoated capillary). The method will be of great value in studies of inhibitors of protein farnesyltransferase *in vitro*.

23.4 METHOD DEVELOPMENT GUIDELINES

Many factors may influence the quality of CE separations and contribute to the resolution of protein modifications. Generally, optimization strategies, precautions, and CE conditions applied to peptides are also applicable to the separation of modified proteins. However, optimum parameters for protein separations must still be determined empirically to some extent. In addition to well-known factors influencing resolution in CE like capillary surface and coating, buffer pH, also effects of different buffer cations and anions, buffer concentration, additives, and chaotropic agents can be utilized to enhance resolution of protein mixtures.

23.4.1 Buffers

The running buffer selection is substantial to the success of any CE separation. Factors being most important are the UV absorbance at low wavelengths, good pH stability, and conductivity. The highest sensitivity detection is achieved at low UV wavelengths (190–220 nm) where peptide bonds have an absorbance maximum. Good sensitivity at these wavelengths requires buffer systems with high-UV transparency. The most frequently used buffers are sodium phosphate, sodium citrate, Tris, and sodium borate. Sodium formate shows optimal compatibility to mass spectrometric analysis and is therefore often used in CE coupled online to MS.

The buffer pH should be stable during a series of analyses. For reproducible analysis times the buffer should be exchanged or remixed every 5–10 runs. Cellulose-based buffers should be freshly prepared at least once a week, depending on the number of runs, and should not(!) be stored in the refrigerator. The quality and purity of all buffer components should be of the highest grade and the buffers should be filtered through 0.45 μm pore size filters before use. A starting buffer, which gives superior results in many cases, is a 100 mM sodium phosphate buffer (pH 2.5) with the addition of 0.02% HPMC. Owing to the excellent UV transparency of the separation buffer, proteins can be detected very sensitively at 200 nm. In case this system does not provide satisfying results, more complex separation conditions must be taken into account according to specific properties of the proteins (e.g., pI, hydrophobicity, etc.).

23.4.2 Capillary Treatment

Using an untreated capillary every 10–15 injections the capillary should be rinsed with water, 0.1 M NaOH, water, 0.5 M H_2SO_4, and finally with the running buffer. Washing should be done for 2 min with each solvent. The same wash, but without running buffer, should be applied at the end of sample analyses. After flushing the capillary with air, it can be stored.

There are many different static wall coatings described in the literature, recently reviewed by Horvath and Dolnik.[8] Static coated capillaries should be treated according to the manufacturer's guidelines. Solvents such as NaOH and H_2SO_4 must not be used.

23.4.3 Method Evaluation

The evaluation of the applicability of a certain CE method for the separation of modified proteins can easily be performed by analyzing acetylated lysozyme generated by the treatment of the protein with acetic anhydride (described in Section 23.3.1.2). An aqueous solution of lysozyme (0.1 mM, 0.5 mL) adjusted to pH 10 by using 0.1 M NaOH is incubated with 4.5 μL acetic anhydride (100 mM in dioxane) for 5 min at room temperature according to Cordova et al.[27] An example using the suggested phosphate buffer with HPMC is shown in Figure 23.6.

23.5 CONCLUDING REMARKS

The analysis of post-translationally modified proteins is due to their biological significance a field of ever increasing importance. For many years, different kinds of PAGE have been the most commonly used tool for studying modified proteins with all the shortcomings of this technique generally known. The advent of capillary electrophoresis provided a promising new tool for their separation. HPCE has made significant advances in recent years and is establishing itself as a rapid, reproducible, sensitive, and highly resolving method. Hyphenated techniques, for example, the combination with MS, will

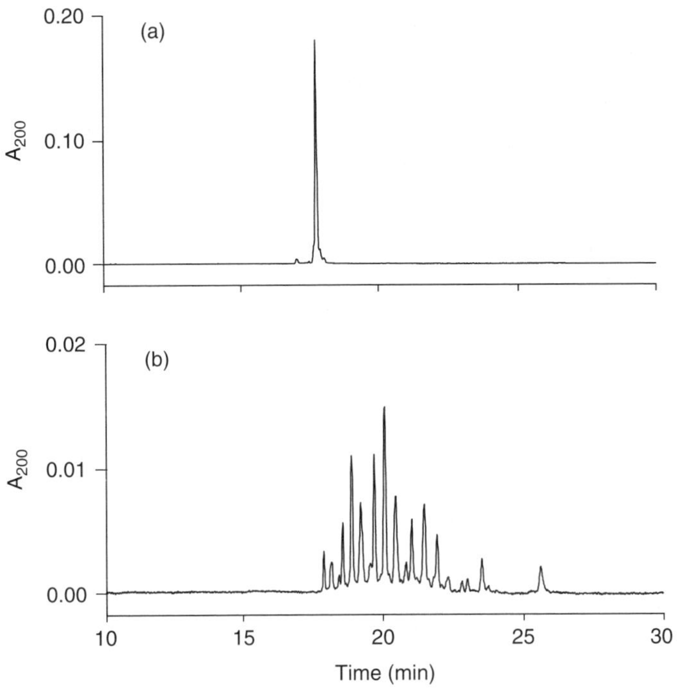

FIGURE 23.6 CE analysis of the lysozyme charge ladder. (a) Untreated lysozyme and (b) charge ladder from acetylation of lysozyme. Conditions were the same as in Figure 23.2.

be able to provide detailed structural information like modification status and sites and will further increase the range of potential CE applications in this field.

ACKNOWLEDGMENTS

This work, as part of the European Science Foundation EUROCORES Programme EuroDYNA, was partly supported by funds from the Austrian Science Foundation (project I23-B03) and by the Jubilee Fund of the Austrian National Bank, grant no. 9319.

REFERENCES

1. Yan, J. X., N. H. Packer, A. A. Gooley, and K. L. Williams, Protein phosphorylation: Technologies for the identification of phosphoamino acids, *J. Chromatogr. A*, 808, 23–41, 1998.
2. Paik, W. K. and S. Kim, *Protein Methylation*, John Wiley & Sons, New York, 1980.
3. Wold, F., *In vivo* chemical modification of proteins (post-translational modification), *Annu. Rev. Biochem.*, 50, 783–814, 1981.
4. Lindner, H. and W. Helliger, Age-dependent deamidation of asparagine residues in proteins, *Exp. Gerontol.*, 36, 1551–1563, 2001.
5. Lindner, H., B. Sarg, B. Hoertnagl, and W. Helliger, The microheterogeneity of the mammalian H1(0) histone. Evidence for an age-dependent deamidation, *J. Biol. Chem.*, 273, 13324–13330, 1998.
6. Fraga, M. F., E. Ballestar, A. Villar-Garea, M. Boix-Chornet, J. Espada, G. Schotta, T. Bonaldi, et al., Loss of acetylation at Lys16 and trimethylation at Lys20 of histone H4 is a common hallmark of human cancer, *Nat. Genet.*, 37, 391–400, 2005.
7. Sarg, B., W. Helliger, B. Hoertnagl, B. Puschendorf, and H. Lindner, The *N*-terminally acetylated form of mammalian histone H1(0), but not that of avian histone H5, increases with age, *ABB*, 372, 333–339, 1999.

8. Horvath, J. and V. Dolnik, Polymer wall coatings for capillary electrophoresis, *Electrophoresis*, 22, 644–655, 2001.
9. Strahl, B. D. and C. D. Allis, The language of covalent histone modifications, *Nature*, 403, 41–45, 2000.
10. Turner, B. M., Histone acetylation and an epigenetic code, *Bioessays*, 22, 836–845, 2000.
11. Jenuwein, T. and C. D. Allis, Translating the histone code, *Science*, 293, 1074–1080, 2001.
12. Lindner, H., W. Helliger, A. Dirschlmayer, H. Talasz, M. Wurm, B. Sarg, M. Jaquemar, and B. Puschendorf, Separation of phosphorylated histone H1 variants by high- performance capillary electrophoresis, *J. Chromatogr.*, 608, 211–216, 1992.
13. Lindner, H., M. Wurm, A. Dirschlmayer, B. Sarg, and W. Helliger, Application of high-performance capillary electrophoresis to the analysis of H1 histones, *Electrophoresis*, 14, 480–485, 1993.
14. Lindner, H., W. Helliger, B. Sarg, and C. Meraner, Effect of buffer composition on the migration order and separation of histone H1 subtypes, *Electrophoresis*, 16, 604–610, 1995.
15. Yoon, S., K. Y. Han, H. S. Nam, L. V. Nga, and Y. S. Yoo, Determination of protein phosphorylation and the translocation of green fluorescence protein-extracellular signal-regulated kinase 2 by capillary electrophoresis using laser induced fluorescence detection, *J. Chromatogr. A*, 1056, 237–242, 2004.
16. Wei, J., L. Yang, A. K. Harrata, and C. S. Lee, High resolution analysis of protein phosphorylation using capillary isoelectric focusing—electrospray ionization—mass spectrometry, *Electrophoresis*, 19, 2356–2360, 1998.
17. Shiraishi, M., R. D. Loutzenhiser, and M. P. Walsh, A highly sensitive method for quantification of myosin light chain phosphorylation by capillary isoelectric focusing with laser-induced fluorescence detection, *Electrophoresis*, 26, 571–580, 2005.
18. Lindner, H., W. Helliger, A. Dirschlmayer, M. Jaquemar, and B. Puschendorf, High-performance capillary electrophoresis of core histones and their acetylated modified derivatives, *Biochem. J.*, 283 (Pt 2), 467–471, 1992.
19. Lindner, H., B. Sarg, C. Meraner, and W. Helliger, Separation of acetylated core histones by hydrophilic-interaction liquid chromatography, *J. Chromatogr. A*, 743, 137–144, 1996.
20. Wiktorowicz, J. E. and J. C. Colburn, Separation of cationic proteins via charge reversal in capillary electrophoresis, *Electrophoresis*, 11, 769–773, 1990.
21. Beattie, J. H., A. M. Wood, and G. J. Duncan, Rat metallothionein-2 contains N(alpha)-acetylated and unacetylated isoforms, *Electrophoresis*, 20, 1613–1618, 1999.
22. Andon, B., J. Barbosa, and V. Sanz-Nebot, Separation and characterization of rabbit liver apothioneins by capillary electrophoresis coupled to electrospray ionization time-of-flight mass spectrometry, *Electrophoresis*, 27, 3661–3670, 2006.
23. Conti, M., C. Gelfi, and P. G. Righetti, Screening of umbilical cord blood hemoglobins by isoelectric focusing in capillaries, *Electrophoresis*, 16, 1485–1491, 1995.
24. Negin, R. S. and J. D. Carbeck, Measurement of electrostatic interactions in protein folding with the use of protein charge ladders, *J. Am. Chem. Soc.*, 124, 2911–2916, 2002.
25. Gao, J., M. Mammen, and G. M. Whitesides, Evaluating electrostatic contributions to binding with the use of protein charge ladders, *Science*, 272, 535–537, 1996.
26. Menon, M. K. and A. L. Zydney, Effect of ion binding on protein transport through ultrafiltration membranes, *Biotechnol. Bioeng.*, 63, 298–307, 1999.
27. Cordova, E., J. Gao, and G. M. Whitesides, Noncovalent polycationic coatings for capillaries in capillary electrophoresis of proteins, *Anal. Chem.*, 69, 1370–1379, 1997.
28. Carbeck, J. D., J. C. Severs, J. Gao, Q. Wu, R. D. Smith, and G. M. Whitesides, Correlation between the charge of proteins in solution and in the gas phase investigated by protein charge ladders, capillary electrophoresis, and electrospray ionization mass spectrometry, *J. Phys. Chem. B*, 102, 10596–10601, 1998.
29. Catai, J. R., T. J. Sastre, P. M. Jongen, G. J. de Jong, and G. W. Somsen, Analysis of recombinant human growth hormone by capillary electrophoresis with bilayer-coated capillaries using UV and MS detection, *J. Chromatogr. B Analyt. Technol. Biomed. Life Sci.*, 852, 160–166, 2007.
30. Won, J. I., R. J. Meagher, and A. E. Barron, Characterization of glutamine deamidation in a long, repetitive protein polymer via bioconjugate capillary electrophoresis, *Biomacromolecules*, 5, 618–627, 2004.

31. Dawson, J. F., M. P. Boland, and C. F. Holmes, A capillary electrophoresis-based assay for protein kinases and protein phosphatases using peptide substrates, *Anal. Biochem.*, 220, 340–345, 1994.
32. Gamble, T. N., C. Ramachandran, and K. P. Bateman, Phosphopeptide isomer separation using capillary zone electrophoresis for the study of protein kinases and phosphatases, *Anal. Chem.*, 71, 3469–3476, 1999.
33. Li, H., H. Y. Wu, Y. Wang, C. E. Sims, and N. L. Allbritton, Improved capillary electrophoresis conditions for the separation of kinase substrates by the laser micropipet system, *J. Chromatogr. B Biomed. Sci. Appl.*, 757, 79–88, 2001.
34. Bonewald, L. F., L. Bibbs, S. A. Kates, A. Khatri, J. S. McMurray, K. F. Medzihradszky, and S. T. Weintraub, Study on the synthesis and characterization of peptides containing phosphorylated tyrosine, *J. Pept. Res.*, 53, 161–169, 1999.
35. Tadey, T. and W. C. Purdy, Capillary electrophoretic resolution of phosphorylated peptide isomers using micellar solutions and coated capillaries, *Electrophoresis*, 16, 574–579, 1995.
36. Sandra, K., J. Van Beeumen, I. Stals, P. Sandra, M. Claeyssens, and B. Devreese, Characterization of cellobiohydrolase I N-glycans and differentiation of their phosphorylated isomers by capillary electrophoresis-Q-Trap mass spectrometry, *Anal. Chem.*, 76, 5878–5886, 2004.
37. Cao, P. and J. T. Stults, Mapping the phosphorylation sites of proteins using on-line immobilized metal affinity chromatography/capillary electrophoresis/electrospray ionization multiple stage tandem mass spectrometry, *Rapid Commun. Mass Spectrom.*, 14, 1600–1606, 2000.
38. Cao, P. and J. T. Stults, Phosphopeptide analysis by on-line immobilized metal-ion affinity chromatography-capillary electrophoresis-electrospray ionization mass spectrometry, *J. Chromatogr. A*, 853, 225–235, 1999.
39. Ganzler, K., N. W. Warne, and W. S. Hancock, Analysis of rDNA-derived proteins and their posttranslational modifications, in *Capillary Electrophoresis in Analytical Biotechnology*, P. G. Rhigetti (Ed.) CRC press, Boca Raton, FL, pp. 185–238, 1996.
40. Berezovski, M., W. P. Li, C. D. Poulter, and S. N. Krylov, Measuring the activity of farnesyltransferase by capillary electrophoresis with laser-induced fluorescence detection, *Electrophoresis*, 23, 3398–3403, 2002.

24 Extreme Resolution in Capillary Electrophoresis: UHVCE, FCCE, and SCCE

Wm. Hampton Henley and James W. Jorgenson

CONTENTS

24.1 INTRODUCTION

The majority of difficult separations can be divided into two main classes: complex mixtures and challenging pairs (Figure 24.1). Complex mixtures, such as proteins and protein digests, demonstrate "random" or broad spectrum behavior with respect to their properties, whereas "well-behaved" analytes such as nucleic acids display predictable behavior determined mainly by the size of the molecule. The analysis of these complex samples requires a separation system with high peak capacity and the ability to cover a broad spectrum of physical properties.

Challenging pairs, such as enantiomers and isotopomers, display properties that are so similar in nature that a separation technique must have extremely high efficiency in order to resolve the analytes. The high efficiency can come at the expense of the ability to analyze a broad spectrum of different analyte properties because the species of interest are confined to closely migrating bands.

Conventional approaches to increasing the resolution for difficult separations in capillary electrophoresis (CE) usually require extensive development of buffers and/or capillary coatings particular to an individual separation. The improvement in resolution is usually the result of exploiting a specific intrinsic property of the analytes. A universal approach to achieving broad spectrum high resolution separations can be achieved using CE with an extremely high applied potential. Separation efficiency increases linearly with applied voltage, and resolution improves with the square root of applied voltage. Ultrahigh voltage CE (UHVCE) uses existing buffer systems and separation conditions but gives the analytes more time to separate at a given electric field strength.

The separation of challenging pairs that are not resolved using traditional methods is often achieved by developing novel pseudo-stationary phases for enantiomers or by utilizing mass spectrometry for analyzing isotopomers. Resolution of these tough pairs can also be achieved using

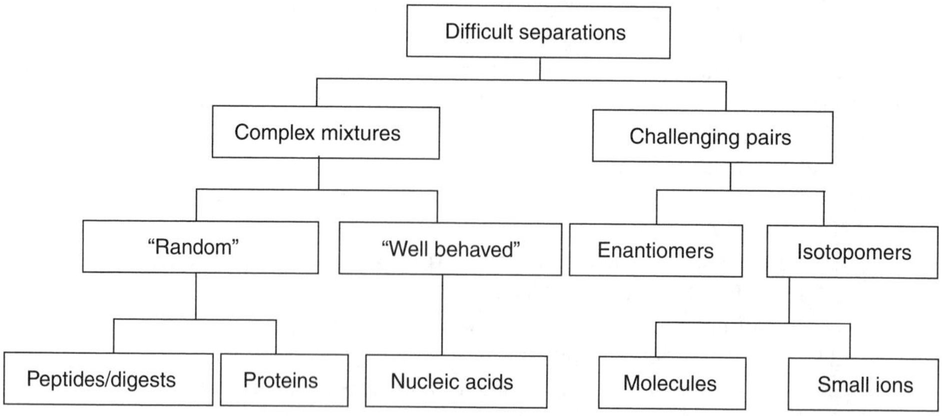

FIGURE 24.1 Dividing difficult separations into several categories can help in the development of separation strategies.

extended separation times under high electric field strength. Because the difficult pairs exist as closely migrating analyte bands, techniques such as flow counterbalanced CE (FCCE) and synchronous cyclic CE (SCCE) can be used. These methods use conventional separation voltages in relatively short capillaries but use procedures to increase the duration in a high electric field.

24.2 ULTRAHIGH VOLTAGE CAPILLARY ELECTROPHORESIS

24.2.1 Background and Theory

Early work with CE demonstrated how higher applied voltages offered vast improvements for electrophoretic separations.[1–5] Separation efficiency shows a linear dependence on applied voltage when analytes are detected very close to the end of the capillary:

$$N \cong \frac{(\mu_{\mathrm{e}} + \mu_{\mathrm{EOF}})\, V}{2D}, \tag{24.1}$$

where N is the efficiency in theoretical plates, μ_{e} is the electrophoretic mobility of the analyte, μ_{EOF} is the mobility of the electroosmotic flow (EOF), V is the applied voltage, and D is the diffusion coefficient of the analyte.[6] Resolution between analytes with electrophoretic mobilities μ_1 and μ_2 show a dependence on the square root of the applied voltage:[7]

$$\mathrm{Rs} = 0.177(\mu_1 - \mu_2)\left(\frac{V}{\overline{D}(\overline{\mu}_{\mathrm{e}} + \mu_{\mathrm{EOF}})}\right)^{1/2}. \tag{24.2}$$

In addition, the speed of a separation can be greatly increased with higher applied voltages. Migration time for an analyte (t_{m}) displays an inverse linear relationship with applied voltage for the same length capillary (L) and separation distance (l):

$$t_{\mathrm{m}} = \frac{lL}{V(\mu_{\mathrm{e}} + \mu_{\mathrm{EOF}})}. \tag{24.3}$$

The flat electrokinetic flow profile suggests that as long as Joule heating can be prevented, there is no theoretical limit to the improvements to be gained by increasing the applied voltage. In practice, several factors combined to limit the applied voltage used in early instrumentation to approximately 30 kV.

High electric field strengths can lead to resistive or Joule heating of the buffer solution within the capillary. A detailed discussion of the effects of Joule heating with voltages up to 60 kV has been described by Palonen et al.[8] In the simplest of terms, the power (P) running through the capillary is given by

$$P = iV = \frac{V^2}{R}, \tag{24.4}$$

where R is the resistance of the capillary. Since it is desirable to have the greatest possible voltage, the resistance of the capillary must be large enough to keep the power below about 1 W per meter of capillary length.

Initial experiments conducted at voltages in excess of 30 kV resulted in broken capillaries. A quick calculation of the radial electric field at the silica/buffer interface (Figure 24.2) using the following equation explains why this occurs:[9]

$$E_{\mathrm{radial}} = \frac{V_{\mathrm{i}} - V_{\mathrm{o}}}{\ln(b/r_{\mathrm{c}})Kr_{\mathrm{c}}}, \tag{24.5}$$

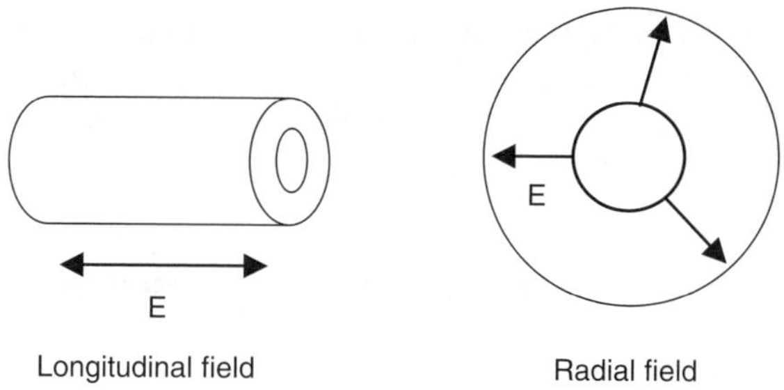

FIGURE 24.2 Longitudinal electric fields are used to separate analytes in CE. Radial fields also exist through the capillary wall, and at high voltages, they can destroy the capillary via dielectric breakdown.

where V_i is the voltage inside the capillary, V_o is the voltage of a nearby conductor distance b from the center of the capillary, r_c is the length of the inner radius of the capillary, and K is the dielectric constant of the capillary wall (~3.9 for fused silica). For a capillary 50 μm in diameter, 1 m from the nearest grounded conductor, with 100 kV applied, the radial electric field strength is about 100 million V/m. Radial electric fields greater than the dielectric strength of fused silica (~15 million V/m) damage the fused silica wall and cause defects in the structure leading to fragmentation.

Early efforts using a van de Graff generator[10] at ~200 kV and a thick dielectric coating such as a sleeve of Teflon® tubing around the capillary were unsuccessful. In another attempt, a plastic tube surrounding the capillary was filled with isopropyl alcohol.[11] In theory, both the capillary and the tube filled with alcohol can be treated as long resistors, and therefore the potential will drop linearly down the length of the capillary. The potential drop in the outer tube should match the potential drop in the capillary and therefore reduce the radial field through the capillary wall. Unfortunately, the capillaries still degraded in a very short time.

The first capillary shielding system that could prevent dielectric breakdown of the fused-silica capillary was demonstrated by Hutterer and Jorgenson.[12–15] The capillary was placed inside a series of metal rings charged to potentials closely matching those inside the capillary (Figure 24.3). The electric potential from the rings counteracts the radial field from the capillary and shields the capillary from damage. The high voltage direct current (HVDC) was generated using a 26-stage Cockcroft–Walton voltage multiplier[16] with 2.4 nF HV capacitors. The voltages needed to charge the rings were supplied by different stages of the multiplier. The entire instrument was submerged in a tank containing transformer oil to provide electrical insulation.

This instrument showed the expected improvements in efficiency and resolution for samples of peptides via CE,[7,12] hyaluronic acid via capillary gel electrophoresis (CGE),[12,14] and oligosaccharide mixtures via micellular electrokinetic capillary chromatography.[12,15] Application of up to 120 kV was possible, and no degradation of the capillary was observed. Figure 24.4 shows electropherograms of a cytochrome *c* digest separated on the same capillary using 28 kV and 120 kV.[12] It is easy to see the improvements in resolution and the decrease in separation time when higher voltages are applied.

The current state-of-the-art instrumentation uses larger capacitors (10 nF) to create a 100-fold Cockcroft–Walton voltage multiplier capable of reaching potentials as high as 410 kV.[6] A total of 25 shielding rings keep the potential difference between the capillary and shielding system at less than 20 kV. The benchtop instrument runs in air, using plastic dielectric material instead of the tank of transformer oil. The HV polarity can easily be reversed, allowing both positive and negative HV separations. The temperature of the instrument can be controlled to within a few degrees over a range of ambient temperature up to 60°C. It can be easily moved and coupled to different detection systems including laser-induced fluorescence (LIF) detection, UV absorbance detection, and electrospray ionization-mass spectrometry (ESI-MS). Separations of peptides, proteins, and nucleic acids using the instrument with these detectors are described in Sections 24.2.2.4, 24.3.3.3, and 24.4.3.3.

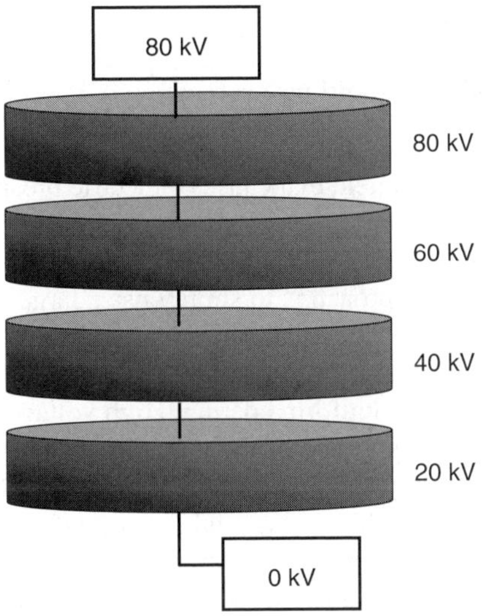

FIGURE 24.3 Charged metal shielding rings can be used to reduce the radial electric field strength. Closely matching the potential charge on the ring to that within the capillary prevents dielectric breakdown and allows operation at very high voltages. (Reprinted from Henley, W. H., PhD Dissertation, University of North Carolina, Chapel Hill, 2005. Copyright 2005. With permission.)

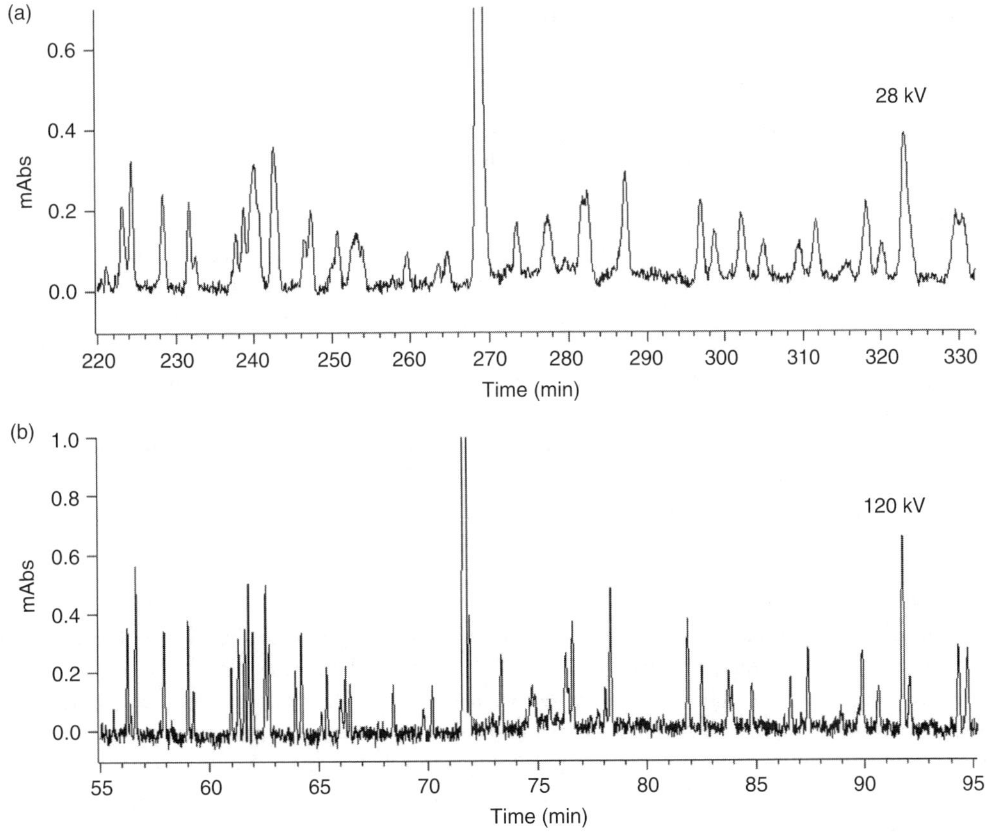

FIGURE 24.4 Separations of cytochrome *c* tryptic digest using the same capillary at conventional (top) and UHV conditions. (Example: Reprinted from Hutterer, K. M., Doctorial Thesis. University of North Carolina, Chapel Hill, 2000. Copyright 2000. With permission.)

24.2.2 Practical Implementation

24.2.2.1 High Voltage Capillary Shielding System

The first step in implementing UHVCE is to devise a capillary shielding system adequate for the desired separation voltage. The radial field that causes dielectric breakdown of the capillary must be dealt with in some manner. The dielectric breakdown strength of silica is approximately 15 MV/m. Equation 24.4 can be used to show that small diameter capillaries have a large radial electric field at relatively modest voltages. One approach is to simply use a large i.d. capillary with a thick wall. This may work for relatively low voltages, but not without band broadening and heat dissipation problems inherent to large bore capillaries. Another approach uses a conductive, but high resistance, material to surround the capillary in a uniform coating. High voltage is applied to the capillary and to the surrounding material. If the resistance of the material is equal along its length, the potential drop should mirror that of the capillary and thus neutralize the radial electric field through the wall of the capillary. Attempts using isopropyl alcohol in a tube surrounding the capillary resulted in failure after a relatively short time.[11] Nevertheless, such a system should be possible.

The most suitable method found so far is the use of charged metal rings to surround the capillary and neutralize the radial field (Figure 24.3). Briefly, the capillary is surrounded by a series of metal rings charged to a potential that roughly matches those inside the capillary. Long lengths of capillary can be equally spaced within the rings by winding them onto a suitable spool. This method has been successfully used without any noticeable degradation of the capillary.

24.2.2.2 Power Supply and Development

The development of the power supply for UHVCE requires several considerations. Voltages of the magnitude used for UHVCE can easily arc through the air and "crawl" along insulating surfaces for several feet. Safety is paramount, and a grounded metal shell or Faraday cage should be used to protect the operator from electrocution and to protect sensitive nearby equipment from electromagnetic pulses caused by stray electric arcs. The supply must generate high voltage with sufficient current to drive the separation and charge the shielding system. Loss of charge from corona and stray capacitance must be prevented if the full potential of the power supply is to be realized.

The generation of high voltage can be accomplished in many ways. Perhaps the simplest method is the van de Graff generator.[10] Briefly, charge is pumped from one roller to anther roller held inside a metal sphere via a belt of dielectric material. The van de Graff generator is a constant current source, and therefore the voltage is limited by the loss of charge to the surrounding air and resistive load. Submerging the generator in transformer oil can dramatically reduce charge loss and improve the efficiency of the generator. Since the voltage is generated in a single step, charging metal shield rings to different potentials to shield the capillary may not be practical.

If a single commercial high voltage power supply is used, a string of high voltage resistors set up to divide the voltage will be needed to bias the shielding rings. Such a system would be most practical for voltages up to about 100 kV, and even then require careful insulation of the resistors and high voltage wires.

The Cockcroft-Walton voltage multiplier has been used to generate the high voltage for the two UHVCE systems developed thus far.[6,7,12,14] The instrumentation has already been thoroughly detailed elsewhere, and so will only briefly be discussed here. The multiplier is a convenient and inexpensive solution to several problems surrounding UHVCE. It produces successively higher d.c. voltages as the multiplier chain is extended, providing multiple voltage sources to bias the shielding rings. The electronics are simple and can be easily contained within modestly sized shielding rings so that dielectric stresses on the electronic components themselves can be minimized (Figure 24.5). The efficiency of the multiplier decreases as the number of stages is increased. This can be mediated somewhat by using large capacitors in the first stages and decreasing the capacitance of each successive stage. Using fast recovery diodes can prevent excessive losses from reverse current leakage. The

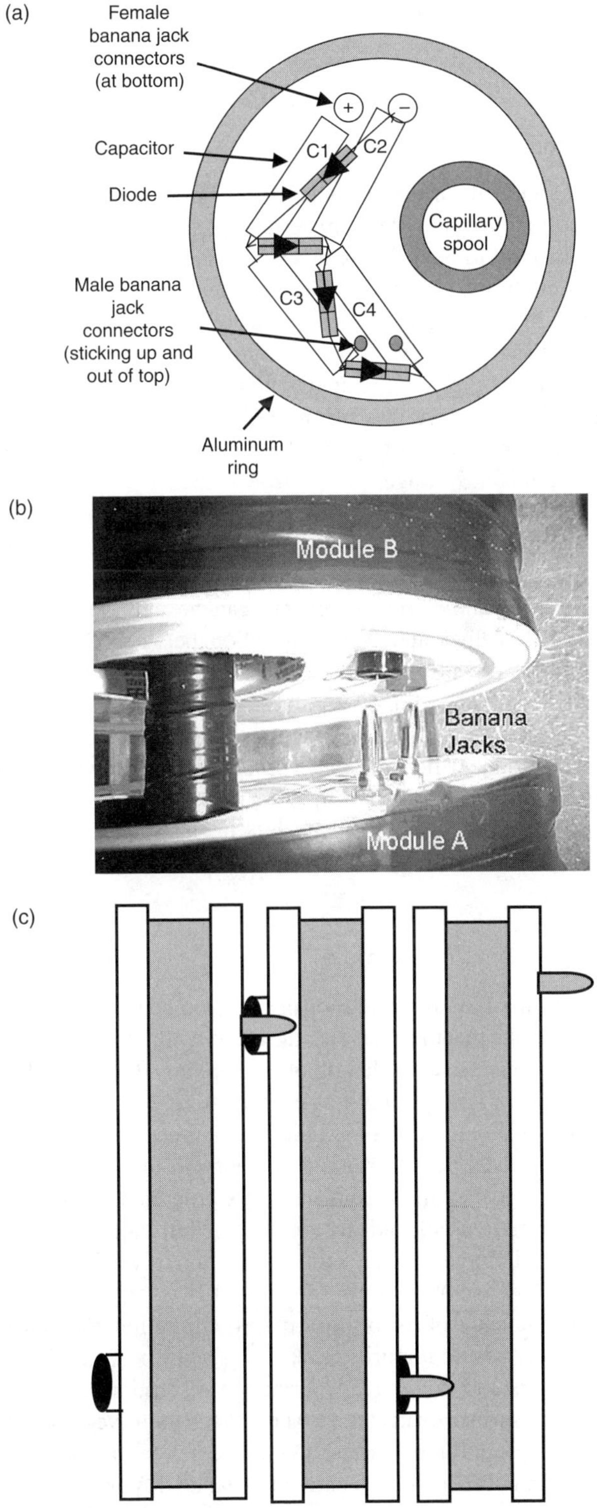

FIGURE 24.5 (a) A schematic representation of the electronics in the individual modules that make up the Cockcroft–Walton voltage multiplier and shielding rings. The capillary spool can be seen in the photo (b), as well as the connections between modules (c). (Reprinted from Henley, W. H., PhD Dissertation, University of North Carolina, Chapel Hill, 2005. Copyright 2005. With permission.)

a.c. input frequency should be carefully selected for maximum efficiency of the Cockcroft-Walton generator.

24.2.2.3 HV Insulation

When the electric field at a metal surface becomes greater in magnitude than the dielectric strength of the surrounding material, ionization of that material will occur. This phenomenon is called corona when it occurs in air. It is undesirable because it is a source of potential voltage loss and can lead to an electric arc. The strength of the electric field is dependent on the radius of curvature of the metal shielding ring. For every 30 kV of potential, the radius of curvature must increase 1 cm to prevent corona. Potentials of 500 kV require a radius of curvature of over 33 cm. Such large diameters can be avoided if material of sufficient dielectric strength and thickness is used to surround a more modest diameter. The first UHVCE instrument was surrounded by a tank of transformer oil. The oil proved difficult to work in. It absorbed moisture from the atmosphere and became slightly conductive, reducing the maximum attainable voltage.

The second UHVCE instrument[6] used thick coatings of plastics for dielectric insulation (Figure 24.5). All metal surfaces were painted with vinyl coatings. The electronic components were potted in thermoplastic polyethylene commonly available as "hot melt" glue. This material is inexpensive, has good dielectric properties, can be applied in thick coatings, and can be easily reworked with a scalpel and heat. Insulation between the shielding rings was provided by polyethylene caps. The insulation from the surrounding air was provided by an acrylic tube with many layers of polyester film. The polyester film had an exceptionally large dielectric strength ($\sim 3 \times 10^6$ V/cm). Every effort was taken to prevent an ionizable path to ground. The high voltage end of the instrument housing was completely sealed with polyester/styrene casting resin except for the tiny injection port. The injection port was plugged with a polyetheretherketone (PEEK) injection rod (Figure 24.6) covered in silicone vacuum grease to exclude air from the seal.

24.2.2.4 Modes of Operation and Examples

Open tube UHVCZE can be used for the separation of small molecules, peptides, proteins, and protein digests.[6,7,12] Increasing the time that the analytes migrate will improve resolution and efficiency. One method of achieving this is to increase the length of the capillary, but when the maximum attainable voltage is limited, the electric field strength will be relatively low in long capillaries. UHVCE can improve these separations by maintaining a high field strength in a long capillary. Figure 24.7 shows two separations of a model peptide sample in a 566-cm-long capillary. The sample was first separated at 28 kV (close to the maximum conventional CE applied voltage) resulting in an electric field strength of only 50 V/cm. Leu-enk had an efficiency of 1 million theoretical plates and required approximately 328 min to migrate to the detector. Using the UHVCE power supply at 330 kV, the electric field strength was 580 V/cm. Leu-enk showed a 10-fold efficiency improvement at higher voltage, resulting in an efficiency of 10 million plates. Migration time was reduced to 28 min. Resolution between the analytes and impurity peaks can also been seen.

In open tube UHVCE, it is important to reduce analyte adsorption as much as possible. Analytes that adsorb to the fused-silica surface do not tend to show much improvement as the length of capillary and applied voltage are increased. Figure 24.8 shows a model protein separation that demonstrated a small amount of adsorption in a 50 cm long separation. Using UHV in a 501 cm capillary, resolution and efficiency are actually reduced, and migration time is proportionally longer.

Capillaries coated with aminopropyl species have been used to reduce the adsorption of peptides and proteins. UHVCE separations of peptides and proteins using these coating have been reported

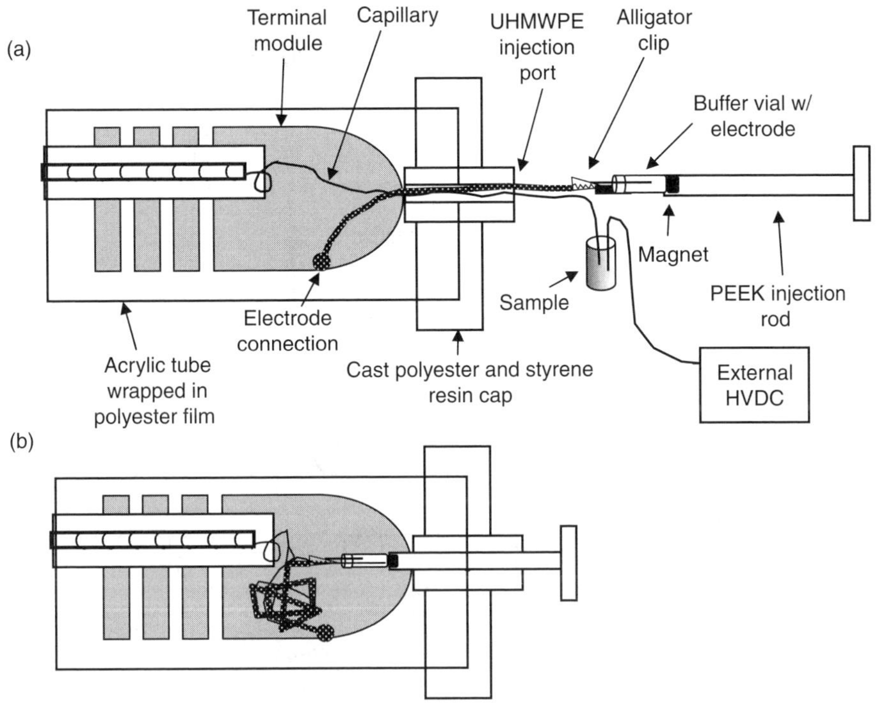

FIGURE 24.6 Schematic diagram showing the operation of the injection port for sampling at low voltage. (Reprinted from Henley, W. H., PhD Dissertation, University of North Carolina, Chapel Hill, 2005. Copyright 2005. With permission.)

with UV absorbance and ESI-MS.[6] Large improvements were seen in resolution and efficiency using UHVCE compared with shorter capillaries run at the same field strength at lower voltages.

If the capillary is filled with a suitable sieving matrix, polymers such as nucleic acids[6] and hyaluronic acid[12,14] can be separated using UHV capillary gel electrophoresis (UHVCGE). Figure 24.9 shows four separations of MegaBASE™ 4 Color Sequencing Standard using different lengths of 75μm i.d., 360μm o.d. acrylamide-coated capillary filled with MegaBACE™ Long Read Matrix (Amersham Biosciences). The applied voltage ranges from −9 kV in the 60 cm capillary to –83 kV in the 5.5 m capillary to maintain a constant field strength of 150 V/cm. The time has been rescaled for the longer separations by a factor of X (given in the legend) so that the resolution improvements at higher applied voltages can be easily seen.

Neutral analytes have been separated using UHV micellular electrokinetic chromatography (UHVMEKC).[12,15] This was demonstrated by the separation of oligosaccharide mixtures in a surfactant-containing buffer. Higher resolution was achieved using UHV than standard running voltages, and plate counts over 1 million plates were seen.

24.2.2.5 Limitations

Despite the advantages offered by UHVCE, several problems still limit its application. The lack of available instrumentation and unfamiliarity with voltages in excess of 30 kV keep many potential applications from being explored. The dangers involved with working at such high voltages are more perceived than real, and safety can be assured by using a grounded Faraday cage with interlocks designed to protect the operator from high voltage discharges. Also, the stored energy in most high voltage systems is usually very small, much less than 50 J for the current UHVCE system.

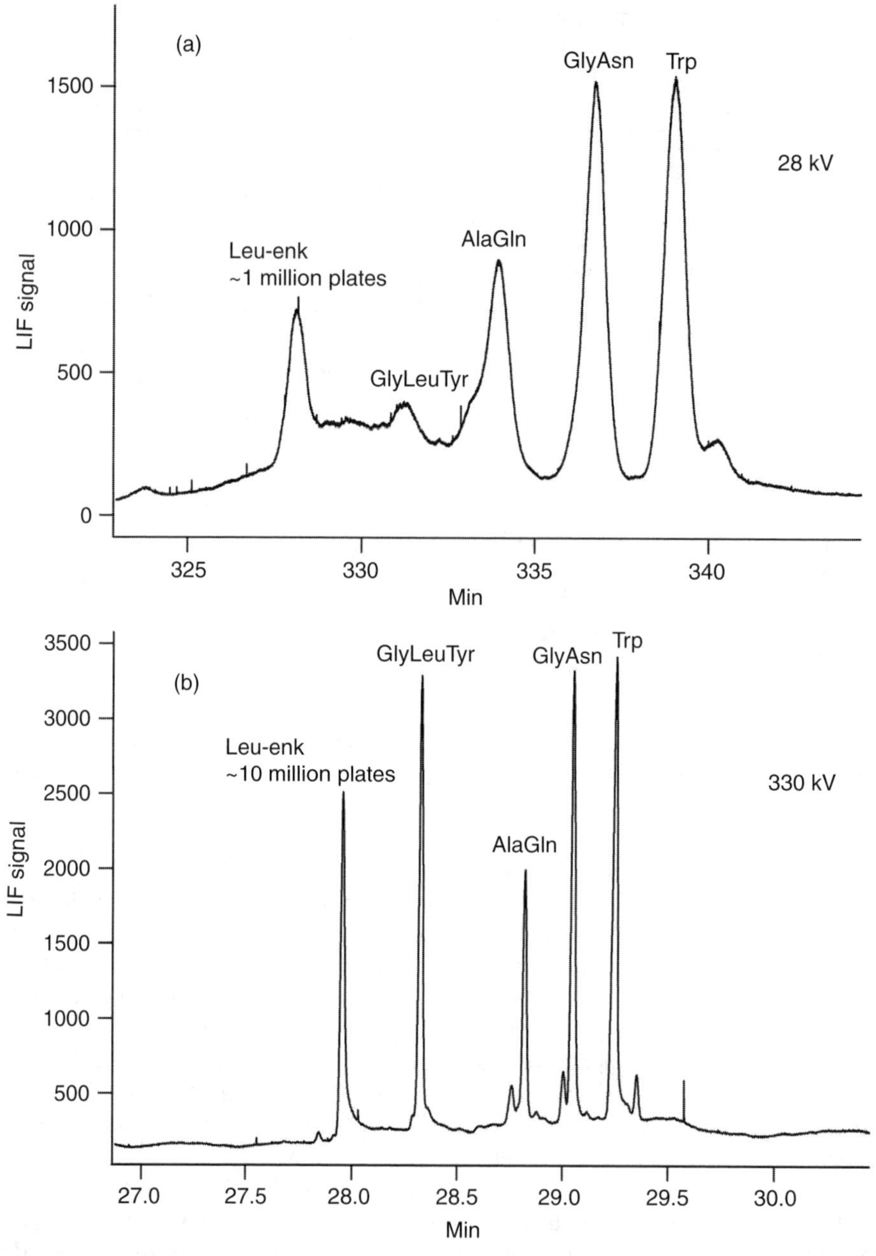

FIGURE 24.7 Separations of a model peptide sample run on the same capillary at conventional and UHV. The 10-fold increase in electric field strength improves efficiency and reduces separation time 10-fold. Resolution increase can be clearly seen by the closely migrating impurity peaks. (Reprinted from Henley, W. H., PhD Dissertation, University of North Carolina, Chapel Hill, 2005. Copyright 2005. With permission.)

24.2.2.6 Methods Development Guidelines

Part of the advantage of UHVCE is that it allows the use of known separation systems that work well for similar analytes but are unable to completely resolve the analytes of interest using conventional voltages and capillary lengths. Several factors relating to the buffer system and capillary should be taken into consideration before attempting to use a separation system for UHVCE.

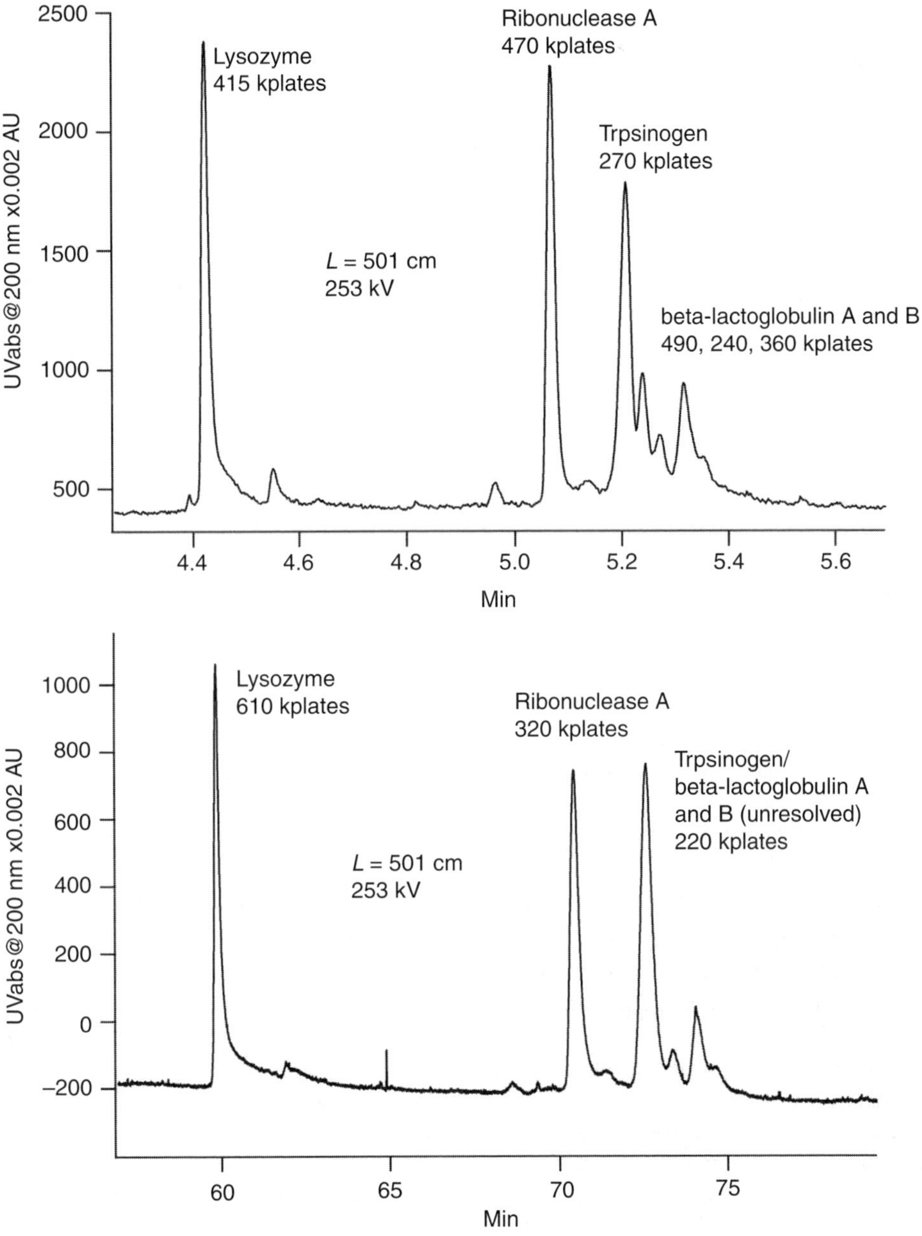

FIGURE 24.8 Separation improvements are not realized in longer capillaries at higher voltages if analyte adsorption occurs. Shown here is a separation of proteins in short (a) and long (b) capillaries under the same electric field strength. Efficiency and resolution are actually reduced at UHV conditions due to analyte adsorption over the long capillary wall. (Reprinted from Henley, W. H., PhD Dissertation, University of North Carolina, Chapel Hill, 2005. Copyright 2005. With permission.)

24.2.2.7 Buffer Selection

When selecting a buffer for UHVCE, the conductivity of the buffer must be as low as possible so that the electric field strength can be high without unacceptable Joule heating. The maximum field strength for a buffer/capillary system can easily be found by generating a current to field strength

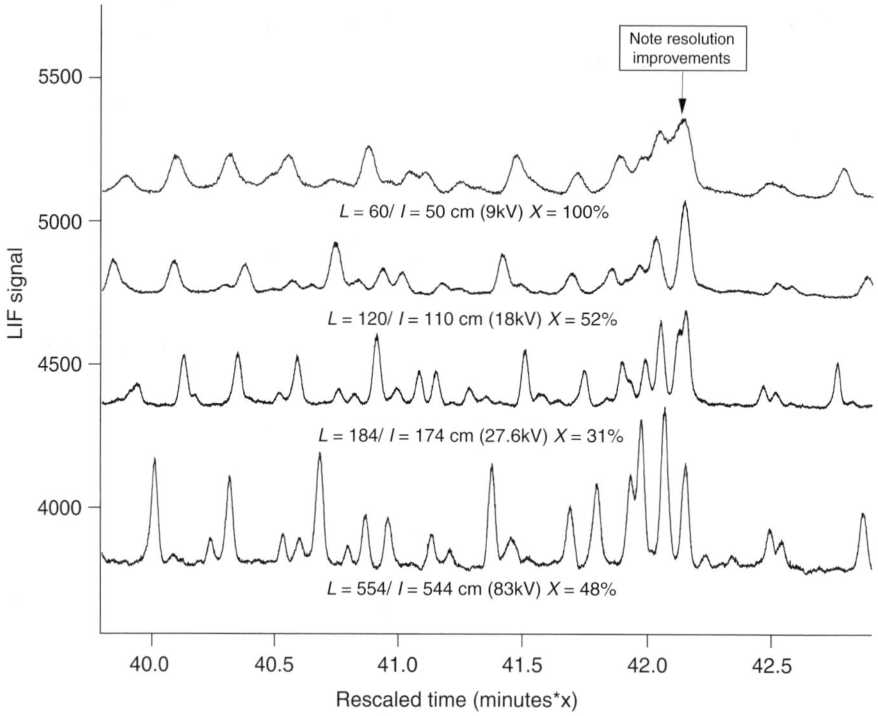

FIGURE 24.9 Separations of DNA sequencing standard rescaled in the time domain by factor X for comparison of resolution improvements at higher applied voltages. Longer lengths (L) of capillary were used to maintain a constant 150 V/cm field strength. (Reprinted from Henley, W. H., PhD Dissertation, University of North Carolina, Chapel Hill, 2005. Copyright 2005. With permission.)

plot using a conventional power supply in a short piece of capillary. The onset of appreciable Joule heating can be found at the field strength where the plot begins to show nonlinear behavior.

High concentrations of salt are used in some buffers to prevent adsorption of analyte molecules (peptides and proteins[17] in particular) to the capillary wall. Zwitterionic molecules contain an equal number of positive and negative charges at a pH equal to their isoelectric point. They can be used to raise the ionic strength of the buffer without increasing the conductivity greatly. In UHVCE applications, they can sometimes be substituted for more conductive buffers to allow the use of higher field strengths while maintaining sufficient ionic strength to reduce analyte adsorption. It is important to note that the pI of the zwitterionic molecule may be several pH units different from their pK_a values. In cases where this occurs, they cannot be relied upon to buffer the solution effectively.

High concentrations of ions with small radii of hydration such as H_3O^+, K^+, and OH^- should be avoided. These small ions are rapidly transported within the linear electric field of the capillary, resulting in high electric currents. For this reason, ions with lower mobility such as Li^+ and CH_3COO^- should be substituted wherever possible.[18]

The buffer capacity can become important in separations of extended duration. For example, a CGE separation requiring 30 min on a 30 cm capillary will require 300 minutes on a 300 cm capillary if the same electric field strength is used. The 10-fold increase in analysis time results in 10-fold the amount of electrolysis products that must be dealt with by the buffer. The moles of OH^- at the cathode can be calculated from the electrophoretic current (i):

$$it\frac{1}{F} = \text{molOH}^-, \tag{24.6}$$

where t is time in seconds and F is Faraday's constant. A significant pH shift can cause problems with reproducibility. Large capacity buffer vials can be used if the instrumentation has enough available space.

Buffer systems that do not adequately address problems of analyte adsorption will not be compatible with UHVCE in long capillaries.[6] Buffers that contain ions with significantly different mobilities from the analyte ions will also result in electrodispersion that will be exacerbated by long separation distances. Both of these problems can be avoided by examining the shape of the analyte bands in short capillaries at lower voltages. Peaks that show fronting or tailing indicate potential problems, whereas highly Gaussian peaks indicate compatibility with UHVCE.

24.2.2.8 Capillary Selection

The length of the capillary should be determined by the desired degree of improvement over a known, lower voltage, separation. For example, if twice the resolution of a known system using 30 kV in a 50 cm capillary is desired, then 120 kV should be used in a capillary 2 m long. This separation will also require fourfold the time.

Joule heating can be avoided by simply reducing the power consumption of the capillary. Reducing the capillary diameter by half decreases the current fourfold. For the separation described above, a 50 cm length of capillary with half the diameter could be used at 120 kV without excessive Joule heating. In this case, the separation would require fourfold *less* time.

The inner diameter of the capillary cannot be decreased too greatly if UV absorbance is used for detection. Below ~50 μm, the pathlength is too short for sensitive detection. ESI-MS also requires a certain flow rate for optimal detection that may not be achievable in very narrow capillaries. LIF detection is well suited for use with narrow capillaries, but not every analyte can be labeled with fluorescent tags.

24.2.2.9 Sample Separation Strategies

Unlike techniques that are limited to a narrow range of analyte properties, UHVCE is well suited for the analysis of complex samples containing analytes with a broad spectrum of properties (Figure 24.1). These samples include, among others, cell lysates, protein digests, or DNA sequencing preparations. The UHVCE analysis of samples should ideally start with a fairly concentrated sample containing low amounts of salt. High sample ionic strength can result in electrodispersion of the sample when used with low conductivity buffers. Dilute samples can be problematic if poor detection limits require excessively large sample injections.[19] This may or may not be solved using sample stacking techniques.

The best solution found thus far for the preventing peptide and protein samples from adsorbing to the capillary wall in UHVCE is electrostatic repulsion. This can be accomplished in bare silica capillaries by using very basic pH buffers where the deprotonated silanol groups and the majority of peptides and proteins are negatively charged. This method is not directly compatible with positive mode ESI-MS, but may work for UV absorbance detection or LIF.[17]

Various coatings that impart a positive charge to the surface can be used for low pH separations. Covalently bound coatings such as aminopropyltrimethoxysilane have been used with good success for UV detection and ESI-MS of peptides.[20,21] Other coating schemes based on the same principles have been used to varying degrees of success, and diisopropylaminopropylethyoxysilane has shown remarkable stability to degradation from hydrolysis.[6,22] Cationic polymers such as PolyE-323[23,24] have been used to great effect for CE/ESI-MS for proteins and peptides. These coating are easily applied to long capillaries and stable in buffers with organic modifiers.

The analysis or sequencing of DNA in CE and its advantages over standard slab gels has been well documented.[25–40] Preliminary study of UHVGCE for DNA separations indicates that it may be best suited for high-resolution separations of relatively short sequences of DNA.[6]

Polysaccharides and other complex carbohydrates can be difficult to analyze for several reasons. The lack of charged moieties requires the use of a sieving gel or pseudo-stationary phase. Sieving gels can be quite useful for highly branched structures found in some of these analytes. UHVCGE or UHVMEKC has been shown to be useful for high-resolution analysis of these samples.[12,14,15]

24.3 FLOW COUNTERBALANCED CAPILLARY ELECTROPHORESIS

24.3.1 BACKGROUND

The effectiveness of capillary electrophoretic separations can be greatly affected by the magnitude and direction of the bulk fluid flow. As pointed out by Jorgenson and Lukacs,[4] analyte bands that migrate in the direction opposite to the EOF spend a greater amount of time in the electric field and thus are more affected than analytes that migrate with the EOF. Direct manipulation of the EOF or application of hydrodynamic (pressure driven) flow can be used to retard the migration of analytes to improve efficiency and resolution. Resolution improvement by the direct control of the EOF through the manipulation of radial electric fields has been shown at low pH and low ionic strength.[41] Cheng et al.[42] noted the potential of a counter flow to improve the resolution of analytes in a paper investigating the effect of a constant applied pressure from a sheath flow cuvette on separations of amino acids during CE separations. Pressure driven flows have also been used for the preconcentration of samples for capillary isotachophoresis with long preconcentration times in short capillaries.[43–49] Reduction of the EOF using surfactants and polymers to increase the viscosity of background electrolyte near the wall has also been demonstrated for resolution improvement.[50–55] Lucy and McDonald[51] used this technique to resolve the major isotopes of chlorine using CE. Several preparative scale purification techniques using a counter flow have been reported for enantiomers and other closely related species.[56–63] Another study used a counter flow to prevent buffer components that improve selectivity but increase the UV absorbance background signal from reaching the detector.[64]

A practical system for analytical scale high-resolution separations was first demonstrated by Culbertson and Jorgenson.[65] Narrow diameter capillaries and LIF detection were used to resolve analytes with extremely small differences in mobility such as peptide isomers. Misleveling of the buffer reservoirs, using a siphoning action to provide a counter flow in large diameter capillaries, was also reported with modest resolution improvements.[66]

Band broadening caused by the counter flow limited high-resolution FCCE to narrow bore capillaries and LIF detection for over a decade. Recently, the use of large diameter capillaries packed with bare silica particles to flatten the parabolic flow profile of the counter flow has been demonstrated.[67] Packed capillary FCCE should not be confused with capillary electrokinetic chromatography (CEC). Packing material used in FCCE serves no chromatographic purpose and does not act as a stationary phase. The purpose of the packing material is simply to prevent band broadening in large bore capillaries so that UV absorbance detection can be used to detect unlabeled analytes.

24.3.2 THEORETICAL ASPECTS

The power of FCCE lies in the ability to maintain analyte bands within a high electric field strength for a long period of time. A counter flow is applied that pushes the bulk solution in the direction opposite to the analyte's migration, extending the duration of the separation. For a given length (L) of capillary with applied voltage (V), the electric field strength (E) is simply

$$E = \frac{V}{L}. \tag{24.7}$$

The time (t_m) required for an analyte band to migrate a distance equal to the length of the capillary (L) can be determined from the following equation:

$$t_{\mathrm{m}} = \frac{L}{E(\mu_{\mathrm{e}} + \mu_{\mathrm{EOF}})}, \tag{24.8}$$

where μ_e is the analyte's electrophoretic mobility and μ_{EOF} is the mobility of the EOF. Suppose that a counter flow is applied opposite to the direction of the analyte's net migration so that the analyte band now requires fourfold the time to transverse the same length of capillary. The electric field strength, length of capillary, and mobilities of the analyte band and EOF remain unchanged, but now the effective distance (L_{eff}) traveled by the analyte band is equal to $4L$. Neglecting any band broadening, an equivalent separation without a counter flow would require a capillary four times as long. In order to obtain the same electric field strength, four times the voltage would have to be applied. This can be thought of as the effective voltage (V_{eff}) that can be determined from the known electric field strength:

$$E = \frac{V}{L} = \frac{V_{\mathrm{eff}}}{L_{\mathrm{eff}}}. \tag{24.9}$$

Since the resolution between two analyte bands scales with the square root of the applied voltage, fourfold the effective voltage would generate twofold the resolution. For example, consider two analyte bands that transverse a 30 cm capillary in an average time of 30 s with 30 kV applied. If the resolution between the bands is 0.5, then a capillary 120 cm long with 120 kV applied would be needed to obtain a resolution of 1.0. The same resolution can be obtained using the 30 cm capillary with 30 kV applied if a counter flow is applied so that the analytes transverse the same distance in 120 s.

EOF is generated at the wall of the capillary, resulting in a flat flow profile that limits band broadening to that caused by longitudinal diffusion. This broadening can be expressed as a variance (σ_{B}^2) using Einstein's equation:

$$\sigma_{\mathrm{B}}^2 = 2Dt, \tag{24.10}$$

where D is the diffusion coefficient of the analyte and t is time. The application pressure results in a hydrodynamic flow. Flow velocity is greatest in the center of the capillary due to drag at the capillary wall and a parabolic flow profile is the result. Analyte bands diffuse laterally across this flow profile and broaden to a degree proportional to its magnitude (Taylor dispersion). Culbertson and Jorgenson [65] derived an equation for the degree of band broadening for FCCE in an open tube. In addition, the maximum allowable diameter of the capillary was determined from the following expression:

$$\sigma_{\mathrm{c}}^2 = \frac{d_{\mathrm{c}}^2 v_{\mathrm{pa}}^2 t}{96D} \leq 2Dt = \sigma_{\mathrm{B}}^2, \tag{24.11}$$

where the broadening of the analyte bands from hydrodynamic flow is expressed as a variance, σ_{C}^2, which should be less than or equal to the band broadening caused by longitudinal diffusion alone. The diameter of the open capillary (d_{C}) can be calculated from the average velocity of the hydrodynamic flow (v_{pa}) required to counterbalance the electrokinetic migration of the analyte band and the diffusion coefficient of the analytes (D). For the analytes investigated, open tubular capillaries 5–10 μm were deemed acceptable.[65,68] Narrow bore capillaries have a limited optical pathlength and LIF detection is usually required for sensitive detection.

Band broadening from hydrodynamic flow in packed capillaries is independent of capillary diameter, allowing the use of capillaries with a pathlength large enough for UV absorption detection.

A similar equation for band broadening in packed capillary FCCE can be derived, as can a similar method of calculating the required particle diameter (d_p) to minimize band broadening.[6,67] Briefly, the total broadening is expressed as a variance (σ^2_{total}), which is the sum of the contributions of longitudinal diffusion and hydrodynamic flow:

$$\sigma^2_{total} = \sigma^2_B + \sigma^2_C = 2Dt + \frac{d^2_p v^2_{pa} t_b}{10D}, \tag{24.12}$$

where t_b is the time during which the counter flow is applied. Using the same strategy as expression 11, the maximum particle diameter can be calculated. The efficiency for FCCE in packed capillaries can be found using the following equation:

$$N = \frac{10D(tv_{ek})^2}{20D^2t + d^2_p v^2_{pa} t_b}, \tag{24.13}$$

where v_{ek} is the average electrokinetic velocity for the analyte bands and N is the efficiency in theoretical plates.[6,67] High efficiencies can be most easily obtained by using small diameter particles and long separation times.

24.3.3 Practical Implementation

24.3.3.1 Instrumentation

The instrumentation required to perform FCCE is relatively simple. The power supply can be any commercially available HVDC power source, but an ideal power supply would have dual polarity, 0–30 kV at several 100 μA, with safety interlocks and computer control capabilities. The maximum voltage needed is determined by the desired electric field strength and length of the capillary. In order to allow for some margin of error in control of the counter flow, the minimum length of capillary should be at least three to four times the physical width of the desired separation window. These margins allow for the analyte bands to be maintained within the inner half of the capillary so that they are not pushed out on either side of the capillary, with detection occurring in the center of the capillary.

The source of the counter flow, or pumping system, is determined by the velocity of the analyte band and the backpressure required to reverse or stop its forward migration. Where the analyte bands migrate slowly in an open tube, the pumping system may be as simple as a misleveling of the buffer vials. Rapidly migrating analyte bands in narrow open capillaries may require a compressed gas driven system such as that shown in Figure 24.10.[65,68] Packed capillaries require much greater pressures and so an high-performance liquid chromatography (HPLC) pump or syringe pump capable of generating the desired fluid velocity must be used (Figure 24.11).[6,67] If an HPLC pump is used, some consideration should be given regarding the volumetric flow rate needed to maintain a stable pressure. Typically, flow rates for HPLC columns range in the mL or μL per minute range. The flow rate for FCCE will be much smaller, typically nL/min. The use of a "splitter" capillary to divert the majority of the flow volume back to the pump's buffer reservoir can save many liters of running buffer. This may be especially important for buffers containing expensive reagents. Syringe pumps have limited capacity and the flow rate will therefore determine the maximum separation time. One notable exception to this is the Haskel International Inc. (Burbank, CA) DSHF-300 Air Driven Fluid Pump. This is a syringe-style hydraulic pump capable of producing up to 50,000 psi of liquid pressure via a 300-fold amplification of applied gas pressure. The advantages of this pump include its stable applied pressure over a volumetric flow rate ranging from no flow at all to a few mL per minute. It has a low stoke volume of a few mL, but it automatically recycles and refills itself at the end of a piston stoke, maintaining a fairly constant pressure with built in check valves.

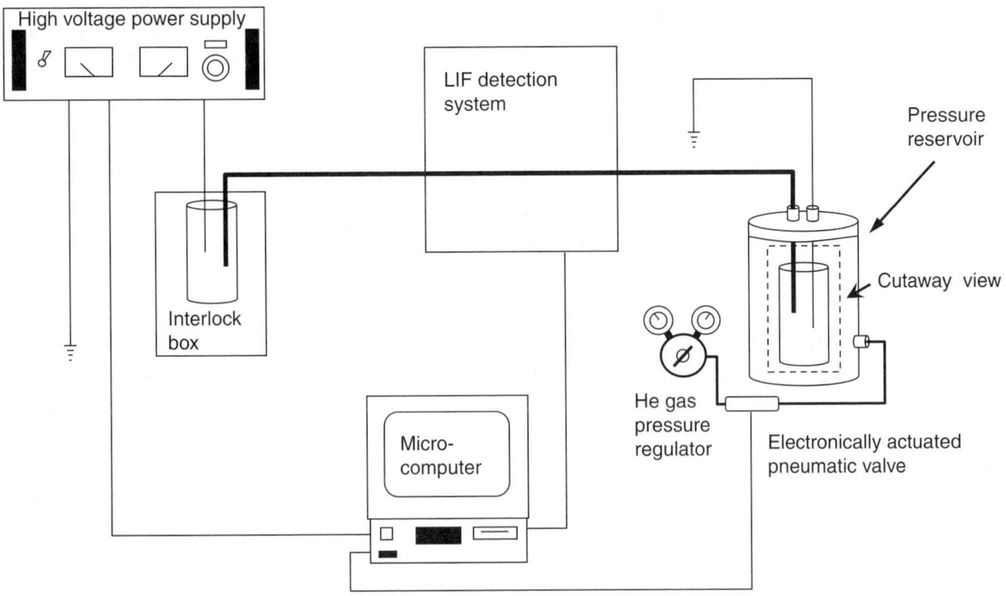

FIGURE 24.10 Open tube FCCE using a gas pressure driven pumping system. (Reprinted from Culbertson, C. T. and Jorgenson, J. W., *Analytical Chemistry* 1994, 66, 955–962. Copyright 1994 American Chemical Society. With permission.)

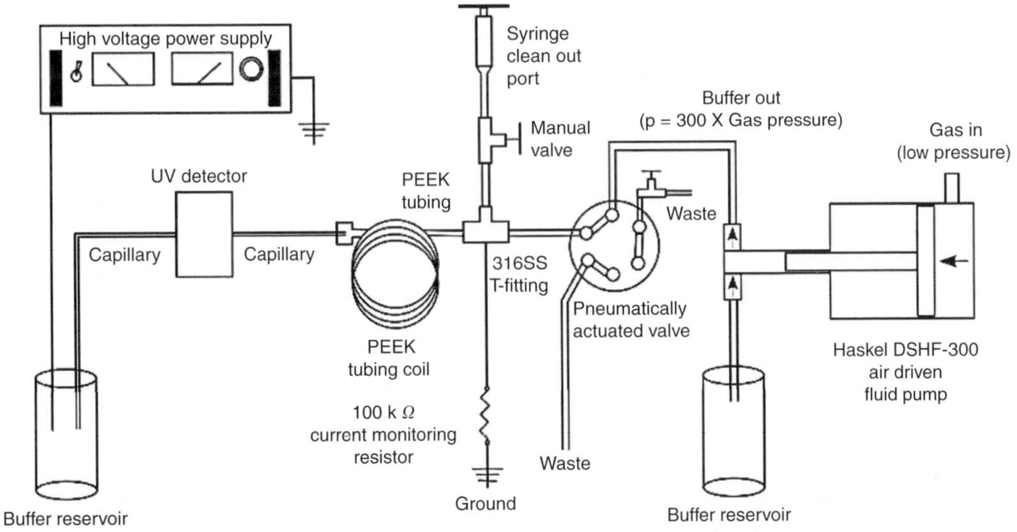

FIGURE 24.11 Packed capillary FCCE using high pressure from a hydraulic pump controlled by a pneumatically actuated valve. (Reprinted from Henley, W. H., et al., *Analytical Chemistry* 2005, 77, 7024–7031. Copyright 2005 American Chemical Society. With permission.)

Ideally, the velocity of the counter flow should be exactly equal and opposite to the velocity of the analyte band so that band broadening from pressure driven flow is minimized. The practical implementation of such a system requires a detector capable of imaging the entire width of the separation window such as a linear diode array or charge-coupled device (CCD) camera.

Single point detectors using UV absorbance or LIF detection require that the analyte bands pass though the detection point in order for data to be collected. Therefore, analytes are first allowed to migrate past the detection point electrokinetically. Then a pressure great enough to push them backwards past the detector is applied via a computer-controlled valve. Once the analytes have been sufficiently pushed back, the pressure is released and the cycle is repeated. The electrophoretic voltage is maintained during the entire course of the separation, so the analytes are continuously separating electrokinetically, even while being pushed backwards by the counter flow. An alternative scheme using constant pressure with intermittently applied voltage would only resolve the analytes while the voltage was applied.

Valves for controlling the back pressure must actuate quickly, be rated to the desired pressure, and ideally be computer controlled. As long as the pressure remains constant, dead volume is usually not a major concern. If the valve is constructed from metal, it must be electronically isolated from the electrophoretic system or else electrolysis processes may corrode and destroy the valve, in addition to ruining the separation. Several suitable valves are manufactured by Valco Instrument Co. Inc. (VICI) (Houston, TX).

When making the connection between the capillary and pumping system, it is very important to consider the flow of electrophoretic current and the effect it may have on components of the system. Figure 24.12 shows two ways of connecting a packed separation capillary to the high-pressure pumping system. A stainless steel Swagelok® compression fitting forms a union between the fused-silica capillary and a piece of PEEK tubing. The configuration in Figure 24.12a results in electrolysis products entering the capillary and spoiling the separation. Even though the metal fitting is electrically floating, it is not isolated from the electrophoretic current. Current flows as ions through the capillary and through the PEEK tubing, but the highly conductive metal fitting offers a low resistance path for electrons to travel. The result is that electrolytic processes occur at the surface of the metal near the exit of the capillary and the entrance to the PEEK tubing. An easy way to avoid this is to simply push the capillary into the PEEK tubing so that the metal fitting is truly electrically isolated from the electrophoretic current (Figure 24.12b).

24.3.3.2 Capillary Packing

Packing the capillary with nonporous bare silica can greatly reduce the amount of band broadening observed at similar counter flow linear velocities. Figure 24.13 shows the temporal variance of an

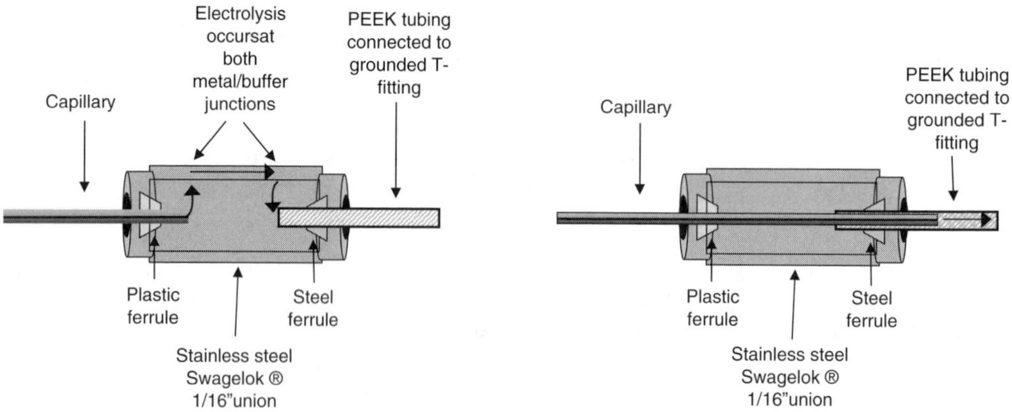

FIGURE 24.12 In order to prevent electrolysis products from entering the capillary at the connection of the capillary to high-pressure counter flow system, metal surfaces must be floating and isolated from the path of the electrophoretic current. (Reprinted from Henley, W. H., PhD Dissertation, University of North Carolina, Chapel Hill, 2005. Copyright 2005. With permission.)

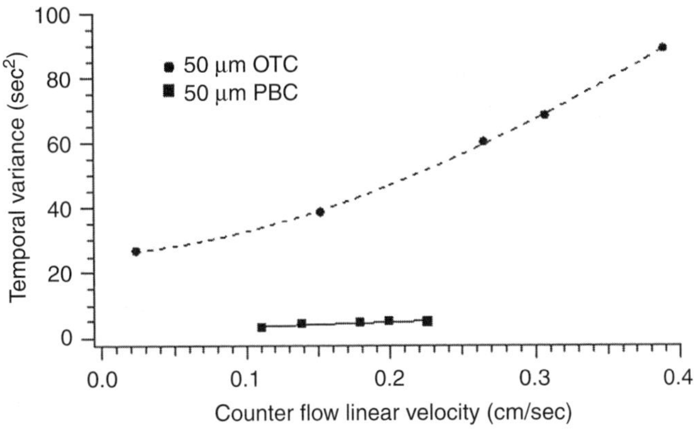

FIGURE 24.13 A comparison between the band broadening seen in a pack capillary and open tubular capillary under similar counter flow linear velocities shows much less broadening in the packed capillary. (Reprinted from Henley, W. H., et al., *Analytical Chemistry* 2005, 77, 7024–7031. Copyright 2005 American Chemical Society. With permission.)

R-(−)-epinephrine peak using a 50 μm open tubular capillary (OTC) and a 75 μm capillary packed with 4.1 μm nonporous bare silica particles. The diameters of the capillaries were chosen so that they would have approximately the same dead volume and electrophoretic current. It is easy to see from the plot of temporal variance versus counter flow linear velocity that much less broadening is seen in the packed capillary as compared with the large diameter open tube.

The preparation of the packed capillary for FCCE using UV absorbance detection is not difficult, but it differs in several aspects from the methods used to prepare capillary HPLC or UHPLC columns.[69] For chromatography, capillaries are typically run with mobile phases flowing in the same direction as the capillary was packed. For FCCE, the alternative application and release of pressure flowing opposite to the flow of the EOF causes large voids to develop within the packed bed. Consolidation of the capillary can be achieved via sonication under hydrodynamic flow.[70,71] Briefly, dry silica particles are tapped into one end of an OTC. Heat from an electric arc is used to sinter them together and to the capillary wall. Pressure (up to 50,000 psi) from packing bomb is used to force a ~10 mg/mL slurry of packing material into the open end of the capillary. After the bed is packed, an inlet frit is sintered at the head of the bed using a heated tungsten wire. The pressure is released, and the capillary is cut to length. The capillary, except for the fritted ends, is placed into a water bath and sonicated for ~5 min. Pressure is then applied to the capillary in the direction opposite to that in which it was originally packed. A ~5–10% reduction in the length of the packed bed should be observed. A new frit is then sintered using a heated tungsten wire while the capillary is still under pressure. Sonication can be repeated with pressure applied from the opposite end if further consolidation is desired. Figure 24.14 shows how the consolidation of the packing material reduces band broadening caused by the counter flow. The same capillary was used for measurements of an *R*-(–)-epinephrine peak using 3.1 μm nonporous bare silica packing.[6,67]

24.3.3.3 Modes of Operation and Examples

Open tube FCCE can be performed in capillaries of relatively large diameter if the analyte bands migrate slowly and only a modest improvement in resolution is desired. Misleveling of buffer vials can provide a simple way of improving the resolution for analyte bands that are almost, but not quite resolved under standard conditions.[66] Band broadening will occur but usually not to a great extent over short durations.

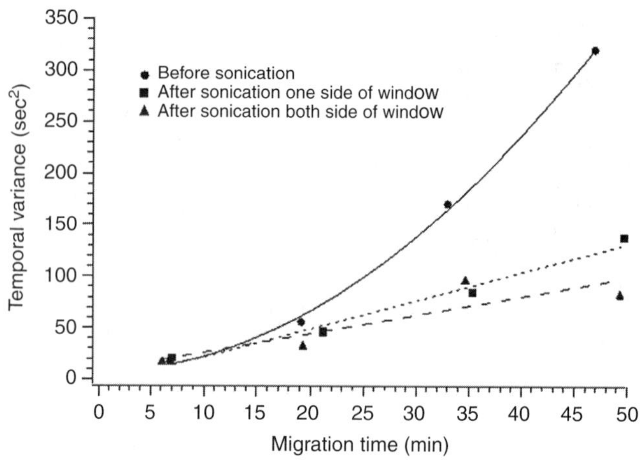

FIGURE 24.14 Reducing the interparticle porosity of the capillary packing can decrease the amount of band broadening caused by the counter flow. Data taken from migration of *R*-(−)-epinephrine using 3.1 μm nonporous bare silica packing. (Reprinted from Henley, W. H., et al., *Analytical Chemistry* 2005, 77, 7024–7031. Copyright 2005 American Chemical Society. With permission.)

Open tube FCCE in narrow capillaries using zone electrophoresis is useful for resolving analytes that can be tagged with fluorescent labels for LIF detection. Culbertson[72] demonstrated resolution of the peptide fragment YAGAVVNDL and YAGAVVNDI (Figure 24.15), which only have an electrokinetic mobility difference of 1.0033.

Open tube FCCE using micellular electrokinetic chromatography (MEKC) can be used to obtain high resolution for charged or neutral analytes. High-resolution separations of TRITC-labeled phenylalanine with different numbers of hydrogen atoms replaced with deuterium atom showed resolution between analytes with differences of a single neutron (0.16% mass difference). The relative levels of deuterium substitution could also be determined from the peak areas. The ability of FCCE with MEKC to separate these isotopomers is believed to be based on slight differences in hydrophobicity caused by the influence of the heavier deuterium atom.[68]

Packed capillary FCCE can be used with larger diameter capillaries with sufficient pathlength for sensitive UV/Vis absorption detection. UV detection allows analysis of nonfluorescent analytes and sample impurities. Most organic and some inorganic compounds absorb UV or visible light to some degree, and therefore packed capillary FCCE can be used for a much larger range of analytes than open tubular FCCE.

Separations of racemic samples of several basic and acidic pharmaceuticals using packed capillary FCCE have been reported previously.[6,67] Figure 24.16 shows the separation of the enantiomers of fenoprofen using β-cyclodextrin as the chiral pseudo-stationary phase. The top portion of the figure (a) shows the entire separation. The analyte peak is allowed to electrokinetically migrate pass the detector (forward pass) and then pushed backwards under hydrodynamic flow (reverse pass). Seven cycles with forward passes (F1–F7) and six reverse passes (R1–R6) can be seen in Figure 24.16a. Figure 24.16b shows the first electrokinetic pass in front of the detector. This part of the electropherogram represents the resolution (0.34) and efficiency (21,000 plates) that would result from traditional CE under these separation conditions. Figure 24.16c shows the final pass in front of the detector, where the analyte bands have broadened by a small degree, but are approaching baseline resolution with greatly increased separation efficiency (320,000 plates). Table 24.1 gives a list of acidic and basic pharmaceutical compounds separated with packed capillary FCCE. Electrokinetic only migration times, total number of passes, along with resolution and efficiency improvements are given.

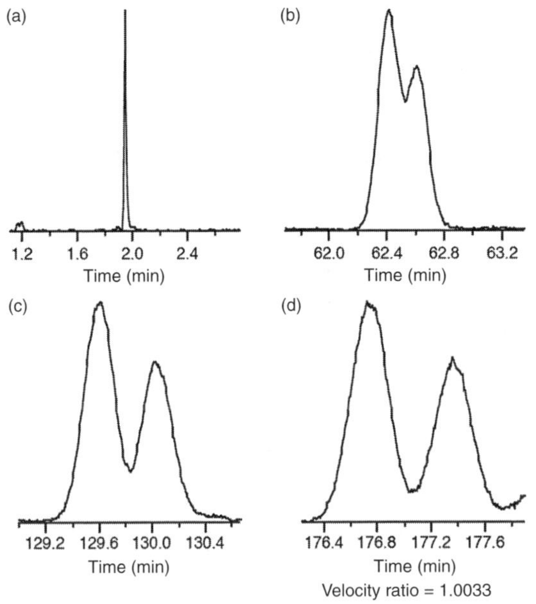

FIGURE 24.15 Open tube FCCE with LIF detection showing the resolution of YAGAVVNDL and YAGAVVNDI. (Reprinted from Culbertson, C.T., p. 264, 1996, University of North Carolina, Chapel Hill, NC. Copyright 1996. With permission.)

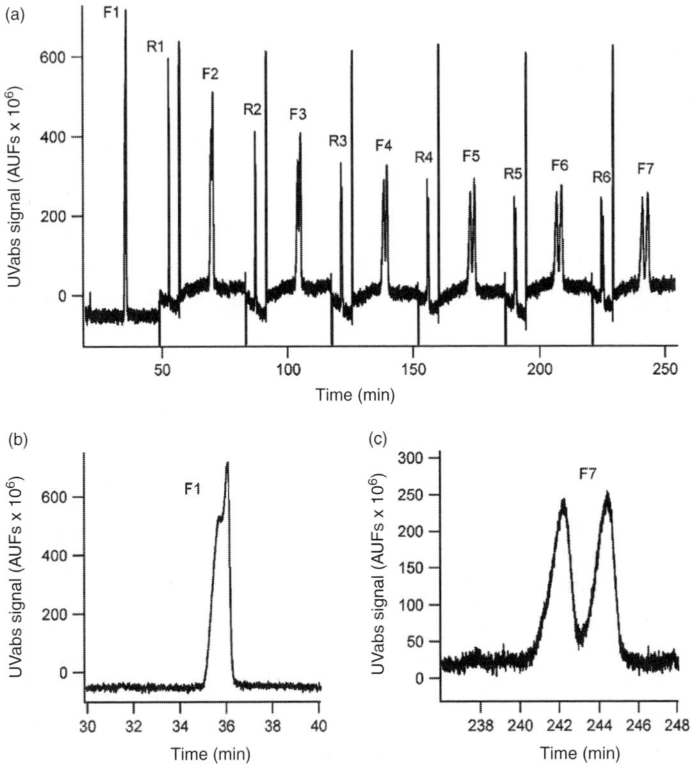

FIGURE 24.16 Packed capillary FCCE electropherogram showing the separation of fenoprofen enantiomers using β-cyclodextrin as a chiral selector. (Reprinted from Henley, W. H., et al., *Analytical Chemistry* 2005, 77, 7024–7031. Copyright 2005 American Chemical Society. With permission.)

TABLE 24.1
Racemic Mixtures Resolved Using FCCE in Packed Capillaries Using β-Cyclodextrin

Compound	Migration Time (min) for 1st Forward Pass	Total Run Time (min)	Total No. of Forward Passes	Degree of Resolution at 1st Pass/at Last Pass	Efficiency of 1st Pass/Last Pass (103 Plates)
Epinephrine[a]	40.9	329	8	0.42/1.6	31/213
Norepinephrine[a]	66.7	275	7	0.36/1.02	14/77
Synephrine[a]	117	138	2	0.988/1.02	15/25
Norphenylephrine HCl[a]	62.1	147	4	1.13/1.76	35/68
Phenylpropanolamine[a]	71	195	5	0.21/0.77	9/41
Octopamine[a]	61.07	403	10	ND/0.55	8/45
Chlorpheniramine maleate[a]	32.8	118	3	2.12/3.35	90/240
Doxylamine[a]	23.5	202	6	1.59/3.38	120/820
Ibuprofen[b]	34.1	259	9	0.496/1.08	44/550
Ketoprofen[b]	39.8	688	22	ND/0.88	10/340
Fenoprofen[b]	35.7	242	7	0.34/0.88	21/380

ND = none detected.

[a]Basic chiral analytes were separated in 50 mM phosphoric acid, 10 mM β-cyclodextrin 0.5% triethylamine.
[b]Acidic compounds were analyzed using a 100 mM MES, 10 mM β-cyclodextrin buffer system.

Source: Reprinted from Henley, W. H., et al., *Analytical Chemistry* 2005, 77, 7024–7031. Copyright 2005 American Chemical Society. With permission.

Several isotopomer separations have been reported for packed capillary FCCE. Deuterium-substituted phenylalanine showing no resolution during the first forward pass was almost baseline resolved after an 11 h separation time.[67] Figure 24.17 shows a packed capillary FCCE separation of bromine-79 and bromine-81 using UV absorbance detection. After 6 h of run time (Figure 24.17b), a distinct notch is seen in the bromide ion peak. The separation was maintained for over 60 forward passes requiring over 1100 min, and clear separation of the isotopes can easily be seen.[6,67]

Before attempting MEKC in packed capillaries, an examination of the properties of the micelles under high pressures should be considered. Unpublished work suggests that at the high pressures required for reversing the analyte bands, the properties of the micelles may change, resulting in an unstable system.

24.3.4 Methods Development Guidelines

24.3.4.1 Handling Extended Separation Times

The high resolution and efficiency obtained in FCCE separations comes at the expense of extended separation time. Although the buffer within the capillary is replenished by the counter flow,[60] care should be paid to ensure sufficient buffer capacity for reservoirs with fixed volume. Measurements of the electrophoretic current can be used to calculate the expected pH change in each buffer reservoir for a given separation time. In addition, steps should be taken to prevent electrolysis products from entering the capillary. Figure 24.11 shows a long coil of PEEK tubing connected between the capillary and the grounded electrode. This coil has sufficient length and volume to prevent gas bubbles or UV absorbing electrolysis products from entering the separation capillary for many hours under typical counter flow rates.

Automation of the valve control using a computer program is not only useful for ensuring that the analyte band stays within the capillary but also lets the operator leave the instrument running

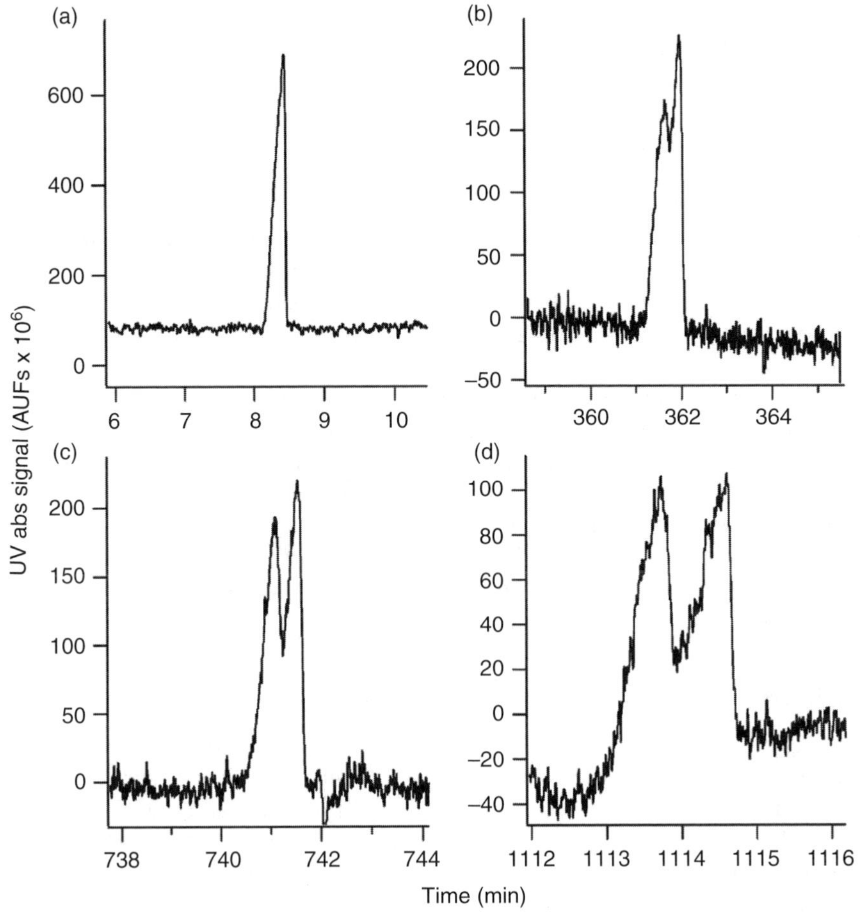

FIGURE 24.17 A separation of bromine isotopes using packed capillary FCCE. (Reprinted from Henley, W. H., et al., *Analytical Chemistry* 2005, 77, 7024–7031. Copyright 2005 American Chemical Society. With permission.)

autonomously for hours at a time. There are many different ways to automate the control of the system, but a few basic principles should be followed for reliable operation. After injection of the sample, the electrokinetic migration velocity of the analyte band can be determined by the time it takes to travel the distance to the detector. After allowing the band to travel some distance beyond the detector, the counter flow can be applied (while maintaining the applied voltage). The time required for the analyte band to be pushed back in front of the detector can be used to calculate the analyte band velocity under hydrodynamic flow. These two parameters can be used to calculate the position of the analytes at any given point in the FCCE separation cycle. The progress of the separation should be checked every few hours to correct any timing errors. Adjustments will have to be made for long separations due to variations in pump pressure, EOF velocity, and buffer viscosity caused by temperature changes or buffer depletion.

24.3.4.2 Band Broadening Prevention

One of the main advantages of FCCE is that published buffer conditions that may not completely resolve certain analytes in a traditional CE system can be used to fully resolve the analytes by giving them more time to separate. This can save method development time when analyzing a variety of compounds with similar characteristics. The main limitation on the timescale of an FCCE experiment is the degree to which band broadening reduces the analyte signal strength, and therefore the counter flow velocity should be kept as low as possible.

Adsorption of the analyte to the capillary wall or bare silica packing material must also be prevented to avoid signal loss. Electrostatic repulsion works well for negatively charged species in basic buffer with bare capillaries. Amine or other basic analytes can be analyzed at low pH in bare capillaries if a competitive inhibitor such as triethylamine is added to the running buffer. The use of dynamic coatings such as polymers or surfactants in the running buffer to reduce adsorption of analytes is probably best suited to open tubular FCCE as increased viscosity may result in high back pressures in packed capillaries. Permanent, covalently bound coating should be used only after their long-term stability has been confirmed, and analyte band migration time should be closely monitored.

Electrodispersion can be a major source of band broadening in FCCE due to the long separation times. Close matching of the electrophoretic mobilities of the analyte bands and running buffer can reduce electrodispersion greatly. Gaussian-shaped peaks usually indicate a compatible sample/buffer whereas fronted or tailed peaks indicate potential problems. Furthermore, the ionic strength of the running buffer should be closely matched to that of the sample.

In addition, in packed capillary FCCE, care must be taken to ensure that the packing material is fully consolidated. Voids in the packed bed will dramatically increase the observed broadening as seen in Figure 24.14.

24.3.4.3 Determining the Best Separation Strategy

The first thing to consider when designing an FCCE separation strategy for a particular sample is the desired method of detection. Both open tubular FCCE and packed capillary FCCE have their advantages and drawbacks. Open tube FCCE uses a slightly less complicated instrumental setup (Figure 24.10) than packed capillary FCCE (Figure 24.11). While not difficult, the equipment and skills needed to pack capillaries with bare silica particles can represent a considerable investment. The high sensitivity of LIF detection makes open tube FCCE the method of choice for analytes that can be easily labeled with fluorescent tags such as peptides and protein fragments.

FCCE in packed capillaries can be used for many more types of analytes. Direct or indirect UV/Vis absorbance detection can be used to detect everything from monoatomic ions like Br^- to basic and acidic pharmaceuticals.

Pseudo-stationary phases can be used to improve the resolving power of either type of FCCE. MEKC works well in narrow diameter open tubes for separating neutral analytes based on their hydrophobicity differences. Chiral selectors such as cyclodextrins can be used with either method. Other methods of detection such as noncontact conductivity and modes such as CEC have yet to be explored but may prove useful in the future.

24.4 SYNCHRONOUS CYCLIC CAPILLARY ELECTROPHORESIS

24.4.1 BACKGROUND

Synchronous cyclic CE, as the name implies, uses a series of electrodes placed at regular intervals around a closed loop to "chase" analyte bands around a continuous separation channel. Voltage switching is synchronized to maintain analytes of a particular mobility within the separation channel for a long period of time.[73] A similar concept had been applied in chromatography using two chromatographic columns and timed valve changes to continuously separate analyte bands.[74] The problems connecting separation channels on the extremely small scale of capillary electrophoretic systems prohibited the implementation of SCCE until several years later.[75]

24.4.1.1 Microchip SCCE

Burggraf et al.[75–77] reported the first practical SCCE system using microfluidic technology. Photolithographic techniques can be used to cheaply manufacture microfluidic structures with channels one to several hundred microns in dimension. The interconnection of these channels can be used to create networks with negligible dead volumes. The first pattern used for SCCE was four overlapping, 2-cm channels arranged in a square. An additional channel intersecting the middle of a main channel was used to perform electrokinetic injections. Several other designs have since been reported by Manz's group.[78,79] Figure 24.18 shows the layout for an SCCE that was used to separate fluorescently label amino acids and human urine.[79]

24.4.1.2 Capillary SCCE

The use of polyimide coated fused-silica capillaries for SCCE with low dead volume connections was first reported by Zhao et al. in 1999.[80,81] Fused-silica capillaries offered several advantages over the microchip-base SCCE systems. The capillaries provided longer separation distances for the analyte bands. While the geometry of the connections on a microfluidic device cannot be easily changed during an experiment, the use of adjustable gaps between the capillaries provides very low dead volume. In addition, conventional UV-Vis absorption detection was feasible for SCCE for the first time using capillaries 50 μm in diameter. A basic schematic representation of the instrumental setup is shown in Figure 24.19.[80]

Another device has been reported, the "electrophoretron" that used two capillaries with different polarity surface treatments. The capillaries were connected in a loop and a single power supply was used to apply a fixed potential across the two capillaries. The different polarity surface treatments cause the EOF to flow in a continuous loop around the capillaries. Injected analytes would continuously travel around the loop under the influence of the EOF.[82] Such a device may prove useful for CEC separations of neutral analytes. However, any separation of charged species in one capillary will be almost negated in the second capillary due to the change in the polarity of the electric field.[73]

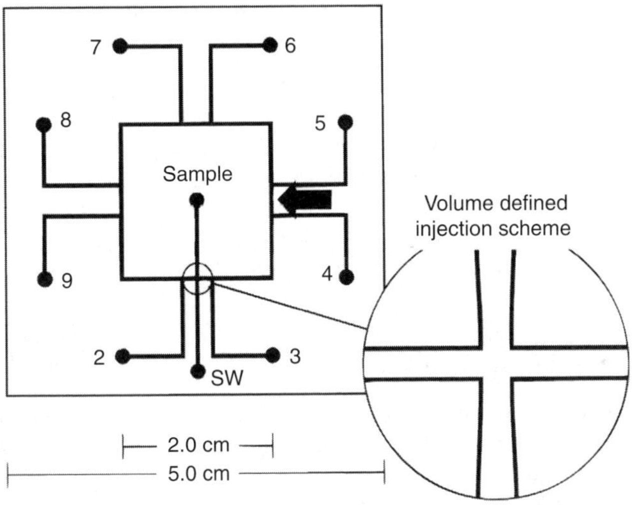

FIGURE 24.18 Microchip-base SCCE system used with MEKC to separate FITC-labeled amino acids and human urine components. (Reprinted from von Heeren, F., et al., *Analytical Chemistry* 1996, 68, 2044–2053. Copyright 1996 American Chemical Society. With permission.)

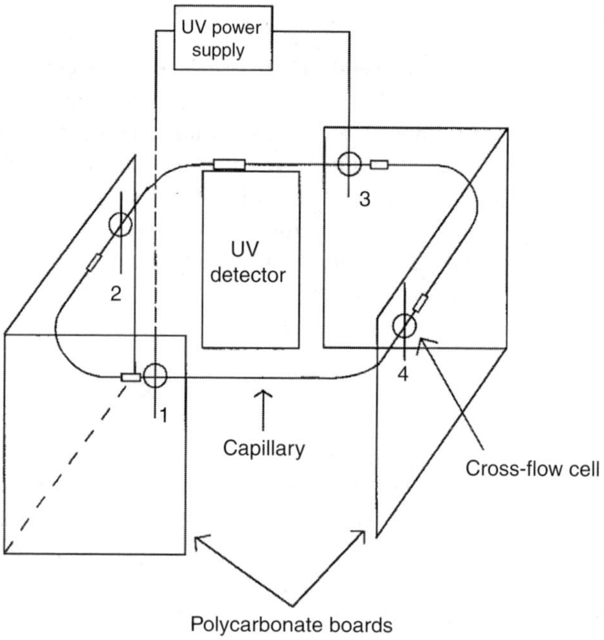

FIGURE 24.19 Schematic representation of capillary-based SCCE system used to resolve isotopomers and enantiomers. (Reprinted from Zhao, J., et al., *Journal of Microcolumn Separations* 1999, 11, 431–437. Copyright 1999 John Wiley & Sons, Inc. With permission.)

24.4.2 Theoretical Aspects

24.4.2.1 Absolute versus "Effective" Voltage

The electric field strength is determined by the applied potential difference divided by the distance over which it is applied. SCCE is similar to FCCE in that the effective separation voltage can be much higher than the actual, physically applied voltages. For a single cycle, the effective voltage is twice that of the applied voltage. For an analyte band transversing n cycles, the effective voltage (V_{eff}) is

$$V_{\text{eff}} = 2nV_{\text{applied}}, \tag{24.14}$$

where V_{applied} is the applied voltage. Using Equation 24.2, an equation for the expected resolution between two analyte bands for a given applied voltage can be derived:

$$\text{Rs} = 0.177(\mu_1 - \mu_2)\left(\frac{2nV_{\text{applied}}}{\overline{D}((\overline{\mu}_{\text{e}} + \mu_{\text{EOF}}}\right)^{1/2}. \tag{24.15}$$

24.4.2.2 Band Broadening Mechanisms

There are several sources of band broadening in SCCE to consider. Longitudinal diffusion occurs in all CE separations. As seen in Equation 24.10, it is mainly governed by separation time and the analyte's diffusion coefficient.

Turns in the separation channel create another source of band broadening.[83–87] As the analyte bands migrate around a turn, the portion of the band on the inside of the turn travels a shorter distance than the portion of the band on the outside of the turn. In addition, the electric field is concentrated near the wall of the inside of the turn, increasing the rate of band broadening. Taylor dispersion then smoothes out the distorted band shape. The dispersion created by a turn in the separation channel can be described as a variance

$$\sigma^2_{\text{turn}} = \frac{(\Delta l)^2}{X} = \frac{(2\theta w\,(1 - \exp(-t_{\text{D}}/t_{\text{t}})))^2}{X}. \qquad (24.16)$$

where Δl is the difference in length traveled around the inside and outside of the turn, θ is the angle subtended in radians, w is the width at the top of the separation channel, t_D is transverse diffusion equilibrium time, t_t is the turn transit time, and X is a constant $\cong$12 when t_D is large.[83] The smaller radius of curvature found in microchip-based channels exacerbate this phenomenon, which is typically not a concern in capillary-based SCCE.

Broadening at the connections (corners on microchip-based SCCE) between the separation channels is another important source of band broadening and sample loss in SCCE. The total broadening can be expressed as a variance, σ^2_{total}

$$\sigma^2_{\text{total}} = \sum_{i=\text{sides}} \sigma_i^2 + \sum_{j=\text{corners}} \sigma_j^2 = \sigma^2_{\text{migr.dist}} + n\sigma^2_{\text{corner}}, \qquad (24.17)$$

where n is the number of corners the band travels through, $\sigma^2_{\text{migr.dist}}$ is the variance from the total migration distance, and σ^2_{corner} is the broadening caused each time a band travels through a corner or connection of channels.[75]

24.4.3 Practical Implementation

24.4.3.1 Power Supply and Voltage Switching

Power supplies for microchip-based SCCE and fused-silica capillary-based SCCE mainly differ in the number of power supplies and in the magnitude of the applied voltage. The small size found in most microchip-based SCCE systems limits the separation channel length and so relatively small potential differences are needed to generate high electric field strengths (~2 kV/cm). In contrast, capillary-based SCCE use approximately 50 cm lengths of capillary, and so power supplies generating up to 30 kV and a few hundred microamperes are typically used. The configuration of the separation channels determines the number of HV electrodes needed. The actual number of HV supplies needed can be reduced if relays are used to switch the potential between different electrodes. The fused-silica capillary-based SCCE instrumentation reported used a single 30 kV HV supply with four HVDC relays.[80] The first chip-based SCCE system used four HV power supplies at 2.5 kV and eight HV relays to control the voltage switching.[75,76] Later work with chip-based SCCE used six 10 kV HV supplies and thirteen HV relays.[79] Different geometries with shorter capillary channels were also studied. A triangular arrangement resulted in the fewest HV supplies needed.[78]

24.4.3.2 Connection of Fused-Silica Capillaries

While the connection of the separation channels used in microchip-based SCCE can be manufactured with almost no dead volume using photolithographic techniques, connection of fused-silica capillaries is challenging. Zhao et al.[80,81] report the use of a controllable gap using a cross flow connection and a tight fitting Teflon sleeve (Figure 24.20). The gap between the capillaries offers several advantages over other methods. In order to allow the flow of electrophoretic current, the gap is opened, which also allows fresh buffer to flow into the capillary network and old buffer to be removed. While the analyte bands migrate through the gap, the gap is tightly closed, reducing the dead volume dramatically. This system can be implemented with solenoid, piezoelectric, hydraulic, or even manual actuation. Careful preparation of the capillaries is required to ensure a tight, low loss

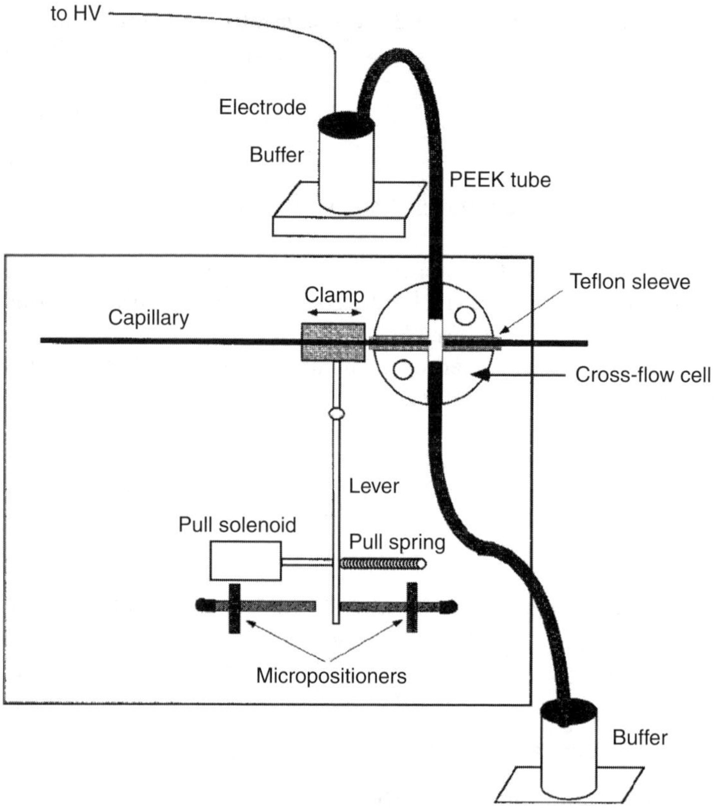

FIGURE 24.20 Schematic diagram detailing the connection of the capillaries and the controlled gap for capillary-based SCCE. (Reprinted from Zhao, J., et al., *Journal of Microcolumn Separations* 1999, 11, 431–437. Copyright 1999 John Wiley & Sons, Inc. With permission.)

fit. Capillaries were cut to precisely equal lengths so that *E* fields and flow velocities would be equal in each length of the four capillaries. The ends of the capillaries were polished flat using a jeweler's lathe and fine grit sandpaper to ensure a perfectly flat mating surface.[80]

24.4.3.3 Modes of Operation and Examples

Capillary zone electrophoresis with microchip-based SCCE has been used to separate fluorescent dye from degradation products[88] and to separate fluorescein isothiocyanate (FITC) labeled amino acids.[78] Capillary zone electrophoresis (CZE) in fused-silica capillaries has been used to separate the racemic mixtures of (α-hydroxybenzyl)methyltrimethylammonium and (2-hydroxy-1-phenyl)ethyltrimethylammonium with β-cyclodextrin as the chiral pseudo-stationary phase.[81] L-Phenylalanine and L-phenylalanine-ring-D5 (the hydrogens in the aromatic ring were substituted with deuterium)[81] and another separation of the closely related amino acids phenylalanine and tyrosine have been resolved using CZE with capillary-based SCCE.[80]

MEKC has been implemented in both microchip-based SCCE separations and capillary-based SCCE separations. Figure 24.21a shows a microchip-based SCCE separation using MEKC of FITC-labeled amino acids.[79] The same chip was also used to resolve FITC-labeled components of human urine using MEKC in addition to performing an MEKC-based immunoassay of serum theophylline levels.[79] MEKC using capillary-based SCCE has resulted in separations with as many as 100 million theoretical plates in 15 h of run time. A separation of L-phenylalanine and L-phenylalanine-ring-D5 is

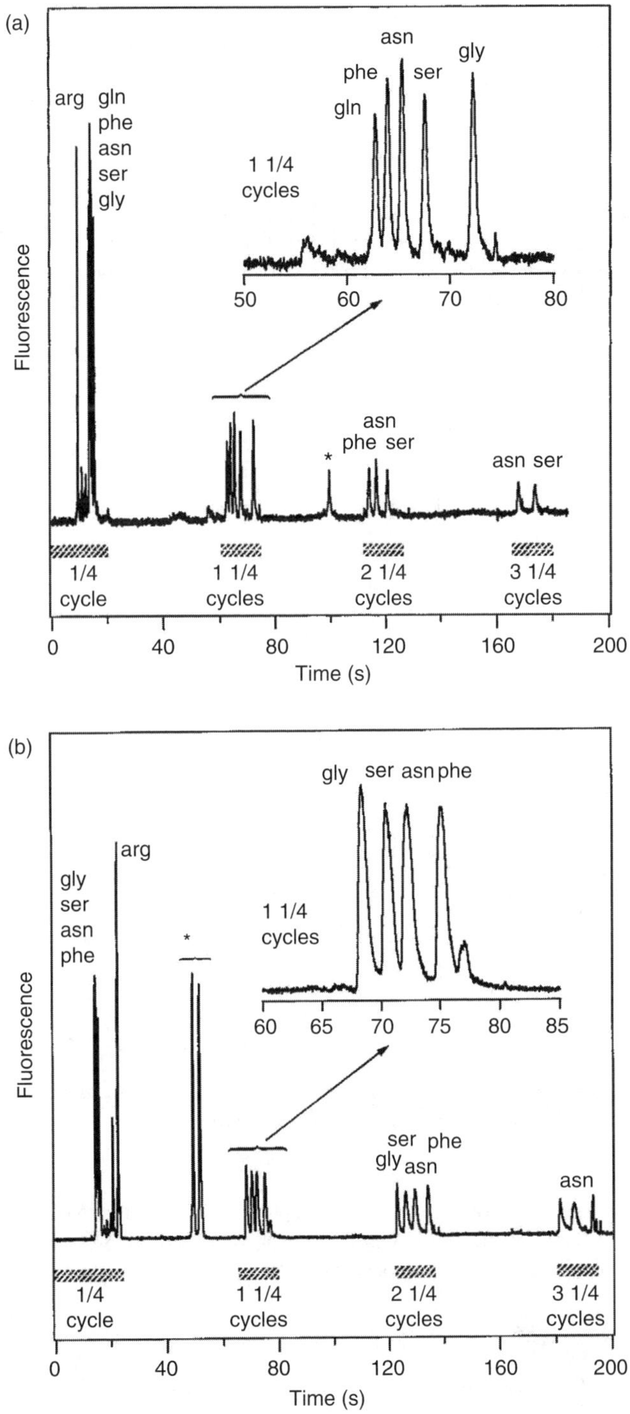

FIGURE 24.21 (a) MEKC separation of FITC-labeled amino acids using microchip-based SCCE. (b) CGE separation of FITC-labeled amino acids using the same microchip-based SCCE device. (Reprinted from von Heeren, F., et al., *Analytical Chemistry* 1996, 68, 2044–2053. Copyright 1996 American Chemical Society. With permission.)

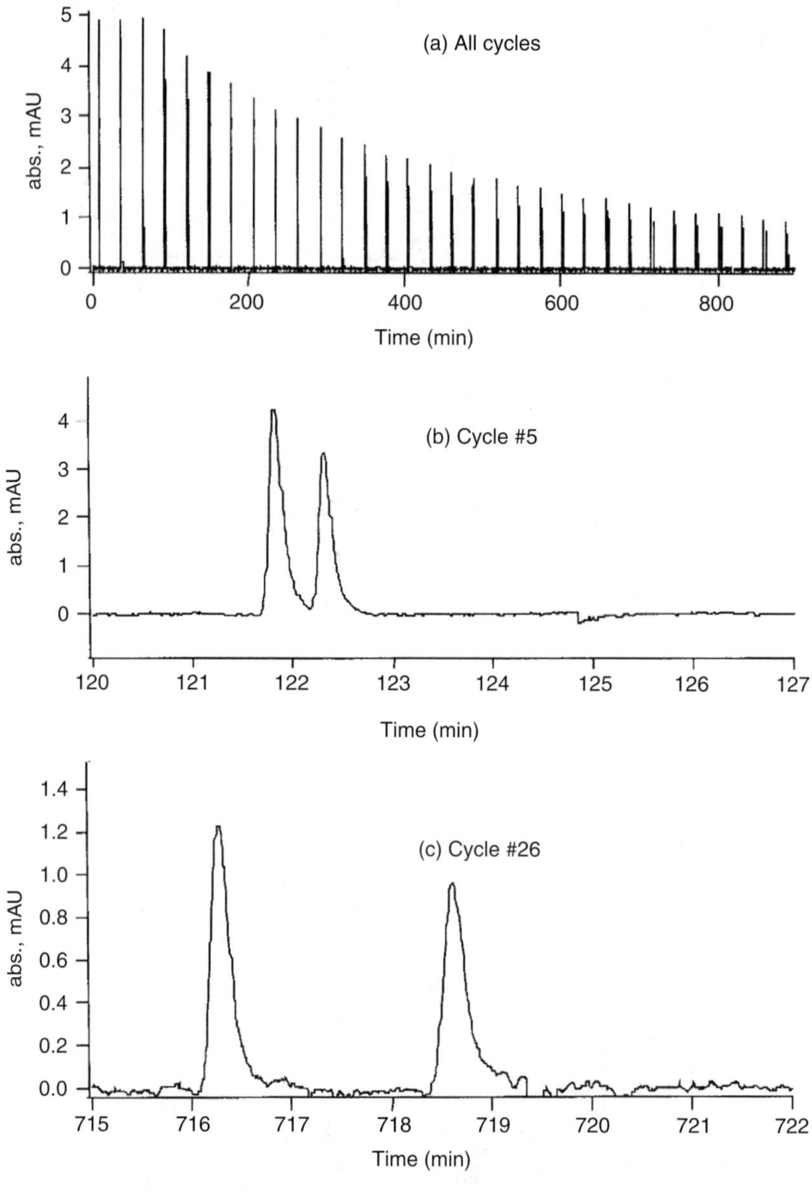

FIGURE 24.22 MEKC separation of phenylalanine and phenylalanine-ring-D5 using capillary-based SCCE. (Reprinted from Zhao, J. and Jorgenson, J. W., *Journal of Microcolumn Separations* 1999, 11, 439–449. Copyright 1999 John Wiley & Sons, Inc. With permission.)

shown in Figure 24.22. These two analytes have only a 0.4% mobility difference and were separated in only 14 h.[81]

CGE separations have been reported using microchip-based SCCE. Figure 24.21b shows an SCCE with CGE separation of FITC-labeled amino acids.[79] SCCE with CGE has also been used with another microchip design for the separation of a PhiX174/Hae III DNA ladder. Separations with difference synchronization times were used to isolate different DNA fragments.[78]

24.4.3.4 Methods of Detection

Laser-induced fluorescence detection is a logical choice for high-resolution SCCE when performed on microfluidic chip-based systems. High laser power can result in photobleaching, decreasing the analyte signal every time it passes through the beam.[75,76,79]

UV absorbance can easily be done on capillary-based systems using capillaries at least 50–75 μm in diameter. It is a universal method of detection that can be used with high sensitivity with most analytes. The short pathlength provide by the shallow channels make UV absorbance detection difficult to implement on microfluidic systems. In addition, microchip-based systems fabricated from glass instead of fused silica absorb significant amounts of UV, further complicating its implementation. In future experiments, detection methods may include direct conductivity measurements (microchip-based SCCE) and contactless conductivity methods (capillary-based SCCE).

24.4.3.5 Limitations

Theoretically, the ability of SCCE to separate analytes is limited only by longitudinal diffusion. In practice, broadening from the dead volume connecting the separation capillaries and adsorption of the analytes to the capillary wall ultimately limit the efficiency of the separation. Microchip-based SCCE also loses efficiency from broadening around tight turns.

SCCE, similar to FCCE, is a method in which each cycle reduces the range of mobilities that can be analyzed. The result is that the peak capacity of the separation will be reduced during each subsequent cycle. This effect is most notable when short channels and low voltages are used.[78]

The switching of potentials in SCCE causes analytes of the selected mobility to continuously travel around the separation loop in the same direction. Analytes that migrate at higher mobilities are lost at the corners and flushed into the waste reservoirs. Analytes of lower mobility that do not make it into the next separation channel before the potentials are switched will travel in the reverse direction. This can cause data analysis problems for complex samples when these backward migrating peaks pass in front of the detector again. This has been observed in several microchip-based separations including the SCCE separation of fluorescein isothiocyanate and its degradation products,[88] the MEKC and CGE with SCCE separations of FITC-labeled amino acids and human urine,[79] and SCCE of double-stranded DNA in a sieving matrix.[78] While this phenomenon is not much of a problem with relatively simple samples, data analysis may be confusing and complicated for complex samples such as cell lysates.

24.4.4 Methods Development Guidelines

24.4.4.1 Handling Extended Separation Times

The selection of the buffer components is very important in SCCE. Just as in FCCE and UHVCE, careful attention must be paid toward the amount of analyte adsorption to the channel walls. At extended separation times, losses from adsorption will add to reduce the signal to noise value greatly. Although the separation channels are flushed with fresh buffer during the SCCE separation, buffering capacity must be fairly high for long separation times if the reservoir size is small. This is a bigger problem with microchip-based SCCE where typical reservoir volumes are less than 1 mL. Capillary-based SCCE typically uses 20 mL buffer reservoirs and buffering capacity problems are usually not observed.

Similar to FCCE, automation of SCCE voltage switching can greatly improve the reliability and usability of the SCCE system. Zhao et al.[80] report the use of software that determined the rising and falling signal of the detected analyte above a predetermined threshold value during each cycle and recalculated switching times to ensure that the analyte bands stayed within the separation channels.

Longer separation channels have a few advantages over shorter channels when separation times are long. Longer channels allow longer separation distances without as much broadening and sample loss associated with channel connections and corners. In addition, longer separation channels maintain a higher peak capacity during extended run times, and the analyte bands can broaden considerably without being cut off at the edges.

24.4.4.2 Band Broadening Prevention

Band broadening prevention strategies for SCCE are similar to those used in UHVCE and FCCE. Known formulas can be used for buffer components if little peak tailing is seen under standard CE separations. Electrostatic repulsion from channel walls that share the same electrical charge as the analytes can reduce broadening greatly.

The electrophoretic mobilities of the buffer ions should closely match those of the sample ions to prevent broadening from electrodispersion. Similarly, the ionic strength of the buffer should be greater than or equal to that of the sample.

Several strategies for reducing dead volume at the channel connections have been reported. The use of T-shaped junctions at the middle of the separation channels (Figure 24.19) has been employed in an effort to reduce losses at the corner connections of the separation channels. Experimental observation indicated that broadening was more significant than that seen in previous designs with reservoir connections at the corners.[79] Recently, Manz et al.[78] have reported the use of side channels that are much shallower (~1/8th depth) than the separation channel in an effort to reduce losses at the connections of the channels.

The use of narrower rounded turns for microchip-based SCCE is reported for reducing band broadening at the corners.[78] This approach has been shown to reduce the broadening seen for turns in other microfluidic chip designs.[83,86,87,89,90] Other, more exotic approaches to reducing dispersion at the turns of channels on microfluidic devices have been reported. One technique using a pulsed UV laser to modify the surface of plastic chips at the turn to increase the EOF by up to 4%. This technique was shown to reduce band broadening at the turns.[91]

24.4.4.3 Sample Separation Strategies

Amino acids and peptides are easily labeled with fluorescent dyes. CZE, CGE, or MEKC in capillary or microchip-based SCCE can typically resolve these analytes. Microchip-based SCCE offers higher sensitivity for very small samples, and higher electric field strengths can be used without Joule heating problems. Capillary-based SCCE offers a much higher peak capacity and the use of UV absorbance detection of unlabeled analytes.

Isotopomers are best separated using capillary-based SCCE due to the extremely long separation times required. The components of the buffer must be carefully selected so that their ionic strength and ion mobility are very close to that of the analytes and adsorption to the capillary wall is prevented. MEKC can be used in some cases to separate isotopomers more quickly than CZE, most likely due to small changes in hydrophobicity caused by deuterium substitution.

Resolution of the enantiomers in a racemic mixture will require the use of a chiral selector such as a cyclodextrin. Fluorescently labeled analytes can be detected at very low concentration using microchip-based SCCE, and unlabeled analytes can be easily detected with UV absorption using capillary-based SCCE. The long separation times available with capillary-based SCCE allows the use of less than optimal chiral resolving agents, whereas a more selective chiral selector is needed in microchip-based SCCE.

24.5 CONCLUDING REMARKS

The three techniques presented in this chapter demonstrate different approaches to improving the resolution and efficiency of CE separations. All three methods work by increasing the time and distance that the analyte bands electrophoretically migrate under a high electric field strength, but the different ways in which each method accomplishes this goal determine their useful application.

UHVCE uses the direct application of extremely high voltages on long capillaries, and it allows the examination of analytes that span the entire range of mobilities found within the sample. Samples containing complex mixtures of proteins, peptides, nucleic acids, and many others display a wide range of mobilities and UHVCE is well suited to their analysis.

FCCE and SCCE use conventional voltages on relatively short capillaries, but they maintain analytes with a narrow range of mobilities within the capillaries for very long periods of time. Because high-resolution analysis is limited to such a narrow range of mobilities, FCCE and SCCE are best suited for the analysis of difficult to resolve pairs such as closely related peptides, enantiomers, and isotopomers. The ultimate utility of SCCE and FCCE depends on the ability of the experimenter to eliminate sources of band broadening. The amount of broadening observed in SCCE depends mainly on the design of the instrumentation. Band broadening in FCCE is inherent to the magnitude of the counter flow required to reverse the migration of the analyte bands.

The analysis of extremely complex samples such as cell lysates can be simplified through a combination of these methods. Preliminary high-efficiency, high-resolution analysis with UHVCE can be used to elucidate areas of interest within the sample. After the mobilities of the analytes of interest have been determined, FCCE or SCCE can then be used for much more detailed analysis of particular bands. The analysis of complicated biological samples remains an area where the implementation of these techniques can provide new insight.

ACKNOWLEDGMENT

National Science Foundation Predoctoral Fellowship Program.

REFERENCES

1. Jorgenson, J. W. and Lukacs, K. D., Capillary zone electrophoresis. *Journal of Chromatography Library* 1985, 30 (Microcolumn Sep.), 121–131.
2. Jorgenson, J. W. and Lukacs, K. D., Capillary zone electrophoresis. *Science (Washington, DC, United States)* 1983, 222, 266–272.
3. Jorgenson, J. W. and Lukacs, K. D., Free-zone electrophoresis in glass capillaries. *Clinical Chemistry (Washington, DC, United States)* 1981, 27, 1551–1553.
4. Jorgenson, J. W. and Lukacs, K. D., Zone electrophoresis in open-tubular glass capillaries. *Analytical Chemistry* 1981, 53, 1298–1302.
5. Lukacs, K. Theory, instrumentation, and applications of capillary zone electrophoresis. PhD Dissertation, University of North Carolina, Chapel Hill, 1983.
6. Henley, W. H. Two distinct approaches for increasing the resolution and efficiency of capillary electrophoretic separations. PhD Dissertation, University of North Carolina, Chapel Hill, 2005.
7. Hutterer, K. M. and Jorgenson, J. W., Ultrahigh-voltage capillary zone electrophoresis. *Analytical Chemistry* 1999, 71, 1293–1297.
8. Palonen, S., Jussila, M., Porras, S. P., Hyotylainen, T., and Riekkola, M. L., Extremely high electric field strengths in non-aqueous capillary electrophoresis. *Journal of Chromatography A* 2001, 916, 89–99.
9. Halliday, D., Resnick. R., and Krane, K. S., *Physics*. John Wiley & Sons, Inc., New York, 1992.
10. Van de Graaff, R. J., Compton, K. T., and Van Atta, L. C., The electrostatic production of high voltage for nuclear investigations. *Physical Review* 1933, 43, 149–157.

11. Kraft, E. M. S., Development of capillary zone electrophoresis at 100,000 to 150,000 volts. M.S. Thesis, University of North Carolina, Chapel Hill, NC, 1996.
12. Hutterer, K. M., Ultra high voltage capillary electrophoresis. Doctorial Thesis, University of North Carolina, Chapel Hill, NC, 2000.
13. Hutterer, K. M. and Dolnik, V., Capillary electrophoresis of proteins 2001–2003. *Electrophoresis* 2003, 24, 3998–4012.
14. Hutterer, K. M. and Jorgenson, J. W., Separation of hyaluronic acid by ultrahigh-voltage capillary gel electrophoresis. *Electrophoresis* 2005, 26, 2027–2033.
15. Hutterer, K. M., Birrell, H., Camilleri, P., and Jorgenson, J. W., High resolution of oligosaccharide mixtures by ultrahigh voltage micellar electrokinetic capillary chromatography. *Journal of Chromatography B* 2000, 745, 365–372.
16. Verma, R., Shyam, A., and Nair, L., Development of battery powered 100 kV dc power supply. *Review of Scientific Instruments* 2006, 77 (10, Pt. 1), 106104/1–106104/4.
17. Gordon, M. J., Lee, K. J., Arias, A. A., and Zare, R. N., Protocol for resolving protein mixtures in capillary zone electrophoresis. *Analytical Chemistry* 1991, 63, 69–72.
18. Duso, A. B. and Chen, D. D. Y., Proton and hydroxide ion mobility in capillary electrophoresis. *Analytical Chemistry* 2002, 74, 2938–2942.
19. Huang, X., Coleman, W. F., and Zare, R. N., Analysis of factors causing peak broadening in capillary zone electrophoresis. *Journal of Chromatography* 1989, 480, 95–110.
20. Moseley, M. A., Deterding, L. J., Tomer, K. B., and Jorgenson, J. W., Determination of bioactive peptides using capillary zone electrophoresis/mass spectrometry. *Analytical Chemistry* 1991, 63, 109–114.
21. Moseley, M. A., Jorgenson, J. W., Shabanowitz, J., Hunt, D. F., and Tomer, K. B., Optimization of capillary zone electrophoresis/electrospray ionization parameters for the mass spectrometry and tandem mass spectrometry analysis of peptides. *Journal of the American Society for Mass Spectrometry* 1992, 3, 289–300.
22. Kirkland, J. J., Glajch, J. L., and Farlee, R. D., Synthesis and characterization of highly stable bonded phases for high-performance liquid chromatography column packings. *Analytical Chemistry* 1989, 61, 2–11.
23. Hardenborg, E., Zuberovic, A., Ullsten, S., Soderberg, L., Heldin, E., and Markides, K. E., Novel polyamine coating providing non-covalent deactivation and reversed electroosmotic flow of fused-silica capillaries for capillary electrophoresis. *Journal of Chromatography A* 2003, 1003, 217–221.
24. Ullsten, S., Zuberovic, A., Wetterhall, M., Hardenborg, E., Markides, K. E., and Bergguist, J., A polyamine coating for enhanced capillary electrophoresis-electrospray ionization-mass spectrometry of proteins and peptides. *Electrophoresis* 2004, 25, 2090–2099.
25. Albarghouthi, M. N. and Barron, A. E., Polymeric matrices for DNA sequencing by capillary electrophoresis. *Electrophoresis* 2000, 21, 4096–4111.
26. Albarghouthi, M. N., Buchholz, B. A., Doherty, E. A. S., Bogdan, F. M., Zhou, H., and Barron, A. E., Impact of polymer hydrophobicity on the properties and performance of DNA sequencing matrices for capillary electrophoresis. *Electrophoresis* 2001, 22, 737–747.
27. Albarghouthi, M. N., Buchholz, B. A., Stein, T. M., and Barron, A. E., PolyDuramide: A novel, hydrophilic, self-coating polymer matrix for DNA sequencing by capillary electrophoresis. *Abstracts of Papers, 222nd ACS National Meeting,* Chicago, IL, United States, August 26–30, 2001, ANYL-119.
28. Buchholz, B. A., Doherty, E. A. S., Albarghouthi, M. N., Bogdan, F. M., Zahn, J. M., and Barron, A. E., Microchannel DNA sequencing matrices with a thermally controlled "viscosity Switch". *Analytical Chemistry* 2001, 73, 157–164.
29. Carrilho, E., Ruiz-Martinez, M. C., Berka, J., Smirnov, I., Goetzinger, W., Miller, A. W., Brady, D., and Karger, B. L., Rapid DNA sequencing of more than 1000 bases per run by capillary electrophoresis using replaceable linear polyacrylamide solutions. *Analytical Chemistry* 1996, 68, 3305–3313.
30. Dolnik, V., DNA sequencing by capillary electrophoresis (review). *Journal of Biochemical and Biophysical Methods* 1999, 41, 103–119.
31. Dolnik, V., Gurske, W. A., and Padua, A., Galactomannans as a sieving matrix in capillary electrophoresis. *Electrophoresis* 2001, 22, 707–719.
32. Heller, C., Separation of double-stranded and single-stranded DNA in polymer solutions. Part 2. Separation, peak width, and resolution. *Electrophoresis* 1999, 20, 1978–1986.

33. Salas-Solano, O., Carrilho, E., Kotler, L., Miller, A. W., Goetzinger, W., Sosic, Z., and Karger, B. L., Routine DNA sequencing of 1000 bases in less than one hour by capillary electrophoresis with replaceable linear polyacrylamide solutions. *Analytical Chemistry* 1998, 70, 3996–4003.
34. Slater, G. W., Desruisseaux, C., Hubert, S. J., Mercier, J.-F., Labrie, J., Boileau, J., Tessier, F., and Pepin, M. P., Theory of DNA electrophoresis: A look at some current challenges. *Electrophoresis* 2000, 21, 3873–3887.
35. Song, L., Liang, D., Chen, Z., Fang, D., and Chu, B., DNA sequencing by capillary electrophoresis using mixtures of polyacrylamide and poly(*N*,*N*-dimethylacrylamide). *Journal of Chromatography A* 2001, 915, 231–239.
36. Song, L., Liang, D., Kielescawa, J., Liang, J., Tjoe, E., Fang, D., and Chu, B., DNA sequencing by capillary electrophoresis using copolymers of acrylamide and *N*,*N*-dimethyl-acrylamide. *Electrophoresis* 2001, 22, 729–736.
37. Wei, W. and Yeung, E. S., Improvements in DNA sequencing by capillary electrophoresis at elevated temperature using poly(ethylene oxide) as a sieving matrix. *Journal of Chromatography B: Biomedical Sciences and Applications* 2000, 745, 221–230.
38. Zhou, H., Miller, A. W., Sosic, Z., Buchholz, B., Barron, A. E., Kotler, L., and Karger, B. L., DNA Sequencing up to 1300 bases in two hours by capillary electrophoresis with mixed replaceable linear polyacrylamide solutions. *Analytical Chemistry* 2000, 72, 1045–1052.
39. Huber, C. G. and Oberacher, H., Analysis of nucleic acids by on-line liquid chromatography-mass spectrometry. *Mass Spectrometry Reviews* 2002, 20, 310–343.
40. Shibata, K., Itoh, M., Aizawa, K., Nagaoka, S., Sasaki, N., Carninci, P., Konno, H., et al., RIKEN integrated sequence analysis (RISA) system—384-format sequencing pipeline with 384 multicapillary sequencer. *Genome Research* 2000, 10, 1757–1771.
41. Culbertson, C. T. and Jorgenson, J. W., Increasing the resolving power of capillary electrophoresis through electroosmotic flow control using radial fields. *Journal of Microcolumn Separations* 1999, 11, 167–174.
42. Cheng, Y. F., Wu, S., Chen, D. Y., and Dovichi, N. J., Interaction of capillary zone electrophoresis with a sheath flow cuvette detector. *Analytical Chemistry* 1990, 62, 496–503.
43. Williams, B. A. and Vigh, G., The use of hydrodynamic counterflow to improve the resolution and detection of the minor component in the capillary electrophoretic analysis of enantiomers. *Enantiomer* 1996, 1, 183–191.
44. Bier, M., Twitty, G. E., and Sloan, J. E., Recycling isoelectric focusing and isotachophoresis. *Journal of Chromatography* 1989, 470, 369–376.
45. Everaerts, F. M., Vacik, J., Verheggen, T. P. E. M., and Zuska, J., Displacement electrophoresis. Experiments with counterflow of electrolyte. *Journal of Chromatography* 1970, 49, 262–268.
46. Everaerts, F. M., Vacik, J., Verheggen, T. P. E. M., and Zuska, J., Isotachophoresis. Experiments with electrolyte counterflow. *Journal of Chromatography* 1971, 60, 397–405.
47. Everaerts, F. M., Verheggen, T. P. E. M., and Van de Venne, J. L. M., Isotachophoretic experiments with a counter flow of electrolyte. *Journal of Chromatography* 1976, 123, 139–148.
48. Reinhoud, N. J., Tjaden, U. R., and van der Greef, J., Automated isotachophoretic analyte focusing for capillary zone electrophoresis in a single capillary using hydrodynamic back-pressure programming. *Journal of Chromatography* 1993, 641, 155–162.
49. Reinhoud, N. J., Tjaden, U. R., and van der Greef, J., Strategy for setting up single-capillary isotachophoresis-zone electrophoresis. *Journal of Chromatography A* 1993, 653, 303–312.
50. Clark, S. L. and Remcho, V. T., Electrochromatographic retention studies on a flavin-binding RNA aptamer sorbent. *Analytical Chemistry* 2003, 75, 5692–5696.
51. Lucy, C. A. and McDonald, T. L., Separation of chloride isotopes by capillary electrophoresis based on the isotope effect on ion mobility. *Analytical Chemistry* 1995, 67, 1074–1078.
52. Soga, T., Inoue, Y., and Ross, G. A., Analysis for halides, oxyhalides and metal oxoacids by capillary electrophoresis with suppressed electroosmotic flow. *Journal of Chromatography A* 1995, 718, 421–428.
53. Terabe, S., Yashima, T., Tanaka, N., and Araki, M., Separation of oxygen isotopic benzoic acids by capillary zone electrophoresis based on isotope effects on the dissociation of the carboxyl group. *Analytical Chemistry* 1988, 60, 1673–1677.

54. Yeung, K. K. C. and Lucy, C. A., Isotopic separation of [14N]- and [15N]aniline by capillary electrophoresis using surfactant-controlled reversed electroosmotic flow. *Analytical Chemistry* 1998, 70, 3286–3290.
55. Yeung, K. K. C. and Lucy, C. A., Ultrahigh-resolution capillary electrophoretic separation with indirect ultraviolet detection: Isotopic separation of [14N]-and [15N]ammonium. *Electrophoresis* 1999, 20, 2554–2559.
56. Brewer, A. K., Madorsky, S. L., and Westhaver, J. W., The concentration of K39 and K41 by balanced ion migration in a counterflowing electrolyte. *Science (Washington, DC, United States)* 1946, 104, 156–157.
57. Chankvetadze, B., Burjanadze, N., Bergenthal, D., and Blaschke, G., Potential of flow-counterbalanced capillary electrophoresis for analytical and micropreparative separations. *Electrophoresis* 1999, 20, 2680–2685.
58. Ivory, C. F., Preparative free-flow electrofocusing in a vortex-stabilized annulus. *Electrophoresis* 2004, 25, 360–374.
59. McLaren, D. G. and Chen, D. D. Y., A quantitative study of continuous flow-counterbalanced capillary electrophoresis for sample purification. *Electrophoresis* 2003, 24, 2887–2895.
60. McLaren, D. G. and Chen, D. D. Y., Continuous electrophoretic purification of individual analytes from multicomponent mixtures. *Analytical Chemistry* 2004, 76, 2298–2305.
61. Thome, B. and Ivory, C. F., Continuous fractionation of enantiomer pairs in free solution using an electrophoretic analog of simulated moving bed chromatography. *Journal of Chromatography, A* 2002, 953, 263–277.
62. Thome, B. M. and Ivory, C. F., Development of a segmented model for a continuous electrophoretic moving bed enantiomer separation. *Biotechnology Progress* 2003, 19, 1703–1712.
63. Tracy, N. I. and Ivory, C. F., Preparative isoelectric focusing of proteins using binary buffers in a vortex-stabilized, free-flow apparatus. *Electrophoresis* 2004, 25, 1748–1757.
64. Kutter, J. and Welsch, T., The effect of electroosmotic and hydrodynamic flow profile superposition on band broadening in capillary electrophoresis. *Journal of High Resolution Chromatography* 1995, 18, 741–744.
65. Culbertson, C. T. and Jorgenson, J. W., Flow counterbalanced capillary electrophoresis. *Analytical Chemistry* 1994, 66, 955–962.
66. Iwata, T. and Kuroshu, Y., Enhancement of the sensitivity and resolution by flow-counterbalanced capillary electrophoresis. *Analytical Sciences* 1995, 11, 131–133.
67. Henley, W. H., Wilburn, R. T., Crouch, A. M., and Jorgenson, J. W., Flow counterbalanced capillary electrophoresis using packed capillary columns: Resolution of enantiomers and isotopomers. *Analytical Chemistry* 2005, 77, 7024–7031.
68. Culbertson, C. T. and Jorgenson, J. W., Separation of fluorescently derivatized deuterated isotopomers of phenylalanine using micellar electrokinetic chromatography and flow counterbalanced micellar electrokinetic chromatography. *Journal of Microcolumn Separations* 1999, 11, 175–183.
69. MacNair, J. E., Lewis, K. C., and Jorgenson, J. W., Ultrahigh-pressure reversed-phase liquid chromatography in packed capillary columns. *Analytical Chemistry* 1997, 69, 983–989.
70. Shalliker, R. A., Broyles, B. S., and Guiochon, G., Evaluation of the secondary consolidation of columns for liquid chromatography by ultrasonic irradiation. *Journal of Chromatography, A* 2000, 878, 153–163.
71. Tong, D., Bartle, K. D., Clifford, A. A., and Edge, A. M., Theoretical studies of the preparation of packed capillary columns for chromatography. *Journal of Microcolumn Separations* 1995, 7, 265–278.
72. Culbertson, C. T. New methods for increasing the resolving power and lowering the UV absorbance detection limits in capillary electrophoresis. PhD Thesis, University of North Carolina, Chapel Hill, 1996.
73. Eijkel, J. C. T., van den Berg, A., and Manz, A., Cyclic electrophoretic and chromatographic separation methods. *Electrophoresis* 2004, 25, 243–252.
74. Ramsteiner, K. A., Systematic approach to column switching. *Journal of Chromatography* 1988, 456, 3–20.

75. Burggraf, N., Manz, A., De Rooij, N. F., and Widmer, H. M., Synchronized cyclic capillary electrophoresis: A novel concept for high-performance separations using low voltages. *Analytical Methods & Instrumentation* 1993, 1, 55–59.
76. Burggraf, N., Manz, A., Effenhauser, C. S., Verpoorte, E., de Rooij, N. F., and Widmer, H. M., Synchronized cyclic capillary electrophoresis—A novel approach to ion separations in solution. *Journal of High Resolution Chromatography* 1993, 16, 594–596.
77. Burggraf, N., Manz, A., Verpoorte, E., Effenhauser, C. S., Widmer, H. M., and de Rooij, N. F., A novel approach to ion separations in solution: Synchronized cyclic capillary electrophoresis (SCCE). *Sensors and Actuators, B: Chemical* 1994, 20, 103–110.
78. Manz, A., Bousse, L., Chow, A., Metha, T. B., Kopf-Sill, A., and Parce, J. W., Synchronized cyclic capillary electrophoresis using channels arranged in a triangle and low voltages. *Fresenius′ Journal of Analytical Chemistry* 2001, 371, 195–201.
79. von Heeren, F., Verpoorte, E., Manz, A., and Thormann, W., Micellar electrokinetic chromatography separations and analyses of biological samples on a cyclic planar microstructure. *Analytical Chemistry* 1996, 68, 2044–2053.
80. Zhao, J., Hooker, T., and Jorgenson, J. W., Synchronous cyclic capillary electrophoresis using conventional capillaries: System design and preliminary results. *Journal of Microcolumn Separations* 1999, 11, 431–437.
81. Zhao, J. and Jorgenson, J. W., Application of synchronous cyclic capillary electrophoresis: Isotopic and chiral separations. *Journal of Microcolumn Separations* 1999, 11, 439–449.
82. Choi, J. G., Kim, M., Dadoo, R., and Zare, R. N., Electrophoretron: A new method for enhancing resolution in electrokinetic separations. *Journal of Chromatography, A* 2001, 924, 53–58.
83. Culbertson, C. T., Jacobson, S. C., and Ramsey, J. M., Dispersion sources for compact geometries on microchips. *Analytical Chemistry* 1998, 70, 3781–3789.
84. Fu, L.-M., Yang, R.-J., and Lee, G.-B., Analysis of geometry effects on band spreading of microchip electrophoresis. *Electrophoresis* 2002, 23, 602–612.
85. Griffiths, S. K. and Nilson, R. H., Design and analysis of folded channels for chip-based separations. *Analytical Chemistry* 2002, 74, 2960–2967.
86. Wang, Y., Lin, Q., and Mukherjee, T., System-oriented dispersion models of general-shaped electrophoresis microchannels. *Lab on a Chip* 2004, 4, 453–463.
87. Yao, Z.-H., Yoder, G. L., Culbertson, C. T., and Ramsey, J. M., Numerical simulation of dispersion generated by a 180 Deg turn in a microchannel. *Chinese Physics (Beijing, China)* 2002, 11, 226–232.
88. Manz, A., Verpoorte, E., Effenhauser, C. S., Burggraf, N., Raymond, D. E., and Widmer, H. M., Planar chip technology for capillary electrophoresis. *Fresenius' Journal of Analytical Chemistry* 1994, 348, 567–571.
89. Culbertson, C. T., Jacobson, S. C., and Ramsey, J. M., Microchip devices for high-efficiency separations. *Analytical Chemistry* 2000, 72, 5814–5819.
90. Ramsey, J. D., Jacobson, S. C., Culbertson, C. T., and Ramsey, J. M., High-efficiency, two-dimensional separations of protein digests on microfluidic devices. *Analytical Chemistry* 2003, 75, 3758–3764.
91. Johnson, T. J., Ross, D., Gaitan, M., and Locascio, L. E., Laser modification of preformed polymer microchannels: Application to reduce band broadening around turns subject to electrokinetic flow. *Analytical Chemistry* 2001, 73, 3656–3661.

25 Separation of DNA for Forensic Applications Using Capillary Electrophoresis

Lilliana I. Moreno and Bruce McCord

CONTENTS

25.1 INTRODUCTION

Forensic science is defined as the application of science to the law. It is the goal of the forensic scientist to identify and compare physical evidence and use the resulting observations to aid in solving criminal or civil matters. From its inception, forensic scientists have recognized the potential of capillary electrophoresis (CE) as a useful tool to assist in the analysis of a wide array of trace evidence samples [1,2]. This is particularly true for applications in forensic biology. In this discipline, biological fluids (blood, hair, semen) left behind at a crime scene are probed to establish the *essential facts* of the crime. Sample analysis is performed via extraction and analysis of the genetic material within these samples. The key to the procedure is targeting specific locations in the genome containing polymorphisms (different allelic forms) that permit differentiation between individuals. The statistical probability of an individual inheriting any given polymorphism can then be used to assess the evidence and compare it with known samples from victims and suspects.

In the early years of DNA analysis, the separation and sizing of DNA was accomplished by slab gel electrophoresis. However, since 1998 CE has gradually taken over slab gel techniques in this field because of its ease of automation, minimal sample consumption (which is of utmost importance in the field), and its high-throughput capabilities.

25.2 BACKGROUND

Although seldom used for genotyping in current forensic applications, slab gels set the precedent for the development of CE. Slab gel systems had an advantage in forensic analysis in that multiple samples could be run simultaneously, allowing easy comparisons and rapid analysis. The ability to compare different samples on the same gel was particularly useful in early procedures involving restriction fragment length polymorphisms (RFLPs) and multilocus probes. With the RFLP techniques, DNA fragment sizes ranged up to 20,000 bases and results were based on the presence or absence of bands within different size ranges. Separations took place on large format agarose gels and genetic loci were detected using Southern blotting with radioactive probes. As a result, a complete analysis could take several weeks because of the time involved in repetitive DNA transfer, hybridization, and exposure to radioactive film [3].

With the advent of the polymerase chain reaction (PCR), there was a complete paradigm shift. The PCR permits a few copies of a DNA sequence to be amplified to millions of copies [4], thus increasing sensitivity and eliminating the need for radioactive probes. DNA fragment sizes were reduced to a few hundred bases, and the total time for an analysis dropped to a single day. With PCR, sample quantities were sufficient for detection via simple silver staining processes; however due to the smaller size of the alleles, higher resolution acrylamide gels were required. As the exceptional capabilities of the PCR in DNA typing became apparent, law enforcement personnel began to take advantage of the expanded speed and sensitivity that the new technique had to offer. Consequently, an increasing number of evidence samples began to be submitted to laboratories for testing.

Therein arose a new concern. How were laboratories going to be able to handle the bulk of samples received in a timely manner? An analytical technique was needed that could provide as good or better results than those obtained via slab gel systems in a standardized and automated fashion. This technique was to be CE.

The principle governing both slab gel and CE techniques is the same: the separation of a series of analytes based on size selective sieving through a gel or polymer network under the influence of an electric field. The slab gel matrix permits a size-dependent separation of fragments since larger DNA fragments contain an invariant charge-to-size ratio due to their sugar-phosphate backbone. However, slab gels require manual operations such as the preparation of the gel and the loading of samples. Automated CE systems eliminated these tedious tasks and substituted replaceable, entangled polymer solutions for the rigid cross-linked acrylamide gels. Also, the improved heat dissipation of the fused-silica capillary permits much higher voltages to be used, reducing the run time of the analysis [5]. Laser-induced fluorescence detection also improved throughput by allowing multiwavelength analysis [6]. Soon, the forensic science community became interested in this robust technology and began to adopt it for a variety of analyses. In particular, CE systems can be used for DNA quantification, genotyping, sequencing, and mutation detection.

25.3 THEORETICAL ASPECTS

25.3.1 The Capillary and the Sieving Matrix

In 1988, it was shown that DNA single nucleotide separation could be achieved using cross-linked polyacrylamide gel filled capillaries coupled with UV/Vis detection [7]. These chemical gels provided size selective separation of DNA molecules based on their ability to migrate through transient pores created in the polymer matrix. Unfortunately, cross-linked gel filled capillaries had a limited lifetime due to the ready formation of voids in the gel during electrophoresis [8]. There were also concerns with carryover from one run to the next, an important point when precious evidentiary samples were being run. The solution to this problem was the development of CE systems with entangled polymer buffers such as polyacrylamide, hydroxyethylcellulose, and polyethylene oxide [9]. With proper control of polymer molecular weight and concentration, these buffers provided equivalent separation to cross-linked gels with the added advantage that they permitted refilling and reuse of the capillary. This property allowed the same capillary to be used up to 100 times before replacement.

An additional problem with these novel sieving polymers was the effect of the electroosmotic flow (EOF) on migration time reproducibility. For optimum results, the polymer network must not only separate the DNA fragments but also eliminate wall effects such as adsorption and osmotic flow. This is an important issue as the primary function of the method is to precisely determine the size of the DNA fragments. Variations in migration time due to slight differences in EOF can negatively affect estimation of fragment size. For this reason, linear polydimethyl acrylamide "POP" became the polymer of choice for these separations because of its relatively low viscosity and its capability to eliminate EOF [10,11].

Another critical issue in the development of the separation was the choice of buffer composition. The optimal buffer should produce low and stable currents and be able to separate DNA under both denaturing—where the formation of secondary structure of the sample is halted by denaturing agents such as heat, formamide, or urea [12,13]—or nondenaturing conditions—where secondary structure formation is desired to study subtle differences in DNA sequences [14,15].

For the majority of genotyping applications such as sequencing and fragment length determination, denaturing conditions are preferred as resolution is superior and there is a predictable relationship between fragment size and migration time. Buffers such as trishydroxymethylaminomethane (Tris) and *N-tris*[hydroxymethyl]methyl-3-aminopropanesulfonic acid (TAPS) are commonly used as they produce low currents and can buffer at physiological pH. A commonly used buffer for sequencing and genotyping applications consists of 100 mM TAPS at pH 8.0 with 5% pyrolidinone and 8 M urea as denaturants [11]. Four to six percent polydimethyl acrylamide at a controlled molecular weight is added to sieve the DNA and eliminate EOF. Uncoated fused-silica capillaries of 50 μm are typically used as this diameter provides a good compromise between resolution, sensitivity, and resistance to clogging.

25.3.2 Injection

For most applications in forensic DNA analysis, electrokinetic injection is used. This injection mode provides improved peak shape and intensity as a result of field amplified sample stacking. The stacking process results when the DNA sample is prepared in low ionic strength solutions. The application of an electric field accelerates the DNA to the sample/buffer interface where it concentrates because of the drop in field strength at that point. Because this process is highly dependent on the ionic strength of the sample solution, great care must be taken to keep salt concentrations low. Unfortunately PCR amplifications occur at relatively high-salt concentrations (70–100 mM KCl) and various buffers are often added to the stabilize DNA for long-term storage. Thus, the sample must be diluted or dialyzed before injection as smaller buffer ions have higher electrophoretic mobility and interfere with the injection of the DNA.

Sample injection is an important issue in forensic analysis as there should be a semi-quantitative relationship between peak height and sample concentration. This relationship helps the analyst assess the quality of the sample and its preparation. It also helps define the relative level of different contributors to a mixture. The removal of salts through dialysis or the unintentional addition of higher quantities of salt can produce peak intensities that are less representative of sample quantity and affect the aforementioned relationship.

In a typical injection, 1–2 μL of a completed PCR is diluted in 10–20 μL of formamide. Water can also be used but the formamide produces complete denaturation of the sample without further processing. High purity, low conductance formamide must be used to avoid loss of peak intensity or other injection artifacts.

25.3.3 Detection

Laser-induced fluorescence is the most common method of detection for DNA analyses due to its high sensitivity and multiplex capabilities. Typically argon-ion lasers are used with excitation wavelengths of 488 and 514 nm. Detection occurs using a charge-coupled device (CCD) that collects the fluorescence emission produced by the various dyes bound to the DNA molecules. There are three basic methods for detection of the DNA using these dyes: intercalation, amplification with dye-labeled PCR primers, or incorporation of dye-labeled bases into the DNA sequence during replication. For single channel detection of native DNA, intercalating dyes produce excellent results [16]. These dyes may simply be added to the CE buffer and the DNA is labeled as it moves through the gel toward the detector. For genotyping of denatured DNA, dye-labeled primers are used [17] while dye-labeled bases are used in DNA sequencing [18].

The specific dyes used for genotyping and DNA sequencing are designed to simultaneously absorb at a single laser wavelength but emit at a variety of different wavelengths. Rhodamine and fluorescein derivatives are commonly used. Using these dyes, multiple loci can be amplified and genotyped without interfering with each other by simply labeling each set of PCR reactants with a different dye. Current commercial systems have the capability to detect as many as five different dyes simultaneously on a single CE capillary. Special software is used to eliminate problems with dye overlap by applying virtual filters and various calibration procedures [17].

25.3.4 Size Estimation

As mentioned previously, the primary function of the separation in genetic analysis is the estimation of fragment size. Multichannel fluorescence systems typically reserve one channel for use as an internal standard. The internal standard consists of multiple peaks throughout the size range of the analysis and is used by the computer to permit a precise estimate of unknown fragments in the other dye channels. Computer algorithms can then be used to produce a size estimation of the unknown DNA fragments in the other dye lanes. The samples can be further processed by comparison of these data with an external standard that consists of all known mutations. Using both sets of standards, it is possible to produce size estimates with a precision of better than 0.17 base pairs (bp) making the system capable of distinguishing single base differences at sizes up to 350 bp [19]. It is important to note at this point that the size estimates produced by these techniques can be influenced by temperature or sequence effects, and thus careful control of temperature and denaturant concentration is important in order to maintain precision [20,21].

25.4 PRACTICAL APPLICATIONS IN FORENSIC BIOLOGY

The major application of CE in forensic biology is in the detection and analysis of short tandem repeats (STRs). STR markers are preferred because of the powerful statistical result that is possible with these markers and the large databases that exist for convicted offenders' profiles. Other related applications include the analysis of haploid markers in the Y chromosome and in mitochondrial DNA (mtDNA). Nonhuman DNA testing can also be performed depending on the circumstances of the case. The techniques involved include genotyping, DNA sequencing, and mutation detection.

25.4.1 STRs

The analysis of STR loci in DNA is the most common method for the determination of human identity and can indisputably distinguish between two or more unrelated individuals if sufficient loci can be detected [22]. STR loci occur in noncoding regions of the human genome and consist of short segments of DNA 2–7 bp in length such as AATG, which are repeated consecutively multiple times. The number of repeats at a given locus can vary between individuals and there is a statistical probability that a given individual will have a set number of repeats at a particular STR locus. It has been estimated that over 100,000 STR loci exist in the human genome and research continues in an effort to determine their exact function [23]. Among this large number of STRs, the forensic community in the United States has established a set of 13 loci (Table 25.1) [24,25] that can be used to develop a genetic profile for the identification of individuals in criminal casework. To process the results from each analysis, large database known as Combined DNA Index System (CODIS) has been set up. This database stores profiles from convicted offenders and unsolved casework. Similar databases have been set up in Europe, Japan, and other countries. The information in these databases can be used to detect and apprehend serial offenders by permitting rapid exchange of information between crime laboratories [26–28].

STRs can be targeted for PCR amplification by preparing primer sequences that bind to more conserved sequences flanking the variable STR regions. The chemistry and protocols necessary for identifying each set of STRs are included in one or a combination of kits provided by companies such as Applied Biosystems and Promega. These kits (Table 25.2) have been subjected to strict validation processes to ensure the quality of the data [19,29–31]. The kits permit multiplex PCR amplification of up to 16 loci (including the sex determination gene) simultaneously from a single sample (Figure 25.1). The different loci included in the kit contain multiple alleles and are separated by size and dye label. The repeat motifs are 4–5 bases in length and the motif sequence can repeat itself up to 51 times (Table 25.1). To help define the size and migration time of each known allele an external standard known as an allelic ladder is run subsequent to each set of samples and used to more precisely define the identity of each peak (Figure 25.2).

TABLE 25.1
Thirteen STR Markers Commonly Used for Forensic DNA Analyses in the United States

Marker (Locus)	Repeat Motif	Allele Range
CSF1PO	TAGA	6–16
FGA	CTTT	15–51.2
THO1	TCAT	3–14
TPOX	GAAT	6–13
vWA	[TCTG][TCTA]	10–24
D3S1358	[TCTG][TCTA]	9–20
D5S818	AGAT	7–16
D7S820	GATA	6–15
D8S1179	[TCTA][TCTG]	8–19
D13S317	TATC	5–15
D16S539	GATA	5–15
D18S51	AGAA	7–27
D21S11	[TCTA][TCTG]	24–38

Source: Adapted from Butler, J.M., *Forensic DNA Typing: Biology, Technology, and Genetics of STR Markers*, 2nd ed., Academic Press, San Diego, CA, 2005. With permission.

TABLE 25.2
Commercially Available Human STR Amplification Kits

Kit Name	Target Loci	Discrimination Power
(A) Applied Biosystems, Foster City, CA		
AmpFLSTR® Profiler®	D3S1358, vWA, FGA, THO1, TPOX, CSF1PO, D5S818, D13S317, D7S820, and Amelogenin	1:3.6 × 10^9
AmpFL®SEFiler™	D2S1338, D3S1358, D8S1179, D16S539, D18S51, D19S433, D21S11, SE-33, FGA, vWA, and Amelogenin	
AmpFL® Cofiler®	CSF1PO, D16S539, THO1, Amelogenin, TPOX, D3S1358, and D7S820	1:8.4 × 10^5
AmpFL Profiler Plus	D3S1358, D5S818, D7S820, D8S1179, D13S317, D18S51, D21S11, FGA, vWA, and Amelogenin	1:9.6 × 10^{10}
AmpFL® SGM Plus®	D2S1338, D3S1358, D8S1179, D16S539, D18S51, D19S433, D21S11, THO1, FGA, vWA	1:3.3 × 10^{12}
AmpFL® Identifiler®	CSF1PO, D3S1358, D5S818, D7S820, D8S1179, D13S317, D16S539, D18S51, D21S11, vWA, FGA, THO1, TPOX, D2S1338, D19S433, and Amelogenin	1:2.1 × 10^{17}
AmpFL® Green I	THO1, TPOX, Amelogenin, and CSF1PO	1:410
AmpFL® Blue	D3S1358, vWA, FGA	1:5000
(B) Promega Corporation, Madison, WI		
PowerPlex® 16 System	Penta E, D18S51, D21S11, THO1, D3S1358, FGA, TPOX, D8S1179, vWA, Amelogenin, Penta D, CSF1PO, D16S539, D7S820, D13S317, and D5S818	1:1.8 × 10^{17}

Source: Applied Biosystems, Foster City, CA (www.appliedbiosystems.com) and Promega Corporation, Madison, WI (www.promega.com).

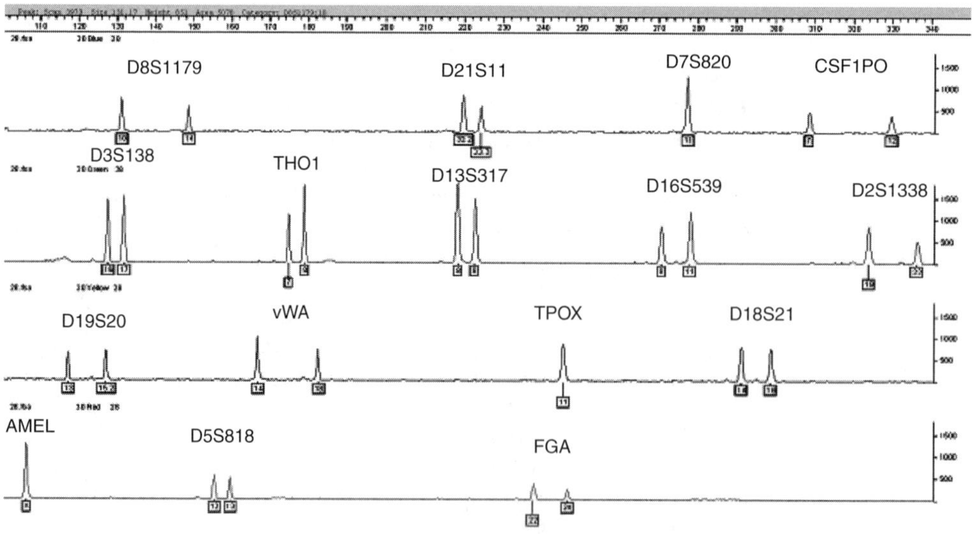

FIGURE 25.1 AmpF*l*STR Identifiler™16-plex amplification results. The electropherogram shows 15 STR loci as well as the amelogenin (sex determining) marker. Individual STR loci are separated by amplicon size and dye label. Each row represents a different dye marker: 1, the 6-FAM labeled loci; 2, VIC labeled loci; 3, NED labeled loci; 4, PET labeled loci; an internal standard labeled in rox is also run simultaneously but is not shown. (Courtesy of Ada Nuñez, Florida International University.)

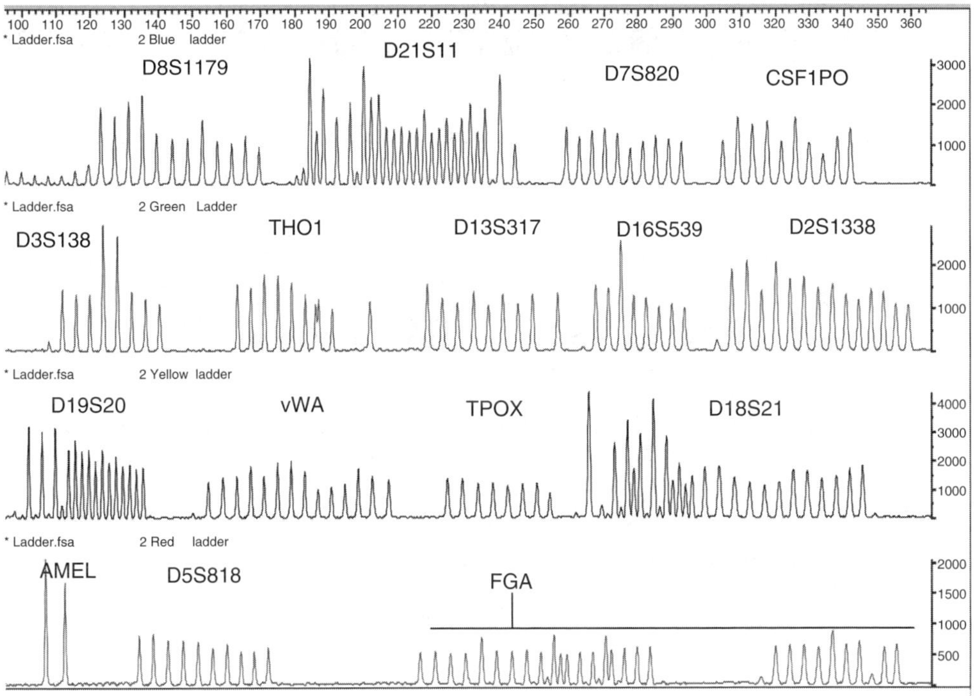

FIGURE 25.2 AmpF/STR™Identifiler kit allelic ladder. A separation of the most common alleles for each STR loci. The STR analysis software utilizes this ladder as an external standard to assign the correct alleles to evidence samples such as is illustrated in Figure 25.1. Notice the presence of two base variant alleles (from the normal four base STR repeat) at the D21, D18, D19, and FGA loci. There is also a one base variant allele located in the THO1 locus. (Courtesy of Ada Nuñez, Florida International University.)

One of the major issues in the separation of these large multiplex sets by CE is the potential presence of variant alleles that differ from the standard 4 base repeat unit by 1 or 2 bp. Single base resolution and high precision are necessary over the range of fragment lengths up to 350 bp in order to reliably detect these variant alleles and distinguish them from artifacts, spikes, and noise.

Another issue that is unique to this application is the role of the PCR in defining system sensitivity. Amplification of DNA concentrations less than 100 pg (about 17 cells) can produce stochastic intensity fluctuations leading to peak imbalance and occasional loss of signal (allele dropout). Low level mixtures may also be present in the electropherogram further complicating the interpretation of the data [32,33]. The ability to produce clear and unambiguous electropherograms is critical in criminal casework since DNA evidence may be the only information tying the suspect to the crime. Loss of peak intensity can complicate data interpretation. As a result, rigid rules have been developed for interpretation of signals below an intensity threshold that can be defined by the ability to reproducibly amplify DNA and detect it at a level above the system's limit of quantification [34].

For mixtures such as those often encountered in sexual assault cases, special techniques have been developed for isolation of male DNA (sperm heads) from female epithelial cells [35]. These procedures aid the interpretation of the data by removing most of the female contribution to the profile. This process, known as differential extraction, can be performed by selective digestion of the epithelial cells followed by isolation of undigested sperm heads through centrifugation. The sperm fraction is further digested and extracted using dithiothreitol (DTT). However in certain situations such as the case with vasectomized males or multiple assailants, isolation of individual profiles is more difficult. In these cases careful analysis of the results may be necessary to reach

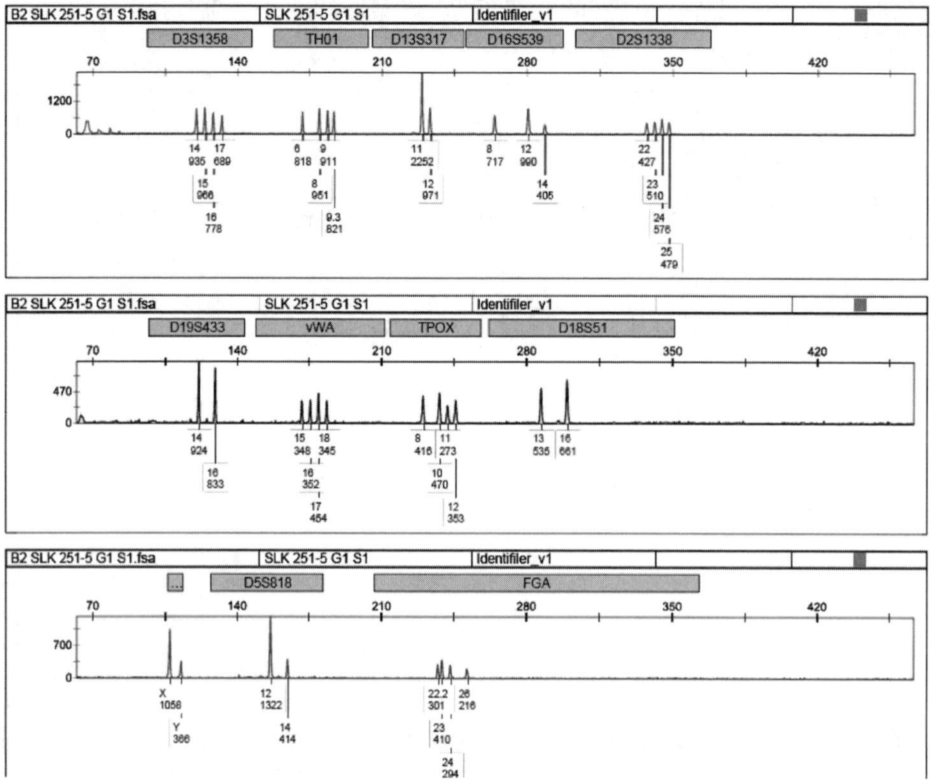

FIGURE 25.3 Mixed male/female DNA Profile. The amplification of a 1:1 mixture of male and female DNA using the Identifiler™STR multiplex. Note the presence of 2, 3, or 4 alleles at each locus indicating the presence of 2-, 1-, or 0-shared alleles. Also, notice the 3:1 ratio of X to Y sex typing alleles in the third panel due to the mixture of XX and XY alleles. Analysis performed on an ABI310 Genetic Analyzer. (Courtesy of Stephanie King and George Duncan, Broward County Sheriff's Office Crime Laboratory.)

a conclusion. Figure 25.3 illustrates the electropherogram of a mixed profile. Note that the sample can be determined to be a mixture of male and female DNA based on the relative peak areas of the Amelogenin sex marker. In this case there is a 3/1 ratio of X- to Y-chromosome, suggesting a 1/1 mixture of male (XY) to female (XX) DNA.

Once the separation is complete, the data are analyzed using specialized software that assigns an allele repeat number to each peak based on the alleles identified by the previously run allelic ladder. Statistical analyses can be performed using relevant population frequencies to determine the overall probability of another individual having an identical DNA profile [36]. Because the different genetic loci used in determining the profile are inherited independently, the individual probabilities of having a particular set of alleles at one locus can be multiplied together for all loci producing a random match probability for the 13 CODIS loci of 1×10^{-15} [24]. For all intents and purposes, the data from a single source profile such as that shown in Figure 25.1 can be used to provide identification of a suspect [37]. Analysis of more complex samples such as those containing multiple contributors (Figure 25.3) or related individuals is less straightforward and requires in-depth statistical analysis [38–41].

25.4.2 Mini STRs

Large STR multiplex sets are used for most forensic evidence samples. However, biological fluids or materials left behind in a crime scene may be degraded because of exposure to a variety of harsh

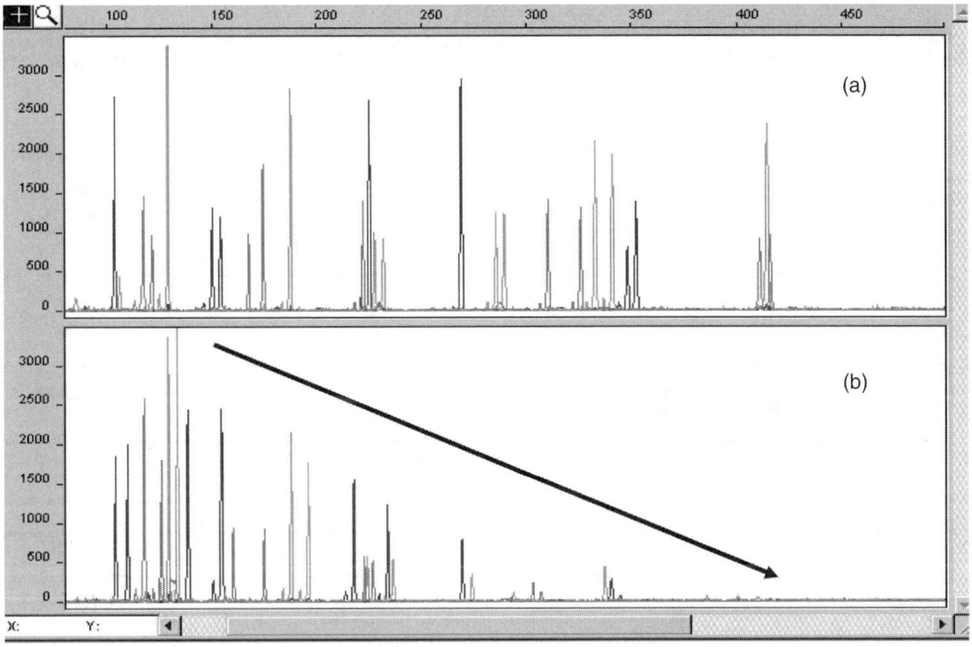

FIGURE 25.4 **(See color insert following page 810.)** A comparison between a 9947 control DNA sample (a) and a degraded DNA sample (b) extracted from a recovered bone fragment. There is an evident loss of signal at the larger size loci in sample b (arrow) due to sample degradation. The sample is extracted and amplified DNA prepared using the Promega Powerplex STR multiplex kit and consists of 16 separate genetic loci labeled with three different dyes and separated via capillary gel electrophoresis using the ABI 310 genetic analyzer. (Courtesy of Kerry Opel, International Forensic Research Institute, Florida International University.)

conditions and/or due to the presence of various contaminants that have been mixed with the sample [42,43]. In both cases, the result may be a partial DNA profile (Figure 25.4) in which some alleles are missing from the profile or are present below a laboratory's interpretational threshold [44,45]. Characteristically, the larger alleles lose intensity due to increasing decomposition of the template into smaller fragments. The resulting partial electropherograms are far less definitive than a full profile since fewer loci are available for statistical analysis. This leads to an increase in the probability of finding a random unrelated individual in the general population with a matching profile.

In these situations, one approach has been to perform further testing using mtDNA sequencing. Since there are multiple copies of mitochondrion in each cell, the likelihood of obtaining a result is greatly increased when compared to nuclear DNA. However, mtDNA analysis involves difficult and expensive analytical procedures, and because it is inherited solely through the female line, maternally related individuals will all have the same profile and statistical results are much less conclusive.

In recent years, investigators driven by this issue have developed a viable alternative—the use of mini STRs. Mini STRs are reduced size STR amplicons that can be obtained by redesigning the amplification primers in such a way that they bind closer to the STR repeat regions [46–48]. To keep the fragment size as short as possible only one or two STR loci can be used in each dye lane. These shorter amplicons can be detected when the original template is too fragmented to properly amplify (Figure 25.5).

Many of these STR markers targeted by modified primers are the same as those already established for the CODIS database [46]. Concordance studies between the mini STR and the commercially available STR primers have been performed and some discrepancies (0.2%) have been observed in the allele calls. These, however, can be explained by the fact that the primers from the mini STRs bind at different locations than those of the original STRs, thus insertions/deletions can affect primer

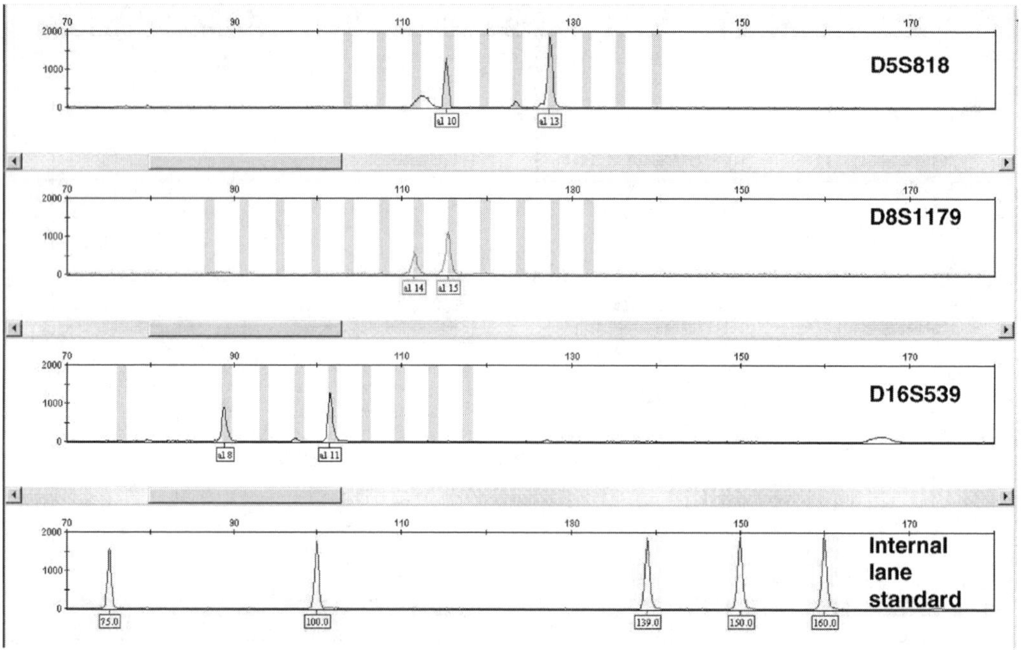

FIGURE 25.5 Mini STR amplification. The electropherogram illustrates the analysis of an extracted and amplified bone sample. The electropherogram consists of four panels showing three STR loci and an internal lane standard run simultaneously and detected with four different dyes. Unlike the larger multiplexes illustrated in previous figures, only one locus appears in each dye lane in an effort to keep the amplified products as short as possible. (Courtesy of Kerry Opel, International Forensic Research Institute, Florida International University.)

binding resulting in an apparent discrepancy in allele size [49]. This problem is being addressed through alternative primer designs, and researchers are continuing to identify new mini STRs to increase the information content of STR assays [50].

25.4.3 Mitochondrial DNA Analysis

In situations in which minimal DNA can be recovered from a sample due to severe degradation or lack of recoverable DNA, mtDNA can be exploited due to its relatively small size and presence in multiple copies within the cell [51]. The mitochondrion is a self-replicating organelle that is able to synthesize its own DNA [52]. The DNA is circular and contains 16,569 bp [53]. It contains both coding and noncoding regions, and it is in the noncoding hypervariable regions, HVI and HVII, where information relevant to forensic information is found [18,54]. These hypervariable regions have high-mutation rates that enable them to be a useful tool for human genetic analysis [55]. Owing to its small size, mtDNA does not contain STRs or other repetitive elements and instead analysis relies on variations in sequences [51,56]. Mutation detection takes place through the analysis of single nucleotide polymorphisms (SNPs); deletions, additions, and substitutions of various nucleotides in HVI and HVII. The treatment the samples receive before being loaded into the CE system is also different from that of the STRs. Because these samples are generally present at low copy number and are highly degraded, extensive use of control samples with strict isolation protocols is necessary in order to avoid cross contamination and maintain data reliability.

The samples are first amplified and the resultant native, unlabeled PCR products are quantified using CGE or microfluidic CGE with fluorescent intercalating dyes to determine input levels and

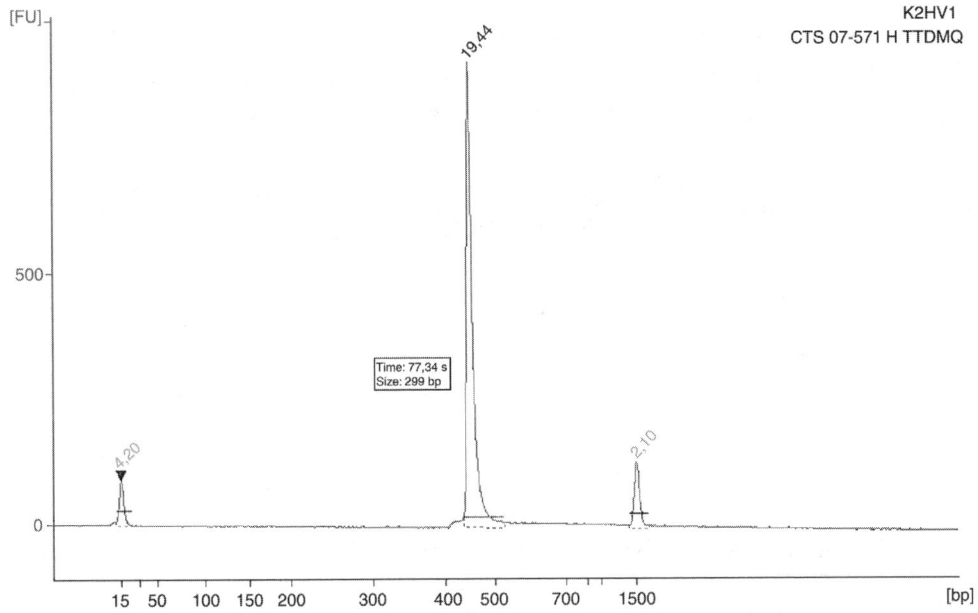

FIGURE 25.6 Quantitative analysis of PCR-amplified mtDNA by microfluidic CE. Determination of the overall quality and quantity of the DNA before sequencing is important to assure high-quality results. The figure illustrates the detection of an mtDNA amplicon from the HVI region located between two internal standards used for sizing. The analysis is performed on an extracted blood stain by an Agilent 2100 bioanalyzer using fluorescent detection with an intercalating dye. (Courtesy of the DNA Analysis Unit II at the FBI Laboratory, Quantico, VA.)

sample quality for the subsequent sequencing reaction [16,57] (Figure 25.6). Sequencing reactions are then performed using the PCR template, and products are separated using denaturing CGE. The results are compared to a reference sequence to catalog the specific point mutations that are present [58] (Figure 25.7). The statistical analyses of these data are not as definitive as that obtained from STR typing since mitochondria are transmitted directly from mothers to their progeny and there is no admixing or shuffling of genetic information such as occurs in meiosis. Thus mtDNA analysis cannot produce the high-statistical certainty of identification produced in STR typing and instead is most useful as a means for maternal lineage determination [59,60]. However, in situations in which the only evidence available is badly degraded or where only a few cells containing DNA are present (such as the situation in which a single shed hair is recovered), mtDNA may be the only way useful genetic information can be obtained [51,61–63].

25.4.4 Y-STRs

As with mtDNA, which provides maternal lineage, the Y-chromosome can be used to provide paternal lineage. The first Y-STR was discovered in 1992 [64] Since then, a number of additional Y-STRs have been validated for use in the forensic field [65,66] and several commercial multiplex amplification kits are now available (Table 25.3). Initially used for paternity testing and rape case scenarios [67,68], Y-STRs are now being used to aid in missing persons investigations [69], genealogical research, and evolutionary studies [70–72]. Y-chromosome markers are particularly useful in the detection of small amounts of male DNA in the presence of an overwhelming abundance of female DNA. This type of STR loci are most useful when trying to isolate the male fraction from a DNA mixture (Figure 25.8) [73,74]. .

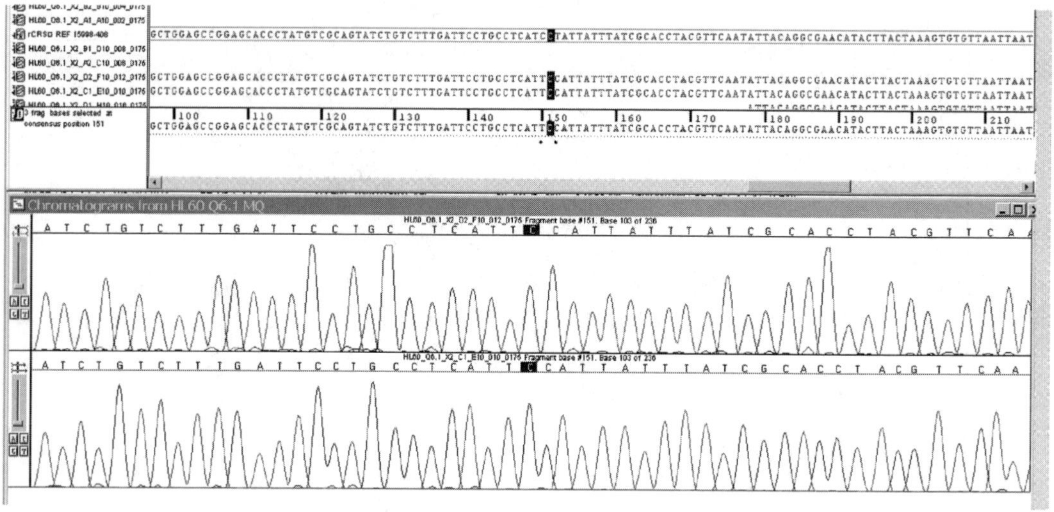

FIGURE 25.7 **(See color insert following page 810.)** Mitochondrial DNA profile of an HVII sequence from a hair shaft. A repetitive analysis of the same hair sample is used to compare and align the sequences. The nuclear DNA in hair is often badly degraded and difficult to amplify. mtDNA provides an alternative procedure that can provide a DNA haplotype for forensic DNA profiling based on point mutations detected in the sequence when compared to a reference sample. Analysis performed on an ABI 310 Genetic Analyzer. (Courtesy of the DNA Analysis Unit II at the FBI Laboratory, Quantico, VA.)

TABLE 25.3
Validated Y-STR Loci for Forensic Casework

Locus	Repeat	Size
DYS393	AGAT	108–132
DYS392	TAT	236–263
DYS391	TCTA	275–295
DYS389I	(TCTG)(TCTA)	239–263
DYS389II	(TCTG)(TCTA)	353–385
Y-GATA A7.2	TAGA	174–190
DYS438	TTTTC	203–233
DYS385	GAAA	252–300
DYS19	TAGA	242–254
DYS425	TGT	104–110
DYS388	ATT	119–131
DYS390	(TCTA)(TCTG)	200–251
DYS439	AGAT	242–258
DYS434	ATCT	106–116
DYS437	TCTA	188–192
Y-GATA C.4	TATC	250–269
Y-GATA A7.1	ATAG	104–112
Y-GATA H.4	TAGA	130–143

Source: Adapted from Daniels, D.L., Hall, A.M., and Ballantyne J., *J Forensic Sci*, 43, 668, 2004. With permission.

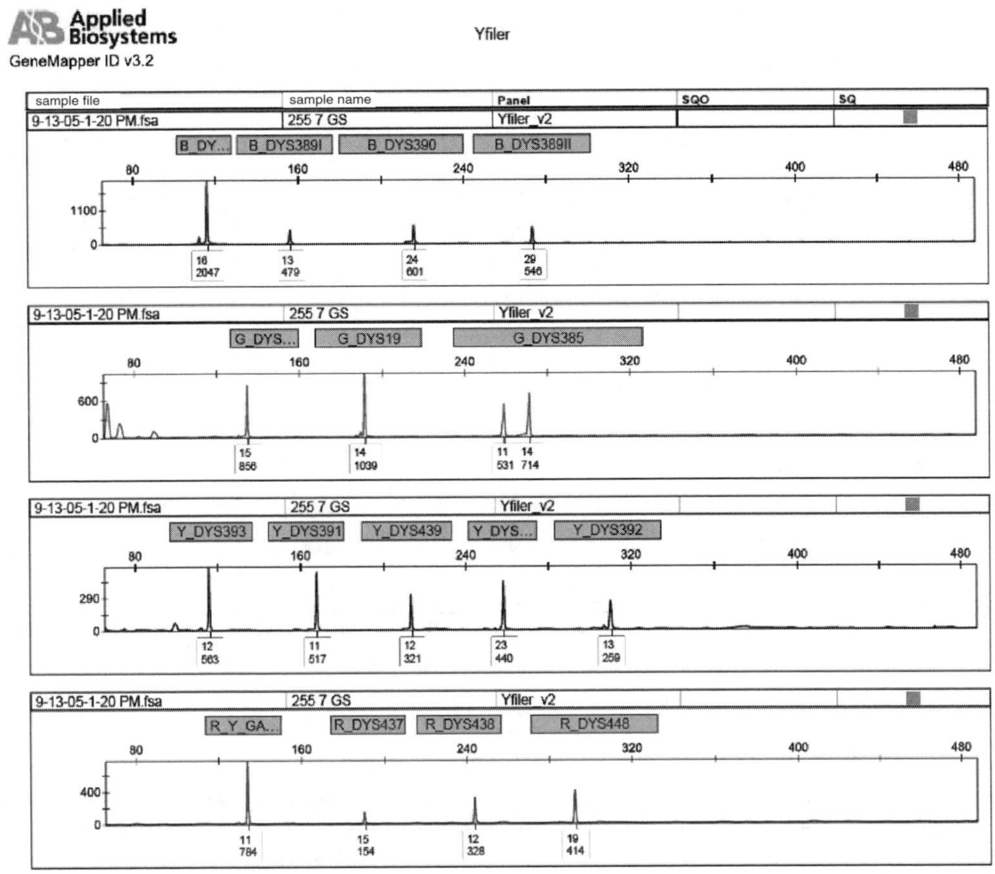

FIGURE 25.8 A Y-STR profile of the mixture in Figure 25.3 using the ABI Y Filer™Y chromosomal STR multiplex. Note that the DY385 locus has two alleles due to a duplication of the sequence on the Y chromosome. Analysis performed using an ABI 310 Genetic Analyzer. (Courtesy of Stephanie King and George Duncan, Broward County Sheriff's Office Crime Laboratory.)

An example of such a scenario might be a fingernail scraping from a female victim or the detection of a mixed DNA profile from the handle of an automobile. The profile depicted in Figure 25.9a results from an extract taken from a knife handle that includes DNA from the male suspect along with two female profiles. Use of the Y profile (Figure 25.9b) permits isolation of just male DNA, although it should be noted that all of the suspect's male relatives from his paternal lineage would also have the same Y-haplotype. Statistical analysis of the result involves performing an estimate of the frequency of a given profile in a database. Frequency estimates for Y profiles are much less specific than those obtained with autosomal STRs as individual allele frequencies cannot be multiplied together. Nevertheless, the data provide important information regarding the potential placement of an individual at a crime scene.

25.4.5 Single Nucleotide Polymorphisms

The various types of point mutations detected via mtDNA sequencing can also be targeted in nuclear DNA. SNPs are particularly useful in ethnicity testing and in the analysis of highly degraded or compromised samples [57,75,76]. In the human genome SNPs occur every 1000–2000 bp, thereby

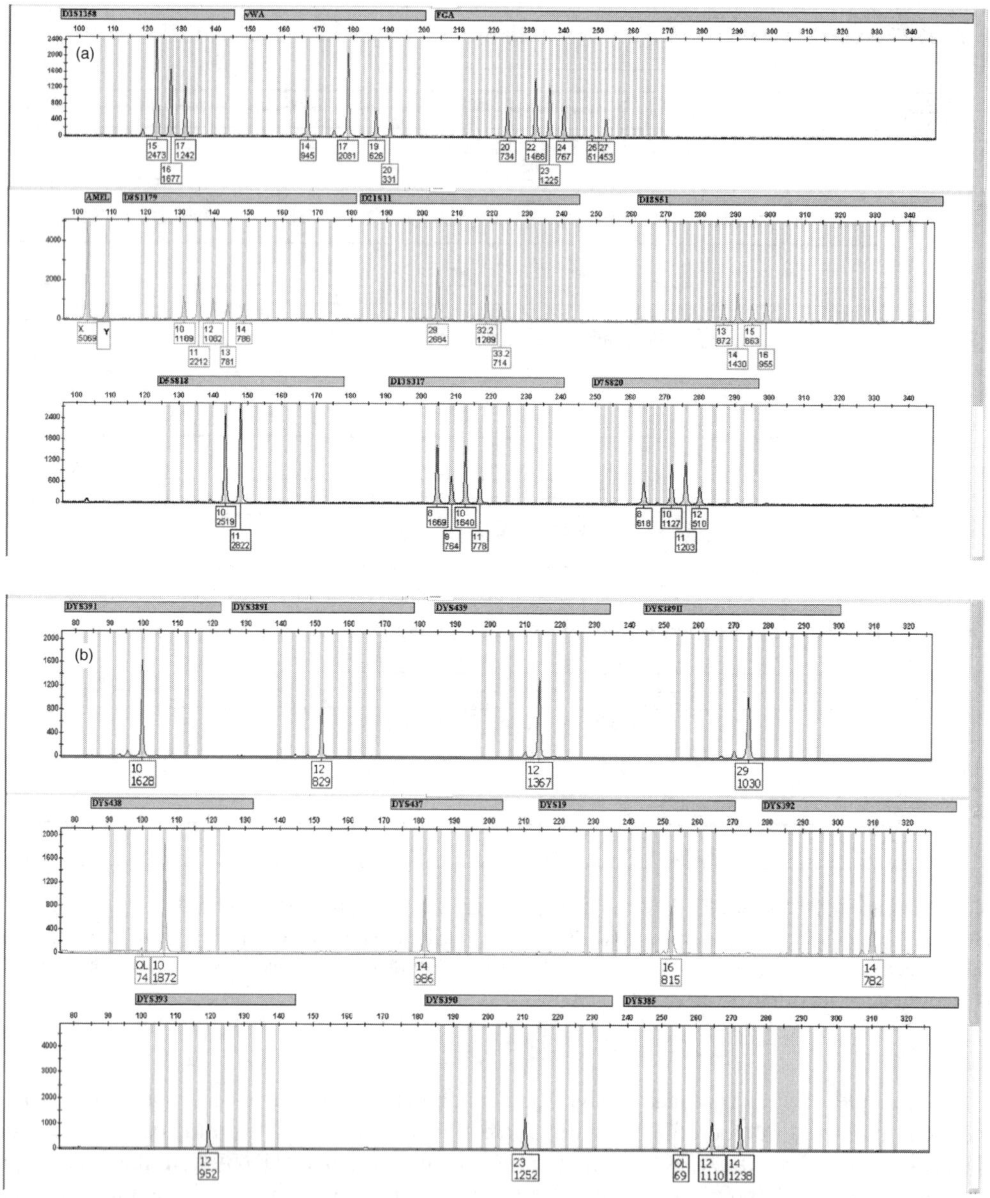

FIGURE 25.9 A comparison of (a) a mixed DNA profile (two females and one male contributor) followed by (b) the Y-STR profile of the same sample. The sample was collected with a single surface swab (moistened with DI water) from a knife handle (homicide case). Amplification was performed using PowerPlex®Y with 1.2 ng and Profiler Plus™with 0.8 ng of template DNA. "The male component DNA obtained from the knife handle and the suspect has the same Y-haplotype; therefore, the suspect could not be eliminated as the source of the male DNA in the mixture. The results also indicate the presence of only one male contributor in the DNA extract. It should be noted that all of the suspect's male relatives from his paternal lineage would also have the same Y-haplotype." (Courtesy of DNA Unit—Orange County Sheriff-Coroner Department, CA.)

accounting for about 90% of genetic variation [75,77,78]. Unlike tandem repeats, SNPs are found both in the coding and noncoding regions of the DNA molecule. They can play an important role in the field of forensics as they have, in contrast to STRs, lower mutation rates and may eventually be linked to physical features such as hair color, stature, and skin shade [79,80]. Unfortunately, SNP systems have limited numbers of alleles and therefore many more loci are required to perform a full genetic profile than is the case with STRs. In fact, about 50 SNPs would require analysis to achieve the same level of statistical uniqueness developed with 13 STRs [78]. In addition, SNPs cannot easily be used in mixture studies as fragment sizes overlap making it difficult to isolate multiple contributors to a profile. SNPs can be detected using a single base primer extension assay in which a special primer is designed to target a known SNP location [81]. A polymerase is then added and used to probe the site of the polymorphism (the next unincorporated base pair) through the addition of a terminal fluorescently labeled dideoxynucleotide (ddNTP). The SNP assay can be multiplexed by simultaneously labeling multiple numbers of these loci and using changes in primer length to permit all locations to be detected simultaneously. Like mtDNA sequencing and mini STRs analysis, SNPs can be valuable in the recovery of information from degraded DNA. Figure 25.10 illustrates the comparison of results from the standard STR typing of a degraded DNA sample followed by SNP typing of that same DNA extract. The SNP amplicons are much shorter and provide additional genetic information to assist in the identification of the partial STR profile.

25.4.6 Mutation Detection

Although more commonly used in oncology and detection of genetic diseases [82–84], mutation detection techniques can provide useful information in forensic analysis. Sequence polymorphisms resulting from one or more base pair changes in PCR-amplified DNA fragments can be exploited through differences in melting points, heteroduplex formation, or conformational variations. A variety of different CE procedures exist to detect these differences, including single-stranded conformational polymorphism (SSCP), heteroduplex polymorphism, and constant denaturant CE. Using these techniques, DNA fragments of the same or similar length that would otherwise comigrate can be differentiated based on mobility differences resulting from the effect of temperature, denaturant concentration, or rehybridization effects. These types of mobility assays have been used in paternity disputes [85] and in ABO allele discrimination [86]. These techniques are relatively inexpensive to perform and can be highly sensitive to slight differences in sequence [87] making them highly useful in the detection of previously unknown mutational events.

25.4.7 Nonhuman DNA

Nonhuman DNA can provide crucial information in a variety of crime scenarios [88]. Items found at a crime scene, such as animal hair, plant material, fungal spores, and soils, may all contain recoverable DNA. Genetic markers such as STRs have been isolated from these types of materials and a number of different loci have been validated for use in criminal casework [89,90]. For other items such as microbial DNA, soil and plant materials, sequence information may not exist, and various exploratory procedures must be used to recover genetic information.

25.4.7.1 Animal DNA

Feline and canine STR analyses have been used in forensic casework [91–94] and specific kits have been developed to determine the genetic fingerprint of both animals [95,96]. Canine genetic material has also been analyzed via CE using mtDNA sequencing [97]. As with human samples, canine DNA can help link a suspected animal to crimes such as bite attacks or mauling. Similarly, it can also be used to exclude the animal from implication in attacks or unlawful deaths [98]. More commonly, however, dog hair and other animal hair has been found associated with victims at crime scenes [94,99]. Figure 25.11 illustrates an electropherogram of a suspect dog hair that might be used to associate an animal with the victim of an abduction. Genetic analysis of this material provides further information on the circumstances of the crime and can connect the victim to a particular

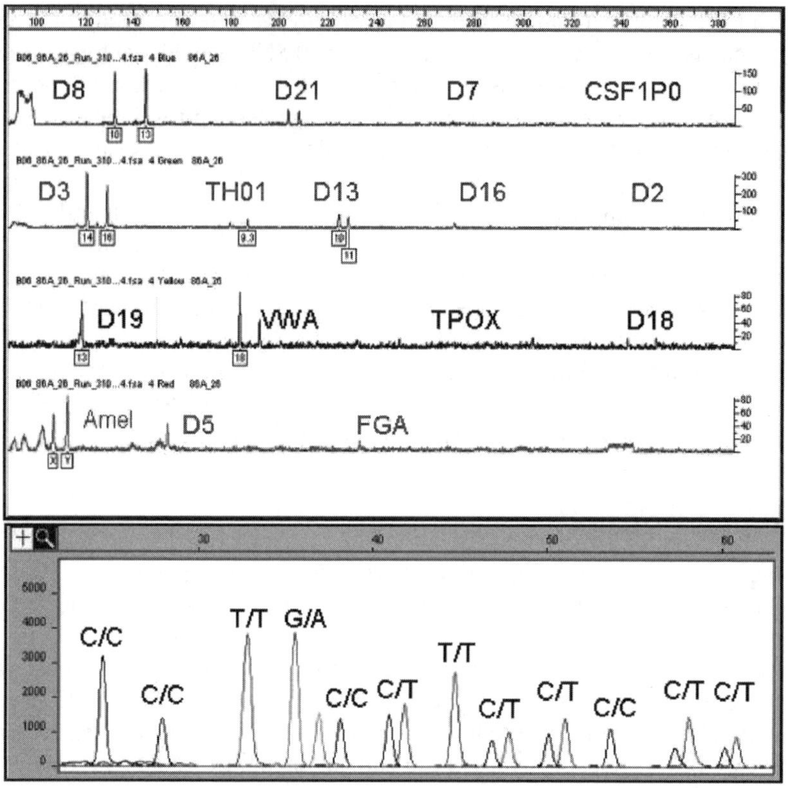

FIGURE 25.10 (**See color insert following page 810.**) The recovery of information from degraded DNA using the Identifiler™STR multiplex and an SNP multiplex, both analyzed using the ABI 310 genetic analyzer. The top electropherogram depicts a degraded sample amplified via STRs. The bottom profile depicts a SNP multiplex on the same sample. The STR profile is blank for many larger alleles such as D7 CSF, D16, D2, TPOX, D18, and FGA. The smaller sized SNP fragments permit recovery of additional genetic information from the sample. (From Vallone, P.M., Decker, A.E, and Butler, J.M., The evaluation of Autosomal SNP assays, *Chemical Science and Technology Laboratory, National Institute of Standards and Technology*, Div.831.)

location or suspect. Although more commonly used for lineage analysis, other genetic material such as equine and bovine DNA could also be used in criminal casework [100].

25.4.7.2 Botanical DNA

Plant DNA analysis has been increasing over the past couple of years because of its ability to pinpoint the origin of drug-related plant material [101,102]. Plant STRs are not yet well characterized; however, other molecular methods can be combined with capillary gel electrophoresis to establish the identity of plant species [103]. These techniques include amplified fragment length polymorphism (AFLP), random amplification of polymorphic DNA (RAPD), and other similar random primer annealing techniques used for mutation detection.

AFLP is a molecular tool in which the DNA is cut by a combination of restriction enzymes for subsequent amplification utilizing relatively nonspecific primers that are tagged with a fluorescent dye [104,105]. These primers will anneal only to the fraction of the cut fragments that contain a short complimentary sequence, leaving behind those that do not. Since DNA sequences vary, the cut

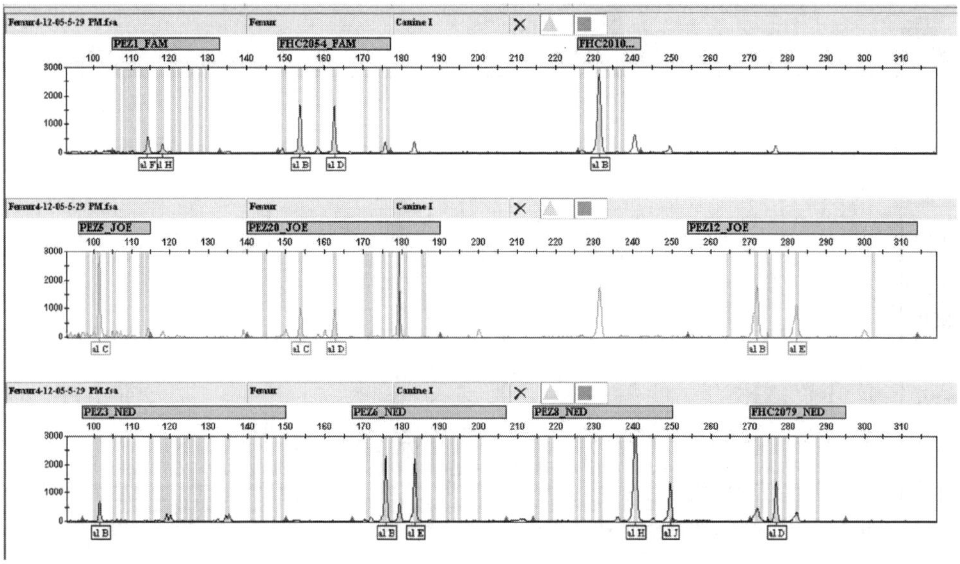

FIGURE 25.11 Canine STR analysis. Ten canine markers were successfully amplified and analyzed after extracting DNA from a single dog hair. (Courtesy of Lilliana Moreno, Forensic DNA Profiling Facility, Florida International University.)

sites produced by the selected restriction enzymes will vary between the different plant materials submitted for analysis. The result is a set of DNA fragments of different sizes and intensities that can be related to a particular plant cultivar. The specific fragments can be isolated and detected by capillary gel electrophoresis. Various computer software are then used to define and catalog the differences between samples. AFLP has proven to be a particularly valuable tool for establishing plant identification and for tracing plant material back to its original source [106]. Figure 25.12 illustrates the analysis of a seized marijuana sample using AFLP with multichannel fluorescent CE detection.

25.4.7.3 Microbial DNA

Recent events involving the use of microbes as bioterrorists' weapons have intensified research in the developing field of microbial forensics. This new field couples law enforcement efforts with existing procedures to identify patterns in disease outbreaks, determine the pathogen involved, control its spread, and trace the microorganism back to its source [107]. A variety of techniques are currently being used as tools for the classification and identification of microorganisms and these procedures can also be used to assist investigators in the detection of criminal acts. While most of the reported incidents to date involve hoaxes with nonpathogenic material, the consequences of such an attack require that laboratories be properly prepared to meet the threat. In addition, these same techniques can also be used to examine trace evidence such as soils left behind following a criminal act [108,109]. A number of molecular techniques for microbial detection and analysis are listed below.

Terminal restriction fragment length polymorphism (TRFLP) uses end-labeled primers that will bind to specific primer sequences used when amplifying DNA by PCR [110,111]. After digestion of the PCR products with restriction enzymes, the samples are loaded into the CE system for separation.

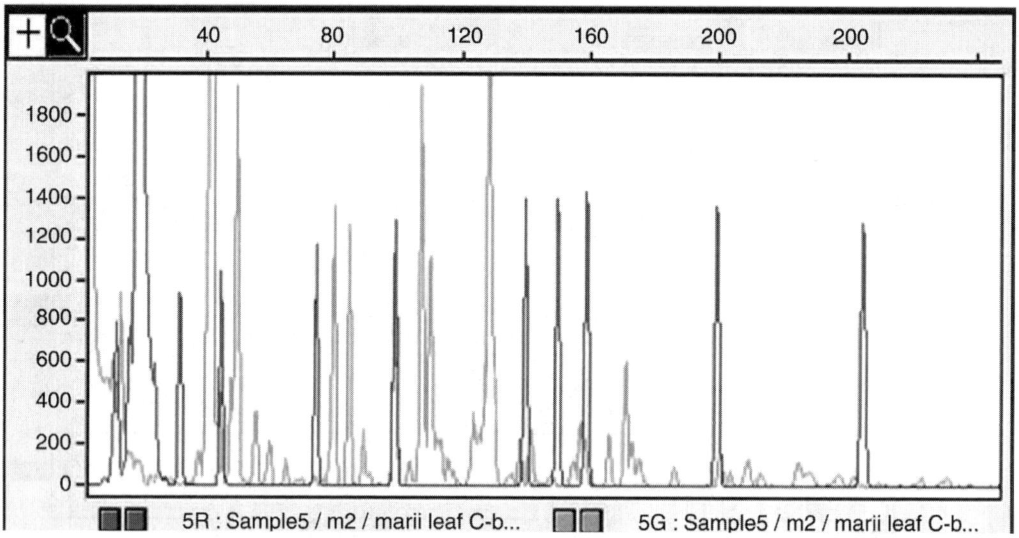

FIGURE 25.12 Genetic analysis of a marijuana sample using AFLP. The sample consists of amplified DNA fragments characteristic of the plant overlaid on top of an internal lane standard used to size the individual fragments. Analysis performed on an ABI 310 genetic analyzer. (Courtesy of Ira Lurie, Yin Shen, and Bruce McCord.)

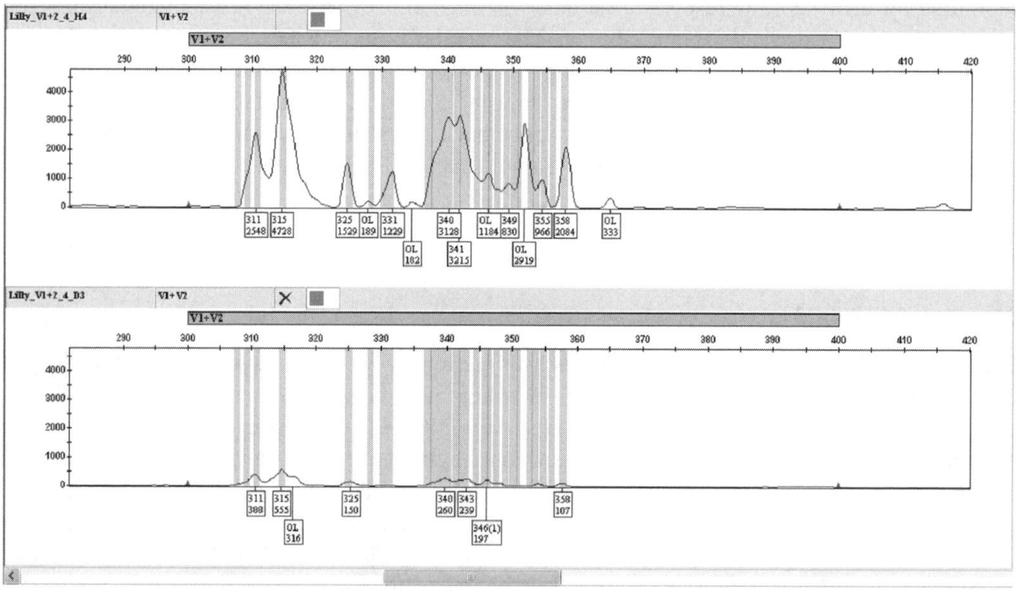

FIGURE 25.13 Soil microbial DNA profile. The 16S rRNA gene hypervariable region V1_V2 was amplified using ALH technique. The different DNA fragments identified in the figure depict the different microbes present in the soil. These data can then be used to aid in the identification of soils associated with a crime scene and to help answer important questions such as has a body been moved from its current location. Analysis performed using an ABI 310 genetic analyzer. (Courtesy of Lilliana Moreno, Forensic DNA Profiling Facility, Florida International University.)

This method has been extensively used in conjunction with the small ribosomal subunit (16S rRNA) to establish differences between microbial entities [112,113] and was the technique first used in the analysis of microbial communities for forensic purposes [108]. Subsequent researchers in microbial forensics examined a method known as amplicon length heterogeneity (ALH), which is commonly used in microbial ecology. ALH bases its profiles on differences in length within select hypervariable domains of the 16S rRNA genes. DNA extracts from different microbial communities are amplified using universal primers that bind outside of the variable region and are capable of hybridizing to the majority of target organisms. The output is a pattern of peaks that provide, just like STRs, a unique pattern that is used to differentiate microbial communities (Figure 25.13). Unlike TRFLP, this highly reproducible method does not require restriction endonucleases but is based on the natural variation in sequence lengths of specific regions within the gene [114,115]. Because of its ubiquitous nature, soil can be a valuable evidence in crime scene investigations; however, it is seldom used because the physical analysis is complicated and requires experts in the field of geology. A novel approach utilizing ALH-PCR to discriminate between soils for forensic purposes has been developed and tested and provides a promising foundation for the future of the application [109].

25.5 CONCLUSIONS

Capillary electrophoresis technology has become an indispensable tool for forensic scientists in the biology field since it is able to provide valuable information to aid in the process of law enforcement. The primary application of the technique is in the qualitative analysis of STRs. Isolation of STR mixtures is also possible using relative peak heights. Other applications of CE include quantitative analysis of PCR products, mtDNA sequencing, and mutation detection for the analysis of plant and bacterial DNA. Based on the performance of the methods illustrated above, it is reasonable to expect future researchers and practitioners to continue working to exploit the capabilities of this robust scientific technique and its application to criminal investigations.

ACKNOWLEDGMENTS

The authors would like to thank John Butler, Dee Mills, Alice Isenberg, Kate Theisen, Pete Vallone, Ed Buse, Ada Nuñez, Kerry Opel, Ira Lurie, Stephanie King, Yin Shen, and George Duncan for their contributions to this work. Major funding from the National Institute of Justice is also gratefully acknowledged. Points of view in the document are those of the authors and do not necessarily represent the official view of the U.S. Department of Justice.

REFERENCES

1. Tagliaro, F. and Smith, F., Forensic capillary electrophoresis, *Trends Anal Chem*, 15, 513, 1996.
2. Northrop, D., McCord, B., and Butler, J., Forensic applications of capillary electrophoresis, *J Capillary Electrophor*, 1, 158, 1994.
3. Barron, A., Sunada, W., and Blanch, H., Capillary electrophoresis of DNA in uncrosslinked polymer solutions: Evidence for a new mechanism of DNA separation, *Biotechnol Bioeng*, 52, 259, 1996.
4. Arnheim, N., Polymerase chain reaction strategy, *Annu Rev Biochem*, 61, 131, 1992.
5. Luckey, J. and Smith, L., Advances in DNA sequencing technology, *Proc SPIE*, 1891, 21, 1993.
6. Fregeau, C. and Fourney, R., DNA typing with fluorescently tagged short tandem repeats: A sensitive and accurate approach to human identification, *BioTechniques*, 15, 100, 1993.
7. Cohen, A., Najarian, D., Paulus, A., et al., Rapid separation and purification of oligonucleotides by high-performance capillary gel electrophoresis, *Proc Natl Acad Sci USA*, 85, 9660, 1988.
8. Kasper, T., Melera, M., Gozel, P., et al., Separation and detection of DNA by capillary electrophoresis, *J Chromatogr*, 458, 303, 1988.

9. Albarghouthi, M., Buchholz, B., Huiberts, P., et al., Poly-*N*-hydroxyethylacrylamide (polyDuramide): A novel, hydrophilic, self-coating polymer matrix for DNA sequencing by capillary electrophoresis, *Electrophoresis*, 23, 1429, 2002.
10. Tseng, W. and Chang, H., A new strategy for optimizing sensitivity, speed and resolution in capillary electrophoretic separation of DNA, *Electrophoresis*, 22, 763, 2001.
11. Mandabhushi, R., Separation of 4-color DNA sequencing extension products in noncovalently coated capillaries using low viscosity polymer solutions, *Electrophoresis*, 19, 224, 1998.
12. Rocheleau, M., Grey, R., Chen, D., et al., Formamide modified polyacrylamide gels for DNA sequencing by capillary gel electrophoresis, *Electrophoresis*, 13, 484, 1992.
13. Guttman, A. and Cooke, N., Denaturing capillary gel electrophoresis, *Am Biotechnol Lab*, 9, 10, 1991.
14. Brining, S., Chen, N., Yi, D., et al., Gel electrophoretic distinction between toxic and nontoxic forms of beta-amyloid (1-40), *Electrophoresis*, 20, 1398, 1999.
15. Ozawa, S., Sugano, K., Sonehere, T., et al., High resolution for single-strand conformation polymorphism analysis by capillary electrophoresis, *Anal Chem*, 76, 6122, 2004.
16. Butler, J., McCord, B., Jung, J., et al., Quantitation of polymerase chain reaction products by capillary electrophoresis using laser fluorescence, *J Chromatogr B*, 658, 271, 1994.
17. Butler, J., Buel, E., Crivellente, F., et al., Forensic DNA typing by capillary electrophoresis using the ABI Prism 310 and 3100 genetic analyzers for STR analysis, *Electrophoresis*, 25, 1397, 2004.
18. Wilson, M., Stoneking, M., Holland, M., et al., Guidelines for the use of mitochondrial DNA sequencing in forensic science, *Crime Lab Digest*, 20, 68, 1993.
19. LaFountain, M., Schwartz, M., Svete, P., et al., TWGDAM validation of the AmpFlSTR profiler Plus and AmpFlSTR Cofiler STR multiplex systems using capillary electrophoresis, *J Forensic Sci*, 46, 1191, 2001.
20. Hartzell, B., Graham, K., and McCord, B., Response of short tandem repeat systems to temperature and sizing methods, *Forensic Sci Int*, 133, 228, 2003.
21. Nock, T., Dove, J., McCord, B., et al., Temperature and pH studies of short tandem repeat systems using capillary electrophoresis at elevated pH, *Electrophoresis*, 22, 755, 2001.
22. Budowle, B., Chakraborty, R., Carmody, G., et al., Source attribution of a forensic DNA profile, *Forensic Sci Comm*, 2, 1, 2000.
23. Moxon, E. and Wills, C., DNA microsatellites: Agents of evolution? *Sci Am*, 280, 94, 1999.
24. Butler, J., *Forensic DNA Typing: Biology, Technology, and Genetics of STR Markers*, 2nd edn. Academic Press, San Diego, CA, 2005.
25. www.fbi.gov, accessed 12/2006.
26. Budowle, B., Moretti, T. R., Niezgoda, S. J., et al., CODIS and PCR-based short tandem repeat loci: Law enforcement tools., *Forensic Science Research and Training Center, FBI Academy*, 73–77.
27. Ruitberg, C. M., Reeder, D. J., and Butler, J. M., STRBase: A short tandem repeat DNA database for the human identity testing community, *Nucleic Acids Res*, 29, 320, 2001.
28. http://www.FBi.gov/hg/lab/codis/program.htm, accessed 8/21/07.
29. Cotton, E., Allsop, R., Guest, J., et al., Validation of the AMPFlSTR SGM plus system for use in forensic casework, *Forensic Sci Int*, 112, 151, 2000.
30. Krenke, B., Tereba, A., Anderson, S., et al., Validation of a 16-locus fluorescent multiplex system, *J Forensic Sci*, 47, 773, 2002.
31. Hobson, D., Smerick, J., Brown, A., et al., A summary of the validation of the STR analysis by capillary electrophoresis: development of interpretation guidelines for the Profiler Plus and Cofiler systems for use in forensic casework, in *Proceedings from the 10th International Symposium on Human Identification*, Promega Corp, Coronoado Springs (1999).
32. Gill, P., Sparkes, R., Pinchin, R., et al., Interpreting simple STR mixtures using allele peak areas, *Forensic Sci Int*, 91, 41, 1997.
33. Ladd, C., Lee, H., Yang, N., et al., Interpretation of complex forensic DNA mixtures, *Croat Med J*, 42, 244, 2001.
34. Gill, P., Whitaker, J., Flaxman, C., et al., An investigation of the rigor of interpretation rules for STRs derived from less than 100 pg of DNA, *Forensic Sci Int*, 112, 17, 2000.
35. Gill, P., Jeffreys, A., and Werrett, D., Forensic application of DNA "fingerprints," *Nature*, 318, 577, 1985.

36. Weir, B., The second national research council report on forensic DNA evidence, *Am J Hum Genet*, 59, 497, 1996.
37. Sajantila, A. and Budowle, B., Identification of individuals with DNA testing, *Ann Med*, 23, 637, 1991.
38. Fung, W. and Hu, Y., Interpreting DNA mixtures based on the NRC-II recommendation 4.1, *Forensic Sci Comm*, 2, 1–11, 2000.
39. Weir, B., Triggs, C., Starling, L., et al., Interpreting DNA mixtures, *J Forensic Sci*, 42, 213, 1997.
40. Clayton, T., Whitaker, J., Sparkes, R., et al., Analysis and interpretation of mixed forensic stains using DNA STR profiling, *Forensic Sci Int*, 91, 55, 1998.
41. Evett, I., Gill, P., and Lambert, J., Taking account of peak areas when interpreting mixed DNA profiles, *J Forensic Sci*, 43, 62, 1998.
42. Ludes, B., Pfitzinger, H., and Mangin, P., DNA fingerprinting from tissues after variable postmortem periods, *J Forensic Sci*, 38, 686, 1993.
43. Smith, S. and Morin, P., Optimal storage conditions for highly dilute DNA samples: A role for trehalose as a preserving agent, *J Forensic Sci*, 50, 1101, 2005.
44. Whitaker, J., Clayton, T., Urquhart, A., et al., Short tandem repeat typing of bodies from a mass disaster: High success rate and characteristic amplification patterns in highly degraded samples, *BioTechniques*, 18, 670, 1995.
45. Anderson, T., Ross, J., Roby, R., et al., A validation study for the extraction and analysis of DNA from human nail material and its application to forensic casework, *J Forensic Sci*, 44, 1053, 1999.
46. Butler, J. M., Shen, Y., and McCord, B. R., The development of reduced size STR amplicons as tools for analysis of degraded DNA, *J Forensic Sci*, 48, 1054, 2003.
47. Wiegand, P. and Kleiber, M., Less is more—Length reduction of STR amplicons using redesigned primers, *Int J Legal Med*, 114, 285, 2001.
48. Tsukada, K., Takayanagi, K., Asamura, H., et al., Multiplex short tandem repeat typing in degraded samples using newly designed primers for the TH01, TPOX, CSF1PO, and vWA loci, *Legal Med*, 4, 239, 2002.
49. Drabek, J., Chung, D., Butler, J., et al., Concordance study between miniplex assays and a commercial STR typing kit, *J Forensic Sci*, 49, 859, 2004.
50. Coble, M. and Butler, J., Characterization of new miniSTR loci to aid analysis of degraded DNA, *J Forensic Sci*, 50, 43, 2005.
51. Bender, K., Schneider, P., and Rittner, C., Application of mtDNA sequence analysis in forensic casework for the identification of human remains, *Forensic Sci Int*, 113, 103, 2000.
52. Saccone, C., Gissi, C., Lanave, C., et al., Evolution of the mitochondrial genetic system: An overview, *Gene*, 261, 153, 2000.
53. Anderson, S., Bankier, A., Barrell, B., et al., Sequence and organization of the human mitochondrial genome, *Nature*, 290, 457, 1981.
54. Holland, M. and Parsons, T., Mitochondrial DNA sequence analysis—Validation and use for forensic casework, *Forensic Sci Rev*, 11, 21, 1999.
55. Jazin, E., Soodyall, H., Jalonen, P., et al., Mitochondrial mutation rate revisited: Hot spots and polymorphisms, *Nat Genet*, 18, 109, 1998.
56. Wilson, M., DiZinno, J., Polanskey, D., et al., Validation of mitochondrial DNA sequencing for forensic casework analysis, *Int J Legal Med*, 108, 68, 1995.
57. Kline, M., Vallone, P., Redman, J., et al., Mitochondrial DNA typing screens with control region and coding region SNPs, *J Forensic Sci*, 50, 377, 2005.
58. Isenberg, A. and Moore, J., Mitochondrial DNA analysis at the FBI laboratory, *Forensic Sci Comm*, 1, 1999.
58. Cann, R., Stoneking, M., and Wilson, A., Mitochondrial DNA and human evolution, *Nature*, 325, 31–36 1987.
59. Baasner, A., Schafer, C., Juge, A., et al., Polymorphic sites in human mitochondrial DNA control region sequences: Population data and maternal inheritance, *Forensic Sci Int*, 98, 169, 1998.
60. Budowle, B., Allard, M., Wilson, M., et al., Forensics and mitochondrial DNA: Applications, debates, and foundations, *Annu Rev Genom Hum G*, 4, 119, 2003.
61. Stone, A., Starrs, J., and Stoneking, M., Mitochondrial DNA analysis of the presumptive remains of Jesse James, *J Forensic Sci*, 46, 173, 2001.

62. vonWurmb-Schwark, N., Harbeck, M., Wiesbrock, U., et al., Extraction and amplification of nuclear and mitochondrial DNA from ancient and artificially aged bones, *Legal Med*, 5, 169, 2003.
63. Roewer, L. and Epplen, J., Rapid and sensitive typing of forensic stains by PCR amplification of polymorphic simple repeat sequences in case work, *Forensic Sci Int*, 53, 163, 1992.
64. Daniels, D., Hall, A., and Ballantyne, J., SWGDAM developmental validation of a 19-locus Y-STR system for forensic casework, *J Forensic Sci*, 49, 1, 2004.
65. Hall, A. and Ballantyne, J., The development of an 18-locus Y-STR system for forensic casework, *Anal Bioanal Chem*, 376, 1234, 2003.
66. Kayser, M. and Sajantila, A., Mutations at Y-STR loci: Implications for paternity testing and forensic analysis, *Forensic Sci Int*, 118, 116, 2001.
67. Corach, D., Filgueira, R., Marino, M., et al., Routine Y-STR typing in forensic casework, *Forensic Sci Int*, 118, 131, 2001.
68. Huffine, E., Crews, J., Kennedy, B., et al., Mass identification of persons missing from the break-up of the former Yugoslavia: Structure, function, and role of the International Commission of Missing Persons, *Croat Med J*, 42, 271, 2001.
69. Butler, J., Recent developments in Y-short tandem repeat and Y-single nucleotide polymorphism analysis, *Forensic Sci Rev*, 15, 91, 2003.
70. Hedman, M., Pimenoff, V., Lukka, M., et al., Analysis of 16 Y STR loci in the Finnish population reveals a local reduction in the diversity of male lineages, *Forensic Sci Int*, 142, 37, 2004.
71. Silva, D., Carvalho, E., Costa, G., et al., Y-chromosome genetic variation in Rio De Janeiro population, *Am J Hum Biol*, 18, 829, 2006.
72. Cerri, N., Ricci, U., Sani, I., et al., Mixed stains from sexual assault cases: Autosomal or Y-chromosome short tandem repeats? *Croat Med J*, 44, 289, 2003.
73. Hanson, E. and Ballantyne, J., A highly discriminating 21 locus Y-STR "megaplex" system designed to augment the minimal haplotype loci for forensic casework, *J Forensic Sci*, 49, 40, 2004.
74. Budowle, B., SNP typing strategies, *Forensic Sci Int*, 146, 139, 2004.
75. Vallone, P., Decker, A., and Butler, J., Allele frequencies for 70 autosomal SNP loci with US Caucasian, African American, and Hispanic samples, *Forensic Sci Int*, 149, 279, 2005.
76. Stoneking, M., Single nucleotide polymorphisms: From the evolutionary past, *Nature*, 409, 821, 2001.
77. Venter, J. C., Adams, M. D., Myers, E. W., et al., The sequence of the human genome, *Science*, 291, 1304, 2001.
78. Amorim, A. and Pereira, L., Pros and cons in the use of SNPs in forensic kinship investigation: A comparative analysis with STRs, *Forensic Sci Int*, 150, 17, 2005.
79. Payseur, B. and Cutter, A., Integrating patterns of polymorphism at SNPs and STRs, *Trends Genet*, 22, 424, 2006.
80. Syvanen, A., Accessing genetic variation: Genotyping single nucleotide polymorphisms, *Nat Rev Genet*, 2, 930, 2001.
81. Tian, H., Jaquins-Gerstl, A., Munro, N., et al., Single-strand conformation polymorphism analysis by capillary and microchip electrophoresis: A fast, simple method for detection of common mutations in BRCA1 and BRCA2, *Genomics*, 63, 25, 2000.
82. Atha, D., Kasprzak, W., O'Connell, C., et al., Prediction of DNA single-strand conformation polymorphism: Analysis by capillary electrophoresis and computerized DNA modeling, *Nucleic Acids Res*, 29, 4643, 2001.
83. Orita, M., Iwahana, H., Kanazawa, H., et al., Detection of polymorphisms of human DNA by gel electrophoresis as single-strand conformation polymorphisms, *Proc Natl Acad Sci USA*, 86, 2766, 1989.
84. Fukuda, M. and Tamaki, Y., Subtyping of D11S488 STR alleles by single-strand conformation polymorphism (SSCP) analysis in two cases of disputed parentage, *Nihon Hoigaku Zasshi*, 52, 42, 1998.
85. Tsai, L., Kao, L., Chang, J., et al., Rapid identification of the ABO genotypes by their single-strand conformation polymorphism, *Electrophoresis*, 21, 537, 2000.
86. Sunnucks, P., Wilson, A., Beheregaray, L., et al., SSCP is not so difficult: The application and utility of single-stranded conformation polymorphism in evolutionary biology and molecular ecology, *Mol Ecol*, 9, 1699, 2000.

87. Day, A., Nonhuman DNA testing increases DNA's power to identify and convict criminals, *Silent Witness*, 6, 2001.
88. Hellmann, A., Rohleder, U., Eichmann, C., et al., A proposal for standardization in forensic canine DNA typing: Allele nomenclature of six canine-specific STR loci, *J Forensic Sci*, 51, 274–281, 2006.
89. Eichmann, C., Berger, B., and Parson, W., A proposed nomenclature for 15 canine-specific polymorphic STR loci for forensic purposes, *Int J Legal Med*, 118, 249, 2004.
90. Sensabaugh, G. and Kaye, D., Non-human DNA evidence, *Jurimetrics J*, 38, 1, 1998.
91. Eichmann, C., Berger, B., Steinlechner, M., et al., Estimating the probability of identity in a random dog population using 15 highly polymorphic canine STR markers, *Forensic Sci Int*, 151, 37, 2005.
92. Halverson, J. and Basten, C., A PCR multiplex and database for forensic DNA identification of dogs, *J Forensic Sci*, 50, 352, 2005.
93. State of Washington v. Kenneth J. Leuluaialii and Geroge J. Tuilefano, in *No. 43507-8-I consolidated with No. 437-1-I*, Court of Appeals of the State of Washington, 2003.
94. Butler, J., David, V., O'Brien, S., et al., The MeowPlex: A new DNA test using tetranucleotide STR markers for the domestic cat, *Profiles DNA*, 5, 7, 2002.
95. www.appliedbiosystems.com, accessed 12/2006.
96. State of California v. David Alan Westerfield, in *Case No. SCD165805*, Superior Court of California, County of San Diego Central Division, 2002.
97. Braunner, P., Reshef, A., and Gorski, A., DNA profiling of trace evidence—Mitigating evidence in a dog biting case, *J Forensic Sci*, 46, 1232, 2001.
98. Menotti-Raymond, M., David, V., and O'Brien, S., Pet cat hair implicates murder suspect, *Nature*, 386, 774, 1997.
99. Giovambattista, G., Ripoli, M., Liron, J., et al., DNA typing in a cattle stealing case, *J Forensic Sci*, 46, 1484, 2001.
100. Gilmore, S., Peakall, R., and Robertson, J., Short tandem repeat (STR) DNA markers are hypervariable and informative in *Cannabis sativa*: Implications for forensic investigations, *Forensic Sci Int*, 131, 65, 2003.
101. Hsieh, H., Hou, R., Tsai, L., et al., A highly polymorphic STR locus in *Cannabis sativa*, *Forensic Sci Int*, 131, 53, 2003.
102. Yoon, C., Forensic science: Botanical witness for the prosecution, *Science*, 260, 894, 1993.
103. Vos, P., Hogers, R., Bleeker, M., et al., AFLP: A new technique for DNA fingerprinting, *Nucleic Acids Res*, 23, 4407, 1995.
104. Coyle, H., Ladd, C., Palmbach, T., et al., The green revolution: Botanical contributions to forensics and drug enforcement, *Croat Med J*, 42, 340, 2001.
105. Bless, C., Palmeter, H., and M.M., Identification of acer rubrum using amplified fragment length polymorphism, *J Forensic Sci*, 51, 31, 2006.
106. Budowle, B., Genetics and attribution issues that confront the microbial forensics field, *Forensic Sci Int*, 146, 185, 2004.
107. Horswell, J., Cordiner, S., Maas, E., et al., Forensic comparison of soils by bacterial community DNA profiling, *J Forensic Sci*, 47, 350, 2002.
108. Moreno, L., Mills, D., Entry, J., et al., Microbial metagenome profiling using amplicon length heterogeneity-polymerase chain reaction proves more effective than elemental analysis in discriminating soil specimens, *J Forensic Sci*, 51, 2006.
109. Marsh, T., Terminal restriction fragment length polymorphism (T-RFLP): An emerging method for characterizing diversity among homologous populations of amplification products, *Curr Opin Microbiol*, 2, 323–327, 1999.
110. Clement, B., Kehl, L., DeBord, K., et al., Terminal restriction fragment patterns (TRFPs), a rapid, PCR-based method for the comparison of complex bacterial communities, *J Microbiol Meth*, 31, 135, 1998.
111. Chan, O. C., Yang, X., Fu, Y., et al., 16S rRNA gene analyses of bacterial community structures in the soils of evergreen broad-leaved forests in south-west China, *FEMS Microbiol Ecol*, 58, 247, 2006.
112. Kvist, T., Ahring, B., Lasken, R., et al., Specific single-cell isolation and genomic amplification of uncultured microorganisms, *Appl Microbiol Biotechnol*, 74, 926–935, 2006.

113. Mills, D., Fitzgerald, K., Litchfield, C., et al., A comparison of DNA profiling techniques for monitoring nutrient impact on microbial community composition during bioremediation of petroleum-contaminated soils, *J Microbiol Meth*, 54, 57, 2003.
114. Dunbar, J., Ticknor, L., and Kuske, C., Assessment of microbial diversity on four southwestern United States soils by 16S rRNA gene terminal restriction fragment analysis, *Appl Environ Microbiol*, 66, 2943, 2000.

26 Clinical Application of CE

Zak K. Shihabi

CONTENTS

26.1 INTRODUCTION

The movement of soil colloidal particles was the first description of electrophoresis as early as 1809. However, Arne Tiselius (~1937) was the first to construct a successful instrument useful for the separation of serum protein by electrophoresis using the boundary separation principle. Because of the clinical significance of this type of separation, many improvements and refinements followed, such as utilizing paper, cellulose acetate, gel, and more recently capillaries in order to speed up and better separate (into distinct zones) the different proteins. The electric current can be utilized in the clinical applications to accomplish not just separation but other tasks:

1. It can be utilized to move fluids (through the electroosmotic flow, EOF). This feature is very useful for analysis in the microchip and in capillary electrochromatography (CEC) where the additions of pumps to move and mix fluids are not feasible.
2. It can be used to concentrate dilute compounds directly during the electrophoretic step (through stacking) as it will be discussed later. This feature is very important for analysis of clinical compounds present in biological fluids at very low concentration.
3. It can be used to separate, quantify, and identify different compounds in clinical samples using different electrophoretic principles.

Capillary electrophoresis (CE) is a general analytical technique for separation and quantification of a wide variety of molecules including those of clinical interest, utilizing narrow bore capillaries under high voltage. It separates various compounds not only based on charge but also based on size, hydrophobicity, and stereospecificity as discussed in Chapters 1, 2, 19. The flexibility of this technique stems from its ability to incorporate easily in the separation buffer different additives, that can interact with some of the analytes relative to others to alter their velocity so as to achieve the desired separation. These additives give the CE great ability to separate numerous clinical compounds in the same instrument using nonexpensive capillaries. Thus, most of the clinical tests can be adapted to CE; however in practice, some are better suited than others for analysis by this method.

The clinical/biological field and pharmaceutical industry are the main areas that are benefiting most from the application of the CE. Most of the clinical tests can be analyzed by CE as well by other techniques, such as high-performance liquid chromatography (HPLC) or slab gel electrophoresis (SGE). However, the CE offers certain advantages for clinical analyses notably a high plate number, a characteristic that is useful when dealing with complex samples containing numerous compounds such as serum or urine. The CE also offers rapid analysis time and a low cost per test with full automation. These characteristics were the driving forces for adopting the CE for completing the human genome sequence project.

However, the CE in clinical analysis requires more thoughtful planning to achieve a good separation compared with other methods. For example, many steroids cannot be separated in free zone electrophoresis. Many of these are neutral or weakly charged compounds and migrate with EOF. They separate much better by the addition of a micellar compound, such as sodium dodecyl sulfate (SDS) to achieve separation by micellar electrokinetic capillary chromatography (MEKC). The CE has some challenges such as poor sensitivity of detection, problems with the sample matrix, and adsorption to the capillary walls as it will be discussed later here and Chapter 13. However, with careful planning these obstacles can be resolved. Many applications can fall under the umbrella of clinical applications; however, tests that have routine applications in the clinical field will be discussed more in detail in this chapter.

26.1.1 Special Aspects of CE in Clinical Analysis

The clinical field is a very vast one encompassing molecules of different physical and chemical characteristics. The sizes vary from small ions such as Na and K to very large ones such as protein and DNA. It encompasses compounds that are very abundant in concentration such as albumin

(~30,000 mg/L) to compounds very low in concentration such as prostate-specific antigen (~4 μg/L). It encompasses compounds with different solubility and hydrophobicity. Thus, the analysis of these different compounds requires different strategies and represents different degrees of difficulties. Some analyses work well and easily with capillary zone electrophoresis (CZE) (e.g., serum proteins), size separation (e.g., DNA) while others are difficult to adapt to any form of CE. The lower the concentration is the more difficult the analysis becomes.

26.1.2 Advantages of CE in the Clinical Field

The main advantage of the CE in clinical analysis is the flexibility of the separation so that the same instrument can be utilized to separate numerous biological compounds. It is easy to add different additives to the separation buffer to induce a change in the velocity of some of the compounds leading to a better separation. The principle of the separation can easily be changed from free solutions to hydrophobicity separation (MEKC), size separation, chiral, isofocusing to suit a particular group of compounds as described under Chapters 1, 2. The second important advantage is the high plate number in CE generated in the capillary due to the flat profile, compared with the laminar flow in HPLC, without dealing with the high pressure or the high cost of the HPLC packed column. A third feature is the ability to perform separations without the need to use large volumes of organic solvents. Organic solvents, which are used often in HPLC, are becoming more expensive to purchase and more difficult, under many state laws, to store and dispose of. This eventually may compel shifting of many separations from HPLC to the CE.

26.1.3 Limitations of CE in the Clinical Field

On the other hand, CE suffers from a few problems. Unlike HPLC, CE is greatly affected by sample matrix (i.e., salts and proteins) [1–4], which are very high in biological samples. Another major problem in clinical analysis is the suboptimal detection sensitivity. The third problem is sample interaction with the capillary walls. To utilize the CE successfully for practical separation of clinical samples it is important to understand how these factors affect the CE and the different maneuvers needed to overcome them.

26.2 PRACTICAL ASPECTS OF CE IN CLINICAL ANALYSIS

26.2.1 Background

Many applications fall under the umbrella of clinical applications. For the sake of simplicity, these can be divided into two general areas. One is the research, which is performed occasionally for gaining basic scientific information. The other area is the routine analysis for patient care and diagnosis such as detection of monoclonal gammopathies to detect specific tumors. This test is performed often. In research, the need is for the flexibility and versatility with good separation with emphasis on detection of compounds at low concentration. In routine work, the emphasis is more on good reproducibility (precision) with minimum amount of analytical steps. The different forms of CE can fulfill both goals. For example, the ability to add different additives to the buffer allows for great selectivity and versatility, for example, from separation based on charge to separation based on hydrophobicity, or to size. On the other hand, the MEKC form of the CE offers ability to simply inject the sample directly on the capillary without any treatment.

26.2.2 Sample Matrix Effects

Many compounds can be analyzed easily from pure standard solutions, tablets, or from samples with clean matrix. However, the analysis of the same compounds from biological sources poses many more difficulties. This is because of the effects of both the high salt and high protein content of the biological samples on the separation in CE.

Although the sample, in most instances, constitutes a very small portion of the overall volume in the capillary once injected (<1%), the matrix of the sample has profound effects in CE. This is due to two main factors: contribution of the sample to the total current conductance and also due to its interaction with the capillary walls (especially proteins). On the basis of how the sample is prepared and how the separation buffer is selected, sample matrix effects can be either favorable or detrimental to the analysis [5]. Peak shape, separation, quantification, and reproducibility are affected greatly by sample matrix.

A simple, but limited, solution to overcome the problems of high salt and protein content is to dilute the sample. However, this not only lowers the salt and the protein content but also decreases the concentration of the analytes of interest too!

26.2.3 High Salt Content in the Sample

Conductivity differences between the analyte zones and the pure background electrolyte (BGE) zones can lead to local electric field strength differences that, in turn, can distort the shape of the analyte bands yielding asymmetric peaks and result in reduced separation efficiencies [1,3,5]. Sample desalting is not easy to perform. If it is necessary, this can be accomplished through special columns or after sample extraction. However, samples with high salt content can be analyzed directly provided the separation buffers have high ionic strength [3]. These aspects will be discussed more in detail later on.

26.2.4 High Protein Content in the Sample

High protein content in the sample can mask the absorption of the compounds of interest especially those with low concentration. Proteins can bind to many molecules and alter their migration. They can also cause peaks asymmetry [1,5]. However, more importantly they adsorb to the capillary walls and change the zeta potential and the characteristics of the capillary. Notably, the reproducibility suffers greatly from protein adsorption. High protein content in the sample eventually ruins the capillary because of their adsorption to the walls. Since proteins tend to adsorb more on the capillary inlet, cutting off a few millimeters of an old capillary inlet sometimes can restore back the separation. Excess protein such as in serum samples can be removed by deproteinization before the CE step. If the molecule of interest is a small to medium in size (<5000 Da) acetonitrile (2 V acetonitrile: 1 V of serum) is an effective means to remove the protein while providing ability to concentrate (stacking) the analyte as described below.

If the molecule of interest is neutral or weakly charged then the sample can be analyzed directly without removing the protein or any treatment using MEKC. Analysis by MEKC can tolerate proteins since the surfactants in the MEKC solubilize them.

26.2.5 Wide Variability in Concentration

The recent interest in proteomics placed great emphasis on analyzing proteins and peptides that are present in very low concentration. Serum and many biological fluids contain numerous different proteins (~10,000) the majority of these are in very low concentration. However, at the same time, very few proteins (mainly albumin, globulins, and transferrin) are present in these fluids in very large concentrations. If these abundant proteins are not removed first, they make the analysis very difficult because they overload the capillary. Several methods can be used to remove these proteins and enrich the ones of interest; some are commercially available as kits. In general, several separation or binding steps are used to remove or enrich these proteins (e.g., size exclusion, anionic exchange chromatography, hydrophobic chromatography, and solid-phase ligands binding) [6,7]. For analysis of small molecules, sample deproteinization is necessary but easy to perform. This removes the proteins but leaves the small molecules in the supernatant.

26.2.6 Precision

In general, but not in all cases, the HPLC tends to give slightly better reproducibility than the CE. The reproducibility in CE depends greatly on the capillary surface and to some extent on the number of steps for sample preparation. A thorough wash with diluted sodium hydroxide or phosphoric acid removes the adsorbed proteins from the capillary surface and improves the reproducibility. Furthermore, the reproducibility improves by employing internal standards, peak area rather than peak height, effective mobility rather than migration time, and frequent calibration [8]. The MEKC can decrease or eliminate the extra steps for sample preparation. It allows direct injection of serum on the capillary thus indirectly improving precision.

26.2.7 Coated versus Noncoated Capillary

A noncoated (untreated) capillary is sufficient for most routine CE analysis, especially when it is used with high ionic strength buffers. It is less expensive and offers good precision. However, for the analysis of proteins and peptides present in low concentrations, a coated capillary gives a much better theoretical plate number for proteins [9]. This is true of the capillaries and also for microchips [10]. Such capillaries are also important for performing isofocusing and isotachophoresis (ITP) to eliminate the EOF and are more suitable for the detecting microheterogenity. The coating can be covalently (permanently) bound or hydrodynamically by adding the coating material to the buffer. The walls can be modified to possess either positive or negative charges or neutral. Linear polyacrylamide was the first polymer used for permanent coating but it was not very stable. Other coatings were prepared with other polymers such as dextran, fluorinated aromatic hydrocarbons, and hexamethyl disiloxane poly(dimethyl acrylamide). Some of the covalently coated capillaries have a very short life. Several dynamic coatings have been described, for example, Polybrene-poly(vinyl sulfonate), Polybrene, poly(methoxyethoxyethyl)-ethylenimine, poly(diallyldimethylammonium chloride), and starch derivatives [11–13]; some of these are commercially available as kits. An interesting dynamic coating (adsorbed surfactant) was described based on washing the capillary with special surfactant such as dimethylditetradecylammonium bromide and didodecyldimethylammonium bromide. The coating binds and remains tightly bound enough for next several samples eliminating analytes adsorption [14].

26.2.8 Stacking of Compounds of Clinical Interest

Because of the need for better sensitivity of detection in CE, sample concentration is crucial for the widespread and practical use of this technique. Sample concentration can be accomplished by several physical means outside the capillary, such as by liquid- and solid-phase extraction; but more easily in CE by concentration on the capillary directly (stacking). Stacking is a general term referring to the contiguous concentrated zones resembling a "stack of coins."

Several methods for staking including the theoretical aspects have been described [15–18]. Few of these methods are general while others are specific and more suitable for certain types of compounds. In most of the stacking techniques, discontinuous buffers of different kinds are employed as the basic means for altering the charge of the analytes or the field strength of the sample zone [15]. Thus, the same analyte molecules present at the different edges of the band move at different velocities in such a way that leads to sample concentration. Buffer discontinuity can be brought about simply by altering the sample conductivity or pH so as to be different from that of the separation buffer. Sometimes this can occur unintentionally [16].

Two general types of stacking are used often in CE for clinical compounds. The first is the high field strength stacking. In this method, the sample is diluted in the same separation buffer but at 10 times dilution. Thus, the sample molecules are subjected to higher field strength than that of the electrophoresis buffer. Thus, they accelerate in this region but slow at the interface of the sample and the buffer. This method can be used for large and small molecules. The second type is acetonitrile stacking. Mixing acetonitrile with the sample (2:1 v/v) is used mainly to remove proteins. However, the presence of acetonitrile in the sample (not in the buffer), has several important additional advantages. For example, (1) it counteracts the deleterious effects of salts, (2) it yields better stacking for small molecules than that obtained in dilute buffers, and (3) it allows larger volumes of sample (in some cases half of the capillary volume) to be injected. The overall effect is an increased sensitivity of small and medium molecules (<5000 Da). The stacking occurs in both the hydrodynamic and the electromigration injection. The sodium chloride present in serum at about 150 mmol/L together with the acetonitrile used in the deproteinization both bring about 10–30 times sample concentration. The mechanism of the stacking is pseudotransient ITP [19]. Thus, biological samples having high salt and protein content are well suited to CZE analysis after treatment with acetonitrile.

ITP [15], especially transient ITP, are used for small and large molecules stacking. In this technique, the sample is mixed with appropriate leading and terminating ions [16] as described in Chapter 13. This technique is used to selectively enrich low-abundance compounds (e.g., peptides in the absence of any bioaffinity interactions) [20]. This method is more difficult in practice, but it offers the benefits of speed and very high concentration.

Neutral molecules analyzed by MEKC require special and more difficult stacking methods [18]. They concentrate based on stacking either by having regions of difference in the electric field or by sweeping, that is, picking and accumulating of analytes by the pseudostationary phase that penetrates the sample [18].

26.2.9 Compounds Difficult to Analyze by CE

In spite of few attempts, glycoproteins, carbohydrates, and lipids are difficult to analyze by CE. Carbohydrates and glycoproteins are important in the immunological recognition, control, and attack of pathogens and in protein folding. Carbohydrates do not have strong absorption in the UV and most do not have strong charges. They are analyzed at very high pH and require labeling mostly with fluorescent reagents. The analysis of carbohydrates by CE-MS has been reviewed [21].

Lipids are important since they are part of the cell membrane and the lipoproteins have important diagnostic values in coronary heart disease. Lipids too do not have strong absorption and lack strong charges; in addition to that, they are hydrophobic. Lipoproteins have been analyzed by MEKC [22]. CZE was used to separate the isoforms of low-density lipoprotein (LDL) particles in human serum based on charge/volume ratios of the particles. LDL, dense LDL, were analyzed by CE [23]. A CE-MS method incorporating SDS for separating very-low-density lipoprotein (VLDL), LDL, high-density lipoprotein (HDL) has been described [24].

26.2.10 Compounds Suited for Analysis by CE

Some of the clinical compounds are better suited than others for separation by this technique. In general, polar compounds possessing a strong charge and present in high concentration are most suited to analysis by CE (e.g., proteins, hemoglobins and their chains, some peptides, and drugs). Enzymes and DNA are also suited for analysis by CE even though they are not present in high concentration. The DNA is amplified first and then analyzed elegantly by CE using sensitive detection fluorescent dyes while enzymes are analyzed through the accumulation of products (catalytic activity on the substrates). Following sections discusses those clinical compounds that have been well studied by CE and analyzed often for patient care.

26.3 PROTEINS

Proteins are important not only for their function but also for their diagnostic significance. Proteins maintain structural integrity and perform different functions such as catalytic, hormonal, or receptors. The function of many of the proteins of clinical interest remains not well understood nevertheless these proteins remain to be important. The CE is useful for quantification, purity check, and detection of microheterogeneity of proteins.

Proteins and peptides are composed of amino acids, which are zwitterions carrying both positive and negative charges. Thus proteins possess different isoelectric points. These proteins acquire a charge based on the buffer they are dissolved in and move toward the opposite electrode depending on the net charge. Usually, slightly alkaline buffers (pH > 8.0) are chosen for separation of proteins. Under these conditions, the majority of proteins are negatively charged moving toward the anode but they are pulled toward the cathode by the EOF. They can be also separated at acidic buffer conditions (pH < 3.0) (i.e., carrying a positive charge). However, because of the absence of the EOF at this pH, the separation requires a much longer time. Different proteins tend to behave differently in CZE depending on their net charge. For example, basic proteins tend to bind to the capillary wall and give distorted peak shape especially when they are in low concentration. To improve their separation by CE, different additives, high salts, or coated capillaries are used to decrease the binding to the walls.

The advantages of CE over agarose gel (AG) electrophoresis for analysis of proteins are the speed, automation, small sample volume, and avoidance of many staining/destaining steps. This led some companies to design instruments dedicated only to protein analysis by CE. Furthermore, other manufacturers designed special CE instruments to perform capillary isoelectric focusing (CIEF) with absorption imaging detectors in order to focus, concentrate, and separate better the different proteins as discussed in Chapter 19. These instruments can detect protein microheterogeneity better than the common CZE instruments. Here are some clinical applications of proteins and peptides measurement by CE.

26.3.1 Serum Proteins

The interest in serum proteins stems from their diagnostic significance. Few thousands of different proteins are present in the serum arising from the different cells and tissues; however, mostly at very low level—way below the CE detection limits. A few of these proteins are present at high enough concentration to be detected directly by CE. These proteins separate into a few bands (~5–12), which are not pure or single proteins but a group of several proteins. Serum proteins are analyzed routinely on a daily basis in most large hospitals to detect several disorders such as renal failure and infections; but most importantly, monoclonal gammopathies.

In CZE, serum proteins have been separated using different buffers (e.g., Tris and Tricine), but mostly borate, with pH of 8–11 [22,25–29]. Serum protein separation can be completed by CE in about 2–10 min in contrast to 1–2 h for agarose electrophoresis (AG) (Figure 26.1). The correlation coefficient between CE and AG for the separated bands is good [30–32]. Some commercial instruments use multicapillaries of narrow diameter (25 μm) to increase the throughput of the analysis. The narrow capillaries produce better resolution than the wider capillaries with a much shorter migration time [28]. This is true for all CE separations.

26.3.2 Immunofixation (Immunosubtraction)

Serum immunoglobulins are composed of heavy and light chains and classified based on their reaction with specific antibodies into the classes of IgG, IgA, IgM, IgD, and IgE. These can be secreted in high concentration as some of clones of cells become malignant. Multiple myeloma, Waldenstrom's disease, and light chain disease all represents different malignancies of plasma cells and can cause increased levels of these proteins (paraproteins). The identification of the type of

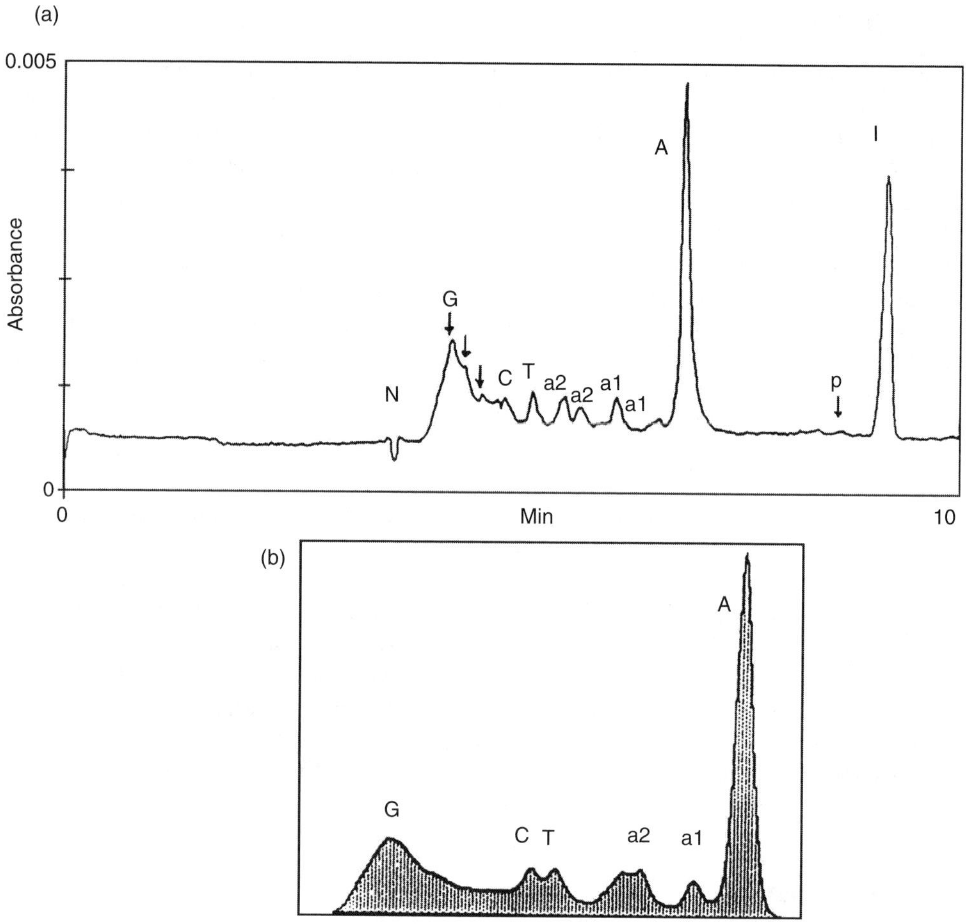

FIGURE 26.1 Comparison of the CZE to agarose gel (AG) electrophoresis for a patient with a streptococcal infection: (a) CZE and (b) AG. (I: internal standard; P, prealbumin; A, albumin; a1, α_1 globulin; a2, α_2 globulin; T, transferrin; C, complement; G, γ-globulins; N, neutral compounds). Arrows in the γ-region indicate oligobands. (From Shihabi, Z., *Electrophoresis*, 17, 1607, 1996. With permission.)

these different paraproteins is important for the proper patient treatment. These paraproteins are detected and classified with immunofixation, a laborious procedure performed in gel electrophoresis. Immunofixation is an important part [32] of working up an unknown monoclonal band. CZE has been adapted to perform the immunofixation method based on reacting serum proteins with specific antibodies. The Antigen-antibody complex can be separated from the free antibody or antibodies based on binding to a solid matrix or simple separation in the capillary. The sample is assayed before and after the binding [30,32]. The difference between the two "immunosubtraction" represents the specific type of the monoclonal abnormal serum protein. This simple method has been shown to be reliable.

26.3.3 Cryoglobulins

Cryoglobulins analysis is an important but unfortunately a neglected test in clinical practice because of the difficulty of the test. Cryoglobulins are special type of immunoglobulins that reversibly precipitate from serum at cold temperatures [33]. They can be associated with skin lesions (purpura), glomerulonephritis, and peripheral neuropathy, or malignancy [8]. These are divided into two general categories. First, a small fraction (about 5–10%) that are pure monoclonal antibodies (type I) and which representing malignancies. The second type, which is the majority ~95%, are immune complexes (mixed cryoglobulins) representing immune stimulation due to infection or presence of autoantibodies. The latter group is further divided based on if the cryoprecipitate contains a monoclonal rheumatoid factor (type II) or polyclonal rheumatoid factor (type III) immunoglobulins [33]. The mixed cryoglobulins can precipitate in the different tissues of the body such as the kidney and the extremities causing vasculitis.

Cryoglobulins are detected, quantified, and phenotyped all simultaneously by precipitating an aliquot of the serum at 4°C, centrifuging, and dissolving the precipitate in a buffer followed by electrophoresis using the same conditions as those for serum proteins (Figure 26.2) [34]. Cryoglobulins are well suited for analysis by CE. The main advantages are the higher sensitivity, the use of small volumes of serum, speed (15 min vs. 2 h), and improved quantification compared to the AG method. We have used the CE for routine cryoglobulins analysis for our patients for over a decade with good success [21].

26.3.4 Urinary Proteins

Usually urine proteins are present at about 10–100 times lower concentrations compared with serum but in the presence of numerous interfering UV-absorbing compounds. This renders urinary proteins to be more difficult to measure directly by CE when compared with serum. The majority of urine samples require concentration (based on the protein content) before the CE analysis. They are concentrated through special commercial "membrane concentrators" followed by washing with saline solution or using chromatographic column [32] to decrease the interfering UV-absorbing materials. The urine contains several proteins of clinical interest, especially Bence-Jones proteins, which are important for detecting plasma cell malignancies and for diagnosis of nephropathy. Very few urine samples that are very high in proteins can be analyzed directly without any preparation. The same buffers and conditions for serum proteins are basically used for the analysis of urine protein [35].

26.3.5 Cerebrospinal Fluid Proteins

The main clinical significance of cerebrospinal fluid (CSF) protein electrophoresis is for the detection of the oligoclonal bands, which are present in multiple sclerosis in the gamma region. Similar to urine, proteins in the CSF are present fluid in very low concentration (100 times less than serum). For the majority of the samples, a 10- to 20-fold concentration is preferred before analysis by CE (by the same membrane concentrators used for urine). CSF protein separation can be accomplished in less than 10 min with CE versus 2 h for AG with the ability to detect oligoclonal banding by this technique [36].

26.4 ENZYMES

Since enzymes are essentially proteins, they can be measured in CE by direct light absorbency like other proteins or by their enzymatic activity. The CE offers versatility for enzyme measurement: the enzyme itself, the substrates, or the products (Figure 26.3) all can be measured in CE. Most of the enzymes in biological fluids are present in very low concentration so that they cannot be determined by direct light absorbency. The catalytic activity is much more sensitive and more versatile. Catalytic activity is more suited for enzymes with low activity because the reaction product can be amplified with time. The catalytic activity can be measured in several ways: (A) incubation in the capillary that is used as a microreactor [37–39]; (B) online, postcapillary reaction (more difficult); and (C) incubation outside the capillary that is more common. If a long incubation step is needed then it is more convenient to perform the incubation outside the instrument. After the incubation step, the reaction is stopped preferably by the addition of acetonitrile (rather than acid) (Figure 26.3), which

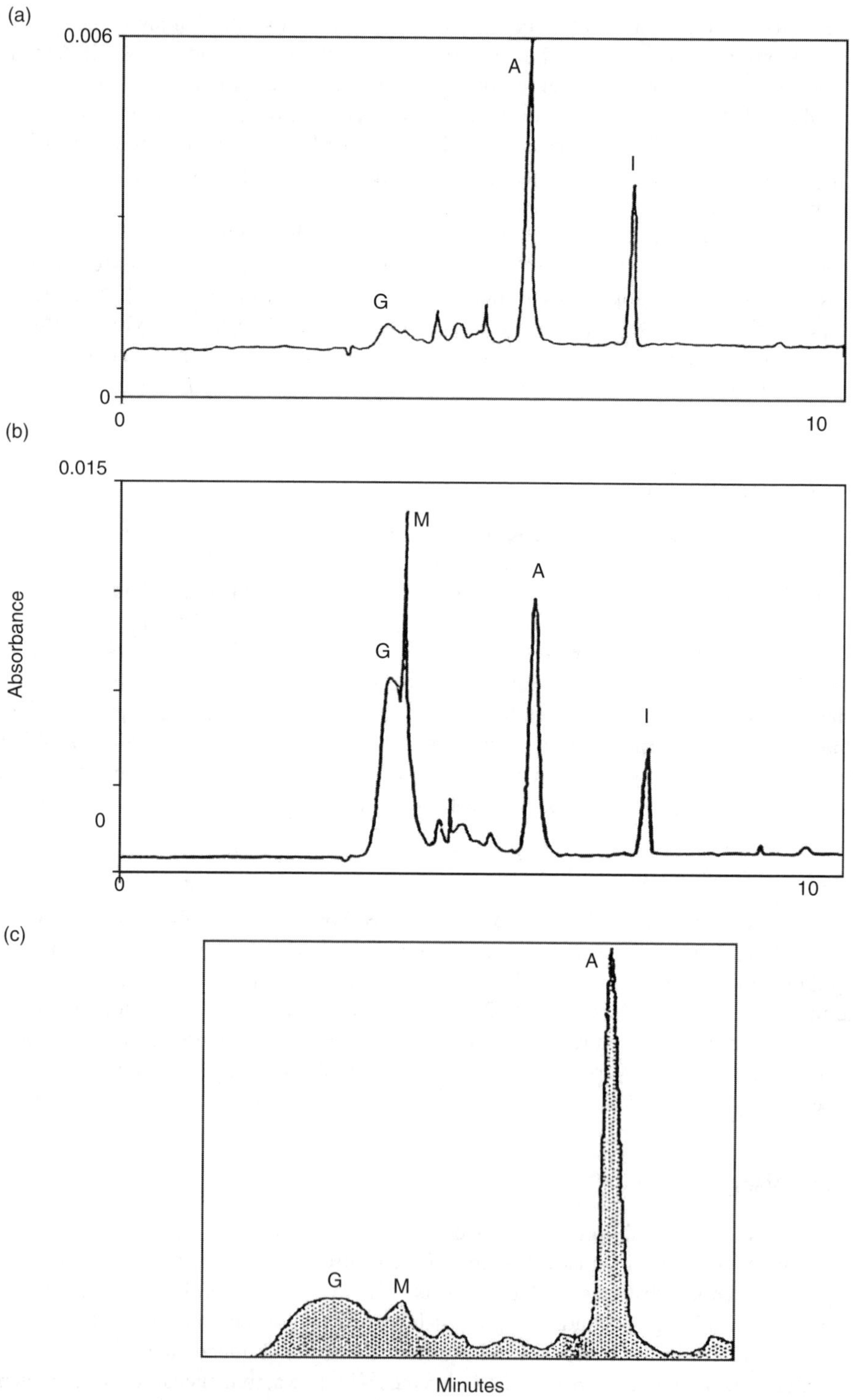

FIGURE 26.2 Serum electrophoresis of a patient with type II cryoglobulins (650 mg/L): (a) serum CZE, (b) Cryoglobulins by CZE, and (c) cryoglobulins by AG (see legend of Figure 26.1). (From Shihabi, Z., *Electrophoresis*, 17, 1607, 1996. With permission.)

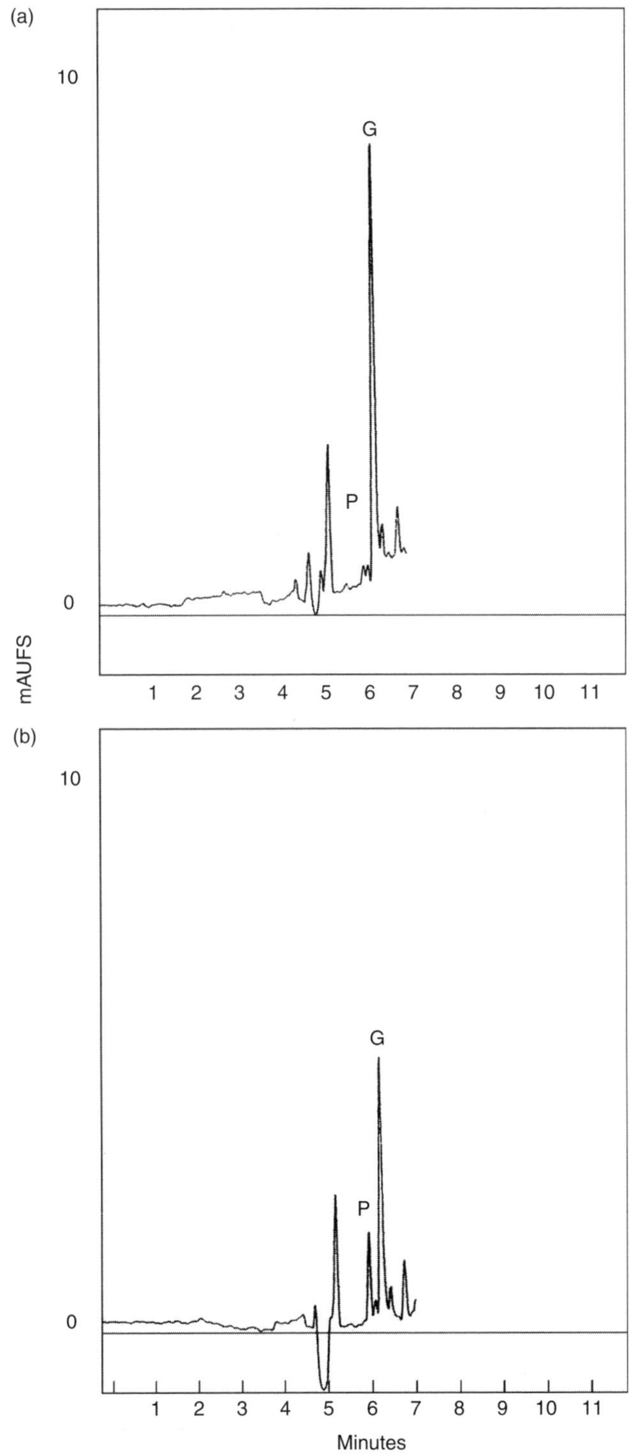

FIGURE 26.3 Glutathione transferase activity: rat heart tissue homogenate 25 μL were mixed with 50 μL of 1-Cl-2-dinitrobenzene (0.2 mg/mL, pH 6.6) and 25 μL of reduced glutathione (2 mg/mL) and incubated for 10 min at 37°C. The reaction is stopped with 200 μL acetonitrile, mixed, and centrifuged. (a) At 0 min and (b) at 10 min of incubation (P, product of the reaction; G, glutathione peak). (Separation buffer: 250 mM borate, 50 mM Tris, pH 8.0 at 10 kV, 214 nm, 20 s injection on a capillary 30 cm × 50 μm (ID).

can remove the proteins, decreasing the UV absorption, and concentrate the analyte (by stacking) at the same time.

Proteolytic enzymes and those enzymes with low activity or with expensive substrates are well suited for analysis by CE in this manner. The proteolytic enzymes can be detected simply based on their absorption in the UV range in CE. Several enzymes have been analyzed by CE, such as chloramphenicol acetyl transferase [40], glutathione peroxidase, glutathione transferase

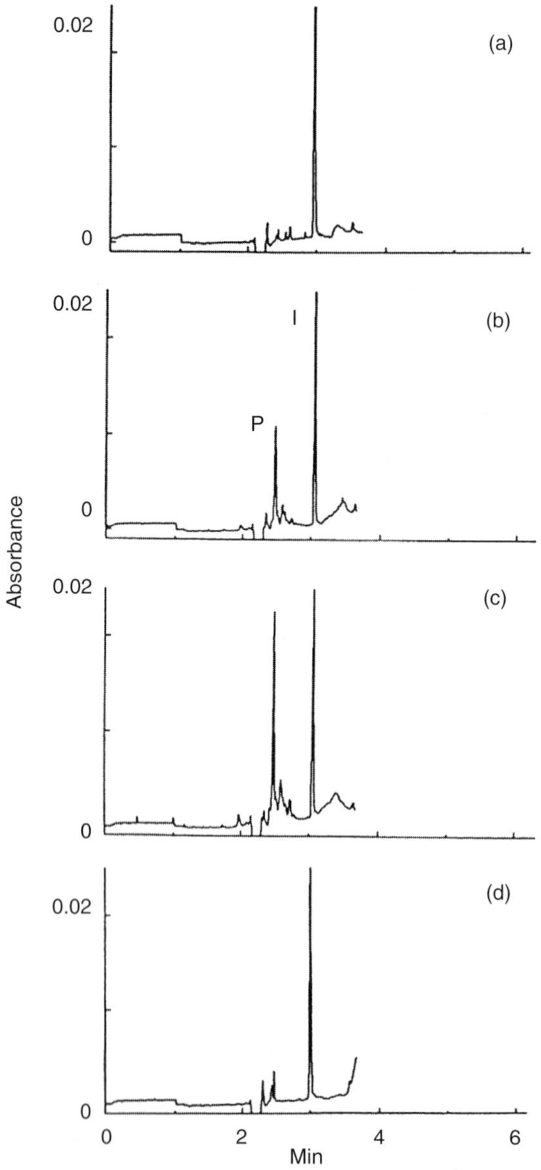

FIGURE 26.4 Enzymatic activity of breast tumor homogenate activity for Cathepsin D (106 pmol/mg protein) at different periods of incubation: (a) 0 min; (b) 10 min; (c) 20 min; and (d) at 20 min in the presence of pepstatin. (P, split peptide; I, Iothalamic acid). (From Shihabi, Z.K. and Kute, T.E., *J. Chromatogr. B*, 683, 125, 1996. With permission.)

(Figure 26.3), ornithine transcarbamylase [40], angiotension converting enzyme [41], Cathepsin D (Figure 26.4) [42], and elastase [43].

The advantages of CE for analysis of enzymes are the use of small volumes, versatility, and ability to avoid the extra steps of indicator reactions. We found in some instances, for example, analysis of glutathione transferase it is easier to assign enzymatic activity units (IU) based on the CE because both the substrate and the products can be monitored at the same wavelength. Unfortunately, researchers did not take full advantage of the CE for enzyme analysis. In practice, kinetic spectrophotometric methods remain to be most widely used for routine work while the CE is reserved for those difficult and specialized tests.

26.5 HEMOGLOBIN

26.5.1 Hemoglobin Variant

Hemoglobin (Hb) carries the vital oxygen to the different tissues. Its concentration is used clinically as an indicator of the different types of anemia. Hemoglobin is a tetramer composed of 2α chains and 2β chains. The β chain is more susceptible than the α chain for amino acid substitution (mutations), which results in different variants often present in special populations. Most of the variants are harmless and of research interest. However, few such as Hb S are associated with severe anemia, decreased capacity to carry oxygen, and altered red blood cell shape. The most encountered variants of hemoglobin are A, F, S, C, and E. Hemoglobin electrophoresis is carried out for two purposes: detection of Hb variants and detect the presence of thalassemias (decreased synthesis of one of the Hb chains).

Because of the small charge difference of the isoelectric point (pI), Hb variants do not separate well by CZE. For good separation of hemoglobin variants by CZE, a high buffer concentration, a narrow capillary (20–30 μm) (i.d.), and coated (hydrodynamically treated) capillary are chosen. Tris, Tricine, and arginine buffers at pH 8–8.4 give a good separation [27,44,45] (Figure 26.5), which resembles very closely to that of the alkaline separation by AG.

Although CE instruments are not well designed for CIEF, many variants can be separated better by this technique. In addition to the common variants, G Philadelphia, A2, and Bart's can all be separated by CIEF [46–48]. The separation by CIEF compares well with gel isoelectric focusing and with HPLC. The variants have also been analyzed by both CE and CIEF equipped with special absorption imaging detectors. These types of detection devices eliminate the extra steps needed to move the peaks, after the focusing step, to the detector and can simultaneously detect several capillaries with better precision and faster results than CE instruments. HB A_2, which is increased in β-thal, is better quantified by CE compared with AG electrophoresis [44]. Hemoglobin analysis by CE has been reviewed recently [49–51].

26.5.2 Hemoglobin A_{1C}

Hemoglobin HbA_{1C} is very important clinically because it is used to follow the control of blood glucose over extended periods of time in patients with diabetes. It measures the amount of glucose bound to hemoglobin (glycated Hb). Unlike blood glucose, HbA_{1C} is more stable and represents the average glucose in the past 3 months (the average life of the red blood cell). Clinical labs perform this test routinely on a daily basis because of the widespread of incidence of diabetes and the need to modify the glucose level or modify the patient treatment. A dedicated HPLC instrument solely for this test is sold commercially. However, the cost/test is relatively expensive. For analysis of this hemoglobin by CE, a coated capillary either dynamically [52,53] or permanently is required [54]. Both CZE [52,53] and isoelectric focusing [54,55] are used for its analysis.

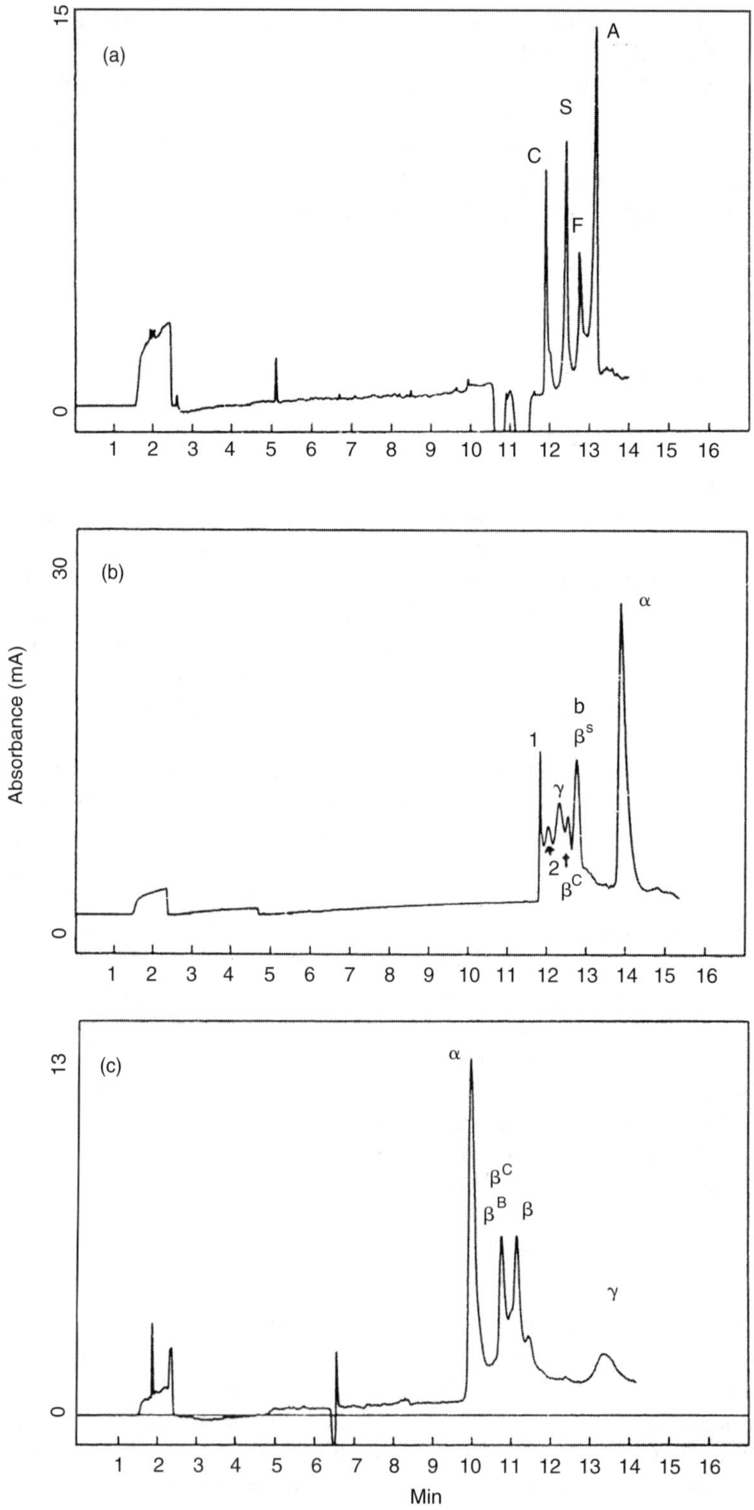

FIGURE 26.5 Separation of a mixture of the common Hb variants using the same diluted sample and same CE conditions for the intact Hb molecules and the HB chains (4 s injection, 214 nm). (a) Intact Hb molecule; (b) Hb chains under acidic conditions; and (c) Hb chains under basic conditions. Peak 1, albumin; peak 2, unknown. (From Shihabi, Z.K. and Hinsdale, M.E., *Electrophoresis*, 26, 581, 2005. With permission.)

26.5.3 Globin Chains of Hemoglobin

The globin chains are useful for investigating the presence of thalassemias that lead to decreased globin synthesis (either α or β chain). This could be minor (without much clinical consequences) or major (severe or life threatening) symptoms, depending on the number of genes involved. The deviation of the ratio of α/β in the patient from that of the normal (~1:1) indicates the presence of thalassemia. The thalassemia represents a decrease in the synthesis of either α or β chains. The α-thalassemia is common in people of the African origin while the β-thalassemia is common in those of Mediterranean sea origin. These chains are analyzed by cellulose acetate, which is a laborious technique requiring extraction with acetone, denaturation by high concentration of urea, and strip staining. Few CE methods used similar protocol to the cellulose acetate methods with analysis in phosphate buffer either at pH 11.8 or at pH 2.5–4.5 [56,57]. We simplified the CE by using direct Hb hemolysates without extraction or denaturation simply by analyzing the diluted red blood cells at high alkaline pH 12.6 [58] or low acidic pH 2.15 (Figure 26.5). These buffers with high or low pH induce the chain denaturation and at the same time separate the heme and serum proteins from the chains. Thus method is extremely simple and very rapid since it does not require sample treatment, staining or destaining. Furthermore, the common variants of the β chains, such as βS, βC, and βE, are also separated from each other. Thus, this method gives further confirmation of the Hb variants separation by CZE.

26.6 PEPTIDES AND POLYPEPTIDES

Peptides and polypeptides are similar to protein in structure but smaller in size. Many of the peptides are biologically active compounds with different functions, such as adrenocorticotropic hormone (ACTH), angiotension, substance P, or glutathione. Most of these peptides are present at very low concentration. However, peptides can also arise from the digestion of purified proteins *in vitro* especially as a step during protein characterization and can be present in high concentration.

After sequencing the human genome, there is a great interest in peptide analysis as a means to identify those proteins coded by the different genes discovered recently. Peptide analysis by CE can be used for quality control or purity check in the pharmaceutical industry. In this respect, the CE is well suited for this purpose. After protein hydrolysis, the different peptides can be separated by HPLC or CE and analyzed by the mass spectra. On the other hand, peptides can be present naturally in different biological fluids such as in serum or spinal fluid. In most of the biological fluids, they are usually present in low concentration among high concentration of interfering substances. Thus, the analysis becomes much more difficult. Sample concentration before or on the capillary becomes vital for the analysis. We used acetonitrile to concentrate some peptides, for example, glutathione, insulin chains (Figure 26.6), and enkaphlin [59,60]. However, for those peptides in very low concentration they require much more concentration either by other types of stacking [61] or by chromatography [62]. For better analysis, coated capillaries are preferred [9,10]. Analysis of peptide has been reviewed recently [63,64].

26.7 CE AND PROTEOMICS

Proteins are synthesized as a result of the transcription and translation steps of the information stored in the DNA. However, because of post-translational modification, during development and disease, proteins primary structure can be modified greatly and so the function. In cells or tissues, there are usually a large number of proteins that vary widely in their physicochemical properties, including molecular weight, pI, solubility, and folding. Many factors that affect the synthesis can result in many subtle variations, for example, due to sugar binding or SH interaction, which can account for the wide variations between different individuals.

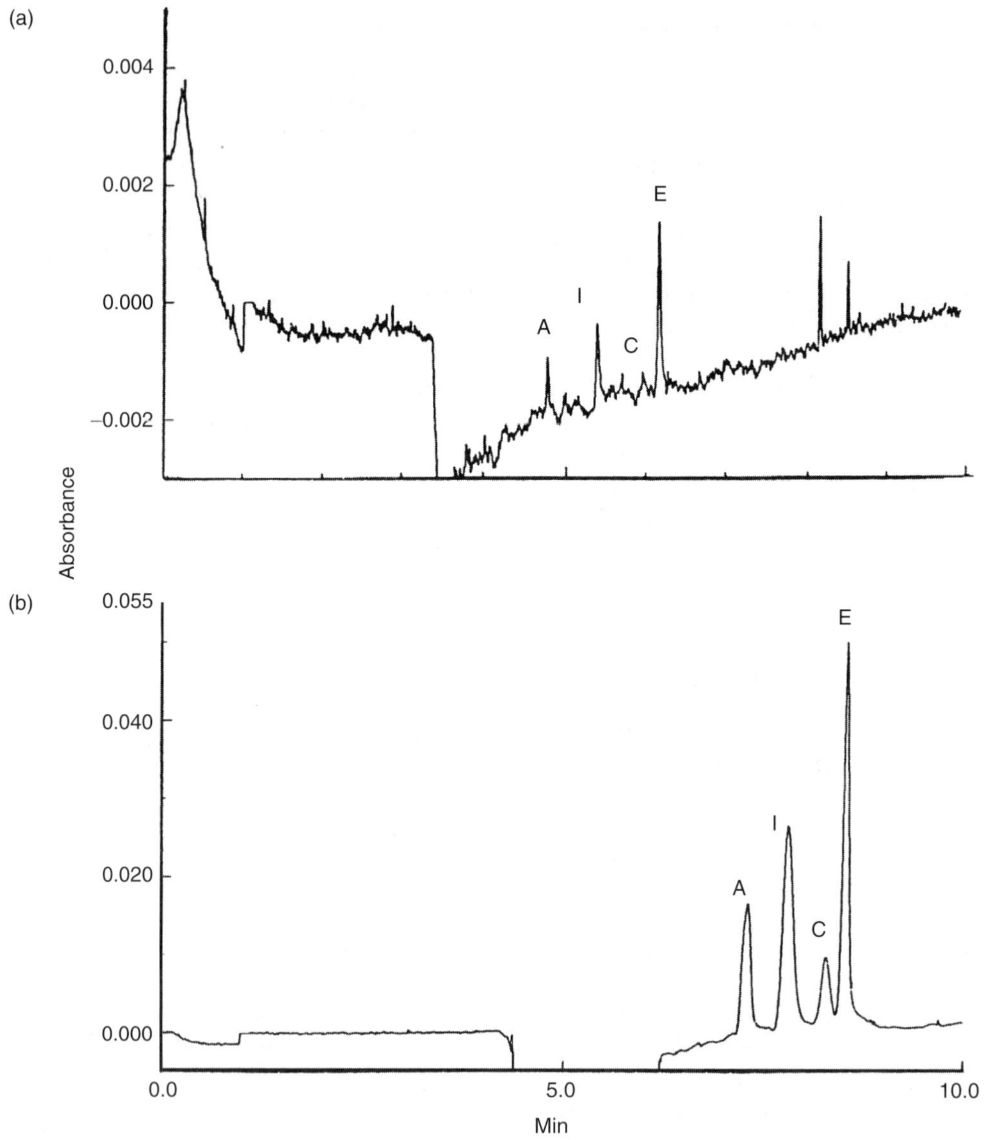

FIGURE 26.6 Effect of stacking for some natural peptides on sensitivity of detection (A, angiotensin; I, insulin B chain; C, impurity in the insulin B chain; and E, Leu-enkephalin; (a) at 1.5% loading of the capillary and (b) at 30% (stacking in acetonitrile). (From Shihabi, Z.K., *J. Chromatogr. A*, 1996, 744, 231. With permission.)

At the present time, proteomics is an area far from the routine work; however, this probably will change in the near future because of the importance and the accelerated research. We have witnessed the rapid change from research to routine work for DNA analysis. DNA fragments now are routinely analyzed for viral detection; and for leukemia's typing. Finding proteins (biomarkers) that play key roles in the development, malfunctions of cells or tissues is a helpful for early detection and diagnosis of diseases as well as searching for new targets for production of new drugs. There is a great interest in identification and quantification of multiple proteins, including those in very low concentration that constitute or control a particular biological process. Thus, techniques providing high sensitivity and high peak capacity are still greatly demanded for the analysis of biological samples.

At the present time, most of the studies for proteomics employ 2D gel or HPLC-MS. The CE-MS has some promise in this area because of the speed and high plate number [65]. However, Huang et al. [66] have summarized some of the problems involved with separation of protein by CE as part of the proteomics "poor reproducibility, low-sample loading capacity, and low throughput due to ineffective interfaces between the separation and MS systems." Size separation is often used in protein characterization in gel electrophoresis. This separation works very well in CE for the DNA strands; however, it is more difficult for serum proteins. It is based on adding polymers such as different cellulose derivatives, dextrans, or linear polyacrylamide to the buffers to retard the migration of the large peptides. The proteins have to be denatured and carry the same charge (in SDS) [67]. Several companies offer special kits suitable for this purpose.

Several pure proteins have been purified and checked for their purity, microheterogeneity, or diagnostic significance by CE. In this case, coated capillaries are preferred for this separation. Many of the proteins can be analyzed by CE; however, sensitive detectors such as fluorescence are necessary [68–70]. Examples of proteins studied by CE separately or in a profile are transferrin isoforms, which are important as markers of alcoholism [71], α-1 antitrypsin [72], recombinant human erythropoietin glycoforms that stimulates erythopiosis [73], plasminogen tissue activator [74], prions [75], urothelial carcinoma proteins in urine [76,77], and numerous urinary proteins [78].

26.7.1 Nucleic Acids (DNA)

This is an area that represents the best application of CE in the clinical/biological field as discussed in Chapters 12, 16 and 25.

26.8 SMALL MOLECULE ANALYSIS

26.8.1 Drug Analysis

Therapeutic drug monitoring is an expanding area for both the clinical and the pharmaceutical industry as discussed in Chapter 4.

26.8.2 Endogenous Compounds

Many metabolites both ionizable and nonionizable can be measured by different forms of CE such as CZE, MEKC, and chirality. Metabolites with strong UV absorption such as nucleotides, phenolic amino acids, and their metabolites are easy to measure by CE. However, some of these require concentration and clean up before the CE step [5]. This can be achieved by traditional concentrating methods, such as solid phase and solvent extraction; or by concentration on the capillary (stacking). Examples of small molecules that have been analyzed by CE are nucleotides [79,80], amino acids, catecholamines [81–83], and sugars [84,85]. Below is a more detailed discussion of some of these compounds.

26.8.3 Amino Acids

Amino acid analysis can be performed for detecting all the amino acids or for few specific ones. Regardless of the type of method, analysis of amino acids is a difficult task because the majority of the amino acids lack a strong chromophore and they resemble each other in structure, which makes the separation very difficult. Thus, they need extra reaction steps to be derivatized pre- or postseparation. Furthermore, physiological fluids like serum contain many interfering compounds such as peptides and the uncommon amino acids. Amino acid analysis can be requested for a wide variety of purposes such as clinical (diagnostic), nutritional, or for basic structure determination.

1. *Specific amino acids*: Analysis of a few or specific amino acids is useful and easy to perform. This often requested in the detection of inborn errors of metabolism (phenylketonuria or maple syrup disease). In these cases, the abnormal amino acids are present in high concentration that makes the analysis simple. Several amino acids have been determined by CE for this purpose, such as tyrosine, proline, and phenylalanine.
2. *Amino acids arising from protein hydrolysate*: This analysis is used often in food industry as well in basic biochemistry. It is important in determination of the structure of protein and in the assessment of the nutritional value of different proteins. Here the analysis of about 20 amino acids is sufficient. This type of analysis is relatively more difficult than that for a single amino acid.
3. *Quantification of free amino acids in serum, urine, or other physiological fluids*: The separation here is complicated by the presence of many interfering substances, such as small peptides and uncommon amino acids. This analysis is requested often for screening purposes.

Traditionally amino acid analysis is carried out by specialized column chromatography instruments (amino acid analyzers). These instruments are quite expensive. More recently, general HPLC with specialized columns have been adapted to run this analysis. Many workers have attempted analysis of amino acids by CE using different pre- and postreactions to enhance their detection. Both free CZE and MEKC have been used with different degrees of success similar to that of amino acids detection by HPLC. The separation of amino acids is better achieved in coated capillaries and also better with MEKC. Derivatization, especially with fluorescent agents, offers much better sensitivity. Several additive agents such as urea, cyclodextrin, and tetrabutyl ammonium salts improve the separation. Chiral amino acid separation based on MEKC has been developed to analyze and quantify both D- and L-amino, which can be useful for detecting bacterial contamination [86,87]. On-capillary derivatization of amino acids in serum with laser-induced fluorescence (LIF) detection has also been reported [88]. Amino acid by CE has been reviewed [89–92]. The analysis of amino acids from biological fluids by CE without interference and with good reproducibility remains a challenge. A dedicated CE instrument for amino acids separation might someday be commercially available.

26.8.4 Ion Analysis

Many organic and inorganic ions are components of the biological fluids and cells. Most of these have clinical and physiological importance. A change in their normal level is associated with different disorders. Inorganic ions in the serum are important for maintaining cell viability, muscle excitability, osmotic pressure, and regulating the pH. Thus they prevent muscle, renal, and heart malfunctions. Organic acids are intermediates in the metabolism of many compounds in the cell. Clinically they are important for detection of inborn errors of metabolism, infection, and different metabolic disorders, such as diabetes. Some of the ions are not easy to measure regardless of the method. These ions resemble each other, lack strong absorbency and may be present in low concentration. Depending on the type of the ion, they are analyzed by several techniques [e.g., HPLC, automated clinical analyzer, gas chromatography (GC), atomic absorption, or mass spectra]—all technically very involved. Many inorganic ions are measured routinely in clinical labs. In practice, the common inorganic ions, such as Na, K, Ca, can be measured more conveniently and rapidly in the clinical labs by ion-selective electrodes built into highly specialized instruments.

The CE is quite suited for analysis of these ions too. Because of their relative charge to the small mass, they tend to migrate rapidly in this technique giving fast separation with very high theoretical plate numbers at a low cost per test. Both cations and anions can be analyzed in the same run. The separation can be based on simple free solution CE or based on a suitable chelating additive. In general, cations are measured in a low pH electrolyte containing a UV active species (imidazole or benzylamine), while anions are measured after reversing the EOF and also after adding a high mobility UV active species.

Organic acids are more difficult to measure compared with inorganic ions. Usually, these are measured by the GC or HPLC. However, CE offers speed, precision, and specificity over other methods. Many of these compounds have been measured by CE directly, or by indirect UV absorbency after the addition of a UV-absorbing compound such as benzoate, naphthalene, sulfonate, imidazole, or benzylamine. For example, oxalate and citrate (Figure 26.7), which are important in stone formation, have been measured after urine dilution by both direct and indirect UV detection [93–95]. Lactate, pyruvate, ascorbate, and oxalate were measured by CE in serum and in cerebrospinal fluids of patients in ~10 min [96,97]. Methylmalonic acid, which is a sensitive measure of vitamin B_{12} deficiency, preceding any clinical symptoms or changes in the serum level, has been determined in urine by CE after sample extraction and concentration [65,98]. Some of the uncommon ions such as nitrite and nitrate can be measured easily with CE (Figure 26.8) [99]. Plasma NO_2 and NO_3 were analyzed after sample dilution using absorbance at 214 nm [100] and after serum deproteinization with acetonitrile (Figure 26.8). We found the CE is much faster and less expensive for nitrate analysis when compared with the enzymatic methods. CE has been applied for the determination of arsenic acid and its related compounds using indirect detection in ammonium formate buffer [101]. Few

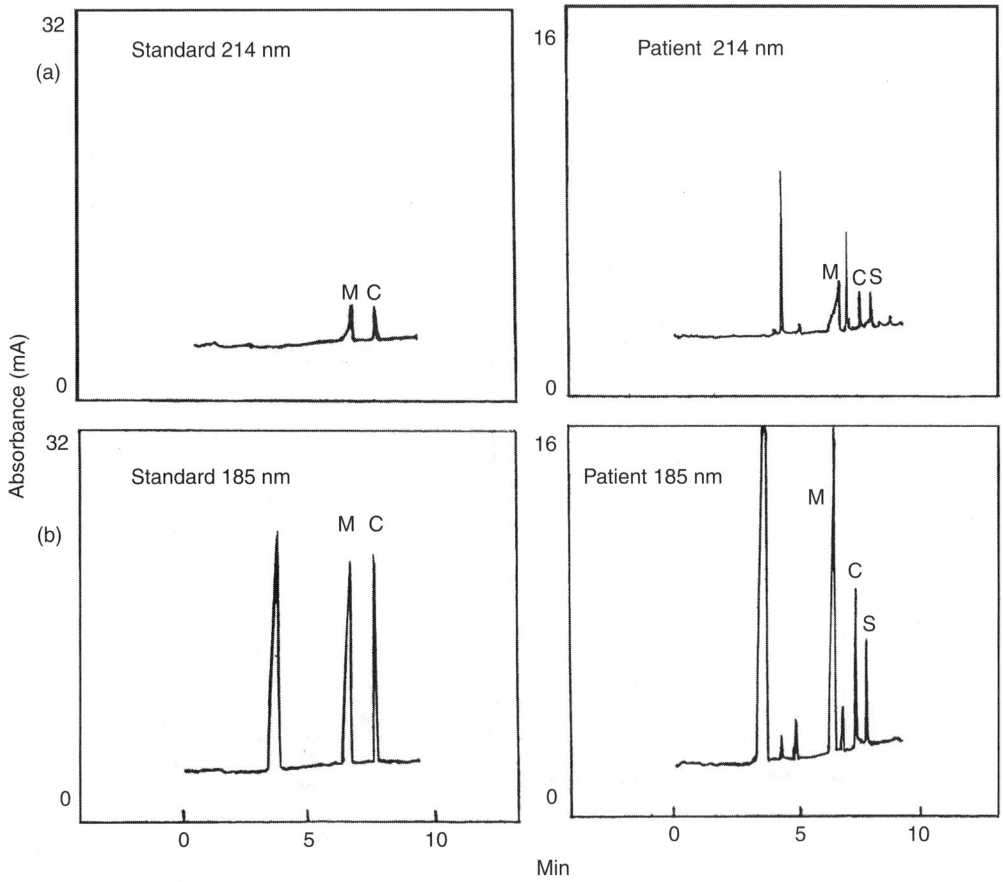

FIGURE 26.7 Effect of the wavelength on citrate detection by CE. (a) Standard of malonic acid (500 mg/L) and standard of citric acid (1000 mg/L) with detection at 214 nm; and patient urine at 214 nm; (b) same as top but detection at 185 nm (C, citric acid; S, Succinic acid; M, malonic acid. (From Shihabi, Z., et al., *J. Liq. Chromatogr.*, 24, 3197, 2001. With permission.)

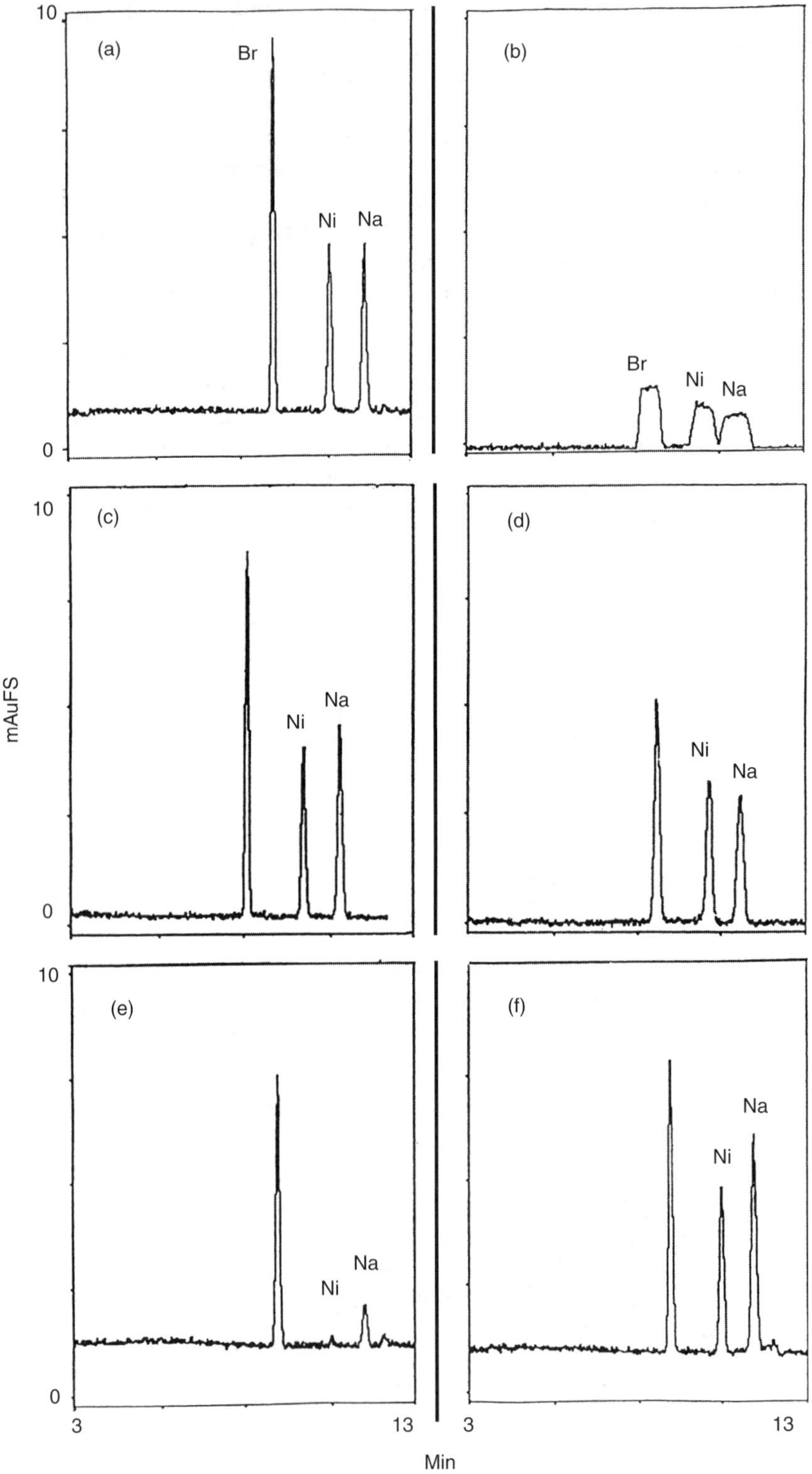

FIGURE 26.8 Nitrite (Ni) and nitrate (Na) (6 mg/L), analysis by CE, added to (a) water without acetonitrile, (b) 0.9% NaCl without acetonitrile, (c) 66% acetonitrile in water, and (d) 66% acetonitrile in 0.9% saline. (e) Serum from an individual with low levels of nitrite and nitrate (for spiking), deproteinized with acetonitrile (66% acetonitrile final concentration), and (f) the same serum spiked with nitrite and nitrate (6 mg/L, 66% acetonitrile final concentration). (From Friedberg, M. A., et al., *J. Chromatogr. A*, 781, 491, 1997.)

ions, such as Fe, Ni, Zn, citrate, and oxalate, have been determined in urine based on transient ITP [102]. Hydrogen peroxide that is involved in many biological reactions has been measured after acetonitrile stacking [103].

26.9 CONCLUDING REMARKS

The CE can be adapted to the separation of numerous compounds of clinical interest with some compounds being more suited than others. It offers different advantages in the clinical field notably, speed, high plate number, automation, and low operating cost. It is a good complement to other separation techniques. At the same time, the CE has few limitations, such as poor detection limits and the effect of the sample matrix on the separation. Understanding the principles behind these limitations allows better strategies to encounter them. Biological samples usually have a high content of protein and salts with a very wide range of concentrations for many compounds. This requires special attention to how the samples can be treated, how the separation buffer is chosen, and how the capillary is treated. The practical aspects and the special conditions for the separation of biological samples are described. The CE has matured enough that there are several companies putting on the market specialized fully automated instruments dedicated to special functions or special clinical compounds such as serum protein separation, DNA analysis, or isofocusing of proteins. Numerous applications of CE in the clinical field are described especially in the area of proteins, hemoglobin, enzymes, and ions. Since the operating costs are much less than that of the HPLC, and it does not require any appreciable amounts of organic solvents, application of CE in the clinical field is expected to expand. The CE-MS is being refined for applications in the area of proteomics. There remains a need for simpler methods of isofocusing in capillaries and simpler methods for size separation for proteins. Further researches concerning amino acid analysis, dyes that bind to proteins to improve the detection limits and improve the specificity are needed. The analysis of proteins by CE probably will expand in the future as the separation and detection of microheterogeneity are improved.

REFERENCES

1. Shihabi, Z.K. Effect of sample composition on electrophoretic migration application to hemoglobin analysis by capillary electrophoresis and agarose electrophoresis. *J. Chromatogr. A*, 1027, 179, 2004.
2. Foret, F., Klepárník, K., Gebauer, P., and Boèek, P. Ionic boundaries in biological capillary electrophoresis. *J. Chromatogr. A*, 1053, 43, 2004.
3. Garcia, L.L., and Shihabi, Z.K. Sample matrix effects in capillary electrophoresis: I. Basic considerations. *J. Chromatogr. A*, 652, 465, 1993.
4. Shihabi, Z.K. Sample matrix effects in capillary electrophoresis: II. Acetonitrile deproteinization. *J. Chromatogr. A*, 652, 471, 1993.
5. Shihabi, Z.K. *Sample Preparation and Stacking for Capillary Electrophoresis. Encyclopedia of Chromatography*, 2nd ed. (Cazes, J., ed.), Taylor & Francis, New York, NY, 2005, 1490–1498.
6. Fortis, F., Guerrier, L., Righetti, P.G., Antonioli, P., and Boschetti, E. A new approach for the removal of protein impurities from purified biologicals using combinatorial solid-phase ligand libraries. *Electrophoresis*, 27, 3018, 2006.
7. Schappler, J., Guillarme, D., Prat, J., Veuthey, J.-L., and Rudaz, S. Enhanced method performances for conventional and chiral CE-ESI/MS analyses in plasma. *Electrophoresis*, 27, 1537, 2006.
8. Shihabi, Z.K. Capillary electrophoresis for the determination of drugs in biological fluids. *Handbook of Analytical Separation* (Elsevier), 5, 77, 2004.
9. Dolnik, V., Capillary electrophoresis of proteins 2003–2005. *Electrophoresis*, 27, 126, 2006.
10. Liu, J., and Lee, M.L. Permanent surface modification of polymeric capillary electrophoresis microchips for protein and peptide analysis. *Electrophoresis*, 27, 3533, 2006.
11. Cordova, E., Gao, J., and Whitesides, G.M. Noncovalent polycationic coatings for capillaries in capillary electrophoresis of proteins. *Anal. Chem.*, 69, 1370, 1997.

12. Catai, J.R., Tervahauta, H.A., de Jong G.J., and Somsen, G.W. Noncovalently bilayer-coated capillaries for efficient and reproducible analysis of proteins by capillary electrophoresis. *J. Chromatogr. A*, 1083, 185, 2005.
13. Sakai-Kato, K., Kato, M., Nakajima, T., Toyo'oka, T., Imai, K., and Utsunomiya-Tate, N. Cationic starch derivatives as dynamic coating additives for protein analysis in capillary electrophoresis. *J. Chromatogr. A*, 1111, 127, 2006.
14. Yassine, M.M., and Lucy, C.A. Preparative capillary zone electrophoresis using a dynamic coated wide-bore capillary. *Electrophoresis*, 27, 3066, 2006.
15. Shihabi, Z.K. Stacking and discontinuous buffers in capillary zone electrophoresis. *Electrophoresis*, 21, 2872, 2000.
16. Krivankova, L., Pantuckova, P., Gebauer, P., Bocek, P., Caslavska, J., and Thormann, W. Chloride present in biological samples as a tool for enhancement of sensitivity in capillary zone electrophoretic analysis of anionic trace analytes. *Electrophoresis*, 24, 505, 2003.
17. Shihabi, Z.K. Stacking in capillary zone electrophoresis. *J. Chromatogr. A*, 902,107, 2000.
18. Shihabi, Z.K. Transient pseudo-isotachophoresis for sample concentration in capillary electrophoresis. *Electrophoresis*, 23, 1612, 2002.
19. Kim, J.B., and Terabe, S. On-line sample preconcentration techniques in micellar electrokinetic chromatography. *J. Pharm. Biomed. Anal.*, 30, 1625, 2003.
20. An, Y., Cooper, W.J., Balgley, M.B., and Lee, C.S. Selective enrichment and ultrasensitive identification of trace peptides in proteome analysis using transient capillary isotachophoresis/zone electrophoresis coupled with nano-ESI-MS. *Electrophoresis*, 27, 3503, 2006.
21. Campa, C., Coslovi, A., Flamigni, A., and Rossi, M. Overview on advances in capillary electrophoresisass spectrometry of carbohydrates: A tabulated review. *Electrophoresis*, 27, 2027, 2006.
22. Lehmann, R., Liebich, H., Grübler, G., and Voelter, W. Capillary electrophoresis of human serum proteins and apolipoproteins. *Electrophoresis*, 16, 998, 1995.
23. Liu, M., McNeal, C.J., and Macfarlane, R.D. Charge density profiling of circulating human low-density lipoprotein particles by capillary zone electrophoresis, *Electrophoresis*, 25, 2985, 2004.
24. Macfarlane, R.D., Bondarenko, P.V., Cockrill, S.L., Cruzado, I.D., Koss, W., McNeal, C.J., Spiekerman, M., and Watkins, L.K. Development of a lipoprotein profile using capillary electrophoresis and mass spectrometry. *Electrophoresis*, 18, 1796, 1997.
25. Gordon, M.J., Lee, K.J., Arias, A.A., and Zare, R.N. Protocol for resolving protein mixtures in capillary zone electrophoresis. *Anal. Chem.*, 63, 69, 1991.
26. Chen, F.T., Liu, C.M. Hsieh, Y.Z., and Sternberg, J.C. Capillary electrophoresis—A new clinical tool. *Clin. Chem.*, 37, 14, 1991.
27. Chen, F.T., and Sternberg, J. Characterization of proteins by capillary electrophoresis in fused-silica columns: Review on serum protein analysis and application to immunoassays. *Electrophoresis*, 15, 13, 1994.
28. Chen, F.A. Rapid protein analysis by capillary electrophoresis. *J. Chromatogr.*, 559, 445–453, 1991.
29. Dolnik, V. Capillary zone electrophoresis of serum proteins: Study of separation variables. *J. Chromatogr.*, 709, 99–110, 1995.
30. Roudiere, L., Boularan, A.M., Bonardet, A., Vallat, C., Cristol, J.P., and Dupuy, A.M. Evaluation of a capillary zone electrophoresis system versus a conventional agarose gel system for routine serum protein separation and monoclonal component typing. *Clin. Lab.*, 52, 19, 2006.
31. Louahabi, A., Philippe, M, Lali, S., Wallemacq, P., and Maisin, D. Evaluation of a new Sebia kit for analysis of hemoglobin fractions and variants on the capillary system. *Clin Chem. Lab. Med.*, 44, 340, 2006.
32. Mussap, M., Ponchia, S., Zaninotto, M., Varagnolo, M., and Plebani, M., Evaluation of a new capillary zone electrophoresis system for the identification and typing of Bence Jones Protein. *Clin. Biochem.*, 39, 152, 2006.
33. Shihabi, Z. Cryoglobulins: An important but a neglected clinical test. *Ann. Clin Lab Sci.*, 36, 395, 2006.
34. Shihabi, Z. Analysis and general classification of serum cryoglobulins by capillary zone electrophoresis. *Electrophoresis*, 17, 1607, 1996.
35. Friedberg, M.A. and Shihabi, Z.K. Urine protein analysis by capillary electrophoresis. *Electrophoresis*, 18, 1836, 1997.

36. Sanders, E., Katzmann, J.A., Clark, R., Oda, R.P., Shihabi, Z., and Landers, J.P. Development of capillary electrophoresis as an alternative to high resolution agarose electrophoresis for the diagnosis of multiple sclerosis. *Clin. Chem. Lab Med.*, 37, 37,1999.
37. Bao, J., and Regnier, F.E. Ultramicro enzyme assays in a capillary electrophoretic system. *J. Chromatogr.*, 608, 217–224, 1992.
38. Wu, D., and Regnier, F.E. Native protein separations and enzyme microassays by capillary zone and gel electrophoresis. *Anal. Chem.*, 65, 2029–2035, 1992.
39. Miller, K.J., Leesong, I., Bao, J., Regnier, F.E., and Lytle, F.E. Electrophoretically mediated microanalysis of leucine aminopeptidase in complex matrices using time-resolved laser-induced fluorescence detection. *Anal. Chem.*, 65, 3267, 1993.
40. Landers, J.P., Schuchard, M.D., Subramaniam, M., Sismelich, T.P., and Spelsberg, T.C. High-performance capillary electrophoretic analysis of chloramphenicol acetyl transferase activity. *J. Chromatogr.*, 603, 247, 1992.
41. Shihabi, Z.K. Analysis of angiotension-converting enzyme by capillary electrophoresis. *J. Chromatogr. A*, 853, 185, 1999.
42. Shihabi, Z.K., and Kute, T.E. Analysis of cathepsin D from breast tissues by capillary electrophoresis. *Chromatogr. B*, 683, 125, 1996.
43. Viglio, S., Zanaboni, G., Luistetti, M., Cetta, G., Guglielminetti, M., and Iadarola, P. Micellar electrokinetic chromatography: A convenient alternative to colorimetric and high performance liquid chromatographic detection to monitor protease activity. *Electrophoreis*, 19, 2083, 1998.
44. Shihabi, Z.K., Hinsdale, M.E., and Daugherty, H.K., Jr. Hemoglobin A2 quantification by capillary zone electrophoresis. *Electrophoresis*, 21, 749, 2000.
45. Cotton, F., Lin, C., Fontaine, B., Gulbis, B., Janssens, J., and Vertongen, F. Evaluation of a capillary electrophoresis method for routine determination of hemoglobins A2 and F. *Clin. Chem.*, 45, 237, 1999.
46. Mario, N., Baudin, B., Aussel, C., and Giboudeau, J. Capillary isoelectric focusing and high-performance cation-exchange chromatography compared for qualitative and quantitative analysis of hemoglobin variants. *Clin. Chem.*, 43, 2137, 1997.
47. Jenkins, M.A. and Ratnaike, S. Capillary isoelectric focusing of haemoglobin variants in the clinical laboratory. *Clin. Chim. Acta*, 289, 121, 1999.
48. Hempe, J.M., and Craver, R.D. Quantification of hemoglobin variants by capillary isoelectric focusing. *Clin. Chem.*, 40, 2288, 1994.
49. Wang, J., Zhou, S., Huang, W., Liu, Y., Cheng, C., Lu, X., and Cheng, J. CE-based analysis of hemoglobin and its applications in clinical analysis, *Electrophoresis*, *27*, 3108, 2006.
50. Jenkins, M., and Ratnaike, S. Capillary electrophoresis of hemoglobin. *Clin. Chem. Lab Med.*, 41, 747, 2003.
51. Jabeen, R., Payne, D., Wiktorowicz, J., Mohammad, A., and Petersen, J. Capillary electrophoresis and the clinical laboratory. *Electrophoresis*, 27, 2413, 2006.
52. Gerritsma, J., Sinnige, D., Drieze, C., Sittrop, B., Houtsma, P., Hulshorst-Jansen, N., and Huisman, W. Quantitative and qualitative analysis of haemoglobin variants using capillary zone electrophoresis. *Ann. Clin. Biochem.*, 37, 380, 2000.
53. Siren, H., Laitinen, P., Turpeinen, U., and Karppinen, P. Direct monitoring of glycohemoglobin A1c in the blood samples of diabetic patients by capillary electrophoresis. Comparison with an immunoassay method. *J. Chromatogr. A*, 979, 201, 2002.
54. Yao, X.-W., and Regnier, F.E. Polymer- and surfactant-coated capillaries for isoelectric focusing. *J. Chromatogr.*, 632, 185, 1993.
55. Vincenzi, J.A., and Franco, M.M. Novel approach for the analysis of glycated hemoglobin using capillary isoelectric focusing with chemical mobilization. *J. Chromatogr. B*, 785, 285, 2003.
56. Ong, C.-N., Liau, L.S., and Ong, H.Y. Separation of globins using free zone capillary electrophoresis. *J. Chromatogr.*, 576, 346, 1992.
57. Ferranti, P., Malorni, A., Pucci, P., Fanali, S., Nardi, A., and Ossicini, L. Capillary zone electrophoresis and mass spectrometry for the characterization of genetic variants of human hemoglobin. *Anal. Biochem.*, 194, 1, 1991.
58. Shihabi, Z.K., and Hinsdale, M.E. Simplified hemoglobin chain detection by capillary electrophoresis. *Electrophoresis*, 26, 581, 2005.

59. Shihabi, Z.K. Peptide stacking by acetonitrile-salt mixtures for capillary zone electrophoresis. *J. Chromatogr. A*, 744, 231, 1996.
60. Shihabi, Z.K., Hinsdale, M.E., and Cheng, C.P. Analysis of glutathione by capillary electrophoresis based on sample stacking. *Electrophoresis*, 22, 2351, 2001.
61. Yang, Y., Boysen, R.I., and Hearn, M.T. Optimization of field-amplified sample injection for analysis of peptides by capillary electrophoresis-mass spectrometry. *Anal. Chem.*, 78, 4752, 2006.
62. Sandra, K., Lynen, F., Devreese, B., Van Beeumen, J., and Sandra, P. On-column sample enrichment for the high-sensitivity sheath-flow CE-MS analysis of peptides. *Anal. Bioanal. Chem.*, 385, 671, 2006.
63. Stutz, H. Advances in the analysis of proteins and peptides by capillary electrophoresis with matrix-assisted laser desorption/ionization and electrospray-mass spectrometry detection. *Electrophoresis*, 26, 1254, 2005.
64. Kasicka, V. Recent developments in capillary electrophoresis and capillary electrochromatography of peptides. *Electrophoresis*, 27, 142, 2006.
65. Fliser, D., Wittke, S., and Mischak, H. Capillary electrophoresis coupled to mass spectrometry for clinical diagnostic purposes *Electrophoresis*, 26, 2708, 2005.
66. Huang, Y.-F., Huang, C.-C., Hu, C.-C., and Chang, H.-T. Capillary electrophoresis-based separation techniques for the analysis of proteins. *Electrophoresis*, 27, 3503, 2006.
67. Manabe, T., Oota, H., and Mukai, J. Size separation of sodium dodecyl sulfate complexes of human plasma proteins by capillary electrophoresis employing linear polyacrylamide as a sieving polymer. *Electrophoresis*, 19, 2308, 1998.
68. Yan, W., Sloat, A.L., Yagi, S., Nakazumi, H., and Colyer, C.L. Protein labeling with red squarylium dyes for analysis by capillary electrophoresis with laser-induced fluorescence detection. *Electrophoresis*, 27, 1347, 2006.
69. Sluszny, C., He, Y., and Yeung, E.S. Light-emitting diode-induced fluorescence detection of native proteins in capillary electrophoresis. *Electrophoresis*, 26, 4197, 2005.
70. Lacroix, M., Poinsot, V., Fournier, C., and Couderc, F. Laser-induced fluorescence detection schemes for the analysis of proteins and peptides using capillary electrophoresis. *Electrophoresis*, 26, 2608, 2005.
71. Giordano, B.C., Muza, M., Trout, A., and Landers, J.P. Dynamically-coated capillaries allow for capillary electrophoretic resolution of transferrin sialoforms via direct analysis of human serum. *J. Chromatogr. B*, 742, 79, 2000.
72. Lupi, A., Viglio, S., Luisetti, M., Gorrini, M., Coni, P., Faa, G., Cetta, G., and Iadarola, P. Alpha1-antitrypsin in serum determined by capillary isoelectric focusing. *Electrophoresis*, 21, 3318, 2000.
73. Sanz-Nebot, V., Benavente, F., Vallverdu, A., Guzman, N.A., and Barbosa, J. Separation of recombinant human erythropoietin glycoforms by capillary electrophoresis using volatile electrolytes. Assessment of mass spectrometry for the characterization of erythropoietin glycoforms, *Anal. Chem.*, 75, 5220, 2003.
74. Thorne, J.M., Goetzinger, W.K., Chen, A.B., Moorhouse, K.G., and Karger, B.L. Examination of capillary zone electrophoresis, capillary isoelectric focusing and sodium dodecyl sulfate capillary electrophoresis for the analysis of recombinant tissue plasminogen activator. *J. Chromatogr. A*, 744, 155, 1996.
75. Schmerr, M.J., Jenny, A.L., Bulgin, M.S., Miller, J.M., Hamir, A.N., Cutlip, R.C., and Goodwin, K.R. Prions have been studied by capillary electrophoresis [Use of capillary electrophoresis and fluorescent labeled peptides to detect the abnormal prion protein in the blood of animals that are infected with a transmissible spongiform encephalopathy]. *J Chromatogr. A*, 853, 207, 1999.
76. Theodorescu, D., Fliser, D., Wittke, S., Mischak, H., Krebs, R., Walden, M., Ross, M., Eltze, E., Bettendorf, O., Wulfing, C., and Semjonow, A. Pilot study of capillary electrophoresis coupled to mass spectrometry as a tool to define potential prostate cancer biomarkers in urine. *Electrophoresis*, 26, 2797, 2005.
77. Theodorescum, D., Wittke, S., Ross, M.M., Walden, M., Conaway, M., Just, I., Mischak, H., and Frierson, H.F. Urothelial carcinoma was identified by CE_MS [Discovery and validation of new protein biomarkers for urothelial cancer: a prospective analysis]. *Lancet Oncol.*, 7, 230, 2006.

78. Ru, Q.C., Katenhusen, R. A, Zhu, L.A., Silberman, J., Yang, S., Orchard, T.J., Brzeski, H., Liebman, M., and Ellsworth, D.L. Proteomic profiling of human urine using multi-dimensional protein identification technology. *J. Chromatogr. A*, 1111, 166, 2006.
79. Geldart, S.E. and Brown, P.R. Analysis of nucleotides by capillary electrophoresis. *J. Chromatogr. A*, 828, 317, 1998.
80. Liu, C.C., Huang, J.S., Tyrrell, D.L., and Dovichi, N.J. Capillary electrophoresis-electrospray-mass spectrometry of nucleosides and nucleotides: Application to phosphorylation studies of anti-human immunodeficiency virus nucleosides in a human hepatoma cell line. *Electrophoresis*, 26, 1424, 2006.
81. Arcibal, I.G., Santillo, M.F., and Ewing, A.G. Recent advances in capillary electrophoretic analysis of individual cells. *Anal. Bioanal. Chem.*, 387, 51, 2007.
82. Weng, Q., Xu, G., Yuan, K., and Tang, P. Determination of monoamines in urine by capillary electrophoresis with field-amplified sample stacking and amperometric detection. *J. Chromatogr. B*, 835, 55, 2006.
83. Peterson, Z.D., Collins, D.C., Bowerbank, C.R., Lee, M.L., and Graves, S.W. Determination of catecholamines and metanephrines in urine by capillary electrophoresis-electrospray ionization-time-of-flight mass spectrometry. *J. Chromatogr. B*, 776, 221, 2002.
84. Chen, G., Zhang, L., and Zhu, Y. Determination of glycosides and sugars in Moutan Cortex by capillary electrophoresis with electrochemical detection. *J. Pharm. Biomed. Anal.*, 41, 129, 2006.
85. Yang, Y., Breadmore, M.C., and Thormann, W. Analysis of the disaccharides derived from hyaluronic acid and chondroitin sulfate by capillary electrophoresis with sample stacking. *J. Sep. Sci.*, 28, 2381, 2005.
86. Carlavilla, D., Moreno-Arribas, M.V., Fanali, S., and Cifuentes, A. Chiral MEKC-LIF of amino acids in foods: Analysis of vinegars. *Electrophoresis*, 27, 2551, 2006.
87. Chen, F., Zhang, S., Qi, L., and Chen, Y. Chiral capillary electrophoretic separation of amino acids derivatized with 9-fluorenylmethylchloroformate using mixed chiral selectors of β-cyclodextrin and sodium taurodeoxycholate. *Electrophoresis*, 27, 2896, 2006.
88. Veledo, M.T., de Frutos, M., and Diez-Masa, J.C. On-capillary derivatization and analysis of amino acids in human plasma by capillary electrophoresis with laser-induced fluorescence detection: Application to diagnosis of aminoacidopathies. *Electrophoresis*, 27, 3101, 2006.
89. Poinsot, V., Lacroix, M., Maury, D., Chataigne, G., Feurer, B., and Couderc, F. Recent advances in amino acid analysis by capillary electrophoresis. *Electrophoresis*, 27, 176, 2006.
90. Poinsot, V., Bayle, C., and Couderc, F. Recent advances in amino acid analysis by capillary electrophoresis. *Electrophoresis*, 24, 4047, 2003.
91. Prata, C., Bonnafous, P., Fraysse, N., Treilhou, M., Poinsot, V., and Couderc, F. Recent advances in amino acid analysis by capillary electrophoresis. *Electrophoresis*, 22, 4129, 2001.
92. Smith, J.T. Recent advancements in amino acid analysis using capillary electrophoresis. *Electrophoresis*, 20, 3078, 1999.
93. Holmes, R.P. Measurement of urinary oxalate and citrate by capillary electrophoresis and indirect ultraviolet absorbance. *Clin. Chem.*, 41, 1297, 1995.
94. Wildman, B.J., Jackson, P.E., Jones, W.R., and Alden, P.G. Analysis of anion constituents of urine by inorganic capillary electrophoresis. *J. Chromatogr.*, 546, 459, 1991.
95. Shihabi, Z., Holmes, R., and Hinsdale, M. Urinary citrate analysis by capillary electrophoresis, *J. Liq. Chromatogr.*, 24, 3197, 2001.
96. Dolnik, V. and Dolnikova, J. Capillary zone electrophoresis of organic acids in serum of critically ill children. *J. Chromatogr. A*, 716, 269, 1995.
97. Hiraoka, A., Akai, J., Tominaga, I., Hattori, M., Sasaki, H., and Arato, T. Capillary zone electrophoretic determination of organic acids in cerebrospinal fluid from patients with central nervous system diseases. *J. Chromatogr. A*, 680, 243, 1994.
98. Marsh, D.B., and Nutall, K.L. Serum methylmalonic acid by capillary zone electrophoresis using electrokinetic injection and indirect photometric detection. *J. Capillary Electrophor.*, 2, 63, 1995.
99. Friedberg, M.A., Hinsdale, M.E., and Shihabi, Z.K. Analysis of nitrate in biological fluids by capillary electrophoresis. *J. Chromatogr. A*, 781, 491, 1997.

100. Leone, A.M., Francis, P.L., Rhodes, P., and Moncada, S. A rapid and simple method for the measurement of nitrite and nitrate in plasma by high performance capillary electrophoresis. *Biochem. Biophys. Res. Comm.*, 200, 951, 1994.
101. Kitagawa, F., Shiomi, K., and Otsuka, K. Analysis of arsenic compounds by capillary electrophoresis using indirect UV and mass spectrometric detections. *Electrophoresis*, 27, 2233, 2006.
102. Timerbaev, A.R., and Hirokawa, T. Recent advances of transient isotachophoresis-capillary electrophoresis in the analysis of small ions from high-conductivity matrices. *Electrophoresis*, 27, 323, 2006.
103. Shihabi, Z. Direct analysis of hydrogen peroxide by capillary electrophoresis. *Electrophoresis*, 27, 4215, 2006.

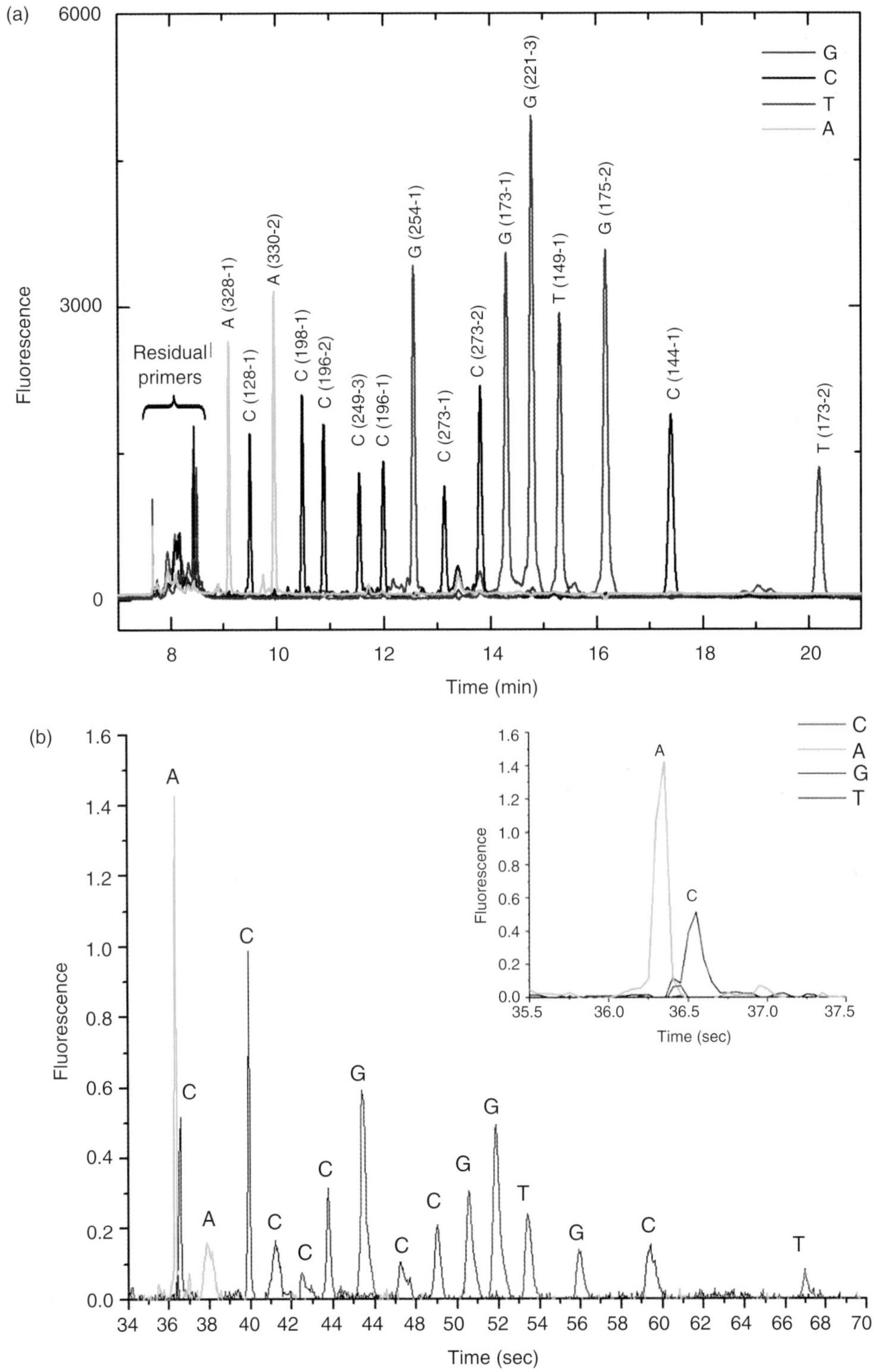

FIGURE 12.10 Electropherograms of multiplexed SBE-FSCE separation of 16 loci in a wild-type *p*53 gene. (a) CE separation in 36 cm capillary (47 cm total length, 50 μm inner diameter) in 1× TTE (89 mM Tris, 89 mM TAPS, 2 mM EDTA), 7 M urea and 0.5% (v/v) POP-6 for dynamic coating. (b) Separation on 8 cm microchip in 1× TTE (49 mM Tris, 49 mM TAPS, 2 mM EDTA) with 7 M urea on chip coated with Poly-*N*-hydroxyethylacrylamide. Both separations were performed at 55°C. (Reprinted with permission from Meagher, R. J., et al. *Analytical Chemistry*, 79, 1848–1854, 2007.)

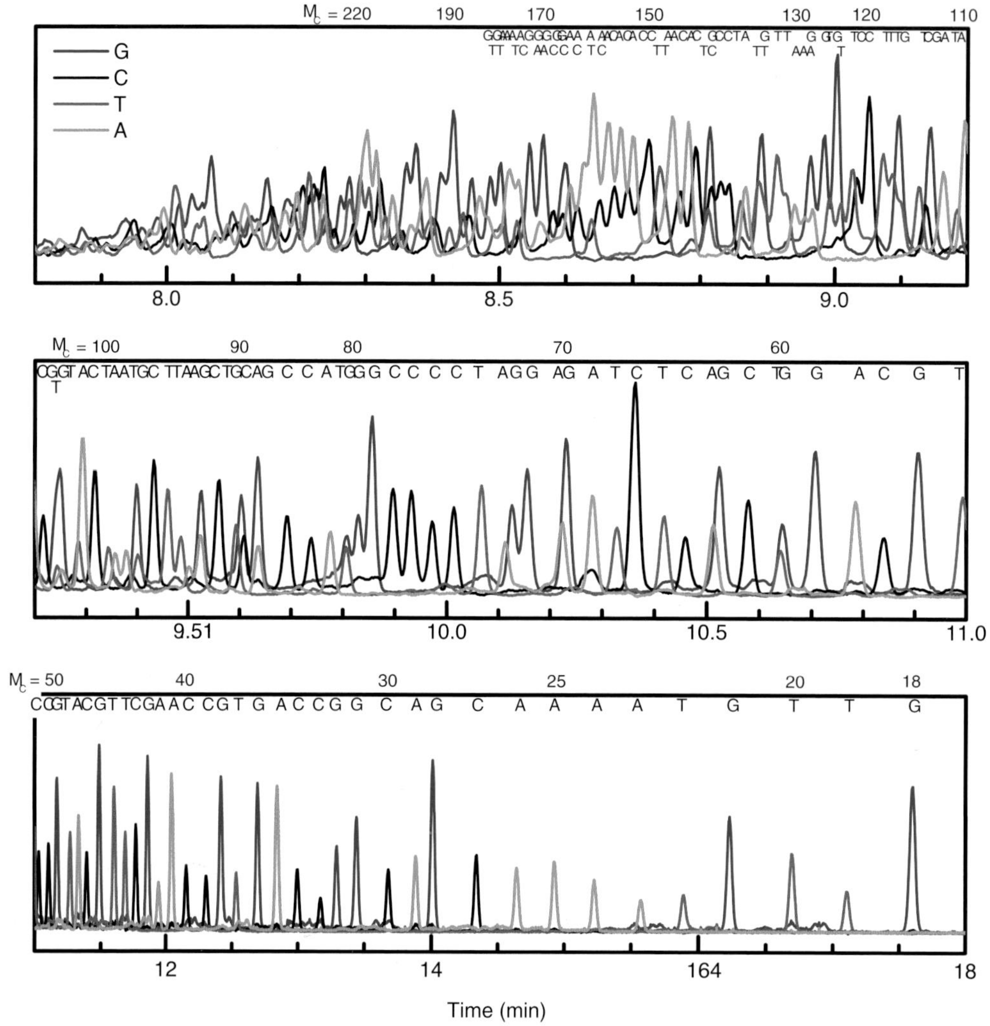

FIGURE 12.11 DNA sequencing analysis of M13mp18 ssDNA in free-solution with genetically engineering protein polymer drag-tag in 36 cm capillary (47 cm total length, 50 μm inner diameter) with 312 V/cm field strength and electrokinetic injection of 1 kV for 20 s. Capillaries were filled with 1× TTE (89 mM Tris, 89 mM TAPS, 2 mM EDTA) with 7 M urea and 0.5% (v/v) POP-6 to dynamically coat the walls of the capillary. (Reprinted with permission from Meagher, R. J., et al. In preparation 2007.)

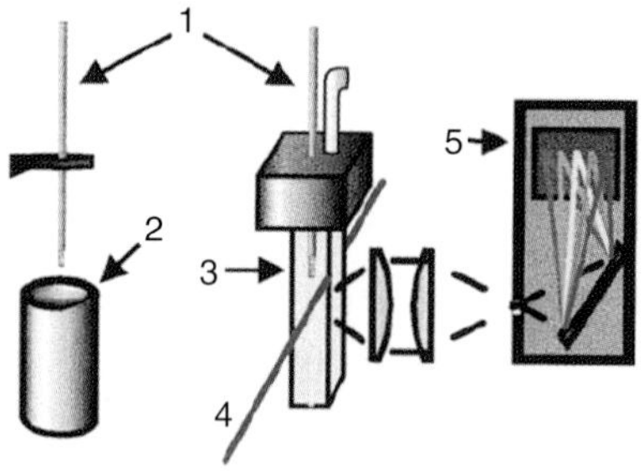

FIGURE 14.4 CE system with simultaneous multiwavelength fluorescence detection. Cellular samples were electrokinetically drawn into a capillary (1) from a steel microvial (2). The analytes were separated in the capillary, and eluted into a square-cuvette sheath-flow cell (3) where fluorescence was induced with a laser beam (4). The fluorescent light was collected orthogonally and directed into a spectrograph and CCD where a complete spectrum was obtained for each separated analyte (5). (From Fuller, R.R. et al., *Neuron*, 20, 173, 1999. With permission.)

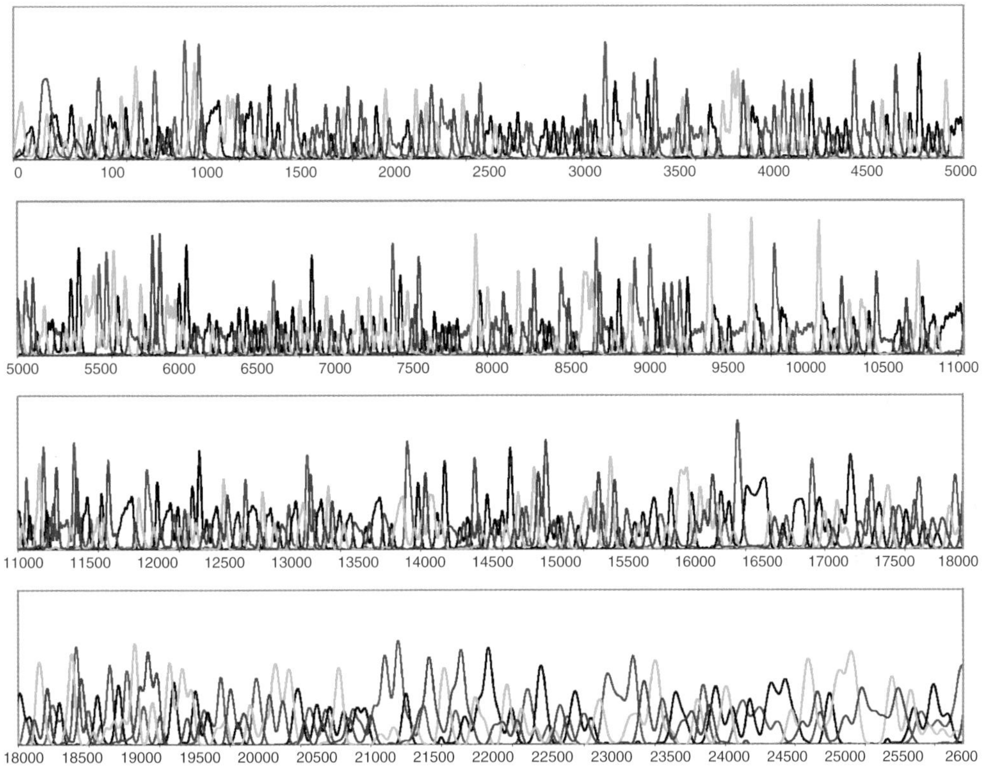

FIGURE 16.7 Analyzed four-color dye-terminator sequencing from a CEQ analysis system (Beckman Coulter): 1009 bases called to 98% accuracy.

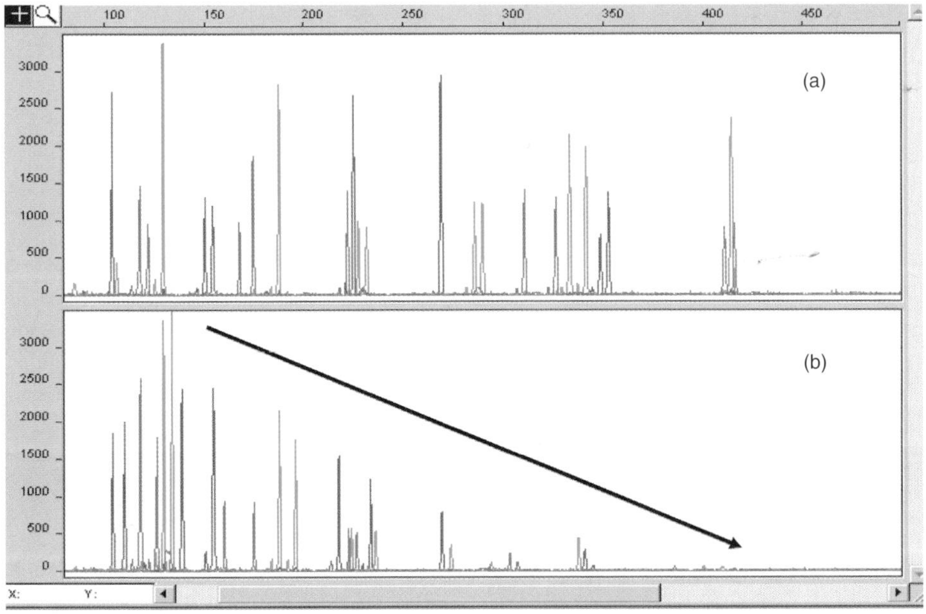

FIGURE 25.4 A comparison between a 9947 control DNA sample (a) and a degraded DNA sample (b) extracted from a recovered bone fragment. There is an evident loss of signal at the larger size loci in sample b (arrow) due to sample degradation. The sample is extracted and amplified DNA prepared using the Promega Powerplex STR multiplex kit and consists of 16 separate genetic loci labeled with three different dyes and separated via capillary gel electrophoresis using the ABI 310 genetic analyzer. (Courtesy of Kerry Opel, International Forensic Research Institute, Florida International University.)

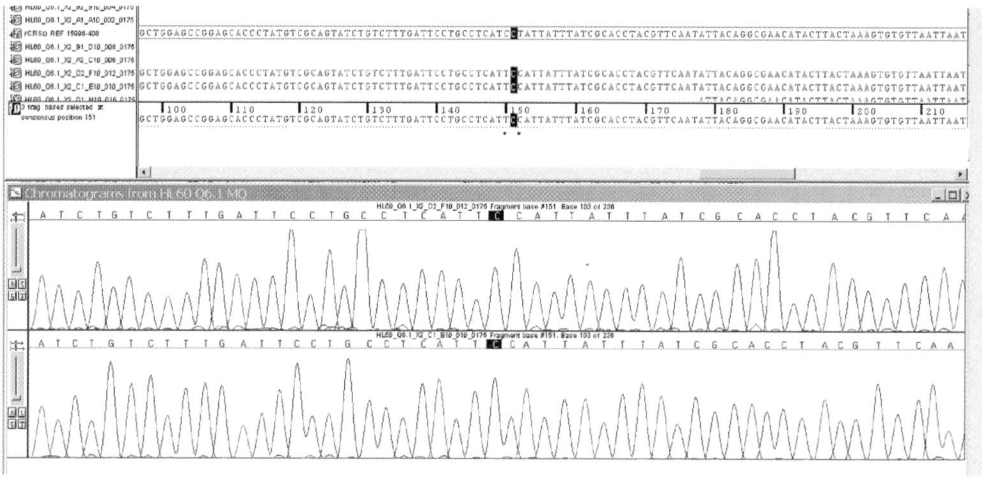

FIGURE 25.7 Mitochondrial DNA profile of an HVII sequence from a hair shaft. A repetitive analysis of the same hair sample is used to compare and align the sequences. The nuclear DNA in hair is often badly degraded and difficult to amplify. mtDNA provides an alternative procedure that can provide a DNA haplotype for forensic DNA profiling based on point mutations detected in the sequence when compared to a reference sample. Analysis performed on an ABI 310 Genetic Analyzer. (Courtesy of the DNA Analysis Unit II at the FBI Laboratory, Quantico, VA.)

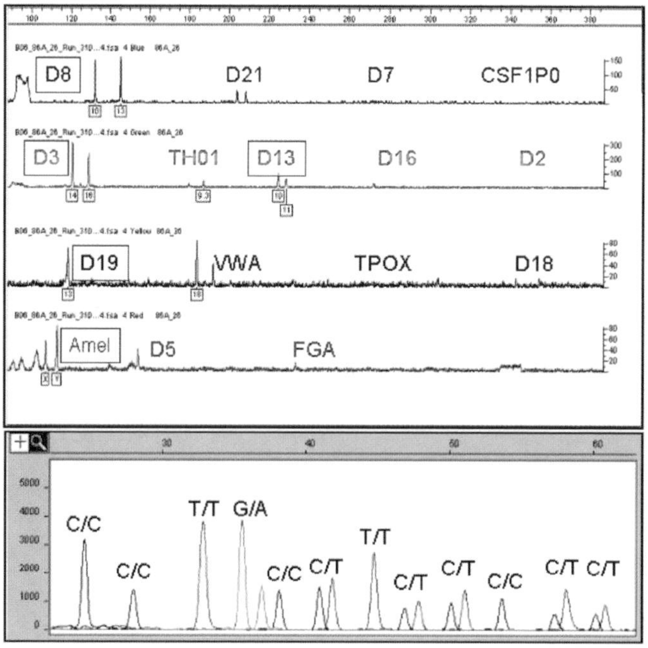

FIGURE 25.10 The recovery of information from degraded DNA using the Identifiler™STR multiplex and an SNP multiplex, both analyzed using the ABI 310 genetic analyzer. The top electropherogram depicts a degraded sample amplified via STRs. The bottom profile depicts a SNP multiplex on the same sample. The STR profile is blank for many larger alleles such as D7 CSF, D16, D2, TPOX, D18, and FGA. The smaller sized SNP fragments permit recovery of additional genetic information from the sample. (From Vallone, P.M., Decker, A.E., and Butler, J.M., The evaluation of Autosomal SNP assays, *Chemical Science and Technology Laboratory, National Institute of Standards and Technology*, Div. 831.)

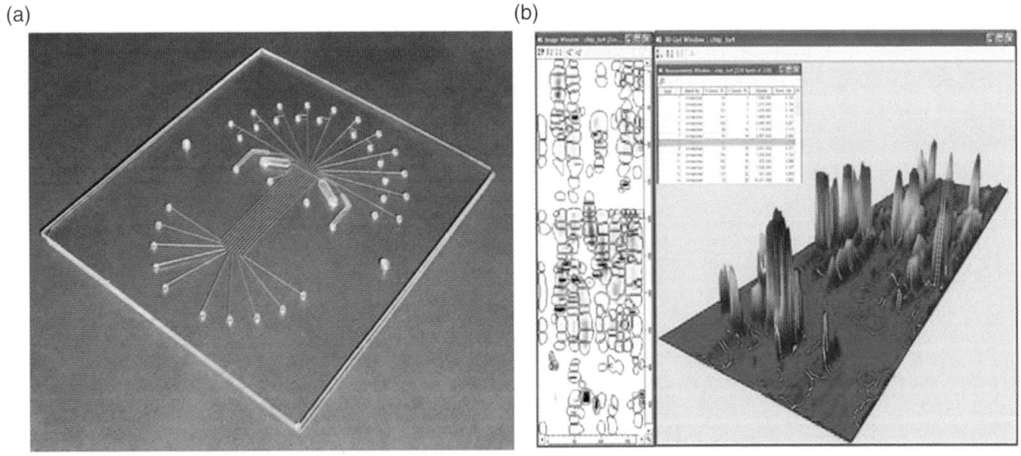

FIGURE 33.3 (a) Microfluidic PMMA chip for 2-D IEF/SDS–CGE protein separations and (b) pseudo-gel image resulting from a yeast cell lysate separation.

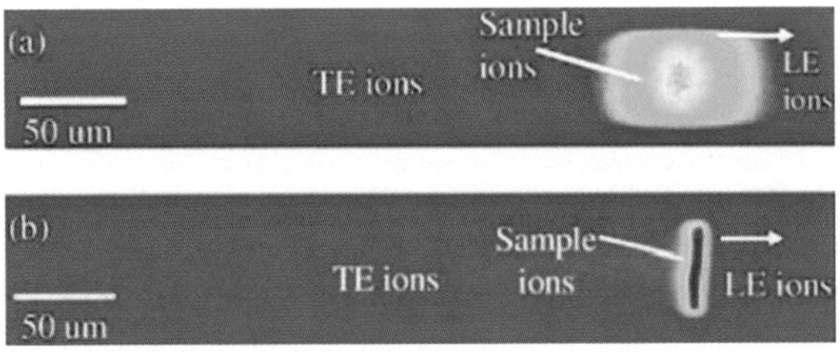

FIGURE 38.14 CCD camera images of on-chip sample peaks of AlexaFluor 488 at the LE/TE interface in two different ITP experiments. In (a) there is finite (nonuniform) EOF and the sample peak streamwise dimension is on the order of channel width or larger. In (b) EOF is suppressed and the sample is concentrated in narrower zone (~5 μm) at relatively high electric field. While Taylor dispersion based analysis is probably applicable in the first case, more comprehensive modeling is required for case (b).

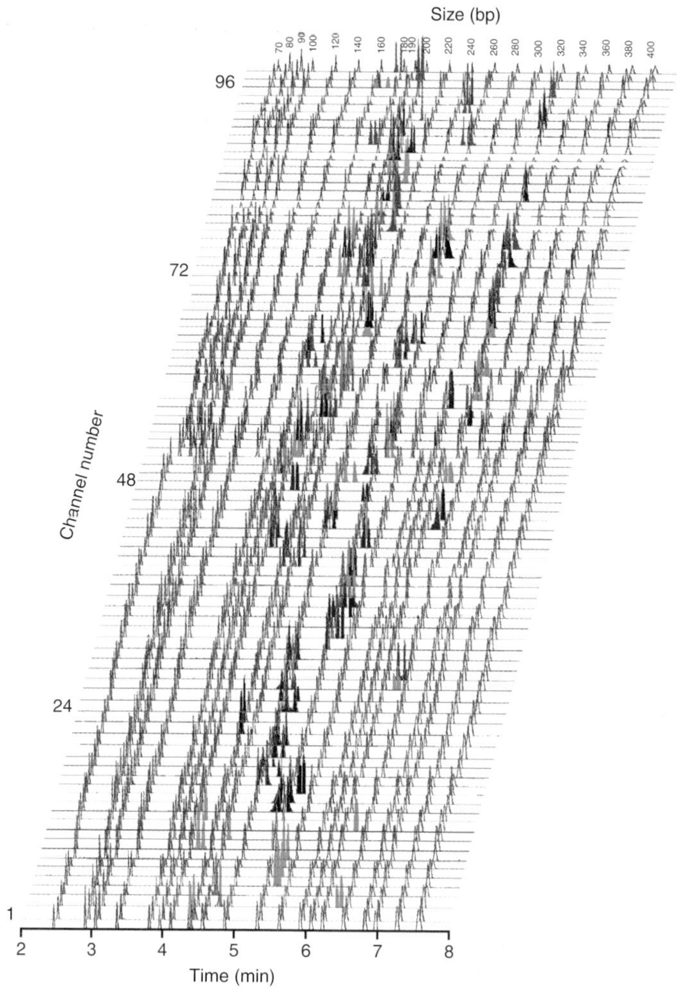

FIGURE 46.7 Electropherograms from a multiplexed short tandem repeat sizing separation on a 150-mm diameter, 96-channel μCAE device. Separations are successful in all lanes and sizing against DNA standards (blue and red) was accurate to within 1% of known values. (Reproduced from Medintz, I.L. et al., *Clin. Chem.*, 47, 1614, 2001. With permission.)

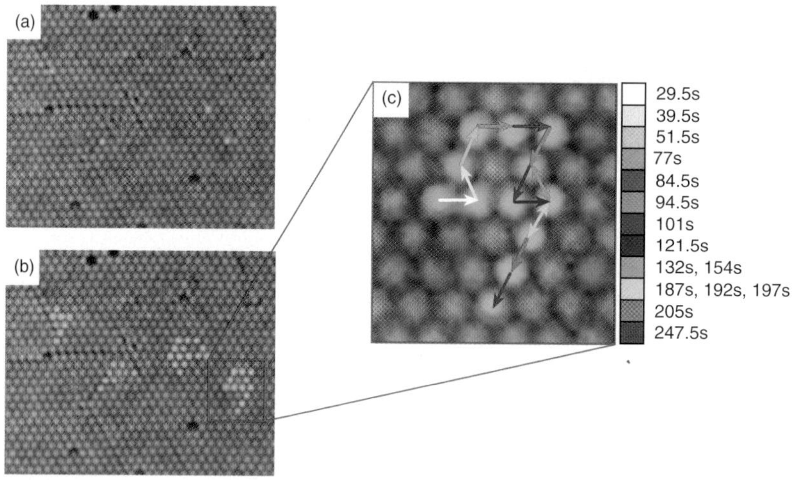

FIGURE 54.12 DNA dynamics in a polymer matrix formed from colloidal templating in the absence of an electric field. (a) At a given instance in time, four DNA molecules are localized in different pores. (b) Over time, the DNA explores the local pores. (c) The jumping between pores is random in both direction and the waiting time between jumps. (Reprinted from Nykypanchuk, D., et al., *Science*, 297, 987, 2002. With permission.)

27 Solid-Phase Microextraction and Solid Phase Extraction with Capillary Electrophoresis and Related Techniques

Stephen G. Weber

CONTENTS

27.1 INTRODUCTION

Improving the concentration detection limit is the goal of much analytical research and development. While miniaturization leading to lower mass detection limits is by no means trivial, it is at least conceptually obvious how to proceed. On the other hand, approaches to better concentration detection limits often require considerable creativity. In some cases, better detectors offer lower detection limits. Because properties of molecules differ, it is difficult to be globally accurate in a statement of relative merits of detectors, but experience shows that the inherent detection limits for compounds that are well suited for each detector are fluorescence < electrospray ionization mass spectrometry (ESI-MS) ~ electrochemistry < optical absorbance. It is worth noting that low detection limit is correlated with the selectivity of the detector. In fact, in real samples of sufficient complexity (are there any real samples that are not complex?), Nagels[1–4] has shown that detection limits correlate with the selectivity of the detector. The reason that detection limits correlate with the selectivity of the detector is that the limitation in real analyses is interferences from chemical constituents of the sample. These may be perceptible peaks in an electropherogram, or they may be a "baseline" of poorly resolved, multiple unknown components that vary from sample to sample. When this is the case, attempts to improve the detection limit by increasing the concentration of all the components in a sample will fail.

The high-resolution capability of capillary electrophoresis (CE) coupled with recent progress in comprehensive two-dimensional separations give a glimmer of hope that some day all components of a sample will be resolved, and then detection limits will be dictated by the inherent detector properties. If this is the case, then the act of concentrating all of the components in a sample will lead to better detection limits. A consideration of the peak capacity problem, however, tells us that we have a long way to go to realize this goal. We must, therefore, consider ways to improve analytical methods based on CE and related techniques for real samples knowing that there are more components in a sample that our most well-designed separations system can resolve. What is required is a method that concentrates selectively the desired analytes, and does not concentrate, or better yet, that dilutes undesired components of a sample.

Microextractions have the capability to concentrate selectively compounds with particular chemical properties, and to reject solutes with other particular chemical properties. Microextractions function on a small volume scale. As the volumes involved in microextraction and CE can be (but are not necessarily) the same, and given the capability of microextraction to concentrate analytes selectively, microextractions are a natural choice for improving virtually any CE (micellar electrokinetic chromatography, MEKC; capillary electrochromatography, CEC; etc.) method.

27.2 MICROEXTRACTION APPROACHES

This chapter will describe two approaches to microextraction: solid-phase extraction (SPE) and solid-phase microextraction (SPME). It is not a comprehensive review. Rather certain key elements of each approach and how it interfaces to the separations step will be described. As SPE is more intuitively understood, and SPME perhaps less so, there is a focus on SPME. SPME has been reviewed from several different perspectives.[5–17]

SPME was first developed in Pawliszyn's laboratory and applied to gas chromatography (GC).[18] The technique has been commercialized and is available through Supelco (now part of the Sigma/Aldrich company in the United States). The technique has some marvelous attributes—it is more than a different way to carry out a separation. Figure 27.1 shows in schematic form the basic idea as originally conceived. An SPME probe is simply an inert rod or fiber (e.g., an optical fiber) coated with any of a number of types of materials such as the highly polar poly(acrylic acid) or the highly nonpolar poly(dimethylsiloxane). The probe tip can be immersed directly in the sample or head space sampling may be preferred. The most common application is to gas chromatography (GC). The probe is inserted into the injection port where the high temperature and gas flow desorb and carry away the analytes. The commercial device incorporates the SPME probe into a syringe barrel. This both protects it and makes the injection into a GC seamless. Recently, a commercial device, also from Supelco, has become available for interfacing with high-performance liquid chromatography (HPLC).

One property of SPME that distinguishes it from other types of extraction is the phase ratio (organic or extracting phase volume/sample volume). The phase ratio in SPME can be very small (e.g., 1/1000). When the phase ratio is small, then the number of moles of an analyte extracted can be small in comparison to the total number of moles in the sample. Unlike in bulk extraction where most of the analyte can be extracted from a sample, in SPME it is possible that a negligible fraction of the analyte in a sample is extracted. Because of this, the quantity extracted will reflect the "free

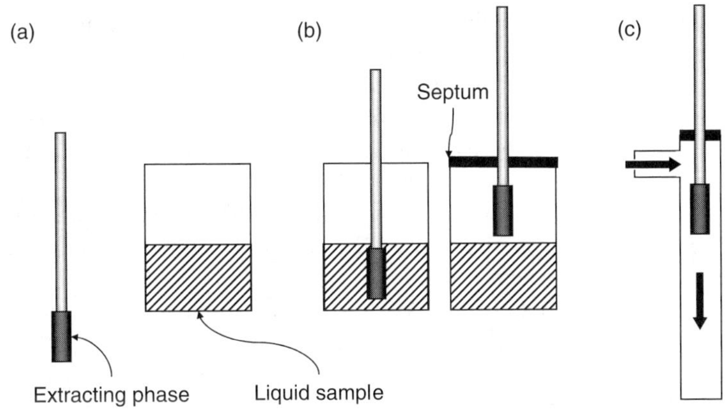

FIGURE 27.1 Scheme of SPME. (a) An SPME probe with an extracting phase as a film on the end of a rod or fiber, and a liquid sample in a container. (b) Extraction by immersion of the probe in a sample or by headspace sampling. (c) Desorption in the inlet of a GC column.

concentration" of the analyte, not the "total concentration." For example, it is possible to determine total drug concentration from blood serum using a bulk extraction. On the other hand, an SPME experiment can be configured so that the "free" drug concentration is measured. In a sense SPME is to bulk extractions what pH measurements are compared to titrations with base.

Another remarkable property of SPME can be realized with volatile solutes. It turns out to be faster (and more selective) to extract volatile solutes with headspace extraction rather than by immersing the SPME probe into the sample. Gaining speed from headspace extraction seems counterintuitive — after all, headspace analysis involves an extra step (sample—vapor phase—extracting phase vs. sample—extracting phase). The reason that headspace can be faster (more rapid progress toward equilibrium; more analyte extracted in a given time) is that the rate-limiting step in the extraction is typically solute getting out of the sample. The flux of analyte in this case is related to a mass transport coefficient (that depends on diffusion and fluid flow if present) for the analyte in the sample and also to the surface area through which the analyte flux goes. The probe tip-sample surface area is the area across which the rate-limiting flux occurs in the direct immersion case. In the case of headspace analysis, the sample-vapor phase surface area is the area across which analyte flux occurs. The latter area is far larger than the former, so headspace extraction proceeds faster than direct immersion.

Another advantage of SPME over bulk extractions especially and over SPE to a lesser degree is the variety of extraction phases that can be made, and the ease with which phases can be developed and used in research laboratories. In bulk extraction, there are certainly differences among organic solvents in their ability to extract analytes. However, using one organic solvent in preference to another is not a powerful method to control the extraction selectivity. In SPE, chromatographic particles are used. Here, there is a variety of choices. However, preparation of a new phase is time consuming at best, and difficult to do and difficult to analyze at worst. In contrast, making a film of polymer or other type of material on a surface can be very easy. Thickness can be controlled by simple means using dip coating.

SPME does have drawbacks. One potential drawback is the equilibration time. Probes typically equilibrate with sample on the timescale of tens of minutes to an hour or more. In many applications, it is not necessary to wait for equilibrium, thus the extraction step need not take a long time. However, by making the extraction yield depend on kinetic as well as equilibrium phenomena, there will be more matrix effects to consider. Things like viscosity, percent solids, and the presence of species that can foul the probe surface have an influence on a kinetically limited extraction, whereas they do not on an equilibrium extraction. SPME is ideally suited to GC. The fit with liquid-based techniques is not as simple. Desorption must be into a solution or liquid film requiring some thinking about the chemistry, while in GC, there is nothing really to think about except the details: elevating the temperature in the gas phase will release the extracted analytes into the column. Desorption in CE (and HPLC) often involves a solvent that can itself dissolve in the film, so desorption can be faster than adsorption—on the minutes timescale. On the other hand, it takes some effort to mate a purely organic extract with CE.

Finally, there are some secondary considerations in SPME that, if properly handled, can improve analysis. SPME probes are reusable, which is great from a cost and environmental burden perspective, but as a consequence carryover is a concern. Washing (or heating for volatile solutes) is essential between applications. Modifying the sample, for example, by adding salt, can influence analyte yield. Unlike in SPE or bulk extraction, it is not necessary to choose between a polar and a nonpolar phase. Phases with both properties, e.g., Carboxen/PDMS exists. Carboxen is a porous carbon that adsorbs polar and nonpolar solutes, and PDMS is quite nonpolar.

SPE is conceptually simpler than SPME. Sample is passed through a packed bed. The bed strips the analyte from the sample. A desorbing solvent is then passed through the bed and passed on for analysis. Commonly used phases are C18 and polystyrene/divinylbenzene (PS/DVB), but others more selective are also used as will be discussed below. It is difficult to generalize, but despite the nominally similar size of the objects used for SPME and SPE (mm length scale, 100 μm diameter scale) the phase ratio (extracting phase/sample) is larger in SPE than SPME because of the porosity of most materials used in SPE. Consequently, it is not typically the case that the extraction reveals the "free" concentration of an analyte. Also because of the porosity and thinness of the extracting layer, for example, in a C18 material, the presence of the support (nominally the same in an optical fiber-based SPME and a silica-based SPE) can be more evident in SPE. Some skill is required to prepare CE- or capillary HPLC-scale packed beds. There are practical advantages of SPE, too. Packed beds are a natural match to flowing fluid, so adsorption and desorption are easily carried out by solvent switching. There is a considerable history of adsorbent selectivity that can be applied to new problems. As the typical particles are small, adsorbent layers are thin and mass transport outside of the particles is rapid because the fluid flow augments diffusion, equilibration times (equivalent roughly to the C term in the van Deemter equation) are short.

The defining features of SPME are mostly film-based, and the film volume is markedly smaller than the sample volume. SPE is typically particle-based with a larger phase ratio. However, these terms are not precise, and they are used differently in the literature. Also, there are some approaches that simply do not fall easily into one category or the other. This writer is not enthusiastic about creating unique terms for even finer distinctions between techniques, so this will be avoided here. The terms SPME and SPE will be used generally, though not exclusively, as the original authors used them. When all else fails, or for general comments, the abbreviation SP(M)E will be used.

SPME has been applied to CE and related techniques, first by Li and Weber[19] offline and then by Nguyen and Luong[20] with online back extraction both in 1997. The approach possesses a number of potential advantages, not the least of which is the obvious applications to the chip format in which film and gel (as in sol-gel) formation are possible. Another advantage is that desorption can often be followed by stacking or a variant of it obviating the problem of bandspreading induced by the desorption step. The size scale of CE and SPME are also similar, so the fit is natural from a conceptual perspective. On the other hand, there are engineering difficulties that include among other things how to manage the many fluids required (adsorption, desorption, separation, column rinsing, SPME rinsing), and how to incorporate fluid flow and electrokinetic flow. There are several very useful reviews cited above that are considerably broader than the current overview that provides examples of SPME-CE and related techniques. The remainder of this chapter draws from the primary literature in the hope of presenting the sort of detailed understanding that makes this approach to analytical chemistry understandable, and allows the novice to anticipate problems and recognize opportunities.

27.3 PREPARATION OF SPME PROBES

27.3.1 COMMERCIAL SPME PROBES

Several workers use commercial SPME probes in conjunction with CE and related techniques. In all cases, probes were conditioned both before first use and between uses. Poly acrylate (PA) probes were conditioned in 50:50 methanol/water (30 min)[21] or the desorption solvent, for example, acetonitrile.[22] A carboxen/PDMS probe was cleverly conditioned by exposing it to GC inlet conditions (300°C for 2 h).[23] The solutes in this case were halophenols. Obviously, this treatment would not be useful for ionic compounds, for example, cations in a PA phase (although if the cation were a low molecular weight protonated nitrogen base, it may work). Thirty minutes in methanol for a PDMS/divinyl benzene (DVB) probe proved suitable.[24] Other workers[25] used a 30 min initial conditioning for four different types of probes, and used a 20 min conditioning between analyses. The theme is clear. A 30 min exposure to a suitable solvent will work to condition SPME probes for use in CE both before and between analyses.

27.3.2 FILMS AND FIBERS

Others interested in the development of SPME *per se* prepare their own probes. It is not necessary to start from scratch. One group[20] purchased a particular optical fiber from Polymicro and used dimethylformamide (DMF) to srip off a layer of nylon. This revealed a layer of PDMS underneath. Of course, PDMS is a well-studied and understood extraction phase, so this is a very clever route to something home made, and thus suited to a particular experiment, but commercially available. Another approach uses a differently coated optical fiber. The polyimide coating on an optical fiber was removed using 98% sulfuric acid at 120°C for 15 h. (*Note*: Hot, concentrated sulfuric acid is extremely corrosive and must be treated with utmost respect. Use the absolute minimum quantity to carry out the task. Use an apparatus in which both the containers of sulfuric acid and the fiber are held in place reliably and securely.) Following this treatment, the clean surface could react with aminopropyltriethoxysilane (APTES) to form an amine (ammonium in neutral or acidic solutions) surface. The surface could also be exposed to a solution of pyrrole and ammonium persulfate in 1:1 water/isopropanol for forming a poly(pyrrole) surface.[26] When interfacing with CE, handling the probe is a problem. Typically, CE does not involve a device for injection, rather the injection is made directly into the separation capillary with no other mechanical device (even if a mechanical device such as an autosampler is used, at most the mechanical assistance in the injection *per se* is something to pierce the septum enclosing the sample). Thus, unlike HPLC and GC, interfacing requires creativity. This topic will be taken up below. Here, though, it is appropriate to describe an approach that permits both modification of a surface and adaptation of the probe to a device. Whang and Pawliszyn[27] developed a method to create a probe suitable for CE, and that fits into the commercial, syringe-like holding apparatus (see Figure 27.2). The probe thus formed is placed into the syringe-like apparatus that is commercially available after the probe with which it came is removed. While the field of SPME has been dominated by a handful of phases spanning a wide range of polarity, there are other phases with excellent properties. Poly vinylchloride (PVC), appropriately plasticized, is the basis for ion-selective electrodes. In a sense, SPME experiments have been done on this material for decades, but without that explicit purpose in mind. Valenta et al.[28,29] has described the fact that the polarity of plasticized PVC depends on the plasticizer. Zhang et al. has demonstrated that the polarity also depends on the ratio of plasticizer to polymer.[30] Li and Weber[19] have used this material in SPME-CE. The probes are based on a stainless steel rod. A primer coating of poly(vinylchloride-*co*-vinyl alcohol-*co*-maleic acid) permits a coating of plasticized PVC (dip coat from tetrahydrofuran). The probes are reusable. The polymer also plays a role in the selectivity with of extractions based on molecular recognition. A barbiturate receptor in plasticized PVC extracts barbiturates more selectively the lower the polarity of the plasticizer.

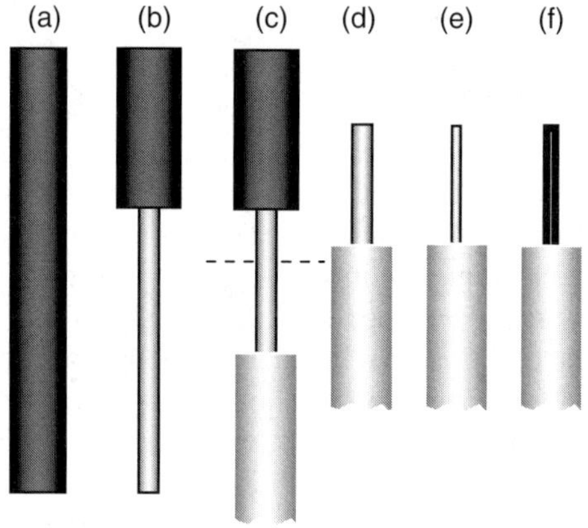

FIGURE 27.2 Preparation of an SPME probe for CE.[27] (a) A 245 μm diameter PA coated optical fiber 3 cm long. (b) Strip the PA from 2 cm. (c) Glue a 10 cm long piece of fused silica capillary (100 μm ID, 245 μm OD) to the distal 0.8 cm of the exposed 2 cm. (d) Cut at a point 1.0 cm from the end of the capillary. (e) Etch with HF (*Note*: use in a hood. Breathing HF is harmful.) to about 40 μm diameter. (f) Dip coat X3 in 3% poly(acrylate).

Porous materials can be used for SPME. Carboxen particles have been mentioned above. A monolithic porous polymer has been prepared for SPME. Wei et al.[31] made a porous polymer monolith with monomers methacrylic acid, ethylene glycol dimethacrylic acid, and a toluene porogen. The material was prepared inside a 250 μm capillary, which had been previously derivatized with a methacrylic acid-containing silane. Following polymerization at 60°C for 16 h, the monolith was washed with methanol, a popular choice for conditioning. Application (which is described more below) uses flow through the medium, so perhaps this is more properly called micro-SPE. However, the authors call it SPME, and it is an interesting approach, so it is being described here. Zhou et al.[32] used molecular recognition to assist in the SPME. A conjugate of the well-known small molecule host, calix-[4]-arene, with a silane yields a reactive ethoxy silane that is functionalized with the calixarene. A sol of this and tetraethoxysilane with a porogen (-OH terminated silicone oil) was prepared. A fiber was dipped into the sol several times, then the sol was given time to react to form the gel. Control probes were made in which the calixarene-modified silane was not present. These probes were used for head space extractions, then back extraction and analysis with CE. The calixarene probe was more effective for propranalol than the control.

27.3.3 Packed Beds

Packed beds, as mentioned above, are sometimes described as SPME, and indeed, it may be accurate—that is the extracting phase may reach or approach equilibrium with a sample without extracting a significant amount of analyte on average from the sample passed through the packed bed. Typically, a small piece of a filter material is used as a "frit" to block particle flow at the boundary between a capillary holding the particles and another capillary.[33] If the particles are large enough with respect to the downstream capillary diameter, then there is no need for a frit or other device to contain the particles. In Reference 34, 10 μm particles were introduced into a 180 μm capillary as a very dilute suspension at modest pressure. The bed was formed where the 180 μm capillary met a restrictor that was 50 μm in diameter.

27.4 INTERFACING SP(M)E WITH CE AND RELATED TECHNIQUES

27.4.1 General Considerations

Interfacing SPME with CE is not as straightforward as interfacing SPME with GC or as interfacing SPE with CE. Because of the difficulty of interfacing, a lot of work has been done with offline "interfacing." The general issues faced when considering interfacing (in general) are as follows:

Preconcentration: A good extraction/back extraction cycle not only results in a cleaner sample, but analytes are concentrated in the process. Obviously, this argues for using the smallest possible back extraction solvent volume. If this volume needs to be manipulated by hand, that is, if it is not a volume resulting from a programmed mechanical device or by switching on a chip, then the minimum volume of the back extraction solvent will probably be about 1 μL.

Kinetics: An effective extraction/back extraction cycle is fast. Slow extraction or back extraction requires that the designer of the system be cognizant of the details. Slow extraction can obviously lead to poor yield. Operating in a kinetic regime means that many more parameters influence the outcome in comparison to operating with the extraction at equilibrium. Slow back extractions have the same problem. In addition, a slow back extraction online can add to bandspreading.

Fluid management: Depending on the design of a system there may be five different fluids needed for an analysis: sample, back extraction solution, running buffer/electrolyte, extraction phase rinsing, and column rinsing/conditioning solution. These liquids need to be managed appropriately for a successful outcome of a single analysis and for good reproducibility.

On balance, the design of a system is a trade-off between the complexity of mechanization/automation versus the savings of time (and improvement in reproducibility) once the effort to automate/mechanize has been made. Because the effort to create online systems is considerable, there has been significant effort in offline systems. Both offline and online systems are covered below. The section on "online" systems includes methods that have at least one online step.

27.4.2 Offline

Both in-sample and headspace approaches have been used in conjunction with CE. The forward extraction, as is typical for SPME, can take an hour or more to equilibrate. The yield of the extraction is improved by the addition of salt (typically NaCl). As the probe is intended to reach equilibrium with the sample and not extract a large fraction of the moles of analyte in the sample, the volume of sample can be immaterial. The first SPME-CE paper used offline extraction and back extraction.[19] Management of the back extraction solution was made easier by keeping it in a piece of Teflon®. The general scheme is shown in Figure 27.3.

Using such a system, the back extraction solution can be kept to a few microliters. In most instances, however, the back extraction is into a much larger volume of 20–200 μL.[21–24,31,32] In most cases, the back extraction medium is not highly ionic. This permits stacking, electric field-induced concentration at the interface between the relatively high conductivity running buffer and the relatively low conductivity back extraction solvent. The option to stack means that it is less urgent to use a small volume of back extraction solvent.

In most of the articles cited above, the authors measured the desorption kinetics. These turn out to be remarkably consistent—with about 90% recovery (from the probe) in minutes (~3–10 min). This is interesting, because at least two modes of back extraction are used. In one, a solvent more "organic" than water is used (e.g., methanol or acetonitrile). On the other hand, the ionic nature of

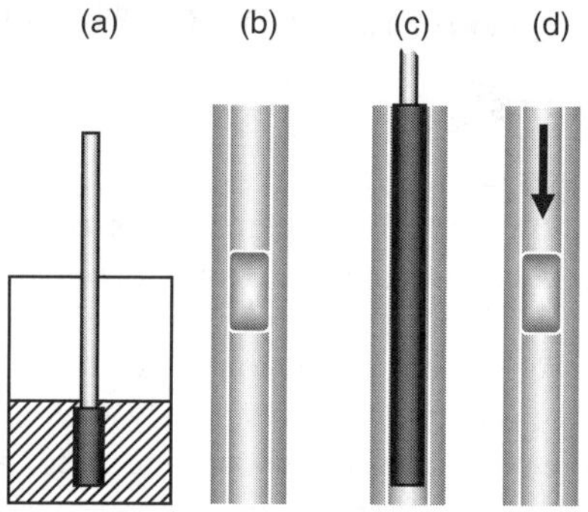

FIGURE 27.3 Off-line extraction and back extraction with back extraction solution management.[19,21] (a) extraction with SPME (dark portion = extracting phase, diagonal hatching = sample). (b) Separately, inject a small volume (few μL of back extraction solvent into a Teflon® tube sized to have a diameter just larger than the diameter of the SPME probe. (c) Insert the probe into the back extraction solution-containing probe. The solution will spread between the wall of the tube and the extracting phase. (d) Following the back extraction, remove the SPME probe. Using an empty syringe, move the droplet, which re-forms when the probe is removed, at will.

the analyte is switched by changing the pH of the back extraction solvent from the value existing in the sample.

There is one report of an offline SPME (or perhaps more properly SPE) extraction system for CE analysis in which the offline extraction is automated.[31] The extraction medium is a monolith placed in the loop position of a six-port loop injector. An injector loop upstream contains a large loop with sample. The sample from the loop in the upstream injector is passed through the extraction medium for a controlled time at a controlled flow rate. Following a rinse, the back extraction solvent is then pumped through the monolith into a collection vial. This system has the advantage of the reproducibility of an automated system, and simultaneously avoids the complexity of the interfacing process.

27.4.3 Online

When the separation medium is online, it may be within the separation path or not. Figure 27.4 shows the two most common arrangements. One pair of arms (horizontal or vertical) is for sample management, while the other pair (vertical or horizontal) is for the separation. The tubing need not be of the same diameter in each arm or branch. If the SP(M)E medium is in the center of the cross arrangement (Figure 27.4a), then obviously the SP(M)E medium is "online" during the separation. Elution from the SP(M)E material occurs in the separation column. This places severe demands on the effectiveness of the back extraction and the volume of back extraction solvent. Of course, stacking may be used to address problems of having an initial back extraction zone that is too large in many cases. The back extraction must be complete in the arrangement, too. If it is not, the peaks will tail, or worse, the separation will be impossible. If the SP(M)E medium is in one branch of the cross, then the demands on the back extraction are not as severe. Incomplete back extraction is permitted. A flow-gate type of injection is possible to assure narrow zones. On the other hand, the back extraction solution can be directed onto the separation column if stacking, sweeping, or other online reconcentration is possible. Variations on these themes exist as well.

This sort of cross arrangement has been used by several workers. Peptides were trapped, separated by CE, and detected by ESI-MS using an arrangement as shown in Figure 27.4b.[34] This paper demonstrated the creation of the online SP(M)E in a PDMS chip. Once every 10 s a 1 s injection was made from the back extraction solution. It appeared that the peptides eluted from the SP(M)E

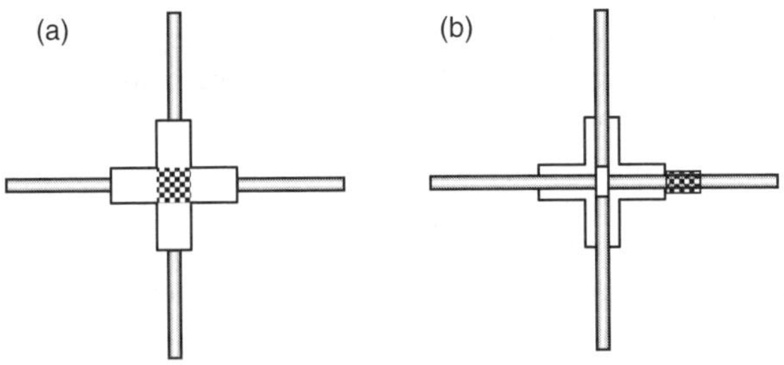

FIGURE 27.4 (a) A cross arrangement for online extraction and back extraction with the extraction medium centered in the cross. (b) Reaction medium in one arm.

[very high surface area hypercrosslinked poly(styrene)] over about a 3 min period. In a similar way (Figure 27.4b), a short capillary packed with small fibers (Zylon) was used for the extraction of tricyclic antidepressants.[35] Samples could be driven through the extraction medium at a high flow rate (80 μL/min). Back extraction volumes varied, but were in the range of 2 μL, of which 4 nL were injected onto the CE column. An alternative arrangement similar to Figure 27.4b has only three arms. The "waste" line for the extraction and the running buffer source (positive electrode in "normal" polarity) capillary are the same.[36] This leads to interesting possibilities for flow control, as pressure flow going through the SP(M)E medium splits—a predictable amount going to waste, and a predictable amount going onto the separation column.

A cross like Figure 27.4a has been used for affinity capture by Guzman.[37] In this case, each of the four arms is outfitted with a valve as well, so the flow is completely controllable. As mentioned above, however, it is important to desorb the analytes into a small volume and rapidly. Guzman uses 100 nL of desorption solution (pH 3.4 glycine buffer or acetonitrile) to dissociate the antibody–analyte complexes. A cross with medium centered in it was fabricated on a chip.[38] In this paper, the goal was manufacturing, not applying, the SP(M)E. A small packed bed served as the hydrophobic medium. Laser-induced fluorescence was used for detection of the "analyte," fluorescein. Interestingly, the beam was split so that the luminescence from the fluorescein in the extraction chamber could be seen. Evidence of rapid and complete desorption was obtained in this way.

It is possible to construct a system in which there is only one pair of "arms"—that is with an extraction medium and a separation capillary in series. In an example of this, Yates and coworkers[33] extracted proteins and then desorbed them into a zone of methanol introduced with a 45 s electrokinetic injection.

The previous works have been based on beds of particles or fibers. Films, that is, with a construction similar to commercially available SPME probe tips, can be used online. The first publication in this area was half online. The sample extraction was carried out offline, but desorption and analysis was managed online.[20] In this and other papers, the extraction fiber is constructed as in Figure 27.2. A fiber with analytes adsorbed is inserted into the injection end of the capillary. The injection process involves introducing not sample but a desorption solution. The flow may be stopped to allow for desorption (recall that it can take several minutes), then restarted for separations. In a similar vein, Whang and Pawliszyn[27] created a system in which fibers with extracted analytes were moved to a capillary for desorption/analysis. In this case, though, the desorption solution is injected into the capillary before inserting the SPME fiber into the end of the capillary. Also in this case, the fiber is attached to a syringe for ease of handling. While it is an advantage in some sense, the use of a low (organic/aqueous) phase ratio does limit the number of moles extracted from a sample. As this places limitations on the follow-up analysis (mass detection limit), there is some advantage to keeping the SPME tip "long." The "active" extracting tips are typically ~1 cm in length. One may wonder about the effect of creating an annular space filled with analyte on bandspreading in CE. Stoyanov and Pawliszyn[39] have shown that the problem is not too large. Because the cross-sectional area of the annular space created by inserting the probe into the injection end of the capillary is small compared with the full cross-sectional area of the capillary, the field is higher in the annular space than the capillary. This requires that the velocity of the analytes slows down when they reach the full capillary (having come from the annular space). The result is a compression of sorts that helps keep the injection-induced bandspreading to a minimum.

Another way to fight the potential bandspreading from the desorption/injection process is to perform isoelectric focusing rather than C(Z)E.[26] The focusing effect of the separation technique undoes any bandspreading resulting from the desorption injection process.

Despite clever approaches, the online desorption/injection from SPME is not a "natural" fit. An approach that holds promise for macromolecular analytes is to use microdialysis as an intermediate step. Liu and Pawliszyn[40] achieved this by back extracting and moving analytes electrokinetically into the lumen of a microdialysis probe, then inject from the lumen into the separation capillary. The beauty of this is that the sample stays in the (small) microdialysis probe volume, but electrolyte and current can pass through the walls of the microdialysis probe.

Some general practical comments are in order. Care must be taken to make good butt connections when coupling capillaries. Band spreading can result from poor connections. Many applications of SP(M)E-CE are partly offline, or involve physically working with capillaries. Also, solvents of various sorts are used for desorption. Therefore, there is abundant opportunity for the formation of bubbles. It is a good idea at some point in the process to use pressure to flush the system before using electrokinetic forces for separations. Many publications discuss moving fluid where desired using both pressure and electroosmotic flow. Because flow control generally needs to be quantitative, it is often necessary to have valves or pressurized electrolyte containers to facilitate predictable fluid flow.

27.5 RANGE OF APPLICABILITY

The range of analytes accessible to SP(M)E/CE is wide. Drugs, pollutants (and very hydrophobic ones like polyaromatic hydrocarbon, PAHs), proteins/peptides have all been determined using the methods described herein. There does not seem to be any sacrifice in separation modes, either. CE, capillary isoelectric focusing, MEKC, nonaqueous CE can all function with some form of SP(M)E. Methods of detection are similarly broad, encompassing at least the three most common: optical absorbance, laser-induced fluorescence, and MS. It is perhaps foolish to attempt to generalize, but it does seem that an improvement in concentration detection limits of about 100-fold can be realized through SPME in conjunction with CE.

27.6 CHALLENGES FOR THE FUTURE

27.6.1 Selectivity

Many analyses are for a few similar molecules, or a class of molecules. In such cases, selectivity is highly desirable in the extraction. Selectivity in SPME has been reviewed.[41] Information is also available at the Supelco website (this author is not and has never been funded by Supelco, and this author has no financial interest in Supelco) http://www.sigmaaldrich.com/Brands/Supelco_Home/Spotlights/SPME_central.html.

Ultimately, molecular selectivity is desirable. Calix-[4]-arenes improved the extraction of propranolol. Antibodies can be used for affinity separations. Li et al.[42] have demonstrated selectivity for barbiturates using a specially designed receptor molecule. The need for more selectivity is made clear by the following common observation. Recall that SPME extractions can take a long time to reach equilibrium. Is it worthwhile to wait for equilibrium? It can be as mentioned above, it should be more reproducible. But interestingly, for many samples extracted by direct immersion of the probe into the sample there is no detection limit advantage to longer extractions because interferences are extracted along with analytes. More research and development on selective phases is certainly needed.

27.6.2 Desorption/Injection

There remains, despite some very clever approaches, a need for a seamless union of SPME and CE. This may ultimately come with lab-on-a-chip developments, as the goal of much of this sort of research is complete analysis. In other words, rather than the necessary voltages, pumps, valves, and pressure sources being a bother (as it is to a worker using commercial CE, commercial SPME, commercial valves, etc.) it is a research opportunity for the courageous chemists and engineers designing chips. This prognosis, then, points out a need for the development of extraction phases (with selectivity, see above) that can be emplaced in the microfabrication process. This is an important area for future work.

27.6.3 Smaller Probes

There is room to make smaller, thinner probes to manage microanalysis. Interfacing them brings new challenges, but also may obviate some problems. Smaller, thinner phases with faster equilibration times may improve the desorption/injection process.

27.7 CONCLUDING REMARKS

The combination of SP(M)E and CE is a good idea. CE suffers from poor concentration detection limits; microextractions provide for preconcentration. There is apparently no inherent sacrifice that must be made in separations mode or detection mode in order to make SP(M)E work with CE and related techniques. The biggest advantage occurs when the extraction is fairly selective: then the act of concentrating the analytes (and not concentrating interferences) improves the concentration detection limit for the method. Examples of where this is the case would be in head space extractions, and extractions based on selective or affinity phases. The engineering of the extraction and back extraction online is still awkward, but improving.

ACKNOWLEDGMENTS

The author is grateful to the NSF Division of Chemistry for funding. Contribution from the Department of Chemistry, University of Pittsburgh, Pittsburgh, PA.

REFERENCES

1. Creten, W. L. and Nagels, L. J., Computation of determination limits for multicomponent chromatograms, *Analytical Chemistry*, 59, 822, 1987.
2. Nagels, L. J., Creten, W. L., and Parmentier, F., Statistical model for organic chromatographic trace analysis of complex samples. A case study: Plant extracts, *International Journal of Environmental Analytical Chemistry*, 25, 173, 1986.
3. Nagels, L. J., Creten, W. L., and Van Haverbeke, L., Determination limits in high-performance liquid chromatography of plant phenolic compounds with an ultraviolet detector, *Analytica Chimica Acta*, 173, 185, 1985.
4. Nagels, L. J. and Creten, W. L., Quantitative evaluation of chromatographic analysis of complex mixtures by establishing limits of determination, *Analytica Chimica Acta*, 169, 299, 1985.
5. Andreu, V. and Pico, Y., Determination of pesticides and their degradation products in soil: Critical review and comparison of methods, *Trends in Analytical Chemistry*, 23, 772, 2004.

6. Fritz, J. S. and Macka, M., Solid-phase trapping of solutes for further chromatographic or electrophoretic analysis, *Journal of Chromatography A*, 902, 137, 2000.
7. Jinno, K., Sawada, H., and Han, Y., Drug analysis by capillary electrophoretic methods, *Biomedical Chromatography*, 12, 126, 1998.
8. Liu, Z. and Pawliszyn, J., Online coupling of solid-phase microextraction and capillary electrophoresis, *Journal of Chromatographic Science*, 44, 366, 2006.
9. Mile, B., Chemistry in court, *Chromatographia*, 62, 3, 2005.
10. Pawliszyn, J., New developments and applications of solvent-free sampling and sample preparation technologies for the investigation of living systems, *Australian Journal of Chemistry*, 56, 155, 2003.
11. Pawliszyn, J., *Applications of Solid Phase Microextraction*, Royal Soc. Chem., Cambridge, 1999.
12. Pawliszyn, J., *Comprehensive Analytical Chemistry, Volume 37*. (Sampling and Sample Preparation for Field and Laboratory: Fundamentals and New Directions in Sample Preparation.), Elsevier Science B.V., Amsterdam, The Netherlands, 2002.
13. Pawliszyn, J., Special Issue: *Microextraction, Part I*. [In: *Journal of Chromatography Science*; 2006, 44(6)], Preston Publications, Niles, IL, 2006.
14. Pawliszyn, J. B., *Solid Phase Microextraction: Theory and Practice*, VCH, New York, NY, 1997.
15. Saito, Y. and Jinno, K., Miniaturized sample preparation combined with liquid phase separations, *Journal of Chromatography, A*, 1000, 53, 2003.
16. Theodoridis, G., Koster, E. H. M., and de Jong, G. J., Solid-phase microextraction for the analysis of biological samples, *Journal of Chromatography, B: Biomedical Sciences and Applications*, 745, 49, 2000.
17. Vas, G. and Vekey, K., Solid-phase microextraction: A powerful sample preparation tool prior to mass spectrometric analysis, *Journal of Mass Spectrometry*, 39, 233, 2004.
18. Belardi, R. P. and Pawliszyn, J. B., The application of chemically modified fused silica fibers in the extraction of organics from water matrix samples and their rapid transfer to capillary columns, *Water Pollution Research Journal of Canada*, 24, 179, 1989.
19. Li, S. and Weber, S. G., Determination of barbiturates by solid-phase microextraction and capillary electrophoresis, *Analytical Chemistry*, 69, 1217, 1997.
20. Nguyen, A.-L. and Luong, J. H. T., Separation and determination of polycyclic aromatic hydrocarbons by solid phase microextraction/cyclodextrin-modified capillary electrophoresis, *Analytical Chemistry*, 69, 1726, 1997.
21. Fan, X. and Deng, Y., Separation and identification of aromatic acids in soil and the Everglades sediment samples using solid-phase microextraction followed by capillary zone electrophoresis, *Journal of Chromatography, A*, 979, 417, 2002.
22. Jinno, K., Han, Y., Sawada, H., and Taniguchi, M., Capillary electrophoretic separation of toxic drugs using a polyacrylamide-coated capillary, *Chromatographia*, 46, 309, 1997.
23. Kannamkumarath, S. S., Wuilloud, R. G., Jayasinghe, S., and Caruso, J. A., Fast speciation analysis of iodophenol compounds in river waters by capillary electrophoresis-inductively coupled plasma-mass spectrometry with off-line solid-phase microextraction, *Electrophoresis*, 25, 1843, 2004.
24. Hernandez-Borges, J., Cifuentes, A., Garcia-Montelongo, F. J., and Rodriguez-Delgado, M. A., Combining solid-phase microextraction and on-line preconcentration-capillary electrophoresis for sensitive analysis of pesticides in foods, *Electrophoresis*, 26, 980, 2005.
25. Hernandez-Borges, J., Rodriguez-Delgado, M. A., Garcia-Montelongo, F. J., and Cifuentes, A., Highly sensitive analysis of multiple pesticides in foods combining solid-phase microextraction, capillary electrophoresis-mass spectrometry, and chemometrics, *Electrophoresis*, 25, 2065, 2004.
26. Liu, Z. and Pawliszyn, J., Coupling of solid-phase microextraction and capillary isoelectric focusing with laser-induced fluorescence whole column imaging detection for protein analysis, *Analytical Chemistry*, 77, 165, 2005.
27. Whang, C.-W. and Pawliszyn, J., Solid phase microextraction coupled to capillary electrophoresis, *Analytical Communications*, 35, 353, 1998.
28. Valenta, J. N., Sun, L., Ren, Y., and Weber, S. G., Solvatochromic study of poly(vinyl chloride) plasticizers and their solutions in chloroform: application to phenobarbital partitioning and molecular recognition of phenobarbital, *Analytical Chemistry*, 69, 3490, 1997.
29. Valenta, J. N. and Weber, S. G., Molecular recognition of phenobarbital in plasticizers. Equilibrium investigations on the solubility of the barbiturate artificial receptor and its binding to phenobarbital in plasticizers, *Journal of Chromatography, A*, 722, 47, 1996.

30. Zhang, X., Zhao, H., Chen, Z., Nims, R., and Weber, S. G., Effect of polymer concentration on partitioning and molecular recognition in plasticized poly(vinyl chloride), *Analytical Chemistry*, 75, 4257, 2003.
31. Wei, F., Fan, Y., Zhang, M., and Feng, Y.-Q., Poly(methacrylic acid-ethylene glycol dimethacrylate) monolith in-tube solid-phase microextraction applied to simultaneous analysis of some amphetamine derivatives in urine by capillary zone electrophoresis, *Electrophoresis*, 26, 3141, 2005.
32. Zhou, X., Li, X., and Zeng, Z., Solid-phase microextraction coupled with capillary electrophoresis for the determination of propranolol enantiomers in urine using a sol-gel derived calix[4]arene fiber, *Journal of Chromatography, A*, 1104, 359, 2006.
33. Tong, W., Link, A., Eng, J. K., and Yates, J. R., III, Identification of proteins in complexes by solid-phase microextraction/multistep elution/capillary electrophoresis/tandem mass spectrometry, *Analytical Chemistry*, 71, 2270, 1999.
34. Dahlin, A. P., Bergstroem, S. K., Andren, P. E., Markides, K. E., and Bergquist, J., Poly(dimethylsiloxane)-based microchip for two-dimensional solid-phase extraction-capillary electrophoresis with an integrated electrospray emitter tip, *Analytical Chemistry*, 77, 5356, 2005.
35. Jinno, K., Kawazoe, M., Saito, Y., Takeichi, T., and Hayashida, M., Sample preparation with fiber-in-tube solid-phase microextraction for capillary electrophoretic separation of tricyclic antidepressant drugs in human urine, *Electrophoresis*, 22, 3785, 2001.
36. Tempels, F. W. A., Wiese, G., Underberg, W. J. M., Somsen, G. W., and de Jong, G. J., On-line coupling of size exclusion chromatography and capillary electrophoresis via solid-phase extraction and a Tee-split interface, *Journal of Chromatography, B: Analytical Technologies in the Biomedical and Life Sciences*, 839, 30, 2006.
37. Guzman, N. A., Improved solid-phase microextraction device for use in on-line immunoaffinity capillary electrophoresis, *Electrophoresis*, 24, 3718, 2003.
38. Tuomikoski, S., Virkkala, N., Rovio, S., Hokkanen, A., Siren, H., and Franssila, S., Design and fabrication of integrated solid-phase extraction-zone electrophoresis microchip, *Journal of Chromatography, A*, 1111, 258, 2006.
39. Stoyanov, A. V. and Pawliszyn, J., CE in a nonuniform capillary modulated by a cylindrical insert, and zone-narrowing effects during sample injection, *Analytical Chemistry*, 75, 3324, 2003.
40. Liu, Z. and Pawliszyn, J., Microdialysis hollow fiber as a macromolecule trap for on-line coupling of solid phase microextraction and capillary electrophoresis, *Analyst (Cambridge, United Kingdom)*, 131, 522, 2006.
41. Li, S. and Weber, S. G., in *Applications of Solid Phase Microextraction*, J. Pawliszyn, Ed., Royal Society of Chemistry, Cambridge, 49, 1999.
42. Li, S., Sun, L., Chung, Y., and Weber, S. G., Artificial receptor-facilitated solid-phase microextraction of barbiturates, *Analytical Chemistry*, 71, 2146, 1999.

28 CE-SELEX: Isolating Aptamers Using Capillary Electrophoresis

Renee K. Mosing and Michael T. Bowser

CONTENTS

28.1 INTRODUCTION

Since their discovery in 1990, aptamers have gained increasing attention due to their high affinity and specificity for a wide range of target molecules.[1–5] While aptamer applications are growing exponentially, the time and difficulty of the process used to isolate aptamers are limiting widespread adoption. Recently, the high resolving power of capillary electrophoresis (CE) has been used to select high-affinity aptamers.[6,7] This offers a number of advantages over conventional selection protocols due to the dramatic enrichment rate and selection stringency made possible by CE. These advantages result in a significantly faster and simpler process allowing high-affinity aptamers to be obtained in several days rather than the weeks to months typical of pre-existing procedures. This chapter discusses the significance of CE-SELEX and describes details of the selection process.

28.2 BACKGROUND

The word aptamer comes from the Latin root "aptus" meaning "to fit." This is appropriate since aptamers are single-stranded deoxyribonucleic acid (ssDNA) or ribonucleic acid (RNA) sequences that fold into unique structures allowing them to bind target molecules with high affinity and selectivity. Aptamers are typically 70–120 bases long, containing primer sequences on both ends to facilitate polymerase chain reaction (PCR) amplification. As shown in Figure 28.1, depending on their sequence, single-stranded nucleic acids can fold into a variety of loops, hairpins, and bulges to generate a wide range of structures. The plethora of structures available make nucleic acids an attractive combinatorial library since sequences can be isolated with affinity for virtually any molecular target.

While the large number of structures available in a library of even relatively short nucleic acids allow aptamers to be isolated for a wide range of targets, it also raises the question of how to identify high-affinity sequences out of so many possibilities. This was addressed in 1990 with the introduction of a process for isolating aptamers that has been referred to as SELEX (Systematic Evolution of Ligands by EXponential enrichment),[8] *in vitro* selection[9] or *in vitro* evolution.[10] The process has been described in detail in a number of excellent reviews.[1–5] As illustrated in Figure 28.2, a structurally diverse library containing approximately 10^{13}–10^{15} ssDNA or RNA molecules is incubated with the target molecule. The nucleic acids contain a 30–80 base random sequence region flanked by two primer regions. Sequences with affinity for the target are separated from nonbinding sequences using filtration, affinity chromatography, or panning. Binding sequences are PCR amplified, purified, and made single stranded to generate a new nucleic acid pool suitable for further rounds of enrichment. This process is continued until the pool converges on a collection of sequences with high affinity for the target. Typically, 8–15 selection cycles are required to generate a pool containing a significant abundance of high-affinity aptamers. Individual aptamers are then cloned and sequenced from this final pool for further characterization.

Aptamers have been obtained for hundreds of targets.[1–5] Low nanomolar to picomolar dissociation constants are typical for large protein targets. Aptamers of therapeutic interest have been isolated for a number of drug targets including HIV components,[11,12] the influenza virus surface

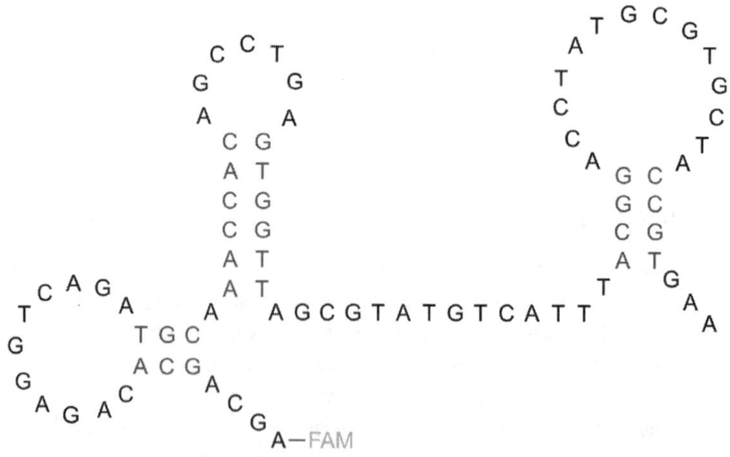

FIGURE 28.1 Secondary structure of a DNA aptamer selected using CE-SELEX to have affinity for neuropeptide Y. (From Mendonsa, S. D., Bowser, M. T., *J. Am. Chem. Soc.*, 127, 9382–9383, 2005.)

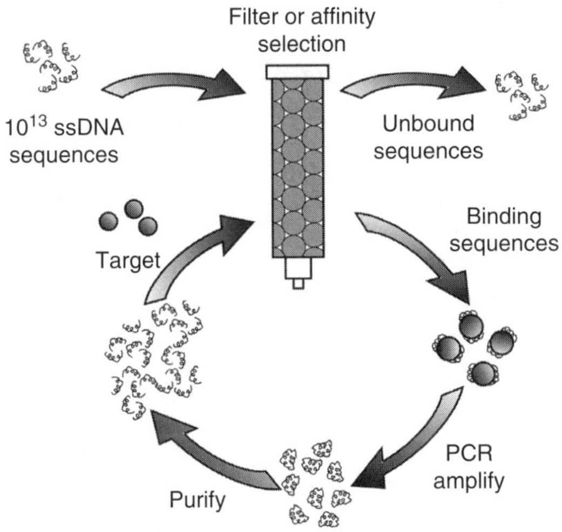

FIGURE 28.2 Schematic representation of the SELEX process. A random sequence ssDNA library is incubated with the target molecule. Binding sequences are selected using a filter or affinity separation, PCR amplified, purified, and made single stranded, generating a new pool suitable for further cycles of enrichment. High-affinity aptamers are typically obtained after 8–15 rounds of selection.

glycoprotein hemagglutinin,[13,14] and the amyloid conformer of the prion protein.[15,16] Macugen, a treatment for age-related macular degeneration (AMD), is the first aptamer-based drug to gain Food and Drug Administration (FDA) approval and is now available to the public.[17] There are a number of other promising therapeutic aptamers in clinical trials.[18]

Aptamers have an even more diverse role in diagnostics.[19,20] Highly specific aptamers have been used as affinity probes in CE[21–23] and proteomic microarrays.[24–28] Aptamers have been incorporated into fluorescent switches and beacons.[29,30] They have been incorporated into chromatography[31–33] and capillary electrochromatography[34–40] stationary phases. They have even been used as the sensing element of cantilever-based detectors.[41] Clearly, the combination of high affinity and selectivity provided by aptamers can be applied to a wide range of analytical techniques.

The high affinity and selectivity of aptamers draw obvious comparisons with antibodies.[19] Although similar in many respects, aptamers hold several advantages. Once their sequence is known, aptamers are cheaply synthesized with little variation between batches. Modifications such as fluorescent tags and non-natural bases can be easily incorporated into the synthesis. Isolating aptamers does not require animals. Aptamers can therefore be obtained for molecules that are toxic or do not stimulate an immune response. Aptamers are not restricted to use in physiological conditions. The structure and affinity of aptamers can be easily manipulated by changing buffer conditions, heat, pH, and so forth. Aptamers will often refold into their original structure once returned to nondenaturing conditions. The relatively small size of aptamers allows them to penetrate tissues more easily, which is attractive in therapeutic applications. Furthermore, their smaller size allows tighter packing and higher loading capacities when attached to stationary surfaces in chromatography or microarrays.

28.3 THEORY

28.3.1 CE-SELEX

Recently, the high resolution power of CE has been used to isolate high-affinity aptamers (see Table 28.1 and references therein). This process has been referred to as CE-SELEX.[6,7] The CE-SELEX process is illustrated in Figure 28.3. The target molecule is incubated with a nucleic

TABLE 28.1
Comparison of Aptamers Selected Using CE-SELEX with Those Selected Using Conventional SELEX Protocols

	CE-SELEX		SELEX	
Target	K_d **(nM)**	**Cycles**	K_d **(nM)**	**Cycles**
IgE	23 ± 12[7]	4	6[53]	15
HIVRT	0.18 ± 0.07[11]	4	1[54]	12
Neuropeptide Y	300 ± 200[42]	4	370[55,*]	12
PFTase	0.5[45]	1	n/a	
mutS	3.6 ± 0.5[56]	3	n/a	
Kinase C-δ	122[57]	9	n/a	—
Ricin	58 ± 19[44]	4	105 ± 41[44]	9

*RNA aptamer.

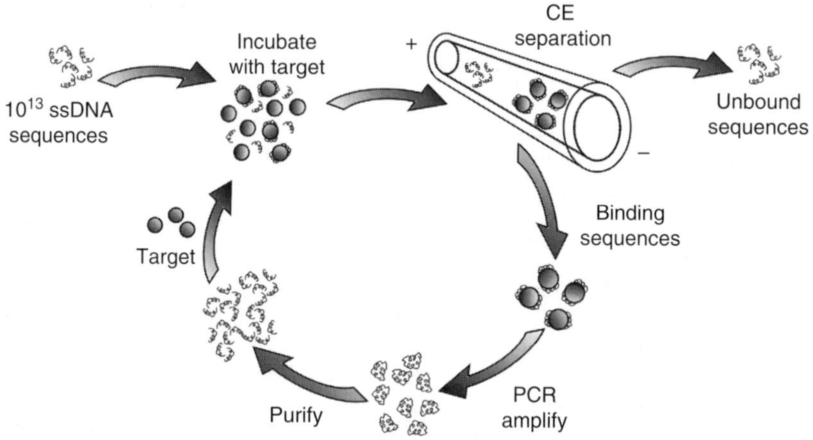

FIGURE 28.3 Schematic representation of the CE-SELEX process. A random sequence ssDNA library is incubated with the target molecule. Binding sequences are separated from nonbinding sequences using CE, PCR amplified, purified, and made single stranded, generating a new pool suitable for further cycles of enrichment. High-affinity aptamers are typically obtained after 2–4 rounds of selection.

acid library. Several nanoliters of this incubation mixture are injected onto a capillary and separated using free zone CE. Nonbinding oligonucleotides migrate through the capillary with the same mobility, regardless of their length or sequence. Complexing the target changes the size and/or charge of binding sequences causing them to migrate as a separate fraction. These binding sequences are collected as a separate fraction at the outlet of the capillary. As in conventional SELEX selections, the fraction containing binding sequences is amplified, purified, and made single stranded for further rounds of enrichment. The process is repeated until no further improvement in affinity is observed between rounds.

28.3.2 Library Size

In general, as many independent sequences as possible should be included in the initial library. This maximizes the probability that high-affinity aptamers are present in the initial pool. Presumably, sequences with the highest affinity for the target are characterized by more complex, and

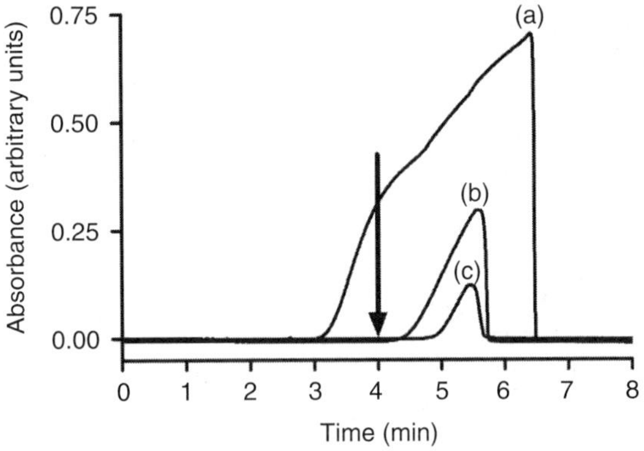

FIGURE 28.4 Effect of injection size on the peak shape of a 2.5 mM ssDNA library separated using CE. Injection parameters were (a) 1 psi, 5 s; (b) 1 psi, 3 s; and (c) 0.5 psi, 3 s. Using a 1 psi, 3 s injection, 1.8×10^{13} sequences were loaded onto the capillary. The arrow shows the position on the electropherogram where collection of binding sequences was stopped. CE conditions: TGK buffer, pH 8.3, UV detection at 254 nm, 50.2 cm, 50 μm i.d., 360 μm o.d. capillary, 30 kV separation voltage. (From Mosing, R. K. et al., *Analytical Chemistry*, 77, 6107–6112, 2005.)

therefore rare structures than moderate binders with lower selectivity. The number of possible independent sequences in an oligonucleotide library is given by 4^n, where n is the number of bases in the random region. Total coverage of all potential sequences becomes quickly unmanageable. For example, a moderate length library containing a 60 base random region contains over 10^{36} independent sequences. In conventional selections, 10^{13}–10^{15} sequences are often used in the initial round of selection as a compromise between an adequate coverage of sequence space and practical constraints. CE selection puts further constraints on library size. Injection volume in CE is limited to several nL. Considering this, the concentration of the library must be in the low mM range to inject 10^{13} sequences onto the capillary. This high concentration of ssDNA introduces significant band broadening, making separation of aptamers from nonbinding sequences more difficult (see Figure 28.4). In some cases, the concentration of the library must be lowered to ensure that bound sequences are well separated from nonbinding sequences. Therefore, choosing the optimum library size in CE-SELEX is a trade-off between maximizing sequence diversity and achieving adequate resolution to separate aptamers from nonbinding sequences.

28.3.3 Sequence Length

Nucleic acid libraries are typically made up of 70–120 base ssDNA or RNA molecules. These single-stranded oligonucleotides contain a 30–80 base random sequence region flanked by two primer regions. The random region should contain at least 30 bases so that it is long enough to form common structural motifs such as hairpins, bulges, pseudoknots, and g-quartets. Longer sequences add diversity to the pool by providing more positions along the sequence where these motifs can be located. The upper limit to sequence length is often determined by synthetic limitations. Yield and purity drop off dramatically during synthesis of oligonucleotides longer than 100 bases.

28.3.4 Modified Nucleic Acid Libraries

Fluorescently labeled libraries can be used to facilitate detection during selections. Laser-induced fluorescence (LIF) detection is orders of magnitude more sensitive than UV-absorbance detection making it much easier to reliably observe peaks for the unbound sequences during selections. Fluorescent labels are easily incorporated at the 5′ end of ssDNA during initial synthesis. Fluorescent labels can be incorporated into sequences generated during the selection process by using a 5′ labeled forward primer during PCR.

If the final application of the aptamer includes attaching a fluorescent label (e.g., an affinity sensor), it is advantageous to incorporate the label during selection. Postselection modification of the aptamer could modify its structure, and consequently affect affinity or selectivity.

28.3.5 Target Concentration

An advantage of CE-SELEX is the flexibility to manipulate the selection stringency by varying the target concentration in the incubation mixture. Target concentrations as low as 1 pM have successfully been used in CE-SELEX selections.[6] Considering the injection volume this resulted in 10^{13} ssDNA molecules competing for binding sites on 10,000 target molecules. This level of stringency is difficult to obtain using conventional macroscale separations. More recent selections have used higher target concentrations.[7,11,42] There is some concern that extreme competition promotes the selection of aptamers that bind target sites that do not optimally bind nucleic acids.

28.3.6 Capillary Inner Diameter

One of the easiest ways to increase the injection volume, and therefore number of sequences, is to increase the inner diameter of the CE separation capillary. This increase in loading capacity must be balanced with the potential for Joule heating. Joule heating is generally avoided in CE due to the additional band broadening it causes. This is not the major concern in CE-SELEX since peaks are already broad due to destacking caused by the ionic strength of the nucleic acid library. In CE-SELEX, the concern is that Joule heating can change or even eliminate the structures that allow aptamers to bind their targets. Melting temperatures as low as 40°C are not uncommon for some of the structural motifs found in aptamers. Therefore, strict temperature control and elimination of Joule heating are critical for selections to be successful.

28.3.7 Capillary Length

Capillary length should be optimized in much the same way as in typical CE separations. An additional factor to consider is that longer capillaries force the complex to stay intact for a longer period of time before collection. Note that even if equilibrium is reached during incubation before injection, as soon as the separation begins the target–aptamer complex migrates into buffer that does not contain the uncomplexed target. Therefore, the off-rate must be on the order of minutes for the complex to migrate through the capillary intact. This can be used to increase the stringency of the selection. Longer capillaries will promote selection of aptamers with slow off rates, which generally have higher affinity for the target.

28.3.8 Capillary Coatings

Coatings are typically used in CE to prevent analyte interactions with the capillary wall.[43] These wall interactions can give rise to severe peak tailing, degrading resolution. In CE-SELEX, interactions between the oligonucleotide library and the capillary wall are minimal since both are negatively charged. Interactions between the target and the wall are more common. These should be avoided since it is unclear how adsorption onto the capillary surface will affect binding sites on the target. Typically, a CE analysis of the target alone is performed before attempting a selection. If a symmetrical peak is observed, this indicates that interactions between the target and capillary wall are minimal. Tailing would suggest that wall interactions are taking place and that buffer conditions or the capillary surface should be modified.

28.3.9 Buffer Composition

There are a number of important considerations when choosing the incubation and separation buffers. The end use of the aptamer is most important when choosing the incubation buffer. The selection protocol enriches sequences that bind best under these specific conditions. There is no guarantee that the selected aptamers will bind the target under different conditions. Therefore, the incubation buffer should mimic the conditions that the aptamer will be used in as closely as possible. The separation buffer should be similar to the incubation buffer, taking constraints of the CE separation into consideration. High ionic strength buffers give rise to Joule heating and should be avoided. Divalent ions such as Ca^{2+} and Mg^{2+} interact strongly with the fused-silica walls of uncoated capillaries, affecting electroosmotic flow (EOF). The incorporation of buffer additives to modify the separation should also be avoided since this introduces the possibility of selecting aptamers with affinity for the additive, not the target.

28.3.10 Negative Selections

Great care must be taken when performing conventional filter or chromatography-based SELEX to avoid isolating aptamers with affinity for the stationary surfaces to which they are exposed to during selection. Negative selections are often necessary to prevent this. Before incubation with the target the library is passed through a filter or the raw stationary support to eliminate sequences that have any background affinity for these stationary surfaces. While effective, this procedure decreases the diversity of the nucleic acid pool before selection even begins.

Negative selections have not proven to be necessary in CE-SELEX selections. No aptamers have been identified that exhibit affinity for the capillary surface. This has been equally true for selections performed using uncoated or neutrally coated capillaries.[6,7,11,42] This greatly simplifies the SELEX procedure and removes one of the significant pitfalls of aptamer selection.

28.4 PRACTICAL APPLICATION

Mendonsa et al.[6,7] were the first to isolate aptamers using CE-SELEX. They used CE to isolate aptamers with low nM dissociation constants for immunoglobulin E (IgE) after as few as two rounds of selection. This was a significant improvement over conventional selections, which typically take 8–15 rounds. Since then CE-SELEX has been used to isolate aptamers for a range of targets including human immunodeficiency virus reverse transcriptase (HIV-RT),[11] neuropeptide Y (NPY),[42] ricin,[44] protein famesyltransferase (PFTase),[45] and MutS[46] (see Table 28.1). Berezovski et al.[45] were able to isolate aptamers with subnanomolar dissociation constants for PFTase after a single round of CE-SELEX selection, demonstrating the extreme enrichment provided by CE.

Aptamers selected using CE-SELEX exhibit similar if not better affinity for their targets than aptamers selected using conventional methods (see Table 28.1). Mosing et al.[11] used CE-SELEX to isolate aptamers with dissociation constants for HIV reverse transcriptase as low as 180 pM, fivefold better than aptamers selected using nitrocellulose filtration. Tang et al.[44] directly compared CE-SELEX with conventional selections by isolating aptamers for ricin using both methods. They found that after four rounds of CE-SELEX 87.2% of the nucleic acid pool bound ricin while even after nine rounds of conventional selection only 38.5% of the pool bound target.

Target size is a potential limitation to CE-SELEX. The target must be large enough to significantly shift the mobility of the aptamer upon binding for fraction collection to be successful. To test this limitation, Mendonsa and Bowser[6] used CE-SELEX to isolate aptamers for NPY. NPY is a 36 amino acid peptide with a molecular weight of 4272 g/mol, significantly smaller than the ssDNA in the nucleic acid library (~25 kDa). Aptamers with 300 nM dissociation constants for NPY were isolated, similar to RNA aptamers previously selected using conventional methods. These results demonstrated for the first time that CE-SELEX can be used to select aptamers for targets smaller than the library itself. It is important to note that the target did not need to be attached to a stationary phase or a large affinity tag as is necessary to perform conventional SELEX selections against small molecules.

28.5 METHODS DEVELOPMENT GUIDELINES

28.5.1 Identification of the Collection Window

For CE-SELEX experiments to be successful, there must be adequate separation between the aptamer–target complex and nonbinding sequences. Initial experiments should be performed to assess the probability that resolution will be sufficient and determine the migration window of the analyte–target complex. This is complicated by the low abundance of binding sequences in early rounds of selection and the low concentration of target used in selections. Both these factors prevent direct detection of aptamer–target complex during selections. Before attempting selections, the nucleic acid library and target should be analyzed by CE separately to determine their migration times. In almost all cases the aptamer–target complex will migrate at a time between the target and unbound oligonucleotides. For large targets it can be assumed that the complex will migrate closer to the free target peak. Note that in almost every case the aptamer complex will have a mobility that is less negative than the unbound sequences (see Figure 28.5). It is difficult to imagine many targets with a more negative charge density than ssDNA or RNA that would give rise to an aptamer–target complex with a more negative mobility than unbound oligonucleotides. If the target and unbound sequences are not separated then it is unlikely that the aptamer–target complex will be sufficiently resolved for successful fraction collection. In this case separation buffer, voltage, injection time, capillary coatings, capillary length, and/or inner diameter should be modified to achieve resolution.

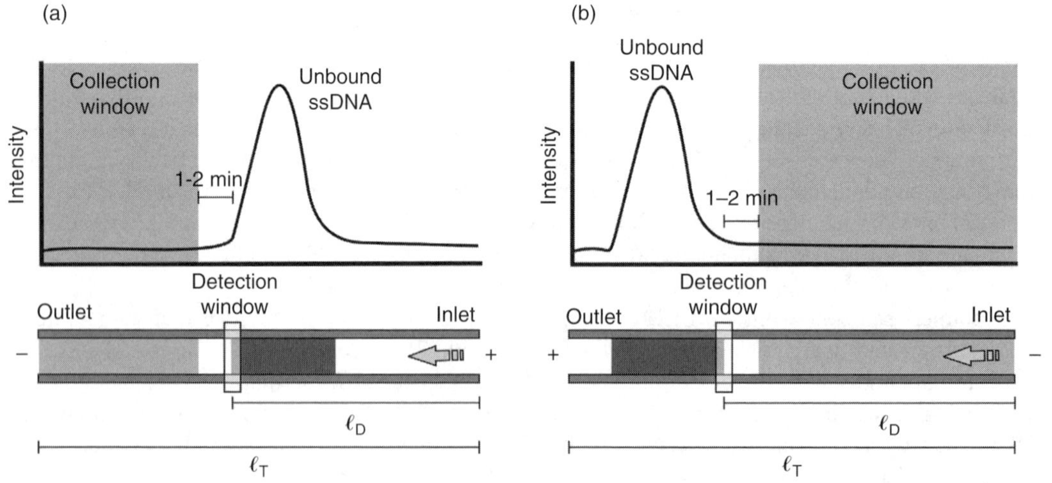

FIGURE 28.5 Schematics illustrating the procedure for fraction collection in CE-SELEX. (a) When using an uncoated capillary that generates EOF the aptamer–target complex will migrate earlier than the unbound ssDNA. Equation 28.2 should be used to calculate the time required for an analyte that reached the detection window 1–2 min before the leading edge of the unbound ssDNA peak to reach the outlet of the separation capillary. (b) When using a coated capillary that does not generate EOF the aptamer–target complex will migrate after the unbound ssDNA. Equation 28.2 should be used to calculate the time required for an analyte that reached the detection window 1–2 min after the trailing edge of the unbound ssDNA peak to reach the outlet of the separation capillary.

28.5.2 Fraction Collection

Once an adequate separation is achieved, and the migration window of the aptamer–target complex has been identified, CE-SELEX selections can begin. Before selection the nucleic acid library is heated to 72°C for 2 min and allowed to cool to room temperature. This ensures that all sequences fold into their stable room temperature conformations. The library is then incubated with the target (50–500 pM) for 20 min at room temperature to allow binding to occur. It is important to keep the target concentration lower than that of the nucleic acid pool to promote competition for binding sites. An aliquot of the incubation mixture is injected onto the CE capillary and the separation voltage is applied.

The strategy for fraction collection is shown in Figure 28.5. Figure 28.5a shows a typical selection using an uncoated capillary in the presence of EOF. EOF pulls the negatively charged nucleic acid library toward the cathode. The aptamer–target complex is generally less negative and therefore will migrate earlier than the unbound sequences. All sequences that migrate before the unbound nucleic acid peak should therefore be collected. Sequences are collected during the CE separation directly into the outlet vial containing a small volume (~50 μL) of separation buffer. The volume of buffer in the collection vial should be small to facilitate PCR amplification after fractions have been collected. To account for run-to-run variability, the collection window should be set to end a certain period of time (e.g., 1–2 min) before the leading edge of the unbound sequences migrates off the end of the capillary. This time can be adjusted based on the resolution anticipated between the aptamer–target complex and the unbound sequences. As shown in Figure 28.5a, the on column detection of CE facilitates this fraction collection strategy. As there is a length of capillary after the detection window, it takes a certain period of time for analytes to reach the outlet of the capillary after they have been detected. The time that an analyte will reach the outlet of the capillary (t_{out}) is given by

$$t_{out} = \frac{\ell_T}{\ell_D}(t_{det}), \tag{28.1}$$

where t_{det} is the time required for the analyte to reach the detector, ℓ_T is the total length of the capillary, and ℓ_D is the capillary length to the detector. To account for variation in migration time between CE separations the separation is monitored until the front of the peak corresponding to the unbound sequences is observed. Equation 28.1 can then be used to calculate the time it would take an analyte that reached the detector 1–2 min earlier to reach the outlet of the capillary. The separation is stopped at this time. The vial at the capillary outlet is removed and any sequences remaining on the capillary are washed to waste using a pressure rinse.

Figure 28.5b illustrates the fraction collection procedure when using a coated capillary that eliminates EOF. The polarity of this separation has been reversed since in the absence of EOF nucleic acids will migrate toward the anode. The aptamer–target complex will migrate after the unbound oligonucleotides if it is less negative. In this case, the separation is monitored until the end of the peak corresponding to the unbound sequences is observed at the detector. Equation 28.1 is used to estimate the time it would take an analyte that reached the detector 1–2 min later than the trailing end of this peak to reach the outlet of the separation capillary. At this time the separation voltage is turned off and the vial at the capillary outlet is replaced with a collection vial containing a small volume (~10–50 μL) of PCR buffer. A pressure rinse is used to collect any sequences remaining on the capillary into this vial.

28.5.3 PCR Amplification

After fraction collection, binding sequences are PCR amplified using standard procedures. Care must be taken during this amplification step due to the extremely low concentration of ssDNA collected. The precise melting, annealing, and extension temperatures are determined by the lengths of the primers and the sequences in the library. The reverse primer should be biotinylated at the 5′ end to facilitate the purification procedure (described in the following section). If the initial library was fluorescently labeled, the forward primer should be similarly labeled at the 5′ end.

Note that PCR facilitates release of even high-affinity aptamers from their targets. The first step of the PCR process is a melting stage where the sequences are brought to ~90°C. This melting eliminates all secondary structure from the ssDNA, removing the aptamer's affinity for the target. Once released, the target is quickly diluted in the PCR buffer, making reformation of the aptamer–target complex unlikely.

Successful PCR amplification can be confirmed using standard gel electrophoresis protocols (2% agarose, ethidium bromide). A single solid band of the proper length with little smearing should be observed. Careful controls should be carried out to ensure that the amplified DNA is from the collection and not the result of contamination.

28.5.4 Purification

PCR amplification generates double-stranded DNA (dsDNA). ssDNA must be recovered for binding sequences to fold into their proper binding structures. Complementary sequences are biotinylated during PCR to facilitate ssDNA recovery. The dsDNA is passed through a streptavidin column. After a short incubation period, the column is rinsed with streptavidin binding buffer (10 × 500 μL, 10 mM Tris, 50 mM NaCl, and 1 mM ethylenediamine tetraacetic acid, EDTA) to remove excess PCR reagents. The column is then heated to 37°C in the presence of 0.15 M NaOH to separate the dsDNA without disrupting the biotin–streptavidin complex. The single-stranded forward sequences are eluted from the column while the complementary sequences remain attached to the stationary phase through the biotin–streptavidin interaction.

The ssDNA is immediately neutralized (0.15 M acetic acid) to avoid degradation: 40 μL of 3 M sodium acetate and 1 mL of 100% ice-cold ethanol are added to initiate ethanol precipitation to further purify and concentrate the ssDNA. The solution is incubated at –80°C for 1 h to ensure complete precipitation. The solution is centrifuged at 14,000 rpm for 15 min at 4°C to pellet the ssDNA precipitate. The supernatant is discarded leaving approximately 50 μL at the bottom of the vial. The following steps are repeated three times to wash the ssDNA pellet:

1. Add 1 mL of ice cold 70:30 ethanol/water
2. Centrifuge 14,000 rpm for 25 min at 4°C
3. Discard the supernatant leaving approximately 50 μL of solution remaining in the bottom of the vial

The samples are then dried in a speed vac at 60°C for 25 min or until dry. The resulting DNA pellet is resuspended in 30 μL of buffer, generating a new ssDNA pool suitable for further rounds of enrichment. In the next selection cycle 10–15 μL is used. The remaining portion is used to assess the affinity of the pool for the target. An aliquot is also stored for further characterization or cloning and sequencing once the selection process is complete.

28.5.5 Dissociation Constant Estimation

The affinity of the nucleic acid pool for the target is measured after every round of selection to monitor the progress of the enrichment. Affinity of the library for the target should increase as the selection progresses. Selection cycles should continue until no further improvement in affinity is observed between rounds. Dissociation constants in the low nM to pM range are typically observed at the completion of the CE-SELEX process.

Dissociation constants can be estimated using affinity capillary electrophoresis (ACE).[47] In these experiments, a constant concentration of the nucleic acid pool is titrated with increasing concentrations of target. Use of fluorescently labeled sequences is recommended since LIF detection allows detection of much lower concentrations. After incubation, each solution is analyzed using CE. The intensity of the peak corresponding to the unbound sequences decreases as the target concentration increases and the equilibrium is shifted toward the aptamer–target complex. If the condition that the nucleic acid concentration is much lower than the target concentration is met, the dissociation constant (K_d) can be determined by fitting the peak intensities to the following equation:

$$\frac{I_0 - I}{I_0} = \frac{c}{K_d + [\text{target}]}, \quad (28.2)$$

where I_0 is the height of the unbound nucleic acid peak in the absence of target, I is the height of the unbound nucleic acid peak in the presence of target, c is a constant, and [target] is the concentration of the target.

28.5.6 Cloning and Sequencing

Once the affinity of the nucleic acid pool has plateaued, individual sequences can be cloned and sequenced from the pool for further characterization. The final pool is PCR amplified for eight cycles to ensure that all sequences are double stranded. Primers should not be 5′ labeled with biotin or a fluorescent tag since this will prevent ligation of the sequences into the cloning vector. dsDNA is ligated into a pGEM vector and transfected into DH5 *Escherichia coli* and colonies are raised.[7,11] Plasmids from approximately 30 (or more) colonies are chosen at random and the sequences of individual clones are determined using the T7 promoter sequence.

28.5.7 Aptamer Characterization

Once the sequences of individual aptamers are obtained, it is relatively inexpensive to have sufficient quantities for further characterization synthesized. In conventional selections, sequence patterns or motifs are commonly observed in the aptamers recovered from cloning. This suggests that the selection has converged on certain sequence elements that are important for binding the target. Software programs such as ClustalW are readily available for identifying these motifs in large numbers of sequences.[48] Only the random region of the sequences should be entered when using these programs. Otherwise, the homology of the primer regions will dominate the analysis. Contrary to conventional selections, sequence motifs are rare in aptamers selected using CE-SELEX.[6,7,11,42] This suggests that the heterogeneity of the pool remains high even after the selection has converged on a collection of high-affinity aptamers.

Common structural elements can be identified using mFold, a program that predicts secondary structure of ssDNA or RNA.[49] For this analysis, it is necessary to include the primer regions since they most likely contribute to the overall structure of the nucleic acid. This analysis is useful in identifying common structural elements in aptamers even if they are generated by differing sequences.

Sequencing and synthesis allow the affinities of individual aptamers to be determined. Dissociation constants can be estimated using ACE as described above. Affinity of the aptamer should be compared with that of the unselected library or a nonsensical sequence such as polyT as a control experiment. Selectivity should be assessed by measuring the affinity of the aptamer for molecules similar to the target. Affinity should be measured under a range of conditions and in the presence of potential interferents depending on whether the aptamer is to be used as a therapeutic or diagnostic agent. Activity assays should be performed if the aptamer was designed to inhibit the function of the target to determine if the aptamer interferes with the active site.

28.6 ADVANTAGES OF CE-SELEX

CE-SELEX provides a number of benefits over traditional SELEX methods. The high resolving power and efficiency of CE generates high rates of enrichment allowing aptamers to be isolated after 2–4 rounds of CE-SELEX selection. Conventional selections typically require 8–15 selection cycles, a process that can take weeks to months to complete. CE-SELEX performs selections in free solution, dramatically reducing the potential for nonspecific interactions with stationary surfaces. CE-SELEX generates a large number of independent aptamer sequences, producing many lead compounds for therapeutic or diagnostic applications. The abundance of high-affinity sequences in the final nucleic acid pool is nearly 100%. In comparison, it is not uncommon for half of the sequences selected using conventional techniques to not demonstrate affinity for the target. Finally, CE-SELEX is much more compatible with libraries that incorporate non-natural nucleic acids to improve aptamer stability. These non-natural nucleic acids often contain hydrophobic functional groups, which promote nonspecific interactions with hydrophobic filter materials, such as nitrocellulose, commonly used in conventional selections.

28.7 FUTURE WORK

CE-SELEX has been used successfully to isolate aptamers for a variety of targets (see Table 28.1). While these initial proofs of concept experiments are promising, a more detailed study of the selection process is now needed. CE-SELEX has made the selection process dramatically faster. This will allow fundamental experiments assessing the effect of variables such as library concentration, target concentration, and incubation time to be performed for the first time. The 4–6 weeks that are necessary to complete a conventional selection have made these experiments unfeasible until now.

CE-SELEX opens the door to further advances in automating the selection process. Several researchers have developed automated SELEX instruments based on conventional selection protocols.[50–52] These experiments have demonstrated that automation greatly improves the throughput of aptamer selection. The electrophoretic selection of CE-SELEX suggests that development of an automated microfluidic SELEX device may be possible. Considering the dramatic improvements in the time required for separation, amplification, and purification, which can be obtained by performing these processes on the microscale, it is not difficult to imagine a fully automated microfluidic device capable of performing the entire SELEX process in as little as several hours.

REFERENCES

1. Bunka, D. H. J., Stockley, P. G., Aptamers come of age—At last, *National Reviews Microbiology*, 4, 588–596, 2006.
2. Klug, S. J., Famulok, M., All you wanted to know about SELEX, *Molecular Biology Reports*, 20, 97–107, 1994.
3. Famulok, M., Mayer, G., Blind, M., Nucleic acid aptamers from selection *in vitro* to applications *in vivo*, *Accounts of Chemical Research*, 33, 591–599, 2000.
4. Osborne, S. E., Ellington, A. D., Nucleic acid selection and the challenge of combinatorial chemistry, *Chemical Reviews*, 97, 349–370, 1997.
5. Forst, C. V., Molecular evolution: A theory approaches experiments, *Journal of Biotechnology*, 64, 101–118, 1998.
6. Mendonsa, S. D., Bowser, M. T., *In vitro* evolution of functional DNA using capillary electrophoresis, *Journal of American Chemical Society*, 126, 20–21, 2004.
7. Mendonsa, S. D., Bowser, M. T., *In vitro* selection of high-affinity DNA ligands for human IgE using capillary electrophoresis, *Analytical Chemistry*, 76, 5387–5392, 2004.

8. Tuerk, C., Gold, L., Systematic evolution of ligands by exponential enrichment: RNA ligands to bacteriophage T4 DNA polymerase, *Science*, 249, 505–510, 1990.
9. Ellington, A. D., Szostak, J. W., *In vitro* selection of RNA molecules that bind specific ligands, *Nature*, 346, 818–822, 1990.
10. Joyce, G. F., Amplification, mutation, and selection of catalytic RNA, *Gene*, 82, 83–87, 1989.
11. Mosing, R. K., Mendonsa, S. D., Bowser, M. T., Capillary electrophoresis-SELEX selection of aptamers with affinity for HIV-1 reverse transcriptase, *Analytical Chemistry*, 77, 6107–6112, 2005.
12. Dey, A. K., Khati, M., Tang, M., Wyatt, R., Lea, S. M., James, W., An aptamer that neutralizes R5 strains of human immunodeficiency virus type 1 blocks gp120-CCR5 interaction, *Journal of Virology*, 79, 13806–13810, 2005.
13. Jeon, S. H., Kayhan, B., Ben-Yedidia, T., Arnon, R., A DNA aptamer prevents influenza infection by blocking the receptor binding region of the viral hemagglutinin, *Journal of Biological Chemistry*, 279, 48410–48419, 2004.
14. Misono, T. S., Kumar, P. K., Selection of RNA aptamers against human influenza virus hemagglutinin using surface plasmon resonance, *Analytical Biochemistry*, 342, 312–317, 2005.
15. Proske, D., Blank, M., Buhmann, R., Ansgar, R., Aptamers—Basic research, drug development, and clinical applications, *Applied Microbiology and Biotechnology*, 69, 367–374, 2005.
16. Rhie, A., Kirby, L., Sayer, N., Wellesley, R., Disterer, P., Sylvester, I., Gill, A., Hope, J., James, W., Tahiri-Alaoui, A., Characterization of 2′-fluoro-RNA aptamers that bind preferentially to disease associated conformations of prion protein and inhibit conversation, *Journal of Biological Chemistry*, 278, 39697–39705, 2003.
17. Siddiqui, M. A. A., Keating, G. M., Pegaptanib: In exudative age-related macular degeneration, *Drugs*, 65, 1571–1577, 2005.
18. Lee, J. F., Stovall, G. M., Ellington, A. D., Aptamer therapeutics advance, *Current Opinion in Chemical Biology*, 10, 282–289, 2006.
19. Jayasena, S. D., Aptamers: An emerging class of molecules that rival antibodies in diagnostics, *Clinical Chemistry*, 45, 1628–1650, 1999.
20. Tombelli, S., Minunni, M., Mascini, M., Analytical applications of aptamers, *Biosensors and Bioelectronics*, 20, 2424–2434, 2005.
21. German, I., Buchanan, D. D., Kennedy, R. T., Aptamers as ligands in affinity probe capillary electrophoresis, *Analytical Chemistry*, 70, 4540–4545, 1998.
22. Haes, A. J., Giordano, B. C., Collins, G. E., Aptamer-based detection and quantitative analysis of ricin using affinity probe capillary electrophoresis, *Analytical Chemistry*, 78, 3758–3764, 2006.
23. Pavski, V., Le, X. C., Detection of human immunodeficiency virus type 1 reverse transcriptase using aptamers as probes in affinity capillary electrophoresis, *Analytical Chemistry*, 73, 6070–6076, 2001.
24. Collett, J. R., Cho, E. J., Ellington, A. D., Production and processing of aptamer microarrays, *Methods*, 37, 4–15, 2005.
25. Cho, E. J., Collett, J. R., Szafranska, A. E., Ellington, A. D., Optimization of aptamer microarray technology for multiple protein targets, *Analytica Chimica Acta*, 564, 82–90, 2006.
26. Bock, C., Coleman, M., Collins, B., et al., Photoaptamer arrays applied to multiplexed proteomic analysis, *Proteomics*, 4, 609–618, 2004.
27. Golden, M. C., Collins, B. D., Willis, M. C., Koch, T. H., Diagnostic potential of PhotoSELEX-evolved ssDNA aptamers, *Journal of Biotechnology*, 81, 167–178, 2000.
28. Petach, H., Gold, L., Dimensionality is the issue: Use of photoaptamers in protein microarrays, *Current Opinion in Biotechnology*, 13, 309–314, 2002.
29. Matthew Levy, S. F. C. A. D. E., Quantum-dot aptamer beacons for the detection of proteins, *ChemBioChem*, 6, 2163–2166, 2005.
30. Cao, Z., Suljak, S. W., Tan, W., Molecular beacon aptamers for protein monitoring in real-time and in homogeneous solutions, *Current Proteomics*, 2, 31–40, 2005.
31. Brumbt, A., Ravelet, C., Grosset, C., Ravel, A., Villet, A., Peyrin, E., Chiral stationary phase based on a biostable L-RNA aptamer, *Analytical Chemistry*, 77, 1993–1998, 2005.
32. Ravelet, C., Grosset, C., Peyrin, E., Liquid chromatography, electrochromatography and capillary electrophoresis applications of DNA and RNA aptamers, *Journal of Chromatography A*, 1117, 1–10, 2006.

33. Ravelet, C., Boulkedid, R., Ravel, A., Grosset, C., Villet, A., Fize, J., Peyrin, E., A L-RNA aptamer chiral stationary phase for the resolution of target and related compounds, *Journal of Chromatography A*, 1076, 62, 2005.
34. Kotia, R. B., Li, L., McGown, L. B., Separation of nontarget compounds by DNA aptamers, *Analytical Chemistry*, 72, 827–831, 2000.
35. Vo, T. U., McGown, L. B., Selectivity of quadruplex DNA stationary phases toward amino acids in homodipeptides and alanyl dipeptides, *Electrophoresis*, 25, 1230–1236, 2004.
36. Charles, J. A. M., McGown, L. B., Separation of Trp-Arg and Arg-Trp using G-quartet-forming DNA oligonucleotides in open-tubular capillary electrochromatography, *Electrophoresis*, 23, 1599–1604, 2002.
37. Rehder, M. A., McGown, L. M., Open-tubular capillary electrophoresis electrochromatography of bovine beta-lactoglobulin variants A and B using an aptamer stationary phase, *Electrophoresis*, 22, 3759–3764, 2001.
38. Rehder-Silinski, M. A., McGown, L. B., Capillary electrochromatographic separation of bovine milk proteins using a G-quartet DNA stationary phase, *Journal of Chromatography A*, 1008, 233, 2003.
39. Vo, T. U., McGown, L. B., Effects of G-quartet DNA stationary phase destabilization on fibrinogen resolution in capillary electrochromatography, *Electrophoresis*, 27, 749–756, 2006.
40. Deng, Q., German, I., Buchanan, D., Kennedy, R. T., Retention and separation of adenosine and analogous by affinity chromatography with an aptamer stationary phase, *Analytical Chemistry*, 73, 5415–5421, 2001.
41. Navani, N. K., Li, Y., Nucleic acid aptamers and enzymes as sensors, *Current Opinion in Chemical Biology*, 10, 272, 2006.
42. Mendonsa, S. D., Bowser, M. T., *In vitro* selection of aptamers with affinity for neuropeptide Y using capillary electrophoresis, *Journal of American Chemical Society*, 127, 9382–9383, 2005.
43. Judit Horvath, V. D., Polymer wall coatings for capillary electrophoresis, *Electrophoresis*, 22, 644–655, 2001.
44. Tang, J., Xie, J., Shao, N., Yan, Y., The DNA aptamers that specifically recognize rixin toxin are selected by two *in vitro* selection methods, *Electrophoresis*, 27, 1303–1311, 2006.
45. Berezovski, M., Drabovich, A., Krylova, S. M., Musheev, M., Okhonin, V., Petrov, A., Krylov, S. N., Nonequilibrium capillary electrophoresis of equilibrium mixtures: A universal tool for development of aptamers, *Journal of the American Chemical Society*, 127, 3165–3171, 2005.
46. Drabovich, A. P., Berezovski, M., Okhonin, V., Krylov, S. N., Selection of smart aptamers by methods of kinetic capillary electrophoresis, *Analytical Chemistry*, 78, 3171–3178, 2006.
47. Østergaard, J., Heegaard, N. H. H., Bioanalytical interaction studies executed by preincubation affinity capillary electrophoresis, *Electrophoresis*, 27, 2590–2608, 2006.
48. Chenna, R., Sugawara, H., Koike, T., Lopez, R., Gibson, T. J., Higgins, D. G., Thompson, J. D., Multiple sequence alignment with the clustal series of programs, *Nucleic Acids Research*, 31, 3497–3500, 2003.
49. Zuker, M., Mfold web server for nucleic acid folding and hybridization prediction, *Nucleic Acids Research*, 31, 3406–3415, 2003.
50. Cox, J. C., Hayhurst, A., Hesselberth, J., Bayer, T. S., Georgiou, G., Ellington, A. D., Automated selection of aptamers against protein targets translated *in vitro*: From gene to aptamer, *Nucleic Acids Research*, 30, e108, 2002.
51. Cox, J. C., Rudolph, P., Ellington, A. D., Automated RNA selection, *Biotechnology Progress*, 14, 845–850, 1998.
52. Hybarger, G., Bynum, J., Williams, R. F., Valdes, J. J., Chambers, J. P., A microfluidic SELEX prototype, *Analytical and Bioanalytical Chemistry*, 384, 191–198, 2006.
53. Wiegand, T. W., Williams, P. B., Dreskin, S. C., Jouvin, M. H., Kinet, J. P., Tasset, D., High-affinity oligonucleotide ligands to human IgE inhibit binding to Fce receptor I, *Journal of Immunology*, 157, 221–230, 1996.
54. Schneider, D. J., Feigon, J., Hostomsky, Z., Gold, L., High-affinity ssDNA inhibitors of the reverse transcriptase of type 1 human immunodeficiency virus, *Biochemistry*, 34, 9599–9610, 1995.
55. Proske, D., Hofliger, M., Soll, R. M., Beck-Sickinger, A. G., Famulok, M., A Y2 receptor mimetic aptamer directed against neuropeptide Y, *Journal of Biological Chemistry*, 277, 11416–11422, 2002.

56. Drabovich, A., Berezovski, M., Krylov, S. N., Selection of smart aptamers by equilibrium capillary electrophoresis of equilibrium mixtures, *Journal of the American Chemical Society*, 127, 11224–11225, 2005.
57. Mallikaratchy, P., Stahelin, R. V., Cao, Z., Tan, W., Selection of DNA ligands for protein kinase C-omega, *Chemical Communications*, 3229–3231, 2006.

29 Microfluidic Technology as a Platform to Investigate Microcirculation

Dana M. Spence

CONTENTS

29.1 INTRODUCTION

Of the numerous applications of capillary- or microflow-based analyses, especially those involving biological assays, there perhaps is none more fitting than employing microflow techniques to study the circulatory system. In fact, the dimensions of the microcirculation so closely match those of microbore capillary tubing and the channels in microfluidic devices, it is rather amazing that such analytical tools and devices have not been in use to study the events in the bloodstream to a greater extent. Here, a brief introduction to the microcirculation is given such that the nonexpert is able to understand better why microflow techniques are an excellent tool to study *in vivo* processes that occur in the bloodstream. Next, real examples will be given in order to demonstrate how such devices are effective as tools to increase our understanding of the processes occurring in the bloodstream.

29.2 BACKGROUND

One of the key features of employing microbore tubing in a microflow system or a microfluidic device to study events occurring in the microcirculation is that the studies can be performed in a controlled environment. That is, *in vivo*, the circulation is very dynamic. Not only do vessel diameters change (due to vasodilation) but also do the linear rates of the bloodstream, the viscosity of the blood, and hematocrit.[1] There are also slight changes in pH,[2] oxygen tensions,[3] and concentrations of molecules in the bloodstream. Finally, there are also changes that occur at the cellular level that can affect the overall cellular function. Common examples are cell aging and cellular oxidant status that can affect the cell's membrane properties.[4]

The use of a microflow system enables quantitative determinations to be performed in an environment where the many variables mentioned above can be controlled.[5] An excellent example is the linear rate of the blood flow. *In vivo*, it has been shown recently that red blood cells (RBCs) have the ability to control vessel diameter due to their ability to release adenosine triphosphate (ATP) upon shear-induced deformation.[6–8] This ATP is a recognized stimulus of nitric oxide (NO) production in certain cell types including the endothelium[9] and platelets.[10] NO is a potent vasodilator, resulting in the increase of vessel diameter and an increase in blood flow.[11–13] The ability to study the effect of linear rate on the release of ATP from RBCs would be difficult in an animal model because, as the ATP is released (resulting in subsequent NO-induced vasodilation), the vessel diameter would increase, thus changing the overall linear rate of the blood flow through the vessel. By performing such studies on a controlled *in vitro* platform (microbore tubing), the ATP can be measured rather easily via chemiluminescence, but the linear rate will remain constant (because microbore tubing does not dilate).

Another advantage of performing measurements involving the bloodstream using microflow technologies is that blood is a complex sample and, as such, often requires some type of preparation prior to the measurement portion of the analysis. Examples of such preparation schemes using microfluidic devices have been described for automated, high-speed genotyping from droplets of blood[14] and separation of plasma from a flowing blood sample.[15] There also exists numerous examples of cell sorting on microfluidic devices.

Finally, and what may prove in the immediate near future to be the most important aspect of investigating the bloodstream with microflow techniques is the finding that many processes in the bloodstream are actually activated due to the flow itself.[8,16,17] It has long been accepted that endothelium-derived NO is actually stimulated due to a shear stress placed upon the endothelium.[11–13] However, it is also becoming increasingly clear that other events in the microcirculation are also stimulated due to the forces placed upon certain cells when flowing through open tubes (vessels). These same forces can be applied to cells in an *in vitro* format in a microflow system and the resultant cellular response can be quantitatively measured or monitored.[18] In order to better understand the importance of flow for the induction of many cellular events *in vivo*, it is necessary to understand the properties of blood flow *in vivo*.

29.2.1 Components of the Circulation Investigated with Microfluidic Techniques

The microcirculation is generally considered to be the blood flow through vessels having diameters that are less than 100 μm. Specifically, this set of vessels consists of capillaries, arterioles, and venuoles. There do exist other classifications within this range of vessels (e.g., meta-arterioles are often defined as those vessels whose diameters range between 10 and 25 μm). However, capillaries can be defined to have diameters below 25 μm while arteriole diameters range between 25 and 100 μm.[1] However, there exist many different cell types that are determinants of microcirculatory maintenance. There traditional roles and potential new roles being that are being discovered through microanalytical techniques follow.

29.2.1.1 Erythrocytes

Many researchers consider the microcirculation to be a very important component of the circulatory system due to its 60,000 miles of architecture (vessels) that deliver the different cell types in the

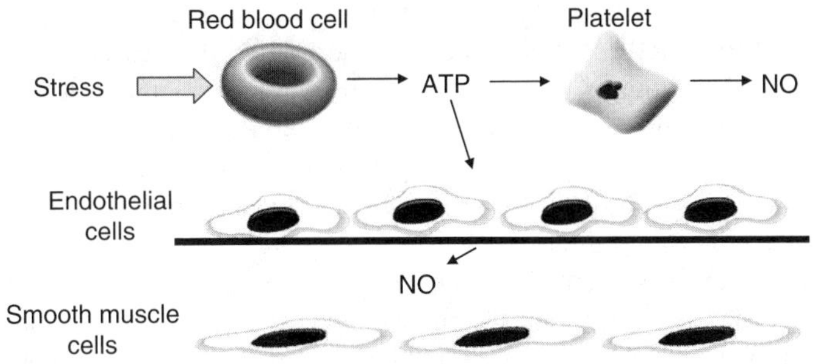

FIGURE 29.1 Proposed mechanism for the involvement of RBCs and platelets as determinants of vascular caliber in the microcirculation.

bloodstream to the tissues and organs in the human body. Of these cell types, erythrocytes or red blood cells (RBCs) are deemed important due to their classic task of delivering oxygen to the various tissues and organs *in vivo*. RBCs are generally considered as one of the more nonexciting cell types in the circulation. They do not possess a nucleus, mitochondria, or any other key type of organelle. Recently, however, RBCs have gained some attention in the literature due to their association with NO, both as a carrier of the molecule[19,20] and as a potential determinant in NO production.[7,21]

RBCs, when traversing microvascular beds such as in the pulmonary circulation, are subjected to mechanical deformation. Previously, it was reported that, in the isolated perfused rabbit lung, RBCs obtained from either rabbits or healthy humans were a required component of the perfusate in order to demonstrate flow-induced endogenous NO synthesis in the pulmonary circulation. Importantly, it was reported that the property of these RBCs that was responsible for the stimulation of NO synthesis was their ability to release ATP in response to mechanical deformation. Moreover, it was reported that the release of ATP from RBCs of rabbits and humans increased as the degree of deformation increased.[6,22,23]

These reports suggested a novel mechanism for the control of vascular caliber (Figure 29.1). In this construct, as the RBC is increasingly deformed by increments in the velocity of blood flow through a vessel and/or by reductions in vascular diameter, it releases ATP that stimulates endothelial synthesis of NO resulting in relaxation of vascular smooth muscle and, thereby, an increase in vascular caliber. This vasodilation results in a decrease in vascular resistance as well as a decrease in the stimulus for RBC deformation and ATP release. Indeed, it has been proposed that RBC-derived ATP contributes to vascular resistance in both the pulmonary and systemic circulation.

The finding that ATP is released from RBCs in response to mechanical deformation suggests that variables such as vascular diameter and flow rate may be important determinants of ATP release as these cells traverse the intact circulation. An understanding of the contribution of these parameters individually and in combination to ATP release from RBCs in the microcirculation, and in particular in resistance vessels, is of paramount importance for the comprehension of the role of the RBC as a determinant of vascular caliber.

29.2.1.2 Platelets

Other formed elements in the bloodstream that are considered crucial are the platelets. Platelets are non-nucleated cells in the bloodstream that are approximately 2 μm in diameter. The main role of platelets involves hemostasis, the prevention of excessive blood loss following vascular injury. The actions of platelets take place in a dynamic environment: flowing blood. As such, the larger and more numerous RBCs occupy the axial portion of the vessel forcing platelets toward the peripheral endothelial lining.[24] Platelets normally circulate without adhering to undisturbed vascular endothelium. Upon vascular insult, subendothelial collagen is exposed that acts as the primary stimulus for platelet activation. However, several other endogenous agonists also exist such as thrombin, ADP, thromboxane A_2, serotonin, and epinephrine, which also promote platelet activation.[25] Platelet activation initiates a change in the platelet shape, thus promoting adhesion to the vascular walls and the subsequent recruitment of additional platelets. NO has been widely shown to mediate this process by activating soluble guanylate cyclase (GC) which initiates a protein kinase G (PKG) dependent pathway.[26] In this construct, NO becomes an important determinant in platelet adhesion *in vivo*.[10] If uncontrolled, these adhered platelets become a major constituent of thrombus formation and subsequent vessel blockage.

Recently, platelets have been reported to have unique properties in disease states other than cardiovascular problems. For example, a procedure used in multiple sclerosis, plasma replacement therapy, involves removing such formed elements as RBCs and leukocytes from the patient's plasma and then replenishing new plasma with the previously removed cells. The plasma contains the platelets so, in addition to obtaining new plasma, the patient also receives new platelets. Interestingly, the platelets of people with multiple sclerosis have been shown to be more susceptible to hyperactivity, often aggregating more easily than the platelets of healthy people that do not have multiple sclerosis.[27] The importance of this platelet activity is that platelets are known to create NO upon activation and people with multiple sclerosis have been shown to have high levels of NO and NO metabolites in their cerebral spinal fluid and urine.[28,29] This trait of platelets (prone to activation and aggregation) is also known to exist for patients that have cystic fibrosis.[30] Finally, patients with diabetes also have platelets that are known to be activated more readily than healthy, nondiabetic patients.[31] Exemplifying the importance and complexity of the relationship between platelets and NO is that NO has the ability to inhibit platelet activation.[10]

Interestingly, platelets have the ability to create bioavailable NO in the bloodstream in multiple ways other than by activation. For example, NO can be synthesized inside of the platelet by one of multiple isoforms of NOS[32,33] or from the denitrosation of *S*-nitrosothiols (*S*-nitrosoglutathione or *S*-nitrosoalbumin) enzymatically[34] or by copper-containing proteins.[35]

29.2.1.3 Endothelial Cells

Endothelial cells form a single layer within the blood vessels comprising the circulatory system. They separate the flowing blood within the lumen from the vessel wall. While they are present in both arteries and capillaries they have significantly different properties depending on the vessel size. Endothelial cells are involved in many biological roles some of which include vasoconstriction, blood pressure, inflammation, and transfer of materials into and out of the circulatory system. Research of vascular disease has significantly increased the culturing of endothelial cells that are dominantly obtained from the bovine aorta, bovine adrenal capillaries, rat brain capillaries, human umbilical veins as well as other sources. Once cultured, a simple protocol for cell dissociation involving trypsin can be followed to subculture the cells, reseeding them for future divisions or integrating them into microfluidic devices for *in vivo* mimic studies.

29.3 THEORETICAL ASPECTS

29.3.1 Unique Flow Properties of the Bloodstream

Collectively, the formed elements that comprise blood in the microcirculation have the properties of a non-Newtonian fluid. As such, flow properties encountered in open tubes are not necessarily found when the fluid contains components such as RBCs or platelets. Many properties of blood flow in a vessel *in vivo* are difficult to mimic in a section of microbore tubing or a channel in a microfluidic device. For example, the overall pressure drop across an open tube can be described by the expression shown in the following equation.

$$q = \frac{\Delta P r^4 \pi}{8\ \mu L}$$

Here, the overall flow q is dependent on variables such as the pressure applied across the tube (ΔP), the radius (r), length (L) of the tube, and the viscosity (μ) of the solution. Unfortunately, this equation must be modified somewhat because blood is a non-Newtonian fluid having some pseudoelasticity. Therefore, at similar applied pressures, blood will flow faster than a typical Newtonian fluid. Furthermore, *in vivo*, the vessel diameter (due to dilation and constriction of the resistance vessels) is dynamic and constantly changing. Moreover, the viscosity of blood is very difficult to measure because the hematocrit of blood changes depending on the diameter of the vessel in which it flows. Because of this, blood viscosity *in vivo* is often reported as an "apparent" viscosity. Finally, in the bloodstream, blood flow is somewhat pulsatile, especially in the venous side of the circulation. In sum, it is difficult to know the exact linear rate of the blood flow or a pressure drop at any one point in the microcirculation.

Another aspect of blood flow *in vivo*, and one that plays a major role in such circulation complications such as atherosclerosis and stroke, is that cells will partition themselves into certain areas of the flowing stream. For example, RBCs will typically travel in the middle of the open tube where the convective forces are at a minimum. However, platelets, being smaller in size, then become forced to flow more toward the wall. In fact, a close examination of the components of blood at the wall of the vessel or tube will show that there is actually a cell-free layer (often called the skimming layer) along the wall. This phenomenon was first noticed in the 1920s by Fahraeus and Lindqvist today bears the name of their discoverers.[1]

Interestingly, it may also be a key to a potential new role for the RBC, namely, the ability to participate in the control of vascular caliber in the intact circulation. Furthermore, it is these flow properties that enable such formed elements as the platelets to participate in the repair of damaged endothelium.

29.4 APPLICATIONS

29.4.1 Disease Biomarkers

Although tissue analysis is an outstanding point to begin for many experiments in search of biomarkers, the bloodstream represents one of the more accessible tissues from which to extract samples. An example of microfluidic technology helping to determine such a biomarker was recently reported by Carroll, where the amount of ATP released from RBCs that were exposed to shear-induced mechanical deformation.[4] Specifically, RBCs obtained from healthy control subjects were pumped through microbore tubing having diameters that approximated those of resistance vessels *in vivo*. The ATP released from these deformed RBCs was measured using the well-known luciferin chemiluminescence assay for ATP. When compared to RBCs obtained from patients having type II diabetes, the control patients release approximately two times the amount of ATP (190 ± 10 nMvs. 90 ± 10 nM, respectively) as the RBCs obtained from the patients with type II diabetes (Figure 29.2). Such a screen could be an important marker for adult onset diabetes because one could screen the ATP release, archiving the results for the patient, and subsequently monitor any decrements on perhaps an annual basis.

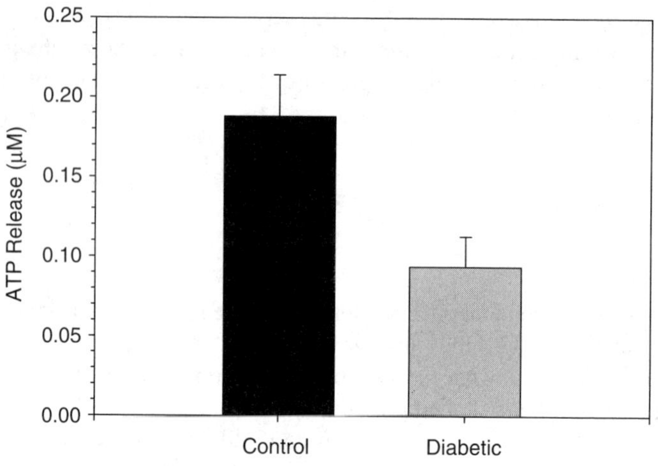

FIGURE 29.2 Determination of ATP release from the RBCs of diabetic ($n = 7$) and nondiabetic ($n = 7$) patients is shown. The RBCs were pushed through microbore tubing that had an inside diameter of 50 μm. The error bars represent standard errors about the mean. The values shown in the figure differ significantly ($p < 0.001$).

29.4.2 High-Throughput Drug Efficacy Investigations

A microfluidic device has been utilized to explore the mechanism by which vasodilation occurs and the effects of iloprost, a known vasodilator. Iloprost is a second generation prostacyclin analog that has been shown to directly affect vasorelaxation *in vivo* by increasing the amount of ATP released from RBCs transversing the circulatory system. ATP is then free to stimulate the production of NO from endothelial cells resulting in the production of guanosine monophosphate, which is known to relax smooth muscle cells surrounding resistance vessels. This study incorporates the methodology of the previous research and examines the effect of iloprost on the release of ATP from RBCs when forced through a microchannel to mimic the circulation system. An irreversibly sealed poly(dimethyl)siloxane (PDMS) microfluidic device was fabricated from a 4 in. silicon wafer master that was obtained using standard negative photoresist lithographic processes (Figure 29.3a). The channel dimensions measured approximately 100 μm wide by 100 μm deep and were patterned from a negative photomask. This T-channel chip was irreversibly sealed to another PDMS chip absent of any channels after perforating sample introduction and waste ports. A displacement syringe pump and Tygon tubing was used to introduce samples via hypodermic steel tubing inserted into the channels. A 7% RBC solution treated and left untreated with iloprost was mixed with a luciferin/luciferace solution to examine the iloprost stimulated, RBC-derived ATP. The chemiluminescence for all samples and standards was measured in real time by a photomultiplier tube at a rate of 10 measurements per second for 30 s using software written in-house. The results from this study are shown in Figure 29.3b.

29.4.3 Measuring Cell-to-cell Communication

Another advantage of microflow techniques to investigate the events occurring in the microcirculation is those involving multiple cell types in a controlled environment. For example, the ability to place endothelial cells in a microfluidic device and monitor the ability of platelets to adhere to these endothelial cells would be beneficial in studies pertaining to vessel blockage due to thrombus formation. Quite often, a thrombus is formed due to platelets, macrophages, and even RBCs adhering to the endothelial wall under certain conditions. The ability to quantitatively monitor those

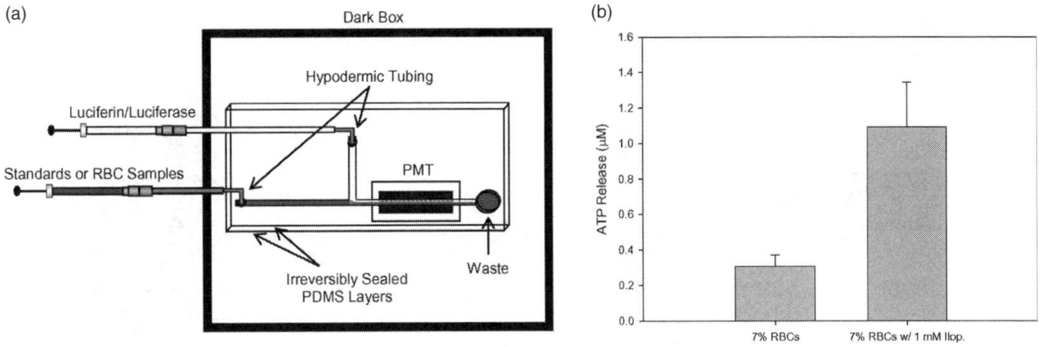

FIGURE 29.3 (a, b) Determination of ATP release from RBCs incubated with iloprost, a stable analog of prostacyclin, in a microfluidin format.

molecules involved in the thrombus formation or those molecules that prevent such adhesion events would be useful to the biomedical community. Figure 29.4a shows an endothelium coating in the channel of a microfluidic device.[9] The dimensions of this channel approximate those of resistance vessels *in vivo*. Figure 29.4b shows activated platelets flowing over an endothelium. Importantly, a close observation reveals that, in certain areas, platelet recruitment is occurring; that is, upon activation, platelets become somewhat sticky and adhere to each other in addition to the endothelial wall. In such a scenario, these platelets become determinants in the mechanism of vessel blockage. The ability to create a multicellular composition in a controlled environment will facilitate the investigations of thrombus formation, occurrence of stroke, and studies involved in discovering new thrombolytic materials.

29.5 METHODOLOGY

29.5.1 Cell Harvesting

One of the inherent challenges for performing studies involving the circulation is that, due to an absence of a nucleus, RBCs and platelets cannot be cultured. Therefore, to work with a meaningful sample of RBCs, platelets, plasma, and so forth, one has to gather these cells from a mammalian subject prior to completing the analysis. The RBCs and platelets are both obtained from the whole blood of the animal or human. This means that approval from either an animal investigation committee or a human investigation committee is required before such harvesting can proceed. In addition, if the harvesting will involve a nonsurvival procedure (with an animal subject), it is likely that certain drugs will be employed that may be controlled substances or require a special license in order to be used in the procedure. Of course, the appropriate training will also be required prior to any harvesting procedures.

Once the necessary approvals are in place, the harvesting of the RBCs and platelets can proceed. This procedure for removal of cells from rabbits has been outlined in many publications. Rabbits can be anesthetized with ketamine (8 mL/kg, i.m.) and xylazine (1 mg/kg, i.m.) followed by pentobarbital sodium (15 mg/kg i.v.). A cannula is then placed in the trachea so that the animals can be ventilated with room air. A catheter is then placed into a carotid artery for administration of heparin and for phlebotomy. After heparin (500 units, i.v.), animals are exsanguinated and the whole blood collected in a 50 mL tube. Depending on the animal size, varying amounts of blood can be collected.

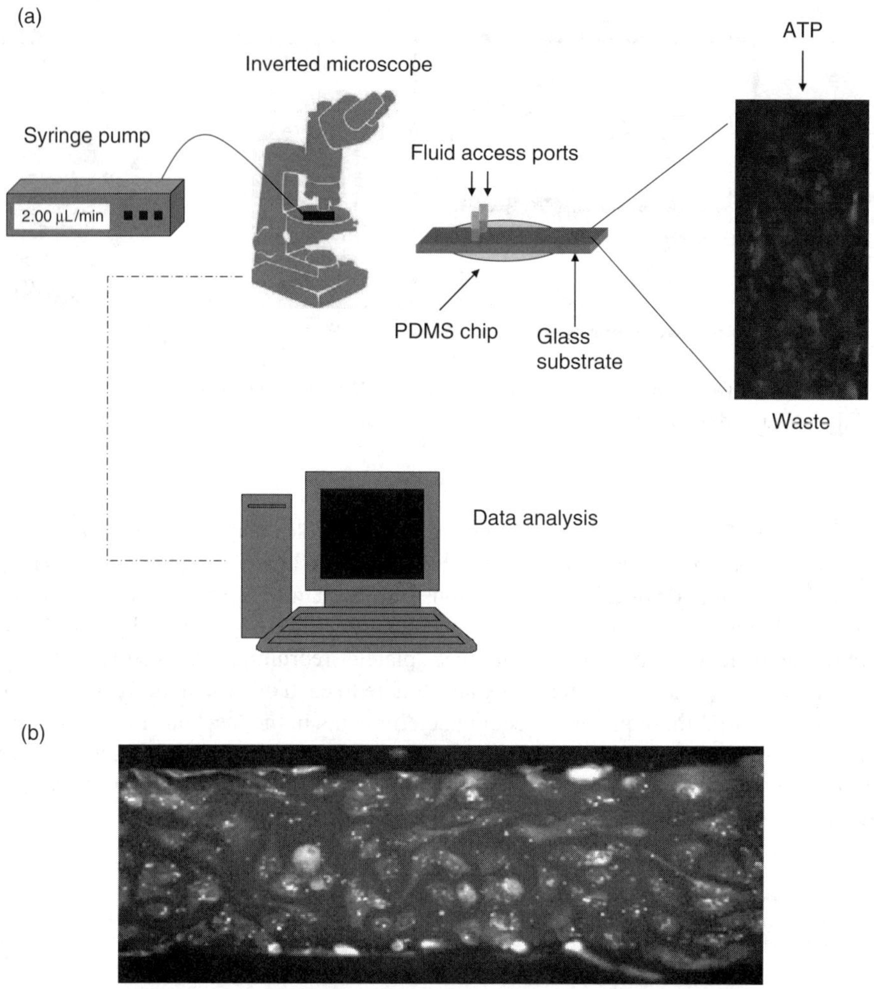

FIGURE 29.4 (a) A PDMS chip was fabricated to contain 100 μm × 100 μm × 2 cm channels. bPAECs were cultured into the microfluidic channels, and the chip was reversibly sealed to a glass substrate, making sure to align the channel and the fluid access port. Solutions were perfused across the cells (2 μL min^{-1}) using 75 μm microbore tubing to connect the syringe pump to the microfluidic adaptor on the fluid access port. Fluorescence measurements were acquired using an Olympus IX71M microscope equipped with an electrothermally cooled CCD (Orca, Hamamatsu) and MicroSuite software (Olympus). (b) Platelets are shown adhering to an immobilized endothelium, even in the presence of flow. Such devices may be used in the future to investigate thrombus formation or other vaso-occlusive mechanisms.

29.5.2 Sample Preparation

Once the RBCs or platelets are harvested from the subject, there still exists some sample preparation prior to the measurement portion of the analysis. For example, in order to purify platelets, the collected blood is centrifuged at 500 *g* at 37°C for 10 min. The platelet-rich plasma (PRP) is decanted for the subsequent isolation of platelets. Platelets are then isolated from the PRP by centrifugation (15 min at 2000 *g* at 37°C) and washed three times in Tyrode-albumin solution (pH 7.4). The first wash contains heparin (2 U/mL) and apyrase (1 U/mL); the second wash contains only apyrase (1 U/mL); and the third wash contains Tyrode's solution without apyrase and heparin.

In order to purify the RBCs, the plasma and buffy coat (white cells) are removed for further purification (see above). RBCs are then resuspended and washed three times in a physiological salt solution [PSS; in mM, 4.7 KCl, 2.0 $CaCl_2$, 140.5 NaCl, 12 $MgSO_4$, 21.0 tris(hydroxymethyl)aminomethane, 11.1 dextrose with 5% bovine serum albumin (final pH 7.4)]. Cells should generally be prepared and studied on the day of use within 8 h of removal from animal or human subjects.

In addition to the cell purification, it is also important to remember that, *in vivo*, RBCs exist at various hematocrits throughout the vasculature, depending on the size of the vessel. For example, it is generally accepted that the average hematocrit of RBCs *in vivo* is approximately 40–45%. However, in resistance vessels, this value can decrease to levels as small as single-digit percentiles.

Endothelial cells are, in some ways, easier to work with than the RBCs and platelets because they can be grown in standard tissue culture flasks. However, even though these cells may be easier to obtain (committee approval is not required), preparing them for a quantitative investigation in a microflow system is challenging because one must immobilize these cells prior to investigation. For example, Kotsis has demonstrated the ability to culture endothelial cells from bovine pulmonary artery on the walls of a microbore tubing. Although these cells responded to an ATP bolus by producing NO, the preparation of the cell microflow reactors was tedious and met with a low rate of success. More recently, immobilizing these cells to the walls of a channel in a microfluidic device has been reported by multiple groups. Immobilizing the cells in the microfluidic device is advantageous because the chip device is planar (similar to standard tissue culture flasks), the channels are accessible thus allowing media to flow over the cells, and the cells can be easily monitored using standard microscopy (Figure 29.4a). There also exist different methods for cell immobilization.

Spence et al.[36] employed a method by which endothelial cells were vacuum delivered into a channel that was coated with fibronectin. Li et al.[37] described the use of a "trough" method for immobilizing a dopaminergic cell model. This method appears to be useful for cell types that have a tendency to cluster. Other methods have been described as well. In most cases, hydrodynamic pumping schemes have been employed for such immobilization determinations.[38]

29.5.3 Detection

Detection of the molecules produced, consumed, and secreted by the cells described here is challenging for two main reasons. First, the cell is dynamic and constantly tries to maintain balance. As such, molecules concentrations or speciation are usually changing. Second, the matrix in which these measurements are typically performed is very complex. Thus, the technique of choice needs to have some built-in feature that enables the analyst to overcome the matrix. To date, a variety of measurements have been employed to learn more about the roles of the cells in the microcirculation. Specifically, fluorescence, chemiluminescence, and amperometry have all been used extensively. Not surprisingly, all three of these detection schemes are readily employed in capillary electrophoresis-based determinations. Therefore, many of the measurements employ technology from the CE field. However, due to the cell matrix complexity, techniques are required to overcome potential interferents. For example, Kovarik et al.[39] employed a Nafion coating over a micromolded ink electrode for selectivity in detecting dopamine in the presence of an anion interferent (ascorbate). For similar reasons, Ku[40] employed the classic method of multiple standard additions to quantitatively determine the amount of NO released from activated platelets in a flowing stream.

29.6 FUTURE ASPECTS

The concept of performing studies involving the circulation in an environment that closely mimics blood vessels and its components is relatively new. Therefore, there has not been a great deal of studies involving actual blood flow in microflow systems. However, as pointed out in a recent review by Martin, Root, and Spence, the number of papers published over the past decade involving biological cells and microfluidic devices is growing at a very rapid rate. One of the obstacles in performing cellular studies with cells typically found in the circulation is that many of these cells cannot be cultured; that is, they must be harvested. One of the initial steps in any analysis is obtaining the sample and obtaining the sample is not easy in these studies. However, the potential range of studies that still need to be investigated are immense.

One potential area where microflow studies of the circulation will be of major benefit involves studies involving glycosylation products. In a recent report, it was stated that nearly 80% of all proteins in the bloodstream were glycosylated. With such products being implicated heavily in such diseases as diabetes and cancer, it would seem logical to employ the systems described here with other separation schemes, detection schemes, or protocols in general to determine various biomarkers in a system that more readily mimics the circulation *in vivo*.

As eluded to above, one of the key features that this technology offers is the ability to perform important investigations early. For example, most biomarker detection is successful on plasma, serum, or other tissues after the disease of interest has already taken hold of the patient. By monitoring the bloodstream, one may be able to find certain biomarkers before the formation of tumors, vessel blockages, organ damage, and so forth. Moreover, it may also be possible to perform biomarker studies not only on patients already inflicted with a disease but also to perform such studies on healthy patients, creating an archive of their health and subsequently monitoring the decline in certain proteins, peptides, or metabolites.

REFERENCES

1. Enderle, J., Blanchard, S., Bronzino, J. *Introduction to Biomedical Engineering*; Academic Press, San Diego, CA, 2000.
2. Bergfeld, G. R., Forrester, T. Release of ATP from human erythrocytes in response to a brief period of hypoxia and hypercapnea, *Cardiovascular Research*, 1992, *26*, 40–47.
3. Olearczyk, J. J., Stephenson, A. H., Lonigro, A. J., Sprague, R. S. Heterotrimeric G protein Gi is involved in a signal transduction pathway for ATP release from erythrocytes, *American Journal of Physiology* 2004, *286*, H940–H945.
4. Carroll, J., Raththagala, M., Subasinghe, W., Baguzis, S., Oblak, T. D. A., Root, P., Spence, D. An altered oxidant defense system in red blood cells affects their ability to release nitric oxide-stimulating ATP, *Molecular BioSystems*, 2006, *2*, 305–311.
5. Martin, R. S., Root, P. D., Spence, D. M. Microfluidic technologies as platforms for performing quantitative cellular analyses in an *in vitro* environment, *Analyst (Cambridge, United Kingdom)*, 2006, *131*, 1197–1206.
6. Sprague, R. S., Olearczyk, J. J., Spence, D. M., Stephenson, A. H., Sprung, R. W., Lonigro, A. J. Extracellular ATP signaling in the rabbit lung: Erythrocytes as determinants of vascular resistance, *American Journal of Physiology–Heart and Circulatory Physiology*, 2003, *285*, H693–H700.
7. Sprague, R. S., Ellsworth, M. L., Stephenson, A. H., Lonigro, A. J. ATP: The red blood cell link to NO and local control of the pulmonary circulation, *American Journal of Physiology*, 1996, *271*, H2717–H2722.
8. Sprague, R. S., Ellsworth, M. L., Stephenson, A. H., Lonigro, A. J. Increases in flow rate stimulate adenosine triphosphate release from red blood cells in isolated rabbit lungs, *Experimental Clinical Cardiology*, 1998, *3*, 73–77.
9. Oblak, T. D. A., Root, P., Spence, D. M. Fluorescence monitoring of ATP-stimulated, endothelium-derived nitric oxide production in channels of a poly(dimethylsiloxane)-based microfluidic device, *Analytical Chemistry*, 2006, *78*, 3193–3197.
10. Freedman, J. E., Loscalzo, J., Barnard, M. R., Alpert, C., Keaney, J. F., Michelson, A. D. Nitric oxide released from activated platelets inhibits platelet recruitment, *The Journal of Clinical Investigation*, 1997, *100*, 350–356.
11. Ignarro, L. J., Buga, G., Dhaudhuri, G. EDRF generation and release from perfused bovine pulmonary artery and vein, *European Journal of Pharmacology*, 1988, *149*, 79–88.

12. Moncada, S., Palmer, R., Higgs, E. A. Biosynthesis of nitric oxide from L-arginine, *Biochemical Pharmacology*, 1989, *38*, 1709–1715.
13. Palmer, R., Ferrige, M. J., Moncada, S. Nitric oxide release accounts for the biological activity of endothelium-derived relaxation factor, *Nature*, 1987, *327*, 524–526.
14. Easley, C. J., Karlinsey, J. M., Bienvenue, J. M., Legendre, L. A., Roper, M. G., Feldman, S. H., Hughes, M. A., et al. A fully integrated microfluidic genetic analysis system with sample-in-answer-out capability, *Proceedings of the National Academy of Sciences of the United States of America*, 2006, *103*, 19272–19277.
15. Yang, S., Uendar, A., Zahn, J. D. A microfluidic device for continuous, real time blood plasma separation. *Lab on a Chip*, 2006, *6*, 871–880.
16. Sprague, R. S., Ellsworth, M. L., Stephenson, A. H., Kleinhenz, M. E., Lonigro, A. J. Deformation-induced ATP release from red blood cells requires cystic fibrosis transmembrane conductance regulator activity, *American Journal of Physiology*, 1998, *275*, H1726–H1732.
17. Sprague, R. S., Stephenson, A. H., Ellsworth, M. L., Keller, C., Lonigro, A. J. Impaired release of ATP from red blood cells of humans with primary pulmonary hypertension. *Exp Biol Med*, 2001, *226*, 434–439.
18. Sprung, R. J., Sprague, R. S., Spence, D. M. Determination of ATP release from erythrocytes using microbore tubing as a model of resistance vessels *in vivo*, *Analytical Chemistry*, 2002, *74*, 2274–2278.
19. Gow, A. J., Stamler, J. S. Reactions between nitric oxide and haemoglobin under physiological conditions, *Nature*, 1998, *391*, 169–173.
20. Stamler, J. S., Jia, L., Eu, J. P., Mcmahon, T. J., Demchenko, I. T., Bonaventura, J., Gernert, K., Piantadosi, C. A. Blood flow regulation by *S*-nitrosohemoglobin: A dynamic activity of blood involved in vascular control, *Science*, 1997, *276*, 2034–2037.
21. Ellsworth, M. L., Forrester, T., Ellis, C. G., Dietrich, H. H. The erythrocyte as a regulator of vascular tone, *American Journal of Physiology*, 1995, *269*, H2155–H2161.
22. Sprague, R. S., Stephenson, A. H., Dimmitt, R. A., Weintraub, N. A., Branch, C. A., McMurdo, L., Lonigro, A. J. Effect of L-NAME on pressure-flow relationships in isolated rabbit lungs: Role of red blood cells, *American Journal of Physiology–Heart and Circulatory Physiology*, 1995, *269*, H1941–H1948.
23. Fischer, D. J., Torrence, N. J., Sprung, R. J., Spence, D. M. Determination of erythrocyte deformability and its correlation to cellular ATP release using microbore tubing with diameters that approximate resistance vessels *in vivo*. *Analyst*, 2003, *128*, 1163–1168.
24. Radomski, M. W., Radomski, A. S. In *Vascular Endothelium in Human Physiology and Pathophysiology*, Vallance, P., Webb, D., Eds., Academic Publication, Dordrecht, The Netherlands, 1999.
25. Packham, M. A. Role of platelets in thrombosis and hemostasis, *Canadian Journal of Physiology and Pharmacology*, 1994, *72*, 278–284.
26. Morrell, C. N., Matsushita, K., Chiles, K., Scharpf, R. B., Yamakuchi, M., Mason, R. J., Bergmeier, W., et al. Regulation of platelet granule exocytosis by *S*-nitrosylation, *Proceedings of the National Academy of Sciences of the United States of America*, 2005, *102*, 3782–3787.
27. Neu, I. S., Prosiegel, M., Pfaffenrath, V. Platelet aggregation and multiple sclerosis, *Acta Neurologica Scandinavica*, 1982, *66*, 497–504.
28. Giovannoni, G., Heales, S. J. R., Land, J. M., Thompson, E. J. The potential role of nitric oxide in multiple sclerosis, *Multiple Sclerosis*, 1998, *4*, 212–216.
29. Smith, K. J., Lassmann, H. The role of nitric oxide in multiple sclerosis, *Lancet Neurology*, 2002, *1*, 232–241.
30. O'Sullivan, B. P., Linden, M. D., Frelinger, A. L., III, Barnard, M. R., Spencer-Manzon, M., Morris, J. E., Salem, R. O., Laposata, M., Michelson, A. D. Platelet activation in cystic fibrosis, *Blood*, 2005, *105*, 4635–4641.
31. Coppola, L., Verrazzo, G., La Marca, C., Ziccardi, P., Grassia, A., Tirelli, A., Giugliano, D. Effect of insulin on blood rheology in non-diabetic subjects and in patients with Type 2 diabetes mellitus, *Diabetic Medicine: A Journal of the British Diabetic Association*, 1997, *14*, 959–963.
32. Sase, K., Michel, T. Expression of constitutive endothelial nitric oxide synthase in human blood platelets, *Life Sciences*, 1995, *57*, 2049–2055.

33. Muruganandam, A., Mutus, B. Isolation of nitric oxide synthase from human platelets, *Biochimica et Biophysica Acta*, 1994, *1200*, 1–6.
34. Root, P., Sliskovic, I., Mutus, B. Platelet cell-surface protein disulfide-isomerase mediated *S*-nitrosoglutathione consumption, *Biochemical Journal*, 2004, *382*, 575–580.
35. Gordge, M. P., Hothersall, J. S., Neild, G. H., Dutra, A. A. N. Role of a copper(I)-dependent enzyme in the anti-platelet action of *S*-nitrosoglutathione, *British Journal of Pharmacology*, 1996, *119*, 533–538.
36. Spence, D. M., Torrence, N. J., Kovarik, M. L., Martin, R. S. Amperometric determination of nitric oxide derived from pulmonary artery endothelial cells immobilized in a microchip channel, *Analyst*, 2004, *129*, 995–1000.
37. Li, M., Spence, D. M., Martin, R. S. A microchip-based PC 12 cell reactor: Determination of immobilization conditions and amperometric detection of catecholamines released upon stimulation with calcium, *Electroanalogy*, 2005, *17*, 1171–1180.
38. Khademhosseini, A., Langer, R., Borenstein, J., Vacanti, J. P. Microscale technologies for tissue engineering and biology, *Proceedings of the National Academy of Sciences of the United States of America*, 2006, *103*, 2480–2487.
39. Kovarik, M. L., Torrence, N. J., Spence, D. M., Martin, R. S. Fabrication of carbon microelectrodes with a micromolding technique and their use in microchip-based flow analyses, *Analyst (Cambridge, United Kingdom)*, 2004, *129*, 400–405.
40. Ku, C. J. *Analytical Chemistry*, 2007, *79*, in press.

30 Capillary Electrophoresis Applications for Food Analysis

Belinda Vallejo-Cordoba and
María Gabriela Vargas Martínez

CONTENTS

30.1 INTRODUCTION

The usefulness and importance of capillary electrophoresis (CE) in food analysis is now well recognized and established, as shown with many applications in the study of substances of food concern, ranging from naturally ocurring compounds to contaminants. This fact is corroborated by more than 650 papers published in the past 10 years,[1] including several review papers[2–4] and a handbook for method development in food analysis.[5] Also, the scope of applications is broad in terms of molecular size of food components, ranging from the analysis of small molecules such as organic acids (OA) or amino acids (AA) to the analysis of large biomolecules such as proteins, carbohydrates, or DNA.[1] An exhaustive review examining the latest developments in the application of capillary electromigration methods for the analysis of food components, as well as for the investigation of food interactions and food processing was recently published.[2] In addition, the analysis of organic contaminants in foods with emphasis on procedures for extracting and concentrating compounds such as pesticides, antibiotics, biological toxins, and pathogens was also reviewed.[3] Another important area in food analysis where CE offers good potential is food authentication. Several examples of food authentication were summarized in two review papers.[4,6] Consequently, the aim of this chapter is not a comprehensive review of the literature as this has been done in other works, but to present selected applications of CE that are most relevant to the analysis of different food components and food quality.

30.2 APPLICATIONS

30.2.1 CARBOHYDRATES

The analysis of carbohydrates by CE has been reviewed by several authors as scientific papers[1,7–9] or chapters of books.[5,10] In fact, Chapter 7 by Khandurina in this handbook deals specifically with this topic. Carbohydrates have traditionally been classified by food researchers into sugars and polysaccharides, although mixtures of them such as glucose syrups are also used, taking advantage of their respective characteristics. Sugars are utilized for their sweetening power, preservative action (osmotic pressure), and crystallinity in foodstuffs; polysaccharides provide foodstuffs with texture, body, and colloidal properties.

Currently, the assay of carbohydrates is a very important criterion for the quality control of drinks and foodstuffs, especially in dietary products. Assay of carbohydrates is also very important in the fields of monitoring of food-labeling claims, analysis of sweeteners, bulking agents, and fat substitutes as well as for establishing authenticity and in the fermentation monitoring in the production of alcoholic beverages.

A representative selection of applications is shown in Table 30.1, which illustrates the wide choice of food sample types that have been analyzed, with attention being paid to the CE mode, its running conditions, the derivatization technique and its conditions, as well as the detection method utilized.[11–26] Since carbohydrates lack both a charge and a strong ultraviolet (UV) chromophore, several derivatization techniques have been described during the development of the CE methodologies.[11,27] Wang et al.[11] reported an excellent detection limit of 420 amol for a standard mixture of monosaccharides previously derivatized with 3-aminophthalhydrazide and an online chemiluminescence detector with hexacyanoferrate as the postcapillary chemiluminescence reagent. A more conventional approach is the derivatization with a UV chromophore, although this does not give such good sensitivity as the chemiluminescence detection. Several chromophores have been utilized to derivatize carbohydrates such as 2-aminobenzoic acid,[12] 6-aminoquinoline,[13] 8-aminonaphthyl-1,3,6-trisulfonic acid,[14] 9-aminopyrene-1,4,6-trisulfonic acid (APTS),[19,23] *p*-nitroaniline,[22] and *p*-aminobenzoic acid.[26] While these methods lead to improved sensitivity and resolution, the complexity of derivatization limits its use.

Alternatively, methods for the analysis of underivatized carbohydrate have been developed. These methods include the use of high alkaline electrolyte to ionize the carbohydrates and make them suitable for indirect UV detection.[15–17,20,24,28–30] These techniques enabled to analyze acidic, neutral, and amino sugars without derivatization. Soga and Serwe[18] reported the separation of the standard mixtures of 28 carbohydrates (Figure 30.1)[18] including monosaccharides, disaccharides, acidic sugars, amino sugars, and sugar alcohols using indirect detection at high pH with 20 mM 2,6-pyridinedicarboxylic acid (PDC) for indirect UV detection and 0.5 mM cetyltrimethylammonium hydroxide (CTAH) for reversing the direction of the electroosmotic flow (EOF). The detection limits for fructose, glucose, and sucrose were in the range from 12 to 16 mg/L with pressure injection of 50 mbar for 6 s. The reader is referred to two additional reviews on carbohydrate analysis that have recently been published.[31,32]

30.2.2 AMINO ACIDS AND BIOGENIC AMINES

An overwhelming majority of foods contain AA, either in the free form (e.g., fruit juice) or in the form of protein (partially hydrolyzed or intact). Proteins are polymers of AA, and as such, represent the principal source of dietary AA for humans when enzymatically digested to liberate their constituents AA. Consequently, the determination of the AA content in food is important in a number of applications that include food as the sole source of nutrition (e.g., infant formula), prescribed fortified nutritional products (e.g., product enriched with glutamine), the verification of the absence of an specific AA in certain inborn errors of metabolism (e.g., phenylalaline in phenylketoneurea), as well as for regulatory concerns.

TABLE 30.1
Methods for the Analysis of Carbohydrate Samples by CE

Type of Food	Detection Method	Sample Derivatization	CE Mode	Run Buffer	CE Conditions	References
Standard mixture of monosaccharides	Online chemiluminescence	Precapillary derivatization with 3-aminophthal-hydrazine and postcapillary reaction with 25 mM hexacyanoferrate in 3 M NaOH	CZE	200 mM borate (pH 10), 100 mM hydrogen peroxide	107 cm × 75 μm i.d. fused-silica capillary, 24 kV, 25°C	11
Mono- and disaccharides	UV absorbance 214 nm	Mono- and disaccharides derivatized with 2-aminobenzoic acid	CZE	50 mM sodium phosphate, pH 7	70 cm × 50 μm i.d. fused-silica capillary, 20 kV, 25°C	12
Peanut fungal pathogens and Baker's yeast	UV absorbance 254 nm	*N*-Acetylglucosamine and glucose derivatized with 6-aminoquinoline prior to analysis	CZE	100 mM sodium phosphate monobasic (pH 7)	80 cm × 20 μm i.d. fused-silica capillary, 18 kV	13
Cereals/corn	UV absorbance 235 nm	Reducing sugars derivatized with 8-aminonaphthyl-1,3,6-trisulfonic acid	CZE	250 mM phosphoric acid titrated pH 4.5 with triethylamine	67 cm × 50 μm i.d. fused-silica capillary, 20 kV, 30°C	14
Monosaccharides in orange juice	Indirect UV at 570 nm	Underivatized	CZE	70 mM NaOH, 2 mM naphthol blue-black, and 80 mM sodium phosphate (pH 12.5)	23 cm × 50 μm i.d. fused-silica capillary, 9 kV, 25°C	15
Monosaccharides in soft drinks, isotonic beverages, fruit juice, and sugarcane spirits	Indirect contactless conductivity detection	Underivatized	CZE	10 mM NaOH, 4.5 mM Na_2HPO_4, 200 μm CTAB	44 cm × 20 μm i.d. × 375 μm o.d. fused-silica capillary 2 kV, 30°C	16
Carbohydrates in alcoholic drinks/wine	ESI-MS	Underivatized	CE-MS	300 mM diethylamine (DEA). Sheath liquid: 2-propanol/water (80/20% v/v) with 0.25% DEA at 4 μL/min	70 cm × 50μm i.d. fused-silica capillary, 20 kV, 25°C	17

Continued

TABLE 30.1 (Continued)

Type of Food	Detection Method	Sample Derivatization	CE Mode	Run Buffer	CE Conditions	References
Mono- and disaccharides, amino sugars, and sugar alcohols in orange juice, apricot, sake, yogurts	Indirect detection 350 nm	Underivatized	CZE	20 mM 2,6-PDC with 0.5 mM CTAH (pH 12.1)	112.5 cm × 50 μm i.d. fused-silica capillary, −25 kV, 20°C	18
Oligo- and monosaccharides in enzymatic polysaccharide digestion	Capillary array electrophoresis-LIF	Derivatized with 0.2 M 9-aminopyrene-1,4,6-trisulfonic acid trisodium (APTS) in 15% acetic acid and 2 M $NaBH_3CN$ in tetrahydrofuran	CZE	25 mM Lithium acetate (pH 5.0)	62 cm × 75 μm i.d. fused-silica capillary, 15 kV, 27 or 30°C	19
Mix of carbohydrates	Frequency based electrochemical	Underivatized	CZE	0.1 M NaOH running electrolyte	Poly(dimethylsiloxane)/glass microchip with a channel 20-μm wide and 7-μm high. 50-μm wide copper electrode vs. Ag/AgCl	20
Carbohydrates in plants (lupine sample)	UV detection 220 nm	Derivatization with 12.5 μL of 0.15 M tryptamine dissolved in 10% propanol and heated at 90°C for 10 min + 4.5 μL sodium cyanoborohydride (0.3 g/mL), 90°C for 60 min	MEKC	100 mM Sodium tetraborate decahydrate, 35 mM cholic acid, and 2% 1-propanol, pH 9.7	760 mm × 0.05 mm i.d. fused-silica capillary, 30 kV, 30°C	21
Carbohydrates in infant and milk powder, rice syrup, and cola drink	UV-visible on-column LED 406 nm	Derivatization with 50 μL of glacial acetic acid, 100 μL of 0.036 M *p*-nitroaniline in methanol and 75 μL of 0.080 M $NaBH_3CN$	CZE	0.17 M boric acid (pH 9.7)	72 cm × 50 μm i.d. fused-silica capillary, −20 kV, 27 or 30°C	22

Polysaccharides, *iota*, *kappa*, and *lambda* carrageenans	LIF	Derivatization with 2 μL of 75% v/v acetic acid, 2 μL 9-aminopyrene-1,4,6-trisulfonic acid (APTS) reagent (0.2 M in 15% v/v acetic acid) and 2 μL of 1 M sodium cyanoborohydride in THF	CZE	25 mM ammonium acetate (pH 8.0)	60/67 cm × 50 μm i.d. polyvinyl alcohol coated capillary, 30 kV, 50°C	23
Polysaccharides	Indirect DAD 210 nm	Underivatized	CEC	1% DMSO in 20 mM Tris, pH 8/ACN	Sulfonated polystyrene/DVB stationary phase. 300 Å, 45 cm (53 cm total) × 100 μm, 10 kV, 20°C	24
		Derivatized with phenyisocyanate	CEC	THF with 2% water	Mixed Bed X400-PLGel C (40:60), 25(33) cm, 75 μm; 25 kV; 20°C	24
Carbohydrates in sake	Indirect UV	Underivatized	CZE	20 mM PDC, 0.5 mM CTAH (pH 12.1)	112.5 cm × 50 μm i.d. fused-silica capillary, 30 kV, 15°C	25
Glucose, maltose, and maltotriose in nonalcoholic drinks/beverages	UV detection 280 nm	Precolumn derivatization 20 mg of $NaBH_3CN$ in 1 mL of methanolic solution of 250 mM *p*-aminobenzoic acid (PABA) and 20% AcOH	CZE	20 mM sodium tetraborate (pH 10.2)	57 cm × 75 μm i.d. fused-silica capillary, 20 kV, 25°C	26

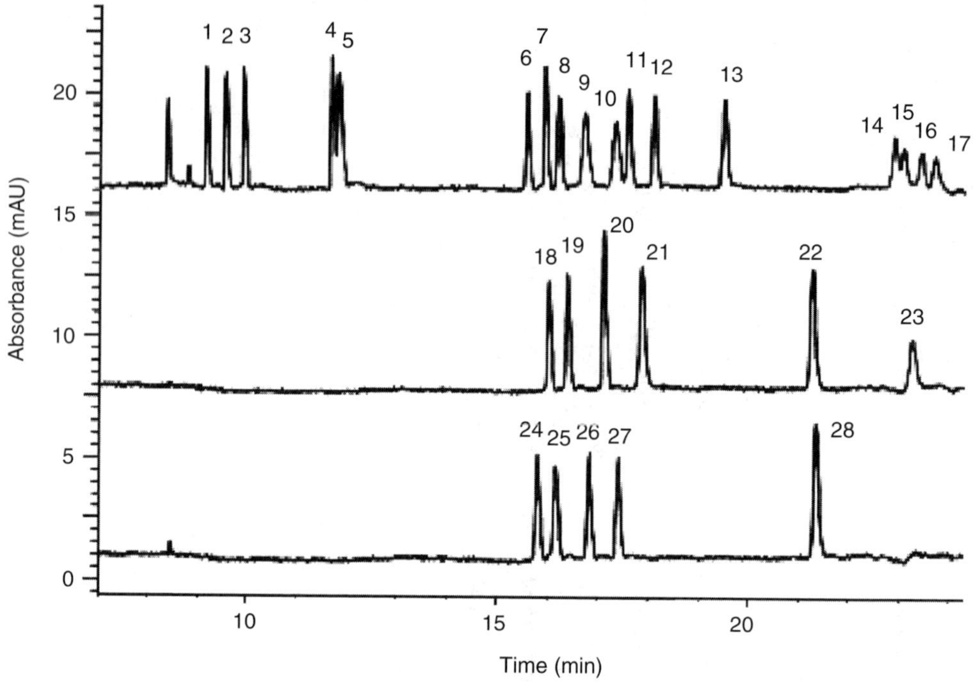

FIGURE 30.1 Separation of 28 standard carbohydrates by CE. Peak assignment: (1) mannuronic acid, (2) glucuronic acid, (3) galacturonic acid, (4) *N*-glycolylneuraminic acid (NGNA), (5) *N*-acetylneuraminic acid (NANA), (6) ribose, (7) mannose, (8) xylose, (9) glucosamine, (10) glucose, (11) galactosamine, (12) galactose, (13) fucose, (14) mannitol, (15) sorbitol, (16) xylitol, (17) inositol, (18) fructose, (19) rhamnose, (20) lactulose, (21) lactose, (22) sucrose, (23) galactitol, (24) *N*-acetylmannosamine, (25) *N*-acetylglucosamine, (26) *N*-acetylgalactosamine, (27) arabinose, and (28) raffnose, 200 mg/mL each. Experimental conditions: capillary, fused-silica 50 μm ID × 112.5 cm (104 cm effective length); background electrolyte, 20 mM PDC, pH 12.1; injection, 6 s at 50 mbar; temperature, 20°C; indirect detection signal 350 nm, reference 275 nm. (Reprinted from Soga, T., and Serwe, M., *Food Chem.*, 69, 339–344, 2000. With permission from Elsevier Science-NL.)

Biogenic amines (BA) are bacterial degradation products of AA found in virtually all foods. They are also normal constituents in fermented foods, such as cheese, wine, and beer. Since excessive dietary intake of BA can lead to adverse physiological effects, such as migraine headaches (due to their toxicity), and because their presence is used as indicators of the degree of freshness or spoilage of food, it is important to analyze BA content for food quality. Histamine, putrescine, cadaverine, tyramine, tryptamine, β-phenylethylamine, spermine, and spermidine are considered to be the most important BA occurring in foods. The reader is referred to several reviews[1,7–10] that summarize the CE methodologies that have been developed for the determination of BA[33–39] and AA,[40–42] as well as some book chapters on the same topic.[5,10,32] The most significant reports and its separation conditions are given in Table 30.2.

BA is challenging in that it displays poor chromophore and fluorophore properties. A similar situation exists with AA because very few of them have a chromophoric moiety that allows for facile detection after separation; consequently, derivatization procedures to facilitate their detection are usually required. Fortunately, the presence of common functional groups (amine and carboxylic acid) chemical handles for derivatization schemes that render the BA and AA amenable to spectrophotometric and fluorimetric detection after separation.

TABLE.30.2
Methods for the Detection of AA and BA Separated by CE

Type of Food	Detection Method	Sample Derivatization	CE Mode	Run Buffer	CE Conditions	References
AA growth medium	LIF	Derivatized with 3-(4-carboxybenzoyl)-2-quinoline-carboxaldehyde before analysis	MEKC	6.25 mM borate (pH 9.66), 150 mM SDS, 10 mM tetra-hydrofuran	107 cm × 75 μm i.d. fused-silica capillary, 24 kV, 25°C	43
Standards of AA	CCD spectrometer	Derivatized with FITC and 5-iodoacetamido fluorescein (5-IAF)	CZE	20 mM borate buffer at pH 10	Fused-silica capillary, 50 mm i.d., 360 μm o.d., length, 50 cm inner surface modified with γ-glycidoxypropyltrimethoxysilane (GOPTMS), aplied voltage 17 kV	44
AA in alcoholic drinks/beer	Electrochemical	Derivatized with naphthalene-2,3-dicarboxaldehyde before analysis	CZE	Borate (pH 9.48)	80 cm × 20 μm i.d. fused-silica capillary, 18 kV	45
AA in beer and yeast	Contactless conductivity	Underivatized	CZE	2.3 M acetic acid, pH 2.1, 0.1% w/w hydroxyethylcellulose	80 cm × 50 μm i.d. fused-silica capillary, 30 kV, 25°C	46
Complex mixture of AA and blood	ESI-MS	Underivatized	CE-MS	Background electrolyte of 1 M formic acid	130 cm × 20 μm i.d. fused-silica capillary, 30 kV	47
AA standard mixture	ESI-MS	Underivatized	CE-MS	1 M Formic acid solution	100 cm × 50 μm i.d. fused-silica capillary, 30 kV, 20°C	48
Standard solution of AA and catecholamine	LIF	Derivatized with 5 mM NDA and 43 mM KCN	CZE	125 mM Borate bufffer (pH 8.7)	65 cm × 50 μm i.d. fused-silica capillary, 10 kV/7 min, 20 kV/8 min	49
AA in soy sauce	UV-VIS at 214 nm	Underivatized	CZE	50 mM Phosphate buffer (pH 2.5)	Uncoated fused silica column, i.d. 50 μm, 150 cm (effective length 37.8 cm), 15 kV, 25°C	50

Continued

TABLE.30.2 (Continued)

Type of Food	Detection Method	Sample Derivatization	CE Mode	Run Buffer	CE Conditions	References
BA (histamine, tryptamine, phenylethylamine, tyramine, agmatine, ethanolamine, serotonin, cadaverine, and putrescine) in red wines	DAD	Derivatization with 1,2-naphthoquinone-4-sulfonate	CZE	40 mM aqueous sodium tetraborate solution, pH 10.5, 2-propanol (25%, v/v)	67 cm × 75 μm i.d. × 375 μm o.d. fused-silica capillary, 30 kV, 25°C	51
BA in red and white wines	ESI-MS	Underivatized	CE-MS	25 mM citric acid (pH 2.0)	75 cm × 50 μm i.d. × 360 μm o.d. fused-silica capillary,13 kV, 20°C	52
BA in fish and wine	LIF	Derivatized with OPA	CZE	50 mM borate buffer (pH 9.0) 20 mM MβCD, 25 mM SβCD 10% ethanol	47 cm × 50 μm i.d. × 360 μm o.d. fused-silica capillary, 25 kV, 25°C	53
Standards mix of AA	DAD at 205 nm	Underivatized	MEKC	20 mmol/L sodium borate buffer (pH 9.3) 140 mmol/L SDS	60 cm × 75 μm i.d. × 365 μm o.d. etched bare fused-silica capillary, 20 kV, 25°C	54
Standards of AA	UV detection at 200 nm	Underivatized	CZE	100 mM sodium phosphate (pH 2.0)	57 cm × 50 μm i.d. fused-silica capillary, 20 kV, 20°C	55
Standards of AA	Fluorescence	Derivatized with fluorescamine 2.5 mM	CZE	50 mM borate buffer (pH 9.5)	70 cm × 50 μm i.d. × 375 μm o.d. fused-silica capillary, 20 kV	56
AA in infant food	ESI-MS	Underivatized	CE-MS	300 mM formic acid or 100 mM triethylamine	70 cm × 50 μm i.d. fused-silica capillary, 20 kV	57
BA and AA in beer	LIF	Derivatized with 10 μL of 1×10^{-3} mol/L SAMF, 75 μm H_3BO_3 buffer, pH 8.0	MEKC	25 mM boric acid running buffer (pH 9.0), containing 24 mM SDS, 12.5% v/v AcN	57 cm × 75 μm i.d. uncoated fused-silica capillary, 22.5 kV, 25°C	58
Standards of BA	Indirect fluorescence	Underivatized	Microchip-CE CZE	30 mM phosphate (pH 9.4) 3% 2-propanol, 500 μM Rhodamine 110, pH 2.2	Separation channel, 40 μm × 10 μm × 16 mm, 2000 V	59
AA in beer	Conductivity	Underivatized	CZE	50 mM AMP/10 mM CAPS, pH 10.8	60 cm × 25 μm i.d. × 360 μm o.d. fused-silica capillary, 25 kV	60

BA (histamine, putrescine, cadaverine, tyramine, spermidine, spermine, agmatine) in fish, cheese, meat products, vegetarian products	DAD at 210, 214, and 320 nm	Underivatized	CZE	20 mM sodium citrate buffer, 25 mM phosphate buffer, both BGE pH 2.5 and 6.5	56 cm × 50 μm i.d. fused-silica capillary, 10/25 kV, 25/35°C	61
AA (cysteine, cysteinyl-glycine, γ-glutamyl-cysteine, glutathione, and (γ-glutamyl-cysteinyl) 2-glycine (PC_2) in Durum wheat seeds	LIF	Derivatized with 400 μM 5-bromomethylfluorescein (5-BMF)	CZE	50 mM boric acid, pH 9.7–10.6 400 μM 5-BMF, 1 mM EDTA 1 μM GSH	57 cm × 75 μm i.d. fused-silica capillary, 30 kV, 30°C	62
Standards of aromatic	Electrochemical	Underivatized	Microchip-CZE	30 mM acetate buffer solution (pH 4.5)	Chip 86 × 20 × 2 mm × 75-mm long separation channel, 5-mm long injection channel, 1500 V	63
Stock solutions of AA	UV detection at 200 nm	Derivatized with phthalic anhydride 10 mM on-column	CZE	20 mM phosphate buffer, pH 10	40.5 cm × 50 μm i.d., fused-silica capillary, 24 kV, 25°C	64
BA (trimethylamine, putrescine, cadaverine, spermine, tryptamine, spermidine, phenylethylamine, tyramine) in fish, meat, and sausage	UV detection at 214 nm	Derivatized with 800 μL of a 1% (w/v) Dns-Cl in acetone solution	CZE	6 mM copper sulfate, 6 mM 18-crown-6-ether, and 4 mM formic acid (pH 2.7)	70 cm × 75 μm i.d. × 375 μm o.d. capillary tubing, 20 kV, 30°C	65
Mixture of BA standards (histamine, tyramine, cadaverine, and putrescine)	Spectrophotometric at 420 nm	5 mM 4-(4-sulfo-phenylazo)-1-hydroxy-2-naphthaldehyde sodium salt (AZO_2)	MEKC	10 mM phosphate buffer, pH 7.8, 70% v/v ethanol, 0.3 mM Brij 35, 10 mM Na_2SO_4	48.5 cm × 50 μm i.d. fused-silica capillary, 20 kV, 35°C	66
AA in plant seeds	PDA	Underivatized	CZE MEKC	75 mM phosphoric acid (pH 1.85) 50 mM NaH_2PO_4 100 mM SDS (pH 7.0)	570 mm × 50 μm i.d., uncoated fused-silica capillary, 25 kV, 15°C (CZE); 25 kV, 20°C (MEKC)	67

Continued

TABLE.30.2 (Continued)

Type of Food	Detection Method	Sample Derivatization	CE Mode	Run Buffer	CE Conditions	References
Standard mixtures of nine AA	UV-visible detection	In-column derivatization with OPA (50 mM) in borate buffer (pH 9.5, 50 mM)	CEC	100 mM phosphate buffer (pH 4.0)	Proline-coated column 75 cm × 75 μm i.d., –15 kV	68
BA in a sample of Thai fish sauce	Fluorescence emission, 488 nm excitation	On-chip derivatization with 25 μM dichlorotriazine fluorescein (DTAF)	Microchip CE	80 mM borate buffer (pH 9.2.)	Window of 40 μm × 40 μm	69
Mixture of AA (L-threonine and L-tyrosine) standards	Chemiluminescence detection (CL) with a on column coaxial flow detection interface	Underivatized	pCEC	MeOH/phosphate 20/80 (pH 8.0), 5 mM Cu(II) + 1×10^{-5} mol/L luminol pump flow rate, 0.01 mL/min; CL reagents, H_2O_2 + NaOH media; CL reagent flow rates, 1.3 μL/min	*In situ* polymerization monolithic stationary phase (30 cm); capillary, 75 μm i.d. × 375 μm o.d., −4 kV; back pressure, 500 psi	70
BA (histamine, tyramine, putrescine, and tryptamine) in red wine	UV spectrophotometer	Derivatized with 10 mM OPA and 10 mM NAC and DNS-OH	MEKC	25 mM sodium borate buffer (pH 10.0); 25 mM sodium SDS; 5% v/v methanol	Channel widths/lengths of reaction chambers 120 μm × 2.3 mm (PRSM 1), 200 μm × 1.8 mm (PRSM 2), and 50 μm × 5.6 mm (PRSM 3). Glass microchips of soda lime silicate, 75 mm × 25 mm × 1 mm	71
AA in sake	Indirect UV	Underivatized	CZE	20 mM 2,6-pyridinedicarboxylic acid (PDC), 0.5 mM CTAH (pH 12.1)	112.5 cm × 50 μm i.d. fused-silica capillary, –30 kV, 15°C	72
AA in beer	UV at 210 and 270 nm	Underivatized	MEKC	25 mM $Na_2B_4O_7$ (pH 10.5) and 110 mM SDS	57 cm × 75 μm i.d. fused-silica capillary	73
Histamine in tuna fish samples	DAD	Underivatized	CZE	50 mmol/L phosphate (pH 2.5) 0.05% hydroxyethylcellulose	50 cm × 75 μm i.d. fused-silica capillary, 23°C	76

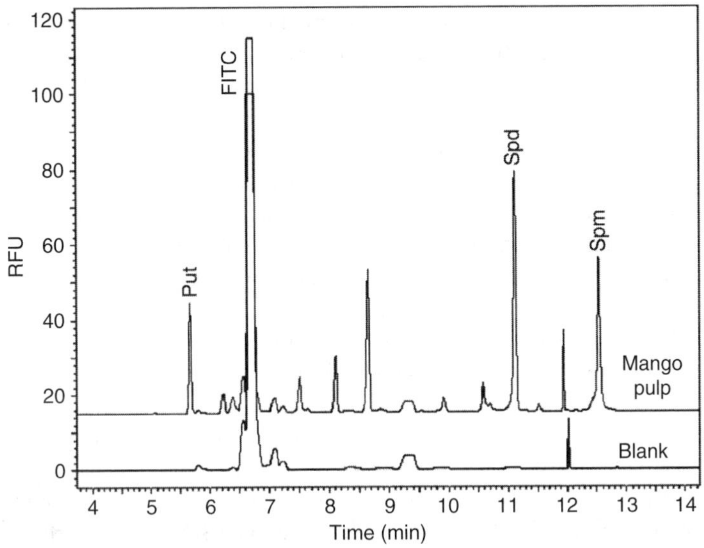

FIGURE 30.2 Electropherograms of the polyamines standards spiked in Manila Mangos pulp and its blank. Conditions: Pre-column derivatization of the sample with FITC, 20 mM borate buffer, pH 9.5, 50 mM SDS, 30 kV, 25°C. Put = putrescine, Spd = spermidine and Spm = spermine.

Many reported methods for BA and AA involve pre- or postcolumn (or capillary) derivatization of these compounds before detection using 3-(4-carboxybenzoyl)-2-quinoline-carboxaldehyde,[43] 5-(4,6-dichloro-s-triazin-2-ylamino) fluorescein,[44] 1,2-naphthoquinone-4-sulfonate,[45] *o*-Phthalaldehyde (OPA),[53,68,71] fluorescamine,[56] and many other procedures reported in Table 30.2. OPA is disadvantageous in that it reacts only with primary amines, and the fluorescent derivatives are associated with significant instability. Dabsyl- and dansylchloride[65] are better in this respect as they react with both primary and secondary amino groups, and provide stable derivatives. Indirect UV,[72] indirect fluorescence detection,[59] conductivity,[46,60] and electrochemical detection [45,63] have been utilized after CE separation as well as mass spectrometry (MS).[47,48,52,57] Kvasnicka et al.[74] developed a direct, sensitive, and rapid ($<$15 min) CE method with conductometric detection for the determination of BAs in food products (salami, cheese, wine, and beer). Vargas et al.[75] reported the use of fluorescein isothiocyanate isomer 1 (FITC) for the derivatization of polyamines present in manila mangoes. The separation by CE with detection by laser-induced fluorescence (LIF) was critical to studying the levels of polyamines in Mexican manila mango and its relationship to storage at chilling temperature (Figure 30.2).

Nineteen AA were detected in beer samples derivatized with napthalene 23-dicarboxaldehyde (NDA) and cyanide (CN^-) by capillary zone electrophoresis (CZE) with electrochemical detection.[45] Under the optimum conditions, the limits of detection (LODs) for individual AA were between 84 and 893 amol. An interesting method recently developed by Ruiz-Jiménez and Luque de Castro[65] reported the determination of BA in nine solid food samples using a full-automated method based on pervaporation coupled online with CE and indirect UV detection. The pervaporator allowed leaching, formation of the volatile analytes, and their removal by evaporation and diffusion through a membrane. The isolated analytes were injected online into the CE system while the solid matrix remained in the pervaporator. With this approach, BA have been determined in fish, meat, and sausage, with LODs ranging between 0.2 and 0.6 μg/mL.

30.2.3 Vitamins

The fat-soluble vitamins include vitamins A, D, E, K, and the carotenoids, which are precursors of vitamin A. The water-soluble vitamins are vitamins C, B_1, B_2, B_3, B_6, folacine, B_{12}, biotin, and pantothenic acid. Vitamin assay in foods are carried out for a variety of purposes: (1) to assure compliance with contract specifications and nutrient labeling regulation; (2) to provide quality assurance for supplement products; (3) to provide data for food composition tables; (4) to study changes in vitamin content attributable to food processing, packaging, and storage; and (5) to assess the effects of geographical, environmental, and seasonal conditions. Several scientific reviews[77–78] and book chapters[6,10,32] have been published for the analysis of vitamins by CE, and some separation details and the associated references can be seen in Table 30.3. CE methods for the determination of vitamins in food are limited to the determination of vitamin C in fruit,[79,80] vegetables,[80,81,88] and beverages,[82–89] niacin in the range of foods,[90–93] and thiamine in the samples of meat[94] and milk.[95]

As detailed in Chapter 3 by Terabe, micellar electrokinetic chromatography (MEKC) is a useful technique in the retention analysis of water-soluble compounds. The separation and analysis of lypophilic analytes, however, may be difficult in MEKC due to the strong affinity of lypophilic compounds to the micelle resulting in long separation times and poor resolution. An interesting approach for the simultaneous analysis of water- and fat-soluble vitamins by microemulsion electrokinetic chromatography (MEEKC) was proposed by Sanchez.[96] The separation of both water- and fat-soluble vitamins (B_1, B_2, B_3, B_6, B_{12}, C, A palmitate, D, E acetate, and K) was obtained when the microemulsion was prepared with sodium dodecyl sulfate (SDS) as the surfactant, octane as the nonpolar modifier, butanol as the cosurfactant, and propanol as the second cosurfactant. Complete separation of all vitamins was carried out within 55 min; however, this approach was tested only in multivitamin formulation.

The electrophoretic profile of flavins in commercially available baker's yeast was reported by Cataldi et al.[99] An assay for riboflavin in common natural products was developed using LIF detection; benefiting from its intrinsic fluorescent nature, flavins could be selectivity detected at very low concentrations using a middle basic running electrolyte (i.e., phosphate buffer at pH 9.8). In addition, a high sensitivity methodology was obtained with a minimal sample preparation by Aturki et al., who reported a rapid method for the separation and detection of tocopherols (TOHs) in vegetable oils by capillary electrochromatography (CEC) (Figure 30.3).[101] The method proposed was faster than the conventional chromatographic separation, with complete resolution of four TOHs achieved in less than 2.5 min with LODs of 1.25 mg/mL.

30.2.4 Organic Acids and Inorganic Ions

Organic acids are among the most frequently assayed substances in food, occurring naturally in a variety of plant and animal substrates. OA present in food originate from biochemical processes, either from their addition or from the activity of some microorganisms (particularly yeasts and bacteria), and are important contributors to the sensory properties of foods. According to the regulations set forth by the Food and Drug Administration (FDA), OA can be used as acidulants (e.g., citric, fumaric, malic, and sorbic acid), antimicrobial additives (e.g., propionic acid), and sequestrants (e.g., tartaric acid).[105] On the other hand, inorganic ions (IA) in food are important from a health-related viewpoint. Cations such as Na^+ and K^+ are essential for the maintenance of a proper electrolyte balance, while excessive Na^+ levels is associated with high blood pressure. The divalent cations, Mg^{2+} and Ca^{2+}, are important for bone growth and are regulated in infant formulas. Other IA are used in food as additives, including nitrite as a preservative and color enhancer in meat, bromate to improve the strength of flour and reduce fermentation time, and sulfite as antimicrobial growth factor in food.

A scan of recent literature on the topic clearly reveals that CE is increasingly used for the analysis of OA and IA in food. These reports are summarized in Table 30.4, and excellent reviews about this topic can be found in the literature.[106,107] The detectors most frequently used for the OA analysis by CE are direct UV at 206–220 nm, and conductivity; while for IA, indirect UV and conductivity detection are the most often reported. Cortacero-Ramírez et al.[109] applied direct UV detection to analyze 19 OA in beer samples. The inner surface of the capillary was dynamically coated to facilitate a fast anodic EOF with the addition of a polycation (hexadimetrine bromide,

TABLE.30.3
Methods for the Detection of Vitamins Separated by CE

Type of Food	Detection Method	Sample Derivatization	CE Mode	Run Buffer	CE Conditions	References
Ascorbic acid in vegetables/blue berries	PDAD 254 nm	Underivatized	MEKC	0.05 M sodium deoxycholate, 0.01 M $Na_2B_4O_7$, 0.01 M KH_2PO_4, pH 8.6	56 cm × 50 μm i.d. light path fused-silica column, 25 kV	80
Niacin in cereals, meat, fruit, vegetables, and selected foods	PDAD 254 nm	Underivatized	CZE	15% CH_3CN, 85% 0.01 M KH_2PO_4, 0.01 M Na_2HPO_4 (pH 7)	56 cm × 50 μm i.d. light path fused-silica column, 20 kV	91
Ascorbic acid in spinach, watermelon, potato, and tomato	DAD spectrophotometric	Underivatized	CZE	60 mM sodium dihydrogen phosphate, 60 mM NaCl, 0.0001% HDM (pH 7)	33.5 cm × 50 μm i.d. fused-silica capillary, −15 kV, 23°C	97
L-Ascorbic acid and D-isoascorbic acid in lemon, pineapple, sunkist, spinach	DAD 265 nm	Underivatized	CZE	0.2 M borate buffer (pH 9.0)	57 cm × 75 μm i.d. fused-silica capillary, 30 kV, 25°C	98
Alcoholic drinks/beer	UV absorbance, 210 and 270 nm	Underivatized	MEKC	25 mM $Na_2B_4O_7$ (pH 10.5) and 110 mM SDS	57 cm × 75 μm i.d. fused-silica capillary	73
Riboflavins vitamers in vegetables, wheat flours, and tomatoes and baker's yeasts	LIF at 442 nm	Underivatized	CZE	30 mM phosphate buffer, pH 9.8	84 cm × 75 μm i.d. fused-silica capillary, 30 kV, 15°C	99
Alcoholic drinks/wine	LIF at 442 nm	Underivatized	CZE	30 mM phosphate buffer, pH 9.8	84 cm × 75 μm i.d. fused-silica capillary, 30 kV, 15°C	100

Continued

TABLE.30.3 (Continued)

Type of Food	Detection Method	Sample Derivatization	CE Mode	Run Buffer	CE Conditions	References
Vegetables oils/virgin olive, hazelnut, sunflower, and soybean	DAD at 205 nm	Underivatized	CEC	Methanol and acetonitrile (50/50 v/v) containing 0.01% ammonium acetate	33 cm × 75 μm i.d. × 375 μm o.d. fused-silica capillary, partially packed with ChromSpher C_{18} (3 μm), −25 kV, 20°C	101
Ascorbic acid in soft drinks	Conductivity detection	Underivatized	CE	10 mM histidine/0.135 mM tartaric acid, 0.1 mM CTAB, pH 6.5, 0.025% HP-β-CD	60 cm × 50 μm i.d. × 360 μm o.d. fused-silica capillary, −15 kV	102
			Microchips-CE	The same condition except, 0.06% HP-β-CD, 0.125 mM CTAB	PMMA microchips 90 × 6 mm, with a pair of antiparallel orientated electrodes (1 mm × 1.4 cm), and with a gap of 0.5 mm	
Riboflavin, flavin mononucleotide in wines, milk, yoghurt, and raw eggs	LIF	Underivatized	CZE	30 mM phosphate buffer, pH 9.8	92 cm × 75 μm i.d. fused-silica capillary, 30 kV, 15°C	103
Vitamin C in *Lupinus albus* L. var. *Multolupa*	UV detection	Underivatized	MEKC	1.08 g of deoxycholate in 50 mL of a 1:1 mixture of 0.02 M borate/phosphate, pH 8.6	47 cm × 75 μm i.d. fused-silica capillary, 18 kV, 28°C	104

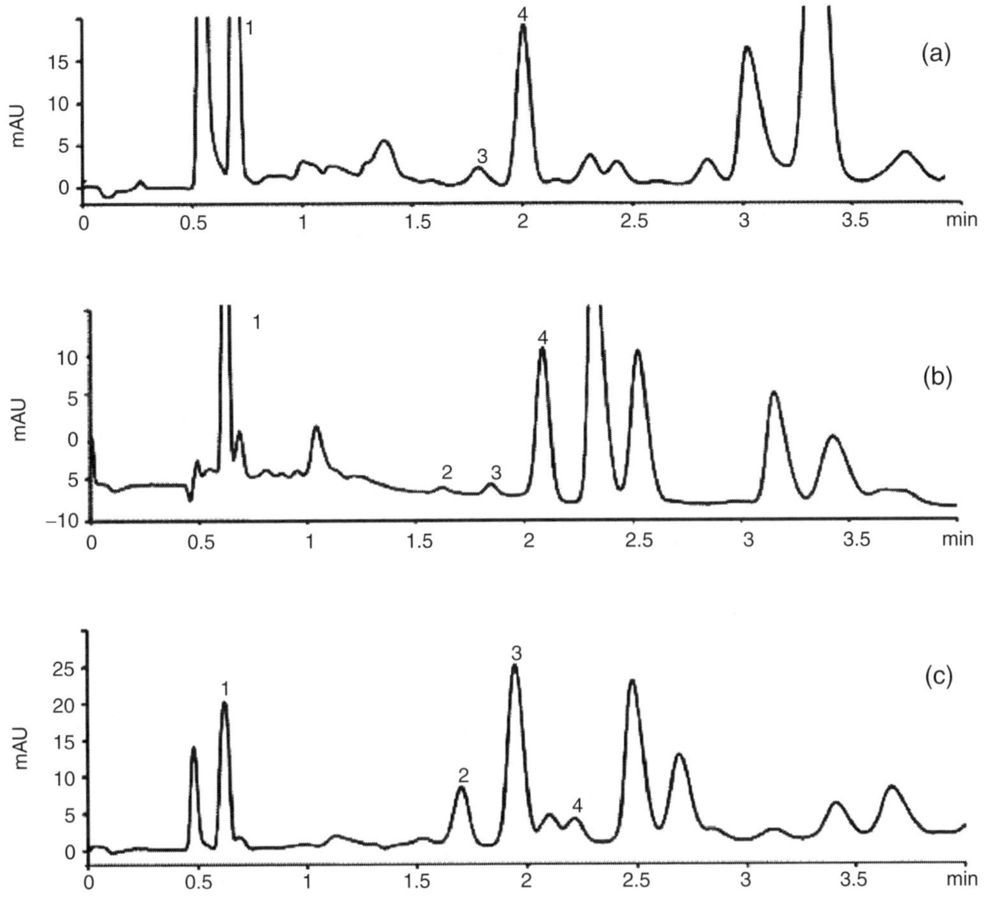

FIGURE 30.3 CEC analysis of the TOHs content in different vegetable oils. (a) Virgin olive oil: (b) Sunflower oil: (c) Soybean oil. 1, BHT; 2, δ-TOH; 3, γ-TOH; 4, α-TOH.: 1, BHT; 2, d-TOH; 3, g-TOH; 4, a-TOH. Oil samples were diluted 5 times (10 times for sunflower oil) with the mobile phase and injected in the CEC system Butylated hydroxytoluene (BHT) was added as antioxidant to prevent the loss of TOHs. (Reprinted from Aturki, Z., et al., *Electrophoresis*, 26, 798–803, 2005. With permission from WILEY-VCH Verlag GmbH & Co.)

HDB) to the electrolyte. This resulting method was very useful because of the short time (22 min) for the analysis of 19 compounds, which makes this method suitable for screening in the brewing industry.

Recently, Suárez-Luque et al.[111] developed a method for the analysis of chloride, nitrate, sulfate, phosphate, and formic acid in honey samples by CE with indirect UV detection. The separation was achieved using 2 mM potassium dichromate and 0.05 mM tetraethylenepentamine (TEPA), pH 4.00. The detection limit was in the range between 0.03 and 20 mg/kg of the IA.

30.2.5 Toxins, Contaminants, Pesticides, and Residues

Safety of food is a basic requirement of food quality. "Food safety" implies absence or acceptable and safe levels of contaminants, adulterants, naturally occurring toxins, or any other substance that may make food injurious to health on an acute or chronic basis. Most countries, therefore, have established official tolerance levels for chemical additives, residues, toxins, and contaminants in food products. Table 30.5 summarizes several CE reports that can be found in the literature that deal with the analysis of toxins, contaminants, pesticides, and residues. A number of reviews appearing

TABLE.30.4
Methods for the Analysis of OA and IA by CE

Application	Detection Method	Sample Derivatization	CE Mode	Run Buffer	CE Conditions	References
Tartrate in spiked sample matrices; cocoa, jam, lemonade, sugar syrup, Madeira cake, and digestive biscuit	Direct UV at 214 nm	Underivatized	CZE	200 mM phosphate buffer (pH 7.5)	Neutral polyacrylamide coated capillary, 50-μm i.d., 66 cm total length (56-cm effective length) and, −14 kV, 22°C	108
OA in malt beer	Direct UV at 210 nm	Underivatized	CZE	50 mM sodium phosphate (pH 8), 0.001% HDB, and 25% 2-propanol	57 cm × 75 μm i.d. fused-silica capillary, −15 kV, 23°C	109
OA in milk powder, Cheddar cheese, and plain liquid yogurt	Indirect UV at 200 nm	Underivatized	CZE	4.4 mM KHP and 0.27 mM CTAB (pH 11.2)	105 cm × 75 μm i.d. fused-silica capillary, −20 kV, 30°C	110
IA (chloride, nitrate, sulfate, phosphate) and formic acid in honey	Indirect UV at 254 nm.	Underivatized	CZE	2 mM potassium dichromate, 0.05 mM tetraethylene-pentamine (TEPA), pH 4.0	60 cm × 75 μm i.d. fused-silica capillary, −27 kV, 25°C	111
Citric, isocitric, tartaric, and malic acids in natural and commercial orange juices	Direct UV	Underivatized	CZE	200 mM phosphate buffer (pH 7.50)	57 cm × 50 μm i.d. fused-silica capillary, −14 kV, 25°C	112
OA in coffee	Direct UV at 200 nm	Underivatized	CZE	500 mM phosphate buffer (pH 6.25), 0.5 mM cetyltrimethylammonium bromide (CTAB)	57 cm × 50 μm i.d. fused-silica capillary, −10 kV, 25°C	113
OA (tartaric, malic, lactic, succinic, and acetic acids) in white and red port wines	Indirect UV at 200 nm	Underivatized	CZE	5 mM 2,6-PDC, 0.5 mM CTAB (pH 5.0)	78 cm × 75 μm i.d. fused-silica capillary, −10 kV, 18°C	114
Acetic, citric, fumaric, lactic, malic, oxalic, succinic, and tartaric acids, nitrate and sulfite ions in white and red wines	Direct UV at 200 nm	Underivatized	CZE	200 mM phosphate buffer (pH 7.50)	50 cm × 50 μm i.d. neutral coated capillary, −14 kV, 20°C	115
OA (oftartaric, malic, acetic, succinic, and lactic) in wine	Direct UV, 185 nm Indirect UV, 254 nm	Underivatized	CZE	3 mM phosphate with 0.5 mM MTAB (pH 6.5) 7 mM phthalic acid, 2 mM MTAB, 5% v/v methanol (pH 6.1)	60 cm × 75 μm i.d. fused-silica capillary, 15 kV, 25°C	116

IA and OA in sake	Indirect UV at 200 nm	Underivatized	CZE	20 mM 2,6-PDC, 0.5 mM CTAH, pH 12.1	112.5 cm × 50 μm i.d. fused-silica capillary, −30 kV, 15°C	72
Gibberellic acid in fermentation broth and commercial products	UV detection at 214 nm	Underivatized	MEKC	25 mM disodium tetraborate, pH 9.2, 100 mM SDS	57 cm × 50 μm i.d. uncoated fused-silica capillary, 30 kV, 25°C	117
IA (NH_4^+, K^+, Ca^{2+}, Na^+, Mg^{2+}, Cl^+, NO_3^+, SO_4^{2+}) in alcoholic and nonalcoholic beverages	Conductivity	Underivatized	μ-chip CE	10.5 mM His, 50 mM acetic acid, 2 mM 18-crown-6, pH 4.10 20 mM MES/His, pH 6.0 10 mM His, 7 mM glutamic acid, pH 5.75	Microchip 85 mm total length, 4 kV	118
Inorganic and organic acid anions in beverage drinks	Indirect UV at 240 nm	Underivatized	CZE	3 mmol/L 1,3,5-benzenetricarboxylic acid (BTA), 15 mmol/L tris(hydroxymethyl)aminomethane 1.5 mmol/L tetraethylenepentamine (TEPA), pH 8.4	65 cm × 50 μm i.d. fused-silica capillary, −25 kV	119
Citric and lactic acid in food and beverages	Conductivity	Underivatized	μ-chip-CE	10 mM MES/His, 2 mM 18-crown-6 (pH 6) 10 mM 2-(cyclohexylamino) ethanesulfonic acid (CHES), 6 mM arginine electrolyte, pH 9 10 mM 3-(cyclohexylamino)-1-propanesulfonic acid (CAPS), 10 mM arginine buffer, pH 10	Microchip 85 mm length, 3 kV	120
Malic acid in apple juice	PDA	Underivatized	CZE	1 mM CuSO4 +1 mM L-tartaric acid, pH 5.1	56 cm × 50 μm i.d. polyvinyl alcohol-coated capillary, −20 kV, 30°C	121

TABLE.30.5
Methods for the Detection of Toxins, Contaminants, Pesticides, and Residues Separated by CE

Type of Food	Detection Method	Sample Derivatization	CE Mode	Run Buffer	CE Conditions	References
Acidic pesticides (o-phenylphenol, ioxynil, haloxyfop, acifluorfen, picloram) in apple, grapes, oranges, tomatoes	UV ES-SM	Underivatized	CE-UV CE-MS	4 mM ammonium formate-formic acid (pH 3.1), 32 mM ammonium formate/acid formic buffer (pH 3.1)	60 cm × 75 μm i.d. fused-silica capillary, 25 kV, 25/15°C	127
AA herbicides (glufosinate and aminomethylphosphonic acid, the major metabolite of glyphosate) in agricultural soil samples	LIF	Derivatized with sulfoindo-cyanine succinimidyl ester (Cy5) and 1-ethyl-1-[5-(*N*-succinimidyl-oxycarbonyl) pentyl]-tetramethyl-indodi-carbocyanine (NIR-641)	CZE	(Cy5) 50 mM borate, 15% ACN (pH 8.0) (NIR-641) 150 mM borate, 15% ACN (pH 9.5)	57 cm × 50 μm i.d. fused-silica capillary, 25 kV, 25°C	128
Aniline metabolites (3-chloroaniline,3-chloro-4-hydroxyaniline and 3-chloro-4-methoxyaniline) of chlorpropham in potato	LIF	Derivatized with 150 μL of 5 mM 5-(4,6-dichloro-*s*-triazin-2-ylamino) fluorescein (DTAF)	MEKC	20 mM boric acid, 30 mM SDS, and 10 mM Triton X-100 (pH 9.5)	57 cm × 50 μm i.d. × 375 μm o.d. fused-silica capillary, 25 kV, 25°C	129
Aflatoxins B_1, B_2, and G_1 and the cholera toxin stock solution	MPE-Fluorescence	Underivatized	CZE	20 mM Tris, 10 mM carboxymethyl-β-cyclodextrin buffer (pH 7.5)	20.4 cm × 2.1 μm i.d. capillary	130
Insecticide (pirimicarb) and fungicide (azoxystrobin) in soil and tomato	DAD	Underivatized	CEC	Acetonitrile/5 mM aqueous Tris 40/60 v/v (pH 8.6)	33 cm × 100 μm ID × 375 μm OD polyimide coated fused-silica capillary, 30 kV, 30°C	131
Substituted urea pesticides in orange and tomato	DAD	Underivatized	MEKC	4 mM borate and 35 mM SDS (pH 9.0)	60 cm × 75 μm i.d. polyimide-coated fused-silica capillary, 25 kV, 25°C	132

Pesticides (aldicarb, carbofuran, isoproturon, chlorotoluron, metolachlor, mecoprop, dichlorprop, MCPA, 2,4-D, methoxychlor, TDE, DDT, dieldrin, and DDE) in drinking water	UV detection	Underivatized	MEKC	50 mM SDS, 10 mM borate buffer, 15 mM β-CD, 22% acetonitrile, pH 9.6	52 cm × 50 μm i.d. fused-silica capillary, 25 kV	133
Bacterial endotoxins (lipopolysaccharides) in protein complexation	UV absorption	Underivatized	CZE	0.05 M Tris–HCl buffer, pH 8.5	28 cm × 50 μm i.d. fused-silica capillary, 25 kV, 10°C	134
Fungicide validamycin A in rice plants	Indirect UV detection	Underivatized	CZE	10 mM aminopyrine, 2 mM ethylenediaminete-traacetic acid (pH 5.2)	60 cm × 75 μm i.d. fused-silica capillary, 15 kV, 25°C	135
Herbicide, glyphosate (standard)	ESI-MS	Underivatized	CE-MS	Ammonium formate buffer (pH 2.5)	75 cm × 50 μm i.d. fused silica capillary column, –15 kV, 25°C	136
Pesticides (pyrimethanil, pyrifenox, cyprodinil, cyromazine, and pirimicarb) in grapes and orange juice	UV and MS	Underivatized	CE-MS	0.3 M ammonium acetate/acetic acid (pH 4.0)	57 cm × 50 μm i.d. fused-silica capillary, 25 kV, 25°C	137
Pesticides (carbendazim , simazine, atrazine, propazine and ametryn, diuron and linuron, carbaryl and propoxur), in drinking water and carrot	DAD at 200 nm	Underivatized	MEKC	20 mmol/L phosphate buffer (pH 2.5), 25 mmol/L SDS, 10% methanol	58.3 cm × 75 μm i.d. × 375 μm o.d. fused-silica capillary, –25 kV, 25°C	138
Antibiotics [Quinolone residues (danofloxacin, enro-floxacin, flumequine, ofloxacin, and pipemidic acid)] in chicken and fish	ESI-MS	Underivatized	CE-MS	60 mM $(NH_4)_2CO_3$ (pH 9.2)	75 cm × 75 μm i.d. × 375 μm o.d. fused-silica capillary, 20 kV, 20°C	139

Continued

TABLE.30.5
(Continued)

Type of Food	Detection Method	Sample Derivatization	CE Mode	Run Buffer	CE Conditions	References
Pesticides (acrinathrin, bitertanol, cyproconazole, fludioxonil, flutriafol, myclobutanil, pyriproxyfen, and tebuconazole) in lettuce, tomato, grape, and strawberry	DAD	Underivatized	MEKC	6 mM sodium tetraborate decahydrate, 75 mM of cholic acid sodium solution (pH 9.2)	57 cm × 75 μm i.d., uncoated fused-silica capillary, 30 kV, 25°C	140
Pesticides (cyprodinil, cyromazine, pyrifenox, pirimicarb, and pyrimethanil) in water, apple, and orange juice	UV	Underivatized	CZE	0.4 mM cetyltrimethylammonium chloride (CTAC), 0.4 M acetic acid (pH 4), 5% v/v 2-propanol	57 cm × 50 μm ID, bare fused-silica capillary, 22 kV, 25°C	141
Pesticide residues (dinoseb, pirimicarb, procymidone, pyrifenox, pyrimethanil, and thiabendazole) in peaches and nectarines	DAD 214 nm ESI-MS	Underivatized	CZE-UV CZE-MS	0.3 M ammonium acetate (pH 4.0), 10% methanol	57 cm × 75 μm i.d. × 375 μm o.d. fused-silica capillary, 30 kV, 25°C	142
Fungicide residues (procymidone and thiabendazole) in apples, grapes, oranges, pears, strawberries, and tomatoes	ESI-MS	Underivatized	CE-MS	12 mM ammonium formate, 20 mM formic acid (pH 3.5), 2% methanol	150 cm × 75 μm i.d. × 375 μm o.d. fused-silica capillary, 30 kV, 25°C	143
Pesticides and metabolites (naphthalene acetamide, carbaryl, 1-naphthol, thiabendazole, and carbendazime) in cucumbers	UV-visible	Underivatized	MEKC	30 mM ammonium chloride/ammonia buffer (pH 9.0), 15 mM SDS	57 cm × 75 μm i.d., 375 μm o.d. fused-silica capillary, 29 kV, 25°C	144
Amino phosphonic acid herbicides (model mixture)	LIF	In-capillary derivatization with 5-(4,6-dichloro-*s*-triazin-2-ylamino) fluorescein (DTAF)	MEKC	50 mM boric acid, 30 mM Brij-35, adjusted to pH 9.5	57 cm × 50 μm i.d. fused-silica capillary, 5 kV, 35°C	145

Penicillin V and related substances in mixture of a real fermentation broth	UV ESI-MS	Underivatized	CZE NACE-MS	20 mM ammonium acetate (unadjusted, pH 6.5), 20 mM ammonium acetate in acetonitrile/MeOH (60/40 v/v)	70 cm × 50 μm i.d. × 360 μm o.d. fused-silica capillary, −20 kV, 25°C	146
Seven paralytic shellfish toxins (PSTs), namely decarbamoylsaxitoxin (dcSTX), saxitoxin (STX), neosaxitoxin (NEO), gonyautoxin-2 (GTX-2), gonyautoxin-3 (GTX-3), gonyautoxin-1 (GTX-1), and gonyautoxin-4 (GTX-4)	UV at 200 nm	Underivatized	ITP-CE	LE: morpholine buffer (pH 5.0) TE: 10 mM formic acid, (pH 2.7) CZE: morpholine buffer (pH 5.0)	57-cm-long fused-silica capillary with polyacrylamide (effective length of 51.8 cm), 75 μm, 14 kV, 20°C	147
Antibiotics residues (amoxicillin, doxycycline hydrochloride, streptomycin sulfate, thiamphenicol, florphenicol, nifursol, enrofloxacin, ciprofloxacin, norfloxacin) from poultry and porcine tissues	UV	Underivatized	CZE	Different concentration of borate + phosphate buffer (30 mM NaH_2PO_4)	47, 57, 67, and 77 cm × 75 or 100 μm i.d. × 375 μm o.d. fused-silica capillary, 25 kV, 25°C	148
Quinolones [enrofloxacin (ENR), ciprofloxacin (CPR), danofloxacin (DAN), difloxacin (DIF), marbofloxacin (MAR), flumequine (FLU), and oxolinic acid (OXA)] in pig kidney tissue	DAD	Underivatized	NACE	20 mM ammonium acetate 0.004% polycation hexadimethrine bromide (HDB), 4% acetic acid (pH 5.4) in methanol/acetonitrile (50:50 v/v)	64.5 cm × 75 μm i.d. fused-silica capillary, 30 kV, 25°C	149

Continued

TABLE.30.5
(Continued)

Type of Food	Detection Method	Sample Derivatization	CE Mode	Run Buffer	CE Conditions	References
E. coli in meat	LIF	Underivatized	CE	4.5 mM Tris, 4.5 mM boric acid, and 0.1 mM EDTA) polyduramide (0.1% w/v)	27 cm × 75 μm i.d. fused-silica capillary, 370 V/cm	150
Six bacterial contamination in corn flakes, milk, baby food, juice, and frankfurter	UV	Underivatized	CZE	25 mM phosphate buffer (pH 7.0) + 25 μM calcium chloride + 35 μM myoinositol hexakisphosphate.	47 cm × 75 μm i.d. fused-silica capillary, 15 kV, 20°C	151
Glycoalkaloids in potato	MS-MS	Underivatized	NACE-MS	90:10 v/v of MeCN-MeOH containing 50 mM ammonium acetate and 1.2 M acetic acid	80 cm × 50 μm i.d. × 375 μm o.d. fused-silica capillary, 25.5 kV, 20°C	152
Eight colorants in (orange, apple, and grape) soft drinks (grape, pineapple, and peach), jellies, and (apple) milk beverages	PAD	Underivatized	CE	20 mM NaOH, 15 mM disodium tetraborate (borax) pH 10.0, 7 mM β-CD	50.2 cm × 50 μm i.d. fused-silica capillary, 25 kV, 25°C	153
Nine sulfonamides in meat (animal food)	UV visible detection	Underivatized	CZE	45 mM sodium phosphate, 10% methanol (pH 7.3)	64.5 cm × 75 μm i.d. fused-silica capillary, with bubble cell, 25 kV, 27°C	154

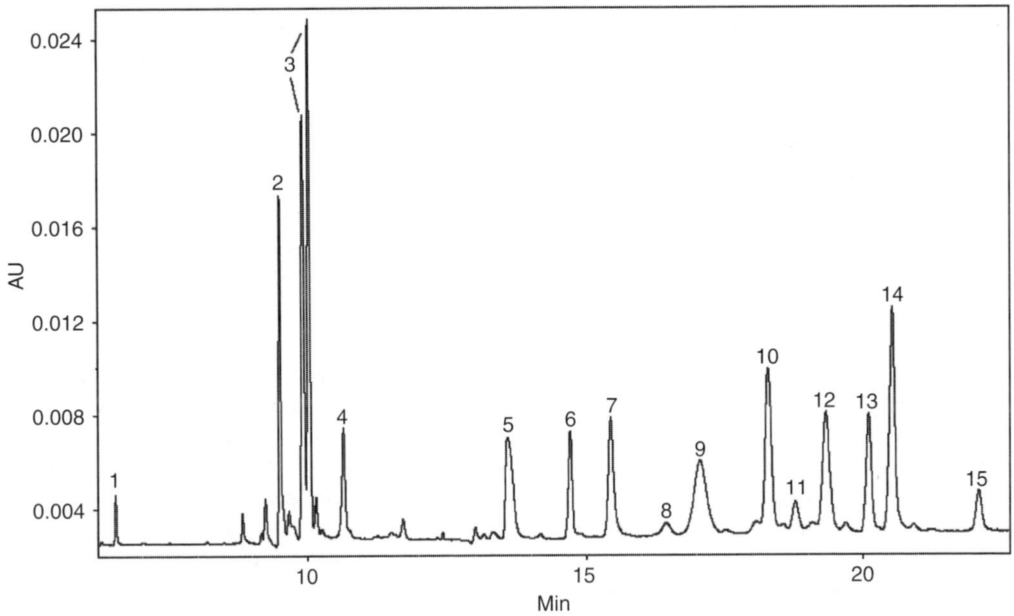

FIGURE 30.4 Electropherograms of a bovine milk sample spiked with 15 antibiotic standard mixture. Conditions: 16 ppm antibiotics, 100 mM borate buffer pH 8.0, with the addition of 5 mM α-CD and 5 mM HPα-CD, 30 kV, 25°C, detection 200 nm. (1) Procaine, (2) trimethoprim, (3) spyramicin, (4) tylosin, (5) enrofloxacin, (6) cloramphenicol, (7) tetraciclina, (8) oxytetracyclin, (9) clortetraciclin, (10) amoxicillin, (11) cefalexine, (12) ampicillin, (13) dicloxacillin, (14) oxacillin, and (15) penicillin G.

in the literature over the past half decade provide an excellent encapsulation of publications on these topics.[122–126]

Several drugs are frequently fed to domestic cattle because modern intensive animal breeding demands permanent suppression of diseases caused by viruses, bacteria, protozoa, and/or fungi. Therefore, residues of these drugs can be found in foods of animal origin such as milk, eggs, and meat. Kowalski et al.[148] reported the development and validation of CZE methodologies for eight antibiotics and one coccidiostatic (nifursol) residues in poultry and porcine tissues. They proposed a simplified clean-up procedure, including deproteinization by acetonitrile and liquid-liquid extraction with ethyl acetate, for drug substances at concentrations below 20 mg/kg in a variety of food types. Castillo and Vargas[155] reported a CE method for the simultaneous analysis of different kinds of antibiotic residues in bovine milk, which would allow for the routine monitoring of the presence of these compounds for quality control (Figure 30.4).[155]

On the other hand, contaminants are substances that have not been intentionally added to food. These substances may be present in food as a result of the various stages of their production, packaging, transport, or holding, or might result from environmental contamination. In this area, Palenzuela et al.[151] proposed an excellent CE method for the detection of bacterial contamination. The method was based on the interaction of ions with biocolloids that allows for their reliable separation of eight different types of bacteria by CE in only 25 min—a dramatic reduction in the analysis time resulted (7 h of enrichment vs. the 24–48 h typically required by culturing methods) compared with classical microbiological analyses.

In the field of pesticide residues in food, CE continues to gain ground and garner significant attention. However, one of the main drawbacks of CE is its low sensitivity in the monitoring of pesticide residue. Therefore, the use of different preconcentration strategies before the separation [e.g., solid-phase extraction (SPE), liquid-liquid extraction, etc.] or different stacking procedures in the capillary column have been applied and recently reviewed.[156,157] Da Silva et al.[138] demonstrated the usefulness of sample stacking techniques for pesticide analysis with the separation of nine pesticides in complex matrices such as fruits and vegetables. Online preconcentration of the analytes during MEKC separation was established, providing LODs of 2.5 mg/L for all analytes in carrots. Moreover, the use of SPE, in combination with an online preconcentration strategy, allowed for the determination of pesticides at the 0.1 mg/L level in drinking water. Other toxic compounds such as mycotoxins[130] and bacterial endotoxins,[134] colorants,[153] and sulfonamine residues[154] have also been analyzed and reported using CE methodologies.

30.2.6 Analysis of Phenolic Compounds in Food

Phenolic compounds (PCs) are a group of naturally occurring secondary metabolites present in plants, all of which possess one common structural feature—a phenol moiety, and behave as excellent antioxidants due to the reactivity of this group. PCs are of interest due to their potential contribution to the taste (astringency, bitterness, and sourness) and formation of off-flavor in food, including tea, coffee, and various fruits juices, during storage. Currently, PCs are divided into two types: simple phenols and polyphenols. Phenols are a widely found group of secondary plant metabolites (e.g., benzoic and cinnamic acid derivatives), and polyphenols have at least two phenol subunits—flavonoids are an example of a common polyphenol in plants. Some recent application of CE to the analysis of PCs in tea, wines, and other foods are summarized in Table 30.6. It should be noted that there is no need to derivatize samples of PCs because they are aromatic and, therefore, show intense absorption in the UV region. Several reviews have been published that can be used as a guide for the analysis of these compounds in food by CE.[158–161] In a recent review, Li et al.[160] discussed the strategies that have been used during the optimization of CZE for analysis of phytochemical bioactive compounds, and proposed the use of multivariate experimental designs to simplify this task. Blasco et al.[162] recently proposed a new and attractive route for fast, simultaneous detection of prominent natural antioxidants, including phenolics and nonphenolic antioxidants, using a CE microchip with electrochemical detection via a glassy carbon electrode. Separation of the pair of standard compounds (+)-catechin and ascorbic acid, (+)-catechin and rutin, as well as arbutin and phlorizdin was carried out in less than 400 s and later applied for the determination of these compounds in pear juice. An interesting method for the simultaneous separation and determination of selected PCs in six different collections of *Hypericum perforatum* (St. John's wort) was proposed by Hamoudová et al.[163] This method used an online combination of CZE with capillary ITP (CE-ITP) allowing detection limit as low as 60 ng/mL and total analysis time of less than 35 min.

Using a different mode of CE, MEEKC, and MEKC, Huang et al.[181] separated 13 PCs. The authors compared both methodologies and demonstrated that both MEEKC and MEKC methods possess the ability to analyze PCs in different food samples. The same authors also reported a method for the separation of the nine flavonoids most often found in grape wine, namely, apigenin, baicalein, naringenin, luteolin, hesperetin, galangin, kaempferol, quercetin, and myricetine, using a 35 mM borax, pH 8.9, 16.8 kV of applied voltage with an analysis time of within 16 min.[182] A high-pressure liquid chromatography (HPLC) method was also developed for the same group of compounds and later compared with the CZE. They concluded that both strategies were useful for the analysis of PCs. An electropherogram of the flavonoid separation can be seen in Figure 30.5.[182]

30.2.7 Chiral Analysis of Food Compounds

Most of the major nutritional organic components of food and beverage are chiral, including proteins, AA, carbohydrates, fats, and some vitamins. Many flavor and fragrance components of food and beverage are chiral as well. The significance of the enantiomeric composition of food components is that they can have a dramatic effect on aroma, taste, and nutritional value. Chiral separations for food components are useful for evaluating age, treatment, and storage effects; for control or

TABLE.30.6
Methods for the Analysis of Phenolic Compounds by CE

Application	Detection Method	CE Mode	Run Buffer	CE Conditions	References
Polyphenolic fraction of extra-virgin olive oil	UV absorption	CZE	Buffer solution 45 mM of sodium tetraborate pH 9.3	50 μm i.d., 47 cm of total length (40 cm to the detector) with a detection window of 100 × 200 μm	162
PCs in hypericum perforatum (St. John's wort)	ITP conductivity detector CZE UV at 270 nm and conductivity	CE-ITP	LE of 10 mM HCl with Tris 0.2% 2-HEC, 20% v/v methanol pH* 7.20 TE 50 mM H_3BO_3 pH* 8.2, 20% v/v methanol CZE 50 mM Tris buffer + 25 mM MOPSO + 65 mM boric pH* 8.3, 20% v/v methanol	ITP-fluorinated ethylene propylene copolymer (FEP), 9.0 cm × 60.8 mm i.d. CZE-FEP, 16 cm × 60.3 mm i.d.	163
PCs in olive oil	ESI-MS UV at 200 nm	CZE-ESI-MS	60 mM NH_4OAc at pH 9.5 with 5% of 2-propanol	Fused-silica capillary 50 μm i.d. (7 cm effective length for the UV detector and 100 cm total length for the MS detection	164
PCs in chess (*Bromus inermis* L.)	UV at 254 nm	CZE	20 mM sodium tetraborate (pH 9.2) with 5% v/v methanol	53.5 cm fused-silica capillary (45 cm effective length), 75 μm i.d., 365 μm o.d.	165
Rutin and quercetin in plants	Electrochemical	MEKC	20 mM borate (pH 8.8), 40 mM SDS, 10% acetonitrile	50 cm × 50 μm i.d. fused-silica capillary, 12 kV	166
Quercetin, rutin, kaempferol, catechin, gallic acid in plants	UV at 270 nm	CZE	100 mM borate (pH 10)	51 cm × 50 μm i.d. fused-silica capillary, 15 kV, 32°C	167
Procyanidins after thiolysis	UV absorption at 214 nm	MEKC	50 mM phosphate (pH 7), 40 mM sodium cholate, 10 mM SDS	47 cm × 50 μm i.d. fused-silica capillary, 15 kV, 25°C	168
Catechins in green tea	UV absorption at 230 nm	MEEKC	Microemulsion of 50 mM phosphate (pH 2.5), SDS, *n*-heptane, 2-hexanol, or cyclohexanol	24 cm × 50 μm i.d. fused-silica capillary, 10 kV, 40°C	169, 170

Continued

TABLE.30.6
(Continued)

Application	Detection Method	CE Mode	Run Buffer	CE Conditions	References
Catechins in green tea	UV absorption at 200 nm	MEKC	20 mM phosphate + 50 mM borate + 200 mM SDS (3:1:2), (pH 7)	47 cm × 50 μm i.d. fused-silica capillary, 30 kV, 29°C	171, 172
Theaflavin composition of black tea	UV at 380 nm	NACE	71% v/v acetonitrile, 25% v/v methanol, 0.1 M potassium hydroxide, 4% v/v glacial acetic acid, 90 mM ammonium acetate	40 cm × 50 μm i.d. fused-silica capillary, 22.5 kV, 18.5°C	173, 174
Resveratrol in wines, herbs, and health food	Electrochemical	CZE	100 mM borate (pH 9.24)	65 cm × 25 μm i.d. fused silica capillary, 30 kV	175
Resveratrol and piceid in red wine	DAD	MEKC	20 mM sodium tetraborate, 25 mM polyethylene glycol 400, 25 mM SDS, 10% methanol	57 cm × 75 μm i.d. fused-silica capillary, 28 kV, 25°C	176
Anthocyanins in wine	Visible at 599 nm	CZE	50 mM borate (pH 8.4), 15% methanol	46 cm × 75 μm i.d. fused-silica capillary, 25 kV, 10°C	177
Resveratrol, catechin, rutin, quercetin, myricetin, caffeic acid, chlorogenic acid, gallic acid in plants	UV at 240 nm	CZE	25 mM borate (pH 9.4)	75 cm × 50 μm i.d. fused-silica capillary, 18 kV	178
Chlorogenic acid, ferulic acid, vanillic acid, caffeic acid, catechol in coffee extracts	DAD	CZE	50 mM borate (pH 9.5)	48.5 cm × 50 μm i.d. fused-silica capillary, 20 kV, 25°C	179
Phenolic acids (derivatives of benzoic and cinnamic acids)	Contactless conductivity detection	CE	150 mM 2-amino-2-methylpropanol (pH 11.6), 30 mM MES/His, 30 mM CTAB, 10% methanol (pH 6)	48.5 cm × 50 μm i.d. × 50 μm o.d. fused silica capillary, 15 kV, 25°C	180
	UV visible	CEC	AMPD, 300 mM 2-amino-2-methylpropanediol; AMP, 150 mM 2-amino-2-methylpropanol; MA, 50 mM methylamine	Optical fibers 300 μm i.d., 330 μm clad diameter, 360 μm o.d.	

Phenolic compounds (syringic acid, *p*-coumaric acid, vanillic acid, caffeic acid, gallic acid, 3,4-dihydroxybenzoic acid, 4-hydroxybenzoic acid, (+)-catechin, (−)-epigallocatechin, (−)-epicatechin gallate, (−)-epigallocatechin gallate, (−)-epicatechin, and (−)-gallocatechin) in teas and grapes	UV-Vis detector	MEKC	2.89% (w/v) SDS, 1.36% (w/v) heptane, 7.66% (w/v) cyclohexanol, 2% (w/v) acetonitrile, and 86.1% (v/v) phosphate solution of pH 2.0 (25 mM)	48.5 cm × 50 μm i.d. fused-silica capillary, −20 to −27 kV, 25-35°C	181
		MEKC	2.89% (w/v) SDS, 2% (w/v) methanol, and 95.1% (v/v) phosphate solution of pH 2.0 (25 mM)		
Flavonoids (apigenin, baicalein, naringenin, luteolin, hesperetin, galangin, kaempferol, quercetin, and myricetine) in grape wine	DAD	CZE	35 mM Borax (pH 8.9)	70 cm × 75 μm i.d. fused-silica capillary	182
Flavonoids (quercetin and isorhamnetin)	UV-visible detection	CZE-ITP	LE 10 mM HCl, Tris, 20% CH_3OH (pH 7.2) TE 50 mM H_3BO_3, 20% CH_3OH (pH 8.2) BGE 25 mM MOPS, 50 mM Tris, 55 mM H_3BO_3 (pH 8.36)	Sampling, 800 V for 80 s; Stacking, 3000 V (1000 V/cm) for 24 s; Separation, 2700 V (675 V/cm)	183
Phenolic acids in olive oils	UV detection	CZE	25 mM sodium borate (pH 9.6)	57 cm × 75 μm i.d. × 375 μm o.d. fused-silica capillary, 25 kV, 25°C	184
Flavonoids and phenolic compounds (ferulic acid, apigenin, luteolin, rosmarinic acid, and caffeic acid) in *Perilla frutescens* L.	Electrochemical	CE	100 mmol/L borate buffer (pH 8.7)	75 cm × 25 μm i.d. × 360 μm o.d. fused-silica capillary, 18 kV, 20°C	185

Continued

TABLE. 30.6 (Continued)

Application	Detection Method	CE Mode	Run Buffer	CE Conditions	References
Flavonoids and phenolic acids in red wine	Conductivity	CZE-ITP	LE: 10 mM HCl of pH* 7.2 with Tris, 20% v/v methanol. TE: 50 mM boric acid of pH* 8.2, 20% v/v methanol. CZE: 25mM β-hydroxy-4-morpholino propanes ulfonic acid (MOPSO), 50 mM Tris, 15mM boric acid and 5 mM β-CD of pH* 8.5, 20% v/v methanol	ITP-Pre-separation capillary 9 cm × 0.8 μm ID, CZE-16 cm × 0.3 μm ID, 25°C	186
Flavonoids and phenolic acids (trans-resveratrol, cinnamic acid, ferulic acid, *p*-coumaric acid, quercetin, and morin) in berries	UV detection	CZE	35 mM sodium tetraborate (pH 9.3) containing 5% v/v methanol	55 cm × 50 μm ID fused-silica capillary, 20 kV, 25°C	187
Catechins and theaflavins (+)-catechin, catechin gallate, (−)-epicatechin, epicatechin-3-gallate, epigallocatechin, epigallocatechin-3-gallate, theaflavin, theaflavin-3-monogallate, theaflavin-3′-monogallate and theaflavin-3,3′-gallate in green and black teas	UV at 205 nm	CE	800 mL of 500 mM boric acid (pH 7.2), 200 mL of 100 mM a KH_2PO_4 (pH 4.5), 450 mL of 20 mM β-CD and 550 mL of acetonitrile	Light path capillary 40 cm × 50 μm i.d. 25 kV, 30°C	188
Flavonoids and isoflavonoids in beer	UV at 210 and 270 nm	MEKC	25 mM $Na_2B_4O_7$ (pH 10.5) and 110 mM SDS	57 cm × 75 μm i.d. fused-silica capillary	73
Phenolic oligomers [(+)-catechin and (−)-epica-techin] in soaking water from lentils, white beans and black beans, and in food by-products (almond peels)	DAD detection	MEKC	50 mM boric acid/sodium tetraborate, 100 mM SDS (pH 9 and 8.2)	27 cm × 50 μm i.d. fused-silica capillary, 10 kV, 25°C	189
Phenolic acids in red wine	Amperometric	Microchip CE	15 mM borate buffer (pH 9.5), 10% of methanol	Microchip 74 mm total length 2000 V	190

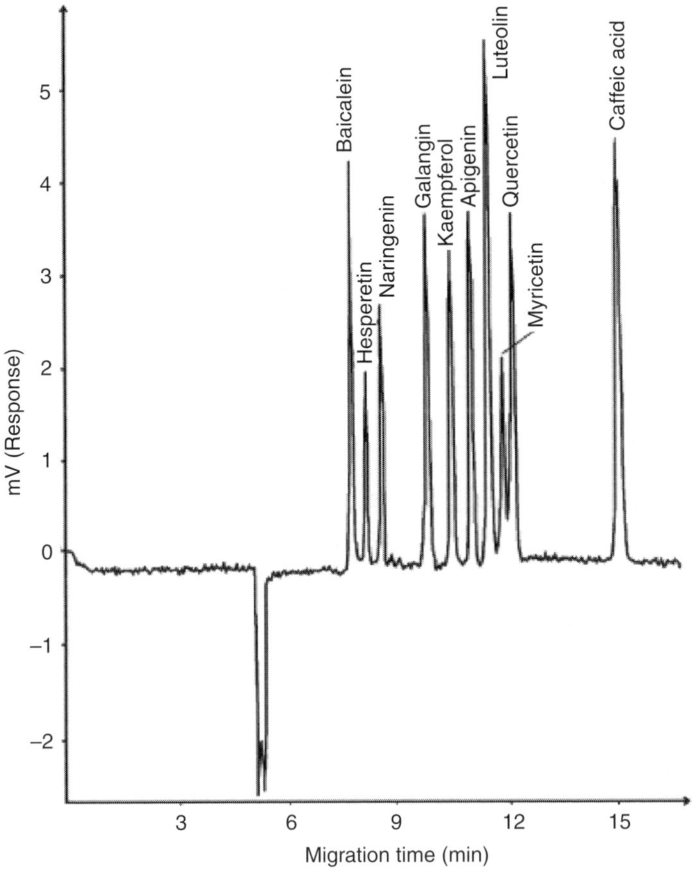

FIGURE 30.5 Electropherogram of the flavonoids separation. Conditions: 35 mM borax, pH 8.9, and applied field strength of 240 V/cm on a fused capillary of 70 cm (effective length: 45 cm) × 75 μm. (Reprinted from Wang, S. P. and Huang, K.-J., *J. Chromatogr. A*, 1032, 273–279, 2004. With permission from Elsevier Science-NL.)

monitoring fermentation processes; to identify adulterated products; to understand and control flavors and fragrances; and for fingerprint complex mixtures.

AA in food naturally occurs as the L-AA form. However, the D-enantiomers can be found in food that have been exposed to microbial activity (fermentation, aging, etc.), highly processed food (those that have been exposed to extremes of pH, heat, etc.), and food that naturally contains D-AA. Excellent reviews have been published that deal with chiral separation by CE[193–196,198]and specifically those separations that are applied to food analysis.[191,192,197] Table 30.7 summarizes the most recently published scientific paper on chiral compounds present in food and beverages.

Nouadje et al.[199] proposed a method some years ago using β-cyclodextrins (β-CDs) and SDS to determine L- and D-AA previously derivatized with FITC. A number of AA, including L-proline, L-aspartic acid, D-Asp, L-serine, L-asparagine, L-glutamic acid, D-Glu, L-alanine, L-arginine, D-Arg, and the nonchiral γ-amino-*n*-butyric acid (GABA), were analyzed in orange juice. The buffer consisted of 100 mM sodium tetraborate, 30 mM SDS, and 20 mM β-CD at pH 9.4, with an applied voltage of 23 kV. It was shown that L-Arg, L-Asp, and GABA were the most important compounds for the classification of commercial orange juices to provide useful information about quality and the efficacy of processing. Simo et al.[200] reported the use of stepwise discriminating analysis for 26 standard samples of commercial orange juices (i.e., nectars, orange juices reconstituted from

TABLE. 30.7
Methods for the Detection Chiral Analysis of Food Compounds Separated by CE

Type of Food	Detection Method	Sample Derivatization	CE Mode	Run Buffer	CE Conditions	References
Chiral AA content in orange juices	LIF	Derivatized with 200 m 3.75 mM FITC	MEKC	100 mM sodium tetraborate 20 mM β-CD, 30 mM SDS (pH 9.4)	57 cm × 50 μm i.d. fused-silica capillary, 23 kV, 25°C	200
Chiral AA in vinegars	LIF	Derivatized with 200 mL of 3.75 mM fluorescein isothiocianate (FITC) Dextrin 10, β-CD, hydroxypropyl-β-CD, and dimethyl-β-CD	MEKC	100 mM sodium tetraborate, 30 mM SDS, 20 mM β-CD (pH 9.7)	57 cm × 50 μm i.d. fused-silica capillary, 15 kV, 20°C	202
Fungicide (Imazalil Residue) in orange	Direct detection at 200 nm	Underivatized	CE	5 mM ammonium dihydrogen phosphate + 50 mM phosphate buffer (pH 3.0), 4 mM 2-hydroxypropyl-β-cyclodextrin	64.5 cm × 75 μm i.d., fused-silica bubble cell capillary, 25 kV, 20°C	203
Fungicide (vinclozolin) in wine	Direct detection at 203 nm	Underivatized	MEKC	20 mM phosphate, 5 mM borate buffer (pH 8.5), 50 mM γ-CD, 100 mM SDS	64.5 cm × 75 μm i.d. fused-silica capillary, 20 kV, 20°C	204
Biogenic DL-amino acids (Ala, Phe, Tyr, Ser, Cys, Met, Val, Leu, Ile, Thr, His, Pro, Trp, Arg, Lys, Glu, Gln, Asp and Asn) in solid-phase peptide synthesis (SPPS)	UV detection at 200 nm	Derivatized with 1 mL of 2 mM *N*-fluorenylmethoxy carbonyl-L-alanyl *N*-carboxyanhydride (FMOC-L-Ala-NCA) in acetone	CZE MEKC	20 mM borate buffer (pH 9.2) (0, 4, 10 mM γ-CD) 20 mM borate buffer (pH 9.2), (0, 10, 20, 80 mM) SDS	73 cm × 75 μm i.d. open-tubular silica capillary, 15 kV	205
AA enantiomers	UV absorption	Derivatized with 9-fluoreny methyl chloroformate (FMOC) *β*-CD and sodium taurodeoxy cholate (STDC) as selectors	CCE	150 mM borate and 18% v/v isopropanol (pH 8.0), 30 mM β-CD and 30 mM STDC	60 cm × 75 μm i.d. fused-silica capillary, 15 kV, 25°C	206
Chiral analysis of pantothenic acid in a soft drink	Direct UV at 200 nm	Underivatized	CZE	60 mM phosphate buffer (pH 7.0) + 60 mM 2-hydroxypropyl-β-cyclodextrin and 10% (v/v) methanol	Fused-silica bubble cell capillary of 56 cm, 375 mm i.d., 20 kV at 15°C	207

Monosaccharides (mannose, galactose, fucose, glucose, xylose, and arabinose)	DAD-UV	Derivatized with 1-phenyl-3-methyl-5-pyrazolone (PMP)	CZE	200 mM borate and 200 mM (S)-3-amino-1,2-pro-panediol (SAP), pH 9.2	80.5 cm × 50 μm i.d. fused-silica capillary, 20 kV, 10–50°C	208
Standards of α-aminoisobutyric acid (AIB) and isovaline	UV detection at 492 nm LIF 488 nm excitation, 520 nm emission	Derivatives of chloroethylnitro-sourea of benzyloxycarbonyllysine (*tert*-butyl ester) (CENU-Lys)	MEKC	20 mM borate buffer (pH 9.0) + 20 mM Brij 58 5 mM carbonate buffer (pH 10.3) + 20 mM Brij 58	Fused-silica capillary, 75 μm i.d., 20 kV, 25°C UV: 64.5 cm total length (56.5 cm to detector) LIF: 60 cm total length (55 cm to the detector)	209
Halogenated AA	DAD	Underivatized L-4-hydroxyproline, L-histidine (HP-L-4-Hypro) and (HO-L-4-Hypro) as chiral selectors	Ligand exchange CE (LE-CE)	20 mM HO-L-4-Hypro, 10 mM Cu(II) sulfate, in 5 mM phosphoric acid solution, pH 4.5, injection, 10 mbar 61 s; 10 kV	38.5 cm × 50 μm i.d. fused-silica capillary, 5–20 kV, 25°C	210
Flavanone-7-*O*-glycosides (naringin, prunin, narirutin, hesperidin, neohesperidin, and eriocitrin) in lemon juice	DAD	Underivatized	CCE	0.2 M borate buffer (pH 10.0), 5 mM γ-CD as chiral selector	77 cm × 75 μm i.d. × 375 μm o.d. fused silica capillary, 30 kV, 25°C	211
DL-Lactic acid yoghurts, and beverages (wine, sake, beer, and a soft drink)	Direct detection at 200 nm	Underivatized	CZE	90 mM phosphate buffer (pH 6.0), 240 mM 2-hydroxypropyl-β-cyclodextrin	50 cm × 50 μm i.d. (PVA)-coated bubble cell capillary, −30 kV, 16°C	212
DL-tartaric acid in grape juices, wines, soft drinks, sakes, cooking sakes, jams, candies, tablet candies, and pickles	Direct detection at 250 nm	Underivatized	LE-CE	1 mM copper(II) sulfate 10 mM D-quinic acid (pH 5.0)	48.5 cm × 50 μm i.d. (PVA)-coated bubble cell capillary, −15 kV, 30°C	213

Continued

TABLE. 30.7 (Continued)

Type of Food	Detection Method	Sample Derivatization	CE Mode	Run Buffer	CE Conditions	References
Dansyl AA and phenoxy acid herbicides	UV detection	Underivatized	CEC	AA: 1.6 mM Sodium phosphate composed of 20% v/v acetonitrile and 80% v/v aqueous phosphate buffer (pH 5.5) Herbicides: 2 mM sodium phosphate composed of 60% v/v acetonitrile and 40% v/v aqueous phosphate buffer (pH 6.0)	Capillary column: 27 cm × 100 μm i.d. × 360 μm i.d. fused-silica tubing, 15 kV	214
AA and nonprotein AA	LIF	Derivatized with 4-fluoro-7-nitro-2, 1, 3-benzoxadiazole (NBD-F)	CEC	P1: 5 μm silica particles modiofied with (*S*)-*N*-3, 5-dinitrobenzoyl-1-naphthylglicine P2: 5 μm silica particles modiofied with (*S*)-*N*-3, 5-dinitrophenyl aminocarbonyl-valine 5 mM phosphate buffer (pH 2.5) acetonitrile 30:70	Capillary column 30 cm × 75 μm i.d., 0.83 kV/cm 0.50 kV/cm	215
Standards of AA	LIF	Derivatized with FITC	MEKC	100 mM borate buffer (pH 9.5) 30 mM SDS β- and γ-CD as chiral selectors	74.4/99 cm × 50 μm i.d. fused-silica capillary, 22 kV, 25°C	216
Standards of aliphatic amines	Conductivity	Underivatized	On-chip CE	0.5 M acetic acid 5 mM of DM-β-CD 5 mM of (+)-18C_6 H_4 (pH 2.45)	60 cm × 50 μm i.d. × 375 μm o.d. fused-silica capillary, 15 kV, 22°C (PMMA), 90 mm × 16 mm	217
Standards of AA	UV-visible	Underivatized	CZE	20 mM ammonium acetate (pH 5) 2.5 mM vancomycin	35 cm × 50 μm i.d. × 375 μm o.d. fused-silica capillary, −15 kV, 20°C	218

Aromatic AA and L-mimosine in extracts of plant seeds and nonprotein AA	UV detection at 200 nm	Underivatized	CZE	60 mM phosphoric acid buffer, pH 2.5 containing 175 mM HP-α-CD 60 mM phosphoric acid buffer (pH 2.5) containing 2.3 mM HS-γ-CD	502 mm × 50 μm i.d. fused-silica capillary, 25.1 kV, 20°C, 25.1 kV, 18°C	219
Aromatic AA (phenylalanine, tyrosine, and tryptophan)	DAD	Underivatized	CZE	20 mM phosphate (pH 2.0), 0.5 mM S-β-CD, and 0.55% dextran sulfate	55 mm × 25 μm i.d. fused-silica capillary, 30 kV, 25°C	220
Flavonoids medicarpin and vestitone from transgenic legumes extracts	UV visible	Underivatized	CZE	2 mM hidroxypropyl-β-ciclodextrin 20 mM hidroxypropyl-γ-ciclodextrin 25 mM borate (pH 10.0), 10% v/v methanol	50 cm × 50 μm i.d. × 375 μm o.d. fused-silica capillary, 15 kV, 25°C	221
Standards mixture of AA	UV at 254 nm	Underivatized	CZE	100 mmol/L MES + 10 mmol/L His (pH 5.2) + 20 mmol/L MA-β-CD + 2 mmol/L γ-CD	Polyacrylamide coated capillary 80 cm total length (54 cm to the detector) × 75 μm i.d., 24 kV	222
Chiral AA in orange juices and orange concentrates	LIF	Derivatized with FITC	MEKC	100 mM sodium tetraborate, 30 mM SDS (pH 9.4), 20 mM β-CD	50 cm × 50 μm i.d. fused-silica capillary, 20 kV, 15°C	223

concentrates, and pasteurized orange juices not from concentrates) and selected L-Arg, L-Asp, and GABA as the most important variable L-AA to differentiate the samples. With the use of β-CDs as the chiral selector to separate the L-AA, it was possible to classify correctly the 100% of the samples.

Carlavilla et al.[202] recently published a chiral MEKC-LIF procedure that allows fast and sensitive analysis of the 11 main L- and D-AAs typically found in vinegars. The separation of the 11 compounds could be achieved in a relatively short time (less than 20 min) with a LOD down to 16.6 nM. It was shown that using these profiles resulted in a very useful strategy for characterizing vinegars.

Gel-Moreto et al.[211] reported, for the first time, the complete separation of the 2*R*- and 2*S*-diastereomer of six selected flavanone-7-*O*-glycosides (naringin, prunin, narirutin, hesperidin, neohesperidin, and eriocitrin). CE measurements were done using 0.2 M borate buffer, 5 mM γ-CD, pH 10.0, and later applied for the separation of 2*S*- and 2*R*-diastereomers of eriocitrin and hesperidin in lemon juice. Another application of chiral analysis in food by CE was published by Liu et al.[201] They developed a method for the determination of 1-aminocyclopropane-1-carboxylic acid (ACC) in apple tissues based on the derivatization of ACC with 3-(2-furoyl) quinoline-2-carboxaldehyde (FQ) and the use of CE-LIF (488 nm). Optimal conditions for the separation of ACC were obtained using 20 mM borate buffer, pH 9.35, with 40 mM SDS and 10 mM Brij 35, allowing for the analysis of ACC in crude apple extracts to be done with an impressive LOD of 10 nM. Finally, Cinquina et al.[76] reported the determination of histamine in tuna fish samples. Histamine is a degradation product of the bacterial decarboxylation of the AA histidine, which is present at substantial concentration in fish tissues. A CE method and an HPLC method with diode array detection were compared showing that both techniques were associated with correlation coefficients that exceeded 0.999, and LODs of 1 and 2 mg/kg for HPLC, and 0.5 and 1 mg/kg for CE.

30.2.8 Proteins and Peptides

Proteins are one of the main components of foods and possibly one of the most challenging to analyze, not only because of the heterogeneity of the fractions but also because of their tendency to aggregate. Thus, it is not surprising that CE has evolved as an alternative technique to the classical gel electrophoresis for the analysis of proteins and peptides with the advantage that it allows for online detection and automation. In fact, the analysis of food proteins is one of the main applications of CE in food analysis, with most focus on the analysis of dairy proteins, followed closely by cereal proteins.[2] The third category includes other food proteins such as those from meat, eggs, and others.[2] Table 30.8 summarizes some selected applications for the analysis of food proteins by CE. For a comprehensive review, including applications up to 1999, the reader is referred to Frazier et al.[5] and to Cifuentes[2] for a more updated review including applications from June 2002 to June 2005. For specific reviews on food proteins of animal origin, the review by Recio et al.[224] describes CE methods of analysis for milk, egg, meat, and fish proteins and peptides, with special emphasis on dairy products. In addition, various hurdles associated with food analysis technology, such as the assessment of technological processes, quality, and authenticity control of animal foods, were considered.[224] Similarly, Bean and Lookhart[225] offer a specific review covering methods and applications for the separation of three major groups of food proteins: meat, dairy, and cereal proteins.

The first prerequisite for the analysis of proteins is their solubilization from the food matrix. Nonliquid food sources usually require homogenization before extraction to allow efficient recovery of proteins from the material.[226] For this purpose, milling, blending, homogenizing, and the use of ultrasound (sonication) are common methods. The homogenization step may be combined with extraction of the proteins using suitable extraction buffers, which may give a selective extraction of specific groups of proteins, often combining precipitation, filtration, and dialysis steps.[226] Defining a strategy for isolation of a protein or a group of proteins from a specific food matrix will usually encompass a choice of method chosen on the basis of known characteristics of the protein(s) that

TABLE. 30.8
Food Proteins Analyzed by CE

Application	Detection Method	Sample Pretreatment	CE Mode	Run Buffer	References
Caseins and whey proteins	UV at 214 nm	Milk, diluted (1:5) with 10 mM phosphate buffer (4.8 M urea, 0.2% Tween 20, pH 2.5), heating 5 min at 40°C	CZE	50 mM phosphate buffer (4 M urea, 0.1% Tween 20, pH 2.5), coated capillary	230
	UV at 214	Milk, diluted (1:5) with 5 mM sodium citrate (5 mM DTT, 6 M urea, pH 8.0)	CZE	10 mM phosphate or citrate buffer (0.05% MHEC, 6 M urea, pH 2.5), coated capillary	231
	UV at 214	Milk, diluted (1:5) with 167 mM Tris, 42 mM MOPS, 67 mM EDTA, 17 mM DTT, 10 M urea	CZE	20 mM sodium citrate buffer (0.05% MHEC, 6 M urea, pH 3.0), coated capillary	232
Caseins and whey proteins, denatured β-lactoglobulin, para-κ-casein	UV at 214	Milk, cheese, 167 mM Tris, 42 mM MOPS, 67 mM EDTA, 17 mM DTT, 8 M urea, 0.5 g/L MHEC	CZE	0.48 mM citric acid – 13.6 mM trisodium citrate in 4.8 M urea (pH 2.3), coated capillary	233
Caseins and whey proteins	UV at 214	Purified proteins, denatured by addition of 10% SDS and 7% DTT, heating 3 min at 100°C	MEEKC	3 mM sodium borate, 8.2 mM SDS (pH 9.5), uncoated capillary	234
Whey proteins, separation of β-lactoglobulin variants A and B	UV at 214 nm	Milk, whey proteins prepared by casein precipitation diluted (1:4) with 8.25 mM borate buffer, 0.1% Tween 20, pH 8.0	CZE	0.05 M borate buffer, Tween 20 0.1%, pH 8.0, coated capillary	235
Cereal storage proteins	UV at 200 nm	Various cereals, extraction with 50% 1-propanol, glutenins with 50% 1-propanol, 1% DTT	CZE	50 mM iminodiacetic acid IDA, 20% AcN, 0.05% HPMC	246
Sarcoplasmic/myofibrillar meat proteins	UV at 214	Meat, sarcoplasmic proteins extracted in bidistilled deionized water, myofibrillar extracted in 0.6 M NaCl/0.01 M phosphate buffer, 0.5% polyphosphates, pH 6.0, Biorad SDS sample buffer	MEKC	SDS-CE, nm, Bio-Rad SDS run buffer	247
Lysozyme from egg white	MS	Meat adulterated with chicken egg white	CE-MS	75 mM ammonium acetate/acetic acid, pH 5.5, coated capillary	249

include molecular weight or isoelectric point; however, some general strategies as outlined by Sorensen et al. [226] may be useful as a starting point before performing a CE analysis.

Similarly, the separation of proteins by CE is highly dependent on the characteristics of the proteins, especially the isoelectric points of the proteins and also information on the differences between the proteins to be separatedare important.[226] The most important point to consider in the choice of buffer systems used in analyzing proteins is how to prevent or control the adsorption of proteins to the capillary wall.[226] Adsorption of the proteins to the silica surface of the capillary wall is the main reason for the observed efficiency loss, poor reproducibility in migration times, and low protein recovery.[2] This problem may be solved by the use of coated capillaries.[226,227,228]A coating polymer was demonstrated to provide good separations of acidic and basic proteins, including proteins from whey.[227] Coating regeneration was achieved by flushing the capillary between injections with an aqueous solution of the polymer.[227] Alternatively, the ionization of the silanol groups can be suppressed by working with background electrolytes (BGE) with low pH values or with a buffer pH above the pI of the sample protein and the use of dynamic coatings.[226,228]A comprehensive review of different dynamic and static capillary coatings for reducing protein wall adsorptions offered different alternatives for solving the problem.[228]

The separation of dairy proteins by CE has been generally carried out by CZE and has been exhaustively covered in several review papers,[2,224,225,229] thus Table 30.8 only presents the key methodologies that offer the reader an overview of their most distinctive features. Basically, dairy protein analysis has been performed in whole milk for the simultaneous determination of caseins and whey proteins, or in fractions isolated from milk after casein precipitation. The first approach being used when the quantitative determination of the major proteins is required for the calculation of casein/whey protein ratios or for authentication purposes where an analysis of the whole protein profile is required. In both cases, accurate quantitative data must be derived. However, few studies have addressed the analysis of both groups of proteins in a single run by presenting quantitative data based on calibration curves constructed with analytical standards and good recovery of all proteins from milk samples.

The first method reported to provide quantitative recovery of major whey and casein proteins in a single run was by Vallejo-Cordoba (Table 30.8).[230] With this method,[230] sample preparation was critical for the quantitative recovery of all proteins in their native state; while in most other methodologies, sample preparation included a strong denaturing reducing buffer.[231,232,233] Sample and run buffer contained the minimum concentration of urea (4 M) that allowed good casein solubility. Keeping urea to a minimum concentration was essential to prevent β-lactoglobulin denaturation and to minimize urea crystallization.[230] In addition, the nonionic detergent, Tween, was also used to help with maintaining caseins in solution. A typical electropherogram of whey proteins and caseins in fresh milk separated in a coated capillary at pH of 2.5 is shown in Figure 30.6.

Several groups have applied the method of de Jong et al.[231] with modifications to monitor milk proteins; these conditions not only allowed the separation of the individual milk proteins but also some of their genetic variants.[7,232,233] Although the original method offered by de Jong et al.[231] was the first method described for simultaneously determining casein and whey proteins, quantitative data were not presented. The original method of de Jong[231] included a reducing denaturing sample buffer at pH 8.5 and a running buffer at pH 2.5 or 3.0 containing a cellulose polymer. To minimize protein and capillary wall interactions, separation was carried out in a coated capillary. However, this method was later modified to minimize protein absorption by optimizing the sample and running buffer (Table 30.8).[232,233]

Finally, a third method[234] was based on MEKC, where proteins were separated after complete denaturation with SDS and DL-dithiothreitol in uncoated capillaries at pH 9.5. Although the method had the advantage of being very rapid (separation completed in less than 90 s), it was not quantitative (Table 30.8). The second approach reported for the analysis of dairy proteins was the analysis of the whey fraction after casein precipitation. Unlike the methods described earlier, separations were carried out in uncoated capillaries using polymeric additives, a high ionic strength, and high pH

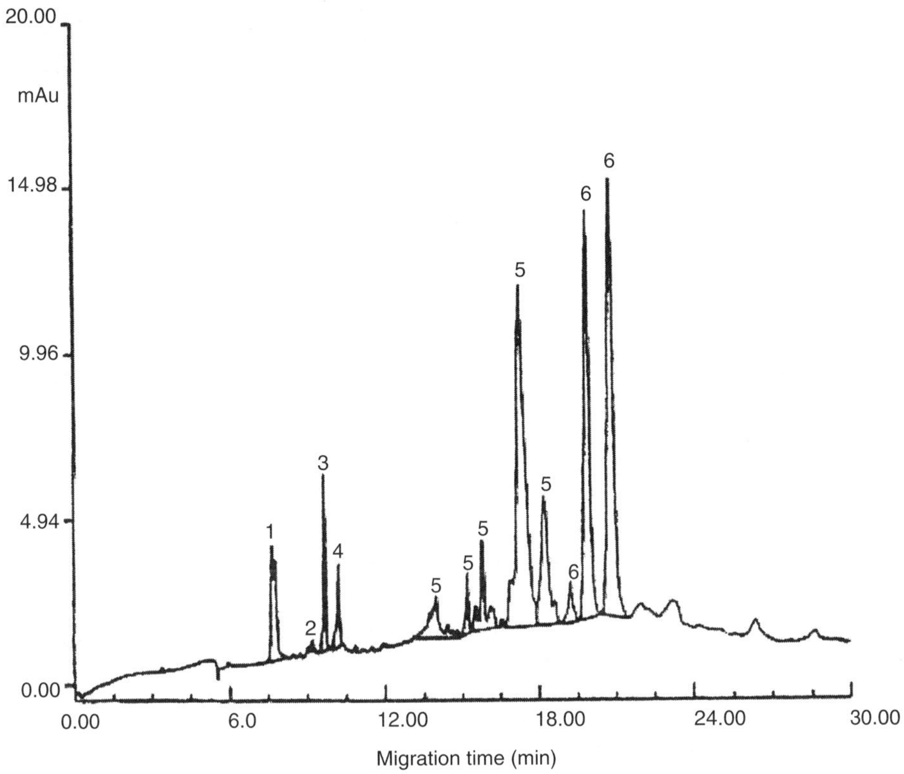

FIGURE 30.6 Typical electropherogram of proteins in fresh milk. (1) β-lactoglobulin, (2) bovine serum albumin, (3) α-lactalbumin, (4) conalbumin (added internal standard), (5) α-casein and (6) β-casein (From Vallejo-Cordoba, B. *J. Cap. Elec.*, 4, 219, 1997. With permission.)

separation buffer, which collectively reduced adsorption of proteins to the capillary wall.[229] Several examples of this strategy were reviewed;[224,229] however, few of these methods were demonstrated to be quantitative.[229] To this end, Olguin-Arredondo and Vallejo-Cordoba[235] presented a method that allowed the quantification of genetic variants of β-lactoglobulin (Table 30.8). Excellent resolution for variants A and B was observed under the established conditions (Figure 30.7).

Most applications for the analysis of dairy proteins are based on the methods described earlier (Table 30.8). However, since UV detection was used, the detection sensitivity of these methods is usually limited to proteins in the μM range.[236] Thus, to overcome this drawback, an on-capillary derivatization method with the fluorogenic reagent 3-(2-furoyl)quinoline-2-carboxaldehyde (FQ) and LIF was developed for the determination of dairy proteins.[236] This method allowed the determination of trace amounts of β-lactoglobulin in hypoallergenic infant formulas.[237] CE was used for the determination of added rennet casein and caseinate to processed cheeses[238]and for the simultaneous quantitative determination of bovine, ovine, and caprine casein fractions in Mexican unripened cheese (Panela).[239] CE was also useful for characterizing milk protein hydrolysates to investigate the molecular basis for differences in bioactivity,[240] for monitoring proteolysis of casein in packaged pasteurized milk during refrigerated storage in relation to hygienic and microbiological characteristics of starting raw milk,[4] and for peptide mapping in the investigation of milk protein genetic variants.[242]

Cereal proteins are important not only for their nutritional quality but also for their functional role in foods. The analysis of cereal proteins by CE has been covered in several reviews.[228,243–245] Cereal proteins are complex mixture of proteins that are often difficult to solubilize and separate. Because of this, a wide range of analytical techniques including CZE or sodium SDS-CE have

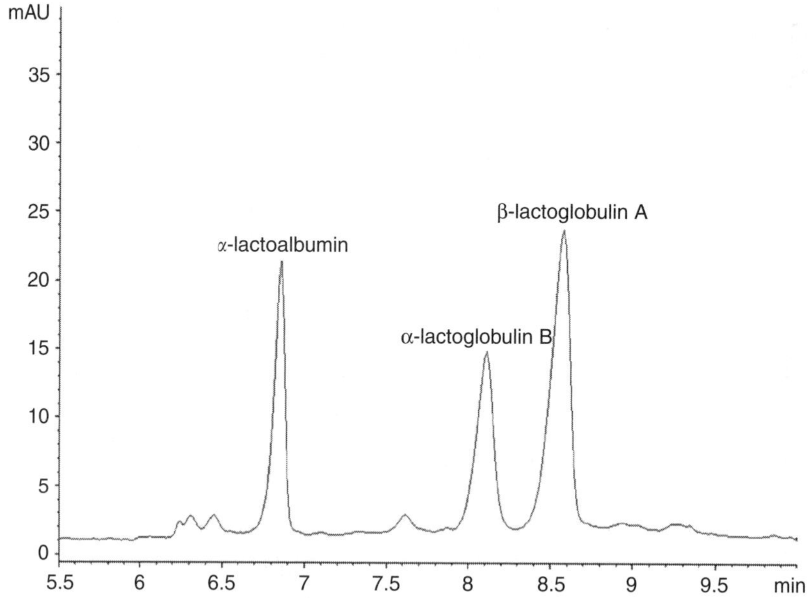

FIGURE 30.7 Separation of β-lactoglobulin variants A and B in cow's milk. (From Olguin-Arredondo, H. and Vallejo-Cordoba, B., *J. Elec. Microchip Tech.*, 6, 145, 1999. With permission.)

been used.[243] These reviews also present the applications of CE for cultivar identification, classification, and prediction of quality.[243–245] Since the objective of most cereal protein applications was to qualitatively characterize protein profiles, most method development efforts were toward improving resolution, reproducibility, and speed of the analysis. A good example was an improved method that allowed rapid (2–8 min) separations of grain proteins for several cereals (wheat, oats, rice, barley, and rye) with high resolution and reproducibility using isoelectric buffers (Table 30.8).[246] These buffers allow the use of extremely high voltages, which in turn produces very rapid separations and resolution.[246] Also, it was shown that acetonitrile worked as well as urea in solubilizing maize and sorghum proteins, the most hydrophobic storage proteins. In addition, some cellulose polymer modifier was added to the buffer to prevent protein–capillary wall interactions.[246]

CE has also been successfully applied to the study of muscle proteins, and some of these applications have been recently reviewed.[6,224,244] Methods for the determination of muscle proteins were based on CZE, SDS-CE, or isoelectric focusing (CEIF).[6,224,244] Meat species identification was carried out by analyzing sarcoplasmic or myofibrillar proteins by a replaceable polymer-filled SDS-CE method (Table 30.8).[247] However, the analysis of sarcoplasmic protein profiles allowed better differentiation among beef, pork, and turkey meat (Figure 30.8).[247] The importance of sample preparation in the established method was highlighted since sarcoplasmic proteins extracted by simply homogenizing meat with cold bidistilled water were most useful for meat species identification when protein profiles were examined by linear discriminant analysis.[248] On the other hand, myofibrillar proteins extracted with 0.6 M NaCl/0.01 M phosphate buffer with 0.5% polyphosphates (pH 6.0) were not useful for raw meat species identification, although they may be of importance in the identification of heat-processed meats.[248]

The analyses of egg proteins by CZE, CIEF, or MEKC were reviewed by Recio et al.[224]In addition, an interesting application of CE coupled to mass spectrometry (CE-MS) was the detection of lysozyme from chicken and turkey egg white.[249] Since one of the problems of protein separations was the possible interaction with the capillary wall, a polymer coating compatible with CE-MS was developed, and the usefulness of this method was shown by the analysis of a minced meat extract containing chicken egg white as adulterant (Table 30.8).[249]

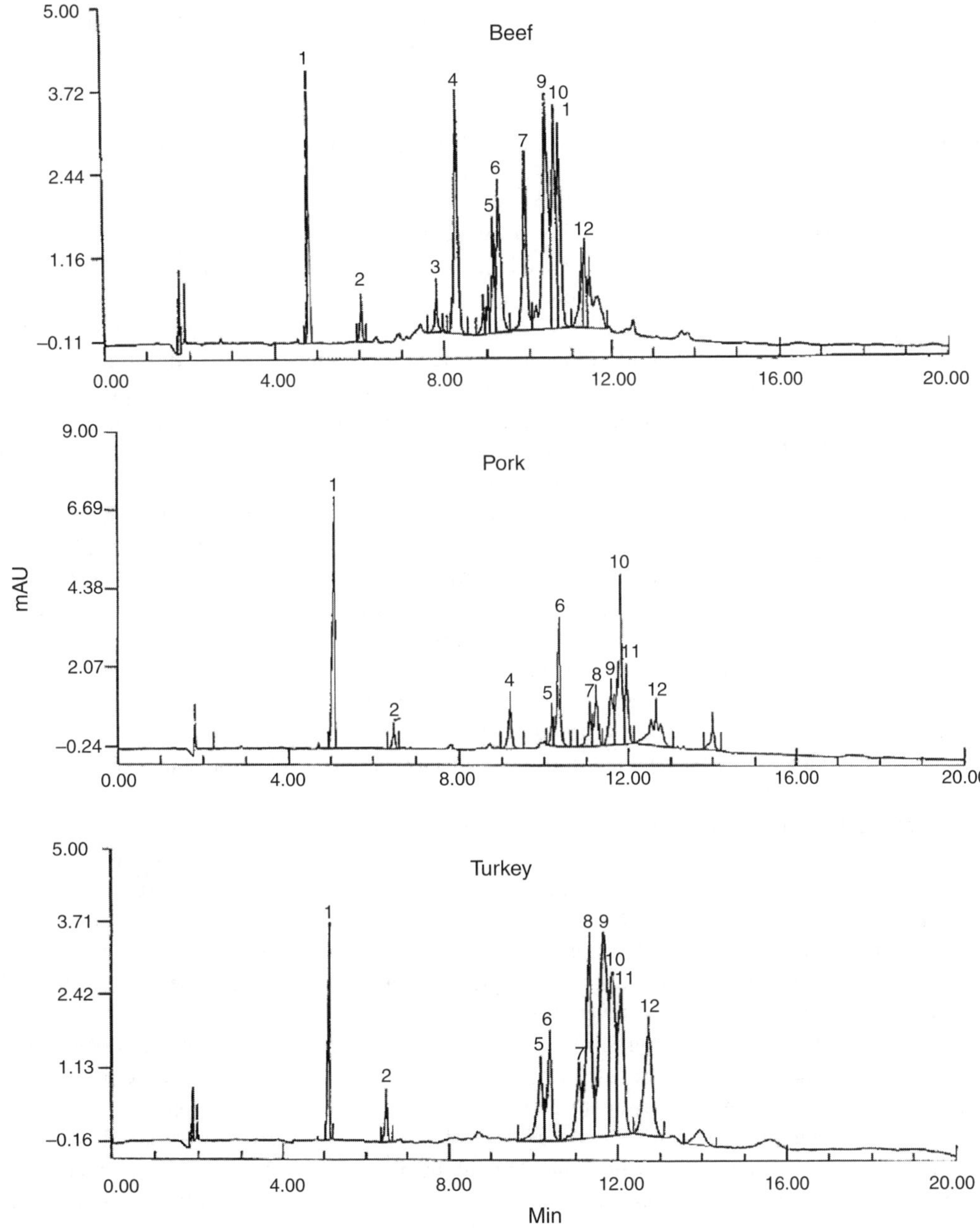

FIGURE 30.8 CE-SDS electropherograms of sarcoplasmic proteins extracted from beef, pork and turkey meat. (From Cota-Rivas, M. and Vallejo-Cordoba, B., *J. Cap. Elec.*, 4, 197, 1998. With permission.)

30.2.9 DNA AND MICROCHIPS

Unlike food proteins, nucleic acids have no nutritional value but are characteristic of the various biological components in complex products. Thus, the analysis of nucleic acids in food allows control laboratories to determine the presence or absence of certain ingredients in complex products or the identification of specific single food components.[250] These analyses were based on nucleic acid probes, including the polymerase chain reaction (PCR), which made the detection of minute amounts of nucleic acids and their sequence determination possible.[250] Also, as DNA is more thermostable than many proteins, analyses using nucleic acids are less liable to be disrupted by processing of foods. Thus, nucleic-acid-based technologies are developing rapidly and the use of suitable methods by food control laboratories has the potential to greatly simplify methods for food authentication.[251] The use of molecular techniques and CE has tremendously facilitated analytical procedures since this combination has the benefit of high specificity and sensitivity of molecular techniques coupled with the high resolving power and automation of CE.[2]

DNA-based analysis methods are highly dependent on the DNA extraction and purification techniques. In particular, the application of molecular methods to food samples requires stringent extraction and purification strategies, which ensure efficient recovery of nucleic acid and removal of the numerous compounds inhibiting PCR assay.[252] A comparison of DNA extraction methods for food analysis was recently reported that highlighted the efficiency of two different commercial kits.[252] Also, four different DNA extraction methods from maize flour in genetically modified organisms (GMOs) in foods were compared and the SDS/proteinase K method was chosen as the most convenient.[253] Thus, these studies highlighted the need for suitable DNA extraction methods to obtain highly purified nucleic acids without inhibitors before PCR-CE can be carried out.

PCR-based techniques combined with CE separations has resulted in powerful methods in two specific areas of food analysis, namely, food authentication and microbiology. PCR-CE applications for meat authentication were reviewed by Vallejo-Cordoba et al.,[6] while applications including other foods were reviewed by Cifuentes[2] and Kvasnicka.[4] An overview of the combined use of PCR and CE for the detection of transgenic foods, meat species identification, and microbial analysis discussing advantages and drawbacks of these combined techniques was presented.[254] Specific applications of food authentication consisted in the detection of GMOs[255] and the identification of meat species[256] or in beef sexing;[257] while food microbiology applications included the detection of food-borne pathogens,[258] food-spoilage bacteria,[259] and the characterization of toxigenic fungi in dry, cured meat.[260]

Commercial use of transgenic plants and other GMOs has raised several ideological and ethical issues in recent years. Therefore, this has created a demand for analytical methods that can detect and quantify the amount of GMO in foods.[254] A method that combines PCR and CGE-LIF was developed for the quantification of genetically modified Bt maize in foods.[255] The method developed was based on the coamplification of specific DNA maize sequences with internal standards using QC-PCR.[255] Different DNA isolation and amplification techniques, which are being used for detecting GMOs in foods, addressing quantitative aspects were reviewed.[261]

In the past, species identification in muscle foods routinely involved the detection of species-specific proteins when attempting to discern the origins of the material, but they were not without problems. The processing of foods by heating can cause denaturation of the proteins under study and, in addition, protein expression is usually tissue dependent. Thus, study in the past decade has seen DNA replace protein in species identification owing to the protein's instability at high temperatures, and to the fact that the DNA's structure is conserved within all tissues of an individual.[251] To this end, a PCR-RFLP method that used fluorescence sensor CE for DNA fingerprinting of pork, goat, and beef generated by restriction enzyme digestion following a fluorescent-labeling PCR amplification was reported.[256] This method was based on the amplification of the mitochondrial 12S rRNA gene with a unique primer pair and incorporated a fluorescent-labeling nucleotide with sufficient specificity to detect the three animal species. The method could reliably identify pork, goat, and beef and semiquantify any of these meats when they were present in meat mixtures at levels less than 1%.[256]

Although QC-PCR by CE for the determination of meat species has not been reported, this methodology was useful for quantitatively analyzing GMOs in foods. An alternative method, real time PCR (RT–PCR), is gaining popularity over QC-PCR for the quantification of GMOs in food samples, although these methods are still under development for the simultaneous detection of several transgenes. Also, the interlaboratory reproducibility of RT–PCR was very low since %RSD values of 40% were reported.[255] In addition, although instrumentation required for RT–PCR or QC-PCR-CE may be costly for most food laboratories, CE is a more versatile technique because a wider range of analytes may be determined with the same instrumentation. Thus, QC-PCR-CE offers good potential for the quantitative determination of meat species or soy in heat-processed products.

Several of these food DNA-based analyses were recently translated to microchips as shown from the applications reviewed by Cifuentes.[2] An exemplary application of microchips was an improved PCR-RFLP method for fish species identification.[262] The objective of the improved method was to replace the gel-electrophoretic steps for fragment separation, detection, and analysis by employing a chip-based CE system (Agilent 2100 Bioanalyser) to analyze PCR-RFLP fingerprints. The use of this system allowed simultaneous postrestriction digestion analysis of 12 samples in under 40 min, which is a considerable time reduction over conventional gel-based methods. In addition, repeatability was less than 3%, allowing species identification without the need to run reference materials with every sample. Using DNA admixtures, discrimination of 5% salmon DNA in trout DNA was detected.[262] This improved lab-on-a-chip method was applied to the development of PCR-RFLP profiles on a commercial microchip electrophoresis instrument that can be used to identify a range of white fish species without the need for concurrent analysis of reference materials. The method was applied to a range of products and subjected to an interlaboratory study carried out by five UK food control laboratories. One hundred percent correct classification of single species samples and six of nine admixture samples was achieved by all the laboratories. The results indicated that the fish species identification could be carried out using a database of PCR-RFLP profiles without the need for reference materials.[263]

30.2.10 Food Additives

Food additives are substances added to food with the sole purpose of preserving it or improving its flavor and appearance. The extended use of additives has made necessary the development of new analytical procedures that are able to characterize them to ensure that food manufacturers comply with labeling regulations. A comprehensive review was reported by Boyce,[264] focusing on antioxidants, preservatives, colorings, and sweeteners added to food. In addition, an exhaustive survey of capillary electromigration methods to analyze natural antioxidants was presented together with some discussion of the use of these substances as functional foods.[161] Most recently, an updated overview on this subject was reported by Cifuentes.[2] In general, CZE and MEKC methods with UV detection were used for the analysis of food additives. A very useful application that demonstrated the versatility of MEKC for the simultaneous rapid separation of preservatives, sweeteners, preservatives, and colors in soft drinks was reported by Frazier et al.[265] Minimum sample preparation was required since soft drinks were only subjected to degassing by sonication. Resolution between all additives was achieved within a 15-min run using carbonate buffer containing an optimum SDS concentration at the micellar phase. By using the UV-visible range (190–600 nm), the identity of sample components was confirmed by spectral matching relative to standards.[265] Food additives that were used in this analysis included a range of seven synthetic food colors (quinoline yellow, sunset yellow FCF, carnosine, ponceay 4R, brilliant blue FCF, green S, and black PN), three artificial sweeteners (acefulfame K, aspartame, and saccharin) and two preservatives (benzoic and sorbic acid). Although this example required minimum sample preparation, sample matrix interference is directly related to food complexity. The analysis of eight colorants in milk beverages required prior sample pretreatment with a polyamide SPE column.[266] Baseline resolution of the carminic acid, a natural colorant, and seven synthetic dyes was achieved within 9 min by using a tetraborate buffer contaning β-CD as additive. Also, the recoveries of the eight food colorants from milk beverages were better than 85% with detection levels of less than 0.5 μg/mL.[266]

On the other hand, the development of a MEEKC allowed the determination of the same food colorants with most instances not requiring sample pretreatment. The effects of SDS surfactant, organic modifier (acetonitrile), cosurfactant, and oil were examined to optimize the separation. A highly efficient MEEKC separation method, where the eight colorants were separated with baseline resolution within 14 min, was achieved by using a microemulsion solution of pH 2.0 containing 3.31% SDS, 0.81% octane, 6.61% 1-butanol, and 10% acetonitrile.[267] To this end, an electrolyte composed of tetraborate (TBS), Brij 35, and acetonitrile (ACN) was optimized for the separation of 11 dyes allowed in Brazil providing baseline separation in less than 9 min. The optimization procedure that provided baseline separation in less than 9 min was aided by a factorial design.[268]

Other applications of CE to analyze food additives include the determination of vitamin C and preservatives (benzoate and sorbate) by both conventional CE and microchip electrophoresis with capacitively coupled contactless conductivity detection.[269] The separation was optimized by adjusting the pH value of the buffer and the use of hydroxypropyl-β-CD (HP-β-CD) and CTAB as additives. For conventional CE, optimal separation conditions were achieved in a histidine/tartrate buffer at pH 6.5, containing 0.025% HP-β-CD and 0.25 mM CTAB with a LOD ranging from 0.5 to 3 mg/L, whereas a histidine/tartrate buffer with 0.06% HP-β-CD and 0.25 mM CTAB gave a LOD ranging from 3 to 10 mg/mL. By using a microchip electrophoresis format, a considerable reduction of analysis time was accomplished.[269]

Another useful application of CE is the determination of the preservative nitrate and nitrite ions in meat products.[270] It is known that nitrite causes methaemoglobinaemia and, with secondary and tertiary amines, yields the carcinogenic nitrosoamines. Owing to these toxic effects, it is important to develop new analysis methods for the simultaneous determination of the two anions reducing the matrix effect in meat samples. The method developed was based on the separation of the two anions in a capillary coated with polyethyleneimine (PEI). Since PEI is a cationic polymer, the EOF is reversed over a wide pH range and the fast separation of anions is achieved without the addition of any electroosmotic modifier to the Tris separation buffer at pH 7.5. The LODs of the method for nitrite and nitrate were 0.10 μg/mL and 0.09 μg/mL, respectively.[270]

Because of its aromatic properties, vanillin is the most widely used flavoring material in foods and beverages.[271] Vanillin is the main aromatic compound in natural vanilla, but it can also be synthesized from low-cost materials such as 2-methoxyphenol, eugenol, and lignin. On the other hand, ethylvanillin, which is commonly used as another synthetic compound, has much more flavoring strength than vanillin and is used in formulation and imitation products. Thus, a CE method was developed for the determination of vanillin, ethylvanillin, 2-methoxyphenol, and 2-ethoxyphenol, simultaneously in cocoa drink samples by using a photodiode array detector.[270] Separation was carried out in phosphate buffer containing CTAH at pH 10 with 10% acetonitrile. Vanillin and related compounds were determined in 7 min, with the LOD at 1.6 μg/mL. Mean recoveries were 96.3–103.8%. With this method, it was possible to determine vanillin and ethylvanillin, which were originally contained as flavoring in cocoa drinks, but it was also found that vanillin and ethylvanillin were metabolized, respectively, to 2-methoxyphenol and 2-ethoxyphenol, by growth of *Bacillus firmus* causing an off-flavor.[271]

Artificial sweeteners are added to foods as a sugar substitute, particularly to low-calorie foods. Sucralose is one of these artificial sweeteners that was determined by CE in different food matrices.[272] A CE method was optimized chemometrically for the quantification of sucralose from different food matrices. Separation from food matrix components was obtained in a dinitrobenzoic acid/sodium hydroxide background electrolyte with a pH of 12.1 and detection was achieved at 238 nm by indirect UV. The method allows the detection of sucralose at >30 mg/kg making it suitable for implementation of the recently amended "Sweetners for use in foodstuffs" European Directive.[272]

Antioxidants are often added to food to retard oxidative processes that cause food off-flavors. The most commonly permitted additives are butylated hydroxyanisole (BHA), butylated hydroxytoluene (BHT), tert-butylhydroquinone (BHQ), and the esters of gallic acid. Some of the applications of CE to synthetic antioxidants were reviewed by Boyce.[264] A recently reported method was the analysis of alkyl gallates and nordihydroguaguaiaretic acid (NDGA) by a microchip MEKC with pulsed amperometric detection. Simultaneous separation of the antioxidants propyl gallate (PG), octyl gallate (OG), lauryl gallate (LG), and NDGA was carried out in borate buffer, pH 9.7, containing SDS. The measured detection limits were in the range of 2-6 fmol of the analyte.[273]

30.2.11 Food Quality

The versatility of CE makes it a very useful method for monitoring food quality from its different aspects, namely, nutritional, sanitary, compositional (including authenticity), and sensorial. Also, these technological processes such as heat treatment, roasting, freezing, drying, and so forth may affect different aspects of food quality, starting with changes in compositional and nutritional quality. While some recent applications of CE for evaluating food quality has been reviewed by Cifuentes,[2] the CE analysis of dairy proteins to determine the effect of different processing conditions, to follow proteolysis in different types of cheese, or to assess authenticity[7] represent, by far, the largest number of CE applications in food quality assessment. A comprehensive review of these applications has been presented by Recio et al.[224]

Capillary electrophoresis has been very useful for studying the effect of heat treatment on dairy proteins, particularly in the study of the interaction between carbonyl-containing compounds such as reducing sugars with amino groups. These types of interactions, known as Maillard reaction, have been found to be very important in food science and medicine. The Maillard reaction results in the formation of large protein aggregates as well as low molecular weight compounds known as Maillard reaction products (MRP) that impart various flavor, aroma, and color characteristics to foods. Thus, the formation of MRP not only affects food compositional quality but also impacts nutritional and sensorial aspects of quality. In this sense, CE methods were reported that allowed the determination of a broad spectrum of compounds ranging from large molecular weight aggregates, such as lactosylated milk proteins,[274] to small molecular weight compounds, such as furosine.[275] Deterioration due to Maillard reaction of milk powder upon storage was monitored by CE showing a native and a modified fraction of β-lactoglobulin.[274] The method used for the analysis of milk proteins was the one previously reported by Recio and Olieman.[232] In comparison to other methods, the CE method proved to be rapid, easy to execute, and sensitive for monitoring early deterioration of milk powders during storage due to Maillard reaction. The determination of the ratio of unmodified whey proteins to total whey proteins by CE was considered a suitable method to evaluate the loss in nutritional value of milk powders with reference to protein quality.[274]

A new CE method was established for the quantitative determination of furosine in dairy products. Sample preparation consisted of drying hydrolyzed samples, redissolving them in NaOH, and purifying them by SPE. The electrophoretic separation was carried out in an uncoated capillary using phosphate buffer containing the additive hexadecyl trimethylammonium bromide (HDTAB). The LOD for furosine was 0.5 ppm, a concentration that corresponds to 4.5 mg/mL of protein in milk samples. Electropherograms showing very well-defined peaks for furosine in UHT and evaporated milks are depicted in Figure 30.9. As expected, higher furosine concentration was determined for evaporated milk since it is the most severely heat-treated product. Thus, furosine determination by CE method may be used to assess the extent of protein damage caused by heating and as a useful indicator of the intensity of food processing conditions that cause the deterioration of the nutritive value.[275]

Another heat-induced marker formed as a result of the Maillard reaction is hydroxymethylfurfural (HMF). A MEKC method was developed for the determination of HMF in milk-based formulas by using an uncoated capillary and phosphate buffer (pH 7.5) containing SDS.[276] A similar compound, furfural, is also formed as a result of Maillard reaction and could be detected in distilled agave drinks such as Tequila and Bacanora by CE method. Furfural is formed during the cooking of agave heads before juice extraction, fermentation, and distillation. A MEKC procedure was carried out with phosphate buffer containing SDS at pH 7.5.[277] Typical electropherograms showing the furfural peak in Bacanora (Figure 30.10a) and Bacanora spiked with the analytical standard (Figure 30.10b) are shown. In addition, some CE-MS and CE-MS/MS applications for the study of Maillard reaction products were reviewed by Simo et al.[249]

Food authenticity is yet another food quality control parameter in addition to the parameters described earlier. Food fraud may involve substitution or addition of ingredients of inferior value

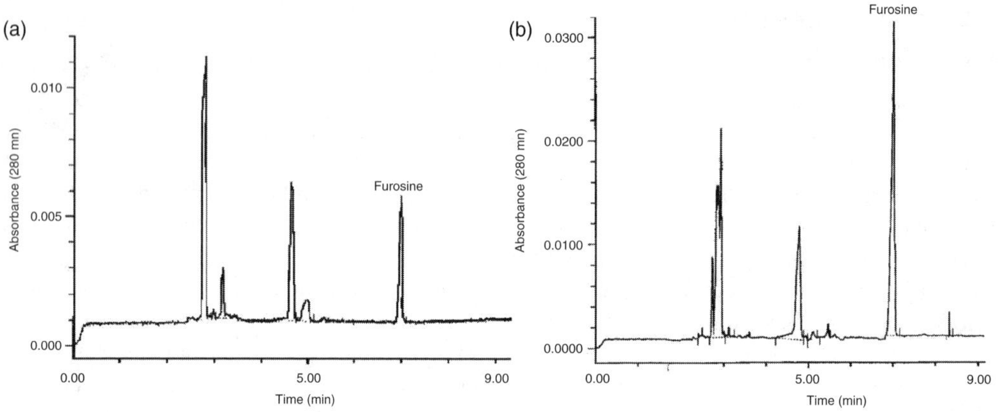

FIGURE 30.9 Typical electropherograms of milk showing the furosine peak. (a) UHT milk containining 269 mg of furosine/100 g protein, (b) Evaporated milk containing 730 mg of furosine/100 g protein. (From Vallejo-Cordoba, B., et al., *J. Agric. Food Chem.*, 52, 5787, 2004. With permission.)

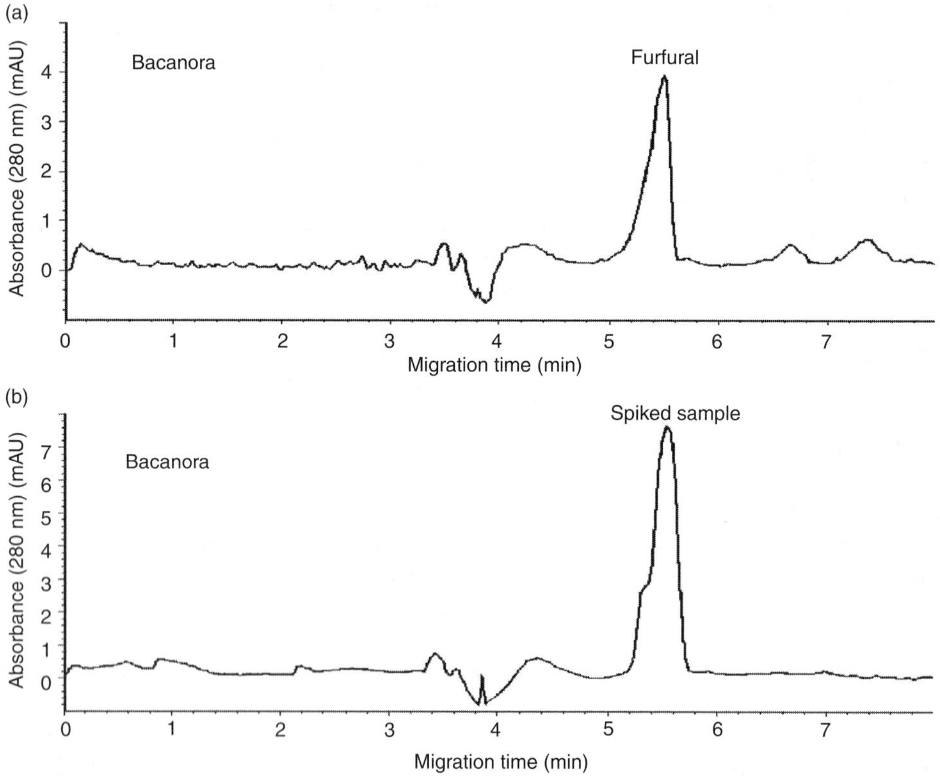

FIGURE 30.10 Typical electropherograms of Bacanora (a), and Bacanora spiked with the analytical standard (b). (From Vallejo-Cordoba, B. and Gonzalez-Cordova, A. F., in *Hispanic Foods, Chemistry and Flavor*, ACS Series, 946, Tunick, M. H., and González de Mejia, E. Ed., 2007, p. 153. With permission.)

and/or quality required for normal or accepted processing or product formulation.[6] A comprehensive review of CE applications in food authenticity was offered by Kvasnicka.[4] Authenticity issues dealing with meat and meat products have been covered by Vallejo-Cordoba et al.,[6] and applications dealing with dairy products authenticity were reviewed by Recio et al.[224] Detailed knowledge on raw materials constituting foods is essential, and this is especially true for the diverse foods and feeds made from cereal grains. Varietal identity and ability to predict dough properties of wheat consignments are largely dependent on protein composition. Whereas DNA analysis indicates genotype, protein composition provides information on variety and likely processing parameters. In this sense, the Australian industry has standard methods of protein and DNA analysis involving gel electrophoresis, CE, and most recently, Lab-on-a-chip equipment for protein analysis and microarrays for DNA composition.[278] The treatment of foods can be another issue in food authenticity; for example, meat may have been previously frozen and falsely represented as fresh, especially with fresh or chilled meat associated with higher prices than their frozen counterparts. Thus, fraudulent marketers may thaw frozen meat and sell it as chilled or fresh. Methods for the differentiation between fresh and frozen/thawed meat were based on the release of enzymes. Thus, the application of electrophoretically mediated microanalysis (EMMA) to determine β-hydroxyacyl CoA-dehydrogenase (β-HADH) activity in a model system and in meat samples was reported. The enzymatic assay and the separation of the reaction products were carried out in an uncoated capillary using a plug–plug reaction mode at variable potential. The NAD^+ produced by β-HADH activity in juice extracted from previously frozen meat was at least three times the concentration of that produced in fresh meat (Figure 30.11).[279]

The most important issue in food quality analysis is food safety. Therefore, the development of rapid methods for the identification and quantification of bacterial contamination in foods is of utmost importance. Thus, a CE method with UV detection was proposed for the identification and quantification of bacterial contamination in food samples. The proposed method allowed for the effective separation of eight different types of bacteria in only 25 min. Electrophoretic resolution was improved by using cations in phosphate buffer (pH 7.0) that interacted with the bacterial surfaces changing its electrical properties and electrophoretic mobility. The validity of the method was established by comparison with the standard plate counting method,[151] where bacterial cells were separated as

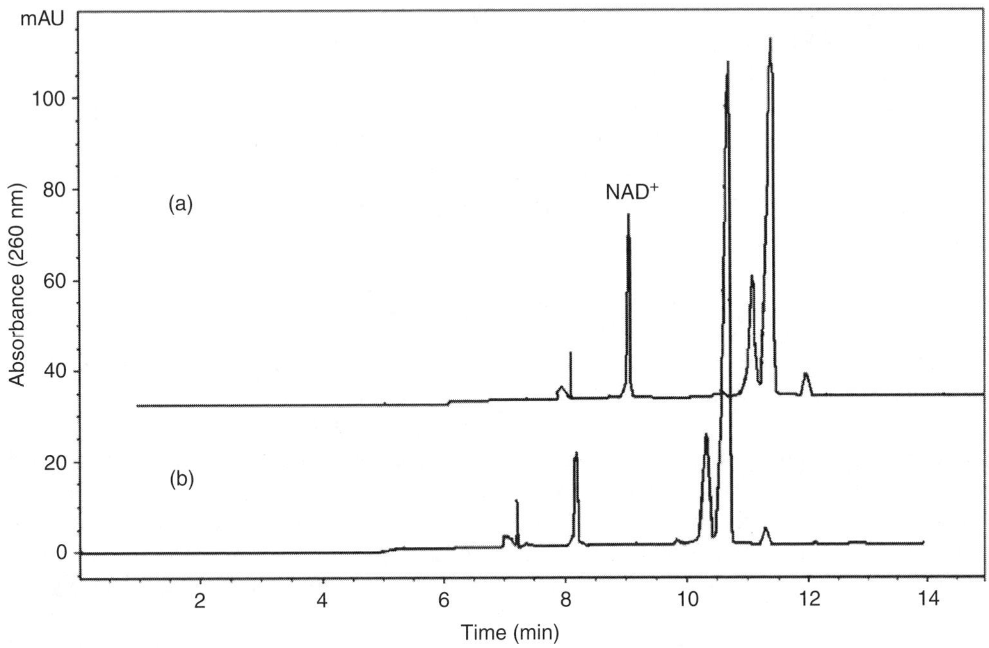

FIGURE 30.11 Electropherogram showing β-HADH activity by EMMA in extracted juice from beef. (a) Frozen thawed and (b) fresh. (From Vallejo-Cordoba, B., et al., *J. Cap. Elec. Microchip Tech.*, 8, 81, 2003. With permission.)

"biocolloids," which under the influence of an electric field, move at a velocity that is proportional to their electrophoretic mobility. This means that under appropriate experimental conditions, different microorganisms exhibit a differential electrophoretic mobility.[151] Another approach for the identification of foodborne pathogens or food-spoilage bacteria consisted in multiplex-PCR-CGE-LIF procedures.[258,259] The main advantage of these techniques over conventional microbiological techniques is that it is possible to obtain fast and specific identification of microorganisms at the molecular/genetic level. One of the PCR-CGE-LIF protocols allowed for the simultaneous detection and differentiation of the genera *Leuconostoc* and *Carnobacterium*,[258] while a second protocol allowed the detection of *Staphylococcus aureus*, *Listeria monocytogenes*, and *Salmonella* spp.[259]

The determination of food fatty acids is important from the standpoint of food quality composition since fatty acid composition is required in nutritional labeling. Also, the quantification of *trans*-fatty acids, which are formed during oil hydrogenation, is particularly important since new regulations require their labeling to allow the consumer to make informed choices on their diet. Thus, a novel CE methodology using UV indirect detection at 224 nm for the analysis of *trans*-fatty acids in hydrogenated oils was reported. The electrolyte consisted of phosphate buffer (pH 7.0) containing sodium dodecylbenzenesulfonate, Brij 35, octanol, and acetonitrile, where, under optimized conditions, 10 fatty acids including C:12, C:13 (internal standard), C:14, C:16, C:18, C18:1c, C18:1t, C18:2cc, C18:2tt, and C18:3ccc were baseline separated in less than 12 min. The proposed method was used to monitor the formation of *trans*-fatty acids during hydrogenation of Brazilnut oil.[280] Similarly, the determination of short-chain free fatty acids, which are associated with lipolized flavors, is important for good sensory quality. However, the quantification of individual free fatty acids (FFA) in dairy products is particularly complicated, since FFA represent less than 0.5% of the total fat. FFA with chain lengths from 2 to 20 carbon atoms are present, and short-chain FFA with fewer than eight carbon atoms are extremely volatile. Because of these problems, the methods most frequently described for the analysis of FFA in milk and dairy products involve fat extraction, separation of FFA from the bulk of the extracted lipid, and injection of the fraction containing FFA directly in the gas chromatograph, or fat extraction, derivatization, and gas chromatography of the methyl esters formed.[282] To circumvent these problems, an attractive alternative was the use of CE, particularly in a MEKC mode. To this end, the release of short-chain free fatty acids was monitored from milk fat during hydrolysis with lipase by using MEKC.[282] Sample buffer containing CD and methanol allowed for the solubilization of FFA, followed by separation in an uncoated capillary with Tris buffer containing *p*-anisate as chromophore and trimethyl β-CD at pH 8.0. Indirect UV detection at 270 nm was used[282] and excellent separation was observed for FFA C_4-C_{12} in a FFA standard mixture (Figure 30.12a) or in milk fat lipolyzed with lipase (Figure 30.12b).

30.3 CONCLUDING REMARKS AND OUTLOOK

The versatility of CE in food analysis was clearly shown by the numerous, wide ranging methods and applications presented. Moreover, the efficiency and rapidity of CE separations were the main advantages that make this technique an attractive alternative in food analysis laboratories. CE is still far from being as well established as chromatographic techniques, although it certainly has gained popularity as an alternative to conventional gel electrophoresis for the analysis of food proteins, most likely due to the quantitative nature of CE, high resolving power, and speed of analysis. However, for the analysis of trace component determination for an organic contaminant, low sensitivity and low reproducibility of quantitative analysis are issues that need to be addressed. Perhaps the greatest potential of CE was found in food diagnostics for quality assurance and authenticity, with most recent approaches involving Lab-on-chip equipment. Authenticity issues such as meat and fish species, GMOs, and cereal variety identification and quantification are challenging problems where DNA-based CE methods in microchip-based formats offer opportunities for development. Similarly, DNA-based CE methods for microorganism identification is another niche that may be explored.

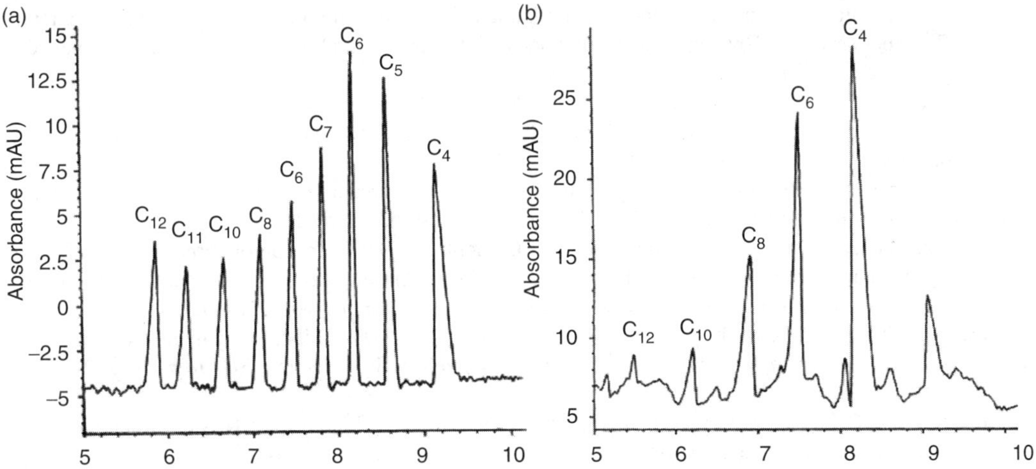

FIGURE 30.12 Typical electropherograms of free fatty acids. (a) Standard mixture (100 ppm of each), (b) milk fat after a 30 min hydrolysis with lipase. (From Vallejo-Cordoba, B., et al., *J. Cap. Elec.*, 5, 11, 1998. With permission.)

ACKNOWLEDGMENTS

M.G. Vargas would like to thank the National Autonomous University of Mexico (Universidad Nacional Autónoma de México, UNAM, FES-Cuautitlán, Depto de Química) for time and resources to complete this work. Author B. Vallejo-Cordoba would like to acknowledge the Research Center for Food and Development (Centro de Investigación en Alimentación y Desarrollo, A.C., CIAD, A.C.) for their financial assistance, Dr. James Landers for his kind invitation to make this contribution, and Dr. Norberto Guzman for sharing his enthusiasm on CE technology.

REFERENCES

1. Castañeda, G., Rodríguez-Flores, J., and Ríos, A., Analytical approaches to expanding the use of capillary electrophoresis in routine food analysis, *J. Sep. Sci.*, 28, 915, 2005.
2. Cifuentes, A., Recent advances in the application of capillary electromigration methods for food analysis, *Electrophoresis*, 27, 283, 2006.
3. Juan-García, A., Font, G., and Picó, Y., Determination of organic contaminants in food by capillary electrophoresis, *J. Sep. Sci.*, 28, 793, 2005.
4. Kvasnicka, F., Capillary electrophoresis in food authenticity, *J. Sep. Sci.*, 28, 813, 2005.
5. Frazier, R. A., Ames, J. M., and Nursten, H. E., *Capillary electrophoresis for food analysis: Method development*, The Royal Society of Chemistry, Cambridge, U.K., 2000, p. 67.
6. Vallejo-Córdoba, B., González-Córdoba, A.F., Mazorra-Manzano, M.A., and Rodríguez-Ramírez, R., Capillary electrophoresis for the analysis of meat authenticity, *J. Sep. Sci.*, 28, 826, 2005.
7. Frazier, R. A., Ames, J. M., and Nursten, H. E., The development and application of capillary electrophoresis methods for food analysis, *Electrophoresis*, 20, 3156, 1999.
8. Frazier, R. A., Recent advances in capillary electrophoresis methods for food analysis, *Electrophoresis*, 22, 4197, 2001.
9. Frazier, R. A., and Papadopoulou, A., Recent advances in the application of capillary electrophoresis for food analysis, *Electrophoresis*, 24, 4095, 2003.
10. Sørensen, H., Sørensen, S., Bjergegaard, C., and Michaelsen, S., *Chromatographic and Capillary Electrophoresis in Food Analysis*, The Royal Society of Chemistry, Cambridge, U.K., 1999.
11. Wang, X., Wang, Q., Chen, Y., and Han, H., Determination of carbohydrates as their 3-aminophthalhydrazide derivatives by capillary zone electrophoresis with on-line chemiluminescence detection, *J. Chromatogr. A*, 992, 181, 2003.

12. He, L., Sato, K., Abo, M., Okubo, A., and Yamazaki, S., Separation of saccharides derivatized with 2-aminobenzoic acid by capillary electrophoresis and their structural consideration by nuclear magnetic resonance, *Anal. Biochem.*, 314, 128, 2003.
13. Zhang, M., Melouk, H. A., Chenault, K., and El Rassi, Z., Determination of cellular carbohydrates in peanut fungal pathogens and Baker's yeast by capillary electrophoresis and electrochromatography, *J. Agric. Food Chem.*, 49, 5265, 2001.
14. Doner, L. W., Johnston, D. B., and Singh, V., Analysis and properties of arabinoxylans from discrete corn wet-milling fiber fractions, *J. Agric. Food Chem.*, 49, 1266, 2001.
15. Cabálková, J., Žídková, J., Pribyla, L., and Chmelík, J., Determination of carbohydrates in juices by capillary electrophoresis, high-performance liquid chromatography, and matrix-assisted laser desorption/ionization-time of flight-mass spectrometry, *Electrophoresis*, 25, 487, 2004.
16. Carvalho, A. Z., da Silva, J. A. F., and do Lago, C. L., Determination of mono- and disaccharides by capillary electrophoresis with contactless conductivity detection, *Electrophoresis*, 24, 2138, 2003.
17. Klampfl, C. W., and Buchberger, W., Determination of carbohydrates by capillary electrophoresis with electrospray-mass spectrometric detection, *Electrophoresis*, 22, 2737, 2001.
18. Soga, T., and Serwe, M., Determination of carbohydrates in food samples by capillary electrophoresis with indirect UV detection, *Food Chem.*, 69, 339, 2000.
19. Khandurina, J., Anderson, A. A., Olson, N. A., Stegel, J. T., and Guttman, A., Large-scale carbohydrate analysis by capillary array electrophoresis: Part 2. Data normalization and quantification, *Electrophoresis*, 25, 3122, 2004.
20. Hebert, N. E., Kuhr, W. G., and Brazill, S. A., Microchip capillary electrophoresis coupled to sinusoidal voltammetry for the detection of native carbohydrates, *Electrophoresis*, 23, 3750, 2002.
21. Andersen, K. E., Bjergegaard, C., and Sørensen, H., Analysis of reducing carbohydrates by reductive tryptamine derivatization prior to micellar electrokinetic capillary chromatography, *J. Agric. Food Chem.*, 51, 7234, 2003.
22. Momenbeik, F., Johns, C., Breadmore, M. C., Hilder, E. F., Macka, M., and Haddad, P. R., Sensitive determination of carbohydrates labelled with *p*-nitroaniline by capillary electrophoresis with photometric detection using a 406 nm light-emitting diode, *Electrophoresis*, 27, 4039, 2006.
23. Mangin, C. M., Goodall1, D. M., and Roberts, M. A., Separation of ι-, χ- and λ-carrageenans by capillary electrophoresis, *Electrophoresis*, 22, 1460, 2001.
24. Mistry, K., Krull1, I., and Grinberg, N., Size-exclusion capillary electrochromatographic separation of polysaccharides using polymeric stationary phases, *Electrophoresis*, 24, 1753, 2003.
25. Cortacero-Ramírez, S., Segura-Carretero, A., Cruces-Blanco, C., Hernáinz-Bermúdez de Castro, M., and Fernández-Gutiérrez, A., Analysis of carbohydrates in beverages by capillary electrophoresis with precolumn derivatization and UV detection, *Food Chem.*, 87, 471, 2004.
26. Campa, C., Coslovi1, A., Flamigni1, A., and Rossi1, M., Overview on advances in capillary electrophoresis-mass spectrometry of carbohydrates: A tabulated review, *Electrophoresis*, 27, 2027, 2006.
27. Martínez-Montero, C., Rodríguez-Dodero, M. C., Guillen-Sánchez, D. A., and Barroso, C. G., Analysis of low molecular weight carbohydrates in food and beverages: A review, *Chromatographia*, 59, 15, 2004.
28. Klockow, A., Paulus, A., Figueiredo, V., Amado, R., and Widmer, H. M., Determination of carbohydrates in fruit juices by capillary electrophoresis and high-performance liquid chromatography, *J. Chromatogr. A*, 680, 187, 1994.
29. Vorndran, A. E., Oefner, P. J., Scherz, H., and Bonn, G. K., Indirect UV detection of carbohydrates in capillary zone electrophoresis, *Chromatographia*, 33, 163, 1992.
30. Soga, T., and Heiger, D. N., Simultaneous determination of monosaccharides in glycoproteins by capillary electrophoresis, *Anal. Biochem.*, 261, 73, 1998.
31. Suzuki, S., and Honda S., Miniaturization in carbohydrate analysis, *Electrophoresis*, 24, 3577, 2003.
32. Legaz, M. E., Pedrosa, M.M., and Vicente, C., Carbohydrates, polymeric sugars and their constituents, in *Advanced Chromatographic and Electromigration Methods in Bioscience*, Deyl, S., Milsik, I., Tagliaro, F., and Tesarova, E., Ed. *J. of Chromatogr.* 60, 264, 1998.
33. Chiu, T. C., Lin, Y. W., Huang, Y. F., and Chang, H. T., Analysis of biologically active amines by CE, *Electrophoresis*, 27, 4792, 2006.

34. Pais, P., and Knize, M.G., Chromatographic and related techniques for the determination of aromatic heterocyclic amines in foods, *J. Chromatogr. B*, 747, 139, 2000.
35. Önal, A., A review: Current analytical methods for the determination of biogenic amines in foods, *Food Chem.*, 108, 1475, 2007.
36. Oguri, S., Electromigration methods for amino acids, biogenic amines and aromatic amines, *J. Chromatogr. B*, 747, 1, 2000.
37. Kovács, Á., Simon-Sarkadi, L., and Ganzler, K., Determination of biogenic amines by capillary electrophoresis, *J. Chromatogr. A*, 836, 305, 1999.
38. Liu, X., Yang, L. X., and Lu, Y. T., Determination of biogenic amines by 3-(2-furoyl) quinoline-2-carboxaldehyde and capillary electrophoresis with laser-induced fluorescence detection, *J. Chromatogr.A*, 998, 213, 2003.
39. Rodríguez, I., Lee, H. K., and Li, S. F. Y., Microchannel electrophoretic separation of biogenic amines by micellar electrokinetic chromatography, *Electrophoresis*, 20, 118, 1999.
40. Molnar-Perl, I., Role of chromatography in the analysis of sugars, carboxylic acids and amino acids in food, *J. Chromatogr. A*, 891, 1, 2000.
41. Poinsot, V., Lacroix, M., Maury, D., Chataigne, G., Feurer, B., and Couderc, F., Recent advances in amino acid analysis by capillary electrophoresis, *Electrophoresis*, 27, 176, 2006.
42. Simó, C., Barbas, C., and Cifuentes, A., Capillary electrophoresis-mass spectrometry in food analysis, *Electrophoresis*, 26, 1306, 2005.
43. Ummadi, M., and Weimer, B. C., Use of capillary electrophoresis and laser-induced fluorescence for attomole detection of amino acids, *J. Chromatogr. A*, 964, 243, 2002.
44. Kostal, V., Zeisbergerova, M., Hrotekova, Z., Slais, K., and Kahle V., Miniaturized liquid core waveguide-based fluorimetric detection cell for capillary separation methods: Application in CE of amino acids, *Electrophoresis*, 27, 4658, 2006.
45. Dong, Q., Jin, W., and Shan, J., Analysis of amino acids by capillary zone electrophoresis with electrochemical detection, *Electrophoresis*, 23, 559, 2002.
46. Coufal, P., Zuska, J., van de Goor, T., Smith, V., and Gaš, B., Separation of twenty underivatized essential amino acids by capillary zone electrophoresis with contactless conductivity detection, *Electrophoresis*, 24, 671, 2003.
47. Schultz, C. L., and Moini, M., Analysis of underivatized amino acids and their d/l-enantiomers by sheathless capillary electrophoresis/electrospray ionization-mass spectrometry, *Anal. Chem.*, 75, 1508, 2003.
48. Soga, T., and Heiger, D.N., Amino acid analysis by capillary electrophoresis electrospray ionization mass spectrometry, *Anal. Chem.*, 72, 1236, 2000.
49. Siri, N., Lacroix, M., Garrigues, J. C., Poinsot, V., and Couderc, F., HPLC-fluorescence detection and MEKC-LIF detection for the study of amino acids and catecholamines labelled with naphthalene-2,3-dicarboxyaldehyde, *Electrophoresis*, 27, 4446, 2006.50.
50. Lioe, H. N., Apriyantono, A., Takara, K., Wada, K., Naoki, H., and Yasuda, M., Low molecular weight compounds responsible for savory taste of indonesian soy sauce, *J. Agric. Food Chem.*, 52, 5950, 2004.
51. García-Villar, N., Saurina, J., and Hernández-Cassou, S., Capillary electrophoresis determination of biogenic amines by field-amplified sample stacking and in-capillary derivatization, *Electrophoresis*, 27, 474, 2006.
52. Santos, B., Simonet, B. M., Ríos, A., and Valcárcel, M., Direct automatic determination of biogenic amines in wine by flow injection-capillary electrophoresis-mass spectrometry, *Electrophoresis*, 25, 342, 2004.
53. Male, K. B., and Luong, J. H. T., Derivatization, stabilization and detection of biogenic amines by cyclodextrin-modified capillary electrophoresis–laser-induced fluorescence detection, *J. Chromatogr. A*, 926, 309, 2001.
54. Qu, Q., Liu, Y., Tang, X., Wang, C., Yang, G., Hu1, X., and Yan, C., Etched bare fused-silica capillaries for online preconcentration of amino acids in CE, *Electrophoresis*, 27, 4500, 2006.
55. Lee, J. H., Choi, O. K., Jung, H. S., Kim, K. R., and Chung, D. S., Capillary electrophoresis of nonprotein and protein amino acids without derivatization, *Electrophoresis*, 21, 930, 2000.
56. Zhan, W., Wang, T., and Li, S. F. Y., Derivatization, extraction and concentration of amino acid and peptides by using organic/aqueous phases in capillary electrophoresis with fluorescence detection, *Electrophoresis*, 21, 3593, 2000.

57. Klampfl C. W., and Ahrer, W., Determination of free amino acids in infant food by capillary zone electrophoresis with mass spectrometric detection, *Electrophoresis*, 22, 1579, 2001.
58. Cao, L., Wang, H., and Zhang, H., Analytical potential of 6-oxy-(*N*-succinimidyl acetate)-9-(2′-methoxycarbonyl) fluorescein for the determination of amino compounds by capillary electrophoresis with laser-induced fluorescence detection, *Electrophoresis*, 26, 1954, 2005.
59. Beard, N. P., and de Mello, A. J., A polydimethylsiloxane/glass capillary electrophoresis microchip for the analysis of biogenic amines using indirect fluorescence detection, *Electrophoresis*, 23, 1722, 2002.
60. Tanyanyiwa, J., Schweizer, K., and Hauser, P. C., High-voltage contactless conductivity detection of underivatized amino acids in capillary electrophoresis, *Electrophoresis*, 24, 2119, 2003.
61. Lange, J., Thomas, K., and Wittmann, C., Comparison of a capillary electrophoresis method with high performance liquid chromatography for the determination of biogenic amines in various food samples, *J. Chromatogr. B*, 779, 229, 2002.
62. Hart, J. J., Welch, R. M., Norvell, W. A., and Kochian, L.V., Measurement of thiol-containing amino acids and phytochelatin (PC2) via capillary electrophoresis with laser-induced fluorescence detection, *Electrophoresis*, 23, 81, 2002.
63. Shin, D., Tryk, D. A., Fujishima, A., Muck Jr., A., Chen, G., and Wang, J., Microchip capillary electrophoresis with a boron-doped diamond electrochemical detector for analysis of aromatic amines, *Electrophoresis*, 25, 3017, 2004.
64. Zhang, Y., and Gomez, F. A., On-column derivatization and analysis of amino acids, peptides and alkylamines by anhydrides using capillary electrophoresis, *Electrophoresis*, 21, 3305, 2000.
65. Ruiz-Jiménez, J., and Luque de Castro, M. D., Pervaporation as interface between solid samples and capillary electrophoresis, Determination of biogenic amines in food, *J. Chromatogr. A*, 1110, 245, 2006.
66. Driouich, R., Takayanagi, T., Oshima, M., and Motomizu, S., Synthesis of water-soluble phenylazosalicylaldehyde analogues and their application to capillary electrophoretic determination of primary amines, *Electrophoresis*, 27, 3460, 2006.
67. La, S., Kim, A., Kim, J. H., Choi, O. K., and Kim, K. R., Profiling and screening analysis of 27 aromatic amino acids by capillary electrophoresis in dual modes, *Electrophoresis*, 23, 1080, 2002.
68. Lin, C. C., and Liu, C. Y., Proline-coated column for the capillary electrochromatographic separation of amino acids by in-column derivatization, *Electrophoresis*, 25, 3216, 2004.
69. Beard, N. P., Edel, J. B., and de Mello, A. J., Integrated on-chip derivatization and electrophoresis for the rapid analysis of biogenic amines, *Electrophoresis*, 25, 2363, 2004.
70. Lin, Z., Xie, Z., Lu, H., Lin, X., Wu, X., and Chen, G., On-column coaxial flow chemiluminescence detection for underivatized amino acids by pressurized capillary electrochromatography using a monolithic column, *Anal. Chem.*, 78, 5322, 2006.
71. Ro, K. W., Lim, K., Kim, H., and Hahn, J. H., Poly(dimethylsiloxane) microchip for precolumn reaction and micellar electrokinetic chromatography of biogenic amines, *Electrophoresis*, 23, 1129, 2002.
72. Soga, T., and Imaizumi, M., Capillary electrophoresis method for the analysis of inorganic anions, organic acids, amino acids, nucleotides, carbohydrates and other anionic compounds, *Electrophoresis*, 22, 3418, 2001.
73. Cortacero-Ramírez, S., Segura-Carretero, A., Cruces-Blanco, C., Hernáinz-Bermúdez de Castro, M., and Fernández-Gutiérrezet, A., Direct multicomponent analysis of beer samples constituents using micellar electrokinetic capillary chromatography, *Electrophoresis*, 25, 1867, 2004.
74. Kvasnicka, F., and Voldrich, M., Determination of biogenic amines by capillary zone electrophoresis with conductometric detection, *J. Chromatogr. A*, 1103, 145, 2006.
75. Vargas M. G., Muñoz Ramírez, E., and Trejo Márquez, A., Development of analytical methods for the determination of polyamines by MECK using LIF detection, *Presented at the 12th Latin-American Symposium on Biotechnology, Biomedical, Biopharmaceutical and Industrial Applications of Capillary Electrophoresis and Microchip Technology*, Queretaro, Mexico, 2–5 December, 2006.
76. Cinquina, A. L., Longo, F., Cali, A., De Santis, L., Baccelliere, R., and Cozzani, R., Validation and comparison of analytical methods for the determination of histamine in tuna fish samples, *J. Chromatogr. A*, 1032, 79, 2004.

77. Ong, C. P., Ng, C. L., Lee, H. K., and Li, S. F. Y., Separation of water- and fat-soluble vitamins by micellar electrokinetic chromatography, *J. Chromatogr. A*, 547, 419, 1991.
78. Craige Trenerry, V., The application of capillary electrophoresis to the analysis of vitamins in food and beverages, *Electrophoresis*, 22, 1468, 2001.
79. Chiari, M., Nesi, M., Carrea, G., and Righetti, P. G., Determination of total vitamin C in fruits by capillary zone electrophoresis, *J. Chromatogr. A*, 645, 197, 1993.
80. Thompson, C. O., and Trenerry, V. C., A rapid method for the determination of total L-ascorbic acid in fruits and vegetables by micellar electrokinetic capillary chromatography, *Food Chem.*, 53, 43, 1995.
81. Panak, K., Giorgieri, S. A., Ruiz, O., and Diaz, L. E., Postharvest analysis of vitamin C and inorganic cations in lettuce by CZE, *J. Capil. Electrophor.*, 5, 59, 1998.
82. Boyce, M. C., and Spickett, E. E., Separation of food grade antioxidants (synthetic and natural) using mixed micellar electrokinetic capillary chromatography, *J. Agric. Food. Chem.*, 47, 1970, 1999.
83. Buskov, S., Moller, P., Sorensen, H., Sorensen, J. C., and Sorensen, S., Determination of vitamins in food based on supercritical fluid extraction prior to micellar electrokinetic capillary chromatographic analyses of individual vitamins, *J. Chromatogr. A*, 802, 233, 1998.
84. Begona, B. M., and van de Werken, G., Determination of green and black tea composition by capillary electrophoresis, *J. High Resolut. Chromatogr.*, 22, 225, 1999.
85. Schiewe, J., Mrestani, Y., and Neubert, R., Application and optimization of capillary zone electrophoresis in vitamin analysis, *J. Chromatogr. A*, 717, 255, 1995.
86. Koh, E. V., Bissell, M. G., and Ito, R. K., Measurement of vitamin C by capillary electrophoresis in biological fluids and fruit beverages using a stereoisomer as an internal standard, *J. Chromatogr. A*, 633, 245, 1993.
87. Lin Ling, B., Baeyens, W. R. G., Van Ackers, P., and Dewaele, C., Determination of ascorbic acid and isoascorbic acid by capillary zone electrophoresis: Application to fruit juices and to a pharmaceutical formulation, *J. Pharm. Biomed. Anal.*, 10, 717, 1992.
88. Choi, O., and Jo, J., Determination of L-ascorbic acid in foods by capillary zone electrophoresis, *J. Chromatogr. A*, 781, 435, 1997.
89. Marshall, P. A., Trenerry, V. C., and Thompson, C. O., The determination of total ascorbic acid in beers, wines, and fruit drinks by micellar electrokinetic capillary chromatography, *J. Chromatogr. Sci.*, 33, 426, 1995.
90. Ward, C. M., Trenerry, V. C., and Pant, I., The application of capillary electrophoresis to the determination of total niacin in concentrated yeast, *Food Chem.*, 58, 185, 1997.
91. Ward, C. M., and Trenerry, V. C., The determination of niacin in cereals, meat and selected foods by capillary electrophoresis and high performance liquid chromatography, *Food Chem.*, 60, 667, 1997.
92. Diaz-Pollan, C., and Vidal-Valverde, C., Niacin determination in legumes by capillary electrophoresis (CE). Comparison with high performance liquid chromatography (HPLC), *J. High Resolut. Chromatogr.*, 21, 81, 1998.
93. Windahl, K. W., Trenerry, V. C., and Ward, C. M., The determination of niacin in selected foods by capillary electrophoresis and high performance liquid chromatography: Acid extraction, *Food Chem.*, 65, 263, 1998.
94. Vidal-Valverde, C., and Diaz-Pollan, C., Optimization analysis by capillary electrophoresis of thiamine in meat: Comparison with high perfomance liquid chromatography, *Eur. Food Res. Technol.*, 209, 355, 1999.
95. Vidal-Valverde, C., and Diaz-Pollan, C., Comparison of capillary electrophoretic and high performance liquid chromatographic thiamin determination in milk, *Milchwissenschaft*, 55, 307, 2000.
96. Sanchez, J. M., and Salvado, V., Comparison of micellar and microemulsion electrokinetic chromatography for the analysis of water- and fat-soluble vitamins, *J. Chromatogr. A*, 950, 241, 2002.
97. Herrero-Martinez, J. M., Simo-Alfonso, E. F., Ramis-Ramos, G., Deltoro, V. I., Calatayud, A., and Barreno, E., Simultaneous determination of L-ascorbic acid, glutathione, and their oxidized forms in ozone-exposed vascular plants by capillary zone electrophoresis, *Environ. Sci. Technol.*, 34, 1331, 2000.

98. Liao, T., Jiang, C. M., Wu, M. C., Hwang, J. Y., and Chang, H. M., Quantification of L-ascorbic acid and total ascorbic acid in fruits and spinach by capillary zone electrophoresis, *Electrophoresis*, 22, 1484, 2001.
99. Cataldi, T. R. I., Nardiello, D., Carrara, V., Cirielloa, R., and De Benedetto G. E., Assessment of riboflavin and flavin content in common food samples by capillary electrophoresis with laser-induced fluorescence detection, *Food Chem.*, 82, 309, 2003.
100. Cataldi, T. R. I., Nardiello, D., Scrano, L., and Scopa, A., Assay of riboflavin in sample wines by capillary zone electrophoresis and laser-induced fluorescence detection, *J. Agric. Food Chem.*, 50, 6643, 2002.
101. Aturki, Z., D'Orazio, G., and Fanali, S., Rapid assay of vitamin E in vegetable oils by reversed-phase capillary electrochromatography, *Electrophoresis*, 26, 798, 2005.
102. Law, W. S., Kubáň, P., Hong Zhao, J., Li, S. F. Y., and Hauser, P. C., Determination of vitamin C and preservatives in beverages by conventional capillary electrophoresis and microchip electrophoresis with capacitively coupled contactless conductivity detection, *Electrophoresis*, 26, 4648, 2005.
103. Cataldi, T. R. I., Nardiello, D., De Benedetto, G. E., and Bufo, S. A., Optimizing separation conditions for riboflavin, flavin mononucleotide and flavin adenine dinucleotide in capillary zone electrophoresis with laser-induced fluorescence detection, *J. Chromatogr. A*, 968, 229, 2002.
104. Frias, J., Miranda, M. L., Doblado, R., and Vidal-Valverde, C., Effect of germination and fermentation on the antioxidant vitamin content and antioxidant capacity of *Lupinus albus* L. var. Multolupa, *Food Chem.*, 92, 211, 2005.
105. Aurant, L., Woods, A., and Wells, M., Food laws and regulation. In *Food Composition and Analysis*, Van Nostrand Reinhold, New York, 1987, p. 14.
106. Galli, V., Garcia, A., Saavedra, L., and Barbas, C., Capillary electrophoresis for short-chain organic acids and inorganic anions in different samples, *Electrophoresis*, 24, 1951, 2003.
107. Cortacero-Ramirez, S., Hernainz-Bermudez de Castro, M., Segura-Carretero, A., and Cruces-Blanco, C., Analysis of beer components by capillary electrophoretic methods, *Trends Anal. Chem.*, 22, 440, 2003.
108. Vickers, P. J., Braybrook, J., Lawrence, P., and Gray, K., Detecting tartrate additives in foods: Evaluating the use of capillary electrophoresis, *J. Food Composition Anal.*, 20, 252, 2007.
109. Cortacero-Ramírez, S., Segura-Carretero, A., Hernáinz-Bermúdez de Castro, M., and Fernández-Gutiérrez, A., Determination of low-molecular-mass organic acids in any type of beer samples by coelectroosmotic capillary electrophoresis, *J. Chromatogr. A*, 1064, 115, 2005.
110. Izco, J. M., Tormo, M., and Jimenez-Flores, R., Development of a CE method to analyze organic acids in dairy products: Application to study the metabolism of heat-shocked spores, *J. Agric. Food Chem.*, 50, 1765, 2002.
111. Suárez-Luque, S., Mato, I., Huidobro, J. F., Simal-Lozano, J., and Sancho, M. T., Capillary zone electrophoresis method for the determination of inorganic anions and formic acid in honey, *J. Agric. Food Chem.*, 54, 9292, 2006.
112. Saavedra, L., Rupérez, F. J., and Barbas, C., Capillary electrophoresis for evaluating orange juice authenticity: A study on Spanish oranges, *J. Agric. Food Chem.*, 49, 9, 2001.
113. Galli, V., and Barbas, C., Capillary electrophoresis for the analysis of short-chain organic acids in coffee, *J. Chromatogr. A*, 1032, 299, 2004.
114. Esteves, V. I., Lima, S. S. F., Lima, D. L. D., and Duarte, A. C., Using capillary electrophoresis for the determination of organic acids in Port wine, *Anal. Chim. Acta*, 513, 163, 2004.
115. Saavedra, L., and Barbas, C., Validated capillary electrophoresis method for small-anions measurement in wines, *Electrophoresis*, 24, 2235, 2003.
116. Castiñeira, A., Peña, R. M., Herrero, C., and Garci⊥'ıa-Martín, S., Analysis of organic acids in wine by capillary electrophoresis with direct UV detection, *J. Food Comp. Anal.*, 15, 319, 2002.
117. Nhujak, T., Srisa-Art, M., Kalampakorn, K., Tolieng, V., and Petsom, A., Determination of gibberellic acid in fermentation broth and commercial products by micellar electrokinetic chromatography, *J. Agric. Food Chem.*, 53, 1884, 2005.
118. Kubáň, P., and Hauser, P. C., Application of an external contactless conductivity detector for the analysis of beverages by microchip capillary electrophoresis, *Electrophoresis*, 26, 3169, 2005.
119. Fung, Y. S., and Lau, K. M., Analysis of organic acids and inorganic anions in beverage drinks by capillary electrophoresis, *Electrophoresis*, 24, 3224, 2003.

120. Tanyanyiwa, J., and Hauser, P.C., High-voltage capacitively coupled contactless conductivity detection for microchip capillary electrophoresis, *Anal. Chem.*, 74, 6378, 2002.
121. Yamamoto, A., Akiba, N., Kodama, S., Matsunaga, A., Kato, K., and Nakazawa, H., Enantiomeric purity determination of malic acid in apple juices by multibeam circular dichroism detection, *J. Chromatogr. A*, 928, 139, 2001.
122. Menzinger, F., Schmitt-Kopplin, Ph., Freitag, D., and Kettrup, A., Analysis of agrochemicals by capillary electrophoresis, *J. Chromatogr. A*, 891, 45, 2000.
123. Weinberger, R., Capillary electrophoresis of venoms and toxins, *Electrophoresis*, 22, 3639, 2001.
124. Hernández-Borges, J., Rodríguez-Delgado, M. Á., García-Montelongo, F. J., and Cifuentes A., Chiral analysis of pollutants and their metabolites by capillary electromigration methods, *Electrophoresis*, 26, 3799, 2005.
125. Ahmed, F. E., Analyses of pesticides and their metabolites in foods and drinks, *Trends Anal. Chem.*, 20, 649, 2001.
126. Eash, D. T., and Bushway, R. J., Herbicide and plant growth regulator analysis by capillary electrophoresis, *J. Chromatogr. A*, 880, 281, 2000.
127. Rodríguez, R., Mañes, J., and Picó, Y., Off-line solid-phase microextraction and capillary electrophoresis mass spectrometry to determine acidic pesticides in fruits, *Anal. Chem.*, 75, 452, 2003.
128. Orejuela, E., and Silva, M., Rapid and sensitive determination of phosphorus-containing amino acid herbicides in soil samples by capillary zone electrophoresis with diode laser-induced fluorescence detection, *Electrophoresis*, 26, 4478, 2005.
129. Orejuela, E., and Silva, M., Rapid determination of aniline metabolites of chlorpropham in potatoes by micellar electrokinetic chromatography using negative-charged mixed micelles and laser-induced fluorescence detection, *Electrophoresis*, 26, 2991, 2005.
130. Wei, J., Okerberg, E., Dunlap, J., Ly, C., and Shear, J. B., Determination of biological toxins using capillary electrokinetic chromatography with multiphoton-excited fluorescence, *Anal. Chem.*, 72, 1360, 2000.
131. Cooper, P. A., Jessop, K. M., and Moffatt, F., Capillary electrochromatography for pesticide analysis: Effects of environmental matrices, *Electrophoresis*, 21, 1574, 2000.
132. Rodríguez, R., Picó, Y., Font, G., and Mañes, J., Determination of urea-derived pesticides in fruits and vegetables by solid-phase preconcentration and capillary electrophoresis, *Electrophoresis*, 22, 2010, 2001.
133. Fung, Y. S., and Mak, J. L. L., Determination of pesticides in drinking water by micellar electrokinetic capillary chromatography, *Electrophoresis*, 22, 2260, 2001.
134. Kilár, A., Kocsis, B., Kustos, I., Kilár, F., and Hjertén, S., CE to monitor endotoxins by protein complexation, *Electrophoresis*, 27, 4188, 2006.
135. He, J., Chen, S. W., Ruan, L. F., Cao, L. L., Yao, J., and Yu, Z. N., Determination of the fungicide validamycin A by capillary zone electrophoresis with indirect UV detection, *J. Agric. Food Chem.*, 51, 7523, 2003.
136. Safarpour, H., and Asiaie, R., Determination of glyphosate as cross-contaminant in a commercial herbicide by capillary electrophoresis-electrospray ionization-mass spectrometry, *Electrophoresis*, 26, 1562, 2005.
137. Hernández-Borges, J., Rodríguez-Delgado, M. Á., García-Montelongo, F. J., and Cifuentes A., Highly sensitive analysis of multiple pesticides in foods combining solid-phase microextraction, capillary electrophoresis-mass spectrometry, and chemometrics, *Electrophoresis*, 25, 2065, 2004.
138. Da Silva C. L., de Lima, E. C., and Tavares, M. F. M., Investigation of preconcentration strategies for the trace analysis of multiresidue pesticides in real samples by capillary electrophoresis, *J. Chromatogr. A*, 1014, 109, 2003.
139. Juan-García, A., Font, G., and Picó, Y., Determination of quinolone residues in chicken and fish by capillary electrophoresis-mass spectrometry, *Electrophoresis*, 27, 2240, 2006.
140. Juan-García A., Picó, Y., and Font, G., Capillary electrophoresis for analyzing pesticides in fruits and vegetables using solid-phase extraction and stir-bar sorptive extraction, *J. Chromatogr. A*, 1073, 229, 2005.

141. Hernández-Borges, J., Cifuentes, A., García-Montelongo, F. J., and Rodríguez-Delgado, M. Á., Combining solid-phase microextraction and online preconcentration-capillary electrophoresis for sensitive analysis of pesticides in foods, *Electrophoresis*, 26, 980, 2005.
142. Juan-Garciá, A., Font, G., and Picó, Y., Quantitative analysis of six pesticides in fruits by capillary electrophoresis-electrospray-mass spectrometry, *Electrophoresis*, 26, 1550, 2005.
143. Rodríguez, R., Pico, Y., Font, G., and Mañes, J., Analysis of thiabendazole and procymidone in fruits and vegetables by capillary electrophoresis–electrospray mass spectrometry, *J. Chromatogr. A*, 949, 359, 2002.
144. Segura Carretero, A., Cruces-Blanco, C., Cortacero Ramírez, S., Carrasco Pancorbo, A., and Fernandez Gutierrez, A., Application of micellar electrokinetic capillary chromatography to the analysis of uncharged pesticides of environmental impact, *J. Agric. Food Chem.*, 52, 5791, 2004.
145. Molina, M., and Silva, M., In-capillary derivatization and analysis of amino acids, amino phosphonic acid-herbicides and biogenic amines by capillary electrophoresis with laser-induced fluorescence detection, *Electrophoresis*, 23, 2333, 2002.
146. Hilder, E. F., Klampfl, C. W., Buchberger, W., and Haddad, P. R., Comparison of aqueous and nonaqueous carrier electrolytes for the separation of penicillin V and related substances by capillary electrophoresis with UV and mass spectrometric detection, *Electrophoresis*, 23, 414, 2002.
147. Wu, Y., Ho, A. Y. T., Qian, P-Y., Leung, K. S-Y., Cai, Z., and Lin, J.-M., Determination of paralytic shellfish toxins in dinoflagellate *Alexandrium tamarense* by using isotachophoresis/capillary electrophoresis, *J. Separation Sci.*, 29, 399, 2006.
148. Kowalski, P., Olędzka, I., and Lamparczyk, H., Capillary electrophoresis in analysis of veterinary drugs, *J. Pharm. Biomed. Anal.*, 32, 937, 2003.
149. Hernández, M., Borrull, F., and Calull, M., Using nonaqueous capillary electrophoresis to analyze several quinolones in pig kidney samples, *Electrophoresis*, 23, 506, 2002.
150. Kourkine, I. V., Ristic-Petrovic, M., Davis, E., Ruffolo, C. G., Kapsalis, A., and Barron, A. E., Detection of *Escherichia coli* O157:H7 bacteria by a combination of immunofluorescent staining and capillary electrophoresis, *Electrophoresis*, 24, 6551, 2003.
151. Palenzuela, B., Simonet, B. M., García, R. M., Ríos, A., and Valcárcel, M., Monitoring of bacterial contamination in food samples using capillary zone electrophoresis, *Anal. Chem.*, 76, 3012, 2004.
152. Bianco, G., Schmitt-Kopplin, P., De Benedetto, G., Kettrup, A., and Cataldi, T. R. I., Determination of glycoalkaloids and relative aglycones by nonaqueous capillary electrophoresis coupled with electrospray ionization-ion trap mass spectrometry, *Electrophoresis*, 23, 2904, 2002.
153. Huang H. Y., Chiu, C. W., Sue, S. L., and Cheng, C. F., Analysis of food colorants by capillary electrophoresis with large-volume sample stacking, *J. Chromatogr. A*, 995, 29, 2003.
154. Soto-Chinchilla, J. J., García-Campaña, A. M., Gámiz-Gracia, L., and Cruces-Blanco, C., Application of capillary zone electrophoresis with large-volume sample stacking to the sensitive determination of sulfonamides in meat and ground water, *Electrophoresis*, 27, 4060, 2006.
155. Castillo, M. A., and Vargas, M. G., Development of capillary electrophoretic methods for bovine milk quality control, *Presented at the 12th Latin-American Symposium on Biotechnology, Biomedical, Biopharmaceutical and Industrial Applications of Capillary Electrophoresis and Microchip Technology*, Queretaro, Mexico, 2-5 December, 2006.
156. Vargas, M. G., Trejo-Márquez, M. A., and Rodríguez, F., Analysis of herbicides by capillary electrophoresis, In *Applications of Analytical Chemistry in Environmental Research*, Special Review Books, Research Singpost, Trivandrum, Kerala, India, 2005, ISBN: 81-303-0057-8.
157. Hernández-Borges, J., Frías-García, S., Cifuentes, A., and Rodríguez-Delgado, M. A., Pesticide analysis by capillary electrophoresis, *J. Sep. Sci.*, 27, 947, 2004.
158. Da Costa, C. T., Horton, D., and Margolis, S. A., Analysis of anthocyanins in foods by liquid chromatography, liquid chromatography–mass spectrometry and capillary electrophoresis, *J. Chromatogr. A*, 881, 403, 2000.
159. Gu, X., Chu, Q., O'Dwyer, M., and Zeece, M., Analysis of resveratrol in wine by capillary electrophoresis, *J. Chromatogr. A*, 881, 471, 2000.
160. Li, P., Li, S. P., and Wang, Y. T., Optimization of CZE for analysis of phytochemical bioactive compounds, *Electrophoresis*, 27, 4808, 2006.
161. Herrero, M., Ibáñez, E., and Cifuentes, A., Analysis of natural antioxidants by capillary electromigration methods, *J. Sep. Sci.*, 28, 883, 2005.

162. Blasco, A. J., Barrigas, I., González, M. C., and Escarpa, A., Fast and simultaneous detection of prominent natural antioxidants using analytical microsystems for capillary electrophoresis with a glassy carbon electrode: A new gateway to food environments, *Electrophoresis*, 26, 4664, 2005.
163. Hamoudová, R., Pospíšilová, M., and Spilková, J., Analysis of selected constituents in methanolic extracts of *Hypericum perforatum* collected in different localities by capillary ITP-CZE, *Electrophoresis*, 27, 4820, 2006.
164. Štěrbová, D., Včèek, J., and Kubáň, V., Capillary zone electrophoretic determination of phenolic compounds in chess (*Bromus inermis* L.) plant extracts, *J. Sep. Sci.*, 29, 308, 2006.
165. Carrasco-Pancorbo, A., Gómez-Caravaca, A. M., Cerretani, L., Bendini, A., Segura-Carretero, A., and Fernández-Gutiérrez, A., A simple and rapid electrophoretic method to characterize simple phenols, ligands, complex phenols, phenolic acids, and flavonoids in extravirgin olive oil, *J. Sep. Sci.*, 29, 2221, 2006.
166. Li, X., Zhang, Y., and Yuan, Z., Separation and determination of rutin and quercetin in the flowers of *Sophora japonica* L. by capillary electrophoresis with electrochemical detection, *Chromatographia*, 55, 243, 2002.
167. Suntornsuk, L., Kasemsook, S., and Wongyai, S., Quantitative analysis of aglycone quercetin in mulberry leaves (*Morus alba* L.) by capillary zone electrophoresis, *Electrophoresis*, 24, 1236, 2003.
168. Herrero-Martinez, J. M., Ràfols, C., Rosés, M., Bosch, E., Lozano, C., and Torres, J. L., Micellar electrokinetic chromatography estimation of size and composition of procyanidins after thiolysis with cysteine, *Electrophoresis*, 24, 1404, 2003.
169. Pompanio, R., Gotti, R., Santagati, N. A., and Cavrini, V., Analysis of catechins in extracts of *Cistus* species by microemulsion electrokinetic chromatography, *J. Chromatogr. A*, 990, 215, 2003.
170. Pompanio, R., Gotti, R., Luppi, B., and Cavrini, V., Microemulsion electrokinetic chromatography for the analysis of green tea catechins: Effect of the cosurfactant on the separation selectivity, *Electrophoresis*, 24, 1658, 2003.
171. Bonoli, M., Pelillo, M., Toschi, T. G., and Lercker, G., Analysis of green tea catechins: Comparative study between HPLC and HPCE, *Food Chem.*, 81, 631, 2003.
172. Bonoli, M., Colabufalo, P., Pelillo, M., Toschi, T. G., and Lercker, G., Fast determination of catechins and xanthines in tea beverages by micellar electrokinetic chromatography, *J. Agric. Food Chem.*, 51, 1141, 2003.
173. Wright, L. P., Pieter Aucamp, J., and Apostolides, Z., Analysis of black tea theaflavins by nonaqueous capillary electrophoresis, *J. Chromatogr. A*, 919, 205, 2001.
174. Wright, L. P., Mphangwe, N. I. K., Nyirenda, H. E., and Apostolides, Z., Analysis of the theaflavin composition in black tea (*Camellia sinensis*) for predicting the quality of tea produced in Central and Southern Africa, *J. Sci. Food Agric.*, 82, 517, 2002.
175. Gao, L., Chu, Q., and Ye, J., Determination of trans-Resveratrol in wines, herbs and health food by capillary electrophoresis with electrochemical detection, *Food Chem.*, 78, 255, 2002.
176. Brandolini, V., Maietti, A., Tedeschi, P., Durini, E., Vertuani, and Manfredini, S., Capillary electrophoresis determination, synthesis, and stability of resveratrol and related 3-O-β-D-glucopyranosides, *J. Agric. Food Chem.*, 50, 7407, 2002.
177. Saenz-Lopez, R., Fernandez-Zurbano, P., and Tena, M. T., Development and validation of a capillary zone electrophoresis method for the quantitative determination of anthocyanins in wine, *J. Chromatogr. A*, 990, 247, 2003.
178. Vaher, M., and Koel, M., Separation of polyphenolic compounds extracted from plant matrices using capillary electrophoresis, *J. Chromatogr. A*, 990, 225, 2003.
179. Del Castillo, M. D., Ames, J. M., and Gordon, M. H., Effect of roasting on the antioxidant activity of coffee brews, *J. Agric. Food Chem.*, 50, 3698, 2002.
180. Kubáň, P., Štěrbová, D., and Kubáň, V., Separation of phenolic acids by capillary electrophoresis with indirect contactless conductometric detection, *Electrophoresis*, 27, 1368, 2006.
181. Huang, H.-Y., Lien, W-C., and Chiu, C.-W., Comparison of microemulsion electrokinetic chromatography and micellar electrokinetic chromatography methods for the analysis of phenolic compounds, *J. Sep. Sci.*, 28, 973, 2005.
182. Wang, S. P., and Huang, K.-J., Determination of flavonoids by high-performance liquid chromatography and capillary electrophoresis, *J. Chromatogr. A*, 1032, 273, 2004.

183. Ma, B., Zhou, X., Wang, G., Huang, H., Dai, Z., Qin, J., and Lin, B., Integrated isotachophoretic preconcentration with zone electrophoresis separation on a quartz microchip for UV detection of flavonoids, *Electrophoresis*, 27, 4904, 2006.
184. Carrasco Pancorbo, A., Cruces-Blanco, C., Segura Carretero, A., and Fernandez Gutierrez, A., Sensitive determination of phenolic acids in extra-virgin olive oil by capillary zone electrophoresis, *J. Agric. Food Chem.*, 52, 6687, 2004.
185. Peng, Y., Ye, J., and Kong, J., Determination of phenolic compounds in *Perilla frutescens* L. by capillary electrophoresis with electrochemical detection, *J. Agric. Food Chem.*, 53, 8141, 2005.
186. Hamoudová, R., Urbánek, M., Pospíšilová, M., and Polášek. M., Assay of phenolic compounds in red wine by on-line combination of capillary isotachophoresis with capillary zone electrophoresis, *J. Chromatogr. A*, 1032, 281, 2004.
187. Ehala, S., Vaher, M., and Kaljurand, M., Characterization of phenolic profiles of Northern European Berries by capillary electrophoresis and determination of their antioxidant activity, *J. Agric. Food Chem.*, 53, 6484, 2005.
188. Lee, B. L., and Ong, C. N., Comparative analysis of tea catechins and theaflavins by high performance liquid chromatography and capillary electrophoresis, *J. Chromatogr. A*, 881, 439, 2000.
189. Cifuentes, A., Bartolomé, B., and Gómez-Cordovés, C., Fast determination of procyanidins and other phenolic compounds in food samples by micellar electrokinetic chromatography using acidic buffers, *Electrophoresis*, 22, 1561, 2001.
190. Scampicchio, M., Wang, J., Mannino, S., and Prakash Chatrathi, M., Microchip capillary electrophoresis with amperometric detection for rapid separation and detection of phenolic acids, *J. Chromatogr. A*, 1049, 189, 2004.
191. Barbas, C., and Saavedra, L., Chiral analysis of aliphatic short chain organic acids by capillary electrophoresis, *J. Sep. Sci.*, 25, 1190, 2002.
192. Simo, C., Rizzi, A., Barbas, C., and Cifuentes, A., Chiral capillary electrophoresis-mass spectrometry of amino acids in foods, *Electrophoresis*, 26, 1432, 2005.
193. Ward, T. J., Chiral Separations, *Anal. Chem.*, 78, 3947, 2006.
194. Van Eeckhaut, A., and Michotte, Y., Chiral separations by capillary electrophoresis: Recent developments and applications, *Electrophoresis*, 27, 2880, 2006.
195. Vespalec R., and Boček, P., Chiral separations in capillary electrophoresis, *Chem. Rev.*, 100, 3715, 2000.
196. Gübitz, G., and Schmid, M. G., Advances in chiral separation using capillary electromigration techniques, *Electrophoresis*, 27, 1, 2006.
197. Simó, C., Barbas, C., and Cifuentes, A., Chiral electromigration methods in food analysis, *Electrophoresis*, 24, 2431, 2003.
198. Gübitz, G., and Schmid, M. G., Recent advances in chiral separation principles in capillary electrophoresis and capillary electrochromatography, *Electrophoresis*, 23, 3981, 2004.
199. Nouadje, G., Nertz, M., and Courderc, F., Study of the racemization of serine by cyclodextrin-modified micellar electrokinetic chromatography and laser-induced fluorescence detection, *J. Chromatogr. A*, 716, 331, 1995.
200. Simo, C., Martin-Alvarez, P. J., Barbas, C., and Cifuentes, A., Application of stepwise discriminant analysis to classify commercial orange juices using chiral micellar electrokinetic chromatography-laser induced fluorescence data of amino acids, *Electrophoresis*, 25, 2885, 2004.
201. Liu, X., Li, D. F., Wang, Y., and Lu, Y. T., Determination of 1-aminocyclopropane-1-carboxylic acid in apple extracts by capillary electrophoresis with laser-induced fluorescence detection, *J. Chromatogr. A*, 1061, 99, 2004.
202. Carlavilla, D., Moreno-Arribas, M. V., Fanali, S., and Cifuentes, A., Chiral MEKC-LIF of amino acids in foods: Analysis of vinegars, *Electrophoresis*, 27, 2551, 2006.
203. Kodama, S., Yamamoto, A., Ohura, T., Matsunaga, A., and Kanbe, T., Enantioseparation of imazalil residue in orange by capillary electrophoresis with 2-hydroxypropyl-â-cyclodextrin as a chiral selector, *J. Agric. Food Chem.*, 51, 6128, 2003.
204. Kodama, S., Yamamoto, A., Saitoh, Y., Matsunaga, A., Okamura, K., Kizu, R., and Hayakawa, K., Enantioseparation of vinclozolin by β-cyclodextrin-modified micellar electrokinetic chromatography, *J. Agric. Food Chem.*, 50, 1312, 2002.

205. Pumera, M., Flegel, M., Lepša, L., and Jelínek, I., Chiral analysis of biogenic DL-amino acids derivatized by urethane-protected α-amino acid *N*-carboxyanhydride using capillary zone electrophoresis and micellar electrokinetic chromatography, *Electrophoresis*, 23, 2449, 2002.
206. Chen, F., Zhang, S., Qi, L., and Chen, Y., Chiral capillary electrophoretic separation of amino acids derivatized with 9-fluorenylmethylchloroformate using mixed chiral selectors of β-cyclodextrin and sodium taurodeoxycholate, *Electrophoresis*, 27, 2896, 2006.
207. Kodama, S., Yamamoto, A., and Matsunaga, A., Direct chiral resolution of pantothenic acid using 2-hydroxypropyl-β-cyclodextrin in capillary electrophoresis, *J. Chromatogr. A*, 811, 269, 1998.
208. Kodama, S., Aizawa, S.-I., Taga, A., Yamashita, T., and Yamamoto, A., Chiral resolution of monosaccharides as 1-phenyl-3-methyl-5-pyrazolone derivatives by ligand-exchange CE using borate anion as a central ion of the chiral selector, *Electrophoresis*, 27, 4730, 2006.
209. Vandenabeele-Trambouze, O., Albert, M., Bayle, C., Couderc, F., Commeyras, A., Despois, D., Dobrijevic, M., and Grenier Loustalot, M.-F., Chiral determination of amino acids by capillary electrophoresis and laser-induced fluorescence at picomolar concentrations, *J. Chromatogr. A*, 894, 259, 2000.
210. Koidl, J., Hödl, H., Schmid, M. G., Pantcheva, S., Pajpanova, T., and Gübitz, G., Chiral separation of halogenated amino acids by ligand-exchange capillary electrophoresis, *Electrophoresis*, 26, 3878, 2005.
211. Gel-Moreto, N., Streich, R., and Galensa, R., Chiral separation of six diastereomeric flavanone-7-*O*-glycosides by capillary electrophoresis and analysis of lemon juice, *J. Chromatogr. A*, 925, 279, 2001.
212. Kodama, S., Yamamoto, A., Matsunaga, A., Soga, T., and Minoura, K., Direct chiral resolution of lactic acid in food products by capillary electrophoresis, *J. Chromatogr. A*, 875, 371, 2000.
213. Kodama, S., Yamamoto, A., Matsunaga, A., and Hayakawa, K., Direct chiral resolution of tartaric acid in food products by ligand exchange capillary electrophoresis using copper (II)–D-quinic acid as a chiral selector, *J. Chromatogr. A*, 932, 139, 2001.
214. Zhang, M., and El Rassi, Z., Enantiomeric separation by capillary electrochromatography II. Chiral separation of dansyl amino acids and phenoxy acid herbicides on sulfonated silica having surface-bound hydroxypropyl-β-cyclodextrin, *Electrophoresis*, 21, 3135, 2000.
215. Kato, M., Dulay, M. T., Bennett, B., Chen, J.-R., and Zare, R. N., Enantiomeric separation of amino acids and nonprotein amino acids using a particle-loaded monolithic column, *Electrophoresis*, 21, 3145, 2000.
216. Jin, L. J., Rodriguez, I., and Li, S. F. Y., Enantiomeric separation of amino acids derivatized with fluoresceine isothiocyanate isomer I by micellar electrokinetic chromatography using b- and g-cyclodextrins as chiral selectors, *Electrophoresis*, 20, 1538, 1999.
217. Gong, X. Y., and Hauser, P. C., Enantiomeric separation of underivatized small amines in conventional and on-chip capillary electrophoresis with contactless conductivity detection, *Electrophoresis*, 27, 4375, 2006.
218. Fanali, S., Crucianelli, M., De Angelis, F., and Presutti, C., Enantioseparation of amino acid derivatives by capillary zone electrophoresis using vancomycin as chiral selector, *Electrophoresis*, 23, 3035, 2002.
219. La, S., Ahn, S., Kim, J.-H., Goto, J., Choi, O.-K., and Kim, K.-R., Enantioseparation of chiral aromatic amino acids by capillary electrophoresis in neutral and charged cyclodextrin selector modes, *Electrophoresis*, 23, 4123, 2002.
220. Zakaria, P., Macka, M., and Haddad, P. R., Optimisation of selectivity in the separation of aromatic amino acid enantiomers using sulfated β-cyclodextrin and dextran sulfate as pseudostationary phases, *Electrophoresis*, 25, 270, 2004.
221. Allen, D. J., Gray, J. C., Payva, N. L., and Smith, J. T., An enantiomeric assay for the flavonoids medicarpin and vestitone using capillary electrophoresis, *Electrophoresis*, 21, 2051, 2000.
222. Mikuš, P., Kaniansky, D., and Fanali, S., Separation of multicomponent mixtures of 2,4-dinitrophenyl labelled amino acids and their enantiomers by capillary zone electrophoresis, *Electrophoresis*, 22, 470, 2001.
223. Simó, C., Barbas, C., and Cifuentes, A., Sensitive micellar electrokinetic chromatography-laser-induced fluorescence method to analyze chiral amino acids in orange juices, *J. Agric. Food Chem.*, 50, 5288, 2002.

224. Recio, I., Ramos, M., and López-Fandiño, R., Capillary electrophoresis for the analysis of food proteins of animal origin, *Electrophoresis*, 22, 1489, 2001.
225. Bean, S. R., and Lookhart, G. L., High-performance capillary electrophoresis of meat, dairy and cereal proteins, *Electrophoresis*, 22, 4207, 2001.
226. Sorensen, H., Sorensen, S., Bjergegaard, C., and Michaelsen, S., Protein purification and analysis, in *Chromatrography and Capillary Electrophoresis in Food Analysis, RSC Food Analysis Monographs*, Belton, P.S., Ed., Norwich, U.K., 1999, p. 315.
227. Gonzalez, N., Elvira, C., San Román, J., and Cifuentes, A., New physically coated adsorbed polymer coating for reproducible separations of basic and acidic proteins by capillary electrophoresis, *J. Chromatogr. A*, 1012, 95, 2003.
228. Dolnik, V., Capillary electrophoresis of proteins 2003-2005, *Electrophoresis*, 27, 126, 2006.
229. Recio, I., Amigo, L., Lopez-Fandiño, R., Assessment of the quality of dairy products by capillary electrophoresis of milk proteins, *J. Chromatogr. B*, 697, 231, 1997.
230. Vallejo-Cordoba, B., Rapid separation and quantification of major caseins and whey proteins of bovine milk by capillary electrophoresis, *J. Cap. Elec.*, 4, 219, 1997.
231. de Jong, N., Visser, S., and Olieman, C. Determination of milk proteins by capillary electrophoresis, *J. Chromatogr. A*, 652, 207, 1993.
232. Recio, I., and Olieman, C., Determination of denatured serum proteins in the casein fraction of heat-treated milk by capillary zone electrophoresis, *Electrophoresis*, 17, 1228, 1996.
233. Miralles, B., Rothbauer, V., Manso, M., Amigo, l., Krause, I., and Ramos, M., Improved method for the simultaneous determination of whey proteins, caseins, and para-k-casein in milk and dairy products by capillary electrophoresis. *J. Chromatogr. A*, 915, 225, 2001.
234. Fairise, J. F., and Cayot, P., New ultrarapid method for the separation of milk proteins by capillary electrophoresis, *J. Agric. Food Chem.*, 46, 2628, 1998.
235. Olguin-Arredondo, H., and Vallejo-Cordoba, B., Separation and determination of β-lactoglobulin variants A and B in cow's milk by capillary free zone electrophoresis. *J. Cap. Elec. and Microchip Tech.*, 6, 145, 1999.
236. Veledo, M. T., Frutos, M., and Diez-Masa, J. C., Development of a method for quantitative analysis of the major whey proteins by capillary electrophoresis with online capillary derivatization and laser-induced fluorescence detection, *J. Sep., Sci.*, 28, 935, 2005.
237. Veledo, M. T., Frutos, M., and Diez-Masa, J. C., Analysis of trace amounts of bovine β-lactoglobulin in infant formulas by capillary electrophoresis with on-capillary derivatization and laser-induced fluorescence detection, *J. Sep., Sci.*, 28, 941, 2005.
238. Miralles, B., Krause, I., Ramos, M., and Amigo, L., Comparison of capillary electrophoresis and isoelectric focusing for analysis of casein/caseinate addition in processed cheeses, *Int. Dairy J.*, 16, 1448, 2006.
239. Rodriguez-Nogales, J. M., and Vazquez, F., Application of electrophoretic and chemometric analysis to predict the bovine, ovine and caprine milk percentages in Panela cheese, an unripened cheese, *Food Control*, 18, 580, 2007.
240. Otte, J., Shalaby, S. M., Zakora, M., Pripp, A. H., El-Shabrawy, S. A., Angiotensin-converting enzyme inhibitory activity of milk protein hydrolysates: Effect of substrate, enzyme and time of hydrolysis, *Int. Dairy J.*, 17, 488, 2007.
241. De Noni, I., Pellegrino, L., Cattaneo, S., Resmini, P., HPLC of proteose peptones for evaluating ageing of packaged pasteurized milk. *Int. Dairy J.*, 17, 12, 2007.
242. Olguin-Arredondo, H., Vallejo-Cordoba, B., and Gonzalez-Cordova, A. F., Micropreparative separation, fractionation, an peptide mapping of β-lactoglobulin A and B variants by capillary free zone electrophoresis, *J. Cap. Elec. Microchip Tech.*, 9, 2005.
243. Lookhart, G., Bean, L., and Scott, R., Capillary electrophoresis of cereal proteins: An overview, *J. Cap. Elec. Microchip Technol.*, 9, 23, 2004.
244. Lookhart, G., and Bean, L., High performance capillary electrophoresis of meat, dairy, and cereal proteins, *Electrophoresis*, 22, 4207, 2001.
245. Bean, S. R., Bietz, J. A., and Lookhart, G., High performance capillary electrophoresis of cereal proteins, *J. Chromatogr. A*, 814, 25, 1998.

246. Bean, S. R., Bietz, J. A., and Lookhart, G., Ultrafast capillary electrophoretic analysis of cereal storage proteins and its applications to protein characterization and cultivar differentiation, *J. Agric. Food Chem.*, 48, 344, 2000.
247. Cota-Rivas, M., and Vallejo-Cordoba, B., Capillary electrophoresis for meat species differentiation, *J. Cap. Elec.*, 4, 195, 1998.
248. Vallejo-Cordoba, B., and Cota-Rivas, M., Meat species identification by linear discriminant analysis of capillary electrophoresis protein profiles, *J. Cap. Elec.*, 5, 171, 1998.
249. Simo, C., Elvira, C., Gonzalez, N., San Roman, J., Barbas, C., and Cifuentes, A., Capillary electrophoresis-mass spectrometry of basic proteins using a new physically adsorbed polymer coating. Some applications in food analysis, *Electrophoresis*, 25, 2056, 2004.
250. Meyer, R., and Candrian, U., PCR-based DNA analysis for the identification and characterization of food components, *Lebensm-Wiss u. Technol.*, 29, 1, 1996.
251. Lockley, A. K., and Bardsley, R. G., DNA-based methods for food authentication. *Trends Food Sci. Technol.*, 11, 67, 2000.
252. Di Pinto, A., Forte, V. T., Guastadisegni, M. C., Martino, C., Schena, F. P., and Tantillo, G. A., Comparison of DNA extraction methods for food analysis, *Food Control*, 18, 76, 2007.
253. Garcia-Cañas, V., González, R., and Cifuentes, A., Detection of genetically modified maize by polymerase chain reaction and capillary gel electrophoresis with UV detection and laser-induced fluorescence, *J. Agric. Food Chem.*, 50, 1016, 2002.
254. Garcìa-Cañas, V., Gonzalez, R., and Cifuentes, A., The combined use of molecular techniques and capillary electrophoresis in food analysis, *Trends Anal. Chem.*, 23, 637, 2004.
255. Garcia-Cañas, V., Cifuentes, A., and Gonzalez, R., Quantitation of transgenic maize using double quantitative competitive polymerase chain reaction and capillary gel electrophoresis laser induced fluorescence, *Anal. Chem.*, 76, 2306, 2004.
256. Sun, Y. L., and Lin, C. S., Establishment and application of a fluorescent polymerase chain reaction restriction fragment length polymorphism (PCR-RFLP), method for identifying porcine, caprine, and bovine meats, *J. Agric. Food Chem.*, 51, 1771, 2003.
257. Zeleny, R., Bernreuther, A., Schimmel, H., and Pauwels, J., Evaluation of PCR-based beef sexing methods, *J. Agric. Food Chem.*, 50, 4169, 2002.
258. Alarcon, B., Garcia-Cañas, V., Cifuentes, A., González, R., and Aznar, R., Simultaneous and sensitive detection of three foodborne pathogens by multiplex PCR, capillary gel electrophoresis, and laser-induced fluorescence, *J. Agric. Food Chem.*, 52, 5583, 2004.
259. Garcia-Cañas, V., Macian, M. C., Chenoll, E., Aznar, R., González, R., and Cifuentes, A., Detection and differentiation of several food-spoilage lactic acid bacteria by multiplex polymerase chain reaction, capillary gel electrophoresis, and laser-induced fluorescence, *J. Agric. Food Chem.*, 52, 5583, 2004.
260. Martin, A., Jurado, M., Rodriguez, M., Nuñez, F., and Cordoba, J. J., Characterization of molds from dry cured meat products and their metabolites by micellar electrokinetic capillary electrophoresis and random amplified polymorphic DNA PCR, *J. Food Prot.*, 67, 2234, 2004.
261. Garcìa-Cañas, V., Cifuentes, A., and Gonzalez, R., Detection of genetically modified organisms in foods by DNA amplification techniques, *Crit. Rev. Food Sci. Nutr.*, 44, 425, 2004.
262. Doodley, J. J., Sage, H. D., Brown, H. M., and Garrett, S. D., Improved fish species identification by use of lab-on-a-chip technology, *Food Control*, 16, 601, 2005.
263. Doodley, J. J., Sage, H. D., Clarke, M. A. L., Brown, H. M., and Garrett, S. D., Fish species identification using PCR-RFLP analysis and lab-on-a-chip capillary electrophoresis. Application to detect white fish species in food products and an interlaboratory study, *J. Agric. Food Chem.*, 53, 3348, 2005.
264. Boyce, M. C., Determination of additives in food by capillary electrophoresis, *Electrophoresis*, 22, 1447, 2001.
265. Frazier, R. A., Inns, E. L., Dossi, N., Ames, J. M., and Nursten, H. E., Development of a capillary electrophoresis method for the simultaneous analysis of artificial sweetners, preservatives and colors in soft drinks. *J. Chromatogr. A*, 876, 213, 2000.
266. Huan, H. Y., Shih, Y. C., and Chen, Y. C., Determining eight colorants in milk beverages by capillary electrophoresis, *J. Chromatogr. A*, 959, 317, 2002.
267. Huan, H. Y., Chuang, C. L., Chiu, C. W., and Chung, M. C., Determination of food colorants by microemulsion electrokinetic chromatography, *Electrophoresis*, 26, 867, 2005.

268. Jager, A. V., Tonin, F. G., and Tavares, M. F. M., Optimizing the separation of food dyes by capillary electrophoresis, *J. Sep. Sci.*, 28, 957, 2005.
269. Law, W. S., Kuba, P., Zhao, J. H., Li, S. F. Y., and Hauser, P. C., Determination of vitamin C and preservatives in beverages by conventional capillary electrophoresis and microchip electrophoresis with capacitively coupled contactless conductivity detection, *Electrophoresis*, 26, 4648, 2005.
270. Oztekin, N., Nutku, M. S., and Erim, F. B., Simultaneous determination of nitrite and nitrate in meat products and vegetables by capillary electrophoresis, *Food Chem.*, 76, 103, 2002.
271. Ohashi, M., Omae, H., Hashida, M., Sowa, Y., and Imai, S., Determination of vanillin and related flavor compounds in cocoa drink by capillary electrophoresis, *J. Chromatogr. A*, 1138, 262, 2007.
272. McCourt, J., Stroka, J., and Anklam, E., Experimental design-based development and single laboratory validation of a capillary zone electrophoresis method for the determination of the artificial sweetener sucralose in food matrices, *Anal. Bioanal. Chem.*, 382, 1269, 2005.
273. Ding, Y., Mora, M. F., and Garcia, C. D., Analysis of alkyl gallates and nordihydroguaiaretic acid using plastic capillary electrophoresis—microchips, *Analytic Chimica Acta*, 561, 126, 2006.
274. De Block, J., Merchiers, M., Mortier, L., Brraekman, A., Ooghe, W., and Van Renterghem, R., Monitoring nutritional quality of milk powders: Capillary electrophoresis of the whey protein fraction compared with other methods, *International Dairy Journal*, 13, 87, 2003.
275. Vallejo-Cordoba, B., Mazorra-Manzano, M. A., and Gonzalez-Cordova, A. F., New capillary electrophoresis method for the determination of furosine in dairy products, *J. Agric. Food Chem.*, 52, 5787, 2004.
276. Morales, F., and Jimenez-Perez, S., Hydroxymethylfurfural determination in infant milk-based formulas by micellar electrokinetic capillary chromatography. *Food Chem.*, 72, 525, 2001.
277. Vallejo-Cordoba, B., and Gonzalez-Cordova, A. F., Latest advances in the chemical characterization of Mexican distilled beverages: Tequila, Mezcal, Bacanora and Sotol, In *Hispanic Foods, Chemistry and Flavor*, ACS Series, 946, Tunick, M.H., and González de Mejia, E., Ed., 2007, 153.
278. Wrigley, C. W., Bate, I. L., Uthayakumaran, S., and Rathmell, W. G., Modern approaches to food diagnostics for grain quality assurance. *Food Australia*, 58, 538, 2006.
279. Vallejo-Cordoba, B., Mazorra-Manzano, M. A., and Gonzalez-Cordova, A. F., Determination of β-hydroxyacyl CoA-dehydrogenase activity in meat by electrophoretically mediated microanalysis, *J. Cap. Elec. Micro. Tech.*, 8, 81, 2003.
280. de Oliveira, M. A. L., Solis, V. E. S., Gioielli, L. A., Polakiewicz, B., and Tavares, M. F. M., Method development for the analysis of transfatty acids in hydrogenated oils by capillary electrophoresis, *Electrophoresis*, 24, 1641, 2003.
281. Vallejo-Cordoba, B., Mazorra-Manzano, M. A., and Gonzalez-Cordova, A. F., Determination of short-chain free fatty acids in lipolyzed milk fat by capillary electrophoresis, *J. Cap. Elec.*, 5, 11, 1998.

31 Separation Strategies for Environmental Analysis

Fernando G. Tonin and Marina F. M. Tavares

CONTENTS

31.1 INTRODUCTION

In the past 25 years, capillary electrophoresis (CE) has conquered a solid position in the scientific community, supported by vast literature compiling the intricacy of its theoretical aspects and the diversity of its applications (Chapter 1 by Landers brings an introductory panel on CE technology). Several relevant characteristics of CE such as high resolution, high efficiency, and speed of analysis in addition to excellent mass sensitivity, low consumption of reagents, and small sample volumes are in perfect tuning with the demands of environmental analyses. Moreover, whenever limits of detection (LOD) are favorable, the direct injection of complex samples is possible without being detrimental to the column integrity, which in turn contributes to preclude time-consuming sample preparation, typical of environmental procedures. Another particularity of the CE technique is the variety of separation mechanisms that can be practiced in a single capillary column, allowing the simultaneous assessment of distinct classes of pollutants in a given sample.

31.2 SEPARATION STRATEGIES

The CE determination of broad classes of atmospheric pollutants and water and soil contaminants has been reviewed regularly.[1–7] This chapter focuses on separation strategies for environmental analysis involving the two most commonly used CE modes: capillary zone electrophoresis (CZE) and electrokinetic chromatography (EKC).

31.2.1 CZE

CZE separations are based solely on the differences in the electrophoretic mobilities of *charged species*, either in aqueous or nonaqueous media (this latter often referred to as nonaqueous capillary electrophoresis, NACE). In CZE, the migration of a species within the capillary column is the net result of mass transport phenomena and chemical equilibria. Two modes of migration are possible, that is, under suppressed electroosmotic flow (EOF), achieved at low pH buffers or by the use of surface modified capillaries, and in the presence of EOF; in the latter, two possibilities arise: separations under co- and counter-EOF, depending on the relative mobility of the analyte and EOF itself. With the proper control of electrolyte composition (buffer type regarding both co- and counterions, buffer pH and concentration, as well as additives), the analyte mobility can be altered. Flow characteristics are also dependable on the electrolyte composition as well as on the capillary surface condition.

Organic solvents are among the most commonly used additives in CZE. Several other additives (complexing agents to discriminate metal cations, quaternary alkylammonium salts as flow reversers in the separation of small anions, and neutral cyclodextrins [CDs] as chiral selectors) are listed in the literature as modifiers in CZE separations. In contrast, separations with charged CDs or mixtures of neutral and charged CDs are often classified as EKC methodologies (*vide* next section). Considering that low-molecular mass additives represent in fact borderline cases between CZE and EKC systems, arbitrary assumptions will be made in this chapter with the purpose of classifying the literature methodologies. Cyclodextrins either neutral, charged, or mixtures of both will be considered secondary phases in EKC separations whereas other complexing agents (ethylenediaminetetraacetic acid [EDTA], *o*-phenanthroline, crown ethers, etc.) will be classified as additives of CZE electrolytes.

Quaternary alkylammonium salts, used as flow reverser additives in the separation of small anions, will fall technically into the EKC definition, since the surfactant micellization is anticipated due to the electrolyte ionic strength in which the separation is conducted (*vide* section on small ions analysis).

CZE is, therefore, suitable for the determination of explicitly charged small ions, whose mobilities already differ by some extent or might be modified by additives, as well as ionizable organic environmental pollutants (for instance, compounds containing carboxylic acid, phenol, or amine functionalities). Cations are generally separated under co-EOF whereas anions with low to moderately low mobility are separated counter-electroosmotically. Examples of the CZE analysis of small ions in methanol (MeOH)-modified borate electrolytes include the speciation of mercury [inorganic Hg(II), methyl-, ethyl-, and phenylmercury] in water and dogfish muscle[8] and the direct ultraviolet (UV) determination of anions (Br^-, I^-, NO_2^-, $S_2O_3^{2-}$, CrO_4^{2-}, NO_3^-, SCN^-, $Fe(CN)_6^{4-}$, MoO_4^{2-}, and WO_4^{2-}) in effluents of a power plant.[9] For ionizable compounds, the analyte acquires an electrophoretic *effective mobility* (summation of ionic species mobilities weighted by the species availability at a given pH). The selective inspection of 21 aromatic amines in groundwater and soil sample using low pH phosphate buffer in capillaries modified by 1,3-aminopropane is a representative example of the CZE mode for ionizable compounds.[10] Another example is the determination of the priority phenols in spiked wastewater samples using ammonium acetate buffer modified by *N*-methyl formamide/acetonitrile (ACN) mixtures.[11]

31.2.2 EKC

Electrokinetic chromatography separations are those put into practice with electrolytes containing a *secondary phase* that migrate with a particular velocity distinct from that of the analyte. The fundamental requirement for EKC separations is that either the analyte or the secondary phase must be charged under the electrolyte medium pH. As pointed out previously, for charged analytes forming adducts with small molecules, there is a blurred distinction between CZE and EKC separation principles.

31.2.2.1 MEKC

Micellar electrokinetic chromatography (MEKC) is a particular EKC mode where the secondary phase is composed by micellized surfactant (MEKC is discussed in detail in Chapter 3 by Terabe). Solute differential retention occurs as a result of a partition mechanism between a dispersed phase defined by the total volume of micelles and the remaining aqueous phase. MEKC modes of elution comprise normal, restricted, and reversed MEKC, based on the relative migration of the analyte and secondary phase apparent velocity.

The anionic surfactant sodium dodecylsulfate (SDS) is by far the most commonly used surfactant in micellar separations. Examples of the use of simple buffered/SDS systems in environmental applications include the simultaneous analysis of 10 *N*-methylcarbamate pesticides and their hydrolytic phenolic metabolites in river, well, and pond water (pH 8 phosphate/borate buffer/SDS),[12] and the analysis of insecticides (imidacloprid and its metabolite 6-chloronicotinic acid) in air samples collected from a greenhouse cropped with tomatoes (pH 8.5 ammonium chloride/ammonia buffer/SDS).[13]

The use of SDS in MEKC poses two important shortcomings regarding the nature of the compounds eligible for separation: hydrophilic compounds, especially anionic in character (electrostatic repulsion), are poorly retained even at high SDS concentrations, whereas hydrophobic compounds are highly retained even at low SDS concentrations and little selectivity is, therefore, provided.

The alternative use of cationic surfactants (quaternary alkylammonium salts) is recommended to increase retention of highly hydrophilic compounds. Cationic surfactants reverses the EOF. As the cationic micelles migrate electrophoretically against the anodic EOF, an extended migration window results. An example of MEKC separation using cationic surfactants is the analysis of various *s*-triazines, including five chloro-, three methoxy-, and five alkylthio-*s*-triazines in tetradecyltrimethylammonium bromide (TTAB) electrolytes.[14]

For the MEKC analysis of highly hydrophobic compounds, several strategies have been proposed: the use of organic additives (short-chain alcohols, acetonitrile, tetrahydrofuran, dioxane, and dimethylsulfoxide are among the most studied), and the addition of large amounts of urea and glucose have been reported. A few examples of solvent-modified electrolytes within the environmental context include (i) separation of seven polynuclear aromatic hydrocarbons (PAHs) in borate/SDS/ACN for inspection of deliberately contaminated soils submitted to biological decontamination process and spent machine oil;[15] (ii) determination of 11 triazines in groundwater using as optimized electrolyte, borate/phosphate/SDS/1-propanol;[16] (iii) the analysis of 21 mono- and dinaphthalenesulfonates, as well as their hydroxy- and amino-derivatives in river water, with borate/SDS/ACN;[17] and (iv) determination of homologues and isomers of linear alkylbenzenesulfonates (LAS) in household products and sewage sludge samples using borate/phosphate/ACN electrolytes.[18]

Other strategies to cope with hydrophobic compounds include the addition of neutral surfactants or neutral CDs to SDS micelle systems, use of bile salts, and by exploring solvophobic association with tetrahexyl- or tetraheptylammonium ions (added as perchlorate salt [THxAP] or bromide salt [THpAB]) or dioctylsulfosuccinate (DOSS).

In the separation of aromatic sulfonates, 10 electrolyte compositions were investigated with low and high pH buffers containing SDS, Brij35, and octylamine. Twenty-one compounds could be separated and the contents of aromatic sulfonates in an industrial effluent were determined.[19] Other examples of mixed micelle systems applied to compounds of environmental relevance include (i) the analysis of eight azo dyes, mono-, and disulfonated compounds with Brij35/SDS electrolytes;[20] (ii) the separation of 16 arylamine isomers with Tween80/SDS/cholate systems (lake water near industrial area);[21] (iii) the analysis of eight aromatic compounds (phenol derivatives and PAH) with Tween20/SDS electrolytes;[22] and (iv) the separation of six phenylureas and chlorsulfuron pesticides with PEG400/SDS systems.[23]

Although the target function of neutral CDs is to exert chiral selection in the EKC separation of optical isomers, they have often been used as auxiliary complex ligands as a means of improving resolution of closely eluting achiral positional and structural related compounds or to reduce significantly apparent retention factors. The separation of seven positional and structural naphthalenesulfonate isomers (pH 3.0 phosphate buffer/β-CD)[24] and five 2,4-dinitrophenylhydrazine (DNPH)-aldehyde derivatives in vehicular emission (pH 9.0 borate buffer/SDS/β-CD)[25] are examples of neutral CD-mediated separations.

Studies involving bile salts in EKC separations include (i) determination of *s*-triazines and quats in well water samples;[26] (ii) separation of the 16 priority pollutant PAHs for inspection in ambient air samples;[27] (iii) characterization of the electrophoretic behavior of 56 aromatic compounds (phenoxy acid herbicides, phenylalkanoic acids, aromatic carboxylic acids, aromatic sulfonic acids, azo and other dyes, and nitrogen-containing aromatic acids) for further assessment of extraction and sample clean-up procedures using spiked water and soils;[28] and (iv) recovery of synthetic dyes from spiked water and soil matrices.[29] Another interesting application of MEKC with diverse bile salts is the prediction of ecotoxicity of aromatic compounds.[30]

The use of carboxylic and dicarboxylic acids, SDS, bile salts, organic solvents, and alkylammonium ions was explored to study the separation of LAS homologues and positional isomers,[31,32] as well as alkylether sulfate oligomers.[33] The MEKC separation of mixtures of the surfactant classes coconut diethanolamide, cocamido propyl betaine, and alkylbenzene sulfonate was studied in either low pH phosphate or high pH borate or dipentylamine buffers containing as surfactants deoxycholate or SDS, organic solvents (methanol, acetonitrile, *n*-propanol, and *n*-butanol), and anionic solvophobic agents (DOSS, fatty acids).[34]

The use of DOSS/ACN mixtures in the separation of 9-fluoroenylmethyl chloroformate (FMOC)-derivatized anilines was investigated for inspection of lake water.[35] While the organic modifier allowed the solubilization of the hydrophobic solutes and maintained the DOSS surfactant in its monomeric form by inhibiting micellization, the DOSS surfactant associated with the FMOC anilines to a varying degree led to their differential migration and separation. Seven conventional chromatographic ion-pairing agents including tetraalkylammonium series and alkanesulfonic series were tested comparatively in a recent study involving the separation of 13 PAH in methanolic/ACN matrices.[36]

Alternative and more complex surfactants continue to be explored as a resourceful option for MEKC separations. *In situ* generated micelles, which are anionic complexes formed by alkyl- or steroidal-glycoside surfactants and borate ions, the use of sodium 10-undecylenate (SUA) and sodium 10-undecylsulfate (SUS) oligomers as well as surfactants composed of two ionic groups and two liphophilic chains, such as sodium 5,12-*bis*(dodecyloxymethyl)-4,7,10,13-(tetraoxa)-1,16-hexadecanedisulfonate (DBTD), bilayered aggregates such as vesicles and liposomes, and bilayer micelles are a few examples.

In situ generated micelles have been applied to the inspection of aniline pesticidic metabolites in lake water.[37] The separation of 16 PAH in SUA oligomer electrolytes was reported.[38] Creosote-contaminated soil samples were extracted by accelerated solvent extraction using methylene chloride–acetone mixtures. The extracts were further fractioned by gel permeation chromatography before analysis. The EKC chromatogram of a creosote-contaminated soil fraction shows the resolution of at least 50 peaks. The separation of the 11 priority phenols in river and sea water[39] was demonstrated in MEKC with DBTD surfactants, whereas examples of the use of liposomes as carriers include the separation of benzene derivatives and phenols.[40]

High-molecular mass surfactants such as butyl acrylate-butyl methacrylate-methacrylic acid copolymer sodium salts, starburst dendrimers, poly(amidoamines), and diaminobutane-based poly(propyleneimine) as well as cationic polyelectrolytes (ionenes) had all been presented as successful secondary phases for aromatic compounds. The determination of 10 nitrophenols in glycine buffers modified by β-CD (0–10 mmol L^{-1}) and polyvinylpyrrolidone (PVP) (0.5–2.5% w/v) is an example of application of polymer-based electrolytes to rain, tap, and process water.[41]

31.2.2.2 Other EKC Modes

31.2.2.2.1 CD-EKC and CDCD-EKC

Single negatively and positively charged CDs (e.g., sulfobutyl ether β-CD, SB-β-CD, sulfated CD, carboxymethylated β-CD, and methylamino substituted β-CD) or even mixtures of neutral and charged CDs are frequently employed in EKC separations as secondary phases. The separation mechanism is simply defined by host–guest interactions once micellized surfactants are absent. The enantiomeric separation of polychlorinated biphenyls using mixtures of several neutral and charged CD derivatives is a fine example of the EKC with charged CDs impact in environmental analysis.[42] Other examples include the use of a mixture of SB-β-CD and the neutral methyl-β-CD in borate buffer for the analysis of 16 PAH in contaminated soils extracted by supercritical CO_2[43] and the use of SB-β-CD in the analysis of 25 chlorinated and substituted phenolic compounds (including the 11 priority phenols).[44] A complexation model was used for investigating the effect of pH and CD concentration on the electrophoretic mobility. The latter method was applicable for quantifying the level of pentachlorophenol in contaminated soil samples.

31.2.2.2.2 MEEKC

Microemulsion electrokinetic chromatography (MEEKC) is a relatively new technique that accomplishes electrokinetic separations using buffers containing surfactant coated oil droplets (Chapter 4 by Altria and coworkers presents further details on the MEEKC technique). The potential of MEEKC for the separation of priority endocrine disrupting compounds in industrial and domestic wastewater treatment effluents and sludges has been investigated.[45] Using reverse migrating microemulsion, that is, negative polarity at the electrode inlet and a pH 2.8 phosphate buffer containing octane, butanol, and SDS, further modified by propanol, the separation of the breakdown products of alkylphenolic detergents, a few synthetic estrogens, and the plastic monomer bisphenol A was demonstrated.

31.2.2.2.3 Calixarene EKC

Calixarenes are conical-shaped macrocyclic oligomers whose inner cavity can accommodate several guest molecules and, therefore, can serve as a viable option to replace CDs in EKC separations. Environmental examples of the use of calixarene as EKC secondary phases include the separation of chlorinated phenols, benzenediols, and toluidines.[46] Sulfonated calixarenes have also been used as chromophores in the indirect detection of aliphatic amines.[47]

31.2.2.2.4 Suspension EKC

The use of suspended chromatographic silica-based particles in SDS buffers was first demonstrated as secondary phases in the EKC separations of phenol derivatives.[48] Applications of polymer and silica-based particles to several other pollutants have been reviewed recently.[49]

31.2.2.2.5 Ion-exchange EKC

The use of polymer ion additives such as poly(diallyldimethylammonium chloride) (PDDAC) and (diethylamino)ethyldextran (DEAE-dextran) has been successfully proposed to promote dynamic interactions with certain moderately hydrophobic solutes such as naphthalenesulfonate and naphthalenedisulfonate isomers.[50] The fundamentals and scope of ion-exchange EKC with ionic polymers for the separation of small ions of potential importance in the environmental context have been reviewed.[51]

31.3 SENSITIVITY ENHANCEMENT STRATEGIES

The small pathlength defined by the capillary internal diameter coupled with the need of a small injection volume to preserve the high-resolution features of CE place a strong demand on the detection capability. This is especially problematic for absorbance detectors (LOD in the order of 10^{-5} to 10^{-6} mol L^{-1}), considered unsuitable for many environmental applications, where trace level occurrence or matrix complexity issues are generally of concern.

Several approaches to enhance sensitivity in CE have been devised: some are based on sample preconcentration procedures, others rely on special cell geometries and alternative detectors. Both areas have been subjects of intense investigation as described in Chapters 9 by Sweedler and coworkers, 13 by Burgi and Giordano, 14 by Ewing and coworkers, and 27 by Weber.

31.3.1 Preconcentration Schemes

The sensitivity enhancement strategies based on sample manipulation are further subdivided into offline and online procedures (Chapter 13 by Burgi and Giordano). Liquid–liquid extraction (LLE) with a variety of solvents of selected properties and solid-phase extraction (SPE) (Chapter 27 by Weber) with a large assortment of chromatographic stationary phases of distinct chemistries are among the most often performed offline strategies for environmental applications as reviewed recently.[4,5] An interesting example of homogeneous LLE is the pH-induced phase separation of water samples (rain, river, and spring water) treated with perfluorooctanic acetate ($PFOA^-$) surfactant in different water-miscible solvents for the preconcentration of five PAH.[52] At the optimal conditions (above 99% recovery), a 40 mL sample aliquot-containing THF and $PFOA^-$ was treated with HCl to separate 30 μL of sedimented phase. The sedimented liquid phase was then mixed with 30 μL DMSO before analysis. Limits of quantification (LOQ) in the range 10^{-10} to 10^{-9} mol L^{-1} and enrichment factors up to 125,000-fold were obtained. An example of SPE is the use of C_{18}-bonded silica and polystyrene–divinylbenzene (PS–DVB) disks for the enrichment of four triazines and three degradation products of atrazine in drinking and well water samples with recoveries better than 93%

and LOD in the range of 0.02–0.30 mg L^{-1}.[53] Several other LLE and SPE procedures are compiled in the application tables (Section 31.4).

Online preconcentration strategies involve the insertion of large sample volumes in the capillary and can generally be classified into two categories. One category involves the criterious manipulation of the analyte electrophoretic velocity. A collection of strategies grouped by the name of stacking and transient isotachophoresis (tITP) belong to this category. Field-amplified sample stacking of arsenic species in environmental reference materials[54] and organonitrogen pesticides in drainage water[55] as well as the tITP preconcentration of iodide in seawater samples[56] are fine examples of these procedures.

The other category explores the analyte ability of interacting with secondary phases. Stacking of micelles and sweeping are representative of this group (fundamental aspects of these strategies were introduced in Chapter 3 by Burgi and Giordano). Stacking of micelles was employed to preconcentrate dioxin-related compounds[57] and in the analysis of nonsteroidal anti-inflammatory drugs in mineral water,[58] whereas sweeping was employed for the online concentration of bisphenol A and alkylphenols.[59] Stacking of micelles and sweeping were contrasted in the analysis of phenylurea herbicides in tap and pond water[60] and in the multiclass pesticide analysis of drinking water.[61]

Combinations of online procedures are often employed in environmental applications. Examples include the analysis of trace metal ions in factory wastewater by a combination of cation-selective exhaustive injection (CSEI) (a form of stacking, CZE format) and sweeping.[62] In this latter case, sweeping was promoted by dynamic complexation with EDTA, used as carrier. The same CSEI principle has also been applied to the analysis of quaternary ammonium herbicides[63] and environmentally relevant aromatic amines[64] in water, both followed by sweeping with SDS.

Hyphenation of automatic continuous flow systems (such as SPE, dialysis, gas diffusion, evaporation, direct leaching, etc.) to CE and the coupling of automatic sample preparation devices into commercial CE equipments have been devised as a means to simplification and miniaturization of analytical procedures. An automatic online SPE device for the multiresidue extraction of seven pesticides has been described.[65] Four river samples were spiked with the test mixture at three different levels presenting recoveries from 90% to 114%.

An elegant example of SPE method based on ion-exchange retention was used for inline preconcentration of inorganic anions.[66] A single capillary containing a preconcentration zone (adsorbed layer of cationic latex particles) and a separation zone (fused-silica modified by adsorption of a cationic polymer) was used. Analytes were retained in the preconcentration zone and eluted isotachophoretically into the separation zone by means of an eluotropic gradient. This approach was used to determine nitrate in Antarctic ice cores at the 2.2–11.6 μg L^{-1} level.

31.3.2 Alternative Detection Schemes

For the absorbance-based detectors, alternative cell geometries have been designed to extend the optical pathlength such as the Z-shaped cells, bubble cells, rectangular cells, and multireflection capillaries.[6] Although manufacturing and manipulation of these special cells can pose a few operational problems, sensitivity enhancement of tenths of fold have been reported. The determination of the herbicide metribuzin and its major conversion products in soil is an example of the use of bubble cells,[67] whereas the UV-detection of derivatized carbonyl compounds in rain is an example of the use of Z-shaped cells[68] in environmental analysis.

Alternative detection schemes for environmental analysis such as those based on laser-induced fluorescence (LIF), chemiluminescence, conductivity, amperometry, and mass spectrometry (MS) have been reported.

31.3.2.1 Fluorescence Detection

Some pollutants are natively fluorescent such as PAH and metabolites as inspected in biota.[69] Others must be derivatized to fluoresce such as aniline species metabolized from pesticides[70] and low-molecular mass amines labeled as fluorescein isothiocyanate (FITC) derivatives for further inspection of atmospheric aerosol samples reaching 50 pg mL^{-1} (10^{-9} mol L^{-1}) level.[71]

An interesting variant for labeling procedures is the so-called *in-capillary derivatization* reactions, which explores differences in the electrophoretic mobilities to merge distinct zones of analyte and labeling reagent under the effect of an electric field. Molina and Silva[72] explored the possibilities of LIF detection for in-capillary derivatization of nitrogen compounds (herbicides and biogenic amines).

Indirect fluorescence detection is another possibility for nonfluorescent compounds. In this particular case, a fluorophore co-ion is part of the electrolyte system. The simultaneous determination of cations (EDTA complexes) and anions in pond water by indirect fluorescence detection based on the displacement of fluorescein is an example.[73] The use of fluorescein at pH 9.5 as electrolyte was also reported in the determination of anionic organophosphorus herbicides [glufosinate, glyphosate, and its major metabolite aminomethylphosphonic acid (AMPA)] in groundwater,[74] with LOD in the μmol L^{-1} range.

31.3.2.2 Chemiluminescence Detection

Biogenic amines, labeled with *N*-(4-aminobutyl)-*N*-ethylisoluminol) (ABEI) in the presence of *N*,*N*-disuccinimidyl carbonate (DSC) in methanolic medium, were determined in lake water sample by an EKC method using online chemiluminescence detection.[75] Parameters affecting the separation process (pH, SDS concentration, applied voltage, and injection time) and chemiluminescence detection sensitivity [H_2O_2 concentration in the buffer and the postcolumn reagents $K_3Fe(CN)_6$ and NaOH concentrations] were examined in detail. Detection limits in the 1.2×10^{-10} to 3.5×10^{-8} mol L^{-1} range were obtained.

31.3.2.3 Electrochemical Detection

Among the electrochemical detection schemes for CE, conductometric methods have gained great visibility especially for the determination of the high-equivalent conductivity inorganic species. Conductivity detection in CE is performed either by galvanic contact or in a contactless mode (CCD). The galvanic contact mode is further subdivided into on-capillary detection without decoupling and off-capillary, after grounding the separation applied voltage. Most contactless methods use oscillometric techniques. Wet deposition in the 1983–2006 period and related atmospheric chemistry in the metropolitan area of Sao Paulo, Brazil, using CZE-CCD methods has been investigated.[76,77] Electrolytes composed of MES-His (pH 6.2) modified by 18-crown-6 ether for the cation analysis (ammonium, magnesium, sodium, calcium, and potassium), and cetyltrimethylammonium bromide (CTAB) for the anion analysis (sulfate, nitrate, chloride, acetate, and formate) were employed. Limits of detection spanned from 10 to 100 μg L^{-1}. An interesting approach employing dual-opposite end injection and CZE-CCD to determine cations and anions in the same run has been devised.[78] A single electrolyte containing all necessary modifiers was employed (MES/His containing 18-crown-6 ether and CTAB) allowing simultaneous detection of 13 anions and cations in less than 6 min with detection limits in the range of 7–30 μg L^{-1} (except fluoride and phosphate). The method was applied to the inspection of ions in rain, surface, and drainage water and the results were contrasted with ion chromatography showing excellent agreement.

Despite the high sensitivity of amperometric detection, its use in environmental applications has not been exploited accordingly. Nevertheless, a few examples include the MEKC analysis of herbicides in tap water[79] and the CZE and MEKC analysis of aromatic amines in contaminated water samples.[80]

Electrochemical detection of 19 chlorophenols in river samples was performed using a graphite-epoxy working electrode at a potential of 800 mV versus Ag/AgCl.[81] A palladium metal union was used to decouple the electric field from the electrochemical cell and a compensating pressure was applied during the EKC analysis. Detection limits from one to three orders of magnitude lower than UV-detection were achieved (10 $\mu g\,L^{-1}$).

31.3.2.4 Mass Spectrometry Detection

Mass spectrometry (MS) is one of the most powerful detection techniques employed in environmental screening due to the intrinsic structural information and high sensitivity it provides. With the advent of effective interfaces, soft ionization modes (electrospray ionization, ESI, and atmospheric pressure chemical ionization, APCI) and increased affordability of CE-MS instruments, the hyphenated technique has become widespread.

In CE-MS, the use of NACE or mixed aqueous-organic solvent systems is advantageous as nonvolatile buffers are usually avoided. CZE-MS with volatile buffers has been employed in a variety of environmental applications such as the inspection of drug residues in river water (pH 5.1 ammonium acetate),[82] detection of explosives (nitroaromatic and cyclic nitramine compounds) in soil and marine sediment samples (ammonium acetate at pH 6.9 modified by SB-β-CD),[83] and the identification of reactive vinylsulfone and chlorotriazine dyes in spent dyebaths and municipal wastewater treatment plant receiving dyehouse effluents (ammonium acetate at pH 9 modified by 40% ACN).[84] Other examples of volatile electrolyte systems include the use of a pH 3.0 formic acid/ammonium formate system in 50% methanol for the CZE-ESI (sheath liquid: 9:1 MeOH:20 $mmol\,L^{-1}$ acetic acid)-MS (ion trap) analysis of quaternary ammonium herbicides in contaminated irrigation water and spiked mineral water,[85] and pH 9.1 ammonium acetate/ammonium hydroxide/MeOH or pH 11 ammonium hydroxide systems by CZE-ESI (50% isopropanol)-MS (ion trap) for the inspection of methoxy phenols and aromatic acids in biomass burning aerosol samples.[86] When concentrations are rather high CE-MS can still be used as a means of confirming the pollutant presence. An example of such approach includes the determination of LAS in wastewater from treatment plants and coastal waters of Cadiz (Spain) receiving untreated domestic wastewaters by CZE-UV (ammonium acetate at pH 5.6 containing 30% isopropanol) followed by confirmation of the $[M\text{-}H]^-$ ions by ESI (sheath liquid: 80% isopropanol and 0.1% ammonia)-MS (quadrupole).[87]

Particularly in MEKC separations, the presence of high amounts of SDS in the buffer is detrimental causing low ionization efficiency. A number of approaches have been presented to overcome this shortcoming such as the use of high-molecular mass surfactants and anodically migrating micelles as well as partial filling techniques. Examples of this latter approach have been compiled within the pesticide analysis context.[88,89] More recently, technological improvements in the interfacing systems were shown to be more tolerant to the usage of nonvolatile buffers as demonstrated by the MEKC-ESI-MS separation of triazines[90] and the MEKC-APCI-MS analysis of aromatic amines and alkylphthalates.[91]

31.4 REPRESENTATIVE APPLICATIONS

The literature comprises a larger number of methodologies employing electromigration principles, diverse preconcentration strategies, and a variety of detection schemes for pollutant standards combined as test mixtures and for real samples. Tables 31.1 through 31.9 compile representative applications of CZE and EKC methods to the most important classes of pollutants in different environmental compartments.

31.4.1 Pesticides

Pesticide is a generic term used to describe compounds employed in the control, prevention, and elimination of plagues that attack crops and herds. Pesticides are associated with their persistence and toxicity and due to their widespread use in agriculture, these compounds are important sources of environmental contamination.

TABLE 31.1
Selected Applications of Electroseparation Methods for Pesticides in Environmental Samples

Compounds	CE Mode	Matrix	Preconcentration Procedure (Recovery)	Optimal Electrolyte	Detection (LOD)	References
Multiclass pesticides						
Amitrole, atrazin-2-hydroxy in 8 min	CZE	WATER River water Drinking water	—	20 mmol L^{-1} phosphate buffer (pH 3.2)	UV, 200 nm 90–120 μg L^{-1}	98
Acifluorfen, pentachlorophenol, 2,4-DB acid, dinoseb, 2,4,5-TP, 2,4,5-T acid, MCPP acid, dichloroprop, MCPA acid, bentazon, 2,4-D, picloram, 3,5-dichlorobenzoic acid, chloramben, dicamba, 4-nitrophenol in 40 min	CZE	WATER Drinking water	Spiked sample 91–124%	5 mmol L^{-1} NH_4Ac, 40% isopropanol (pH 10.0)	MS, single quadrupole 8–250 μg L^{-1}	104
Atrazine, simazine, paraquat, diquat in 3 min	EKC	WATER Well water	SPE C_{18} 80–95%	10 mmol L^{-1} tetraborate, 25 mmol L^{-1} SDS, 40 mmol L^{-1} perchlorate, 15% ACN (pH 9.3)	220, 254, 300 nm, 150 μm extended optical path capillary 0.6–1.9 μg L^{-1}	26
Carbendazim, simazine, atrazine, propazine, ametryn, diuron, linuron, carbaryl, propoxur in 6 min	EKC	WATER Drinking water (and carrots)	SW, SRMM, SRW SPE C_{18}, NH_2	20 mmol L^{-1} phosphate, 25 mmol L^{-1} SDS, 10% MeOH (pH 2.5)	UV, 220 nm 0.1 μg L^{-1}	61
Fenuron, simazine, atrazine, carbaryl, ametryn, prometryn, terbutryn in 12 min	EKC	WATER River water	online SPE C_{18} 90–114%	10 mmol L^{-1} phosphate, 60 mmol L^{-1} SDS, 8% ACN (pH 9.5)	UV, 226 nm 0.01–0.03 μg mL^{-1}	65
Simazine, aziprotryne, hexazinone, diuron in 10 min	EKC	WATER River water Well water	SPE DVN-VP 41–109%	20 mmol L^{-1} borate, 8.5 mmol L^{-1} SDS (pH 8.30)	UV, 215, 240 nm 0.02–0.17 ng L^{-1}	124
Pyrethroids						
Pyrethrin esters (pyrethrin, cinerin and jasmolin, I and II) in 25 min	EKC	Pyrethrum extract	—	25 mmol L^{-1} Tris, 30 mmol L^{-1} SDS, 25% ACN (pH 9)	UV, 254 nm 1.1–14.1 mg L^{-1}	109

Triazines						
Hydroxyatrazine, desisopropylhydroxyatrazine, desethylhydroxyatrazine, ameline in ca. 40 min	CZE	WATER Tap water	SPE Amberchrom resins 20–95% 4% for ammeline	80 mmol L^{-1} acetate, 62 mmol L^{-1} phosphate (pH 4.7)	UV, bubble cell, 205 nm	100
Atrazine, simazine, propazine, ametryne, prometryne, terbutryne in ca. 9.3 min	CZE	WATER River water Drinking water	SPE Oasis HLB 83–114%	7.5 mmol L^{-1} $HClO_4$, 17 mmol L^{-1} SDS in 50:50 MeOH–ACN mixture (nonaqueous)	UV, 214 nm 2.1–3.4 μg L^{-1} (CE-UV) 0.01–0.05 μg L^{-1} (SPE-CE-UV)	105
Desethylatrazin-2-hydroxy, simazine, prometon, atrazine, simetryn, ametryn, propazine, prometryn, trietazine, terbutylazine, terbutryn in 30 min	EKC	WATER Ground water	SPE C_{18} 52–87%	24 mmol L^{-1} borate, 18 mmol L^{-1} phosphate, 25 mmol L^{-1} SDS, 5% 1-propanol (pH 9.5)	UV, 214 nm 0.05 μg L^{-1}	16
Atrazine, simazine, propazine, prometryn, hydroxyatrazine, deisopropylatrazine, deethylatrazine in 7 min	EKC	WATER Drinking water Well water	SPE 2 PS-DVB disks 73.5–102.4%	10 mmol L^{-1} borate, 60 mmol L^{-1} SDS, 20% MeOH (pH 9.2)	220 nm, 200 μm extended optical path capillary 0.02–0.30 mg L^{-1}	53
Metribuzin, bromacil, terbacil, hexazinone, triadimefon, DEET in 8 min	EKC	WATER Drainage water from highway	SPE C_{18} disks 85%	12.5 mmol L^{-1} borate, 50 mmol L^{-1} SDS (pH 9.0)	UV, 210 nm 0.8 μg L^{-1}	55
Hydroxyatrazine, hydroxyterbutylazine, deethylhydroxyatrazine, deisopropylhydroxyatrazine ameline in 9.5 min	EKC	WATER Tap water Ground water	SPE LiChrolut EN 43.8–93.4% ameline is not recovered	30 mmol L^{-1} borate, 30 mmol L^{-1} SDS (pH 9.3)	UV, 210 nm 0.2–0.5 mg L^{-1} (CE-UV) 0.1–0.25 μg L^{-1} SPE-CE-UV	99
Atrazine, desethylatrazine, desisopropylatrazine, hydroxyatrazine, chloro-, hydroxy-degradation products in 29 min	EKC	HUMIC ACID Solutions	—	15 mmol L^{-1} tetraborate, 60 mol L^{-1} SDS, 10% MeOH (pH 9.3)	UV, 210 nm 2–4 mg L^{-1}	119
Hexazinone, metabolites C, A1, E, B, D in 30 min	EKC	WATER Ground water	SPE graphitized nonporous carbon 30–120%	50 mmol L^{-1} SDS, 12 mmol L^{-1} phosphate, 10 mmol L^{-1} borate, 15% MeOH (pH 9.0)	220, 225, 230, 247 nm <0.38 mg L^{-1}	120

Continued

TABLE 31.1
(Continued)

Compounds	CE Mode	Matrix	Preconcentration Procedure (Recovery)	Optimal Electrolyte	Detection (LOD)	References
Carbamates						
Aminocarb, propoxur, carbofuran, carbaryl, methiocarb, metabolites: 4-dimethylamino-3-methylphenol, 2-isopropoxyphenol, 2,3-dihydro-2,2-dimethyl-7-benzofuranol, 1-naphthol, and 4-methylthio-3,5-xylenol in 20 min	EKC	WATER River water Well water Pond water	SPE PS-DVB 82.2–104.8%	45 mmol L^{-1} borate/phosphate, 40 mmol L^{-1} SDS (pH 8.0)	UV, 202, 214 nm 22–85 ng L^{-1}	12
Urea herbicides						
Monuron, isoproturon, diuron in 16 min	EKC	WATER Tap water Pond water	SW, SRMM, SRW SPE C_{18}	50 mmol L^{-1} SDS, 50 mmol L^{-1} phosphoric acid, 15 mmol L^{-1} γ-CD	UV, 244 nm tested z-shaped cell 1 μg L^{-1}	60
Chlorsulfuron, chlorimuron, metsulfuron in 20 min	EKC	SOIL	LLE-SPE C_{18} >80%	30 mmol L^{-1} borate, 80 mmol L^{-1} SDS, 14% MeOH, 20% isopropanol (pH 7.0)	UV, 214 nm 10 μg L^{-1}	114
Monuron, linuron, diuron, isoproturon, monolinuron in 15 min	EKC	WATER Drinking water	SPE C_{18} 80.2–94.9%	4 mmol L^{-1} tetraborate, 12 mmol L^{-1} phosphate, 30 mmol L^{-1} SDS (pH 7)	UV, 244 nm 0.1 mg L^{-1}	121
Organophosphorus						
Glufosinate, glyphosate, aminomethylphosphonic acid in ca. 8 min	CZE	WATER Ground water	—	1 mmol L^{-1} fluorescein (pH 9.5 adjusted with sodium hydroxide)	LIF, 488 nm 10 mW, argon-ion laser 0.6–1.7 μmol L^{-1}	74
Glyphosate in 12 min	CZE	WATER Lake water	—	5 mmol L^{-1} NH_4Ac (pH 2.8 adjusted with acetic acid) (CTAB rinsing before analysis)	ESI-CNLSD 0.2 μg mL^{-1}	101

Glyphosate, glufosinate, aminomethylphosphonic acid in 10 min	CZE	WATER River water	Spiked sample 95.9–104.1%	40 mmol L^{-1} NH_4Ac (pH 9.0)	ICP-MS 0.11–0.19 mg L^{-1}	103
Glyphosate, aminomethylphosphonic acid in ca. 4 min	EKC	WATER River water Tap water	Strong anion-exchange resin 50–99%	50 mmol L^{-1} phthalate, 0.5 mmol L^{-1} TTAB (pH 7.5)	UV, indirect detection, 254 nm 60–85 ng L^{-1}	118
Quaternary ammonium herbicide						
Paraquat, diquat, difenzoquat in ca. 17.5 min	EKC	WATER Tap water	CSEI–SW	50 mmol L^{-1} phosphate, 80 mmol L^{-1} SDS, 20% ACN (pH 2.5)	UV, 220 nm, 255 nm 0.075–1 μg L^{-1}	63
Paraquat, diquat, difenzoquat, chlormequat mepiquat in 17 min	CZE	WATER Contaminated Irrigation water Mineral water	—	200 mmol L^{-1} formic acid–ammonium formate buffer, 50% MeOH (pH 3.0)	ESI-IT-MS 0.5–2.5 mg L^{-1} (hydrodynamic injection) 1–10 μg L^{-1} (electrokinetic injection)	85
Triazolopyrimidine sulfonanilide						
Flumetsulam, florasulam, cloransulam-methyl, diclosulam, metosulam in 6.5 min	EKC	SOIL	SPE-FESI C_{18} 50–84%	11 mmol L^{-1} formic acid, 16 mmol L^{-1} $(NH_4)_2CO_3$, 2.5 mmol L^{-1} α-CD, 0.00042% HDB (pH 7.6)	UV, 205 nm 18–34 μg kg^{-1}	116
Chloroacetanilide herbicide						
Metolachlor stereoisomers and two metabolites (ethane sulfonic acid, oxanilic acid) in ca. 24.5 min	EKC	WATER Ground water	SPE C_{18} >90%	75 mmol L^{-1} borate, 5% (w/v) γ-CD, 20% MeOH (pH 9.2)	Diodo array detector 5 mg L^{-1}	115
Triazinone herbicide						
Metribuzin and its major conversion products, deaminometribuzin, diketometribuzin, deaminodiketometribuzin in ca. 7 min	EKC	SOIL	Sonication SPE LiChrolut EN 78.3–99.2%	10 mmol L^{-1} NH_4Ac buffer, 100 mmol L^{-1} SDS (pH 10)	UV, 220 nm, 260 nm 19–23.4 μg kg^{-1}	67

Continued

TABLE 31.1
(Continued)

Compounds	CE Mode	Matrix	Preconcentration Procedure (Recovery)	Optimal Electrolyte	Detection (LOD)	References
Chlorophenoxycarboxylic acid						
2,4-dichlorophenoxyoleic, 2,4-dichlorophenoxypropionic, 2,4,5-trichlorophenoxyacetic, 2,4-dichlorophenoxyacetic, phenoxyacetic, 2,4-dichlorophenol (product of their decomposition) in ca. 11 min	EKC	WATER Tap water River water Drinking water Well water	SPE Diapak S16 12.6–84.7%	10 mmol L^{-1} borate buffer, 2 mmol L^{-1} β-CD (pH 9.2)	UV, 205 nm 0.5–1 μg L^{-1}	117
Nicotinoids						
Imidacloprid, metabolite 6-chloronicotinic acid in 6 min	EKC	AIR greenhouse air	SPE amberlite 85–92%	NH_4Cl/NH_3 buffer at 15 mmol L^{-1}, 60 mmol L^{-1} SDS (pH 8.5)	UV, 227, 270 nm 0.71–1.18 ng L^{-1}	13

FESI, field enhanced sample injection; SW, sweeping; SRMM, stacking with reverse migrating micelles; SRW, stacking with reverse migrating micelles and a water plug; SPE, solid-phase extraction; C_8, octa-, C_{18}, octadecyl-, NH_2, amino-bonded silica; PS-DVB, poly(styrene-divinylbenzene); DVN-VP, poly(divinylbenzene-*N*-vinylpyrrolidone); SLM, supported liquid membrane; LLE, liquid–liquid extraction; HDB, hexadimethrine bromide; CD, cyclodextrin; ESI-CNLSD, electrospray condensation nucleation light scattering detector; ICP-MS, inductively coupled plasma-mass spectrometry; ESI-IT-MS, electrospray ionization ion trap-mass spectrometry; CSEI, cation-selective exhaustive injection; ASE, accelerated solvent extraction; TTAB, tetradecylytrimethylammonium bromide; CTAB, cetyltrimethylammonium bromide; LIF, laser-induced fluorescence; MeOH, methanol; ACN, acetonitrile.

Pesticides encompass a large assortment of organic compounds, several of which are positional, geometrical, and optical isomers, with differing degrees of ionization, polarity, and water solubility. Many exhibit spectrochemical, electrochemical, or other functional properties suitable for detection. Unless the pesticide class is explicitly charged, single- and multiclass pesticide analyses are usually assessed by EKC modes. Several excellent review articles have covered the literature on pesticide analysis by CE in the past 5 years.[92–96]

The analytical feasibility of electromigration methods for pesticides has been endorsed by fundamental studies with standards and analyses of spiked and real samples. An example of fundamental study is the determinations of pK_a (CZE: citrate/phosphate buffers and carbonate buffers covering the pH range from 2.05 to 12.45) and pI (CIEF kit from Beckman Coulter) of 12 hydroxy-*s*-triazines standards.[97]

The majority of CZE methodologies for pesticides employs simple buffers at varying pH,[70,98–100] volatile electrolytes[101,102] (usually aiming at CZE-MS applications),[103] solvent modified buffers,[85,104,105] and electrolytes for special detector schemes (indirect fluorescence detection, for instance).[74] Table 31.1 compiles the details of a few CZE methodologies applied to pesticide analyses, classifying the application by the pesticide class and the environmental compartment assessed.

For the EKC mode, pesticide standards are either combined in multiresidue mixtures[106–108] or differentiated by classes: pyrethroids,[109] *s*-triazines,[14,90,99,110] carbamates,[88,111] phenoxyacids,[112] quaternary ammonium salts,[63,113] phosphonic acids,[72] and urea-derived pesticides.[114] MEKC methodologies include buffered systems containing SDS,[99,107] cationic,[14,110,111] or nonionic surfactants[72] and additives such as organic solvents[63,106,113,114] and CDs for chiral separations.[106] The use of poly-SUS for pyrethoids,[109] DOSS for herbicide mixtures,[108] and alkylglycoside chiral surfactants for phenoxy acids[112] have also been reported. Several examples of EKC methodologies applied to single class[12,13,16,53,55,60,63,67,109,114–121] and multiclass[26,61,65,122–124] pesticide residues in real samples are detailed in Table 31.1. Figure 31.1 illustrates the electromigration separation of herbicides.

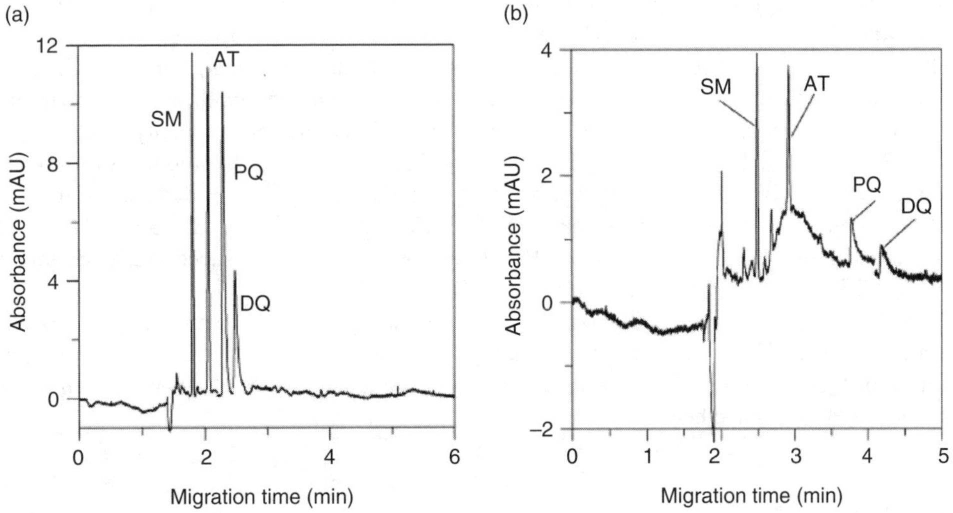

FIGURE 31.1 EKC separation of triazines and quats herbicides. (a) Standard solution and (b) spiked well water. Electrolyte: 10 mmol L^{-1} sodium tetraborate, 25 mmol L^{-1} SDS, 15% ACN, and 40 mmol L^{-1} sodium perchlorate. Other conditions: +30 kV; direct UV detection at 220 nm. Peak labels: simazine (SM), atrazine (AT), paraquat (PQ), and diquat (DQ). (Modified from M.I. Acedo-Valenzuela et al., *Anal. Chim. Acta*, 519, 65–71, 2004. With permission.)

31.4.2 PAH, PCB, AND PCDD

Polynuclear aromatic hydrocarbons are listed by the U.S. Environmental Protection Agency (U.S. EPA) and European Community as priority pollutants due to their mutagenicity and carcinogenicity. PAHs can be formed during natural processes or emitted by anthropogenic activity.

PAHs are characterized by a multitude of structurally similar congeners with inherently neutral and highly hydrophobic character. Therefore, they are not suitable to CZE analysis and as a result, only EKC-based methodologies have been devised in the literature. PAHs have been used in test mixtures as model compounds to introduce a plethora of EKC approaches. MEKC separations include diverse surfactant types (SDS,[69,125] tetradecylsulfate,[126] dodecylbenzenesulfonate,[127] cationic surfactants,[128] bile salts[129]) as well as commonly used additives (urea,[125,130] organic solvents,[125–130] and neutral CDs).[69,125] Other strategies invoke solvophobic association with ThxAP, THpAP,[131] or DOSS[130] in buffered electrolytes as well as charge–transfer interactions with pyrylium salts in nonaqueous media.[132] The use of mixed CD systems (CDCD-EKC), containing both noncharged and negatively charged CDs (SB-β-CD and carboxymethyl-β-CD), has also been demonstrated.[133] The applicability of more elaborate secondary phases such as SUA[134] and SUS[135] oligomers, in combination with neutral CDs,[136] polymeric surfactants (11-acrylamidoundecanoate,[137] sulfated siloxanes[138]), starburst dendrimers,[139] fullerenes,[140] and silica-based particle suspensions[141] have all been demonstrated by PAH test mixtures.

Polychorinated biphenyl (PCB) mixtures were used extensively in the past as coolant fluids in power transformers and capacitors. PCBs were widely used because of their higher stability, but are persistent environmental contaminants due to careless disposal practices, leakage, or accidents. Similar to PAHs, PCBs are neutral, highly hydrophobic, and present a large number of congeners, thus demanding EKC methodologies.

The separation of PCB congeners has been addressed by SDS/neutral CD containing electrolytes,[142,143] modified by organic solvents[144] or urea[142,143] and mixtures of bile salts.[145] CD-EKC and CDCD-EKC modes with a large assortment of ionic CDs and modifiers are often employed for the chiral discrimination of PCB racemates.[42,146] The use of polymeric surfactants such as polysodium undecyl sulfate (poly-SUS), in acetonitrile[147] and its valinate form (poly-*D*-SUV) in combination with hydroxypropyl-γ-CD, methanol, and urea[148] has also been reported.

The term "dioxin" comprises a group of 75 polychlorinated dibenzo-*p*-dioxin (PCDD), and 135 polychlorinated dibenzofuran (PCDF), whose effects on human health and the environment include dermal toxicity, immunotoxicity, reproductive effects and teratogenicity, endocrine disrupting effects, and carcinogenicity. PCDD and PCDF are produced as byproducts of a myriad of processes, including bleaching of pulp, incineration of garbage, recycling of metals, and in the production of common solvents. Fires of many kinds, including forest fires and those in incinerators, also release dioxins into the environment. PCDD congeners have been successfully separated by MEKC in SDS buffers,[149] modified by urea and CDs.[150,151]

Table 31.2 compiles a few applications of EKC methodologies to PAH in real samples. For example, soil,[15,38,43] water,[52,152] air,[27] and other matrices.[125] Figure 31.2 illustrates the electromigration separation of PAH homologs. To date, no environmental applications of electromigration methods for PCBs and PCDDs have been reported.

31.4.3 PHENOLS

Several phenols and their derivatives (chloro- and nitrophenols) are priority polluting substances that are widespread in the environmental aquatic compartments. They originate as byproducts from the coal and oil industry and also due to pesticide and drug decay.

Phenols are UV-absorbing compounds with weakly acidic OH groups ($pK_a \approx 9$), which, upon dissociation at high pH electrolytes, generate anionic species, making them suitable for CZE analysis. However, when a large number of phenolic compounds and derivatives must be assessed

TABLE 31.2
Selected Applications of Electroseparation Methods for PAH in Environmental Samples

Compounds	CE Mode	Matrix	Preconcentration Procedure (Recovery)	Optimal Electrolyte	Detection (LOD)	References
7 PAH homologs in 10 min	EKC	SOIL spiked heath sand	LLE cyclohexane (48–90%)	8.5 mmol L^{-1} borate, 85 mmol L^{-1} SDS, 50% ACN (pH 9.9)	UV, 280 nm 10 mg L^{-1}	15
16 PAH homologs in 35 min	EKC	AIR ambient air	Air sample collected on polyurethane foams	0.1 mol L^{-1} phosphate, 0.1 mol L^{-1} borate, 50 mmol L^{-1} STDC, 30% acetone	UV, 214 nm 3–25 ng mL^{-1}	27
13 PAH homologs in 15 min	EKC	WATER aqueous standard mixture extracted by SPME	SPME Silica impregnated with polydimethylsiloxane	50 mmol L^{-1} NH_4Ac, 100 mmol L^{-1} THA^+ in 100% MeOH	UV, 254 nm	36
16 PAH homologs in 21 min	EKC	SOIL contaminated with creosote	ASE 50:50 CH_2Cl_2-acetone, 150°C, 2500 psi, 10 min followed by HPGPC	20% THF, 0.00625 mol L^{-1} OSUA	UV, 214 nm	38
11 PAH homologs in 14 min	EKC	SOIL wood preserving lot	SFE CO_2 at 400 atm, 400 mL/min, 120°C, 20 min; sample collected in CH_2Cl_2 and diluted in MeOH/water	25 mmol L^{-1} SB-β-CD, 20 mmol L^{-1} M-β-CD, 50 mmol L^{-1} borate	LIF, bubble cell, 325 nm, 2.5 mW HeCd laser 0.9–21.7 μg L^{-1}	43
5 PAH homologs in 30 min	EKC	WATER rain water river water spring water	LLE perfluoro surfactants in THF (>99%)	0.2 mol L^{-1} $PFOS^-$, 50% DMSO, 0.1 mol L^{-1} H_3PO_4	UV, 280, 333 nm 10^{-10} to 10^{-9} mol L^{-1}	52
5-Hydroxy-PAH homologs in 20 min	EKC	ISOPODS hepatopancreas FLATFISH bile	PROTEOLYSIS protease K in Tris buffer pH 9, 37°C, 18 h	30 mmol L^{-1} borate, 60 mmol L^{-1} SDS, 12.5 mmol L^{-1} γ-CD (pH 9.0)	LIF Nd-YAG, 266 nm	125

ASE, accelerated solvent extraction; HPGPC, high-performance gel permeation chromatography; OSUA, oligomers of sodium undecylenic acid; SFE, supercritical fluid extraction; SDβCD, sulfobutyl ether β-cyclodextrin; MβCD, methyl β-cyclodextrin; $PFOS^-$, perfluorooctanic sulfate; DMSO, dimethyl sulfoxide; THA^+, tetrahexylammonium; SPME, solid-phase microextraction; STDC, sodium taurodeoxycholate; LIF, laser-induced fluorescence; LLE, liquid–liquid extraction.

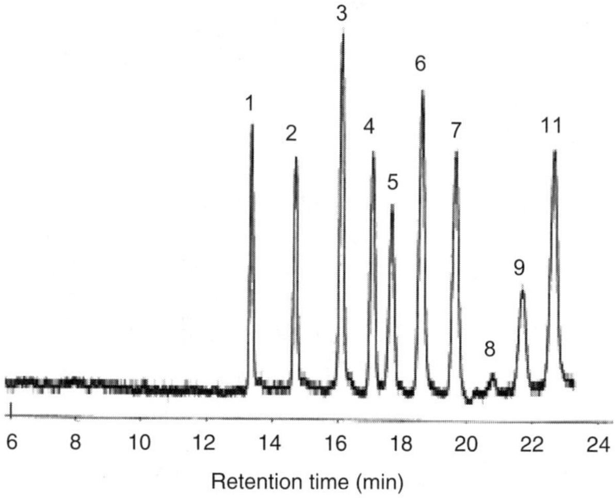

FIGURE 31.2 EKC separation of PAHs. Electrolyte: 10 mmol L^{-1} H_3PO_4 and 70 mmol L^{-1} sodium *n*-tetradecylsulfate in 75:25 methanol:H_2O. Other conditions: −20 kV, direct UV-detection at 254 nm. Peak labels: benzo[*a*]perylene (1), perylene (2), benzo[*a*]anthracene (3), pyrene (4), 9-methylanthracene (5), anthracene (6), fluorene (7), naphthalene (8), and benzophenone (9). (Modified from J. Li and J.S. Fritz, *Electrophoresis*, 20, 84–91, 1999. With permission.)

simultaneously (chlorophenol congeners, for instance), EKC methodologies should be selected as a means to increasing selectivity. Nevertheless, CZE has, by far, received more attention for real sample applications.

Literature proposed CZE methods for phenols and derivatives using test mixtures based on aqueous buffered systems (phosphate–borate[153] and borate[154]), volatile electrolytes (ammonium hydrogencarbonate,[155] diethylmalonic acid/dimethylamine in isopropanol[156] and L-cysteic acid, 3-amino-1-propanesulfonic acid, aminomethanesulfonic acid, and diethylmalonic acid[157]), and non-aqueous media (ammonium acetate in ACN/acetic acid in MeOH;[158] acetate, bromide, chloride, and malonate in ACN; and diprotic acids/tetrabutylammonium hydroxide[159] and maleate in MeOH[160]).

Environmental assessment using CZE methodologies includes pressurized hot water extraction of phenols from sea sand and soil,[161] chlorophenols[162] and priority phenols[11] in wastewater, nitrophenols[163] and iodophenol[164] in river waters, and methoxyphenols in biomass burning.[86] Details of these methodologies are compiled in Table 31.3.

The impact of different surfactants (SDS,[22] DOSS,[165] CTAB[166] and hexadimethrine bromide, HDB,[167] bile salts[29,30]), nonionic[168] and mixed micelles,[22,169] and additives (neutral[170] and anionic CDs,[44] tetraalkylammonium salts,[171] organic solvents[29,165]) in EKC separations has been demonstrated with phenol test mixtures. In addition, phenols have been chosen to introduce the applicability of more exotic EKC secondary phases such as SDS modified by bovine serum albumin,[172] water-soluble calixarene,[173] starburst dendrimers,[174] cationic replaceable polymeric phases,[175] ionenes,[176] amphiphilic block copolymers,[177] polyelectrolye complexes,[178] and liposome-coated capillaries.[179]

The separation of phenols of environmental interest[180] as well as the sources and transformations of chlorophenols in the natural environment[181] have been revised. Examples of the investigation of phenols by EKC methodologies in aquatic systems,[39,41,81] soil,[44] and gas phase[182] are compiled in Table 31.3. Figure 31.3 illustrates the electromigration separation of phenols by both CZE and EKC modes.

31.4.4 Amines

Amines are organic bases that are usually present in biological materials (biogenic amines), processed foods and beverages (of concern are the nitrosamines in fried bacon, smoked/cured meat and fish,

TABLE 31.3
Selected Applications of Electroseparation Methods for Phenols in Environmental Samples

Compounds	CE Mode	Matrix	Preconcentration Procedure (Recovery)	Optimal Electrolyte	Detection (LOD)	References
26 Priority phenols in 40 min	CZE	WATER Waste water	SPE PSDVB spiked sample 63.3–113.4%	150 mmol L^{-1} NH_4Ac in 75:25 *N*-methylformamide-ACN mixture (non-aqueous)	UV, 280 nm 28–629 μg L^{-1}	11
11 Priority phenols in 12 min	EKC	WATER River water Sea water	SPE spiked sample 74.2–106%	50 mmol L^{-1} phosphate, 25 mmol L^{-1} tetraborate-phosphate, 7.5 mmol L^{-1} DBTHS (pH 7)	UV, 214 nm 28.1–215 nmol L^{-1}	39
10 Nitrophenols in 20 min	EKC	WATER Rain water Tap water Process water	ITP spiked sample	50 mmol L^{-1} glycine, 0.2% m-HEC, 2.5% PVP (pH 9.1)	UV, 254 nm 19–80 μg L^{-1}	41
25 Phenols in 20 min	EKC	SOIL Certified soil Reference standard	—	50 mmol L^{-1} phosphate, 1 mmol L^{-1} SB-β-CD (pH 7.5)	UV, 214 nm 0.05–0.33 mg L^{-1}	44
19 Chlorophenols in ca. 35 min	EKC	WATER River water	SPE PSDVB 81–116%	50 mmol L^{-1} ACES, 22 mmol L^{-1} SDS (pH 6.1)	ELECTROCHEMICAL graphite-epoxy electrode versus Ag/AgCl 0.07–0.2 μg L^{-1}	81
12 Substituted methoxy phenols and aromatic acids in 9.5 min	CZE	AEROSOL Biomass burning Aerosol	Berner type impactor filter (140–420 nm, 50% cut off)	(1) 20 mmol L^{-1} NH_4Ac, 10% MeOH (pH 9.1) (2) 1 mol L^{-1} NH_4OH (pH 11)	ESI-IT-MS (1) 0.1–1.0 μmol L^{-1} (2) 0.3–0.7 μmol L^{-1}	86

Continued

TABLE 31.3
(Continued)

Compounds	CE Mode	Matrix	Preconcentration Procedure (Recovery)	Optimal Electrolyte	Detection (LOD)	References
Phenol, 3-methylphenol, 4-chloro-3-methylphenol, 3,4-dichlorophenol in 6 min	CZE	SOIL Soil and sea sand mixture Sea sand	PHWE 77.7–105.2% (soil sample) 77.7–98.4% (sea sand sample)	30 mmol L^{-1} CHES (pH 9.7)	UV, 220 nm 270–410 $\mu g\,L^{-1}$	161
17 Chlorophenols in 15 min	CZE	WATER Waste water	—	10 mmol L^{-1} phosphate buffer, 40% acetone (pH 8.23)	UV, 214 nm —	162
o-Nitrophenol, *m*-nitrophenol, *p*-nitrophenol in 11 min	CZE	WATER River water	Spiked sample 95.6–100.9%	20 mmol L^{-1} borate-carbonate, 10% MeOH (pH 9.4)	UV, 191 nm 10.2–40.6 $\mu mol\,L^{-1}$	163
2-Iodophenol, 4-iodophenol, 2,4,6-triiodophenol in 6.6 min	CZE	WATER River water	SPME CAR-PDMS ca. 100%	20 mmol L^{-1} CAPS (pH 11.0)	ICP-MS 2.4–3.9 $\mu g\,L^{-1}$ (CE-ICP-MS) 0.03–0.04 $\mu g\,L^{-1}$ (SPME-CE-ICP-MS)	164
12 Chloro and nitrophenols in 10 min	EKC	AIR	Loop-supported liquid film	25 mmol L^{-1}, borate 10 mmol L^{-1} phosphate, 10 mmol L^{-1} SDS (8.86)	UV, 205 nm 3.5–17 $\mu g\,L^{-1}$	182

DBTHS, disodium 5,13-*bis*(dodecyloxymethyl-4,7,11,14-tetraoxa-1,17-heptadecanedisulfonate; ITP, isotachophoresis; m-HEC, methyl-hydroxyethylcellulose; PVP, polyvinylpirrolidone; PSDVB, polystyrene-divinylbenzene copolymer; SB-β-CD, sulfobutylether-β-cyclodextrin; CAR-PDMS, carboxen-poly(dimethylsiloxane); ESI-IT-MS, electrospray ionization ion trap-mass spectrometry; CHES, 2-(*N*-cyclohexylamino)-ethanesulphonic acid; CAPS, 3-(cyclohexylamino)-1-propanesulfonic acid; PHWE, pressurized hot water extraction; ICP-MS, inductively coupled plasma-mass spectrometry; SPME, solid-phase microextraction; SPE, solid-phase extraction.

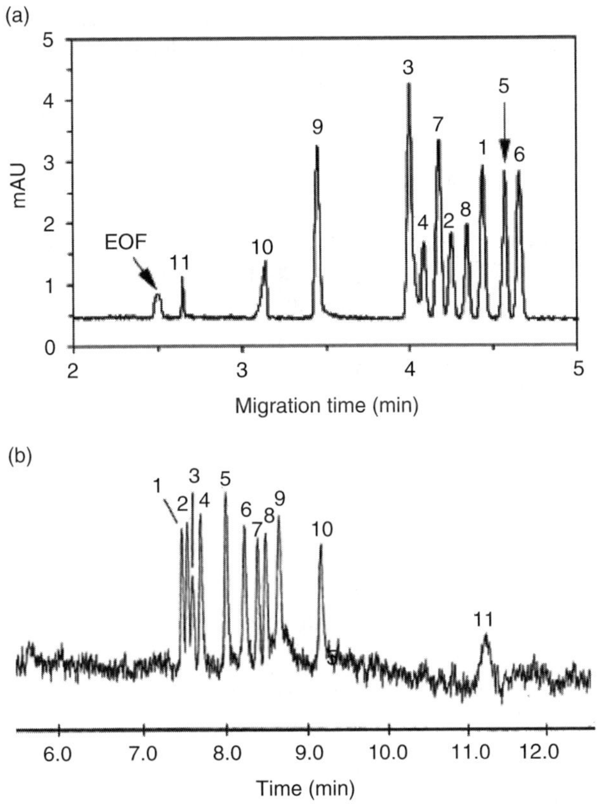

FIGURE 31.3 (a) CZE and (b) EKC separation of EPA phenols. Electrolytes: (a) 20 mmol L^{-1} NaH_2BO_3 at pH 10.00; (b) 20 mmol L^{-1} phosphate, 8% 2-butanol, 0.001% HDB, pH 11.95. Other conditions: (a) +20 kV; direct UV-detection at 254 nm; (b) +25 kV; direct UV-detection at 210 nm. Peak labels: (a) 2,4-dinitrophenol (1), 2-methyl-4,6-dinitrophenol (2), pentachlorophenol (3), 2,4,6-trichlorophenol (4), 4-nitrophenol (5), 2-nitrophenol (6), 2,4-dichlorophenol (7), 2-chlorophenol (8), 4-chloro-3-methylphenol (9), phenol (10), and 2,4-dimethylphenol (11). (b) 2-nitrophenol (1), 2-chlorophenol (2), 2,4,6-trichlorophenol (3), phenol (4), 4-nitrophenol (5), 2,4-dinitrophenol (6), 4-chloro-3-methylphenol (7), 2,4-dichlorophenol (8), 2-methyl-4,6-dinitrophenol (9), 2,4-dimethylphenol (10), and pentachlorophenol (11). (Modified from (a) I. Canals et al., *Anal. Chim. Acta*, 458, 355–366, 2002 and (b) P. Kubáň et al., *J. Chromatogr. A*, 912, 163–170, 2001. With permission.)

canned sausage, etc.), as well as environmental samples. Several amines are strongly toxic and suspected carcinogens. Major sources of amines in the environment include effluents or byproducts of several chemical industry sectors such as oil refining, synthetic polymers, adhesives, rubber tyre manufacturing, leather tanning, pharmaceuticals, pesticide production, and explosives.

31.4.4.1 Aliphatic Amines

The analysis of small aliphatic amines by electromigration methods faces a few challenges. Although readily protonated at low pH electrolytes, the lack of chromophore precludes direct UV-detection in both CZE and EKC mode unless derivatizing schemes are devised or alternative detectors are selected. A further complication of the EKC mode derives from the hydrophilicity and polar characteristic of aliphatic amines, which usually impair strong interaction with commonly used micellized surfactants, demanding alternative secondary phases or more elaborate electrolytes.

Indirect UV-detection is a feasible option and particularly interesting for the CZE determination of primary to quarternary amines without derivatization. Imidazole, *N*-ethylbenzylamine, and benzyltriethylammonium chloride have been proposed as electrolyte chromophoric constituents.[183] The indirect UV-detection of aliphatic amines in ambient air has been performed by CZE in imidazole-based buffer modified by ethanol and EDTA.[184] By replacing imidazole with ammonium, the electrolyte became applicable to MS detection (these details and other CZE methodologies are reported in Table 31.4).

EKC separations of aliphatic amines include electrolyte systems composed of several surfactants (SDS,[185] cholate,[186] Brij 35[72]) modified by certain additives (urea,[185] neutral CDs,[185] organic solvents[75,186]), mixed CDs,[187] and more unusual secondary phases [resorcarene-octacarboxylic acid,[188] calixarene,[47] poly(sodium 4-styrenesulfonate), PSSS[189]]. Derivatization is performed when UV (*o*-phthaldialdehyde, OPA,[185,187] as derivatizing agents) or LIF detection [3-(2-furoyl)quinoline-2-carboxaldehyde,[186] 5-(4,6-dichloro-s-triazin-2-ylamino)fluorescein (DTAF)[72] as labeling agents] is designed.

Examples of the EKC determination of aliphatic amines in water compartments[75] and atmospheric aerosol samples[71] are compiled in Table 31.4.

31.4.4.2 Aromatic Amines

Direct UV spectrophotometric detection of aromatic amines from azo dye reduction in textile industry wastewaters[190] and electrophoretic methods for biogenic and aromatic amines[191] have been the subject of recent reviews. In addition, environmental applications of mutagenic heterocyclic amines have recently been revised.[192]

CZE analysis of aromatic amines include the use of simple buffers at varying pH (acetate,[80] phosphate,[10,193] and borate[70]) with direct UV,[10] electrochemical,[80] or fluorescence[70,80,193] detection.

EKC methodologies for aromatic amines are based on electrolytes containing SDS,[64,91,194,195] mixtures of SDS and nonionic surfactants,[196] and bile salts[29] as well as modifiers (tetraalkylammonium salts,[195] urea,[64] organic solvents,[29,64,91,194,196] and CDs[197]). Combinations of anionic soluble polymers[198] or crown ether[199] and CDs have also been reported. Despite being chromophoric, alternative detectors for aromatic amines have been proposed (CE-ESI-MS,[91] electrochemical,[80] and fluorescence[80] detection).

Examples of EKC methodologies for aromatic amines in water[21,35,37,80] are compiled in Table 31.4. Figure 31.4 illustrates the electromigration separation of aliphatic and aromatic amines.

31.4.5 Carbonyls

Low-molecular mass carbonyls are among the most abundant and ubiquitous volatile organic compounds in the atmosphere. They are produced from industrial activity and incomplete combustion of fossil fuels and biomass. Many aldehydes are also emitted indoors (plastic, foam insulation, lacquers, etc.). As a source of free radicals, aldehydes play an important role in the ozone formation, in urban smog events, as well as in the photochemistry of the unpolluted troposphere. Aldehydes are recognized irritants of the eye and respiratory tract, and often, carcinogenic and mutagenic characteristics are also attributed to them.

To implement CZE and EKC methodologies for the determination of carbonyl compounds, structure-related issues must be addressed. Aldehydes and ketones of environmental importance are essentially neutral molecules, interact poorly with micellar phases, and present no chromophoric moieties. Two methodological approaches have been presented: either generation of charged adducts or generation of large UV-absorbing or fluorescence derivatives. In the former, the charged adducts can be detected directly (if aromatic) or indirectly (if aliphatic). The separation of anionic aldehyde-bisulfite adducts is a fine example of the first approach.[25] The second approach comprises the derivatization of the molecules to generate either neutral or charged UV-absorbing or fluorescent

TABLE 31.4
Selected Applications of Electroseparation Methods for Amines in Environmental Samples

Compounds	CE Mode	Matrix	Preconcentration Procedure (Recovery)	Optimal Electrolyte	Detection (LOD)	References
Aliphatic amines						
Methylamine, dimethylamine, diethylamine, dipropylamine, piperidine, pyrrolidine, morpholine in 10 min	EKC	AIR Particulate aerosol	FITC derivatization	20 mmol L^{-1} borate, 20% acetone, 5 mmol L^{-1} DM-β-CD	LIF, 488 nm Ar laser, 520 nm emission 10^{-9} mol L^{-1}	71
Diaminopropane, putrescine, cadaverine, diaminohexane in 7.5 min	EKC	WATER Lake water	ABEI-DSC derivatization 92.8–106.8%	10 mmol L^{-1} borate, 100 mmol L^{-1} H_2O_2, 80 mmol L^{-1} SDS (pH 9.3)	Chemiluminescence, postcolumn reagents: 3 mmol L^{-1} $K_3Fe(CN)_6$, 0.8 mol L^{-1} NaOH. $(3.5–12) \times 10^{-8}$ mol L^{-1}	75
Methylamine, ethylamine, propylamine, butylamine, dimethylamine, diethylamine, cadaverine, putrescin, spermidine in 9.5 min	CZE	AIR Metal working Fluid aerosols and ambient air	Quartz filter — SPE poly(acrylate-methacrylate) copolymer 95–99%	20 mmol L^{-1} imidazole, 4 mmol L^{-1} EDTA, 25% ethanol (pH 2.5 with acetic acid)	UV, indirect detection, 214 nm 2 μg mL^{-1}	184
Aromatic amines						
21 Aromatic amines in 25 min	CZE	WATER SOIL	LLE/SPE methylene chloride/ 20:80 (w/w) keto derivatized–underivatized PSDVB 47–97%	50 mmol L^{-1} NaH_2PO_4, 7 mmol L^{-1} 1,3-diaminopropane, (pH 2.35 with H_3PO_4)	UV, 280 nm 0.06–1.8 mg L^{-1}	10
2-Toluidine, 4-toluidine, 1-naphthylamine, 2-naphthylamine, 2-aminobiphenyl, 4-aminobiphenyl, 2-methoxy-5-methylaniline, 4-methoxy-2-methylaniline, 2-chloroaniline, 4-chloroaniline, 2,4,5-trimethylaniline, 2,4,6-trimethylaniline, 2-anisidine, 4-anisidine, 2,4-xylidine, 2,6-xylidine in 50 min	EKC	WATER Waste water	SPE/LLE (cation exchange resin; tertiary butyl methyl ether) 78.4–94.4%	50 mmol L^{-1} phosphate—adjusted to pH 8.5 with 100 mmol L^{-1} borate, 300 mmol L^{-1} SDS, 10 mmol L^{-1} cholic acid, 10 mmol L^{-1} Tween 80	UV, 214 nm 0.263–9.525 μg mL^{-1}	21

Continued

TABLE 31.4
(Continued)

Compounds	CE Mode	Matrix	Preconcentration Procedure (Recovery)	Optimal Electrolyte	Detection (LOD)	References
4-Chloroaniline, 4-bromoaniline, 3,4-dichloroaniline, 3-chloroaniline, 3-chloro-4-methylaniline, 4-isopropylaniline, aniline, 3-methylaniline, acetophenone, propiophenone, butyrophenone, valerophenone, hexanophenone, heptanophenone in 26 min	EKC	WATER lake water	FMOC derivatization (anilines)	35 mmol L^{-1} DOSS, 8 mmol L^{-1} borate, 40% ACN (pH 8.5)	UV, 214 nm LIF, 266 nm solid-state UV laser, 310 nm emission 5.7×10^{-8} to 4.9×10^{-7} mol L^{-1}	35
4-Chloroaniline, 4-bromoaniline, 3,4-dichloroaniline, 3-chloroaniline, 3-chloro-4-methylaniline, 4-isopropylaniline, 3-methylaniline, aniline in 15 min	EKC	WATER Tap water Lake water	FITC derivatization	400 mmol L^{-1} borate, 40 mmol L^{-1} OG (pH 9.0)	LIF, 488 nm Ar laser, 520 nm emission 10^{-10} mol L^{-1}	37
p-Toluidine, aniline, *p*-ethoxyaniline, *p*-phenylenediamine in 4 min	CZE	WATER Lake water	FITC derivatization	15 mmol L^{-1} borate (pH 9.5)	Multiphoton-excited fluorescence detection, diode laser, 808 nm 1.25–2.60 μmol L^{-1}	70
1,3-Phenylenediamine, 2-methoxy aniline, 4-ethoxyaniline, 4,4′-diaminobiphenyl, 2-methylaniline, 2,4- dimethylaniline, 2-ethylaniline, 2,6-dimethylaniline in 4 min	EKC	WATER Surface water near textile and leather industries	Fluorescamine derivatization	5 mmol L^{-1} tetraborate, 4.5 mmol L^{-1} boric acid, 20 mmol L^{-1} SDS (pH 9)	Fluorescence, mercury–xenon lamp, 495 nm emission 1 μg L^{-1}	80

ABEI, *N*-(4-aminobutyl)-*N*-ethylisoluminol); DSC, *N*,*N*-disuccinimidyl carbonate; FITC, fluorescein isothiocyanate; OG, *n*-octylglucopyranoside; FMOC, 9-fluoroenylmethyl chloroformate; DOSS, dioctyl sulfosuccinate; EDTA, ethylenediaminetetraacetic acid; PSDVB, polystyrene-divinylbenzene copolymer; DMβCD, di-methyl β-cyclodextrin; SPE, solid-phase extraction; LLE, liquid–liquid extraction; LIF, laser-induced fluorescence.

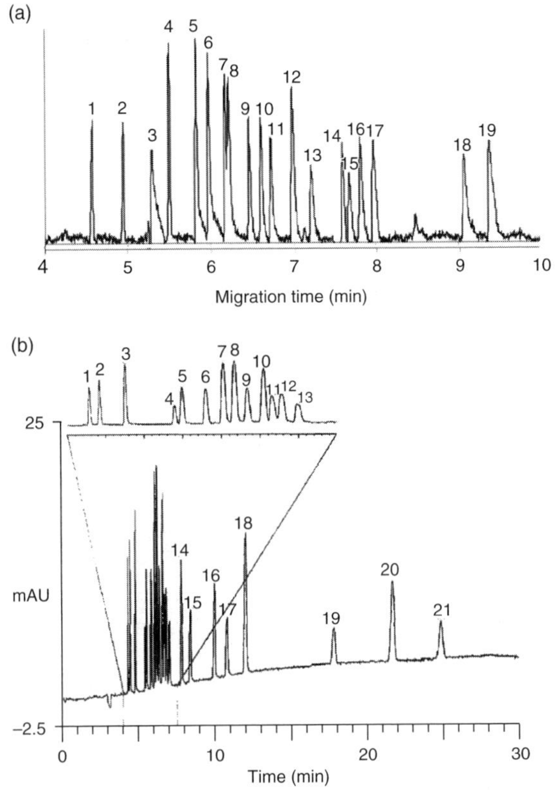

FIGURE 31.4 Separation of (a) aliphatic and (b) aromatic amines. Electrolytes: (a) 20 mmol L^{-1} imidazole-acetate at pH 3.5, 4 mmol L^{-1} EDTA and 25% ethanol; (b) 50 mmol L^{-1} phosphate buffer at pH 2.35, 7 mmol L^{-1} 1,3-diaminopropane. Other conditions: (a) +20 kV; indirect UV-detection at 214 nm; (b) +30 kV applied voltage; direct UV-detection at 280 nm. Peak labels: (a): methylamine (1), dimethylamine (2), Na^+ (3), ethylamine (4), cadaverine (5), putrescine (6), propylamine (7), methanolamine (8), morpholine (9), diethylamine (10), butylamine (11), isopropanolamine (12), piperazine (13), 2-(2-aminoethoxy)-ethanol (14), 2-amino-1-butanol (15), diethanolamine + methyldiethanolamine (16), amino-2-methyl-1-propanol (17), amino-2-ethyl-1,3-propandiol (18), and spermidine (19); (b): pyridine (1), *p*-phenylenediamine (2), benzidine (3), *o*-toluidine (4), aniline (5), *N,N*-dimethylaniline (6), *p*-anisidine (7), *p*-chloroaniline (8), *m*-chloroaniline (9), ethylaniline (10), α-naphthylamine (11), diethylaniline (12), *N*-(1-naphthyl)ethylenediamine (13), 4-aminophenazone (14), *o*-chloroaniline (15), 3,4-dichloroaniline (16), 3,3′-dichlorobenzidine (17), 2-methyl-3-nitroaniline (18), 2,4-dichloroaniline (19), 2,3-dichloroaniline (20), and 2,5-dichloroaniline (21). (Modified from (a) A. Fekete et al., *Electrophoresis*, 27, 1237–1247, 2006 and (b) A. Cavallaro et al., *J. Chromatogr. A*, 709, 361–366, 1995. With permission.)

derivatives, usually assessed by EKC methods. Several derivatizing reagents have been proposed in environmental applications: DNPH,[25,68,200,201] 3-methyl-2-benzothiazoline hydrazinone (MBTH),[202] 5-(dimethylamino)-naphthalene-1-sulfohydrazide (dansylhydrazine, DNSH),[68,203] and 4-hydrazinobenzoic acid (HBA).[204]

Table 31.5 compiles the details of the electromigration methodologies applied to carbonyls in the environment. Figure 31.5 illustrates the electromigration separation of carbonyl compounds as bisulfite adducts and DNSH derivatives.

31.4.6 Small Ions and Organometallic Compounds

Aspects of the determination of small ions and organometallic compounds by CE methodologies as well as elemental speciation have been extensively covered in the literature by excellent reviews

TABLE 31.5
Selected Applications of Electroseparation Methods for Carbonyl Compounds in Environmental Samples

Compounds	CE Mode	Matrix	Preconcentration Procedure (Recovery)	Optimal Electrolyte	Detection (LOD)	References
Acetone, acetaldehyde, propionaldehyde, benzaldehyde, formaldehyde, methylglyoxal in 10.5 min	CZE	WATER Rain water	Dansylhydrazine derivatization	5 mmol L^{-1} phosphate, 10 mmol L^{-1} tetraborate, 20% ACN (pH 8.0)	UV, 218 nm Z-shaped flow cell, 170–300 nmol L^{-1}	68
9 DNSH derivatives including isomers and impurities in 9 min	CZE	AIR Indoors Outdoors	C_{18} catridges impregnated with dansylhydrazine/trichloroacetic acid/MeOH	20 mmol L^{-1} phosphate buffer (pH 7.02)	UV, 214 nm 1.1–9.5 μg L^{-1} LIF, He/Cd 325 nm excitation, 520 emission 0.29–5.3 μg L^{-1}	203
4 HBA derivatives in 6 min	CZE	AIR Indoors	C_{18} catridges impregnated with HBA	0.040 mol L^{-1} tetraborate (pH 9.3)	UV, 290 nm 2.7–8.8 ng L^{-1}	204
Formaldehyde, acetaldehyde, propionaldehyde, acrolein, benzaldehyde (bisulfite derivatives) in ca. 7.5 min	EKC	AIR Vehicular emission	IMPINGER bisulfite derivatization	10 mmol L^{-1} 3,5 dinitrobenzoic acid, 0.2 mmol L^{-1} CTAB (pH 4.5)	UV, indirect detection, 254 nm 10–40 μg L^{-1}	25
5 DNPH derivatives in 14 min	EKC	AIR Vehicular emission	IMPINGER 0.05 g DNPH in 50 mL 2 mol L^{-1} HCl LLE in chloroform	20 mmol L^{-1} borate, 50 mmol L^{-1} SDS, 15 mmol L^{-1} β-CD	UV, 360 nm 0.2–2.0 mg L^{-1}	25
3 DNPH derivatives in 8 min	EKC	GAS Stack gas from an organic plant	IMPINGER 1.56 g DNPH in 500 mL 2 mol L^{-1} HCl LLE in CS_2	0.02 mol L^{-1} borate, 0.05 mol L^{-1} SDS (pH 9)	UV, 214 nm 2.0 mg L^{-1}	200
4 DNPH derivatives in 20 min	EKC	WATER River water	Spiked sample 97–102%	0.02 mol L^{-1} borate-phosphate, 0.05 mol L^{-1} SDS (pH 9)	UV, 360 nm 0.05 mg L^{-1} (formaldehyde) 0.08 mg L^{-1} (acetaldehyde)	201
5 MBTH derivatives in 10 min	EKC	AIR Indoors	IMPINGER 0.05% MBTH stacking with salt	20 mmol L^{-1} tetraborate, 50 mmol L^{-1} SDS (pH 9.3)	UV, 216 nm 0.54–4.0 μg L^{-1}	202

Obs LODs refer to each single aldehyde, not the derivative; DNPH, 2,4-dinitrophenylhydrazine; DNSH, 5-dimethylaminonaphthalene-1-sulfohydrazide; HBA, 4-hydrazinobenzoic acid; MBTH, 3-methyl-2-benzothiazoline hydrazinone; LLE, liquid–liquid extraction; CTAB, cetyltrimethylammonium bromide.

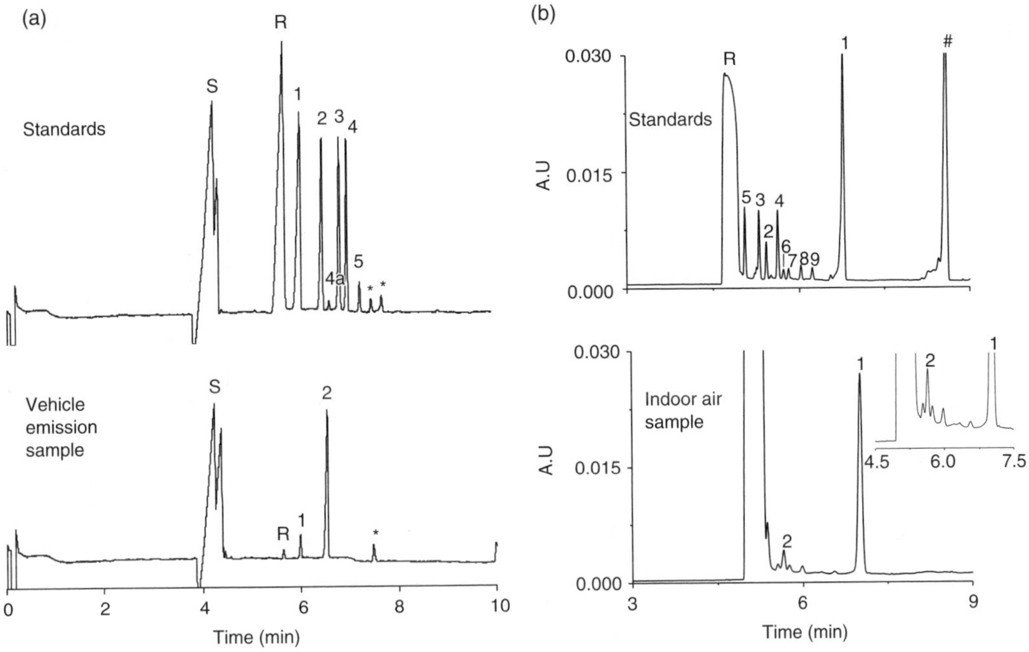

FIGURE 31.5 Separation of carbonyl compounds as (a) bisulfite adducts and (b) DNSH derivatives. Electrolytes: (a) 10 mmol L^{-1} 3,5-dinitrobenzoic acid at pH 4.5 containing 0.2 mmol L^{-1} CTAB; (b) 20 mmol L^{-1} phosphate buffer at pH 7.02. Other conditions: (a) -10 kV; indirect UV detection at 254 nm; (b) $+20$ kV applied voltage; direct UV detection at 214 nm. Peak labels: (a): formaldehyde (1), acetaldehyde (2), propionaldehyde (3), acrolein (4), and benzaldehyde (5), impurities (*) system peak (S), excess reagent (R); (b): formaldehyde (1), acetaldehyde (2 and 7), propionaldehyde (3 and 6), acrolein (4, 8, and 9), and acetone (5), impurities (#), excess reagent (R). (Modified from (a) E.A. Pereira et al., *J. AOAC Int.*, 82, 1562–1570, 1999 and (b) E.A. Pereira et al., *J. Chromatogr. A*, 979, 409–416, 2002. With permission.)

and key articles. A comprehensive review on applications of CE to the analysis of inorganic species including organometallic compounds in diverse environmental matrices (drinking, mineral, surface, and ground waters; rainwater; snow; seawater; brine and waste waters; aerosol; and others) has been compiled.[205] The determination of inorganic ions in environmental aquatic samples of high salinity, particularly seawater, has been revised.[206]

31.4.6.1 Inorganic Cations

The CZE determination of alkali, alkaline earth, and transition metal cations may be conducted under indirect UV-detection in electrolytes composed of cationic chromophores (protonated imidazole, aromatic amines, pyridine derivatives at moderately low pH electrolytes) and weak complexing agents (α-hydroxyisobutyric acid, HIBA, or other mono- and dicarboxylic acids), eventually modified by solvents. The simultaneous separation of 16 metal ions for inspection of river samples is an example of this approach.[207] Direct UV-detection at low wavelengths (190 nm) using nonabsorbing electrolytes (pH 10 borate buffer) is a possibility for absorbing cations such as ammonium, as determined in river and sewage samples.[208] Alternatively, UV-absorbing chelating (EDTA)[62] or ion-pair forming agents (*o*-phenanthroline)[209–211] can be selected for the determination of alkaline earth and transition metal ions. Figure 31.6 illustrates the electromigration separation of small cations.

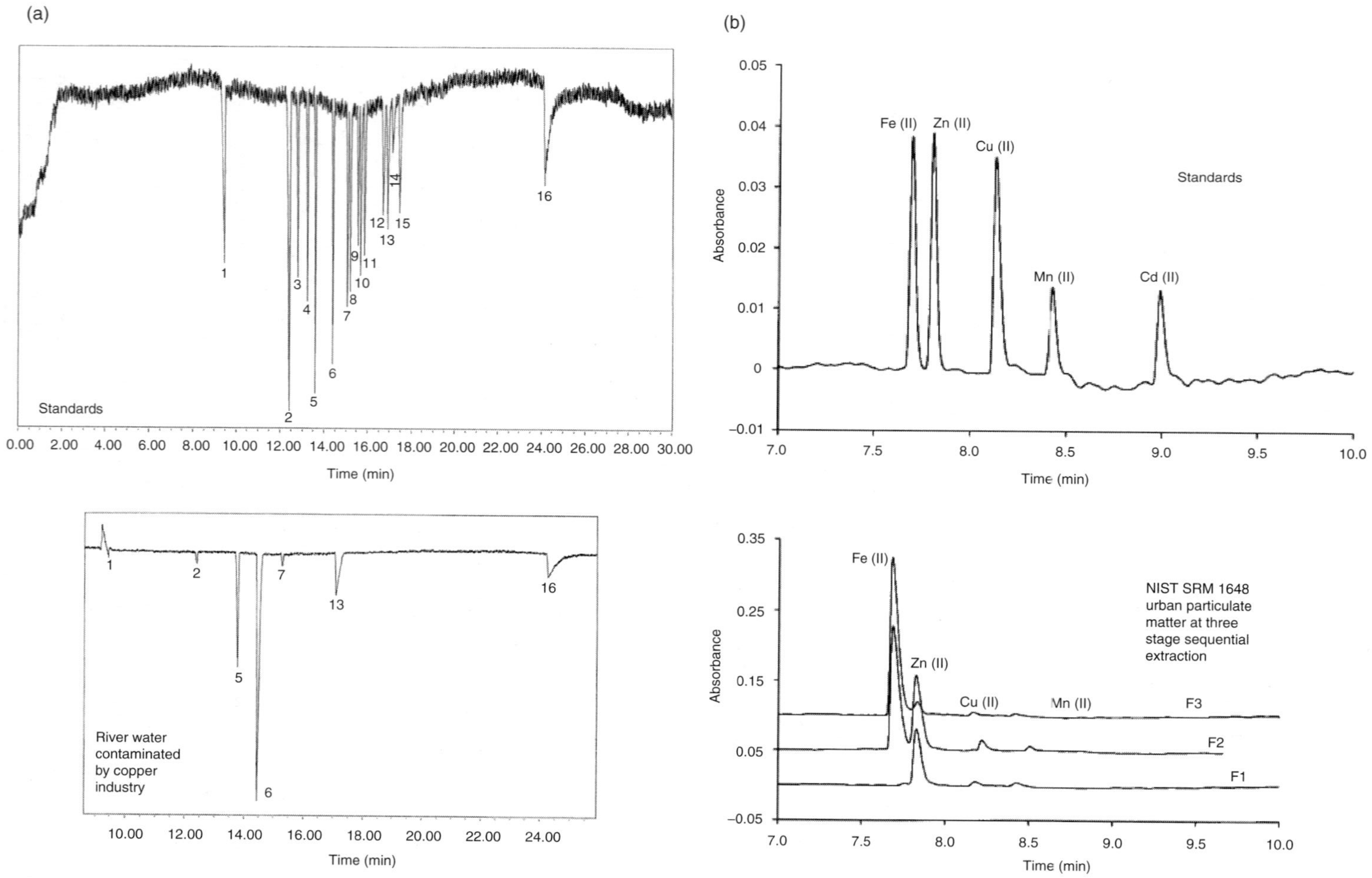

FIGURE 31.6 Separation of metal ions. Electrolytes: (a) 10 mmol L^{-1} 4-aminopyridine and 6.5 mmol L^{-1} HIBA at pH 4.5; (b) 200 mmol L^{-1} ammonium acetate at pH 5.5, 0.5 mmol L^{-1} 1,10-phenanthroline, 10 mmol L^{-1} hydroxylamine, and 20% acetone. Other conditions: (a) +25 kV; indirect UV detection at 214 nm; (b) +14 kV applied voltage; direct UV-detection at 226 nm. Peak labels: (a): potassium (1), sodium (2), barium (3), strontium (4), calcium (5), magnesium (6), manganese (7), lithium (8), iron (9), cobalt (10), cadmium (11), niquel (12), zinc (13), lead (14), chromium (15), and copper (16). (Modified from (a) N. Shakulashvili et al., *J. Chromatogr. A*, 895, 205–212, 2000 and (b) E. Dabek-Zlotorzynska et al., *Anal. Bioanal. Chem.*, 372, 467–472, 2002. With permission.)

31.4.6.2 Inorganic Anions

Similar to the inorganic cations, the determination of inorganic anions and those derived from low-molecular mass carboxylic acids are commonly approached by indirect UV-detection. However, here the electrolyte chromophore must be anionic (chromate,[212,213] benzoate derivatives,[214] naphthalene derivatives[215,216] among others[217,218]) and an EOF reverser (long-chain tetraalkylammonium quaternary salts) is usually employed. It is worth mentioning that many methodologies for inorganic anions and even metal and nonmetal elements determined as oxoanions or anionic complexes that are reported in the literature as CZE separations are, in fact, EKC separations. This is due to the electrolyte ionic strength that forces micellization of the cationic surfactant used as flow reverser.[219] Direct UV-detection at low wavelengths using nonabsorbing electrolytes (phosphate, borate, chloride, formate/chloride buffers modified by solvents, and cationic surfactants) is also a possibility for UV-absorbing anions such as iodide,[56,220,221] iodate,[222] bromide,[223] bromate,[224] nitrate,[225] and nitrite[226] as well as mixtures of anions[9,66,227] and aromatic acids.[228] Figure 31.7 illustrates the electromigration separation of small anions.

31.4.6.3 Simultaneous Detection of Cations and Anions

An interesting strategy based on dual-opposite end sample injection that accomplishes the simultaneous separation of small cations and anions within a single run was demonstrated in the analysis

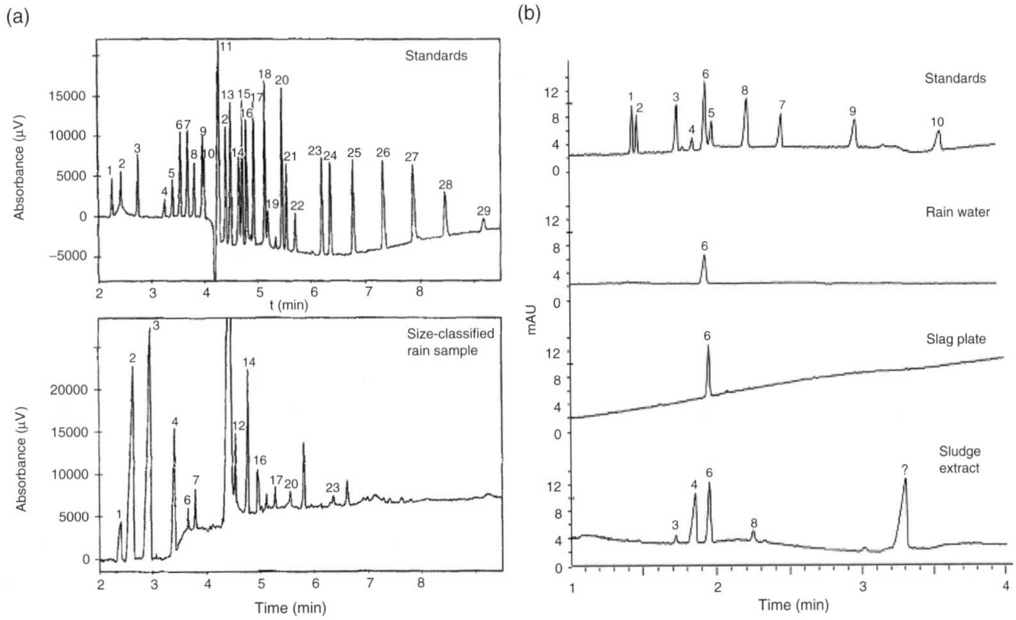

FIGURE 31.7 Separation of inorganic and organic anions. Electrolytes: (a) 7.5 mmol L^{-1} salicylic acid, 15 mmol L^{-1} Tris, 400 μmol L^{-1} DoTAH, 1050 μmol L^{-1} Ca^{2+}, and 600 μmol L^{-1} Ba^{2+}; (b) 50 mmol L^{-1} tetraborate at pH 9.3, containing 5% methanol. Other conditions: (a) −28 kV; indirect UV-detection at 232 nm; (b) −23 kV applied voltage; direct UV-detection at 200 nm. Peak labels: (a): chloride (1), nitrate (2), sulfate (3), formate (4), fumarate (5), malonate (6), succinate (7), maleate (8), glutarate (9), methanesulfonate (10), carbonate/adipate (11), malate (12), pimelate (13), acetate (14), suberate (15), oxalate (16), azelate (17), sebacate (18), glyoxylate (19), propionate (20), methacrylate (21), lactate (22), butyrate (23), hydroxy-*iso*-butyrate (24), valerate (25), capronate (26), enanthate (27), caprylate (28), and pelargate (29); (b): bromide (1), iodide (2), nitrite (3), thiosulfate (4), chromate (5), nitrate (6), $Fe(CN)_6^{4-}$ (7), SCN^- (8), MoO_4^{2-} (9), and WO_4^{2-} (10). (Modified from (a) A. Mainka et al., *Chromatographia*, 45, 158–162, 1997 and (b) M.I. Turnes-Carou et al., *J. Chromatogr. A*, 918, 411–421, 2001. With permission.)

of environmental water samples with indirect fluorescence[73] and contactless conductometric[78] detection.

31.4.6.4 Speciation and Organometallic Compounds

Speciation of chromium (histidine/acetic acid;[229] acetate buffer/EDTA[230]), mercury (borate/MeOH),[8] tin (pyridine/CTAB),[231] lead (SDS/β-CD),[232] arsenic (borate,[233] phosphate or phosphate/borate,[54] phosphate/TTAB,[234] phosphate and phosphate/TTAB[235]), sulfur (phosphate/TTAB/ACN),[236] and selenium (histidine/acetic acid,[237] SDS/β-CD[232]) by CE methods have been reported. Other examples include the analysis of beryllium (as an acetylacetone complex) in digested airborne dust[238] and the determination of Fe(II)-, Cu(I)-, Ni(II)-, Pd(II)-, and Pt(II)-cyano complexes and nitrate from leaching solutions of automobile catalytic converters.[239]

Metals including As, Co, Fe, Mn, Ni, and V are found at concentrations of 1–1000 ppm in crude oils. About 27–100% of the total metals in crude oil occur as organometallic porphyrin complexes. These petroporphyrins, derived from heme or chlorophyll, usually contain Ni(II) or V(IV)O and are distinguished by various ring types (Etio, DPEP, Rhodo analogs) and different substituents (H, *n*-alkyl). Petroporphyrin separations are of interest in geochemical sciences, environmental monitoring, and process control. Separation of petroporphyrin models [Ni(II) and V(II)O Etio I and Octaethyl type porphyrins] was approached by MEKC in electrolytes of varied composition: 15.8–22.5 mmol L^{-1} borate buffer containing 15–42 mmol L^{-1} SDS and 0–30% acetone, methylethylketone, acetonitrile, or methanol in a 6–30 min window.[240]

Several examples of the aforementioned methodologies and their application in environment compartments are compiled in Table 31.6.

31.4.7 Explosives and Warfare Residues

31.4.7.1 Explosives

A SB-β-CD-assisted EKC method for the determination of cyclic nitramine explosives and related degradation intermediates and the 14 EPA listed explosives (borate/SDS electrolyte) has been described.[241] A volatile electrolyte composed of SB-β-CD modified ammonium acetate buffer was selected for the EKC-MS detection of nitroaromatic and cyclic nitramine compounds in soil and marine sediment,[83] as detailed in Table 31.7. The use of phosphate/SDS electrolytes was reported in the separation of the 14 listed nitramine and nitroaromatic explosives for the analysis of extracts of high explosives such as C-4, tetrytol, and detonating cord.[242]

The simultaneous detection of small cations (ammonium, sodium, potassium, calcium, magnesium, and strontium) and anions (bromide, chloride, nitrite, nitrate, sulfate, perchlorate, thiocyanate, and chlorate) from low explosives in postblast residue using an elaborate electrolyte composed of a cationic chromophore and modifiers (imidazole/HIBA/18-crown-6 ether/ACN), an anionic chromophore (1,3,6-naphthalenesulfonic acid) and flow reversal agent (tetramethylammonium hydroxide) has been presented.[243]

31.4.7.2 Warfare Residues

A CZE method (borate buffer at pH 9.5) with LIF detection (He-Cd laser, 325 nm excitation; 450 nm emission) that simultaneously examines thiols (OPA derivatives) and cyanide (derivatized by OPA-taurine) aiming at monitoring for the potential contamination of drinking water supplies by chemical warfare (CW) nerve agents was described.[244] Detection limits of 9.3 μg L^{-1} for cyanide and from 1.8 to 89 μg L^{-1} for the thiols were obtained.

Table 31.7 compiles the details of CZE[245,247] and EKC[247,248] methodologies for inspection of contaminated water and soil for alkylphosphonic acids, the breakdown products of organophosphorus

TABLE 31.6
Selected Applications of Electroseparation Methods for Small Ions and Organometallic Compounds in Environmental Samples

Compounds	CE Mode	Matrix	Preconcentration Procedure (Recovery)	Optimal Electrolyte	Detection (LOD)	References
Small ions						
Br^-, I^-, NO_2^-, $S_2O_3^{2-}$, CrO_4^{2-}, NO_3^-, $Fe(CN)_6^{4-}$, SCN^-, MoO_4^{2-}, WO_4^{2-} in 3.5 min	CZE	WATER rain water river water drinking water industrial residual water	—	50 mmol L^{-1} tetraborate 5% MeOH (pH 9.3)	UV, 200 nm 0.02–0.1 mg L^{-1}	9
Lactate, butyrate, salicylate, propionate, phosphate, formate, citrate, acetate, Ba^{2+}, Ca^{2+}, Mg^{2+}, Ni^{2+}, Cu^{2+} in ca. 7 min	CZE	WATER pond water	—	5 mmol L^{-1} borate buffer, 5 μmol L^{-1} fluorescein (pH 9.2)	Fluorescence, indirect detection, (LED-induced fluorescence) 3.7–14.6 μmol L^{-1}	73
NH_4^+, K^+, Ca^{2+}, Na^+, Mg^{2+}, Li^+ (internal standard) in ca. 3.8 min	CZE	WATER rain water	—	20 mmol L^{-1} MES, 20 mmol L^{-1} His, 0.2 mmol L^{-1} 18-crown-6 (pH 6.2)	Contactless conductometric detection 9.4–24 μg L^{-1}	77
K^+, Na^+, Ba^{2+}, Sr^{2+}, Ca^{2+}, Mg^{2+}, Mn^{2+}, Li^+, Fe^{2+}, Co^{2+}, Cd^{2+}, Ni^{2+}, Zn^{2+}, Pb^{2+}, Cr^{3+}, Cu^{2+} in 24 min	CZE	WATER river water	—	10 mmol L^{-1} 4-aminopyridine, 6.5 mmol L^{-1} HIBA (pH 4.5 adjusted with sulfuric acid)	UV, indirect detection, 214 nm 92–454 μg L^{-1}	207
Fe^{2+}, Cu^{2+}, Zn^{2+}, Mn^{2+}, Cd^{2+} in 9 min	CZE	AIR airborne particulate	Sequential extraction 78–87%	200 mmol L^{-1} NH_4Ac (pH 5.5), 0.5 mmol L^{-1} 1,10-phenanthroline, 10 mmol L^{-1} hydroxylamine hydrochloride, acetone 20%	UV, 226 nm 29–142 μg L^{-1}	210

Continued

TABLE 31.6 (Continued)

Compounds	CE Mode	Matrix	Preconcentration Procedure (Recovery)	Optimal Electrolyte	Detection (LOD)	References
Zn^{2+}, Cu^{2+}, Co^{2+}, Fe^{2+}, Cd^{2+} in ca. 7 min	CZE	AIR Respirable, fine, and Coarse air Particulate	Andersen sampler, filters sonication	30 mmol L^{-1} hydroxylamine hydrochloride, 0.1 mmol L^{-1} 1,10-phenanthroline, 1% MeOH (pH 3.7)	UV, 265 nm 0.5–3 μg L^{-1}	211
C6–C12 perfluorinated carboxylic acids in 20 min	CZE	WATER Lake water River water Tap water	—	50 mmol L^{-1} TRIS, 7 mmol L^{-1} 2,4-DNBA, 50% MeOH (pH 9.0)	UV, indirect detection, 270 nm 0.6–2.4 mg L^{-1}	214
Bromide in 6 min	CZE	WATER Ground water Surface water	Spiked sample 61.6–102%	5 mmol L^{-1} formic acid, 42 mmol L^{-1} NaCl (pH 3.5)	UV, 200 nm 0.1 mg L^{-1}	223
Potassium in ca. 11 min	CZE	WATER Ground water Surface water	Spiked sample 73.4–97.5%	50 mmol L^{-1} 18-crown-6 10 mmol L^{-1} imidazole (pH 4.5)	UV, indirect detection, 214 nm 0.5 mg L^{-1}	223
Bromate in 4 min	CZE	WATER Tap water River water	Spiked sample 100–110%	10 mmol L^{-1} phosphate, 10 mmol L^{-1} Na_2SO_4. (pH 3.2)	UV, 193 nm 1 μg L^{-1}	224
Nitrite, nitrate in ca. 7 min	CZE	WATER Sea water	tITP	SE: artificial sea water with pH adjusted to 3.0 using phosphate buffer TE: 600 mmol L^{-1} acetate (pre-rinse with 0.1 mmol L^{-1} DDAB)	UV, 210 nm 2.7–3.0 μg L^{-1} as nitrogen	226
Ammonium in ca. 8 min	CZE	WATER River water SEWAGE	—	20 mmol L^{-1} borate (pH 10 adjusted with 1 mol L^{-1} sodium hydroxide)	UV, 190 nm 0.24 mg L^{-1} (as nitrogen)	228

3,4,5,-Trimethoxybenzoic acid, 4-hydroxyphenylacetic acid, salicylic acid, ferulic acid, *p*-coumaric acid, vanillic acid, 4-hydroxybenzoic acid in 11 min	CZE	WATER Lake water	SPE C_{18} 58–108%	13 mmol L^{-1} tetraborate (pH 9.7)	UV, 214 nm	228
Iodide in ca. 35 min	EKC	WATER Sea water	Spiked sample tITP 99.6–100.4%	SE: 0.5 mol L^{-1} NaCl, 25 mmol L^{-1} CTAC (pH 2.4) TE: 500 mmol L^{-1} MES (pH 6.0)	UV, 226 nm 0.6 μg L^{-1}	56
NO_3^-, Cl^-, SO_4^{2-}, formate, acetate in ca. 3.7 min	EKC	WATER Rain water	—	20 mmol L^{-1} MES, 20 mmol L^{-1} His, 0.2 mmol L^{-1} CTAB (pH 6.2)	Contactless conductometric detection 48–115 μg L^{-1} (inorganic ions)	77
Ammonium, potassium, sodium, calcium, magnesium, manganese, lithium, chloride, nitrite, nitrate, sulfate, fluoride, phosphate in 5.5 min	EKC	WATER Rain water Surface water Drainage water	—	20 mmol L^{-1} MES/His 11.5 mmol L^{-1} 18-crown-6, 10 μmol L^{-1} CTAB (pH 6.0)	Contactless conductometric detection 10–250 μg L^{-1}	78
F^-, Cl^-, NO_2^-, SO_4^{2-}, NO_3^-, HCO_3^- in ca. 0.7 min	EKC	WATER Snow	Spiked sample 91–104%	6.0 mmol L^{-1} chromate, 2.5 mmol L^{-1} CTAB, ACN 3.6%, (pH 9.5)	UV, indirect detection, 254 nm 0.03–0.2 mg L^{-1}	213
Oxalate, glycolate, malonate, pyruvate, formate, suberate, malate, acetate, succinate, glyoxylate, phthalate, lactate, methanesulfonate, propionate, glutarate, benzoate in 8.5 min	EKC	AEROSOL Atmospheric aerosol GAS Vehicle emission	47-mm PTFE filters (aerosol) KOH-coated quartz fiber (vehicle emission)	4 mmol L^{-1} NDC, 14.5 mmol L^{-1} *Bis–Tris*, 0.2 mmol L^{-1} TTAB (pH 6.2)	UV, indirect detection, 214 nm 3–10 μg L^{-1}	215
Chloride, nitrate, sulfate, formate, fumarate, malonate, succinate, maleate, glutarate, methanesulfonate, carbonate/adipate, malate, pimelate, acetate, suberate, oxalate, azelate, sebacate, glyoxylate, propionate, methacrylate, lactate, butyrate, hydroxy-iso-butyrate, valerate, capronate, enanthate, caprylate, pelargate in 9.5 min	EKC	WATER Size classified raindrops	freezing in liquid nitrogen (Guttalgor method)	7.5 mmol L^{-1} salicylic acid, 15 mmol L^{-1} Tris, 400 μmol L^{-1} DoTAH, 1050 μmol L^{-1} Ca^{2+}, 600 μmol L^{-1} Ba^{2+} (pH 6.2)	UV, 232 nm Z-shaped flow cell, 90–600 nmol L^{-1}	217

Continued

TABLE 31.6
(Continued)

Compounds	CE Mode	Matrix	Preconcentration Procedure (Recovery)	Optimal Electrolyte	Detection (LOD)	References
Cl^-, NO_3^-, SO_4^{2-}, ClO_4^-, F^- in 2.5 min	EKC	WATER Tap water River water	—	0.25 mmol L^{-1} CTAC, MeOH 30%, 5 mmol L^{-1} chromate (pH 8.0)	UV, indirect detection, 254 nm 0.09–0.23 mg L^{-1}	218
Iodide in ca. 33 min	EKC	WATER Sea water	tITP	SE: 0.5 mol L^{-1} NaCl, 25 mmol L^{-1} CTAC (pH 2.4 adjusted with HCl) TE: 0.5 mol L^{-1} MES (pH 6.0 adjusted with NaOH)	UV, 226 nm 0.2 μg L^{-1}	221
Iodide, iodate in ca. 33 min	EKC	WATER Sea water	—	SE: 0.5 mol L^{-1} NaCl, 12.5 mmol L^{-1} CTAC (pH 2.4 adjusted with HCl) TE: 0.5 mol L^{-1} MES (pH 6.5)	UV, 210 nm, 226 nm 0.23–10 μg L^{-1} (tITP)	222
Nitrate in ca. 7.6 min	EKC	WATER Sea water	—	0.1 mol L^{-1} sodium phosphate, 0.15 mol L^{-1} DDAPS (pH 6.2)	UV, 210 nm 35 μg L^{-1}	225
Nitrate, nitrite, bromide, iodide in ca. 5.8 min	EKC	WATER Sea water	—	(1) 10 mmol L^{-1} DDAPS, 10 mmol L^{-1} phosphate (pH 8.0) (2) 30 mmol L^{-1} DDAPS, 10 mmol L^{-1} phosphate (pH 8.0)	UV, 214 nm 5–11 μmol L^{-1}	227
Organometallic						
Methylmercury, ethylmercury, phenylmercury, mercury (II), as cysteine complexes in 14 min	CZE	WATER Lake water River water	Spiked sample 86.6–111%	100 mmol L^{-1} boric acid, 12% MeOH (pH 9.1)	Atomic fluorescence spectrometry 6.8–16.5 μg L^{-1} (as Hg)	8
Arsenite, arsenate, dimethylarsinate, monomethylarsonate in ca. 19 min	CZE	Environmental reference material	LV-FASI	20 mmol L^{-1} phosphate, 10 mmol L^{-1} borate (pH 9.28)	UV, 200 nm 24–93 μg L^{-1}	54

Cr(III), Cr(IV) in 3 min	CZE	WATER Rinse water from the galvanic industry	—	4.5 mmol L^{-1} His (pH 3.40 adjusted with acetic acid)	Contactless conductometric detection 10–39 μg L^{-1}	229
Cr(III), Cr(IV) in ca. 4.5 min	CZE	WATER Tap water Well water River water Waste water	Spiked sample 98–103%	20 mmol L^{-1} acetate buffer, 1 mmol L^{-1} EDTA (pH 4.7)	Chemiluminescence (luminol) 1–8 pmol L^{-1}	230
Se(IV), Se(VI) in ca. 8 min	CZE	SOIL	Sonication with water 90–101%	8.75 mmol L^{-1} His (pH 4.0 adjusted with acetic acid)	Contactless conductometric detection 7.5–190 μg L^{-1}	237
Trimethyllead, triethyllead, diphenylselenide, phenylselenyl in 20 min	EKC	WATER Drainage water	LLE chloroform 80–104%	50 mmol L^{-1} SDS, 5 mmol L^{-1} β-CD (pH 6.0)	UV, 210 nm 26–67 mg L^{-1}	232
Arsenite, arsenate, dimethyl arsenic acid in 4 min	EKC	WATER Ground water	FASI	10 mmol L^{-1} phosphate buffer, 0.35 mmol L^{-1} TTAB (pH 9.0)	UV, 185 nm 0.1–0.5 mg L^{-1}	234
AsO_2^{2-}, AsO_4^{2-}, dimethylarsinic acid in 2 min	EKC	WATER Ground water	FASI	20 mmol L^{-1} phosphate, 0.75 mmol L^{-1} TTAB (pH 9.0)	UV, 185 nm 1 μmol L^{-1}	235
Be, as diacetylacetonate-beryllium in 15 min	EKC	AIR Digested airborne Dust (model sample)	—	0.1 mol L^{-1} TRIS-nitrate, 50 mmol L^{-1} SDS (pH 7.8)	UV, 295 nm 1 mg L^{-1}	238
Fe(II)-, Cu(I)-, Ni(II)-, Pd(II)-, and Pt(II)-cyano complexes, nitrate in 20 min	EKC	CATALYTIC CONVERTER RESIDUE leaching solutions	alkaline NaCN solutions	20 mmol L^{-1} phosphate, 100 mmol L^{-1} NaCI, 3 mmol L^{-1} NaCN, 1.2 mmol L^{-1} TBAB, 40 μmol L^{-1} TTAB (pH 11)	UV, 208 nm 11–60 μg L^{-1}	239

FASI, field-amplified sample injection; SPE, solid-phase extraction; LV-FASI, large-volume field amplified stacking injection; TBAB, tetrabutylammonium bromide; TTAB, tetradecylytrimethylammonium bromide; CTAB, cetyltrimethylammonium bromide; NDC, 2,6-naphthalenedicarboxylic acid; *Bis Tris*, 2,2-*Bis*(hydroxymethyl)-2,29,20-nitrilotriethanol; DoTAH, dodecyltrimethylammoniumhydroxide; CTAC, cetyltrimethylammonium chloride; 2,4-DNBA, 2,4-dinitrobenzoic acid; SE, separation electrolyte; TE, terminating electrolyte; tITP, transient isotachophoresis; DDAB, dilauryldimethylammonium bromide; DDAPS, 3-(n,N-dimethyldodecylammonium)propane sulfonate; HIBA, α-hydroxyisobutyric acid; MES, 2-[morphine]ethanesulphonic acid; LLE, liquid–liquid extraction.

TABLE 31.7
Selected Applications of Electroseparation Methods for Explosives and Warfare Residues in Environmental Samples

Compounds	CE Mode	Matrix	Preconcentration Procedure (Recovery)	Optimal Electrolyte	Detection (LOD)	Referencess
Methylphosphonic acid, ethylphosphonic acid, propylphosphonic acid, *N*-butylphosphonic acid, phenylphosphonic acid, ethyl methylphosphonic acid, ethyl methylthiophosphonic acid, isopropyl methylphosphonic acid, pinacolyl methylphosphonic acid, dimethyl phenylphosphonate in 16 min	CZE	WATER SOIL	SPE Bond-Elut SCX	50 mmol L^{-1} NH_4Ac (pH 9.0 adjusted by ammonia solution)	FPD, (P) mode, 0.1–0.5 μg L^{-1}	245
Methylphosphonic acid, ethylphosphonic acid, propylphosphonic acid, phenylphosphonic acid, isopropylphosphonic acid, methyl ethylphosphonic acid, ethyl methylphosphonic acid, ethyl ethylphosphonic acid, methyl propylphosphonic acid, propyl methylphosphonic acid in 15 min	CZE	SOIL	PHWE 150 bar, 100°C SPE cation exchange cartridge	15 mmol L^{-1} NH_4Ac (pH 8.8 adjusted by ammonia solution)	ESI-IT-MS 5 μg L^{-1}	246
2,4,6-Trinitrotoluene, 1,3,5-trinitrobenzene, hexahydro-1,3,5-trinitro-1,3,5-triazine, octahydro-1,3,5,7-tetranitro-1,3,5,7-tetrazocine, 2,4,6,8,10,12-hexanitro-2,4,6,8,10,12-hexaazaisowurtzitane in 16 min	EKC	SOIL Contaminated soil SEDIMENT Marine sediment	—	10 mmol L^{-1} SB-β-CD, 10 mmol L^{-1} NH_4Ac (pH 6.9)	ESI-IT-MS 0.025–0.5 mg L^{-1}	83
Methylphosphonic acid and its monoacid/monoalkyl esters (ethyl, isopropyl, and pinacolyl) in 10 min	EKC	WATER Surface water Ground water SOIL	—	30 mmol L^{-1} L-His, 30 mmol L^{-1} MES, 0.7 mmol L^{-1} TTAOH, 0.03% wt Triton X-100	Conductivity	247
Methylphosphonic acid and its monoacid/monoalkyl esters (ethyl, isopropyl, and pinacolyl) in 3 min	EKC	WATER Tap water Ground water Artificial sea water SOIL	SPE On-Guard-Ba On-Guard-Ag On-Guard-H 90–110%	200 mmol L^{-1} boric acid, 10 mmol L^{-1} phenylphosphonic acid, 0.03% wt Triton X-100, 0.35 mmol L^{-1} DDAOH (pH 4.0)	UV, indirect detection, 210 nm 1–2 μg L^{-1} (water samples) 25–50 μg L^{-1} (aqueous leachates of soil)	248

TTAOH, tetradecyltrimethylammonium hydroxide; ESI-IT-MS, electrospray ionization ion trap mass spectrometry; SB-β-CD, sulfobutylether-β-cyclodextrin; PHWE, pressurized hot water extraction; FPD, flame photometric detector; SPE, solid-phase extraction; MES, 2-[morphine]ethanesulphonic acid; DDAOH, didodecyldimethylammonium hydroxide.

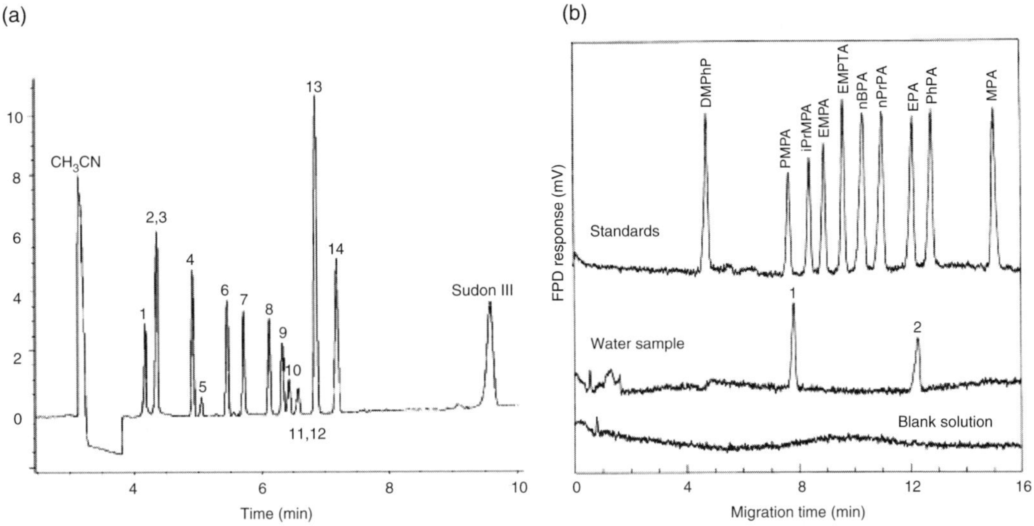

FIGURE 31.8 Separation of explosives (a) and warfare residues (b). Electrolytes: (a) 12 mmol L^{-1} borate at pH 9 and 50 mmol L^{-1} SDS; (b) 50 mmol L^{-1} ammonium acetate at pH 9.0. Other conditions: +30 kV, direct UV-detection; (b) +30 kV; make up: 0.5% formic acid; flame photometric detector. Peak legends: (a): octahydro-1,3,5,7-tetranitro-1,3,5,7-tetrazocine (1); hexahydro-1,3,5-trinitro-1,3,5-triazine (2); 1,3,5-trinitrobenzene (3); 1,3-dinitrobenzene (4); nitrobenzene (5); 2,4,6-dinitrotoluene (6); *N*-2,4,6-tetranitro-*N*-methylaniline (7); 2,4-dinitrotoluene (8); 2,6-dinitrotoluene (9); 2-nitrotoluene (10); 3-nitrotoluene (11); 4-nitrotoluene (12); 2-amino-4,6-dinitrotoluene (13); and 4-amino-2,6-dinitrotoluene (14). (b): methylphosphonic acid (MPA), ethylphosphonic acid (EPA), *n*-propylphosphonic acid (nPrPA), *n*-butylphosphonic acid (nBPA), phenylphosphonic acid (PhPA), ethylmethylphosphonic acid (EMPA), ehtylmethylthiophosphonic acid (EMPTA), isopropylmethylphosphonic acid (iPrMPA), pinacolylmethylphosphonic acid (PMPA), and dimethylphenylphosphonate (DMPhP); (1) and (2) are suspected contaminants of water sample. (Modified from (a) C.A. Groom et al., *J. Chromatogr. A*, 999, 17–22, 2003 and (b) E.W.J. Hooijschuur et al., *J. Chromatogr. A*, 928, 187–199, 2001. With permission.)

nerve agents. The unusual detection system for CE (flame photometric detector)[245] with an ability to detect alkylphosphonic acids a fold below the required 1 μg mL^{-1} level is worth mentioning here.

Figure 31.8 illustrates the electromigration separation of EPA explosives and warfare residues.

31.4.8 Aromatic Sulfonates

Aromatic sulfonates and their amino- and hydroxy-derivatives are produced on a large scale in the chemical industry. Although the acute toxicity and the risk of bioaccumulation appear to be small, they became persistent and widespread environmental pollutants due to their high mobility in the aquatic compartments and limited biodegradability.

The anionic charge of sulfonates within the entire practical pH range of electromigration separations and the UV-absorbing capability of the aromatic moiety make CZE-based methodologies of aromatic sulfonates feasible. However, improved selectivity and resolution of positional isomers have been sought by means of interactions with CDs,[249] nonionic,[250] and cationic[250] surfactants as well as mixed micelles.[251] An overview of the electrophoretic methods for the determination of benzene- and naphthalenesulfonates in water samples has been presented in the literature.[252]

Details of a CZE methodology for the separation of 14 aromatic sulfonates in river water[253] and a CZE-MS evaluation of 22 compounds in influent and effluent samples of a wastewater treatment plant[254] are organized in Table 31.8. With the idea that phytoremediation could be used as an alternative means of wastewater treatment of recalcitrant compounds, a simple borate buffer-based

TABLE 31.8
Selected Applications of Electroseparation Methods for Aromatic Sulfonates, Dyes, and Surfactants in Environmental Samples

Compounds	CE Mode	Matrix	Preconcentration Procedure (Recovery)	Optimal Electrolyte	Detection (LOD)	References
Aromatic sulfonates						
14 Aromatic sulfonates of a wide range of different structure in 9.5 min	CZE	WATER River water Contaminated seepage water	SPE C_{18} LiChrolut EN 72–132%	25 mmol L^{-1} sodium borate (pH 9.3)	UV, 210 nm fluorescence, Xe lamp 0.2 mg L^{-1} (CE) 0.1 μg L^{-1} (SPE-CE)	253
13 Aromatic sulfonates of a wide range of different structure in ca. 20 min	CZE	WATER Waste water treatment	SPE C_{18} LiChrolut EN >50%	5 mmol L^{-1} NH_4Ac (pH 10.5)	ESI-MS 0.1–0.4 mg L^{-1} (CE-MS) 0.1 μg L^{-1} (SPE-CE-MS)	254
Anthraquinone-1-sulphonic acid, anthraquinone-2-sulphonic acid, anthraquinone-1,5-disulphonic acid, anthraquinone-2,6-disulphonic acid, anthraquinone-1,8-disulphonic acid in 7.5 min	CZE	WATER Hydroponic media PLANT EXTRACTS Plant cultivated in spiked hydroponic media	Aqueous extraction (plant samples) direct injection (water samples)	20 mmol L^{-1} tetraborate (pH 9.3)	UV, 254 nm	255
21 Naphthalenesulfonate derivatives in 28 min	EKC	WATER River water	SPE	50 mmol L^{-1} borate, 100 mmol L^{-1} SDS (pH 8.7)	UV, 230 nm 20 μg L^{-1}	17
21 Naphthalenesulfonate derivatives in 30 min	EKC	WATER Industrial effluent	Sample was diluted 1/1000 and filtered	25 mmol L^{-1} tetraborate, 75 mmol L^{-1} Brij 35, 5 mmol L^{-1} octylamine (pH 9.0)	UV, 220 nm 0.5–3.2 mg L^{-1}	19
56 Aromatic organic acids were evaluated; 2,4-dichlorophenoxyacetic acid, 2,4,5-trichlorophenoxyacetic acid, orange II, trypan blue, 2,4,5-trichlorophenoxypropionic acid, 3-nitrobenzoic acid, 4-chlorobenzensulfonic acid (recovery studies) in 20 min	EKC	WATER Deionized water SOIL	Soxhlet (soil), SPE cation exchange 38–101% (water) 26.5–94.4% (soil)	50 mmol L^{-1} borate, 100 mmol L^{-1} cholate (pH 8.3)	UV, 214 nm	28

Surfactants						
4 LAS (C10, C11, C12, C13) in 45 min	CZE	WATER Coastal marine water Waste water Treatment plants	SPE (C18 and SAX) 94–98% (marine water) SPE (Isolute ENV+) 77–93% (wastewater)	50 mmol L^{-1} NH_4Ac, 30% isopropanol (pH 5.6)	UV, 214 nm 1 μg L^{-1} (SPE-CE-UV) ESI-MS (qualitative confirmation)	87
19 LAS isomers (commercial formulation) in 45 min	EKC	WATER Industrial Waste water SEWAGE SLUDGE	LLE, MeOH/NaOH (44.6–96.2%) SPE, C18 and SAX (33.1–75.7%)	10 mmol L^{-1} phosphate, 40 mmol L^{-1} SDS, 30% ACN (pH 6.8)	UV, 200 nm 4 mg L^{-1}	18
Dyes						
5 Vinylsulfone and chlorotriazine reactive dyes in 34 min	CZE	WATER Waste water	SPE C_{18} 81–121%	5 mmol L^{-1} NH_4Ac (pH 9.0)	ESI-MS single quadrupole 23–42 μg L^{-1}	84
Eosin, fluorescein (internal standard) in 5.3 min	CZE	WATER Ground water	Charcoal adsorption	20 mmol L^{-1} borate (pH 9.2)	LIF, 514.5 nm Ar/Kr ion laser 1 nmol L^{-1}	259
Sulfonated azo dyes: Acid Blue 113, Acid Red 73, Acid Red 13, Mordant Yellow 8, Acid Red 1, Acid Red 14, Acid Red 9, Acid Yellow 23 in 16 min	EKC	WATER Waste water	SPE Isolute ENV 54–81%	9.5 mmol L^{-1} NH_4Ac, 0.1% Brij 35 (pH 9)	UV, 214 nm 19–230 mg L^{-1}	20
Monosulfonated dyes: cresol red, acid blue, acid orange, tropaeolin, nuclear fast red, orange II, acid red in 27 min	EKC	WATER Spiked water SOIL	SPE ion-pair (water) C_{18} (soil) 42.5–107% (water) 17.9–105% (soil)	100 mmol L^{-1} cholate, 10% acetone (pH 8.35)	UV, 214 nm	29

SPE, solid-phase extraction; ESI- MS, electrospray ionization mass spectrometry; LIF, laser-induced fluorescence; LLE, liquid–liquid extraction; LAS, linear alkylbenzenesulfonates.

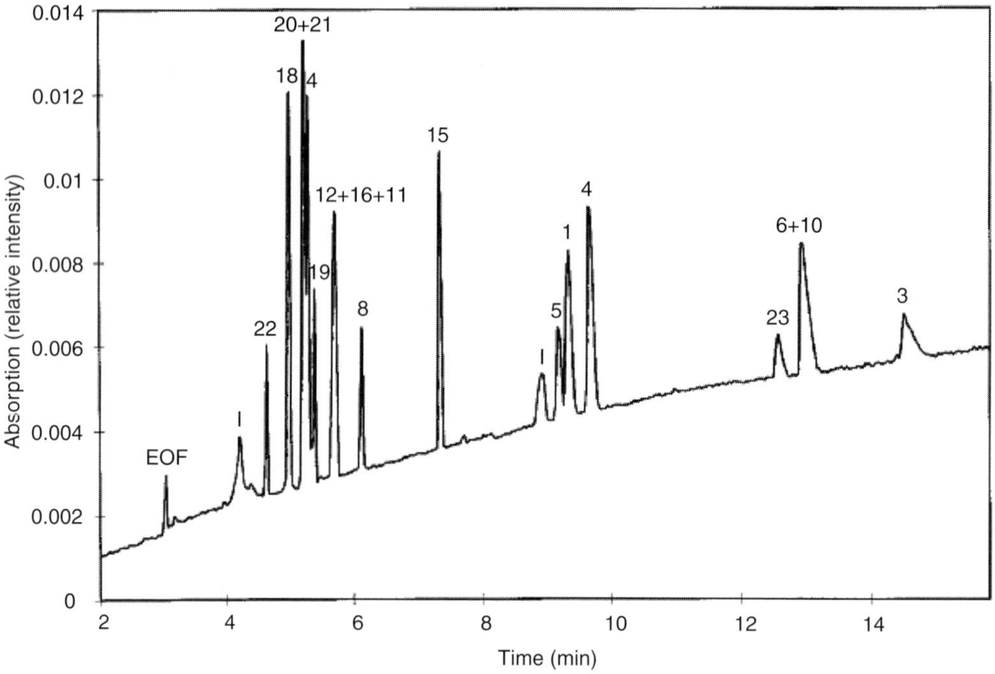

FIGURE 31.9 Separation of aromatic sulfonates. Electrolyte: 12 mmol L^{-1} ammonium acetate at pH 10. Other conditions: +20 kV; direct UV detection at 214 nm. Peak legend: 2-amino-1,5-NDS (1); 1,3,6-NTS (2); 1,3-BDS (3); 1,5-NDS (4); 2,6-NDS (5); 1-OH-3,6-NDS (6); 1-amino-5-NS (7); BS (8); 1-amino-4-NS (9); 2-OH-3,6-NDS (10); 1-OH-6-amino-3-NS (11); 3-nitro-BS (12); 1-amino-6-NS (13); 4-methyl-BS (14); 1-OH-4-NS (15); 4-chloro-BS (16); 2-amino-1-NS (17); 1-amino-7-NS (18); 4-chloro-3-nitro-BS (19); 1-NS (20); 2-NS (21); diphenylamine-4-sulfonate (22); and 8-NH_2-1-OH-3,6-NDS (23). *Abbreviations*: NS, naphthalenesulfonate; NDS, naphthalenedisulfonate; NTS, naphthalenetrisulfonate; BS, benzenesulfonate; BDS, benzenedisulfonate; and OH, hydroxy. (Modified from R. Loos et al., *J. Mass Spectrom.*, 35, 1197–1206, 2000. With permission.)

CZE methodology was used to monitor sulfonated anthraquinones uptake in hydrophonic media, in which plants, bacteria, and algae had grown for 6 weeks.[255]

EKC methodologies for aromatic sulfonates in industrial effluents,[19] mono- and dinaphthalenesulfonates as well as hydroxy- and amino-derivatives in river water,[17] and phenoxyacetic acids in spiked water and soils[28] are compiled in Table 31.8. Figure 31.9 illustrates the electromigration separation of aromatic sulfonates.

31.4.9 Surfactants

Anionic surfactants such as alkanesulfonates, alkyl sulfates, and LAS are water-soluble, surface-active materials that are consumed in large quantities in industrial and commercial formulations. Thus, their disposal may impact water reservoirs. LAS surfactants, which are used commercially, are quite complex mixtures containing several homologs and positional isomers. The degradation rates and toxicity of LAS depend on the alkyl chain length and the position of the phenyl ring.

Since LAS are chromophoric-charged compounds, they are amenable to CZE[18,87] determination. EKC mode is preferred when homologue discrimination or isomeric distribution is requested[31,32] or mixture of surfactant classes are screened.[33,34]

LAS were determined in influent and effluent samples of wastewater treatment plants and coastal waters receiving untreated domestic effluents using a volatile buffer-based CZE method (UV and

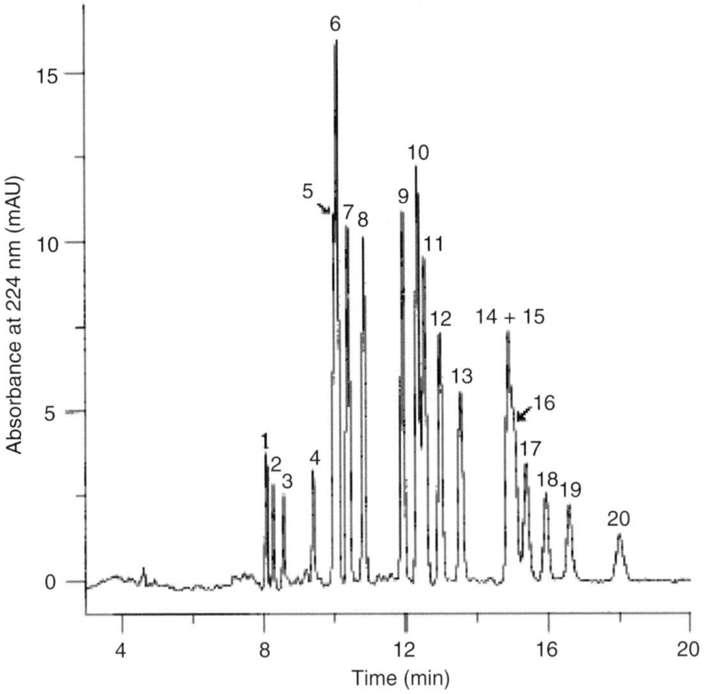

FIGURE 31.10 Separation of LAS surfactants. Electrolyte: 10 mmol L^{-1} sodium phosphate at pH 6.8, 40 mmol L^{-1} SDS and 30% acetonitrile. Other conditions: +15 kV applied voltage. Peak legend: positional isomers: 5-C10 (1), 4-C10 (2), 3-C10 (3), 2-C10 (4), 6-C11 (5), 5-C11 (6), 4-C11 (7), 3-C11 (8), 2-C11 (9), 6-C12 (10), 5-C12 (11), 4-C12 (12), 3-C12 (13), 2-C12 (14), 7-C13 (15), 6-C13 (16), 5-C13 (17), 4-C13 (18), 3-C13 (19), and 2-C13 (20); first number represents the attachment point of the phenyl ring to the carbon chain. (From J.M. Herrero-Martínez et al., *Electrophoresis*, 24, 681–686, 2003. With permission.)

MS detection).[87] Phosphate and borate buffers containing acetonitrile were used to discriminate LAS homologues in sewage sludge, whereas unmodified buffers were applied to wastewater samples.[18] Isomeric separation was only possible using electrolytes with high contents of SDS and acetonitrile as organic modifier. Table 31.8 compiles the details of the aforementioned methodologies. Figure 31.10 illustrates the electromigration separation of LAS surfactants.

31.4.10 Dyes

Synthetic dyes, including azo compounds, are widely used as coloring agents in a variety of products such as textiles, paper, leather, gasoline, and foodstuffs. Synthetic dyes persist even after conventional water treatment procedures due to their hydrophilic character and high solubility (they usually bear carboxylic or sulfonic acid groups in their structure) and, therefore, can be distributed in the environment from urban and industrial wastewater.

CZE separation of synthetic dyes has been approached by simple (borate and citrate)[256] and volatile buffers (ammonium acetate)[257] modified by solvents as well as nonaqueous systems (ammonium acetate/acetic acid in MeOH).[258] Environmental applications of CZE methodologies include the analysis of spent dyebaths and wastewater samples[257] and the monitoring of groundwater migration, where eosin was used as a fluorescent tracer (details in Table 31.8).[259]

EKC separations of synthetic dyes include buffered SDS systems[260] and polymeric electrolytes,[261] modified by organic solvents.[260] Examples of the determination of azo dyes, mono- and disulfonated compounds in water samples,[20] as well as synthetic dyes in spiked water and soil matrices[29] have been compiled in Table 31.8. Figure 31.11 illustrates the electromigration separations of synthetic dyes.

31.4.11 Endocrine Disruptors and Pharmaceuticals

Recently, it has been established that certain synthetic organic chemicals affect the reproductive health of higher organisms by contributing to infertility in various ways and even increasing the rate of cancer in reproductive organs. These chemicals have been termed "environmental estrogens" due to their disrupting effects on the endocrine system of hormone production and transmission. In addition to organochlorine insecticides, PCBs and dioxins (covered in Section 31.4.2), phenolic compounds such as bisphenol A—a widely used substance that is polymerized industrially into polycarbonate and nonylphenol—a product of the breakdown of large surfactants used in spermicides, some plastics and plasticizers such as phthalic acid esters, as well as pharmaceuticals such as synthetic estrogens belong to the list of chemicals suspected to have endocrine-disrupting effects.

31.4.11.1 Phenolic Compounds

The CZE separation of halogenated phenolic and bisphenolic compounds from 25 potentially interfering phenolic derivatives for inspection in water, sludge, and sediments has been attempted[262–264] using aqueous (borate buffer at pH 9.4) and nonaqueous (borate buffer in MeOH) systems (details in Table 31.9).

MEKC separation and online concentration of bisphenol A and alkyl phenols has been approached in a series of experiments using SDS and other alkyl chain anionic surfactants, bile salts, and TTAB in organic solvent and CD-modified buffers.[59,265,266]

The EKC determination of several target endocrine disruptors in water[267,268] and biosolids (sewage sludge and treated sludge)[269] has been detailed in Table 31.9. The potential of MEEKC for the separation of priority endocrine-disrupting compounds in wastewater samples has been investigated.[45]

31.4.11.2 Phthalate Esters

Excessive use of phthalate esters in industrial applications, mainly as plasticizers, is the main cause of their persistent presence in consumer goods and in the environment. The economic and social interest in the control of phthalate esters and the availability of analytical methodologies for areas such as environmental and food analyses have been discussed.[270]

31.4.11.3 Pharmaceutical Residues

The occurrence and fate of pharmaceuticals from various prescription classes and related metabolites and medicinal products for veterinary use in aquatic environments as well as their removal is one of the emerging issues in environmental chemistry.[271] More than 80 different prescribed pharmaceuticals, metabolites, and veterinary drugs have been detected in aquatic environment (e.g., sewage influent and effluent samples, surface and groundwater, and even drinking water), as reported by studies carried out in several European countries, United States, Canada, and Brazil.[271]

Tetracyclines,[272] nonsteroidal anti-inflammatory,[58,273] antidepressants,[274] and mixtures of acidic drugs[82,275] as well as veterinary drug[276] residues have been determined by CZE (citric acid/citrate, borate, borate in MeOH, phosphate/MeOH, ammonium formate/formic acid/ACN, ammonium acetate, ammonium acetate/acetic acid/MeOH) and EKC (SDS/pH 2.5 phosphate buffer/ACN) methodologies in water samples from influent and effluent of sewage treatment plants as well as wastewater, river, surface and groundwater as compiled in Table 31.9. Estrogens have been receiving increased attention due to the already mentioned possible interference with the reproductive

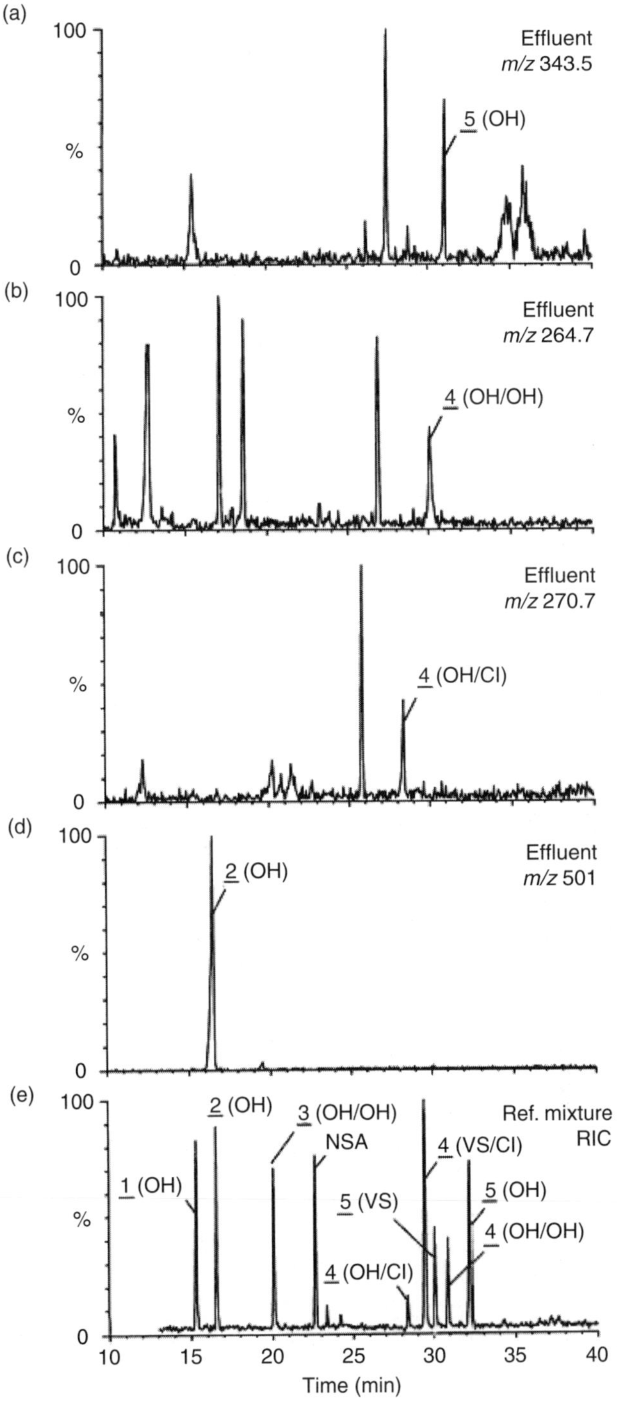

FIGURE 31.11 Identification of synthetic dyes in a sewage extract effluent (mass traces a–d) and a reference mixture (e) by CE-MS. Electrolyte: 5 mmol L^{-1} ammonium acetate with 40% acetonitrile. Other conditions: +30 kV applied voltage; sheath liquid: 80:20 isopropanol:water. *Abbreviations*: OH, hydroxy form; Cl, chloride form; VS, vinylsulfone; NSA, naphthalene sulfonic acid; RIC, reconstructed ion chromatogram. (From T. Poiger et al. *J. Chromatogr. A*, 886, 271–282, 2000. With permission.)

TABLE 31.9
Selected Applications of Electroseparation Methods for Endocrine Disruptors and Pharmaceuticals in Environmental Samples and Miscellaneous

Compounds	CE Mode	Matrix	Preconcentration Procedure (Recovery)	Optimal Electrolyte	Detection (LOD)	References
Endocrine disruptors						
2,4,6-Tribromophenol, pentabromophenol, tetrabromobisphenol A, tetrachlorobisphenol A, 2,6-Dibromophenol in 28 min	CZE	WATER River water Waste water	SPE-LVSEP Polystyrene–divinylbenzene 95.5–105.8% (river water) 73–94% (Waste water)	20 mmol L^{-1} tetraborate in MeOH (non-aqueous) (pH 9.4)	UV, 210 nm, 230 nm 0.4–1.7 $\mu g L^{-1}$	262
4-Bromo-3-methylphenol, 2-bromo-4-methylphenol, pentabromophenol, 2,4,6-tribromophenol, 2,4-dibromophenol, 2-bromophenol, 2,6-dibromophenol, tetrabromobisphenol A, tetrachlorobisphenol A in ca. 8 min	CZE	WATER Waste water	SPE PS–DVB	20 mmol L^{-1} sodium tetraborate (pH 9.6)	UV, 210 nm	263
2,6-Dibromophenol, 2,4,6-tribromophenol, tetrabromobisphenol A, pentabromophenol, tetrachlorobisphenol A in 28 min	CZE	SEWAGE SEDIMENT River sediment Marine sediment	MSPD-LVSEP Florisil 75.7–106.4%	20 mmol L^{-1} tetraborate in MeOH (nonaqueous) (pH 9.4)	UV, 210 nm, 230 nm 2.7–5.3 $\mu g L^{-1}$	264
Octylphenol, nonylphenol in 10 min	EKC	WATER Waste water treatment effluents and sludges	SPE C_{18} spiked sample 25.6–50.8%	25 mmol L^{-1} phosphate, 200 mmol L^{-1} SDS, 900 mmol L^{-1} butanol, 80 mmol L^{-1} heptane, 20% propanol (pH 2)	UV, 214 nm 50 mg L^{-1} (tested)	45
17β-Estradiol, diethylstilbestrol, ethynylestradiol, octylphenol, nonylphenol, bisphenol A in 15 min	EKC	WATER River water	Spiked sample analytes solubilized in 10% ACN:90% buffer	100 mmol L^{-1} phosphate, 12.5% ACN, 25 mmol L^{-1} SDS, 1 mmol L^{-1} HP-β-CD (pH 1.8)	UV, 214 nm low mg L^{-1} range	267

Estriol, phenol, trichlorophenol, bisphenol A, pentachlorophenol, butylphenol, estrone, β-estradiol, diethylstilbestrol, ethinylestradiol, nonylphenol in 25 min	EKC	WATER River water	Spiked sample	20 mmol L^{-1} CAPS, 25 mmol L^{-1} SDS, 15% ACN (pH 11.5)	UV, 200 nm 2.0–7.4 mg L^{-1}	268
Estrone, β-estradiol, and ethynylestradiol in 10 min	EKC	WATER Spiked water	SPE C18; sweeping >96%	30 mmol L^{-1} phosphoric acid, 80 mmol L^{-1} SDS, 20% MeOH	UV, 214 nm 0.16–0.30 nmol L^{-1}	278
Pharmaceutical residues						
Clofibric acid, naproxen, bezafibrate, diclofenac, ibuprofen, mefenamic acid in ca. 20 min	CZE	WATER River water	LLE/SPE hexane/MTBE (50:50)/Bondesil ODS 43.1–98%	20 mmol L^{-1} NH_4Ac (pH 5.1 adjusted with acetic acid)	ESI-MS single quadrupole 25–59 μg L^{-1}	82
Tetracycline, oxytetracycline, doxycycline in 17.5 min	CZE	WATER Surface water	Online SPE STRATA-X 94–106%	50 mmol L^{-1} citric acid (pH 2.5 adjusted with HCl 1 mol L^{-1})	UV, 260 nm 1.6–2 μg L^{-1}	272
Acetylsalicylic acid, ibuprofen, fenoprofen, naproxen, diclofenac, ketoprofen in 26 min	CZE	WATER Bottled water	SPE-LVSEP LiChrolut RP-18 64–95%	30 mmol L^{-1} tetraborate, 70% MeOH (pH 9.8)	UV, 214 nm 2–9 ng L^{-1}	273
Fluoxetine, venlafaxine, citalopram, sertraline, paroxetine, clomipramine, trazodone in 17 min	CZE	WATER River water SEWAGE	SPE hydrophilic divinylbenzene 85–99%	1.5 mol L^{-1} formic acid, 50 mmol L^{-1} ammonium formate, 15% ACN	ESI-TOF-MS 13–53 μg L^{-1} (SPE-CE-MS)	274
Clofibric acid, naproxen, bezafibrate in 15 min	CZE	WATER Surface water SEWAGE	SPE LiChrolut RP-18 70–89% (surface water) LLE/SPE hexane/MTBE (50:50)/LiChrolut RP-18 50–58% (sewage)	20 mmol L^{-1} NH_4Ac, 60% MeOH (pH 4.5 adjusted with acetic acid)	ESI-MS single quadrupole 100 ng L^{-1}	275

Continued

TABLE 31.9
(Continued)

Compounds	CE Mode	Matrix	Preconcentration Procedure (Recovery)	Optimal Electrolyte	Detection (LOD)	References
Sulfapyridine, sulfamethazine, sulfamerazine, sulfamether, sulfadiazine, sulfadimethoxine, sulfamethoxazole, sulfachlorpyridazine, sulfamethizole in 18 min	CZE	WATER Ground water	SPE-LVSS Oasis HLB 75.6–100.3%	45 mmol L^{-1} sodium phosphate, 10% MeOH (pH 7.3)	UV, 265 nm 2.59–22.95 μg L^{-1}	276
Diclofenac, ibuprofen, fenoprofen, naproxen, ketoprofen in 35 min	EKC	WATER Mineral water	SPE-SRMM, SRMM-ASEI, FESI-RMM LiChrolut RP-18 70–100%	25 mmol L^{-1} phosphate, 75 mmol L^{-1} SDS, 40% ACN (pH 2.5 adjusted with HCl)	UV, 214 nm 0.07–0.23 μg L^{-1} (SRMM) 0.050–0.195 μg L^{-1} (SRMM-ASEI) 0.7–1.6 μg L^{-1} (FESI-RMM)	58
Miscellaneous						
Anatoxin-a, microcystin-LR, cylindrospermopsin in 3 min	CZE	WATER Water bloom Crude extracts	Centrifugation, freezing–thawing, filtering	25 mmol L^{-1} sodium tetraborate (pH 9.3)	UV, 230, 240, 278 nm 0.89–2.77 mg L^{-1}	286
Anatoxin-a, microcystin-LR, cylindrospermopsin in 9 min	EKC	WATER Water bloom Crude extracts	Centrifugation, freezing–thawing, filtering	25 mmol L^{-1} sodium tetraborate, 100 mmol L^{-1} SDS (pH 9.3)	UV, 230, 240, 278 nm 0.73–2.81 mg L^{-1}	286
Pollen allergens and organic pollutants (nonspecified); 20 compounds in 13 min	EKC	AIR Airborne dust samples	LLE acetone followed by acidic and alkaline aqueous solutions	20 mmol L^{-1} TRIS, 5 mmol L^{-1} H_3PO_4, 50 mmol L^{-1} SDS (pH 8.7)	UV, 206 nm	288

HP-β-CD, hydroxypropyl-β-cyclodextrin; CAPS, cyclohexylamino-1-propanesulfonic acid; LVSS, large-volume sample stacking; SRMM, stacking with reverse migrating micelles; SRMM–ASEI, stacking with reverse migrating micelles–anion selective exhaustive injection; FESI–RMM, field-enhanced sample injection with reverse migrating micelles; LVSEP, large-volume sample stacking using the electroosmotic flow pump; ESI-TOF-MS, electrospray ionization time of flight mass spectrometry; ESI-MS, electrospray ionization mass spectrometry; MSPD, matrix solid-phase dispersion; PS–DVB, polystyrene–divinylbenzene; LLE, liquid–liquid extraction; SPE, solid-phase extraction.

system of man and animals. A few applications have been reported (Table 31.9).[277,278] Figure 31.12 illustrates the electromigration separation of endocrine disruptors and pharmaceuticals.

31.4.12 MISCELLANEOUS

31.4.12.1 Humic Substances

CZE of natural organic matter[279] and separation methods for humic substances[280] have been subjects of recent reviews. CZE and EKC separations (borate/TRIS/EDTA modified by CDs/SDS) and

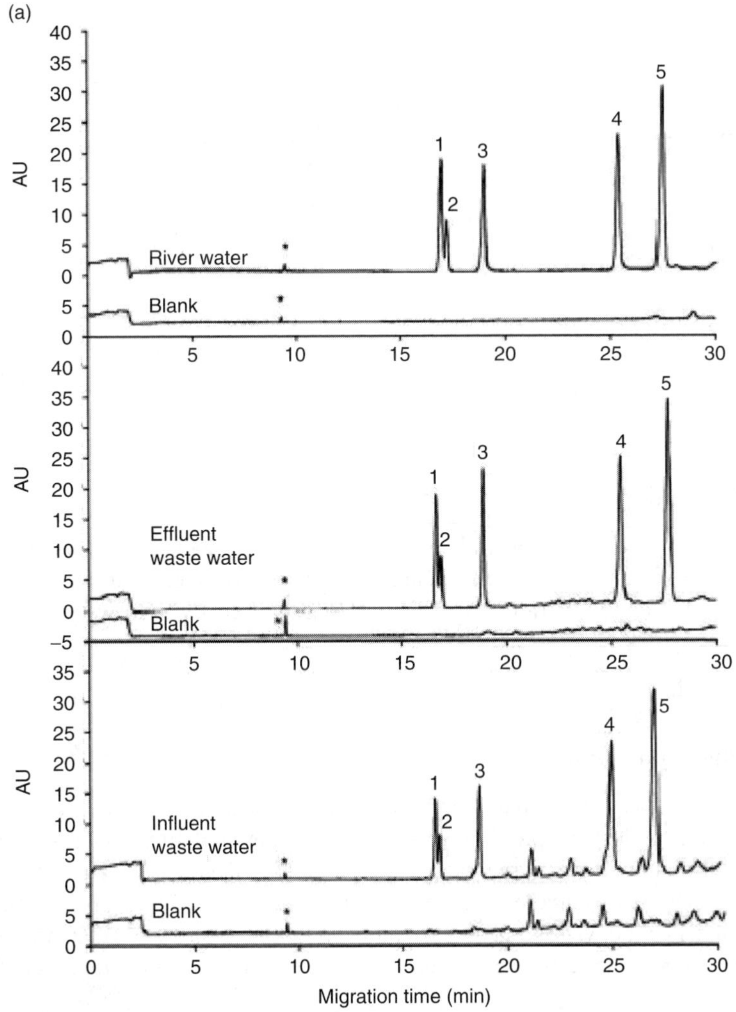

FIGURE 31.12 Separation of (a) endocrine disruptors and (b) veterinary drug residues. Electrolytes: (a) 20 mmol L^{-1} sodium tetraborate at pH 9.4; (b) 45 mmol L^{-1} sodium phosphate at pH 7.3 with 10% methanol. Other conditions: (a) −30 kV; direct UV-detection at 210 nm; (b) +25 kV; direct UV-detection at 265 nm. Peak legends: (a): 2,4,6-tribromophenol (1), pentabromophenol (2), 2,6-dibromophenol (3), tetrabromobisphenol A (4), and tetrachlorobisphenol A (5). (b): Sulfapyridine (SPD), sulfamethazine (SMZ), sulfamerazine (SMR), sulfamether (SMT), sulfadiazine (SDZ), sulfadimethoxine (SDM), sulfamethoxazole (SMX), sulfachloropyridazine (SCP), sulfamethizole (SMI), and *p*-aminobenzoic acid (PABA). (Modified from (a) E. Blanco et al., *J. Chromatogr. A*, 1071, 205–211, 2005 and (b) J.J. Soto-Chinchilla et al., *Electrophoresis*, 27, 4360–4368, 2006. With permission.)

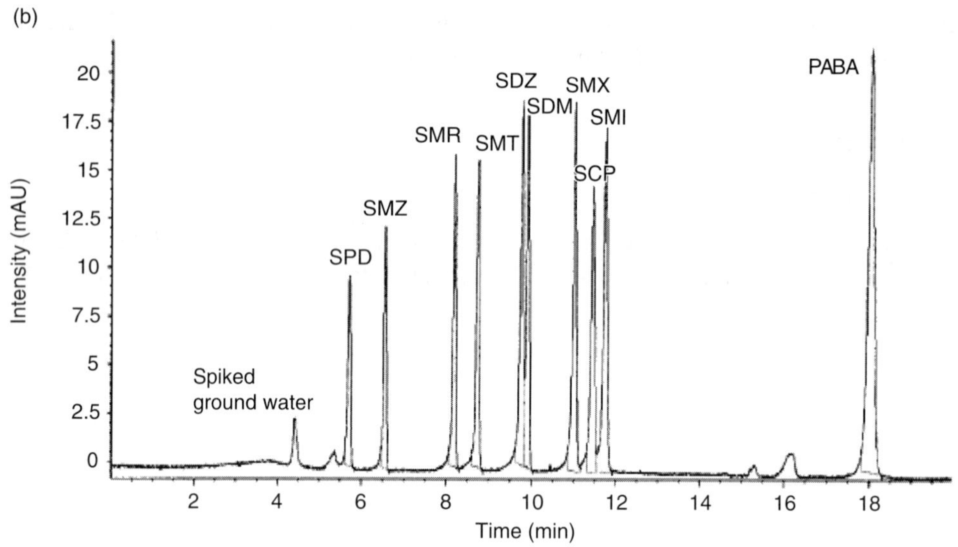

FIGURE 31.12 (Continued)

characterization (TOF-MS) of humic acids isolated from Antarctica soil[281] as well as interactions of organic matter and environmental pollutants[282,283] were reported (details in Table 31.9).

31.4.12.2 Algal Toxins

Methods for determining microcystins and microcystin-producing cyanobacteria have been reviewed.[284] Details of CZE and EKC analyses of cyanobacterial toxins in environmental samples[285–287] are compiled in Table 31.9. Figure 31.13 presents the separation of cyanobacterial toxins.

31.4.12.3 Other Applications

CZE and MEKC methods were used to monitor extraction procedures involving pollen allergens and organic pollutants from dust samples collected before, during, and after pollen seasons at different locations (car-traffic tunnel in Prague and a metro station in Paris) using air-filtration devices.[288] Water and acetic acid extracts were analyzed by CZE using acetic acid as background electrolyte whereas water and alkaline water–SDS-buffer extracts were analyzed by MEKC in Tris-phosphate–SDS electrolytes. Significant differences were found in the profiles of dust extracts from different origins.

Drinking water and humidity condensate samples collected from U.S. Space Shuttle and the Russian Mir Space Stations are analyzed routinely at the NASA–Johnson Space Center as a means of verifying water quality and monitoring the environment of the spacecraft.[289] Anions and cations were determined by ion chromatography whereas carboxylates and amines were determined by CE (phthalate/TTAB for carboxylates and imidazole/HIBA for amines). Results showed that Shuttle water is of distilled quality whereas Mir-recovered water contains various levels of minerals. Organic ions were rarely detected in potable water samples but were present in humidity condensates.

Table 31.9 compiles the details of the previously cited methodologies.

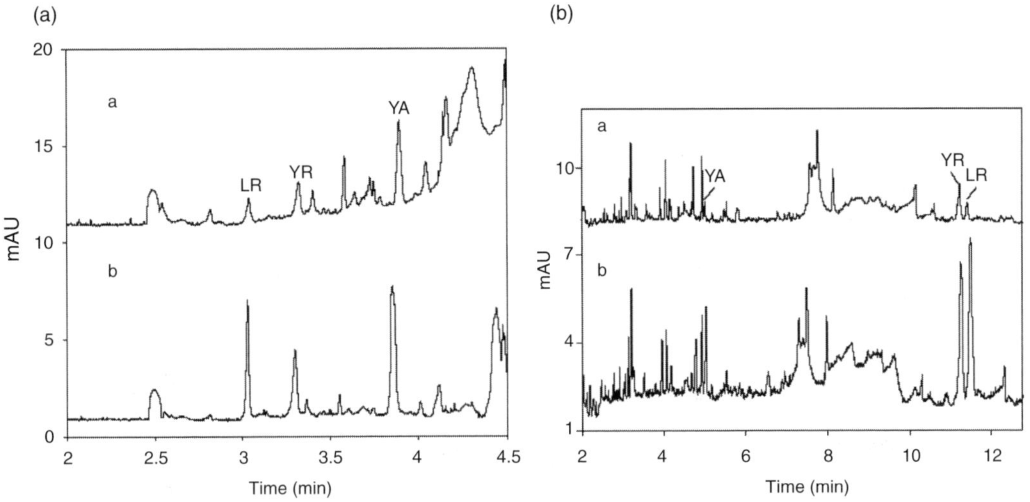

FIGURE 31.13 (a) CZE and (b) MEKC separation of cyanobacterial toxins. Electrolytes: (a) 25 mmol L^{-1} borate at pH 9.3; (b) 25 mmol L^{-1} borate at pH 9.3 containing 75 mmol L^{-1} SDS. Other conditions: (a) and (b) +25 kV; direct UV-detection at 238. Peak legends: microcystin YA (YA), microcystin YR (YR), and microcystin LR (LR); a: water bloom sample; and b: spiked sample. (Modified from G. Vasas et al. *J. Biochem. Biophys. Methods*, 66, 87–97, 2006. With permission.)

31.5 A METHOD DEVELOPMENT GUIDE FOR ENVIRONMENTAL ANALYSIS

Figure 31.14 depicts a simplified diagram that can be useful during the first stages of method development for environmental analysis. In CE, method development relies strongly on the knowledge of the analyte structural features. For neutral compounds (for instance, many classes of pesticides, PAH, PCB, dioxins, derivatized carbonyls, etc.), the obvious choice is the selection of an EKC protocol. As a first attempt, MEKC in borate/SDS electrolytes is usually recommended. Depending on the result of the exploratory run, several additives can be further tested as a means to modulating the solute–micelle interaction. Other EKC modes are also feasible options, especially when SDS fails in promoting separation, which is the case of either highly hydrophilic or highly hydrophobic compounds. Occasionally, EKC has been the technique of choice for certain classes of ionizable pollutants, such as phenol derivatives and amines, simply because the separation of a large number of positional isomers and conformers is attempted and extra selectivity is required.

Method development procedures for charged compounds, either organic pollutants with ionizable functional groups within the CE operational pH range (phenols, amines, carboxylic acids, many dyes, pharmaceuticals, etc.) or explicitly ionic compounds such as small ions, sulfonates, LAS surfactants, and so forth, will follow two distinct routes in the diagram of Figure 31.14, depending on the compound's UV-absorbing characteristics (only UV-absorbance detectors were considered here because they are practically a part of all commercially available CE equipments).

If the pollutant is charged and exhibits UV-absorbing properties, the CZE mode is readily recommended. For basic pollutants, moderately low to low pH buffers are indicated and the analyte migrates coelectroosmotically as a cation (protonated species) whereas for acidic pollutants, high pH buffers will promote the analyte dissociation and it migrates counterelectroosmotically as an anion. In both cases, buffer pH and concentration are the variables to optimize before the addition of any modifiers is considered.

If the pollutant is charged but does not present a chromophoric group, either a cationic (separation of cation) or an anionic (separation of anions) chromophore must be added to the electrolyte and

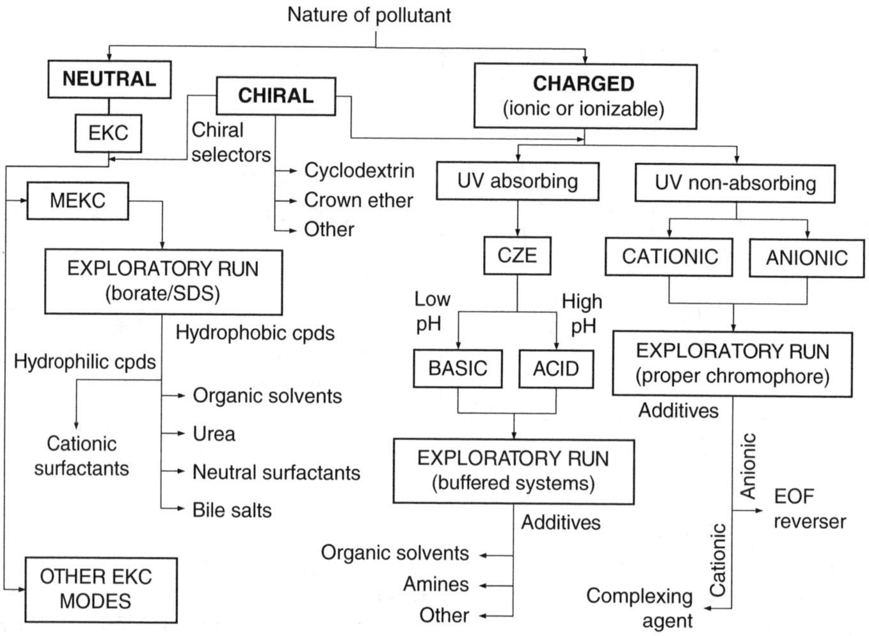

FIGURE 31.14 A method development guide for environmental analysis.

indirect detection is performed. For the cations, complexing agents will further improve resolution of similar mobility species. In the case of nonabsorbing anions, a judicious choice of both co- and counterions is recommended, the co-ion due to peak symmetry considerations while the counterion for signal enhancement. Furthermore, an EOF modifier (alkyl quaternary ammonium salt) is usually selected to expedite the separation.

Finally, in the electrolyte design during chiral pollutants method development, a chiral selector must always be considered. Cyclodextrins are among the most commonly used chiral additives and can be employed in both EKC and CZE methodologies, depending on the nature of the pollutant.

31.6 CONCLUDING REMARKS

On the basis of the diversity of applications presented in this chapter, it becomes clear that CE has found a niche among separation techniques for environmental analysis. Future directions will likely include refinement of preconcentration strategies and further detector improvements to achieve the desirable low-concentration limits of detection. In addition, with the consolidation of CE-MS technology, more robust methods are likely to emerge with enhanced sensitivity and superior selectivity, improving further the acceptance of CE in the routine determination of pollutants in real samples.

ACKNOWLEDGMENTS

The authors wish to acknowledge the Conselho Nacional de Desenvolvimento Científico e Tecnológico (CNPq) and the Fundação de Amparo à Pesquisa do Estado de São Paulo (Fapesp) of Brazil for financial support (Fapesp 04/08503-2; 04/08931-4) and fellowships (CNPq 306068/2003-6).

REFERENCES

1. E. Dabek-Zlotorzynska and V. Celo, Recent advances in capillary electrophoresis and capillary electrochromatography of pollutants, *Electrophoresis*, 27, 304–322, 2006.
2. E. Dabek-Zlotorzynska, H. Chen and L. Ding, Recent advances in capillary electrophoresis and capillary electrochromatography of pollutants, *Electrophoresis*, 24, 4128–4149, 2003.
3. I. Ali, V.K. Gupta and H.Y. Aboul-Enein, Chiral resolution of some environmental pollutants by capillary electrophoresis, *Electrophoresis*, 24, 1360–1374, 2003.
4. D. Martínez, M.J. Cugat, F. Borrul and M. Calull, Solid-phase extraction coupling to capillary electrophoresis with emphasis on environmental analysis, *J. Chromatogr. A*, 902, 65–89, 2000.
5. M.C. Bruzzoniti, C. Sarzanini and E. Mentasti, Preconcentration of contaminants in water analysis, *J. Chromatogr. A*, 902, 289–309, 2000.
6. D. Martínez, F. Borrull and M. Calull, Strategies for determining environmental pollutants at trace levels in aqueous samples by capillary electrophoresis, *Trends Anal. Chem.*, 18, 282–291, 1999.
7. G.W. Sovocool, W.C. Brumley and J.R. Donnelly, Capillary electrophoresis and capillary electrochromatography of organic pollutants, *Electrophoresis*, 20, 3297–3310, 1999.
8. X.-P. Yan, X.-B. Yin, D.-Q. Jiang and X.-W. He, Speciation of mercury by hydrostatically modified electroosmotic flow capillary electrophoresis coupled with volatile species generation atomic fluorescence spectrometry, *Anal. Chem.*, 75, 1726–1732, 2003.
9. M.I. Turnes-Carou, P. López-Mahía, S. Muniategui-Lorenzo, E. Fernández-Fernández and D. Prada-Rodríguez, Capillary zone electrophoresis for the determination of light-absorbing anions in environmental samples, *J. Chromatogr. A*, 918, 411–421, 2001.
10. A. Cavallaro, V. Piangerelli, F. Nerini, S. Cavalli and C. Reschiotto, Selective determination of aromatic amines in water samples by capillary zone electrophoresis and solid-phase extraction, *J. Chromatogr. A*, 709, 361–366, 1995.
11. S. Morales and R. Cela, Highly selective and efficient determination of US Environmental Protection Agency priority phenols employing solid-phase extraction and non-aqueous capillary electrophoresis, *J. Chromatogr. A*, 896, 95–104, 2000.
12. M. Molina, D. Peréz-Bendito and M. Silva, Multi-residue analysis of *N*-methylcarbamate pesticides and their hydrolytic metabolites in environmental waters by use of solid-phase extraction and micellar electrokinetic chromatography, *Electrophoresis*, 20, 3439–3449, 1999.
13. A. Segura-Carretero, C. Cruces-Blanco, S. Pérez Duran and A. Fernández Gutiérrez, Determination of imidacloprid and its metabolite 6-chloronicotinic acid in greenhouse air by application of micellar electrokinetic capillary chromatography with solid-phase extraction, *J. Chromatogr. A*, 1003, 189–195, 2003.
14. C.-E. Lin, C.-C. Hsueh, T.-Z. Wang, T.-C. Chiu and Y.-C. Chen, Migration behavior and separation of s-triazines in micellar electrokinetic capillary chromatography using a cationic surfactant, *J. Chromatogr. A*, 835, 197–207, 1999.
15. O. Brüggemann and R. Freitag, Determination of polycyclic aromatic hydrocarbons in soil samples by micellar electrokinetic capillary chromatography with photodiode-array detection, *J. Chromatogr. A*, 717, 309–324, 1995.
16. S. Frías, M.J. Sánchez and M.A. Rodríguez, Determination of triazine compounds in ground water samples by micellar electrokinetic capillary chromatography, *Anal. Chim. Acta*, 503, 271–278, 2004.
17. S.J. Kok, E.H.M. Koster, C. Gooijer, N.H. Velthorst and U.A.Th. Brinkman, Separation of twenty-one naphthalene sulfonates by means of capillary electrophoresis, *J. High Resol. Chromatogr.*, 19, 99–104, 1996.
18. K. Heinig, C. Vogt and G. Werner, Determination of linear alkylbenzenesulfonates in industrial and environmental samples by capillary electrophoresis, *Analyst*, 123, 349–353, 1998.
19. S. Angelino, A. Bianco Prevot, M.G. Genaro and E. Pramauro, Ion-interaction high-performance chromatography and micellar electrokinetic capillary chromatography: Two complementary techniques for the separation of aromatic sulfonated compounds, *J. Chromatogr. A*, 845, 257–271, 1999.
20. E.R. Cunha and M.F. Alpendurada, Comparison of HPLC and micellar electrokinetic chromatography in determination of sulfonated azo dyes in waste water, *J. Liq. Chromatogr. Rel. Technol.*, 25, 1835–1854, 2002.
21. R.J.G. Jeevan, M. Bhaskar, R. Chandrasekar and G. Radhakrishan, Analysis of arylamine isomers by micellar electrokinetic chromatography, *Electrophoresis*, 23, 584–590, 2002.

22. M. Wang, D. Wu, Q. Yao and X. Shen, Separation and selectivity in micellar electrokinetic chromatography using sodium dodecyl sulfate micelles or Tween 20-modified mixed micelles, *Anal. Chim. Acta*, 519, 73–78, 2004.
23. L. Song, Q. Ou, W. Yu and G. Li, Separation of 6 phenylureas and chlorsulfuron standards by micellar, mixed micellar and microemulsion electrokinetic chromatography, *J. Chromatogr. A*, 699, 371–382, 1995.
24. M.H. Chen and W.H. Ding, Separation and migration behavior of positional and structural naphthalenesulfonate isomers by cyclodextrin-mediated capillary electrophoresis, *J. Chromatogr. A*, 1033, 167–172, 2004.
25. E.A. Pereira, M.F.M. Tavares and A.A. Cardoso, Alternative methodologies for the determination of aldehydes by capillary electrophoresis, *J. AOAC Int.*, 82, 1562–1570, 1999.
26. M.-I. Acedo-Valenzuela, T. Galeano-Días, N. Mora-Díez and A. Silva-Rodríguez, Determination of neutral and cationic herbicides in water by micellar electrokinetic capillary chromatography, *Anal. Chim. Acta*, 519, 65–71, 2004.
27. E. Dabek-Zlotorzynska and E.P.C. Lai, Separation of polynuclear aromatic hydrocarbons by micellar electrokinetic capillary chromatography using sodium taurodeoxycholate modified with organic solvents, *J. Cap. Elec.*, 3, 31–35, 1996.
28. W.C. Brumley and C.M. Brownrigg, Electrophoretic behavior of aromatic-containing organic acids and the determination of selected compounds in water and soil by capillary electrophoresis, *J. Chromatogr.*, 646, 377–389, 1993.
29. W.C. Brumley, C.M. Brownrigg and A.H. Grange, Capillary liquid-chromatography-mass spectrometry and micellar electrokinetic chromatography as complementary techniques in environmental analysis, *J. Chromatogr. A*, 680, 635–644, 1994.
30. J.M. Bermúdez-Saldaña, M.A. García, M.J. Medina-Hernández and M.L. Marina, Micellar electrokinetic chromatography with bile salts for predicting ecotoxicity of aromatic compounds, *J. Chromatogr. A*, 1052, 171–180, 2004.
31. J.M. Herrero-Martínez, E.F. Simó-Alfonso and G. Ramis-Ramos, Separation of homologues and isomers of linear alkylbenzenesulfonates by capillary electrophoresis with sodium dodecyl sulfate, carboxylic acids and bile salts, *Electrophoresis*, 24, 681–686, 2003.
32. J.M. Herrero-Martínez, E.F. Simó-Alfonso and G. Ramis-Ramos, Separation and determination of homologues of linear alkylbenzenesulfonates by nonaqueous capillary zone electrophoresis using alkylammonium salts in ethanol, *Electrophoresis*, 22, 2017–2024, 2001.
33. V. Bernabé-Zafón, S. Ortega-Gadea, E.F. Simó-Alfonso and G. Ramis-Ramos, Characterization and quantitation of mixtures of alkyl ether sulfates and carboxylic acids by capillary electrophoresis with indirect photometric detection, *Electrophoresis*, 24, 2805–2813, 2003.
34. V. Bernabé-Zafón, J.R. Torres-Lapasió, S. Ortega-Gadea, E.F. Simó-Alfonso and G. Ramis-Ramos, Capillary electrophoresis enhanced by automatic two-way background correction using cubic smoothing splines and multivariate data analysis applied to the characterisation of mixtures of surfactants, *J. Chromatogr. A*, 1065, 301–313, 2005.
35. W. Wall, K. Chan and Z. El Rassi, Electrically driven microseparation methods for pesticides and metabolites. VI. Surfactant-mediated electrokinetic capillary chromatography of aniline pesticidic metabolites derivatized with 9-fluoroenylmethyl chloroformate and their detection by laser-induced fluorescence, *Electrophoresis*, 22, 2320–2326, 2001.
36. J.T. Koch, B. Beam, K.S. Phillips and J.F. Wheeler, Hydrophobic interaction electrokinetic chromatography for the separation of polycyclic aromatic hydrocarbons using non-aqueous matrices, *J. Chromatogr. A*, 914, 223–231, 2001.
37. W. Wall and Z. El Rassi, Electrically driven microseparation methods for pesticides and metabolites. V. Micellar electrokinetic capillary chromatography of aniline pesticidic metabolites derivatized with fluorescein isothiocyanate and their detection in real water at low levels by laser-induced fluorescence, *Electrophoresis*, 22, 2312–2319, 2001.
38. T.W. Moy, P.L. Ferguson, A.H. Grange, W.H. Matchett, V.A. Kelliber, W.C. Brumley, J. Glassman and J.Q. Farley, Development of separation systems for polynuclear aromatic hydrocarbon environmental contaminants using micellar electrokinetic chromatography with molecular micelles and free zone electrophoresis, *Electrophoresis*, 19, 2090–2094, 1998.

39. H. Harino, S. Tsunoi, T. Sato and M. Tanaka, Applicability of micellar electrokinetic chromatography with a double-chain surfactant having two sulfonate groups to the determination of pollutant phenols in water, *Anal. Sci.*, 16, 1349–1351, 2000.
40. S.K. Wiedmer, J. Hautala, J.M. Holopainen, P.K.J. Kinnunen, and M.-L. Riekkola, Study on liposomes by capillary electrophoresis, *Electrophoresis*, 22, 1305–1313, 2001.
41. D. Kaniansky, E. Krčmová, V. Madajová, M. Masár, J. Marák and F.I. Onuska, Determination of nitrophenols by capillary zone electrophoresis in a hydrodynamically closed separation compartment, *J. Chromatogr. A*, 772, 327–337, 1997.
42. C. García-Ruiz, Y. Martín-Biosca, A.L. Crego and M.L. Marina, Rapid enantiomeric separation of polychlorinated biphenyls by electrokinetic chromatography using mixtures of neutral and charged cyclodextrin derivatives, *J. Chromatogr. A*, 910, 157–164, 2001.
43. R.S. Brown, J.H.T. Luong, O.H.J. Szolar, A. Halasz and J. Hawari, Cyclodextrin-modified capillary electrophoresis: Determination of polycyclic aromatic hydrocarbons in contaminated soils, *Anal. Chem.*, 68, 287–292, 1996.
44. C.A. Groom and J.H.T. Luong, Sulfobutylether-beta-cyclodextrin-mediated capillary electrophoresis for separation of chlorinated and substituted phenols, *Electrophoresis*, 18, 1166–1172, 1997.
45. B. Fogarty, E. Dempsey and F. Regan, Potential of microemulsion electrokinetic chromatography for the separation of priority endocrine disrupting compounds, *J. Chromatogr. A*, 1014, 129–139, 2003.
46. D. Shobat and E. Grushka, Use of calixarenes to modify selectivities in capillary electrophoresis, *Anal. Chem.*, 66, 747–750, 1994.
47. L. Arce, A. Segura-Carretero, A. Rios, C. Cruces, A. Fernández and M. Valcárcel, Use of calixarene compounds as selectivity modifiers in capillary electrophoresis separations, *J. Chromatogr. A*, 816, 243–249, 1998.
48. K. Bächmann, B. Göttlicher, I. Haag, K.I. Han, W. Hensel and A. Mainka, Capillary electrokinetic chromatography with a suspension of chromatographic particles, *J. Chromatogr. A*, 688, 283–292, 1994.
49. B. Göttlicher and K. Bächmann, Application of particles as pseudo-stationary phases in electrokinetic chromatography, *J. Chromatogr. A*, 780, 63–73, 1997.
50. S. Terabe and T. Isemura, Ion-exchange electrokinetic chromatography with polymer ions for the separation of isomeric ions having identical electrophoretic mobilities, *Anal. Chem.*, 62, 650–652, 1990.
51. J.S. Fritz, M.C. Breadmore, E.F. Hilder and P.R. Haddad, Use of ionic polymers as stationary and pseudo-stationary phases in the separation of ions by capillary electrophoresis and capillary electrochromatography, *J. Chromatogr. A*, 942, 11–32, 2002.
52. Y. Takagai and S. Igarashi, Homogeneous liquid-liquid extraction and micellar electrokinetic chromatography using sweeping effect concentration system for determination of trace amounts of several polycyclic aromatic hydrocarbons, *Anal. Bioanal. Chem.*, 373, 87–92, 2002.
53. E. Turiel, P. Fernández, C. Pérez-Conde and C. Cámara, Trace-level determination of triazines and several degradation products in environmental waters by disk solid-phase extraction and micellar electrokinetic chromatography, *J. Chromatogr. A*, 872, 299–307, 2000.
54. P. Zhang, G. Xu, J. Xiong, Y. Zheng, Q. Yang and F. Wei, Determination of arsenic species by capillary electrophoresis with large-volume field-amplified stacking injection, *Electrophoresis*, 22, 3567–3572, 2001.
55. Y. He and H.-K. Lee, Combination of solid-phase extraction and field-amplified concentration for trace analysis of organonitrogen pesticides by micellar electrokinetic chromatography, *Electrophoresis*, 18, 2036–2041, 1997.
56. T. Hirokawa, T. Ichihara, K. Ito and A.R. Timerbaev, Trace ion analysis of seawater by capillary electrophoresis: Determination of iodide using transient isotachophoresis preconcentration, *Electrophoresis*, 24, 2328–2334, 2003.
57. K. Otsuka, H. Hayashibara, S. Yamauchi, J.P. Quirino and S. Terabe, Highly sensitive micellar electrokinetic chromatographic analysis of dioxin-related compounds using on-line concentration, *J. Chromatogr. A*, 853, 413–420, 1999.

58. A. Macià, F. Borrull, M. Calull and C. Aguilar, Different sample stacking strategies to analyze some nonsteroidal anti-inflammatory drugs by micellar electrokinetic capillary chromatography in mineral waters, *J. Chromatogr. A*, 1117, 234–245, 2006.
59. S. Takeda, A. Omura, K. Chayama, H. Tsuji, K. Fukushi, M. Yamane, S.-I. Wakida, S. Tsubota and S. Terabe, Separation and on-line concentration of bisphenol A and alkylphenols by micellar electrokinetic chromatography with anionic surfactant, *J. Chromatogr. A*, 1014, 103–107, 2003.
60. J.P. Quirino, N. Inoue and S. Terabe, Reversed migration micellar electrokinetic chromatography with off-line and on-line concentration analysis of phenylurea herbicides, *J. Chromatogr. A*, 892, 187–194, 2000.
61. C.L. Silva, E.C. Lima and M.F.M. Tavares, Investigation of preconcentration strategies for the trace analysis of multiresidue pesticides in real samples by capillary electrophoresis, *J. Chromatogr. A*, 1014, 109–116, 2003.
62. K. Isoo and S. Terabe, Analysis of metal ions by sweeping via dynamic complexation and cation-selective exhaustive injection in capillary electrophoresis, *Anal. Chem.*, 75, 6789–6798, 2003.
63. O. Núñez, J.B. Kim, E. Moyano, M.T. Galcerán and S. Terabe, Analysis of the herbicides paraquat, diquat and dibenzoquat in drinking water by micellar electrokinetic chromatography using sweeping and cation selective exhaustive injection, *J. Chromatogr. A*, 961, 65–75, 2002.
64. J.P. Quirino, Y. Iwai, K. Otsuka and S. Terabe, Determination of environmentally relevant aromatic amines in the ppt levels by cation selective exhaustive injection-sweeping-micellar electrokinetic chromatography, *Electrophoresis*, 21, 2899–2903, 2000.
65. P. Hinsmann, L. Arce, A. Ríos and M. Valcárcel, Determination of pesticides in water by automation online solid-phase extraction-capillary electrophoresis, *J. Chromatogr. A*, 866, 137–146, 2000.
66. M.C. Breadmore, A.S. Palmer, M. Curran, M. Macka, N. Avdalovic and P.R. Haddad, On-column ion-exchange of inorganic anions in open tubular capillary electrochromatography with elution using transient-isotachophoresis gradients. 3. Implementation and method development, *Anal. Chem.*, 74, 2112–2118, 2002.
67. J.F. Huertas-Pérez, M. del Olmo Iruela, A.M. García-Campaña, A. González-Casado and A. Sánchez-Navarro, Determination of the herbicide metribuzin and its major conversion products in soil by micellar electrokinetic chromatography, *J. Chromatogr. A*, 1102, 280–286, 2006.
68. A. Mainka and K. Bächmann, UV detection of derivatized carbonyl compounds in rain samples in capillary electrophoresis using sample stacking and a Z-shaped flow cell, *J. Chromatogr. A*, 767, 241–247, 1997.
69. J. Kuijt, C. García-Ruiz, G.J. Stroomberg, M.L. Marina, F. Ariese, U.A.Th. Brinkman and C. Gooijer, Laser-induced fluorescence detection at 266 nm in capillary electrophoresis. Polycyclic aromatic hydrocarbon metabolites in biota, *J. Chromatogr. A*, 907, 291–299, 2001.
70. S. Chen, Y. Xu, Y. Bi, W. Du and B.-F. Liu, Analysis of environmental pollutants metabolized from pesticides using capillary electrophoresis with multiphoton-excited fluorescence detection, *Talanta*, 70, 63–67, 2006.
71. E. Dabek-Zlotorzynska and W. Maruszak, Determination of dimethylamine and other low-molecular mass amines using capillary electrophoresis with laser-induced fluorescence detection, *J. Chromatogr. B*, 714, 77–85, 1998.
72. M. Molina and M. Silva, In-capillary derivatization and analysis of amino acids, amino phosphonic acid-herbicides and biogenic amines by capillary electrophoresis with laser-induced fluorescence detection, *Electrophoresis*, 23, 2333–2340, 2002.
73. S.-J. Chen, M.-J. Chen and H.-T. Chang, Light-emitting diode-based indirect fluorescence detection for simultaneous determination of anions and cations in capillary electrophoresis, *J. Chromatogr. A*, 1017, 215–224, 2003.
74. S.Y. Chang and C.-H. Liao, Analysis of glyphosate, glufosinate and aminomethylphosphonic acid by capillary electrophoresis with indirect fluorescence detection, *J. Chromatogr. A*, 959, 309–315, 2002.
75. Y.-M. Liu and J.-K. Cheng, Separation of biogenic amines by micellar electrokinetic chromatography with on-line chemiluminescence detection, *J. Chromatogr. A*, 1003, 211–216, 2003.
76. F.R. Rocha, J.A.F. da Silva, C.L. do Lago, A. Fornaro and I.G.R. Gutz, Wet deposition and related atmospheric chemistry in the Sao Paulo metropolis, Brazil; Part 1. Major inorganic ions in rainwater

as evaluated by capillary electrophoresis with contactless conductivity detection, *Atmos. Environ.*, 37, 105–115, 2003.
77. A. Fornaro and I.G.R. Gutz, Wet deposition and related atmospheric chemistry in the Sao Paulo metropolis, Brazil; Part 2. Contribution of formic and acetic acids, *Atmos. Environ.*, 37, 117–128, 2003.
78. P. Kubáň, B. Karlberg, P. Kubáň and V. Kubáň, Application of a contactless conductometric detector for the simultaneous determination of small anions and cations by capillary electrophoresis with dual-opposite end injection, *J. Chromatogr. A*, 964, 227–241, 2002.
79. M. Chicharro, A. Zapardiel, E. Bermejo and A. Sánchez, Simultaneous UV and electrochemical determination of the herbicide asulam in tap water samples by micellar electrokinetic capillary chromatography, *Anal. Chim. Acta*, 469, 243–252, 2002.
80. A. Asthana, D. Bose, A. Durgbanshi, S.K. Sanghi and W.Th. Kok, Determination of aromatic amines in water samples by capillary electrophoresis with electrochemical and fluorescence detection, *J. Chromatogr. A*, 895, 197–203, 2000.
81. M. van Bruijnsvoort, S.K. Sanghi, H. Poppe and W.Th. Kok, Determination of chlorophenols by micellar electrokinetic chromatography with electrochemical detection, *J. Chromatogr. A*, 757, 203–223, 1997.
82. W. Ahrer, E. Scherwenk and W. Buchberger, Determination of drug residues in water by the combination of liquid chromatography or capillary electrophoresis with electrospray mass spectrometry, *J. Chromatogr. A*, 910, 69–78, 2001.
83. C.A. Groom, A. Halasz, L. Paquet, S. Thiboutot, G. Ampleman and J. Hawari, Detection of nitroaromatic and cyclic nitramine compounds by cyclodextrin assisted capillary electrophoresis quadrupole ion trap mass spectrometry, *J. Chromatogr. A*, 1072, 73–82, 2005.
84. T. Poiger, S.D. Richardson and G.L. Baughman, Identification of reactive dyes in spent dyebaths and wastewater by capillary-electrophoresis-mass spectrometry, *J. Chromatogr. A*, 886, 271–282, 2000.
85. O. Núñez, E. Moyano and M.T. Galceran, Capillary electrophoresis-mass spectrometry for the analysis of quaternary ammonium herbicides, *J. Chromatogr. A*, 974, 243–255, 2002.
86. Y. Iinuma and H. Herrmann, Method development for the analysis of particle phase substituted methoxy phenols and aromatic acids from biomass burning using capillary electrophoresis/electrospray ionization mass spectrometry (CE/ESI/MS), *J. Chromatogr. A*, 1018, 105–115, 2003.
87. J. Riu, P. Eichhorn, J.A. Guerrero, Th.P. Knepper and D. Barceló, Determination of linear alkylbenzenesulfonates in wastewater treatment plants and coastal waters by automated solid-phase extraction followed by capillary electrophoresis-UV detection and confirmation by capillary electrophoresis-mass spectrometry, *J. Chromatogr. A*, 889, 221–229, 2000.
88. M. Molina, S.K. Wiedmer, M. Jussila, M. Silva and M.L. Riekkola, Use of partial filling technique and reverse migrating micelles in the study of *N*-methylcarbamate pesticides by micellar electrokinetic chromatography-electrospray ionization mass spectrometry, *J. Chromatogr. A*, 24, 191–202, 2001.
89. F. Menzinger, P. Schmitt-Kopplin, M. Frommberger, D. Freitag and A. Kettrup, Partial-filling micellar electrokinetic chromatography and nonaqueous capillary electrophoresis for the analysis of selected agrochemicals, *Fresenius J. Anal. Chem.*, 371, 25–34, 2001.
90. Y.-R. Chen, M.-C. Tseng, Y.-Z. Chang and G.-R. Her, A low-flow CE/electrospray ionization MS interface for capillary zone electrophoresis, large-volume sample stacking, and micellar electrokinetic chromatography, *Anal. Chem.*, 75, 503–508, 2003.
91. K. Isoo, K. Otsuka and S. Terabe, Application of sweeping to micellar electrokinetic chromatography-atmospheric pressure chemical ionization-mass spectrometric analysis of environmental pollutants, *Electrophoresis*, 22, 3426–3432, 2001.
92. O.H.J. Szolar, Environmental and pharmaceutical analysis of dithiocarbamates, *Anal. Chim. Acta*, 582, 191–200, 2007.
93. L. Gámiz-Gracia, A.M. García-Campaña, J.J. Soto-Chinchilla, J.F. Huertas-Pérez and A. González-Casado, Analysis of pesticides by chemiluminescence detection in the liquid phase, *Trends Anal. Chem.*, 24, 927–942, 2005.
94. V. Andreu and Y. Picó, Determination of pesticides and their degradation products in soil: Critical review and comparison of methods, *Trends Anal. Chem.*, 23, 772–789, 2004.
95. J. Hernández-Borges, S. Frías-García, A. Cifuentes and M.A. Rodrígues-Delgado, Pesticide analysis by capillary electrophoresis, *J. Sep. Sci.*, 27, 947–963, 2004.

96. Y. Picó, R. Rodriguez and J. Mañes, Capillary electrophoresis for the determination of pesticide residues, *Trends Anal. Chem.*, 22, 133–151, 2003.
97. Ph. Schmitt, T. Poiger, R. Simon, D. Freitag, A. Kettrup and A.W. Garrison, Simultaneous determination of ionization constants and isoelectric points of 12 hydroxy-*s*-triazines by capillary zone electrophoresis and capillary isoelectric focusing, *Anal. Chem.*, 69, 2559–2566, 1997.
98. M. Chicharro, A. Zapardiel, E. Bermejo and M. Moreno, Determination of 3-amino-1,2,4-triazole (amitrole) in environmental waters by capillary electrophoresis, *Talanta*, 59, 37–45, 2003.
99. R. Loos and R. Niessner, Analysis of atrazine, terbutylazine and their *N*-dealkylated chloro and hydroxyl metabolites by solid-phase extraction and gas chromatography mass spectrometry and capillary electrophoresis-ultraviolet detection, *J. Chromatogr. A*, 835, 217–229, 1999.
100. H. Stutz, K. Pittertschatscher and H. Malissa, Jr., Capillary zone electrophoresis determination of hydroxymetabolites of atrazine in potable water using solid-phase extraction with amberchrom resins, *Mikrochim. Acta*, 128, 107–117, 1998.
101. J. You, M. Kaljurand and J.A. Koropchak, Direct determination of glyphosate in environmental waters using capillary electrophoresis with electrospray condensation nucleation light scattering detection, *Intern. J. Environ. Anal. Chem.*, 83, 797–806, 2003.
102. J. Hernández-Borges, F.J. García-Montelongo, A. Cifuentes and M.A. Rodríguez-Delgado, Determination of herbicides in mineral and stagnant waters at ng/L levels using capillary electrophoresis and UV detection combined with solid-phase extraction and sample stacking, *J. Chromatogr. A*, 1070, 171–177, 2005.
103. R.G. Wuilloud, M. Shah, S.S. Kannamkumarath and J.C. Altamirano, The potential of inductively coupled plasma-mass spectrometric detection for capillary electrophoretic analysis of pesticides, *Electrophoresis*, 26, 1598–1605, 2005.
104. X. Song and W.L. Budde, Determination of chlorinated acid herbicides and related compounds in water by capillary electrophoresis-electrospray negative ion mass spectrometry, *J. Chromatogr. A*, 829, 327–340, 1998.
105. R. Carabías-Martínez, E. Rodríguez-Gonzalo, J. Dominguez-Álvarez and J. Hernández-Méndez, Determination of triazine herbicides in natural waters by solid-phase extraction and non-aqueous capillary zone electrophoresis, *J. Chromatogr. A*, 869, 451–461, 2000.
106. N.V. Komarova and L.A. Kartsova, Factors responsible for the electrophoretic behavior of carboxylic acid and triazine derivatives under conditions of capillary zone electrophoresis and micellar electrokinetic chromatography, *J. Anal. Chem.*, 59, 662–668, 2004.
107. M. Chicharro, A. Zapardiel, E. Bermejo and A. Sanchez, Multiresidue analysis of pesticides in environmental waters by capillary electrophoresis using simultaneous UV and electrochemical detection, *Electroanalysis*, 16, 311–318, 2004.
108. R. Carabías-Martínez, E. Rodríguez-Gonzalo, P. Revilla-Ruiz and J. Domínguez-Alvarez, Solid-phase extraction and sample stacking-micellar electrokinetic capillary chromatography for the determination of multiresidue of herbicides and metabolites, *J. Chromatogr. A*, 990, 291–302, 2003.
109. C.W. Henry III, S.A. Shamsi and I.M. Warner, Separation of natural pyrethrum extracts using micellar electrokinetic chromatography, *J. Chromatogr. A*, 863, 89–103, 1999.
110. C.-E. Lin, Y.-C. Liu, T.-Y. Yang, T.-Z. Wang and C.-C. Yang, On-line concentration of *s*-triazine herbicides in micellar electrokinetic chromatography using a cationic surfactant, *J. Chromatogr. A*, 916, 239–245, 2001.
111. Y.S. Wu, H.K. Lee and S.F.Y. Li, A fluorescence detection scheme for capillary electrophoresis of *N*-methylcarbamates with on-column thermal decomposition and derivatization, *Anal. Chem.*, 72, 1441–1447, 2000.
112. A. Karcher and Z. El Rassi, Electrically driven microseparation methods for pesticides and metabolites: IV. Effects of the nature of fluorescent labels on the enantioseparation of pesticides and their degradation products by capillary zone electrophoresis with UV and laser-induced fluorescence detection, *Electrophoresis*, 21, 2043–2050, 2000.
113. O. Núñez, E. Moyano, L. Puignou and M.T. Galcerán, Sample stacking with matrix removal for the determination of paraquat, diquat and difenzoquat in water by capillary electrophoresis, *J. Chromatogr. A*, 912, 353–361, 2001.
114. G. Dinelli, A. Vicari and A. Bonetti, Separation of sulfonylurea metabolites in water by capillary electrophoresis, *J. Chromatogr. A*, 700, 195–200, 1995.

115. C. Klein, R.J. Schneider, M.T. Meyer and D.S. Aga, Enantiomeric separation of metolachlor and its metabolites using LC-MS and CZE, *Chemosphere*, 62, 1591–1599, 2006.
116. J. Hernández-Borges, F.J. García-Montelongo, A. Cifuentes and M.A. Rodríguez-Delgado, Analysis of triazolopyrimidine herbicides in soils using field-enhanced sample injection-coelectroosmotic capillary electrophoresis combined with solid-phase extraction, *J. Chromatogr. A*, 1100, 236–242, 2005.
117. N.V. Komarova and L.A. Kartsova, Determination of herbicides of the chlorophenoxycarboxylic acid type in natural and drinking water by capillary zone electrophoresis, *Russian J. Appl. Chem.*, 76, 238–243, 2003.
118. M. Corbera, M. Hidalgo, V. Salvado and P.P. Wieczorek, Determination of glyphosate and aminomethylphosphonic acid in natural water using capillary electrophoresis combined with enrichment step, *Anal. Chim. Acta*, 540, 3–7, 2005.
119. H. Prosen, M. Guček and L. Zupančič-Kralj, Optimization of liquid chromatography and micellar electrokinetic chromatography for the determination of atrazine and its first degradation products in humic waters without preparation, *Chromatographia*, 60(Suppl. 60), S107–S112, 2004.
120. D.T. Kubilius and R.J. Bushway, Determination of hexazinone and its metabolites in groundwater by capillary electrophoresis, *J. Chromatogr. A*, 793, 349–355, 1998.
121. M.B. Barroso, L.N. Honda and G. Morovjan, Multiresidue analysis of phenylurea herbicides in environmental and drinking water by capillary electrophoresis, *J. High Resol. Chromatogr.*, 22, 171–176, 1999.
122. J.L. Jarman, W.J. Jones, L.A. Howell and A.W. Garrison, Application of capillary electrophoresis to study the enantioselective transformation of five chiral pesticides in aerobic soil slurries, *J. Agric. Food Chem.*, 53, 6175–6182, 2005.
123. R. Rodríguez, Y. Picó, G. Font and J. Mañes, Analysis of post-harvest fungicides by micellar electrokinetic chromatography, *J. Chromatogr. A*, 924, 387–396, 2001.
124. M. Chicharro, A. Zapardiel, E. Bermejo, J.A. Perez and A. Moreno, Multiresidue analysis of *S*-triazines herbicides in environmental waters by micellar electrokinetic capillary chromatography, *J. Liq. Chromatogr. Rel. Technol.*, 24, 461–478, 2001.
125. X. Xu and R.J. Hurtubise, Separation of polycyclic aromatic hydrocarbon metabolites by gamma-cyclodextrin-modified micellar electrokinetic chromatography, *J. Liq. Chromatogr. Rel. Technol.*, 23, 1657–1670, 2000.
126. J. Li and J.S. Fritz, Nonaqueous media for separation of nonionic organic compounds by capillary electrophoresis, *Electrophoresis*, 20, 84–91, 1999.
127. G. Kavran and F.B. Erim, Separation of polycyclic aromatic hydrocarbons with sodium dodecylbenzenesulfonate in electrokinetic chromatography, *J. Chromatogr. A*, 949, 301–305, 2002.
128. G.K. Belin, E.B. Erim and F. Gulacar, Capillary electrokinetic separation of polycyclic aromatic hydrocarbons using cetylpyridinium bromide, *Polycyclic Arom. Compds.*, 24, 343–352, 2004.
129. H. Ren, X. Li, M. Qi, C. Stathakis and N.J. Dovichi, Capillary electrophoretic separation of polynuclear aromatic hydrocarbons using sodium cholate in mixed aqueous-organic buffers, *J. Chromatogr. A*, 817, 307–311, 1998.
130. J.H.T. Luong, The combined effect of acetonitrile and urea on the separation of polycyclic aromatic hydrocarbons using dioctyl sulfosuccinate in electrokinetic chromatography, *Electrophoresis*, 19, 1461–1467, 1998.
131. Y. Shi and J.S. Fritz, Capillary zone electrophoresis of neutral organic molecules in organic–aqueous solution, *J. High Resol. Chromatogr.*, 17, 713–718, 1994.
132. J.L. Miller, M.G. Khaledi and D. Shea, Separation of hydrophobic solutes by nonaqueous capillary electrophoresis through dipolar and charge-transfer interactions with pyrylium salts, *J. Microcolumn Sep.*, 10, 681–685, 1998.
133. A.-L. Nguyen and J.H.T. Luong, Separation and determination of polycyclic aromatic hydrocarbons by solid-phase microextraction/cyclodextrin-modified capillary electrophoresis, *Anal. Chem.*, 69, 1726–1731, 1997.
134. C.P. Palmer, M.Y. Khaledi and H.M. McNair, A monomolecular pseudostationary phase for micellar electrokinetic capillary chromatography, *J. High Resol. Chromatogr.*, 15, 756–762, 1992.
135. D. Norton and S.A. Shamsi, Separation of methylated isomers of benzo[*a*]pyrene using micellar electrokinetic chromatography, *Anal. Chim. Acta*, 496, 165–176, 2003.

136. C. Akbay, S.A. Shamsi and I.M. Warner, Separation of monomethyl-benz[*a*]antracene isomers using cyclodextrin-modified electrokinetic chromatography, *J. Chromatogr. A*, 910, 147–155, 2001.
137. C. Fujimoto, Y. Fujise and S. Kawaguchi, Macromolecular surfactant as a pseudo-stationary phase in micellar electrokinetic capillary chromatography, *J. Chromatogr. A*, 871, 415–425, 2000.
138. D.S. Peterson and C.P. Palmer, An anionic siloxane polymer as a pseudostationary phase for electrokinetic chromatography, *Electrophoresis*, 21, 3174–3180, 2000.
139. N. Tanaka, T. Fukutome, K. Hosoya, K. Kimata and T. Anaki, Polymer-supported pseudo-stationary phase for electrokinetic chromatography. Electrokinetic chromatography in a full range of methanol–water mixtures with alkylated starburst dendrimers, *J. Chromatogr. A*, 716, 57–67, 1995.
140. J.M. Treubig Jr. and P.R. Brown, Novel approach to the analysis and use of fullerenes in capillary electrophoresis, *J. Chromatogr. A*, 873, 257–267, 2000.
141. B. Göttlicher and K. Bachmann, Investigation of the separation efficiency of hydrophobic compounds in suspension electrokinetic chromatography, *J. Chromatogr. A*, 768, 320–324, 1997.
142. I. Benito, J.M. Saz, M.L. Marina, J. Jiménez-Barbero, M.J. González and J.C. Díez-Masa, Micellar electrokinetic capillary chromatographic separation of polychlorinated biphenyl congeners, *J. Chromatogr. A*, 778, 77–85, 1997.
143. M.L. Marina, I. Benito, J.C. Díez-Masa and M.J. González, Separation of chiral polychlorinated biphenyls by micellar electrokinetic chromatography using β- and γ-cyclodextrin mixtures in the separation buffer, *J. Chromatogr. A*, 752, 265–270, 1996.
144. W.C. Lin, F.C. Chang and C.H. Kuei, Separation of atropisomeric polychlorinated biphenyls by cyclodextrin modified micellar electrokinetic chromatography, *J. Microcolumn Sep.*, 11, 231–238, 1999.
145. A.L. Crego, M.A. García and M.L. Marina, Enantiomeric separation of chiral polychlorinated biphenyls by micellar electrokinetic chromatography using mixtures of bile salts and sodium dodecyl sulphate with and without gamma-cyclodextrin in the separation buffer, *J. Microcolumn. Sep.*, 12, 33–40, 2000.
146. C. García-Ruiz, A.L. Crego and M.L. Marina, Comparison of charged cyclodextrin derivatives for the chiral separation of atropisomeric polychlorinated biphenyls by capillary electrophoresis, *Electrophoresis*, 24, 2657–2664, 2003.
147. S.H. Edwards and S.A. Shamsi, Micellar electrokinetic chromatography of polychlorinated biphenyl congeners using a polymeric surfactant as the pseudostationary phase, *J. Chromatogr. A*, 903, 227–236, 2000.
148. S.H. Edwards and S.A. Shamsi, Chiral separation of polychlorinated biphenyls using a combination of hydroxypropyl-gamma-cyclodextrin and a polymeric chiral surfactant, *Electrophoresis*, 23, 1320–1327, 2002.
149. A. Hilmi, J.H.T. Luong and A.-L. Nguyen, Applicability of micellar electrokinetic chromatography to kinetic studies of photocatalytic oxidation of dibenzo-*p*-dioxin, *Chemosphere*, 36, 3113–3117, 1998.
150. J. Grainger, P.C. McClure, A. Liu, et al., Isomer identification of chlorinated dibenzo-*p*-dioxins by orthogonal spectroscopic and chromatographic techniques, *Chemosphere*, 32, 13–23, 1996.
151. S. Terabe, Y. Miyashita, O. Shibata, E.R. Barnhart, L.R. Alexander, D.G. Patterson, B.L. Karger, K. Hosoya and N. Tanaka, Separation of highly hydrophobic compounds by cyclodextrin-modified micellar electrokinetic chromatography, *J. Chromatogr. A*, 516, 23–31, 1990.
152. J.T. Koch, B. Beam, K.S. Phillips and J.F. Wheeler, Hydrophobic interaction electrokinetic chromatography for the separation of polycyclic aromatic hydrocarbons using nonaqueous matrices, *J. Chromatogr. A*, 914, 223–231, 2001.
153. C.-E. Lin, C.-C. Hsueh, W.-C. Lin and C.-C. Chang, Migration behavior and separation of trichlorophenols by capillary zone electrophoresis, *J. Chromatogr. A*, 746, 295–299, 1996.
154. I. Canals, E. Bosch and M. Roses, Prediction of the separation of phenols by capillary zone electrophoresis, *Anal. Chim. Acta*, 458, 355–366, 2002.
155. S. Takeda, Y. Tanaka, M. Yamane, Z. Siroma, S.-I. Wakida, K. Otsuka and S. Terabe, Ionization of dichlorophenols for their analysis by capillary electrophoresis-mass spectrometry, *J. Chromatogr. A*, 924, 415–420, 2001.
156. O. Jáuregui, E. Moyano and M.T. Galceran, Capillary electrophoresis-electrospray ion-trap mass spectrometry for the separation of chlorophenols, *J. Chromatogr. A*, 896, 125–133, 2000.
157. O. Jáuregui, L. Puignou and M.T. Galceran, New carrier electrolytes for the separation of chlorophenols by capillary electrophoresis, *Electrophoresis*, 21, 611–618, 2000.

158. S. Morales and R. Cela, Capillary electrophoresis and sample stacking in non-aqueous media for the analysis of priority pollutant phenols, *J. Chromatogr. A*, 846, 401–411, 1999.
159. S.P. Porras, R. Kuldvee, S. Palonen and M.-L. Riekkola, Capillary electrophoresis of methyl-substituted phenols in acetonitrile, *J. Chromatogr. A*, 990, 35–44, 2003.
160. S.P. Porras, R. Kuldvee, M. Jussila, S. Palonen and M.-L. Riekkola, Capillary zone electrophoresis of phenol and methylphenols at high pH in methanol, *J. Sep. Sci.*, 26, 857–862, 2003.
161. J. Kronholm, P. Revilla-Ruiz, S.P. Porras, K. Hartonen, R. Carabías-Martínez and M.-L. Riekkola, Comparison of gas chromatography-mass spectrometry and capillary electrophoresis in analysis of phenolic compounds extracted from solid matrices with pressurized hot water, *J. Chromatogr. A*, 1022, 9, 2004.
162. S. Fu, F. Li, S.G. Chu and X.B. Xu, The determination of chlorophenols in waste water by capillary zone electrophoresis with an organic modifier, *Chromatographia*, 56, 69–72, 2002.
163. X. Guo, Z. Wang and S. Zhou, The separation and determination of nitrophenol isomers by high-performance capillary zone electrophoresis, *Talanta*, 64, 135–139, 2004.
164. S.S. Kannamkumarath, R.G. Wuilloud, S. Jayasinghe and J.A. Caruso, Fast speciation analysis of iodophenol compounds in river waters by capillary electrophoresis-inductively coupled plasma-mass spectrometry with off-line solid-phase microextraction, *Electrophoresis*, 25, 1843–1851, 2004.
165. W. Wall and Z. El Rassi, Electrically driven microseparation methods for pesticides and metabolites. Part VII: Capillary electrophoresis and electrochromatography of underivatized phenol pesticidic metabolites. Preconcentration and laser-induced fluorescence detection of dilute samples, *J. Sep. Sci.*, 25, 1231–1244, 2002.
166. A. Zemann and D. Volgger, Separation of priority pollutant phenols with coelectroosmotic capillary electrophoresis, *Anal. Chem.*, 69, 3243–3250, 1997.
167. S.M. Masselter and A.J. Zemann, Influence of organic solvents in coelectroosmotic capillary electrophoresis of phenols, *Anal. Chem.*, 67, 1047–1053, 1995.
168. M.R.N. Monton, J.P. Quirino, K. Otsuka and S. Terabe, Separation and on-line preconcentration by sweeping of charged analytes in electrokinetic chromatography with nonionic micelles, *J. Chromatogr. A*, 939, 99–108, 2001.
169. M.R.N. Monton, K. Otsuka and S. Terabe, On-line sample preconcentration in micellar electrokinetic chromatography by sweeping with anionic-zwitterionic mixed micelles, *J. Chromatogr. A*, 985, 435–445, 2003.
170. G. Jeevan, M. Bhaskar, R. Chandrasekar and G. Radhakrishnan, Separation of harmful chlorophenols by cyclodextrin-assisted capillary electrokinetic chromatography, *J. Sep. Sci.*, 25, 1143–1146, 2002.
171. P. Kuban, M. Berg, C. García and B. Karlberg, On-line flow sample stacking in a flow injection analysis-capillary electrophoresis system: 2000-fold enhancement of detection sensitivity for priority phenol pollutants, *J. Chromatogr. A*, 912, 163–170, 2001.
172. L. Gaillon, S. Cozette, J. Lelievre and R. Gaboriaud, New pseudo-stationary phases for electrokinetic capillary chromatography. Complexes between bovine serum albumin and sodium dodecyl sulfate, *J. Chromatogr. A*, 876, 169–182, 2000.
173. T. Zhao, X.B. Hu, J.K. Cheng and X.R. Lu, P-sulfonic calix[4]arene as running buffer additive in electrokinetic chromatography, *J. Liq. Chromatogr. Rel. Technol.*, 21, 3111–3124, 1998.
174. A.L. Gray and J.T. Hsu, Novel sulfonic acid-modified Starburst dendrimer used as a pseudostationary phase in electrokinetic chromatography, *J. Chromatogr. A*, 824, 119–124, 1998.
175. B. Maichel, B. Potocek, B. Gaš and E. Kenndler, Capillary electrokinetic chromatography with polyethyleneimine as replaceable cationic pseudostationary phase. Influence of methanol and acetonitrile on separation selectivity, *J. Chromatogr. A*, 853, 121–129, 1999.
176. A.V. Pirogov and O.A. Shpigun, Application of water-soluble polymers as modifiers in electrophoretic analysis of phenols, *Electrophoresis*, 24, 2099–2105, 2003.
177. T. Nakamura, A. Ohki, M. Mishiro, O. Tsuyashima and S. Maeda, Aggregate of amphiphilic block copolymer as a pseudo-stationary phase in capillary electrophoresis, *Anal. Sci.*, 15, 879–883, 1999.
178. A.V. Pirogov, A.V. Shpak and O.A. Shpigun, Application of polyelectrolyte complexes as novel pseudo-stationary phases in MEKC, *Anal. Bioanal. Chem.*, 375, 1199–1203, 2003.

179. J.T. Hautala, S.K. Wiedmer and M.-L. Riekkola, Anionic liposomes in capillary electrophoresis: Effect of calcium on 1-palmitoyl-2-oleyl-sn-glycero-3-phosphatidylcholine/phosphatidylserine-coating in silica capillaries, *Anal. Bioanal. Chem.*, 378, 1769–1776, 2004.
180. A.L. Crego and M.L. Marina, Capillary zone electrophoresis versus micellar electrokinetic chromatography in the separation of phenols of environmental interest, *J. Liq. Chrom. Rel. Technol.*, 20, 1–20, 1997.
181. M. Czaplicka, Sources and transformations of chlorophenols in the natural environment, *Sci. Total Environ.*, 322, 21–39, 2004.
182. S. Kar and P.K. Dasgupta, Measurement of phenols on a loop-supported liquid film by micellar electrokinetic chromatography and direct UV detection, *J. Chromatogr. A*, 379, 379–387, 1996.
183. W.H. Matchett and W.C. Brumley, Preconcentration of aliphatic amines from water determined by capillary electrophoresis with indirect UV detection, *J. Liq. Chromatogr. Rel. Technol.*, 20, 79–100, 1997.
184. A. Fekete, M. Frommberger, G. Ping, M.R. Lahaniatis, J. Lintelman, J. Fekete, I. Gebefugi, A.K. Malik, A. Kettrup and P. Schmitt-Kopplin, Development of a capillary electrophoretic method for the analysis of low-molecular-weight amines from metal working fluid aerosols and ambient air, *Electrophoresis*, 27, 1237–1247, 2006.
185. W. Maruszak, Analysis of aliphatic amines by micellar electrokinetic chromatography, *J. High Resol. Chromatogr.*, 22, 126–128, 1999.
186. X. Liu, L.-X. Yang and Y.-T. Lu, Determination of biogenic amines by 3-(2-furoyl)quinoline-2-carboxaldehyde and capillary electrophoresis with laser-induced fluorescence detection, *J. Chromatogr. A*, 998, 213–219, 2003.
187. K.B. Male and J.H.T. Luong, Derivatization, stabilization and detection of biogenic amines by cyclodextrin-modified capillary electrophoresis-laser-induced fluorescence detection, *J. Chromatogr. A*, 926, 309–317, 2001.
188. A. Bazzanella, H. Mörbel, K. Bächmann, R. Milbradt, V. Böhmer and W. Vogt, Highly efficient separation of amines by electrokinetic chromatography using resorcarene-octacarboxylic acids as pseudostationary phases, *J. Chromatogr. A*, 792, 143–149, 1997.
189. B.A. Musial, M.N. Martin and N.D. Danielson, Effect of an anionic polymer on the separation of cationic molecules by capillary electrophoresis with conductivity detection, *J. Sep. Sci.*, 25, 311–318, 2002.
190. H.M. Pinheiro, E. Toraud and O. Thomas, Aromatic amines from azo dye reduction: Status review with emphasis on direct UV spectrophotometric detection in textile industry wastewaters, *Dyes Pigments*, 61, 121–139, 2004.
191. S. Oguri, Electromigration methods for amino acids, biogenic amines and aromatic amines, *J. Chromatogr. B*, 747, 1–19, 2000.
192. H. Kataoka, Methods for the determination of mutagenic heterocyclic amines and their applications in environmental analysis, *J. Chromatogr. A*, 774, 121–142, 1997.
193. S.-A. Leung and A.J. de Mello, Electrophoretic analysis of amines using reversed-phase, reversed-polarity, head-column field-amplified sample stacking and laser-induced fluorescence detection, *J. Chromatogr. A*, 979, 171–178, 2002.
194. G. Taibi, M.R. Schiavo, P.C. Rindina, R. Muratore and C.M.A. Nicotra, Micellar electrokinetic chromatography of polyamines and monoacetylpolyamines, *J. Chromatogr. A*, 921, 323–329, 2001.
195. C.M. Knapp and J.J. Breen, Effects of tetraalkylammonium salts on the micellar electrokinetic chromatography of aniline and substituted anilines, *J. Chromatogr. A*, 799, 289–299, 1998.
196. C.-E. Lin, C.-C. Chen, H.-W. Chen, H.-C. Huang, C.-H. Lin and Y.-C. Liu, Optimization of separation and migration behavior of chloropyridines in micellar electrokinetic chromatography, *J. Chromatogr. A*, 910, 165–171, 2001.
197. S.D. Mendonsa and R.J. Hurtubise, Capillary electrophoresis (CE) methods for the separation of carcinogenic heterocyclic aromatic amines, *J. Liq. Chrom. Rel. Technol.*, 22, 1027–1040, 1999.
198. P. Zakaria, M. Macka and P.R. Haddad, Mixed-mode electrokinetic chromatography of aromatic bases with two pseudo-stationary phases and pH control, *J. Chromatogr. A*, 997, 207–218, 2003.
199. D.W. Armstrong, L.W. Chang and S.S.C. Chang, Mechanism of capillary electrophoresis enantioseparations using a combination of an achiral crown ether plus cyclodextrins, *J. Chromatogr. A*, 793, 115–134, 1998.

200. S.-L. Zhao, T.-Y. Dai, Z. Liu, F.-S. Wei, H.-F. Zou and X.-B. Xu, Determination of lower aliphatic carbonyl compounds in stack gas as their 2,4-dinitrophenylhydrazones by micellar electrokinetic chromatography, *Chemosphere*, 35, 2131–2136, 1997.
201. S. Takeda, S. Wakida, M. Yamane and K. Higashi, Analysis of lower aliphatic-aldehydes in water by micellar electrokinetic chromatography with derivatization to 2,4-dinitrophenylhydrazones, *Electrophoresis*, 15, 1332–1334, 1994.
202. E.A. Pereira, A.A. Cardoso and M.F.M. Tavares, Determination of low-aliphatic aldehydes indoors by micellar electrokinetic chromatography using sample dissolution manipulation for signal enhancement, *Electrophoresis*, 24, 700–706, 2003.
203. E.A. Pereira, E. Carrilho and M.F.M. Tavares, Laser-induced fluorescence and UV detection of derivatized aldehydes in air sample using capillary electrophoresis, *J. Chromatogr. A*, 979, 409–416, 2002.
204. E.A. Pereira, M.O.O. Rezende and M.F.M. Tavares, Analysis of low molecular weight aldehydes in air samples by capillary electrophoresis after derivatization with 4-hydrazinobenzoic acid, *J. Sep. Sci.*, 27, 28–32, 2004.
205. K. Fukushi, S. Takeda, K. Chayama and S.-I. Wakida, Application of capillary electrophoresis to the analysis of inorganic ions in environmental samples, *J. Chromatogr. A*, 834, 349–362, 1999.
206. A.R. Timerbaev and K. Fukushi, Analysis of seawater and different highly saline natural waters by capillary zone electrophoresis, *Marine Chem.*, 82, 221–238, 2003.
207. N. Shakulashvili, T. Faller and H. Engelhardt, Simultaneous determination of alkali, alkaline earth and transition metal ions by capillary electrophoresis with indirect UV detection, *J. Chromatogr. A*, 895, 205–212, 2000.
208. K. Fukushi, H. Ito, K. Kimura, K. Yokota, K. Saito, K. Chayama, S. Takeda and S.-I. Wakida, Determination of ammonium in river water and sewage samples by capillary zone electrophoresis with direct UV detection, *J. Chromatogr. A*, 1106, 61–66, 2006.
209. E. Dabek-Zlotorzynska, M. Kelly, H. Chen and C.L. Chakrabarti, Application of capillary electrophoresis combined with a modified BCR sequential extraction for estimating of distribution of selected trace metals in PM2.5 fractions of urban air particulate matter, *Chemosphere*, 58, 1365–1376, 2005.
210. E. Dabek-Zlotorzynska and R. Aranda-Rodriguez, Development and validation of capillary electrophoresis for the determination of selected metal ions in airborne particulate matter after sequential extraction, *Anal. Bioanal. Chem.*, 372, 467–472, 2002.
211. Y.-S. Fung and H.-S. Tung, Capillary electrophoresis for trace metal ion analysis on environmental studies, *Electrophoresis*, 20, 1832–1841, 1999.
212. A. Gáspár and E. Dudás, Application of internal universal calibration for determination of fully dissociated species in capillary electrophoresis using indirect UV detection and electrokinetic injection, *J. Chromatogr. A*, 1110, 254–260, 2006.
213. E. Yang and Z. Zhang, Simultaneous and fast detection of anions in snow using short tube by capillary zone electrophoresis, *Intern. J. Environ. Anal. Chem.*, 82, 353–360, 2002.
214. L. Wójcik, K. Korczak, B. Szostek and M. Trojanowicz, Separation and determination of perfluorinated carboxylic acids using capillary zone electrophoresis with indirect photometric detection, *J. Chromatogr. A*, 1128, 290–299, 2006.
215. E. Dabek-Zlotorzynska, M. Piechowski, M. McGrath and E.P.C. Lai, Determination of low-molecular-mass carboxylic acids in atmospheric aerosol and vehicle emission samples by capillary electrophoresis, *J. Chromatogr. A*, 910, 331–345, 2001.
216. E. Dabek-Zlotorzynska, R. Aranda-Rodrigues and L. Graham, Capillary electrophoresis determinative and GC-MS confirmatory method for water-soluble organic acids in airborne particulate matter and vehicle emission, *J. Sep. Sci.*, 28, 1520–1528, 2005.
217. A. Mainka, P. Ebert, M. Kibler, T. Prokop, B. Tenberken and K. Bächmann, Development of new methods for the analysis of carboxylic acids and carbonyl compounds in size classified raindrops by CE for application in modeling atmospheric processes, *Chromatographia*, 45, 158–162, 1997.
218. A.G. Diress and C.A. Lucy, Study of the selectivity of inorganic anions in hydro-organic solvents using indirect capillary electrophoresis, *J. Chromatogr. A*, 1085, 155–163, 2005.
219. J.E. Melanson, N.E. Baryla and C.A. Lucy, Dynamic capillary coatings for electroosmotic flow control in capillary electrophoresis, *Trends Anal. Chem.*, 20, 365–374, 2001.

220. V. Martínez, N. García, I. Antigüedad, R.M. Alonso and R.M. Jiménez, Capillary electrophoresis as a useful tool for the analysis of chemical tracers applied to hydrological systems, *J. Chromatogr. A*, 1032, 237–242, 2004.
221. K. Ito, T. Ichibara, H. Zhuo, K. Kumamoto, A.R. Timerbaev and T. Hirokawa, Determination of trace iodide in seawater by capillary electrophoresis following transient isotachophoretic preconcentration. Comparison with ion chromatography, *Anal. Chim. Acta*, 497, 67–74, 2003.
222. Z. Huang, K. Ito, A.R. Timerbaev and T. Hirokawa, Speciation studies by capillary electrophoresis—Simultaneous determination of iodide and iodate in seawater, *Anal. Bioanal. Chem.*, 378, 1836–1841, 2004.
223. S. Rovio, M. Mäntynen and H. Sirén, Determination of bromide and potassium in saline groundwaters by capillary electrophoresis without prior dilution, *Appl. Geochem.*, 19, 1331–1337, 2004.
224. T. Takayanagi, M. Ishida, J. Mbuna, R. Driouich and S. Motomizu, Determination of bromate ion in drinking water by capillary zone electrophoresis with direct photometric detection, *J. Chromatogr. A*, 1128, 298–302, 2006.
225. C. Tu and H.K. Lee, Determination of nitrate in seawater by capillary zone electrophoresis with chloride-induced sample self-stacking, *J. Chromatogr. A*, 966, 205–212, 2002.
226. K. Fukushi, Y. Nakayama and J.-I. Tsujimoto, Highly sensitive capillary zone electrophoresis with artificial seawater as the background electrolyte and transient isotachophoresis as the on-line concentration procedure for simultaneous determination of nitrite and nitrate in seawater, *J. Chromatogr. A*, 1005, 197–205, 2003.
227. M.A. Woodland and C.A. Lucy, Altering the selectivity of inorganic anion separations using electrostatic capillary electrophoresis, *Analyst*, 126, 28–32, 2001.
228. Y. Deng, X. Fan, A. Delgado, C. Nolan, K. Furton, Y. Zuo and R.D. Jones, Separation and determination of aromatic acids in natural water with preconcentration by capillary zone electrophoresis, *J. Chromatogr. A*, 817, 145–152, 1998.
229. P. Kubáň, P. Kubáň and V. Kubáň, Speciation of chromium (III) and chromium (VI) by capillary electrophoresis with contactless conductometric detection and dual opposite end injection, *Electrophoresis*, 24, 1397–1403, 2003.
230. W.-P. Yang, Z.-J. Zhang and W. Deng, Simultaneous, sensitive and selective online chemiluminescence determination of Cr(III) and Cr(VI) by capillary electrophoresis, *Anal. Chim. Acta*, 485, 169–177, 2003.
231. F. Han, J.L. Fasching and P.R. Brown, Speciation of organotin compounds by capillary electrophoresis using indirect ultraviolet absorbance detection, *J. Chromatogr. A*, 669, 103–112, 1995.
232. C.L. Ng, H.K. Lee and S.F.Y Li, Determination of organolead and organoselenium compounds by micellar electrokinetic chromatography, *J. Chromatogr. A*, 652, 547–553, 1993.
233. G. Forte, M. D'Amato and S. Caroli, Capillary electrophoresis speciation analysis of various arsenical compounds, *Microchem. J.*, 79, 15–19, 2005.
234. K.F. Akter, Z. Chen, L. Smith, D. Davey and R. Naidu, Speciation of arsenic in ground water samples: A comparative study of CE-UV, HG-AAS and LC-ICP-MS, *Talanta*, 68, 406–415, 2005.
235. Z.L. Chen, J.-M. Lin and R. Naidu, Separation of arsenic species by capillary electrophoresis with sample-stacking techniques, *Anal. Bioanal. Chem.*, 375, 679–684, 2003.
236. Z. Chen and R. Naidu, Separation of sulfur species in water by co-electroosmotic capillary electrophoresis with direct and indirect UV detection, *Intern. J. Environ. Anal. Chem.*, 83, 749–759, 2003.
237. P. Kubáň, P. Kubáň and V. Kubáň, Rapid speciation of Se(IV) and Se(VI) by flow injection-capillary electrophoresis system with contactless conductivity detection, *Anal. Bioanal. Chem.*, 378, 378–382, 2004.
238. M. Takaya, Development of an analytical method for beryllium in airborne dust by micellar electrokinetic chromatography, *J. Chromatogr. A*, 850, 363–368, 1999.
239. M. Aguillar, A. Farran and V. Martí, Capillary electrophoretic determination of cyanide leaching solutions from automobile catalytic converters, *J. Chromatogr. A*, 778, 397–402, 1997.
240. J.M. Robert and C.D. Spinks, Separation of metallated petroporphyrin models using micellar electrokinetic capillary chromatography, *J. Liq. Chrom. Rel. Technol.*, 20, 2979–2995, 1997.

241. C.A. Groom, A. Halasz, L. Paquet, P. D'Cruz and J. Hawari, Cuclodextrin-assisted capillary electrophoresis for determination of the cyclic nitramine explosives RDX, HMS and CL-20. Comparison with high-performance liquid chromatography, *J. Chromatogr. A*, 999, 17–22, 2003.
242. S.A. Oehrle, Analysis of nitramine and nitroaromatic explosives by capillary electrophoresis, *J. Chromatogr. A*, 745, 233–237, 1996.
243. K.G. Hopper, H. LeChair and B.R. McCord, A novel method for analysis of explosives residue by simultaneous detection of anions and cations via capillary electrophoresis, *Talanta*, 67, 304–312, 2005.
244. C.L. Copper and G.E. Collins, Separation of thiol and cyanide hydrolysis products of chemical warfare agents by capillary electrophoresis, *Electrophoresis*, 25, 897–902, 2004.
245. E.W.J. Hooijschuur, A.E. Kientz and U.A.Th. Brinkman, Application of microcolumn liquid chromatography and capillary electrophoresis with flame photometric detection for the screening of degradation products of chemical warfare agents in water and soil, *J. Chromatogr. A*, 928, 187–199, 2001.
246. M. Lagarrigue, A. Bossée, A. Bégos, A. Varenne, P. Gareil and B. Bellier, Separation and identification of isomeric acidic degradation products of organophosphorus chemical warfare agents by capillary electrophoresis-ion trap mass spectrometry, *J. Chromatogr. A*, 1137, 110–118, 2006.
247. T.E. Rosso and P.C. Bossle, Capillary ion electrophoresis screening of nerve agent degradation products in environmental samples using conductivity detection, *J. Chromatogr. A*, 814, 125–134, 1998.
248. A.-E.F. Nassar, S.V. Lucas and L.D. Hoffland, Determination of chemical warfare agent degradation products at low-part-per-billion levels in aqueous samples and sub-part-per-million levels in soils using capillary electrophoresis, *Anal. Chem.*, 71, 1285–1292, 1999.
249. M.-H. Chen and W.-H. Ding, Separation and migration of positional isomers and structural naphthalenesulfonate isomers by cyclodextrin-mediated capillary electrophoresis, *J. Chromatogr. A*, 1033, 167–172, 2004.
250. M.J. Cugat, F. Borrull and M. Calull, Separation of aromatic sulfonate compounds by coelectroosmotic micellar electrokinetic chromatography, *Chromatographia*, 50, 229–234, 1999.
251. M.J. Cugat, F. Borrull and M. Calull, Comparative study of capillary zone electrophoresis and micellar electrokinetic chromatography for the separation of twelve aromatic sulphonate compounds, *Chromatographia*, 46, 204–208, 1997.
252. M.J. Cugat, F. Borrull and M. Calull, An overview of electrophoretic methods for the determination of benzene- and naphthalenesulfonates in water samples, *Trends Anal. Chem.*, 20, 487–499, 2001.
253. R. Loos and R. Niessner, Analysis of aromatic sulfonates in water by solid-phase extraction and capillary electrophoresis, *J. Chromatogr. A*, 822, 291–303, 1998.
254. R. Loos, J. Riu, M.C. Alonso and D. Barceló, Analysis of polar hydrophilic aromatic sulfonates in waste water treatment plants by CE/MS and LC/MS, *J. Mass Spectrom.*, 35, 1197–1206, 2000.
255. S. Aubert and J.-P. Schwitzguébel, Capillary electrophoretic separation of sulphonated anthraquinones in a variety of matrices, *Chromatographia*, 56, 693–697, 2002.
256. L. Farry, S.A. Oxspring, W.F. Smyth and R. Marchant, A study of the effects of injection mode, on-capillary stacking and off-line concentration on the capillary electrophoresis limits of detection for four structural types of industrial dyes, *Anal. Chim. Acta*, 349, 221–229, 1997.
257. T. Poiger, S.D. Richardson and G.L. Baughman, Analysis of anionic metallized azo and formazan dyes by capillary electrophoresis-mass spectrometry, *J. Chromatogr. A*, 886, 259–270, 2000.
258. A.R. Fakhari, M.C. Breadmore, M. Macka and P.R. Haddad, Non-aqueous capillary electrophoresis with red light emitting diode absorbance detection for the analysis of basic dyes, *Anal. Chim. Acta*, 580, 188–193, 2006.
259. W. Brumley and J.F. Farley, Determining eosin as a groundwater migration tracer by capillary electrophoresis/laser-induced fluorescence using a multiwavelength laser, *Electrophoresis*, 24, 2335–2339, 2003.
260. S.M. Burkinshaw, D. Hinks and D.M. Lewis, Capillary zone electrophoresis in the analysis of dyes and other compounds employed in the dye-manufacturing and dye-using industries, *J. Chromatogr. A*, 640, 413–417, 1993.
261. P. Blatny, C.-H. Fisher, A. Rizzi and E. Kenndler, Linear polymers applied as pseudo-phases in capillary zone electrophoresis of azo compounds used as textile dyes, *J. Chromatogr. A*, 717, 157–166, 1995.

262. E. Blanco, M.C. Casais, M.C. Mejuto and R. Cela, Analysis of tetrabromobisphenol A and other phenolic compounds in water samples by non-aqueous capillary electrophoresis coupled to photodiode array ultraviolet detection, *J. Chromatogr. A*, 1071, 205–211, 2005.
263. E. Blanco, M.C. Casais, M.C. Mejuto and R. Cela, Comparative study of aqueous and non-aqueous capillary electrophoresis in the separation of halogenated phenolic and bisphenolic compounds in water samples, *J. Chromatogr. A*, 1068, 189–199, 2005.
264. E. Blanco, M.C. Casais, M.C. Mejuto and R. Cela, Approaches for the simultaneous extraction of tetrabromobisphenol A, tetrachlorobisphenol A and related phenolic compounds from sewage sludge and sediment samples based on matrix solid-phase dispersion, *J. Chromatogr. A*, 78, 2772–2778, 2006.
265. S. Takeda, A. Omura, K. Chayama, H. Tsuji, K. Fukushi, M. Yamane, S.-I. Wakida, S. Tsubota and S. Terabe, Separation and on-line concentration of bisphenol A and alkylphenols by micellar electrokinetic chromatography with cationic surfactant, *J. Chromatogr. A*, 979, 425–429, 2002.
266. S. Takeda, S. Ilida, K. Chayama, H. Tsuji, K. Fukushi and S. Wakida, Separation of bisphenol A and three alkylphenols by micellar electrokinetic chromatography, *J. Chromatogr. A*, 895, 213–218, 2000.
267. F. Regan, A. Moran, B. Fogarty and E. Dempsey, Novel modes of capillary electrophoresis for the determination of endocrine disrupting chemicals, *J. Chromatogr. A*, 1014, 141–152, 2003.
268. B. Fogarty, F. Regan and E. Dempsey, Separation of two groups of oestrogen mimicking compounds using micellar electrokinetic chromatography, *J. Chromatogr. A*, 895, 237–246, 2000.
269. F. Regan, A. Moran, B. Fogarty and E. Dempsey, Development of comparative methods using gas chromatography-mass spectrometry and capillary electrophoresis for determination of endocrine disrupting chemicals in bio-solids, *J. Chromatogr. B*, 770, 243–253, 2002.
270. A. Goméz-Hens and M.P. Aguilar-Caballos, Social and economic interest in the control of phthalic acid esters, *Trends Anal. Chem.*, 22, 847–857, 2003.
271. T. Heberer, Occurrence, fate and removal of pharmaceutical residues in the aquatic environment: A review of recent research data, *Toxicol. Lett.*, 131, 5–17, 2002.
272. L. Nozal, L. Arce, B.M. Simonet, A. Rios and M. Valcárcel, Rapid determination of trace levels of tetracyclines in surface water using a continuous flow manifold coupled to a capillary electrophoresis system, *Anal. Chim. Acta*, 517, 89–94, 2004.
273. A. Macià, F. Borrull, C. Aguilar and M. Calull, Improving sensitivity by large-volume sample stacking using the electroosmotic flow pump to analyze some nonsteroidal anti-inflammatory drugs by capillary electrophoresis in water samples, *Electrophoresis*, 24, 2779–2787, 2003.
274. M. Himmelsbach, W. Buchberger and C.W. Klampfl, Determination of antidepressants in surface and waste water samples by capillary electrophoresis with electrospray ionization mass spectrometric detection after preconcentration using off-line solid-phase extraction, *Electrophoresis*, 27, 1220–1226, 2006.
275. A. Macià, F. Borrull, M. Calull and C. Aguilar, Determination of some acidic drugs in surface and sewage treatment plant waters by capillary electrophoresis-electrospray ionization-mass spectrometry, *Electrophoresis*, 25, 3441–3449, 2004.
276. J.J. Soto-Chincilla, A.M. García-Campaña, L. Gámiz-Gracia and C. Cruces-Blanco, Application of capillary zone electrophoresis with large-volume sample stacking to the sensitive determination of sulfonamides in meat and ground water, *Electrophoresis*, 27, 4360–4368, 2006.
277. J.-B. Kim, K. Otsuka and S. Terabe, On-line sample concentration in micellar electrokinetic chromatography with cationic micelles in a coated capillary, *J. Chromatogr. A*, 912, 343–352, 2001.
278. H. Harino, S. Tsunoi, T. Sato and M. Tanaka, Applicability of micellar electrokinetic chromatography to the analysis of estrogens in water, *Fresenius J. Anal. Chem.*, 369, 546–547, 2001.
279. P. Schmitt-Kopplin and J. Junkers, Capillary zone electrophoresis of natural organic matter, *J. Chromatogr. A*, 998, 1–20, 2003.
280. P. Janoš, Separation methods in the chemistry of humic substances, *J. Chromatogr. A*, 983, 1–18, 2003.
281. D. Gajdošová, K. Novotná, P. Prošek and J. Havel, Separation and characterization of humic acids from Antarctica by capillary electrophoresis and matrix-assisted laser desorption ionization time-of-flight mass spectrometry. Inclusion complexes of humic acids with cyclodextrins, *J. Chromatogr. A*, 1014, 117–127, 2003.
282. M.L. Pacheco, E.M. Pena-Méndez and J. Havel, Supramolecular interaction of humic acids with organic and inorganic xenobiotics studied by capillary electrophoresis, *Chemosphere*, 51, 95–108, 2003.

283. H. Prosen, S. Fingler, L. Zupančič-Kralj and V. Drevenkar, Partitioning of selected environmental pollutants into organic matter as determined by solid-phase microextraction, *Chemosphere*, 66, 1580–1589, 2007.
284. L.N. Sangolkar, S.S. Maske and T. Chakrabarti, Methods for determining microcystins (peptide hepatotoxins) and microcystin-producing cyanobacteria, *Water Res.*, 40, 3485–3496, 2006.
285. G. Vasas, D. Szydlowska, A. Gáspár, M. Welker, M. Trojanowicz and G. Borbély, Determination of microcystins in environmental samples using capillary electrophoresis, *J. Biochem. Biophys. Methods*, 66, 87–97, 2006.
286. G. Vasas, A. Gáspár, C. Páger, G. Surányi, C. Máthé, M.M. Hamvas and G. Borbely, Analysis of cyanobacterial toxins (anatoxin-a, cylindrospermopsin, microcystin-LR) by capillary electrophoresis, *Electrophoresis*, 25, 108–115, 2004.
287. A. Gago-Martínez, J.M. Leão, N. Piñeiro, E. Carballal, E. Vaquero, M. Nogueiras and J.A. Rodríguez-Váquez, An application of capillary electrophoresis for the analysis of algal toxins from the aquatic environment, *Intern. J. Environ. Anal. Chem.*, 83, 443–456, 2003.
288. P. Sázelová, V. Kašiška, S. Koval, A. Prusík and G. Peltre, Evaluation of the efficiency of extraction of ultraviolet-absorbing pollen allergens and organic pollutants from airborne dust samples by capillary electrophoresis, *J. Chromatogr. B*, 770, 303–311, 2002.
289. D. Horta, P.D. Mudgett, L. Ding, M. Drybread, J.R. Schultz and R.L. Sauer, Analysis of water from the Space Shuttle and Mir Space Station by ion chromatography and capillary electrophoresis, *J. Chromatogr. A*, 804, 295–304, 1998.

Part IIIA

Microchip-Based: Core Methods and Technologies

32 Cell Manipulation at the Micron Scale

Thomas M. Keenan and David J. Beebe

CONTENTS

32.1 INTRODUCTION

The advent of microfabrication technology and its more recent application to the field of biology has greatly enhanced our ability to engineer *in vitro* cell culture environments. Cells and their sensing elements are several hundred nanometers to tens of micrometers (microns) in size, and thus interact within a microscale environment, or "microenvironment." Unlike conventional cell culture methods, microfabrication-based methods allow manipulation of the physical, chemical, and electrical properties of the cellular microenvironment at the microscale. As a result of this enhanced functionality, microengineered culture environments are becoming more widely used in the biological research community for a variety of applications, including investigating fundamental questions in biology, enriching specific cell types from mixed populations, and examining the response of cells to novel chemical compounds. Microengineered culture environments impose unique requirements on analysis tools for detecting and characterizing changes in cell behavior or physiology in response to specific environmental perturbations. Optimizing existing analysis tools for use with microengineered culture environments would greatly enhance the utility of microfabrication-based methods for biological studies. In this chapter, we provide a broad overview of the most common and recent microfabrication-based methods for controlling the position of cells within a culture environment, and the properties of the cellular microenvironment.

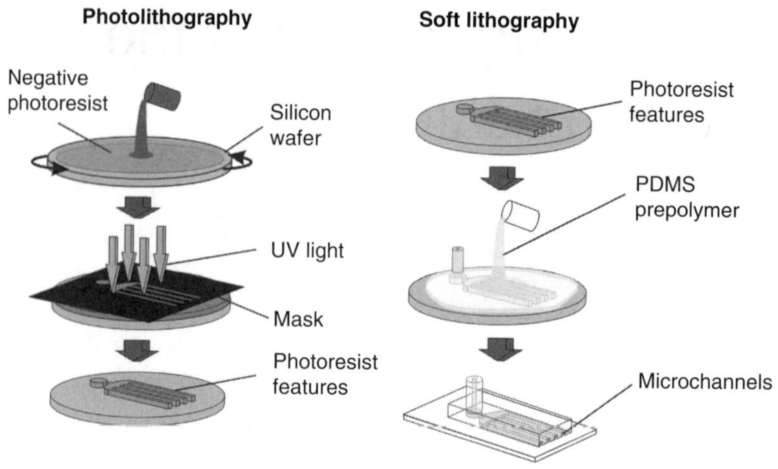

FIGURE 32.1 Photolithography and soft lithography. Photolithography selective polymerizes a photoactive polymer coating using UV light filtered by a patterned mask to create microscale features of defined dimensions. Soft lithography utilizes the photolithographically defined substrates to mold polymer replicas of the microscale features. (Reprinted and adapted from Li, N., et al., *Crit Rev Biomed Eng*, 31, 423. Copyright 2003. With permission from Begell House, Inc.)

32.2 MICROFABRICATION TOOLS

Many microfabrication tools have been developed over the past 50 years to create more intricate and advanced microelectronics. The ability of these tools to precisely control the size and shape of polymers, metals, and other materials at scales as small as 1 μm, make them extremely adept at creating unique cell culture microenvironments. Two processes called "photolithography" and "soft lithography" are the most commonly used for biological studies.[1,2] Photolithography begins by coating a substrate with a thin layer of photosensitive polymer called a "photoresist" (Figure 32.1). The thickness of the photoresist layer is controlled by spinning the substrate at a defined speed and length of time. Thinner layers are produced by spinning the wafer faster and longer. The coated substrate is then placed in contact with a thin sheet of glass or plastic that has user-defined opaque and transparent regions, called a "mask." Ultraviolet (UV) light is then shone through the mask onto the coated substrate causing exposed regions to degrade (positive photoresist) or polymerize (negative photoresist). The exposed substrate is then placed in a chemical solution called "developer," which removes the degraded or unpolymerized photoresist leaving user-defined features patterned on the substrate. Soft lithography uses these patterned substrates, called "masters," to mold elastomeric polymer replicas. The most commonly used elastomer is poly(dimethyl siloxane) (PDMS), due to its biocompatibility, transparency, and high-fidelity replica molding properties.

32.3 CONTROLLING CELL POSITION IN 3D CULTURES

The ability to control where a cell is positioned relative to its physical environment, other cells, or regions with particular chemical identities is useful for studying cell exploration and migration in the presence of various physical and chemical cues, investigating specific cell–cell or cell–substrate interactions, characterizing autocrine or paracrine signaling, or for automating cell analysis. There is an extensive body of literature describing methods to control the position of cells on two-dimensional (2D) substrates.[1,3,4] For some applications 2D culture systems are ideal; however, for many others they are not. Although invaluable in developing our current understanding of many biological phenomena, 2D culture systems are artificial environments not representative of *in vivo* culture

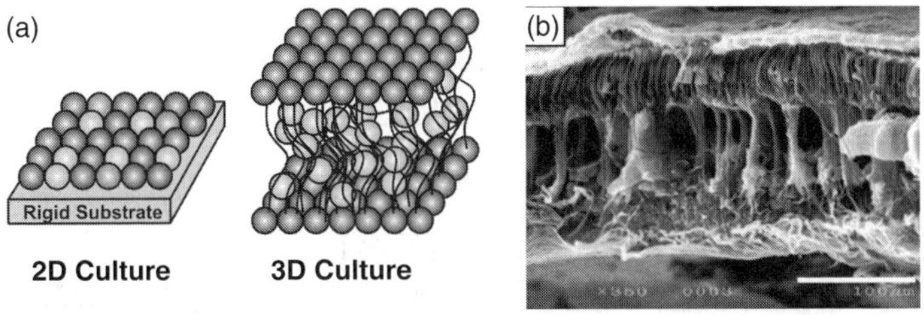

FIGURE 32.2 Comparison of 2D vs. 3D culture architectures. (a) 2D cultures consist of patterned or random cells (shaded spheres) on a rigid or semirigid substrate. 2D cultures can offer cells a richer environment with cell organization in three dimensions and incorporation of extracellular matrix components (black lines) with entrained growth factors and signaling molecules. (b) Scanning electron micrograph (SEM) of the deep-sea gulper shark retina. (Photo courtesy of Jill Olin, Hofstra University. With permission.)

environments except in a few distinct cases (e.g., endothelial or epithelial cells). Nearly all cells found *in vivo* are completely surrounded by an intricate and highly organized arrangement of extracellular matrix proteins, other cells, and a rich milieu of soluble biomolecules distributed throughout the extracellular fluid. Cell–cell contact and the juxtaposition of different cells or cell types relative to one another are essential components of normal biological function. Cells communicate in a variety of ways including integrin and gap junction signaling, or by secreting signaling proteins into the extracellular environment that act on the cell itself (i.e., autocrine signaling) or on neighboring cells (i.e., paracrine signaling). Two-dimensional cell cultures cannot fully replicate these intricate and complex environments (Figure 32.2) and as a result may be limited in the insight and information they can provide about normal and abnormal cell physiology. Three-dimensional (3D) culture architectures better simulate complex *in vivo* environments and may be able to provide more comprehensive and relevant information about cell responses to specific environmental factors. Because of the thorough description of 2D patterning methods in existing literature and the advantages of 3D culture systems, we will focus our discussion on efforts to control cell position in 3D culture architectures.

32.3.1 Physical Entrapment

The simplest method of patterning cells in a 3D culture system is to create voids on the surface of a biocompatible gel (Figure 32.3). Voids can be created by degrading select regions of the gel,[5] embossing the gel with patterns defined by a microfabricated stamp,[6] or by molding the gel on a microfabricated surface.[7,8] Cell patterning is achieved by seeding cells at random and rinsing away those that have not fallen into the voids created in the gel. The voids can subsequently be refilled with new gel to provide full encapsulation. Physical entrapment methods can be used with virtually any soft or moldable polymer; however, its reliance on conventional photolithography allows cell patterning in only two dimensions.

32.3.2 Photopatterning

Cells can also be patterned in 3D culture constructs using a process called "photopatterning." Photopatterning (Figure 32.4) utilizes photoactive hydrogels that become cross-linked or polymerized when exposed to light.[9] Cells are mixed with prepolymerized gel solution and exposed to UV light through a mask, similar to photolithography. Regions exposed to UV light polymerize and trap the cells contained within. The unpolymerized solution can be rinsed away, replaced with acellular gel solution, and flood exposed with UV light to provide a blank background. Alternatively, solutions containing other cell types can be photopatterned relative to the first pattern to create a wide variety of coculture architectures. Photopatterning is a very effective means of patterning cells in 3D cultures,

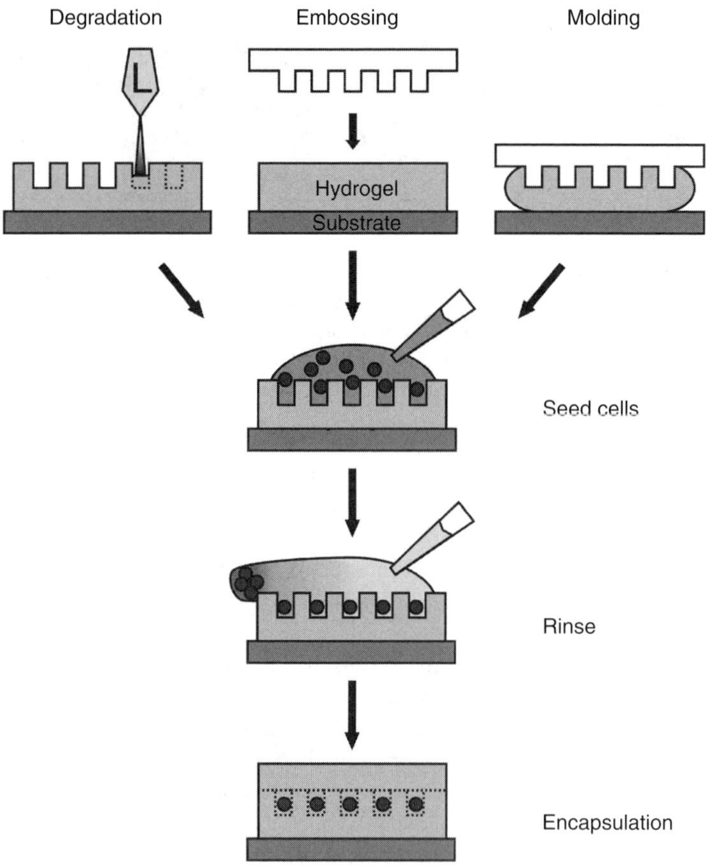

FIGURE 32.3 Physical entrapment cell patterning. Cells can be patterned in 3D hydrogel cultures by creating voids using laser (L) degradation, embossing, and molding. Cells are seeded in the voids and those that do not can be rinsed away. Addition of more hydrogel allows full 3D encapsulation.

even though cell position can only be controlled in two dimensions. However, its use of UV light to initiate polymerization may have adverse effects on cell viability and function. UV light can damage deoxyribonucleic acid (DNA) and creates polymer or cross linker radicals that may directly damage cells.

32.3.3 Dielectrophoresis

A phenomenon called *dielectrophoresis* has been used to pattern a variety of cell types on 2D substrates[10,11] and more recently in 3D culture constructs.[9] Unlike electrophoresis where charged species move in an applied electric field due to Coulombic forces ($F = qE$), dielectrophoresis capitalizes on the ability of a cell to become *polarized* when placed in an electric field. Dielectrophoresis is most often used in conjunction with alternating current (AC) electric fields since AC fields eliminate electrophoretic movement, and have less physiological impact on cells than direct current (DC) fields.[11] When a cell is placed in an AC field, the magnitude and polarity of the induced dipole depend on the frequency of the applied field and the conductivities of the cell and the surrounding medium, described by the equation

$$\underline{p}(r) = 4\pi\varepsilon_{\mathrm{m}}R^3\left(\frac{\underline{\varepsilon}_{\mathrm{c}} - \underline{\varepsilon}_{\mathrm{m}}}{\underline{\varepsilon}_{\mathrm{c}} + 2\underline{\varepsilon}_{\mathrm{m}}}\right), \tag{32.1}$$

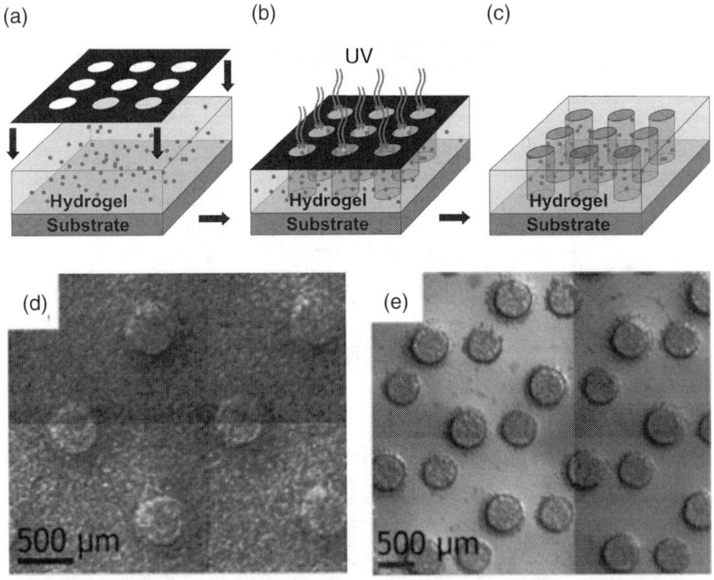

FIGURE 32.4 Photopatterning. (a) Cells mixed with prepolymerized hydrogel solutions can be selectively illuminated with UV light through the use of an applied photomask (shown in black). (b) Regions exposed to UV become polymerized and entrap the cells. (c) The unpolymerized hydrogel can be rinsed away, replaced with acellular prepolymerized hydrogel solution, and flood exposed with UV to provide a blank background. (d) Photopatterning has been used to pattern 3T3 fibroblasts in PEG hydrogels against a blank background. (e) Aligned with other photopatterned cell types for coculture experiments. (Adapted from Albrecht, D. R., et al., *Lab Chip*, 5, 111, 2005. With permission from The Royal Society of Chemistry.)

where $\underline{p}(r)$ is the induced dipole, R is the radius of the cell, $\underline{\varepsilon}$ is the complex permittivity ($\underline{\varepsilon} = \varepsilon + \sigma/(j\omega)$) of the cell (c) or medium (m), σ is the conductivity of the cell or medium, $j = \sqrt{-1}$, ω is the angular frequency of the applied electric field, and $E(r)$ is the applied electric field.

If cells are placed in a nonuniform AC field, the spatial field gradient imparts different amounts of force on each half of the cell dipole (Figure 32.5a). When the complex permittivity of the cell is greater than that of the surrounding medium, the cell forms a positive dipole that causes it to move toward the electrode with higher field strength in what is known as "positive dielectrophoresis" (pDEP). If cell permittivity is less than that of the surrounding medium, a negative dipole is formed resulting in repulsion away from the electrode of higher field strength, or "negative dielectrophoresis" (nDEP). By engineering the shape and position of the electrodes, spatial field gradients can be created that localize cells to precise, user-defined locations (Figure 32.5b). The magnitude and direction of cell movement can be tuned by adjusting the conductivity of the surrounding medium or the frequency of the applied AC field.

Since dielectrophoresis relies on the geometry of the electrodes to create the nonuniform electric fields, the method has been limited to the creation of 2D cell patterns within 3D culture constructs (Figure 32.5c).[9] True 3D cell patterning would require more complex electrode geometries than those employed to date. Although dielectrophoresis has been used to levitate cells[12–14] and could be used for 3D patterning, the method has not been utilized solely or in combination with any other methods to pattern cells in all three dimensions within a 3D culture construct.

Because many biological hydrogels (e.g., collagen, fibrin, Matrigel™) have conductivities similar enough to cells to make dielectrophoretic cell patterning difficult, synthetic hydrogels such as poly(ethylene glycol) (PEG) are most commonly used.[9] PEG hydrogels are biocompatible, but are devoid of the numerous ligands and other supportive signals that cells normally receive from the extracellular environment. Although there are numerous chemistries that can be incorporated into PEG

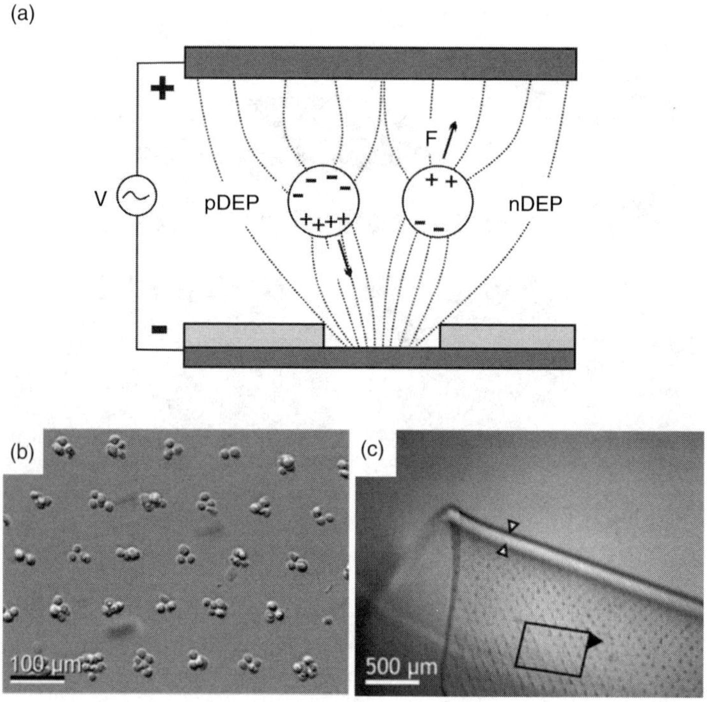

FIGURE 32.5 Dielectrophoresis. (a) Nonuniform magnetic fields place different amounts of force on cells with induced dipoles, causing them to move toward (pDEP) or away from (nDEP) high field densities. (b) Dielectrophoresis has been used to pattern 3T3 fibroblasts by controlling the shape and position of the electrodes. (c) Dielectrophoresis has also been used to pattern 3T3 fibroblasts in 3D PEG hydrogels. (Adapted from Albrecht, D. R., et al., *Lab Chip*, 5, 111, 2005. With permission from The Royal Society of Chemistry.)

hydrogels,[15] the inability to fully recapitulate the natural environment may confound experimental observations. Other confounding variables may be contributed by localized heating from the applied electric field, which could damage cells or activate intracellular signaling cascades. The electric field can also perturb the electrical state of the cell membrane and the function of membrane-embedded ion channels.

32.3.4 Inkjet Printing

Inkjet printing of biological solutions has been used to pattern a variety of biological molecules on 2D substrates[16] and could prove to be an efficient method for patterning cells in engineered 3D culture architectures. Biological inkjet printers are adapted conventional inkjet printers[17,18] that deposit one or more biological solutions in user-defined patterns under the control of a computer. The computer controls the location of fluid ejection by rastering the printer head over the substrate or by moving the substrate relative to a stationary printer head. The solutions are ejected from the microfluidic channels and nozzles that constitute the inkjet printer heads using either heat-based or piezoelectric fluid displacement. Heat-based fluid displacement, also known as "bubble jet" printing, utilizes a heating element to vaporize a portion of the ink and create a bubble that ejects fluid from the nozzle and onto the substrate (Figure 32.6a). When the bubble collapses it creates a vacuum that pulls more ink into the printer head. Piezoelectric fluid displacement uses a piezoelectric transducer to physically displace the fluid within the microchannel when a voltage is applied, and eject the fluid out of the nozzle and onto the substrate (Figure 32.6b). Addition of a z-stepper motor allows 3D

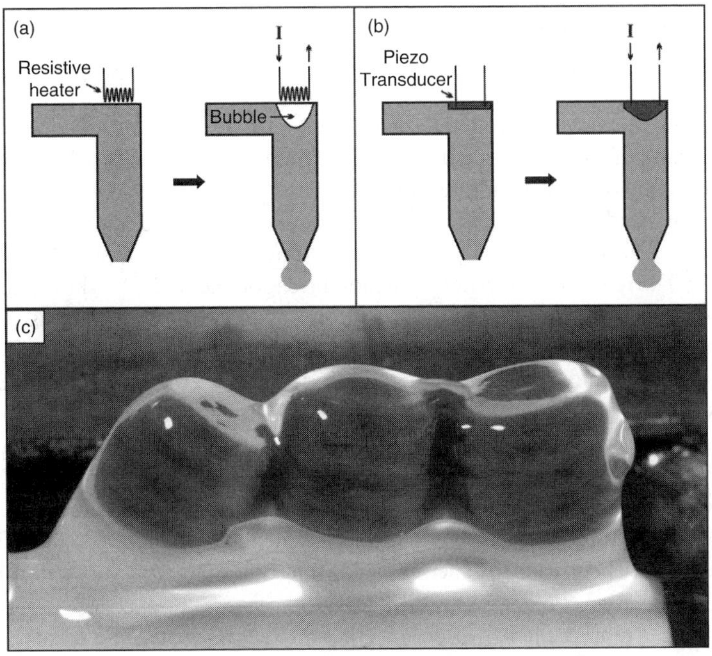

FIGURE 32.6 Inkjet printing. (a) Bubble jet printer heads use resistive heating elements to vaporize the ink and create a gas bubble that displaces fluid out of the nozzle. (b) Piezoelectric printer heads use a piezoelectric transducer to mechanically displace fluid out of the nozzle. (c) Bovine aortic endothelial cells were inkjet printed on an alginate-coated scaffold to create tubes 4 mm in diameter. (Reprinted from Varghese, D., et al., *J Thorac Cardiovasc Surg*, 129, 2, 470. Copyright 2005. With permission from American Association for Thoracic Surgery.)

cell culture constructs to be created using a layer-by-layer process. This method has been shown to effectively create 3D cultures of neuronal cell lines in fibrin gels[19] and a variety of cells in alginate gels[18] (Figure 32.6c). Because of the use of microfluidic channels in inkjet printer heads, care must be taken to choose microchannel dimensions that do not shear damage cells or other biological molecules within the printing solution. Minimization of shear effects often comes at the expense of drop size resolution, which limits how close regions of different composition can be placed relative to one another. In addition, the heating of biological ink when using bubble jet biological inkjet printers may damage cells or cause denaturation of proteins.

32.4 ENGINEERING THE CELLULAR MICROENVIRONMENT

The ability to control the physical, chemical, and electrical properties of the *in vitro* cell culture environment will perhaps be the single greatest contribution of microfabrication methods to the field of biology. Engineering the cellular microenvironment can be used to create more physiologically relevant culture conditions, study biological phenomena (e.g., growth, development, response to damage, death), or to direct the behavior of cells toward specific industrial outcomes (e.g., tissue engineering, chemical manufacturing). There are three ways in which the cellular microenvironment can be altered—mechanically, chemically, or electromagnetically. Although most methods intend to alter the cellular microenvironment in only one of these ways, it is often difficult to avoid secondary perturbations to the cellular environment via one or both of the other modalities. Here, we review the major methods for altering the microenvironment using microfabrication technology.

32.4.1 Mechanical Microenvironment

32.4.1.1 Static Manipulations

Mechanical manipulations of the cell microenvironment can be either static or dynamic. Static manipulations consist of modifications to the physical environment with which the cell interacts. Altering the topography of the cell culture substrate is useful for understanding how cell growth and proliferation is influenced by topographical or morphological cues, and how cells explore or migrate in specific physical environments. Many studies have shown that topographical cues alone can alter cell behavior.[20–22] The recent development of nanofabrication methods[23–27] has enabled the creation of substrate topographies similar in size to those cells encounter *in vivo*, such as the protein fibers that comprise the extracellular matrix (Figure 32.7).

A variety of different microfabrication methods have been used to create complex topographies on cell culture substrates. Many of these studies[20] require expensive fabrication tools and complex methods and will not be discussed further here. The most common method of adding topography to cell culture substrates is by using photolithography to either add features to the substrate or mask the substrate for selective chemical or plasma etching. Alternatively, cell culture substrates can be molded[28–30] or embossed[31] directly from photolithographically defined substrates. Photolithography and soft lithography have been used to generate a wide range of cell culture substrate topographies[28,32] including replicas of cells[33] (Figure 32.8).

More complex 3D topographies can be generated using 3D patterning techniques.[34] The simplest method of generating 3D topographies is employing multiple iterations of photolithography (Figure 32.9a). Unfortunately, multilayer photolithography requires precise alignment of each subsequent layer and may result in poor feature resolution and fidelity. Generating multiple level features in a single photolithography step can be accomplished using gray-scale photolithography (Figure 32.9b). Gray-scale photolithography uses positive photoresists and masks printed with different gray levels[35] (i.e., transparency) or masks made of microfluidic channels filled with dye.[36] Although gray-scale photolithography is effective at making features of different heights, it cannot make more complex features such as overhangs, closed loops, or hollow objects. These types of complex 3D structures require methods such as multiphoton polymerization.[37–39] Multiphoton polymerization uses long wavelength lasers, where each photon has half the energy necessary to activate the photoinitiator in a photoactive polymer. When two beams are timed so that two long wavelength photons arrive at a single photoinitiator molecule at the same time, the photoinitiator absorbs both photons and is excited to its active state causing localized polymerization. By moving the focal point of the two incident laser beams, complex 3D structures can be constructed (Figure 32.9c). These features can subsequently be combined with soft lithography to form more intricate 3D structures than those that can be produced with standard photolithography (Figure 32.9d–e).[39]

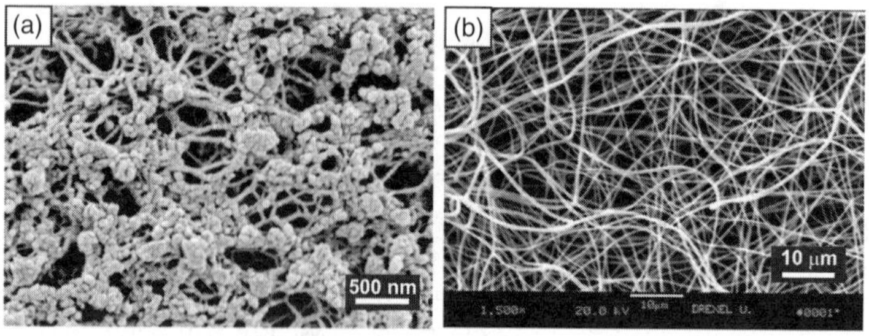

FIGURE 32.7 Natural versus synthetic substrates. (a) SEM of the basement membrane of the rhesus macaque urothelium. (From Figure 2 in Abrams, G. A., et al., *Urol Res*, 31, 341, 2003. Copyright 2003. With permission from Springer Science and Business Media.) (b) Electrospun poly (lactide-co-glycolide) (PLGA) fibers (scale bar = 10 μm). (Adapted from Li et al., *J Biomed Mat Res*, 60, 4, 613, 2002. With permission.)

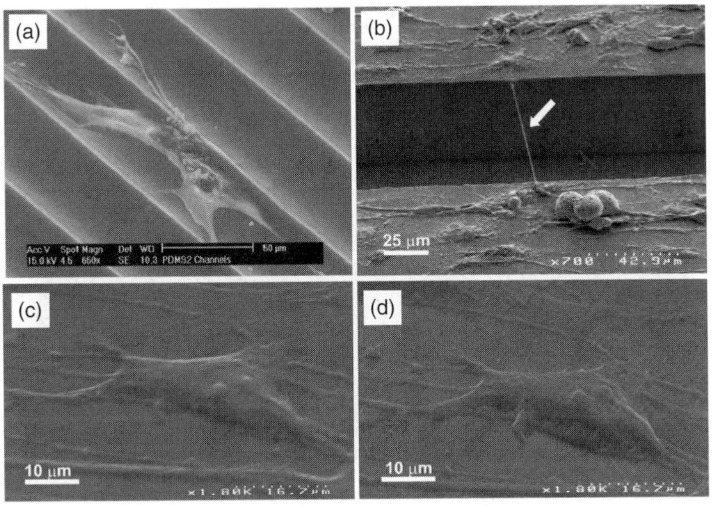

FIGURE 32.8 Photo-/soft lithography generated topographies. (a) Bone marrow derived connective tissue progenitor cells growing on PDMS channels molded from photolithographically defined masters. (From Figure 5b in Mata, A., et al., *Biomed Microdev*, 4, 4, 267, 2002. Copyright 2002. With permission from Springer Science and Business Media.) (b) Neurons growing on PDMS ridges with neurite bridge (arrow) spanning the gap. (Reprinted from Goldner, J. S., et al., *Biomaterials*, 27, 460. Copyright 2006. With permission from Elsevier.) (c) Fixed rat Schwann cells were used to create (d) PDMS replicas that were subsequently found to guide neurite extension when seeded with dorsal root ganglion neurons (not shown). (Reprinted from Bruder, J. M., et al., *Langmuir*, 22, 20, 8266. Copyright 2006. With permission from American Chemical Society.)

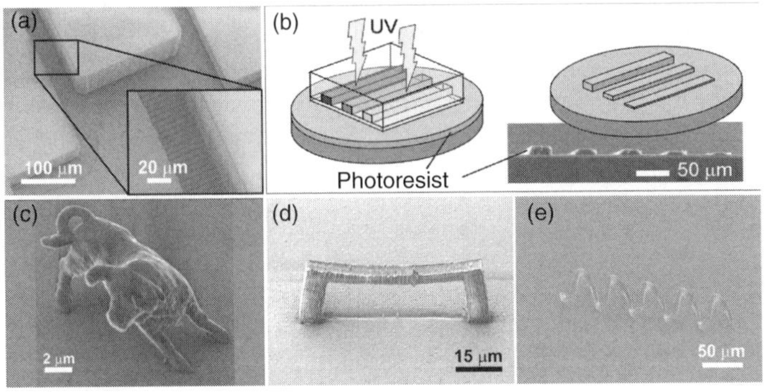

FIGURE 32.9 3D topographies. (a) A three-level master made with multiple iterations of photolithography. Three mesas, with the center mesa taller than the other two, are connected by ridges 1 × 1 μm in cross section (inset). (b) Multiple level features fabricated with gray-scale photolithography and microfluidic masks. (Adapted from Chen, C., et al. *Proc Natl Acad Sci USA*, 100, 1499, 2003. With permission.) (c) A 10-μm long, 7-μm tall bull made with two-photon photopolymerization in urethane acrylate polymer. (Reprinted from Kawata, S., et al., *Nature*, 412, 6848, 697, Copyright 2001. With permission from Macmillan Publishers Ltd.) (d–e) 3D PDMS structures molded from masters fabricated with multiphoton photopolymerization. (Adapted from LaFratta, C. N., et al., *Proc Natl Acad Sci USA*, 103, 23, 8589, 2006. With permission.)

32.4.1.2 Dynamic Manipulations

Dynamic manipulations of the mechanical microenvironment consist of active elements that impose mechanical force on cells. Many cell types experience mechanical forces *in vivo*. Chondrocytes and osteoblasts within skeletal and load bearing connective tissues endure large compressive forces

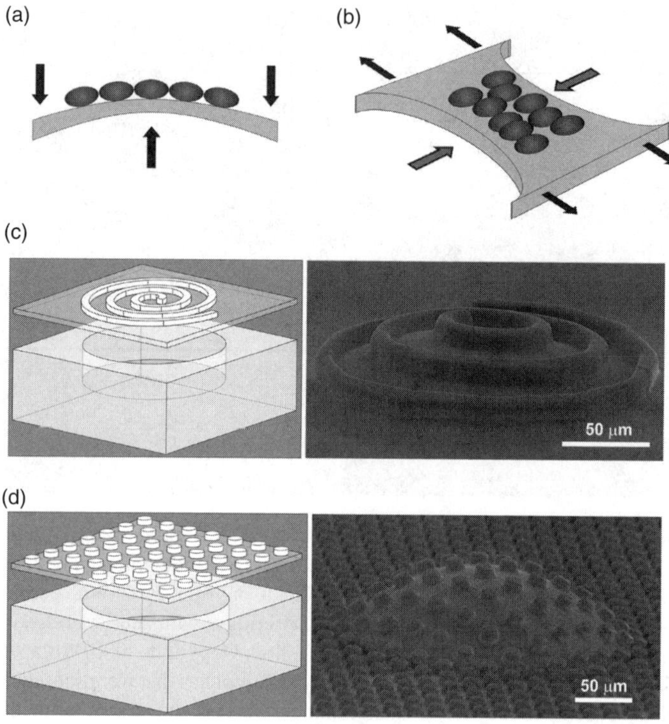

FIGURE 32.10 Deformable substrates. (a) Force (black arrows) can be used to bend the cell culture substrate and mechanically strain attached cells. (b) The cell culture substrate can also be stretched, but due to the Poisson effect this results in accessory stresses and strains perpendicular to the direction of applied force (gray arrows), preventing mechanical loads from being truly uniaxial. (c–d) Microfabrication technology enhances the ways in which cells can be mechanically strained. Cells could be cultured on thin PDMS membranes that are patterned with a variety of unique topographies, and then mechanically strained when the membrane is deflected by pressurizing underlying microfluidic channels. (Adapted from Hoffman, J. M., et al., *Adv Mater*, 16, 23, 2201, 2004. With permission.)

during locomotion. Osteoblasts and endothelial cells experience large shear forces generated by fluid flow within the osteon and vasculature, respectively. The ability to culture cells in environments where different forms of mechanical loadings can be imposed provides invaluable insight into cell physiology.

A variety of methods have been developed to apply mechanical forces to cells.[40] One very common method is to culture cells on a deformable substrate that can be bent (Figure 32.10a) or stretched (Figure 32.10b). Transducers provide controlled loading with user-defined strains and strain rates, or cyclic loading with user-controlled frequency and amplitude. A recent advance in microfabrication technology now allows cells to be grown on microtopographies while being dynamically loaded. The method developed by Folch and coworkers[41] employs thin PDMS membranes less than 20 μm thick that are patterned with complex substrate topographies (Figure 32.10c–d). The membranes are bonded to microfluidic channels connected to pressure or vacuum sources, allowing the membranes to be deflected dynamically. The resulting concave or convex surfaces mechanically strain cells interacting with the unique substrate topography.

Fluidic shear forces are commonly generated *in vitro* using cone and plate viscometers.[42] A cone and plate viscometer (Figure 32.11a) consists of a cone that rotates in close proximity to a flat substrate. Constant shear stress is created in the region between the cone and the substrate by engineering the taper of the cone to offset the influence of the increasing tangential velocity that

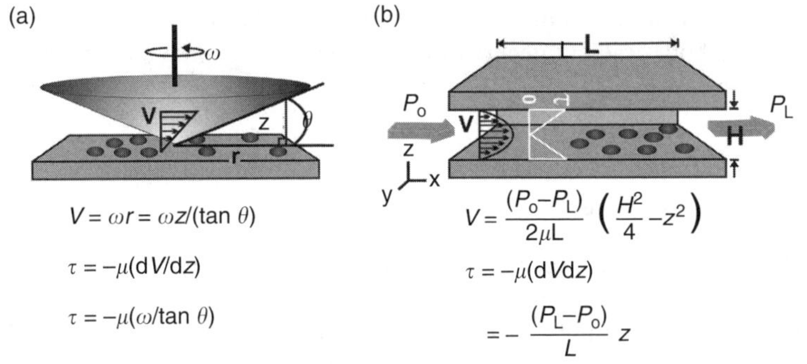

FIGURE 32.11 Generating shear forces. (a) Cone and plate viscometers create constant shear stress throughout the volume between the cone and substrate. The shear stress (τ) is independent of the radius (r) or distance from the cell culture surface (z), and dependent on the angular velocity (ω) of the cone and the angle of the taper (θ). (b) Microfluidic channels have at least one dimension less than 1 mm creating a laminar flow regime restriction that creates steady velocity profiles and shear stresses along the direction of fluid flow. The well-characterized flow regime allows mathematical calculation of both the velocity profile (V) and the shear stress distribution across the channel (white graph perpendicular to fluid flow direction) using only the microchannel dimensions (L and H) and the applied pressure gradient (P_O and P_L). The negative sign convention indicates that shear forces push down on the cells.

accompanies increasing radius on the shear rate. Cells cultured on the flat substrate are exposed to constant shear forces and can be examined for behavioral or physiological effects.

Microfabrication technology provides a complementary method to expose cells to specific shear stress environments. Straight microfluidic channels (Figure 32.11b) with highly precise dimensions and controlled fluid flow rates have been used to study the effects of shear stress on cells.[43,44] Microfluidic channels have one or more dimensions less than 1 mm, which forces flowing liquids to be confined to laminar flow regimes.[45] Laminar fluid flow is characterized by a parabolic velocity profile that does not change downstream of a short flow stabilization region near the entrance of the microchannel. The steady-state parabolic velocity profile produces constant shear stresses in the direction of fluid flow. The velocity profile can be calculated from the fluid channel dimensions and the applied pressure gradient, and be used to calculate the shear stress throughout the microchannel. Because the velocity profiles near the junction of two microchannel walls can be more complex to predict, cells are usually cultured near the middle of microchannels that are much wider than they are tall.

32.4.2 Chemical Microenvironment

Chemical manipulation of the cellular microenvironment encompasses chemical modification of the physical components with which the cell comes into contact (i.e., substrate, extracellular matrix, etc.), and the distribution of soluble chemical species in the culture medium in which the cells are grown. Numerous methods have been developed to modify the chemistry of cell culture substrates.[1,46] Here we focus our discussion on different microfluidic methods for controlling the distribution of soluble chemical species in the cellular microenvironment. For this application, microfluidic devices offer unique advantages due to their restriction to laminar fluid flow.[45] In addition to having a constant, parabolic velocity profile, laminar fluid flow is also characterized by more linear streamlines that follow the contours of the microchannel. The fluid does not flow in circuitous, disorganized paths as it does in turbulent flow. The lack of eddies and convective mixing results in a system where chemical species can only mix via diffusion. This limitation allows creation of predictable chemical gradients, which is not possible in conventional cell culture systems. The lack of convective mixing also allows tight control over the location of different flowing streams within the microchannel, which has proven useful for exposing specific regions of cells to defined chemical environments. Here, we will discuss both gradient generators and subcellular fluidic compartmentalization using microfluidic channels.

32.4.2.1 Gradient Generators

Gradients of diffusible biological molecules play critical roles in many biological phenomena including development, cancer, inflammation, and wound healing. A wide variety of methods have been developed to expose cells to soluble chemical gradients.[47–52] Unfortunately, all of these methods offer little temporal or spatial control over the gradient, and in many cases provide no way to quantify the gradient that each cell perceives.

A device we will refer to as the "Premixer Device" (Figure 32.12a)[53] has been used to expose a variety of cells to biochemical gradients.[54–56] The device generates steady-state, user-controlled gradients using constant fluid flow. The device splits and recombines inlet fluids in an upstream microfluidic mixer to generate a variety of different gradient shapes (Figure 32.12b). The use of transparent PDMS to fabricate the device and the closed channel geometry allows direct quantification of the gradient to which cells are exposed using a conventional wide field fluorescence microscope. The only disadvantage of the device is that it exposes cells to fluid flow and shear forces that can destroy autocrine or paracrine signaling, activate intracellular signaling cascades, and alter cell migration.[57]

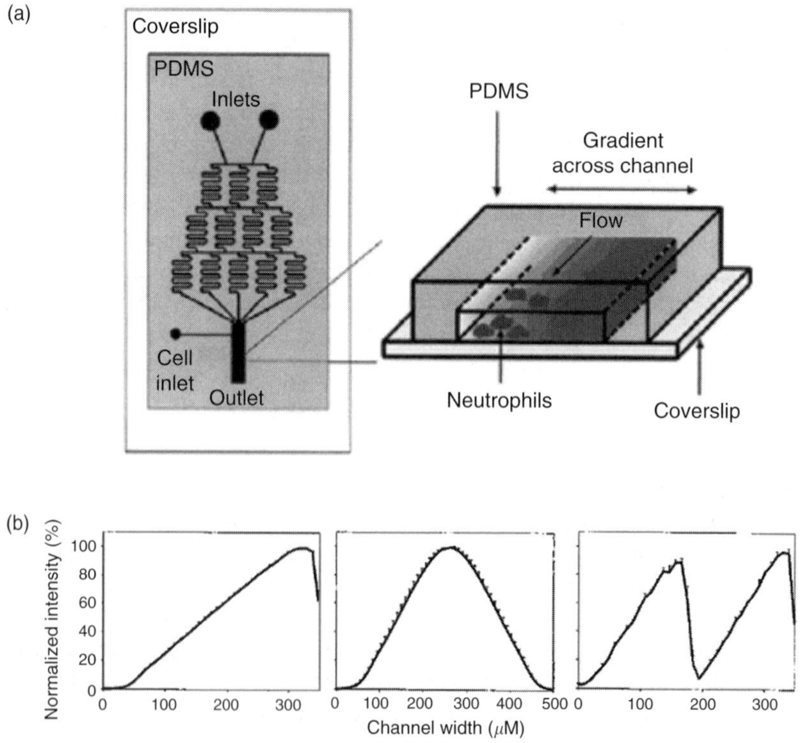

FIGURE 32.12 Premixer device. (a) An upstream microfluidic mixer splits and recombines fluids to expose cells to user-defined gradients under constant fluid flow. (b) The premixer can be designed to generate a wide variety of gradient shapes. (Adapted from Jeon, N. L., et al., *Nat Biotechnol*, 20, 8, 826. Copyright 2002. With permission from Macmillan Publishers Ltd.)

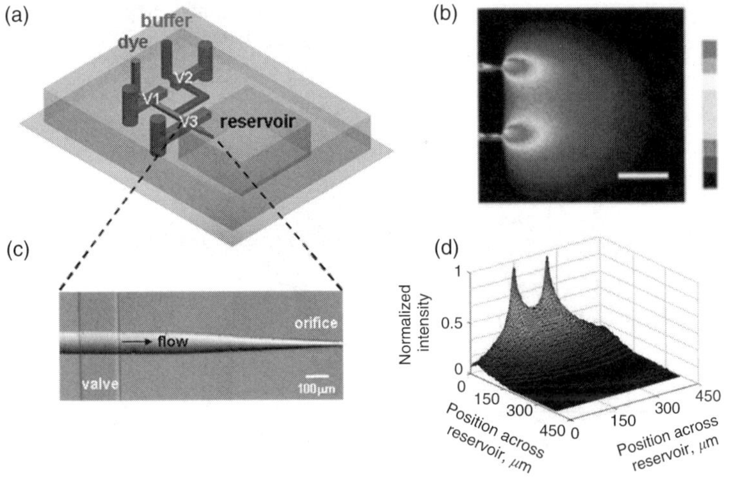

FIGURE 32.13 Microfluidic multi-injector. (a) 3D schematic representation of the device shows two small-diameter channels under the control of microfluidic valves that pneumatically eject fluid into a cell culture reservoir to form soluble molecule gradients. (b) Top view of the device in operation forming gradients of fluorescein isothiocyanate (FITC)-conjugated dextran. (c) 3D plot of the fluorescence intensity within the cell culture reservoir. (Adapted from Chung, B. G., et al., *Lab Chip*, 6, 6, 764, 2006. Reproduced with permission from The Royal Society of Chemistry.)

Another recently developed microfluidic device replicates a method developed by Gunderson and Barrett[49] in a microfluidic platform. In the Microfluidic Multi-injector[58] (Figure 32.13a), fluids are ejected pneumatically out of a pair of small orifices into a larger cell culture reservoir, under the control of independent microfluidic valves. Once the fluid is ejected into the reservoir, it forms a diffusive gradient (Figure 32.13b–c). Although the device generates gradients without appreciable fluid flow in the cell culture reservoir, as evidenced by a symmetric gradient, it cannot generate steady-state gradients or the complex gradients of the Premixer Device.

A microfluidic generator called the "Microjets Device" (Figure 32.14a) generates stable gradients on open surfaces without exposing cells to appreciable fluid flow.[59] The open architecture allows efficient nutrient and gas exchange, and the lack of flow minimizes shear forces and facilitates autocrine and paracrine signaling. Like the previously described device, the Microjets Device generates gradients by pneumatically ejecting fluids out of small orifices, which minimize convective flow and allow a diffusive gradient to form in the cell culture area. However, the Microjets Device uses two opposed arrays of small orifices delivering different concentrations of biochemical factor to generate steady-state chemical gradients. In addition, dynamic changes to the slope and position of the gradient can be accomplished by adjusting the air pressures driving the fluid out of each orifice array (Figure 32.14b–c). The Microjets Device is not capable of generating the complex gradients possible with the Premixer Device nor can the gradient be visualized with a conventional wide field fluorescence microscope. The architecture of the device requires confocal microscopy to evaluate the gradient to which the cells are exposed.

32.4.2.2 Subcellular Chemical Compartmentalization

The laminar flow restriction and small dimensions of microfluidic devices have been exploited to expose different portions of a single cell to different chemical environments. A method developed by Whitesides and coworkers[60,61] utilized hydrodynamic focusing[62] on a microfluidic platform to selectively label portions of a cell with different dyes (Figure 32.15). Laminar fluid flow confined the dye solution to particular regions of the channel despite being in contact with streams containing

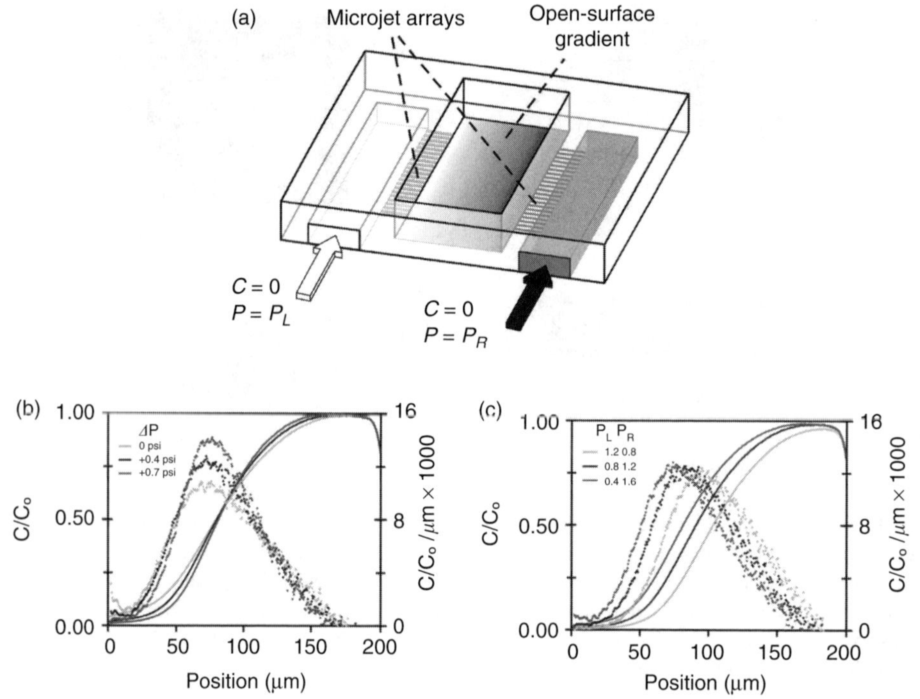

FIGURE 32.14 Microjets device. (a) 3D schematic representation of the device showing opposed arrays of small orifices (i.e., small microfluidic channels) creating gradients in an open cell culture area. (b) Plot of the concentration profile (solid lines, left axis) and profile derivatives (dots, right axis) for three different pressure conditions illustrating how increasing the driving pressure delivered to the respective orifice arrays by the same magnitude causes an increase in gradient slope with little change in the gradient position. (c) Equal magnitude driving pressure offsets cause the gradient to shift position within the cell culture reservoir but does not affect the gradient shape. (Reprinted from Thomas M., et al., *Appl Phys Lett*, 89, 114103, 2006. Copyright 2006. With permission from American Institute of Physics.)

no dye. Cells within the microchannel were thus labeled only in the specific subcellular regions exposed to a particular dye. Microfluidic hydrodynamic focusing has since been used to examine the dynamics of epidermal growth factor (EGF) receptor signaling[63] and the effects of focal application of agrin in stimulating *in vitro* neuromuscular junction formation.[64,65]

Microfluidic channels can also be used to physically compartmentalize portions of a cell. Jeon and coworkers[66] replicated a Campenot[67] chamber using microfluidic channels to provide subcellular isolation of central nervous system growth cones from their cell bodies. Neurons seeded in a central microchannel were allowed to extend axons through narrow restrictions to another large microchannel (Figure 32.16). The restrictions prevent appreciable fluid connections from being established between the two microchannels allowing the growth cones to be exposed to a different chemical environment than the cell bodies.

32.4.3 Electrical Microenvironment

Several important cell types are sensitive to the electrical characteristics of the cellular microenvironment. Neurons, cardiac myocytes, and retinal cells all generate and can be stimulated with electrical impulses. Because of the importance of these cell types, various methods have been devised to record or stimulate electrical activity within cells cultured *in vitro*. Traditionally, the electrical activity of cells has been recorded or stimulated by simply placing electrodes in the

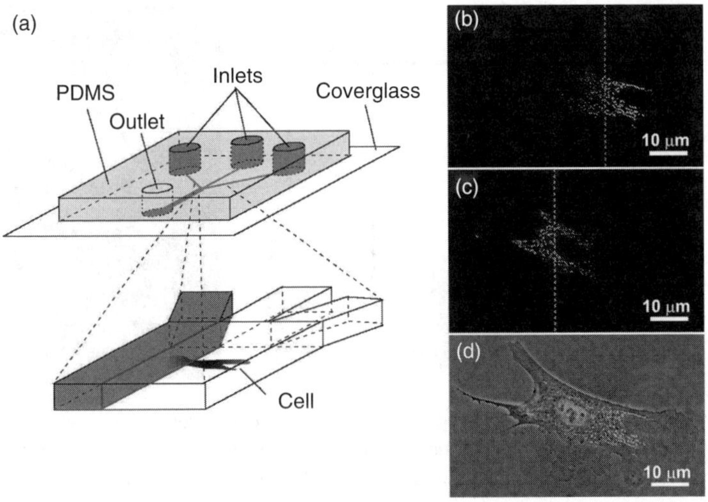

FIGURE 32.15 Hydrodynamic focusing. (a) A 3D schematic representation of the device shows how a fluid stream containing dye was confined by adjacent flow streams to expose only certain subcellular regions of the cell to the dye. (b) A bovine capillary endothelial cell labeled on one side with Mitotracker Green, and (c) on the other side with FM Mitotracker Red CM-H_2XRos. (d) Phase micrograph of the cell overlayed with fluorescence images from (b) and (c). (Adapted from Takayama, S., et al., *Nature*, 411, 1016, Copyright 2001. With permission from Macmillan Publishers Ltd.)

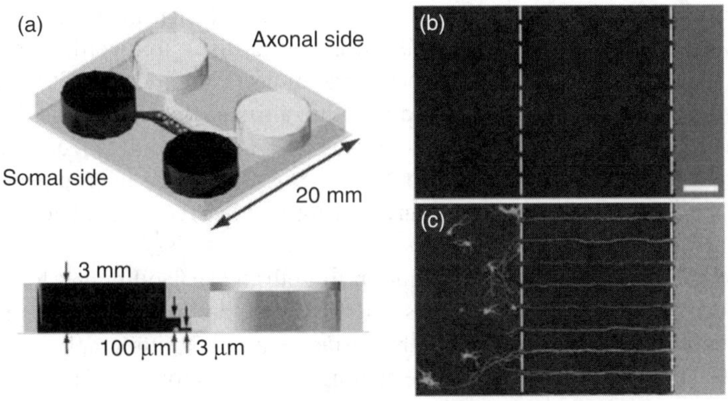

FIGURE 32.16 Physical compartmentalization. (a) 3D schematic representation and 2D cross section of the device shows how neurons loaded in the somal side of the device (black) are unable to cross through the narrow constriction, unlike their axons, which can extend into the axonal side (white). (b) Texas red dextran injected in the axonal side (right) does not cross into the somal side showing fluidic isolation between the compartments. (c) When green cell tracker dye is loaded into the axonal side it retrograde labels the axons and somas of the neurons. (Adapted from Taylor, A. M., et al., *Nat Methods*, 2, 8, 599. Copyright 2005. With permission from Macmillan Publishers Ltd.)

culture medium, or alternatively, using a process known as "patch clamping" (Figure 32.17a). Patch clamping uses glass capillaries that are heated and then pulled at a defined speed to form a drawn tip approximately 1 μm in diameter. The drawn capillary, or micropipette, is then filled with a conductive electrolyte solution and a metal electrode, and brought to the surface of a cell using a micromanipulator. In this configuration the micropipette can be used to record the extracellular electrical activity or stimulate cells. To measure single cell transmembrane ion currents, a small hole is ripped in the cell membrane by placing the micropipette tip in contact with the cell membrane and

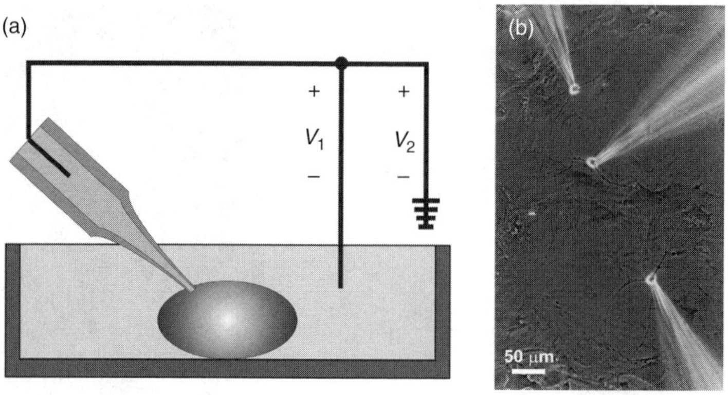

FIGURE 32.17 Micropipette electrical recording/stimulation. (a) Pulled glass pipettes are used to measure the electrical activity of cells across the cell membrane (V_1) by accessing the intracellular fluidic environment, or extracellularly (V_2) by simply placing the micropipette near the cell. (b) A major limitation of micropipette-based electrical recording/stimulation is the inability to record from more than a few cells at a time due to space limitations of the micropipette set up. (Adapted Fitzsimonds, R. M., et al., *Nature*, 388, 439. Copyright 1997. With permission from Macmillan Publishers Ltd.)

gently pulling suction in the capillary. The cell membrane is pulled into the micropipette forming a high-resistance seal, and ruptured within the micropipette to provide fluidic, and thus electrical, access to the intracellular environment.

Although the quality of the recordings or stimulations is excellent using micropipettes, the method suffers significantly from its lack of scalability. It is very difficult to record or stimulate electrical activity in more than just a few cells due to space limitations (Figure 32.17b). Trying to comprehensively characterize the electrical activity of cells within a tissue section or large dissociated culture is not possible using individually placed micropipettes. It is also difficult to record from or stimulate cells for more than a few hours due to biofouling or clogging of the micropipette tip by proteins and cell components, and the instability of any seal formed between the micropipette and the cell membrane.

Microelectrode arrays (MEA) offer a solution to both the scalability and long-term recording limitations of micropipette electrodes. MEAs (Figure 32.18) are made by depositing a conductive metal on a cell culture substrate, patterning the surface with photolithography, and using the photoresist pattern as a mask for subsequent metal etching. After selective metal etching, the photoresist is stripped and the resulting metal electrodes are connected to a multichannel recorder or stimulator. Dissociated cells or tissue sections are then seeded on the MEA and can be recorded from or stimulated electrically. Micron-scale control over the size, shape, and position of the metal electrodes allows many electrodes to be placed on the cell culture substrate in a variety of configurations. Because MEAs are integrated into the cell culture substrate, they offer the ability to record from cell cultures for long periods of time.

MEAs have been used primarily to record electrical activity in tissues and cultured cells,[68] although they can also be used to electrically stimulate cells[69,70] or other functions such as electroporate dissociated cell cultures with precise spatiotemporal control.[71] MEAs are extremely useful for providing a comprehensive electrical profile of an entire cell culture, or for increasing experimental throughput by simultaneously recording from or stimulating multiple cells.[68] Although the quality of the electrical recordings of MEAs is not as good as patch clamping, it is comparable with or better than other extracellular recording methods. Effective recording or stimulation of electrically active cells using MEAs requires close contact between the cell and the electrode. Large cell-electrode distances will make action potentials difficult to detect and electrical activity difficult to stimulate due to decreasing field strength with distance from the cell or electrode, respectively. To

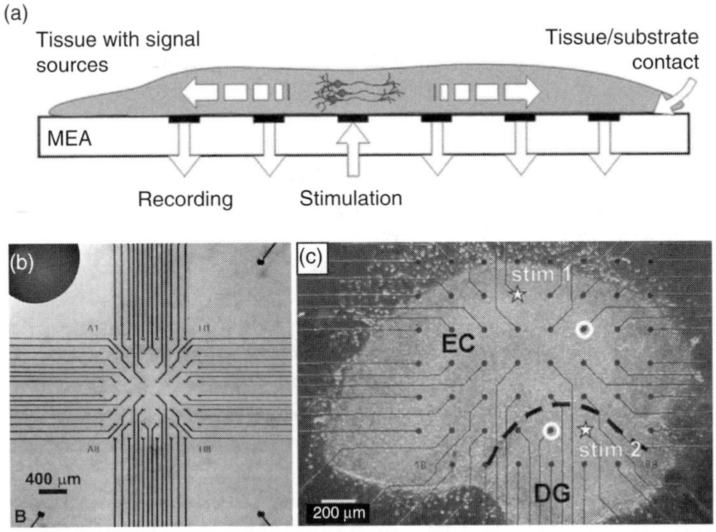

FIGURE 32.18 Microelectrode arrays. (a) MEAs can be used to record or stimulate electrical activity in dissociated cell cultures, engineered 3D cultures, or whole tissue explants. (b) MEAs are constructed using metal deposition and photolithography resulting in intricate, user-defined electrode configurations. (c) Brain tissue explant cultured on a MEA after 7 days *in vitro*. (Adapted from Stett, A., et al., *Anal Bioanal Chem*, 377, 486, 2003. Copyright 2003. With permission from Springer Science and Business Media.)

encourage close contact between cells and electrodes, MEAs are often patterned with adhesive proteins such as poly(lysine) and/or extracellular matrix proteins, or adhesion-promoting self-assembled monolayers using microstamping or other patterning techniques.[1]

MEAs have an added advantage in that they are compatible with and can be incorporated into many other methods discussed in this chapter to address more complex questions. For example, microfluidic channels constructed around an MEA[72] would allow the electrical activity of cells or whole tissue sections to be recorded or stimulated while the cells are exposed to unique chemical, mechanical, or thermal[73] environments. Cells could be exposed to chemical gradients, regulated shear forces, or disparate fluidic environments for many days while under constant electrical monitoring or stimulation.

32.5 SUMMARY

Microfabrication technology has provided a plethora of tools and methods to engineer the position and microenvironment of cells *in vitro*. The unprecedented level of control over the mechanical, chemical, and electrical nature of the cellular microenvironment allows investigation of questions not addressable with conventional tools and methods. The unique insight into normal and abnormal cell behavior afforded by microfabricated tools and methods may one day lead to cures for injuries and diseases, and the ability to direct cell growth and behavior for tissue engineering or industrial applications.

Although microfabrication technology has greatly enhanced the type of complex environments in which cells can be grown and studied, there is a large need for the development of complementary analysis methods to detect and characterize changes in cell response. Existing analysis methods, some of which are covered in the other chapters of this book, are often not adaptable to or optimized for use with microfabricated cell culture tools or environments. The development of novel methods that integrate engineered cell culture environments with complementary analysis tools would greatly accelerate our understanding of normal and abnormal cell behavior.

REFERENCES

1. Folch, A. and Toner, M., Microengineering of cellular interactions, *Annu Rev Biomed Eng*, 2, 227, 2000.
2. Li, N., Tourovskaia, A., and Folch, A., Biology on a chip: Microfabrication for studying the behavior of cultured cells, *Crit Rev Biomed Eng*, 31, 423, 2003.
3. Falconnet, D., Csucs, G., Grandin, H. M., and Textor, M., Surface engineering approaches to micropattern surfaces for cell-based assays, *Biomaterials*, 27, 3044, 2006.
4. Yap, F. L. and Zhang, Y., Protein and cell micropatterning and its integration with micro/nanoparticles assembly, *Biosens Bioelectron*, 22, 775, 2007.
5. Liu, Y., Sun, S., Singha, S., Cho, M. R., and Gordon, R. J., 3D femtosecond laser patterning of collagen for directed cell attachment, *Biomaterials*, 26, 4597, 2005.
6. Suh, K. Y., Seong, J., Khademhosseini, A., Laibinis, P. E., and Langer, R., A simple soft lithographic route to fabrication of poly(ethylene glycol) microstructures for protein and cell patterning, *Biomaterials*, 25, 557, 2004.
7. Tsang, V. L. and Bhatia, S. N., Three-dimensional tissue fabrication, *Adv Drug Deliv Rev*, 56, 1635, 2004.
8. Zhu, A., Chen, R., and Chan-Park, M. B., Patterning of a random copolymer of poly[lactide-co-glycotide-co-(epsilon-caprolactone)] by UV embossing for tissue engineering, *Macromol Biosci*, 6, 51, 2006.
9. Albrecht, D. R., Tsang, V. L., Sah, R. L., and Bhatia, S. N., Photo- and electropatterning of hydrogel-encapsulated living cell arrays, *Lab Chip*, 5, 111, 2005.
10. Gray, D. S., Tan, J. L., Voldman, J., and Chen, C. S., Dielectrophoretic registration of living cells to a microelectrode array, *Biosens Bioelectron*, 19, 1765, 2004.
11. Voldman, J., Electrical forces for microscale cell manipulation, *Ann Rev Biomed Eng*, 8, 425, 2006.
12. Kaler, K. V. I. S. and Jones, T. B., Dielectrophoretic spectra of single cells determined by feedback-controlled levitation, *Biophys J*, 57, 173, 1990.
13. Fuhr, G., Arnold, W. M., Hagedorn, R., Muller, T., Benecke, W., Wagner, B., and Zimmermann, U., Levitation, holding, and rotation of cells within traps made by high-frequency fields, *Biochim Biophys Acta*, 1108, 215, 1992.
14. Hakoda, M., Hachisu, T., Wakizaka, Y., Mii, S., and Kitajima, N., Development of a method to analyze single cell activity by using dielectrophoretic levitation, *Biotechnol Prog*, 21, 1748, 2005.
15. Roberts, M. J., Bentley, M. D., and Harris, J. M., Chemistry for peptide and protein PEGylation, *Adv Drug Deliv Rev*, 54, 459, 2002.
16. Boland, T., Xu, T., Damon, B., and Cui, X., Application of inkjet printing to tissue engineering, *Biotechnol J*, 1, 910, 2006.
17. Pardo, L., Wilson, W. C., and Boland, T. J., Characterization of patterned self-assembled monolayers and protein arrays generated by the ink-jet method, *Langmuir*, 19, 1462, 2003.
18. Varghese, D., Deshpande, M., Xu, T., Kesari, P., Ohri, S., and Boland, T., Advances in tissue engineering: Cell printing, *J Thorac Cardiovasc Surg*, 129, 470, 2005.
19. Xu, T., Gregory, C. A., Molnar, P., Cui, X., Jalota, S., Bhaduri, S. B., and Boland, T., Viability and electrophysiology of neural cell structures generated by the inkjet printing method, *Biomaterials*, 27, 3580, 2006.
20. Flemming, R. G., Murphy, C. J., Abrams, G. A., Goodman, S. L., and Nealey, P. F., Effects of synthetic micro- and nano-structured surfaces on cell behavior, *Biomaterials*, 20, 573, 1999.
21. Stevens, M. M. and George, J. H., Exploring and engineering the cell surface interface, *Science*, 310, 1135, 2005.
22. Abrams, G. A., Murphy, C. J., Wang, Z. Y., Nealey, P. F., and Bjorling, D. E., Ultrastructural basement membrane topography of the bladder epithelium, *Urol Res*, 31, 341, 2003.
23. Chen, Y. and Pepin, A., Nanofabrication: Conventional and nonconventional methods, *Electrophoresis*, 22, 187, 2001.
24. Murugan, R. and Ramakrishna, S., Nano-featured scaffolds for tissue engineering: A review of spinning methodologies, *Tissue Eng*, 12, 435, 2006.
25. Norman, J. and Desai, T., Methods for fabrication of nanoscale topography for tissue engineering scaffolds, *Ann Biomed Eng*, 34, 89, 2006.
26. Pham, Q. P., Sharma, U., and Mikos, A. G., Electrospinning of polymeric nanofibers for tissue engineering applications: A review, *Tissue Eng*, 12, 1197, 2006.

27. Li, W. J., Laurencin, C. T., Caterson, E. J., Tuan, R. S., and Ko, F. K., Electrospun nanofibrous structure: A novel scaffold for tissue engineering, *J Biomed Mat Res*, 60, 613, 2002.
28. Mata, A., Boehm, C., Fleischman, A. J., Muschler, G., and Roy, S., Analysis of connective tissue progenitor cell behavior on polydimethylsiloxane smooth and channel micro-textures, *Biomed Microdev*, 4, 267, 2002.
29. Martin, C. R. and Aksay, I. A., Microchannel molding: A soft lithography-inspired approach to micrometer-scale patterning, *J Mat Res*, 20, 1995, 2005.
30. Li, N. Z. and Folch, A., Integration of topographical and biochemical cues by axons during growth on microfabricated 3-D substrates, *Exp Cell Res*, 311, 307, 2005.
31. Narasimhan, J. and Papautsky, I., Polymer embossing tools for rapid prototyping of plastic microfluidic devices, *J Micromech Microeng*, 14, 96, 2004.
32. Goldner, J. S., Bruder, J. M., Li, G., Gazzola, D., and Hoffman-Kim, D., Neurite bridging across micropatterned grooves, *Biomaterials*, 27, 460, 2006.
33. Bruder, J. M., Monu, N. C., Harrison, M. W., and Hoffman-Kim, D., Fabrication of polymeric replicas of cell surfaces with nanoscale resolution, *Langmuir*, 22, 8266, 2006.
34. Geissler, M. and Xia, Y. N., Patterning: Principles and some new developments, *Adv Mat*, 16, 1249, 2004.
35. Henke, W., Hoppe, W., Quenzer, H. J., Staudtfischbach, P., and Wagner, B., Simulation and process design of gray-tone lithography for the fabrication of arbitrarily-shaped surfaces, *Jpn J Appl Phys Part 1–Regular Papers Short Notes & Review Papers*, 33, 6809, 1994.
36. Chen, C., Hirdes, D., and Folch, A., Gray-scale photolithography using microfluidic photomasks, *Proc Natl Acad Sci USA*, 100, 1499, 2003.
37. Kawata, S., Sun, H. B., Tanaka, T., and Takada, K., Finer features for functional microdevices—Micromachines can be created with higher resolution using two-photon absorption, *Nature*, 412, 697, 2001.
38. Baldacchini, T., LaFratta, C. N., Farrer, R. A., Teich, M. C., Saleh, B. E. A., Naughton, M. J., and Fourkas, J. T., Acrylic-based resin with favorable properties for three-dimensional two-photon polymerization, *J Appl Phys*, 95, 6072, 2004.
39. LaFratta, C. N., Lit, L. J., and Fourkas, J. T., Soft-lithographic replication of 3D microstructures with closed loops, *Proc Natl Acad Sci USA*, 103, 8589, 2006.
40. Brown, T. D., Techniques for mechanical stimulation of cells *in vitro*: A review, *J Biomech*, 33, 3, 2000.
41. Hoffman, J. M., Shao, J., Hsu, C. H., and Folch, A., Elastomeric Molds with tunable microtopography, *Adv Mat*, 16, 2201, 2004.
42. Duerrschmidt, N., Stielow, C., Muller, G., Pagano, P. J., and Morawietz, H., NO-mediated regulation of NAD(P)H oxidase by laminar shear stress in human endothelial cells, *J Physiol*, 576 (Pt 2), 557, 2006.
43. Frame, M. D., Chapman, G. B., Makino, Y., and Sarelius, I. H., Shear stress gradient over endothelial cells in a curved microchannel system, *Biorheology*, 35, 245, 1998.
44. Bransky, A., Korin, N., Nemirovski, Y., and Dinnar, U., An automated cell analysis sensing system based on a microfabricated rheoscope for the study of red blood cells physiology, *Biosens Bioelectr*, 22, 165, 2006.
45. Bird, R. B., Stewart, W. E., and Lightfoot, E. N., *Transport Phenomena*, John Wiley & Sons, Inc., 1960.
46. Shin, H., Jo, S., and Mikos, A. G., Biomimetic materials for tissue engineering, *Biomaterials*, 24, 4353, 2003.
47. Boyden, S., The chemotactic effect of mixtures of antibody and antigen on polymorphonuclear leucocytes, *J Exp Med*, 115, 453, 1962.
48. Zigmond, S. H., Ability of polymorphonuclear leukocytes to orient in gradients of chemotactic factors, *J Cell Biol*, 75 (Pt 1), 606, 1977.
49. Gunderson, R. W. and Barrett, J. N., Neuronal chemotaxis: Chick dorsal-root axons turn toward high concentrations of nerve growth factor, *Science*, 206, 1079, 1979.
50. Zicha, D., Dunn, G. A., and Brown, A. F., A new direct-viewing chemotaxis chamber, *J Cell Sci*, 99 (Pt 4), 769, 1991.
51. Foxman, E. F., Campbell, J. J., and Butcher, E. C., Multistep navigation and the combinatorial control of leukocyte chemotaxis, *J Cell Biol*, 139, 1349, 1997.

52. Chen, H., He, Z., Bagri, A., and Tessier-Lavigne, M., Semaphorin–neuropilin interactions underlying sympathetic axon responses to class III semaphorins, *Neuron*, 21, 1283, 1998.
53. Jeon, N. L., Dertinger, S. K. W., Chiu, D. T., Choi, I. S., Stroock, A. D., and Whitesides, G. M., Generation of solution and surface gradients using microfluidic systems, *Langmuir*, 16, 8311, 2000.
54. Jeon, N. L., Baskaran, H., Dertinger, S. K. W., Whitesides, G. M., Van de Water, L., and Toner, M., Neutrophil chemotaxis in linear and complex gradients of interleukin-8 formed in a microfabricated device, *Nat Biotechnol*, 20, 826, 2002.
55. Chung, B. G., Flanagan, L. A., Rhee, S. W., Schwartz, P. H., Lee, A. P., Monuki, E. S., and Jeon, N. L., Human neural stem cell growth and differentiation in a gradient-generating microfluidic device, *Lab Chip*, 5, 401, 2005.
56. Saadi, W., Wang, S. J., Lin, F., and Jeon, N. L., A parallel-gradient microfluidic chamber for quantitative analysis of breast cancer cell chemotaxis, *Biomed Microdev*, 8, 109, 2006.
57. Walker, G. M., Sai, J., Richmond, A., Stremler, M., Chung, C. Y., and Wikswo, J. P., Effects of flow and diffusion on chemotaxis studies in a microfabricated gradient generator, *Lab Chip*, 5, 611, 2005.
58. Chung, B. G., Lin, F., and Jeon, N. L., A microfluidic multi-injector for gradient generation, *Lab Chip*, 6, 764, 2006.
59. Keenan, T. M., Hsu, C. H., and Folch, A., Microfluidic "jets" for generating steady-state gradients of soluble molecules on open surfaces, *Appl Phys Lett*, 89, 114103, 2006.
60. Takayama, S., Ostuni, E., LeDuc, P., Naruse, K., Ingber, D. E., and Whitesides, G. M., Subcellular positioning of small molecules, *Nature*, 411, 1016, 2001.
61. Takayama, S., Ostuni, E., LeDuc, P., Naruse, K., Ingber, D. E., and Whitesides, G. M., Selective chemical treatment of cellular microdomains using multiple laminar streams, *Chem Biol*, 10, 123, 2003.
62. Spielman, L. and Goren, S. L., Improving resolution in Coulter counting by hydrodynamic focusing, *J Colloid Interface Sci*, 26, 175, 1968.
63. Sawano, A., Takayama, S., Matsuda, M., and Miyawaki, A., Lateral propagation of EGF signaling after local stimulation is dependent on receptor density, *Dev Cell*, 3, 245, 2002.
64. Tourovskaia, A., Figueroa-Masot, X., and Folch, A., Differentiation-on-a-chip: A microfluidic platform for long-term cell culture studies, *Lab Chip*, 5, 14, 2005.
65. Tourovskaia, A., Kosar, T. F., and Folch, A., Local induction of acetylcholine receptor clustering in myotube cultures using microfluidic application of agrin, *Biophys J*, 90, 2192, 2006.
66. Taylor, A. M., Blurton-Jones, M., Rhee, S. W., Cribbs, D. H., Cotman, C. W., and Jeon, N. L., A microfluidic culture platform for CNS axonal injury, regeneration and transport, *Nat Methods*, 2, 599, 2005.
67. Campenot, R. B., Local control of neurite development by nerve growth factor, *Proc Natl Acad Sci USA*, 74, 4516, 1977.
68. Stett, A., Egert, U., Guenther, E., Hofmann, F., Meyer, T., Nisch, W., and Haemmerle, H., Biological application of microelectrode arrays in drug discovery and basic research, *Anal Bioanal Chem*, 377, 486, 2003.
69. Jimbo, Y., Kasai, N., Torimitsu, K., Tateno, T., and Robinson, H. P. C., A system for MEA-based multisite stimulation, *IEEE Trans Biomed Eng*, 50, 241, 2003.
70. Shenai, M. B., Putchakayala, K. G., Hessler, J. A., Orr, B. G., Banaszak Holl, M. M., and Baker, J. R., Jr., A novel MEA/AFM platform for measurement of real-time, nanometric morphological alterations of electrically stimulated neuroblastoma cells, *IEEE Trans Nanobiosci*, 3, 111, 2004.
71. Jain, T. and Muthuswamy, J., Microsystem for transfection of exogenous molecules with spatio-temporal control into adherent cells, *Biosens Bioelectron*, 22, 863, 2007.
72. Pearce, T. M., Williams, J. J., Kruzel, S. P., Gidden, M. J., and Williams, J. C., Dynamic control of extracellular environment in *in vitro* neural recording systems, *IEEE Trans Neural Syst Rehabil Eng*, 13, 207, 2005.
73. Pearce, T. M., Wilson, J. A., Oakes, S. G., Chiu, S. Y., and Williams, J. C., Integrated microelectrode array and microfluidics for temperature clamp of sensory neurons in culture, *Lab Chip*, 5, 97, 2005.

33 Multidimensional Microfluidic Systems for Protein and Peptide Separations

Don L. DeVoe and Cheng S. Lee

CONTENTS

33.1 INTRODUCTION

Since the concept of micro total analysis systems (μTAS) was first proposed,[1] the field has advanced rapidly, with ongoing developments promising to profoundly revolutionize modern bioanalytical platforms and methodologies. Whether termed μTAS, lab-on-a-chip, or microfluidics, the technologies that define the field offer important innovations capable of transforming the ways in which bioanalytical techniques are performed. For example, reduced size and power requirements can enable improved portability and higher levels of integration, with lower per-unit cost for disposable applications. Low volume fluid control enabled by microfluidics allows smaller dead volumes and reduced sample consumption, while many pumping methods including capillary action and electroosmotic flow scale favorably in these systems, enabling precise and valveless flow control at the microscale. Similarly, thermal time constants tend to be extremely small due to large surface area to volume ratios inherent in these systems, reducing the onset of significant Joule heating during electrokinetic separations and thus allowing higher separation voltages and shorter analysis times with equivalent or better separation resolution for complex mixtures in an integrated format.

A significant advantage of microfluidics is that lithographic and replication-based fabrication techniques readily lend themselves to the formation of complex systems, providing a path for effective manufacturing of highly parallel analytical tools. Specifically, we consider here the promise that microfluidics technology holds for realizing robust multidimensional separations of proteins and peptides in an integrated platform. Assuming the separation techniques used in a given two-dimensional

(2-D) system are fully orthogonal, that is, the two separation techniques are based on independent physicochemical properties of analytes, the overall peak capacity is the product of the peak capacities of the individual one-dimensional (1-D) methods.[2] As a result, multidimensional systems are of great interest for the analysis of complex mixtures.[3] Nowhere is the demand for effective analysis of complex matrices greater than in the field of proteomics. A major challenge faced in proteomics analysis is the vast number of proteins present in a typical biological sample, together with the large variation of protein relative abundances (typically greater than 6 orders of magnitude,[4] and higher than 10 orders of magnitude for the case of blood plasma[5]), which ultimately dictate the need for high resolution separations to be performed on protein or peptide samples before detection by mass spectrometry (MS). Thus, multidimensional separation technologies capable of reducing dynamic range and enhancing detection sensitivity without substantially sacrificing analytical throughput are highly desirable for many proteomic applications.

While intact protein analysis via 2-D polyacrylamide gel electrophoresis (2-D PAGE)[6] and bottom-up peptide analysis combining multidimensional chromatography with tandem MS analysis[7,8] remain the workhorse technologies for most modern proteomic studies, ongoing improvements in microfluidics for sample preparation and separations in proteomics are beginning to reveal their potential impact on the field. In this chapter, microfluidic platforms for protein and peptide analysis employing multidimensional electrophoretic separations and combined electrophoretic and chromatographic separations are discussed.

33.2 MICROFLUIDIC PLATFORMS FOR MULTIDIMENSIONAL SEPARATIONS

33.2.1 Time-Multiplexed Separations

The majority of reported multidimensional microfluidic systems employ serial separations that are time multiplexed. In this approach, fractions from the first separation dimension are sampled sequentially, with each fraction separated in series within the second dimension while the first-dimension separation continues. In one of the first demonstrations of this approach, Rocklin et al.[9] performed 2-D separations of peptide mixtures in a microfluidic device using micellar electrokinetic chromatography (MEKC) and capillary zone electrophoresis (CZE) as the first and second dimensions, respectively. A schematic representation of their platform, fabricated in a glass substrate, is shown in Figure 33.1a. The MEKC separation was performed within a 65-mm-long serpentine channel located between the upper and lower channel intersection points shown in the figure, with a 10-mm-long CZE separation performed past the lower channel intersection. With a time scale for CZE on the order of a few seconds, substantially faster than the MEKC separation, large numbers of fractions could be sampled from the first dimension and separated by CZE with an overall analysis time under 10 min. Because MEKC is a transient electrokinetic separation, only a small portion (~10%) of the analyte in the first dimension could be sampled into the CZE microchannel, since no fractions could be sampled while the second dimension separation was being performed. Despite this sample loss issue, a revised chip design was later reported for peptide separations with a peak capacity greater than 4000, such as the comparative separation of tryptic digest from bovine and human hemoglobin shown in Figure 33.1b revealing distinct peptide variations between these samples.[10]

Similar approaches based on time multiplexing have been employed by a number of researchers. For example, Gottschlich et al.[11] used a spiral shaped glass channel coated with a C_{18} stationary phase for performing chromatographic separation of trypsin-digested peptides. By providing a cross interface, peptides eluted from an MEKC[9] or reversed-phase[11] chromatography channel were sampled and rapidly separated by CZE in a short glass microchannel. Herr et al.[12] similarly coupled isoelectric focusing (IEF) with serial CZE for 2-D separations of model proteins, a concept that

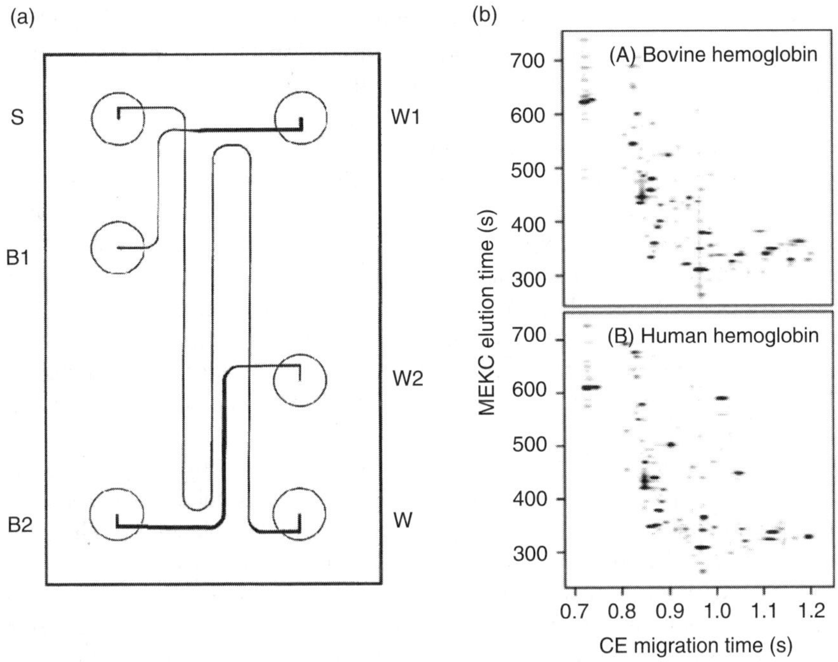

FIGURE 33.1 Time-multiplexed MEKC/CZE peptide separation chip schematic (From Rocklin, R. D. et al., *Anal. Chem.*, 2000, *72*, 5244–5249.) and (b) pseudo-gel views for chip-based separations of human and bovine hemoglobin. (From Ramsey, J. D. et al., *Anal. Chem.*, 2003, *75*, 3758–3764.)

was later implemented in a poly(dimethylsiloxane) (PDMS)-based microfluidic system using elastomeric valving to isolate the separation dimensions.[13] In addition, Shadpour and Soper[14] developed a microfluidic chip combining sodium dodecyl sulfate (SDS) gel electrophoresis with MEKC, again with multiple fractions sampled serially from the gel electrophoresis separation for time-multiplexed fractionation by MEKC in the second dimension.

In each of the preceding examples, the separations were time multiplexed, with rapid second dimension separations chosen to enable repeated sampling of analyte eluted from a slower first dimension separation. Such time-multiplexed separations have also been employed in multidimensional capillary separations, for example, through the combination of a first dimension chromatographic separation together with a second dimension separation based on CZE,[15] with low dead-volume switching valves used to couple the separation dimensions. For many applications, multidimensional microfluidic systems employing time-multiplexed separations may not offer compelling advantages over their capillary analogs, but rather introduce additional complexities, costs, and performance limitations, which may be avoided through the use of more traditional capillary technologies. On the other hand, unique benefits exist for applications that can leverage the small footprints and integrated nature of these microfluidic platforms, particularly when coupled with on-chip sample preparation.

33.2.2 Spatially Multiplexed Separations

In contrast to time multiplexing, high-throughput multidimensional separations may also be performed using spatial multiplexing, with an array of second dimension microchannels used to simultaneously separate analyte sampled as multiple fractions from the entirety of a first dimension separation channel. Using this approach relaxes the requirement for the second dimension separation to be substantially faster than the first dimension, opening the door to a wider range of potential separation modes. Furthermore, spatial multiplexing enables complete sampling of the first dimension separation, even for transient separation modes, thereby preventing sample loss due to a duty cycle mismatch between the dimensions. The use of spatial multiplexing also leverages a significant advantages of microfluidics, namely the ability to combine large numbers of channels in a single chip without the need for complex fluidic interconnects between the channels.

The concept of spatially-multiplexed microfluidic separations was first embodied in a microfabricated quartz device proposed by Becker et al.[16] which contained a single channel for a first dimension and an array of 500 parallel channels with submicron dimensions positioned orthogonally to the first dimension. While no separations were actually performed in this device, the concept was later extended and validated by several groups. For example, Chen et al.[17] described a microfluidic 2-D capillary electrophoresis platform based on a reconfigurable system of six individual PDMS layers. The chip consisted of a 2.5-cm-long microchannel for performing IEF, with an intersecting array of parallel 6-cm-long microchannels for SDS–PAGE. This six-layer PDMS microfluidic device required the alignment, bonding, removal, realignment, and rebonding of various combinations of the six layers to perform a full 2-D protein separation.

A fully integrated system developed by Li et al.[18] combined IEF with SDS gel electrophoresis of intact proteins in a single polycarbonate chip containing 10 parallel second dimension microchannels (Figure 33.2). By combining both separation dimensions in a single rigid polymer chip, dispersion of analyte during the transfer between the dimensions was minimized. Furthermore, different separation media were employed in each of the dimensions, allowing sample mobility to be independently tailored for rapid free-solution IEF separations and high resolution SDS-capillary gel electrophoresis (CGE) separations. By using a replaceable polyethylene oxide (PEO) gel in the SDS–CGE channels, the interface between the separation media also provided for sample stacking during the transfer of focused proteins bands into the second dimension. Using this system, a comprehensive 2-D protein separation was completed in less than 10 min, with the majority of time consumed in the required SDS–protein complexation reaction. A peak capacity of $\sim$170 in the second dimension of size-based separation was estimated from a measured bandwidth of 150 μm over a 2.5 cm channel length. Because the separation mechanisms in IEF and SDS–CGE were completely orthogonal, the overall peak capacity was estimated to be 1700 (10 fractions from IEF $\times$ 170 from SDS–CGE). Improvements in peak capacity were proposed through the use of longer channels during the size-based separation, and by increasing the density of microchannels in the array to increasing the number of IEF fractions analyzed during the size-based separation. Owing to the use of parallel

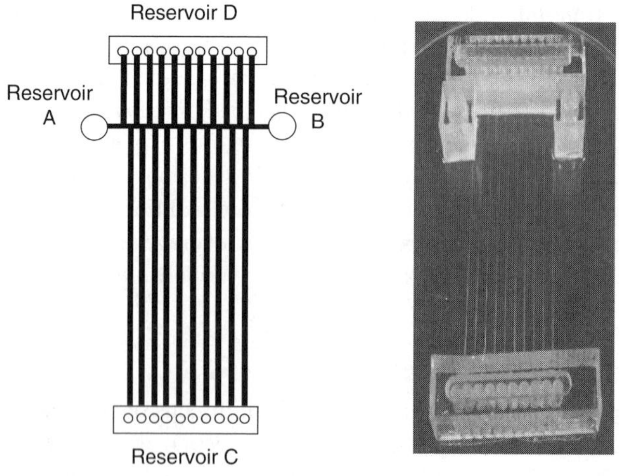

FIGURE 33.2 Schematic representation and photograph of a 2-D IEF/SDS–CGE microfluidic chip containing 10 spatially-multiplexed channels for SDS–CGE separation. (From Li, Y. et al., *Anal. Chem.*, 2004, *76*, 742–748.)

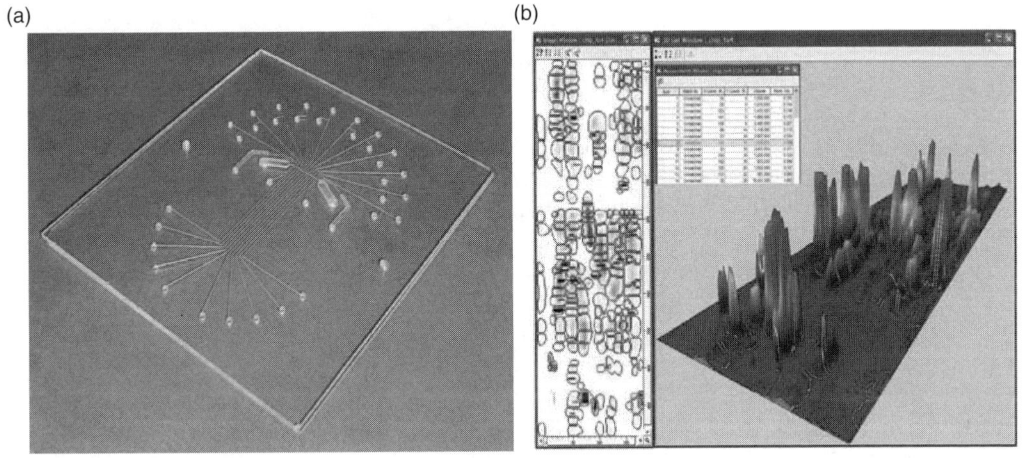

FIGURE 33.3 **(See color insert following page 810.)** (a) Microfluidic PMMA chip for 2-D IEF/SDS–CGE protein separations and (b) pseudo-gel image resulting from a yeast cell lysate separation.

separations in the second dimension, there is no accompanying increase in the analysis time. Further improvements in separation resolution were later achieved through the use of individual reservoirs for the injection of PEO gel and SDS solutions into the chip, and through electrical isolation of each second dimension channel. A revised chip design is shown in Figure 33.3a, with an example separation of intact proteins from yeast cell lysate depicted in Figure 33.3b.

Ivory and coworkers[19] demonstrated a multistage separation platform based solely on multiple IEF separations, with an initial separation performed in a straight channel using broad-range ampholytes, followed by a second separation using a set of orthogonal microchannels that terminated in reservoirs containing ampholytes covering more narrow pI ranges for higher focusing resolution. By providing multiple sets of the orthogonal channels, progressively narrower pI ranges were established within the main separation channel. Alternately, multiple parallel narrow-range IEF separations could be performed using the apparatus. A three-stage transient IEF separation was demonstrated in this system, and additional stages were proposed. Multidimensional microfluidic separation platforms with even higher dimensionality have been proposed, such as a concept for a four-dimensional (4-D) platform combining IEF, isotachophoresis (ITP), SDS–PAGE, and reversed-phase liquid chromatography (RPLC)[20] (Figure 33.4). It is not yet clear whether significant advantages can be realized by such high-dimensional fractionation, since analyte loss, dispersion, and dilution may ultimately prevent effective downstream detection (e.g., by MS), in particular given the relatively low loading capacities typical of many microfluidic systems. Regardless, this 4-D concept suggests an intriguing direction for new platforms and multidimensional separation modalities that can be enabled by the unique advantages offered by integrated microfluidics technology.

33.2.3 Interfacing Multidimensional Separations with MS

Mass spectrometry is an essential tool for the characterization of biomolecules, revealing the charge-to-mass ratio of analyte molecules following ionization. There is a strong need for effective MS in the analysis of proteins and peptides, where MS coupling is required to provide sufficient mass resolution and sequence information for modern proteomic analyses. Unlike deoxyribonucleic acid (DNA), no techniques exists for the direct amplification of proteins, and thus detection sensitivity is paramount for the identification of low abundance species from limited samples. The matter is further complicated by the enormous dynamic range of protein abundance in complex samples. For

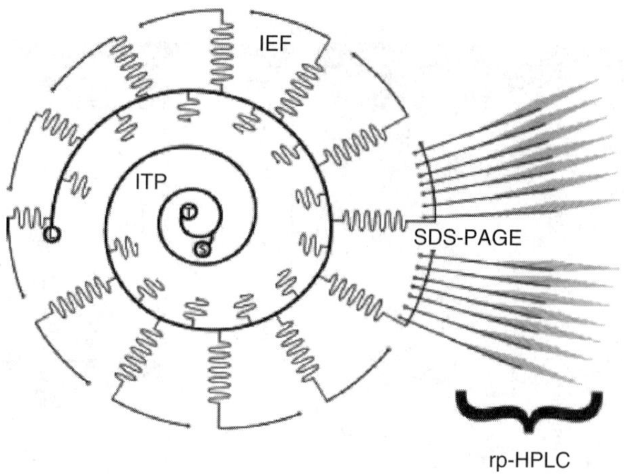

FIGURE 33.4 Conceptual layout of a spatially-multiplexed 4-D microfluidic system combining IEF, ITP, SDS–PAGE, and RPLC. (From Ivory, C. F., *Electrophoresis*, 2007, *28*, 15–25.)

multidimensional microfluidic technologies to substantially impact the field of proteomics, effective methods for MS interfacing are a necessity.

The integration of microfluidic systems and MS received substantial attention following demonstration of the first microfluidic electrospray ionization (ESI)–MS interfaces in 1997 by the groups of Karger and coworkers,[21] Ramsey and Ramsey,[22] and Aebersold and coworkers.[23] Extensive studies on ESI–MS interfacing for coupling microfluidics to MS have been reported, with online ESI–MS offering a simple approach for directly interfacing microfluidic analyses with MS, while providing good sensitivity and mass accuracy, and reproducible signals for effective analyte quantification. A number of reviews have appeared in recent years, which illuminate this topic. Oleschuk and Harrison[24] offer an early summary of microfluidic ESI–MS interfacing, and Sung et al.[25] provide a more recent review addressing this area. A further review by Limbach and Meng[26] in 2002 updates the developments in microfluidics for ESI–MS, and reviews by Figeys and Pinto,[27] Lion et al.,[28] and Marko-Varga et al.[29] focus on microfabricated systems in proteomics, including summaries of relevant microfluidic-to-MS interfaces. A more detailed discussion of microfluidic ESI–MS interfacing is provided in Chapter 53 by Lazar.

While ESI–MS is well suited as an interface between serial time-multiplexed microfluidic systems and MS, as reviewed by Foret and Kusý,[30] interfaces based on matrix-assisted laser desorption/ionization (MALDI) offer several important advantages for spatially-multiplexed (i.e., parallel analysis) microfluidics. Unlike online ESI–MS, MALDI–MS is an offline soft ionization method, which is typically applied to the analysis of solid phase analyte co-crystallized with an energy-absorbing matrix material on the surface of a supporting target plate. In comparison with ESI–MS, MALDI–MS tends to be more tolerant of salts and other sample contaminants, offers excellent detection limits, and produces mass spectra which are relatively simple to interpret due to the absence of multiple charge states. Although generally limited to the analysis of higher molecular weight species due to interference from matrix components below ~500 Da, MALDI–MS can be applied to a wide range of sample mass, with good sensitivity above 300 kDa, compared with less than 100 kDa for ESI–MS.

From the point of view of multidimensional microfluidic systems, the off-line nature of MALDI–MS analysis represents a key advantage. First, MALDI–MS allows on-chip sample processing steps to be decoupled from back-end MS analysis. This is necessary when, for example, the time scales for biomolecular separations and online MS data acquisition are incompatible. More importantly for the present discussion, off-line MALDI–MS analysis provides a method for coupling simultaneous parallel on-chip analyses with MS detection, where ESI–MS from multiple microchannels is simply not practical or even feasible. It should be noted that although MALDI–MS is a serial process once the sample plate has been prepared, its high duty cycle enables high throughput for large numbers of deposited samples. For reviews of work in this area, Foret and Preisler[31] presented an early summary of microfabricated systems for MALDI–MS interfacing, and DeVoe and Lee[32] recently reviewed the current state of microfluidic interfacing with MALDI–MS. Additional discussion of microfluidic MALDI–MS interfacing may be found in Chapter 49 by Laurell.

For microfluidic systems that employ parallel chromatography channels, MALDI–MS interfacing is particularly straightforward. The concept of interfacing multiplexed capillary chromatography separations with MALDI–MS for high-throughput proteome analysis was first described by Aebersold and coworkers[33] using a system comprising four parallel capillary chromatography columns coupled with MALDI–MS/MS. In their system, a network of capillaries connected together with microjunctions was used to interface a single gradient pump to four parallel separation columns, with eluent from each column deposited by direct mechanical spotting onto a MALDI target at a flow rate of 1 μL/min using a commercial fraction collector. Flow rate variations of ±5–6% were observed between the columns, presumably due to variations in hydrodynamic flow resistance resulting from inhomogeneity of the packed silica beads. In a recent demonstration by Knapp and coworkers,[34] a multichannel cyclic olefin copolymer (COC) chip was developed for parallel nanoflow RPLC analysis. Chip effluent was deposited onto a MALDI target through an electrically mediated deposition technique,[35] which was shown to be robust for the case of a two-channel chromatography chip.

While electrospray[36] and pulsed electric field[35] deposition has been demonstrated from multichannel chips, there are several disadvantages associated with the use of electric fields to assist sample deposition. In addition to increased system complexity, the reliability and repeatability of droplet deposition degrade as the number of parallel channels is increased. In contrast, the direct mechanical spotting strategy described by Aebersold and coworkers[33] in a capillary format is well suited for multichannel microfluidic chips employing a chromatographic separation in the second dimension. For example, the eight-channel COC chip shown in Figure 33.5 contains 100-μm wide and 60-μm deep channels, with a single fluid port connected to all eight channels through a symmetric on-chip splitter. The end of the chip was cut with a high-speed semiconductor dicing saw to expose the channel exits. Using a syringe pump, a panel of nine model peptides were deposited onto 64 target spots (8 channels × 8 spotting events) on a custom MALDI target, with a target deposition volume of 150 nL per spot corresponding to sample amounts ranging from 25 to 75 fmol for each peptide. Using a Kratos Amixa MALDI-time-of-flight (TOF) instrument, the resulting spectra showed good uniformity, with an average relative standard deviation (RSD) of 44% across all peptides.

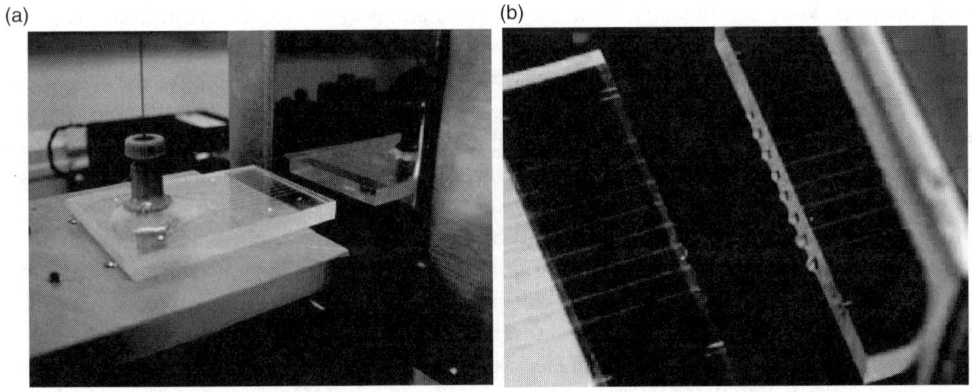

FIGURE 33.5 (a) Photo of an eight-channel COC chip aligned to a MALDI target mounted on a three-axis robotic positioner and (b) close up of chip exit showing droplet formation.

This compares well with an RSD of 50% in MALDI signal intensities reported for multiple peptide spots deposited from a single capillary using an optimized matrix seed layer technique.[37] Further improvements in deposition volume uniformity were realized for an eight-channel chip containing a methacrylate monolith developed for reversed-phase chromatography in the second separation dimension, with negligible variations at flow rates down to 200 nL/min per channel.

33.3 METHODS DEVELOPMENT GUIDELINES

33.3.1 Fabrication of Spatially Multiplexed Microfluidic Chips

There are a number of practical issues that must be addressed to realize functional multidimensional microfluidic separation platforms. For devices based on spatially-multiplexed microfluidics, challenges exist even at the fabrication level, since producing microfluidic substrates with large numbers of parallel channels can be difficult. For example, when rapid prototyping methods based on hot embossing of thermoplastics such as polymethyl methacrylate (PMMA) or COC are used for substrate patterning, high frictional forces imposed during demolding of the plastic from dense microchannel arrays can lead to distortion of channel geometry, substrate warping, template damage, and in extreme cases prevent demolding from occurring entirely. It can also be difficult to achieve sufficient adhesion strength when bonding chips containing closely-spaced microchannels due to the reduced bond surface area. While these problems can be alleviated by employing low aspect ratio and tapered channel geometries, limiting the overall chip real estate occupied by the channels, and by keeping "reasonable" spacing between adjacent channels, such solutions can often be in conflict with desired performance goals for the system. The appropriate channel spacing must be evaluated on the basis of the particular fabrication methods, materials, and channel dimensions being employed, and weighed against potential performance losses resulting from a compromised design.

33.3.2 Electrical and Hydrodynamic Crosstalk

Another fundamental challenge results from electrical and hydrodynamic crosstalk between multiple interconnected channels that are inherent in spatially-multiplexed systems. To illustrate this point, consider the schematic representation of the 2-D IEF/SDS–CGE protein separation chip previously depicted in Figure 33.2a. This chip contains 11 parallel channels above the horizontal IEF channel, and 10 parallel channels below. As shown in the figure, each of these channels terminate in one of two common reservoirs, which connect the distal ends of the upper and lower channel arrays. In practice, this configuration results in a resistive network that shunts electrical current through the reservoirs during IEF separation, thereby producing an uneven electric field distribution within the IEF channel. More significantly, both analytes and ampholytes (for the creation of pH gradient in the IEF channel) can be unintentionally injected into the upper and lower channel networks by electrophoresis, resulting in potentially severe sample loss, and preventing establishment of the desired pH gradient for the present example of IEF as the first separation dimension. Issues with crosstalk can also manifest during hydrodynamic sample injection, since the resistive network introduced by the interconnected SDS–CGE channels applies equally to hydrodynamic and electrical current shunts.

One approach to solve this problem would be to integrate flow control elements such as elastomeric valves into the system, but the additional complexity and inherent fabrication constraints make this solution less than ideal for real-world applications. Another obvious approach for eliminating crosstalk would be to provide individual reservoirs for all channels within the second dimension array, thereby isolating each channel both electrically and fluidically. As an example, this approach was chosen for the IEF/SDS–CGE device shown in Figure 33.3. However, there are practical limitations to this solution. As the number of reservoirs increases, so does the challenge of achieving efficient interfacing between off-chip plumbing and on-chip channels. For applications requiring

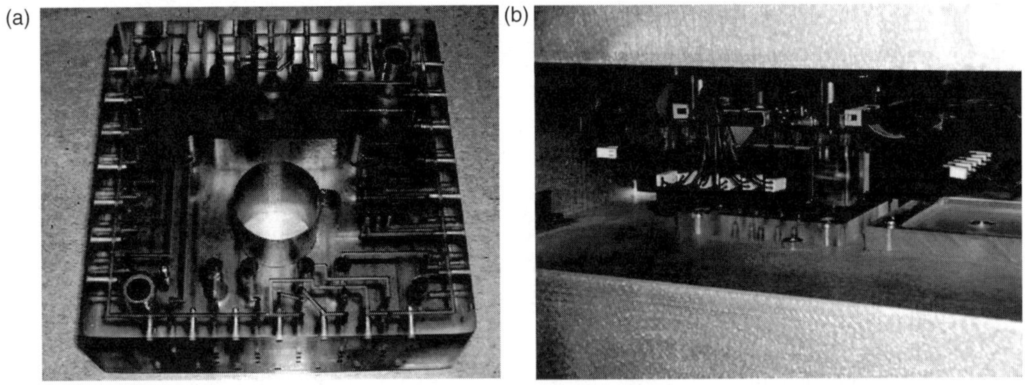

FIGURE 33.6 (a) Manifold interface for providing fluidic interconnects between on-chip microchannels and off-chip pumps, valves, and electrodes and (b) manifold with integrated valves positioned above an IEF/SDS–CGE chip during testing.

pressure injection of sample, buffer, or other solutions into the device, on-chip connections are generally required at each reservoir (with the potential exception of waste reservoirs) to provide interfacing to pumps, valves, and/or electrodes, leading to a complex overall interface. For applications in which simple open reservoirs are suitable, for example, when electrokinetic sample injection is favored over pressure injection, the need for large numbers of interconnects to off-chip components may be alleviated, but additional challenges are introduced by bulk flow induced by hydrodynamic pressure gradients resulting from uneven fluid heights among large numbers of reservoirs in the chip. Furthermore, each added reservoir represents an additional point for bubble formation and entrapment, reducing the robustness of the microfluidic system.

There are several solutions that may be used to address this interface challenge. For the case of the relatively simple IEF/SDS–CGE chip previously described in Figure 33.3a, 38 individual reservoirs were ultimately required for effective operation of the system, with a manifold interface used to simplify the fluidic and electrical interface. The manifold consisted of a machined PMMA block as shown in Figure 33.6, with O-ring compression seals and integrated valves, which provided leak-free hydrodynamic pressure control for chip preparation and sample introduction, and electrodes for electrokinetic control. The use of manifolds for microfluidic interfacing can substantially reduce the time and effort required for chip preparation and testing by eliminating the need for discrete fluidic or electrical ports to be added to each on-chip reservoir. Another solution is to reduce the required number of reservoirs by connecting limited numbers of channels to common fluid ports for off-chip access. Consider the example of a chip design comprising 32 second dimension channels shown in Figure 33.7a. Each set of eight adjacent channels within the upper and lower arrays are connected to a single reservoir, substantially reducing the total number of required ports. Although channels within each group remain subject to crosstalk during the first dimension separation, the overall current uniformity is greatly improved by segregating each group from the others as revealed in Figure 33.7b.

33.3.3 Sample Dispersion and Loss at Channel Intersections

When spatial or time multiplexing is employed as a multidimensional separation strategy with a first dimension based on an electrokinetic separation mode, an inherent difficulty arises from the need for the first dimension separation channel to be intersected by one of more second dimension channels. Regardless of whether the first dimension separation is operated in a transient (e.g., CZE) or steady-state (e.g., IEF) mode, electric field lines extending into the intersecting channels result in dispersion of sample out of the first dimension channel, and ultimately to sample loss as diffusion

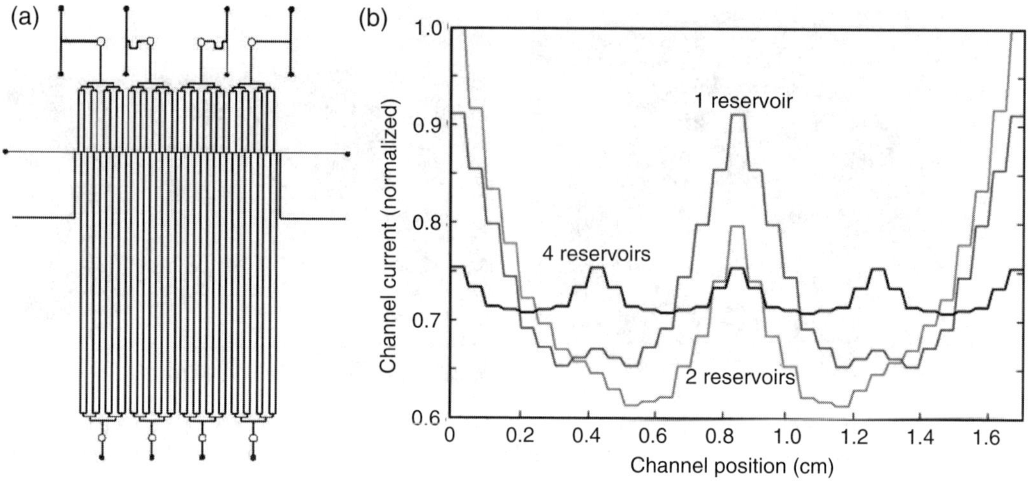

FIGURE 33.7 (a) Schematic representation of 2-D separation chip containing 32 second dimension microchannels, with groups of 8 second microchannels terminating at one of four common reservoirs and (b) model results showing relative current uniformity within the first dimension channel when using 1, 2, and 4 reservoirs.

allows sample to travel beyond the influence of the radiating electric field lines. This is of particular concern when analyte passing the intersections is in a concentrated band, since a substantial portion of the analyte may be lost into the interconnecting channel, while the remaining portion suffers from broadening after passing the intersection.

Li et al.[18] addressed this issue by integrating multiple separation media, with relatively low resistivity ampholyte and buffer in the first separation dimension for IEF and higher resistivity PEO gel used for the second dimension for CGE. As a result, the electric field remained largely constrained within the first dimension microchannel during IEF, and minimal sample dispersion was observed. In addition to constraining the electric field, low diffusion in the gel allowed dispersed analyte to be mobilized back into the first dimension channel without suffering a high degree of loss. As discussed previously, another benefit of this approach is that the multiple separation media further enabled effective sample stacking as focused analyte was electrokinetically transferred to the second separation dimension. Cui et al.[38] proposed a solution based on the integration of addressable electrodes near each channel intersection to control the local current streamlines and thereby limit sample dispersion and loss. This is a more general approach that may be applied to a range of multidimensional systems. A potential disadvantage is that additional current is injected by each electrode, leading to current and electric field gradients within the first dimension separation channel, which can affect separation performance.

33.3.4 Inhibiting Bulk Flow in Multidimensional Microchannel Networks

An important practical issue that affects many microfluidic systems, and multidimensional microfluidics in particular, is unwanted hydrodynamic flow induced by on-chip pressure gradients. Owing to the complexity of the microchannel networks and relatively large numbers of reservoirs or other off-chip interconnects required to operate multidimensional separation chips, there is a large potential for pressure gradients to be induced by sources such as gravity (due to uneven fluid heights in the reservoirs), variations in capillary forces at the reservoirs, or trapped air bubbles within the reservoirs or microchannel networks. In the authors' experience, fully eliminating these pressure gradient sources in complex microfluidic devices is rarely feasible, even when employing automated manifold interfaces, taking great care in chip preparation, and allowing pressure gradients to equilibrate over long periods of time. A more effective approach is to minimize hydrodynamic flow through the use of media, which inhibit bulk fluid motion, either by increasing the effective viscosity of fluids or by generating a physical barrier to flow. For example, Cui et al.[19] added a 2.5% solution of methylcellulose to all reservoirs in a multidimensional IEF chip to limit the intrusion of reservoir solutions into the separation channels, while the replaceable PEO gel used as a second dimension separation medium by Li et al.[18] served a similar purpose. In general, multidimensional systems employing gel-based separations are likely to suffer from less bulk flow than those using free solutions, as are systems employing chromatographic media that require high-pressure gradients to generate bulk flow.

33.4 CONCLUDING REMARKS

The development of novel platforms for multidimensional separations of proteins and peptides remains an area of great interest throughout the analytical community. Microfluidic technologies offer real advantages over capillary-based systems in this area, particularly for the analysis of limited samples. Despite the promise and numerous examples of multidimensional microfluidic platforms for proteomic analysis, a number of challenges remain before these systems can become competitive with modern capillary systems such as 2-D liquid chromatography system for the analysis of complex substrates. The authors believe that as multidimensional microfluidic systems continue to evolve, the benefits of the technology will become increasingly evident. Specific areas that require attention include improved methods for fluidic interfacing and on-chip valving, which do not substantially impact system complexity or cost. Another important area of research currently being addressed by a number of groups is the development microfluidic systems with novel coupled separation dimensions that may offer improvements in separation efficiency. Interfacing spatially-multiplexed systems with MS also remains a significant challenge, and although a variety of solutions have been proposed and explored more efforts are needed to achieve parity with serial capillary ESI–MS.

New development efforts in the area of time-multiplexed multidimensional separations may also provide real-world benefits for proteomic analysis. One of the difficulties with time multiplexing relates to sample loss resulting from the inability to fully sample all separated analyte bands within the first dimension microchannel. When a transient separation is employed in the first dimension, sample loss is inherent in this configuration due to the inability to collect fractions while the second dimension separation is proceeding. The difficulties with sample loss imposed by the serial nature of the second separation dimension could potentially be alleviated through the use of on-chip trap columns to store fractions before injection into the second dimension channel, providing parity between the desired rate of fraction acquisition and the duty cycle of the second dimension separation. Alternately, a combined time- and spatial-multiplexing scheme could be envisioned, wherein multiple second dimension separation channels sequentially sample the first dimension.

Clearly, there are research issues that remain to be addressed before multidimensional microfluidics is commercially competitive against more traditional multidimensional capillary systems. However, with the demand for higher analytical throughput and smaller sample quantities in proteomics continuing to increase, there is an open opportunity for novel multidimensional platforms based on microfluidics to meet this demand. Beyond the concepts described in this chapter such as rapid and parallel separations, zero dead-volume flow control, and effective multichannel MS interfacing, further functionalities are also of great interest, for example, the combination of cell lysis, protein preconcentration, and digestion before multidimensional separation. Such technologies, already demonstrated for related lab-on-a-chip applications, offer an intriguing glimpse of what could ultimately be possible in a fully integrated system for protein or peptide analysis.

REFERENCES

1. Manz, A., Graber, N., Widmer, H. M., *Sens. Actuat.* 1990, *B1*, 244–248.
2. Giddings, J. C., *J. High Res. Chromatogr.* 1987, *10*, 319–323.
3. Giddings, J. C., *United Separation Science*, Wiley, New York, 1991.
4. Corthals, G. L., Wasinger, V. C., Hochstrasser, D. F., Sanchez, J.-C., *Electrophoresis* 2000, *21*, 1104–1115.
5. Jacobs, J. M., Adkins, J. N., Qian, W.-J., Liu, T., Shen, Y., Camp, D. G. I., Smith, R. D., *J. Prot. Res.* 2005, *4*, 1073–1085.
6. Rabilloud, T., *Anal. Chem.* 2000, *72*, 48A–55A.
7. Gygi, S. P., Rist, B., Griffin, T. J., Eng, J., Aebersold, R., *J. Prot. Res.* 2002, *1*, 47–54.
8. Peng, J., Elias, J. E., Thoreen, C. C., Licklider, L. J., Gygi, S. P., *J. Prot. Res.* 2003, *2*, 43–50.
9. Rocklin, R. D., Ramsey, R. S., Ramsey, J. M., *Anal. Chem.* 2000, *72*, 5244–5249.
10. Ramsey, J. D., Jacobson, S. C., Culbertson, C. T., Ramsey, J. M., *Anal. Chem.* 2003, *75*, 3758–3764.
11. Gottschlich, N., Jacobson, S. C., Culbertson, C. T., Ramsey, J. M., *Anal. Chem.* 2001, *73*, 2669–2674.
12. Herr, A. E., Molho, J. I., Drouvalakis, K. A., Mikkelsen, J. C., Utz, P. J., Santiago, J. G., Kenny, T. W., *Anal. Chem.* 2003, *75*, 1180–1187.
13. Wang, Y.-C., Choi, M. H., Han, J., *Anal. Chem.* 2004, *76*, 4426–4431.
14. Shadpour, H., Soper, S. A., *Anal. Chem.* 2006, *78*, 3519–3527.
15. Evans, C. R., Jorgenson, J. W., *Anal. Bioanal. Chem.* 2004, *378*, 1952–1961.
16. Becker, H., Lowack, K., Manz, A., *J. Micromech. Microeng.* 1998, *8*, 24–28.
17. Chen, X., Wu, H., Mao, C., Whitesides, G. M., *Anal. Chem.* 2002, *74*, 1772–1778.
18. Li, Y., Buch, J. S., Rosenberger, F., DeVoe, D. L., Lee, C. S., *Anal. Chem.* 2004, *76*, 742–748.
19. Cui, H., Horiuchi, K., Dutta, P., Ivory, C. F., *Anal. Chem.* 2005, *77*, 7878–7886.
20. Ivory, C. F., *Electrophoresis* 2007, *28*, 15–25.
21. Xue, Q., Foret, F., Dunayevskiy, Y. M., Zavracky, P. M., McGruer, N. E., Karger, B. L., *Anal. Chem.* 1997, *69*, 426–430.
22. Ramsey, R. S., Ramsey, J. M., *Anal. Chem.* 1997, *69*, 1174–1178.
23. Figeys, D., Ning, Y., Aebersold, R., *Anal. Chem.* 1997, *69*, 3153–3160.
24. Oleschuk, R. D., Harrison, D. J., *Trends Anal. Chem.* 2000, *19*, 379–388.
25. Sung, W.-C., Makamba, H., Chen, S.-H., *Electrophoresis* 2005, *26*, 1783–1791.
26. Limbach, P. A., Meng, Z., *Analyst* 2002, *127*, 693–700.
27. Figeys, D., Pinto, D., *Electrophoresis* 2001, *22*, 208–216.
28. Lion, N., Rohner, T. C., Dayon, L., Arnaud, I. L., Damoc, E., Youhnovski, N., Wu, Z. Y., et al., *Electrophoresis* 2003, *24*, 3533–3562.
29. Marko-Varga, G., Nilsson, J., Laurell, T., *Electrophoresis* 2004, *25*, 3479–3491.
30. Foret, F., Kusý, P., *Electrophoresis* 2006, *27*, 4877–4887.
31. Foret, F., Preisler, J., *Proteomics* 2002, *2*, 360–372.
32. DeVoe, D. L., Lee, C. S., *Electrophoresis* 2006, *27*, 3559–3568.
33. Lee, H., Criffin, T. J., Gygi, S. P., Rist, B., Aebersold, R., *Anal. Chem.* 2002, *74*, 4353–4360.
34. Ro, K. W., Liu, J., Knapp, D. R., *J. Chromatogr. A* 2006, *1111*, 40–47.
35. Ericson, C., Phung, Q. T., Horn, D. M., Peters, E. C., Fitchett, J. R., Ficarro, S. B., Salomon, A. R., Brill, L. M., Brock, A., *Anal. Chem.* 2003, *75*, 2309–2315.
36. Wang, Y., Zhou, Y., Balgley, B., Cooper, J., Lee, C. S., DeVoe, D. L., *Electrophoresis* 2005, *26*, 3631–3640.
37. Onnerfjord, P., Ekstrom, S., Bergquist, J., Nilsson, J., Laurell, T., Marko-Varga, G., *Rapid Comm. Mass Spectrom.* 1999, *13*, 315–322.
38. Cui, H., Huang, Z., Dutta, P., Ivory, C. F., *Anal. Chem.* 2007, *79*, 1456–1465.

34 Microchip Immunoassays

Kiichi Sato and Takehiko Kitamori

CONTENTS

34.1 INTRODUCTION

It is predicted that an enormous number of analyses will be required in the future in our daily lives, especially in the clinical and biochemical fields. Rapid progress in molecular biology and biochemistry has brought about a deluge of useful and important information. From these achievements, a significant number of genetic mutations have been elucidated, and a number of biomarkers concordant with disease have been established. These are not only useful for revealing the presence and, potentially, staging the disease but also for the a priori prediction of patient predisposition to diseases. By simple calculation, the number of required diagnostic analyses is thought to be the product of the population size and the number of indices. To perform all of these analyses, it is clear that significant investment of funds, time, and resources will be required, which is very difficult to realize with conventional analytical methods. One form of evolving technology, microfluidic devices or microchips can be a powerful solution to this problem.

Microchip devices for chemical and biochemical analyses have been greatly advanced owing to the progress of microfabrication techniques. Microchemical systems using these devices have attracted much attention, not only from scientists and engineers but also from the clinical, forensic, and biomedical research communities. Thus far, most studies describing microchip-based analytical systems have focused on deoxyribonucleic acid (DNA) analysis by microchip electrophoresis with laser-induced fluorescence detection. These microchip-based electrophoretic systems have great advantages in some applications, especially in the clinical diagnostics and molecular biological fields. Other analytical methods must also be performed for select applications, which involve several distinct biochemical processes. To realize process functionality in these complicated systems, it is necessary to utilize the chemical properties and potential of molecules involved in an effective way.

Immunoassay, founded by Rosalyn Yalow [1] in the late 1950s in the form of the "radioimmunoassay," is still today one of the most powerful and important analytical methods used in clinical diagnoses and biochemical studies, primarily because of its extremely high selectivity and sensitivity. The antibody-based method used today has two generic classifications: homogeneous and heterogeneous. In heterogeneous immunoassays, the antibody is immobilized on a solid surface, while homogeneous assays take place in a solution phase. Integration of both types of immunoassay into microchips has been accomplished with reasonable success by numerous research groups. Most of the homogeneous immunoassays carried out in the microchip format have been based on a microchip electrophoresis [2,3], where the benefit of the chip-based integration was primarily the reduction in separation time. In heterogeneous microchip immunoassay, however, a number of other advantages exist—and this represents the focus of this chapter.

Enzyme-linked immunosorbent assay (ELISA) or other immunosorbent assay systems, in which antigen or antibodies are fixed on a solid surface, are applicable to many analytes with high sensitivity. These are used practically in many fields including clinical diagnoses and life science researches. The conventional heterogeneous immunoassay, however, requires a relatively long assay time and involves troublesome liquid-handling procedures and large quantities of expensive antibody reagents. Moreover, realization of point-of-care (POC) testing is difficult with the conventional immunoassay, since rather large devices are necessary for automated practical diagnosis systems. To overcome these drawbacks, a microchip-based system is effective. The integration of analytical systems into a microchip should bring about enhanced reaction efficiency, a reduced assay time, simplified procedures, and a lowered consumption of samples, reagents, and energy.

The main reason of the long assay time in the conventional heterogeneous assay is that the reaction efficiency is very poor. This is because the reactions occur only on the solid surface. Moreover, it takes long time to complete the reaction due to the long molecular diffusion time. In a microchip, since it is easy to reduce the diffusion distance and increase the surface area-to-volume ratio, the reaction time can be reduced to several minutes rather than hours or days.

A major source of the loss of reproducibility in the conventional assay is human error, because the assay procedure contains many troublesome manual operations. In the microchip format, all procedures can be controlled by pumps and valves that are regulated by a computer. Automation is very favorable without the need for any large-scale robotic equipment.

Antibodies, enzymes, and other reagents used for immunoassay are expensive, hence, when used in the conventional method causes a large problem. Moreover, in many cases, samples are very precious or only a very small amount of samples can be obtained. Therefore, any reduction of the consumption in the microchip systems is welcomed.

34.2 BASICS OF IMMUNOASSAY MICROCHIPS

Microfluidic devices for heterogeneous immunoassays consist of microchannels for transporting solutions, reaction solid phase, and detection area. There are several important factors to develop a high performance system.

Microchip substrate is open to a variety of materials. Silicon wafer is a good material to build up microstructures if fabrication facilities are available. Glass has good chemical and optical properties, and some polymers are cost effective for mass production. The surface treatment of the microchannel is critically important for all materials. This is because of the nonspecific adsorption of the analytes and antibodies to the channel wall that will result in considerable analytical error. It is very important to modify the surface with some blocking reagents or other materials to prevent protein adsorption before experiments. Therefore, we must choose a material of which, surface chemistry is well understood.

Reaction solid phase is the core of the immunoassay microchip where the primary antibody (capture antibody) is immobilized on the surface. There are some formats for the immobilization (Figure 34.1). The simplest method is a fixation of the antibody to the channel wall [4]. Several strategies have been used to adsorb or fix antibodies on the surface (e.g., direct adsorption), covalent bonding to react with function groups on the surface, and microcontact printing. While direct adsorption is common on hydrophobic polymer surfaces, covalent bonding by silanization reagents is used for silicon and glass surface. Since the amount and activity of the immobilized antibody is strongly dependent on the immobilization method, study of the fixation method is important [5].

Another type of solid support for capture antibody is microbeads. Beads coated with antibodies are packed in the microchannel with a dam or cage structure. Bead format is considerably superior to

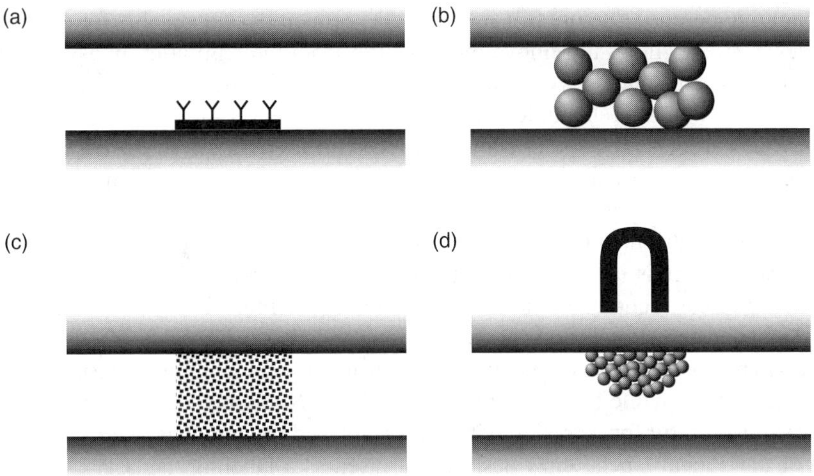

FIGURE 34.1 Formats of the reaction solid phase for microchip immunoassay. (a) channel wall, (b) packed microbeads, (c) monolithic structure, and (d) magnetic beads.

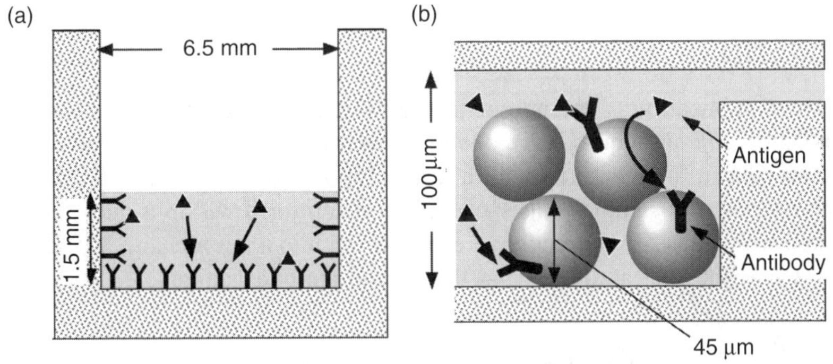

FIGURE 34.2 Schematic illustrations of the antigen–antibody reaction. (a) Microtiter plate and (b) microchip.

the wall surface immobilization methods. Microbeads bring extremely large surface area-to-volume ratio and decrease the size of the liquid phase (i.e., decreasing the diffusion distance). These effects are useful to reduce the assay time dramatically. An example is shown in Figure 34.2.

The surface area-to-volume ratio of a 50-μL solution in a microtiter plate well (0.65 mm in diameter) was estimated to be 13 cm^{-1}, whereas that of the microchannel (11 beads in 100 μm × 100 μm × 200 μm channel space) was 480 cm^{-1}. Therefore, the surface area-to-volume ratio the microchannel was 37 times larger than that of the microtiter plate, and the reaction rate may be increased because of this larger reaction field.

In the case of the conventional microtiter plate assay, a 1.5-mm movement would be necessary for the most distant located antibody molecule to react with an antigen fixed on the surface of the well, since the liquid depth was 1.5 mm. On the other hand, the liquid phase of the microchannel filled with polystyrene beads was much smaller. The maximum distance from an antibody molecule to the reaction-solid surface will not exceed 20 μm. Because the diffusion time is proportional to the squares of the diffusion distance, the diffusion time of the antibody molecule to the antigen in the microchip would be more than 5600 times shorter than in the conventional method [6].

The monolithic porous structure has also the same properties and suitable for the reaction solid phase. However, it is difficult to remove the monolith from the chip after an assay so the chip is then disposed. Hence, this happens even though the change of the beads is very easy. Therefore, the monolithic format is not favorable unless the cost of the chip is very low.

Magnetic fine particles are another choice for the solid support. These particles can be captured in the channel easily with a magnet without any microstructures inside. In many cases, however, the particles are gathered very tightly and reagents are unable to go through them. That limits the number of particles contributed to the reaction.

There are several kinds of detection methods for microchip-based heterogeneous immunoassays. The most popular technique is a fluorescent detection. Fluorescent molecules are used as a labeling material and detection is done at the reaction solid phase. A combination of a laser-induced fluorescence microscope, a highly sensitive detector, and a good labeling material may bring very sensitive determination, while the beads may scatter light. Nonfluorescent material, colloidal gold nanoparticle, is also useful for detection with a laser-induced thermal lens microscope (TLM).

Instead of the optical labels, enzymes are also used as a labeling material [7]. ELISA is one of the most popular techniques in the conventional immunoassay and suitable in the microchip format. By using a fluorogenic reaction, the fluorescent signal can be taken at the downstream microchannel without interruption by microbeads. In the case of chromogenic reaction, a TLM is useful [8].

34.3 APPLICATIONS

An example of the application of the microchip-based immunosorbent assay is shown below. The human carcinoembryonic antigen (CEA), one of the most widely used tumor markers for serodiagnosis of colon cancer, was assayed with a microchip-based system [9]. An ultratrace amount of CEA dissolved in serum samples was successfully determined within a short time. Polystyrene beads pre-coated with anti-CEA antibody were introduced into a microchannel. Then a serum sample containing CEA, the first antibody, and the second antibody conjugated with colloidal gold,

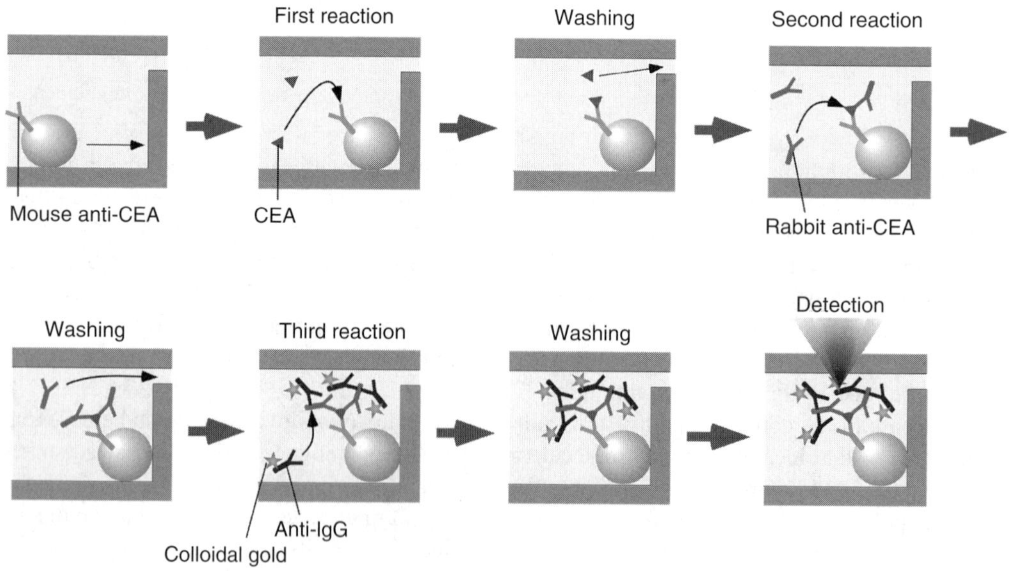

FIGURE 34.3 Schematic illustrations of the microchip-based immunosorbent assay using microbeads.

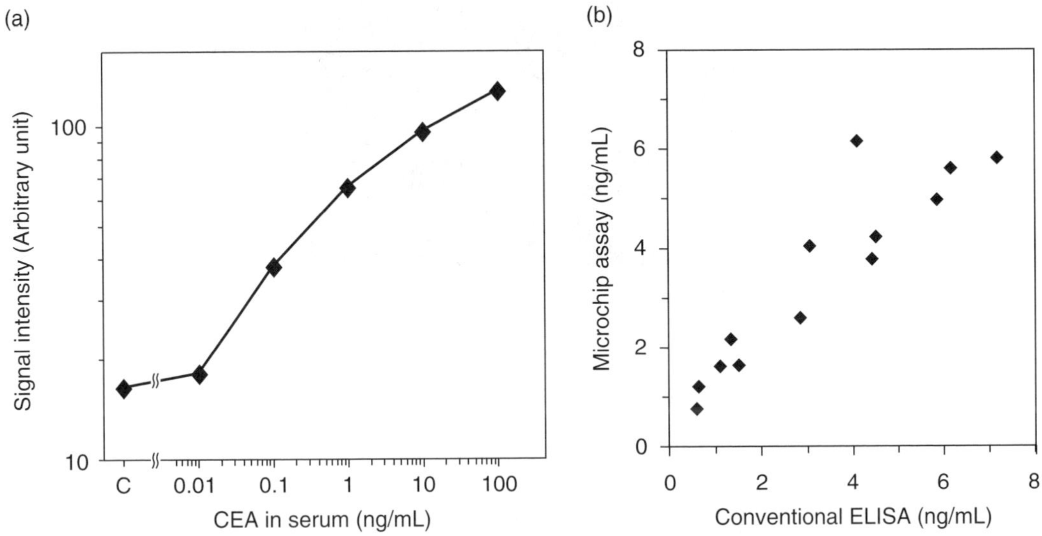

FIGURE 34.4 Determination of a tumor marker, human CEA. (a) Calibration curve for CEA in human sera. (b) Correlation between the conventional ELISA and the microchip-based immunoassay.

were reacted successively (Figure 34.3). The resulting antigen–antibody complex, fixed on the bead surface, was detected using a TLM. A highly selective and sensitive determination of an ultratrace amount of CEA in human sera was made possible by a sandwich immunoassay system that requires three antibodies for an assay. A detection limit in the orders of magnitude lower than the conventional ELISA was achieved (Figure 34.4). Moreover and pertinent to the clinical application of this technology, the analysis of serum samples from 13 patients showed a high correlation ($r = .917$) with the conventional ELISA. This integration reduced the time necessary for the antigen–antibody reaction to ~1%, thus shortening the overall analysis time from 45 h to 35 min. This microchip-based diagnostic system might be the first micrototal analysis system (μTAS) to show a practical usefulness for clinical diagnoses with short analysis times, high sensitivity, and easy procedures.

In numerous microchip systems, researchers have shown that multiplexed analysis can be achieved by increasing the number of channels available for parallel processing of numerous samples simultaneously for protein crystallization [10] and genetic [11] analysis. This type of higher integration has been achieved for multiplexed immunoassay analysis capable of processing several samples simultaneously [12]. In this integrated system, the chip had branching multichannels connected to four reaction and detection regions; thus, the system could process four samples at a time with only a single fluid pumping unit (Figure 34.5). To show utility, interferon gamma was assayed by a three-step sandwich immunoassay with the system coupled to a TLM as a detector. The biases of the signal intensities obtained from each channel were within 10%, with percent coefficient of variance (CVs) were almost the same level as the single straight channel assay. The total assay time for all four samples was 50 min when compared with 35 min for one sample in the single channel assay; hence, a higher throughput was realized with the branching structure chip. The simultaneous assay of many samples may also be achieved by simply arraying many channels in parallel on a chip. This approach, however, requires many pumps and capillary connections, and high integration seems to be difficult. On the other hand, a microchip with branching microchannels seems to be suitable for carrying out the simultaneous assay, since the numbers of pumps and capillary connections required for the system can be minimized.

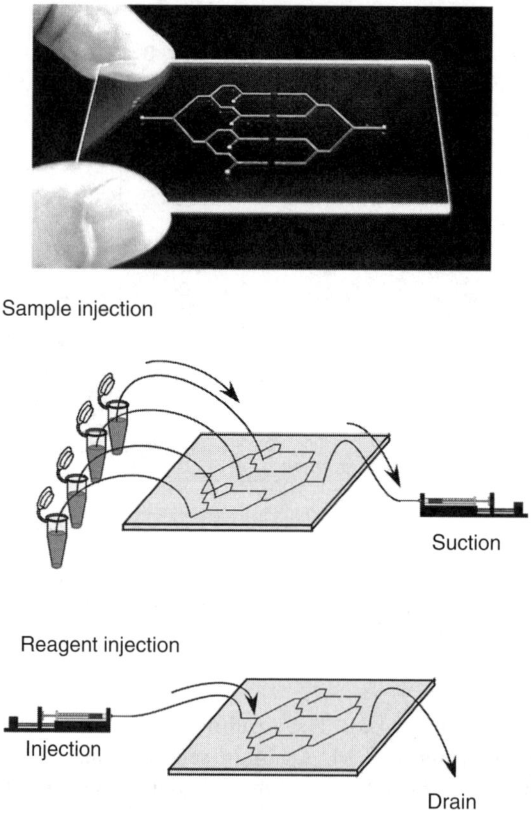

FIGURE 34.5 Immunoassay chip with a branching microchannels for simultaneous assay.

34.4 FUTURE DIRECTIONS

It is clear that much research has been done in this area, with the general sense that the microfluidic immunoassay can be superior to conventional immunoassay systems. It is interesting that, despite the large potential market size, microchip-based immunoassay systems have not yet been commercialized. One of the main reasons is the difficulty associated with creating a truly high-throughput assay without interfacing with automated fluidic control systems. While one microchip immunoassay can be completed in 30 min with manual operation, 96 assays can be completed within several hours in the conventional format. Therefore, it is clear that parallelization and automation will be required to realize the full potential. However, this will require the development and perfection of compact, cheap, precise, and automatic fluidic controller systems (including pumps, valves, and mixers). Precise control of submicroliter solutions without dead volume is a technical challenge and it is difficult to realize it only with inexpensive devices—however, such developments are underway in numerous labs. In addition, assay cost presents a problem, largely due to the cost of a 96-well microtiter plate at less than \$2 dollars, while microchip costs could approach \$100 dollars. Clearly, developing a cost-effective mass production approach will be a necessity. A candidate design for the automated and compact system might be a compact disk (CD) format. It is very easy to array more than 96 microchannels radially on a CD and all solutions/reagents can be controlled by centrifugal force without any pumping. The structured portable CD player or CD drive in a laptop PC might be a good analyzer as it can control the rotation speed precisely and has optical readout and could accommodate fluorescent detection. Ultimately, development of a functional, easy-to-use, and cost-effective system is a key to the acceptance and dissemination of such technology.

REFERENCES

1. Yalow, R. S., Berson, S. A. *J. Clin. Invest.* 1960, *39*, 1157–1175.
2. Koutny, L. B., Schmalzing, D., Taylor, T. A., Fuchs, M. *Anal. Chem.* 1996, *68*, 18–22.
3. Chiem, N. H., Harrison, D. J. *Clin. Chem.* 1998, *44*, 591–598.
4. Bernard, A., Michel, B., Delamarche, E. *Anal. Chem.* 2001, *73*, 8–12.
5. Yakovleva, J., Davidsson, R., Lobanova, A., Bengtsson, M., Eremin, S., Laurell, T., Emneus, J. *Anal. Chem.* 2002, *74*, 2994–3004.
6. Sato, K., Tokeshi, M., Odake, T., Kimura, H., Ooi, T., Nakao, M., Kitamori, T. *Anal. Chem.* 2000, *72*, 1144–1147.
7. Eteshola, E., Leckband, D. *Sens. Actuators B-Chem.* 2001, *72*, 129–133.
8. Sato, K., Yamanaka, M., Hagino, T., Tokeshi, M., Kimura, H., Kitamori, T. *Lab Chip* 2004, *4*, 570–575.
9. Sato, K., Tokeshi, M., Kimura, H., Kitamori, T. *Anal. Chem.* 2001, *73*, 1213–1218.
10. Hansen, C.L., Skordalakes, E., Berger, J.M., Quake, S.R. *Proc. Natl Acad. Sci. USA*. 2002, 99, 16531–16536.
11. Yeung, S.H., Greenspoon, S.A., McGuckian, A., Crouse, C.A., Emrich, C.A., Ban, J., Mathies, R.A. *J. Forensic Sci.* 2006, 51, 740–747.
12. Sato, K., Yamanaka, M., Takahashi, H., Tokeshi, M., Kimura, H., Kitamori, T. *Electrophoresis* 2002, *23*, 734–739.

35 Solvent Extraction on Chips

Manabu Tokeshi and Takehiko Kitamori

CONTENTS

35.1 INTRODUCTION

Solvent extraction is a method capable of separating compounds based on their solution preferences for two different immiscible liquids. In other words, it is the extraction of a substance from one liquid phase into another liquid phase. Solvent extraction is a fundamental technique in chemical/biological laboratories and chemical industries, where it is carried out in the macroscale using separatory funnels. The advantage of doing a solvent extraction in the microspace on a chip is the scale merit of using small dimensions, that is, a large specific interface area (the interface-to-volume ratio) and a short diffusion distance, which together results in a short diffusion time. High speed and high performance solvent extraction systems are possible on chips, without the need for mechanical stirring, mixing, or shaking, all of which are necessary for the conventional method.

This chapter focuses on the analytical aspects of solvent extraction on microchips. Although there are several reports in the literature on the application of chemical reactions,[1–7] the topics dealt with here are limited only to application using parallel (side-by-side) flow schemes inside microchannels. Other examples, such as using a porous membrane,[8] metal mesh,[9] or droplets[10] lie outside the scope of this chapter.

35.2 COCURRENT SOLVENT EXTRACTION

35.2.1 SIMPLE APPROACHES

In order to perform an on-chip solvent extraction, a chip with at least two inlets for two immiscible liquids is required. However, the difficulty associated with its use, and the resultant extraction efficiency depends on the shape and dimensions of the channel, as well as other variables. Glass is the substrate usually used when performing solvent extraction using chips due to its chemical inertness toward organic solvents. If a polymer is to be used as the substrate, chemical modification of the channel is typically required.

The first solvent extraction carried out using a two-phase parallel flow in a microchannel was demonstrated by Tokeshi et al.[11] using the typical extraction set up shown in Figure 35.1. An aqueous solution of Fe-bathophenanthrolinedisulfonic acid complex and a chloroform solution of capriquat were introduced into the microchannel, using syringe pumps at a constant flow rate to establish a parallel two-phase flow in the microchannel (Figure 35.2). In the case of ordinary solvent extraction using a separatory funnel, the two solutions in the separatory funnel are separated vertically by the difference in their specific gravities. Generally, in a microchannel, the liquid–liquid interface does not have this vertical arrangement, instead the flows are positioned parallel to the sidewalls

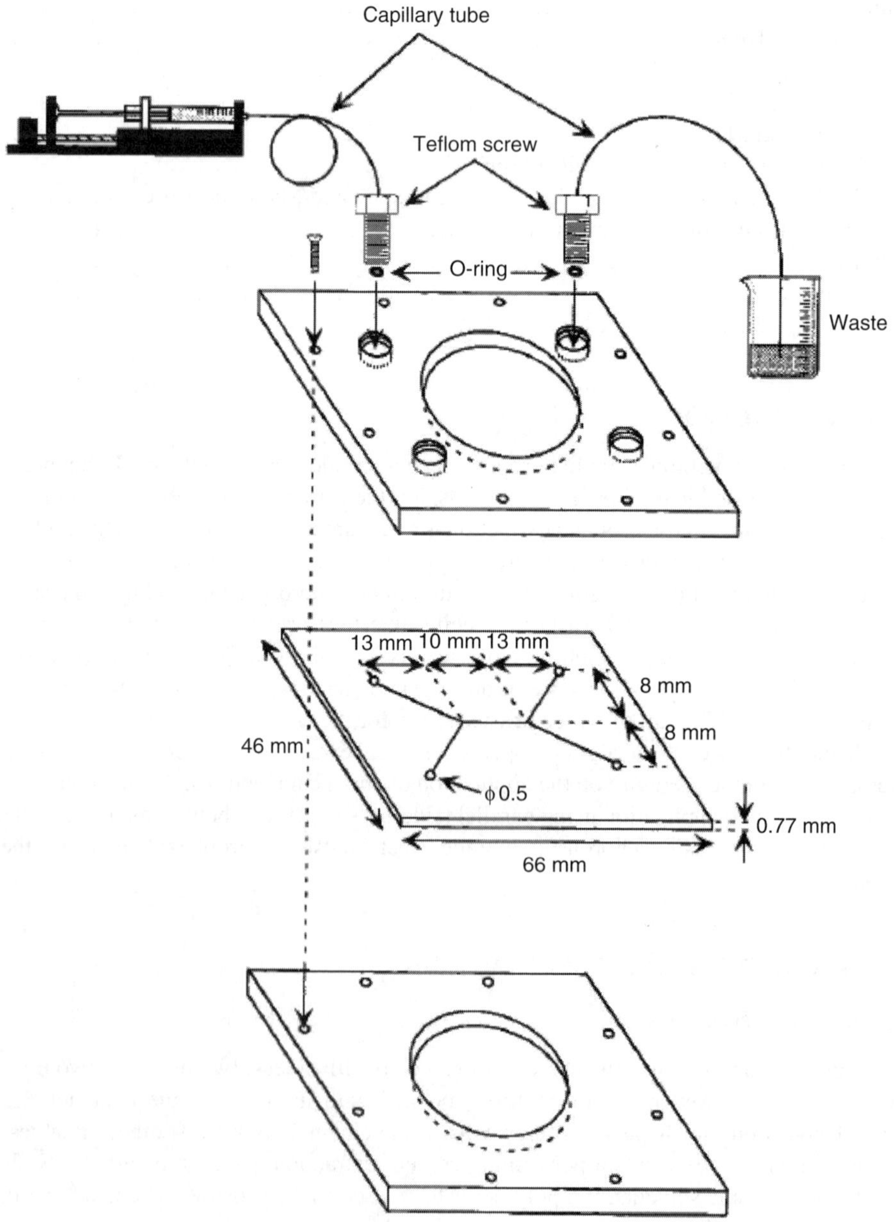

FIGURE 35.1 Schematic illustration of experimental setup. (From Tokeshi, M. et al., *Anal. Chem.*, 72, 1711, 2000.)

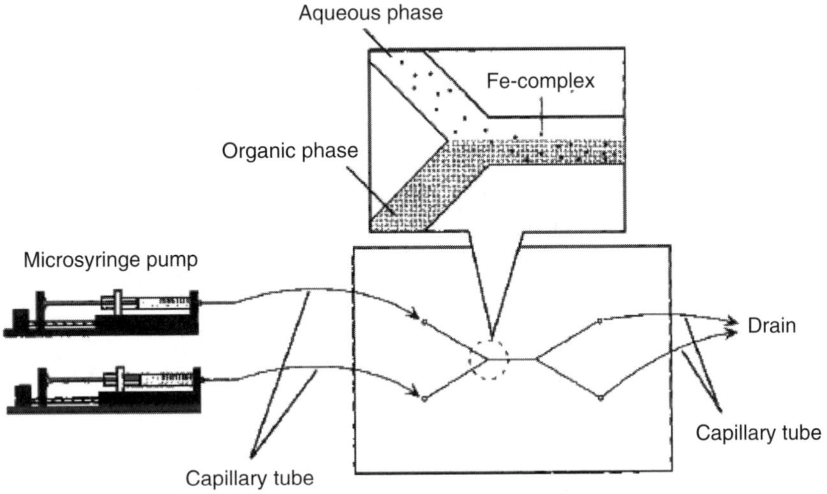

FIGURE 35.2 Schematic illustration of microextraction system. (From Tokeshi, M. et al., *Anal. Chem.*, 72, 1711, 2000.)

of the microchannel. This is due to the fact that the interfacial tension and friction force are much stronger than the force of gravity in the microspace provided by the microchannel.[12] However, this also depends on the microchannel geometry—for example, the properties of the wall surface (hydrophobicity, hydrophilicity, wettability), the properties of liquids (viscosity, interfacial tension), and so forth.[13] The time taken for the extraction in the 250 μm (w) × 100 μm (d) microchannel of Figure 35.2 was determined to be 45 s—this roughly coincides with the molecular diffusion time. The extraction time was also at least one order of magnitude shorter when compared with the conventional extraction time using a separatory funnel and mechanical mixing (shaking). In a similar manner, the Ni-dimethylglyoxime complex,[14] Co-2-nitroso-5-dimethylaminophenol (DMAP) complex,[15] and Al-2,2′-dihydroxyazobenzen (DHAB) complex[16] were all analyzed by performing an extraction at a two-phase interface, water/chloroform, water/toluene, and water/1-butanol, respectively. The determination of radioactive nuclides, such as uranium, was carried out by the extraction of U(IV) from an HNO_3 aqueous solution to 30% (70% *n*-dodecane) and 100% tri-*n*-butyl phosphate phase (TBP).[17] In the conventional method, 100% TBP was not used as an extraction solvent because it is impossible to get a separation between these two phases due to the fact that the specific gravity of the two liquids is almost identical. Alternatively, with the chip, the extraction of U(IV) can be achieved by using a two-phase aqueous/100% TBP flow because specific gravity is not a dominant variable. Moreover, using a three-phase flow, it is possible to obtain a rapid solvent extraction as illustrated in the following example, in which the extraction solvent *m*-xylene, was sandwiched by two aqueous solutions of Co-2-nitroso-DMAP complex to form a microchannel-based three-phase flow (Figure 35.3).[12] The concentration of Co-2-nitroso-DMAP complex in the *m*-xylene phase was determined by using thermal lens microscopy (TLM)[18] and corresponded well with the time taken for the extraction process. The dependence of the TLM signal on the distance x from the junction is shown in the graph of Figure 35.3. The extraction process reached equilibrium after only a distance of 1.5 cm, indicating that the extraction equilibrium was attained within 3 s of contact. The extraction efficiency can, however, be improved by using specially designed microchannels, for example, an asymmetrical zigzag-side-walled microchannel[19] or a microchannel, which has intermittent partition walls in the center of the confluent part.[20]

Uniquely, Minagawa et al.[21] demonstrated the integration of two chemical processes on a chip: a chelating reaction and solvent extraction. The layout of the microchannels fabricated on the glass plates is shown in Figure 35.4 and a schematic illustration of the molecular behavior demonstrated

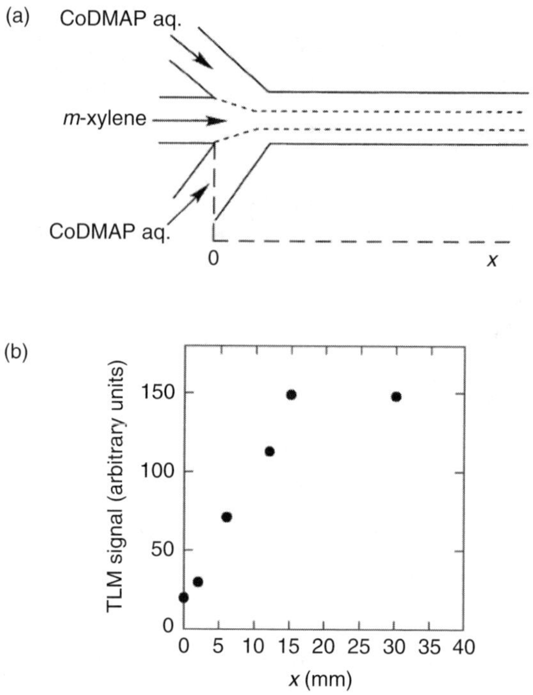

FIGURE 35.3 (a) Schematic illustration of multilayer extraction system. (b) Position dependence of thermal lens signal intensity on distance x. (From Hibara, A. et al., *Anal. Sci.*, 17, 89, 2001.)

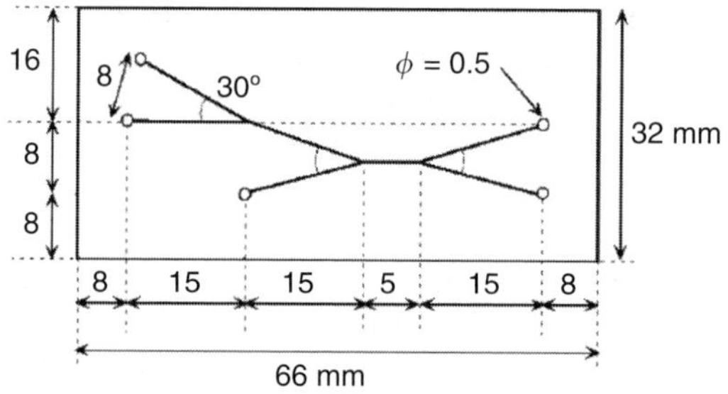

FIGURE 35.4 Layout and dimensions of microchip. Microchannel was 100-μm deep and 250-μm wide. (From Minagawa, T. et al., *Lab Chip*, 1, 72, 2001.)

in the microchannels is given in Figure 35.5. First, the *m*-xylene and the 2-nitroso-1-naphthol (NN) aqueous solutions were introduced into the two inlets of the first Y-shaped microchannel by the first syringe pump at a constant flow rate. These two liquids met at the intersection point, and a parallel two-phase flow, that is, an organic/aqueous interface, was formed in the microchannel. In this region, the NN in the aqueous phase could not be extracted into the *m*-xylene, whether or not the liquids were flowing. Next, the sample solution of cobalt ion [Co(II)] was introduced into the third inlet of the second Y-shaped microchannel using a second syringe pump at the same constant flow rate. Therefore, a three-phase parallel flow was formed upstream of the intersection point of the second Y-shaped structure. There was almost no interdiffusion in the three fluids under the flow condition, although the lack of interdiffusion depended on the flow rate. However, under these experimental conditions, interdiffusion can never occur. Thus, the chelating

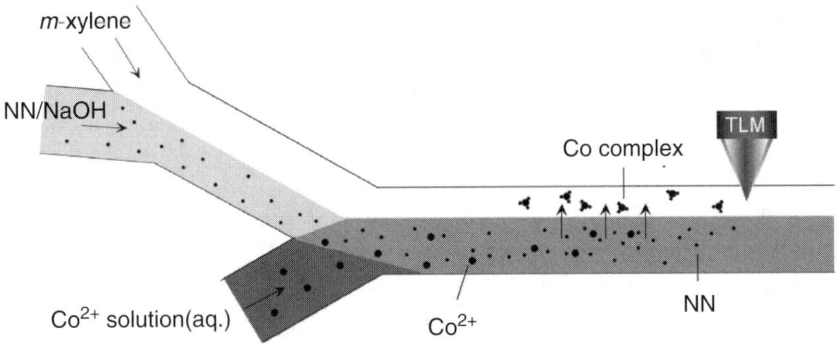

FIGURE 35.5 Schematic illustration of the chelating reaction and solvent extraction in microchannel. (From Minagawa, T. et al., *Lab Chip*, 1, 72, 2001.)

reaction of Co(II) with the NN and the extraction of Co chelates did not occur under flow. When the flow was stopped, the NN solution and the Co(II) sample solution promptly mixed, and Co(II) reacted with NN. Then, the reaction product, Co chelate in the aqueous phase was extracted into the organic phase. By using on-chip detection with a thermal lens microscope, troublesome operations such as the phase separation necessary for the conventional system can be avoided.

35.2.2 Ion Sensing

The liquid microspace provides a short diffusion distance, and large specific interfacial area, of the liquid–liquid interface. As a result several novel and attractive analytical features arise including extremely fast ion sensing, and a need for only an ultra small volume of reagent solution. In contrast to the slow response time of a standard ion-selective optode, where the response time is basically governed by the slow diffusion of ionic species in the viscous polymer membrane, that of the on-chip ion-sensing system is clearly faster due to the short diffusion distance and low viscosity of organic solution. The ion-pair extraction scheme is an established methodology employed for ion-selective optodes, which provide highly selective optical ion determination for various kinds of ions, using a single lipophilic pH indicator dye and a highly selective neutral ionophore. However, exploitation of an ion-pair extraction reaction and chip technology provide attractive advantages that would not be achieved by conventional ion sensors. The advantages of on-chip ion-sensing systems are as follows:

1. Reaction time (response time) can be dramatically reduced by the fast molecular transport achievable in a microspace.
2. Since the required reagent solution volume is extremely small ($\sim$100 nL), a fresh organic phase can be used in every measurement. Subsequently, response degradation caused by the leaching of ion-sensing components, a typical problem of ion-selective optodes, would not need to be taken into consideration. This merit is directly reflected in the excellent reproducibility of the response during continuous measurements and the effective reduction of the amount of expensive reagents consumed in one measurement.
3. Ion determination can essentially be carried out by detection of the protonation/deprotonation process of a single lipophilic anionic dye in the organic phase. Therefore, no special color-responsive chelating reagents are required for the measurement of another kind of analyte ion. One only need to alter the neutral ionophore for different ion-selective ionophores. From the viewpoint of optical instrumentation, this merit is

quite important; there is no need to change the excitation source to match the excitation wavelength of the different chelating reagents.

4. Highly selective ionophores or carriers, for various kinds of ions developed for application to ion-selective electrodes, are commercially available and can be used without any chemical modification.

Figure 35.6 shows a schematic illustration of the experimental setup and ion-pair extraction scheme.[22] The organic solution containing a neutral ionophore, a lipophilic pH indicator dye, and an aqueous solution containing sample ion (K^+ or Na^+) were independently introduced into the microchannel to form an organic/aqueous interface. Then determination of the ion was done using a thermal lens microscope positioned downstream, where the ion-pair extraction occurred in the organic phase under continuous-flow conditions. The response time and minimal required volume of reagent for the on-chip ion-sensing system were about 8 s and 125 nL, respectively.

In other work, a sequential multi-ion-sensing system using a single chip was successfully identified by expanding the concept of on-chip ion sensing.[23] Figure 35.7 shows the basic concept for this chip-based multi-ion sensor. Different organic phases containing the same lipophilic pH indicator dyes, but different ionophores were introduced sequentially into the microchannel by the on–off switching of syringe pumps. In this case, having each organic phase contain different ionophores avoids contamination. The aqueous sample solution containing the different ions is introduced from the other inlet, to form a parallel two-phase flow with the intermittently pumped

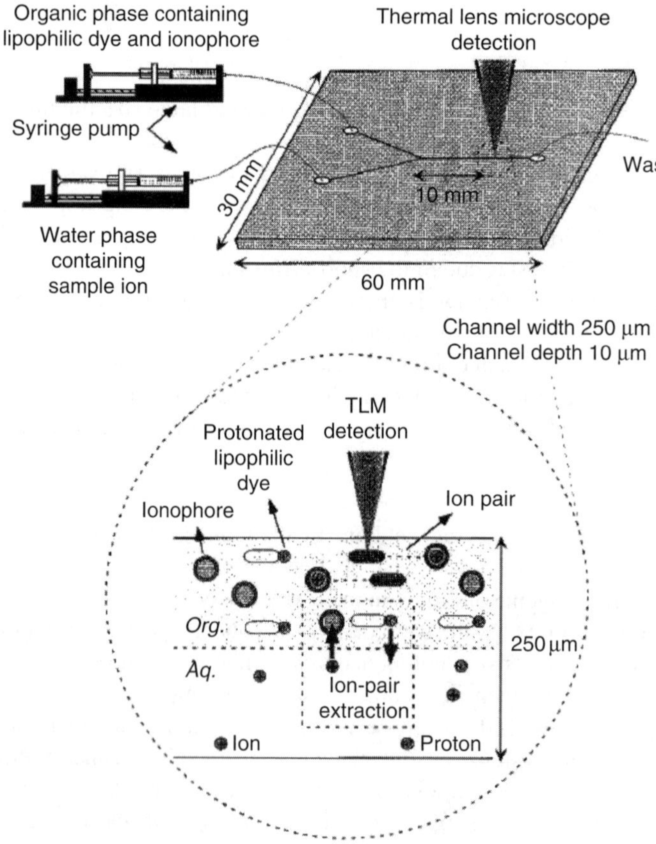

FIGURE 35.6 Schematic illustration of experimental setup and ion-pair extraction in microchannel. (From Hisamoto, H. et al., *Anal. Chem.*, 73, 1382, 2001.)

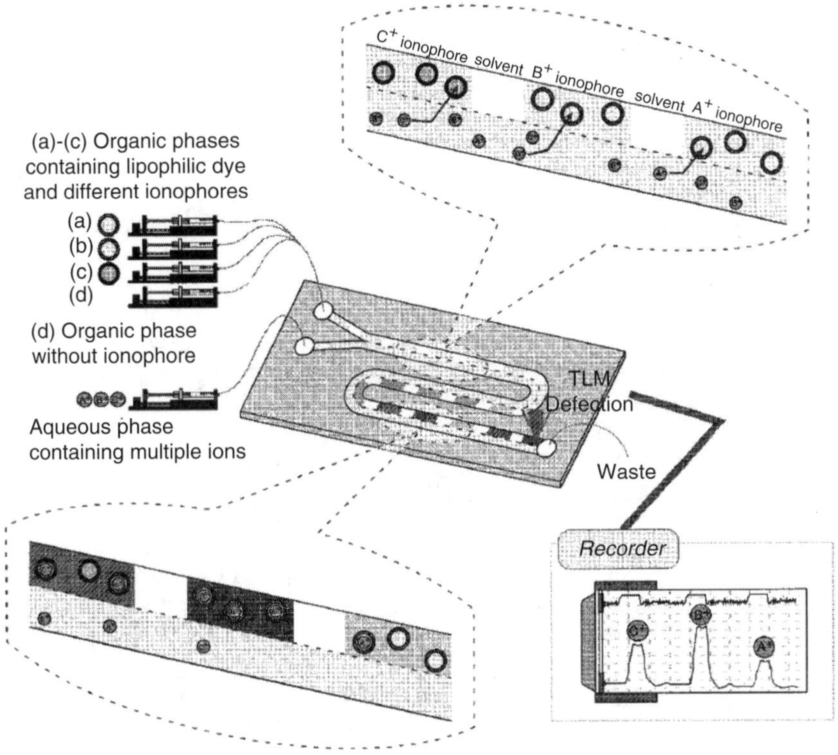

FIGURE 35.7 Concept of sequential ion-sensing system using single microchip. (From Hisamoto, H. et al., *Anal. Chem.*, 73, 5551, 2001.)

organic phases. The selective ion-pair extraction reaction proceeds during flow; thus, different ions can be selectively extracted into different organic phases, depending on the selectivity of the neutral ionophores contained in the respective organic phases. Downstream in the flow, the ion-pair extraction reaction becomes equilibrated; thus, downstream detection of the color change of the organic phase, allowed for sequential and selective multi-ion sensing in the single aqueous sample solution containing multiple ions. In this case, valinomycin and 2,6,13,16,19-pentaoxapentacyclo-$[18.4.4.4.^{7,12}0.^{1,20}0^{7,12}]$dotriacontane (DD16C5)—known to exhibit high selectivity when used in conventional ion sensors—were used as highly selective potassium and sodium ionophores, respectively. Three types of aqueous sample solutions were analyzed with the system: a buffer solution containing 10^{-2} M K^+, 10^{-2} M Na^+, or both ions. When the aqueous phases containing a single type of ion were used, selective extractions occurred in each case; that is, potassium ions were extracted only for an organic phase segment containing valinomycin, and sodium ions were extracted only for that containing DD16C5. When the aqueous phase containing both ions was examined, both ions were independently extracted into different organic phases, depending on the nature of the ionophores in the respective organic phase (Figure 35.8). The minimum volume of single organic phase needed to obtain an equilibrium response without dilution by cross dispersion of two organic phases was ~500 nL in the system, indicating that the required amounts of expensive reagents in one measurement could be reduced to a few nanograms.

35.2.3 Liquid Membrane

A liquid membrane, composed of an organic phase with two aqueous phases, has a wide variety of industrial and analytical applications, including separation, concentration, and removal of

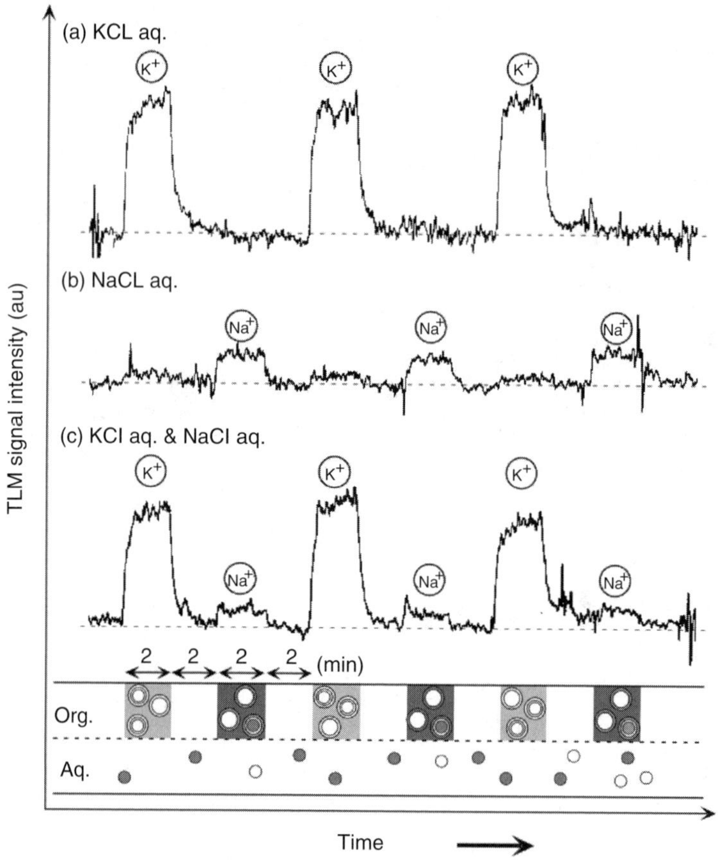

FIGURE 35.8 Response profiles for different aqueous solutions obtained by intermittent pumping of organic phase: (a) Aqueous solution containing 10^{-2} M KCl, (b) aqueous solution containing 10^{-2} M NaCl, and (c) aqueous solution containing both 10^{-2} M KCl and 10^{-2} M NaCl. (From Hisamoto, H. et al., *Anal. Chem.*, 73, 5551, 2001.)

analytes from wastewater, environmental, and biomedical samples. Surmeian et al.[24] demonstrated molecular transport on a chip using a stable three-layer flow membrane system, water/organic and solvent/water. Under continuous-flow conditions, the analyte (methyl red) was rapidly extracted across the microchannel from the donor to the acceptor phase through the organic phase (cyclohexene). The thickness of the organic phase, sandwiched by the two aqueous phases, was ~64 μm, and it was considered a thin liquid organic membrane. Permeability studies identified the effects of molecular diffusion, layer thickness, and organic solvent-water partition coefficient on the molecular transport. In the chip, complete equilibration was achieved in several seconds, in contrast to a conventional apparatus, where it would require tens of minutes for the comparable extraction. Maruyama et al.[25] showed another selective separation by using a three-layer flow membrane, water/*n*-heptane/water, to separate yttrium ions within a few seconds. In studies such as these, the formation of a stable liquid/liquid interface inside the microchannel under flow conditions is very important. The use of the microchannel structures and surface modification are particularly effective for stabilization of the liquid/liquid interface inside the microchannel.[7,20,24–28]

35.2.4 Continuous Flow Chemical Processing

The integration of chemical processes on a chip by using continuous-flow chemical processing (CFCP) in combination with microunit operations, such as solvent extraction, phase separation, and

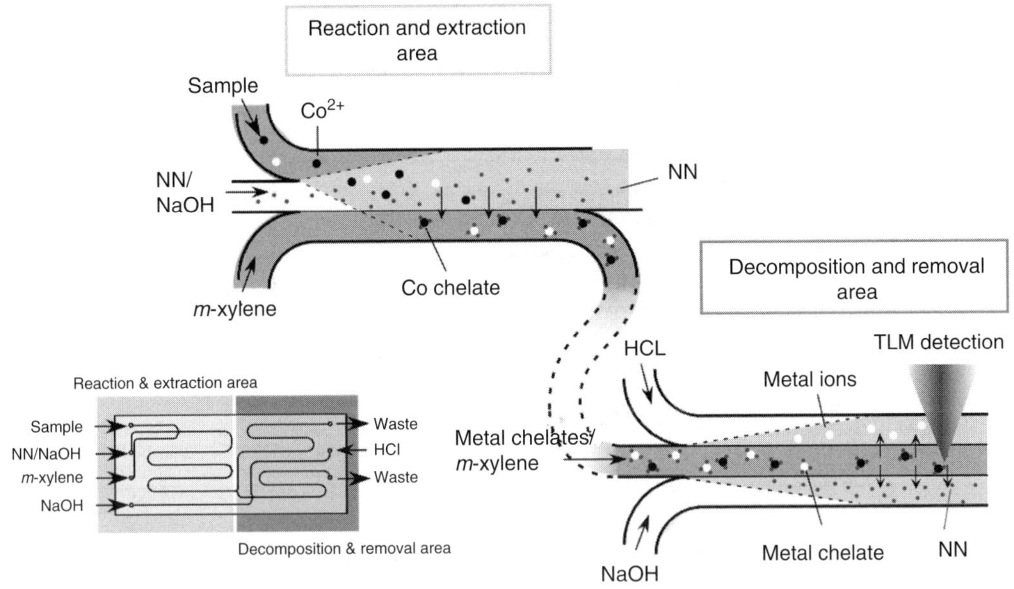

FIGURE 35.9 Schematic illustration of Co(II) determination by using CFCP.

so forth, under continuous flow conditions is a strategy used for the construction of real micro-total analysis systems.[27–30] The integration of a Co(II) wet analysis provides a good example of CFCP,[27] and Figure 35.9 schematically illustrates this analysis. The microchip consists of two different areas: the former is the reaction and extraction area, and the latter is the washing, that is, decomposition and removal, area. In the former area, a sample solution containing Co(II) ions, a NN solution and *m*-xylene, are introduced at a constant flow rate through three inlets using syringe pumps. The three liquids meet at the intersection point, and a parallel two-phase flow, consisting of an organic/aqueous interface, forms in the microchannel. The chelating reaction of Co(II) and NN and extraction of the resulting Co(II) chelate proceed as the reacting mixture flows along the microchannel. Since the NN reacts with the coexisting metal ions [such as Cu(II), Ni(II), and Fe(II)], these coexisting metal chelates are also extracted into *m*-xylene. Therefore, a postextraction washing process is necessary for the decomposition and removal of the undesired coexisting metal chelates. The coexisting metal chelates decompose when they make contact with hydrochloric acid, and the metal ions are moved into an HCl solution (back extraction). The decomposed chelating reagent, NN, is dissolved in a sodium hydroxide solution. In contrast to the coexisting metal chelates, the Co chelate is stable in HCl and NaOH solutions, and remains. In the latter (washing) area, the *m*-xylene phase containing Co chelates and the coexisting metal chelates from the former (reaction and extraction) area is interposed between the other two inlets at a constant flow rate. Then, the three-phase flow, HCl/*m*-xylene/NaOH, forms in the microchannel. The decomposition and removal of the coexisting metal chelates proceed along the microchannel in the same manner as described above. Finally, the target chelates in *m*-xylene are detected downstream. Cobalt in an admixture sample was successfully determined with a rapid analysis time of less than 1 min. The advantages of this approach, compared with a conventional method, are the simplified nature of the procedure and avoiding troublesome operations. In the conventional method, the acid and alkali solutions cannot be used simultaneously, and alternative washing procedures must be repeated several times. The same effect can be obtained by using three-phase flow in the microchannel. In a subsequent paper, Kikutani et al.[31] expanded the concept of CFCP to integrate four parallel analyses of Co(II) and Fe(II) on a chip in a three-dimensional microchannel network.

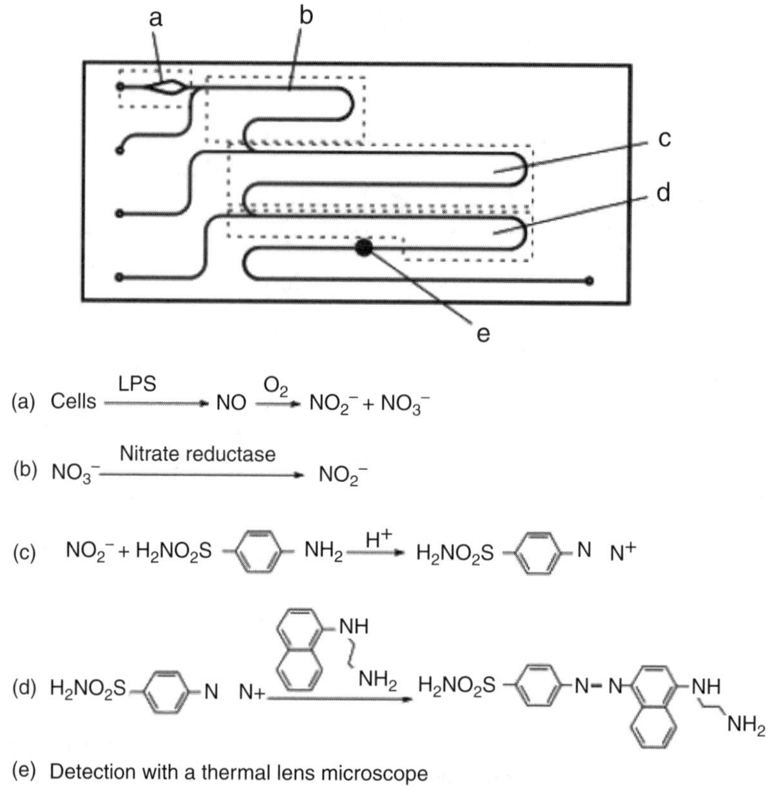

FIGURE 35.10 Chemical processes carried out in microchip for bioassay of macrophage-stimulating agent. (From Tokeshi, M. et al., *Anal. Chem.*, 74, 1565, 2002.)

Goto et al.[32] demonstrated the integration of a bioassay system that illustrated all processes required for a bioassay, that is, cell culture, chemical stimulation of cells, chemical and enzymatic reactions, and detection, on a chip using CFCP (Figure 35.10). By using the temperature control device, spatial temperature control of the system was possible, with areas on the chip maintained at different temperatures. Nitric oxide released from macrophage-like cells stimulated by lipopolysaccharide was successfully monitored by this system. The total assay time was reduced from 24 to 4 h, and the detection limit for nitric oxide was improved from 1×10^{-6} m to 7×10^{-8} M compared with the conventional batch methods. Moreover, the system could monitor a time course of the release, which is difficult to measure by conventional methods.

Smirnova et al.[33] demonstrated the determination of the insecticide, carbaryl, using a two-chip system. The first chip (for the hydrolysis of carbaryl) had a simple Y-shaped channel while the second chip (for the diazo coupling reaction between hydrolyzed products and 2,4,6-trimethylaniline)—the extraction required special channel shapes with a partial surface—modification obtained by using capillary-restricted modification (CARM) (Figure 35.11).[34] Determination of carbaryl pesticide in water with sufficient sensitivity was carried out with an analysis time of 8 min. In a similar manner, Honda et al.[35] developed a combination of a tube-type enzyme-immobilized microreactor and a microextractor with partial surface modification to produce optically pure amino acids.

35.2.5 Sample Preparation for Gas Chromatography and Mass Spectroscopy

As the successful applications of a commercially available liquid chromatography (LC) chip for mass spectroscopy (MS) have been made apparent, the development of sample preparation chips for other

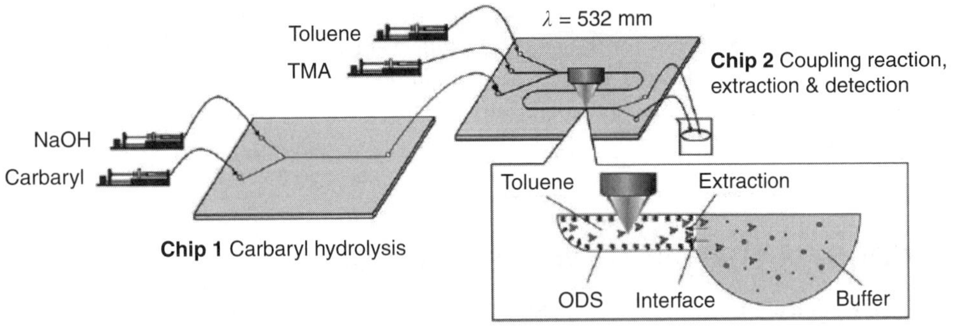

FIGURE 35.11 Concept of integration of carbaryl determination onto microchips. Target carbaryl was hydrolyzed in chip 1 to the product 1-naphthol. In chip 2,1-naphthol was coupled with diazonium ion and was extracted into the organic phase. (From Smirnova, A. et al., *Anal. Chim. Acta*, 558, 69, 2006.)

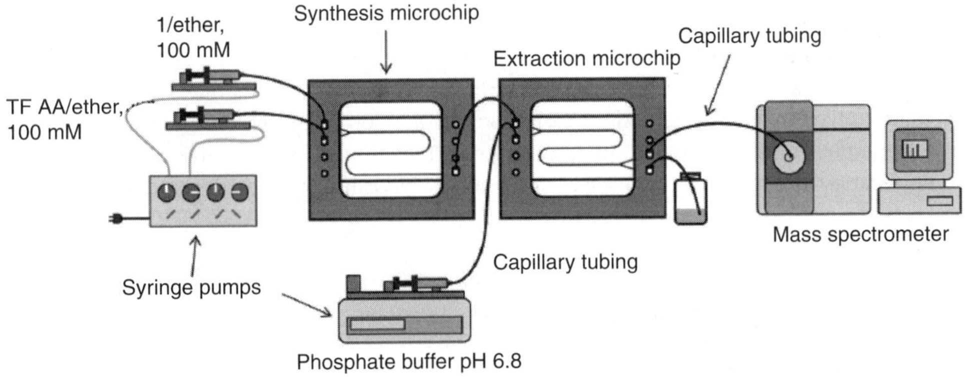

FIGURE 35.12 Schematic illustration of synthesis microchip-extraction microchip-ESI-MS. (From Takahashi, Y. et al., *J. Mass Spectrom. Soc. Jpn.*, 54, 19, 2006.)

conventional analytical systems, such as gas chromatography (GC), are also strongly desirable. As described above, microfluidic chips have great potential to integrate chemical processing such as sample preparation (see Chapter 44 by Biemvenue). Recently, several groups have developed the sample preparation chips for GC and GC–MS, and MS analyses.

Miyaguchi et al.[36] developed a solvent extraction chip for the off-chip GC analysis of amphetamine-type stimulants (a class of illegal drugs) in urine. Microchannels, partially fluoroalkylated by the CARM method,[34] were employed for stabilizing the 1-chlorobutane/alkalize urine interface and obtaining a GC-amenable fraction. As a practical demonstration, methoxyphenamine hydrochloride was administered to three healthy volunteers, and the concentration of methoxyphenamine in their urine determined. This showed the potential of chip-based sample preparation to contribute to the rapid automated analysis in forensic toxicology. Similarly, Xiao et al.[37] carried out solvent extraction of ephedrine using a polydimethylsiloxane (PDMS)/glass chip that had been obtained by surface modification of half of the glass wall, with octadecyltrichlorosilane. On-chip extraction of hydrocarbons from a North Sea oil for GC-MS analysis was also reported.[38]

Figure 35.12 shows a schematic diagram of an online high-throughput detection system for a reaction product synthesized by a microreactor.[39] The system has a synthesis chip that a microreactor for the synthesis of 2,2,2-trifluore-*N*-phenetyl acetamide (TPA), an extraction chip for the purification of TPA from the reaction mixture, and an electrospray ionization mass spectrometer (ESI-MS) for detection. In this system, the extraction chip is placed between the microreactor and the ESI-MS to reduce the ionization hindrances for the reaction products. This placement makes the detection of the reaction product possible under any reaction conditions. The system as described here may be useful for the optimization of reaction conditions and the screening of valuable chemicals such as precursors of pharmaceuticals.

35.3 COUNTER-CURRENT SOLVENT EXTRACTION

From the viewpoint of recovery efficiency, cocurrent solvent extraction as described above, can reach a theoretical plate number of only unity. In contrast to cocurrent solvent extraction, a higher theoretical plate number is expected using counter-current solvent extraction. However, counter-current flow in a microchannel cannot be easily formed because of interfacial tension and viscous force of both (aqueous and organic) phases. In an ordinary microchannel, counter-current flow cannot occur because the two phases collide, and high shear stress at the liquid/liquid interface causes breakup. To form counter-current flow, the aqueous solution must flow along one side of the channel, and the organic solution flow along the other, without breakup. Aota et al.[40] successfully developed a counter-current flow system, which was obtained by selectively modifying the lower half of a microchannel with a hydrophobic group (Figure 35.13). Using this system, they realized that a theoretical plate number of 4.6 was needed for the extraction of a Co(II)-2-nitroso-dimethylaminopyridine (DMAP) complex in an aqueous-toluene counter-current flow. Counter-current flow is expected to be applicable to enrichment processes for various environmental analyses and biomolecular separations.

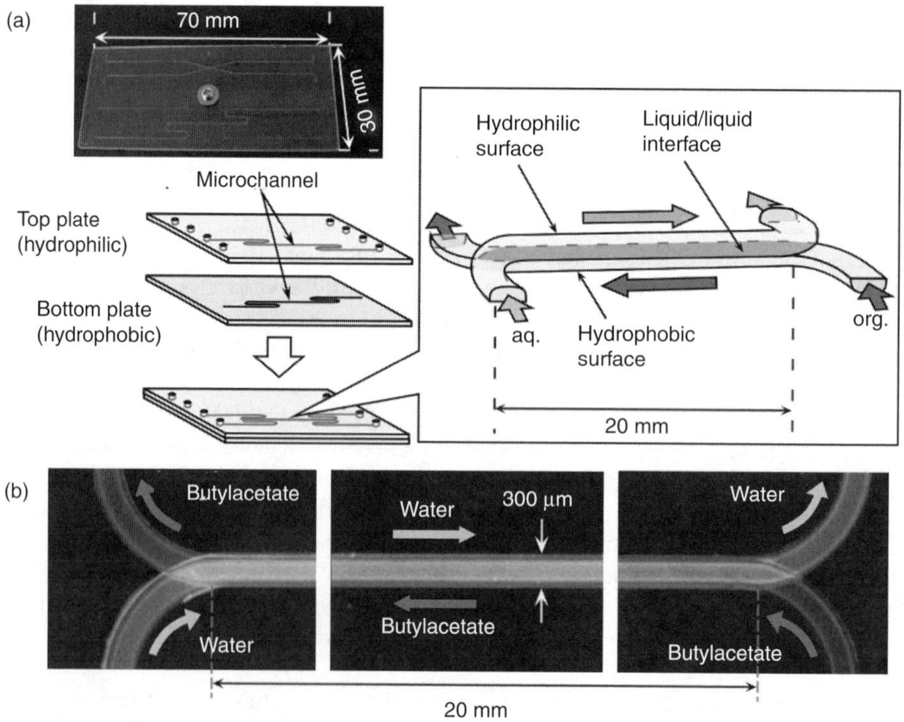

FIGURE 35.13 (a) Photograph of microchip and (b) schematic illustration of counter-current extraction system. (From Aota, A. et al., *Angew. Chem. Int. Ed.*, 46, 878, 2007.)

35.4 FUTURE DIRECTIONS

The development and application of chip-based solvent extraction is gradually expanding as shown in this chapter. With methods for surface modification, the liquid–liquid interface in a microchannel can be stabilized, and so special techniques for solvent extraction are not required. Therefore, since solvent extraction is one of the basic chemical processes, it is expected that it will be used more often in the future.

In the next 5 years, from a practical viewpoint, the combination of solvent extraction chips with conventional analytical apparatuses, such as GC, LC, MS, and so forth, and the integration of solvent extraction and other chemical processes will become more and more important. From the viewpoint of the basic science of two-phase and multi-phase flows, the flow itself near the liquid–liquid interface is very interesting. In fact, chaotic vortices and transient vortex-like flows have been observed around the liquid–liquid interface in the counter-current flow.[41]

ACKNOWLEDGMENTS

We would like to thank our coworkers whose work has been referenced in this chapter. Kanagawa Academy of Science and Technology and The Ministry of Education, Culture, Sports, Science and Technology of Japan are acknowledged for financial support.

REFERENCES

1. Hisamoto, H., Saito, T., Tokeshi, M., Hibara, A. and Kitamori, T., Fast and high conversion phase-transfer synthesis exploiting the liquid–liquid interface formed in a microchannel chip, *Chem. Commun.*, 2662, 2001.
2. Ueno, M., Hisamoto, H., Kitamori, T. and Kobayashi, S., Phase-transfer alkylation reactions using microreactors, *Chem. Commun.*, 936, 2003.
3. Shaw, J., Nudd, R., Naik, B., Turner, C., Rudge, D., Benson, M. and Garman, A., Liquid/liquid extraction systems using micro-contactor arrays, *Proc. μTAS 2000 Symposium*, Enschede, The Netherlands, 371, 2000.
4. Ehrfeld, W., Hessel, V. and Löwe, H., *Microreactors*, Wiley-VCH, Weinheim, 2000, Chapter 5.
5. Kikutani, Y., Horiuchi, T., Uchiyama, K., Hisamoto, H., Tokeshi, M. and Kitamori, T., Glass microchip with three-dimensional microchannel network for 2 × 2 parallel synthesis, *Lab Chip*, 2, 188, 2002.
6. Kikutani, Y., Hibara, A., Uchiyama, K., Hisamoto, H., Tokeshi, M. and Kitamori, T., Pile-up glass microreactors, *Lab Chip*, 2, 193, 2002.
7. Maruyama, T., Uchida, J., Ohkawa, T., Futami, T., Katayama, K., Nishizawa, K., Sotowa, K, Kobota, F., Kamiya, N. and Goto, M., Enzymatic degradation of *p*-chlorophenol in a two-phase flow microchannel system, *Lab Chip*, 3, 308, 2003.
8. Cai, Z.-X., Fang, Q., Chen, H.-W. and Fang, Z.-L., A microfluidic chip based liquid–liquid extraction system with microporous membrane, *Anal. Chim. Acta*, 556, 151, 2006.
9. Wenn, D. A., Shaw, J. E. A. and Machenzie, B., A mesh microcontactor for 2-phase reactions, *Lab Chip*, 3, 180, 2003.
10. Chen, H., Fang, Q., Yin, X.-F. and Fang, Z.-L., Microfluidic chip-based liquid-liquid extraction and preconcentration using a subnanoliter-droplet trapping technique, *Lab Chip*, 5, 719, 2005.
11. Tokeshi, M., Minagawa, T. and Kitamori, T., Integration of a microextraction system on a glass chip: Ion-pair solvent extraction on Fe(II) with 4,7-diphenyl-1,10-phenanthrolinedisulfonic acid and tri-*n*-octylmethylammonium chloride, *Anal. Chem.*, 72, 1711, 2000.
12. Hibara, A., Tokeshi, M., Uchiyama, K., Hisamoto, H. and Kitamori, T., Integrated multilayer flow system on a microchip, *Anal. Sci.*, 17, 89, 2001.
13. Kuban, P., Berg, J. and Dasgupta, P. K., Vertically stratified flows in microchannels. Computational simulations and applications to solvent extraction and ion exchange, *Anal. Chem.*, 75, 3549, 2003.
14. Sato, K., Tokeshi, M., Sawada, T. and Kitamori, T., Molecular transport between two phases in a microchannel, *Anal. Sci.*, 16, 455 2000.

15. Tokeshi, M., Minagawa, T. and Kitamori, T., Integration of a microextraction system: solvent extraction of Co-2-nitroso-5-dimethylaminophenol complex on a microchip, *J. Chromatogr. A*, 894, 19, 2000.
16. Kim, H.-B., Ueno, K., Chiba, M., Kogi, O. and Kitamura, N., Spatially-resolved fluorescence spectroscopic study on liquid/liquid extraction processes in polymer microchannels, *Anal. Sci.*, 16, 871, 2000.
17. Hotokezaka, H., Tokeshi, M., Harada, M., Kitamori, T. and Ikeda, Y., Development of the innovative nuclide separation system for high-level radioactive waste using microchip-extraction behavior of metal ions from aqueous to organic phase in microchannel, *Prog. Nucl. Energy*, 47, 439, 2005.
18. Kitamori, T., Tokeshi, M., Hibara, A. and Sato, K., Thermal lens microscope and microchip chemistry, *Anal. Chem.*, 76, 52A, 2004.
19. Ueno, K., Kim, H.-B. and Kitamura, N., Channel shape effects on the solution-flow characteristics and liquid/liquid extraction efficiency in polymer microchannel chips, *Anal. Sci.*, 19, 391, 2003.
20. Maruyama, T., Kaji, T., Ohkawa, T., Sotowa, K., Matsushita, H., Kubota, F., Kamiya, N., Kusakabe, K. and Goto, M., Intermittent partition walls promote solvent extraction of metal ions in a microfluidic device, *Analyst*, 129, 1008, 2004.
21. Minagawa, T., Tokeshi, M. and Kitamori, T., Integration of a wet analysis system on a glass chip: determination of Co(II) as 2-nitroso-1-naphthol chelates by solvent extraction and thermal lens microscopy, *Lab Chip*, 1, 72, 2001.
22. Hisamoto, H., Horiuchi, T., Tokeshi, M., Hibara, A. and Kitamori, T., On-chip integration of neutral ionophore-based ion pair extraction reaction, *Anal. Chem.*, 73, 1382, 2001.
23. Hisamoto, H., Horiuchi, T., Uchiyama, K., Tokeshi, M., Hibara, A. and Kitamori, T., On-chip integration of sequential ion-sensing system based on intermittent reagent pumping and formation of two-layer flow, *Anal. Chem.*, 73, 5551, 2001.
24. Surmeian, M., Slyadnev, M. N., Hisamoto, H., Hibara, A. and Kitamori, T., Three-layer flow membrane system on a microchip for investigation of molecular transport, *Anal. Chem.*, 74, 2014, 2002.
25. Maruyama, T., Matsushita, H., Uchida, J., Kubota, F., Kamiya, N. and Goto, M., Liquid membrane operations in a microfluidic device for selective separation of metal ions, *Anal. Chem.*, 76, 4495, 2004.
26. Hibara, A., Nonaka, M., Hisamoto, H., Uchiyama, K., Kikutani, Y., Tokeshi, M. and Kitamori, T., Stabilization of liquid interface and control of two-phase confluence and separation in glass microchips by utilizing octadecylsilane modification of microchannels, *Anal. Chem.*, 74, 1724, 2002.
27. Tokeshi, M., Minagawa, T., Uchiyama, K., Hibara, A., Sato, K., Hisamoto, H. and Kitamori, T., Continuous-flow chemical processing on a microchip by combining microunit operations and a multiphase flow network, *Anal. Chem.*, 74, 1565, 2002.
28. Kikutani, Y., Ueno, M., Hisamoto, H., Tokeshi, M. and Kitamori, T., Continuous-flow chemical processing in three-dimensional microchannel network for on-chip integration of multiple reaction in a combinatorial mode, *QSAR Comb. Sci.*, 24, 742, 2005.
29. Sato, K., Hibara, A., Tokeshi, M., Hisamoto, H. and Kitamori, T., Integration of chemical and biochemical analysis systems into a glass microchip, *Anal. Sci.*, 19, 15, 2003.
30. Tokeshi, M., Kikutani, Y., Hibara, A., Sato, K., Hisamoto, H. and Kitamori, T., Chemical process on microchips for analysis, synthesis and bioassay, *Electrophoresis*, 23, 3583, 2003.
31. Kikutani, Y., Hisamoto, H., Tokeshi, M. and Kitamori, T., Micro wet analysis system using multi-phase laminar flows in three-dimensional microchannel network, *Lab Chip*, 4, 328, 2004.
32. Goto, M., Sato, K., Murakami, A., Tokeshi, M. and Kitamori, T., Development of a microchip-based bioassay system using cultured cells, *Anal. Chem.*, 77, 2125, 2005.
33. Smirnova, A., Mawatari, K., Hibara, A., Proskurnin, M. A. and Kitamori, T., Micro-multiphase laminar flow for the extraction and detection of carbaryl derivative, Anal. *Chim. Acta*, 558, 69, 2006.
34. Hibara, A., Iwayama, S., Matsuoka, S., Ueno, M., Kikutani, Y., Tokeshi, M. and Kitamori, T., Surface modification method of microchannels for gas-liquid two phase flow in microchips, *Anal Chem*, 77, 943, 2005.
35. Honda, T., Miyazaki, M., Yamaguchi, Y., Nakamura, H. and Maeda, H., Integrated microreaction system for optical resolution of racemic amino acids, *Lab Chip*, 7, 366, 2007.
36. Miyaguchi, H., Tokeshi, M., Kikutani, Y., Hibara, A., Inoue, H. and Kitamori, T., Microchip-based liquid-liquid extraction for gas-chromatography analysis of amphetamine-type stimulants in urine, *J. Chromatogr. A*, 1129, 105, 2006.

37. Xiao, H., Liang, D., Liu, G., Guo, M., Xing, W. and Cheng, J., Initial study of two-phase laminar flow extraction chip for sample preparation for gas chromatography, *Lab Chip*, 6, 1067, 2006.
38. Bowden, S. A., Monaghan, P. B., Wilson, R., Parnell, J. and Cooper, J. M., The liquid–liquid diffusive extraction of hydrocarbons from a North Sea oil using a microfluidic format, *Lab Chip*, 6, 740, 2006.
39. Takahashi, Y., Sakai, R., Sakamoto, K., Yoshida, Y., Kitaoka, M. and Kitamori, T., On-line high-throughput ESIMS detection of a reaction product using synthesis and extraction microchips, *J. Mass Spectrom. Soc. Jpn.*, 54, 19, 2006.
40. Aota, A., Nonaka, M., Hibara, A. and Kitamori, T., Countercurrent laminar microflow for highly efficient solvent extraction, *Angew. Chem. Int. Ed.*, 46, 878, 2007.
41. Aota, A., Hibara, A. and Kitamori, T., Dependence of the number of theoretical plates of micro counter-current extraction of flow rate, *Proc. μTAS 2005 Symposium*, Boston, USA, 118, 2005.

36 Electrophoretic Microdevices for Clinical Diagnostics

Jerome P. Ferrance

CONTENTS

36.1 INTRODUCTION

Microfluidic devices are changing the way analytical procedures are performed. The speed that they bring to an analysis, the small sample size and limited reagent use, integration of multiple processing steps into a single unit, as well as the possibility for on-site analysis, all point to changes in the way analytical methods are to be used in the relatively near future. Nowhere is this going to be more important than in the field of clinical analysis. Microfluidic devices in the doctor's office, at the patient's bedside, or at any point-of-care will be as common and easy to use as the blood glucose and home pregnancy testing devices currently available. In the operating room, where rapid analysis while a patient is still in surgery can provide information to the surgeon regarding the extent of damage, to the emergency room, where rapid feedback could identify agents that might cause an epidemic, or are the result of a terrorist attack, the clinical analysis methods provided by microchips will be invaluable. Developments in microfluidics methods directed toward the clinical laboratory have been progressing over the past few years, and fully integrated devices are now being reported. This chapter will review what has been achieved, and look to what is still needed to move this technology into the mainstream of clinical diagnostics.

36.2 BACKGROUND

Not long after the early development of microchips for electrophoretic separations of DNA, the potential of these devices for clinical applications was realized.[1–3] While microchip electrophoresis remains a major area in clinical analysis to which microchip processing is applied, implementation of additional processing and analysis steps on microchips over the past few years is also providing new devices designed to replace existing methods utilized in the clinical laboratory.[4–6] The goal of many of these efforts is to eliminate the need for sample transportation to and processing in a central laboratory, by bringing the microfluidic technology directly to the point of care.[7] A report sponsored by the National Cancer Institute discusses the development of point-of-care systems for cancer diagnosis and prognosis, in which integrated microfluidic systems play an essential role.[8]

The purpose of this chapter is to explore clinical analyses that have been transferred to an electrophoretic microchip, along with some of the microchip preparatory methods that are now being integrated with an electrophoretic analysis step to provide lab-on-a-chip type devices. There are a vast number of clinical analyses that are now being transferred to microdevices, but the scope of this chapter is limited to only those that have an electrophoretic component in accord with the nature of this volume. A number of clinical analysis methods simply take advantage of the microfluidic architecture in microchips to decrease sample size and reaction times without requiring a separation step. The most common example of these is microarrays, recently reviewed by Situma et al.,[9] that utilize oligonucleotide to detect the presence of specific DNA sequences.[10–14] Similar types of immunoassays for clinical detection of specific proteins have also been developed on microfluidic devices by immobilizing antibodies within microchannels;[15,16] while these are not included in this chapter, a number of reviews[17,18] as well as an earlier Chapter 34 in this book by Sato and Kitamori cover these types of microchip immunoassays with applications to clinical analyses in more detail. Protein immunoassays on microchips that incorporate a separation step are included in this chapter along with analyses for specific ions,[19] glucose,[20] pH,[21] enzymatic activity,[21,22] and other types of bioaffinity interactions.[23]

The use of mass spectroscopy in clinical analysis is increasing, but is not yet widely employed, thus electrophoretic microchips designed for interfacing with MS detection are also not included here; these devices are covered in Chapter 49 by Laurell for coupling with matrix-assisted laser desorption/ionization-mass spectrometry (MALDI-MS) detection, and Chapter 53 by Lazar for electrospray-MS. A number of microchip applications also relate to the trapping, growth, and analysis of cells on microdevices, but these are not methods utilized in clinical analysis and thus are not included. The reader is directed to a recent review of these techniques by El-Ali et al.[24] as well as in Chapter 32 by Keenan and Beebe. Even with these limitations, the number of clinical methods that have been transferred to microchips is impressive and a comprehensive review is not possible. A number of reviews on microchip separations for clinical analyses have been written,[25–29] the most recent of which reported on microchip separations utilized for detection of cancer, cardiovascular disease, renal disease, neurological disease, immune disorders, diabetes, hereditary disorders, thyroid functioning, and infectious agents in body fluids.[29]

36.3 THEORETICAL ASPECTS

Microchip electrophoretic separations themselves and the theory behind them have been covered in Chapters 33 by DeVoe and Lee and 55 by Mahmoudian et al. This is also true of sample processing methods on microdevices, which have been covered in Chapter 43 by Bienvenue and Landers, Chapter 50 by van Midwoud and Verpoorte. The theory and engineering behind integration of multiple processes on a single device, in which both sample pretreatment and analysis steps are performed without removal of the sample from the device, are also explored in Chapter 43 by Bienvenue and Landers, as well as Chapter 40 by Easley and Landers describing valves to control flow within and between processes. This chapter seeks to show how the methods and theory discussed in these other chapters are being applied to real problems in clinical analysis. In addition, some of the issues involved in developing microdevices for utilization in the clinical laboratory not covered elsewhere will be described.

36.4 PRACTICAL APPLICATIONS

36.4.1 DNA ANALYSES

The ability to use laser-induced fluorescence (LIF) for detection allowed DNA separations on microchips to advance more rapidly than separations of other types of clinically relevant compounds. Chapter 6 by Szántai and Guttman reviewed the use of capillary electrophoresis of DNA for clinical diagnostic purposes, and any of those methods can easily be transferred to the microchip platform. Most often, clinical microchip DNA diagnostic devices utilize separations of specific fragments of DNA amplified using the polymerase chain reaction (PCR). Nucleic acid targets of clinical interest for amplification include mutations associated with particular diseases, the presence of exogenous DNA or RNA from pathogens, or over expressed DNA indicating the presence of a single clonal variation. In the coming genetic medicine revolution, where treatment options are also based on particular DNA sequences within a patient's genome, the use of microchip DNA separations will expand even further in the clinical laboratory.

36.4.1.1 Genotyping and Mutation Detection

One of the most common methods for detection of mutations in PCR-amplified DNA fragments is single-stranded conformational polymorphism (SSCP) analysis. Tian et al.[30] reported microchip separations for the detection of common mutations in *BRCA1* and *BRCA2*, two breast cancer susceptibility genes specific for the Ashkenazi Jewish population. Separations were performed in 140 s, fourfold faster than the CE-based assay, with profiles from the wild-type and mutant alleles easily distinguishable. As was important in the development of all of these clinical microchip methods, no loss of diagnostic ability was seen in moving to the microfluidic analysis method. This method was further pursued by Kang et al.[31] for rapid detection of a point mutation in the obesity gene. In addition to the SSCP technique, heteroduplex analysis (HDA) is a second method utilized for mutation analysis that has been transferred to microchips. Hestekin et al.[32] used a combination of SSCP and HDA separations on a microchip to evaluate mutations in specific exons in the *p*53 gene, a gene for a transcription factor important in regulating the cell cycle and preventing tumor growth. Figure 36.1 shows microchip separations of fragments generated from samples of a wild type and an exon 8 mutant of the *p*53 gene. Tian et al.[33] also utilized HDA on a microchip, combining it with allele-specific PCR to identify additional mutations in the *BRCA* genes as well as the *PTEN* tumor suppressor gene. The clinical potential for all of these electrophoretic methods based on migration differences due to the presence of mutations was recently reviewed by Hestekin and Barron.[34]

The mutations for which the above methods are most applicable are simple deletion and insertion mutations affecting only a few bases. Substitution mutations are difficult to monitor in this way,[30] but represent a significant class of DNA mutations called single nucleotide polymorphisms (SNP), which will play a significant role in the area of personalized medicine. Woolley et al.[35] performed SNP detection in *HLA-H*, a marker gene for hemochromatosis, on a 12-channel microchip using selective restriction digestion of a PCR-amplified product followed by traditional DNA sizing electrophoresis. Shi et al.[36] utilized this same method on a 96-channel radial array microchip for SNP detection in the gene for methylenetetrahydrofolate reductase, the protein product of which regulates folate and methionine metabolism. This same 96-channel microchip design was used by Medintz et al.[37] to evaluate a polymorphism in the *HFE* gene resulting in hereditary hemochromatosis using

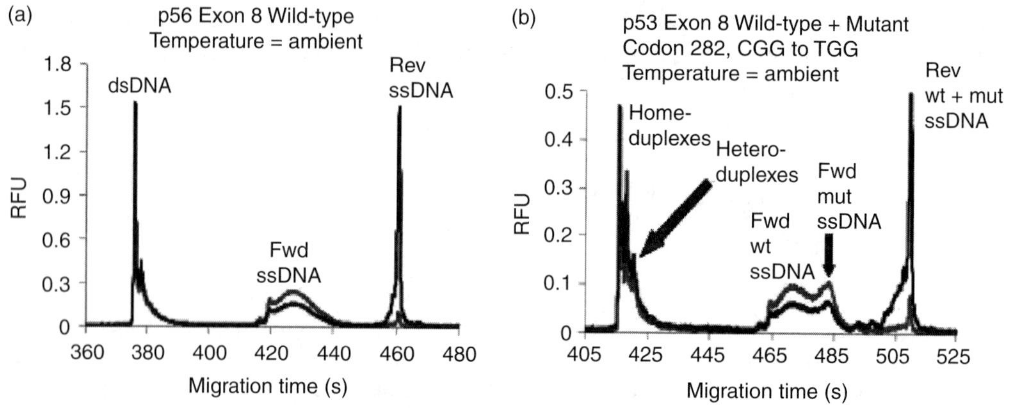

FIGURE 36.1 Electropherograms showing analysis of *p53* exon 8 amplicons for mutation detection by tandem SSCP/HA. dsDNA peaks are identified by the overlap of the FAM (forward strand) and JOE (reverse strand) dyes. ssDNA peaks are identified by the predominance of one fluorescent dye. Peaks due to the mutation are indicated by the arrows. Separation conditions: ambient temperature, 0.1% PHEA dynamically coated channel, 350–450 V/cm applied electric field strengths. (Reprinted from Hestekin, C. N., et al., *Electrophoresis*, 27, 3823, 2006. With permission.)

allele-specific primers to generate different sized products. Figure 36.2 shows the separation and detection of both homozygote and heterozygote mutations in the S62C variants of the *HFE* gene on a microchip. The authors extended this work to simultaneously detect three SNP mutations within this same gene based on the sizes of the DNA fragments produced.[38]

Simple size-based electrophoretic separations in microchips can also be employed for insertion and deletion mutations, which modify the size of the PCR-amplified DNA fragments. Sung et al.[39] utilized this method on a plastic microchip to evaluate fragile X alleles that have multiple repeat units of a trimeric CGG repeat. A tetranucleotide repeat in microsatellite alleles, related to hypercholesterolemia in families, was investigated by Cantafora et al.[40] using the Agilent 2100 Bioanalyzer, a commercially available microchip electrophoresis instrument capable of analyzing up to 12 samples in 30 min. This same instrumentation was used in the evaluation of deletion and duplication mutations associated with Duchenne muscular dystropy in two papers by Ferrance et al.,[41,42] which looked at a total of 13 loci in the dystrophin gene using two PCR amplifications and electrophoretic separations. A total of 50 samples were evaluated in the initial work, with all 35 samples with mutations being identified.[41] Figure 36.3 shows the traditional Southern blot analysis versus the microchip separation for the detection of a particular deletion. The subsequent work went on to establish normal ranges for each exon product using 40 control samples, which were then used to correctly evaluate patient samples for the presence of mutations.[42]

The Bioanalyzer was also used in two studies by Sohni et al.[43,44] that looked at both variable number of tandem repeat and deletion mutations in genes of clinical interest. The first study investigated polymorphisms in the cytokine genes *IL-1RN* and *CCR5*, both of which are important in immune system functioning.[43] The second study investigated nucleotide repeats in the promoter region of the inducible NO_2 synthase gene and the endothelial NO_2 gene, both associated with complications in diabetes, and a deletion/insertion mutation in the angiotensin-converting enzyme gene associated with NO_2 activity.[44] In addition to the Agilent instrument, Hitachi also developed a commercial microchip electrophoresis instrument (SV 1210) both of which were utilized by Zhang et al.[45] to evaluate gene fragments generated from 3 μL of blood in less than 20 min using a DNA extraction disk and a capillary PCR instrument.

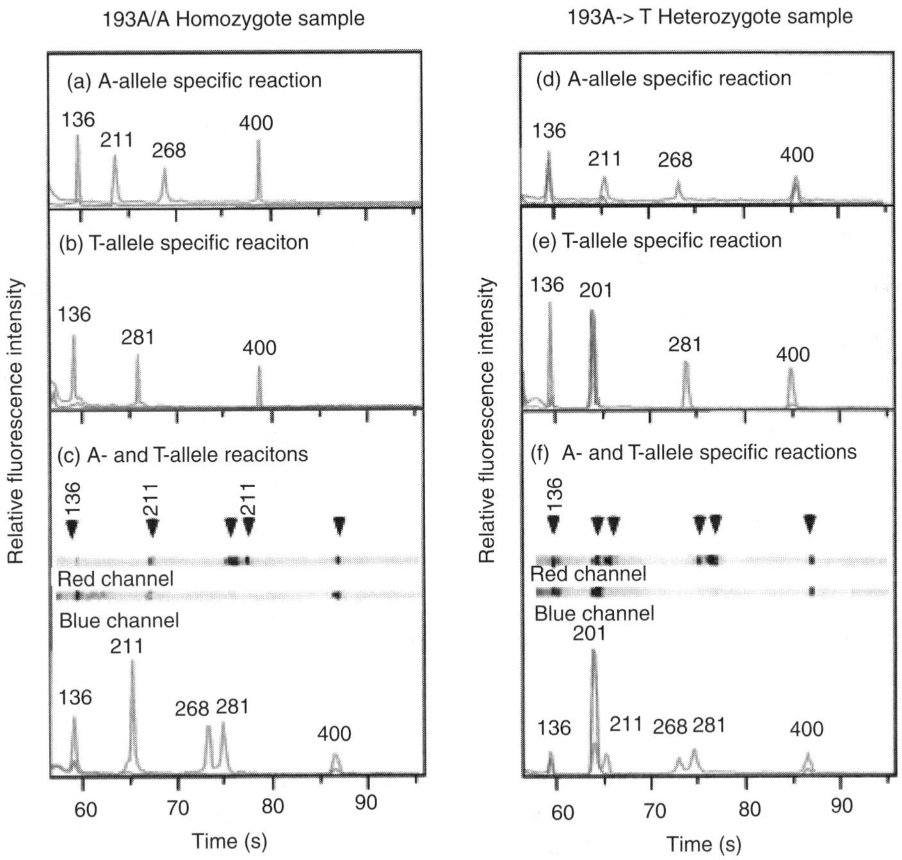

FIGURE 36.2 Detection of a heterozygous mutation (right panel) in the *HFE* gene using allele-specific PCR with two-color (red/blue) detection. (a and d) Separation of the A-allele specific amplicon (211 bp) from the 136 bp and 400 bp internal standards (red channel); (b and e) separation of T-allele specific amplicon (201 bp) from the same samples (blue channel); (c and f) separation of both amplicons showing only the A-allele fragment in the homozygous sample but both fragments in the heterozygous mutant sample. Numbers above the peaks indicate fragment length (bp). (Reprinted from Medintz, I., et al., *Electrophoresis*, 21, 2352, 2000. With permission.)

36.4.1.2 Other Nucleic Acid Analyses

Mutation detection and genotyping represent a significant portion of clinical DNA analyses, but other DNA separations of clinical interest have also been transferred to microchips. In lymphoproliferative disorders, a single clonal variation is often overexpressed in T- or B-cells resulting in a loss of heterogeneity in the immune system. Munro et al.[46] translated slab gel separations of PCR-amplified fragments, traditionally used to detect these disorders, to a microchip, effectively detecting clonal populations in the variable region of the T-cell receptor-γ gene and the immunoglobulin heavy chain gene. Again, no loss of diagnostic information was seen in the transition to the microchip separation system, in fact, a patient diagnosed as not having a lymphoma using the gel-based method was re-evaluated based on the microchip results and found to be in an early stage of the disease. Quantification of mRNA species has also been proposed as a clinical method, and reverse transcriptase-PCR (RT-PCR) products from renal biopsies have also been separated on microchips as a diagnostic method.[47]

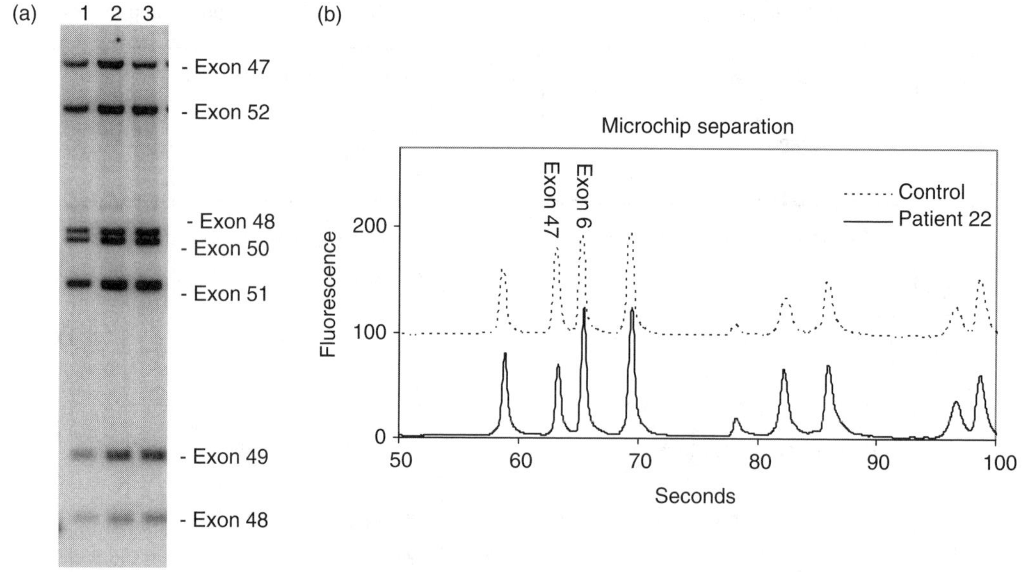

FIGURE 36.3 (a) Southern blot results obtained using DNA digested with HindIII and probed with cDMD 8. Lane 1: normal male control; lane 2: normal female control; lane 3: female patient with a deletion in exon 47. (b) Microchip electropherograms of a multiplex amplification for a control and a female patient with a deletion in exon 47.

The presence of exogenous DNA or RNA in a patient is also of clinical interest, as this can be used to indicate the presence of a bacterial or viral pathogen. Höfgartner et al.[48] used a single-channel glass microchip for the detection of herpes simplex virus (HSV) in cerebrospinal fluid (CSF) through separation of PCR-amplified products. The microchip-based electrophoresis method allowed for rapid diagnosis of herpes simplex encephalitis (HSE), in less than 100 s, compared to 18 h for identical results obtained by hybridization. Chen et al.[49] were able to detect viral RNA from the hepatitis C virus through RT-PCR amplification with product identification performed using an electrophoretic separation on a plastic microdevice. This group went on to further quantify the amount of viral DNA present using a competitive PCR procedure, with product analysis again on the plastic microchips.[50] The Agilent Bioanalyzer also provides an efficient method for the separation of PCR products generated from exogenous DNA in CSF to show the presence of viral infections. Separations showing the presence of infections from a number of viral pathogens are shown in Figure 36.4. The Agilent instrument was also used to evaluate repetitive sequence PCR products from *Mycobacterium* species by Goldberg and coworkers.[51] This kind of genotypic analysis can be used to identify particular strains, track tuberculosis transmission in outbreaks, and identify cross contamination in the laboratory.

36.4.2 Protein Analysis

Electrophoretic analysis of proteins was one of the first methods transitioned to capillaries in clinical laboratories—this is highlighted in Chapter 2 by Hempe. Unfortunately, protein analysis has not made the same transition to microchips. This is due to the difficulties in implementing a ultraviolet (UV) absorbance method on microchips, the method normally utilized for proteins detection on CE. A number of fluorescent methods have been employed both for standard separations as well as for immunoassay analyses that utilize separations. Two-dimensional separations have also been performed on microchips,[52] as reviewed in Chapter 33 by Lee, and these might have clinical significance at some point in the future.

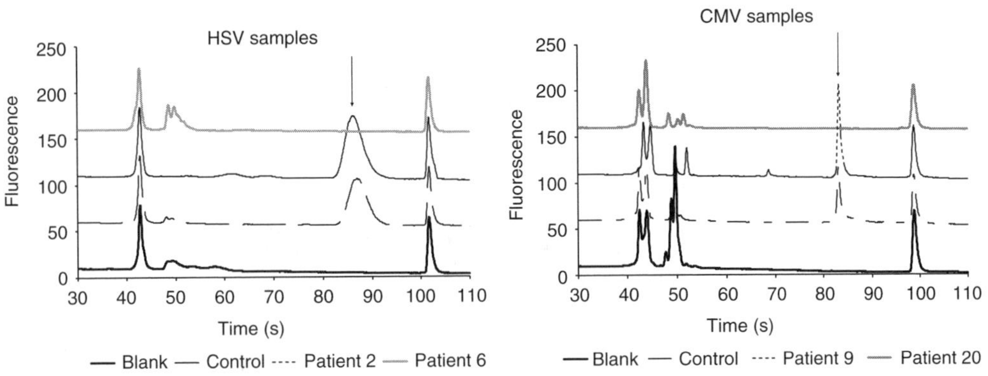

FIGURE 36.4 Electropherograms of PCR-amplified cerebral spinal fluid samples from patients with suspected infectious agents. (a) Samples from patients possibly infected with HSV; clinical diagnosis by traditional means: patient 2—infected HSV type 1, patient 6—negative. (b) Samples from patients possibly infected with cytomegalovirus; clinical diagnosis by traditional means: patient 9—infected, patient 20—negative.

36.4.2.1 Protein Separations

Human serum protein quantification is an important method for determining the relative abundances of albumin, IgG, and other important serum proteins. Colyer et al.[53] investigated serum proteins through microchip separations of fluorescently labeled proteins, showing that quantification was possible even after labeling. Real samples, however, could not be evaluated because of poor sensitivity. Callewaert et al.[54] also utilized a fluorescent labeling scheme to tag N-glycan components in serum, but was able to separate these real clinical samples on a microchip; a typical separation is shown in Figure 36.5, in which the difference between serum from a patient with cirrhosis and one with chronic hepatitis is illustrated. Separation of lipoproteins in serum on a microchip was reported by Weiller et al.[55] who utilized NBD as a fluorescent label to allow detection of both high- and low-density lipoproteins, important risk factors for cardiac disease. This group further refined the work to allow analysis of the low-density lipoprotein (LDL) components to show the presence of different forms of LDL in serum samples.[56]

A number of other covalent and dynamic fluorescent labeling schemes have now been proposed for the detection of proteins on microchips, all of which could be applied for clinical protein analysis.[57–60] Bio-Rad has adopted an on-column dynamic labeling scheme in their commercial microchip electrophoresis instrument, which Chan and Herold[61] utilized to evaluate microalbuminuria, a marker for cardiovascular disease and nephropathy in diabetic patients. Through the use of an SDS-based separation, these experiments can provide quantitative and qualitative information about the proteins. However, the use of covalent and dynamic labeling of native proteins must be evaluated for each system studied as fluorescent tags are not incorporated homogeneously across proteins, and are known to be influenced by the hydrophobicity and secondary structure of the proteins. This heterogeneity affects both the electrophoretic mobility and fluorescence response of each protein differently, and thus must be accounted for in the results obtained from any clinical protein separation. Giordano et al.[60] developed a partially denaturing technique for fluorescent labeling, to minimize preferential binding, in which SDS was added to the sample, but no heat denaturation step was employed. Figure 36.6 shows the microchip protein profiling of human sera, using this method for the detection of gammopathies. The separations show clear spikes in the protein profiles of severely affected patient sera, but detection of more subtle changes was not evaluated.

UV detection on microchips has not been completely bypassed, but it does require quartz microdevices, as the normal glass and plastic microchips absorb too much UV radiation in the wavelengths of interest. Zhuang et al.[62] have developed quartz microchips for protein detection, and

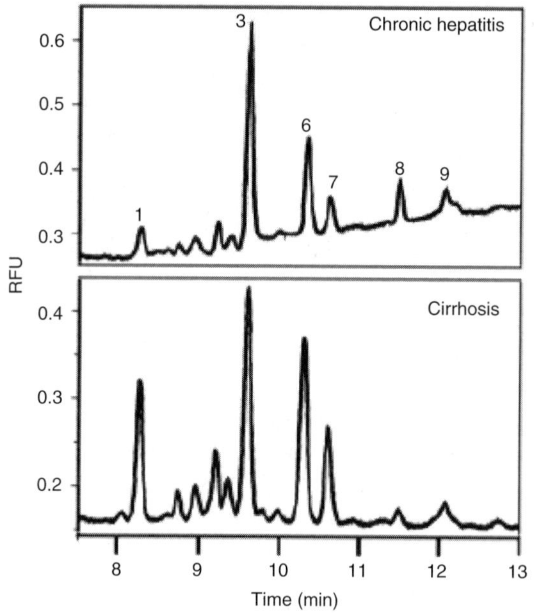

FIGURE 36.5 Profiling of two serum samples: one from a noncirrhotic chronic hepatitis patient and one from a patient with cirrhosis. Log (peak 7/peak 8) is diagnostic for liver cirrhosis, and these two peaks are well resolved in the microfluidics profiles. (Reprinted from Callewaert, N., et al., *Electrophoresis*, 25, 3128, 2004. With permission.)

have utilized them for electrophoretic analysis of urinary proteins of clinical interest.[63] This group has also now applied quartz microchips to the analysis of serum lipoproteins[64] without the need for fluorescent tagging as discussed above.[55] UV detection is also covered in a review by Wang et al.[65] who detail the use of electrophoretic microchips for the clinical analysis of hemoglobin, with various detection methods.

36.4.2.2 Microchip Immunoassays

A number of microchip immunoassays have been developed that utilize an on-chip reaction followed by a separation component to separate free antigen from that bound to the antibody—if a labeled antigen is used—or to separate free antibody from the antigen/antibody complex—if labeled antibodies are used. Schmalzing et al.[66] utilized the latter method to demonstrate an immunoassay for T4 in serum for clinical evaluation of thyroid function on fused-silica microchips. The immunoassay used a competitive format with labeled antigens competing against antigens in the serum for antibody-binding sites; the amount of free labeled antigen was used to quantify the amount of T4 in the samples. The same group also reported a microchip electrophoretic immunoassay for cortisol in serum using a similar method,[67] which could quantify cortisol in serum within the clinical range without extraction from the sample or other preparation steps. Qiu and Harrison[68] were able to incorporate a calibration step into their immunoassay microchip that will be important for eventual implementation in clinical analyses. Using a microchip with multiple inlets, they could control the amount of antibody mixed with labeled antigen, in this case bovine serum albumin (BSA). While this antigen is not clinically relevant, the ability to calibrate the microchip response, as shown in Figure 36.7, will be important for protein quantification.

In addition to fluorescent-based assays, electrochemical detection has also been utilized for microchip immunoassays. Wang et al.[69] initially utilized an alkaline phosphatase-labeled antibody,

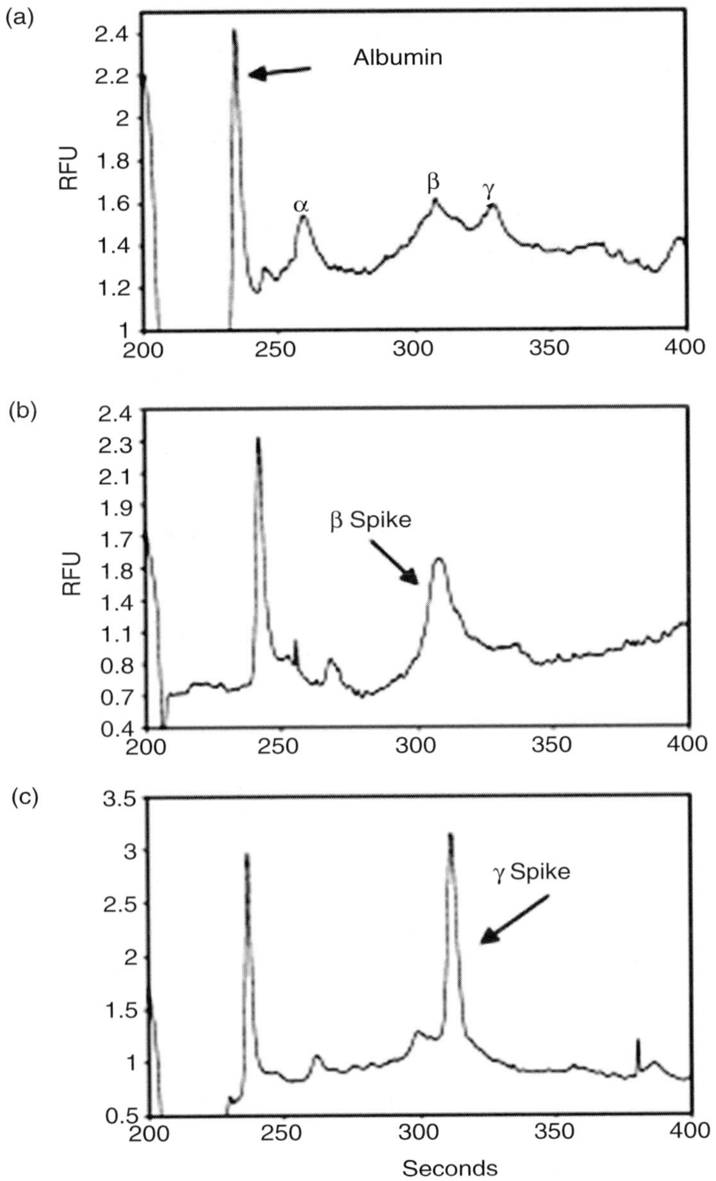

FIGURE 36.6 Comparison of CZE analysis of partially denatured human sera using run buffer containing 0.04% NanoOrange dye. Sera were diluted 1:500 in 0.5% SDS, with 1% dye included in the sample buffer. Separation conditions: 100 mM borate, 3 mM diaminobutane, pH 8.5. (Reprinted from Giordano, B. C., et al., *Anal. Chem.*, 76, 4705, 2004. With permission.)

which was mixed with the antigen (mouse IgG), then separated on-chip to resolve the free antibody from the complex. Using a post column reaction of the alkaline phosphatase to produce 4-aminophenol, they were able to monitor the products using electrochemical detection. This same group also used ferrocene tagged antibodies, simply measuring the reduction of the ferrocene in the free antibody and the complex using amperometric detection.[70] Measurement of both IgG and T3 antigens were shown in this format. More recently, Herr et al.[71] showed fluorescent microchip immunoassays for both tetanus antibody and tetanus toxin, utilizing fluorescently labeled tetanus toxin C-fragment and a fluorescent marker as an internal standard for quantification. The microchip separation provided a direct immunoassay for the presence of tetanus antibody, and a competitive

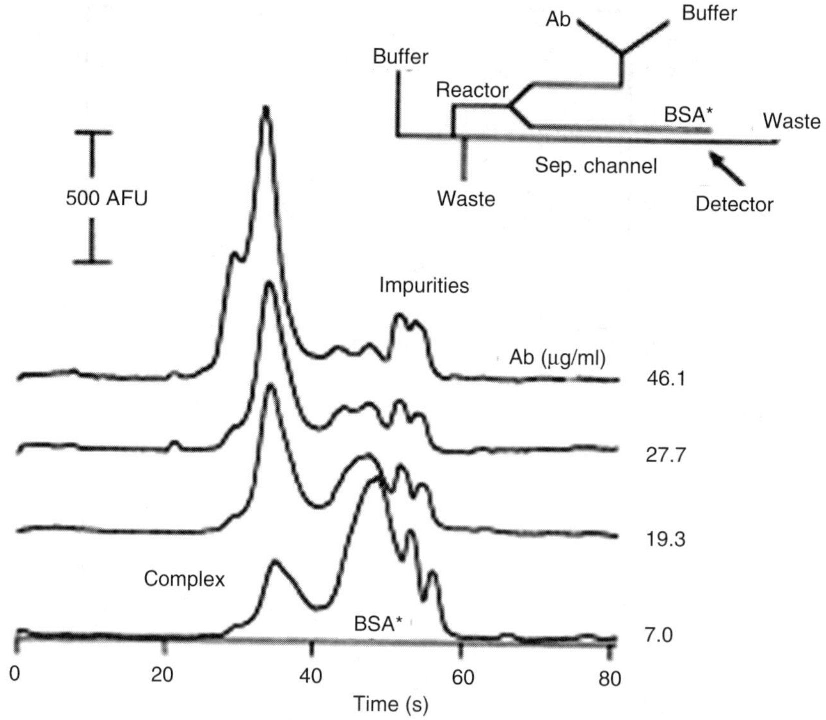

FIGURE 36.7 Traces from an on-chip calibration series. Intensity changes as a function of the amount of anti-BSA delivered to the reaction, the complex peak increases in intensity while the labeled BSA peak decreases. At the highest antibody concentration impurities can be seen in the BSA peak region. (Reprinted from Qiu, C. X., and Harrison, D. J., *Electrophoresis*, 22, 3949, 2001. With permission.)

assay for the presence of tetanus toxin could be used to generate a dose-response curve as shown in Figure 36.8; the microchip design, however, did not include an integrated incubation step.

36.4.3 Additional Clinically Relevant Analyses

In addition to DNA and proteins, a number of other important clinical markers have been evaluated using electrophoretic separations on microchips. Fanguy and Henry[72] quantified uric acid in urine using a microchip electrophoretic separation with electrochemical detection. This work was extended to include three markers of renal function normally measured in clinical assays: uric acid, creatinine, and creatine.[73] Detection of glucose has also been performed directly from human plasma on a microchip, with no need for sample pretreatment. Du et al.[74] utilized a poly(dimethylsiloxane) (PDMS) microchip containing a copper electrode for electrocatalytic oxidation of the glucose as the detection method. Additional investigations involving the analysis of carbohydrates by microchip electrophoresis, a few of which are of interest for clinical analyses, are covered in a review by Suzuki and Honda.[75] This review covers not only electrochemical detection methods but also reports on the use of fluorescent and UV absorbing tags; the authors suggest, however, that the need for a derivatization step eliminates the rapid separation advantage provided by the microchip analysis.

A number of small molecules of interest in clinical analyses have also been analyzed using microchip-based separations. Zhao et al.[76] utilized a microchip separation for detection on the antibiotic lincomycin directly from urine samples. This method, also employing electrochemical detection, allowed lincomycin analysis in under 40 s. Crevillen et al.[77] utilized a similar method for

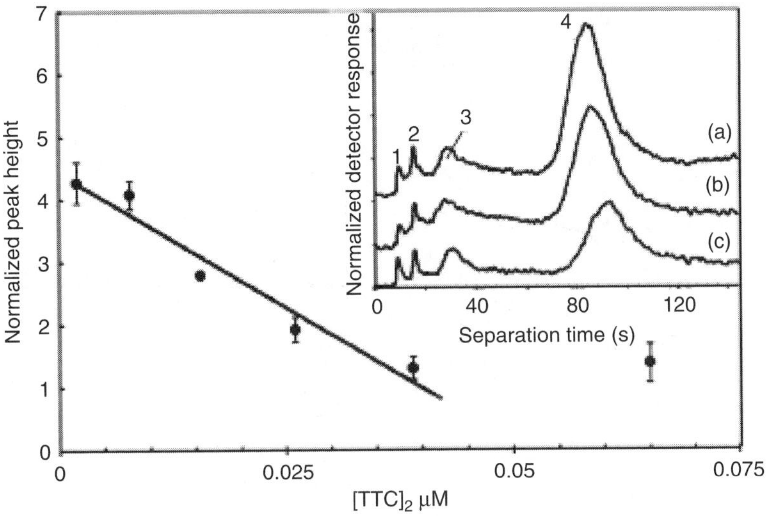

FIGURE 36.8 Dose–response curve for the tetanus toxin C-fragment using a microchip competitive immunoassay. Symbols show the normalized response (peak height of complex/peak height of free dye) with error bars indicating the standard error in the measurements ($n = 3$–5). A correlation coefficient of .95 was obtained for the linear fit through the points indicated. The inset shows electropherograms with TTC concentrations of (a) 2.0, (b) 7.8, and (c) 15.6 nM, with a labeled TTC* concentrations of 13.0 nM and 6.0 nM anti-TTC. Peaks 1 and 2 are due to free dye with peak 2 used for normalization; peak 3 is free TTC*, and peak 4 is the complex. Separation conditions: 300 V/cm, total length 6.1 cm, and length to detector 6 mm. (Reprinted from Herr, A. E., et al., *Anal. Chem.*, 77, 585, 2005. With permission.)

separation and detection of hydroquinone and hydroquinone glycoside, important in clinical analyses because of their presence in pharmaceuticals and natural remedies. They were able to separate and detect these components using different injection methods, and also applied the microchip method to the analysis of urine samples. Figure 36.9 shows application of this method to a variety of sample types. Electrochemical detection was also utilized on microchips for the separation of six organic acids (fumaric, citric, succinic, pyruvic, acetic, and lactic acids) and three cations (K^+, Na^+, and Li^+).[78] This microchip used a four-electrode system with no direct contact between the electrodes and the separation channel, but was not tested with real samples.

Munro et al.[79] showed separation and detection of amino acids on microchips using an indirect fluorescence detection method. Figure 36.10 shows application of this method to urine samples with no pretreatment other than dilution in the appropriate separation buffer. Abnormal amounts of amino acids can easily be detected in the two patient samples compared to the healthy control sample. An absorbance detection based approach was utilized for the clinical analysis of calcium ion in serum, which is important in the regulation of a number of physiological processes.[80] Beads with an immobilized calcium reactive dye were placed into the detection region, and the samples mobilized past the beads using electrophoretic flow. While a true separation was not intended, the interference of magnesium ions was significantly less than seen in pressure flow devices,[19] since Mg^{2+} moves through the channel faster than Ca^{2+} allowing less interaction time with the beads.

36.4.4 Sample Preparation

There are a large number of biological sample processing steps that have now been translated to microchip methods and been shown to provide the same results, while utilizing smaller amounts of sample and less reagents, thus reducing both cost and waste generated. At the same time, the miniaturization of these processes has often produced a significant reduction in the time required,

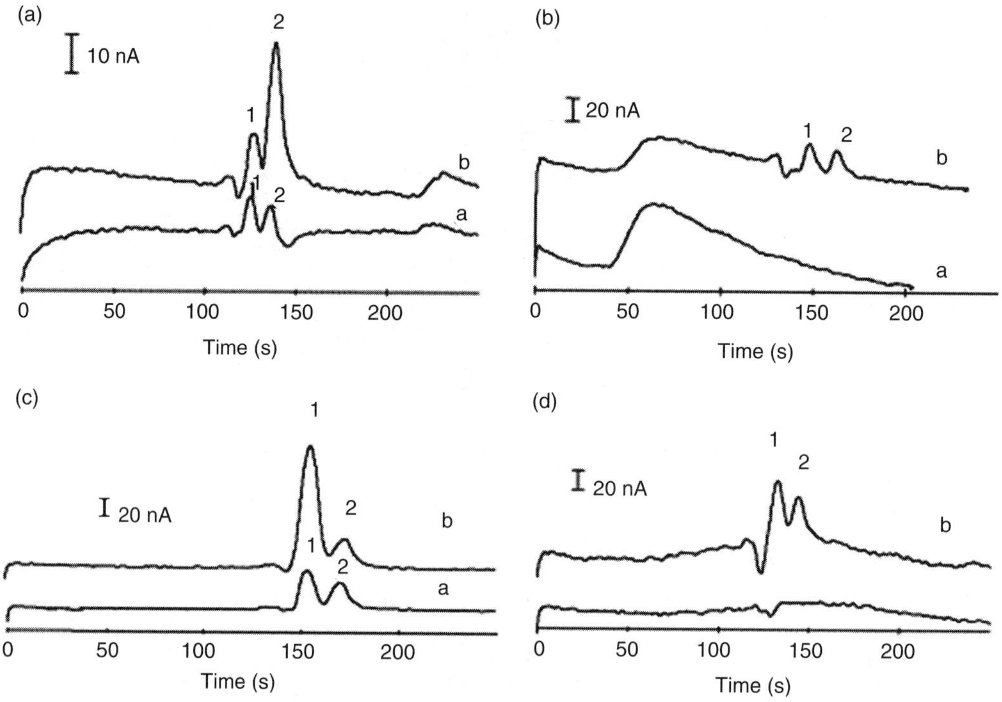

FIGURE 36.9 Electropherograms of (a) the nutraceutical HMP, (b) urine, (c) a cosmetic formulation, and (d) a wine sample. Peak labels correspond to (1) hydroquinone and (2) arbutin. The (a) traces are unspiked sample (b and d) or spiked with 150 ppm of (a) hydroquinone and (c) arbutin. The (b) traces are spiked with 150 ppm of hydroquinone and 125 ppm of arbutin. (Reprinted from Crevillen, A. G., et al., *Anal. Chim. Acta*, 562, 137, 2006. With permission.)

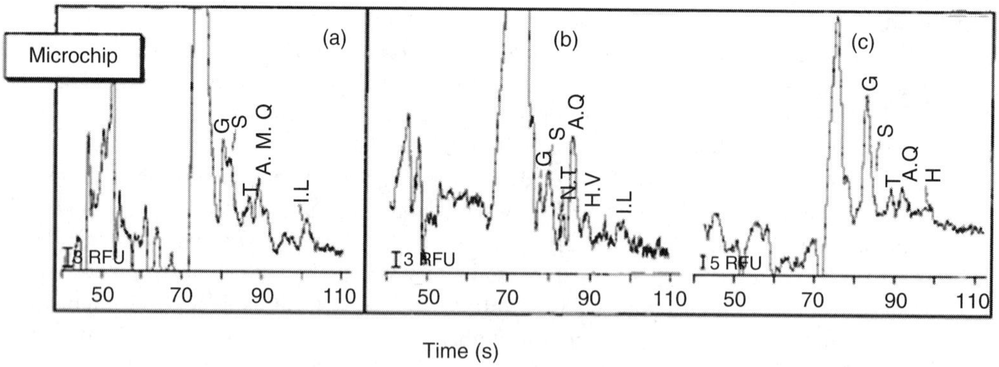

FIGURE 36.10 Microchip electrophoretic separations of (a) normal and (b and c) abnormal urine samples. Urine samples were diluted 1:10 in water with 0.5 mM fluorescein, 1 mM sodium carbonate and 0.2 mM CTAOH. Separation conditions: 1.0 mM sodium carbonate, 0.5 mM fluorescein, and 0.2 mM CTAOH, pH 10.3; l_{eff} 5.5 cm, 15-s injection at 417 V/cm (reversed polarity), 183 V/cm separation voltage. *Abbreviations*: G, glycine; S, serine; N, asparagine; T, threonine; A, alanine; M, methionine; Q, glutamine; H, histidine; V, valine; L, leucine; I, isoleucine. (Reprinted from Munro, N. J., et al., *Anal. Chem.*, 72, 2765, 2000. With permission.)

by reducing diffusional lengths, thermocycling times, and column volumes. This chapter will not address all of these processes, but again will be limited to those processes directly related to clinical analysis and those that would be utilized in conjunction with a subsequent electrophoretic separation for analysis. Chapter 43 by Bienvenue and Landers, Chapter 50 by van Midwoud and Verpoorte in this volume address additional methods and provide more in-depth coverage of the microchip sample preparation literature. In addition, early work on sample preparation methods utilized for clinical molecular diagnostics that have been translated to microfluidic formats is covered in a review by Huang et al.[81] These methods include cell sorting/selection, which can be used for clinical diagnostics but are not covered here, as well as nucleic acid purifications, and nucleic acid amplifications methods.

36.4.4.1 Enzymatic Reactions

Enzymatic amplification of specific fragments of a nucleic acid template, particularly through the use of the PCR, is one of the most widely translated methods to microchips associated with a subsequent electrophoretic analysis. A number of microdevices and thermocycling methods have been described and are summarized in a recent review by Zhang et al.[82] covering the basic designs and practical applications of microchip PCR. Not all of the microchip PCR developments showed direct clinical utility, and thus are not included here, but most of the methods could theoretically be utilized for clinical analysis. In addition to the traditional PCR-based DNA amplification, amplifications from RNA using RT to first generate a cDNA are also of clinical interest, as well as other enzymatic methods, which are not as widely utilized.

Most clinically relevant microchip PCR amplifications have been utilized for the detection of infectious agents. Mikhailovich et al.[83] showed allele-specific amplifications of mycobacterium tuberculosis to evaluate rifampicin resistance as a method for monitoring the use of this drug for treatment. They compared this method to a hybridization method, providing similar results but greatly decreasing the analysis time. Yang et al.[84] utilized a plastic microchip to test the sensitivity and specificity of detection of a K12-specific fragment from *Escherichia coli*. They were able to detect as low as 10 *E. coli* cells with this method in the presence of 2% blood, and could do multiplex PCR to show specificity. For the detection of hepatitis B, a common infectious disease and blood borne pathogen, Cho et al.[85] conducted a large-scale study of a microchip PCR method for the virus in real clinical samples. The authors utilized a real-time detection method, employing a miniature fluorescence detection system using light emitting diodes and photodiode detectors, to monitor amplification of sequences specific for the hepatitis B virus, thus a separation was not needed. The study, however, evaluated specificity, sensitivity, cross reactivity, reproducibility, and limits of detection in a large-scale clinical evaluation of the method ($n = 563$) to show the applicability of microchip methods for routine clinical analysis.

Real-time detection was also used for an isothermal amplification procedure termed nucleic acid sequence-based amplification, for the detection of artificial human papilloma virus sequences in SiHa cells.[86] This method was not applied to real samples, but demonstrates the clinical applicability of additional microchip enzymatic amplification methods. Not all clinical microchip PCR amplifications looked for infectious agents, however. Cheng et al.[87] showed a microchip PCR method for multiplex amplifications of dystrophin gene fragments for the detection of mutations associated with muscular dystrophy.

Other types of enzymatic reactions have also been implemented on microchips. Using PCR-amplified DNA as the template, Lou et al.[88] utilized ligase chain reaction to identify a mutation in the *NOD2/CARD15* gene, a marker for inflammatory bowel disease, using DNA from blood of healthy volunteers. The reaction used the ligase enzyme to join short oligonucleotides sequences when the corresponding complementary wild-type or mutant DNA template was present. Digestion of DNA on a microchip using restriction enzymes has also been reported, but at this time, this has not yet been utilized for analyses relevant to clinical work.[89] Likewise, digestions of proteins have been shown on microchips, but have not been applied for clinical analyses.[90,91] A number of these enzymatic reactions and others have been transferred to microchips, but are covered within Section 36.4.5 below since they have been directly coupled to the subsequent electrophoretic separation on the same microchip.

36.4.4.2 Purification and/or Concentration

A number of diverse methods fall under this category with respect to clinical analysis. One important process in clinical analyses is the isolation of plasma from whole blood, which has been performed on a microchip using a transverse flow microfilter design.[92] Utilizing capillary action for the separation, both the plasma and the cell components of the blood were available on-chip for additional processing. Processing of blood to separate the leukocytes from other components has also been shown on a microchip for clinical applications requiring removal of erythrocytes.[93] This microchip was designed to mimic intrinsic blood flow processes, such as leukocyte margination, to separate the two cell types and allow collection of the white cells. Another interesting technique applied to serum and urine samples is laminar flow diffusion, which takes advantage of the laminar flow and short diffusion lengths provided by microchips.[94] By flowing the sample in parallel with a water stream, low molecular weight components are extracted into the water phase, which can then be collected in a separate stream from the original sample. While used only for spectroscopic detection in the original work, this could be implemented for the removal of small molecular weight components for analysis, such as urea and creatine from urine. Additional details on this process can be found in Chapter 35 on liquid/liquid extraction by Tokeshi and Kitamori.

One of the most widely investigated purification methods on microchips is DNA extraction, which has been shown using a number of different phases and microchip designs, but only those methods that have been shown to deal with real samples such as whole blood are of clinical interest. Breadmore et al.[95] capture DNA directly from blood utilizing silica beads immobilized in place with a silica sol–gel matrix, showing that it was possible to perform PCR amplification on the eluted genomic DNA. Chung et al.[96] used beads with organic surface groups for capture of DNA on a poly(methyl methacrylate) (PMMA) microchip. The beads were immobilized to the surface of the PMMA through the organic surface groups, and the DNA containing solution passed back and forth over the immobilized beads. DNA from *E. coli* cells could be extracted and amplified from serum and whole blood, with the immobilized beads providing significantly better extraction efficiency than free beads as the number of *E. coli* in the blood sample decreased.

Rather than packing beads into a microchannel, Wu et al.[97] utilized the liquid precursor of the silica sol-gel to form a monolithic matrix in a microchannel for DNA extraction. This solid phase provided extraction efficiency with good reproducibility using standards, but extraction efficiency fell off quickly with whole blood when the devices were used repeatedly. This is not necessarily a problem, in that these devices for clinical applications would be designed for only a single use to prevent cross contamination of samples. Figure 36.11 shows the protein and DNA profiles generated on the sol-gel microchip during purification of DNA directly from lysed blood. While the sol-gel matrix was easier to generate in the channel, the liquid precursor filled all of the channels. Wen et al.[98] solved this problem through the use of a photopolymerizable matrix, which they coated with a sol-gel precursor to increase the DNA-binding capacity. For whole blood application, they found that the proteins competed for binding sites, and significantly decreased the capacity of the monolith for DNA. A dual phase microchip was developed using an initial C18 phase to capture the proteins from the lysed blood as the DNA passed through for capture on the monolith. Volumes as large as 10 μL of blood could be processed on this device with good extraction efficiencies.

In addition to proteins competing for binding sites in the DNA extraction process, removal of proteins from the solid phase is normally performed using an alcohol-based wash step to prevent protein coelution with the DNA and possible interference with the subsequent processing steps. As reported by Legendre et al.[99] this wash step causes its own problems in that isopropanol inhibits PCR, which is often the next process on an integrated device. Wen et al.[98] were able to eliminate this

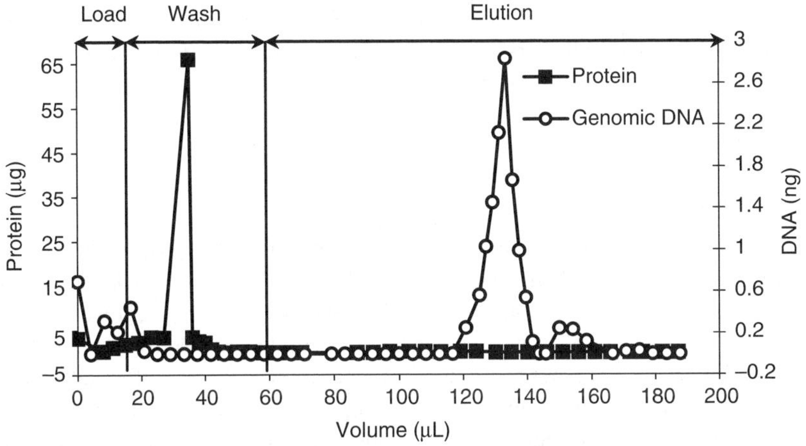

FIGURE 36.11 Protein and DNA extraction profiles from a lysed human blood sample loaded onto a sol-gel filled microchip. Protein (closed squares) is removed in the wash step and DNA (open circles) is released during the elution step.

step using the dual phase method, but other approaches have also been investigated. Witek et al.[100] utilized a photoactivated polycarbonate (PPC) microfluidic device for DNA extraction from *E. coli*, which incorporated an ethanol wash step, but the device could be dried before elution of the DNA to eliminate inhibition from the wash buffer. Larger elution volumes were required, however, and this work did not address DNA purification from human whole blood. An amine-coated microchip that had lower protein binding was used in work by Nakagawa et al.[101] to extract DNA from whole blood. No wash step was required, avoiding the use of PCR-inhibiting reagents, but the amine phase showed only moderate efficiency. A competing approach was developed by Cao et al.[102] using the protonated amine groups on chitosan at a low pH to capture the negatively charged DNA. The DNA was easily released at a high pH at which the amine groups became neutral. Very little protein was found to bind to the chitosan phase, thus a wash step was also not required in this method. Using the chitosan technology, a microchip capture device could be constructed simply by immobilizing the chitosan directly on the surface of a bifurcated channel structure within a microchip.

While DNA represents a major focus of purification on microchips, capture of RNA has also been explored. Kokoris et al.[103] utilized a silica membrane immobilized in a microchip for RNA purification. The "lab card" device provided RNA suitable for reverse transcriptase amplification with a yield and quality better than that obtained using traditional RNA extraction methods. A number of protein preconcentration or extraction microchip devices have also been developed, using either a general hydrophobic capture phase to capture all proteins, or an antibody or ligand-based phase for specific capture of a protein of interest. Most of these are designed for use with MS analysis of the collected products, however, and are covered in Chapter 53 by Lazar. One exception is the use of a phosphocholine ligand capture phase for selective capture of C-reactive protein (CRP), of interest as a cardiac and inflammation marker.[104] This work required subsequent electrophoretic separation to quantify the CRP because of capture of additional phosphocholine-binding proteins in the extraction; quantitative results generated from the microchip extraction process agreed well with current CRP analysis method results.

36.4.5 Chip-Based Integration of Sample Preparation Steps

The ultimate motivation for transfer of electrophoretic separations from a capillary to a microchip was the ability to perform these additional processing steps in the microchip format, which was not as easily achieved in a capillary system. This provides the microchip the advantage of sequential integration of a number of processing steps directly with the electrophoretic analysis, and many of the integrated devices that resulted are of interest in clinical analyses.[105] As with microchip electrophoresis itself, much of this work has focused on DNA for genetic analysis[106] (see Chapter 43 by Bienvenue and Landers), but a number of integrated protein microchips have also been reported, along with multiprocess assays for small molecules.

36.4.5.1 DNA Amplification

A significant number of groups have reported on integration of the PCR process directly with the subsequent electrophoretic separation on the same microchip.[87,107–115] As with the microchip PCR for clinical applications, most of the integrated microchip work focused on detection of infectious agents, starting in 1996 with Woolley et al.[107] who showed amplification and electrophoresis of a fragment of salmonella DNA. Figure 36.12 shows the microchip separation of the amplified salmonella fragment along with a co-injection of the amplified product with a size standard for peak identification. Waters et al.[108,109] incorporated cell lysis, and amplified and separated fragments from multiple regions of the *E. coli* genome in their integrated device. Bacterial detection of both *E. coli* 0157 and salmonella was performed by Koh et al.[113] in a plastic integrated device fabricated in cyclic olefin copolymer (COC). Limits of detection as low as six copies were possible. Lagally et al.[114] also evaluated an integrated microchip for pathogen detection, amplifying genes from both *E. coli* and *Staphylococcus aureus*. Limits of detection in this system were as low as two to three bacterial cells. Easley et al.[115] developed an integrated microchip that utilized valves for pressure-based injections from the PCR chamber into the separation channel for the detection of salmonella DNA. This allowed dilution

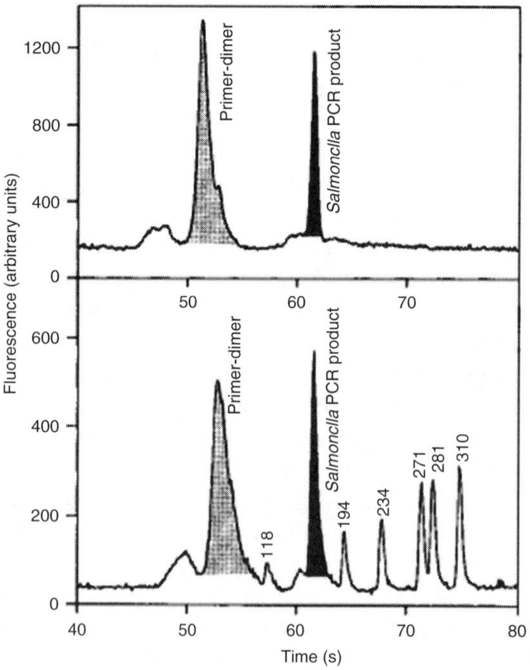

FIGURE 36.12 Integrated PCR-electrophoresis microdevice assay for the detection of salmonella DNA. (a) Microchip separation of the PCR product was performed immediately following a 39-min amplification in the integrated microdevice. Primer-dimer (light gray peak) appears at 51 s and the PCR product (dark gray peak) appears at 61 s. (b) Sizing of the *Salmonella* product (1:100 dilution) in a separate microchip using a ΦX174 HaeIII digest DNA standard (1 ng/μL). Total analysis time for the salmonella sample using the integrated microdevice was less than 45 min. (Reprinted from Woolley, A. T., et al., *Anal. Chem.*, 68, 4081, 1996. With permission.)

of the PCR product with water to provide a stacking effect as it was co-injected with a DNA size standard for accurate sizing.

For the detection of a severe acute respiratory syndrome coronavirus, Zhou et al.[116] used an integrated microchip to perform RT-PCR along with electrophoretic sizing. Nasopharyngeal swab samples were utilized in the analysis with the microchip method showing infection in 17 of 18 patient samples; this was better than the conventional RT-PCR method that identified only 12 infected patients. Pal et al.[117] incorporated an additional reaction step in their integrated microchip for the identification of hemagglutinin A subtypes of influenza virus. Identification was based on restriction fragments, thus an on-chip enzymatic digestion was performed after the PCR. Separation of the digested products allowed successful discrimination of two influenza strains. In more recent work, Kaigala et al.[118] used a PCR electrophoresis microchip to assess the risk of nephropathy in renal transplant patients. The analysis could distinguish between low, medium, and high BK viral loads, indicating patients at risk for complications; detection of as low as two viral copies was possible. Easley et al.[119] went further, incorporating a DNA extraction step with the PCR and separation steps to show a fully-integrated analysis following off-chip cell lysis. They reported on analysis of mouse blood for the detection of *Bacillus anthracis* infections showing detection in asymptomatic mice from less than 1 μL of blood samples. A similar sized sample of nasal aspirate was used to show the presence of *Bordetella pertussis* in a patient confirmed to have whooping cough. Figure 36.13a shows a picture of the integrated microchip utilized in this work, Figure 36.13b shows the IR-mediated thermocycling profile on the device, and Figure 36.13c shows consecutive injections and separations of the amplified product for the *B. anthracis* analysis.

For detection of mutations, Cheng et al.[87] used an integrated PCR-microchip electrophoresis device for a multiplex amplification to identify deletions in the dystrophin gene, with successful amplification and resolution of all of the products. Taylor et al.[120] detected mutations in mitochondrial DNA, which are indicated in a wide range of human disease states, utilizing a different set of on-chip processes. DNA was digested using a restriction enzyme, then denatured and reannealed to perform a HDA in the separation step for the identification of mutations.

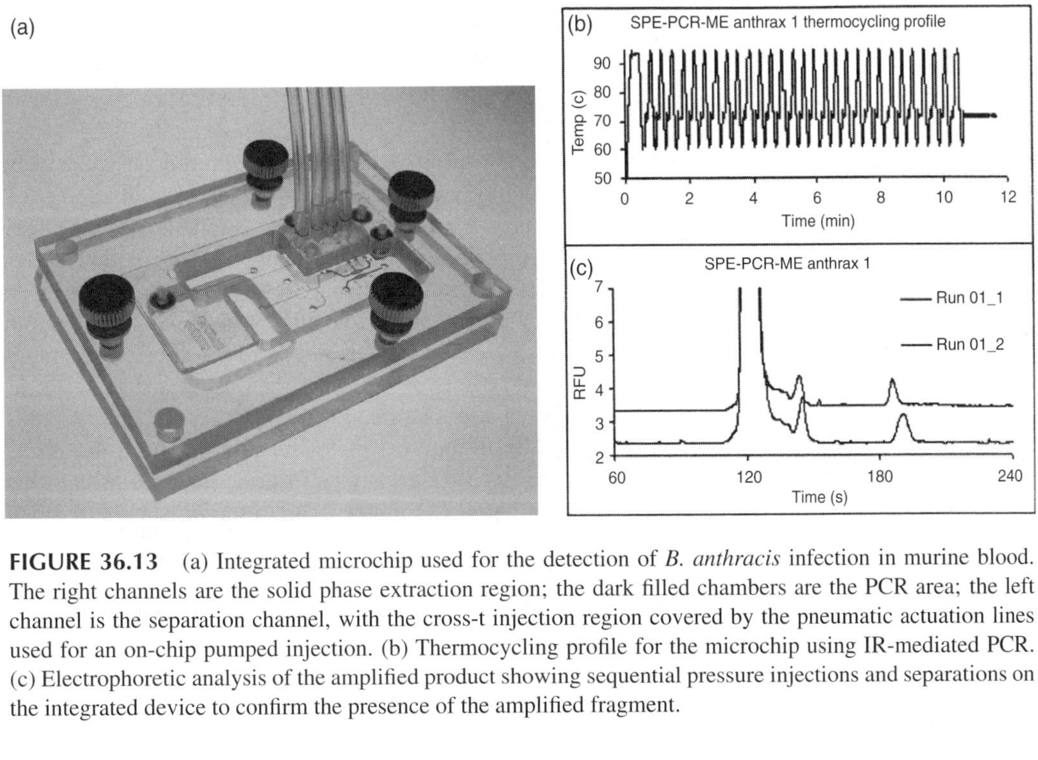

FIGURE 36.13 (a) Integrated microchip used for the detection of *B. anthracis* infection in murine blood. The right channels are the solid phase extraction region; the dark filled chambers are the PCR area; the left channel is the separation channel, with the cross-t injection region covered by the pneumatic actuation lines used for an on-chip pumped injection. (b) Thermocycling profile for the microchip using IR-mediated PCR. (c) Electrophoretic analysis of the amplified product showing sequential pressure injections and separations on the integrated device to confirm the presence of the amplified fragment.

36.4.5.2 Onboard Sample Preparation for Other Analytes

Integrated microdevices have also been developed for protein analysis, most often using MS detection, as in the discovery of new biomarkers useful for clinical analysis.[121] Gottschlich et al.[122] showed early integration on a microchip, combining a protein digestion step directly with an electrophoretic separation of the products, but often only a preconcentration step is integrated with the on-chip separation. Membranes have been formed in a cross channel for concentrating proteins during the injection step using both laser patterning[123] and photopolymerization.[124] While these methods might be applicable to clinical analysis, no clinical sample was evaluated. Both concentration and selective protein removal was shown on a microchip that integrated two electrophoretic channels with built-in electrochemical detection.[125] The first channel was used for sample preconcentration utilizing an isotachophoresis method, followed by a normal CZE separation; a myoglobin sample could be concentrated by greater than 60-fold in this system. Selective removal of proteins from a peptide sample was also shown in conjunction with desalting of the sample for subsequent analysis of the peptides.

In addition to proteins, small molecules have also been analyzed on integrated microchips. Analyses of three markers of renal function, uric acid, creatinine, and creatine, along with *p*-aminohippuric acid were investigated using integrated microchips by Wang and Chatrathi.[126] This work utilized fluid control to mix in creatinase and creatininase enzymes with the samples for on-chip enzymatic reactions before the electrophoresis. The unreacted acid species were separated from the neutral hydrogen peroxide, generated in the two enzymatic reactions, with detection using an amperometric method. To individually quantify the creatine and creatinine concentrations in a sample, the creatininase was simply eliminated from the reaction. A similar method was employed by Wang et al.[127] for the measurement of uric acid integrated with measurements of ascorbic acid, glucose, and acetaminophen on a single device. This work utilized the integrated microchip to mix the sample with glucose oxidase for generating hydrogen peroxide from the glucose present. Again, electrophoretic separation from the acidic species was possible with amperometric detection used for quantification. Acetaminophen, another neutral species, which migrated with the hydrogen peroxide, was detected by analyzing the sample with and without glucose oxidase to determine the difference in detector response when the neutral components passed the detector.

36.5 METHOD DEVELOPMENT GUIDELINES

Implementation of microchip methods for clinical analysis has thus far been almost exclusively confined to the research laboratory. This is changing with the availability and ease of use of the commercially available microchip electrophoresis instrumentation. These instruments provide both quantification and size determination of DNA or protein samples, thus could be immediately implemented into any clinical laboratory. Microdevices for sample preparation steps, at least for DNA processing, are currently under commercial development, but have not yet been introduced. Sample extractions and concentrations on microdevices are easily implemented in any laboratory, simply by placing the appropriate solid phase matrix into a microchip rather than using a suspension or large column, or centrifugal tube format. The only issue is fabrication or purchase of the microchips that have the appropriate sized channel and a constriction or frit for retention of the solid phase within the channel. Interfaces for the connection of syringe pumps to flow material through the microchip column have been described in the literature,[95] and microchip reservoirs/connectors are also commercially available from IDEX® Corporation.

In-house development of a microchip PCR amplification system is a bit more challenging than the DNA extraction system, but a noncontact IR heating design has been described in the literature.[128] As an easier approach, simply putting the microfluidic device into a conventional thermocycler also works, and Legendre et al.[99] showed that it was possible to integrate microchip DNA extraction with the PCR amplification using this technique. Integration of capture and separations, or enzymatic or labeling reactions with the electrophoretic separation, are also possible in many laboratories, again obtaining or fabricating the specifically designed microchips will be a significant part of the processes. Development of fully-integrated devices, however, is at this point beyond the realm of most laboratories that do not have experience with microfluidic design and utilization.

In addition to simply designing and fabricating the devices and instrumentation to use microchip methods, a number of issues also have to be dealt with that are more specific for clinical application of these devices. The use of microfluidic devices with biological samples provides some important differences relative to the use of these devices for standards or "real samples" that have been cleaned up or spiked to greatly increase the concentration of the compound of interest. The proteins, lipids, and DNA mixture found in any biological sample have numerous charged species, as well as both hydrophobic and hydrophilic compounds. Thus, surfaces in microchips formed from almost any substrate will attract or interact with some component in the sample. In addition, there are issues with the sheer mass of material, such as proteins that can compete nonspecifically with an assay, the particulate matter left by cell lysis that can clog the microstructures, and the viscosity of biological samples that make it hard to pump material through the microdevice. The development of microchips for clinical applications then must look beyond the initial testing of the microchip assays using standards, and show that the assay, method, and microchip are rugged and reliable enough for use with real samples at the required concentrations, which is not always the case.[2]

Often, the primary issue when implementing on-chip clinical analyses is biofouling, where proteins or lipids bind to the microchip surface interfering with the assay being performed on the device; this is particularly an issue in microchips because of the large surface-to-volume ratio presented by these devices. One of the biggest clinical applications in which this is important is PCR amplification of DNA, in which binding of *Taq* polymerase and Mg^{2+} to the microchip surface can inhibit the reaction.[129,130] A review by Kricka and Wilding[131] detailed the various methods that had been utilized for passivating microchip surfaces to prevent PCR inhibition, and this group has shown increased efficiency for PCR through coating selection.[132] Preventing reaction components from binding to the surface can also be accomplished by preventing the components in the aqueous solutions from reaching the surfaces by surrounding the aqueous phase with an organic phase such as a fluorocarbon oil, as reported by Roach et al.[133]

Of more interest is some recent work that looks at the modification of the microchip surfaces to resist nonspecific adsorption in polymeric devices, since these represent the future of inexpensive, disposable microchips that will be needed for clinical analyses. Bi et al.[134] utilized an acrylate copolymer to modify the surface of PMMA devices to exhibit poly(ethylene glycol) (PEG) on the surface. PEG is known to prevent protein adsorption, and these microdevices were utilized in electrophoretic separations of proteins, with good theoretical efficiencies. Application of these devices directly to serum or plasma should be possible with no biofouling of the separation channel surface. This same group was also able to generate a phospholipid coating on the surface of PMMA microchip channels as a biomimetic surface to prevent protein binding.[135] This coating was also evaluated for the electrophoretic separation of proteins, showing stable electrophoretic mobility, with little nonspecific adhesion of proteins or platelets when serum or plasma was utilized in the device. For PDMS devices, Wu et al.[136] grafted hydrophilic polymers onto the surface to create a surface with a strongly suppressed electroosmotic flow (EOF), and significantly decreased adsorption of BSA and lysozyme than the native PDMS surface. These microchips were also utilized for electrophoretic separations of protein, both native and denatured, as well as peptides.

In addition to adsorption, issues to be addressed for the routine use of microchips for clinical analysis include the development of lower cost, compact instrumentation for detection on microchips. Shrinivasan et al.[137] detailed the development of a miniaturized LIF detector applicable for fluorescent detection of DNA on microchips, which could also be applied to proteins with the necessary fluorescent tags. By replacing the argon-ion laser with a diode laser, the cost and power requirements were significantly decreased, with only a small loss in sensitivity. Integration of fluid flow and mixing components into the microchips must also be addressed. Kim et al.[138] describe a microchip for blood typing that incorporates flow splitting designs, micromixers, and microfilters directly fabricated into an injection molded COC device.[139] This group also developed a COC microchip for partial oxygen, glucose, and lactate levels in blood that incorporates an on-chip power source for fluid manipulation. Vestad et al.[140] addressed the need to reduce the associated hardware for flow control through microchips through the use of capillary pumping, allowing surface tension to pull fluid through their devices. Flow path can be controlled through the use of elastomeric valves, similar to those utilized in the integrated devices. Again, the goal is to eliminate the need for external power sources in the instrumentation for fluid flow control, to simplify the instrumentation and make it cost effective and compact enough for use in point-of-care applications.

36.6 CONCLUDING REMARKS

There are a number of clinical areas in which microchip devices will be implemented over the next decade. For DNA, the switch will actually be away from the devices described here to a large extent as real time-PCR analysis becomes the predominant method for detecting and quantifying specific DNA sequences. Real-time PCR takes advantage of the amplification method while eliminating the separation step to reduce the complexity of the analysis and the microchip design. These devices will be implemented both for mutation detection and infectious agent analysis. For proteins, the separation will remain an integral part of the analysis, thus microchip separations will expand to form a larger part of the clinical diagnostic arena. Most of these devices will utilize MS detection, however, unless new breakthroughs in protein labeling or UV detection on microchips are seen. Electrophoretic microchips for small molecule detection will play some part in clinical analysis, but individualized instrumentation specific for a particular analyte or group of analytes may be the final format.

For this to become reality, reductions in the costs of both the microdevices and the instrumentation will be needed to move these methods into routine clinical laboratory use. On the instrumentation side, replacing research designs with commercial grade instruments, with simple user interfaces and integrated components for sample and reagent loading, alignment, and thermal processing must be completed. Initial commercial microchip electrophoresis instruments show that this is possible, but more demand from the clinical community may be needed to show the potential for additional microchip instrumentation to commercial entities. The microchips themselves must also be cost effective for the analyses being performed, with disposable devices required to prevent contamination or false positive and false negative results. The cost of a PCR microchip, even a plastic and disposable one, will never be as inexpensive as a PCR tube, thus additional advantages must be provided by the microchip methods to justify their implementation. These can include reduced reagent consumption in the microchip, or incorporation of additional processing steps such as the DNA extraction, but some added benefit must be present in the microchips if they are to be accepted and used by the clinical community.

In terms of integrating additional steps onto a microdevice, issues related to fluid flow control, either through on-chip or off-chip valving or flow metering, still need to be more fully addressed. Handling of the nanoliter volumes utilized in microchips presents additional challenges for instrumentation development, particularly with reagent addition and mixing with components already on a microdevice. Further work is also needed in the area of on-chip storage of reagents, possibly already lyophilized in the necessary compartments, or sealed off for storage but easily available on-chip when needed. Development of integrated microchips will continue, however, and these will begin to focus more on the actual end applications now that the concepts have been shown using standards and simulants. More specific clinical analyses will be translated to microchips over the next few years, and the complexity of the designs and instrumentation will determine how quickly they will be transitioned into the clinical laboratory.

REFERENCES

1. Burtis, C. A., Converging technologies and their impact on the clinical laboratory, *Clin. Chem.*, 42, 1735, 1996.
2. Colyer, C. L., Tang, T., Chiem, N., and Harrison, D. J. V., Clinical potential of microchip capillary electrophoresis systems, *Electrophoresis*, 18, 1733, 1997.
3. Kricka, L. J., Miniaturization of analytical systems, *Clin. Chem.*, 44, 2008, 1998.
4. Tudos, A. J., Besselink, G. A. J., and Schasfoort, R. B. M., Trends in miniaturized total analysis systems for point-of-care testing in clinical chemistry, *Lab Chip*, 1, 83, 2001.
5. Schulte, T. H., Bardell, R. L., and Weigl, B. H., Microfluidic technologies in clinical diagnostics, *Clin. Chim. Acta*, 321, 1, 2002.
6. Verpoorte, E., Microfluidic chips for clinical and forensic analysis, *Electrophoresis*, 23, 677, 2002.
7. Pugia, M. J., Blankenstein, G., Peters, R. P., Profitt, J. A., Kadel, K., Willms, T., Sommer, R., Kuo, H. H., and Schulman, L. S., Microfluidic tool box as technology platform for hand-held diagnostics, *Clin. Chem.*, 51, 1923, 2005.
8. Soper, S. A., Brown, K., Ellington, A., Frazier, B., Garcia-Manero, G., Gau, V., Gutman, S. I., et al., Point-of-care biosensor systems for cancer diagnostics/prognostics, *Biosens. Bioelec.*, 21, 1932, 2006.
9. Situma, C., Hashimoto, M., and Soper, S. A., Merging microfluidics with microarray-based bioassays, *Biomol. Eng.*, 23, 213, 2006.
10. Aytur, T., Foley, J., Anwar, M., Boser, B., Harris, E., and Beatty, P. R., A novel magnetic bead bioassay platform using a microchip-based sensor for infectious disease diagnosis, *J. Immunol. Meth.*, 314, 21, 2006.
11. Soper, S. A., Hashimoto, M., Situma, C., Murphy, M. C., McCarley, R. L., Cheng, Y. W., and Barany, F., Fabrication of DNA microarrays onto polymer substrates using UV modification protocols with integration into microfluidic platforms for the sensing of low-abundant DNA point mutations, *Methods*, 37, 103, 2005.
12. Peytavi, R., Raymond, F. R., Gagne, D., Picard, F. J., Jia, G., Zoval, J., Madou, M., Boissinot, K., Boissinot, M., Bissonnette, L., Ouellette, M., and Bergeron, M. G., Microfluidic device for rapid (<15 min) automated microarray hybridization, *Clin. Chem.*, 51, 1836, 2005.
13. Rubina, A. Y., Pan'kov, S. V., Dementieva, E. I., Pen'kov, D. N., Butygin, A. V., Vasiliskov, V. A., Chudinov, A. V., Mikheikin, A. L., Mikhailovich, V. M., and Mirzabekov, A. D., Hydrogel drop microchips with immobilized DNA: Properties and methods for large-scale production, *Anal. Biochem.*, 325, 92, 2004.
14. Sohni, Y. R., Cerhan, J. R., and O'Kane, D., Microarray and microfluidic methodology for genotyping cytokine gene polymorphisms, *Human Immunol.*, 64, 990, 2003.
15. Honda, N., Lindberg, U., Andersson, P., Hoffman, S., and Takei, H., Simultaneous multiple immunoassays in a compact disc-shaped microfluidic device based on centrifugal force, *Clin. Chem.*, 51, 1955, 2005.
16. Rossier, J. S., and Girault, H. H., Enzyme linked immunosorbent assay on a microchip with electrochemical detection, *Lab Chip*, 1, 153, 2001.
17. Dupuy, A. M., Lehmann, S., and Cristol, J. P., Protein biochip systems for the clinical laboratory, *Clin. Chem. Lab. Med.*, 43, 1291, 2005.
18. Jia, H. X., Wu, Z. Y., and Fang, Z. L., Microfluidic chip-based immunoassay, *Chin. J. Anal. Chem.*, 33, 1489, 2005.
19. Malcik, N., Ferrance, J. P., Landers, J. P., and Caglar, P., The performance of a microchip-based fiber optic detection technique for the determination of Ca^{2+} ions in urine, *Sens. Act. B–Chem.*, 107, 24, 2005.
20. Srinivasan, V., Pamula, V. K., and Fair, R. B., An integrated digital microfluidic lab-on-a-chip for clinical diagnostics on human physiological fluids, *Lab Chip*, 4, 310, 2004.
21. Morimoto, K., and Suzuki, H., Micro analysis system for pH and protease activities with an integrated sample injection mechanism, *Biosens. Bioelec.*, 22, 86, 2006.
22. Upadhyay, S., Ohgami, N., Kusakabe, H., and Suzuki, H., Electrochemical determination of gamma-glutamyl transpeptidase activity and its application to a miniaturized analysis system *Biosens. Bioelec.*, 21, 1230, 2006.
23. Guijt, R. M., Baltussen, E., and van Dedem, G. W. K., Use of bioaffinity interactions in electrokinetically controlled assays on microfabricated devices, *Electrophoresis*, 23, 823, 2002.

24. El-Ali, J., Sorger, P. K., and Jensen, K. F., Cells on chips, *Nature*, 442, 403, 2006.
25. Jin, L. J., Ferrance, J., and Landers, J. P., Miniaturized electrophoresis: An evolving role in laboratory medicine, *Biotechniques*, 31, 1332, 2001.
26. McGlennen, R. C., Miniaturization technologies for molecular diagnostics, *Clin. Chem.*, 47, 393, 2001.
27. Oin, J. H., Fung, Y. S., and Lin, B. C., DNA diagnosis by capillary electrophoresis and microfabricated electrophoretic devices, *Exp. Rev. Molec. Diag.*, 3, 387, 2003.
28. Mastrangelo, C. H., Burns, M. A., and Burke, D. T., Microfabricated devices for genetic diagnostics, *Proc. IEEE*, 86, 1769, 1998.
29. Li, S. F. Y., and Kricka, L. J., Clinical analysis by microchip capillary electrophoresis, *Clin. Chem.*, 52, 37, 2006.
30. Tian, H. J., Jaquins-Gerstl, A., Munro, N., Trucco, M., Brody, L. C., and Landers, J. P., Single-strand conformation polymorphism analysis by capillary and microchip electrophoresis: A fast, simple method for detection of common mutations in BRCA1 and BRCA2, *Genomics*, 63, 25, 2000.
31. Kang, S. H., Jang, S., and Park, S. K., Single-strand conformation polymorphism analysis by microchip electrophoresis for the rapid detection of point mutation in human obesity gene, *Bull. Korean Chem. Soc.*, 27, 1346, 2006.
32. Hestekin, C. N., Jakupciak, J. P., Chiesl, T. N., Kan, C. W., O'Connell, C. D., and Barron, A. E., An optimized microchip electrophoresis system for mutation detection by tandem SSCP and heteroduplex analysis for p53 gene exons 5–9, *Electrophoresis*, 27, 3823, 2006.
33. Tian, H. J., Brody, L. C., Fan, S. J., Huang, Z. L., and Landers, J. P., Capillary and microchip electrophoresis for rapid detection of known mutations by combining allele-specific DNA amplification with heteroduplex analysis, *Clin. Chem.*, 47, 173, 2001.
34. Hestekin, C. N., and Barron, A. E., The potential of electrophoretic mobility shift assays for clinical mutation detection, *Electrophoresis*, 27, 3805, 2006.
35. Woolley, A. T., Sensabaugh, G. F., and Mathies, R. A., High-speed DNA genotyping using microfabricated capillary array electrophoresis chips, *Anal. Chem.*, 69, 2181, 1997.
36. Shi, Y. N., Simpson, P. C., Scherer, J. R., Wexler, D., Skibola, C., Smith M. T., and Mathies, R. A., Radial capillary array electrophoresis microplate and scanner for high-performance nucleic acid analysis, *Anal. Chem.*, 71, 5354, 1999.
37. Medintz, I., Wong, W. W., Sensabaugh, G., and Mathies, R. A., High speed single nucleotide polymorphism typing of a hereditary haemochromatosis mutation with capillary array electrophoresis microplates, *Electrophoresis*, 21, 2352, 2000.
38. Medintz, I., Wong, W. W., Berti, L., Shiow, L., Tom, J., Scherer, J., Sensabaugh G., and Mathies, R. A., High performance multiplex SNP analysis of three hemochromatosis-related mutations with capillary array electrophoresis microplates, *Genome Res.*, 11, 413, 2001.
39. Sung, W. C., Lee, G. B., Tzeng, C. C., and Chen, S. H., Plastic microchip electrophoresis for genetic screening: The analysis of polymerase chain reactions products of fragile X (CGG)n alleles, *Electrophoresis*, 22, 1188, 2001.
40. Cantafora, A., Blotta, I., Bruzzese, N., Calandra, S., and Bertolini, S., Rapid sizing of microsatellite alleles by gel electrophoresis on microfabricated channels: Application to the D19S394 tetranucleotide repeat for cosegregation study of familial hypercholesterolemia, *Electrophoresis*, 22, 4012, 2001.
41. Ferrance, J., Snow, K., and Landers, J. P., Evaluation of microchip electrophoresis as a molecular diagnostic method for Duchenne muscular dystrophy, *Clin. Chem.*, 48, 380, 2002.
42. Ferrance, J. P., Wu, Q. R., Giordano, B., Hernandez, C., Kwok, Y., Snow, K., Thibodeau, S., and Landers, J. P., Developments toward a complete micro-total analysis system for Duchenne muscular dystrophy diagnosis, *Anal. Chim. Acta*, 500, 223, 2003.
43. Sohni, Y. R., Cerhan, J. R., and O'Kane, D., Microarray and microfluidic methodology for genotyping cytokine gene polymorphisms, *Human Immunol.*, 64, 990, 2003.
44. Sohni, Y. R., Burke, J. P., Dyck, P. J., and O'Kane, D. J., Microfluidic chip-based method for genotyping microsatellites, VNTRs and insertion/deletion polymorphisms, *Clin. Biochem.*, 36, 35, 2003.
45. Zhang, L. H., Dang, F. Q., Kaji, N., and Baba, Y., Fast extraction, amplification and analysis of genes from human blood, *J. Chrom. A*, 1106, 175, 2006.

46. Munro, N. J., Snow, K., Kant, E. A., and Landers, J. P., Molecular diagnostics on microfabricated electrophoretic devices: From slab gel- to capillary- to microchip-based assays for T- and B-cell lymphoproliferative disorders, *Clin. Chem.*, 45, 1906, 1999.
47. Cohen, C. D., Frach, K., Schlondorff, D., and Kretzler, M., Quantitative gene expression analysis in renal biopsies: A novel protocol for a high-throughput multicenter application, *Kidney Inter.*, 61, 133, 2002.
48. Höfgartner, W. T., Huhmer, A. F. R., Landers, J. P., and Kant, J. A., Rapid diagnosis of herpes simplex encephalitis using microchip electrophoresis of PCR products, *Clin. Chem.*, 45, 2120, 1999.
49. Chen, Y. H., Wang, W. C., Young, K. C., Chang, T. T., and Chen, S. H., Plastic microchip electrophoresis for analysis of PCR products of hepatitis C virus, *Clin. Chem.*, 45, 1938, 1999.
50. Young, K. C., Lien, H. M., Lin, C. C., Chang, T. T., Lee, G. B., and Chen, S. H., Microchip and capillary electrophoresis for quantitative analysis of hepatitis C virus based on RT-competitive PCR, *Talanta*, 56, 323, 2002.
51. Cangelosi, G. A., Freeman, R. J., Lewis, K. N., Livingston-Rosanoff, D., Shah, K. S., Milan, S. J., and Goldberg, S. V., Evaluation of a high-throughput repetitive-sequence-based PCR system for DNA fingerprinting of *Mycobacterium tuberculosis* and *Mycobacterium avium* complex strains, *J. Clin. Microbiol.*, 42, 2685, 2004.
52. Ramsey, J. D., Jacobson, S. C., Culbertson, C. T., and Ramsey, J. M., High-efficiency, two-dimensional separations of protein digests on microfluidic devices, *Anal. Chem.*, 75, 3758, 2003.
53. Colyer, C. L., Mangru, S. D., and Harrison, D. J., Microchip-based capillary electrophoresis of human serum proteins, *J. Chromatogr. A*, 781, 271, 1997.
54. Callewaert, N., Contreras, R., Mitnik-Gankin, L., Carey, L., Matsudaira, P., and Ehrlich, D., Total serum protein N-glycome profiling on a capillary electrophoresis-microfluidics platform, *Electrophoresis*, 25, 3128, 2004.
55. Weiller, B. H., Ceriotti, L., Shibata, T., Rein, D., Roberts, M. A., Lichtenberg, J., German, J. B., de Rooij, N. F., and Verpoorte, E., Analysis of lipoproteins by capillary zone electrophoresis in microfluidic devices: Assay development and surface roughness measurements, *Anal. Chem.*, 74, 1702, 2002.
56. Ceriotti, L., Shibata, T., Folmer, B., Weiller, B. H., Roberts, M. A., de Rooij, N. F., and Verpoorte, E., Low-density lipoprotein analysis in microchip capillary electrophoresis systems, *Electrophoresis*, 23, 3615, 2002.
57. Yao, S., Anex, D. S., Caldwell, W. B., Arnold, D. W., Smith, K. B., and Schultz, P. G., SDS capillary gel electrophoresis of proteins in microfabricated channels, *Proc. Natl. Acad. Sci. USA*, 96, 5372, 1999.
58. Liu, Y. J., Foote, R. S., Jabobson, S. C., Ramsey R. S., and Ramsey, J. M., Electrophoretic separation of proteins on a microchip with noncovalent postcolumn labeling, *Anal. Chem.*, 72, 4606, 2000.
59. Bousse, L., Mouradian, S., Minalla, A., Yee, H., Williams, K., and Dubrow, R., Protein sizing on a microchip, *Anal. Chem.*, 73, 1207, 2001.
60. Giordano, B. C., Jin, L. J., Couch, A. J., Ferrance, J. P., and Landers, J. P., Microchip laser-induced fluorescence detection of proteins at submicrogram per milliliter levels mediated by dynamic labeling under pseudonative conditions, *Anal. Chem.*, 76, 4705, 2004.
61. Chan, O. T. M., and Herold, D. A., Chip electrophoresis as a method for quantifying total microalbuminuria, *Clin. Chem.*, 52, 2141, 2006.
62. Zhuang, G. S., Liu, J., Jia, C. P., Jin, Q. H., Zhao, J. L., and Wang, H. M., Manufacture of reversible electrophoresis chip and its application to protein detection, *Acta Chim. Sinica*, 63, 1003, 2005.
63. Zhuang, G. S., Liu, J., Jia, C. P., Jin, Q. H., Wang, H. M., Yang, M. S., and Zhao, J. L., Capillary electrophoresis microchip assay of clinical urinary proteins, *Acta Chim. Sinica*, 64, 229, 2006.
64. Zhuang, G. S., Jin, Q. H., Liu, J., Cong, H., Liu, K. D., Zhao, J. L., Yang, M. S., and Wang, H. M., A low temperature bonding of quartz microfluidic chip for serum lipoproteins analysis, *Biomed. Microdevice*, 8, 255, 2006.
65. Wang, J., Zhou, S., Huang, W., Liu, Y., Cheng, C., Lu, X., and Cheng, J., CE-based analysis of hemoglobin and its applications in clinical analysis, *Electrophoresis*, 27, 3108, 2006.
66. Schmalzing, D., Koutny, L. B., Taylor, T. A., Nashabeh, W., and Fuchs, M., Immunoassay for thyroxine (T4) in serum using capillary electrophoresis and micromachined devices, *J. Chromatogr. B*, 697, 175, 1997.

67. Koutny, L. B., Schmalzing, D., Taylor, T. A., and Fuchs, M., Microchip electrophoretic immunoassay for serum cortisol, *Anal. Chem.*, 68, 18, 1996.
68. Qiu, C. X., and Harrison, D. J., Integrated self-calibration via electrokinetic solvent proportioning for microfluidic immunoassays, *Electrophoresis*, 22, 3949, 2001.
69. Wang, J., Ibanez, A., Chatrathi, M. P., and Escarpa, A., Electrochemical enzyme immunoassays on microchip platforms, *Anal. Chem.*, 73, 5323, 2001.
70. Wang, J., Ibanez, A., and Chatrathi, M. P., Microchip-based amperometric immunoassays using redox tracers, *Electrophoresis*, 23, 3744, 2002.
71. Herr, A. E., Throckmorton, D. J., Davenport, A. A., and Singh, A. K., On-chip native gel electrophoresis-based immunoassays for tetanus antibody and toxin, *Anal. Chem.*, 77, 585, 2005.
72. Fanguy, J. C., and Henry, C. S., The analysis of uric acid in urine using microchip capillary electrophoresis with electrochemical detection, *Electrophoresis*, 23, 767, 2002.
73. Garcia, C. D., and Henry, C. S., Direct detection of renal function markers using microchip CE with pulsed electrochemical detection, *Analyst*, 129, 579, 2004.
74. Du, Y., Yan, J. L., Zhou, W. Z., Yang, X. Y., and Wang, E. K., Direct electrochemical detection of glucose in human plasma on capillary electrophoresis microchips, *Electrophoresis*, 25, 3853, 2004.
75. Suzuki, S., and Honda, S., Miniaturization in carbohydrate analysis, *Electrophoresis*, 24, 3577, 2003.
76. Zhao, X. C., You, T. Y., Qiu, H. B., Yan, J. L., Yang, X. R., and Wang, E. K., Electrochemiluminescence detection with integrated indium tin oxide electrode on electrophoretic microchip for direct bioanalysis of lincomycin in the urine, *J. Chromatogr. B*, 810, 137, 2004.
77. Crevillen, A. G., Barrigas, I., Blasco, A. J., Gonzalez, M. C., and Escarpa, A., Microchip-electrochemistry route for rapid screening of hydroquinone and arbutin from miscellaneous samples: Investigation of the robustness of a simple cross-injector system, *Anal. Chim. Acta*, 562, 137, 2006.
78. Laugere, F., Guijt, R. M., Bastemeijer, J., van der Steen, G., Berthold, A., Baltussen, E., Sarro, P., van Dedem, G. W. K., Vellekoop, M., and Bossche, A., On-chip contactless four-electrode conductivity detection for capillary electrophoresis devices, *Anal. Chem.*, 75, 306, 2003.
79. Munro, N. J., Huang, Z., Finegold, D. N., and Landers, J. P., Indirect fluorescence detection of amino acids on electrophoretic microchips, *Anal. Chem.*, 72, 2765, 2000.
80. Caglar, P., Tuncel, S. A., Malcik, N., Landers, J. P., and Ferrance, J. P., A microchip sensor for calcium determination, *Anal. Bioanal. Chem.*, 386, 1303, 2006.
81. Huang, Y., Mather, E. L., Bell, J. L., and Madou, M., MEMS-based sample preparation for molecular diagnostics, *Anal. Bioanal. Chem.*, 372, 49, 2002.
82. Zhang, C. S., Xu, J. L., Ma, W. L., and Zheng, W. L., PCR microfluidic devices for DNA amplification, *Biotech. Adv.*, 24, 243, 2006.
83. Mikhailovich, V. M., Lapa, S. A., Gryadunov, D. A., Strizhkov, B. N., Sobolev, A. Y., Skotnikova, O. I., Irtuganova, O. A., et al., Detection of rifampicin-resistant *Mycobacterium tuberculosis* strains by hybridization and polymerase chain reaction on a specialized TB-microchip, *Bull. Exp. Biol. Med.*, 131, 94, 2001.
84. Yang, J. N., Liu, Y. J., Rauch, C. B., Stevens, R. L., Liu, R. H., Lenigk, R., and Grodzinski, P., High sensitivity PCR assay in plastic micro reactors, *Lab Chip*, 2, 179, 2002.
85. Cho, Y. K., Kim, J., Lee, Y., Kim, Y. A., Namkoong, K., Lim, H., Oh, K. W., et al., Clinical evaluation of micro-scale chip-based PCR system for rapid detection of hepatitis B virus, *Biosens. Bioelec.*, 21, 2161, 2006.
86. Gulliksen, A., Solli, L. A., Drese, K. S., Sorensen, O., Karlsen, F., Rogne, H., Hovig, E., and Sirevag, R., Parallel nanoliter detection of cancer markers using polymer microchips, *Lab Chip*, 5, 416, 2005.
87. Cheng, J., Waters, L. C., Fortina, P., Hvichia, G., Jacobson, S. C., Ramsey, J. M., Kricka, L. J., and Wilding, P., Degenerate oligonucleotide primed polymerase chain reaction and capillary electrophoretic analysis of human DNA on microchip-based devices, *Anal. Biochem.*, 257, 101, 1998.
88. Lou, X. J., Panaro, N. J., Wilding, P., Fortina, P., and Kricka, L. J., Mutation detection using ligase chain reaction in passivated silicon-glass microchips and microchip capillary electrophoresis, *Biotechniques*, 37, 392, 2004.
89. Easley, C. J., Legendre, L. A., Roper, M. G., Wavering, T. A., Ferrance, J. P., and Landers, J. P., Extrinsic Fabry-Perot interferometry for noncontact temperature control of nanoliter-volume enzymatic reactions in glass microchips, *Anal. Chem.*, 77, 1038, 2005.

90. Peterson, D. S., Rohr, T., Svec, F., and Frechet, J. M. J., Enzymatic microreactor-on-a-chip: Protein mapping using trypsin immobilized on porous polymer monoliths molded in channels of microfluidic devices, *Anal. Chem.*, 74, 4081, 2002.
91. Jin, L. J., Ferrance, J., Sanders, J. C., and Landers, J. P., A microchip-based proteolytic digestion system driven by electroosmotic pumping, *Lab Chip*, 3, 11, 2003.
92. Crowley, T. A., and Pizziconi, V., Isolation of plasma from whole blood using planar microfilters for lab-on-a-chip applications, *Lab Chip*, 5, 922, 2005.
93. Shevkoplyas, S. S., Yoshida, T., Munn, L. L., and Bitensky, M. W., Biomimetic autoseparation of leukocytes from whole blood in a microfluidic device, *Anal. Chem.*, 77, 933, 2005.
94. Mansfield, C. D., Man, A., Low-Ying, S., and Shaw, R. A., Laminar fluid diffusion interface preconditioning of serum and urine for reagent-free infrared clinical analysis and diagnostics, *Appl. Spec.*, 59, 10, 2005.
95. Breadmore, M. C., Wolfe, K. A., Arcibal, I. G., Leung, W. K., Dickson, D., Giordano, B. C., Power, M. E., Ferrance, J. P., Feldman, S. H., Norris, P. M., and Landers, J. P., Microchip-based purification of DNA from biological samples, *Anal. Chem.*, 75, 1880, 2003.
96. Chung, Y. C., Jan, M. S., Lin, Y. C., Lin, J. H., Cheng, W. C., and Fan, C. Y., Microfluidic chip for high efficiency DNA extraction, *Lab Chip*, 4, 141, 2004.
97. Wu, Q. R., Bienvenue, J. M., Hassan, B. J., Kwok, Y. C., Giordano, B. C., Norris, P. M., Landers, J. P., and Ferrance, J. P., Microchip-based macroporous silica sol-gel monolith for efficient isolation of DNA from clinical samples, *Anal. Chem.*, 78, 5704, 2006.
98. Wen, J., Guillo, C., Ferrance, J. P., and Landers, J. P., DNA extraction using a tetramethyl orthosilicate-grafted photopolymerized monolithic solid phase, *Anal. Chem.*, 78, 1673, 2006.
99. Legendre, L. A., Bienvenue, J. M., Roper, M. G., Ferrance, J. P., and Landers, J. P., A valveless microfluidic sample preparation device for DNA extraction and amplification of DNA from nanoliter volume samples using conventional instrumentation, *Anal. Chem.*, 78, 1444, 2006.
100. Witek, M., Llopis, S. D., Wheatley, A., McCarley, R. L., and Soper, S. A., Purification and preconcentration of genomic DNA from whole cell lysates using photoactivated polycarbonate (PPC) microfluidic chips, *Nucleic Acids Res.*, 34, e74/1, 2006.
101. Nakagawa, T., Tanaka, T., Niwa, D., Osaka, T., Takeyama, H., and Matsunaga, T., Fabrication of amino silane-coated microchip for DNA extraction from whole blood, *J. Biotechnol.*, 116, 105, 2005.
102. Cao, W. D., Easley, C. J., Ferrance, J. P., and Landers, J. P., Chitosan as a polymer for pH-induced DNA capture in a totally aqueous system, *Anal. Chem.*, 78, 7222, 2006.
103. Kokoris, M., Nabavi, M., Lancaster, C., Clemmens, J., Maloney, P., Capadanno, J., Gerdes, J., and Battrell, C. F., Rare cancer cell analyzer for whole blood applications: Automated nucleic acid purification in a microfluidic disposable card, *Methods*, 37, 114, 2005.
104. Roper, M. G., Frisk, M. L., Oberlander, J. P., Ferrance, J. P., McGrory, B. J., and Landers, J. P., Extraction of C-reactive protein from serum on a microfluidic chip, *Anal. Chim. Acta*, 569, 195, 2006.
105. Liu, Y., Garcia, C. D., and Henry, C. S., Recent progress in the development of μTAS for clinical analysis, *Analyst*, 128, 1002, 2003.
106. Lagally, E. T., and Soh, H. T., Integrated genetic analysis microsystems, *Crit. Rev. Solid State Mat. Sci.*, 30, 207, 2005.
107. Woolley, A. T., Hadley, D., Landre, P., deMello, A. J., Mathies, R. A., and Northrup, M. A., Functional integration of PCR amplification and capillary electrophoresis in a microfabricated DNA analysis device, *Anal. Chem.*, 68, 4081, 1996.
108. Waters, L. C., Jacobson, S. C., Kroutchinina, N., Khandurina, J., Foote, R. S., and Ramsey, J. M., Microchip device for cell lysis, multiplex PCR amplification, and electrophoretic sizing, *Anal. Chem.*, 70, 158, 1998.
109. Waters, L. C., Jacobson, S. C., Kroutchinina, N., Khandurina, J., Foote, R. S., and Ramsey, J. M., Multiple sample PCR amplification and electrophoretic analysis on a microchip, *Anal. Chem.*, 70, 5172, 1998.
110. Cheng, J., Waters, L. C., Fortina, P., Hvichia, G., Jacobson, S. C., Ramsey, J. M., Kricka, L. J., and Wilding, P., Degenerate oligonucleotide primed polymerase chain reaction and capillary electrophoretic analysis of human DNA on microchip-based devices, *Anal. Biochem.*, 257, 101, 1998.

111. Khandurina, J., McKnight, T. E., Jacobson, S. C., Waters, L. C., Foote, R. S., and Ramsey, J. M., Integrated system for rapid PCR-based DNA analysis in microfluidic devices, *Anal. Chem.*, 72, 2995, 2000.
112. Lagally, E. T., Simpson, P. C., and Mathies, R. A., Monolithic integrated microfluidic DNA amplification and capillary electrophoresis analysis system, *Sens. Act. B–Chem.*, 63, 138, 2000.
113. Koh, C. G., Tan, W., Zhao, M. Q., Ricco, A. J., and Fan, Z. H., Integrating polymerase chain reaction, valving, and electrophoresis in a plastic device for bacterial detection, *Anal. Chem.*, 75, 4591, 2003.
114. Lagally, E. T., Scherer, J. R., Blazej, R. G., Toriello, N. M., Diep, B. A., Ramchandani, M., Sensabaugh, G. F., Riley, L. W., and Mathies, R. A., Integrated portable genetic analysis microsystem for pathogen/infectious disease detection, *Anal. Chem.*, 76, 3162, 2004.
115. Easley, C. J., Karlinsey, J. M., and Landers, J. P., On-chip pressure injection for integration of infrared-mediated DNA amplification with electrophoretic separation, *Lab Chip*, 6, 601, 2006.
116. Zhou, Z. M., Liu, D. Y., Zhong, R. T., Dai, Z. P., Wu, D. P., Wang, H., Du, Y. G., Xia, Z. N., Zhang, L. P., Mei, X. D., and Lin, B. C., Determination of SARS-coronavirus by a microfluidic chip system, *Electrophoresis*, 25, 3032, 2004.
117. Pal, R., Yang, M., Lin, R., Johnson, B. N., Srivastava, N., Razzacki, S. Z., Chomistek, K. J., et al., An integrated microfluidic device for influenza and other genetic analyses, *Lab Chip*, 5, 1024, 2005.
118. Kaigala, G. V., Huskins, R. J., Preiksaitis, J., Pang, X. L., Pilarski, L. M., and Backhouse, C. J., Automated screening using microfluidic chip-based PCR and product detection to assess risk of BK virus-associated nephropathy in renal transplant recipients, *Electrophoresis*, 27, 3753, 2006.
119. Easley, C. J., Karlinsey, J. M., Bienvenue, J. M., Legendre, L. A., Roper, M. G., Feldman, S. H., Hughes, M. A., et al., A fully-integrated microfluidic genetic analysis system with sample in-answer out capability, *Proc. Natl. Acad. Sci. USA*, 103, 19272, 2006.
120. Taylor, P., Manage, D. P., Helmle, K. E., Zheng, Y., Glerum, D. M., and Backhouse, C. J., Analysis of mitochondrial DNA in microfluidic systems, *J. Chromatogr. B*, 822, 78, 2005.
121. Fortier, M. H., Bonneil, E., Goodley, P., and Thibault, P., Integrated microfluidic device for mass spectrometry-based proteomics and its application to biomarker discovery programs, *Anal. Chem.*, 77, 1631, 2005.
122. Gottschlich, N., Culbertson, C. T., McKnight, T. E., Jacobson, S. C., and Ramsey, J. M., Integrated microchip-device for the digestion, separation and postcolumn labeling of proteins and peptides, *J. Chromatogr. B*, 745, 243, 2000.
123. Song, S., Singh, A. K., and Kirby, B. J., Electrophoretic concentration of proteins at laser-patterned nanoporous membranes in microchips, *Anal. Chem.*, 76, 4589, 2004.
124. Hatch, A. V., Herr, A. E., Throckmorton, D. J., Brennan, J. S., and Singh, A. K., Integrated preconcentration SDS-PAGE of proteins in microchips using photopatterned cross-linked polyacrylamide gels, *Anal. Chem.*, 78, 4976, 2006.
125. Silvertand, L. H. H., Machtejevas, E., Hendriks, R., Unger, K. K., van Bennekom, W .P., and de Jong, G. J., Selective protein removal and desalting using microchip CE, *J. Chromatogr. B*, 839, 68, 2006.
126. Wang, J., and Chatrathi, M. P., Microfabricated electrophoresis chip for bioassay of renal markers, *Anal. Chem.*, 75, 525, 2003.
127. Wang, J., Chatrathi, M. P., Tian, B. M., and Polsky, R., Microfabricated electrophoresis chips for simultaneous bioassays of glucose, uric acid, ascorbic acid, and acetaminophen, *Anal. Chem.*, 72, 2514, 2000.
128. Easley, C. J., Legendre, L. A., Landers, J. P., and Ferrance, J. P., Rapid DNA amplification in glass microdevices, in *Microchip Capillary Electrophoresis: Methods and Protocols*, C.S. Henry, ed., Humana Press, Totowa, NJ, chap. 15, 2006.
129. Panaro, N. J., Lou, X. J., Wilding, P., Fortina, P., and Kricka, L. J., Surface effects on PCR reactions in multichip microfluidic platforms, *Biomed. Microdev.*, 6, 75, 2004.
130. Krishnan, M., Burke, D. T., and Burns, M., A polymerase chain reaction in high surface-to-volume ratio SiO_2 microstructures, *Anal. Chem.*, 76, 6588, 2004.
131. Kricka, L. J., and Wilding, P., Microchip PCR, *Anal. Bioanal. Chem.*, 377, 820, 2003.
132. Lou, X. J., Panaro, N. J., Wilding, P., Fortina, P., and Kricka, L. J., Increased amplification efficiency of microchip-based PCR by dynamic surface passivation, *Biotechniques*, 36, 248, 2004.

133. Roach, L. S., Song, H., and Ismagilov, R. F., Controlling nonspecific protein adsorption in a plug-based microfluidic system by controlling interfacial chemistry using fluorous-phase surfactants, *Anal. Chem.*, 77, 785, 2005.
134. Bi, H. Y., Meng, S., Li, Y., Guo, K., Chen, Y. P., Kong, J. L., Yang, P. Y., Zhong, W., and Liu, B. H., Deposition of PEG onto PMMA microchannel surface to minimize nonspecific adsorption, *Lab Chip*, 6, 769, 2006.
135. Bi, H., Zhong, W., Meng, S., Kong, J., Yang, P., and Liu, B., Construction of a biomimetic surface on microfluidic chips for biofouling resistance, *Anal. Chem.*, 78, 3399, 2006.
136. Wu, D. P., Zhao, B. X., Dai, Z. P., Qin, J. H., and Lin, B. C., Grafting epoxy-modified hydrophilic polymers onto poly(dimethylsiloxane) microfluidic chip to resist nonspecific protein adsorption, *Lab Chip*, 6, 942, 2006.
137. Shrinivasan, S., Norris P. M., Landers, J. P., and Ferrance, J. P., A low-cost, low-power consumption, miniature laser-induced fluorescence system for DNA detection on a microfluidic device, *Clin. Lab. Med.*, 27, 173, 2007.
138. Kim, D. S., Lee, S. H., Ahn, C. H., Lee, J. Y., and Kwon, T. H., Disposable integrated microfluidic biochip for blood typing by plastic microinjection moulding, *Lab Chip*, 6, 794, 2006.
139. Ahn, C. H., Choi, J. W., Beaucage, G., Nevin, J. H., Lee, J. B., Puntambekar, A., and Lee, J. Y., Disposable smart lab on a chip for point-of-care clinical diagnostics, *Proc. IEEE*, 92, 154, 2004.
140. Vestad, T., Marr, D. W. M., and Oakey, J., Flow control for capillary-pumped microfluidic systems, *J. Micromech. Microeng.*, 14, 1503, 2004.

37 Advances in Microfluidics: Development of a Forensic Integrated DNA Microchip (IDChip)

Katie M. Horsman and James P. Landers

CONTENTS

37.1 INTRODUCTION

The past half decade has seen deoxyribonucleic acid (DNA) analysis performed on microdevices become more accepted and much more routine. Only recently, however, have analytical chemists addressed the specific needs of forensic science to allow the forensics community to harness these advances in microfluidics. The research focus of a number of groups has led to the development of microfluidic systems specifically tailored to forensic applications. As a result, we are beginning to witness the validation of microfluidic devices in forensic laboratories, and their widespread adoption in the coming years can be anticipated. Microfluidics has the potential to advance forensic DNA analysis in ways that were unforeseen at the time that restriction fragment length polymorphisms (RFLPs) were adopted. It will introduce a rapid, automatable technology that will enable our severely backlogged crime laboratories to process casework more efficiently.

In many respects, the forensics community has been awaiting the advent of microdevices. Microfluidic chips have been proven to carry out processes more efficiently than their macroscale counterparts (e.g., polymerase chain reaction or PCR and electrophoretic separations), thus, considerably decreasing sample processing time.[1,2] The current advances in integration of multiple processes on a single device serve to further enhance the processing speed.[3–5] Single-process devices, such as those for DNA purification as well as multiprocess devices, will significantly reduce the opportunity for laboratory sources of sample contamination by completing the sample preparation in a closed system and, thus, eliminating the multiple tube transfers and additional handling. Forensics, perhaps more than any other field, stands to benefit from this inherently closed-system design. The ease of automation of the microchip methods is advantageous to the forensics community in the same way it will benefit the clinical community—by decreasing technician time and, thus, dramatically decreasing costs associated with DNA analysis. This decrease in sample processing time and costs could, in turn, have a dramatic impact on the DNA analysis backlog in crime laboratories. Although yet to be fully realized, it is reasonable to expect a concomitant improvement in limit of detection by the transfer to the microscale format from the macroscale counterpart. Therefore, we can expect to see successful DNA typing in cases where limited or no profile was previously obtained with conventional methods.

Conventional forensic DNA analysis involves DNA extraction and purification, quantification, and amplification, followed by separation of PCR products, detection, and data analysis. In sexual assault casework, differential extraction, a cell sorting process by which DNA is differentially extracted from sperm and vaginal epithelial cells (or other nonsperm cell types), must be utilized to obtain enriched fractions of male and female DNA. In this chapter, we will discuss the specific development of these techniques on the microscale for application to forensic human DNA analysis.

37.2 DIFFERENTIAL EXTRACTION

Differential extraction is employed by forensic DNA analysts to obtain enriched fractions of male and female DNA from sexual assault evidence. This evidence, typically obtained from a vaginal swab during evidence collection at the emergency room, contains sperm cells from the perpetrator often in an overwhelming number of vaginal epithelial cells from the victim. Although the victim's DNA profile can often be obtained by other means, a "clean" profile of the perpetrator is sought for prosecution. As a result, Gill[6] developed differential extraction, which exploits the differential stability of the nuclei of sperm and vaginal epithelial cells to obtain enriched male and female DNA fractions. The mixed cell sample is first extracted using mild conditions (Gill Buffer [10 mM trizma, pH 8, 10 mM EDTA, 0.1 M NaCl, and 2% SDS] and 20 μg/mL proteinase K, incubate for 2+ h at 56°C), the swab removed, and sample centrifuged. The supernatant contains female DNA (from lysed vaginal epithelial cells) while the pellet contains unlysed nuclei from sperm heads. After extensive washing of the sperm cell pellet, incubation under harsh extraction conditions [Gill buffer containing 40 mM dithiothreitol (DTT) and 20 μg/mL proteinase K] results in the lysis of the sperm cell nuclei.

In the translation of this process to the microscale, several approaches have been developed. The conventional method has not been directly translated to the chip, most likely due to the inherent need for centrifugation. While centrifugation has been demonstrated on microdevices,[7] it is not in widespread use, requires specialized chip design, and the development of methods for washing the sperm cell pellet in order to obtain results comparable to the macroscale method.

One microscale method developed for differential extraction involves a filter-based system and lysis using acoustic energy.[8] The sample is first infused over a filter (size and material not indicated) in which the sperm cells (~4–6 μm diameter) pass through unimpeded and the much larger epithelial cells (~50 μm diameter) are retained. The DNA is then extracted using ultrasonic disruption of the cells. Although it is too early to gauge the success of this method, filtration has been explored on the macroscale for this application without widespread success.

Another method for the analysis of sexual assault evidence exploits the differential physical and chemical properties of sperm and epithelial cells to result in a cell separation, from which DNA can be extracted from each population independently.[9] In this simple method, the epithelial cells settle to the bottom of the microchip inlet reservoir more rapidly than the sperm cells (as a result of their size and density). By subsequently invoking flow in the microchannel, sperm cells are swept from the inlet to the outlet reservoir, where they can be either collected for subsequent analysis or packed up against a silica bead/sol-gel bed for DNA extraction (Figure 37.1a). Efficient separation of a mixture of sperm and epithelial cells has been demonstrated via short tandem repeat

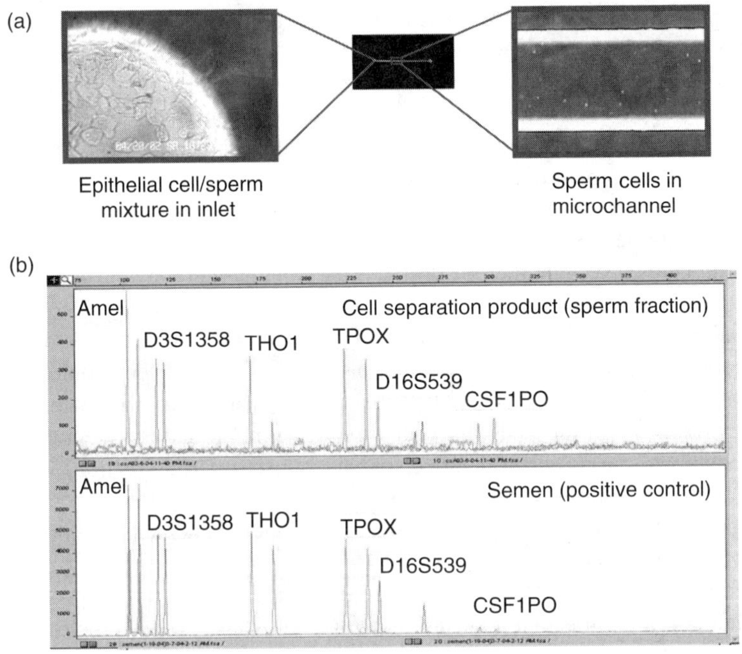

FIGURE 37.1 Separation of sperm and vaginal epithelial cells on a glass microdevice. (a) Photographs of the straight-channel microdevice (center), the inlet reservoir (containing a mixture of sperm and vaginal epithelial cells), and the microchannel (containing sperm cells) during cell separation. In this method, the cells were separated based on their physicochemical differences, resulting in differential sedimentation of the cell types. Upon application of a negative pressure at the outlet reservoir, sperm cells were swept into the microchannel, whereas the epithelial cells adsorbed to the glass substrate of the inlet reservoir. (b) Forensic STR profile obtained from the material in the outlet reservoir following cell separation. This compares favorably with the semen donor's profile and indicates an efficient separation from the vaginal epithelial cell DNA. (Adapted and reproduced from Horsman et al., *Anal Chem* 2005, 77, 742–749. Copyright 2005. With permission from American Chemical Society.)

(STR) profiling (Figure 37.1b), and, in addition, cell separation with integrated DNA extraction has also been demonstrated.[10] A drawback associated with this method is the sample volume that can be loaded into the chip. The proof-of-principle work demonstrated a sample load that was in the 10–50 μL range, where biological material from vaginal swabs are typically eluted in 500 μL or more. Because the number of sperm cells on a given swab is unknown (and, thus, the amount of DNA available for subsequent processing), the device must be capable of accepting the entire volume from the evidentiary sample. In addition, this method represents a divergence from the conventional macroscale method, assuming that both cell types exist on and are eluted from the swab intact. Any epithelial cells that have lysed in storage (due to dehydration) will yield free female DNA that co-migrates to the outlet reservoir with the sperm cells, thus contaminating the male fraction. While this method has shown promise for use in forensic DNA analysis, it did not meet the requirements for accommodating larger sample volumes.

An interesting approach reported for obtaining purified male and female fractions on microdevices involves acoustic differential extraction (ADE)[11] and exploits microacoustic transducer developments (see Chapter 44 by Laurell for details). This method involves the elution of biological material from the vaginal swab under mild lysis conditions, resulting in a mixture of sperm cells and epithelial cell lysate. The sample is then infused in the microdevice, and sperm cells trapped in a monolayer above an ultrasonic transducer while free DNA (from the lysed epithelial cells) flows through the system unretained. Flow of the epithelial cell lysate is directed to the outlet reservoir using laminar flow valving[12] while the sperm cells are retained by the transducer. Upon deactivation of the ultrasound, the sperm cells are released and directed to the sperm cell outlet. Highly enriched fractions of male and female DNA have been shown to be obtained using this method. In addition, this approach addresses the shortcoming of the aforementioned cell sorting method, in that it is capable of readily processing large volume samples, although cell sorting integrated with DNA extraction is under development but has not yet been demonstrated.

All of these methods are amenable to eventual integration with downstream sample preparation steps, such as the solid-phase extraction of DNA, for automated processing of samples in a format that is a closed, fully contained system, thereby diminishing contamination by outside sources. However, none of the methods described here identify sperm cells specifically. Before DNA analysis, analysts conduct a time-consuming "sperm search" using light microscopy to identify sperm morphologically and gauge the relative abundance. While the "sperm search" is currently considered a separate entity, any method that *could* incorporate the positive identification of sperm cells, would be attractive to the forensic laboratories and eliminate another time-consuming analysis step. Although not on a microchip, one method that has been detailed for this is laser microdissection, which has recently been used to identify and capture sperm cells from a membrane-coated microscope slide.[13,14] While effective, particularly for evidence containing low cell numbers, this method is not likely to be used for high-throughput casework.

While several microscale alternatives to differential extraction have been proposed and demonstrated, one school of thought is that none are likely to significantly impact the forensic community without direct integration with DNA extraction.

37.3 DNA EXTRACTION

DNA extraction and purification on microdevices has gained widespread use for clinical and forensic applications, and has been reviewed in detail in Chapter 43 by Bienvenue. On the macroscale, DNA purification is completed using organic extraction (historically), or, more commonly, solid phases such as QIAamp and DNA IQ. The development of DNA extraction methods on the microscale has largely focused on the use of silica-based solid phases,[15–19] although purification has also been recently demonstrated using an ion-exchange resin in a microdevice.[20] Any microscale DNA extraction method must demonstrate high efficiency of extraction comparable to or better than the currently available methods (e.g., ~46% for whole blood[21] using QIAamp). In addition, the extraction bed must be of relatively high capacity to enable binding of sufficient DNA for downstream processing, and the extraction must be capable of accepting large volumes (e.g., up to 500 μL) in a timely manner.

Most solid-phase DNA extraction techniques involve loading the DNA onto the solid phase using a chaotropic salt (e.g., Guanidine HCl), washing with ethanol to remove proteins, and eluting the DNA with water or other solvent. A more detailed explanation of these methods can be found in Chapter 43, entitled "Integrating Sample Preparation Into Multifluidic Devices."

Christel et al.[16] reported the first microdevice-based solid-phase extraction of DNA in 1999 using an etched silica solid phase consisting of silicon pillars. This work reported extraction efficiencies of approximately 50% from prepurified DNA. As a result, the extension of this method to forensic samples, which are often low-copy number samples, seems unlikely. Higher extraction efficiencies have been demonstrated using silica bead and sol-gel-based solid phases. These phases are added to the microchannel as liquid suspensions, thus, circumventing the complex etching required for the pillar design described previously. Silica beads and sol-gel have been demonstrated individually[3,22] and in tandem[17] where the sol-gel acts as a glue holding the beads in place. In both methods, DNA purification has been demonstrated using blood and other biological fluids common to forensic analyses with high extraction efficiencies (up to 60%) and high capacities (~20 ng, 1.5 cm extraction bed) comparable to the macroscale counterpart such as the QIAamp kit. In a clear demonstration of the application of this work to forensic analyses, on-chip sperm cell lysis and DNA extraction using a silica bead solid phase was reported by the addition of 40 mM DTT to the 6 M guanidine lysis solution.[18] The purified DNA was shown to be amenable to downstream (off-chip) processing including STR amplification and separation of PCR products.

DNA purification on microdevices, as currently performed, appears to meet the needs of the forensics community. That is, extraction efficiencies and capacities comparable to the conventional methods have been demonstrated. Most notably, these methods typically result in elution volumes on the order of 10 μL or less, negating the need for an additional concentration step (e.g., Microcon) and, thus, further diminishing DNA loss and additional processing time. Because the commercially available kits accept limited volumes of DNA, concentration of the eluted DNA is commonly required in forensic samples in order to maximize the amount of template DNA added to the STR amplification. In addition, because the target amount of template DNA added to an STR reaction is important in order to minimize possible PCR artifacts such as allelic dropout, off-scale alleles, and so forth, DNA quantification following extraction is necessary.

37.4 DNA QUANTIFICATION

Unlike DNA diagnostics with clinical samples, where the amount of DNA in a volume of blood or other biological fluid can be estimated with little consequence, postpurification DNA quantification is of the utmost importance in forensic sample analysis. The target mass of DNA required for effective use of the commercially available STR kits is $\sim$1 ng. When a reduced mass of DNA is added to the PCR, stochastic effects that could include "allelic dropout," may make interpretation difficult. Alternatively, an excess of DNA in the PCR ($>$5 ng) can result in off-scale peaks, "pull-up" (color bleeding usually due to off-scale peaks), increased levels of stutter (small peaks typically one repeat unit shorter than the STR allele, believed to be due to slippage of the primer or template during replication) or other artifacts in the subsequent electrophoretic separation.

Until recently, DNA quantification in forensic analyses has most commonly involved the use of slot-blot methods. In the past few years, the forensics community has shifted focus to the use of quantitative PCR as the preferred method for quantification. Both methods allow for human-specific DNA quantification, although qPCR has significantly lower limits of detection and gives an indication of the PCR-amplifiability of the DNA as well. Recently, multiplex qPCR assays have been developed to simultaneously determine the total amount of genomic and mitochondrial DNA,[23–25] total genomic and male DNA,[26,27] or assess the extent of DNA degradation for specific forensic applications.[28,29] Because of the utility of qPCR to the forensic community, we focus on the development of qPCR on microdevices here.

Although real-time PCR has been demonstrated on microdevices, no qPCR has been reported to date. That is, fluorescence detection during the accumulation of PCR products throughout cycling has been demonstrated. However, no standard curve was simultaneously amplified to allow for DNA quantification based on the signal generated. Real-time PCR has been demonstrated by Lin et al.[30] in 25 μL reactions using fluorogenic probes and the SYBR Green I intercalating dye, respectively. In both cases, multiplex PCRs that include generation of a standard curve for quantification have not been described. In 2003, Quake and coworkers[31] reported the development of picoliter volume PCR chambers on a microfluidic device for parallel real-time PCR assays. Although not demonstrated, the authors indicate that the eight flow channels can be used for the generation of a standard curve and, thus, the chip can be used for quantitative PCR.

While DNA quantification remains the least-developed sample processing step on microchip for forensic DNA analysis, it is anticipated that extension of the reported methods to qPCR will be forthcoming. Any such qPCR method should be directly applicable to forensic DNA quantification by use of the appropriate primers and probes described in the literature for the macroscale counterpart. Following successful demonstration of qPCR on-chip, integration of this method with DNA extraction and DNA amplification would be the next microfluidic challenge; however, recent advances in this arena suggest this is possible.[32–34] In addition, development of valving or other means of metering the appropriate volume of purified DNA into the amplification reaction must be carried out. This then results in the need for a storage mechanism on chip or off chip for the unused extracted DNA, which is derived from the original evidence sample. This material must be saved for subsequent reprocessing, permanent storage, or use by opposing legal council.

37.5 DNA AMPLIFICATION

DNA amplification on microdevices has been demonstrated routinely, and Roper et al.[35] have not only provided an extensive review of the literature on this topic (also see Chapter 43 by Bienvenue) but also provided a scoring method for device comparison. However, it is noteworthy that few examples of the complex PCR associated with forensic DNA typing have been demonstrated on chip. Forensic DNA typing most commonly involves the amplification of STRs via PCR using multiplex PCRs involving up to 16 primer pairs for coamplification of STRs from multiple loci. Primers are tagged with fluorescent tags (typically one of three or four fluorophores) for discrimination of loci. Commercially available kits have been developed, are the industry standard, and are used by every forensic DNA laboratory in the United States. Of primary interest to the forensics community in the development of a microfluidic device for PCR is the limit of detection for the amplification fragments (low-copy number amplifications), unbiased amplification, limiting the DNA template required for effective amplification, total reaction time, the ability to perform multiplex amplifications (as in the commercially available kits), and multiple chamber amplification (to amplify >1 sample at a time). Of course, any microdevice method must show that the PCR amplification product is comparable to that obtained on the macroscale, permitting any future results to be related to those previously obtained and are contained in the national DNA database.

Microchip DNA amplification, to date, has already demonstrated the potential for low limits of detection, with amplification demonstrated from single-copy sources.[36] The useful limit of detection (so as to prevent stochastic effects in forensic analyses) with microchip PCR is yet to be determined. However, the reduced reaction volume inherent to the microchip supports the expectation of lower limits of detection. Fast thermocycling has been demonstrated by a number of microchip PCR designs and methods with amplification rates approaching the biological limit of the processivity of *Taq* polymerase, that is, with 20–25 cycles in under 5 min.[1,2] This fast thermocycling has not yet been applied to STR amplifications, although preliminary data have shown a reduction in total amplification time by almost half.[37] Therefore, it is not unreasonable to anticipate that the approximately 3-h conventional thermocycling protocol will be reduced to 45 min or less. There are a few demonstrations of STR amplifications in microdevices. Liu et al.[4] have demonstrated the amplification of a mini Y-STR multiplex, Bienvenue et al.[5] and Legendre et al.[37] have demonstrated amplification using the AmpFlSTR COfiler and Profiler Plus kits, and Schmidt et al.[40] demonstrated microchip amplification using the PowerPlex 16 STR kit. Recently, amplification of mitochondrial DNA has been reported on a glass microchip.[41] Following amplification, a subsequent mitochondrial sequencing reaction was performed. These data indicated amplification with template amounts as low as 1 pg, where product was not detected by the conventional method. Large masses of mitochondrial DNA (500 pg–1 ng) resulted in some nonspecific product formation when amplified in the low volume reaction. Table 37.1 summarizes the characteristics of these reported microchip forensic DNA amplifications. In addition to multiproduct amplification in each chamber (corresponding to the multiple loci), simultaneous amplification in multiple chambers will likely be needed to compete with the conventional thermocyclers. This has been shown by Waters et al.[42] (four chambers), Legendre et al.[37] (three reactions performed—although four chambers are fabricated in the device) (Figure 37.2), Lutz-Bonengel et al.[41] (60 reaction spots), and Schmidt et al.[40] (23 reactions

TABLE 37.1
Forensic DNA Amplifications on Microdevices

Author (Citation)	Reaction Volume	Reaction Time	Heating Method	Amplification Demonstrated	Integrated?	Multichamber?
Liu et al.[31]	160 nL	64 min (35 cycles)	Resistive	Mini Y-STR multiplex incl. Amelogenin	Yes, to microchip electrophoresis	No
Legendre et al.[37]	200 nL	100 min (28 cycles)	Block (conventional)	COfiler, Profiler Plus	No	Yes, 4
Schmidt et al.[40]	1 μL	210 min (32 cycles)	Block (conventional)	PowerPlex 16	No	Yes, 60
Bienvenue et al.[5]	1.2 μL	180 min[a] (28 cycles)	Block (conventional)	COfiler, Profiler Plus	Yes, to DNA purification	Yes, 4
Lutz-Bonengel et al.[41]	1 μL	180 min[b] (32 cycles)	Block (conventional)	Mitochondrial	No	Yes, 60

[a]This reaction time is reported for the PCR amplification and the DNA extraction in total.
[b]This reaction time is reported for the PCR amplification only, not the sequencing step, which requires approximately an additional 220 min.

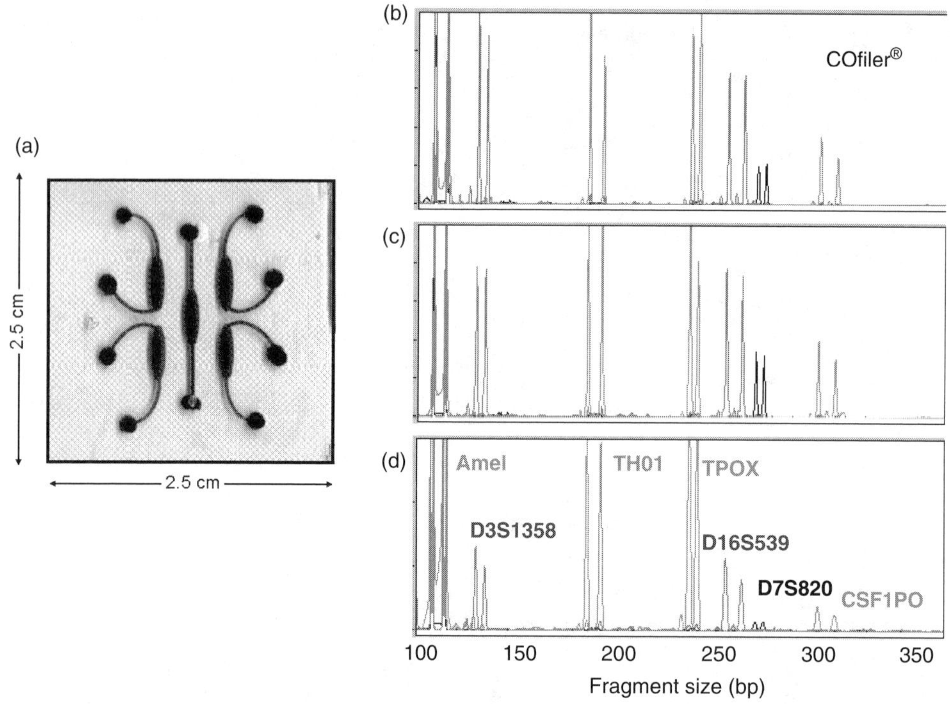

FIGURE 37.2 Multiple simultaneous STR amplifications on a single glass microdevice. (a) Photograph of the microdevice for multiple chamber PCR microdevice. The chip contains five chambers: one for temperature control (center) and four surrounding PCR chambers for parallel amplification. Amplification of the COfiler multiplex STR loci was performed simultaneously in three of the four reaction chambers on this microdevice using less than 1 ng template DNA. (Adapted and reproduced from Legendre, L. et al. Multiplex microchip PCR for STR Analysis. Poster presented at The *15th International Symposium on Human Identification*, 2004. With permission.) The resulting STR profiles (following conventional DNA separation) are shown in panels B–D, indicating amplification in each chamber with no apparent PCR artifacts that would complicate analysis of the DNA profile. (Figures 37.2B through 37.2D are courtesy of L. Legendre.)

performed—although the chip contains 60 reaction spots) (Figure 37.3). Multiple PCR chambers per device will enable high-throughput processing but also allow analysts to simultaneously process the necessary controls (reagent blanks, positives, negatives).

The application of microfabricated devices to the amplification of STR fragments in forensic DNA typing provides numerous advantages over the conventional methodology—specifically, lower limits of detection,[36] less reagent and sample consumption, enhanced processing speed,[1] and so forth. However, it also brings along issues that are not currently encountered with conventional methodology. For example, the STR amplifications on microchip discussed above involve total reaction volumes that range from 160 nL to 1.2 μL. Thus, the DNA added to the reaction must be contained in a fraction of a microliter, rather than the common 10–200 μL obtained from a Qiagen purification and then used for a conventional 25 μL amplification reaction. Consequently, the DNA must be concentrated to a greater extent than needed for conventional amplification. As a result, for microchip PCR to be utilized in forensic laboratories it will most likely require direct integration with microchip DNA extraction, where the DNA is routinely eluted in microliter volumes. This need to amplify the entire mass of DNA and, thus, concentrate the purified DNA to a high degree, is most critical in cases where low starting copies of DNA template are available. Bienvenue et al.[5] have demonstrated DNA purification and PCR amplification on the same microdevice, carrying out microliter-scale DNA extraction as reported previously in a glass

(a)

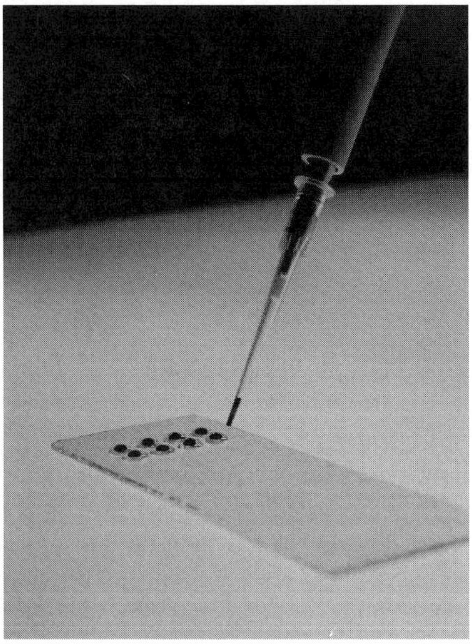

(b)

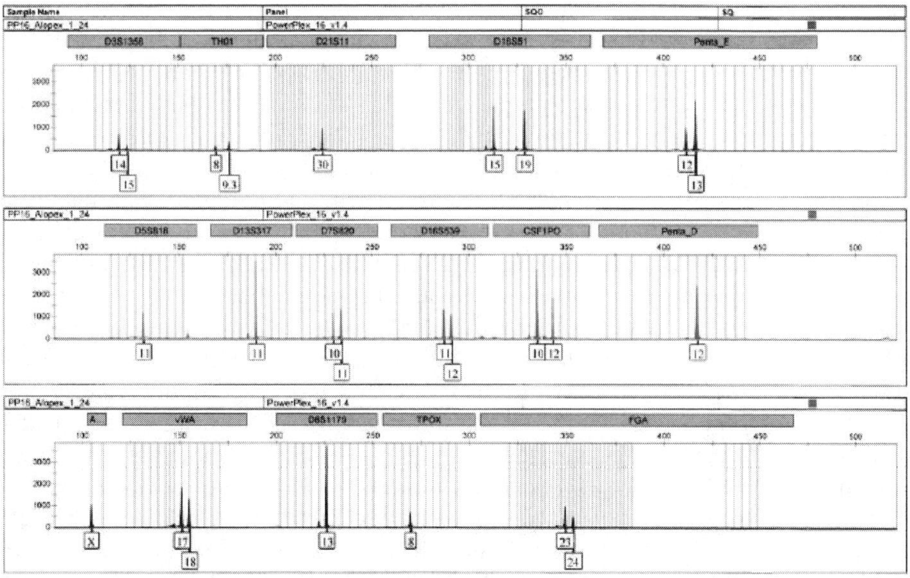

FIGURE 37.3 (a) Photograph of a glass microchip containing 60 reaction spots, each with a hydrophilic surface surrounded by a hydrophobic ring. Each spot is loaded with 0.5 μL PCR mix and 0.5 μL of DNA template and covered with 5 μL oil to decrease evaporation and contamination. Thus, the total reaction volume on this device is 1 μL. The chip was thermocycled on conventional instrumentation using an *in situ* adapter. (Photo courtesy of Schmidt, U.) (b) A complete STR profile generated by PowerPlex 16 amplification of 32 pg genomic DNA on the chip shown in (a). Allele names are identified in the gray boxes above each set of peaks. Repeat numbers for the corresponding alleles are indicated below each peak. (Adapted and reproduced from Schmidt, U. et al., *Int J Legal Med* 2006, 120, 42–48. Copyright 2005. With permission from Springer-Verlag.)

microdevice.[17,22,43] However, in a unique approach, the DNA was eluted from the solid phase using PCR master mix and retained in a chamber. Subsequent placement of the device into a conventional (block) thermocycler allowed for amplification of the STR fragments in the Cofiler and Profiler Plus kits. These results are significant as they represent the first work demonstrating the potential for interfacing microfluidic DNA extraction technology with conventional PCR instrumentation. Of course, for application to forensic casework, the integration of a DNA quantification step will also be necessary.

37.6 SEPARATION OF PCR PRODUCTS

As described in detail in Chapter 25 by McCord, forensic DNA profiling is routinely accomplished by capillary electrophoresis (CE) in forensic laboratories. Recently, microchip electrophoresis is being evaluated for use in casework by the *Virginia Department of Forensic Science* and the *Forensic Science Service* (U.K.). While DNA separations on microdevices are becoming more commonplace, separations for forensic DNA analysis have been less-frequently demonstrated. DNA separations for forensic analyses, like DNA sequencing, require single base pair resolution and multicolor fluorescence detection (see Chapters 6, 27, and 44). In addition, the STR separation must be rapid (less than 30 min) to compete with the current capillary-based method. For nondisposable, single-process devices, the sieving matrix must be of relatively low viscosity to enable loading and replacement of the matrix after each separation (as in the capillary) to prevent carryover.

Table 37.2 provides some of the pertinent details associated with the STR separations carried out on microdevices to date. All of these STR separations have utilized long-read polyacrylamide (LPA) or polydimethylacrylamide (PDMA) [the main component of the commercially available polyolefin plastomer polymers, known as "POP" polymers]; POP-4 (where the number dictates the %PDMA) is the polymer used for capillary STR separations on ABI instruments. Acceptable polymers for these separations must be capable of single base pair resolution at low temperatures (preferably, room temperature) with short effective separation length (L_{eff} – distance from injection point to detection point) and of relatively low viscosity to enable replacement and ease of filling the microchannel. Overall, it appears that LPA is capable of providing the requisite resolution with microchannels having an L_{eff} of 11.5 cm,[44] although shorter distances are unlikely to be adequate with the current methodology.[45] In general, the separations require 25 or more minutes,[46–48] which is comparable to that obtained on the conventional capillary instrumentation. One embodiment[46,48], however, utilizes a 96-microchannel device (Figure 37.4) and can accomplish 96 simultaneous separations, resulting in greater throughput capabilities than the commonly encountered 4- and 16-capillary conventional CE instruments. An example of data obtained on this microdevice is shown in Figure 37.5. POP-4 has also been used for the separation of STR fragments on microdevices, where the separation of an Identifiler allelic ladder was accomplished with an L_{eff} of only 8 cm and in under 12 min.[49] This separation is distinct in that the optical requirements for the detection system are relatively simple—containing only a laser and a single photomultiplier tube (PMT), because of the electronic filtering capabilities of an inline acousto-optic tunable filter (AOTF) to select the desired emission wavelengths sequentially (see Chapter 45 by Karlinsey). The value of the AOTF stems from the ease with which four-color detection can be expanded to five or more colors, which may be optimal for portable systems. One of the benefits of microchip electrophoresis reported by Goedecke et al.[47] (Figure 37.6) was the improved data quality and stability when compared to the commercially available capillary array instruments. This advantage arises from the enhanced thermal and mechanical stability of the separation channel, as well as the improved heat-sinking inherent to glass microchips.

As mentioned earlier, devices similar to those described in Schmalzing et al.[50] and Yeung et al.[48] are currently undergoing evaluation in the United States and the United Kingdom. It is anticipated

TABLE 37.2
STR Separations on Microdevices

Author (Citation)	Sample Treatment	Microchip Substrate	Separation Channel (Effective Length, cm)	Sieving Matrix	Channel Coating	Separation Field Strength	STR Kit Utilized	Separation Time (min)	Resolution of 9.3/10 (THO1) Demonstrated?	Multichannel?	Detection (Color(s))
Schmalzing et al.[52]	Diluted in buffer, heat denatured, snap cooled	Glass (fused silica)	2.6	4% LPA in one time TBE/3.5 M urea/30% formamide	Modified Hjerten	200 V/cm	CTTv	<2	No	No	1
Schmalzing et al.[50]	Diluted in H_2O, heat denatured, snap cooled	Glass (fused silica)	11.5	4% LPA in one time TBE/3.5 M urea/30% formamide, 50°C	Modified Hjerten	200 V/cm	CTTv	<10	Yes	No	2
Medintz et al.[53]	Desalted (Qiagen), resuspended in 0.5 × TE, diluted in formamide, heat denatured, snap cooled	Glass (borofloat)	5.5	Long-read LPA (Amersham), 40°C	Modified Hjerten	[a]	(ET dyes)	<8	Yes	Yes, 96	4

Continued

TABLE 37.2
(Continued)

Author (Citation)	Sample Treatment	Microchip Substrate	Separation Channel (Effective Length, cm)	Sieving Matrix	Channel Coating	Separation Field Strength	STR Kit Utilized	Separation Time (min)	Resolution of 9.3/10 (THO1) Demonstrated?	Multichannel?	Detection (Color(s))
Mitnik et al.[44]	Heat denatured, snap cooled	Glass (borosilicate)	11.5	4% LPA in one time TTE/7 M urea, 50°C	Modified Hjerten	180 V/cm	PowerPlex 16	<35	Yes	No	4
Shi and Anderson[54]	Heat denatured, snap cooled	Plastic (polyolefin)	4.5	4% LPA in one time TTE/7 M urea, 35°C	2% p(DMA/DEA)	165 V/cm	CTTv	<10	Yes	No	2
Goedecke et al.[55]	Combined, purified (GFX spin column), resuspended/diluted in water, heat denatured, snap cooled	Glass (borosilicate)	20	4% LPA in one time TTE + 7 M urea, 50°C	Modified Hjerten	180 V/cm	PowerPlex 16	40	Yes	Yes, 16	4
Crouse et al.[46]	Diluted in formamide, heat denatured, snap cooled	Glass (borofloat)	17–18	Long-read LPA (Amersham)	Modified Hjerten	[b]	PowerPlex 16	<25	Yes	Yes, 96	4

Shi[56]	Heat denatured, snap cooled	Plastic (polycycloolefin)	10	3% LPA in one time TTE/7 M urea, 35°C	2% p(DMA/DEA)	150 V/cm	CTTv and Profiler Plus	<18	Yes	No	4
Yeung et al.[48]	Diluted in formamide, heat denatured, snap cooled	Glass (borofloat)	15.9	Long-read LPA, five times TTE, 67°C	Modified Hjerten	150 V/cm	PowerPlex 16 and Profiler Plus	<22 and <17	Yes	Yes, 96	4
Karlinsey and Landers[49]	Dilute in formamide, spin filter (Microcon 30), heat denatured, snap cooled	Glass (borofloat)	8	POP-4 (Applied Biosystems)	Modified Hjerten	250 V/cm	Identifiler	18	No	No	5

[a] The separation channels are not straight and, therefore, the field strength is not uniform. Voltage was applied as follows: +1300 V at the anode reservoir, +200 V at the cathode reservoir, +325 V at the sample reservoir, and +325 V at the waste reservoir.

[b] The separation channels are not straight and, therefore, the field strength is not uniform. Voltage was applied as follows: +2500 V at the anode reservoir, +0 V at the cathode reservoir, +180 V at the sample reservoir, and +200 V at the waste reservoir.

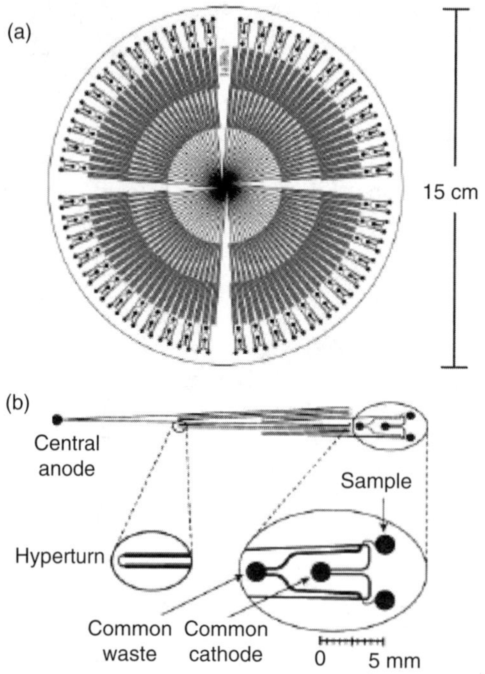

FIGURE 37.4 (a) Design for the 96-channel μCAE microdevice by Mathies and coworkers. The glass microdevice is approximately the dimensions of a compact disk. The microchannel structure consists of 48 doublet structures, each containing two electrophoresis lanes that share common cathode and waste reservoirs. (b) Magnified view of the doublet structure, highlighting the "hyperturn," which prevents band-broadening dispersion as the DNA molecules traverse the turn, thereby enabling high-resolution DNA separations in a compact device. (Adapted and reproduced from Yeung, S. H. et al., *J Forensic Sci* 2006, 51, 740–747. Copyright 2006. With permission from Blackwell Publishing, Inc.)

that microdevices for the separation of STR products will appear in casework soon. Because little advantage has been gained in processing time with these devices, as compared to the conventional multiple-capillary CE instrumentation, there is little impetus for the forensic community to implement this separation-only microdevice. While not used in forensic laboratories, there are commercial 96-capillary CE instruments capable of the same throughput as the 96-microchannel microcapillary array electrophoresis (μCAE) device in a similar time frame. One would anticipate more widespread adoption by forensic laboratories either following a reduction in the separation time or, better still, when the integration of μPCR and DNA separation has been achieved in an effective and robust manner. Liu et al.[4] have demonstrated an instrument capable of μPCR of a mini-Y STR multiplex and separation of the PCR product in under 64 min.

37.7 IMPACT

Advancements in forensic DNA analysis methodology have, historically, focused on alterations in the basic science permitting individualization of biological material. In its origins, RFLP was used to generate the DNA profile. However, the focus then shifted to the use of variable number of tandem repeats (VNTRs) and, eventually, to STRs. It is the STRs that are currently being used to build the national and international DNA databases (containing DNA of convicted offenders and from unsolved cases) and are, therefore, unlikely to be supplanted in the near future. Microdevices do not represent a shift in the basic science, but rather the technology platform on which the STR profiles

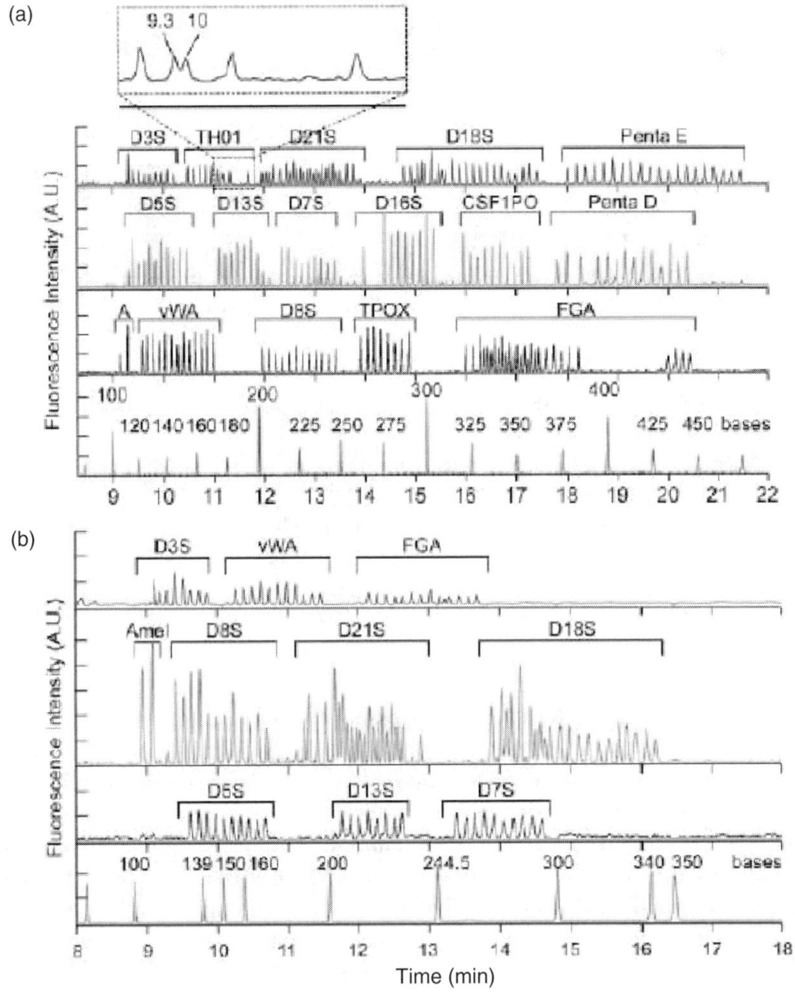

FIGURE 37.5 Electropherograms of (a) Promega PowerPlex 16 allelic ladder and (b) Applied Biosystems AmpFlSTR Profiler Plus allelic ladder obtained on the μCAE microdevice shown in Figure 37.4. (Reproduced from Yeung. S. H. et al., *J Forensic Sci* 2006, 51, 740–747. Copyright 2006. With permission from Blackwell Publishing, Inc.)

are obtained much like the witnessed transitions from slab gels to the capillary format for the DNA separations. While the effectiveness of this technology platform and its inherent ability to enhance throughput might be considered revolutionary, it will not likely be viewed as a "paradigm shift" in the same way that the introduction of PCR was to forensic DNA analysis. Instead, it may be viewed as a necessary overhaul of the technology which, ultimately, will lead to better, more efficient processing of DNA evidence. Because the underlying principles of the DNA analysis have not been changed, merely the technological platform, its introduction, and acceptance by the courts should be relatively straightforward. With similar validation studies as those completed when shifting from slab gels to CE, their acceptance is not anticipated to be threatened.

As a result of the shift from the conventional methodology and technology to the microchip format, it is possible that DNA testing in criminal laboratories will undergo a "make-over," giving it a very different look than it has today. While the criminalistics on the *front-end* (e.g., determining which samples should be tested) will remain largely unchanged, there will be dramatically less analyst

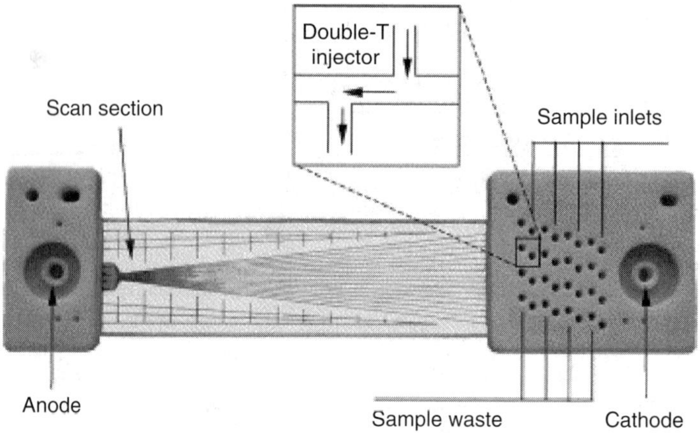

FIGURE 37.6 Microdevice for STR separations, capable of 16 simultaneous separations. The device is constructed of two glass layers with interface boards at each end. The left interface board contains the anode buffer reservoir (2 mL). The right interface board contains the cathode buffer reservoir (2 mL), 16 sample inlets (70 μL), and 16 sample waste reservoirs (70 μL). The entire device measures approximately 5-cm wide and 25-cm long, and the separation channels have an effective length of 20 cm. (Adapted and reproduced from Goedecke, N. et al., *J Chromatogr A* 2006, 1111, 206–213. Copyright 2006. With permission Elsevier.)

intervention in the analysis during the sample processing due to the automation. Thus, the job of the analyst will necessarily shift from carrying out the entire analysis to only the criminalistics on the *front-end* and the interpretation of the STR profile results produced by the instrument. While this will increase the throughput of forensic DNA laboratories, it will, in most laboratories, not decrease the number of analysts employed. Instead, the number of samples submitted to and processed by laboratories is expected to increase. In this way, DNA testing in cases where it is not currently routine (e.g., robbery) in all jurisdictions may become more commonplace. While a major advantage of microfluidic devices in forensic analyses is the concomitant automation anticipated, automation alone is not the only advantage of the microdevice, as robotic methods are also able to achieve more automated analyses. Microdevices allow the distinct advantage of integrating multiple sample processes (such as DNA extraction, quantification, PCR, and separation) on a single platform. This allows for DNA analysis in a closed system and, therefore, less opportunity for contamination by the analyst with tube transfers and sample handling. In addition, many microfluidic processes have been shown to be more efficient than their macroscale counterparts such as DNA amplification, which typically requires 2–3 h conventionally, but can be completed in an order-of-magnitude less time on the microscale with concomitant decreases in sample and reagent consumption. Decreased sample consumption in forensic casework, where there is often not abundant sample at the outset, is highly valued. This might be expected to lead to the ability to obtain an STR profile where one cannot be obtained with conventional methodology and will, likely, be a major driving force for the transition to microfabricated technology in forensic laboratories.

In recent years, we have witnessed the incorporation of microdevices into the U.S. and U.K. crime laboratories. Currently, only separation of PCR products is completed on these instruments, with limited advantage over the current conventional methodology, CE. However, the major advantage of microdevice technology is in the ability to incorporate multiple sample processing steps on a single platform. Thus, the true value of the microchip in forensic laboratories will be realized when multiple processes are integrated onto a single automated device, thus, decreasing analyst time as well as total processing time and reagent consumption. Since microchip electrophoresis is the most mature of the microchip processes described above, it is of no surprise that this would be the first process to be introduced into the forensic laboratory. The more unseasoned microchip processes such as PCR and

DNA extraction are expected to follow suit in due time. The significant advantages of incorporating these devices will, undoubtedly, drive their introduction into forensic laboratories.

37.8 CONCLUDING REMARKS

The microfluidics community has successfully demonstrated microscale methods for cell sorting/differential extraction, DNA extraction, PCR amplification, and electrophoretic separation of STR products. Most of these individual microscale processes meet needs and desires of the forensics community by demonstrating efficient extraction of DNA, rapid thermocycling, and single base pair resolution (with multicolor detection) of STR fragments. However, quantitative PCR has yet to be demonstrated on chip. Several examples of cell sorting (for use in sexual assault evidence) have been demonstrated, although much work remains to show that efficiency and sensitivity of these methods are comparable to the conventional differential extraction and, thus, merit implementation. Most notably, many of these methods fall short addressing the macro-to-micro volume connection where the macroscale sample size (mL) must be interfaced with the small volumes (nL to μL) of the microscale environment. With forensic samples involving large volumes of solution that need to be processed, volume reduction need to be invoked on chip in order to obtain sufficient DNA for STR typing. Therefore, the *front-end* processes, such as cell sorting and DNA extraction, must be capable of processing large sample volumes in order to compete with their macroscale counterparts. Developments like the ADE offer the potential to address this issue.[51] In addition, with the maturity of the microfluidics field, as with any other analytical development, one can anticipate further advances in the automation of these processes. This automation, although not yet demonstrated for all processes, provides a significant advantage over the conventional processes used in the financially strapped forensic laboratories.

Also, the concept of single-process versus integrated (multiprocess) microdevices has been discussed. Single-process devices are designed with higher-throughput capabilities to process multiple samples simultaneously. However, these devices are capable of only one step of the analysis process. Integrated devices are typically lower-throughput but feature a totally integrated analysis in a "sample-in-answer-out" fashion. For example, a microliter of blood could be introduced to the inlet of the device and all sample processing steps occur on the device, resulting in a DNA profile detected at the outlet. While both of these designs have promise in forensic DNA analysis, their adoption may be highly dependent on the intricacies of the sample type and the application. For samples such as those of convicted felons to be added to the national DNA databank, high-throughput single-process devices may be the microchip of choice. These devices can be easily designed with throughput in mind, such as the 96-channel microdevice for STR separations. Alternatively, casework samples may be better suited for an integrated device that is all inclusive and would not require sample handling after introduction of the sample to the inlet reservoir. Ultimately, the choice of microchip format most appropriate in forensic laboratories will need to be driven by the forensics community, as many caveats to DNA typing are not intuitive to the analytical chemists and engineers developing microfluidic devices.

REFERENCES

1. Easley, C., Humphrey, J., Landers, J. P., *J. Micromech. Microeng.* 2007, 17, 1758–1766.
2. Hashimoto, M., Chen, P. C., Mitchell, M. W., Nikitopoulos, D. E., Soper, S. A., Murphy, M. C. *Lab Chip* 2004, 4, 638–645.
3. Easley, C. J., Karlinsey, J. M., Bienvenue, J. M., Legendre, L. A., Roper, M. G., Feldman, S. H., Hughes, M. A., Hewlett, E. L., Merkel, T. J., Ferrance, J. P., Landers, J. P. *Proc Natl Acad Sci USA* 2006, 103, 19272–19277.
4. Liu, P., Seo, T. S., Beyor, N., Shin, K. J., Scherer, J. R., Mathies, R. A. *Anal Chem* 2007, 79, 1881–1889.
5. Bienvenue, J. M., Legendre, L. A., Ferrance, J. P., Landers, J. P. in preparation.

6. Gill, P., Jeffreys, A. J., Werrett, D. J. *Nature* 1985, 318, 577–579.
7. Duffy, D. C., Gillis, H. L., Lin, J., Sheppard, N. F., Kellogg, G. J. *Anal Chem* 1999, 71, 4669–4678.
8. Devitt, A. J., Aflatooni, N., Vinas, M., Loh, N., Pourahmadi, F., Yuan, R., and Northrup, M. A. *Proceedings of the International Symposium on Miniaturized Total Analysis Systems*, Boston, MA, USA, October 9, 2005, pp. 7–9.
9. Horsman, K. M., Barker, S. L., Ferrance, J. P., Forrest, K. A., Koen, K. A., Landers, J. P. *Anal Chem* 2005, 77, 742–749.
10. Horsman, K. M., Barker, S. L. R., Bienvenue, J. M., Voorhees, J. C., Blaiser, K., Schroeder, B., Koen, K. A., Weingart, G., Landers, J. P., Ferrance, J. P. *Special Publication—Royal Soc Chem* 2004, 297, 288–290.
11. Horsman, K. M., Evander, M., Easley, C. J., Nilsson, J., Laurell, T., Landers, J. P. *Micro Total Analysis Systems 2006, Proceedings of the mTAS 2006 Symposium*, 10th, Tokyo, Japan, 2006, 2, 1055–1057.
12. Blankenstein, G., Scampavia, L., Branebjerg, J., Larsen, U. D., Ruzicka, J. *Proceedings of the 2nd International Symposium on Miniaturized Total Analysis Systems* μ*TAS96*, Basel, Switzerland, 1996, 82–84.
13. Di Martino, D., Giuffre, G., Staiti, N., Simone, A., Le Donne, M., Saravo, L. *Forensic Sci Int* 2004, *146 Suppl*, S151–S153.
14. Sanders, C. T., Sanchez, N., Ballantyne, J., Peterson, D. A. *J Forensic Sci* 2006, 51, 748–757.
15. Cady, N. C., Stelick, S., Batt, C. A. *Biosens Bioelectron* 2003, 19, 59–66.
16. Christel, L. A., Petersen, K., McMillan, W., Northrup, M. A. *J Biomech Eng-Trans ASME* 1999, 121, 22–27.
17. Breadmore, M. C., Wolfe, K. A., Arcibal, I. G., Leung, W. K., Dickson, D., Giordano, B. C., Power, M. E., Ferrance, J. P., Feldman, S. H., Norris, P. M., Landers, J. P. *Anal Chem* 2003, 75, 1880–1886.
18. Bienvenue, J. M., Duncalf, N., Marchiarullo, D., Ferrance, J. P., Landers, J. P. *J Forensic Sci* 2006, 51, 266–273.
19. Chung, Y. C., Jan, M. S., Lin, Y. C., Lin, J. H., Cheng, W. C., Fan, C. Y. *Lab Chip* 2004, 4, 141–147.
20. Cao, W., Easley, C. J., Ferrance, J. P., Landers, J. P. *Anal Chem* 2006, 78, 7222–7228.
21. Wen, J., Guillo, C., Ferrance, J. P., Landers, J. P. *Anal Chem* 2006, 78, 1673–1681.
22. Wolfe, K. A., Breadmore, M. C., Ferrance, J. P., Power, M. E., Conroy, J. F., Norris, P. M., Landers, J. P. *Electrophoresis* 2002, 23, 727–733.
23. Alonso, A., Martin, P., Albarran, C., Garcia, P., Garcia, O., de Simon, L. F., Garcia-Hirschfeld, J., Sancho, M., de La Rua, C., Fernandez-Piqueras, J. *Forensic Sci Int* 2004, 139, 141–149.
24. Andreasson, H., Gyllensten, U., Allen, M. *Biotechniques* 2002, 33, 402–404, 407–411.
25. Timken, M. D., Swango, K. L., Orrego, C., Buoncristiani, M. R. *J Forensic Sci* 2005, 50, 1044–1060.
26. Horsman, K. M., Hickey, J. A., Cotton, R. W., Landers, J. P., Maddox, L. O. *J Forensic Sci* 2006, 51, 758–765.
27. Nicklas, J. A., Buel, E. *J Forensic Sci* 2006, 51, 1005–1015.
28. Swango, K. L., Hudlow, W. R., Timken, M. D., Buoncristiani, M. R. *Forensic Sci Int* 2007, 170, 35–45.
29. Swango, K. L., Timken, M. D., Chong, M. D., Buoncristiani, M. R. *Forensic Sci Int* 2006, 158, 14–26.
30. Lin, Y. C., Li, M., Chung, M. T., Wu, C. Y., Young, K. C. *Sensors Materials* 2002, 14, 199–208.
31. Liu, J., Enzelberger, M., Quake, S. *Electrophoresis* 2002, 23, 1531–1536.
32. Unger, M. A., Chou, H. P., Thorsen, T., Scherer, A., Quake, S. R. *Science* 2000, 5463, 113–116.
33. Grover, W. H., Skelley, A. M., Liu, C. N., Lagally, E. T., Mathies, R. A. *Sensors Actuat B—Chem* 2003, 59, 315–323.
34. Karlinsey, J. M., Monahan, J., Marchiarullo, D. J., Ferrance, J. P., Landers, J. P. *Anal Chem* 2005, 77, 3637–3643.
35. Roper, M. G., Easley, C. J., Landers, J. P. *Anal Chem* 2005, 77, 3887–3893.
36. Lagally, E. T., Medintz, I., Mathies, R. A. *Anal Chem* 2001, 73, 565–570.
37. Legendre, L. L., Bienvenue, J. M., Ferrance, J. P., Landers, J. P. *15th International Symposium on Human Identification*, Phenix, AZ, USA, 2004.
38. Schmidt, U., Lutz-Bonengel, S., Weisser, H. J., Sanger, T., Pollak, S., Schon, U., Zacher, T., Mann, W. *Int J Legal Med* 2006, 120, 42–48.
39. Lutz-Bonengel, S., Sanger, T., Heinrich, M., Schon, U., Schmidt, U. *Int J Legal Med* 2007, 121, 68–73.

40. Waters, L. C., Jacobson, S. C., Kroutchinina, N., Khandurina, J., Foote, R. S., Ramsey, J. M. *Anal Chem* 1998, 70, 5172–5176.
41. Wu, Q., Bienvenue, J. M., Hassan, B. J., Kwok, Y. C., Giordano, B. C., Norris, P. M., Landers, J. P., Ferrance, J. P. *Anal Chem* 2006, 78, 5704–5710.
42. Mitnik, L., Carey, L., Burger, R., Desmarais, S., Koutny, L., Wernet, O., Matsudaira, P., Ehrlich, D. *Electrophoresis* 2002, 23, 719–726.
43. Schmalzing, D., Adourian, A., Koutny, L., Ziaugra, L., Matsudaira, P., Ehrlich, D. *Anal Chem* 1998, 70, 2303–2310.
44. Crouse, C. A., Yeung, S., Greenspoon, S., McGuckian, A., Sikorsky, J., Ban, J., Mathies, R. *Croat Med J* 2005, 46, 563–577.
45. Goedecke, N., McKenna, B., El-Difrawy, S., Gismondi, E., Swenson, A., Carey, L., Matsudaira, P., Ehrlich, D. J. *J Chromatogr A* 2006, 1111, 206–213.
46. Yeung, S. H., Greenspoon, S. A., McGuckian, A., Crouse, C. A., Emrich, C. A., Ban, J., Mathies, R. A. *J Forensic Sci* 2006, 51, 740–747.
47. Karlinsey, J. M. *Advances and Applications of Microfluidic Analysis Systems*, University of Virginia, Charlottesville, VA, 2007.
48. Schmalzing, D., Koutny, L., Chisholm, D., Adourian, A., Matsudaira, P., Ehrlich, D. *Anal Biochem* 1999, 270, 148–152.
49. Petersson, F., Nilsson, A., Holm, C., Jonsson, H., Laurell, T. *Analyst* 2004, 129, 938–943.
50. Schmalzing, D., Koutny, L., Adourian, A., Belgrader, P., Matsudaira, P., Ehrlich, D. *Proc Natl Acad Sci USA* 1997, 94, 10273–10278.
51. Medintz, I. L., Berti, L., Emrich, C. A., Tom, J., Scherer, J. R., Mathies, R. A. *Clin Chem* 2001, 47, 1614–1621.
52. Shi, Y., Anderson, R. C. *Electrophoresis* 2003, 24, 3371–3377.
53. Goedecke, N., McKenna, B., El-Difrawy, S., Carey, L., Matsudaira, P., Ehrlich, D. *Electrophoresis* 2004, 25, 1678–1686.
54. Shi, Y. *Electrophoresis* 2006, 27, 3703–3711.

38 Taylor Dispersion in Sample Preconcentration Methods

Rajiv Bharadwaj, David E. Huber, Tarun Khurana, and Juan G. Santiago

CONTENTS

38.1 INTRODUCTION

Sample preconcentration methods, including stacking and focusing techniques, enable high-sensitivity detection by increasing analyte ion concentration. Sample preconcentration also leads to a decrease in the axial dimension of sample plugs (i.e., peak widths) and this can improve resolution in electrophoretic separations. In this chapter, we describe the basic principles and limitations of electrokinetic sample stacking and focusing techniques. We highlight the importance of Taylor dispersion in determining the efficiency of preconcentration methods. Sample preconcentration techniques can be broadly classified into two types:

1. In an increase in sample ion concentration occurs due to a local decrease in the magnitude of the drift velocity of an ion. For drift velocity vector field $\bar{v}_{\mathrm{d}}$, stacking occurs when

locally $\nabla \cdot \bar{v}_d < 0$. These gradients are typically limited to a small region relative to the length of a separation channel. Ions enter the region, are stacked, leave the region, and subsequently disperse and decrease in concentration. After stacking, peak (spatial) variance tends to grow linearly in time. Stacking includes field-amplified sample stacking (FASS) and its derivative techniques.

2. In focusing, the condition $\nabla \cdot \bar{v}_d < 0$ is typically satisfied throughout a large extent of the channel and there is, additionally, a focal point, $\bar{x}_{foc}$ (or focal region), for the solute. There is a reference frame where the continuous vector field $\bar{v}_d$ changes sign at $\bar{x}_{foc}$. In this frame, the opposing signs of $\bar{v}_d$ drive the initially widely distributed sample to $\bar{x}_{foc}$, where sample ions accumulate. For a finite amount of sample, sample peak width reaches a steady-state value determined by the opposing effects of the $\bar{v}_d$ gradient and dispersion. Focusing includes isoelectric focusing (IEF) and thermal gradient focusing (TGF) where $\bar{x}_{foc}$ is fixed in space; and also includes isotachophoresis (ITP) where $\bar{x}_{foc}$ propagates at a wave speed set by the bulk flow and the ion mobilities in the system.

Leveraging heterogeneous buffer systems to effect changes in sample drift velocity is a common feature of electrophoretic preconcentration techniques. The sample and background electrolyte (BGE) can differ in ionic strength and ionic makeup (e.g., ion mobility or valence). The theoretical concentration enhancement of most of these techniques has been well known for many decades. For example, the maximum concentration enhancement in FASS is predicted to be equal to the ratio of the (relatively high) conductivity of the BGE to the (relatively low) conductivity of the sample solution. This ratio determines the low-to-high analyte drift velocity ratio. In ITP, the theoretical drift velocity ratio and associated prediction of concentration enhancement is a function of the leading ion concentration and various ionic mobilities. Several numerical simulation tools that can predict the concentration enhancement in various electrophoretic preconcentration techniques are available. One fairly comprehensive and useful simulation tools is Simul, which is available free download on the Web [1]. Simul solves one-dimensional multispecies electromigration–diffusion phenomena (e.g., for electromigration in long thin tubes of constant area). The code includes ionization equilibria of weak electrolytes, calculation of local pH fields, and the dependence of ionic mobilities on ionic strength and pH. Given initial conditions and ion information (e.g., fully ionized mobility, pK_a values, diffusivity), the code predicts the evolution of ionic species in the presence of an electric field. Simul is, however, limited to one dimension and does not include flow-induced dispersion caused by nonuniform electroosmosis or body forces due to conductivity gradients.

In practice, both molecular diffusion and advective dispersion (i.e., Taylor dispersion [2]) limit the achievable concentration enhancement of preconcentration techniques. In some techniques, such as TGF, an external-pressure-driven velocity field is integral to the focusing effect and dispersive effects are inherent in the system. In other techniques such as FASS, external-pressure-driven flows are minimal (e.g., due to unwanted hydrostatic pressure heads in end-channel reservoirs). However, despite efforts to suppress electroosmotic flow (EOF), electrophoresis in capillaries is typically accompanied by at least trace EOF. The heterogenous nature of buffers creates gradients in EOF mobility. These heterogeneities include ion density and ion mobilities and may also include temperature gradients, valence (via chemical reactions), and perhaps permittivity and viscosity. More importantly, variations in local ionic conductivity (and local permittivity) affect strong variations in local electric field. Both these effects (gradients in EOF mobility and gradients in field) create nonuniform EOF that leads to the generation of internal pressure gradients. Secondary flows generated by these pressure gradients tend to disperse analyte plugs and reduce the preconcentration effect.

In this chapter, we will identify the role and the scaling of dispersive effects in determining the ultimate limits of sample preconcentration techniques. Where possible, we will offer simple quantitative theory for estimating the effects of dispersion, and in all cases we will provide scaling arguments that can help guide empirical optimization of preconcentration techniques. First, we will explain the basic physics of Taylor dispersion and provide reduced order models to qualitatively and quantitatively describe it. Second, we will present simple relations to estimate the importance of Taylor dispersion in relation to other peak broadening mechanisms in capillary electrophoresis (CE) systems. These relations are useful in, for example, empirically identifying optima in applied field and conductivity ratios. Finally, we will briefly describe three examples of preconcentration techniques as instructive "case studies" in which dispersion plays an important role in determining concentration enhancement.

38.2 BACKGROUND

Dispersion, the tendency for ordered molecules to decrease gradients and local concentration, is caused by both molecular diffusion and nonuniform bulk liquid motion. High dispersion rates may be advantageous for mixing and chemical reactions, but are undesirable in separation and purification applications. For separations, minimizing dispersion improves resolution and sensitivity [3] and yields improved dynamics for concentration and purification [4]. As a consequence, the physical processes that lead to dispersion have been a subject of intense interest for more than a century. In recent years, the development of the concept of the micro-total analysis system (μTAS) or "labs on a chip" has motivated further exploration of dispersion in microchannel flows.

In the field of CE, researchers are most familiar with dispersion caused by molecular diffusion and Taylor dispersion [2]. In 1953, G. I. Taylor demonstrated that, under certain conditions, the cross-sectional average of the unsteady, three-dimensional concentration field within a channel evolves as a one-dimensional convective–diffusion equation with a coefficient analogous to a modified diffusivity. Here, the advective dispersion in the axial direction is balanced by spanwise and depthwise (e.g., radial in a cylindrical tube) diffusion, causing peak variance to increase linearly in time, with the characteristic slope determined by an effective dispersion coefficient. Taylor further demonstrated that given enough time, theoretically, all solute plugs flowing within a channel ultimately reach this limit. Subsequently, Aris [5] proved that the (now named) Taylor regime could be unified with the pure diffusive regime by using an effective dispersion coefficient, which was the sum of the molecular diffusivity and the Taylor dispersion coefficient. The effective dispersion concept has proved to be extremely useful and has been extended to other geometries [6], generalized using alternative analyses [7,8], and extended to apply to EOF [9] and electrophoresis in nanochannels [10,11].

38.3 THEORETICAL ASPECTS

The nonuniform velocity distribution over the cross section of a capillary or a microchannel leads to distortion of the sample plug. This velocity-induced distortion generates a radial concentration gradient of the analyte. Molecular diffusion acts to reduce this radial concentration gradient. The two effects in tandem determine the effective dispersion rate of the sample plug. The increase in variance can be quantified in terms of an effective dispersion coefficient, D_{eff}. In this section, we will describe the details of calculating this effective diffusion coefficient for a sample velocity profile. This section is targeted at readers who may want to understand the mathematical details behind calculation of D_{eff}. Practitioners who are not interested in these details may skip to the next section without loss of continuity. The next section describes simple relations to estimate the role of dispersion in peak broadening in CE.

38.3.1 TAYLOR DISPERSION ANALYSIS VIA AREA AVERAGING

We derive the Taylor dispersion equation using an area averaging approach as described by Stone and Brenner [8]. We consider axisymmetric flow in a cylindrical capillary of radius "a," but will later discuss the application of these principles to other cross sections (such as typical wet etched

"D" profiles). We consider a nonuniform (in x) drift velocity to account for electrophoresis, and the scaling of streamwise coordinates by a characteristic length $U_p t_c$, where U_p is the magnitude of the deviation of axial velocity from the area-averaged velocity and t_c is the characteristic time for stacking or focusing. This approach can be applied to stacking and focusing. The following equations describe the convective dispersion of a charged species in a parabolic velocity field:

$$\frac{\partial C_i}{\partial t} + \frac{\partial}{\partial x}\left(u_{ti}(x,r)\,C_i\right) = D_i\left(\frac{1}{r}\frac{\partial}{\partial r}\left(r\frac{\partial C_i}{\partial r}\right) + \frac{\partial^2 C_i}{\partial x^2}\right); \quad i = 1:N,$$
$$\frac{\partial C_i}{\partial r} = 0 \quad \text{at } r = a, 0, \tag{38.1}$$

where C_i is the concentration of a dilute and, in general, charged solute of interest. In general electrokinetic flow physics, applied electric fields couple with conductivity gradients and lead to the generation of net charge in the bulk liquid. However, this charge density is negligible compared to the total background ion concentration. We can, therefore, assume that the solution is approximately electrically neutral everywhere outside of the electrical double layer, so that $\sum_{i=1}^{N} z_i C_i \approx 0$, where z_i is the valence. This approximation is discussed in detail elsewhere [12–14].

In Equation 38.1, we have made the typical "nearly parallel flow" approximation common in lubrication analysis that the flow is nearly parallel to the axis in this long thin tube [15]. This approximation holds as long as axial gradients in electroosmotic mobility and conductivity scale is $1/\sigma$, where the characteristic plug width of the solute, σ, is such that $\sigma \gg a$. This condition is commonly met in sample stacking and focusing techniques, although it may not be met in, for example, high field regimes of ITP in the so-called peak mode. This and other limitations are discussed later.

We consider the velocity field bounded by a cylindrical slip surface, which excludes a thin electric double layer (EDL) [16]. The velocity scale u_t includes the effects of local pressure gradients, electroosmosis, and electrophoresis and can be expressed as follows:

$$u_{ti}(x,r,t) \cong u_p(x,r,t) + U_{\text{eph},i}(x,t) + U_{\text{eof}}(x,t), \tag{38.2}$$

where $u_p(x,r,t)$ is the axial velocity component due to a local pressure gradient, $\partial p(x,t)/\partial x$, and equal to $-a^2(\partial p/\partial x)(1 - r^2/a^2)/4\eta$, where η is dynamic viscosity.

$U_{\text{eph},i}(x,t)$ is an axial electrophoretic drift due to an approximately purely axial electric field, $E(x,t)$, and can be written as $z_i F \mu_{\text{eph},i} E(x,t)$, where z_i, $\mu_{\text{eph},i}$, and F are the valence, electrophoretic mobility, and Faraday's constant [12]. $U_{\text{eof}}(x,t)$ is the (exactly axial) velocity component due to electroosmosis, which can be expressed as $zF\mu_{\text{eof}}E(x,t)$, where μ_{eof} is electroosmotic mobility and $E(x,t)$ is the axial field component at the slip plane (near the wall). We also consider changes slow enough such that fluid inertia and charge relaxation are negligible. See Storey et al. [17] and Lin et al. [18] for more detailed discussion of quasi-steady electromigration, electroosmosis, and pressure-driven flow in long thin channels.

We define $U_e(x,t) = U_{\text{eph},i}(x,t) + U_{\text{eof}}(x,t)$ as the sum of the electroosmotic and approximately parallel electrophoretic velocities. Note that $\nabla\cdot(U_{\text{eof}} + u_p) = \partial(U_{\text{eof}} + u_p)/\partial x = 0$. The convective diffusion equation then becomes

$$\frac{\partial C_i}{\partial t} + (U_{e,i} + u_p)\frac{\partial C_i}{\partial x} + \frac{\partial U_{\text{eph},i}}{\partial x}C_i = D_i\left(\frac{1}{r}\frac{\partial}{\partial r}\left(r\frac{\partial C_i}{\partial r}\right) + \frac{\partial^2 C_i}{\partial x^2}\right). \tag{38.3}$$

The dependent variables are expressed in terms of cross-sectional averages and deviations:

$$\begin{aligned} C_i(x,r,t) &= \langle C_i\rangle(x,t) + C_i'(x,r,t), \\ u(x,r,t) &= \langle u_p\rangle(x,t) + \langle U_{e,i}\rangle(x,t) + u_p'(x,r,t) = U_p + U_{e,i} + u_p. \end{aligned} \tag{38.4}$$

The cross-sectional average is defined as $\langle \cdots \rangle = 1/\pi a^2 \int_0^a 2\pi r(\cdots)\mathrm{d}r$. Note u_p' is $U_p(1-2r^2/a^2)$, where $U_p = -\pi a^4(\partial p/\partial x)/8\eta$ is the bulk (area-averaged) velocity due to pressure gradients. We can interpret primed quantities as describing the concentration field in a frame moving with the area-averaged velocity of the solute. Substituting these definitions into Equation 38.1:

$$\frac{\partial \langle C_i\rangle}{\partial t} + \frac{\partial C_i'}{\partial t} + (U_{e,i} + \langle u_p\rangle)\frac{\partial \langle C_i\rangle}{\partial x} + (U_{e,i} + \langle u_p\rangle)\frac{\partial C_i'}{\partial x} + u_p'\frac{\partial \langle C_i\rangle}{\partial x} + u_p'\frac{\partial C_i'}{\partial x}$$
$$+ \frac{\partial U_{\mathrm{eph},i}}{\partial x}\langle C_i\rangle + \frac{\partial U_{\mathrm{eph},i}}{\partial x}C_i' = D_i\left(\frac{1}{r}\frac{\partial}{\partial r}\left(r\frac{\partial C_i'}{\partial r}\right) + \frac{\partial^2\langle C_i\rangle}{\partial x^2} + \frac{\partial^2 C_i'}{\partial x^2}\right). \tag{38.5}$$

Subject to

$$\frac{\partial C_i'}{\partial r} = 0 \quad \text{at } r = a, 0.$$

Note that $\langle U_{\mathrm{eph},i}\rangle = U_{\mathrm{eph},i}$ and $U'_{\mathrm{eph},i} = 0$. The boundary conditions (BC) reduce to this since $\partial\langle C_i\rangle/\partial r$ is exactly zero by definition, $\langle C_i\rangle$ being only a function of x. Next, we take a cross-sectional average of Equation 38.5:

$$\frac{\partial \langle C_i\rangle}{\partial t} + \underbrace{\left\langle\frac{\partial C_i'}{\partial t}\right\rangle}_{=0 \text{ by def}} + (U_{e,i} + \langle u_p\rangle)\frac{\partial \langle C_i\rangle}{\partial x} + \underbrace{(U_{e,i} + \langle u_p\rangle)\left\langle\frac{\partial C_i'}{\partial x}\right\rangle}_{=0 \text{ by def}} + \underbrace{\left\langle u_p'\right\rangle\frac{\partial \langle C_i\rangle}{\partial x}}_{=0 \text{ by def}}$$
$$+ \left\langle u_p'\frac{\partial C_i'}{\partial x}\right\rangle + \frac{\partial U_{\mathrm{eph},i}}{\partial x}\langle C_i\rangle + \underbrace{\left\langle\frac{\partial U_{\mathrm{eph},i}}{\partial x}C_i'\right\rangle}_{=0 \text{ by def}} = D_i\left(\underbrace{\left\langle\frac{1}{r}\frac{\partial}{\partial r}\left(r\frac{\partial C_i'}{\partial r}\right)\right\rangle}_{\text{evaluate using BC}} + \frac{\partial^2\langle C_i\rangle}{\partial x^2} + \underbrace{\left\langle\frac{\partial^2 C_i'}{\partial x^2}\right\rangle}_{=0 \text{ by def}}\right), \tag{38.6}$$

where the notes "=0 by def" and "evaluate using BC" denote a quantity zero by definition and a term that can be evaluated with the BCs. Evaluating the first term on the right-hand side:

$$\left\langle\frac{1}{r}\frac{\partial}{\partial r}\left(r\frac{\partial C_i'}{\partial r}\right)\right\rangle = \frac{1}{\pi a^2}\int_0^a 2\pi\frac{\partial}{\partial r}\left(r\frac{\partial C_i'}{\partial r}\right)\mathrm{d}r = \frac{2}{a^2}\int_0^a \mathrm{d}\left(\mathrm{r}\frac{\partial C_\mathrm{i}'}{\partial \mathrm{r}}\right) = r\left.\frac{\partial C_i'}{\partial r}\right|_a - r\left.\frac{\partial C_i'}{\partial r}\right|_0 = 0.$$

(From BC, $\partial C_i'/\partial r$ is zero at both $r = 0$ and $r = a$.) Collecting nonzero terms in Equation 38.6:

$$\frac{\partial\langle C_i\rangle}{\partial t} + \left(U_{e,i} + \langle u_p\rangle\right)\frac{\partial\langle C_i\rangle}{\partial x^2} + \frac{\partial U_{\mathrm{eph},i}}{\partial x}\langle C_i\rangle = D_i\left(\frac{\partial^2\langle C_i\rangle}{\partial x^2}\right) - \left\langle u_p'\frac{\partial C_i'}{\partial x}\right\rangle. \tag{38.7}$$

The cross-correlation term is placed on the right-hand side as it acts as a source of dispersion (as does diffusion). We now derive an expression for C_i'. To do so, subtract Equation 38.7 from Equation 38.5:

$$\frac{\partial C_i'}{\partial t} + u_p'\frac{\partial\langle C_i\rangle}{\partial x} + \left(U_{e,i} + \langle u_p\rangle\right)\frac{\partial C_i'}{\partial x} + u_p'\frac{\partial C_i'}{\partial x} + \frac{\partial U_{\mathrm{eph},i}}{\partial x}C_i'$$
$$= D_i\left(\frac{1}{r}\frac{\partial}{\partial r}\left(r\frac{\partial C_i'}{\partial r}\right) + \frac{\partial^2 C_i'}{\partial x^2}\right) + \left\langle u_p'\frac{\partial C_i'}{\partial x}\right\rangle. \tag{38.8}$$

Scale the variables as follows: $r^* = r/a$, $x^* = x/\sigma$, $C_i'^* = C_i'/C_{io}'$, $\langle C_i\rangle^* = \langle C_i\rangle/C_{io}$, $t^* = t/t_c$, and $\langle u_p\rangle^* = \langle u_p\rangle/U_p = 1$, $u_p'^* = u_p'/U_p$, and $\langle U_{e,i}\rangle^* = \langle U_{e,i}\rangle/(U_{\text{eph},i}(x,t) + U_{\text{eof}}(x,t)) = 1$. We choose both C_{io}' and C_{io} as we will use the ratio C_{io}'/C_{io} as a smallness parameter in our scaling arguments.

We scale axial gradients using the characteristic width of dispersion σ. For techniques where variance width of the sample plug increases approximately linearly in time, we can interpret σ as the characteristic length scale of advective dispersion, $U_p t_c$. Here, t_c is the time of observation or time between injection and detection. In focusing techniques where solute dispersion length scales reach a steady value, σ is an inherent length scale associated with the focusing stacking dynamics. For example, in ITP stacking or TGF of finite injection volumes, sample widths eventually reach a steady-state value. In these techniques, the axial concentration and velocity gradients of the peak are determined by the interplay between electromigration (which provides the focusing fluxes) and diffusion and dispersion (which limit the amount of stacking). The inherent length scale of dispersion, σ, is then an internal length scale (i.e., determined by the specific condition). For such cases, we will assume $\sigma \gg a$ with the understanding that the analysis is checked for self consistency and is only valid if it indeed predicts long thin solute plugs.

Equation 38.8 can then be written as

$$\left(\frac{C_{o,i}'}{t_c}\right)\frac{\partial C_i^*}{\partial t^*} + \left(\frac{U_p C_{o,i}}{\sigma}\right)u_p'^*\frac{\partial \langle C_i\rangle^*}{\partial x^*} + \left(\frac{C_{o,i}'}{\sigma}\right)(U_p + U_{e,i} + U_p u_p'^*)\frac{\partial C_i^*}{\partial x^*} + \left(\frac{U_p C_{o,i}'}{\sigma}\right)\frac{\partial U_{\text{eph},i}}{\partial x^*}C_i'$$
$$= \left(\frac{D_i C_{o,i}'}{a^2}\right)\left(\frac{1}{r^*}\frac{\partial}{\partial r^*}\left(r^*\frac{\partial C_i^*}{\partial r^*}\right)\right) + \left(\frac{D_i C_{o,i}'}{\sigma^2}\right)\frac{\partial^2 C_i'^*}{\partial x^{*2}} + \left(\frac{U_p C_{o,i}'}{\sigma}\right)\left\langle u_p'^*\frac{\partial C_i'^*}{\partial x^*}\right\rangle. \quad (38.9)$$

Next, multiply this equation by $a^2/(D_i C_{o,i})$.

$$\underbrace{\left(\frac{a^2 C_{o,i}'}{D_i t_c C_{o,i}}\right)}_{\text{order } \varepsilon^2}\frac{\partial C_i'^*}{\partial t^*} + \underbrace{\left(\frac{a^2 U_p}{\sigma D_i}\right)}_{\text{order } \varepsilon}u_p'^*\frac{\partial \langle Ci\rangle^*}{\partial x^*} + \underbrace{\left(\frac{a^2 U_p C_{o,i}'}{\sigma D_i C_{o,i}}\right)}_{\text{order } \varepsilon^2}\left(1 + \frac{U_{e,i}}{U_p} + u_p'^*\right)\frac{\partial C_i'^*}{\partial x^*} + \underbrace{\left(\frac{a^2 UpC_{o.i}'}{\sigma D_i C_{o,i}}\right)}_{\text{order } \varepsilon^2}$$
$$\times \frac{\partial U_{\text{eph},i}'}{\partial x}C_i' = \underbrace{\left(\frac{C_{o,i}'}{C_{o,i}}\right)}_{\text{order } \varepsilon}\left(\frac{1}{r^*}\frac{\partial}{\partial r^*}\left(r^*\frac{\partial C_i'^*}{\partial r^*}\right)\right) + \underbrace{\left(\frac{a^2 C_{o,i}'}{\sigma D_i C_{o,i}}\right)}_{\text{order } \varepsilon^2}\frac{\partial^2 C_i'^*}{\partial x^{*2}} + \underbrace{\left(\frac{a^2 U_p C_{o,i}'}{\sigma D_i C_{o,i}}\right)}_{\text{order } \varepsilon^2}\left\langle u_p'^*\frac{\partial C_i'^*}{\partial x^*}\right\rangle.$$
(38.10)

The notes below the various terms denote the order of magnitude. Of interest are the long times relative to streamwise transport so that $t_c \gg a^2/D_i$. In other words, the plug is long compared to the radius of the capillary such that $\sigma \gg a$ and the smallness parameter, ε, is

$$\varepsilon \sim \left(\frac{a}{\sigma}\right)^2 \sim \frac{a^2}{D_i t_c} \ll 1.$$

For the third term on the left-hand side, we assume that the displacement time, $\sigma/(U_p + U_{e,i}(x))$, is much longer than the radial diffusion time, a^2/D_i. For TGF, $U_p + U_{e,i}(x) = 0$ and the term in question is zero. For FASS and ITP, the solute plug velocity, $U_p + U_{e,i}$, is finite and this assumption implies the diffusion time, a^2/D_i, must be smaller than the time required for the plug to move a characteristic distance σ. This condition should hold for FASS where typically σ is significantly larger than a. For ITP, we will restrict our analysis to the cases where this third term is negligible. In comparing the second and third terms on the left-hand side, we see this assumption allows $[U_p + U_{e,i}]/U_p$ to be somewhat larger than unity (or smaller), but not so large that $[U_p + U_{e,i}]C_{o,i}'/(U_p C_{o,i})$ is order unity. This seems reasonable for ITP plugs with interface axial lengths, σ, of order a and larger. (The challenges of analyzing Taylor dispersion in ITP are discussed further below.) The current assumptions lead to the following:

$$\left(\frac{a^2 U_p}{\sigma D_i}\right) u_p'^* \frac{\partial \langle C_i \rangle^*}{\partial x^*} \approx \left(\frac{C'_{o,i}}{C_{o,i}}\right)\left(\frac{1}{r^*}\frac{\partial}{\partial r^*}\left(r^* \frac{\partial C_i'^*}{\partial r^*}\right)\right). \tag{38.11}$$

Or, in dimensional form:

$$u_p' \frac{\partial \langle C_i \rangle}{\partial x} \approx D_i \left(\frac{1}{r}\frac{\partial}{\partial r}\left(r \frac{\partial C_i'}{\partial r}\right)\right). \tag{38.12}$$

Perturbations in the axial velocity must be balanced by radial diffusion. Substituting for u_p'

$$U_p\left(1 - \frac{2r^2}{a^2}\right)\frac{\partial \langle C_i \rangle}{\partial x} \approx D_i \left(\frac{1}{r}\frac{\partial}{\partial r}\left(r \frac{\partial C_i'}{\partial r}\right)\right).$$

Multiply both sides by r and integrate in r

$$U_p \frac{\partial \langle C_i \rangle}{\partial x}\left(\frac{r^2}{2} - \frac{r^4}{2a^2}\right) \approx D_i \left(r \frac{\partial C_i'}{\partial r}\right) + C_1.$$

Apply BC at $r = a$: $C_1 = 0$. Divide both sides by r (and D) and integrate again:

$$C_i'(r,x,t) = C_i'(0,x,t) + \frac{U_p(x,t)\, a^2}{4D_i}\frac{\partial \langle C_i \rangle}{\partial x}\left[\frac{r^2}{a^2} - \frac{r^4}{2a^4}\right]. \tag{38.13}$$

where $C_i'(0,x,t)$ is the constant of integration obtained from the BC at $r = 0$. Now, we can evaluate the cross term in Equation 38.7. Take the x-derivative of Equation 38.13,

$$\frac{\partial C_i'(r,x,t)}{\partial x} = \frac{\partial C_i'(0,x,t)}{\partial x} + \frac{\partial}{\partial x}\left(U_p \frac{\partial \langle C_i \rangle}{\partial x}\right)\frac{a^2}{4D_i}\left[\frac{r^2}{a^2} - \frac{r^4}{2a^4}\right].$$

multiply this by u' (from Equation 38.4), and integrate over the cross section:

$$\left\langle u' \frac{\partial C_i'}{\partial x}\right\rangle = -\frac{\partial}{\partial x}\left(\frac{a^2 U_p^2}{48 D_i}\frac{\partial \langle C_i \rangle}{\partial x}\right). \tag{38.14}$$

Substitute this into Equation 38.7, which was the area average of the originally decomposed convective diffusion equation, to yield

$$\frac{\partial \langle C_i \rangle}{\partial t} + (U_{e,i} + \langle u_p \rangle)\frac{\partial \langle C_i \rangle}{\partial x} + \frac{\partial U_{\text{eph},i}}{\partial x}\langle C_i \rangle = D_i\left(\frac{\partial^2 \langle C_i \rangle}{\partial x^2}\right) + \frac{\partial}{\partial x}\left(\frac{a^2 U_p^2}{48 D_i}\frac{\partial \langle C_i \rangle}{\partial x}\right). \tag{38.15}$$

To review, the main assumptions are that the velocity and electric fields are nearly parallel and that $a^2/(D_i t_{\text{obs}}) \ll 1$ and $a/\sigma \ll 1$.

We rewrite Equation 38.15 as

$$\frac{\partial \langle C_i \rangle}{\partial t} + (U_{\text{eph},i} + U_{\text{eof}} + U_p)\frac{\partial \langle C_i \rangle}{\partial x} + \frac{\partial U_{\text{eph},i}}{\partial x}\langle C_i \rangle = \frac{\partial}{\partial x}\left(D_{\text{eff},i}\frac{\partial \langle C_i \rangle}{\partial x}\right), \tag{38.16}$$

where $D_{\text{eff},i}(x,t) = D_i + a^2U_p(x,t)^2/(48D_i) = D_i(1 + Pe_i^2/48)$. Here Pe_i is a Peclet number defined as U_pa/D_i. We do not know of analytical solutions to the nontrivial cases where $U_{\text{eph},i}$, U_{eof}, and U_p indeed vary in x (as in the case of nonuniform EOF velocities), and so this requires numerical solutions. However, we argue that significant insight can be gleaned from this equation, and this insight is valuable in optimizing sample preconcentration processes.

The previously mentioned equation shows that the area-averaged solute plugs will travel along the capillary with a nonuniform wave velocity of the form $U_{\text{eph},i} + U_{\text{eof}} + U_p$, as expected. The local dispersion coefficient, $D_{\text{eff},i}$, is strictly a function of the pressure-driven flow component, which has nonzero radial gradients, and has the same form as the well-known Taylor–Aris solution [5], but with axial and temporal variations in the dispersion coefficient. The focusing effect of electrophoresis is captured by the electrophoresis term $\langle C_i\rangle \partial U_{\text{eph},i}/\partial x$. This analysis captures the interplay between dispersion and preconcentration forces in electrophoretic stacking methods. $D_{\text{eff},i}$ acts to increase sample plug variance (in the absence of stacking fluxes, variance scales as $D_{\text{eff},i}t$, where t is time). This dispersion is countered by gradients of ion drif velocity that generate a negative value of $\langle C_i\rangle \partial U_{\text{eph},i}/\partial x$. For a positive x-direction electric field, preconcentration (local increase in $\langle C_i\rangle$) occurs for a cation when $U_{\text{eff},i}$ has negative axial gradient, $\partial U_{\text{eph},i}/\partial x < 0$ (and where $\partial U_{\text{eph},i}/\partial x > 0$ for an anion). Opposite signs for $\partial U_{\text{eph},i}/\partial x$ cause so-called electromigration dispersion and associated reductions in $\langle C_i\rangle$ [19]. As described earlier, focusing occurs when we have a frame of reference in which $\partial U_{\text{eph},i}/\partial x$ is accompanied by a change in sign of the electrophoretic velocity, and solute is driven toward this focal point from large regions of the channel. [U_{eof} and U_p cannot cause preconcentration since $\nabla \cdot (U_{\text{eof}} + U_p)$ is identically zero.] The relative strength of stacking and focusing versus dispersion is determined by the relative strength of $\partial U_{\text{eph},i}/\partial x$ and $D_{\text{eff},i}$; more simply, $\partial U_{\text{eph},i}/\partial x$ acts to reduce σ while $D_{\text{eff},i}$ acts to increase it. This balance determines the maximum achievable concentration increase, and can determine the resolution of simultaneous preconcentration and separation. This concept is further developed in the next section.

As mentioned earlier, Taylor-type dispersion analyses have been extended to other geometries including flow in rectangular channels of varying aspect ratio and channels with the characteristic "D" shape of isotropic chemical etching [6,20]. In general, these analyses lead to an effective dispersion coefficient of the form $D(1 + \kappa Pe^2)$, where κ takes into account geometric dependences. Table 38.1 list sample κ values for several flow geometries. This elegant and general form of the dispersion coefficient implies that the insights gained by the area-averaging are generally applicable to dispersion in a wide range of geometries.

Often, microchannels are etched with finite (and sometimes high) width-to-depth aspect ratios, w/h, with typical values ranging from 2 to 20. For such channels, the time scale for diffusion across

TABLE 38.1
Taylor Dispersion κ Values for Various Cross-sectional Geometries

Cross-Sectional Geometry	κ
Ellipse	$\frac{1}{192}\left(\frac{24 - 24e^2 + 5e^4}{24 - 12e^2}\right)$; $e = \sqrt{1 - d^2/W^2}$; where d and W are the minor and major axes, respectively
Rectangle	$\frac{1}{210}\left(\frac{8.5W^2}{d^2 + 2.4dW + W^2}\right)$; where d and W are the channel depth and width, respectively
Cylinder	$\frac{1}{48}$

the channel depth, $\tau_h \sim h^2/D_i$ is short compared to across the channel width $\tau_w \sim w^2/D_i$. Ajdari et al. [20] points out that dispersion in shallow-channels with smooth spanwise height distributions is controlled by the product κPe_w^2, where $Pe_w = wU/D_i$. In wide, shallow channels dispersion, owing to spanwise (width direction) velocity gradients, occurs at rates that are not negligible compared to spanwise diffusion. Statistical sampling of solute molecules along the spanwise direction is, therefore, less efficient and leads to increased dispersion over that of an idealized, infinitely wide channel with the same depth. A key consequence of the analysis is that the largest cross-section dimension controls the time scale to reach the Taylor dispersion limit. For arbitrarily shaped channels, the criteria for achieving Taylor dispersion are then modified to $t_c \gg w^2/D_i$ (where w is the largest cross section scale) and $\sigma \gg a$, where again σ is the characteristic axial dimension of the solute (or interface region) of interest.

38.3.2 Scaling Relations to Estimate Dispersion Effects

In the presence of axial gradients in ionic strength (and conductivity) or pH, both electric field and electroosmotic mobility will vary along the axis of the capillary or microchannel [21]. As discussed earlier, this leads to the generation of internal pressure gradients to satisfy the continuity equation (i.e., mass conservation). In the absence of an external, applied pressure difference, u_p is generated strictly by gradients in EOF mobility $U_{\text{eof}} = \mu_{\text{eof}}(x)E(x)$. For this case, we can estimate the relative importance of Taylor dispersion to molecular diffusion by performing a simple scaling analysis. This exercise highlights the importance of judicious choices of system parameters to minimize dispersion. These choices include separation voltage, EOF suppression strategy, and channel shape and dimensions.

We present an example scaling analysis for dispersion and optimum electric field in a single-interface FASS problem. (The scaling of pressure velocity magnitudes versus local electric field and electroosmotic mobility will also hold for other single-interface preconcentration methods such as the ITP method demonstrated by Jung et al. [22]). This simple case yields approximations as with closed-form analytical expressions. Obtaining estimates for the electroosmotic velocity is the first step. This can be achieved in at least two ways: direct experimental measurement (e.g., in calibration experiments) and theoretical estimates. EOF velocities can be measured under various conditions using the current monitoring method [23], micron-resolution particle image velocimetry measurements [24], or neutral marker tracking [25]. If these measurements are difficult to obtain, EOF velocity can be estimated using various models for the electrical double layer. For thin EDLs, the Helmholtz–Smoluchowski model yields a simple relation [26]:

$$u_{\text{EOF}} = -\frac{\varepsilon\zeta E}{\eta}, \tag{38.17}$$

where ε is the permittivity constant and ζ is the zeta potential [27]. The zeta potential is a function of ionic strength and pH of the electrolyte. Various models and correlations are available in the literature to estimate the zeta potential for various substrates [28–30]. In the presence of axial gradients in conductivity, electric field will also vary along the axis of the channel. Electric field variation can be estimated by invoking conservation of current and Gauss's law. For example, consider the two-zone preconcentration problem in a cylindrical tube as depicted in Figure 38.1. After negligible charge relaxation time scales [13], the following relations hold true [31]:

$$\begin{aligned} \sigma_{e1}E_1 &\cong \sigma_{e2}E_2, \\ E_1xL + E_2(1-x)L &= V, \end{aligned} \tag{38.18}$$

where σ_e refers to electrical conductivity and x is the fraction of channel filled with σ_{e1}. These equations can be solved simultaneously to obtain E_1 and E_2. Upon deriving the estimates for the

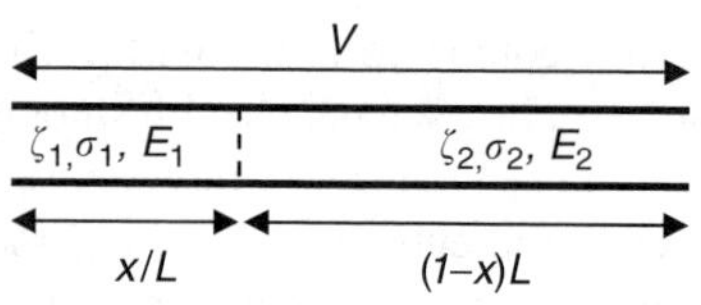

FIGURE 38.1 Schematic representation of a capillary with axial gradient in conductivity, zeta potential, and electric field. The interface is located at distance x/L from the end of the channel (i.e., the applied potential), where x is the fraction of the length L occupied by the liquid 1. The potential drop across the capillary is V.

local electric fields and zeta potentials, the EOF mismatch-induced pressure-driven component of the velocity field can be obtained. To do so, we invoke the continuity equation:

$$U_{\text{EOF},1} + U_{p1} = U_{\text{EOF},2} + U_{p2}, \tag{38.19}$$

where U_{p1} and U_{p2} are related by following:

$$\frac{U_{p1}}{U_{p2}} = -\frac{(1-x)}{x}. \tag{38.20}$$

Finally, the maximum pressure-driven component of the velocity field is (assuming $x < 0.5$):

$$U_{p1} = (U_{\text{EOF},2} - U_{\text{EOF},1})(1-x). \tag{38.21}$$

Burgi and Chien [31] have derived similar relations for estimating internal pressure gradients and for predicting optimum conductivity ratio for FASS. Now, the dispersion coefficient is estimated:

$$D_{\text{eff}} = D + \frac{a^2 U_{p1}^2}{48D} = D + \frac{a^2}{48D}(U_{\text{EOF},2} - U_{\text{EOF},1})^2(1-x)^2. \tag{38.22}$$

Furthermore,

$$U_{\text{EOF},2} - U_{\text{EOF},1} = \varepsilon\frac{(\zeta_1 - \gamma\zeta_2)}{\eta}E_1, \tag{38.23}$$

where $\gamma = \sigma_1/\sigma_2$. Noting $E_1 = E_0/(x + \gamma(1-x))$, where $E_0 = V/L$ is the nominal electric field, we have

$$D_{\text{eff}} = D + \frac{a^2\varepsilon^2 E_0^2}{48D\eta^2}(\zeta_1 - \gamma\zeta_2)\frac{(1-x)^2}{x + \gamma(1-x)}. \tag{38.24}$$

We now assume that we are concerned with a stacking technique such as FASS where variance increases in time. For such a technique, the peak variance due to molecular diffusion and Taylor dispersion scales is

$$\sigma_{\text{total}}^2 \sim 2D_{\text{eff}}\,\tau_{\text{obs}}. \tag{38.25}$$

Note that D_{eff} varies in space and time as per equation (38.24). However, for scaling purposes the maximum dispersion coefficient can be used. The observation time, τ_{obs}, is a complex quantity to determine, depending on the electrophoretic, electroosmotic, and induced pressure-driven velocities.

A closed form solution may be found in some simple scenarios, but in all cases the time scales linearly with E_0, such that $\tau_{obs} \sim L/\mu_{eff}E_0$, where μ_{eff} is an effective species mobility. Thus,

$$\sigma^2_{total} \sim \left[\frac{a^2\varepsilon^2}{24D\eta^2}(\zeta_1 - \gamma\zeta_2)^2 \frac{(1-x)^2}{x+\gamma(1-x)}\frac{L}{\mu_{eff}}\right]E_0 + \left[2D\frac{L}{\mu_{eff}}\right]\frac{1}{E_0}. \tag{38.26}$$

In this relation, the first term on the right-hand side is the contribution of Taylor dispersion to the overall band broadening. The second is due to molecular diffusion. A key point is that the contribution to the variance due to Taylor dispersion increases with nominal electric field, E_0, whereas variance due to molecular diffusion decreases with electric field, $1/E_0$. This is a consequence of the U_p^2 and therefore, E_0^2 scaling of D_{eff} on the one hand, and the L/u and therefore, $1/E_0$ scaling of time. Therefore, there exists an optimum electric field that will minimize the peak broadening. This optimum electric field can be derived by minimizing the variance with respect to field to obtain

$$E_{optimum} \sim \sqrt{\frac{48D^2\eta^2(x+\gamma(1-x))}{a^2\varepsilon^2(\zeta_1-\gamma\zeta_2)^2(1-x)^2}}. \tag{38.27}$$

As we discuss later in this chapter, techniques achieving a steady-state dispersion-limited width typically also have optimum electric fields that reduce dispersion. At negligible Taylor dispersion conditions (i.e., low electric fields), the sample axial dimension is determined by a balance between electrophoretic focusing fluxes and molecular diffusion. Plug axial dimension, therefore, scales as D/E_0 (see Equations 38.40, 38.41, 38.44, and associated discussions). At higher electric fields, Taylor dispersion becomes important and axial dimensions scale as D_{eff}/E_0. At sufficiently high fields, the Taylor dispersion term, which scales as $a^2\varepsilon^2E_0^2/(48D\eta^2)$, dominates and then interface width again scales as E_0.

In all cases, the optimum field for minimization of Taylor dispersion is a function of sample (through D), channel geometry, solution viscosity, and zeta potential value for a given conductivity ratio and solution permittivity. Finally, we note the earlier analysis neglects the effects of Joule heating [32,33] on dispersion. Joule heating places additional constraints on the optimum field as temperature rise is a function of both conductivity distribution and pore diameter. We, however, stress the main purpose of the analysis is to argue that all preconcentration phenomena have optimum fields and that they are often limited by the dispersion associated with even residual EOF.

38.4 PRACTICAL APPLICATIONS

38.4.1 Field-Amplified Sample Stacking

38.4.1.1 Introduction

Field-amplified sample stacking is a fairly widely applicable method of achieving increased sensitivity for capillary and on-chip assays in a scheme that is easily integrated with electrophoretic separation techniques [4,34–43]. FASS is typically used as a preconcentration step that occurs before the electrophoretic separation of analyte ions.

The transport phenomena associated with FASS (as in all preconcentration methods) are, in general, a complex coupling of convective–diffusion, electrostatics, and electrokinetics along with the unsteady effects associated with the response of the electrical double layer to varying bulk ion concentrations. The detailed understanding of the process dynamics is important for optimization of high-sensitivity systems. The effects of EOF on preconcentration and separation are very important to studies of FASS as even slight EOF couples with axial conductivity gradients to generate internal

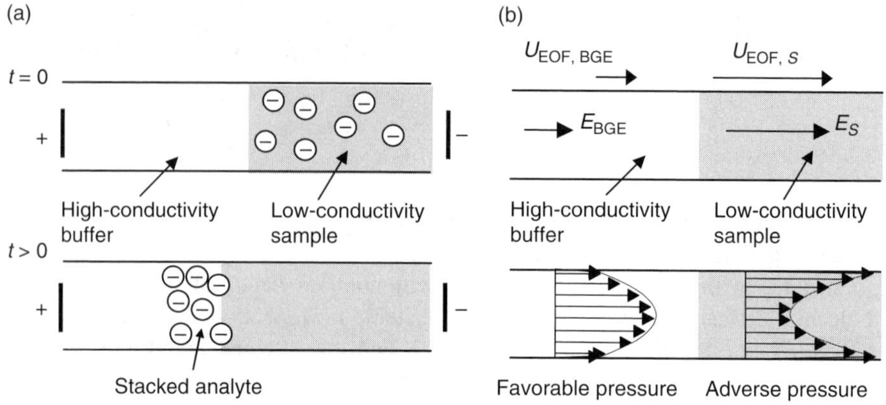

FIGURE 38.2 (a) Schematic diagram showing FASS of anionic species in the absence of EOF. A gradient in the BGE ion concentration is established. The sample is in a region of locally low conductivity. Upon application of an electric field, the axial gradient in conductivity results in an electric field gradient. Since area-averaged current density is uniform along the axis of the channel, the low conductivity section is a region of high electric field, and the region of high conductivity contains relatively low electric field. As sample ions exit the high field/high electrophoretic velocity region and enter the low velocity region, they locally accumulate and increase in concentration. (b) Stacking in the presence of EOF. Gradients in conductivity generate axial variation in electric field and electroosmotic mobility. The system generates internal pressure gradients, which tend to disperse the sample.

pressure gradients. These internal pressure gradients disperse the sample and thereby, limit the practically achievable concentration enhancement.

38.4.1.2 Basic Theory and Implementation

The principle behind FASS is shown schematically in Figure 38.2a. An axial gradient in ionic conductivity (and therefore electric field) is achieved by preparing the sample in an electrolyte solution of lower concentration than the BGE. Upon application of an axial potential gradient, the sample region acts as a high electrical resistance zone in series with the rest of the channel and a locally high electric field is generated within the sample zone. Under the influence of electric field, sample ions migrate from the high to low drift velocity region. This leads to a local accumulation or "stacking" of sample ions near the interface between high and low conductivity regions. This stacking increases sample concentration and results in an increased signal. The process depicted in Figure 38.2a is for an idealized case where diffusion and advection are neglected. The maximum concentration enhancement is given by a conservation of species at the interface

$$\frac{C_{\text{Stacked}}}{C_{\text{Initial}}} = \frac{E_S}{E_{\text{BGE}}} = \gamma, \tag{38.28}$$

where E_S and E_{BGE} are the electric field in the sample and the BGE regions, respectively. In essence, FASS relies strictly on electric field gradients generated by spatial variations in ion density. This makes FASS unlike other preconcentration techniques (such as ITP and IEF), which require more specific buffer chemistries and ion mobilities. For example, FASS can be implemented with the same type of ions in the BGE and sample zones. As such, FASS provides flexibility in the choice of BGE buffer chemistry and can nearly always be performed under well-controlled pH conditions.

Figure 38.2b shows a more realistic system where finite EOF is present. The gradient in the electrolyte concentration required for stacking leads to a gradient in electric field and electroosmotic mobility. This causes a mismatch of electroosmotic velocity and hence generation of a pressure

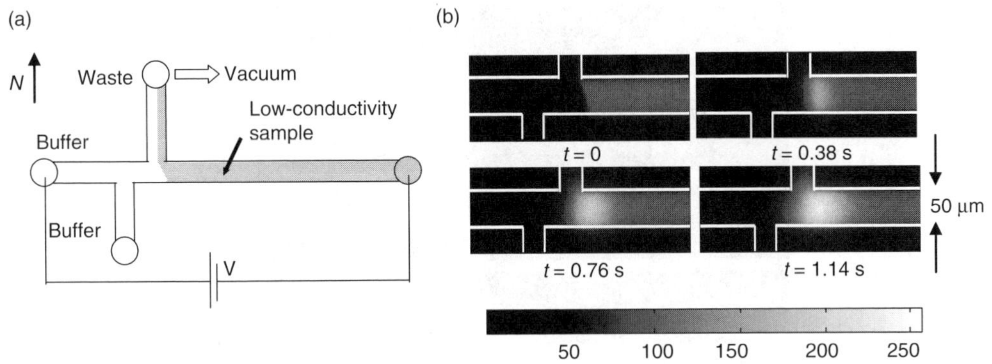

FIGURE 38.3 Single interface stacking experiments. (a) Schematic diagram of a microchip. Width and centerline depth of channels were 50 μm and 20 μm, respectively. (b) Epifluorescence CCD images showing establishment of initial condition for conductivity gradient and subsequent stacking across interface. The sample was anionic 17 μm bodipy dye and the buffer was HEPES at pH = 7.

gradient (consistent with the continuity constraint). The pressure gradient tends to disperse the concentration fields and thereby lower the efficiency of stacking.

FASS has been applied in a variety of assay formats in both capillaries and microchips. The configurations can be broadly classified as (1) single interface configuration and (2) finite-plug configuration. The single interface configuration includes techniques such as field-amplified sample injection (FASI) [36,37,44], large volume sample stacking (LVSS) [38], and Head-Column FASS [45]. These techniques involve a single interface between the sample region and BGE. In these techniques, the volume of sample loaded into the capillary or the microchannel often exceeds the total volume of (initially low concentration) sample loaded into the capillary, or the microchannel can exceed the total volume of the capillary. In contrast, the finite-plug technique involves sample zones of fixed size. The sample zone size can be defined by duration of sample injection by either electrokinetic flows or by hydrodynamics flow. In microchannel networks, the sample size can be accurately controlled by chip geometry, for example, by using pinched-injection or staggered-T injection [46–48].

Figure 38.3 shows schematic representation of microchip-based single-interface FASS system. The interface between high and low conductivity buffer regions is generated by applying a vacuum at the north reservoir. Once a buffer–buffer interface is established, the vacuum is released and an axial electric field is applied. Upon application of an axial electric field along the west-to-east direction, sample stacks at the interface between buffer streams. Figure 38.3b shows images of the stacking process at selected times. In Figure 38.4, instantaneous images of stacking process are shown for a case where EOF was not suppressed. Since the EOF velocity in all regions is greater than the negative electrophoretic velocity of fluorescein dye, the stacked region moves in the direction of EOF. The images clearly show the favorable pressure-gradient-induced curvature of the stacked ions on the downstream (left-hand) side of the interface. As described earlier, these pressure gradients act to disperse sample and reduce the efficiency of FASS. From Figures 38.3b and 38.4, the efficacy of the EOF-suppression method is apparent as the conductivity interface is nearly stationary and there is negligible pressure-induced curvature of stacked analyte. Figure 38.5 shows the temporal development of the spatial concentration distribution of sample ions. The peak intensity increases roughly exponentially at first and then saturates at a maximum achievable concentration enhancement of γ.

38.4.1.3 Dispersion Theory

In most electrophoretic experiments, the quantity of practical interest is the cross-sectional area-averaged concentration distribution of sample ions. This quantity is, for example, proportional to the

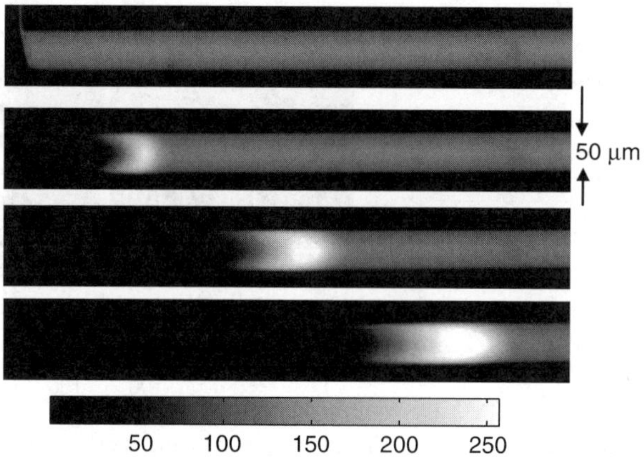

FIGURE 38.4 CCD images of on-chip FASS in an untreated channel with significant EOF mobility. The images clearly show the development of stacked fluorescein in the favorable pressure gradient region (i.e., the high conductivity region). The sample was 25 μm fluorescein dye dissolved in 5 mM Borate buffer. The BGE for this particular experiment was 25 mM Borate buffer (pH = 9.2). The electric field in the sample region was 50 V/cm and $\gamma = 4$.

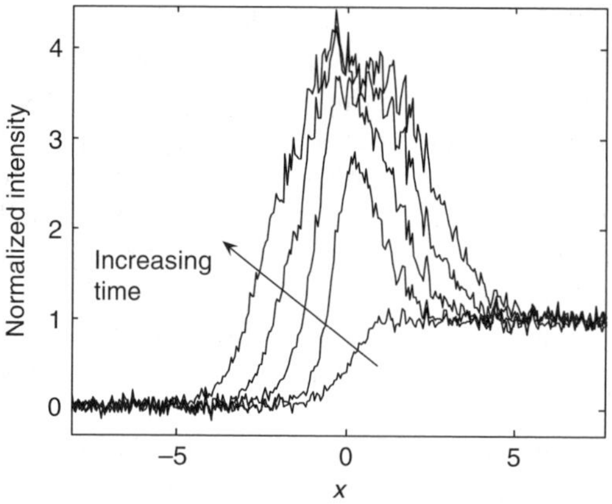

FIGURE 38.5 Measured axial intensity profiles for sample ions. The profiles were obtained by averaging two-dimensional image intensity data along the width of the channel. The applied nominal electric field was 588 V/cm and $\gamma = 4$. Time between individual profiles was 0.15 s.

measured signal intensity of line-of-sight optical integrators such as pointwise fluorescence detectors, transmitted-mode absorption detectors, and width-averaged electrophoregrams from charge-coupled device (CCD) arrays. This signal determines the key detectability constraints of electrophoretic separations [3]. As described earlier, Taylor dispersion analysis allows us to develop cross-sectional area-averaged transport equations. Such models provide useful insight into the physics of the process and lead to the identification of key parameters that can be used to develop optimization strategies for FASS experiments.

The FASS model requires description of electromigration, diffusion, and advection of sample ions as well as BGE ions. The general system of equations is highly coupled and nonlinear and, therefore, difficult to solve. However, the concentration of sample ions is much smaller than the buffer ions (typically μM sample ions concentration or less versus order 1 mM buffer ion concentrations). Therefore, we can decouple the buffer and sample ion concentration fields. Using this approach, Bharadwaj and Santiago [4] have developed a dynamic model for FASS in a flat-plate

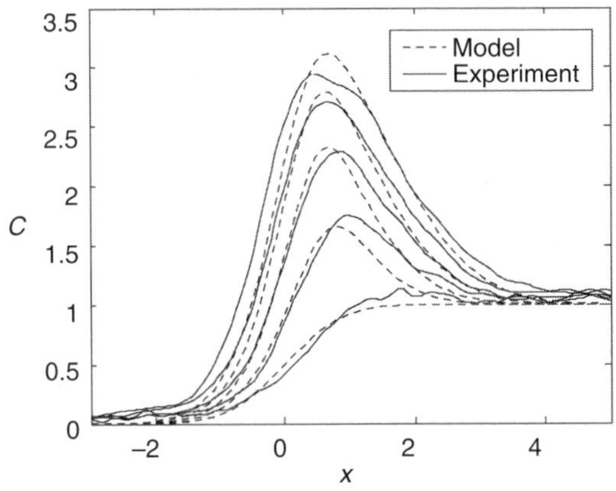

FIGURE 38.6 Comparison of model prections and measured concentration profiles. $\gamma = 4$, $E_0 = 379$ V/cm, and time-between-frames was 76 ms. The model parameters are $Pe_e = 55$, $\alpha = 0.23$, $\beta = 0.28$, $\delta = 1.27$.

geometry. The analysis provides the following equation for the cross-sectional area-averaged sample ion distribution, C_S:

$$\frac{\partial C_S}{\partial t} + \alpha \langle u \rangle \frac{\partial C_S}{\partial x} = D \frac{\partial^2 C_S}{\partial x^2} \frac{1}{Pe_e} + \alpha^2 \beta^2 Pe_e \frac{8g(x,t)}{105D} \frac{\partial}{\partial x} \left(g(x,t) \frac{\partial C_S}{\partial x} \right) - z_S \mu_S \frac{\partial (C_S E)}{\partial x}. \quad (38.29)$$

The parameters governing this system of equations are

$$Pe_e = \frac{E_0 \mu_{\text{eph}} F s_S}{D_S}; \quad \alpha = \frac{-\varepsilon_0 \varepsilon_r \varsigma / \eta}{\mu_{\text{eph}}} = \frac{\mu_{\text{eof}}}{\mu_{\text{eph}}}; \quad \beta = \frac{d}{s_S}; \quad \text{and} \quad \delta = \frac{s_B}{s_S}.$$

Pe_e is the electric Peclet number, expressed as the ratio of diffusion time to electromigration time; α is the ratio of electroosmosis to electrophoretic mobility; β is the ratio of channel width to characteristic length scale for the initial sample ion concentration distribution; and δ is the ratio of the length scale of the initial BGE and sample ion concentration gradients. Here, d is the channel depth, s_S and s_B are the initial sample and BGE ion concentration gradients, and E_0 is the nominal electric field. The function $g(x,t)$ in the advective dispersion term accounts for the axial variation in pressure-driven velocity profile (cf. Figure 38.2).

Figure 38.6 shows comparisons between dispersion model predictions and experimentally measured concentration profiles. Experiments are shown for $\gamma = 4$ and 9 and E_S values of 379 and 588 V/cm. There is a quantitative agreement between measured area-averaged concentration profiles and the model prediction throughout the time of observation. The model predictions were obtained by solving Equation 38.29. As shown in Figure 38.3b, for times approaching 1 s, the region of high area-averaged concentration becomes two-dimensional as it enters the staggered-T injection region of the system, which cannot be captured by one-dimensional model. The dispersion model is able to capture important features such as the development of peak width and the temporal growth of the maximum concentration. Also, the model describes convective–electromigration–diffusion dynamics in a purely two-dimensional flow in a wide, shallow channel (neglecting the influence of side walls). In reality, however, the microchannels in the experiment have a D-shape characteristic of an isotropic etch with a width to maximum depth ratio of 2.5. This assumption may be improved in future refinements of the dispersion model. For example, Dutta and Leighton [6] have investigated the effect of isotropic-etched microchannel geometries on the dispersion coefficient for simple pressure-driven flows. Their analysis shows, in the Taylor dispersion limit (ruled by the widest channel dimension), the dispersion coefficients for the D-shaped channels can be three to four times

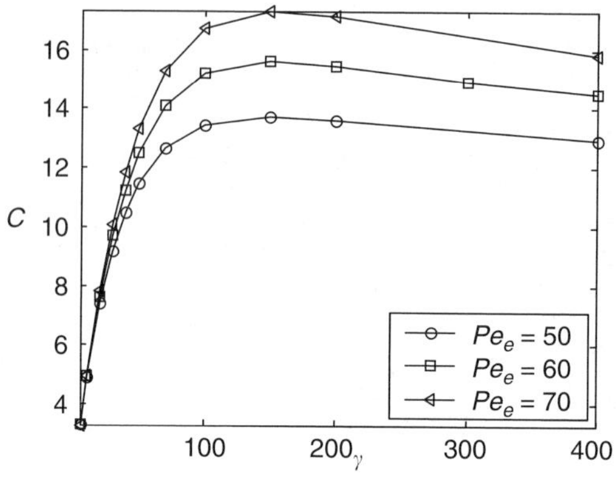

FIGURE 38.7 Optimum value of γ for fixed stacking times of 1 s. The parameters for the dispersion model are: $\alpha = 0.5$, $\beta = 2$, and $\delta = 1$. For a given analysis time and fixed values of Pe_e, α, β, and δ, there is a unique value of γ that provides maximum concentration enhancement.

larger than those predicted by simple two-dimensional analysis. Such advective–diffusion effects would be most important in flow for large γ values (associated with larger internally generated pressure gradients) and the large Peclet numbers associated with high electric fields. Another possible refinement of the model would include dispersive effects due to Joule heating [49], which should also lead to reduction in the rate of concentration increase.

38.4.1.4 Performance and Guidelines

In this section, we summarize some parametric and optimization results from dispersion models. First, we consider the effect of conductivity ratio, γ, on stacking efficiency. Figure 38.7 shows clearly that there is an optimum value of γ for a given set of parameters and analysis time. In these plots, the analysis time is fixed as we are interested in FASS as a preconcentration step before electrophoretic separation. These fixed-time comparisons help to determine the time needed to achieve adequate concentration enhancement and the initial condition of the subsequent uniform-conductivity separation process. These model predictions were generated with a parametric variation study using the numerical model described earlier. This result is in contrast with the ideal concentration factor described earlier, which shows that increasing γ always increases the concentration enhancement (Equation 38.28). This in an important feature of the dispersion analysis as it gives experimentalists a method of choosing values of γ to yield optimal signal detections.

The existence of an optimal value of γ can be better understood by considering the scaling of the parameters of interest. Equation 38.28 shows that the maximum sample concentration is proportional to γ:

$$C_{C,\max} \propto \gamma. \tag{38.30}$$

In contrast, the ratio of the EOF velocities in the low-conductivity region to the value in the high-conductivity region scales is

$$\frac{U_{\mathrm{EOF},S}}{U_{\mathrm{EOF},B}} \sim \frac{\zeta_S E_S}{\zeta_B E_B} \sim \gamma^{1+n}. \tag{38.31}$$

The parameter n refers to the ratio of the zeta potential in the sample and the BGE regions. Typical values of this parameter range between 0.2 and 0.3 [50,51], so that the advective dispersion effects of mismatched slip velocities is negligible for low γ but dominates at high γ. An analogous scaling

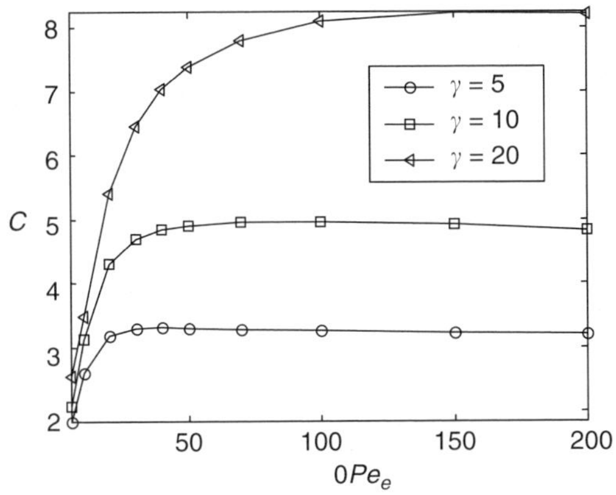

FIGURE 38.8 Optimum value of Pe_e for a fixed stacking time of 1 s. The parameters for the dispersion model are $\alpha = 0.5$, $\beta = 2$, and $\delta = 1$. At low Pe_e, diffusive dispersion dominates and concentration enhancement suffers. At high Pe_e, advective dispersion is dominant and again concentration enhancement suffers. For fixed γ and analysis time, there is a unique, optimal Pe_e (e.g., an optimal electric field for a given channel system) which results in maximum concentration increase.

observation was made by Burgi and Chien [31]. They discuss the existence of an optimum γ using simple scaling arguments. They developed an algebraic model for the long-time behavior of a finite-length sample plug variance as a function of γ, using a one-dimensional Taylor dispersion approximation. In contrast to their model, the dynamic model described here allows quantitative prediction of both temporal and spatial development of the sample ion, BGE ions, and electric field profiles.

Another important parameter determining the convective dispersion and hence, rate of concentration increase in FASS is α, or the ratio of electroosmotic and electrophoretic mobilities. For a typical value of electrophoretic mobility (e.g., 3E−8 m^2 V^{-1} s^{-1}), α is approximately equal to 2 for glass microchips. Therefore, the dispersion dynamics of untreated glass chips are well in the advection regime. However, at least an order of magnitude reduction in electroosmotic mobility is possible by dynamic surface coatings using neutral water-soluble polymers [52]. It is, therefore, interesting to experimentalist to quantify the importance of suppressing EOF in determining maximum achievable concentration increases. Bharadwaj and Santiago [4] show that, even for the case of a 10-fold decrease in electroosmotic mobility ($\alpha = 0.2$), there is significant convective dispersion. The preconcentration time required to reach the maximum concentration enhancement can be as much as 50% longer for $\alpha = 0.2$ case as compared to a case where there is no EOF. This result has important consequences in the design of microchip-based FASS systems because a slower rate of concentration increase can adversely affect both the amount of sample required per separation and the throughput of the device. To increase the rate of concentration enhancement, the electric field and Peclet number should be increased.

Figure 38.8 describes the effect of Peclet number, $Pe_e = E_0\mu_{\mathrm{eph}}F_{SS}/D_S$, on maximum concentration achievable for a given analysis time. Initially, increases in Pe_e are favorable for concentration enhancement. This is due to a reduced contribution of molecular diffusion, which scales as Pe_e^{-1}. However, as Pe_e is increased further, the concentration increase slows down and, past a critical Pe_e, the achievable concentration begins to decrease. The latter effect is due to the aforementioned fact that convective dispersion increases with increasing Pe_e. Equation 38.29 shows the dispersion term scales as $\alpha^2\beta^2 Pe_e$. There is, therefore, an optimum value of Pe_e (e.g., an optimum electric field for a given process and geometry) for a given analysis time and fixed values of γ, α, and β. Note that the optimum value of Pe_e may be somewhat smaller in practice than that predicted by our dispersion model since we do not account for the effects of Joule heating [49]. Joule heating is proportional to the square of local electric field and is expected to be important for very high field strengths and relatively large channels.

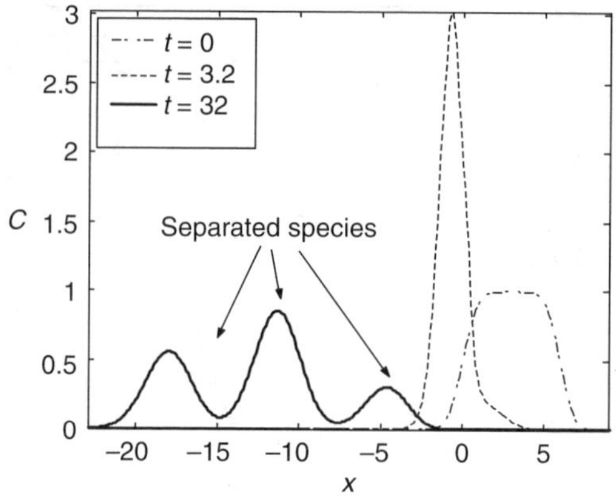

FIGURE 38.9 Stacking and separation dynamics of three negatively charged sample species. The model parameters are $Pe_e = 40$, $\gamma = 5$, $\beta = 2$, $\alpha = 0.05$, $\delta = 1$, and $h = 6$. The dimensionless electrophoretic mobilities of the species are 1, 2, and 3, respectively. The initial dimensionless concentrations of the sample species are 1/6, 1/2, and 1/3, respectively.

Finally, Figure 38.9 shows model results for sample stacking and separation dynamics of a finite injection volume (initially approximately six channel depths wide) of three anionic sample ions. Initially, there is rapid stacking (accumulation) of the sample ions as they exit the low conductivity region and enter the high conductivity region. Once sample ions enter the high conductivity region, sample stacking ends and ions are subsequently electrophoretically separated into three distinct peaks. The dispersion model can be used to optimally design FASS-based electrophoretic separation systems for the analysis of multiple sample species. For example, the model predictions can guide the location of detector and width of initial sample plug to ensure adequate signal-to-noise ratio (SNR) and resolution.

38.4.2 Temperature Gradient Focusing

38.4.2.1 Introduction

So far, we have alluded to a number of preconcentration techniques that are of interest to the CE community and, in the previous section, we examined FASS in detail. In this section, we consider dispersion as it relates to focusing techniques. The quintessential focusing technique is IEF [53–55]. In IEF, charged species migrate through a pH gradient under the influence of an electric field until they reach their isoelectric, or pI, point. At the pI, a species becomes net neutral and ceases to migrate. While this is the most common focusing technique, it is not the most general from the perspective of dispersion analysis and scaling. Consequently, we choose microfluidic temperature gradient focusing (TGF) as it represents an excellent case study in dispersion and provides a more general model. The approach and discussions presented can be extended to the analysis of Taylor dispersion in IEF.

TGF is a form of electric field gradient focusing, where a temperature-induced gradient in electric field helps produce the gradient in electrophoretic velocity required for focusing (in contrast to IEF, where the focusing effect is caused by the changing charge on an ampholyte). TGF was first described in a seminal paper by Ross and Locascio [56], where they demonstrated the successful focusing of charged fluorescent dyes, amino acids, green fluorescent protein, DNA, and polystyrene particles, illustrating the general utility of TGF. TGF has subsequently been extended to DNA hybridization assays and single nucleotide polymorphism detection [57], as well as the detection of chiral enantiomers [58]. Focusing of neutral and ionic hydrophobic analytes (e.g., coumarin)

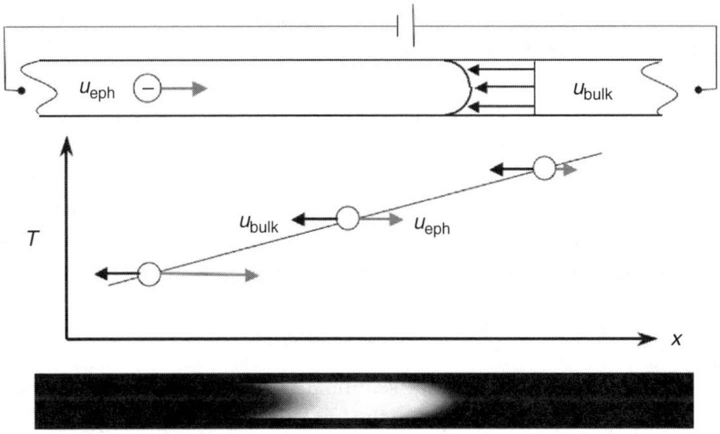

FIGURE 38.10 Schematic of TGF process with advective dispersion. An electrophoretic velocity is countered by an opposing liquid flow, composed of both pressure-driven and electroosmototic flow (top). A temperature gradient is applied to a microchannel, inducing a gradient in the electrophoretic velocity of an analyte. The analyte focuses where the electrophoretic and convective ("bulk") fluxes sum to zero (middle). Both molecular diffusion and advective dispersion broaden the band about the focus point. The bottom image shows Bodipy focused in a 20 by 200 μm wide channel with an applied electric field and temperature gradient of 60 V/mm and 10°C/mm, respectively.

has also been achieved [59,60] using TGF combined with micellar electrokinetic chromatography (MEKC). The capabilities of TGF allow it to be used in analytical (i.e., detection and separation) and preparative (i.e., concentration and purification) applications. In both applications, minimization of dispersion is an important design goal.

We have already discussed the basics of dispersion, noting how decreased dispersion improves resolution and sensitivity in separation applications [61], and also yields improved dynamics for concentration and purification applications [4]. However, there are some key differences to consider when comparing focusing techniques such as TGF with other techniques. We describe the basics of TGF theory, implementation details, and the modifications to Taylor dispersion required for TGF. Finally, we present tips for empirical optimization of TGF preconcentration factors and resolution.

38.4.2.2 Basic Theory and Implementation

Temperature gradient focusing focuses charged species by balancing an axially varying electrophoretic flux with a counterflow, causing species to focus at locations where their net fluxes sum to zero (Figure 38.10) [56].

$$\langle u_{bulk} \rangle + u_{eph}(x_{foc}) = 0, \tag{38.32}$$

where x_{foc} is the focus location for the species in question and "bulk" refers to the net effect of both pressure-driven flow and electroosmosis. Note that, unlike IEF, the electrophoretic velocity of the charged species is always nonzero, so a counterflow, and its associated dispersive effects, is always present and significant. In practice, the counterflow is provided by a combination of flow due to electroosmosis and flow caused by an externally applied pressure difference. Variation in electrophoretic flux is accomplished by applying a temperature gradient along the axis of the microchannel and by using a buffer with a temperature-dependent ionic strength (e.g., due to temperature-dependent dissociation of a weak electrolyte).

Within the channel, local electric field is inversely proportional to conductivity, which in turn is a function of local viscosity and ion density. Using the convention of Ross et al. [56], we write the conductivity as $\sigma(T) = \mu_0\sigma_0/(\mu f)$, where μ_0 and σ_0 are the buffer viscosity and conductivity at a defined reference temperature, $\mu(T)$ is the viscosity, and $f(T)$ is a nondimensional function incorporating any remaining conductivity dependencies (primarily the change in ionic strength). A similar decomposition is applied to the electrophoretic mobility, yielding $\nu_{\text{eph}} = \nu_{0,\text{eph}}\mu_0/(\mu f_{\text{eph}})$, where $\nu_{0,\text{eph}}$ is the analyte's electrophoretic mobility at the reference temperature, and $f_{\text{eph}}(T)$ accounts for any other temperature dependencies. The usefulness of this decomposition becomes apparent when we assume uniform current, I, use Ohm's law to determine the local electric field, and solve for the electrophoretic velocity:

$$u_{\text{eph}} = \nu_{0,\text{eph}} E_0 f / f_{\text{eph}}, \tag{38.33}$$

where $E_0 = I/A\sigma_0$ and A is the channel cross-sectional area. Most analyses assume $f_{\text{eph}}(T)$ remains near unity, which is equivalent to assuming that the ion has a charge and drag coefficient independent of temperature. In this case, we see the electrophoretic velocity is a function only of the temperature, through f. In addition, we note the rate of (global) mass accumulation in the entire volume of a channel due to electrophoretic focusing is proportional to the difference in f on the boundaries. However, for a differential volume or for linear $f(T(x))$, the focusing is proportional to the gradient of f.

TGF has been implemented in a variety of microfluidic formats, including channels imprinted in polymer substrates and embedded capillaries [56,62]. In most cases, the temperature gradient is imposed by mounting the channel across temperature-regulated blocks, such that the temperature gradient is established via thermal conduction in the gap between the blocks. Figure 38.11 shows

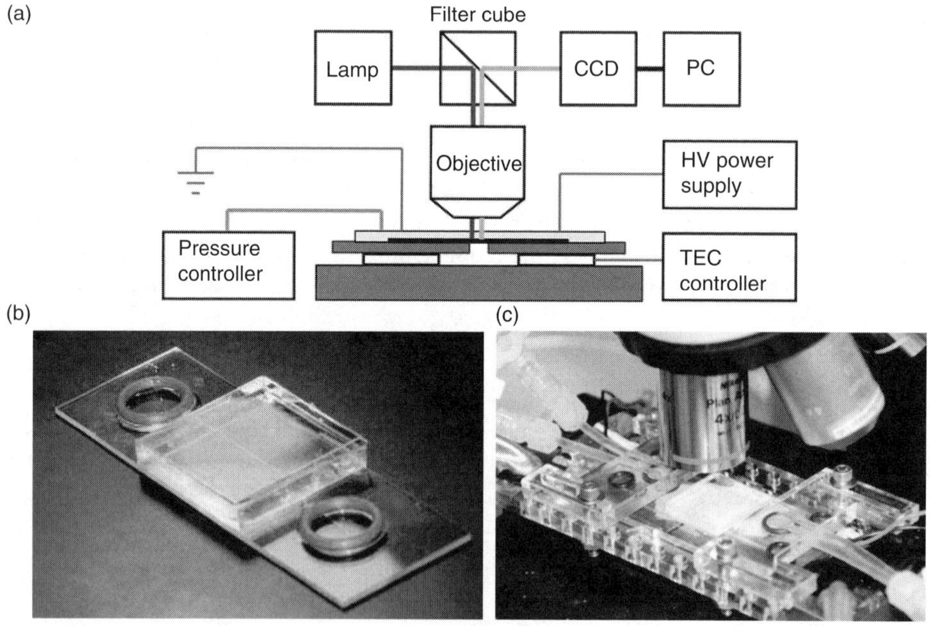

FIGURE 38.11 Control schematic and images of TGF fixture and capillary assembly. The TGF assembly (a) provides the thermal, fluidic, and electrical interface to the microchannel. A temperature gradient is established across a gap between two copper plates, each heated or cooled by a thermoelectric (Peltier) device. Pressure control is accomplished by adjusting the relative heights of two external reservoirs. The capillary assembly (b) consists of a 20 by 200 μm rectangular glass borosilicate capillary that spans the distance between two O-ring reservoirs and underneath an insulating PDMS block. In the assembly photo (c), the capillary assembly is mounted on the TGF fixture. The encapsulating PDMS block is located directly below the objective, and the fluidic manifolds are to the left and right of the block [62].

one such implementation. Alternatively, Joule heating has also been used to form the temperature gradient independent of external fixtures [56,63].

38.4.2.3 Dispersion Theory

To analyze dispersion in TGF, we must modify our previous derivation to account for changes due to the temperature gradient. This includes an axially varying electrophoretic velocity and diffusivity. In its most general form, the transport equation becomes

$$\frac{\partial c}{\partial t} + \mathbf{u}_{\text{bulk}} \cdot \nabla c + \nabla \cdot \left(\mathbf{u}_{\text{eph}} c\right) = \nabla \cdot \nabla \left(Dc\right), \tag{38.34}$$

where c is the concentration of the sample analyte and D is the analyte's molecular diffusivity. Note the placement of the electrophoretic velocity and diffusivity within the gradient operators. In particular, the placement of the diffusivity within the second gradient operator reflects the use of the Fokker–Planck diffusivity law, $\mathbf{J} = -\nabla(Dc)$. Although it is common practice to use Fick's law for the diffusive flux (yielding the traditional $\nabla \cdot D\nabla c$), Fick's law strictly applies only to diffusion with homogeneous D [64]. To recover the more familiar diffusion representation, we differentiate, yielding the terms, $D\nabla c$ and $c\nabla D$. The latter term thus represents a flux due to a diffusivity-induced velocity, ∇D, which is typically small in comparison to the electrophoretic velocity.

For a purely axial temperature gradient, if the cross section mean decomposition is again performed, a new mean concentration transport equation is derived

$$\frac{\partial \langle c \rangle}{\partial t} + \langle u_{\text{bulk}} \rangle \frac{\partial \langle c \rangle}{\partial x} = \frac{\partial}{\partial x}(\langle u_{\text{eph}} \rangle \langle c \rangle) = \frac{\partial}{\partial x}\left(D\frac{\partial \langle c \rangle}{\partial x} - \langle u'_{\text{bulk}} c' \rangle\right). \tag{38.35}$$

The deviation transport equation again reduces to a form analogous to Equation 38.11 following our scaling arguments, although with one exception, the characteristic time scale for TGF is no longer the observation time, but is instead the focusing time for a nondiffusive particle approaching its focal point, τ_{foc}. (See solution for form of τ_{foc}.) Note that if any of these assumptions are violated, then new dispersion mechanisms arise. For example, if τ_{foc} is of order a^2/D or less, then ballistic dispersion [20] can become significant. The correlation term on the right-hand side of Equation 38.35 once again corresponds to the advective dispersion arising from transverse variations in the axial velocity. To evaluate this term, we need an expression for u'.

An analytical solution for u' was derived by Huber and Santiago [62]. Using a decomposition on the Helmholtz–Smolukowski equation similar to that performed by Ross for the electrophoretic velocity, they determined the nonuniform electroosmotic slip velocity to be

$$u_{eo} = v_{eo,0} E_0 f(T) g(T), \tag{38.36}$$

where v_{eo} is the electroosmotic mobility, $g(T) \equiv v_{eo}(T)\mu(T)/v_{eo,0}\mu_0 = \varepsilon(T)\zeta(T)/\varepsilon_0\zeta_0$, ε is the permittivity, ζ is the zeta potential, and the subscript zero indicates a value at the reference temperature [16]. By extending the lubrication flow solution of Ghosal [65] to include variable viscosity and the nonuniform electroosmotic slip velocity of Equation 38.3, Huber and Santiago derived an analytical expression for the bulk velocity, from which the deviation velocity was determined:

$$u'_{\text{bulk}} = U_p \left(\frac{1}{2} - \frac{3y^2}{2a^2}\right), \tag{38.37}$$

$$U_p = \frac{a^2}{3\{\mu\}} \frac{\Delta P}{L_{\text{ch}}} + v_{eo,0} E_0 \frac{\{\mu f g\}}{\{\mu\}} - v_{eo,0} E_0 f g, \tag{38.38}$$

where a is the channel half-height, L_{ch} is the length of the channel, ΔP the applied pressure difference, and the curved brackets indicate an axial mean over the length of the channel. We see the flow field is a superposition of a uniform electroosmotic component and a parabolic (in y) pressure-driven component (for flow between parallel plates). The two are linked through continuity, so as one decreases, the other must increase. Thus, U_p, the pressure-driven flow component, contains both the externally applied pressure gradient and the internally generated pressure gradient, which results from the local slip velocity deviating from the axial average. Note that the axially averaged terms are uniform and constant, and f and g are functions of T; therefore, the velocity profile varies in x in response to axial temperature changes.

We analyze a flat-plate geometry again for simplicity and as an approximation to the full focusing problem. This implies, we are interested in characteristic focusing times much longer than that for molecular diffusion along the depth but not for spanwise diffusion along the width of the channel. This implies that our analysis is an estimate of the maximum concentrations achieved over these relatively short focusing times (along the centerline of the channel) and does not account for the finite amount of dispersion introduced by spanwise velocity gradients. Improvements on this model are discussed in the following text.

Given u' and c', the advective dispersion term may be determined, which yields

$$\frac{\partial \langle c\rangle}{\partial t} + \langle u_{\text{bulk}}\rangle \cdot \frac{\partial \langle c\rangle}{\partial x} + \frac{\partial}{\partial x}(\langle u_{\text{eph}}\rangle \langle c\rangle) = \frac{\partial}{\partial x}\left(D_{\text{eff}}\frac{\partial \langle c\rangle}{\partial x}\right), \tag{38.39}$$

$$D_{\text{eff}} = D\left(1 + \frac{2}{105}\frac{U_p^2 a^2}{D^2}\right). \tag{38.40}$$

This 1-D convection–diffusion equation has features of the Taylor–Aris dispersion equation. However, in contrast to Taylor–Aris, here D_{eff} is a function of temperature and the axial coordinate, as it depends on both D and U_p. When f is a linear function of the axial dimension, Equation 38.39 can be solved in closed form subject to the form of D_{eff}. If D_{eff} is uniform, the solution is a Gaussian with peak variance $\sigma^2 = 2D_{\text{eff}}\tau_{\text{foc}}$, where $\tau_{\text{foc}} = 1/2E_0 v_0 |\mathrm{d}f/\mathrm{d}x|$. If D_{eff} is also a linear function of x, the solution is

$$c = c_0 \exp\left\{-\frac{x/L - \ln(1 + x/L)}{\sigma^2/L^2}\right\}, \tag{38.41}$$

where c_0 is the peak height and $L = D_{\text{eff}}(0)/(\mathrm{d}D_{\text{eff}}/\mathrm{d}x)$ [62]. Although unusual in form, this solution produces Gaussian-like peaks with a skew that grows with the slope of D_{eff}. For more general cases, the equation can be solved numerically.

Figure 38.12 shows sample full-field fluorescence images of focused Bodipy proprionic acid in an applied temperature gradient of 10°C/mm and electric fields from 1 to 215 V/mm. The two cases illustrate different dispersion regimes. The images on the left show focusing in the molecular diffusion dominated regime. Note the lack of spanwise curvature in the peaks and the near inverse square root dependence of peak width on electric field (dotted line). The images on the right-hand side show focusing experiments where the Taylor dispersion criteria are violated about $E_0 = 40$ V/mm and there is enhanced advective (or ballistic) dispersion [20]. The solid line is a theoretical prediction for the peak width based on an heuristic ballistic dispersion model [66]. In both cases the direction of electrophoretic flux is left to right, while the bulk flow is right to left, driven by electroosmosis. The experimental and theoretical peak widths agree closely with some deviation at low fields, where the focused peak remains slightly "over-focused" as field was decreased, having had insufficient time to diffuse out to full width.

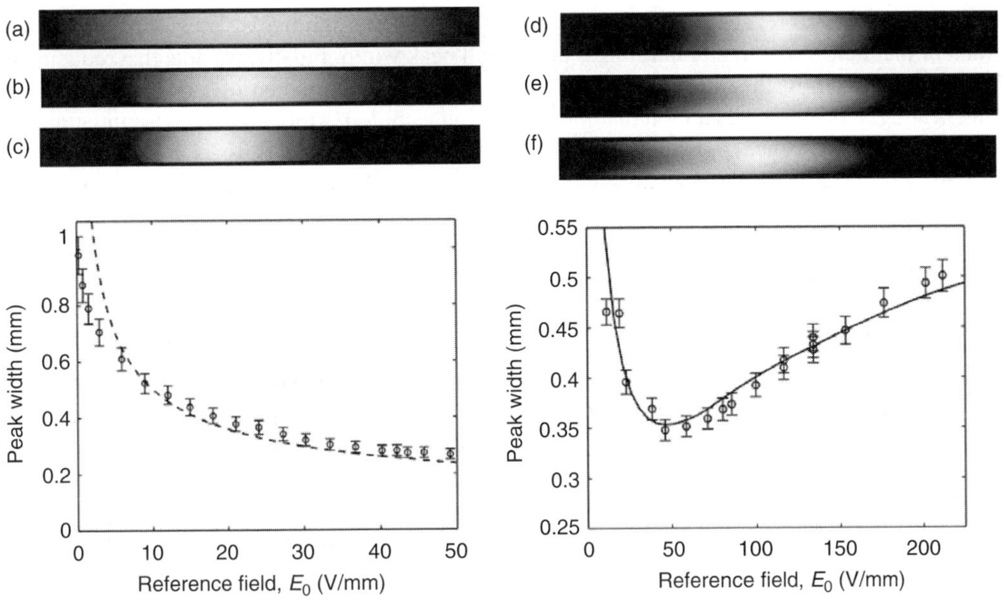

FIGURE 38.12 Bodipy peak images and width as function of reference field. The images show focused Bodipy dye within a 20 × 200 μm rectangular capillary, while the bottom figures plot the peak widths versus electric field as determined by the fitting of Gaussian profiles to each peak. The applied temperature gradient was 10°C/mm and the applied electric fields ranged from 1 to 215 V/mm. The left hand side shows results for focusing in the molecular diffusion dominated regime. The current-normalized fields, E_0, were (a) −1.5, (b) −15, and (c) −30 V/mm. The right hand gives results which violate the Taylor scaling arguments above 40 V/mm and thus feature enhanced ballistic dispersion. (Note the curvature in the images of the peaks.) The fields were (d) −38, (e) −81, and (f) −212 V/mm [62,66].

38.4.2.4 Performance and Guidelines

As we have described, TGF has successfully demonstrated the simultaneous concentration and separation of a wide range of species in a variety of implementations. Key figures of merit for TGF as a separation modality are the concentration factor and peak capacity. Static TGF, where the externally applied pressure remains constant, has demonstrated concentration factors in excess of 20,000 but is limited to peak capacities <10 [56]. As a result, there has been recent work to develop a dynamic form of TGF called scanning TGF [67], where the externally applied pressure is varied with time. This technique allows higher peak capacity and tunable resolution and concentration by adjusting the rate at which the procedure scans through the applied pressures.

In implementation, the complexities of TGF extend beyond that due to advective dispersion and encompass system design details such as device geometry, temperature control, and buffer selection. In static TGF, larger temperature gradients produce faster focusing and sharper peaks but reduce resolution [56]. The latter may be addressable using scanning TGF, but particular attention must be paid to the temperatures within the system because analytes of interest (e.g., enzymes) may be extremely sensitive to temperature and Joule heating may adversely influence the temperature gradients. In order to produce desired temperature profiles, designers must consider conductive and convective heat transfer within the channel, Joule heating due to the electric field, and heat transfer from the channel to its fixture and the environment. While it is possible to model these effects, in most cases, empirical studies will be required to validate models and refine optimizations.

Optimization of TGF preconcentration and (static) resolution is similar to other sample preconcentration methods. As described earlier, the contribution of molecular diffusion to the total peak variance scales as $1/E$, while the Taylor dispersion coefficient scales as U_p^2 and therefore, E^2. Sample peak variance, therefore, scales as $1/E$ at low fields and as E at high fields. Thus, there will (again) be an optimal field strength that minimizes the overall peak width. Finally, we note that reducing the characteristic width of the channel simultaneously decreases the amount of Taylor dispersion and Joule heating. However, we caution that most detection techniques rely on depthwise integration of a signal (e.g., fluorescence), which then scales as the channel height, h. Thus, TGF system designers must also optimize the channel width subject to their desired detection limits.

38.4.3 Isotachophoresis

38.4.3.1 Introduction

Isotachophoresis [68] is an electrophoretic preconcentration and separation technique that utilizes a heterogenous buffer system of disparate electrophoretic mobilities. Typically, a plug containing sample ions to be focused and separated is introduced between the leading and trailing electrolyte (LE and TE) whose mobilities are respectively higher and lower than any of the mobilities of sample ions. Under the influence of an electric field, the sample ions separate and redistribute themselves in contiguous zones in order of reducing mobility starting from LE to TE (each is focused into its respective, mobility-dependent focal point). At steady state, these focused sample zones migrate at a same speed as the leading zone, hence the name "iso-tacho-phoresis."

ITP, also known as displacement electrophoresis, was first performed in capillary tubes by Everaerts et al. [69] for the separation of strong anions using a thermocouple detector. Since then, ITP has been used for the analysis of various important chemical and biological species such as amino acids [70], peptides [71], nucleotides [72], proteins [73,74], heavy metal ions [75], and other organic/inorganic ions [76,77] on a variety of detection platforms such as UV absorbance, conductivity and fluorescence detection. Over the past 15 years, ITP has been used as a preconcentration technique in conjunction with CE [78]. This mode of ITP, referred to as transient isotachophoresis (tITP), has been implemented on microchip platform in the recent years to achieve improved sensitivity [22,79,80].

38.4.3.2 Basic Theory and Implementation

ITP leverages differences in the mobility of sample ions to create a segregation of species on the basis of mobility. ITP separation results in a system of contiguous sample zones, sandwiched between leading and trailing electrolytes, migrating at identical speeds. The inherent preconcentration effect of ITP maintains sharp concentration boundaries between adjacent sample zones. In the absence of dispersion due to (radially) nonuniform bulk flow, the thickness of these boundaries is governed by the balance of electromigration and diffusion flux, as we shall discuss later. Away from this diffused boundary region, the sample concentration is uniform and can be obtained in each zone using species conservation equations and the electroneutrality condition [81]. Consider a simple model ITP system, shown in Figure 38.13, consisting of a plug of sample ions (X_i) injected between the leading electrolyte (LE) and trailing electrolyte (TE), and common counterion (A) present everywhere. For now, we assume EOF is fully suppressed to negligible levels. After sufficient focusing time, the various zones in the system will develop locally uniform concentrations (concentration "plateaus" around their respective focal points) where diffusive fluxes are locally insignificant. For this long-time condition, the species conservation equations can be simplified to obtain the well-known Kohlrausch regulating function (KRF) [82] given by

$$\left(\sum_i \frac{C_i}{\mu_{\mathrm{eph},i}}\right)_x = f(x), \tag{38.42}$$

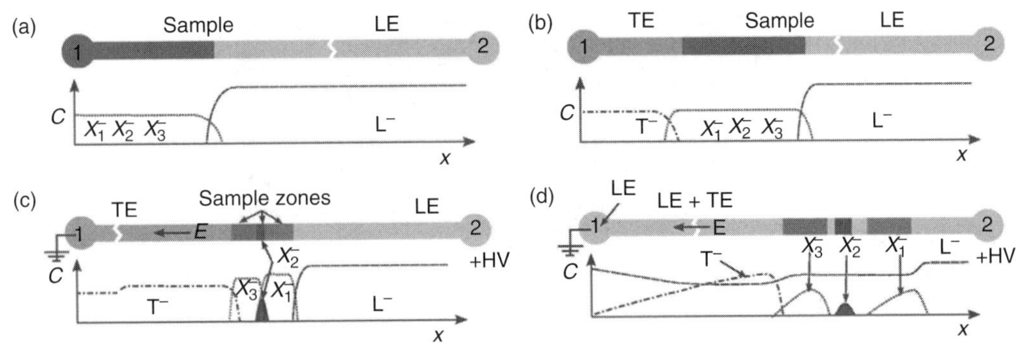

FIGURE 38.13 A schematic of the steps involved in a typical anionic ITP experiment is shown. (a) First, the capillary/microchannel is filled with the leading electrolyte using pressure driven flow. Sample electrolyte is introduced in well 1 and the capillary/channel is partially filled with this sample electrolyte by applying vacuum on well 2. (b) Well 1 is emptied and filled with the trailing electrolyte which is then drawn into the capillary/channel. (c) Next, high voltage is applied across the capillary and an electric field is setup inside the capillary/channel initiating separation of sample zones and achieving ITP condition. Sample ions X_1 and X_3 are present in high initial concentration and hence their zones form plateau shaped peaks and achieve the concentration as required by KRF. X_2 appears as a peak in the diffused interface between X_1 and X_3 zones and has a much lower concentration than predicted by KRF. (d) Finally, well 1 is emptied and is filled with LE. The LE ions overspeed the TE and sample ions and "break" the ITP mode into CE mode. The sample peaks separate in the CE mode and also disperse due to electromigration dispersion.

where C_i and $\mu_{\text{eph},i}$ are the concentration and the mobility of species i, at axial location x in the channel. The constant (in time) function $f(x)$ is governed by the initial condition. This relation can be stated as follows: The sum of concentration-to-mobility ratio of all species at a given location in the channel (relative to the channel wall at negligible electroosmosis conditions) remains invariant with time.

Using the KRF, we can arrive at the following expression to obtain the concentration adjustment of an ITP zone assuming no dispersions and bulk fluid velocity [83]:

$$C_X = C_L \frac{\mu_X}{\mu_L} \frac{\mu_A + |\mu_L|}{\mu_A + |\mu_X|}, \tag{38.43}$$

where the adjusted sample concentration, C_X, is strictly a function of the LE concentration and electrophoretic mobilities in the nondiffuse one-dimensional model. Note that KRF may be used to estimate sample concentration in an ITP zone only if the species are strong ions (fully ionized). Further, the species are present in sufficiently high initial concentration to form concentrated regions with finite-width, but locally uniform concentrations (a plateau in analyte peaks, e.g., samples X_1 and X_3 in Figure 38.13c).

For low initial concentrations (or short preconcentration times), the ITP zone width can be on the order of the diffusion width between zones (e.g., sample X_2 in Figure 38.13c). In that case, as pointed out by Svoboda [84], analyte width is a function of both initial concentration and injected plug length (see also the quantitative data of Jung et al. [22]). This regime of ITP has been called the spiked mode [85] and here, the analyte zone appears as a spike between adjacent zones rather than a plateau. One often encounters this spiked mode ITP in trace analyte detection and separation. In this regime, Taylor dispersion is a critical factor in determining the maximum concentration enhancement achieved. As a result, the sample peak widths (in spike mode) and interfaces between ITP zones can often be greater than peak widths predicted by diffusion alone. In our work, this factor is also reflected in the rate of growth of sample peaks. Despite this, there has been very little work in the analysis and modeling of Taylor dispersion in ITP.

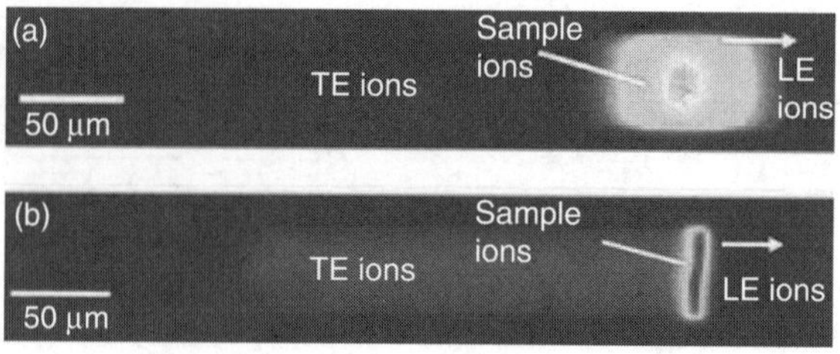

FIGURE 38.14 **(See color insert following page 810.)** CCD camera images of on-chip sample peaks of AlexaFluor 488 at the LE/TE interface in two different ITP experiments. In (a) there is finite (nonuniform) EOF and the sample peak streamwise dimension is on the order of channel width or larger. In (b) EOF is suppressed and the sample is concentrated in narrower zone (~5 μm) at relatively high electric field. While Taylor dispersion based analysis is probably applicable in the first case, more comprehensive modeling is required for case (b).

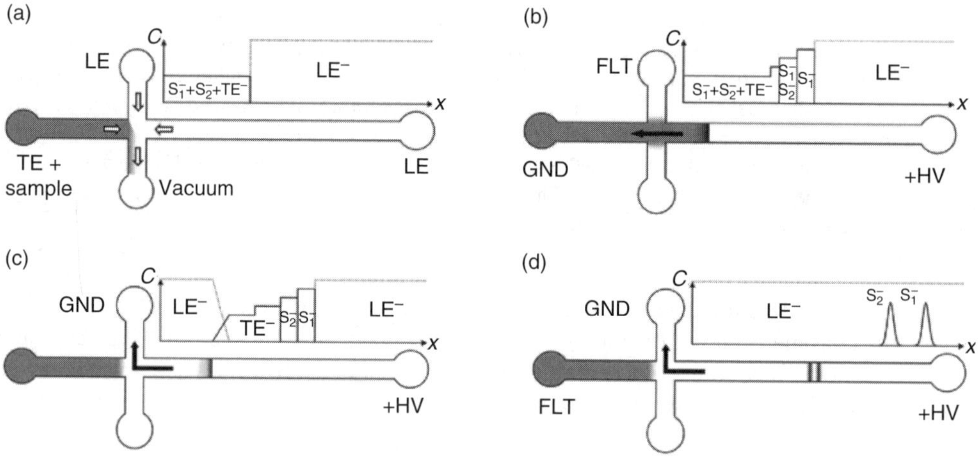

FIGURE 38.15 Schematic of ITP/CE assay protocol in a microchip. Configurations of co-ions are also shown at each step. (a) The north and the south reservoirs are filled LE, and the west reservoir is filled with a mixture of TE and sample. TE/LE boundary is formed by applying vacuum at the south reservoir. White arrows show the direction of pressure-driven flows. (b) ITP preconcentration is initiated by applying high voltage and ground at the east and west reservoirs, respectively. The black arrow denotes the direction of electric field. Sample anions electromigrate toward the anode as EOF is suppressed. The early stage of ITP preconcentration results in a partial separation (i.e., moving boundary electrophoresis). (c) The field is switched toward the north reservoir to inject LE ions behind the sample and initiate CE. ITP preconcentration continues until LE ions overtake the TE and sample ions. (d) Separation of samples occurs further downstream where sample ions electromigrate in nearly homogeneous LE electrolyte (remnant of TE not shown).

Example visualizations of ITP-focused sample peaks in a 50 μm wide by 20 μm deep (isotropically etched) microchannel are shown in Figure 38.14. The top image shows a typical ITP sample peak in a channel with finite EOF and Taylor dispersion. Interface lengths are on the order of or larger than the characteristic channel cross section dimensions. In this experiment, the ITP stacking occurs over a long duration (~5 min) such that the peak has a substantial axial dimension compared to the channel width and has clearly dispersed edges. The area-averaging dispersion analysis and scaling presented earlier should apply here. Figure 38.14b shows early stage of an ITP plug created on-chip with high-quality EOF suppression (using poly-*N*-hydroxyethylacrylamide [PHEA] coating on a borosilicate glass wall [22]), high electric field (~1000 V/cm in the TE), and high LE concentration (~1 M). The injection protocol used here is shown in Figure 38.15a,b. Here, the sample peak is a narrow concentration "shock wave" with extremely high electric field and concentration gradients. Accurate prediction of the focusing dynamics of such a peak will probably require fairly comprehensive two- and three-dimensional models.

The coupling of ITP with CE is shown in Figures 38.13d and 38.15c–d. This coupling is referred to as transient ITP (tITP). tITP is typically achieved by replacing the TE ions with LE ions, which interrupts the ITP process. After sufficient sample preconcentration via ITP (as in Figure 38.13c), LE ions are introduced behind the contiguous sample zones. This is accomplished by either manually exchanging the buffer in well 1 (e.g., with a hand pipette, as shown in Figure 38.13c) or by using a side channel of a microchip to inject LE ions electrokinetically behind the train of sample zones (Figure 38.15c,d) [22,86]. The LE ions enter the sample zones faster. Sample zone concentrations reduce in order to satisfy the KRF-type regulation, and sample plug typically forms long tails as shown. This form of dispersion is termed as electromigration dispersion (EMD), and has been extensively investigated [87–89]. Once the ITP process is interrupted, the samples separate via standard mobility-based CE separation.

38.4.3.3 Dispersion Theory

The isotachophoretic boundary between two adjacent zones, under constant current condition and in the absence of bulk flow, assumes a constant width governed by the balance of electromigration and dispersion fluxes. For negligible electroosmosis (and negligible Taylor dispersion), the dispersion is determined by diffusion alone. Analytical solution to the concentration of the species in this diffused boundary, for a three-component fully ionized system, has been presented by Saville et al. [90]. The characteristic length-scale, δ, of the ITP boundary in this case is given by

$$\delta = \frac{\mu_L \mu_T}{\mu_L - \mu_T} \frac{C_L}{j} \left(\frac{k_B T}{e} \right), \tag{38.44}$$

where μ_L and μ_T are electrophoretic mobilities of LE and TE, respectively, C_L is the concentration of leading ion, j is the current density, and $k_B T/e$ is the thermal voltage. For negligible Taylor dispersion (e.g., low fields), peak axial dimension scales as D/E (since diffusivity is proportional to mobility by the Nernst–Einstein relation). Typical concentration profiles of LE, TE, and counterion in an ITP boundary (with molecular diffusion alone contributing to dispersion) are shown in Figure 38.16.

The case of nonsuppressed EOF conditions is much more complex. For even trace EOF mobilities, there will be a mismatch in electroosmotic mobility and electric field in the adjacent zones. This results in a mismatch in the electroosmotic velocity and hence generation of internal pressure gradients as discussed earlier. These pressure gradients result in increased width of the interface due to dispersion and have detrimental effects on preconcentration of trace samples via ITP.

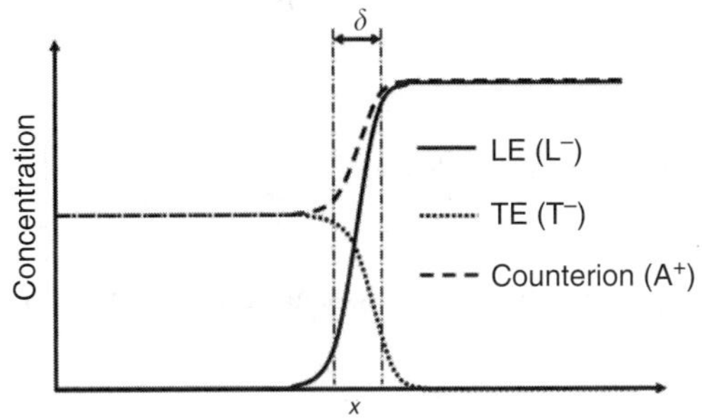

FIGURE 38.16 Schematic of the distribution of ions in the diffused interface between leading and trailing electrolyte in an ITP system. δ is the characteristic length scale of this diffused boundary, obtained from the balance of electromigration and diffusion fluxes.

Saville [91] investigates the effect of electroosmosis on the interface between two ITP zones. To our knowledge, this is the only published Taylor dispersion analysis for ITP. Saville's analysis is for a simplified problem. He neglects the influence of electrolyte concentration and pH on electroosmotic mobility. Similar to the analyses presented earlier, a purely axial flow velocity is split into the bulk area average velocity $\langle u \rangle$ and a deviation velocity, $u'_p = u - \langle u \rangle$. The pressure-driven flow velocity component due to a mismatch in electroosmotic slip velocity (alone) is $u'_p = 2(\mu_{\text{eof}} E - \{\mu_{\text{eof}} E\})(1 - 2r^2/a^2)$. While simplifying approximations, Saville assumes μ_{eof} is uniform, and the local electric field governing EOF (but not the electric field governing electrophoresis) is approximately uniform in x. For example, assuming the TE voltage drop is much larger than that of the LE; and assuming solute plug is approximately halfway along the length of the channel, our expression reduces to $\mu_{\text{eof}} E_{\text{TE}}(x)(1 - 2r^2/a^2)$. Saville writes this as

$$u'_p = \mu_{\text{eof}} E_0 (1 - 2r^2/a^2), \tag{38.45}$$

where E_0 is assumed constant and uniform. Saville further expands μ_{eof} as $-\varepsilon\zeta E/\eta$ (where ε is the permittivity, μ is the viscosity of the liquid, and ζ is the zeta potential) but we shall retain the more compact form here. In a frame of reference moving at $\langle u_{\text{bulk}} \rangle = U_{\text{EOF}} + U_p$, the species conservation equation for any species i can be expressed as

$$\frac{\partial C_i}{\partial t} + u'_p \frac{\partial C_i}{\partial x} = \frac{\partial}{\partial x}\left[-\mu_{\text{eph},i} C_i E_x(x,r) + D_i \frac{\partial C_i}{\partial x}\right] + \frac{1}{r}\frac{\partial}{\partial r} r \left[-\mu_{\text{eph},i} C_i E_r(x,r) + D_i \frac{\partial C_i}{\partial r}\right], \tag{38.46}$$

where $\mu_{\text{eph},i}$ is the electrophoretic mobility (m^2/V/s), E is the electric field (V/m), and C_i is the concentration of the species i. The species relations are constrained by the electroneutrality approximation expressed as

$$\sum_i z_i C_i = 0. \tag{38.47}$$

Next, the conservation equations are transformed to the frame of reference moving with the interface at the speed U_{ITP} ($U_{\text{ITP}} = \mu_{\text{L}} E_{\text{LE}}$, where E_{LE} is the electric field strength in the LE region) and subsequently nondimensionalized to obtain the following dimensionless equations:

$$\begin{aligned} &-C_i^* + Pe(1 - 2r^{*2})\frac{\partial C_i^*}{\partial x^*} \\ &= \frac{\partial}{\partial x^*}\left[-\mu^*_{\text{eph},i} C_i^* E_x^* + \mu^*_{\text{eph},i}\frac{\partial C_i^*}{\partial x^*}\right] + \lambda^2 \frac{1}{r^*}\frac{\partial}{\partial r^*} r^* \left[-\mu^*_{\text{eph},i} C_i^* E_r^* + \mu^*_{\text{eph},i}\frac{\partial C_i^*}{\partial r^*}\right] \end{aligned} \tag{38.48}$$

and

$$\sum_i z_i C_i^* = 0. \tag{38.49}$$

In arriving at the previous equation, the following scaling has been used:

$$C_i^* = C_i/C_A, \quad \mu^*_{\text{eph},i} = \mu_{\text{eph},i}/\mu_{\text{eph},A}, \quad x^* = x/l, \quad r^* = r/a, \quad E_x^* = E_x \mu_{\text{eph},A} l/D_A$$

and

$$E_r^* = E_r \mu_{\text{eph},A} a/D_A.$$

Here, the following dimensionless parameters appear: Peclet number $Pe = \mu_{\text{eof}} E_0 l / D_A$ and the aspect ratio $\lambda = l^2/a^2$, where l is the electrical length scale obtained by balancing diffusion and electromigration flux, $l = D_A/U_{\text{ITP}}$.

The typical value of Peclet number is ~10 indicating that convection plays an important role in determining the shape of the interface. By performing asymptotic expansion on the concentration and potential terms in the species conservation equation and area-averaging the equation over the cross section of the capillary, Saville obtains the following:

$$-\frac{\partial \langle C_i^* \rangle}{\partial x^*} = \frac{\partial}{\partial x^*}[-\mu^*_{\text{eph},i}\langle C_i^* \rangle \langle E_x^* \rangle] + \left(\mu^*_{\text{eph},i} + \frac{(Pe)^2}{48\mu^*_{\text{eph},i}\lambda^2}\right)\frac{\partial^2 \langle C_i^* \rangle}{\partial x^{*2}}. \quad (38.50)$$

Reverting back to dimensional form, we write

$$-U_{\text{ITP}}\frac{\partial \langle C_i \rangle}{\partial x} = \frac{\partial}{\partial x}[-\mu_{\text{eph},i}\langle C_i \rangle \langle E_x \rangle] + \left(D_i + \frac{(\mu_{\text{eof}} E_0)^2 a^2}{48 D_i}\right)\frac{\partial^2 \langle C_i \rangle}{\partial x^2}. \quad (38.51)$$

Hence, we arrive at a expression similar to the one derived in Taylor dispersion using area-averaging section, with

$$D_{\text{eff},i} = \left(D_i + \frac{(\mu_{\text{eof}} E_0)^2 a^2}{48 D_i}\right).$$

Note that Equation 38.49 follows from our more general Equation 38.16 (e.g., transform the equation to the U_{ITP} frame of reference and assume a uniform constant value of D_{eff}). Saville presents numerical solutions of his model for the case where $\lambda = 1$, $\mu^*_{\text{eph,LE}} = 1$, $\mu^*_{\text{eph,TE}} = 0.5$, and $Pe = 10$. The model shows interface width (e.g., the interface between two adjacent species) as a strong function of Pe (and E_0). In the conditions Saville modeled, Taylor dispersion increases the effective dispersion coefficient up to ~18-fold over diffusion alone.

Despite the dearth of work on ITP Taylor dispersion, there is fairly clear experimental evidence that Taylor dispersion often limits maximum achievable concentration increase. Next, we present experimental studies of Taylor dispersion in ITP and present an empirical optimization of ITP that minimizes dispersion to maximize preconcentration factor. We studied the effects of Taylor dispersion on the width of single interface ITP systems shown schematically in Figure 38.15a-b, under constant (in time) current conditions. In a typical ITP experiment, the LE contains high concentration of Cl^- (100 mM to 1 M), the TE contains $HEPES^-$ (5–100 mM) and the sample ions (Alexa Fluor 488) in trace concentration (100 aM to 1 nM). We setup an interface between the LE and TE using pressure-driven flow, and ITP is initiated by applying high voltage (~3 kV) across this interface. Under the influence of the applied electric field, sample ions overspeed TE ions in the TE zone and stack at the LE/TE interface. Figure 38.17 shows a typical result for the width (along the axial direction) of the focused sample plug as it propagates down the channel toward the anode. Data are shown for four values of applied current in this 50 μm wide by 20 μm deep borosilicate glass microchannel (Mycralyne Inc.). Here, EOF was suppressed by adding polyvinyl pyrrolidone (PVP) (0.2% w/w) to the LE and TE. In the absence of Taylor dispersion (but including the effects of diffusion), we expect the sample plug width to remain constant. However, the sample width clearly grows nearly linearly in space. This behavior of sample plug width is observed even in ITP experiments with high degree of EOF suppression. We hypothesize that this type of growth is due to a mismatch in the EOF between the LE and TE region, which induces Taylor dispersion. This experimental evidence suggests that the interface dimension in ITP is often controlled by Taylor dispersion.

Analytical and even numerical computations of ITP processes including diffusion and Taylor dispersion are difficult. However, the analyses presented in this chapter can be used as a guide for

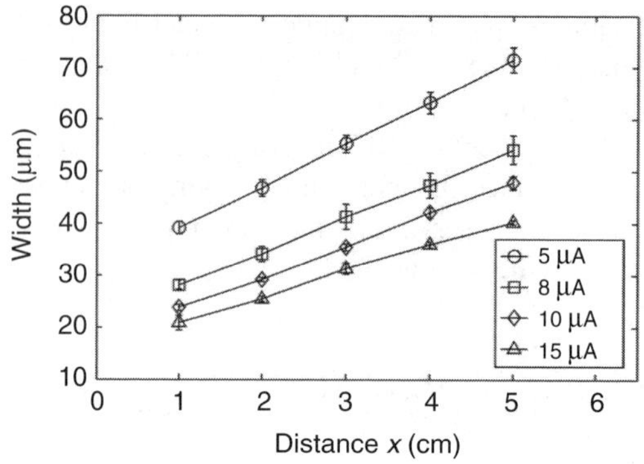

FIGURE 38.17 Plot of the width of the sample plug (10 nM Alexa Fluor 488) focused at the interface of LE (750 mM Tris–Hcl) and TE (25 mM Tris-Hepes) at various locations downstream in a 50 μm wide (20 μm deep) microchannel. Each data set represents five realizations of a constant current ITP experiment.

empirical optimizations of the ITP process. Key parameters influencing the sample preconcentration achieved are the concentrations of LE, TE, and the initial sample concentration; the local applied electric field; the initial conductivity gradient of the initial condition, which partly determines time to reach steady state; and the degree to which EOF is suppressed.

Jung et al. [22] performed a systematic variation of LE, TE, and sample concentration and measured the effect of these variations on concentration increase, CI (CI = $C_{\text{sample,final}}/C_{\text{sample}}$). In each case, we ensure that the ITP zones reached a fully preconcentrated state (i.e., maximum focusing) by verifying the transients of the preconcentration procedure using full-field imaging at low magnifications. In all cases, ITP zones reached a quasi-steady concentration, but we emphasize this does not imply that the KRF analysis is valid (since plateau-shaped peaks were not achieved [84]). We suppressed EOF (to minimize Taylor dispersion) by adding 0.1% (w/w) poly-*N*-hydroxyethylacrylamide (PHEA) to all electrolytes to suppress EOF. We tried several suppression strategies including poly-ethyPEO coating and PVP, but PHEA yielded the lowest electroosmotic mobility (as verified by current monitoring measurements [23]). For convenience in working with relatively high voltages (up to 3 kV), we used constant voltage control in this optimization of preconcentration.

Effects of LE (NaCl) concentration, C_{LE}, TE (HEPES) concentration, C_{TE} and initial concentration of sample (Alexa Fluor 488), $C_{\text{S,initial}}$ on sample preconcentration are summarized in Figure 38.18. C_{LE} was varied from 10 mM to 1 M to study its effect on maximum focused sample concentration, $C_{\text{S,final}}$ and concentration increase, CI (Figure 38.18a). The 5 mM HEPES TE solution contained 1 nM Alexa Fluor 488 as a sample. The focused sample concentration is nearly directly proportional to the concentration of LE, as expected from a one-dimensional nondispersive model (i.e., KRF theory). However, the nondispersive model drastically underpredicts the proportionality constant; the measured focused sample concentrations are 3500- to 7900-fold less than that predicted by Equation 38.43 despite all cases reaching fully-focused state. This gross difference between KRF theory and experiments is because the sample is in a "smeared" region of locally varying conductivity and electric field, as dictated by the effects of diffusion and Taylor dispersion.

Figure 38.18a also shows the effect of C_{TE} on ITP preconcentration, where C_{TE} were varied from 1 to 100 mM. The LE was fixed at 1 M NaCl, and $C_{\text{S,initial}}$ was fixed at 1 nM. The KRF model suggests that focused sample concentration is not a function of $C_{\text{S,initial}}$ or C_{TE}. However, the measurements show that $C_{\text{S,final}}$ increased for lower C_{TE} (i.e., as conductivity ratio increases). High LE-to-TE conductivity ratios (associated with low TE concentrations) increase the electric fields in the TE and focused sample zones. High electric field leads to fast-focusing dynamics and high electric Peclet numbers ($U_{\text{eph}}a/D$, where a is the characteristic channel scale) and thereby, high $C_{\text{S,final}}$, as the preconcentration process is less susceptible to dispersion. Higher electric fields would

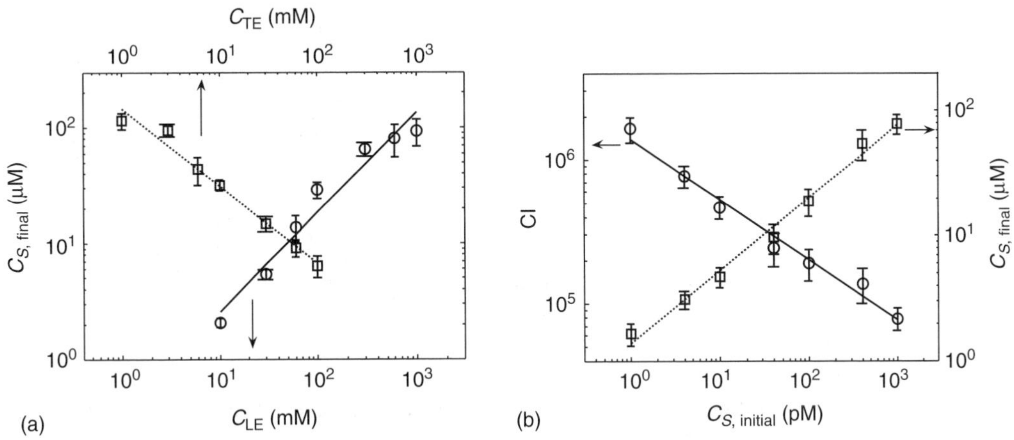

FIGURE 38.18 Parametric variations of initial concentration profile. The nominal applied field was 220 V/cm. CCD viewing area was centered 30 mm downstream of the channel intersection. (a) Maximum focused sample concentration, $C_{S,\text{final}}$, versus LE concentration, C_{LE} and TE concentration, C_{TE}. For variation of C_{LE}, sample analyte and TE were respectively 1 nM alexa fluor 488 and 5 mM HEPES. The regression coefficient, R^2, is 0.95. For variation of C_{TE}, sample analyte and LE were respectively 1 nM alexa fluor 488 and 1 M NaCl. $R^2 = 0.97$. (b) CI and $C_{S,\text{final}}$ versus initial sample concentration, $C_{S,\text{initial}}$. TE and LE were 5 mM HEPES and 1 M NaCl, respectively. The regression coefficients for CI and $C_{S,\text{final}}$ are respectively 0.97 and 0.98.

eventually lower maximum focusing due to the effects of dispersion (e.g., due to Taylor dispersion or Joule heating or both).

Next, initial sample concentrations, $C_{S,\text{initial}}$ were varied from 1 pM to 1 nM as shown in Figure 38.18b. The LE and TE were, respectively, fixed at 1 M NaCl and 5 mM HEPES. The LE-to-TE conductivity ratio was kept constant at 1.3×10^3 in an effort to decouple the dependence of the TE zone electric field on this conductivity ratio. The data shows CI increases as $C_{S,\text{initial}}$ decreases. This trend is roughly consistent with the KRF model. However, the dependence of CI on $C_{S,\text{initial}}$ is weaker than the inversely proportional dependence predicted by the simple model (CI changes just under two orders of magnitude while $C_{S,\text{initial}}$ changes three orders of magnitude). This discrepancy is also apparent in the measurements of the maximum concentration, $C_{S,\text{final}}$. The experimental data show $C_{S,\text{final}}$ as a linear function of (although not directly proportional to) $C_{S,\text{initial}}$, which is not attributable to changes in local field in the TE zone. We again hypothesize that this discrepancy between the simple model and observations is due directly to the effects of dispersion. Dispersed interfaces of finite width cause a final sample concentration to be a function of initial sample concentration.

38.4.3.4 Performance and Guidelines

These experimental parametric studies yield important insight into key ITP stacking parameters and suggest strategies for optimizing ITP in practice. Suppression of EOF to minimize Taylor dispersion is a key component. Also important are high LE concentration and low initial sample concentration to maximize achievable concentration increase, and the implementation of a single-column ITP configuration (where initially, there is a single interface between the LE and the TE/sample mixture) to inject a large effective sample width. As shown in Figure 38.18b, these simple strategies derived from earlier scaling arguments achieve ITP preconcentration with final-to-initial sample concentration ratios exceeding one million.

An example implementation of these strategies is shown in Figure 38.19. These are results of a tITP assay (again, ITP followed by injection of LE ions on the cathode side of the TE to interrupt ITP and initiate CE), which uses the single interface ITP configuration described schematically in Figure 38.15. The figure shows two example separations of 100 aM (100 attomolar) concentrations

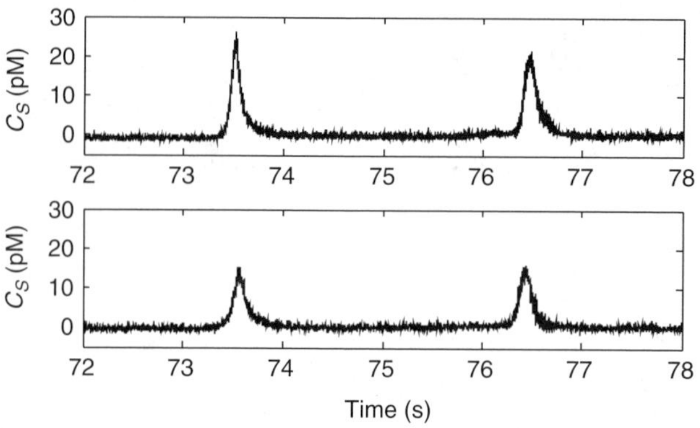

FIGURE 38.19 Two 100 aM sample electropherograms of ITP/CE separation of Alexa Fluor 488 (the peaks near 73.5 s) and Bodipy (peaks near 76.5 s). A glass microchip (microchannel cross-sectional dimensions are 50 μm wide and 20 μm deep) and 60× water immersion objective (N.A. = 0.9) were used. The detector was located 30 mm downstream of the injection region.

each of Alexa Fluor 488 and Bodipy, detected 30 mm after injection with 40 s ITP preconcentration under a nominal field of 220 V/cm. LE and TE were 600 mM NaCl and 5 mM HEPES, respectively. The concentrations measured after injection, ITP preconcentration, and separation are, respectively, 21 and 16 pM for Alexa Fluor 488 and bodipy, as averaged across five realizations. This experiment achieves a concentration increase of 2.1×10^5 fold relative to the initial sample concentration of 100 aM. This 100 aM sensitivity is, to our knowledge, the highest ever reported sensitivity for an electrophoresis experiment.

38.5 CONCLUDING REMARKS

Dispersion effects arising from nonuniform fluid motion have detrimental effects on CE sensitivity and resolution. Quantification and minimization of dispersion effects is especially important for sample stacking and focusing techniques. In preconcentration methods, EOF of heterogeneous buffer systems give rise to internal pressure gradients and strong dispersive forces. Generalized Taylor dispersion analysis is a powerful approach for quantifying these effects. More importantly, scaling, analytical solutions, and numerical solutions of Taylor dispersion provide unique insight. This can guide the choices of separation voltage, EOF suppression strategy, and channel shape, width, and depth for a given target analyte and buffer chemistry. The goal of such models and understanding is the optimization of preconcentration efficiency and resolution across a broad range of physical regimes and techniques.

Further improvements in the predictive capability of the stacking and focusing models presented here can be made by accounting for the 3D effects of typical D-shaped, wet-etched channel cross sections. More sophisticated dispersion analyses of the unsteady three-dimensional velocity field should also be carried out. This is particularly important for the high electric field regimes of TGF and ITP, where the sample plug widths become on the order of, or smaller than, the channel width. More comprehensive models should aid in the systematic design and optimization of CE separations under a wide range of conditions.

REFERENCES

1. Vlastimil Hruska, M.J., Bohuslav Gas, Simul 5—Free dynamic simulator of electrophoresis. *Electrophoresis*, 2006, 27: 984–991.

2. Taylor, G., Dispersion of soluble matter in solvent flowing slowly through a tube. *Proc. Roy. Soc. A*, 1953, 225: 186–203.
3. Bharadwaj, R., Santiago, J.G., and Mohammadi, B., Design and optimization of on-chip capillary electrophoresis. *Electrophoresis*, 2002, 23: 2729–2744.
4. Bharadwaj, R. and Santiago, J.G., Dynamics of field-amplified sample stacking. *J. Fluid Mech.*, 2005, 543: 57–92.
5. Aris, R., On the dispersion of a solute in a fluid flowing through a tube. *Proc. Roy. Soc. London, Ser. A*, 1956, 235: 67–77.
6. Dutta, D. and Leighton, D.T., Dispersion reduction in pressure-driven flow through microetched channels. *Anal. Chem.*, 2001, 73: 504–513.
7. Brenner, H., A general theory of Taylor dispersion phenomena. *PCH PhysicoChem. Hydrodyn.*, 1980, 1: 91–123.
8. Stone, H.A. and Brenner, H., Dispersion in flows with streamwise variations of mean velocity: Radial flow. *Ind. Eng. Chem. Res.*, 1999, 38: 851–854.
9. Griffiths, S.K. and Nilson, R.H., Hydrodynamic dispersion of a neutral nonreacting solute in electroosmotic flow. *Anal. Chem.*, 1999, 71: 5522–5529.
10. Pennathur, S. and Santiago, J.G., Electrokinetic transport in nanochannels. 1. Theory. *Anal. Chem.*, 2005, 77: 6772–6781.
11. Griffiths, S.K. and Nilson, R.H., Charged species transport, separation, and dispersion in nanoscale channels: Autogenous electric field-flow fractionation. *Anal. Chem.*, 2006, 78: 8134–8141.
12. Probstein, R.F., *Physicochemical Hydrodynamics: An Introduction*, 2nd ed. 1994, New York: John Wiley & Sons, p. 400.
13. Lin, H., Storey, B.D., Oddy, M.H., Chen, C.-H., and Santiago, J.G., Instability of electrokinetic microchannel flows with conductivity gradients. *Phys. Fluids*, 2003, 16: 1922–1935.
14. Chen, C.H., et al., Convective and absolute electrokinetic instability with conductivity gradients. *J. Fluid Mech.*, 2005, 524: 263–303.
15. Deen, W.M., *Analysis of Transport Phenomena*. 1998, New York: Oxford University Press.
16. Santiago, J., Electroosmotic flows in microchannels with finite inertial and pressure forces. *Anal. Chem.*, 2001, 73: 2353–2365.
17. Storey, B.D., et al., Electrokinetic instabilities in thin microchannels. *Phys. Fluids*, 2005, 17: 1922–1935.
18. Lin, H., et al., Instability of electrokinetic microchannel flows with conductivity gradients. *Phys. Fluids*, 2004, 16: 1922–1935.
19. Gas, B. and Kenndler, E., Dispersive phenomena in electromigration separation methods. *Electrophoresis*, 2000, 21: 3888–3897.
20. Ajdari, A., Bontoux, N., and Stone, H.A., Hydrodynamic dispersion in shallow microchannels: The effect of cross-sectional shape. *Anal. Chem.*, 2006, 78: 387–392.
21. Anderson, J.L. and Idol, W.K., Electroosmosis through pores with nonuniformly charged walls. *Chem. Eng. Comm.*, 1985, 38: 93–106.
22. Jung, B., Bharadwaj, R., and Santiago, J.G., On-chip million fold sample stacking using single interface isotachophoresis. *Anal. Chem.*, 2006, 78: 2319–2327.
23. Huang, X.H., Gordon, M.J., and Zare, R.N., Current-monitoring method for measuring the electroosmotic flow-rate in capillary zone electrophoresis. *Anal. Chem.*, 1988, 60: 1837–1838.
24. Santiago, J.G., Wereley, S.T., Meinhart, C.D., Beebee, D.J., and Adrian, R.J., A particle image velocimetry system for microfluidics. *Exp. Fluids*, 1998, 25: 316–319.
25. Devasenathipathy, S. and Santiago, J.G., Electrokinetic flow diagnostics, in *Micro- and Nano-Scale Diagnostic Techniques*. 2002, Ed. K.S. Breuer, New York: Springer Verlag.
26. Probstein, R.F., *Physicochemical Hydrodynamics*. 1994, New York: John Wiley & Sons.
27. Hunter, R.J., *Zeta Potential in Colloid Science*. 1981, London: Academic Press.
28. Kirby, B.J. and Hasselbrink, E.F., Zeta potential of microfluidic substrates: 2. Data for polymers. *Electrophoresis*, 2004, 25: 203–213.
29. Kirby, B.J. and Hasselbrink, E.F., Zeta potential of microfluidic substrates: 1. Theory, experimental techniques, and effects on separations. *Electrophoresis*, 2004, 25: 187–202.
30. Yao, S., et al., Porous glass electroosmotic pumps: Design and experiments. *J. Coll. Interf. Sci.*, 2003, 268: 143–153.

31. Burgi, D.S. and Chien, R.L., Optimization of sample stacking for high performance capillary electrophoresis. *Anal. Chem.*, 1991, 63: 2042–2047.
32. Xuan, X.C. and Li, D.Q., Analytical study of Joule heating effects on electrokinetic transportation in capillary electrophoresis. *J. Chromatogr. A*, 2005, 1064: 227–237.
33. Tang, G.Y., et al., Modeling of electroosmotic flow and capillary electrophoresis with the joule heating effect: The Nernst–Planck equation versus the Boltzmann distribution. *Langmuir*, 2003, 19: 10975–10984.
34. Burgi, D.S., Large-volume stacking of anions in capillary electrophoresis using an electroosmotic flow modifier as a pump. *Anal. Chem.*, 1993, 65: 3726–3729.
35. Burgi, D.S. and Chien, R.L., Optimization in sample stacking for high-performance capillary electrophoresis. *Anal. Chem.*, 1991, 63: 2042–2047.
36. Chien, R.L. and Burgi, D.S., Field amplified sample injection in high-performance capillary electrophoresis. *J. Chromatogr.*, 1991, 559: 141–152.
37. Chien, R.L. and Burgi, D.S., Field amplified polarity-switching sample injection in high-performance capillary electrophoresis. *J. Chromatogr.*, 1991, 559: 153–161.
38. Chien, R.L. and Burgi, D.S., Sample stacking of an extremely large injection volume in high-performance capillary electrophoresis. *Anal. Chem.*, 1992, 64: 1046–1050.
39. Chien, R.L., Sample stacking revisted: A personal perspective. *Electrophoresis*, 2003, 24: 486–497.
40. Chien, R.L. and Burgi, D.S., Field amplified sample injection in high-performance capillary electrophoresis. *J. Chromatogr.*, 1991, 559: 141–152.
41. Chien, R.L. and Burgi, D.S. On-column sample concentration using field amplification in CZE. *Anal. Chem.*, 1992, 64: A489–A496.
42. Jung, B., Bharadwaj, R., and Santiago, J.G., Thousand-fold signal increase using field amplified sample stacking for on-chip electrophoresis. *Electrophoresis*, 2003, 24: 3476–3483.
43. Jacobson, S. and Ramsey, J. Microchip electrophoresis with sample stacking. *Electrophoresis*, 1995, 16: 481–486.
44. Chien, R.L., Mathematical modeling of field amplified sample injection in high performance capillary electrophoresis. *Anal. Chem.*, 1991, 63: 2866–2869.
45. Zhang, C.X. and Thormann, W., Head-column field-amplified sample stacking in binary system capillary electrophoresis. 2. Optimization with a preinjection plug and application to micelllar electrokinetic chromatography. *Anal. Chem.*, 1998, 70: 540–548.
46. Jacobson, S.C., Ermakov, S.V., and Ramsey, J.M., Minimizing the number of voltage sources and fluid reservoirs for electrokinetic valving in microfluidic devices. *Anal. Chem.*, 1999, 71: 3273–3276.
47. Jacobson, S.C., et al., High-speed separations on a microchip. *Anal. Chem.*, 1994, 66: 1114–1118.
48. Jacobson, S.C., et al., Effects of injection schemes and column geometry on the performance of microchip electrophoresis devices. *Anal. Chem.*, 1994, 66: 1107–1113.
49. Grushka, E., McCormick, R.M., and Kirkland, J.J., Effect of temperature-gradients on the efficiency of capillary zone electrophoresis separations. *Anal. Chem.*, 1989, 61: 241–246.
50. Kirby, B.J., Zeta potential of microfluidic substrates: 1. Theory, experimental techniques, and effects on separations. *Electrophoresis*, 2004, 25: 187–202.
51. Yao, S., Hertzog, D.E., Zeng, S., Mikkelsen, J.C., and Santiago, J.G., Porous glass electroosmotic pumps: Design and experiments. *J. Coll. Interf. Sci.*, 2003, 268: 143–153.
52. Preisler, J. and Yeung, E.S., Characterization of nonbonded poly(ethylene oxide) coating for capillary electrophoresis via continuous monitoring of electroosmotic flow. *Anal. Chem.*, 1996, 68: 2885–2889.
53. Hjerten, S. and Zhu, M.D., Adaptation of the equipment for high-performance electrophoresis to isoelectric-focusing. *J. Chromatogr.*, 1985, 346: 265–270.
54. Righetti, P.G., *Isoelectric Focusing: Theory, Methodology, and Applications*, 1st ed. 1983, New York: Amsterdam, xv, 386 pages.
55. Herr, A.E., et al., On-chip coupling of isoelectric focusing and free solution electrophoresis for multidimensional separations. *Anal. Chem.*, 2003, 75: 1180–1187.
56. Ross, D. and Locascio, L.E., Microfluidic temperature gradient focusing. *Anal. Chem.*, 2002, 74: 2556–2564.
57. Balss, K.M., et al., DNA hybridization assays using temperature gradient focusing and peptide nucleic acids. *J. Am. Chem. Soc.*, 2004, 126: 13474–13479.

58. Balss, K.M., et al., Simultaneous concentration and separation of enantiomers with chiral temperature gradient focusing. *Anal. Chem.*, 2004, 76: 7243–7249.
59. Balss, K.M., et al., Micellar affinity gradient focusing: A new method for electrokinetic focusing. *J. Am. Chem. Soc.*, 2004, 126: 1936–1937.
60. Kamande, M.W., et al., Simultaneous concentration and separation of coumarins using a molecular micelle in micellar affinity gradient focusing. *Anal. Chem.*, 2007, 79: 1791–1796.
61. Ghosal, S., Electrokinetic flow and dispersion in capillary electrophoresis. *Annu. Rev. Fluid Mech.*, 2006, 38: 309–338.
62. Huber, D.E. and Santiago, J.G., Taylor-Aris dispersion in temperature gradient focusing. *Electrophoresis*, 2007, 28: 2333–2344.
63. Kim, S.M., et al., Low-power concentration and separation using temperature gradient focusing via Joule heating. *Anal. Chem.*, 2006, 78: 8028–8035.
64. Van Milligen, B.P., et al., On the applicability of Fick's law to diffusion in inhomogeneous systems. *Eur. J. Phys.*, 2005, 26: 913–925.
65. Ghosal, S., Lubrication theory for electro-osmotic flow in a microfluidic channel of slowly varying cross-section and wall charge. *J. Fluid Mech.*, 2002, 459: 103–128.
66. Huber, D.E., *Transport and Dispersion in Microfluidic Temperature Gradient Focusing*, Ph.D Thesis, 2006, Stanford University.
67. Hoebel, S.J., et al., Scanning temperature gradient focusing. *Anal. Chem.*, 2006, 78: 7186–7190.
68. Everaerts, F.M., Beckers, J.L., and Verheggen, T.P.E.M., *Isotachophoresis: Theory, Instrumentation, and Applications*. 1976, New York: Amsterdam.
69. Martin, A.J.P. and Everaert, F.M., Displacement electrophoresis. *Anal. Chim. Acta*, 1967, 38: 233.
70. Everaerts, F.M. and Van der Put, A.J.M., Isotachophoresis—Separation of amino acids. *J. Chromatogr.*, 1970, 52: 415.
71. Holloway, C.J. and Pingoud, V., The analysis of amino acids and peptides by isotachophoresis. *Electrophoresis*, 1981, 2: 127–134.
72. Bruchelt, G.G., Niethammer, D.D., and Schmidt, K.K.H., Isotachophoresis of nucleic acid constituents. *J. Chromatogr.*, 1993, 618: 57–77.
73. Delmotte, P., Analysis of complex protein mixtures by capillary isotachophoresis. *Sci. Tools*, 1977, 24: 33.
74. Schmitz, G. and Möllers, C., Analysis of lipoproteins with analytical capillary isotachophoresis. *Electrophoresis*, 1994, 15: 31–39.
75. Everaerts, F.M., et al., Determination of heavy metals by isotachophoresis. *J. Chromatogr. A*, 1985, 320: 263–268.
76. Sollenberg, J., Analytical isotachophoresis in biological monitoring of exposure to industrial chemicals. *J. Chromatogr. A*, 1991, 545: 369–374.
77. Gebauer, P. and Bocek, P., Recent progress in capillary isotachophoresis. *Electrophoresis*, 2000, 21: 3898–3904.
78. Krivankova, L., Gebauer, P., and Bocek, P., Some practical aspects of utilizing the online combination of isotachophoresis and capillary zone electrophoresis. *J. Chromatogr. A*, 1995, 716: 35–48.
79. Wainright, A., et al., Sample preconcentration by isotachophoresis in microfluidic devices. *J. Chromatogr. A*, 2002, 979: 69–80.
80. Xu, Z.Q., et al., Performance of electrokinetic supercharging for high-sensitivity detection of DNA fragments in chip gel electrophoresis. *Electrophoresis*, 2004, 25: 3875–3881.
81. Mosher, R.A., Saville, D.A., and Thormann, W., *The Dynamics of Electrophoresis*. 1992: VCH Weinheim New York.
82. Kohlrausch, F., Ueber concentrations-Verschiebungen durch Electrolyse in Inneren Von Losungen and losungsgemischen. *Ann. Phys. (Leipzig)*, 1897, 62: 209–239.
83. Martin, A.J.P. and Everaert, F.M., Displacement electrophoresis. *Proc. Roy. Soc. Lond. Ser. A—Math. Phys. Sci.*, 1970, 316: 493.
84. Svoboda, M. and Vacik J., Capillary electrophoresis with ultra violet detection some quantitative aspects, *Journal of Chromatography A*, 1976, 119: 539–547.
85. Nagyova, I. and Kaniansky, D., Discrete spacers for photometric characterization of humic acids separated by capillary isotachophoresis. *J. Chromatogr. A*, 2001, 916: 191–200.

86. Wainright, A., et al., Preconcentration and separation of double-stranded DNA fragments by electrophoresis in plastic microfluidic devices. *Electrophoresis*, 2003, 24: 3784–3792.
87. Thormann, W., Description and detection of moving sample zones in zone electrophoresis: Zone spreading due to the sample as a necessary discontinuous element. *Electrophoresis*, 1983, 4: 383–390.
88. Beckers, J.L., Steady-state models in electrophoresis: From isotachophoresis to capillary zone electrophoresis. *Electrophoresis*, 1995, 16: 1987–1998.
89. Hjertén, S.S., Zone broadening in electrophoresis with special reference to high-performance electrophoresis in capillaries: An interplay between theory and practice. *Electrophoresis*, 1990, 11: 665–690.
90. Saville, D.A. and Palusinski, O.A., Theory of electrophoretic separations. 1. Formulation of a mathematical-model. *Aiche J.*, 1986, 32: 207–214.
91. Saville, D.A., The effects of electroosmosis on the structure of isotachophoresis boundaries. *Electrophoresis*, 1990, 11: 899–902.

39 The Mechanical Behavior of Films and Interfaces in Microfluidic Devices: Implications for Performance and Reliability

Matthew R. Begley and Jennifer Monahan

CONTENTS

39.1 INTRODUCTION

As electrophoretic separations move beyond traditional capillary systems to include microfabricated "lab-on-a-chip" technology, researchers must consider the mechanical response of devices, as dictated by the materials and methods involved in channel fabrication. Microfluidic chips created using planar lithography (e.g., [1–4]) enable increasingly complex separations, such as multidimensional chromatography [5–9] and on-chip sample preparation [10,11]. Unfortunately, lithographic techniques generate open channels, which require a subsequent bond step to produce an enclosed capillary network suitable for electrophoretic separations. In contrast to tubular capillaries comprised of a single material piece, the bonded interface of a microchip separation channel introduces a new failure mechanism, which limits electrophoretic or chromatographic methods that rely on pressure (i.e., capillary electrokinetic chromatography (CEC) [12–15], microchip isoelectric focusing (mIEF) [16], high ionic strength buffers [17], or polymer-based separations [18]).

Relatively recent additions of external and/or integrated features that control fluid motion (e.g., valves [19–21]) further increase the need to understand and quantify the mechanics controlling interface strength. This is especially true as hybrid chips pairing different types of materials (e.g., glass and polydimethylsiloxane (PDMS)) become increasingly prevalent; the bonding of surfaces with elastic and thermal expansion mismatch can induce significant stresses, which influence the performance and reliability of the final device. This chapter is designed to provide separation scientists with an introduction to interface and film mechanics that can be used to promote successful microfluidic design and fabrication.

Interface stability affects several aspects of lab-on-chip development. Successful device fabrication requires consistent bonding between different layers, often of different materials, and this remains among the most challenging obstacles to device development. Figure 39.1 provides schematic illustrations of several features/geometries that play increasingly important roles in microfluidic separation devices. Figure 39.1a is a top view of a typical multilayer assembly; a layer with lithographically patterned channels is "sealed" by bonding additional layers. The interface formed by the layers is typically weaker than the layers themselves, such that interface failure defines device reliability. Figure 39.1b is a side view of the interface failure shown in Figure 39.1a; the corner formed by bonding a capping layer to the layer with etched channels serves as the initiation site for failure. Device performance may further include deformable elements interacting with adjacent elements (e.g., valving). Such dynamic elements may require controlled debonding (e.g., valve release [19–21]) or site-specific bonding for ideal performance (e.g., check valves [22,23]). Figures 39.1c and 39.1d illustrate the side views of two common multilayer valve assemblies; the clear layer is a deformable element that is pushed or pulled via external pressure actuation to allow (Figure 39.1c) or prevent (Figure 39.1d) fluid motion in through the channels.

Thus, device reliability for many systems is dictated by interface stability, which in turn dictates the maximum allowable conditions and pressures. The use of pressure in microfluidic chips naturally introduces a competition between pressure limitations due to interface strength and that required for operation. Consider two examples: (1) filling a microchip channel with a viscous separation polymer (e.g., for DNA sequencing [24]) and (2) creating an integrated valve that is actuated using pressure [19–21]. Both require elevated pressures, which must be balanced against the possibility of breaking the chip apart. In designing features for these diverse devices, two fundamental questions arise:

I. What are the physical parameters that affect interface stability?
II. How do geometry, material properties and device pressures govern performance and reliability?

The principle motivation for this chapter is to establish the framework to predict device performance and reliability a priori, by detailing the mechanics underlying the behavior of typical geometries.

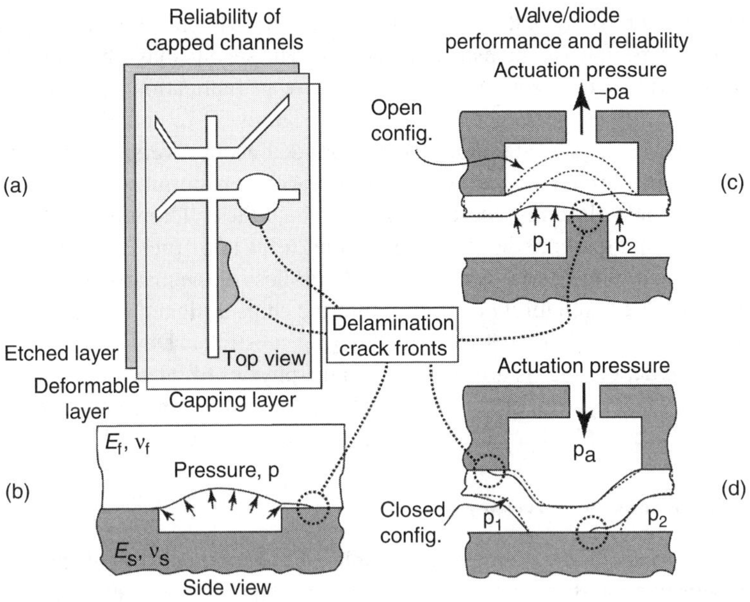

FIGURE 39.1 Schematic illustrations of common microfluidic features utilized in microchip separation devices: (a) top view of a multilayer assembly of microchannels formed by capping a layer with etched channels, (b) side view of micropatterned channel with capping film layer, (c) side view of a vacuum-driven three-layer negative pressure valve, (d) side view of a positive pressure valve.

To complement this, we describe basic materials characterization experiments used to determine the relevant parameters. This combination can be used to optimize proven device configurations and develop new strategies to improve fabrication, performance and reliability.

39.2 BACKGROUND

Microfluidic separation devices have increased in complexity beyond the original cross-t design of Manz and coworkers (e.g., [25,26]) to include a wide variety of designs with intricate intersecting channels. The bulk of these devices are fabricated in glass or PDMS. Several reviews (and previous chapters in this book) discuss the methods used in the fabrication of microfluidic devices [27,28]. Nearly all are based on planar lithography and "soft" lithographic techniques to produce channels embedded in the surface of a substrate (e.g., [29,30]). To create an enclosed microcapillary network suitable for electrophoretic separations, these channels must be bonded to a cover plate(s) as shown in Figure 39.1a. The bonding of two different layers creates an interface that is typically weaker than either layer; hence, the "seam" that runs along the edge of the channel acts as an initiation site for chip failure. This is shown schematically in Figure 39.1a (top view) and Figure 39.1b (cross-section view). Interface failure usually corresponds to device failure, as fluid from the channel is lost to the interface "crack"; with sufficient pressure (or volume injected into the channel), the entire interface may debond.

Here we will focus on two material systems common to chip based electrophoretic separations. The first, *glass-on-glass* devices, are common for "cross-t" and gated electrokinetic injections. Such devices depend on a high temperature anneal to facilitate bonding between the cover plate and substrate with etched channels [27,28]. The successful high temperature anneal of two pieces of glass requires flat, defect free surfaces and the uniform application of pressure during fabrication. In many ways, successful bonding of *glass-on-glass* devices is considered a "black art": while strong bonds produce robust chips that are ideal for pressurized applications, fabrication yield can be depressingly low, especially in the hands of a novice. The second class of chips, *glass-on*-PDMS hybrid devices, are widely utilized because of their ease in fabrication/assembly and to produce integrated elastomeric valves, such as those shown in Figures 39.1c and 39.1d. Depending on the application, glass-PDMS devices are either weakly bonded using reversible van der Waals forces, or more strongly bonded through oxidative modification of the material surfaces prior to bonding.

The physical demands placed on these devices can be quite different due to the rigid nature of glass–glass interfaces and extreme deformation (arising from large pressures relative to the elastic modulus) that may occur with PDMS hybrid chips. Regardless of the material(s) or fabrication techniques, the performance and reliability of all microfluidic chips designed for CE separations depend upon controlling the interface bond between neighboring substrates. Device operation (either interface failure or desired debonding) is governed by (1) the physics of interface debonding and (2) the relationship between loading (e.g., pressure and/or temperature) and chip deformation. The two are strongly interrelated, because both are essentially governed by energy stored by elastic deformation.

39.2.1 The Mechanics of Interface Failure

The mechanics of interface failure, while often complex, has been firmly established in the context of thin film multilayers commonly found in microelectronics and protective coatings. This large body of work serves as the foundation for the present chapter; it is succinctly summarized by the review article by Hutchinson and Suo [31] and covered in great detail in the book by Freund and Suresh [32]. We present the most basic framework needed to quantify the performance of microfluidic devices, and summarize results for geometries and materials common to such devices. The focus is on prediction of debonding initiated from sharp changes in geometry, as illustrated schematically in Figure 39.2.* For these systems, the act of debonding or delaminating under pressure is simplified by considering the process as the formation and/or propagation of a crack at the interface.

As a first approximation for analyzing interfaces, an elementary "strength of materials" approach is to determine the maximum stress in the component and then compare its value to the strength of the material (or interface), as measured via a conventional uniaxial tension test. Unfortunately, this approach is not applicable in the presence of material discontinuities, such as an interface that is only partially bonded or a corner formed by bonding two materials (see Figure 39.2a). If one obtains a complete solution for the stress distribution, the stress field diverges to infinity as the discontinuity is approached: that is, the stress diverges to infinity as distance from the crack tip becomes smaller and smaller. This is shown schematically in Figure 39.2b (refer to log-log scale on the right-hand side). Hence, the discontinuity represents a *singularity* in the stress distribution. This nonphysical result (i.e., an infinite stress at a crack tip) is a consequence of assuming elastic behavior applies at

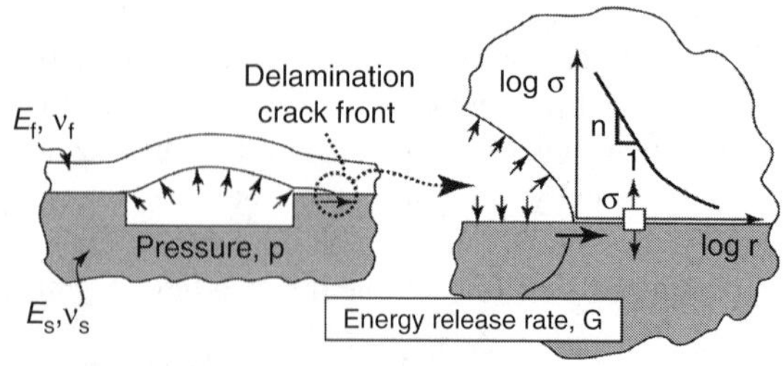

FIGURE 39.2 Schematic illustration of the stress distribution at the tip of a delamination crack: the presence of a stress singularity necessitates an energy-based approach.

* "Delamination" is commonly used to describe the debonding of thin films or bonded stacks of thin films (i.e., multilayers); hence, in the present context, debonding and delamination refer to the same phenomena.

an arbitrarily small length-scale. To get around these conceptual difficulties, *linear elastic fracture mechanics* (LEFM) theory [33] is a very successful approach to predicting crack growth without resorting to molecular simulations. That is, one can successfully predict failure using macroscopic continuum solutions. It is interesting to note that for materials with limited mechanisms for nanoscale deformation,* continuum elasticity solutions are applicable to impressively small scales, that is, several nanometers. Hence, it is not surprising that they can serve as the basis for predicting failure in predominantly elastic systems.

With the exception of an infinitely long material interface, any mathematically sharp geometry change introduces a singularity, for example, right-angled corners, the edge of a debonded region, or material interfaces that intersect a free edge. The form of this singularity depends on the angle of the junction (e.g., the 90° corner junction shown in Figure 39.2a), as well as the elastic mismatch between the materials comprising the junction (e.g., [34–36]). At the moment, the mechanics of crack stability are considerably more established than that required to predict debond initiation from a bimaterial corner junction. That is, the criteria needed to predict the advance of a sharp interface delamination crack is more clearly defined than that needed to predict its initiation from a right-angled corner. Debond initiation from blunt corners is a critical problem that requires continued study.† However, as luck would have it, when the film and substrate have very large elastic mismatch—such as for PDMS-*glass* systems—it is reasonable to assume that the substrate is "rigid." In this limit, the corner initiation problem reduces to that of an interface crack: the entire span shown in Figure 39.2a corresponds to the "crack," and well-established interface fracture mechanics is applicable. This is because the substrate does not deform, such that it is immaterial whether or not the substrate geometry forms a perfect 90° corner or is partially debonded as shown in Figure 39.2a.

In systems where the elastic moduli of the layers are comparable—critically, glass-glass or any homogenous system—one must distinguish between the corner initiation geometry and the partially debonded region shown in Figure 39.2a. This is because the deformation of the substrate lying between the original corner and leading edge of the debond region (i.e., crack tip) may influence the stress distribution at the edge of the crack tip. When the distance between the corner and the crack tip is sufficiently large, the influence of geometry of the channel (and hence corner) can be neglected and one can again treat the entire debonded length as a crack between geometrically uniform layers. This limit is reached when the distance between the corner and debond edge is several times greater than the film thickness (or thickness of the more compliant layer) (see [37] for the related study of delamination from vertical edges of a two layer system). The fact that even short cracks (measured from the edge of the corner to the debond crack tip) might lead to undesirable behavior—such as fluid or pressure loss—highlights the need for additional mechanics analyses of the initiation and short-crack propagation problems for elastically similar materials. Such analyses are beyond the scope of this introduction, and hence, we must assume here that the debond crack tip is located at least several film thicknesses from the corner.

The stress singularity implies that stress-based failure criteria are not feasible and one must adopt energy frameworks to describe the conditions that lead to debonding.‡ The central concept that is that the crack front advances when the energy available to drive the crack reaches a critical value, often referred to as the interface toughness. The energy available to drive the crack, or energy release rate (ERR), is dictated by the potential energy of the system. This is comprised of the strain energy stored in the deformed structure and the work done by applied loads. The critical value of the ERR

---

* Glass is a classic example: atomic bonds can stretch (elasticity) or rupture, but typically do not reform. In metals, atomic bonds can "slide" and reform, a phenomenon described by dislocation theory; this gives rise to energy dissipation, which makes it much more difficult to propagate a crack in metals.

† Significant inroads into the corner initiation problem have been established (e.g., [34–36]; these studies, however, require more extensive discussion of mechanics concepts that are beyond the scope of this work.

‡ Or, one can use equivalent stress-based frameworks that deal with the nature of the singularity rather than a "maximum stress": these approaches use the stress intensity factor, which is essentially the amplitude of the singularity (see Section 39.3.1.)

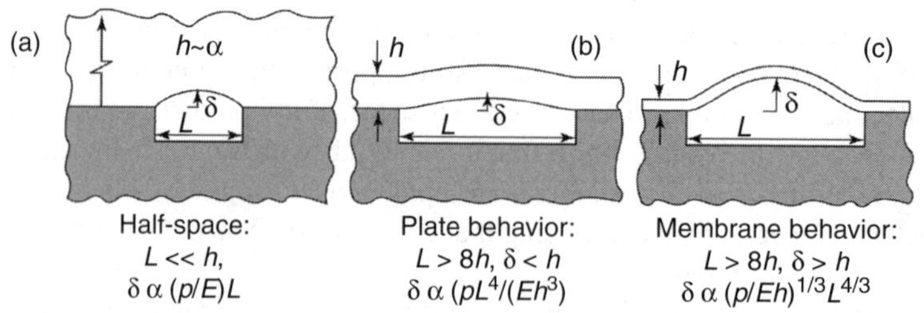

FIGURE 39.3 Schematic illustration of three types of film response: (a) elastic half-space behavior, dominated by bulk compression of the film, (b) plate behavior, dominated by bending deformation of the film, and (c) membrane behavior, dominated by axial stretching of the film.

(which has units of N/m or J/m^2) is the energy per unit area needed to create new crack surface and is fundamentally dictated by the bonding at the interface. As such, it is often referred as the adhesion energy. This chapter outlines the mechanics needed to compute the ERR for various combinations of load and geometry, as well as that needed to design and interpret experiments for extracting the interface toughness.

39.2.2 Overview of the Mechanics Describing Deformable Plates/Films

The relationships between geometry, external loading (i.e., pressures) and deformation has been an area of intense study in the solid mechanics community for more than century; hence, our goal is to provide a condensed overview of the theory and solutions that apply specifically to microfluidic applications. A number of classical mechanics results are relevant to microfluidic devices and the trick is to determine which is applicable. The key consideration is the "aspect ratio" of the device that relates the pressurized span, L, to the film thickness, h, as shown in Figure 39.3. Three different types of deformation "modes" are amenable to closed-form analytical solutions. Figure 39.3a depicts small aspect ratios L/h: very thick structures respond as *half-spaces*, in that the deformation does not depend on the film thickness, which is taken to be semi-infinite. Figure 39.3b depicts moderate to large aspect ratios L/h: moderately thin structures respond as *plates*, in that bending deformation of the film dominates film stretching. Figure 39.3c depicts large aspect ratios L/h: very thin structures respond as *membranes*, in that axial stretching deformation of the film dominates the effects of bending.*

The scaling relationships shown in Figure 39.3 apply regardless of the out-of-plane geometry (e.g., elliptical or rectangular), which only affects a constant prefactor. In this chapter, we review the relevant analytical solutions that bracket the range of behaviors, as well as numerical solutions that can be used for intermediate aspect ratios. The focus is on providing guidelines that enable chip designers to identify appropriate solutions for a given geometry and pressure magnitude.

39.3 THEORETICAL ASPECTS

Since a truly comprehensive treatment of the mechanics relevant to microfluidic devices is beyond the scope of a single chapter, we provide a basic introduction that adopts several simplifications. First and foremost, we assume that materials are linearly elastic; while, this is obviously objectionable for elastomers at large strains, it allows for closed-form analytical solutions that can be used to guide initial device design and broadly assess reliability. It is important to note that even when

* It should be noted that many refer to flexible thin films (commonly elastomers) as "membranes," regardless of their dimensions, presumably because they are easily deformed by relatively low pressures (relative to atmospheric pressure). Here, we adopt the classical mechanics definition of *membrane*, that is, a stretch-dominated structure (as opposed to bending-dominated).

deformation is large (i.e., displacements are significant compared to film thickness), strains may still be small. Moreover, the loss of accuracy due to linearization of material response is oftentimes not significant for strains less than ~30% (e.g., [38]). Secondly, we assume that interface debonding occurs at an asymptotically sharp crack front, as opposed to a bimaterial corner defined by a finite angle. This approximation has no impact on glass–PDMS interfaces due to the elastic mismatch; for interfaces formed between similar materials, this approximation is valid when the debond crack tip is sufficiently far from corners, other crack tips and so on. Finally, we assume that the interface is sufficiently weak to ensure that delamination cracks propagate at the interface, that is, interface delaminations will not "kink" and turn to run in the materials on either side of the interface.

39.3.1 Interface Fracture Mechanics

Interfacial stability is predicted using the ERR, which is defined as the mechanical energy that is released during an incremental increase in crack length. Crack advance occurs when the ERR reaches a critical value, often referred to as the interface toughness, denoted here as Γ_i. Thus, the crack growth criterion is $G \geq \Gamma_i$. The ERR is defined as [33]

$$G \equiv -\frac{\partial U}{\partial A}, \tag{39.1}$$

where A is the area of the new surface formed by crack advance (dictated by the geometry of the crack front) and U is the mechanical energy of the system. U is a function of the current crack size, loads such as applied pressure or residual stress and the elastic properties of the system. When the crack propagates with a straight crack front, the ERR can be written interms of the crack length and the associated length of the crack front. The ERR thus has units of N/m, or J/m^2 and hence, the interface toughness has units of energy per unit area; for this reason, the toughness is often referred to as the critical adhesion energy, or simply the adhesion energy. The interface toughness is governed by the nature of bonding at the interface and should be considered a material property to be measured for each system and fabrication approach.

For bimaterial interfaces, the interface toughness is also a function of the *mode-mixity*, which describes the relative amounts of crack tip opening, sliding and tearing deformation [31,32]. Figure 39.4 illustrates the definitions of the three fundamental modes of crack growth. For straight crack fronts with no deformation gradients in the direction parallel to the crack front, there is no out-of-plane tearing and mode III is identically zero. The stress intensity factors K_I and K_{II} relate to the contributions of opening and sliding displacements, respectively and hence are outcomes of the solution to the crack tip elasticity problem. In essence, the stress intensity factors describe the amplitude of the stress singularity shown schematically in Figure 39.2b. As such, they are dependent on the applied load, crack geometry and elastic properties on either side of the interface. It is important to note that bending of thin films near crack tips can introduce a significant mode II component; thus, even though pressurized crack faces may seem to be "pure mode I," the deformation that involves relative sliding of the material near the interface crack tip, which implies that mode II is present.

It has been demonstrated that any two-dimensional bimaterial problem can be uniquely described in terms of the two Dundur's parameters, defined as

$$\alpha = \frac{\bar{E}_f - \bar{E}_s}{\bar{E}_f + \bar{E}_s}, \quad \beta = \frac{E_f\,(1+\nu_s)\,(1-2\nu_s) - E_s\,(1+\nu_f)\,(1-2\nu_f)}{E_f\,(1+\nu_s)\,(1-\nu_s) + E_s\,(1+\nu_f)\,(1-\nu_f)}, \tag{39.2}$$

where $\bar{E} = E/\left(1-\nu^2\right)$ and the subscripts f and s refer to the film and substrate, respectively. That is, the behavior of any interface crack with identical Dundur's parameters is independent of individual elastic properties of the film and substrate. Dundur's parameters for several material systems relevant to microfluidic systems are shown in Table 39.1.

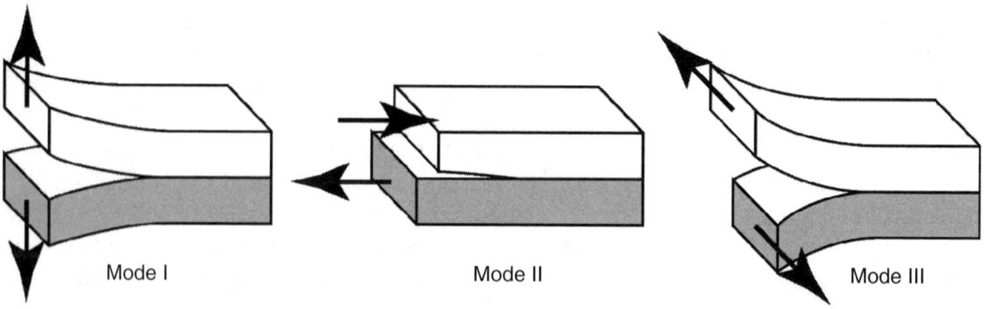

FIGURE 39.4 Modes of crack loading: for the scenarios considered here, mode III is assumed to be zero due to the absence of deformation gradients parallel to the crack front.

TABLE 39.1
Dundur's Parameters for Several Interfaces Relevant to Microfludic Devices

Materials Forming the Interface				Dundur's Parameters	
Material #1	***E, v***	**Material #2**	***E, v***	α	β
Elastomer	1 MPa, 0.5	Glass	70 GPa, 0.2	~ -1	1.4×10^{-5}
Elastomer	1 MPa, 0.5	Polymer	3 GPa, 0.4	~ -1	1.2×10^{-4}
Polymer	3 GPa, 0.4	Glass	70, GPa, 0.2	–0.91	–0.28
Identical	—	Identical	—	0	0

Note: Interchanging materials #1 and #2 simply changes sign of Dundur's parameters.

Here, we limit our attention to scenarios where $\beta \sim 0$, which essentially eliminates combinations of materials with comparable—*but not identical*—stiffness. Note that β is identically zero for identical materials. When $\beta \sim 0$ the crack tip fields are analogous to those in a homogenous material (i.e., the crack tip stresses scale as $\sigma_{ij} \sim 1/\sqrt{r}$) and K_I and K_{II} dictate the contributions of opening and sliding displacements, respectively (or normal and shear stresses). If $\beta \neq 0$, the opening and sliding modes are coupled and crack advance criteria require additional parameters [31,32].

The simplification that $\beta = 0$ implies that crack advance can be predicted in the context of a mixed-mode interface toughness, that is,

$$\Gamma_{\mathrm{i}} = f(\psi), \quad \text{where } \psi \equiv \tan^{-1}\left(\frac{K_{\mathrm{II}}}{K_{\mathrm{I}}}\right), \tag{39.3}$$

where K_{I} and K_{II} are computed from the elasticity solution to the given crack problem.

Since interface toughness depends strongly on the amount of mode-mixity, the ERR alone is not sufficient to predict delamination; one must also compute the phase angle from the relevant elasticity solution and have measurements that describe $\Gamma_{\mathrm{i}} = f(\psi)$.

However, measurements on a variety of systems have illustrated that the minimum interface toughness, $\Gamma_{\mathrm{i}}^{\mathrm{min}}$, is obtained for pure mode I. That is, the measured toughness for cracks dominated by mode II ($\psi \to \pi/2$) is generally much higher. Hence, assuming $\psi \sim 0$ and adopting the crack advance criterion $G \geq \Gamma_{\mathrm{i}}\,(\psi = 0) = \Gamma_{\mathrm{i}}^{\mathrm{min}}$ is generally conservative, in that it underestimates crack stability. Moreover, experiments on interfaces between glassy polymers (e.g., PMMA) and glass suggest $\Gamma_{\mathrm{i}} \approx \Gamma_{\mathrm{i}}(\psi = 0)$ for $-15° < \psi < 60°$ [31,32].

If the interface toughness is known, the prediction of debonding thus essentially boils down to computing the potential energy of the system, U and computing its derivative with respect to crack length. The potential energy of the system is computed from the solution to the solid mechanics problem at hand; this is described in the next section. For simple geometries, this can be done in closed-form, as illustrated in Section 39.4.2. For more complex geometries, this is typically done with commercial finite element analysis (FEA) software and many codes have a built-in option for doing so.

39.3.2 Potential Energy and Pressure–Deflection Relationships

The relationship between loads, geometry and deformation is governed by the solution of partial differential equations (PDEs) that are derived using three fundamental relationships: (1) equilibrium conditions, expressed as linear PDEs involving up to six stress components, (2) strain-displacement compatibility conditions, expressed as differential equations involve up to six strain components and three displacements and (3) a constitutive description, expressed as an algebraic equation relating the six stress and six strain components. The governing equations are completely linear for linear elastic materials that experience relatively small displacements. For large deflections, the strain-displacement relationships are nonlinear, such that numerical solutions are typically needed. The deformation of linearly elastic membranes is a notable exception, in that several important nonlinear solutions (that account for large-displacements) can be derived in closed form.

The stresses and displacements at the boundaries of the deformable domain (applied or enforced) dictate the *boundary conditions*, which effectively determine the particular solution to the PDEs. If the problem is truly three-dimensional (because geometry and/or loads vary in all three directions) then numerical solutions are needed for all but the most elementary example cases. FEA is the de facto choice for such scenarios and commercial codes abound; ABAQUS and ANSYS (software programs) are arguably the most common. An example of a simple geometry that is nevertheless a truly three-dimensional problem is a film capping an elliptical hole, that is, an elliptical plate or membrane.

If the geometry and loads do not vary in the third dimension, the derivatives with respect to the third coordinate are identically zero and the problem is inherently two-dimensional. Important examples include long, straight-walled channels, or axisymmetric structures such as circular plates: one can obtain the relevant solution by analyzing a two-dimensional "slice" taken from the full structure. For simple boundary conditions, such scenarios are highly amenable to analytical solutions. This is the focus of the remainder of this section.

39.3.2.1 Small Deformation Plate Behavior

Consider a film to a pressure distribution shown in Figure 39.5a. It is assumed that the variation in displacement in the z-direction is zero, that is, $\varepsilon_z = 0$: this is referred to as *plane-strain deformation*. (Note that this does not imply $\sigma_z = 0$.) First, consider platelike behavior, wherein the bending strain

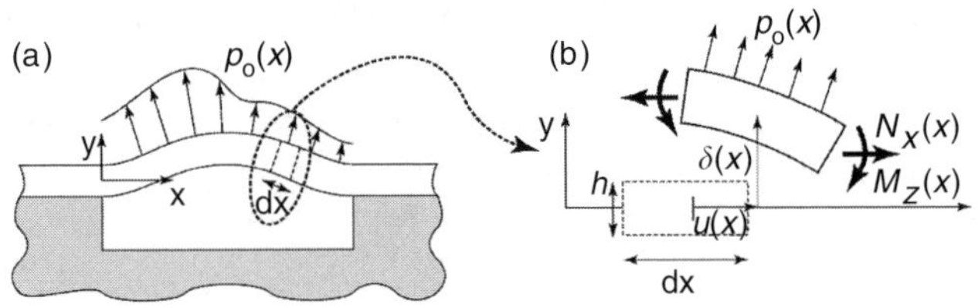

FIGURE 39.5 (a) Schematic illustration of film subjected to an arbitrary pressure distribution, (b) a differential element of the film showing the variables used for strain–displacement and equilibrium equations.

in the film dominates axial stretching. For small deformations relative to the film thickness, the only relevant strain-displacement relationship is:

$$\varepsilon_{\mathrm{XX}}(x, y) = \delta''(x) \cdot y, \tag{39.4}$$

where $\delta''(x)$ is the second derivative of the vertical deflection with respect to the x-coordinate (i.e., the local curvature) and y is the distance from the centerline of the film.

The net internal moment acting at any cross-section (about the z-axis) is given by

$$M(x) = \int_{-h/2}^{h/2} \sigma_{\mathrm{XX}}(x, y) \cdot y \cdot dy = \frac{\bar{E}h^3}{12}\delta''(x), \tag{39.5}$$

where $\sigma_{\mathrm{XX}} = \bar{E}\varepsilon_{\mathrm{XX}}$ is used as the constitutive law. $\bar{E} = E/(1 - \nu^2)$ is the plane-strain modulus, where ν is the Poisson's ratio of the film. This assumes that the in-plane stress σ_{xx} dominates the vertical stress σ_{yy} through most of the film thickness; that is, the structure translates the vertical pressure into a linear bending stress. Vertical equilibrium of the differential element shown in Figure 39.5b dictates that the moment distribution, vertical displacements and external pressure related as follows:

$$M''(x) = \frac{\bar{E}h^3}{12}\delta^{IV}(x) = -p_{\mathrm{o}}(x), \tag{39.6}$$

where $p_{\mathrm{o}}(x)$ is the resultant pressure acting on the film (i.e., the superposition of pressure on top and bottom). Thus, by combining a kinematic relationship between strain and displacement, a constitutive law and equilibrium, a differential equation is obtained for the displacements of the film in terms of the applied load.

One can integrate this result for any pressure distribution: the four integration constants are determined by the boundary conditions at either end. Force balance applied to the differential element of Figure 39.5b also dictates that the resultant vertical shear force is given by $V(x) = M'(x) = \left(\bar{E}h^3/12\right)\delta'''(x)$. This completes the possible boundary conditions that can be used to solve for unknown constants: for example, a clamped or bonded film implies zero displacement and slope, that is, $\delta(x = 0) = \delta'(x = 0) = 0$ and $\delta(x = L) = \delta'(x = L) = 0$. Or, for free ends (i.e., cantilevers), the net moment and shear force acting at a free end is zero, such that $\delta''(x = L) = \delta'''(x = L) = 0$.

For the case of a uniform applied pressure (i.e., $p_{\mathrm{o}}(x) = p_{\mathrm{o}}$), the total potential energy of the system (for a slice across of the channel of width b) is given by

$$U = b\int_0^L \int_{-h/2}^{h/2} 1/2\sigma_{xx}(x, y)\,\varepsilon_{xx}(x, y) - bp_{\mathrm{o}}\int_0^L \delta(x)\,\mathrm{d}x, \tag{39.7}$$

where L is the length of the span that can deform. The first term is the strain energy in the deflected film and the second is the external work done by the applied pressure. Note that the last term is equivalent to the pressure times the volume of the bulge: this observation becomes important in calculation the ERR for bulge tests under fixed volume (see Section 39.4.2). Thus, for any uniform film that is subjected to constant pressure and dominated by bending deformation, the potential energy is given by

$$U = b\int_0^L \left[\frac{\bar{E}h^3}{24}\left(\delta''(x)\right)^2 - p_{\mathrm{o}}\delta(x)\right]\mathrm{d}x. \tag{39.8}$$

This result can be used with Equation 39.1 to determine the ERR for scenarios involving small deflections; Section 39.4.2.1 provides an illustration.

39.3.2.2 Large Deformation Membrane Behavior

Similar elementary results can be derived for deformation in the membrane limit, that is, when the centerline stretch of the film dominates bending. In this scenario, the strain-displacement relationship is

$$\varepsilon_{\mathrm{xx}}(x) = u'(x) + 1/2\left(\delta'(x)\right)^2, \tag{39.9}$$

where $u(x)$ is the displacement in the x-direction (see Figure 39.5). One can illustrate via energy minimization that the strain and hence the resultant stress $\sigma_{xx}(x) = \bar{E}\varepsilon_{\mathrm{xx}}(x) = \sigma_{\mathrm{o}}$, is uniform along the membrane.* This stress is an unknown that must be determined. Additionally, the energy minimization implies the following equilibrium condition:

$$h\sigma_{\mathrm{o}}\delta''(x) = -p(x). \tag{39.10}$$

Again, the solution by direction integration involves two constants; these constants, as well as the unknown membrane stress, are dictated by the displacements of the end points. Noting that the membrane stress is spatially uniform and integrating Equation 39.9 yields

$$u(L) = \frac{\sigma_{\mathrm{o}}L}{\bar{E}} + \frac{1}{2}\int_0^L \left(\delta'(x)\right)^2 \mathrm{d}x. \tag{39.11}$$

For uniform applied pressure, the total potential energy of the membrane system is given by

$$U = b\int_0^L \left(\frac{h\sigma_{\mathrm{o}}^2}{2\bar{E}} - p_{\mathrm{o}}\delta(x)\right)\mathrm{d}x = \frac{bhL\sigma_{\mathrm{o}}^2}{2\bar{E}} - bp_{\mathrm{o}}\int_0^L \delta(x)\,\mathrm{d}x. \tag{39.12}$$

Once again, the last term represents the pressure times the volume of the bulge, which is relevant to interface fracture toughness measurements conducted by injecting a fixed volume (see Section 39.4.2.2).

39.4 PRACTICAL APPLICATIONS AND DEVELOPMENT GUIDELINES

39.4.1 DEBONDING OF MULTILAYERS FROM INTERFACE FLAWS AT EDGES

This section addresses interface failure between two elastic plates bonded together, as shown in Figure 39.6. Stresses are generated when the plates have different initial curvatures and/or different thermal expansion properties; the resulting stored elastic energy promotes debonding. It is assumed that two plates with different curvatures bond together after they have been pressed completely flat and their interfaces have been aligned. Put another way, the material on each side of the interface experiences relative sliding prior to bonding. This is an important assumption, because it implies that each layer will be stressed in the bonded state *even when their initial curvature, elastic properties*

* One can derive equivalent expressions using force balances: however, the finite angles that are retained in membrane theory (e.g., $\cos\beta \approx 1 - (1/2)\,\beta^2$) make such derivations cumbersome.

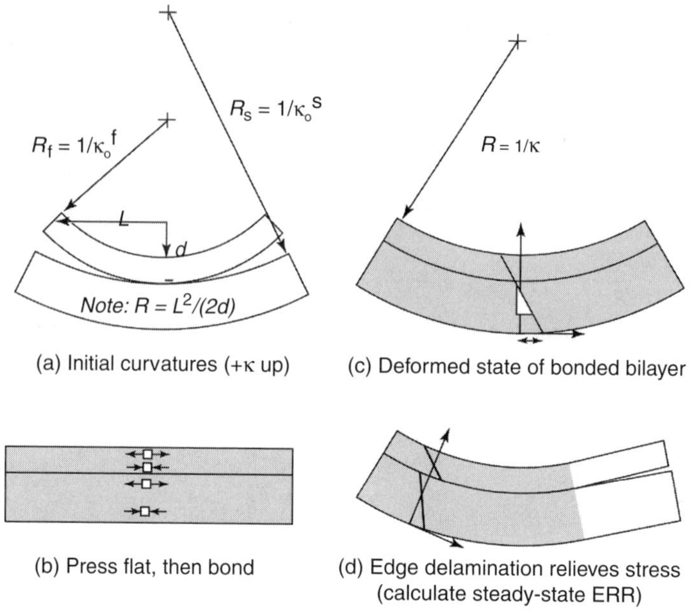

FIGURE 39.6 Geometry and variables used in the analysis of bilayer deformation and debonding: (a) initial configuration of curved plates, (b) it is assumed the plates are pressed flat prior to bonding, (c) resulting curved bilayer with continuous total strain distribution, and (d) schematic illustration of edge debonding and stress state in bonded region.

and thermal expansion coefficients are identical. This is a consequence of the fact that "opposite" faces are bonded: for example, if both plates have the same curvature, the compression side of the top plate will be bonded to the tensile side of the bottom plate (see Figure 39.6b).

The *mechanical* strain distributions in the thickness direction in each plate can be expressed as*

$$\varepsilon_{xx}^{f}(y) = \varepsilon_o - \kappa \cdot y + \kappa_o^{f}(y - h_s - h_f/2) - \theta_f, \tag{39.15a}$$

and

$$\varepsilon_{xx}^{s}(y) = \varepsilon_o - \kappa \cdot y + \kappa_o^{s}(y - h_s/2) - \theta_s. \tag{39.15b}$$

Here, κ_o^i is the initial curvature of the *i*th layer (i.e., the layer traces a circular arc with radius $R_o^i = 1/\kappa_o^i$) and the subscripts *f* and *s* refer to the film and substrate, respectively. Positive curvature is taken as concave up: if the curvature of the two layers opposes one another, the individual layer curvatures are of opposite sign. y is the distance from the bottom of the substrate. The thermal strain in the *ith* layer is defined as: $\theta_i \equiv \alpha_i\left(T_i - T_o^i\right)$, where α_i is the coefficient of thermal expansion and T_i is the current temperature. T_o^i is the reference temperature at which there is no thermal strain, that is, the temperature at which the layer is fabricated. ε_o is the axial stretch along the bottom of the substrate (i.e., $y = 0$) and κ is the curvature of the bilayer after bonding; hence, ε_o and κ are the solution variables of interest.

* The total strain is simply $\varepsilon_{tot}(x) = \varepsilon_o - \kappa(x) \cdot y$: one subtracts the thermal strains from the total strain to get the mechanical strain, that is, the strain used to determine stresses via $\sigma = \bar{E}\varepsilon$.

Absent any external loading on the bonded bilayer, the net resultant axial force, N and net resultant moment, M, are zero. This implies

$$N = \int_0^{h_s} \sigma_{xx}^s(y)\ dy + \int_{h_s}^{h_s+h_f} \sigma_{xx}^f(y)\ dy = 0, \tag{39.16}$$

and

$$M = \int_0^{h_s} \sigma_{xx}^s(y) \cdot y\ dy + \int_{h_s}^{h_s+h_f} \sigma_{xx}^f(y) \cdot y\ dy = 0 \tag{39.17}$$

Substituting $\sigma_{xx}^i(y) = \bar{E}_i \varepsilon_{xx}^i(y)$ and performing the integration yield two linear equations for the unknowns ε_o and κ. The stored elastic energy of the bilayer can be calculated by integrating the strain energy density per unit volume, $1/2\, \sigma_{xx}^i \varepsilon_{xx}^i$, throughout the volume, using the results for ε_o and κ with Equation (39.15); integration over thickness yields the strain energy per unit area in the plane of the bilayer.

The ERR for an edge crack such as that shown in Figure 39.6d can be trivially calculated using the elastic energy. Here, we assume that interface failure relieves all stresses in the layers, that is, they return to their undeformed configuration for portions of the plate that are no longer bonded.* When the length of the interface crack greatly exceeds either layer thickness, edge effects are negligible; this case is referred to as the *steady-state* ERR, because the ERR is independent of flaw size. The steady-state ERR is the maximum possible—that is, for shorter interface flaws the ERR is smaller. The steady-state ERR is simply the change in strain energy per unit crack advance, that is, $G = (U_{\text{bonded}} - U_{\text{cracked}})/(b\Delta a) = U_{\text{bonded}}/(b\Delta a)$. This result can be written as†

$$G = \bar{E}_s h_s f_{sf} + \bar{E}_f h_f f_{fs}, \tag{39.18a}$$

where

$$f_{ij} = \frac{g_{ij}}{24} h_j^2 \left[\kappa_i^2 - 2\kappa_i\kappa_j + 4\kappa_j^2\right] + 6h_j\kappa_j\left[h_i\kappa_j + 2(\theta_s - \theta_f)\right] + 3\left[h_i\kappa_j + 2(\theta_s - \theta_f)\right]^2 \tag{39.18b}$$

and

$$g_{ij} = \frac{\bar{E}_i h_i}{\bar{E}_j h_j} + 4 + 6\frac{h_j}{h_i} + 4\left(\frac{h_j}{h_i}\right)^2 + \left(\frac{h_j}{h_i}\right)^3 \frac{\bar{E}_j}{\bar{E}_i}. \tag{39.18c}$$

Note that the ERR is symmetric with respect to the film and substrate, that is, one can arbitrarily label the top or bottom film as the substrate.

This broadly general result has interesting and important implications, even when the two plates are these same material and thickness. For example, consider two identical plates with the same initial curvature: the ERR is

$$G = \frac{\bar{E}h^3}{12R_o^2}\ \text{(curvature opposing)} \quad \text{or} \quad G = \frac{\bar{E}h^3}{16R_o^2}\ \text{(curvature aligned)} \tag{39.19}$$

* In some instances, the plates in the debonded region remain subject to out-of-plane constraints, such that they are not entirely stress-free: this requires slightly different algebra (see Reference 31).

† Note that the Einstein summation convention does *NOT* apply.

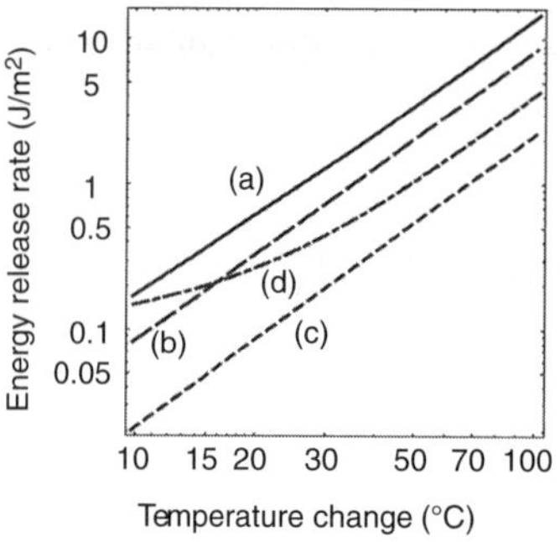

Substrate, Film	(a) Glass, polymer	(b) Polymer, polymer	(c) Glass, elastomer	(d) Glass, glass
E_s, α_s, h_s (GPa, ppm/°C, mm)	70, 6, 1	4, 40, 2	70, 6, 10	70, 7, 2
E_f, α_f, h_f (GPa, ppm/°C, mm)	4, 40,1	4, 40, 2	0.001, 300, 5	70, 7, 2
$\Delta T_f, \Delta T_s$(°C)	same: temp change after bonding	0, ΔT_f, temp, difference of bonding	same: temp change after bonding	0, ΔT_f, temp, difference of bonding
R_s, R_f (meters)	20, 20	20, −20	0, 0	20, −20
$\Gamma_i(\Psi=0°)$	~2–5 J/m²	~2–20 J/m²	~0.05–1 J/m²	~0.1–5 J/m²
$\Gamma_i(\Psi=90°)$	~10–50 J/m²	??	~5–30 J/m²	~5–30 J/m²

FIGURE 39.7 Examples of energy release rates for several different bilayer scenarios: (a) glassy polymer film on glass, (b) two glassy polymers bonded at different temperatures, and (c) an elastomer film bonded to a glass substrate, and (d) two glass slides bonded at different temperatures.

These somewhat surprising results imply that aligning initially curved plates may not be remarkably effective in suppressing delamination. It should be noted that if the initial curvatures are aligned, the final curvature of the plate will be twice the initial curvature: the internal moments generated by the internal stresses introduced during pressing are complementary. Conversely, as one would expect, the final curvature of the bilayer is zero if the initial curvatures are opposed and identical.

As anyone in the microelectronics community can attest to, the true headaches begin when two different materials with different thermal expansion coefficients are bonded together, often times at different temperatures due to the fabrication process. In this case, the ERR arising from bonding perfectly flat plates is given by

$$G = \frac{1}{2g_{\mathrm{fs}}}\bar{E}_{\mathrm{f}}h_{\mathrm{f}}\,(\theta_{\mathrm{s}}-\theta_{\mathrm{f}})^2\left(1+\left(\frac{\bar{E}_{\mathrm{f}}h_{\mathrm{f}}^3}{\bar{E}_{\mathrm{s}}h_{\mathrm{s}}^3}\right)\right). \qquad (39.20)$$

The implications of these results for several scenarios relevant to microfluidic devices are illustrated in Figure 39.7, which depicts the ERR for bilayer debonding as a function of temperature change. Approximate numbers for interface toughness are also listed, based on an attempt to roughly summarize findings from a diverse range of interface studies. It is worth noting that the ERR for all cases is in the ballpark of the pure mode I interface toughness; this indicates that many common fabrication sequences inherently introduce stresses that are "large," in that the device will be near the threshold for debonding. This rather thumbnail sketch illustrates the difficulty in identifying highly repeatable bonding processes. It must be noted, however, that many of the scenarios considered in Figure 39.7 will not necessarily delaminate; the presence of mode II deformation, for which the interface toughness is considerably larger (see table in Figure 39.7), implies greater stability. Nevertheless, it is clear that many common bilayer scenarios are not "far" from debonding, highlighting the need for more precise and systematic characterization of such systems.

39.4.2 Debonding of Pressurized Films Covering Long Channels

Figure 39.8a depicts film debonding due to pressurizing a sealed cavity; the width (span) of the cavity is denoted as L. Recall from Section 39.2.1 that this length should be taken as the channel width plus any debonded region next the interior edge of the film/substrate interface (see Figure 39.2b). For PDMS-glass systems, this inclusion of a debonded region next to the channel involves no significant approximation, because the substrate is effectively rigid and the geometry of the unbonded region is not relevant. For glass–glass systems, the geometry of the unbonded substrate is relevant if the distance from the edge of the channel (see Figure 39.2b) to the crack tip is greater than several

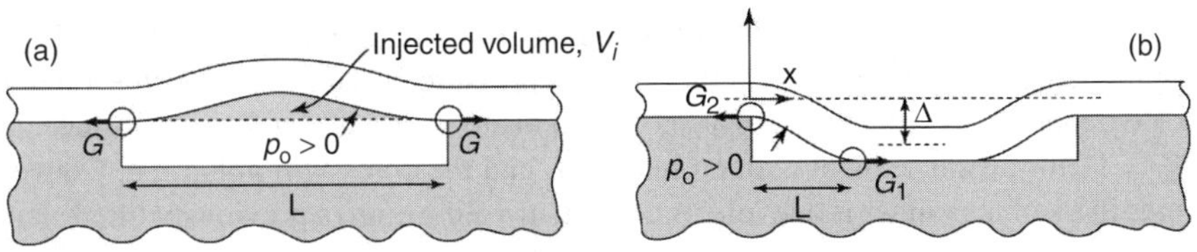

FIGURE 39.8 Side views of geometries relevant to debonding along channels: (a) edge debonding due to internal pressure, (b) debonding to release a "stuck" film.

times the film thickness. In the following, we assume that L refers to the total length of the debonded region, with the understanding that this must extend beyond the edge of the channel when considering glass-glass systems.

This section considers plate and membrane behavior, as opposed to scenarios where the film is thick enough to be considered an elastic half-space (as in Figure 39.2a). Thus, the applications in Section 39.4.2 thus correspond directly to the theory described in Section 39.3.2. The analysis of two relatively thick films (compared to crack length) requires numerical solutions not discussed in Section 39.3.2 and hence is discussed in Section 39.4.3.1 in conjunction with pressure-deflection relationships.

39.4.2.1 Plate Behavior: $L >\sim 10$ h and $\delta_{max} < h$

Figure 39.8b illustrates a pressurized film that adheres to an adjacent layer: this is a more general case of the simple span shown in Figure 39.8a. The film is subjected to a uniform pressure and is deflected at one end (from its initially stress-free position). Solution to Equation 39.6 subject to the boundary conditions $\delta(0) = \delta'(0) = \delta'(L) = 0$ and $\delta(L) = -\Delta$ yield the following deflection profile for the film

$$\delta(x) = \frac{p_o L^4}{2\bar{E}h^3}\left(\frac{x}{L}\right)^2\left(1-\left(\frac{x}{L}\right)\right)^2 - \Delta\left(\frac{x}{L}\right)^2\left(3-2\left(\frac{x}{L}\right)\right), \tag{39.21}$$

where p_o is the resultant pressure, that is, the superposition of that acting on the top and bottom of the film. Plugging this into Equation 39.8, the ERR is

$$G = \frac{p_o^2 L^4}{24\bar{E}h^3}\left(1 \pm 6\frac{\Delta\bar{E}h^3}{p_o L^4}\right)^2. \tag{39.22}$$

Despite the fact that the pressure seems to act purely to "open the crack," it should be noted that there is a mode II component arising from bending of the film at the edge of the delamination: the phase angle is $\psi = -45°$. This highlights the need for mixed-mode interface toughness measurements for material systems relevant to microfluidic systems.

When utilizing Equation 39.22, one must determine the proper sign of the operator in parenthesis. This operation depends on the stress condition at the interface. For instance, if the displacement of the film places the interface in tension (e.g., G_1 in Figure 39.8b where the film is debonding from the bottom of the channel), then a positive sign should be used in Equation 39.22. In this scenario, there is a driving force for debonding even when there is no internal pressure. Conversely, if one was to consider debonding at the upper left edge of the channel—labeled G_2—the displacement of the film acts to close an interface crack at the edge of the channel: for this scenario, the negative sign is used, since the displacement decreases the driving force. One may use the same reasoning to determine the sign of the contribution of the film displacement for interfaces created by the bonding of a third layer on top of the film.

With $\Delta = 0$, the result corresponds to debonding from an interface crack along the edge of a long, straight channel (Figure 39.8a). If the internal pressure is held constant, the ERR increases as the crack extends (i.e., L increases). This corresponds to *unstable crack growth*: once debonding initiates, $G > \Gamma_i$ for all subsequent debond lengths L and the crack will not arrest. Conversely, the crack extends in a stable manner if one injects a constant volume into a previously filled channel. The injected volume will initially be accommodated by the deflection of the film, as shown in Figure 39.8a. Integration of Equation 39.21 (i.e., the displaced profile of the film) yields the relationship between injected volume, V_i and the resulting pressure. This result is then used with Equation 39.8. Since the second term in Equation 39.8 refers to the pressure times the injected volume, it is constant with respect to crack length and it does not enter into the ERR calculation. Thus, the derivative of the potential energy is equal to the derivative with respect to the strain energy, that is, the first term only. This yields

$$G = \frac{7\bar{E}h^3 \left(15V_i\right)^2}{8bL^6}, \tag{39.23}$$

where b is the length of the channel (i.e., the distance in the z-direction over which the injected volume is distributed). This result indicates that crack extension due to injected volume will be stable, in that the ERR will decrease as the crack extends (L increases). This has important implications for both design (i.e., devices that operated under fixed volume conditions), as well as bulge testing to infer interface toughness (see Section 39.4.6.2).

39.4.2.2 Membrane Behavior: $L > \sim 10$ h and $\delta_{max} > h$

For films that experience deflections that are much greater than the film thickness, the membrane models of Section 39.3.2.2 are appropriate. For a membrane covering a long channel of width L, the solution is found by integrating Equations 39.10 and 39.11 and imposing the boundary conditions $\delta\,(x = 0) = \delta\,(x = L) = 0$ and $u\,(x = 0) = u\,(x = L) = 0$. The results are

$$\sigma_o = \frac{\bar{E}}{2 \cdot 3^{1/3}} \left(\frac{p_o L}{\bar{E}h}\right)^{2/3}, \quad \text{(membrane stress)} \tag{39.24}$$

and

$$\delta\,(x) = L\left(\frac{3p_o L}{\bar{E}h}\right)^{1/3} \left(1 - \frac{x}{L}\right)\left(\frac{x}{L}\right). \quad \text{(membrane deflection)} \tag{39.25}$$

The ERR for constant pressure in the channel can be calculated from Equation 39.12: this yields

$$G = \frac{7p_o L}{8}\left(\frac{p_o L}{9\bar{E}h}\right)^{1/3}. \tag{39.26}$$

Once again, one can solve for the ERR that arises due to constant volume injection, using Equation 39.25 to calculate the volume for given pressure. In calculating the potential energy used in the ERR calculation, the second term (pressure times injected volume) is dropped. The result is

$$G = 126\bar{E}h\left(\frac{V_i}{bL^2}\right)^4. \tag{39.27}$$

Under constant volume injection, the ERR again falls as the crack extends (i.e., L increases), implying stable crack extension under constant volume injection.

TABLE 39.2
Summary of Pressure–Deformation Relationships

	Half-Space		Plate		Membrane	
	δ_{max}	Volume	δ_{max}	Volume	δ_{max}	Volume
Channel	$\frac{pL}{E}$	$\frac{\pi L\delta_{max}}{4}$	$\frac{P_0L^4}{32\bar{E}h^3}$	$\frac{8L\delta_{max}}{15}$	$\frac{L}{4}\left(\frac{3p_0L}{\bar{E}h}\right)^{1/3}$	$\frac{2L\delta_{max}}{3}$
Circle or Ellipse	$\frac{4p_0a}{\pi\bar{E}}$	$\frac{2\pi a^2\delta_{max}}{3}$	$\frac{3pa^4/2\bar{E}h^3}{\left(3+2(a/b)^2+3(a/b)^3\right)}$	$\frac{\pi ab\delta_{max}}{3}$	$h\left(\frac{3(1-\nu)p_0a^4}{8Eh^4}\right)^{1/3}$	$\frac{\pi a^2\delta_{max}}{2}$
			Ellipse: $b \geq a$	Ellipse: $b \geq a$		

39.4.3 Pressure–Deflection Relationships for Valve Design

The relationship between pressure, deformation and geometry plays a critical role in microchip design, as it essentially dictates the actuation pressure of valves and the fluidic capacitance introduced by deformable channels. Table 39.2 summarizes several classical results for simple geometries.

39.4.3.1 Long Capped Channels: Plane-Strain Deformation

Figure 39.9 presents the film displacements resulting from pressurizing a cavity, for small pressures (such that deformations are small compared to the film thickness) *and small spans that are comparable to the film thickness*. Results are shown for an elastomer film bonded to a glass substrate: due to the extreme compliance of the elastomer, the substrate is effectively rigid. The curves in Figure 39.9 were generated via numerical FEA of the two-dimensional, plane-strain geometry shown in the inset. Such techniques are required when the deformation of the film is not dominated by bending, as is the case for films that are of comparable thickness to the span.

For very thick films, the behavior is well described by results that assume the film behaves as a semi-infinite half-space. In this limit, the geometry corresponds to that of a pressurized crack, such that the inner surface of the film displaces according to the well-known analytical result (e.g., [33])

$$\delta(x) = \delta_\infty\sqrt{1-\left(\frac{2x}{L}\right)^2}, \tag{39.28}$$

the maximum surface deflection is given by

$$\delta_\infty = \frac{pL}{\bar{E}}. \tag{39.29}$$

Note that this analytical result is independent of the film thickness, which is a consequence of invoking the assumption that the film behaves as a half-space. The maximum displacement of the inner surface, δ_∞, is used to normalize the displacement results in Figure 39.9; hence, the curves asymptote to unity for small aspect ratios (note $a/h = L/2h$).

Figure 39.9 illustrates that for shorter spans, the deflection of the top of the film will be considerably smaller than that of the bottom; this has significant implications for the actuation pressure of thick elastomer films used as valves. Monitoring the deflection of the top surface enables use of the

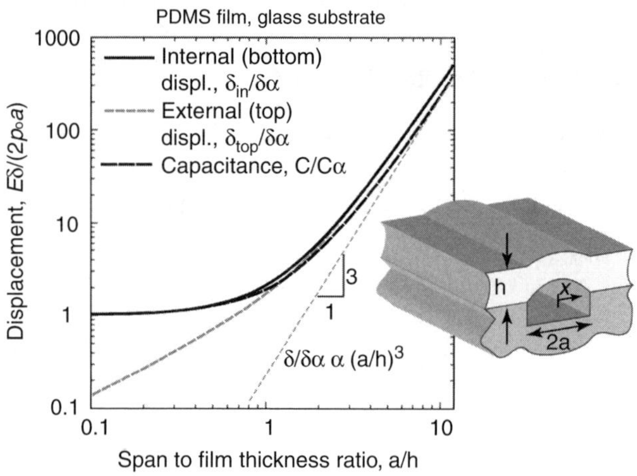

FIGURE 39.9 Film deformation as a function of span to thickness ratio for elastomer films bonded to glass substrates, for *long, straight-walled channels* (plane-strain); results illustrate the transition from the half-space solution (where bending is negligible) to the plate solution (dominated by bending).

film as a pressure sensor to quantify the internal cavity pressure, or direct measurement of the fluidic capacitance of the channel. For larger spans where bending dominates, the deflection of the top (or outer) surface is identical to the bottom surface. When bending dominates, the film deflection is accurately described by the solution for a clamped plate. The deflections of the plate are given by Equation 39.21 with $\Delta = 0$.

The capacitance of the deformable film (discussed more fully in Section 39.4.5) is also shown: this is essentially the channel volume introduced by deformation of the film and is calculated by integrating the film's spatial displacement profile obtained from the numerical analysis. These results are normalized by the volume obtained by integration of Equation 39.28. The capacitance results in Figure 39.9 can be reinterpreted as the ERR for debonding, with the crack length defined according to the convention discussed at the outset of Section 39.4.2. This reinterpretation involves replacing the normalized capacitance with a normalized ERR, G/G_o, where G_o is the ERR for a pressurized crack between two half-spaces: this quantity is given in Table 39.5. The coincidence of the normalized capacitance and that of the normalized ERR is not exact (i.e., there is no obvious analytical reason the two should coincide, at least not obvious to us), but rather an observation based on numerical results. The error is less than ~20% over the range shown in Figure 39.9.

For films that are sufficiently thin to allow for bending, a central question is whether or not pressures are large enough to induce large deflections, in which case one must account for axial stretching of the film (along its centerline). At sufficiently low pressure or small deflections, the deflection varies linearly with pressure (i.e., plate behavior). At large pressures or large deflections, the film deflections scale with $p^{1/3}$. This is referred to as membrane behavior. The transition between plate and membrane behavior can be illustrated using a closed-form solution that implicitly defines the deflection of the film for a given pressure. The solution that describes the applied pressure and the deflection in terms of the axial tension developed in the membrane is [32]

$$\frac{p_o L^4}{32\bar{E}h^4} = \lambda = \frac{\hat{\sigma}^2 \sinh\sqrt{\hat{\sigma}}}{12\left[(24+4\hat{\sigma})\cosh 2\sqrt{\hat{\sigma}} - 18\sqrt{\hat{\sigma}}\sinh 2\sqrt{\hat{\sigma}} - 24 - 16\hat{\sigma}\right]}, \tag{39.30}$$

$$\delta_{\max} = 12h\lambda\left(\frac{\sqrt{\hat{\sigma}} - 2\tanh\left(\frac{\sqrt{\hat{\sigma}}}{2}\right)}{\hat{\sigma}^{3/2}}\right), \tag{39.31}$$

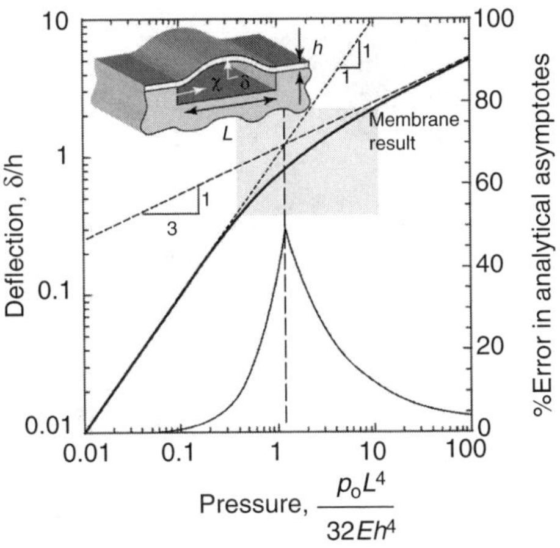

FIGURE 39.10 Deflection–pressure relationship for films over *long, rectangular channels* (plane-strain deformation) illustrating the transition from plate-to-membrane behavior. The shaded region represents combinations of pressures and displacements for which simple analytical solutions have greater than 10% error.

where λ represents the normalized (dimensionless) pressure and is the normalized axial tension in the film: $\hat{\sigma} = 3NL^2/\bar{E}h^3$ where N is the axial resultant (force per unit width of the film). Clearly, the deflection for a given applied pressure can found by numerical solution to Equation 39.5 and substitution into Equation 39.6. However, it is often simpler to generate the load–displacement relationship via the parametric plotting features of commercial codes such as *Mathematica*. The outcome of this procedure is illustrated in Figure 39.10.

For sufficiently large pressure, one obtains the classical membrane solution, described by Equation 39.25. The shaded region in Figure 39.5 represents combinations of pressures and displacements for which simple analytical solutions have greater than 10% error.

39.4.3.2 Circular Films

Similar behavior is observed for circular films, although a closed-form expression for the pressure–deflection relationship in the transition region is not possible. Figure 39.11 illustrates the transition from plate-to-membrane behavior as the pressure applied to a circular film is increased. It should be noted that for circular (i.e., axisymmetric) films, the dependence on Poisson's ratio is slightly different from that of the plane-strain case. Strictly speaking, the pressure–deflection relationship shown in Figure 39.11 depends on the Poisson's ratio; however, the dependence is rather weak and tabulated elsewhere [39]. The asymptotic limits at low and high pressure (i.e., plate and membrane, respectively) are listed in Table 39.2; note that those listed inside Figure 39.11 are for $\nu = 1/2$.

39.4.4 Design of Check Valves for Specific Adhesion Energy and/or Dimensions

The results of the previous sections can be used to design check valves that control fluid motion in powerful ways. Consider the geometry shown in Figure 39.12 [40]: a wall of thickness w separates two fluidic channels within a single layer. At low pressure, the check valve closes the channel by seating on this wall. When a critical pressure is reached, the valve film debonds from its seat and allows fluid to pass from one channel to the other. Assuming displacements are much smaller than the film thickness (such that plate behavior applies) the critical pressure required to release the check

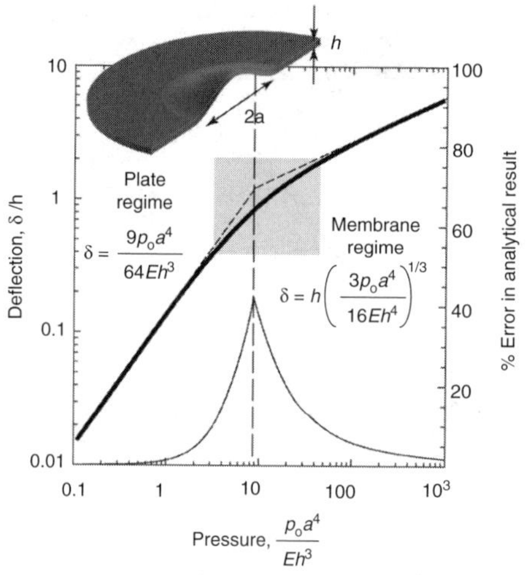

FIGURE 39.11 Deflection–pressure relationship for *circular* films with $v = {}^1\!/_2$, illustrating the transition from plate-to-membrane behavior. The shaded region represents combinations of pressures and displacements for which analytical solutions have greater than 10% error.

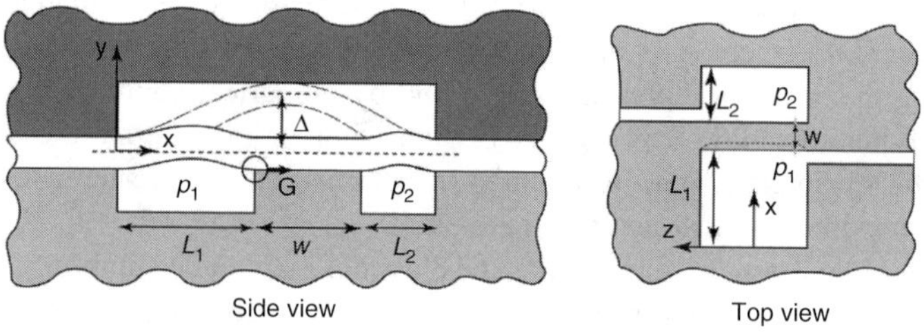

FIGURE 39.12 Schematic diagram of a fluidic check valve: by modulating the dimensions (and/or interface adhesion energy), one can design valves that allow flow in only one direction, and achieve the desired deflection after debonding at the design pressure.

valve and allow for fluid flow is dictated by Equation 39.22 with $\Delta = 0$: that is,

$$p_1^c = \frac{\sqrt{24\bar{E}h^3\Gamma_i}}{L_1^2}, \tag{39.32}$$

where Γ_i is the adhesion energy of the interface formed by the film and the valve seat. Clearly, one may control the relative pressures to allow for flow from left-to-right (or vice versa) by suitable choice of the channel width, L. The ratio of activation pressures for forward (1 to 2) and reverse (2 to 1) directions thus scales as

$$\frac{p_1^c}{p_2^c} = \left(\frac{L_2}{L_1}\right)^2. \tag{39.33}$$

Above the critical pressure, the valve debonds, which connects the two fluid channels. The valve displaces according to Equation 39.21 with $L = L_1 + L_2 + w$ and $\Delta = 0$.

Suppose that in the "open" position, the desired film displacement is Δ; presumably this is chosen to achieve a specific flow rate at the activation pressure of the valve. Alternatively, Δ may represent the maximum allowable displacement of the valve film to prevent adhesion with overlying layers, as shown in Figure 39.12. Thus, one can identify the combinations of material properties and dimensions that result in a valve that is either closed or open to a prescribed clearance by setting $p_1^c = p_{\max}(\Delta)$. This combination is

$$\frac{\Gamma_i}{\bar{E}h} = \frac{\left(\frac{\Delta}{L_1}\right)^2\left(\frac{h}{L_1}\right)^2}{\left(1+\frac{L_2}{L_1}+\frac{w}{L_1}\right)^8}. \tag{39.34}$$

In ideal world, one might modulate the adhesion energy of the valve seat to achieve the desired performance for a specific set of dimensions. This would likely be difficult as the chemistry of the fluidic environment, which obviously depends on the application, often affects interface toughness. However, one can measure the interface toughness for a given application and then choose the dimensions of the chambers to satisfy Equation 39.34: the result is a check valve that actuates to a specific clearance for a given interface condition.

One can naturally derive similar design guidelines for chamber dimensions comparable to the film thickness (i.e., the elastic half-space regime) or valve displacements greater than the film thickness (i.e., the membrane regime).

39.4.5 Fluidic Capacitance

Any feature of the fluidic network whose internal volume changes with pressure acts as a fluidic capacitor, because those volume changes imply that additional mass will be "stored" inside the feature. This behavior can be exploited to alter the characteristic timing of flow inside a fluidic "circuit" (e.g., [41]). *Assuming density changes are negligible*, the mass flow rate can be written as

$$\hat{q} = \frac{\mathrm{d}m}{\mathrm{d}t} = \rho\frac{\mathrm{d}}{\mathrm{d}t}\left(V_o + \frac{\mathrm{d}V}{\mathrm{d}\,(\Delta p)}\mathrm{d}(\Delta p)\right), \tag{39.35}$$

where V_0 is the initial fixed volume of the element, ρ is the mass density and $\Delta p = p_{\text{in}} - p_{\text{out}}$ is the difference between pressure inside and outside the element. Obviously, the volume flow rate (at constant density) is simply $\hat{q} = q/\rho$. In terms of volume flow rate,

$$q = \frac{\mathrm{d}V}{\mathrm{d}p}\cdot\frac{\mathrm{d}p}{\mathrm{d}t} \equiv C\dot{p}, \tag{39.36}$$

where it should be noted $\mathrm{d}V_o/\mathrm{d}t = 0$ by definition. Clearly, the flow rate is defined as positive towards the capacitance element inside the chamber: when the pressure inside is greater, the film deflects outward and the capacitor stores more fluid (positive flow rate). Hence, for any enclosed volume, the fluidic capacitance is defined as the change in volume with respect to the difference between internal and external pressure.

Table 39.2 lists the pressure-volume relationships for various geometries: the fluidic capacitance is found simply by differentiating with respect to pressure. For small deformations, volume varies linearly with applied pressure, such that the capacitance is not a function of the pressure: it merely defines the proportionality between increases in pressure and increases in "stored" mass. For such cases, the fluid circuit analysis is linear, because flow rate and pressure drops are related via linear expressions. For large deformations (i.e., the membrane limit), the fluidic capacitance is a function of the pressure: this implies that the fluidic circuit behavior will be nonlinear. Obviously, once the fluidic resistance, capacitance and inductance have been identified via geometry (and material properties), complicated networks can be analyzed using commercially available circuit analysis software such as SPICE [42].

39.4.6 Material Property Measurements

This section outlines several approaches to characterizing the elastic modulus of thin films and the interface toughness when bonded to substrates. Naturally, there is an elastic enormous variety of test configurations that can be used; attention here is limited to those with specimen fabrication that is similar to that used for typical microfluidic devices. The text by Freund and Suresh [32] describes additional test approaches, particularly those pertaining to quantifying the effects of mode-mixity on interface toughness.

39.4.6.1 Indentation Modulus Measurements

As evident from the calculations in previous sections, the elastic modulus of a material must be known if a researcher hopes to predict interface toughness for a fabricated device. An increasingly prevalent method to characterize modulus is instrumented indentation, wherein an indenter tip of known shape is pressed into a surface: see Figure 39.13a. The plane-strain modulus of the material, that is, $\bar{E} = E/\left(1 - \nu^2\right)$, is extracted from the measured force–displacement relationship. The test method is particularly attractive because it requires minimal sample preparation and mounting: one must merely ensure that a nominally flat film can be attached to substrate that can be inserted into the apparatus. This makes it ideal for testing thin films, as samples can be deposited via spin casting, chemical vapor deposition and so on. Moreover, the dynamic ranges of force and displacement measurement continue to expand, enabling tests on a broad range of film thickness. Conventional micro-indenters typically apply forces in the Newton range and are capable of measuring displacements in the > 10 micron range. Nanoindentation systems can accurately apply forces beneath a milli-Netwon and measure displacements with nanometer precision over a broad dynamic range spanning tens of microns. Such systems are increasingly common and are becoming the de facto choice for thin film modulus measurements, due to sophisticated and automated test control and interpretation.* The book chapter by Hay and Pharr [43] provides an excellent, detailed overview of the instrumentation and theory involved in indentation testing.

Typical load–displacement curves obtained from instrumented indentation experiments are shown in Figure 39.13b. Results are shown for silica (E = 72 GPa, ν = 0.2) and polystyrene (PS, E = 2 GPa, ν = 0.4). The load–displacement curve for PDMS (E = 1.5 MPa, $\nu \sim 0.5$) involve forces that are too low to appear on the scale used in Figure 39.13b, due to the fact its elastic modulus is three orders of magnitude lower than polystyrene.

Depending on the material being tested, the loading path is a complicated function of elastic and inelastic deformation (such as creep or plasticity). Conversely, the initial portion of the unloading curve is dominated by purely elastic deformation; it is this portion of the measurement that is used to extract the modulus. The initial slope of the unloading curve is often referred to as the contact stiffness and is denoted here as S. The postprocessing method for determining S by curve-fitting the unloading portion of the curve is described in detail in [43]. The basic relationship that is used is

$$\frac{EE_{\mathrm{i}}}{\left(1 - \nu^2\right) E_{\mathrm{i}} + \left(1 - \nu_{\mathrm{i}}^2\right) E} = \frac{S}{\sqrt{A\left(\delta_{\mathrm{c}}\right)}}, \tag{39.37}$$

where E_{i} is the modulus of the indenter and S is the measured stiffness (i.e., slope of the initial portion of the unloading curve). A is the area function of the indenter tip: this function describes the relationship between contact depth, δ_{c}, and the projected contact area between the indenter tip and surface (see Figure 39.13a).

* Indentation testing can also be performed with an atomic force microscope, which has pico-Newton force and angstrom displacement resolution at the lower end; the draw back is that the dynamic displacement range is limited to several microns. Moreover, on older systems, the control software, as well as that needed for test interpretation, may not be readily available.

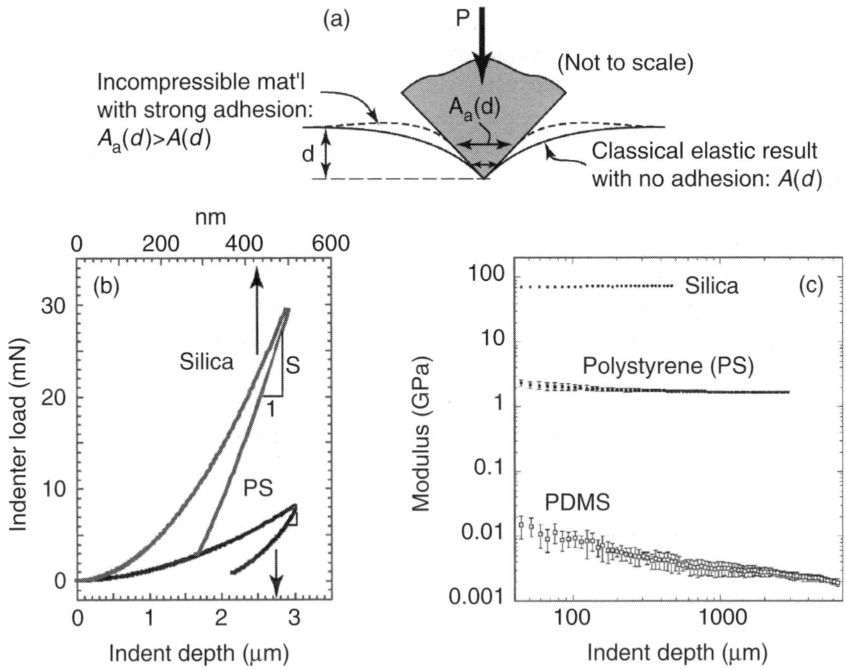

FIGURE 39.13 Typical measurements obtained using a commercial nanoindenter: (a) schematic of the contact area, (b) typical load–displacement measurements, and (c) inferred modulus as a function of penetration depth for several materials.

The accuracy of the test method relies critically on the accurate determination of the area function, which varies with tip geometry. At the scale of many nanoindentation experiments, minute differences in tip geometry—such as those caused by wear induced by repeated experiments—can have a large effect on the area function. For this reason, the area function is typically measured by indenting a known material to several depths and computing the function from Equation 39.37. This experimentally determined tip function is then used to interpret further tests on unknown materials. Fused silica (or quartz) is commonly used to "calibrate the tip," that is, determine the area tip function, because its material response is nearly perfectly elastic and it does not exhibit adhesion with diamond indenter tips.

Provided one is testing relatively stiff material such as glassy polymers, ceramics and metals, the choice of indenter tip geometry is largely one of convenience; by far the most common is a diamond that has been cleaved on crystallographic planes to generate a well-defined tip profile. The resulting shape is that of a pyramid with 65.3° face angle and a three-sided base, known as a Berkovich tip. Other common indenter tips include Vickers (a pyramid with a square base), spheres, cones and even cube-corners (i.e., the corner of a cube that is lopped off and mounted such that the direction of indentation aligns with the diagonal of the cube). It should be emphasized that the actual geometry differs from these nominal shapes at small length scales; for example, a Berkovich tip will typically have a rounded tip with a radius of ~100–300 nm after hundreds of indents on stiff materials such as ceramic or metals. This implies that careful tip calibration is critical for indents whose maximum penetration is smaller than several microns.

Modern hardware and testing protocols are capable of superimposing a small oscillatory component to the apply load. This effectively introduces a series of unloading cycles that can be used to extract the modulus; this technique is commonly referred to as the dynamic contact method (DCM) and is a standard option in commercial nanoindentation systems. The result is that one can measure the modulus of the material as a function of indentation depth in one rapid test.* Examples of data obtained from such nanoindentation testing are shown in Figure 39.13c for several materials: the results represent the average of a dozen tests, with the error bars indicating standard deviation. The results for the ceramic (silica) and glassy polymer (polystyrene) are typical of relative stiff materials whose modulus is greater than 1 GPa: for penetration depths greater than one hundred nanometers, the modulus is independent of probe depth. Moreover, for these stiff materials there is very little scatter. This indicates that the tests successfully probed the "bulk" material properties and were not unduly influenced by surface phenomena.

In contrast, the results in Figure 39.13c for the elastomer (PDMS) exhibit larger scatter and an elastic modulus that depends on depth. The indentation testing of elastomers is obviously relevant to microfluidic devices; unfortunately, such materials introduce a host of challenges that continue to be addressed by the mechanics community. First and foremost, such materials exhibit strong adhesion with indenter tips; adhesion obfuscates the inferred modulus by artificially increasing the contact area for a given depth (see Figure 39.13a). This increases the slope of the unloading curve and one infers a larger modulus than reality (see Figure 39.13c: the average modulus for strains less than 3% is 1.5 MPa). Second, such materials exhibit nonlinear stress–strain behavior, which strictly speaking, violates the models derived earlier. Even though the material is purely elastic (does not experience permanent deformation), this leads to ambiguity regarding the imposed strain level at which a modulus is extracted. The most desirable strategy that avoids these problems is to use a flat-ended punch; in this case, provided the indenter is not unloaded too far, adhesion plays no role during unloading because the contact area remains in compression and does not change.

Alternatively, one can indent a freestanding film (i.e., a blanket film covering a hole or microfluidic chamber) using an instrumented probe, as shown in Figure 39.14. With careful choice of dimensions, the measured load-deflection response is dominated by plate bending or membrane stretching, in which case analytical solutions are accurate (e.g., [38]). Adhesion is less of a factor, because the load–deflection response is dominated by behavior outside the contact area. In many regards, the instrumentation can be easier to set up; one can simultaneously measure load and displacement with the same probe (as opposed to pressure loading in "bulge tests," (Section 39.4.6) where film deflection must be monitored separately from the applied pressured). This has been illustrated for macroscale elastomer films in the membrane regime (PDMS: $\sim$100 μm thick and 5 cm in diameter [38,44]). Similar tests in the plate regime have been demonstrated for ultrathin glassy polymers that cover microfabricated holes ($\sim$750 nm thick PMMA films over holes 60 μm in diameter: [45]), using a commercial nanoindenter as the instrumented probe.

39.4.6.2 Bulge Testing to Determine Modulus and Interface Toughness

The configurations shown in Figures 39.3, 39.10, and 39.11 can be used to probe mechanical properties of the system by monitoring film displacement while modulating the pressure: this is typically referred to as a "bulge test" (e.g., [31,32,46–48]). If the delamination crack front can be monitored while the pressure (or injected volume) is modulated, the interface toughness can also be determined. There are two design considerations for the test: (1) the range of strain imposed during the test and (2) the nature of the load–deflection relationship as dictated by the film thickness-to-span ratio. The former determines whether or not the strains are sufficiently small to allow for a linear elastic approximation and to prevent film rupture. Table 39.3 summarizes the relationships between applied pressure and the maximum strain in the film for various configurations. To address the second consideration, the film thickness and span is ideally such the range of applied pressure enables the accurate application of a simple closed-form analytical expression, that is, plate solutions or membrane solutions.

* Or more importantly, one can extract an accurate area tip function by averaging the results of dozens of indents.

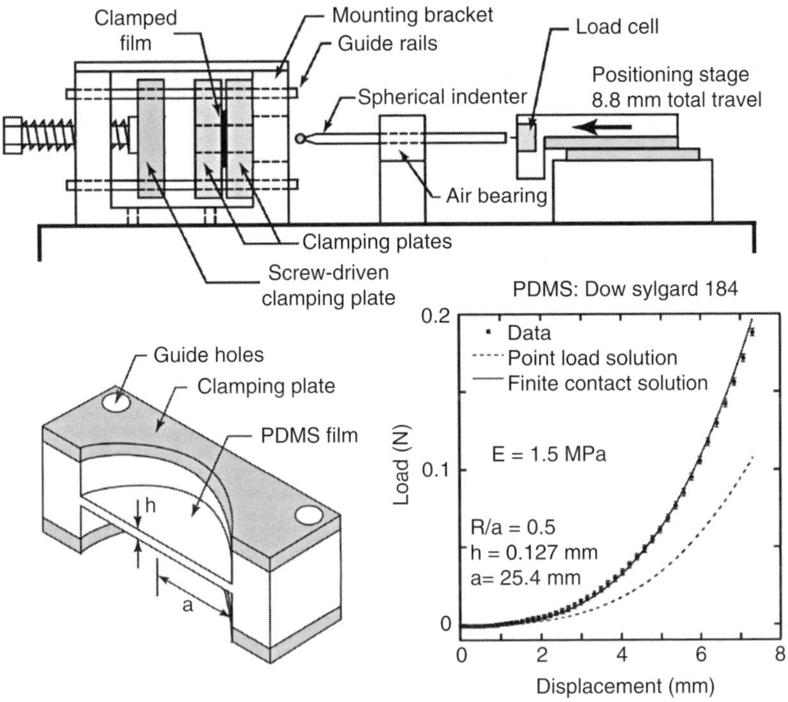

FIGURE 39.14 Schematic illustration of deflection test on freestanding PDMS film (Dow Sylgard 184 with a 10:1 mixing ratio) [38].

TABLE 39.3
Estimate for Maximum Strain in the Film

Long Channel: Plane Strain		Circular	
Plate	**Membrane**	**Plate**	**Membrane**
$\dfrac{p_0 L^2}{4\bar{E}h^2}$	$\dfrac{1}{6}\left(\dfrac{3p_0 L}{\bar{E}h}\right)^{2/3}$	$\dfrac{3p_0 a^2}{4\bar{E}h^2}$	$\left(\dfrac{(1-\nu)p_0 a}{2Eh}\right)^{2/3}$

Table 39.4 summarizes simple rules of thumb for designing microfluidic systems developed from the pressure-strain solutions presented in Section 39.4. Based on the resolution of the available instrumentation and anticipated modulus, these recommendations can be adjusted for film thickness and span as required. The broad generalizations of Table 39.4 can be specified more quantitatively using the results shown in Figures 39.9 through 39.11; that is, critical values of pressure and deflection can be identified such that deflections are within a specified percentage of the analytical solutions.

The bulge test described above can be used to characterize the interface as well as the film itself. Table 39.5 summarizes the ERR (G) for the half-space, plate and membrane limits, in terms of both applied pressure and injected volume. The same limits identified in Table 39.4 can be used to establish the limits of accuracy. The use of injected volume is preferred because the crack front advances in a stable manner: this implies that a single film can be used to make multiple measurements. In order to extract interface toughness, one must monitor the "crack length," that is, the width of the debonded (or bulging) region. For transparent films, the gap between the bulging film and substrate creates enough contrast to monitor the position of the crack front using optical methods. Depending on the film thickness, optical methods are often sufficient even for opaque films, as the edge of the bulging region

TABLE 39.4
Design Guidelines for Accurate Closed-Form Pressure-Deflection Solutions

Long Channel: Plane Strain		Circular ($v = 1/2$)	
Plate	**Membrane**	**Plate**	**Membrane**
$\delta_{\max} < h/3$	$\delta_{\min} > 2h$	$\delta_{\max} < h/3$	$\delta_{\min} > 2h$
$p_o^{\max} < 0.8\bar{E}\left(\frac{h}{a}\right)^4$	$p_o^{\min} > 30\bar{E}\left(\frac{h}{a}\right)^4$	$p_o^{\max} < 3\bar{E}\left(\frac{h}{a}\right)^4$	$p_o^{\min} > 40\bar{E}\left(\frac{h}{a}\right)^4$

TABLE 39.5
Summary of Energy Release Rates

	Half-Space	Plate		Membrane	
	Pressure	**Pressure**	**Volume Injection**	**Pressure**	**Volume Injection**
Channel	$\frac{\pi p_o^2 L}{4\bar{E}}$ (*)	$\frac{p_o^2 L^4}{24\bar{E}h^3}$	$\frac{3\bar{E}h}{2}\left(\frac{h}{L}\right)^2\left(\frac{10V_i}{bL^2}\right)^2$	$\frac{7p_o L}{8}\left(\frac{p_o L}{9\bar{E}h}\right)^{1/3}$	$126\bar{E}h\left(\frac{V_i}{bL^2}\right)^4$
Circle	$\frac{2p_o^2 a}{\pi\bar{E}}$ (*)	$\frac{3p_o^2 a^4}{32\bar{E}h^3}$	$\frac{24\bar{E}h}{\pi^2}\left(\frac{h}{a}\right)^2\left(\frac{V_i}{a^3}\right)^2$	$\frac{5}{18}\left(\frac{3(1+v)p_o^4 a^4}{\bar{E}h}\right)^{1/3}$	$\frac{112\bar{E}h}{(1+v)}\left(\frac{V_i}{a^3}\right)^4$

(∗) These results assume the substrate is rigid compared to the film: if the film and substrate have identical elastic properties, these results should be multiplied by two.

is clearly visible (particularly when using white light interferometry). Alternatively, if the volume of injected material is known, the crack length can be inferred from the maximum deflection of the film. The toughness can be determined by least squares fit to the measured $L = f\ (V_i)$ relationship, using $G = \Gamma_i$ as the unknown fitting parameter. Monitoring small deflections generally ensures that the linearly elastic material assumption is valid; however, this also implies small injected volumes that may be difficult to achieve. Thus, one should design the test—that is, adjust film thickness and span—to lie in the appropriate regime of behavior based on the resolution limits of the volume injection and displacement monitoring.

39.4.6.3 Peel Tests to Determine Interface Toughness

Perhaps the most straightforward test to determine interface toughness is the peel test, wherein one measures the applied force required to peel a film from the substrate. The test is illustrated in Figure 39.15 and has the advantage of measuring at least part of the dependence of the interface toughness on mode-mixity. The ERR for this configuration is [32,49]

$$G = \sigma_a h(1 - \cos\theta) + \frac{\sigma_a^2 h}{2\bar{E}}, \tag{39.38}$$

where σ_a is the stress applied to the film at the end. (Note that this simply the force applied to the film divided by the cross sectional area of the strip being pulled.) The test involves mixed-mode loading due to bending of the film near the edge of the debonded region, except for the case where $\theta = 0°$ that corresponds to pure mode II. For compliant films on much stiffer substrates (i.e., $\alpha \approx -1$) the

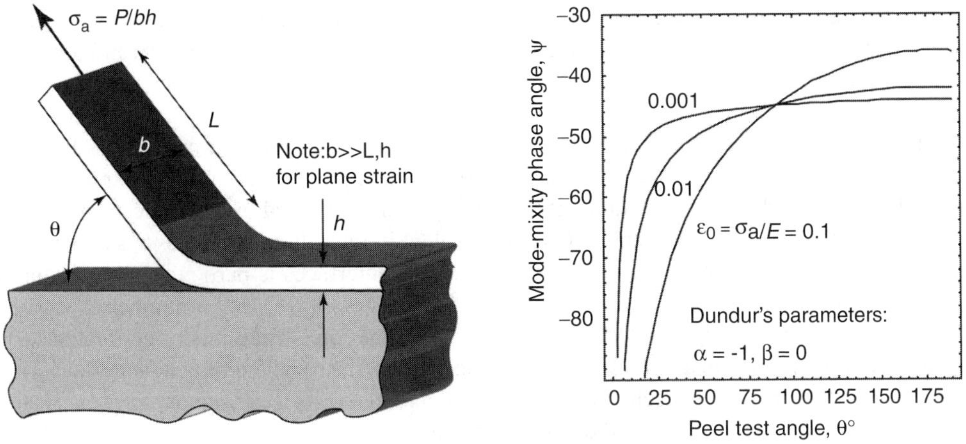

FIGURE 39.15 Schematic of the peel test, and associated mode-mixity phase angle for the case of a flexible film on a rigid substrate; mode II arises from bending near the debond edge.

phase angle of the test ψ is given by

$$\psi = -\tan^{-1}\left(\frac{\cos\theta - \sqrt{\frac{2(1-\cos\theta)}{\varepsilon_o} + \sin^2\theta}}{-\cos\theta + \sqrt{\frac{2(1-\cos\theta)}{\varepsilon_o} + \sin^2\theta}}\right), \tag{39.39}$$

where $\varepsilon_o = \sigma_a/\bar{E}$ is the level of strain in the film far from the debond edge. Care should be taken to ensure that the material is appropriately modeled using linear elasticity for the strain level at debonding.

The phase angle (Ψ) is shown in Figure 39.15 as a function of peel test angle for several values of applied strain. In all cases, the phase angle is negative due to a negative mode II component: thus, the test is only capable of measuring mixed-mode interface toughness in this regime. When the film is pulled vertically (i.e., $\theta = 90°$), the phase angle is $\psi = -45°$ regardless of the applied strain level: this is identical to the bulge test in the plate regime.

39.5 FUTURE PROSPECTS AND CONCLUDING REMARKS

The convergence of microfabrication technology developed for microelectronics, microelectromechanical systems (MEMS) and lab-on-chip technology is steadily progressing, and creating rapidly expanding opportunities for new types of miniaturized chemical analysis. It seems obvious that future of microfluidic systems lies in hybrid/composite devices, which incorporate a wide range of materials and geometric features and hence, fabrication techniques. Future chips will undoubtedly involve integrated metallic electrodes, piezoelectric materials for actuation, micropatterned polymers for chemical sensing and so on. The successful realization of such complex microfluidic devices relies critically on the translation of thin film mechanics to those material systems that differ from typical semiconductor and metallic devices. The theoretical and experimental approaches to predicting the mechanical response of multilayer systems outlined here provide a well-established foundation that should be applied to these material systems.

The increasing use of polymers in microfluidic devices, especially elastomers, is a significant departure from microelectronic devices that tend to rely on comparatively stiff ceramics and metals. Hence, there is a critical need for sustained and systematic experimental and theoretical characterization of polymer/ceramic and polymer/metal interfaces. The utility of such studies will be strongly influenced by the successful partnering of chemists and engineers; it is clear that relevant material systems must be defined by those with an understanding of chemical compatibility and device performance requirements, while those familiar with mechanical integrity offer an efficient route to characterization.

A critical area that needs analysis is the initiation of interface failure at the corners formed by bonding capping layers over patterned channels. While there have been significant inroads to this problem from the mechanics community, it has largely been motivated by microelectronic devices. As such, the analysis of geometry and materials prevalent in microfluidic devices has received only cursory treatment. Similarly, the behavior of PDMS/glass interfaces is particularly important, due to their ubiquitous utilization in emerging devices. While there have been a number of studies in this regard, the understanding of interface toughness as a function of both mode-mixity and surface treatment requires significant additional study. Related to this, the role of nonlinear material behavior and large strain deformation in mechanics of interface delamination is largely unexplored, yet is likely to play an important role in successful device fabrication.

NOMENCLATURE/GLOSSARY

- a Radius of a circular span, used in expressions to describe the behavior of a deformable film bonded over a circular hole.
- b The length of the crack front that experiences debonding, often taken as unity.
- α, β Dundur's parameters: dimensionless parameters that functions of the elastic properties of two materials on either side of an interface, which play a critical role in interface mechanics.
- α_i Coefficient of thermal expansion of the ith layer, used to include the effects of thermal strains introduced during deposition.
- C Fluidic capacitance, that is, volume of fluid stored in a deformable element per unit pressure.
- δ deflection of a film or half-space in the direction normal to the interface (or axis of the film), may appear as a function of position along the film axis (i.e., $\delta(x)$).
- Δ The maximum or requisite displacement of a valve film, which generally is dictated by the fluidic channel depth.
- ε Strain, usually refers to the direct strain (aligned with the major axis) of a deformable film.
- ε_o The axial extensional strain resulting from bonding two layers.
- E_i Elastic modulus (Young's modulus) of layer "i": if there is no subscript, then this refers to the modulus of the film (as the substrate is assumed to be rigid).
- $\bar{E}_{\mathrm{i}}$ The plane-strain elastic modulus of the layer "i": $\bar{E} = E/(1 - r^2)$
- G The energy release rate, or "crack driving force," which is used to predict crack stability.
- Γ_i The critical value of the energy release rate at which crack extension occurs, often referred to as the fracture toughness, or adhesion energy: the subscript indicates that the quantity refers to the interface toughness (as opposed to a bulk material).
- h_i Thickness of layer "i": if there is no subscript, then this refers to the Poisson's ratio of the film (as the substrate is assumed to be rigid).
- κ The curvature of a bilayer created by bonding together two layers.
- L Total span (or crack length) across a long, rectangular channel.
- M Resultant moment in a thin film.
- N Axial force resultant in a thin film.
- p_o The *net* pressure acting inside a cavity; that is, the sum of pressures acting on both sides of the film: assumed to be spatially uniform.
- P, S Indentation load and stiffness (i.e., the slope of the load–displacement curve).
- T_i, T_i^o Temperature and reference temperatures, respectively, of the ith layer: the reference temperature is defined as that at which there is no thermal strain, typically the fabrication temperature of the layer.
- θ_i $= \alpha_i \left(T_i - T_i^{\mathrm{o}}\right)$, the thermal strain in the ith layer.

σ Stress.

ψ Phase angle describing mode-mixity of crack driving force: describes the relative amounts of mode I (opening) and mode II (sliding) deformation at a crack tip.

U Potential energy of the system, comprised of the strain energy of a deformed solid and the work done by applied loads.

ν_i Poisson's ratio of layer "i": if there is no subscript, then this refers to the Poisson's ratio of the film (as the substrate is assumed to be rigid).

REFERENCES

1. Ruska, W.S. *Microelectronic Processing: An Introduction to the Manufacture of Integrated Circuits*, McGraw-Hill: New York, 1987.
2. Morgan, D.V. and Board, K. *An Introduction to Semicondutor Microtechnology*, 2nd edn. John Wiley & Sons: New York, 1990.
3. Madou, M. *Fundamentals of Microfabrication*, CRC Press: Boca Raton, 1997.
4. van Zant, P. *Microchip Fabrication: A Practical Guide to Semiconductor Processing*, McGraw-Hill: New York, 2000.
5. Becker, H., Lowack, K., and Manz, A. "Planar quartz chips with submicron channels for two dimensional capillary electrooresis applications," *Journal of Micromechanics and Microengineering* 1998, 8, 24–28.
6. Rocklin, R.D., Ramsey, R.S., and Ramsey, J.M. "A microfabricated fluidic device for performing two-dimensional liquid-phase separations," *Analytical Chemistry* 2000,72, 5244–5249.
7. Gottschlich, N., Jacobson, S.C., Culbertson, C.T., and Ramsey, J.M. "Two dimensional electrochromatography/capillary electrophorsis on a microchip," *Analytical Chemistry* 2001, 73, 2669–2674.
8. Ramsey, J.D., Jacobson, S.C., Culbertson, C.T., and Ramsey, J.M. "High-efficiency, two-dimensional separations of protein digests on microfluidic devices," *Analytical Chemistry* 2003, 75, 3758–3764.
9. Wang, Y.C., Choi, M.N., and Han, J.Y. "Two-dimensional protein separation with advanced sample and buffer isolation using microfluidic valves," *Analytical Chemistry* 2004, 76, 4426–4431.
10. Wu, H.K, Wheeler, A., and Zare, R.N. "Chemical cytometry on a picoliter-scale integrated microfluidic chip," *Proceedings of the National Academy of Sciences* 2004, 101351, 12809–12813.
11. Roper, M.G., Shackman, J.G., Dahlgren, M., and Kennedy, R.T., "Microfluidic chip for continuous monitoring horome secretion from live cells using an electrophoresis-based immunoassy," *Analytical Chemistry* 2003, 75, 4711–4717.
12. Oleschuk, R.D., Shultz-Lockyear, L.L., Ning, Y.B., and Harrison, D.J. "Trapping of bead-based reagents within microfluidic systems: On-Chip solid-phase extraction and electrochromatography," *Analytical Chemistry* 2000, 72, 585–590.
13. Jemere, A.B., Oleschuk, R.D., Ouchen, F., Fajuyigbe, F., and Harrison, D.J. "An integrated solid-phase extraction system for sub-picomolar detection," *Electrophoresis* 2002,23, 3537–3544.
14. Stachowiak, T.B., Svec, F., and Frechet, J.M.J. "Chip electrocrmatography," *Journal of Chromatography A* 2004, 1044, 97–111.
15. Peterson, D.S. "Solid supports for micro analytical systems," *Lab on a Chip* 2005, 5, 132–139.
16. Guillo, C., Karlinsey, J.M., and Landers, J.P. "On-chip pumping for pressure mobilization of the focused zones following microchip isoelectric focusing," *Lab on a Chip* 2007, 7, 112–118.
17. Satow, T., Machida, A., Funakushi, I.C., and Palmieri, R. "Effects of the sample matrix on the separation of peptides by high performance capillary electrophoresis," *Journal of High Resolution Chromatography* 1991, 14, 276.
18. Svec, F. "Preparation and HPLC applications of rigid macroporous organic polymer monoliths," *Journal of Separation Science* 2004, 27, 747–766.
19. Thorsen, T., Maerkl, S.J., and Quake, S.R. "Microfluidic large-scale integration," *Science* 2002, 298, 580–584.
20. Fu, A.Y., Chou, H.P., Spence, C., Arnold, F.H., and Quake, S.R. "An integrated microfabricated cell sorter," *Analytical Chemistry* 2002, 74, 2451–2457.

21. Grover, W.H., Skelley, A.M., Liu, C.N., Lagally, E.T., and Mathies, R.A. "Monolithic membrane valves and diaphragm pumps for practical large-scale integration into glass microfluidic devices." *Sensors and Actuators B—Chemical*, 2003, 89, 315–323.
22. Jeon, N.L., Chiu, D.T., Wargo, C.J., Wu, H., Choi, I.S., Anderson, J.R., and Whitesides, G.M. Design and fabrication of integrated passive valves and pumps for flexible polymer 3-dimensional microfluidic systems, *Biomedical Microdevices*, 2002, 4, 117–121.
23. Easley, C.J., Karlinsey, J.M., Leslie, D.C., Begley, M.R., and Landers, J.P. Directional and frequency-dependent flow control in microfluidic circuits using passive elastomeric components, *Proceedings of the mTAS 2006 Conference, 10th International Conference on Miniaturized Systems for Chemistry and the Life Sciences (Tokyo)*, 2006, 2, 1064–1068.
24. Buchholz, B.A., Doherty, E.A.S., Albarghouthi, M.N., Bogdan, F.M., Zahn, J.M., and Barron, A.E., "Microchannel DNA sequencing matrices with a thermally controlled "viscosity switch," *Analytical Chemistry*, 2001, *v*73, 157–164.
25. Manz, A., Harrison, D.J., Verpoorte, E., Fettinger, J.C., Paulus, A., Ludi, H., and Widmer, H.M., "Planar Chips Technology for Miniaturization and Integration of Seperation Techniques into Monitoring Systems, CE on a chip" *Journal of Chromatography*, 1992, 593, 253–258.
26. Harrison, D.J., Manz, A., Fan, Z.H., Ludi, H., and Widmer, H.M., "Capillary Electrophoresis and Sample Injection Systems Integrated on a Planar Glass Chip," *Analytical Chemistry*, 1992, 64, 1926–1932.
27. Geschke, O., Klank, H., and Telleman, P., Ed. *Microsystem Engineering of Lab-on-a-Chip Devices*, Wiley VCH: Weinheim, 2004.
28. Li, Paul C.H. *Microfluidic Lab-On-a-Chip for Chemical and Biological Analysis and Discovery*, CRC Press Taylor & Francis: Boca Raton, 2006.
29. Xia, Y. and Whitesides, G.M. "Soft lithography," *Angewandte Chemie, International Edition* 1998, 37, 550–575.
30. Xia, Y., and Whitesides, G.M. "Soft lithography," *Annual Review of Materials Science* 1998, 28, 153–184.
31. Hutchinson, J.W. and Suo, Z. "Mixed mode cracking in layered materials." In *Advances in Applied Mechanics*, Hutchinson, J.W., Wu, T.Y. Eds. Academic Press: London 1992, 29, 63–191.
32. Freund, L.B. and Suresh, S. *Thin Film Materials: Stress, Defect Formation, and Surface Evolution*, Cambridge University Press: Cambridge, 2003.
33. Kanninen, M.F. and Popelar, C.H., *Advanced Fracture Mechanics*, Oxford University Press, New York, 1985.
34. Mohammed, H. and Leichti, K.M., "Cohesive zone modeling of crack nucleation at bimaterial corners," *Journal of the Mechanics and Physics of Solids*, 2000, 48, 735–764.
35. Labossiere, P.E.W., Dunn, M.L., and Cunningham, S.J., "Application of bimaterial interface corner failure mechanics to silicon/glass anodic bonds," *Journal of the Mechanics and Physics of Solids* 2002, 50, 405–433.
36. Reedy, Jr., E.D., "Strength of butt and sharp cornered joints," *Comprehensive Adhesion Science*, Elsevier Press, Amsterdam. 2001.
37. Ambrico, J.M. and Begley, M.R., "The role of flaw geometry in film delamination from two-dimensional interface flaws along free edges," *Engineering Fracture Mechanics* 2003, 70, 1721–1736.
38. Scott, O.N., Begley, M.R., Komaragiri, U., and Mackin, T.J. "Indentation of freestanding elastomer films using spherical indenters," *Acta Materialia* 2005, 52, 4877–4885.
39. Komaragiri, U., Begley, M.R., and Simmonds, J.G. The mechanical response of freestanding circular elastic films under point and pressure loads, *Journal of Applied Mechanics* 2005, 72, 203–212.
40. Easley, C.J., Leslie, D.L., Landers, J.P., Utz, M., and Begley, M.R. "Design of microfluidic diodes and check valves," to be published, 2007.
41. Easley, C.J., Leslie, D.C., Karlinsey, J.M., Begley, M.R., and Landers, J.P. "Microfluidic waveform shaping with passive elastomeric components," to be published, 2007.
42. Roberts, G. and Sedra, A., *SPICE*, Oxford University Press, New York, 1996.
43. Hay, J.L. and Pharr, G.M. "Instrumented indentation testing," chapter in *ASM Handbook Volume 8: Mechanical Testing and Evaluation*, 10th edn., H. Kuhn and D. Medlin, Ed., ASM International, Materials Park, OH, 2000, pp. 232–243.

44. Begley, M.R. and Mackin, T.J. "Spherical indentation in the membrane regime," *Journal of the Mechanics and Physics of Solids* 2004, 52, 2005–2023.
45. Maner, K.C., Begley, M.R., and Oliver, W.C. "Nanomechanical testing of circular freestanding polymer films with sub-micron thickness," *Acta Materialia* 2004, 52, 5451–5460.
46. Dannenberg, H. "Measurement of Adhesion by a blister method" *Journal of Applied Polymer Science*, 1961, 5, 125.
47. Malyshev, D.M. and Salganik, R.L., *International Journal of Fracture Mechanics*, 1965, 1, 11.
48. Cotterell, B. and Chen, Z. "The blister test—Transition from plate to membrane behavior for elastic material," *International Journal of Fracture* 1997, 86, 191–198.
49. Williams, J.G. "Energy release rates for the peeling of flexible membranes and the analysis of blister tests," *International Journal of Fracture* 1997, 87, 265–288.
50. Thouless, M.D. and Jensen, H.M. Elastic fracture mechanics of the peel-test geometry, *Journal of Adhesion* 1992, 38, 185–197.

40 Practical Fluid Control Strategies for Microfluidic Devices

Christopher J. Easley and James P. Landers

CONTENTS

40.1 MICROFLUIDIC FLOW CONTROL INTRODUCTION

Maintaining precise control over fluid within confined microfluidic networks has arguably become one of the major hurdles in realizing the true potential of microfluidic science. The nearly 20-year-old promise[1] of integrating multiple processes into single devices—whether customized as miniature factories for synthetic purposes or as microscale total analysis systems (μ-TAS) for sample processing and analysis—had fallen short of expectations until the recent implementation of practical and robust microfluidic valving systems.[2,3] With the basic valving systems now in place, there is enormous potential for many unique experiments relevant to a wide range of disciplines from fundamental physics to applied biology. Furthermore, the development of passive components to complement these active valves should prove to enhance control even further, while simultaneously reducing peripheral instrumentation.

This chapter is not intended to be a comprehensive review of microfluidic valving. On the contrary, we intend to provide a summary of the two most widely utilized valving strategies developed in the past decade.[2,3] Our approach is based on the practicality of implementing these types of valves.

In particular, we propose that the fabrication simplicity and compatibility of active elastomeric membrane valves[2,3] are key factors in their robustness and usefulness, a notion that is supported by the recent surge in related publications.[4–15] The various microelectromechanical systems (MEMS) approaches to microfluidic valving are typically limited by operational complexity and fabrication,[5] thus they are not discussed here; interested readers are referred to a recent review by Oh and Ahn.[16]

Throughout this chapter, flow control techniques that require an external (off-chip) stimulus are considered *active*, while those requiring no external stimulus are considered *passive*. The two dominant active valving techniques are discussed in Section 40.2 along with a simplified manual alternative, while some of the recent passive flow control methods are reviewed in Section 40.3. Various applications of the active valving systems are discussed in Section 40.4, which is followed by a section including guidelines that provide the reader with a "beginner's knowledge" of implementing these types of valves (Section 40.5). Although this chapter places more emphasis on active valving for microfluidic flow control, we conclude (Section 40.6) by stating the importance of conducting further research into practical passive flow control methods that should allow more widespread use of microfluidic devices in general.

40.2 ACTIVE MICROFLUIDIC VALVING

Depending on the particular device, microfluidic volumes can range from picoliters to microliters, with typical total solution volumes in the nanoliter range. Before the development of robust valves that were simple to fabricate, first by the Quake and coworkers[2] and then by the Mathies and coworkers,[3] it was nearly impossible to reliably control these small volumes of fluid using practical fabrication methods. The central theme of these two approaches is the use of a flexible poly(dimethylsiloxane) (PDMS) membrane as a pneumatically actuated material to produce on-chip valves, or fluidic switches. While it should be noted that PDMS is not the only elastomeric membrane amenable to such application, it is the one that has been extensively used thus far. The use of PDMS in microfluidics was pioneered by the Whitesides and coworkers,[17,18] who developed the soft-lithography techniques used by Quake and coworkers[2] to fabricate their valves. The material is well suited for optical interrogation due to its transparency in the visible spectral region,[17] and it is also amenable to the integration of electrochemical detection,[19] making it well suited for analytical purposes. Included in this section are summaries of the architectures and functionalities of the normally open Quake valves[2] and the normally closed Mathies valves,[3] which are easily operated by computer control, followed by a brief description of a low-power, manually operated alternative proposed by the Whitesides and coworkers.[20]

40.2.1 NORMALLY OPEN[21] VALVES USING SOFT LITHOGRAPHY

Figure 40.1a and b illustrates the basic architecture of valves developed by the Quake and coworkers.[2,22] These normally open, pressure-actuated valves are fabricated in two configurations, push-down[2] or push-up[22] valves. The push-down valves (Figure 40.1a) require approximately an order of magnitude larger pneumatic actuation pressure to close than do the push-up valves (Figure 40.1b). Furthermore, the fluidic channels in the push-down configuration are restricted to a depth of less than ~20 μm, while the push-up valves will function with much deeper channels while simultaneously requiring lower pressures. A key feature of these types of valves is the essentially zero dead volume, as can be inferred from the illustration of pressure-actuated closing of the fluid channel (Figure 40.1c). With this advantage, the valves can be arranged into high-density architectures ($\geq$30 valves mm^{-2}),[23] as shown in Figure 40.1d and e. With at least three valves coupled in series, peristaltic pumps can be created,[2] allowing the precise metering of solutions within the device at flow rates up to 2.35 nL s^{-1}. Figure 40.1e shows a magnified view of Figure 40.1d, with labels for the peristaltic pumps and cell growth chambers used in this bacterial chemostat study.[23]

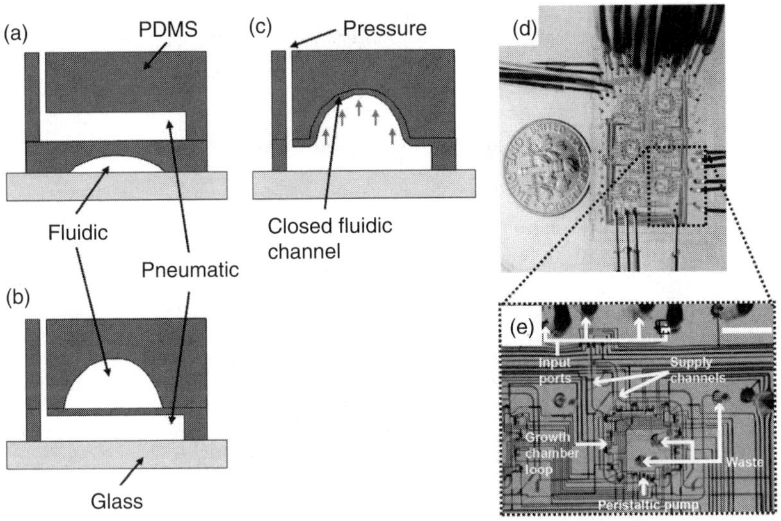

FIGURE 40.1 Basic architecture of Quake's normally open valves. (a) Push-down configuration, which is typically limited to fluidic channels <20 μm. (b) Push-up configuration, which requires approximately an order of magnitude less pressure to actuate than the push-down valves and can be used with deeper channels. (Adapted from Studer, V., et al. *J. Appl. Phys.*, 2004, 95, 393–398.) (c) Illustration of pressure-actuated closing of the fluid channel using a push-up valve. (d) High-density valve arrangements were shown in a microfluidic chemostat for programmed population control of bacteria. (e) Magnified view of device in (d). (From Balagadde, F. K., et al. *Science*, 2005, *309*, 137–140. With permission from American Association for Advancement of Science.)

Since their development, these valves have been proven functional through their utilization in a wide variety of applications such as microfluidic large-scale integration,[4] protein crystallization,[24] nucleic acid processing,[6] and multistep radiolabel synthesis.[9] These applications will be discussed further in Section 40.4. One disadvantage of these valves is that, in order to actuate a large number of individually addressable valves, there must be an equal number of external pneumatic controls (typically computer-controlled solenoid valves). In this situation, the controlling instrumentation begins to overwhelm the microscale nature of the device. Although multiplexers have been developed,[4] these pose restrictions on individual addressability of the valves. In other words, a continuous external pressure is required on each valve to maintain actuation, thus a latchable fluidic control structure has yet to be developed in this valve configuration.

40.2.2 Normally Closed[21] Valves Using Hybrid Devices

Alternatively, the valves developed by the Mathies and coworkers[3] are fabricated in a normally closed architecture, as shown in Figure 40.2a, and are actuated open by vacuum, with the option of pressure to promote closing. Rather than patterning the PDMS by soft-lithography, these valves are defined by patterned glass layers (chemically etched), which sandwich a PDMS membrane. The fact that the majority of the channel architecture can be made from glass is advantageous, parleying the rigidity and chemical resistivity of the glass substrate with the system, while maintaining valving capabilities. These valves can be fabricated in either three-layer (Figure 40.2a) or four-layer (Figure 40.2b) architectures, in which the four-layer design minimizes contact with PDMS surfaces that have a known incompatibility with certain solvents and analytes.[25] Figure 40.2c illustrates the actuation of the three-layer valves, where a vacuum is applied to the pneumatic control chamber to open the fluid path below; this valving action is mechanistically equivalent to the four-layer valve actuation. Either type of valve can be arranged into a diaphragm pumping configuration (at least three valves), with reported flow rates up to 380 nL s^{-1},[3] approximately two orders of magnitude larger

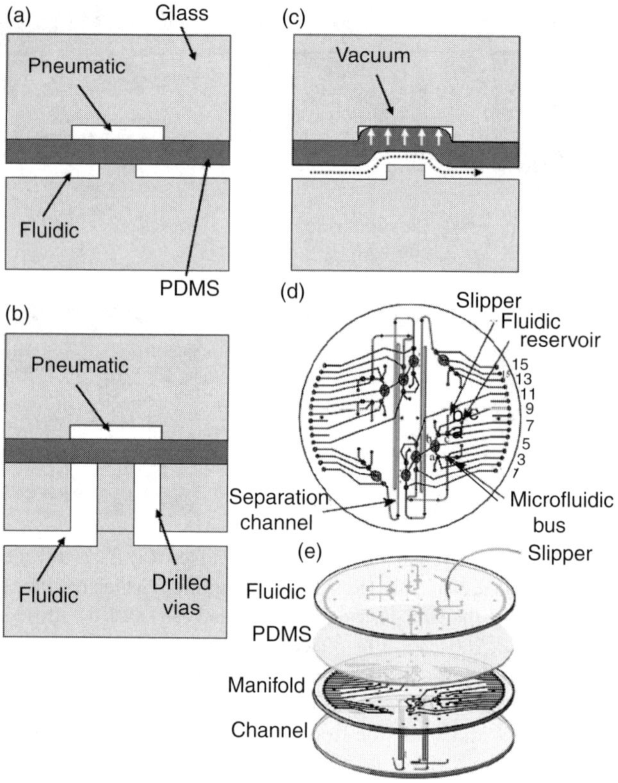

FIGURE 40.2 Basic architecture of Mathies' normally closed valves. (a) The three-layer configuration in PDMS-glass hybrid devices and (b) the four-layer configuration (used to limit PDMS-solution contact) can be (c) actuated open to fluid flow by applying a vacuum to the pneumatic chamber at the valve seat. (Adapted from Grover, W. H., Skelley, A. M., Liu, C. N., Lagally, E. T., Mathies, R. A. *Sens. Actuat. B*, 2003, 89, 315–323.) (d) An example of the assembled four-layer device that was used to detect evidence of life in extreme environments in the MOA. (e) The preassembled MOA device. (From Skelley, A. M., et al. *Proc. Natl. Acad. Sci. USA*, 2005, 102, 1041–1046. Copyright 2005. With permission from National Academy of Science, USA.)

than the Quake pumps. An illustration of an exemplary four-layer device is shown in Figure 40.2d (assembled) and Figure 40.2e (pre-assembly). This device was utilized for amino acid detection in extreme environments with the long-term goal of searching for evidence of life on Mars.[8,14] Since these valves were developed three years after the Quake valves, they have been utilized in fewer applications to date. However, recent reports describe the use of these valves for DNA computing,[26] pressure injections for electrophoresis chips,[10,27] DNA sequencing,[28] serial dilution circuits,[29] and fully integrated genetic analysis devices.[11] A disadvantage of these valves is that the dead volume is not negligible, which is also related to the lower achievable valve density compared to the Quake valves, although the valves can be designed for <10 nL dead volume[3] and have shown capable of injection of samples that are submicroliter in volume.[10,27]

An interesting modification of this valve configuration was recently made by the Mathies group, in which the researchers showed that it was possible to fabricate "latching" structures that no longer required a continuous external stimulus.[12] Vacuum pulses of 120 ms were sufficient to latch the valves for up to 2 min. This was an important advance in flow control technology, because it demonstrated that $2^{(n-1)}$ valves could be controlled with only n control lines. Although the authors showed just five inputs to control 16 valves, it should be possible to extrapolate the concept, for example, to control 1024 valves with only 11 external control lines, thereby greatly reducing the complexity of the external hardware associated with valve control.

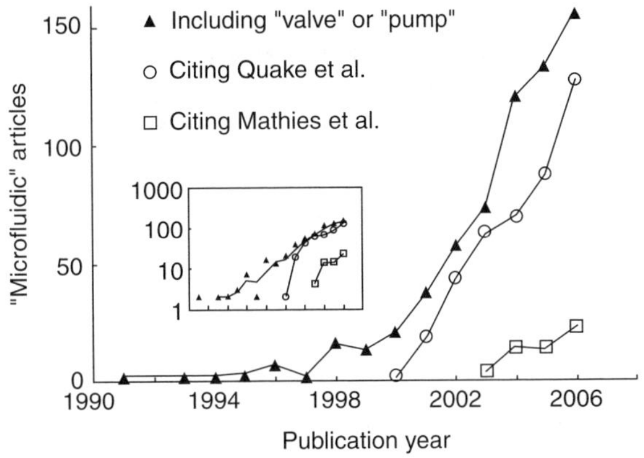

FIGURE 40.3 The ISI Web of Science database was searched with the subject keywords of "microfluid* AND (valv* OR pump*)" through December 2006. These results were plotted as a function of publication year (solid triangles), and the totals were further limited to those which cite the initial valve descriptions by the Quake (open circles) or Mathies (open squares) group. The surge of publication since the introduction of the Quake valves is evident in either the linear plot or the inset logarithmic plot. Ninety-seven percent of the total "microfluidic valve or pump" publications in 2006 cited either the Quake or Mathies. Key reasons for the success of these valves are simplicity of fabrication, valve actuation robustness, and design scalability.

40.2.3 Robust and Scalable Nature of Elastomeric Valving

The Quake[2] and Mathies[3] valve configurations can arguably be considered as the most robust that have been developed to date. This argument is supported by the surge in publications that have referenced these valves. Using the ISI Web of Science database,[30] a search was performed including the terms "microfluid* AND (valv* OR pump*)" through December 2006. These results were plotted as a function of publication year (Figure 40.3, solid triangles), and the resulting publications were further limited to those that cite the initial valve descriptions by the Quake (open circles) or Mathies (open squares) group. It is clear from Figure 40.3 that the number of publications including "valves" or "pumps" began to surge after the introduction of Quake's valves, a fact that is visually confirmed in the inset logarithm-scale plot. Although the surge could have occurred due to general scientific progress alone, this notion is refuted by the fact that an average of 76% of all the publications referenced either Quake or Mathies since 2001, with 97% in 2006. While this representation of the literature is not infallible, Figure 40.3 demonstrates the general importance of these elastomeric membrane valves. The key reasons for the success of these types of valves are the simplicity with which they can be fabricated, the valve actuation robustness, and the design scalability. The keys to "fabrication simplicity" are that the fabrication methods necessary for either configuration[2,3] are essentially no more complex than typical microdevice fabrication,[17,31] and that they have been shown to be highly scalable, for example, with $\geq$30 valves mm^{-2}.[4,5]

40.2.4 Manually Operated Torque Valves

For certain applications, portability and power savings outweigh the need for highly scalable, computer-controlled valving. It should be clear to those experienced in elastomeric valving on nontrivial chip-based architectures that the external hardware also becomes nontrivial in both size and complexity. A useful alternative was presented by Whitesides and coworkers,[20] in which torque-actuated valves were used to collapse PDMS channels, thereby closing the channels to fluid flow. The mechanistic action of these valves was similar to the Quake valves,[2] in that a pressure was applied above a thin layer of PDMS to collapse a fluidic channel. However, since the torque-actuated valves (coined as TWIST valves) relied on a miniature machine screw to collapse the channels, they did not require power to maintain the valves in either the open (flow on) or closed (flow off) state indefinitely, thus functioning as latchable fluidic switches. Moreover, these valves were also shown to be amenable to settings between "on" and "off," thereby functioning as variable fluidic resistors. The machine screws were fixed into recessed regions of the PDMS devices through a simple photochemical curing process that embedded them into polyurethane, which simultaneously bonded with the PDMS. These devices were used to carry out sandwich immunoassays.[20] The TWIST valves have since been proven useful for the development of entirely portable microfluidic systems capable of flow rate control, sample introduction, filtration, mixing, and bubble generation.[32]

40.3 PASSIVE MICROFLUIDIC FLOW CONTROL

Although active valves have clearly been proven as effective fluid control elements, the ultimate usefulness of these valves is inherently limited by the pneumatic interfacing problem. As demands for functionality increase, the number of valves must also increase; therefore, the controlling instrumentation often begins to overwhelm the microscale nature of the device. In this section, several methods for passive fluid control are summarized. As before, we do not intend to provide a comprehensive review of passive flow control, only a sampling of the more popular methods. Although there is limited review literature on this subject, the interested reader is referred to an informative focus article on capillarity by Eijkel and van den Berg,[33] who refer to passive flow control methods as "a set of everyday forces that are 'always there' . . . gravity, suction, and the capillary force . . . ." Whenever possible, these and other passive flow control strategies should be utilized as alternatives to active valving, and active valves should be used only when necessary. Since passive components require no external stimulus, careful management of active and passive methods—working in concert—should offer the most optimal balance of instrumentation and space.

40.3.1 FLUIDIC RESISTORS FOR CONTROLLED FLOW-SPLITTING

A fundamental mode of design-driven passive flow control in microfluidic devices is to utilize differential fluidic flow resistances to direct the majority of flow through the least resistant path. This straightforward, yet powerful, technique is illustrated in Figure 40.4. When a pressure greater than atmospheric pressure is applied to the inlet channel of a device, and this channel is split, by design, into two exit channels, the flow rate of fluid through these channels depends on their relative flow resistance values. The flow resistance, in turn, increases with an increase in length or a decrease in cross-sectional area of the channel. In fact, Harrison and coworkers[34] have outlined a simple method to calculate flow resistance values, and they showed that highly accurate flow splitting could be achieved simply by designing the channel layout accordingly. The researchers started with the expression for the average linear flow rate, U, in a rectangular channel

$$U = \frac{wd}{\eta}\frac{\Delta P}{L}F, \tag{40.1}$$

where w, d, and L are the half-width, half-depth, and length of the microchannel, respectively; η is the viscosity of the fluid; and ΔP is the pressure difference along the channel. The value F refers to a geometric form factor that depends on the ratio d/w (when $d \leq w$), which can be calculated by

$$F = \frac{w}{3d} - \frac{64w^2}{\pi^5 d^2}\sum_{n=0}^{\infty}\frac{\tan h\left[\frac{(2n+1)\pi d}{2w}\right]}{(2n+1)^5}. \tag{40.2}$$

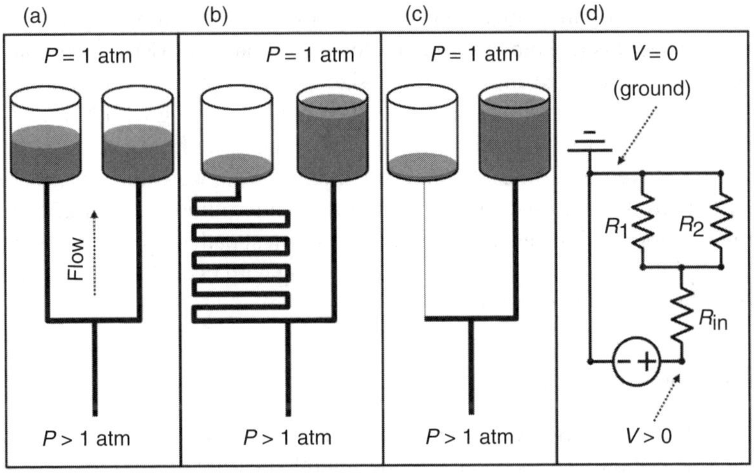

FIGURE 40.4 Output fluid levels at an arbitrary time point illustrate the effects of length and cross-sectional area of opposing exit channels. (a) Conceptual device designed for equal flow resistance in both exit paths, forcing the output flow rates to be equal upon application of a pressure gradient. Other devices were designed for a 10-fold larger flow resistance in the left exit path using (b) increased channel length or (c) decreased cross-sectional area, thereby forcing the output flow rates in their right exit paths to be 10-fold larger than the opposing (left) paths. (d) This behavior can be modeled using simple electrical circuit diagrams as analogs to fluidic networks.

Equation 40.1 was then manipulated to give the pressure difference as follows:

$$\Delta P = \frac{Q}{A}\left(\frac{\eta L}{wdF}\right) = Q\left(\frac{4\eta L}{(wd)^2 F}\right), \tag{40.3}$$

where Q is the volumetric flow rate and A is the cross-sectional area of the rectangular channel (note that $U = Q/A$). The right-hand side of Equation 40.3 represents the Ohm's law ($\Delta V = IR$) equivalent for microfluidics, in which ΔP is analogous to the voltage drop and Q is analogous to current flow through a wire, meaning that the term in parentheses on the right in Equation 40.3 represents the fluidic resistance,

$$R = \frac{4\eta L}{(wd)^2 F}. \tag{40.4}$$

In the device of Figure 40.4a, fluidic flow rates through each exit channel are equal due to their identical flow resistance values. However, in the devices of Figure 40.4b and c, the left exit channels were designed to be 10-fold more resistant to flow than the opposing (right) exit channel. This was accomplished in the Figure 40.4b device by simply designing a 10-fold longer channel in the left exit path, while the Figure 40.4c device accomplished the identical goal by reducing the cross-sectional area of the left exit path. Essentially independent of the magnitude of the input pressure, the flow rates through these devices is 10-fold larger in the right exit path (by design), as illustrated by the fluid levels at an arbitrary time point shown in Figure 40.4a–c. This flow behavior can be accurately modeled using the analogy to Ohm's law in electrical circuits, via Equation 40.3. The electrical circuit equivalent to the devices in Figure 40.4 is shown in Figure 40.4d, where $R_1 = R_2$ in Figure 40.4a device, while $R_1 = 10\,R_2$ in the Figure 40.4b and c device.

Unfortunately, even though control of relative flow resistances within channel networks is straightforward, it appears to be somewhat overlooked by much of the microfluidic research community. Through more judicious design constraints, researchers could benefit greatly from this simple method. Harrison and coworkers[34] utilized relative flow resistances to reproducibly sample small volumes for electrophoresis injections from a much larger sample introduction channel. Landers and coworkers also used this technique for microchip electrophoresis (ME) injections from on-chip diaphragm pumps,[10,11] where the injection plug size was minimized by carefully designed flow splitting ratios at the injection cross-tee. Lam et al.[35] combined valving with flow resistances, in this case to produce a digitally variable fluidic resistor with 16 possible flow resistance values. A different approach was employed by Whitesides and coworkers,[36] who were able to show analog/parallel computing capability using shortest-path determinations in microfluidic mazes. City maps were used as mask patterns to fabricate channel networks, and the shortest path between two points in the "city" were determined rapidly, suggesting that parallel computing with microfluidics is a viable alternative for some computationally difficult problems.

40.3.2 Gravimetric Control

When possible, flow control based on gravity alone can provide a simple, yet useful, alternative to active valving. These siphoning effects can provide smooth pressure profiles based on Equation 40.5. The pressure difference between input and output reservoirs, ΔP, can be described by

$$\Delta P = \rho g L \tag{40.5}$$

with a working fluid of density ρ and height L, where g is the gravitational acceleration constant. Although it is difficult to generate large pressures using this method, it can be sufficient for many applications using channels with sufficiently large cross-sectional area (small fluidic resistance). For example, Du et al.[37] utilized gravity-driven flows for an automated flow injection analysis system for on-chip core-waveguide spectrometric detection of the complexation of *o*-phenanthroline with Fe(II). Highly precise flow control was achieved by Whitesides and coworkers,[38] who utilized reservoirs on motorized stages to adjust the L term of Equation 40.5, providing switching between forward and reverse flow. This gravity-based flow was used to control droplet fluidics for encryption and decryption of signals coded in the intervals between droplets. Disadvantages of this method include pressure variability over time due to evaporation or source depletion, difficulty in flow switching, and interference from feedback pulses based on device compliance.[39]

40.3.3 Capillarity

In the aforementioned focus article by Eijkel and van den Berg,[33] the authors highlight the fact that capillary forces scale better with miniaturization when compared to other passive forces such as gravity, suction, centrifugation, or evaporation. To this end, Walker and Beebe[40] showed that useful pumping work could be performed in microfluidic systems by simply controlling the shape of droplets at fluidic input ports. They showed that the highest pressure attainable for a given port radius is a hemispherical drop with a radius equal to that of the port. A flow rate of 1.25 $\mu L\ s^{-1}$ was demonstrated using only a 0.5 μL drop of water, and the method was shown to be strong enough to pump against gravitational potential. In other words, based on the Young–Laplace equation, the smaller droplet possessed a higher internal pressure than the larger drop, thus was capable of pumping against a much larger droplet in an opposing reservoir simply based on the shape of the droplet at the air–water interface[40] (Figure 40.5). The change in volume with respect to time was described in the following equation:

$$\frac{dV}{dt} = \frac{1}{R}\left(\rho g L - \frac{2\gamma}{r}\right), \tag{40.6}$$

where r is the radius of the small spherical "pumping" drop, R is the microchannel resistance (see Equation 40.4), and γ is the surface free energy of the liquid. The pressure created by the larger outlet reservoir drop of density ρ and height L is represented by the gravimetric term ρgL (see

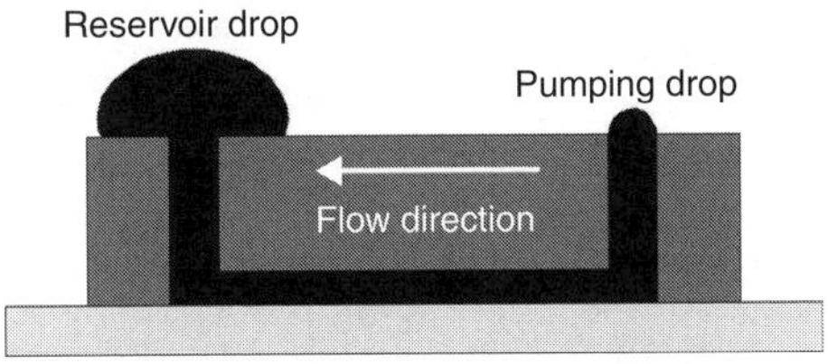

FIGURE 40.5 Capillarity was shown to be a more dominant force at the microscale by a passive pumping method. The small "pumping" drop (right) was designed to provide enough capillary force to overcome the gravimetric force from the large "reservoir" drop (left), thus the flow progressed from right to left in the figure (see Equation 40.6). (Adapted from Walker, G. M., Beebe, D. J. *Lab Chip*, 2002, 2, 131–134.)

Equation 40.6), which opposes the capillary pressure but is typically much smaller in magnitude in microscale systems. This technique was shown by Walker and Beebe[40] to be useful for injecting a plug within a microchannel for electrophoresis applications.

Eijkel and van den Berg[33] also noted that the capillary pressure can be tuned by changing the contact angle or the device geometry. When pumping in this a constant cross section, the meniscus within the channel proceeds with the square root of time due to the linear increase in flow resistance as the channel fills. Different geometries and cross sections can be used to control the relative wetting of channels within a microfluidic network, providing more precise control over these autonomous systems. Kim et al.[41] showed this method to be capable of passive control of two merging laminar streams whose widths could be tuned based on flow resistance of downstream channels. Capillarity can also be used to control sample and sieving polymer loading for DNA separations, as shown by Ono and Fujii.[42] This method allows for DNA injection and separation with a simple two-electrode system.

40.3.4 Passive Structural Components

Instead of merely exploiting microscale physical phenomena using simple straight-channel patterns with constant cross section, it is certainly feasible to make use of well-developed microfabrication methods for introducing flow control functionality into individual components. Although various MEMS approaches have been utilized for this purpose (as reviewed by Oh and Ahn[16]), many of these techniques require complex fabrication and will not be discussed here. This section focuses on more practical passive components that can be fabricated in PDMS or PDMS-glass hybrid devices, requiring minimal fabrication complexity in addition to the typical microchannel assembly methods.

Passive flow rectifiers analogous to electrical diodes have been developed by several groups.[43–45] In a single channel, these components allow larger flow rates in the "forward" direction over the "reverse" direction. The flow rectifiers developed by the Landers and coworkers[45,46] provide the most facile fabrication of any reported to date (mask design shown in Figure 40.6a, with magnified view of rectifier inset; fluidic layer in black, pneumatic in gray). These components can be fabricated alongside the normally closed valves/pumps developed by the Mathies and coworkers,[3] with no additions to the fabrication steps other than the essentially negligible additions to the mask design process. As shown in Figure 40.6b, the rectifiers[45,46] were shown to eliminate negative flow pulses that are inherent to the on-chip diaphragm pumps (flow rate of a diaphragm pump, gray trace; rectified flow, black trace). This behavior was shown to mimic the behavior of an electrical half-wave rectifier circuit (Figure 40.6c and d).

Easley et al.[45,46] also provided proof-of-principle data for discrete components referred to as "fluidic capacitors." While the concept of fluidic capacitance is by no means a new concept, little work has been done to exploit this behavior in microfluidic networks. To the knowledge of the authors, only recently has there been a report that discusses the fluidic compliance of an entire microchip made up of a flexible polymer, PDMS.[39] In this work, which was concurrent with the work by Easley

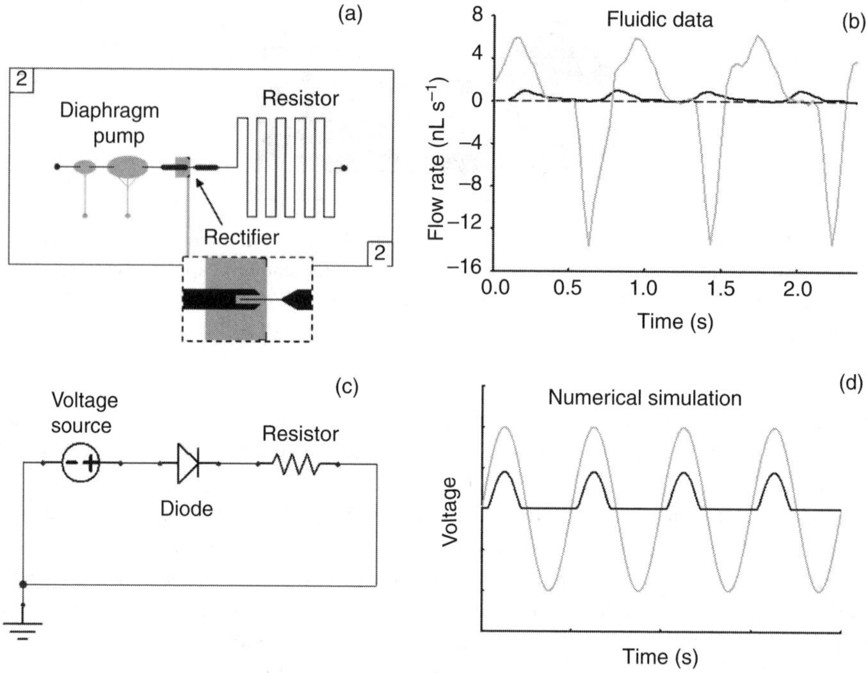

FIGURE 40.6 Fluidic flow rectifiers that require no additional fabrication steps above the steps required to construct the Mathies valves. (a) Mask design of a fluidic half-wave rectifier, including a diaphragm pump, flow rectifier (with inset image), and a fluidic resistor for flow visualization. Fluidic layer is in black, while the pneumatic layer is in gray. (b) Flow rate profiles from these rectifiers (black trace) revealed their success in eliminating negative flow pulses that are inherent to the on-chip diaphragm pumps (gray trace). The fluidic flow rate profiles matched qualitatively with the behavior of an analogous electrical circuit, (c) a half-wave rectifier using a voltage source in series with a single diode, where the (d) voltage profiles of the pump (gray trace) and the rectified flow (black trace) are shown here. (Adapted from Easley, C. J., et al. Micro-Total-Analysis-Systems, *10th International Conference on Miniaturized Systems for Chemistry and Life Sciences*, Tokyo, Japan, November 2006.)

et al., Beebe and coworkers showed that the dynamic characteristics of the entire system were similar to a lowpass filter in electrical circuits. However, there were no attempts to exploit these effects using individual microfluidic components. Easley et al. proposed the use of discrete components to control the fluidic capacitance of individual fluidic networks within the same device.[45,46] This treatment was based on the electrical analogy, where the volumetric flow rate, Q, through microchannels in the presence of compliant membranes is dependent on the time derivative of the pressure, P, through the channel as below:

$$Q = C\frac{\mathrm{d}P}{\mathrm{d}t}, \tag{40.7}$$

where C is the fluidic capacitance in units of mm^3 kPa^{-1}, representing the volume stored in the compliant membrane per applied pressure. This way, the frequency response of a first-order system could be characterized based on similar equations. For example, a low-pass or high-pass filter could be characterized by its cutoff frequency, f_0, at which the power output of a circuit is reduced to one-half of its maximum value. Using the fluidic capacitance and resistance analogies presented above, the fluidic cutoff frequency can be calculated by

$$f_0 = \frac{1}{2\pi RC}. \tag{40.8}$$

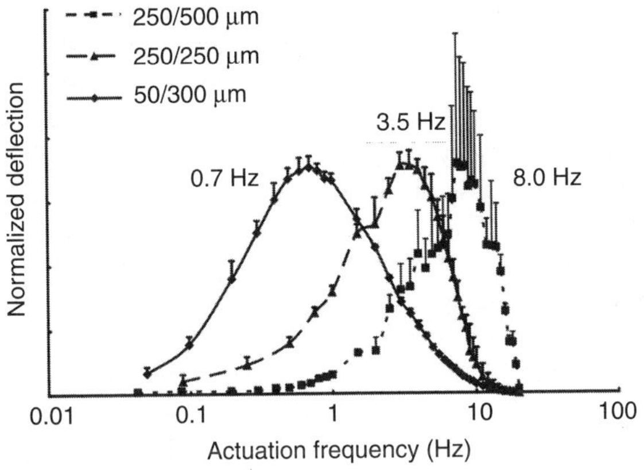

FIGURE 40.7 Characteristic frequencies could be shifted by an order of magnitude by locally altering PDMS membrane thicknesses ("fluidic capacitors"). Different curves represent different combinations of the thicknesses of two fluidic capacitor components in the same fluidic network. These fluidic bandpass filters provide the proof-of-concept of a new paradigm in microfluidic flow control, where actuation frequency could be used to passively control relative flow rates.

This equation was also deduced by Beebe and coworkers.[39] In fact, Easley et al.[45,46] have recently shown that multiple fluidic capacitor components could be combined to produce actuation-frequency-dependent control over flow rates through microfluidic networks. These fluidic bandpass filters with tunable fundamental frequencies, which should provide a new paradigm in flow control, are currently in development. As shown in Figure 40.7, the fundamental frequencies of the fluidic bandpass filters could be shifted by an order of magnitude, from 0.7 to 8.0 Hz, by simply altering the thicknesses of the deflectable PDMS layers (fluidic capacitors) at discrete locations within the network. Combinations of membrane thicknesses between 50 and 500 μm were used in this work (see legend in Figure 40.7).

Finally, Kartalov et al.[13] recently presented a straightforward method for creating microfluidic vias, which allowed solutions to be passed between multilayers of channel networks. The fabrication of these vias was consistent with the multilayer soft-lithography methods[17,18] used by Quake and coworkers.[2] A novel, passive flow shaping component, coined an "autoregulator," was enabled using these vias. The autoregulator was shown to provide the fluidic equivalent to an electrical current source, in which the output flow rate was essentially independent of input pressure above a certain threshold pressure.[13] These components were also shown to possess flow rectification character. With further development of the types of passive components outlined in this section—ones that are complimentary to the now well-developed valving techniques—the true potential of microfluidic flow control could be realized.

40.4 APPLICATIONS OF ACTIVE VALVING SYSTEMS

As illustrated by Figure 40.3, microfluidic valving techniques have proven highly useful in recent years. The Quake valves, in particular, have been put to use for a wide variety of functionalities from synthesis to genetic screening. Hansen et al.[24] developed a method for the screening of conditions for protein crystallization based on free-interface diffusion. The researchers were able to accurately meter solutions on the picoliter scale in a highly integrated and parallel fashion. A spectrum of screening conditions, covered by 144 parallel reactions, was carried out with each using only 10 nL of protein sample, to allow growth of diffraction-quality crystals. The device was shown to outperform conventional techniques, while using a mass of protein sample that was smaller by two orders of magnitude. More recent work has shown this technique to be useful for crystallizing challenging proteins by using knowledge of phase behavior to generate more rational screening tests.[47] Balagadde et al.,[23] with the device shown in Figure 40.1d and e, utilized the Quake valves to provide long-term programmed population control of bacterial in a microchemostat. A feedback mechanism, based on bacterial quorum sensing, was used to regulate the cell density of extremely small cell populations with single cell resolution over hundreds of hours. Lee et al.[9] fabricated a "miniature chemical factory" for multistep synthesis of a radiolabeled molecular imaging probe, 2-deoxy-2-[^{18}F]fluoro-D-glucose, in an integrated microfluidic device. The device was able to carry out five sequential processes: [^{18}F]fluoride concentration, water evaporation, radiofluorination, solvent exchange, and hydrolytic deprotection. This automated method provided high radiochemical yield and purity with shorter synthesis time relative to conventional automated synthesis, and the products were used successfully for positron emission tomography (PET) imaging in mice.

The Mathies valves have also found practical use for a variety of applications. Indeed, the slight chronological shift associated with the development of the Mathies valving system compared to the Quake system (Figure 40.3) suggests that the Mathies valves will perhaps adopt a similar trend. Skelley et al.[8] have developed a Mars Organic Analyzer (MOA) (device schematic shown in Figure 40.2d and e), which was capable of detecting a molecular signature of life (homochiral amino acids) in the Atacama Desert, Chile, and in the Panoche Valley, CA—two of the driest locations on Earth. The integrated and portable MOA system consisted of a microfabricated capillary electrophoresis instrument for sensitive amino acid biomarker analysis, along with the necessary high voltage power supplies, pneumatic controls, and fluorescence detection optics, and was capable of amino acid detection in mineral deposits as low as 70 parts per trillion. This system has since been shown to be capable of analyzing a wide variety of fluorescamine-labeled amine-containing biomarker compounds, including amino acids, mono and diaminoalkanes, amino sugars, nucleobases, and nucleobase degradation

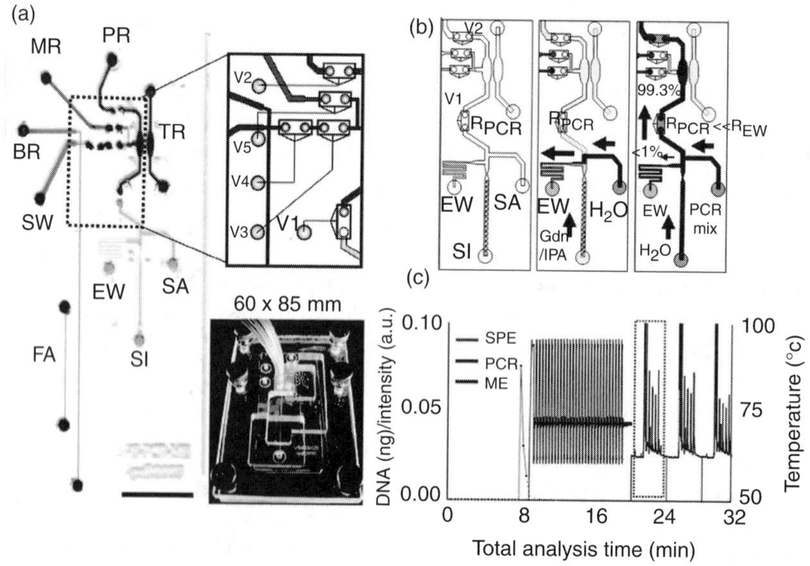

FIGURE 40.8 A clever combination of the Mathies valves and passive flow control methods allowed full integration of sample cleanup and analysis steps directly from blood or nasal aspirates, resulting in (a) a microfluidic genetic analysis (MGA) device. (b) The combination of valving, laminar flow, and differential channel resistances was used to prevent fouling of PDMS valves with incompatible solvents when transitioning from SPE to the PCR. (c) SPE and PCR were then coupled to ME using valve-based pumping to inject the purified DNA sample. The presence of *Bacillus anthracis* could be detected directly from murine blood in less than 24 min, and multiple injections were allowed using the valves to confirm the diagnoses. (From Easley, C. J., et al. *Proc. Natl. Acad. Sci. USA*, 2006, 103, 19272–19277. Copyright 2006. With permission from National Academy of Science, USA.)

products.[14] The Mathies group has also applied their valves to DNA sequencing on a microfluidic format,[28] integrating thermal cycling, sample purification, and capillary electrophoresis.

The Landers and coworkers[27] first utilized the Mathies valves to show their utility for pressure injection on a microfluidic format, then applied this technology to the integration of DNA amplification (polymerase chain reaction, PCR) with electrophoretic separation.[10] As shown in Figure 40.8, this technology was combined with upstream sample cleanup (solid-phase extraction, SPE) of crude clinical samples such as blood and nasal aspirates, providing fully integrated genetic analysis in <30 min.[11] The device architecture is shown in Figure 40.8a, flow control steps between SPE and PCR are illustrated in Figure 40.8b, and the results of a <24 min detection of *Bacillus anthracis* directly from the blood of an asymptomatic mouse are shown in Figure 40.8c. This work provided true sample-in, answer-out analyses of complex samples, as predicted by Manz et al.[1] over a decade ago. The work was carried out on a relatively simple device design, for the authors utilized a creative combination of active and passive flow control methods to interface the SPE and PCR steps (Figure 40.8c), where laminar flow and relative flow resistance values were exploited to prevent incompatible solvents from contacting the PDMS valves. Most recently, Guillo et al.[15] used these valves for pressure mobilization of focused protein zones following isoelectric focusing (IEF) to provide a robust microchip alternative to conventional IEF.

40.5 VALVE FABRICATION AND IMPLEMENTATION GUIDELINES

This section provides brief summaries of fabrication methods and instrumentation requirements associated with the microfluidic, pneumatic valving techniques outlined above. This section is meant to provide a general guide for fabrication. For more detailed descriptions of chip fabrication methods, see Chapter 10 by Legendre and Landers. Readers interested in implementing these valving techniques should refer to the original publications[2,3,22] for details and/or clarity. The organization of this section parallels that of Section 40.2 in that the normally open and normally closed valve configurations are separately described.

40.5.1 NORMALLY OPEN VALVES USING SOFT LITHOGRAPHY

In order to achieve either the push-down[2] or push-up[22] versions of the Quake valves, as shown in Figure 40.1a and b, standard multilayer soft-lithography equipment is necessary. For either valve configuration, typically a layer of SU-8 (negative) or AZ (positive) photoresist is spun onto two silicon wafers to the desired thicknesses. One wafer is then exposed to UV radiation using the fluidic photomask, while the other is exposed using the pneumatic photomask. The fluidic layer master is fabricated out of AZ photoresist, which can then be reflowed with heat to give a rounded channel profile, and the pneumatic layer is fabricated out of SU-8 photoresist. The methods deviate at this point, for the push-down valves (Figure 40.1a) require the fluidic layer to be below the pneumatic layer, while the push-up valves (Figure 40.1b) require the opposite. A thin PDMS film (10–100 μm) is then spun onto the lower layer's master (whether fluidic or pneumatic), with the thickness of the PDMS above the master determining the valving membrane thickness. A thick film of PDMS (1–10 mm) is simply poured onto the opposing layer's master, at which time the two layers are cured, then the upper layer is peeled from the master, aligned with the lower layer, and mated. In order to mate the two layers, it is often preferable to prepare the thin PDMS film with an excess of elastomer base ($\sim$20:1, base:curing agent), while the thick film is prepared with an excess of curing agent ($\sim$5:1, base:curing agent).[2] This way, the two layers can be essentially fused into one by heating postassembly at 60–80°C for 1–2 h. Finally, the PDMS device is peeled from the master, access holes are punched, and the device is completed by either mating with another (unpatterned) layer of PDMS or a glass slide.

Note that the push-down valves (Figure 40.1a) require approximately an order of magnitude larger pneumatic actuation pressure to close than do the push-up valves (Figure 40.1b) due to the preferential geometry of the push-up configuration.[22] Furthermore, the fluidic channels in the push-down configuration are restricted to a depth of less than ~20 μm, while the push-up valves will function with much deeper channels while simultaneously requiring lower pressures. However, in cases where sample contact with a glass substrate (or some other material mated to PDMS) is necessary, such as high-resolution confocal microscopy, the push-down valves must be used due to the lower position of the fluidic layer. On the other hand, the recently developed microfluidic via technology[13] may provide a remedy to this problem in some cases.

40.5.2 Normally Closed Valves Using Hybrid Devices

When a normally open configuration is not sufficient, or when glass fluidic devices are preferred over PDMS devices, the Mathies valve configuration provides a useful alternative.[3] These valves also have two configurations, the three-layer (Figure 40.2a) or four-layer (Figure 40.2b) mode. When PDMS material compatibility is not an issue, the three-layer configuration should be adequate. However, if certain reagents are incompatible, the four-layer configuration can be used instead to minimize PDMS/solution contact. With either configuration, the photolithographic patterning is carried out in a glass substrate rather than PDMS. The PDMS portion of these devices can simply be purchased in thin sheets, and slices can be cut to the appropriate size. Glass slides can be purchased with prespun photoresist, and fluidic/pneumatic layers can be etched into these slides using standard wet etching methods with hydrofluoric acid (HF). (*Warning: Exposure to HF can produce harmful health effects that may not be immediately apparent, such as destruction of deep tissue and even bone. Extreme care should be taken when working with these solutions.*) For the three-layer configuration, the PDMS layer (valving membrane) is then sandwiched between the aligned fluidic and pneumatic layers. With the four-layer configuration, access holes must be drilled into another glass slide, in alignment with the valves (two holes per valve), and this slide must be thermally bonded to the fluidic layer.

Although the fabrication of these valves is relatively simple, common problems associated with these valves include delamination of the PDMS layer from the glass layers at higher fluidic or pneumatic pressures, resulting in sample or solvent leakage. In addition, the dead volume of these valves is not negligible, and care must be taken to avoid bubble trapping within the valve seats, especially in the four-layer configuration (Figure 40.2b). Then again, there appears to be no evidence of an upper-limit restriction on fluid channel depth, for the valve actuation dimensions are separate from those of the fluid channel.

40.5.3 Instrumentation Requirements

The common theme in these microfluidic valving systems is the requirement of a vacuum or pressure pump/compressor that is used to actuate the pneumatic lines that are interfaced with the on-chip valves. These pneumatic lines, in turn, are usually controlled individually using computer-operated solenoid valves with the corresponding circuitry and manifolds. Various tubing can be used to connect each solenoid valve to the corresponding microfluidic interface. Interfacing to the device can be accomplished through a variety of means, which are not discussed here. Interested readers should refer to the cited works that are particularly relevant to their own research.

40.6 CONCLUDING REMARKS

In writing this chapter, the intent of the authors was to provide the reader with a rudimentary knowledge of the practical aspects associated with elastomeric valving technology—one that could be directly implemented in the laboratory. At this point, the reader should have a basic understanding of the mechanics of actuation for the two featured valving methods,[2,3] as well as an understanding of how innovative design of passive components can considerably improve flow control with or without valves in place. It should be noted that for many microfluidic applications, the exquisite control of small volumes that is attainable with these valves and components is unparalleled. Thus, researchers should not be hesitant to implement these techniques in their laboratories. In the years to come, once microfluidic researchers have a set of tools comparable to the various active and passive circuit components available to electrical engineers, many previously undiscovered avenues of experimentation will undoubtedly be revealed.

REFERENCES

1. Manz, A., Graber, N., Widmer, H. M. *Sens. Actuat. B*, 1990, 1, 244–248.
2. Unger, M. A., Chou, H.-P., Thorsen, T., Scherer, A., Quake, S. R. *Science*, 2000, 288, 113–116.
3. Grover, W. H., Skelley, A. M., Liu, C. N., Lagally, E. T., Mathies, R. A. *Sens. Actuat. B*, 2003, 89, 315–323.
4. Thorsen, T., Maerkl, S. J., Quake, S. R. *Science*, 2002, 298, 580–584.
5. Hong, J. W., Quake, S. R. *Nat. Biotechnol.*, 2003, 21, 1179–1183.
6. Hong, J. W., Studer, V., Hang, G., Anderson, W. F., Quake, S. R. *Nature Biotechnology* 2004, 22, 435–439.
7. Lagally, E. T., Scherer, J. R., Blazej, R. G., Toriello, N. M., Diep, B. A., Ramchandani, M., Sensabaugh, G. F., Riley, L. W., Mathies, R. A. *Anal. Chem.*, 2004, 76, 3162–3170.
8. Skelley, A. M., Scherer, J. R., Aubrey, A. D., Grover, W. H., Ivester, R. H., Ehrenfreund, P., Grunthaner, F. J., Bada, J. L., Mathies, R. A. *Proc. Natl. Acad. Sci. USA*, 2005, 102, 1041–1046.
9. Lee, C. C., Sui, G., Elizarov, A., Shu, C. J., Shin, Y. S., Dooley, A. N., Huang, J., et al. *Science*, 2005, 310, 1793–1796.
10. Easley, C. J., Karlinsey, J. M., Landers, J. P. *Lab Chip*, 2006, 6, 601–610.
11. Easley, C. J., Karlinsey, J. M., Bienvenue, J. M., Legendre, L. A., Roper, M. G., Feldman, S. H., Hughes, M. A., et al. *Proc. Natl. Acad. Sci. USA*, 2006, 103, 19272–19277.
12. Grover, W. H., Ivester, R. H., Jensen, E. C., Mathies, R. A. *Lab Chip*, 2006, 6, 623–631.
13. Kartalov, E. P., Walker, C., Taylor, C. R., Anderson, W. F., Scherer, A. *Proc. Natl. Acad. Sci. USA*, 2006, 103, 12280–12284.
14. Skelley, A. M., Cleaves, H. J., Jayarajah, C. N., Bada, J. L., Mathies, R. A. *Astrobiology*, 2006, 6, 824–837.
15. Guillo, C., Karlinsey, J. M., Landers, J. P. *Lab Chip*, 2007, 7, 112–118.
16. Oh, K. W., Ahn, C. H. *J. Micromech. Microeng.*, 2006, 16, R13–R39.
17. Duffy, D. C., McDonald, J. C., Schueller, O. J. A., Whitesides, G. M. *Anal. Chem.*, 1998, 70, 4974–4984.
18. McDonald, J. C., Duffy, D. C., Anderson, J. R., Chiu, D. T., Wu, H., Schueller, O. J., Whitesides, G. M. *Electrophoresis*, 2000, 21, 27–40.
19. Garcia, C. D., Henry, C. S. *Anal. Chem.*, 2003, 75, 4778–4783.
20. Weibel, D. B., Kruithof, M., Potenta, S., Sia, S. K., Lee, A., Whitesides, G. M. *Anal. Chem.*, 2005, 77, 4726–4733.
21. Herein, "normally open valves" are defined as valves open to fluid flow in the nonactuated, resting state. Conversely, "normally closed valves" are defined as valves closed to fluid flow in the nonactuated, resting state. In both cases, the "resting state" refers to the state requiring no power to maintain.
22. Studer, V., Hang, G., Pandolfi, A., Ortiz, M., Anderson, W. F., Quake, S. R. *J. Appl. Phys.*, 2004, 95, 393–398.
23. Balagadde, F. K., You, L., Hansen, C. L., Arnold, F. H., Quake, S. R. *Science*, 2005, 309, 137–140.
24. Hansen, C. L., Skordalakes, E., Berger, J. M., Quake, S. R. *Proc. Natl. Acad. Sci. USA*, 2002, 99, 16531–16536.
25. Lee, J. N., Park, C., Whitesides, G. M. *Anal. Chem.*, 2003, 75, 6544–6554.
26. Grover, W. H., Mathies, R. A. *Lab Chip*, 2005, 5, 1033–1040.
27. Karlinsey, J. M., Monahan, J., Marchiarullo, D. J., Ferrance, J. P., Landers, J. P. *Anal. Chem.*, 2005, 77, 3637–3643.
28. Blazej, R. G., Kumaresan, P., Mathies, R. A. *Proc. Natl. Acad. Sci. USA*, 2006, 103, 7240–7245.

29. Paegel, B. M., Grover, W. H., Skelley, A. M., Mathies, R. A., Joyce, G. F. *Anal. Chem.*, 2006, 78, 7522–7527.
30. The Thomson Corporation. *ISI Web of Science*, 2007, Citation Database.
31. Manz, A., Harrison, D. J., Verpoorte, E. M. J., Fettinger, J. C., Paulus, A., Luedi, H., Widmer, H. M. *J. Chromatogr.*, 1992, 593, 253–258.
32. Garstecki, P., Fuerstman, M. J., Fischbach, M. A., Sia, S. K., Whitesides, G. M. *Lab Chip*, 2006, 6, 207–212.
33. Eijkel, J. C., van den Berg, A. *Lab Chip*, 2006, 6, 1405–1408.
34. Attiya, S., Jemere, A. B., Tang, T., Fitzpatrick, G., Seiler, K., Chiem, N., Harrison, D. J. *Electrophoresis*, 2001, 22, 318–327.
35. Lam, E. W., Cooksey, G. A., Finlayson, B. A., Folch, A. *Appl. Phys. Lett.*, 2006, 89, 164105.
36. Fuerstman, M. J., Deschatelets, P., Kane, R., Schwartz, A., Kenis, P. J. A., Deutch, J. M., Whitesides, G. M. *Langmuir*, 2003, 19, 4714–4722.
37. Du, W.-B., Fang, Q., He, Q.-H., Fang, Z.-L. *Anal. Chem.*, 2005, 77, 1330–1337.
38. Fuerstman, M. J., Garstecki, P., Whitesides, G. M. *Science*, 2007, 315, 828–832.
39. Kim, D., Chesler, N. C., Beebe, D. J. *Lab Chip*, 2006, 6, 639–644.
40. Walker, G. M., Beebe, D. J. *Lab Chip*, 2002, 2, 131–134.
41. Kim, S. J., Lim, Y. T., Yang, H. S., Shin, Y. B., Kim, K., Lee, D. S., Park, S. H., Kim, Y. T. *Anal. Chem.*, 2005, 77, 6494–6499.
42. Ono, K., Fujii, T. *Micro Total Analysis Systems*, San Diego, CA, USA, 2005, p. 271.
43. Adams, M. L., Johnston, M. L., Scherer, A., Quake, S. R. *J. Micromech. Microeng.*, 2005, 15, 1517.
44. Jeon, N. L., Chiu, D. T., Wargo, C. J., Wu, H., Choi, I. S., Anderson, J. R., Whitesides, G. M. *Biomed. Microdev.*, 2002, 4, 117–121.
45. Easley, C. J., Karlinsey, J. M., Leslie, D. C., Begley, M. R., Landers, J. P. Micro-Total-Analysis-Systems, *10th International Conference on Miniaturized Systems for Chemistry and Life Sciences*, Tokyo, Japan, November 2006.
46. Easley, C. J. Ph.D. Dissertation, University of Virginia, Charlottesville, VA, USA, 2006.
47. Anderson, M. J., Hansen, C. L., Quake, S. R. *Proc. Natl. Acad. Sci. USA*, 2006, 103, 16746–16751.

41 Low-Cost Technologies for Microfluidic Applications

Wendell Karlos Tomazelli Coltro and Emanuel Carrilho

CONTENTS

41.1 INTRODUCTION

The amazing research field related to miniaturized systems has provided a true revolution in analytical and bioanalytical sciences in the past two decades. An exponential increase in the number of publications has been observed since the concept of a micrototal analysis system (μTAS) was first reported in the early 1990s. Short analysis time, reduced reagent consumption, low waste production, and unprecedented device portability are key features of a μTAS. These, combined with the integration of other analytical procedures, provide advantages that have stimulated the scientific community to explore the value of executing analysis in the microworld.[1–4] Since the first report in literature,[5,6] a large number of materials and microfabrication techniques have been theorized, investigated, and developed in microfabrication sciences.[1–4]

Standard techniques based on photolithographic processes have been extensively used to fabricate micromachined systems, and the details associated with these techniques can be found in Chapter 10 by Legendre. The initial development of μTAS was focused on glass or quartz substrates to create microdevices. Both materials imparted the advantages associated with low electric conductivity and high thermal conductivity, thus allowing the use of high voltages, providing good transparency for optical detection and most importantly, well-developed surface chemistry.[5,6] However, the conventional microfabrication processes are laborious, time consuming, and present a relatively high cost per device.

Furthermore, the standard processes can require some sophisticated instrumentation, particularly if located in costly cleanroom environments, some of which are not always readily accessible to the researchers. For this reason, the quest for simple and low-cost procedures is a constant focus for all the scientific community.

As a result, the introduction of new types of materials has offered new ways for the fast, low-cost prototyping of microdevices of the disposable nature. In this light, the inexpensive nature of certain polymeric materials presents unique opportunities for microchip fabrication, and there is a significant interest in the development and use of these as substrates. One of the primary advantages of most polymeric devices is that they can be fabricated outside of the clean room environment.[2–4,7] The emerging use of polymers can be attributed largely to their enhanced biocompatibility, great flexibility, reduced cost, and easy processing.[7,8] Various polymeric materials have been used to create microfluidic devices including poly(methyl methacrylate) (PMMA),[8,9] poly(carbonate),[10] photoresists (e.g., SU-8),[11] Zeonor 1020[12] and, obviously, poly(dimethylsiloxane) (PDMS).[13,14] A variety of techniques have been utilized for the manufacture of polymeric microstructures including laser ablation, X-ray lithography, injection molding, and imprinting from masters templates.[2–4] Although these techniques are simpler, in some ways, than those used in glass and quartz μchip fabrication, they still require sophisticated instrumentation for the micromachining and production of the molding template.[2–4]

This chapter presents some detailed fundamentals of low-cost technologies that can be used to produce microchannels, microelectrodes as well as integrated systems using the low-cost instrumentation commonly found chemical/biochemical laboratories. The main goal of this chapter is to enable the reader with rapid prototype capabilities for lab-on-a-chip fabrication without a significant investment in specific instrumentation, cleanroom facilities or time.

All fabrication techniques presented here are based on the use of toner from office laser printers, which is used to print a layout on a transparency film or a wax paper sheet. The transparency film is essentially used as a poly(ethylene terephthalate) (PET) substrate for polyester-toner (PT) microchannels preparation. The wax paper is used as an initial surface onto which the pattern is printed—the patterns are subsequently transferred either to a recordable compact disks (CD-R) or a glass surface. In either case, this transference is carried out with heat and pressure using the type of heat transfer machine that is used for T-shirt transfer machine or, in more sophisticated form, a tool similar to that used for hot embossing.

41.2 FUNDAMENTAL ASPECTS

41.2.1 A Brief History

The potential for the use of toner in microfabrication technology has a short but promising history that originated at the beginning of this century. Tan et al.,[15] introduced a simple method for the fabrication of PDMS microchips by using a photocopier to produce a high-relief master on polyester film. This low-cost master was employed to produce channels in PDMS using a replica molding process. This pioneering work, using toner-assisted microfabrication science, yielded channels in a PDMS substrate in less than 1.5 h.

do Lago et al.[16] proposed a direct-printing process for fast and direct production of microfluidic devices at a very low cost. In this process, a laser printer is used to selectively deposit a toner layer on a polyester film, which is subsequently laminated against a blank polyester film (single toner layer, STL) or against a mirrored image of the layout (double toner layer, DTL). Compared to the proposed technique by Tan et al.,[15] the direct-printing process makes possible the fabrication of microchips in a matter of minutes, not hours. Furthermore, tens of devices can be simultaneously printed over a single transparency sheet. This simple and inexpensive technology has been used to fabricate electrophoresis microchips coupled to end-channel amperometric[17–19] and contactless conductivity detection.[16,20] Electrospray[16] and mixing[21] microdevices have also been proposed by using do Lago's process, as well as the production of printed masters for prototyping of PDMS microchips.[22]

Daniel and Gutz[23] have used this direct-printing process for quick production of both single or multiple coplanar gold electrodes and microfluidic flow cells, as well as microfluidic cells with interdigitated gold electrode array.[24] In their work, the layouts were laser printed on a wax paper, and thermally transferred onto the gold sputtered CDtrodes' surface. This same technology was used to fabricate disposable twin gold electrodes for amperometric detection in conventional capillary electrophoresis[25] (CE) and microchip electrophoresis.[20] Lowinsohn et al.[26] have recently reported the fabrication and characterization of inexpensible gold-disk CDtrodes using toner masks, while obtaining highly reproducible electrode area. The relative standard deviation (RSD) for the voltammetric response of 10 different CDtrodes was below 1% for a 2 mmol L^{-1}solution of $K_4Fe(CN)_6$.

Recently, do Lago et al.[27] proposed a new process to fabricate glass-toner microdevices. In this new approach, laser printers do not print the layout directly over the glass surface. Alternatively, the layout is to be first printed on a wax paper and then transferred onto glass surface by heating under pressure. In fact, following the printing step, each layout can be cut out and positioned (like a decal) over a glass wafer for thermal transference. In a manner similar to the steps used for preparing CDtrodes,[17,23–26] heating under low pressure allows transfer of the toner from wax paper to the glass surface. These glass-toner devices present a specific functional advantage in that higher electroosmotic flow (EOF) is generated than in PT devices, primarily because of the greater silanol content in the bottom and top surfaces of the channels. Using the same approach, free-flow electrophoresis in glass-toner microchips have also been recently proposed by the same group.[28]

Liu et al.[21] proposed the use of the direct-printing process to fabricate passive micromixers in microfluidic devices. Micromixers were projected by applying a gray-scale coloring of the graphic software. This gray-scale effect provides the presence of toner particles into channel that act as obstacles to the flow stream for advection mixing, allowing these particles to be used as efficient mixing elements. Vullev et al.[29] used this concept of the direct-printing process to print positive-relief masters on smooth substrates. In fact, although the authors have not referenced previously reported works,[16,17,22] in this "nonlithographic" process, the masters were printed over a polyester film just as previously proposed for PT microdevices fabrication[16,17] as well as for the masters production for rapid prototyping of PDMS channels.[22]

41.2.2 The Working Principle

It is well known that UV photolithography is a standard protocol to transfer an image from a mask to a planar surface of a substrate such as glass, silicon, or quartz (see Legendre Chapter 10 for details). In this case, the layout drawn on a mask is exposed to UV-radiation and patterned on a photosensitive layer (photoresist). After light patterning step, a wet chemical or dry etching step is often employed to create channels in the substrate surface after removing the photoresist layer. In this step, the resulting channel can have an isotropic or anisotropic profile, depending on both the material and etch solution characteristics. The bonding of these etched channels is often performed by a thermal bonding step, which requires elevated temperature. Even though this high-end technology allows researchers to develop well-defined microstructures with desired aspect ratios, these steps are laborious, time consuming, and require expensive equipment, thus driving the search for alternative methods for fabrication with inherent simplicity and lower cost.

To the best of our knowledge, the direct-printing process proposed by do Lago et al.[16] is the simplest, easiest, fastest, and least expensive microfabrication process of all those described in the literature. Basically, this process can be explained by the working theory of a laser printer. When you design a layout by using graphic software, this layout is sent to a laser printer that selectively deposits a toner layer over a surface (paper or transparency film). The graphic software should be able to prepare the layouts on a 1:1 scale and for their printing without geometrical aberrations.

Toner is a complex powder deposited by a laser printer to form an image (in our case the channel walls). Laser printers force toner powder to form the desired image on the polyester film and as

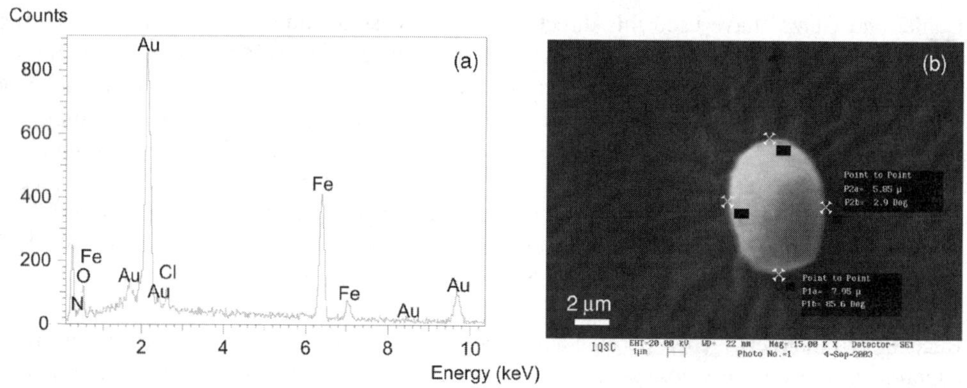

FIGURE 41.1 Typical EDX spectrum of (a) toner sample and (b) toner particle deposited by laser printer. (Reprinted from Coltro, W.K.T. *Fabrication and Evalution of Electrophoresis Microdevices with Electrochemistry Detection*, Sao Carlos, 2004, 126 p. Dissertation (Master in Science). Institute of Chemistry at Sao Carlos, University of Sao Paulo. Online available at: http://www.teses.usp.br/)

a final step, the toner image is melted onto the surface. In this process, the toner started off as a powder, passed through a fluid, and ended up as a solid structure bonded to the transparency sheet or to wax paper surface. At the chemical level, toner is constituted mainly of iron oxide (50–55% w/w) and a polymeric resin (45–50% w/w). Other pigments or ingredients can be added according to the manufacturer. Figure 41.1a shows a typical spectrum obtained by energy dispersive X-ray (EDX) spectroscopy for a toner sample—the spectrum reveals the presence of iron and oxygen. The gold that is present results from the metal deposition for a previous scanning electron microscopy (SEM) analysis, presented in Figure 41.1b. In the EDX spectrum, it can also be seen that chlorine is present, probably the result of residual $FeCl_3$. Interestingly, the polymeric resin is composed of two components: PMMA (ca. 20%) and poly(styrene) (ca. 80%).[16,17] The other material used in this microchip preparation, the PET film, is coated with a thin layer of silica with a thickness that varies depending on the manufacturer. X-ray photoelectronic spectroscopy (XPS) experiments showed the presence of silicon (ca. 0.6% estimated as SiO_2) over the polyester film.[18] While the magnitude of EOF on glass or PDMS chips basically correlates with the extent of deprotonation of the silanol groups, this is not trivial on PT chips. Here, the parameters affecting EOF in this hybrid, highly complex system are numerous and not well understood, since the bottom and the top of channel are made up of silica doped PET and the channel walls are made up of toner. However, the low abundance of silica in the channel explains the low EOF found in PT microchips, which is typically 10 times lower than glass and PDMS devices.[17]

As described previously,[16,17] the limitation of this low-cost fabrication technology is related to the laser-printer resolution. Theoretically, a laser printer with 600 dots per inch (dpi) resolution could be used to print 50-μm width channels because such resolution (600 dpi) is about 42 μm. However, toner layer is composed of smooth particles that are of irregular round or elliptical shape, with dimensions around 6 × 8 μm (see Figure 41.1b). Experimentally, it is possible to produce channels with 50-μm width, however, with limitations such in the channel definition and possible appearance of edge regularity. Furthermore, the presence of some toner particles in narrow channels at the mentioned dimension (50 μm) can result in partially or completely obstructed channels. The RSD for channels with 50-μm width is around 25% with the high RSD inherently related to the printing process. For channels having a width in the 150 μm to 1 mm range, this parameter is decreased to 10% and 3%, respectively ($n = 10$).

While glass, silicon, or quartz channels are often thermally bonded at high temperatures (550–650°C), the PT channels are bonded by a thermal lamination step in a standard office laminator. The lamination should be carried out at temperature high enough to provide a desired seal between the toner and the polyester (STL), or to other toner layer (DTL). For STL channels, the printed channel is bonded against a blank transparency (i.e., a transparency cover without toner layer), while for DTL channels, the printed layout should be firstly aligned with its mirror image, and subsequently laminated together. During the lamination step, the toner layer binds by fusion on the blank transparency or against another toner layer providing a stable sealing. After the bonding process, a monolithic structure of polyester and toner results with channels and reservoirs produced where no toner was deposited.

41.2.3 Required Instrumentation

The direct-printing process does not require any sophisticated instrumentation. Basically, two pieces of equipment are required as fundamental tools: a computer equipped with graphic software (e.g., Corel Draw, AutoCad, Macromedia FreeHand, or Microsoft Power Point) and a laser printer (although different brands provide different results as will be discussed later). Accessories such as an office laminator, a heating press (T-shirt transfer machine), and a paper driller represent the low-cost tools we have used. Finally, polyester films (transparency films or overhead projector films), wax paper sheets, a commercially available toner cartridge, microscope slides (glass) and CDs are the consumables, since they are used to produce microchips, microelectrodes, and integrated systems at very low cost.

For the sake of comparison, the acquisition of this standard office equipment costs less than 3000 U.S. dollars. In contrast, the cost of a standard microfabrication laboratory including medium-size cleanrooms, photo exposure devices, and spinners, for example, is close to one million U.S. dollars. Even though a Class 100 laminar flow cleanhoods are substantially less expensive, the total amount involved is still elevated. In addition to hardware costs, the comparative price of consumables for the low-cost and standard techniques is staggering. While a toner cartridge, polyester films, CDs, and wax paper can be obtained for a few hundred dollars, the purchase of photoresists, developer, and remover solutions, as well as wafers (e.g., glass and silicon substrates) may amount to several thousand dollars. Furthermore, the mask price used in the UV-exposure step can range from $20 to $300, depending on the chosen technology.

41.3 THE DIRECT-PRINTING PROCESS

The direct-printing technology has been used for several applications including electrophoretic microchips,[16–20] CDtrodes[23–26] for amperometric detection in conventional capillary[25] and microchip electrophoresis,[17] microfluidic cells with interdigitated Au–CDtrodes,[23,24] electrospray,[16] and micromixers[21] devices. Electrophoresis microchips with integrated electrodes for contactless conductivity detection,[20] glass-toner chips,[27] microchip free-flow electrophoresis,[28] as well as glass-etched channels[32] have recently been proposed using the direct-printing technique in one or more stages during the device preparation.

41.3.1 PT Microchips Fabrication

The PT microchip fabrication is depicted on Figure 41.2, where the production of both (A) STL and (B) DTL channels are shown. For both types channel microfabrication, the first step is related to the projection of the desired layout using any applicable graphic software such as Corel Draw, AutoCAD, and Macromedia Freehand. The drawn layout is sent to the laser printer that prints the image on a transparency film (I) carefully positioned on the printer tray. In this procedure, the white color is used, during the drawing step, for the regions where microfluidic channels and reservoirs (and consequently the channel width) are desired (II). On the other hand, the toner layer not only defines the channel walls but, consequently, also the channel depth. The thickness of the toner layer deposited on polyester surface depends of three main factors: (1) printing mode, (2) printer

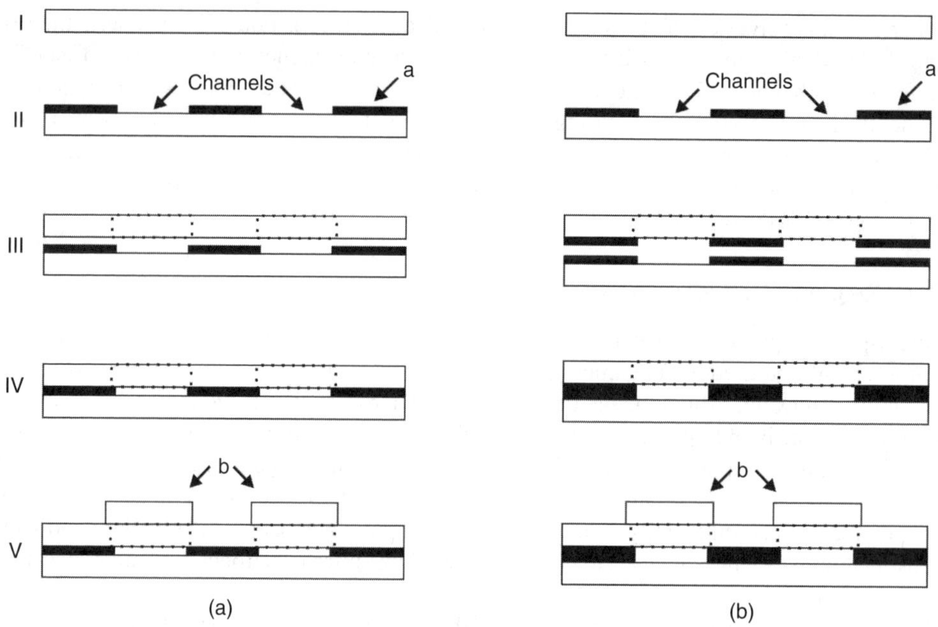

FIGURE 41.2 Step-by-step procedure of the direct-printing process for (a) STL and (b) DTL. (I) Polyester film; (II) printed layout over the PET film; (III) printed layout with aligned cover [blank polyester in (a), or printed mirror image in (b)] containing holes for accessing the microfluidic network; (IV) laminated STL and DTL chips; and (V) microdevices with solution reservoirs. In this scheme, the toner layer and the reservoirs glued on polyester film are depicted by A and B, respectively.

resolution, and (3) printer brand. Using a Hewlett Packard LaserJet printer and working at vectorial, best quality printing mode, the thickness is around 7 ± 1 μm. Minor differences have been observed in the channels defined by using laser printers with 600 or 1200 dpi resolution. The main observed difference is that, for 1200 dpi, the toner layer is more compact than for 600 dpi, which will be relevant only in toner-based lithography.[32] Significant alterations will appear when using high-resolution laser printers (≥3000 dpi) similar to photoplotters commonly employed to print masks for photolithography.

For the simplest case shown at Figure 41.2, the STL channel, the printed channels are laminated against a blank polyester film (containing no toner), where the STL channel is defined solely by one toner layer. Consequently, the channel depth is close to toner layer thickness (i.e., 7 μm). The DTL channels, require the use of two toner layers, which ultimately yields an improved aspect ratio. However, since office laser printers were not projected to have to meet the needs of the analytical microfabrication community, perfect alignment of both layers in repetitive printing steps is almost impossible. The strategy found for do Lago et al.[16] was to make these DTL channels by printing the desired layout and its mirror image simultaneously on the same polyester film. Then, both structures were aligned on top of each other for a DTL device.

Following the printing step, but before the lamination step, it is necessary to create access holes to the microfluidic network on PT chips, these can simply be made by using an office paper driller. This step is obviously much simpler (and faster) than preparing holes in glass channels with specialized drilling tools (see Chapter 10 by Legendre). The bonding step of the PT channels is provided by thermal lamination, but only after the printed channel and its cover (with holes) have been carefully aligned (III). For DTL channels, special attention should be given to this step because the alignment between the printed layout and its mirror image needs to be exact in order to avoid channel obstruction during the lamination step. The use of alignment markers or guidelines on the layout optimizes repeatability in chip-to-chip fabrication.

The lamination step, typically taking 40–50 s for each microdevice, however, the temperature at which this is carried out represents an important variable. The optimum temperature is around the melting point of the toner (i.e., 120°C) if the working temperature is below this value, repetitive lamination steps (two or three times) will be required in order to avoid fluid leakage. Since the transparency film inherently is hydrophilic in nature, it is necessary to contain the buffer solutions in reservoirs; otherwise, the fluid will spread out over the surface via capillarity. The reservoirs can be prepared by gluing small cylinders (V), constructed from the base of 100 μL pipette tips or poly vinyl chloride tubes onto the polyester surface using a bicomponent epoxy resin. It is strongly recommended that the internal diameter (i.d.) of these reservoirs be larger than the diameter of the access holes (access to the microfluidic network) in order to avoid blockage of the channel entrance. The glue curing time for the preparation of reservoirs consumes the longest portion of the fabrication time (ca. 10 min) in this low cost process. Likewise, a silicon or glass wafer can be explored to micromachine several layouts simultaneously, a polyester sheet can also be used to prepare many more devices in a same printing step. This advantage comes from the fact that a letter or A4 size sheet has a total area four times larger than the area of a standard wafer (25 in.^{2}).

As described here, the direct-printing process is highly dependent on the laser-printer resolution. Consequently, in this respect, the definition of the PT channels is worse when compared to glass and PDMS chips produced by conventional photolithography. Figure 41.3 shows three micrographs obtained by SEM depicting the surface of channels typically obtained in PT (Figure 41.3a), the edge from channel to the bulk toner (Figure 41.3b), as well as a cross-sectional profile (Figure 41.3c) of two channels. The presence of many toner particles inside the channel as well as a significant roughness to toner wall is illustrated by Figure 41.3a and b. Figure 41.4 presents two examples of electrophoresis microchips fabricated in PT using DTL channels format. For capacitively coupled contactless conductivity detection (C^4D) measurements, the PT chips have been projected with a double-T format injection channel,[16,20] while for amperometric detection, a simple cross-channel has provided satisfactory results.[17–19]

41.3.2 Microelectrode Preparation

The toner material and the direct-printing approach may not only be used for PT microchannel fabrication but also to produce a great variety of features, devices, or elements for analytical microsystems. For example, other essential elements for electrophoresis microchips are the

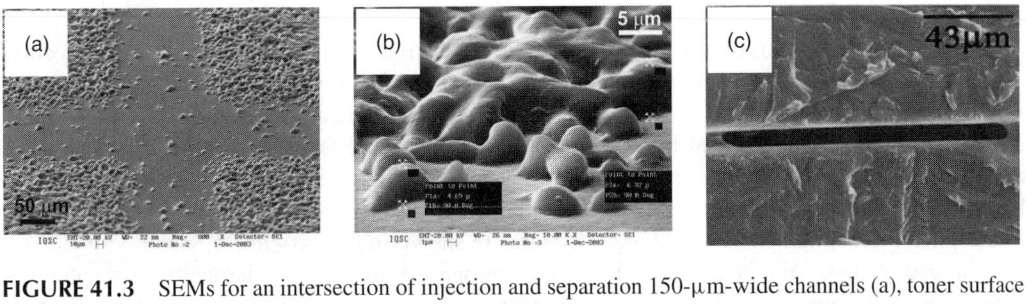

FIGURE 41.3 SEMs for an intersection of injection and separation 150-μm-wide channels (a), toner surface at channel wall (b), and a transversal cut of a DTL channel (c). (Images (a) and (b) were reprinted from Coltro, W.K.T. *Fabrication and Evalution of Electrophoresis Microdevices with Electrochemistry Detection*, Sao Carlos, 2004, 126 p. Dissertation (Master in Science). Institute of Chemistry at Sao Carlos, University of Sao Paulo. Online available at: http://www.teses.usp.br/) and image (c) with permission from He, F.-Y., et al., *Anal. Bioanal. Chem.*, 382, 192, 2005 (c). With permission.)

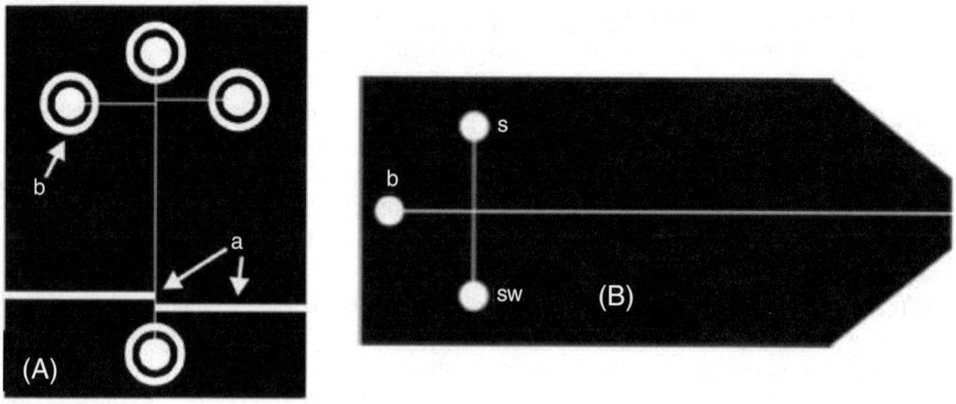

FIGURE 41.4 Examples of electrophoresis microchip fabricated in PT. (A) Microdevice with cupper tape electrodes (a) positioned externally over the channel for C^4D measurements. Point b is a representation for the solution reservoirs. (Reprinted from do Lago, C.L., Silva, H.D.T., Neves, C.A., Brito-Neto, J.G.A. and Fracassi da Silva, J.A., *Anal. Chem.*, 75, 3853, 2003. With permission.) (B) Microdevice layout for end-channel amperometric detection. S, SW, and b represent, sample, sample waste, and buffer reservoirs, respectively. (Reprinted from do Lago, C.L., et al., *Anal. Chem.*, 75, 3853, 2003. and Coltro, W.K.T., et al., *Electrophoresis*, 25, 3832, 2004. With permission.)

electrodes. The preparation of microelectrodes can also benefit from this direct-printing process, that is, electrodes for both electrokinetic control and electrochemical (EC) detection. Simultaneously to release of this technology, Daniel and Gutz[23] described an innovative method to manufacture gold microelectrodes using commercial recordable CD as metal source.[31] In this method, depicted in Figure 41.5, the direct-printing process is used to print the electrode design over a wax paper (a), of the same type used as a support for commercially available adhesive labels. Then, this printed image is thermally transferred onto the CD gold layer (b), from which the protective film was previously removed with HNO_3.[31] Commonly, the thermal transference is carried out in a thermal press at 120°C for 90 s under a pressure of approximately 0.1–0.5 MPa. Following the toner transference, the wax paper is easily removed from CD surface, similarly to a decal step. The toner-free gold areas are etched away (c) by a short exposure to an iodide/iodine solution. After this etching step, the toner is selectively removed with the help of a cotton swab soaked with acetonitrile, exposing the underlying gold layer (d) and, thus, producing the electrodes.[23] An amplified image from a dual electrode is presented in Figure 41.5e.

Richter et al.[25] have used this process to produce electrodes (see Figure 41.5) for EC detection in CE. These Au–CDtrodes were used first in a home-made CE system, in which an electrophoretic separation of iodide, ascorbic acid, dipyrone, and acetaminophen was successfully performed. Au–CDtrodes were later applied to an electrophoresis microchip fabricated with PT,[17] where the effectiveness of the proposed system was demonstrated with a separation of iodide and ascorbic acid. A series of 10 repetitive injections obtained in a conventional CE–EC system and one electropherogram obtained in miniaturized system are shown in Figure 41.6.

The use of these Au–CDtrodes in both conventional capillary and microchip electrophoresis presents some advantages. The first is that the low cost of electrode fabrication, considering that ca. 50 electrodes can be prepared from a single CD, aids in the quest of having cost-effective, disposable, single-use devices. The second is that, for both systems, the electrode replacement can be made quickly requiring only a new adjustment between the electrode and the capillary/channel extremity. A third advantage involves the use of the dual electrodes (see Figure 41.5) to simultaneously detect

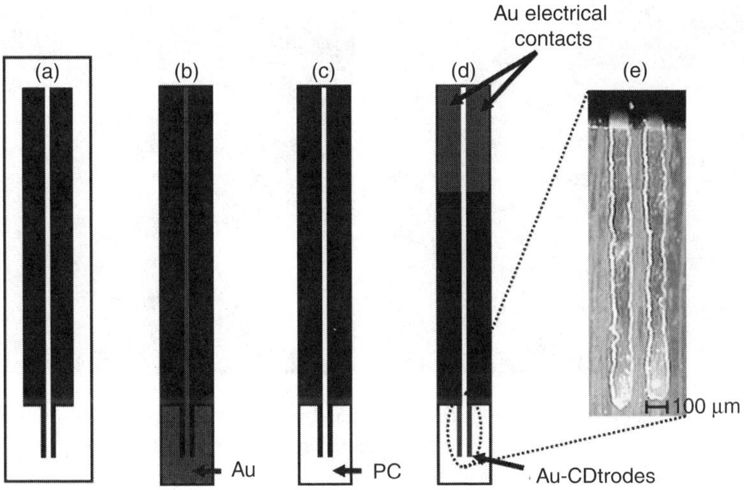

FIGURE 41.5 Au–CDtrode microfabrication process. (a) Drawing and printing of electrode layout on a wax paper; (b) thermal transference of the desired layout over a Au–CD surface; (c) Au etching step with iodide/iodine solution; (d) dual microelectrodes obtained after toner removal; and (e) amplified image from a dual Au–CDtrode with ~100-μm wide and gap ~100 μm. Black and gray colors indicate the apparent toner and the Au surface, respectively. (Adapted and reprinted from Coltro, W.K.T., et al., *Electrophoresis*, 25, 3832, 2004. With permission.)

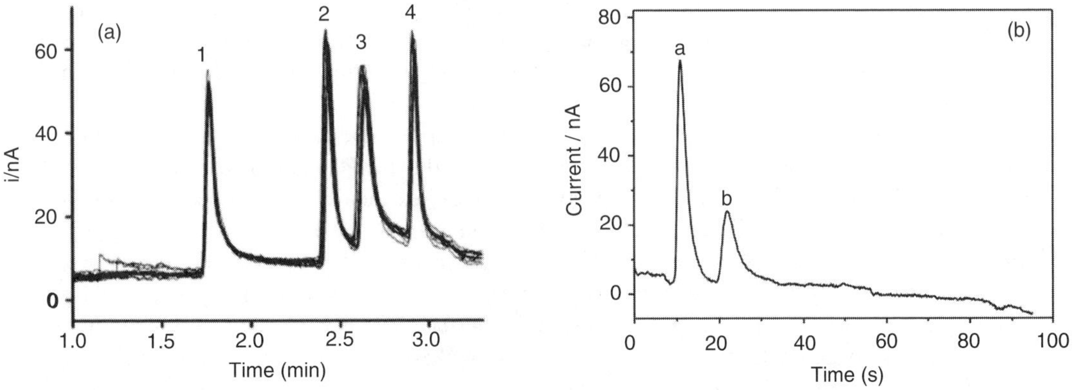

FIGURE 41.6 Electropherograms obtained for conventional capillary (A) and microchip electrophoresis (B) coupled with end-column and end-channel amperometric detection, respectively. In (A) the sample is constituted of iodide (1), ascorbic acid (2), dipyrone (3), and acetophen (4). In (B) the sample is composed of iodide (a) and ascorbic acid (b). (Reprinted from Richter, E.M., et al., *Electrophoresis*, 25, 2965, 2004 and Coltro, W.K.T., et al., *Electrophoresis*, 25, 3832, 2004. With permission.)

both the oxidation and the reduction reaction at the electrodes. Since amperometry is a selective method, the dual detection can provide additional information. In addition to these advantages, the production of electrodes can be improved by using a high-resolution laser printer. With printer of higher resolution, narrower electrodes can be prepared, thus decreasing the peak broadening related to diffusion at electrode surface.

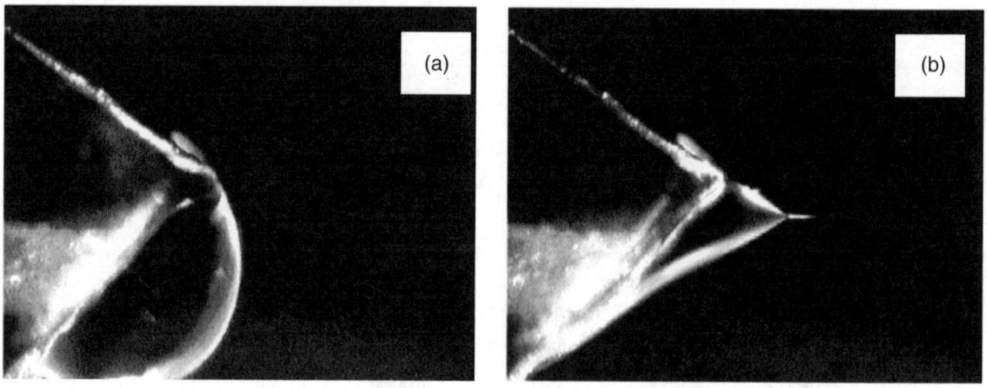

FIGURE 41.7 Examples of electrospray devices fabricated in PT. (a) A drop of KCl solution at the channel outlet and (b) electrospray tip with Taylor's cone generated by electrostatic field. (Reprinted from do Lago, C.L., et al., *Anal. Chem.*, 75, 3853, 2003. With permission.)

41.3.3 Fabrication of Other Microanalytical Devices

The major contribution of this direct-printing process is that it is not restricted to the examples described above. One possibility that has not completely been explored yet is the use of microelectrospray devices fabricated in PT. The generation of Taylor's cone from a microchannel outlet on a PT chips was shown by do Lago et al.[16] Owing to the hydrophilic nature the substrate, the edge of the device was dipped in a silicone solution to impart enough hydrophobic character to the external surface to prevent the aqueous solution from spreading. The electrospray phenomenon on the PT device was observed[16] for a solution drop hanging at the wedge-cut tip (Figure 41.7), but the coupling with mass spectrometry has not yet been shown and is currently under investigation in our laboratory.

The concept of μTAS suggests that several analytical steps should be integrated on the same chip. Different analytical operations have already been integrated using standard techniques and popular substrates.[33–37] For PT chips, this task is not trivial and requires creativity. As shown in Figure 41.3a, the presence of toner particles in the channel is a limitation associated with the direct-printing process. And while this may be viewed as disadvantageous, these toner particles can be exploited as obstacles to create convective mixing, or to adsorb analytes/reagents. In other words, toner particles could act as filter, preconcentrator, microreactor (the enhanced surface area is particularly enable with these latter two points), or obstacles that induce mixing. It is well established that using photolithographic techniques it is possible to create obstacles that have an ordered configuration inside the channel. In the direct-printing process, toner particle clusters can be generated by using a gray-scale effect from graphic software such as Corel Draw. This gray-scale effect allows the deposition of toner particles inside the channel, potentially creating obstacles for flow or turbulent mixing—although it remains to be seen whether this can be accomplished in an accurate way. Theoretically, it is possible to apply gray-scale effects ranging from 0% to 100%, with the lowest percentage providing no toner and 100% will generate a channel fully loaded with toner particles. Exemplary micrographs for the gray-scale effect applied to PT microchannels are depicted in Figure 41.8. Three different stages of this effect (0%, 50%, and 90%) are identified in a 150-μm wide channel in Figure 41.8a, where it is possible to observe that the channel is almost totally obstructed at 90% gray. In previous experiments,[30,38] it has been estimated that beyond 80% of gray-scale effect, the channel is partially obstructed. Figure 41.8b shows this effect applied to the 500-μm wide channel where Liu et al.[21] were able to show that this effect could be applied to create passive micromixers in these low-cost devices.

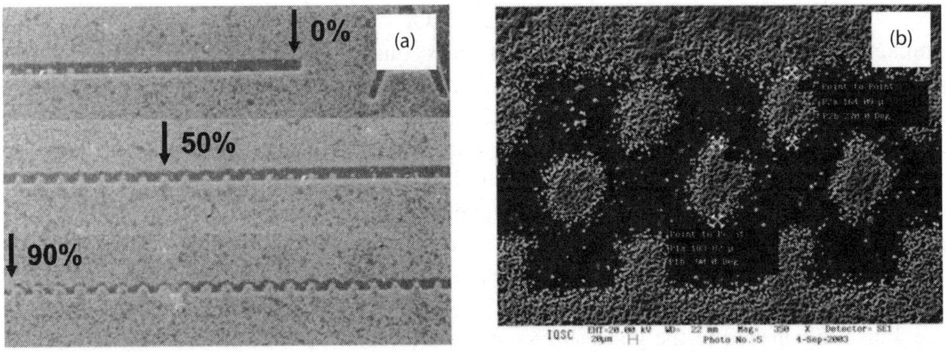

FIGURE 41.8 SEM images showing a gray-scale effect applied to PT channels: (a) 150-μm wide channel with a gradient of gray-tone applied to different points over the channel and (b) 500-μm wide channel with 80% of gray-tone applied to the entire channel. (Reprinted from Coltro, W.K.T. *Fabrication and Evaluation of Electrophoresis Microdevices with Electrochemistry Detection*, Sao Carlos, 2004, 126 p. Dissertation (Master in Science). Institute of Chemistry at Sao Carlos, University of Sao Paulo. Online available at: http://www.teses.usp.br/)

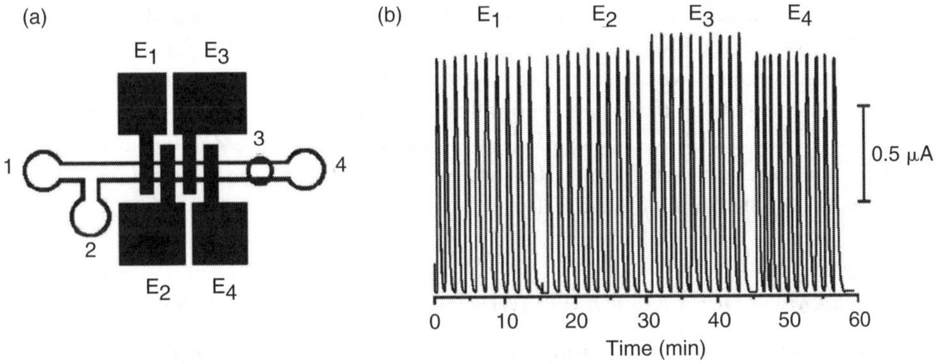

FIGURE 41.9 Example of electrodes produced by direct-printing process. (a) A schematic view of the microfluidic device with four interdigitated Au–CDtrodes. (1) and (4) inlet and outlet reservoirs, respectively; (2) and (3) reservoirs for reference electrodes; E_{1-4}, gold working electrodes. (b) amperometric response at all four electrodes for a $K_4Fe(CN)_6$ sample introduced into microchannel by FIA. (Reprinted from Daniel, D. and Gutz, I.G.R., *Talanta*, 68, 429, 2005. With permission.)

Daniel and Gutz[24] have also proposed the fabrication of microfluidic flow injection cells with interdigitated array of gold electrodes. Figure 41.9 presents a schematic diagram of this microdevice (a) as well as a sequence of amperometric response to a solution of $K_4Fe(CN)_6$ exposed to all four interdigitated gold electrodes in microfluidic flow cell (1.4 cm length × 0.1 cm width × 19 μm depth) (b). The microfluidic channel was prepared by thermal transference of three toner layers over the polycarbonate CD base, with the bonding provided by a second polycarbonate slice. The fluid transport was carried out with a peristaltic pump for flow injection analysis (FIA) applications.

41.3.4 Fabrication of Glass-Toner Microchips and Toner-Mediated Lithography

Polyester-toner devices present advantages such as low cost per device, as well as easy and speed of fabrication. In addition, dozens of devices can be prepared in parallel using a single transparency film.[16,17,30] As mentioned earlier, the magnitude of EOF on PT devices is 10 times lower than that in common microfabrication substrates such as glass, quartz, and PDMS.[17,30] While the EOF in these popular materials can be raised by chemical or physical modification of the surface (activation by oxygen plasma or sodium hydroxide), similar treatment of PT chips does not yield the same enhancement. The use of sodium hydroxide is not recommended because it attacks the toner surface rapidly (in a few minutes), even at low concentration. However, the low EOF can certainly be an advantage for the analysis of analytes with similar electrophoretic mobilities. One strategy to obtain a simple microdevice but with greater EOF is the fabrication of glass-toner microchips.[27]

The toner layer cannot be printed over a glass surface since this substrate is not an acceptable substrate for commercial laser printers. Alternatively, the layout can be printed on a wax paper first, and then transferred to the glass surface using a thermal transfer press methodology already described. In this case, one or more toner layers can be sequentially transferred onto glass surface increasing the aspect ratio. The sealing of these channels is also provided by using thermal press, in which a glass cover binds to upper toner layer under heating and pressure. Recently, this technology has been used to fabricate microchips for free-flow electrophoresis.[28]

The problems related to toner porosity and weaker adherence of the toner to the glass surface, compared to PT devices, reveals a drawback of these glass-toner devices. In recent studies,[32] we improved the adherence of the toner to the glass surface allowing the exploration of the toner layers as layout masks for preparing glass channels by wet chemical etching (toner-based lithography). In this case, glass channels are obtained without the use of any photolithographic step during the process and the costs related to it. The toner possesses limited chemical resistance to hydrofluoric acid solution, which can be overcome by multilayer stacking. However, the resistance is substantial enough to create channels up to 40-μm deep in less than 10 min. Figure 41.10 presents two SEM images, where it was possible to observe a cross-intersection (Figure 41.10a) and a transversal section of the bonded channel (Figure 41.10b) but, as expected, the channel did not present a perfect definition due to the limitation of the direct-printing process.

41.3.5 Fabrication of PT Microchips with Integrated Electrodes

One of the most exciting applications using the direct-printing process is related to the fabrication of PT microchips with integrated electrodes for C^{4}D.[20] The C^{4}D system is an universal detection mode that has several advantages when coupled with conventional CE or microchip electrophoresis: (i) the electronic circuitry is simple, inexpensive, and decoupled from the high-voltage used for separation; (ii) the formation of bubbles at the electrodes is avoided; and (iii) electrochemical modification

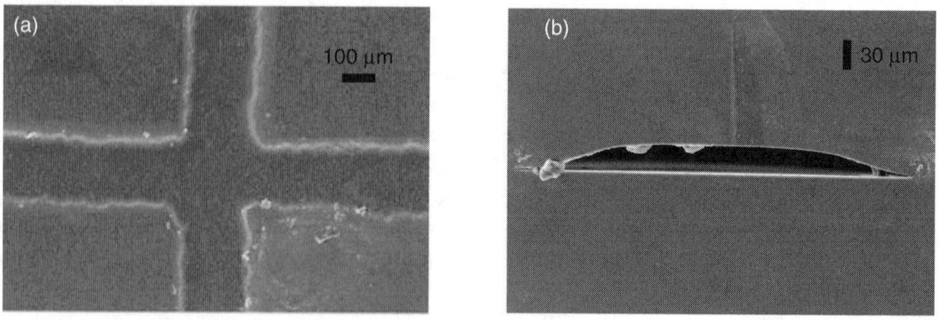

FIGURE 41.10 SEM images showing the glass channel obtained by toner-mediated lithographic process. (a) Intersection of injection and separation 200-μm-wide microchannels and (b) sectional profile of a bonded glass–glass microchannel. (Reprinted from Coltro, W.K.T., Piccin, E., Fracassi da Silva, J.A., do Lago, C.L. and Carrilho, E., *Lab Chip*, 7, 931, 2007. Reproduced by permission of The Royal Society of Chemistry.)

or degradation of the electrode surface is prevented, thereby, allowing a wide variety of electrode materials and background electrolytes.[39,40]

The detection electrodes for C^4D are often prepared by manually gluing copper stripes over the separation channel. This manual electrode positioning limits the repeatability on the chip-to-chip preparation, in which the accurate width, gap, and position over the separation channel are essential parameters. To minimize this problem, the integration of microfabricated electrodes on polyester films was a successful alternative. The electrodes for C^4D were prepared using toner masks for metal deposition via sputtering. Briefly, the electrode geometry was drawn and printed over a polyester film, followed by the sputtering of a thin layer, of titanium/aluminum over the printed electrode mask. The resulting electrodes were obtained by soaking the substrate in acetonitrile and dissolving the underlying toner layer. As the toner layer was removed with the solvent, the sputtered Ti/Al could be lifted off, and the electrode material remained anchored to polyester substrate. Note that aluminum was chosen due to its lower cost when compared to gold and platinum. The integration of these electrodes with PT channels was also carried by thermal lamination while maintaining the electrodes carefully positioned upwards, that is, isolated from contact with the solution in the channel by the thickness of the polyester film (100 ± 10 μm). The final C^4D-PT chip carries the detection electrodes on the outside of the top polyester film and the microfluidic network is directly printed on the bottom polyester film.

Figure 41.11a presents one layout of this integrated system as well as a typical electropherogram obtained on this device (Figure 41.11b). In addition to enhanced chip-to-chip repeatability, the use of integrated electrodes improved the signal-to-noise ratio. The use of toner masks for preparation of electrodes can be also extended to other EC modes such as amperometry, voltammetry, and potentiometry. Furthermore, the use of copper tapes as electrodes is limited to the classical format (antiparallel design), while the use of toner masks allows one to investigate different electrode design for C^4D.

41.4 CONCLUDING REMARKS

As seen in this brief chapter, the direct-printing process, based on laser printing of layouts on polyester films or wax paper, has the potential to become a powerful technology for the rapid prototyping of microfluidic devices at very low cost, and even a source of low-cost production of disposable devices. This is supported by the fact that the required instrumentation is commonly found at offices and chemistry laboratories. Besides the typical injection and separation channels for electrophoresis, this technology has shown that mixing, preconcentration, clean-up, reactor devices,

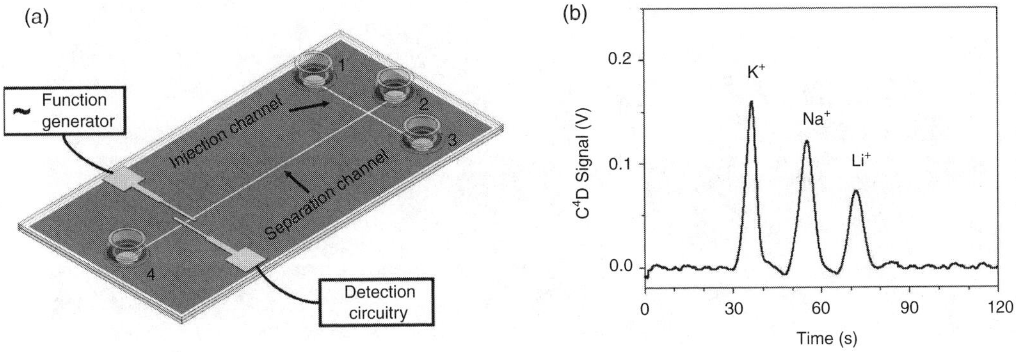

FIGURE 41.11 Representation of a (a) PT microdevice with integrated electrodes for C^4D and (b) a typical electropherogram obtained for a mixture of inorganic cations (25 μM each) in this device. (Reprinted from Coltro, W.K.T., Fracassi da Silva, J.A. and Carrilho, E., unpublished material.)

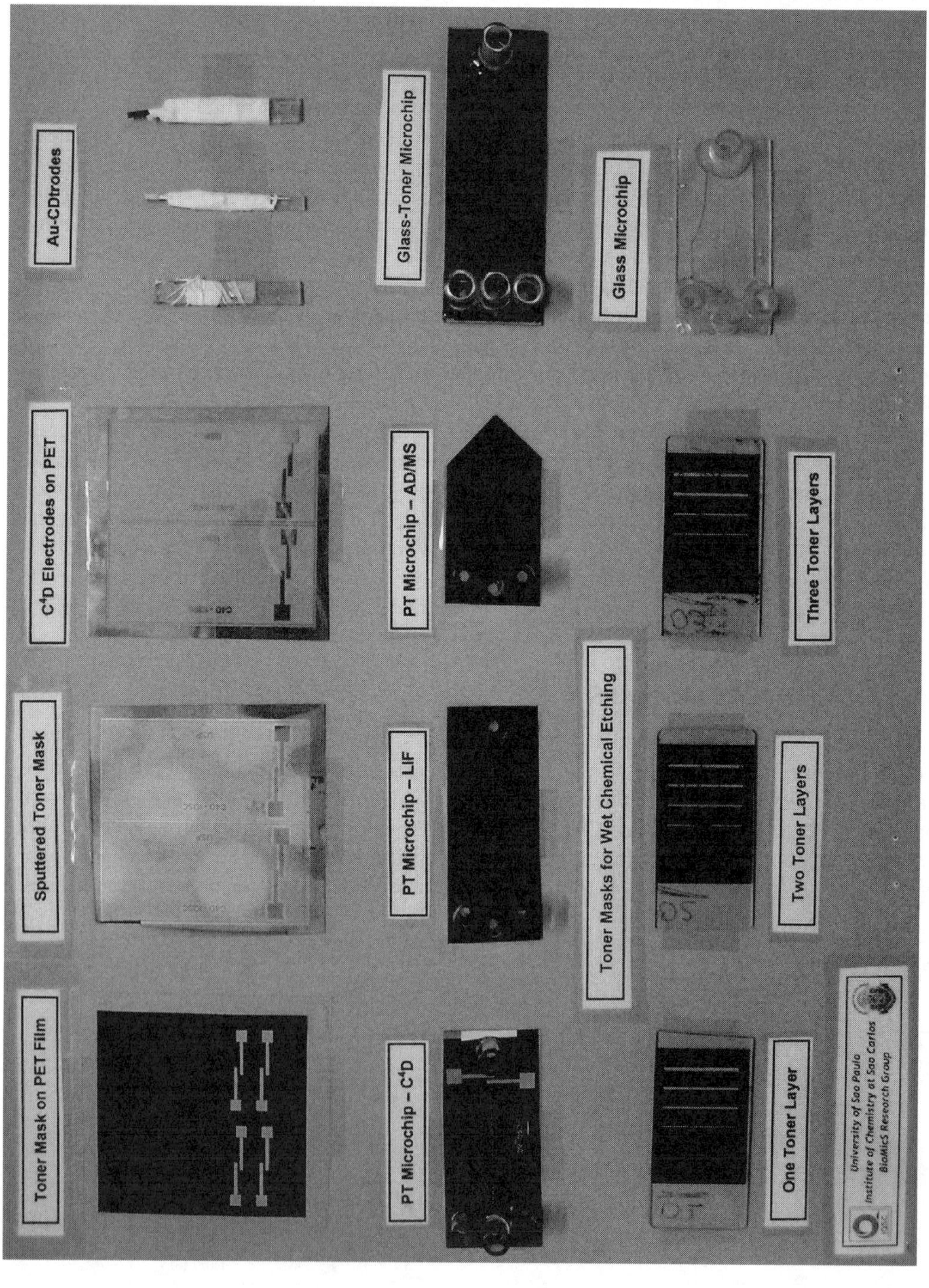

FIGURE 41.12 Examples of devices, elements, and features that can be achieved using the toner-based technology.

as well as electrodes can also be incorporated into PT channel design using graphic resources and toner masks. Currently, our main goal is focused in the development of a working μTAS by exploring just the direct-printing process. Furthermore, the versatility of the direct-printing process has been demonstrated with microelectrodes for amperometric detection in conventional CE and microchip electrophoresis, microfluidic cells with interdigitated electrodes, electrospray, micromixers, and free-flow electrophoresis devices, as well as glass-toner and glass-etched microdevices that can be fabricated and explored for specific applications.

Overall, as presented here in some detail, this technology presents the opportunity to all scientists in any laboratory interested in lab-on-a-chip technology, requiring only common tools, easily found in offices/general laboratories. To conclude this chapter, we present in Figure 41.12 a portfolio of many devices prepared using toner technology.

ACKNOWLEDGMENTS

We would like to thank the scholarship granted from Fundação de Amparo à Pesquisa do Estado de São Paulo (FAFESP—grant 04/01525-0) to W.K.T. Coltro and research fellowship granted from Conselho Nacional de Desenvolvimento Científico e Tecnológico (CNPq) (grant 304100/2005-6) to E. Carrilho. CNPq is also acknowledge for financial support (grants 477982/2003-5 and 478467/2006-0).

REFERENCES

1. Reyes, D.R., Iossifidis, D., Auroux, P.A. and Manz, A., *Anal.Chem.*, 74, 2623, 2002.
2. Auroux, P.A., Iossifidis, D., Reyes, D.R. and Manz, A., *Anal.Chem.*, 74, 2627, 2002.
3. Vilkner, T., Janasek, D. and Manz, A., *Anal.Chem.*, 76, 3373, 2004.
4. Dittrich, P.S., Tachikawa, K. and Manz, A., *Anal. Chem.*, 78, 3887, 2006.
5. Manz, A., Miyahara, Y., Miura, J., Watanabe, Y., Miyagi, H. and Sato, K., *Sens. Actuators B*, 1, 249, 1990.
6. Manz, A., Graber, N. and Widmer, H.M., *Sens. Actuators B*, 1, 244, 1990.
7. Becker, H. and Locascio, L.E., *Talanta*, 56, 267, 2002.
8. Kelly, R.T. and Woolley, A.T., *Anal. Chem.*, 75, 1941, 2003.
9. Galloway, M., Stryjewski, W., Henry, A., Ford, S.M., Llopis, S., McCarley, R.L. and Soper, S.A., *Anal. Chem.*, 74, 2407, 2002.
10. Liu, Y.J., Ganser, D., Schneider, A., Liu, R., Grodzinski, P. and Kroutchinina, N., *Anal. Chem.*, 73, 4196, 2001.
11. Jackman, R.J., Floyd, T.M., Ghodssi, R., Schmidt, M.A. and Jensen, K.F., *J. Micromech. Microeng.*, 11, 263, 2001.
12. Kameoka, J., Craighead, H.G., Zhang, H.W. and Henion, J., *Anal. Chem.*, 73, 1935, 2001.
13. Duffy, D.C., McDonald, J.C., Schuller, O.J.A. and Whitesides, G.M., *Anal. Chem.*, 70, 4974, 1998.
14. McDonald, J.C. and Whitesides, G.M., *Acc. Chem. Res.*, 35, 491, 2002.
15. Tan, A., Rodgers, K., Murrihy, J.P., O'Mathuna, C., Glennon, D., *Lab Chip*, 1, 7, 2001.
16. do Lago, C.L., Silva, H.D.T., Neves, C.A., Brito-Neto, J.G.A. and Fracassi da Silva, J.A., *Anal. Chem.*, 75, 3853, 2003.
17. Coltro, W.K.T., Fracassi da Silva, J.A., Silva, H.D.T., Richter, E.M., Furlan, R., Angnes, L., do Lago, C.L., Mazo, L.H. and Carrilho, E., *Electrophoresis*, 25, 3832, 2004.
18. He, F.-Y., Liu, A.-L., Yuan, J.-H., Coltro, W.K.T., Carrilho, E. and Xia, X.-H., *Anal. Bioanal. Chem.*, 382, 192, 2005.
19. Liu, A.-L., He, F.-Y., Lu, Y. and Xia, X.-H., *Talanta*, 68, 1303, 2006.
20. Coltro, W.K.T., Fracassi da Silva, J.A. and Carrilho, E., unpublished results.
21. Liu, A.-L., He, F.-Y., Wang, K., Zhou, T., Lu, Y. and Xia, X.-H., *Lab Chip*, 5, 974, 2005.
22. Bao, N., Zhang, Q., Xu, J.-J. and Chen, H.-Y., *J. Chromatogr. A*, 1089, 270, 2005.
23. Daniel, D. and Gutz, I.G.R., *Electrochem. Commun.*, 5, 782, 2003.

24. Daniel, D. and Gutz, I.G.R., *Talanta*, 68, 429, 2005.
25. Richter, E.M., Fracassi da Silva, J.A., Gutz, I.G.R., do Lago, C.L. and Angnes, L., *Electrophoresis*, 25, 2965, 2004.
26. Lowinsohn, D., Richter, E.M., Angnes, L. and Bertotti, M., *Electroanalysis*, 18, 89, 2006.
27. do Lago, C.L., Neves, C.A., de Jesus, D.P., Silva, H.D.T., Brito-Neto, J.G.A. and Fracassi da Silva, J.A., *Electrophoresis*, 25, 3825, 2004.
28. de Jesus, D.P., Blanes, L. and do Lago, C.L., *Electrophoresis*, 27, 4935, 2006.
29. Vullev, V.I., Wan, J., Heinrich, V., Landsman, P., Bower, P.E., Xia, B., Millare, B. and Jones, G., *J. Am. Chem. Soc.*, 128, 16062, 2006.
30. Coltro, W.K.T. *Fabrication and Evalution of Electrophoresis Microdevices with Electrochemistry Detection*, Sao Carlos, 2004, 126 p. Dissertation (Master in Science). Institute of Chemistry at Sao Carlos, University of Sao Paulo. Online available at: http://www.teses.usp.br/
31. Angnes, L., Richter, E.M., Augelli, M.A. and Kume, G.M., *Anal. Chem.*, 72, 5503, 2000.
32. Coltro, W.K.T., Piccin, E., Fracassi da Silva, J.A., do Lago, C.L. and Carrilho, E., *Lab Chip*, 7, 931, 2007.
33. Easley, C.J., Karlinsey, J.M., Bienvenue, J.M., Legendre, L.A., Roper, M.G., Feldman, S.H., Hughes, M.A., et al., *Proc. Natl. Acad. Sci. USA*, 103, 19272, 2006.
34. Skelley, A.M., Scherer, J.R., Aubrey, A.D., Grover, W.H., Ivester, R.H.C., Ehrenfreund, P., Grunthaner, F.J., Bada, J.L. and Mathies, R.A., *Proc. Natl. Acad. Sci. USA*, 102, 1041, 2005.
35. Burns, M.A., Johnson, B.N., Brahmasandra, S.N., Handique, K., Webster, J.R., Krishnan, M., Sammarco, T.S., et al., *Science*, 282, 484, 1998.
36. Liu, R.H., Yang, J., Lenigk, R., Bonanno, J. and Grodzinski, P., *Anal. Chem.*, 76, 1824, 2004.
37. Unger, M.A., Chou, H.-P., Thorsen, T., Scherer, A. and Quake, S.R., *Science*, 288, 113, 2000.
38. Coltro, W.K.T. and Carrilho, E., *Evaluation of Printed Patterns in Polyester Films as a Preconcentration Microdevice for Separation Microchips*, presented at the 9th Latin American Symposium on Biotechnology, Biomedical, Biopharmaceutical and Industrial Applications of Capillary Electrophoresis and Microchip Technology, Mexico City, 2003.
39. Brito-Neto, J.G.A., Fracassi da Silva, J.A., Blanes, L. and do Lago, C.L., *Electroanalysis*, 17, 1198, 2005.
40. Brito-Neto, J.G.A., Fracassi da Silva, J.A., Blanes, L. and do Lago, C.L., *Electroanalysis*, 17, 1207, 2005.

42 Microfluidic Reactors for Small Molecule and Nanomaterial Synthesis

Andrew J. deMello, Christopher J. Cullen, Robin Fortt, and Robert C.R. Wootton

CONTENTS

42.1 INTRODUCTION

In 1828, when attempting to prepare ammonium cyanate from silver cyanide and ammonium chloride, Friedrich Wöhler accidentally synthesized urea [1]. Contemporary wisdom at the time held that organic compounds could only be created from a "vital force," which existed within living organisms. In a letter to Jöns Jakob Berzelius, Wöhler's excitement was apparent; "I can no longer, so to speak, hold my chemical water and must tell you that I can make urea without needing a kidney, whether of man or dog; the ammonium salt of cyanic acid is urea." Although Wöhler's discovery was not the first synthesis of an organic compound it sparked huge interest in making organic compounds from nonliving substances and marked the beginning of organic chemistry as an academic and industrial discipline.

Since that time, organic chemistry (the study of the structure, properties, composition, reactions, and synthesis of carbon-containing compounds) has flourished and is of vital importance to the pharmaceutical, chemical, cosmetic, petrochemical, and textile industries. For the vast majority of the time, since (and for centuries before) Wöhler's synthesis of urea, the chemist's toolkit has predominantly consisted of macroscopic components fabricated from glass. Examples of such components include round-bottomed flasks, test tubes, distillation columns, reflux condensers, Erlenmeyer flasks, drying tubes, separation funnels, recrystallization tubes, and burettes. Despite the enormous advances that have been made in experimental, mechanistic, and theoretical organic chemistry over the past 150 years it is noteworthy that the basic experimental techniques and associated laboratory equipment remain largely unaffected. Perhaps this is unsurprising, since standard laboratory glassware provides a fitting environment in which to perform the vast majority of synthetic transformations. Glass as a generic material is robust, possessing good chemical inertness. It also exhibits high thermal conductivity, low electrical conductivity, and good transparency within the visible region of the electromagnetic spectrum. This means that standard glassware can be used to process most chemical reagents under a wide range of temperatures and pressure. A more pragmatic reason for the adoption of macroscale glassware in organic synthesis relates to the fact that chemists as individuals feel comfortable performing reactions in environments, which they can easily manipulate, control, and observe. Nevertheless, although a chemist may prefer to perform a reaction within a round bottom flask, at the molecular scale it makes little difference if a reaction is processed within a volume of 100 mL or 100 pL. What is more important is that the "ideal" chemical reactor should provide an environment in which chemical state functions are precisely controllable, allowing the rapid synthesis of a desired product in high yield. To this end, over the past decade, the application of micromachining techniques cultivated within the semiconductor and microelectronics industries have allowed the creation of a new instrumental platform able to efficiently process and analyze molecular reactions on the micron to nanometer scale. This chapter aims to provide an introduction to the concept of reaction miniaturization. The theory of miniaturization is discussed and followed by a practical assessment of the use of microfluidic systems in synthetic chemistry.

42.2 THEORETICAL BACKGROUND

42.2.1 FLUID FLOW ON THE MICROSCALE

The foremost reason why microfluidic systems provide attractive environments in which to perform synthetic chemistry lies in the dependency of fluid flow on scale. Put simply, fluid handling within microfluidic environments differs markedly from typical macroscale flow [2]. A variety of phenomena manifest themselves upon moving from the macroscale to the microscale, which in turn have a profound effect on both the efficiency and manner in which chemical reactions proceed.

A chemical reaction is initiated by bringing the required reactants into intimate contact. Efficient execution of this process often defines the efficiency of the reaction and, therefore, requires considered examination when designing a reaction system. The physical transport of a component along a concentration gradient by molecular diffusion and turbulent convection is known as mass transfer. The transport of mass through an interface between two media or phases of the same medium is extremely important, since chemical reactions are normally coupled to the mass transfer efficiency. The mass transfer coefficient is used to quantify the efficiency of mixing in macrofluidic systems and is given by

$$K_c = \frac{N_A}{C_s - C_B}, \tag{42.1}$$

where K_c is the mass transfer coefficient (m s^{-1}), N_A the molar flux (mol m^{-2} s^{-1}), C_s the concentration at the phase boundary (kg m^{-3}), and C_B the concentration in bulk solution (kg m^{-3}).

A conspicuous effect of reactor miniaturization is that fluid properties become increasingly controlled by viscous forces rather than inertial forces as reaction volumes are reduced [3]. In macroscale environments, reagents are brought into intimate contact by turbulent flow. In such a regime, fluid elements (of varying sizes) travel randomly and create a mixture that is highly segregated with a finely dispersed structure. These eddies are a product of areas of circulatory motion in the fluid created by the inertial forces of flowing liquids. The mixture that is produced, although finely dispersed, is still heterogeneous at the molecular level and mixing of these fine fluid elements is subsequently achieved through the random, diffusive motion of the component molecules. The degree of turbulence in a system depends on its scale and the density and viscosity of the fluid. For example, in large systems such as fluidic jet streams, where turbulence is high, fluid elements can move large distances and therefore mixing is much faster than mixing by molecular diffusion alone.

Fluid flow in the absence of turbulence is described as being laminar. In such a situation, viscous forces dominate inertial forces and eddies are no longer present. Laminar fluid flow can be conceptualized as consisting of concurrently flowing parallel streams with little or no mixing between streams. Since the same viscous forces that restrain the fluid streams in laminae also dampen out irregularities and streamlines, laminar flow is not disturbed by obstacles and instead maintains a smooth path. With the removal of the highly segregated structure created by eddies, mixing must be achieved by molecular diffusion alone.

The Reynolds' number is a dimensionless number used in fluid dynamics to determine dynamic similitude [4] and is generically defined as

$$Re = \frac{v\delta}{U}, \tag{42.2}$$

where v defines the mean fluid velocity, δ is a characteristic dimension of the reactor and U defines the kinematic viscosity of the fluid. Significantly, the Reynolds' number provides a measure of the ratio of inertial forces to viscous forces and is most commonly used to assess whether flow will be laminar or turbulent. In simple terms, viscosity describes the resistance of a fluid to deformation under shear stress and encourages fluid to flow in parallel layers, with no disruption between the layers (laminar flow), while inertia describes the tendency of the fluid to resist acceleration and thus counteract laminar flow. For microfluidic systems, such as capillaries or blood vessels in the body or the microchannel networks common to microfluidic systems, Reynolds' numbers are typically much less than 10^2. This represents a situation where the flow regime may be considered essentially laminar and contrasts with macroscale pipes or reactors (exhibiting Reynolds' numbers in excess of 10^4), where flow regimes are almost always turbulent. This behavior has a direct consequence on mixing within microfluidic systems. Before a reaction between two reagents can occur, intimate contact between the component molecules must be realized through mixing. Mixing within microfluidic systems in its simplest manifestation is achieved by bringing together pure fluid component streams within channels having characteristic cross-sectional dimensions measured in tens of microns. Since low Reynolds' and Peclet numbers (which measure the relative importance of mass transport due to convection with respect to diffusional mass transport) are typical of such environments, mixing can only be accomplished through diffusion, rather than the fast convective processes that dominate in turbulent systems. Thus under these conditions, two or more distinct fluid streams moving in the same channel do not develop turbulence at the interface between them, or at the interface with the channel walls and the only mechanism of mixing of their components is diffusion across the former interface.

Diffusive mixing efficiencies can be assessed using the Fourier number, which is defined by

$$F_{\text{o}} = \frac{Dt}{l^2}, \tag{42.3}$$

where D is the molecular diffusion coefficient, t is the contact time, and l is a characteristic length over which diffusion occurs. Broadly speaking, adequate mixing of fluidic streams occurs for Fourier numbers of 0.1, with complete mixing defined by Fourier numbers greater than 1 [5]. Equation 42.3 illustrates that diffusive-mediated mixing timescales increase with the characteristic dimension of the reactor. Consequently, although mixing via diffusion is highly inefficient for reactors with characteristic dimensions greater than 1 mm, when diffusion distances are reduced below 100 μm mixing times can become small. For example, a small molecule such as fluorescein in water exhibits a mixing time of approximately 90 ms across a diffusion distance of 10 μm.

42.2.2 Mixing and Chemical Reactions

The effect of mixing on the extent of a chemical reaction and the resulting product distribution is crucial when designing microfluidic reactors. It is generally recognized that first-order irreversible reactions are unaffected by local turbulent mixing, but are controlled by the residence time of the reaction and the conversion of such systems can therefore be easily calculated. However, in the case of fast reactions, where two or more reagents are initially present in separate streams, reaction rarely occurs throughout the whole volume uniformly. As the mixture is initially highly segregated with a heterogeneous and dispersed structure the reaction is stifled until mixing occurs. In this situation, the rate of reaction is no longer related to the rate constant but instead is limited by the rate of diffusion. In the case of fast reactions, where a single reaction product is produced, the yield can be regarded as a direct measure of the degree of mixing in the reaction vessel. In the case of fast reactions with two or more products, the product distribution has similar connotations. In this way, the product distribution can be manipulated by increasing or decreasing the amount of segregation or the efficiency of mixing within the reactor. The relationship between the rate of a reaction and the rate of mixing in a reaction falls into one of three main categories. These are the chemical regime, the diffusional regime and the mixed chemical/diffusional regime. The interaction between chemical reactions and fluid dynamics is described by the Damkoehler number, which represents the contribution of these two systems [6]. The Damkoehler number can be represented in either of two forms, *NDaI* and *NdaII*, where *NDaI* is a ratio of the chemical reaction rate to the bulk mass flow rate and *NDaII* is the ratio of the chemical reaction rate to the molecular diffusion rate. In the chemical regime, the mixing time is fast in comparison to the reaction rate and is represented by a Damkoehler number that is close to zero. In this situation, the chemical reaction is slow compared to the mass transfer and the reaction system can be regarded as an unreactive fluid. That is to say that mixing is complete before significant amounts of products are present and once mixed, the reaction proceeds uniformly throughout the reaction volume. In the case of a competitive-consecutive reaction as detailed in Scheme 1.1, the smallest amount of secondary products is formed. In the diffusion regime (where $NDa \approx \infty$), the chemical reaction is very fast and thus the reaction is limited by the rate of mixing. In this condition, the area available for reaction is reduced from the volume of the entire reaction vessel at $NDa \approx 0$ to a plane between reacting streams at which each reagent has zero concentration. When a chemical reaction occurs, microscopic concentration gradients are introduced increasing segregation, which in turn leads to a decreased level of mixedness. The reaction rate is then independent of the rate constant and the formation of secondary products in this situation is the greatest for the reaction system detailed in Scheme 1.1. In a mixed chemical/diffusional regime ($NDa = 1$), the greatest interaction between chemical reactions and fluid dynamics occurs and the product distribution depends on both the chemical

$$\mathrm{A} + \mathrm{B} \xrightarrow{k_1} \mathrm{R}$$

$$\mathrm{R} + \mathrm{B} \xrightarrow{k_2} \mathrm{S}$$

Scheme 1.1 The competitive reaction of the product R with the initial reagent B leads to the formation of by-product S and a reduction in the overall yield of R.

factors, (e.g., rate constants) and also on the diffusional factors (e.g., efficiency of mixing). In this situation, the amount of secondary product formed lies between the extremes of the chemical and diffusional regimes. Some practical examples, where micromixing has a profound effect on the product are fast consecutive competing reactions, polymerizations and precipitations. In competitive-consecutive reactions, a reaction occurs between two reagents (A and B) to produce a product (R) at rate k_1. The product in this situation is then able to react with one or more of the initial starting reagents in a competitive manner at rate k_2 to form a by-product (S), leading to a reduction in the initial product (Scheme 1.1). The competitive reaction of the product R with the initial reagent B leads to the formation of by-product S and a reduction in the overall yield of R In the case of fast competitive-consecutive reactions k_1 is much larger than k_2 and k_2 is still large. If a small amount of B is mixed with a large excess of A, R is instantaneously formed. If the fluid is completely mixed, R and B are dispersed throughout the reaction volume, B is consumed by the excess of A and R is preserved. If a small amount of B is mixed with a large excess of A and the fluid is a segregated mixture, R stays at the fluid interface of B. In the presence of B, R is immediately converted to S, the concentration of which then may be used as a measurement of segregation or unmixedness in the vessel.

42.3 MICROFABRICATED REACTION SYSTEM

42.3.1 Fabrication of Microfluidic Reactors

Over the past decade a diversity of microfluidic systems have been designed for the rapid mixing of pure fluid components. Microfluidic devices can be manufactured using a diversity of fabrication techniques originally developed in the semiconductor and microelectronics industries. Since most system features are relatively large (>1 μm), fabrication is straightforward and can be achieved using well-established lithographic and microstructuring methods. Although, extensive details of common fabrication methods can be found in excellent articles elsewhere [7], a brief description of a common protocol for the fabrication of planar glass (or silicon) microfluidic reaction systems is provided (Figure 42.1).

In the simplest case, a fluidic channel pattern is structured within a planar substrate using a combination of photolithography, wet-etching, and bonding. Initially, a photosensitive polymer resist is deposited on to a glass substrate and then exposed to ultraviolet (UV) radiation through a mask (which defines the fluidic topography). Subsequent development in a solvent allows the removal of portions of photoresist, which have been exposed (or unexposed) to radiation, leaving a polymerized resist pattern with high chemical resistance in some areas and exposed substrate in others. Etching of the substrate material then allows the two-dimensional resist pattern to be transferred to the substrate material. Typically, wet etching protocols, involving the use of aqueous etchants, such as HF, HNO_3, KOH, and tctramcthyl ammonium hydroxide, are used to create three dimensional structures in the substrate material. The final stage in the fabrication of a basic microfluidic reaction system involves the assembly of the structured substrate with an unstructured coverplate (typically glass or silicon) to form the enclosed chip structure. Anodic bonding is used to bond glass and silicon substrates, while for glassy materials thermal bonding (450–900°C) provides the simplest way to assemble substrates.

42.3.2 Fluid Motivation

An important factor to consider when using microfluidic reactors to perform chemical synthesis is the method by which reagents are moved through the microdevice. Normally, pressure-driven flow using infusion pumps is used to force fluid through microchannels. The wide volumetric flow rate ranges (pL/min to mL/min) and applicability to all solvent systems makes hydrodynamic pumping the most versatile and popular method of flow motivation. In addition, electroosmotic pumps can be used to generate fluid flow by application of a potential difference through a conductive solvent [8].

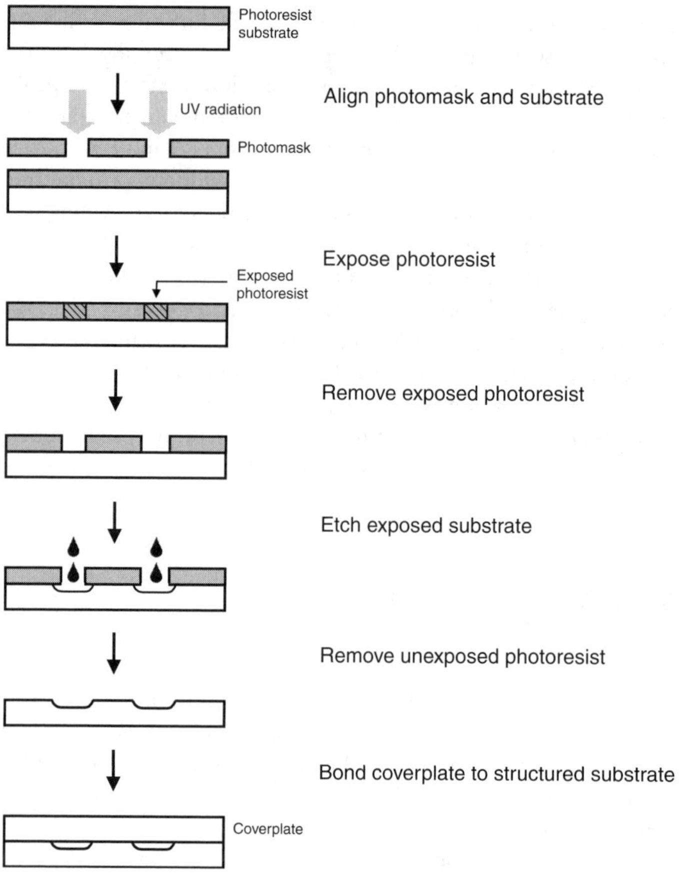

FIGURE 42.1 Schematic representation of component processes in the fabrication of a basic glass microfluidic reactor using photolithography, wet-chemical etching, and thermal bonding.

The primary advantage of such an approach is that it creates a flat flow profile, and thus allows definition of precise flow rates and narrow residence time distributions. Nevertheless, electroosmotic flow (EOF) pumps are severely limited in their widespread application to molecular synthesis due to a need for a conductive solvent and the fact that varying electrophoretic mobilities of reagents and products leads to time-dependent concentration gradients within the reactor that can degrade performance.

42.3.3 Mixing Modalities

A wide range of microfluidic systems have been designed to efficiently mix pure fluid components [9]. Broadly, all of these systems can be classified as being either passive or active in operation. Passive mixers rely on the geometric properties of the channel shape and fluidic streams to achieve mixing, whereas active mixers rely on time-dependent perturbations of fluid flow to achieve mixing. In general, passive systems are simple to fabricate and integrate with other functional components, while active systems require the integration of external actuators and involve more complex fabrication techniques. Passive mixing is realized via a diversity of mechanisms including serial or parallel flow lamination, chaotic advection, and injection of substreams of one component into a primary stream of another component and microdroplet formation. Similarly, active mixers exploit actuation via pressure, electrohydrodynamic, dielectrophoretic, acoustic, thermal, and electrokinetic mechanisms.

Detailed reviews of the structure, function, and mode of operation of microfluidic mixers are provided elsewhere.

Of particular interest in relation to reactive microfluidic systems have been recent developments in the use of chaotic advection to accelerate mixing. In many situations, practical limitations (such as minimum feature dimensions) mean that basic flow lamination is inefficient at generating high degrees of mixedness within short times. Rapid mixing with low reagent consumption is, however, readily achievable using chaotic advection [10]. Chaotic advection may enhance mixing in laminar flow systems by continuously "stretching" and "refolding" concentrated solute volumes, thereby creating an exponential decrease in striation thickness. In microfluidic systems, this can be achieved by introducing obstacles within channels or by modifying channel geometries. Chaotic advection has been shown to generate efficient mixing within both high and low Reynolds' number regimes. For example, zigzag-shaped channels have been shown to cause chaotic advection at high Reynolds' numbers by recirculation around turns [11], while generation of chaotic flow at low Reynolds' numbers has been achieved through the use of grooves on channel surfaces [12].

The ability to achieve rapid mixing of fluids while minimizing the deleterious effects of dispersion is a significant challenge when creating synthetic systems. Put simply, dispersion acts to increase residence time distributions within continuous-flow [13], which in turn causes significant variation in the yield, efficiency and product distribution. Although chaotic flow can be used to reduce dispersion in microfluidic systems, localization of reagents within discrete droplets has been shown to be effective at entirely eliminating this phenomenon [14]. Indeed, a number of recent studies have exploited the formation of droplets with microfluidic channels to perform a variety of synthetic processes [15]. In these reports, droplets are made to spontaneously form when multiple lamina streams of aqueous reagents are injected into an immiscible carrier fluid. The formed picoliter droplets are isolated from channel surfaces and other droplets, with each one acting as an individual reaction vessel. Importantly, the use of winding channel geometries is highly efficient in generating chaotic mixing within droplets and allows rapid and dispersion free mixing.

42.4 APPLICATION OF MICROFLUIDIC REACTORS IN SYNTHETIC CHEMISTRY

42.4.1 Key Benefits of Microfluidic Systems in Synthesis

Small molecule synthesis forms the basis of the pharmaceutical and fine chemical industries. These industries have to address sometimes competing concerns: the synthesis of high purity compounds and the synthesis of large quantities of a compound. Microfluidic systems have been employed to address these concerns and show many advantages over existing techniques. Many of these advantages are case specific, but nevertheless certain key areas can be considered.

From a synthetic point of view, the essential features of the microscale are functions of mass and energy transfer. The rapidity and efficiency of mixing is often one of the limiting factors on the rate of reaction for chemical processes. The other commonly found limit is the rate at which heat can be transferred out of (or less commonly into) the reaction vessel. As we have seen previously, mixing on the microscale is a rapid and controlled diffusive process, limited only by the diffusion coefficients of the solutes themselves. Given small enough diffusive distances, such as can be found in multilaminar microfluidic mixers, homogeneity can be achieved faster than is possible in macroscale systems, and the mixing is easily quantifiable [16]. Mass transfer between phases in microfluidic reactors is also rapid, though more complex than the simple case for diffusive mixing. The flow regimes within microreactors are normally strictly laminar and so readily quantifiable into flow modes. Even so, the prediction of flow modes is complex and strongly dependent on local conditions such as solvent viscosity and phase compressibility. Irrespective of this, work on microfluidic multiphase reactions has shown that an acceleration of the rate of mass transfer of at least two orders of magnitude is experienced in microfluidic systems compared to commonly quoted laboratory scale reactions [17]. Studies on slug flow reaction systems have demonstrated convincing benefits to employing microreactors for phase-transfer reactions due to the enhanced rate of transfer between phases in these systems [18].

Heat transfer is limited by a "cube/square" relationship. Heat is generated throughout the reaction volume, but can only be removed at the reactor surface. From this point of view, it can be argued that the traditional spheroidal reaction vessel is perhaps the worst configuration imaginable for effective heat transfer from the vessel to the surroundings. Given the very high surface area-to-volume ratios present in microfluidic reactors these systems represent a very efficient heat transfer system indeed. The limitation on the system in these devices is often found to be the intrinsic conduction properties of the reactor materials themselves, rather than the heat transfer between the reaction medium and the vessel wall [19]. The precise control of local thermal and concentration gradients possible within microfluidic reactors enable much greater control of synthetic processes. Many systems have a narrow optimal range of conditions for efficient and specific synthesis and benefit from the greater control of these factors inherent in micron scale reactors. In general, it is fair to say that different areas within the chemical and pharmaceutical industries require different approaches within the broader theme of microfluidic flow systems. These requirements are briefly discussed below.

42.4.2 Fields of Application

42.4.2.1 High-Throughput Synthesis

In high-throughput synthesis, the chief problems that have to be overcome are the time it takes to perform individual syntheses and the purity of the products. The volume of product produced is not of primary concern in most cases because the target compounds synthesized are destined for lead screening, a process requiring milligram amounts or sometimes less. The chief aim within this area is to produce as many pure compounds as possible within the shortest possible space of time with high target purity. The compound libraries thus generated are then evaluated in a process often completely divorced from the original synthesis. Traditional approaches toward library generation have relied upon two divergent paradigms: solid phase synthesis and automation.

Solid phase chemistry has been the technique of choice for research into high-throughput synthesis toward the generation of large compound libraries. In this approach, the synthetic targets are attached by a linker group to insoluble functionalized, polymeric material, allowing them to be readily separated from excess reagent, soluble reaction by-product or solvents. Such an approach involves immobilization of substrates on a polymeric resin or beads using linker groups followed by a synthetic sequence and finally a cleavage step to obtain the free product.

This approach suffers from problems related to the influence of the support on reaction chemistries and the optimization of solid-supported chemistries, which may have completely different kinetics from their solution-phase analogues. Difficulties exist in assessing product purity or structure of compounds in the solid phase, often leading to expensive spectroscopic solutions. Working in the solid phase also limits the range of chemistry, which can be applied to a synthetic problem, not least because many solid phase approaches are so dependent upon amide bond formation. A further issue arises from the need for the perfect linker: uncleaved in normal reaction conditions but perfectly cleaved when desired, preferably without a footprint in the target molecule.

The above limitations with solid phase chemistries have necessitated the pursuit of solution-phase chemistries for library generation. The advantages of solution-phase syntheses are legion, but the major benefits include; unlimited numbers and types of reaction, lower requirements of solvents and reagents compared to solid phase syntheses and that such chemistries can be developed and monitored with relative ease. A conventional solution-phase chemical synthesis involves the use of arrays of sub millimeter wells as discrete reaction vessels to which the reagents are delivered using automated robotic systems. Although such systems have been successful, the batch nature of

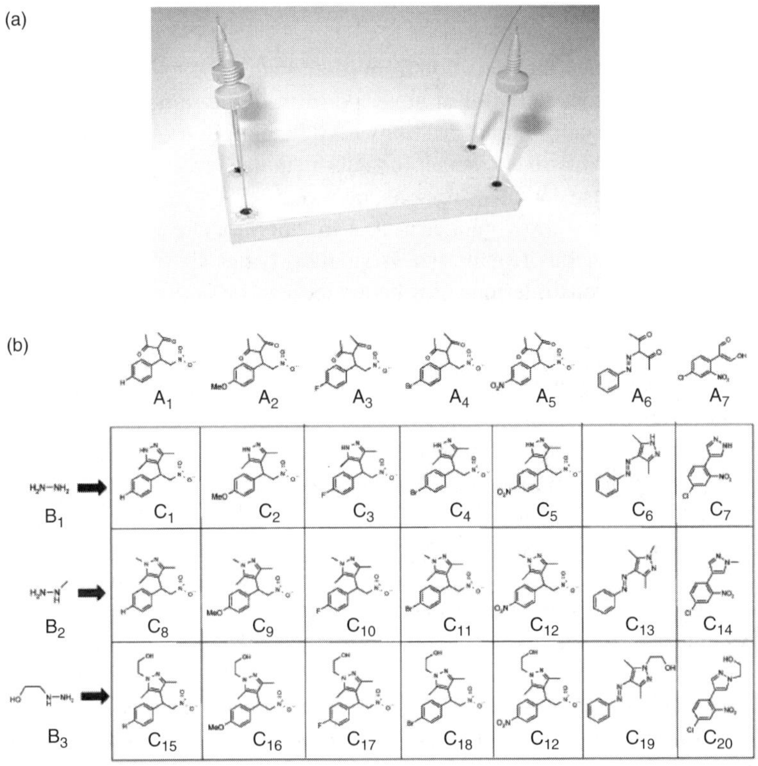

FIGURE 42.2 (a) Continuous-flow glass microfluidic reactor used for generating combinatorial libraries. (b) The formation of a 7 × 3 pyrazole library was programmed and executed in the automated platform using the device in part (a). (Taken from Garcia-Egido, E., et al., *Lab on a Chip*, 2003, **3**: 73–76.)

the general approach is not ideal for efficient process optimization and high-throughput chemical processing.

The use of microfluidic systems in high-throughput chemistry offers great advantages. Reactions can be run in the solution phase, while still allowing rapid reaction throughput and with no detectable cross-contamination. For example, work by Mitchell et al. [20] using an integrated continuous-flow reaction/detection system (expressed as a logic gate) showed that the rapid generation of a small library of compounds via a multicomponent reaction could be accomplished with ease. Using such an approach, valuable information about reaction intermediates can be gathered in real time with the entire process being readily controlled by simple automation protocols. Garcia-Egido et al. [21] also showed that using slugs of reagents as separate reaction vessels small libraries of thiazoles could be generated then purified and analysed by in-line LC-MS (Figure 42.2), with a noticeable improvement in turnover compared to bulk systems and no detectable cross-contamination. Indeed utilizing similar systems, the synthesis of libraries of cycloadducts [22], pyrazoles [21], nitrostilbene esters [23], and thioethers [24] has been accomplished.

Although the bulk of these systems represent only one synthetic step, multistep combinatorial processes have been undertaken, notably by Sakai et al. [25]. High-throughput synthesis on the microfluidic scale has been proven to be advantageous in terms of the available chemistries, product purity and, ease of separation. Table 42.1 lists most of the successfully advantageous examples of synthesis performed within microfluidic reactors and recorded within the literature. One other advantage that has been demonstrated is that the same chemistry used to produce the sample can be used to produce large quantities of product. This multiscale nature of microfluidic reactors is one of enormous potential value to the chemical and pharmaceutical sector.

42.4.2.2 Multiscale Synthesis

In traditional synthetic schemes, the transition from small scale production to large-scale production is a troublesome one. Reactions will almost never perform in the same way on a production scale as they do on a laboratory scale. Some reactions may be simply too dangerous to perform on a large scale, especially if they involve a dangerous intermediate, or a strong exotherm. Scale-up of a chemical process can, for these and other reasons, take years.

Continuous-flow processes allow greater access to laboratory reactions than batch processes. This is because the instantaneous reactor size is smaller, better simulating laboratory conditions and the assay of any dangerous intermediates generated can be kept small. Microfluidic reaction systems offer one superlative advantage over other methodologies: that of scale-out. Scale-out (or numbering up) is a simple proposition. If one microfluidic reactor produces a small quantity of product, when more is desired the number of reactors and not the scale of the reactor, is increased. Because microfluidic devices are easily mass produced, this system of parallel reactors is relatively cheap. Furthermore, because of the continuous-flow nature of the reaction environment a surprisingly large amount of product can be produced in a short space of time and the reaction conditions within the production reactor exactly mimic those of the research reactor. In other words, no scale-up development work is necessary. An excellent example of this approach is in the pioneering work undertaken by R.D. Chambers and coworkers in the use of elemental fluorine for fluorinations.

Direct fluorination is a process that is generally viewed as unscaleable. The highly exothermic nature of the system ensures that large-scale reactions are problematic and the use of fluorine gas in large assay is in itself undesirable. Early work on microfluidic fluorinations showed that the direct reaction was comparable in selectivity to the industrial Schiemann process [26] and that the system could be applied to a wide range of compounds [27]. In a series of papers, Chambers et al. [28,29] showed that by the simple numbering up of reactor channels a commercially viable reactor could be constructed with no loss of reaction efficiency or selectivity.

The numbering up of microfluidic systems is a powerful tool in the rapid development of synthetic processes from the laboratory to fine scale and beyond and is only beginning to show its versatility [30] (Figure 42.3).

42.4.2.3 High-Selectivity Synthesis

The highly ordered mass and heat transfer processes within microreactors often produce very selective reactions. As will be seen later this has implications for nanomaterial synthesis, but even within well-known reactions this selectivity can be important. Burns and Ramshaw [31] showed that nitration processes within microfluidic systems produce cleaner products than bulk scale systems. Similarly high yields and low by-product formation has been reported in on-chip peptide formation [32]. This is almost certainly attributable to the thermal flatness found within microfluidic channels and the ordered and predictable mixing within the system. It appears that reactions within the microfluidic regime are cleaner and often quicker, than their bulk equivalents.

42.4.2.4 Catalytic Systems

The use of microfluidic systems for catalytic reactions has several advantages. The high surface area-to-volume ratios encountered on the microscale allow for good contact between heterogeneous catalysts and the reaction medium, and the thermal flatness eliminates hotspot formation. This has made microfluidic systems a valuable tool for process development in systems dependent upon a limited range of catalyst temperatures for optimal operation. For example, Worz and coworkers [33] at BASF showed that by using this approach, the efficiency of their catalytic process was increased and the reliability of the catalyst improved. Worz [33] states that the use of this technology represented a considerable saving in development time. For heterogeneous catalysis, many systems have been

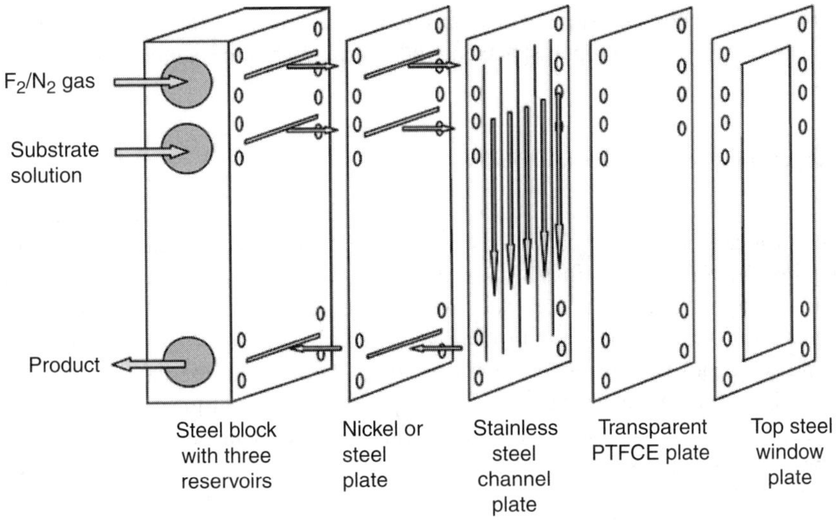

FIGURE 42.3 A modular microreactor system for the direct fluorination of ethyl acetoacetate by fluorine gas. The design allows for multiple channels to be supplied from single reservoir sources and a multichannel device to be constructed in a facile manner from a disposable channel plate. (Taken from Chambers, R.D., et al., *Lab on a Chip,* 2005, 5: 191–198. With permission.)

used from packed beds [17] through solid supports [34] to derivitized channel walls [35–37], each with advantages for particular catalytic systems. Indeed the breadth of application for microfluidic reaction systems in catalytic process development is hard to overestimate.

42.5 REAL WORLD APPLICATIONS

Real world synthetic problems are rarely solved in one step, but are more normally addressed by a combination of factors leading to complex solutions. The "plug and play" nature of microfluidic components makes them a flexible tool in problems of this nature, even when the whole process in question cannot be transferred onto a single microfluidic device. The following section outlines some illustrative examples of such applications.

42.5.1 POINT OF USE MANUFACTURE

Synthesis of short half-life radiotracers positron emission tomography (PET) is an emerging field within radiography. The technique utilizes positron-emitting radiotracers to trace metabolic activity directly within the body. The chief disadvantage of this methodology is the need for the tracers themselves: the short half-lives dictate that the compounds must be synthesized close to the point of use and the fiercely radioactive nature of the isotopes used entails prodigious shielding. The general approach toward these syntheses is to use automated robotic synthesis stations, often custom built, to perform the necessary synthesis and processing within a shielded compartment known as a hot cell. Space within these cells is at a premium and the number of automated stations per cubic metre is the major constraint on the procedure.

Recent work by Jeffery et al. [71] showed that many of the standard synthetic operations that are necessary in the synthesis of $^{[18]}$FDG (a major PET adiotracer) could be incorporated into microfluidic devices. Moreover, Lu et al. [72] showed that the methodology was also applicable to $^{[11]}$C compounds. In an elegant recent study, Lee and coworkers [73] showed that all of the major processes in $^{[18]}$F radiolabeling tracer compounds could be integrated onto a single device. Such advances are revolutionizing the field, with continuous-flow syntheses on microfluidic devices being more efficient in terms of throughput, space and indeed energy requirements than their conventional counterparts.

42.5.2 Measurement of Chemical Phase Space

All chemical processes are optimized to some degree. This entire procedure can be thought of as finding a set of localized optimal conditions within a reaction's output. The most common way of quantifying this form of relationship is by using phase space, an n-dimensional notional space, where each input to or output from a reaction is dealt with as a separate dimension. Using this model allows an intuitive handling of data. The main problem with applying the phase space model to reactions in the real world is the lack of data. When each experiment takes 3 h to run the number of data points generated will be quite low unless the optimization process has an unlimited budget. Automation of kinetic and calorimetric data gathering has made some improvements and allows the use of Design of Experiments algorithms, which can help to find minima quickly. However, using microfluidic devices it is possible to speed up the data gathering process. Studies by Le Bars and coworkers [74] has shown that calorimetric readings can be taken in continuous flow in a microfluidic device. Furthermore, in-line monitoring using, inter alia, NMR spectroscopy allows real-time evaluation of reaction conditions [75]. In addition, a useful demonstration of the benefits of microfluidic devices for reaction optimisation was presented by Ratner et al. [68] who investigated the influence of systematic variations in reaction time and temperature on glycosylation reactions (Figure 42.4).

In a series of studies our group demonstrated that not only could reactions be monitored on-line [76] and these data used for very rapid optimization [77], but also that this information could be used to plot local phase space in detail rarely attempted before. Several thousand data points per minute can be generated using appropriate in-line sampling and control techniques, allowing detailed phase space surfaces to be plotted [78]. This represents an increase in speed of over four orders of magnitude when compared to conventional methods.

42.5.3 Microfluidic Systems for Nanomaterial Production

As discussed, microfluidic systems are well suited for performing reagent mixing in a rapid and controllable manner. This combined with control over other variables, such as reactor temperature, concentration gradients and pressure dictate that continuous-flow processing on the microscale can be used to synthesize species of specific yet variable characteristics. Perhaps the most interesting demonstration of this feature is in the synthesis of nanomaterials (such as nanoparticles, nanotubes, and nanorods). The physical characteristics of nanocrystallites are primarily determined by quantum confinement effects with properties such as the optical band gap often differing considerably from the bulk material. Consequently, they are seen as customized precursors for functional materials in applications such as biological sensing and optoelectronics [79]. Since the electronic and optical properties of these materials are ultimately controlled by the physical dimensions of the crystallites, there is considerable interest in processing routes that yield nanoparticles of well-defined size and shape. Synthetic approaches for nanomaterial synthesis involve particle growth on an atom-by-atom or molecule-by-molecule basis until the desired size is achieved [80]. Such growth takes place spontaneously in super-saturated solutions and has been successfully used to create crystallites of well-defined size and shape. "Bottom-up" or synthetic approaches are attractive due to their versatility and ease of use; however, standard syntheses rarely yield product size distributions better than $\pm 5\%$. This means that it is almost always necessary to use some form of post-treatment (including electrophoresis, chromatography, precipitation and photocorrosion) to extract particles of well-defined size. Although nanoparticles with extremely narrow size distributions can ultimately be extracted, the starting point for all such methods is a polydisperse sample and thus product yields and conversions are necessarily low.

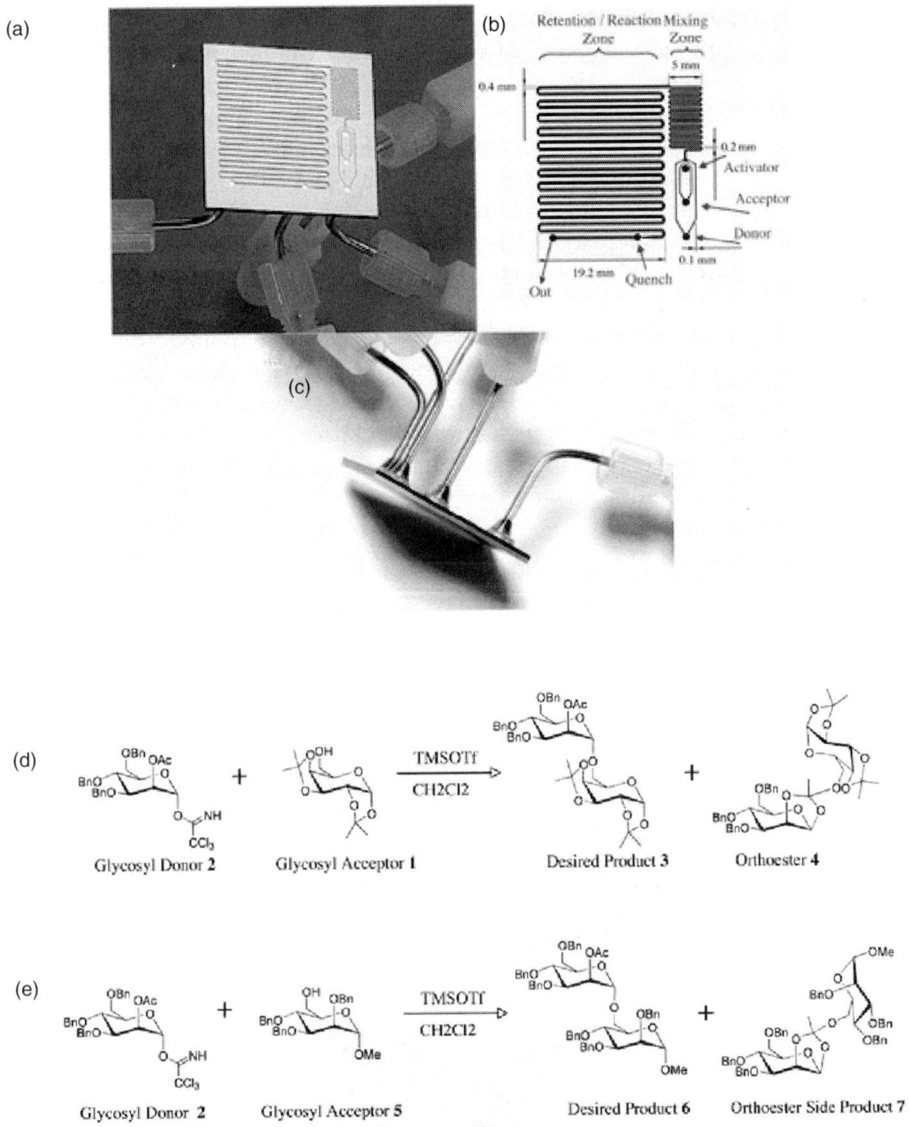

FIGURE 42.4 (a) Silicon microfluidic reactor for performing glycosylation reactions over a wide range of experimental conditions. (b) Schematic representation of microfluidic circuit, which comprises three primary inlets, a mixing and reaction zone, a secondary inlet for quenching reagent and an outlet for analysis/collection. (c) Side view of microreactor. (d and e) Glycoside coupling reactions. (Taken from Ratner, D.M., et al., *Chemical Communications*, 2005, 578–580. With permission.)

At a fundamental level monodisperse nanoparticle populations can be created by ensuring that initial nucleation of solute molecules (to form "seed" particles) occurs on a timescale, which is rapid compared with the growth process (in which the seed particles confine dissolved solutes) [81]. Moreover, nuclei formation and growth should occur within an environment in which state functions are precisely controlled throughout. When these circumstances are not met, the size of nuclei and the particle growth rates vary according to location and result in a wide distribution of particle sizes. Unsurprisingly, a number of recent studies have demonstrated the effectiveness of continuous-flow microfluidic reactors in performing controlled nanoparticle synthesis [82]. These systems generally utilize simple flow regimes, whereby pure fluid component streams are mixed within low Reynolds' number regimes. Variation in reaction residence times and reagent concentrations can be used control average particle size, while product polydispersity is minimized through reduction in residence time distributions and precise control of chemical state functions. For example, silver, cobalt, copper, cadmium sulphide [83,84], cadmium selenide [14,85], gold [86,87], palladium [88], titania [89,90], and CdSe-ZnS [91] core-shell have all been produced within a range of microfluidic reactors. In all cases low-polydispersity nanoparticles of varying size could be synthesized with space-time yields significantly higher than corresponding macroscale approaches.

More recent studies have addressed the issue of minimizing particle size distributions through the development of segmented flow reactors. For example, Ismagilov and coworkers [92] have reported multistep chemical production of CdS and CdS/CdSe core-shell nanoparticles within a microdroplet-based reactor. Importantly, such an approach allowed for millisecond time control and also enabled multiple reactions to be initiated by flowing additional reagent streams directly into individual droplets. In addition, Chan et al. [93] have reported the use of microfluidic droplet reactors for the high-temperature synthesis of CdSe nanoparticles, while Yen et al. [94] have a used gas–liquid segmented flow reactor containing multiple temperature zones (Figure 42.5) for the synthesis of high quality CdSe quantum dots. In all studies, enhanced mixing and reduced residence time distributions provide the driving force for improvements in reaction yield and size distribution.

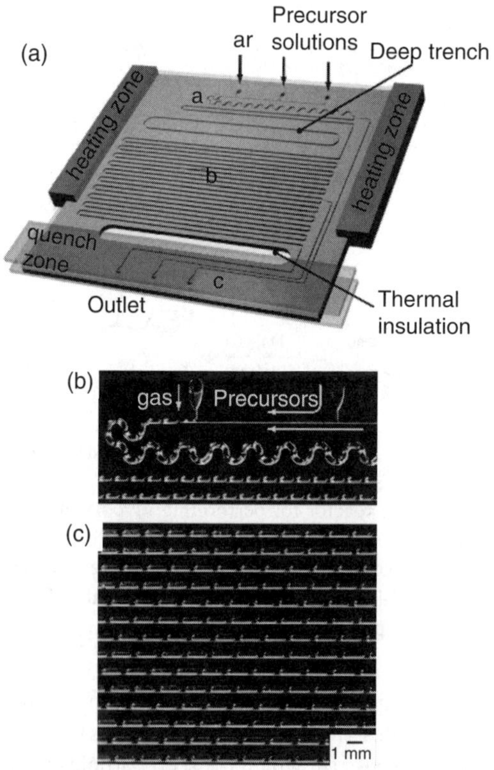

FIGURE 42.5 Microfluidic reactor for nanoparticle production. (a) Schematic illustrating the key components of the reactor. The reactor performs precursor mixing (section a), controlled particle growth (section b) and reaction quenching (section c). A halo etch region allows localization of temperature zones for reaction (>260°C) and quenching (<70°C). (b) Photograph of heated inlets and (c) photograph of main channel section. Red segments show the reaction solution, with dark segments defining argon gas. (Taken from Brian, K.H., et al., *Angewandte Chemie-International Edition in English*, 2005, 44, 5447–5451. With permission)

TABLE 42.1
Diversity of small-molecule syntheses performed within microfluidic reactors.

Reaction	Product(s)	Conversion	Comments	Year/Reference
Bayer–Villiger oxidation	Lactones	<99%	Increased reactivity and selectivity	2006/[38]
Baylis–Hillman reaction	Baylis–Hillman adducts	<95%	C–C bond forming	2006/[39]
Autocatalytic nitration	Nitrophenols	65–76%	Autocatalysis	2005/[40]
Catalytic dehydrogenation	Toluene	88%	Gas-solid heterogenous catalysis	2005/[41]
Suzuki–Miyaura coupling	Functionalised biphenols	<76%	Hydrodynamic pumping	2005/[42]
Grignard exchange reaction	Pentafluorobenzene	92%	Formation of Grignard reagents	2005/[43]
Friedel–Crafts aminoalkylation	1-(*N*-Butyl-*N*-methoxycarbonyl aminomethyl)-2,4,6-trimethoxybenzene	96%	Improved selectivity with highly reactive substrates	2005/[44]
Imine hydrogenation	Amines	Quantitative		2005/[45]
Stereoselective alkylation	C–C bond	41% (ee 82%)	Using chiral auxiliary	2004/[46]
Hydrogenations	Alkenes	Quantitative	Triphase reaction system immobilized Palladium	2004/[47]
Michael addition	1,3-Diketones	95%	Stopped flow technique	2002/[48]
Suzuki coupling	Functionalized biphenols	70–100%	Microwave assisted	2004/[49]
Heterocyclic synthesis	1,2-Azoles	98–100%	Significant improvement on batch	2004/[50]
Fluorination	Fluorinated diketones	70%	Safety enhancement	1999/[27]
Esterification	Benzoic acid esters, phenyl esters	100%	EOF	2003/[51]
Photo-oxygenation	Ascaridole	85%	Safety enhancement	2002/[52]
Sandmeyer reaction	Diazonium intermediates	Various	Control over unstable intermediate	2003/[53]
Carbamate synthesis	Methyl carbamates	91%	Control over exothermic conditions	2004/[54]
Ring expansion (6–7 membered ring)	*N-tert*-Butoxycarbonyl-5-ethoxycarbonyl-4-perhydroazepinone	89%	Safety and control of reactive ethyl diazoacetate reagent	2004/[54]
Nitration	Nitrobenzene	65%		2001/[55]
Wittig reaction	Alkene C=C bond formation	39–59%		2001/[23]
Peptide synthesis	Peptides	100%	Quantitative yield in 20 min	2001/[32]
Aldol reaction	C–C bond via enolate	100%	Enolate generated via silyl enol ethers	2001/[56]
Enamine formation	Enamines	42%	Formation using DCC	2001/[57]

Continued

TABLE 42.1
(Continued)

Reaction	Product(s)	Conversion	Comments	Year/Reference
Ugi 4 component coupling	α-Dialkylacetamide	—	Detection of reaction products and intermediates	2001/[58]
Dehydration of alcohols	Alkene	85–95%	Sulfated Zirconia catalyst	2000/[59]
Kumada–Corriu coupling	Coupling of aryl halide and Grignard reagent	60%	Nickel chelated to immoblilsed salen ligand	2001/[60]
Knoevenagel condensation	Condensation followed by hetero Diels–Alder reaction forming cycloadducts	50–68%	2 × 2 combinatorial array produced four products	2002/[61]
Phase transfer diazo coupling		~100%	Increased surface area, rapid diffusive phase transfer	2001/[62]
Phase transfer alkylations	Alkylated β-keto esters	71–96%	Increased surface area, rapid diffusive phase transfer	2003/[63]
Hantzsch synthesis	2-Aminothiazoles	58–99%	Yields improved over bulk—heated under EOF	2002/[64]
Knorr reaction	Pyrazoles	52–99%	Sequential synthesis of 7 × 3 library	2003/[21]
Acid catalyzed esterification	Pyrenebutyric acid ethyl ester	83%	Surface area related rate enhancement	2003/[65]
Swern oxidations	Carbonyl compounds	81–100%	Room temperature conversions—Batch Swern usually below −50°C	2005/[66]
Photochlorination	1-Chloromethyl-2,4-diisocyanatobenzene	55%	80% selectivity (big incresase over batch)	2002/[67]
Glycosylations	Glycosidic link for oligosaccharide assembly	—	Small quantity of glycosylating agent required	2005/[68]
Heck coupling reaction	Ethyl cinnamate	99%	Using a modular microreaction system	2005/[69]
Horner–Wadsworth–Emmons	Olefins	91%	Using a modular microreaction system	2005/[69]
Diels–Alder cycloaddition	Cycloadduct	100%	Using a modular microreaction system	2005/[69]
Henry reaction	2-Nitro-1-phenylethanol	76%	Nitroaldol addition	2005/[69]
Reimer–Tiemann formylation	Formylated β-napthol	10%	Yields similar to bulk, allowed study of temperature effects	2005/[70]

42.6 CONCLUSIONS

The nature of the microfluidic environment allows for a control over physical properties such as energy transfer, mass transfer and mixing that is difficult to achieve using other approaches. The consequences of this for the synthetic chemist are that processes, which are controlled in large part by these properties, can be exploited to produce a range of products that are otherwise inaccessible. The control on the microscale leads to faster, cleaner and more specific reactions. In addition to this, the nature of the relationship between surface and bulk properties in the microfluidic environment ensures that reactions involving rapid exchanges of energy become manageable in a way that is not possible in larger reaction vessels.

Microfluidic approaches also show strong advantage in high-throughput systems, whether these be for catalyst screening or for combinatorial synthesis. The rapidity of data generation, which has already revolutionized the science of genomics, is also a major bonus in the application of microfluidics to these synthetic problems.

REFERENCES

1. Wöhler F., *Grundriß der Organischen Chemie.* 1848, Berlin: Duncker und Humblot.
2. Purcell, E.M., Life at low Reynolds number. *American Journal of Physics,* 1977, **45**: 3–11.
3. Atencia, J. and D.J. Beebe, Controlled microfluidic interfaces. *Nature*, 2005, **437**: 648–655.
4. Rott, N., Note on the history of the Reynolds number. *Annual Review of Fluid Mechanics*, 1990, **22**: 1–11.
5. Crank, J., *The Mathematics of Diffusion.* 2 edn. 1975, Oxford, England: Oxford University Press.
6. Chakraborty, S. and V. Balakotaiah, Low-dimensional models for describing mixing effects in laminar flow tubular reactors. *Chemical Engineering Science*, 2002, **57**: 2545–2564.
7. Kovacs, G.T.A., K. Petersen, and M. Albin, Silicon Micromachining. *Analytical Chemistry*, 1996, **68**: A407–A412.
8. Feng, X.Z., S.J. Haswell, and P. Watts, Organic synthesis in microreactors. *Current Topics in Medicinal Chemistry*, 2004, **4**: 707–727.
9. Nguyen, N.T. and Z. Wu, Micromixers—A review. *Journal of Micromechanics and Microengineering*, 2005, **15**: R1–R16.
10. Ottino, M., The mixing of fluids. *Scientific American*, 1989, **260**: 56–67.
11. Mengeaud, V., J. Josserand, and H.H. Girault, Mixing processes in a zigzag microchannel: Finite element simulations and optical study. *Analytical Chemistry*, 2002, **74**: 4279–4286.
12. Stroock, A.D., et al., Chaotic mixer for microchannels. *Science*, 2002, **295**: 647–651.
13. Krishnadasan, S., et al., On-line analysis of CdSe nanoparticle formation in a continuous-flow chipbased microreactor. *Journal of Materials Chemistry*, 2004, **14**: 2655–2660.
14. Bringer, M.R., et al., Microfluidic systems for chemical kinetics that rely on chaotic mixing indroplet. *Philosophical Transactions of the Royal Society of London A*, 2004, **362**: 1087–1104.
15. Song, H., D.L. Chen, and R.F. Ismagilov, Reactions in droplets in microfluidic channels. *Angewandte Chemie-International Edition*, 2006, **45**: 7336–7356.
16. Ehrfeld, W., et al., Characterization of mixing in micromixers by a test reaction: Single mixing units and mixer arrays. *Industrial and Engineering Chemical Research*, 1999, **38**: 1075–1082.
17. Losey, M.W., M.A. Schmidt, and K.F. Jensen, Microfabricated multiphase packed-bed reactors: Characterization of mass transfer and reactions. *Industrial and Engineering Chemistry Research*, 2001, **40**: 2555–2562.
18. Burns, J.R. and C. Ramshaw, The intensification of rapid reactions in multiphase systems using slug flow in capillaries. *Lab on a Chip*, 2001, **1**: 10–15.
19. Rebrov, E., M.H.J.M. de Croon, and J.C. Schouten, Design of a microstructured reactor with integrated heat-exchanger for optimum performance of a highly exothermic reaction. *Catalysis Today*, 2001, **69**: 183–192.
20. Mitchell, M.C., V. Spikmans, and A. De Mello, Microchip-based synthesis and analysis: Control of multicomponent reaction products and intermediates. *Analyst*, 2001, **126**: 24.
21. Garcia-Egido, E., et al., Synthesis and analysis of combinatorial libraries performed in an automated micro reactor system. *Lab on a Chip*, 2003, **3**: 73–76.

22. Fernandez-Suarez, M., S.Y.F. Wong, and B.H. Warrington, Synthesis of a three-member array of cycloadducts in a glass microchip under pressure driven flow. *Lab on a Chip*, 2002, **2**: 170–174.
23. Skelton, V., et al., The preparation of a series of nitrostilbene ester compounds using micro reactor technology. *Analyst*, 2001, **126**: 7–10.
24. Jonsson, D., B.H. Warrington, and M. Ladlow, Automated flow-through synthesis of heterocyclic thioethers. *Journal of Combinatorial Chemistry*, 2004, **6**: 584–595.
25. Sakai, R., et al., On-line MS detection for a multi-step combinatorial synthesis system., in *μTAS 2004, Proceedings of the 8th International Conference on Miniaturised Systems for Chemistry and the Life Sciences*, T. Laurell, et al., Ed. 2004, Royal Society of Chemistry, Cambridge, UK, p. 96–98.
26. Jahnisch, K., et al., Direct fluorination of toluene using elemental fluorine in gas/liquid microreactors. *Journal of Fluorine Chemistry*, 2000, **105**: 117–128.
27. Chambers, R.D. and R.C.H. Spink, Microreactors for elemental fluorine. *Chemical Communications*, 1999, 883.
28. Chambers, R.D., et al., Gas–liquid thin film microreactors for selective direct fluorination. *Lab on a Chip*, 2001, **1**: 132–137.
29. Chambers, R.D., et al., Elemental fluorine—Part 16. Versatile thin-film gas–liquid multi-channel microreactors for effective scale-out. *Lab on a Chip*, 2005, **5**: 191–198.
30. Commenge, J.-M. and M. Matlosz, From process miniaturisation to structured multiscale design: the innovative, high-performance chemical reactors of tomorrow. *Chimia*, 2002, **56**: 654.
31. Burns, J.R. and C. Ramshaw, Development of a microreactor for chemical production. *Transactions of the Institute of Chemical Engineering*, 1999, **77**: 206–211.
32. Watts, P., et al., The synthesis of peptides using micro reactors. *Chemical Communications*, 2001, 990–991.
33. Worz, O., et al., Microreactors, a new efficient tool for optimum reactor design. *Chemical Engineering Science*, 2001, **56**: 1029–1033.
34. Wiles, C., P. Watts, and S.J. Haswell, The use of solid-supported reagents within EOF-based microreactors., in *μTAS 2004, Proceedings of the 8th International Conference on Miniaturised Systems for Chemistry and the Life Sciences*, T. Laurell, et al., Ed. 2004, Royal Society of Chemistry, Cambridge, UK, p. 104–106.
35. Wootton, R.C.R. and A.J. deMello, A one-step protocol for the chemical derivatisation of glass microfluidic devices. *Lab on a Chip*, 2006, **6**: 471–473.
36. Kobayashi, J., Y. Mori, and S. Kobayashi, Triphase hydrogenation reactions utilizing palladiumimmobilized capillary column reactors and a demonstration of suitability for large scale synthesis. *Advanced Synthesis and Catalysis*, 2005, **347**: 1889–1892.
37. Wan, Y.S.S., et al., TS-1 zeolite microengineered reactors for 1-pentene epoxidation. *Chemical Communications*, 2002, 878–879.
38. Mikami, K., et al., Nanoflow microreactor for dramatic increase not only in reactivity but also inselectivity: Baeyer-Villiger oxidation by aqueous hydrogen peroxide using lowest concentration of a fluorous lanthanide catalyst. *Journal of Fluorine Chemistry*, 2006, **127**: 592–596.
39. Acke, D.R.J. and C.V. Stevens, Study of the Baylis-Hillman reaction in a microreactor environment: First continuous production of Baylis-Hillman adducts. *Organic Process Research and Development*, 2006, **10**: 417–422.
40. Ducry, L. and D.M. Roberge, Controlled autocatalytic nitration of phenol in a microreactor. *Angewandte Chemie-International Edition*, 2005, **44**: 7972–7975.
41. Roumanie, M., et al., Design and fabrication of a structured catalytic reactor at micrometer scale: Example of methylcyclohexane dehydrogenation. *Catalysis Today*, 2005, **110**: 164–170.
42. Phan, N.T.S., J. Khan, and P. Styring, Polymer-supported palladium catalysed Suzuki-Miyaura reactions in batch and a mini-continuous flow reactor system. *Tetrahedron*, 2005, **61**: 12065–12073.
43. Wakami, H. and J. Yoshida, Grignard exchange reaction using a microflow system: From bench to pilot plant. *Organic Process Research and Development*, 2005, **9**: 787–791.
44. Nagaki, A., et al., Control of extremely fast competitive consecutive reactions using micromixing. Selective Friedel-Crafts aminoalkylation. *Journal of the American Chemical Society*, 2005, **127**: 11666–11675.
45. Saaby, S., et al., The use of a continuous flow-reactor employing a mixed hydrogen-liquid flow stream for the efficient reduction of imines to amines. *Chemical Communications*, 2005, 2909–2911.

46. Wiles, C., et al., Stereoselective alkylation of an Evans auxiliary derivative within a pressure-driven micro reactor. *Lab on a Chip*, 2004, **4**: 171–173.
47. Kobayashi, J., et al., A microfluidic device for conducting gas-liquid-solid hydrogenation reactions. *Science*, 2004, **304**: 1305–1308.
48. Wiles, C., et al., 1,4-Addition of enolates to a,b-unsaturated ketones within a micro reactor. *Lab on a Chip*, 2002, **2**: 62–64.
49. He, P., S.J. Haswell, and P.D.I. Fletcher, Microwave heating of heterogeneously catalysed Suzuki reactions in a micro reactor. *Lab on a Chip*, 2004, **4**: 38–41.
50. Wiles, C., et al., The application of microreactor technology for the synthesis of 1,2-azoles. *Organic Process Research and Development*, 2004, **8**: 28–32.
51. Wiles, C., et al., Solution phase synthesis of esters within a micro reactor. *Tetrahedron*, 2003, **59**: 10173–10179.
52. Wootton, R.C.R., R. Fortt, and A.J. de Mello, A microfabricated nanoreactor for safe, continuous generation and use of singlet oxygen. *Organic Process Research and Development*, 2002, **6**: 187–189.
53. Fortt, R., R.C.R. Wootton, and A.J. de Mello, Continuous-flow generation of anhydrous diazonium species: Monolithic microfluidic reactors for the chemistry of unstable intermediates. *Organic Process Research and Development*, 2003, **7**: 762–768.
54. Zhang, X.N., S. Stefanick, and F.J. Villani, Application of microreactor technology in process development. *Organic Process Research and Development*, 2004, **8**: 455–460.
55. Doku, G.N., et al., Electric field-induced mobilisation of multiphase solution systems based on the nitration of benzene in a micro reactor. *Analyst*, 2001, **126**: 14–20.
56. Wiles, C., et al., The aldol reaction of silyl enol ethers within a micro reactor. *Lab on a Chip*, 2001, **1**: 100–101.
57. Sands, M., et al., The investigation of an equilibrium dependent reaction for the formation of enamines in a microchemical system. *Lab on a Chip*, 2001, **1**: 64–65.
58. Mitchell, M.C., V. Spikmans, and A.J. de Mello, Microchip-based synthesis and analysis: Control of multicomponent reaction products and intermediates. *Analyst*, 2001, **126**: 24–27.
59. Wilson, N.G. and T. McCreedy, On-chip catalysis using a lithographically fabricated glass microreactor—the dehydration of alcohols using sulfated zirconia. *Chemical Communications*, 2000, 733–734.
60. Haswell, S.J., B. O'Sullivan, and P. Styring, Kumada–Corriu reactions in a pressure-driven microflow reactor. *Lab on a Chip*, 2001, **1**: 164–166.
61. Fernandez-Suarez, M., S.Y.F. Wong, and B.H. Warrington, Synthesis of a three-member array of cycloadducts in a glass microchip under pressure driven flow. *Lab on a Chip*, 2002, **2**: 170–174.
62. Hisamoto, H., et al., Fast and high conversion phase-transfer synthesis exploiting the liquid–liquid interface formed in a microchannel chip. *Chemical Communications*, 2001, 2662–2663.
63. Ueno, M., et al., Phase-transfer alkylation reactions using microreactors. *Chemical Communications*, 2003, 936–937.
64. Garcia-Egido, E., S.Y.F. Wong, and B.H. Warrington, A Hantzsch synthesis of 2-aminothiazoles performed in a heated microreactor system. *Lab on a Chip*, 2002, **2**: 31–33.
65. Brivio, M., et al., Surface effects in the esterification of 9-pyrenebutyric acid within a glass micro reactor. *Chemical Communications*, 2003, 1924–1925.
66. Kawaguchi, T., et al., Room-temperature swern oxidations by using a microscale flow system. *Angewandte Chemie-International Edition*, 2005, **44**: 2413–2416.
67. Ehrich, H., et al., Application of microstructured reactor technology for the photochemical chlorination of alkylaromatics. *Chimia*, 2002, **56**: 647–653.
68. Ratner, D.M., et al., Microreactor-based reaction optimization in organic chemistry glycosylation as a challenge. *Chemical Communications*, 2005, 578–580.
69. Snyder, D.A., et al., Modular microreaction systems for homogeneously and heterogeneously catalyzed chemical synthesis. *Helvetica Chimica Acta*, 2005, **88**: 1–9.
70. Iles, A., R. Fortt, and A.J. de Mello, Thermal optimisation of the Reimer-Tiemann reaction using thermochromic liquid crystals on a microfluidic reactor. *Lab on a Chip*, 2005, **5**: 540–544.
71. Jeffery, N.T., et al., Radiochemistry on microfluidic devices: proof of principle. *Supplement to the Journal of Nuclear Medicine: Society of Nuclear Medicine*, 2004, 145.

72. Lu, S.Y., et al., Syntheses of C-11- and F-18-labeled carboxylic esters within a hydrodynamicallydriven micro-reactor. *Lab on a Chip*, 2004, **4**: 523–525.
73. Lee, C.-C., et al., Multistep synthesis of a radiolabeled imaging probe using integrated microfluidics. *Science*, 2005, **310**: 1793–1796.
74. Le Bars, J., et al., A scale-transparent reaction calorimetric assay for rapid catalyst selection. *Advanced Synthetic Catalysis*, 2001, **343**: 207.
75. Wensink, H., et al., Real-time monitoring of chemical reactions inside a micro NMR chip, in *μTAS 2004, Proceedings of the 8th International Conference on Miniaturised Systems for Chemistry and the Life Sciences*, T. Laurell, et al., Ed. 2004, Royal Society of Chemistry, Cambridge, UK, p. 192–194.
76. Fortt, R., R.C.R. Wootton, and A.J. deMello, Continuous-flow generation of anhydrous diazonium species: monolithic microfluidic reactors for the chemistry of unstable intermediates. *Organic Process Research and Development*, 2003, **7**: 762–768.
77. Leung, S.-A., et al., A method for rapid reaction optimisation in continuous-flow microfluidic reactors using online Raman spectroscopic detection. *The Analyst*, 2005, **130**: 46–51.
78. Cullen, C.J., et al. Rapid phase space surface generation using an integrated microfabricated device reaction detection system and automated control, in *Micro Total Analytical Systems 2006: The 10th International Conference on Miniaturized Systems for Chemistry and Life Sciences*, 2006, Tokyo: Society for Chemistry and Micro-Nano Systems.
79. Alivisatos, A.P., Semiconductor clusters, nanocrystals, and quantum dots 10.1126/ science.271.5251.933. *Science*, 1996, **271**: 933–937.
80. Murray, C.B., C.R. Kagan, and M.G. Bawendi, Synthesis and characterization of monodisperse nanocrystals and close-packed nanocrystal assemblies. *Annual Review of Materials Science*, 2000, **30**: 545–610.
81. LaMer, V.K. and R.H. Dinegar, Theory, production and mechanism of formation of monodispersed hydrosols. *Journal of the American Chemical Society*, 1950, **72**: 4847–4854.
82. Mello, J.D. and A.D. Mello, FocusMicroscale reactors: nanoscale products. *Lab on a Chip*, 2004, **4**: 11N–15N.
83. Edel, J.B., et al., Microfluidic routes to the controlled production of nanoparticles. *Chemical Communications*, 2002, 1136–1137.
84. Hung, L.-H., et al., Alternating droplet generation and controlled dynamic droplet fusion in microfluidic device for CdS nanoparticle synthesis. *Lab on a Chip*, 2006, **6**: 174–178.
85. Yen, B.K.H., N.E. Stott, K.F. Jensen, and M.G. Bawendi, A continuous-flow microcapillary reactor for the preparation of a size series of CdSe nanocrystals. *Advanced Materials*, 2003, **15**: 1858–1862.
86. Wagner, J., et al., Generation of metal nanoparticles in a microchannel reactor. *Chemical Engineering Journal*, 2004, **101**: 251–260.
87. Shalom, D., et al., Synthesis of thiol functionalised gold nanoparticles using a continuous flow microfluidic reactor. *Materials Letters*, 2007, **61**: 1146–1150.
88. Song, Y., C.S.S.R. Kumar, and J. Hormes, Synthesis of palladium nanoparticles using a continuous-flow polymeric micro reactor. *Journal of Nanoscience and Nanotechnology*, 2004, **4**: 788–793.
89. Wang, H., et al., Preparation of titania particles utilizing the insoluble phase interface in a microchannel reactor. *Chemical Communications*, 2002, 1462–1463.
90. Cottam, B.F., et al., Accelerated synthesis of titanium oxide nanostructures using microfluidic chips. *Lab On a Chip*, 2007, **7**: 167–169.
91. Wang, H., et al., Continuous synthesis of CdSe-ZnS composite nanoparticles in a microfluidic reactor. *Chemical Communications*, 2004, 48–49.
92. Shestopalov, I., J.D. Tice, and R.F. Ismagilov, Multi-step synthesis of nanoparticles performed on millisecond time scale in a microfluidic droplet-based system. *Lab on a Chip*, 2004, **4**: 316–321.
93. Chan, E.M., A.P. Alivisatos, and R.A. Mathies, High-temperature microfluidic synthesis of CdSe nanocrystals in nanoliter droplets. *Journal of the American Chemical Society*, 2005, **127**: 13854–13861.
94. Brian, K.H., A.G. Yen, M.A. Schmidt, K.F. Jensen, and M.G. Bawendi, A microfabricated gas–liquid segmented flow reactor for high-temperature synthesis: the case of CdSe quantum dots. *Angewandte Chemie-International Edition in English*, 2005, **44**: 5447–5451.

Part IIIB

Microchip-Based: Specialized Methods and Technologies

43 Sample Processing with Integrated Microfluidic Systems

Joan M. Bienvenue and James P. Landers

CONTENTS

43.1 INTRODUCTION

As the genetic bioanalytical community continues to search for ways to improve DNA analysis for human identification, pathogen detection, and disease diagnosis, affecting more rapid, efficient, and timely results, the development of new analysis platforms becomes paramount.[1,2] Microfluidic systems have become increasingly attractive analytical tools for applications in many fields. Polymerase chain reaction (PCR)[3–6] and high-resolution DNA separations[7–13] are now readily carried out on-chip, as well as microfluidic purification of DNA or a variety of applications, including those in the clinical, biohazardous, and forensic sectors.[14–19] With successful microchip adaptation of these processes now commonplace, research focus has shifted toward the integration of these methods and with other sample processing steps (cell lysis and sorting, DNA quantitation)—the first step toward creation of a standalone device with full-genetic profiling capabilities. Due to the multistep nature of the DNA analysis process, careful consideration of solution compatibility and chemistry, fluidic interfacing and device engineering, as well as computer control and automation must be undertaken for seamless integration of these sample processing technologies. That is, the firmware, hardware, and "chemware" must all be carefully considered and optimized to create a multi-component, multi-functional design that can accommodate the complex process of genetic analysis. As a result, much attention is now being paid to device design and concept, interfacing diverse and complex chemical analyses, and computer-controlled automation, and, as a result, multi-component, microfluidic sample processing is now becoming a reality.

Integrated microfluidic platforms offer unique solutions to many of the problems currently facing genetic analysis for numerous applications. First, traditional genetic analyses necessitate multiple, time-consuming sample processing steps, often requiring different instrumentation and sample handling steps for each process, creating ample opportunity for contamination or loss of sample during transfer.[1] Integrated microfluidic systems put forward the opportunity to have automated sample handling, reducing the user intervention and sample manipulation, as well as instrumentation required to perform the analysis.[1,2] In addition to reducing the time, cost, and handling required to accomplish these processing steps, a reduction in the amount of sample required for analysis is inherent to integrated microfluidic systems. By reducing sample and reagent consumption, a more efficient analysis is achieved. Consequently, a shift to microfluidic technology for genetic analysis will be impelled by the benefits of faster, more efficient, more automated and integrated sample processing that produces timely and accurate results with the reduced possibility of contamination and/or loss of sample.

Although microchips have been used for a variety of applications, this chapter is not a comprehensive review of this diverse field; rather, it will focus on the development of the "chemware" required for microfluidic sample processing for genetic analysis. In particular, the discussion will center on nucleic acid purification and PCR amplification, with highlights of efforts to develop other on-chip sample processing techniques, such as sample cleanup, cell sorting and lysis, and DNA quantitation, both alone and as components of integrated multi-process devices. Particular attention will be paid to the development of solid-phase purification of DNA and RNA in microfluidic systems due to the lack of review publications on this subject. The development of microfluidic hardware will also be addressed; however, the preponderance of this chapter will devoted to the enhancement and integration of sample processing from a chemware perspective. In addition, this chapter will detail the development of integrated microfluidic processing devices for multiple genetic applications, as well as the utilization of microchips to enable fast, reproducible sample handling for human identification, rapid disease diagnosis, and pathogen detection.

43.2 DNA PURIFICATION

As the microfluidic community looks to develop systems capable of performing a full genetic evaluation, purified DNA or RNA is essential for most of these analyses. PCR-based methods for detection of disease and infection, as well as for human identification and other specific genetic analyses, have become the standard for many applications. In order for PCR to work efficiently and effectively, however, the DNA or RNA template used in the reaction must be free of contaminants that will inhibit the polymerase or interfere with other reagents necessary for the reaction to proceed. Consequently, a DNA purification step is vital to ensure that the starting template for PCR is free of contaminants and suitable for this enzymatic amplification. Thus, the inclusion of a DNA purification step in any functional genetic analysis micrototal analysis system (μTAS) is imperative. DNA purification (as opposed to cell isolation) prior to amplification provides many advantages, including, most importantly, the removal of sample PCR-inhibitors that can include cellular constituents (either proteins or lipids) in cell lysates, such as hemoglobin,[20] and an as yet unidentified factor in eosinophils that specifically inhibits reverse transcription-PCR.[21] In addition to these endogenous contaminants and inhibitors, in forensic science, the removal of environmental exogenous impurities such as humic acid from soil[22] and pollen,[23] is also vital, as the presence of these contaminants will also prevent successful amplification. Effective removal of both of these external and sample-based inhibitors is imperative to affect robust and efficient downstream PCR.

In addition, because many benefits associated with performing PCR in microdevices are achieved only when the volume of the PCR chamber is reduced, the use of a concentrating method for DNA isolation is not only advantageous, but also necessary, especially for samples that contain a low number of DNA template starting copies. Solid phase extraction (SPE) methods, such as those described herein, are not only effective at extracting highly purified DNA for subsequent analysis, but can also function as a concentrating step prior to PCR amplification. This concentration effect is not only beneficial for interfacing microfluidic techniques, but also could allow for a more successful PCR amplification in cases where low starting copy numbers are present, by providing the sample to be amplified in a smaller volume. Consequently, DNA purification, both as a means to remove endogenous and exogenous contaminants and as a concentration step, is a critical component of microfluidic sample processing.

Although purification of DNA from biological samples has been accomplished using a variety of methods (see Rudi et al.[24] for a more comprehensive review), more recently developed conventional and commercially available techniques for DNA purification exploit silica SPE methods, these methods reducing the time required for the extraction while maintaining recovery, sample purity, and integrity. These techniques typically rely on a three-step bind/wash/elute protocol to purify DNA from interfering proteins and cellular debris. In addition to enabling faster sample preparation, these protocols are more easily translated into microdevice formats; as a consequence, solid phases such as silica beads or sol-gels, which can be easily packed into microdevices to create a SPE bed or column for DNA purification, have become some of the more common phases utilized for microscale purification, as will be discussed in the next section.[14,15,18,19,25–27] In addition, a variety of novel silica solid phases can be created in microdevices during the fabrication process[16,17] that are suitable for sample purification. All of these phases will be highlighted in detail in the text.

Silica-based methods such as those referenced above typically rely on DNA adsorption to the silica phase in the presence of a chaotropic agent, followed by elution of proteins with an isopropanol or ethanol wash, and subsequent DNA elution in water or buffer. DNA, a strong polyelectrolyte carrying two negative charges per base pair at most pHs, has a large negative surface charge. In addition, the surface of silica is also negatively charged, due to weakly acidic silanol groups, with an average pK_a ranging from 5 to 7.[28] This heterogenous, negatively charged silica surface makes the net electrostatic repulsion of the fixed charges on the DNA and the silica surfaces strongly disfavor adsorption at low ionic strength.[28]

In high ionic strength environments, however, the situation is markedly different. The dissolution of the chaotropic agent, guanidine hydrochloride (GuHCl) in water, is an endothermic process with a positive entropy change.[28] One molecule of GuHCl will bind, on average, 4.5 molecules of water (maximum six), which enables the effective dehydration of the DNA molecule and the surface of the silica at high concentrations.[29] Decreasing the water activity in solution through the addition of a chaotropic salt results in a decrease in solvent-accessible surface area, a loss of water bound, and an increase in entropy of the solution. In addition, the loss of water at high ionic strength conditions greatly reduces the electrostatic penalty for placing the negatively charged DNA adjacent to the negatively charged silica surface. Finally, when the pH of the solution is lowered from 8 to 5 (a pH $\sim$ 6 is typically utilized in many silica-based purifications), there is an increase surface hydroxyl groups on the silica phase, thus increasing the ability of the silica to form hydrogen bonds with the DNA in solution, and a decrease in free hydroxyls that diminishes solution competition with DNA for these hydrogen-bonding sites on the silica surface.[28] These combined effects are the major driving force behind DNA adsorption to silica: an increase in the entropy of the water molecules released from the DNA and silica surface and the reduction of the negative potential at the silica surface by the pH of the solution. In combination, they allow for the binding of nucleic acids to silica and the subsequent purification of nucleic acids from high ionic strength (6–8 M) chaotropic solutions.

The first reference of a true microchip-based, silica solid phase extraction exploiting this purification technique was published by Christel et al.[17] in 1999. In this research, a microdevice containing silica pillars with high surface-area-to-volume ratios to increase DNA adsorption was designed for the purification and concentration of DNA for PCR amplification. This represented the first microchip-based DNA purification accomplished; however, the fabrication of this device relied on a complex reactive ion etching technique, which limits the potential utility of this format. In addition, capture efficiencies reported with this device using prepurified stock lambda DNA were only 50%, considerably lower than what would be expected (upward of 80% for other silica-based methods) for a sample that did not contain proteins or other cellular debris (such as blood) limiting its utility for low copy number samples where effective retention and release of DNA is imperative. This device, however, provided the first example of a truly miniaturized DNA purification system and, consequently, set the stage for other microfluidic, silica-based extraction methods. Cady et al.[16] employed

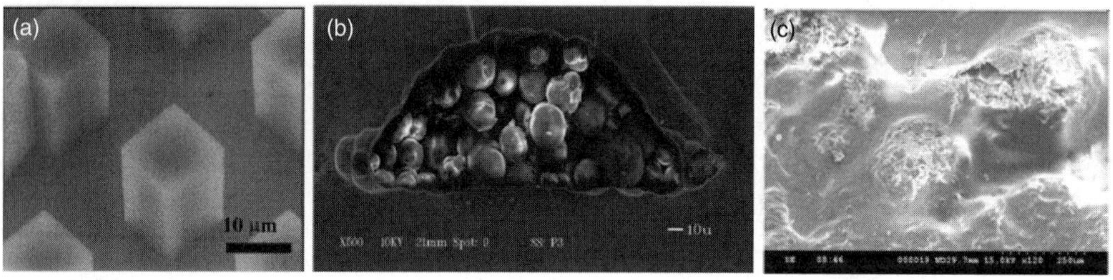

FIGURE 43.1 Examples of solid phases used for DNA purification in microdevices. (a) Scanning electron micrograph of silica pillars microfabricated into a DNA extraction microdevice. The spacing between pillars was 10 μm and the depth and height of the pillars was adjusted between 20 and 50 μm. (Figure adapted and reproduced from Cady, N. C., et al., *Biosens. Bioelectron.*, 2003, 19, 59–66. Copyright 2003. With permission from Elsevier.) (b) Micrograph (500 times) of a microchannel packed with silica particles immobilized with sol-gel resulting in a high-surface area silica solid phase for DNA purification. (Figure adapted and reproduced from Breadmore, M. C., et al., *Anal. Chem.*, 2003, 75, 1880–1886. Copyright 2003. With permission from American Chemical Society.) (c) Photomicrograph of a microchannel containing immobilized beads for DNA extraction. (Figure adapted and reproduced from Chung, Y. C., et al., *Lab Chip*, 2004, 4, 141–147. Copyright 2004. With permission from Royal Society of Chemistry.)

a similar design concept (depicted in Figure 43.1a), fabricating a device containing high surface area pillars, with the goal of providing a system with higher binding capacity (>200 ng) the capable of handling large input volumes. It was, however, difficult to effectively evaluate this device design for sample preparation as no extraction efficiencies were reported in this manuscript. In addition, while effort toward the development of devices capable of extracting DNA from large volume samples is certainly needed, any purification method that will be effective in a microfluidic system must also yield concentrated DNA (in a small elution volume) amenable to downstream processes (e.g., PCR, where typical reaction volumes range from nanoliters to picoliters). With DNA eluted from the solid phase in an unusually large volume (250 μL) in this work, the likelihood that such a method could be effectively integrated with other microchip processes is limited. Devices that contain pillars created during the fabrication process do, however, represent one potential silica-based design that could provide reproducible and robust microscale solid-phase purifications of nucleic acids.

In a more direct translation of current macroscale, silica-based, SPE protocols other, microchip-based purification systems have focused on utilizing a packed silica-bead bed or silica sol-gel matrix solid phase for purification. This type of extraction was first miniaturized in a capillary format, to demonstrate the utility of the proposed method in the microscale, by Tian et al.,[25] who utilized a 500 nL capillary-based chamber packed with silica particles to establish that PCR-amplifiable DNA (with 80–90% of proteins removed during the load and wash steps) could be obtained from white blood cells with high extraction efficiencies (70%). This demonstrated the feasibility of incorporating such silica-based column purification methods into microfabricated devices and the effectiveness of such methods for the purification of DNA from a wide variety of biological species (white blood cells, cultured cells, and whole blood).

This same microscale extraction technique was extrapolated to the microchip by Wolfe et al.,[18] who evaluated a variety of silica and silica bead/sol-gel matrices for microchip DNA purification. In this work, a potential problem associated with using silica beads or particles in a microdevice was highlighted, that is, the tendency of these particles to pack tightly under flow during repetitive use, thus affecting the reproducibility of flow and the repeatability of the extractions. Recently, however, it has been suggested that if the devices are single use (which would be typical of devices designed for many applications), packed silica-bead solid phases are acceptable purification phases for DNA or RNA extraction, as demonstrated in Easley et al.,[30] Bienvenue et al.,[31] and Hagan et al.,[32] and highlighted elsewhere in this chapter. Alternatively, the use of sol-gels (liquid colloidal suspensions of silica-based materials that can be acid or base catalyzed to gel in place), as described by Wolfe et al.,[18] have been demonstrated as efficient, reusable solid phases. These solutions are simply flowed into microchambers and, by controlling the catalyzed reaction, allowed to gel and form a porous DNA extraction bed with enough surface area for the binding of DNA, as demonstrated by Wu et al.,[19] who utilized this phase to extract DNA from bacterial (anthrax), viral (varicella zoster and herpes simplex), and human (blood) sources, with greater than 65% extraction efficiency demonstrate from blood. In a more recent translation of this work, Wen et al.[27] have employed the use of a photopolymerizable sol-gel monolith to extract DNA in a capillary-based system, which was further extrapolated to a microchip-based extraction. The photopolymerization step allows for easy and precise formation of the solid phase within the microdevice, without the use of retaining weirs or other microfabricated features, making it attractive for use in integrated systems. In addition, this solid phase has recently been incorporated into a novel two-stage microdevice that was developed for DNA extraction from blood—a C18 reverse phase column for protein capture (Stage 1) in series with a monolithic column for DNA extraction (Stage 2). This device has a high capacity for DNA in blood (>240 ng) and was found to achieve ~70% extraction efficiency, with effective removal of contaminating proteins. Consequently, sol-gels, like column-based solid phases, have been effectively employed as reproducible phases for DNA purification in microfluidic systems.

In addition to being used as solid phase for DNA purification, the sol-gel extraction medium can also be used as a "glue" to immobilize a silica-bead phase,[33] thus maintaining a hybrid, reproducible extraction column from run-to-run. This hybrid sol-gel/bead solid phase (depicted in Figure 43.1b) was evaluated extensively by Breadmore et al.,[15] who optimized flow rates and loading pH to effect a sample purification in 15 min from bacterial sources (anthrax and salmonella) and whole blood, with amplifiability of the resultant purified DNA demonstrated. In addition, intra- and interdevice reproducibility was demonstrated, with extraction efficiencies as high as 79% reported. The utility of this hybrid solid phase and extraction protocol was also demonstrated for DNA purification from sperm cells, with a view to the analysis of sexual assault evidence,[14] as described in Chapter 37. These results demonstrated the potential of microchip-based extraction methods for forensic analysis. In addition, this group has utilized silica beads alone as a purification phase, both for the purification of DNA[30,31] and RNA.[32] The purification of RNA from biological samples remains a challenging and often overlooked sample processing step. The work of Hagan et al.[32] represents the first example of a silica-based purification of RNA from crude samples, those of both clinical (cells lines of alveolar rhabdomyosarcoma (ARS) tumors) and forensic (semen stains) interest. The only other instance of a microfluidic purification of RNA present in the research to date was published by the Quake group[34] and is discussed later in this section.

Another unique silica-based approach to microscale DNA extraction currently under development utilizes a serpentine channel design, combined with an immobilized silica-bead solid phase and fluidic oscillation. This method, developed by Chung et al.,[35] relies on silica beads immobilized on the plasma-oxidized surface of the polymethylmethacrylate (PMMA) channels, instead of a packed-silica solid phase, as depicted in Figure 43.1c. Following bead immobilization, the solutions required for DNA binding, purification, and release are flowed back-and-forth through the device. This fluidic oscillation over the immobilized phase results in marked improvement of recovery and extraction efficiency over the same extraction methods with free beads. This method represents yet another variation of silica-based purifications that has been accomplished in microfluidic systems, exploiting previously optimized chemistries. In summary, the development of macroscale, commercial, silica SPE protocols has enabled the facile translation of DNA, and now RNA, extraction into microfluidic systems for a variety of applications.

Although silica-based purifications of DNA and RNA are, by far, the most common microfluidic methods for isolation of nucleic acids, there have also been other solid phases and methods explored for DNA purification that deserve mention. Like more recently developed commercial technologies for nucleic acid purifications, pH-induced methods for extraction have also been miniaturized. Cao et al.[26] have utilized chitosan-coated silica beads that extract DNA at pH 5 and, subsequently, release the DNA into solution at pH 9. With this purification technique, the PCR inhibitory reagents typically utilized in silica-based purifications (isopropanol, ethanol, GuHCl, etc.) are not necessary, thus eliminating a major source of inhibitory contamination. Extraction efficiencies upward of 92% from biological samples including blood were reported and the purified nucleic acids were suitable for PCR amplification. Nakagawa et al.[36] have also reported a novel purification method for DNA using a microchannel coated with 3-aminopropyltriethoxysilane (APTES) or 3-[2-(2-aminoethylamino)-ethylamino]-propyltrimethoxysilane (AEEA) to introduce amine groups on the surface. Relying on surface electrostatic interactions between amine groups and DNA, this method also relies on a pH change (from 7.5 to 10.6) to elute DNA from the device and these researchers were able to recover 27–40% of the DNA, which was PCR-amplifiable, from a whole blood sample.

Quake and coworkers[34] have utilized an affinity column purification of both DNA and RNA in microfluidic devices, packing a microfluidic channel with derivatized polymer magnetic beads. Cell lysates can then be passed over the beads, the DNA or RNA retained on the column, and eluted in a wash step. With this system and a moderately abundant target (zinc finger OZF), sensitivity of detection on the order of 2 to 10 cells was obtained. In addition, the device was also utilized for DNA purification, with PCR-amplifiable DNA successfully isolated from *Escherichia coli* (*E. coli*) culture samples, demonstrating that the device design and method were versatile enough to accommodate both nucleic acid types. Additionally, Witek et al.[37] have utilized a photoactivated polycarbonate (PPC) device for isolation of DNA from both cell lysates and whole blood. The PPC chip was fabricated by exposing pristine polycarbonate surfaces to ultraviolet (UV) radiation that resulted in the formation of surface carboxylate groups through a photo-oxidation reaction. Subsequently, DNA can be precipitated onto the activated polycarbonate using a PEG/NaCl ethanol buffer, as seen in Figure 43.2, the proteins rinsed away, and the purified DNA eluted with deionized water. Extraction efficiencies of this device with prepurified DNA were upward of 85% and the DNA purified from blood was also suitable for PCR amplification. Finally, in a slightly different approach, Lee et al.[38] as described in more detail in a latter section of this chapter, have developed a way to use carboxyl-terminated beads to isolate DNA from lysed sample. By selectively binding and removing the proteins in the sample using these magnetic beads, the resultant purified DNA is left behind in solution and is PCR-amplifiable. This unique approach lends itself well to single-chamber purification/amplification sample processing devices, as will be described. Each of the solid-phase purification methods described here, although not as thoroughly characterized as their silica-based counterparts, exploits unique chemistries to purify DNA from interfering species and each may be appropriate in the development of diverse biological genetic analyses.

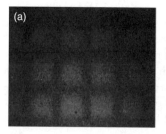

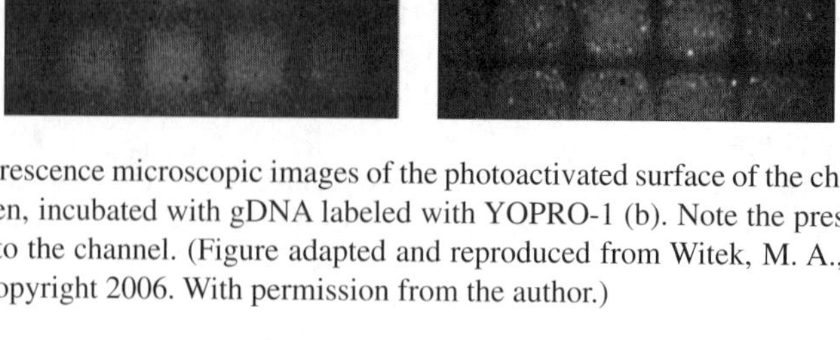

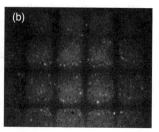

FIGURE 43.2 Fluorescence microscopic images of the photoactivated surface of the channel incubated without gDNA (a) and then, incubated with gDNA labeled with YOPRO-1 (b). Note the presence of fluorescently labeled DNA bound to the channel. (Figure adapted and reproduced from Witek, M. A., et al., *Nucleic Acids Res.*, 2006, 34, 74. Copyright 2006. With permission from the author.)

43.3 PCR AMPLIFICATION

Polymerase chain reaction, an enzymatic process in which a specific region of DNA is replicated repeatedly to yield many copies of a particular sequence, was first described by Mullis et al.[39] in 1985. PCR is, theoretically, an exponential amplification of a target DNA sequence, that is, for every cycle completed, the number of target fragments doubles, such that after 30 cycles, approximately 30 billion copies of the target region of the DNA template have been created.[39] This amplification process relies on the heating and cooling (thermocycling) of samples using precise thermal control, typically through three temperatures, to allow for denaturing of the DNA (94°C), specific primer annealing (~60°C), and extension of the PCR fragment (~72°C). Thermocycling for PCR amplification can be accomplished in a number of ways. Conventional systems use thermal blocks that are heated and cooled to, in turn, heat and cool the reaction tubes that contain the reaction solutions. This indirect heating, combined with the large reaction volumes typically required (10–50 μL), limits cycling rates and results in 1–2 h amplification times for many clinical analyses and over 3 h for standard forensic amplifications.

Miniaturization of the reaction vessel provides one mechanism for decreasing the overall amplification time; by increasing the surface area-to-volume ratio, as accomplished in microfluidic devices, homogeneous solution temperatures can be achieved more rapidly, thus reducing the time required for thermocycling (for a more thorough discussion of microfluidic PCR systems, see the following reviews).[40,41] In addition, microfluidic-based miniaturization allows for a number of different heating methods to be exploited. Many groups[3,6,38,42–56] have chosen to utilize direct contact methods for microchip thermocycling, similar to those used to accomplish conventional thermocycling. With these methods, the microdevice is either in direct contact with a heating element or a heating element is fabricated into the system. With noncontact methods, the solution in the chamber is heated directly, independent of the chamber itself, to perform thermocycling and affect a more rapid analysis, as pioneered by Landers and coworkers[4,5,57] and demonstrated by Ahn and coworkers.[58]

Contact methods for PCR thermocycling are, by in large, the most common techniques used for amplification in microfluidic systems. By utilizing microdevices and contact heating, a number of groups[42,44–52] have significantly reduced the volume of sample and reagents required for amplification (on the nanoliter to picoliter range) and the importance of accomplishing PCR amplification in reduced volumes cannot be underestimated. In addition to potentially reducing the time required to perform amplification and the reduction in cost associated with the reduction in reagents used, the volume of solution generated is amenable to upstream and downstream processing and analysis components. To date, Quake and coworkers[45] have accomplished the lowest volume PCR amplifications, performing 450 pL RT–PCR on a 72-chamber chip. Although this paper has set the lower limit of reaction volume, speed of analysis was not the major focus and thermocycling was performed by placing the entire device in a conventional thermocycler. Consequently, even with a reduced reaction volume ideally leading to extremely fast thermocycling times, over an hour was still needed for RT–PCR amplification of a 240 bp fragment of human β-actin RNA. Although not all of the nanoliter–picoliter amplification devices cited previously accomplished rapid thermocycling, the reduction in volume permitted by these microprocessing systems will be an important component of successfully integrated microfluidic systems.

In addition to reducing the sample and reagent volume necessary to perform PCR amplification, multiple groups[5,6,38,53–56,59] have focused on reducing the time required to carry out thermocycling for PCR. Once again, using contact methods for heating in one of the fastest amplifications accomplished on-chip to-date, Hashimoto et al.[6] used a flow-through device consisting of discrete temperature zones (depicted in Figure 43.3) produced by electrical resistance heaters to amplify a 500 bp region of λ-phage DNA in 1.7 min and a 997 bp fragment of the same target in 3.2 min. It was determined, however, that the starting copy number of DNA will affect the speed at which PCR amplification can be performed and with a linear velocity of 10 mm/s, the lowest DNA starting copy to provide amplified signal in 20 cycles was 1×10^7 copies. Consequently, without refinement, low copy number samples would be difficult to process with this system. Obeid et al.[54] also described

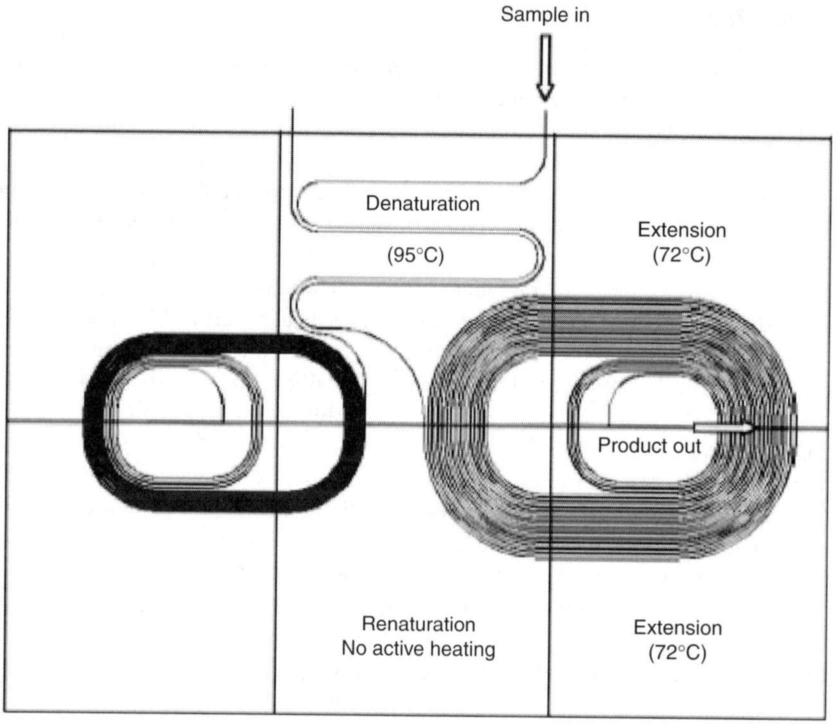

FIGURE 43.3 Schematic views of the flow-through PCR device layout and model. The device on the right was used in the experiments described and had 50-mm wide channels separated by 250 mm. The device on the left had 50-mm wide channels separated by 50 mm and was not used in the research presented, however, it illustrates the versatility of the flow-through design. (Figure adapted and reproduced from Hashimoto, M., et al., *Lab Chip*, 2004, 4, 638–645. Copyright 2004. With permission from Royal Society of Chemistry.)

a flow-through amplification system that used four heating blocks, three for PCR amplification and the fourth for reverse transcription, which was utilized to perform 30 cycles of PCR in 6 min. Once again, however, the lower limit of starting copies required for this device was 6.25×10^6, making usage of this device problematic for situations where low template copy numbers are suspected. The amplification was, however, accomplished in 700 nL, making this device not only capable of rapid PCR but also lower volume amplification.

Finally, using infrared-mediated heating, a noncontact method has been developed to accomplish rapid PCR amplification. With this method, near-infrared (IR) radiation is utilized to selectively heat the reaction solution through excitation of vibrational modes of water molecules.[4,5,57] By doing so, the time required to heat the solution is dramatically decreased, resulting in a concomitant decrease in amplification time. In addition, by etching away the glass surrounding the PCR chamber, the thermal mass of the glass can be decreased, contributing to faster cycling rates. This modified glass device and indirect heating method has been utilized to perform rapid amplification, accomplishing 30 cycles in less than 5 min.[60] This technique has also been utilized to perform RT–PCR in a device with two IR-heating regions: one for the RT incubation and a second for PCR thermocycling.[58] The further development of this and other rapid thermocycling techniques presented herein represents a considerable step forward in the development of rapid, microfluidic PCR amplification and, combined with the reduced volume techniques described previously, provides a significant advancement of current technology. In addition, by executing low-volume, rapid thermocycling techniques, microfluidic PCR amplification will integrate more seamlessly with other microscale sample processing and genetic analysis steps, helping to facilitate the ultimate goal of an integrated total analysis system.

43.4 CELL SORTING, CELL LYSIS, AND DNA QUANTITATION

The selective isolation of different cell populations can be an important step in sample processing for a number of different genetic analyse. As described in Chapter 37, separation of sperm cells and vaginal epithelial cells is an important step for the effective analysis of sexual assault evidence and clinically, cell-sorting techniques have typically involved the isolation of specific cell types (i.e., cancer cells) from mixtures of other interfering species. A variety of methods can be exploited to separate different cell types, including dielectrophoresis, separation on the basis of the different physical and chemical properties of cell types (see Chapter 37), and acoustic differential separations. A number of groups have utilized microscale dielectrophoresis, the manipulation of particles in nonuniform electric fields, for separating different cell types (for comprehensive reviews on detailing the use of dielectrophoresis for cell sorting, see References 61 and 62). Typically, for these methods, electrodes are fabricated into the microdevice and differences in the dielectric makeup of the cells are exploited to allow for separation when these particles are placed in an electric field. Consequently, dielectrophoresis has become a noninvasive way to isolate different cell populations that is easily miniaturized in microfluidic systems.[61,62]

In addition, as described in detail in Chapter 37, cell populations that have inherent physical differences (such as sperm and vaginal epithelial cells) can be separated by simply exploiting these innate dissimilarities.[63] Also exploiting the size variation of different cell types to promote their facile separation, Yuen et al.[64] have utilized a weir-filter design to effectively isolate white blood cells (12–15 μm) from red blood cells (7–8 μm), as will be described later in this chapter. In addition, Quake and coworkers[65,66] have developed a pressure-driven cell sorter that can selectively isolate fluorescently-labeled cell populations. Austin and coworkers[67] have translated magnetic-activated cell sorting (MACS) to microscale using antibody-coated magnetic beads in a high magnetic field; however, to date there has been no published data showing the successful sorting of cells with this method. The same group,[68,69] however, has utilized deterministic lateral displacement to isolate red and white blood cells, and as well as platelets, from whole blood and for total cell removal from whole blood for plasma preparation. Finally, the development of microfluidic acoustic technology has enabled the separation of cells for a number of applications[70–74] including its employment for differential extraction (see Chapter 37). Selective cell sorting, although not the focus of this chapter, is a technique that lends itself well to microfluidics and provides for efficient sample enrichment prior to other sample processing techniques.

The lysis of cells on-chip (either after a cell-sorting step or from mixed sample sources) will, undoubtedly, be an important step in the development of sample processing microfluidic devices for total sample handling of crude biofluids. There are many examples in the literature that describe the lysis of cells in microfluidic systems and that can be easily divided into two categories: (1) those that utilize a chemical-based lysis and (2) those that employ a physical breakdown of the cell. Chemical cell lysis has been demonstrated using a variety of buffers and reagents[14,75–78] and typically involves flowing the particular lysing reagent on-chip to mix with the cells of interest. Physical breakdown of the cell can be caused by a number of different methods that disrupt the cellular structure and release the nucleic acids into solution. Electroporation,[79,80] electric field,[81–84] thermal lysis,[85–88] sonication,[89] laser/vibrationally induced,[38] and mechanical lysis with beads[90] or barbs[91] are all methods that have been utilized to rupture cells on chip. Successful release of cellular constituents through effective, complete, and rapid cell lysis is essential for downstream sample processing steps, such as SPE and PCR. Consequently, although often overlooked, microscale cell lysis is an indispensable component of any microfluidic sample processing systems.

Finally, on-chip DNA quantitation will undoubtedly be an important feature of microfluidic sample processing for a number of different applications where the amount of input DNA for downstream genetic analysis is crucial. As discussed more extensively in Chapter 37, quantitative PCR (qPCR) methods in microfluidic systems have not been demonstrated to date. However, there are a number of publications describing real-time PCR in microdevices and it is likely only a matter of time before these methods will be extended to include qPCR (for a more detailed discussion of real-time PCR, see Chapter 37).

In addition to on-chip qPCR, another approach being developed to accomplish quantitation involves a more direct translation of current macroscale approaches to the microscale. Using a fluorescence-based assay, a valved device is used to generate a calibration curve, which is used to quantitate an unknown sample with on-chip mixing of the DNA with an intercalating dye.[92] This method was designed with the goal of integrating this technology with microscale solid phase purification techniques, for post-purification/pre-PCR quantitation, making it a potentially facile inclusion in sample preparatory genetic analysis chips. Although less focus has been paid to developing microfluidic techniques for on-chip determination of DNA concentration, this sample processing step is no less a necessity than the other more routinely miniaturized techniques and will be an important inclusion in the development of genetic μTAS for a number of applications.

The discussion of on-chip sample processing presented here is by no means exhaustive, rather it is meant to highlight the progress made in miniaturizing the main genetic processing steps required to affect a total analysis. The methods described represent some of the first and most recent examples of microchip-based sample preparation for downstream sample processing and a major step toward the development of microfluidic systems capable of accepting crude samples for DNA analysis. As advancement of these technologies has continued, the focus has shifted from basic method development of these and other individual processing techniques to the integration of several sample treatment steps in multi-process devices.

43.5 INTEGRATION OF MICROFLUIDIC HARDWARE: DESIGN, ENGINEERING, AND FLUIDIC CONTROL

Although this chapter highlights sample processing in microfluidic systems in both stand-alone, single process devices and integrated multi-process systems (i.e., the development of the "chemware" to support miniaturized genetic analysis), it is of critical importance to note that the successful integration of such diverse chemistries would not be functionally possible without the concomitant advancement of microfluidic hardware. Advances at the device level have included innovative engineering to allow for elegant chip fluidic control through valving[48,93–98] and novel device designs,[30,34,99–102] as well as fluidic modeling of these systems where the surface-area-to-volume ratios are radically different from the macroworld (see Chapter 40 by Easley). The complexity and functionality of integrated microfluidic devices, as depicted in Figure 43.4, has made rapid sample processing and analytics for genetic typing possible. Without these advancements, the integrated "chemware" that will be detailed in the remainder of this chapter would simply not be feasible and consequently, it is worth nothing these achievements and recognizing that it is these innovations that have permitted the functional microfluidic integration of genetic analysis.

43.6 INTEGRATED SAMPLE PROCESSING WITHOUT ONLINE DETECTION

Recent literature has provided descriptions of several devices that have integrated sample processing steps, such as on-chip cell lysis, cell selection, DNA purification, and PCR amplification. In one of

FIGURE 43.4 (Continued) (c) An integrated microfluidic device designed by the Burns group. (*Top*) Schematic representation of the device, containing multiple different liquid entry channels ("L," sample, PCR reagents and RD reagents), several metering channels, drop mixing intersections, a sealed PCR chamber, an open RD chamber, individually controlled valves, and an electrophoresis channel. (*Bottom*) Photograph of an assembled device (1.5 cm by 1.6 cm), which can control discrete liquid drops that are 100–240 nL, with fluidic channel dimensions of 200–600-μm wide and 50-μm deep. (Figure adapted and reproduced from Pal, R., et al., *Lab Chip*, 2005, 5, 1024–1032. Copyright 2005. With permission from Royal Society of Chemistry.)

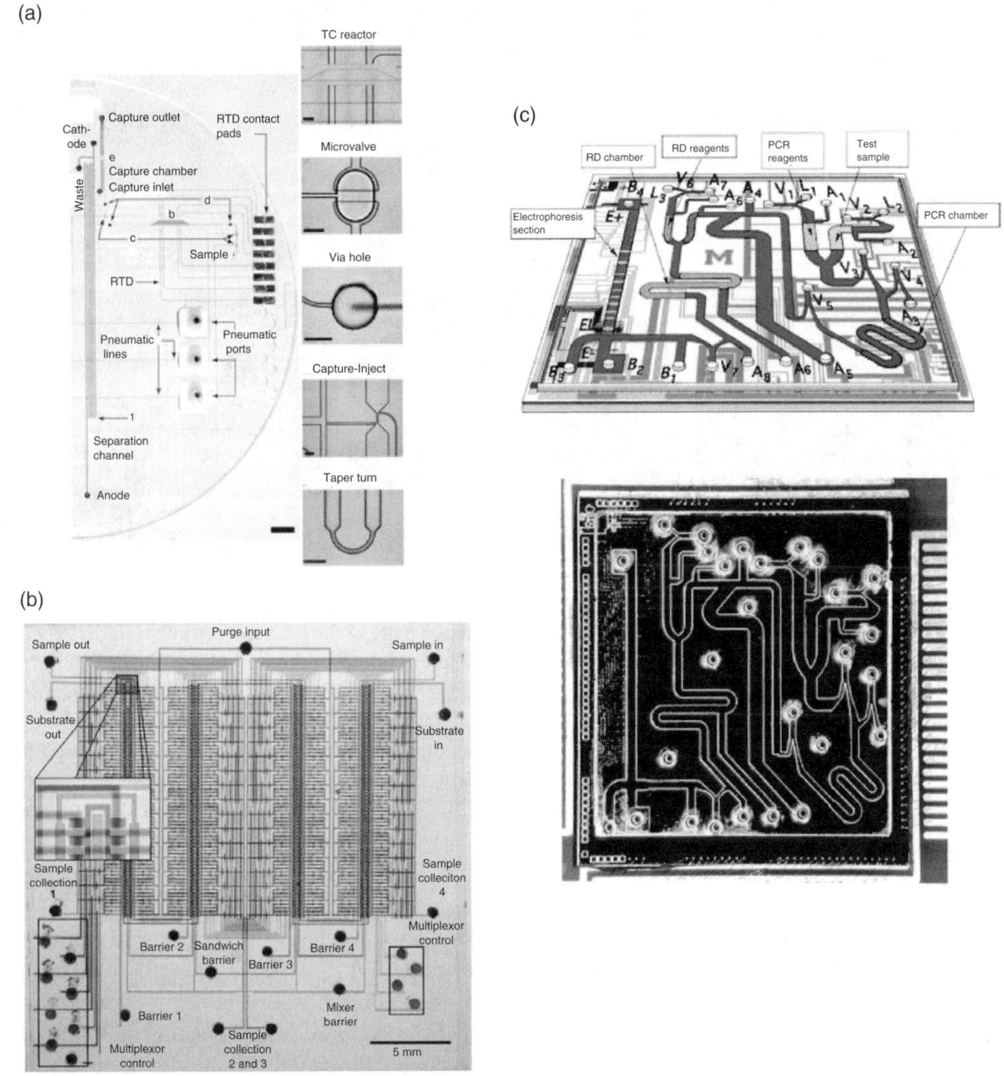

FIGURE 43.4 (a) The bioprocessor developed by the Mathies and coworkers. A photograph of the microdevice, showing one of two complete nucleic acid processing systems is shown (left), with sequencing reagent, capture gel, separation gel, and pneumatic channels. (Scale bar, 5 mm). The pictures on the right show multiple different components of the device, including (from *top* to *bottom*): a 250-nL thermal cycling reactor with RTDs (Scale bar, 1 mm), a 5-nL displacement volume microvalve, a 500-μm-diameter via hole, a capture chamber and cross injector, and a 65-μm-wide tapered turn (Scale bars, 300 μm). (Figure adapted and reproduced from Blazej, R. G., et al., *Proc. Natl. Acad. Sci. USA*, 2006, 103, 7240–7245. Copyright 2006. With permission from National Academy of Sciences.) (b) An optical micrograph of a nanofluidic system developed by the Quake and coworkers that can be used for parallelized high-throughput screening of fluorescence-based single-cell assays. The various inputs have been loaded with food dyes to show the channels and subelements of the fluidic logic. This chip has 2056 valves, which are used to manipulate 256 compartments containing bacterial cells expressing an enzyme of interest (or a library of mutants of that enzyme) that are combined on a pairwise basis with 256 other compartments containing a fluorogenic substrate used to assay for a desired activity. Cells that display a particularly interesting activity can be selected and recovered from the chip using valve-based addressing of the compartments. (Figure adapted and reproduced from Hong, J. W., et al., *Nat. Biotechnol.*, 2003, 10, 1179–1183. Copyright 2003. With permission from Nature Publishing Group.) **(Caption continued opposite on page 1216.)**

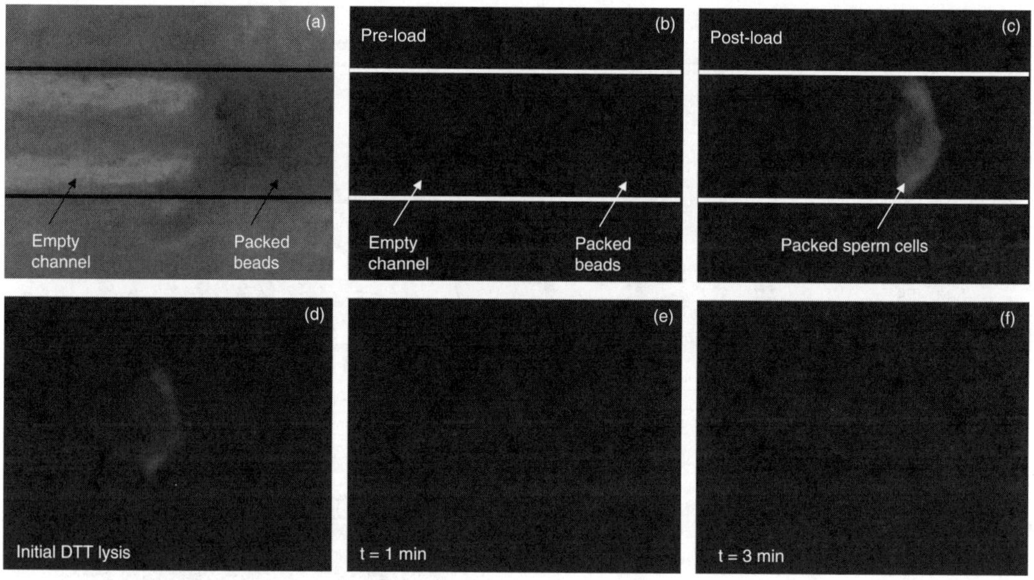

FIGURE 43.5 On-chip lysis of sperm cells integrated with microchip-based DNA extraction. All panels show a section of the microchip extraction channel (~1 mm) at the front edge of the silica bed by either light (Panels a and b) or fluorescence (Panels c through f) microscopy. No inherent fluorescence was seen in the channel when irradiated with 480–550 nm light (b), before cell loading. DEAD Red™-labeled sperm cells were flowed into the channel and packed up against the beads, while the silica bed front was visualized (Panel c). After flow of the dual-lysis/loading buffer was subsequently initiated (Panel d), the fluorescence decreased over the next few minutes in (Panels e and f) and was essentially nonexistent after 5 min (data not shown). (Figure adapted and reproduced from Bienvenue, J. M., et al., *J. Forensic Sci.*, 2006, 51, 266–273. Copyright 2006. With permission from Blackwell Publishing.)

the more simplistic integrations of two sample processing steps, cell lysis and nucleic acid extraction were demonstrated in the same microdevice by Bienvenue et al.,[14] who developed a method for loading whole sperm cells into a device containing a hybrid silica-bead/sol-gel extraction column. The cells packed against the leading edge of the silica-packed column, penetrating only slightly into the solid phase, as depicted in Figure 43.5. Following the packing step (Figure 43.5a through c), a dual cell lysis/DNA loading buffer was flowed over the packed cells and column (Figure 43.5d through f), lysing them and releasing their contents into solution for subsequent binding to the silica bed. Purification was then carried out as would typically be accomplished for silica-based methods, with an isopropanol wash to remove interfering species, and finally elution with water. The resultant eluate was subjected to forensic short tandem repeat (STR) amplification and the genetic profile compared to that obtained using standard conventional silica methods. The microchip-obtained profile was consistent with the profile generated using the kit-based purification, demonstrating that not only was the dual-process device as least as functional as the open-system conventional methods, but also that this closed microfluidic system have potential as a forensic sample processing tool, where prevention of contamination is of paramount importance.

In 2001, Yuen et al.[64] published a report describing an integrated system that performed cell isolation and direct PCR amplification. Using weir-type filter constructed of silicon that spanned the microfluidic channel to perform the isolation, blood (<3 μL) was flowed into the channel. Owing to their larger size (12–15 μm in diameter as opposed to 7–8 μm in red blood cells), the white blood cells were trapped between the top of the weir and the bottom of the microdevice cover plate. Held there, the PCR master mix was flowed into the chamber and the device subjected to thermocycling (and thermal lysis of the cells) using a copper block heater fabricated into a plexiglass holder. With this system, successful cell isolation and PCR amplification of a 236 bp fragment of the human coagulation Factor V was accomplished; however, the authors note that the presence of too many white blood cells resulted in significant inhibition of amplification, most likely due to the inhibitory factor in eosinophils.[21] These results suggested that although the cell separation and direct PCR method presented was a functional means to obtain sequence-specific genetic results, the inclusion of a true DNA purification step will still be essential for further development of such a technique.

In addition to this work, Hong et al.[34] described a nanoliter-scale processor that performs cell isolation, cell lysis, and nucleic acid purification on a single platform. In their work, a device was designed that contained 100 μm wide fluidic channels, in which three holding chambers (lysis buffer, cell, and bead) were created using valve isolation. The various reagents required for each different step of the analysis were flowed into the chip through different inlets, with cells mobilized in the device and retained in cell holding chambers. Following cell retention, the valve between the cell chamber and the lysis buffer chamber was opened, allowing lysis buffer to mix with and lyse the cells. The resultant corresponding cell lysate was then flowed over an on-chip mRNA affinity column (as described previously in the extraction section of this chapter), during which the desired mRNA was retained on the beads. The beads were then removed from the device and used directly in the off-chip RT–PCR amplification. This device and method were utilized to isolate mRNA from samples containing single cells, with two different targets (β-actin, a high-abundance transcript and zinc finger OZF, a moderate-abundance transcript). This report also described the development of a DNA purification chip, a device capable of cell lysis and affinity purification. Unique to this device was the use of novel channel designs and flow patterns, as well as the use of rotary mixers that provide comprehensive on-chip mixing of cells and lysate—a process that would require consume hours to complete relying on diffusive mixing. Once again, the beads were removed from the device subsequent to DNA capture, and the amplification of the gene encoding prelipin peptidase-dependent protein (*ppdD*) was accomplished off-chip. Importantly, these two devices demonstrated the feasibility of incorporating multiple sample processing steps in microfluidic systems for the preparation of both DNA and RNA for genetic interrogation, successfully integrating cell lysis and nucleic acid purification in devices with unique architectures and precision fluidic control elements.

Collectively, these reports highlight the importance of integrating multiple sample processing steps into single multi-purpose microdevices. By incorporating on-chip cell lysis or cell sorting with either purification or amplification, more efficient processing was accomplished in closed, microfluidic systems, reducing not only the opportunity for contamination and the degree of user intervention required, but also the volume of sample and reagents consumed. Although each of these designs requires further refinement and inclusion of additional sample processing and/or analytical steps before they can be considered complete, through these examples, great progress has been made toward the development of a lab-on-a-chip, μTAS device capable of total sample preparation and analysis.

The integration of DNA purification and PCR amplification remains a formidable and much-overlooked undertaking. As argued earlier, purification of DNA before PCR amplification is imperative for a variety of reasons—yet there are only a few reports of integrated DNA purification and amplification devices in the literature to-date.[31,38,103,104] The inherent incompatibility of the reagents (chaotropic salts, organics, etc.) necessary to perform many types of nucleic acid purifications (particularly silica-based nucleic acid isolation) with the polymerase chain reaction, make fluidic coupling of these two dramatically and inherently different processes a challenging task. Legendre et al.[103] circumvent these problems with a valveless chip design that used the previously described silica-bead hybrid solid phase and combined with a conventional block thermocycler for PCR . Using a dual syringe system, reagents from the load and wash step of this silica-based purification were simply passed through the PCR domain and out of the device. DNA could then be eluted from the SPE domain directly into the PCR chamber, while master mix containing the reagents necessary for PCR was flowed in through a side channel for subsequent amplification, removing the

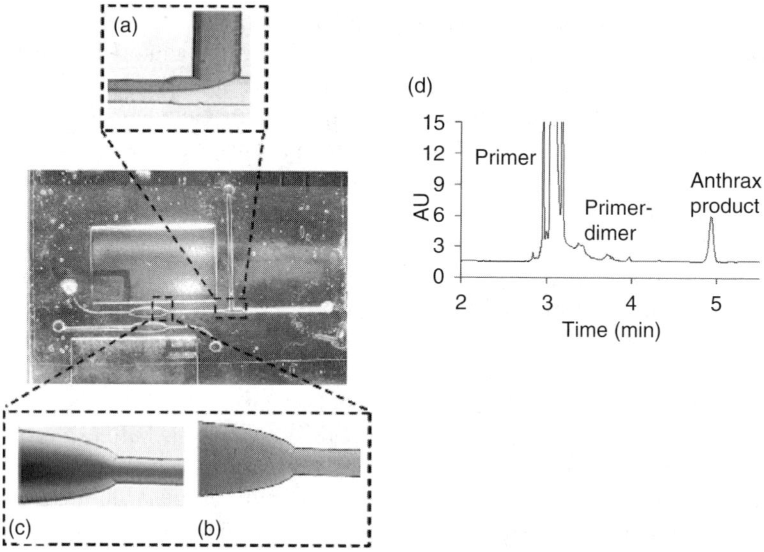

FIGURE 43.6 Microfluidic mixing in the integrated purification/amplification device. (a) Blue dye was flowed through the side channel, while yellow food dye was flowed through the SPE bed. Note the laminar flow. (b) After entrance into the PCR chamber, blue and yellow dyes were only mixed in the center of the channel, due to diffusion. (c) Following the initial denaturation step of thermocycling, green color was observed in the PCR chamber indicating adequate mixing of the two streams. (d) Electropherogram showing the results of application of the integrated DNA purification/amplification device to biowarfare agent detection. Anthrax spores on a nasal swab were eluted in lysis buffer and loaded onto the device for purification of DNA, integrated with IR-mediated PCR. The capillary electropherogram depicts successful amplification of a 211 bp product peak of *B. anthracis* and this total sample processing was accomplished in the total analysis time less than 23 min. (Figure adapted and reproduced from Legendre, L. A., et al., *Anal. Chem.*, 2006, 78, 1444–1451. Copyright 2006. With permission from American Chemical Society.)

offending contaminating solvents from the device. Co-mobilization of the reagents delivered through two separate syringes allowed for 1:1 mixing of the eluting DNA and the PCR reagents in the PCR chamber (as depicted in Figure 43.6a through c), where the mixture was could be thermocycled for DNA amplification using a conventional thermocycler. This dual-process microdevice was utilized for the successful purification of DNA from a diversity of biofluids, including buccal cells, blood, and semen followed by target-specific amplification.[103] In addition, this microdevice was also utilized to accomplish the microfluidically integrated purification and targeted DNA amplification of sequences from *Bacillus anthracis* (*B. anthracis*) isolated from a mock nasal swab with a total analytical time of less than 23 min,[103] as depicted in Figure 43.6d, using the IR-mediated, noncontact thermocycling method described earlier in this chapter. This work represented the initial efforts to integrate DNA purification and PCR amplification, with a view toward the development of a sample processing device capable of handling of a wide array of input samples for a wide variety of applications.

A similar device design was also utilized to explore the integration of DNA purification and multiplexed amplifications for forensic analysis.[31] In this work, device design and capacity were further explored and methodologies to incorporate the commercially-available reagents commonly utilized for forensic analyses were investigated. The challenges associated with microfluidic integration for forensic analysis are complicated by the need to develop systems that utilize established, validated, and commercially available reagents/protocols, as described in Chapter 37. As a consequence, microfluidic systems must be developed that interface seamlessly with existing, validated, and court-tested reagents, protocols, and technology, providing a substantial decrease in reagents and sample consumed by the analysis, as well as cost per sample, and significant improvements in time to result and prevention of contamination. These changes will provide the motivation to move beyond previously established techniques and promote the inclusion of microfluidic systems in general casework analysis. The simple device described earlier was used to demonstrate the feasibility of microfluidic sample processing in a closed system, utilizing common benchtop equipment and commercially-available, kit-based reagents. It was utilized to purify DNA from semen (a commonly encountered biofluid in sexual assaults) and perform integrated, multiplexed amplification using the standardized kit-based reagents employed for conventional forensic genetic interrogation, as highlighted in Chapter 37.

Each of the devices described in this section has functionally integrated two or more sample processing steps. Using novel chip designs, multiple sample processing steps (including cell sorting, cell lysis, DNA purification, and PCR amplification) could be accomplished in an integrated system, providing more automated and more efficient sample handling. The basic integration of these steps has served as a catalyst to the development of fully integrated systems with both sample processing and analytical capabilities. The functional development of these integrated chemware systems has and will continue to impel, the evolution of μTAS for genetic analysis.

43.7 INTEGRATED-SAMPLE PROCESSING WITH ONLINE DETECTION

It is important to distinguish integrated-sample processing devices from devices that have incorporated both sample processing and an endpoint analysis step. All of the devices highlighted in the previous section are capable of sample manipulation, but do not have an online detection step in their microfluidic architecture. The following section will detail μTAS that are capable of both sample handling and analytical detection for genetic analysis. Building in complexity, these devices contain the hardware and "chemware" necessary to perform integrated total analysis, providing a genotypic "answer" following sample handling, and will be discussed in detail in the next section. In addition, although the focus of this chapter is on the development of sample processing, both in stand-alone microfluidic devices and integrated processing systems, it is important to note that the integration of PCR amplification and microchip electrophoresis has been accomplished by a number of groups.[42,99,100,105–116] Integration of PCR with electrophoretic analysis is an important and essential step toward the development of μTAS devices for genetic interrogation; however, we have chosen to focus this chapter on the integration of multiple sample processing steps, leaving the integration of sample processing and analytical steps for future discussion, except in the rare cases where an analytical step was incorporated with multi-process sample handling. Consequently, detailed explanations of integrated PCR-(ME) devices will not be explored.

In one of the earliest examples of a device capable of performing multiple, integrated sample processing steps followed by analysis, Waters et al.[88] described a device in which that represents one of the first examples of integrated sample processing combined with an analytical step to affect a more complete analysis in a closed microfluidic system. In this particular device, thermal cell lysis, followed by direct PCR amplification and electrophoretic sizing was accomplished. With this method, whole *E. coli* cells were loaded into the device along with PCR master mix. Following loading, the entire device was placed into a conventional benchtop thermocycler and subjected to a typical PCR amplification. During the initial 4-min heating step (94°C), the cells were lysed, releasing their DNA into solution, which could then be amplified during thermocycling, ultimately followed by electrophoretic sizing using microchip electrophoresis. Although this method was successfully employed to perform integrated cell lysis, amplification, and electrophoretic analysis, without the inclusion of a DNA purification step in this method, this device would be limited to use with samples that do not contain significant concentrations of PCR inhibitors. It does, however, represent one of the first examples of integrated sample processing, combined with an analytical step, to affect a more complete analysis in a closed microfluidic system.

As a more complex example of integrated sample processing, Lagally et al.[117] developed a device capable of dielectrophoretic cell sorting and on-chip cell lysis, followed by sequence-specific

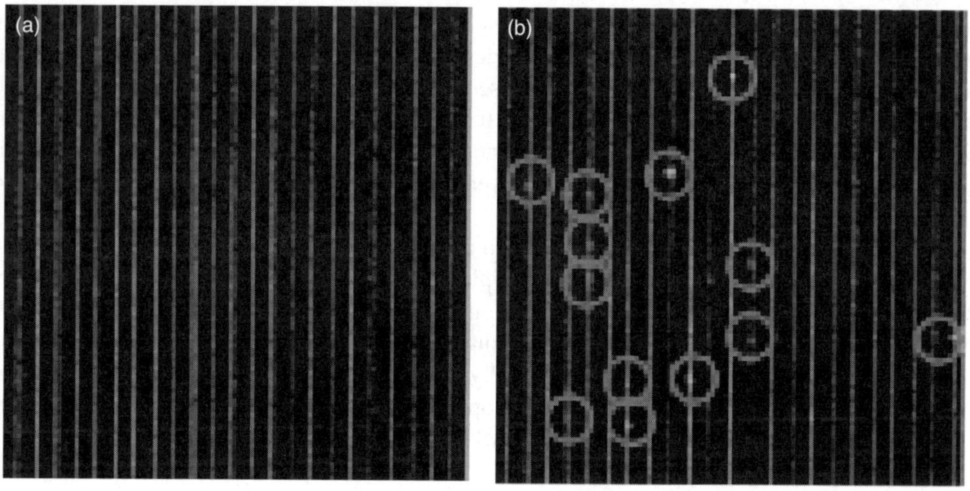

FIGURE 43.7 On-chip trapping of cells was accomplished using the integrated microfluidic system described by Lagally et al. (a) An epifluorescence image taken after 10 min of cell flow over the interdigitated microelectrodes without DEP voltage applied. Nonspecific binding does not occur and no cells are visible. (b) The same microelectrodes after 10 min of flow with 7 V applied at 1 kHz, as pictured, cells are captured by DEP onto the electrodes where they can be further processed. (Figure adapted and reproduced from Lagally, E. T., et al., *Lab Chip*, 2005, 5, 1053–1058. Copyright 2005. With permission from Royal Society of Chemistry.)

hybridization. This device was used to affect a 160-fold increase in cell concentration by trapping and concentrating the cells in a continuous flowing stream (100 μL/h for 10 min) using positive dielectrophoresis in a 100 nL plug and an on-chip valving system. Following cell trapping (as depicted in Figure 43.7), the lysis buffer and molecular beacon are added to the cells and detection of *E. coli* MC1061 cells was accomplished via the sequence-specific hybridization of the beacon to rRNA of the bacteria with fluorescence measured by laser scanning confocal microscopy. By using high concentrations of guanidine thiocyanate (4 M), not only was cell lysis promoted, but also a more rapid hybridization of target DNA to the molecular beacon probe was achieved (immediate binding versus upward of 40 min when placed in TE buffer), due to a decrease in the melting temperature of the nucleic acid/molecular beacon probe hybrids. Detection of as few as 25 cells in less than 30 min was accomplished in this manner, making this device a rapid total analysis tool capable of taking in crude sample, performing integrated sample processing, followed by an analytical/detection step to produce sequence-specific results.

Yeung et al.[118] have described a device for integrated thermal cell lysis, specific target DNA isolation, followed by PCR amplification and hybridization detection. In this work, intact cell were broken down by thermal lysis (at 90°C) in the presence of biotinylated genome capture probes. Following cell lysis, the temperature of the chamber was lowered to 50°C, where the probes bound to their target sequences and, following the addition of avidin-coated magnetic particles, specific genome sequences were isolated from cellular debris and interfering proteins by washing away the unbound species. The captured DNA was then be amplified using asymmetric PCR and the resultant amplicon hybridized to detection electrodes, labeled with gold nanoparticles, and detected by electrocatalytic silver deposition and electrochemical silver dissolution. With this method, two different bacterial species (*E. coli* and *Bacillus subtilis* (*B. subtilis*) could be detected in a mixed sample, however, the device was only evaluated with cultured cells, so its performance with more complex biofluids is unknown at this time.

In another example of integrated sample processing, Lee et al.[38] accomplished cell lysis, DNA purification, and PCR amplification for pathogen detection using a single-chamber device. In this work, the authors developed a Laser-Irradiated Magnetic Bead System (LIMBS) for cell lysis and

DNA purification from pathogens. With this method, pathogen and carboxyl-terminated magnetic beads were loaded into the microchip and placed into a chip guide module. A high-power laser beam (808 nm) was then applied to the microdevice while it was simultaneously vibrated to facilitate lysis of the cells. In addition to promoting cell lysis, when the laser was fired, the proteins were adsorbed to the surface of the magnetic beads, while the DNA remained free in solution. Following removal of the beads, the resultant DNA-containing solution was suitable for real-time PCR amplification. Although extraction efficiencies with this method were not reported, this single-chamber protocol was utilized to successfully lyse *E. coli* and Gram-positive bacterial cells, as well as hepatitis B virus mixed with human serum, with a higher efficiency of DNA release using the LIMBS method than that of other conventional cell lysis methods (e.g., boiling). In addition, successful lysis, purification, and real-time PCR amplification was performed in this single-chamber device using cultured *E. coli* cells, demonstrating the potential of this method for pathogen detection.

To date, only a few examples can be found in the literature describing a singular microfluidic chip capable of accepting crude sample lysate (from complex biofluids), purifying and extracting the DNA, accomplishing a target-specific amplification, and then performing online genetic analysis. In 2004, Liu et al.[85] described an integrated microfluidic system that performed multiple sample processing steps and integrated detection of target amplicon. In this system, sample and immunomagnetic capture beads were loaded into a chamber for incubation. Following successful isolation of the target cells, a washing step was performed, followed by loading of the chamber with the necessary PCR reagents. Subsequently, thermal lysis and PCR thermocycling was carried out, followed by incubation with and acoustic mixing of the amplicon for a hybridization reaction. Following immobilization of the DNA, ferrocene-labeled signaling probes hybridized with the target DNA, were bound to the immobilized probes, and could be detected on-chip. The isolation and detection of *E. coli* from a mock *E. coli*/rabbit blood mixture, as well as single-nucleotide polymorphism analysis directly from diluted blood were demonstrated. This device is the earliest example of a fully functional microanalytical system capable of accepting crude sample and producing a genotypic readout, establishing that multifunctional, microfluidic devices could function as viable automated genetic analysis systems.

More recently, Cady et al.[104] have described a method for integrating sample processing with detection, by developing a device capable of silica-based purification using the silica-coated pillar design described earlier in this chapter, integrated with online, real-time PCR. With this device, a silica-based purification was achieved for the pathogenic bacterium *Listeria monocytogenes*—a rod-shaped bacterium that is the causative agent in listeriosis, a serious infection caused by eating contaminated food—followed by integrated real-time amplification using SYBR Green, in an average of 45 min. In addition, this integrated system could detect as few as 104–107 cells, making this a sensitive screening technique that combines both sample processing and analytical analysis in a simple two-component device.

Finally, late in 2006, Easley et al.[30] described a valved microdevice with discreet functional domains: a silica-bead-based purification chamber, fluidically integrated to a PCR chamber suitable for IR-mediated thermocycling and connected to an electrophoretic domain, where separation and detection of amplicon was accomplished. This microgenetic analysis (MGA) device was the first of its kind that was capable of accepting sample as crude as whole blood and performing complete sample processing and genotypic analysis. As depicted in Figure 43.8, the device was evaluated with blood drawn from a mouse infected with *B. anthracis* before prior to the onset of symptoms. Purification of nucleic acids from less than 1 μL of this sample was accomplished in less than 10 min, followed by target-specific amplification in 11 min. Utilizing the sophisticated valve system, amplicon could then be repeatedly co-injected with sizing standard (both via pressure) for separation, detection, and accurate sizing using microchip electrophoresis in less than 3 min, making the total time required for this analysis less than 24 min.

In addition, the versatility of the same microdevice and method was demonstrated by application to the complete genetic analysis of 1 μL of nasal aspirate from a patient symptomatic of whooping cough. The presence of *Bacillus pertussis* (*B. pertussis*) was confirmed by the amplification of a

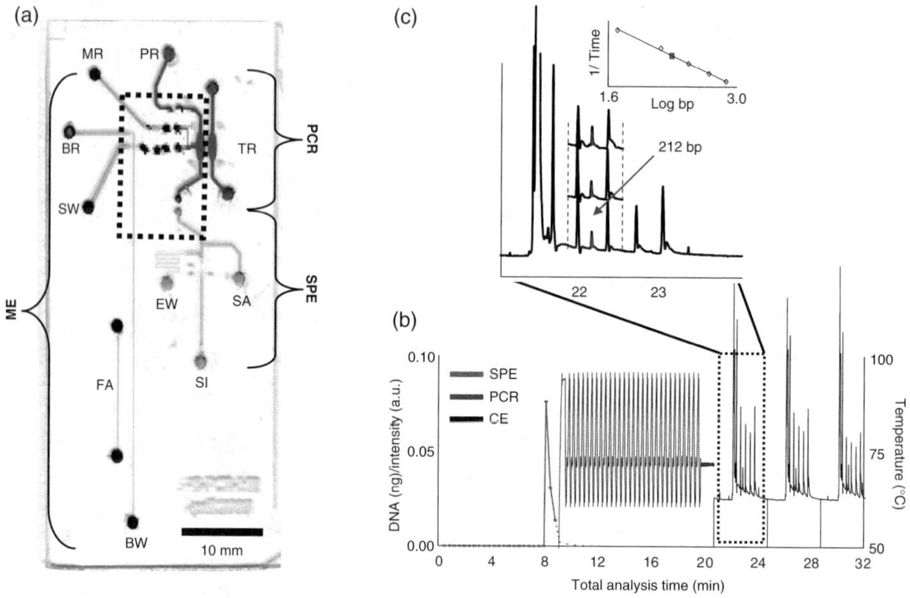

FIGURE 43.8 The integrated μTAS described by Easley et al. used for the integrated-sample processing and detection of *B. anthracis* in murine blood. (a) The IgA device with dyes placed in the channels for visualization. Domains for DNA extraction, PCR amplification, injection, and separation are connected through a network of channels and vias. SPE reservoirs are labeled for sample inlet (SI), sidearm (SA), and extraction waste (EW). Injection reservoirs are labeled for PCR reservoir (PR), marker reservoir (MR), and sample waste (SW). Electrophoresis reservoirs are labeled for buffer reservoir (BR) and buffer waste (BW). Additional domains patterned onto the device include the temperature reference (TR) chamber and fluorescence alignment (FA) channel. The flow control region is outlined by a dashed box. Device dimensions are 30 × 63.5 mm, with a total solution volume less than 10 mL. (a) Detector responses during all three stages of sample processing and analysis are portrayed in terms of total analysis time. The SPE trace (green) was taken from an offline DNA extraction of the same murine sample and is representative of the total DNA concentration observed in a typical extraction. The temperature and fluorescence intensity represent online data, with a total analysis time of less than 24 min. Three sequential injections and separations were carried out to ensure the presence of amplified product. (c) Magnified view of the first separation shown in (b), with the product peak marked and sized. The second and third runs are overlaid with the time axis cropped. The inset plot shows the sizing curve of inverse migration time versus log (base pairs) with both the sizing standard peaks (open diamonds) and product (red square) plotted for all three runs shown in (a) (error bars included). From this data, the product was sized as 212 ± 3 bp. (Figure adapted and reproduced from Easley, C. J., et al., *Proc. Natl. Acad. Sci. USA*, 2006, 103, 19272–19277. Copyright 2006. With permission from National Academy of Sciences.)

fragment of the *IS481* repeated insertion sequence, as confirmed by sequencing of the amplicon. With a total analysis time again of less than 24 min, this application further demonstrated the versatility and robustness of the method and system, thus highlighting the potential of this type of total analysis device for application to rapid, point-of-care testing for a variety of applications. Like the work presented by Liu et al.,[85] this device integrated multiple sample processing steps in a single microfluidic device. However, improving upon this earlier work, Easley et al. were able to establish a rapid analysis (<24 min versus 3.5 h) and utilize two crude input samples (infected murine blood from asymptomatic mice and human nasal aspirate versus a mock rabbit blood/bacteria mixture). This μTAS and the other integrated devices described here represent the first in what will surely be a long line of increasingly more sophisticated integrated sample processing and analytical devices for genetic interrogation and total genetic profiling.

43.8 CONCLUDING REMARKS

The path to creating integrated microdevices for bioanalysis is not only riddled with cutting-edge scientific challenges, it is ripe with cutting-edge engineering challenges as well. The μTAS field today has reached a milestone, solving or circumventing many of these critical challenges, both with hardware and "chemware," to generate the integrated multi-functional microdevices described herein. This has required clever design and novel approaches to integrating the necessary the hardware, firmware, and "chemware" into the microdevice in a functional manner. Compartmentalization for isolation of different chemistries in different functional domains, as described in the examples presented herein represents a breakthrough in terms of functional bandwidth for microfluidic devices. In addition, the development of precision fluidic control components and novel device designs has enabled the advancement of integrated chemistries; providing the necessary hardware to support the "chemware" as an essential component of integrated microfluidics.

The successful development of integrated microfluidic systems for genetic analysis represents an important approach to the improvement of analytical technology for many applications (e.g., forensics, clinical analysis). The advancement of reproducible, optimized methods for microchip DNA purification, PCR amplification, and other sample pretreatment techniques has enabled the subsequent development of integrated sample processing systems; without consistent, reliable standalone microfluidic processing, successful and consistent integrated analyses are an impossibility. The work presented here describes the development of these sample processing devices, highlights the flexibility of microfluidic systems for a accomplishing a wide variety of genetic analysis applications, and the underscores importance of the inclusion of pretreatment steps in the development of μTAS.

Future work with integrated sample processing microdevices will likely focus on the development of systems capable of total sample manipulation, with the versatility to handle multiple input samples. In addition, new generation devices will likely be able to perform sample-dependent processing, that is, the device will have many different fluidic handling steps incorporated, and the steps applied in the processing can be tailored to the type of input sample. Automation of these devices will be essential and the integration of these sample processing techniques with microfluidic analytical technologies will be imperative. The continued successful development of a genetic μTAS will hinge on sustained effort directed toward the advancement of enhanced microfluidic sample processing and integrated handling systems.

REFERENCES

1. Butler, J. M. *Forensic DNA Typing, Biology and Technology Behind STR Markers*, 2nd ed., Academic Press, San Diego, CA, 2001.
2. Rathore, A., and Guttman, A., Eds. *Electrokinetic Phenomena*, Marcel Dekker, Inc., New York, 2004.
3. Kopp, M. U., Mello, A. J., and Manz, A. *Science*, 1998, 280, 1046–1048.
4. Huhmer, A. F., and Landers, J. P. *Anal. Chem.*, 2000, 72, 5507–5512.
5. Giordano, B. C., Ferrance, J., Swedberg, S., Huhmer, A. F., and Landers, J. P. *Anal. Biochem.*, 2001, 291, 124–132.
6. Hashimoto, M., Chen, P. C., Mitchell, M. W., Nikitopoulos, D. E., Soper, S. A., and Murphy, M. C. *Lab Chip*, 2004, 4, 638–645.
7. Gawron, A. J., Martin, R. S., and Lunte, S. M. *Eur. J. Pharm. Sci.*, 2001, 14, 1–12.
8. Jin, L. J., Ferrance, J., and Landers, J. P. *Biotechniques*, 2001, 31, 1332–1335, 1338–1340, 1342, *passim*.
9. Zhang, L., Dang, F., and Baba, Y. *J. Pharm. Biomed. Anal.*, 2003, 30, 1645–1654.
10. Khandurina, J., and Guttman, A. *Curr. Opin. Chem. Biol.*, 2003, 7, 595–602.
11. Chen, L., and Ren, *J. Comb. Chem. High Throughput Screen.*, 2004, 7, 29–43.
12. Dolnik, V., and Liu, S. *J. Sep. Sci.*, 2005, 28, 1994–2009.
13. Li, S. F., and Kricka, L. J. *Clin. Chem.*, 2006, 52, 37–45.

14. Bienvenue, J. M., Duncalf, N., Marchiarullo, D., Ferrance, J. P., and Landers, J. P. *J. Forensic. Sci.*, 2006, 51, 266–273.
15. Breadmore, M. C., Wolfe, K. A., Arcibal, I. G., Leung, W. K., Dickson, D., Giordano, B. C., and Power, M. E., et al. *Anal. Chem.*, 2003, 75, 1880–1886.
16. Cady, N. C., Stelick, S., and Batt, C. A. *Biosens. Bioelectron.*, 2003, 19, 59–66.
17. Christel, L. A., Petersen, K., McMillan, W., and Northrup, M. A. *J. Biomech. Eng-Trans ASME*, 1999, 121, 22–27.
18. Wolfe, K. A., Breadmore, M. C., Ferrance, J. P., Power, M. E., Conroy, J. F., Norris, P. M., and Landers, J. P. *Electrophoresis*, 2002, 23, 727–733.
19. Wu, Q., Bienvenue, J. M., Hassan, B. J., Kwok, Y. C., Giordano, B. C., Norris, P. M., Landers, J. P., and Ferrance, J. P. *Anal. Chem.*, 2006, 78, 5704–5710.
20. Rolfs, A., Schuller, I., Finckh, U., and Weber-Rolfs, I. *PCR: Clinical Diagnostics and Research*, Springer-Verlag, Berlin, 1992.
21. Hamalainen, M. M., Eskola, J. U., Hellman, J., and Pulkki, K. *Clin. Chem.*, 1999, 45, 465–471.
22. Miller, D. N., Bryant, J. E., Madsen, E. L., and Ghiorse, W. C. *Appl. Environ. Microbiol.*, 1999, 65, 4715–4724.
23. Wilson, I. G. *Appl. Environ. Microbiol.*, 1997, 63, 3741–3751.
24. Rudi, K., and Jakobsen, K. S. *Methods Mol. Biol.*, 2006, 345, 23–35.
25. Tian, H., Huhmer, A. F., and Landers, J. P. *Anal. Biochem.*, 2000, 283, 175–191.
26. Cao, W., Easley, C. J., Ferrance, J. P., and Landers, J. P. *Anal. Chem.*, 2006, 78, 7222–7228.
27. Wen, J., Guillo, C., Ferrance, J. P., and Landers, J. P. *Anal. Chem.*, 2006, 78, 1673–1681.
28. Melzak, K., Sherwood, C., Turner, R., and Haynes, C. *J. Coll. Interface Sci.*, 1996, 181, 635–644.
29. Mason, P. E., Neilson, G. W., Enderby, J. E., Saboungi, M. L., Dempsey, C. E., MacKerell, A. D., Jr., and Brady, J. W. *J. Am. Chem. Soc.*, 2004, 126, 11462–11470.
30. Easley, C. J., Karlinsey, J. M., Bienvenue, J. M., Legendre, L. A., Roper, M. G., Feldman, S. H., and Hughes, M. A., et al. *Proc. Natl. Acad. Sci. USA*, 2006, 103, 19272–19277.
31. Bienvenue, J. M., Legendre, L. A., Ferrance, J. P., and Landers, J. P. *J. Forensic Sci.*, 2007, Submitted.
32. Hagan, K., Bienvenue, J. M., Ferrance, J. P., and Landers, J. P. *Anal. Chem.*, 2007, Submitted.
33. Kato, M., Dulay, M. T., Bennett, B. D., Quirino, J. P., and Zare, R. N. *J. Chromatogr. A*, 2001, 924, 187–195.
34. Hong, J. W., Studer, V., Hang, G., Anderson, W. F., and Quake, S. R. *Nat. Biotechnol.*, 2004, 22, 435–439.
35. Chung, Y. C., Jan, M. S., Lin, Y. C., Lin, J. H., Cheng, W. C., and Fan, C. Y. *Lab Chip*, 2004, 4, 141–147.
36. Nakagawa, T., Tanaka, T., Niwa, D., Osaka, T., Takeyama, H., and Matsunaga, T. *J. Biotechnol.*, 2005, 116, 105–111.
37. Witek, M. A., Llopis, S. D., Wheatley, A., McCarley, R. L., and Soper, S. A. *Nucleic Acids Res.*, 2006, 34, e74.
38. Lee, J. G., Cheong, K. H., Huh, N., Kim, S., Choi, J. W., and Ko, C. *Lab Chip*, 2006, 6, 886–895.
39. Mullis, K., Faloona, F., Scharf, S., Saiki, R., Horn, G., and Erlich, H. *Cold Spring Harb. Symp. Quant. Biol.*, 1986, 51 (Pt 1), 263–273.
40. Zhang, C., Xu, J., Ma, W., and Zheng, W. *Biotechnol. Adv.*, 2006, 24, 243–284.
41. Roper, M. G., Easley, C. J., and Landers, J. P. *Anal. Chem.*, 2005, 77, 3887–3893.
42. Lagally, E. T., Scherer, J. R., Blazej, R. G., Toriello, N. M., Diep, B. A., Ramchandani, M., Sensabaugh, G. F., Riley, L. W., and Mathies, R. A. *Anal. Chem.*, 2004, 76, 3162–3170.
43. Krishnan, M., Agrawal, N., Burns, M. A., and Ugaz, V. M. *Anal. Chem.*, 2004, 76, 6254–6265.
44. Liu, J., Hansen, C., and Quake, S. R. *Anal. Chem.*, 2003, 75, 4718–4723.
45. Marcus, J. S., Anderson, W. F., and Quake, S. R. *Anal. Chem.*, 2006, 78, 956–958.
46. Lee, D. S., Park, S. H., Yang, H., Chung, K. H., Yoon, T. H., Kim, S. J., Kim, K., and Kim, Y. T. *Lab Chip*, 2004, 4, 401–407.
47. Matsubara, Y., Kerman, K., Kobayashi, M., Yamamura, S., Morita, Y., Takamura, Y., and Tamiya, E. *Anal. Chem.*, 2004, 76, 6434–6439.
48. Pal, R., Yang, M., Johnson, B. N., Burke, D. T., and Burns, M. A. *Anal. Chem.*, 2004, 76, 3740–3748.

49. Gulliksen, A., Solli, L., Karlsen, F., Rogne, H., Hovig, E., Nordstrom, T., and Sirevag, R. *Anal. Chem.*, 2004, 76, 9–14.
50. Guttenberg, Z., Muller, H., Habermuller, H., Geisbauer, A., Pipper, J., Felbel, J., Kielpinski, M., Scriba, J., and Wixforth, A. *Lab Chip*, 2005, 5, 308–317.
51. Liu, C. N., Toriello, N. M., and Mathies, R. A. *Anal. Chem.*, 2006, 78, 5474–5479.
52. Koh, C. G., Tan, W., Zhao, M. Q., Ricco, A. J., and Fan, Z. H. *Anal. Chem.*, 2003, 75, 4591–4598.
53. Lee, T. M., Carles, M. C., and Hsing, I. M. *Lab Chip*, 2003, 3, 100–105.
54. Obeid, P. J., Christopoulos, T. K., Crabtree, H. J., and Backhouse, C. J. *Anal. Chem.*, 2003, 75, 288–295.
55. Woolley, A. T., Hadley, D., Landre, P., deMello, A. J., Mathies, R. A., and Northrup, M. A. *Anal. Chem.*, 1996, 68, 4081–4086.
56. Chen, J., Wabuyele, M., Chen, H., Patterson, D., Hupert, M., Shadpour, H., Nikitopoulos, D., and Soper, S. A. *Anal. Chem.*, 2005, 77, 658–666.
57. Oda, R. P., Strausbauch, M. A., Huhmer, A. F., Borson, N., Jurrens, S. R., Craighead, J., Wettstein, P. J., Eckloff, B., Kline, B., and Landers, J. P. *Anal. Chem.*, 1998, 70, 4361–4368.
58. Lee, S., Kim, S.-W., and Ahn, C. H., μTAS, Tokyo, Japan 2006.
59. Krishnan, M., Ugaz, V. M., and Burns, M. A. *Science*, 2002, 298, 793.
60. Easley, C., Humphrey, J., and Landers, J. P. *J. Micromech. Microeng.*, 2007, 17, 11–19.
61. Hughes, M. P. *Electrophoresis*, 2002, 23, 2569–2582.
62. Gonzalez, C. F., and Remcho, V. T. *J. Chromatogr. A*, 2005, 1079, 59–68.
63. Horsman, K. M., Barker, S. L., Ferrance, J. P., Forrest, K. A., Koen, K. A., and Landers, J. P. *Anal. Chem.*, 2005, 77, 742–749.
64. Yuen, P. K., Kricka, L. J., Fortina, P., Panaro, N. J., Sakazume, T., and Wilding, P. *Genome Res.*, 2001, 11, 405–412.
65. Fu, A. Y., Chou, H. P., Spence, C., Arnold, F. H., and Quake, S. R. *Anal. Chem.*, 2002, 74, 2451–2457.
66. Fu, A. Y., Spence, C., Scherer, A., Arnold, F. H., and Quake, S. R. *Nat. Biotechnol.*, 1999, 17, 1109–1111.
67. Berger, M., Castelino, J., Huang, R., Shah, M., and Austin, R. H. *Electrophoresis*, 2001, 22, 3883–3892.
68. Davis, J. A., Inglis, D. W., Morton, K. J., Lawrence, D. A., Huang, L. R., Chou, S. Y., Sturm, J. C., and Austin, R. H. *Proc. Natl. Acad. Sci. USA*, 2006, 103, 14779–14784.
69. Huang, L. R., Cox, E. C., Austin, R. H., and Sturm, J. C. *Science*, 2004, 304, 987–990.
70. Kumar, M., Feke, D. L., Belovich, J. M. *Biotechnol. Bioeng.*, 2005, 89, 129–137.
71. Jonsson, H., Nilsson, A., Petersson, F., Allers, M., and Laurell, T. *Perfusion*, 2005, 20, 39–43.
72. Petersson, F., Nilsson, A., Holm, C., Jonsson, H., and Laurell, T. *Analyst*, 2004, 129, 938–943.
73. Petersson, F., Nilsson, A., Holm, C., Jonsson, H., and Laurell, T. *Lab Chip*, 2005, 5, 20–22.
74. Wang, Z., Grabenstetter, P., Feke, D. L., and Belovich, J. M. *Biotechnol. Prog.*, 2004, 20, 384–387.
75. Irimia, D., Tompkins, R. G., and Toner, M. *Anal. Chem.*, 2004, 76, 6137–6143.
76. Sethu, P., Anahtar, M., Moldawer, L. L., Tompkins, R. G., and Toner, M. *Anal. Chem.*, 2004, 76, 6247–6253.
77. Di Carlo, D., Ionescu-Zanetti, C., Zhang, Y., Hung, P., and Lee, L. P. *Lab Chip*, 2005, 5, 171–178.
78. Heo, J., Thomas, K. J., Seong, G. H., and Crooks, R. M. *Anal. Chem.*, 2003, 75, 22–26.
79. Lu, H., Schmidt, M. A., and Jensen, K. F. *Lab Chip*, 2005, 5, 23–29.
80. Lu, K. Y., Wo, A. M., Lo, Y. J., Chen, K. C., Lin, C. M., and Yang, C. R. *Biosens. Bioelectron.*, 2006, 22, 568–574.
81. Culbertson, C. T. *Methods Mol. Biol.*, 2006, 339, 203–216.
82. Wang, H. Y., Bhunia, A. K., and Lu, C. *Biosens. Bioelectron.*, 2006, 22, 582–588.
83. McClain, M. A., Culbertson, C. T., Jacobson, S. C., Allbritton, N. L., Sims, C. E., and Ramsey, J. M. *Anal. Chem.*, 2003, 75, 5646–5655.
84. Gao, J., Yin, X. F., and Fang, Z. L. *Lab Chip*, 2004, 4, 47–52.
85. Liu, R. H., Yang, J., Lenigk, R., Bonanno, J., and Grodzinski, P. *Anal. Chem.*, 2004, 76, 1824–1831.
86. He, Y., Zhang, Y. H., and Yeung, E. S. *J. Chromatogr. A*, 2001, 924, 271–284.
87. El-Ali, J., Gaudet, S., Gunther, A., Sorger, P. K., and Jensen, K. F. *Anal. Chem.*, 2005, 77, 3629–3636.
88. Waters, L. C., Jacobson, S. C., Kroutchinina, N., Khandurina, J., Foote, R. S., and Ramsey, J. M. *Anal. Chem.*, 1998, 70, 158–162.

89. Belgrader, P., Hansford, D., Kovacs, G. T., Venkateswaran, K., Mariella, R., Jr., Milanovich, F., Nasarabadi, S., Okuzumi, M., Pourahmadi, F., and Northrup, M. A. *Anal. Chem.*, 1999, 71, 4232–4236.
90. Kim, J., Hee Jang, S., Jia, G., Zoval, J. V., Da Silva, N. A., and Madou, M. J. *Lab Chip*, 2004, 4, 516–522.
91. Di Carlo, D., Jeong, K. H., and Lee, L. P. *Lab Chip*, 2003, 3, 287–291.
92. Shrinivasan, S., Norris, P. M., Landers, J. P., Ferrance, J. P. *Clin. Lab. Med.*, 2007, 27(1) 173–181.
93. Grover, W. H., Ivester, R. H., Jensen, E. C., and Mathies, R. A. *Lab Chip*, 2006, 6, 623–631.
94. Grover, W. H., and Mathies, R. A. *Lab Chip*, 2005, 5, 1033–1040.
95. Skelley, A. M., Scherer, J. R., Aubrey, A. D., Grover, W. H., Ivester, R. H., Ehrenfreund, P., Grunthaner, F. J., Bada, J. L., and Mathies, R. A. *Proc. Natl. Acad. Sci. USA*, 2005, 102, 1041–1046.
96. Unger, M. A., Chou, H. P., Thorsen, T., and Scherer, A., Quake, S. R. *Science*, 2000, 288, 113–116.
97. Melin, J., and Quake, S. R. *Annu. Rev. Biophys. Biomol. Struct.*, 2007, 36, 213–231.
98. Thorsen, T., Maerkl, S. J., and Quake, S. R. *Science*, 2002, 298, 580–584.
99. Blazej, R. G., Kumaresan, P., and Mathies, R. A. *Proc. Natl. Acad. Sci. USA*, 2006, 103, 7240–7245.
100. Pal, R., Yang, M., Lin, R., Johnson, B. N., Srivastava, N., Razzacki, S. Z., Chomistek, K. J., et al. *Lab Chip*, 2005, 5, 1024–1032.
101. Paegel, B. M., Emrich, C. A., Wedemayer, G. J., Scherer, J. R., and Mathies, R. A. *Proc. Natl. Acad. Sci. USA*, 2002, 99, 574–579.
102. Hong, J. W., and Quake, S. R. *Nat. Biotechnol.*, 2003, 21, 1179–1183.
103. Legendre, L. A., Bienvenue, J. M., Roper, M. G., Ferrance, J. P., and Landers, J. P. *Anal. Chem.*, 2006, 78, 1444–1451.
104. Cady, N. C., Stelick, S., Kunnavakkam, M. V., and Batt, C. A. *Sens. Actuat. B: Chem.*, 2005, B107, 332–341.
105. Burns, M. A., Johnson, B. N., Brahmasandra, S. N., Handique, K., Webster, J. R., Krishnan, M., Sammarco, T. S., et al. *Science*, 1998, 282, 484–487.
106. Dunn, W. C., Jacobson, S. C., Waters, L. C., Kroutchinina, N., Khandurina, J., Foote, R. S., Justice, M. J., Stubbs, L. J., and Ramsey, J. M. *Anal. Biochem.*, 2000, 277, 157–160.
107. Hataoka, Y., Zhang, L., Mori, Y., Tomita, N., Notomi, T., and Baba, Y. *Anal. Chem.*, 2004, 76, 3689–3693.
108. Hong, J. W., Fujii, T., Seki, M., Yamamoto, T., and Endo, I. *Electrophoresis*, 2001, 22, 328–333.
109. Khandurina, J., McKnight, T. E., Jacobson, S. C., Waters, L. C., Foote, R. S., and Ramsey, J. M. *Anal. Chem.*, 2000, 72, 2995–3000.
110. Lagally, E. T., Simpson, P. C., and Mathies, R. A. *Sens. Actuat. B: Chem.*, 2000, B63, 138–146.
111. Lagally, E. T., Medintz, I., and Mathies, R. A. *Anal. Chem.*, 2001, 73, 565–570.
112. Lagally, E. T., Emrich, C. A., and Mathies, R. A. *Lab Chip*, 2001, 1, 102–107.
113. Liu, C. N., Toriello, N. M., Maboudian, R., and Mathies, R. A. *Special Publication—Royal Soc. Chem.*, 2004, 297, 297–299.
114. Rodriguez, I., Lesaicherre, M., Tie, Y., Zou, Q., Yu, C., Singh, J., Meng, L. T., et al. *Electrophoresis*, 2003, 24, 172–178.
115. Waters, L. C., Jacobson, S. C., Kroutchinina, N., Khandurina, J., Foote, R. S., and Ramsey, J. M. *Anal. Chem.*, 1998, 70, 5172–5176.
116. Woolley, A. T., Hadley, D., Landre, P., deMello, A. J., Mathies, R. A., and Northrup, M. A. *Anal. Chem.*, 1996, 68, 4081–4086.
117. Lagally, E. T., Lee, S. H., and Soh, H. T. *Lab Chip*, 2005, 5, 1053–1058.
118. Yeung, S. W., Lee, T. M., Cai, H., and Hsing, I. M. *Nucleic Acids Res.*, 2006, 34, e118.

44 Cell and Particle Separation and Manipulation Using Acoustic Standing Waves in Microfluidic Systems

Thomas Laurell and Johan Nilsson

CONTENTS

44.1 INTRODUCTION

There is an urgent need to find means to efficiently separate and manipulate cells and particles in microfluidic chips because conventional mechanical separation concepts are not readily amenable to downscaling as the channel dimensions and the particle sizes are approaching each other and thus, physical restrictions in the flow paths tend to cause channel blockage. Therefore, noncontact modes to induce forces on particles are of interest. Recently, ultrasonic standing waves have proven beneficial in the quest for techniques that generate relatively large forces, which simultaneously are compatible with microchip technologies and system integration.

44.2 BACKGROUND

Manipulation and spatial control of particulate matter in microspace are gaining interest within the microtechnology community. This has become especially attractive when combining microfluidic structures with physical concepts that induce controlled forces on particles inside microchannels. Modern lithographic techniques enable the design and fabrication of static microstuctures that allow mechanical separation and capturing of cells and particles in streaming microsystems. On the other hand, concepts that allow noncontact cell separation and capturing offer an enhanced functionality in terms of nonclogging continuous flow separation with low mechanical stress, that is, low shear rate and low surface interaction. The literature is well provided with mechanical filter concepts, ranging from conventional slits[1,2] for size exclusion separation to advanced Patchinco-style single-cell capture devices.[3] Recent developments by the Seki[4] group demonstrate elegant microscale solutions to laminar-flow-based size selection of particles and cells that, for example, enable clear discrimination of erythrocytes from leukoytes.[5] A similar approach was pioneered by the Austin group[6,7] demonstrating the deterministic lateral displacement separation device that enables discrimination of particles with size differences of tens of micrometers. Common to these concepts are all the uses of sieving features or laminar flow splitting schemes in a continuous flow.

In contrast, techniques that induce an external force on a particle in a microfluidic environment, not necessarily requiring a continuous flow mode for stable operation, are gaining attention. The use of an optically induced force, optical tweezers, for spatial control of cells under microscope surveillance has since long been a valuable research tool[8,9] and is now available as commercial instrumentation. Likewise, the use of the dielectrophoretic properties of cells and particles have been widely researched and utilized in the development of microsystems for particle and cell separation and trapping.[10–14] More recent developments have taken dielectrophoresis electrodes to a single-sided chip layout, thereby, vastly reducing complexity in dielectrophoretic chip designs.[15,16] Further work by Voldman[17] has also demonstrated particle separation based on combined physical features such as iso-dielectrophoresis where lateral ion gradients orthogonal to the induced dielectrophoretic force serve as the basis for spatially defining a particle's zero dielectrophoretic force position in a channel.

With the advent of biofunctionalized magnetic microparticles, magnetophoretic separation has also recently emerged as a potential mode of advanced particle separation. Pamme and Manz[18] have reported fundamental developments in microchip-integrated particle separation based on the net force acting on the particles streaming in a flow orthogonal to the direction of the applied magnetic field. This principle was later also employed to the magnetic bead-based extraction of mouse macrophages and human ovarian cancer cells.[19]

An alternative to dielectrophoretic or magnetic force manipulation of microparticles in suspension is the use of acoustic standing wave forces, which also have served as the basis for particle and cell manipulation in microfluidic devices. Historical work by Kundt[20] demonstrated early on the ability to move cork particles in a standing wave pattern. The fundamental acousto-physical principles have been well researched and described in the literature by King[21] and Gorkov.[22]

Normally, particles are trapped in the pressure node of the acoustic standing wave and standing wave patterns can thus be utilized to concentrate and aggregate particles in the pressure nodes. Numerous applications have utilized the aggregation effect, for example, demonstrating yeast cells concentration,[23,24] blood plasma clarification,[25,26] hybridoma cell aggregation,[27] and cell retention in general.[28] These are all examples of systems developed as macroscale devices in stagnant systems where aggregated cells are collected at the bottom of a chamber. Efforts to realize continuous flow separation systems based on acoustic focusing of particles and cells have, for example, been reported by Yasuda et al.[29] and Benes et al.[30] The true benefits of continuous flow separation becomes evident as the scale is reduced and the flow becomes truly laminar at low Reynolds numbers, reducing influence on separation from unsteady flow or turbulence. This fact clearly speaks for acoustic separation to be most efficiently applied in combination with microfluidic-based devices in a laminar flow mode, serving as a non-fluid-contact mode of controlling and separating microparticles.

When reducing the dimensions, that is, going into the microscale domain with channel dimensions smaller than 500 μm, inherent benefits in terms of an increased primary acoustic radiation force is obtained due to the higher resonance frequency of the resonator.

With the progress within the lab-on-a-chip field, the need for new modes of controlling cells and particles in a gentle but precise way is highly sought and acoustics combined with microfluidics is now emerging as one strong candidate in this respect. This chapter will cover recent developments within microchip-based cell and particle separation and manipulation. Applications in continuous-flow blood component separation and blood purification will be outlined along with new modes of cell trapping and in-chip manipulation.

44.3 THEORY

44.3.1 STANDING WAVE FORCES

44.3.1.1 Primary Axial Radiation Force

The primary axial radiation force (PRF) acting on a spherical particle in an acoustic standing wave is defined by

$$F_{\mathrm{PRF}} = \frac{-\pi \cdot P_0^2 \cdot V \cdot \beta_0}{2\lambda} \cdot \sin\left(\frac{4\pi \cdot z}{\lambda}\right) \cdot \left(\frac{(5\rho_{\mathrm{p}} - 2\rho_0)}{(2\rho_{\mathrm{p}} + \rho_0)} - \frac{\beta_{\mathrm{p}}}{\beta_0}\right)$$

$$= \frac{-\pi \cdot P_0^2 \cdot V \cdot \beta_0}{2\lambda} \cdot \Phi \cdot \sin\left(\frac{4\pi \cdot z}{\lambda}\right) \tag{44.1}$$

$$\Phi = \left(\frac{(5\rho_{\mathrm{p}} - 2\rho_0)}{(2\rho_{\mathrm{p}} + \rho_0)} - \frac{\beta_{\mathrm{p}}}{\beta_0}\right), \tag{44.2}$$

where P_0 is the applied acoustic pressure amplitude, V defines the particle volume, β_{p} is the compressibility and ρ_{p} is the density of the particle in a fluid with compressibility β_0 and density ρ_0, λ is the acoustic standing wavelength, and z is the position along the wave propagation axis. Φ is commonly denoted as the acoustic contrast factor. Force equation, Equation 44.1, was presented by Gould and Coakley[31] and is based on the theory on acoustic forces on small particles in a fluid presented by Gorkov.[22] Basic theory on the interaction between particles and acoustic pressure fields assuming a rigid sphere and a friction-less liquid was presented by King.[21] Later, Yosioka[32] extended the theory to also encompass compressible particles.

44.3.1.2 Primary Lateral Acoustic Radiation Force

The acoustic forces acting on the particles based on the theory by Gorkov are described in a three-dimensional (3D) space using the negative gradient of a 3D potential function given by the standing wave field. Equation 44.1 determines the axial force but there are also lateral/radial forces acting on the particles from radial gradients in the standing wave field that will force the particles toward the central axis. For particles with density and compressibility higher than the surrounding medium, the PRF will be directed toward the nearest pressure node in the standing wave fields. When performing acoustic focusing of particles in a continuous flow the particles will be retained in the pressure node, that is, being trapped there as long as the drag force on the particle/cluster from the surrounding medium is lower than the acoustic force. Normally, the axial force is larger than the radial force.

44.3.1.3 Secondary Forces

As particles gather and the interparticle distances become small, the influence of scattered sound from neighbouring particles increases, which in turn gives rise to an interparticle acoustic force that is commonly named secondary forces, or Bjerknes forces named after Bjerknes.[33] He first described the theory behind this phenomena, Equation 44.3, where a is the radius of the particle, d is the distance between the particles, and θ is the angle between the centre line of the particles and the direction of propagation of the incident acoustic wave.

$$F_{\mathrm{B}}(x) = 4\pi a^6 \left[\frac{(\rho_{\mathrm{p}} - \rho_0)^2 (3\cos^2\theta - 1)}{6\rho_0 d^4} v^2(x) - \frac{\omega^2 \rho_0 (\beta_{\mathrm{p}} - \beta_0)^2}{9d^2} p^2(x) \right] \quad (44.3)$$

As an acoustic standing wave starts to act on a particle suspension, particles are driven toward the standing wave nodal plane by the primary radiation force in an initial stage. As the particle density in the nodal plane increases, the average interparticle distance becomes small and thus the influence of the secondary force increases. This can be observed as particles moving toward one another, forming clusters of densely packed particles. If monodispersed particles are used, nice periodic hexagonal particles patterns are formed in the primary focal nodal plane, Figure 44.1.

44.3.2 Microscale Acoustic Standing Wave Separators

Acoustic standing wave separators are commonly designed as so-called layered resonators, where the boundaries of a resonance chamber is defined at one side by a transducer (or a coupling layer

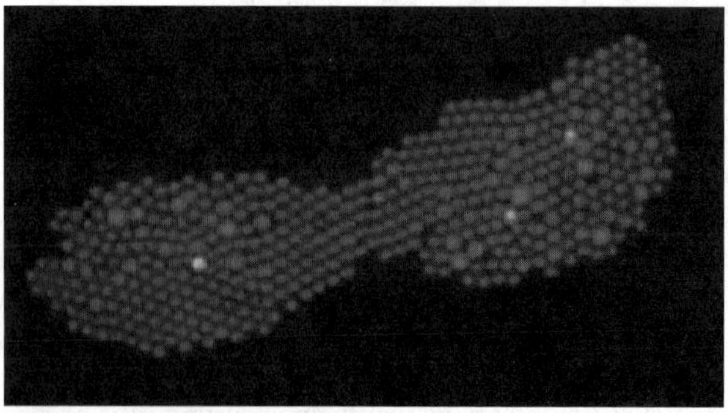

FIGURE 44.1 Fluorescently labeled polystyrene particles gathered in the pressure nodal plane (in plane with the image focus) of an acoustic standing wave by the primary acoustic radiation force in a 70 μm deep microchannel. The microbeads are also clustered in a dense hexagonal pattern by Bjerknes forces.

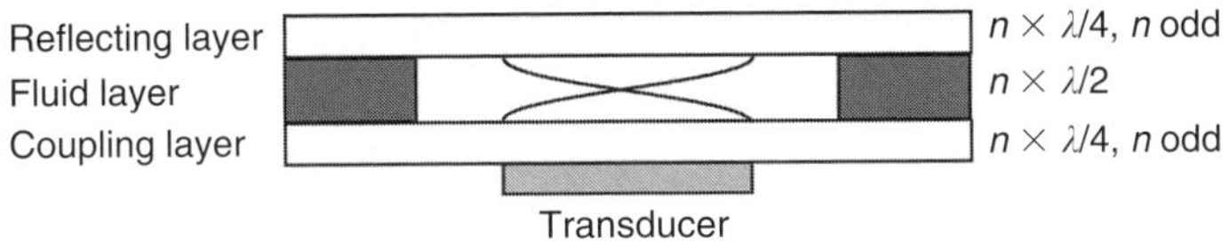

FIGURE 44.2 Schematic of a layered resonator design.

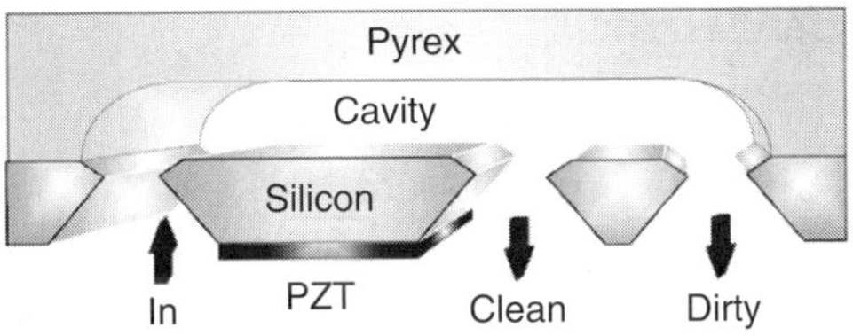

FIGURE 44.3 A microfabricated layered resonator for continuous flow particle enrichment. (Reproduced from Harris N.R., et al., *Sensors and Actuators B: Chemical*, 95, 425–434, 2003. With permission from Elsevier.)

linked to a transducer) and a highly reflecting surface at the opposing side (Figure 44.2) generating an acoustic standing wave pressure node in the centre of the fluid chamber.

The common design typically encompasses a quarter wavelength ($n \times \lambda/4$; n = odd numbers) coupling layer, a half wavelength fluid layer, and a quarter wavelength ($n \times \lambda/4$; n = odd numbers) reflecting layer. The reflecting layer can alternatively be defined by a second acoustic element. In this case, the phase shift between the two transducers can be tuned to position the standing wave pressure node at an arbitrary position along the direction of the wave propagation.

These types of layered separators have been researched by several groups and were applied to various particle and cell separation applications. A microfluidic separation device was later developed by Hawkes and Coakley[34] by electrodischarge machining of stainless steel films to fabricate microchannels in a high Q-value material. The sidewalls were subsequently fine polished to obtain a good acoustic reflecting surface and the component was assembled as a layered resonator device. The separator was realized both in the form of an H-separator as presented by Benes et al.[35] and as a Y-shaped separator. This rather complicated procedure of fabricating a qualitative microscale resonator clearly speaks for the transition into microfabrication technology for the acoustic separator fabrication.

An early initiative to utilize microfabrication technology in this respect was presented by Harris et al.[36,37] who employed both anisotropically wet-etched silicon and isotropically etched glass that were anodically bonded together, forming a separation device (Figure 44.3). An single inlet was employed via the silicon chip and a flow-through cavity etched in the opposing acoustically reflecting glass chip forming a resonator between the transducer surface and the glass recess and focused the incoming particles into a sheet. The device had two outlets via the silicon chip, which allowed to balance the flows at the two outlets such that the particle-enriched fraction was collected at one of the outlets and the clear fraction at the other.

A common drawback with layered resonators is that visual inspection generally is not possible as the reflecting glass layer serves as the observation window and thus, cells and particles are focused in the plane of the observing window.

An optional mode of microscale resonator design, the Lund method, was proposed by Nilsson et al.,[38] describing a microchip with a rectangular-shaped acoustic focusing channel that was anisotropically etched in standard ⟨100⟩ silicon. The end of the separation channel was provided with a trifurcation flow splitting outlet. The chip was sealed by a anodic bonding of a glass lid. Figure 44.4a shows a schematic cross-section of the separation chip and Figure 44.4b shows the

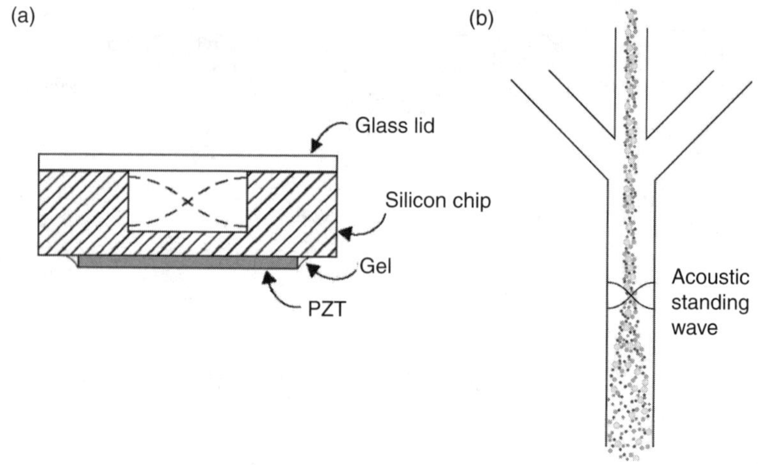

FIGURE 44.4 Schematic view of an acoustic particle concentrator according to the Lund method, enabling online visual quality control of the particles focusing. (a) Cross-section of the Lund method acoustic focusing chip. (b) Schematic top view of the Lund method acoustic focusing chip.

top view of the chip in operation. The almost perfect vertical silicon channel sidewalls served as the acoustic reflecting surfaces, thus providing a standing wave nodal plane orthogonal to the visual inspection window, which offers direct feedback on the performance of the separation.

As the chip is actuated from underneath and an acoustic standing wave is induced in the flow channel since the excitation frequency is set to match the $\lambda/2$ resonance criterion of the channel width. It should be noted that the obtained standing wave in the flow channel is orthogonal to the incident wave propagation direction. This fact enables acoustic control of particle and cell displacement in the lateral direction, in plane with the microchip, which in turns opens the route to simple integration with other downstream acoustic separations steps or other microfluidic unit operations.

The following section of this chapter will outline several applications of using the Lund method of ultrasonic standing wave particle/cell focusing, where the feature of visual control is of outmost importance.

44.4 PRACTICAL APPLICATIONS

44.4.1 Binary Modes of Acoustic Separation

44.4.1.1 Acoustic Manipulation Based on Density Properties

Acoustic differentiation of cells and particles based on the primary acoustic radiation force equation, Equation 44.1, inherently opens the route to design acoustic microparticle and cell separation systems. The most straightforward system configuration is to make a binary separation (separation of mixed particles into one or two populations) where the primary radiation force has different signs for, for example, two population of particles. This situation is obtained when the acoustic contrast factor, Φs in Equation 44.2 has different signs for the two particle populations to be separated. The decisive factors for this are the density ratio and the compressibility ratio for the particle versus the carrier solution. Most commonly, a sign shift in the acoustic contrast factor can be obtained by a proper selection of carrier fluid density versus the density of the particle to be separated. A natural system that nicely illustrates this can be seen in the focusing of lipid emulsions in milk into the pressure antinodes along the sidewalls of the acoustic resonator rather than into the central pressure node. Figure 44.5 shows milk streaming at the outlet of a Lund-separator without (44.5a) and with (44.5b)

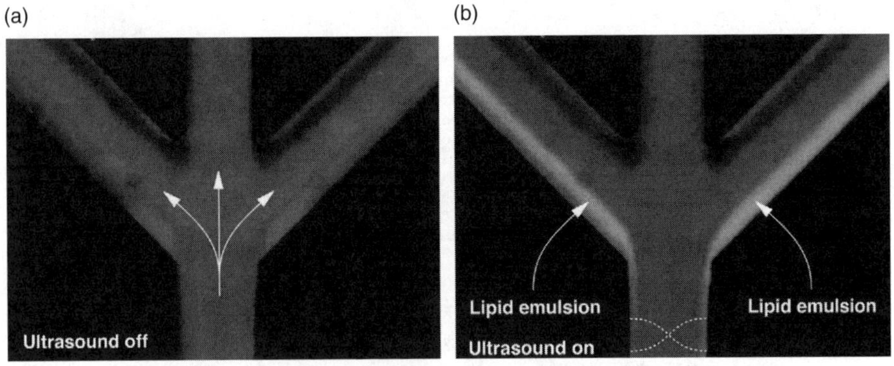

FIGURE 44.5 (a) Milk streaming through a Lund-separator without ultrasonic actuation. (b) Lipid emulsion in milk focused in the acoustic standing wave pressure antinodes along the sidewalls of a Lund-separator.

the ultrasonic actuation at 2 MHz. Once the lipid emulsion is focused along the channel sidewalls, the laminar flow directs the lipid emulsion to the side outlets at the channel trifurcation.

44.4.1.2 Carrier Media Density Manipulation

In situations where particles are difficult to separate owing to similar size and acoustic contrast factors, it may still be possible to accomplish a good separation of the species by adapting the density of the carrier media such that the acoustic contrast factor displays a sign shift for one of the particles to separate. As tabular data on cell or particle density many a times are lacking, one way to find the correct carrier media density for a qualitative separation is to perform a series of centrifugation runs in which the carrier medium density is titrated. The buffer density at which the particle populations are differentiated into the sedimentation pellet and the supernatant, respectively, defines the buffer for a successful acoustic separation. Petersson et al.[39] demonstrated this approach when failing to separate 3 μm red coloured polystyrene particles (density 1.05 g/cm^3) from 3 μm white polymethyl methacrylate (PMMA) particles (density 1.22 g/cm^3) suspended in distilled water. By adding CsCl (0.22 mg/mL) to the aqueous carrier, the medium density was adjusted to 1.16 g/cm^3, which is between the densities of the two particle types to be separated. Centrifugation experiments clearly confirmed this situation, collecting the white PMMA particles in the sedimentation pellet and red particles were recovered in the supernatant. When running this buffer system in the acoustic separator, 88% of the polystyrene particles were recovered in the side outlet and 96% of the PMMA particles were collected from the centre outlet.

These findings clearly identify binary acoustic separations as a chip-integrated modality matching density media centrifugations. Centrifugation is a major workhorse in all bio-oriented labs and the vast variety of centrifugation kits using adapted density media for extracting cells or biological particles can now be directly applied to acoustic separation systems.

44.4.1.3 Acoustic Manipulation Based on Frequency Switching

An optional mode of performing a binary particle separation utilizes the mobility of the particle species to be separated. The motion of each particle is defined by its individual primary acoustic radiation force and the opposing hydrodynamic drag force, Equation 44.4, where v is the fluid velocity, μ is the fluid viscosity, and D_p is the particle diameter:

$$F = 3 \cdot \pi \cdot \mu \cdot D_\mathrm{p} \cdot v \tag{44.4}$$

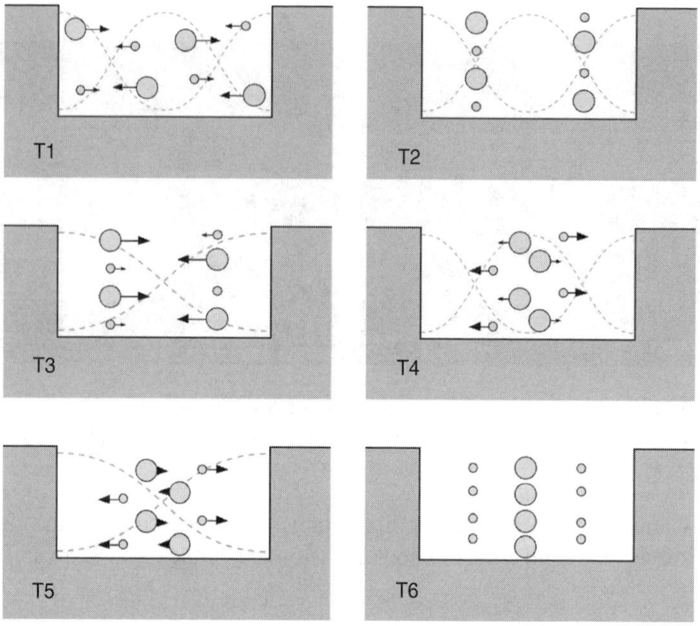

FIGURE 44.6 Particle separation by means of resonance frequency switching where separation is performed based on the different migration velocities of differently sized particles. By accurate tuning of the frequency, switching size discrimination to the first harmonic and the fundamental resonance nodes are obtained.

The net force acting on the particle defines a unique travelling speed in the acoustic standing wave field for each particle type. If the actuation frequency is switched from the fundamental channel resonance ($\lambda/2$) to a λ resonance criterion before the particles have reached their equilibrium position in the channel centre, the particles will start to migrate to the pressure nodes defined at the $\lambda/4$ position from each channel wall. By repeating the frequency switching, a situation can be accomplished where particles of different sizes can be separated. This is accomplished due to the particle-specific primary acoustic radiation force and the varying magnitude of the standing wave field with respect to the spatial position in the channel. Figure 44.6 gives a schematic representation of the separation of particles by means of acoustic resonance frequency switching. The schematic depicts six sequential time points during the acoustic switching process.

At T1, the channel is actuated in a single wavelength (first harmonic of the resonator) standing wave resonance at 4 MHz, driving particles to the pressure nodal planes $\lambda/4$ from each sidewall, T2. When switching to the fundamental resonance (2 MHz), T3, the particles will migrate toward the pressure nodal plane in the channel centre. Larger particles move faster due to a higher net force. If the frequency is switched back at the right moment to a double node resonance mode, T4, the larger particles will, at the switching occasion, be located close to the centre of the channel, where the axial PRF is at its minimum in the first harmonic resonance. The smaller particles, on the other hand, will have moved only a small distance from their original position and start to move back to the position seen at time T2 under the influence of a higher radiation force. After switching frequency again, T5, the larger particles will be closer to the centre than at T2 while the smaller will be closer to their original double node position. If the switching continues, the larger particles will end up in the center of the resonator channel and the smaller ones one-quarter of the channel width from each sidewall. If this is performed in a continuous flow mode in a Lund-separator as presented by Siversson et al.,[40] the switching system can be tuned such that the two particle types are aligned in laminar flow lanes that can be diverted to different outlets at the end of the separation channel. Figure 44.7a shows the outlet of a Lund separator operated in frequency switching mode. The particle suspension enters the

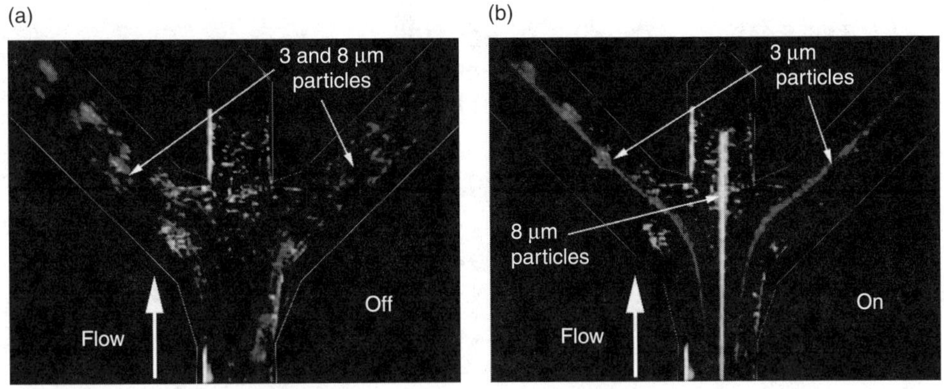

FIGURE 44.7 (a) Microscope image of the situation when the ultrasound is turned off. Both 3 μm polystyrene and 8 μm PMMA particles exit through the side outlets. (Reproduced from, *Chemical Society Reviews*, 36, 492–506, 2007. With permission from The Royal Society of Chemistry.) (b) Microscope image of the situation when the ultrasound is turned on. The 8 μm particles are focused in the fundamental resonance pressure node and exit through the centre outlet while the 3 μm particles are gathered in the first harmonic pressure nodes and exit through the side outlets. The fundamental resonance and the first harmonic were typically active for 800 and 200 μs, respectively (total flow rate: 90 μL/min). (Reproduced from *Chemical Society Reviews*, 36, 492–506, 2007. With permission from The Royal Society of Chemistry.)

separation channel via two side inlets where the particles are seen laminated against the sidewalls. In Figure 44.7b, the acoustic switching is in operation showing 8 μm PMMA beads being routed to the center outlet and 3 μm polystyrene beads directed to the side outlets.

44.4.2 Applications of Ultrasonic Standing Wave Microresonators to Blood Component Handling

44.4.2.1 Lipid Microemboli Separation

The Lund-method has been developed to address several clinical needs in blood component handling where the direct visual access to the quality of separation has been instrumental for the platform development. In thoracic surgery, a substantial loss of blood commonly occurs resulting in shed blood being recovered from the chest cavity for retransfusing the blood to the patient. As the shed blood is contaminated by tissue debris from the surgery as well as triglycerides, leaking from adipose tissue undergoing surgery, the blood has to be purified before returning the blood to the patient. Normally a mechanical filter with cutoff of about 40 μm serves as the main purification step. Centrifugal instrumentation developed for this purpose is also available. In spite of these measures, a substantial amount of triglycerides in the form of lipid microdroplets still prevail in the retransfused blood. An increasing awareness now pinpoints lipid microemboli as an important source of cognitive dysfunction in conjunction with major surgery where autotransfusion of shed blood is standard. The lipid microemboli that cause cognitive dysfunction are in the size range of approximately 8 μm and larger. As neither mechanical filters nor centrifugation eliminate lipid microemboli from shed blood sufficiently well, new means of separation are needed. In view of these needs, acoustic separation has recently emerged as a new potential and mild way of purifying shed blood. The components to separate are erythrocytes and triglyceride microdroplets and as discussed earlier, a correctly tuned acoustic standing wave binary separation mode should be feasible. When analyzing fundamental acousto-physical data for the two components, it is evident that the two should be possible to separate in a standard acoustic resonator operated at its fundamental resonance criterion (Table 44.1). The negative acoustic contrast factor, Φ, for triglycerides and the corresponding positive factor

TABLE 44.1
Acousto-Physical Data of Relevant Blood Constituents, Showing a Sign Shift in the Acoustic Contrast Factor, Φ, for Erythrocytes versus Triglycerides

	Density (kg/m^3)	Compressibility (ms^2/kg)	Velocity (m/s)	Φ
Triglycerids	913	5.34E−10	1435	−0.31
Erythrocytes	1100	3.48E−10	1616	0.29
Plasma	1010	4.40E−10	1500	

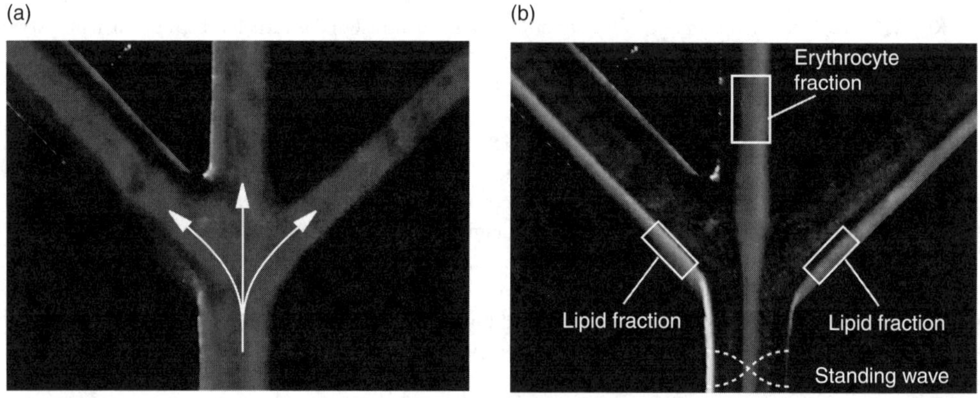

FIGURE 44.8 Proof of principle of lipid particle separation from blood. Milk mixed with blood is shown in the perfused nonactuated chip, left, and the same chip is shown when actuated at the fundamental resonance frequency, 2 MHz, of the resonator channel. The erythrocyte fraction is clearly seen exiting via the center outlet and the lipid fraction (white streaks along the sidewalls) in the side channels.

for erythrocytes shows that erythrocytes should be possible to collect at the center outlet of the a Lund-separator and the lipid microemboli in the side channels.

A proof of concept of this hypothesis was performed on a mixture of blood and milk, where the lipid emulsion in milk modeled the triglyceride suspension in real blood.[41] Figure 44.8 shows a binary separation of erythrocytes from the milk lipid fraction in a mixture of blood and milk. The corresponding studies on blood and triglycerides were reported by Petersson et al.[42] demonstrating triglyceride microdroplet removal from blood at efficiencies as high as 95%. Figure 44.9 shows the triglyceride elimination efficiency as measured on a radioactively labeled triglyceride emulsion (0.3 μm droplet size) spiked at 1% level in blood. The separation was performed at a flow rate of 300 μL/min.

Clinical requirements on a separation system for autolog blood washing and transfusion request throughputs of at least 1000 mL/h. In view of the limited throughput of the reported acoustic lipid elimination microsystem, a clinial implementation of this separation strategy seems, at a first glance, not feasible. However, when considering the potential of microfabrication to generate large arrays of identical acoustic separation channels, this can very well be accomplished. A clear benefit of the Lund-method is that multiple channels may be actuated simultaneously by the same transducer, which reduces the requirements on the actuation unit in an instrumental setup for clinical purposes.

An effort to demonstrate the possibility of upscaling the throughput was presented by Jönsson et al.[43] Eight parallel separation channels, 350 μm wide and 125 μm deep each, were connected in

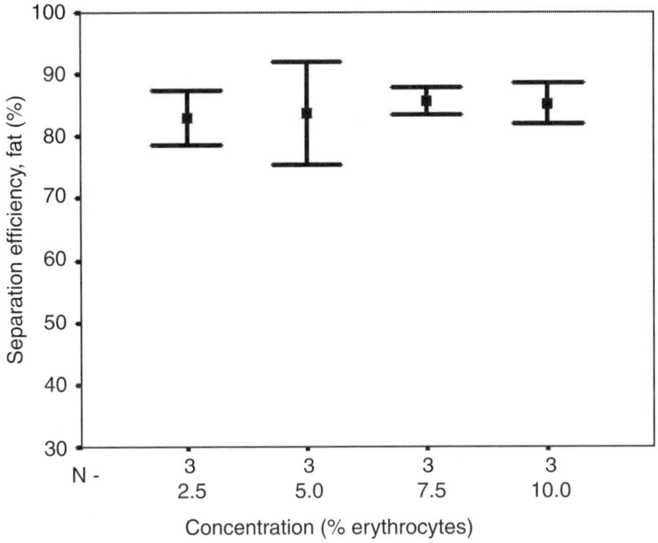

FIGURE 44.9 Acoustic separation of lipid particles from blood at varying erythrocyte concentrations. Flow rate = 300 μL/min. (Reproduced from Petersson F., et al., *The Analyst*, 129, 938–943, 2004. With permission from Royal Society of Chemistry.)

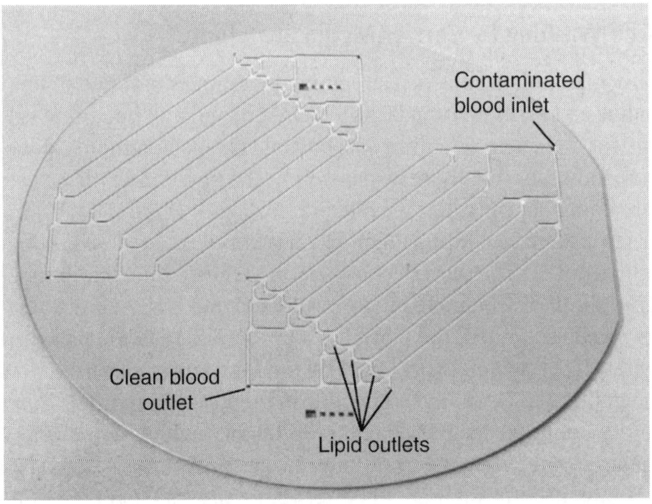

FIGURE 44.10 To increase the throughput of the acoustic separation device, eight parallel channels were connected in a bifurcation network. The image shows two eight channel separators on a 3″ silicon wafer.

a microfluidic bifurcation structure such that the chip was supplied with a single inlet for blood, a single outlet for purified blood, and a set of outlets at the chip backside for the triglyceride fraction. The separator was operated at a flow rate of 60 mL/h. Triglyceride elimination was recorded to range up to 90%. It should also be noted that the processing of the blood in the acoustic resonator did not induce any measurable hemolysis above the background already present in the inlet blood fraction. Figure 44.10 shows an early version of two eight-channel separator structures connected in bifurcation structures on a 3″ silicon wafer.

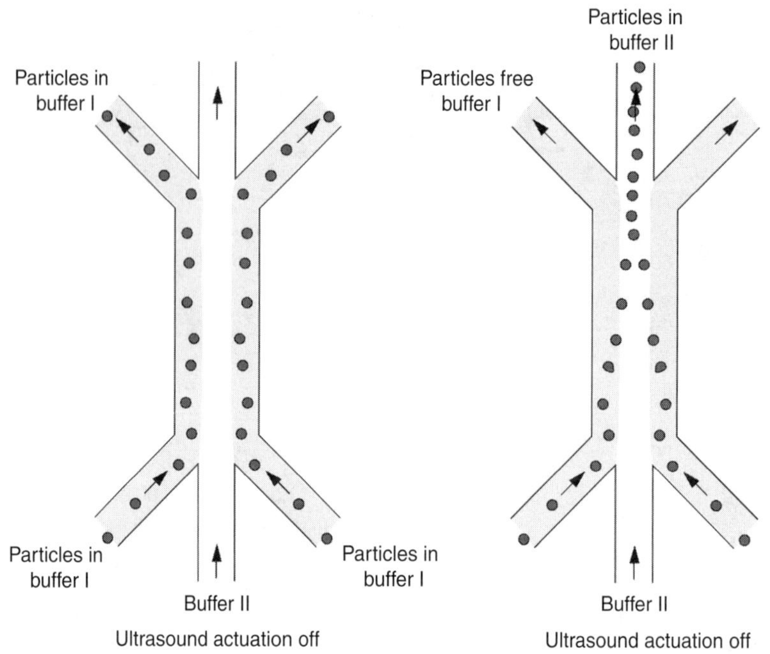

FIGURE 44.11 Acoustic switching of particles between two buffer systems.

44.4.2.2 Blood Cell Washing by Carrier Media Switching

The ability to spatially manipulate cells in a streaming medium as outlined in this chapter opens new routes to perform online cell manipulation tasks. A fundamental unit operation is to be able to change the buffer conditions for a particle or cell in suspension. This is commonly done by centrifugation, removal of the supernatant, and finally resuspension in the new buffer. These steps can be directly implemented in a streaming acoustic half wavelength resonator (Figure 44.11) by allowing particles suspended in buffer I to enter a separation channel via two symmetrical side inlets to a main channel that is provided with buffer II. The particle suspension is thus laminated in its original buffer fluid along the sidewalls of the flow channel. Figure 44.11 schematically shows this setting and as the ultrasonic standing wave is activated, the particles start to migrate to the pressure node in the centre of the channel (Figure 44.11), which is occupied by the buffer II stream.

This mode of operation has been investigated using the Lund-separator to wash blood contaminated by inflammatory components and coagulation factors, which is the case for drainage blood collected from patients in the intensive care units in the postsurgery state. Normally, this blood is centrifuged and retransfused to the patients after removing the supernatant and resuspension in plasma replacement fluid. Petersson et al.[44] demonstrated an acoustic resonator microchip with a channel configuration as outlined in Figure 44.11. Figure 44.12 shows a close up of the outlet from such a separation chip where a model system with white polyamide particles, 5 μm, are suspended in a buffer together with the colour compound, Evans blue. The nonactuated chip, Figure 44.12a, shows the particles exiting the chip together with the original blue coloured buffer via the side outlets. As the acoustic actuation is turned on, the particles are immediately transferred into the uncoloured buffer stream in the centre of the channel, see Figure 44.12b.

Blood washing using the proposed particle switching strategy demonstrated wash efficiencies up to 95%. To meet the needs for throughput, the same approach to parallel channel design as proposed for the lipid microemboli removal chip can be implemented. Switching of particles between different media is a powerful contact-free microfluidic unit operation for cell manipulation on chip where, for example, protocols for compound testing and receptor activation ranges can be titrated.

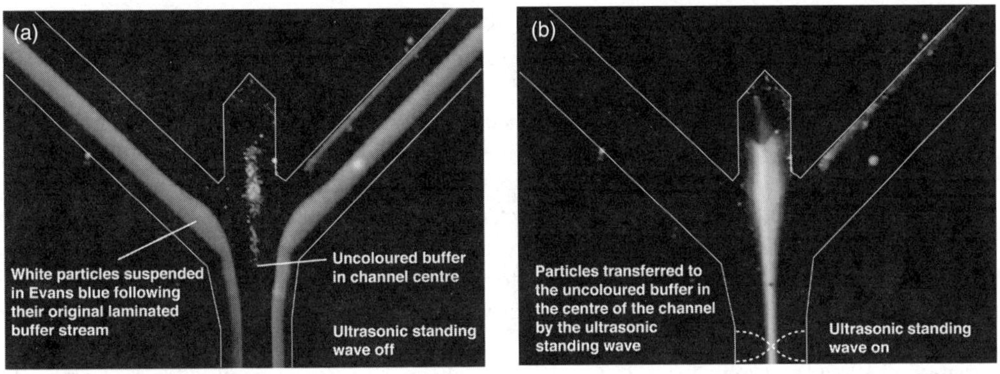

FIGURE 44.12 (a) White particles suspended in a blue colour compound, Evans blue, are seen to follow their original buffer stream to the side outlets of the separation chip. As the chip is actuated at the resonance frequency, 2 MHz, of the resonator channel, the particles are forced by the axial primary radiation force into the central uncoloured buffer stream and exits the chip via the centre outlet (b).

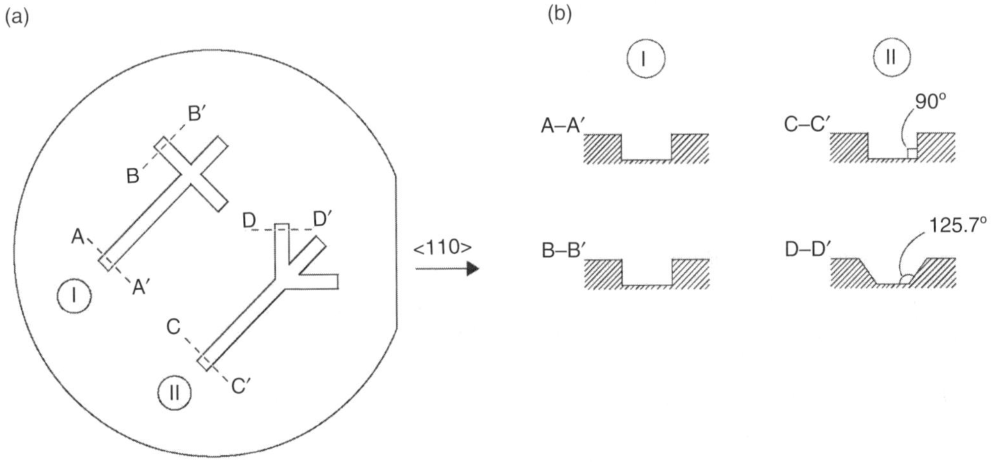

FIGURE 44.13 Schematic of the layout for a separation chip fabricated in standard <110>- oriented silicon with either x-shaped (I) or y-shaped (II) channel outlet. (a) Mask alignment. Reproduced from Nilsson A., et al., *Lab on a Chip*, 4, 131–135, 2004. With permission from Royal Society of Chemistry. (b) Channel cross-sections. Reproduced from Nilsson A., et al., *Lab on a Chip*, 4, 131–135, 2004. With permission from Royal Society of Chemistry.

44.4.3 Guidelines to a Standard Lund-Separator

A separation chip, according to the Lund-method, is most easily made by anisotropic etching of a ⟨100⟩-oriented silicon wafer. The channel mask should be aligned 45° offset to the ⟨110⟩-cut phase of the wafer, Figure 44.13a, to accomplish the desired rectangular channel cross-section, Figure 44.13b.

After performing the photolithography and etching of the 1 μm surface oxide on the wafer in buffered hydrofluoric acid (HF), the flow channel is etched in KOH (potassium hydroxide, Merck KGaA, 64271 Darmstadt, Germany), 40 g in 100 mL deionised water to the desired channel width and depth. Normal etch rates are approximately 1 μm/min. After terminating the anisotropic etching, the remaining surface oxide is stripped in HF, washed and blown dry. Inlet and outlet holes should be etched from the backside of the wafer to provide an inspection area at the channel side, free from connecting tubings. The chip is sealed by anodic bonding of a Pyrex glass lid to the channel side of the chip. Finally, silicone rubber tubings are glued to the backside of the wafer with standard silicone rubber elastomer.

The ultrasonic transducer was a PZ26 piezoelectric ceramic disc (Ferroperm Piezoceramics A/S, Kvistgard, Denmark). The tranducer should be designed to operate in the region of the resonance frequency of the resonator channel.

The piezoelectric element is glued to the backside of the chip with epoxy (2 Ton Clear Epoxy, ITWDevcon, Danvers, MA, USA). Alternatively, a small aliquot of ultrasonic coupling gel can be clamped between the transducer and the resonator chip (Figure 44.4a). Normal actuation voltages range between 5 and 20 V for performing a standard acoustic particles-focusing experiment. A power amplifier may be required for the driving of the piezoelectric element. Flow rates are normally in the range of 10–200 μL/min for a 350 × 125 μm channel (width × depth). Fluid is preferably aspirated through the chip, that is, syringe pumps are connected to the chip outlets, drawing fluid from reservoirs in which the inlet tubing is immersed.

44.4.4 Acoustic Trapping

As have been mentioned earlier, trapping of particles will occur in an acoustic standing wave field if the retaining forces from the acoustic field are larger than the drag forces exerted from a surrounding fluid flow (Equations 44.1 and 44.4). The standing wave field may either be generated along the fluid flow lines for longitudinal trapping or transversal to the fluid flow lines for lateral trapping. According to the theory given earlier in this chapter, the trapping force is stronger in the longitudinal case compared to the lateral for a given acoustic intensity.

44.4.4.1 Longitudinal Trapping

An example of a system using longitudinal trapping is shown in Figure 44.14. An 8.5 MHz focused PZT-transducer, 20-mm in diameter, is used together with a molybdenum reflector to generate a standing wave inside a quartz capillary.[45] Figure 44.15 shows trapped 4.7 μm latex spheres inside a 75 μm i.d. capillary. The standing wave pattern is clearly visible after the particles have been trapped. The system was used to demonstrate size selective separation, where 4.7 μm latex particles were retained in the capillary while the majority of 3.0 μm particles passed through the trap. This method generates several trapping points along the capillary and it is quite difficult to selectively trap particles in a certain position. The advantage is that the primary axial radiation force used will generate a stronger trapping force for a given acoustic intensity, as discussed earlier. Alternatively, the acoustic intensity can be decreased while maintaining trapping.

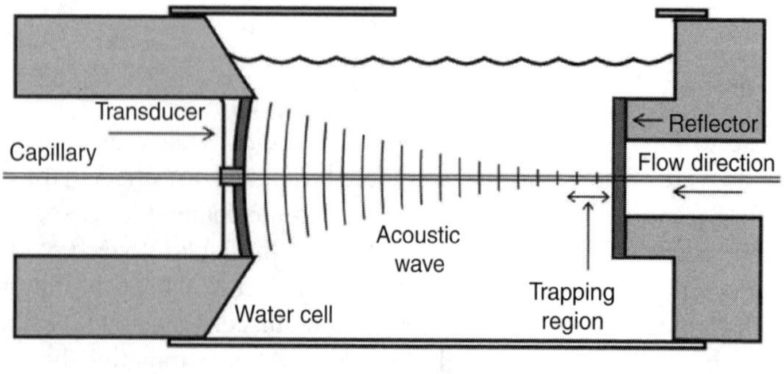

FIGURE 44.14 Experimental setup for the longitudinal capillary ultrasonic trap. (Reproduced from Wiklund M., et al., *Ultrasonics*, 41, 329, 2003. With permission.)

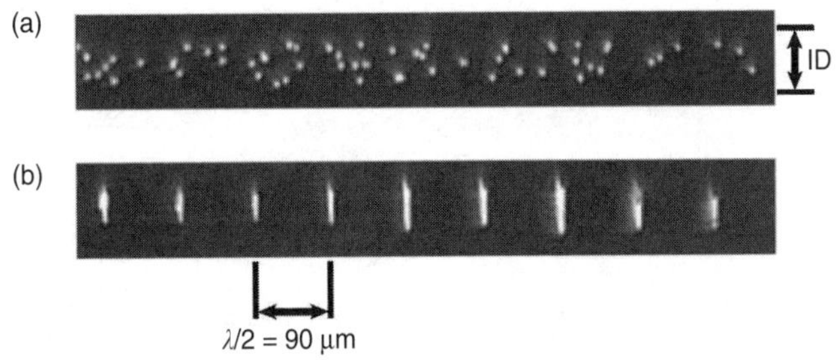

FIGURE 44.15 Fluorescence images of 4.7 μm latex particles inside the 75 μm I.D. capillary without (a) and with (b) activated transducer. (Reproduced from Wiklund M., et al., *Ultrasonics*, 41, 329, 2003. With permission.)

The same group has also presented an alternative way to couple acoustic energy into a microfluidic chip using a flat transducer and a refractive element placed on top of the chip.[46] This setup is used in combination with dielectrophoretic (DEP) forces with an outlook to combined manipulation of bioparticles or individual cells using the two forces simultaneously.

44.4.4.2 Lateral Trapping

In the case of lateral trapping, the standing wave is applied perpendicular to the fluid flow through the trapping device. A multilayer resonant structure is commonly used to generate the standing wave (cf. Figure 44.2). Basically, two different transducer configurations have been presented in the literature. The transducer is either connected via an intermediate coupling layer to the trapping cavity or is in direct contact with it. The trapping cavity is normally sealed with a glass lid acting as both acoustic reflector and inspection window. The trapping cavity height should be a multiple of $\lambda/2$, depending on the number of desired trapping positions. The glass resonator should be an odd number of $\lambda/4$ in thickness and if an intermediate coupling layer is used between the transducer and the trapping cavity, this is normally chosen to have a thickness of $\lambda/4$.

A transducer with an intermediate coupling layer was used in a setup by Bazou et al. (Figure 44.16).[47] A circular resonator was designed for cell manipulation experiments. A 1.5 MHz disc transducer, 12 mm in diameter, was glued to a steel coupling layer ($3\lambda/4$ thick) forming the bottom of the resonator. The resonator cavity (height: 0.5 mm) was sealed by a 1 mm quartz glass lid and the device was driven at 1.57 MHz, corresponding to the resonance frequency of the resonator cavity. The cell suspension was filled into the resonator and the fluid flow was stopped as the ultrasound was activated. The cells moved into the resonator nodal plane within 1 s (Figure 44.17).

Wiklund and coworkers[48] used a similar setup with a glass-liquid-glass resonator activated by a 5 mm in diameter external transducer (Figure 44.18). The channel height ($\lambda/2$, 260 μm) was defined by a polydimethylsiloxane (PDMS) spacer. The glass coupling layer and the glass reflector were both chosen to have a thickness of $\lambda/4$, 550 μm. The system was run at a frequency close to 3 MHz and was used for investigation of adherent COS-7 cell viability after being exposed to the ultrasound in the trap under continuous perfusion with cell medium. No adverse effect was found for exposure times up to 75 min.

44.4.4.3 Integrated Transducers

Trapping devices may also be designed without the coupling layer between the transducer and the resonance chamber, that is, the transducer is in direct contact with the liquid in the chamber. This has been demonstrated by, for example, Spengler et al.[49] A 5 × 15 mm transducer was used in contact with a 5 mm wide channel formed by a 0.8 mm polytetrafluoroethylene (PTFE) spacer. The channel height corresponded to one wavelength, that is, two pressure node planes were generated

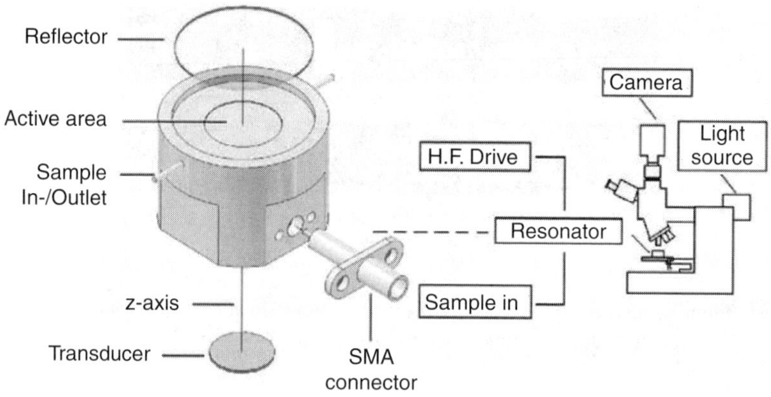

FIGURE 44.16 Acoustic resonator for cell trapping experiments. Its main components were a 1.5 MHz disk transducer that was glue-attached to a steel acoustic coupling layer, a sample volume, and glass acoustic reflector. The thicknesses of the different layers were selected to give a highly resonant system. (Reproduced from Bazou D., et al., *Ultrasound in Medicine and Biology*, 31, 423, 2005. With permission.)

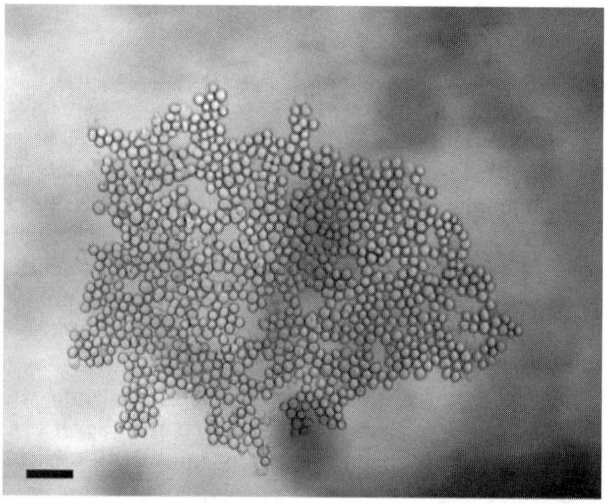

FIGURE 44.17 Neural cells suspended in the ultrasound trap 30 min after initiation of ultrasound. Scale bar is 70 μm. (Reproduced from Bazou D., et al., *Ultrasound in Medicine and Biology*, 31, 423, 2005. With permission.)

when driven at 1.93 MHz. The channel was closed by a 0.75 mm thick quartz glass lid acting as an acoustic reflector.

A similar configuration using miniaturized transducers was demonstrated by Lilliehorn et al.[50] Transducer elements measuring 0.8 × 0.8 mm were mounted on a printed circuit board (PCB) and cast in epoxy that was polished down to the upper surface of the transducers (Figure 44.19). A cavity was drilled in the PCB underneath the transducer to ensure air backing of the transducer, which is important for the operation of the resonant system with a minimum off loss (high Q-value). The flow channels ($\lambda/2$ height) were structured on a soda-lime glass plate using a photosensitive polymer, SU-8, for defining the walls. The glass plate acted as a reflector and the thickness was 1.55 mm that corresponded to an odd number of $\lambda/4$. The system was operated at around 10 MHz. A higher frequency is advantageous to use since the forces acting on the particles will be larger for a given acoustic intensity (Equation 44.1). When studying the resulting trapping behavior of the device, it was

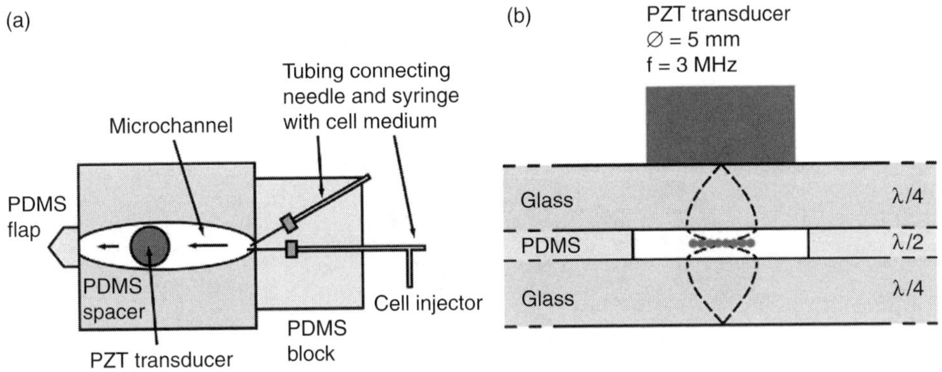

FIGURE 44.18 (a). Top view of the glass–PDMS–glass microfluidic chip with an integrated PDMS block containing two separate inlets with tubing and needles. (b). Cross-section of the microfluidic chip showing the circular 3-MHz PZT transducer and the three-layered structure. (Reproduced from Hultstrom J., et al., *Ultrasound in Medicine and Biology*, 33, 145, 2007. With permission.)

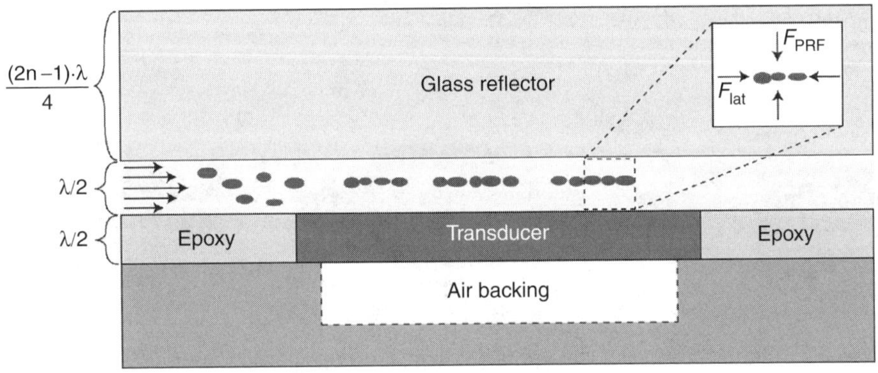

FIGURE 44.19 A side view schematic of the microfluidic trapping device using miniature transducers. The acoustic forces trap the particles in clusters in the center of the channel as illustrated in the insert. (Reproduced from Evander M., et al., *Analytical Chemistry*, 79, 2984–2991, 2007. With permission.)

seen that the 5 μm polyamide particles were trapped in several clusters above the transducer surface in the centre of the channel, as indicated in Figure 44.19. This was theoretically and experimentally shown to be a result of the nearfield pattern in the acoustic intensity generated by the transducer (Figure 44.20). The acoustic field varies spatially across the transducer surface since the wavelength of the acoustic signal is much shorter than the dimensions of the transducer, which results in an interference pattern.

44.4.4.4 Bioassays, Dynamic Arraying

The intention of the device presented by Lilliehorn et al.[50] was to utilize the miniaturized transducers in combination with the laminar flow conditions in microfluidic systems to create a dynamic arraying device holding several individually controlled transducers as envisioned in Figure 44.21.[51] Here, an array of 16 transducers are loaded with different functionalized beads in a first step. The beads are acoustically trapped in the positions given by the transducers. In a second step, samples are supplied through the orthogonal channels and the response is, for example, read by fluorescence. The transducers are then deactivated and the beads are flushed out of the device and the whole procedure is repeated again for a new set of beads and samples.

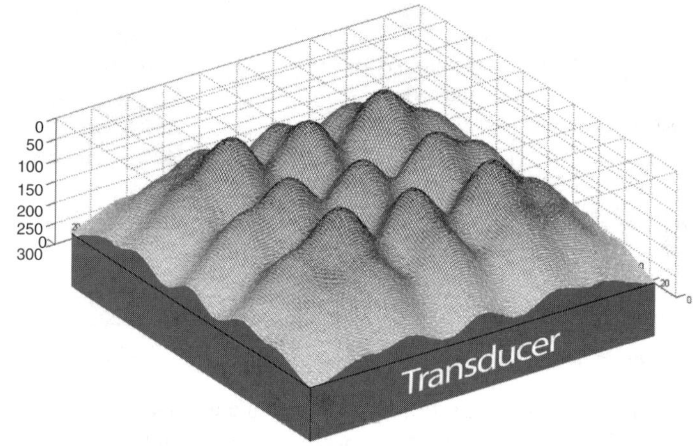

FIGURE 44.20 Spatial variation of the acoustic field across the transducer surface. The actual trapping positions are given by the near field pressure distribution as shown in the 3D-image. Particles will be trapped in clusters at the local pressure minima, indicated by the peaks in the figure. (Reproduced from Evander M., et al., *Analytical Chemistry*, 79, 2984–2991, 2007. With permission.)

FIGURE 44.21 Illustrations of the concept of dynamic arraying showing insertion of the solid phase of different specificity (a) through inlets (A, B...X), trapping of bead clusters using the ultrasonic transducer array (b) and perfusion of sample (c) through inlets (A, B...Y) followed by fluorescence read-out. (Reproduced from Lilliehorn T. et al., *Sensors and Actuators B*, 106, 851, 2005. With permission.)

A simplified model system is shown in Figure 44.22 using three (550 × 550 μm) transducers. A channel structure (width 600 μm, height 61 μm) incorporating hydrodynamic focusing of the sample inlet was fabricated by wet etching of glass. Beads were trapped from a continuous flow and moved between the different transducers by activating them in a sequence. A simple bioassay was performed by trapping 6.7 μm biotinylated polystyrene beads and perfusing them with fluorescently tagged avidin through the orthogonal channels (marked "Analyte inlet" in Figure 44.22).

44.4.4.5 Cell Trapping and Culturing

There is a clear trend today within the bioanalytical and biomedical fields toward more frequent use of cell-based studies. The dimensions of microfluidic systems are well matched to meet the demand on cell-based systems. Still, new methods are needed that can efficiently handle and manipulate cells in those formats. Examples have already been given in this chapter where acoustic forces are used to trap and manipulate cells. The device in Figure 44.22 has been further developed for use in cell-based bioassays.[52] The temperature characteristics of the device have been examined to be able to control the temperature during the cell experiments. The major source of heat in the acoustic resonance systems presented here is the power dissipation in the transducer itself. The power dissipation follows a

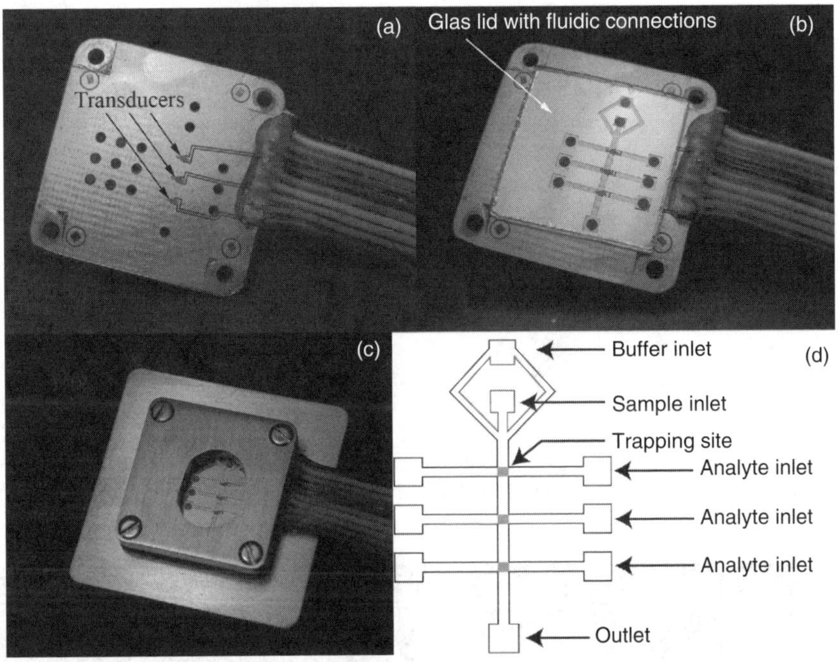

FIGURE 44.22 The microfluidic acoustic resonator is based on a PCB with three miniature PZT-transducers, 550 × 550 μm (a). The PCB provides fluidic and electric connections to the transducers. A glass lid with microfluidic channels placed over the PCB defines the resonator cavity over each transducer (b), and the entire assembly is fixed by a brass holder (c). A schematic of the channels with the transducers, that is, trapping sites, marked with grey (d). (Reproduced from Evander M., et al., *Analytical Chemistry*, 79, 2984–2991, 2007. With permission.)

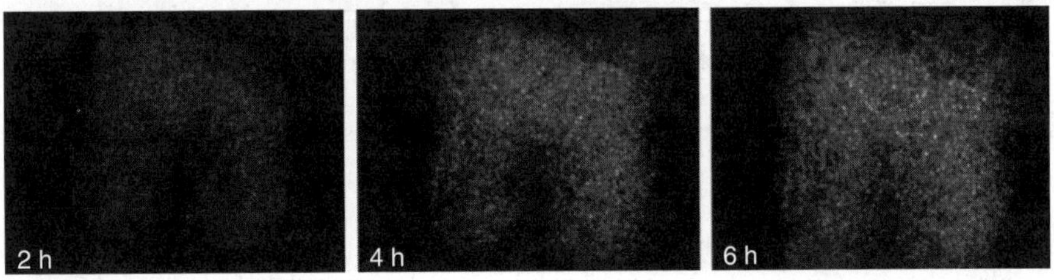

FIGURE 44.23 Growth of yeast cells trapped in the acoustic device while being perfused with cell medium. The images show the increase of the number of cells in the cell cluster after 2, 4, and 6 h of cultivation. (Reproduced from Evander M., et al., *Analytical Chemistry*, 79, 2984–2991, 2007. With permission.)

quadratic dependence on the drive voltage. Experiments have shown that it is possible to select a drive voltage amplitude that gives both good trapping conditions and an appropriate temperature profile for cell-based experiments. Figure 44.23 shows the results from growing yeast cells in the trapping device. The cells were cultivated for 6 h under continuous perfusion of cell medium. Images were recorded every hour to follow the growth of the cells.

44.4.4.6 Cell Enrichment

An interesting application of the acoustic trap is to use it for enrichment of particles/cells from a dilute sample. The origin may be a rare event experiment or a sample where everything but

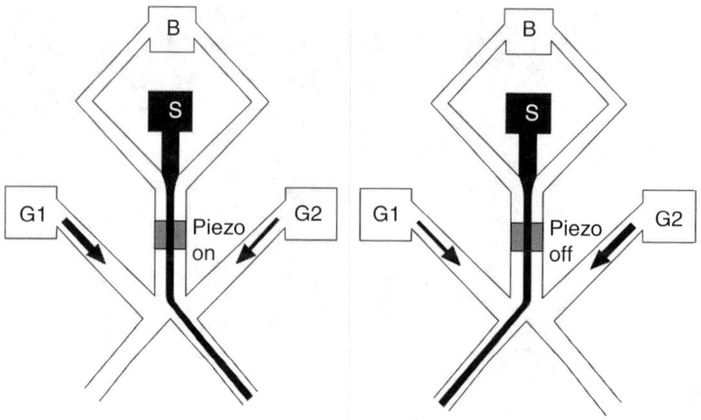

FIGURE 44.24 Channel structure with the sample inlet, S, and the hydrodynamically focusing buffer inlet, B, together with with the valve-less switching inputs G1 and G2.

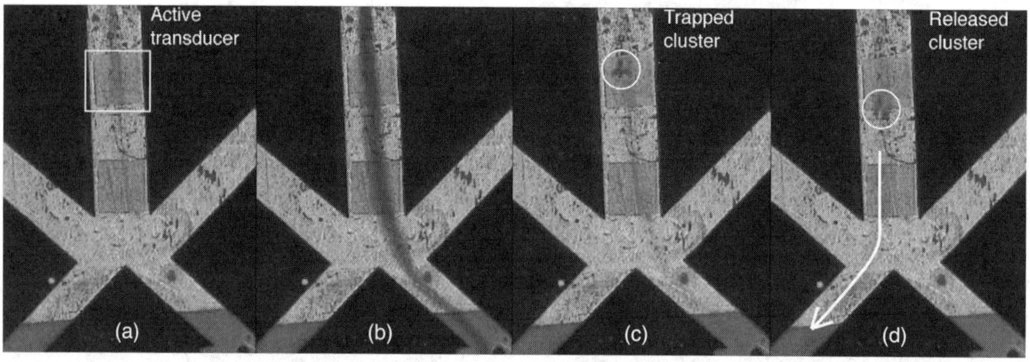

FIGURE 44.25 Principle of cell enrichment using the acoustic trapping. (a) The tranducer is activated, (b) the sample is supplied, (c) the particles/cells in the sample liquid are trapped at the transducer, and (d) the switching flow is switched and the transducer is deactivated. The trapped particles/cells are released to the exit down left.

the cells/particles of interest have been lysed or dissolved. This has been demonstrated using the transducer plate shown in Figure 44.22a together with new design of the glass channel system, shown in Figure 44.24. Hydrodynamic focusing is used for the sample inlet (Inlet B and S) to ensure high trapping efficiency.[52] Two inlets have been added (G1 and G2) for valveless switching of the flow between the two outlets below. If the flow in G1 is significantly higher than the flow in G2, the sample/buffer flow will be directed to the right outlet and vice versa. The channel dimensions at the transducer were 900 μm width and 70 μm height. The system was operated at 10.8 MHz.

The operation of the device is shown in Figure 44.25 using polystyrene beads in a dye as model sample fluid. The transducer is activated and the hydrodynamically focused sample flow is supplied. The switching flows (inputs G1 and G2) are set for guiding the sample flow to the right outlet after passing the transducer. The particles/cells in the sample flow meeting the conditions for trapping are retained at the transducer in a cluster (Figure 44.25c). After optional washing of the cluster, the flows in inlets G1 and G2 are switched to address the left output and the transducer is deactivated. The trapped cluster is released and supplied to the output for further processing.

This system has been used for sample preparation for enhanced analysis of sexual assault evidence.[53] A sample containing sperm cells and lysed female epithelial cells was infused into the device during 5 min. The sperm cells were trapped at the transducer. After finished infusion, the cells were washed with PBS for another 5 min, then released and collected at the outlet for qPCR analysis. The original sample contained 5% male fraction and 95% female fraction. After processing in the trapping device, the collected liquid at the female outlet (down right in Figure 44.25) contained 0.5% male fraction and 99.5% female, while the collected liquid at the male outlet (down left in Figure 44.25) contained 85% male fraction and 15% female fraction, that is, a 17-fold enrichment of the male fraction.

44.5 CONCLUSIONS

Acoustic standing wave manipulation of particles in microfluidic systems offers the implementation of a wide range of particle/cell handling unit operations without introducing any moving parts, where binary separations, buffer media switching, particles sizing, cell enrichment, and cell trapping can be accomplished. Cell manipulation in microfluidic systems using ultrasonic standing waves have proven to be a mild and yet robust technique well suited for performing chip-integrated cell handling tasks. In contrast to dielectrophoretic and optical tweezer technology, higher throughputs are offered in acoustically controlled systems. The low device complexity and the fact that changes in ionic strength of the carrier buffer does not affect the acoustic forces induced on the cells to be handled make acoustic separators an especially attractive platform for cell handling in their native body fluids and in long-term experiments with, for example, gradient perfusion systems.

With the rapid development of cell-based microfluidic systems in the lab-on-a-chip field, ultrasonic standing wave manipulation and control of cells in microspace can be anticipated to play an increasing role in future microfluidic systems.

REFERENCES

1. Kittisland G., Stemme G., and Nordén B., A submicro particle filter in silicon, *Sensors and Actuators*, A21–A23, 904–907, 1990.
2. Brody J.P., Osborn T.D., Forster F.K., and Yager P., A planar microfabricated fluid filter, *Sensors and Actuators A*, 54, 704–708, 1996.
3. Di Carlo D., Wu L.Y., and Lee L.P., Dynamic single cell culture array, *Lab Chip*, 6, 1445–1449, 2006.
4. Yamada M. and Seki M., Microfluidic particle sorter employing flow splitting and recombining, *Analytical Chemistry*, 78, 1357–1362, 2006.
5. Yamada M. and Seki M., Microfluidic device for continuous and hydrodynamic separation of blood cells, in *Micro Total Analysis Systems, Proceedings of μTAS 2006 Conference*. T. Kitamori, H. Fujita, and S. Hasabe (Eds.), Society for Chemistry and Micro-Nano Systems, Tokyo, Japan, 1052–1054, 2006.
6. Huang L.R., Cox E.C., Austin R.H., and Sturm J.C., Continuous particle separation through deterministic lateral displacement, *Science*, 304, 987–990, 2004.
7. Davis J.A., Inglis D.W., Morton K.J., Lawrence D.A., Huang L.R., Chou S.Y., Sturm J.C., and Austin R.H., Deterministic hydrodynamics: Taking blood apart, *Proceedings of the National Academy of Sciences of the United States of America*, 103, 14779–14784, 2006.
8. Ashkin A., Dziedzic J.M., Bjorkholm J.E., and Chu S., Observation of a single-beam gradient force optical trap for dielectric particles, *Optics Letters*, 11, 288–290, 1986.
9. Block S.M., Making light work with optical tweezers, *Nature*, 360, 493–495, 1992.
10. Huang Y. and Pethig R., Electrode design for negative dielectrophoresis, *Measurement Science and Technology*, 2, 1142–1146, 1991.
11. Markx G.H., Talary M.S., and Pethig R., Separation of viable and nonviable yeast using dielectrophoresis, *Journal of Biotechnology*, 32, 29–37, 1994.
12. Schnelle T., Hagedorn R., Fuhr G., Fiedler S., and Muller T., Three-dimensional electric field traps for manipualtion of cells—calculation and experimental verification, *Biochimica et Biophysica Acta*, 1157, 127–140, 1993.

13. Fiedler S., Shirley S.G., Schnelle T., and Fuhr G., Dielectrophoretic sorting of particles and cells in a microsystem, *Analytical Chemistry*, 70, 1909–1915, 1998.
14. Cummings E.B. and Singh A.K., Dielectrophoresis in microchips containing arrays of insulating posts: Theoretical and experimental results, *Analytical Chemistry*, 75, 4724–4731, 2003.
15. Rosenthal A. and Voldman J., Dielectrophoretic traps for single-particle patterning, *Biophysical Journal*, 88, 2193–2205, 2005.
16. Rosenthal A., Taff B.M., and Voldman J., Quantitative modeling of dielectrophoretic traps, *Lab on a Chip*, 6, 508–515, 2006.
17. Vahey M.D. and Voldman J., Iso-dielectric separation: A new method for the continuous-flow screening of cells, in *Micro Total Analysis Systems, Proceedings of μTAS 2006 Conference*. T. Kitamori, H. Fujita, and S. Hasabe (Eds.), Society for Chemistry and Micro-Nano Systems, Tokyo, Japan, 1052–1054, 2006.
18. Pamme N. and Manz A., On-chip free-flow magnetophoresis: Continuous flow separation of magnetic particles and agglomerates, *Analytical Chemistry*, 76, 7250–7256, 2004.
19. Pamme N. and Wilhelm C., Continuous sorting of magnetic cells via on-chip free-flow magnetophoresis, *Lab on a Chip*, 6, 974–980, 2006.
20. Kundt A. and Lehmann O., Ueber longitudinale Schwingungen und Klangfiguren in cylindrischen Flüssigkeitssäulen, *Annalen der Physik und Chemie*, 153, 1, 1874.
21. King L.V., On the acoustic radiation pressure on spheres, *Proceedings of the Royal Society of London*, A147, 212–240, 1934.
22. Gorkov L.P., On the forces acting on a small particle in an acoustical field in an ideal fluid, *Soviet Physics Doklady*, 6, 773, 1962.
23. Limaye M.S., Hawkes J.J., and Coakley W.T., Ultrasonic standing wave removal of microorgansims from suspension in small batch systems, *Journal of Microbiological Methods*, 27, 211, 1996.
24. Hawkes J.J. and Coakley W.T., A continuous flow ultrasonic cell-filtering method, *Enzyme and Microbial Technology*, 19, 57, 1996.
25. Cousins C.M., Holownia P., Hawkes J.J., Limaye M.S., Price C.P., Keay P.J., and Coakley W.T., Plasma preparation from whole blood using ultraound, *Ultrasound in Medicine and Biology*, 26, 881, 2000.
26. Peterson S., Perkins G., and Baker C., Development of an ultrasonic blood separator, *Proceedings of the IEEE Eighth Annual Conference of the Engineering in Medicine and Biology Society*, 154–156, 1986.
27. Trampler F., Sonderhoff S.A., Pui P.W.S., Kilburn D.G., and Piret J.M., Acoustic cell filter for high-denisty perfusion culture of hybridoma cells, *Biotechnology*, 12, 281–284, 1994.
28. Doblhoffdier O., Gaida T., Katinger H., Burger W., Groschl M., and Benes E., A novel ultrasonic resonance field device for the retention of animal cells, *Biotechnology Progress*, 10, 428–432, 1994.
29. Yasuda K., Umemura S., and Takeda K., Concentration and fractionation of small particles in liquid by ultrasound, *Japanese Journal of Applied Physics Part 1—Regular Papers Short Notes & Review Papers*, 34, 2715, 1995.
30. Benes E., Groschl M., Nowotny H., Trampler F., Keijzer T., Bohm H., Radel S., et al., Ultrasonic separation of suspended particles, *Ultrasonics Symposium, 2001 IEEE*, 1, 649, 2001.
31. Gould R.K. and Coakley W.T., The effects of acoustic forces on small particles in suspension. In *Proceeding of the 1973 Symposium of Finite Amplitude Wave Effects in Fluids*, Bjørno L. (Ed.), IPC, Guilford, UK, pp. 252–257.
32. Yosioka K. and Kawasima Y., Acoustic radiation pressure on a compressible sphere, *Acustica*, 5, 167–173, 1955.
33. Bjerknes V.F.K., in "*Die Kraftfelder*," F. Vieweg, (Ed.), Braunschweig, Germany, 1909.
34. Hawkes J.J. and Coakley W.T., Force field particle filter, combining ultrasound standing waves and laminar flow, *Sensors and Actuators B: Chemical*, 75, 213, 2001.
35. Benes E., Groschl M., Nowotny H., Trampler F., Keijzer T., Bohm H., Radel S., et al., Ultrasonic separation of suspended particles, *Ultrasonics Symposium, 2001 IEEE*, 1, 649–659, 2001.
36. Harris N., Hill M., Shen Y., Townsend R.J., Beeby S., and White N., A dual frequency, ultrasonic, microengineered particle manipulator, *Ultrasonics*, 42, 139, 2004.
37. Harris N.R., Hill M., Beeby S., Shen Y., White N.M., Hawkes J.J., and Coakley W.T., A silicon microfluidic ultrasonic separator, *Sensors and Actuators B: Chemical*, 95, 425, 2003.

38. Nilsson A., Petersson F., Jönsson H., and Laurell T., Acoustic control of suspended particles in micro fluidic chips, *Lab on a Chip*, 4, 131–135, 2004.
39. Petersson F., Åberg L., and Laurell T., Acoustic separation of particles with similar acoustic properties by means of medium density manipulation, in *Micro Total Analysis Systems, Proceedings of μTAS 2006 Conference*. T. Kitamori, H. Fujita, and S. Hasabe (Eds.), Society for Chemistry and Micro-Nano Systems, Tokyo, Japan, 1052–1054, 2006.
40. Siversson C., Petersson F., Nilsson A., and Laurell T., Acoustic particle sizing in microchannels by means of ultrasonic frequency switching, in *Micro Total Analysis Systems, Proceedings of μTAS 2004*. T. Laurell, J. Nilsson, J. D., Harrison, K. Jensen, and J. P. Kutler (Eds.), The Royal Society for Chemistry, Cambridge, UK, 330–332, 2004.
41. Petersson F., Nilsson A., Holm C., Jönsson H., and Laurell T., Separation of lipids from blood ultilizing ultrasonic standing waves in microfluidic channels, *The Analyst*, 129, 938–943, 2004.
42. Petersson F., Nilsson A., Holm C., Jönsson H., and Laurell T., Continuous separation of lipid particles from erythrocytes by means of laminar flow and acoustic standing wave forces, *Lab on a Chip*, 5, 20–22, 2005.
43. Jönsson H., Holm C., Nilsson A., Petersson F., and Laurell T., Separation by ultrasonic standing waves can ameliorate brain damage after cardiac surgery, *Annals of Thoracic Surgery*, 78, 1572–1578, 2005.
44. Petersson F., Nilsson A., Holm C., Jönsson H., and Laurell T., Carrier medium exchange through ultrasonic particle switching in microfluidic channels, *Analytical Chemistry*, 77, 1216–1221, 2005.
45. Wiklund M., Spégel P., Nilsson S., and Hertz H.M., Ultrasonic-trap-enhanced selectivity in capillary electrophoresis, *Ultrasonics*, 41, 329, 2003.
46. Wiklund M., Günther C., Lemor R., Jäger M., Fuhr G., and Hertz H.M., Ultrasonic standing wave manipulation technology integrated into a dielectrophoretic chip, *Lab on a Chip*, 6, 1537, 2006.
47. Bazou D., Kuznetsova L.A., and Coakley W.T. Physical environment of 2-D animal cell aggregates formed in a short pathlength ultrasound standing wave trap, *Ultrasound in Medicine and Biology*, 31, 423, 2005.
48. Hultstrom J., Manneberg O., Dopf K., Hertz H.M., Brismar H., and Wiklund M., Proliferation and viability of adherent cells manipulated by standing-wave ultrasound in a microfluidic chip. *Ultrasound in Medicine and Biology*, 33, 145, 2007.
49. Spengler J.F., Jekel M., Christensen K.T., Adrian R.J., Hawkes J.J., and Coakley W.T., Observation of yeast cell movement and aggregation in a small-scale MHz-ultrasonic standing wave field, *Bioseparation*, 9, 329, 2000.
50. Lilliehorn T., Simu U., Nilsson M., Almqvist M., Stepinski T., Laurell T., Nilsson J., and Johansson S., Trapping of microparticles in the near field of an ultrasonic transducer. *Ultrasonics*, 43, 293, 2005.
51. Lilliehorn T., Nilsson M., Simu U., Johansson S., Almqvist M., Nilsson J., and Laurell T., Dynamic arraying of microbeads for bioassays in microfluidic channels, *Sensors and Actuators B*, 106, 851, 2005.
52. Evander M., Johansson L., Lilliehorn T., Piskur J., Lindvall M., Johansson S., Almqvist M., Laurell T., and Nilsson J., Noninvasive acoustic cell trapping in a microfluidic perfusion system for online bioassays. *Analytical Chemistry*, 79, 2984–2991, 2007.
53. Evander, M., Using acoustic differential extraction to enhance analysis of sexual assault evidence on a valveless glass microdevice, in *Micro Total Analysis Systems, Proceedings of μTAS 2006 Conference*. T. Kitamori, H. Fujita, and S. Hasebe (Eds.), CHEMINAS, Tokyo, Japan, 1055, 2006.

45 Optical Detection Systems for Microchips

James M. Karlinsey and James P. Landers

CONTENTS

45.1 INTRODUCTION

Several detection techniques adapted for the capillary format have been applied to fluidic microchips, including optical, electrochemical, and mass spectrometric detection. While the aim of this chapter is not to review all of these various techniques, interested readers are directed to several recent discussions presented in the literature.[1–6] Instead, the focus will be on optical methods, which will be described briefly along with some examples—the reader is also referred to the complimentary chapter on capillary-based systems by Sweedler in Chapter 9. Because of its broad application in microchip electrophoresis, a significant amount of discussion will be paid to laser-induced fluorescence (LIF) detection, which has played a key role in many of the microchip advancements achieved to date. This discussion will include a practical guide to assembling and aligning an LIF detection setup. Other optical techniques presented include ultraviolet (UV) absorption, chemiluminescence (CL) and electrochemiluminescence (ECL), Raman spectroscopy, thermal lens microscopy (TLM), surface plasmon resonance (SPR), and refractive index (RI) detection.

45.2 OPTICAL DETECTION SYSTEMS

The optical detection methods applied to microchip-based analysis are, for the most part, similar to those employed in capillary electrophoresis (CE). The main differences, however, are that UV absorbance is not as common and that the microchip format creates the possibility of incorporating additional functionality during the fabrication steps (especially in plastic devices). Each of the optical methods described here includes specific requirements, such as the nature of the analyte and the microchip substrate, and these must be taken into consideration when choosing and implementing a system.

45.2.1 LASER-INDUCED FLUORESCENCE

Despite the increase in alternative methods of optical detection, laser-induced fluorescence (LIF) remains the most widely applicable detection method in microchip electrophoresis systems. There are several reasons why LIF is ideally suited to the microchip format; these include (1) its high sensitivity, (2) its compatibility with microchannel dimensions, (3) its fast response time, (4) its noncontact interrogation, and (5) its addressable nature. When Harrison and coworkers[7] initially miniaturized a CE channel onto a planar glass chip in 1992, the fluorescent signals obtained for fluorescein and calcein dyes were used to evaluate the feasibility of performing CE separations on a microchip (Figure 45.1). Since then, LIF detection had been used in many different applications on-chip including single molecule detection and DNA sequencing. A more thorough examination of various aspects of LIF detection, including setup and alignment of a confocal system, can be found later in this chapter.

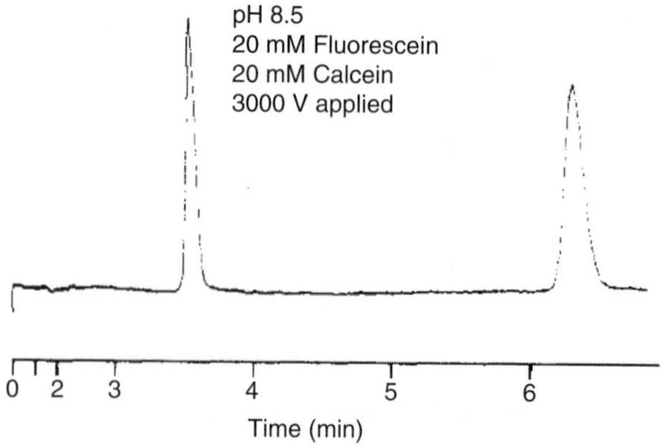

FIGURE 45.1 First published microchip electropherogram using LIF. A mixture of fluorescein and calcein (concentrations shown) were injected for 30 s across a separation channel and subsequently electrophoresed past the detector. (Reprinted from Harrison, D.J., et al., *Analytical Chemistry*, 64, 1926, 1992. Copyright 1992. With permission from American Chemical Society.)

45.2.2 Absorbance

Although it is a universal detector in CE and high pressure liquid chromatography (HPLC), absorbance detection has not found the same success in microchip applications. This is primarily due to the short optical path lengths (i.e., channel depths) and the difficulty in coupling light into and out of the microchip, resulting in detection limits several orders of magnitude higher than LIF. One of the earliest examples of UV detection on-chip was the imaging of an isoelectric focusing (IEF) channel with a charge-coupled device (CCD) camera. Sample was loaded onto a quartz chip and excited at 280 nm using a Xe lamp. With the benefit of the concentrating effect of the IEF step, the detection limit of myoglobin was found to be 30 $\mu g \cdot mL^{-1}$.[8] In most microchip applications, however, there is no concentration step and the short optical path lengths present a challenge. An initial attempt to increase the path length involved the patterning of a silicon flow cell that achieved path lengths up to 5 mm using successive reflections in a channel 50 μm deep. However, there was significant amount of loss at each surface.[9] This was addressed by another group that fabricated a multireflection absorbance cell in a glass microchip using a three-mask process to deposit aluminum mirrors above and below the flow channel. With entrance and exit apertures patterned into the mirrors, optical path lengths of 50–272 μm were obtained in the multireflection absorbance cell (compared to channel depths of 10–30 μm), increasing the sensitivity 5- to 10-fold over single-pass devices.[10] Another approach exploited total internal reflection (TIR) to form a liquid-core waveguide by sandwiching Teflon tubing embedded in a polydimethylsiloxane (PDMS) block between two silica plates containing fluidic channels. By increasing the optical path length to 5 mm while maintaining a detection volume less than 1 μL, the limit of detection (LOD) for crystal violet was found to be ~1 μM.[11]

The different approaches for coupling light into the absorbance cell primarily involve optical fibers and planar waveguides. Using fibers alone, a separation of peptides was shown by Jindal and Cramer[12] using on-chip electrochromatography with a UV absorbance detector. Optical fibers for excitation and collection were positioned above and below a hybrid quartz/PDMS device, which featured an optical path length of only 23 μm, yet resulted in a LOD of 167 μM for thiourea. A similar approach involved embedding a ball lens in a layer of PDMS that sealed against a glass microchannel. An optical fiber brought incident 668 nm light into the lens, which then collected the reflected light from a layer of aluminum patterned on the other side of the channel. Calcium levels were monitored through a reaction with arsenazo III, and a detection limit of 85 μM was found for calcium before applying the assay to clinical urine samples.[13] The alternative approach for coupling light with the absorbance cell was taken by Kutter's group, where fiber couplers and planar waveguides were fabricated into the microchip along with the fluidic channels. Using deep-reactive ion etching (DRIE) to form high-aspect ratio channels in silicon, an effective optical path length of 1.2 mm was obtained in a U-shaped channel (Figure 45.2). By eliminating propagation loss along the waveguides, the detection limit for paracetamol at 254 nm was found to be 3 $\mu g \cdot mL^{-1}$, at present the most sensitive waveguide device in the literature, and a three-component separation was shown on the device.[14] The U-shaped channel design was initially presented by Liang and Harrison[15] in a microchip that featured a dual absorbance and fluorescence cell, with 488 nm light brought into the device with an optical fiber. This work provided an excellent comparison between UV and LIF detection on microchip, as detection limits were found to be 6 μM (for fluorescein isothiocyanate) and 3 nM (for fluorescein), respectively.

45.2.3 Chemiluminescence

Chemiluminescence refers to the release of light energy from a chemical reaction. This is based on the type of reaction first apparent to scientists in the form of "cold" light from fireflies. The firefly's glow mechanism, which hinges on the oxidation of firefly luciferin, is incredibly efficient, as 80% of reacting molecules present generate a photon of light. The inherent benefit of a CL system is that no

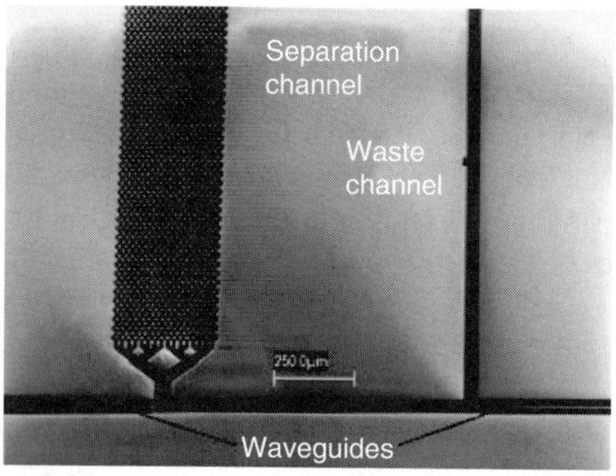

FIGURE 45.2 Scanning electron micrograph of the U-shaped detection region of a microchip designed for absorbance detection using integrated waveguides. (Reprinted from Mogensen, K.B., et al., *Electrophoresis*, 25, 3792, 2004. With permission.)

excitation source is required, making it ideally suited to a miniaturized analysis system. Xu et al.[16] used this feature, along with the benefit of the reduced reagent consumption in a microreactor chip, to monitor chromium III from aqueous samples. A detection limit of 100 nM was obtained in the continuous flow system, where the CL yield was observed to benefit from the short (<1 s) mixing time. Liu et al.[17] examined two model CL systems (the ion-catalyzed luminol-peroxide reaction and the dansyl species conjugated peroxalate-peroxide reaction) using three different designs in PDMS microchips, and sub-μM concentrations were detected. Importantly, the identification of three metal cations in the first system and the chiral recognition of dansyl phenylalanine enantiomers in the second was shown in <1 min, demonstrating the utility of CL detection coupled with a separation mode. Some other analytes examined on microchip using CL detection include dopamine and catechol (detection limits of 20 and 10 μM, respectively),[18] glucose (detection limit of 10 μM),[19] and the cancer marker, *immunosuppressive acidic protein* (detection limit of 100 nM).[20]

45.2.4 Electrochemiluminescence

Electrochemiluminescence detection is similar to CL, except that the luminescent reaction is effected by electrochemical stimulation. Along with the advantages of CL, the stimulation reaction allows the time and position of the light-emitting reaction to be controlled. $Ru(bpy)_3^{2+}$ is one of the most efficient and thoroughly examined ECL molecules and, while the reaction scheme may change depending on the analyte, the excited product is always $Ru(bpy)_3^{2+*}$, which emits light at a maximum intensity at 620 nm when returning to its ground state.[21] In 1998, a fully integrated probe was designed that featured a gold interdigitated microelectrode array (IDA) and a photodiode detector on a silicon microchip to perform ECL. The device was applied to the model system of $Ru(bpy)_3^{2+}$ and tripropylamine, and a 500 nM detection limit was obtained for the ruthenium complex.[22] L'Hostis et al.[23] fabricated an ECL detector using a hybrid silicon and SU-8 photoresist device. The detector (with a platinum IDA) was initially evaluated using a reaction of codeine and $Ru(bpy)_3^{2+}$ (detection limit of 100 μM), and it was then incorporated into an enzymatic microreactor (with a carbon IDA) to assay glucose (detection limit of 50 μM). Applying an ECL detector to a micellar electrokinetic chromatographic (MEKC) separation on a glass microchip, a U-shaped floating platinum electrode was placed across the separation channel to perform both direct and indirect detection.[24] Applying ECL to clinical samples, the antibiotic lincomycin was analyzed using a microchip CE system with an integrated indium tin oxide (ITO) working electrode (Figure 45.3). Without pretreatment, lincomycin was analyzed in urine in <40 s with a detection limit of 9.0 μM.[25] A benefit of using ECL is that

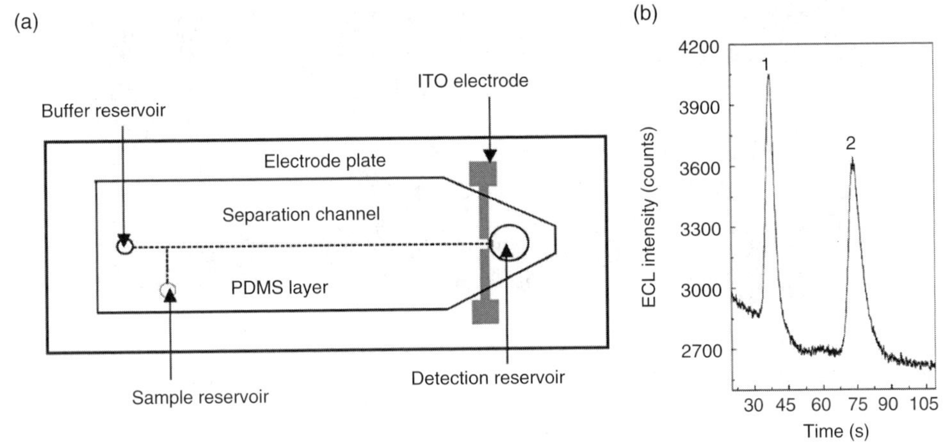

FIGURE 45.3 Microchip design and sample data using ECL detection. (a) The microchip was fabricated in PDMS using a glass mold and the ITO electrodes were patterned with an etch step. (b) The separated mixture contains 100 μM each of lincomycin (1) and clindamycin phosphate (2), which were detected using $Ru(bpy)_3^{2+}$. (Reprinted from Zhao, X., et al., *J Chromatography B*, 810, 137, 2004. With permission.)

it can be combined with EC detection, as demonstrated by Qiu et al.[26] in the determination of the neurotransmitter dopamine and several pharmaceuticals.

45.2.5 Raman Spectroscopy

Raman spectroscopy is similar to IR spectroscopy, but superior in microchip systems because water is virtually Raman transparent. Similar to LIF, sample is typically excited with a laser and spectra can be collected from a very small volume. A real benefit of Raman spectroscopy is the ability to monitor structural changes or the progress of a chemical reaction on chip. Pan and Mathies[24] used a chip-based flow experiment to examine the chromophore structure of rhodopsin photointermediates as well as changes in protein–chromophore interactions in a glass microchip at room temperature. Fletcher et al.[27] used a Raman microscopic spectrometer to observe the formation of ethyl acetate from ethanol and acetic acid in a glass microreactor (Figure 45.4). Using an excitation wavelength of 780 nm, it was shown that the spectral intensities of Raman bands specific to each species were proportional to the concentration present in the microreactor at a given time, providing the time dependence of the product formation. A similar system was used to optimize the catalytic oxidation of isopropyl alcohol to acetone in a continuous flow microreactor.[28] Ramser et al.[29] utilized the microchip environment to trap a single red blood cell and follow the dynamics of an oxygenation cycle after environmental stimulation using Raman spectroscopy in real-time. Of the various optical methods presented in this chapter, Raman spectroscopy possesses the greatest promise as a microchip detection technique, especially considering the signal enhancement (up to 10^{14})[30] that can be obtained performing surface-enhanced Raman spectroscopy (SERS) on a surface that has been coated with silver or gold.

45.2.6 Thermal Lens Microscopy

The Kitamori group has pioneered the use of TLM in microchip systems, where both an excitation and a probe beam are focused into a liquid sample—an example of its utility is provided with immunoassay in Chapter 34. The energy of the excitation beam is absorbed by the sample species and results in a localized temperature increase that affects the refractive index (RI) within the medium. The probe beam, which is selected such that there is no absorption, is subject to the "thermal lens

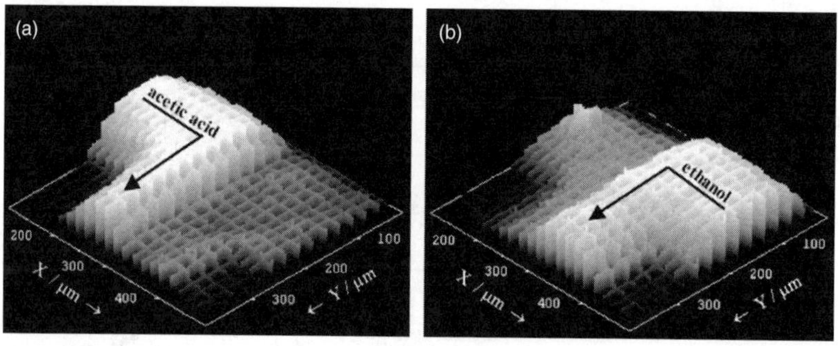

FIGURE 45.4 Three-dimensional plots of Raman intensity where two laminar flow streams come together. The region was probed for specific bonds to identify (a) acetic acid at 893 cm^{-1} and (b) ethanol at 882 cm^{-1}. (Reprinted from Fletcher, P.D.I., et al., *Electrophoresis*, 24, 3239, 2003. With permission.)

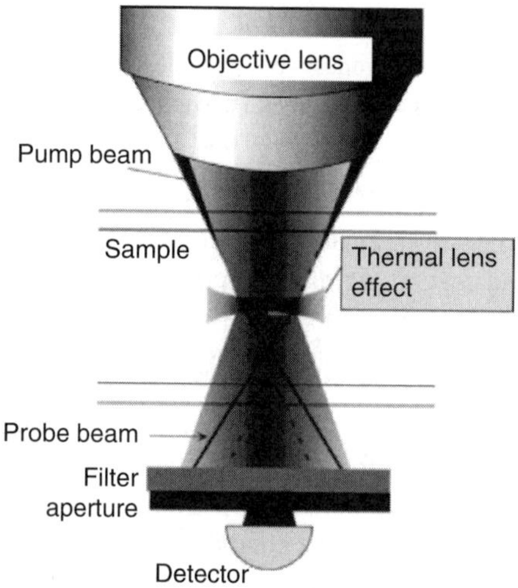

FIGURE 45.5 Schematic illustration of the principle of the thermal lens effect in TLM. (Reprinted from Tokeshi, M., et al., *Electrophoresis*, 24, 3583, 2003. With permission.)

effect" that results from the change in RI due to the increased temperature (Figure 45.5). Because the probe beam is modulated by RI changes, the resulting change in intensity can be recorded with a photodetector. The sensitivity of this detection method has been reported at subsingle-molecule detection for a nonfluorescent species. To demonstrate this, a test sample of Pb(II) octaethylporphyrin was prepared in benzene and analyzed in a 100 μm deep quartz microchannel. The concentrations used corresponded to 0.4–3.4 molecules with a detection limit of 0.34 (and an average temperature rise for a single molecule of 3.1 μK).[31] Another impressive demonstration of the TLM technique was in the two-dimensional imaging of a single cell to monitor the release of cytochrome *c* during the apoptosis process with a spatial resolution of ~1 μm.[32] A review of the advances made by the Kitamori group can be found in the literature,[33] but it should be noted that another group has recently developed a portable thermal lens spectrometer and shown a 30 nM limit of detection for xylenecyanol in a fused silica microchip.[34]

45.2.7 Surface Plasmon Resonance

Surface plasmon resonance is a surface-sensitive method for chemical sensing based on RI changes on the surface of a metal film. Variations in light intensity reflected from the back of the film are detected, thus, no labeling is required. Because the SPR signal is strongest when measuring near the sensor surface, it is often used to detect the surface binding characteristics of biomolecules. Yager and coworkers[35] developed a special microscope for performing SPR detection, and applied it to characterize protein absorption on the microchannel surface of Mylar devices. An amplification technique was then developed using a precipitating enzyme to increase the signal reflectivity by up to 70%.[36] Furuki et al.[37] sputtered gold onto glass before sandwiching a UV resin between glass layers to define a microchannel. The gold surface was then chemically modified with a photobiotin layer to promote the specific adsorption of avidin at a concentration of 25 $\mu g{\cdot}mL^{-1}$, probing with a 670 nm laser diode. PDMS microchips have also been used for SPR, and have been demonstrated for protein binding to a gold surface in both flow-through microchip (using a commercial SPR detector)[38] and microarray[39] formats. A review of SPR technology and applications can be found elsewhere.[40]

45.2.8 Refractive Index Detection

Refractive index detection is based on the factor by which light is slowed down (relative to vacuum) as it travels through a medium. RI is sensitive to changes in temperature, pressure, and flow rate, and, as previously discussed, plays a role in other detection techniques. Before those methods, however, a holographic RI detector was applied to the separation of carbohydrates on chip.[41] In the setup, an incident laser beam passed through a holographic optical element that divided the beam into two. One beam served as a reference and passed solely through the glass substrate while the probe beam passed through the solution in the separation channel. A photodiode array was placed to collect the fringe pattern from the interfering beams, and this pattern changed whenever an analyte passed through the separation channel (Figure 45.6a). Using this system, a proof of principle separation of three sugars was performed in less than 17 s, although their concentration was high (33 mM). An alternative method was demonstrated by Costin and Synovec,[42] who detected the angular deflection of a diode laser beam incident on adjacent laminar flow streams. The beam entered the PDMS microchannel orthogonal to the direction of flow and the diffusion gradient of the two streams, and the resulting angle of the beam corresponded to a RI gradient (Figure 45.6b). The device was evaluated as a molecular weight sensor for poly(ethylene glycol) solutions and was found be sensitive to changes in RI of 4.5×10^{-6}.

45.3 FLUORESCENCE DETECTION

The main considerations when performing LIF detection include the target analyte, the instrumentation, and the microchip, and these will be discussed in the following sections. Some advances in microchip-based LIF will also be presented, including multichannel and multicolor detection, single molecule detection, and integrated optics.

45.3.1 Fluorescent Molecules

While one of the limitations of LIF detection is that few target molecules exhibit native fluorescence, especially biologically relevant proteins and DNA of clinical interest, several different labeling approaches both on- and off-chip have been demonstrated. For the former, the microchip fabrication steps provide the ability to incorporate additional structures into the design, in many cases without adding more steps (or cost) to the fabrication process. Examples include the additional channels and reaction chambers used to perform both precolumn[43] and postcolumn[44] labeling of amino acids.

45.3.1.1 Native Fluorescence

There are few examples in the microchip literature that feature analytes of interest exhibiting native fluorescence, such as the phycobiliproteins present in cyanobacteria and some algae.[45,46] Recently,

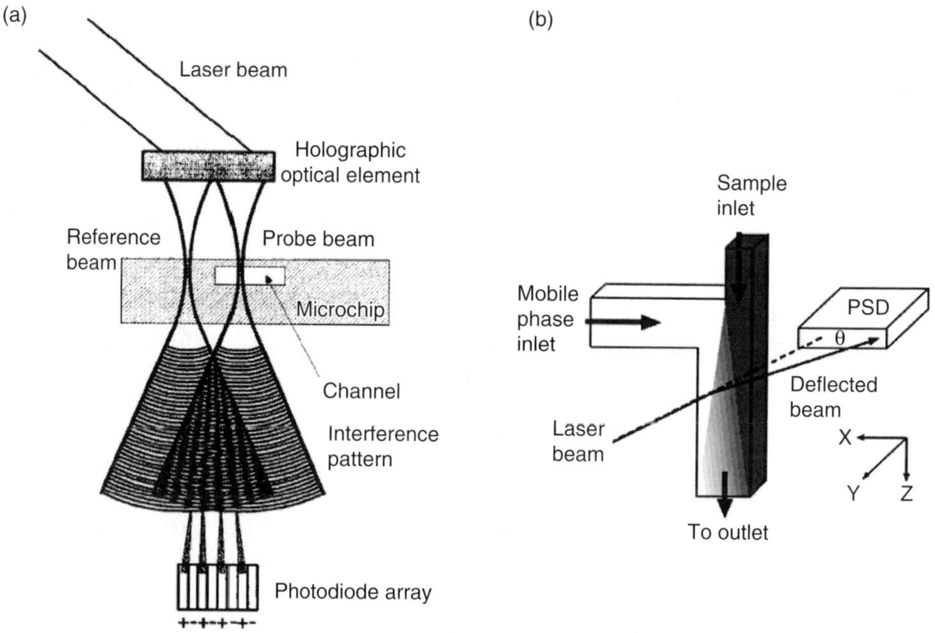

FIGURE 45.6 Different experimental approaches for RI detection. (a) Principle of holographic RI detection, where a reference and probe beam interact to produce an interference pattern that changes when an analyte passes through the channel. (From Burggraf, N., et al., *Analyst*, 123, 1443, 1998. Reproduced by permission of The Royal Society of Chemistry.) (b) Illustration of angular detection as a gradient forms between two different laminar flow streams to create an RI gradient. (Reprinted from Costin, C.D. and Synovec, R.E., *Analytical Chemistry*, 74, 4558, 2002. Copyright 2002 American Chemical Society. With permission.)

however, deep UV fluorescence detection has been demonstrated on microchip using 266 nm excitation (using a frequency quadrupled Nd:YAG laser). Hellmich et al.[47] demonstrated UV-LIF of the amino acid tryptophan (detection limit of 17 μM) and a protein mixture of 500 μM lysozyme C and 125 μM avidin in hybrid PDMS/glass devices. With a similar system, Schulze et al.[48] demonstrated UV-LIF of several different proteins including a mixture of lysozyme (900 nM), trypsinogen (500 nM), and chymotrypsinogen (500 nM) using fused silica devices.

Still, the majority of literature describing naturally fluorescent molecules analyzed on-chip deal primarily with novel device, separation, and detection technologies. In these studies, fluorescein and rhodamine dyes are most common because they can be readily obtained in various derivatized forms.[49] For example, our group used a combination of fluorescein and carboxy-X-rhodamine dyes in an electrophoresis experiment to evaluate an on-chip pressure injection,[50] and later added rhodamine 6G to the test mix to evaluate a scanning multicolor detection platform.[51] Also, fluorescent dyes can be used to perform indirect fluorescent detection of nonfluorescent species, as demonstrated by Munro et al.[52] by adding fluorescein to the run buffer to analyze amino acids in urine. Sirichai and de Mello[53] also added fluorescein to the run buffer to perform quantitative analysis on a commercial photographic developer solution and reported detection limits of 5 $\mu g{\cdot}mL^{-1}$. Recently, deep UV fluorescence detection has been demonstrated on microchip using 266 nm excitation (using a frequency quadrupled Nd:YAG laser). Hellmich et al.[47] demonstrated UV-LIF of the amino acid tryptophan (detection limit of 17 μM) and a protein mixture of 500 μM lysozyme C and 125 μM avidin in hybrid PDMS/glass devices. With a similar system, Schulze et al.[48] demonstrated UV-LIF of several different proteins (Figure 45.7) including a mixture of lysozyme (900 nM), trypsinogen (500 nM), and chymotrypsinogen (500 nM) using fused silica devices. This technique is promising because it offers the benefits of LIF detection while enabling native detection of biomolecules,

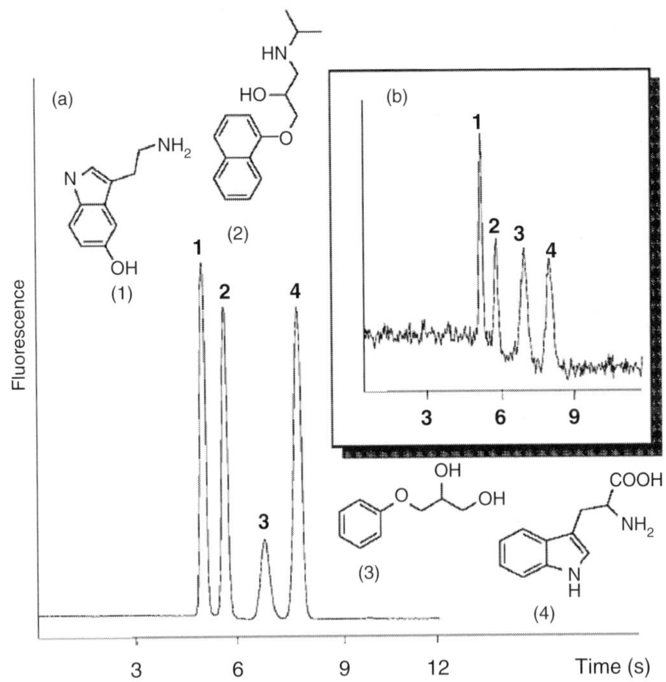

FIGURE 45.7 Peaks collected using deep UV-LIF detection. The electropherogram in the foreground represents a mixture of (1) serotonin, (2) propranolol, (3) 3-phenoxy-1,2-propanediol, and (4) tryptophan, each at a concentration of 40 $\mu g \cdot mL^{-1}$. The same mixture is shown in the inset with concentrations closer to their limits of detection. (Reprinted from Schulze, P., et al., *Analytical Chemistry*, 77, 1325, 2005. Copyright 2005 American Chemical Society. With permission.)

however, there are greater demands on the optics and microchip substrate to provide high UV transmittance with low autofluorescence.

45.3.1.2 Protein Labeling

As mentioned earlier, the ability to integrate precolumn and postcolumn labeling of amino acids brought an increased functionality to the microchip format that is not readily achieved on capillary alone. The initial report of a postcolumn reactor contained a mixing tee after the separation column where an additional reagent reservoir, separated from the reservoirs used for the injection and separation steps, electrokinetically introduced the fluorescent tag *o*-phthaldialdehyde (OPA). The OPA then mixed with the partially separated amino acids by diffusion (Figure 45.8a) in the reaction column as the separation continued.[44] With the precolumn reactor, the analytes and OPA were introduced into the same chamber prior from separate reservoirs before diffusing into the separation channel in a gated injection scheme (Figure 45.8b).[43] Both integrated techniques, performed in glass devices, enabled the labeling and separation to occur in very small volumes (<1 nL) in a matter of seconds, demonstrating the benefits of using microfluidic analysis systems. The same group later used a postcolumn reactor to label proteins noncovalently with a NanoOrange dye, which offered fast reaction kinetics and high fluorescence yield.[54] This dye also proved effective for dynamic on-column labeling, resulting in a detection limit of 500 $ng \cdot mL^{-1}$ for bovine serum albumin.[55] Another precolumn labeling reaction was demonstrated on a glass/PDMS device using dichlorotriazine fluorescein (DTAF), a derivatizing agent for biogenic amines. In addition to fast reaction kinetics (<60 s for on chip mixing and reaction), detection limits were reduced to 1 nM.[56]

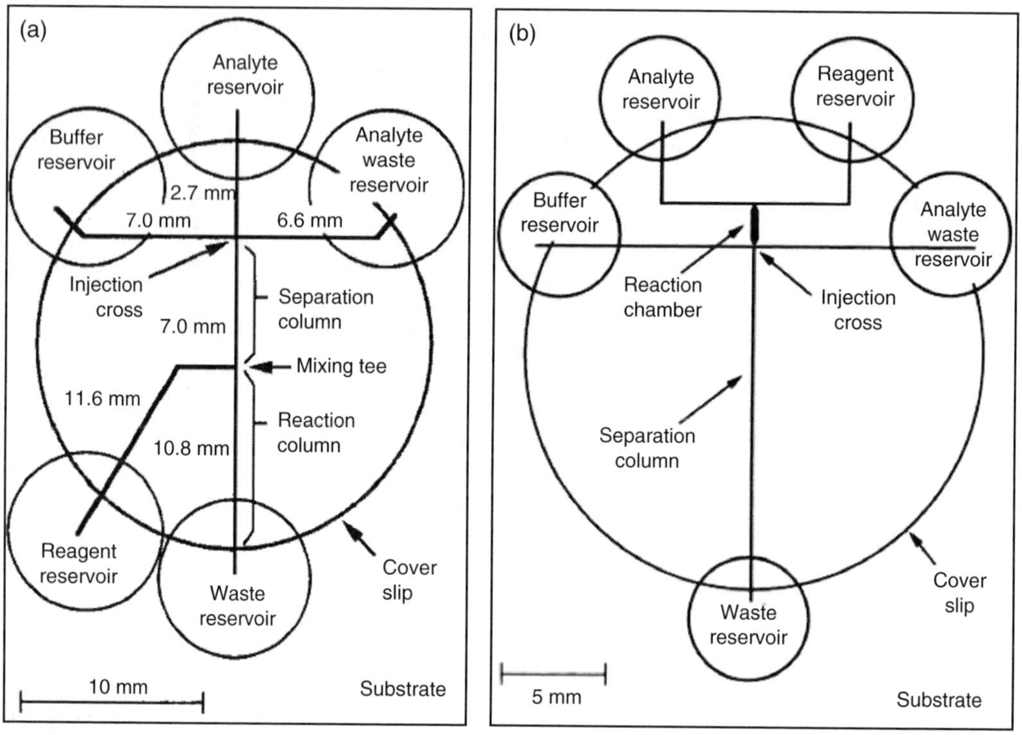

FIGURE 45.8 Integrated reactors for amino acid labeling. (a) Schematic of the microchip layout for post-column labeling, incorporating a reactor column after the separation. (Reprinted from Jacobson, S.C., et al., *Analytical Chemistry*, 66, 3472, 1994. Copyright 1994 American Chemical Society. With permission.) (b) Schematic of the layout for precolumn labeling, incorporating a reactor chamber prior to injecting sample into the separation channel for analysis. (Reprinted from Jacobson, S.C., et al., *Analytical Chemistry*, 66, 4127, 1994. Copyright 1994 American Chemical Society. With permission.)

45.3.1.3 DNA Labeling

Laser-induced fluorescence is the most common detection method in DNA sizing and sequencing applications, on both capillary and microchip, owing to the different ways to incorporate fluorophores. In double-stranded DNA (dsDNA) separations, intercalating dyes are typically added to the separation matrix. Upon binding the DNA strand, the dye exhibits a significant increase in fluorescence yield over unbound dye. While a wide variety of nucleic acid stains are available, depending on the application,[49] the protocol for dsDNA separations in our lab involves adding the monomeric cyanine dye, YO-PRO®-1 iodide (Invitrogen), to the sieving matrix at a 0.1% v/v concentration.[57] For single-stranded DNA (ssDNA) separations, fluorescently tagged primers are incorporated into the target DNA by the polymerase chain reaction (PCR) method. This allows multiple fluorescent labels to be incorporated in a single application step and is commonly used in four- and five-color sequencing and genotyping applications. Derivatization chemistries enable conjugations of most dyes to oligonucleotides,[49] and fluorescent labels can often be added (for a fee) when ordering DNA primers. In many commercial multiplexed PCR amplification kits (i.e., Applied Biosystems AmpFlSTR® kits) proprietary dyes are used, but the Mathies group has designed and synthesized fluorescent oligonucleotide primers using the principle of fluorescence resonance energy transfer (FRET),[58] and have used their dyes in sequencing[59] and genotyping[60] applications on microchip.

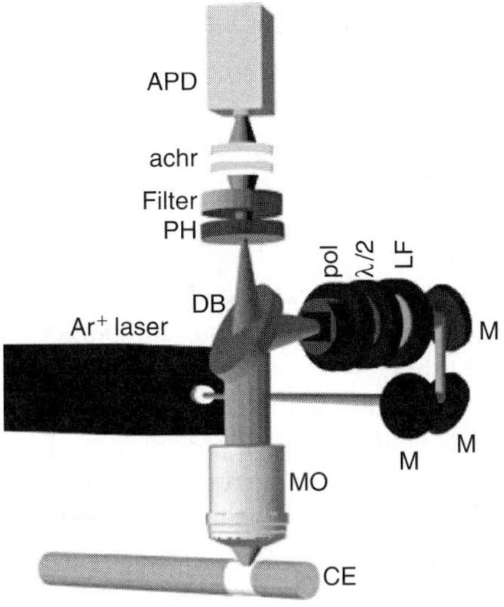

FIGURE 45.9 Schematic of a confocal LIF detection system with source, optics, and detector shown. The optics include mirrors (M), laser line filter (LF), half-wave plate (λ/2), polarizer (pol), dichroic beamsplitter (DB), microscope objective (MO), pinhole (ph), filter, and achromat lenses (achr). The source shown is an argon ion laser, and the detector is an avalanche photodiode (APD). While the electrophoresis channel shown here is in a capillary (CE), the system could be readily applied to a microchip. (Reprinted from Johnson, M.E. and Landers, J.P., *Electrophoresis*, 25, 3515, 2004. With permission.)

45.3.2 LIF Instrumentation

A typical method for performing LIF detection on a microchip is to bring the excitation light into the microchannel orthogonal to the device plane, which is relatively simple to align, and to then focus the beam into the channel using a microscope objective or some combination of achromat lenses. LIF emission is then collected with a separate objective or lens combination, spatially and spectrally filtered, and recorded on a photodetector. Confocal detection systems are quite popular in the literature, first demonstrated by Guzman and coworkers[61] for LIF detection on a capillary in 1991, where the focusing and collection optics are the same and the excitation and emission wavelengths are separated with a dichroic beamsplitter (Figure 45.9). Variations in LIF detection setup exist, however, and He et al.[62] present a detailed illustration of their laser-induced detection system where the excitation is brought into the microchannel from an angle. A recent review on the fundamentals and practice of LIF detection in microanalytical systems presents a thorough examination of ultrasensitive systems (sub-pM concentration).[63] While the goal of this discussion is not necessarily to achieve ultrasensitivity, the principles and considerations are very much the same.

45.3.2.1 Sources

The most commonly used excitation sources for LIF detection on microchip are helium-neon and argon ion gas lasers, both of which feature various emission lines. Typical laser powers range from 0.5 to 50 mW. Benefits of these lasers include their beam quality and long lifetimes, but they are still costly and bulky relative to the microchip scale. There have been other reports of excitation sources, including dye lasers (i.e., rhodamine 6G), metal-vapor lasers (i.e., HeCd), and solid-state lasers (i.e., Nd:YAG), and semiconductor lasers (i.e., laser diodes), but these still present size and cost issues. What is promising is the recent use of light-emitting diodes (LEDs) for LIF detection because, in addition to being small and cheap, they can be incorporated into the devices themselves.

It is important to note that there are several probes and dye labels available that are ideally suited to the different laser excitation wavelengths.[49]

45.3.2.2 Detectors

Photomultipliers tubes (PMTs) remain a common choice for detection. PMTs consist of a photocathode, a series of dynodes, and an anode. Incident photons strike the photocathode surface, causing electrons to be emitted by the photoelectric effect. These electrons are accelerated toward the first dynode, which is held at a positive potential, and collide to release additional electrons. This process is repeated several times, with each dynode held at a higher potential, until the final accumulation of cascading electrons (multiplying a signal by orders of magnitude) reach the anode. This results in a sharp current pulse indicating that photons have been detected by the PMT. While PMTs are quite common as photodetectors for microchip LIF, some of the other detectors that have been used include avalanche photodiodes (APDs) and CCD arrays. While the former is most often used to achieve ultrasensitive detection,[63] the latter is typically used in whole channel imaging and multidimensional applications, which will be described next.

45.3.3 Multidimensional Approaches

45.3.3.1 Multichannel Detection

While several different glass types have been reported in the literature for microchip separations, they are all compatible with LIF detection due to their optical transparency. This is significant because it creates the opportunity to interrogate any region of the microchip, offering more flexibility than the capillary format. Because the glass capillaries used in CE are coated with a polymer layer that must be removed to expose the glass window for excitation, there is inherent limitation to where detection can be performed. In addition to performing LIF detection at a single location, several techniques have been shown for multipoint (within a single channel) or multichannel detection. The former includes the incorporation of waveguide structures into the microchip design to create several excitation points along a single channel, which was used to perform velocity measurements of particles flowing a microchannel.[64] Several approaches for multipoint detection have been demonstrated for microchip IEF as an alternative to mobilizing resolved peaks past a fixed detection point.[65] Similar to the approach for detecting IEF by UV absorbance,[8] Yao et al.[66] collected the fluorescence signal from an IEF separation using a CCD array for whole column imaging. Interestingly, the excitation source was also an array, featuring organic LEDs. As opposed to imaging the entire IEF channel, Raisi et al.[67] designed a scanning fluorescence detector based on a computer-controlled translation stage with a single PMT and were able to scan the entire separation channel every 9 s to collect temporal and spatial measurements. It is ironic this approach parallels that devised by Hjerten[68] and reviewed more than half a century ago. Another scanning approach using a single PMT is presented using an acoustooptic deflector (AOD) system (described later) to sweep the excitation beam across an IEF channel during the focusing step.[69]

Considering the capabilities of current capillary instrumentation, which routinely handle bundles of capillary at a time, the ability to increase the number of parallel analysis channels and, therefore, sample throughput is arguably one of the biggest challenges (and successes) in the microchip field. Modest increases in throughput have been achieved featuring both moving and nonmoving parts for beam displacement to address multiple channels. Cheng et al.[70] used a mirror attached to a galvano-scanner to raster an excitation beam across a six-channel device designed to perform independent immunoassays simultaneously (Figure 45.10). The galvano-scanner changed its position in an arc in response to a change in current. Huang et al.[71] used an AOD to change the diffraction angle of an incident laser beam to address an eight-channel device designed for parallel DNA separations. The AOD featured no moving parts, relying solely on the interaction of the laser beam and an acoustic

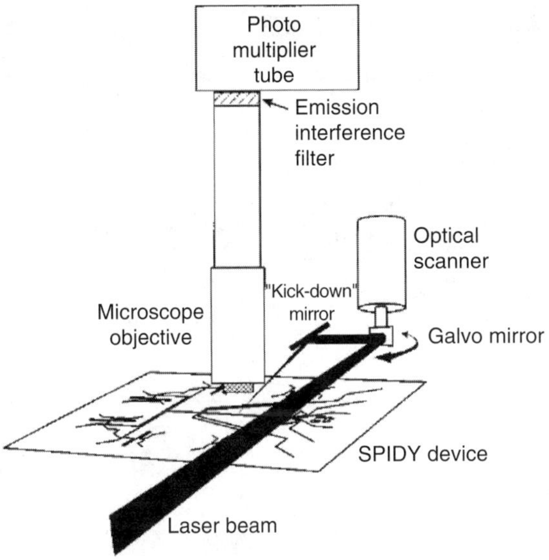

FIGURE 45.10 Multichannel approach for LIF detection on a microchip. A mirror attached to a galvano-scanner was used to position the excitation beam so that all eight channels on the device (named SPIDY) could be addressed to monitor parallel immunoassays. (Reprinted from Cheng, S.B., et al., *Analytical Chemistry*, 73, 1472, 2001. Copyright 2001 American Chemical Society. With permission.)

wave propagating through the crystal to scan across the channels, and exhibited fast response times (~200 ns). Most impressive, however, are the advances made by the Mathies group with their radial devices for DNA analysis, which have achieved parallel processing of 96[72] and 384[73] samples. Using circular devices, with analysis channels extending from the center, the detection points can be kept close together, and a circular scan path enables high scan rates and positional accuracy. A description of the radial scanner, which includes a rhomb prism at the top of a hollow rotating shaft, can be found in the literature.[74]

45.3.3.2 Multicolor Detection

In multicolor LIF experiments, the most common technique for sampling at multiple emission wavelengths is to add additional photodetectors along with the appropriate dichroic beamsplitters to separate the fluorescent signal. This is the approach taken by several groups to perform multi-DNA sequencing[75] and genotyping[76] on microchip. Despite the existence of commercial capillary instruments that utilize a spectrograph to disperse the emission wavelengths onto a CCD array, this technique has seldom been reported in the literature for microchip LIF detection. However, Backhouse et al.[77] used a diffraction grating to disperse the LIF emission onto a CCD array for four-color sequencing, and Simpson et al.[78] used a transmission imaging spectrograph with a wide imaging area to disperse LIF emission from a two-color genotyping experiment run in 48 parallel channels onto a CCD array. In the CCD systems, the diffracted emission wavelengths are displaced and then binned in rows or columns of CCDs. Recently, a system was developed to complement the microchip platform by featuring a single acoustooptic tunable filter (AOTF) and a single PMT.[51] The filter behaves similar to the previously described AOD, except that when an acoustic wave traveling through the crystal at a specific frequency interacts with the light, only select wavelength bands are diffracted by the crystal and collected by the PMT. Capable of addressing several wavelengths with only a single photodetector (19 are shown in Figure 45.11), an AOTF-based detection system has

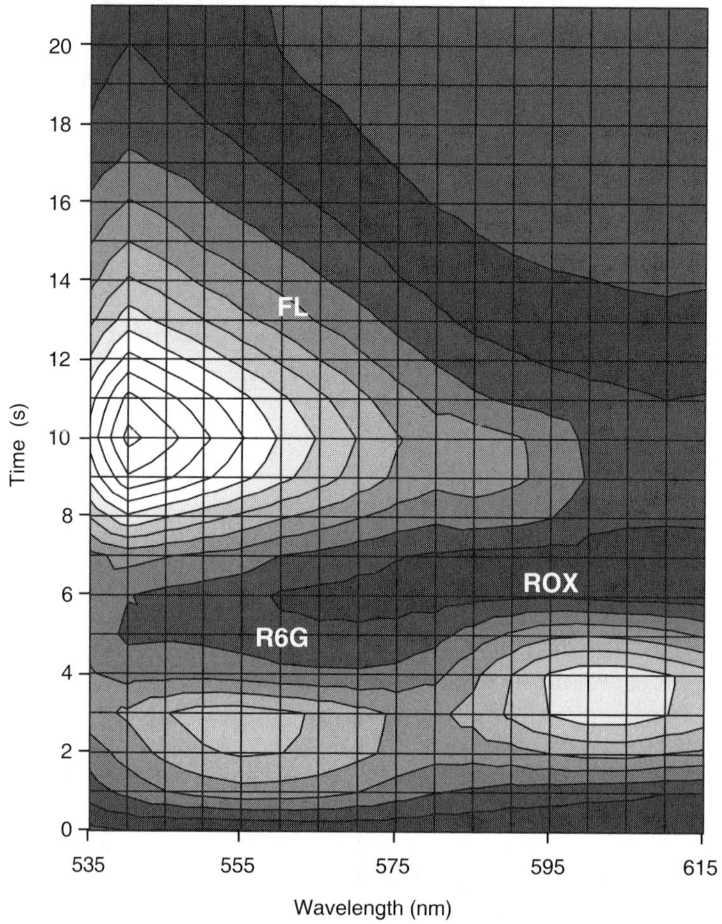

FIGURE 45.11 Three-dimensional multicolor data collected with an AOTF. A mixture of three dyes (FL: fluorescein, R6G: rhodamine 6G, and ROX: carboxy-X-rhodamine) was injected and separated on chip while scanning the AOTF through 19 wavelengths. (Adapted from Karlinsey, J.M. and Landers, J.P., *Analytical Chemistry*, 78, 5590, 2006.)

been shown for four- and five-color genotyping applications on microchip.[79] In all of these cases, the appropriate optics are placed after the pinhole in the confocal detection setup.

45.3.4 Single Molecule Detection

The earliest report of single-molecule detection on microchip was in 1998, with concentration detection limits of 1.7 pM for rhodamine 6G and 8.5 pM rhodamine B separated in 10 μm deep glass channels in less than 35 s. The fluorescent molecules were excited with the 514 nm line from an argon ion laser at 540 μW, a 100× objective was used in a confocal setup, and a single-photon avalanche diode was used to collect the signal.[80] Shortly afterward, single-molecule detection was applied to DNA separations, focusing the sample through a 1 μm diameter focused laser beam. Using the 488 nm line from an argon ion laser and a 100× objective in a confocal microscope, thiazole orange intercalator was excited to detect single DNA molecules during the separation of a 100–1000 bp DNA sizing ladder.[81] Foquet et al.[82] patterned nanochannels in fused silica, creating channel volumes ~100 times smaller than the observation volumes obtained using conventional confocal optics. This enabled single-fluorophore detection to be achieved at a higher concentration

(2 nM) than typically reported (pM). Although the examples presented here involve LIF detection, single-molecule detection using TLM has also been demonstrated (see earlier).[31]

45.3.5 Integrated Optics

Recently, several groups have integrated optical components into their devices, and some of the more impressive examples are presented here. In some cases, no additional fabrication steps are required, as the optics are patterned along with the fluidic features. After pioneering microchip fabrication in PDMS, the Whitesides group assembled an integrated fluorescence detection system including an optical fiber embedded in the patterned fluidic layer to couple excitation light into the channel, a microavalanche photodiode (μAPD) embedded in a second PDMS layer, and a colored polycarbonate filter placed between the two layers (Figure 45.12). The device was then used to detect proteins and small molecules separated on the chip, and a detection limit of ~25 nM was obtained for fluorescein solution.[83] A microoptical system was designed in glass that involved the additional patterning of microlens arrays and aperture arrays deposited on both sides of the chip to enable the detection of a 3.3 nM dye solution, comparing favorably with confocal systems.[84] Polystyrene chips were fabricated by Maims et al.[85] with an embossed diffractive element to obtain fluorescence and absorption spectra using laser and LED sources. One research group integrated an optical waveguide network with fluidic microchannels to perform particle sorting.[86] Optical trapping was performed by focusing a femtosecond pulsed laser into fused silica to form the waveguide, and the

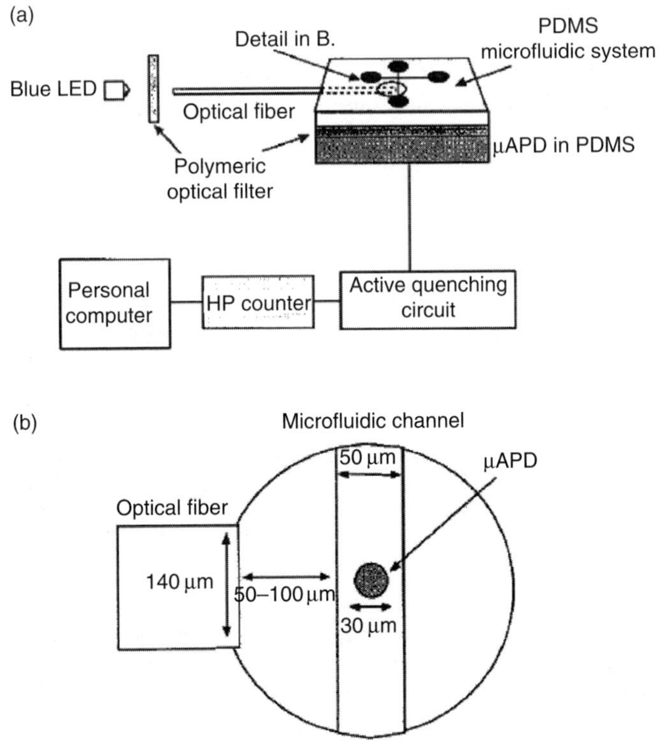

FIGURE 45.12 Diagrams for a PDMS chip containing integrated optics. (a) The microchip consists of three layers containing the fluidics, a filter, and an APD detector. Light from an LED source is coupled in the microchip through an optical fiber embedded in the fluidic layer. (b) Top-down view of the microchip that shows the spatial relationship between the channel, detector, and optical fiber. (Reprinted from Chabinyc, M.L., et al., *Analytical Chemistry*, 73, 4491, 2001. Copyright 2001 American Chemical Society. With permission.)

trapped particles (exhibiting different fluorescence signals) were then sorted into different flow paths. Miniaturized spectrometers have been assembled by combining molded elastomeric microchannels with filtered silicon detector arrays, and a solution of Bromophenol Blue was detectable down to the sub-μM range by monitoring its absorption at 588 nm.[87] Thin-film photoelements have been incorporated into PDMS devices to serve as both excitation sources and detectors. Using a confocal detection setup, a thin-film polyfluorene LED was used to excite fluorescein and carboxyfluorescein dyes in an electrophoresis experiment, with a detection limit of 1 μM.[88] A thin-film organic diode was implemented as a photodetector for a CL assay on chip, with a preliminary detection limit of ~1 mM.[89]

45.4 PRACTICAL APPLICATION OF LIF

The following sections are designed to assist someone interested in assembling a microchip LIF detection setup, or merely to help reinforce some of the considerations for those already collecting data. It is by no means the only way to perform microchip LIF, and it is probably not the best. However, one should be able to routinely achieve 1–100 nM detection limits on the system described herein using a fluorescein dye in a typical CE buffer system.

45.4.1 Experimental LIF Setup

The confocal epifluorescent detection scheme we use is common among many who use this detection mode[90] and features a cube-and-rail assembly system, specifically the microbench system from LINOS Photonics (Milford, MA) on an optical breadboard to maintain proper alignment of components (Figure 45.13). The excitation source is a multiline argon ion gas laser (model Reliant 150 m, Laser Physics, West Jordan, UT) that features user-selectable wavelengths (457, 488, and 514 nm)

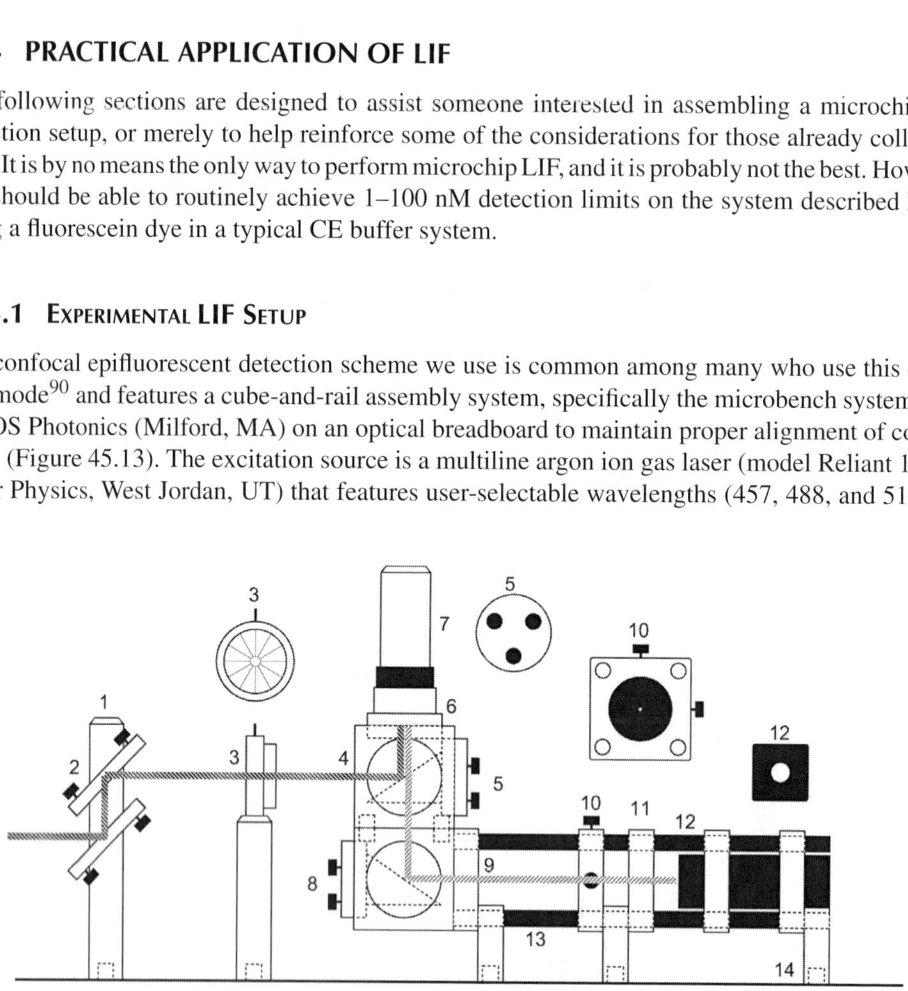

FIGURE 45.13 Confocal epifluorescent detection setup. Laser excitation (dark shading) enters the cube assembly from the steering mirrors and is reflected off of the dichroic beamsplitter before being focused into a microchannel. LIF emission (light shading) is then collected with the objective, passed through the beamsplitter and off a mirror before focusing onto a pinhole. The light passing though the pinhole and subsequent filter is finally collected by the photodetector. The microchip and microchip stage, which should feature translation in the *z*-axis and at least one of the *x* or *y*-axes, have been left out of the diagram. Frontal views are provided of some components when their features are not clear in the side-view. (1: mounting post, 2: steering mirror, 3: iris diaphragm, 4: cube assembly, 5: dichroic beamsplitter on 45° adapter (with three positioning knobs), 6: objective adapter, 7: microscope objective, 8: mirror on 45° adapter, 9: 90° adapter plate, 10: pinhole holder with *x*-*y* positioning, 11: emission filter holder, 12: PMT detector, 13: rail, 14: rail mount.)

and can be operated in multiline mode up to 150 mW. The excitation wavelength enters normal to one of the cubes and is incident on a dichroic beamsplitter tilted at a 45° angle. The dichroic is selected with a particular cutoff wavelength, below which light is reflected and above which light is transmitted. Reflected excitation light is directed into the back end of a microscope objective where it is then focused into the microchannel. The fluorescence emission that is collected by the same objective is then transmitted through the dichroic beamsplitter and focused onto a pinhole placed at the appropriate distance from the objective. The pinhole serves as the spatial filter in the image plane of the objective and defines the collection volume. An appropriate emission filter is placed after the pinhole, and the filtered emission light is then incident on the fluorescence detector where signal is collected for analysis. Both the dichroic beamsplitters and emission filters we use are obtained from Omega Optical (Brattleboro, VT), and the pinhole is from National Aperture (Salem, NH).

In a true confocal system, a point source for excitation is imaged onto the sample and the resulting fluorescence is imaged onto a spatial filter, although most microchip systems use a laser beam as the source. Several discussions regarding the selection of optics, including the objective and pinhole, can be found in the literature.[63,90] The standard microchip LIF setup used in our lab features a 488 nm excitation beam diameter of 2 mm and a 16× magnification objective (model 04 OAS 012, Melles Griot) with a focal length of 10.8 mm and a numerical aperture (NA) of 0.32, yielding a theoretical spot size of 3.4 μm. Aberrations, however, such as those arising from RI mismatches at the air/glass and glass/buffer interfaces, enlarge the effective spot diameter, but our channel widths are typically 30 μm across at their narrowest point (due to the narrowest linewidths we can pattern on our photomasks). Assuming a typical isotropic etch depth of 40 μm, a hemispherical channel is formed with a bottom width of 30 μm (the initial line width on the photomask) and a top width of 110 μm (the initial line width in addition to the etch in each direction). Imaging the channel onto a 1 mm pinhole using the 16× objective, therefore, probes a 62.5 μm spot, which is adequate for sampling from the middle of the channel. When increased sensitivity is desired, a 40× objective (model 440864, Zeiss) with an NA of 0.6 and a correction ring to adjust for cover glass thickness is used, experimentally increasing the signal almost two-fold. By switching objectives, the excitation spot size increases to 7 μm and the collection spot decreases to 25 μm, and so the channel is probed more efficiently.

45.4.2 Alignment of the Optics

Alignment of the optical system is critical to the LIF detection, and a technique for setting up and/or aligning a cube-and-rail system is presented here. While this technique is not the only approach, it is relatively straightforward and easy to adapt for various applications. Before alignment, it is helpful to assemble the cube system containing the dichroic beamsplitter and mirror cubes (both containing adjustable 45° adapters), objective adapter, emission filter holder, 90° adapter plate, pinhole holder with x-y positioning, and appropriate rails and rail mounts.

1. The excitation beam must first be aligned parallel to the optical table or breadboard. A way to insure that the height is maintained along the table is to position an iris diaphragm at an appropriate height (i.e., where the light exits the laser), and alternately move it along the beam path from a point close to the laser output to a point as far away as possible, while still on the table. This works best if the beam path follows the predrilled holes on the optical table. Depending on the laser system, this may involve leveling a box enclosure or adjusting the mounts on a tube enclosure.
2. Once the beam is parallel, a set of two steering mirrors are used in tandem to position the beam at a height that matches the center of the top cube containing the dichroic beamsplitter. The mirrors should be mounted at 45° angles on mirror mounts with adjustable pitch and should be placed so that the first mirror reflects the excitation beam perpendicular to the table and the second reflects the beam parallel to the table, with the beam incident at

the desired height. The steering mirrors are then adjusted to align the beam parallel to the table, again using the iris diaphragm. The iris should be centered at the height of the center of the top cube and moved alternately between two holes on the optical table, one close to the steering mirrors and one further away. The pitch on the initial mirror mount should be adjusted while the iris is located close to the steering mirrors, and the pitch on the second mirror should be adjusted after moving the iris further away. This step is repeated until the beam is parallel at the appropriate height to enter the cube. The iris diaphragm can then be fixed in place before the entrance of the cube assembly with the iris diameter adjusted slightly smaller than the beam.

3. The excitation beam should be normal to the cube assembly, and this is tested by placing a microscope slide across the face of the cube. A mirror held against the slide should reflect the excitation beam back through the iris. Once the cube assembly is positioned, it should be fixed to the table. By maintaining an excitation path parallel to the holes in the optical table, the cube assembly should already be centered using the appropriate rail mounts.
4. With the microchip objective removed, the dichroic beamsplitter should be adjusted so that the excitation beam exits the assembly perpendicular to the table. This will put the dichroic at the desired 45°. The position of the beam on the dichroic beamsplitter should then be checked using a mirror above the objective adapter to direct the excitation beam back out of the cube assembly. This is easily observed with the iris still in place. If the beam is not centered, the dichroic mount can be moved back and forth along the beam path. This may require the use of a washer between the cube and the mount to back it out sufficiently.
5. The alignment of the dichroic and the microscope objective must then be checked. The best way to check this is to place a card with a hole in it above the beam exiting the top of the cube assembly, leaving enough space for the objective to be put in place. After lining the hole up with the beam, the objective should be placed in the adapter. With proper alignment, the beam exiting the objective will be centered on the hole. If the beam does not line up, the dichroic adapter should be adjusted accordingly. After this is done, step 4 should be repeated and the alignment of the microscope objective should be rechecked.
6. Once the dichroic beamsplitter is positioned, the mirror adapter in the bottom cube should be adjusted to direct the emission beam parallel to the table as it exits the cube assembly. Because of the filtering behavior of the dichroic, this step is most easily performed with a multiline laser (with a mirror above the objective adapter and the microscope objective removed), but a concentrated solution of fluorophore can also be used to generate a strong signal for alignment (with the objective in place and the beam incident on a filled microchannel). Since the dichroic is aligned, the beam incident on the mirror is normal to the table and so the mirror should be adjusted for a 45°. This is done by setting an iris diaphragm to the height of the center of the bottom cube, and placing it alternately close to the assembly and further away, while adjusting the adapter knobs until the beam is parallel. Because the 45° adapter has three adjustable knobs, only two should be adjusted at a time to ensure that all three are used to align the exit beam.
7. The *x*–*y* positioner with the pinhole should then be placed at the appropriate distance from the shoulder of the microscope objective (the top of the objective adapter). This distance is the tube length (typically 160 mm, specified on the objective housing) minus 10 mm (that accounts for the placement of the eyepiece in finite-corrected light microscopes). Once the pinhole is in position (in this case, 150 mm from the shoulder of the objective) a fluorophore-filled microchannel should be placed above the objective with the focused excitation beam in the channel. Adjusting the *z*-axis should increase or decrease the size of the focused emission beam on the pinhole. The *x*- and *y*-axes should be adjusted on the positioner to align the pinhole. The easiest way to do this is to adjust the *z*-axis until

the outer fringes of the focused emission beam are visible around the pinhole, allowing it to be centered easily.

8. After all of the key components are aligned, the emission filter and detector, typically a PMT, can be placed in the setup. Depending on the detector, additional optics, such as focusing lenses or a spectrograph, may be placed after the pinhole. The setup should then be covered to reduce background signal from stray light. The main source of background signal is scattered excitation light, which can be blocked with a notch filter. This type of filter is very expensive, however, and should only be necessary for ultrasensitive fluorescence detection experiments.

The benefit of using fluorescence detection is that it is path length independent and, therefore, the microchannel dimensions (which commonly range from 5 to 200 μm) do not limit the effectiveness of the detection. However, it is important that the probe volume is located within the channel. While the lateral position of the focused beam is readily identified by eye, observing the sidewalls of the channel that scatter the light and provide an overhead image like a contour map, the depth of the focus within the channel is more difficult to discern. The most sensitive alignment in the z-axis is achieved by observing a fluorescent signal while adjusting the z-position. In some applications, such as when using a DNA intercalating dye that exhibits a weak fluorescent signal when unbound, there is enough signal present to maximize the emission signal. However, in most cases, the separation channel is filled with run buffer or polymer matrix that does not exhibit a fluorescence signal. Instead of optimizing the z-position by adding a fluorophore to the analysis channel, and therefore contaminating the detection region, it is recommended that a separate alignment channel, characterized by the same width and etch depth as the separation channel, be included in the microchip design. Once the z-position of the microchip stage is tuned using the alignment channel, the chip can then be repositioned with the beam focused in the analysis channel.

45.4.3 Evaluation of the Detection System

One of the more problematic issues when working with glass microchips is the glass–glass bonding step, usually performed by cleaning the bonding surfaces and thermally annealing at high temperatures (>600°C) between ceramic plates with some added weight to generate pressure. Not only is this a time-consuming step (typically performed overnight), but it can also introduce optical defects in the glass due to the combination of high temperature and pressure. It is not unusual to observe "frosting" in the glass following one or more bonding cycles. In regions of the chip where the glass is frosted or clouded, it is difficult to both focus the excitation beam and collect the fluorescence emission. It is recommended, following thermal bonding, that the analysis channels in a microchip be inspected using the LIF detection system before running any samples, in order to identify appropriate windows for interrogation.

When evaluating a detection setup, it is most efficient to do so in the context of a given set of experimental conditions (i.e., device dimensions, buffer system, etc.). These conditions can be standardized within a research group or lab setting for comparison between users and instruments, as well as to evaluate any changes to a system. There may be times, however, when a specific application (i.e., DNA sequencing) merits its own evaluation of the LIF setup. In these cases, it is important to maintain similar conditions in evaluating the detection as will be implemented in the analysis. For example, an off-chip desalting of PCR product before being loaded into the sample well on a microchip will have a significant effect on the amount of sample that gets injected into the separation channel. In this case, the change is not made to the detection setup, but the fluorescence emission collected from the sample will be very different compared to a test sample that may not have been desalted.

45.5 CONCLUDING REMARKS

Using primarily LIF detection, the microchip community has achieved the analysis speeds and reagent reductions proposed some 15 years ago, and the process of miniaturizing and integrating optical components onto and around the microchip has already begun. While a universal detector is yet to be developed, researchers have increased the capability of LIF detection systems by integrating additional complexity on the devices themselves, incorporating multiple reaction steps as well as optics. At the same time, the number of microchip analysis techniques is steadily growing. According to SciFinder (which offers access to the Chemical Abstracts database), the number of publications per year regarding microchip detection techniques doubled in the years 2003–2004 and is currently holding steady. Using the optical methods presented in this chapter along with various other forms of chip-based analysis, including electrochemical detection, mass spectroscopy, and even nuclear magnetic resonance, it will be the responsibility of the microchip community to find new applications and elucidate more processes in the future as we increase our ability to control solution flow and mass transfer on the micro- and nanoscale.

REFERENCES

1. Auroux, P.A., Iossifidis, D., Reyes, D.R., and Manz, A., Micro total analysis systems. 2. Analytical standard operations and applications, *Analytical Chemistry*, 74, 2637–2652, 2002.
2. Uchiyama, K., Nakajima, H., and Hobo, T., Detection method for microchip separations, *Analytical and Bioanalytical Chemistry*, 379, 375–382, 2004.
3. Vilkner, T., Janasek, D., and Manz, A., Micro total analysis systems. Recent developments, *Analytical Chemistry*, 76, 3373–3385, 2004.
4. Mogensen, K.B., Klank, H., and Kutter, J.P., Recent developments in detection for microfluidic systems, *Electrophoresis*, 25, 3498–3512, 2004.
5. Dittrich, P.S., Tachikawa, K., and Manz, A., Micro total analysis systems. Latest advancements and trends, *Analytical Chemistry*, 78, 3887–3907, 2006.
6. Viskari, P.J. and Landers, J.P., Unconventional detection methods for microfluidic devices, *Electrophoresis*, 27, 1797–1810, 2006.
7. Harrison, D.J., Manz, A., Fan, Z.H., Ludi, H., and Widmer, H.M., Capillary electrophoresis and sample injection systems integrated on a planar glass chip, *Analytical Chemistry*, 64, 1926–1932, 1992.
8. Mao, Q.L. and Pawliszyn, J., Demonstration of isoelectric focusing on an etched quartz chip with UV absorption imaging detection, *Analyst*, 124, 637–641, 1999.
9. Verpoorte, E., Manz, A., Ludi, H., Bruno, A.E., Maystre, F., Krattiger, B., Widmer, H.M., Vanderschoot, B.H., and Derooij, N.F., A silicon flow cell for optical-detection in miniaturized total chemical-analysis systems, *Sensors and Actuators B—Chemical*, 6, 66–70, 1992.
10. Salimi-Moosavi, H., Jiang, Y.T., Lester, L., McKinnon, G., and Harrison, D.J., A multireflection cell for enhanced absorbance detection in microchip-based capillary electrophoresis devices, *Electrophoresis*, 21, 1291–1299, 2000.
11. Duggan, M.P., McCreedy, T., and Aylott, J.W., A noninvasive analysis method for on-chip spectrophotometric detection using liquid-core waveguiding within a 3D architecture, *Analyst*, 128, 1336–1340, 2003.
12. Jindal, R. and Cramer, S.M., On-chip electrochromatography using sol-gel immobilized stationary phase with UV absorbance detection, *Journal of Chromatography A*, 1044, 277–285, 2004.
13. Malcik, N., Ferrance, J.P., Landers, J.P., and Caglar, P., The performance of a microchip-based fiber optic detection technique for the determination of Ca^{2+} ions in urine, *Sensors and Actuators B—Chemical*, 107, 24–31, 2005.
14. Mogensen, K.B., Eriksson, F., Gustafsson, O., Nikolajsen, R.P.H., and Kutter, J.P., Pure-silica optical waveguides, fiber couplers, and high-aspect ratio submicrometer channels for electrokinetic separation devices, *Electrophoresis*, 25, 3788–3795, 2004.

15. Liang, Z.H., Chiem, N., Ocvirk, G., Tang, T., Fluri, K., and Harrison, D.J., Microfabrication of a planar absorbance and fluorescence cell for integrated capillary electrophoresis devices, *Analytical Chemistry*, 68, 1040–1046, 1996.
16. Xu, Y., Bessoth, F.G., Eijkel, J.C.T., and Manz, A., Online monitoring of chromium(III) using a fast micromachined mixer/reactor and chemiluminescence detection, *Analyst*, 125, 677–683, 2000.
17. Liu, B.F., Ozaki, M., Utsumi, Y., Hattori, T., and Terabe, S., Chemiluminescence detection for a microchip capillary electrophoresis system fabricated in poly(dimethylsiloxane), *Analytical Chemistry*, 75, 36–41, 2003.
18. Su, R.G., Lin, J.M., Qu, F., Chen, Z.F., Gao, Y.H., and Yamada, M., Capillary electrophoresis microchip coupled with online chemiluminescence detection, *Analytica Chimica Acta*, 508, 11–15, 2004.
19. Xu, Z.R. and Fang, Z.L., Composite poly(dimethylsiloxane)/glass microfluidic system with an immobilized enzymatic particle-bed reactor and sequential sample injection for chemiluminescence determinations, *Analytica Chimica Acta*, 507, 129–135, 2004.
20. Tsukagoshi, K., Jinno, N., and Nakajima, R., Development of a micro total analysis system incorporating chemiluminescence detection and application to detection of cancer markers, *Analytical Chemistry*, 77, 1684–1688, 2005.
21. Gerardi, R.D., Barnett, N.W., and Lewis, S.W., Analytical applications of *tris*(2,2′-bipyridyl)ruthenium(III) as a chemiluminescent reagent, *Analytica Chimica Acta*, 378, 1–41, 1999.
22. Fiaccabrino, G.C., de Rooij, N.F., and Koudelka-Hep, M., On-chip generation and detection of electrochemiluminescence, *Analytica Chimica Acta*, 359, 263–267, 1998.
23. L'Hostis, E., Michel, P.E., Fiaccabrino, G.C., Strike, D.J., de Rooij, N.F., and Koudelka-Hep, M., Microreactor and electrochemical detectors fabricated using Si and EPON SU-8, *Sensors and Actuators B—Chemical*, 64, 156–162, 2000.
24. Pan, D.H. and Mathies, R.A., Chromophore structure in lumirhodopsin and metarhodopsin I by time-resolved resonance Raman microchip spectroscopy, *Biochemistry*, 40, 7929–7936, 2001.
25. Zhao, X.C., You, T.Y., Qiu, H.B., Yan, J.L., Yang, X.R., and Wang, E.K., Electrochemiluminescence detection with integrated indium tin oxide electrode on electrophoretic microchip for direct bioanalysis of lincomycin in the urine, *Journal of Chromatography B—Analytical Technologies in the Biomedical and Life Sciences*, 810, 137–142, 2004.
26. Qiu, H.B., Yin, X.B., Yan, J.L., Zhao, X.C., Yang, X.R., and Wang, E.K., Simultaneous electrochemical and electrochemiluminescence detection for microchip and conventional capillary electrophoresis, *Electrophoresis*, 26, 687–693, 2005.
27. Fletcher, P.D.I., Haswell, S.J., and Zhang, X.L., Monitoring of chemical reactions within microreactors using an inverted Raman microscopic spectrometer, *Electrophoresis*, 24, 3239–3245, 2003.
28. Leung, S.A., Winkle, R.F., Wootton, R.C.R., and deMello, A.J., A method for rapid reaction optimisation in continuous-flow microfluidic reactors using online Raman spectroscopic detection, *Analyst*, 130, 46–51, 2005.
29. Ramser, K., Enger, J., Goksor, M., Hanstorp, D., Logg, K., and Kall, M., A microfluidic system enabling Raman measurements of the oxygenation cycle in single optically trapped red blood cells, *Lab on a Chip*, 5, 431–436, 2005.
30. Moskovits, M., Surface-enhanced Raman spectroscopy: a brief retrospective, *Journal of Raman Spectroscopy*, 36, 485–496, 2005.
31. Tokeshi, M., Uchida, M., Hibara, A., Sawada, T., and Kitamori, T., Determination of subyoctomole amounts of nonfluorescent molecules using a thermal lens microscope: Subsingle molecule determination, *Analytical Chemistry*, 73, 2112–2116, 2001.
32. Tamaki, E., Sato, K., Tokeshi, M., Sato, K., Aihara, M., and Kitamori, T., Single-cell analysis by a scanning thermal lens microscope with a microchip: Direct monitoring of cytochrome c distribution during apoptosis process, *Analytical Chemistry*, 74, 1560–1564, 2002.
33. Tokeshi, M., Kikutani, Y., Hibara, A., Sato, K., Hisamoto, H., and Kitamori, T., Chemical processing on microchips for analysis, synthesis, and bioassay, *Electrophoresis*, 24, 3583–3594, 2003.
34. Mawatari, K., Naganuma, Y., and Shimoide, K., Portable thermal lens spectrometer with focusing system, *Analytical Chemistry*, 77, 687–692, 2005.

35. Munson, M.S., Hasenbank, M.S., Fu, E., and Yager, P., Suppression of nonspecific adsorption using sheath flow, *Lab on a Chip*, 4, 438–445, 2004.
36. Hasenbank, M.S., Fu, E., and Yager, P., Lateral spread of an amplification signal using an enzymatic system on a conductive surface, *Langmuir*, 22, 7451–7453, 2006.
37. Furuki, M., Kameoka, J., Craighead, H.G., and Isaacson, M.S., Surface plasmon resonance sensors utilizing microfabricated channels, *Sensors and Actuators B—Chemical*, 79, 63–69, 2001.
38. Wheeler, A.R., Chah, S., Whelan, R.J., and Zare, R.N., Poly(dimethylsiloxane) microfluidic flow cells for surface plasmon resonance spectroscopy, *Sensors and Actuators B—Chemical*, 98, 208–214, 2004.
39. Kanda, V., Kariuki, J.K., Harrison, D.J., and McDermott, M.T., Label-free reading of microarray-based immunoassays with surface plasmon resonance imaging, *Analytical Chemistry*, 76, 7257–7262, 2004.
40. Homola, J., Yee, S.S., and Gauglitz, G., Surface plasmon resonance sensors: review, *Sensors and Actuators B—Chemical*, 54, 3–15, 1999.
41. Burggraf, N., Krattiger, B., de Mello, A.J., de Rooij, N.F., and Manz, A., Holographic refractive index detector for application in microchip-based separation systems, *Analyst*, 123, 1443–1447, 1998.
42. Costin, C.D. and Synovec, R.E., A microscale-molecular weight sensor: Probing molecular diffusion between adjacent laminar flows by refractive index gradient detection, *Analytical Chemistry*, 74, 4558–4565, 2002.
43. Jacobson, S.C., Hergenroder, R., Moore, A.W., and Ramsey, J.M., Precolumn reactions with electrophoretic analysis integrated on a microchip, *Analytical Chemistry*, 66, 4127–4132, 1994.
44. Jacobson, S.C., Koutny, L.B., Hergenroder, R., Moore, A.W., and Ramsey, J.M., Microchip capillary electrophoresis with an integrated postcolumn reactor, *Analytical Chemistry*, 66, 3472–3476, 1994.
45. Colyer, C., Kinkade, C., Viskari, P., and Landers, J., Analysis of cyanobacterial pigments and proteins by electrophoretic and chromatographic methods, *Analytical and Bioanalytical Chemistry*, 382, 559–569, 2005.
46. Karlinsey, J.M., Phycobiliprotein Analysis on Microchip, *Advances and Applications of Microfluidic Analysis Systems*, 2007, University of Virginia: Charlottesville, VA.
47. Hellmich, W., Pelargus, C., Leffhalm, K., Ros, A., and Anselmetti, D., Single cell manipulation, analytics, and label-free protein detection in microfluidic devices for systems nanobiology, *Electrophoresis*, 26, 3689–3696, 2005.
48. Schulze, P., Ludwig, M., Kohler, F., and Belder, D., Deep UV laser-induced fluorescence detection of unlabeled drugs and proteins in microchip electrophoresis, *Analytical Chemistry*, 77, 1325–1329, 2005.
49. Haugland, R.P., *The Handbook: A Guide to Fluorescent Probes and Labeling Technologies*, 10th ed. 2005: Invitrogen Corp.
50. Karlinsey, J.M., Monahan, J., Marchiarullo, D.J., Ferrance, J.P., and Landers, J.P., Pressure injection on a valved microdevice for electrophoretic analysis of submicroliter samples, *Analytical Chemistry*, 77, 3637–3643, 2005.
51. Karlinsey, J.M. and Landers, J.P., Multicolor fluorescence detection on an electrophoretic microdevice using an acoustooptic tunable filter, *Analytical Chemistry*, 78, 5590–5596, 2006.
52. Munro, N.J., Huang, Z.L., Finegold, D.N., and Landers, J.P., Indirect fluorescence detection of amino acids on electrophoretic microchips, *Analytical Chemistry*, 72, 2765–2773, 2000.
53. Sirichai, S. and de Mello, A.J., A capillary electrophoresis microchip for the analysis of photographic developer solutions using indirect fluorescence detection, *Analyst*, 125, 133–137, 1999.
54. Liu, Y.J., Foote, R.S., Jacobson, S.C., Ramsey, R.S., and Ramsey, J.M., Electrophoretic separation of proteins on a microchip with noncovalent, postcolumn labeling, *Analytical Chemistry*, 72, 4608–4613, 2000.
55. Giordano, B.C., Jin, L.J., Couch, A.J., Ferrance, J.P., and Landers, J.P., Microchip laser-induced fluorescence detection of proteins at submicrogram per milliliter levels mediated by dynamic labeling under pseudonative conditions, *Analytical Chemistry*, 76, 4705–4714, 2004.
56. Beard, N.P., Edel, J.B., and deMello, A.J., Integrated on-chip derivatization and electrophoresis for the rapid analysis of biogenic amines, *Electrophoresis*, 25, 2363–2373, 2004.

57. Sanders, J.C., Breadmore, M.C., Kwok, Y.C., Horsman, K.M., and Landers, J.P., Hydroxypropyl cellulose as an adsorptive coating sieving matrix for DNA separations: Artificial neural network optimization for microchip analysis, *Analytical Chemistry*, 75, 986–994, 2003.
58. Ju, J.Y., Kheterpal, I., Scherer, J.R., Ruan, C.C., Fuller, C.W., Glazer, A.N., and Mathies, R.A., Design and synthesis of fluorescence energy-transfer dye-labeled primers and their application for DNA-sequencing and analysis, *Analytical Biochemistry*, 231, 131–140, 1995.
59. Berti, L., Xie, J., Medintz, I.L., Glazer, A.N., and Mathies, R.A., Energy transfer cassettes for facile labeling of sequencing and PCR primers, *Analytical Biochemistry*, 292, 188–197, 2001.
60. Medintz, I.L., Berti, L., Emrich, C.A., Tom, J., Scherer, J.R., and Mathies, R.A., Genotyping energy-transfer-cassette-labeled short-tandem-repeat amplicons with capillary array electrophoresis microchannel plates, *Clinical Chemistry*, 47, 1614–1621, 2001.
61. Hernandez, L., Escalona, J., Joshi, N., and Guzman, N., Laser-induced fluorescence and fluorescence microscopy for capillary electrophoresis zone detection, *Journal of Chromatography*, 559, 183–196, 1991.
62. He, B., Tait, N., and Regnier, F., Fabrication of nanocolumns for liquid chromatography, *Analytical Chemistry*, 70, 3790–3797, 1998.
63. Johnson, M.E. and Landers, J.P., Fundamentals and practice for ultrasensitive laser-induced fluorescence detection in microanalytical systems, *Electrophoresis*, 25, 3513–3527, 2004.
64. Mogensen, K.B., Kwok, Y.C., Eijkel, J.C.T., Petersen, N.J., Manz, A., and Kutter, J.P., A microfluldic device with an integrated waveguide beam splitter for velocity measurements of flowing particles by Fourier transformation, *Analytical Chemistry*, 75, 4931–4936, 2003.
65. Guillo, G., Karlinsey, J.M., and Landers, J.P., On-chip pumping for pressure mobilization of the focused zones following microchip isoelectric focusing, *Lab on a Chip*, 7, 112–118, 2007.
66. Yao, B., Yang, H.H., Liang, Q.L., Luo, G., Wang, L.D., Ren, K.N., Gao, Y.D., Wang, Y.M., and Qiu, Y., High-speed, whole-column fluorescence imaging detection for isoelectric focusing on a microchip using an organic light emitting diode as light source, *Analytical Chemistry*, 78, 5845–5850, 2006.
67. Raisi, F., Belgrader, P., Borkholder, D.A., Herr, A.E., Kintz, G.J., Pourhamadi, F., Taylor, M.T., and Northrup, M.A., Microchip isoelectric focusing using a miniature scanning detection system, *Electrophoresis*, 22, 2291–2295, 2001.
68. Hjerten, S., Free zone electrophoresis, *Chromatographic Reviews*, 9, 122–219, 1967.
69. Sanders, J.C., Huang, Z.L., and Landers, J.P., Acousto-optical deflection-based whole channel scanning for microchip isoelectric focusing with laser-induced fluorescence detection, *Lab on a Chip*, 1, 167–172, 2001.
70. Cheng, S.B., Skinner, C.D., Taylor, J., Attiya, S., Lee, W.E., Picelli, G., and Harrison, D.J., Development of a multichannel microfluidic analysis system employing affinity capillary electrophoresis for immunoassay, *Analytical Chemistry*, 73, 1472–1479, 2001.
71. Huang, Z., Munro, N., Huhmer, A.F.R., and Landers, J.P., Acoustooptical deflection-based laser beam scanning for fluorescence detection on multichannel electrophoretic microchips, *Analytical Chemistry*, 71, 5309–5314, 1999.
72. Simpson, P.C., Roach, D., Woolley, A.T., Thorsen, T., Johnston, R., Sensabaugh, G.F., and Mathies, R.A., High-throughput genetic analysis using microfabricated 96-sample capillary array electrophoresis microplates, *Proceedings of the National Academy of Sciences of the United States of America*, 95, 2256–2261, 1998.
73. Emrich, C.A., Tian, H.J., Medintz, I.L., and Mathies, R.A., Microfabricated 384-lane capillary array electrophoresis bioanalyzer for ultrahigh-throughput genetic analysis, *Analytical Chemistry*, 74, 5076–5083, 2002.
74. Shi, Y.N., Simpson, P.C., Scherer, J.R., Wexler, D., Skibola, C., Smith, M.T., and Mathies, R.A., Radial capillary array electrophoresis microplate and scanner for high-performance nucleic acid analysis, *Analytical Chemistry*, 71, 5354–5361, 1999.
75. Liu, S.R., Shi, Y.N., Ja, W.W., and Mathies, R.A., Optimization of high-speed DNA sequencing on microfabricated capillary electrophoresis channels, *Analytical Chemistry*, 71, 566–573, 1999.

76. Schmalzing, D., Koutny, L., Chisholm, D., Adourian, A., Matsudaira, P., and Ehrlich, D., Two-color multiplexed analysis of eight short tandem repeat loci with an electrophoretic microdevice, *Analytical Biochemistry*, 270, 148–152, 1999.
77. Backhouse, C., Caamano, M., Oaks, F., Nordman, E., Carrillo, A., Johnson, B., and Bay, S., DNA sequencing in a monolithic microchannel device, *Electrophoresis*, 21, 150–156, 2000.
78. Simpson, J.W., Ruiz-Martinez, M.C., Mulhern, G.T., Berka, J., Latimer, D.R., Ball, J.A., Rothberg, J.M., and Went, G.T., A transmission imaging spectrograph and microfabricated channel system for DNA analysis, *Electrophoresis*, 21, 135–149, 2000.
79. Karlinsey, J.M., Multicolor Short Tandem Repeat (STR) Analysis on Microchip, *Advances and Applications of Microfluidic Analysis Systems*, 2007, University of Virginia: Charlottesville, VA.
80. Fister, J.C., Jacobson, S.C., and Ramsey, J.M., Ultrasensitive cross correlation electrophoresis on microchip devices, *Analytical Chemistry*, 71, 4460–4464, 1999.
81. Haab, B.B. and Mathies, R.A., Single-molecule detection of DNA separations in microfabricated capillary electrophoresis chips employing focused molecular streams, *Analytical Chemistry*, 71, 5137–5145, 1999.
82. Foquet, M., Korlach, J., Zipfel, W.R., Webb, W.W., and Craighead, H.G., Focal volume confinement by submicrometer-sized fluidic channels, *Analytical Chemistry*, 76, 1618–1626, 2004.
83. Chabinyc, M.L., Chiu, D.T., McDonald, J.C., Stroock, A.D., Christian, J.F., Karger, A.M., and Whitesides, G.M., An integrated fluorescence detection system in poly(dimethylsiloxane) for microfluidic applications, *Analytical Chemistry*, 73, 4491–4498, 2001.
84. Roulet, J.C., Volkel, R., Herzig, H.P., Verpoorte, E., de Rooij, N.F., and Dandliker, R., Performance of an integrated microoptical system for fluorescence detection in microfluidic systems, *Analytical Chemistry*, 74, 3400–3407, 2002.
85. Maims, C., Hulme, J., Fielden, P.R., and Goddard, N.J., Grating coupled leaky waveguide micro channel sensor chips for optical analysis, *Sensors and Actuators B—Chemical*, 77, 671–678, 2001.
86. Applegate, R.W., Squier, J., Vestad, T., Oakey, J., Marr, D.W.M., Bado, P., Dugan, M.A., and Said, A.A., Microfluidic sorting system based on optical waveguide integration and diode laser bar trapping, *Lab on a Chip*, 6, 422–426, 2006.
87. Adams, M.L., Enzelberger, M., Quake, S., and Scherer, A., Microfluidic integration on detector arrays for absorption and fluorescence microspectrometers, *Sensors and Actuators A—Physical*, 104, 25–31, 2003.
88. Edel, J.B., Beard, N.P., Hofmann, O., deMello, J.C., Bradley, D.D.C., and deMello, A.J., Thin-film polymer light emitting diodes as integrated excitation sources for microscale capillary electrophoresis, *Lab on a Chip*, 4, 136–140, 2004.
89. Hofmann, O., Miller, P., Sullivan, P., Jones, T.S., deMello, J.C., Bradley, D.D.C., and deMello, A.J., Thin-film organic photodiodes as integrated detectors for microscale chemiluminescence assays, *Sensors and Actuators B—Chemical*, 106, 878–884, 2005.
90. Ocvirk, G., Tang, T., and Harrison, D.J., Optimization of confocal epifluorescence microscopy for microchip-based miniaturized total analysis systems, *Analyst*, 123, 1429–1434, 1998.

46 Microfabricated Electrophoresis Devices for High-Throughput Genetic Analysis: Milestones and Challenges

Charles A. Emrich and Richard A. Mathies

CONTENTS

46.1 INTRODUCTION

The first microfabricated separation system introduced by Harrison et al. [1,2] highlighted the unique marriage of photolithography with analytical chemistry to perform ultrafast, high-resolution separations of subnanoliter volume samples. The lab-on-a-chip concept that grew from these early papers leveraged two important aspects of microfabrication. First, the ability to integrate complicated electrical and mechanical structures directly with the separation channels enables prepurification sample preparation, a concomitant reduction in manual sample handling steps and costly reagent volumes, and the eventual miniaturization of the "lab" for point-of-analysis studies. The second advantage that microfabrication adds is the ability to create arbitrarily complex arrays of channels on the same substrate. The photolithographic techniques used are equivalent regardless of whether the desired channel geometry is simple or complex, in sharp contrast to conventional capillary arrays where each additional capillary exponentiates the complexity and decreases robustness. Thus, microfabricated capillary electrophoresis (CE) systems are ideally suited to perform high-throughput analysis.

The first microfabricated capillary array electrophoresis (μCAE) device appeared in 1997 [3] featuring 12 separation channels interrogated by a scanning laser-induced fluorescence (LIF) detector.

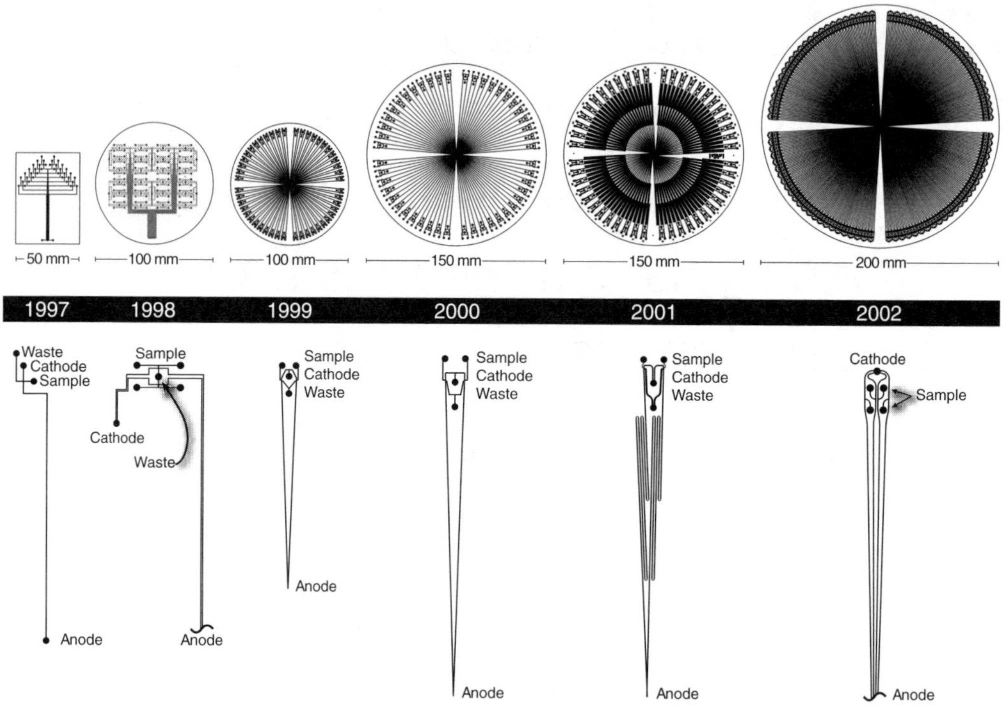

FIGURE 46.1 The evolution of high-throughput μCAE devices. The earliest devices were fabricated on large microscope slides and used rectilinear arrays of channels, which terminated at a common anode. More complex rectilinear arrays were clearly not scalable and were replaced by modern radial arrays. Radial channel arrays are simple to design and operate and are easily scalable by increasing the wafer diameter.

A race for extending and maximizing the density of channel arrays quickly followed. As Figure 46.1 illustrates, arrays grew in channel density first to 48, then 96, and ultimately to 384 channels [4]. To accommodate the growing number of channels, the "microchips" on which they were fabricated also grew from the size of microscope slides to window-size plates half a meter in length [5]. Much of the growth in channel number and areal density was facilitated by an early shift from rectilinear to radial arrays of channels [6] and the simultaneous development of rotary confocal detection systems.

Despite the great increase in throughput provided by μCAE systems, they are only half way to displacing conventional separation formats. While the resolution and sequencing read lengths of μCAE systems are on par with their conventional counterparts and the time required for similar separations is shorter, they are much closer to a capillaries-on-a-chip than they are to labs-on-a-chip. The next step in the evolution of lab-on-a-chip devices for high-throughput analysis is the integration of sample preparation components before separation. DNA amplification processes such as the polymerase chain reaction (PCR) and Sanger cycle sequencing are the most attractive sample preparation steps and have been integrated with CE exploiting the development of on-chip temperature and valve control elements [7]. On-chip extraction techniques have also been recently demonstrated for purification of DNA directly from clinical samples that often contain an array of inhibitory background molecules [8]. Sample cleanup steps have also been developed to desalt and concentrate Sanger extension fragments before electrokinetic injections, greatly increasing the performance of high-resolution DNA-sequencing separations [9]. The integration of these elements into multichannel arrays has advanced with the recent demonstration of four-channel PCR-CE systems [10,11]. Total integration of sample preparation techniques with on-chip CE analysis came of age in 2006 with the demonstration of single-channel systems for DNA sequencing [12] and infectious disease typing [13]. Clearly, the successful implementation of integrated components is the next step in μTAS lab-on-a-chip systems.

The demand for high-throughput electrophoretic analysis, which grew exponentially after the birth of PCR in 1988 [14] and with the sequencing of the human genome [15] has only increased in recent years. New metagenomic tests are being used to sample the genomes of environmental specimens [16], mitochondrial barcodes are now used to catalog the world's animal species [17], and increasingly DNA evidence has become the gold standard for human identification in criminal investigations. Taken further, human identification by electrophoresis can be used beyond the crime lab in cases of parental determination and to identify victims of mass disasters [18].

DNA electrophoretic techniques are also becoming increasingly relevant not only in the understanding of cancer but also, enabled by recent breakthroughs in so-called targeted therapies, in charting a course for its treatment. The Lazarus-like response of certain patients to the lung cancer drug "gefitinib" was explained by mutations in the growth factor receptor EGFR, which were found by DNA sequencing [19]. More recently, the identification of a metagene construct for evaluating the severity of nonsmall cell lung cancers promises a more accurate determinant of individual patient therapy than the traditional staging system [20]. Because of the genetic variability of both cancers and patients, it is likely that the need for high-throughput genetic screens will only increase in the future. Thus, it is imperative that electrophoretic techniques be pushed toward higher-throughput, lower cost, and smaller size to truly enable the coming generation of personalized medicine. Such developments in integrated μCE systems incorporating sample purification, amplification, and analysis with robust temperature and fluidic control have recently begun to emerge [21] and will provide an effective foundation on which to build future bioanalysis devices.

46.2 BACKGROUND AND EVOLUTION μCAE SYSTEMS

The first high-throughput arrayed CE microdevice was demonstrated 5 years after the first single-channel analysis system and consisted of 12 channels with linear scanning detection [3]. The design of the 12-channel μCAE device presented in Figure 46.1 is a modification of the original cross-injector design set out by Harrison et al. [22] and proven by Woolley and Mathies [23] for CE separations of DNA. The injectors of the various channels are arrayed to minimize the dimensions of the chip (maximizing channel density) while leaving sufficient space between reservoirs to maintain electrical and sample isolation. Each of the 12 channels terminates at a common anode reservoir located just below the detection point, for a total of $3N+1$ drilled reservoirs where N is the number of samples. Interrogation of the channels is performed by linearly translating the μCAE device below a laser confocal fluorescence point detection system.

The 12-channel μCAE devices were fabricated on glass microscope slides by using photolithography and wet chemical etching to 8-μm depths. The etch depth of the channels, in this case, was limited because during photolithography the photoresist was patterned directly on the surface of the wafers without the use of a thin-film "hard" mask layer. Photoresists typically exhibit poor adhesion to hydrophilic surfaces such as glass, resulting in significant undercutting of the photoresist during wet chemical etching steps. This undercutting introduces defects in the etched surface, especially with long etches, and significantly reduces the resolution of the photolithographic pattern.

The 12-channel μCAE device was successfully demonstrated by genotyping twelve individuals for mutations in the human *HLA-H* gene in a restriction fragment length polymorphism (RFLP) assay. Samples were noncovalently labeled with DNA intercalating dyes and run simultaneously with DNA standards for accurate sizing. The 0.5% hydroxyethylcellulose (HEC) separation matrix has low viscosity facilitating its loading into the shallow channels. Despite the differing channel routing in the μCAE device, the electropherograms presented in Figure 46.2 display little channel-to-channel variation, achieved resolution of better than 10 bp, and were complete in 160 s.

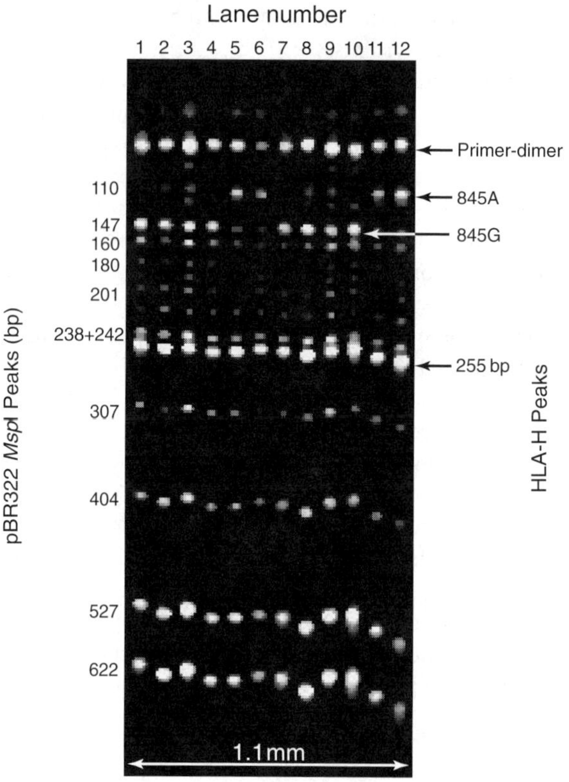

FIGURE 46.2 Electropherograph of a HLA-H genotyping separation using the 12-channel μCAE device. Fluorescent labeling is achieved by the noncovalent intercalating dyes thiazole orange and butyl TOTIN. Separations were monitored by linearly translating the device under a laser confocal detection system focused just upstream of the common anode shared by all channels. A pBR *Msp*I digest is added as a sizing standard. (Reproduced from Woolley, A.T. et al., *Anal. Chem.*, 69, 2181, 1997. With permission.)

46.2.1 Advancing to 96 Samples

The next significant step in the evolution of μCAE devices required transitioning from glass microscope slides to larger, 100-mm diameter wafers made from Borofloat glass, which is more compatible with microfabrication processes because of its low sodium content and also exhibits lower autofluorescence, improving detection limits. A 48-channel μCAE device built on the larger wafers is presented in Figure 46.3 and features 96 sample reservoirs, pairs of which share the forty-eight 10-cm long separation channels [24]. The 48-channel μCAE chips were microfabricated using an amorphous silicon hard-mask deposited by plasma-enhanced chemical vapor deposition. The addition of this hard-mask step was critical to achieving high-fidelity microfabrication of 21-μm deep channels by wet chemical etching. These deeper channels permit higher-viscosity sieving matrices to be used, which enable higher resolution separations. Interrogation of the channels is performed by a linear galvoscanner to successfully sample each channel at a rate of 40 Hz.

The 48-channel μCAE device employs 96 sample reservoirs, 6 cathode reservoirs, 24 waste reservoirs, and a common anode reservoir. This design reduces the number of necessary reservoirs to $5/4N+7$, which is close to the theoretical minimum of $N+3$ for the case where the anode, cathode, and waste reservoirs are all shared. This minimization of total reservoirs is important because the reservoirs are the largest consumer of area on the chip surface. Not only are they 1.25-mm in diameter, but positional error in drilling places limits on the reservoir density. The 48-channel μCAE chip was

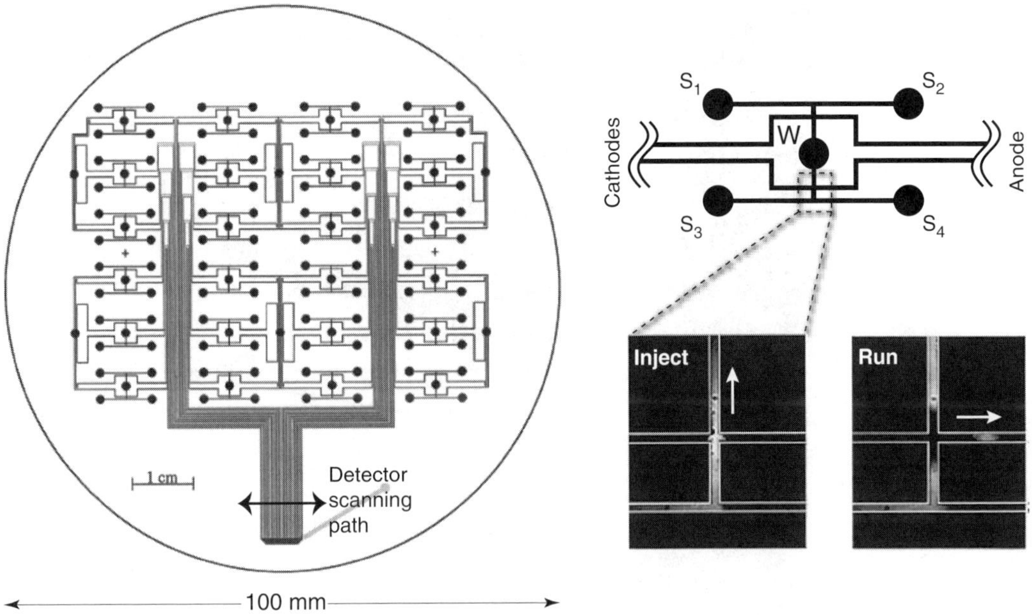

FIGURE 46.3 Design of the 48-channel μCAE chip. The unique cross-injector design allowed two sample reservoirs (S) to utilize the same separation channel in succession. The total number of reservoirs necessary for operation is greatly reduced by grouping waste (W) and cathode reservoirs. Each channel has a 10-cm effective separation length terminating in a common anode. Detection is achieved using a galvoscanner along the indicated linear path. (Adapted from Simpson, P.C. et al., *Proc. Natl. Acad. Sci. USA*, 95, 2256, 1998. With permission.)

demonstrated by genotyping 96 individuals for the C282Y mutation in the human *HFE* gene using a PCR-RFLP assay. Each separation channel in the 48-channel device is used to serially analyze two separate samples. Figure 46.4 presents the results from such a double-injection run of 96 *HFE* genotyping samples. Electrophoresis is complete within 8 min for both sets of samples but significant variation in peak retention times is observed, possibly as a result of channel geometry or variation of electrode position in the reservoirs.

The rectilinear layout of this chip enables quick sample loading via multichannel pipettors, but this design paradigm scales poorly for higher channel density arrays. Similarly, the integration of sample preparation components is clearly difficult for such designs because of the limited area around each sample reservoir. Another concern is that while channel lengths are equalized in this design, they still incorporate right-angle turns that were later revealed to be a significant source of dispersion in on-chip electrophoretic separations [25].

46.2.2 Transition to Radial Formats

Further improvements in high-resolution, high-throughput μCAE necessitated a paradigm shift from rectilinear to radial arrays of channels and a concomitant advancement in detection technologies. The first radial μCAE device [6] arrayed 96 separation channels around a central common anode on a 100-mm diameter glass substrate as presented in Figure 46.5. Pairs of channels are grouped into doublet structures that share common cathode and waste reservoirs. The design of this radial μCAE device offers significant advantages over previous rectilinear arrays. First, all the channels have equivalent lengths and geometries, negating any variability in electrophoretic conditions that might be present in irregular channels. Second, the radial array paradigm is perfectly complimentary to the circular wafers used in microfabrication processes, maximizing utilization of existing wafer

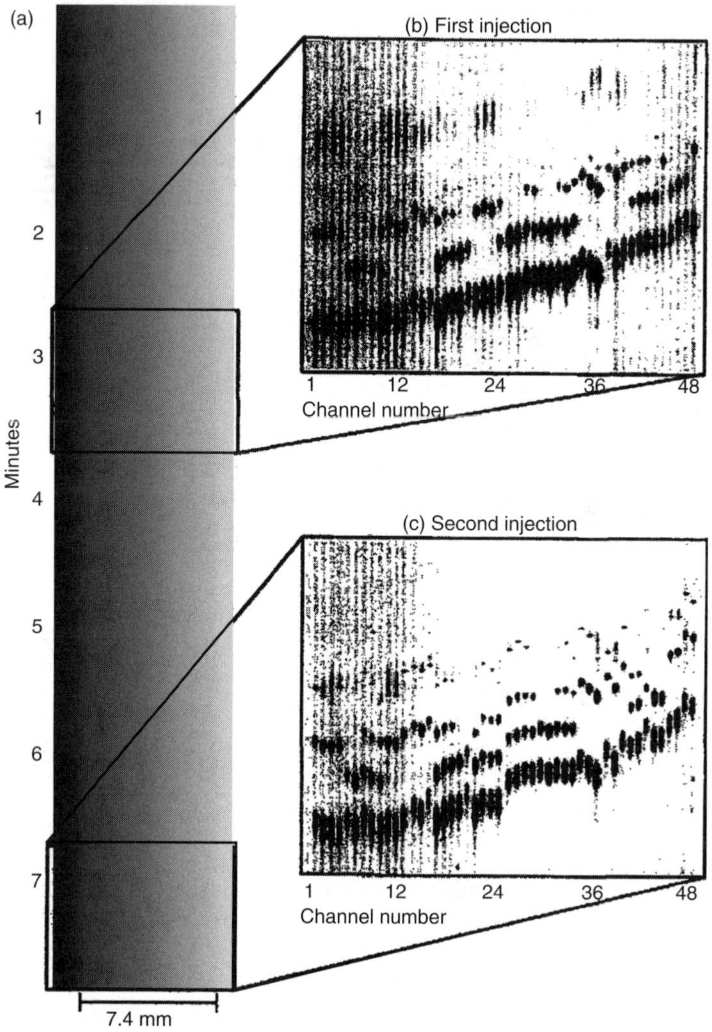

FIGURE 46.4 Electrophoresis pseudo gel image from a genotyping separation performed on the 48-channel μCAE device. The 96 samples are analyzed in two successive runs, successfully discriminating between full-length 167-bp fragments and 111 or 140-bp digested fragments. (Reproduced from Simpson, P.C. et al., *Proc. Natl. Acad. Sci. USA*, 95, 2256, 1998. With permission.)

area by positioning the drilled reservoirs at the edge of the wafer where the per-channel-area is greatest. Positioning the reservoirs at the periphery of the wafer in a radially symmetric fashion also allows simple construction of electrode arrays to address each reservoir. Third, the separation channels in a radial array μCAE device can be straight between the injection point and the detection zone. While longer separation channels generally yield higher resolution than shorter channels, any turns in those channels can act as a source of dispersion in the bands. Fourth, radial arrays of channels are clearly the best design for scaling to ultrahigh-throughput analysis devices, as wafer sizes >100-mm diameter are readily available and the tools for processing of wafers up to 300-mm diameter are well developed and commonplace in the semiconductor industry. Thus, increasing the number of channels can be accomplished by increasing wafer size to accommodate the necessary number of sample, cathode, and waste reservoirs. Further, integration of sample preparation steps into a μCAE design becomes viable when the sample reservoirs are at the periphery of the analysis device where there is more available surface area. Components of such an integrated device can then be incorporated by placing them distal to the sample reservoirs, and the number of preseparation steps will be limited only by the ultimate size of the substrate.

The shift to radial channel arrays also necessitated a shift in detection methodology from linear scanning to rotary scanning as presented in Figure 46.6. The current generation rotary scanner directs excitation from an Ar^+ laser up through a hollow rotating shaft atop which sits a rhomb prism and objective that respectively deflect the laser 1-cm off the axis of rotation and focus the beam into the microchannels. Fluorescence is collected along the same path as excitation, passing through a dichroic beamsplitter and into a four-color confocal photomultiplier tube (PMT) detector array. In addition, the Berkeley rotary confocal fluorescence scanner is an inverted optical system, which puts all the optical elements on the underside of the chip and all electrical connections on the opposite, top side of the chip, thereby maximizing flexibility of the chip design.

The performance of the 96-channel, 100-mm diameter μCAE device was demonstrated by simultaneously genotyping 96 individuals for the C677T mutation in the human *MTHFR* gene [6]. Electrophoresis was performed at 200 V/cm through a 1.0% w/v HEC sieving matrix and used dimeric intercalating dyes for noncovalent fluorescent labeling. Separations on the 3.5-cm long channels were complete within 120 s, corresponding to a sample throughput of approximately 0.5 samples/s. The utility of this radial design was further advanced by increasing the diameter of the substrate from 100 to 150 mm. By increasing the wafer size, the effective separation length of each channel was increased from 3.5 to 5.5 cm, which increased analysis time, but also increased the ultimate resolution of the μCAE device. This increase in resolution enabled more demanding analyses such as the short tandem repeat typing assays commonly used in forensic identification, which require near single-base resolution. Medintz et al. [26] performed the first μCAE short tandem repeat (STR) separations using a covalent fluorescent multiplexing scheme, demonstrating extremely high-resolution separations in under 8 min with $<1\%$ sizing variance. Results from a single multiplex STR genotyping separation are presented in Figure 46.7. The improvement in resolution over previous separations can be attributed not only to the increase in effective separation length from 3.5 to 5.5 cm, but also to the use of a high-performance, denaturing linear polyacrylamide (LPA) separation matrix. The development of a simple and robust high-pressure gel loader and washer by Scherer et al. [27] facilitated the use of high-viscosity LPA in the 150-mm μCAE chip. By applying pressurized helium gas to the sieving matrix, it can be forced into the channels of μCAE device through the common central anode reservoir in under 10 min. After the completion of an electrophoretic run, sieving matrix can then be washed from the radial μCAE chip with water and dried using the same apparatus.

46.2.3 Optimizing Microchip DNA Sequencing

The ultimate test of any electrophoretic separation system is DNA sequencing of Sanger extension fragments. Sequencing separations not only require single-base resolution, but also require that degree of resolution over 2–3 orders of magnitude of sample fragment size. In addition, the longer the DNA sequencing read length, the more value it has during the final sequence assembly step. The read lengths demonstrated by microfabricated DNA separation systems by 2000 had been significantly less than the 500+ base reads from commercial capillary array systems [28,29]. Notable exceptions include straight-channel microdevices with separation lengths from 11 to 40 cm [30,31], which achieved DNA sequencing reads of up to 800 bases. However, the increased size and complexity of fabrication and design of these longer "capillary-on-a-window" style separation devices negated many of the advantages of miniaturization that are the hallmarks of microfabricated separation systems. The obvious solution was to increase the effective separation length of channels on μCAE chips without increasing substrate size by folding the channels back along themselves. However, turns are effective sources of dispersion in μCE separations [25,32,33]. The dispersion is due to the racetrack effect, in which species all traveling at the same velocity will pass through a turn more quickly along the shorter inside track of the turn than along the longer outside track. This effect is

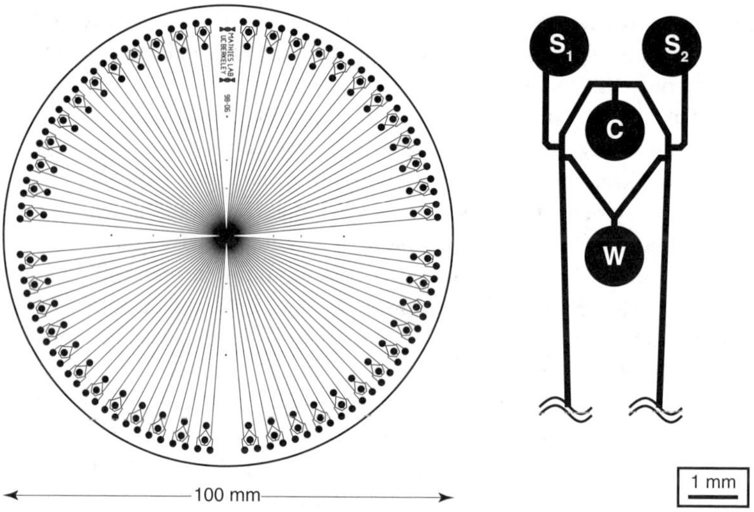

FIGURE 46.5 The first radial array μCAE chip featured 96 sample channels arrayed about a single, common anode on a 100-mm diameter glass substrate. Utilization of space was maximized by placing reservoirs at the periphery of the wafer and grouping pairs of channels into a doublet structure, which shared cathode (C) and waste (W) reservoirs. Detection was performed along a circular path 1 cm from the anode with the scanner described in Figure 46.6. (Adapted from Shi, Y.N. et al., *Anal. Chem.*, 71, 5354, 1999. With permission.)

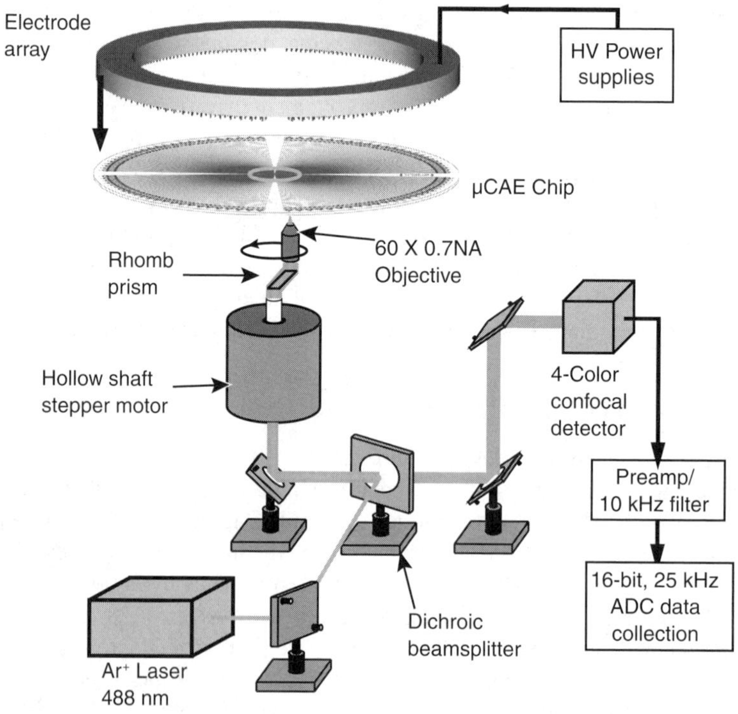

FIGURE 46.6 The Berkeley rotary confocal fluorescence scanner developed by James R. Scherer. Laser excitation is directed up through the hollow drive shaft of a stepper motor, deflected off axis by a rhomb prism, and is then focused with a microscope objective, tracing a 1-cm radius circular detection path on the chip. Fluorescence is collected along the same path and then spatially and spectrally filtered by a four-color PMT array. The inverted nature of this system allows easy access to the upper surface of a μCAE device for electrical, pneumatic, and sample interfacing.

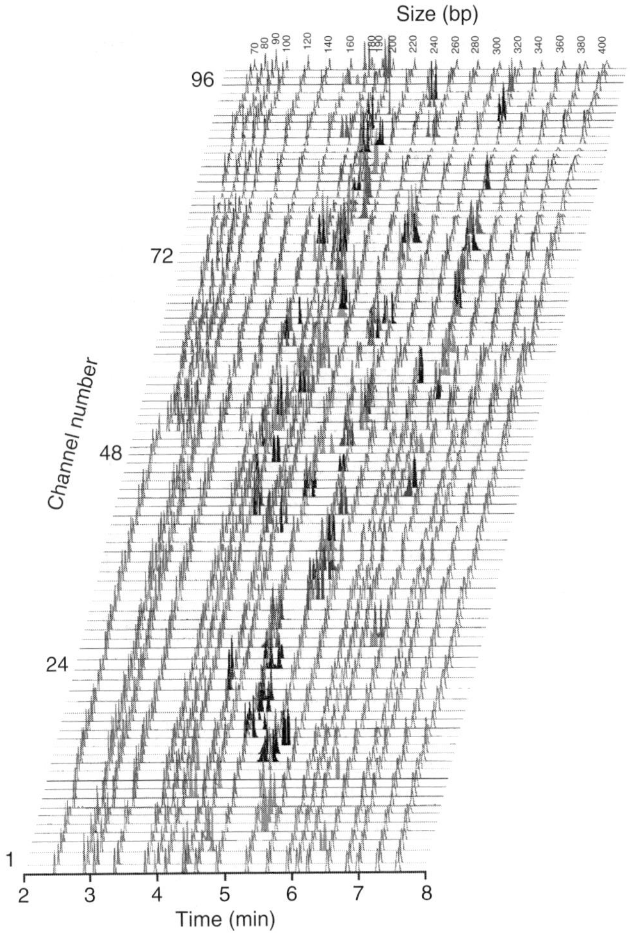

FIGURE 46.7 **(See color insert following page 810.)** Electropherograms from a multiplexed short tandem repeat sizing separation on a 150-mm diameter, 96-channel μCAE device. Separations are successful in all lanes and sizing against DNA standards (blue and red) was accurate to within 1% of known values. (Reproduced from Medintz, I.L. et al., *Clin. Chem.*, 47, 1614, 2001. With permission.)

exacerbated in electrokinetic transport because the shorter inside path also has a lower resistance than the outside path, and will thus experience a higher electric field strength that further increases the difference in transit time between the inside and outside tracks. In an elegant series of experiments, Paegel et al. [34] discovered that the racetrack dispersion of bands passing through turns could be almost eliminated simply by narrowing the width of the channel for the duration of the turn.

The development of these hyperturns was a critical advance toward development of a μCAE system for high-throughput DNA sequencing by enabling the longer separation lengths necessary for high-resolution separations. By folding the separation channels back on themselves with hyperturns it was possible to extend the channels on a 150-mm diameter wafer to a total effective separation length of 15.9 cm (see Figure 46.8). The total combined separation length on the μCAE device presented in Figure 46.8 is 15.25 m, with the channels taking up the majority of the surface area of the wafer. Maximizing the amount of utilized space on the wafer surface was also found to significantly improve the success rate of thermal bonding of μCAE devices for two reasons. The smaller contact area between the wafers will experience greater pressure during bonding and the presence of the channels provides convenient escape routes for gas trapped between the wafers during bonding.

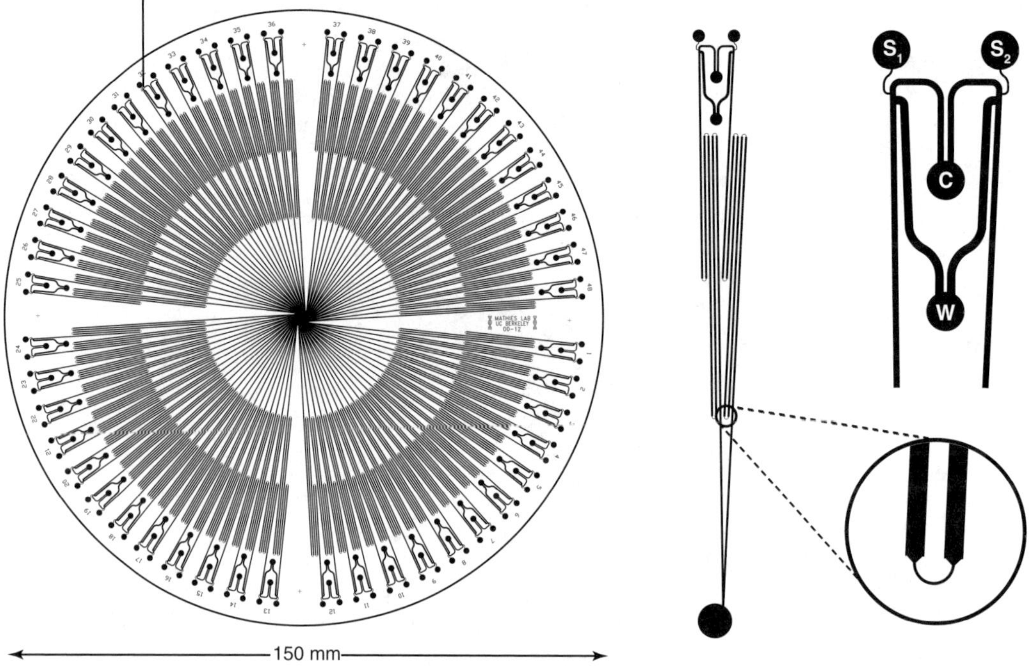

FIGURE 46.8 Design of the 96-channel μCAE sequencer. Channels are grouped into doublets and folded back on themselves twice for an effective separation length of 15.9-cm. Each turn uses the indicated hyperturn geometry to minimize turn-related dispersion. The lengths and widths of channels in the 250-μm offset cross-injector are balanced to provide equal fluidic resistance during gel loading. (Adapted from Paegel, B.M. et al., *Proc. Natl. Acad. Sci. USA*, 99, 574, 2002. With permission.)

The performance of the 96-channel μCAE sequencer was demonstrated by simultaneous electrophoresis of 96 identical *M13mp18* vector DNA sequencing standards as presented in Figure 46.9. Electrophoresis was performed at 240 V/cm through LPA under denaturing conditions resulting in continuous DNA sequencing reads of 430 bases with >99% accuracy in 24 min [35]. This level of accuracy, corresponding to a Phred score of 20 is recognized as the standard metric for determining read length. The longer time required for separations of DNA sequencing fragments is necessary because of the increased length of the channels, the more efficient long-chain sieving matrix, and the upper practical limit placed on the electric field due to the onset of biased repetition of the larger fragments. The longer electrophoresis times necessary for high-resolution separation of DNA sequencing fragments also necessitated an increase in the buffering capacity of the cathode and anode reservoirs. This volume increase was made possible by affixing Plexiglas moats to the surface of the chip, and filling the moats with 3 mL of Tris-TAPS-EDTA run buffer. The added buffer volume was necessary to offset the hydrolytic production of hydroxyl ions at the cathode and protons at the anode reservoirs. These hydrolytic products titrate and then acidify the anode buffer, creating a local pH drop that reduces the net negative charge on the DNA and significantly reduces the overall separation efficiency.

The 96-channel μCAE sequencer proved to be a robust high-resolution separation platform that is easily extended to other applications. Through creative fluorescent multiplexing, Blazej et al. [36] used the μCAE sequencer to mine two human mitochondrial genomes for mutations, analyzing an entire genome versus a reference genome in a single electrophoretic run. More recently, Yeung et al. [37] demonstrated the utility of the μCAE sequencer to high-resolution STR typing separations for forensic DNA identification. The system performed separations with resolution and sensitivity equal

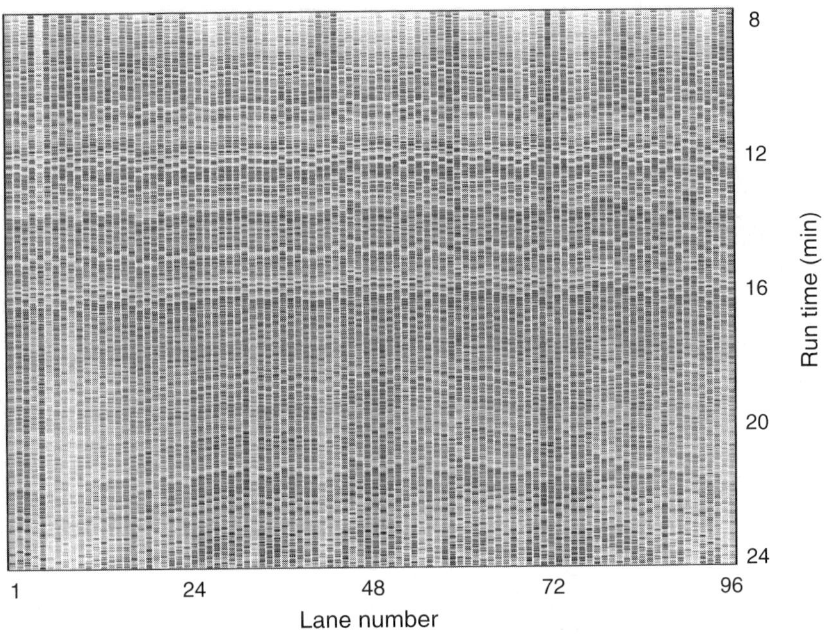

FIGURE 46.9 Results from a DNA sequencing separation of *M13mp18* standards on the 96-channel μCAE device. Average sequencing read lengths of 430 bases of ≥99% accuracy were obtained, and only one of the 96 lanes failed. (Reproduced from Paegel, B.M. et al., *Proc. Natl. Acad. Sci. USA*, 99, 574, 2002. With permission.)

to conventional capillary systems in less than half the time, and proved useful not only for standard sizing ladders but also for actual (nonprobative) casework samples.

46.2.4 To the Ultimate 384 Lane Microdevice

Increasing throughput beyond what was demonstrated for 96-channel μCAE devices require an increase in the wafer diameter. Figure 46.10 presents the design of a 384-lane μCAE device for ultrahigh throughput genotyping [4]. This channel density is close to the practical maximum for radial arrays and the Berkeley rotary confocal fluorescence detection and required the replacement of standard cross-injector with a direct injector. The direct injector used in the 384-lane μCAE bioanalyzer provides significant savings of area by eliminating the waste reservoirs associated with a cross-injector and also groups channels into quartets, which share a common cathode reservoir. Thus, the total number of reservoirs is reduced to $5/4N+1$, but the elimination of the cross-injector comes at a cost of reduced resolution because of the larger direct-injected plug size. Much of this lost resolution is made up by the increased 8.0-cm effective separation length afforded by the larger diameter wafer, achieving theoretical plate numbers of up to 4×10^6 for separations of DNA size standards. The performance of the 384-channel μCAE bioanalyzer was demonstrated by simultaneously genotyping 384 individuals for the H63D mutation in the human *HFE* gene via an RFLP assay. The electrophoretic separation was performed at 260 V/cm and was complete within 325 s for an overall throughput of >1 samples/s. Results of the separation presented in Figure 46.11 reveal excellent reproducibility between channels and the 98.7% success rate of the separations stands as a testament to the robustness of both the radial array concept and the rotary confocal fluorescence scanner.

The design, manufacture, and operation of the 384-lane μCAE bioanalyzer likely represents a practical maximum for throughput in microfabricated electrophoresis systems. The fabrication

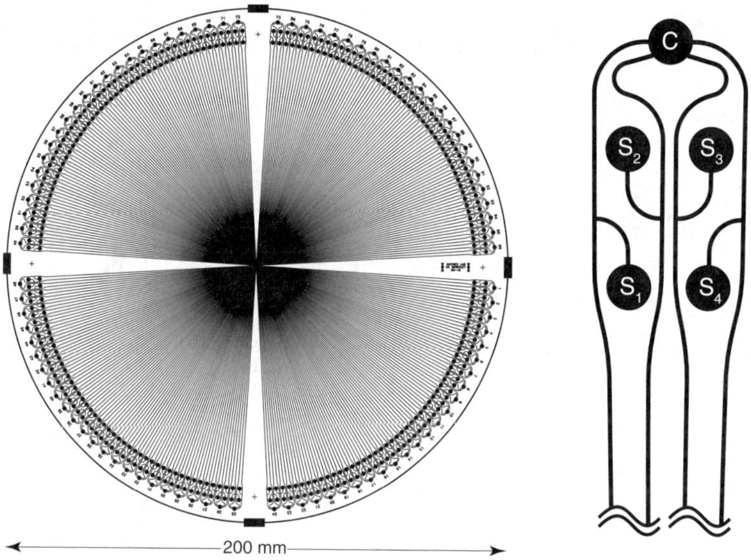

FIGURE 46.10 Design of the 384-lane μCAE bioanalyzer. Channels are grouped into quartets, which share a common cathode reservoir. To save space on the 200-mm diameter substrate, a direct injection scheme is used in place of a cross-injector. The resolution lost by the transition to direct injectors is partially recovered by an increased effective separation length of 8.0 cm. (Adapted from Emrich, C.A. et al., *Anal. Chem.*, 74, 5076, 2002. With permission.)

of larger devices requires significant investments in new tools, specialized and time-consuming hole drilling equipment, and robotics for performing sample loading. A similarly high-throughput system for DNA sequencing was recently demonstrated by Aborn et al. [5] fabricated on 25 × 50 cm rectangular glass substrates. This system also utilizes a rotary confocal fluorescence scanner for detection, but the size and complexity of the μCAE devices necessitates a complex and costly robotic station for sample and buffer loading, gel filling, and cleaning of the wafers. Further, the cross-injector design necessary for impressive, high-resolution sequencing increases the complexity of design of the device enough to require nonuniform separation channel lengths.

Overall, μCAE devices presented in the literature thus far offer only limited advantages over conventional capillary electrophoresis systems for high-throughput bioanalysis: a decadic improvement in speed, and no change in detection limits. Thus, μCAE systems are not as compelling an upgrade as CAE was over slab gel systems (i.e., automatable technology, 10× faster analysis, simplified and more sensitive detection methodology). To offer the same kind of paradigm shift in high-throughput analysis, μCAE systems will need to leverage integration of sample preparation components to fully exploit the advantages in speed, and reductions in both manual sample handling and reagent volumes that microfabricated analysis systems offer.

46.2.5 Integration: The Final Frontier

The cross-injector enables analyses with microfabricated systems an order of magnitude faster than capillary-based systems. This advantage is due to the extremely small sample zones that can be injected (≤1 nL) with the cross-injector, but the consequence of these minute zones is that the cross-injector is extremely inefficient. Samples are typically pipetted at 1–2 μL volumes into the drilled reservoirs on a microdevice. Even when considering large injected sample plugs of 1 nL and a low-volume, pipetted sample (1 μL), the cross-injector is only 0.1% efficient. That is, despite the excellent low-volume fluid-handling promises of microsystems, they still require input volumes

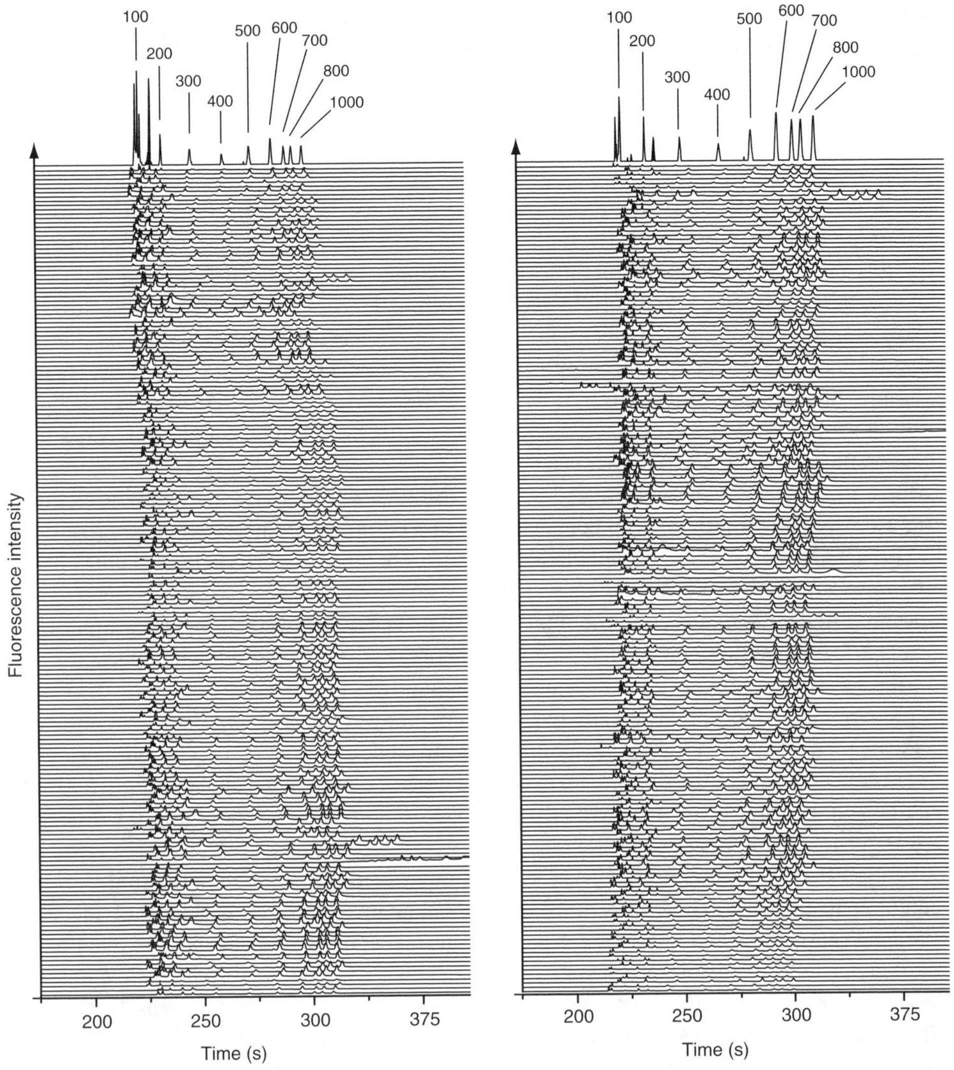

FIGURE 46.11 Results of a single run on the 384-lane μCAE bioanalyzer demonstrating high resolution (4.0×10^6 theoretical plates) and low lane-to-lane variation in peak elution time (overall RSD < 2%). Fragments were typed for the H63D mutation in the human *HFE* gene by RFLP successfully in 379 of the 384 lanes. Electrophoresis was carried out at 260 V/cm through a PDMA sieving matrix and used a fluorescent intercalating dye for labeling. (Adapted from Emrich, C.A. et al., *Anal. Chem.*, 74, 5076, 2002. With permission.)

hundreds or thousands of times larger than the volume of sample that is ultimately analyzed. The scale of inefficiency of the cross-injector should not be underestimated. If slab gel electrophoresis were 0.1% efficient with respect to the analyzed volume of loaded sample, each sample would needed to be loaded in 10 mL aliquots, instead of the 10 μL volumes typically used.

The use of on-chip PCR reactors both decreases the analysis time by taking advantage of the low thermal mass of microfabricated systems and also increases the efficiency of the cross-injector. PCR reactors in microfabricated devices are typically ~250 nL in volume and are directly coupled to the cross-injector [7]. The reduction in initial sample size increases the scale of efficiency 5- to 10-fold when compared to a drilled reservoir, but introduces another drawback: injection bias in high-salt samples. Any electrokinetic injection is biased toward high-concentration, high-mobility species. The injected fraction f of any species i is given by

$$f_i = \frac{c_i \mu_i}{\sum_i c_i \mu_i}, \tag{46.1}$$

where c is concentration and μ is electrophoretic mobility. According to Equation 46.1, high-concentration, high-mobility species, such as salts, will be injected preferentially over scarce, low-mobility DNA. This problem is significant for samples that have been amplified by PCR, which typically contain $\geq$50 mM Cl^-, and $\sim\mu$M dNTPs. Both will be injected preferentially with respect to lower-concentration amplified products. This deconcentration event is the reverse of sample stacking phenomena [38] that are used to concentrate samples before or at the outset [39] of many separations. For example, the high-salt buffer (50 mM Tris–HCl, 100 mM NaCl, 10 mM $MgCl_2$) used in *Mbo*I digestion of samples in [4] necessitated the manual desalting of all 384 samples analyzed to obtain satisfactory injections and separations.

Various approaches have been investigated to mitigate the extreme inefficiency of the cross-injector, most by performing on-chip concentration and desalting of analytes before cross-injection [9,40,41]. Paegel developed an oligonucleotide-capture chamber to purify and concentrating DNA-sequencing extension fragments [9]. This capture system used a LPA gel copolymerized with an Acrydite-modified oligo that is complimentary to a sequence of the product to be captured. Under an applied electric field, amplified products are electrophoresed through the gel, where they are captured by the tethered oligos, allowing other charged species such as salt, excess primers, and dNTPs to pass through into a waste reservoir. The captured products, in this case *M13mp18* sequencing extension products, are then released by heating to $T > T_m$ and injected for CE analysis. This scheme resulted in significant concentration of the sample (up to the μM concentration of the capture oligo) and obviated the time-consuming and labor-intensive ethanol precipitation step commonly used for purification of DNA sequencing fragments.

Recently, Blazej et al. combined this oligo-capture system directly with a thermal-cycling reaction chamber to create the first fully integrated sequencing device. The design of the device is presented in Figure 46.12 and incorporates on-chip thermal control elements [7] and pneumatic fluid control elements [42] to perform a Sanger cycle sequencing reaction, the products of which are then fed into an oligo-capture chamber. Purified extension fragments are then thermally released and electrophoretically separated at 167 V/cm on a 30-cm long folded channel and detected on a four-color fluorescence scanner. The device was able to generate 556 continuous bases of sequence with $\geq$99% accuracy starting from only 1 fmol of template, which is a $>$10$\times$ reduction in the amount of necessary template and reagents for sequencing versus conventional methods. This device achieves at last the vision for integration set forth by Harrison in 1993 [2] by combining microfabricated components in a way that exceed the performance of commercial systems in terms of input volume, reagent consumption, analysis time, and sensitivity.

Adapting similar integrated systems to multichannel array systems has been challenging due to the difficulty inherent in designing robust microfluidic elements that can be controlled with a minimum of complexity. The first steps toward such integrated array CE systems were recently taken by Liu and Toriello, who developed a high-sensitivity PCR-CE microdevice [10,11] capable of high-throughput genetic analysis with very low limits of detection. The microdevice, presented in Figure 46.13, consists of a "standard" four-layer glass sandwich structure that incorporates thin-film metal heaters and temperature sensors optimized for maximal thermal uniformity and scalability with 380-nL PCR reactors. The four-layer structure consists of a PDMS membrane layer (0.25 mm thick) sandwiched between a pair bonded of glass wafers for CE and a glass wafer for pneumatic actuation (each 1.1 mm thick). This structural paradigm allows for high-resolution CE separations to take place in all-glass channels and allows fluidic routing paths to cross the separation channels, greatly extending design flexibility. Pneumatic control is afforded by PDMS membrane valves that

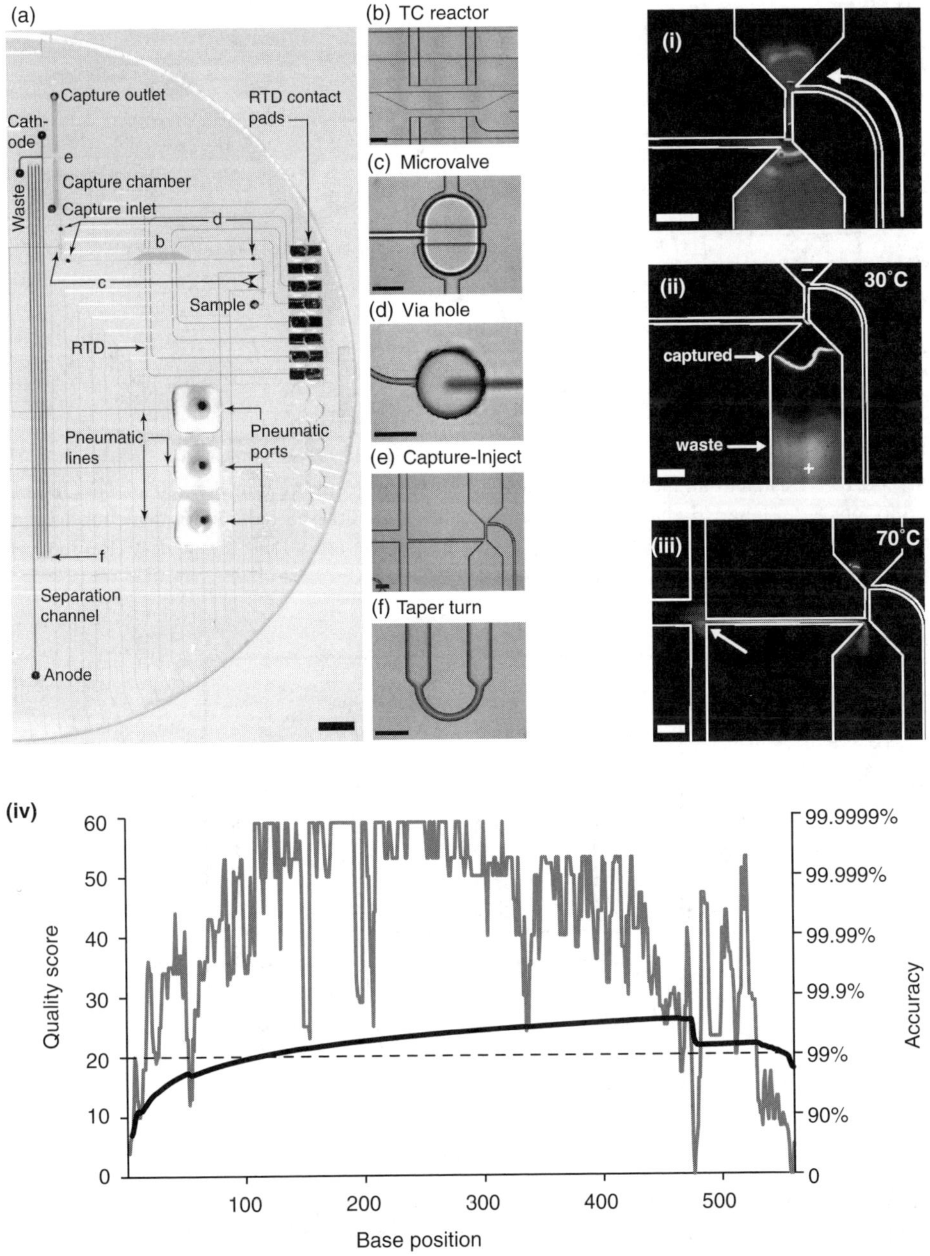

FIGURE 46.12 Design and operation of a fully integrated high-resolution DNA sequencing microdevice. (a–f) Thermal control elements are embedded together with a pneumatically sealed chamber for Sanger cycle sequencing. Product fragments are then directed by pneumatic pumping (i) to a separate chamber for purification by affinity capture (ii), injected with a cross-injector (iii) and then separated on a 30-cm long folded channel. (iv) Sequencing reads of at least 99% accuracy, corresponding to a Phred score of 20, were reproducibly obtained to lengths of >500 bases. (Reproduced from Blazej, R.G. et al., *Proc. Natl. Acad. Sci. USA*, 103, 7240, 2006. With permission.)

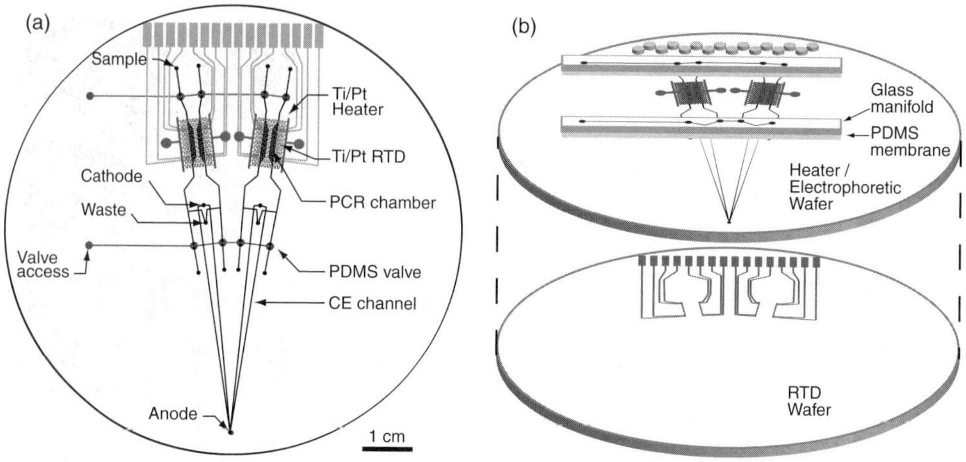

FIGURE 46.13 Design of an integrated PCR-CE array system featuring four PCR chambers (380 nL each) coupled to four 5-cm long CE channels. Individual elements are arrayed radially for detection with a rotary LIF scanner. (b) Exploded view of the four-layer structure of the array PCR-CE device. (Reproduced from Liu, C.N., et al., *Anal. Chem.*, 78, 5474, 2006. With permission.)

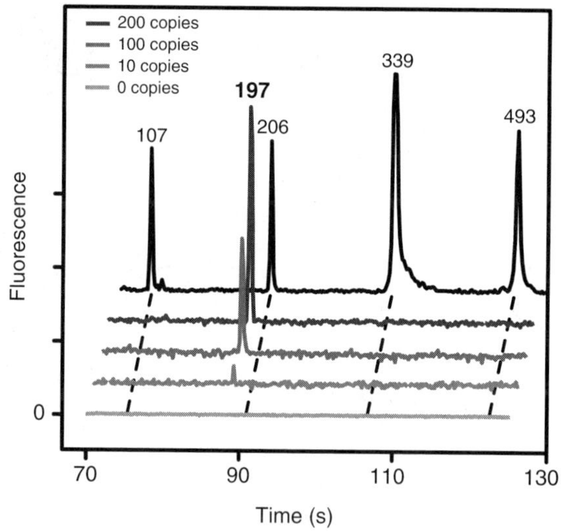

FIGURE 46.14 Limits of detection of the array-PCR device shown in the previous figure demonstrate that successful amplification and CE analysis can be performed starting with as few as 10 template copies in the 380-nL PCR reactors. The total analysis time for this four-layer integrated analysis device, including 30 cycles of PCR was less than 30 min. (Reproduced from Liu, C.N., et al., *Anal. Chem.*, 78, 5474, 2006. With permission.)

are actuated through a glass manifold. Both the manifold and the PDMS membrane are replaceable so that intrarun carryover of DNA can be eliminated. The adoption of a modular valve design that allows disassembly was critical to the development of a reproducible, reusable, and robust array PCR-CE system. Each of the reactors and CE channels in this microdevice are arrayed radially for detection with the Berkeley rotary scanner and to enable the facile radial layout scaling. The performance of the microdevice was demonstrated in two sets of experiments, the first of which performed amplification on both purified DNA templates and from whole bacterial cells [10] to a limit of 10 initial template copies in only 27 min (Figure 46.14). The second set demonstrated the utility of the array-PCR microdevice for RNA analysis by reverse-transcription analysis of bacteria with subcellular detection limits and analysis of RNA splice site variations associated with human breast cancer [11]. Both sets of experiments demonstrated excellent reproducibility and low limits of detection. Moreover, these devices present a scalable layout that forms the foundation of next-generation integrated μCAE devices for high-throughput bioanalysis.

46.3 CONCLUSIONS AND PROSPECTS

The last decade has seen the birth and growth of μCAE systems from simple rectilinear to complex radial designs. The radial array paradigm and rotary LIF scanners enabled μCAE systems to reach a practical limit of nearly 400 channels with analysis times shorter than their conventional CAE counterparts, in some cases reaching throughputs of >1 sample per second. Further increases in throughput and continued growth in the capabilities of μCE systems will be realized by integrating sample preparation components directly with CE analysis. The first step toward integrated sample preparation components with high-resolution separations has already been taken, marking significant progress along the path toward true lab-on-a-chip systems. Among the most critical steps to be made are the development of modular functional components that are easily combined and incorporated into μCE systems. Thus executed, the development of μCAE systems in the coming decade will be more fruitful than the last.

ACKNOWLEDGMENTS

We thank all the members of the Mathies group for their support, and Robert Blazej and Stephanie Yeung for assistance in figure preparation. This work was supported by NIH grant HG003583 and the donors to the UC Berkeley Center for Analytical Biotechnology.

REFERENCES

1. Harrison, D.J. et al., Capillary electrophoresis and sample injection systems integrated on a planar glass chip. *Anal. Chem.*, 64, 1926, 1992.
2. Harrison, D.J. et al., Micromachining a miniaturized capillary electrophoresis-based chemical analysis system on a chip. *Science*, 261, 895, 1993.
3. Woolley, A.T., Sensabaugh, G.F., and Mathies, R.A., High-speed DNA genotyping using microfabricated capillary array electrophoresis chips. *Anal. Chem.*, 69, 2181, 1997.
4. Emrich, C.A. et al., Microfabricated 384-lane capillary array electrophoresis bioanalyzer for ultrahigh-throughput genetic analysis. *Anal. Chem.*, 74, 5076, 2002.
5. Aborn, J.H. et al., A 768-lane microfabricated system for high-throughput DNA sequencing. *Lab Chip*, 5, 669, 2005.
6. Shi, Y.N. et al., Radial capillary array electrophoresis microplate and scanner for high-performance nucleic acid analysis. *Anal. Chem.*, 71, 5354, 1999.
7. Lagally, E.T., Emrich, C.A., and Mathies, R.A., Fully integrated PCR-capillary electrophoresis microsystem for DNA analysis. *Lab Chip*, 1, 102, 2001.
8. Breadmore, M.C. et al., Microchip-based purification of DNA from biological samples. *Anal. Chem.*, 75, 1880, 2003.
9. Paegel, B.M., Yeung, S.H.I., and Mathies, R.A., Microchip bioprocessor for integrated nanovolume sample purification and DNA sequencing. *Anal. Chem.*, 74, 5092, 2002.
10. Liu, C.N., Toriello, N.M., and Mathies, R.A., Multichannel PCR-CE microdevice for genetic analysis. *Anal. Chem.*, 78, 5474, 2006.
11. Toriello, N.M., Liu, C.N., and Mathies, R.A., Multichannel reverse transcription-polymerase chain reaction microdevice for rapid gene expression and biomarker analysis. *Anal. Chem.*, 78, 7997, 2006.
12. Blazej, R.G., Kumaresan, P., and Mathies, R.A., Microfabricated bioprocessor for integrated nanoliter-scale Sanger DNA sequencing. *Proc. Natl. Acad. Sci. USA*, 103, 7240, 2006.

13. Easley, C.J. et al., A fully integrated microfluidic genetic analysis system with sample-in-answer-out capability. *Proc. Natl. Acad. Sci. USA*, 103, 19272, 2006.
14. Saiki, R.K. et al., Primer-directed enzymatic amplification of DNA with a thermostable DNA-polymerase. *Science*, 239, 487, 1988.
15. Lander, E.S. et al., Initial sequencing and analysis of the human genome. *Nature*, 409, 860, 2001.
16. Venter, J.C. et al., Environmental genome shotgun sequencing of the Sargasso Sea. *Science*, 304, 66, 2004.
17. Hebert, P.D.N. et al., Biological identifications through DNA barcodes. *Proc. R. Soc. Lond. Ser. B-Biol. Sci.*, 270, 313, 2003.
18. Brenner, C.H. and Weir, B.S., Issues and strategies in the DNA identification of World Trade Center victims. *Theor. Popul. Biol.*, 63, 173, 2003.
19. Lynch, T.J. et al., Activating mutations in the epidermal growth factor receptor underlying responsiveness of non-small-cell lung cancer to gefitinib. *N. Engl. J. Med.*, 350, 2129, 2004.
20. Potti, A. et al., A genomic strategy to refine prognosis in early-stage non-small-cell lung cancer. *N. Engl. J. Med.*, 355, 570, 2006.
21. Dittrich, P.S., Tachikawa, K., and Manz, A., Micro total analysis systems. Latest advancements and trends. *Anal. Chem.*, 78, 3887, 2006.
22. Harrison, D.J. et al., Rapid separation of fluorescein derivatives using a micromachined capillary electrophoresis system. *Anal. Chim. Acta*, 283, 361, 1993.
23. Woolley, A.T. and Mathies, R.A., Ultra-high-speed DNA fragment separations using microfabricated capillary array electrophoresis chips. *Anal. Chem.*, 91, 11348, 1994.
24. Simpson, P.C. et al., High-throughput genetic analysis using microfabricated 96 sample capillary array electrophoresis microplates. *Proc. Natl. Acad. Sci. USA*, 95, 2256, 1998.
25. Culbertson, C.T., Jacobson, S.C., and Ramsey, J.M., Dispersion sources for compact geometries on microchips. *Anal. Chem.*, 70, 3781, 1998.
26. Medintz, I.L. et al., Genotyping energy-transfer-cassette-labeled short-tandem-repeat amplicons with capillary array electrophoresis microchannel plates. *Clin. Chem.*, 47, 1614, 2001.
27. Scherer, J.R. et al., High-pressure gel loader for capillary array electrophoresis microchannel plates. *Biotechniques*, 31, 1150, 2001.
28. Woolley, A.T. and Mathies, R.A., Ultra-high-speed DNA-sequencing using capillary electrophoresis chips. *Anal. Chem.*, 67, 3676, 1995.
29. Liu, S.R. et al., Optimization of high-speed DNA sequencing on microfabricated capillary electrophoresis channels. *Anal. Chem.*, 71, 566, 1999.
30. Koutny, L. et al., Eight hundred base sequencing in a microfabricated electrophoretic device. *Anal. Chem.*, 72, 3388, 2000.
31. Salas-Solano, O. et al., Optimization of high-performance DNA sequencing on short microfabricated electrophoretic devices. *Anal. Chem.*, 72, 3129, 2000.
32. Jacobson, S.C. et al., Precolumn reactions with electrophoretic analysis integrated on a microchip. *Anal. Chem.*, 66, 4127, 1994.
33. Molho, J.I. et al., Optimization of turn geometries for microchip electrophoresis. *Anal. Chem.*, 73, 1350, 2001.
34. Paegel, B.M. et al., Turn geometry for minimizing band broadening in microfabricated capillary electrophoresis channels. *Anal. Chem.*, 72, 3030, 2000.
35. Paegel, B.M. et al., High-throughput DNA sequencing with a microfabricated 96-lane capillary array electrophoresis bioprocessor. *Proc. Natl. Acad. Sci. USA*, 99, 574, 2002.
36. Blazej, R.G., Paegel, B.M., and Mathies, R.A., Polymorphism ratio sequencing: A new approach for single nucleotide polymorphism discovery and genotyping. *Genome Res.*, 13, 287, 2003.
37. Yeung, S.H.I. et al., Rapid and high-throughput forensic short tandem repeat typing using a 96-lane microfabricated capillary array electrophoresis microdevice. *J. Forensic Sci.*, 51, 740, 2006.
38. Mikkers, F.E.P., Everaerts, F.M., and Verheggen, T., Concentration distributions in free zone electrophoresis. *J. Chromatogr.*, 169, 1, 1979.
39. Laemmli, U.K., Cleavage of structural proteins during assembly of head of bacteriophage-T4. *Nature*, 227, 680, 1970.

40. Foote, R.S. et al., Preconcentration of proteins on microfluidic devices using porous silica membranes. *Anal. Chem.*, 77, 57, 2005.
41. Hatch, A.V. et al., Integrated preconcentration SDS–PAGE of proteins in microchips using photopatterned cross-linked polyacrylamide gels. *Anal. Chem.*, 78, 4976, 2006.
42. Grover, W.H. et al., Monolithic membrane valves and diaphragm pumps for practical large-scale integration into glass microfluidic devices. *Sens. Actuator B-Chem.*, 89, 315, 2003.

47 Macroporous Monoliths for Chromatographic Separations in Microchannels

Frantisek Svec and Timothy B. Stachowiak

CONTENTS

47.1 INTRODUCTION AND BACKGROUND

History of chromatographic separation processes is closely related to particulate packings for more than a century. Tsvett[1] used powdered calcium carbonate to achieve the separation of chlorophylls. Separation of biomacromolecules in both low-pressure size-exclusion and ion-exchange chromatographic modes became available after introduction of stationary phases based on modified natural polysaccharides in the 1950s.[2–4] High-performance liquid chromatography (HPLC) commenced in the 1960s partly because bonded porous silica particles were available.[5,6] The shape of these particles has changed from relatively large irregular pieces to current regular beads, the size of which can now be even less than 2 μm. The porous structure and surface chemistry also have evolved to reach current high standards. Typically, particulate separation media are prepared in larger batches and a small portion of these porous beads with well-defined size, shape, porosity, and chemistry are packed into columns that are subsequently used for chromatographic separations. Despite the many advantages that led to their widespread use, columns packed with typical particulate materials also have some limitations. The necessity of packing procedure, the slow diffusional mass transfer of high molecular weight solutes into the stagnant phase present in the pores of the beads, and the large void volume between the packed particles, are the most important and best-known limitations of the present day column technologies.[6,7]

Separation scientists realized that some of these restrictions could be eliminated by using separation media with limited discontinuity. For example, R. Synge[8] envisioned "a continuous block of porous material" and its use for the separation. Unfortunately, suitable materials were not available and his idea could not be materialized.[9] The first experimental attempt to create via polymerization a "single piece" of gel-like separation medium that did not contain interparticular voids date back to the late 1960s.[10] Disappointingly, this hydrogel did not perform well. In contrast, the permeability of monolithic open-pore polyurethane foams introduced slightly later was excellent.[11,12] However, less suitable chromatographic characteristics of these materials prevented their broader use in both liquid and gas chromatography. Thus, macroporous discs[13,14] and compressed soft polyacrylamide gels placed in a cartridge or column[15] represented the first successful demonstrations of media that exhibited no interstitial porosity. These elegant approaches have recently been described in detail in a series of excellent review articles.[16–19]

In the early 1990s, yet another category was developed.[20] These rigid macroporous organic polymer monoliths were formed by a very simple *in situ* "molding" process in which a liquid mixture of monomers and solvents was polymerized under carefully controlled conditions and immediately used within a closed tube or similar container. Many review articles describing various aspects of these materials have been published during the years since their inception.[19,21–28]

Porous inorganic materials, in general, and porous silica in particular, are very popular supports widely used in chromatography. Therefore, development of silica monoliths followed closely to that of the organic polymers.[29–31] Detailed accounts of these materials have also been published several times.[32–35] Separations using these monolithic separation media have also been treated theoretically.[36–41]

In contrast to columns, the microfabricated devices feature a network of microchannels etched in glass or imprinted in a polymer plate that are designed to enable much smaller sample volumes to be analyzed at an increased speed and permitting a large number of analysis to be performed simultaneously, thus increasing the overall throughput. Microfluidic analytical devices with open channel geometry are currently most frequently used to achieve the desired functions. Open channels are best suited for separation systems in which no interactions with functionalities located at the solid surface are required or are even undesirable, such as in electrophoresis.

In contrast, the mode of operation of a number of other modules in microfluidic devices, such as reactors, solid-phase extraction, and separation units require interactions of molecules in solution with functionalities attached to a solid support. Obviously, this support can itself be the channel wall but, in this case, the overall surface-to-volume ratio is very small. This ratio can be increased by filling the channel with a solid phase, and the monolithic materials represent one of the options to fill the channel with a suitable stationary phase.

47.2 PRACTICAL APPLICATIONS

47.2.1 REVERSED-PHASE HPLC

According to International Union of Pure and Applied Chemistry (IUPAC),[42] chromatography is a physical method of separation in which the components to be separated are distributed between two phases, one of which is stationary while the other moves in a definite direction. HPLC is then defined as "an analytical separation technique used to detect and quantitate analytes of interest in more or less complex mixtures and matrices" that uses "elevated pressures to force the liquid through the bed of the stationary phase."[43] The bed is most often located in a column. Since almost four decades of its inception, HPLC technique has advanced and there is enormous literature available on this technique.

In contrast, very small activity can be observed in the field of HPLC on chip. The simplest approach to microfluidic HPLC devices that mimic typical chromatographic columns include channels packed with chromatographic beads.[44–54] However, an early attempt by Ocvirk et al.[44] to pack channel on a silicon microchip with beads was not very successful due to technical difficulties, and the 20-mm long microfluidic HPLC "column" generated only 200 theoretical plates (4000 plates/m).

Using a different approach, Oleschuk et al.[55] succeeded in packing octadecyl silica beads from a side channel into a specifically designed chamber of the microchip.

Although packing is a typical approach for the preparation of HPLC columns, filling channels with porous beads is difficult to accomplish due to the need of frits or other retaining structures and the difficulty associated with efficient packing of the channels. Therefore, for a long time, microfluidic separation modules packed with particles did not match the performance of their counterparts, such as standard packed columns. Only recently, Agilent Laboratories developed a microfluidic system for HPLC coupled with mass spectrometry (HPLC-MS). Although characteristics such as column efficiency are not shown, their separations appear acceptable.[56]

Another reason for the limited interest in HPLC-like separations on chip appears to be the complexity of plumbing since an external pump typical of standard HPLC instrumentation has to be included. The role of the chip then degrades to a capillary-like column and the advantages attributed to microfluidic devices vanish. In order to avoid attachment of the chip to a bulky mechanical pump, Penrose et al.[57] used centripetal force to propel the mobile phase through a bed of particles packed in a channel. Since no detector was added to the system, the readout was a scan of the bed after visualizing the bands in ultraviolet (UV) light.

An alternative option to packed channels is the use of monolithic materials, which may have many of the same benefits as packed beds, including high surface area and easily controlled surface chemistry. However, a distinct advantage of monoliths is the ability to prepare them easily and rapidly via polymerization of liquid precursors within the channels of the microdevice without the need for any retaining structures. Despite the popularity of monolithic capillary columns for separations of a variety of low and high molecular weight compounds in HPLC mode,[19,26,58] their first application in microfluidic chips dates back only to 2005.

Several modules had to be developed before a functional microfluidic HPLC system could be assembled. In a series of papers, the Sandia National Laboratory group demonstrated high-pressure connectors for introduction of liquids, fabrication of integrated polymer checking microvalve elements, on-chip high-pressure picoliter injector, and eventually a complete HPLC chip.[59–62]

Their plastic connectors linking the chip with the external pump via common capillary fittings were glued to the fluid access holes. Although this approach is not ideal for mass production due to the extensive skilled manual work needed to obtain reasonable yield of useful chips, the best units withstood pressures up to 8.5 MPa. Certainly, interconnects integrated into the top part of the chip represent a better solution to the problem,[63] but these would be very difficult to implement with glass chips.

The flow control elements or valves developed in Sandia were plugs of nonporous monolithic polymer prepared from fluorinated acrylates via UV-initiated polymerization directly in one of the channels. The reason for using the fluorinated monomers is to manage the surface energy, decrease their friction in channel, enable actuation at a low pressure, as well as to avoid both swelling and shrinking in solvents typically used in reversed-phase separations such as water and acetonitrile. The friction could be further decreased by modification of the valve channel with fluorinated alkylsilane.[61] However, the latter may be counterproductive in systems used for the separation of proteins and peptides because they tend to adsorb on highly hydrophobic surfaces.

The high-pressure picoliter injector also consists of a moving monolithic plug similar to that used for valving, thus significantly reduced in size.[60] The injection volume is controlled by pressure difference between sample and mobile phase channels and by time. For example, a 0.25 s long injection afforded a sample volume of 180 pL.

The fused-silica chip designed for the separation of peptides and proteins is shown in Figure 47.1.[62] It consists of a sample channel, an injector valve, a mobile phase channel, a separation channel accommodating the monolithic column, and a miniaturized fluorescence detector. The fluids are driven through the channels by syringe pumps. The 17-mm long monolithic column was prepared by *in situ* UV-initiated polymerization developed previously by our group.[64,65] A solution of monomers comprising stearyl acrylate, 1,6-hexanediol diacrylate, and tetrahydrofurfural

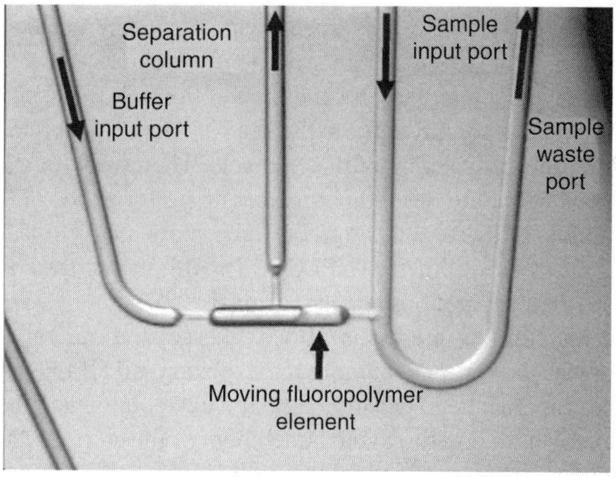

FIGURE 47.1 Four-port valve allowing rapid sample changes. Sample is changed by flushing at low pressure when the sample waste port is open. Sample is injected by closing sample waste port and pressurizing the sample line. (Reprinted with permission from Reichmuth, D. S., et al., *Anal. Chem.*, 77, 2997, 2005. Copyright 2005 American Chemical Society.)

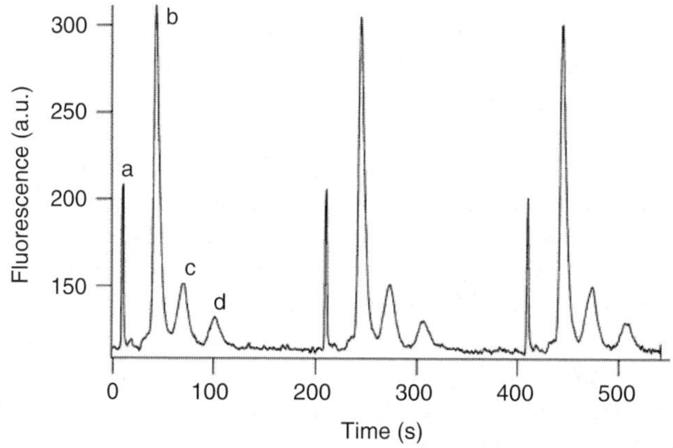

FIGURE 47.2 Repeated isocratic HPLC separations of proteins on chip. (Reprinted with permission from Reichmuth, D. S., et al., *Anal. Chem.*, 77, 2997, 2005. Copyright 2005 American Chemical Society.) Conditions: lauryl methacrylate monolith; mobile phase 24% acetonitrile + 0.16% heptafluorobutyric acid in 5 mmol/L phosphate buffer (pH 2.0); Injections 6,4 nL, 750 ms, pressure 0.2 MPa. Peaks: free dye (a), insulin (b), antibiotin (c), α-lactalbumin (d).

acrylate in a mixture of acetate buffer solution (pH 5.0) and methoxyethanol serving as porogens was polymerized using azobisisobutyronitrile (AIBN) as photoinitiator.

First, excellent control and repeatability of the injection volumes was confirmed with over 100 injections of a peptide mixture consisting of glycine-tyrosine, valine-tyrosine-valine, methionine enkephalin, leucine enkephalin, and angiotensin II. The overall relative standard deviation for peak area and retention time was 3.9% and 3.6%, respectively. Chromatographic performance of the chip in reversed-phase mode was then tested with a mixture of proteins (insulin, antibiotin, and α-lactalbumin). Repeated separations of these compounds are shown in Figure 47.2.[62]

This relatively simple system demonstrates the potential of a microfluidic chip for separations in HPLC mode. It was certainly not ideally suited for the separations of more complex mixtures since it was only used in isocratic conditions and the several minutes long retention times are too long for the desired high throughput systems. However, the column efficiency of 25,000 plates/m found for α-lactalbumin using less-than-ideal conditions is remarkably good.

Moschou et al.[66] used the compact disc microfluidic platform that is spun to allow the centripetal force to achieve pumping of the mobile phase through the channel. They incorporated a poly(2-hydroxyethyl methacrylate-co-ethylene dimethacrylate-co-[2-(methacryloyloxy) ethyl] trimethylammonium chloride) monolith in the channel of a poly(dimethylsiloxane) (PDMS) chip. The unusual feature of this preparation is the use of irradiation with microwaves in a common domestic microwave oven to initiate the polymerization reaction, which is then completed in only 4.5 min. This is a significantly shorter time compared to 20 h required to obtain monolith using thermally initiated polymerization of the same mixture. This monolith was then used for the purification of enhanced green fluorescent protein in ion-exchange mode.

Despite the high popularity of silica-based monoliths in HPLC, use of this chemistry in microfluidic chips is scarce. Ishida et al.[67] etched a 42-cm long, 400 μm wide, and 30 μm deep serpentine channel in glass substrate. In order to avoid the "race track" effect, the turns were tapered to a width of only 50 μm. Four syringes were attached to the chip through glued outlet capillaries (Figure 47.3). The separation channel was filled by drawing the sol solution prepared by adding tetramethoxysilane

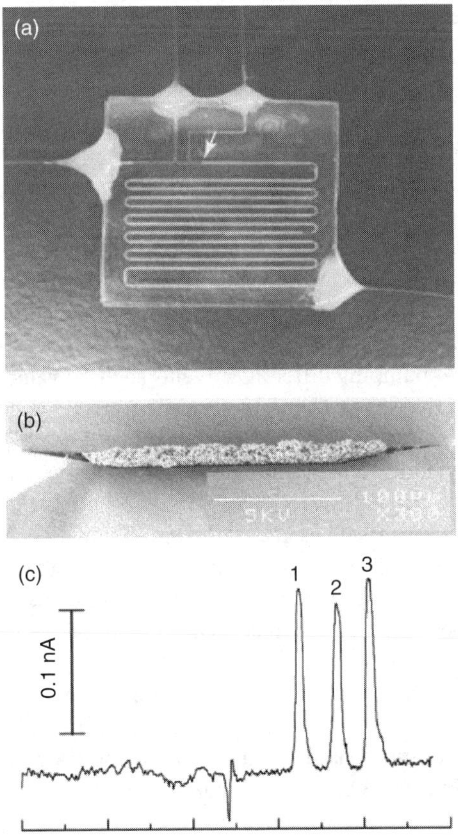

FIGURE 47.3 Fabricated LC chip with a 45-cm long serpentine channel. (Reprinted with permission from Ishida, A., et al., *J. Chromatogr. A*, 1132, 90, 2006. Copyright 2004 Elsevier B.V.) (a) Photograph of the assembled chip. The arrow indicates the top end of ODS monolithic silica column. (b) SEM image of the cross-section of the separation channel filled with monolithic silica. (c) Chromatographic separation of catechin (1), epigallocatechin gallate (2) and epicatechin (3) on chips with tapered turns. Conditions: sample concentration, 1 mmol/L, mobile phase methanol–water (3:2, v/v), 0.02 mol/L phosphate buffer (pH 2.0), flow rate, 300 nL/min; applied potential, +400 mV versus Ag/AgCl.

to poly(ethylene glycol) (PEG) containing urea and acetic acid and the contents heated to 40°C for 24 h. The generated gel was then brought to a temperature of 120°C for 3 h and after cooling, the resulting silica monolith was washed with water and methanol. The porous silica was dried at 60°C for 10 h, followed by heating at 200°C for 50 h. After cooling to the room temperature, pores within the monolith were modified by flushing it with a toluene solution of octadecyltrichlorosilane at 100°C for 12 h. As apparent from the description of the preparation, this process is significantly more complex than the simple photoinitiated polymerization affording organic polymer monoliths and indicates why these monoliths are more common stationary phases in microfluidic HPLC. Figure 47.3 shows separation of three catechins using the silica-based monolithic stationary phase and electrochemical detection. Clearly, with a smallest height equivalent to theoretical plate (HETP) of 60 μm representing about 16,500 plates/m, this separation does not match that of typical monolithic silica columns prepared in capillary.

A variety of detectors have been developed that are useful for monitoring separations on chip. Most of the detectors are based on electrochemical and optical concepts.[68] For example, electrochemical detection is successful in detection of electroactive compound (*vide supra*).[67] The very popular laser-induced fluorescence (LIF) is sensitive enough to detect molecules containing a suitable chromo-phore. Since only a small number of compounds include this feature, most of them must be first modified with a fluorescent label. While this may not be difficult for certain type of molecules, labeling of proteins and peptides possessing a plurality of reactive groups is a problem since their separation may then also occur according to the number of attached labels. In addition, knowledge of the accurate molecular mass is required in fields such as proteomics, glycomics, metabonomics, and so forth. The optical and electrochemical detection methods do not provide this type of information. Therefore, several chip designs were developed that include both monolithic column and an interface that enable connection to mass spectrometer (MS).[69,70]

In one of the early implementations, the chip features shown in Figure 47.4 were fabricated on silicon wafer using SU-8-based multilayer technology.[71] The channels were then covered with a glass wafer. Standard fused-silica capillaries coated with polyimide were glued to the channel ends to enable connection with the nanoflow pump and injector at one end and the coupling to MS at the other end. This chip withstands an internal pressure of up to 5 MPa without visible delamination and no damage was observed after pumping different solvents such as water, acetonitrile, and acetic acid through the channel for several hours. The polymerization mixture consisting of lauryl methacrylate, ethylene dimethacrylate, cyclohexanol, ethylene glycol, and azobisiso butyronitrile (AIBN) was injected in the channel and irradiated through a mask to achieve photopolymerization only in the desired 22-mm-long part of the channel. This device afforded certain separations of peptides originating from digested cytochrome C as also shown in Figure 47.4.

In the following chapter, this group used a monolith in an improved chip design provided with a nib fabricated on the bottom silicon plate to enhance the nano-ESI interface.[72] In contrast to the previous system, less-hydrophobic butyl methacrylate instead of the lauryl derivative was used to prepare the monolith. No real separation of a peptide mixture was attempted and the performance of this system was only demonstrated with desalination of the peptide sample before MS analysis.

Oleschuk and coworkers[73] prepared monoliths in a glass chip with semicircular channel cross-section. They used slightly modified polymerization mixture developed by Peters et al.[74] consisting of butyl acrylate, ethylene dimethacrylate, 2-acrylamido-2-methyl-1-propanesulfonic acid, and photoinitiator benzoin methyl ether all dissolved in porogenic solvent comprising water, 1-propanol, and 1,4-butanediol. The device was made by patterning a plug of the monolith in the center of a doubled unit followed by dicing it in two parts to obtain chips with a monolith reaching to the channel exit. In contrast to open channel, implementation with the monolith inside afforded a stable electrospray at a wide range of flow rates (50–500 nL/min). The best results were obtained at lower flow rates of 50–100 nL/min. Figure 47.5 compares mass spectra of leucine enkephalin obtained from both open and monolithic formats, and demonstrates significantly impaired spray stability and sensitivity found for the open channel device. The positive effect of the monolith was ascribed to multiple Taylor

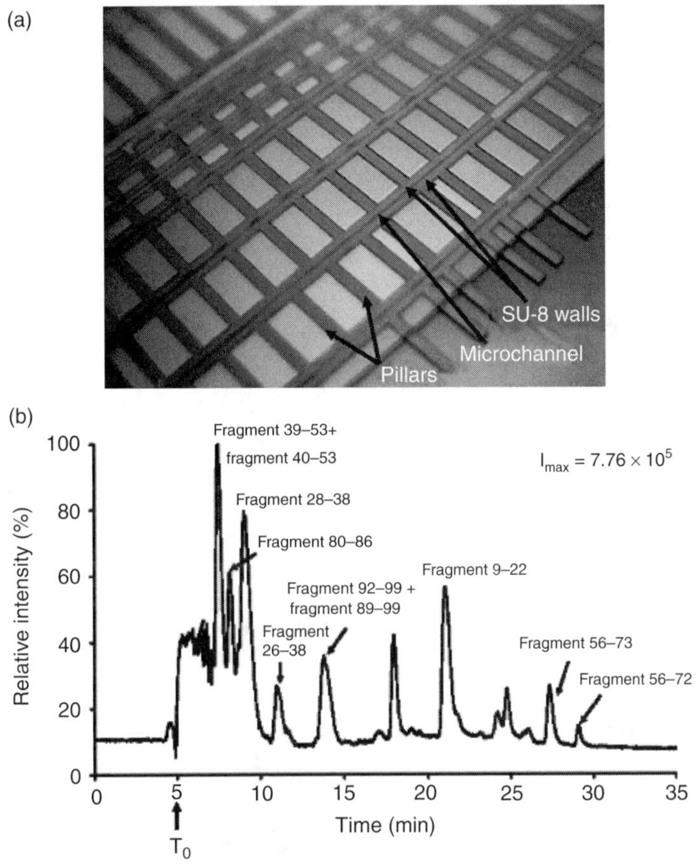

FIGURE 47.4 Photograph of a microsystem fabricated via the multilayered structure route (top) and on-chip HPLC separation of cytochrome c digest using a lauryl methacrylate-based monolithic column (bottom). (Reprinted with permission from Le Gac, S., et al., *J. Chromatogr. B*, 808, 3, 2004. Copyright 2004 Elsevier B.V.) Conditions: sample concentration 800 fmol, mobile phase A, 5% acetonitrile in 0.1% aqueous formic acid, mobile phase B, 95% acetonitrile in 0.1% aqueous formic acid, gradient 5–50% B in A in 30 min, 50–95% B in A in 1 min, flow-rate 200 nL/min.

cones originating from numerous large pores at the face of the monolithic material that generated the mist. The multiple paths also minimized clogging and the hydrophobic nature of the monolith avoided droplet spreading at the emitter.

A 70-fold enhancement in stability of the spray and sensitivity of detection has also been demonstrated with plastic chips containing 5-mm long monolithic sprayer in the zero dead volume emitting tip.[75] The microfluidic devices were fabricated from three different polymeric substrates including poly(methyl methacrylate) (PMMA), cyclic olefin copolymer (COC), and PDMS. The channel surface had to be first modified by photografting to afford covalent attachment of the monolith to the channel walls.[76,77] The monolith was then patterned in the desired location via photopolymerization of a mixture consisting of lauryl acrylate, ethylene dimethacrylate, benzoin methyl ether, methanol, and 2-propanol. PDMS did not appear as a suitable substrate for the formation of monoliths, most likely due to its high permeability for oxygen,[78] which is a known inhibitor of free-radical polymerization reactions. PMMA was not ideal either since it tends to dissolve in the polymerization mixture, thus causing distortion of the channel shape.[75] In contrast, COC has an excellent UV transparency required for photoinitiation of the polymerization reaction within the chip; it does not dissolve in the polymerization mixture or in typical chromatographic solvents, and enables

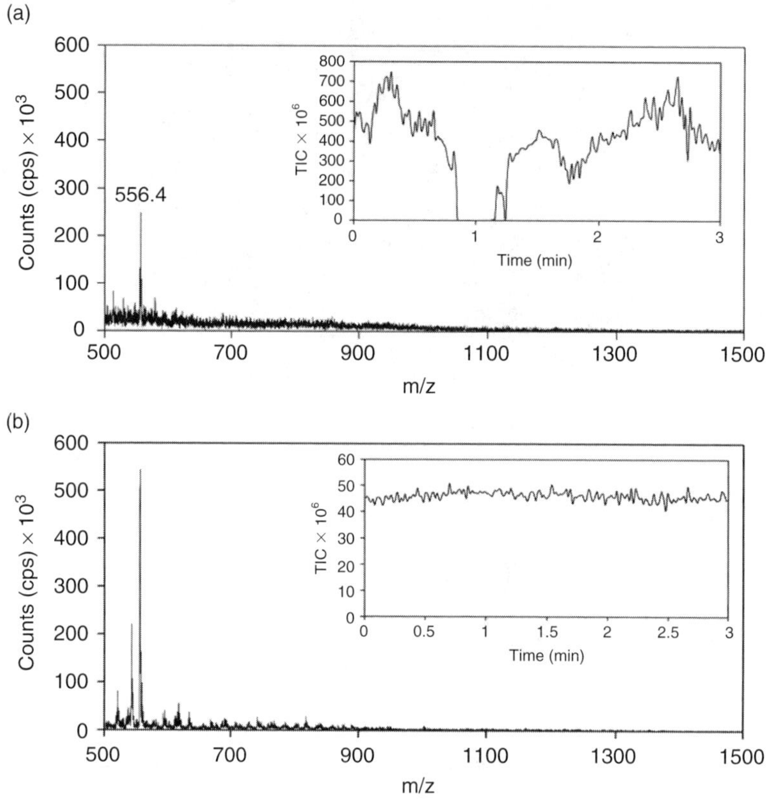

FIGURE 47.5 Representative mass spectrum of leucine-enkephalin obtained using chip with open channel emitter (a) and porous polymer monolith-assisted electrospray process from a microdevice at a constant infusion (b). (Reprinted with permission from Koerner, T. and Oleschuk, R. D., *Rapid Comm. Mass Spectr.*, 19, 3279, 2005. Copyright 2005 J. Wiley.) Conditions: peptide solution 1.0 μmol/L in 50% aqueous acetonitrile with 0.5% acetic acid, flow rate of 100 nL/min, applied voltage of 3.5 and 3.0 kV, respectively.

in situ formation of excellent monoliths.[63,79] The hydrophobic monolith/emitter was also used for preconcentration and desalting of myoglobin. A 225 nL plug of the protein solution in ammonium acetate buffer containing 450 pmol of myoglobin was pumped through a 50-mm long poly(lauryl acrylate-co-ethylene dimethacrylate) monolith, the channel was then rinsed with water, and the protein released and sprayed using aqueous acetonitrile. This process afforded an excellent MS spectrum.[75]

Knapp and coworkers[80] approached the electrospraying from the COC chip containing a monolith differently. First, they used the channel opening on the edge of the device as the emitter. Similar to observations of Bedair and Oleschuk,[75] they also found hydrophobicity of the native COC insufficient to maintain stable electrospray ionization (ESI) from the channel opening on the edge. Therefore, they used radiofrequency (RF) plasma-assisted chemical vapor deposition (CVD) with octafluorocyclobutane to form a thin hydrophobic fluorocarbon layer around the channel exit. Unfortunately, this coating was quickly damaged by occasional corona discharge. The problem was then solved by adding a thin layer of perfluoroalkyl ether Krytox 1625 over the fluorocarbon-coated surface. The octafluorocyclobutane fluorocarbon covalently attached to the COC surface serves as an anchor for the perfluoroalkyl ether coating. This flexible layer has the ability to repair arcing damage and keep the hydrophobic surface around the edge exit. The high voltage for ESI was applied through the carbon ink line. Lifetime of the resulting microfluidic emitter was found to exceed 20 h. Despite all the success, the fabrication of the spray appears to be rather complex.

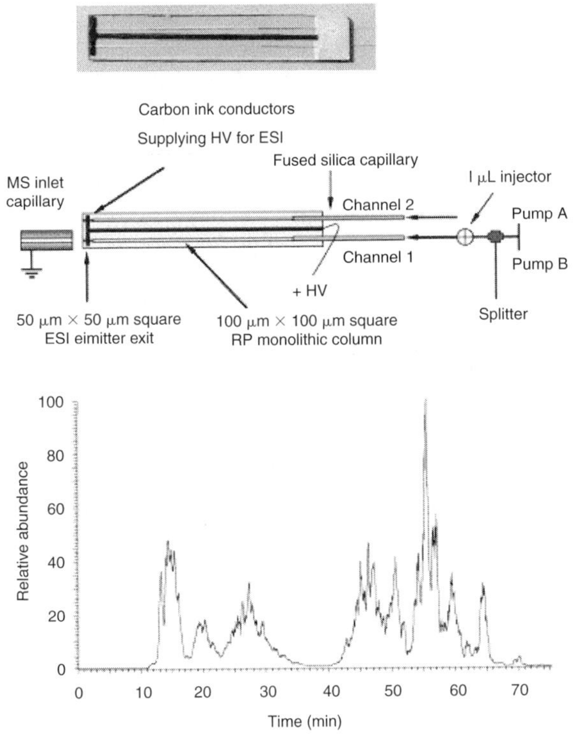

FIGURE 47.6 Schematic diagram of the microfluidic device fabricated from cyclic olefin copolymer containing poly(ethylhexyl methacrylate-co-ethylene dimethacrylate) monolith (top) and chromatogram of tryptic digested bovine serum albumin in the reversed-phase mode. Conditions: digest sample 5 pmol/μL, mobile phase A 5% acetonitrile in 0.1% aqueous TFA, mobile phase B 70% acetonitrile in 0.1% aqueous TFA, gradient 0% B for 5 min, then from 0% to 70% B in 60 min, flow rate 300 nL/min.

To avoid formation of voids at the monolith–channel wall interface, the inert COC was photografted[78] with trimethylolpropane triacrylate to afford polymerizable double bonds enabling covalent anchoring of the monolith. The channel was then filled with a polymerization mixture consisting of ethylhexyl methacrylate, ethylene dimethacrylate, 1-decanol, and photoinitiator 2,2-dimethoxy-2-phenylacetophenone and polymerization initiated by UV light through a mask. The 55-mm long monolith having a cross-section of about 50 × 50 μm ended 3 mm before the exit, thus adding a postcolumn dead volume of only 7.5 nL.

The performance of the device was demonstrated with the separation of a tryptic digest of bovine serum albumin (BSA) in the reversed-phase mode using a gradient of acetonitrile in aqueous solution of trifluoroacetic acid as the mobile phase.[80] The separated peptides were detected by mass spectrometer. Figure 47.6 shows both the device and separation of a sample containing 5 pmol of peptides. Comparison with a database revealed an excellent amino acid sequence coverage of 70%.

Similar chip design was also used for the offline interface with matrix-assisted laser desorption/ionization time-of-flight mass spectrometry (MALDI-TOF-MS).[81] In contrast to inline ESI, MALDI provides for time flexibility. Separated compounds are collected on the plate and archived, thus enabling reanalyzing the sample later if so desired. Also, MS or MS/MS performance can be optimized independently in this implementation. An important part of the system is the noncontact electric field-driven deposition system mounted on an x,y,z stage, enabling computer-controlled deposition of sample spots across the plate.

The chip similar to that shown in Figure 47.6 included two parallel circular channels with a diameter of 100 μm, of which both ends were provided with connecting capillaries, leaving 6-cm long

TABLE 47.1
Comparison of Sequence Coverage for BSA Acquired by MALDI-MS and Microchip LC-MALDI-MS[a]

		Microchip LC-MALDI-MS	
	MALDI-MS	Column A	Column B
No. of detected masses	21	64	67
No. of identified peptides	13	37	39
Sequence coverage (%)	23	58	60

[a] Database: SwissProt. Search tool: MS-Fit; peptide mass tolerance: ±1 Da.

space for the monolithic column. The wall modification and preparation of the poly(ethylhexyl methacrylate-co-ethylene dimethacrylate) monolith was identical with that shown earlier.[80] The ultimate test of the performance of the system included separation of a digest of three proteins—BSA, myoglobin, and cytochrome c. The sequence coverage obtained using the microchip with two monolithic columns run in parallel is shown in Table 47.1 and compared with MALDI-MS analysis data obtained for nonfractionated sample. Clearly, the separation on chip containing monolith affords significantly better results.

47.2.2 Affinity Chromatography

This chromatographic mode represents a simple on–off technique. The compound of interest is fished-out from a more complex mixture via a very selective interaction with a ligand immobilized on solid support. Once the entire sample is processed through the column and followed by washing, the captured compound is released using a strong eluent.[82] Thanks to the simple binary separation mechanism, plate counts are not important and the development of devices mostly focuses on binding capacity and selectivity.

Li and Lee[83] used a very simple microfluidic glass chip containing two crossing channels. The length of the main separation channel was 2 cm. Using a mask-enabling irradiation of only a part of the channel, they prepared 2- and 10-mm long poly(glycidyl methacrylate-cotrimethylolpropane trimethacrylate) monoliths using photopolymerization procedure developed earlier.[64] They then reacted epoxy groups of the monolith with an excess of ethylenediamine in methanol by pumping this solution through the channel at a flow rate of 1.0 μL/min for 10 min, followed by stopping the flow and letting the reaction continue for another 6 h at a temperature of 70°C. The last step was a reaction with a reactive dye, Cibacron-blue-3G-A, to afford specific ligands. The dye was attached to the solid support, after its solution in aqueous methanol was pumped through the monolith to fill completely the pores and left to react at 60°C for 3 h.

The function of the device containing 2-mm long monolith was first tested on capture of fluorescein isothiocyanate (FITC)-labeled lysozyme from its mixture with cytochrome c. The capture was monitored via an increase in fluorescence of the monolith. The ability of the monolith to selectively capture lysozyme from the mixture was validated by MALDI-TOF mass spectrometry. Since the Cibacron-blue-3G-A ligand has a strong affinity for albumins, the ultimate function was successfully demonstrated with capture of albumin from human cerebrospinal fluid.[83]

Lectins represent a group of affinants that are very specific for interactions with sugars and were also used for the bioaffinity separations of glycoproteins in microfluidic chip.[84] First, 0.5-mm long monolith was prepared by UV-initiated photopolymerization in a vinylized glass chip using a mixture consisting of glycidyl methacrylate, ethylene dimethacrylate, cyclohexanol, dodecanol, and 2,2-dimethoxy-2-phenylacetophenone (photoinitiator). The movement of sample and other fluids

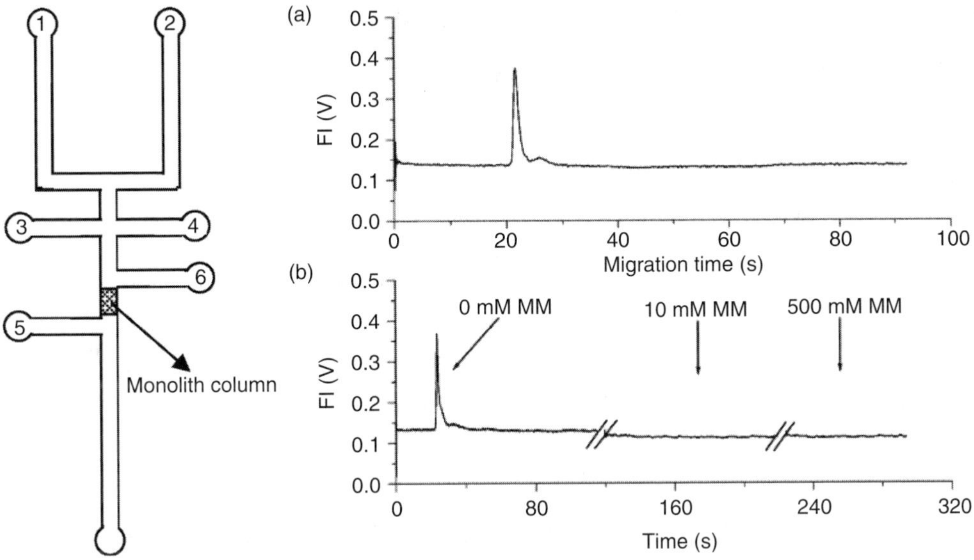

FIGURE 47.7 Layout of the microfluidic chip including running buffer reservoir (1), eluent buffer reservoir (2), sample reservoir (3), sample waste reservoir (4), washing reservoirs (5,6) and waste reservoir (7) and electropherograms of human serum albumin detected above (a) and below the affinity monolith column with immobilized *Pisum sativum* agglutinin (b). (Reprinted with permission from Mao, X., et al., *Anal. Chem.*, 76, 6941, 2004. Copyright 2004 American Chemical Society.)

between the seven reservoirs located at the ends of seven channels comprising the chip shown in Figure 47.7 was achieved by voltage. The pores of the monolith were filled with a solution of lectin *Pisum sativum* agglutinin that was allowed to react for 16 h at 30°C. A solution of fluorescently labeled glycoproteins, turkey ovalbumin, chicken ovalbumin, and ovomucoid was then driven through the monolith. Only the proteins with terminal mannosyl residues or *N*-acetylchitobiose fucose structures were adsorbed while other glycoforms were washed off by the running buffer. The captured proteins were then released from the matrix in fractions according to their different affinity toward the immobilized lectin using a mobile phase with an increasing concentration of displacing sugar, methyl-α-D-mannopyranoside, and detected via LIF. For example, glycoproteins from egg white with different glycan structures were separated into fractions according to the strength of their affinity toward immobilized lectin (Figure 47.7). This study demonstrated that the microfluidic chip-based lectin affinity chromatography of glycoforms significantly reduces the analysis time. The whole analytical process could be completed in less than 6 min, whereas it needs at least 4 h using the traditional affinity chromatography.

So far, only a very limited number of reports on affinity chromatography on chip are available. Interestingly, all monoliths in these publications were prepared from the same pair of monomers, glycidyl methacrylate and ethylene dimethacrylate. The epoxide functionality of the former is being used for the attachment of ligands in affinity chromatography since the 1970s.[85] Although quite reliable, the direct reaction of nucleophilic groups of proteins with epoxides is slow and very long reaction times are required. This feature may not be desirable for immobilization of sensitive ligands. Alternative multistep immobilization approaches that include reaction with a diamine or hydrolysis of epoxide to diol, its oxidation to aldehyde using periodate, followed by fast immobilization of the ligand and stabilization of the imine via hydrogenation using a hydride are rather tedious.[86] Therefore, we introduced azlactone chemistry, originally advocated by 3M company,[87] in the field of monoliths.[88] This functionality reacts with proteins in a single step and significantly faster as demonstrated on the fabrication of enzymatic microreactor on chip for peptide mapping.[89]

47.2.3 Separations in Electrochromatographic Mode

47.2.3.1 Concept of Electrochromatography

In contrast to mechanical pumping typical of HPLC discussed earlier, which relies on pumps with moving parts, electroendoosmotic flow (EOF) is generated within a stationary system by applying an electrostatic potential across the entire length of a device. The potential advantages of the flat flow profile generated by EOF in chromatography were recognized by Pretorius as early as in 1974.[90] Although the first electrochromatographic separations in packed capillary columns (CEC) were demonstrated at the beginning of 1980s,[91] serious technical difficulties resulted in limited interest, until a revival of interest in capillary electrochromatography (CEC) occurred in the mid 1990s arising from the search for new miniaturized separation methods with vastly enhanced efficiencies and peak capacities.

Although not completely correct, CEC is often presented as a hybrid method that combines the capillary column format and EOF typical of high-performance capillary electrophoresis (HPCE) with the use of a solid stationary phase and a separation mechanism based on specific interactions of solutes with a stationary phase characteristic of HPLC. Therefore, CEC is most commonly implemented by means typical of both HPLC (packed columns) and HPCE (electrophoretic instrumentation). To date, both columns and instrumentation developed specifically for CEC remain scarce.

CEC packing plays a dual role: in addition to providing sites for the required interactions as in HPLC, they must also be involved in EOF. This realization of the basic difference between HPLC and CEC has stimulated the development of both specific particulate packings having properties tuned for the needs of CEC as well as alternative column technologies.

The successful use of packed capillary columns in CEC separations has already been demonstrated in numerous papers and monographs.[92–94] The preparation of these columns includes two key steps: the fabrication of retaining frits within a capillary and the packing of small diameter particles into narrow-bore tubes. Both these steps require considerable experimental skills and experience in order to obtain stable columns with reproducible properties. Although packing is currently a well-established technique for the production of CEC columns, reproducibility of this procedure remains problematic.

Technical difficulties associated with packed columns have spurred the development of various alternative approaches. For example, Regnier's group developed systems with features microfabricated within the channel that does not fit the traditional definitions of either open channel or packed bed columns.[95–101] The microfabricated channels contain so-called collocated monolith support structures (COMOSS), shown in Figure 47.8, which are essentially an array of tightly packed posts. The size and shape of the posts, as well as the dimensions of the channels between them, can be varied as desired. Although the support structures are called directly fabricated "particles," the resulting column is referred to as "a bundle of interconnecting capillaries with frequent mixing nodes." COMOSS have numerous advantages, including the uniformity and regularity of the support particles, the high level of control over the channel dimensions and geometry, and the ability to control the extent of mixing in the column.[100]

Yet another broadly explored option is columns containing monolithic *in situ* prepared separation media adopted from a concept developed for HPLC columns.[25,92,102] As a result of their unique properties, monolithic materials have attracted considerable attention. Perhaps the most appealing aspect of the monolithic materials is their ease of preparation. The simple polymerization process performed directly within the confines of a capillary avoids the problems related to both frit formation and packing. In addition, columns of virtually any length and shape are easily accessible. The polymerization mixture may also be prepared using a wide variety of monomers, allowing a nearly unlimited choice of both matrix and surface chemistries. This flexibility enables the easy tailoring of both the interactions that are required for specific separation modes and the level of EOF generated by the support. Finally, the control that can be exerted over the polymerization process enables the

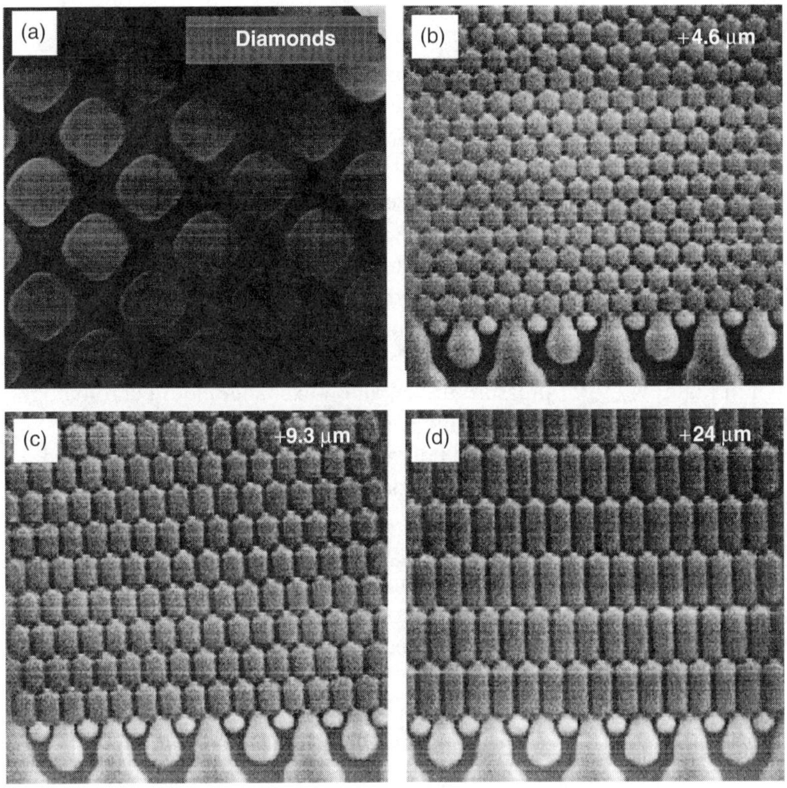

FIGURE 47.8 SEM micrographs of a section of a diamond COMOSS column (a) and its extended hexagonal modifications with the extension lengths of 4.6 (b), 9.3 (c), and 24 μm (d). (Reprinted with permission from Slentz, B. E., et al., *J. Sep. Sci.*, 25, 1011, 2002. Copyright 2007 Elsevier B.V.).

facile optimization of the porous properties of the monolith that directly affect the flow rate and chromatographic efficiency of the system.

Electrochromatography includes features of both electrophoresis and liquid chromatography. Therefore, electrochromatographic separations are achieved as a result of differences in both electrophoretic mobility (for ionized analytes) and specific interactions of solutes with the stationary phase. This combination of separation mechanisms enables efficient electrochromatographic separations of both ionizable and neutral compounds.

One of the major advantages of electrochromatography compared to HPLC is the ease with which it can be miniaturized to a microfluidic format and makes it most frequently used mode of separation in chips. The separation in electrochromatographic mode is achieved by simply applying an electric field across the channel. Since the fluid flow is driven by EOF, the separation device does not need to include mechanical pumps or valves.[103] In addition, the flat flow profile generated by EOF minimizes dispersion of analyte bands during their passage through the stationary phase and allows very high plate counts to be achieved. Since EOF is largely independent of channel or particle size, monolithic stationary phases with small pores can be used, thus facilitating solute mass transfer without generating the large pressure drops associated with classical pressure-driven HPLC. As a result, electrochromatography on chip is an attractive choice for "lab-on-a-chip" separations due to its excellent separation performance and relatively easy implementation. The topic of chip electrochromatography was summarized partly or completely in several very good review articles.[104–113] Our own recent review[114] also became the major source for writing this section.

47.2.3.2 Acrylamide-Based Monoliths

Despite a significant number of studies concerning capillary electrochromatography using monoliths in capillaries, the first monolithic material prepared in channel of a very simple chip was demonstrated by Hjertén's group only in 2000.[115] This first polymer-based monolithic stationary phase was prepared in 30.6-cm long and approximately 40-μm wide and 20-μm deep semicircular serpentine microchannels etched into quartz substrates and sealed with a "roof" formed by 4-μm thick layer of silicon dioxide shown in Figure 47.9. Monolithic stationary phase for reversed-phase electrochromatography was obtained by polymerization of methacrylamide, isopropyl acrylamide, piperazine diacrylamide (crosslinker), and vinylsulfonic acid mixtures in an aqueous buffer solution using the common redox initiating system ammonium peroxosulfate/tetramethylethylenediamine (TEMED). In addition, similar polymerization mixtures were used to create polymer monoliths in straight channel shown in Figure 47.9 for pressure-driven anion-exchange liquid chromatography with the exception that isopropyl acrylamide and vinylsulfonic acid were used instead of diallyldimethylammonium chloride. Since polymerization commences once the initiator is added, channels have to be filled immediately after the mixture is prepared. Before formation of the polymer monolith, the channels were treated with [3-(methacryloyloxy)propyl] trimethoxysilane to provide them with pendant methacryloyl groups that are incorporated during polymerization into the growing polymer monolith, thus covalently attaching it to the channel walls.

The location of the polymer bed could be controlled to a limited extent by using fluid plugs containing PEG, TEMED, and ammonium peroxodisulfate, but no monomers. These fluid plugs displaced the polymerization mixture to keep certain sections of the microchannel clear of monolith. The plugs also helped to create sharp ends to the monolith. The completed chips were attached to a special clamp apparatus for fluid interfacing and operated using a highly modified CE instrument capable of UV-detection.

The performance of the acrylamide-based monolithic stationary phase in the chip was compared to that in a fused-silica capillary containing monolith prepared from the same mixture. Not surprisingly, van Deemter plots were virtually identical for both devices since the channel etched in quartz, which did not have on-chip injection arms, closely resembled the capillary. Neutral alkyl phenones and positively charged antidepressants were separated in the serpentine channel in 18 and 9 min, respectively. A faster separation was demonstrated with a rapid separation of uracil, phenol, and benzyl alcohol in a Y-shaped channel (Figure 47.9) containing only a 1.8-cm long monolith. The separation was completed in less than 20 s, with all three peaks eluting within a 7 s window. A column efficiency calculated for unretained peak of uracil was 6300 plates, which represents 350,000 plates/m.

Overall, this implementation using the expensive and difficult-to-fabricate quartz microchips does not represent a significant advantage compared to typical CEC in much cheaper fused-silica capillaries placed in standard instruments. Also, carrying out experiments required laborious manual changing of contents of the inlet reservoir to achieve the sample injection.

Later, the same group developed what they call a hybrid microdevice as an alternative to quartz microchips.[116] This device consisted of a short, fused-silica capillary column containing a 2.8-cm long polymer gel prepared from acrylamide, 2-acrylamido-2-methyl-1-propanesulfonic acid (AMPS), and allyl-β-cyclodextrin as crosslinker. This capillary was set in a groove on a polyvinyl chloride support plate that included sample reservoirs, electrodes, and a slit for "on-tube" UV-detection. In addition, a ball lens was mounted beneath the detection slit to focus the UV beam. Gel electrophoresis and electrochromatography were then performed in this system. For example, several alkyl phenones were separated in the electrochromatographic mode in less than 200 s. The suggested separation mechanism was related to interactions between the hydrophobic cavity of the β-cyclodextrin and the analytes. However, no direct evidence for this mechanism was presented.

Claimed advantages of this hybrid device compared to traditional microchips include lower cost in terms of materials and preparation, UV-detection with no need for fluorescent labels, the ability to

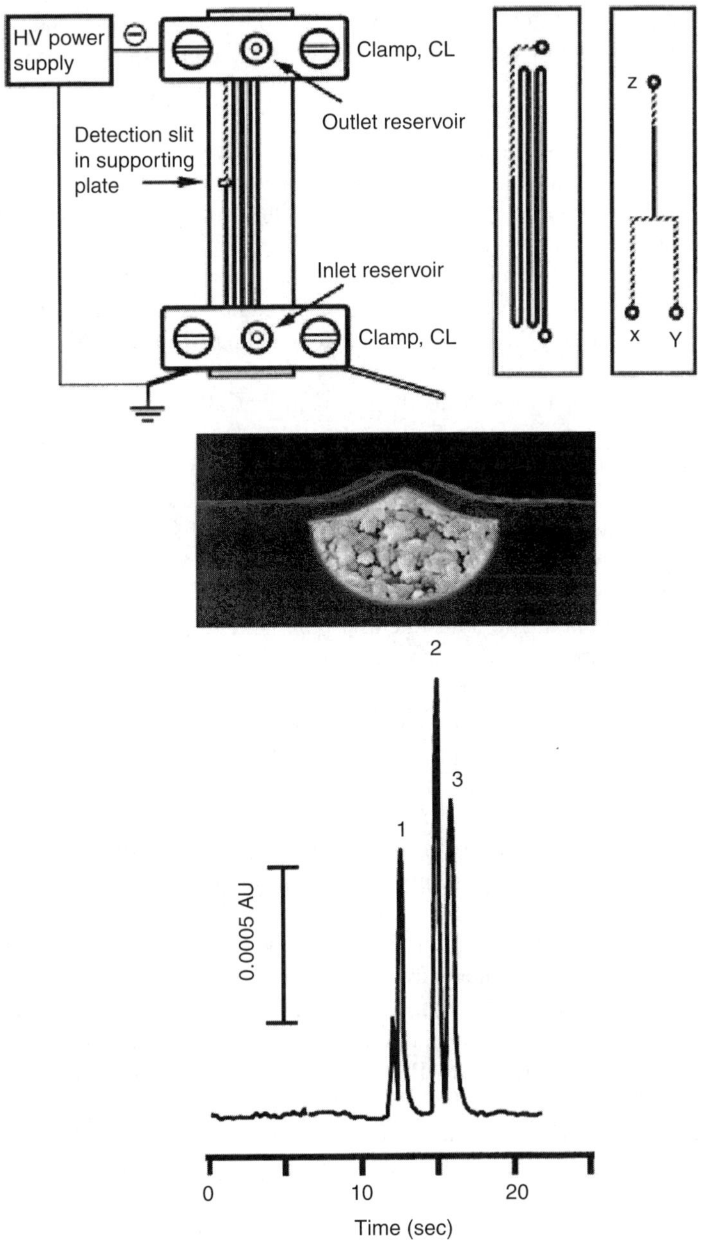

FIGURE 47.9 Chip layout, cross-section of the monolith, and fast anion-exchange chromatography of uracil (1), phenol (2), and benzyl alcohol (3) using the Y-shaped chip format. (Reprinted with permission from Ericson, C., et al., *Anal. Chem.*, 72, 81, 2000. Copyright 2000 Elsevier B.V.) Conditions: straight monolithic column 18-mm long, mobile phase 5 mmol/L sodium phosphate buffer (pH 7.4) containing 15% acetonitrile, applied voltage, 2.4 kV (500 V/cm), EOF 2.1 mm/s, UV detection at 254 nm, electrokinetic injection 100 V for 2 s.

change the separation column by simply switching the capillary, and the ability to directly utilize the existing body of experience available for capillary separations. However, this approach is limited to very simple geometries and is less suitable for the more complex layouts typical of micro total analysis system (μTAS).

Szekely and Freitag[117] fabricated separation chips from glass with the channels sawed with a 70-μm resinoid blade. While this approach only enables the fabrication of straight channels, it is much faster and less complex than methods based on photolithography and etching, and does not require work in the clean room. A multifunctional connection and support unit containing optical probes, reservoirs, as well as high-pressure and high-voltage connectors, constituted the interface to the macro world. To demonstrate their possible application in voltage-driven electrochromatography, a monolithic stationary phase was prepared using a mixture of piperazine diacrylamide, *N*,*N*-dimethylacrylamide, sodium vinyl sulfonate, and ammonium sulfate dissolved in phosphate buffer. The polymerization reaction was initiated by the addition of TEMED and ammonium persulfate dissolved in water. A charge-coupled device (CCD) camera serving as a fluorescence detector and a plug of fluorescein moving through the 52-mm long, 70-μm deep, and 80 μm wide semicircular channel driven by voltage demonstrated functionality of the chip.

Latter, they added another module to their system that allowed setting and maintaining the flow rate in microfluidic systems using EOF to drive the mobile phase.[118] The sensing principle is claimed to be compatible with any type of microanalytical device in which the mobile phase resides in a reservoir. The relatively simple unit consists of a light-emitting diode, fiber optics, and a photomultiplier. This sensor optically records the speed of sinking of the liquid level in the reservoirs. Readings are taken with a frequency of 1000 Hz allowing measurements of the liquid level in intervals less than 0.4 s. On the basis of these measurements, they could maintain a flow rate at a preset value via automatic adjustments in the field strength applied across the channel. The concept was again demonstrated on stabilized flow through an acrylamide-based monolith in the channel. Since the reservoir appears to be open, evaporation of the liquid can also occur. The observed flow rate would then be higher than the actual value.

Using this system, the same group also studied effects of electrochemical processes on the electrodes located in reservoirs of chips containing acrylamide-based monolith.[119] For example, the pH in the cathodic reservoir containing 50 mmol/L potassium chloride changed after 4 min at 4000 V from 7.58 to 8.49. Use of phosphate buffer pH 9.2 helped to stabilize the pH and the increase was only 0.21 pH units after 1 min. Surprisingly, these changes did not affect EOF generated by the ionizable functionalities of the monolith.

Zeng et al.[120] fabricated PDMS chips including two crossed channels using typical soft lithographic replication process. They advocated extended irradiation of the chip leading to formation of hydroxyl functionalities at the channel surface. Following modification with [3-(methacryloyloxy)propyl] trimethoxysilane, they drew in the separation channel a solution of acrylamide, methylenebisacrylamide, AMPS, allyl substituted γ-cyclodextrine (γ-CD), TEMED, and ammonium persulfate in borate buffer. The γ-CD moieties then enabled enantioseparation of labeled derivatives of amino acids. Although some separation in electrochromatographic mode could be observed, the chromatographic evaluation of results was very poor to make it worth of significant attention.

47.2.3.3 Acrylate- and Methacrylate-Based Monoliths

The early monolithic columns were prepared by thermally initiated polymerization. Although useful for larger scale columns, this technique is not ideally suited for the fabrication of monoliths within microfluidic devices. Therefore, in the late 1990s, we introduced polymerization initiated by irradiation with UV light, which is better suited for the preparation of monoliths in capillaries and microchips.[64,121] Later, we refined this approach to enable both preparation of the monolith and its *in situ* surface functionalization.[77–79,122–126] The significant advantage of photoinitiation is the ability to photolithographically define the size and position of the monolith using a mask through which the liquid polymerization mixture in the channel is irradiated. This process was demonstrated with the preparation of a variety of methacrylate-based monolithic porous polymer stationary phases. The typical polymerization mixtures contained monomers, initiator, and porogenic solvents.

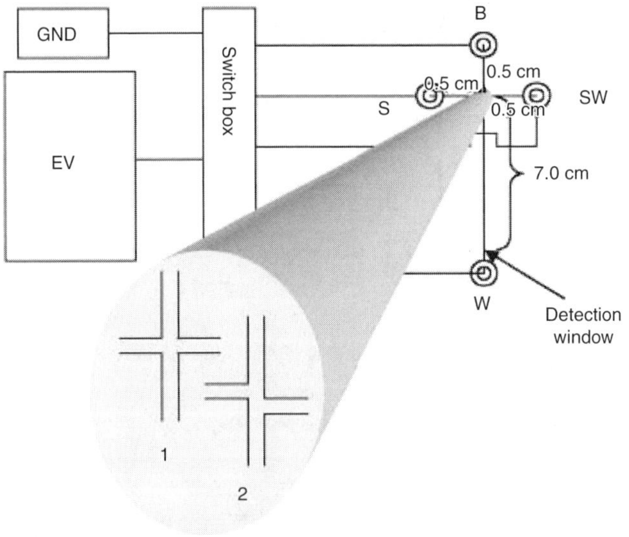

FIGURE 47.10 Chip filled with monolith connected to switch box and power. (Reprinted with permission from Fintschenko, Y., et al., *Fresenius' J. Anal. Chem.*, 371, 174, 2001. Copyright 2001 Springer Verlag.) (1) T injector, (2) offset T-injector. Chip dimensions: 8.5 × 2.0 cm, channel cross section 50 μm wide, 25 μm deep, separation channel 7.5 cm, total length of monolith to separation window 7 cm. Reservoirs: B running buffer, W waste, S sample, SW sample waste. Injection and run voltages 2 kV.

For example, the early monomer mixtures contained crosslinker such as ethylene dimethacrylate and trimethylolpropane trimethacrylate, monovinyl monomers such as butyl methacrylate, and AMPS. The initiator was originally 2,2′-azobisisobutyronitrile. However, the use of aromatic phenone initiators later enabled significant acceleration of the polymerization process such that it was completed in several minutes as opposed to more than 12 h. A large number of various porogenic solvents was also tested.[127]

Microfluidic electrochromatography with polymer-based monolithic stationary phases prepared *in situ* by photoinitiation was again studied extensively in Sandia National Laboratories. Although the early studies primarily focused on the development of acrylate-based polymer monoliths in capillaries, initial test separations were also carried out in glass-/fused-silica microchips with a simple double-T cross layout (Figure 47.10).[128–130] Both positively and negatively charged monoliths were prepared. Positively charged monoliths were found to be useful for separations at low pH since they afforded high EOF and, at the same time, prevented undesired Coloumbic interactions with positively charged compounds, such as amines. A typical polymerization mixture comprised crosslinker 1,3-butanediol diacrylate, butyl acrylate or combinations of butyl acrylate with lauryl acrylate, ionizable monomer (AMPS for negatively charged monolith or [2-(acryloyloxy)-ethyl] trimethylammonium methyl sulfate for positively charged monoliths), AIBN (photoinitiator), and adhesion promoter (z-6030, Dow Chemicals). The porogenic solvent contained ethanol, acetonitrile, and a phosphate buffer. The surface of channels in glass microchips was silanized using treatment with [3-(methacryloyloxy)propyl] trimethoxysilane.[131] The 50-μm wide and 25-μm deep microchannels were filled with the liquid polymerization mixture using capillary action and monoliths were prepared by exposure to UV light for 10–30 min. The ionized functionalities on the monolith and aqueous buffer present in the porogenic solvent enabled the use of EOF, rather than pressure-driven flow, to purge the unreacted components from the porous material after polymerization. Figure 47.11 demonstrates the separation of three bioactive peptides, fluorescently labeled with naphthalene-2,3-dicarboxaldehyde (NDA) using a 7-cm long monolithic stationary phase, affording efficiencies of 105–5800 plates

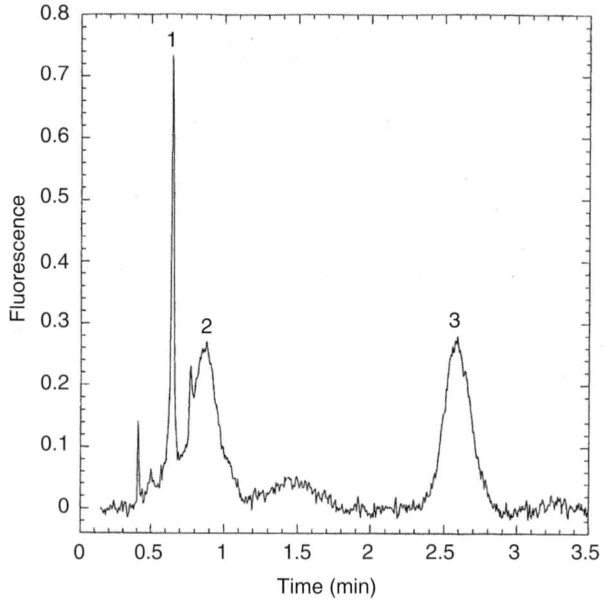

FIGURE 47.11 Electrochromatographic separation of NDA-labeled papain inhibitor GGYR (1), α-casein (fragment 90–95) RYLGYL (2), and Ile-angiotensin III, RVYIHPI (3) using negatively charged lauryl methacrylate-based monolithic stationary phase in a glass chip. (Reprinted with permission from Shediac, R., et al., *J. Chromatogr. A*, 925, 251, 2001. Copyright 2001 Elsevier B.V.) Conditions: mobile phase 35:65 acetonitrile–12.5 mmol/L phosphate buffer (pH 7.0), LIF detection at 413 nm, field strength 1.2 kV/cm, channel dimensions: 25-μm deep, 50-μm wide, separation channel length 8.0 cm; injection arms 1.0 cm each; length to detector 7.0 cm.

(1500–83,000 plates/m). The performance was limited by undesirable electrostatic interactions of the peptides with the stationary phase.[129]

The alkyl acrylate-based monoliths were also used for the separation of polycyclic aromatic hydrocarbons (PAHs). Good performance was demonstrated on the separation of three neutral PAHs with efficiencies of 8400–13,900 plates (140,000–231,000 plates/m).[128] In a different study, the number of PAH analytes was increased to 13.[130] In this work, the monolith-free window required for detection was created by decomposition of the monolith using the energy of a frequency-doubled argon ion laser, which was part of the LIF detector. The authors claim that irradiation for 10–20 min while electroosmotically pumping buffer solution through the device was sufficient to completely remove the monolith from the desired area. Chips containing polymer monolith in two different configurations were prepared and evaluated. In the first device, all the channels were completely filled with porous polymer monolith. The second then had the monolith located only in a specific area of the chip using polymerization initiated through a mask. This photopatterned monolith filled all but the injection arms of the device. Interestingly, its performance was poorer than that of the former, featuring efficiencies of 100,000–150,000 plates/m as opposed to 180,000–200,000 plates/m for the chip completely filled with monolith. Heterogeneous regions in the patterned polymer, specifically at the ends of the monolith, and/or zeta potential mismatch between the sections of the chip with and without monolith were blamed for the poorer performance. The latter could have introduced parabolic flow, which would have led to band broadening and decreased efficiency. In addition to comparing their separation performance, the operational limitations and the reproducibility of the two types of monolith devices were also investigated. For example, low resistance to flow in the device with open injection arms allowed running buffer to flow into the fluid reservoirs and dilute the sample during pinched separation runs. Similar problem was observed by Jacobson et al.[132] On the other hand,

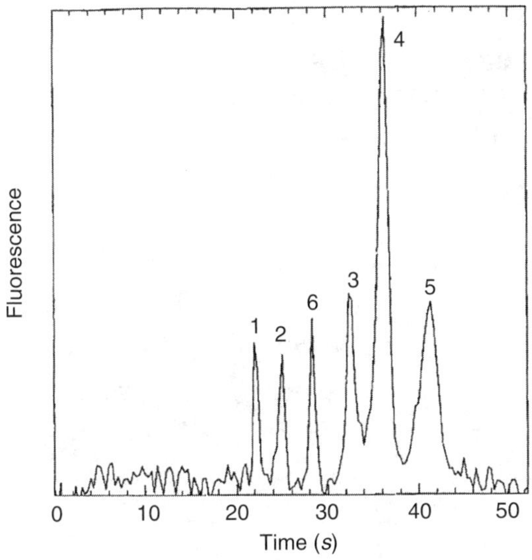

FIGURE 47.12 Reversed-phase electrochromatography of NDA-labeled peptides papain inhibitor (1), proctolin (2), opioid peptide (α-casein fragment 90–95) (3), Ile-angiotensin III (4), angiotensin III (5), and GGG (6) in a microchip containing negatively charged lauryl acrylate monolith. (Reprinted with permission from Throckmorton, D. J., et al., *Anal. Chem.*, 74, 784, 2002. Copyright 2002 American Chemical Society.) Conditions: mobile phase 30:70, acetonitrile-25 mmol/L borate buffer (pH 8.2) containing 10 mmol/L octane sulfonate, LIF detection at 413 nm, field strength 770 V/cm (5 kV).

injection times as long as 5 min or more were required to ensure uniform sample injections in the devices with monolith-filled sample arms. This study also highlighted the importance of carefully sealing fluid reservoirs during chip operation. Using unsealed reservoirs, 95% reductions in signal peak heights were observed over the course of just five separations due to dissolved oxygen that quenched the fluorescence of the PAHs. In addition, variations in retention times were rather high, with deviations up to 20%. Thus, further work was needed to improve the overall performance of these chips.

Microchips filled with acrylate-based polymer monolith were later used to separate bioactive peptides and amino acids.[133] Citing decreased performance due to zeta potential mismatch between open and monolith-filled channels, all the channels of the microchip were filled with monolith except for a small detection window. NDA-labeled peptides were separated with excellent efficiencies in fused-silica chips with channels either 25-μm deep and 90-μm wide or 40-μm deep and 130-μm wide. The mobile phase contained octane sulfonate, apparently acting as an ion-pairing agent for the charged peptides.[134] The effect of voltage on the peptide separation was evaluated in a range of 1–5 kV. The best performance occurred at 2 kV (~300 V/cm), affording efficiencies of 6350–30,000 plates/chip (127,000–600,000 plates/m). However, nearly baseline resolution was achieved even at the highest field strengths (770 V/cm) and the separation was completed in impressive 45 s (Figure 47.12). In contrast to the previous report,[130] these new devices featured an improved reproducibility with relative standard deviation (RSD) values for the retention times of approximately 2%. Since glass-based microchips are costly, this report also describes a procedure for their recycling via thermal incineration of the monolith. However, our results with recycled glass chips did not match those obtained with the new chips.

Lazar et al.[134] have recently demonstrated the first example of interfacing chip electrochromatography with mass spectrometric detection. The glass microchip contained a methacrylate-based monolith as well as an integrated capillary tip for ESI. The straight 130-μm wide and 50-μm deep main channel of the chip contained a 5 to 6-cm long separation section that led directly to the ESI

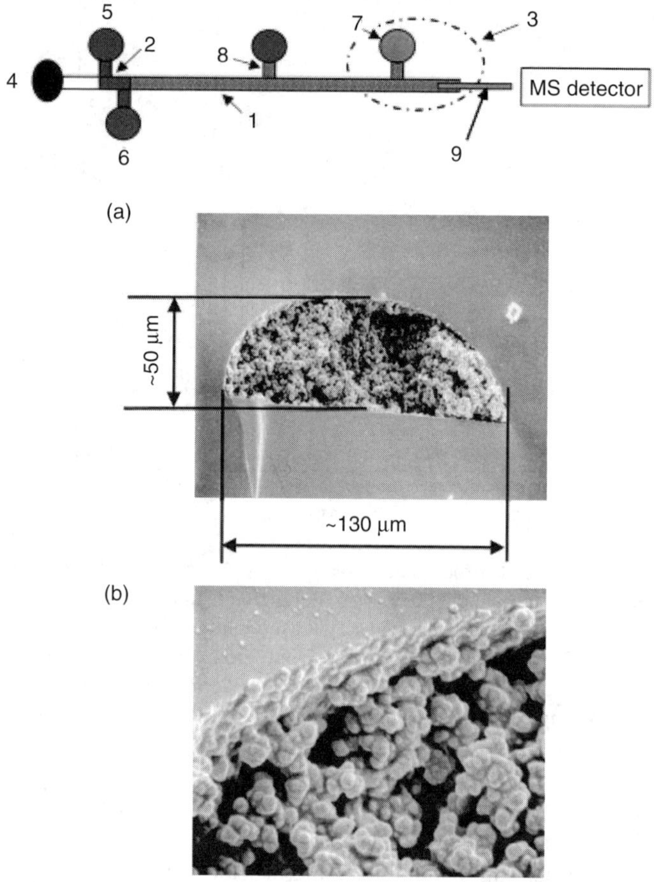

FIGURE 47.13 Schematic diagram of CEC system integrated on microchip: separation channel containing poly(glycidyl methacrylate-co-methyl methacrylate-co-ethylene dimethacrylate) monolith (1), double-T injector (2), ESI source (3), eluent reservoir (4), sample inlet reservoir (5), sample waste reservoir (6), eluent waste reservoir that houses the porous glass gate (7), side channel for flushing the monolith (8), and ESI emitter (9). SEM micrographs of channel cross-section with the monolith (a) and enlarged view of the monolith structure near the channel wall (b). (Reprinted with permission from Lazar, I. M., et al., *Electrophoresis*, 24, 3655, 2003. Copyright 2003 Elsevier B.V.)

emitter as shown in Figure 47.13. A 5-mm long trench was etched into the cover plate before the chip was bonded to allow insertion of the emitter tip into the end of the separation channel. Further, the chip design included an additional fluid reservoir near the end of the separation channel. This reservoir contained a conductive, semipermeable glass disk, which allowed voltage to be applied to the fluid in the channel but prevented bulk fluid movement. With this design, electroosmotically driven fluid in the main channel was forced out through the ESI capillary tip. Similar to the Sandia approach,[130,133] all of the microchannels were completely filled with monolith to ensure uniform EOF throughout the chip. The monolithic stationary phase was prepared by photoinitiated polymerization of mixtures consisting of glycidyl methacrylate, methyl methacrylate, and ethylene dimethacrylate in the presence of formamide and 1-propanol as porogenic solvents and 2,2-dimethoxy-2-phenyl-acetophenone as the initiator. Subsequently, the monolith was incubated with *N*-ethylbutylamine overnight at 70°C. The amine reacted with the epoxy groups of glycidyl methacrylate on the monolith, forming tertiary amine functionalities and rendering the surface positively charged. This chemistry was chosen in order to minimize electrostatic interactions between the monolith and the positively charged peptides

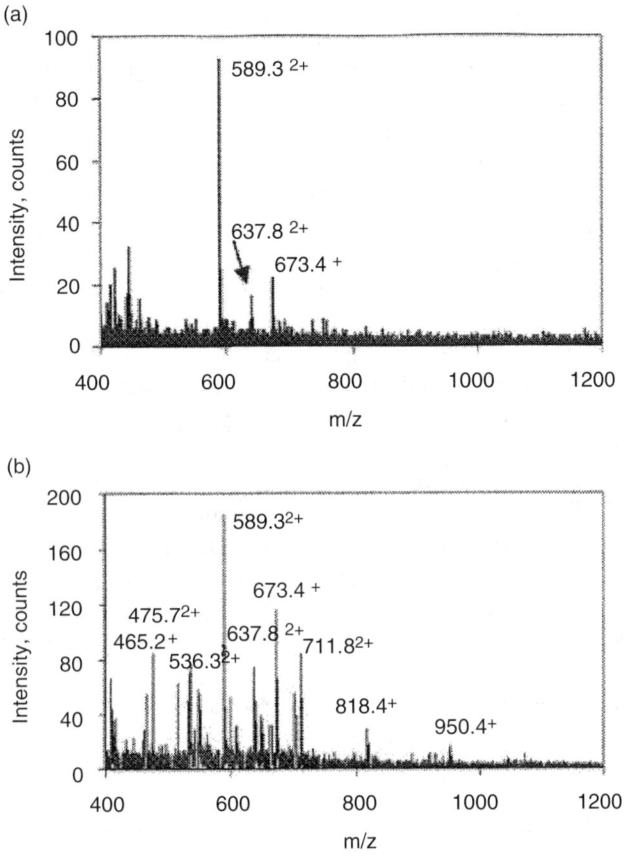

FIGURE 47.14 Time-of-flight mass spectra of peptides prepared via tryptic digestion of bovine hemoglobin. (Reprinted with permission from Lazar, I. M., et al., *Electrophoresis*, 24, 3655, 2003. Copyright 2003 Elsevier B.V.) (a) Peptides separated in the CEC system with monolithic column shown in Figure 47.13, (b) unseparated peptides detected in an infusion experiment.

of bovine hemoglobin tryptic digest. Although it would also be possible to separate mixtures of peptides on negatively charged monoliths such as those prepared by copolymerization of AMPS, this approach would require the addition of an ion-pairing agent in the mobile phase. This approach would be counterproductive since most ion-pairing agents are incompatible with ESI mass spectrometry.

The positively charged surface of the monolith drives the electroosmotic flow to run from cathode to anode. This EOF was opposite to the direction of electrophoretic migration of the analyzed peptides and increased the time they spent in the relatively short column. Separations of the peptide mixture using mass spectrometric detection afforded a sequence coverage of 70–80% and efficiencies of 3000–4000 plates. While the tryptic digest was not completely separated on the monolithic stationary phase, the mass spectra were greatly simplified compared to a spectrum of the original peptide mixture (Figure 47.14). In addition, multiplexed devices with 3–4 operating lines were also constructed.

Performing injections proved to be problematic for this system. While the positively charged monolith and "reversed" EOF improved separation performance, they also significantly enhanced electrophoretic discrimination during injection. In addition, despite the tuned chemistry, the analytes interacted with the amine functionalities and neutralized them, completely suppressing the EOF in the injection arms after multiple injections. This prevented long-term use of the device. Use of disposable, perhaps plastic, chips appears to be a simple solution to this problem.

47.2.3.4 Silica-Based Monoliths

Monolithic silica is the most recently introduced class of stationary phases for electrochromatography on microchips. They benefit from their very high surface area, adjustable pore size, and controllable surface chemistry. Functionalities required for the separation in reversed-phase mode are typically incorporated onto the silica monolith by including an appropriate silicon alkoxide in the precursor mixture or by silanization of the surface after the monolith has been formed. Monolithic silica-based columns are generally known to exhibit superior performance in HPLC separations of small molecules, which is attributed to the presence of mesopores entailing large surface areas.[135]

Silica monoliths in capillaries are usually prepared via acid-catalyzed condensation of alkoxysilanes followed by heat treatment.[92] Thus, the spatial definition of such a monolith is difficult.[136–138] This is why use of silica-based monoliths in chips is scarce. In one example, Zare and coworkers[139,140] at Stanford has developed a new method for the preparation of photopatternable sol-gel monoliths. They mixed a partially gelled solution of 3-methacryloxypropyl-trimethoxysilane (MPTMS) with a solution of photoinitiator in toluene and immediately filled the channels with this mixture via capillary action. Exposure to UV light through a photomask initiated a free-radical polymerization of the methacrylate vinyl groups of MPTMS. The resulting monolith is thus a hybrid of both silica and polymer monoliths. The electrochromatographic performance of a 4.7-cm long monolith prepared in this way in a glass chip with 35-μm deep and 90-μm wide channels was demonstrated with the separation of two neutral coumarin dyes shown in Figure 47.15. Although a baseline resolution was achieved, only modest efficiencies 280 and 940 plates (6000–20,000 plates/m) were found for the two peaks.

In another approach, Breadmore et al.[141,142] developed silica-based monolithic stationary phases suitable for microchip electrochromatography. They modified the procedure originally developed

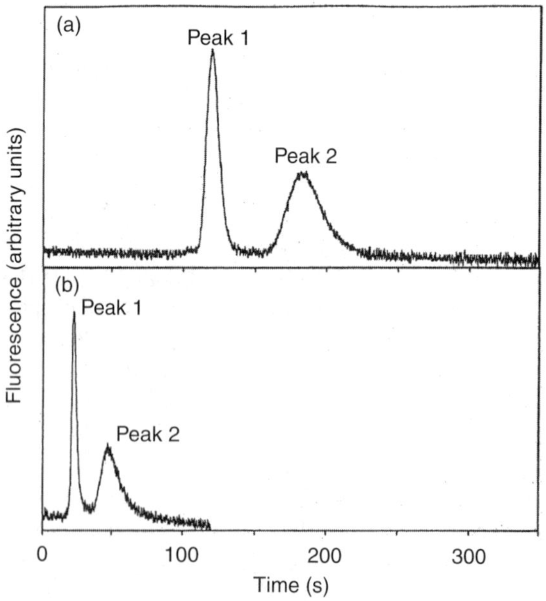

FIGURE 47.15 Electrochromatograms showing the separation of a mixture of Coumarin 314 (peak 1) and Coumarin 510 (peak 2). (Reprinted with permission from Morishima, K., et al., *J. Sep. Sci.*, 25, 1226, 2002. Copyright 2002 J. Wiley.) Conditions: sample concentration 0.6 mg/mL, mobile phase 1/3/6 (v/v/v) 50 mmol/L ammonium acetate (pH 6.5) (10%), water (30%), acetonitrile (60%), voltage 2.0 kV for (a) and 2.5 kV for (b), LIF detection at 420 nm, peaks detected on open channel adjacent to 4.7-cm long monolithic section (a), and on-column 1.2 cm from the top of the monolith (b).

by Tanaka and coworkers.[143] First, an unmodified silica monolith was formed by hydrolytic condensation of tetramethyl orthosilicate in the presence of PEG. This water-soluble polymer induced a phase separation during gelation, resulting in a monolith with well-controlled porous properties. Varying molecular weight of PEG significantly changed the resulting structure of the monolith and its chromatographic performance. The resulting monoliths were then functionalized using two different procedures. First, cationic polymer poly(diallyldimethylammonium chloride) (PDDAC) was dynamically adsorbed from its solution pumped through the monolith to make it suitable for anion-exchange electrochromatography.[141] To maintain the coating during separations, PDDAC was also included in the mobile phase. In the second approach, the silica monolith was coated with multiple, alternating layers of polymer electrolytes of opposite charges forming polyelectrolyte multilayer.[142] The cationic layers in the multilayer coating were made from PDDAC, while anionic layers consisted of either dextran sulfate or poly(styrene sulfate). This system afforded very stable and reproducible EOF. In addition, the multilayers could be used for multiple runs in ion-exchange and reversed-phase modes without the need for recoating. The use of coatings has several advantages over both modification with functional silanes and the use of functional silica precursors. For example, the silica monolith must be completely dried before modification by silanization. This drying process can lead to shrinkage or even cracking of the monolith. Next, although addition of functional alkoxysilanes to the monolith reaction mixture creates a functional monolith in a single step, the sol-gel process is more complex, and any attempt to adjust the composition of the monolith mixture may require reoptimization of the sol-gel process. With coatings, the monolith can be functionalized without the need for drying and, further, the monolith formulation is decoupled from the coating formulation. However, application of this approach has yet to be demonstrated in microchips.

47.3 CONCLUDING REMARKS

The field of chromatographic separations on chip using channels containing a stationary phase is still in its infancy; although significant effort has already been dedicated to development and application of a variety of stationary phases. Obviously, implementations involving monolithic separation media are likely to continue to grow. Their major advantage is the formation from liquid precursors that are easy to fill the channel and simple to polymerize *in situ*. The focus must also shift toward improving their separation performance. Current chips do not by far match selectivities and efficiencies of their older relatives—capillary columns.

In addition to serving as the stationary phases, monolithic materials are also finding numerous other applications in the microfluidic world.[144] Their use in on-chip solid-phase extraction and preconcentration,[28] as supports for immobilization of enzymes to form enzymatic microreactors for protein mapping,[27] static mixers,[145] and valves[146,147] represent just a few examples of modules in the microfluidic toolbox and further growth is inevitable.

However, to make the microfluidic devices affordable, it is imperative to switch from expensive chips made from glass or quartz to inexpensive, mass-produced plastic devices. Although a certain success with plastic devices has been demonstrated, it is clear that further development is needed.

Obviously, much remains to be done on the road leading to the creation of more complex analytical devices called lab-on-a-chip or total microanalytical systems (μ-TAS) and the monoliths are expected to help to achieve this goal.

ACKNOWLEDGMENTS

Support of this work by a grant of the National Institute of Biomedical Imaging and Bioengineering, National Institutes of Health (EB-006133) is gratefully acknowledged. Work at the Molecular Foundry was supported by the Director, Office of Science, Office of Basic Energy Sciences, Division of Materials Sciences and Engineering, of the U.S. Department of Energy under Contract No. DE-AC02-05CH11231.

REFERENCES

1. Tsvett, M. S., Adsorption analysis and chromatographic method. The use of it in the chemistry of chlorophyll, *Ber. Deut. Botan. Gessel.*, 24, 322, 1906.
2. Porath, J. and Flodin, P., Gel filtration: a method for desalting and group separation, *Nature*, 193, 1657, 1959.
3. Hjertén, S., Chromatographic separation according to size of macromolecules and cell particles on columns of agarose suspensions, *Arch. Biochem. Biophys.*, 99, 466, 1962.
4. Sober, H. A., Chromatography of proteins on cellulose ion-exchangers 5829, *J. Am. Chem. Soc.*, 76, 1711, 1954.
5. Snyder, L. R. and Kirkland, J. J. *Introduction to Modern Liquid Chromatography*, 2nd ed., Wiley: New York, 1979.
6. Unger, K. K. *Packings and Stationary Phases in Chromatographic Techniques,* Dekker: New York, 1990.
7. Martin, A. J. P. and Synge, R. L. M., A new form of chromatogram employing two liquid phases. I. Theory. II Aplication to the microdetermination of the higher monoamino acids in proteins, *Biochem. J.*, 35, 1358, 1941.
8. Mould, D. L. and Synge, R. L. M., Electrokinetic ultrafiltration analysis of polysaccharides. A new approach to the chromatography of large molecules, *Analyst*, 77, 964, 1952.
9. Mould, D. L. and Synge, R. L. M., Separations of polysaccharides related to starch by electrokinetic ultrafiltration in collodion membranes, *Biochem. J.*, 58, 571, 1954.
10. Kubin, M., Spacek, P., and Chromecek, R., Gel permeation chromatography on porous poly(ethylene glycol methacrylate), *Coll. Czechosl. Chem. Commun.*, 32, 3881, 1967.
11. Ross, W. D. and Jefferson, R. T., *In situ* formed open-pore polyurethane as chromatography support, *J. Chrom. Sci.*, 8, 386, 1970.
12. Hansen, L. C. and Sievers, R. E., Highly permeable open pore polyurethane columns for liquid chromatography, *J. Chromatogr.*, 99, 123, 1974.
13. Tennikova, T. B., Svec, F., and Belenkii, B. G., High performance membrane chromatography. A novel method of protein separation, *J. Liquid Chromatogr.*, 13, 63, 1990.
14. Svec, F. and Tennikova, T. B., High performance membrane chromatography: highly efficient separation method for proteins in ion exchange, hydrophobic interaction, and reversed phase modes, *J. Chromatogr.*, 646, 279, 1993.
15. Hjertén, S., Liao, J. L., and Zhang, R., High-performance liquid chromatography on continuous polymer beds, *J. Chromatogr.*, 473, 273, 1989.
16. Josic, D. and Strancar, A., Application of membranes and compact porous units for the separation of biopolymers, *Ind. Eng. Chem. Res.*, 38, 333, 1999.
17. Liao, J. L., Continuous bed for conventional column and capillary column chromatography, *Adv. Chromatogr.*, 40, 467, 2000.
18. Strancar, A., Podgornik, A., Barut, M., and Necina, R., Short monolithic columns as stationary phases for biochromatography, *Adv. Biochem. Eng. Biotechnol.*, 76, 49, 2002.
19. Svec, F., Tennikova, T. B., and Deyl, Z., *Monolithic Materials: Preparation, Properties, and Applications,* Elsevier: Amsterdam, 2003.
20. Svec, F. and Fréchet, J. M. J., Continuous rods of macroporous polymer as high performance liquid separation media, *Anal. Chem.*, 54, 820, 1992.
21. Svec, F. and Fréchet, J. M. J., New designs of macroporous polymers and supports: from separation to biocatalysis, *Science*, 273, 205, 1996.
22. Svec, F. and Fréchet, J. M. J., Molded separation media: an inexpensive, efficient, and versatile alternative to packed columns for the fast HPLC separation of peptides, proteins, and synthetic oligomers and polymers, *Macromol. Symp.*, 110, 203, 1996.
23. Svec, F. and Fréchet, J. M. J., Molded rigid monolithic porous polymers: an inexpensive, efficient, and versatile alternative to porous beads for the design of materials with high flow characteristics for numerous applications, *Ind. Eng. Chem. Res.*, 36, 34, 1999.
24. Peters, E. C., Svec, F., and Fréchet, J. M. J., Rigid macroporous polymer monoliths, *Adv. Mater.*, 11, 1169, 1999.
25. Svec, F., Recent developments in the field of monolithic stationary phases for capillary electrochromatography, *J. Sep. Sci.*, 28, 729, 2005.

26. Svec, F., Organic polymer monoliths as stationary phases for capillary HPLC, *J. Sep. Sci.*, 27, 1419, 2004.
27. Svec, F., Less common applications of monoliths: I. Microscale protein mapping with proteolytic enzymes immobilized on monolithic supports, *Electrophoresis*, 27, 947, 2006.
28. Svec, F., Less common applications of monoliths: II. Preconcentration and solid phase extraction, *J. Chromatogr. B*, 841, 52, 2006.
29. Tanaka, N., Ishizuka, N., Hosoya, K., Kimata, K., Minakuchi, H., Nakanishi, K., and Soga, N., Octadecylsilylated porous silica rod for reversed-phase liquid chromatography, *Kuromatogurafi*, 14, 50, 1993.
30. Fields, S. M., Silica xerogel as a continuous column support for high-performance liquid chromatography, *Anal. Chem.*, 68, 2709, 1996.
31. Minakuchi, H., Nakanishi, K., Soga, N., Ishizuka, N., and Tanaka, N., Octadecylsilylated porous silica rods as separation media for reversed-phase liquid chromatography, *Anal. Chem.*, 68, 3498, 1996.
32. Cabrera, K., Wieland, G., Lubda, D., Nakanishi, K., Soga, N., Minakuchi, H., and Unger, K. K., Silicarod™—A new challenge in fast high-performance liquid chromatography separations, *Trends Anal. Chem.*, 17, 50, 1998.
33. Cabrera, K., Lubda, D., Eggenweiler, H. M., Minakuchi, H., and Nakanishi, K., A new monolithic-type HPLC column for fast separations, *J. High Resolut. Chromatogr.*, 23, 93, 2000.
34. Tanaka, N., Nagayama, H., Kobayashi, H., Ikegami, T., Hosoya, K., Ishizuka, N., Minakuchi, H., Nakanishi, K., Cabrera, K., and Lubda, D., Monolithic silica columns for HPLC, micro-HPLC, and CEC, *J. High Resolut. Chromatogr.*, 23, 111, 2000.
35. Tanaka, N., Kobayashi, H., Ishizuka, N., Minakuchi, H., Nakanishi, K., Hosoya, K., and Ikegami, T., Monolithic silica columns for high-efficiency chromatographic separations, *J. Chromatogr. A*, 965, 35, 2002.
36. Meyers, J. J. and Liapis, A. I., Network modeling of the convective flow and diffusion of molecules adsorbing in monoliths and in porous particles packed in a chromatographic column, *J. Chromatogr. A*, 852, 3, 1999.
37. Leinweber, F. C. and Tallarek, U., Chromatographic performance of monolithic and particulate stationary phases: Hydrodynamics and adsorption capacity, *J. Chromatogr. A*, 1006, 207, 2003.
38. Paces, M., Kosek, J., Marek, M., Tallarek, U., and Seidel-Morgenstern, A., Mathematical modelling of adsorption and transport processes in capillary electrochromatography: Open-tubular geometry, *Electrophoresis*, 24, 380, 2003.
39. Vervoort, N., Gzil, P., Baron, G. V., and Desmet, G., A correlation for the pressure drop in monolithic silica columns, *Anal. Chem.*, 75, 843, 2003.
40. Vervoort, N., Gzil, P., Baron, G. V., and Desmet, G., Model column structure for the analysis of the flow and band-broadening characteristics of silica monoliths, *J. Chromatogr. A*, 1030, 177, 2004.
41. Nischang, I., Chen, G., and Tallarek, U., Electrohydrodynamics in hierarchically structured monolithic and particulate fixed beds, *J. Chromatogr. A*, 1109, 32, 2006.
42. Ettre, L., IUPAC nomenclature for chromatography, *Pure Appl. Chem.*, 65, 872, 1993.
43. Neue, U. *HPLC Columns: Theory, Technology, and Practice,* Wiley: New York, 1997.
44. Ocvirk, G., Verpoorte, E., Manz, A., Grassbauer, M., and Widmer, H. M., HPLC partially integrated onto a silicon chip, *Anal. Meth. Instrument.*, 2, 74, 1995.
45. Oleschuk, R. D., Shultz-Lockyear, L. L., Ning, Y. B., and Harrison, D. J., Trapping of bead-based reagents within microfluidic systems: On-chip solid-phase extraction and electrochromatography, *Anal. Chem.*, 72, 585, 2000.
46. Ghitun, M., Bonneil, E., Fortier, M. H., Yin, H., Killeen, K., and Thibault, P., Integrated microfluidic devices with enhanced separation performance: application to phosphoproteome analyses of differentiated cell model systems, *J. Sep. Sci.*, 29, 1539, 2006.
47. Pregibon, D. C., Toner, M., and Doyle, P. S., Magnetically and biologically active bead-patterned hydrogels, *Langmuir*, 22, 5122, 2006.
48. Zaytseva, N. V., Montagna, R. A., and Baeumner, A. J., Microfluidic biosensor for the serotype-specific detection of dengue virus RNA, *Anal. Chem.*, 77, 7520, 2005.
49. Xie, J., Miao, Y., Shih, J., Tai, Y. C., and Lee, T. D., Microfluidic platform for liquid chromatography-tandem mass spectrometry analyses of complex peptide mixtures, *Anal. Chem.*, 77, 6947, 2005.

50. Ramsey, J. D. and Collins, G. E., Integrated microfluidic device for solid-phase extraction coupled to micellar electrokinetic chromatography separation, *Anal. Chem.*, 77, 6664, 2005.
51. Peterson, D. S., Solid supports for microanalytical systems, *Lab Chip*, 5, 132, 2005.
52. Yin, H., Killeen, K., Brennen, R., Sobek, D., Werlich, M., and Van de Goor, T., Microfluidic chip for peptide analysis with an integrated HPLC column, sample enrichment column, and nanoelectrospray tip, *Anal. Chem.*, 77, 527, 2005.
53. Lettieri, G. L., Dodge, A., Boer, G., de Rooij, N. F., and Verpoorte, E., A novel microfluidic concept for bioanalysis using freely moving beads trapped in recirculating flows, *Lab Chip*, 3, 34, 2003.
54. Losey, M. W., Schmidt, M. A., and Jensen, K. F., Microfabricated multiphase packed-bed reactors: Characterization of mass transfer and reactions, *Ind. Eng. Chem. Res.*, 40, 2555, 2001.
55. Oleschuk, R. D., Shultz-Lockyear, L. L., Ning, Y. B., and Harrison, D. J., Trapping of bead-based reagents within microfluidic systems: On-chip solid-phase extraction and electrochromatography, *Anal. Chem.*, 72, 585, 2000.
56. Yin, H., Killeen, K., Brennen, R., Sobek, D., Werlich, M., and Van de Goor, T., Microfluidic chip for peptide analysis with an integrated HPLC column, sample enrichment column, and nanoelectrospray tip, *Anal. Chem.*, 77, 527, 2005.
57. Penrose, A., Myers, P., Bartle, K., and McCrossen, S., Development and assessment of miniaturized centrifugal chromatograph for reversed phase separations in microchannels, *Analyst*, 129, 709, 2004.
58. Svec, F., Preparation and HPLC applications of rigid macroporous organic polymer monoliths, *J. Sep. Sci.*, 27, 747, 2004.
59. Hasselbrink, E. F., Shepodd, T. J., and Rehm, J. E. High-pressure microfluidic control in lab-on-a-chip devices using mobile polymer monoliths. *Anal. Chem.*, 74, 4913, 2002.
60. Reichmuth, D. S., Shepodd, T. J., and Kirby, B. J., On-chip high-pressure picoliter injector for pressure-driven flow through porous media, *Anal. Chem.*, 76, 5063, 2004.
61. Kirby, B. J., Reichmuth, D. S., Renzi, R. F., Shepodd, T. J., and Wiedenman, B. J., Microfluidic routing of aqueous and organic flows at high pressures: Fabrication and characterization of integrated polymer microvalve elements, *Lab Chip*, 5, 184, 2005.
62. Reichmuth, D. S., Shepodd, T. J., and Kirby, B. J., Microchip HPLC of peptides and proteins, *Anal. Chem.*, 77, 2997, 2005.
63. Mair, D. A., Geiger, E., Pisano, A. P., Svec, F., and Fréchet, J. M. J., Injection molded microfluidic chips featuring integrated interconnects, *Lab Chip*, 6, 1346, 2006.
64. Viklund, C., Ponten, E., Glad, B., Irgum, K., Horsted, P., and Svec, F., Molded macroporous poly(glycidyl methacrylate-co-trimethylolpropane trimethacrylate) materials with fine controlled porous properties: Preparation of monoliths using photoinitiated polymerization, *Chem. Mater.*, 9, 463, 1997.
65. Yu, C., Svec, F., and Frechet, J. M. J., Towards stationary phases for chromatography on a microchip: Molded porous polymer monoliths prepared in capillaries by photoinitiated *in situ* polymerization as separation media for electrochromatography, *Electrophoresis*, 21, 120, 2000.
66. Moschou, E. A., Nicholson, A. D., Jia, G., Zoval, J. V., Madou, M. J., Bachas, L. G., and Daunert, S., Integration of microcolumns and microfluidic fractionators on multitasking centrifugal microfluidic platforms for the analysis of biomolecules, *Anal. Bioanal. Chem.*, 385, 596, 2006.
67. Ishida, A., Yoshikawa, T., Natsume, M., and Kamidate, T., Reversed-phase liquid chromatography on a microchip with sample injector and monolithic silica column, *J. Chromatogr. A*, 1132, 90, 2006.
68. Xu, X., Li, L., and Weber, S. G., Electrochemical and optical detectors for capillary and chip separations, *Trends Anal. Chem.*, 26, 68, 2007.
69. Yang, Y., Li, C., Kameoka, J., Lee, K. H., and Craighead, H. G., A polymeric microchip with integrated tips and in situ polymerized monolith for electrospray mass spectrometry, *Lab Chip*, 5, 869, 2005.
70. Yang, Y., Li, C., Lee, K. H., and Craighead, H. G., Coupling on-chip solid-phase extraction to electrospray mass spectrometry through an integrated electrospray tip, *Electrophoresis*, 26, 3622, 2005.
71. Le Gac, S., Carlier, J., Camart, J. C., Cren-Olive, C., and Rolando, C., Monoliths for microfluidic devices in proteomics, *J. Chromatogr. B*, 808, 3, 2004.

72. Carlier, J., Arscott, S., Thomy, V., Camart, J. C., Cren-Olive, C., and Le Gac, S., Integrated microfabricated systems including a purification module and an on-chip nano electrospray ionization interface for biological analysis, *J. Chromatogr. A*, 1071, 213, 2005.
73. Koerner, T. and Oleschuk, R. D., Porous polymer monolith assisted electrospray from a glass microdevice, *Rapid Comm. Mass Spectr.*, 19, 3279, 2005.
74. Peters, E. C., Petro, M., Svec, F., and Fréchet, J. M. J., Molded rigid polymer monoliths as separation media for capillary electrochromatography 1. Fine control of porous properties and surface chemistry, *Anal. Chem.*, 70, 2288, 1998.
75. Bedair, M. F. and Oleschuk, R. D., Fabrication of porous polymer monoliths in polymeric microfluidic chips as an electrospray emitter for direct coupling to mass spectrometry, *Anal. Chem.*, 78, 1130, 2006.
76. Hu, S., Ren, X., Bachman, M., Sims, C. E., Li, G. P., and Allbritton, N. L., Surface-directed, graft polymerization within microfluidic channels, *Anal. Chem.*, 76, 1865, 2004.
77. Stachowiak, T. B., Rohr, T., Hilder, E. F., Peterson, D. S., Yi, M., Svec, F., and Fréchet, J. M. J., Fabrication of porous polymer monoliths covalently attached to the walls of channels in plastic microdevices, *Electrophoresis*, 24, 3689, 2003.
78. Rohr, T., Ogeltree, D. F., Svec, F., and Frechet, J. M. J., Surface functionalization of thermoplastic polymers for the fabrication of microfluidic devices by photoinitiated grafting, *Adv. Funct. Mater.*, 13, 265, 2003.
79. Rohr, T., Hilder, E. F., Donovan, J. J., Svec, F., and Fréchet, J. M. J., Photografting and the control of surface chemistry in porous polymer monoliths, *Macromolecules*, 36, 1677, 2003.
80. Liu, J., Ro, K. W., Nayak, R., and Knapp, D. R., Monolithic column plastic microfluidic device for peptide analysis using electrospray from a channel opening on the edge of the device, *Int. J. Mass Spectr.*, 259, 65, 2007.
81. Ro, K. W., Liu, J., and Knapp, D. R., Plastic microchip liquid chromatography-matrix-assisted laser desorption/ionization mass spectrometry using monolithic columns, *J. Chromatogr. A*, 1111, 40, 2006.
82. Turkova, J. *Bioaffinity Chromatography,* Elsevier: Amsterdam, 1993.
83. Li, C. and Lee, K. H., Affinity depletion of albumin from human cerebrospinal fluid using Cibacron-blue-3G-A-derivatized photopatterned copolymer in a microfluidic device, *Anal. Biochem.*, 333, 381, 2004.
84. Mao, X., Luo, Y., Dai, Z., Wang, K., Du, Y., and Lin, B. C., Integrated lectin affinity microfluidic chip for glycoform separation, *Anal. Chem.*, 76, 6941, 2004.
85. Turkova, J., Blaha, K., Malanikova, M., Vancurova, J., Svec, F., and Kalal, J., Glycidyl methacrylate gels with epoxide groups as a support for immobilization of enzymes in pH range 2–12, *Biochim. Biophys. Acta*, 524, 162, 1978.
86. Petro, M., Svec, F., and Fréchet, J. M. J., Immobilization of trypsin onto "molded" macroporous poly(glycidyl methacrylate-co-ethylene dimethcrylate) rods and use of the conjugates as bioreactors, *Biotech. Bioeng.*, 49, 355, 1996.
87. Heilmann, S. M., Rasmussen, J. K., and Krepski, L. R., Chemistry and technology of 2-alkenyl azlactones, *J. Polym. Sci. A, Polym. Chem.*, 39, 3655, 2001.
88. Xie, S., Svec, F., and Fréchet, J. M. J., Monolithic poly(2-vinyl-4,4-dimethylazlactone-co-acrylamide-co-ethylene dimethacrylate) support for design of high throughput biorectors, *Polym. Prepr.*, 38, 211, 1997.
89. Peterson, D. S., Rohr, T., Svec, F., and Fréchet, J. M. J., Enzymatic microreactor-on-a-chip: Protein mapping using trypsin immobilized on porous polymer monoliths molded in channels of microfluidic devices, *Anal. Chem.*, 74, 4081, 2002.
90. Pretorius, V., Hopkins, B. J., and Schieke, J. D., Electroosmosis. A new concept for high-speed liquid chromatography, *J. Chromatogr.*, 99, 23, 1974.
91. Jorgenson, J. W. and Lukacs, K. D., Electroosmosis. A new concept for high speed liquid chromatography, *J. Chromatogr.*, 218, 209, 1981.
92. Svec, F. and Deyl, Z., *Capillary Electrochromatography*, Elsevier: Amsterdam, 2001.
93. Krull, I. S., Stevenson, R., Mistry, K., and Schwarz, M. E. *Capillary Electrochromatography and Pressurized Flow Capillary Electrochromatography: An Introduction*, HNB Publishing: New York, 2000.

94. Svec, F., Capillary electrochromatography: A rapidly emerging separation method, *Adv. Biochem. Eng. Biotechnol.*, 76, 1, 2002.
95. He, B., Tait, N., and Regnier, F., Fabrication of nanocolumns for liquid chromatography, *Anal. Chem.*, 70, 3790, 1998.
96. He, B. and Regnier, F., Microfabricated liquid chromatography columns based on collocated monolith support structures, *J. Pharm. Biomed. Anal.*, 17, 925, 1998.
97. He, B., Ji, J. Y., and Regnier, F. E., Capillary electrochromatography of peptides in a microfabricated system, *J. Chromatogr. A*, 853, 257, 1999.
98. Slentz, B. E., Penner, N. A., Lugowska, E., and Regnier, F., Nanoliter capillary electrochromatography columns based on collocated monolithic support structures molded in poly(dimethyl siloxane), *Electrophoresis*, 22, 3736, 2001.
99. Slentz, B. E., Penner, N. A., and Regnier, F. E., Capillary electrochromatography of peptides on microfabricated poly(dimethylsiloxane) chips modified by cerium(IV)-catalyzed polymerization, *J. Chromatogr. A*, 948, 225, 2002.
100. Slentz, B. E., Penner, N. A., and Regnier, F., Geometric effects of collocated monolithic support structures on separation performance in microfabricated systems, *J. Sep. Sci.*, 25, 1011, 2002.
101. Slentz, B. E., Penner, N. A., and Regnier, F. E., Protein proteolysis and the multidimensional electrochromatographic separation of histidine-containing peptide fragments on a chip, *J. Chromatogr. A*, 984, 97, 2003.
102. Hilder, E. F., Svec, F., and Fréchet, J. M. J., Development and application of polymeric monolithic stationary phases for capillary electrochromatography, *J. Chromatogr. A*, 1044, 3, 2004.
103. Seiler, K., Fan, Z. H. H., Fluri, K., and Harrison, D. J., Electroosmotic pumping and valveless control of fluid-flow within a manifold of capillaries on a glass chip, *Anal. Chem.*, 66, 3485, 1994.
104. Regnier, F. E., He, B., Lin, S., and Busse, J., Chromatography and electrophoresis on chips: Critical elements of future integrated, microfluidic analytical systems for life science, *Trends Biotech.*, 17, 101, 1999.
105. Bruin, G. J. M., Recent developments in electrokinetically driven analysis on microfabricated devices, *Electrophoresis*, 21, 3931, 2000.
106. Kutter, J. P., Current developments in electrophoretic and chromatographic separation methods on microfabricated devices, *Trends Anal. Chem.*, 19, 352, 2000.
107. Szumski, M. and Buszewski, B., State of the art in miniaturized separation techniques, *Crit. Revs. Anal. Chem.*, 32, 1, 2002.
108. de Mello, A., On-chip chromatography: The last twenty years, *Lab Chip*, 2, 48N, 2002.
109. Mistry, K., Krull, I., and Grinberg, N., Capillary electrochromatography: An alternative to HPLC and CE, *J. Sep. Sci.*, 25, 935, 2002.
110. Lion, N., Rohner, T. C., Dayon, L., Arnaud, I. L., Damoc, E., Youhnovski, N., Wu, Z. Y., et al., Microfluidic systems in proteomics, *Electrophoresis*, 24, 3533, 2003.
111. Kasicka, V., Recent advances in capillary electrophoresis and capillary electrochromatography of peptides, *Electrophoresis*, 24, 4013, 2003.
112. Bedair, M. and El Rassi, Z., Recent advances in polymeric monolithic stationary phases for electrochromatography in capillaries and chips, *Electrophoresis*, 25, 4110, 2004.
113. Pumera, M., Microchip-based electrochromatography: Designs and applications, *Talanta*, 66, 1048, 2005.
114. Stachowiak, T. B., Svec, F., and Fréchet, J. M. J., Chip electrochromatography, *J. Chromatogr. A*, 1044, 97, 2004.
115. Ericson, C., Holm, J., Ericson, T., and Hjertén, S., Electroosmosis- and pressure-driven chromatography in chips using continuous beds, *Anal. Chem.*, 72, 81, 2000.
116. Vegvari, A. and Hjertén, S., A hybrid microdevice for electrophoresis and electrochromatography using UV detection, *Electrophoresis*, 23, 3479, 2002.
117. Szekely, L. and Freitag, R., Fabrication of a versatile microanalytical system without need for clean room conditions, *Anal. Chim. Acta*, 512, 39, 2004.
118. Szekely, L. and Freitag, R., Module for real-time noninvasive control of the electroosmotic flow in microfluidic systems, *Anal. Chim. Acta*, 539, 165, 2005.

119. Szekely, L. and Freitag, R., Study of the electroosmotic flow as a means to propel the mobile phase in capillary electrochromatography in view of further miniaturization of capillary electrochromatography systems, *Electrophoresis*, 26, 1928, 2005.
120. Zeng, H. L., Li, H. F., and Lin, J. M., Chiral separation of dansyl amino acids by PDMS microchip gel monolithic column electrochromatography with [gamma]-cyclodextrin bonded in polyacrylamide, *Anal. Chim. Acta*, 551, 1, 2005.
121. Yu, C., Svec, F., and Fréchet, J. M. J., Towards stationary phases for chromatography on a microchip: Molded porous polymer monoliths prepared in capillaries by photoinitiated in situ polymerization as separation media for electrochromatography, *Electrophoresis*, 21, 120, 2000.
122. Svec, F., Yu, C., Rohr, T., and Fréchet, J. M. J., Design of a toolbox for fabrication of analytical microfluidic systems using porous polymer monoliths, in *Micro Total Analysis Systems 2001*, Ramsey, J. M., van den Berg, A., Eds., Kluwer Acad. Publ.: Dordrecht, 2001.
123. Svec, F., Fréchet, J. M. J., Hilder, E. F., Peterson, D. S., and Rohr, T. in *Micro Total Analysis Systems 2002*, Baba, Y., van den Berg, A., Eds., Kluwer Academic Publishers: Dordrecht, 2002.
124. Stachowiak, T. B., Svec, F., and Fréchet, J. M. J., Patternable Protein resistant surfaces for multifunctional microfluidic devices via surface hydrophilization of porous polymer monoliths using photografting, *Chem. Mater.*, 18, 5950, 2006.
125. Hilder, E. F., Svec, F., and Fréchet, J. M. J., Shielded stationary phases based on porous polymer monoliths for capillary electrochromatography of highly basic biomolecules, *Anal. Chem.*, 76, 3887, 2004.
126. Pucci, V., Raggi, M. A., Svec, F., and Fréchet, J. M. J., Monolithic columns with a gradient of functionalities prepared via photoinitiated grafting for separations using capillary electrochromatography, *J. Sep. Sci.*, 27, 779, 2004.
127. Yu, C., Xu, M., Svec, F., and Fréchet, J. M. J., Preparation of monolithic polymers with controlled porous properties for microfluidic chip applications using photoinitiated free radical polymerization, *J. Polym. Sci. A, Polym. Chem.*, 40, 755, 2002.
128. Ngola, S. M., Fintschenko, Y., Choi, W. Y., and Shepodd, T. J., Conduct-as-cast polymer monoliths as separation media for capillary electrochromatography, *Anal. Chem.*, 73, 849, 2001.
129. Shediac, R., Ngola, S. M., Throckmorton, D. J., Anex, D. S., Shepodd, T. J., and Singh, A. K., Reversed-phase electrochromatography of amino acids and peptides using porous polymer monoliths, *J. Chromatogr. A*, 925, 251, 2001.
130. Fintschenko, Y., Choi, W. Y., Ngola, S. M., and Shepodd, T. J., Chip electrochromatography of polycyclic aromatic hydrocarbons on an acrylate-based UV-initiated porous polymer monolith, *Fresenius' J. Anal. Chem.*, 371, 174, 2001.
131. Courtois, J., Szumski, M., Bystroem, E., Iwasiewicz, A., Shchukarev, A., and Irgum, K., A study of surface modification and anchoring techniques used in the preparation of monolithic microcolumns in fused silica capillaries, *J. Sep. Sci.*, 29, 14, 2006.
132. Jacobson, S. C., Hergenroder, R., Koutny, L. B., Warmack, R. J., and Ramsey, J. M., Effects of injection schemes and column geometry on the performance of microchip electrophoresis devices, *Anal. Chem.*, 66, 1107, 1994.
133. Throckmorton, D. J., Shepodd, T. J., and Singh, A. K., Electrochromatography in microchips: Reversed-phase separation of peptides and amino acids using photopatterned rigid polymer monoliths, *Anal. Chem.*, 74, 784, 2002.
134. Lazar, I. M., Li, L. J., Yang, Y., and Karger, B. L., Microfluidic device for capillary electrochromatography-mass spectrometry, *Electrophoresis*, 24, 3655, 2003.
135. Tanaka, N., Kobayashi, H., Ishizuka, N., Minakuchi, H., Nakanishi, K., Hosoya, K., and Ikegami, T., Monolithic silica columns for high-efficiency chromatographic separations, *J. Chromatogr. A*, 965, 35, 2002.
136. Gottschlich, N., Jacobson, S. C., Culbertson, C. T., and Ramsey, J. M., Two-dimensional electrochromatography/capillary electrophoresis on a microchip, *Anal. Chem.*, 73, 2669, 2001.
137. Ericson, C., Holm, J., Ericson, T., and Hjertén, S., Electroosmosis- and pressure-driven chromatography in chips using continuous beds, *Anal. Chem.*, 72, 81, 2000.
138. Xiong, L. and Regnier, F. E., Channel-specific coatings on microfabricated chips, *J. Chromatogr. A*, 924, 165, 2001.

139. Dulay, M. T., Quirino, J. P., Bennett, B. D., Kato, M., and Zare, R. N., Photopolymerized sol-gel monoliths for capillary electrochromatography, *Anal. Chem.*, 73, 3921, 2001.
140. Morishima, K., Bennett, B. D., Dulay, M. T., Quirino, J. P., and Zare, R. N., Toward sol-gel electrochromatographic separations on a chip, *J. Sep. Sci.*, 25, 1226, 2002.
141. Breadmore, M. C., Shrinivasan, S., Wolfe, K. A., Power, M. E., Ferrance, J. P., Hosticka, B., Norris, P. M., and Landers, J. P., Towards a microchip-based chromatographic platform. Part 1: Evaluation of sol-gel phases for capillary electrochromatography, *Electrophoresis*, 23, 3487, 2002.
142. Breadmore, M. C., Shrinivasan, S., Karlinsey, J., Ferrance, J. P., Norris, P. M., and Landers, J. P., Towards a microchip-based chromatographic platform. Part 2: Sol-gel phases modified with polyelectrolyte multilayers for capillary electrochromatography, *Electrophoresis*, 24, 1261, 2003.
143. Ishizuka, N., Minakuchi, H., Nakanishi, K., Soga, N., Nagayama, H., Hosoya, K., and Tanaka, N., Performance of a monolithic silica column in a capillary under pressure-driven and electrodriven conditions, *Anal. Chem.*, 72, 1275, 2000.
144. Ro, K. W., Nayak, R., and Knapp, D. R., Monolithic media in microfluidic devices for proteomics, *Electrophoresis*, 27, 3547, 2006.
145. Rohr, T., Yu, C., Davey, M. H., Svec, F., and Fréchet, J. M. J., Porous polymer monoliths: Simple and efficient mixers prepared by direct polymerization in the channels of microfluidic chips, *Electrophoresis*, 22, 3959, 2001.
146. Yu, C., Mutlu, S., Selvaganapathy, P., Mastrangelo, C. H., Svec, F., and Fréchet, J. M. J., Flow control valves for analytical microfluidic chips without mechanical parts based on thermally responsive monolithic polymers., *Anal. Chem.*, 75, 1958, 2003.
147. Luo, Q., Mutlu, S., Gianchandani, Y. B., Svec, F., and Fréchet, J. M. J., Monolithic valves for microfluidic chips based on thermoresponsive polymer gels, *Electrophoresis*, 24, 3694, 2003.

48 Microdialysis and Microchip Systems

Barbara A. Fogarty, Pradyot Nandi, and Susan M. Lunte

CONTENTS

48.1 INTRODUCTION

Microdialysis is a generic sampling technique that can be used for the continuous monitoring of small molecules present in aqueous samples. It has found application in a number of areas such as *in vivo* monitoring, process analytical chemistry and environmental sampling. In microdialysis, sampling is accomplished through the diffusion of analyte molecules across a semipermeable membrane that is built into the probe. Sampling is achieved by pumping a solution through the probe that is similar in composition to that of the matrix being sampled. A concentration gradient of the analyte is generated between the perfusate and the surrounding medium, causing analytes to diffuse into the probe. They are then delivered by the syringe pump to a fraction collector or an appropriate analysis system. It is also possible to deliver compounds into the aqueous environment outside the probe by adding the compound of interest to the perfusate.

A generalized schematic illustrating a typical sampling and analysis setup is shown in Figure 48.1. Most commonly, dialysate samples are collected and analyzed offline using techniques such as liquid chromatography (LC) coupled to optical detection modes (UV, fluorescence), electrochemical detection or mass spectrometry (MS). Alternatively, online analysis can be performed by directly coupling the microdialysis probe to flow-through sensors or separation-based analytical systems. The most commonly used separation techniques for the analysis of such microdialysis samples are LC

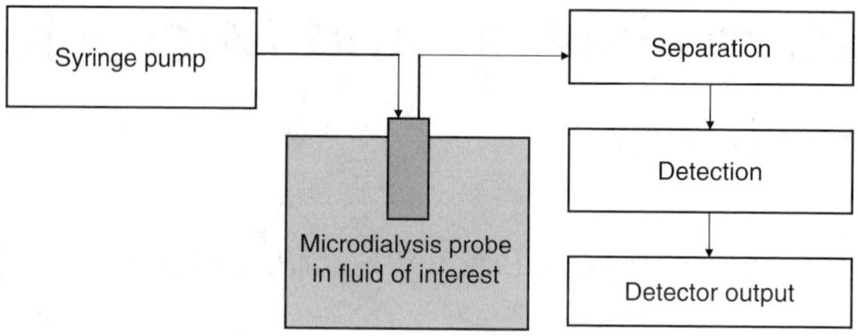

FIGURE 48.1 Typical microdialysis sampling and analysis setup.

and capillary electrophoresis (CE). These have been combined with electrochemical or fluorescence detection for near real-time monitoring applications.

More recently, microchip-based analytical systems have proven to be an attractive alternative for the analysis of small volume-limited samples. Microchip systems also enable faster separations, can be made portable, and are less expensive than conventional instrumentation. The integration of microdialysis sampling with microchip analysis systems has the potential for achieving very fast analysis times, thereby maximizing temporal resolution. This chapter will discuss the theoretical aspects of microdialysis sampling and considerations for the coupling of the sampling technique to microchip systems and will conclude with future directions of this technology.

48.2 BACKGROUND

Microdialysate samples have been analyzed using a variety of nonseparation-based analytical techniques including immunoassay, biosensors, and MS [1–4]. The main limitation to the use of these methods is that they are typically restricted to the measurement of a single analyte. For more complex samples, the detection of multiple substances is usually necessary. In this case, the dialysate sample is normally analyzed by conventional chromatographic or electrophoretic separation methods employing optical, electrochemical, or mass spectrometric modes of detection [5].

When developing an assay based on microdialysis sampling, several parameters must be taken into consideration. These include the concentration range of the analyte to be sampled, the temporal resolution desired, and the volume of sample that is available. Low flow rates generally yield a higher recovery of analyte but also generate much smaller samples per unit time. Electrophoresis-based methods require nanoliter (nL) to picoliter (pL) volumes of sample, which can lead to good temporal resolution but also exhibit reduced sensitivity for mass sensitive detectors. Larger volumes can be introduced with chromatographic separation methods but frequently at the expense of temporal resolution.

The direct coupling of microdialysis to separation systems yields a system that is capable of continuous monitoring of the analyte(s) of interest. Microdialysis sampling has been coupled online to conventional separation systems such as LC and CE [6–8]. As samples are protein-free, they can be delivered and injected directly into the analytical system. A major advantage of online analysis is that it obviates the need for the manipulation of small volume samples ($<1\ \mu L$) and thus reduces the chance for sample loss and contamination as well as improving temporal resolution. The interface that physically couples the microdialysis system to the separation system, however, can be challenging to implement. The tubing and connectors associated with such an interface can lead to an increase in dead volume in the system that may lead to a delayed response in online monitoring experiments as well as causing band broadening.

An important consideration for online coupling of microdialysis sampling and conventional LC separation is the time required for analysis and the relatively high sample volume requirements

(5–10 μL) that can limit temporal resolution. CE offers the advantage of small sample volumes requirements (nL), highly efficient separations, and fast analysis times. However, in the case of electrophoretic separation systems, it is necessary to isolate the microdialysis probe and sample body (e.g., an animal in animal sampling experiments) from the high field strengths used for separation. In addition, when electrochemical detection is used, care must be taken to isolate the potentiostat from the separation voltage to prevent grounding through the detector or animal.

The first reported online coupling of microdialysis with conventional CE-laser-induced fluoroscence (LIF) was reported by Hogan et al. [8], in 1994. The system achieved separation of an antineioplastic agent and its major metabolite in less than 60 s. The overall temporal resolution was 90 s. This system utilized a 60 nL injection valve and a capillary-based interface for delivering the sample to the CE system. The interface between the microdialysis and electrophoresis systems could also be used to introduce derivatizing reagents for chemical labeling of biological analytes; these reagents are often needed for fluorescence and electrochemical detection [9]. Using a capillary-based system with very short capillaries, Kennedy's group has been able to perform extremely fast separations (12 s) of amino acid neurotransmitters [10]. It is anticipated that the advent of microchip technology will facilitate separations on the subsecond scale.

While the vast majority of microdialysis sampling to date has been achieved in conjunction with conventional systems, microdialysis has also been coupled online with biosensors and, more recently, to microchip-based separation devices. These devices can be fabricated using low cost materials and are amenable to mass production [11]. The smaller dimensions of microchip systems minimize sample and reagent volume requirements and allow placement of the analysis system closer to the aqueous sample or animal being sampled. Due to the planar nature of microchip devices, additional sample handling procedures such as preconcentration, mixing, extraction, or derivatization can also be carried out on the same platform [12–15]. Considerations for coupling microdialysis to microchip instrumentation systems will be discussed in Section 48.5.

48.3 THEORETICAL ASPECTS

Microdialysis sampling involves the recovery of low molecular weight analytes following their diffusion across a semipermeable membrane [16]. The membrane is housed in hollow tubing connected to an inlet and outlet. A solution similar in ionic strength to that of the sample matrix is perfused through the tubing, and the analytes recovered represent some fraction of the original analyte concentrations in the area sampled (Figure 48.2).

Analyte recovery is dependent on characteristics of the probe (length, pore size, chemical composition) and physicochemical properties of the sampled analyte(s) (coefficient of diffusion, hydrophobicity) as well as the rate of perfusion across the probe. The length of the probe is primarily correlated to the membrane surface area that is available for diffusion—a greater surface area facilitates higher sample recovery. The pore size of the semipermeable membrane determines the molecular weight cutoff and, hence, can be customized depending on application as well as the analyte(s) of interest. With regard to the perfusion fluid, lower flow rates can allow greater sample recoveries; however, longer perfusion times are needed if higher sample volumes (μL–mL) are required for analysis. An alternative strategy to improve analyte recovery involves the addition of chemical additives such as cyclodextrins or osmotic agents to the perfusate [17,18].

Microdialysis can achieve both sample recovery and cleanup before analysis. The pore size of the microdialysis membrane can be tailored for selective exclusion of macromolecules such as proteins and enzymes, thus producing cleaner samples with reduced possibility of enzymatic degradation of the sample.

Selection of a suitable microdialysis probe will be dictated by the target of interest and sample matrix being sampled. As far as the types of probe are concerned, probes for brain sampling are usually rigid and shorter in length (cannula probes) compared to probes for subcutaneous or organ

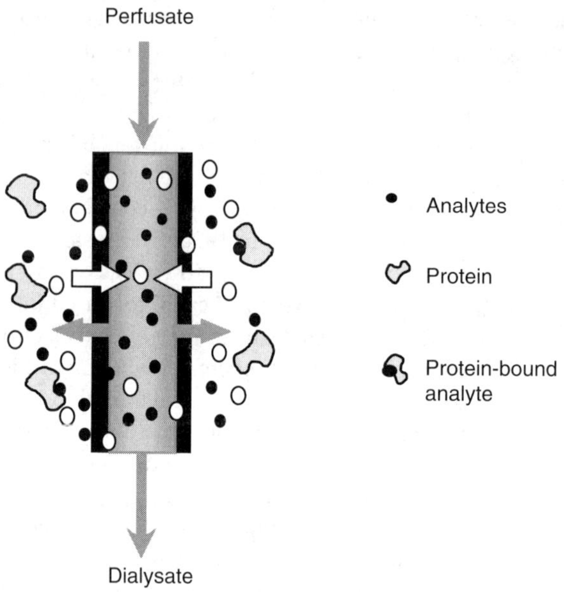

FIGURE 48.2 Microdialysis sampling.

implantation (linear probes), which are typically longer and flexible [19]. It should be kept in mind that in the case of *in vivo* microdialysis experiments (e.g., brain, muscle) the probe samples only the extracellular environment, not the intracellular fluid. It also does not measure analyte that is protein bound. Once the microdialysis system has been optimized, attention must focus on the analysis of recovered samples.

48.4 PRACTICAL APPLICATIONS

A distinct advantage of microdialysis is that it allows continuous sampling with no net fluid loss; as a result, it has been used most extensively for *in vivo* sampling. Sampling can be performed in awake, freely moving animals over several days with the animal acting as its own control. Microdialysis has been used for recovery of endogenous compounds such as amino acids, catecholamines, and peptides [20,21]. It has been applied extensively to the study of neurotransmitters [22], for pharmacokinetic modeling of drug substances [23], and for drug delivery [24]. Microdialysis has also been used to monitor blood gases [25]. As a generic sampling method, microdialysis has also found application in the areas of biological, environmental, and biofermentation monitoring [26,27].

Flow-through microchip systems incorporating an integrated membrane [28,29] or coupled to an external probe have been used for monitoring biomolecules such as glucose and lactate [30]. These systems incorporated electrochemical sensors for analyte detection. An online flow-through biosensor has been demonstrated for the *in vivo* determination of glucose, choline, and glutamate in a freely moving rat dosed with aspartate, an agonist of the N-methyl D-aspartate (NMDA) receptor [31]. An electrode array was used for discrimination of individual analytes.

Microdialysis sampling has also been coupled to separation-based microchip devices. External probes have been used in microchip electrophoresis for the monitoring of enzyme reactions [32] and have been applied to the *in vivo* analysis of glutamate in the striatum of an anesthetized rat [33]. Considerations for coupling microdialysis sampling to flow-through and separation-based microchip devices are similar and will now be discussed in more detail, with reference to specific examples from the literature.

48.5 METHOD DEVELOPMENT GUIDELINES

48.5.1 Microchip Fabrication

The substrate for the microfluidic device should be selected with consideration of the end application. Substrates used to fabricate the microchip device should not interact with target analytes, and must be compatible with the detection method employed (i.e., should not exhibit background fluorescence, BGF.). For the analysis of nonpolar compounds, it should be kept in mind that substrates such as poly(dimethyl)siloxane (PDMS) can adsorb hydrophobic analytes such as peptides and proteins. Plasma oxidation or treatment of the surface can sometimes be useful to minimize these interactions [34,35]. For perfusates containing organic solvents, compatibility with polymer substrates can also be an issue. Substrates to be used for the fabrication of electrophoresis-based separation devices should be capable of supporting a stable electroomostic flow (EOF). The use of a low cost material and standard processing procedures can permit mass fabrication of devices.

One of the first demonstrations of coupling microdialysis to a flow-through miniaturized analysis system was reported by Steinkuhl et al. [36] for glucose monitoring. The system was designed and fabricated using a full wafer process and used glass and silicon substrates. Device components were small enough to be housed in a wearable device. The system was tested *in vitro* with human serum and standard solutions of glucose with a stability of 5% achieved over 5 days.

48.5.2 Microdialysis Sampling

48.5.2.1 Choice of Membrane

Selection of the dialysis membrane will be dictated by the target analytes and sample matrix. There are several types of commercial probes available for biological monitoring applications. The most fundamental of these probe types are based on length and molecular weight cutoff. Accessories such as guide cannulae, tubing connectors, and so forth compatible with probe dimensions must also be selected. Both integrated membranes and external probes have been used in conjunction with microchip devices. When external probes and tubing are connected to microchip platforms, sample can be delivered directly to the microchip device. This has been demonstrated with both flow-through and separation-based devices. Dialysis membranes should be conditioned with perfusion fluid before sampling commences.

48.5.2.2 Integrated Membranes

Interfacing microchip devices to the macro world is a challenging task [37]. Microdialysis may be coupled to microchip systems through integrated membranes built into the device or by coupling external probes to the device. The integration of membranes into microchip devices has recently been reviewed [38]. Dialysis membranes have been integrated into flow-through microchip devices by clamping or bonding between channel layers. Cellulose membranes have been sandwiched between two acrylic plates for blood monitoring (Figure 48.3) [28].

Another approach involved the use of a carbonate membrane bonded to SU8 channels with a glass layer used to seal the device [29]. The system used a stacked approach with a dialysis membrane sandwiched between an SU8 perfusion channel and a poly(dimethylsiloxane) flow channel (Figure 48.4). Bonding the membrane reduced the chance of leakage and resulted in a more robust device. A fluorescent dye and glucose solution were used for optical characterization of the system. A commercially available glucose sensor was used to determine the concentration of glucose in the dialysate. A recovery of 80% was obtained at flow rate of 1.5 μL/min.

Dialysis membranes can also be fabricated *in situ* on microchip devices. This was achieved by Song et al. [14] who filled glass (fused silica) microchannels with monomer solutions and initiated polymerization by exposure to a UV laser. The membranes produced could be used between

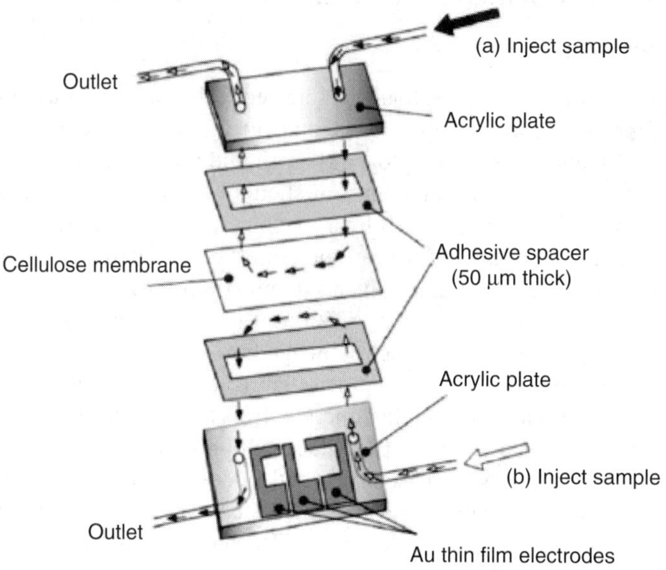

FIGURE 48.3 Schematic representation of fluidic device with integration of cellulose membrane. (Reproduced from Kurita, R., et al., *Biosens. Bioelectron.*, 21, 1649, 2006. With permission.)

pH 2.7 and 9, and minimal protein adhesion was observed. The system was designed for desalting of protein samples. The advantage of using photopatterning is that the molecular weight cutoff of the membrane can be tailored to suit the application.

48.5.2.3 Interfacing to External Probes

For *in vivo* or *in vitro* monitoring, external probes can be used and implanted in tissue or a solution of interest. The microdialysis system is then interfaced to the microchip device for detection (flow-through systems) or separation of recovered analytes. Where possible, connectors should be fabricated as part of the device to minimize the number of external fittings required and associated dead volumes and band broadening. For interfaces attached using adhesives (e.g., epoxy quick dry), care must be taken to ensure that no material leaches into the device as this can lead to contamination and interferences in detection. A thorough characterization of the interface is necessary to ensure that there is no sample leakage, which can result in a variation in perfusion flow rates through the probe. When electrophoresis is used as the separation mechanism for *in vivo* monitoring applications, the interface between the microdialysis setup and the microchip may also need to be grounded to achieve isolation of the animal from the high voltage. Specific approaches to coupling external probes to separation-based devices will be discussed in Section 48.5.4.

48.5.3 Detection

Integration of a dialysis membrane or coupling to a microdialysis probe can help to exclude many of the larger molecules such as proteins that are typically responsible for fouling the sensor elements of microchip analysis systems. The nature of the analysis will dictate whether coupling to a flow-through or separation-based sensor must be achieved and the kind of detection elements to be employed (optical, electrochemical, etc.). One of the issues with microchip devices is the very small

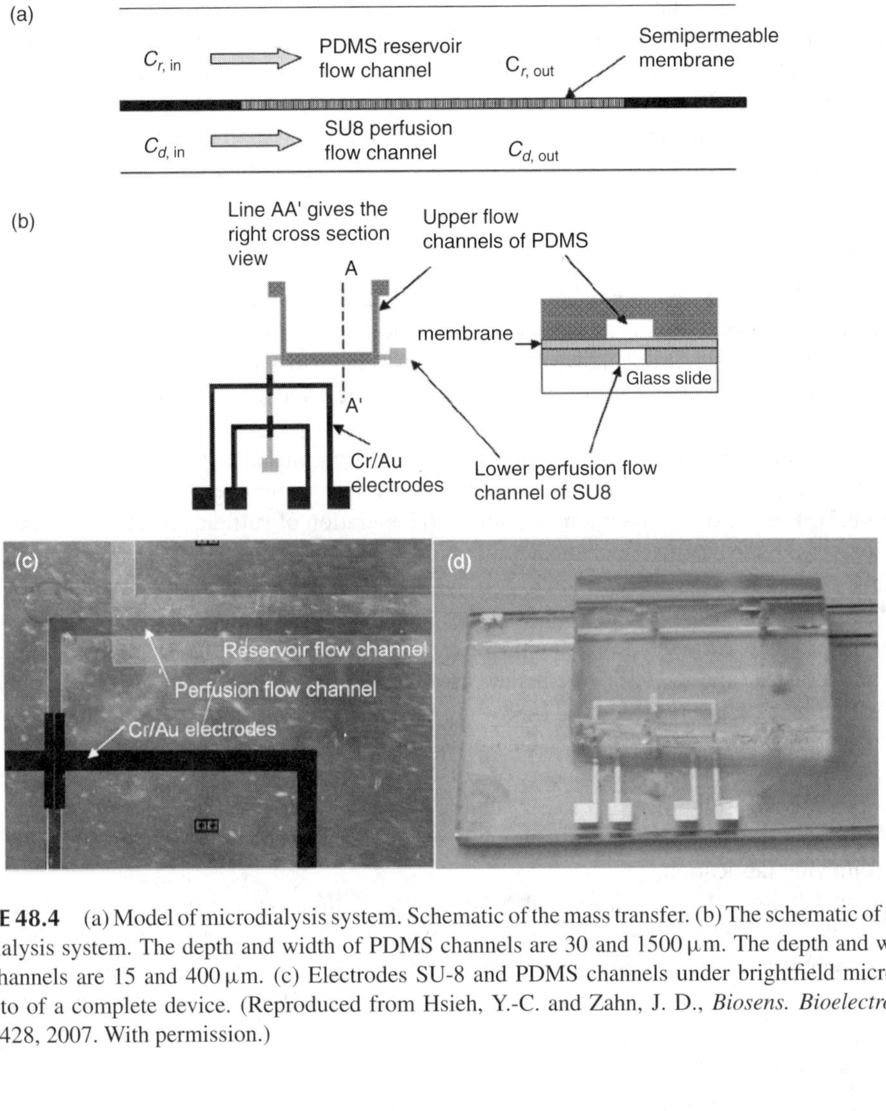

FIGURE 48.4 (a) Model of microdialysis system. Schematic of the mass transfer. (b) The schematic of stacked microdialysis system. The depth and width of PDMS channels are 30 and 1500 μm. The depth and width of SU-8 channels are 15 and 400 μm. (c) Electrodes SU-8 and PDMS channels under brightfield microscopy. (d) Photo of a complete device. (Reproduced from Hsieh, Y.-C. and Zahn, J. D., *Biosens. Bioelectron.*, 22, 2422–2428, 2007. With permission.)

volumes (nL–pL) of sample that are injected, which can impact the sensitivity of the analysis [39]. While miniaturized optical components and electrode materials can be integrated onto microchip platforms, sensitive off-chip detection systems (LIF; MS) may also be used.

If a single analyte is to be detected, a simple flow-through microdialysis/microchip system may suffice. If multianalyte determination is needed, a separation-based microchip device can allow resolution and detection of several analytes in a single sample (Section 48.5.4). In flow-through devices, the perfusate is directed to an array of sensors, which may be also be modified to allow detection of different analytes and improve detection sensitivity.

Detection of lactate in whole blood has been achieved using a miniaturized flow-through device with gold film working electrodes modified with lactate oxidase and a poly(vinylpyridine) mediator containing horseradish peroxidase. The analysis results were achieved in 2 min without the need for centrifugal separation or serum separation. A flow-through potentiometric sensor based on semipermeable dialysis tubing in a silicon device has also been described [40]. The system consisted of a microdialysis probe, a sensor array, and a calibration element. The *in vitro* sensor was able to detect a standard solution of potassium as part of an integrated microdialysis system. The sampling time was 60 s with a flow rate of 2 μL/min.

Impedance electrodes have also been investigated as detectors for microchip microdialysis applications. An on-chip microdialysis system with inline sensing electrodes for impedance detection was developed [41]. Cr/Au electrodes were used to determine the electrical resistance of changes in the concentration of phosphate-buffered saline (PBS) solutions that were used to characterize the system. The system monitored concentration changes with a 210-s system response delay. The lag time was attributed to dead volume in the tubing between the syringe pumps and the microsystem.

48.5.4 Separation

There have only been a few reports of coupling microdialysis to microchip-based separation systems. Ideally, the microchip system should allow the injection of discrete sample plugs from a continuously flowing stream of dialysate without disturbing the separation element of the analysis. This allows maximal temporal resolution and limits the effect of perfusion flow rate on system performance.

The first reported coupling of microdialysis to a separation-based microchip device was achieved by Huynh et al. [32], who coupled an external commercially available probe to a glass microchip (Figure 48.5). The use of a glass microchip allowed generation of sufficient EOF to gate the sample stream away from the separation channel until an injection was required. This can be more difficult to achieve with polymer substrates such as poly(dimethylsiloxane) where the magnitude of EOF supported by the substrate is often considerably less than that of glass. The device was used to monitor the products formed during an enzyme assay in near real time. Laser-induced fluorescence detection was used, and device performance as a function of flow rate and applied voltage was characterized using a fluorescent dye. The resulting system was then used to monitor the progress of an enzyme reaction (Figure 48.6). Some optimization of the setup was needed to balance the perfusion flow rate with the requirements of the separation system. Using perfusion flow rates of 1μ L/s, a temporal resolution of 30 s was achieved with a system lag time of ~6.5 min. This was attributed to dead volume due to the microchip and associated connectors that were added to the device following fabrication.

On-chip labeling of analytes may also be needed if fluorescence or electrochemical detection is employed. This has been demonstrated with flow-gated design using naphthalene 2,3-dicarboxaldehye and 2-mercaptoethanol [33,42]. Labeling reagents were added to the buffer reservoir and on-column labeling was achieved (Figure 48.7). Using this approach, the

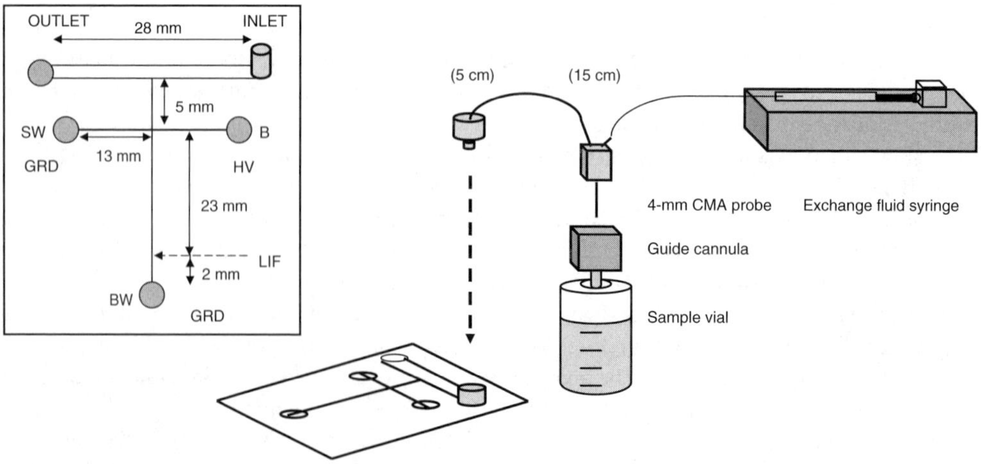

FIGURE 48.5 Layout and schematic of microdialysis/microchip CE system. (Reproduced from Huynh, B. H., et al., *Anal. Chem.*, 76, 6440, 2004. With permission.)

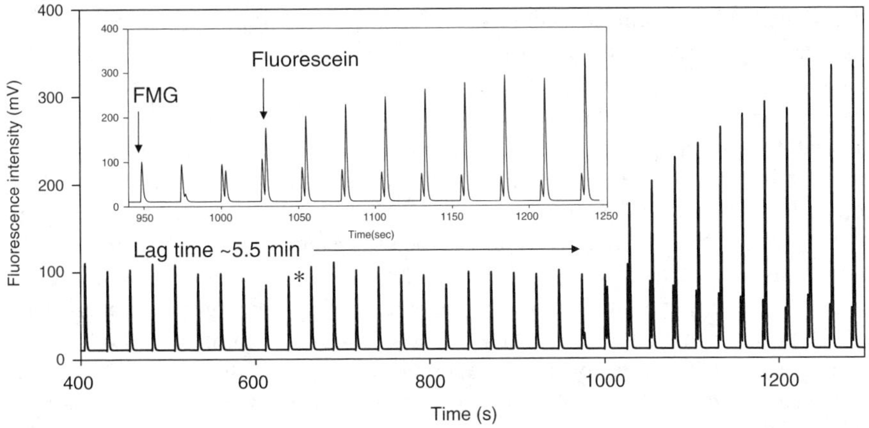

FIGURE 48.6 Near real-time monitoring of an enzyme reaction using microdialysis/microchip CE system. (Reproduced from Huynh, B. H., et al., *Anal. Chem.*, 76, 6440, 2004. With permission.)

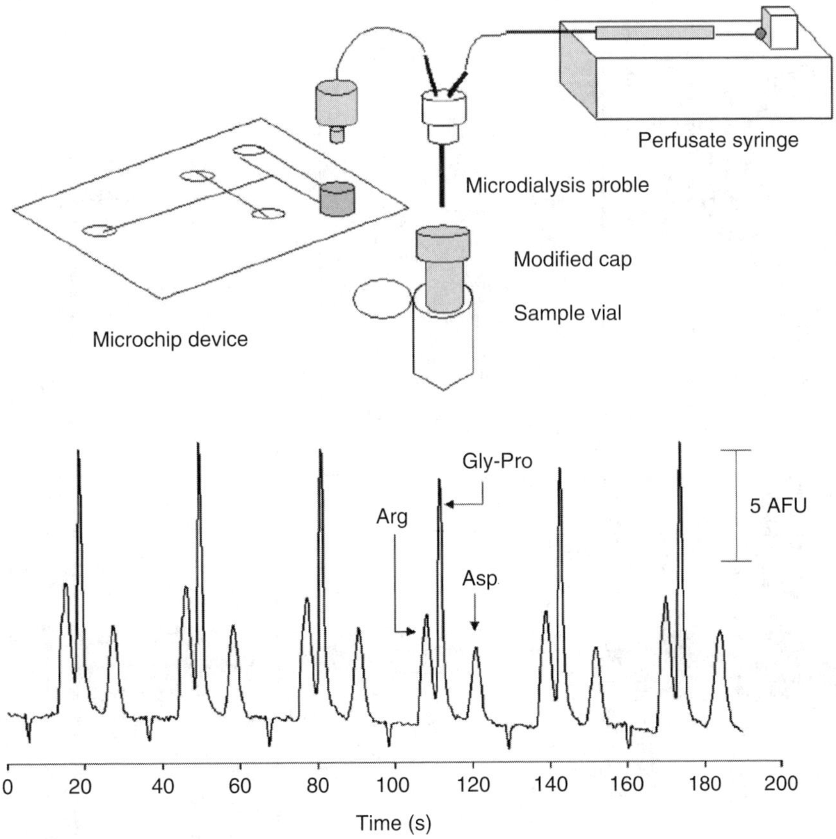

FIGURE 48.7 (a) Schematic of microdialysis/microchip setup and (b) continuous separation of three-component mixture sampled through external microdialysis probe. (Reproduced from Huynh, B. H., et al., *J. Pharm. Biomed. Anal.*, 42, 529, 2006. With permission.)

microdialysis/microchip system can achieve a high degree of process integration with sample preparation, delivery, derivatization, separation, and detection all accomplished on the one platform. Multiple separations of a three-component mixture that was sampled through the external microdialysis probe and fluorescently labeled on-chip have been reported.

An alternative approach by Li et al. [43] examined the use of integrated polymer valves to achieve discrete injection of microdialysate samples for separation on chip. Polymer valves have been incorporated in microchip devices previously [44]; however, this was the first report of integrating these valves into an electrophoretic microchip system. While fabrication of microchips incorporating polymer valves is not trivial, the injector design allows discrete injection of microdialysate samples for separation on chip without the need for flow-gating. External pneumatic operation of a polymer valving channel causes it to expand and seal off the fluidic channel until an injection is required. The injector design reported by Li et al. was further optimized to include a push-back channel and avoid sample leakage following injection. The device was then coupled to an external microdialysis probe and used to monitor a recovery and separation of a fluorescein dye mixture (Figure. 48.8). An injection frequency of 43 Hz was reported with a system lag time of only 14 s; this illustrates the potential of the system for applications requiring high temporal resolution.

The first application of a separation-based microchip/microdialysis system to the *in vivo* monitoring of a live animal has been reported [33]. The device design (Figure 48.9) incorporated a

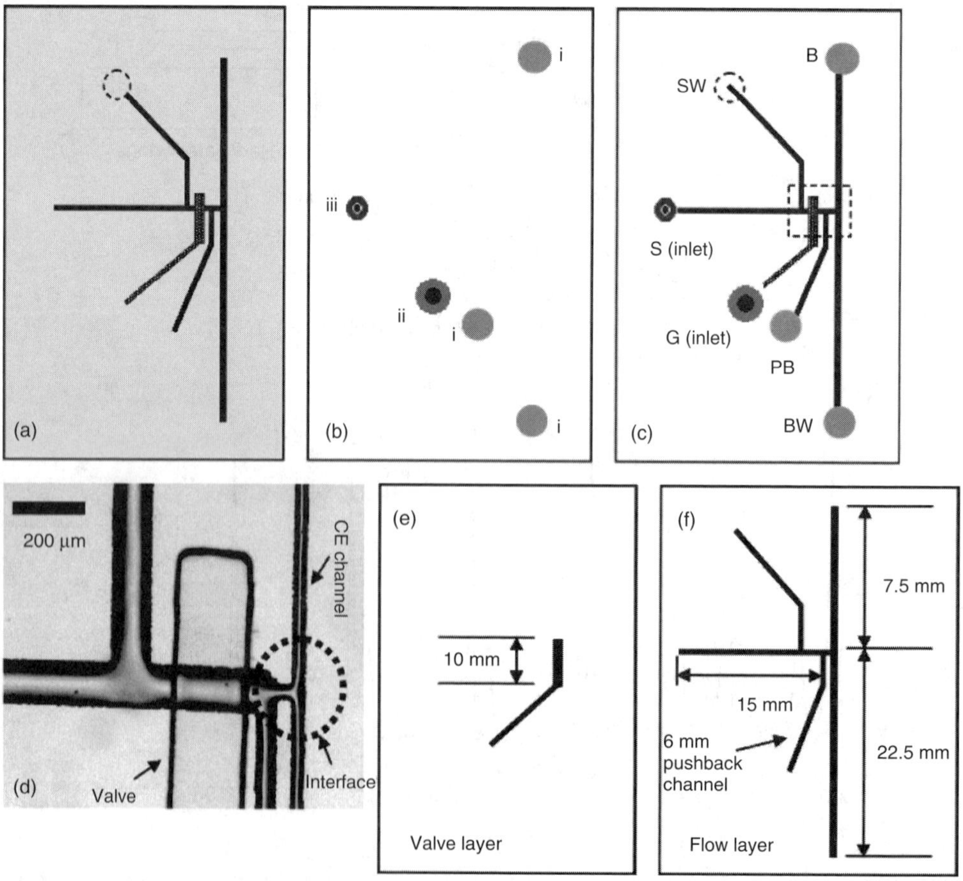

FIGURE 48.8 Assembled microchip with integrated polymer valves. (Reproduced from Li, M. W., et al., *Anal. Chem.*, 78, 1042, 2006. With permission.)

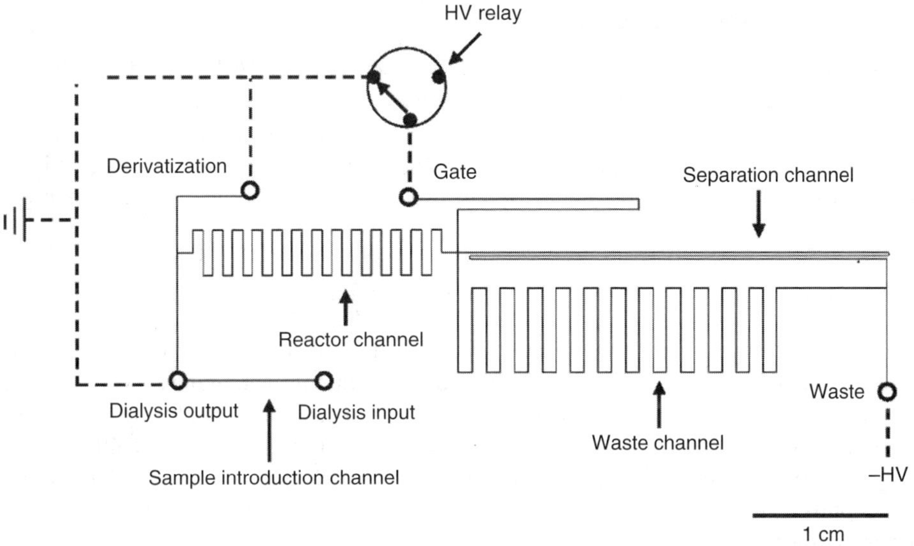

FIGURE 48.9 Layout and schematic of microdialysis/microchip CE system for online monitoring of neurotransmitters. (Reproduced from Sandlin, Z. D., et al., *Anal. Chem.*, 77, 7702, 2005. With permission.)

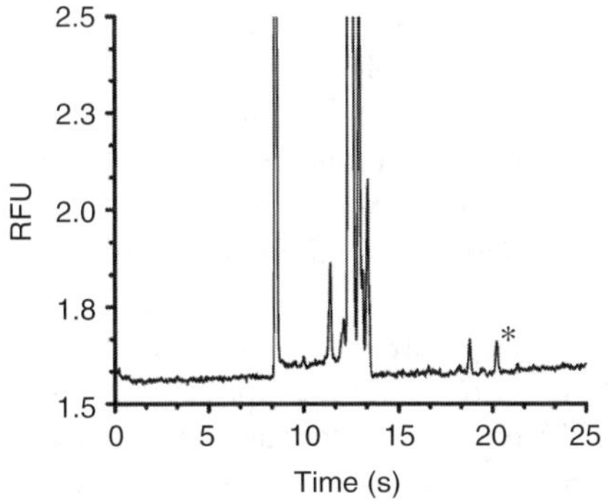

FIGURE 48.10 Typical electropherogram obtained *in vivo* from the rat striatum using microchip/microdialysis system indicates the peak for glutamate. (Reproduced from Sandlin, Z. D., et al., *Anal. Chem.*, 77, 7702, 2005. With permission.)

sample introduction channel for the dialysate, a precolumn reactor for derivatization, a flow-gated injector, and a separation channel. Laser-induced fluorescence detection was used. The device was applied to the analysis of glutamate following an infusion of the glutamate uptake inhibitor L-*trans*-pyrolidine-2,4-dicarboxylic acid into the striatum of an anesthetized rat (Figure 48.10). Two microchip layouts were examined, with the second approach achieving separations of selected neurotransmitters within 60 s.

External microdialysis probes fabricated in-house were used in these experiments. Initial studies found that the addition of ethylenediamine tetraacetic acid (EDTA) to the derivatizing reagent was necessary to prevent divalent cations (Mg^{2+} and Ca^{2+}) present in the cerebral spinal fluid from decreasing the EOF. Using the device, separations could be carried out at 1.8 min intervals; however, the effective temporal resolution was estimated to be between 2 and 4 min due to delay times attributed to the dead volume in the system. Increases in perfusion flow rates led to a decreased delay time but reduced recovery through the probe was observed.

48.5.5 Integrated System Control

Control of discrete system elements, which allows added flexibility for sampling, is important to consider. An integrated computer-controlled system was demonstrated for a microdialysis probe coupled online to a flow-though system for lactate and glucose monitoring in serum samples [30]. The system comprised a microdialysis probe and a micromachined silicon stack along with a miniaturized flow cell and sensor array. The integrated system was computer controlled with a response time of 4 min.

48.5.6 Validation

In most cases, the investigator should compare the results of the microchip analysis with those an offline standard technique (CE; high-performance liquid chromatography, HPLC) for method validation. Control microdialysate samples must be collected, particularly for *in vivo* experiments that involve injection of a compound or drug. Such control samples can help identify or eliminate peaks from the microchip analysis. In many cases it is useful to incorporate an internal standard into the sampling process to correct for any changes in the recovery, reaction chemistry or with the on-chip injection process.

48.6 CONCLUDING REMARKS

Microdialysis is a well-established sampling technique that has found application in many areas, most notably in biological monitoring. Advances in fabrication technologies have led to the development of innovative microchip technologies suitable for real-world monitoring applications. Microdialysis/microchip systems offer several advantages, including lower costs, faster analysis for improved temporal resolution, and simplified connections and fittings for device fabrication and operation. While the performance of conventional bench-top systems is currently superior to that of the separation-based devices, initial reports clearly demonstrate the potential of coupling microdialysis to microchip platforms.

Future developments will see the optimization of device design and the investigation of alternative materials for microchip, electrode, and membrane fabrication. It is envisioned that further advances in fabrication and integration procedures will allow the development of implantable/wearable microdialysis/microchip systems for personal or on-animal monitoring. Integration of the separation-based systems with powerful detection techniques such as MS will further improve the detection capability of these systems for biological, pharmaceutical, and environmental monitoring.

ACKNOWLEDGMENTS

This research was supported by research grants from the National Institutes of Health (R01 NS042929-04) and from the National Science Foundation (CHE-0111618). Support in the form of a predoctoral fellowship from the American Heart Association for Pradyot Nandi is also gratefully acknowledged. The authors would also like to thank Nancy Harmony for her assistance in the preparation of this manuscript.

REFERENCES

1. Hanes, S. D. and Herring, V. L., Gentamicin enzyme-linked immunosorbent assay for microdialysis samples, *Therap. Drug Monitor.*, 23, 689, 2001.
2. Fujii, T., et al., Effects of physostigmine and calcium on acetylcholine efflux from the hippocampus of freely moving rats as determined by *in vivo* microdialysis and a radioimmunoassay, *Neurosci. Lett.*, 289, 181, 2000.

3. Moscone, D., et al., Rapid determination of lactulose in milk by microdialysis and biosensors, *Analyst (Cambridge, United Kingdom)*, 124, 325, 1999.
4. Weiss, D. J., et al., Applications of microdialysis/mass spectrometry to drug discovery, in *The Mass Spectrometer in Drug Discovery: A Practical Guide to Instrumentation, Experimental Strategies and Applications*, Rossi, D. T. and Sinz, M. W. eds., Chapter 12, Marcel Dekker, New York, 2001, pp. 377–397.
5. Khandelwal, P., et al., Nanoprobe NMR spectroscopy and *in vivo* microdialysis: new analytical methods to study brain neurochemistry, *J. Neurosci. Method*, 133, 181, 2004.
6. Scott, D. O., et al., *In vivo* microdialysis sampling coupled to liquid chromatography for the study of acetaminophen metabolism, *J. Chromatogr.*, 506, 461, 1990.
7. Ruiz-Jimenez, J. and Luque de Castro, M. D., Coupling microdialysis to capillary electrophoresis, *Trend Analyt. Chem.*, 25, 563, 2006.
8. Hogan, B. L., et al., Online coupling of *in vivo* microdialysis sampling with capillary electrophoresis, *Anal. Chem.*, 66, 596, 1994.
9. Zhou, S. Y., et al., Continuous *in vivo* monitoring of amino acid neurotransmitters by microdialysis sampling with online derivatization and capillary electrophoresis separation, *Anal. Chem.*, 67, 594, 1995.
10. Lada, M. W., et al., High temporal resolution monitoring of glutamate and aspartate *in vivo* using microdialysis online with capillary electrophoresis with laser-induced fluorescence detection, *Anal. Chem.*, 69, 4560, 1997.
11. Whitesides, G. M., The origins and the future of microfluidics, *Nature (London, United Kingdom)*, 442, 368, 2006.
12. Wu, Q., et al., Microchip-based macroporous silica sol-gel monolith for efficient isolation of DNA from clinical samples, *Anal. Chem.*, 78, 5704, 2006.
13. Roper, M. G., et al., Advances in polymerase chain reaction on microfluidic chips, *Anal. Chem.*, 77, 3887, 2005.
14. Song, S., et al., Microchip dialysis of proteins using *in situ* photopatterned nanoporous polymer membranes, *Anal. Chem.*, 76, 2367, 2004.
15. Dodge, A., et al., PDMS-based microfluidics for proteomic analysis, *Analyst (Cambridge, United Kingdom)*, 131, 1122, 2006.
16. Torto, N., et al., Microdialysis sampling challenges and new frontiers, *LC-GC (Europe)*, 14, 536, 2001.
17. Trickler, W. J. and Miller, D. W., Use of osmotic agents in microdialysis studies to improve the recovery of macromolecules, *J. Pharm. Sci.*, 92, 1419, 2003.
18. Khramov, A. N. and Stenken, J. A., Enhanced microdialysis recovery of some tricyclic antidepressants and structurally related drugs by cyclodextrin-mediated transport, *Analyst*, 124, 1027, 1999.
19. Davies, M. I. and Lunte, C. E., Microdialysis sampling coupled online to microseparation techniques, *Chem. Soc. Rev.*, 26, 215, 1997.
20. Beyer, C. E., et al., Comparison of the effects of antidepressants on norepinephrine and serotonin concentrations in the rat frontal cortex: an *in vivo* microdialysis study, *J. Psychopharmacol. (London, United Kingdom)*, 16, 297, 2002.
21. Dawson, L. A., et al., Rapid high-throughput assay for the measurement of amino acids from microdialysates and brain tissue using monolithic C18-bonded reversed-phase columns, *J. Chromatogr., B: Analyt. Technol. Biomed. Life Sci.*, 807, 235, 2004.
22. Watson, C. J., et al., *In vivo* measurements of neurotransmitters by microdialysis sampling, *Anal. Chem.*, 78, 1391, 2006.
23. de Lange, E. C. M., et al., Methodological issues in microdialysis sampling for pharmacokinetic studies, *Adv. Drug Del. Rev.*, 45, 125, 2000.
24. Hocht, C., et al., Application of microdialysis for pharmacokinetic-pharmacodynamic modelling, *Expert Opin. Drug Discovery*, 1, 289, 2006.
25. Cooney, C. G., et al., Optical pH, oxygen and carbon dioxide monitoring using a microdialysis approach, *Sensors and Actuators, B: Chemical*, B69, 183, 2000.
26. Miro, M. and Frenzel, W., Implantable flow-through capillary-type microdialyzers for continuous *in situ* monitoring of environmentally relevant parameters, *Anal. Chem.*, 76, 5974, 2004.

27. Wei, M.-C., et al., Determination of organic acids in fermentation products of milk with high performance liquid chromatography/on-lined micro-dialysis, *Chromatographia*, 54, 601, 2001.
28. Kurita, R., et al., Miniaturized one-chip electrochemical sensing device integrated with a dialysis membrane and double thin-layer flow channels for measuring blood samples, *Biosens. Bioelectron.*, 21, 1649, 2006.
29. Hsieh, Y.-C. and Zahn, J. D., Glucose recovery in a microfluidic microdialysis biochip, *Sensors and Actuators, B: Chemical*, B107, 649, 2005.
30. Dempsey, E., et al., Design and development of a miniaturized total chemical analysis system for online lactate and glucose monitoring in biological samples, *Anal. Chim. Acta*, 346, 341, 1997.
31. Shi, G., et al., On-line biosensors for simultaneous determination of glucose, choline, and glutamate integrated with a microseparation system, *Electrophoresis*, 24, 3266, 2003.
32. Huynh, B. H., et al., On-line coupling of microdialysis sampling with microchip-based capillary electrophoresis, *Anal. Chem.*, 76, 6440, 2004.
33. Sandlin, Z. D., et al., Microfluidic electrophoresis chip coupled to microdialysis for *in vivo* monitoring of amino acid neurotransmitters, *Anal. Chem.*, 77, 7702, 2005.
34. Roman, G. T. and Culbertson, C. T., Surface engineering of poly(dimethylsiloxane) micro fluidic devices using transition metal sol-gel chemistry, *Langmuir*, 22, 4445, 2006.
35. Vickers, J. A., et al., Generation of hydrophilic poly(dimethylsiloxane) for high-performance microchip electrophoresis, *Anal. Chem.*, 78, 7446, 2006.
36. Steinkuhl, R., et al., Microdialysis system for continuous glucose monitoring, *Sensors and Actuators, B: Chemical*, B33, 19, 1996.
37. Roddy, E. S., et al., Sample introduction techniques for microfabricated separation devices, *Electrophoresis*, 25, 229, 2004.
38. de Jong, J., et al., Membranes and microfluidics: a review, *Lab on a Chip*, 6, 1125, 2006.
39. Pasas, S., et al., Detection on microchips: principles, challenges, hyphenation, and integration, in *Separation Methods in Microanalytical Systems*, Kutter, J. P. and Fintschenko, Y. eds., Chapter 9, CRC Press LLC, Boca Raton, FL, 2006, pp. 433–497.
40. Pijanowska, D. G., et al., A flow-through potentiometric sensor for an integrated microdialysis system, *Sensors and Actuators, B: Chemical*, B103, 350, 2004.
41. Hsieh, Y.-C. and Zahn, J. D., On-chip microdialysis system with flow-through sensing components, *Biosens. Bioelectron.*, 22, 2422–2428, 2007.
42. Huynh, B. H., et al., A microchip electrophoresis device with on-line microdialysis sampling and on-chip sample derivatization by naphthalene 2,3-dicarboxaldehyde/2-mercaptoethanol for amino acid and peptide analysis, *J. Pharm. Biomed. Anal.*, 42, 529, 2006.
43. Li, M. W., et al., Design and characterization of poly(dimethylsiloxane)-based valves for interfacing continuous-flow sampling to microchip electrophoresis, *Anal. Chem.*, 78, 1042, 2006.
44. Unger, M. A., et al., Monolithic microfabricated valves and pumps by multilayer soft lithography, *Science (Washington, DC)*, 288, 113, 2000.

49 Microfluidic Sample Preparation for Proteomics Analysis Using MALDI-MS

Simon Ekström, Johan Nilsson, György Marko-Varga, and Thomas Laurell

CONTENTS

49.1 INTRODUCTION

Explosive progress in the last decade in the field of genomics has generated new insight and an enormous amount of data concerning the codes and functions of different genes. Proteomics constitutes the huge corresponding effort to systematically unravel the functions of the different proteins encoded in the genome. Identification of proteins and studies of protein expression, interaction, and post-translational modifications (PMTs) are imperative for understanding biological systems. This has led to an increased interest in the development of microanalysis systems that allows proteins to be subjected to mass screening.

This chapter will briefly cover some aspects of microfluidic sample preparation techniques in the field of proteomics with a special focus on applications using matrix-assisted laser desorption ionization-mass spectrometry (MALDI-MS) for the analysis readout. The reader is referred to Chapter 54 by Lazar as a complement to this chapter. Owing to the fact that the amount of available sample in protein analysis often is very limited, the use of microfluidic structures in combination with MALDI-MS has many great advantages, such as the ability to handle small amounts of sample, miniaturized sample handling that reduces sample losses, rapid digestion of proteins, effective solid-phase extraction (SPE) and easy automation.

49.2 BACKGROUND

49.2.1 MINIATURIZATION OF ANALYTICAL SYSTEMS

Miniaturization of analytical systems has been a long-standing priority and one of the first examples of pioneering work in chip-based miniaturization is the gas chromatograph developed by Terry et al. at Stanford in 1975.[1] In the late 1980s, the term micrototal analysis system (μTAS) was introduced by Michael Widmer and colleagues at Ciba Geigy in Switzerland, a concept followed up later in 1993 when chip-based capillary electrophoresis of amino acids was presented.[2] In the last decade, an ever increasing number of applications have been developed for microfluidic devices (e.g., in protein analysis).[3–11] genomic analysis,[12–14] forensic science,[15] environmental analysis,[16] cell handling,[17–21] clinical analysis,[22–24] combinatorial chemistry,[25] microreactors,[26–28] enzymatic assays,[29–31] and for fundamental science studies of microfluidics.[32–34]

The main driving force behind the increasing interest for miniaturization in proteomics is the need to integrate and automate analysis to facilitate increased speed and reduced costs per analysis. The benefits of miniaturization arise from the increased reaction kinetics in small volumes and the possibility to perform sample handling procedures at a high speed in micro-/nanoliter systems. Reduced dead volumes in integrated systems allow for increased system performance. Extreme surface-to-volume ratios (80000 m^{-1}) compared to conventional microtiterplates (500 m^{-1}) enables elution of samples in very small volumes[4] (e.g., after SPE). The art of microfluidics is utilizing the scaling laws for new effects and better performance. This is not always a straightforward process as the behavior of fluids at the microscale can be significantly different from generic behavior in macroscale. Factors such as surface tension and frictional forces give a liquid behavior governed by capillary forces, with a laminar flow as viscosity and not inertia dominates fluid behavior. Other aspects that can be used to advantage when constructing microfluidic systems are low thermal mass and efficient mass transport, for example, if a channel is reduced in width by a factor 10 the mean diffusion time for an analyte across the channel is reduced by a factor 100. All microfluidic structures inherently benefit from the lower sample and reagent consumption and the high degree of parallelity that can be achieved. For additional user-friendliness, microfluidic structures should be manufactured as disposables when it is possible.

49.2.2 PROTEOMICS

The idea of mapping all the human proteins was presented in 1982, with the human protein index suggested by Anderson and Anderson.[35] At that time, technology for this purpose was rather limited, but with time and technology development, more and more scientists became aware of the gains of such a project. The term proteome was introduced in 1994 by the Australian researchers, Keith Williams and Marc Wilkins, at the 1994 Siena meeting[36] ("2-D Electrophoresis: From Protein Maps to Genomes" September 5–7, 1994). The proteome is defined as the complete set of proteins from the information encoded in a genome that can be expressed and modified by a cell, tissue, or organism. The number of transcribed genes in the human genome is currently believed to be around 20,000–25,000, although this number is fiercely debated and might soon change. With regard to the number of PMTs, the 20,000–25,000 genes are estimated to give rise to approximately 1 million different protein molecules.

Protein technology is also inherently more complex than DNA-based technology. Not only is the basic alphabet bigger and more heterogeneous, genes can also be variously spliced[37] resulting in numerous different protein products from a single stretch of DNA. In addition, mRNA editing is relatively common, leading to modified messages and corresponding protein products.[38] There are also more than 200 different PMTs by which proteins can be modified after they have been synthesized. Proteins once synthesized can be cleaved, to eliminate signal sequences, transit or propeptides, and initiator methionines. There are examples of proteins that have been found to possess more than 1000 different isoforms.[39] The proteome analysis projects have no equivalent to the high-throughput sequencers, which have served the genome projects so well. Neither is it possible to amplify the proteins by any analog to the polymerase chain reaction (PCR) technique used for DNA amplification, leaving only the proteins isolated from the original sample available for analysis. While the genome is virtually static, at least in the case of DNA, the proteome continuously changes its state. In principle, therefore, an infinite number of proteomes can be assigned to a given genome. Adding to the complexity is also the huge dynamic range of the proteome;[40] on average, a human cell contains about 10,000 different proteins present at concentration ranges from 10^1 to 10^{12}. Finding a specific low-abundance molecule in such a complex mixture is more difficult than having to seek a certain person out of the entire population of Earth.

The huge complexity of proteomic analysis makes it a technically challenging goal. The technologies required to separate large numbers of proteins, to identify them, and to study their modifications are by no means straightforward or homogenous,[41–44] but as proteomics research evolves so does the technology used in this field. Explosive progress the last decades in the fields of protein science, bioinformatics, and cell and molecular biology have resulted in an increased demand for development of instrumental technology that allows identification and detailed structural studies of proteins to be automatically performed at high speed and with a high sensitivity. Miniaturization can, with advantage, be implemented by applying microfabrication techniques. Although, it should be mentioned that a lot of miniaturization in the field of proteomics have been accomplished by using conventional means (e.g., using capillaries).

Despite the many advantages of miniaturization there are some concerns. For instance, if the speed of analysis is limited in a preceding or subsequent step, it is not necessarily a huge gain for the overall analysis time if the application of miniaturization allows for one rapid sample preparation step. In cases where the sample/analytes are not limited, it might not be necessary to use miniaturization to facilitate analysis. Another difficulty is that miniaturized sample preparation has the inherent characteristic that small sample volumes have to be used; while this has some advantages, it includes the disadvantage that fewer analyte molecules are available for detection. Consider a scenario where a sample containing an analyte at a concentration of 1 pM is to be analyzed with a miniaturized device that allows for loading of 1 nL sample. Disregarding any sample losses or dilution in the analysis device, there will be ~600 molecules (1 zeptomole) available for detection. Although there are detection techniques that can facilitate single molecule detection,[45–50] this makes analysis more difficult than required if a more concentrated sample could be applied.

To counter the effect of small sample loads in miniaturized analytical systems, a huge number of methods for preconcentration of analytes have been deployed.[51] These include numerous "electrokinetic concentration" processes such as isotachophoresis,[52,53] field-amplified sample stacking,[54] and electrokinetic trapping.[55,56] Chromatographic SPE methods have shown to provide 1000-fold preconcentration before electrochromatography[57,58] and micellar electrokinetic chromatography (MEKC).[59] Membrane preconcentration have also been adapted to microfluidics with success.[60–63] A recently presented technique used an integrated nanofluidic filter to accomplish electrokinetic trapping[64] with concentration factors as high as 10^6–10^8. Examples of other techniques for concentration of analytes are liquid–liquid extraction,[65–67] microdialysis,[68] and immunoaffinity capture.[69–71]

49.2.3 Mass Spectrometry for Analysis and Identification of Proteins

In the late 1980s, two technology breakthroughs in soft desorption/ionization of large biomolecules occurred with the introduction of electrospray ionization (ESI)[72,73] and laser desorption ionization.[74,75] These methods have paved the way for the intense use of MS for analysis of biomolecules that we see today, and were awarded with a Nobel prize in 2002. The generic mass spectrometer consists of an ion source, an analyzer and a detector (Figure 49.1). The mass spectra generated provide the mass of the analyzed molecule and in the case of a MS/MS experiment, the masses of the breakdown products of a selected analyte.

When the mass of a protein or peptide in a mass spectra does not immediately give the identity of the analyte, it was quickly understood that if a protein was cleaved in a predicable manner, the peptides generated from a given protein would form a fingerprint for the protein.[76] This protein fingerprint could then be compared to *in silico* patterns of protein sequences generated from databases. In 1993, this technique of peptide mass fingerprinting (PMF) was independently described by a number of groups.[77–81] The most commonly used proteolytic enzyme for generation of this fingerprint is trypsin, which cleaves specifically at the C-terminal end of arginine and lysine (Figure 49.2). Owing to its ease of use PMF analysis with MALDI-MS is still the most common way of identifying proteins separated by two-dimensional gel electrophoresis (2-DE).[82]

As an alternative to PMF, proteins can be identified by MS/MS of peptides generated from the proteins. This identification method was introduced in 1994 by Mathias Mann[83] (Sequence Tags) and John Yates[84] (cross-correlation). In an MS/MS experiment a certain ion is selected in MS mode to undergo fragmentation in the gas phase. Peptides fragment in a sequence-specific manner, yielding a MS/MS spectrum that reflects its amino acid composition. The fragmentation of the peptide is commonly brought about by collision-induced dissociation (CID), but many other forms of fragmentation can also be used, for example, PSD fragmentation with MALDI,[85,86] or Electron Capture Dissociation (ECD). The usefulness of ESI-MS/MS took a further leap with the sensitivity increasing and sample-saving nanospray. The ability to perform sensitive MS/MS analysis of peptides is the basis for the "shotgun"/multidimensional protein identification technology proteomics (MudPIT).

From the beginning, MALDI was commonly coupled with time-of-flight (TOF) analyzers and ESI with ion trap or quadrupole analyzers.[87] Modern mass spectrometers are evolving at a rapid pace and numerous different configurations exist on the market today. Both MALDI and ESI ion sources can be combined with a number of analyzers, for example, triple quadrupole (TQ), quadrupole TOF (Q-TOF), ion traps (IT), quadrupole ion traps (QIT), or Fourier transform ion-cyclotron resonance analyzers (FT-ICR). The new technologies in MS, such as linear ion traps,[88] orbitrap,[89,90] ion mobility,[91,92] MALDI TOF-TOF,[93] and FT-ICR-MS[94,95] provide ever-increasing sensitivity and speed of analysis. The fact is that MS has matured so that it is no longer the most limiting factor in

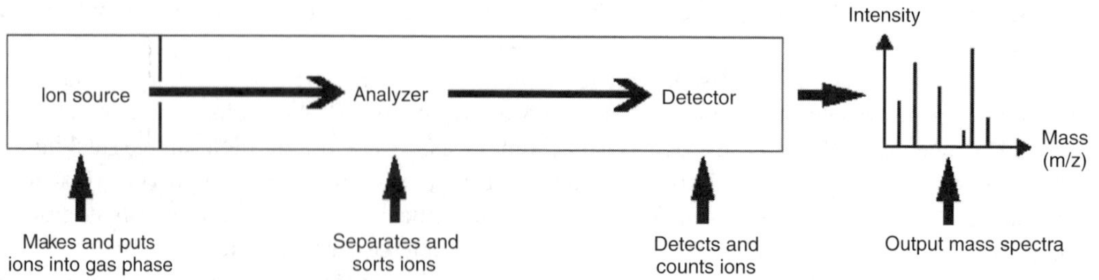

FIGURE 49.1 Diagram of a generic mass spectrometer. Sample analytes are transposed to ions in the ion source and separated in the analyzer. The ions hit the detector and a mass spectra is recorded, where the intensity of the ions are plotted against their mass.

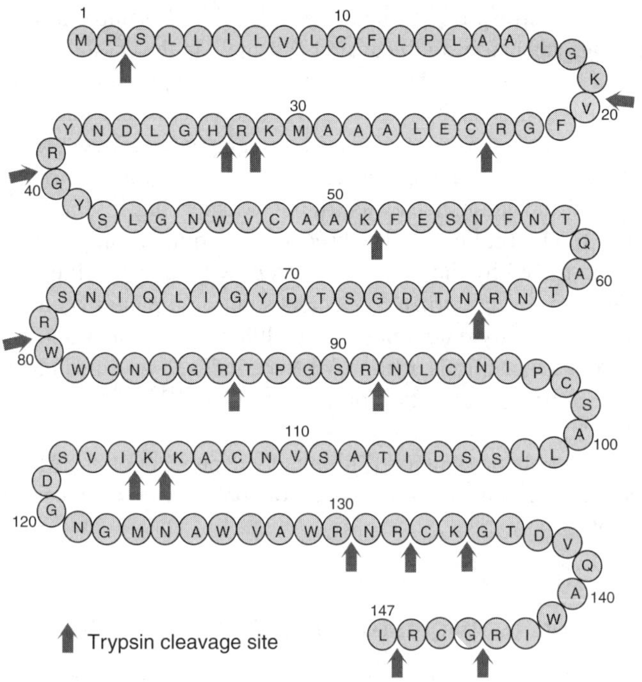

FIGURE 49.2 Amino acid sequence of lysozyme and trypsin cleavage sites (carboxy terminal to lys and arg). The specificity of the trypsin digestion results in a "fingerprint" for the protein when analyzed by MS.

proteomics. The difficulties of analyzing low-abundance proteins are currently more limited by the sample preparation.

49.2.4 Sample Preparation for MALDI-MS

The laser desorption technique almost exclusively used today is the MALDI technique developed by Franz Hillenkamp and Michael Karas.[74,96,97] The sample is dispersed in a large excess of matrix material, which will strongly absorb the incident laser light. The matrix also serves to isolate sample molecules in a chemical environment that enhances the probability of ionization without fragmentation. Short pulses of laser light focused (10–200 μm spot diameter) on the sample spot will cause the sample and matrix to volatilize. The analyte ions formed are then separated according to mass in the analyzer (commonly a TOF) and subsequently recorded as an electrical signal as they arrive at the detector.

The most important step in MALDI-MS, is finding the optimal matrix-sample preparation method.[98] The first reports in the field all use the conventional "dried-droplet" (DD) method,[99] where analyte and matrix solutions are premixed in a 1:1 ratio and subsequent deposition of approximately 1 μL volumes onto a metal target; the spot is left to dry and crystallize within a few minutes. However, this method often leads to sample inhomogeneties, that is, the crystal formation originating from the rim of the droplet results in crystals of varying size and density. The final sample surface has spots with little or no crystals at some locations, while a dense crystal layer normally forms at the rim. Since several laser scans have to be averaged in order to obtain a mass spectrum with a good signal-to-noise ratio, there is a major interest in maximizing the shot-to-shot reproducibility. Both inhomogeneous analyte incorporation into the matrix crystal structure[100] and variation of the analyte/matrix ratio as a result of the crystallization process[101] are believed to contribute to the poor reproducibility. Improved reproducibility have been reported using crystal crushing,[102] spin-coated drying,[103] stirring,[104] aerospray,[105] electrospray,[106] and recrystallization[107] sample preparation methods. The presence of microcrystals can cause significant enhancement in the crystal growth rate[108] and hence result in smaller crystal sizes. A method that has been shown to provide good reproducibility is fast evaporation method.[99] An approach that uses fast evaporation to produce microcrystals followed by a second layer of matrix to obtain a homogeneous crystal layer have been presented by Dai et al.[109] A similar approach to seed-layer method was developed in our group for high reproducibility and high-throughput purposes and automated by use of microdispensing.[110] It is also possible to first add the matrix to the MALDI target followed by subsequent addition of sample to the dried matrix. This is very conveniently facilitated by the thin-layer method where nitrocellulose is included into the matrix solution.[111] The thin-layer matrix a convenient solution for automated peptide analysis as the samples can be deposited directly onto a predeposited thin-layer matrix surface in nanovials. This method was later improved by Miliotis et al.[112,113] Recent developments in MALDI matrix-sample preparation are direct desorption/ionization on porous silicon,[114] carbon nanotube matrix,[115] and liquid ionic matrix.[116,117]

The optimal matrix-sample preparation is ultimately dependent on the MALDI instrument used and the nature of the sample. Thus, the optimal sample preparation for a certain application has to be deduced empirically.

49.2.5 Microdispensing Techniques for MALDI-MS Sample Preparation

The most well known and successful application of microfluidics is inkjet printing.[118] Sample volumes in the nanoliter–picoliter range are very difficult to handle by conventional pipetting and in this context, droplet dispensers for sample handling are finding increased use in life science[119] and chemistry.[120] Therefore special tools need to be developed (Figure 49.3).

The development is fueled by rising demands for speed in high-throughput screening systems and the overall trend toward miniaturized analysis systems. The technology has evolved from the conventional inkjet printing technology, where droplets are generated from an enclosed volume of liquid. The use of droplet dispensers have been presented in a number of applications for biological sample handling,[120–125] for example, for sample introduction in CE,[126] fabrication of sensors,[127] and sample deposition in MALDI.[128–130] An example of a commercial dispenser for proteomics applications is the Chemical inkjet Printer (Chip) marketed by Shimadzu Corporation (Kyoto, Japan), as tool for on-membrane protein digestion.[131,132] The limitation of these dispensers is that they are of the single-ended type where the sample is either filled from a stationary reservoir or aspirated through the orifice of the dispenser.

FIGURE 49.3 Pictures illustrating the difficulties of liquid deposition in nanovials using conventional techniques (right pipetting and left capillary deposition). Nanovial dimensions are 400 × 400 μm.

To handle the many physically diverse sample solutions that are encountered in the field of proteomics, a piezo-actuated flow-through microdispenser developed within our group[133] has found many uses for depositing sample volumes ranging from 65 pL to over 300 nL. The dispensing principle is based on a piezo-ceramic element, which elongates when a voltage pulse is applied, thus creating a pressure pulse that results in droplet ejection at the orifice. A schematic cross-section of a dispenser structure is illustrated in Figure 49.4. A raised pyramid-shaped nozzle ensures stable droplet directivity and flow-through capability allows for online coupling to high-resolution liquid chromatography (LC), as have been implemented by Miliotis et al.[112,113,134]

During the years, a number of improvements of the piezoelectric microdispenser have been made in order to shrink internal volumes and obtain optimal operational characteristics. In addition, a number of new design solutions have been tested; an example is the compound microdispenser[135] that was developed to avoid crystal formation at the nozzle orifice, which will have an adverse effect, on the droplet formation. This compound dispenser, when adapted for gas-flow-guided dispensing (Figure 49.5) was also found to offer potential improvements for the situation where piezoelectric microdispensing was coupled to online LC separations.[136] Especially in situations where the composition of the liquid varies with time, as the changes in surface tension of the dispensed liquid could lead to decreased droplet stability, directionality (formation of satellite droplets that may have a different trajectory) and in the worst case disrupted dispensing.

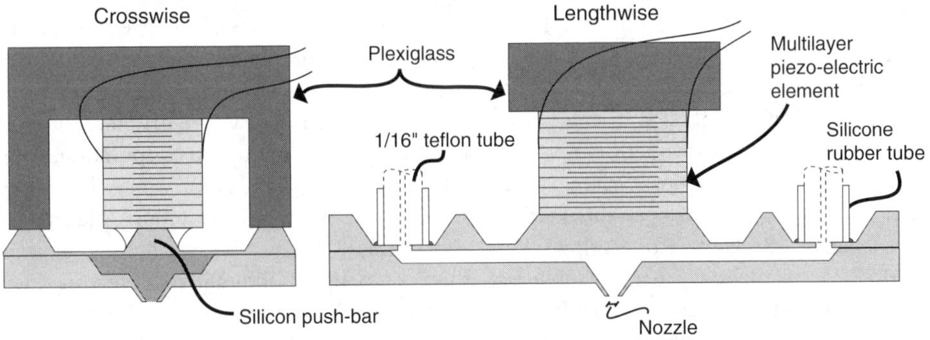

FIGURE 49.4 Schematic cross-section of a microdispenser. (Reproduced with permission from Miliotis, T., Kjellstrom, S., Nilsson, J., Laurell, T., Edholm, L. E., and Marko-Varga, G., *Journal of Mass Spectrometry*, 35, 369–377, 2000, copyright 2000, John Wiley & Sons.)

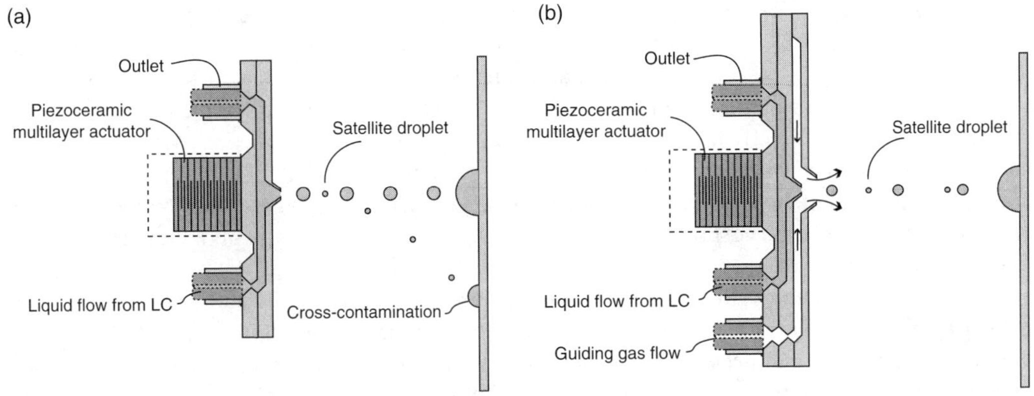

FIGURE 49.5 Droplet ejection with satellite droplets from an ordinary flow-through dispenser (Figure 49.7a) and from a gas-flow guided dispenser (Figure 49.7b). The gas flow is achieved by adding a second nozzle side with a single inlet for gas to the dispenser depicted in Figure 49.4.

49.2.6 Microspot Techniques for MALDI-MS Sample Preparation

It is a well-known fact that a small spot size is beneficial to MALDI analysis.[137,138] As MALDI-MS desorbs/ionizes the analytes with a laser having a focal spot size of 50–200 μm, it follows that the analyte density in the deposited sample spot is critical for the final sensitivity. To increase the sensitivity and reproducibility in MALDI-MS, different microspot preparation techniques have been developed. There are two main ways to accomplish a microspot. It can be done either by having a moiety on the target plate that confines the sample or by use of a specialized deposition of the sample. Examples of methods that can be applied to ensure a small spot are drop-on-demand dispensing,[128,139–141] hydrophilic anchor point on a hydrophobic target[142] or deposition on a hydrophobic target,[143] capillary deposition,[144–147] electrowetting-on-dielectric (EWOD)-based deposition,[148] wall-less sample preparation,[149] and electrospray deposition.[150]

49.2.7 Solid-Phase Extraction as Sample Preparation Before MALDI-MS

Despite the high sensitivity and the relatively high tolerance for contaminants of MALDI-MS, successful analysis often requires pure and concentrated analytes. One of the most robust and convenient techniques for sample clean up is SPE. A study of the literature will reveal numerous strategies that can be used to combine MALDI-MS with miniaturized SPE. They can be summarized as (i) micropipette tips packed with SPE media; (ii) microtiterplates with integrated SPE; (iii) MALDI target plates for on-target (on-probe) purification of samples and (iv) microfabricated on-chip-based SPE technologies. The micropipette tip-based SPE is probably the easiest to implement, the first examples used ordinary pipette tips packed with beads.[151,152] More recently, reports of similar SPE tips using monolith media[153] and pieces of SPE membranes[154] have been published. Currently, there are several suppliers of different microextraction tips, for example, ZipTip™ (Millipore Corporation, Billerica, MA, USA), SuproTip (The Nest Group, Inc, Southborough, MA, USA), Omix pipette tips (Varian, Inc. Paulo Alto, CA, USA), Eppendorf PerfectPure Tips (Eppendorf AG, Hamburg, Germany), and Agilent Cleanup C18 Pipette Tips (Agilent, Palo Alto, CA, USA). The advantage of the micropipette tip-based SPE is that it is easy to automate using ordinary pipette robotics and relatively inexpensive, with a cost per sample of 2$–3$. The drawbacks are that there are many sample transfers, where sample can be lost and the requirement of a low flow resistance (large pores) in the tips leads to a slow mass transfer of analyte to the SPE media and tip-to-tip variations. Thus, several aspiration dispensing cycles are needed and recovery (i.e., sensitivity) has been questioned for the ZipTips when small elution volumes are used.[155,156]

The microtiterplate-based technologies such as the ZipPlate (Millipore Corporation, Billerica, MA, USA) and the device presented by Nissum et al.[157] where the SPE media is integrated at the bottom of each well (300 nL in the case of ZipPlate) offers a more an elegant solution. Here, sample losses due to unnecessary sample transfers are minimized and this is claimed to result in higher sensitivities. Standard pipetting robotics can be used to perform most of the sample transfers and the sample can be eluted directly onto a MALDI-MS target placed under the device. Although, special adapters have to be made in order to facilitate the final elution onto the MALDI target. These devices are relatively new and there is a lack of published data on their performance, but the methodology is sound and as long as the cost per analysis is kept low the use of these type of devices can become as common as the tip-based approach.

On-target purification (SPE) has been accomplished by modifying the surface of the MALDI target to promote binding of the analytes and allowing contaminants to be washed away.[158–160] There are many examples of surfaces and strategies that can be used for on-target cleanup, as described in a review by Xu et al.[161] Examples of commercial on-target SPE technology are the protein chip used for SELDI analysis[162] (Ciphergen, Fremont, CA, USA) and MassPrep PROtarget (Waters, Milford, MA, USA). Just as the microtip and microtiterplate-based SPE standard robotics can be used, here the final elution step is avoided as elution is done by addition of the matrix solution to the spot (i.e., the recovery should be 100% unless some of the analyte is lost during washing). Sample losses due to many sample transfers are potential drawbacks of SPE. It will also become important to avoid spreading of the sample spot when the matrix is added to the spot as sensitivity will suffer with increasing spot size. Limited capacity of a "planar" surface another concern that might be of less importance when the goal is purification/concentration of peptides from in-gel digest. But when the goal is biomarker discovery where the sample contains many different species of a broad concentration range, this capacity deficit becomes more of a concern.

The coupling of microfabricated on-chip SPE devices to MALDI-MS has previously been presented by our group.[163–167] Another approach to on-chip SPE—the microfluidic compact disk Gyro Lab MALDI SP1 (Gyros AB, Sweden)—has been presented and commercialized by Gyros AB, where centrifugal force is used to transport liquids for the purpose of SPE before MALDI-MS.[168] EWOD[148] and electrocapture[55] are techniques that can be used instead of SPE for purification and concentration before MALDI-MS.

49.2.8 Separation Techniques Interfaced to MALDI-MS

Since its introduction in 1975 by Klose[169] and O'Farrell,[170] 2-DE has been the most commonly used protein separation technique.[171] A number of developments have extended the usefulness of 2-DE, such as narrow range immobilized pH gradients (IPGs),[172] and 2-D difference gel electrophoresis[173] (DIGE). High resolution and the ability to get a visual representation of the intact proteins and their isoforms are the advantages of 2-DE. Among the difficulties with 2-DE is the separation of membrane proteins, separation of alkaline proteins, and the difficulty of identifying low-abundance proteins from the separation. Owing to the offline nature of 2-DE (i.e., intact proteins in-gel), identification of the gel-separated proteins are with advantage done by MALDI-MS or MS/MS.

An increasingly popular approach in proteomics is to cleave (usually with a proteolytic enzyme) the proteins to peptides, which are separated by LC, followed by MS/MS protein identification of the peptides.[174] The separation is usually done by combining two or more orthogonal chromatographic steps such as a first step of chromatofocusing, size exclusion, ion exchange, hydrophobic interaction, or affinity, followed by a last step of reversed-phase LC, which has the best compatibility with MS. By using ESI MS/MS the separation is analyzed online and this is the most common method.[175–177] Alternatively, the separation is fractionated directly onto a MALDI target as discrete spots.[112,113,134,178–180] While multidimensional separations coupled to ESI MS (MS/MS) have a greater throughput compared to protein identification with MALDI-MS it is not clear that either technique has a clear advantage. Rather it seems that the two techniques are complementary.

Capillary electrophoresis has also been applied to separate peptides before MS analysis with very high separation efficiency and several review publications can be recommended.[181–183] LC is more commonly used in conventional proteomics than CE because of the rather limited capacity of CE. In microfluidics, CE is the most well-developed separation technique.[184–187]

Capillary electrochromatography[188] (CEC) where a stationary phase is introduced in the separation channel, is another option for on-chip separation of biomolecules. This has been achieved by using chemically derivatized walls,[189,190] packed beads in microchannels,[57,191] *in situ* polymerization to obtain monolithic stationary phases;[58,192] alternatively, molded collocated monolithic support structures (COMOSS) can be used to increase the surface-to-volume ratio.[193,194] Pressure-driven LC has been more rarely used in microfluidics; but in principle, many of the microstructures used for CEC or SPE could be applied to LC and coupled offline to MALDI-MS. The increased use of LC in proteomics is now starting to reflect in microfluidics[195–197] and Agilent (Agilent, Palo Alto, CA, USA) is currently marketing a high-performance liquid chromatography (HPLC) chip[198] that integrates trapping, separation column, and an ESI nozzle.

There are also numerous different integrated microfluidic systems published in the literature, but to describe all of them here would neither be possible nor do them justice. The coupling of microfluidics to ESI MS has been extremely successful since the first examples where published[199–201] and the developments have been described in several reviews.[202,203]

Integrated microfluidic systems capable of protein digestion have been presented in many configurations,[56,204–212] and will continue to be a goal in the development of microfluidic proteomics technologies.

49.3 PRACTICAL APPLICATIONS

49.3.1 Application of Microdispensing and Nanovials for MALDI-MS Sample Preparation

The use of microdispensing in combination with nanovials has a number of advantages, as it combines both a moiety and a deposition technique for achieving the MALDI microspot. The combination of microdispensing and micromachined silicon nanovials allows for enrichment of the analyte in the nanovial by making multiple depositions. This "spot-on-a-chip" enrichment (Figure 49.6) has been shown to provide a signal amplification of at least 10–50 times compared to ordinary sample preparations.[213] The linearity of the enrichment effect was demonstrated and the signal amplification technique was applied for analysis of relevant biological samples. The cytokine IL-8, involved in inflammatory responses, was not detectable with the standard dried-droplet sample preparation at a 100 nM concentration; but by using the "spot-on-a-chip" enrichment, the IL-8 peaks were evident in the MALDI-spectrum. Figure 49.7 illustrates the sensitivity gain provided by sample enrichment through microdispensing.

The use of nanovials to confine the sample spots to a size approximately matching that of the laser focal point is also a major advantage in the automation process since searching for high intensity locations ("hot-spots") can be avoided.

The silicon nanovial MALDI target arrays used in our laboratory are usually recirculated because using them as disposables are considered too expensive. Therefore, an effort was made to manufacture low-cost polymeric high-density nanovial MALDI targets, which could be used as disposables and

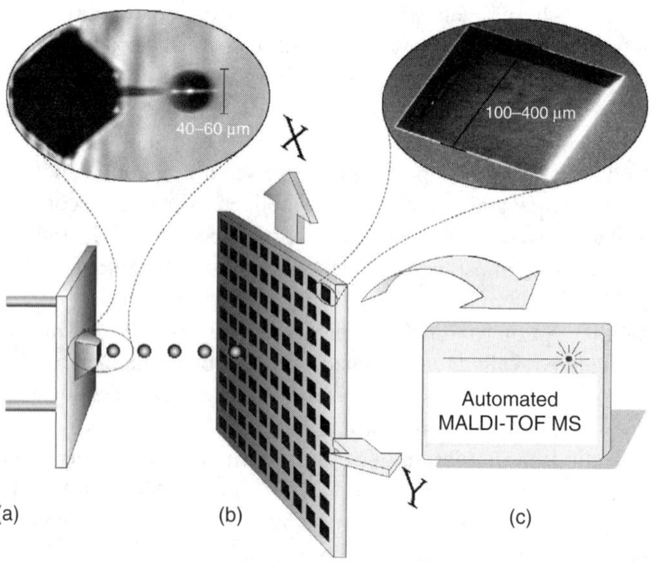

FIGURE 49.6 The "spot-on-a-chip" platform: (a) the piezo-actuated microdispenser (b) heated high-density MALDI-target plate, and (c) automated MALDI-TOF MS. (Reproduced with permission from Ekstrom, S., Ericsson, D., Önnerfjord, P., Bengtsson, M., Nilsson, J., Marko-Varga, G., and Laurell, T., *Analytical Chemistry*, 73, 214–219, 2001, copyright 2001, American Chemical Society.)

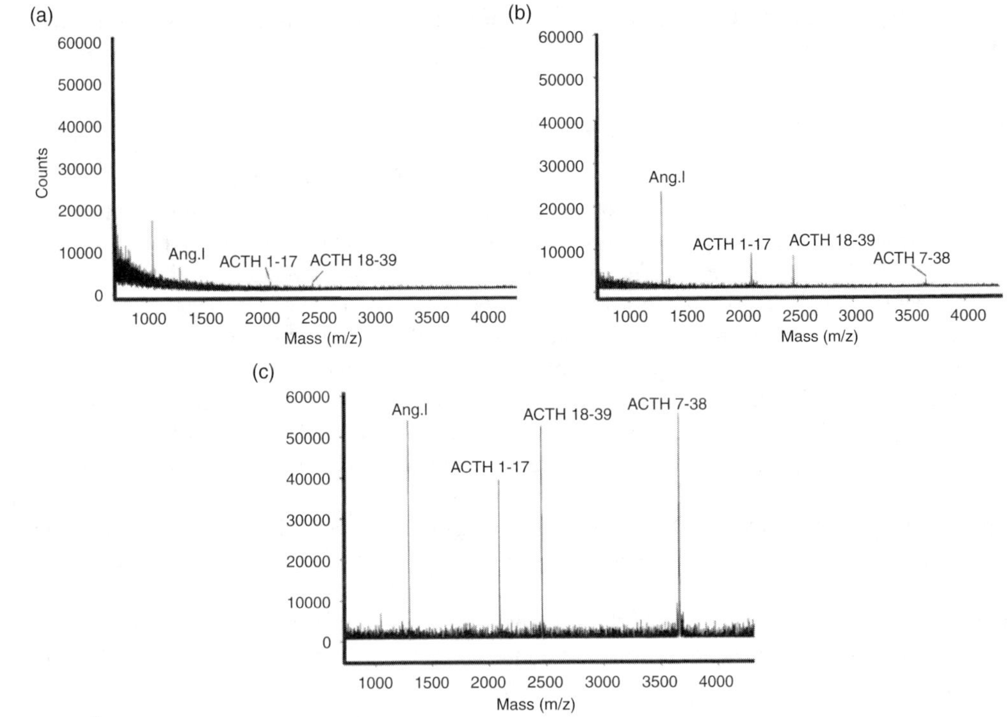

FIGURE 49.7 Effect of enrichment when analyzing a calibration mixture consisting of four peptides, angiotensin I, ACTH (1–17 clip), ACTH (18–39 clip) and ACTH (7–38 clip), at a concentration of 5 nM. In Figure 49.13a 100 droplets (50 amol) have been deposited and three of the four peptides can be detected. In Figure 49.13b 2000 droplets (1 amol) have been deposited and finally in Figure 49.13c a total of 8000 droplets, corresponding to an absolute amount of 4 fmol have been deposited and the saturation level of the detector was reached. Matrix: 2,5-dihydroxybenzoic acid. (Reproduced with permission from Ekstrom, S., Ericsson, D., Onnerfjord, P., Bengtsson, M., Nilsson, J., Marko-Varga, G., and Laurell, T., *Analytical Chemistry*, 73, 214–219, 2001, copyright 2001, American Chemical Society.)

FIGURE 49.8 Pictures of high-density polymeric nanovial MALDI target plates. (Reproduced with permission from Ekstrom, S., Nilsson, J., Helldin, G., Laurell, T., and Marko-Varga, G., *Electrophoresis*, 22, 3984–3992, 2001, copyright 2001, John Wiley & Sons.)

allow for archiving of samples[214,215] (Figure 49.8). These disposable high-density MALDI target plate were made either from polymethylmethacrylate (PMMA) and polycarbonate (PC) using precision micromilling and cold embossing to produce nanovials with 0.3–0.5 mm diameter (Figure 49.8). The polymeric high-density nanovial MALDI target plates was found to provide similar performance as silicon nanovial arrays and an extension of the in-vial chemistry that facilitates identification of phosphorylated peptides was applied in these nanovial arrays.

49.3.2 Miniaturized Enzymatic Digestion of Proteins for MALDI-MS

Owing to the complex nature of proteins and the fact that MS has a lower limit-of-detection (LOD) for peptides than for proteins, most protein identifications are made from tryptic peptides. This imposes a limitation as trypsin proteolysis is concentration dependent, that is, there has to be enough protein available in the sample for effective proteolytic digestion to take place.[216] Similar to most other proteases, trypsin has Michaelis-Menten constants, K_m in the low μM range. So the problem of insufficient digestion becomes apparent when trying to digest low nM concentrations of protein. It should be noted that low nM concentrations is to be considered relatively high in the field of proteomics. It is well known that identification of proteins from 2-DE is limited by this, where spots containing <100 fmol of protein are very difficult to identify. This also affects the gel-free proteomics approaches (shotgun) where the proteins are present in largely varying amounts and here the repeatability of the digestion also becomes important.[217] Some methods of solving the problems of insufficient digestion are scaling-up/concentrating the protein sample, the introduction of a more effective way of cleaving protein to peptides (preferably in a specific manner), or performing the digestion in a smaller volume. Scaling-up the sample is the earliest method, but this is not always possible with limited samples. While the proteomics community is waiting for a more effective digestion protocol to be invented, the only remaining solution is to shrink the volumes while increasing protein concentration. By performing the digestion in a concentrated low microliter or better nanoliter volume, the enzyme kinetics becomes more favorable as the enzyme substrate ratio allows operation at favorable K_m.

In the late 1990s, a number of microfluidic structures have been developed at our department (Electrical Measurements, Lund Institute of Technology; Lund University, Sweden) including the porous silicon immobilized enzyme reactor (IMER) and the piezoelectric dispenser. The optimal porosity for IMER and the manufacturing procedure for this porous layer had been determined for glucose monitoring systems in our earlier work. It was anticipated that flow-through digestion of proteins could be facilitated by immobilizing trypsin in the IMER. Initial applications of piezoelectric dispensing for MALDI-MS sample deposition had been very promising.[140,218] The sensitivity gain obtained by confining the MALDI spots in small area spots had been demonstrated in the literature,[137] and nanovials[138,139] seemed a particularly convenient method of confining the MALDI sample spot. With the aforementioned technologies at hand, it was decided to investigate the possibility of creating a proteomics workstation. This effort resulted in an integrated microanalytical platform for automated identification of proteins. The silicon micromachined analytical tools, (i.e., μ-chip IMER, piezoelectric microdispenser, and high-density nanovial target plates, are the corner stones in the system (Figure 49.9). The μ-chip IMER provided online enzymatic digestion of protein samples within 1–3 min and the microdispenser enabled subsequent online picoliter sample preparation in a high-density format. Interfaced to automated MALDI-TOF MS, this system provided a highly efficient platform that could analyze 100 protein samples in 3.5 h.

The trypsin μ-chip IMER was later improved, with respect to porosity and geometry, to allow for 12 s digestion of β-casein in a study by Bengtsson et al.[219] An excellent review on the subject of immobilized enzyme reactors and their integration to other analytical devices have been written by Massolini et al.[220]

49.3.3 In-Nanovial Chemistry

The integrated microanalytical platform for automated identification of proteins described earlier was found lacking sensitivity when samples containing <500 nM were to be digested. There were also concerns about the possibility of cross-contamination arising from the flow-through

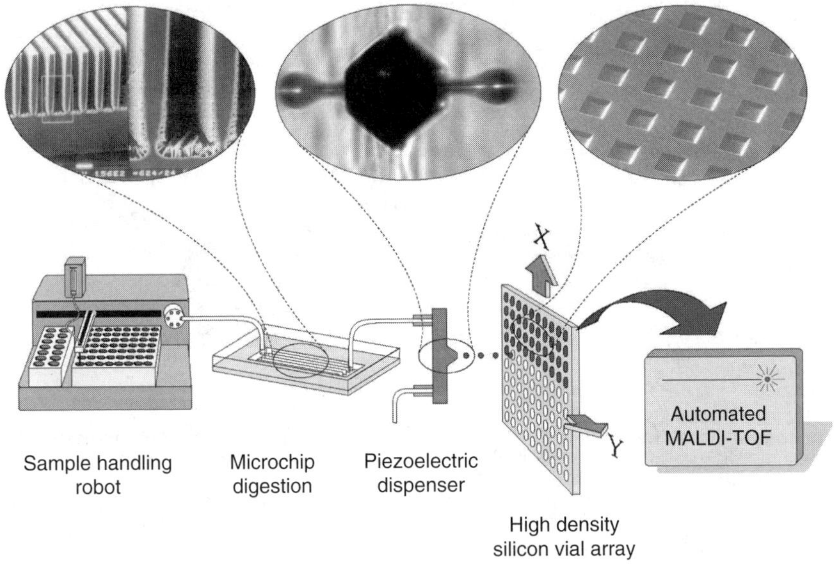

FIGURE 49.9 The integrated analysis system presented in its different parts: automated sample pretreatment and injection; microchip IMER, photo insert shows a SEM picture of the lamella structure with the porous layer; the microdispenser used to deposit sample into microvials; shallow nanovials ($300 \times 300 \times 20$ μm) on the MALDI-target plate; and automated MALDI-TOF MS analysis. (Reproduced with permission from Ekström, S., Önnerfjord, P., Bengtsson, M., Nilsson, J., Laurell, T., Marko-Varga, G., *Anal. Chem.*, 72, 286–293, 2000, copyright 2000 American Chemical Society.)

IMER methodology. This was addressed by performing the digestion directly in the nanovials,[221] thus excluding the IMER for enzymatic digestion. The in-vial digestion protocol is described in Figure 49.10. The in-vial chemistry allowed for a high degree of sample preconcentration, of low volume samples (1.8–8 nL digestion volume) for reduced sample loss; improved Michaelis-Menten kinetics and enhanced sensitivity allowed enzymatic digestion of proteins in concentrations (low nM), which is normally difficult to process. The in-vial digestion technology was later improved by the use of porosified nanovial arrays allowing for improved and faster digestion.[222]

49.3.4 In-Vial Identification of Post-Translational Modifications

Proteomic analysis of PMTs is of paramount interest as these are used to modulate the activity of most eukaryote proteins.[223] The PMTs can serve as molecular switches for enzyme activation, signaling processes, and subcellular localization. There are over 200 different PMTs that can be encountered. Examples of important PMTs are phosphorylation,[224,225] glycosylation,[226–229] and ubiqination.[230] Microfluidic structures can be applied to facilitate identification of these highly important PMTs. By adapting the sample preparation method developed for in-vial digestion, phosphorylated peptides can be identified. This was accomplished by first depositing the sample in two adjacent nanovials and subsequently treated one of the samples with the enzyme alkaline phosphatase; phosphopeptides treated in this manner will show a mass shift of –80 Da. The in-vial chemistry and dispensing allowed detection of phosphopeptides at low fmol amounts (Figure 49.11). Miniaturized phosphopeptide analysis has also been demonstrated using compact disk (CD)-based microfluidics with immobilized metal affinity capture and alkaline phosphatase treatment.[231]

49.3.5 Microfabricated Protein Arrays for MALDI-MS Analysis

Microarray-based proteomics, where antibodies or other molecules that have been arrayed onto a surface to selectively capture proteins, has been presented and covered in a number of reviews.[232–235]

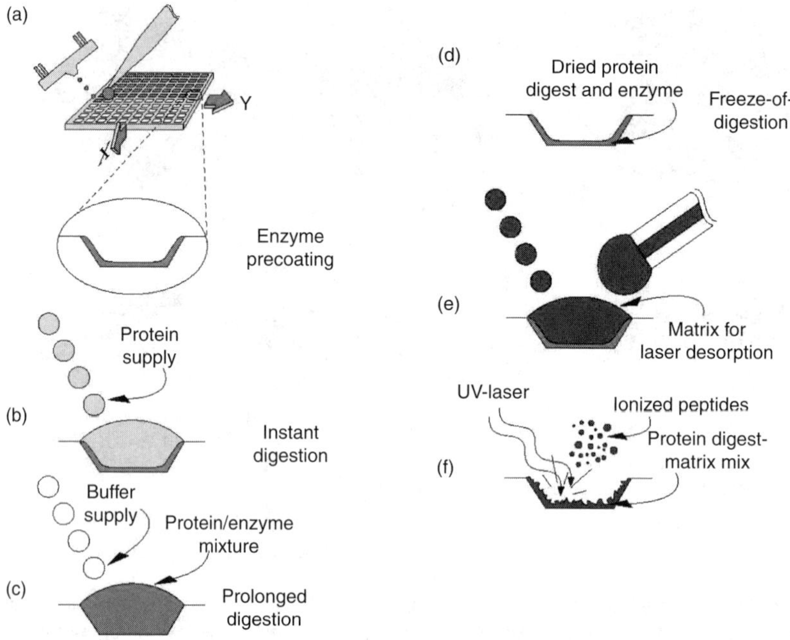

FIGURE 49.10 Step-by-step illustration of the in-vial digestion protocol. (a) Enzyme precoating of the nanovial array by dispenser or pipette. (b) Dispensing of protein sample-instant digestion begins. (c) Dispensing of digestion buffer or water-prolonged digestion. (d) Digestion stops when let to dry-freeze-of-digestion. (e) MALDI-matrix deposition by dispensing (i) (left) or GELoader-tip (ii) capillary-and-pump (iii) (right). (f) MALDI-TOF MS analysis. (Reproduced with permission from Ericsson, D., Ekstrom, S., Nilsson, J., Bergquist, J., Marko-Varga, G., and Laurell, T., *Proteomics*, 1, 1072–1081, 2001, copyright 2001, John Wiley & Sons.)

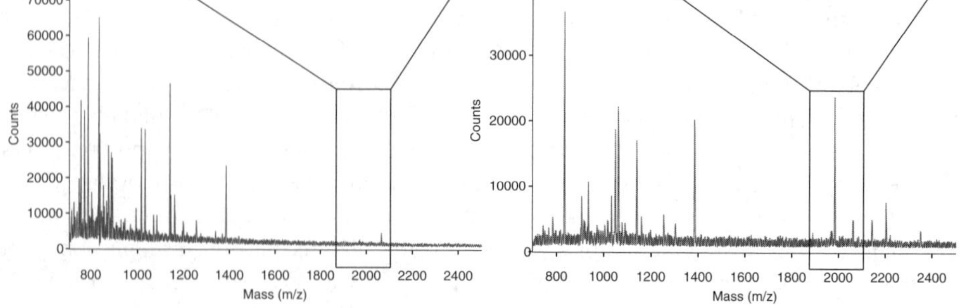

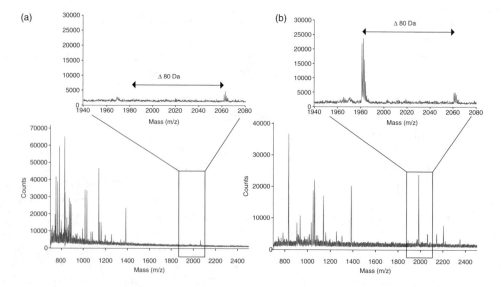

FIGURE 49.11 In-vial dephosphorylation of β-casein. (a) Mass spectra obtained from the untreated β-casein and (b) mass spectra obtained from a nanovial containing <10 fmol β-casein after microdispensing of alkaline phosphatase (0.1 unit/μL) in 50 nM NH_4HCO_3, pH 7, 8 at 6 Hz for 60 s (corresponding to 36 nL alkaline phosphatase). During this treatment the entire target plate was heated to 40°C. The deposited enzyme was then allowed to dry for 5 min followed by microdispensing of 10 nL matrix. The marked peak in (b) shows the dephosphorylated peptide that is generated by the loss of a phosphate group from the marked peak in (a) by the enzymatic treatment. (Reproduced with permission from Ekstrom, S., Nilsson, J., Helldin, G., Laurell, T., and Marko-Varga, G., *Electrophoresis*, 22, 3984–3992, 2001, copyright 2001, John Wiley & Sons.)

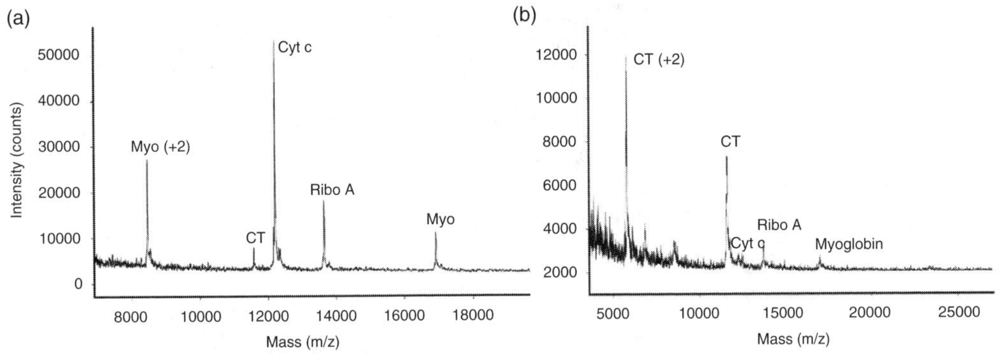

FIGURE 49.12 Mass spectra generated using a micro chip coated with a scFv antibody fragment (CT-17) specific for choleratoxin. The chip was first allowed to react with a protein mixture before being washed. Mass spectroscopy spectra were performed before (a) and after washing (b). (Reproduced with permission from Borrebaeck, C. A. K., Ekstrom, S., Hager, A. C. M., Nilsson, J., Laurell, T., and Marko-Varga, G., *Biotechniques*, 30, 1126, 2001, copyright 2001, Eaton Publishing.)

The MALDI-MS "spot-on-a-chip" sample preparation platform has been applied to protein array experiments, by microdispensing of a recombinant single-chain variable fragment (scFv) antibody specific for choleratoxin onto a nitrocellulose-coated nanovial array chip[236] (Figure 49.12). This concept has later been considerably expanded by use of porous silicon as an arraying substrate by Ressine et al.[237–239] The implementation of microfluidic array-based antibody assays to MALDI-MS has a huge potential both as a biomarker discovery tool and for in-depth investigation of the biological mechanisms of known antigens.

49.3.6 Application of Solid-Phase Extraction Sample Preparation before MALDI-MS

To make any MALDI microspot technique (e.g., the "spot-on-a-chip" method) to work optimal, the sample has to be devoid of contaminants. Purification and enrichment of proteins and peptides is conveniently performed by SPE. Previously, standard commercial SPE options for miniaturized SPE[240] or tips packed with beads[151,152] were used for purification and enrichment before the "spot-on-a-chip" sample preparation. These methods lacked the desired sensitivity, reproducibility and integration possibilities for large scale proteome analysis. To enable integrated sensitive, high-throughput SPE, a silicon microfabricated extraction chip (SMEC) was developed. The manufacturing and investigation of this silicon microextraction chip for sample cleanup and trace enrichment of peptides was investigated using both standard samples and samples derived from 2-DE separations.[164] The SMEC structure used a weir structure to trap reversed-phase chromatography media (Poros™ R2 beads) that facilitated a flow-injection-based methodology for sample purification/enrichment of contaminated and dilute samples before the MALDI-MS analysis (Figure 49.13).

The SMEC design was also optimized with regard to microfluidic properties by replacing the weir with a grid structure, in another publication (not included in this thesis) by Bergkvist et al.[163] This improved design (Figures 49.14 and 49.15) was evaluated both numerically and experimentally.

The intended use of the SMEC chip was to enable SPE followed, by "spot-on-a-chip" sample preparation and integration with microdispensing. Despite considerable efforts to realize this, it could not be implemented satisfactorily. The problems were inherent in the microfluidic design. The pressure-driven flow-injection-based methodology for SPE (Figure 49.13) was negatively affected by air-bubble formation, due to the high back-pressure across the bead-trapping structure. These air bubbles made integration with the microdispenser awkward, since they were trapped in the microdispenser and provided a compliance that absorbed the pressure pulse generated by

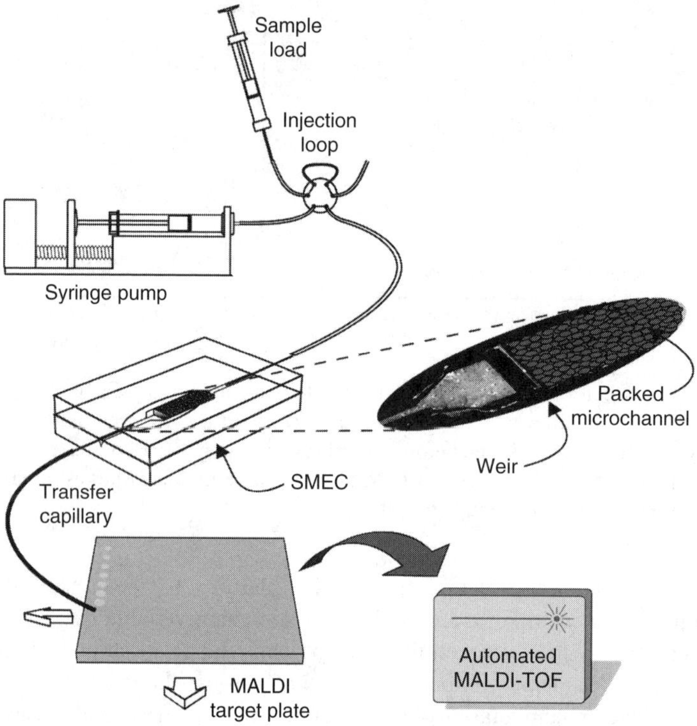

FIGURE 49.13 The instrumental setup used for the SMEC sample cleanup/enrichment. (Reproduced with permission from Ekstrom, S., Malmstrom, J., Wallman, L., Lofgren, M., Nilsson, J., Laurell, T., and Marko-Varga, G., *Proteomics*, 2, 413–421, 2002, copyright 2002, John Wiley & Sons.)

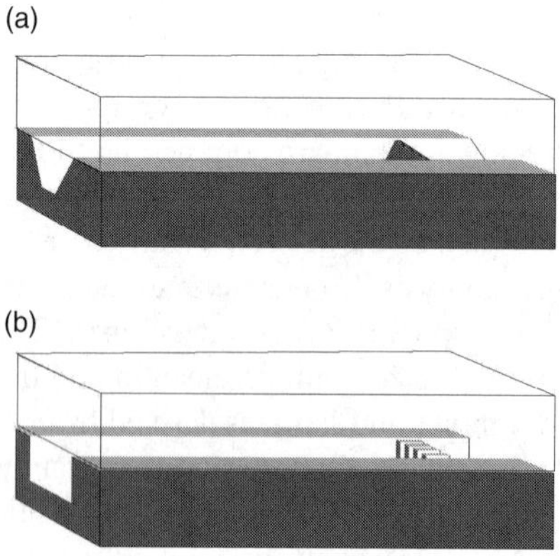

FIGURE 49.14 Schematic drawings of the (a) weir-SMEC and (b) the improved grid-SMEC structures. (Reproduced with permission from Bergkvist, J., Ekstrom, S., Wallman, L., Lofgren, M., Marko-Varga, G., Nilsson, J., and Laurell, T., *Proteomics*, 2, 422–429, 2002, copyright 2002, John Wiley & Sons.)

the piezo-actuation (Figure 49.16a). This could be minimized by degassing all solutions used in the system. A microdispenser with integrated SPE bead trapping was manufactured and tested (Figure 49.16b) but long-term operation was still a problem. Furthermore, aspects of carryover from the solid-phase bed were not readily solved in such a setup and repeated use would inherently alter the binding capacity and flow resistance of the packed bed with time. If the microextraction

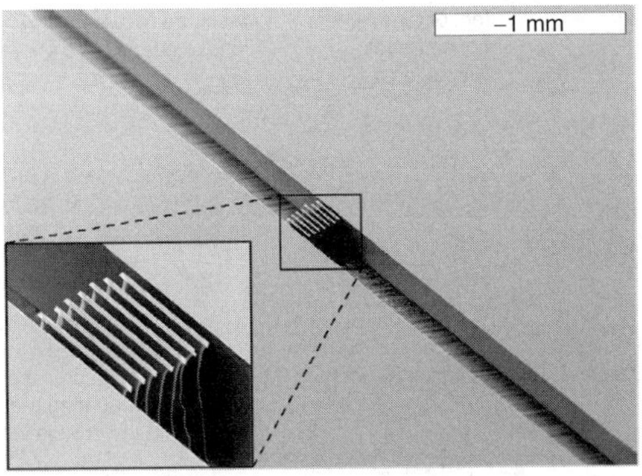

FIGURE 49.15 SEM photograph of the improved grid-SMEC structure. (Reproduced with permission from Bergkvist, J., Ekstrom, S., Wallman, L., Lofgren, M., Marko-Varga, G., Nilsson, J., and Laurell, T., *Proteomics*, 2, 422–429, 2002, copyright 2002, John Wiley & Sons.)

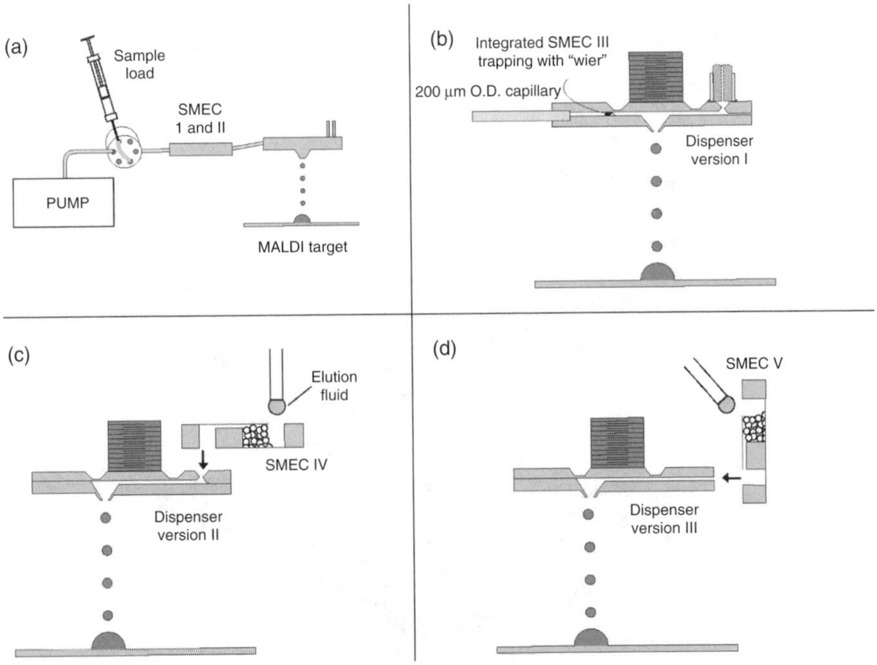

FIGURE 49.16 Schematic pictures of some different designs that were investigated in the study, (a) and (b) used the FIA methodology that later was abandoned. The first generation of the modular capillary filling SPE system (c) having the fluid inlet on the top of the dispenser had problems with capillary filling being erratic and the second generation (d) having the inlet on the side of the dispenser chip, in the central pin.

chip has to be exchanged for each sample in a FIA setup, automation efficiency is severely affected and the fluidic operations are also inherently time consuming.

A more attractive approach would be to enable a parallel mode operation, while avoiding pressure-driven flow. To accomplish this, it was decided to develop a modular capillary force-driven system. This was not a straightforward task and a number of designs (shown in Figure 49.16) had to be manufactured and tested before arriving at the design presented (Figure 49.16d).

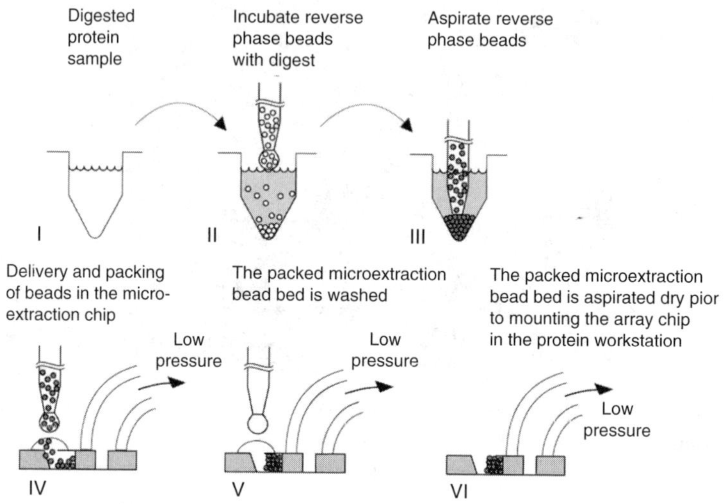

FIGURE 49.17 Process steps in preparing microextraction array. (Reproduced with permission from Wallman, L., Ekstrom, S., Magnusson, M., Bolmsjo, G., Olsson, M., Nilsson, J., Marko-Varga, G., Laurell, T., *Measurement Science & Technology*, 17(12), 3147–3153, 2006, copyright 2006 Iop Publishing.)

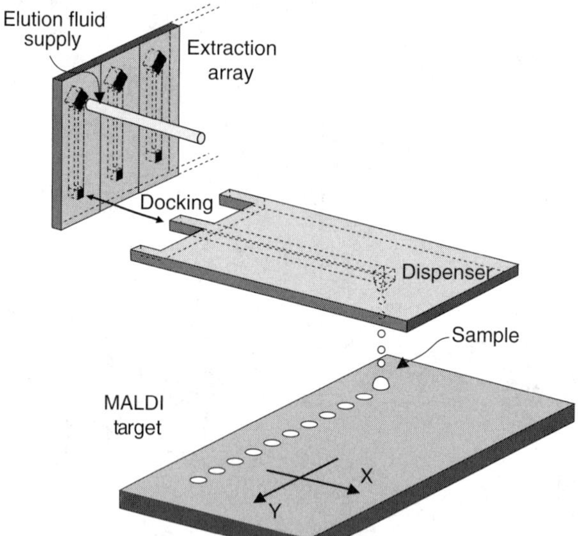

FIGURE 49.18 Operational setup of the final microsystem. The dispenser is docked to one outlet of the extraction array and elution fluid is applied at the SMEC inlet. The eluted analyte is subsequently dispensed onto the MALDI target. The total time for one cycle (one position) was less than 1 min. (Reproduced with permission from Wallman, L., Ekstrom, S., Marko-Varga, G., Laurell, T., and Nilsson, J., *Electrophoresis*, 25, 3778–3787, 2004, copyright 2004, John Wiley & Sons.)

The final design presented shown in Figure 49.16d was a solid-phase microextraction system that performed the sample enrichment and cleanup with the packed microbead bed in an array chip format, offline, and using low pressure to aspirate the sample as well as the washing fluid through the chip (Figure 49.17).

The SPE array chip is then transferred to a robotic station where displacement of the sample from the solid phase is performed by adding a droplet of elution fluid to the extraction chip inlet, which subsequently is drawn by capillary forces through the packed bed to the chip outlet. A newly designed microdispenser with a capillary inlet is docked to the microextraction chip outlet, whereby capillary forces pull the eluted sample automatically into the dispenser. The elution front contains the enriched sample, which thus can be transferred by microspotting onto a MALDI target (Figure 49.18). This system avoided the drawbacks of the flow-injection-based methodology and in an evaluation study, it outperformed two commercially available SPE techniques.[166]

The modular capillary force-driven system resulted in an automated workstation for SPE before MALDI-MS analysis. While miniaturization has the prerequisites for optimizing the SPE process and provides the desired parallelization and sensitivity, it can often lead to quite complicated design solutions that are expensive produce, and requires specialized robotics for sample delivery/transfers since interfacing the microworld to the macroworld is not a trivial task in the laboratory. As an alternative to the modular capillary force-driven SPE system, a less complex solution was developed. The least complex solution would be to integrate the SPE with the MALDI target. Solutions for on-target SPE have been presented in literature and are commercially available. Other options are the microtiterplate-based SPE as presented by Nissum et al.[157] and the ZipPlate. On the basis of this knowledge and our previous efforts, it followed that a miniaturized device that incorporated the vacuum-driven flow-through approach of the microtiterplate SPE devices with a minimum of sample transfers and the final elution as a small analyte spot on the same device would have great potential for SPE before MALDI-MS. The first design of such a device utilized a sieve structure for the bead trapping. These initial experiments showed huge promise, but the eluted MALDI spot became large (>2 mm); elution was difficult to control and beads were not securely trapped. Therefore a new design comprising of a single outlet that retained the beads either due to the beads being larger than the outlet or by the key-stone effect.[241] This device was named Integrated Selective Enrichment (ISET). The ISET serves as both a sample treatment device and MALDI-MS target, and contains an

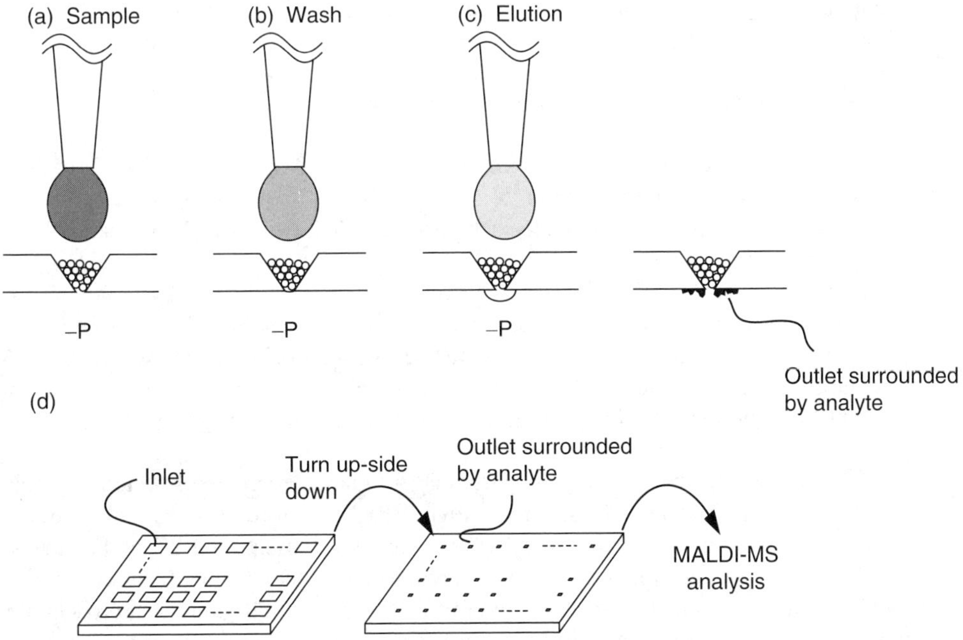

FIGURE 49.19 Workflow when using the ISET, (a) sample, (b) wash, and (c) elution is drawn through the media by a applying a vacuum underneath the ISET. After elution, the analytes forms a spot (0.5–1 mm) on the analysis zone surrounding the outlet. After elution the ISET is turned up side down and the analyte positions are subjected to analysis (d). (Reproduced with permission from Ekstrom, S., Wallman, L., Hok, D., Marko-Varga, G., and Laurell, T., *Journal Proteome Research*, 5, 1071–1081, 2006, copyright 2006, American Chemical Society.)

array of 96 perforated nanovials, which can be filled with approximately 40 nL of reversed-phase beads. The workflow for ISET sample preparation is described in Figure 49.19. Liquid transport is achieved by placing the ISET in a vacuum fixture. Samples can then be loaded either by incubating the beads in the sample solution and subsequently transfer the beads to the perforated nanovials or the sample solutions can be drawn through array positions prefilled with beads. After capture, contaminants are removed by drawing a washing solution through each position; in the case of SPE, the final elution can then be performed with acetonitrile solution including the matrix. MALDI spot sizes of ~500 μm are achieved by control of the elution volume and the applied vacuum. The spot size can be further reduced, with improved sensitivity, by making the analysis zone on the ISET hydrophobic. The ISET sample preparation minimizes the number of sample transfers and the total surface area available for undesired adsorption of the analytes in order to provide high-sensitivity analysis.

Examples of ISET sample preparation protocols used for SPE are given below.

Protocol for Transfer of Analytes while Being Bound to Beads

1. 1 μL of a suspension of Poros R2 beads in 50% ACN/0.1%TFA was added to the sample (acidified with TFA) while stored in an external container (e.g., a microtiterplate). The analytes (peptides) were allowed to bind for at least 30 min.
2. The beads, with the captured analytes, were transferred to the ISET by aspiration of 5 μL sample/beads from the bottom of the sample container and loaded into the perforated nanovials under maximum vacuum (−600 to −800 mbar as measured at the gauge).
3. Wash with 2 × 3 μL, 0.1% TFA, under maximum vacuum, followed by removing the ISET from the vacuum fixture and rinsing the entire backside with Milli-Q water (i.e. ultrapure water) (MQ).
4. Elution of the analytes onto the backside of the ISET with 2 × 0.3 μL, 50% ACN/0.1%TFA containing 1 mg/mL of cyano-4-hydroxy-cinnamic acid and ACTH 18–39/μL (internal calibrant). The elution was done at a lower vacuum (approximately –100 mbar).

Protocol for ISET Sample Preparation with Prefilled Beads

1. Each perforated nanovial was filled with approximately 40 nL Poros R2 beads in 50% ACN/0.1%TFA, at maximum vacuum.
2. 5 μl of acidified sample was applied to each position and drawn through the bead volume by applying maximum vacuum.
3. Wash with 2 × 3 μL, 0.1% TFA, under maximum vacuum, followed by removing the ISET from the vacuum fixture and rinsing the entire backside with MQ.
4. Elution of the analytes onto the backside of the ISET with 2 × 0.3 μL, 50% ACN/0.1%TFA containing 1 mg/ml of cyano-4-hydroxy-cinnamic acid and ACTH 18–39/μL (internal calibrant). The elution was done at a lower vacuum (approximately −100 mbar).

The ISET technology has been successfully applied for characterization of proteins coisolated by affinity chromatography of prostate-specific antigen (PSA) from human seminal fluid[165] and subjected to a more thorough study of the performance, when used for SPE of peptide samples, including a favorable performance comparison with some of the commercially available SPE techniques.[167] The main advantage of the ISET platform is the generic microfluidic approach to sample cleanup and enrichment in a high-throughput format with a very high performance. In addition, the platform is a very cost-effective alternative to existing sample preparation strategies. The ISET platform can be used to address several needs in the field of MS, based on life science research and circumvents many of the problems associated with miniaturized SPE, as the uncomplicated design minimizes sample transfers and makes custom robotics unnecessary.

49.4 CONCLUSIONS

When introduced the word proteomics was used to describe the ambitious goal of cataloging the proteins in different organisms. Today there are many more meanings and the field of proteomics has been differentiated considerably, with subdivisions such for, structural proteomics, clinical proteomics, and so forth.

In addition to the proteomics techniques described here, which by no account can be claimed to be representing all the flavors of proteomics, there are numerous others. Examples of novel technologies that are highly interesting but not mentioned earlier is protein–protein interaction analysis by the yeast two-hybrid system,[242,243] and tandem affinity purification (TAP).[244] Currently strong emerging field of proteomics is clinical proteomics and biomarker analysis (*Journal of Proteome Research*, Special Biomarker Issue, Vol. 4: August 2005), where the goal is to discovery proteins with medical relevance and their pathophysiological implications. An interesting MALDI-MS technique for biomarker discovery is the tissue imaging methodology developed by Caprioli and coworkers.[245]

Despite the fact that mRNA and protein expression have been shown to differ from predicted values,[246] it is also foreseeable that the information about the proteome obtainable from the transcriptome, that is, by looking at the mRNA present via cDNA microarray analysis, will be expanded as systems biology matures.[247]

The shear amount of methodologies in the field of proteomics can be grasped by the fact that a topic search using the search words such as "proteomics or proteome or proteomic" in the Web of Science® database returns 12081 publications (*search done 2005-10-27*). The huge amount of publications does by no means imply that there is no room for improvement. The application of microfluidics to proteomics analysis is still in its infancy and as the detection technologies becomes even more sensitive the ability to handle minute sample amounts will become paramount for continued progress.

The development of new technology and methodology in both proteomics and microfluidics, as well as an increased merging of the two fields, guarantees that the future holds bright promise for new routes. This will lead to improved bioanalytical performance and pave the way to better understanding of biological processes and disease pathogenesis.

REFERENCES

1. Terry, S. C. and Jerman, J. H., A Miniature Gas-chromatograph Utilized in a Portable Gas-analysis System, *Abstracts of Papers of the American Chemical Society* 185 (MAR), 123-Anyl, 1983.
2. Harrison, D. J., Fluri, K., Seiler, K., Fan, Z. H., Effenhauser, C. S., and Manz, A., Micromachining a miniaturized capillary electrophoresis-based chemical-analysis system on a chip, *Science*, 261, 895–897, 1993.
3. Lion, N., Rohner, T. C., Dayon, L., Arnaud, I. L., Damoc, E., Youhnovski, N., Wu, Z. Y., Roussel, C., Josserand, J., Jensen, H., Rossier, J. S., Przybylski, M., and Girault, H. H., Microfluidic systems in proteomics, *Electrophoresis*, 24, 3533–3562, 2003.
4. Lion, N., Reymond, F., Girault, H. H., and Rossier, J. S., Why the move to microfluidics for protein analysis?, *Current Opinion in Biotechnology*, 15, 31–37, 2004.
5. Erickson, D. and Li, D. Q., Integrated microfluidic devices, *Analytica Chimica Acta*, 507, 11–26, 2004.
6. Lagally, E. T. and Mathies, R. A., Integrated genetic analysis microsystems, *Journal of Physics D–Applied Physics*, 37, R245–R261, 2004.
7. Vilkner, T., Janasek, D., and Manz, A., Micro total analysis systems. Recent developments, *Analytical Chemistry*, 76, 3373–3385, 2004.
8. Lee, S. J. and Lee, S. Y., Micro total analysis system (micro-TAS) in biotechnology, *Applied Microbiology and Biotechnology*, 64, 289–299, 2004.
9. Laurell, T., Nilsson, J., and Marko-Varga, G., The quest for high-speed and low-volume bioanalysis, *Analytical Chemistry*, 77, 264a–272a, 2005.
10. Marko-Varga, G. A., Nilsson, J., and Laurell, T., New directions of miniaturization within the biomarker research area, *Electrophoresis*, 25, 3479–3491, 2004.

11. Marko-Varga, G., Nilsson, J., and Laurell, T., New directions of miniaturization within the proteomics research area, *Electrophoresis*, 24, 3521–3532, 2003.
12. Auroux, P. A., Koc, Y., deMello, A., Manz, A., and Day, P. J. R., Miniaturised nucleic acid analysis, *Lab on a Chip*, 4, 534–546, 2004.
13. Paegel, B. M., Blazej, R. G., and Mathies, R. A., Microfluidic devices for DNA sequencing: Sample preparation and electrophoretic analysis, *Current Opinion in Biotechnology*, 14, 42–50, 2003.
14. Kelly, R. T. and Woolley, A. T., Microfluidic systems for integrated, high-throughput DNA analysis, *Analytical Chemistry*, 77, 96a–102a, 2005.
15. Carey, L. and Mitnik, L., Trends in DNA forensic analysis, *Electrophoresis*, 23, 1386–1397, 2002.
16. Saito, Y., Nakao, Y., Imaizumi, M., Morishima, Y., Kiso, Y., and Jinno, K., Miniaturized solid-phase extraction as a sample preparation technique for the determination of phthalates in water, *Analytical and Bioanalytical Chemistry*, 373, 81–86, 2002.
17. Kiermer, V., FACS-on-a-chip, *Nature Methods*, 2, 91, 2005.
18. Johann, R. and Renaud, P., A simple mechanism for reliable particle sorting in a microdevice with combined electroosmotic and pressure-driven flow, *Electrophoresis*, 25, 3720–3729, 2004.
19. Wang, M. M., Tu, E., Raymond, D. E., Yang, J. M., Zhang, H. C., Hagen, N., Dees, B., Mercer, E. M., Forster, A. H., Kariv, I., Marchand, P. J., and Butler, W. F., Microfluidic sorting of mammalian cells by optical force switching, *Nature Biotechnology*, 23, 83–87, 2005.
20. Yao, B., Luo, G. A., Feng, X., Wang, W., Chen, L. X., and Wang, Y. M., A microfluidic device based on gravity and electric force driving for flow cytometry and fluorescence activated cell sorting, *Lab on a Chip*, 4, 603–607, 2004.
21. Andersson, H. and van den Berg, A., Microfluidic devices for cellomics: A review, *Sensors and Actuators B–Chemical*, 92, 315–325, 2003.
22. Schulte, T. H., Bardell, R. L., and Weigl, B. H., Microfluidic technologies in clinical diagnostics, *Clinica Chimica Acta*, 321, 1–10, 2002.
23. Liu, Y., Garcia, C. D., and Henry, C. S., Recent progress in the development of mu TAS for clinical analysis, *Analyst*, 128, 1002–1008, 2003.
24. Verpoorte, E., Microfluidic chips for clinical and forensic analysis, *Electrophoresis*, 23, 677–712, 2002.
25. Watts, P. and Haswell, S. J., Microfluidic combinatorial chemistry, *Current Opinion in Chemical Biology*, 7, 380–387, 2003.
26. Krenkova, J. and Foret, F., Immobilized microfluidic enzymatic reactors, *Electrophoresis*, 25, 3550–3563, 2004.
27. Fletcher, P. D. I., Haswell, S. J., Pombo-Villar, E., Warrington, B. H., Watts, P., Wong, S. Y. F., and Zhang, X. L., Micro reactors: Principles and applications in organic synthesis, *Tetrahedron*, 58, 4735–4757, 2002.
28. Yoshida, J., Suga, S., and Nagaki, A., Selective organic reactions using microreactors, *Journal of Synthetic Organic Chemistry Japan*, 63, 511–522, 2005.
29. Hadd, A. G., Jacobson, S. C., and Ramsey, J. M., Microfluidic assays of acetylcholinesterase inhibitors, *Analytical Chemistry*, 71, 5206–5212, 1999.
30. Moser, I., Jobst, G., Svasek, P., Varahram, M., and Urban, G., Rapid liver enzyme assay with miniaturized liquid handling system comprising thin film biosensor array, *Sensors and Actuators B–Chemical*, 44, 377–380, 1997.
31. Wang, J., On-chip enzymatic assays, *Electrophoresis*, 23, 713–718, 2002.
32. Gao, P., Le Person, S., and Favre-Marinet, M., Scale effects on hydrodynamics and heat transfer in two-dimensional mini and microchannels, *International Journal of Thermal Sciences*, 41, 1017–1027, 2002.
33. Jones, T. B., Wang, K. L., and Yao, D. J., Frequency-dependent electromechanics of aqueous liquids: Electrowetting and dielectrophoresis, *Langmuir*, 20, 2813–2818, 2004.
34. Kamholz, A. E. and Yager, P., Molecular diffusive scaling laws in pressure-driven microfluidic channels: Deviation from one-dimensional Einstein approximations, *Sensors and Actuators B–Chemical*, 82, 117–121, 2002.
35. Anderson, N. G. and Anderson, L., The human protein index, *Clinical Chemistry*, 28 (Pt 2), 739–748, 1982.

36. Wilkins, M. R., Sanchez, J. C., Gooley, A. A., Appel, R. D., Humphery Smith, I., Hochstrasser, D. F., and Williams, K. L., Progress with proteome projects: Why all proteins expressed by a genome should be identified and how to do it, *Biotechnology and Genetic Engineering Reviews*, 13, 19–50, 1996.
37. Black, D. L., Mechanisms of alternative pre-messenger RNA splicing, *Annual Review Biochemistry*, 72, 291–336, 2003.
38. Schaub, M. and Keller, W., RNA editing by adenosine deaminases generates RNA and protein diversity, *Biochimie*, 84, 791–803, 2002.
39. Ullrich, B., Ushkaryov, Y. A., and Sudhof, T. C., Cartography of neurexins: More than 1000 isoforms generated by alternative splicing and expressed in distinct subsets of neurons, *Neuron*, 14, 497–507, 1995.
40. Anderson, N. L. and Anderson, N. G., The human plasma proteome—History, character, and diagnostic prospects, *Molecular and Cellular Proteomics*, 1, 845–867, 2002.
41. Wang, H. and Hanash, S., Intact-protein based sample preparation strategies for proteome analysis in combination with mass spectrometry, *Mass Spectrometry Review*, 24, 413–426, 2005.
42. Lambert, J. P., Ethier, M., Smith, J. C., and Figeys, D., Proteomics: From gel based to gel free, *Analytical Chemistry*, 77, 3771–3787, 2005.
43. Patterson, S. D., Proteomics: Evolution of the technology, *Biotechniques*, 35, 440–444, 2003.
44. Patterson, S. D. and Aebersold, R. H., Proteomics: The first decade and beyond, *Nature Genetics*, 33, 311–323, 2003.
45. Dittrich, P. S. and Manz, A., Single-molecule fluorescence detection in microfluidic channels—The Holy Grail in μTAS?, *Anal Bioanalytical Chemistry*, 382, 1771–1782, 2005.
46. Koo, T. W., Chan, S., and Berlin, A. A., Single-molecule detection of biomolecules by surface-enhanced coherent anti-Stokes Raman scattering, *Optics Letters*, 30, 1024–1026, 2005.
47. Karhanek, M., Kemp, J. T., Pourmand, N., Davis, R. W., and Webb, C. D., Single DNA molecule detection using nanopipettes and nanoparticles, *Nano Letters*, 5, 403–407, 2005.
48. Koopman, M., Cambi, A., de Bakker, B. I., Joosten, B., Figdor, C. G., van Hulst, N. F., and Garcia-Parajo, M. F., Near-field scanning optical microscopy in liquid for high resolution single molecule detection on dendritic cells, *FEBS Letters*, 573, 6–10, 2004.
49. Twerenbold, D., Gerber, D., Gritti, D., Gonin, Y., Netuschill, A., Rossel, F., Schenker, D., and Vuilleumier, J. L., Single molecule detector for mass spectrometry with mass independent detection efficiency, *Proteomics*, 1, 66–69, 2001.
50. Fredriksson, S., Gullberg, M., Jarvius, J., Olsson, C., Pietras, K., Gustafsdottir, S. M., Ostman, A., and Landegren, U., Protein detection using proximity-dependent DNA ligation assays, *Nature Biotechnology*, 20, 473–477, 2002.
51. Lichtenberg, J., de Rooij, N. F., and Verpoorte, E., Sample pretreatment on microfabricated devices, *Talanta*, 56, 233–266, 2002.
52. Wainright, A., Williams, S. J., Ciambrone, G., Xue, Q., Wei, J., and Harris, D., Sample pre-concentration by isotachophoresis in microfluidic devices, *Journal of Chromatography A*, 979, 69–80, 2002.
53. Kautz, R. A., Lacey, M. E., Wolters, A. M., Foret, F., Webb, A. G., Karger, B. L., and Sweedler, J. V., Sample concentration and separation for nanoliter-volume NMR spectroscopy using capillary isotachophoresis, *Journal of American Chemical Society*, 123, 3159–3160, 2001.
54. Jung, B., Bharadwaj, R., and Santiago, J. G., Thousandfold signal increase using field-amplified sample stacking for on-chip electrophoresis, *Electrophoresis*, 24, 3476–3483, 2003.
55. Astorga-Wells, J., Jornvall, H., and Bergman, T., A microfluidic electrocapture device in sample preparation for protein analysis by MALDI mass spectrometry, *Analytical Chemistry*, 75, 5213–5219, 2003.
56. Astorga-Wells, J., Bergman, T., and Jornvall, H., Multistep microreactions with proteins using electrocapture technology, *Analytical Chemistry*, 76, 2425–2429, 2004.
57. Oleschuk, R. D., Shultz-Lockyear, L. L., Ning, Y. B., and Harrison, D. J., Trapping of bead-based reagents within microfluidic systems: On-chip solid-phase extraction and electrochromatography, *Analytical Chemistry*, 72, 585–590, 2000.

58. Yu, C., Davey, M. H., Svec, F., and Frechet, J. M. J., Monolithic porous polymer for on-chip solid-phase extraction and preconcentration prepared by photoinitiated *in situ* polymerization within a microfluidic device, *Analytical Chemistry*, 73, 5088–5096, 2001.
59. Ramsey, J. D. and Collins, G. E., Integrated microfluidic device for solid-phase extraction coupled to micellar electrokinetic chromatography separation, *Analytical Chemistry*, 77, 6664–6670, 2005.
60. Rohde, E., Tomlinson, A. J., Johnson, D. H., and Naylor, S., Protein analysis by membrane preconcentration capillary electrophoresis: Systematic evaluation of parameters affecting preconcentration and separation, *Journal of Chromatography B–Analytical Technologies in the Biomedical and Life Sciences*, 713, 301–311, 1998.
61. Khandurina, J., Jacobson, S. C., Waters, L. C., Foote, R. S., and Ramsey, J. M., Microfabricated porous membrane structure for sample concentration and electrophoretic analysis, *Analytical Chemistry*, 71, 1815–1819, 1999.
62. Foote, R. S., Khandurina, J., Jacobson, S. C., and Ramsey, J. M., Preconcentration of proteins on microfluidic devices using porous silica membranes, *Analytical Chemistry*, 77, 57–63, 2005.
63. Song, S., Singh, A. K., and Kirby, B. J., Electrophoretic concentration of proteins at laser-patterned nanoporous membranes in microchips, *Analytical Chemistry*, 76, 4589–4592, 2004.
64. Wang, Y. C., Stevens, A. L., and Han, J. Y., Million-fold preconcentration of proteins and peptides by nanofluidic filter, *Analytical Chemistry*, 77, 4293–4299, 2005.
65. Weigl, B. H. and Yager, P., Microfluidics—Microfluidic diffusion-based separation and detection, *Science*, 283, 346–347, 1999.
66. Chen, H., Fang, Q., Yin, X. F., and Fang, Z. L., Microfluidic chip-based liquid–liquid extraction and preconcentration using a subnanoliter-droplet trapping technique, *Lab on a Chip*, 5, 719–725, 2005.
67. Fang, Q., Chen, H., and Cai, Z. X., Stopped-flow liquid–liquid extraction on microfluidic chips, *Chemical Journal of Chinese Universities–Chinese*, 25, 261–263, 2004.
68. Xiang, F., Lin, Y. H., Wen, J., Matson, D. W., and Smith, R. D., An integrated microfabricated device for dual microdialysis and on-line ESI ion trap mass spectrometry for analysis of complex biological samples, *Analytical Chemistry*, 71, 1485–1490, 1999.
69. Thomas, D. H., Rakestraw, D. J., Schoeniger, J. S., Lopez-Avila, V., and Van Emon, J., Selective trace enrichment by immunoaffinity capillary electrochromatography on-line with capillary zone electrophoresis—Laser-induced fluorescence, *Electrophoresis*, 20, 57–66, 1999.
70. Guzman, N. A., Improved solid-phase microextraction device for use in on-line immunoaffinity capillary electrophoresis, *Electrophoresis*, 24, 3718–3727, 2003.
71. Guzman, N. A. and Phillips, T. M., Immunoaffinity CE for proteomics studies, *Analytical Chemistry*, 77, 60a–67a, 2005.
72. Whitehouse, C. M., Dreyer, R. N., Yamashita, M., and Fenn, J. B., Electrospray interface for liquid chromatographs and mass spectrometers, *Analytical Chemistry*, 57, 675–679, 1985.
73. Fenn, J. B., Mann, M., Meng, C. K., Wong, S. F., and Whitehouse, C. M., Electrospray ionization for mass spectrometry of large biomolecules, *Science*, 246, 64–71, 1989.
74. Karas, M. and Hillenkamp, F., Laser desorption ionization of proteins with molecular masses exceeding 10,000 Daltons, *Analytical Chemistry*, 60, 2299–2301, 1988.
75. Tanaka, K., Waki, H., Ido, Y., Akita, S., Yoshida, Y., and Yoshida, T., Protein and polymer analyses up to m/z 100,000 by laser ionization time-of-flight mass spectrometry, *Rapid Communication Mass Spectrometry*, 2, 151–153, 1988.
76. Henzel, W. J., Watanabe, C., and Stults, J. T., Protein identification: The origins of peptide mass fingerprinting, *Journal of the American Society for Mass Spectrometry*, 14, 931–942, 2003.
77. James, P., Quadroni, M., Carafoli, E., and Gonnet, G., Protein identification by mass profile fingerprinting, *Biochemical Biophysical Research Communication*, 195, 58–64, 1993.
78. Mann, M., Hojrup, P., and Roepstroff, P., Use of mass spectrometric molecular information to identify proteins in sequence databases, *Biological Mass Spectrometry*, 22, 338–345, 1993.
79. Hezel, W. J., Billeci, T. M., Stults, J. T., Wong, S. C., Grimely, C., and Watanabe, C., Identifying proteins from two-dimensional gels by molecular mass searching of peptide fragments in protein sequence databases, *Proceedings of the National Academy Science USA*, 90, 5011–5015, 1993.
80. Pappin, D. J. C., Hojrup, P., and Bleasby, A. J., Rapid identification of proteins by peptide-mass fingerprinting, *Current Biology*, 3, 327–332, 1993.

81. Yates, J., Speicher, S., Griffin, P. R., and Hunkapiller, T., Peptide mass maps: A highly informative approach to protein identification, *Analytical Biochemistry*, 214, 397–408, 1993.
82. Roepstorff, P., MALDI-TOF mass spectrometry in protein chemistry, *Exs*, 88, 81–97, 2000.
83. Mann, M. and Wilm, M., Error-tolerant identification of peptides in sequence databases by peptide sequence tags, *Analytical Chemistry*, 66, 4390–4399, 1994.
84. Eng, J. K., Mccormack, A. L., and Yates, J. R., An approach to correlate tandem mass-spectral data of peptides with amino-acid-sequences in a protein database, *Journal of the American Society for Mass Spectrometry*, 5, 976–989, 1994.
85. Kaufmann, R., Spengler, B., and Lutzenkirchen, F., Mass spectrometric sequencing of linear peptides by product-ion analysis in a reflectron time-of-flight mass spectrometer using matrix-assisted laser desorption ionization, *Rapid Communication Mass Spectrometry*, 7, 902–910, 1993.
86. Noble, D., MALDI-TOF pulses ahead, *Analytical Chemistry*, 67, 497A–501A, 1995.
87. Chernushevich, I. V., Loboda, A. V., and Thomson, B. A., An introduction to quadrupole-time-of-flight mass spectrometry, *Journal of Mass Spectrometry*, 36, 849–865, 2001.
88. Douglas, D. J., Frank, A. J., and Mao, D., Linear ion traps in mass spectrometry, *Mass Spectrometry Review*, 24, 1–29, 2005.
89. Hardman, M. and Makarov, A. A., Interfacing the orbitrap mass analyzer to an electrospray ion source, *Analytical Chemistry*, 75, 1699–1705, 2003.
90. Hu, Q. Z., Noll, R. J., Li, H. Y., Makarov, A., Hardman, M., and Cooks, R. G., The orbitrap: A new mass spectrometer, *Journal of Mass Spectrometry*, 40, 430–443, 2005.
91. Tempez, A., Ugarov, M., Egan, T., Schultz, J. A., Novikov, A., Della-Negra, S., Lebeyec, Y., Pautrat, M., Caroff, M., Smentkowski, V. S., Wang, H. Y. J., Jackson, S. N., and Woods, A. S., Matrix implanted laser desorption ionization (MILDI) combined with ion mobility-mass spectrometry for bio-surface analysis, *Journal of Proteome Research*, 4, 540–545, 2005.
92. Tang, K., Li, F., Shvartsburg, A. A., Strittmatter, E. F., and Smith, R. D., Two-dimensional gas-phase separations coupled to mass spectrometry for analysis of complex mixtures, *Analytical Chemistry*, 77, 6381–6388, 2005.
93. Medzihradszky, K. F., Campbell, J. M., Baldwin, M. A., Falick, A. M., Juhasz, P., Vestal, M. L., and Burlingame, A. L., The characteristics of peptide collision-induced dissociation using a high-performance MALDI-TOF/TOF tandem mass spectrometer, *Analytical Chemistry*, 72, 552–558, 2000.
94. Marshall, A. G., Hendrickson, C. L., and Jackson, G. S., Fourier transform ion cyclotron resonance mass spectrometry: A primer, *Mass Spectrometry Review*, 17, 1–35, 1998.
95. Bogdanov, B. and Smith, R. D., Proteomics by FTICR mass spectrometry: Top down and bottom up, *Mass Spectrometry Reviews*, 24, 168–200, 2005.
96. Karas, M., Bachmann, D., and Hillenkamp, F., Influence of the wavelength in high-irradiance ultraviolet-laser desorption mass-spectrometry of organic-molecules, *Analytical Chemistry*, 57, 2935–2939, 1985.
97. Karas, M., Bachmann, D., Bahr, U., and Hillenkamp, F., Matrix-assisted laser desorption of non-volatile compounds, *International Journal of Mass Spectrometry Ion Processes*, 78, 53–68, 1987.
98. Kussmann, M. and Roepstorff, P., Sample preparation techniques for peptides and proteins analyzed by MALDI-MS, *Methods in Molecular Biology*, 146, 405–424, 2000.
99. Vorm, O., Roepstorff, P., and Mann, M., Improved resolution and very high sensitivity in MALDI-TOF of matrix surfaces made by fast evaporation, *Analytical Chemistry*, 66, 3281–3287, 1994.
100. Beavis, R. C. and Bridson, J. N., Epitaxial protein inclusion in sinapinic acid crystals, *Journal of Physics D: Applied Physics*, 26, 442–447, 1993.
101. Gusev, A. I., Wilkinson, W. R., Proctor, A., and Hercules, D. M., Improvement of signal reproducibility and matrix/comatrix effects in MALDI analysis, *Analytical Chemistry*, 67, 1034–1041, 1995.
102. Xiang, F. and Beavis, R. C., A method to increase contaminant tolerance in protein matrix-assisted laser desorption/ionization by the fabrication of thin protein-doped polycrystalline films, *Rapid Communication Mass Spectrometry*, 8, 199–204, 1994.
103. Perera, I. K., Perkins, J., and Kantartzoglou, S., Spin-coated samples for high resolution matrix-assisted laser desorption/ionization time-of-flight mass spectrometry of large proteins, *Rapid Communication Mass Spectrometry*, 9, 180–187, 1995.

104. Westman, A., Demirev, P., Huth-Fere, T., Bielawski, J., and Sundqvist, B. U. R., Sample exposure effects in matrix-assisted laser desorption-ionization mass spectrometry of large biomolecules, *International Journal of Mass Spectrometry Ion Proceedings*, 130, 107–115, 1994.
105. Castro, J. A., Köster, C., and Wilkins, C., Matrix-assisted laser desorption/ionization of high-mass molecules by Fourier-transform mass spectrometry, *Rapid Communication Mass Spectrometry*, 6, 239–241, 1992.
106. Axelsson, J., Hoberg, A.-M., Waterson, C., Myatt, P., Shield, G. L., Varney, J., Haddleton, D. M., and Derrick, P. J., Improved reproducibility and increased signal intensity in matrix-assisted laser desorption/ionization as a result of electrospray sample preparation, *Rapid Communication Mass Spectrometry*, 11, 209–213, 1997.
107. Allwood, D. A., Perera, I. K., Perkins, J., Dyer, P. E., and Oldershaw, G. A., Preparation of "near" homogeneous samples for the analysis, *Applied Surface Science*, 103, 231–244, 1996.
108. Matsuoka, M. and Eguchi, N., Growth of *m*-chloronitrobenzene crystals in the prescence of microcrystals, *Journal of Physics D: Applied Physics*, 26, B162–B167, 1993.
109. Dai, Y., Whittal, R. M., and Li, L., Confocal fluorescence microscopic imaging for investigating the analyte distribution in MALDI matrixes, *Analytical Chemistry*, 68, 2494–2500, 1996.
110. Önnerfjord, P., Ekström, S., Bergquist, J., Nilsson, J., Laurell, T., and Marko-Varga, G., Homogeneous sample perparation for automated high throughput analysis with MALDI-TOF MS, *Rapid Communication Mass Spectrometry*, 13, 315–322, 1999.
111. Shevchenko, A., Wilm, M., Vorm, O., and Mann, M., Mass spectrometric sequencing of proteins from silver-stained polyacrylamide gels, *Analytical Chemistry*, 68, 850–858, 1996.
112. Miliotis, T., Kjellstrom, S., Nilsson, J., Laurell, T., Edholm, L. E., and Marko-Varga, G., Capillary liquid chromatography interfaced to matrix-assisted laser desorption/ionization time-of-flight mass spectrometry using an on-line coupled piezoelectric flow-through microdispenser, *Journal of Mass Spectrometry*, 35, 369–377, 2000.
113. Miliotis, T., Kjellstrom, S., Onnerfjord, P., Nilsson, J., Laurell, T., Edholm, L. E., and Marko-Varga, G., Protein identification platform utilizing micro dispensing technology interfaced to matrix-assisted laser desorption ionization time-of-flight mass spectrometry, *Journal of Chromatography A*, 886, 99–110, 2000.
114. Thomas, J. J., Shen, Z. X., Crowell, J. E., Finn, M. G., and Siuzdak, G., Desorption/ionization on silicon (DIOS): A diverse mass spectrometry platform for protein characterization, *Proceedings of the National Academy of Sciences of the USA*, 98, 4932–4937, 2001.
115. Pan, C. S., Xu, S. Y., Zou, H. F., Guo, Z., Zhang, Y., and Guo, B. C., Carbon nanotubes as adsorbent of solid-phase extraction and matrix for laser desorption/ionization mass spectrometry, *Journal of the American Society for Mass Spectrometry*, 16, 263–270, 2005.
116. Zabet-Moghaddam, M., Heinzle, E., Lasaosa, M., and Tholey, A., Pyridinium-based ionic liquid matrices can improve the identification of proteins by peptide mass-fingerprint analysis with matrix-assisted laser desorption/ionization mass spectrometry, *Analytical and Bioanalytical Chemistry*, 384, 215–224, 2006.
117. Armstrong, D. W., Zhang, L. K., He, L., and Gross, M. L., Ionic liquids as matrixes for matrix-assisted laser desorption/ionization mass spectrometry, *Analytical Chemistry*, 73, 3679–3686, 2001.
118. Petersen, K. E., Fabrication of an integrated, planar silicon ink-jet structure, *IEEE Transaction on Electron Devices*, ED-26, 1918, 1979.
119. Wallace, D. B., Ink-jet based fluid microdispensing in biochemical applications, *Lab Automation News*, 1, 6–9, 1996.
120. Lemmo, A. V., Rose, D. J., and Tisone, T. C., Inkjet dispensing technology: Applications in drug discovery, *Current Opinion Biotechnology*, 9, 615–617, 1998.
121. Howard, E. I. and Cachau, R. E., Ink-jet printer heads for ultra-small-drop protein crystallography, *Biotechniques*, 33, 1302–1306, 2002.
122. Roda, A., Guardigli, M., Russo, C., Pasini, P., and Baraldini, M., Protein microdeposition using a conventional ink-jet printer, *Biotechniques*, 28, 492–496, 2000.
123. Limpanuphap, S. and Derby, B., Manufacture of biomaterials by a novel printing process, *Journal of Material Science Materials in Medicine*, 13, 1163–1166, 2002.

124. Xu, T., Petridou, S., Lee, E. H., Roth, E. A., Vyavahare, N. R., Hickman, J. J., and Boland, T., Construction of high-density bacterial colony arrays and patterns by the ink-jet method, *Biotechnology Bioengineering*, 85, 29–33, 2004.
125. Schober, A., Günther, R., Schwienhorst, A., Döring, M., and Lindemann, B. F., Accurate high-speed liquid handling of very small biological samples, *Biotechniques*, 15, 324–329, 1993.
126. Sziele, D., Brüggeman, O., Döring, M., Freitag, R., and Schügerl, K., Adaption of a microdrop injector to sampling in capillary electrophoresis, *Journal of Chromatography A*, 669, 254–258, 1994.
127. Nilsson, S., Lager, C., Laurell, T., and Birnbaum, S., Thin-layer immunoaffinity chromatography with bar code quantitation of C-reactive protein, *Analytical Chemistry*, 67, 3051–3056, 1995.
128. Allmaier, G., Picoliter to nanoliter deposition of peptide and protein solutions for matrix-assisted laser desorption/ionization mass spectrometry, *Rapid Communication Mass Spectrometry*, 11, 1567–1569, 1997.
129. Little, D. P., Cornish, T. J., and O'Donnell, M. J., MALDI on a chip: Analysis of arrays of low-femtomole to subfemtomole quantities of synthetic oligonucleotides and DNA diagnostic products dispensed by a piezoelectric pipet, *Analytical Chemistry*, 69, 4540–4546, 1997.
130. Meier, M. A., de Gans, B. J., van den Berg, A. M., and Schubert, U. S., Automated multiple-layer spotting for matrix-assisted laser desorption/ionization time-of-flight mass spectrometry of synthetic polymers utilizing ink-jet printing technology, *Rapid Communication Mass Spectrometry*, 17, 2349–2353, 2003.
131. Sloane, A. J., Duff, J. L., Wilson, N. L., Gandhi, P. S., Hill, C. J., Hopwood, F. G., Smith, P. E., Thomas, M. L., Cole, R. A., Packer, N. H., Breen, E. J., Cooley, P. W., Wallace, D. B., Williams, K. L., and Gooley, A. A., High throughput peptide mass fingerprinting and protein macroarray analysis using chemical printing strategies, *Molecular and Cellular Proteomics*, 1, 490–499, 2002.
132. Ohtsu, I., Nakanisi, T., Furuta, M., Ando, E., and Nishimura, O., Direct matrix-assisted laser desorption/ionization time-of-flight mass spectrometric identification of proteins on membrane detected by Western blotting and lectin blotting, *Journal of Proteome Research*, 4, 1391–1396, 2005.
133. Laurell, T., Wallman, L., and Nilsson, J., Design and development of a silicon microfabricated dispenser for on-line picolitre sample handling, *Journal Micromechanical Microengineering*, 9, 369–376, 1999.
134. Miliotis, T., Marko-Varga, G., Nilsson, J., and Laurell, T., Development of silicon microstructures and thin-film MALDI target plates for automated proteomics sample identifications, *Journal of Neuroscience Methods*, 109, 41–46, 2001.
135. Nilsson, J., Bergkvist, J., Ekstrom, S., and Laurell, T., *Micro Total Analysis Systems 2001*, Kluwer Academic Publishers, London, 2001.
136. Ekström, S. Bergkvist, J., Laurell, T., Nilsson, J., *Micro Total Analysis Systems 2002*, Kluwer Academic Publishers, London, 2002.
137. Zhang, H., Andren, P. E., and Caprioli, R. M., Micro-preparation procedure for high-sensitivity matrix-assisted laser desorption ionization mass spectrometry, *Journal Mass Spectrometry*, 30, 1768–1771, 1995.
138. Jespersen, S., Niessen, W. M. A., Tjaden, U. R., Vandergreef, J., Litborn, E., Lindberg, U., and Roeraade, J., Attomole detection of proteins by matrix-assisted laser-desorption ionization mass-spectrometry with the use of picolitre vials, *Rapid Communications in Mass Spectrometry*, 8, 581–584, 1994.
139. Little, D. P., Cornish, T. J., O'Donnell, M. J., Braun, A., Cotter, R. J., and Koster, H., MALDI on a chip: Analysis of arrays of low femtomole to subfemtomole quantities of synthetic oligonucleotides and DNA diagnostic products dispensed by a piezoelectric pipet, *Analytical Chemistry*, 69, 4540–4546, 1997.
140. Onnerfjord, P., Nilsson, J., Wallman, L., Laurell, T., and Marko-Varga, G., Picoliter sample preparation in MALDI-TOF MS using a micromachined silicon flow-through dispenser, *Analytical Chemistry*, 70, 4755–4760, 1998.
141. Ekström, S., Ericsson, D., Önnerfjord, P., Bengtsson, M., Nilsson, J., Laurell, T., and Marko-Varga, G., Signal amplification using "spot-on-a-chip" technology for the identification of proteins via MALDI-TOF MS, *Analytical Chemistry*, 73, 214–219, 2001.
142. Gobom, J., Schuerenberg, M., Mueller, M., Theiss, D., Lehrach, H., and Nordhoff, E., Alpha-cyano-4-hydroxycinnamic acid affinity sample preparation. A protocol for MALDI-MS peptide analysis in proteomics, *Analytical Chemistry*, 73, 434–438, 2001.

143. Hung, K. C., Ding, H., and Guo, B. C., Use of poly(tetrafluoroethylene)s as a sample support for the MALDI-TOF analysis of DNA and proteins, *Analytical Chemistry*, 71, 518–521, 1999.
144. Whittal, R. M., Keller, B. O., and Li, L., Nanoliter chemistry combined with mass spectrometry for peptide mapping of proteins from single mammalian cells, *Analytical Chemistry*, 70(24), 5344–5347, 1998.
145. Keller, B. O., Wang, Z. P., and Li, L., Low-mass proteome analysis based on liquid chromatography fractionation, nanoliter protein concentration/digestion, and microspot matrix-assisted laser desorption ionization mass spectrometry, *Journal of Chromatography B–Analytical Technologies in the Biomedical and Life Sciences*, 782, 317–329, 2002.
146. Rejtar, T., Hu, P., Juhasz, P., Campbell, J. M., Vestal, M. L., Preisler, J., and Karger, B. L., Off-line coupling of high-resolution capillary electrophoresis to MALDI-TOF and TOF/TOF MS, *Journal of Proteome Research*, 1, 171–179, 2002.
147. Stupak, J., Liu, H. Z., Wang, Z. P., Brix, B. J., Fliegel, L., and Li, L., Nanoliter sample handling combined with microspot MALDI-MS for detection of gel-separated phosphoproteins, *Journal of Proteome Research*, 4, 515–522, 2005.
148. Wheeler, A. R., Moon, H., Bird, C. A., Loo, R. R. O., Kim, C. J., Loo, J. A., and Garrell, R. L., Digital microfluidics with in-line sample purification for proteomics analyses with MALDI-MS, *Analytical Chemistry*, 77, 534–540, 2005.
149. Bogan, M. J. and Agnes, G. R., Wall-less sample preparation of mu m-sized sample spots for femtomole detection limits of proteins from liquid based UV-MALDI matrices, *Journal of the American Society for Mass Spectrometry*, 15, 486–495, 2004.
150. Hensel, R. R., King, R. C., and Owens, K. G., Electrospray sample preparation for improved quantitation in matrix-assisted laser desorption/ionization time-of-flight mass spectrometry, *Rapid Communication Mass Spectrometry*, 11, 1785–1793, 1997.
151. Annan, R. S., Mculty, D. E., and Carr, S. A., *Proceedings of the 44th ASMS Conference on Mass Spectrometry and Allied Topics*, Portland, OR, 1996, p. 702, 1996.
152. Kussmann, M., Nordhoff, E., Rahbek-Nielsen, H., Haebel, S., Rossel-Larsen, M., Jakobsen, L., Gobom, J., Mirgorodskaya, E., Kroll-Kristensen, A., Palm, L., and Roepstorff, P., Matrix-assisted laser desorption/ionization mass spectrometry sample preparation techniques designed for various peptide and protein analytes, *Journal of Mass Spectrometry*, 32, 593–601, 1997.
153. Miyazaki, S., Morisato, K., Ishizuka, N., Minakuchi, H., Shintani, Y., Furuno, M., and Nakanishi, K., Development of a monolithic silica extraction tip for the analysis of proteins, *Journal of Chromatography A*, 1043, 19–25, 2004.
154. Rappsilber, J., Ishihama, Y., and Mann, M., Stop and go extraction tips for matrix-assisted laser desorption/ionization, nanoelectrospray, and LC/MS sample pretreatment in proteomics, *Analytical Chemistry*, 75, 663–670, 2003.
155. Stewart, I. I., Thomson, T., and Figeys, D., O-18 Labeling: A tool for proteomics, *Rapid Communications in Mass Spectrometry*, 15, 2456–2465, 2001.
156. Larsen, M. R., Cordwell, S. J., and Roepstorff, P., Graphite powder as an alternative or supplement to reversed-phase material for desalting and concentration of peptide mixtures prior to matrix-assisted laser desorption/ionization-mass spectrometry, *Proteomics*, 2, 1277–1287, 2002.
157. Nissum, M., Schneider, U., Kuhfuss, S., Obermaier, C., Wildgruber, R., Posch, A., and Eckerskorn, C., In-gel digestion of proteins using a solid-phase extraction microplate, *Analytical Chemistry*, 76, 2040–2045, 2004.
158. Warren, M. E., Brockman, A. H., and Orlando, R., On-probe solid-phase extraction MALDI-MS using ion-pairing interactions for the cleanup of peptides and proteins, *Analytical Chemistry*, 70, 3757–3761, 1998.
159. Brockman, A. H., Shah, N. N., and Orlando, R., Optimization of a hydrophobic solid-phase extraction interface for matrix-assisted laser desorption/ionization, *Journal of Mass Spectrometry*, 33, 1141–1147, 1998.
160. Worrall, T. A., Cotter, R. J., and Woods, A. S., Purification of contaminated peptides and proteins on synthetic membrane surfaces for matrix-assisted laser/desorption ionization mass spectrometry, *Analytical Chemistry*, 70, 750–756, 1998.
161. Xu, Y., Bruening, M. L., and Watson, J. T., Non-specific, on-probe cleanup methods for MALDI-MS samples, *Mass Spectrometry Review*, 22, 429–440, 2003.

162. Tang, N., Tornatore, P., and Weinberger, S. R., Current developments in SELDI affinity technology, *Mass Spectrometry Reviews*, 23, 34–44, 2004.
163. Bergkvist, J., Ekstrom, S., Wallman, L., Lofgren, M., Marko-Varga, G., Nilsson, J., and Laurell, T., Improved chip design for integrated solid-phase microextraction in on-line proteomic sample preparation, *Proteomics*, 2, 422–429, 2002.
164. Ekstrom, S., Malmstrom, J., Wallman, L., Lofgren, M., Nilsson, J., Laurell, T., and Marko-Varga, G., On-chip microextraction for proteomic sample preparation of in-gel digests, *Proteomics*, 2, 413–421, 2002.
165. Ekstrom, S., Wallman, L., Malm, J., Becker, C., Lilja, H., Laurell, T., and Marko-Varga, G., Integrated selective enrichment target—A microtechnology platform for matrix-assisted laser desorption/ionization-mass spectrometry applied on protein biomarkers in prostate diseases, *Electrophoresis*, 25, 3769–3777, 2004.
166. Wallman, L., Ekstrom, S., Marko-Varga, G., Laurell, T., and Nilsson, J., Autonomous protein sample processing on-chip using solid-phase microextraction, capillary force pumping, and microdispensing, *Electrophoresis*, 25, 3778–3787, 2004.
167. Ekstrom, S., Wallman, L., Hok, D., Marko-Varga, G., and Laurell, T., Miniaturized solid-phase extraction and sample preparation for MALDI MS using a microfabricated integrated selective enrichment target, *Journal Proteome Research*, 5, 1071–1081, 2006.
168. Gustafsson, M., Hirschberg, D., Palmberg, C., Jornvall, H., and Bergman, T., Integrated sample preparation and MALDI mass spectrometry on a microfluidic compact disk, *Analytical Chemistry*, 76, 345–350, 2004.
169. Klose, J., Protein mapping by combined isoelectric focusing and electrophoresis of mouse tissues. A novel approach to testing for induced point mutations in mammals, *Humangenetik*, 26, 231–243, 1975.
170. O'Farrell, P. H., High resolution two-dimensional electrophoresis of proteins, *Journal of Biological Chemistry*, 250, 4007–4021, 1975.
171. Van den Bergh, G. and Arckens, L., Recent advances in 2D electrophoresis: An array of possibilities, *Expert Review of Proteomics*, 2, 243–252, 2005.
172. Bjellqvist, B., Ek, K., Righetti, P. G., Gianazza, E., Gorg, A., Westermeier, R., and Postel, W., Isoelectric focusing in immobilized pH gradients: Principle, methodology and some applications, *Journal of Biochemical and Biophysical Methods*, 6, 317–339, 1982.
173. Unlu, M., Morgan, M. E., and Minden, J. S., Difference gel electrophoresis: A single gel method for detecting changes in protein extracts, *Electrophoresis*, 18, 2071–2077, 1997.
174. Lambert, J. P., Ethier, M., Smith, J. C., and Figeys, D., Proteomics: From gel based to gel free, *Analytical Chemistry*, 77, 3771–3787, 2005.
175. Appella, E., Padlan, E. A., and Hunt, D. F., Analysis of the structure of naturally processed peptides bound by class I and class II major histocompatibility complex molecules, *Exs*, 73, 105–119, 1995.
176. Yates, J. R., 3rd, McCormack, A. L., Schieltz, D., Carmack, E., and Link, A., Direct analysis of protein mixtures by tandem mass spectrometry, *Journal of Protein Chemistry*, 16, 495–497, 1997.
177. Wolters, D. A., Washburn, M. P., and Yates, J. R., An automated multidimensional protein identification technology for shotgun proteomics, *Analytical Chemistry*, 73, 5683–5690, 2001.
178. Chen, H. S., Rejtar, T., Andreev, V., Moskovets, E., and Karger, B. L., High-speed, high-resolution monolithic capillary LC-MALDI MS using an off-line continuous deposition interface for proteomic analysis, *Analytical Chemistry*, 77, 2323–2331, 2005.
179. Lee, H., Griffin, T. J., Gygi, S. P., Rist, B., and Aebersold, R., Development of a multiplexed microcapillary liquid chromatography system for high-throughput proteome analysis, *Analytical Chemistry*, 74, 4353–4360, 2002.
180. Li, N., Shaw, A. R. E., Zhang, N., Mak, A., and Li, L., Lipid raft proteomics: Analysis of in-solution digest of sodium dodecyl sulfate-solubilized lipid raft proteins by liquid chromatography-matrix-assisted laser desorption/ionization tandem mass spectrometry, *Proteomics*, 4, 3156–3166, 2004.
181. Simpson, D. C. and Smith, R. D., Combining capillary electrophoresis with mass spectrometry for applications in proteomics, *Electrophoresis*, 26, 1291–1305, 2005.
182. Schmitt-Kopplin, P. and Englmann, M., Capillary electrophoresis-mass spectrometry: Survey on developments and applications 2003–2004, *Electrophoresis*, 26, 1209–1220, 2005.

183. Stutz, H., Advances in the analysis of proteins and peptides by capillary electrophoresis with matrix-assisted laser desorption/ionization and electrospray-mass spectrometry detection, *Electrophoresis*, 26, 1254–1290, 2005.
184. Bousse, L., Cohen, C., Nikiforov, T., Chow, A., Kopf-Sill, A. R., Dubrow, R., and Parce, J. W., Electrokinetically controlled microfluidic analysis systems, *Annual Review of Biophysics and Biomolecular Structure*, 29, 155–181, 2000.
185. Bruin, G. J. M., Recent developments in electrokinetically driven analysis on microfabricated devices, *Electrophoresis*, 21, 3931–3951, 2000.
186. Cooper, J. W., Wang, Y. J., and Lee, C. S., Recent advances in capillary separations for proteomics, *Electrophoresis*, 25, 3913–3926, 2004.
187. Bandilla, D. and Skinner, C. D., Capillary electrochromatography of peptides and proteins, *Journal of Chromatography A*, 1044, 113–129, 2004.
188. Stachowiak, T. B., Svec, F., and Frechet, J. M. J., Chip electrochromatography, *Journal of Chromatography A*, 1044, 97–111, 2004.
189. Kutter, J. P., Jacobson, S. C., and Ramsey, J. M., Solid phase extraction on microfluidic devices, *Journal of Microcolumn Separations*, 12, 93–97, 2000.
190. Gottschlich, N., Jacobson, S. C., Culbertson, C. T., and Ramsey, J. M., Two-dimensional electrochromatography/capillary electrophoresis on a microchip, *Analytical Chemistry*, 73, 2669–2674, 2001.
191. Jemere, A. B., Oleschuk, R. D., and Harrison, D. J., Microchip-based capillary electrochromatography using packed beds, *Electrophoresis*, 24, 3018–3025, 2003.
192. Shediac, R., Ngola, S. M., Throckmorton, D. J., Anex, D. S., Shepodd, T. J., and Singh, A. K., Reversed-phase electrochromatography of amino acids and peptides using porous polymer monoliths, *Journal of Chromatography A*, 925, 251–263, 2001.
193. He, B., Tait, N., and Regnier, F., Fabrication of nanocolumns for liquid chromatography, *Analytical Chemistry*, 70, 3790–3797, 1998.
194. Slentz, B. E., Penner, N. A., Lugowska, E., and Regnier, F., Nanoliter capillary electrochromatography columns based on collocated monolithic support structures molded in poly(dimethyl siloxane), *Electrophoresis*, 22, 3736–3743, 2001.
195. O'Neill, A. P., O'Brien, P., Alderman, J., Hoffman, D., McEnery, M., Murrihy, J., and Glennon, J. D., On-chip definition of picolitre sample injection plugs for miniaturised liquid chromatography, *Journal of Chromatography A*, 924, 259–263, 2001.
196. Shintani, Y., Hirako, K., Motokawa, M., Iwano, T., Zhou, X., Takano, Y., Furuno, M., Minakuchi, H., and Ueda, M., Development of miniaturized multi-channel high-performance liquid chromatography for high-throughput analysis, *Journal of Chromatography A*, 1073, 17–23, 2005.
197. Reichmuth, D. S., Shepodd, T. J., and Kirby, B. J., Microchip HPLC of peptides and proteins, *Analytical Chemistry*, 77, 2997–3000, 2005.
198. Yin, N. F., Killeen, K., Brennen, R., Sobek, D., Werlich, M., and van de Goor, T. V., Microfluidic chip for peptide analysis with an integrated HPLC column, sample enrichment column, and nanoelectrospray tip, *Analytical Chemistry*, 77, 527–533, 2005.
199. Ramsey, R. S. and Ramsey, J. M., Generating electrospray from microchip devices using electroosmotic pumping, *Analytical Chemistry*, 69, 1174–1178, 1997.
200. Xue, Q. F., Foret, F., Dunayevskiy, Y. M., Zavracky, P. M., McGruer, N. E., and Karger, B. L., Multichannel microchip electrospray mass spectrometry, *Analytical Chemistry*, 69, 426–430, 1997.
201. Figeys, D., Gygi, S. P., McKinnon, G., and Aebersold, R., An integrated microfluidics tandem mass spectrometry system for automated protein analysis, *Analytical Chemistry*, 70, 3728–3734, 1998.
202. Sung, W. C., Makamba, H., and Chen, S. H., Chip-based microfluidic devices coupled with electrospray ionization-mass spectrometry, *Electrophoresis*, 26, 1783–1791, 2005.
203. Zhang, S. and Van Pelt, C. K., Chip-based nanoelectrospray mass spectrometry for protein characterization, *Expert Review of Proteomics*, 1, 449–468, 2004.
204. Slentz, B. E., Penner, N. A., and Regnier, F. E., Protein proteolysis and the multi-dimensional electrochromatographic separation of histidine-containing peptide fragments on a chip, *Journal of Chromatography A*, 984, 97–107, 2003.

205. Li, J. J., LeRiche, T., Tremblay, T. L., Wang, C., Bonneil, E., Harrison, D. J., and Thibault, P., Application of microfluidic devices to proteomics research—Identification of trace-level protein digests and affinity capture of target peptides, *Molecular and Cellular Proteomics*, 1, 157–168, 2002.
206. Peterson, D. S., Rohr, T., Svec, F., and Frechet, J. M. J., Enzymatic microreactor-on-a-chip: Protein mapping using trypsin immobilized on porous polymer monoliths molded in channels of microfluidic devices, *Analytical Chemistry*, 74, 4081–4088, 2002.
207. Jin, L. J., Ferrance, J., Sanders, J. C., and Landers, J. P., A microchip-based proteolytic digestion system driven by electroosmotic pumping, *Lab on a Chip*, 3, 11–18, 2003.
208. Peterson, D. S., Rohr, T., Svec, F., and Frechet, J. M. J., Dual-function microanalytical device by *in situ* photolithographic grafting of porous polymer monolith: Integrating solid-phase extraction and enzymatic digestion for peptide mass mapping, *Analytical Chemistry*, 75, 5328–5335, 2003.
209. Gao, J., Xu, J. D., Locascio, L. E., and Lee, C. S., Integrated microfluidic system enabling protein digestion, peptide separation, and protein identification, *Analytical Chemistry*, 73, 2648–2655, 2001.
210. Wang, C., Oleschuk, R., Ouchen, F., Li, J. J., Thibault, P., and Harrison, D. J., Integration of immobilized trypsin bead beds for protein digestion within a microfluidic chip incorporating capillary electrophoresis separations and an electrospray mass spectrometry interface, *Rapid Communications in Mass Spectrometry*, 14, 1377–1383, 2000.
211. Cooper, J. W., Chen, J. Z., Li, Y., and Lee, C. S., Membrane-based nanoscale proteolytic reactor enabling protein digestion, peptide separation, and protein identification using mass spectrometry, *Analytical Chemistry*, 75, 1067–1074, 2003.
212. Qu, H. Y., Wang, H. T., Huang, Y., Zhong, W., Lu, H. J., Kong, J. L., Yang, P. Y., and Liu, B. H., Stable microstructured network for protein patterning on a plastic microfluidic channel: Strategy and characterization of on-chip enzyme microreactors, *Analytical Chemistry*, 76, 6426–6433, 2004.
213. Ekstrom, S., Ericsson, D., Onnerfjord, P., Bengtsson, M., Nilsson, J., Marko-Varga, G., and Laurell, T., Signal amplification using "spot on-a-chip" technology for the identification of proteins via MALDI-TOF MS, *Analytical Chemistry*, 73, 214–219, 2001.
214. Ekstrom, S., Nilsson, J., Helldin, G., Laurell, T., and Marko-Varga, G., Disposable polymeric high-density nanovial arrays for matrix assisted laser desorption/ionization-time of flight-mass spectrometry: II. Biological applications, *Electrophoresis*, 22, 3984–3992, 2001.
215. Marko-Varga, G., Ekstrom, S., Helldin, G., Nilsson, J., and Laurell, T., Disposable polymeric high-density nanovial arrays for matrix assisted laser desorption/ionization-time of flight-mass spectrometry: I. Microstructure development and manufacturing, *Electrophoresis*, 22, 3978–3983, 2001.
216. Quadroni, M. and James, P., Proteomics and automation, *Electrophoresis*, 20, 664–677, 1999.
217. Hagman, C., Ramstrom, M., Jansson, M., James, P., Hakansson, P., and Bergquist, J., Reproducibility of tryptic digestion investigated by quantitative Fourier transform ion cyclotron resonance mass spectrometry, *Journal of Proteome Research*, 4, 394–399, 2005.
218. Onnerfjord, P., Ekstrom, S., Bergquist, J., Nilsson, J., Laurell, T., and Marko-Varga, G., Homogeneous sample preparation for automated high throughput analysis with matrix-assisted laser desorption/ionisation time-of-flight mass spectrometry, *Rapid Communications in Mass Spectrometry*, 13, 315–322, 1999.
219. Bengtsson, M., Ekstrom, S., Marko-Varga, G., and Laurell, T., Improved performance in silicon enzyme microreactors obtained by homogeneous porous silicon carrier matrix, *Talanta*, 56, 341–353, 2002.
220. Massolini, G. and Calleri, E., Immobilized trypsin systems coupled on-line to separation methods: Recent developments and analytical applications, *Journal of Separation Science*, 28, 7–21, 2005.
221. Ericsson, D., Ekstrom, S., Nilsson, J., Bergquist, J., Marko-Varga, G., and Laurell, T., Downsizing proteolytic digestion and analysis using dispenser-aided sample handling and nanovial matrix-assisted laser/desorption ionization-target arrays, *Proteomics*, 1, 1072–1081, 2001.
222. Finnskog, D., Jaras, K., Ressine, A., Malm, J., Marko-Varga, G., Lilja, H., and Laurell, T., High-speed biomarker identification utilizing porous silicon nanovial arrays and MALDI-TOF mass spectrometry, *Electrophoresis*, 27, 1093–1103, 2006.
223. Mann, M. and Jensen, O. N., Proteomic analysis of post-translational modifications, *Nature Biotechnology*, 21, 255–261, 2003.

224. Loyet, K. M., Stults, J. T., and Arnott, D., Mass spectrometric contributions to the practice of phosphorylation site mapping through 2003: A literature review, *Molecular Cell Proteomics*, 4, 235–245, 2005.
225. Mukherji, M., Phosphoproteomics in analyzing signaling pathways, *Expert Review of Proteomics*, 2, 117–128, 2005.
226. Zhang, H., Yi, E. C., Li, X. J., Mallick, P., Kelly-Spratt, K. S., Masselon, C. D., Camp, D. G., Smith, R. D., Kemp, C. J., and Aebersold, R., High throughput quantitative analysis of serum proteins using glycopeptide capture and liquid chromatography mass spectrometry, *Molecular and Cellular Proteomics*, 4, 144–155, 2005.
227. Sprung, R., Nandi, A., Chen, Y., Kim, S. C., Barma, D., Falck, J. R., and Zhao, Y. M., Tagging-via-substrate strategy for probing O-GlcNAc modified proteins, *Journal of Proteome Research*, 4, 950–957, 2005.
228. Dube, D. H. and Bertozzi, C. R., Glycans in cancer and inflammation. Potential for therapeutics and diagnostics, *Nature Reviews Drug Discovery*, 4, 477–488, 2005.
229. Zala, J., Mass spectrometry of oligosaccharides, *Mass Spectrometry Reviews*, 23, 161–227, 2004.
230. Kirkpatrick, D. S., Denison, C., and Gygi, S. P., Weighing in on ubiquitin: The expanding role of mass-spectrometry-based proteomics, *Nature Cell Biology*, 7, 750–757, 2005.
231. Hirschberg, D., Jagerbrink, T., Samskog, J., Gustafsson, M., Stahlberg, M., Alvelius, G., Husman, B., Carlquist, M., Jornvall, H., and Bergman, T., Detection of phosphorylated peptides in proteomic analyses using microfluidic compact disk technology, *Analytical Chemistry*, 76, 5864–5871, 2004.
232. Lueking, A., Cahill, D. J., and Mullner, S., Protein biochips: A new and versatile platform technology for molecular medicine, *Drug Discovery Today*, 10, 789–794, 2005.
233. Sheehan, K. M., Calvert, V. S., Kay, E. W., Lu, Y. L., Fishman, D., Espina, V., Aquino, J., Speer, R., Araujo, R., Mills, G. B., Liotta, L. A., Petricoin, E. F., and Wulfkuhle, J. D., Use of reverse phase protein microarrays and reference standard development for molecular network analysis of metastatic ovarian carcinoma, *Molecular and Cellular Proteomics*, 4, 346–355, 2005.
234. Sheehan, K. M., Fishman, D. A., Liotta, L. A., Petricoin, E. F., and Wulfkuhle, J. D., Mapping the molecular network of metastatic ovarian carcinoma: Theranostics using proteomics, *Journal of the Society for Gynecologic Investigation*, 12, 285a, 2005.
235. Lopez, M. F. and Pluskal, M. G., Protein micro- and macroarrays: Digitizing the proteome, *Journal of Chromatography B–Analytical Technologies in the Biomedical and Life Sciences*, 787, 19–27, 2003.
236. Borrebaeck, C. A. K., Ekstrom, S., Hager, A. C. M., Nilsson, J., Laurell, T., and Marko-Varga, G., Protein chips based on recombinant antibody fragments: A highly sensitive approach as detected by mass spectrometry, *Biotechniques*, 30, 1126, 2001.
237. Steinhauer, C., Ressine, A., Marko-Varga, G., Laurell, T., Borrebaeck, C. A. K., and Wingren, C., Biocompatibility of surfaces for antibody microarrays: Design of macroporous silicon substrates, *Analytical Biochemistry*, 341, 204–213, 2005.
238. Finnskog, D., Ressine, A., Laurell, T., and Marko-Varga, G., Integrated protein microchip assay with dual fluorescent- and MALDI read-out, *Journal of Proteome Research*, 3, 988–994, 2004.
239. Ressine, A., Ekstrom, S., Marko-Varga, G., and Laurell, T., Macro-/nanoporous silicon as a support for high-performance protein microarrays, *Analytical Chemistry*, 75, 6968–6974, 2003.
240. Pluskal, M. G., Microscale sample preparation—The demand for microscale analysis has stimulated the development of new solid phase formats in sample preparation, *Nature Biotechnology*, 18, 104–105, 2000.
241. Ishihama, Y., Rappsilber, J., Andersen, J. S., and Mann, M., Microcolumns with self-assembled particle frits for proteomics, *Journal of Chromatography A*, 979, 233–239, 2002.
242. Miller, J. and Stagljar, I., Using the yeast two-hybrid system to identify interacting proteins, *Methods Molecular Biology*, 261, 247–262, 2004.
243. Causier, B., Studying the interactome with the yeast two-hybrid system and mass spectrometry, *Mass Spectrometry Review*, 23, 350–367, 2004.
244. Puig, O., Caspary, F., Rigaut, G., Rutz, B., Bouveret, E., Bragado-Nilsson, E., Wilm, M., and Seraphin, B., The tandem affinity purification (TAP) method: A general procedure of protein complex purification, *Methods*, 24, 218–229, 2001.
245. Caldwell, R. L. and Caprioli, R. M., Tissue profiling by mass spectrometry: A review of methodology and applications, *Molecular Cell Proteomics*, 4, 394–401, 2005.

246. Gygi, S. P., Rist, B., Gerber, S. A., Turecek, F., Gelb, M. H., and Aebersold, R., Quantitative analysis of complex protein mixtures using isotope-coded affinity tags, *Nature Biotechnology*, 17, 994–999, 1999.
247. Greenbaum, D., Colangelo, C., Williams, K., and Gerstein, M., Comparing protein abundance and mRNA expression levels on a genomic scale, *Genome Biology*, 4, 117, 2003.

50 Implementing Sample Preconcentration in Microfluidic Devices

Paul M. van Midwoud and Elisabeth Verpoorte

CONTENTS

50.1 INTRODUCTION

The capillary format has proven to be ideal for separations requiring an applied electric field. The micrometer-dimensioned capillary cores facilitate the analysis of smaller, nanoliter-sized samples, which in turn means higher separation resolution and much shorter separation times. Reduction of column diameters means a commensurate reduction in detection volumes, however, making detection of low analyte concentrations even more challenging as numbers of analyte species available for detection are decreased. Sample preconcentration methods are thus required to compress more analytes into smaller volumes to improve detection limits.

This report gives an overview of the sample preconcentration techniques which have to date been implemented in microfluidic devices. Some of these techniques originate from the conventional capillary electrophoresis (CE) and high performance liquid chromatography (HPLC) world and are quite recognizable despite incorporation into a miniaturized format. Other techniques are brand new, having been enabled by the access to micro- and nanofluidics that new fabrication technologies provide us. Interestingly, recent reviews that appeared while this chapter was in preparation categorize available on-chip preconcentration methods in much the same manner as we do, in terms of electrokinetic preconcentration, surface extraction, and similar nanofilter and membrane approaches,[1,2] The reader is advised to consult these and other papers for more comprehensive overviews of what has appeared in the literature over the past few years. We have opted to focus more on describing the mechanisms underlying various methods, and how these methods have been realized practically in terms of device. This chapter does not, however, provide an in-depth "how-to" instruction manual for implementing these techniques in microfluidic devices. However, we hope that the information given here will be enough for the reader to be able to decide what approach may be best suited to his or her problem.

50.2 ELECTROKINETIC PRECONCENTRATION TECHNIQUES

Fortunately, applied electric fields can be very effectively exploited not only for separation but also for on-line sample preconcentration in capillaries, as evidenced by a number of well-established electrokinetic preconcentration techniques for CE (see, for instance, Chapters 13 and 19 of this book for more details).

The microfluidic chip format may not directly resemble the capillary format, but certainly the microchannels that make up microfluidic networks have inner dimensions that are comparable to capillaries. Electric fields can thus be used just as effectively in microchips to perform CE and similar types of separations. Generation of EOF has also proven to be the most popular mechanism to date by which fluids are moved through microchannels. On the flip side, detection of low concentrations in small volumes is as great or an even greater issue in chips. As a result, sample preconcentration methods for electrokinetically driven microdevices have received significant attention in the past few years. The transfer of existing electrokinetic preconcentration techniques to the microchip format is an obvious approach, given the similarity in system inner dimensions and the fact that CE had already been successfully realized in chips. In addition, some new approaches have been developed in the last decade, enabled by micro- and nanotechnologies. This section will investigate further how a number of these techniques have been implemented in a microfluidic format.

50.2.1 FIELD-AMPLIFIED SAMPLE STACKING (FASS)

Field-amplified sample stacking (FASS) is a preconcentration technique which is commonly used in CE. The technique was first mentioned by Mikkers et al.[3] in 1979 and has been intensively studied by Burgi and Chien et al.[4–6] In 1995, Jacobson and Ramsey[7] reported the first FASS experiment on a microchip.

To achieve a preconcentration of the sample with FASS, the sample is diluted in a buffer which has a much lower conductivity (σ) than the surrounding run buffer used for the separation. In FASS, the high- and low-conductivity buffers differ only in concentration (which is related to the ionic strength and thus the conductivity), to overcome problems due to unwanted electrodispersion, which could occur when two buffers with different chemical compositions are introduced.

The introduction of a low-conductivity sample plug into higher-conductivity run buffer results in the local variation of the applied electric field, with a higher electric field in the sample segment than in run buffer segments. The relationship between the electric field and the conductivity is given by

$$E = \frac{I}{A\sigma} \tag{50.1}$$

where I is the electric current running through the channel, A is the channel cross-sectional area, and σ is the conductivity of the buffer. Since the current running through any given section of the

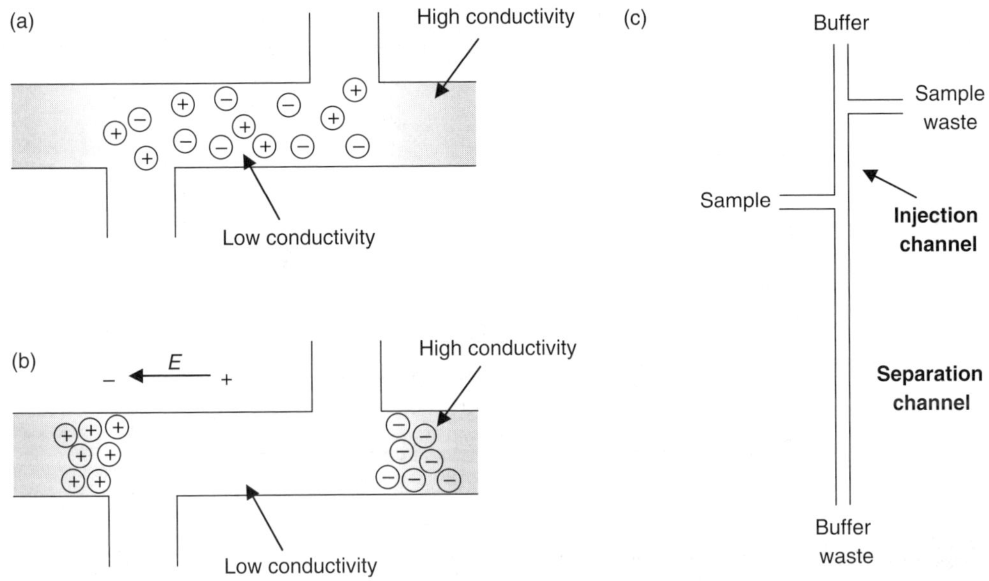

FIGURE 50.1 (a, b) Schematic view of on-chip FASS and (c) the general chip layout used for FASS. (a) Anionic and cationic sample ions are introduced into the injection region (b) and on application of an electric field along the separation channel, sample ions exit the low-conductivity region and enter the high-conductivity region.

microchannel will be the same, a decrease of σ in a certain segment will result in a larger electrical field in that segment in relation to the rest of the system.

The FASS stacking phenomenon is schematically shown in Figure 50.1a and b, in a sample plug that has been formed at a double-T injector. Since the electrophoretic velocity of charged molecular species is proportional to the applied electrical field (see Chapter 1), species migrate faster in higher applied electric fields. In the amplified electric field within the sample plug, both cations and anions migrate with increased velocity, the former toward the negative electrode and the latter toward the positive electrode (see Figure 50.1b). When the ions encounter a sudden drop in electric field strength at the sample-run buffer boundaries, their electrophoretic velocity decreases sharply and they are focused into narrow zones at the boundaries of the sample plug. These zones subsequently migrate into the surrounding run buffer and are separated by CE (not shown). As Figure 50.1b shows, an initial separation of cations from anions is achieved in the stacking process.

There are several different strategies for realizing sample stacking on microfluidic devices. The simplest involves the injection of a sample dissolved in low-conductivity buffer into a high-conductivity buffer stream, employing either volume-defined injection[8–12] or gated injection.[7,13] A commonly used channel geometry for volume-defined injection is shown in Figure 50.1c. Known as a "double-T" injector device,[14] the injection loop that defines the volume to be analyzed is the channel located between the sample and sample-waste reservoirs, labeled "injection channel" in this figure. The injection channel is filled with sample diluted in a buffer with low conductivity by applying an electric field between the sample and sample-waste reservoirs. The analytes are focused and separated in the separation channel by applying a potential between the buffer and buffer-waste reservoirs. A number of different variations of this type of injection geometry have been employed for FASS on chip.[8–10,12] Gated injection is carried out in so-called "cross layouts," at the intersection of two channels laid out in the form of a cross. The volume of sample injected is defined by the amount of time the sample is allowed to enter the separation channel.[7]

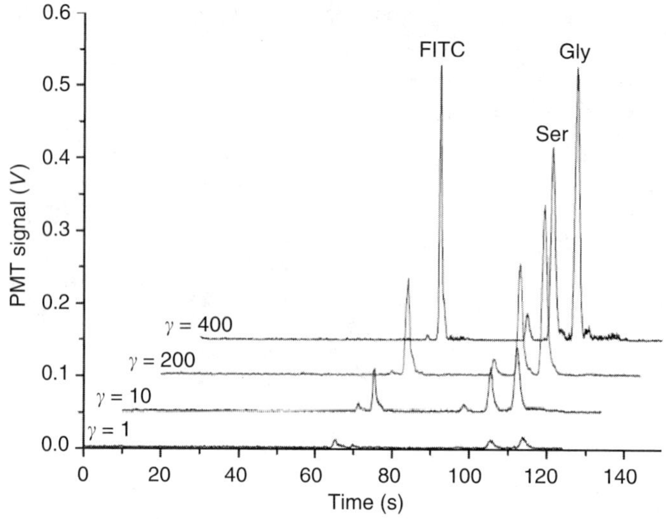

FIGURE 50.2 Comparison of microchip CE of two fluorescently labeled amino acids (10 μM Ser and Gly) with ($\gamma > 1$) and without ($\gamma = 1$) sample stacking. Separation conditions: separation length, 7.5 cm; electric field, 460 V/cm; run buffer, 32 mM carbonate at pH 9.6. (From Lichtenberg, J., et al., *Electrophoresis*, 22, 258, 2001. With permission.)

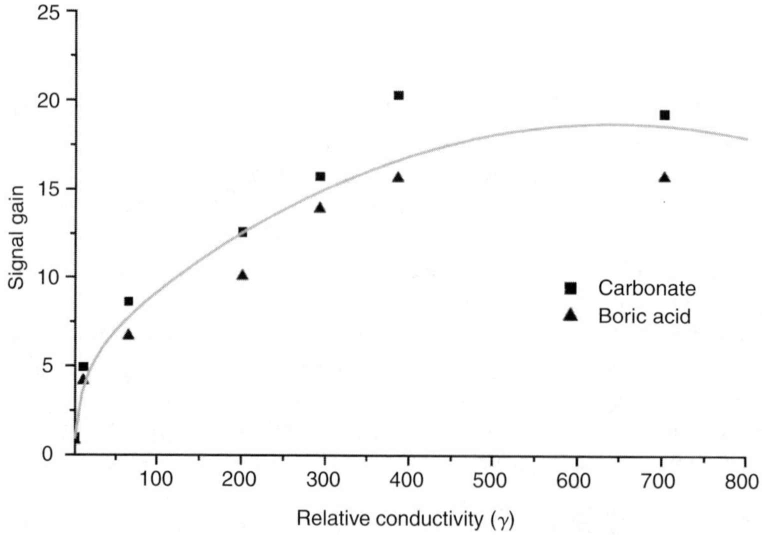

FIGURE 50.3 Signal gain (peak height) as a function of relative conductivity for a high-conductivity (32 mM carbonate, pH 9.6) and medium-conductivity (100 mM Tris/20 mM boric acid, pH 9.1) buffer. Analyte: glycine. Experimental conditions as in Figure 50.2. (From Lichtenberg, J., et al., *Electrophoresis*, 22, 258, 2001. With permission.)

Ideally, the stacking effect is only due to the conductivity differences between the sample and run buffers. This means that the relative conductivity is the most important factor for FASS. The relative conductivity (γ) is given by

$$\gamma = \frac{c_r}{c_s} = \frac{\sigma_r}{\sigma_s} = \frac{E_s}{E_r} \tag{50.2}$$

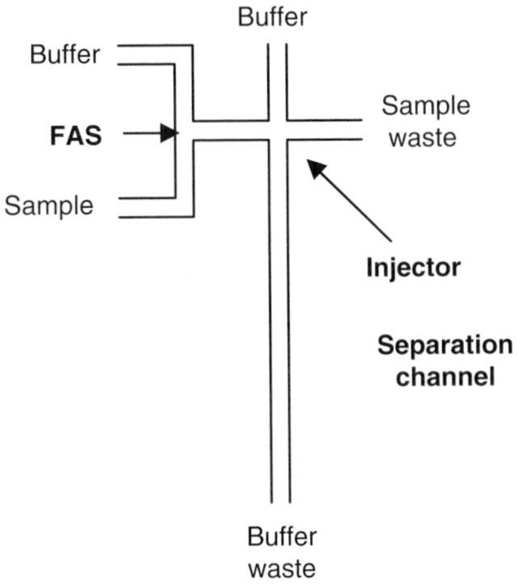

FIGURE 50.4 Schematic of FASS on a microchip demonstrated by Yang and Chien.[16]

where c_r and c_s are the buffer concentrations for the run buffer and the sample buffer, respectively. σ_r and σ_s are the conductivities of the two buffers and E_s and E_r are the electric field strengths in the two buffers. According to the theory, the peak height of the analytes should increase by a factor γ and the peak width should decrease by a factor γ. Unfortunately, this is not fully true, as Figures 50.2 and 50.3 show. The rate at which peak height increases as a function of γ is relatively constant up to $\gamma = 10$ or so, in accordance with other reports.[7,13] However, for γ greater than 10, this rate decreases steadily, until it levels off at values of γ of 400 or so in this case.[8] This is a result of the generation of additional flow effects when conductivity differences are large, due to the electroosmotic flow (EOF) in the sample buffer being higher than in the run buffer. Due to this mismatch of EOFs, a backpressure will be generated in the channel as the sample plug attempts to displace the slower moving buffer both in front and behind it. The appearance of this backpressure results in dispersion of the stacked zone, and a loss of peak height as a result.[4,15] This means there is a trade-off between the optimum relative conductivity and the appearance of peak broadening. The effect of relative conductivity on the signal gain is discussed by Lichtenberg et al.[8] for FASS on a microchip, and by Burgi and Chien[5] for FASS in conventional CE.

Yang and Chien[16] have reported FASS on a microchip with a 100-fold increase in signal concentration. In the described experiment, the walls of the channels were coated with an acrylamide polymer layer to eliminate EOF. This coating minimizes the problems associated with the mismatch of EOF and results in a concentration boundary, which is stationary except for dispersion caused by diffusion. An optimized chip layout used by Yang and Chien for static sample stacking is given in Figure 50.4. The steps required to achieve stacking are depicted in Figure 50.5. Interesting in this example is that flow control in the channels is accomplished by controlling the pressures rather than the voltages applied at all the reservoirs, using an instrument especially designed for this purpose. Yang and Chien demonstrated a 100-fold increase in signal at a theoretical relative conductivity of 200. Yang and Chien provide a thorough "how-to" description of the conditions used for their experiments and details of chip operation in Reference 16.

The stacking approaches described up to this point are all based on the preconcentration of analytes in relatively small volumes of sample. For trace analysis applications, such small samples often will not contain enough analyte molecules for detection, even after stacking. Several researchers have, therefore, explored how to introduce and stack larger samples in microfluidic devices. Lichtenberg et al.[8] reported a field-amplified stacking method with a long channel for large-volume stacking. A schematic layout of the chip along with a depiction of its operation are shown in Figure 50.6.

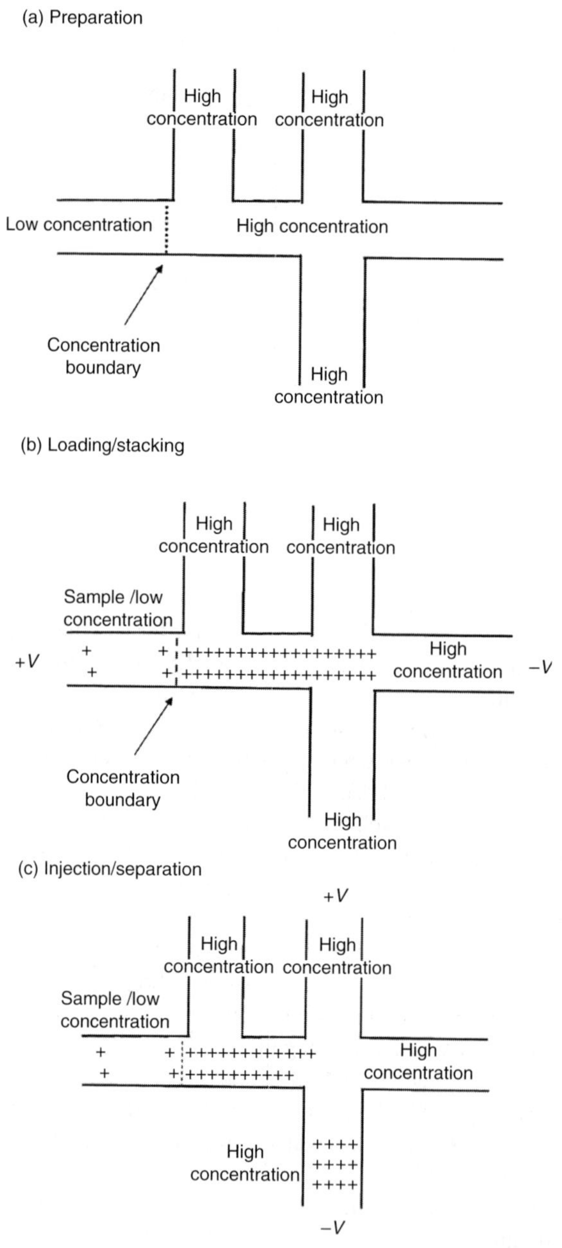

FIGURE 50.5 Operation of FASS device in which EOF has been eliminated through deposition of a polymer coating onto microchannel walls. In Step (a), a concentration boundary is formed between high- and low-conductivity solutions by filling the various channels with the appropriate solutions. In Step (b), a voltage is applied in the horizontal direction to load sample from the left. Stacking occurs at the concentration boundary, as shown. In Step (c), a voltage is applied along the vertical (separation) channel. A well-defined portion of the stacked sample plug is injected in the downward direction and separated. (From Yang et al., *J. Chromatog. A*, 924, 155, 2001. With permission.)

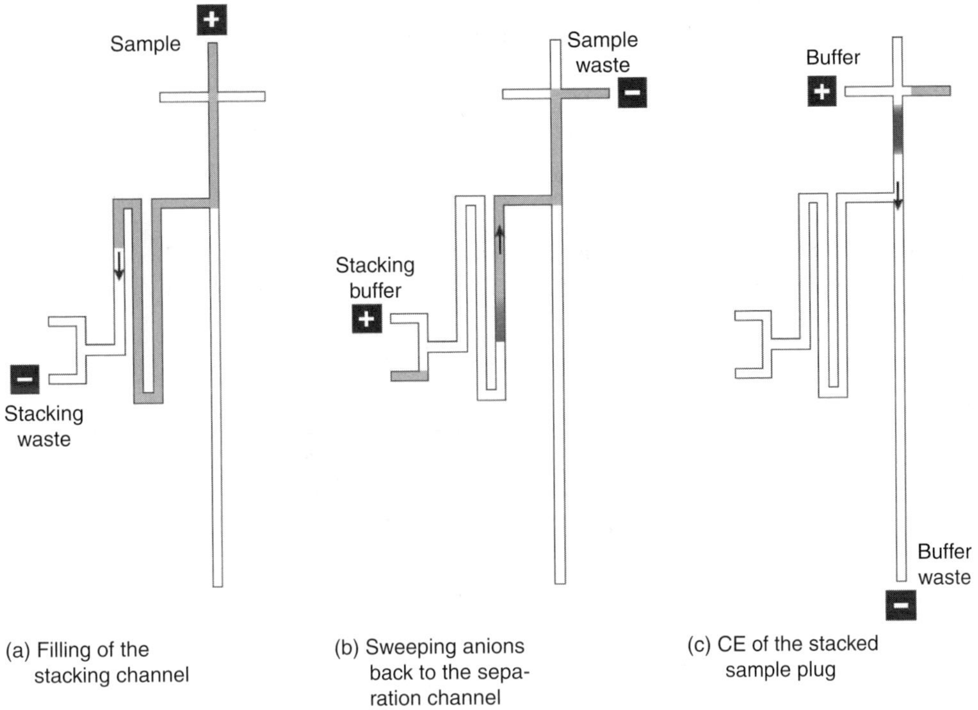

FIGURE 50.6 Schematic diagram of a chip layout with a long preconcentration ("stacking") side channel (7.5 cm) used by Lichtenberg et al.[8] The operation of the device is also outlined in this figure. All flows were controlled electrokinetically. In this particular example, anions are stacked at the stacking buffer-sample boundary, as this boundary is pushed back toward the injection intersection. (From Lichtenberg et al., *Electrophoresis*, 22, 258, 2001. With permission.)

By applying a voltage between the sample inlet and the stacking waste, the long stacking channel is filled with sample diluted in a low-conductivity buffer. When the channel is filled with sample, the electric field is reversed, and a potential difference is applied between the stacking buffer and the sample waste. The high-conductivity stacking buffer ensures that the anions are concentrated in the stacking channel. With this setup, a much higher sample volume can be injected, so more analytes can be focused and a higher sensitivity achieved. The focused sample plug is brought into the injection channel and, by applying a voltage between the buffer and the buffer waste, injected into the separation channel.

One possible limitation with this setup is that only cations or anions can be injected into the separation channel. However, this could also be an advantage if only one of the two is of interest. With this particular chip, a 65-fold increase in sample concentration can be obtained with an injection plug length of 7.5 cm. The effect of injection plug length on sensitivity is also reported by Burgi et al.[5] for FASS in conventional CE and by Beard et al.[11] for FASS on chip.

In the examples discussed earlier, conductivity gradients between sample and run buffers are established through use of electrokinetic flows to bring the different solutions into contact with one another for subsequent stacking. In addition to the adverse pressure-driven flow (PF) effects that can arise, dispersion of the stacked zones can also be caused by electrokinetic flow instabilities that occur in electrokinetically driven flow systems with high-conductivity gradients. Excessive dispersion at buffer–buffer interfaces can result. To overcome this problem, Jung et al.[12] recently reported a new chip layout incorporating a porous polymer plug between buffers of differing conductivity. The layout of the chip is shown in Figure 50.7. The high flow resistance of the plug allows the introduction

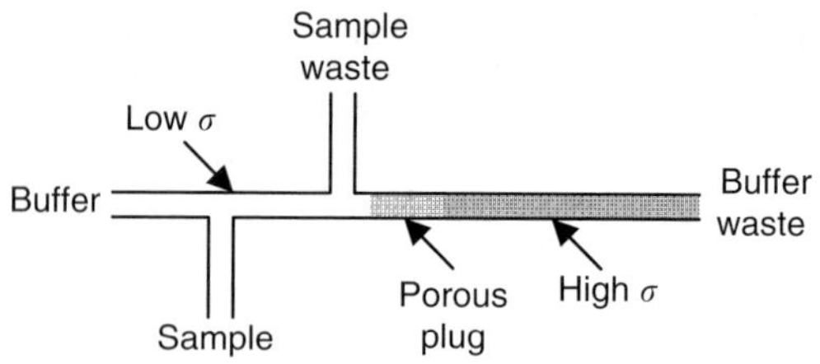

FIGURE 50.7 Schematic view of the chip layout that is used by Jung et al. [12]. The porous plug is made by photopolymerization of methacrylates and has a median pore diameter of about 4.6 μm. Length of the plug, 500 μm.

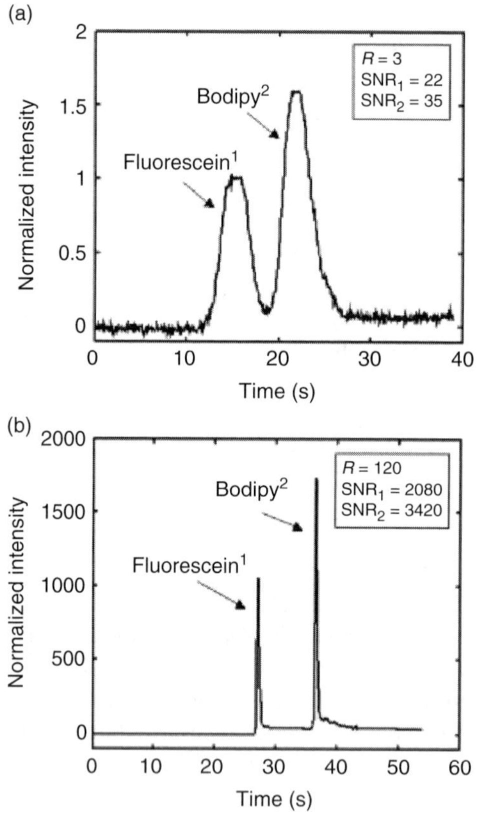

FIGURE 50.8 Electropherograms of fluorescein and BODIPY separations. (a) Electropherogram of analytes in a CE separation and detection performed without stacking: conductivity ratio, γ, of unity. (b) Stacked CE electropherogram for a conductivity ratio, γ, of 1290. The differences in electromigration times are due to the effects of FASS. The signal increase is 1100-fold for the stacked case, and resolution increases from 3 to 120. (From Jung et al., *Electrophoresis*, 24, 3476, 2003. With permission.)

of the required buffers on either side of the plug using syringe pumps without any mixing, while electrical contact is maintained between the different solutions. In this way, a conductivity gradient may be established without the use of electrical fields and the attendant dispersion at the buffer–buffer interface. Volume-defined injection of sample in a low-conductivity buffer can then be performed electrokinetically, with subsequent stacking and separation in the horizontal separation channel.

Furthermore, sample dispersion during the FASS-CE process is reduced by coating microchannel walls with methylcellulose to suppress EOF. The preconcentration factors attainable using this approach are excellent, with a concentration increase of 1100-fold reported for two fluorescent dyes (Figure 50.8).

Field-amplified sample injection (FASI) is a variation of FASS which can lead to very high-sensitivity enhancements on the order of several thousand in conventional CE.[17] To perform FASI of anionic analytes, a plug of very low–conductivity solution such as water is introduced between the run buffer and the sample, which is injected last at the capillary inlet. The application of a reversed polarity voltage (EOF toward inlet) then results in sample anions rapidly migrating through the water to stack at the buffer interface, while water is pumped out of the capillary at the inlet. Subsequent reversal of the polarity allows the stacked analytes to be carried down the capillary and separated, to be detected near the outlet. FASI has recently been successfully implemented in microfluidic devices by Gong et al.[17] and Shiddiky et al.[18] The former group managed signal gains between 90 and 160 for fluorescent dyes using a very simple cross layout. The latter group combined a FASI step (signal enhancement of about 1000) with a preceding FASS step (signal enhancement of 5) to obtain a total gain of 5200 for detection of trace phenolics by micellar electrokinetic chromatography (MEKC).

Field-amplified stacking methods are only applicable to charged species. However, stacking of neutral species is possible if a micelle-containing run buffer is employed. Termed *sweeping*, this involves the injection of an analyte zone into a column of separation buffer containing a pseudo-stationary phase (micelles). On application of the electric field, the micelles are swept through the analyte zone, collecting analyte as they go, and resulting in a unique focusing effect. Charged species can also be preconcentrated in this way. When first described,[19] preconcentration factors of several thousand-fold were achieved. In a recent variation of sweeping, sensitivity increases approaching 1,000,000-fold have been reported.[20]

A few examples of sweeping implemented on microchips have appeared in the literature. One approach, reported by the Landers and coworkers[21,22] for both capillaries and microchips, involves the addition of salt (e.g., NaCl) to the sample matrix to increase its conductivity to levels 2–3 times higher than the run buffer. The latter typically contains negatively charged micelles of sodium dodecyl sulphate (SDS) or sodium cholate. The result is a substantial drop in electric field strength at the sample plug-run buffer interfaces, and a field amplification effect analogous to FASS. In this case, it is the negatively charged micelles, with electrophoretic mobilities counter to the EOF that stack at the cathodic interface of the sample plug and the buffer. Analyte is collected and preconcentrated by the micelles as it is electroosmotically driven through this interfacial zone. The micelle zone is also effectively "locked" in place up against the higher concentration chloride, until this latter zone has diffused sufficiently to let cholate anions enter. Hence, a sharp interface is maintained for analyte stacking, which makes this type of stacking very attractive for electrokinetic injection of the sample. This technique lends itself very well to chip-based separation, with 20-fold peak height improvement being observed for an 80-s injection of a solution containing 67 nM BODIPY (Figure 50.9).[21] Sera et al.[23] have also reported increased sensitivity for the detection of fluorescent dyes using a sweeping technique, with signal gains of 90-fold for rhodamine B and 1500-fold for cresyl fast violet. More recently, Kitagawa et al.[24] reported the combination of sweeping with very sensitive thermal lens microscopy to achieve increases in sensitivity on the order of 10^5 to 10^6.

50.2.2 Isotachophoresis (ITP)

Isotachophoresis (ITP) was, like FASS, originally developed for CE, but it relies on zero EOF. ITP is a quantitative analytical separation technique in its own right, with a wide range of applications. It has also been successfully applied in conjunction with CE for both sample preconcentration and removal of highly abundant charged species, by coupling the ITP and CE columns. The first

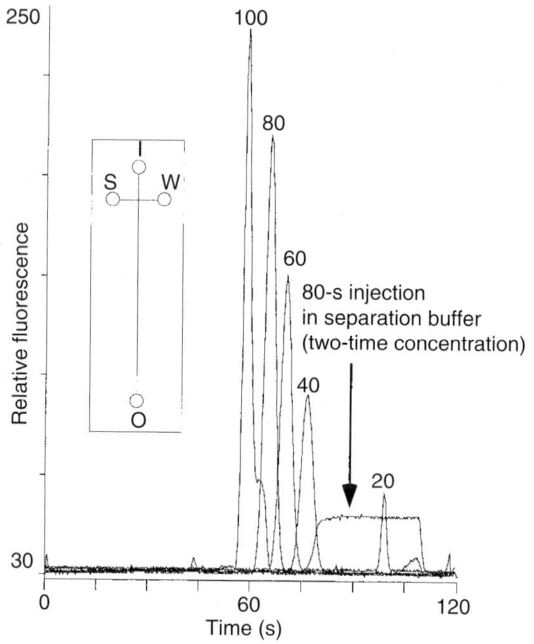

FIGURE 50.9 Electrokinetic stacking injection on a microchip. Sample: 67 nM BODIPY (a neutral fluorescent dye) in 150 mM NaCl. Run buffer: 80 mM sodium cholate, 10% ethanol, 5 mM tetraborate, pH 9. Sample was first injected by applying an electric field of 183 V/cm between the sample reservoir (S) and the outlet (O) for various periods of time as indicated in the figure. Separation was then carried out by applying 366 V/cm between the inlet (I) and outlet (O). An 80-s injection of 134 nM BODIPY in run buffer is included to show nonstacking injection. (From Palmer et al., *Anal. Chem.*, 73, 725, 2001. With permission.)

microfluidic device for ITP separation was reported by Walker et al.,[25] who demonstrated separation of the herbicides paraquat and diquat at submicromolar levels using Raman spectroscopic detection.

ITP extends the stacking concept of FASS to ternary buffer systems. In ITP, the sample is injected between a leading buffer (LE) and a terminating buffer (TE). The mobility of ions in the LE is larger than the mobility of the fastest electrolyte present in the sample, and the mobility of the ions in the TE is lower than the mobility of the slowest electrolyte present in the sample. Owing to the different buffer electrolyte mobilities, a mobility gradient is established between the two buffers. Applying a voltage results in separation of the different electrolytes in the sample between the LE and the TE in order of descending mobilities. Owing to a self-focusing mechanism governed by the Kohlrausch regulation function,[26,27] sharp analyte zones are achieved. This mechanism is based on the fact that ions diffusing out of an analyte zone are always recaptured due to the different field strengths in the neighboring zones. To illustrate this, consider an analyte ion that diffuses out of the front end of an analyte zone. It experiences a sudden drop of the electric field strength as the higher mobility ions in the neighboring zone provide less resistance to current flow. The escaped ion, therefore, slows down and is quickly reabsorbed into its particular analyte zone. The same reasoning is also valid for the back end of the analyte zone. The zone following the analyte zone contains lower-mobility ions, resulting in a higher electric field. Analyte ions diffusing into this zone will exhibit increased velocity back toward their own zone. The focusing effect in ITP for negatively charged analytes is schematically shown in Figure 50.10.

To perform ITP in the conventional format, three different solutions must be introduced sequentially into a fused-silica capillary, making ITP a technique that requires specialized instrumentation and expertise. Researchers have also successfully constructed coupled-column arrangements with capillaries to combine ITP and CE, but again, the attachment of capillaries to one another with

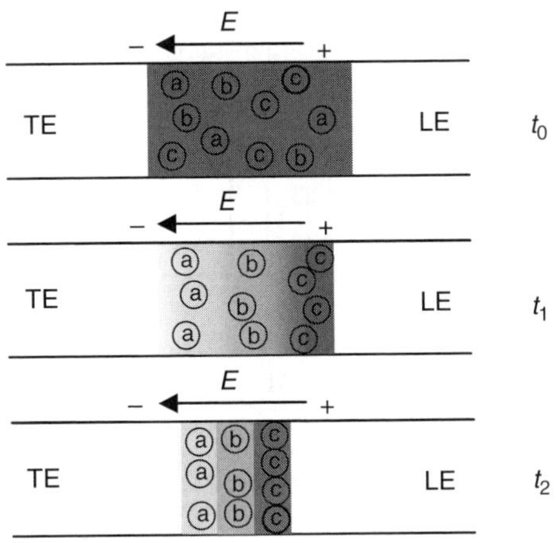

FIGURE 50.10 Schematic of an isotachophoresis sample stacking process for negatively charged analytes. The sample is introduced between the terminating electrolyte and the leading electrolyte (t_0). By applying a potential difference over the sample, the analytes start to migrate through the sample plug (t_1). At a certain point (t_2), small zones containing analytes with different electrophoretic mobilities have formed.

minimum dead volumes to form a robust network is not trivial. In contrast, the interconnected, branched microchannel networks of microfluidic devices are easily fabricated using microtechnologies, making this an attractive alternative for facilitated implementation of ITP and ITP-CE systems. While the first example of on-chip ITP employed only a single serpentine microchannel of uniform cross-section,[25] later examples that appeared starting in 2000 boasted more sophisticated designs. Kaniansky et al.,[28] Grass et al.,[29] and Bodor et al.[30,31] realized ITP on-chip with two coupled separation columns with on-chip conductivity detectors. An example of one of these devices is shown in Figure 50.11, along with an isotachopherogram of a sample containing 14 different anions. With this chip, it was also possible to extract a sample fraction containing focused low-concentration analytes from the ITP zone system and electrophoretically transfer them to the second column for separation by CE. In the process, high-concentration matrix constituents were left behind in the ITP region. This makes the technique very amenable for trace analysis in complex mixtures, where analytes present in high concentration would otherwise mask analytes present in low concentration.

The preconcentration factors or gains in signal that can be achieved with on-chip ITP can be quite spectacular, as will be discussed later. For a model ITP system, which consists of leading electrolyte ion (LE^-), analyte ion (S^-), trailing electrolyte ion (TE^-), and a common leading counterion (LC^+), an expression for the concentration adjustment between two zones in ITP can be derived as[32,33]

$$C_{S,\text{final}} = C_{LE} \frac{v_S\,(v_{LE} + v_{LC})}{v_{LE}\,(v_S + v_{LC})} \tag{50.3}$$

where $C_{S,\text{final}}$ is the adjusted sample concentration and v is the linear velocity. From the derived equation, it can be seen that sample concentration, $C_{S,\text{final}}$, is proportional to the concentration of leading electrolyte. This effect was investigated by Jeong et al.,[32] who demonstrated a threefold improvement of the peak height when increasing the leading electrolyte concentration from 100 to 250 mM. The resolution between analytes for the experiment described by Jeong et al.[34] was reduced, however. The peak height increases with increasing LE concentration, but the time needed for efficient stacking also increases.

Jeong et al.[32] described an improvement in signal of 500-fold, using ITP as a stacking method to preconcentrate simple fluorescein electrolytes using 250 mM NaCl as leading electrolyte.

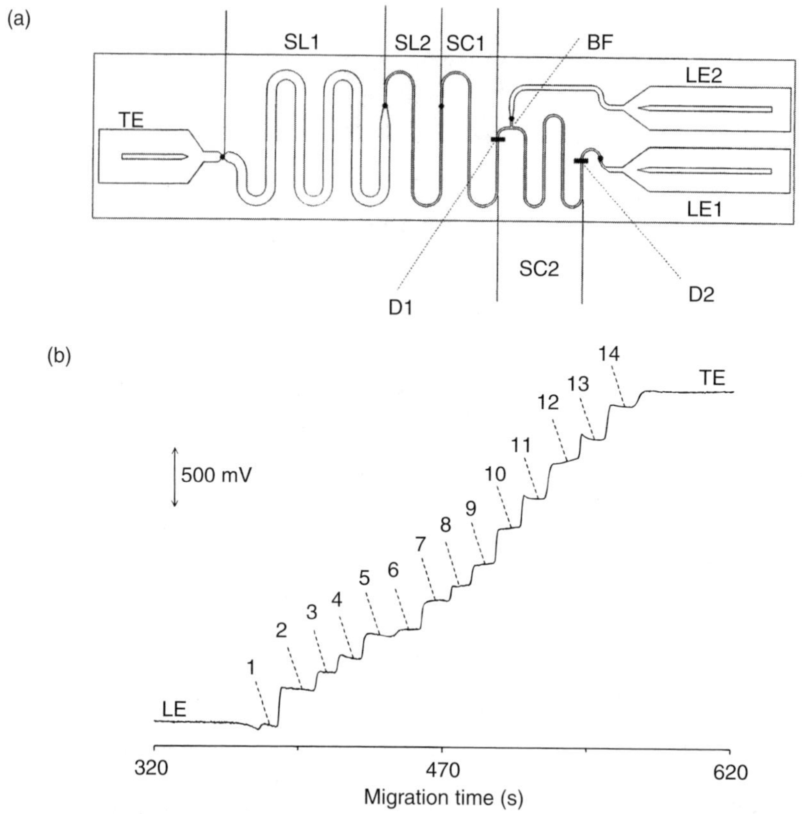

FIGURE 50.11 (a) Schematic diagram of the PMMA ITP microchip used by Kaniansky et al.[28] and described in Reference 29. Chip dimensions: 2 cm × 7 cm. TE: terminating buffer reservoir (10 μL); SL1: large sample loop (15 μL); SL2: small sample loop (1.2 μL); SC1: preseparation column (1.2 μL); SC2: analytical column (1.2 μL); BF: bifurcation block; LE1, LE2: leading buffer reservoirs (15 μL); D1: conductivity detector 1; D2: conductivity detector 2 (both these detectors consist of integrated thin-film Pt electrodes). There are a number of points in the system, indicated by vertical lines in the diagram, at which the different solutions required can be introduced to the various sample loops and separation channels using syringes or syringe pumps. (From Grass et al., *Sens. Act. B.*, 72, 249, 2001. With permission.) (b) An isotachopherogram for the separation of a 14-component model mixture of anions using concentration-cascade ITP. After an initial separation in SC1, the components were introduced to SC2 for a second ITP separation under different conditions, using D2 to monitor the separation. The injected sample contained the anions at 200 μmol/L concentrations. Zone assignments: (LE) leading anion (chloride); (1) chlorate; (2) methanesulfonate; (3) dichloroacetate; (4) phosphate; (5) citrate; (6) isocitrate; (7) glucuronate; (8) β-bromopropionate; (9) succinate; (10) glutarate; (11) acetate; (12) suberate; (13) propionate; (14) valerate; (TE) terminating anion (capronate). (From Kaniansky et al., *Anal. Chem.*, 72, 3596, 2000. With permission.)

Wainright et al.[35] also showed a decrease in the limit of detection by a factor of 400 for samples containing fluorescent molecules having a range of electrophoretic mobilities. The chip designs used in these examples are similar to that in Figure 50.11a. In contrast, Jung et al.[33] demonstrated a million-fold increase in sample concentration using ITP performed in a simple cross layout of microchannels made in glass. The operation of this device, which also includes CE separation of the concentrated analytes, is shown in Figure 50.12. Figure 50.13 compares a CE separation of 100 nM concentrations of two fluorescent dyes with an ITP-CE separation of the same dyes but at 1 pM concentrations. The concentration increase and signal increase for Alexa Fluor 488 (first peak) are 6.4×10^4 and 4.5×10^5, respectively. The best performance in this particular study was obtained for

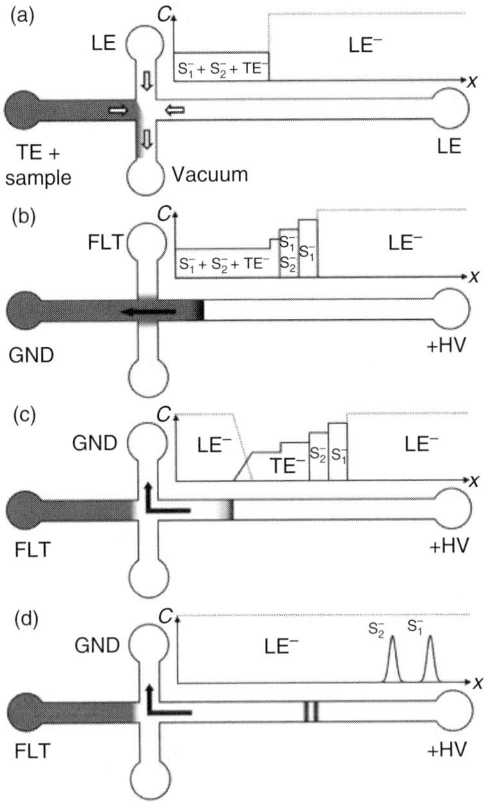

FIGURE 50.12 (a) Schematic of ITP/CE assay protocol. Configurations of co-ions are also shown at each step. (a) The upper and lower reservoirs are filled with LE, and the left reservoir is filled with a mixture of TE and sample. TE/LE boundary is formed by applying vacuum at the lower reservoir. White arrows show the direction of PF. (b) ITP stacking is initiated by applying high voltage and ground at the right and left reservoirs, respectively. The black arrow denotes the direction of electric field. (c) The field is switched toward the upper reservoir to inject LE ions behind the sample and initiate CE. ITP stacking continues until LE ions overtake the TE and sample ions. (d) Separation of samples occurs further downstream where sample ions electromigrate in nearly homogeneous LE electrolyte. (From Jung et al., *Anal. Chem.*, 78, 2319, 2006. With permission.)

an initial fluorescent dye concentration of 100 fM, for which a concentration increase of 2 million was observed, with a signal-to-noise ratio greater than 11. The same researchers later developed a high-sensitivity detection system in conjunction with this microfluidic ITP-CE approach, which allowed the detection of two fluorescent dyes in a sample having an initial concentration of 100 aM (10^{-18} M). With detected concentrations on the order of 15–20 pM, the concentration increase achieved by ITP was 2.1×10^5 fold.[36]

Isotachophoresis as a sample preconcentration method is very useful for concentrating analytes in biological samples. A similar approach as described by Jeong et al.[37] was used for capillary gel electrophoresis of DNA fragments on microchips. Preconcentration of double-stranded DNA fragments based on isotachophoresis with a zone electrophoretic separation step has also been reported by Wainright et al.[38] HCl-imidazole buffer was used as leading electrolyte and HEPES-imidazole buffer as terminating buffer. An increase in detection sensitivity of 40-fold compared to standard chip zone electrophoresis was observed. The advantage of isotachophoresis, in contrast to field-amplified stacking, is the option of analyzing samples with a high salt concentration. However, it is not possible

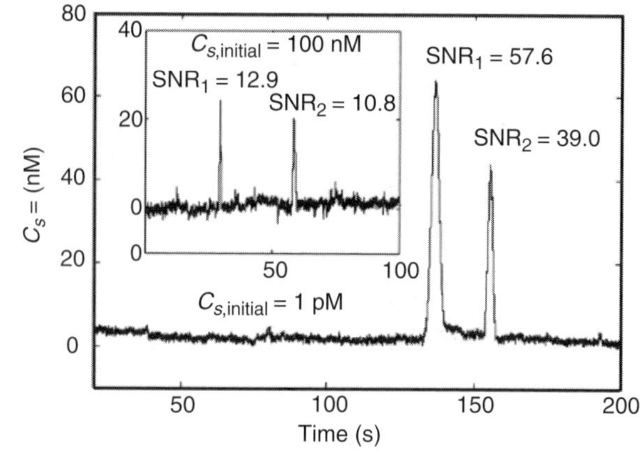

FIGURE 50.13 Comparison of ITP/CE and CE separations of two fluorescent dyes Alexa Fluor 488 (first peak) and BODIPY (second peak). Inset electropherogram shows separation (with no stacking) of 100 nM initial concentrations of Alexa Fluor 488 and BODIPY. 5 mM HEPES was used as a background electrolyte and the applied electric field was 280 V/cm. For ITP/CE mode (main plot), initial concentrations of Alexa Fluor 488 and BODIPY were both 1 pM in 5 mM HEPES (i.e., diluted by a factor of 1×10^5). Here, LE was 1 M NaCl and the nominal applied field was 220 V/cm. The concentration increase and signal increase for Alexa Fluor 488 (first peak) are 6.4×10^4 and 4.5×10^5, respectively. (Reprinted from Jung et al., *Anal. Chem.*, 78, 2319, 2006. With permission.)

to concentrate anions and cations in the same run, and it is not possible to concentrate neutral species in their native form.

50.2.3 Isoelectric Focusing (IEF)

One of the ways in which amphoteric molecules differ from one another is their isoelectric point (pI), the pH at which overall molecular charge is zero. In isoelectric focusing (IEF), a pH gradient is generated in a channel or capillary in either gel or free-solution form, and sample analytes are driven through this gradient by an applied electric field. The anode is located at the acidic end of the gradient and the cathode at the basic end. In most cases, the pH ranges from 3 to 10, a range that is established by filling the channel with an acidic compound (so-called anolyte) and a basic compound (so-called catholyte). Application of an electric field results in the formation of different buffer regions in the system. The analytes will travel through the capillary and become focused at the location in the gradient where the pH equals their pI. A molecule with a net positive charge, for example, will migrate in the increasingly basic gradient and gradually become less charged, until the point where the local pH corresponds to the pI of the molecule. At this point, the overall molecular charge is zero, and the molecule will cease to migrate in the applied electric field.[39] Since the majority of biological samples are amphoteric molecules, IEF is suitable for stacking many types of biological analytes. Chapter 19 in this book provides an excellent discussion on IEF in capillary systems.

IEF is the conventional method for protein separation in two-dimensional (2D) polyacrylamide gel electrophoresis and is considered one of the most powerful techniques available for separating proteins. However, IEF in gels is a time-consuming technique. Hjertén and Zhu,[40] therefore, adapted IEF to fused-silica capillaries, so-called capillary IEF (cIEF), to minimize analysis times. No gels are used in cIEF; instead, the capillary is filled with ampholytes in free solution to create the pH gradient. This technique is also called liquid-phase IEF. Additional advantages of cIEF are the potential for system automation and the possibility of on-line detection.

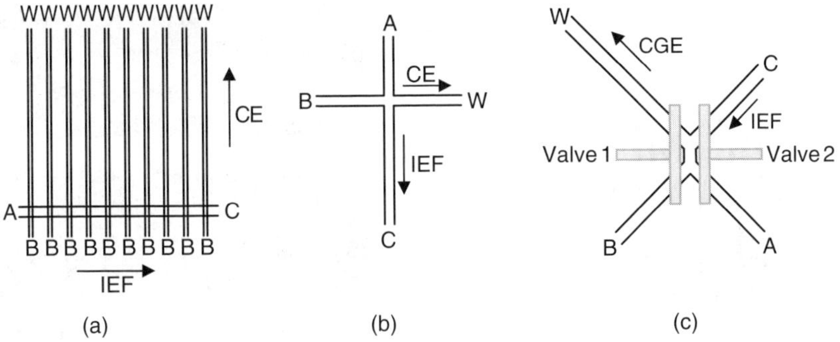

FIGURE 50.14 Different chip designs for IEF coupled to CE. The chip with gel-filled microchannel used by Chen et al.[45] (a), the chip used by Herr et al.[46] (b), and the chip used by Wang et al.[47] (c). The first dimension (IEF) extends from reservoir A (anolyte) to reservoir C (catholyte) and the second dimension from reservoir B (buffer) to reservoir W (waste).

Conventional IEF and cIEF are generally known as analytical separation techniques. However, IEF can also be used as a stacking or preconcentration technique, particularly when coupled to SDS-gel electrophoresis for protein analysis. Two-dimensional gel electrophoresis (2D-GE) has become one of the workhorses for the rapidly growing field of proteomics. The use of IEF as the first dimension provides typical preconcentration factors of between 50 and 100, in addition to excellent separation resolution.[41] The miniaturization and integration of 2D protein separations into microfluidic devices has attracted much attention recently, as such platforms would provide a significant reduction in analysis times. Column switching is also achieved much more easily in these devices. However, the goal of 2D-GE has proven to be elusive, due in part to the need for multiple separation media, which must be localized in different regions of a device.

There are a few early examples of IEF performed in microfluidic devices, including IEF of green fluorescent protein (GFP) in a plastic microchannel,[42] continuous concentration of bacteria in a microfluidic flow cell,[43] and microchannel IEF for protein separation and the study of protein–protein interactions.[44] However, the first published example of preconcentration of protein mixtures with IEF in gel-filled microchannels coupled to SDS-gel electrophoresis was reported more recently, by Chen et al. in 2002.[45] The device, formed in the elastomer, poly(dimethylsiloxane) (PDMS), consisted of several layers, containing both in-plane and vertical microchannels to form a 3D microfluidic system. A schematic diagram of the system layout is shown in Figure 50.14a. As this figure shows, IEF is carried out in a single channel first, to form a series of focused, separated zones. These zones are then transferred simultaneously to an array of microchannels, which are positioned in another level of the device and perpendicular to the IEF channel. A series of vertical holes ensures analyte transfer from one level to the next. Each zone then undergoes further separation by SDS-GE in an individual channel. In this example, the IEF gel was embedded by applying vacuum to one end of the PDMS microchannel to draw gel solution from a reservoir at the other end. Unfortunately, no factors for sample enhancement were given.

The device by Chen et al. required performing the IEF separation in one PDMS device first, then disassembling this to remove the enclosed IEF channel for reassembly in a second PDMS device also containing the microchannels for SDS-GE. Li et al.[41] reported a simpler device for 2D-GE having a layout again like that shown in Figure 50.14a, but formed by imprinting a single network of microchannels into a polycarbonate (PC) layer. This layer was sealed by a second PC layer into which a series of holes had been made to align with the microchannels. One of the main challenges with this device was controlling the local composition of media in the IEF and SDS-GE channels, which needed to be different for these two operations. The absence of integrated valves to close off one channel system from the other meant that it became more difficult to introduce a medium into one system without affecting the other. Two approaches were found to resolve this issue, however.

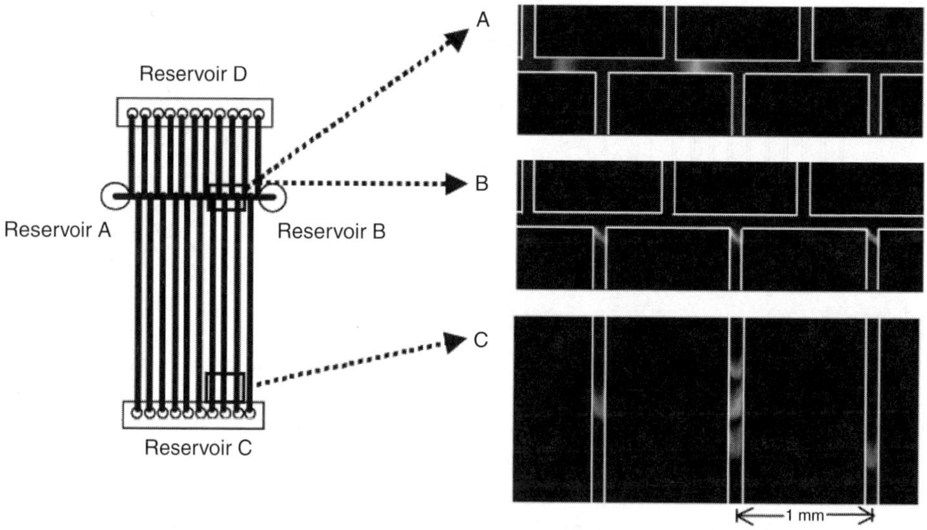

FIGURE 50.15 Fluorescent images of on-chip 2D separation of five model proteins using multiple separation media. (a) Non-native IEF with focusing order of (i) actin; (ii) bovine serum albumin, ovalbumin, and trypsin inhibitor; and (iii) parvalbumin from left to right; (b) electrokinetic transfer of focused proteins; (c) SDS-gel electrophoresis. (From Li et al., *Anal. Chem.*, 76, 742, 2004. With permission.).

One involved simply using the same gel medium for both separations so that the entire chip was filled initially with the same gel and solution phase. The IEF solution was replaced after separation by flushing the SDS-GE channels with the SDS-containing buffer required for this separation. The second approach involved filling the entire chip with gel medium, followed by introduction of IEF solution medium into the IEF channel, thereby replacing the gel in this region. Figure 50.15 shows the layout of channels in the PC chip on the left. The images on the right show images of IEF focusing (panel A), electrokinetic transfer of proteins (panel B), followed by SDS-GE separation of the transferred proteins (panel C).

Herr et al.[46] reported the coupling of liquid-phase IEF with free-solution CE in an acrylic device using a simple cross layout of microchannels. The IEF channel was filled with a liquid pH gradient, by filling the buffer reservoir of the cathode with sodium hydroxide (pH 10) and the buffer reservoir of the anode with phosphoric acid (pH 3). The sample components were separated by their isoelectric points in the IEF channel and were mobilized to an intersection by EOF, where fluidic volumes were electrokinetically sampled into the CE channel. With this chip, an increase in sample concentration of 70-fold was obtained in a separation time of less than 1 min. A schematic view of the chip is given in Figure 50.14b. Images showing IEF of proteins and the intermittent sampling of the resulting zones for CE separation in the second dimension are shown in Figure 50.16.

A third approach for coupling IEF with capillary gel electrophoresis has been reported by Wang et al.[47] and is given in Figure 50.14c. Microfluidic valves based on a thin deformable PDMS membrane and pneumatic actuation[48] were integrated into this device. These valves could be operated in such a way as to physically separate the two separation systems and prevent mixing between the different buffers used. The sample was first focused in the IEF channel. When the analyte of interest entered the transfer region, both valves were closed in order to trap the analytes. After opening valve 1, the trapped analytes were guided into and separated in the gel electrophoresis channel. With the two valves, very controllable flows were obtained. The advent of integrated valving has meant new options for researchers developing microfluidic applications in which procedures that are incompatible from a solution or medium point-of-view must be coupled with one another.

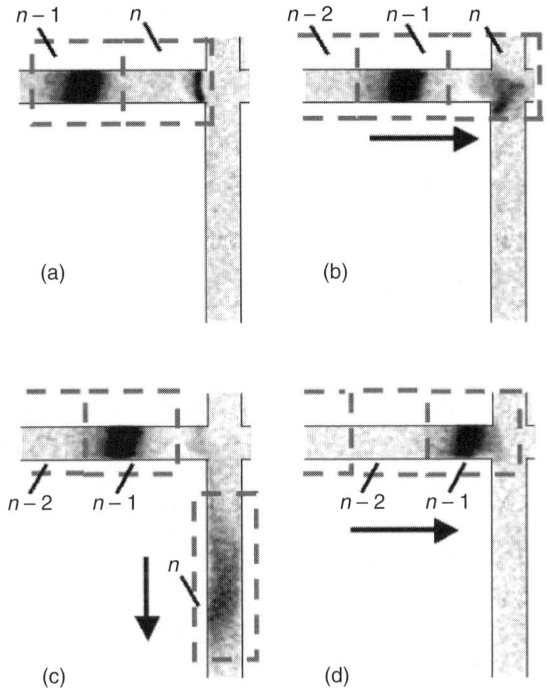

FIGURE 50.16 CCD images during species sampling. (a) Species are focused by IEF in the first dimension (dark bands in horizontal channel). Simultaneously, the bands are mobilized by EOF toward the catholyte reservoir (not visible on the right). (b) Once a fluid volume of interest, n, reaches the microchannel intersection, all applied potentials are switched to floating. (c) High voltage is then applied at the upper reservoir while the lower reservoir is grounded. This initiates protein separation in the second dimension by CE. (d) On completion of the CE separation, IEF/EOF is reinitiated causing sample species to refocus and the next fluidic volume $(n - 1)$ to migrate to the intersection. (From Herr et al., *Anal. Chem.*, 75, 1180, 2003. With permission.)

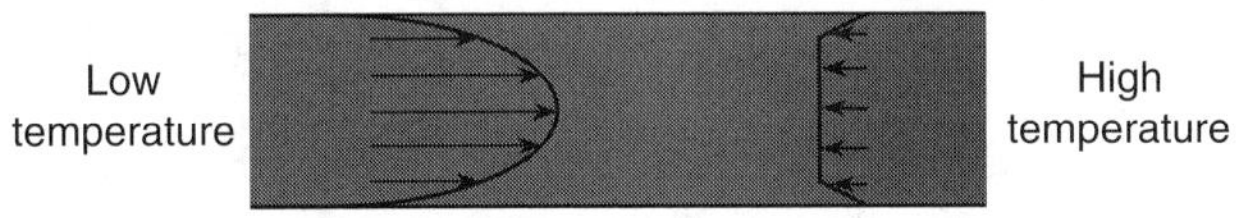

FIGURE 50.17 Schematic representation of the concentration of ionic species in temperature gradient focusing.[49]

50.2.4 Temperature Gradient Focusing (TGF)

Temperature gradient focusing (TGF) is a relatively new technique for the concentration and separation of ionic species in microchannels or capillaries (Figure 50.17). The first TGF on chip was reported by Ross and Locascio in 2002.[49] In TGF, one end of the channel is heated while the other end is cooled, resulting in a temperature difference along the channel.

Since the conductivity of a buffer is temperature-dependent, a conductivity gradient is established in the channel due to the temperature difference. The relationship between the conductivity and the temperature can be written as

$$\sigma(T) = \frac{\eta(20)\sigma_0}{\eta(T)f(T)} \tag{50.4}$$

where T is the temperature, σ_0 is the conductivity at $T = 20°C$, $\eta(20)$ is the viscosity at $T = 20°C$, $\eta(T)$ is the temperature-dependent viscosity, and $f(T)$ is a function that accounts for all other potential temperature-dependent factors besides viscosity that could affect conductivity, such as ionic strength. $f(T)$ is normalized to $f(20) = 1$.

Because the conductivity is related to the electric field (see Equation 50.1), the electric field is also temperature-dependent. Using Equations 50.1 and 50.4, the electric field dependence on the temperature can be written as

$$E(T) = \frac{I\eta(T)f(T)}{A\sigma_0\eta(20)} \equiv E_0\frac{\eta(T)f(T)}{\eta(20)} \quad (50.5)$$

where E_0 is the electric field at $T = 20°C$.

The electrophoretic mobility of a charged analyte is also dependent on viscosity, which is dependent on temperature, as given in Equation 50.6. The relationship between the viscosity and the temperature is given by

$$\mu_{EP}(T) = \frac{\eta(20)\mu_{EP,0}}{\eta(T)f_{EP}(T)} \quad (50.6)$$

where $\mu_{EP,0}$ is the electrophoretic mobility at $T = 20°C$, and $f_{EP}(T)$ is a function that accounts for all other temperature-dependent factors besides viscosity that affect mobility. $f_{EP}(T)$ is normalized so that $f_{EP}(20) = 1$.

Combining Equations 50.5 and 50.6 results in the velocity of an analyte (ν_{EP}):

$$\nu_{EP}(T) = E_0\mu_{EP,0} \cdot \frac{f(T)}{f_{EP}(T)} \quad (50.7)$$

According to Walden's rule, the conductivity is inversely proportional to the viscosity. This means that the temperature dependence of the electric field and electrophoretic mobility tend to cancel each other out (see Equation 50.7), except for buffers where $f(T)/f_{EP}(T)$ is not constant. Most buffers and analytes have constant $f_{EP}(T)$, so buffers with a temperature-dependent $f(T)$ have to be chosen to achieve a velocity gradient.

Ross and Locascio showed that buffers made from Tris and boric acid have a temperature-dependent $f(T)$, due to the ionic strength being temperature dependent. This means that an electrophoretic velocity gradient can be achieved in the channel when a temperature gradient is applied. Concentration of the analytes is achieved by balancing the electrophoretic mobility of analytes (μ_{EP}) against a bulk flow of buffer solution (μ_{Buf}), which is shown in Figure 50.17. The average bulk mobility of the buffer can be driven by using an applied pressure gradient, electroosmosis, or a combination of the two.

The analytes will focus at point (μ_T), where the sum of the electrophoretic mobility of the analyte (μ_{EP}) and the bulk mobility of the buffer (μ_{Buf}) equals zero:

$$\mu_T = \mu_{EP} + \mu_{Buf} = 0 \quad (50.8)$$

Achieving high-resolution separation in CE requires long channels or capillaries. In contrast, equivalent resolution can be achieved with TGF with much shorter microchannels (a few millimeters). For this reason, TGF is very suitable for integration into microfluidics.

Ross and Locascio[49] reported the first microfluidic device for temperature gradient focusing. They used the Tris/borate buffer due to the non-constant $f(T)$. Copper blocks were used at the ends of the microchannels to realize the temperature gradient. With this setup, a 10,000-fold increase in analyte concentration was achieved after 100 min of focusing for a carboxylic acid solution. After 1 min, the degree in focusing ranged from 2- to 100-fold, depending on the analyte mobility, applied voltage, and temperature gradient.

Ross and Locascio also demonstrated a more elegant way to apply a temperature gradient within a microchannel, namely a gradient generated by internal Joule heating. This means that the applied electric field, which drives the focusing, also generates the temperature gradient. Change in resistance within the channel is a result of changing the dimensions of the channel so that the diameter of the channel at one end is smaller than at the other end.[50] The resistance of the fluid (R) is given by

$$R = \frac{\rho L}{A} \tag{50.9}$$

where ρ is the solution density, L, the length of the column, and A the cross-sectional area. It can be concluded from Equation 50.9 that when the cross-sectional area of the channel increases (increasing channel diameter), the resistance of the fluid decreases. The resistance is related to the heat liberated, according to Joule's law:

$$Q = I^2 R t \tag{50.10}$$

where Q is the amount of heat liberated, I is the current, R is the resistance, and t is the time. Equations 50.9 and 50.10 show that when the cross-sectional area of the channel decreases, the resistance increases and the heat liberated also increases. This increase in resistance is due to the increase in current density. A schematic view of the chip used by Ross and Locascio is shown in Figure 50.18.

The analytes are focused in the temperature gradient. With this chip (only 1.7 cm long), a 300-fold increase in concentration was obtained for a carboxylic acid solution.

Balss et al.[51] reported the concentration of enantiomers with TGF using two copper blocks for the temperature gradient. The concentration of small molecules, including fluorescent dyes, DNA, proteins, particles, and whole cells has also been demonstrated using the TGF technique.[49,51,52] Balss et al.[53] combined TGF with MEKC to obtain so-called micellar affinity gradient focusing (MAGF) for focusing neutral analytes. Although the hydrophobic interiors of the micelles interact with neutral analytes by partitioning, micelles are charged, which means they can be focused in the temperature gradient based on their electrophoretic mobility.

50.2.5 Flow-Induced Electrokinetic Trapping (FIET)

The use of recirculating flows for the preconcentration of samples in microfluidic devices is, like TGF and nanofluidic filtering (see later), a relatively new approach (early 2000s). The technique, which has been termed flow-induced electrokinetic trapping (FIET), involves the concentration and

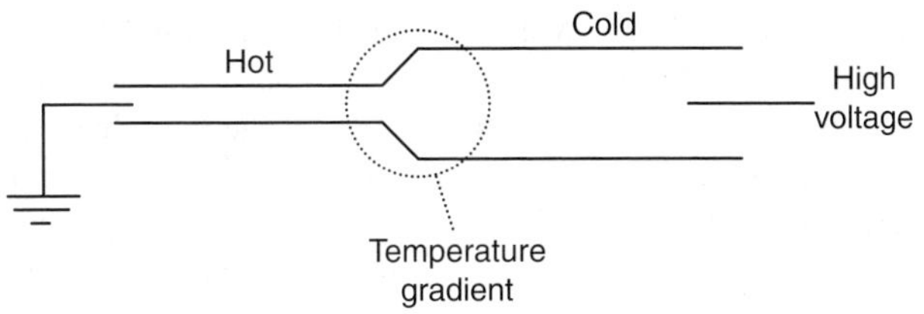

FIGURE 50.18 Schematic view of the chip used for TGF generated by internal Joule heating. (Adapted from Ross et al., *Anal. Chem.*, 74, 2556, 2002.)

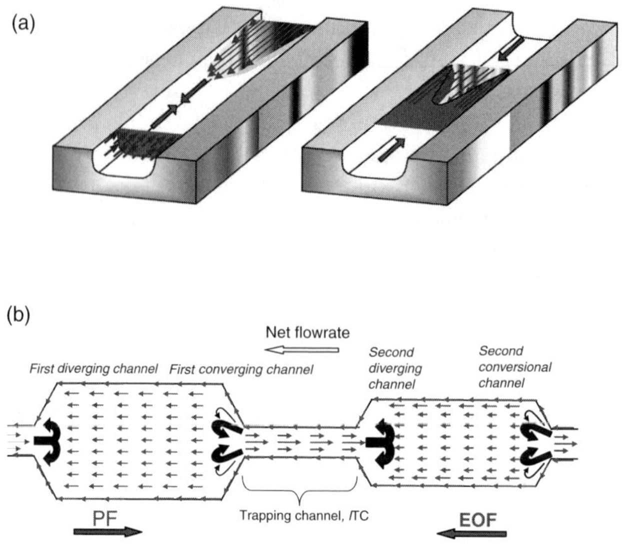

FIGURE 50.19 (a) Schematic showing how bidirectional flow is generated in a microchannel. (b) Schematic diagram of a FIET device. (From Lettieri et al., *Lab Chip.*, 3, 34, 2003. With permission.)

manipulation of particles using fluid flows only. Electrophoretic migration of charged particulate species is species-dependent, dictated by the species' electrophoretic mobility, $\mu_{ep,part}$, which is a direct function of its electrostatic surface properties. If a charged particle is to be captured and accumulated in an applied electric field, conditions that result in a zero net force being applied on the particle need to be established. In FIET, this is accomplished by applying counteracting EOF and PF in a microchannel. The technique relies on the different flow profiles exhibited by EOF and PF (Figure 50.19a).

The microchannel geometry used is easily made by microfabrication and consists of a narrow channel (<100 μm), which opens up into two tapered elements at either end (Figure 50.19b). When EOF and PF co-exist, the resulting velocity profile is simply the sum of the velocity profiles for the two flows (Figure 50.19a).[54–56] The observed ratio of flows required for trapping is such that bidirectional flow is established in the narrow channel segment. PF prevails down the channel core in one direction with EOF along the channel walls in the other. The tapered elements are thus described as being either diverging (PF exits the narrow channel) or converging (PF enters the narrow channel). The geometry of the diverging and converging elements results in the establishment of well-defined electric field and pressure gradients in these regions. The flow velocity gradients thus created are manifested as recirculating flow patterns. Particles within the microfluidic device are drawn toward these gradients and are caught in the rotating flows, resulting in the formation of concentrated particle zones. Although each zone of trapped particles is stationary, the particles themselves continue to follow rotating streamlines. A number of different types of 1- and 2-μm polymer microspheres have been successfully captured in this device, as shown in Figure 50.20.

A couple of features about FIET are worth noting. First, as alluded to earlier, the conditions of electric field E and pressure under which a particle is trapped are dictated by its electrophoretic velocity. In other words, FIET can be tuned to the $\mu_{\mathrm{ep,part}}$ of the particle of interest. Second, while bidirectional flow exists over a range of electric field strengths at any given applied pressure, there is only one value of electric field strength at which total net flow is equal to zero. Thus, particles can be trapped under conditions of nonzero net flow. This makes the FIET system unique, as it allows concentration of particles in a stationary zone under conditions of net flow. This offers the possibility to process the trapped particles through perfusion with different reagents, and to flush away untrapped species.

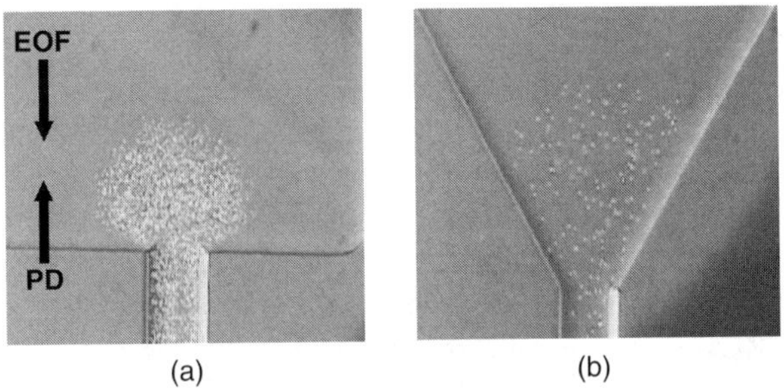

FIGURE 50.20 These photos show trapping of Protein-A coated fluorescent polymer microspheres (1 μm) at diverging channel elements. Boundary conditions are the same for both diverging channels, that is, an average electric field of $E = 200$ V/cm and PD = 7 mbar (7 cm of buffer).

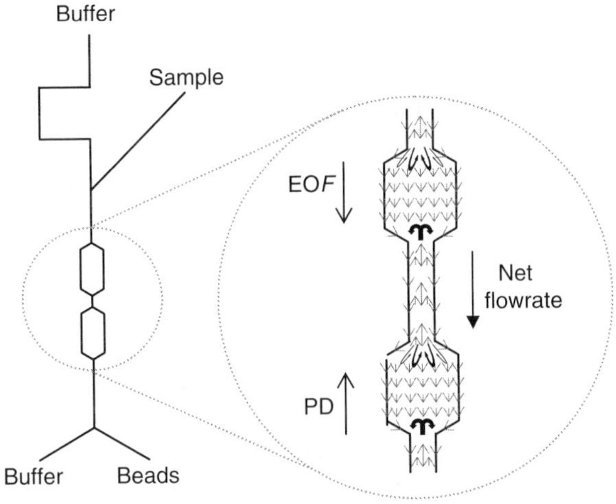

FIGURE 50.21 Schematic of a device layout used by Lettieri et al.[57–59] and an exploded view of the microfluidic vortices when the overall velocity is close to zero.

A schematic view of one of the devices used by Lettieri et al.[57–59] for the recirculating flow is shown in Figure 50.21. In this particular embodiment, the beads are introduced into the system from the lower right reservoir by applying a pressure difference using a narrow column of buffer solution several centimeters high. A voltage is applied between the buffer reservoir at the top and the one at the bottom, resulting in an EOF. When the overall velocity of the beads is close to zero, the beads are trapped in the small central channel and are subjected to recirculation. Lettieri et al. demonstrated the potential of FIET for bead-based assays by perfusing trapped negatively charged streptavidin-coated beads with a solution containing fluorescein-labeled biotin and monitoring the development of a fluorescent signal within the bead cluster.

Lettieri et al.[58,60] also demonstrated the recirculation and concentration of double-stranded λDNA in a trapping channel. The device used for the experiment was similar to the one presented in Figure 50.21. This means that DNA can be extracted and concentrated using the FIET principle, extending the applicability of this technique to biological "particles". This type of approach could be a nice alternative for many microfluidic applications now requiring solid-phase extraction (SPE). Because trapping occurs in free solution, there is no need for a SPE step of any kind. This simplifies the transfer of material from the FIET component to a subsequent unit operation in a microfluidic device, as it obviates the need to remove material from a surface. FIET is just one example of how

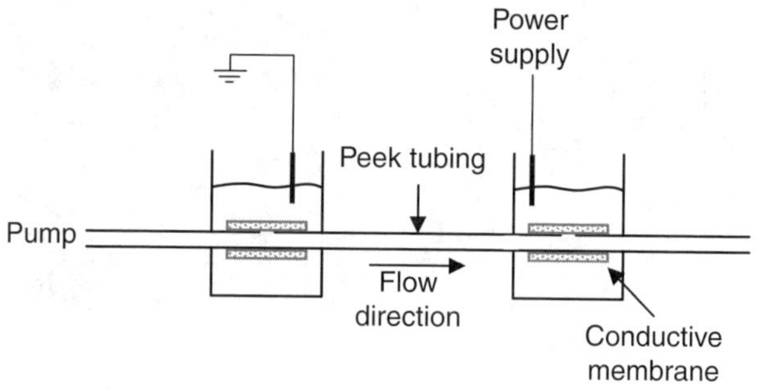

FIGURE 50.22 Schematic representation of the device used for preconcentration by Astorga-Wells. (Adapted from Astorga-Wells et al., *Anal. Chem.*, 76, 2425, 2004.)

the unique flow properties at the micro- and nanometer scale can be exploited to perform certain sample handling operations in completely new ways.

Astorga-Wells et al.[61,62] demonstrated a similar trapping principle for the preconcentration of peptides and proteins for conventional liquid-based separation systems. A schematic representation of the microfluidic device used for the preconcentration is given in Figure 50.22.

Since poly(etheretherketone) PEEK tubing is used (wall is not charged), no EOF is obtained. The proteins and peptides were introduced into the PEEK tubing by PF. When a voltage was applied over the PEEK tubing, the charged species having an electrophoretic velocity higher than the velocity of the flow will be retained. Neutral species and species with a lower electrophoretic velocity will not be retained. By decreasing the applied voltage over the tubing, the charged species are eluted out of the PEEK tubing depending on their electrophoretic velocity. The device described by Astorga-Wells is not yet implemented on a microchip, but could easily be done.

Other examples of the use of gradients of counteracting electroosmotic and pressure-driven forces to preconcentrate small molecules have been reported. Hori et al.[63] demonstrated this principle of analyte concentration in a conventional flow system in 1993, concentrating small molecules and DNA by factors of 10 to 47. Ista et al.[64] recently demonstrated this approach in a microfluidic device for small molecules.

50.3 SOLID-PHASE EXTRACTION (SPE)

In pressure-driven chromatography, the technique most used for sample concentration is solid-phase extraction (SPE). The first experimental applications of SPE were reported five decades ago in conjunction with the first attempts to perform trace analysis of organic compounds in water.[65] The number of citations for the term "SPE in pharmaceutical chemistry" has increased tremendously since the late 1980s.[65]

SPE is an extraction method that uses a solid phase and a liquid phase to isolate and preconcentrate analytes of interest from a solution. The technique is based on partitioning of the analyte between the liquid phase and the extraction material (solid phase), with the mechanism of interaction depending on the extraction material. SPE is very popular, not only because it concentrates the sample, but can also be used to remove the analyte from interfering molecules if the interfering molecules do not show interactions with the SPE material. Chapter 27 of this book is devoted to SPE and its cousin, solid-phase microextraction.

There are three possibilities for the implementation of SPE in microfluidic devices, namely, (1) to fill a channel with particles that serve as the extraction material, (2) coat the channel wall with the extraction material, and (3) fill a channel with a polymeric rod (monolithic phase) extraction material.

50.3.1 Particles

While approaches such as FIET may replace the way we perform extraction and preconcentration in the future, it is clear that the general availability of a large number of optimized extraction phases will ensure that SPE is not abandoned any time soon in microfluidics. However, using particulate phases requires developing good protocols for packing them into microchannels. Irregular packing can occur when small microchannels are packed with particles having diameters not much less than the channels themselves. SPE particles are often quite small, about 3–5 μm in diameter. However, if they are packed into a channel having a depth of only 20 μm, it is clear that not more than four particles fit across the channel. Voids in the packing formed particularly at the wall can result in bandbroadening during elution.

Particles also need to be retained locally in a microchannel, particularly if the on-chip SPE process is to be coupled with a separation directly in the device. This generally requires that some kind of porous physical barrier or frit be formed in a microchannel, often a nontrivial process. Ramsey and Collins[66] reported the fabrication of methacrylate frits in microchip devices made of glass. The channel was filled with a polymer solution containing a monomer, cross-linker, photoinitiator, and an EOF promoter. Polymerization was achieved by a photopolymerization process started on exposure of the polymer solution to ultraviolet (UV) radiation. To polymerize only a small part of the channel, two pieces of black electrical tape were used to mask the channel except for a small slit (~200 μm wide). After polymerization, the device was flushed thoroughly, and the channel was packed by pressure with octadecyl-derivatized silica beads. The formation of silicate frits in microchannels to aid in the generation of EOF for better flow control in glass-based microreactors has been described by Christensen et al.[67] and Skelton et al.[68] The main problem generally in fabricating frits for retaining packing material is the poor reproducibility, an issue that several authors have discussed.[69–71]

Oleschuk et al.[72,73] showed that it is possible to create a structure within the microchip to retain the particles by using differential etching instead of frits. Two photomasks were used to create the device shown in Figure 50.23. The first was used to perform a shallow etch of the tops of the weirs (walls over which solution enters the chamber) and the second to etch the channels. The space between the top of the weirs and the cover plate was only 1 μm, which is less than the diameter of the individual particles (1.5–4 μm). This means the particles were unable to cross the weir. The particles were introduced into the chamber using an additional channel. In Figure 50.23, a cross-section of the packed chamber and the channel layout with the bead introduction channel are shown. To achieve a uniformly packed chamber, particle slurry was introduced by applying a high voltage between the bead introduction channel and the sample inlet and outlet. The EOF obtained transferred the neutral

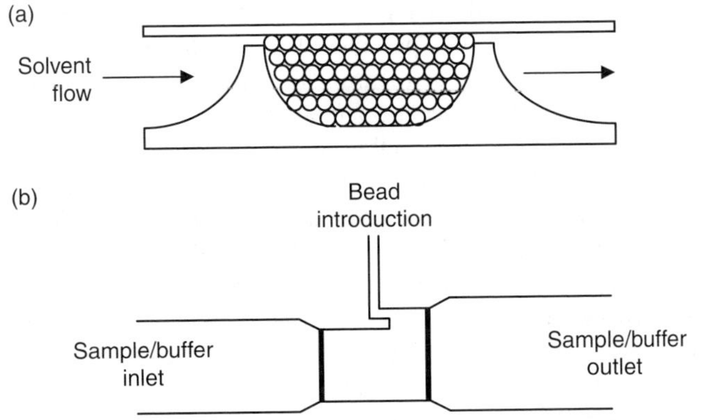

FIGURE 50.23 (a) Schematic view of cross-section of packed chamber without the need for frits and (b) top view of channel layout used by Oleschuk et al. (From Oleschuk et al., *Anal. Chem.*, 72, 585, 2000. With permission.)

(a)

(b)

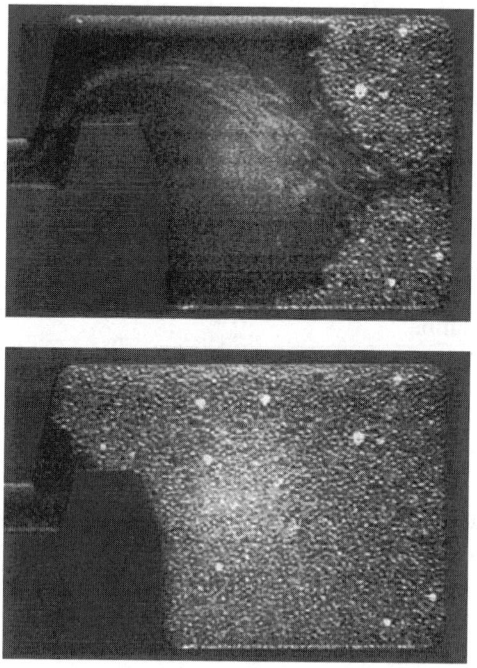

FIGURE 50.24 Images of packing the chamber (a) at an initial stage and (b) after it is completely filled with beads. (From Oleschuk et al., *Anal. Chem.*, 72, 585, 2000. With permission.)

particles to the chamber. The bead introduction channel is hook-shaped at the chamber entrance. This results in uniform chamber packing and keeps the particles in the chamber after packing. It also minimizes backflow into the bead introduction channel.

An advantage of the chip described by Oleschuk et al.[72] is the possibility to remove the beads packed in the chamber and replace them with other beads having a different partitioning mechanism. If the chamber is not tightly packed, the beads can be removed and repacked just by reversing the voltages between the sample inlet and outlet channels and the bead introduction channel. The two images in Figure 50.24 show how the chamber is packed electrokinetically. With this chip, a concentration enhancement of up to 500-fold is observed for the nonpolar fluorescent dye, BODIPY.

Another way to pack channels with particles without using frits is to make use of the so-called "keystone effect." Lord et al.[74] showed that tapering conventional capillaries to an inner diameter of approximately 10 μm was sufficient to trap particles of 3 μm. Mayer et al.[75] also used the "keystone effect" to pack capillaries for capillary electrochromatography (CEC). Introducing particles toward the taper by applying a pressure on the inlet and vacuum on the outlet reservoir results in an increase in density of the particles at the taper. Due to this density increase, the particles aggregate without requiring a physical barrier or filter. The first particles that are aggregated act as the "keystones," blocking the other particles and allowing the packed section to grow. Ceriotti et al.[76] adapted the "keystone effect" to microfluidic devices for CEC applications. The "keystone effect" and the channel dimensions used by Ceriotti et al. are shown in Figure 50.25.

Surprisingly, the C18 packings made by Ceriotti et al. remained in place when solvent was flushed through them, even though there was no taper or frit at the end of the microchannel to retain the packing. It was surmised that a heat treatment was sufficient to stabilize the packing. This treatment consisted of filling the channel with water and placing the chip in an oven overnight at 115°C. Due to this heating, the outer surface of the particles is believed to dissolve in water, forming a saturated solution of polysilic acids, which redeposits as silica between the particles on cooling. This interparticle bonding was proposed by Adam et al.[77]

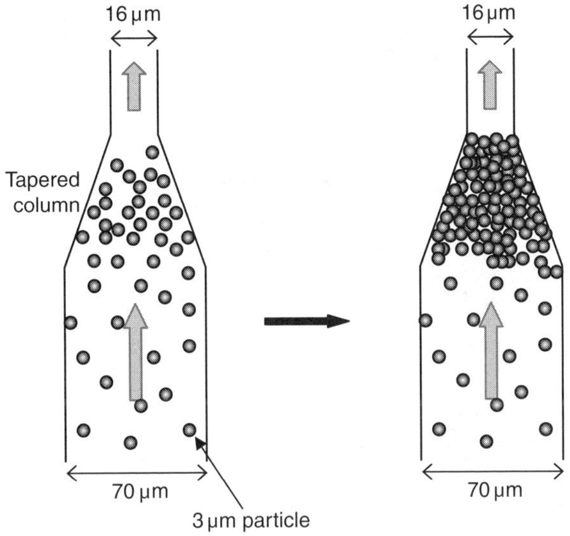

FIGURE 50.25 Schematic view of the "keystone effect." (From Ceriotti et al., *Anal. Chem.*, 74, 639, 2002. With permission.)

The concentration of the particles in the suspension is critical. At high concentrations, the particles aggregate before the tapered column and at low concentrations the particle density at the tapered column is too low, resulting in particles entering the injection channel (the channel with an internal diameter of 16 μm).

There have been no articles published to date reporting the use of the "keystone effect" for packing channels for sample preconcentration, although this should be possible.

The first commercially available chip with sample enrichment column for the analysis of biological samples has been available since early 2005 from Agilent Technologies (Palo Alto, California, USA). The chip integrates an HPLC column, sample enrichment column, and nanoelectrospray tip. The sample and mobile phase are introduced with an external nanopump. The enrichment column and analytical column are packed using a modified version of the method described by Smith and Evans.[78] The particles (5 μm) are kept in the trap column by closing it with a frit at one end. There are five microchannels with an internal diameter smaller than the particle size at the other end, which also act as a frit. With this setup, it is possible to elute the analytes from the packing material by using a backward flush method. More details about the chip layout and the working of the chip are extensively described in References 79–81.

50.3.2 Wall Coating

Kutter et al.[82] reported a microfluidic device made of glass with C18-coated channels. The channels of the chip were modified using octadecyltrimethoxysilane in dried toluene at room temperature.[83] The trimethoxysilane reacts with the silanol groups of the glass resulting in covalently bonded octadecyltrimethoxysilane on the wall of the channel. An 80-fold increase in concentration was obtained for a coumarin dye. In Figure 50.26, an example is given of the preconcentration of an 8.7 nM coumarin dye C460, which was preconcentrated for 160 s. A borate buffer with 15% acetonitrile was used for sample loading and a borate buffer with 60% acetonitrile was used for elution.

The chip described by Kutter et al. has limited sample trapping capacity. The phase ratio of the stationary phase bonded to the inner wall is low, and not all of the sample may be extracted using

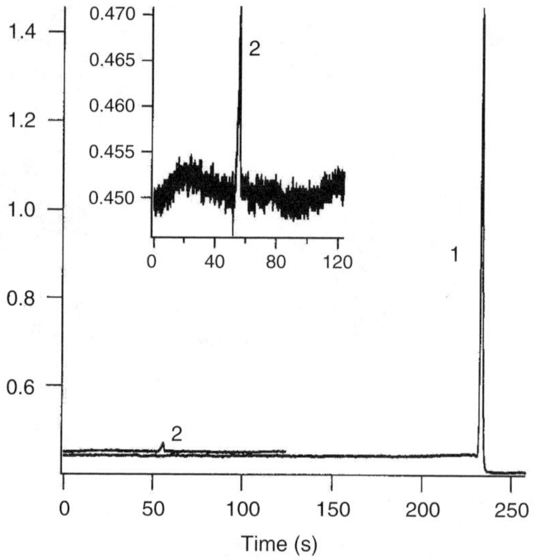

FIGURE 50.26 Chromatograms of (1) 160-s injection with concentration and (2) 1-s injection without concentration. (From Kutter et al., *J. Microcol. Sep.*, 12, 93, 2000. With permission.)

this type of approach. Despite their high surface-to-volume ratios, the available surface area for extraction can be significantly increased in microchannels through packing with microspheres.

50.3.3 Polymers

To increase the surface-to-volume ratio in the preconcentration channel without the need for particles, the channel can be filled with a polymeric rod. These are formed by *in situ* polymerization, during which the polymer material also reacts with the wall of the channel. As a result, no frits are needed to hold the material in place. Columns filled with a polymeric rod, so-called monolithic columns, were originally developed for conventional liquid chromatography. They are made by sol-gel technology,[84] which enables the formation of a highly porous material containing macropores and mesopores in its structure.[85] The use of a monolithic phase circumvents the problems encountered when packing a column with particles.

The polymerization of monolithic material in-chip has been described by several authors. In most cases, the monolithic phases were prepared by photoinitiated polymerization[86] of methacrylate mixtures introduced to the microchannels. Once the microchannels are filled with polymer solution, the surface of the chip is covered with a mask having an open window to expose a specific part of the channel to UV light. The microchip is then covered with some kind of foil or tape, except for the open window, and after illuminating the chip with UV light, the polymerization reaction starts. The advantage of using photoinitiated polymerization is that the polymerized plug can be localized to any desired location on chip. The dimensions of the plug or rod can also be precisely defined.

The feasibility of synthesizing monolithic material in PDMS,[87] glass,[88,89] and polymeric substrates[90,91] has been shown. However, PDMS is not so suited for the analysis of nonpolar analytes. The analytes are absorbed onto and into the PDMS, which results in poor recovery and low enrichment. Without surface modification, PDMS is only suited for polar samples such as genomic DNA.[87,92] In Figure 50.27, a SEM image of monolithic material inside a Zeonor polymer microchannel is shown.[90]

Yu et al. described the integration of ion-exchange and hydrophobic concentrators based on a methacrylate-based monolithic phase in a chip made of glass.[88] With this device, they were able to concentrate a hydrophobic tetrapeptide, Coumarin 519, and green fluorescent protein (GFP) by a factor of 1000.

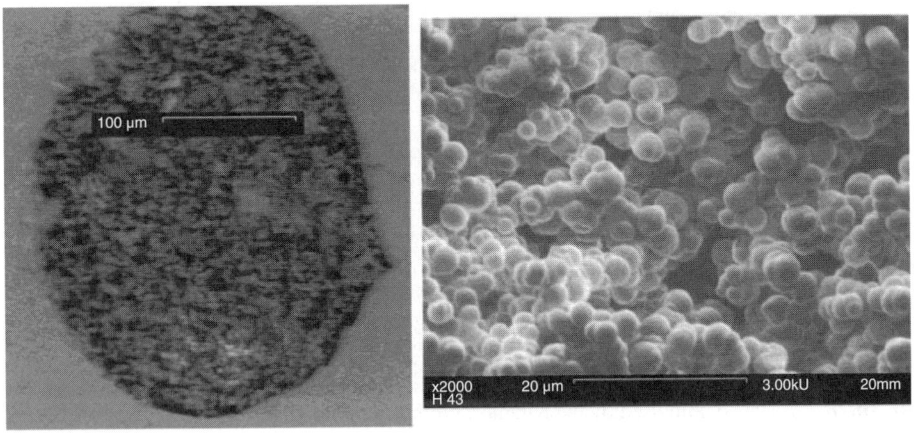

FIGURE 50.27 SEM images showing a cross-section of monolithic material inside a Zeonor polymer chip described by Tan et al. (From Tan et al., *Anal. Chem.*, 75, 5504, 2003. With permission.)

The extraction efficiency for nucleic acids using monolithic silica material in microfluidic devices is not nearly as good as what can be achieved using silica beads.[93] However, with monolithic material, better repeatabilities are obtained. For these reasons, Karwa et al.[87] and Wolfe et al.[93] combined the two materials. Both groups used a sol-gel to immobilize silica beads in a channel. This approach has already proven its usefulness in CEC.[94] The channel is packed by a two-step process, where the particles are first introduced into the channel and then held in place with a polymeric phase. This packing technique provided excellent efficiency, repeatability, and stability. In fact, Wolfe et al. found that the extraction efficiency for a λDNA sample improved to a value of about 70%, compared to 57% for silica beads alone, and 20–30% for the sol-gel matrices tested. Due to the presence of the particles in the column, matrix cracking caused by internal pressure differences within the pores of the sol-gel matrix is reduced.

50.4 LIQUID–LIQUID EXTRACTION (LLE)

Liquid–liquid extraction (LLE) is among the oldest sample extraction and preconcentration techniques available in analytical chemistry. LLE is a method whereby two immiscible phases, generally an organic solvent and an aqueous solution, are brought into contact in order to extract one or more analytes from one phase into the other. If the receiving phase has a smaller volume than the donor phase, preconcentration can be effected. The separation mechanism is, like SPE, based on partitioning. At equilibrium, the partition coefficient of analyte i (K_i) in a two-phase system is given by

$$K_i = \frac{C_{i,\text{organic\_phase}}}{C_{i,\text{sample}}} \tag{50.11}$$

where $C_{i,\text{organic\_phase}}$ is the concentration of extracted analyte in the extraction solvent and $C_{i,\text{sample}}$ the concentration of analyte in the sample at equilibrium. The use of microfluidic devices is attractive for LLE, since the contact area between phases can be enlarged to facilitate a more rapid equilibrium of analyte partitioning between phases.

Tokeshi et al.[95] reported the first two-phase microextraction system on a microchip in 2000. An Fe(II)-complex was extracted from an aqueous solution into a chloroform solution with a simple Y-shaped chip (see Figure 50.28). The solvents were introduced with two syringe pumps, and a parallel two-layer flow was formed in the microchannel because the surface tension and frictional forces are much stronger than that due to the specific gravity. The extraction of Fe(II)-complex from aqueous

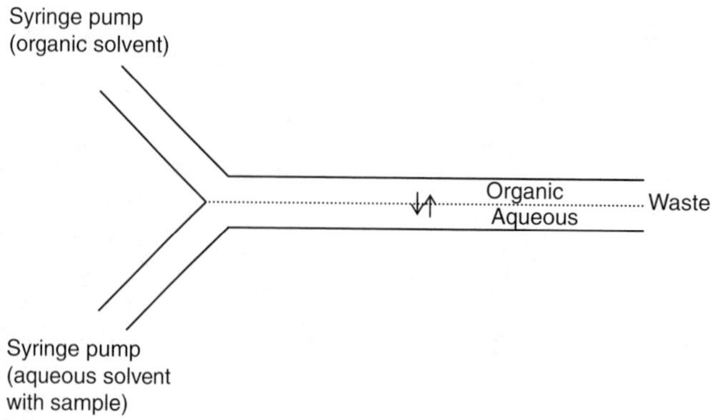

FIGURE 50.28 Y-shaped chip used by the Kitamori group for micro-LLE.[95–98,101]

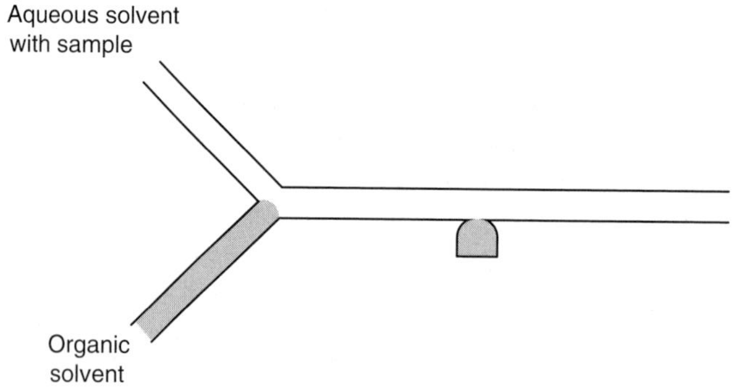

FIGURE 50.29 Chip with recess used for micro-LLE by Chen et al.[102]

solution took place where the two-layer flow was formed. Later on, Kitamori and coworkers showed more applications for their chip[96–99] and they have even modified the chip to create a three-layer flow system.[100,101] However, these devices were not intended to be used specifically for sample extraction and purification, but to study the molecular transport of analytes through an organic membrane. In this section, we will briefly focus on the implementation of LLE in microchips specifically for sample enrichment. Readers are referred to Chapter 36, which details the efforts made by Kitamori and coworkers to incorporate LLE into microfluidic devices for different purposes.

Chen et al.[102,103] realized a chip for LLE employing a stationary organic solvent and flowing aqueous sample. The chip was Y-shaped, similar to the chip used by Kitamori and coworkers, but the solvents were introduced by differential hydrostatic pressure created by varying liquid levels in the reservoirs. The chip, made of glass, was first filled with an organic solvent by filling the organic solvent reservoir and keeping the other reservoirs empty. Then the sample reservoir was filled with aqueous sample, which produced a flow of this solution through the channel. The organic solvent was rinsed out of the sample solvent channel, but remained in the organic solvent channel due to the pressure introduced by the organic solvent volume in the reservoir. This resulted in a stable interface between the aqueous and organic phases at the exit of the organic solvent inlet channel (see Figure 50.29, left half of structure). The sample, butyl-rhodamine B (BRB) was extracted at this point from the aqueous solvent into the organic solvent.[103] A preconcentration factor of 40 was obtained and the relative standard deviation was only 2.7% for this method.

More effective for LLE proved to be the introduction of a single microrecess in the channel wall for organic solvent trapping.[102] This ensured that droplets of organic solvent could be trapped and held in place as aqueous solution flowed past.[102] Different recess geometries were studied, and the chip with the recess that resulted in the best preconcentration factor is depicted in Figure 50.29.

The chip works the same as the chip that was used for BRB extraction.[103] It was first filled with an organic solvent, which filled the recess, and then with aqueous solution. The organic solvent was rinsed out of the sample solvent channel by the aqueous solution, but stayed in the recess where the extraction then took place. After the extraction, the organic solvent with analytes trapped in the recess was flushed out of the chip by passing acetone through the solvent channel. An enrichment factor of about 5000 was achieved in a recess of ~750 nL, but at the expense of a very long extraction time of 50 min. In a much smaller recess (~140 pL), an enrichment factor of 250 was achieved within 5 min. This enrichment factor was not only achieved by extraction of the analytes from the aqueous solvent into the organic solvent, but was also due to dissolution of organic solvent in the flowing aqueous sample. During the extraction, a part of the stagnant organic solvent dissolved into the aqueous solvent, resulting in a decrease in organic solvent volume and higher enrichment factors. This chip was, unfortunately, not coupled to a separation channel.

Cai et al.[104] recently published an example of LLE on-chip based on the principle of wetting-film extraction. Wetting-film extraction was introduced by Ruzicka, Christian, and coworkers in 1996.[105] The technique is based on the film-forming characteristics of an organic solvent on hydrophobic walls. In conventional wetting-film extraction, Teflon tubing is used as the support for the wetting film. An organic solvent is passed through a tubing or channel, and due to the hydrophobic wall, a uniform organic film is formed on the wall. When passing an aqueous sample through the tubing or channel, the hydrophobic analytes will extract from the aqueous liquid into the organic liquid. The extracted analytes are then eluted with an organic liquid.

Cai et al. used a polymeric chip (polycarbonate), because the native surface is an excellent support for creating organic-wetting films. The organic liquid and the aqueous sample were sequentially injected using a gravity pump. Butanol was used as the organic wetting film and as eluting solvent, and BRB diluted in sodium hydroxide was used as sample (pH 13.0). An enrichment factor of 24 was obtained. With this chip, an RSD of only 1.5% was obtained for eleven BRB injections at 10 μM.

50.5 NANOPOROUS MEMBRANES AND FILTERS

50.5.1 MEMBRANE FILTRATION

The first membrane filtration was done by Dr. Sourirajan in 1960 in his laboratory at UCLA (http://www.gewater.com/library/tp/698_Membranes_the.jsp). Since then, many scientists have been developing new membranes that can specifically separate particles from solution or other particles in solution.[106] Membranes are semipermeable barriers which are used to separate analytes based on their size, making membrane filtration a size-exclusion technique. Membranes have pores with specific size ranges, and molecules and particles having diameters that are bigger than the membrane pores cannot pass. In contrast, molecules and particles with smaller diameters can pass through the membrane.

With membrane filtration techniques, analytes can be concentrated on the membranes if the molecular diameter is bigger than that of the pores of the membrane. Analytes are dissolved in a buffer containing ions that are smaller than the pores of the membrane. The buffer ions pass through the membrane, and the analytes with sizes bigger than the membrane pores are concentrated at the membrane. In theory, the degree of concentration is only limited by the ratio of the initial sample volume to the surface of the membrane, and the solubility limit of the samples.

The first microfluidic chip with a porous membrane for the preconcentration of a biological sample was presented by Khandurina et al. in 1999.[107] In this experiment, they demonstrated preconcentration of DNA samples. The channels were etched into a glass substrate and bonded to a cover plate using a silicate solution as an adhesive. The chip layout used in this experiment is presented in Figure 50.30.

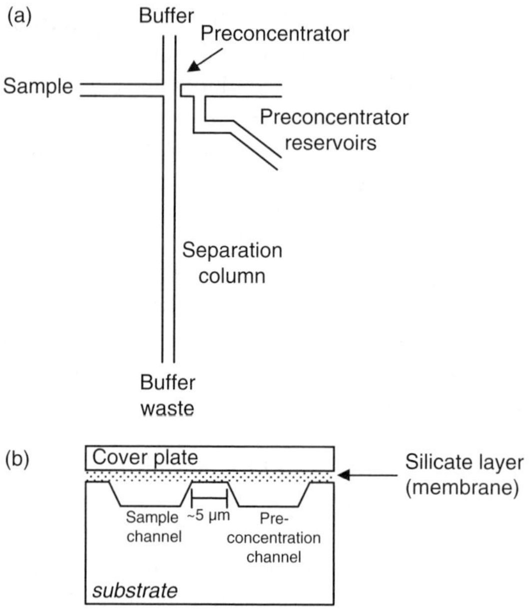

FIGURE 50.30 (a) Schematic of the chip layout used by Khandurina et al.[107] (b) Schematic cross-sectional view through the preconcentrator.

As can be seen in Figure 50.30, the preconcentration channel and the sample channel are not directly connected to each other, but are separated by a distance of about 3–5 μm (see Figure 50.30b). Electrical contact between the two separated channels is maintained through the porous, semipermeable silicate membrane. The DNA sample is loaded by applying a voltage between the sample inlet and the preconcentrator reservoirs. The buffer ions pass through the silicate membrane, but the large DNA molecules accumulate at the preconcentrator due to their hindered transport through the membrane. After preconcentration, the voltage is applied between the buffer and buffer waste, and the DNA molecules are injected onto the separation column.

With this chip, DNA can be concentrated by a factor of up to two orders of magnitude. Khandurina et al.[108] also reported an integrated system for polymerase chain reaction (PCR)-based analysis on a microchip with this preconcentration technique. The chip layout was similar to that depicted in Figure 50.30, with an additional Peltier thermoelectric element integrated for the DNA amplification. With this chip, a 25-fold preconcentration of the DNA sample due to the semipermeable silicate membrane was observed.

Foote et al.[109] recently published an article where they have used the same chip as described by Khandurina et al.[107] for the preconcentration of proteins. However, they observed distortions of peak shape and loss of resolution for some sample components. This could be due to slow desorption of sample from the membrane surface or changes in buffer composition in the injection channel and adjoining portions of the separation channel during preconcentration. To overcome these problems, they changed the chip layout, as shown in Figure 50.31.

The samples used in this experiment were fluorescein-labeled proteins. The sample was concentrated by applying a voltage between the sample inlet and the preconcentrator reservoirs. The proteins were first accumulated at the preconcentrator. After concentration, the sample was loaded into the injector region by applying a voltage between the sample-inlet and the sample-waste reservoirs, and injected by applying a voltage between the buffer and buffer-waste reservoirs. With this chip, good peak shapes were obtained, as seen in Figure 50.32. Peak resolutions improved as a function of concentration time up to the 8-min point, after which time resolution started

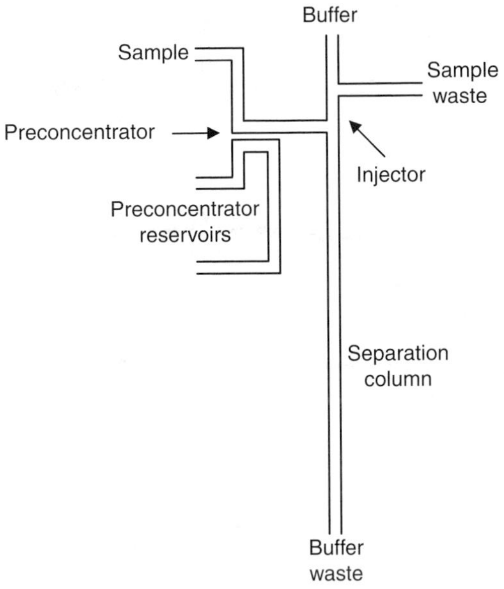

FIGURE 50.31 Schematic of chip layout used by Foote et al.[109] for side channel preconcentration.

to deteriorate again. Although these effects were not fully understood, it appeared to be due to time-dependent changes in the sample buffer. However, the stacking was more complete at longer preconcentration times. At 8 min concentration time, a maximum signal enhancement of a factor of approximately 600 was obtained. As Figure 50.33 shows, the rate at which peak height increases for a protein grows as a function of concentration time. This is most likely due to the fact that there was a delay between application of the loading potential and the arrival of the sample proteins at the preconcentrator. Proteins had to migrate through gel-filled channels in order to get to this element.

Another approach to use the membrane filtration technique in microchips is to integrate the membrane into the separation channel, as Song et al.[110] and Hatch et al.[111] have demonstrated. The chip layout used by Song et al. is given in Figure 50.34a.

Song et al.[110] used a zwitterionic polymer membrane (molecular weight cutoff <5700), which was integrated by using photoinitiated polymerization. The sample was loaded onto the bottom left part of the membrane by applying a voltage between the sample and sample waste. By applying a voltage between the buffer and buffer waste, the concentrated sample could be removed from the membrane again, separated, and detected. Song et al. showed a concentration increase by two orders of magnitude for both BSA and phosphorylase *b* with a 100-s concentration time (see Figure 50.35).

Hatch et al.[111] have also integrated a membrane (polyacrylamide) into the injection channel by photopolymerization. With this chip, a concentration increase of over 100-fold was observed for four disease biomarkers (IL-6, IL-1β, TNFα, and CRP) in saliva.

Membrane filtration is an effective technique for the preconcentration of biological macromolecules like proteins and DNA. However, relatively low enrichment factors are obtained compared to LLE, SPE, and TGF. Concentration enrichments with a factor of about 100 were shown for both proteins and DNA molecules by several groups. A problem that has to be taken into account is the probable adsorption of the macromolecules on the membrane surface. This could result in tailing in the injection channel and/or in a low recovery of the molecules. Another issue that has to be taken into account is the good probability that concentration increase is not a linear function of time, as observed by Khandurina et al.[107] and Foote et al.[109]

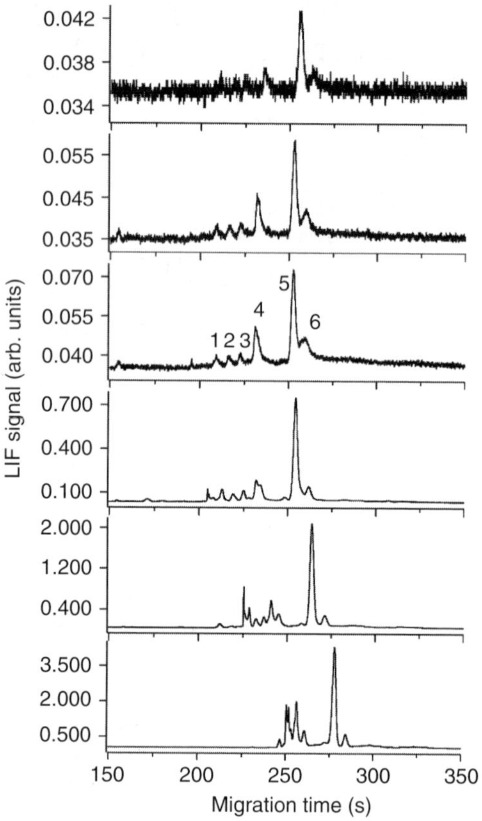

FIGURE 50.32 CGE separations of fluorescent high-molecular-weight protein standards following preconcentration in the sample channel. Starting with the top electropherogram, preconcentration times were 0, 1, 2, 4, 6, and 8 min. Peak numbers: (1) trypsin inhibitor, (2) carbonic anhydrase, (3) alcohol dehydrogenase, (4) bovine serum albumin, (5) β-galactosidase, and (7) myosin. (From Foote et al., *Anal. Chem.*, 77, 57, 2005. With permission.)

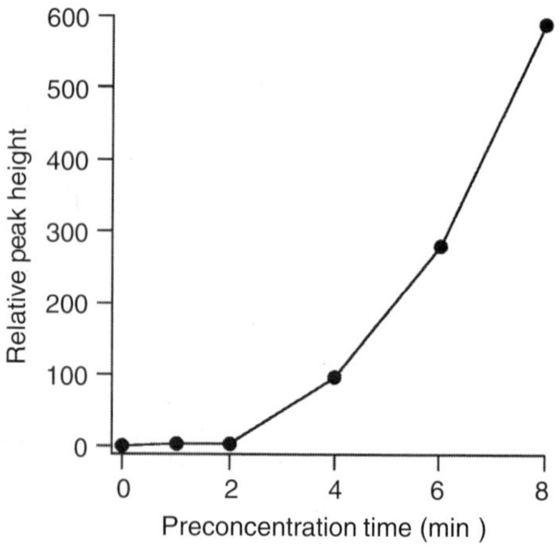

FIGURE 50.33 Relative peak height of β-galactosidase (peak 5) versus preconcentration time from data in Figure 50.32. (From Foote et al., *Anal. Chem.*, 77, 57, 2005. With permission.)

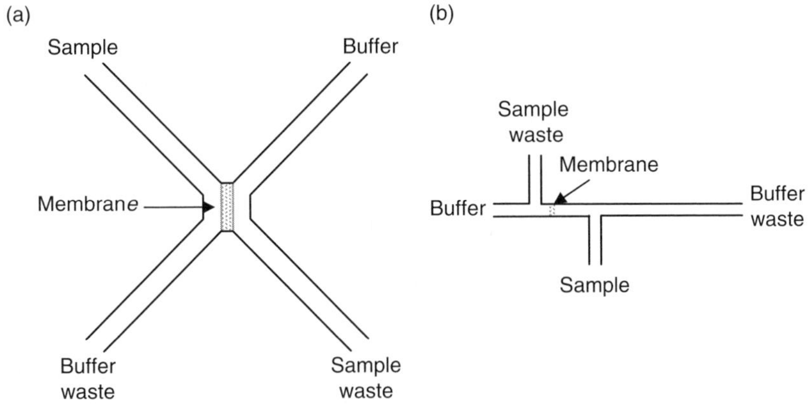

FIGURE 50.34 (a) Schematic layout of the chip with integrated membrane in the separation channel used by Song et al.[110] and (b) the chip layout used by Hatch et al.[111]

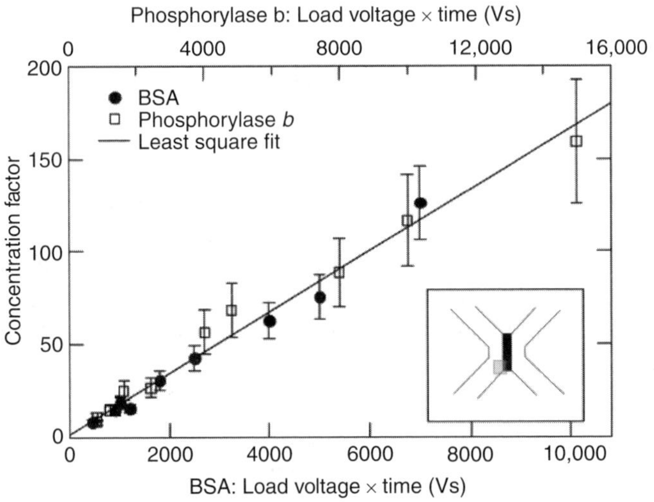

FIGURE 50.35 Linear increase in the concentration of both BSA and phosphorylase *b* with the voltage-time product. The curve is linear for over two orders of magnitude increase in concentration factor. (From Song et al., *Anal. Chem.*, 76, 4589, 2004. With permission.)

50.5.2 Nanofluidic Filters

Nanofluidic filtering is based on an electrokinetic trapping mechanism enabled by recent advances in nanochannel fabrication. A nanofluidic filter consists of two microfluidic channels bridged by buffer-filled nanochannels which act like an ion-selective membrane. The technique is a relatively new technique which has a demonstrated potential for sample enrichment. A schematic chip layout of a nanofluidic filter is given in Figure 50.36.

The substrate most often used for making nanofluidic filter chips is glass. Glass surfaces are negatively charged, due to the presence of deprotonated silanol groups. Positively charged counterions from the buffer form an electrical double-layer at this surface. When the ionic strength of the buffer is decreased, the thickness of the double-layer is increased because the square root of the buffer concentration is inversely proportional to the diffuse double-layer thickness. With buffer

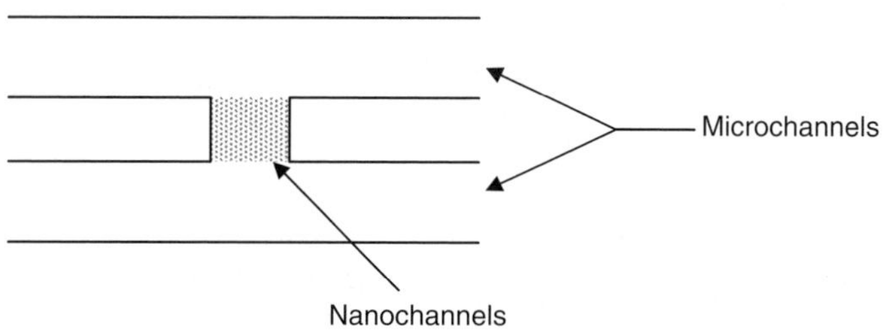

FIGURE 50.36 Schematic chip layout of a nanofluidic filter.

concentrations of 10^{-2}–10^{-6} M, double layers in the range of 3–300 nm are obtained.[112] This means that when the diameter of the nanochannel is small enough and buffer concentration is low enough, a double-layer overlap can occur in the nanochannel. The cation concentration (A^+) will be greater than the anion concentration (B^-) in the nanochannel, due to the slight excess of cations present in the electrical double-layer. In other words,

$$[A^+] = \alpha \cdot [B^-] \tag{50.12}$$

where α is greater than 1. In the case of a positively charged nanochannel wall, a negatively charged double layer is established and the anion concentration will exceed that of the cations in the nanochannel upon double-layer overlap. In this situation, α is less than 1.

A difference in the bulk conductivity of the micro- and nanochannels results from the double-layer overlap, even if they are filled with the same buffer. At low buffer concentrations, the interface conductivity (at the nanochannel) starts to dominate over the conductivity of the buffer ions. This results in a current being carried by different ratios of cations to anions in a nanochannel compared to a microchannel. Assuming that A^+ and B^- have equal concentrations in the microchannel, the relationship between the ion fluxes in the microchannel (J_m^A) and nanochannel (J_n^A) for the cations can be described as[112]

$$J_m^A = \left(\frac{\mu_A + \mu_B/\alpha}{\mu_A + \mu_B}\right) \cdot J_n^A \tag{50.13}$$

where μ_A is the electrophoretic mobility of the cations and μ_B the electrophoretic mobility of the anions. When the nanochannel is negatively charged, α is greater than one (see Equation 50.12). This means, $J_m^A < J_n^A$.

The microchannel takes away cations on the cathode side of the nanochannel and the nanochannel provides cations. However, because $J_m^A < J_n^A$, an enrichment of cations is obtained at the cathode side of the nanochannel. On the anode side of the nanochannel, the opposite is observed and a decrease in cations is obtained, so-called ion-depletion. The ion fluxes for anions (B) can be described as[112]

$$J_m^B = \left(\frac{\mu_B + \alpha \cdot \mu_A}{\mu_A + \mu_B}\right) \cdot J_n^B \tag{50.14}$$

The wall of the nanochannel is negatively charged, thus $\alpha > 1$, which results in $J_m^B < J_n^B$. The microchannel provides anions on the cathode side of the nanochannel and the nanochannel takes away anions at the cathode side. Again, because $J_m^B < J_n^B$, an enrichment of anions as well as cations is observed at the cathode side. A depletion of anions occurs on the anode side of the nanochannel. This means that on the cathode side of the nanochannel, enrichment of cations and anions is obtained, while on the anode side, a depletion of cations and anions results.

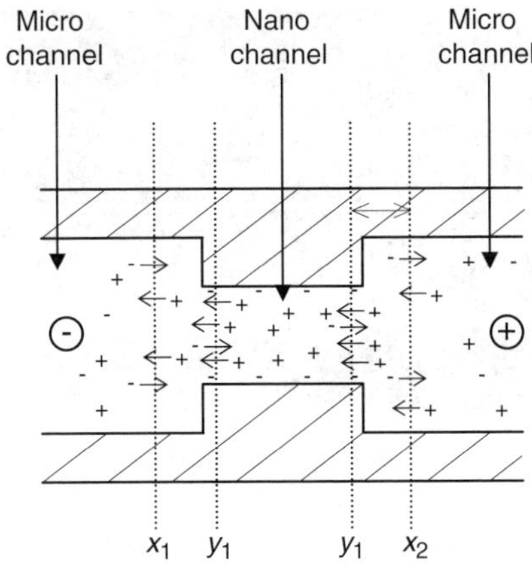

FIGURE 50.37 Schematic representation of the ion-enrichment and ion-depletion effect. (From Pu et al., *Nano Letters*, 4, 1099, 2004. With permission.)

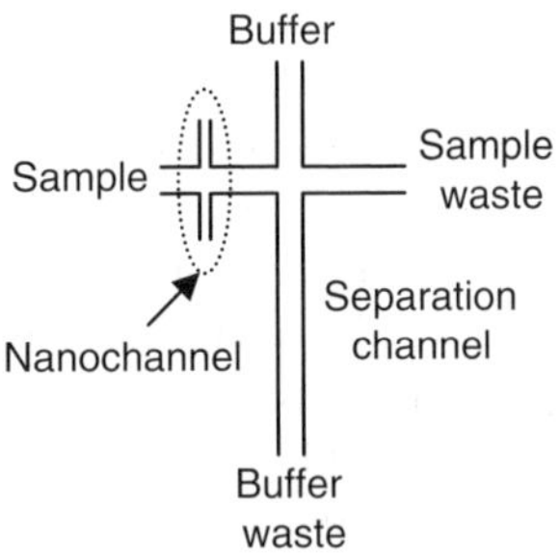

FIGURE 50.38 The chip with nanofluidic filter, which is used by Petersen and Ramsey (From Petersen, N. J. and Ramsey, J. M., in *Micro Total Analysis Systems (μ TAS)*, Boston, 2005, pp. 1252–1254.)

When the wall of the nanochannel is positively charged, the double-layer overlap in the nanochannel is due to the negatively charged buffer ions, and $\alpha < 1$. In this case, the enrichment of ions is obtained on the anode side and depletion on the cathode side. This ion-enrichment and depletion effect is schematically shown in Figure 50.37, where $\mu_A = \mu_B$ and $\alpha = 3$.

Figure 50.37 has four cross-sectional planes, x_1, y_1, y_2, and x_2, respectively. The two x-cross-sectional planes represent the current passing through the microchannels and the two y-planes represent the current through the nanochannel. x_1 and y_1 are at the cathode end and x_2 and y_2 are at the anode end. At both x-planes, where no double-layer overlap exists, the current is carried by two cations and two anions. At both y-planes, where the interface conductivity starts to dominate the bulk conductivity, the current is carried by three cations and one anion. This results in an ion-enrichment at the cathode side and an ion-depletion at the anode side for both cations and anions.

Pu et al.[112] demonstrated this ion-enrichment or ion-depletion effect for a solution containing either negatively charged fluorescein or positively charged rhodamine-6G. Both species were enriched at the cathode end and depleted at the anode end of the nanochannels. Unfortunately, no enrichment factors were given.

Petersen and Ramsey[113] demonstrated a signal enhancement of a factor 100 within 3 min for ionic analytes. The chip layout that was used is given in Figure 50.38, while an experiment showing the concentration of a positively charged fluorescent dye is shown in the fluorescence image of Figure 50.39.

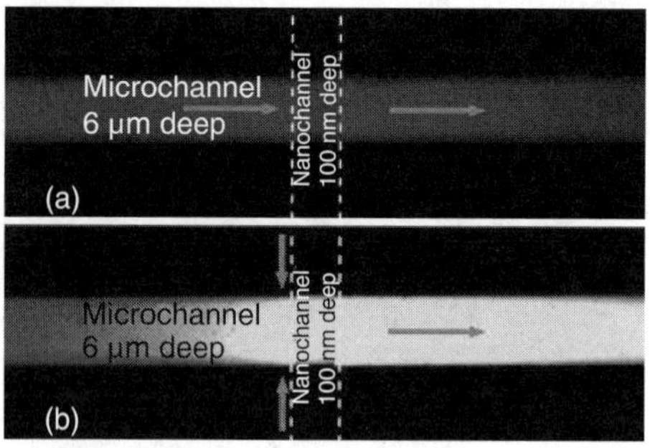

FIGURE 50.39 Concentration of the positively charged rhodamine-6G at a micro–nanochannel junction. All channels and reservoirs were initially filled with 5 μM rhodamine-6G solution. (a) Potential applied through the microchannel and (b) potential applied between nanochannels (positively biased) and right-hand microchannel (negatively biased). (From Petersen, N. J., et al., in *Micro Total Analysis Systems (μ TAS)*, Boston, 2005, 1252. With permission.)

Wang et al.[114] demonstrated the use of a nanofluidic filter for the preconcentration of biological samples. They demonstrated a concentration enrichment factor of 10^6–10^7 for GFP within 1 h with the chip, which is schematically shown in Figure 50.36. The filter collects sample molecules from a relatively large volume (few μL) and concentrates them into small volumes (pL–nL). Much higher enrichment factors are obtained compared to small analytes. This is probably due to electrostatic exclusion of the proteins from the nanochannel, resulting in a concentration effect before the nanochannel interface. Buffer ions, on the other hand, can pass through the nanochannel.[115] This means the whole sample gets concentrated before the nanofluidic filter. Wang et al.[114,116,117] also demonstrated the enrichment of peptides with a factor of 10^7.

The use of nanofluidic filters in microchips for the preconcentration of peptides and proteins is one of the best chip-based preconcentrators reported so far. Small ion samples could also be concentrated with this nanofluidic filter. The integration of the nanochannel is very easy, because it does not need special buffers or any physical barriers for sample enrichment. Long times are required, however, to achieve high enrichment factors (30 min to 1 h).

50.6 CONCLUSIONS

Microfluidics is a rapidly growing discipline being driven by new technological applications in the medical, material, and chemical sciences, to name but a few. There are a number of advantages when using microfluidics technology, including dramatically reduced analysis times and reagent/sample consumption, and the potential for engineering high-throughput systems and systems for cell-based analysis. However, system miniaturization also implies reduced detection volumes, which in turn often results in poor detection sensitivity. A built-in preconcentration technique that compresses the analytes of interest into a smaller volume is often desirable.

The preconcentration techniques that have to date been most used in microfluidic devices have been presented in the preceding pages. Most of the techniques were originally developed for conventional CE or liquid chromatography and have (more or less simply) been adapted to microchips. FASS, ITP, and IEF are examples of these types of techniques. A few new techniques have been developed for the preconcentration of samples on microfluidic devices. TGF and nanofluidic filtering are both examples of techniques that have been developed solely for microfluidic chips. High enrichment factors can be achieved with both techniques, on the order of 10,000 and 1,000,000, respectively. Table 50.1 summarizes the top performers in each device category with respect to preconcentration factor and time.

TABLE 50.1
Different Preconcentration Techniques with Their Preconcentration Factors

Technique	FASS	ITP	IEF	TGF	SPE	LLE	Membranes	Nanofluidic filter
Reference	12	33	46	49	72	102	110	114
Separation	CE	CE	CE	No	CEC	No	CE	CE
Time	≪1 s	40 s	10 s	100 min	< 5 min	5 min	100 s	40 min
Preconcentration	1100-fold	2,000,000-fold	70-fold	10,000-fold	500-fold	250-fold	100-fold	10^7-fold
Analyte	BODIPY Fluorescein	Alexa Fluor 488	Recombinant GFP Q Biogene FITC-labeled dextran FITC-labeled ovalbumin	Carboxylic acid solution	BODIPY 493/503	BRB	BSA Phosphorylase *b*	GFP

It is worth noting that in the studies described in this chapter, only fluorescent dyes or simple, fluorescently labeled biological molecules have been analyzed for the most part. Few real physiological or biological samples were actually tested. To investigate the suitability of these preconcentration techniques, they must ultimately be applied to the biological sample matrices that are of actual interest in proteomics, genomics, and the other life sciences. Thus, some interesting challenges remain for researchers in this area, despite the significant advances that have been made to date.

REFERENCES

1. Breadmore, M. C., Recent advances in enhancing the sensitivity of electrophoresis and electrochromatography in capillaries and microchips, *Electrophoresis*, 28, 254–281, 2007.
2. Song, S. and Singh, A. K., On-chip sample preconcentration for integrated microfluidic analysis, *Analytical and Bioanalytical Chemistry*, 384, 41–43, 2006.
3. Mikkers, F. E. P., Everaerts, F. M., and Verheggen, T. P. E. M., High-performance zone electrophoresis, *Journal of Chromatography*, 169, 11–20, 1979.
4. Chien, R.-L. and Helmer, J. C., Electroosmotic properties and peak broadening in field-amplified capillary electrophoresis, *Analytical Chemistry*, 63, 1354–1361, 1991.
5. Burgi, D. S. and Chien, R.-L., Optimization in sample stacking for high-performance capillary electrophoresis, *Analytical Chemistry*, 63, 2042–2047, 1991.
6. Burgi, D. S. and Chien, R.-L., Application of sample stacking to gravity injection in capillary electrophoresis, *Journal of Microcolumn Separations*, 3, 199–202, 1991.
7. Jacobson, S. C. and Ramsey, J. M., Microchip electrophoresis with sample stacking, *Electrophoresis*, 16, 481–486, 1995.
8. Lichtenberg, J., Verpoorte, E., and De Rooij, N. F., Sample preconcentration by field amplification stacking for microchip-based capillary electrophoresis, *Electrophoresis*, 22, 258–271, 2001.
9. Zhang, L. and Yin, X. F., Field amplified sample stacking coupled with chip-based capillary electrophoresis using negative pressure sample injection technique, *Journal of Chromatography A*, 1137, 243–248, 2006.
10. Mourzina, Y., Kalyagin, D., Steffen, A., and Offenhausser, A., Electrophoretic separations of neuromediators on microfluidic devices, *Talanta*, 70, 489–498, 2006.
11. Beard, N. P., Zhang, C.-X., and deMello, A. J., In-column field-amplified sample stacking of biogenic amines on microfabricated electrophoresis devices, *Electrophoresis*, 24, 732–739, 2003.
12. Jung, B., Bharadwaj, R., and Santiago, J. G., Thousand-fold signal increase using field-amplified sample stacking for on-chip electrophoresis, *Electrophoresis*, 24, 3476–3483, 2003.
13. Kutter, J. P., Ramsey, R. S., Jacobson, S. C., and Ramsey, J. M., Determination of metal cations in microchip electrophoresis using on-chip complexation and sample stacking, *Journal of Microcolumn Separations*, 10, 313–319, 1998.
14. Effenhauser, C. S., Manz, A., and Widmer, H. M., Glass chips for high-speed capillary electrophoresis separations with submicrometer plate heights, *Analytical Chemistry*, 65, 2637–2642, 1993.
15. Herr, A. E., Molho, J. I., Santiago, J. G., Mungal, M. G., and Kenny, T. W., Electroosmotic capillary flow with nonuniform zeta potential, *Analytical Chemistry*, 72, 1053–1057, 2000.
16. Yang, H. and Chien, R.-L., Sample stacking in laboratory-on-a-chip devices, *Journal of Chromatography A*, 924, 155–163, 2001.
17. Gong, M. J., Wehmeyer, K. R., Limbach, P. A., Arias, F., and Heineman, W. R., On-line sample preconcentration using field-amplified stacking injection in microchip capillary electrophoresis, *Analytical Chemistry*, 78, 3730–3737, 2006.
18. Shiddiky, M. J. A., Park, H., and Shim, Y. B., Direct analysis of trace phenolics with a microchip: In-channel sample preconcentration, separation, and electrochemical detection, *Analytical Chemistry*, 78, 6809–6817, 2006.
19. Quirino, J. P. and Terabe, S., Exceeding 5000-fold concentration of dilute analytes in micellar electrokinetic chromatography, *Science*, 282, 465–468, 1998.

20. Quirino, J. P. and Terabe, S., Approaching a million-fold sensitivity increase in capillary electrophoresis with direct ultraviolet detection: Cation-selective exhaustive injection and sweeping, *Analytical Chemistry*, 72, 1023–1030, 2000.
21. Palmer, J., Burgi, D. S., Munro, N. J., and Landers, J. P., Electrokinetic injection for stacking neutral analytes in capillary and microchip electrophoresis, *Analytical Chemistry*, 73, 725–731, 2001.
22. Palmer, J., Munro, N. J., and Landers, J. P., A universal concept for stacking neutral analytes in micellar capillary electrophoresis, *Analytical Chemistry*, 71, 1679–1687, 1999.
23. Sera, Y., Matsubara, N., Otsuka, K., and Terabe, S., Sweeping on a microchip: Concentration profiles of the focused zone in micellar electrokinetic chromatography, *Electrophoresis*, 22, 3509–3513, 2001.
24. Kitagawa, F., Tsuneka, T., Akimoto, Y., Sueyoshi, K., Uchiyama, K., Hattori, A., and Otsuka, K., Toward million-fold sensitivity enhancement by sweeping in capillary electrophoresis combined with thermal lens microscopic detection using an interface chip, *Journal of Chromatography A*, 1106, 36–42, 2006.
25. Walker, P. A., Morris, M. D., Burns, M. A., and Johnson, B. N., Isotachophoretic separations on a microchip. Normal Raman spectroscopy detection, *Analytical Chemistry*, 70, 3766–3769, 1998.
26. Everaerts, F. M., Beckers, J. L., and Verheggen, T. P. E. M., *Isotachophoresis: Theory, Instrumentation, and Applications*, New York, 1976.
27. Mosher, R. A., Saville, D. A., and Thormann, W., *The Dynamics of Electrophoresis*, 1 ed., Weinheim, 1992.
28. Kaniansky, D., Masár, M., Bielciková, J., Iványi, F., Eisenbeiss, F., Stanislawski, B., Grass, B., Neyer, A., and Jöhnck, M., Capillary electrophoresis separations on a planar chip with the column-coupling configuration of the separation channels, *Analytical Chemistry*, 72, 3596–3604, 2000.
29. Grass, B., Neyer, A., Jöhnck, M., Siepe, D., Eisenbeiss, F., Weber, G., and Hergenroder, R., A new PMMA-microchip device for isotachophoresis with integrated conductivity detector, *Sensors and Actuators B-Chemical*, 72, 249–258, 2001.
30. Bodor, R., Madajová, V., Kaniansky, D., Masár, M., Jöhnck, M., and Stanislawski, B., Isotachophoresis and isotachophoresis-zone electrophoresis separations of inorganic anions present in water samples on a planar chip with column-coupling separation channels and conductivity detection, *Journal of Chromatography A*, 916, 155–165, 2001.
31. Bodor, R., Zúborová, M., Ölvecká, E., Madajová, V., Masár, M., Kaniansky, D., and Stanislawski, B., Isotachophoresis and isotachophoresis-zone electrophoresis of food additives on a chip with column-coupling separation channels, *Journal of Separation Science*, 24, 802–809, 2001.
32. Jeong, Y., Choi, K., Kang, M. K., Chun, K., and Chung, D. S., Transient isotachophoresis of highly saline samples using a microchip, *Sensors and Actuators B*, 104, 269–275, 2005.
33. Jung, B., Bharadwaj, R., and Santiago, J. G., On-chip million-fold sample stacking using transient isotachophoresis, *Analytical Chemistry*, 78, 2319–2327, 2006.
34. Krivánková, L., Gebauer, P., Thormann, W., Mosher, R. A., and Bocek, P., Options in electrolyte systems for on-line combined capillary isotachophoresis and capillary zone electrophoresis, *Journal of Chromatography*, 637, 119–135, 1993.
35. Wainright, A., Williams, S. J., Ciambrone, G., Xue, Q., Wei, J., and Harris, D., Sample preconcentration by isotachophoresis in microfluidic devices, *Journal of Chromatography A*, 979, 69–80, 2002.
36. Jung, B. G., Zhu, Y. G., and Santiago, J. G., Detection of 100 aM fluorophores using a high-sensitivity on-chip CE system and transient isotachophoresis, *Analytical Chemistry*, 79, 345–349, 2007.
37. Xu, Z. Q., Hirowaka, T., Nishine, T., and Arai, A., High-sensitivity capillary gel electrophoretic analysis of DNA fragments on an electrophoresis microchip using electrokinetic injection with transient isotachophoretic preconcentration, *Journal of Chromatography*, 990, 53–61, 2003.
38. Wainright, A., Nguyen, U. T., Bjornson, T., and Boone, T. D., Preconcentration and separation of double-stranded DNA fragments by electrophoresis in plastic microfluidic devices, *Electrophoresis*, 24, 3784–3792, 2003.
39. Laas, T., Olsson, I., and Söderberg, L., High-voltage isoelectric focusing with pharmalyte: Field strength and temperature distribution, zone harpening, isoelectric spectra, and pI determinations, *Analytical Biochemistry*, 101, 449–461, 1980.
40. Hjertén, S. and Zhu, M.-D., Adaption of the equipment for high-performance electrophoresis to isoelectric focusing, *Journal of Chromatography*, 346, 265–270, 1985.

41. Li, Y., Buch, J. S., Rosenberger, F., DeVoe, D. L., and Lee, C. S., Integration of isoelectric focusing with parallel sodium dodecyl sulfate gel electrophoresis for multidimensional protein separations in a plastic microfluidic network, *Analytical Chemistry*, 76, 742–748, 2004.
42. Xu, J. D., Locascio, L., Gaitan, M., and Lee, C. S., Room-temperature imprinting method for plastic microchannel fabrication, *Analytical Chemistry*, 72, 1930–1933, 2000.
43. Cabrera, C. R. and Yager, P., Continuous concentration of bacteria in a microfluidic flow cell using electrokinetic techniques, *Electrophoresis*, 22, 355–362, 2001.
44. Tan, W., Fan, Z. H., Qiu, C. X., Ricco, A. J., and Gibbons, I., Miniaturized capillary isoelectric focusing in plastic microfluidic devices, *Electrophoresis*, 23, 3638–3645, 2002.
45. Chen, X., Wu, H., Mao, C., and Whitesides, G. M., A prototype two-dimensional capillary electrophoresis system fabricated in poly(dimethylsiloxane), *Analytical Chemistry*, 74, 1772–1778, 2002.
46. Herr, A. E., Molho, J. I., Drouvalakis, K. A., Mikkelsen, J. C., Utz, P. J., Santiago, J. G., and Kenny, T. W., On-chip coupling of isoelectric focusing and free solution electrophoresis for multidimensional separations, *Analytical Chemistry*, 75, 1180–1187, 2003.
47. Wang, Y.-C., Choi, M. H., and Han, J., Two-dimensional protein separation with advanced sample and buffer isolation using microfluidic valves, *Analytical Chemistry*, 76, 4426–4431, 2004.
48. Unger, M. A., Chou, H. P., Thorsen, T., Scherer, A., and Quake, S. R., Monolithic microfabricated valves and pumps by multilayer soft lithography, *Science*, 288, 113–116, 2000.
49. Ross, D. and Locascio, L. E., Microfluidic temperature gradient focusing, *Analytical Chemistry*, 74, 2556–2564, 2002.
50. Ross, D., Gaitan, M., and Locascio, L. E., Temperature measurement in microfluidic systems using a temperature-dependent fluorescent dye, *Analytical Chemistry*, 73, 4117–4123, 2001.
51. Balss, K. M., Vreeland, W. N., Phinney, K. W., and Ross, D., Simultaneous concentration and separation of enantiomers with chiral temperature gradient focusing, *Analytical Chemistry*, 76, 7243–7249, 2004.
52. Balss, K. M., Ross, D., Begley, H. C., Olsen, K. G., and Tarlov, M. J., DNA hybridization assays using temperature gradient focusing and peptide nucleic acids, *Journal of the American Chemical Society*, 126, 13474–13479, 2004.
53. Balss, K. M., Vreeland, W. N., Howell, P. B., Henry, A. C., and Ross, D., Micellar affinity gradient focusing: A new method for electrokinetic focusing, *Journal of the American Chemical Society*, 126, 1936–1937, 2004.
54. Li, H. and Gale, R. J., Hydraulic and electroosmotic flow through silica capillaries, *Langmuir*, 9, 1150–1155, 1993.
55. Datta, R. and Kotamarthi, V. R., Electrokinetic dispersion in capillary electrophoresis, *AIChE J*, 36, 916–926, 1990.
56. Rice, C. L. and Whitehead, R., Electrokinetic flow in a narrow cylindrical capillary, *Journal of Physical Chemistry*, 69, 4017–4024, 1965.
57. Lettieri, G.-L., Dodge, A., Boer, G., De Rooij, N. F., and Verpoorte, E., A novel microfluidic concept for bioanalysis using freely moving beads trapped in recirculating flows, *Lab on a Chip*, 3, 34–39, 2003.
58. Lettieri, G.-L., *A Novel Microfluidic Concept for Bioanalysis Using Controlled Recirculating Flows*, University of Neuchatel, 2004.
59. Lettieri, G.-L., De Rooij, N. F., and Verpoorte, E., Microfluidic device for bioanalysis using freely moving beads trapped in a recirculating flow, in *Micro Total Analysis System (μTAS)*, Nara, 2002, pp. 630–632.
60. Lettieri, G.-L., Ceriotti, L., De Rooij, N. F., and Verpoorte, E., On-chip DNA trapping and preconcentration employing recirculating flow devices, in *7th International Conference on Miniaturized Chemical and Biochemical Analysis Systems*, California, USA, 2003, pp. 737–740.
61. Astorga-Wells, J., Bergman, T., and Jörnvall, H., Multistep microreactions with proteins using electrocapture technology, *Analytical Chemistry*, 76, 2425–2429, 2004.
62. Astorga-Wells, J., Vollmer, S., Tryggvason, S., Bergman, T., and Jornvall, H., Microfluidic electrocapture for separation of peptides, *Analytical Chemistry*, 77, 7131–7136, 2005.

63. Hori, A., Matsumoto, T., Nimura, Y., Ikedo, M., Okada, H., and Tsuda, T., Electroconcentration by using countercurrent due to pressurized flow and electrophoretic mobility, *Analytical Chemistry*, 65, 2882–2886, 1993.
64. Ista, L. K., Lopez, G. P., Ivory, C. F., Ortiz, M. J., Schifani, T. A., Schwappach, C. D., and Sibbett, S. S., Microchip countercurrent electroseparation, *Lab on a Chip*, 3, 266–272, 2003.
65. Simpson, N. J. K., *Solid-Phase Extraction: Principles, Techniques, and Applications*, 1st ed., Marcel Dekker, New York, 2000.
66. Ramsey, J. D. and Collins, G. E., Integrated microfluidics device for solid-phase extraction coupled to micellar electrokinetic chromatography separation, *Analytical Chemistry*, 77, 6664–6670, 2005.
67. Christensen, P. D., Johnson, S. W. P., McCreedy, T., Skelton, V., and Wilson, N. G., The fabrication of microporous silica structures for microreactor technology, *Analytical Communications*, 35, 341–343, 1998.
68. Skelton, V., Greenway, G. M., Haswell, S. J., Styring, P., Morgan, D. O., Warrington, B. H., and Wong, S. Y. F., The generation of concentration gradients using electroosmotic flow in micro reactors allowing stereoselective chemical synthesis, *Analyst*, 126, 11–13, 2001.
69. Van den Bosch, S. E., Heemstra, S., Kraak, J. C., and Poppe, H., Experiences with packed capillary electrochromatography at ambient pressure, *Journal of Chromatography A*, 755, 165–177, 1996.
70. Colón, L. A., Reynolds, K. J., Alicea-Maldonado, R., and Fermier, A. M., Advances in capillary electrochromatography, *Electrophoresis*, 18, 2162–2174, 1997.
71. Boughtflower, R. J., Underwood, T., and Paterson, C. J., Capillary electrochromatography – some important considersations in the preparation of packed capillaries and the choice of mobile phase bufffers, *Chromatographia*, 40, 329–335, 1995.
72. Oleschuk, R. D., Shultz-Lockyear, L. L., Ning, Y., and Harrison, D. J., Trapping of bead-based reagents within microfluidic systems: On-chip solid-phase extraction and electrochromatography, *Analytical Chemistry*, 72, 585–590, 2000.
73. Jemere, A. B., Oleschuk, R. D., Ouchen, F., Fajuyigbe, F., and Harrison, D. J., An integrated solid-phase extraction system for subpicomolar detection, *Electrophoresis*, 23, 3537–3544, 2002.
74. Lord, G. A., Gordon, D. B., Myers, P., and King, B. W., Tapers and restrictors for capillary electrochromatography and capillary electrochromatography-mass spectrometry, *Journal of Chromatography A*, 768, 9–16, 1997.
75. Mayer, M., Rapp, E., Marck, C., and Bruin, G. J. M., Fritless capillary electrochromatography, *Electrophoresis*, 20, 43–49, 1999.
76. Ceriotti, L., De Rooij, N. F., and Verpoorte, E., An integrated fritless column for on-chip capillary electrochromatography with conventional stationary phases, *Analytical Chemistry*, 74, 639–647, 2002.
77. Adam, T., Unger, K. K., Dittmann, M. M., and Rozing, G. P., Towards the column bed stabilization of columns in capillary electroendosmotic chromatography immobilization of microparticulate silica columns to a continuous bed, *Journal of Chromatography A*, 887, 327–337, 2000.
78. Smith, N. W. and Evans, M. B., The analysis of pharmaceutical compounds using electrochromatography, *Chromatographia*, 38, 649–657, 1994.
79. Yin, H., Killeen, K., Brennen, R., Sobek, D., Werlich, M., and Van de Goor, T., Microfluidic chip for peptide analysis with an integrated HPLC column, sample enrichment column, and nanoelectrospray tip, *Analytical Chemistry*, 77, 527–533, 2005.
80. Fortier, M.-H., Bonneil, E., Goodley, P., and Thibault, P., Integrated microfluidic device for mass spectrometry-based proteomics and its application to biomarker discovery programs, *Analytical Chemistry*, 77, 1631–1640, 2005.
81. Rozing, G., Trends in HPLC column formats—Microbore, nanobore, and smaller, *LC–GC Europe*, 16, 14–19, 2003.
82. Kutter, J. P., Jacobson, S. C., and Ramsey, J. M., Solid phase extraction on microfluidic devices, *Journal of Microcolumn Separations*, 12, 93–97, 2000.
83. Kutter, J. P., Jacobson, S. C., Matsubara, N., and Ramsey, J. D., Solvent-programmed microchip open-channel electrochromatography, *Analytical Chemistry*, 70, 3291–3297, 1998.
84. Malik, A., Advances in sol-gel based columns for capillary electrochromatography: Sol-gel open-tubular columns, *Electrophoresis*, 23, 3973–3992, 2002.

85. Rieux, L., Niederländer, H. A. G., Verpoorte, E., and Bischoff, R. P. H., Silica monolithic columns: Synthesis, characterisation and applications to the analysis of biological molecules, *Journal of Separation Science*, 28, 1628–1641, 2005.
86. Rohr, T., Ogletree, D. F., Svec, F., and Fréchet, J. M. J., Surface functionalization of thermoplastic polymers for the fabrication of microfluidic devices by photoinitiated grafting, *Advanced Functional Materials*, 13, 264–270, 2003.
87. Karwa, M., Hahn, D., and Mitra, S., A sol-gel immobilization of nano and micron size sorbents in poly(dimethylsiloxane) (PDMS) microchannels for microscale solid phase extraction (SPE), *Analytical Chimica Acta*, 546, 22–29, 2005.
88. Yu, C., Davey, M. H., Svec, F., and Fréchet, J. M. J., Monolithic porous polymer for on-chip solid-phase extraction and preconcentration prepared by photoinitiated *in situ* polymerization within a microfluidic device, *Analytical Chemistry*, 73, 5088–5096, 2001.
89. Peterson, D. S., Rohr, T., Svec, F., and Fréchet, J. M. J., Dual-function microanalytical device by *in situ* photolithographic grafting of porous polymer monolith: Integrating solid-phase extraction and enzymatic digestion for peptide mass mapping, *Analytical Chemistry*, 75, 5328–5335, 2003.
90. Tan, A., Benetton, S., and Henion, J.D., Chip-based solid-phase extraction pretreatment for direct electrospray mass spectrometry analysis using an array of monolithic column in a polymeric substrate, *Analytical Chemistry*, 75, 5504–5511, 2003.
91. Yang, Y., Li, C., Lee, K. H., and Craighead, H. G., Coupling on-chip solid-phase extraction to electrospray mass spectrometry through an integrated electrospray tip, *Electrophoresis*, 26, 3622–3630, 2005.
92. Sia, S. K. and Whitesides, G. M., Microfluidic devices fabricated in poly(dimethylsiloxane) for biological studies, *Electrophoresis*, 24, 3563–3576, 2003.
93. Wolfe, K. A., Breadmore, M. C., Ferrance, J. P., Power, M. E., Conroy, J. F., Norris, P. M., and Landers, J. P., Toward a microchip-based solid-phase extraction method for isolation of nucleic acids, *Electrophoresis*, 23, 727–733, 2002.
94. Dulay, M. T., Kulkarni, R. P., and Zare, R. N., Preparation and characterization of monolithic porous capillary columns loaded with chromatographic particles, *Analytical Chemistry*, 70, 5103–5107, 1998.
95. Tokeshi, M., Minagawa, T., and Kitamori, T., Integration of a microextraction system on a glass chip: Ion-pair solvent extraction of Fe(II) with 4,7-diphenyl-1,10-phenanthrolinedisulfonic acid and tri-*n*-octylmethylammonium chloride, *Analytical Chemistry*, 72, 1711–1714, 2000.
96. Sato, K., Tokeshi, M., Sawada, T., and Kitamori, T., Molecular transport between two phases in a microchannel, *Analytical Science*, 16, 455–456, 2000.
97. Hisamoto, H., Horiuchi, T., Tokeshi, M., Hibara, A., and Kitamori, T., On-chip integration of neutral ionophore-based ion-pair extraction reaction, *Analytical Chemistry*, 73, 1382–1386, 2001.
98. Hibara, A., Nonaka, M., Hisamoto, H., Uchiyama, K., Kikutani, Y., Tokeshi, M., and Kitamori, T., Stabilization of liquid interface and control of two-phase confluence and separation in glass microchips by utilizing octadecylsilane modification of microchannels, *Analytical Chemistry*, 74, 1724–1728, 2002.
99. Tokeshi, M., Minagawa, T., and Kitamori, T., Integration of a microextraction system Solvent extraction of a Co-2-nitroso-5-dimethylaminophenol complex on a microchip, *Journal of Chromatography A*, 894, 19–23, 2000.
100. Hibara, A., Tokeshi, M., Uchiyama, K., Hisamoto, H., and Kitamori, T., Integrated multilayer flow system on a microchip, *Analytical Sciences*, 17, 89–93, 2001.
101. Surmeian, M., Slyadnev, M. N., Hisamoto, H., Hibara, A., Uchiyama, K., and Kitamori, T., Three-layer flow membrane system on a microchip for investigation of molecular transport, *Analytical Chemistry*, 74, 2014–2020, 2002.
102. Chen, H., Fang, Q., Yin, X.-F., and Fang, Z.-L., Microfluidic chip-based liquid–liquid extraction and preconcentration using a subnanoliter-droplet trapping technique, *Lab on a Chip*, 5, 719–725, 2005.
103. Fang, Q., Chen, H., and Fang, Z.-L., Microfluidic chips systems based on stopped-flow liquid–liquid extraction, in *Micro Total Analysis System (μTAS)*, Boston, 2005, pp. 128–130.
104. Cai, Z., Chen, H., Chen, B., and Huang, C., A gravity driven micro flow injection wetting film extraction system on a polycarbonate chip, *Talanta*, 68, 895–901, 2006.

105. Luo, Y., Al-Othman, R., Ruzicka, J., and Christian, G. D., Solvent extraction-sequential injection without segmentation and phase separation based on the wetting film formed on a Teflon tube wall, *The Analyst*, 121, 601–606, 1996.
106. Van Dijk, J. C., Munneke, B. R., Kramer, B., and Wouters, J. W., Membrane filtration: A realistic option in the field of water supply? *Desalination*, 81, 229–247, 1991.
107. Khandurina, J., Jacobson, S. C., Waters, L. C., Foote, R. S., and Ramsey, J. M., Microfabricated porous membrane structure for sample concentration and electrophoretic analysis, *Analytical Chemistry*, 71, 1815–1819, 1999.
108. Khandurina, J., McKnight, T. E., Jacobson, S. C., Foote, R. S., and Ramsey, J. M., Integrated system for rapid PCR-based DNA analysis in microfluidic devices, *Analytical Chemistry*, 72, 2995–3000, 2000.
109. Foote, R. S., Khandurina, J., Jacobson, S. C., and Ramsey, J. M., Preconcentration of proteins on microfluidic devices using porous silica membranes, *Analytical Chemistry*, 77, 57–63, 2005.
110. Song, S., Singh, A. K., and Kirby, B. J., Electrophoretic concentration of proteins at laser-patterned nanoporous membranes in microchips, *Analytical Chemistry*, 76, 4589–4592, 2004.
111. Hatch, A. V., Herr, A. E., Throckmorton, D. J., Brennan, J. P., Giannobile, W. V., and Singh, A. K., On-chip preconcentration of proteins for picomolar detection in oral fluids, in *Micro Total Analysis System (μTAS)*, Boston, 2005, pp. 1042–1044.
112. Pu, Q., Yun, J., Temkin, H., and Lin, S., Ion-enrichment and ion-depletion effect of nanochannel structures, *Nano Letters*, 4, 1099–1103, 2004.
113. Petersen, N. J. and Ramsey, J. M., Concentration and separation of ionic analytes using nano–microchannel junctions, in *Micro Total Analysis System (μTAS)*, Boston, 2005, pp. 1252–1254.
114. Wang, Y.-C., Stevens, A. L., and Han, J., Million-fold preconcentration of proteins and peptides by nanofluidic filter, *Analytical Chemistry*, 77, 4293–4299, 2005.
115. Plecis, A., Schoch, R. B., and Renaud, P., Ionic transport phenomena in nanofluidics: experimental and theoretical study of the exclusion-enrichment effect on a chip, *Nano Letters*, 5, 1147–1155, 2005.
116. Wang, Y.-C., Tsau, C. H., Burg, T. P., Manalis, S., and Han, J., Efficient biomolecule preconcentration by nanofilter-triggered electrokinetic trapping, in *Micro Total Analysis System (μTAS)*, Boston, 2005, pp. 238–240.
117. Erickson, B. E., Protein preconcentrator, *Analytical Chemistry*, 77, 283A–284A, 2005.

51 Using Phase-Changing Sacrificial Materials to Fabricate Microdevices for Chemical Analysis

Hernan V. Fuentes and Adam T. Woolley

CONTENTS

51.1 INTRODUCTION

For many years, miniaturization has been used to improve separation performance in chemical analysis. Since first demonstrated over 25 years ago,[1,2] capillary electrophoresis (CE) has been found to offer many advantages compared to slab gel electrophoresis, and CE has become a powerful technique for the analysis of complex biological samples due to its high efficiency and ease of automation.[3]

The potential for using silicon-based technologies to fabricate microdevices for chemical analysis was initially demonstrated in 1979 by Terry et al.[4] Over a decade later, the concept of integrated microfluidic devices, or micrototal analysis systems (μTAS), was presented by Manz et al.[5] Chip-based analytical systems offer many advantages over their conventional, larger-size counterparts. Separations in microchips are particularly attractive due to rapid analysis times, parallel runs, high sample throughput, ability to manipulate ultra-small volumes, and potential for integration of sample processing steps in series with separation.[6] Because of its relative simplicity, CE was first demonstrated in a microfabricated format in 1992.[7] Today, microchip electrophoresis (μCE) has become a well-developed technique with numerous analysis applications reported for DNA,[8,9] proteins,[10,11] cells,[12] small molecules,[13,14] explosives,[15] and environmental contaminants,[16] among others.

Microfluidic analysis devices were initially fabricated almost exclusively in glass and silicon, using micromachining techniques such as photolithography and wet chemical etching.[4,17–19] Glass substrates have been implemented widely due to their favorable optical properties, well-known surface chemistry, and established fabrication techniques. However, the construction of glass microchips requires a well-equipped cleanroom and the use of hazardous chemicals such as hydrofluoric acid. In most cases, the bonding step is performed at high temperatures (>600°C) for several hours, making the process slow and impractical for mass production of microdevices.[20] Silicon microfabrication uses almost the same technology and surface chemistry as glass; however, crystalline silicon is brittle and must be handled with care. In addition, silicon is opaque to visible and ultraviolet (UV) light and thus, cannot be used with laser-induced fluorescence (LIF) detection.

These problems with glass and silicon microdevices have led to growing interest in using polymer-based substrates. Indeed, plastic microchips should overcome several limitations of their glass and silicon counterparts, since plastics are easier to manufacture and handle, enabling the development of lower cost and more rugged microfluidic devices.[21] Moreover, polymeric materials possess a broad range of physical and chemical properties, which should enable compatibility with many biological and chemical reagents and assays.[22] A variety of polymers have been used to fabricate microfluidic devices, including poly(methyl methacrylate) (PMMA),[23,24] poly(dimethylsiloxane) (PDMS),[25,26] polycarbonate (PC),[27] polyimide (PI),[28] cyclic olefin copolymer (COC),[29] polystyrene,[30] and polyethylene terephthalate (PET).[31] Finally, polymer manufacturing methods, such as hot embossing[32,33] and injection molding,[34] offer great potential for the mass production of low-cost platforms containing complex microfluidic patterns.

Polymer microchips are expected to contribute strongly in biological analysis in the postgenome era, especially in the field of proteomics.[35,36] Considerable research has already focused on separations, the use of new materials, and transferring conventional analysis methods to a microchip platform, but more work remains to be done to enable the exploitation of the full potential of plastic microdevices. In particular, the use of conventional device-bonding techniques has limited the broad application of polymer microchips.

Thermal bonding is a widely used approach for sealing microfluidic patterns made in polymer substrates.[21–23,27,37–40] The standard thermal bonding method involves heating a patterned substrate and cover layer to near the glass transition temperature of the material.[23,37] Other reported approaches have used heated weights,[41] or bonding in boiling water.[42] These methods are used because they are easy to implement and allow the formation of microchannels with a surface entirely composed of the same material. However, applying elevated temperatures to a microchip can be incompatible with some biological analyses. Moreover, since bonding takes place near the glass transition temperature of the material and under applied pressure, the microstructures can be distorted readily during this step. This issue is especially problematic for small channel dimensions or low-aspect-ratio microchannels, and can affect the yield and reproducibility of device fabrication. In addition, thermal bonding does not provide sufficient device robustness for introducing viscous sieving solutions into polymeric microchips. Therefore, the development of new technologies to enclose polymer microfluidic systems (without using heat) should broaden the range of applications of plastic microchips.

Several studies have been performed to develop alternate sealing methods for plastic microdevices. Thermal lamination[34] is one approach for enclosing microfluidic channels, but this technique leads to channels whose walls are composed of different materials, which results in inhomogeneities in zeta potential and band broadening.[43] Bonding of microchips using an intermediate adhesive layer is easy to implement;[44–47] however, this approach is limited by a high incidence of channel blockage. In addition, the difference in surface composition between the substrate and adhesive has a negative effect on separation efficiency.[44,48] Lai et al.[49] developed a resin–gas injection bonding method, in which blank and patterned substrates were held together, and a monomer solution that filled the gap between the two pieces was cured using UV radiation. This technique allows for surface modification and bonding to be performed simultaneously; however, the monolayer inside the channel must be uniform and stable to prevent irregular electroosmotic flow. Moreover, this method is only applicable to UV-transparent polymers. Chemical bond formation at contact surfaces has been used to seal plastic microdevices;[50–54] however, this approach requires surface modification before bonding[52] or the use of intermediates such as catalysts and initiators.[53,54]

Solvent bonding is an attractive alternative to conventional thermal bonding.[52,55–58] Briefly, polymeric surfaces to be bonded are exposed to an organic solvent, which partially dissolves the polymer at the surface. When two solvent-wetted polymer pieces are brought together under applied pressure, the polymer chains in both surfaces can interdiffuse and entangle, forming a high-strength bond.[59–61] In principle, solvent bonding of polymer substrates should produce robustly enclosed microchannels due to the strong molecular interaction between the surfaces. However, the use of this technique has been limited due to the ease with which microchannels can be blocked by dissolved polymer, or swelling or softening of the substrate.

Several papers have reported solvent-assisted bonding of polymer microdevices at room temperature.[55–58] Liu et al.[62] demonstrated the solvent bonding of PC devices, but at an elevated temperature of approximately 200°C. Recently, Shah et al.[58] reported the enclosure of polymer microfluidic devices by inducing a flow of solvent through a microfabricated channel via capillary action. This approach was simple to implement, but the bond strength of the microdevices was low (80 psi), which would limit applications that require viscous gels or chromatographic stationary phases to be pumped into microchannels under pressure. Therefore, the development of fabrication protocols that more fully realize the advantages of solvent bonding is warranted.

This chapter describes the theory, methodology, and application of a microfabrication process that uses phase-changing sacrificial layers (PCSLs) as intermediates to protect microchannel features during bonding or hydrogel polymerization. We focus on key process details associated with the fabrication of microchips, and the application of PCSL-formed microfluidic devices in CE separations and other electric field-based analysis methods. Finally, we provide a brief overview of potential future trends and applications of PCSL fabrication methods in microfluidics.

51.2 BACKGROUND AND THEORY

In the past 15 years, the use of microfluidic devices for chemical analysis has increased tremendously. Indeed, a broad range of chromatographic and electrophoretic separation methods have been implemented in microchips.[63] However, for widespread utilization of microfabricated devices in analysis applications, particularly in the field of proteomics, further efforts are needed to develop simple fabrication techniques that achieve functional integration of multiple tasks in a single device.[64] In this section, we describe the fabrication of microdevices using sacrificial materials and discuss some of the advantages of this approach over conventional microfabrication methods.

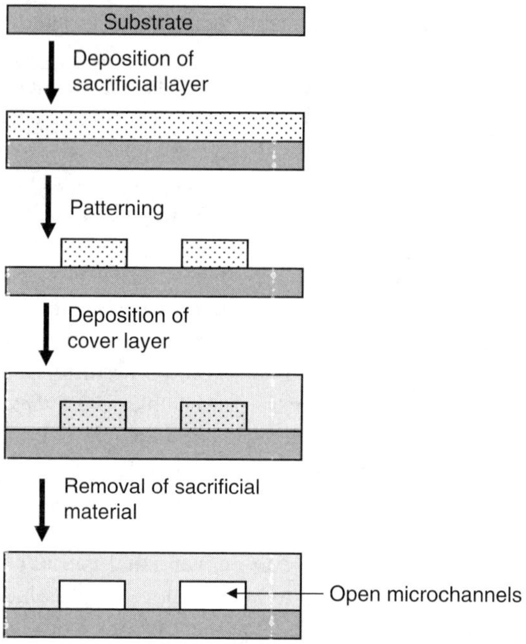

FIGURE 51.1 Schematic diagram of the general processing steps involved in the fabrication of microchannel structures using sacrificial materials.

51.2.1 Sacrificial Materials

Microfabrication using sacrificial layers is well developed in the field of microelectromechanical systems. Reports include the fabrication of micro- and nanomechanical components,[65] electroosmotic micropumps in silicon and glass substrates,[66] and nano-[67] or microchannels[68,69] with potential applications in biology.[70] Unlike bonding protocols, in which a cover plate is affixed to a patterned substrate to seal microchannels, sacrificial layer methods can obviate the bonding step, making this approach very attractive.[71,72]

Figure 51.1 provides a general overview of the process for making microfluidic systems using sacrificial materials. Briefly, a sacrificial layer is deposited on a substrate by spin coating, vapor deposition, or some other method. Then, using photolithography, the sacrificial material is patterned and serves as a temporary "placeholder," which defines the channel geometry. After deposition of a top layer, the sacrificial material is removed, leaving behind open tubular features between the substrate and the cover layer.[71]

There are a variety of materials that can be used as sacrificial cores. Inorganic sacrificial materials include SiO_2[73] and metals such as aluminum,[74] titanium,[75] and nickel.[76] Polymers such as PI,[77] PMMA,[65] PC,[68] and photoresist[78] have also been used as sacrificial materials. After deposition of the cover film, removal of the sacrificial layer can be achieved by dissolution,[79] etching,[66,74,80] or thermal degradation.[81,82] These removal methods each have benefits and drawbacks; selection of the optimal approach is specific to particular combinations of substrate, sacrificial layer, and cover film.[73,83] Recently, Whitesides and coworkers[84] implemented a fabrication method using water-soluble sacrificial cores. Poly(acrylic acid) and dextran proved to be effective sacrificial layers that could be dissolved in water or aqueous NaCl, for making metallic microstructures by nickel electrodeposition.

Importantly, microfabrication using sacrificial cores is a versatile technique. First, there are no limitations in the size or flatness of the channel substrate, since no bonding is involved.[81] Second, the choice of sacrificial material can influence the geometry of the resulting microchannels; for example,

Peeni et al.[74] used aluminum/photoresist, SU-8, and aluminum to obtain semicircular, rectangular, and trapezoidal geometries, respectively. In addition, even though glass[66,74] and silicon[85,86] are the two most common substrates for microfabrication using sacrificial layers, the surface properties of microfluidic vias can be adapted to specific applications by depositing selected materials beneath and on top of the sacrificial layer. Moreover, sacrificial approaches facilitate the formation of three-dimensional fluidic systems through sequential deposition of layers.[67,87] Finally, the fabrication of microchannels *in situ* should enable the integration of fluidic and electronic components in microdevices.

51.2.2 Sacrificial Layer Fabricated Microfluidic Devices

51.2.2.1 Silica and Glass Materials

Sacrificial etching methods have been used widely in the microelectronics industry for years; however, only a few examples of separations in sacrificially formed microfluidic systems have been demonstrated. Craighead and coworkers[88] used thin-film deposition and wet chemical etching to define approximately 100-nm-dimension nanochannels, which were used to measure the electrophoretic mobilities of two different lengths of DNA. More recently, the same group fabricated submicrometer-dimension fluidic channels for the analysis of less than 100 fg of DNA.[80] Hawkins and coworkers fabricated a microdevice using thin-film techniques, and used it to perform single-molecule detection.[89] Woolley and coworkers have shown CE separations of fluorescently labeled amino acids in a 9-mm-long microchannel fabricated by thin-film deposition of silicon dioxide over a photoresist/aluminum sacrificial layer.[74] Subsequent optimization also enabled peptide separations to be carried out in these devices.[90] These studies demonstrate the potential for performing separations in microdevices fabricated using sacrificial layer techniques, but further work remains.

Despite the successful construction of microchips using sacrificial layer methods, several issues must still be addressed with the fabrication process and the resulting devices. First, fabrication requires cleanroom facilities, costly instrumentation for thin-film deposition and patterning, and expertise to use such equipment, all of which make this approach inaccessible to some researchers. In addition, there are several constraints on the types of substrates, sacrificial layers, and overlayers that can be used.[68,85] For example, the sacrificial material must produce a high quality and uniform film of several micrometers thickness.[73] In addition, the sacrificial layer must adhere well to the substrate and not react with the covering film. Ideally, the sacrificial layer will be removed completely, leaving a smooth inner surface in the microchannel. Moreover, neither the substrate nor the cover layer must be affected during the sacrificial layer removal step.[68] Unfortunately, the removal of the sacrificial material can take hours to days, which limits the mass production of microchips by this method. Solubility can be a concern, especially when a polymer is used as the cover layer with a photoresist-based sacrificial material. In this case, the solvent present in the coating polymer solution can attack the sacrificial photoresist and compromise microchannel integrity.[68] Similarly, if a thermally decomposable polymer is used as a sacrificial layer, complete decomposition should occur in a narrow temperature range, which must be compatible with the other materials and components on the substrate.[91]

Structural strength is another important consideration in the fabrication of microchannels for fluid handling. In the sacrificial layer fabrication approach, microchannels are generated by deposition of a thin film on top of a patterned material, and the resulting structures are generally weaker than those made by standard bonding methods. Hubbard et al.[83] investigated possible causes of channel wall failure during fabrication with an aluminum sacrificial layer and silicon dioxide overcoating. Elevated internal pressure from gas generated during the etching of aluminum was found to be the main cause of microchannel wall failure during fabrication. A finite element model showed that channel width and film thickness were the most critical parameters to optimize in avoiding device failure.

In summary, sacrificial fabrication of microfluidic devices in glass and silicon overcomes some limitations of conventional manufacturing methods. Important benefits include the ability to fabricate nanometer- to several micrometer-dimension structures and the elimination of a bonding step. Nevertheless, several problems with sacrificial fabrication have limited the implementation of this technique in making microfluidic devices. First, the fabrication protocols are not generally easy to implement and require sophisticated instrumentation. Second, device materials are limited to glass and silicon, the substrates most compatible with the manufacturing methods. Finally, fabrication yield and reproducibility still need improvement. Thus, further work is essential to develop sacrificial fabrication techniques.

51.2.2.2 Polymeric Materials

We have been interested in applying sacrificial layer methods for making polymer microfluidic systems. We were inspired by work that utilized the solid–liquid phase change of waxes at their melting point, to form actuators such as valves in fluidic microchips.[92–95] This phase transition offers a promising approach for filling (as a liquid) and protecting (as a solid) microchannels during device processing steps.

In brief, PCSL microfabrication involves heating a solid sacrificial material to a temperature above its melting point and using it to fill a microchannel imprinted in a polymeric substrate. After cooling, the sacrificial material solidifies and protects the microchannel from solvent and dissolved polymer during bonding. Next, solvent is spread on top of the PCSL-protected channels, and a polymeric cover layer containing access reservoirs is aligned and held in contact with the imprinted substrate until a robust bond is obtained, resulting in enclosed PCSL-filled channels. Finally, the device is heated above the melting temperature of the sacrificial material, and vacuum is applied to remove the melted PCSL, creating open microchannels. Indeed, a paraffin wax PCSL was recently used to construct microfluidic devices; the combination of PCSLs and solvent bonding enabled polymer microchips to be fabricated and evaluated in μCE analysis of amino acids and peptides.[96]

PCSL microfabrication is an attractive alternative for making robustly enclosed microchannels by solvent bonding, while protecting microchannel integrity using a sacrificial material. This technique does not require sophisticated thin-film deposition instrumentation and should be easy to implement by microfluidics researchers. In the next section, we provide guidelines for the fabrication of polymer microfluidic systems using PCSLs.

51.3 FABRICATION TECHNIQUES AND METHOD DEVELOPMENT

Because of the many advantages of polymer microchips, there is a tremendous need for alternative fabrication processes to allow the realization of their full potential.[64,97] Thus, this section describes the steps to follow to make polymer microdevices using PCSL and solvent-bonding techniques. Because of its good optical properties and well-known patterning procedures, PMMA is used frequently for the fabrication of microfluidic chips.[23,24,34] As a result, the PCSL microfabrication method will be detailed for PMMA microchip fabrication. However, constructing microdevices from different polymers using the PCSL method should be straightforward, if the appropriate combination of substrate, sacrificial material, and solvent are selected. The basic fabrication procedure includes (a) template fabrication, (b) microchannel imprinting, (c) filling channels with PCSL, (d) microdevice bonding, (e) sacrificial layer removal, and (f) microchip evaluation. A schematic overview of the fabrication process is shown in Figure 51.2.

51.3.1 Template Fabrication

Polymer microfluidic systems for chemical analysis are normally fabricated by transferring a patterned microstructure from a template to a substrate.[23] Micromachined silicon templates are often

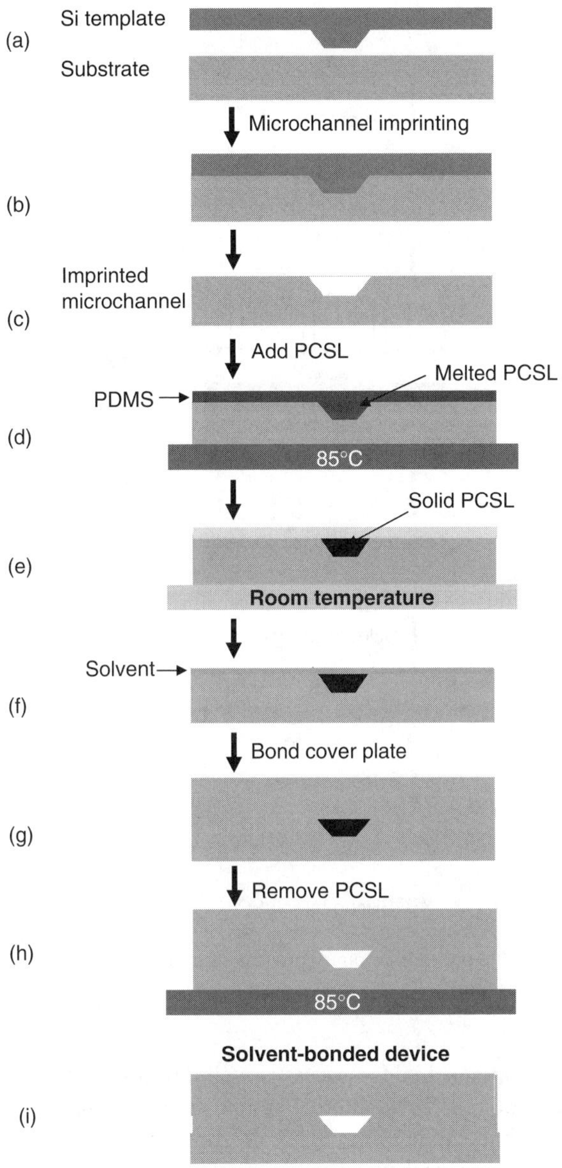

FIGURE 51.2 Fabrication of microchannels in a polymer substrate using PCSLs and solvent bonding (additional details are in the text).

used; these are created by standard photolithographic and etching methods.[42] Figure 51.3 briefly summarizes the steps to transfer a pattern from a computer-designed mask to a silicon wafer. The process starts with a silicon substrate that has an SiO_2 layer formed on the surface. Next, a photosensitive material (photoresist) is spin coated on the wafer, forming a film whose thickness depends on the spin speed and time. Excess solvent in the photoresist is removed by heating the wafer, leaving a solid photosensitive material on the surface, which is then exposed to UV radiation through a patterning mask. The UV light induces a chemical reaction in the photoresist, altering its solubility in a developer solution. After the photoresist pattern is developed, the wafer is subjected to solution- or gas-phase etching to remove the SiO_2 from the areas not protected by the hardened photoresist. Finally, wet chemical or gas-phase etching is used to generate the elevated features, which should be identical to the mask pattern.

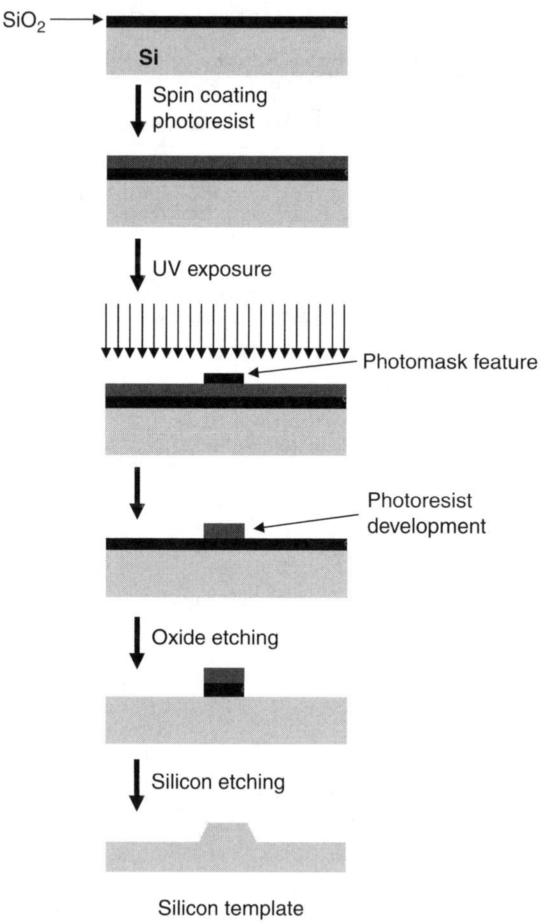

FIGURE 51.3 Schematic of the photolithography process to transfer a pattern to a silicon wafer, forming a template for microchannel imprinting.

Although the silicon micromachining process is well developed, extensive use of costly clean-room instrumentation is required. Thus, alternative and simpler template construction approaches are being pursued. One such method utilizes SU-8, an epoxy-based negative photoresist, which has excellent chemical resistance and mechanical properties. Patterned SU-8 is being applied increasingly in making microstructures for templates in microchip production.[98–100]

51.3.2 Microchannel Imprinting

Using a procedure reported by Locascio and coworkers,[23] elevated features in a Si template can be imprinted into PMMA substrates to form microchannels (see Figure 51.2a through c). Briefly, the silicon template and polymer substrate are sandwiched between glass slides (Figure 51.2b), using two 2″ C-clamps to apply pressure. It is important to use flat and clean substrates to hold the polymer and template together, and to apply pressure evenly to avoid breaking the template. The pieces are heated to 110°C, which is above the glass transition temperature of PMMA. After 10 min, the assembly is removed, the C-clamps are tightened 1/8 turn, and the pieces are returned to the oven. The clamp tightening procedure is repeated two more times for a total imprinting duration of 30 min. This process results in the embedding of the template features into the PMMA (Figure 51.2b). After removal from the oven, the imprinted PMMA (Figure 51.2c) is cooled to room temperature.

51.3.3 Filling Microchannels with PCSL

Before solvent bonding, the microchannels must be filled with PCSL. Some practical considerations for selecting a material for a sacrificial layer are (1) it should have a lower melting point than the substrate itself; (2) it must be resistant to the bonding solvent; (3) it should not react with the polymer substrate; (4) it should form a smooth surface after solidification; and (5) it must be easy to remove from the microchannel after bonding.

The following is a short overview of the process for depositing a PCSL into imprinted features. A PMMA piece having imprinted microchannels (formed as described earlier) is sealed reversibly to an approximately 1-mm-thick PDMS film, which has small holes that correspond to the channel ends. This temporarily enclosed PMMA-PDMS structure is placed on a hotplate at 85°C, heated paraffin wax is added to all but one of the channel access holes in the PDMS film, and vacuum is applied at the unfilled opening, causing liquid PCSL to fill the entire microchannel network (Figure 51.2d). Cooling the assembly to room temperature produces a solid PCSL, protecting the microchannels as indicated in Figure 51.2e.

Occasionally, some PCSL solidifies outside the microchannel region, and this excess wax needs to be removed for proper bonding. We have devised two solutions to the issue of PCSL deposition outside of channels. First, if the paraffin wax PCSL is left in contact with PDMS for 1–2 h, the PCSL outside the channels dissolves into the PDMS.[96] A second solution is to add the liquid paraffin wax to the device without applying vacuum; in this case, the PCSL enters the microchannels by capillary action, and a more uniform filling is observed. Once well-defined microchannels loaded with solid sacrificial material are obtained, the PDMS film is removed carefully, and the PCSL-filled device is ready for bonding.

51.3.4 Microdevice Bonding

To form enclosed microchannels, the PCSL-protected PMMA piece must be bonded to a cover plate having channel access holes. To this end, solvent is added on top of the imprinted substrate (Figure 51.2f), a polymer top piece is aligned with the PCSL-protected microstructure, and applied pressure for approximately 2 min completes microdevice bonding (Figure 51.2g). The solvent bonding step must be performed sufficiently quickly so that the solvent does not evaporate or dissolve the substrate extensively, which could affect channel integrity. In practice, we found that bringing the two PMMA pieces together at an angle allowed air bubbles to escape from the side, leading to higher-quality microdevice bonding. Finally, it is important for the cover PMMA piece to be about the same size as the imprinted substrate.

The amount of solvent added also affects the quality of the bonded devices. Sufficient solvent must be added to cover the entire PCSL-protected surface, without leaving bubbles or unwetted areas. On the other hand, if an excess of solvent is applied, it flows outside the bonding interface and can compromise the optical quality of the surface of the resulting device. To overcome this problem, PDMS films can be placed on the nonbonding sides of the PMMA, to keep the solvent from damaging the microchip exterior surfaces. Using masking tape to protect the PMMA surfaces has also proven to be successful. Experimentally, we found that a 2 × 4 cm PMMA substrate with an approximately 3-cm-long separation channel could be completely bonded with as little as 35 μL of acetonitrile.[96]

51.3.5 Sacrificial Layer Removal

The final step in PCSL fabrication of microfluidic systems is removal of the sacrificial material to yield clean, smooth microchannel inner surfaces. Paraffin wax PCSL is removed by heating the microdevice to 85°C and applying vacuum to a reservoir in the PMMA cover piece to aspirate the liquid PCSL (Figure 51.2h). Nonpolar organic solvents such as hexane or cyclohexane can be used subsequently to dissolve and remove residual sacrificial material from inside the channel. With

PMMA microdevices, complete removal of the PCSL (Figure 51.2i) is verified by an absence of air bubbles in the channels on filling with water.

51.3.6 Microchip Evaluation

Several aspects of the microchips can be characterized after bonding is completed. For instance, the microchannel shape can be indicative of the effectiveness of the sacrificial material in protecting the microchannel during solvent bonding. Edge-on photomicrographs allow measurement of the cross-sectional area of features, both before and after bonding. As an example, Figure 51.4 shows cross-sectional views of a microchannel at several stages of the fabrication process. As is visible in Figure 51.4b, the paraffin wax PCSL appears to undergo a small amount of shrinkage on solidification. Importantly, this minor shrinkage does not compromise microchannel usefulness, even with structures having depths as small as 7 μm.[96]

Another important parameter to evaluate in PCSL-fabricated microchips is the bond strength. Several techniques have been reported to determine the bonding strength of microfluidic chips; for example, by measuring the tensile force in pulling pieces apart[101] or by evaluating the shear force at the bonding interface of chips.[102] These methods normally use epoxy to attach a device to an external force; thus, the bond strength determination may be limited by the strength of the epoxy rather than the microchip itself. We have devised an approach for determining the maximum internal pressure that can be sustained in PCSL solvent-bonded devices. We take an appropriately threaded PMMA piece (which can be connected to a N_2 cylinder via 1/16-in copper tubing) and bond it to a second PMMA substrate utilizing the same conditions (solvent, applied pressure, and time) as for microchips. Once this assembly is connected to a gas cylinder, the applied pressure can be increased until either the

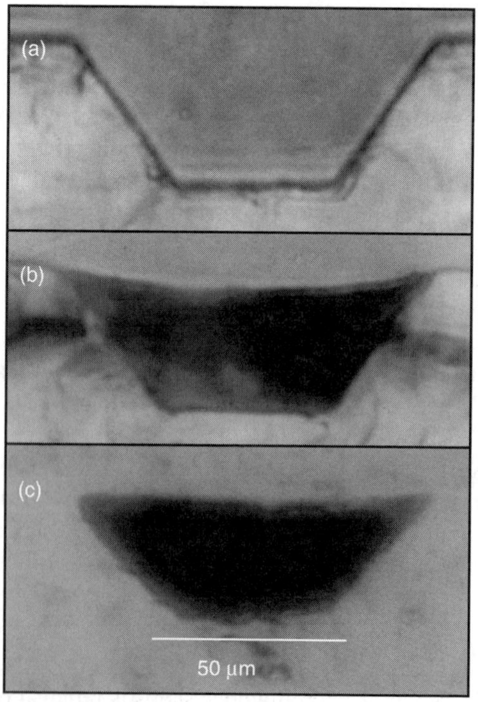

FIGURE 51.4 Edge-on photomicrographs of channel cross-sections at different fabrication stages. (a) Imprinted channel. (b) Imprinted channel filled with PCSL. (c) Bonded device after PCSL removal. (Reprinted from Kelly, R. T., et al., *Anal. Chem.*, 77, 3536, 2005. Copyright 2005. With permission from American Chemical Society.)

device fails or the maximum cylinder value (~2200 psi) is reached. Using this approach, we found that PMMA bonded using acetonitrile and our conditions was stable to at least 2200 psi internal pressure.[96] To more precisely measure the bond strength, one could appropriately connect PMMA substrates to a high-pressure liquid chromatography pump and measure the maximum pressure obtained. As with all high-pressure device failure determination methods, it is necessary to take proper safety precautions to avoid injury from PMMA projectiles that may be produced when failure occurs.

51.4 PRACTICAL APPLICATIONS

51.4.1 MICROCHIP CAPILLARY ELECTROPHORESIS

Capillary electrophoresis in polymer microfluidic devices was first demonstrated 10 years ago, in both PDMS[26] and PMMA.[23] The analysis of amino acids and peptides by μCE is also well established.[42,96,103,104] The operational parameters for injection and separation in μCE are presented in Figure 51.5.[105] First, the microchannels are filled with buffer solution, ensuring that the channels are bubble free. Sample solution is pipetted into the sample reservoir (SR), and a potential is applied to the injection reservoir (IR) for a short duration, while keeping the other three reservoirs grounded to load sample. To separate the small sample plug loaded into the injector, the voltage configuration is switched so a potential that is several fold higher than the injection voltage is applied to the waste reservoir (WR), the buffer reservoir (BR) is kept at ground, and the SR and IR are held at the injection voltage. Detection of separated analytes takes place in the separation channel near the WR. Owing to its high sensitivity and ability to probe small volumes, LIF is the most commonly used detection method in μCE. However, LIF requires that many analytes be labeled fluorescently before analysis, for example, using reagents such as fluorescein isothiocyanate (FITC).[106] In this section, we describe the application of PCSL-formed PMMA microchips to the μCE analysis of amino acids and peptides.

51.4.1.1 Amino Acid and Peptide Analysis

The PCSL techniques described in Section 51.3 were used to fabricate PMMA microchips for μCE analysis. The microchannel lengths from the injection intersection to the BR, SR, and IR were 0.5 cm, and the distance from the injection intersection to the WR was 2.5 cm (see Figure 51.5). The channels had a trapezoidal cross-section and were 40-μm wide at the base and 30-μm deep (see Figure 51.4).[96] Before analysis, the microchannels were filled with 10 mM carbonate buffer

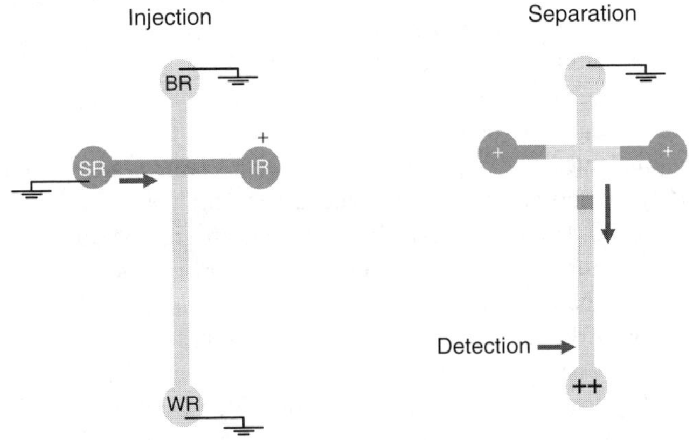

FIGURE 51.5 Schematic representation of the injection and separation steps in μCE (additional details are in the text).

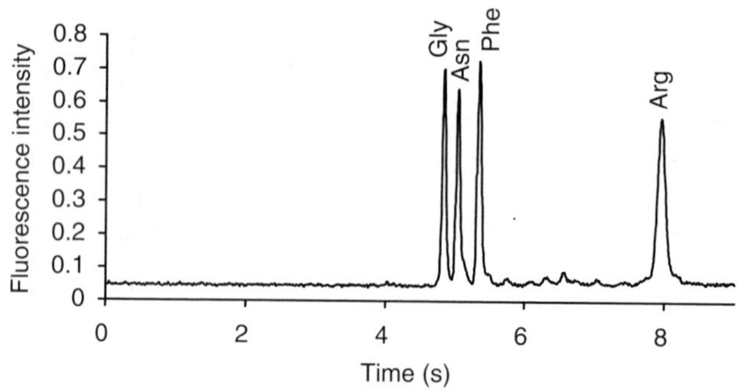

FIGURE 51.6 Separation of FITC-labeled amino acids (75 nM) in 10 mM carbonate buffer, pH 9.2. The injection voltage was +800 V, and the separation voltage was +3.0 kV. (Reprinted from Kelly, R. T., et al., *Anal. Chem.*, 77, 3536, 2005. Copyright 2005. With permission from American Chemical Society.)

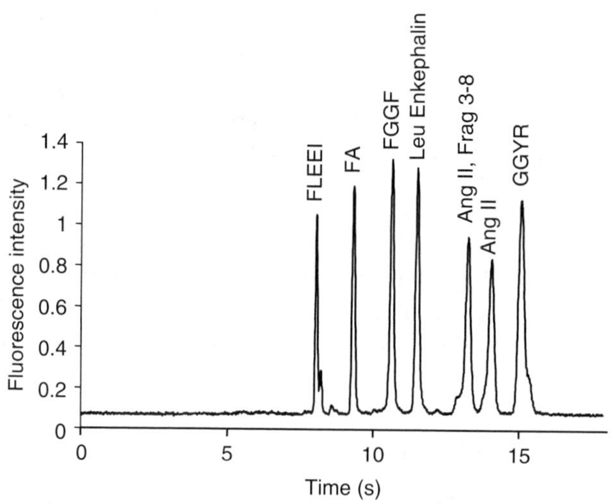

FIGURE 51.7 Separation of 110 nM FITC-labeled peptides; μCE separation conditions were the same as in Figure 51.6. (Reprinted from Kelly, R. T., et al., *Anal. Chem.*, 77, 3536, 2005. Copyright 2005. With permission from American Chemical Society.)

(pH 9.2), containing 0.5% (w/v) hydroxypropyl cellulose (HPC) to minimize electroosmotic flow and prevent analyte adsorption.[107] Buffer solution was pipetted from the SR, and approximately 10 μL of sample solution (amino acid or peptide) was added.

Figure 51.6 shows an electropherogram of four FITC-labeled amino acids separated in a PCSL-formed solvent-bonded microchip. Importantly, these PCSL-fabricated microdevices are able to withstand applied electric fields of at least 1500 V/cm, which is the highest reported field strength for μCE in polymeric devices. The use of high electric fields provided separation efficiencies of over 40,000 theoretical plates, which are comparable to values obtained in glass microchips.[104,108] Figure 51.7 shows an electropherogram of seven FITC-labeled peptides separated in a PMMA microchip constructed using solvent bonding with PCSLs. The separation efficiency in the analysis of FITC-labeled peptides was similar to that for the analysis of amino acids. The ability to carry out high-performance μCE of peptides indicates that PCSL-formed devices have potential for proteomic analysis. In addition, PCSL-fabricated microchips were used to perform more than 300 μCE runs over a 3-month period with no degradation of separation performance. These results demonstrate the excellent potential for using PCSL solvent bonding in the fabrication of robust, low-cost microchips for high quality and rapid electrophoretic analysis of biomolecules.

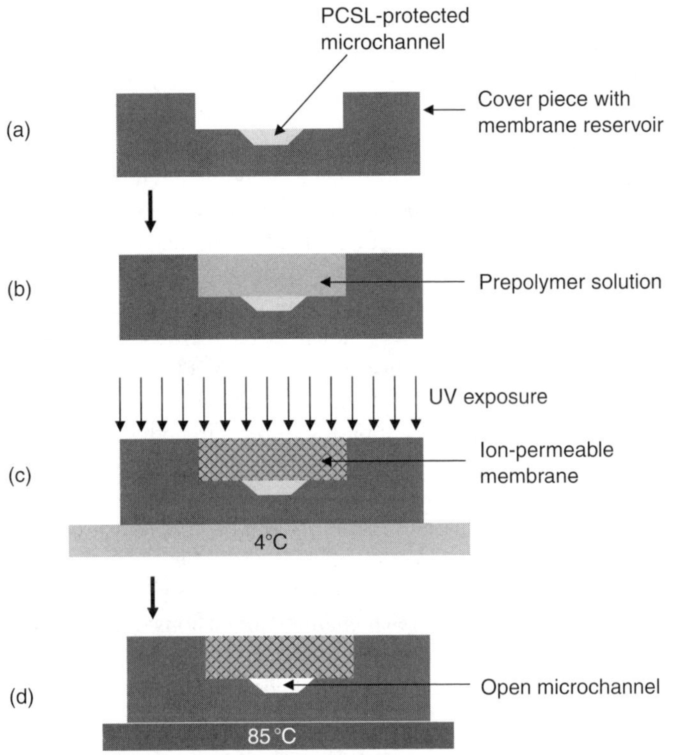

FIGURE 51.8 Schematic diagram of using PCSLs to interface microfluidics with ion-permeable membranes. (a) A PCSL-protected microchannel substrate is bonded to a PMMA cover piece having a membrane reservoir. (b) Prepolymer solution is poured into the membrane reservoir. (c) An ion-permeable hydrogel is photopolymerized. (d) The PCSL is melted and removed from the channel. (Adapted from Kelly, R. T., et al., *Anal. Chem.*, 78, 2565, 2006. Copyright 2006. With permission from American Chemical Society.)

51.4.2 PCSLs for Integrating Membranes with Microfluidics

51.4.2.1 Fabrication

PCSLs offer a straightforward approach for incorporating additional components *in situ* during the fabrication of microchips. For example, Figure 51.8 shows a fabrication scheme for integrating an ion-permeable membrane with a microfabricated channel using PCSLs.[109] First, a PCSL-protected microchannel substrate is bonded to a PMMA cover piece having one or more reservoirs for membrane formation (Figure 51.8a). Next, prepolymer solution is poured into the membrane reservoir and UV polymerized while maintaining the whole assembly at 4°C to avoid melting the PCSL (Figure 51.8b,c). Finally, the microchip is heated to allow facile removal of the liquefied PCSL, leaving a membrane-integrated microchannel in the device (Figure 51.8d). In addition to the characteristics noted in Section 51.3.3, PCSLs for the integration of membranes with microchannels must not react with or dissolve in either the membrane monomer solution or the polymerized hydrogel.

51.4.2.2 Application to Electric Field Gradient Focusing

Electric field gradient focusing (EFGF) is an equilibrium-gradient separation method,[110] in which analytes are focused at different equilibrium positions in an electric field gradient.[111,112] In EFGF, the electrophoretic motion of charged molecules in an electric field gradient in a column is opposed by hydrodynamic flow. The result is that each analyte will focus (stop) at an equilibrium position where the electrophoretic and laminar flow velocities are opposite and equal. The principle of EFGF has been studied theoretically and experimentally.[111–115]

Perhaps the simplest way to do EFGF is to interface an open column of constant cross-sectional area with a semipermeable membrane of changing cross-sectional area.[116] The ion-permeable membrane allows free flow of small buffer ions, but proteins are constrained to the open column, and the changing cross-sectional area of the semipermeable copolymer creates an electric field gradient. Interfacing an open column with an ion-permeable membrane is challenging using conventional methods,[116] but is a task that is ideally suited for PCSL fabrication.

Woolley and coworkers[109] have fabricated EFGF microchips using the PCSL approach. In these devices, the membrane reservoir was a shaped region (for generating an electric field gradient) in the PMMA cover piece, and the open channel in the patterned substrate was protected with PCSL during membrane polymerization. The performance of PCSL-fabricated EFGF microchips was compared to that in capillary-based EFGF devices by analyzing two natively fluorescent proteins, R-phycoerythrin (R-PE) and green fluorescent protein (GFP), in both systems. The results indicated a fourfold reduction in peak width and a threefold improvement in resolution for PCSL-fabricated EFGF systems compared to capillary devices. Moreover, resolution for the analysis of peptides in microchip EFGF devices was comparable to that in μCE, but sample was concentrated 150-fold in the EFGF system.

Using PCSLs to integrate microfabricated channels with ion-permeable membranes improves EFGF experiments. The smaller cross-sectional channel dimensions in the microchip EFGF devices reduce band width and improve the resolution of protein peaks. Importantly, the fabrication protocols are flexible and easy to adapt as needed for different device designs.

51.4.2.3 Application to Protein Preconcentration

The same PCSL fabrication protocols for interfacing microfluidics and membranes in EFGF can also be used for on-chip protein preconcentration and separation.[109,117] A PCSL-filled microchannel imprinted in PMMA was affixed to a cover plate containing a membrane reservoir, and hydrogel was polymerized as described earlier.[109] To concentrate protein samples at the membrane, a voltage was applied between the sample and membrane reservoirs. Concentration factors as high as 10,000-fold for a model protein (R-PE) were observed.[109] Preconcentrated proteins were separated electrophoretically by applying a potential between the membrane reservoir and the reservoir at the end of the channel. A protein mixture containing R-PE and GFP has been concentrated and separated electrophoretically in a membrane-integrated device.[117] These experiments verify the power of the PCSL approach in making microchips for integrated sample enrichment and separation.

51.5 CONCLUDING REMARKS AND FUTURE TRENDS

In this chapter, we have presented phase-changing sacrificial layers as enabling tools for making a variety of microfluidic devices in polymers. Several successful uses of PCSLs in biological analysis are discussed herein, which should motivate the pursuit of new developments and applications of this emerging approach.

In the future, we anticipate the generalization of PCSL fabrication methods to polymers in addition to PMMA. Table 51.1 summarizes the key considerations for the substrate, sacrificial material, and solvent in PCSL fabrication.[118] We have recently obtained promising results in extending the PCSL technique to other polymeric substrates, including PC, PET, and COC.[118]

In addition to the possibility of using PCSLs to construct microfluidics in various polymeric materials, this approach could also facilitate the implementation of new analysis techniques in a microchip format. For example, applications that require high pressures, such as for loading viscous sieving polymers for DNA separation or for packing small-diameter-particle stationary phases for

TABLE 51.1
PCSL Solvent Bonding Method Development Considerations

	Requirements
Substrate	Easily imprinted with template
	Inexpensive
	Seals to elastomer
	Does not interfere with chemical analysis
Sacrificial material	Transitions from solid to liquid near room temperature
	Does not interact with polymer/elastomer
	Appropriate viscosity
	Forms smooth, even channels
	Solubility differs from substrate
Solvent	Dissolves substrate but not sacrificial layer
	Appropriate rate for dissolving substrate

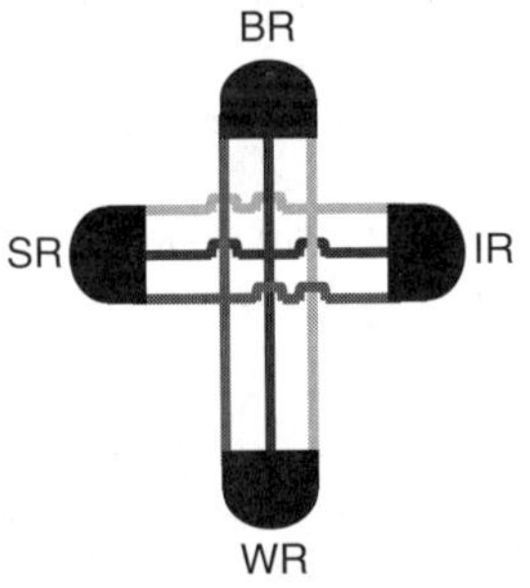

FIGURE 51.9 Layout of a multilayer microdevice for parallel, simultaneous analysis of a single sample in three channels.

chromatography, may become feasible in PCSL-fabricated polymer microchips, because of their high-pressure stability.

51.5.1 Fabrication of Multilayer Microfluidic Arrays

Interestingly, most research efforts with polymer microfluidic structures have focused on applying the same planar two-dimensional layouts of glass microchips in plastic devices. However, multilayer microstructures, wherein multiple channels can cross over one another without contamination, would significantly increase the operational flexibility in miniaturized analysis. Multilayer microfluidics have been fabricated in glass,[119] PDMS,[120,121] and PDMS/glass hybrid microchips.[122] However, this design has not been implemented in easy-to-fabricate thermoplastic (rigid) polymers. Importantly, thermal bonding is problematic in the fabrication of multilayer structures, because the heat and pressure often deform the surface, making it difficult to bond subsequent layers; moreover, microchannels can be constricted or blocked as a result of repeated heating steps. In contrast, PCSL fabrication protocols are simple and straightforward to implement, which would be advantageous for making multilayer polymer microchips. Indeed, using PCSL microfabrication, multiple substrates can be bonded together successfully, without affecting the previous layers.

Important new applications could be developed in multilayer polymer microfluidic systems. For example, multiple replicate analyses of a sample in a single reservoir could be performed in parallel on a single device. Figure 51.9 shows a schematic diagram of a multilayer microdevice with channels crossing over each other. This device could be used to separate a sample in three microchannels using a single input, which would allow parallel, replicate analyses of the same specimen. Additional designs could be formed with parallel channels having different surface properties or separation media, to implement orthogonal separation mechanisms. Such layouts could then be used in parallel analysis of the same sample under different conditions. Finally, multilayer structures with crossing channels would facilitate the integration of end-column fluorescent or other labeling techniques in μCE.

ACKNOWLEDGMENTS

The authors are grateful to the National Institutes of Health (EB006124) for providing financial support to complete this work. We thank Dr. Ryan T. Kelly for training and useful tips on the fabrication of PCSL microdevices, and Bridget Peeni, Tao Pan, Weichun Yang, and Dr. Xiuhua Sun for helpful discussions.

REFERENCES

1. Mikkers, F. E. P., Everaerts, F. M., and Verheggen, T. P. E. M., High-performance zone electrophoresis, *J. Chromatogr.*, 169, 11, 1979.
2. Jorgenson, J. W., and Lukacs, K. D., Zone electrophoresis in open-tubular glass capillaries, *Anal. Chem.*, 53, 1298, 1981.
3. Jorgenson, J. W., and Lukacs, K. D., Capillary zone electrophoresis, *Science*, 222, 266, 1983.
4. Terry, S. C., Jerman, J. H., and Angell, J. B., A gas chromatographic air analyzer fabricated on a silicon wafer, *IEEE Trans. Electron Devices*, ED-26, 1880, 1979.
5. Manz, A., Graber, N., and Widmer, H. M., Miniaturized total chemical-analysis systems: A novel concept for chemical sensing, *Sens. Actuat. B*, 1, 244, 1990.
6. Kutter, J. P., Current developments in electrophoretic and chromatographic separation methods on microfabricated devices, *TrAC*, 19, 352, 2000.
7. Manz, A., Harrison, D. J., Verpoorte, E. M. J., Fettinger, J. C., Paulus, A., Lüdi, H., and Widmer, H. M., Planar chips technology for miniaturization and integration of separation techniques into monitoring systems—Capillary electrophoresis on a chip, *J. Chromatogr.*, 593, 253, 1992.
8. Roper, M. G., Easley, C. J., and Landers, J. P., Advances in polymerase chain reaction on microfluidic chips, *Anal. Chem.*, 77, 3887, 2005.
9. Kelly, R. T., and Woolley, A. T., Microfluidic systems for integrated, high-throughput DNA analysis, *Anal. Chem.*, 77, 96A, 2005.
10. Kato, M., Gyoten, Y., Sakai-Kato, K., Nakajima, T., and Toyo'oka, T., Analysis of amino acids and proteins using a poly(methyl methacrylate) microfluidic system, *Electrophoresis*, 26, 3682, 2005.
11. Roper, M. G., Frisk, M. L., Oberlander, J. P., Ferrance, J. P., McGrory, B. J., and Landers, J. P., Extraction of C-reactive protein from serum on a microfluidic chip, *Anal. Chim. Acta*, 569, 195, 2006.
12. Poulsen, C. R., Culbertson, C. T., Jacobson, S. C., and Ramsey, J. M., Static and dynamic acute cytotoxicity assays on microfluidic devices, *Anal. Chem.*, 77, 667, 2005.
13. Ding, Y., and Garcia, C. D., Determination of nonsteroidal anti-inflammatory drugs in serum by microchip capillary electrophoresis with electrochemical detection, *Electroanalysis*, 18, 2202, 2006.
14. Götz, S., Revermann, T., and Karst, U., Quantitative on-chip determination of taurine in energy and sports drinks, *Lab Chip*, 7, 93, 2007.
15. Pumera, M., Analysis of explosives via microchip electrophoresis and conventional capillary electrophoresis: A review, *Electrophoresis*, 27, 244, 2006.
16. Ding, Y., and Garcia, C. D., Pulsed amperometric detection with poly(dimethylsiloxane)-fabricated capillary electrophoresis microchips for the determination of EPA priority pollutants, *Analyst*, 131, 208, 2006.
17. Harrison, D. J., Manz, A., Fan, Z., Lüdi, H., and Widmer, H. M., Capillary electrophoresis and sample injection systems integrated on a planar glass chip, *Anal. Chem.*, 64, 1926, 1992.
18. Woolley, A. T., and Mathies, R. A., Ultra-high-speed DNA fragment separations using microfabricated capillary array electrophoresis chips, *Proc. Natl. Acad. Sci. USA*, 91, 11348, 1994.

19. Jacobson, S. C., Hergenröder, R., Koutny, L. B., and Ramsey, J. M., High-speed separations on a microchip, *Anal. Chem.*, 66, 1114, 1994.
20. de Mello, A., Plastic fantastic? *Lab Chip*, 2, 31N, 2002.
21. Boone, T. D., Fan, Z. H., Hooper, H. H., Ricco, A. J., Tan, H., and Williams, S. J., Plastic advances microfluidic devices, *Anal. Chem.*, 74, 78A, 2002.
22. Soper, S. A., Ford, S. M., Qi, S., McCarley, R. L., Kelly, K., and Murphy, M. C., Polymeric microelectromechanical systems, *Anal. Chem.*, 72, 642A, 2000.
23. Martynova, L., Locascio, L. E., Gaitan, M., Kramer, G. W., Christensen, R. G., and MacCrehan, W. A., Fabrication of plastic microfluid channels by imprinting methods, *Anal. Chem.*, 69, 4783, 1997.
24. Qi, S., Liu, X., Ford, S., Barrows, J., Thomas, G., Kelly, K., McCandless, A., Lian, K., Goettert, J., and Soper, S. A., Microfluidic devices fabricated in poly(methyl methacrylate) using hot-embossing with integrated sampling capillary and fiber optics for fluorescence detection, *Lab Chip*, 2, 88, 2002.
25. Duffy, D. C., McDonald, J. C., Schueller, O. J. A., and Whitesides, G. M., Rapid prototyping of microfluidic systems in poly(dimethylsiloxane), *Anal. Chem.*, 70, 4974, 1998.
26. Effenhauser, C. S., Bruin, G. J. M., Paulus, A., and Ehrat, M., Integrated capillary electrophoresis on flexible silicone microdevices: Analysis of DNA restriction fragments and detection of single DNA molecules on microchips, *Anal. Chem.*, 69, 3451, 1997.
27. Liu, Y., Ganser, D., Schneider, A., Liu, R., Grodzinski, P., and Kroutchinina, N., Microfabricated polycarbonate CE devices for DNA analysis, *Anal. Chem.*, 73, 4196, 2001.
28. Metz, S., Holzer, R., and Renaud, P., Polyimide-based microfluidic devices, *Lab Chip*, 1, 29, 2001.
29. Kameoka, J., Craighead, H. G., Zhang, H., and Henion, J., A polymeric microfluidic chip for CE/MS determination of small molecules, *Anal. Chem.*, 73, 1935, 2001.
30. Russo, A. P., Apoga, D., Dowell, N., Shain, W., Turner, A. M. P., Craighead, H. G., Hoch, H. C., and Turner, J. N., Microfabricated plastic devices from silicon using soft intermediates, *Biomed. Microdevices*, 4, 277, 2002.
31. Wu, Z., Xanthopoulos, N., Reymond, F., Rossier, J. S., and Girault, H. H., Polymer microchips bonded by O_2-plasma activation, *Electrophoresis*, 23, 782, 2002.
32. Becker, H., and Locascio, L. E., Polymer microfluidic devices, *Talanta*, 56, 267, 2002.
33. Becker, H., and Heim, U., Hot embossing as a method for the fabrication of polymer high aspect ratio structures, *Sens. Actuat. A*, 83, 130, 2000.
34. McCormick, R. M., Nelson, R. J., Alonso-Amigo, M. G., Benvegnu, D. J., and Hooper, H. H., Microchannel electrophoretic separations of DNA in injection-molded substrates, *Anal. Chem.*, 69, 2626, 1997.
35. Lion, N., Reymond, F., Girault, H. H., and Rossier, J. S., Why the move to microfluidics for protein analysis? *Curr. Opin. Biotechnol.*, 15, 31, 2004.
36. Marko-Varga, G., Nilsson, J., and Laurell, T., New directions of miniaturization within the proteomics research area, *Electrophoresis*, 24, 3521, 2003.
37. Galloway, M., Stryjewski, W., Henry, A., Ford, S. M., Llopis, S., McCarley, R. L., and Soper, S. A., Contact conductivity detection in poly(methyl methacrylate) based microfluidic devices for analysis of mono- and polyanionic molecules, *Anal. Chem.*, 74, 2407, 2002.
38. Stachowiak, T. B., Rohr, T., Hilder, E. F., Peterson, D. S., Yi, M., Svec, F., and Fréchet, J. M. J., Fabrication of porous polymer monoliths covalently attached to the walls of channels in plastic microdevices, *Electrophoresis*, 24, 3689, 2003.
39. Vreeland, W. N., and Locascio, L. E., Using bioinspired thermally triggered liposomes for high-efficiency mixing and reagent delivery in microfluidic devices, *Anal. Chem.*, 75, 6906, 2003.
40. Yang, Y., Li, C., Lee, K. H., and Craighead, H. G., Coupling on-chip solid-phase extraction to electrospray mass spectrometry through an integrated electrospray tip, *Electrophoresis*, 26, 3622, 2005.
41. Ford, S. M., Davies, J., Kar, B., Qi, S. D., McWhorter, S., Soper, S. A., and Malek, C. K., Micromachining in plastics using x-ray lithography for the fabrication of microelectrophoresis devices, *J. Biomech. Eng.*, 121, 13, 1999.
42. Kelly, R. T., and Woolley, A. T., Thermal bonding of polymeric capillary electrophoresis microdevices in water, *Anal. Chem.*, 75, 1941, 2003.

43. Liu, Y., Fanguy, J. C., Bledsoe, J. M., and Henry, C. S., Dynamic coating using polyelectrolyte multilayers for chemical control of electroosmotic flow in capillary electrophoresis microchips, *Anal. Chem.*, 72, 5939, 2000.
44. Wu, H., Huang, B., and Zare, R. N., Construction of microfluidic chips using polydimethylsiloxane for adhesive bonding, *Lab Chip*, 5, 1393, 2005.
45. Song, L., Fang, D., Kobos, R. K., Pace, S. J., and Chu, B., Separation of double-stranded DNA fragments in plastic capillary electrophoresis chips by using $E_{99}P_{69}E_{99}$ as separation medium, *Electrophoresis*, 20, 2847, 1999.
46. Satyanarayana, S., Karnik, R. N., and Majumdar, A., Stamp-and-stick room-temperature bonding technique for microdevices, *J. Microelectromech. Syst.*, 14, 392, 2005.
47. Huang, Z., Sanders, J. C., Dunsmor, C., Ahmadzadeh, H., and Landers, J. P., A method for UV-bonding in the fabrication of glass electrophoretic microchips, *Electrophoresis*, 22, 3924, 2001.
48. Bianchi, F., Wagner, F., Hoffmann, P., and Girault, H. H., Electroosmotic flow in composite microchannels and implications in microcapillary electrophoresis systems, *Anal. Chem.*, 73, 829, 2001.
49. Lai, S. Y., Cao, X., and Lee, L. J., A packaging technique for polymer microfluidic platforms, *Anal. Chem.*, 76, 1175, 2004.
50. Chen, G., Li, J., Qu, S., Chen, D., and Yang, P., Low temperature bonding of poly(methylmethacrylate) electrophoresis microchips by *in situ* polymerization, *J. Chromatogr. A*, 1094, 138, 2005.
51. Xu, G., Wang, J., Chen, Y., Zhang, L., Wang, D., and Chen, G., Fabrication of poly(methyl methacrylate) capillary electrophoresis microchips by *in situ* surface polymerization, *Lab Chip*, 6, 145, 2006.
52. Brown, L., Koerner, T., Horton, J. H., and Oleschuk, R. D., Fabrication and characterization of poly(methylmethacrylate) microfluidic devices bonded using surface modifications and solvents, *Lab Chip*, 6, 66, 2006.
53. Fiorini, G. S., Lorenz, R. M., Kuo, J. S., and Chiu, D. T., Rapid prototyping of thermoset polyester microfluidic devices, *Anal. Chem.*, 76, 4697, 2004.
54. Liu, J., Hansen, C., and Quake, S. R., Solving the "world-to-chip" interface problem with a microfluidic matrix, *Anal. Chem.*, 75, 4718, 2003.
55. Griebel, A., Rund, S., Schönfeld, F., Dörner, W., Konrad, R., and Hardt, S., Integrated polymer chip for two-dimensional capillary gel electrophoresis, *Lab Chip*, 4, 18, 2004.
56. Wang, J., Pumera, M., Chatrathi, M. P., Escarpa, A., Konrad, R., Griebel, A., Dörner, W., and Löwe, H., Toward disposable lab-on-a-chip: Poly(methylmethacrylate) microchip electrophoresis device with electrochemical detection, *Electrophoresis*, 23, 596, 2002.
57. Kricka, L. J., Fortina, P., Panaro, N. J., Wilding, P., Alonso-Amigo, G., and Becker, H., Fabrication of plastic microchips by hot embossing, *Lab Chip*, 2, 1, 2002.
58. Shah, J. J., Geist, J., Locascio, L. E., Gaitan, M., Rao, M. V., and Vreeland, W. N., Capillarity induced solvent-actuated bonding of polymeric microfluidic devices, *Anal. Chem.*, 78, 3348, 2006.
59. Glasgow, I. K., Beebe, D. J., and White, V. E., Design rules for polyimide solvent bonding, *Sens. Mater.*, 11, 269, 1999.
60. Harper, C. A., and Petrie, E. M., *Plastic Materials and Processes: A Concise Enclyclopedia*, Wiley Interscience, New York, 2003.
61. Lin, C. B., Lee, S., and Liu, K. S., The microstructure of solvent-welding of PMMA, *J. Adhes.*, 34, 221, 1991.
62. Liu, R. H., Yang, J., Lenigk, R., Bonanno, J., and Grodzinski, P., Self-contained, fully integrated biochip for sample preparation, polymerase chain reaction amplification, and DNA microarray detection, *Anal. Chem.*, 76, 1824, 2004.
63. Dittrich, P. S., Tachikawa, K., and Manz, A., Micro total analysis systems. Latest advancements and trends, *Anal. Chem.*, 78, 3887, 2006.
64. Stone, H. A., and Kim, S., Microfluidics: Basic issues, applications, and challenges, *AIChE J.*, 47, 1250, 2001.
65. Teh, W. H., and Smith, C. G., Fabrication of quasi-three-dimensional micro/nanomechanical components using electron beam cross-linked poly(methyl methacrylate) resist, *J. Vac. Sci. Technol. B*, 21, 3007, 2003.

66. Edwards, J. M., IV, Hamblin, M. N., Fuentes, H. V., Peeni, B. A., Lee, M. L., Woolley, A. T., and Hawkins, A. R., Thin film electroosmotic pumps for biomicrofluidic applications, *Biomicrofluidics*, 1, 014101, 2007.
67. Reano, R. M., and Pang, S. W., Sealed three-dimensional nanochannels, *J. Vac. Sci. Technol. B*, 23, 2995, 2005.
68. Reed, H. A., White, C. E., Rao, V., Allen, S. A. B., Henderson, C. L., and Kohl, P. A., Fabrication of microchannels using polycarbonates as sacrificial materials, *J. Micromech. Microeng.*, 11, 733, 2001.
69. Wu, X., Reed, H. A., Wang, Y., Rhodes, L. F., Elce, E., Ravikiran, R., Shick, R. A., Henderson, C. L., Allen, S. A. B., and Kohl, P. A., Fabrication of microchannels using polynorbornene photosensitive sacrificial materials, *J. Electrochem. Soc.*, 150, H205, 2003.
70. Li, W., Tegenfeldt, J. O., Chen, L., Austin, R. H., Chou, S. Y., Kohl, P. A., Krotine, J., and Sturm, J. C., Sacrificial polymers for nanofluidic channels in biological applications, *Nanotechnology*, 14, 578, 2003.
71. Howe, R. T., Surface micromachining for microsensors and microactuators, *J. Vac. Sci. Technol. B*, 6, 1809, 1988.
72. Harnett, C. K., Coates, G. W., and Craighead, H. G., Heat-depolymerizable polycarbonates as electron beam patternable sacrificial layers for nanofluidics, *J. Vac. Sci. Technol. B*, 19, 2842, 2001.
73. Sharma, S., Buchholz, K., Luber, S. M., Rant, U., Tornow, M., and Abstreiter, G., Silicon-on-insulator microfluidic device with monolithic sensor integration for μTAS applications, *J. Microelectromech. Syst.*, 15, 308, 2006.
74. Peeni, B. A., Conkey, D. B., Barber, J. P., Kelly, R. T., Lee, M. L., Woolley, A. T., and Hawkins, A. R., Planar thin film device for capillary electrophoresis, *Lab Chip*, 5, 501, 2005.
75. Burbaum, C., Mohr, J., Bley, P., and Ehrfeld, W., Fabrication of capacitive acceleration sensors by the LIGA technique, *Sens. Actuat. A*, 27, 559, 1991.
76. Maciossek, A., Löchel, B., Quenzer, H.-J., Wagner, B., Schulze, S., and Noetzel, J., Galvanoplating and sacrificial layers for surface micromachining, *Microelectron. Eng.*, 27, 503, 1995.
77. Bagolini, A., Pakula, L., Scholtes, T. L. M., Pham, H. T. M., French, P. J., and Sarro, P. M., Polyimide sacrificial layer and novel materials for postprocessing surface micromachining, *J. Micromech. Microeng.*, 12, 385, 2002.
78. Song, I.-H., and Ajmera, P. K., Use of a photoresist sacrificial layer with SU-8 electroplating mould in MEMS fabrication, *J. Micromech. Microeng.*, 13, 816, 2003.
79. Chan, K., and Gleason, K. K., Air-gap fabrication using a sacrificial polymeric thin film synthesized via initiated chemical vapor deposition, *J. Electrochem. Soc.*, 153, C223, 2006.
80. Foquet, M., Korlach, J., Zipfel, W., Webb, W. W., and Craighead, H. G., DNA fragment sizing by single molecule detection in submicrometer-sized closed fluidic channels, *Anal. Chem.*, 74, 1415, 2002.
81. Wu, X., Reed, H. A., Rhodes, L. F., Elce, E., Ravikiran, R., Shick, R. A., Henderson, C. L., Allen, S. A. B., and Kohl, P. A., Photoinitiation systems and thermal decomposition of photodefinable sacrificial materials, *J. Appl. Polym. Sci.*, 88, 1186, 2003.
82. Metz, S., Jiguet, S., Bertsch, A., and Renaud, P., Polyimide and SU-8 microfluidic devices manufactured by heat-depolymerizable sacrificial material technique, *Lab Chip*, 4, 114, 2004.
83. Hubbard, N. B., Howell, L. L., Barber, J. P., Conkey, D. B., Hawkins, A. R., and Schmidt, H., Structural models and design rules for on-chip micro-channels with sacrificial cores, *J. Micromech. Microeng.*, 15, 720, 2005.
84. Linder, V., Gates, B. D., Ryan, D., Parviz, B. A., and Whitesides, G. M., Water-soluble sacrificial layers for surface micromachining, *Small*, 1, 730, 2005.
85. Esinenco, D., Psoma, S. D., Kusko, M., Schneider, A., and Muller, R., SU-8 micro-biosensor based on Mach-Zehnder interferometer, *Rev. Adv. Mater. Sci.*, 10, 295, 2005.
86. Nam, W. J., Bae, S., Kalkan, A. K., and Fonash, S. J., Nano- and microchannel fabrication using column/void network deposited silicon, *J. Vac. Sci. Technol. A*, 19, 1229, 2001.
87. Bhusari, D., Reed, H. A., Wedlake, M., Padovani, A. M., Allen, S. A. B., and Kohl, P. A., Fabrication of air-channel structures for microfluidic, microelectromechanical, and microelectronic applications, *J. Microelectromech. Syst.*, 10, 400, 2001.
88. Turner, S. W., Perez, A. M., Lopez, A., and Craighead, H. G., Monolithic nanofluid sieving structures for DNA manipulation, *J. Vac. Sci. Technol. B*, 16, 3835, 1998.

89. Yin, D., Deamer, D. W., Schmidt, H., Barber, J. P., and Hawkins, A. R., Single-molecule detection sensitivity using planar integrated optics on a chip, *Opt. Lett.*, 31, 2136, 2006.
90. Peeni, B. A., Lee, M. L., Hawkins, A. R., and Woolley, A. T., Sacrificial layer microfluidic device fabrication methods, *Electrophoresis*, 27, 4888, 2006.
91. Jayachandran, J. P., Reed, H. A., Zhen, H., Rhodes, L. F., Henderson, C. L., Allen, S. A. B., and Kohl, P. A., Air-channel fabrication for microelectromechanical systems via sacrificial photosensitive polycarbonates, *J. Microelectromech. Syst.*, 12, 147, 2003.
92. Liu, R. H., and Grodzinski, P., Development of integrated microfluidic system for genetic analysis, *J. Microlith. Microfab. Microsyst.*, 2, 340, 2003.
93. Klintberg, L., Svedberg, M., Nikolajeff, F., and Thornell, G., Fabrication of a paraffin actuator using hot embossing of polycarbonate, *Sens. Actuat. A*, 103, 307, 2003.
94. Sethu, P. and Mastrangelo, C. H., Polyethylene glycol (PEG)-based actuator for nozzle-diffuser pumps in plastic microfluidic systems, *Sens. Actuat. A*, 104, 283, 2003.
95. Pal, R., Yang, M., Johnson, B. N., Burke, D. T., and Burns, M. A., Phase change microvalve for integrated devices, *Anal. Chem.*, 76, 3740, 2004.
96. Kelly, R. T., Pan, T., and Woolley, A. T., Phase-changing sacrificial materials for solvent bonding of high-performance polymeric capillary electrophoresis microchips, *Anal. Chem.*, 77, 3536, 2005.
97. Verpoorte, E., Microfluidic chips for clinical and forensic analysis, *Electrophoresis*, 23, 677, 2002.
98. Jackman, R. J., Floyd, T. M., Ghodssi, R., Schmidt, M. A., and Jensen, K. F., Microfluidic systems with on-line UV detection fabricated in photodefinable epoxy, *J. Micromech. Microeng.*, 11, 263, 2001.
99. Huang, F.-C., Liao, C.-S., and Lee, G.-B., An integrated microfluidic chip for DNA/RNA amplification, electrophoresis separation and on-line optical detection, *Electrophoresis*, 27, 3297, 2006.
100. Huang, C.-W, Huang, S.-B, and Lee, G.-B, Pneumatic micropumps with serially connected actuation chambers, *J. Micromech. Microeng.*, 16, 2265, 2006.
101. Ito, T., Sobue, K., and Ohya, S., Water glass bonding for micrototal analysis system, *Sens. Actuat. B*, 81, 187, 2002.
102. Zhuang, G., Jin, Q., Liu, J., Cong, H., Liu, K., Zhao, J., Yang, M., and Wang, H., A low temperature bonding of quartz microfluidic chip for serum lipoproteins analysis, *Biomed. Microdevices*, 8, 255, 2006.
103. Effenhauser, C. S., Bruin, G. J. M., and Paulus, A., Integrated chip-based capillary electrophoresis, *Electrophoresis*, 18, 2203, 1997.
104. Ramsey, J. D., Jacobson, S. C., Culbertson, C. T., and Ramsey, J. M., High-efficiency, two-dimensional separations of protein digests on microfluidic devices, *Anal. Chem.*, 75, 3758, 2003.
105. Jacobson, S. C., Hergenröder, R., Koutny, L. B., Warmack, R. J., and Ramsey, J. M., Effects of injection schemes and column geometry on the performance of microchip electrophoresis devices, *Anal. Chem.*, 66, 1107, 1994.
106. Monnig, C. A., and Jorgenson, J. W., On-column sample gating for high-speed capillary zone electrophoresis, *Anal. Chem.*, 63, 802, 1991.
107. Sanders, J. C., Breadmore, M. C., Kwok, Y. C., Horsman, K. M., and Landers, J. P., Hydroxypropyl cellulose as an adsorptive coating sieving matrix for DNA separations: Artificial neural network optimization for microchip analysis, *Anal. Chem.*, 75, 986, 2003.
108. Rocklin, R. D., Ramsey, R. S., and Ramsey, J. M., A microfabricated fluidic device for performing two-dimensional liquid-phase separations, *Anal. Chem.*, 72, 5244, 2000.
109. Kelly, R. T., Li, Y., and Woolley, A. T., Phase-changing sacrificial materials for interfacing microfluidics with ion-permeable membranes to create on-chip preconcentrators and electric field gradient focusing microchips, *Anal. Chem.*, 78, 2565, 2006.
110. Giddings, J. C., and Dahlgren, K., Resolution and peak capacity in equilibrium-gradient methods of separation, *Sep. Sci.*, 6, 345, 1971.
111. Tolley, H. D., Wang, Q., LeFebre, D. A., and Lee, M. L., Equilibrium gradient methods with nonlinear field intensity gradient: A theoretical approach, *Anal. Chem.*, 74, 4456, 2002.
112. Koegler, W. S., and Ivory, C. F., Focusing proteins in an electric field gradient, *J. Chromatogr. A*, 726, 229, 1996.
113. Ivory, C. F., A brief review of alternative electrofocusing techniques, *Sep. Sci. Technol.*, 35, 1777, 2000.

114. Wang, Q., Tolley, H. D., LeFebre, D. A., and Lee, M. L., Analytical equilibrium gradient methods, *Anal. Bioanal. Chem.*, 373, 125, 2002.
115. Petsev, D. N., Lopez, G. P., Ivory, C. F., and Sibbett, S. S., Microchannel protein separation by electric field gradient focusing, *Lab Chip*, 5, 587, 2005.
116. Humble, P. H., Kelly, R. T., Woolley, A. T., Tolley, H. D., and Lee, M. L., Electric field gradient focusing of proteins based on shaped ionically conductive acrylic polymer, *Anal. Chem.*, 76, 5641, 2004.
117. Li, Y., Membrane-based protein preconcentration microfluidic devices, M.S. Thesis, Brigham Young University, 2006.
118. Draper, M., Solvent bonding of polymeric microdevices for chemical analysis, Honors Thesis, Brigham Young University, 2005.
119. Verpoorte, E., van der Schoot, B. H., Jeanneret, S., Manz, A., Widmer, H. M., and de Rooij, N. F., Three-dimensional micro flow manifolds for miniaturized chemical analysis systems, *J. Micromech. Microeng.*, 4, 246, 1994.
120. Thorsen, T., Maerkl, S. J., and Quake, S. R., Microfluidic large-scale integration, *Science*, 298, 580, 2002.
121. Kuo, T.-C., Cannon, D. M., Jr., Chen, Y., Tulock, J. J., Shannon, M. A., Sweedler, J. V., and Bohn, P. W., Gateable nanofluidic interconnects for multilayered microfluidic separation systems, *Anal. Chem.*, 75, 1861, 2003.
122. Grover, W. H., Skelley, A. M., Liu, C. N., Lagally, E. T., and Mathies, R. A., Monolithic membrane valves and diaphragm pumps for practical large-scale integration into glass microfluidic devices, *Sens. Actuat. B*, 89, 315, 2003.

52 Materials and Modification Strategies for Electrophoresis Microchips

Charles S. Henry, Brian M. Dressen

CONTENTS

52.1 INTRODUCTION

Microchip capillary electrophoresis (MCE), like conventional capillary electrophoresis (CE) and most other chemical separation techniques, is heavily dependent on surface chemistry. This connection is obvious for techniques that use a stationary phase but is also apparent for separation modes like zone electrophoresis where the separation is based on the mobility of the ions. An important component of zone electrophoresis is the electroosmotic flow (EOF), which is a surface-derived phenomenon. Furthermore, the degree of surface hydrophobicity can cause adsorption resulting in band broadening. As a result of the significance of the surface chemistry, it is important to develop an understanding of methods that provide control of surface chemistry in MCE. This problem is more pronounced in MCE than in CE because a wide range of substrate materials are used. The goal of this chapter is to first introduce the reader to the substrate materials and the associated fabrication methods that are common to MCE. Next, general methods for modifying the surface chemistry of different substrate materials are provided. This is not meant to be an all-inclusive listing of methods but instead to provide the reader with general guidelines for controlling the all important surface chemistry of electrophoresis microchips through selection of the substrate material and/or modification of the surface. Finally, a brief summary of surface characterization methods is described, which permit evaluation of different surface modification methods.

52.2 SUBSTRATE MATERIALS

Since the initial report on MCE, a wide range of materials have been used to produce microchips [1]. Early on, all microchips were made from glass or similar materials (quartz, fused silica, etc.) because these materials have very good optical clarity, have well-established fabrication methods, and are chemically similar to the fused silica used with conventional CE. Soon other materials were explored for use with MCE in a search for lower cost substrates, providing a broader range of chemical properties including but not limited to flexibility, the ability to make microchips from multiple materials, and control of surface chemistry through differences in bulk material. In addition, some materials, polymers in particular, are amenable to parallel batch fabrication methods that can reduce the cost of the production in addition to the cost of the materials. Here a summary of each of the major materials used in MCE is provided, along with a description of the fabrication methods appropriate for each type of material.

52.2.1 Silicon-Based Materials

Silicon oxide materials, glass, quartz, and fused silica are arguably the most common substrates used for MCE and will be referred to here as glass for simplicity. Many reasons exist for the use of glass. First, glass is generally chemically stable. Common organic solvents ranging in polarity from methanol to hexanes are completely compatible with the material. Furthermore, aqueous buffers with most pH ranges (2–13) can be used with glass microchips as long as there is no covalent modification to the surface. Second, glass has good optical transmission through a broad range of wavelengths from the UV to the visible. Optical transmission is important for detection whether it is simple absorbance or the more sensitive laser-induced fluorescence (LIF). Third, a large number of fabrication methods preexisted the development of MCE as a result of the need to process silicon-based materials in the integrated circuit industry. These methods can be largely adapted to MCE fabrication. Finally, the material is rigid, making it easy to handle and integrate with external connections to pumps, and so forth. Glass is not without limitations, with fabrication methodology being the primary one. Channels are etched with either caustic hydrogen fluoride (HF) or expensive reaction ion etchers. In both cases, the channel dimensions and shapes are limited. Bonding requires high temperature furnaces, careful temperature programming, and long times (24 h) compared to the polymeric materials in use today. Finally, all of these processes must be repeated for each device produced and must be done in a clean room environment. This adds to the cost of commercially produced microchips and/or limits the availability of this technique to those with access to advanced microfabrication facilities. In the end, it is up to the user to decide whether the advantages of glass devices outweigh the disadvantages.

Figure 52.1 shows the general approach for creation of electrophoresis devices using a glass substrate. The main steps are etching, which define the channel dimensions and shape, and bonding, which adds the fourth wall to the channel to create a sealed capillary. Each of these steps can be accomplished in multiple ways, and therefore careful consideration is needed, particularly for etching.

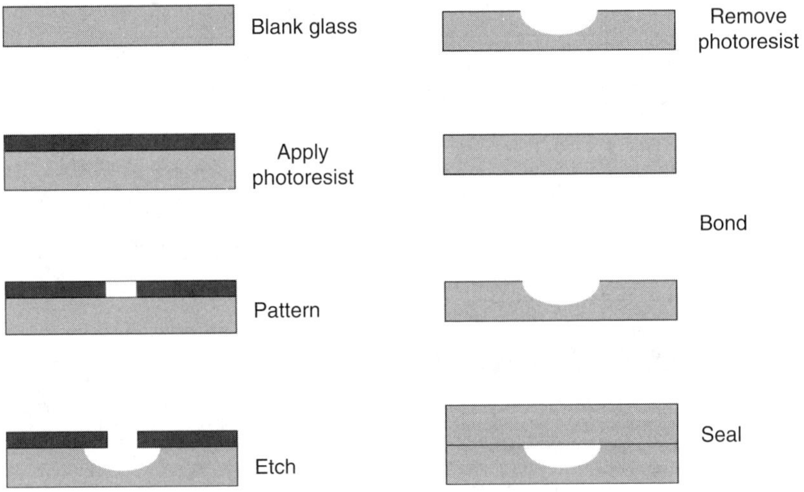

FIGURE 52.1 Schematic showing the fabrication process for glass microchips. The process begins with a blank glass plate. A masking layer (typically a metal plus photoresist) is deposited on top of the glass. The masking layer is patterned using UV light and the exposed metal removed. The underlying glass is etched using a caustic solution to the appropriate depth. The remaining photoresist and metal are removed and the patterned glass substrate is sealed to a second piece of unpatterned glass containing predrilled access holes using thermal methods.

52.2.1.1 Etching

Two general types of etching are used to create glass microchips, wet and dry. Wet etching is by far the most common because it does not require very expensive and complicated reactive ion etching (RIE) instrumentation. Dry etching, on the other hand, can provide better control over channel geometry and aspect ratios as well as give deeper channels. Which method is selected is often dictated by the availability of instrumentation and/or the specific application. For both fabrication methods, the glass is first covered with a thin layer of metal and photoresist. The metal is removed from the surface in the areas where etching will occur using a series of standard photolithography and metal etching steps. Finally, the exposed glass is etched to form three sides of the capillary. For wet etching, the most common etchants are based on HF [either concentrated HF or buffered oxide etch (BOE) that is a mixture of NH_4F and HF]. Wet etching (Figure 52.1) results in a hemicylindrical shape of the channel because etching occurs in all directions (isotropic). This etching mode limits, the aspect ratio of wet-etched glass microchips is limited and most channels are relatively shallow (10–30 μm). In many cases, this is acceptable; however, there are cases where deeper, narrower channels are needed [2]. In the latter case, a dry etching technique that is anisotropic or directional is needed to provide higher aspect ratios. Two dry etching methods, RIE and inductively coupled plasma (ICP) etching, are common in this application (Figure 52.2). Powder blasting has also been used in a limited number of cases, but generally results in rough channel surfaces owing to the etching mechanism. For both RIE and ICP, a cloud of reactive gases (plasma) is generated at reduced pressure using an RF generator and these gases react with the surface. The incoming gas composition can control the selectivity of the etch. For example, CF_4 can etch Si quickly but leave metals relatively untouched. As a result, metal films are traditionally used as etch masks for both processes. In RIE, the substrate to be etched sits on top of one of the electrodes and the gas is driven to the surface, causing reactions that remove surface material (Figure 52.2a). In ICP a plasma is generated using RF frequency, but this reactions occurs above the substrate. The substrate is biased to a potential that causes the reactive gas species to be accelerated toward and react with the surface. Both etching methods provide directional control of etching. As the reactive gases bombard the surface, they react with the first surface they

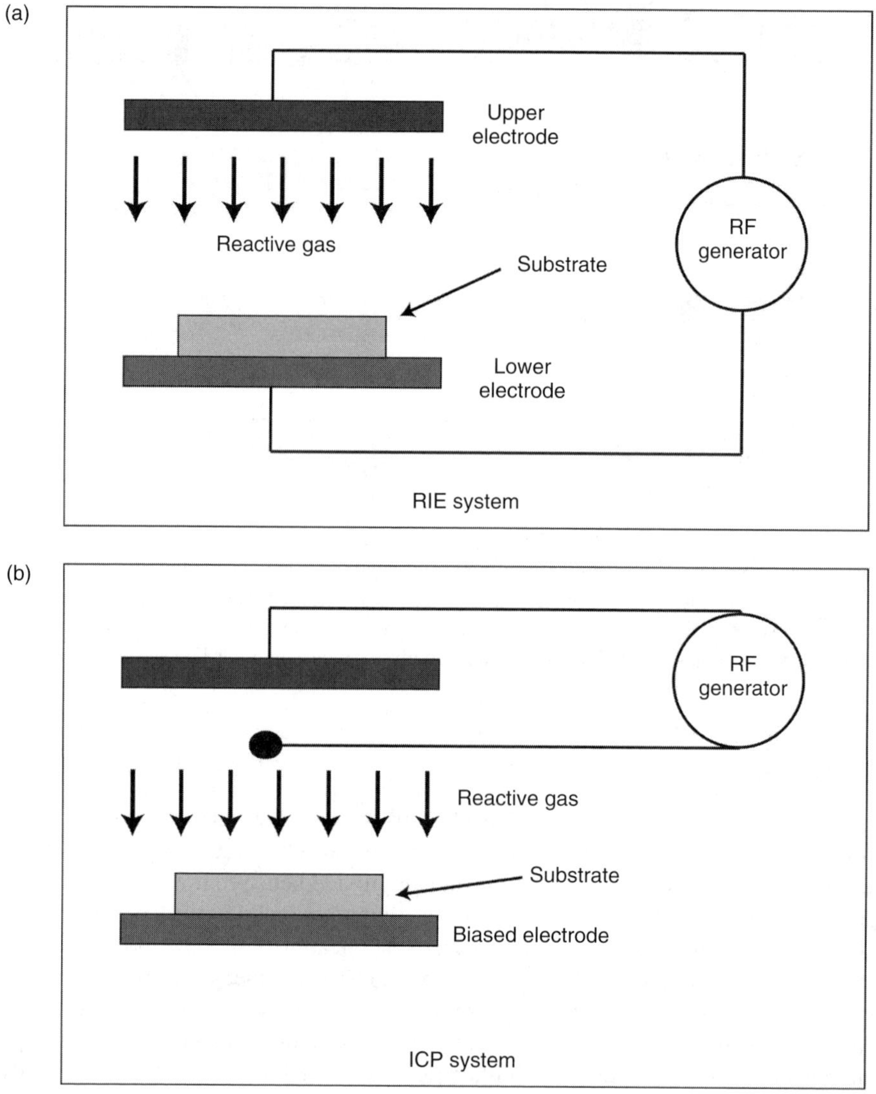

FIGURE 52.2 Schematic drawing of both reactive ion etching (RIE) and inductively coupled plasma (ICP) etching systems. In both cases, RF radiation is used to induce a plasma that is subsequently driven to the surface as a result of the charges on the gases coupled to a potential bias on an electrode underneath the substrate.

come to. Because they are being driven down, perpendicular to the surface, directional etching is achieved with minimal undercutting. In general, ICP etching is preferred for very small features, but RIE is the more common of the two instruments.

52.2.1.2 Bonding

After the channels are etched in a single piece of glass, the patterned substrate must be bonded to a second piece of glass to form the fourth side of the channel. For glass microchips, most bonding is done at elevated temperatures requiring the use of a second piece of glass that has the same thermal expansion coefficient as the etched substrate. After this consideration, bonding is relatively straightforward. Access holes are first drilled in the unpatterned substrate to allow solution to be added to the capillaries. The two substrates are aligned and clamped together after copious washing to remove all particulate contamination. The assembly is subsequently placed in a programmable furnace and brought to a temperature above its glass transition temperature (T_g), usually 600°C–1100°C. At this point, the two glass pieces fuse to form a single integrated device. After sufficient fusion time, the microchips are cooled slowly to room temperature over a period of several hours to prevent cracking due to thermal stress. On the whole, the bonding process can take up to 24 h to complete.

In addition to thermal bonding, several other bonding methods have been reported. The second most common technique is anodic bonding. In anodic bonding, the two substrates are again clamped together after copious washing, but are placed this time between two metal electrodes. The temperature is elevated (400°C) and a high DC voltage (>1 kV) is applied between the two electrodes. Bonding is achieved because sodium ions at the interface of one piece of glass migrate away from the surface, making it highly reactive with the silicon in the second substrate. One of the problems with both thermal and anodic bonding is the harsh conditions required to achieve the desired microchip. Alternative methods have been investigated, which are not as harsh. One good example of this is a photocurable epoxy method developed by the Landers group [3]. In this method, photocurable epoxy is added to special channels that line the microchip. The microchip is exposed to intense UV radiation to cross-link the epoxy, creating a stable bonded microchip.

52.2.2 Polymers

While glass has many benefits for MCE, it is not without its problems, as discussed above. As a result, many investigators have turned to polymers as substitute substrate materials because they are cheaper, easier to fabricate, and have a wide range of surface properties [4]. The first polymeric material used was poly(dimethylsiloxane) or PDMS [5]. After the initial landmark report, many other polymers were demonstrated for use with MCE, including poly(methyl methacrylate) (PMMA), poly(carbonate) (PC), polyester (PE), Teflon, and Parylene C. The monomeric repeat unit for the three most common polymers, PDMS, PMMA, and a typical PC are shown in Figure 52.3. Each material has different properties that make selection application dependent. In general, the three major considerations in selecting a polymer are (1) glass transition temperature of the material, which dictates fabrication methods, (2) the optical transparency and background fluorescence, and (3) the surface chemistry presented by the base material. The following sections describe the fabrication of polymer microchips in detail and are divided by the general methods used for fabrication.

52.2.2.1 Poly(methyl methacrylate) and Poly(carbonate)

Two of the more common polymeric materials for MCE are PMMA and PC. Both of these materials are rigid, have good optical transparency, and can be easily fabricated using one of two methods, hot embossing or injection molding. The general process for both fabrication methods is outlined in Figure 52.4. Hot embossing is the most commonly employed method because it does not require an expensive injection molding machine. Injection molding, however, arguably produces better devices. In both cases, the mold is a critical element and can be made through one of several

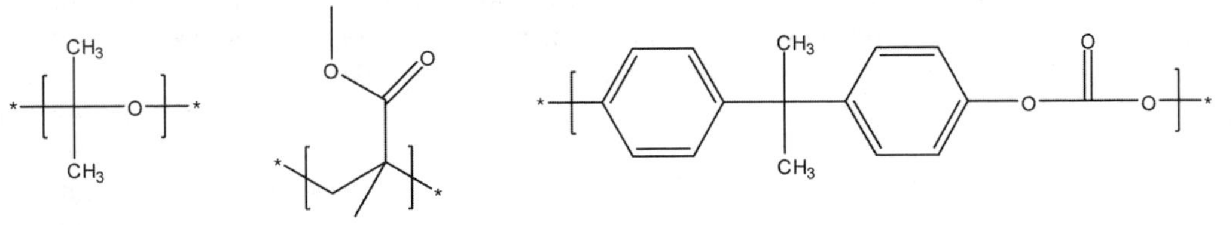

FIGURE 52.3 Comparison of the chemical structures for poly(dimethylsiloxane) (PDMS, left), poly(methyl methacrylate) (PMMA, center), and a typical polycarbonate (PC, right).

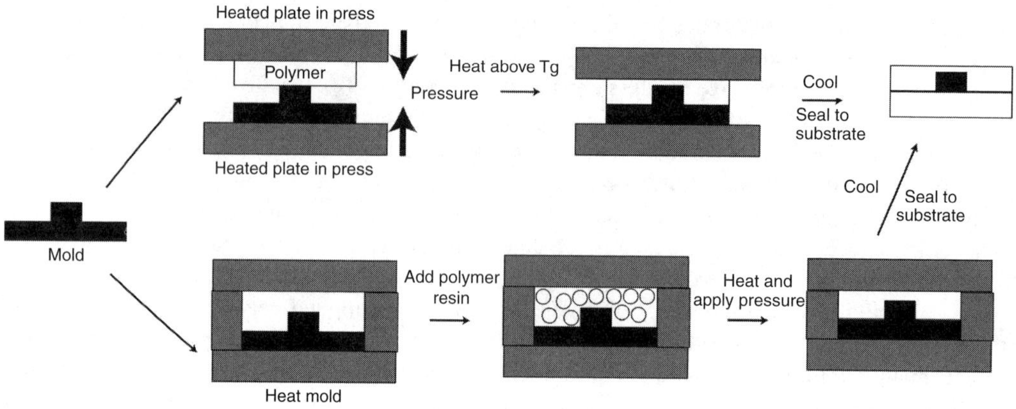

FIGURE 52.4 Comparison of hot embossing (top pathway) and injection molding (bottom pathway) approaches for the fabrication of PMMA or PC microfluidic devices.

different processes. The most common is to generate a mold using single crystal silicon wafers for the template. These molds are easy to make, but are fragile and typically last for 10 molding cycles or less. As a result, other methods have been used to generate metallic molds, usually made from nickel, which are extremely robust and can be used to produce thousands of microchips [6]. Metal molds are normally produced by fabricating an inverse master by normal photolithography. Ni or a similar metal is subsequently electrodeposited on top of this structure after applying an initial metallic seed layer for conductivity. After electrodeposition, the photoresist is released from the metal leaving the completed mold.

52.2.2.2 Poly(dimethylsiloxane)

In 1994, Effenhauser's group reported the use of a commercially available elastomer Sylgard 184 for production of electrophoresis microchips [5] and was followed shortly by a similar application from Whitesides group [7]. Sylgard 184 is a composite material that contains primarily cross-linked PDMS. PDMS rapidly gained favor for production of MCE devices because it is very inexpensive, has good optical clarity into the low UV wavelength range, has very low background fluorescence compared to many polymers, is chemically stable to common aqueous and polar organic solvents, is gas permeable, and is generally biologically compatible. Finally, and arguably most importantly, PDMS uses a simple cast-and-cure molding process to produce complicated microfluidic channels. The molds can be created using a number of methods that varies in complexity from mechanical milling to RIE. The most common method, however, is the use of an epoxy-based photoresist (SU-8). In this fabrication method, a wafer is coated with SU-8, exposed to UV light through a simple transparency, and then developed to remove unexposed material. The remaining photoresist forms the mold upon appropriate curing and can last for upward of a thousand replications if appropriate care is taken. PDMS has been sealed to PDMS, glass, and many other plastics to complete the final microchip with great success. Furthermore, both permanent and reversible seals are possible, depending on the material used for the fourth side of channel and the pretreatment procedures. While PDMS is an attractive material, it has a significant disadvantage for MCE. PDMS is hydrophobic and has been widely used in a noncrosslinked version as a stationary phase in gas chromatography. For MCE, the hydrophobicity causes significant adsorption and therefore, peak tailing. This surface can be rendered hydrophilic for a short period of time using either plasma oxidation or treatment with strong base; however, this treatment is short-lived (often < 30 min) owing to the diffusion of linear oligomers to the channel surface. A second problem caused by the diffusion of the oligomer to the surface is generation of nonuniform EOF. Nonuniform flow creates significant band broadening as well, making PDMS, one of the least attractive materials, which has been used in MCE laboratories, with respect to separation efficiency. Strategies to overcome this problem are addressed in the next section of this chapter.

52.2.2.3 Other Materials

While PC, PMMA, and PDMS are by far the most common polymeric materials used for MCE, several other materials have been used with varying degrees of success. In general, new materials are tested in an effort to combine the fabrication ease of PDMS with the surface and material properties of PMMA, PC, or even glass. This list is also not complete as reports of new materials are published on a regular basis in appropriate journals. Three of the more common examples are thermoset polyester (TPE) [8], parylene C [9], and SU-8 [10]. TPE devices are made from a simple polyester monomer but use a combination of thermal and photo-initiated cross-linking to produce devices from simple molds. The fabrication is akin to that used for PDMS, but produce a rigid polymer microdevice with better surface characteristics. Parylene C is a polymer that is more common to the microelectronics industry and has only recently been used for MCE. Parylene C is deposited using a simple vapor deposition method and forms thin, pinhole-free surfaces. Furthermore, its surface properties are similar to Teflon and therefore it holds significant interest for applications requiring low EOF. Finally, several researchers have reported the use of SU-8 photoresist for creation of microchannels. SU-8 is attractive because devices with a great variation in aspect ratio can be created using very simple photolithography. The devices are also mechanically stable because the base material is an epoxy. The major limitation to SU-8 devices lies in the inability to create microchips that have a uniform surface charge throughout the channel. In current versions of this device, the top and bottom are normally made of glass, while the side walls are the photoresist. The nonuniform zeta potential causes differential flow and therefore band broadening.

52.2.2.4 Summary

MCE devices have been made from a wide variety of materials with greatly varying properties. Selection of an appropriate material is best determined by the specific needs related to detection, cost, available microfabrication facilities, and required separation resolution. In many cases, however, the base material does not provide appropriate chemical properties for the desired application. For this reason, a number of methods for surface modification have been developed to provide chemical control of electrophoresis.

52.3 SURFACE MODIFICATION

Chemically modified surfaces play an important role in many different industries and applications ranging from medical diagnostics and *in vivo* devices to microelectronics. CE has long used chemical modification to control the surface properties of capillaries used for a variety of applications [11]. For example, modern CE-based DNA separations require that the surface charge and therefore the EOF be minimized or eliminated to generate high-resolution separations. In general, there are three methods for surface modification in MCE: covalent, adsorbed, and dynamic coatings. Covalent modification strategies generate new chemical bonds on the surface to change the chemistry, while adsorbed coatings use either electrostatic or hydrophobic interactions to adsorb a polymer onto the surface. Dynamic coatings are different in that the coating agent is in the running buffer and in contact with the surface at all times. A portion of the molecules adsorb to the surface in a dynamic equilibrium to define the chemistry. Surfactants are the most common species for dynamic coatings. In the following sections, a detailed description of the different types of coatings and how they are used is presented.

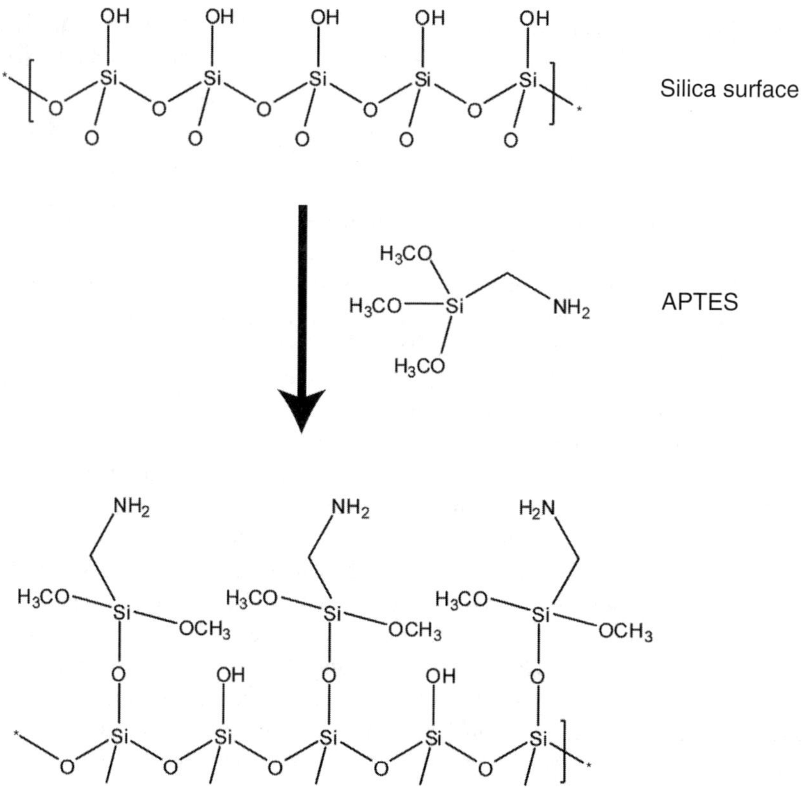

FIGURE 52.5 General reaction schematic for silanizing surfaces. In this example, the silanizing agent is 3-aminopropyltriethoxysilane (APTES) that ultimately presents a primary amine to the surface. Most silanizing agents contain the same triethoxy or trimethoxysilane functional group with a different terminal chain. The terminal chain can be varied to give unique surface properties.

52.3.1 Covalent Methods

Covalent coatings come in two general types, those based on silane chemistry and those based on polymer grafting methods. Silane chemistries are among the oldest and best known coatings for electrophoresis and can be adapted almost directly from conventional CE to MCE for glass microchips. Silanization can also be performed on PDMS since it contains some free SiO_2 functionality. The general schematic for silane modification is shown in Figure 52.5. Generating a silane-coated surface is a relatively time-intensive process, requiring that the surface first be activated with successive washings with NaOH and then HCl, followed by drying at elevated temperature to dehydrate the capillary wall. The reactive silane is added to the capillary and allowed to react for an extended period of time to ensure complete coverage. Silane chemistry is attractive because it is relatively simple and can modify the surface with a wide range of potential functional groups ranging from neutral polymers for reduced EOF to cationic amine coatings for reversed flow applications such as the analysis of anions [12]. Silane chemistry can also be used to deposit a base layer on the surface to eventually grow additional polymer layers for either electrochromatography or to further control flow. Silane chemistry is not without its problems. The most significant concern is the pH stability. Si–O bonds are hydrolyzed at low pH (<3) and moderately to strong alkaline pH (>8) removes the coating from the surface. While this pH range is quite useful for modern liquid chromatography, it excludes an important pH range and group of buffers commonly used with CE. As a result, many groups have focused on adsorbed and dynamic coatings for high pH applications of MCE using glass microchips.

Silanization chemistry works well for silicon-based microchips; however, as mentioned above, there has been a significant shift to the use of polymeric substrate materials in an effort to reduce cost and fabrication complexity. Multiple methods have been used for modifying polymers, including plasma chemistry and surface grafting. Plasma modification is the most common approach and has been applied to many different substrates. Plasmas can be further defined as depositing plasmas or activating plasmas. In depositing plasmas, a thin film is deposited on the surface, typically before the devices are sealed together. Plasma chemistry holds significant, but as of yet, largely unrealized potential because a wide range of monomers can be used to deposit films with a vast array of chemical properties. Activating plasmas tend to be more destructive and work by breaking bonds of the surface generating ionizable functional groups. The most common example of this is the treatment of PDMS with oxygen or air plasmas to generate a hydrophilic surface [7]. Polymer grafting can also be used to modify microchip surfaces. Multiple methods involving UV surface activation have been reported. In one approach, the channel was filled with a solution containing the monomer units, a photoinitiator, and a chain terminator. Upon exposure to UV light, polymer formation was initiated on the surface and continued until the device was removed from the light [13]. In a similar approach, channels were pretreated with the photoinitiator and then dried. The monomer solution was subsequently introduced to the channel and exposed to UV light [14]. Additional methods exist for modification of PMMA and PC. For example, Henry et al. [15] generated a reactive mixture using ethylene or propylenediamine mixed with *n*-butyl lithium to covalently modify PMMA with a primary amine functionality. An alternative approach to generating an anionic surface used a combination of UV radiation and O_3 to convert methoxy groups to carboxylic acids on the surface of PMMA devices [16]. This approach has the advantage of being simple and easy to scale to large numbers of devices. In summary, many methods can be applied to the modification of polymer surfaces for MCE. The choice of the best approach is generally determined by the specific substrate material and desired application.

52.3.2 Adsorbed Coatings

The second major form of coatings in MCE is adsorbed coatings. Adsorbed coatings adhere to the surface through either electrostatic or hydrophobic interactions, are simple to deposit, and can be achieved using a wide range of both neutral and ionic polymers. While adsorbed coatings do not have the same bond strength of a covalently attached film, they are generally stable for at least one day and can be easily regenerated. The major advantage of adsorbed coatings over covalent surface modification strategies, particularly for glass microchips, is the pH sensitivity. Many adsorbed coatings are stable from pH 2 to 13 and even provide a constant EOF in that range. Finally, unlike dynamic coatings, adsorbed coatings do not require any special additives in the run buffer to maintain the surface modification chemistry, allowing the use of a broader range of separation chemistries.

Of the many types of polymers used for adsorbed coatings, neutral polymers are the most common, particularly for the separation of DNA in fragment and sequencing applications. Neutral polymers chemisorb to the surface, and therefore their hydrophobicity and hydrophilicity are important. Hydrophobic polymers tend to adsorb the strongest to the surface of most chip materials, while hydrophilic polymers are the best for reducing the interaction of proteins with the capillary walls. A reasonable compromise can be found with a number of polymers, including hydroxypropylmethylcellulose (HPMC), hydroxyethylcellulose (HEC), poly(vinyl alcohol) (PVA), and poly(*N*,*N*-dimethylacrylamide) (PDMA). Another polymer that is particularly attractive for applications requiring a reduced EOF is poly(ethylene oxide) (PEO). Figure 52.6 shows a series of electropherograms collected using PEO and multilayer PEO-poly(vinyl pyrrolidine) (PVP) coatings for the separation of DNA fragments from ΦX174 [17]. This data shows the difference that coating, including multilayer coating strategies, can have on the ability to resolve different DNA fragments.

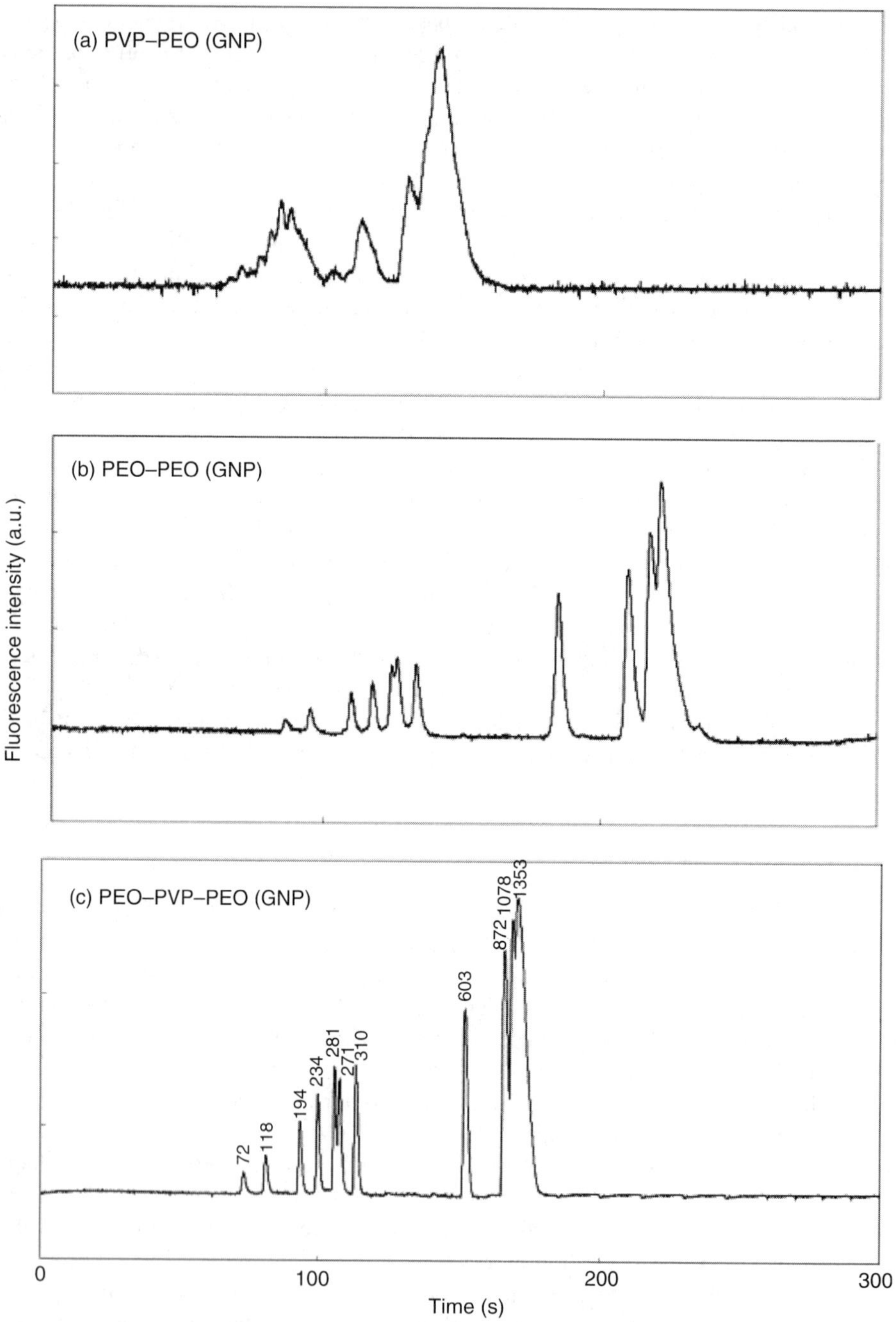

FIGURE 52.6 Comparison of the separation of 10 μg/mL ΦX 174 RF DNA-*Hae*III digest using PMMA chips with three different coatings, using the neutral polymers, PEO and PVP along with gold nanoparticles (GNP) [17].

The second most common type of polymer coatings uses polyelectrolytes. Polyelectrolytes are well known to form somewhat ordered polymer films of varying thickness when deposited on charged surfaces using the simple layer-by-layer deposition method [18]. A schematic drawing of the resulting structures is shown in Figure 52.7. These structures, referred to as polyelectrolyte multilayers (PEMs), can vary in thickness from 1 to 100 nm and exhibit a wide range of chemical properties. They are

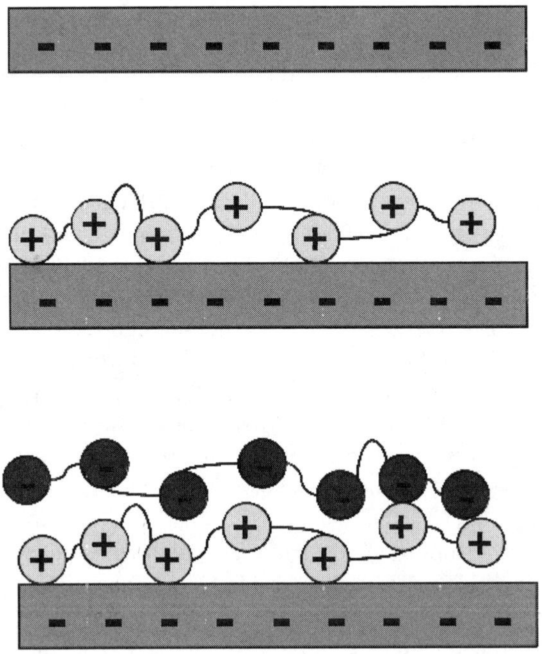

FIGURE 52.7 Schematic drawing showing the buildup of polyelectrolyte layers by the layer-by-layer deposition method in the generation of a polyelectrolyte multilayer structure.

deposited as alternating layers of cationic and anionic polymer. PEMs overcompensate for surface charge by forming a series of loop and tail structures on the surface. The result is that each subsequent PEM layer reverses the surface charge. The number of layers used can range from one or two up to 80–100, depending on the application. Thinner layers are typically used in CE applications, because the films form a chromatographic stationary phase, and thick stationary phases often lead to band broadening. PEM coatings are most commonly used when flow reversal is needed because they provide a relatively simple long-lived coating for this application. A second application of PEMs is to reduce the pH dependence on EOF. Coatings consisting of a strong polyelectrolyte (i.e., one that bears a permanent charge such as poly(styrene sulfonic acid) can give essentially pH-independent EOF because the functional group is charged throughout normal operating pH range. This is a useful characteristic in MCE to improve the reproducibility of separations. Finally, PEM coatings can be used to improve the separation efficiency in MCE separations, particularly for polymer microchips. Polymer microchips are well known to have lower separation efficiency than glass microchips. There are two general causes of this poor performance, use of dissimilar materials, and/or poor surface chemistry. In the case of dissimilar materials where the channel is made from one material and the top and/or bottom is made from a different material, two different ζ potentials are present and thus two different EOFs. In this case, PEM coatings provide a uniform ζ potential and therefore less band broadening. The other case is more difficult. Band broadening in MCE microchips made from the same material can be caused by one of two features, either micro-heterogeneity of the surface charge or inherent hydrophobicity of the material. In either case, coating the channel surface with a PEM can reduce the impact of these problems and improve separation efficiency dramatically.

In general, adsorbed coatings work well for applications requiring reduced EOF. Furthermore, they are attractive because a wide variety of chemistries exist and their deposition is very simple. On the negative side, however, adsorbed coatings have a finite lifetime requiring recoating more frequently than covalent systems.

52.3.3 Dynamic Coatings

The last surface modification strategy for microfluidic devices uses dynamic coatings that are normally composed of surfactants. This process is given its name because the coating material is present in the running buffer and therefore is in a dynamic equilibrium between the capillary surface and the bulk solution. Dynamic coatings are not permanent and are in fact removed by replacing the running buffer with a solution that does not contain the surfactant. There are two molecular categories of surfactants used in dynamic coatings, polymeric and small molecules. Low molecular weight surfactants are the most common, with the compounds shown in Figure 52.8 being the most common. For glass microchips, the surfactants are normally cationic and form a simple bilayer structure that serves two roles. First, the anionic surface is replaced with the cationic surfactant, resulting in reversed EOF. Dynamic coatings using cationic surfactants are arguably the simplest and most common approach to flow reversal in MCE. Second, the bilayer phase can act as a weak stationary phase for electrochromatography. The interactions of surfactants with polymer microchips are different than with glass because of the porosity and difference in surface chemistry of those devices [19,20]. For polymers, both anionic and cationic surfactants have been found to be of use. Cationic surfactants reverse the flow through bilayer formation as noted above. Anionic surfactants, most notably sodium dodecyl sulfate (SDS), can form a tail down monolayer on the surface of some polymers resulting in an increased surface charge and EOF. Support for this binding motif also comes from studies of surfactants that have similar head groups but different tail groups. It was found that surfactants with more hydrophobic tails could remain on the surface much longer than surfactants with shorter tails.

The second group of surfactants that can be used in dynamic coatings are polymeric surfactants. The most common polymers include PEO, poly(ethylene glycol) (PEG), poly(propylene oxide)

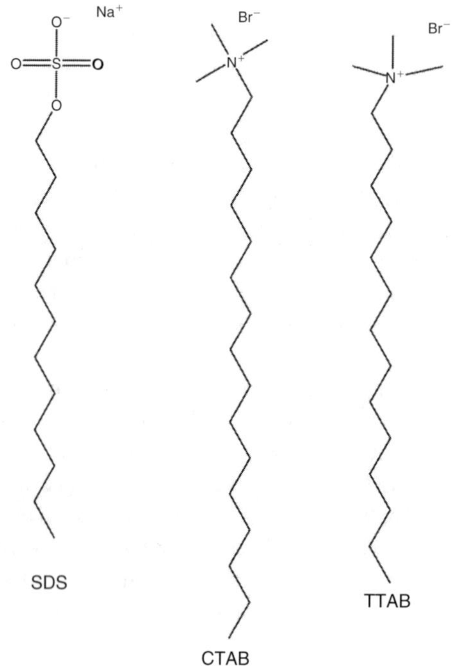

FIGURE 52.8 Three most common ionic surfactants used for coating microfluidic devices. From left to right: sodium dodecyl sulfate (SDS), cetyltrimethyl ammonium bromide (CTAB), and tetradecyltrimethyl ammonium bromide (TTAB).

(PPO), and co- and tri-block variations of these structures. Polymeric surfactants interact with the surface in very different ways than low molecular weight surfactants. In most cases, it is believed that the hydrophobic moiety of the surfactant binds to the surface, exposing the hydrophilic portion to the run buffer. Polymeric surfactants have the advantage in that they can form effective coatings at low concentrations; however, they are typically not as stable as their low molecular weight counterparts.

52.4 SURFACE CHARACTERIZATION METHODS

Finally, it is important to acquaint readers, particularly those more familiar with electrophoretic separations than materials characterization, with techniques that can be used to characterize different substrate materials and coatings. The direct analysis of the interior of capillaries is challenging so most of the methods discussed below are measured on a comparable flat surface. While it can be debated whether this approach is reasonable for traditional fused silica capillaries, the analogy between a flat surface of the material and the surface of the capillary in a microchip device is quite reasonable. The characterization methods that will be discussed here range from measurements of EOF to analysis of surface chemistry using X-ray spectroscopy techniques, with an emphasis on both how the measurement techniques are applied to materials used for electrophoresis and on how the results can be used to improve separation performance.

52.4.1 Physical Characterization

52.4.1.1 Electroosmotic Flow

One of the most important measures of the channel surface chemistry comes from measuring the EOF. EOF is dependent on a number of factors, including surface charge. Furthermore, it is relatively trivial to measure the EOF through one of a number of methods. As a result, EOF measurements provide a convenient method to characterize coatings generated *in situ*. For permanent and adsorbed coatings this is important, but more so for dynamic coatings that are not stable on a surface in the absence of surfactant in the run buffer. EOF measurement provides three valuable pieces of information. First, it provides a measure of the overall surface charge, ζ. Increasing, decreasing, or reversing the surface charge is a common goal of surface modification in MCE. Second, EOF measurements can provide a measure of the coating stability. EOF measurements made on a run-to-run basis are important for dynamic coatings, while EOF measurements made over hours to days are important for determining the stability and even presence of more permanent adsorbed or covalent coatings. Third, measuring EOF as a function of pH can determine acceptable ranges of operation, for a given application.

There are two common methods for measuring EOF as well as a host of additional more difficult, but arguably more useful, methods. The two most common methods are the current monitoring method [21] and the neutral marker method. In the current monitoring method, two different concentrations (ionic strengths) of the same buffer are prepared. The capillary and one reservoir are filled with one buffer and the second reservoir is filled with the second buffer. Voltage is applied to drive the buffer in the second reservoir through the capillary while the separation current is being measured. The current will gradually change as the new buffer replaces the old. The time required to reach a constant current is then used to determine the EOF rate. Alternatively, normal electrophoresis can be performed using a compound that is neutral. The migration time of the neutral compound can be used to determine the EOF. It should be noted that this technique does not work with micellar electrokinetic chromatography owing to partitioning of neutral analytes into charged micelles.

In addition to these two established methods for measuring EOF, several other methods have been published, which are worth mentioning. The first method, periodic flow monitoring, uses a

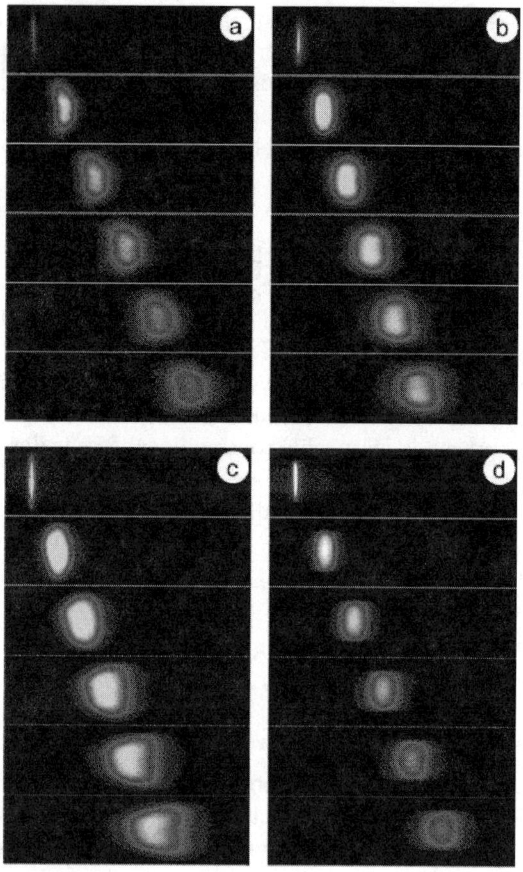

FIGURE 52.9 Images of EOF through various microchannels and capillaries. (a) Acrylic microchannel at 266 V/cm. (b) PDMS microchannel at 248 V/cm. (c) Hybrid acrylic/PDMS microchannel 245 V/cm. (d) Fused-silica capillary at 250 V/cm. In each case, the first image was taken at the time of the UV laser pulse, and the time step between images was 50 ms. Each image is 300 μm by 80 μm [23].

two-spot laser system for periodic photobleaching [22]. This approach allows for measurement of the EOF in near real time during separation experiments and can provide dynamic information on flow during separations. A second method for characterizing EOF was reported by Ross et al. [23]. In this method, the buffer contains a caged fluorescent dye. Caged dyes remain nonfluorescent until hit with a high-intensity laser. Subsequent to this exposure, the dyes become fluorescent and can provide important information on relative EOF in a given channel cross-section. Figure 52.9 shows an image taken from this initial report by Ross et al. showing how EOF dispersion varies with capillary material. The difference in peak width, going from pure to hybrid materials, is significant, and results in a significant reduction in separation efficiency.

52.4.1.2 Contact Angle

A second important measure of coating chemistry is the contact angle. Contact angle measures the relative hydrophobicity/hydrophilicity of a surface and is measured with a simple apparatus known as a goniometer. A simple schematic is shown in Figure 52.10 to provide an idea of how the measurement is achieved and what it means. The contact angle is always measured at the point where all three phases (liquid droplet, solid surface, gas) meet. The lower the contact angle, the more hydrophilic is

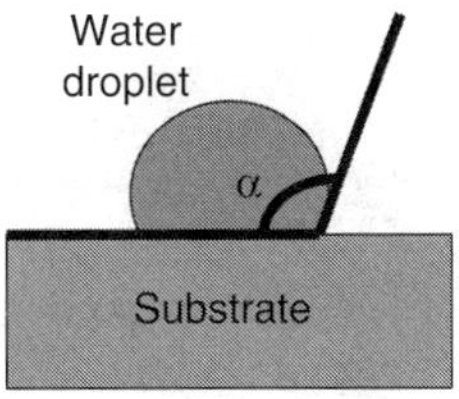

FIGURE 52.10 Simple schematic showing how a contact angle is measured using a goniometer.

the surface in question and the larger the contact angle, the more hydrophobic. For example, clean unmodified glass surfaces commonly have contact angles of $<20°$, while hydrophobic surfaces like unmodified PDMS can have contact angles $>100°$. Different surface treatments can dramatically change the contact angle. For example, treatment of PDMS with a simple oxygen plasma reduces the contact angle to $<60°$ for a short period of time. Contact angle measurements are most useful in MCE for following the conversion of a hydrophobic surface to a hydrophilic one and vice versa. Contact angle measurements should be taken with care, however, as contaminated surfaces can cause inaccurate results.

52.4.1.3 Atomic Force Microscopy

One final physical technique that has been applied to the characterization of MCE materials and coatings is atomic force microscopy (AFM). AFM allows one to profile changes in surface morphology that occur with surface modification. AFM in MCE is most commonly used with polymeric coatings and not with molecular coatings because of the initial surface roughness. Most MCE surfaces are not atomically flat, a requirement for atomic resolution in AFM [24]. The main benefit of using AFM imaging in MCE is to measure surface roughness. Significant surface roughness can cause band broadening and a concomitant loss of separation efficiency.

52.4.2 Chemical Characterization

52.4.2.1 X-Ray Photoelectron Spectroscopy

A final chemical characterization technique that is very useful in MCE is x-ray photoelectron spectroscopy (XPS). XPS measures the energy of photoelectrons emitted from the atom as they undergo excitation and decay as a result of exposure to x-ray radiation. The energy is referred to as the binding energy (BE). Furthermore, because the photoelectrons have to leave the surface, the technique is limited to the measurement of only the outermost layers (<10 nm) of the surface, making it very specific for coatings. XPS can be used for characterization of elemental composition of the surface but arguably its most important use is in the discrimination of specific chemical bonds on a surface. For example, in Figure 52.11, an XPS spectrum for a PDMS surface is shown. The first spectrum is a global scan showing the presence of C, O, and Si as expected. The second spectrum is a high resolution scan of the C1s region. Peak-fitting routines are used to define three different peaks within this system, each corresponding to a different form of carbon. For example, the dominate peak at 284.7 eV is from carbon that is part of the siloxane backbone. The other two peaks come from different forms of carbon that are caused by either the cross-linking process or adsorption of contaminates from the surface. XPS is also a quantitative tool. The peak areas are proportional to the amount of each material present on the surface. This type of information is extremely valuable in defining the chemical content of the surface and has been used with both silicon and polymer-based microchips and coatings.

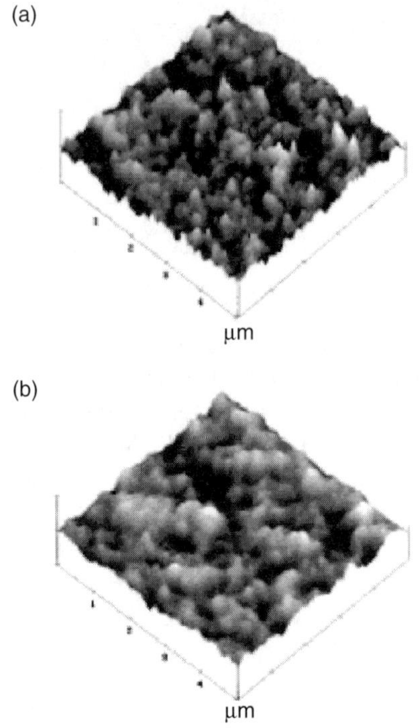

FIGURE 52.11 (a) Low resolution XPS scan of an untreated PDMS surface. (b) High resolution scan of the C1s region of PDMS showing the presence of at least three forms of C with siloxane C being the most dominate.

52.5 SUMMARY

MCE is rapidly growing in importance in modern chemical analysis. As this growth continues, there will be a continuing need to develop cheap, reliable microchip devices for a broad range of applications. Important in this development is the control of the surface chemistry of microfluidic devices. Many approaches can be taken to accomplish this goal, from selecting appropriate substrate materials to modifying existing materials with one from a wide range of coatings. With the number of options available in this area, finding a starting point may seem overwhelming. Given this consideration, the following advice can be given. Start by selecting a microchip material that is readily available. For some scientists with access to clean room facilities, HF etching baths, and furnaces, glass is the best option. For others without access to these kinds of resources, PDMS or one of the other polymers may be the best starting point. The second option to consider is the need for and type of surface modification strategy. Dynamic coatings are the simplest, but may not be compatible with the desired application. Adsorbed coatings are more stable but may not give the desired properties such as removal of surface charge. The main consideration here is the specific application. As is commonly the case, all decisions made regarding coatings and substrate should be made while considering the end application.

REFERENCES

1. Manz, A., Harrison, D. J., Verpoorte, E. M. J., Fettinger, J. C., Paulus, A., Luedi, H., and Widmer, H. M., Planar chips technology for miniaturization and integration of separation techniques into monitoring systems. Capillary electrophoresis on a chip, *J. Chromatogr.*, 593 (1992) 253–258.

2. Culbertson, C. T., Jacobson, S. C., and Ramsey, J. M., Dispersion sources for compact geometries on microchips, *Anal. Chem.*, 70 (1998) 3781–3789.
3. Huang, Z., Sanders, J. C., Dunsmor, C., Ahmadzadeh, H., and Landers, J. P., A method for UV-bonding in the fabrication of glass electrophoretic microchips, *Electrophoresis*, 22 (2001) 3924–3929.
4. Becker, H. and Locascio, L., Polymer microfluidic devices, *Talanta*, 56 (2002) 267–287.
5. Effenhauser, C. S., Bruin, G. J. M., Paulus, A., and Ehrat, M., Integrated capillary electrophoresis on flexible silicone microdevices: analysis of DNA restriction fragments and detection of single DNA molecules on microchips, *Anal. Chem.*, 69 (1997) 3451–3457.
6. Soper, S. A., Ford, S. M., Qi, S., McCarley, R. L., Kelly, K., and Murphy, M. C., Microelectromechnical systems (MEMS) fabricated in polymeric materials: applications in chemistry and life sciences, *Anal. Chem.*, 72 (2000) 545A.
7. Duffy, D. C., McDonald, J. C., Schueller, O. J. A., and Whitesides, G. M., Rapid prototyping of microfluidic systems in poly(dimethylsiloxane), *Anal. Chem.*, 70 (1998) 4974–4984.
8. Fiorini, G. S., Lorenz, R. M., Kuo, J. S., and Chiu, D. T., Rapid prototyping of thermoset polyester microfluidic devices, *Anal. Chem.*, 76 (2004) 4697–4704.
9. Webster, J. R., Burns, M. A., Burke, D. T., and Mastrangelo, C. H., *Micro Electro Mechanical Systems*, pp. 306–310, IEEE Press, Miyazaki, Japan, (2000).
10. Agirregabiria, M., Blanco, F. J., Berganzo, J., Fullaondo, A., Zubiaga, A. M., Mayora, K., and Ruano-Lopez, J. M., SDS-CGE of proteins in microchannels made of SU-8 films, *Electrophoresis*, 27 (2006) 3627–3634.
11. Belder, D. and Ludwig, M., Surface modification in microchip electrophoresis, *Electrophoresis*, 24 (2003) 3595–3606.
12. Horvath, J. and Dolnik, V., Polymer wall coatings for capillary electrophoresis, *Electrophoresis*, 22 (2001) 644–655.
13. Hu, S. W., Ren, X. Q., Bachman, M., Sims, C. E., Li, G. P., and Allbritton, N., Surface modification of poly(dimethylsiloxane) microfluidic devices by ultraviolet polymer grafting, *Anal. Chem.*, 74 (2002) 4117–4123.
14. Hu, S. W., Ren, X. Q., Bachman, M., Sims, C. E., Li, G. P., and Allbritton, N. L., Surface-directed, graft polymerization within microfluidic channels, *Anal. Chem.*, 76 (2004) 1865–1870.
15. Henry, A. C., Tutt, T. J., Galloway, M., Davidson, Y. Y., McWhorter, C. S., Soper, S. A., and McCarley, R. L., Surface modification of poly(methyl methacrylate) used in the fabrication of microanalytical devices, *Anal. Chem.*, 72 (2000) 5331–5337.
16. Zangmeister, R. A. and Tarlov, M. J., UV graft polymerization of polyacrylamide hydrogel plugs in microfluidic channels, *Langmuir*, 19 (2003) 6901–6904.
17. Lin, Y. W. and Chang, H. T., Modification of poly(methyl methacrylate) microchannels for highly efficient and reproducible electrophoretic separations of double-stranded DNA, *J. Chromatogr. A*, 1073 (2005) 191–199.
18. Ai, H., Jones, S. A., and Lvov, Y. M., Biomedical applications of electrostatic layer-by-layer nano-assembly of polymers, enzymes, and nanoparticles, *Cell. Biochem. Biophys.*, 39 (2003) 23–43.
19. Garcia, C. D., Dressen, B. M., Henderson, A., and Henry, C. S., Comparison of surfactants for dynamic surface modification of poly(dimethylsiloxane) microchips, *Electrophoresis*, 26 (2005) 703–709.
20. Ocvirk, G., Munroe, M., Tang, T., Oleschuk, R., Westra, K., and Harrison, D. J., Electrokinetic control of fluid flow in native poly(dimethylsiloxane) capillary electrophoresis devices, *Electrophoresis*, 21 (2000) 107–115.
21. Huang, X., Gordon, M. J., and Zare, R. N., Current monitoring method for measuring the electroosmotic flow rate in capillary electrophoresis, *Anal. Chem.*, 60 (1988) 1837–1838.
22. Pittman, J. L., Henry, C. S., and Gilman, S. D., Experimental studies of electroosmotic flow dynamics in microfabricated devices during current monitoring experiments, *Anal. Chem.*, 75 (2003) 361–370.
23. Ross, D., Johnson, T. J., and Locascio, L. E., Imaging of electroosmotic flow in plastic microchannels, *Anal. Chem.*, 73 (2001) 2509–2515.
24. Chen, L., Ren, J., Bi, R., and Chen, D., Ultraviolet sealing and poly(dimethylacrylamide) modification for poly(dimethylsiloxane)/glass microchips, *Electrophoresis*, 25 (2004) 914–921.

53 Microfluidic Devices with Mass Spectrometry Detection

Iulia M. Lazar

CONTENTS

53.1 INTRODUCTION

The development of microfabricated devices for analytical applications has been pioneered by Manz, Harrison, Widmer, Jacobson, and Ramsey in the early 1990s.[1–9] Since then, significant efforts have been invested to advance the theoretical foundations and the application areas of microfluidics. Under the generic name of "Lab-on-a-Chip" (LOC) or "Micrototal Analysis Systems" (μTAS), microfabricated devices have emerged as powerful and reliable analysis platforms. Key application areas include DNA sequencing/sizing, polymerase chain reaction (PCR) amplification, SNPs Single Nucleotide Polymorphism scoring, one- and two-dimensional (1D/2D) separations, sample dispensing, immunoassays, proteomics, biomedical applications/diagnostics, and biodefense.[10–20] In parallel to microfluidics, mass spectrometry (MS) has evolved into an irreplaceable detection tool for qualitative and quantitative explorations of biological samples. The implementation of novel MS detection strategies on microfluidic chips represents a powerful and attractive alternative to conventional optical detection systems. However, the design of effective interfaces between microfabricated devices and MS instrumentation is not a straightforward process. This chapter is aimed at documenting the challenges and accomplishments related to interfacing microfluidic systems to MS detection, with emphasis on microfluidic capillary electrophoresis (CE) chips. Theoretical considerations are reviewed, and practical aspects of chip-MS interface design, performance, and limitations are described. Relevant applications that demonstrate the power of microfluidic-MS strategies for bioanalytical applications are discussed.

53.2 BACKGROUND

The implementation of the "LOC" concept for bioanalytical and biomedical applications was supported by quality characteristics that are unique to microfabricated devices. Miniaturization enables the development of distinct instrument configurations and novel analytical principles that can significantly reduce analysis times. The miniature format reduces adsorption-related sample losses on the instrumentation surface and is thus ideal for processing small sample amounts. Microfluidic handling operations can be performed accurately, in an automated fashion, and thus facilitate the generation of high-quality data. Microfabrication enables analytical process integration and multiplexing, and consequently high-speed/high-throughput sample analysis. In addition, large-scale parallelization results in the fabrication of inexpensive and disposable platforms that prevent sample contamination and carryover. Ultimately, the microdomain environment enables the emergence of unique physical events. For example, as the size of a device decreases, the surface-to-volume ratio increases; thus, phenomena that are dependent on the largest power of the critical dimension lose their significance, while phenomena that are less depend on size become prevalent.[21] Consequently, surface-driven phenomena are dominating in the microscale world. The use of electroosmotic flow (EOF) for fluidic propulsion on the chip represents a relevant example as it can be effectively generated only in micrometer-sized channels. Major landmarks that demonstrate the power of microfluidic CE platforms include the ability to handle sample volumes as small as 1–10 pL,[4,7] the capability to perform sample manipulations (injection, separation, labeling, and detection) within a few seconds,[8] or ultra high-speed separations within < 1 ms,[22] and the ability to perform high-throughput experiments on multiplexed devices that comprise as many as 384–768 processing lanes.[23,24] For example, the experiment performed by Emrich et al.,[23] that is, DNA sequencing on microfluidic substrates incorporating 384 CE processing lanes that enabled simultaneous genotyping within 325 s, represents a remarkable example of a high-performance microfluidic application that simultaneously demonstrates complexity, speed, and throughput.

Microfluidic platforms comprise a variety of functional elements that perform operations such as pumping, valving, dispensing, labeling, mixing, separation, and detection. Fluidic propulsion is most commonly accomplished by using electrically and pressure-driven forces, especially when sample separations are performed.[13,14] Microchip devices are typically fabricated in glass, quartz, silicon, or polymeric substrates (poly(dimethylsiloxane), PDMS; poly(methylmethacrylate), PMMA; polyimide; polystyrene; polycarbonate; cellulose acetate; poly(ethyleneterephthalate); etc.). The functionality of the chip, the physical and chemical surface properties, the ease of fabrication, and the price are some of the dominant factors that determine the choice of a particular material for chip fabrication.

Microfluidic devices that use electrical fields for fluidic propulsion perform successfully for a variety of electrically driven separations, and thus, present unique opportunities to the analytical and biomedical community. The availability of a broad range of microfluidic functional elements, materials, and processes facilitated the miniaturization of most types of electrically driven separations that use optical and electrochemical detection systems: CE,[3–8] capillary electrochromatography (CEC),[9,25–32] micellar electrokinetic chromatography (MEKC),[33,34] isoelectric focusing (IEF),[35–51] isotachophoresis (ITP),[52–55] and capillary gel electrophoresis (CGE).[56–59] Moreover, a variety of microfluidic 2D separations have been developed, as well, such as MEKC-CE,[60,61] CEC-CE,[62] sodium dodecyl sulfate (SDS)-CGE/MEKC,[63] CGE/temperature gradient gel electrophoresis (TGGE),[64] IEF-CGE,[65,66] and ITP-zone electrophoresis.[67] The recent advent of bioanalytical processes implemented on the chip underline a clear demand for advancing MS detection for high-throughput microfluidic applications. MS combines the benefits of specificity, sensitivity, and resolving power that ultimately result in unambiguous detection capabilities for trace level components. Alternatively, the magnitude of most large-scale proteomics experiments underscore the need for multiplexed and low-cost devices that are capable of fast, sensitive, and reliable analysis. Overall, the data acquired with microfabricated MS devices demonstrate sensitivity, throughput, flexibility, and the great potential of these chips for bioanalytical applications. This chapter is aimed at providing the reader with an update on the status of microfluidic devices with MS detection, particularly on microfluidic CE-MS devices.

53.3 THEORETICAL CONSIDERATIONS FOR DESIGNING MICROCHIP-MASS SPECTROMETRY INTERFACES

53.3.1 CAPILLARY ELECTROPHORESIS RELATED CONSIDERATIONS

Chip Design. Essentially, all theoretical aspects that affect the performance of conventional CE separation techniques also must be considered in the design of microfluidic CE experiments. The chip-MS interface facilitates the removal of the sample from the chip for detection and must be configured adequately to not alter the separation performance. This can be a challenging task, especially for fast CE separations. Typical CE microchips comprise 2–20 cm long × 10–20 μm deep separation channels that operate at field strength of ~300–2000 V/cm. Sample separation can be accomplished within a few seconds or even milliseconds. Two main factors enable the capability to perform high-speed, high-resolution separations on the chip. First, the microfabricated layout enables designs that facilitate the injection of extremely short sample plugs, an essential requirement for high-efficiency CE separations ($\sigma_{\text{inj}}^2 = l_{\text{inj}}^2/12$, where σ represents the contribution of the injection process to peak variance and l represents the injection plug length). Thus, the injection plug length (20–100 μm) can be easily maintained at < 0.1–0.2%[68] of the separation channel length (2–20 cm). Second, the chip itself acts as a good heat sink because of its large mass, and therefore enables the application of high electric field strengths, efficient dissipation of Joule heat, and the generation of a large number of plate counts ($N = \mu V/2D$, where N is the number of theoretical plates, μ is the electrophoretic mobility, V is the voltage, and D is the diffusion coefficient). For example, for a glass microfluidic chip that comprises a 20 μm deep × 25 mm long separation channel, the peak width (s) and volume (nL) of an analyte eluting after 20 s can be anywhere in the range of ~0.3–1 s and 0.3–1 nL, respectively, if efficiencies of 100,000 to 10,000 plates/chip are desired (see theoretical considerations in Reference 68). Consequently, to minimize the detector contribution to peak variance and maintain losses in separation efficiency below 5–10%, the microchip-MS interface related dead volumes should be in the low picoliter volume range (30–100 pL).

Electrical Field Strength. The electrical field strength within each microchannel must be set properly to ensure that the EOFs are balanced in the entire microfluidic network. Channel depths and widths should be maintained small to minimize the effect of undesired parabolic flow profiles that could result due to potential nonuniform EOF in the chip channels. An unbalanced EOF during injection can significantly broaden the initial sample peak width and deteriorate efficiency.

CE Buffer Systems and Channel Coatings. The solution pH and the buffer additives that are used to improve the CE performance, alter the analyte mobility, modify the EOF, reduce analyte-chip interactions, or increase sample solubility must be selected carefully to maintain compatibility with the MS ionization source (see also Section 53.3.2). Ultimately, the choice of a buffer system represents a compromise: (1) To increase the performance, on one hand, CE separations are typically carried out in high-concentration buffers systems (10–200 mM), or surface derivatizations are performed to either alter the EOFs or to suppress the analyte adsorption. To enable MS detection, on the other hand, low concentration (< 20 mM), volatile buffer systems that are compatible with commonly used MS ion sources are preferred. However, such buffer systems can compromise the separation performance. (2) Microfluidic applications are preferentially performed in homogeneous, high pH buffer compositions that promote large EOFs and enable reproducible fluidic manipulations. Not all CE separations, however, can be performed at high pH. (3) Electrospray ion sources operate in a stable manner only in the presence of organic modifiers in the eluent (minimum 10–20%, v/v, CH_3OH or CH_3CN). The addition of organic components to the CE eluents will result in a decrease of EOF. (4) High concentration, nonvolatile detergents and polymeric additives such as the ones used in MEKC, IEF, and ITP separations quickly contaminate the MS ion source and reduce the sensitivity, or even completely suppress the analyte signal. The presence of these additives in the CE eluent considerably limits the practical implementation of these separation techniques with MS detection. (5) Microchip channel coatings must be stable under given experimental conditions and should not bleed out from the channel to generate background ions in the mass spectrometer. Unstable coatings will also affect the EOF magnitude and potentially the stability of the spray. Coating lifetime can represent an issue if the chips are intended for prolonged use but is less of a concern in the case of disposable chips.

CEC Considerations. Complex sample-eluent-stationary phase interactions within a CEC microdevice create additional challenges that interfere with the injection and detection process.[32] The CEC stationary phase must generate EOF and provide for analyte retention. Thus, both charged and hydrophobic moieties must be incorporated in its structure. The charge on the stationary phase (+/–) and the eluent pH will affect the direction of the EOF, the migration of the analytes, and the ionization efficiency. The analyte will interact with the stationary phase through electrostatic and hydrophobic interactions, and its migration through the separation channel will be determined by its electrophoretic mobility and the direction of the EOF. These factors must all be considered and carefully balanced to create optimal conditions for sample separation, for analyte ionization, and for stable EOF generation in the direction of the MS detector.

53.3.2 Mass Spectrometry Detection Related Considerations

Analyzer. The most common mass analyzers used in bioanalysis include ion trap (IT), time-of-flight (TOF), quadrupole (Q), Fourier transform ion cyclotron resonance (FTICR), and hybrid Q-TOF and IT-FTICR instruments. A specific mass analyzer should be chosen according to the sensitivity, mass accuracy, resolution, dynamic range, and data acquisition speed that satisfy the needs of a particular application. Most analyzers today are sensitive enough to enable the detection of low fmol or sub-fmol amounts of sample. Certain applications, such as proteomics, necessitate high mass accuracy, high-resolution characteristics such as displayed by the TOFMS and FTICR-MS detectors, to enable reliable sample identification in complex mixtures. The mass accuracy and resolution of TOF analyzers are in the 5–10 ppm and 10,000–30,000 range, and of the FTICR analyzers in the <1–3 ppm and 100,000–1,000,000 range, respectively. The TOF analyzers present particular advantages for high-speed separations, such as the ones produced on a microfluidic chip, due to the nonscanning characteristics that enable a fast data acquisition process (data generation/storage at > 10 full mass spectra/s). Ion trap instruments that are capable of accumulating ions over prolonged times enable sensitive analysis and MS^n explorations. Quantitation over a broad concentration range can be most reliably accomplished with analyzers that use detectors with a large dynamic range. Slow, scanning mass spectrometers such as the quadrupole instruments provide good performance for large dynamic range quantitation.[69–71]

Ion Sources/Chip-MS Interfaces/Ionization Efficiency. Electrospray ionization (ESI), matrix-assisted laser desorption ionization (MALDI), and atmospheric pressure chemical ionization (APCI) are the most commonly used ionization sources in bioanalysis and have also been implemented on the chip. ESI is a method used to bring ions from the liquid phase to the gas phase (at atmospheric pressure) by a series of processes that involve production of charged droplets at the tip of an ESI capillary, shrinkage of the charged droplets by solvent evaporation and repeated multiple disintegrations, and production of gas-phase ions from very small and highly charged droplets. ESI is a soft ionization method that results in single- and multiple-charged ions by attaching or removing cations (H^+, Na^+, K^+, or NH_4^+, etc.) to and from the analyte molecules. MALDI is an ionization method that involves mixing the sample solution with a volatile matrix (sample/matrix 1:1000), deposition on a target plate, irradiation with a laser pulse, sublimation/expansion of the sample/matrix in the gas phase, and sample ionization by gas-phase proton transfer with the photoionized matrix molecules in the expanding plume. Note that it is the MALDI matrix that absorbs energy at the wavelength of the laser and then transfers it to the analyte molecules. Thus, this process results in a soft ionization mechanism that maintains the analyte molecules intact. APCI produces ions through collisions between the molecules to be analyzed and ionic species already existent in the source (reagent gas) through a variety of mechanisms: proton transfer, hydride abstraction, adduct formation, charge transfer, and so forth.

The chip-MS interface must (1) facilitate efficient sample ionization and transfer from the liquid into the gas phase, (2) ensure that the contribution of the detection process to the peak variance is minimal, and (3) enable easy accommodation of the chip within the mass spectrometer ion source. ESI has found widespread utility especially for the ionization of large biological molecules, as multiple charging brings the analyte *m/z* ratios in the affordable range of most mass spectrometers. Stable and efficient electrospray generation is determined by a number of factors such as sample related parameters (buffer composition, ionic strength, eluent pH, surface tension, viscosity, dielectric strength, and flow rate), ESI emitter size (5–100 μm i.d./10–300 μm o.d.), distance between the ESI emitter and a counter electrode (3–20 mm), and voltage applied to the ESI tip (~700–3000 V).

Low concentration (<20 mM), volatile buffer systems (ammonium bicarbonate or ammonia for high pH and ammonium acetate, acetic acid, or formic acid for low pH separations) in aqueous-organic (H_2O–CH_3OH, H_2O–CH_3CN) eluents of low viscosity and low surface tension are generally recommended for CE-MS applications. The CE eluent pH will affect not only the EOF magnitude and the electrophoretic mobility of the analytes during separation but also the sample protonation/deprotonation equilibria and the ESI efficiency. For example, low pH values will favor the generation of positive ions in the liquid and gas phase, and multiple ion charging. However, the relationship between the concentration of ions present in the solution and those observed in the gas phase is not always straightforward; components with large gas-phase basicity will tend to dominate in the gas phase. Moreover, as there is a competition among ions to occupy the surface of the electrosprayed droplets, in aqueous solutions, the most hydrophobic ions will be likely enriched on the surface of the droplet and will generate more intense signals.

Nano- and micro-ESI sources that operate in the ~50–300 nL/min flow range are an ideal match for typical EOF values generated on the chip (i.e., <300–400 nL/min). The dead volumes associated with nano-ESI sources are minimal. In addition, nano-ESI sources have reasonable tolerance toward buffer composition and concentration, and facilitate sensitive analysis of trace level components (even of unseparated mixtures). Moreover, due to low analyte consumption characteristics, nano-ESI analysis enables extended infusion-MS investigations, increased signal averaging capabilities, and thus the generation of improved quality spectra. However, when high-concentration buffer systems are needed for a CE separation, liquid sheath or liquid junction ESI sources must be used. CE eluent dilution with the liquid sheath or liquid junction solution will inherently result in lower signal intensities, increased dead volumes and band broadening, and ultimately in worse detection limits (10–100 times deterioration is possible).

When the insertion of fine ESI capillary emitters in the chip is contemplated, an important factor to consider is that the small i.d. emitter may act as a flow restrictor on the CE channel, and will result in pressure build-up at the channel terminus. This pressure will generate a laminar backflow through the CE channel that will counteract the EOF. Depending on the microfluidic channel dimensions, and on the length and i.d. of the ESI emitter, the EOF could be suppressed to a level where stable electrospray operation cannot be anymore achieved. For example, a 20 μm i.d $\times$ ~10 mm long ESI emitter was capable of suppressing the EOF generated in a 20 μm deep microchip channel to ~25–40% of its original value.[32]

In parallel with ESI, MALDI has also become a widely used ionization technique for the analysis of biological samples. It is considered a high-throughput MS approach that enables fast, sequential screening of peptide or protein samples deposited into spots on a MALDI plate. In particular, the technique has gained widespread acceptance for the fast analysis of 2D-gel separated protein tryptic digests. As MALDI-MS analysis is performed offline, after the separation has completed, the major concerns related to chip-MALDI-MS interfacing focus on the preservation of separation efficiency during the sample deposition process on a MALDI target plate. While conventional MALDI-MS is performed under high vacuum using TOF analyzers, the recent development of atmospheric/subatmospheric pressure ion sources has enabled the implementation of MALDI sources on a variety of mass analyzers that perform low-energy collision-induced dissociation (CID) for analyte fragmentation and structure elucidation, such as the ion trap and quadrupole. Moreover, recently developed MALDI-TOF/TOF instruments enable analyte fragmentation under high-energy conditions, and the generation of more complex structural information. Thus, the successful merger of a high-throughput analytical sample processing platform, such as a microfluidic chip, and a high-throughput detection strategy, such as a MALDI-MS, will clearly benefit a variety of applications.

Detection Limits. The detection limit of a mass spectrometer is affected by various parameters that include sample complexity, sample ionization efficiency, type of ion source, chip-MS interface design, type of analyzer, scanning or nonscanning operation, data acquisition conditions, speed of analysis, and so forth. The mass spectrometer is a mass sensitive detector that responds to the amount of sample that is reaching the detector. Mass spectra can be summed or averaged over prolonged times to generate a mass spectrum with an acceptable signal/noise (S/N) ratio. The detection limit of a mass spectrometer, per se, can be in the low amol, or even zmol range. For example, the detection of 300–500 zmol of gramicidin S from a 100 nM solution, during an acquisition time of 10 ms with a TOFMS detector, was demonstrated in a chip-MS infusion experiment.[72] A scanning instrument such as an ion trap would have generated similar results in a larger time frame (i.e., 100 ms^{-1} s). From lower concentration solutions, a similar quality spectrum could have been generated by both instruments by prolonged mass spectral averaging.

For practical purposes, it is useful to discuss the detection limit taking into consideration the necessary time to acquire a full mass spectrum. Thus, using a simple nano-ESI infusion experimental setup, detection limits in the range of 10–100 amol of sample delivered to a mass spectrometer during a 1 s acquisition time window, which is sufficient for most instruments to acquire a full mass spectrum, should be relatively easily attainable on a well-tuned instrument. In the context of high-speed separations and microfluidic CE experiments, a 1 s time window is also commensurate with typical peak widths. This is, however, an ideal scenario. One must always distinguish between the amount of sample that is consumed to produce a mass spectrum at a certain S/N ratio, and the amount of sample that is injected into a separation column and that elutes in a few seconds wide peak.

As a result, for the analysis of real world samples that involve the separation of complex mixtures, the possible generation of broad peaks, and the use of different data acquisition conditions, detection limits can be orders of magnitudes worse, i.e., 10–1000 fmol of sample must be loaded on the separation column.

Depending on the objectives of the analysis, the mass spectrometer will be operated in various modes of data acquisition such as MS, tandem MS or MS/MS, selected ion monitoring (SIM), multiple reaction monitoring (MRM), full-scan or limited mass range, data dependent, and so forth. SIM/MRM conditions will improve the detection limits, while data-dependent MS/MS acquisitions will worsen them. The data acquisition/storage speed and the number of averaged spectrum are all factors that will ultimately affect the detection limits.

Dynamic Range/Quantitation. The dynamic range is the ratio of the largest to the smallest detectable signal by a mass spectrometer. The dynamic range of a mass spectrometer can reach values as high as 10^5–10^6, however, the concept is much more complicated. The dynamic range depends on several factors such as the type of the mass spectrometer, the detecting elements (electron multiplier, multichannel plate, and analog versus time-to-digital signal processing), the dynamic range of the ESI source, and so forth. For intraspectrum measurements, the MS dynamic range can reach values of $\sim 10^4$. For interspectrum/overall analysis, slow scanning mass spectrometers such as the quadrupole instruments that utilize electron multipliers and slow, but high dynamic range (16–20 bit) analog-to-digital converters (ADC) provide good performance for quantitation, that is, have a dynamic range of $\sim 10^6$.[69–71] On the other hand, fast, nonscanning TOFMS instruments that utilize electron multipliers or multichannel plates, and high sampling rate, but low dynamic range (8 bit) ADCs or time-to-digital converters, have a low dynamic range and thus require extended spectral summing to improve it.

Speed of Analysis/Data Acquisition/Quantitation. The operational conditions of a microchip-CE system result in fast separations that can be completed within several seconds, and peak widths that can be much less than 1 s. For qualitative purposes, the acquisition of one high-quality mass spectrum per peak could be sufficient for reliable sample identification. However, for quantitative purposes, the number of data acquisition points (i.e., mass spectra) per peak should be >10. Fast, nonscanning TOFMS instruments are particularly useful for such separations. TOF systems that generate 5,000–10,000 spectra/s and store 100–10 spectra/s have the capability to interrogate a narrow peak a sufficient number of times (i.e., 10–20 times) to enable accurate quantitation. In addition, TOF systems record all ions at all times (no loss of information), and enable massive data summing/averaging to improve the detection limits. Tandem MS investigations remain, however, a challenge. Ultimately, for fast separations and complex mixtures, there will be a trade between the data acquisition speed and the achievable detection limits. One should aim for the highest possible separation efficiency that can be sampled appropriately by an MS detector.

Quantitation/Sample Complexity. Complex samples such as cellular extracts and plasma contain proteins that span over abundance ranges of $\sim 10^9$–10^{10}, a much broader range than the typical dynamic range of a mass spectrometer. Low abundant components are generally not detected in such mixtures. In addition, data-dependent acquisition methods will select the most intense peaks in a spectrum for tandem MS investigations and will contribute even more to the loss of low abundant components. Selecting data acquisition parameters that increase the acquisition speed of the mass spectrometer, without significantly compromising spectral quality, could benefit such types of analyses. Prefractionation, affinity enrichment, and removal of the most abundant species are just a few strategies that could help improve the quantitation of such samples.

High-throughput Analysis in the Context of MS Detection. Microfabricated devices are ideal analytical platforms for enabling multiplexed sample processing and high-throughput analysis. Nevertheless, it could be argued that these benefits are not fully exploitable in the context of sequential MS detection. The use of fully integrated and possibly disposable platforms will reduce, however, the entire sample preparation time and effort. On the long run, MALDI-MS, which is already regarded as a high-throughput MS detection approach, and miniaturized and multiplexed mass spectrometers that are currently under development will represent a good match for multiplexed microfabricated devices. Ultimately, it is envisioned that parallelization and large-scale integration will result in the fabrication of inexpensive, disposable platforms that will be easily interfaced to a variety of mass spectrometers.

53.4 MICROCHIP-MASS SPECTROMETER INTERFACES

Microfabricated devices have been interfaced to a variety of mass spectrometers, and a number of bioanalytical applications have been demonstrated.[73–82] Microfluidic platforms typically include the following sample processing steps: infusion, cleanup, preconcentration, digestion, separation, and detection. However, in the context of MS detection, the efforts focused on developing adequate microchip-MS interfaces that enable effective sample ionization from the chip. Since tandem MS capabilities were available until recently only with ESI mass spectrometers, initially, microfluidic ESI sources were developed. Later, technologies that enabled high-throughput laser desorption ionization strategies, in particular MALDI-MS, have been also pursued.

53.4.1 Electrospray Ionization Interfaces

Electrospray ionization from microfluidic devices can be generated directly from the chip surface, from capillary emitters inserted in the chip, or from microfabricated emitters. Liquid sheath, liquid junction, and nano-ESI sources have been implemented on glass, polymeric, or silicon substrates.

A common challenge associated with all microfluidic-MS chips that rely on electrically driven fluidic manipulations is related to the fact that fluidic flows (i.e., EOFs) follow the direction of the electrical potential gradients. In the case of CE/CEC separations, this flow is oriented toward the electrode placed at the separation channel terminus. Thus, the design of the microchip-MS interface must include a functional element that will effectively suppress the eluent flow to the terminal electrode, and will provide for an effective strategy for redirecting the fluid flow to the ESI emitter. This can be a particularly demanding task in the case of CE/CEC microfluidic chips that require very low dead-volume interfaces to not compromise separation efficiency. Several strategies have been used to stabilize the ESI fluid flows: (1) the chip was pressurized by connecting a N_2 cylinder or syringe pump to the sample infusion channel; (2) an external capillary EOF pumping system or a subatmospheric ESI source were connected to the chip to pull the eluent through the CE separation channel; (3) a miniaturized pneumatic nebulizer was fabricated as an integral part of the chip; (4) the ESI emitter was coated with gold and served as a terminal electrode for CE closure and for simultaneous ESI generation; (5) on-chip electroosmotic, electrochemical, and piezo-actuated pumping systems were used to pressurize the fluids on the chip; (6) coated microfluidic channels and porous glass membranes (gates) have been used to suppress the EOF in the direction of the terminal electrode and redirect it to the ESI tip; and (7) the nanospray process itself was used to draw a small fluid flow through the chip. The efforts invested in the development of these microfluidic chip-MS interfaces are reviewed in the following sections of this manuscript.

53.4.1.1 Electrospray Ionization from the Chip Edge

The first attempts to generate ESI from the chip were performed by electrospraying the sample solution directly from the channel terminus at the chip edge (Figure 53.1a).[83–85] Two major, somewhat interrelated, problems were associated with this approach, that is, the electrospray Taylor cone-related dead volumes and the spray instability. Ramsey and Ramsey[85] have estimated that the volume of the Taylor cone at the chip edge is $\sim$12 nL, that is, much broader than are the typical CE peak volumes. The flat edge was not ideal for the establishment of a strong electrical field even if

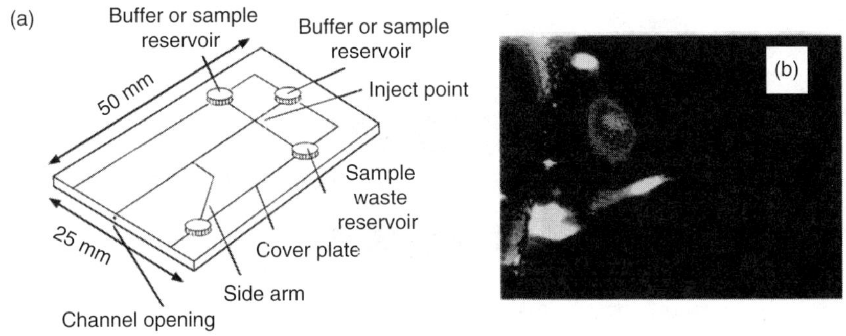

FIGURE 53.1 Electrospray generation from the chip flat edge. (a) Diagram of a microfluidic chip with a coated side arm for EOF reversal. (b) Taylor cone and electrospray generated at the channel opening at 3 kV applied between the microchip and a target electrode. (Reprinted from Ramsey, R.S., Ramsey, J.M., *Anal. Chem.*, 69, 1174–1178, 1997. Copyright 1997. With permission from American Chemical Society.)

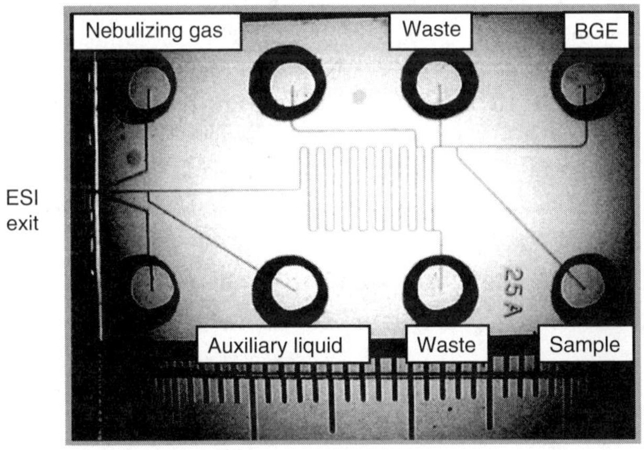

FIGURE 53.2 Microfabricated device with an integrated pneumatic nebulizer. (Reprinted from Zhang, B., et al., *Anal. Chem.*, 71, 3258–3264, 1999. Copyright 1999. With permission from American Chemical Society.)

very large voltages (>3000 V) were applied to the chip, and thus did not facilitate stable and efficient spray generation but rather the build up of a fairly large liquid bubble (Figure 53.1b). Treating the chip surface around the channel exit with a hydrophobic silanizing solution,[83] or using microchips fabricated from polymeric, dielectric, nonwetting materials,[86–88] represent potential solutions that could minimize the growth and spread of the liquid at the channel terminus, minimize dead volumes, and improve spray stability and detection limits. As an alternative, polytetrafluorethylene hydrophobic membranes (50 μm thick, 70% porosity, 0.22 μm average pore size) were used to thermally "seal" the open channels (120 μm × 80 μm × 30 μm) of a polycarbonate chip, and to generate stable spray through the fine membrane pores with minimal lateral dispersion of the Taylor cone.[89]

Samples were typically infused through such devices either by using external syringe pumping[83] or on-chip electroosmotic pumping.[85] In the latter case, the EOF in the main channel was redirected to the chip edge for ESI generation by coating the side-arm channel with a linear polyacrylamide solution for EOF reversal.[85] Alternatively, miniaturized pneumatic nebulizers were integrated on the chip to enable sample infusion and/or CE separations with MS detection (Figure 53.2).[90,91] Common applications of these microfluidic-MS designs included infusion experiments performed from single and multiple channel devices. Relatively good quality mass spectra were produced by infusing 0.1–10 μM peptide/protein solutions.

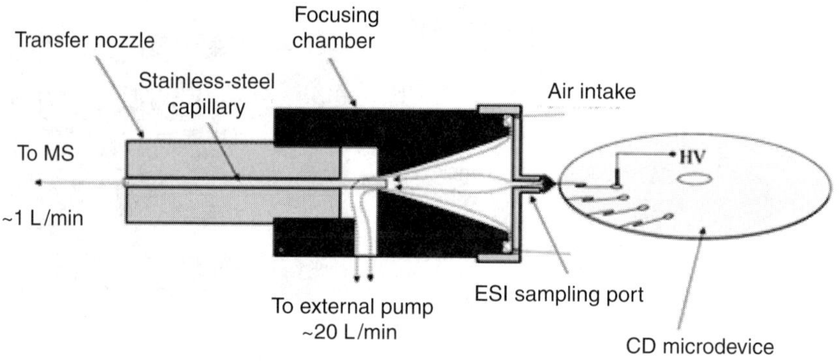

FIGURE 53.3 A microchip-MS interface with a nebulizer that includes an aerodynamic focusing chamber for directing the ESI plume into the mass spectrometer sampling orifice. (Reprinted from Grym, J., et al., *Lab Chip*, 6, 1306–1314, 2006. Copyright 2006. With permission from The Royal Society of Chemistry.)

Recently, a more complex, aerodynamic interface was designed to enable electrospray generation from channel openings at the edge of a polycarbonate multichannel compact disk chip (Figure 53.3).[92] The ion source of an ESI mass spectrometer was fitted with an adapter that assisted the formation and transport of the electrosprayed plume from the microchip device. Electrospray was generated by applying a voltage between the sample wells on the chip and the entrance of the adapter. The flow through the chip (~350 nL/min) and the spray were stabilized by air suction provided by a pump connected to the adapter. The interface was tested with 1–20 μM solutions of peptides and proteins.

53.4.1.2 Electrospray Generation from Capillary Emitters Inserted in the Chip (Sheathless Nano- or Microelectrospray)

Stable and efficient ESI, and thus sensitive MS detection, is most easily accomplished from very sharp emitters (2–10 μm i.d. × 5–20 μm o.d.) fabricated from pulled fused-silica capillaries, that operate at an optimum flow rate of 20–50 nL/min and relatively low ESI on-set voltage (800–1800 V). These are the typical nano-ESI emitters that enable the much desired amol and zmol detection limits. To achieve similar performance, such emitters have been inserted in the chip either through the edge of the chip, directly in the channel terminus,[93–102] or perpendicular to the chip, through an orifice that was communicating with one of the channels on the chip.[72,103] Flow generation and stabilization in these chips was accomplished by various approaches.

Similar to the flat edge designs, simple chips with capillary emitters were used for infusion experiments and their operation relied on external N_2 pressurization[104] or syringe pumping.[102,105] An external EOF pumping approach, implemented by Figeys et al.,[106–109] used an amino-propyl silane activated fused-silica capillary transfer line inserted in the chip for fluidic propulsion. The capillary acted as an external electroosmotic pump to draw the eluent through the chip and transport it to a conventional electrospray source (Figure 53.4). The voltage applied to the capillary terminus that was placed in the ion source was used to promote EOF between the chip and the MS source, and to initiate the ESI process. This approach was demonstrated for infusion experiments and peptide fingerprinting from 3 to 300 nM solutions using single and multichannel devices.

Successful interfacing of microchip-CE to MS was also accomplished by inserting gold (or other conductive material) coated emitters in the chip to simultaneously serve as a CE terminal electrode and an ESI electrode. The ESI emitter actually acted as an extension of the CE separation channel (Figure 53.5).[93] These chips comprised a side channel placed close to the CE channel terminus to enable the infusion (with a syringe pump) of an MS calibrant or CE buffer dilution solution. Gold-coating lifetime and need for a fine fluid flow balance between the chip-EOF and syringe-infused flow are two possible concerns related to this approach. One solution to combining

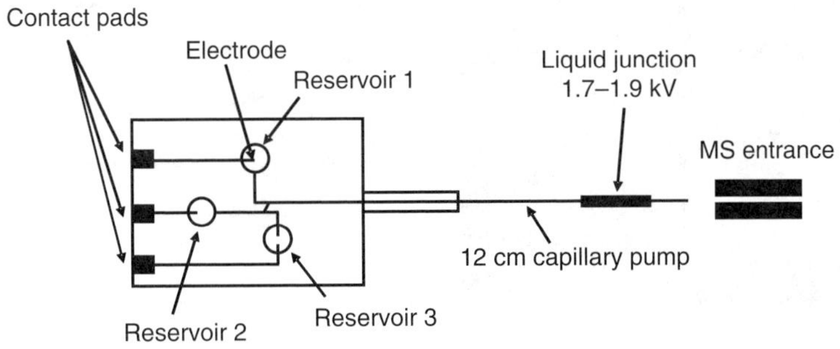

FIGURE 53.4 Diagram of a microfabricated device for MS infusion with an inserted capillary that acts as an electroosmotic pump and transfer line between the chip and ESI interface. (Reprinted from Figeys, D., et al., *Anal. Chem.*, 69, 3153–3160, 1997. Copyright 1997. With permission from American Chemical Society.)

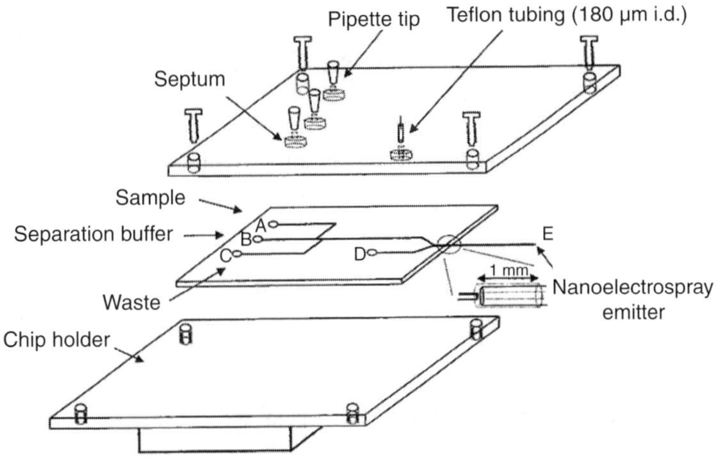

FIGURE 53.5 Schematic diagram of a microchip configuration for CE separation and MS analysis using a disposable nano-ESI capillary emitter. (Reproduced from Li, J., et al., *Anal. Chem.*, 72, 599–609, 2000. Copyright 2000. With permission from American Chemical Society.)

electrokinetic and hydrodynamic flow streams effectively was provided by Razunguzwa et al.,[110]—a hydrodynamic flow restrictor (a multichannel functional element) with increased hydraulic resistance was introduced between the CE channel terminus and the side channel intersection to prevent hydrodynamic backflow through the CE channel. Provided, however, there is sufficient EOF generated in the chip and that the CE buffer is electrosprayable, no dilution solution must be provided to the CE system.[111]

A few alternative strategies have been used to achieve standalone operation of the microchip-ESI interface without external gas or fluid flows supplied to the chip. In the case of microchips equipped with fine nanospray emitters (5 μm i.d.), the electrospray process generated at the emitter tip was capable of sustaining a small, but stable, flow through the microfluidic channel ($\sim$20–30 nL/min).[72] Full mass spectra acquired with a TOFMS at 50–100 spectra/s storage rates enabled the detection of <500 zmol amounts of peptides and proteins from 100 nM solutions. On-chip electroosmotic pumping for EOF generation and ESI stabilization was accomplished in the context of CE and CEC devices by connecting an auxiliary side channel to the main channel via an electrically permeable glass membrane (Figure 53.6).[32,97] The glass membrane had the purpose of enabling the transfer of ions for maintaining CE electrical continuity, but impeding the bulk EOF leakage from the CE channel into the reservoir. Using a similar glass membrane design, a multichannel electroosmotic

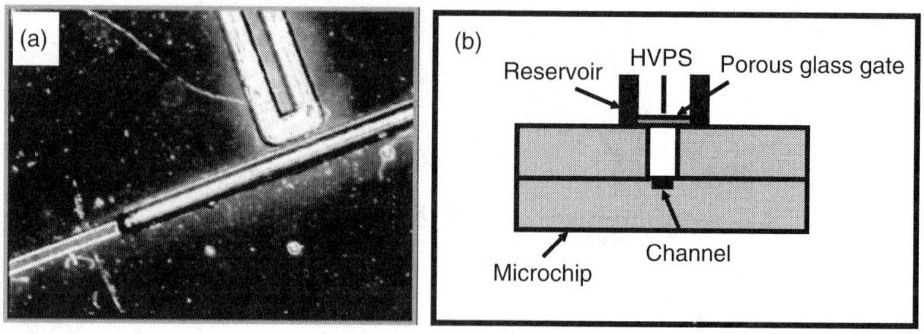

FIGURE 53.6 Electrically permeable glass membranes (porous glass gates). (a) Integrated on the chip, prepared by placing two channels in close proximity to each other (i.e., at ~10–20 μm apart). (Reprinted from Lazar, I.M., et al., *J. Chromatogr. A*, 892, 195–201, 2000. Copyright 2000. With permission from Elsevier.) (b) Exterior to the chip, prepared from a porous (~4 nm pore size) glass disc (4 mm diameter × 0.8 mm thick). (Reprinted from Lazar, I.M., et al., *Electrophoresis*, 24, 3655–3662, 2003. Copyright 2003. With permission from Wiley-VCH.)

pumping system was used to infuse sample or to deliver flow to an on-chip liquid chromatography (LC) system.[112,113] These pumping systems enabled the delivery of 50–400 nL/min flow to the ESI emitter. Alternatively, for similar purposes, the glass membrane was fabricated by etching through one of the chip substrates until generating a ~50 μm thick membrane.[114]

To maintain the separation efficiency and ensure maximum ionization efficiency, the EOF in the CE channel must closely match the optimum electrospray flow of a given ESI emitter. In addition, to minimize dead volumes, the ESI emitters must be precisely aligned with the CE separation channel—a task that can be easily accomplished by etching both chip substrates to provide for a capillary guiding channel.[32,90,97] Somewhat more difficult, but equally effective, channel guides can be drilled in the chip to match the ESI emitter o.d.[95]

53.4.1.3 Liquid Sheath or Liquid Junction ESI Interfaces

Microfluidic chips with inserted ESI emitters were also used in conjunction with liquid sheath and liquid junction interfaces. In one approach, the spraying capillary was inserted in a liquid sheath interface external to chip.[94] Alternatively, a chip-inserted capillary and an auxiliary channel connected to the CE channel terminus were used in a liquid junction interface setup.[96] The mass spectrometer pumping system itself was utilized to vacuum the ESI source and the chip, and to draw fluid flow through the microfluidic system and the chip-MS transfer line. The liquid junction was either an integral part (Figure 53.7)[96] or was external to the chip.[90] The chip-integrated liquid junction comprised a removable electrospray tip inserted in the subatmospheric electrospray interface.[96] By adjusting the pressure in the MS interface, adequate mixing of the CE buffer with the liquid junction solution, and stable sample delivery were accomplished.

Wachs and Deng[115–117] have demonstrated a free-standing liquid junction interface that enabled ESI from the flat edge of a CE microchip. An external pump was used to deliver a suitable solution to the liquid junction element (2–6 μL/min) for collecting and transporting the CE effluent through a pneumatically assisted electrospray needle (Figure 53.8). While liquid sheath and liquid junction interfaces will clearly benefit the implementation of microfluidic CE-MS separations that need high concentration and possibly nonvolatile buffer systems, the large dilution volumes that are added through the interface can significantly deteriorate the detection limits and will necessitate additional means for effective nebulization of the effluent flows.

53.4.1.4 Microfabricated ESI Emitters

As efficient ESI generation from the flat chip surface is difficult to accomplish and the insertion of capillary emitters in the chip is often an impractical process, the fabrication of chip-integrated

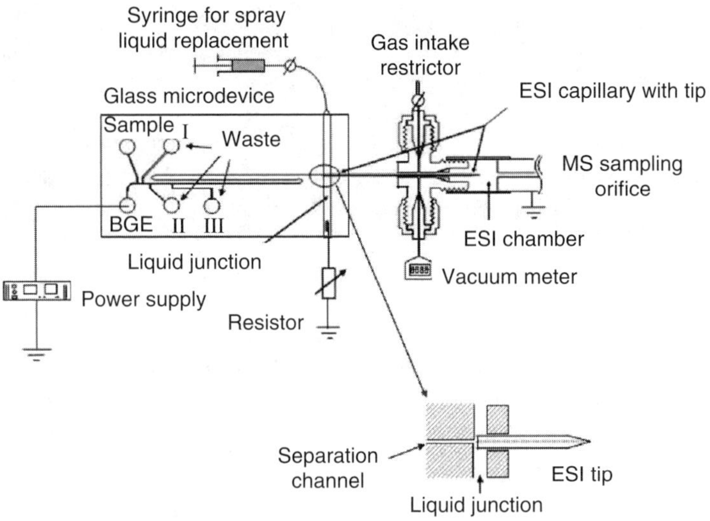

FIGURE 53.7 Microfluidic device with an integrated liquid junction interface and a subatmospheric ESI source for high-performance CE-MS analysis. (Reprinted from Zhang, B., et al., *Anal. Chem.*, 72, 1015–1022, 2000. Copyright 2000. With permission from American Chemical Society.)

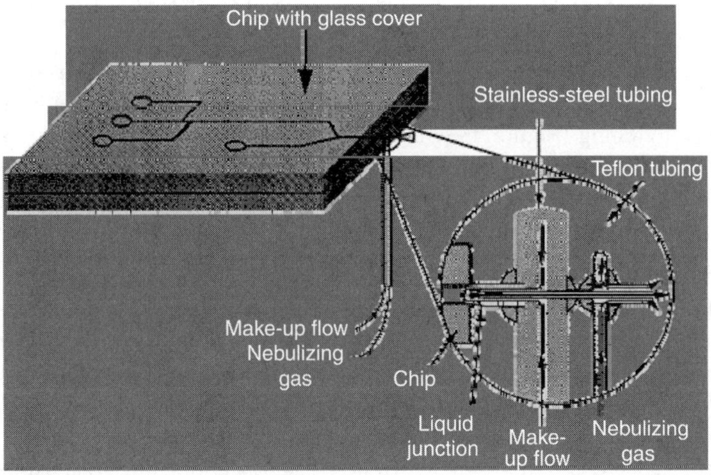

FIGURE 53.8 Schematic diagram of a chip-based CE/MS apparatus with an expanded view of a microliquid junction interface coupled to the exit of the microfluidic channel. (Reprinted from Deng, Y.Z., et al., *Anal. Chem.*, 73, 639–646, 2001. Copyright 2001. With permission from American Chemical Society.)

emitters was actively pursued in the past few years. Complete device integration is highly desirable as it will enable batch generation of multiplexed architectures and improved detection quality and reproducibility. The fabrication of ESI emitters integral to the chip is, however, not a trivial task. A variety of microfabrication techniques have been used to create ESI emitters in polymeric and silicon substrates. Unfortunately, these techniques are less amenable for the generation of high-aspect ratio structures in silica or glass. ESI tips from glass were, however, occasionally fabricated by mechanical machining.[114,118]

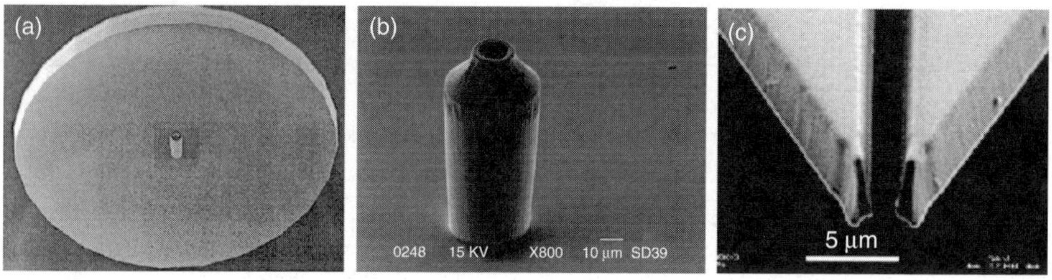

FIGURE 53.9 Silicon-based microfabricated ESI emitters. (a) Silicon emitter fabricated by deep reactive ion etching (10–15 μm orifice diameter). (Reprinted from Schultz, G.A., et al., *Anal. Chem.*, 72, 4058–4063, 2000. Copyright 2000. With permission from American Chemical Society.) (b) Silicon dioxide emitter fabricated by DRIE and oxidation of silicon (10–18 μm diameter orifice). (Reprinted from Griss, P., et al., *J. Micromech. Microeng.*, 12, 682–687, 2002. Copyright 2002. With permission from IOP Publishing Limited.) (c) Polysilicon emitter fabricated by low-pressure chemical vapor deposition, pattern transfer, reactive ion and sacrificial layer etching (1.8 μm × 2 μm and 2.5 μm × 5 μm slot). (Reprinted from Arscott, S., et al., *Sens. Actuat. B*, 106, 741–749, 2005. Copyright 2005. With permission from Elsevier.)

Silicon-Based Emitters. Deep reactive ion etching (DRIE) techniques were used to fabricate silicon ESI nozzles (10 μm i.d., 50 μm depth) with a performance similar to nano-ESI emitters (Figure 53.9a).[119] Later, multiplexed configurations (400 spraying nozzles) on planar silicon wafers were fabricated, as well. The detection of proteins from 10 nM solutions infused at flow rates of 100 nL/min and drugs and metabolites spiked at 2.5 ng/mL in human plasma was demonstrated with these silicon emitters.[119,120] Syringe pumps or high-performance liquid chromatography (HPLC) systems were typically used to infuse the sample through these silicon chips. Alternatively, arrays of tapered SiO_2 needles were prepared by batch silicon etching techniques and oxidation (Figure 53.9b).[121,122] Polysilicon-based emitters were fabricated from two planar, triangular, free-standing cantilevers, by using a combination of low-pressure chemical vapor deposition, pattern transfer, reactive ion etching and sacrificial layer etching (Figure 53.9c).[123] The polysilicon emitters extended 800 μm beyond the edge of the silicon substrate, comprised slot dimensions of 1.8 μm × 2 μm or 2.5 μm × 5 μm, and enabled MS detection from 1 μM peptide solutions by simply applying a small electrospray voltage (700 V) through the rear side of the silicon wafer. The very fine emitter enabled standalone operation with no need for external assistance for spray stabilization.

Polymeric Emitters. A variety of technologies have been pursued to fabricate polymeric ESI emitters and integrate them within polymeric, glass or silicon microfluidic chips. Overall, polymeric microdevices can be more easily and cost-effectively fabricated and thus represent attractive analysis platforms for commercialization purposes. An initial demonstration of polymeric ESI emitters was performed by fabricating hollow needle structures from parylene polymer layers deposited on a silicon substrate (tapered ESI tips with 5 × 10 μm rectangular openings that extend 1 mm beyond the edge of the chip substrate) (Figure 53.10a).[124] Peptide samples (9 μM) were infused by N_2 pressurization at low 35–77 nL/min flow rates. These emitters were later integrated within a microfluidic device that incorporated a LC system driven by on-chip electrochemical pumps.[125,126] Parylene ESI emitters were also fabricated by incorporating a 5 μm thick parylene C film between the substrate and cover plate of a chip prepared by embossing Zeonor 1020R polymer. The parylene film was prepared by lithography and etching into a triangular shape for effective ESI generation (Figure 53.10b).[127] Alternatively, adjustable-gap ESI emitters were fabricated from two triangular-shaped tips mounted in close proximity to each other on a PMMA substrate.[128] The tips were cut out from a 36 μm thick poly(ethyleneterephthalate) film and were hydrophilized by plasma deposition of SiO_2. The gap between the ESI tips was adjustable between 1 and 36 μm, to allow for sample solution pumping at 75–470 nL/min and detection of standard peptides from 0.1 μM solutions.

Polyimide, triangular-shaped ESI emitters prepared by lithography, chemical/plasma etching, lamination bonding, or knife-cutting techniques, were fabricated as an extension of chips made of

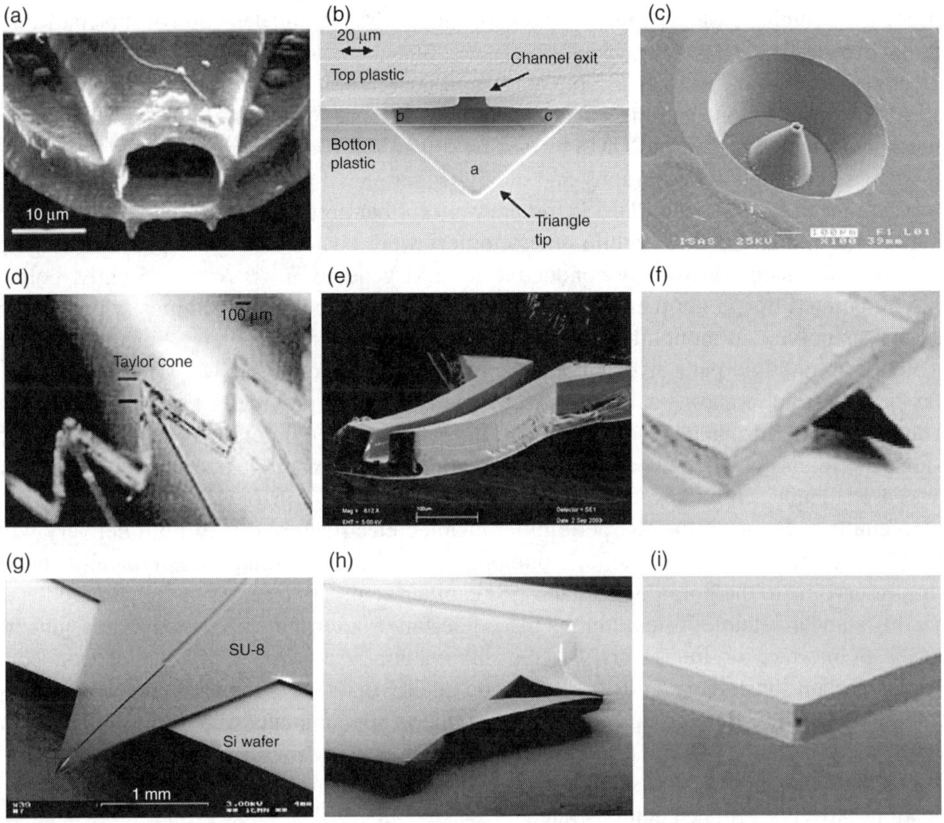

FIGURE 53.10 Polymer-based microfabricated ESI emitters. (a) Parylene emitter fabricated by lithography and wet and plasma etching techniques (5 μm ×10 μm opening). (Reprinted from Licklider, L., et al., *Anal. Chem.*, 72, 367–375, 2000. Copyright 2000. With permission from American Chemical Society.) (b) Parylene C emitter fabricated by lithography and etching (the parylene film lines up with a 10 × 20 μm channel). (Reprinted from Kameoka, J., et al., *Anal. Chem.*, 74, 5897–5901, 2002. Copyright 2002. With permission from American Chemical Society.) (c) PMMA emitter fabricated by micromilling (30 μm opening). (Reprinted from 137. Schilling, M., et al., *Lab Chip*, 4, 220–224, 2004. Copyright 2004. With permission from The Royal Society of Chemistry.) (d) PDMS emitter fabricated by casting and a three-layer photoresist process (30 μm × 50 μm channel opening). (Reprinted from Kim, J.-S., Knapp, D.R., *Electrophoresis*, 22, 3993–3999, 2001. Copyright 2001. With permission from Wiley-VCH.) (e) PDMS emitter fabricated by casting (50 μm channel opening). (Reprinted from Svedberg, M., et al., *Lab Chip*, 4, 322–327, 2004. Copyright 2004. With permission from The Royal Society of Chemistry.) (f) PDMS emitter fabricated by casting and graphite coating (50–60 μm channel opening). (Reprinted from Thorslund, S., et al., *Electrophoresis*, 26, 4674–468, 2005. Copyright 2005. With permission from Wiley-VCH.) (g) SU-8 emitter (8–16 μm slot width) fabricated by UV-photomasking, photoresist development and wafer dicing. (Reprinted from Le Gac, S., et al., *Electrophoresis*, 24, 3640–3647, 2003. Copyright 2003. With permission from Wiley-VCH.) (h) SU-8 emitter fabricated by 2.5D lithography (20 μm slot width). (Reprinted from Le Gac, S., et al., *J. Mass Spectrom.*, 38, 1259–1264, 2003. Copyright 2003. With permission from Wiley-VCH.) (i) SU-8 emitter fabricated by lithography using a three-layer microfabrication process (10 μm × 10 μm to 50 μm × 200 μm channel openings). (Reprinted from Tuomikoski, S., et al., *Electrophoresis*, 26, 4691–4702, 2005. Copyright 2005. With permission from Wiley-VCH.)

copper-coated polyimide printed circuit boards.[129–131] Experiments carried out with these chips relied on the flow that was induced by applying 1600–1800 V between the device and the mass spectrometer and enabled detection limits in the low 40 nM concentration range. Polyimide microfluidic chips that comprise sample enrichment modules, LC channels, and ESI emitters are presently commercialized devices (www.agilent.com) for HPLC applications.[131–133] When connected to bench-top LC pumping systems, these chips enable the separation and detection of sub-fmol amounts of protein digests.

Electrospray emitters were also fabricated from polycarbonate substrates by laser ablation;[134] from PMMA by injection molding,[135] atmospheric molding,[136] micromilling (Figure 53.10c),[137] and mechanical cutting;[138] from PDMS by casting (Figure 53.10d through 53.10f);[139–146] and from SU-8 epoxy resin by photolithography and dicing (Figure 53.10g through 53.10i).[147–152] Some of the polycarbonate, PMMA, and PDMS emitters were further processed by hand polishing, machine milling, or trimming,[135,139–142] while other emitters were coated with gold particles or graphite powder to render them electrically conductive for ESI generation.[135,145,146] Recently, polymeric chips with pointed tips, comprising a microfluidic channel filled at its terminus with a photopatterned porous polymeric monolith prepared from lauryl acrylate-coethylene dimethacrylate, were fabricated. The multiple path, hydrophobic, porous polymer monolithic (PPM) emitter prevented the Taylor cone from spreading at the channel terminus, and thus provided a very stable spray.[153,154]

The ESI-MS performance of most of these microfabricated emitters was tested by performing infusion experiments with 0.1–10 μM peptide/protein solutions. Syringe pumping or N_2 pressurization was used to induce fluid flow through the chips at ~1–20 μL/min, or at <100 nL/min in the case of sharp emitters. Some of the sharper emitter enabled electrospray-driven fluid delivery without external pressurization or electroosmotic pumping.[72,123] The ESI voltage was typically applied to the chip reservoirs, to the gold-coated nano-ESI capillary emitters inserted in the chip, to the interconnecting standard stainless steel unions placed before or after the chip, to the liquid junction or liquid sheath interface, or, interestingly, to the chip material itself that was fabricated from conductive glassy-carbon substrates.[101] Most often, good quality mass spectra were generated from 1 μM concentration samples. Background ions from polymeric contaminants were only seldom reported. However, the fabrication of such emitters, and most importantly their integration within microfluidic chips that perform complex sample processing, continues to remain a challenge and relevant applications have not yet been demonstrated.

53.4.2 Matrix-Assisted Laser Desorption Ionization Interfaces

The development of microfluidic devices with MALDI-MS detection has focused on interfaces that enable off-chip or on-chip MALDI-MS analysis and the microfabrication of silicon/polymeric MALDI target plates.

53.4.2.1 Off-Chip MALDI-MS Detection for Microfluidic Sample Processing

Off-chip sample collection for MALDI-MS detection was essentially accomplished by syringe pumping for sample elution and deposition from the chip onto a standard MALDI target plate, by piezo-actuated sample dispensing from the chip onto the target plate,[155] and by using a rotating ball inlet MALDI-MS interface.[156] The latter two strategies will be discussed. Piezo-actuated silicon flow-through dispensers and nozzles have been developed to reproducibly generate picoliter-sized sample droplets for MALDI-MS analysis (Figure 53.11).[155,157–160] The matrix solution was spotted on the MALDI target simultaneously, in mixture with the sample, or separately, after sample deposition. The droplets were deposited on MALDI targets at a maximum ejection frequency of 500 Hz, and the droplet size could be varied from 30 to 200 pL (depending on nozzle dimensions). The spot size on the MALDI plate was 100–500 μm, a good match for the size of the laser focal point. Pyramid-shaped nozzles that prevented the wetting of the dispensing orifice front surface seemed to improve the directivity of the droplets. Detection limits for peptide/protein samples were in the low amol/fmol range, and multiple depositions on the same spot enabled sample detection from 31 nM solutions. Such a piezoelectric microdispenser with an internal volume of 250 nL (60 pL droplet size) was used for the interfacing of high-resolution LC to MALDI-MS.[157]

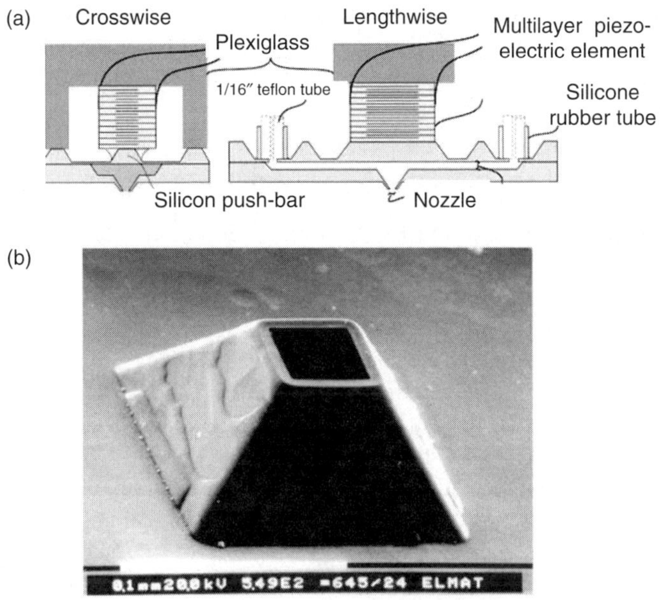

FIGURE 53.11 Piezo-actuated flow-through microdispenser. (a) Schematic diagram of the microdispenser. (b) SEM of a microfabricated dispenser silicon nozzle. (Reprinted from Miliotis, T., et al., *J. Mass Spectrom.*, 35, 369–377, 2000. Copyright 2000. With permission from Wiley-VCH.)

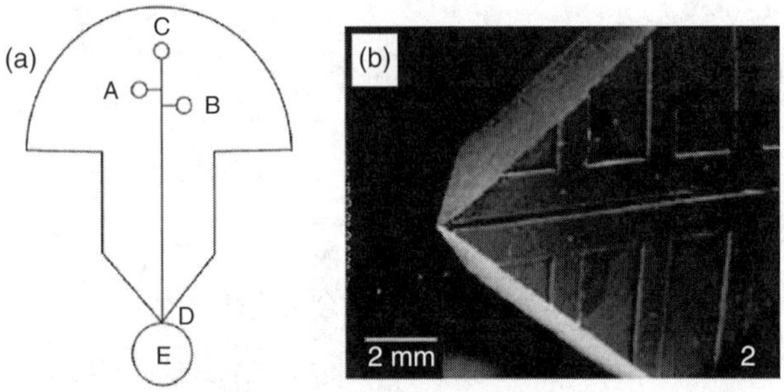

FIGURE 53.12 Microfluidic chip-MALDI-MS rotating ball interface. (a) Schematic diagram of the chip-to-ball direct contact deposition process. (b) SEM image of an embossed microchip tip. (Reprinted from Musiyimi, H.K., et al., *Electrophoresis*, 26, 4703–4710, 2005. Copyright 2005. With permission from Wiley-VCH.)

As an alternative technology, PMMA CE chips fabricated by hot embossing were machined with a triangular tip at the CE channel terminus, to enable MALDI-MS detection by direct contact deposition of the CE effluent onto a rotating ball inlet mass spectrometer (Figure 53.12).[156] The rotating ball acted as a cathode transfer electrode of the sample into the vacuum of the MALDI mass spectrometer, and matrix addition was accomplished in-line, on the surface of the ball. This interface was demonstrated for the CE separation and low fmol MALDI-MS detection of tryptic peptides from 10 nL sample injection volumes.

Microfluidic chips are commonly used for surface plasmon resonance (SPR) biomolecular interaction analysis, as well, and their interfacing to MS, which is highly complementary to SPR detection, was also pursued. MALDI-MS was performed either directly on the SPR chip,[161] or off-chip, either by elution of the binding partners in small volumes and transfer to standard MALDI target plates,[162–164] or by online ESI-MS.[165,166] Alternatively, bifunctional SPR fluid cells with optimized surfaces for SPR and MALDI-MS were fabricated.[167] As a miniaturized SPR sensor has been designed as an online detector for CE,[168] the future integration of microfluidic CE, SPR, and MALDI-MS detection is anticipated to occur.

53.4.2.2 On-Chip MALDI-MS Detection for Microfluidic Sample Processing

MALDI-MS from Open Channels and Reservoirs. Microfluidic CE separations with on-chip MALDI-MS detection were accomplished by performing the separation in the matrix solution in an open CE channel, followed by solvent evaporation and CE channel scanning with the desorption laser beam.[169] The system was demonstrated for the separation and detection of peptide and oligosaccharide samples. Alternatively, CIEF was performed in PMMA or resin chips with imprinted pseudoclosed channels that were covered only during the IEF step. After focusing, the cover plate was removed, matrix solution was deposited on the top of the focused sample with a syringe[170] or by electrospraying,[171] and MALDI-MS detection occurred in the open channel.

MALDI-MS was also implemented as a detection system for monitoring the reaction products of chemical or biochemical reactions in microfluidic reactors. In one approach, a continuous flow glass or silicon microchip that comprised a microdigestion reactor was incorporated into the standard MALDI plate of the MS instrument (Figure 53.13).[172] The vacuum of the MS chamber was used as a driving force for inducing flows in the reaction microchannels. The device was used for organic synthesis and biochemical reactions carried out entirely inside the MALDI-MS vacuum chamber,

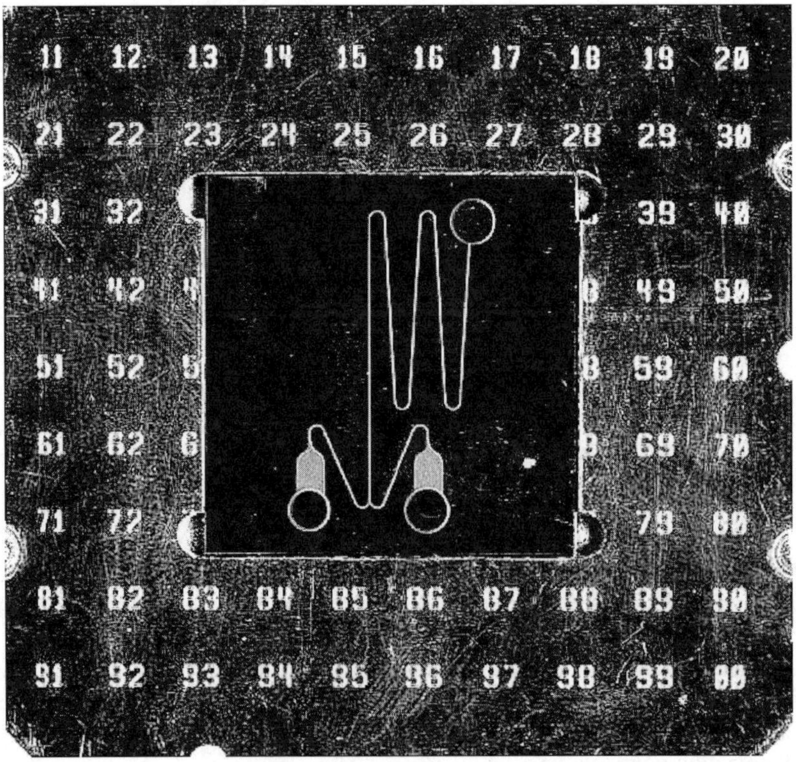

FIGURE 53.13 Microfluidic device integrated into a standard MALDI-TOF target plate. The vacuum of the MALDI-MS vacuum chamber is used as the driving force for inducing fluid flows on the chip (self-activating chip). (Reprinted from Brivio, M., et al., *Anal. Chem.*, 74, 3972–3976, 2002. Copyright 2002. With permission from American Chemical Society.)

and enabled the monitoring of Schiff base reaction products, of polymer separations, oligonucleotide digestions, and peptide sequencing. An improved approach involved the incorporation of a monitoring window along the microfluidic path of a serpentine microreactor chip.[173] The silicon/glass chip contained a rectangular boron doped silicone membrane monitoring window (~2 μm thick, 2.2 mm × 2.5 mm) that was positioned over the microreactor channel. The window was relatively large, to enable laser beam irradiation at an incidence angle of 30°. Fine holes (250 nm × 250 nm or 500 nm × 500 nm) were fabricated in the silicon membrane by a focused ion beam system to allow for ion desorption. The ions were extracted from the channel through the holes by the hitting action of the laser on the silicon window. This chip was used to monitor the reaction products formed in the serpentine microreactor, and opens the way to on-chip kinetic studies with MALDI-MS detection. A more simple approach involved the use of a silicon or glass protein digestion microreactor connected to a reservoir that enabled sample collection, MALDI matrix addition, and MALDI-MS detection from the exposed reservoir. For detection, the chip was simply glued to a standard MALDI target plate.[174]

Centrifugal Compact Disks. A clever microfluidic device that relies on centrifugal fluidic propulsion was developed for sample preparation tasks before MALDI-MS detection, and is presently a commercially available device (www.gyros.com). A high density, fully automated compact disk (CD) microfluidic system contains 96 sample processing lines placed radially on the chip.[175,176] The samples and solvents are loaded in microreservoirs near the center of the disk, the fluids are pumped through microchannels and microchambers to the outer edge of the CD by differential spinning, and the samples are collected in small spots (200 μm × 400 μm) at the end of each working line on the CD. The CD is then inserted into a MALDI mass spectrometer for analysis. These centrifugal CDs can handle sample volumes as low as 1 μL and enable the detection of low 500 amol to 5 fmol sample amounts (Figure 53.14).

Electrowetting-On-Dielectric Actuation Chips (EWOD). A novel microfluidic technique, which relies on electrowetting forces for moving, dispensing, merging, mixing, and reacting droplets at specific locations on an array of electrodes, has been used to perform various sample preparation tasks before MALDI-MS analysis. The technique, also called droplet-based digital microfluidics, was utilized for mixing droplets containing peptides, proteins, matrix, and rinsing solutions, and for generating purified sample/matrix solutions amenable to MALDI-MS detection (Figure 53.15).[177–179] Good quality spectra were generated from ~1 pmol/spot samples (S/N > 10–20).

53.4.2.3 Microfabricated/Functionalized MALDI Target Plates

Micromachined, high-density silicon nanovial target plates that allow for precise sample positioning and confinement, and thus improved MALDI-MS analysis, were fabricated by standard anisotropic wet etching techniques.[180,181] The nanovials had a size of 100 μm × 100 μm to 400 μm × 400 μm, had a pyramidal shape, and were very shallow (i.e., ~20 μm deep). Array densities of up to 4000 nanovials/square inch are achievable with this technology.[182] Piezo-actuated flow-through dispensers were used to deposit sample on such nanovial target plates, and repetitive microdispensing and accurate deposition of 60-pL droplets of a dilute sample to a confined area of 300 μm × 300 μm, allowed for on-spot sample enrichment and 10–50 times signal amplification.[155,158,159] Alternatively, nanovial arrays were fabricated by precision micromilling, cold embossing, and injection molding from PMMA and polycarbonate substrates. The array density was 165 nanovials/cm^2, corresponding to 3300 vials/conventional MALDI plate. Signal intensity from 10 nM peptide solutions, mass resolution, and mass accuracy were similar to that obtained with silicon nanovials.[183,184] These low cost, disposable polymeric nanovial arrays represent ideal platforms for high-throughput proteomic profiling applications.

The functionality of the silicon nanovial target plates was extended by the development of so-called integrated selective enrichment target (ISET) technology. The ISET device contains

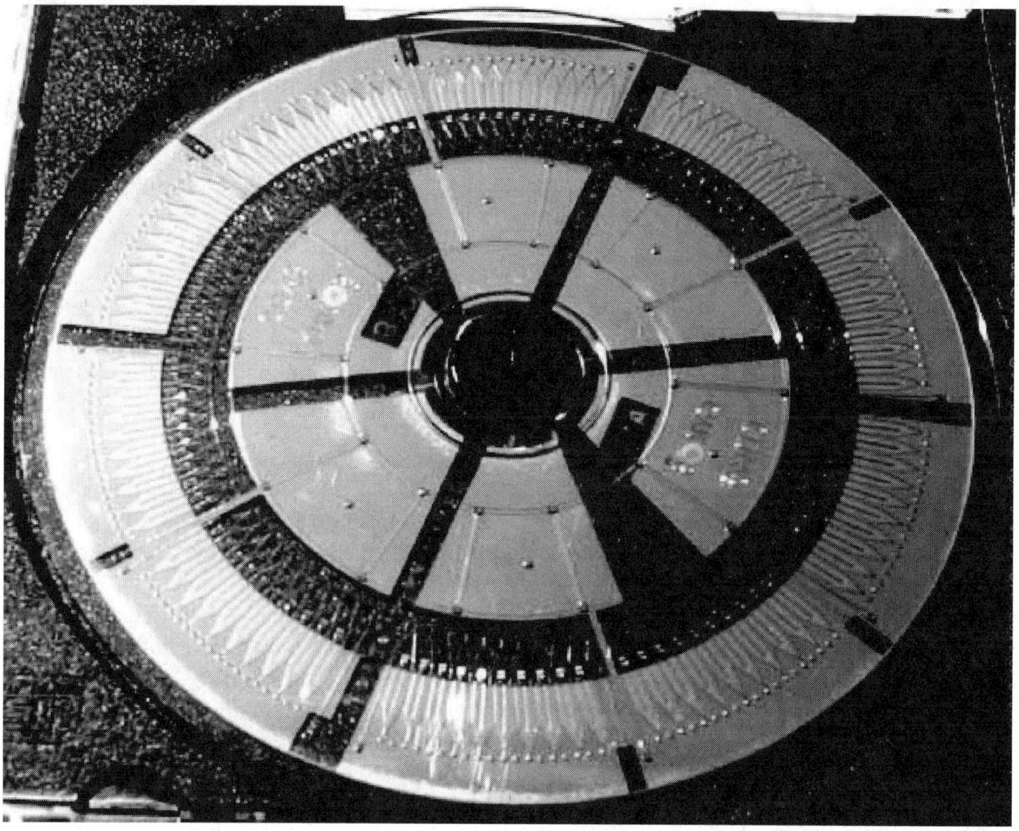

FIGURE 53.14 Centrifugal CD for high-throughput MALDI-MS analysis. (Reprinted from Lazar, I.M., et al., *Mass Spectrom. Rev.*, 25, 573–594, 2006. Copyright 2006. With permission from Wiley-VCH.)

96 perforated nanovials that can be filled with 30–40 nL of reversed-phase beads. The system was used to perform sample pretreatment before MALDI-MS analysis with minimal sample loss due to transfer and surface adsorption.[185] Essentially, Poros R2 beads were used to capture tryptic peptides from a digest solution, were transferred to the front of the ISET device, rinsed by applying negative pressure to the back side of the device through the opening at the bottom of the vial, and eluted with acetonitrile/MALDI matrix to deposit the sample into a ~500 μm spot on the back side of the chip for MALDI-MS detection (Figure 53.16). The ISET device enabled multiple analyses on a limited amount of sample and the identification of proteins isolated by affinity chromatography.

Microfabricated MALDI sample targets were also prepared from PMMA by computer numerical control (CNC) milling. In this case, enzyme functionalized cylindrical posts (360 μm × 360 μm) served as individual sample targets on the MALDI plate and were used for on-probe characterization of nucleic acids.[186] The advantages of this device included reduced sample consumption and handling and reduced analysis times. Alternatively, MALDI targets with twin anchors (400 μm) were prepared from silicon wafers by anisotropic dry etching in an inductively coupled plasma.[187] A pair of elevated anchors with hydrophilic top surface allowed for sample and standard deposition in very close proximity (on the anchors), and thus improved mass calibration with an external standard. The approach actually mimicked an internal standard mass calibration procedure and enabled the analysis of 30 pL sample solutions corresponding to ~1.5 amol of angiotensin peptides.

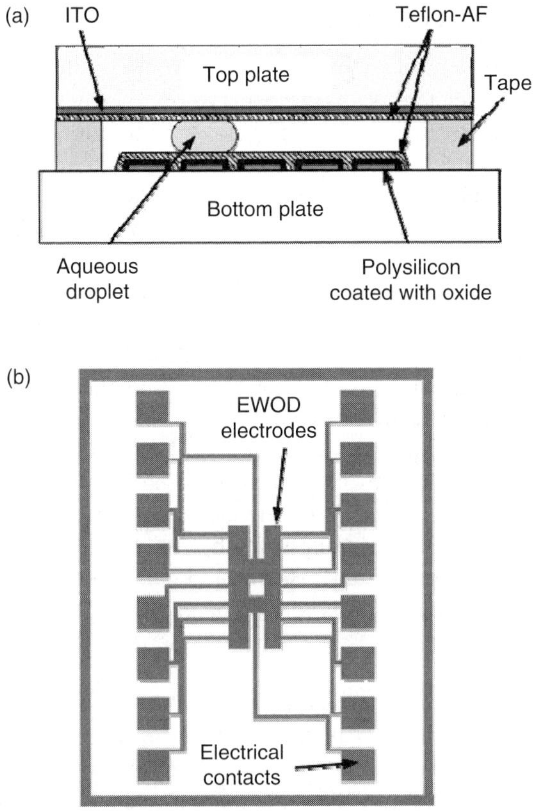

FIGURE 53.15 Schematic diagram of a EWOD device. (a) Side view and (b) top view. The bottom plate served as the base for a pattern of polysilicon EWOD electrodes buried under thermal oxide. A top plate was formed from ITO on glass; both plates were coated with Teflon-AF. The plates were joined with double-sided tape as a spacer. (Reprinted from Wheeler, A.R., et al., *Anal. Chem.*, 76, 4833–4838, 2004. Copyright 2004. With permission from American Chemical Society.)

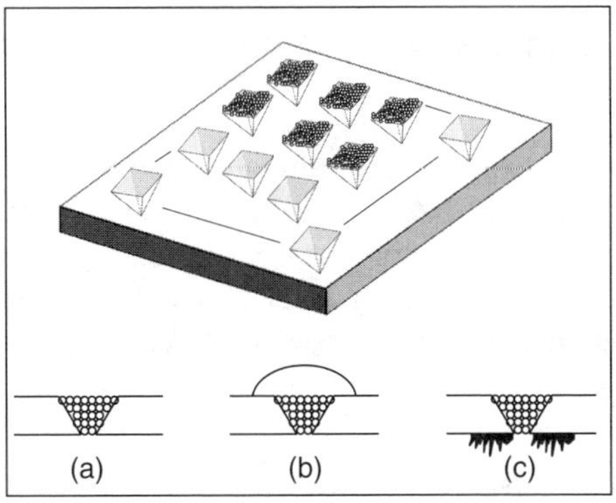

FIGURE 53.16 ISET platform. (a) Nanovial loaded with reversed-phase beads. (b) Elution/matrix mixture applied to the nanovial inlet. (c) Eluted sample crystallized around the nanovial outlet, ready for MALDI-MS. (Reprinted from Ekström, S., et al., *Electrophoresis*, 25, 3769–3777, 2004. Copyright 2004. With permission from Wiley-VCH.)

53.4.3 Other Microchip-MS Devices

While ESI and MALDI are the most commonly used ionization techniques for the MS analysis of biological samples, many other ionization mechanisms do exist and some of them were already implemented on microfluidic devices. APCI and various laser desorption ionization sources were pursued. A miniaturized APCI nebulizer chip, fabricated from silicon and Pyrex glass wafers, was designed to accommodate sample inlet capillaries, a stopper, a vaporizer channel, and a nozzle.[188] The nebulizer chip was used to interface capillary LC to MS, but could be integrated within CE separation chips as well.

Surface-enhanced laser desorption ionization (SELDI) MS is an array-based technology that enables MS detection from functionalized target plates prepared for selective capture and enrichment of proteins from crude biological samples.[189] These functionalized targets are marketed as "protein chips" (www.cipheregn.com) and contain hydrophilic, hydrophobic, cation or anion exchange, enzyme, antibody, and affinity capture spots on organic or inorganic surfaces. The addition of a MALDI matrix is optional depending on the specifics of the application. Alternatively, antibody-coated capture surfaces have been prepared within silicon nanovial target plates or on macroporous silicon targets.[190,191] With the aid of computer algorithms for pattern recognition and differentiation between two groups of subjects, the technique has been demonstrated for biomarker discovery in biological fluids. However, a more intense scrutiny of the SELDI chip approach has established that the technique is not sufficiently sensitive and reproducible to detect very low abundance biomarker components.[192] A recent manuscript reviews the latest developments in SELDI MS.[193]

Surface-assisted laser desorption ionization (SALDI) MS is used as an alternative to MALDI-MS for the analysis of small molecules that are difficult to detect with MALDI because of the interfering low-mass ions from the matrix.[194,195] In SALDI, micrometer-sized inorganic particles with low atomic mass and good stability during desorption such as graphite, nanoparticles, carbon nanotubes, ZnO, TiO2, SiO_2, Si, W, Ag, and so forth, are used as a matrix that facilitates the desorption of analytes. The technique was successfully demonstrated even in its simplest form (i.e., desorption/ionization on porous silicon (DIOS)-MS) for the detection of pharmaceuticals, peptides, and polymers from bare porous silicon surfaces.[196] In this context, the fabrication of ordered silicon nanocavity arrays on Si wafers, by using reactive ion etching and a closely packed SiO_2 nanoparticle array as a mask, was demonstrated for metabolite detection in *Arabidopsis thaliana* root extracts.[197]

53.5 MICROFLUIDIC APPLICATIONS WITH MASS SPECTROMETRY DETECTION

While still in early stages of development, microfluidic devices with MS detection are capable of successfully performing a number of bioanalytical tasks and have been demonstrated in various application areas. In particular, areas requiring capabilities for high-throughput processing of small sample amounts, such as proteomics and drug development, have been pursued. Online and offline sample infusion, preparation, and separation tasks have been implemented on chips interfaced to both ESI and MALDI-MS detectors.

53.5.1 Microfluidic Sample Infusion Analysis

The very first microfluidic-MS applications have involved sample infusion analysis from chips with a flat spraying surface or with inserted ESI capillary emitters. These relatively simple devices were fabricated from glass or polymeric substrates and were interfaced to IT, TOF, or hybrid Q-TOF instruments, and sample infusion was accomplished by connecting syringe pumps, N_2 cylinders, or external EOF pumping capillaries to the chip. Alternatively, infusion experiments have been used to evaluate the effectiveness of ESI-microfabricated emitters. The previous section provides detailed information related to the performance of these chips. These initial single-channel infusion devices were later developed into multiplexed configurations. Completely independent channel arrays connected to multiple ESI emitters,[83,104] or channel arrays connected to a single ESI emitter,[107] were utilized for high-throughput infusion analysis of standard protein digest or yeast proteomic cellular extracts separated on a 2D-gel.[107] Low fmol–pmol sample amounts/spot were detected. At the present time, microfluidic chips with a variety of ESI emitters are used for the infusion analysis of peptides/proteins,[198] glycoconjugates,[199–201] pharmaceuticals,[202] and metabolites.[203]

A multielectrospray device composed of an array of nine microfabricated emitters arranged in a 3×3 configuration, placed ~1.5 mm apart, was used for infusion experiments. For a given flow rate, the device was shown to improve the ESI signal in comparison with conventional-fused capillaries with a single opening.[134] The device was fabricated in polycarbonate substrates by laser etching, and shows potential for applications that need to handle large eluent flow rates.

53.5.2 Microfluidic Sample Preparation

A number of microfluidic applications include the development of sample preparation tasks that are aimed at improving the detection limits. Most commonly, these tasks include processes such as cleanup, desalting, preconcentration, and enzymatic microdigestion.

Sample Cleanup. Sample cleanup and desalting were among the first processes implemented on the chip to improve the detection limits and the quality of MS analysis. Polycarbonate or polyimide chips were devised to enable the incorporation of a sandwiched microdialysis cellulose membrane for sample desalting,[105] or of two microdialysis membranes for the removal of low- and high-molecular weight contaminants.[204] Alternatively, sample desalting was accomplished on poly(vinylidenedifluoride) membranes inserted in the microfluidic sample reservoir,[205,206] or on methacrylate-based monolithic polymeric columns incorporated in SU-8 chips.[207] Such devices enabled the ESI-MS detection of 1.5–5 μM proteins (horse cytochrome c, bovine serum albumin, ubiquitin) from complex matrices containing 100–500 mM NaCl, 100 mM Tris, and 10 mM EDTA, or complex *Escherichia coli* cell lysates.[105] Ultrarapid (tens of milliseconds) desalting of protein solutions was accomplished in a microchannel laminar flow device.[208] The device used a two-layered laminar flow geometry that exploited the differential diffusion of macromolecular analytes and of low molecular weight contaminants.

Recently, EWOD actuation chips were developed into a multiplexed device that was used to simultaneously cleanup four samples. A sequence of seven actuation steps were performed for each sample: (1) generation of sample droplets (0.02–4 μL), (2) transport and drying of sample droplets, (3) generation of rinsing droplets, (4) transport of rinsing droplets to the sample sites for selective dissolution of urea, (5) transport and disposal of the rinsing droplets, (6) generation of MALDI matrix solution droplets, and (7) delivery of matrix droplets to the dried peptide spots.[179]

Sample Preconcentration/Solid-Phase Extraction. A microfluidic device with an external EOF pumping capillary was used as a nanoflow solvent gradient delivery system to a reversed-phase microcolumn to enable peptide sample preconcentration, elution, and detection from 100 pM solutions.[103] The small C18 preconcentration cartridge was placed between the chip and the ESI source of a mass spectrometer. Later, microfluidic designs have integrated the preconcentrator on the chip. For example, trace level peptide samples (2–10 nM) were enriched before MALDI-MS analysis using a microfluidic device that comprised a high-aspect ratio silicon vertical grid structure (silicon microextraction chips, SMEC) that enabled the immobilization of reversed-phase packing materials.[209–211] The need for adequate structures for particle immobilization was avoided by developing porous monolithic structures, polymerized into a miniaturized solid-phase microextraction (SPME) unit, that enabled 10^3-fold peptide/protein sample preconcentration.[212] Multiplexed SPME methacrylate-based monolithic columns, operated by syringe pumps, were integrated within Zeonor polymeric chips, and were used for the extraction and ESI-MS detection of drugs spiked in urine at 0.025 μg/mL level.[213] The implementation of the technology in high-throughput chips comprising 96 or 384 monolithic SPME columns is anticipated.

A variety of microfluidic designs have also been used for affinity enrichment tasks. Methacrylate-based porous monolithic packings, photopolymerized within microfluidic channels and derivatized with Cibacron-blue-3G-A, were used for the capture of lysozyme and albumin from human cerebrospinal fluid.[214] The proteins were eluted with the aid of a syringe pump and analyzed with MALDI-MS. PDMS chips comprising poly(vinylidenefluoride) membranes were used for affinity dialysis/ultrafiltration and concentration of aflatoxin B1 antibody/aflatoxins and phenobarbital antibody/barbiturates. ESI-MS detection sensitivity was improved by ~1–2 orders of magnitude.[215] Affinity capture resins (c-*myc*-specific antibodies immobilized on protein G-Sepharose beads) were also incorporated into complex microfluidic CE-MS platforms to facilitate the trace-level analysis of antigenic peptides spiked at 20 ng/mL into human plasma.[216] In addition, high-throughput parallel screening of 96 samples for tryptic phosphopeptides was made possible by using a microfluidic CD that incorporated a small bed of immobilized metal affinity chromatography (IMAC) beads.[176] Each processing line on the microfluidic CD was devised to perform one or more sample preparation steps: sample preconcentration, affinity selection, desalting, and digestion.

Enzymatic Microdigestion. Several approaches have been taken to perform fast protein digestion (3–12 min) on a chip. Microreactors packed with immobilized trypsin beads,[217,218] PDMS sandwich-type chips incorporating poly(vinylidenefluoride) membranes with immobilized trypsin,[219] or simple chips that relied on a fast liquid-phase digestion protocol[220] were used for this purpose. A representative nanoflow sample processing device included a microchip immobilized enzyme reactor (μ-chip IMER) interfaced to a piezo-actuated flow-through dispenser to deposit the sample from the IMER onto a nanovial-MALDI target plate.[180,221] The IMER was fabricated in silicon and comprised high-aspect ratio, parallel channel structures with porous surface and immobilized proteolytic enzymes, to enable protein digestion in as little as 20 s to 3 min. Alternatively, an IMER chip with immobilized endoglucanases was utilized for the fast hydrolysis (~20–100 times faster than conventional batch hydrolysis) and infusion ESI-MS analysis of methyl cellulose. To speed up the overall analysis, online interfacing of this endoglucanase IMER to LC-ESI-MS and offline interfacing to size exclusion chromatography MALDI-MS, were also performed.[222] Alternatively, a poly(ethyleneterephthalate) chip enabled the multilayer assembly of natural polysaccharides (positively charged chitosan and negatively charged hyaluronic acid) onto the surface of a channel to form a microstructured biocompatible network for enzyme immobilization. Controlled adsorption of trypsin enabled very fast (<5 s) digestion of standard proteins on the chip.[223]

Simple infusion microfluidic chips have been used to quickly (5–15 min) digest proteins for peptide mass fingerprinting.[220] High enzyme:substrate ratios (1:5) were used to perform the tryptic digestion of standard hemoglobin variants from 0.1 to 2 μM (5–10 μL) solutions, or of hemoglobin isolated from ~5 μL human blood samples. Sequence coverage of 70–97% was reported. Alternatively, continuous flow serpentine microreactors, coupled to online HPLC-ESI-MS, were used for the implementation of enzymatic assays for bioactivity screening (enzyme inhibition and substrate conversion).[224] Reaction times were only ~36 s and detection limits were 0.17–2.6 μM, that is, comparable to microtiter plates and bench-top continuous flow assays.

53.5.3 Microfluidic Sample Separations

A variety of electrically driven separation techniques with MS detection have been implemented on the chip, and performance similar to bench-top systems has been demonstrated. Various ESI

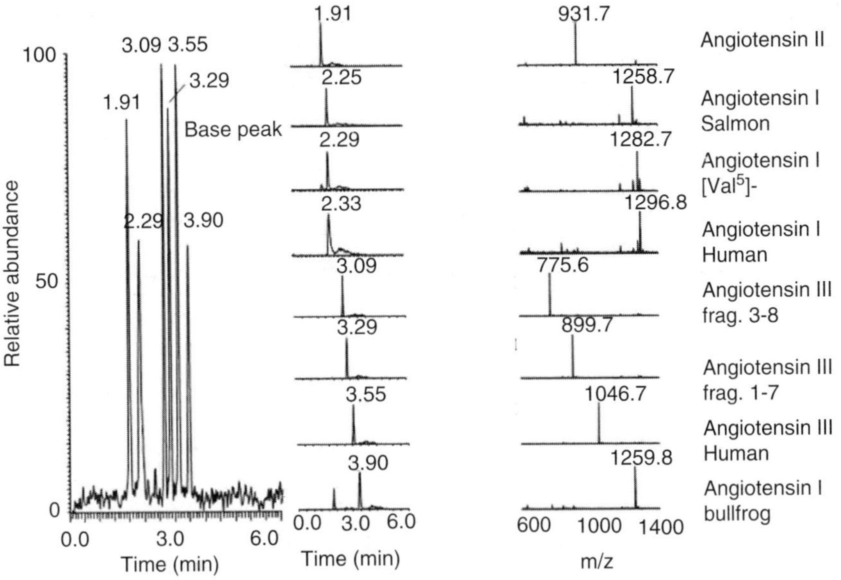

FIGURE 53.17 Microchip-CE/ESI-IT-MS analysis of a mixture of angiotensin peptide (10 μg/mL) using the device shown in Figure 53.7. (Reproduced from Zhang, B., et al., *Anal. Chem.*, 72, 1015–1022, 2000. Copyright 2000. With permission from American Chemical Society.)

and MALDI-MS interfaces have been used in combination with chips that enabled complex sample processing. Relevant examples are discussed in the following text.

Capillary Electrophoresis Chips. Some of the first microchip-CE separations have been performed on glass chips that used on-chip miniaturized pneumatic nebulizers,[90] or on-chip or off-chip liquid junction interfaces coupled to a subatmospheric ESI ion trap instrument.[90,96,225] High-efficiency (~31,000 plates/chip or ~300,000 plates/m) separations on 11 cm long channels, or fast separations on 4.5 cm channels, of peptide/protein samples were achieved (Figure 53.17).[96] Some of these chips were also tested for performing transient isotachophoretic sample preconcentration.[90]

Glass microfluidic chips with inserted nano-ESI emitters, electroosmotic pumping capillaries, or capillary transfer lines inserted in the liquid sheath interface of a Q-TOF instrument, were also used for the CE analysis of standard peptide/protein digests and of gel-isolated proteins from complex cellular extracts (*H. influenza, P. sativum*).[93,95,216,217,226,227] Some of these chips enabled sample enrichment by C18-membrane off-chip preconcentrators, on-chip stacking, or on-chip tryptic digestion with immobilized beads. Typically, CE separations were performed in 1–5 min, and concentration detection limits of 2–40 nM for standard peptides were reached.[93] A relevant proteomic application performed on such a device included the fast (<2 min) CE-MS analysis of 1D-gel-isolated membrane protein digests from *H. influenza*. Approximately 1–2 μg protein extract was applied to the gel and ~25 ng (1–5 μL) was loaded in the chip reservoir (Figure 53.18).[93] An on-chip sample desalting module packed with solid phase extraction (SPE) hyper-crosslinked polystyrene beads, followed by CE separations, was also developed in PDMS substrates for the separation and detection of peptides with ESI-TOFMS.[228]

A free-standing liquid junction interface was coupled to a flat edge glass CE microchip for the analysis of small molecules (drugs, metabolites).[115] It was demonstrated for the detection of recovered carnitine, acylcarnitines, imipramine, and desipramine spiked into urine or plasma at 5–500 μg/mL level. Separations were typically performed in <1 min and intra-assay precisions ranged from 4.1% to 7.3% Relative Standard Deviation RSD.[116,117] A similar device, but fabricated in polymeric Zeonor material, was demonstrated for the analysis of carnitine standards.[229]

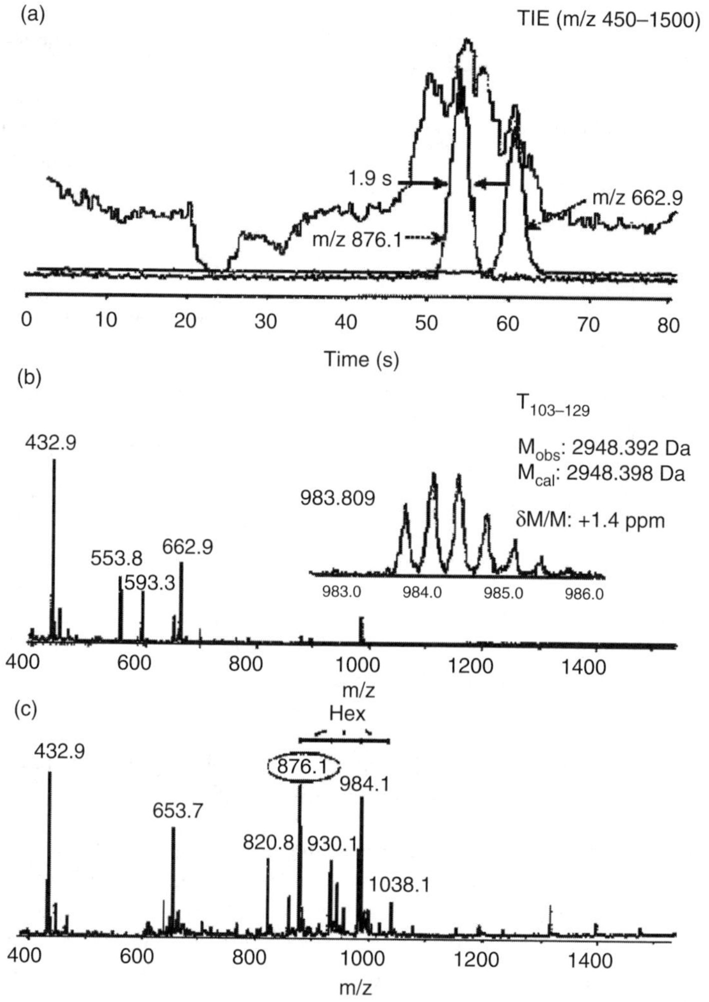

FIGURE 53.18 Microchip-CE/ESI-QqTOF-MS analysis of tryptic peptides from the seed lectin PHA-L using the device shown in Figure 53.5. (Reproduced from Li, J., et al., *Anal. Chem.*, 72, 599–609, 2000. Copyright 2000. With permission from American Chemical Society.)

Microfluidic CE separation chips were also interfaced to MALDI-MS detection. The CE separation of oligosaccharides and peptides (0.5–5 mg/mL) was performed in open CE channels (~250 μm deep) that contained buffer and MALDI matrix. The chips were prepared in glass. After separation, the solvent was evaporated, the chips were placed into a specially designed MALDI source, and the CE channels was scanned with the laser beam.[169] It is anticipated that the CE chip-MALDI-MS protocol could provide a fast and effective alternative for applications that utilize 2D-gel separations followed by MS detection.

Capillary Isoelectric Focusing Chips. Polycarbonate devices comprising a laser-ablated separation channel (50 μm × 30 μm × 16 cm), a poly(ethyleneterephthalate) film sandwiched between the polycarbonate substrates to seal the channels, and a pointed mechanically machined ESI tip were used for the isoelectric focusing of proteins.[230] After protein focusing, the analytes were mobilized into the ESI source by applying a slight pressure at the separation channel inlet, and the spray was stabilized with sheath liquid and sheath gas (N_2) supplied through side channels placed close to the separation channel terminus. Alternatively, PMMA or resin chips were designed to enable

MALDI-MS detection of proteins separated by IEF. The chips contained pseudoclosed channels (i.e., channels that were covered only during the IEF step but not during MALDI) to facilitate the addition of the matrix solution after the focusing step.[170,171] The performance of these microfluidic IEF chips was demonstrated for the detection of standard proteins from ~0.05 to 0.5 mg/mL concentration solutions.

Capillary Electrochromatography Chips. Electrochromatography is an attractive alternative to both CE and LC separations, and since it relies on electrically driven fluid flows, its implementation on microfluidic chips is particularly useful. Recent reviews describe relevant achievements in this area with focus on column technologies.[25,26,154,231] The interfacing of a CEC chip comprising a methacrylate monolithic polymeric separation channel, a double-T injector, and a capillary ESI-MS interface was demonstrated for the separation of $<1\ \mu$M concentration standard protein tryptic digests (Figure 53.19).[32] The ESI interface consisted of a porous glass gate that enabled the transfer of ions to maintain electrical continuity through the channels, but not of bulk flow, which was redirected to the ESI capillary emitter inserted in the chip. Stable spray was generated for hours of chip operation. The dimensions of one CEC analysis line were small enough (6 cm × 0.5 cm) to enable the integration of four CEC channels on a 3″ × 1″ chip, or of eight channels on a 3″ × 2″ chip. More complex, dual function microanalytical devices for SPME and enzymatic digestion, were fabricated by in situ photolithographic grafting of porous polymer methacrylate monoliths in 50 μm capillaries. Transfer of this technology on microfabricated chips will open new avenues for the simple construction of complex μTAS systems.[154]

Liquid Chromatography Chips. Most attempts of implementing LC separations on microfluidic devices involved the fabrication of the separation channel on the chip but made use of external pumping/valving systems to drive the fluids. Initially, the LC analysis of caffeine, with UV and amperometric detection, was demonstrated on a silicon substrate incorporating microfluidic channels of various dimensions (20–200 μm in width, 0.5–10 μm in depth, and 15–50 cm in length) coated with

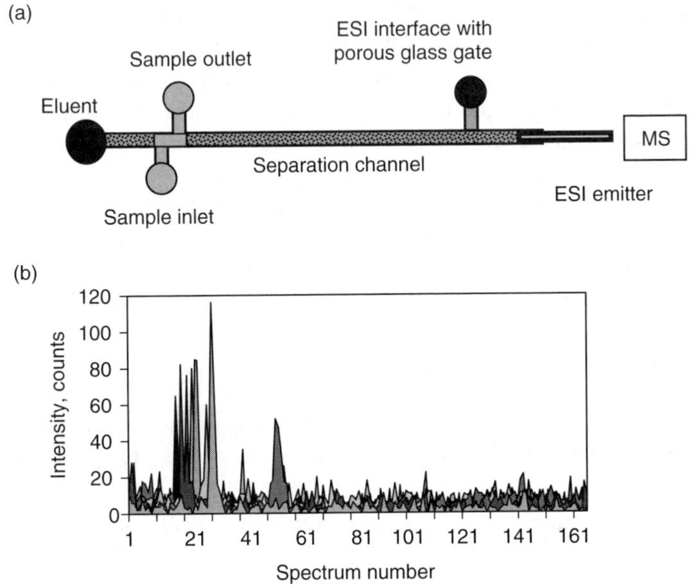

FIGURE 53.19 Microchip-CEC device with a porous polymeric methacrylate-based monolithic stationary phase and ESI-TOFMS detection. (a) Schematic diagram of the CEC device. (b) CEC separation of a protein mixture digest. (Reprinted from Lazar, I.M., et al., *Electrophoresis*, 24, 3655–3662, 2003. Copyright 2003. With permission from Wiley-VCH).

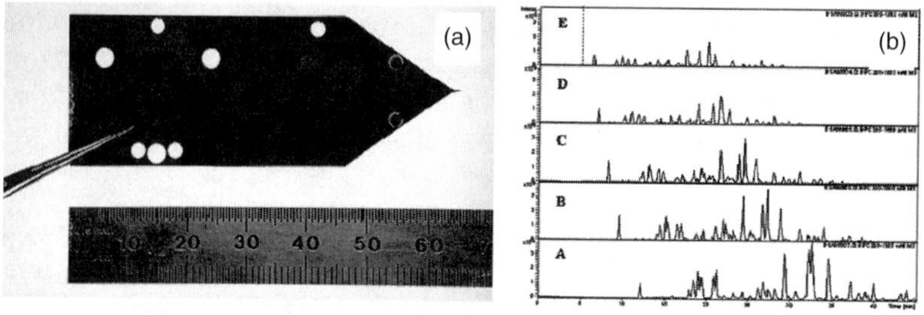

FIGURE 53.20 Microchip-LC driven by a bench-top HPLC pumping system interfaced to ESI-IT-MS. (a) Schematic diagram of a LC polyimide chip comprising laser-ablated channels, ports, frits, and ESI tip. (b) Base peak chromatograms of BSA digest running under different LC flow rate: (A) 100, (B) 150, (C) 200, (D) 300, and (E) 400 nL/min. (Reprinted from Yin, H., et al., *Anal. Chem.*, 77, 527–533, 2005. Copyright 2005. With permission from American Chemical Society.)

n-octyltriethoxysilane.[232] Later, a fused-silica chip that incorporated a laser-polymerized cylindrical fluoropolymer element, which served as a switchable valve for sample injection, was fabricated.[233] Fluidic propulsion, however, was accomplished by connecting the chip to a syringe pump. Recently, a polyimide microfluidic LC chip that comprised laser-ablated separation channels, enrichment columns, ports, frits, and a triangular shape ESI tip, was fabricated (Figure 53.20).[131–133] Sample loading and mobile phase delivery was accomplished by connecting the chip to an external rotary valve, a gradient pump, and a microwell plate autosampler. The chip was demonstrated for the analysis of standard samples and proteins isolated from plasma (1–5 fmol detection limits). Presently, these chips are commercially available analysis platforms (www.agilent.com).

Standalone microfluidic LC devices with ESI-MS detection were fabricated by incorporating the pumping and valving components on glass or silicon chips. In one design, a chip-integrated LC system was devised by developing an on-chip electrochemical pumping system that generated pressure for fluidic propulsion as a result of gas produced by electrolysis (200 psi, 20 nL/min).[125,126] The chip was demonstrated for the LC analysis of standard peptides (Figure 53.21). In another design, a glass microfluidic chip that incorporated a multichannel electroosmotic pumping/valving system (200–400 nanochannels, 1.5–2 μm deep), a packed LC channel, and a capillary ESI-MS interface was developed.[112,113] The pumping/valving system enabled fluidic propulsion and sample loading in electrical field-free regions on the chip. The generated pressures and flow rates (~80 psi, ~80–100 nL/min) were fully consistent with the requirements of bench-top nano-LC platforms that are used routinely in proteomic applications. The system was demonstrated for data-dependent LC-MS/MS analysis of protein mixture digests and protein extracts from the MCF7 breast cancer cell line (Figure 53.22). Two microfluidic LC systems could be integrated on a 1″ × 3″ chip, and under identical experimental conditions, the performance of the microfluidic LC was identical to the one of a commercial μLC instrument. The microfluidic LC system enabled the confident detection ($p < .001$) of 77 proteins by using conventional data filtering parameters, and of five putative cancer biomarkers.

Alternatively, a microfabricated piezo-actuated flow-through dispenser was utilized to interface capillary LC to MALDI/TOF-MS. Accurate deposition of 60-pL sample droplets to a confined area of 300 × 300 μm, and repetitive microdispensing, allowed for on-spot sample enrichment and 10–50 times signal amplification.[157,181] Detection limits were ~100 amol from 1 nM solution.

53.5.4 High-Throughput Sample Processing Chips

Parallel or fast sequential processing was demonstrated with chips that performed both infusion and separation experiments. Initially, Liu et al.[104] has demonstrated high-throughput electrospray-MS

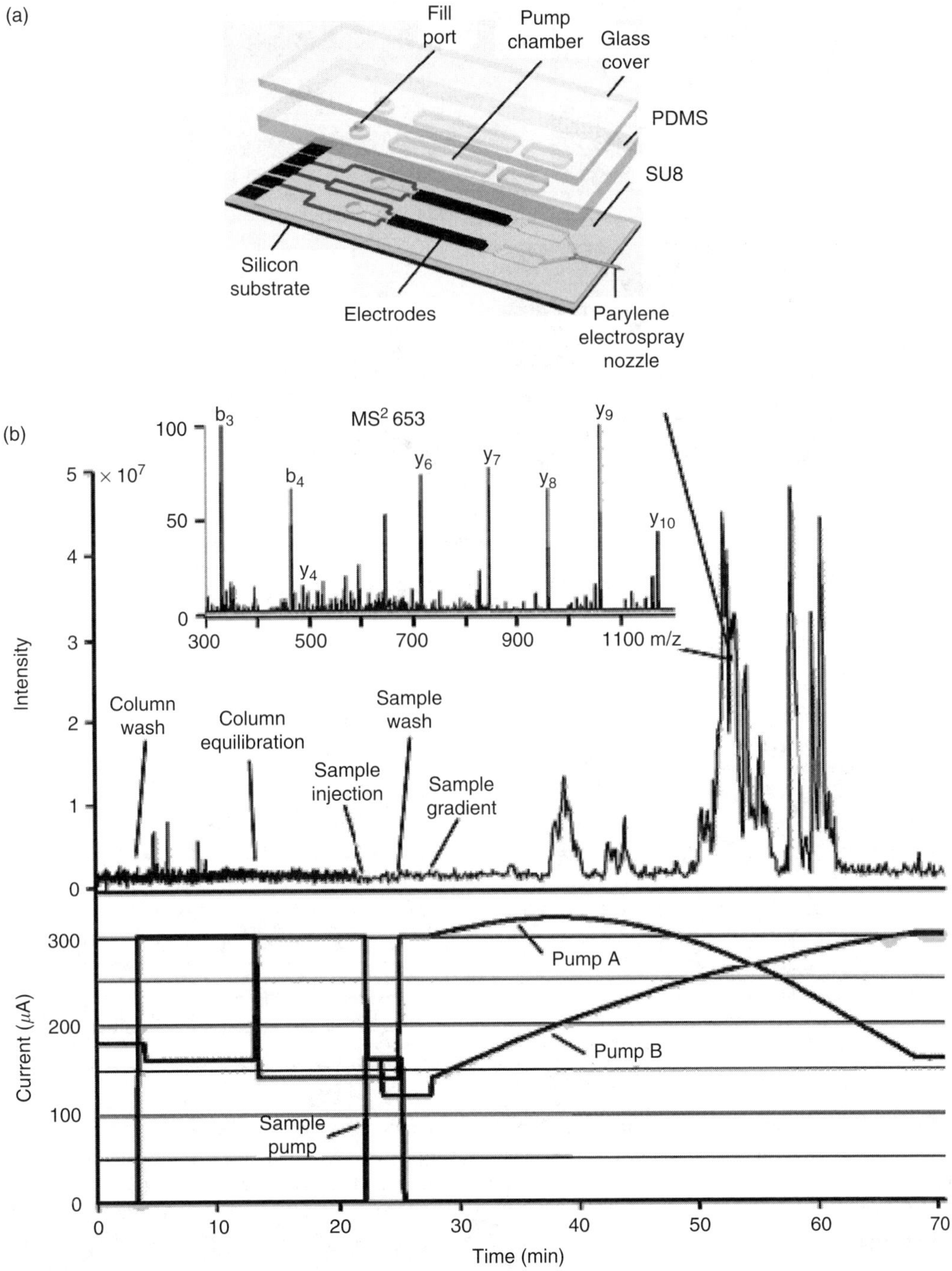

FIGURE 53.21 Microchip-LC driven by an electrochemical pump interfaced to ESI-IT-MS. (a) Schematic representation of an electrochemically driven LC chip. (Reprinted from Xie, J., et al., *Anal. Chem.*, 76, 3756–3763, 2004. Copyright 2004. With permission from American Chemical Society.) (b) Base peak chromatogram of a BSA tryptic peptide mixture (600 fmol injection). Reprinted from Xie, J., et al., *Anal. Chem.*, 77, 6947–6953, 2005. Copyright 2005. With permission from American Chemical Society.)

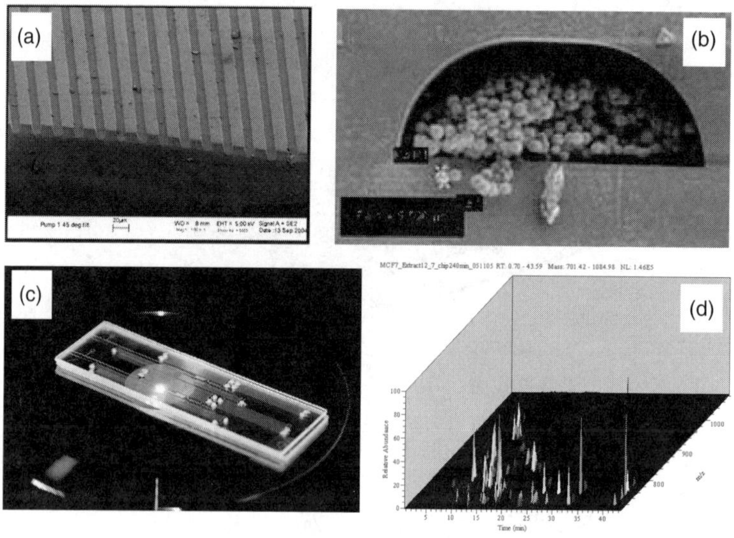

FIGURE 53.22 Microchip-LC driven by a multichannel electroosmotic pump interfaced to ESI-IT-MS. (a) SEM image of the multichannel pumping/valving structure. (b) Microchannel filled with 5 μm particles. (c) Picture of a glass microfluidic chip comprising two fully integrated LC systems. (d) Data-dependent microfluidic LC-MS/MS analysis of a protein digest prepared from the MCF7 breast cancer cell line (SCX fraction eluted with ~50–70 mM NaCl). (Reprinted from Lazar, I.M., et al., *Anal. Chem.*, 78, 5513–5524, 2006. Copyright 2006. With permission from American Chemical Society.)

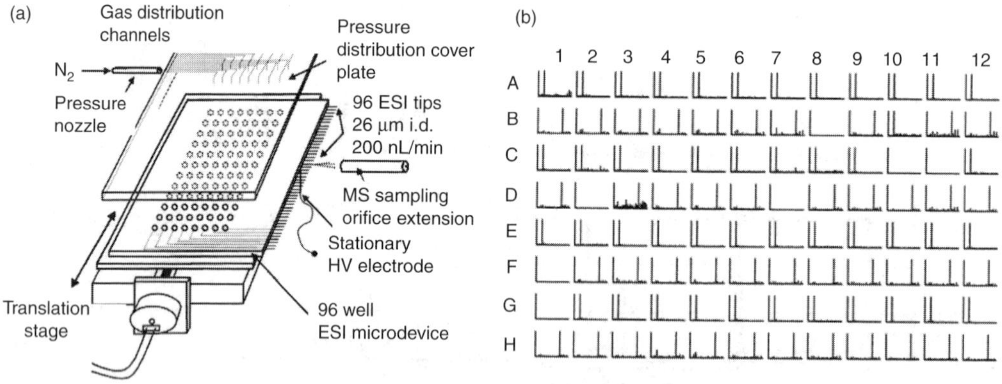

FIGURE 53.23 High-throughput ESI-MS analysis from a microfluidic infusion-MS device. (a) Diagram of a 96-well plate multichannel electropneumatic sample delivery system with multiple ESI emitters. (B) High-throughput ESI-MS analysis (alternate injections) of angiotensin II and III (10 μg/mL). (Reprinted from Liu, H.H., et al., *Anal. Chem.*, 72, 3303–3310, 2000. Copyright 2000. With permission from American Chemical Society.)

from a parallel infusion channel device that comprised 96 channels equipped with an array of 96 capillary emitters. Individual sample wells on the chip were pressurized for infusion-MS with N_2 gas, and the device enabled the fast infusion analysis of 96 peptide samples within 480 s (Figure 53.23). At the present time, silicon ESI chips with 400 emitters are commercialized devices (www.advion.com) that are integrated into an ESI-MS ion source (the NanoMate) that can be mounted on a variety of mass spectrometers. Simple infusion experiments, or pipette tip-based sample desalting followed by infusion-MS, can be reliably performed with this disposable microfabricated device (Figure 53.24).

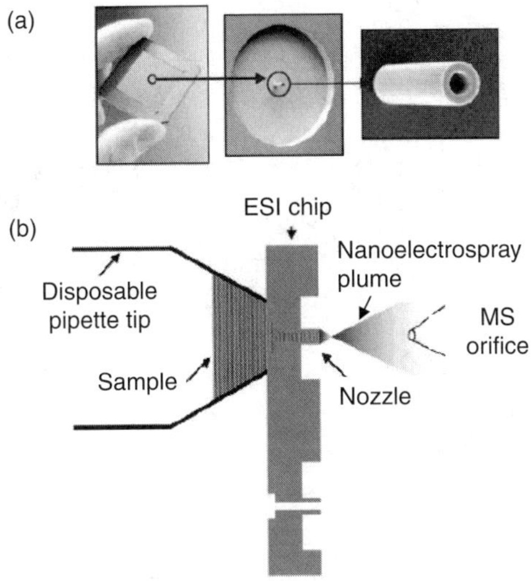

FIGURE 53.24 (a) ESI silicon chip with an array of nano-ESI nozzles. (b) Cross-section of the silicon ESI chip with a disposable pipette tip that presses and seals against the inlet side of the silicon chip. (Reprinted from Zhang, S., et al., *Electrophoresis*, 24, 3620–3632, 2003. Copyright 2003. With permission from Wiley-VCH.)

Some common applications include automated chip-based nano-ESI-MS for rapid identification of proteins separated by 2D-gels,[198] the study of noncovalent protein–protein and protein–ligand interactions,[234] the quantitation of drugs and metabolites in plasma in the low ng/mL range,[202,203] carbohydrate analysis,[199,200] and so forth. In terms of throughput, 40–55 s/sample is achievable with this automated chip-infusion technology.[203] A relevant example involves the fully automated MS characterization of complex carbohydrate systems using a nanoscale liquid delivery system and a chip-based nanoelectrospray silicon emitter array assembly.[200] Recently, these 400 nozzle silicon chips have been incorporated into an analytical platform that performs bench-top LC, fraction collection, and chip-based infusion for MS detection. The system enables extended investigations of sample components/subfractions that are difficult to characterize on a chromatographic time scale.

Alternatively, Foret et al.[225,235] have devised a microfabricated device that incorporated a sample injection loop, a separation channel for CE, IEF, or chromatography, and a liquid junction MS interface. The samples were loaded sequentially from a microtiter plate with the aid of an electropneumatic distributor, and microfluidic manipulations on the chip were accomplished with electrokinetic and pressure fluidic control. The sequential CE-ESI-MS analysis of protein tryptic digests is shown in Figure 53.25.

Li et al.[216,227] have demonstrated a highly integrated analysis system that included a CE separation channel, a microfabricated chamber for sample preconcentration, affinity selection or digestion, and a nano-ESI capillary interface. Sequential sample loading was accomplished with a bench-top autosampler (Figure 53.26). Using peptide mass fingerprinting and online tandem MS, such chips were utilized for the identification of gel-isolated proteins from *Neisseria meningitidis* and prostate cancer cell lines, of affinity selected *c-myc* antigenic peptides spiked into human plasma at 20 ng/mL level, and of IMAC-selected phosphopeptides. Up to 12–30 samples/h were analyzed with such a device, and detection from less than 7 fmol/injection (or 5nM solutions) was achieved. In one example, from a total of 67 tryptic digest bands, only three proteins remained undetected. Sample carryover was estimated to be $< 3\%$.

With MALDI-MS detection, the microfluidic compact disk was demonstrated for the high-throughput analysis of protein digests[175] and phosphorylated peptides.[176] For protein digests,

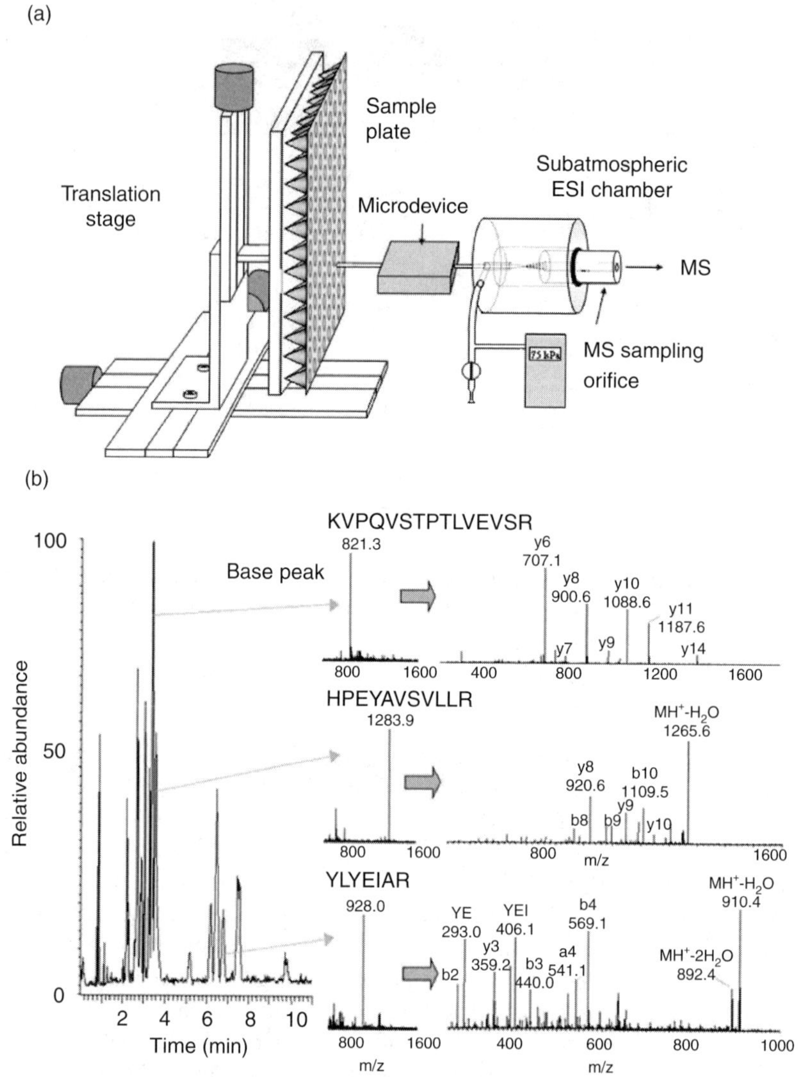

FIGURE 53.25 Microfluidic system for high-throughput separations and MS analysis. (a) Diagram of the total system (automated microwell plate positioning system, sample separation microdevice, and subatmospheric ESI-MS interface). (b) Automated CE-MS/MS analysis of a BSA tryptic digest. (Reproduced from Zhang, B.L., et al., *Anal. Chem.*, 73, 2675–2681, 2001. Copyright 2001. With permission from American Chemical Society.)

96 samples (1 μL) were simultaneously loaded, preconcentrated (on 10 nL nanostructures packed with reversed-phase chromatography particles), desalted, and subsequently eluted onto MALDI target areas (200 μm × 400 μm) to enable detection directly from the CD. Fluidic manipulations through the microstructures on the CD were accomplished using the centrifugal force developed during the rotation of the disk at 1000 rpm. The detection of 50–200 amol amounts of proteolytic peptides was demonstrated. For phosphorylated protein digests, CD nanoscale structures (16 nL) packed with IMAC particles were used to enrich the sample in phosphorylated peptides. Each sample was processed on two lines containing IMAC particles, one with, and one without subsequent

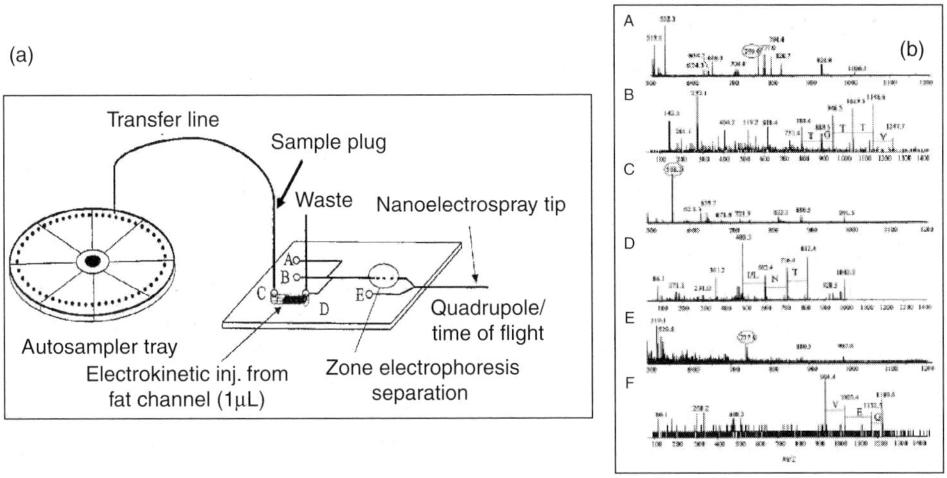

FIGURE 53.26 Microfluidic chip interfaced to an autosampler and to an ESI-TOF-MS detection system. (a) Schematic diagram of the total system. The chip comprises an enzyme microreactor/preconcentrator, a CE separation channel, and a capillary nanoESI-MS interface. (b) Tandem mass spectra of 2D-gel-isolated proteins from *Neisseria meningitidis*. (Reprinted from Li, J., et al., *Proteomics*, 1, 975–986, 2001. Copyright 2001. With permission from Wiley-VCH.)

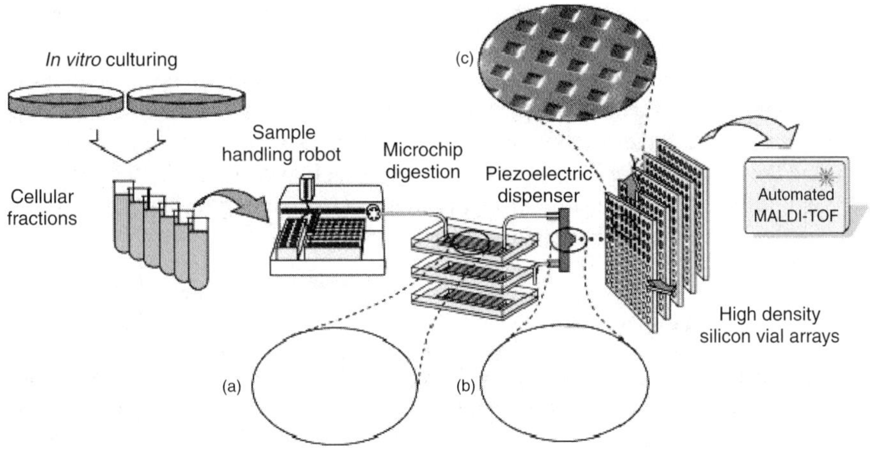

FIGURE 53.27 High-throughput microfabricated platform for protein expression analysis. The platform includes a robotic sample handling system, a porous silicon protease microreactor a flow-through piezoelectric silicon microdispenser, a high-density microchip MALDI target plate (array density 4000 nanovials/in.^{2}), and a MALDI-TOF-MS detector. (Reprinted from Laurell, T., et al., *Trends Anal. Chem.*, 20, 225–231, 2001. Copyright 2001. With permission from Elsevier.)

in situ alkaline phosphatase treatment for peptide dephosphorylation—a process often needed to confirm the presence of a phosphorylation site. Thus, the CD enabled the parallel processing of 48 samples. Detection limits for phosphorylated peptides were ~15 fmol. Alternatively, a μTAS platform that comprised automated sample pretreatment and injection, a silicon microchip IMER, a silicon microdispenser, and a silicon nanovial-MALDI target, enabled automated MALDI-TOF fingerprinting of 100 protein samples within 3.5 h (Figure 53.27).[180]

53.6 METHOD DEVELOPMENT GUIDELINES

53.6.1 Protocol to Evaluate Detection Limits/Ionization Efficiency from the Chip

Up-to-date, the best performance in terms of ESI efficiency is provided by nanospray fused-silica capillaries (~5 μm i.d. × 10–15 μm o.d.). The performance of the chip ESI emitter, per se, could be evaluated by a direct comparison with these capillary emitters. For infusion experiments, the nanospray fused-silica capillary should be connected to a syringe pump and tested at a flow rate of 50–100 nL/min. For example, the infusion of a 0.1 μM solution of a standard peptide mixture at a flow rate of 60 nL/min, should generate mass spectra with S/N > 5–10 when the ion accumulation or spectrum averaging time is lesser or equal to 1 s. Under these conditions, the amount of sample consumed per spectrum and per second, will be 100 amol, an amount that is detectable by most modern mass spectrometers. The peptide solution should be prepared in H_2O : CH_3CN : CH_3COOH, 50:50:0.1, % v/v (CH_3OH could be used instead of CH_3CN). An efficient chip ESI emitter should provide for comparable performance. A microfluidic setup that eliminates the effect of possible contributions from other processes that occur on the chip, will be essential to the success of testing the performance of the chip emitter. Preferably, for this particular experiment, the chip should be connected to a syringe pump such that the flow rate can be precisely set.

For separation chips, a similar experiment should be performed. The injection of a 100 pL plug (the approximate volume of a cross-injector on a chip) of a 10 μM peptide solution (amount injected will be ~1 fmol) should be detectable (S/N > 3) if the peptide is eluting within a lesser or equal to 1 s time window. To clearly define the peak shape for such an experiment, the MS instrument should be capable of acquiring data at a storage rate of 5–10 mass spectra/s.

For on-chip MALDI-MS experiments, the performance of the chip should be similar to a regular MALDI target plate as long as the sample can be deposited in a similar dimension spot and the chip material or eluent additives do not alter or suppress the analyte ionization process. However, microfabricated devices that enable on-spot enrichment should provide for better performance than regular MALDI targets. If the deposition from the chip to a regular target plate is pursued, the effect of the interface design on the separation efficiency should be considered.

53.6.2 Options and Trades

CE versus LC versus CEC. CE will provide for faster separations, while LC will provide for increased sample loading capacity. Complex interactions between the sample, the stationary phase, and the mobile phase can result in a difficult optimization process of CEC-MS methods. LC and CEC do not necessitate high buffer concentrations, thus the ESI signal is not suppressed during analysis. The higher the efficiency of either separation method, the more intense the analyte peaks will be, the better the detection limits, and the more complex samples can be analyzed.

ESI versus MALDI. ESI from the chip will enable chip interfacing to a broader range of mass spectrometers with MS/MS capabilities, and ESI-MS/MS will generally produce more informative peptide fragmentation patterns in comparison to MALDI-MS/MS. This is an outcome of the fact that the ESI process favors multiple charging, while the MALDI process favors single charging of peptides. Singly charged species often do not produced informative fragmentation patterns when subjected to CID, and thus structural information cannot be extracted from the tandem mass spectrum. MALDI ionization, however, provides for simplicity, fast analysis, and high-throughput processing. In addition, high-energy collision fragmentation capabilities, such as the ones available with MALDI-TOF/TOF systems, can provide for more complex structural information.

On-Spot Enrichment versus Separation. On-spot enrichment such as with SELDI-MS is very easy to perform. However, the method is not appropriate for the detection of low abundant components in complex mixtures that are better handled by using enrichment and sample separation steps before MS analysis.

53.6.3 Additional Experimental Considerations and Useful Tips

Detection Limits. The eluent pH will affect the charge state distribution envelope of proteins, that is, low pH values will favor higher charge states. Large MW proteins are capable of acquiring a large number of charges, and thus the ESI signal will be distributed over a larger number of charge states (i.e., ions). This will inherently result in worse detection limits for proteins than for peptides or other small molecules. In addition, the ESI droplet surface gets saturated faster with protein analytes that carry multiple charges, and on the overall, this also results in worse detection levels for protein analytes. For micro-/nano-ESI sources tested with infusion experiments, the ESI signal starts to level out at analyte concentrations higher than 10–30 μM, as at these levels, the ESI droplets become saturated with charged species. Thus, for infusion experiments, the sample should be prepared at lower level concentrations. For MALDI-MS ion sources, a proper choice of matrix components and of their concentration, as well as through mixing of the sample with the matrix, are essential to obtaining good quality signals.

Buffer Systems. Buffer systems should be preferentially volatile and of low concentration. Detergents are particularly detrimental to the ESI process. Volatile buffers should be prepared fresh every day to maintain a stable pH, EOF, and thus stable ion signal. Solvent evaporation, and/or buffer depletion in the reservoirs, will result in a change in pH, EOF, and ESI signal response. Thus, the buffers should be replenished on a regular basis. The EOF must be maintained in the direction of the ESI emitter, especially when nanospray sources are used (i.e., no liquid sheath or liquid junction fluids are delivered to the source).

EOF and Channel Surface Contamination. Flow visualization with fluorescent dyes is very useful for evaluating the direction and magnitude of the EOF and of possible channel surface contamination. Channel coatings must be covalently attached to ensure stable EOF and to prevent bleeding and the generation of background ions in the mass spectrometer. Gluing systems that are used to affix the reservoirs on the chip, or to assemble the chip, can dissolve in the eluent (especially if there are organic constituents), suppress the EOF, and also generate undesirable background ions. Organic contaminants can be potentially burned out from glass chips at 550°C (however, this process will destroy the channel coating), although complete EOF recovery may not be possible.

53.7 CONCLUDING REMARKS

The substantial progress witnessed by microfluidics during the past 18 years demonstrates that the early promises of miniaturization have started to turn into reality. While still in relatively early stages of development, integrated systems with broad functionality deliver today performance similar to or better than the one displayed by conventional instrumentation. Miniaturized systems are capable of performing complex sample preparation tasks and 2D separations. The development of cost-effective, disposable analysis platforms is essential for applications that involve high-throughput analysis of hundreds of microliter-sized samples. Advanced integration with state-of-the-art MS detection will provide a unique and powerful technology that will enable fast, unambiguous detection of trace level components. Continuous improvements in MS instrument sensitivity and the development of miniaturized and possibly multiplexed mass spectrometers will further support the miniaturization trends in bioanalytical instrumentation.

A recent manuscript has compiled a list of over 100 companies that produce and commercialize miniaturized analytical devices.[236] In future, off-the-shelf chips with versatile configurations could become easily available for customary needs. Comprehensive integration of μTAS, microelectromechanical systems (MEMS), and biosensor technologies will benefit practically any field of biological, biomedical, biochemical, or pharmaceutical sciences. It is envisioned that throughput, disposability, and contamination-free analysis will be the key ingredients that will advance these miniaturized systems even in more demanding areas, such as the clinical environment, for high-throughput population screening applications.

ACKNOWLEDGMENTS

This work was supported by NSF grant Career BES-0448840. The author thanks Ms. Jodi Lewis and Ms. Xu Yang for support with the preparation of the manuscript.

REFERENCES

1. Manz, A., Graber, N., Widmer, H.M., Miniaturized total chemical-analysis systems-a novel concept for chemical sensing, *Sens. Actuat. B*, 1, 244–248, 1990.
2. Manz, A., Fettinger, J.C., Verpoorte, E., Ludi, H., Widmer, H.M., Harrison D.J., Micromachining of monocrystalline silicon and glass for chemical-analysis systems-a look into next century technology or just a fashionable craze, *Trends Anal. Chem.*, 10, 144–149, 1991.
3. Manz, A., Harrison, D.J., Verpoorte, E.M.J., Fettinger, J.C., Paulus, A., Lüdi, H., Widmer, H.M., Planar chip technology for miniaturization and integration of separation techniques into monitoring systems. Capillary electrophoresis on a chip, *J. Chromatogr.*, 593, 253–258, 1992.
4. Harrison, D.J., Manz, A., Fan, Z.H., Ludi, H., Widmer, H.M., Capillary electrophoresis and sample injection systems integrated on a planar glass chip, *Anal. Chem.*, 64, 1926–1932, 1992.
5. Harrison, D.J., Fluri, K., Fan, Z., Effenhauser, C.S., Manz, A., Micromachining a miniaturized capillary electrophoresis-based chemical analysis system on a chip, *Science*, 261, 895–897, 1993.
6. Harrison, D.J., Glavina PG, Manz A. 1993. Towards Miniaturized Electrophoresis and Chemical-Analysis Systems on Silicon-an Alternative to Chemical Sensors, *Sens. Actuat. B*, 10, 107–116, 1993.
7. Jacobson, S.C., Hergenroder, R., Koutny, L.B., Warmack, R.J., Ramsey, J.M, Effects of injection schemes and column geometry on the performance of microchip electrophoresis devices, *Anal. Chem.*, 66, 1107–1113, 1994.
8. Jacobson, S.C., Hergenroder, R., Koutny, L.B., Ramsey, J.M., High-speed separations on a microchip, *Anal. Chem.*, 66, 1114–1118, 1994.
9. Jacobson, S.C., Hergenroder, R., Koutny, L.B., Ramsey, J.M., Open-channel electrochromatography on a microchip, *Anal. Chem.*, 66, 2369–2373, 1994.
10. Woolley, A.T., Mathies, R.A., Ultra-high-speed DNA-sequencing using capillary electrophoresis chips, *Anal. Chem.*, 67, 3676–3680, 1995.
11. Jakeway, S.C., de Mello, A.J., Russell, E.L., Miniaturized total analysis systems for biological analysis, *Fresenius J. Anal. Chem.*, 366, 525–539, 2000.
12. Greenwood, P.A., Greenway, G.M., Sample manipulation in micrototal analysis systems, *Trends Anal. Chem.*, 21, 726–740, 2002.
13. Reyes, D.R., Iossifidis, D., Auroux, P.-A., Manz, A., Micro total analysis systems. 1. Introduction, theory, and technology, *Anal. Chem.*, 74, 2623–2636, 2002.
14. Auroux, P.A., Iossifidis, D., Reyes, D.R., Manz, A., Micro total analysis systems. 2. Analytical standard operations and applications. *Anal. Chem.*, 74, 2637–2652, 2002.
15. Becker, H., Gärtner, C., Polymer microfabrication methods for microfluidic analytical applications, *Electrophoresis*, 21, 12–26, 2000.
16. Khandurina, J., Guttman, A., Bioanalysis in microfluidic devices, *J. Chromatogr. A*, 943, 159–183, 2002.
17. Thomas, G.A., Farquar, H.D., Sutton, S., Hammer, R.P., Soper, S.A., BioMEMS using electrophoresis for the analysis of genetic mutations, *Expert Rev. Mol. Diagn.*, 2, 429–47, 2002.

18. Landers, J.P., Molecular diagnostics on electrophoretic microchips, *Anal. Chem.*, 75, 2919–2927, 2003.
19. Weigl, B.H., Bardell, R.L., Cabrera, C.R., Lab-on-a-chip for drug development, *Adv. Drug Deliv. Rev.*, 55, 349–377, 2003.
20. Ziaie, B., Baldi, A., Lei, M., Gu, Y., Siegel, R.A., Hard and soft micromachining for BioMEMS: review of techniques and examples of applications in microfluidics and drug delivery, *Adv. Drug Deliv. Rev.*, 56, 145–172, 2004.
21. Madou, M., *Fundamentals of Microfabrication*, Boca Raton: CRC Press, 1997.
22. Jacobson, S.C., Culbertson, C.T., Daler J.E., Ramsey J.M., Microchip structures for submillisecond electrophoresis, *Anal. Chem.*, 70, 3476–3480, 1998.
23. Emrich, C.A., Tian, H., Medintz, I.L., Mathies, R.A., Microfabricated 384-lane capillary array electrophoresis bioanalyzer for ultrahigh-throughput genetic analysis, *Anal. Chem.*, 74, 5076–5083, 2002.
24. Aborn, J.H., El-Difrawy, S.A., Novotny, M., Gismondi, E.A., Lam, R., Matsudaira, P., Mckenna, B.K., O'Neil, T., Streechon, P., Ehrlich, D.J., A 768-lane microfabricated system for high-throughput DNA sequencing, *Lab Chip*, 5, 669–674, 2005.
25. Stachowiak, T.B., Svec, F., Fréchet, J.M.J., Chip electrochromatography, *J. Chromatogr. A*, 1044, 97–111, 2004.
26. Ro, K.W., Nayak, R., Knapp, D.R., Monolithic media in microfluidic devices for proteomics, *Electrophoresis*, 27, 3547–3558, 2006.
27. Kutter, J.P., Jacobson, S.C., Matsubara, N., Ramsey, J.M., Solvent-programmed microchip open-channel electrochromatography, *Anal. Chem.*, 70, 3291–3297, 1998.
28. He, B., Ji, J., Regnier, F.E., Capillary electrochromatography of peptides in a microfabricated system, *J. Chromatogr. A*, 853, 257–262, 1999.
29. Oleschuk, R.D., Shultz-Lockyear, L.L., Ning, Y., Harrison, D.J., Trapping of bead-based reagents within microfluidic systems: on-chip solid-phase extraction and electrochromatography, *Anal. Chem.*, 72, 585–590, 2000.
30. Ceriotti, L., de Rooij, N.F., Verpoorte, E., An integrated fritless column for on-chip capillary electrochromatography with conventional stationary phases, *Anal. Chem.*, 74, 639–647, 2002.
31. Breadmore, M.C., Shrinivasan, S., Karlinsey, J., Ferrance J.P., Norris, P.M., Landers, J.P., Towards a microchip-based chromatographic platform. Part 2: Sol-gel phases modified with polyelectrolyte multilayers for capillary electrochromatography, *Electrophoresis*, 24, 1261–1270, 2003.
32. Lazar, I.M., Li L.J., Yang, Y., Karger, B.L., Microfluidic device for capillary electrochromatography-mass spectrometry, *Electrophoresis*, 24, 3655–3662, 2003.
33. Moore, A.W., Jacobson, S.C., Ramsey, J. M., Microchip separations of neutral species via micellar electrokinetic capillary chromatography, *Anal. Chem.*, 67, 4184–4189, 1995.
34. Ramsey, J.D., Collins, G.E., Integrated microfluidic device for solid-phase extraction coupled to micellar electrokinetic chromatography separation, *Anal. Chem.*, 77, 6664–6670, 2005.
35. Macounová, K., Cabrera, C.R., Holl, M.R., Yager, P., Generation of natural pH gradients in microfluidic channels for use in isoelectric focusing, *Anal. Chem.*, 72, 3745–3751, 2000.
36. Cabrera, C.R., Finlayson, B., Yager, P., Formation of natural pH gradients in a microfluidic device under flow conditions: model and experimental validation, *Anal. Chem.*, 73, 658–666, 2001.
37. Macounová, K., Cabrera, C.R., Yager, P., Concentration and separation of proteins in microfluidic channels on the basis of transverse IEF, *Anal. Chem.*, 73, 1627–1633, 2001.
38. Raisi, F., Belgrader, P., Borkholder, D.A., Herr, A.E., Kintz, G.J., Pourhamadi, F., Taylor, M.T., Northrop, M.A., Microchip isoelectric focusing using a miniature scanning detection system, *Electrophoresis*, 22, 2291–2295, 2001.
39. Tan, A., Pashkova, A., Zang, L., Foret, F., Karger, B.L., A miniaturized multichamber solution isoelectric focusing device for separation of protein digests, *Electrophoresis*, 23, 3599–3607, 2002.
40. Huang, T., Pawliszyn, J., Microfabrication of a tapered channel for isoelectric focusing with thermally generated pH gradient, *Electrophoresis*, 23, 3505–3510, 2002.
41. Tan, W., Fan, Z.H., Qiu, C.X., Ricco, A.J., Gibbons, I., Miniaturized capillary isoelectric focusing in plastic microfluidic devices, *Electrophoresis*, 23, 3638–3645, 2002.
42. Zilberstein, G.V., Baskin, E.M., Bukshpan, S., Parallel processing in the isoelectric focusing chip, *Electrophoresis*, 24, 3735–3744, 2003.

43. Herr, A.E., Molho, J.I., Drouvalakis, K.A., Mikkelsen, J.C., Utz, P.J., Santiago, J.G., Kenny, T.W., On-chip coupling of isoelectric focusing and free solution electrophoresis for multidimensional separations, *Anal. Chem.*, 75, 1180–1187, 2003.
44. Xu, Y., Zhang, C.-X, Janasek, D., Manz, A., Sub-second isoelectric focusing in free flow using a microfluidic device, *Lab Chip*, 3, 224–227, 2003.
45. Zilberstein, G.V., Baskin, E.M., Bukshpan, S., Korol, L.E., Parallel isoelectric focusing II, *Electrophoresis*, 25, 3643–3651, 2004.
46. Li, Y., Buch, J.S., Rosenberger, F., DeVoe, D.L., Lee, C.S., Integration of isoelectric focusing with parallel sodium dodecyl sulfate gel electrophoresis for multidimensional protein separations in a plastic microfluidic network, *Anal. Chem.*, 76, 742–748, 2004.
47. Cui, H., Horiuchi, K., Dutta, P., Ivory, C.F., Multistage isoelectric focusing in a polymeric microfluidic chip, *Anal. Chem.*, 77, 7878–7886, 2005.
48. Cui, H., Horiuchi, K., Dutta, P., Ivory, C.F., Isoelectric focusing in a poly(dimethylsiloxane) microfluidic chip, *Anal. Chem.*, 77, 1303–1309, 2005.
49. Yao, B., Yang, H., Liang, Q., Luo, G., Wang, L., Ren, K., Gao, Y., Wang, Y., Qiu, Y., High-speed, whole-column fluorescence imaging detection for isoelectric focusing on a microchip using an organic light emitting diode as light source, *Anal. Chem.*, 78, 5845–5850, 2006.
50. Das, C., Fan, Z.H., Effects of separation length and voltage on isoelectric focusing in a plastic microfluidic device, *Electrophoresis*, 27, 3619–3626, 2006.
51. Song, Y.-A., Hsu, S., Stevens, A.L., Han, H., Continuous-flow pL-based sorting of proteins and peptides in a microfluidic chip using diffusion potential, *Anal. Chem.*, 78, 3528–3536, 2006.
52. Veerland, W.N., Williams, S.J., Barron, A.E., Sassi, A.P., Tandem isotachophoresis-zone electrophoresis via base-mediated destacking for increased detection sensitivity in microfluidic systems, *Anal. Chem.*, 73, 3059–3065, 2003.
53. Jeong, Y., Choi, K., Kang, M.K., Chun, K., Chung, D.S., Transient isotachophoresis of highly saline samples using a microchip, *Sens. Actuat. B*, 104, 269–275, 2005.
54. Chen, L., Prest, J.E., Fielden, P.R., Goddard, N.J., Manz, A., Day, P.J.R., Miniaturized isotachophoresis analysis, *Lab Chip*, 6, 474–487, 2006.
55. Janasek, D., Schilling, M., Franzke, J., Manz, A., Isotachophoresis in free-flow using a miniaturized device, *Anal. Chem.*, 78, 3815–3819, 2006.
56. Bousse, L., Mouradian, S., Minalla, A., Yee, H., Williams, K., Dubrow, R., Protein sizing on a microchip, *Anal. Chem.*, 73, 1207–1212, 2001.
57. Jin, L.J., Giordano, B.C., Landers, J., Dynamic labeling during capillary or microchip electrophoresis for laser-induced fluorescence detection of protein–SDS complexes without pre- or postcolumn labeling, *Anal. Chem.*, 73, 4994–4999, 2001.
58. Hatch, A.V., Herr, A.E., Throckmorton, D.J., Brennan, J.S., Singh, A.K., Integrated preconcentration SDS-PAGE of proteins in microchips using photopatterned cross-linked polyacrylamide gels, *Anal. Chem.*, 78, 4976–4984, 2006.
59. Agirregabiria, M., Blanco, F.J., Berganzo, J., Fullaondo, A., Zubiaga, A.M., Mayora, K., Ruano-Lopez, J.M., SDS-CGE of proteins in microchannels made of SU-8 films, *Electrophoresis*, 27, 3627–3634, 2006.
60. Rocklin, R.D., Ramsey, R.S., Ramsey, J. M., A microfabricated fluidic device for performing two-dimensional liquid-phase separations, *Anal. Chem.*, 72, 5244–5249, 2000.
61. Ramsey, J.D., Jacobson, S.C., Culbertson, C.T., Ramsey, J.M., High-efficiency, two-dimensional separations of protein digests on microfluidic devices, *Anal. Chem.*, 75, 3758–3764, 2003.
62. Gottschlich, N., Jacobson, S.C., Culbertson, C.T., Ramsey, J.M., Two-dimensional electrochromatography/capillary electrophoresis on a chip, *Anal. Chem.*, 73, 2669–2674, 2001.
63. Shadpour, H., Soper, S.A., Two-dimensional electrophoretic separation of proteins using poly (methylmethacrylate) microchips, *Anal. Chem.*, 78, 3519–3527, 2006.
64. Buch, J.S., Rosenberger, F., Highsmith, W.E., Kimball, C., DeVoe, D.L., Lee, C.S., Denaturing gradient-based two-dimensional gene mutation scanning in a polymer microfluidic network, *Lab Chip*, 5, 392–400, 2005.

65. Griebel, A., Rund, S., Schonfeld, F., Dorner, W., Konrad, R., Hardt, S., Integrated polymer chip for two-dimensional capillary gel electrophoresis, *Lab Chip*, 4, 18–23, 2004.
66. Usui, K., Hiratsuka, A., Shiseki, K., Maruo, Y., Matsushima, T., Takahashi, K., Unuma, Y., Sakairi, K., Namatame, I., Ogawa, Y., Yokoyama, K., A self-contained polymeric 2-de chip system for rapid and easy analysis, *Electrophoresis*, 27, 3635–3642, 2006.
67. Ölvecká, E., Kaniansky, D., Pollák, B., Stanislawski, B., Separation of proteins by zone electrophoresis on-line coupled with isotachophoresis on a column-coupling chip with conductivity detection, *Electrophoresis*, 25, 3865–3874, 2004.
68. Weinberger, R., *Practical Capillary Electrophoresis*, Academic Press, Inc., 1993.
69. De Hoffmann, E., Stroobant, V., *Mass Spectrometry. Principles and Applications*. 2nd ed., John Wiley & Sons, 2002.
70. Herbert, C.J., Johnstone, R.A.W., *Mass Spectrometry Basics*, Boca Raton: CRC Press, 2003.
71. Gross, J.H., *Mass Spectrometry. A Textbook*, Springer, 2004.
72. Lazar, I.M., Ramsey, R.S., Sundberg, S., Ramsey, J.M., Subattomole-sensitivity microchip nanoelectrospray source with time-of-flight mass spectrometry detection, *Anal. Chem.*, 71, 3627–3631, 1999.
73. Figeys, D., Pinto, D., Proteomics on a chip: promising developments, *Electrophoresis*, 22, 208–216, 2001.
74. Figeys, D., Adapting arrays and lab-on-a-chip technology for proteomics, *Proteomics*, 2, 373–382, 2002.
75. Limbach, P.A., Meng, Z., Integrating micromachined devices with modern mass spectrometry, *Analyst*, 127, 693–700, 2002.
76. Foret, F., Preisler, J., Liquid phase interfacing and miniaturization in matrix-assisted laser desorption/ionization mass spectrometry, *Proteomics*, 2, 360–372, 2002.
77. Marko-Varga, G., Nilsson, J., Laurell, T., New directions of miniaturization within the proteomics research area, *Electrophoresis*, 24, 3521–3532, 2003.
78. Lion, N., Rohner, T.C., Dayon, L., Arnaud, I.L., Damoc, E., Youhnovski, N., Wu, Z.-Y., Roussel, C., Josserand, J., Jensen, H., Rossier, J.S., Przybylski, M., Girault, H.H., Microfluidic systems in proteomics, *Electrophoresis*, 24, 3533–3562, 2003.
79. Marko-Varga, G., Nilsson, J., Laurell, T., New directions of miniaturization within the biomarker research area, *Electrophoresis*, 25, 3479–3491, 2004.
80. Ohnesorge, J., Neusüß, C., Wätzig, H., Quantitation in capillary electrophoresis-mass spectrometry, *Electrophoresis*, 26, 3973–3987, 2005.
81. DeVoe, D.L., Lee, C.S., Microfluidic technologies for MALDI-MS in proteomics, *Electrophoresis*, 27, 3559–3568, 2006.
82. Lazar, I.M., Grym, J., Foret, F., Microfabricated devices: a new sample introduction approach to mass spectrometry, *Mass Spectrom. Rev.*, 25, 573–594, 2006.
83. Xue, Q.F., Foret, F., Dunayevskiy, Y.M., Zavracky, P.M., McGruer, N.E., Karger. B.L., Multichannel microchip electrospray mass spectrometry, *Anal. Chem.*, 69, 426–430, 1997.
84. Xue, Q.F., Dunayevskiy, Y.M., Foret, F., Karger, B.L., Integrated multichannel microchip electrospray ionization mass spectrometry: analysis of peptides from on-chip tryptic digestion of melittin, *Rapid Commun. Mass Spectrom.*, 11, 1253–1256, 1997.
85. Ramsey, R.S., Ramsey, J.M., Generating electrospray from microchip devices using electroosmotic pumping, *Anal. Chem.*, 69, 1174–1178, 1997.
86. Rohner, T.C., Rossier, J.S., Girault H.H., Polymer microspray with an integrated thick-film microelectrode, *Anal. Chem.*, 73, 5353–5357, 2001.
87. Huikko, K., Östman, P, Grigoras, K., Tuomikoski, S., Tiainen, V.-M., Soininen, A., Puolanne, K., Manz, A., Franssila, S., Kostiainen, R., Kotiaho, T., Poly(dimethylsiloxane) electrospray devices fabricated with diamond-like carbon-poly(dimethylsiloxane) coated SU-8 masters, *Lab Chip*, 3, 67–72, 2003.
88. Lozano, P., Martínez-Sánchez, M., Lopez-Urdiales, J.M., Electrospray emission from nonwetting flat dielectric surfaces, *J. Coll. Inter. Sci.*, 276, 392–399, 2004.
89. Wang, Y.-X., Cooper, J.W, Lee, C.S., DeVoe, D. L., Efficient electrospray ionization from polymer microchannels using integrated hydrophobic membranes, *Lab Chip*, 4, 363–367, 2004.

90. Zhang, B., Liu, H, Karger, B.L., Foret, F., Microfabricated devices for capillary electrophoresis-electrospray mass spectrometry, *Anal. Chem.*, 71, 3258–3264, 1999.
91. Brivio, M., Oosterbroeck, R.E., Verboom, W., van den Berg, A., Reinhoudt, D.N., Simple chip-based interfaces for on-line monitoring of supramolecular interactions by nano-ESI MS, *Lab Chip*, 5, 1111–1122, 2005.
92. Grym, J., Otevřel M., Foret, F., Aerodynamic mass spectrometry interfacing of microdevices without electrospray tips, *Lab Chip*, 6, 1306–1314, 2006.
93. Li, J., Kelly, J.F., Chemushevich, I., Harrison, D.J., Thibault, P., Separation and identification of peptides from gel-isolated membrane proteins using a microfabricated device for combined capillary electrophoresis/nanoelectrospray mass spectrometry, *Anal. Chem.*, 72, 599–609, 2000.
94. Li, J., Thibault, P., Bings, N.H., Skinner, C.D., Wang, C., Colyer, C., Harrison, J., Integration of microfabricated devices to capillary electrophoresis-electrospray mass spectrometry using a low dead volume connection: Application to rapid analyses of proteolytic digests, *Anal. Chem.*, 71, 3036–3045, 1999.
95. Bings, N.H., Wang, C., Skinner, C.D., Colyer, C.L., Thibault, P., Harrison, D.J., Microfluidic devises connected to fused-silica capillaries with minimal dead volume, *Anal. Chem.*, 71, 3292–3296, 1999.
96. Zhang, B., Foret, F., Karger, B.L., A microdevice with integrated liquid junction for facile peptide and protein analysis by capillary electrophoresis/electrospray mass spectrometry, *Anal. Chem.*, 72, 1015–1022, 2000.
97. Lazar, I.M., Ramsey, R.S., Jacobson, S.C., Foote, R.S., Ramsey, J.M., Novel microfabricated device for electrokinetically induced pressure flow and electrospray ionization mass spectrometry, *J. Chromatogr. A*, 892, 195–201, 2000.
98. Vrouwe, E.X., Gysler, J., Tjaden, U.R., van der Greef, J., Chip-based capillary electrophoresis with an electrodless nanospary interface, *Rapid Commun. Mass Spectrom.*, 14, 1682–1688, 2000.
99. Meng, Z., Qi, S., Soper, S.A., Limbach, P.A., Interfacing a polymer-based micromachined device to a nanoelectrospray ionization Fourier transform ion cyclotron resonance mass spectrometer, *Anal. Chem.*, 73, 1286–1291, 2001.
100. Sung, W.-C., Huang, S.-Y., Liao, P.-C., Lee, G.-B, Li, C.-W., Chen, S.-H., Poly(dimethylsiloxane)-based microfluidic device with electrospray ionization mass spectrometry for protein identification, *Electrophoresis*, 24, 3648–3654, 2003.
101. Ssenyange, S., Taylor, J., Harrison, D.J., McDermott, M.T., A glassy carbon microfluidic device for electrospray mass spectrometry, *Anal. Chem.*, 76, 2393–2397, 2004.
102. Yang, Y., Li, C., Kameoka, J., Lee, K.H., Craighead, H.G., A polymeric microchip with integrated tips and *in situ* polymerized monolith for electrospray mass spectrometry, *Lab Chip*, 5, 869–876, 2005.
103. Figeys, D., Aebersold, R., Nanoflow solvent gradient delivery from a microfabricated device for protein identifications by electrospray ionization mass spectrometry, *Anal. Chem.*, 70, 3721–3727, 1998.
104. Liu, H.H., Felten, C., Xue, Q.F., Zhang, B.L., Jedrzejewski, P., Karger. B.L., Foret, F., Development of multichannel devices with an array of electrospray tips far high-throughput mass spectrometry, *Anal. Chem.*, 72, 3303–3310, 2000.
105. Xu, N., Lin, Y.H., Hofstadler, S.A., Matson, D., Call, C.J., Smith, R.D., A microfabricated dialysis device for sample cleanup in electrospray ionization mass spectrometry, *Anal. Chem.*, 70, 3553–3556, 1998.
106. Figeys, D., Ning, Y.B., Aebersold, R., A microfabricated device for rapid protein identification by microelectrospray ion trap mass spectrometry, *Anal. Chem.*, 69, 3153–3160, 1997.
107. Figeys, D., Gygi, S.P., McKinnon, G., Aebersold, R., An integrated microfluidics tandem mass spectrometry system for automated protein analysis, *Anal. Chem.*, 70, 3728–3734, 1998.
108. Figeys, D., Lock, C., Taylor, L., Aebersold, R., Microfabricated device coupled with an electrospray ionization quadrupole time-of-flight mass spectrometer: Protein identifications based on enhanced-resolution mass spectrometry and tandem mass spectrometry data, *Rapid Commun. Mass Spectrom.*, 12, 1435–1444, 1998.

109. Figeys, D., Aebersold, R., Microfabricated modules for sample handling, sample concentration and flow mixing: application to protein analysis by tandem mass spectrometry, *J. Biomech. Eng.*, 121, 7–12, 1999.
110. Razunguzwa, T.T., Lenke, J., Timperman, A.T., An electrokinetic/hydrodynamic flow microfluidic CE-ESI-MS interface utilizing a hydrodynamic flow restrictor for delivery of samples under low EOF conditions, *Lab Chip*, 5, 851–855, 2005.
111. Chen, S.-H., Sung, W.-C., Lee, G.-B., Lin, Z.-Y., Chen, P.-W., Liao, P.-C., A disposable poly(methylmethacrylate)-based microfluidic module for protein identification by nanoelectrospray ionization-tandem mass spectrometry, *Electrophoresis*, 22, 3972–3977, 2001.
112. Lazar, I.M., Karger, B.L., Multiple open-channel electroosmotic pumping system for microfluidic sample handling, *Anal. Chem.*, 74, 6259–6268, 2002.
113. Lazar, I.M., Trisiripisal, P., Sarvaiya, H., Microfluidic liquid chromatography system for proteomic applications and biomarker screening, *Anal. Chem.*, 78, 5513–5524, 2006.
114. Yue, G.E., Roper, M.G., Jeffery, E.D., Easley, C.J., Balchunas, C., Landers, J.P., Ferrance, J.P., Glass microfluidic devices with thin membrane voltage junctions for electrospray mass spectrometry, *Lab Chip*, 5, 619–627, 2005.
115. Wachs, T., Henion, J., Electrospray device for coupling microscale separations and other miniaturized devices with electrospray mass spectrometry, *Anal. Chem.*, 73, 632–638, 2001.
116. Deng, Y.Z., Zhang, N.W., Henion, J., Chip-based quantitative capillary electrophoresis/mass spectrometry determination of drugs in human plasma, *Anal. Chem.*, 73, 1432–1439, 2001.
117. Deng, Y.Z., Henion, J., Li, J.J., Thibault, P., Wang, C., Harrison, D.J., Chip-based capillary electrophoresis/mass spectrometry determination of carnitines in human urine, *Anal. Chem.*, 73, 639–646, 2001.
118. Wang, G.-J, Yang, M.-W., Chang, Y.-Z., Dead volume free micro-machined electro-spray chip for mass spectrometry, *Biomed. Microdev.*, 6, 159–164, 2004.
119. Schultz, G.A., Corso, T.N., Prosser, S.J., Zhang, S., A fully integrated monolithic microchip electrospray device for mass spectrometry, *Anal. Chem.*, 72, 4058–4063, 2000.
120. Dethy, J.-M., Ackerman, B.L., Delatour, C., Henion, J.D., Schultz, G.A., Demonstration of direct bioanalysis of drugs in plasma using nanoelectrospray infusion from a silicon chip coupled with tandem mass spectrometry, *Anal. Chem.*, 75, 4058–4063, 2003.
121. Griss, P., Melin, J., Sjodahl, J., Roeraade, J., Stemme, G., Development of micromachined hollow tips for protein analysis based on nanoelectrospray ionization mass spectrometry, *J. Micromech. Microeng.*, 12, 682–687, 2002.
122. Sjödahl, J., Melin, J., Griss, P., Emmer, ?., Stemme, G., Roeraade, J., Characterization of micromachined hollow tips for two-dimensional nanoelectrospray mass spectrometry, *Rapid Commun. Mass Spectrom.*, 17, 337–341, 2003.
123. Arscott, S., Le Gac, S., Rolando, C., A polysilicon nanoelectrospray-mass spectrometry source based on a capillary microfluidic slot, *Sens. Actuat. B*, 106, 741–749, 2005.
124. Licklider, L., Wang, X.Q., Desai, A., Tai, Y.C., Lee, T.D., A micromachined chip-based electrospray source for mass spectrometry, *Anal. Chem.*, 72, 367–375, 2000.
125. Xie, J., Miao, Y., Shih, J., He, Q., Liu, J., Tai, Y.-C., Lee, T.D., An electrochemical pumping system for on-chip gradient generation, *Anal. Chem.*, 76, 3756–3763, 2004.
126. Xie, J., Miao, Y., Shih, J., Tai, Y.-C., Lee, T.D., Microfluidic platform for liquid chromatography-tandem mass spectrometry analysis of complex peptide mixtures, *Anal. Chem.*, 77, 6947–6953, 2005.
127. Kameoka, J., Orth, R., Ilic, B., Czaplewski, D., Wachs, T., Craighead, H.G., An electrospray ionization source for integration with microfluidics, *Anal. Chem.*, 74, 5897–5901, 2002.
128. Ek, P., Sjödahl, J., Roeraade, J., Electrospray ionization from a gap with adjustable width, *Rapid Commun. Mass Spectrom.*, 20, 3176–3182, 2006.
129. Gobry, V., van Oostrum, J., Martinelli, M., Rohner, T.C., Reymond, F., Rossier, J.S., Girault, H.H., Microfabricated polymer injector for direct mass spectrometry coupling, *Proteomics*, 2, 405–412, 2002.
130. Lion, N., Gellon, J.O., Girault, H.H., Flow-rate characterization of microfabricated polymer microspray emitters, *Rapid Commun. Mass Spectrom.*, 18, 1614–1620, 2004.

131. Yin, H., Killeen, K., Brennen, R., Sobek, D., Werlich, M., van de Goor, T., Microfluidic chip for peptide analysis with an integrated HPLC column, sample enrichment column, and nanoelectrospray tip, *Anal. Chem.*, 77, 527–533, 2005.
132. Fortier, M.-H., Bonneil, E., Goodley, P., Thibault, P., Integrated microfluidic device for mass spectrometry-based proteomics and its application to biomarker discovery programs, *Anal. Chem.*, 77, 1631–1640, 2005.
133. Hardoiun, J., Duchateau, M., Joubert-Caron, R., Caron, M., Usefulness of an integrated microfluidic device (HPLC-Chip-MS) to enhance confidence in protein identification by proteomics, *Rapid Commun. Mass Spectrom.*, 20, 3236–3244, 2006.
134. Tang, K., Lin, Y., Matson, D.W., Kim, T., Smith R.D., Generation of multiple electrosprays using microfabricated emitter arrays for improved mass spectrometric sensitivity, *Anal. Chem.*, 73, 1658–1663, 2001.
135. Svedberg, M., Pettersson, A., Nilsson, S., Bergquist, J., Nyholm, L., Nikolajeff, F., Markides, K., Sheathless electrospray from polymer microchips, *Anal. Chem.*, 75, 3934–3940, 2003.
136. Muck, A., Svatos, A., Atmospheric molded poly(methylmethacrylate) microchip emitters for sheathless electrospray, *Rapid Commun. Mass Spectrom.*, 18, 1459–1464, 2004.
137. Schilling, M., Nigge, W., Rudzinski, A., Neyer, A., Hergenroder, R., A new on-chip ESI nozzle for coupling of MS with microfluidic devices, *Lab Chip*, 4, 220–224, 2004.
138. Yuan, C.-H., Shiea, J., Sequential electrospray analysis using sharp-tip channels fabricated on a plastic chip, *Anal. Chem.*, 73, 1080–1083, 2001.
139. Kim, J.-S., Knapp, D.R., Microfabrication of polydimethylsiloxane electrospray ionization emitters, *J. Chromatogr. A*, 924, 137–145, 2001.
140. Kim, J.-S., Knapp, D.R., Microfabricated PDMS multichannel emitter for electrospray ionization mass spectrometry, *J. Am. Soc. Mass. Spectrom.*, 12, 463–469, 2001.
141. Kim, J.-S., Knapp, D.R., Miniaturized multichannel electrospray ionization emitters on poly(dimethylsiloxane) microfluidic devices, *Electrophoresis*, 22, 3993–3999, 2001.
142. Chiou, C.H., Lee, G.B., Hsu, H.T., Chen, P.W., Liao, P.C., Micro devices integrated with microchannels and electrospray nozzles using PDMS casting techniques, *Sens. Actuat. B*, 86, 280–286, 2002.
143. Iannacone, J.M., Jacubowski, J.A., Bohn, P.W., Sweedler, J.V., A multilayer poly(dimethylsiloxane) electrospray ionization emitter for sample injection and online mass spectrometric detection, *Electrophoresis*, 26, 4684–4690, 2005.
144. Svedberg, M., Veszelei, M., Axelsson, J., Vangbo, M., Nikolajeff, F., Poly(dimethylsiloxane) microchip: microchannel with integrated open electrospray tip, *Lab Chip*, 4, 322–327, 2004.
145. Liljegren, G., Dahlin, A., Zettersten, C., Bergquist, J., Nyholm, L., On-line coupling of a microelectrode array equipped poly(dimethylsiloxane) microchip with an integrated graphite electrospray emitter for electrospray ionization mass spectrometry, *Lab Chip*, 5, 1008–1016, 2005.
146. Thorslund, S., Lindberg, P., Andrén P.E., Nikolajeff, F., Bergquist, J., Electrokinetic-driven microfluidic system in poly(dimethylsiloxane) for mass spectrometry detection integrating sample injection, capillary electrophoresis, and electrospray emitter on-chip, *Electrophoresis*, 26, 4674–4678, 2005.
147. Le Gac, S., Arscott, S., Rolando, C., A planar microfabricated nanoelectrospray emitter tip based on a capillary slot, *Electrophoresis*, 24, 3640–3647, 2003.
148. Le Gac, S., Arscott, S., Cren-Olivé, C., Rolando, C., Two-dimensional microfabricated sources for nanoelectrospray, *J. Mass Spectrom.*, 38, 1259–1264, 2003.
149. Arscott, S, Le Gac, S., Druon, C., Tabourier, P., Rolando, C., A micro-nib nanoelectrospray source for mass spectrometry, *Sens. Actuat. B*, 98, 140–147, 2004.
150. Arscott, S, Le Gac, S., Druon, C., Tabourier, P., Rolando, C., A planar on-chip micro-nib interface for nanoESI-MS microfluidic applications, *J. Micromech. Microeng.*, 14, 310–316, 2004.
151. Carlier, J., Arscott, S., Thomy, V., Camart, J.-C., Cren-Olivé, C., Le Gac, S., Integrated microfabricated systems including a purification module and an on-chip nano electrospray ionization interface for biological analysis, *J. Chromatogr. A*, 1071, 213–222, 2005.
152. Tuomikoski, S., Sikanen, T., Ketola, R.A., Kostiainen, R., Kotiaho, T., Franssila, S., Fabrication of enclosed SU-8 tips for electrospray ionization-mass spectrometry, *Electrophoresis*, 26, 4691–4702, 2005.

153. Bedair, M.F., Oleschuk, R.D., Fabrication of porous polymer monoliths in polymeric microfluidic chips as an electrospray emitter for direct coupling to mass spectrometry, *Anal. Chem.*, 78, 1130–1138, 2006.
154. Peterson, D.S., Rohr, T., Svec, F., Fréchet, J.M.J., Dual-function microanalytical device by in situ photolithographic grafting of porous polymer monolith: Integrating solid-phase extraction and enzymatic digestion for peptide mass mapping, Anal. Chem., 75, 5328–5335, 2003.
155. Önnerfjord, P., Nilsson, J., Wallman, L., Laurell, T., Marko-Varga, G., Picoliter sample preparation in MALDI-TOF MS using micromachined silicon flow-through dispenser, *Anal. Chem.*, 70, 4755–4760, 1998.
156. Musiyimi, H.K., Guy, J., Narcisse, D.A., Soper, S.A., Murray, K.K., Direct coupling of polymer-based microchip electrophoresis to on-line MALDI-MS using a rotating ball inlet, *Electrophoresis*, 26, 4703–4710, 2005.
157. Miliotis, T., Kjellstrom, S., Nilsson, J., Laurell, T., Edholm, L.E., Marko-Varga, G., Capillary liquid chromatography interfaced to matrix-assisted laser desorption/ionization time-of-flight mass spectrometry using an on-line coupled piezoelectric flow-through microdispenser, *J. Mass Spectrom.*, 35, 369–377, 2000.
158. Laurell, T., Wallman, L., Nilsson, J., Design and development of a silicon microfabricated flow-through dispenser for on-line picoliter sample handling, *J. Micromech. Microeng.*, 9, 369–376, 1999.
159. Laurell, T., Nilsson, J., Marko-Varga, G., Silicon microstructures for high-speed and high-sensitivity protein identifications, *J. Chromatogr. B*, 752, 217–232, 2001.
160. Laurell, T., Marko-Varga, G., Ekström, S., Bengtsson, M., Nilsson J, Microfluidic components for protein characterization, *Rev. Mol. Biotechnol.*, 82, 161–175, 2001.
161. Krone, J.R., Nelson, R.W., Dogruel, D., Williams, P., Granzow, R., BIA/MS: interfacing biomolecular interaction analysis with mass spectrometry, *Anal. Biochem.*, 244, 124–132, 1997.
162. Sönksen, C.P., Nordhoff, E., Jansson, Ö., Malmqvist, M., Roepstorff, P., Combining MALDI mass spectrometry and biomolecular interaction analysis using a biomolecular interaction analysis instrument, *Anal. Chem.*, 70, 2731–2736, 1998.
163. Gilligan, J.J., Schuck, P., Yergey, A.L., Mass spectrometry after capture and small-volume elution of analyte from a surface plasmon resonance biosensor, *Anal. Chem.*, 74, 2041–2047, 2002.
164. Borch, J., Roepstorff, P., Screening for enzyme inhibitors by surface plasmon resonance combined with mass spectrometry, *Anal. Chem.*, 76, 5243–5248, 2004.
165. Natsume, T., Nakayama, H., Jansson, Ö., Isobe, T., Takio, K., Mikoshiba, K., Combination of biomolecular interaction analysis and mass spectrometric amino acid sequencing, *Anal. Chem.*, 72, 4193–4198, 2000.
166. Sydor, J.R., Scalf, M., Sideris, S., Mao, G.D., Pandey, Y., Tan, M., Mariano, M., Moran, M.F., Nock, S., Wagner, P., Chip-based analysis of protein–protein interactions by fluorescence detection and on-chip immunoprecipitation combined with μLC-MS/MS analysis, *Anal. Chem.*, 75, 6163–6170, 2003.
167. Grote, J., Dankbar, N., Gedig, E., Koenig, S., Surface plasmon resonance/mass spectrometry interface, *Anal. Chem.*, 77, 1157–1162, 2005.
168. Whelan, R.J., Zare, R.N., Surface plasmon resonance detection for capillary electrophoresis separations, *Anal. Chem.*, 75, 1542–1547, 2003.
169. Liu, J., Tseng, K., Garcia, B., Lebrilla, C.B., Mukerjee, E., Collins, S., Smith, R., Electrophoresis separation in open microchannels. A method for coupling electrophoresis with MALDI-MS, *Anal. Chem.*, 73, 2147–2151, 2001.
170. Mok, M.L.-S., Hua, L., Phua, J.B.-C., Wee, M.K.-T., Sze, N.S.-K., Capillary isoelectric focusing in pseudo-closed channel coupled to matrix assisted laser desorption/ionization mass spectrometry for protein analysis, *Analyst*, 129, 109–110, 2004.
171. Guo, X., Chan-Park, M.B., Yoon, S.F., Chun, J.-H., Hua, L., Sze, N.S.-K., UV embossed polymeric chip for protein separation and identification based on capillary isoelectric focusing and MALDI-TOF-MS, *Anal. Chem.*, 78, 3249–3256, 2006.
172. Brivio, M., Fokkens, R.H., Verboom, W., Reinhoudt, D.N., Tas, N.R., Goedbloed, M., van den Berg, A., Integrated microfluidic system enabling (bio)chemical reactions with on-line MALDI-TOF mass spectrometry, *Anal. Chem.*, 74, 3972–3976, 2002.

173. Brivio, M., Tas, N.R., Goedbloed M.H., Gardeniers, H.J.G.E., Verboom, W., Reinhoudt D.N., van den Berg, A., A MALDI chip integrated system with a monitoring window, *Lab Chip*, 5, 378–381, 2005.
174. Sim, T.S., Kim, E.-M., Joo, H.S., Kim, B.G., Kim Y.-K., Application of a temperature-controllable microreactor to simple and rapid protein identification using MALDI-TOF MS, *Lab Chip*, 6, 1056–1061, 2006.
175. Gustafsson, M., Hirschberg, D., Palmberg, C., Jornvall, H., Bergman, T., Integrated sample preparation and MALDI mass spectrometry on a microfluidic compact disk, *Anal. Chem.*, 76, 345–350, 2004.
176. Hirschberg, D., Jagerbrink, T., Samskog, J., Gustafsson, M., Stahlberg, M., Alvelius, G., Husman, B., Carlquist, M., Jornvall, H., Bergman, T., Detection of phosphorylated peptides in proteomic analyses using microfluidic compact disk technology, *Anal. Chem.*, 76, 5864–5871, 2004.
177. Wheeler, A.R., Moon, H., Kim, C.-J., Loo, J.A., Garrell, R.L., Electrowetting-based microfluidics for analysis of peptides and proteins by matrix-assisted laser desorption/ionization mass spectrometry, *Anal. Chem.*, 76, 4833–4838, 2004.
178. Wheeler, A.R., Moon, H., Bird, C.A., Loo, R.R.O., Kim, C.-J., Loo, J.A., Garrell, R.L., Digital microfluidics with in-line sample purification for proteomics analyses with MALDI-MS, *Anal. Chem.*, 77, 534–540, 2005.
179. Moon, H., Wheeler, A.R., Garrell, R.L., Loo, J.A., Kim, C.-J., An integrated digital microfluidic chip for multiplexed proteomic sample preparation and analysis by MALDI-MS, *Lab Chip*, 6, 1213–1219, 2006.
180. Ekström, S., Onnerfjord, P., Nilsson, J., Bengtsson, M., Laurell, T., Marko-Varga, G., Integrated microanalytical technology enabling rapid and automated protein identification, *Anal. Chem.*, 72, 286–293, 2000.
181. Ekström, S., Ericsson, D., Onnerfjord, P., Bengtsson, M., Nilsson, J., Marko-Varga, G., Laurell, T., Signal amplification using "spot on-a-chip" technology for the identification of proteins via MALDI-TOF MS, *Anal. Chem.*, 73, 214–219, 2001.
182. Laurell, T., Nilsson, J., Marko-Varga, G., Proteomics-protein profiling technology: the trend towards a microfabricated toolbox concept, *Trends Anal. Chem.*, 20, 225–231, 2001.
183. Marko-Varga, G., Ekström, S., Helldin, G., Nilsson, J., Laurell, T., Disposable polymeric high-density nanovial arrays for matrix assisted laser desorption ionization-time-of-flight mass spectrometry: I. Microstructure development and manufacturing, *Electrophoresis*, 22, 3978–3983, 2001.
184. Ekström, S., Nilsson, J., Helldin, G., Laurell, T., Marko-Varga, G., Disposable polymeric high-density nanovial arrays for matrix assisted laser desorption ionization-time-of-flight mass spectrometry: I. Biological applications, *Electrophoresis*, 22, 3984–3992, 2001.
185. Ekström, S., Wallman, L., Malm, J., Becker, C., Lilja, H., Laurell, T., Marko-Varga, G., Integrated selective enrichment target-a microtechnology platform for matrix assisted laser desorption ionization-mass spectrometry applied on protein biomarkers in prostate diseases, *Electrophoresis*, 25, 3769–3777, 2004.
186. Berhane, T., Limbach, P.A., Functional microfabricated sample targets for matrix-assisted laser desorption/ionization mass spectrometry analysis of ribonucleic acids, *Anal. Chem.*, 75, 1997–2003, 2003.
187. Sjödahl, J., Kempka, M., Hermansson, K., Thorsén, A., Roeraade, J., Chip with twin anchors for reduced ion suppression and improved mass accuracy in MALDI-TOF mass spectrometry, *Anal. Chem.*, 77, 827–832, 2005.
188. Östman, P., Jäntti, S., Grigoras, K., Saarela, V., Ketola, R. A., Franssila, S., Kotiaho, T., Kostiainen, R., Capillary liquid chromatography-microchip atmospheric pressure chemical ionization-mass spectrometry, *Lab Chip*, 6, 948–953, 2006.
189. Petricoin, E.F., III, Ardekani, A.M., Hitt, B.A., Levine, P.J., Fusaro, V.A., Steinberg, S.M., Mills, G.B., Simone, C., Fishman, D.A., Kohn, E.C., Liotta, L.A., Use of proteomic patterns in serum to identify ovarian cancer, *Lancet*, 359, 572–577, 2002.
190. Borrebaeck, C.A.K., Ekström, S., Malmborg Hager, A.C., Nilsson, J., Laurell, T., Marko-Varga, G., Protein chips based on recombinant antibody fragments: a highly sensitive approach as detected by mass spectrometry, *Biotechniques*, 30, 1126–1131, 2001.
191. Finnskog, D., Ressine, A., Laurell, T., Marko-Varga, G., Integrated protein microchip assay with dual fluorescent- and MALDI read-out, *J. Proteome Res.*, 3, 988–994, 2004.

192. Diamandis, E.P., Mass spectrometry as a diagnostic and cancer biomarker discovery tool, *Mol. Cell. Proteomics*, 3, 367–378, 2004.
193. Tang, N., Tornatore, P., Weinberger, S.R., Current developments in SELDI affinity technology, *Mass Spectrom. Rev.*, 23, 34–44, 2004.
194. Sunner, J., Dratz, E., Chen, Y.-C., Graphite surface-assisted laser desorption ionization time-of-flight mass spectrometry of peptides and proteins from liquid solutions, *Anal. Chem.*, 67, 4335–4342, 1995.
195. Chen, Y.-C., Shiea, J., Sunner, J., Thin-layer chromatography-mass spectrometry using activated carbon, surface-assisted laser desorption/ionization, *J. Chromatogr. A*, 826, 77–86, 1998.
196. Wei, J., Buriak, J.M., Siuzdak, G., Desorption–ionization mass spectrometry on porous silicon, *Nature*, 399, 243–246, 1999.
197. Finkel, N.H., Prevo, B.G., Velev, O.D., He, L., Ordered silicon nanocavity arrays in surface-assisted desorption/ionization mass spectrometry, *Anal. Chem.*, 77, 1088–1095, 2005.
198. Zhang, S., Van Pelt, C.K., Henion, J.D., Automated chip-based nanoelectrospray-mass spectrometry for rapid identification of proteins separated by two-dimensional gel electrophoresis, *Electrophoresis*, 24, 3620–3632, 2003.
199. Zhang, S., Chelius, D., Characterization of protein glycosylation using chip-based infusion nanoelectrospray linear ion trap tandem mass spectrometry, *J. Biomol. Tech.*, 15,120–133, 2004.
200. Zamfir, A., Vakhrushev, S., Sterling, A., Niebel, H.J., Allen, M., Peter-Katalinic, J., Fully automated chip-based mass spectrometry for complex carbohydrate system analysis, *Anal. Chem.*, 76, 2046–2054, 2004.
201. Zamfir, A.D., Lion, N., Vukelić, Ž., Bindilla, L., Rossier, J., Girault, H.H., Peter-Katalinic, J., Thin chip microsprayer system coupled to quadrupole time-of-flight mass spectrometer for glycoconjugate analysis, *Lab Chip*, 5, 298–307, 2005.
202. Leuthold, L.A., Grivet, C., Allen, M., Baumert, M., Hopfgartner, G., Simultaneous selected reaction monitoring, MS/MS and MS3 quantitation for the analysis of pharmaceutical compounds in human plasma using chip-based infusion, *Rapid Commun. Mass Spectrom.*, 18, 1995–2000, 2004.
203. Kapron, J.T., Pace, E., Van Pelt, C.K., Henion, J., Quantitation of midazolam in human plasma by automated chip-based infusion nanoelectrospray tandem mass spectrometry, *Rapid Commun. Mass Spectrom.*, 17, 2019–2026, 2003.
204. Xiang, F., Lin, Y., Wen, J., Matson, D.W., Smith, R.D., An integrated microfabricated device for dual microdialysis and on-line ESI ion trap mass spectrometry for analysis of complex biological samples, *Anal. Chem.*, 71, 1485–1490, 1999.
205. Lion, N., Gobry, V., Jensen, H., Rossier, J.S., Girault, H., Integration of a membrane-based desalting step in a microfabricated disposable polymer injector for mass spectrometric protein analysis, *Electrophoresis*, 23, 3583–3588, 2002.
206. Lion, N., Gellon, J.O., Jensen, H., Girault, H.H., On-chip protein sample desalting and preparation for direct coupling with electrospray ionization mass spectrometry, *J. Chromatogr. A*, 1003, 11–19, 2003.
207. Le Gac, S., Carlier, J., Camart, J.-C., Cren-Olivé, C., Rolando, C., Monoliths for microfluidic devices and proteomics, *J. Chromatogr. B*, 808, 3–14, 2004.
208. Wilson, D.J., Konermann, L., Ultrarapid desalting of protein solutions for electrospray mass spectrometry in a microchannel flow device, *Anal. Chem.*, 77, 6887–6894, 2005.
209. Bergkvist, J., Ekström, S., Wallman, L., Löfgren, M., Marko-Varga, G., Nilsson, J., Laurell, T., Improved chip design for integrated solid-phase microextraction in on-line proteomic sample preparation, *Proteomics*, 2, 422–429, 2002.
210. Ekström, S., Malmström, J., Wallman, L., Löfgren, M., Nilsson, J., Laurell, T., Marko-Varga, G., On-chip microextraction for proteomic sample preparation of in-gel digests. *Proteomics*, 2, 413–421, 2002.
211. Wallman, L., Ekström, S., Marko-Varga, G., Laurell, T., Nilsson, J., Autonomous protein sample processing on-chip using solid-phase microextraction, capillary force pumping, and microdispensing, *Electrophoresis*, 25, 3778–3787, 2004.
212. Yu, C., Davey, M.H., Svec, F., Fréchet, J.M.J., Monolithic porous polymer for on-chip solid-phase microextraction and preconcentration prepared by photoinitiated in situ polymerization within a microfluidic device, *Anal. Chem.*, 73, 5088–5096, 2001.

213. Tan, A., Benetton, S., Henion, J.D., Chip-based solid-phase extraction pre-treatment for direct electrospray mass spectrometry analysis using an array of monolithic columns in a polymeric substrate, *Anal. Chem.*, 75, 5504–5511, 2003.
214. Li, C., Lee, K.H., Affinity depletion of albumin from human cerebrospinal fluid using Cibacron-blue-3G-A-derivatized photopatterned copolymer in a microfluidic device, *Anal. Biochem.*, 333, 381–388, 2004.
215. Jiang, Y., Wang, P.-C., Locascio, L.E., Lee, C.S., Integrated plastic microfluidic devices with ESI-MS for drug screening and residue analysis, *Anal. Chem.*, 73, 2048–2053, 2001.
216. Li, J., LeRiche, T., Tremblay, T.L., Wang, C., Bonneil, E., Harrison, D.J., Thibault, P., Application of microfluidic devices to proteomics research - Identification of trace-level protein digests and affinity capture of target peptides, *Mol. Cel. Proteomics*, 1, 157–168, 2002.
217. Wang, C., Oleschuk, R., Ouchen, F., Li, J.J., Thibault, P., Harrison, D.J., Integration of immobilized trypsin bead beds for protein digestion within a microfluidic chip incorporating capillary electrophoresis separations and an electrospray mass spectrometry interface, *Rapid Commun. Mass Spectrom.*, 14, 1377–1383, 2000.
218. Jin, L.J., Ferrance, J., Sanders, J.C., Landers, J.P., A microchip-based proteolytic digestion system driven by electroosmotic pumping, *Lab Chip*, 3, 11–18, 2003.
219. Gao, J., Xu, J., Locascio, L.E., Lee, C.S., Integrated microfluidic system enabling protein digestion, peptide separation, and protein identification, *Anal. Chem.*, 73, 2648–2655, 2001.
220. Lazar, I.M., Ramsey, R.S., Ramsey, J.M., On-chip proteolytic digestion and analysis using "wrong-way-round" electrospray time-of-flight mass spectrometry, *Anal. Chem.*, 73, 1733–1739, 2001.
221. Bengtsson, M., Ekström, S., Marko-Varga, G., Laurell, T., Improved performance in silicon enzyme microreactors obtained by homogeneous porous silicon carrier matrix, *Talanta*, 56, 341–353, 2002.
222. Melander, C., Momcilovic, D., Nilsson, C., Bengtsson, M., Schagerlöf, H., Tjerneld, F., Laurell, T., Reimann, C.T., Gorton, L., Microchip immobilized enzyme reactors for hydrolysis of methyl cellulose, *Anal. Chem.*, 77, 3284–3291, 2005.
223. Liu, Y., Lu, H., Zhong, W., Song, P., Kong, J., Yang, P., Girault, H.H., Liu, B., Multilayer-assembled microchip for enzyme immobilization as reactor toward low-level protein identification, *Anal. Chem.*, 78, 801–808, 2006.
224. De Boer, A.R., Bruyneel, B., Krabbe, JG., Lingeman, H., Niessen, W.M.A., Irth, H., A microfluidic-based enzymatic assay for bioactivity screening combined with capillary liquid chromatography and mass spectrometry, *Lab Chip*, 5, 1286–1292, 2005.
225. Foret, F., Zhou, H.H., Gangl, E., Karger, B.L., Subatmospheric electrospray interface for coupling of microcolumn separations with mass spectrometry, *Electrophoresis*, 21, 1363–1371, 2000.
226. Li, J., Wang, C., Kelly, J.F., Harrison, D.J., Thibault, P., Rapid and sensitive separation of trace level protein digests using microfabricated devices coupled to a quadrupole-time-of-flight mass spectrometer, *Electrophoresis*, 21, 198–210, 2000.
227. Li, J., Tremblay, T.-L., Wang, C., Attiya, S., Harrison, D.J., Thibault, P., Integrated system for high-throughput protein identification using a microfabricated device coupled to capillary electrophoresis/nanoelectrospray mass spectrometry, *Proteomics*, 1, 975–986, 2001.
228. Dahlin, A.P., Bergström, S.K., Andrén, P.E., Markides, K.E., Bergquist, J., Poly(dimethylsiloxane)-based microchip for two-dimensional solid-phase extraction-capillary electrophoresis with an integrated electrospray emitter tip, *Anal. Chem.*, 77, 5356–5363, 2005.
229. Kameoka, J., Craighead, H.G., Zhang, H., Henion J., A polymeric microfluidic chip for CE/MS determination of small molecules, *Anal. Chem.*, 73, 1935–1941, 2001.
230. Wen, J., Lin, Y.H., Xiang, F., Matson, D.W., Udseth, H.R., Smith, R.D., Microfabricated isoelectric focusing device for direct electrospray ionization-mass spectrometry, *Electrophoresis*, 21, 191–197, 2000.
231. Hilder, E.F., Svec, F., Fréchet, J.M.J., Polymeric monolithic stationary phases for capillary electrochromatography, *Electrophoresis*, 23, 3934–3953, 2002.
232. McEnery, M., Tan, A.M., Alderman, J., Patterson, J., O'Mathuna, S.C., Glennon, J.D., Liquid chromatography on-chip: progression towards a μ-total analysis system, *Analyst*, 125, 25–27, 1999.
233. Reichmuth, D.S., Shepodd, T.J., Kirby, B.J., Microchip HPLC of peptides and proteins, *Anal. Chem.*, 77, 2997–3000, 2005.

234. Keetch, C.A., Hernandez, H., Sterling, A., Baumert, M., Allen, M.H., Robinson, C.V., Use of a microchip device coupled with mass spectrometry for ligand screening of a multi-protein target, *Anal. Chem.*, 75, 4937–4941, 2003.
235. Zhang, B.L., Foret, F., Karger, B.L., High-throughput microfabricated CE/ESI-MS: automated sampling from a microwell plate, *Anal. Chem.*, 73, 2675–2681, 2001.
236. Haber, C., Microfluidics in commercial applications, an industry perspective, *Lab Chip*, 6, 1118–1121, 2006.

54 Nanoscale Self-Assembly of Stationary Phases for Capillary Electrophoresis of DNA

Kevin D. Dorfman and Jean-Louis Viovy

CONTENTS

54.1 INTRODUCTION

As other entries in this handbook have made clear, capillary electrophoresis has led to tremendous improvements in electrophoretic separations, both in terms of separation speed and automation. From the standpoint of speed, the dominant advantage of capillary electrophoresis is the faster dissipation of Joule heat as the surface area to volume ratio increases, permitting much higher applied voltages. Automation has also played a key role, for example in DNA sequencers, by allowing massively parallel operation and high throughput without much user input.

The successes of electrophoresis in narrow bore, fused silica capillaries has spurred further research into miniaturization and integration of capillary electrophoresis into lab-on-a-chip systems. These new devices, fabricated using the same techniques (and concomitant precision) employed in the microelectronics industry, generally consist of a number of interconnected narrow channels. Silicon and glass are common substrates, although the semiconductor properties of silicon make it less attractive for electrophoresis. The field is rapidly moving towards polymeric chips, which are usually formed by polymerizing (molding) the polymer on a templated substrate, itself fabricated in silicon or photoresist. Typical on-chip channel cross-sections range from tens to several hundred microns, with aspect ratios varying from unity to approximately 10.

The capillary electrophoresis techniques discussed elsewhere in this handbook can often be directly translated "on-chip." Chip-based capillary electrophoresis possesses a number of advantages over conventional capillary electrophoresis, most notably (i) the ability to work with channels that are much smaller than what is possible in fused silica (especially as chip fabrication techniques move towards the nanoscale); (ii) the potential to integrate with other upstream operations, such as DNA isolation and polymerase chain reaction (PCR) for bioanalyses; and (iii) the performance of more precise "pinched" or "gated" injections, which increase the resolution versus separation time ratio. One notable downside of chip-based capillary electrophoresis of DNA, especially as the channel dimensions are reduced, is the increased pressure drop needed to load an entangled polymer sieving matrix. While this can be alleviated somewhat through the use of novel thermosensitive polymers [1] that undergo dramatic changes in viscosity with temperature, the use of classical sieving polymers such as linear polyacrylamide will become untenable as the dimensions of the channels are further reduced.

54.2 BACKGROUND

One possible solution to the pressure drop problem is to fabricate the channels and an artificial sieving matrix simultaneously. In principle, such a strategy offers a number of advantages over loading a channel with polymer solution. The channel size can be reduced at will, since a high pressure drop is no longer needed to load a polymeric sieving matrix. At the same time, highly regular sieving matrices can readily be fabricated by lithographic patterning, which should lead to more reproducible separations with reduced band broadening. In the context of materials science, such a lithographic fabrication process is a "top-down" approach—the pattern for the sieving matrix is defined by the user and then etched into the substrate, much in the way a sculptor creates a statue.

The "top-down" approach to fabricating sieving matrices for DNA electrophoresis was pioneered by Volkmuth and Austin [2], who created a hexagonal array of millions of micron-sized posts in silicon. This general principle has since been demonstrated to rapidly separate long DNA [3]. However, these arrays are difficult and time consuming to manufacture, requiring a great deal of specialized equipment. Indeed, the current state-of-the-art quartz post array [3] requires seven clean-room processing steps, including an e-beam lithography step to pattern the nanoscale support. As a result, although microfabricated post arrays offer order-of-magnitude increases in separation efficiency over conventional gel separations, their high cost and easy fouling by contaminants such as dust has prevented their adoption for routine use. This contrasts starkly with the circumstances surrounding the introduction of other revolutionary methods in DNA electrophoresis, such as pulsed-field electrophoresis and capillary sequencers, which found immediate use in the community at large and were quickly commercialized.

The focus of this chapter is on another approach to forming regularly structured stationary phases for capillary electrophoresis: so-called "bottom-up" techniques that take advantage of the transport and thermodynamic properties of small colloidal particles. We will discuss two such methods: (i) assembly of superparamagnetic beads by an external magnetic field and (ii) the convective assembly of colloidal particles in an evaporating water film. Both techniques successfully form DNA separation matrices without the need for a complicated top-down lithography step, and may eventually provide a route for bringing emerging lab-on-a-chip technologies into routine use. However, fabrication by self-assembly is not as well developed as lithography, and suffers from a number of technical difficulties. We will explore both the promise and the limitations of bottom-up approaches to forming stationary phases for capillary electrophoresis, illustrating the potential of this technique and the obstacles that lie ahead in future research.

54.3 THEORY

54.3.1 Self-Assembled Magnetic Arrays

We begin our discussion with the self-assembly of superparamagnetic beads in external magnetic fields. This method has been dubbed "Ephesia," after the Temple of Artemis, one of the seven wonders of the ancient world [4]. The temple, whose roof was supported by a number of columns, was repeatedly destroyed and rebuilt in exactly the same manner throughout history. The magnetic bead system operates in an analogous manner—the array can be repeatedly formed and destroyed by turning the magnetic field on and off.

54.3.1.1 Formation of the Array

The magnetic beads used in the early stages of development of the method were essentially an oil-in-water emulsion, stabilized by surfactant. The oil droplets were relatively monodisperse with an average diameter of 570 nm, and contained small grains of iron oxide that impart superparamagnetic properties to the beads. The technology for manufacturing superparamagnetic nanobeads has advanced significantly since the initial report on their self-assembly processes [5]. Successors of these magnetic emulsions are commercially available from Ademtech SA in Pessac, France: they are now stabilized by replacing the oil with a polymerized material. Indeed, numerous magnetic latexes, differing in composition and size, can be used to assemble arrays, depending on the application (scale, strength). A requirement, however, is that the size distribution of the particles be rather monodisperse. The external magnetic field is supplied by a cooled electromagnet surrounding the microchip [6].

Figure 54.1 [7] illustrates the general principle underlying the formation of the Ephesia array. In the absence of a magnetic field, illustrated in Figure 54.1a, thermal motion leads to the magnetic moments of the grains in the beads being randomly oriented. As a result, the net magnetic moment

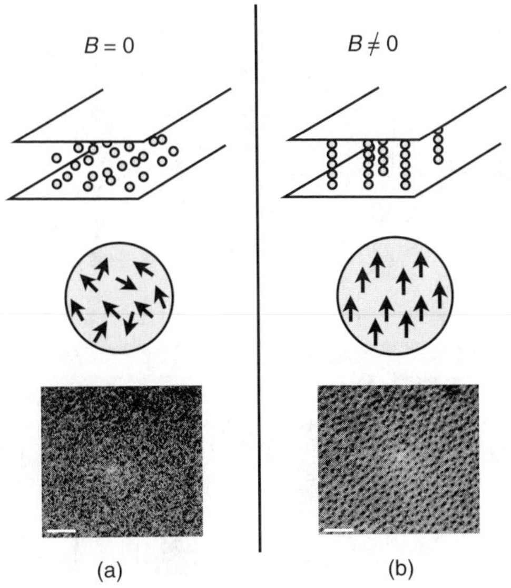

FIGURE 54.1 (a) In the absence of a magnetic field ($B = 0$), the magnetic moments of the iron grains in the bead are randomly oriented and the Brownian motion of the bead center of mass gives rise to the quasiuniform distribution indicated in the lower figure. (b) When the magnetic field is turned on ($B \neq 0$), the magnetic moments align and the beads form columns. (Adapted from Minc, N. Microfluidique et billes magnétique auto-organisées pour les analyses biologiques, Ph.D. Thesis, Université Paris VI, 2005.)

of a given bead vanishes. In this unpolarized state, the beads can be easily injected into a standard poly (dimethylsiloxane) (PDMS) microchannel, where they form the homogeneous suspension depicted at the bottom of Figure 54.1a. When the magnetic field is turned on, the iron grains in the bead become aligned, leading to a net magnetic moment in each bead. This is what is meant by superparamagnetic—the beads only have a magnetic field in the presence of the field (paramagnetism), and the paramagnetic effect occurs because the energy for aligning the local magnetic moments is comparable to thermal energy (the "super" part). Each bead then acts like a "mini-magnet," whereupon nearby beads align their N–S poles and begin to form columns. The column spacing is generally independent of the volume fraction φ of the beads [5]—as the bead concentration increases, the column width increases like $\varphi^{1/2}$ to maintain a constant column spacing [5].

The ability to form columns depends strongly on the strength of the magnetic field. This effect is quantified by the coupling constant [5] (i.e., the ratio of the magnetic and thermal energies). Column structures only form when the coupling parameter is much greater than unity. The change in the suspension morphology with magnetic field is clearly seen in the optical micrographs presented in Figure 54.2 [8]. At weak fields, the beads do not associate and the suspension is homogeneous. As the

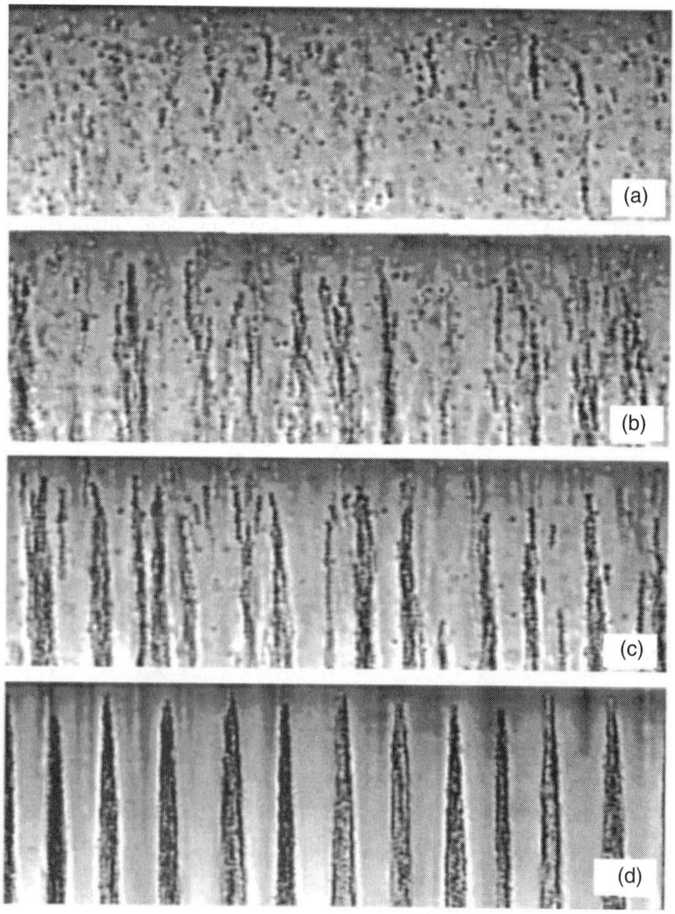

FIGURE 54.2 The growth of the columns depends on the strength of the magnetic field. At a low magnetic fields, (a) there is a coexistence of individual beads and chains. As the magnetic field increases, (b) the distribution shifts towards chains and eventually includes columns, and (c) the columns span the channel. At higher magnetic fields, (d) the columns are stabilized by solid friction with the walls and the structure is a "solid." (Reprinted from Ivey, M., et al., *Phys. Rev. E*, 63, 011403, 2000. With permission.)

field increases, the morphology moves from a mixture of beads and chains towards a solid columnar structure. The columns repel one another because their effective magnetic charges have the same sign at the top and the bottom of the column [8]. The repulsion ultimately leads to the hexagonal structure seen in Figure 54.1.

While the hexagonal structure is the equilibrium state for an infinitely wide (unbounded 2D) magnetic array, electrophoretic applications require channels of finite width. For example, the array in Figure 54.1 typically contains several dozen columns across the channel width. Brownian dynamics simulations [9] and related experiments [10] have shown that the steady-state structure of a confined matrix is a strong function of the wall spacing. When the wall spacing is not commensurate with the crystal lattice spacing, defects form near the wall and propagate into the bulk array [9]. Intuitively, one would expect that perfect arrays would form perfectly flat walls if the gap height h is some integer multiple of the bead radius R. Remarkably, this is not the case [9]. As seen in Figure 54.3, the perfect arrays do not form integer values of h/R. In experiments on real two-dimensional (2D) crystals (formed by sedimenting the beads to the bottom of the channels), imperfections in the PDMS channel walls also play a strong role in determining the bead density across the channel [10]. Although it is useful to think of the imperfections of the array as propagating from a disordered wall "region," it is in fact impossible to unambiguously define the border between the wall region and the bulk crystal [11].

In very large Hele–Shaw cells, at a distance from the walls much larger than the cell thickness, a hexagonal order is obtained only locally, and defects destroy the long-range order. This is expected to arise from the randomness of the interactions between columns and the bottom of the cell. Once a column spans the narrow height of the cell, the solid friction with the floor and ceiling "freezes" the

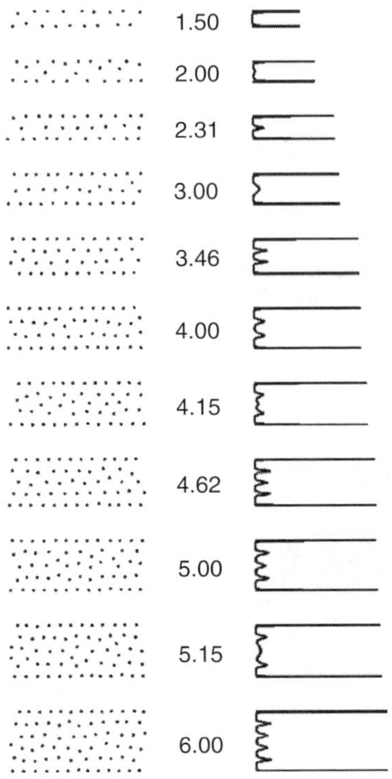

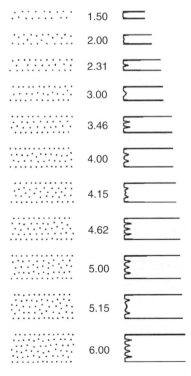

FIGURE 54.3 For finite gap widths h, the distribution of the particles depends on the number of particle radii R between the two walls. Surprisingly, the most uniform density profiles (indicated by the sharpest peaks in the density profiles on the right) do not correspond to integer values of h/R. (Reprinted from Haghgooie, R., and Doyle, P.S., *Phys. Rev. E*, 70, 061408, 2004. With permission.)

column location. In effect, these columns serve as nucleation sites for the growth of the hexagonal crystal. Inasmuch as the position of a given nucleation site is random (with respect to the lattice spacing of nearby crystals formed from other nucleation sites), the hexagonal order emanating from a nucleation site is maintained only until the boundaries of two different nucleated crystals merge.

The column spacing in the array can be controlled by the surfactant used to wash the beads [12]. Figure 54.3 presents three different arrays formed by using Tergitol NP-10, Triton X-405, and Tergitol NP-40 as the stabilizing surfactant. These molecules are amphiphilic, with a hydrophobic head that associates with the oil phase (the bead) and a hydrophilic tail extending into the aqueous phase. As the length of the hydrophilic tail decreases, the steric repulsion energy decreases, leading to denser arrays. The maximum density of the array is limited, however, by an intrinsic instability in ferrofluid suspensions [13]. Indeed, the array depicted at the bottom of Figure 54.4 is close to the highest density that still forms columnar structures in this particular Hele-Shaw cell (of height 12 μm). At higher bead densities, the columnar structure is destabilized and the hexagonal array devolves into a labyrinthine structure.

The self-assembly process possesses an intrinsic limit with respect to the column spacing. Explicitly, the array must be sufficiently concentrated so that the repulsive energy between the different columns is strong enough to enforce the hexagonal structure. The repulsive dipolar energy scales

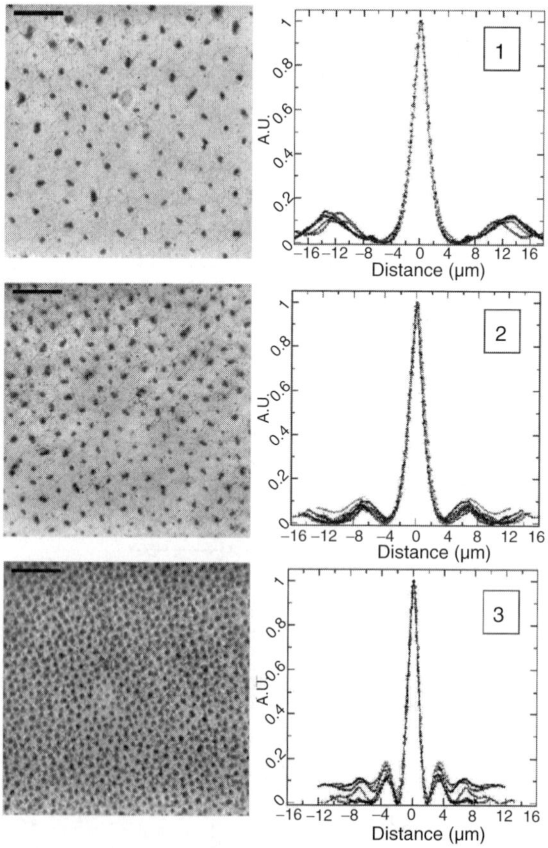

FIGURE 54.4 The density of the array can be altered by changing the length of the surfactant used to stabilize the bead suspension. The pictures on the left side correspond to decreasing hydrophilic arm lengths, from top to bottom. The plots on the right-hand side are cross-sections of the Fourier transforms of the left-hand images. The mean column diameter and spacing, as well as the variance in the spacing, can be computed from these plots [6]. (Reprinted from Minc, N., et al., *Electrophoresis*, 26, 362, 2005. With permission.)

(a) (b)

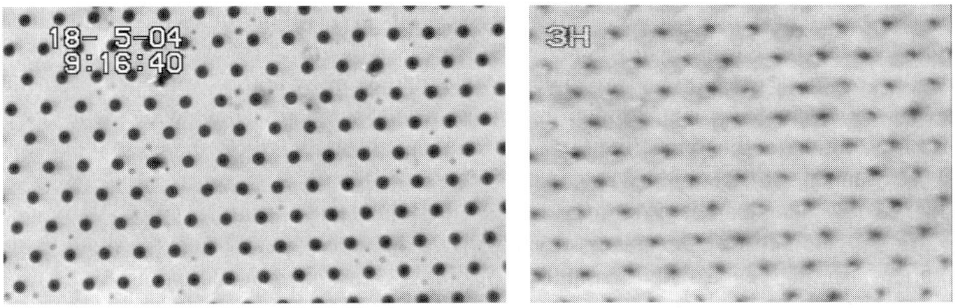

FIGURE 54.5 Highly regular arrays of magnetic bead columns can be produced by patterning the surface with Ni "anchors." (a) Ni spots on the surface, spaced 10 μm apart. (b) Beads that have assembled onto the anchors. (Reprinted from Saliba, A.E., et al., *Proceedings of MicroTAS 2006*, Eds. T. Kitamori, H. Fujita, and S. Hasabe, Tokyo, Japan, Transducers Research Society, 2006.)

like $1/r^3$, which leads to a rapid drop-off in repulsive energy as the array is diluted. This limitation can be side stepped by incorporating magnetic "anchors" into the floor of the channel [14]. The anchors are created by nanopatterning the substrate with nickel spots through e-beam patterning and metal lift-off. The results of this process are illustrated in Figure 54.5. The bead columns faithfully reproduce the nickel surface pattern, without any columns assembling away from the nickel spots. However, the introduction of the nickel pattern eliminates one of the strongest points of the magnetic bead array, namely the fact that it is a bottom-up strategy that does not require any sophisticated lithographic steps. Nevertheless, the nickel patterning is still considerably less complicated than the fabrication required to produce the equivalent post array in quartz [3], and is only necessary for forming very dilute arrays. Another advantage of this "templated" self-assembly is that it allows the formation of structures with an aspect ratio (up to a hundred or more) impossible to reach solely with "top-down" microlithographic techniques. Finally, the cost and complexity of the fabrication of the initial magnetic pattern can be greatly reduced, as compared to the lift-off process, by preparing magnetic plots by microcontact stamping of a magnetic "ink" [15].

54.3.1.2 Separation Resolution

The ability to simply create controlled array structures with a variety of pore sizes (as in Figures 54.4 and 54.5) has led to the development of a number of theoretical models for the transport in such arrays. The first model [4] simply assumed that each time the DNA passes one row of the postarray, it has some probability Π_c of colliding with the post. If there is a collision, the DNA is held up on the post for some average time τ, which depends on the molecular weight and the electric field.

Several simple scaling laws can be readily derived from this Markovian model by assuming that (i) the collision probability Π_c only depends on the density of the array (and not the electric field or the DNA size) and (ii) the trapping time τ varies inversely with the electric field, which is consistent with the "rope-over-pulley" mechanism of escape [16]. In this limit, the mobility μ is independent of the electric field E, while the effective diffusivity (or dispersivity) D^* scales linearly with E. As a result, the separation resolution should be independent of E for a fixed separation length, implying that the separation time in these arrays could be reduced indefinitely by increasing the field. This scaling law is born out for intermediate fields between 20 and 40 V/cm [6], where the model is expected to be valid. The resolution is lower outside of this regime. The loss of resolution at weak fields was subsequently shown to be the result of an increase in D^* due to incomplete stretching of the chain during the unhooking process [17]. The strong field result remains unexplained, although it is speculated [6] that the DNA cannot completely relax between collisions.

This leads to an anomalously long translation time between the collision events, thereby reducing the number of "separation steps" per unit length. Besides this observation, other physical phenomena put a practical limit on the applicable field. First, the surface treatment is never perfect, and it results in a residual electroosmotic flow that can destroy the column array in too strong fields (practically, around 50–100 V/cm). Second, too high fields may result in Joule heating, which is also a cause of loss of resolution (above typically 500 V/cm for channels 50 μm thick; the present experiments [6,12] use channels 10–12 μm thick).

Minc et al. [12] compared this simple model to single-molecule observations of DNA migration in the arrays. The microscale parameters Π_c and τ were measured directly from the single-molecule observations, and the macroscopic parameters μ and D^* were determined from the first-passage time statistics for DNA moving through the viewing area. Although the scaling arguments remain in accord with both the ensemble and single-molecule data, the quantitative agreement between theory and the single-molecule experiments was not satisfactory. Several other transport models also failed to explain the experimental data.

Building upon the insights from the single-molecule experiments [12], a more sophisticated mathematical model was constructed [18]. The model, shown schematically in Figure 54.6, is inspired by the so-called geometration model of gel electrophoresis [19]. The DNA motion is modeled as a three-step process: (i) collision with the post and unraveling into two arms; (ii) a rope-over-pulley disengagement process; and (iii) uniform translation to the next collision, which occurs at some random distance downfield. By making some simple assumptions about the transport processes, most notably by requiring that the probability of a collision in any given row be proportional to the areal post density, and that all DNA motion occurs with the free solution mobility μ_0, the mobility μ, and dispersivity D^* can be computed exactly without any adjustable parameters. The qualitative predictions of the model agree very well with the experiments reported above, and the quantitative agreement is satisfactory given the limited amount of data available [18]. This model was subsequently modified to account for incomplete stretching at weak fields [17], which explains the loss in resolution at weak fields.

54.3.2 Colloidal-Templated Crystals

We now turn our attention to the use of nonmagnetic colloidal particles to form sieving matrix templates. Much in the same way that microelectronics manufacturing has provided the toolbox for creating lab-on-a-chip systems via lithography, photonic crystal production has provided the tools that will be presented in the present section. While we will focus exclusively on the use of colloidal templates to create sieving matrices for DNA electrophoresis, the general methods used here are a burgeoning field of research in materials science and promise to impact a range of applications much broader than capillary electrophoresis. At the same time, the sheer volume of research currently underway into the materials side of the problem promises to greatly improve the variety and regularity of the structures that could ultimately be employed for electrophoresis.

In the present section, we cover the two steps for creating stationary phases from colloidal assembly. In the first step, the colloids are assembled into a regular, close-packed structure through the evaporation of the solvent or a flow-through system. As we will see later, the colloidal crystal, when filled with electrophoresis buffer, can be used directly as a solid support for capillary electrophoresis [20,21]. Alternatively, a polymeric structure can be formed around the bead template. After dissolving the beads, one is left with a regular structure of cavities and pores that can also serve as an electrophoresis support [22].

54.3.2.1 Colloidal Crystals

In the Ephesia system, discussed in Section 54.3.1, an external magnetic field is used to assemble the superparamagnetic particles into an organized array of columns. As illustrated in Figure 54.7,

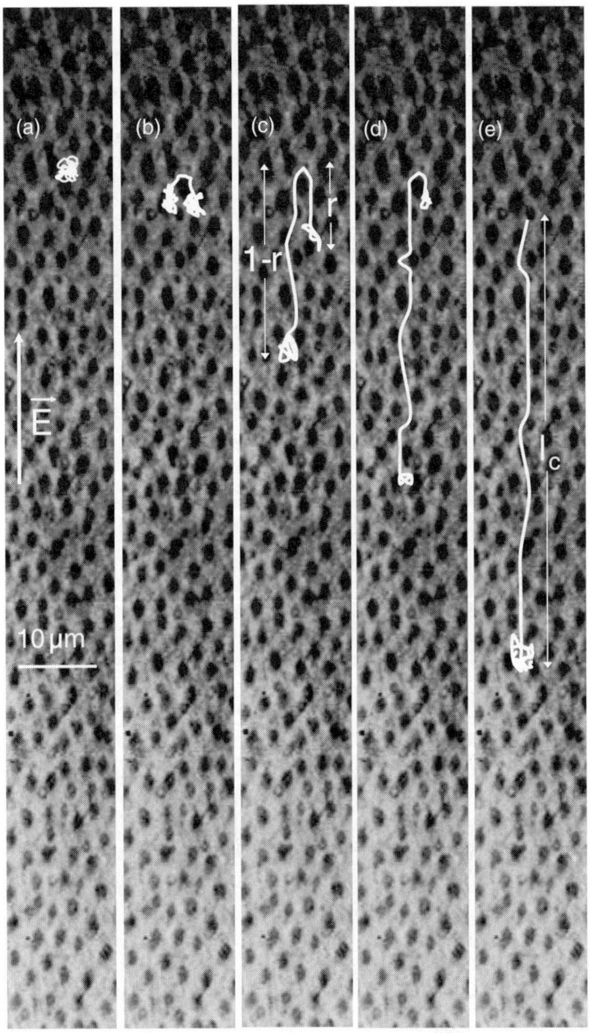

FIGURE 54.6 Schematic of the geometration model of DNA electrophoresis in micropatterned arrays. Representative DNA conformations are superimposed on an image of a self-assembled magnetic bead array. (Reprinted from Minc, N., et al., *Phys. Rev. Lett.*, 94, 198105, 2005. With permission.)

colloidal templating uses the forces exerted by an evaporating liquid film to guide the crystal formation. In the simplest case [23,24], the liquid is evaporated from a flat substrate. The thickness of liquid film is commensurate with the diameter of the colloids, which suppresses Brownian motion of the beads. Once the film evaporates below the level of the colloids, they form a nucleation site for the crystal [24]. This results in a curvature of the water/air interface, whereupon the surface tension and convection in the water phase combine to drive additional colloidal particles towards the nucleation site. An alternate process for driving the crystal formation is to dip the substrate into a suspension of the colloids and slowly withdraw it to form a film, similar to a Langmuir trough [23]. The colloids are then assembled through the convection of the fluid along the meniscus, driven again by the water flux.

The range of particle sizes that can be assembled in this manner is very large, with reports ranging from 200 nm [25–27] to several microns [24,26]. Also note that, although the figure only illustrates the formation of a single layer of colloids, multiple layers are typically formed during the process.

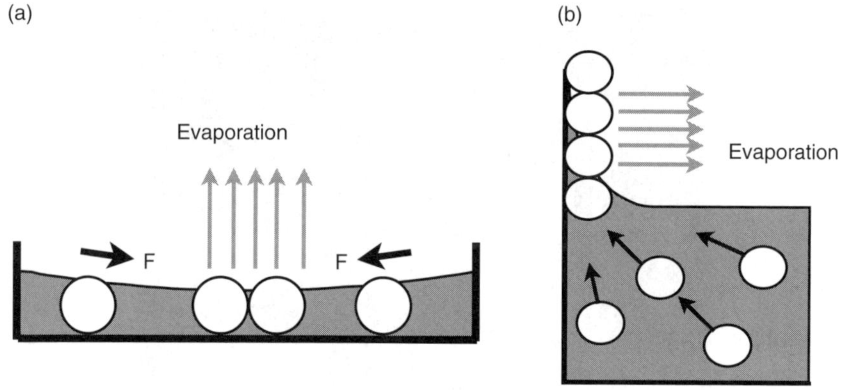

FIGURE 54.7 Mechanism of convective assembly on (a) flat and (b) vertical surfaces. (Adapted from Nagayama, K., *Coll. Surf. A*, 109, 363, 1996.)

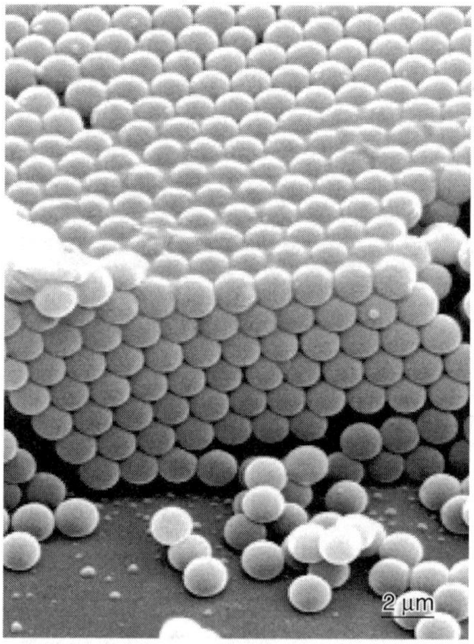

FIGURE 54.8 SEM image of a seven layer colloidal crystal assembled from 396.6 nm particles. (Reprinted from Jiang, P., et al., *Chem. Mater.*, 11, 2132, 1999. With permission.)

For example, the dip coating of a slide will normally form 10–50 layers of colloids [27]. Figure 54.8 illustrates a typical scanning electron micrograph (SEM) picture of a colloidal crystal formed in this manner.

Colloidal crystals possess long-range order, although there are usually defects and stacking faults over the longer length scales needed for electrophoresis [22]. Moreover, since the faults seen in Figure 54.8 are not necessarily colinear, the electric field will not always be aligned with a symmetry axis of the crystal. As we will see shortly, this has important implications for electrophoresis [22]. The presence of domain boundaries is probably the biggest drawback in applications, since the medium does not possess uniform properties over the distances required for DNA separations. Likewise, the boundaries force the DNA to move along the defects until they reach a new, defect-free region [22].

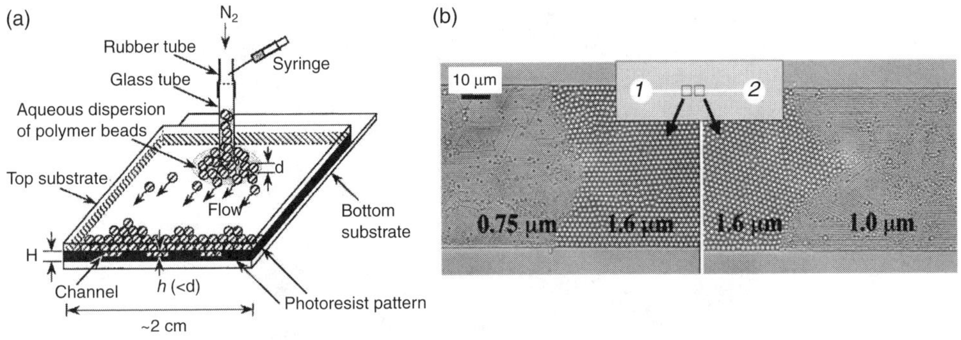

FIGURE 54.9 Two mechanisms for flow-based assembly of colloidal arrays. (a) Assembly through a template weir. (Reprinted from Park, S.H., and Xia, Y., *Chem. Mater.*, 10, 1745, 1998. With permission.) (b) Assembly through permeation-driven flow of water in PDMS. (Reprinted from Randall, G.C., and Doyle, P.S., *Proc. Natl. Acad. Sci. USA*, 102, 10813, 2005. With permission.)

If the colloidal crystal is to be used directly for electrophoresis, it can be sealed on the sides and then covered with a PDMS lid with holes bored for the reservoirs [21]. Similar to the magnetic bead system, a colloidal crystal is a simple method of fabricating a regular nanoscale stationary phase for electrophoresis.

Colloidal crystals can also be assembled through flow-based methods. One of the earliest strategies for flow assembly is depicted in Figure 54.9a [26]. The microfabricated "frame," constructed on the lower substrate in SU-8 resin, has periodic holes to define flow channels. The entire apparatus was covered with an upper substrate and connected to a glass tube. When the colloidal suspension is introduced into the gap between the substrates, the solvent escapes through the gaps in the frame, while the beads remain trapped against the photoresist. This creates a well-defined bead pattern with a height equal to the lower part of the frame.

An alternate flow-based method to form colloidal crystals is to use the permeability of PDMS to form the crystals *in situ* in the channels [28], which could subsequently be used for electrophoresis. The basic strategy is illustrated in Figure 54.9b. The colloidal suspension in the reservoirs creates a pressure head, and the flux of solvent through the walls of the channel leads to a stagnation plane in the center of the channel, with the area-averaged velocity increasing as one moves from the center plane toward the reservoirs. This mode of crystal growth works well in very thin channels; the experimental results in Figure 54.9b used a 2-μm deep, 50-μm wide, 1-cm-long channel. As indicated in the figure, it is possible to create a crystal/crystal interface by seeding each reservoir with different sized colloids.

Finally, if heavy beads such as silica (SG = 2.1) are used, they can be assembled into a regular matrix by sedimentation [20]. One particular advantage of silica beads is that, after assembling them in a glass cell, sucrose can be added to the electrophoresis buffer to closely match the index of refraction of silica. This leads to a transparent material that is ideal for optical detection (at the expense of separation time, since sucrose increases the medium's viscosity, and thus reduces electrophoretic mobility). A second advantage is that some of the well-documented surface treatment strategies against electroosmosis developed for capillary electrophoresis can be directly transposed.

54.3.2.2 Polymer Matrices

Although colloidal crystals can be used directly for electrophoresis [20,21], it is also possible to use the crystals as templates for polymer matrices. In this approach [26,27], the colloidal crystal

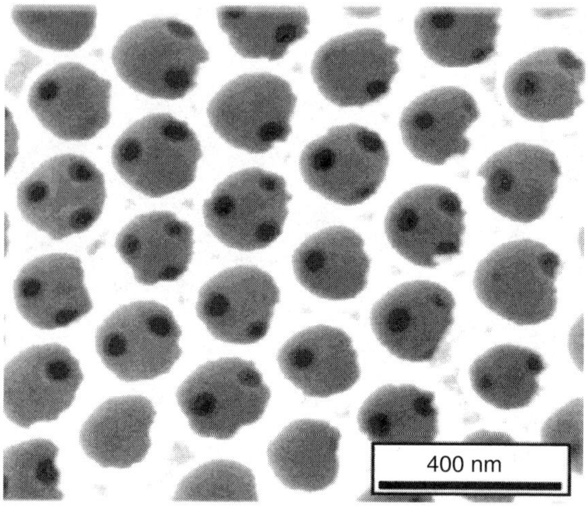

FIGURE 54.10 SEM image of a polymer matrix formed by infusing the monomers into a colloidal crystal, polymerizing the monomers, and then etching away the crystals. The nanopores between the cavities are the result of incomplete wetting of the colloids at their junctions. (Reprinted from Jiang, P., et al., *J. Am. Chem. Soc.*, 121, 11630, 1999. With permission.)

is perfused with a monomer solution that is then polymerized. The solution flows spontaneously into the matrix due to capillary action, although it is suspected that the beads are not completely wetted by the fluid [27]. A large variety of polymers can be used to create the matrix, including polyurethane, polymethyl methacrylate (PMMA), polymalic acid (PMA), polystyrene, and epoxy [27]. For electrophoresis, a mixture of 30% acrylamide monomer and 6% bis-acrylamide was used to create a templated polyacrylamide gel with well-defined pore spaces [22]. Following polymerization, the beads are chemically dissolved. Thus, the key restriction in the choice of polymer matrix is that the polymer must resist degradation by the solvent. For electrophoresis applications [22], the polystyrene beads used to form the matrix were dissolved by toluene, which does not destroy the polyacrylamide. A successful polymerization is easy to see after the toluene is removed; owing to Bragg diffraction of light, the polymer matrix is iridescent.

Figure 54.10 depicts a polystyrene matrix formed by templating of 260.4 nm silica beads and then dissolving the beads with hydrofluoric acid [27]. As is evident from the figure, the polymer matrix consists of an ordered array of cavities with a diameter close to the diameter of the silica beads. The cavities are connected by pores that are typically one-fourth of the bead diameter. The size of the pores can be controlled by changing the viscosity of the monomer solution perfusing the matrix (via the solution temperature), presumably due to a change in the wetting properties and subsequent shrinkage during polymerization [27]. Also note that all cavities are not completely connected, although it is suspected that the network as a whole is simply connected [27].

54.4 APPLICATIONS

54.4.1 Self-Assembled Magnetic Arrays

The first separations of DNA in self-assembled arrays were reported in the literature by Doyle et al. in 2002 [29], although the method appeared in conference proceedings as early as 1997 [30]. In the initial experiments, the magnetic beads were loaded into the channel using a pressure head generated by different fluid heights in the various reservoirs on the chip. While this method will load the beads, much more reproducible arrays are obtained using a pressure control system for the

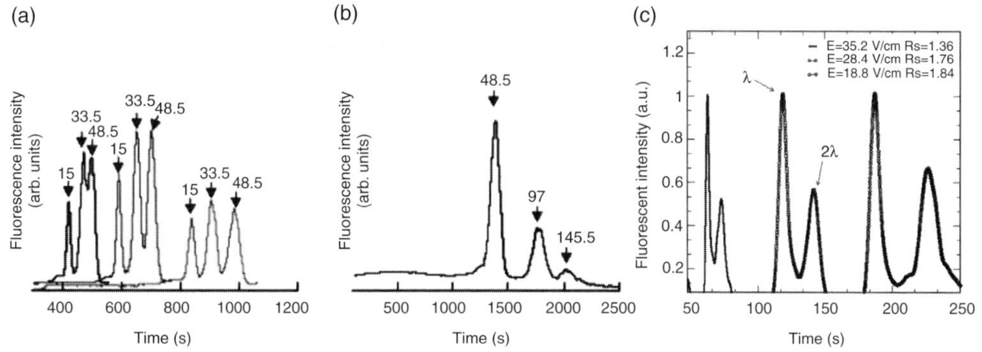

FIGURE 54.11 Electropherogram for separations in magnetic bead arrays. (a) Separation of *XhoI* digests of λ-DNA (48.5 kbp) at 4.8, 7.0, and 10 V/cm. (b) Separation of λ-DNA concatemers at 3.2 V/cm. (Reprinted from Doyle, P.S., et al., *Science*, 295, 2237, 2002. With permission.) (c) Separation of λ-DNA and 2λ-DNA in a more recent experimental setup at different electric field strengths. (Reprinted from Minc, N., et al., *Anal. Chem.*, 76, 3770, 2004. With permission.)

reservoirs [31]. Likewise, the pressure control system eliminates any hydrodynamic flows that might occur from experimental artifacts, such as someone opening the door to the room or nonuniform evaporation from the reservoirs, yielding purely electrophoretic transport through the array. The prototype pressure control system used in experiments [6,12] has since been commercialized by Fluigent.

Following the initial report of this technique [29], Minc et al. [6] made significant technical improvements on the method, which ultimately led to an order-of-magnitude improvement in the separation time and reproducibility. Amongst the technical advances made in this work were (i) the use of the flow control system to load the beads and suppress hydrodynamic flows during the experiment; (ii) an improved electromagnet for forming the array and trapping the structure in the local minimum; and (iii) the use of a pretreatment of the microchannel with a dimethylacrylamide/allyl glycidyl ether copolymer for suppressing electroosmotic flow. In conventional PDMS, this treatment has a limited lifetime, about 90 min of operation, and it must be repeated periodically (for a more detailed account about surface treatments of PDMS and other polymers, see for example Reference 32). The veracity of this new system for DNA electrophoresis has been verified extensively [6].

The key results from the initial separation experiments, using static fluid heads to load the beads, are depicted in Figure 54.11. In each set of experiments, the column spacing was approximately 5.7 μm. Figure 54.11a presents the results for separations between different fragments of a *XhoI* digest of λ-DNA (48.5 kilobase pair or kbp).As is evident from the figure, these DNA fragments can be base-line resolved at 4.8 V/cm after a separation time of approximately 15 min. This compares favorably with traditional pulsed-field gel electrophoresis separation of the same fragments, which requires many hours for an equivalent resolution. The separation time reduced results from the increased post spacing, which breaks the biased reptation with orientation migration mechanism [33], since the DNA can easily coil up into its equilibrium conformation between collisions with the posts. These initial arrays were also able to separate larger DNA, as seen in Figure 54.11b. Two concatemers of λ-DNA and the template were well resolved by an electric field of 3.2 V/cm after approximately 30 min.

The success of the second-generation magnetic bead system [6] is evident from the data presented in Figure 54.11c. In these experiments, the average column diameter is 1.4 μm (indicating aggregation of several single-bead columns) and the average center-to-center spacing is 4.1 μm, implying a pore size of 2.7 μm. These arrays can separate λ-DNA and 2λ-DNA with a resolution greater than unity in several minutes. Indeed, the fastest separation (using an electric field of 35.2 V/cm, an order of magnitude higher than the experiments discussed above) achieves a resolution of 1.36 in only 60 s. The improved performance is the direct result of the technical developments discussed above.

54.4.2 Colloidal Crystals

54.4.2.1 Entropic Trapping

The first application of templated colloidal matrices was a fundamental study of entropic trapping [34]. Entropic trapping is predicted to occur whenever DNA moves through a medium consisting of cavities comparable to the radius of gyration of the DNA that are connected by small pores. The diffusion of the DNA is hindered because it loses entropy during the transit between cavities, since the DNA must uncoil to move through the pores. Theory [35] predicts that, in the entropic trapping limit, the diffusivity of DNA will decay exponentially with molecular weight.

To provide an ordered entropic trapping environment in these experiments, 0.895 μm polystyrene spheres were used to form the crystal and acrylamide was used for the polymer matrix. The resulting structure has a cavity size of 0.92 μm, which is slightly larger than the bead size due to swelling and comparable to the radius of gyration of DNA in the 10 kbp range. The connecting pores are 0.10 μm in radius, which is slightly larger than the DNA persistence length. The matrix was filled with buffer and DNA and then sealed, and no electric field was used.

The dynamics of 7.23 kbp DNA in the array are illustrated in Figure 54.12. The radius of gyration of this size DNA is considerably less than the cavity diameter, allowing them to easily coil therein. At any given instance in time, each DNA molecule occupies a single cavity. The four DNA in Figure 54.12a are indicated by the bright cavities. Over the course of time, as seen in Figure 54.12b, the DNA explore nearby cavities by rapid jumps. These jumps are separated by long periods of time during which the DNA is stationary. The particular trajectory for the right-most DNA is highlighted in Figure 54.12c, where the shading of the lines indicate the broad distribution of waiting times between jumps.

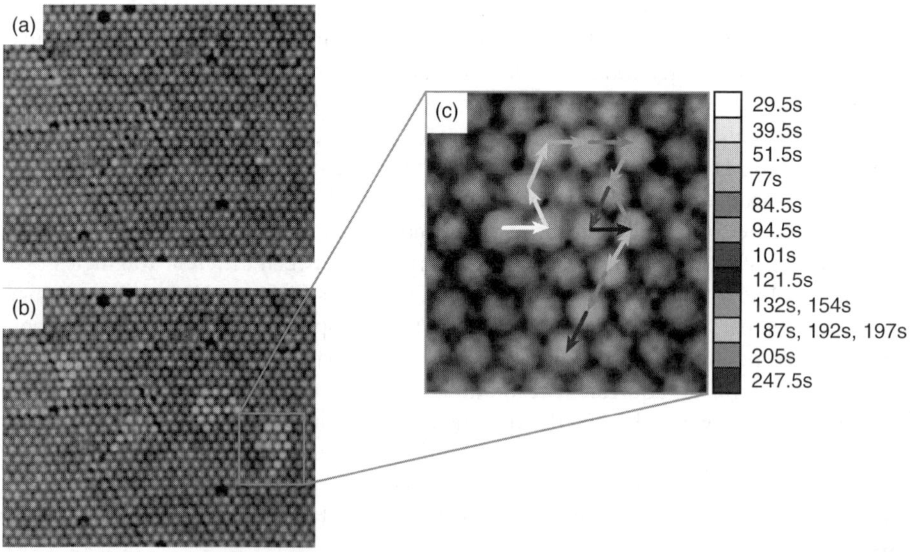

FIGURE 54.12 **(See color insert following page 810.)** DNA dynamics in a polymer matrix formed from colloidal templating in the absence of an electric field. (a) At a given instance in time, four DNA molecules are localized in different pores. (b) Over time, the DNA explores the local pores. (c) The jumping between pores is random in both direction and the waiting time between jumps. (Reprinted from Nykypanchuk, D., et al., *Science*, 297, 987, 2002. With permission.)

The diffusivity of DNA in this matrix, measured by both the jump-time distribution and the growth of the mean-squared displacement, indicates an exponential decay with respect to molecular weight, although the amount of data is very limited. If these data are fit by a power law, one obtains $D \sim N^{-1.1\pm0.7}$. Although this is very close to inverse dependence on molecular weight predicted by the Rouse model, the dynamics are very different. It is clear from these experiments that direct visualization is necessary to determine the true mode of migration [34].

When the radius of gyration exceeds the cavity size, the entropic trapping mechanism breaks down. For example, λ-DNA was observed to move like an inchworm, simultaneously spanning multiple pores [34]. The DNA is not completely extended, however, since it can relax to fill the cavities due to a gain in entropy.

54.4.2.2 DNA Electrophoresis

The first experiments using colloidal templates were performed without the polymerization step, using silica beads with a diameter of 1.04 ± 0.4 μm [20]. The beads were not well-packed over the entire O(cm) region, with monocrystalline regimes periodically interspersed by polycrystalline regimes. Electrophoresis experiments with a range of DNA samples essentially confirmed that the DNA dynamics in this array correspond to the predictions of biased reptation [33]—the mobility scales like $N^{-0.9\pm0.1}$ and the diffusivity scales like $N^{-1.9\pm0.2}$. A reptation plot [36] of the mobility indicates that the occupied pore sizes are consistent with predictions from geometric arguments for the packing of the spheres. Thus, these experiments demonstrated that biased reptation in fact describes the transport of DNA in well-ordered porous media where the pore sizes are much smaller than the radius of gyration of the DNA.

Smaller colloidal crystals, formed by 300 nm silica spheres, have also been used to for DNA electrophoresis [21]. The mean pore size of such an array is around 0.45 nm, considerably smaller that that above and smaller than the persistence length of DNA. Single-molecule observations of λ-DNA moving through this crystal showed that at weak fields (11 V/cm), the DNA moves like an inchworm, occasionally forming the rope-over-pulley shape seen in Figure 54.6. At higher fields, the DNA is stretched and forms strong rope-over-pulley collisions. Thus, it is reasonable to conclude that this smaller silica crystal generates similar dynamics to a quartz nanopillar array [3] without the need for e-beam lithography.

The silica crystal array can also be used for pulsed fields by adding two additional reservoirs to the PDMS cover to create a 120° crossed field. The mobility of different size long DNA (48.5, 23.1, and 9.4 kbp) vary in the pulsed-field array, indicating that they could be separated in such an array if a suitable injection system were employed. The high electric fields and rapid pulse times employed in these experiments are comparable to pulsed-field electrophoresis in the "DNA prism" pillar array [37].

Polymeric matrices formed from a colloidal template have also been used for DNA separations [22]. In contrast to the case in entropic trapping in this type of matrix [34], the electric field compresses the long λ-DNA into a single pore. As illustrated in Figure 54.13, at weak fields the DNA move slowly between the pores, with the direction of the motion biased by the electric field. As the field increases, rope-over-pulley dynamics are observed, whereas at the highest fields the DNA jumps across multiple pores and then coils up. This behavior can only be observed over limited stretches of the medium, since the DNA dynamics change sharply as it tries to transit from one domain to another at the boundaries [22].

The local mobility of DNA in these arrays exhibits a peculiar dependence on molecular weight and electric field, as seen in Figure 54.14. At weak electric fields, the shorter DNA are more mobile. In contrast, there exists a plateau regime at higher electric fields where the longer DNA moves more quickly. The plateau region is of particular interest for applications, since it would allow rapid separations.

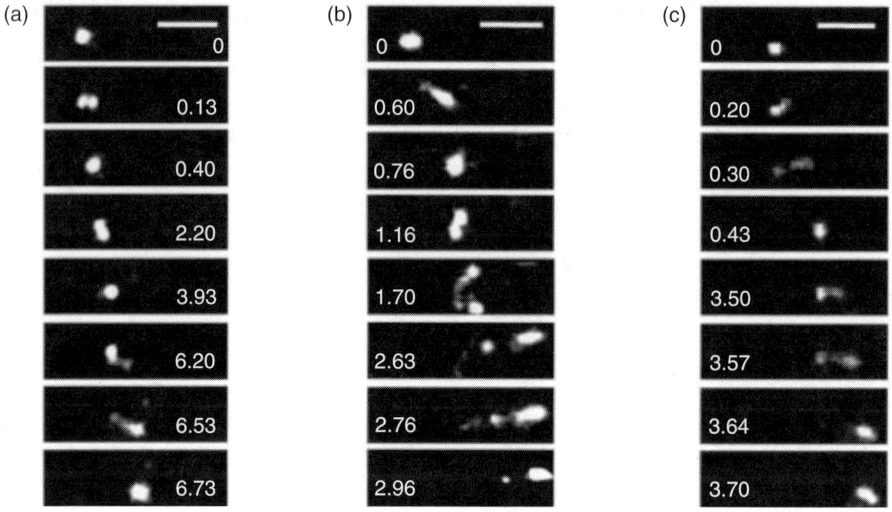

FIGURE 54.13 λ-DNA dynamics in a polymer matrix formed from colloidal templating in the presence of an electric field. (a) Under a weak field (10 V/cm), the DNA move sluggishly between pores, analogous to Figure 54.12. (b) Under a moderate field (25 V/cm), the DNA forms a rope over a pulley, analogous to Figure 54.6. (c) At the highest field (50 V/cm), the DNA jumps over multiple pores. (Reprinted from Zeng, Y., and Harrison, J.D., *Electrophoresis*, 27, 3747, 2006. With permission.)

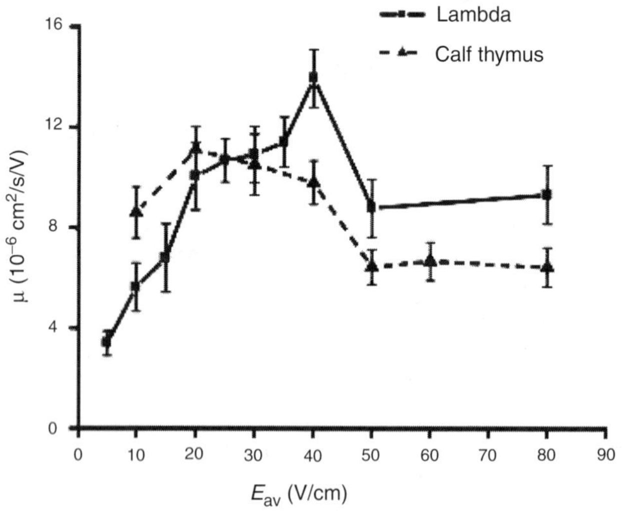

FIGURE 54.14 Electrophoretic mobility of λ-DNA (48.5 kbp) and calf thymus DNA (13 kbp) as a function of electric field in a polymer matrix formed from colloidal templating. (Reprinted from Zeng, Y., and Harrison, J.D., *Proceedings of MicroTAS 2006*, Eds. T. Kitamori, H. Fujita, and S. Hasabe, Tokyo, Japan, Transducers Research Society, 2006. With permission.)

The mobility dependence in Figure 54.14 can be explained qualitatively by a relatively simple model of the transport process [22]. At weak fields, the transport is dominated by entropic trapping, since the change in potential energy for moving in the field is not much greater than the decrease in entropy required to move between the cavities. As a result, the shorter DNA moves more quickly in weak fields, since the entropy loss is lower. The mobility crossover at higher fields is attributed to electrical trapping at the cavity walls. The entropic trapping effect becomes ever less important as the field increases, since the change in potential energy from moving to a new cavity becomes large compared with the entropy cost during transit. However, as noted in section 54.3.2.1, the crystal axes are not necessarily aligned with the field. As a result, if the axes are not aligned with the field, the electrical force pushes the DNA against the cavity walls and flattens it like a pancake. This is a peculiar consequence of the polymer used of creating the "negative" of the bead array, which is a hydrogel that is permeable to ions (and thus to the electric field) but not to DNA. If the "pancake radius" is large enough, then the edges of the pancake will reach the nearby pore entries, whereupon the electric field can pull the DNA out of the cavity. Since larger DNA will have larger pancake radii, they will escape more easily.

The initial experiments discussed above focused on single-molecule observations (i.e., without an injection step), and the matrix was enclosed in a custom electrophoresis cell. A preliminary report [38] using an array of microchannels to sculpt the field [37] indicates that this matrix can be used as an analytical tool. Moreover, pulsed-field separation are possible by mimicking the DNA prism protocol [37,38].

Interestingly, the three-dimensional (3D) network of interconnected cavities described in Reference 22 is a perfect practical realization of the "lakes–straits" model proposed by Zimm [39] as an early model of gel electrophoresis. To be in this regime, a molecule must span several pores, so DNA molecules significantly larger than those used in the experiments described above would be necessary to reach this regime. It would be interesting to check the validity of the lakes–straits model with such molecules.

The above discussions have highlighted the utility of both magnetic arrays and colloidal-templated crystals for DNA electrophoresis. While the variety of solid supports that can be fabricated by these techniques is limited when compared to those available through micro- and nanofabrication, the ease with which these self-assembled arrays can be constructed confers significant advantages for routine use.

While we have only focused here on applications for DNA electrophoresis, colloidal crystals have also been used as a stationary phase for separating hydrophobic dyes and proteins [40]. The arrays may prove especially interesting for electrophoretic protein separations, since the pore size provides a steric constraint, no pressure drop is required for transport through the small pores and the bead surfaces serve as a substrate for reversed-phase adsorption. These arrays thus offer a tunable variety of separation mechanisms. Indeed, proteins with similar hydrophobic groups, which cannot be separated by high performance liquid chromatography (HPLC) due to their similar adsorption properties, are rapidly separated by charge in colloidal arrays [40].

54.5 METHODS DEVELOPMENT GUIDELINES

The array structure (and the formation of the labyrinth) of magnetic arrays on the rate that the magnetic field is applied. In the reported separations [6,12], the magnetic field is applied by ramping the current in the electromagnet to its highest setting as quickly as possible. This traps the bead suspension in a local energy minimum, which may not necessarily correspond to the thermodynamic equilibrium state. As a result, the slow evolution process depicted in Figure 54.2 is not realized and thermodynamic instabilities, such as the labyrinth structure, can be suppressed to a certain extent. Moreover, rapidly applying the magnetic field also appears to increase the reproducibility of the array structure. From a purely theoretical standpoint, this cannot be true, since the thermodynamic minimum should be perfectly reproducible. However, in these experiments, interactions with the wall immobilizing columns through a mechanism resembling solid friction prevents this equilibrium from being reached. Supporting this interpretation, the regularity of the array structure is in general improved if the field is turned on and off several times before the separation to "train" the beads.

Most likely, the initial application of the field brings the beads from an initially random distribution to something close to a local energy minimum. When the field is turned off, Brownian motion allows the beads to perturb their positions. While the magnetic attraction/repulsion is no longer active in the absence of the field, steric effects and van der Waals forces persist, forcing the beads to retain a short-term memory of their previous positions in the presence of the field. If the field is turned back on after a short period of time, the beads now reform the columns, but start from an initial condition closer to the local energy minimum, thereby increasing the probability that they will assemble into a more regular hexagonal structure. Each time the off–on cycle is repeated, the beads reassemble from an initial position closer to the local minimum. This cyclic approach will yield ever more regular arrays until the array is very close to the local energy minimum.

54.6 CONCLUDING REMARKS

In this chapter, we have discussed two different bottom-up approaches to forming stationary phases for capillary electrophoresis: the self-assembly of superparamagnetic beads (the Ephesia system) and colloidal templating. Both methods have proven successful in separating long DNA without the need for a complicated lithographic fabrication step. As a result, they essentially side-step the dedicated clean-room facilities and expensive nanolithography tools needed for comparable top-down approaches and open the possibility of bringing these systems into routine use in biological laboratories. However, as we have seen, the range of structures that can be created through self-assembly is limited. Indeed, for the Ephesia system, it is only possible to create highly open structures by incorporating magnetic anchors into the surface, which itself requires some lithography, and colloidal crystals are plagued by stacking faults at large length scales. Thus, although bottom-up methods offer many exciting possibilities for designing stationary phases for capillary electrophoresis, they are not without their drawbacks.

Given the technological importance of nanofabrication for microelectronics and colloidal assembly for photonic crystals, it is inevitable that (now) conventional lithographic approaches to lab-on-a-chip stationary phases and the newer self-assembled techniques will continue to benefit from advances in related fields. At the present stage, it is difficult to determine which of these approaches will ultimately prevail. If lithographic fabrication costs continue to follow Moore's Law, it is conceivable that methods that are prohibitively expensive today could become affordable in the near future. We suspect, however, that this simple transposition from the microelectronics world will rapidly encounter limitations: biomolecules are by far larger than electrons, and with micron-sized structures, one is already facing limitations associated with the small number of molecules of a given kind contained in the volume of interest. The relative weight of Brownian motion also increases with downscaling. A major element in the success of downscaling in microelectronics, is that it allowed a simultaneous *increase* in the number of components in a microprocessor, and a *decrease* in its footprint, that is, increase the number of processors made in a single step on the same wafer. For the reasons recalled above, this "win-win" strategy will probably not work so well for "lab-on-chips." Likewise, if methods for colloidal templating continue to improve, in particular for generating larger scale and more varied structures, then the ease of this bottom-up strategy could ultimately trump the flexibility of lithography, especially for 3D structures. We anticipate that "hybrid" methods such as the "templated magnetic assembly" seen in Figure 54.5. above, combining relatively simple and inexpensive "top-down" methods to define a bidimensional structure, and "bottom-up" self-assembly to build on it structures too difficult or to expensive to produce by purely top-down approaches, will play an important role in the following years. Other examples, combining microlithography and surface-tension driven self-assembly of nanoparticles, to achieve nanostructures with dimensions much smaller than those of the initial template, can be found, for example in Reference 41.

REFERENCES

1. Buchholz, B.A., Shi, W., and Barron, A.E. Microchannel DNA sequencing matrices with switchable viscosities, *Electrophoresis*, 23, 1398, 2002.
2. Volkmuth, W.D., and Austin, R.A. DNA electrophoresis in microlithographic arrays, *Nature*, 358, 600, 1992.
3. Kaji, N., Tzekua, Y., Ueda, M., Nishimoto, T., Nakanishi, H., Horiike, Y., and Baba, Y. Separation of long DNA molecules by quartz nanopillar chips under a direct current electric field, *Anal. Chem.*, 76, 15–22, 2004.
4. Dorfman, K.D., and Viovy, J.L. Semiphenomenological model for the dispersion of DNA during electrophoresis in a microfluidic array of posts, *Phys. Rev. E*, 69, 011901, 2004.
5. Liu, J., Lawrence, E.M., Wu, A., Ivey, M.L., Flores, G.A., Javier, K., Bibette, and Richard, J. Field-induced structures in ferrofluid emulsions, *Phys. Rev. Lett.*, 74, 2828, 1995.
6. Minc, N., Futterer, C., Dorfman, K., Bancaud, A., Gosse, C., Goubault, C., and Viovy, J.L. Quantitative microfluidic separation of DNA in self-assembled magnetic matrixes, *Anal. Chem.*, 76, 3770, 2004.
7. Minc, N. Microfluidique et billes magnétique auto-organisées pour les analyses biologiques, Ph.D. Thesis, Université Paris VI, 2005.
8. Ivey, M., Liu, J., Zhu, Y., and Cutillas, S. Magnetic-field-induced structural transitions in a ferrofluid emulsion, *Phys. Rev. E*, 63, 011403, 2000.
9. Haghgooie, R., and Doyle, P.S. Structural analysis of a dipole system in two-dimensional channels, *Phys. Rev. E*, 70, 061408, 2004.
10. Haghgooie, R., Li, C., and Doyle, P.S. Experimental study of structure and dynamics in a monolayer of paramagnetic colloids confined by parallel hard walls, *Langmuir*, 22, 3601, 2006.
11. Haghgooie, R., and Doyle, P.S. Structure and dynamics of repulsive magnetorheological colloids in two-dimensional channels, *Phys. Rev. E*, 72, 011405, 2005.
12. Minc, N., Bokov, P., Zeldovich, K.B., Futterer, C., Viovy, J.L., and Dorfman, K.D. Motion of single long DNA molecules through arrays of magnetic columns, *Electrophoresis*, 26, 362, 2005.
13. Rosensweig, R.E. *Ferrohydrodynamics*, Cambridge University Press, New York, 1985.
14. Saliba, A.E., Gosse, C., Minc, N., Roblin, C., Dorfman, K.D., and Viovy, J.L. Ni pattern for guiding magnetic beads self-organization: application to long DNA electrophoresis in dilute arrays, *Proceedings of MicroTAS 2006*, Eds. T. Kitamori, H. Fujita, and S. Hasabe, Tokyo, Japan, Transducers Research Society, 2006.
15. Psychari, E., Saliba, A.E., Futterer, C., Slovakova, M., Goubault, C., and Viovy, J.L. Cell sorting in a microfluidic system with magnetic nanoparticles, *Proceedings of MicroTAS 2005*, Eds. K.F. Jensen, J. Han, D.J. Harrison, and J. Voldman, Boston, MA, USA, Transducers Research Society, 2005.
16. Volkmuth, W.D., Duke, T., Wu, M.C., Austin, R.H. and Szabo, A. DNA Electrodiffusion in a 2D array of posts, *Phys. Rev. Lett.*, 72, 2117, 1994.
17. Dorfman, K.D. DNA electrophoresis in microfluidic post arrays under moderate electric fields, *Phys. Rev. E*, 73, 061922, 2006.
18. Minc, N., Viovy, J.L., and Dorfman, K.D. Non-Markovian transport of DNA in microfluidic post arrays, *Phys. Rev. Lett.*, 94, 198105, 2005.
19. Popelka, S., Kabatek, Z., Viovy, J.L., and Gas, B. Peak dispersion due to geometration motion in gel electrophoresis of macromolecules, *J. Chromatogr. A*, 838, 45, 1999.
20. Meistermann, L., and Tinland, B. DNA electrophoresis in a monodisperse porous medium, *Phys. Rev E*, 62, 4014, 2000.
21. Zhang, H., and Wirth, M.J. Electromigration of single molecules of DNA in a crystalline array of 300-nm silica colloids, *Anal. Chem.*, 77, 1237, 2005.
22. Zeng, Y., and Harrison, J.D. Confinement effects on electromigration of long DNA molecules in an ordered cavity array, *Electrophoresis*, 27, 3747, 2006.
23. Nagayama, K. Two-dimensional self-assembly of colloids in thin liquid films, *Coll. Surf. A*, 109, 363, 1996.
24. Denkov, N.D., Velev, O.D., Kraichevsky, P.A., Ivanov, I.B., Yoshimura, H., and Nagayama, K. Two-dimensional crystallization, *Nature*, 361, 26, 1993.
25. Jiang, P., Bertone, J.F., Hwang, K.S., and Colvin, V.L. Single-crystal colloidal multilayers of controlled thickness, *Chem. Mater.*, 11, 2132, 1999.

26. Park, S.H., and Xia, Y. Fabrication of three-dimensional macroporous membranes with assemblies of microspheres as templates, *Chem. Mater.*, 10, 1745, 1998.
27. Jiang, P., Hwang, K.S., Mittleman, D.M., Bertone, J.F., and Colvin, V.L. Template-directed preparation of macroporous polymers with oriented and crystalline arrays of voids, *J. Am. Chem. Soc.*, 121, 11630, 1999.
28. Randall, G.C., and Doyle, P.S. Permeation-driven flow in poly(dimethylsiloxane) microfluidic devices, *Proc. Natl. Acad. Sci. USA*, 102, 10813, 2005.
29. Doyle, P.S., Bibette, J., Bancaud, A., and Viovy, J.L. Self assembled magnetic matrices for DNA separation chips, *Science*, 295, 2237, 2002.
30. Mayer, P., Bibette, J., and Viovy, J.L. Separation of large DNA using ferrofluid electrophoresis, *Mater. Res. Soc. Symp. Proc.*, 463, 57, 1997.
31. Futterer, C., Minc, N., Bormuth, V., Codarbox, J.H., Laval, P., Rossier, J., and Viovy, J.L. Injection and flow control system for microchannels, *Lab Chip*, 4, 351, 2004.
32. Pallandre, A., de Lambert, B., Attia, R., Jonas, A.M., and Viovy, J.L. Surface treatment and characterization: Perspective to electrophoresis and lab-on-chips, *Electrophoresis*, 27, 584, 2006.
33. Viovy, J.L. Electrophoresis of DNA and other polyelectrolytes: physical mechanisms, *Rev. Mod. Phys.*, 72, 813, 2000.
34. Nykypanchuk, D., Strey, H.H., and Hoagland, D.A. Brownian motion of DNA confined within a two-dimensional array, *Science*, 297, 987, 2002.
35. Baumgartner, A., and Muthukumar, M. A trapped polymer chain in random porous media, *J. Chem. Phys.*, 87, 3082, 1987.
36. Rousseau, J., Drouin, G., and Slater, G.W. Entropic trapping of DNA during gel electrophoresis: effect of field Intensity and gel concentration, *Phys. Rev. Lett.*, 79, 1945, 1997.
37. Huang, L.R., Tegenfeldt, J.O., Kraeft, J.J., Sturm, J.C., Austin, R.H., and Cox, E.C. Separation of 100-kilobase DNA molecules in 10 seconds, *Nat Biotech.*, 20, 1048, 2002.
38. Zeng, Y., and Harrison, J.D. Self-assembled three-dimensional nanofluidic sieves for biomolecule separation, *Proceedings of MicroTAS 2006*, Eds. T. Kitamori, H. Fujita, and S. Hasabe, Tokyo, Japan, Transducers Research Society, 2006.
39. Zimm, B.H. "Lakes-straits" model of field-inversion gel electrophoresis of DNA, *J. Chem. Phys.*, 94, 2187, 1991.
40. Zheng, S., Ross, E., Legg, M.A., Wirth, M.J. High-speed electroseparations inside silica colloidal crystals, *J. Am. Chem. Soc.*, 128, 9016, 2006.
41. Kraus, T., Malaquin, L., Delamarche, E., Schmid, H., Spencer, N.D., and Wolf, A. Closing the gap between self-assembly and microsystems using self-assembly, transfer and integration (SATI) of particles, *Adv. Mater.*, 17, 2438, 2005.

55 Nanoscale DNA Analysis

Laili Mahmoudian, Mohamad Reza Mohamadi, Noritada Kaji, Manabu Tokeshi, and Yoshinobu Baba

CONTENTS

55.1 INTRODUCTION

Microchip electrophoresis has been developed as a promising technology for detection of DNA in samples.[1–3] However, the development of systems for the separation of long DNA strands [greater than several kilo basepairs (kbps)] using commercially available media, has not been successful.[4] In recent years, efforts have been focused on the overcoming the limitations of current DNA electrophoresis methods. Those limitations largely result from an application base that only required the separation of DNA fragments in a narrow size range, one where using fundamental separation modes based on sieving using conventional methods.[5] In conventional gel electrophoresis, DNA molecules move in the electric field based on a size-dependent mobility through pores in a gel matrix.[6] However, length-dependent mobility vanishes for DNA molecules longer than ~40 kbps, primarily due to tendency of DNA molecules to become stretched and oriented to the direction of electric field.[7,8]

Buoyed by the knowledge of micro- and nanofabrication technologies, nanotechnology has been shown to be quite successful for long fragment DNA analysis. To achieve the required efficiency needed for this type of DNA electrophoresis, two directions in the field of nanotechnology have been established: one focused on providing artificial gel structures produced with nanofabrication technology, and the other on nanomaterials for DNA separation.

In this chapter, we discuss the developments and capabilities of nanoscale methods for DNA electrophoresis, and provide some important technical information for each method. Considering the point that the mechanism of DNA separation in each nanotechnique is different (and sometimes unique) from other better-understood separation modes, these will be explained separately.

55.2 NANOFABRICATED STRUCTURES

55.2.1 Obstacle Arrays

The first report on these structures was by Volkmuth in 1992[9] involving a microlithographically constructed obstacle array in SiO_2. Using these arrays, they showed that the DNA molecule could be hooked on the post and DNA fractionation could be done in a direct current (DC) electric field. DNA from *Micrococcus loteus* with an original uniform length of 100 kb (30 μm contour length) was fractionated in a DC field of 1 V/cm (Figure 55.1). Briefly, the highly purified *Micrococcus loteus* was sheared by pipetting leading to a range of fragment sizes, and then introduced to the designed chip; ethidium bromide (EtBr) was used to observe the hooked DNA via epifluorescent video microscopy. Arrays were designed based on unphysical agarose gel, with the effective pore size of 1 μm theoretically corresponding to a physically unstable 0.05% agarose gel. Arrays were fabricated on a standard 3 in. diameter silicon wafer with a 0.5 μm SiO_2 layer created on the surface of the array for optical emission measurement. The arrays consist of a 2.7 × 2.7 mm rectangular containing 2 million posts 0.15 μm high by 1.0 μm in diameter with 2 μm center-to-center spacing. Two different sets of lattices were developed. One set of arrays with the posts on a square lattice, and another set with a square—45°—rotated lattice. For loading the samples, a flat loading area was made at the two ends of the array with a depth equivalent to that of the posts. Field-assisted silicon-glass fusion was utilized to fuse a Pyrex coverslip to the posts. The authors purported that the key to the DNA separation was a dispersion time that correlated with the time required for hooked DNA to thermally free itself from a post.

55.2.1.1 Assymetric Obstacle Courses: Rectification of Brownian Motion

While the separation of DNA based on hooking on micro- or nanoposts has presented an alternative method for gel electrophoresis, it still suffers from the fact that, when an electric field is applied to DNA molecules, different sizes of DNA molecules mobilize at the same speed. To circumvent this problem, an approach was developed by Duke, Austin[10] and Ertas[11] which take advantage of the fact that, while a molecules move, they diffuse at the same time—and at a diffusion rate that is size dependent. In theory, they have shown the possibility of using a two-dimensional obstacle course to sort the fast moving molecules from the slower ones. The elegance of this is that a regular lattice of asymmetric obstacle course, rectifies the lateral Brownian motion of the molecules, so molecules of different size follow different trajectories while they are passing into the device.

Additional effort by Chou et al.[12] applied the above methodology to microfabricate of silicon array. With this device, it was possible to separate a mixture of *Xho*I-cut λ DNA fragments (15 and 33.5 kbp) in the vertical electric field of 1.4 V/cm. They calculated that the DNA fragments moved in trajectories at different angels through the array. Interestingly, molecules that diffuse very slowly

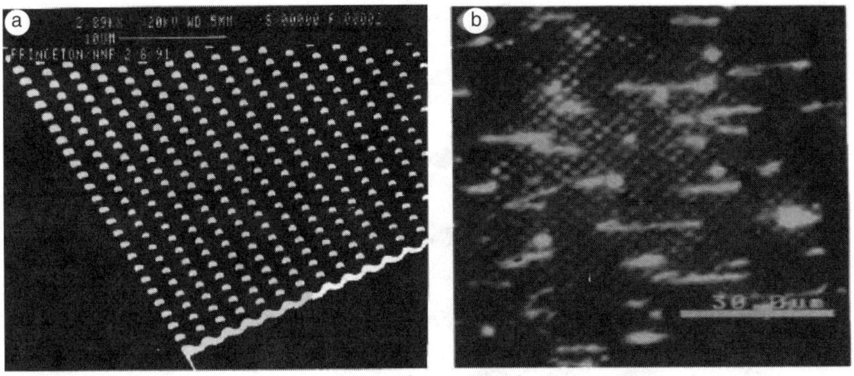

FIGURE 55.1 A scanning electron micrograph of the obstacle course. The obstacles are 0.35 pm high and measure 1.5 × 6.0 pm. The gap between adjacent obstacles is 1.5 pm. An electric field propels the molecules directly through the gaps between the posts with velocity v. Transverse Brownian motion may cause a molecule to skip one channel to the right if it diffuses through displacement a_R, or very rarely, one channel to the left if it diffuses through a_L.

are likely to travel straight without being deflected by the obstacles. In contrast, molecules that diffuse rapidly veer mostly to right in the device. Such a device is advantageous because it provides the possibility for sorting the molecules continuously, in contrast to batch modes in conventional methods. The authors demonstrated that the predicted movement could be done practically with a well-designed device containing 0.5X tris-borate buffer (TBE), 0.1% performance-optimized POP-6 (Perkin Elmer Biosystems, Foster City, California) and 0.1 M dithiothreitol (DTT) as separation buffer. The device contained a lattice of rectangular obstacles with the size 1.5 × 6 μm and an angle of 45° relative to the direction of electric field, designed in a way that allowed for a uniform flow over the whole area of the device. A sacrificial layer was deposited over the silicon nitride floor of the device while the ceiling had a layer of silicon nitride deposited over it using chemical vapor deposition. Small irrigation ducts made it possible to remove the sacrificial layer with a wet etchant (tetramethylammonium hydroxide), followed by sealing the irrigation holes with SiO_2 and etching the loading holes so as to provide a final device that was a quasi-two-dimensional chamber containing patterned silicon nitride obstacles.

A major improvement in fractionation of biomolecules based on rectification of Brownian motion was developed by Oudenaarden.[13] With this technique, originally applied for the analysis of lipid biolayers, a laterally asymmetric diffusion array was fabricated. The separation was done in a direction that was not parallel to the direction of main flow, and lead to the components of the sample to physically discrete positions in the device. This approach brings with it the distinct possibility of inject sample continuously with separation accomplished in a DC electric field.

Cabodi et al.[14] used a laterally asymmetric diffusion array with modified injection strategy termed *out-of-plane sample injection*; enabling separation of diverse biomolecules ranging from long strand DNAs to small proteins. Basically, the laterally asymmetric obstacles rectify the motion to one side. All migrating molecules diffuse laterally according to their diffusion coefficient, with larger molecules diffusing less and traveling straight down in the channel, while smaller molecules, on average, are deflected more extensively than larger ones. Figure 55.2 shows the separation principal of laterally asymmetric diffusion array. Coliphage T2 and T7 were spatially resolved using this device, with a final DNA concentration of 2 μg/mL in 0.5X TBE with 3% β-mercaptoethanol as an antiphotobleaching agent, and YOYO-1 dye was as an intercalating dye. An inverted microscope (Olympus America, Melvilla, New York, USA), illuminated by a 100-W mercury arc lamp was used for DNA observation.

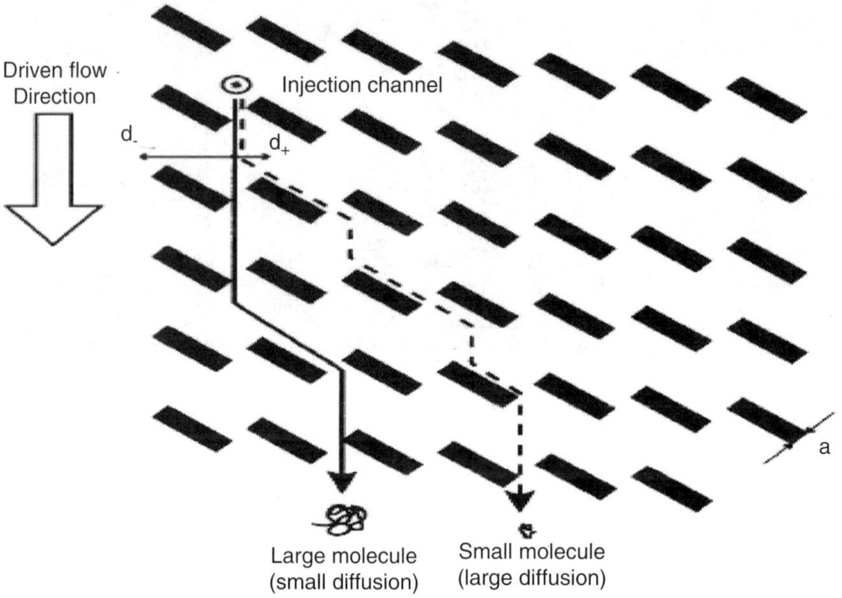

FIGURE 55.2 Principal of separation in asymmetric obstacle array. Basically the latterly asymmetric obstacles rectify the motion to one side. All migrating molecules diffuse laterally according to their diffusion coefficient through the device. Larger molecules diffuse less and travel straight down in the channel and smaller molecules on average are deflected more than larger ones. (Adapted from Cabodi. M, Chen, Y.F, Turner, S.W.P, Craighead. H.G, Austin, R.H, *Electrophoresis*, 2002, 23, 3496–3503. With permission.)

The above technique was further improved by Huang[15] with tilted Brownian ratchet array. Using this design, the flow is tilted relative to vertical axis of the array, since the probability that a molecule could be deflected is greatly increased compared to the previous status (where flow is aligned along the vertical array axis). By this method the separation resolution and speed of DNA separation by the Brownian ratchet array in microchips was increased three and ten times, respectively. A mixture of 48.5 and 164 kb DNA was separated in ~70 min using a 12-mm-long (Figure 55.3).

55.2.2 Entropic Nanotraps

Craighead and coworkers[16] were the first to show that a nanofluidic channel containing entropic traps could be designed for separation of long DNA molecules. Entropic trap array systems include alternating thin and thick regions that correspond to the molecular dimensions of DNA, and were designed for static (DC) electric field and fast separation of large DNA molecules. Figure 55.4 shows the motion of lambda DNA in one of the early designs for an entropic trap device.[16] The fundamental hypothesis describing the function of the device has been described elsewhere.[17] An entropic trap array device designed for the separation of DNA similar to conventional slab gel-pulsed field gel electrophoresis (PFGE), was described by the same group later.[18]

55.2.2.1 Separation Mechanism

In order to fully exploit the power of a microfabricated PFGE system, the principles of this separation should be explained. Basically, the microfabricated entropic trap array system consists of alternating thin and thick regions. With the channel depth of the thin region smaller than the radius of gyration (R_0) of DNA, the system can serve as a molecular sieve. In contrast, the thickness of thick region is larger than R_0, therefore, the DNA can form spherical equilibrium shape in thick region while, in

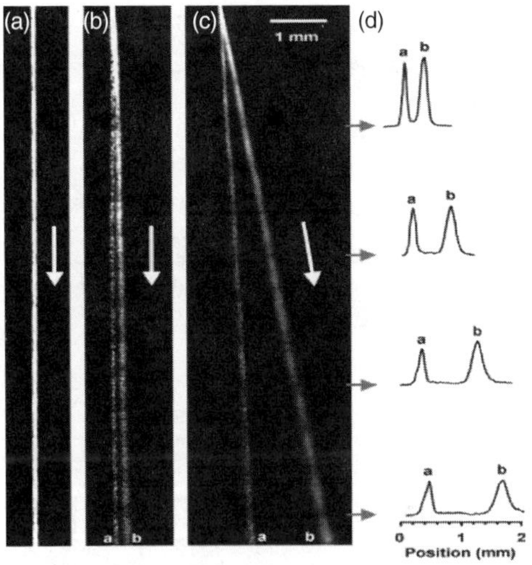

FIGURE 55.3 Basic principle of the Brownian ratchet array. Particles are driven through the array hydrodynamically or electrophoretically (pictured). (a) The rectangular obstacle array prevents particles emerging from gap A and diffusing to the left (1) from reaching gap B– but ratchets particles diffusing to the right (2) to gap B+. (b) Particles of different sizes diffuse to different extents (bell-shaped curves represent lateral distributions of small and large particles), resulting in different probabilities of deflection to B+. The vertical dotted line within the distributions represents the required diffusion for ratcheting to gap B+. (c) The probability of a particle being deflected to B+ is increased by tilting the flow at a small angle with respect to the vertical axis of the array.

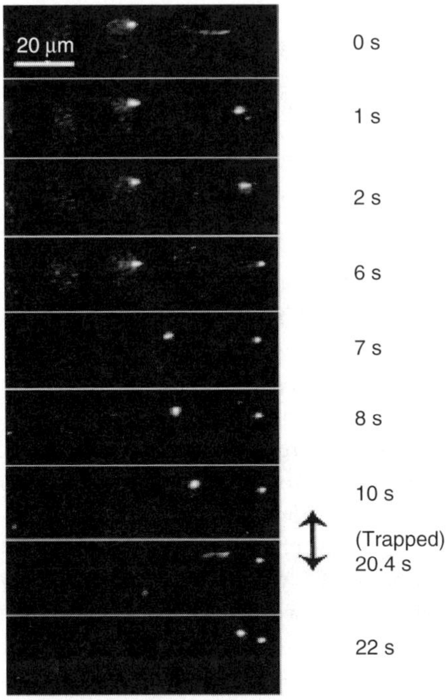

FIGURE 55.4 Sequence of video frames showing the motion of a lambda DNA molecule. Brighter regions are the thick regions of the channel. The left DNA molecule was trapped for 6 s, then escaped the trap and jumped to the next thick region within a second. After traveling the thick region for about 3 s, it was trapped again at the next entropic trap for about 10 s. In this device, *tl*51.4 mm and *ts*590 nm. (Reprinted with permission from Han, J., Craighead, H. G., *J., Vac. Sci. Technol. A*, 1999, 17, 2142–2147. Copyright 1999, American Institute of Physics.)

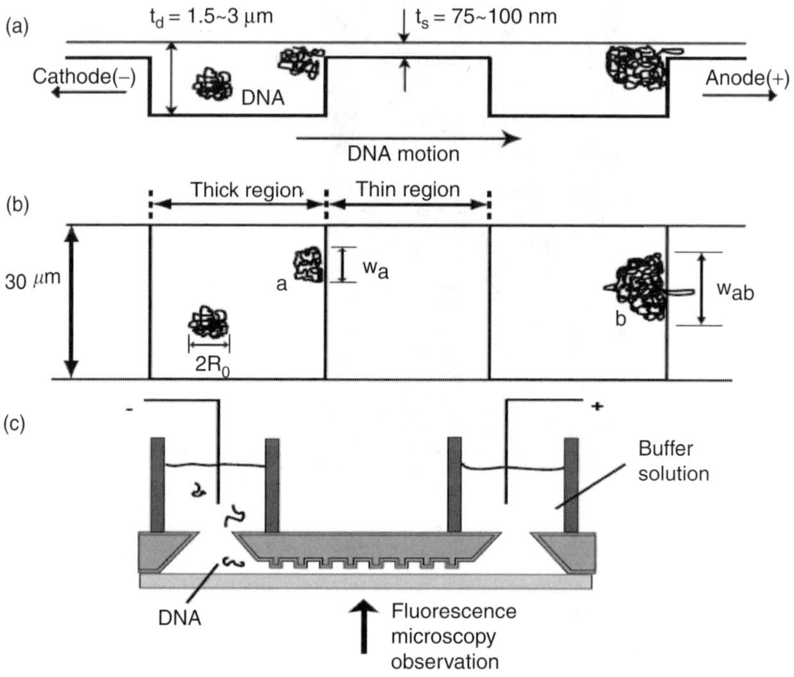

FIGURE 55.5 Nanofluidic separation device with many entropic traps. (a) Cross-sectional schematic diagram of the device. Electrophoresed DNA molecules are trapped whenever they meet a thin region, because their radius of gyration (R_0) is much larger than the thin region depth (here, t_d and t_s are the thick and thin region depths, respectively). (b) Top view of the device in operation. Trapped DNA molecules eventually escape, with a probability of escape proportional to the length of the slit that the DNA molecule covers (w_a and w_b). Larger molecules have a higher escape probability because they cover wider regions of the slit ($w_b > w_a$). (c) Experimental setup. Reservoirs are made at both ends of the channel and filled with DNA solution. (Reprinted from Han, J., Craighead, H. G., *Science*, 2000, 288, 1026–1029. With permission from AAAS.)

thin region, the DNA molecules are deformed. Once the electric field is applied, DNA molecules are temporarily trapped at the entrance to the thin regions, because the relaxed-spherical DNA stretches in electric field facilitating entrance to thin region. This conformation is not entropically preferential, and the DNA tends to escape from the thin region. While the DNA travels in thin and thick regions, a frequent conformational change occurs. Consequently, the DNA mobility will be length dependent. For escape of DNA molecule from the thin region, longer DNA molecules have a larger surface area to contact with the boundary of the thin region, and have a higher probability for escape. As a result, larger DNA molecules move faster than shorter ones (in contrast to normal gel electrophoresis).

The schematic in Figure 55.5 shows the properties and mechanism of separation achievable with a nanofluidic device containing numerous entropic traps for PFGE. Using this system, DNA molecules ranging from 5,000 to ~160,000 base pairs were successfully separated using a 15-mm-long channel. T2 (164 kbp) and T7 (37.9 kbp) DNA, were separated with an electric field of 21 V/cm in 30 min using this microfabricated array system. The authors also showed utility of this device for multiple-channel analysis useful for parallel sample separation.[18] Moreover, the ability of this system to work, (1) with a DC electric field, (2) in the absence of a gel matrix, (3) with miniscule masses of DNA, and finally (4) without the need for high-resolution lithographic techniques, provides distinct advantages for manufacturing integrated devices.

55.2.3 DNA Electrophoresis on Nanopatterned Surface

Nanopatterned surfaces for DNA separation was first introduced several years ago by Young-soo Seo.[18] In earlier works, it has been shown that it is possible to separate DNA on a flat-attractive surface without any topological boundaries or any sieving matrix. The separation mechanism is based on the differences in conformational associated with DNA molecules of different length adsorbed on the surface. This surface-directed separation has opened a new vistas for nonsieving polymer-based DNA separation.

55.2.3.1 Separation Mechanism

On of adsorption of DNA on a flat surface, the balance between the loss of entropy (due to the localization of DNA at the surface) and the energy gain (due to adsorption) determines the resultant conformation of DNA—the resultant conformation may be in the form of loops that extend into solution or "train like" with neighboring segments adsorbed on the surface. In the case of a fixed surface attraction, shorter fragments have higher entropy and have larger number of loops, whereas longer fragments will exploit the energetics achieved upon adsorption and form larger "trains." As a result, in the presence of an electric field applied to the surface, there should be a conformationally based response and, therefore, a length-dependent mobility should result.[19]

The dependence of the separation on the conformational differences emphasizes the need for special care to be taken in the surface design. The sensitivity of conformational changes on a surface can be tuned by increasing the complexity of the design. A design of nanopatterned surface of Ni patches imposed on a Si matrix was developed for DNA electrophoresis.[19] It has been demonstrated that using this method, simultaneous separation of a broad range of DNA sizes, from a few hundred base pairs to genome-size DNA, could be achieved with good resolution. Figure 55.6 shows the schematic of fabrication process and the scanning electron micrograph of surface image of Ni nanopatterned on a Si wafer. The need for only very small amounts of sample and low operating voltages make this technique affordable for integration into chip-based portable devices.

55.2.4 Nanopillars

Micrometer-sized pillars were fabricated and then used for separation of very long DNA molecules over a hundred thousand base pairs.[20,21] As a follow up to this, Baba's group used a new fabrication technique to create a nanopillar chip on a quartz plate.[22] DNA fragments ranging from 1 to 38 kp were rapidly separated as clear bands, within the analysis time of 170 s. Even long DNA molecules that are difficult to separate on conventional gel and capillary electrophoresis (CE) could be separated using the nanopillars. A mixture of λ DNA (48.5 kbp) and T4 DNA (165.6 kbp) was successfully separated by flow through a 380-μm-long nanopillar channel within only 10 s even under a direct current (DC) electric field. Using electron-beam lithography, Ni electroplating, and neutral loop discharge etching, 100–500-nm diameter and 500–5000-nm tall nanopillars were fabricated, and the nanopillar chamber sealed with by quartz coverplate to minimize the loss of resolution that normally accompanies silicon or even silicon-coated insulation materials. Figure 55.7 Shows a schematic representation and a scanning electron micrograph of a microchannel equipped with nanopillar structures fabricated on a quartz glass plate. For analysis, the final DNA concentration was 20–100 μg/mL for DNA separation and $\sim$50 ng/mL for single DNA molecule observation in the 5X TBE electrophoresis buffer (45 mM Tris-borate, 1 mM ethylenediamine tetraacetic acid (EDTA), pH 8.0) containing 4% (v/v) 2-mercaptoethanol to reduce photobleaching. Before electrophoresis, the surface was dynamically coated with 1% poly(vinylpyrrolidone) (M_w 1,300,000, Aldrich) to reduce electroosmotic flow (EOF), and then the solution was substituted with 0.5X TBE electrophoretic

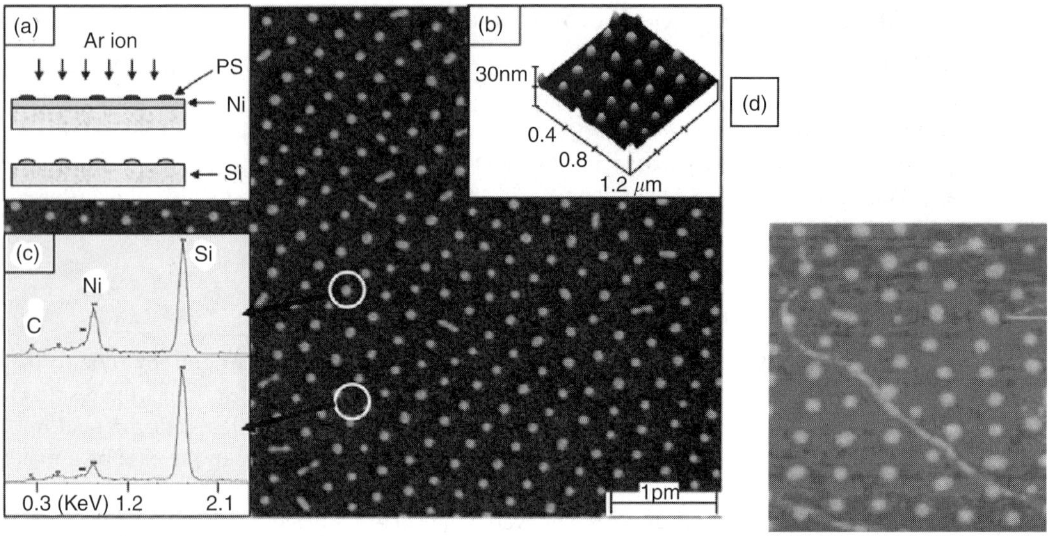

FIGURE 55.6 Surface image of Ni nanopattern on Si wafer using SEM. Inset (a) schematic of etching process, (b) the topography of Ni nanopattern, (c) a fluorescence chemical map of the surface, and (d) scanning probe microscope (SPM) image in air of a λ-DNA adsorbed onto the nanopatterned surface. (Adapted and reprinted with permission from Seo, Y.S., Luo. H, Samuilov. V.A., Rafailovich. M.H., Sokolov. J., Gersappe., Chu. B., *Nano Lett.*, 2004, 4, 659–664. Copyright 2004, American Chemical Society.)

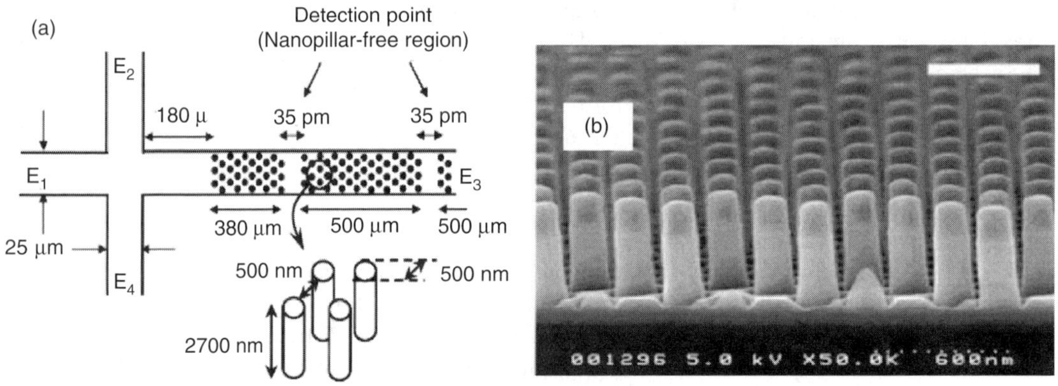

FIGURE 55.7 Nanopillars: (a) Schematic illustration of a microchannel equipped with nanopillar (b) nanopillar structures fabricated on a quartz glass plate before sealing by a cover plate. The nanopillars in this figure are 200 nm wide and 600 nm tall (aspect ratio, 3). (Adapted and reprinted with permission from Kaji. N., Tezuka. Y., Takamura. Y., Ueda. M., Nishimoto. T., Nakanishi. H., Horiike. Y., Baba. Y, *Anal. Chem.*, 2004, 76, 15–22. Copyright 2004, American Chemical Society.)

buffer. DNA separation and a single DNA molecule observation were performed on an inverted microscope.

55.2.4.1 Separation Mechanism

The linear relationship between the logarithmic value of the relative mobilities versus the size of DNA over the range of 1–23 kbp, provided evidence that the electrophoretic behavior of DNA in this size range separated in a nanopillar chip obeyed the Ogston model. In this model, DNA molecules are assumed to be globular in nature, passing through a random network with an average pore size in size range. In contrast to this, T4 DNA shows a cyclic type of motion, and seems to obey a model better described by in "reptation with orientation" regime. Interestingly, single DNA molecule observation in a nanopillar channel revealed that the optimal nanopillar structure induced T4 DNA to form a narrow U-shaped conformation during the electrophoresis, whereas λ DNA maintained a rather spherical conformation.[22]

The most important advantage of nanopillar structure design and fabrication is the capability to modify, with spatial accuracy, particular surfaces in the chip structure. On the other hand, since in separation by nanopillars does not require a DNA sieving matrix, it is more easily implemented in chips designed for totally integrated sample processing, where the different requirements of each domain present challenges. Simplifying the detection domain of multifunction chips will accelerate microfluidic device development for bioanalysis so that the entire analytical process (cell culture, cell selection, DNA extraction, amplification, detection, and collection on a single chip)[22b] overcoming the limitations stemming from injection of high viscosity polymers.

55.3 NANOMATERIALS FOR DNA ANALYSIS

Nanomaterials were developed as an alternative for separation of large DNA fragments. Compared with the difficulties, complexities and costs in the fabrication of micro- and nanofabricated devices, nanomaterials provide a potentially promising system for simple and affordable for lab-on-a-chip technology in medical diagnostics.

55.3.1 Superparamagnetic Particles

Upon applying a magnetic field to superparamagnetic particles with a dielectric constant or magnetic permeability different than that of the surrounding liquid, dipolar interactions are induced within the particles. Provided that these interactions are strong enough, a phase transition of the particle occurs, changing them from liquid to solid. Consequently a column-like structure will appear.[23]

On the basis of this basic theory, self-assembled magnetic particles can be used as a sieving matrix for DNA separation. Doyle et al.[24] have shown that a suspension of superparamagnetic microscale particles can self-assemble into an array with an average interparticle distance of 5.7 μm inside a microchannel device. The device was created using soft lithographic methods that have been previously described by this group.[25] Figure 55.8 shows the columnar structure of magnetic particles as well as the resultant long DNA fragment separation which shows surprisingly good efficiency.

The separation was optimized at an intermediate electric field, where hooking and incomplete extension occurs at the same time. With larger DNA molecules taking on a more extended form, this provides the mobility difference that allows for a size-based DNA separation. Having a low viscosity in the absence of magnetic field, tunable pore size (typically 1–100 μm) and obviating the need for sophisticated microlithography, of the use of suspensions of paramagnetic particles becomes attractive. In addition, they should be potential to the analysis of other analytes like cells, organelles and micro-or nanoparticles.

55.3.2 Nanospheres for DNA Separation

With the development of core-shell type of globular nanoparticles (nanospheres or nanoballs),[26] Tabuchi et al. recently introduced this technology as a potential new type of DNA separation matrix.[27] Nanoballs with low viscosity (0.94 cp at 1.0%; 10 mg/mL) owing to their globular structure, are promising sieving matrixes for long fragment size DNA separation. Using nanospheres, DNA fragments in the range of 100–1000 bp and 1–15 kb were successfully separated with

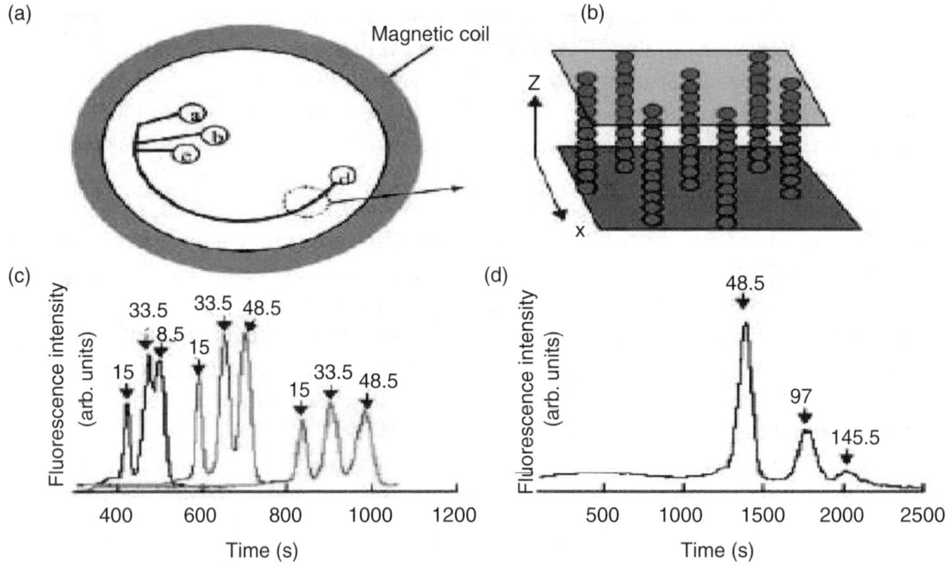

FIGURE 55.8 (a) Schematic of the microchannel and magnetic coil. (b) Columnar structure formed by a suspension of superparamagnetic particles. (c) Fluorescence intensity at 10 mm from the injection zone versus time. Separation of a mixture of λ-phage DNA and λ -DNA digested with *Xho*I (Pharmacia) (the numbers refer to the size of the DNA fragments in kbp) at fields strengths of 4.8 (right), 7 (middle), and 10 left V/cm. (d) Separation of λ-phage DNA concatemers at a field of 3.2 V/cm. Other conditions are identical to those in (c). (Adapted and reprinted from Doyle, P. S., Bibette, J., Bancaud, A., Viovy, J. L., *Science*, 2002, 295, 2237. With permission from AAAS.)

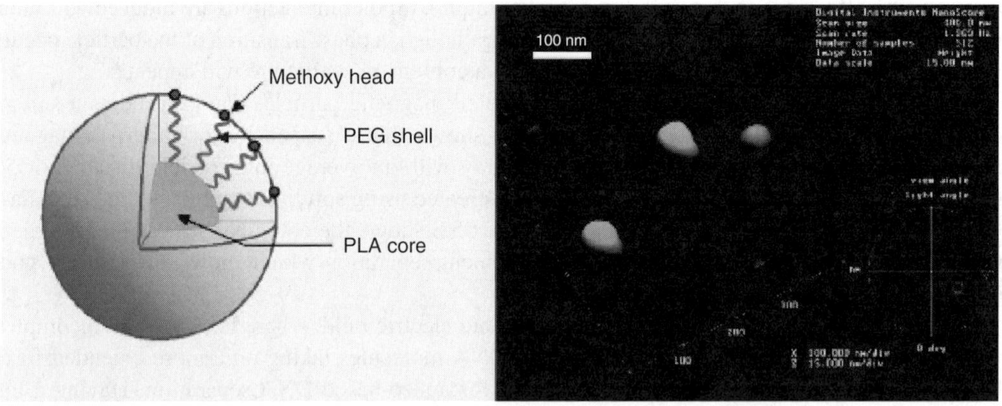

FIGURE 55.9 Schematic of the structure and an atomic force microscope (AFM) image of nanoballs. (From Manz, A., Graber, N., and Widmer, H.M., *Sens. Actuat.* B, 1990, 1, 244–248. Copyright 1990. With permission from Elsevier.)

high resolution and high speed (60 and 100 s, respectively). Figure 55.9 shows the structure of nanospheres.

55.3.2.1 Nanospheres Preparation and Properties

Nanospheres were created by the following procedure: multimolecular micellization, followed by core polymerization of block copolymer of poly(ethylene glycol) (PEG) with poly(lactic acid) processing a methacryloyl group at the polylactide (PLA) chain end (PEG_m-ß-PLA_n-MA_l; Mw(PEG/PLA) = 6100/4000, m ≈ 100, n ≈ 40, l ≈ 70) in an aqueous condition.

Core-shell type nanosphere technology is applicable for separation of wide range of DNA fragments.[27] The PLA segment, forming a spherical core and covered by a relatively high density PEG chains and methacryloyl groups, make stable 30-nm diameter nanospheres. The nanospheres have no surface charge, and a narrow size range and low viscosity. A SV1100 microchip electrophoresis system (Hitachi, Japan) and i-chip 3 microchip (Hitachi, Japan) made of poly(methyl methacrylate) with channels 100-μm wide and 30-mm effective separation channel was used for the DNA analysis. For electrophoresis, all channels were filled with 1% nanosphere solution, sample loaded into the sample well and pressure injected at 1–10 kPa for 1 s using a syringe, with the corresponding well and outlet wells open. Before electrophoretic separation, a second pressure injection (1–10 kPa) for 1 s was applied to the separation channel from the inlet without an electric field. The electrophoresis was then initiated at 220 V/cm with squeezing voltage of 130 V without any pressure.

The double pressurization technique applied for DNA separation was key to achieving high-speed analysis and improved resolution. A primary pressure is needed for sample injection, while the second pressure accelerates the separation speed of the DNA. One of the interesting points in using nanospheres is the focusing effect on the DNA as it migrates in nanosphere solution under an electric field. Observing a single DNA molecule move through a nanoball solution shows that "lumps" of DNA (intrachain aggregated structure) are generated in this condensed environment. The wide range DNA separation ability with high-speed using low viscosity nanoballs provides promising potential for DNA separation in microchannels and could have a widespread application for high-throughout analytical systems in medical diagnostics.

55.3.3 Gold Nanoparticles

Primary trials for manipulation of polymers for long DNA separation have been attempted using ultralow viscosity solutions. Barron et al.[28,29] and Braun et al.[30] have utilized ultradiluted polymer solutions (e.g., down to 0.01%), while Heller et al.[31] and Ren et al.[32] used streptavidin in an end-labeled free solution for CE. The efficiency of gold nanoparticles (GNPs) was shown for the separation of shorter DNA fragments (2–2176 bp),[33] however, recently Huang et al. used GNPs or GNPs mixed with polyethylene oxide (PEO) to demonstrate the effectiveness of nanoparticles for the separation of longer DNA molecules in a low viscosity medium using nanoparticle-filled CE (NFCE).[34] For prevention of aggregation of the GNPs and facilitate a strong interaction with the DNA molecules, the GNPs were noncovalantly bonded to the PEO to make gold nanoparticle/polymer composite (GNPPs).

Separation of high molecular weight DNA with the size range from 8.27 to 48.5 kbp and λ DNA (0.12–23.1 kbp) was accomplished in 6 and 5 min, respectively, using these modified polymers.

55.3.3.1 pH and Salt Dependence of Separation

It has been shown that the resolution of peaks in a GNPP solution depends on pH value, and the concentration ratio of the EtBr and GNPPs. Furthermore, the efficiency of separation is greater if it conducted in the presence of small amounts of salts.

The resolving power of the separation depends on the interactions between the DNA and GNPPs bonded to PEO. A 32-nm GNPPs coated with PEO was reported to have the best resolving power.[34]

55.3.3.2 Separation Mechanism

As described for the transient entanglement mechanism, DNA molecules, in the short term, intertwine with the polymer-adsorbed on the GNPs.[35] There are reasons where slowing the entanglement of DNA fragments may be desired. First, polymer-adsorbed on the GNPs are stiffer and less extended, a character that is dependent on the size of GNPs and the length of the polymer chain. This property causes the polymer-adsorbed nanoparticles to be slightly deformed under the flow.[33,36–38] Second, there is a strong intercalating between the DNA and EtBr adsorbed on the GNPPs. Finally, the DNA migration is affected by the GNPPs (2.0×10^8 g/mol for the 32-nm GNP) which are much heavier than the linear polymer and, to a great extent, slows the DNA migration as well. Therefore it has been proposed that the GNPPs play a similar role to that of streptavidin (that is in providing additional friction) in end-labeled free solution CE.[39,40] Since longer DNA molecules interact with more than one GNPP, the "drag" power would be stronger and, as a result, the mobility of longer DNA is slightly lower than that of shorter ones.

This technique was the first CE method to provided high-speed and separation efficiency for long DNA fragment separation. The potential of polymer-adsorbed nanoparticles in separation science and in the life sciences have been shown in numerous ways. Band broadening depends on the polydispcrsity of thc GNPPs and making more monodispersed GNPPs would certainly offer high resolution.

55.3.4 PEGylated-Latex Mixed Polymer Solution

PEGylated-latex solution mixed with usual water soluble polymer has been shown by Tabuchi et al.[41] to be an effective micro-CE (μ-CE) medium for DNA separation.

A higher separation of DNA fragments from 10 bp to 2 kbp was achieved using the fixed concentrations of 2.25 wt% of 80-nm-PEGylated-latex and 0.49 wt% of hydroxypropyl methycellulose (HPMC) solution under low viscosity of (<5.5 cp).

55.3.4.1 Synthesis of Methoxy-PEG-VB Macromonomer and PEGylated-latex

Synthesis of Methoxy-PEG-VB macromonomer was done by anionic ring-opening polymerization at room temperature and under argon gas[42] and could be accomplished using the following procedure:

1. A mixed solution of 0.16 of mL 2-methoxyethanol (2 mmol) and 5.6 mL (2 mmol) of a 0.35 M potassium naphthalene solution in tetrahydrofolate (THF) should be prepared. Then it should be inserted in a round bottomed-flask containing 30 mL dry hydrofolate (HF) under argon gas and stir for several minutes.
2. Addition of condensed ethylene oxide (11.3 mL, 227 mmol) by a cooled syringe and continuing the stirring for 2 days for polymerization.
3. Addition of 2.8 mL (20 mmol) vinyl benzyl chloride to the polymer and stirring for 24 h at room temperature.
4. Precipitation of the polymer with 100-folds excess of diethyl ether and freeze drying with benzene.

Under these conditions, a 91% yield of polymer could be obtained.

The PEGylated-latex could be obtained by an emulsion polymerization of styrene in presence of PEG macromonomers,[43] and could be done by the following procedure:

1. Dissolving of methoxy-PEG-VB macromonomer (M_n = 5500, 0.75 g) and Potassium peroxodisulfate (KPSO (74 μmol, 20 mg) in 25 mL of deionized water in a round bottomed flask under argon atmosphere.
2. Addition of styrene monomer (0.75 g) and divinyl benzene (76 μmol, 0.01 mL) to the PEG solution and stirring in a mechanical stirrer at 500 rpm for 1 h.
3. Incubation of the suspension in an incubator at 60°C for 24 h.
4. Purification of the latex from water by three-time centrifugation at 150,000 *g* for 15 min.

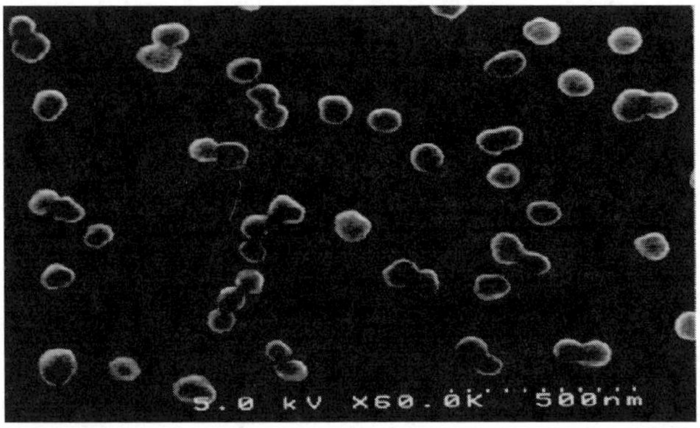

FIGURE 55.10 SEM image of nanosized PEGylated-latex beads. (From Tabuchi, M., Katsuyama, Y., Nogami, K., Nagata, H., Wakuda, K., Fujimoto, M., Nagasaki, Y., Yoshikawa. K., Kataoka, K., Baba, Y., *Lab Chip*, 2005, 5, 199–204. Reproduced by permission of the Royal Society of Chemistry.)

Nanoparticles of PEGylated-latex with the size and polydispersity of 79 nm and $D_w/D_n = 1.23$, respectively, could be prepared by this method. In a similar way, larger PEGylated-latex with the diameter of 193 nm and polydispersity of $D_w/D_n = 1.17$ could be prepared. Figure 55.10 shows the scanning electron micrograph (SEM) image of nanoballs produced by this method.

55.3.4.2 Separation Mechanism

Basically, the enhancement in separation using this system seems to be related to the balance between decreased polymer concentration which provide wider polymer mesh size and the structural obstacles of the particle.

Ultimately, the separation performance depends on the quality, size, and concentration of particles as well as the polymer concentration. The simplicity, low viscosity solution and effective separation of large DNA fragments, are some of advantages of this technique.

55.3.5 Nanogels

Sparsely cross-linked nanogel was developed with covalently linked linear polyacrylamide (%LPA) by the Barron and coworkers.[44] Using nanogel, they achieved a higher-selectivity separation of ssDNA sequencing fragments of longer than 375 bases with reasonable time frames for the analysis. They also produced nanogels with inverse emulsion (water in oil) copolymerization[45,46] of acrylamide and *N*,*N*-methylbisacrylamide (Bis). They purport that the presence of covalent bond enables stabilization of the network and, hence, improves the DNA sequencing read length (~63 bases, 10.4% improvement). In later work, they improved the matrix selectivity via optimization of nanogel cross-linked density with the proper linear polymer.[47] Copolymerization of acrylamide and a low percentage (10^{-4} mol%) of *N*,*N*-methylbisacrylamide (Bis) was utilized to produce a flowable, sparsely cross-linked matrix.

Using this matrix they achieved at 98.5% accuracy for the base calling of 680 bases with 18.7% improvement over the LPA. The power of the system became clarified by wedding this with a high-throughput microfabricated DNA sequencing device. The replaceable, highly entangled nanogel provides a new potential matrix that could match the performance of commercial capillary array electrophoresis (CAE) instruments and chip-based sequencing.

55.3.6 Carbon Nanotubes

Carbon nanotubes (CNs) could be defined as the ultimate fiber formed of perfectly graphitized closed seamless shell[48]—this was discovered in 1991 by Ijima.[49] Xu et al.[48] have verified the enhanced separation of DNA fragments with multiple-wall carbon nanotubes (MWCNs) as buffer additive. An improved resolution of the fragments in a *HaeIII* digest of ϕX174 DNA was observed with an optimized concentration of MWCN. A greater improvement in resolution for separation of 2-Log DNA ladder was shown in the concentrations of MWCN near or below its threshold concentration. They used negatively charged MWCN, which could be suspended in deionized water, and prepared by wet oxidation in concentrated acids.[50] Predetermined amounts of MWCN solution were added to the buffer and stirred for at least 60 min to make a homogenous suspension, which was used as a buffer to dissolve the PVP polymer in it. For this analysis, contactless conductivity detection was used instead of usual ultraviolet (UV) or laser-induced fluorescence (LIF) detection in order to avoid the interference provided by the black color of the CN.

55.3.6.1 Separation Mechanism

Briefly, CNs could form polymer-like networks in buffer.[51] Figure 55.11 shows this network and its corresponding network in PVP solution. On the other hand, CNs can also form DNA-CN hybrids.[52,53] and it is not unreasonable to envision that this adduction might alter DNA fragment mobility. The improved sieving ability generated by adding low concentrations of MWCNs to 1% w/v PVP, might be explained by the fact that large DNA fragments (up to 10 kb) encountering MWCN molecules in a way that is similar to separation of DNA with low concentration polymers (below the entanglement threshold).[33]

The authors concluded that optimized concentrations of MWCNs could be a complementary buffer additive to polymer for CE charged-coupled device (CCD) based DNA separation.

55.4 CONCLUDING REMARKS

Although by introduced techniques, many aspects of DNA migration in different physical or chemical environments has been clarified, still there are some uncertainties in the mechanism, and the behavior of DNA molecules in some environments or on some surfaces is still not well understood.

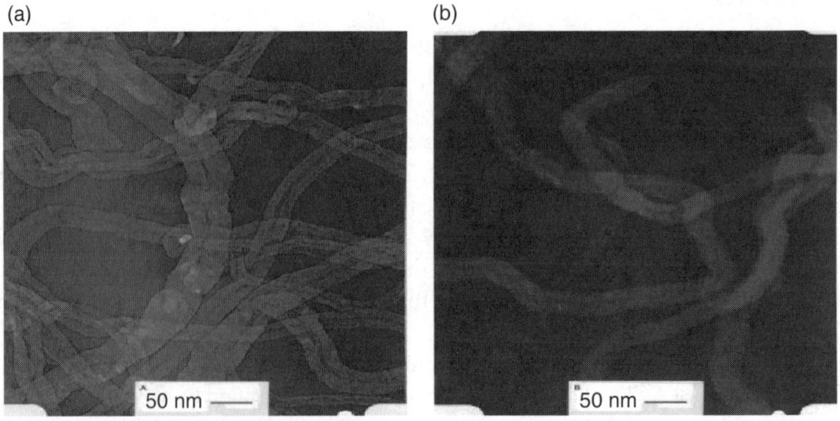

FIGURE 55.11 TEM image of MWCN. (a) 100 ppm MWCN in deionized water. (b) 25 ppm MWCN in 2% w/v PVP solution. (From Xu, Y., Yau Li, S. F., *Electrophoresis* 2006, 27, 4025–4028. With permission.)

New techniques for better understanding of the movement of this macromolecule are essential and nanotechnology is promising filed for this purpose.

REFERENCES

1. Manz, A., Graber, N., Widmer, H. M., *Sens. Actuat. B*, 1990, 1, 244–248.
2. Xu, F., Jabasini, M., Baba, Y., *Electrophoresis*, 2002, 23, 3608–3614.
3. Mohamadi, M. R., Mahmoudian, L., Kaji, N., Tokeshi, M., Chuman, H., Baba, Y., *Nanotoday*, 2006, 1, 38–45.
4. Slater, G. W., Desruisseaux, C., Hubert, S. J., in *Capillary Electrophoresis of Nucleic Acid*, Vol. I (eds. Mitchelson, K. R., Cheng. J.) pp. 27–34 (Humana Press, Totowa, NJ, 2001).
5. Han, J., Craighead, H. G., *Science*, 2000, 288, 1026–1029.
6. Lerman, L. S., Frish, H. L., *Biopolymers*, 1982, 21, 995–997.
7. Duke, T. A. J., Semenov, A. N., Viovy, J. L., *Phys. Rev. Lett.*, 1992, 69, 3260–3263.
8. Duke, T. A. J., Viovy, J. L., Semenov, A. N., *Biopolymers*, 1994, 34, 239–247.
9. Volkmuth, W. D., Austin, R. H., *Nature*, 1992, 358, 600–602.
10. Duke, T. A., J., Austin, R. H., *Phys. Rev. Lett.*, 1998, 80, 1552–1555.
11. Ertas, D., *Phys. Rev. Lett.*, 1998, 80, 1548–1551.
12. Chou, Ch., Bakajin, O., Turner, S. W. P., Duke, T. A. J., Chan, Sh. S., Cox, E. C., Craighead, H. G., Austin, R. H., *Proc. Natl. Adac. Sci. USA*, 1999, 96, 13762–13765.
13. Oudenaarden, A. V., Boxer, S. G., *Science*, 1999, 285, 1046–1048.
14. Cabodi, M., Chen, Y. F., Turner, S. W. P., Craighead, H. G., Austin, R. H., *Electrophoresis*, 2002, 23, 3496–3503.
15. Huang, L. R., Cox, E. C., Austin, R. H., Sturm, J. C., *Anal. Chem.*, 2003, 75, 6963–6978.
16. Han, J., Craighead, H. G., *J. Vac. Sci. Technol. A*, 1999, 17, 2142–2147.
17. Han, J., Turner, S. W., Craighead, H. G., *Phys. Rev. Lett.*, 1999, 83, 1688–1691.
18. Seo, Y., Luo, H., Samuilov, V. A., Rafailovich, M. H., Sokolov, J., Gersappe, D., Chu, B., *Nano Lett.*, 2004, 4, 659–664.
19. Pernodet, N., Samuilov, V., Shin, K., Sokolov, J., Rafailovich, M. H., Gersappe, D., Chu, B., *Phys. Rev. Lett.*, 2000, 85, 5651–5654.
20. Bakajin, O., Duke, T. A., Tegenfeldt, J., Chou, C. F., Chan, S. S., Austin, R. H., Cox, E. C., *Anal. Chem.*, 2001, 73, 6053–6056.
21. Huang, L. R., Tegenfeldt, J. O., Kraeft, J. J., Sturum, J. C., Austin, R. H., Cox, E. C., *Nat. Biotechnol.*, 2002, 20, 1048–1051.
22. Kaji, N., Tezuka, Y., Takamura, Y., Ueda, M., Nishimoto, T., Nakanishi, H., Horiike, Y., Baba, Y., *Anal. Chem.*, 2004, 76, 15–22; Kaji, N., Oki, A., Takamura, Y., Nishimoto, T., Nakanishi, H., Horiike, Y., Tokeshi, M., Baba, Y., Isreal J. Chem., in press.

22b. Easley, C. J., Karlinsey, J. M., Bienvenue, J. M., Legendre, L. A., Roper, M. G., Feldman, S. H., Hughes, M. A., Hewlett, E. L., Merkel, T. J., Ferrance, J. P., Landers, J. P., *Proc. Natl. Acad. Sci. USA*, Dec 19, 2006, 103, 19272–19277.

23. Liu, J., Lawrence, E. M., Wu, A., Ivey, M. L., Flore, G. A., Bibette, J., Richard, J., *Phys. Rev. Lett.*, 1995, 74, 2828–2832.
24. Doyle, P. S., Bibette, J., Bancaud, A., Viovy, J. L., *Science*, 2002, 295, 2237.
25. Mayer, P., Bibette, J., Viovy, J. L., *Mater. Res. Soc. Symp. Proc.*, 1997, 463, 57.
26. Ijima, M., Nagasaki, Y., Okada, T., Kato, M., Kataoka, K., *Macromolecules*, 1999, 32, 1140–1146.
27. Tabuchi, M., Ueda, M., Kaji, N., Yamasaki, N., Nagasaki, Y., Yoshikawa, K., Kataoka, K., Baba. Y., *Nat. Biotech.*, 2004, 22, 337–340.
28. Barron, A. E., Soane, D. S., Blanch, H. W., *J. Chromatogr.*, 1993, 652, 3–16.
29. Barron, A. E., Sunada, W. M., Blanch, H. W., *Electrophoresis*, 1996, 17, 744–754.
30. Braun, B., Blanch, H. W., Prausnitz, J. M., *Electrophoresis*, 1997, 18, 1994–1997.
31. Heller, G. W., Slater, P., Mayer, P., Dovichi, N., Pinto, D., Viovy, J. L., Drouin, G., *J. Chromatogr. A*, 1998, 806, 113–121.
32. Ren, H., Karger, A. E., Oaks, F., Menchen, S., Slater, G. W., Drouin, G., *Electrophoresis*, 1999, 20, 2501–2509.

33. Huang, M. F., Huang, C. C., Chang, H. T., *Electrophoresis*, 2003, 24, 2896–2902.
34. Huang, M. F., Kuo, Y. C., Huang, C. C., Chang, H. T., *Anal. Chem.*, 2004, 76, 192–196.
35. Hubert, S. J., Slater, G. W., Viovy, J. L., *Macromolecules*, 1996, 29, 1006–1009.
36. Nowichi, W., *Macromolecules*, 2002, 35, 1424–1436.
37. Chaplain, V., Janex, M. L., Lafuma, F., Graillat, C., Audebert, R., *Coloid. Polym. Sci.*, 1995, 237, 984–993.
38. Aubouy, M., Raphael, E., *Macromolecules*, 1998, 31, 4357–4363.
39. Mayer, P., Slatter, G. W., Drouin, G., *Anal. Chem.*, 1994, 66, 1777–1780.
40. Vreeland, W. N., Meagher, R. J., Barron, A. E., *Anal. Chem.*, 2002, 74, 4328–4333.
41. Tabuchi, M., Katsuyama, Y., Nogami, K., Nagata, H., Wakuda, K., Fujimoto, M., Nagasaki, Y., Yoshikawa, K., Kataoka, K., Baba, Y., *Lab Chip*, 2005, 5, 199–204.
42. Ito, K., Tsuchida, H., Hayashi, A., Kitano, T., Yamada, E., Matsumoto, T., *Polym. J.* (Tokyo), 1985, 17, 827.
43. Ogawa, R., Nagasaki, Y., Shibata, N., Otsuka, H., Kataoka, K., *Polym. J.* (Tokyo), 2002, 34, 719–726.
44. Doherty, E. A., Kan, C. W., Barron, A. E., *Electrophoresis*, 2003, 24, 4170–4180.
45. Baade, W., Reichet, K. H., *Eur. Polym. J.*, 1984, 20, 505–512.
46. Goetzinger, W., Kotler, L., Carrilho, E., Ruiz-Martinez, M. C., Salas-Solano, O., Karger, B. L., *Electrophoresis*, 1998, 19, 242–248.
47. Doherty, E. A. S., Kan, C. W., Paegel, B. M., Yeung, S. H. I., Cao, S., Mathies, R. A., Barron, A. E., *Anal. Chem.*, 2004, 76, 5249–5256.
48. Xu, Y., Li, S. F., *Electrophoresis*, 2006, 27, 4025–4028.
49. Ijima, S., *Nature*, 1991, 354, 56–58.
50. Chen, G., Zhang, L., Wang, J., *Talanta*, 2004, 64, 1018–1023.
51. Wang, Z. H., Luo, G. A., Chen, J. F., Xiao, S., Wang, Y., *Electrophoresis*, 2003, 24, 4181–4188.
52. Luong, J. H. T., Bouvrette, P., Liu, Y. L., Yang, D. Q., Sacher, E., *J. Chromatogr. A*, 2005, 1074, 187–194.
53. Zheng, M., Jagota, A., Semke, E. D., Diner. B. A. et al., *Nat. Mater*. 2003, 2, 338–342.

Index

A

B

C

D

F

G

H

I

J

K

L

M

N

O

P

Q

R

S

T

U

V

W

X

Y

Z